T0363921

Catalogue of the Cicadoidea

(Hemiptera: Auchenorrhyncha)

Catalogue of the Cicadoidea

(Hemiptera: Auchenorrhyncha)

ALLEN F. SANBORN

With Contributions to the Bibliography by

MARTIN H. VILLET

Amsterdam – Boston – Heidelberg – London – New York – Oxford
Paris – San Diego – San Francisco – Singapore – Sydney – Tokyo
Academic Press is an imprint of Elsevier

Academic Press is an imprint of Elsevier
32 Jamestown Road, London NW1 7BY, UK
225 Wyman Street, Waltham, MA 02451, USA
525 B Street, Suite 1800, San Diego, CA 92101-4495, USA

British Library Cataloguing-in-Publication Data
A catalogue record for this book is available from the British Library

Library of Congress Cataloging-in-Publication Data
A catalog record for this book is available from the Library of Congress

ISBN: 978-0-12-416647-9

For information on all Academic Press publications
visit our website at elsevierdirect.com

Typeset by MPS Limited, Chennai, India
www.adi-mps.com

CONTENTS

INTRODUCTION

This work is the third in a series of catalogues on the Cicadoidea attempting to summarize the published literature of the cicadas. Z.P. Metcalf produced the first catalogue of the Cicadoidea in 1963 as fascicle VIII of the *General Catalogue of the Homoptera* series and included the literature published before 1956. J.P. Duffels and P.A. van der Laan published a second catalogue in 1985 covering the years 1956 to 1980.

This work continues the process of compiling bibliographic information on the cicadas including material published between 1981 and 2010 as well as references located that were not included in Metcalf's or Duffels and van der Laan's bibliographies. The advent of electronic publications and their recent recognition by the International Commission of Zoological Nomenclature as a valid means of disseminating information has meant that only works published electronically before 31 December 2010 are included in this catalogue. These references are cited with a 2010 electronic publication date even if the printed paper version was not published until 2011 and possesses a volume that would only have been published in or after 2011.

Only those references that we were able to obtain and which were verified to have mentioned a genus or species name are included. There are some references that were identified by various computer searches that probably should be included but we were unable to obtain a copy from the author or through various loan mechanisms to verify the mention of a genus or species name. The indexing services often include cicadellids, membracids, fulgorids, etc. when they translate "cicada" so references were only included when we could verify an extant member of either the Cicadidae or Tettigarctidae was mentioned in a reference. Theses and dissertations were not included as many are not officially published documents.

The catalogue is designed to describe the data published on cicadas in the literature. The basic format of the Metcalf and Duffels and van der Laan catalogues has been retained except countries are listed instead of using superscripts with each species. The references are summarized as to the main points discussed in the work along with a listing of any illustrations or tables that discuss the species. General terms were applied, i.e., natural history, when several aspects of the natural history of an organism were discussed in a reference without a specific focus. References written in languages other than Latin characters are also described in general terms since translation services were unavailable. Particular emphasis was made in identifying sonograms and/or oscillograms in references as these have become important tools in the differentiation of species. Any references to the biogeography of a species within a given work are also provided in the catalogue. The general biogeography of a species can be deduced from the localities listed in the references discussing a particular species even though the finding of cryptic species is now clouding the biogeography of some species, e.g. *Cicadetta montana* (Scopoli). Only extant species are included in the list but some primarily paleontological works are included in the bibliography as they compare fossil species to extant forms.

The catalogue is organized to represent the current taxonomy. It has incorporated all taxonomic changes that have occurred through 2012. As a result, the catalogue is also a current species list for the Cicadoidea through the end of 2012. There have been important works published in 2011 and 2012 that changed the higher taxonomy or reassigned species to new genera. These changes were included in the catalogue in an attempt to use and present the current taxonomy and alert individuals that the taxonomy may have changed for a particular species in case a reader was not familiar with the works that modified the taxonomy. A summary of the complete taxonomy is provided at the end of the catalogue.

The taxa are organized into the two families and three subfamilies currently recognized, reduced from the six families and five subfamilies presented by Duffels and van der Laan and highly reorganized from the two families (only one in common) and seven subfamilies presented by Metcalf. Tribes, subtribes, and genera are then listed based on phylogeny using a combination of the previous catalogues, comparative genera used in descriptions of new taxa, and molecular phylogenetic studies that have been published. Species are listed alphabetically within each genus. The species are listed not only with the authority but also the year of publication and placed in parentheses if the taxon was originally described in a different genus following the *Code*. This additional information should facilitate finding original descriptions if needed. The 4th edition of the *International Code of Zoological Nomenclature* was followed in listing species so what was originally a variety in Metcalf may now be listed as a subspecies as varieties published before 1961 became subspecies under the *Code* (Article 45.6.4). Similarly, varieties named after 1960 are considered to be infrasubspecific and not regulated by the *Code* (Article 15.2) and are listed under a nominotypical species in the catalogue unless the taxon has also been used as a subspecies in which case it is listed as a subspecies. If no subspecies was designated in a particular work, the information is listed with the nominotypical species unless it is clear from the location of the study that it would refer to a specific subspecies. Species originally described with diacritic or other marks have been corrected following Article 32.5.2. All species listed in Metcalf and Duffels and van der Laan are included along with all synonymies, including synonymized species, former generic placements and misspellings, even if there were no works discussing the species published between 1981-2010. This provides a complete overview of the extant Cicadoidea.

The genus *Tibicen* is retained in this work. Although there have been many articles written about the status of *Tibicen* (see the treatment for the genus in the catalogue), the petition to the International Commission of Zoological Nomenclature to suppress *Tibicen* (summarized in Melville and Sims 1984) has never been voted upon by the Commission. As a result and regardless of the confusion with the application of the name, *Tibicen* has priority and is conserved here as it was in both the previous catalogues. It was suggested in correspondence with the Commission that another application will need to be submitted to clarify the issue.

The catalogue has listed nomina nuda as outlined in Metcalf and Duffels and van der Laan and more recent literature. Any undetermined species is listed under the genus as originally identified. It is becoming a common trend to identify unknown species as sp. A, sp. B, sp. 1, sp. 2, etc. in published works. These are listed in the catalogue as the original author identified the species in the reference with the author being listed as the first combination of author(s) to mention the taxon. If the species could be identified based on a later description, the data are included with the recognized species. If the undetermined species was compared to a species that was reassigned to a new genus, the undetermined species was also assigned to that genus. It is hoped that complete descriptions will eventually be published for the species (as was done with some nomina nuda from the Duffels and van der Laan catalogue and the more recent literature) so the information can be applied to a particular taxon once it is formed.

The publications are listed in the bibliography alphabetically by author and then chronologically for any given author(s). Letters are added to the publication year to distinguish individual articles published by the same author(s) in a given year. The sequence of lettering does not necessarily correspond to the actual publication dates. The names of periodicals have been spelled out in full to facilitate finding copies of the references.

The catalogue and bibliography could not have been produced without the efforts of many individuals who helped the authors obtain references over the years. Many specialists assisted by providing reprints or copies of their work over the years. Special thanks must be extended to M. Boulard, J.P. Duffels, M. Gogala, M. Hayashi, K.B.R. Hill, Y.J. Lee, M.S. Moulds, F. Mozaffarian, J.A. Quartau, and T. Trilar who provided not only copies of their own work but also helped in obtaining some obscure references. The librarians at Barry University, Rhodes University and the Biology Library of the University of Illinois at Urbana-Champaign filled numerous interlibrary loan requests. J.P. Duffels and C. Simon are also thanked for organizing the cicada symposia at the 10th International Auchenorrhyncha Congress in Cardiff, Wales in 1999 where we were charged with organizing the more recent literature as part of a larger project to compile all cicada literature. It was at this conference that we combined the reference information from our individual libraries to begin what would eventually become the bibliography provided here and the source of information for the catalogue.

Catalogue of the Cicadoidea

SUPERFAMILY CICADOIDEA Westwood, 1840

Family Tettigarctidae Distant, 1905

Tettigarctidae Wooton 1968: 318 (comp. note)
Tettigarctidae Goode 1980: 54 (diversity) Australia
Tettigarctidae Shcherbakov 1981a: 830, 834, 841 (listed, key)
Tettigarctidae Shcherbakov 1981b: 66, 70, 78 (listed, key)
Tettigarctidae Shcherbakov 1982a: 531, 534 (listed, key)
Tettigarctidae Shcherbakov 1982b: 73, 76 (listed, key)
Tettigarctidae Strümpel 1983: 17 (key, comp. note) Australia
Tettigarctidae Whalley 1983: 141 (comp. note)
Tettigarctidae Hayashi 1984: 25–27, Table 1 (comp. note)
Tettigarctidae Boulard 1988f: 51, 64 (key, comp. note)
Tettigarctidae spp. Ewart 1989b: 289 (comp. note)
Tettigarctidae Boulard 1990b: 92 (comp. note)
Tettigarctidae Boulard and Nel 1990: 40–41, 43 (comp. note)
Tettigarctidae Carver 1990: 260 (comp. note)
Tettigarctidae Moulds 1990: 21, 30–31, 45 (described, key, listed, comp. note) Australia
Tettigarctidae Carver, Gross and Woodward 1991: 431, 443 (listed, comp. note)
Tettigarctidae Moulds and Carver 1991: 465–466 (key, characteristics, comp. note) Australia
Tettigarctidae Emeljanov and Kirillova 1991: 799, 811 (comp. note)
Tettigarctidae Carpenter 1992: 221 (key, described, comp. note)
Tettigarctidae Emeljanov and Kirillova 1992: 63, 74 (comp. note)
Tettigarctidae Savinov 1992: 267 (listed)
Tettigarctidae Duffels 1993: 1225–1226 (comp. note)
Tettigarctidae Hayashi 1993: 13 (comp. note)
Tettigarctidae Naumann 1993: 28, 58, 127, 155 (listed) Australia
Tettigarctidae Savinov 1993: 1 (listed)
Tettigarctidae Carver, Gross and Woodward 1994: 327 (listed)
Tettigarctidae Boulard and Mondon 1995: 72 (comp. note)
Tettigarctidae Nel 1996: 84, 88–92 (comp. note)
Tettigarctidae New 1996: 95 (comp. note) Australia
Tettigarctidae Boulard 1997c: 115–117 (comp. note) Australia, Tasmania
Tettigarctidae Chou, Lei, Li, Lu and Yao 1997: 31 (listed)
Tettigarctidae Poole, Garrison, McCafferty, Otte and Stark 1997: 258 (listed)
Tettigarctidae Nel, Zarbout, Barale and Philippe 1998: 594–595, 597, Table 1 (listed, comp. note)
Tettigarctidae Claridge, Morgan and Moulds 1999a: 1831 (comp. note) Australia
Tettigarctidae Hayashi and Moulds 1999: 97, 100 (comp. note)
Tettigarctidae Moulds 1999: 1 (comp. note)
Tettigarctidae Villet and Zhao 1999: 321 (comp. note)
Tettigarctidae Boulard 2000i: 21 (listed)
Tettigarctidae Boulard 2001b: 35–37 (comp. note) Australia, Tasmania
Tettigarctidae Dietrich, Rakitov, Holmes and Black 2001: 296, 300–301, Figs 3–4, Table 1 (comp. note)
Tettigarctidae Sueur 2001: 45, Table 1 (listed)

Catalogue of the Cicadoidea (Hemiptera: Auchenorrhyncha). DOI: http://dx.doi.org/10.1016/B978-0-12-416647-9.00001-2

Tettigarctidae Boulard 2002b: 175 (listed, comp. note)
Tettigarctidae Dietrich 2002: 157–159, Fig 1 (phylogeny, listed, comp. note) Australia
Tettigarctidae Duffels and Turner 2002: 235 (comp. note)
Tettigarctidae Moulds and Carver 2002: 54 (described, listed) Australia
Tettigarctidae Williams 2002: 38 (comp. note) Australia
Tettigarctidae Dietrich 2003: 66–67, 72, Table 1 (listed, comp. note)
Tettigarctidae Holzinger, Kammerlander and Nickel 2003: 475 (comp. note) Australia, Tasmania
Tettigarctidae Shcherbakov 2002: 33 (comp. note)
Tettigarctidae Moulds 2003a: 186–187 (comp. note) Australia
Tettigarctidae Brambila and Hodges 2004: 365 (comp. note)
Tettigarctidae Chen 2004: 38 (listed)
Tettigarctidae Sanborn 2004c: 511 (comp. note) Australia
Tettigarctidae Boulard 2005d: 333 (comp. note)
Tettigarctidae Cryan 2005: 563 (comp. note)
Tettigarctidae Dietrich 2005: 510 (key)
Tettigarctidae Menon 2005: 53 (comp. note)
Tettigarctidae Moulds 2005b: 377, 380, 382, 385–389, 394, 397, 404, 412–415, 426, 430, Fig 31, Table 2 (described, key, history, phylogeny, listed, comp. note) Australia
Tettigarctidae Boulard 2006g: 17 (comp. note)
Tettigarctidae Wang, Ren, Liang, Liu and Wang 2006: 295, 297, Table 1 (listed, comp. note) China
Tettigarctidae Boulard 2007e: 1 (comp. note)
Tettigarctidae Urban and Cryan 2007: 561, Table 1 (listed)
Tettigarctidae Brambila and Hodges 2008: 604 (comp. note)
Tettigarctidae Chawanji, Hodson and Villet 2006: 374 (comp. note)
Tettigarctidae Shiyake 2007: 2, 14 (listed, comp. note)
Tettigarctidae Costa Neto 2008: 453 (comp. note)
Tettigarctidae Sanborn 2008c: 874–875 (described, comp. note)
Tettigarctidae Sanborn 2008d: 3470 (comp. note) Australia
Tettigarctidae Wang, Zhang, Fang and Zhang 2008: 12 (comp. note)
Tettigarctidae Moulds 2009b: 163–164 (comp. note) Australia
Tettigarctidae Santos and Martinelli 2009b: 638 (listed)
Tettigarctidae Shcherbakov 2009: 343–348 (key, listed, comp. note) Australia, Tasmania
Tettigarctidae Wei, Tang, He and Zhang 2009: 337 (comp. note)
Tettigarctidae Cryan and Svenson 2010: 399, 402, Fig 1, Table 1 (phylogeny, comp. note) Australia
Tettigarctidae Phillips and Sanborn 2010: 74 (comp. note) Australia
Tettigarctidae Santos, Martinelli, Maccagnan, Sanborn and Ribeiro 2010: 48 (listed)

Subfamily Tettigarctinae Distant, 1905

Tettigarctinae Zeuner 1944: 110–111, 113 (listed, comp. note)
Tettigarctinae Hayashi 1984: 26–27, Table 1 (comp. note)
Tettigartcinae Popov 1985: 47 (acoustic system, comp. note)
Tettigarctinae Boulard and Nel 1990: 41 (comp. note)
Tettigarctinae Moore 1990: 58 (comp. note)
Tettigarctinae McGavin 1993: 80 (comp. note) Australia
Tettigarctinae Lei, Chou and Li 1994: 54 (comp. note)
Tettigarctinae Nel 1996: 84 (comp. note)
Tettigarctinae Boulard 1997c: 115–117 (comp. note)
Tettigarctinae Chou, Lei, Li, Lu and Yao 1997: 30–36, Figs 5-1–5-3 (phylogeny, key, listed)
Tettigarctinae Boulard 2000i: 21 (listed)
Tettigarctinae Boulard 2001b: 35–37 (comp. note)
Tettigarctinae Sueur 2001: 45, 47, Table 1, Table 3 (listed, diversity, distribution, comp. note)
Tettigarctinae Moulds and Carver 2002: 2 (comp. note)
Tettigarctinae Moulds 2005b: 385–388, 430, Fig 31, Table 2 (described, key, history, phylogeny, listed, comp. note) Australia
Tettigarctinae Shcherbakov 2009: 344, 347, Table 1 (listed, comp. note)

Tribe Tettigarctini Distant, 1905

Tettigarctini Nel 1996: 83 (comp. note)
Tettigarctini Boulard 1997c: 115–117 (comp. note)
Tettigarctini Boulard 2000i: 21 (listed)
Tettigarctini Boulard 2001b: 35–37 (comp. note)
Tettigarctini Sueur 2001: 45, Table 1 (listed)
Tettigarctini Moulds 2005b: 419, 430, Fig 58, Table 2 (phylogeny, listed, comp. note) Australia
Tettigarctini Shcherbakov 2009: 347 (listed)
Tettigarctini Phillips and Sanborn 2010: 74 (comp. note) Australia

Subtribe Tettigarctina Distant, 1905 = Tettigarctaria

Tettigarctaria Boulard 1988f: 53 (comp. note)
Tettigarctaria Nel 1996: 83 (comp. note)

Tettigarctaria Boulard 1997c: 115–117 (comp. note)
Tettigarctaria Boulard 2000i: 21 (listed)
Tettigarctaria Boulard 2001b: 35–37 (comp. note)
Tettigarctaria Moulds and Carver 2002: 54 (listed) Australia
Tettigarctaria Moulds 2005b: 386, 389 (history, listed, comp. note) Australia

***Tettigarcta* White, 1845** = *Tettigareta* (sic) = *Tettigarota* (sic) = *Tettigarta* (sic)
Tettigarcta Zeuner 1944: 111, 113–116 (comp. note)
Tettigarcta Evans 1956: 230, 252, Fig 32 (comp. note)
Tettigarcta Wooton 1968: 318 (comp. note)
Tettigarcta Moore 1981: 167 (comp. note) Australia
Tettigarcta Shcherbakov 1981a: 830, 834, Fig 21 (wing structure, comp. note)
Tettigarcta Shcherbakov 1981b: 66, 70, Fig 21 (wing structure, comp. note)
Tettigarcta Boulard 1982e: 181–182 (comp. note) Australia, Tasmania
Tettigarcta Boulard 1982d: 112 (comp. note)
Tettigarcta Shcherbakov 1982a: 531, Fig 14 (wing structure, comp. note)
Tettigarcta Shcherbakov 1982b: 73, Fig 14 (wing structure, comp. note)
Tettigarcta Strümpel 1983: 17, 117 (listed, comp. note)
Tettigarcta Whalley 1983: 140 (comp. note)
Tettigarcta Boulard 1985e: 1018 (comp. note)
Tettigarcta Boulard 1986c: 191 (comp. note)
Tettigarcta Ewart 1989a: 75 (comp. note) Australia
Tettigarcta Moss 1989: 103, 105 (comp. note) Tasmania
Tettigarcta Boulard 1990b: 88 (comp. note)
Tettigarcta Boulard and Nel 1990: 38, 40–41, 43 (comp. note)
Tettigarcta Moss 1990: 7 (comp. note) Tasmania
Tettigarcta spp. Moulds 1990: 17–19 (comp. note) Australia
Tettigarcta Moulds 1990: 21, 45, 47 (described, communication, diversity, listed, comp. note) Australia, Tasmania
Tettigarcta Carver, Gross and Woodward 1991: 433, 464 (comp. note)
Tettigarcta Moulds and Carver 1991: 465 (comp. note) Australia
Tettigarcta Carpenter 1992: 221 (comp. note) Australia
Tettigarcta Brodsky 1992: 42 (comp. note)
Tettigarcta Brodskiy 1993: 11 (comp. note)
Tettigarcta Duffels 1993: 1226 (comp. note)
Tettigarcta Brodsky 1994: 143 (comp. note)
Tettigarcta Boulard and Mondon 1995: 72 (comp. note)
Tettigarcta Nel 1996: 85, 90–91 (comp. note)
Tettigarcta spp. Nel 1996: 85, 88 (comp. note)
Tettigarcta Boulard 1997c: 116–117 (comp. note)
Tettigarcta spp. Nel, Zarbout, Barale and Philippe 1998: 597 (comp. note)
Tettigarcta Claridge, Morgan and Moulds 1999a: 1831–1833 (song, comp. note) Australia
Tettigarcta Claridge, Morgan and Moulds 1999b: 1 (song, comp. note) Australia
Tettigarcta Hayashi and Moulds 1999: 99 (comp. note)
Tettigarcta Boulard 2000i: 21 (listed)
Tettigarcta Boulard 2001b: 36–37 (comp. note)
Tettigarcta Cooley 2001: 756 (comp. note)
Tettigarcta Moulds and Carver 2002: 54–55 (listed, comp. note) Australia
Tettigarcta Shcherbakov 2002: 33 (comp. note)
Tettigarcta Stölting, Moore and Lakes-Harlan 2002: 4 (vibrational communication, comp. note)
Tettigarcta Dietrich 2003: 72 (comp. note)
Tettigarcta Tishechkin 2003: 144 (comp. note)
Tettigarcta Brambila and Hodges 2004: 365 (comp. note) Australia, Tasmania
Tettigarcta Sanborn 2004c: 511 (comp. note) Australia
Tettigarcta Grimaldi and Engel 2005: 308 (comp. note)
Tettigarcta Moulds 2005b: 388–389, 394, 410, 412–413, 415, 426, 430, Table 2 (type genus of Tettigarctidae, phylogeny, listed, comp. note) Australia
Tettigarcta Boulard 2006g: 164 (comp. note)
Tettigarcta Shiyake 2007: 2 (listed, comp. note)
Tettigarcta Brambila and Hodges 2008: 604 (comp. note) Australia, Tasmania
Tettigarcta Hill 2008: 114 (comp. note) Australia
Tettigarcta Kukalová-Peck 2008: 6 (listed)
Tettigarcta Osaka Museum of Natural History 2008: 133 (comp. note) Australia
Tettigarcta Sanborn 2008c: 874 (comp. note) Australia
Tettigarcta Wang, Zhang, Fang and Zhang 2008: 12 (comp. note)
Tettigarcta Shcherbakov 2009: 343–345, 347, Table 1 (comp. note)
Tettigarcta Wei, Tang, He and Zhang 2009: 337 (comp. note)

***T. crinita* Distant, 1883** = *Tettigarcta criniti* (sic)
Tettigarcta crinita Zeuner 1944: 112–113, Fig 3 (hind wing illustrated, comp. note) Australia
Tettigarcta crinita Evans 1956: 222, 224, 230–231, Figs 24A–B (comp. note)
Tettigarcta crinita Greenup 1964: 22–23 (comp. note) Australia, New South Wales

Tettigarcta crinita Wooton 1968: 319, Fig c, Fig k (wing venation, comp. note)
Tettigarcta crinita Moulds 1983: 429–430 (illustrated, comp. note) Australia, New South Wales, Australian Capital Territory, Victoria
Tettigarcta crinita Moss 1989: 104–105 (comp. note)
Tettigarcta crinita Boulard 1990b: 132–133, Plate XV, Fig 1 (genitalia, comp. note)
Tettigarcta crinita Moss 1990: 7 (comp. note)
Tettigarcta crinita Moulds 1990: 3, 46–51, Figs 21–22, Plate 6, Figs 1a–1b, Plate 24, Fig 7 (illustrated, described, distribution, ecology, type data, listed, comp. note) Australia, Australian National Territory, New South Wales, Victoria
Tettigarcta crinita Moulds and Carver 1991: 465–466, Fig 30.26A (illustrated, comp. note) Australia
Tettigarcta crinita Moulds and Carver 1991: 465–466, Fig 30.26A (illustrated, comp. note) Australia
Tettigarcta crinita Duffels 1993: 1226 (comp. note)
Tettigarcta crinita Naumann 1993: 28, 127, 155 (listed) Australia
Tettigarcta crinita Boer and Duffels 1996b: 309 (distribution, comp. note) Australia
Tettigarcta crinita Moulds 1996b: 93, Plate 68 (illustrated, comp. note) Australia, New South Wales, Victoria
Tettigarcta crinita New 1996: 95 (comp. note) Australia
Tettigarcta crinita Boulard 1997c: 115 (comp. note)
Tettigarcta crinita Nel 1998: 84 (comp. note)
Tettigarcta crinita Claridge, Morgan and Moulds 1999a: 1831–1833, Fig 1 (song, oscillogram, comp. note) Australia, New South Wales
Tettigarcta crinita Hayashi and Moulds 1999: 97–100, Figs 1–12 (eclosion, comp. note) New South Wales
Tettigarcta crinita Boulard 2001b: 35 (comp. note)
Tettigarcta crinita Sueur 2001: 45, Table 1 (listed)
Tettigarcta crinita Dietrich 2002: 159, Fig 2 (illustrated)
Tettigarcta crinita Moulds and Carver 2002: 55 (distribution, ecology, type data, listed) Australia, Australian National Territory, New South Wales, Victoria
Tettigarcta crinita Dietrich 2003: 66, Fig 2(6) (illustrated) Australia
Tettigarcta crinita Moulds 2003a: 187 (comp. note) Australia
Tettigarcta crinita Tishechkin 2003: 144 (comp. note)
Tettigarcta crinita Sanborn 2004c: 511, Fig 171 (illustrated)
Tettigarcta crinita Cryan 2005: 565, 568, Fig 2, Table 1 (phylogeny, comp. note) Australia
Tettigarcta crinita Dietrich 2005: 503, Fig 1A (illustrated) Australia
Tettigarcta crinita Menon 2005: 53 (comp. note) Australia, South Australia
Tettigarcta crinita Moulds 2005b: 380, 382, 388, 397, 399, 406–409, 411, 417–420, Fig 11, Figs 22–24, Figs 46–47, Fig 49, Fig 53, Figs 56–59 (illustrated, phylogeny, listed, comp. note) Australia
Tettigarcta crinita Raven and Yeates 2007: 179 (plant association, comp. note) Australia, Tasmania
Tettigarcta crinita Shiyake 2007: 2, 12, 14, Fig 1 (illustrated, distribution, listed, comp. note) Australia
Tettigarcta crinita Urban and Cryan 2007: 561, 566–567, Figs 3–4, Table 1 (phylogeny, listed) Australia
Tettigarcta crinita Hill 2008: 114 (comp. note) Australia
Tettigarcta crinita Sanborn 2008c: 874, Fig 56 (illustrated, comp. note)
Tettigarcta crinita Moulds 2009b: 164 (comp. note) Australia
Tettigarcta crinita Phillips and Sanborn 2010: 74 (comp. note) Australia

***T. tomentosa* White, 1845** = *Tettigarcta tormentosa* (sic)
Tettigarcta tomentosa Zeuner 1944: 112–114, Fig 2, Figs 5–6 (hind wing illustrated, venation, comp. note) Australia
Tettigarcta tomentosa Moulds 1983: 429 (comp. note) Australia, Tasmania
Tettigarcta tomentosa Strümpel 1983: 17, Fig 29 (illustrated, comp. note)
Tettigarcta tomentosa Boulard 1988f: 53 (type species of Tettigarctidae)
Tettigarcta tomentosa Moss 1989: 104–105 (distribution, comp. note) Tasmania
Tettigarcta tomentosa Moss 1990: 7 (distribution, comp. note) Tasmania
Tettigarcta tomentosa Moulds 1990: 3, 45, 48–51, Plate 6, Figs 2a–2b (type species of *Tettigarcta*, illustrated, described, distribution, ecology, listed, comp. note) Australia, Tasmania
Tettigarcta tomentosa Carver, Gross and Woodward 1991: 432, Fig 30.4F (wings illustrated)
Tettigarcta tomentosa Moulds and Carver 1991: 465–466, Fig 30.26B (illustrated, comp. note) Tasmania
Tettigarcta tomentosa Moulds and Carver 1991: 465–466, Fig 30.26B (illustrated, comp. note) Tasmania

Tettigarcta tomentosa Brodsky 1992: 42, Fig 4B (mesothoracic terga, illustrated, comp. note)
Tettigarcta tomentosa Savinov 1992: 267 (comp. note)
Tettigarcta tomentosa Brodskiy 1993: 11, Fig 4B (mesothoracic terga, illustrated, comp. note)
Tettigarcta tomentosa Duffels 1993: 1226 (comp. note)
Tettigarcta tomentosa Naumann 1993: 58, 127, 155 (listed) Australia
Tettigarcta tomentosa Savinov 1993: 1 (comp. note)
Tettigarcta tomentosa Brodsky 1994: 142–143, Fig 8.10b (mesothorax, illustrated, comp. note)
Tettigarcta tomentosa Carver, Gross and Woodward 1994: 319, Fig 30.4B (wings illustrated)
Tettigarcta tomentosa Boer and Duffels 1996b: 309 (distribution, comp. note) Tasmania
Tettigarcta tomentosa Moulds 1996b: 93 (comp. note) Australia, Tasmania
Tettigarcta tomentosa New 1996: 95 (comp. note) Tasmania
Tettigarcta tomentosa Boulard 1997c: 115–116, Plate III, Figs 1–2 (type species of *Tettigarcta*, Tettigarctaria, Tettigarctini, Tettigarctinae, Tettigarctidae)
Tettigarcta tomentosa Claridge, Morgan and Moulds 1999a: 1831 (comp. note) Tasmania
Tettigarcta tomentosa Boulard 2000i: 21 (listed)
Tettigarcta tomentosa Boulard 2001b: 35–37, Plate III, Figs 1–2 (type species of *Tettigarcta*, Tettigarctaria, Tettigarctini, Tettigarctinae, Tettigarctidae)
Tettigarcta tomentosa Dietrich, Rakitov, Holmes and Black 2001: 296, Table 1 (comp. note) Tasmania
Tettigarcta tomentosa Boulard 2002b: 177, Fig 8 (illustrated, comp. note) Australia, Tasmania
Tettigarcta tomentosa Moulds and Carver 2002: 55 (type species of *Tettigarcta*, distribution, ecology, type data, listed) Australia, Tasmania
Tettigarcta tomentosa Moulds 2003a: 187 (comp. note) Australia
Tettigarcta tomentosa Cryan 2005: 565, 568, Fig 2, Table 1 (phylogeny, comp. note) Australia
Tettigarcta tormentosa (sic) Menon 2005: 53 (comp. note) Australia, Tasmania
Tettigarcta tomentosa Moulds 2005b: 388, 397, 399, 417–420, 426, Figs 56–59, Table 1 (type species of *Tettigarcta*, phylogeny, listed) Australia, Tasmania
Tettigarcta tomentosa Hill 2008: 114–115 (comp. note) Australia
Tettigarcta tomentosa Moulds 2009b: 164 (comp. note) Australia
Tettigarcta tomentosa Cryan and Svenson 2010: 399, 402, Fig 1, Table 1 (phylogeny, comp. note) Australia
Tettigarcta tomentosa Phillips and Sanborn 2010: 74 (comp. note) Australia

Family Cicadidae Latreille, 1802 = Cicadae (sic) = Cicadiae (sic) = Cicididae (sic) = Cicadide (sic) = Plautillidae Distant, 1905 = Plautilidae (sic) = Tettigadidae Distant, 1905 = Tibicinidae Distant, 1905 = Tibiciinidae (sic) = Platypediidae Kato, 1932 = Tibicenidae Van Duzee, 1916 = Ydiellidae Boulard, 1973
Cicadidae Carpenter and Dallas 1867: 186 (comp. note)
Cicadidae Szilády 1870: 119 (listed) Romania
Cicadidae Swinton 1877: 80 (comp. note)
Cicadidae Buchanan 1879: 213 (listed)
Cicadidae Swinton 1879: 80–81 (hearing, comp. note)
Cicadidae Swinton 1880: 233 (comp. note)
Cicadidae Van Duzee 1893: 410 (listed)
Cicadidae Strobl 1900: 201 (listed) Austria
Cicadidae Henshaw 1903: 229 (listed)
Cicadidae Poulton 1907: 360, 362, 371 (predation, listed, comp. note)
Cicadidae Schouteden 1907: 287 (listed)
Cicadidae Surface 1907: 71 (listed, comp. note)
Cicadidae Walker 1920: cxxxi (listed) New Zealand
Cicadidae Costa Lima 1922: 81 (listed) Brazil
Cicadidae Schouteden 1925: 83 (listed) Congo
Cicadidae Faber 1928: 211 (comp. note)
Cicadidae Snodgrass 1930: 182 (comp. note)
Cicadidae Wilson 1930: 65 (comp. note) Florida, Kansas
Cicadidae Weber 1931: 71, 101–102, 135 (listed, comp. note) Germany
Cicadidae Isaki 1933: 380 (listed) Japan
Cicadidae Kamijo 1933: 58, 62, Table 3 (listed, comp. note) Korea
Cicadidae Wu 1935: 1 (listed) China
Cicadidae Steinhaus 1941: 763, Table 1 (listed, bacterial host) Ohio
Cicadidae Tinkham 1941: 165 (comp. note) Texas
Cicadidae Zeuner 1941: 88 (listed)
Cicadidae Zeuner 1944: 110 (listed)
Cicididae (sic) Zeuner 1944: 113 (comp. note)
Cicadidae Lepesme 1947: 168 (listed)
Cicadidae Salmon 1950: 1 (listed) New Zealand
Cicadidae Gómez-Menor 1951: 52, 56, 60 (key, listed, comp. note)
Cicadidae Castellani 1952: 15 (listed) Italy
Cicadidae Wagner 1952: 15 (listed) Italy
Cicadidae Kurokawa 1953: 1, 4–5 (listed, comp. note) Japan
Cicadidae Cho 1955: 237, 256, Table 1 (listed) Korea

Cicadidae Vayssière 1955: 254 (listed)
Cicadidae Evans 1956: 223–224, 252, Fig 32 (comp. note)
Cicadidae Soós 1956: 411 (listed) Romania
Cicadidae Pringle 1957: 25–26, 35, 37, 56, 64, 88, Table 1 (listed, comp. note)
Cicadidae Kim 1958: 94 (listed) Korea
Cicadidae Halkka 1959: 5, 37, Table 1 (listed, comp. note)
Cicadidae Hamann 1960: 166 (lsited)
Cicadidae Eyles 1960: 994 (listed, comp. note) New Zealand
Cicadidae Katsuki 1960: 54 (hearing, comp. note) Japan
Cicadidae Kim 1960: 23 (listed) Korea
Cicadidae Hahn 1962: 8 (listed)
Cicadidae Smit 1964: 137 (listed) South Africa
Cicadidae Cho 1965: 157, 173 (listed) Korea
Cicadidae LeQuesne 1965: 4 (listed, key, comp. note) Great Britain
Cicadidae McCoy 1965: 40 (listed) Arkansas
Cicadidae Meer Mohr 1965: 104 (listed) Sumatra
Cicadidae Chaudhry, Chaudry and Khan 1966: 94 (listed) Pakistan
Cicadidae Dumbleton 1966: 977 (lsited) New Zealand
Cicadidae Evans 1966: 103, 114 (listed, comp. note)
Cicadidae Goodchild 1966: 118 (comp. note)
Cicadidae Synave 1966: 1030 (listed) Senegal
Cicadidae Heinrich 1967: 43–44 (comp. note) Brazil
Cicadidae Wooton 1968: 318 (comp. note)
Cicadidae Weber 1968: 274–275 (listed, comp. note)
Cicadidae Frost 1969: 93 (listed) Florida
Cicadidae Orians and Horn 1969: 937 (listed, predation)
Cicadidae Borror and White 1970: 129, Plate 4 (listed, comp. note)
Cicadidae Crowhurst 1970: 123 (listed, predation, comp. note) New Zealand
Cicadidae Ruffinelli 1970: 3 (listed, comp. note) Uruguay
Cicadidae Dlabola 1971: 382 (listed)
Cicadidae Otero 1971: 116, 144, 175 (listed) Brazil
Cicadidae Popov 1971: 301–302 (listed, comp. note)
Cicadidae Kaltenbach, Steiner and Aschenbrenner 1972: 490 (comp. note) Germany
Cicadidae Meyer-Rochow 1973: 675 (human food, listed) New Guinea
Cicadidae Cullen 1974: 25, 31, 33–34, 36, Fig 21, Table 2, Table 6 (listed)
Cicadidae Bohart and Strange 1976: 313 (parasitism, comp. note) Argentina, Catamarca
Cicadidae Coleman 1977: 109, Table 2 (listed) New Zealand
Cicadidae Johns 1977: 320, 325, Tables 21D–21E (listed) New Zealand
Cicadidae Linnavuori 1977: 65 (listed)
Cicadidae Villiers 1977: 13–14, 17–18, 58–59, 162, 164 (key, comp. note)
Cicadide (sic) Villiers 1977: 19, 21, 25, Fig 5B, Fig 6B, Fig 6D, Fig 8G (comp. note)
Cicadidae Anderson, Ostry and Anderson 1979: 478 (listed) Wisconsin
Tibicinidae Deitz 1979: 25 (listed) New Zealand
Cicadidae Fitzgerald and Karl 1979: 116, Table 2 (listed) New Zealand
Cicadidae Powell and Hogue 1979: 112 (listed, diversity, comp. note) California
Cicadidae Ramos-Elorduy de Conconi and Pino Moreno 1979: 569, Table I (human food, listed) Mexico, Hidalgo
Cicadidae Goode 1980: 54 (diversity) Australia
Cicadidae Dlabola 1981: 198 (listed)
Cicadidae Hamilton 1981: 955, 959, 970–971, Fig 33 (phylogeny, comp. note)
Cicadidae Jankovic and Papovic 1981: 131 (comp. note) Yugoslavia
Cicadidae Lodos and Kalkandelen 1981: 69 (listed) Turkey
Cicadidae O'Connor and Nash 1981: 299 (listed)
Cicadidae Osssiannilsson 1981: 223 (listed)
Cicadidae Pringle 1981: 10–11, Table 1, Fig 12 (listed, comp. note)
Cicadidae Ramos Elorduy de Conconi 1981: 569, Table I (listed) Mexico, Hidalgo
Cicadidae Shcherbakov 1981a: 830, 834, 841 (listed, key)
Cicadidae Shcherbakov 1981b: 66, 70, 78 (listed, key)
Cicadidae Simon and Lloyd 1981: 275 (listed)
Cicadidae Soper 1981: 53, 55 (listed)
Cicadidae Young 1981b: 175 (comp. note)
Tibicinidae Boulard 1982e: 179 (comp. note)
Cicadidae Hayashi 1982b: 187–188 (diversity, distribution) Japan
Cicadidae Hill, Hore and Thornton 1982: 41, 43, 46, 49–50, 159 (listed) Hong Kong
Cicadidae King and Moody 1982: 79, Appendix 3, Table 2 (predation, listed) New Zealand
Cicadidae Koçak 1982: 148 (listed) Turkey
Cicadidae Nast 1982: 310 (listed)
Cicadidae Rosenberg, Ohmart and Anderson 1982: 262, 270 (predation, comp. note) Arizona
Cicadidae Shcherbakov 1982a: 534 (key)
Cicadidae Shcherbakov 1982b: 76 (key)
Cicadidae Young 1982: 103 (listed) Costa Rica

Cicadidae Heath 1983: 12, 14 (comp. note)
Cicadidae O'Connor, Nash and Anderson 1983: 81 (listed)
Cicadidae Panov and Krjutchkova 1983: 147–148 (comp. note) Crimea, Azerbaijan
Cicadidae Ramos 1983: 62 (listed) Dominican Republic
Cicadidae Strümpel 1983: 4, 17–18, 71, 106, 116, 129, 135, 170, Table 5, Table 10 (listed, key, comp. note)
Tibicinidae Strümpel 1983: 4, 17 (listed, key, comp. note)
Platypediidae Strümpel 1983: 17–18 (key, comp. note)
Plautillidae Strümpel 1983: 17–18, 108 (key, comp. note)
Cicadidae Whalley 1983: 139, 142 (listed, comp. note) Britain
Cicadidae Wolda 1983a: 114 (listed) Panama
Cicadidae Young and Josephson 1983a: 183 (comp. note)
Cicadidae Young and Josephson 1983b: 197 (comp. note)
Cicadidae Emmrich 1984: 114 (comp. note)
Cicadidae Hayashi 1984: 25–27, Fig 1a–e, Table 1 (comp. note)
Plautillidae Hayashi 1984: 26, Table 1 (comp. note)
Platypediidae Hayashi 1984: 26, Table 1 (comp. note)
Tettigadidae Hayashi 1984: 26, Table 1 (comp. note)
Cicadidae He 1984: 226 (comp. note) China
Tibicenidae Melville and Sims 1984: 163 (status, history)
Cicadidae Popova and Dubovskaya 1984: 3 (listed) Uzbekistan
Cicadidae Reichhoff-Riehm 1984: 90 (comp. note)
Cicadidae Reis, Souza and Melles 1984: 4, Table 1 (coffee pest, listed) Brazil
Cicadidae Schönefeld and Göllner-Scheiding 1984: 71 (listed) Lombok
Cicadidae Young 1984b: 358 (comp. note)
Cicadidae Young 1984a: 163, 193 (listed) Costa Rica
Cicadidae Claridge 1985: 299, 310 (comp. note)
Tibicinidae Hamilton 1985: 211 (status, comp. note)
Cicadidae Michelsen and Larsen 1985: 548 (listed)
Cicadidae Miller 1985: 18 (listed) California, Channel Islands
Cicadidae Popov 1985: 19, 47 (listed, comp. note)
Cicadidae Popov, Aronov and Sergeeva 1985a: 451, 454 (listed, comp. note)
Cicadidae Popov, Aronov and Sergeeva 1985b: 288, 291 (listed, comp. note)
Cicadidae Sander 1985: 593 (listed) Bulgaria
Cicadidae Theron 1985: 156, 160–161 (diversity, listed, comp. note) South Africa
Plautillidae Boulard 1986c: 203 (stridulatory apparatus, comp. note)
Tibicinidae Boulard, 1986d: 231, 233, 235 (listed)
Platypediidae Boulard 1986d: 224 (comp. note)
Cicadidae Kirillova 1986a: 122 (listed)
Cicadidae Kirillova 1986b: 45 (listed)
Cicadidae Schedl 1986a: 3 (listed, key) Istria
Tibicinidae Schedl 1986a: 3–4 (listed, key) Istria
Cicadidae Davidson and Lyon 1987: 35 (comp. note)
Cicadidae Dlabola 1987: 305 (listed)
Cicadidae Hayashi 1987a: 119–120 (comp. note) Japan
Cicadidae Jamieson 1987: 155–156 (sperm anatomy, comp. note)
Tibicenidae Martinelli and Zucchi 1987b: 469 (listed) Brazil
Cicadidae Martinelli and Zucchi 1987c: 470 (listed) Brazil
Tibicinidae Martinelli and Zucchi 1987c: 470 (listed) Brazil
Cicadidae Moeed and Meads 1987b: 200, Table 1 (listed) New Zealand
Cicadidae Peña 1987: 108–110 (key, described) Chile
Cicadidae Remane 1987: 288, 337 (listed, comp. note) Germany
Cicadidae Seki, Fujishita, Ito, Matsuoka and Tsukida 1987: 102, Table 2 (listed, comp. note)
Cicadidae Toolson 1987: 384 (comp. note)
Cicadidae Villet 1987a: 272, Table 2 (listed) South Africa
Tibicinidae Villet 1987a: 272, Table 2 (listed) South Africa
Cicadidae Villet 1987b: 209 (comp. note)
Cicadidae Walker and Pittaway 1987: 38 (listed)
Cicadidae White 1987: 149, Fig 2 (listed) New Zealand
Cicadidae Williams 1987: 424 (listed)
Cicadidae Wilson 1987: 486 (sugarcane pest, comp. note)
Cicadidae Boulard 1988f: 36, 44, 51, 65–66 (key, comp. note)
Tibicinidae Boulard 1988f: 36, 44–45, 51 (key, comp. note)
Tibicenidae Boulard 1988f: 72 (comp. note)
Plautillidae Boulard 1988f: 51 (key)
Tibicinidae Boulard and Riou 1988: 349 (comp. note)
Cicadidae De Zayas 1988: 11 (key, characteristics, listed) Cuba
Platypediidae Duffels 1988d: 75 (comp. note)
Cicadidae Duffels 1988d: 78 (comp. note)
Cicadidae Gwynne and Schatral 1988: 25, 43 (listed, comp. note)

Cicadidae Lithgow 1988: 65 (listed) Australia, Queensland
Cicadidae Lodos and Kalkandelen 1988: 18 (listed) Turkey
Cicadidae Maes and Téllez 1988: 66, 82 (listed) Nicaragua
Cicadidae Moeed and Meads 1988: 488 (listed) New Zealand
Cicadidae Ohmart, Anderson and Hunter 1988: 94, Table 1 (ecology, listed) Arizona
Cicadidae Ramos 1988: 62–63, Table 4.1 (listed, comp. note) Greater Antilles
Cicadidae Shen 1988: 165 (listed) China
Cicadidae Tishechkin 1988: 7 (listed) Russia
Cicadidae Samson, Evans and Latgé 1988: 40 (listed)
Cicadidae Santos 1988: 27 (listed) Brazil, Paraná
Cicadidae Southcott 1988: 103 (comp. note)
Cicadidae Steward, Smith and Stephen 1988: 348 (listed) Arkansas
Cicadidae Williams and Williams 1988: 3 (listed) Mascarene Islands
Tibicinidae Boulard 1989d: 65 (listed)
Tibicinidae Dietrich 1989: 146, Table 1 (listed)
Cicadidae spp. Ewart 1989b: 289, 301 (comp. note)
Tibicinidae spp. Ewart 1989b: 289 (comp. note)
Tettigadidae spp. Ewart 1989b: 289 (comp. note)
Plautillidae spp. Ewart 1989b: 289 (comp. note)
Platypediidae spp. Ewart 1989b: 289 (comp. note)
Cicadidae Moss 1989: 105 (comp. note)
Cicadidae Noyes and Valentine 1989a: 36 (listed) New Zealand
Cicadidae Popov 1989a: 291 (comp. note)
Cicadidae Popov 1989b: 62 (comp. note)
Cicadidae Schedl 1989: 109 (listed)
Tibicinidae Schedl 1989: 110 (listed)
Tibicinidae Boulard 1990a: 237 (characteristics)
Cicadidae Boulard 1990b: 92, 112, 142–144, 210 (comp. note)
Tibicinidae Boulard 1990b: 92, 142, 209, Fig 12 (comp. note)
Tettigadidae Boulard 1990b: 92 (comp. note)
Plautillidae Boulard 1990b: 143–144, 148, 159 (comp. note)
Tibicinidae Boulard and Nel 1990: 38 (comp. note)
Cicadidae Carver 1990: 260 (comp. note)
Cicadidae Colbourne, Baird, and Jolly 1990: 536–540, Figs 2–3, Fig 4F, Fig 5, Table 1 (predation, comp. note) New Zealand
Cicadidae Duffels 1990b: 65 (listed)
Tibicinidae Duffels 1990b: 65 (listed)
Cicadidae Martin and Simon 1990b: 362, Table 2 (comp. note)
Cicadidae Moulds 1990: 30–31, 45, 52 (described, listed, key, comp. note) Australia
Tibicinidae Moulds 1990: 30 (comp. note)
Tettigadidae Moulds 1990: 30 (comp. note)
Plautillidae Moulds 1990: 30 (comp. note)
Platypediidae Moulds 1990: 30 (comp. note)
Ydiellidae Moulds 1990: 30 (comp. note)
Cicadidae Popov 1990a: 302 (comp. note)
Cicadidae Stephen, Wallis and Smith 1990: 369 (listed) Arkansas
Cicadidae Young 1990: 42 (comp. note)
Cicadidae Young and Kritsky 1990: 25 (comp. note) Indiana
Cicadidae Zaidi, Mohammedsaid and Yaakob 1990: 260–262, 265, 268, Table 1 (listed, comp. note) Malaysia
Tibicinidae Zaidi, Mohammedsaid and Yaakob 1990: 261–262, 265, 268, Table 1 (listed, comp. note) Malaysia
Tibicinidae Boulard 1991e: 259 (comp. note)
Tibicinidae Boulard 1991g: 74 (comp. note)
Cicadidae Emeljanov and Kirillova 1991: 798–800, Fig 3, Table 2 (chromosome number, phylogeny, comp. note)
Cicadidae Carver, Gross and Woodward 1991: 431, 435, 443 (listed, comp. note)
Cicadidae Dolling 1991: 7–8, 64, 83, 143–145, 151, Table 3 (key, listed, comp. note)
Cicadidae Duffels 1991a: 128–129 (comp. note)
Cicadidae Moulds and Carver 1991: 465 (key, described, comp. note)
Cicadidae Liu 1991: 254 (comp. note)
Cicadidae Moulds and Carver 1991: 465–466 (key, composition, comp. note) Australia
Cicadidae Patrick 1991: 3 (listed) New Zealand
Cicadidae Pigott 1991: 1189 (listed)
Cicadidae Popov, Aronov and Sergeeva 1991: 294 (comp. note)
Cicadidae Sanborn 1991: 5753B (comp. note)
Cicadidae Williams and Smith 1991: 289 (comp. note)
Cicadidae Wilson and Hilburn 1991: 415 (listed) Bermuda
Cicadidae Young 1991: 251 (comp. note) Costa Rica
Tibicinidae Boulard 1992a: 119 (comp. note)
Tibicinidae Boulard 1992b: 365 (comp. note)
Tibicinidae Boulard 1992c: 65 (comp. note)
Cicadidae Carpenter 1992: 221 (described, comp. note)
Cicadidae Dean and Milton 1992: 74 (comp. note) South Africa
Cicadidae Emeljanov and Kirillova 1992: 61, 63, 65, Fig 3, Table 2 (chromosome number, phylogeny, comp. note)
Cicadidae Liu 1992: 13 (comp. note)

Cicadidae Messner and Adis 1992: 714–717, 719, Fig 1, Fig 3, Fig 5, Figs 7–8 (cuticle, illustrated, listed, comp. note)
Cicadidae Moalla, Jardak and Ghorbel 1992: 34, 38 (listed, comp. note)
Tibicinidae Moalla, Jardak and Ghorbel 1992: 34 (listed)
Cicadidae Peng and Lei 1992: 97 (listed) China, Hunan
Cicadidae Phillips 1992: 251 (listed)
Cicadidae Popov, Arnov and Sergeeva 1992: 164 (comp. note)
Cicadidae Wheeler, Williams and Smith 1992: 340 (listed)
Cicadidae Wolda and Ramos 1992: 272–273, Tables 17.1–17.2 (listed) Panama
Tibicinidae Wolda and Ramos 1992: 272–273, Tables 17.1–17.2 (listed) Panama
Platypediidae Boulard 1993d: 92 (comp. note)
Tibicinidae Duffels 1993: 1224–1225 (comp. note)
Tettigadidae Duffels 1993: 1225 (comp. note)
Cicadidae Glare, O'Callaghan, and Wigley 1993: 98, 109 (listed) New Zealand
Cicadidae Hayashi 1993: 13 (comp. note)
Cicadidae Helfert and Sänger 1993: 37, 40 (comp. note)
Cicadidae Hogue 1993b: 233–234, 570, Fig 8.6c (illustrated, listed, comp. note)
Cicadidae Kahono 1993: 203, 213, Table 5 (listed, comp. note) Irian Jaya
Tibicinidae Lane 1993: 52 (listed) New Zealand
Cicadidae McGavin 1993: 37, 70, 78, 89–90, 91, 140, 143, 145, 160–161, 177 (wing, illustrated, predation, comp. note)
Cicadidae Naumann 1993: 3–4, 6–9, 12, 15, 19–20, 22–23, 25–26, 34–35, 37, 40–42, 47–48, 50, 52–53, 58, 60–62, 64–65, 67, 70, 72–76, 80–81, 84, 92–93, 95, 100–101, 104, 113, 118, 125, 127, 130, 148 (listed) Australia
Cicadidae Pillet 1993: 100 (listed) Switzerland
Tibicinidae Pillet 1993: 104 (listed) Switzerland
Cicadidae Press and Whittaker 1993: 101, 103, 106 (comp. note)
Cicadidae Remane and Wachman 1993: 75 (key, comp. note)
Tibicinidae Remane and Wachman 1993: 75 (key, comp. note)
Cicadidae Schedl 1993: 797 (listed) Egypt
Tibicinidae Schedl 1993: 798 (listed) Egypt
Cicadidae Wang and Zhang 1993: 188 (listed) China
Tettigadidae Wang and Zhang 1993: 190 (listed) China
Tibicinidae White and Sedcole 1993a: 38 (comp. note) New Zealand
Cicadidae Williams, Smith and Stephen 1993: 1143 (listed)
Cicadidae Wolda 1993: 370–371 (listed) Panama
Tibicinidae Wolda 1993: 371 (listed) Panama
Cicadidae Zaidi 1993: 958–959, Table 1 (listed, comp. note) Borneo, Sarawak
Tibicinidae Zaidi 1993: 958–959, Table 1 (listed, comp. note) Borneo, Sarawak
Cicadidae Carver, Gross and Woodward 1994: 320, 327 (listed, comp. note)
Cicadidae Gogala 1994a: 31–32 (comp. note) Thailand
Cicadidae Lei, Chou and Li 1994: 54 (comp. note)
Cicadidae Maes 1994: 6 (listed, coffee pest) Nicaragua
Cicadidae Nickel 1994: 537 (listed) Germany
Cicadidae Quartau and Rebelo 1994: 137 (listed) Portugal
Cicadidae Shen 1994: 772 (listed) China
Cicadidae Zaidi and Ruslan 1994: 424, 427, 429, Table 2 (comp. note) Malaysia
Tibicinidae Zaidi and Ruslan 1994: 426–427, 429, Table 2 (comp. note) Malaysia
Cicadidae Boulard 1995a: 12–13 (listed, comp. note)
Tibicinidae Boulard 1995a: 12 (listed)
Cicadidae Boulard and Mondon 1995: 65–66, 72 (key, listed, comp. note) France
Tibicinidae Boulard and Mondon 1995: 65–66, 72 (key, listed, comp. note) France
Cicadidae von Dohlen and Moran 1995: 213, Fig 2 (listed, phylogeny, comp. note)
Cicadidae Lee 1995: 17, 21, 51 (listed, comp. note) Korea
Cicadidae Mirzayans 1995: 14 (listed) Iran
Cicadidae Raina, Pannell, Kochansky and Jaffe 1995: 929 (listed) United States
Cicadidae Riede and Kroker 1995: 43 (comp. note) Sabah, Malaysia
Cicadidae Sorensen, Campbell, Gill and Steffen-Campbell 1995: 32, 34, 52 (listed, comp. note)
Cicadidae Van Pelt 1995: 19 (listed) Texas
Cicadidae Wheeler, Schuh and Bang 1995: 123, Table 1 (listed)
Cicadidae Williams and Simon 1995: 269 (listed)
Cicadidae Xu, Ye and Chen 1995: 84 (listed) China, Zhejiang
Cicadidae Yukawa and Yamane 1995: 693, 696 (listed) Indonesia, Krakataus
Cicadidae Zaidi and Ruslan 1995a: 169, 171–172, 174, Tables 1–2 (listed, comp. note) Borneo, Sarawak
Tibicinidae Zaidi and Ruslan 1995a: 169, 171–172, 174, Table 1–2 (listed, comp. note) Borneo, Sarawak

Cicadidae Zaidi and Ruslan 1995b: 218–221, Tables 1–2 (listed, comp. note) Borneo, Sabah
Tibicinidae Zaidi and Ruslan 1995b: 218–221, Tables 1–2 (listed, comp. note) Borneo, Sabah
Cicadidae Zaidi and Ruslan 1995c: 64, 68, 70, Tables 1–2 (listed, comp. note) Peninsular Malaysia
Tibicinidae Zaidi and Ruslan 1995c: 64, 69–70, Tables 1–2 (listed, comp. note) Peninsular Malaysia
Cicadidae Zaidi and Ruslan 1995d: 199–200, 202, Tables 1–2 (listed, comp. note) Borneo, Sabah
Tibicinidae Zaidi and Ruslan 1995d: 199–200, 202, Tables 1–2 (listed, comp. note) Borneo, Sabah
Cicadidae Zborowski and Storey 1995: 87 (illustrated, comp. note)
Cicadidae Boulard 1996c: 96 (characteristics, comp. note)
Cicadidae Boulard 1996d: 114, 115 (listed)
Tibicinidae Boulard 1996d: 115, 146 (listed)
Cicadidae Boulard and Martinelli 1996: 12 (comp. note)
Cicadidae Gogala, Popov and Ribaric 1996: 46 (comp. note)
Tibicinidae Gogala, Popov and Ribaric 1996: 46 (comp. note)
Cicadidae Holzinger 1996: 505, 512 (listed, diversity) Austria
Cicadidae Hoy and Robert 1996: 436, Table 1 (comp. note)
Cicadidae Moore 1996: 221 (listed) Mexico
Cicadidae Nel 1996: 84, 89–92 (comp. note)
Cicadidae New 1996: 88, 95 (listed, comp. note) Australia
Cicadidae Nickel and Sander 1996: 159 (listed) Germany
Cicadidae Pogue 1996: 313–314 (listed, comp. note) Peru
Cicadidae Riede 1996: 80 (listed) Sabah, Malaysia
Cicadidae Zaidi 1996: 100–101, 103, Tables 1–2 (listed, comp. note) Borneo, Sarawak
Tibicinidae Zaidi 1996: 103, Table 2 (listed, comp. note) Borneo, Sarawak
Cicadidae Zaidi and Hamid 1996: 52–55, Tables 1–2 (listed, comp. note) Borneo, Sarawak
Tibicinidae Zaidi and Hamid 1996: 54–55, Table 2 (listed, comp. note) Borneo, Sarawak
Cicadidae Zaidi, Ruslan and Mahadir 1996: 59–60, Table 1 (listed, comp. note) Peninsular Malaysia
Tibicinidae Zaidi, Ruslan and Mahadir 1996: 59–60, Table 1 (listed, comp. note) Peninsular Malaysia
Tibicinidae Boulard 1997a: 180 (comp. note)
Tibicinidae Boulard 1997c: 101, 106 (comp. note)
Tibicenidae Boulard 1997c: 101, 103 (comp. note)
Cicadidae Boulard 1997c: 104–105, 112, 117 (key, comp. note)
Tibiciinidae (sic) Boulard 1997c: 106 (comp. note)
Plautillidae Chou, Lei, Li, Lu and Yao 1997: 30 (listed)
Platypedidae Chou, Lei, Li, Lu and Yao 1997: 30 (listed)
Cicadidae Chou, Lei, Li, Lu and Yao 1997: 357–358 (comp. note)
Cicadidae Holzinger, Frölich, Günthart, Pauterer, Nickel et al. 1997: 49, 61 (listed, diversity, comp. note) Europe
Tibicinidae Holzinger, Frölich, Günthart, Pauterer, Nickel et al. 1997: 49, 61 (listed, diversity, comp. note) Europe
Cicadidae Malaisse 1997: 237, Table 2.11.1 (human food, listed) Zambezi
Cicadidae Matthews and Hildreth 1997: 37 (listed)
Cicadidae Novotny and Wilson 1997: 423, 437 (listed, comp. note)
Tibicinidae Novotny and Wilson 1997: 423, 437 (listed, comp. note)
Cicadidae Poole, Garrison, McCafferty, Otte and Stark 1997: 251, 257–258, 331 (listed, diversity, comp. note) North America
Platypediidae Poole, Garrison, McCafferty, Otte and Stark 1997: 257 (listed) North America
Plautillidae Poole, Garrison, McCafferty, Otte and Stark 1997: 257 (listed) North America
Tettigadidae Poole, Garrison, McCafferty, Otte and Stark 1997: 258 (listed) North America
Tibicinidae Poole, Garrison, McCafferty, Otte and Stark 1997: 258 (listed) North America
Cicadidae Riede 1997a: 276 (comp. note) Sabah, Malaysia
Cicadidae Riede 1997b: 445 (comp. note) Sabah, Malaysia
Cicadidae Schiemenz 1997: 43 (listed) Germany
Cicadidae Shelly and Whittier 1997: 279, Table 16–1 (listed)
Cicadidae Waterhouse 1997: 12, Table 1 (agricultural pests, listed)
Cicadidae Zaidi 1997: 112–113, 115, Tables 1–2 (listed, comp. note) Borneo, Sarawak
Tibicinidae Zaidi 1997: 115, Table 2 (listed, comp. note) Borneo, Sarawak
Cicadidae Zaidi and Ruslan 1997: 219 (listed, comp. note)
Tibicinidae Zaidi and Ruslan 1997: 230 (listed, comp. note)
Cicadidae Henríquez 1998: 86, Fig 68 (illustrated)

Cicadidae Maes 1998: 143 (listed, nymph illustrated, described) Nicaragua
Cicadidae Martinelli, Matuo, Yamada and Malheiros 1998: 134 (listed) Brazil
Cicadidae Schiemenz 1998: 37 (listed) Germany
Cicadidae Schönitzer and Oesterling 1998: 34 (listed) Germany
Cicadidae Zaidi and Ruslan 1998a: 345 (listed)
Tibicinidae Zaidi and Ruslan1998a: 367 (listed)
Cicadidae Zaidi and Ruslan 1998b: 2, 4–5, Tables 1–2 (listed, comp. note) Borneo, Sabah
Tibicinidae Zaidi and Ruslan 1998b: 3–45, Tables 1–2 (listed, comp. note) Borneo, Sabah
Cicadidae Boulard and Riou 1999: 136 (listed)
Tibicinidae Boulard and Riou 1999: 136 (listed)
Cicadidae Claridge, Morgan and Moulds 1999a: 1831 (comp. note)
Cicadidae Hayashi and Moulds 1999: 97, 100 (comp. note)
Cicadidae Holzinger 1999: 431 (listed) Austria
Tibicinidae Holzinger 1999: 431 (listed) Austria
Cicadidae Jamieson, Dallai and Afzelius 1999: 233–236 (sperm anatomy, comp. note)
Cicadidae Lee 1999: 1 (comp. note) Korea
Cicadidae Moulds 1999: 1 (comp. note)
Cicadidae Ohbayashi, Sato and Igawa 1999: 339 (comp. note) Ogasawara (Bonin) Islands
Cicadidae Sanborn 1999a: 35 (listed)
Tibicinidae Sanborn 1999a: 44 (listed)
Platypediidae Sanborn 1999a: 53 (listed)
Cicadidae Sanborn 1999b: 1 (listed) North America
Tibicinidae Sanborn 1999b: 1 (listed) North America
Platypediidae Sanborn 1999b: 2 (listed) North America
Cicadidae Schedl 1999b: 823–824, 830 (listed, comp. note) Israel
Tibicinidae Schedl 1999b: 823, 827, 830 (listed, comp. note) Israel
Cicadidae Villet 1999a: 151 (comp. note)
Cicadidae Villet 1999e: 1 (diversity, comp. note) Africa
Cicadidae Villet and Zhao 1999: 321 (comp. note) China
Cicadidae Zaidi and Ruslan 1999: 2–4, Tables 1–2 (listed, comp. note) Borneo, Sarawak
Tibicinidae Zaidi and Ruslan 1999: 2–3, 5, Tables 1–2 (listed, comp. note) Borneo, Sarawak
Cicadidae Zaidi, Ruslan and Azman 1999: 300 (listed) Borneo, Sabah
Tibicinidae Zaidi, Ruslan and Azman 1999: 318 (listed) Borneo, Sabah
Cicadidae Boulard 2000i: 21, 23, 29, 35, 45, 73 (listed, key) France
Cicadidae Guglielmino, d'Urso and Alma 2000: 163, 167 (diversity, listed, comp. note) Sardinia
Cicadidae Hangay and German 2000: 58 (listed)
Cicadidae Hawkeswood 2000: 419 (listed)
Cicadidae Hua 2000: 60 (listed) China
Tettigadidae Hua 2000: 65 (listed) China
Cicadidae Martin and Webb 2000: 56 (listed) Britain
Cicadidae Maw, Footit, Hamilton and Scudder 2000: 49 (listed)
Cicadidae Moss 2000: 68 (comp. note) Queensland
Cicadidae Noda, Kubota, Miyata and Miyahara 2000: 174 (listed)
Cicadidae Punzo 2000: 97, Table 6 (listed)
Cicadidae Schedl 2000: 257–258, 262 (key, listed, comp. note)
Tibicinidae Schedl 2000: 257–258, 264 (key, listed, comp. note)
Cicadidae Vokoun 2000: 261 (listed) Missouri
Cicadidae Zaidi, Noramly and Ruslan 2000a: 320 (listed, comp. note) Borneo, Sabah
Tibicinidae Zaidi, Noramly and Ruslan 2000a: 320 (listed, comp. note) Borneo, Sabah
Cicadidae Zaidi, Noramly and Ruslan 2000b: 337–338, Table 1 (listed, comp. note) Borneo, Sabah
Tibicinidae Zaidi, Noramly and Ruslan 2000b: 337 (comp. note) Borneo, Sabah
Cicadidae Zaidi, Ruslan and Azman 2000: 198 (listed) Borneo, Sabah
Tibicinidae Zaidi, Ruslan and Azman 2000: 216 (listed) Borneo, Sabah
Cicadidae Boulard 2001b: 25–26, 32, 37 (key, comp. note)
Tibicenidae Boulard 2001b: 21, 24 (comp. note)
Tibicinidae Boulard 2001b: 21, 26 (comp. note)
Tibiciinidae (sic) Boulard 2001b: 26 (comp. note)
Tibicinidae Buckley, Simon, Shimodaira and Chambers 2001: 223 (listed)
Cicadidae Dietrich, Rakitov, Holmes and Black 2001: 296, 300–301, Figs 3–4, Table 1 (comp. note) Illinois
Cicadidae Howard, Moore, Giblin-Davis and Abad 2001: 129, 157 (listed, diversity)
Cicadidae Puissant and Sueur 2001: 429 (comp. note) Corsica
Cicadidae Sanborn 2001b: 449 (listed) El Salvador
Tibicinidae Sanborn 2001b: 449 (listed) El Salvador
Cicadidae Schedl 2001b: 27 (comp. note)
Tibicinidae Schedl 2001b: 27 (comp. note)
Cicadidae Sueur 2001: 36, Table 1 (listed)
Cicadidae Whiles, Callaham, Meyer, Brock and Charleton 2001: 177 (comp. note) Kansas
Cicadidae Yaylayan, Paré, Manti and Bèlanger 2001: 188 (listed)

Cicadidae Yoshizawa and Saigusa 2001: 12 (listed)
Cicadidae Zaidi, Azman and Ruslan 2001a: 112, 115, 117, Table 1 (listed, comp. note) Langkawi Island
Tibicinidae Zaidi, Azman and Ruslan 2001a: 112, 116, 121, Table 1 (listed, comp. note) Langkawi Island
Cicadidae Zaidi, Azman and Ruslan 2001b: 124, 125 (listed, comp. note) Borneo, Sarawak
Cicadidae Zaidi and Nordin 2001: 183, 185, 189, Table 1 (listed, comp. note) Borneo, Sabah
Tibicinidae Zaidi and Nordin 2001: 183, 185, 191, Table 1 (listed, comp. note) Borneo, Sabah
Cicadidae Boulard 2002b: 175, 197 (listed, comp. note)
Platypediidae Boulard 2002b: 205 (comp. note)
Cicadidae Dietrich 2002: 157, 159, Fig 1 (phylogeny, listed, comp. note)
Cicadidae Duffels and Turner 2002: 235 (comp. note)
Tibicinidae Duffels and Turner 2002: 235 (comp. note)
Plautillidae Duffels and Turner 2002: 235 (comp. note)
Platypediidae Duffels and Turner 2002: 235 (comp. note)
Tettigadidae Duffels and Turner 2002: 235 (comp. note)
Cicadidae Ellingson, Anderson and Kondratieff 2002: 284, 287 (comp. note)
Tibicinidae Ellingson, Anderson and Kondratieff 2002: 284, 287 (comp. note)
Cicadidae Jeon, Kim, Tripotin and Kim 2002: 241 (parasitism, comp. note)
Cicadidae Kondratieff, Ellingson and Leatherman 2002: 13, Table 2 (listed) Colorado
Platypediidae Kondratieff, Ellingson and Leatherman 2002: 47 (comp. note)
Tibicinidae Kondratieff, Ellingson and Leatherman 2002: 13, Table 2 (listed) Colorado
Cicadidae Moulds and Carver 2002: 1–3, 54 (described, listed, comp. note) Australia
Plautilidae (sic) Moulds and Carver 2002: 1 (comp. note)
Tibicinidae Moulds and Carver 2002: 1 (comp. note)
Cicadidae Nation 2002: 244 (listed)
Cicadidae Nickel and Remane 2002: 37 (listed) Germany
Cicadidae Paião, Menegium, Casagrande and Leite 2002: S67 (listed)
Cicadidae Perepelov, Burgov and Marymanska-Nadachowska 2002: 217 (karyotype, comp. note) Japan
Cicadidae Picker, Griffith and Weaving 2002: 156 (listed, comp. note) South Africa
Cicadidae O'Geen, McDaniel and Busacca 2002: 1584 (comp. note)
Cicadidae Schedl 2002: 232 (listed, comp. note) Austria
Tibicinidae Schedl 2002: 233 (listed, comp. note) Austria
Cicadidae Stölting, Moore and Lakes-Harlan 2002: 2 (listed)
Cicadidae Williams 2002: 37–38, 156–157 (listed, comp. note) Australia
Tibicinidae Williams 2002: 38 (comp. note) Australia
Cicadidae Zaidi, Nordin, Maryati, Wahab, Norashikin et al. 2002: 2, 8, 10, Tables 1–2 (listed, comp. note) Borneo, Sabah
Tibicinidae Zaidi, Nordin, Maryati, Wahab, Norashikin et al. 2002: 7, 10, Tables 1–2 (listed, comp. note) Borneo, Sabah
Cicadidae Dietrich 2003: 66–67, 72, Table 1 (listed, comp. note) Equals Platypediidae Equals Plautillidae Equals Tettigadidae Equals Tibicinidae
Tibicinidae Dietrich 2003: 72 (comp. note)
Cicadidae Holzinger, Kammerlander and Nickel 2003: 32, 67, 475–476, 478, Fig 9, Fig 259 (phylogeny, key, listed, diversity, key, illustrated, comp. note) Europe
Cicadidae Ibanez 2003: 3–4 (listed, comp. note) France
Tibicinidae Ibanez 2003: 3–4 (listed, comp. note) France
Cicadidae Kubo-Irie, Irie, Nakazawa and Mohri 2003: 990 (comp. note)
Cicadidae Lee and Hayashi 2003a: 152 (listed) Taiwan
Cicadidae Moulds 2003a: 186–187 (comp. note)
Cicadidae Nickel 2003: 70 (listed) Germany
Cicadidae Nickel and Remane 2003: 138 (listed) Germany
Cicadidae Salguero 2003: 30 (listed) Guatemala
Cicadidae Schedl 2003: 423 (comp. note)
Tibicinidae Schedl 2003: 423, 426 (comp. note)
Cicadidae Sueur 2003: 2931 (comp. note)
Cicadidae Tishechkin 2003: 141, 144–146 (listed, comp. note)
Cicadidae Zaidi and Azman 2003: 104, Table 1 (listed, comp. note) Borneo, Sabah
Tibicinidae Zaidi and Azman 2003: 106, Table 1 (listed, comp. note) Borneo, Sabah
Cicadidae Zborowski and Storey 2003: 87 (illustrated, comp. note)
Cicadidae Chen 2004: 38–40 (phylogeny, listed)
Cicadidae Gogala, Trilar, Kozina and Duffels 2004: 2 (listed)

Cicadidae Hirai 2004: 376 (lsited) Japan
Cicadidae Hutchins, Evans, Garrison and Schlager 2004: 269 (listed)
Cicadidae Martinelli 2004: 517 (listed) Brazil
Tibicinidae Martinelli 2004: 517–518 (listed) Brazil
Cicadidae Nuorteva, Lodenius, Nagasawa, Tulisalo and Nuorteva 2004: 52, 54–57, Tables 1–2 (listed, comp. note) Japan
Cicadidae Öztürk, Ulusoy, Erkiliç and S. (Ö.) Bayhan 2004: 5 (listed) Turkey
Cicadidae Pham 2004: 61 (comp. note) Vietnam
Cicadidae Prešern, Gogala and Trilar 2004a: 240 (listed)
Cicadidae Sanborn 2004c: 511 (natural history, comp. note)
Tibicinidae Sanborn 2004c: 511 (comp. note)
Platypediidae Sanborn 2004c: 511 (comp. note)
Tettigadidae Sanborn 2004c: 511 (comp. note)
Plautillildae Sanborn 2004c: 511 (comp. note)
Cicadidae Sanborn, Heath, Heath and Phillips 2004: 66 (listed)
Tibicinidae Sanborn, Heath, Heath, Noriega and Phillips 2004: 282 (comp. note)
Tettigadidae Sanborn, Heath, Heath, Noriega and Phillips 2004: 282 (comp. note)
Cicadidae Sanborn, Heath, Heath, Noriega and Phillips 2004: 282 (comp. note)
Cicadidae Schedl 2004a: 12 (listed) Yemen
Tibicinidae Schedl 2004a: 12 (listed) Yemen
Cicadidae Schedl 2004b: 914 (listed)
Tibicinidae Schedl 2004b: 914 (listed)
Cicadidae Schouten, Duffels and Zaidi 2004: 372–373 (listed) Malayan Peninsula
Tibicinidae Schouten, Duffels and Zaidi 2004: 373, 378, Table 2 (listed) Malayan Peninsula
Cicadidae Stokes and Josephson 2004: 279 (comp. note)
Cicadidae Sueur and Aubin 2004: 217 (comp. note)
Cicadidae Wakakuwa, Ozaki and Aarikawa 2004: 1480 (listed) Japan
Cicadidae Witsack and Nickel 2004: 232 (listed) Germany
Cicadidae Zhao 2004: 471, 477 (listed) China
Cicadidae Boulard 2005b: 112 (comp. note)
Tibicinidae Boulard 2005d: 333 (comp. note) Africa, North America
Cicadidae Boulard 2005d: 332 (comp. note)
Cicadidae Chawanji, Hodson and Villet 2005a: 266–267 (comp. note)
Cicadidae Claridge 2005: I-4 (comp. note)
Cicadidae Cryan 2005: 563 (comp. note)
Cicadidae Dietrich 2005: 510 (key) Equals Platypediidae Equals Plautillidae Equals Tettigadidae Equals Tibicinidae
Cicadidae Ewart 2005b 441 (listed)
Cicadidae Gogala, Trilar and Krpač 2005: 105 (listed) Macedonia
Tibicinidae Gogala, Trilar and Krpač 2005: 113 (listed) Macedonia
Cicadidae Kondratieff, Schmidt, Opler and Garhart 2005: 95, 215 (listed, comp. note)
Cicadidae Lee 2005: 156, 158 (key, listed, comp. note) Korea
Cicadidae Macnamara 2005: 42, 48 (listed) Illinois
Cicadidae Marshall, Cooley, Hill and Simon 2005: S-21 (listed) New Zealand
Cicadidae Moran, Tran and Gerardo 2005: 8805–8806, 8808, Figs 1–2, Fig 4 (comp. note)
Cicadidae Moulds 2005b: 377, 384, 386–389, 396–397, 404, 410, 412–415, 421, 426, 429, 430, 433, Table 2 (described, key, history, phylogeny, listed, comp. note) Australia
Tettigadidae Moulds 2005b: 384–386, 397, 416, 426, Fig 31 (described, key, history, phylogeny, key, listed, comp. note)
Platypediidae Moulds 2005b: 384–389, 393, 396, 416, Fig 31 (described, key, history, phylogeny, listed, comp. note)
Plautillidae Moulds 2005b: 385–388, 396, Fig 31 (described, key, history, phylogeny, listed, comp. note)
Tibicinidae Moulds 2005b: 388–389 (history, listed, comp. note)
Tibicenidae Moulds 2005b: 389 (history, listed, comp. note)
Cicadidae Palmer, Vitelli and Donnelly 2005: 176, Table 1 (listed) Australia, Queensland
Cicadidae Pham 2005a: 216 (comp. note) Vietnam
Cicadidae Pham and Thinh 2005a: 236 (comp. note) Vietnam
Cicadidae Pham and Thinh 2005b: 287 (key, comp. note) Vietnam
Cicadidae Robinson 2005: 208, 217 (listed)
Cicadidae Salazar Escobar 2005: 194 (comp. note) Colombia
Cicadidae Sanborn, Heath, Sueur and Phillips 2005: 191 (comp. note)
Cicadidae Wiesenborn 2005: 656 (listed) Nevada
Cicadidae Yen, Robinson and Quicke 2005a: 198 (listed)
Cicadae (sic) Yen, Robinson and Quicke 2005b: 363, Table 1 (listed)
Cicadidae Zaidi, Azman and Nordin 2005: 27–29, Table 1 (listed, comp. note) Borneo, Sabah
Cicadidae Boulard 2006g: 16 (comp. note)
Cicadidae Chawanji, Hodson and Villet 2006: 374 (comp. note)
Cicadidae Demir 2006: 102 (listed) Turkey
Cicadidae Elliott and Hershberger 2006: 15 (listed)

Cicadidae Marshall 2006: 102–103 (comp. note) North America
Cicadidae Miranda 2006: 384 (listed)
Cicadidae Özdikmen and Kury 2006: 219 (listed)
Cicadidae Pham and Thinh 2006: 525 (listed) Vietnam
Cicadidae Puissant 2006: 19 (listed) France
Cicadidae Sanborn 2006a: 76 (listed)
Cicadidae Sanborn 2006b: 256 (listed)
Cicadidae Simões and Quartau 2006: 52 (listed)
Cicadidae Smith, Kelly and Finch 2006: 1608 (comp. note)
Cicadidae Strauβ and Lakes-Harlan 2006: 2 (listed)
Cicadidae Trilar 2006: 341 (comp. note)
Cicadidae Trilar and Gogala and Szwedo 2006: 313 (listed)
Cicadidae Trilar, Gogala and Popa 2006: 177 (listed)
Tibicinidae Trilar, Gogala and Popa 2006: 177 (listed)
Cicadidae Veiga and Ferrari 2006: 212, Table 1 (listed) Brazil, Amazonia
Cicadidae Vincent 2006: 64 (listed) France
Cicadidae Wang, Ren, Liang, Liu and Wang 2006: 295, 298, Table 1 (listed, comp. note) China
Cicadidae Watanabe and Kobayashi 2006: 224, Table 1 (listed) Japan, Ryukyus
Cicadidae Boulard 2007e: 1, 71–72 (comp. note) Thailand
Cicadidae Demir 2007a: 51 (listed) Turkey
Cicadidae Demir 2007b: 484 (listed) Turkey
Cicadidae De Santis, Medrano, Sanborn and Bolcatto 2007: 2 (listed, characteristics) Argentina
Cicadidae Eaton and Kaufman 2007: 88 (diversity, comp. note) North America
Cicadidae Esson 2007: 21 (comp. note) New Zealand
Cicadidae Ganchev and Potamitis 2007: 288, 304–305, 308, 315, 317, 319–321, Fig 11, Table 1, Tables 7–8 (listed, comp. note)
Cicadidae Nowlin, González, Vanni, Stevens, Fields and Valente 2007: 2174 (listed) North America
Cicadidae Sanborn 2007a: 26 (listed) Venezuela
Cicadidae Sanborn 2007b: 2, 31 (listed) Mexico
Cicadidae Sanborn, Phillips and Sites 2007: 4 (listed) Thailand
Cicadidae Schedl 2007: 153 (listed)
Tibicinidae Schedl 2007: 155 (listed)
Cicadidae Shiyake 2007: 2, 10, 14 (listed, comp. note)
Cicadidae Simões and Quartau 2007: 247–250 (listed)
Cicadidae Söderman 2007: 26 (listed) Finland
Cicadidae Sueur and Puissant 2007a: 127 (comp. note)
Cicadidae Sueur and Puissant 2007b: 55, 61 (comp. note)
Cicadidae Sueur, Vanderpool, Simon, Ouvrard and Bourgoin 2007: 612 (comp. note)
Cicadidae Takakura and Yamazaki 2007: 730 (listed) Japan
Cicadidae Toledo, Marino de Remes Lenicov and López Lastra 2007: 226 (listed)
Cicadidae Trilar and Gogala 2007: 6 (comp. note)
Cicadidae Xia, Bai, Yi, Liu, Chu, Liang et al. 2007: 961 (listed) China
Cicadidae Boulard 2008f: 97 (comp. note) Thailand
Cicadidae Cole 2008: 815 (comp. note)
Cicadidae Costa Neto 2008: 453 (comp. note)
Cicadidae Demir 2008: 475 (listed) Turkey
Tibicinidae Demir 2008: 476 (listed) Turkey
Cicadidae Ewart and Marques 2008: 153 (listed)
Cicadidae Fonseca, Serrão, Pina-Martins, Silva, Mira, et al. 2008: 19, 24–27-30 (listed)
Tibicinidae Fonseca, Serrão, Pina-Martins, Silva, Mira, et al. 2008: 19, 23, 26, 28–29 (listed)
Cicadidae Goble, Price, Barker, and Villet 2008b: 674 (comp. note) Africa
Cicadidae Gogala, Drosopoulos and Trilar 2008a: 91 (comp. note) Greece
Cicadidae Júnior, Almeida, Lima, Nunes, Siqueira et al. 2008: 805, Table 1 (listed) Taiwan
Cicadidae Lee 2008a: 1–3, Table 1 (listed) Vietnam
Cicadidae Marshall 2008: 2785–2786, 2791 (listed, comp. note)
Cicadidae Lee 2008b: 445–446, 468, 462, Table 1 (listed, comp. note) Korea
Cicadidae Moriyama and Numata 2008: 1487 (listed) Japan
Cicadidae Nation 2008: 269 (comp. note)
Cicadidae Nickel 2008: 207 (listed)
Cicadidae Perez-Gerlabert 2008: 202 (listed) Hispaniola
Tibicinidae Perez-Gerlabert 2008: 202 (listed) Hispaniola
Cicadidae Quartau, André, Pinto-Juma, Seabra and Simões 2008: 128 (listed)
Cicadidae Saboori and Lazarboni 2008: 57 (listed) Iran
Cicadidae Sanborn 2008b: 34 (comp. note)
Cicadidae Sanborn 2008c: 874–875 (described, comp. note)
Cicadidae Sanborn 2008d: 3469–3470 (comp. note)
Cicadidae Sanborn 2008f: 686, 688 (listed)
Cicadidae Sanborn, Moore and Young 2008: 3 (listed) Costa Rica
Cicadidae Sanborn, Phillips and Gillis 2008: 5 (listed) Florida
Cicadidae Storm and Whitaker 2008: 353, Table 1 (listed) Indiana
Cicadidae Trilar and Hertach 2008: 185 (listed)

Cicadidae Trilar and Gogala 2008: 29 (listed)
Tibicinidae Trilar and Gogala 2008: 29 (listed)
Cicadidae Vikberg 2008: 60 (listed) Finland
Cicadidae Visser and Ellers 2008: 1316, Table 1 (listed)
Cicadidae Gogala, Drosopoulos and Trilar 2009: 14 (comp. note)
Cicadidae Holliday, Hastings and Coelho 2009: 15 (comp. note)
Cicadidae Ibanez 2009: 208 (listed) France
Cicadidae Lee 2009h: 2617–2618 (listed) Philippines, Palawan
Cicadidae Lee 2009i: 293 (listed) Philippines, Panay
Cicadidae Marshall and Hill 2009: 1 (listed)
Cicadidae Marshall, Hill, Fontaine, Buckley and Simon 2009: 1 (listed)
Cicadidae Moriyama and Numata 2009: 162 (listed) Japan
Cicadidae Moulds 2009a: 504 (listed) Papua New Guinea
Cicadidae Moulds 2009b: 163–164 (comp. note)
Cicadidae Pinto, Quartau, and Bruford 2009: 267, 274 (listed)
Tibicinidae Pinto, Quartau, and Bruford 2009: 267, 274 (listed)
Cicadidae Pham and Yang 2009: 1, 3, 13, 16, Table 2 (listed) Vietnam
Cicadidae Price, Barker and Villet 2009: 618 (comp. note)
Cicadidae Salazar and Sanborn 2009: 270 (listed) Colombia
Cicadidae Sanborn 2009a: 85, 89 (listed, comp. note) Cuba
Cicadidae Sanborn 2009b: 307 (comp. note) Vietnam
Cicadidae Sanborn 2009d: 1 (listed, comp. note) Dominican Republic
Cicadidae Sanborn and Maté-Nankervis 2009: 74, 82, Table 1 (listed, comp. note) Florida
Cicadidae Santos and Martinelli 2009b: 638 (listed)
Cicadidae Shcherbakov 2009: 343–345, 348, Table 1 (comp. note)
Cicadidae Stock, Rivera-Orduño and Flores-Lara 2009: 175 (listed)
Cicadidae Strauβ and Lakes-Harlan 2009: 306 (comp. note)
Cicadidae Wang, Ding, Wheeler, Purcell and Zhang 2009: 1136 (listed) China
Cicadidae Wei, Tang, He and Zhang 2009: 337–338 (listed, comp. note)
Tibicinidae Wei, Tang, He and Zhang 2009: 337 (comp. note)
Platypediidae Wei, Tang, He and Zhang 2009: 337 (comp. note)
Cicadidae Wilson 2009: 35, 48 (listed, comp. note) Fiji
Cicadidae Boulard 2010b: 336 (comp. note)
Cicadidae Cryan and Svenson 2010: 399, 402, Fig 1, Table 1 (phylogeny, comp. note) Australia
Cicadidae Dmitriev 2010: 597 (listed)
Cicadidae Goemans 2010: 3, 5, 7 (listed, comp. note)
Cicadidae Gogala 2010: 82 (listed)
Cicadidae Kemal and Koçak 2010: 1, 3–4, 6 (diversity, listed, comp. note) Turkey
Cicadidae Koçak and Kemal 2010: 135, 148 (listed) Turkey
Cicadidae Lee 2010a: 14–15 (listed, comp. note) Philippines, Mindanao
Cicadidae Lee 2010b: 19–20 (listed, comp. note) Cambodia
Cicadidae Lee 2010c: 1–2 (listed, comp. note) Philippines, Luzon
Cicadidae Lee 2010d: 167 (listed)
Cicadidae Matthews and Matthews 2010: 1084 (listed) Indiana
Cicadidae Moriyama and Numata 2010: 69 (listed) Japan
Cicadidae Mozaffarian and Sanborn 2010: 76 (listed)
Cicadidae Niedringhaus, Biedermann and Nickel 2010: 27 (listed) Luxemburg
Cicadidae Oh, Kim and Kim 2010: 134, Table 1 (listed) Korea
Cicadidae Pham, Thinh and Yang 2010: 63 (listed) Vietnam
Cicadidae Pham and Yang 2010a: 133 (diversity, comp. note) Vietnam
Cicadidae Pham and Yang 2010b: 205 (comp. note) Vietnam
Cicadidae Popple and Emery 2010: 148 (listed) Australia
Cicadidae Puissant and Sueur 2010: 556, 564 (listed)
Cicadidae Saljoqi, Rahimi, Khan and Rehman 2010: 69 (listed, comp. note) Afghanistan
Cicadidae Santos, Martinelli, Maccagnan, Sanborn and Ribeiro 2010: 48, 50 (listed)
Cicadidae Sanborn 2010a: 1579 (listed) Colombia
Cicadidae Sanborn 2010b: 67 (listed)
Cicadidae Trilar and Gogala 2010: 6, 16 (comp. note) Greece
Cicadidae Wei and Zhang 2010: 37–38 (comp. note)

Subfamily Cicadinae Latreille, 1802 = Gaeaninae Distant, 1905 = Daeaninae (sic) = Plautillinae Distant, 1905 = Tibicininae Distant, 1905 = Zammarinae Distant, 1905 = Tibiceninae Van Duzee, 1916 = Platypleurinae Schmidt, 1918 = Platypleurnae (sic) = Lyristinae Gomez-Monor, 1957 = Moaninae Boulard, 1976

Cicadinae Wu 1935: 1 (listed) China
Gaeaninae Wu 1935: 17 (listed) China
Cicadinae Zeuner 1941: 88 (listed)
Platypleurinae Chen 1942: 143 (listed)

Cicadinae Chen 1942: 146 (listed)
Cicadinae Zeuner 1944: 111, 113–115 (listed, comp. note)
Cicadinae Heinrich 1967: 44 (comp. note)
Gaeaninae Heinrich 1967: 44 (comp. note)
Platypleurinae Villiers 1977: 165 (listed, comp. note)
Cicadinae Villiers 1977: 165 (listed, comp. note)
Platypleurinae Medler 1980: 73 (listed) Nigeria
Cicadinae Duffels 1983b: 11 (comp. note)
Tibiceninae Duffels 1983b: 11 (comp. note)
Platypleurinae Strümpel 1983: 18 (comp. note)
Cicadinae Strümpel 1983: 18 (comp. note) Equals Gaeaninae
Moaninae Strümpel 1983: 18 (comp. note)
Tibiceninae Boulard 1984c: 167, 170–171, 174, 176–178 (status)
Cicadinae Boulard 1984c: 178–178 (status, type genus *Cicada*, type species *Cicada orni*)
Gaeaninae Boulard 1984c: 168–169, 179 (type genus *Gaeana*, comp. note)
Platypleurinae Boulard 1984c: 175–176 (type genus *Platypleura*, comp. note)
Cicadinae Hayashi 1984: 26–28, Fig 1a–c, Fig 2, Table 1 (comp. note)
Gaeaninae Hayashi 1984: 26, Table 1 (comp. note)
Platypleurinae Hayashi 1984: 26, 29, Table 1 (listed, comp. note)
Plautillinae Hayashi 1984: 26–27, Fig 1i, Table 1 (comp. note)
Tibiceninae Hayashi 1984: 26, Table 1 (comp. note)
Moaninae Hayashi 1984: 26, Table 1 (comp. note)
Tibiceninae Melville and Sims 1984: 164–165, 181 (status, history)
Lyristinae Melville and Sims 1984: 164–165, 181 (status, history)
Platypleurinae Melville and Sims 1984: 165–166 (status, history)
Cicadinae Boulard 1985d: 212 (status)
Tibiceninae Boulard 1985d: 212 (status)
Cicadinae Boulard 1985e: 1032 (comp. note)
Platypleurinae Boulard 1985e: 1032 (comp. note)
Tibiceninae Hamilton 1985: 211 (status, comp. note)
Cicadinae Moulds 1985: 25 (listed, comp. note)
Tibiceninae Moulds 1985: 26 (comp. note)
Cicadinae Boulard 1986c: 203 (stridulatory apparatus, comp. note)
Gaeaninae Kirillova 1986a: 122 (listed)
Gaeaninae Kirillova 1986b: 45 (listed)
Tibiceninae Kirillova 1986a: 122 (listed)
Tibiceninae Kirillova 1986b: 45 (listed)
Platypleurinae Schedl 1986a: 3–4 (listed, key) Istria
Cicadinae Schedl 1986a: 3–4 (listed, key) Istria
Cicadinae Martinelli and Zucchi 1987b: 469 (listed) Brazil
Platypleurinae Villet 1987b: 209 (comp. note)
Lyristinae Boulard 1988e: 154 (listed)
Tibiceninae Boulard 1988f: 8, 30–35, 72 (comp. note)
Platypleurinae Boulard 1988f: 33–34, 72 (comp. note)
Cicadinae Boulard 1988f: 36, 37, 44, 51 (key, comp. note)
Lyristinae Boulard 1988f: 44–45 (comp. note)
Moaninae Boulard 1988f: 51 (key)
Moaninae Duffels 1988d: 10, 78 (comp. note)
Cicadinae Gwynne and Schatral 1988: 43 (listed)
Tibiceninae Gwynne and Schatral 1988: 43 (listed)
Tibiceninae spp. Ewart 1989b: 289 (comp. note)
Platypleurinae Villet 1989a: 329 (comp. note)
Cicadinae Boulard 1990b: 92, 148, 210 (comp. note)
Lyristinae Boulard 1990b: 92 (comp. note)
Cicadinae Moore 1990: 58 (comp. note)
Lyristinae Moore 1990: 58 (comp. note)
Cicadinae Sueur 2001: 36, 46–47, Tables 1–3 (listed, diversity, distribution, comp. note)
Plautillinae Sueur 2001: 46 (comp. note)
Cicadinae Moulds 1990: 22, 30–31, 52 (listed, key, comp. note) Australia, Tasmania
Gaeaninae Moulds 1990: 30 (comp. note) Australia
Cicadinae Duffels 1991a: 128 (comp. note)
Tibiceninae Duffels 1991a: 128–129 (comp. note)
Gaeaninae Duffels 1991a: 128 (comp. note)
Cicadinae Moulds and Carver 1991: 465, 467 (key, diversity, comp. note) Australia
Cicadinae Moalla, Jardak and Ghorbel 1992: 34, 38 (comp. note)
Platypleurinae Moalla, Jardak and Ghorbel 1992: 34 (comp. note)
Moaninae Duffels 1993: 1223–1224, 1226 (comp. note)
Cicadinae Duffels 1993: 1226 (comp. note)
Tibiceninae Duffels 1993: 1226 (comp. note)
Gaeaninae Duffels 1993: 1226 (comp. note)
Platypleurinae Moore, Huber, Weber, Klein and Bock 1993: 217 (comp. note)
Cicadinae Wang and Zhang 1993: 188 (listed) China
Daeaninae (sic) Wang and Zhang 1993: 190 (listed) China
Cicadinae Lei, Chou and Li 1994: 54 (comp. note)
Plautillinae Lei, Chou and Li 1994: 54 (comp. note)
Lyristinae Boulard 1995c: 163 (comp. note) Equals Platypleurinae
Platypleurinae Boulard 1995c: 163 (comp. note)
Cicadinae Boulard and Mondon 1995: 65 (key) France
Cicadinae Lee 1995: 17, 21–22, 51, 134, 135, 138–139 (listed, key, comp. note) Korea

Platypleurinae Boulard 1996b: 3 (comp. note)
Platypleurinae Boulard 1996c: 95–96 (characteristics, comp. note)
Tibiceninae Boulard 1997c: 83–84, 100–104, 106 (comp. note)
Cicadinae Boulard 1997c: 101, 105–106, 112, 117 (comp. note)
Platypleurinae Boulard 1997c: 101–102, 105, 121 (comp. note)
Lyristinae Boulard 1997c: 104–105 (comp. note)
Cicadinae Chou, Lei, Li, Lu and Yao 1997: 5, 30–36, 106, Figs 5-1–5-3 (phylogeny, key, listed)
Gaeaninae Chou, Lei, Li, Lu and Yao 1997: 4, 30 (listed, comp. note)
Platypleurinae Chou, Lei, Li, Lu and Yao 1997: 30 (listed)
Platypleurnae (sic) Chou, Lei, Li, Lu and Yao 1997: 30 (listed)
Plautillinae Chou, Lei, Li, Lu and Yao 1997: 30–36, Figs 5-1–5-3 (phylogeny, key, listed)
Moaninae Chou, Lei, Li, Lu and Yao 1997: 30 (listed)
Lyristinae Chou, Lei, Li, Lu and Yao 1997: 30 (listed)
Cicadinae Poole, Garrison, McCafferty, Otte and Stark 1997: 251 (listed) North America
Cicadinae Lee 1999: 2 (comp. note) Korea
Cicadinae Moulds 1999: 1 (comp. note)
Zammarinae Moulds 1999: 1 (comp. note)
Platypleurinae Sanborn 1999a: 35 (listed)
Tibiceninae Sanborn 1999b: 1 (listed) North America
Cicadinae Sanborn 1999b: 1 (listed) North America
Cicadinae Villet and Noort 1999: 226, 228–229, Table 12.1 (listed, comp. note)
Cicadinae Villet and Zhao 1999: 321–322 (comp. note) China
Cicadinae Boulard 2000i: 21, 23, 29, 35 (listed, key) France
Tibicenine Boulard 2000i: 20 (comp. note)
Cicadinae Maw, Footit, Hamilton and Scudder 2000: 49 (listed)
Cicadinae Schedl 2000: 262 (listed)
Cicadinae Sueur 2000: 217 (comp. note) Mexico
Cicadinae Boulard 2001a: 139 (comp. note)
Tibiceninae Boulard 2001b: 4–5, 21–24, 26 (comp. note)
Platypleurinae Boulard 2001b: 22–23, 25 (comp. note)
Cicadinae Boulard 2001b: 22, 32, 37 (comp. note)
Lyristinae Boulard 2001b: 25 (comp. note)
Cicadinae Boulard 2001c: 51, 54, 56, 58 (listed)
Cicadinae Puissant and Sueur 2001: 430 (listed) Corsica
Tibiceninae Sanborn 2001b: 449 (listed) El Salvador
Cicadinae Boulard 2002a: 37, 39, 49, 55 (listed)
Cicadinae Boulard 2002b: 195, 197, 204, 207 (comp. note)
Tibiceninae Boulard 2002b: 204 (comp. note)
Cicadinae Duffels and Turner 2002: 235–236 (comp. note) Equals Gaeaninae
Tibiceninae Duffels and Turner 2002: 235–236 (comp. note)
Gaeaninae Duffels and Turner 2002: 236 (comp. note)
Moaninae Duffels and Turner 2002: 236 (comp. note)
Tibiceninae Kondratieff, Ellingson and Leatherman 2002: 13, Table 2 (listed) Colorado
Cicadinae Lee, Choe, Lee and Woo 2002: 3 (comp. note)
Cicadinae Moulds and Carver 2002: 1, 3, 5, 11–12, 14 (listed, comp. note) Australia
Plautillinae Moulds and Carver 2002: 1 (comp. note)
Platypleurinae Moulds and Carver 2002: 1 (comp. note)
Cicadinae Williams 2002: 156–157 (listed) Australia, Queensland, New South Wales, Victoria, South Australia, Tasmania
Cicadinae Boulard 2003a: 98 (listed)
Cicadinae Boulard 2003b: 186 (listed)
Cicadinae Boulard 2003e: 260 (listed)
Cicadinae Boulard 2003c: 172 (listed)
Cicadinae Holzinger, Kammerlander and Nickel 2003: 475, 477 (listed, comp. note) Europe
Cicadinae Lee and Hayashi 2003a: 152 (listed) Taiwan
Cicadinae Moulds 2003a: 187 (comp. note)
Cicadinae Chen 2004: 38–40 (phylogeny, listed)
Plautillinae Chen 2004: 38 (phylogeny, listed)
Cicadinae Pham 2004: 61 (comp. note) Vietnam
Cicadinae Sanborn, Heath, Heath and Phillips 2004: 66 (listed)
Cicadinae Sueur, Puissant, Simões, Seabra, Boulard and Quartau 2004: 178–179 (listed) Portugal
Cicadinae Boulard 2005g: 366 (listed)
Cicadinae Chawanji, Hodson and Villet 2005b: 66 (listed)
Cicadinae Lee 2005: 14, 16, 156 (key, listed, comp. note) Korea
Cicadinae Moran, Tran and Gerardo 2005: 8805, 8808, Fig 1, Fig 4 (comp. note)
Tibiceninae Moran, Tran and Gerardo 2005: 8805, 8808, Fig 1, Fig 4 (comp. note)
Cicadinae Moulds 2005b: 377–381, 383–389, 396, 404, 416, 418, 421, 423, 425–430, Fig 31, Fig 57, Table 2 (described, key, history, phylogeny, listed, comp. note) Equals Gaeaninae Equals Tibiceninae Equals Platypleurinae Australia

Gaeaninae Moulds 2005b: 384–385, Fig 31 (described, key, history, phylogeny, listed, comp. note)
Tibiceninae Moulds 2005b: 384–385, 388–389, 391, 392, 397, 416, Fig 31 (described, key, history, phylogeny, listed, comp. note)
Plautillinae Moulds 2005b: 385–388, 394, 397, 416, 418, 421, 428, Fig 31, Fig 57 (described, key, history, phylogeny, listed, comp. note) Australia
Moaninae Moulds 2005b: 385, 387, 392, 394 (described, key, history, phylogeny, listed, comp. note)
Lyristinae Moulds 2005b: 391 (history, listed, comp. note)
Platypleurinae Moulds 2005b: 385–389, 396–397, Fig 31 (described, key, history, phylogeny, listed, comp. note)
Cicadinae Pham and Thinh 2005a: 238 (listed) Vietnam
Cicadinae Pham and Thinh 2005b: 287–288 (key, comp. note) Equals Tibiceninae Equals Platypleruinae Vietnam
Cicadinae Reynaud 2005: 141 (listed)
Cicadinae Sanborn, Heath, Sueur and Phillips 2005: 191 (comp. note)
Cicadinae Boulard 2006e: 129 (comp. note)
Cicadinae Chawanji, Hodson and Villet 2006: 373–374 (listed, comp. note)
Cicadinae Puissant 2006: 19 (listed) France
Tibiceninae Sanborn 2006a: 76 (listed)
Tibiceninae Sanborn 2006b: 256 (listed)
Cicadinae Trilar 2006: 341 (comp. note)
Cicadinae Boulard 2007a: 93–94 (listed, comp. note)
Cicadinae Boulard 2007e: 86 (comp. note) Thailand
Cicadinae Chawanji, Hodson, Villet, Sanborn and Phillips 2007: 337–338 (listed, comp. note)
Cicadinae Ganchev and Potamitis 2007: 286, 304, Fig 11 (listed)
Tibiceninae Ganchev and Potamitis 2007: 304–305, 319, Fig 11 (listed, comp. note)
Cicadinae Sanborn 2007a: 26 (listed) Venezuela
Cicadinae Sanborn 2007b: 2, 31 (listed) Mexico
Cicadinae Sanborn, Phillips and Sites 2007: 3–4, Table 1 (listed) Thailand
Cicadinae Sueur and Puissant 2007b: 62 (listed) France
Cicadinae Boulard 2008b: 109 (comp. note)
Cicadinae Boulard 2008c: 351–352, 354, 356, 360 (listed, comp. note)
Cicadinae Boulard 2008f: 7, 9, 97 (listed) Thailand
Cicadinae Demir 2008: 475 (listed) Turkey
Cicadinae Lee 2008a: 2–3, Table 1 (listed) Vietnam
Cicadinae Lee 2008b: 446–447, 462, Fig 1, Table 1 (listed, comp. note) Korea
Cicadinae Moulds 2008a: 207–209 (key, listed, comp. note) Australia
Cicadinae Sanborn 2008f: 686, 688 (listed)
Cicadinae Sanborn, Moore and Young 2008: 3 (listed) Costa Rica
Cicadinae Sanborn, Phillips and Gillis 2008: 5 (listed) Florida
Cicadinae Boulard 2009d: 39 (comp. note)
Cicadinae Lee 2009a: 87 (listed)
Cicadinae Lee 2009b: 330 (comp. note)
Cicadinae Lee 2009c: 338 (comp. note)
Cicadinae Lee 2009d: 470 (comp. note)
Cicadinae Lee 2009e: 87 (comp. note)
Cicadinae Lee 2009f: 1487 (comp. note)
Cicadinae Lee 2009g: 306 (comp. note)
Cicadinae Lee 2009h: 2618 (listed) Philippines, Palawan
Cicadinae Lee 2009i: 293 (listed) Philippines, Panay
Cicadinae Lee and Sanborn 2009: 31 (listed)
Cicadinae Moulds 2009b: 164 (comp. note)
Cicadinae Pham and Yang 2009: 1, 5, 13, 16, Table 2 (listed, comp. note) Vietnam
Cicadinae Sanborn 2009a: 89 (listed) Cuba
Cicadinae Sanborn 2009b: 307 (comp. note) Vietnam
Cicadinae Goemans 2010: 5 (listed)
Cicadinae Kemal and Koçak 2010: 3–4 (diversity, listed, comp. note) Turkey
Cicadinae Lee 2010a: 15 (listed) Philippines, Mindanao
Cicadinae Lee 2010b: 20 (listed) Cambodia
Cicadinae Lee 2010c: 2 (listed) Philippines, Luzon
Cicadinae Lee 2010d: 167 (listed)
Cicadinae Lee and Hill 2010: 278, 287, 300, 302 (comp. note)
Cicadinae Mozaffarian and Sanborn 2010: 76 (listed)
Cicadinae Phillips and Sanborn 2010: 74 (comp. note)
Cicadinae Pham, Thinh and Yang 2010: 63 (listed) Vietnam
Cicadinae Pham and Yang 2010a: 133 (diversity, comp. note) Vietnam
Cicadinae Santos, Martinelli, Maccagnan, Sanborn and Ribeiro 2010: 50 (listed)
Cicadinae Sanborn 2010a: 1579 (listed) Colombia
Cicadinae Sanborn 2010b: 67, 74 (listed)
Cicadinae Webi, Ahmed and Rizvi 2010: 28, 33 (comp. note)

Tribe Platypleurini Schmidt, 1918 =
Platypleuraria = Platypleuriti (sic) = Tibicenini Van Duzee, 1916 = Tibicenaria = Lyristini Gomez-Menor, 1957 = Hainanosemiina Kato, 1927 = Hainanosemiaria = Hainanosemiiti (sic)
Platypleurini Chen 1942: 143 (listed)
Platypleurini Medler 1980: 73 (listed) Nigeria
Platypleurini Boulard 1984c: 176–177 (comp. note)
Platypleurini Hayashi 1984: 28 (key)
Tibicenini Kirillova 1986a: 122 (listed)
Tibicenini Kirillova 1986b: 46 (listed)
Tibicenini Chou, Lei and Wang 1993: 82, 85–86 (comp. note) China
Platypleurini Boulard 1985e: 1029, 1032 (comp. note)
Platypleurini Boulard 1986d: 224 (listed)
Platypleurini Kirillova 1986a: 123 (listed)
Platypleurini Kirillova 1986b: 45 (listed)
Platypleurini Villet 1987a: 272, Table 2 (listed) South Africa
Platypleruini Boulard 1988f: 33–34, 72 (comp. note)
Platypleurini Boulard 1990b: 93, 116, 143, 224 (comp. note)
Platypleurini Moulds 1990: 31 (comp. note)
Tibicenini Moulds 1990: 31 (listed) Australia
Platypleuraria Moulds 1990: 31 (comp. note)
Platypleurini Duffels 1991a: 128 (comp. note)
Platypleurini Villet 1994b: 91 (comp. note)
Platypleurini Boulard and Mondon 1995: 65 (key) France
Platypleurini Boer and Duffels 1996b: 302, 330 (distribution, comp. note)
Platypleurini Boulard 1996c: 96 (characteristics, comp. note)
Platypleurini Boulard 1996d: 114, 115 (listed)
Platypleruini Boulard 1997c: 102–103, 105–106, 112, 117, 121 (comp. note)
Platypleuraria Boulard 1997c: 106, 112, 117 (comp. note)
Platypleurini Chou, Lei, Li, Lu and Yao 1997: 106, 167, 298 (key, listed, synonymy, described) Equals Platypleurinae
Platypleurini Novotny and Wilson 1997: 437 (listed)
Platypleurini Boulard 1999a: 78, 81 (listed)
Platypleurini Moulds 1999: 1 (comp. note)
Platypleuraria Sanborn 1999a: 35 (listed)
Platypleurini Villet 1999b: 149 (comp. note)
Platypleurini Villet and Noort 1999: 226, Table 12.1 (comp. note) Tanzania, Kenya, Uganda, South Africa, Botswana, Zimbabwe, Namibia, Angola
Platypleurini Boulard 2000c: 56 (comp. note)
Platypleurini Boulard 2000i: 21, 23. 29 (listed, key)
Platypleurini Boulard 2001a: 131 (comp. note)
Platypleruini Boulard 2001b: 22–23, 25, 32, 37 (comp. note)
Platypleuraria Boulard 2000i: 21 (listed)
Platypleuraria Boulard 2001b: 26, 32, 37 (comp. note)
Platypleurini Boulard 2001c: 51 (listed)
Tibicenini Sanborn 2001b: 449 (listed) El Salvador
Platypleurini Sueur 2001: 36, Table 1 (listed)
Platypleurini Boulard 2002a: 55 (listed)
Platypleurini Boulard 2002b: 197 (comp. note)
Platypleurini Dietrich 2002: 158 (comp. note)
Platypleurini Moulds and Carver 2002: 12 (listed) Australia
Platypleuraria Moulds and Carver 2002: 12 (listed) Australia
Platypleurini Boulard 2003a: 98 (listed)
Platypleurini Boulard 2003b: 186 (listed)
Platypleuraria Boulard 2003b: 186 (listed)
Platypleuraria Boulard 2003c: 172 (listed)
Platypleurini Lee and Hayashi 2003a: 150, 152 (key, listed) Taiwan
Tibicenini Lee and Hayashi 2003a: 150, 152 (key, listed, comp. note) Taiwan
Platypleurini Lee and Hayashi 2003b: 359 (listed) Taiwan
Tibicenini Lee and Hayashi 2003b: 359 (listed) Taiwan
Platypleurini Chen 2004: 39–40 (phylogeny, listed)
Platypleurini Lee and Hayashi 2004: 45 (listed) Taiwan
Tibicenini Lee and Hayashi 2004: 45 (listed) Taiwan
Platypleurini Sueur, Puissant, Simões, Seabra, Boulard and Quartau 2004: 178 (listed) Portugal
Platypleurini Boulard 2005b: 111 (comp. note)
Platypleurini Boulard 2005h: 6 (listed)
Platypleurini Chawanji, Hodson and Villet 2005b: 66 (listed)
Platypleurini Lee 2005: 14, 16, 156–157, 168 (key, listed, comp. note) Korea
Platypleurini Moulds 2005b: 388–389, 391–392, 396, 404, 419, 422–423, 427, 430–432, Fig 58, Fig 60, Table 2 (described, diagnosis, composition, distribution, phylogeny, listed, comp. note) Equals Platypleurina Equals Hainanosemiina Equals Cryptotympanini (partim) Equals Tibiceninae Australia
Tibicenaria Moulds 2005b: 386 (history, listed, comp. note)
Tibicenini Moulds 2005b: 389, 392, 419, Fig 58 (history, phylogeny, listed, comp. note)

Lyristini Moulds 2005b: 390–391, 393 (history, phylogeny, listed, comp. note)
Platypleurina Moulds 2005b: 392, 396, 414, 430, 434, Table 2 (status, listed, comp. note) Equals Platypleuriaria Australia
Hainanosemiina Moulds 2005b: 392, 430, 434, Table 2 (status, listed, comp. note) Equals Hainanosemiaria
Platypleuraria Moulds 2005b: 431 (comp. note)
Platypleurini Pham and Thinh 2005a: 238 (listed) Vietnam
Platypleurini Pham and Thinh 2005b: 287–289 (key, comp. note) Equals Platypleurinae Vietnam
Tibicenini Salmah, Duffels and Zaidi 2005: 15 (comp. note)
Platypleurini Villet, Barker and Lunt 2005: 589 (phylogeny, comp. note) South Africa
Platypleuriti (sic) Villet, Barker and Lunt 2005: 590, Table 1 (listed) South Africa
Hainanosemiiti (sic) Villet, Barker and Lunt 2005: 590, Table 1 (listed) South Africa
Platypleurini Chawanji, Hodson and Villet 2006: 373 (listed)
Platypleurini Trilar 2006: 341 (comp. note)
Platypleurini Boulard 2007d: 494 (listed)
Platypleurini Boulard 2007e: 20, 54 (comp. note) Thailand
Platypleurini Boulard 2007f: 131 (comp. note)
Platypleurini Chawanji, Hodson, Villet, Sanborn and Phillips 2007: 338 (listed)
Platypleurini Price, Barker and Villet 2007: 2575 (distribution, comp. note)
Platypleurini Sanborn, Phillips and Sites 2007: 3–4, 36, Fig 1, Table 1 (listed, distribution) Thailand
Platypleuraria Sanborn, Phillips and Sites 2007: 3, 5, Table 1 (listed) Thailand
Platypleurini Schedl 2007: 153 (listed)
Platypleurini Shiyake 2007: 2, 14 (listed, comp. note)
Platypleuraria Shiyake 2007: 2 (listed, comp. note)
Hainansemiaria Shiyake 2007: 2 (listed, comp. note)
Tibicenini Shiyake 2007: 3, 31 (listed, comp. note)
Platypleurini Boulard 2008f: 7, 15 (listed) Thailand
Platypleurini Boulard 2008c: 361 (listed)
Platypleurini Boulard 2008d: 165 (comp. note)
Platypleurini Lee 2008a: 2–3, Table 1 (key, listed) Vietnam
Platypleurini Lee 2008b: 446, 462, Table 1 (listed, comp. note) Korea
Platypleurini Moulds 2008a: 208–209 (key, comp. note)
Platypleurini Osaka Museum of Natural History 2008: 322 (listed) Japan
Platypleurini Lee 2009h: 2618 (listed) Philippines, Palawan
Platypleurini Lee 2009i: 293 (listed) Philippines, Panay
Platypleurini Pham and Yang 2009: 13, Table 2 (listed) Vietnam
Platypleurini Goemans 2010: 5, 7 (comp. note) Africa
Platypleurini Lee 2010a: 15 (listed) Philippines, Mindanao
Platypleurini Lee 2010b: 20 (listed) Cambodia
Platypleurini Lee 2010c: 2 (listed) Philippines, Luzon
Platypleurini Lee 2010d: 171 (comp. note)
Platypleurini Mozaffarian and Sanborn 2010: 76 (listed)
Platypleurini Phillips and Sanborn 2010: 74 (comp. note)

***Munza* Distant, 1904** = *Numza* (sic)
Munza Hayashi 1984: 34 (comp. note)
Munza Villet 1989b: 54, 63 (listed, comp. note)
Munza spp. Villet 1989b: 59 (comp. note)
Munza Villet 1999b: 153 (comp. note)
Munza Moulds 2005b: 392, 433 (higher taxonomy, listed, comp. note)
Munza Goble, Price, Barker, and Villet 2008a: 673 (phylogeny, comp. note) Africa

***M. basimacula* (Walker, 1850)** = *Platypleura basimacula* = *Numza basimacula* (sic) = *Platypleura reducta* Walker, 1850 = *Platypleura basimacula reducta* = *Poecilopsaltria reducta* = *Poecilopsaltera* (sic) *reducta* = *Munza pygmaea* Jacobi, 1910 = *Munza pygmoea* (sic) = *Platypleura sikumba* Ashton, 1914 = *Munza parva* Villet, 1989 = *Cicada* decora (nec Germar) = *Platypleura decora* (nec Germar) = *Platypleura deusta* (nec Thunburg) = *Platypleura absimilis* (nec Distant)
Munza basimacula Villet 1988: 74, Fig 9 (song, sonogram, comp. note) South Africa
Munza basimaculata Villet 1989b: 63, 65, 69 (comp. note)
Munza parva Villet 1989b: 63, 65–67, Fig 15, Figs 26–27, Fig 31 (n. sp., described, illustrated, habitat, distribution, comp. note) Equals *Cicada decora* (nec Germar) Equals *Platypleura decora* (nec Germar) Equals *Platypleura deusta* (nec Thunburg) Equals *Platyplerua absimilis* (nec Distant) South Africa, Mozambique
Munza parva Villet and Reavell 1989: 334–335, Fig 3 (distribution, habitat, host plants, key, comp. note) South Africa, Mozambique
Platypleura sikuma Villet 1999d: 210, 214 (synonymy, comp. note)
Munza basimacula Villet 1999d: 210, 214 (synonymy, distribution, habitat, comp. note) Equals *Platypleura basimacula* Equals *Platypleura reducta* Equals *Munza pygmaea*

Equals *Platypleura sikuma* Equals *Munza parva* n. syn. Mozambique
Munza pygmaea Villet 1999d: 214 (comp. note)
Munza parva Villet 1999d: 214 (comp. note)
Munza basimacula Villet 1999e: 1 (comp. note) South Africa, Kenya
Munza basimacula Sueur 2001: 37, Table 1 (listed)

***M. furva* (Distant, 1897)** = *Poecilopsalta furva* = *Munza oculata* Jacobi, 1910
Munza furva Evans 1966: 111 (parasitism, comp. note) South Africa
Munza furva Villet 1989b: 59 (comp. note)
Munza furva Malaisse 1997: 237, Table 2.11.1 (human food, listed) Zambezi
Munza furva Huis 2003: 168, Table 1 (predation, comp. note) Democratic Republic of Congo

***M. laticlavia laticlavia* (Stål, 1858)** = *Platypleura laticlavia* = *Munza laticlava* (sic)
Munza laticlavia Villet 1989b: 59 (comp. note)
Munza laticlavia Villet 1999e: 1 (comp. note) South Africa, Kenya
Munza laticlavia Villet, Barker and Lunt 2005: 590–593, Figs 1–5, Table 1 (phylogeny, listed, comp. note) South Africa

***M. laticlavia lubberti* Schumacher, 1913** = *Munza laticlavia lübberti*

***M. laticlavia semitransparens* Schumacher, 1913** = *Munza laticlava* (sic) *semitransparens*

***M. otjosonduensis* Schumacher, 1913**

***M. pallescens* Schumacher, 1913**

M. parva Villet, 1989 to *M. basimacula* (Walker, 1850)

***M. popovi* Boulard, 1975**

***M. revoili* Distant, 1905**

***M. signata* Distant, 1914**

***M. straeleni* Dlabola, 1960**

***M. sudanensis* Distant, 1913**

***M. trimeni* (Distant, 1892)** = *Poecilopsaltria trimeni*
Munza trimeni Gäde and Janssens 1994: 804–807, Fig 2C–D, Table 4 (hormone, comp. note) South Africa
Munza trimeni Ye and Ng 2009: 188 (protein, comp. note)

***M. venusta* Hesse, 1925**

***Soudaniella* Boulard, 1973** = *Soudanielle* (sic)
Soudaniella Medler 1980: 73 (listed) Nigeria
Soudaniella Boulard 1985e: 1019, 1023, 1025 (crypsis, coloration, comp. note) Africa
Soudaniella Boulard 1990b: 167, 175 (comp. note)
Soudaniella Duffels 1991a: 128 (comp. note) Africa
Soudaniella Villet 1994b: 91 (comp. note)
Soudaniella Boulard 1999a: 92 (comp. note)
Soudaniella Boulard 2002b: 202 (comp. note)
Soudaniella Moulds 2005b: 392, 433 (higher taxonomy, listed, comp. note)
Soudaniella Boulard 2006g: 109 (comp. note) Africa
Soudaniella Boulard 2007f: 81, 138, Figs 97–98 (comp. note)
Soudaniella Goble, Price, Barker, and Villet 2008a: 673 (phylogeny, comp. note) Africa

***S. cortustusa* Boulard, 1974** = *Soudanielle* (sic) *cortustusa*
Soudanielle (sic) *cortustusa* Boulard 1984d: 6 (coloration, comp. note)
Soudaniella cortustusa Boulard 1985e: 1023, Plate II, Figs 13–14 (coloration, illustrated, comp. note)
Soudaniella cortustusa Boulard 1990b: 196, 198–199, Plate XXV, Fig 14 (coloration, comp. note) Central African Republic
Soudaniella cortustusa Boulard 1991f: 22 (melanism, comp. note)
Soudaniella cortustusa Boulard 1997b: 22–23, 33, Plate IV, Figs 4–5 (coloration, comp. note)
Soudaniella cortustusa Boulard 2002b: 193, 195, Fig 48 (coloration, illustrated, comp. note) Central African Republic
Soudaniella cortustusa Boulard 2007f: 80, 137, Figs 92–93 (comp. note)

***S. dlabolai* Boulard, 1972** = *Platypleura schoutedeni* Dlabola, 1962

***S. laticeps* (Karsch, 1890)** = *Platypleura laticeps* = *Platypleura (Poecilopsaltria) laticeps*
Soudaniella laticeps Medler 1980: 73 (listed) Nigeria
Soudaniella laticeps Boulard 1999a: 78, 85–86, 96–97, Plate II, Fig 1 (listed, sonogram, illustrated, comp. note) Senegal
Soudaniella laticeps Sueur 2001: 37, Table 1 (listed)
Soudaniella laticeps Boulard 2006g: 131–132, 180 (listed, comp. note)

***S. marshalli* (Distant, 1897)** = *Poecilopsaltria marshalli* = *Platypleura marshalli* = *Platypleura (Poecilopsaltria) marshalli*
Platypleura marshalli Evans 1966: 111 (parasitism, comp. note) South Africa
Platypleura marshalli Sueur 2002b: 122, 129 (listed, comp. note) Equals *Poecilopsaltria marshalli*

***S. melania melania* (Distant, 1904)** = *Soudaniella melania* = *Platypleura (Poecilopsaltria) melania* = *Platypleura melania* = *Platypleura melanaria* (sic) = *Poecilopsaltria melania*
Soudaniella melania Medler 1980: 73 (listed) Nigeria
Soudaniella melania Boulard 1985e: 1039, Plate IV, Fig 28 (coloration, illustrated, comp. note)

Soudaniella melania Boulard 1990b: 178, 180–181, 183, Plate XXI, Fig 6 (illustrated, comp. note)
Soudaniella melania Boulard 1997b: 20, Plate II, Fig 2 (morphology, comp. note)
Soudaniella melania Boulard 2005d: 338 (comp. note) Africa
Soudaniella melania Boulard 2006g: 39–40 (comp. note) Africa
Soudaniella melania Boulard 2007f: 84, Fig 100.2 (illustrated, comp. note)

***S. melania fuscala* Boulard, 1975**

***S. schoutedeni* (Distant, 1913)** = *Platypleura schoutedeni* = *Soudaniella schoutedemi* (sic)

***S. seraphina* (Distant, 1905)** = *Platypleura seraphina*
Soudaniella seraphina Medler 1980: 73 (listed) Nigeria
Soudaniella seraphina Boulard 1990b: 169 (comp. note)
Soudaniella seraphina Boulard 1999a: 78, 86, 88 (listed, sonogram, comp. note) Ivory Coast
Soudaniella seraphina Sueur 2001: 37, Table 1 (listed)
Soudaniella seraphina Boulard 2005d: 344 (comp. note)
Soudaniella seraphina Boulard 2006g: 113, 131–132, 138, 180, Fig 98 (sonogram, listed, comp. note) Ivory Coast

***S. sudanensis* (Distant)** = *Munza sudanensis*

***Esada* Boulard, 1973**
Esada Boulard 1990b: 224 (comp. note)
Esada Moulds 2005b: 392, 433 (higher taxonomy, listed, comp. note)

***E. esa* (Distant, 1905)** = *Platypleura esa* = *Platypleura (Oxypleura) esa*

***Ioba* Distant, 1904** = *Iobao* (sic) = *Lioba* (sic)
Ioba Medler 1980: 73 (listed) Nigeria
Ioba Boulard 1990b: 75, 167, 175 (comp. note)
Ioba Boulard 1995a: 7 (comp. note)
Ioba sp. Malaisse 1997: 237, 241, Tables 2.11.1–2.11.2 (human food, listed) Zambezi
Ioba sp. Malaisse and Parent 1997: 68, Table 2 (nutrient content, listed) Zambezi
Ioba Boulard 1999a: 91 (comp. note)
Ioba Boulard 2002b: 193 (coloration, comp. note) Africa
Ioba Morris 2004: 70 (human food) Malawi
Ioba Moulds 2005b: 392, 433 (higher taxonomy, listed, comp. note)
Ioba Boulard 2006g: 109 (comp. note) Africa
Ioba Schabel 2006: 258 (human food, comp. note) Africa
Ioba Shiyake 2007: 2, 14 (listed, comp. note)
Ioba Moulds 2008a: 208 (comp. note)

***I. bequaerti* Distant, 1913**

***I. horizontalis* (Karsch, 1890)** = *Platypleura horizontalis* = *Platypleura (Poecilopsaltria) horizontalis*
Ioba horizontalis Malaisse 1997: 237, Table 2.11.1 (human food, listed) Zambezi
Ioba horizontalis Sueur 2002b: 122, 129 (listed, comp. note) Equals *Poecilopsaltria horizontalis*
Ioba horizontalis Huis 2003: 168, Table 1 (predation, comp. note) Democratic Republic of Congo

***I. leopardina* (Distant, 1881)** = *Poecilopsaltria leopardina* = *Platypleura (Joba)* (sic) *leopardina* = *Lioba* (sic) *leopardina*
Ioba leopardina Chavanduka 1975: 218, Fig 5 (human food, comp. note) Rhodesia
Ioba leopardina Theron 1985: 156–157, Fig 16.120–16.121, Plate II, Fig 1 (illustrated) South Africa
Ioba leopardina Malaisse 1997: 237, Table 2.11.1 (human food, listed) Zambezi
Ioba leopardina Villet 1999e: 1 (comp. note) South Africa, Kenya
*Ioba leopardina*Villet and Noort 1999: 227–229, Table 12.2 (habitat, listed, comp. note) Tanzania
Ioba leopardina Freytag 2000: 55 (on stamp)
Ioba leopardina Sueur 2002b: 122, 129 (listed, comp. note) Equals *Poecilopsaltria leopardina*
Ioba leopardina Huis 2003: 168, Table 1 (predation, comp. note) Democratic Republic of Congo, Zambia, Zimbabwe
Ioba leopardina Morris 2004: 70, 243 (human food, comp. note) Malawi
Ioba leopardina Shiyake 2007: 2, 12, 14, Fig 2 (illustrated, distribution, listed, comp. note) Africa

***I. limbaticollis* (Stål, 1863)** = *Platypleura limbaticollis* = *Oxypleura limbata* Walker, 1858 (nec *Platypleura limbata* Fabricius, 1775) = *Ioba leopardina* (nec Distant)
Ioba limbaticollis Medler 1980: 73 (listed) Nigeria, Ghana, Zaire
Ioba limbaticollis Boulard 1985e: 1020, 1022, 1038, Plate III, Figs 19–20 (coloration, illustrated, comp. note) Central African Republic
Ioba limbaticollis Boulard 1990b: 188–189, 207, Plate XXIII, Fig 3 (illustrated, comp. note)
Ioba limbaticollis Duffels and van Mastrigt 1991: 174 (predation, comp. note) Central African Republic

Ioba limbaticollis Boulard 1995a: 8 (comp. note)
Ioba limbaticollis Boulard 1999a: 78, 82–83, 88, Plate I, Fig 1 (listed, sonogram, illuatrated, comp. note)
Ioba limbaticollis Sueur 2001: 36, Table 1 (listed)
Ioba limbaticollis Boulard 2002b: 193, 201, 209, Fig 44, sonogram 54b (illustrated, sonogram, comp. note)
Ioba limbaticollis Boulard 2005d: 338, 344 (comp. note) A frica
Ioba limbaticollis Boulard 2006g: 39–40, 131–132, 138, 180, Fig 97 (illustrated, sonogram, listed, comp. note) Africa, Ivory Coast
Ioba limbaticollis Boulard 2007f: 122, 149, Fig 259 (comp. note)

***I. stormsi* (Distant, 1893)** = *Poecilopsaltria stormsi*

***I. veligera* (Jacobi, 1904)** = *Platypleura veligera* = *Platypleura laticollis* Melichar, 1904
Ioba veligera Boulard 1995a: 8 (comp. note)
Ioba veligera Boulard 1999a: 79, 91, 96–97, Plate II, Fig 2 (listed, sonogram, illustrated)
Ioba veligera Sueur 2001: 36, Table 1 (listed)
Ioba veligera Boulard 2006g: 131–132, 150, 180, Fig 110C (illustrated, listed, comp. note)

***Muansa* Distant, 1904**
Muansa Medler 1980: 73 (listed) Nigeria
Muansa Boulard 1985e: 1029–1030 (coloration, comp. note)
Muansa Boulard 1991a: 7–8 (turrets, comp. note)
Muansa Moulds 2005b: 392, 433 (higher taxonomy, listed, comp. note)
Muansa Boulard 2007f: 52 (comp. note)

***M. clypealis* (Karsch, 1890)** = *Platypleura clypealis* = *Platypleura (Poecilopsaltria) clypealis*
Muansa clypealis Medler 1980: 73 (listed) Nigeria, Cameroon, Zaire
Muansa clypealis Strümpel 1983: 57 (comp. note)
Muansa clypealis Boulard 1984d: 6, Plate, Fig 4 (illustrated, nymph behavior, comp. note) Central Africa
Muansa clypealis Boulard 1985e: 1022, 1029, 1046–1047, Plate VIII, Fig 49 (coloration, illustrated, comp. note)
Muansa clypealis Boulard 1988d: 152, 154–155, 158, Fig 12, Fig 18 (wing, pigmentation, leg abnormalities, illustrated, comp. note) Central African Republic
Muansa clypealis Boulard 1990b: 72–75, 79, 81, Plate VI, Figs 1–9 (larvae, turrets, comp. note)
Muansa clypealis Boulard 1991a: 7 (turrets)
Muansa clypealis Boulard 2002b: 207–209, Figs 56–59 (nymph illustrated, comp. note) Central African Republic
Muansa clypealis Rakitov 2002: 127 (comp. note)
Muansa clypealis Boulard 2007f: 131, Fig 21 (illustrated, comp. note)

***Sadaka* Distant, 1904**
Sadaka Medler 1980: 73 (listed) Nigeria
Sadaka 1985c: 175–176 (comp. note)
Sadaka Boulard 1986d: 224 (listed)
Sadaka Boulard 1996d: 124 (listed)
Sadaka Moulds 2005b: 392, 433 (higher taxonomy, listed, comp. note)
Sadaka Moulds 2008a: 208 (comp. note)

***S. aurovirens* Dlabola, 1960**

***S. dimidiata* (Karsch, 1893)** = *Platypleura dimidiata*
Sadaka dimidiata Boulard 1985c: 175–177, 179, 187, Figs 1–6, Figs 36–37 (n. comb., described, neallotype designation, illustrated, comp. note) Equals *Platypleura dimidiata* Cameroon

***S. morini* Boulard, 1985**
Sadaka morini Boulard 1985c: 176–179, 187, Figs 7–11, Figs 38–39 (n. sp., described, illustrated, comp. note) Republic of Congo, Zaire
Sadaka morini Boulard 1988d: 152 (missing organs, comp. note)

***S. radiata centralensis* Boulard, 1986**
Sadaka radiata centralensis Boulard 1986d: 224–225, 238, Table I, Figs 3–4 (n. ssp., described, illustrated, comp. note) Central African Republic, Ivory Coast

***S. radiata radiata* (Karsch, 1890)** = *Platypleura (Oxypleura) radiata* = *Platypleura radiata* = *Sadaka radiata*
Sadaka radiata Schouteden 1925: 83 (listed, comp. note) Congo
Sadaka radiata Medler 1980: 73 (listed) Nigeria, Cameroon, Ivory Coast
Sadaka radiata Boulard 1985c: 46 (comp. note)
Sadaka radiata Boulard 1986d: 224–225, 237 (comp. note) Sierra Léon, Cameroon, Ivory Coast
Sadaka radiata radiata Boulard 1986d: 224, 236–238, Table I, Figs 1–2, Fig 37 (n. comb., neallotype designation, described, illustrated, comp. note) Ivory Coast, Guinea
Sadaka radiata Boulard 1990b: 132–133, 207, Plate XV, Fig 8 (genitalia, comp. note)
Sadaka radiata Boulard 1996d: 114, 124, 126–127, Fig S5 (song analysis, sonogram, listed, comp. note) Ivory Coast
Sadaka radiata Malaisse 1997: 238, Table 2.11.1 (human food, listed) Zambezi
Sadaka radiata Boulard 2000d: 95, 106, Plate O, Fig 1 (song, illustrated, comp. note)
Sadaka radiata Sueur 2001: 37, Table 1 (listed)
Sadaka radiata Huis 2003: 168, Table 1 (predation, comp. note) Democratic Republic of Congo

Sadaka radiata Boulard 2005d: 345–346, Fig 25.42 (sonogram, illustrated, comp. note)
Sadaka radiata Boulard 2006g: 73, 75–76, 85, 100–101, 180, Fig 45, Fig 66 (sonogram, illustrated, listed, comp. note) Ivory Coast

S. sagittifera **Boulard, 1985**
Sadaka sagittifera Boulard 1985c: 178–179, 187, Figs 12–13, Fig 40 (n. sp., described, illustrated, comp. note) Angola

S. virescens **(Karsch, 1890)** = *Platypleura (Oxypleura) virescens* = *Platypleura virescens*
Sadaka virescens Schouteden 1925: 83–84 (listed, comp. note) Congo
Platypleura virescens Boulard 1986d: 224 (type species of *Sadaka*) Angola

Afzeliada **Boulard, 1973**
Afzeliada Medler 1980: 73 (listed) Nigeria
Afzeliada 1985c: 179 (comp. note)
Afzeliada Boulard 1986d: 225 (listed)
Afzeliada sp. Boulard 1986d: 225 (distribution, comp. note) Ivory Coast
Afzeliada sp. Nkouka 1987: 174 (human food) Central Africa
Afzeliada sp. Huis 1996: 7, Table 1 (predation, comp. note) Congo
Afzeliada Chou, Lei, Li, Lu and Yao 1997: 167, 293, 357, Table 6 (key, described, listed, diversity, biogeography, comp. note) China
Afzeliada Villet and Zhao 1999: 322 (comp. note)
Afzeliada sp. Huis 2003: 168, Table 1 (predation, comp. note) Congo
Afzeliada Moulds 2005b: 392, 433 (higher taxonomy, listed, comp. note)
Afzeliada Pham and Thinh 2005a: 238 (listed) Vietnam
Afzeliada Pham and Thinh 2005b: 289 (comp. note) Vietnam
Afzeliada Schabel 2006: 258 (human food, comp. note) Africa
Afzeliada Sanborn, Phillips and Sites 2007: 3–4, Table 1 (listed, comp. note) Thailand
Afzeliada Moulds 2008a: 208 (comp. note)

A. afzelii **(Stål, 1854)** = *Platypleura afzelii* = *Platypleura afzelli* (sic) = *Platypleura (Poecilopsaltria) afzelii* = *Platypleura strumosa* (nec Fabricius) = *Platyplerua aerea* Distant, 1881 = *Platypleura area* (sic) = *Platypleura monodi* Lallemand, 1927
Afzeliada afzelii 1985c: 179 (comp. note)
Platypleura afzelii Boulard 1986d: 224 (type species of *Afzeliada*) Sierra Léon
Platypleura afzellii Chou, Lei, Li, Lu and Yao 1997: 179 (type species of *Afzeliada*)
Afzeliada afzelii Malaisse 1997: 237, Table 2.11.1 (human food, listed) Zambezi
Afzeliada afzelii Huis 2003: 168, Table 1 (predation, comp. note) Democratic Republic of Congo
Platypleura afzellii (sic) Pham and Thinh 2005a: 238 (type species of *Afzeliada*)

A. *badia* (Distant, 1888) to *Platypleura badia*

A. *balachowskyi* Boulard, 1975

A. bernardii **(Boulard, 1971)** = *Platypleura bernardii*
Afzeliada bernardii Medler 1980: 73 (listed) Nigeria
Afzeliada bernardii Boulard 1985e: 1022, 1029, 1038, Plate III, Figs 21–22 (coloration, illustrated, comp. note) Gabon, Central African Republic

A. caluangana **Boulard, 1979**
Afzeliada caluangana 1985c: 179 (comp. note)

A. christinetta **Boulard, 1973** = *Platypleura rutherfordina* Boulard, 1971 nom. nud.
Afzeliada christinetta Boulard 1985e: 1029 (coloration, comp. note)
Afzeliada christinetta Boulard 1986d: 225 (comp. note) Central African Republic

A. circumscripta **(Jacobi, 1910)** = *Platypleura circumscripta* = *Afzeliada circunscripta* (sic)

A. contracta **(Walker, 1850)** = *Oxypleura contracta* = *Platypleura contracta* = *Platypleura (Poecilopsaltria) contracta*
Afzeliada contracta Medler 1980: 73 (listed) Nigeria

A. deheegheri **Boulard, 1975**
Afzeliada deheegheri 1985c: 179 (comp. note)

A. donskoffi **Boulard, 1979** = *A. crassa* Boulard, 1975 nom. nud.
Afzeliada donskoffi Boulard 1985c: 179 (comp. note)

A. duplex **(Dlabola, 1961)** = *Platypleura duplex*
Afzeliada duplex Malaisse 1997: 237, Table 2.11.1 (human food, listed) Zambezi
Afzeliada duplex Huis 2003: 168, Table 1 (predation, comp. note) Democratic Republic of Congo

A. ericina **Boulard, 1973**
Afzeliada ericina Medler 1980: 73 (listed) Nigeria
Afzeliada ericina Boulard 1985e: 1029 (coloration, comp. note)

A. grillotti **Boulard, 1985**
Afzeliada grillotti Boulard 1985c: 179–181, 187, Figs 14–17, Fig 41 (n. sp., described, illustrated, comp. note) Republic of Congo

A. helenae **Boulard, 1975**

A. hyalina **(Distant, 1905)** = *Sadaka hyalina*

A. hyaloptera **(Stål, 1866)** = *Platypleura hyaloptera* = *Platypleura (Oxypleura) hyaloptera*
Afzeliada hyaloptera Boulard 1986d: 225, 236–238, Table I, Fig 38 (distribution, illustrated, comp. note) Ivory Coast, Senegal, Ghana

A. iringana **Boulard, 2012**

A. izzardi (Dlabola, 1960) = *Platypleura izzardi*

A. ladona (Distant, 1919) = *Platypleura ladona* = *Platypleura (Oxypleura) ladona*

A. longula (Distant, 1904) = *Platypleura (Oxypleura) longula* = *Platypleura longula*

A. lusingana Boulard, 1979

A. mikessensis Boulard, 2012

A. passosi Boulard, 1972

A. rutherfordi (Distant, 1883) = *Platypleura rutherfordi* = *Platypleura (Poecilopsaltria) rutherfordi*

Platypleura rutherfordi Schouteden 1925: 83 (listed, comp. note) Congo

Afzeliada rutherfordi Medler 1980: 73 (listed) Nigeria

Platypleura rutherfordi Sueur 2002b: 122, 129 (listed, comp. note)

***Attenuella* Boulard, 1973**

Attenuella Medler 1980: 73 (listed) Nigeria

Attenuella Boulard 1986d: 225 (listed)

Attenuella Novotny and Wilson 1997: 437 (listed)

Attenuella Villet 1999b: 153 (comp. note)

Attenuella Moulds 2005b: 392, 433 (higher taxonomy, listed, comp. note)

Attenuella Shiyake 2007: 2, 14 (listed, comp. note)

A. tigrina (Palisot de Beauvois, 1813) = *Cicada tigrina* = *Platypleura tigrina* = *Ugada tigrina* = *Platypleura attenuata* Distant, 1905 = *Platypleura (Oxypleura) attenuata*

Attenuella tigrina Medler 1980: 73 (listed) Equals *Attenuella attenuata* Nigeria

Cicada tigrina Boulard 1986d: 225 (type species of *Attenuella*)

Attenuella tigrina Boulard 1986d: 225, 236–238, Table I, Fig 39 (distribution, illustrated, comp. note) Ivory Coast

Attenuella tigrina Boulard 1990b: 207 (comp. note)

Attenuella tigrina Boulard 1999a: 78, 81 (listed, sonogram) Ivory Coast

Attenuella tigrina Sueur 2001: 36, Table 1 (listed)

Attenuella tigrina Boulard 2005d: 343 (comp. note)

Attenuella tigrina Boulard 2006g: 72, 78, 180, Fig 39 (sonogram, listed, comp. note) Ivory Coast

Attenuella tigrina 2007: 2, 12, 14, Fig 3 (illustrated, listed, comp. note)

***Severiana* Boulard, 1973**

Severiana Villet 1999b: 153 (comp. note)

Severiana Moulds 2005b: 392, 433 (higher taxonomy, listed, comp. note)

Severiana Goble, Price, Barker, and Villet 2008a: 673 (phylogeny, comp. note) Africa

S. magna Villet, 1999

Severiana magna Villet 1999b: 149–153, Figs 1–9 (n. sp., described, illustrated, distribution, habitat, comp. note) Namibia

S. severini severini (Distant, 1893) = *Poecilopsaltria severini* = *Poecsilopsaltria* (sic) *severini* = *Platypleura severini* = *Platypleura (Poecilopsaltria) severini* = *Platypleura fenestrata* Schumacher, 1913 = *Platypleura (Platypleura) schumacheri* Metcalf, 1955 = *Platypleura schumacheri*

Severiana lindiana Villet 1988: 72, Figs 2a–b (song, sonogram, comp. note) South Africa

Severiana severini Villet 1999b: 153 (comp. note)

Severiana severini Sueur 2001: 37, Table 1 (listed) Equals *Severiana lindiana*

Severiana severini Villet, Barker and Lunt 2005: 590–593, Figs 1–5, Table 1 (phylogeny, listed, comp. note) South Africa

S. severini vitreomaculata (Schumacher, 1913) = *Platypleura fenestrata vitreomaculata* Schumacher, 1913 = *Platypleura fenestrata* var. *vitreomaculata* = *Platypleura (Platypleura) vitreomaculata*

S. similis (Schumacher, 1913) = *Platypleura similis*

Severiana similis Villet 1999b: 153 (comp. note)

***Koma* Distant, 1904** = *Kuma* (sic)

Koma Boulard 1985e: 1023 (coloration, comp. note)

Koma Boulard 1996d: 126 (listed)

Koma Moulds 2005b: 392, 433 (higher taxonomy, listed, comp. note)

Koma Moulds 2008a: 208 (comp. note)

K. bombifrons (Karsch, 1890) = *Platypleura bombifrons* = *Poecilopsaltria bombifrons* = *Kuma* (sic) *bombifrons*

Koma bombifrons Boulard 1990b: 89, 101, 103, Plate X, Figs 10–11 (genitalia, comp. note)

Koma bombifrons Boulard 1996d: 114, 126, 128–129, Fig S6 (song analysis, sonogram, listed, comp. note) Kenya

Koma bombifrons Villet and Noort 1999: 228–229, Table 12.2 (habitat, listed, comp. note) Tanzania

Koma bombifrons Sueur 2001: 36, Table 1 (listed)

Koma bombifrons Sueur 2002b: 122–123, 129 (listed, comp. note) Equals *Poecilopsaltria bombifrons*

Koma bombifrons Boulard 2006g: 180 (listed)

K. intermedia Boulard, 1980

K. semivitrea (Distant, 1914) = *Platypleura semivitrea* = *Koma basilewskyi* Dlabola, 1958

K. umbrosa camusa Boulard, 1980

K. umbrosa umbrosa Boulard, 1980

Systophlochius Villet, 1989 to *Platypleura* Amyot & Audinet-Serville, 1843

S. palochius Villet, 1989 to *Platypleura techowi* Schumacher, 1913

***Platypleura* Amyot & Audinet-Serville, 1843** = *Cicada (Platypleura)* = *Platypleura (Platypleura)* = *Patypleura* (sic) = *Plathypleura* (sic) = *Platypeura* (sic) = *Platypleure* (sic) =

Rlatypleura (sic) = *Poecilopsaltria* Stål, 1866 = *Poicilopsaltria* (sic) = *Paecilopsaltria* (sic) = *Poecilopsaltera* (sic) = *Platypleura (Poecilopsaltria)* = *Systophlochius* Villet, 1989

Platypleura Dallas 1870: 482 (listed)
Poecilopsaltria Dallas 1870: 482 (listed)
Platypleura Distant 1881: ii (comp. note) Madagascar
Platypleura Weber 1931: 101–102, Fig 100 (spiracle illustrated, comp. note)
Platypleura Wu 1935: 1 (listed) Equals *Oxypleura* Equals *Poecilopsaltria* Equals *Dasypsaltria* Equals *Neoplatypleura* China
Platypleura Zeuner 1941: 88 (comp. note)
Platypleura Chen 1942: 143 (listed)
Platypleura Zeuner 1944: 111, 113 (comp. note)
Platypleura Kurokawa 1953: 1 (chromosomes, comp. note) Japan
Platypleura Weber 1968: 274–275, Fig 208c (tracheal system illustrated, comp. note)
Platypleura Medler 1980: 73 (listed) Equals *Poecilopslatria* Nigeria
Platypleura Boulard 1981d: 103 (comp. note) Comoros Islands
Platypleura Gess 1981: 7 (comp. note) South Africa
Platypleura spp. Gess 1981: 8 (comp. note) South Africa
Platypleura Josephson and Young 1981: 220 (comp. note)
Platypleura Shcherbakov 1981a: 830 (listed)
Platypleura Shcherbakov 1981b: 66 (listed)
Platypleura Strümpel 1983: 18, 83, 109, Table 8 (listed, comp. note)
Platypleura Whalley 1983: 141 (comp. note)
Cicada (sic) sp. Ahmad 1984: 270, 271, 273 Fig 3B, 274 Table 3, 275 Fig 5, 276 Table 4, 277 Fig 14, 15 (flight system morphology, wing illustrated) India
Platypleura Boulard 1984c: 175 (type genus of Platypleurinae)
Platypleura Ewing 1984: 228 (comp. note)
Platypleura Hayashi 1984: 28, 29–30, 34 (key, described, synonymy, listed, comp. note) Equals *Oxypleura* Equals *Poecilopsaltria* Equals *Dasypsaltria* Equals *Neoplatypleura*
Poecilopsaltria Hayashi 1984: 31 (comp. note)
Neoplatypleura Hayashi 1984: 31 (comp. note)
Platypleura Smith 1984: 142 (timbal muscle, comp. note)
Platypleura Boulard 1985c: 184 (comp. note)
Platypleura Boulard 1985e: 1023–1024, 1029–1030 (coloration, comp. note)
Platypleura Claridge 1985: 301 (song, comp. note)
Platypleura Popov 1985: 35 (acoustic system, comp. note)
Platypleura sp. Theron 1985: 156–157, Fig 16.120–16.121 (illustrated) South Africa
Platypleura Villet 1987b: 209, 215 (comp. note)
Platypleura Boulard 1988f: 33 (type species of Platypleurinae, comp. note)
Cicada (sic) sp. Byrne, Buchmann and Spangler 1988: 16 (wing, comp. note)
Platypleura Ewart 1989a: 77 (comp. note) South Africa
Platypleura Villet 1989a: 329, 331 (listed, comp. note)
Systophlochius Villet 1989b: 52–53 (n. gen., described, listed, comp. note) South Africa
Platypleura Villet 1989b: 52, 54, 59–60 (comp. note)
Platypleura spp. Villet 1989b: 52, 54, 59 (comp. note)
Poecilopsaltria Villet 1989a: 329 (comp. note)
Platypleura sp. Boulard and Nel 1990: 43 (comp. note)
Platypleura Moulds 1990: 25, 52 (comp. note) Asia, Africa
Platypleura Boulard 1991f: 16 (warning sound, comp. note)
Platypleura Duffels 1991a: 119–120, 128 (comp. note) Africa, Orient
Platypleura spp. Liu 1992: 30 (comp. note)
Platypleura Peng and Lei 1992: 105 (comp. note) China, Hunan
Cicada (sic) spp. Murty, Jamil and Ahmed 1993: 206 (comp. note)
Platypleura Gogala 1994a: 33 (comp. note) Thailand
Platypleura Villet 1994b: 91 (comp. note)
Systophlochius Villet 1994b: 91 (comp. note)
Platypleura Zaidi and Ruslan 1994: 427, Table 2 (comp. note) Malaysia
Platypleura Lee 1995: 17, 22, 52–53, 135 (listed, key, described, ecology, distribution, illustrated, comp. note) Korea
Platypleura spp. Noort 1995: 92 (comp. note) South Africa
Platypleura Zaidi and Ruslan 1995a: 174, Table 2 (listed, comp. note) Borneo, Sarawak
Platypleura Zaidi and Ruslan 1995c: 70, Table 2 (listed, comp. note) Peninsular Malaysia
Platypleura Zaidi and Ruslan 1995d: 202, Table 2 (listed, comp. note) Borneo, Sabah
Platypleura Boer and Duffels 1996b: 302 (comp. note)
Platypleura Boulard 1996d: 128 (listed)
Platypleura Zaidi 1996: 102–103, Table 2 (comp. note) Borneo, Sarawak
Platypleura Zaidi and Hamid 1996: 54–55, Table 2 (listed, comp. note) Borneo, Sarawak
Platypleura Boulard 1997c: 102, 105–106, 117 (type genus of Platypleurini, comp. note)

Platypleura Chou, Lei, Li, Lu and Yao 1997: 167–168, 293, 307, 315, 324, 326, 358, Table 6, Tables 10–12 (key, described, listed, synonymy, diversity, biogeography, comp. note) Equals *Oxypleura* Equals *Poecilopsaltria* Equals *Dasypsaltria* Equals *Neoplatypleura* China

Platypleura Novotny and Wilson 1997: 437 (listed)

Platypleura Villet 1997a: 325 (comp. note) South Africa

Platypleura Zaidi 1997: 115, Table 2 (listed, comp. note) Borneo, Sarawak

Platypleura Lee 1999: 2 (comp. note) Korea

Platypleura Ohbayashi, Sato and Igawa 1999: 340 (parasitism, comp. note)

Platypleura Villet 1999b: 153 (comp. note)

Platypleura Villet and Noort 1999: 227, 229 (comp. note)

Platypleura spp. Villet and Noort 1999: 227–228 (comp. note)

Platypleura Zaidi and Ruslan 1999: 4, Table 2 (listed, comp. note) Borneo, Sarawak

Platypleura sp. Zaidi, Ruslan and Azman 1999: 300 (listed, comp. note) Borneo, Sabah

Platypleura Boulard 2000i: 21 (listed)

Platypleura sp. Zaidi, Ruslan and Azman 2000: 198 (listed, comp. note) Borneo, Sabah

Platypleura Boulard 2001b: 22–26, 37 (type genus of Platypleurini, comp. note)

Platypleura sp. Zaidi and Nordin 2001: 183, 185, 189, 193, Table 1 (listed, comp. note) Borneo, Sabah

Platypleura Boulard 2002b: 193 (coloration, comp. note)

Platypleura spp. Sanborn, Villet and Phillips 2002: 32 (endothermy, comp. note)

Platypleura sp. Zaidi, Nordin, Maryati, Wahab, Norashikin et al. 2002: 2, 8, 10, Tables 1–2 (listed, comp. note) Borneo, Sabah

Platypleura Lee 2003b: 7 (comp. note)

Platypleura Lee and Hayashi 2003a: 153 (key, diagnosis) Taiwan

Platypleura Milius 2003: 408 (comp. note) South Africa

Platypleura spp. Sanborn, Phillips and Villet 2003b: 295 (comp. note) South Africa

Platypleura Sanborn, Villet and Phillips 2003: 305 (thermoregulation, comp. note) South Africa, Eastern Cape

Platypleura spp. Sanborn, Villet and Phillips 2003: 306–308 (comp. note) South Africa

Platypleura sp. Zaidi and Azman 2003: 104, Table 1 (listed, comp. note) Borneo, Sabah

Platypleura Chen 2004: 39–40, 43 (phylogeny, listed)

Platypleura Morris 2004: 70 (human food) Malawi

Platypleura Sanborn, Villet and Phillips 2004a: 819 (comp. note)

Platypleura Schouten, Duffels and Zaidi 2004: 376 (comp. note)

Platypleura Stokes and Josephson 2004: 280 (timbal muscle, comp. note)

Platypleura Boulard 2005d: 345 (comp. note)

Platypleura Lee 2005: 14, 16, 157, 168 (listed, key, comp. note) Korea

Platypleura Moulds 2005b: 388–389, 392, 412–413, 416, 433 (type genus of Platypleurinae, type genus of Platypleruini, higher taxonomy, listed, comp. note)

Platypleura Pham and Thinh 2005a: 238 (listed) Vietnam

Platypleura Pham and Thinh 2005b: 289 (comp. note) Vietnam

Platypleura Villet, Barker and Lunt 2005: 589–593 (diversity, synonymy, comp. note) Equals *Systophlochius* n. syn. South Africa, Kenya, Angola

Systophlochius Villet, Barker and Lunt 2005: 590–593, Figs 1–5, Table 1 (synonymy, listed, comp. note) South Africa

Platypleura spp. Villet, Barker and Lunt 2005: 591 (comp. note)

Platypleura Boulard 2006f: 620 (comp. note) Thailand

Poecilopsaltria Boulard 2006f: 621 (comp. note)

Platypleura Champman 2005: 321 (comp. note)

Platypleura Boulard 2006g: 109, 113–114 (comp. note) Africa

Platypleura Schabel 2006: 258 (human food, comp. note) Africa

Platypleura Boulard 2007e: 35, 47, 77 (comp. note) Thailand

Platypleura Boulard 2007f: 52 (comp. note)

Platypleura Price, Barker and Villet 2007: 2575 (comp. note)

Platypleura Sanborn, Phillips and Sites 2007: 3–4, Table 1 (listed, comp. note) Thailand

Platypleura Shiyake 2007: 2, 14, 18 (listed, comp. note)

Platypleura Sueur and Puissant 2007b: 55 (comp. note)

Platypleura Boulard 2008f: 7, 15, 91 (listed) Thailand

Platypleura (Platypleura) Boulard 2008f: 15 (listed)

Platypleura (Poecilopsaltria) Boulard 2008f: 7, 16 (listed) Thailand

Platypleura (Neoplatypleura) Boulard 2008f: 7, 16 (listed) Thailand

Platypleura Hooper, Hobbs and Thuma 2008: 105–106 (comp. note)

Platypleura Lee 2008a: 2–3, Table 1 (listed) Equals *Poecilopsaltria* Vietnam

Platypleura Lee 2008b: 446, Table 1 (listed, comp. note) Korea
Platypleura Osaka Museum of Natural History 2008: 322 (listed) Japan
Platypleura Lee 2009h: 2618–2619 (synonymy, listed, comp. note) Equals *Poecilopsaltria* Philippines, Palawan
Poecilopsaltria Lee 2009h: 2618 (listed)
Platypleura Lee 2009i: 293–294 (listed, comp. note) Philippines, Panay
Platypleura Pham and Yang 2009: 13, Table 2 (listed) Vietnam
Platypleura Price, Barker and Villet 2009: 618 (comp. note)
Platypleura Tolley, Makokha, Houniet, Swart and Matthee 2009: 120 (comp. note) South Africa
Platypleura Ahmed and Sanborn 2010: 38 (type species, distribution) Africa, Asia, Japan
Platypleura Hayashi, Kamitani and Ohara 2010: 116–117 (comp. note) Ryukyus
Platypleura Lee 2010a: 15 (listed, comp. note) Philippines, Mindanao
Platypleura Lee 2010b: 20 (listed) Cambodia
Platypleura Lee 2010c: 2 (synonymy, listed, comp. note) Equals *Poecilopsaltria* Philippines, Luzon
Poecilopsaltria Lee 2010c: 2 (listed)
Platypleura Mozaffarian and Sanborn 2010: 76 (distribution, listed) Africa, Japan, Iran

***P. adouma* Distant, 1904** = *P. adoreus* (sic)
Platypleura adouma Medler 1980: 73 (listed) Nigeria, Central African Republic, Zaire
Platypleura adouma Whalley 1983: 141–42, Fig 4 (illustrated, comp. note)
Platypleura adouma Boulard 1985e: 1028–1029, 1044–1047, Plate VII, Fig 47, Plate VIII, Fig 49 (coloration, illustrated, comp. note)
Platypleura adouma Nkouka 1987: 174 (human food) Central Africa
Platypleura adouma Villet 1987b: 212 (comp. note) South Africa
Platypleura adouma Boulard 1990b: 188–189, Plate XXIII, Fig 6 (illustrated, comp. note)
Platypleura adouma Huis 1996: 7, Table 1 (predation, comp. note)
Platypleura adouma Villet 1999d: 217 (comp. note) Angola, Congo
Platypleura adouma Huis 2003: 168, Table 1 (predation, comp. note) Democratic Republic of Congo
Platypleura adouma Boulard 2007f: 131, Fig 21 (illustrated, comp. note)

***P. affinis affinis* (Fabricius, 1803)** = *Tettigonia affinis* = *Cicada affinis* = *Poecilopsaltria affinis* = *Platypleura (Poecilopsaltria) affinis*
Tettigonia affinis Dallas 1870: 482 (synonymy) Equals *Poecilopsaltria affinis*
Tettigonia affinis Hayashi 1984: 30 (type species of *Poecilopsaltria*)
Tettigonia affinis Lee 2008a: 3 (type species of *Poecilopsaltria*) India Oriental

***P. affinis distincta* Atkinson, 1884** = *Platypleura nicobarica* Atkinson, 1884 (nec *Platypleura nicobarica* Butler, 1874) = *Poecilopsaltria nicobarica* = *Poecilopsaltria nicobarica* var. *a* Distant, 1889 = *Platypleura (Poecilopsaltria) affinis distincta*

P. albigera Walker, 1850 to *Albanycada albigera*

***P. albivannata* Hayashi, 1974**
Platypleura albivannata Hayashi 1984: 29, 32–34, Figs 14–15, Figs 20–21 (key, described, illustrated, comp. note) Japan
Platypleura albivannata Shiyake 2007: 2, 17–18, Fig 12 (illustrated, distribution, listed, comp. note)
Platypleura albivannata Osaka Museum of Natural History 2008: 322 (listed, acoustic behavior, ecology, comp. note) Japan

***P. andamana* Distant, 1878** = *Platypleura (Platypleura) andamana* = *Platypleura (Poecilopsaltria) andamana* = *Poecilopsaltria andamana* = *Platypleura roepstorffii* Atkinson, 1884
*Poecilopsaltria andamana*Terradas 1999: 48–49, Fig 1 (illustrated, comp. note) Andaman Islands

***P. arabica* Myers, 1928**
Platypleura arabica Al-Azawi 1986: 254 (date palm pest) Qatar, United Arab Emirates
Platypleura arabica Walker and Pittaway 1987: 38–39 (illustrated, natural history, distribution, comp. note) Qatar, United Arab Emirates, Oman
Platypleura arabica Kossenko and Fry 1998: 9–10, Table 4 (predation) Turkmenistan, Uzbekistan
Platypleura arabica Freytag 2000: 55 (on stamp)
Platypleura arabica Gillett and Gillett 2005: 172–173 (illustrated, comp. note) Emirates
Platypleura arabica Schedl 2007: 153–157, Figs 1–2, Plates 1–2 (described, illustrated, ecology, distribution, listed, comp. note) United Arab Emirates, Jordan, Qatar, Oman

***P. argentata* Villet, 1987**
Platypleura argentata Villet 1987a: 272, Table 2 (song intensity, listed, comp. note) South Africa
Platypleura argentata Villet 1987b: 210–214, Figs 1a–b, Fig 2b, Fig 3 (n. sp., described,

illustrated, distribution, habitat, comp. note) South Africa
Platypleura argentata Villet 1988: 72–73, 75–76, Fig 4, Figs 5a–b, Table 1 (song, sonogram, comp. note) South Africa
Platypleura argentata Villet and Reavell 1989: 334–335, Fig 3 (distribution, habitat, host plants, key, comp. note) South Africa
Platypleura argentata Villet 1992: 94–95, Table 1 (acoustic responsiveness, comp. note) South Africa
Platypleura argentata Sueur 2001: 37, Table 1 (listed)
Platypleura argentata Malherbe, Burger and Stephen 2004: 86 (comp. note) South Africa

***P. argus argus* Melichar, 1911**
Platypleura argus Boulard 1985e: 1019–1020, 1034, 1037–1038, Plate I, Fig 4, Plate III, Fig 18, Plate IV, Fig 27 (crypsis, coloration, illustrated, comp. note) Kenya, East Africa
Platypleura argus Boulard 1996d: 114, 128, 130–132, Fig S7 (song analysis, sonogram, listed, comp. note) Kenya
Platypleura argus Boulard 1997b: 20, Plate II, Fig 1 (morphology, comp. note)
Platypleura argus Sueur 2001: 37, Table 1 (listed)
Platypleura argus Boulard 2002b: 193 (coloration, comp. note)
Platypleura argus Boulard 2006g: 180 (listed)
Platypleura argus Boulard 2007f: 84, Fig 100.1 (illustrated, comp. note)

***P. argus marsabitensis* Boulard, 1980** = *Platypleura marsabitensis*
Platypleura argus marsabitensis Boulard 1996d: 114, 131–132, Fig S8 (n. comb., song analysis, sonogram, listed, comp. note) Kenya
Platypleura marsabitensis Boulard 1996d: 131 (synonymy)
Platypleura marsabitensis Villet 1999d: 217 (comp. note)

***P. arminops* Noualhier, 1896** = *Platypleura (Poecilopsaltria) arminops*
Platypleura arminops Boulard 2006f: 621 (comp. note) Thailand
Platypleura (Poecilopsaltria) arminops Boulard 2008f: 16 (biogeography, listed, comp. note) Equals *Platypleura arminops* Thailand
Platypleura arminops Boulard 2008f: 91 (listed) Thailand

***P. assamensis* Atkinson, 1884** = *Platypleura repanda* var. *a* Distant, 1889 = *Platypleura repanda assamensis* = *Platypleura (Poecilopsaltria) assamensis*
Platypleura assamensis Chou, Lei, Li, Lu and Yao 1997: 168, 171–172, 357, Fig 9–41, Plate VII, Fig 76 (key, synonymy, described, illustrated, comp. note) Equals *Platypleura repanda* var. *assamensis* China

***P. auropilosa* Kato, 1940** = *Platypleura auripilosa* (sic)
Platypleura auropilosa Sanborn, Phillips and Sites 2007: 4 (listed, distribution, comp. note) Thailand
Platypleura auropilosa Boulard 2008f: 15, 90–91 (biogeography, listed, comp. note) Thailand, Siam

***P. badia* Distant, 1888** = *Platypleura (Poecilopsaltria) badia* = *Afzeliada badia* = *Platypeura (Neoplatypleura) badia* = *Platypleura fasuayensis* Boulard, 2005
Afzeliada badia Chou, Lei, Li, Lu and Yao 1997: 179–180, 357–358, Fig 9–46 (n. comb. synonymy, described, illustrated, comp. note) Equals *Platypleura badia* China
Afzeliada badia Pham and Thinh 2005a: 238 (synonymy, distribution, listed) Equals *Platypleura badia* Vietnam
Platypleura badia Boulard 2006c: 133 (comp. note)
Platypleura fasuayensis Boulard 2006c: 133–136, Fig B, Figs 2–3 (n. sp., described, illustrated, song, sonogram, comp. note) Thailand
Platypleura (Neoplatypleura) badia Boulard 2007d: 495 (n. comb., synonymy, comp. note) Equals *Afzeliada badia* Equals *Platypleura fasuayensis* Thailand, China,
Afzeliada badia Sanborn, Phillips and Sites 2007: 4, 34 (listed, distribution, comp. note) Thailand, India, Burma, Vietnam, China
Platypleura (Neoplatyplerua) badia Boulard 2008f: 17 (synonymy, biogeography, listed, comp. note) Equals *Platypleura badia* Equals *Afzeliada badia* Equals *Platypleura fasuayensis* n. syn. Thailand, Tenasserim, Burma
Platypleura badia Boulard 2008f: 91 (listed) Thailand
Platypleura badia Lee 2008a: 4 (synonymy, distribution, listed) Equals *Platypleura (Poecilopsaltria) badia* Equals *Afzelidada badia* North Vietnam, China, Yunnan, Thailand, Malaysia, Myanmar, India
Platypleura badia Pham and Yang 2009: 13, Table 2 (listed) Vietnam

***P. basialba* (Walker, 1850)** = *Oxypleura basialba* = *Platypleura (Oxypleura) basialba* = *Platypleura (Poecilopsaltria) basialba*
Platypleura basialba Ahmed and Sanborn 2010: 38 (distribution) Equals *Oxypleura basialba* Pakistan, India

***P. basiviridis* Walker, 1850** = *Poecilopsaltria basiviridis* = *Platypleura (Poecilopsaltria) basiviridis*
Platypleura (Poecilopsaltria) basiviridis Boulard 2008c: 361 (comp. note) India

***P. bettoni* Distant, 1904** = *Platypleura (Poecilopsaltria) bettoni*

P. bouruensis Distant to *Hamza ciliaris* (Linnaeus, 1758)

***P.* "breedeflumensis" Price, Barker & Villet, 2009**

Platypleura "breedeflumensis" Price, Barker and Villet 2009: 618–619, 621–624, Figs 1–2, Fig 3C, Fig 4, Table 1 (illustrated, distribution, phylogeny, comp. note) South Africa

***P. brunea* Villet, 1989** = *Platypleura brunnea* (sic)

Platypleura brunea Villet 1989b: 58–59, 64, 67–68, Fig 7, Fig 19, Fig 29, Figs 37–39 (n. sp., described, illustrated, habitat, song, sonogram, frequency spectrum, distribution, comp. note) South Africa

Platypleura brunea Sueur 2001: 37, Table 1 (listed)

***P.* cf. *brunnea* Phillips, Sanborn & Villet, 2002**

Platypleura cf. *brunnea* Phillips, Sanborn and Villet 2002: 31 (thermal responses) South Africa, Eastern Cape

Platypleura cf. *brunnea* Sanborn, Villet and Phillips 2002: 32 (endothermy, comp. note)

Platypleura cf. *brunnea* Sanborn, Phillips and Villet 2003a: 349, Table 1 (thermal responses, habitat, comp. note) South Africa

Platypleura cf. *brunnea* Sanborn, Villet and Phillips 2003: 305, 307, Table 2 (endothermy, thermoregulation, comp. note) South Africa, Eastern Cape

Platypleura cf. *brunnea* Villet, Sanborn and Phillips 2003a: 2 (endothermy, comp. note) South Africa

Platypleura cf. *brunnea* Sanborn, Villet and Phillips 2004a: 817, 821, Tables 1–2 (thermoregulation, endothermy, habitat, comp. note) South Africa

***P. bufo* (Walker, 1850)** = *Oxypleura bufo* = *Platypleura (Oxypleura) bufo* = *Poecilopsaltria bufo* = *Platypleura (Platypleura) bufo* = *Platypleura (Poecilopsaltria) bufo*

***P. canescens* (Walker, 1870)** = *Oxypleura canescens* = *Platypleura (Oxypleura) canescens* = *Poecilopsaltria canescens*

***P. capensis* (Linnaeus, 1764)** = *Cicada capensis* = *Tettigonia capensis* = *Platypleura stridula* var. *capensis* = *Platypleura stridula capensis* = *Platypleura (Platypleura) stridula capensis* = *Platypleura stridula* var. *b* Stål, 1866

Cicada capensis Villet 1989a: 329 (comp. note)

Cicada stridula var. *capensis* Villet 1989a: 329 (comp. note)

Platypleura stridula var. *capensis* Villet 1989a: 329, 332 (comp. note)

Platypleura capensis Villet 1989a: 329–332, Figs 1C–F, Fig 2B, Fig 3B, Fig 4B, Fig 5B, Table 1 (status, illustrated, song, sonogram, frequency spectrum, habitat, distribution, key, comp. note) South Africa

Platypleura stridula capensis Villet 1989c: 326 (comp. note) South Africa

Platypleura capensis Gäde and Janssens 1994: 803–807, Fig 1A, Figs 2A–B, Fig 3, Tables 1–5 (hormone, comp. note) South Africa

Platypleura capensis Noort 1995: 92–93, Fig 1 (illustrated, comp. note) South Africa

Platypleura capensis Eckert, Gabriel, Birkenbeil, Greiner, Rapus et al. 1996: 402, 406–407, 412, Table 2 (peptide, comp. note) South Africa

Platypleura capensis Socha and Kodrík 1999: 278–279, Fig 1 (comp. note)

Platypleura capensis Sueur 2001: 37, Table 1 (listed)

Platypleura capensis Picker, Griffith and Weaving 2002: 158–159 (described, illustrated, habitat, comp. note) South Africa

Platypleura capensis Phillips, Sanborn and Villet 2002: 31 (thermal responses) South Africa, Eastern Cape

Platypleura capensis Sanborn, Villet and Phillips 2002: 32 (endothermy, comp. note)

Platypleura stridula var. *capensis* Sueur 2002b: 119, 129 (listed, comp. note) Equals *Platypleura capensis*

Platypleura capensis Sanborn, Phillips and Villet 2003a: 348–349, Table 1 (thermal responses, habitat, comp. note) South Africa

Platypleura capensis Sanborn, Phillips and Villet 2003b: 292–293 (comp. note)

Platypleura capensis Sanborn, Villet and Phillips 2003: 305–307, Fig 1, Tables 1–2 (endothermy, thermoregulation, comp. note) South Africa, Eastern Cape

Platypleura capensis Villet, Sanborn and Phillips 2003a: 2–3 (endothermy, comp. note) South Africa

Platypleura capensis Villet, Sanborn and Phillips 2003c: 1443 (comp. note)

Platypleura capensis Gäde, Auerswald, Predel, and Marco 2004: 85 (metabolism, comp. note)

Platypleura capensis Robertson, Villet and Palmer 2004: 463–464, 468, Figs 3a–b, Table, 1(distribution, comp. note)

Platypleura capensis Sanborn, Heath, Heath and Phillips 2004: 70 (listed)

Platypleura capensis Sanborn, Villet and Phillips 2004a: 817–821, Fig 3, Tables 1–2 (thermoregulation, endothermy, habitat, comp. note) South Africa

Platypleura capensis Chawanji, Hodson and Villet 2005a: 258–261, 264–265, Fig 3, Fig 6, Fig 9, Fig 13, Fig 34 (sperm morphology, comp. note) South Africa
Platypleura capensis Gäde 2005: 50 (hormone, comp. note)
Platypleura capensis Sanborn 2005c: 114, Table 1 (comp. note)
Platypleura capensis Villet, Barker and Lunt 2005: 590–593, Figs 1–5, Table 1 (phylogeny, listed, comp. note) South Africa
Platypleura capensis Chawanji, Hodson and Villet 2006: 386–387 (comp. note)
Platypleura capensis Price, Barker and Villet 2007: 2575, 2577, 2580, Table 2 (comp. note) South Africa
Platypleura capensis Yamazaki 2007: 349 (comp. note) South Africa
Platypleura capensis Fullard, Ratcliff, and Jacobs 2008: 242–244, Figs 1–2a (song, oscillogram, comp. note)
Platypleura capensis Lorenz and Gäde 2009: 383 (comp. note)
Platypleura capensis Price, Barker and Villet 2009: 621, Fig 2 (illustrated, phylogeny, comp. note) South Africa
Platypleura capensis Ye and Ng 2009: 188 (protein, comp. note)

***P. capitata* (Olivier, 1790)** = *Cicada capitata* = *Oxypleura capitata* = *Poecilopsaltria capitata* = *Platypleura (Poecilopsaltria) capitata* = *Oxypleura subrufa* Walker, 1850 = *Platypleura subrufa* = *Poecilopsaltria subrufa* = *Paecilopsaltria* (sic) *subrufa*
Platypleura capitata Hanson 1956: 702 (comp. note)
Platypleura capitata Pringle 1957: 26, 55–56, Table 1 (wing beat frequency, muscle kinetics, comp. note)
Platypleura (sic) Pringle 1957: 55, 58 (muscle kinetics, comp. note)
Platypleura capitata Goodchild 1966: 118 (comp. note)
Platypleura capitata Cullen 1974: 34, Table 6 (flight muscles structure, comp. note)
Platypleura capitata Josephson 1981: 24–25, 30–31, 34, Figs 2C–D, Figs 7–8, Fig 10 (timbal muscle structure, comp. note)
Platypleura capitata Josephson and Young 1981: 219–221, 224–233, Figs 1A–1B, Figs 2–6, Figs 7A–C, Tables 1–2 (timbal muscle structure and function, comp. note) Sri Lanka
Platypleura capitata Pringle 1981: 6–7, 10, Fig 8 (oscillogram, timbal muscles, comp. note) Sri Lanka
Platypleura capitata Strümpel 1983: 108 (comp. note)
Platypleura capitata Young and Josephson 1983a: 193 (comp. note)
Platypleura capitata Young and Josephson 1983b: 206 (comp. note)
Platypleura capitata Smith 1984: 142 (timbal muscle, comp. note) Sri Lanka
Platypleura capitata Young and Josephson 1984: 286–287, Fig 1 (comp. note)
Platypleura capitata Claridge 1985: 301 (song, comp. note)
Platypleura capitata Josephson and Young 1985: 204, 206–207 (comp. note)
Platypleura capitata Popov 1985: 37–38, 42, Fig 5 (illustrated, acoustic system, comp. note)
Platypleura capitata Josephson and Young 1987: 993–996, Fig 1, Fig 3 (comp. note)
Platypleura capitata Ewing 1989: 37 (comp. note)
P[latypleura] capitata Yang and Jiang 1994: 258, 260–261 (comp. note)
P[latypleura] capitata Yang, Jiang, Chen and Xu 1994: 217 (comp. note)
P[latypleura] capitata Jiang, Yang, Tang, Xu and Chen 1995b: 229 (song frequency, comp. note)
Platypleura capitata Yang and Jiang 1995: 177 (comp. note)
Platypleura capitata Bennet-Clark and Young 1998: 705 (comp. note)
Platypleura capitata Sanborn 1998: 94 (thermal biology, comp. note)
Platypleura capitata Sueur 2001: 37, Table 1 (listed)
Platypleura capitata Nation 2002: 244, 260 (comp. note)
Platypleura capitata Syme and Josephson 2002: 769, Fig 4A (timbal muscle, comp. note)
Platypleura capitata Srivastava and Bhandari 2004: 1479 (comp. note) India
Platypleura capitata Champman 2005: 321–322, 327 (comp. note)
Platypleura capitata Nation 2008: 264–265 (comp. note)

***P. carlinii* Distant, 1906** = *Platyplerua testacea* de Carlini, 1892 (nec *Zammara testacea* Walker, 1858)

***P. cervina* Walker, 1850** = *Platypleura (Oxypleura) cervina* = *Poecilopsaltria cervina* = *Platypleura straminea* Walker, 1850

***P. cespiticola* Boulard, 2006** = *Platypleura (Poecilopsaltria) cespiticola* = *Platypleura (Poecilopsaltria) cespiticola* var. *fuscalae* Boulard, 2006
Platypleura (Poecilopsaltria) cespiticola Boulard 2006f: 620–630, Figs 1–2, Figs 3a–3c, Figs 4–10 (n. sp., described, illustrated, song, sonogram, comp. note) Thailand

Platypleura (Poecilopsaltria) cespiticola var. *fuscalae* Boulard 2006f: 620, 624, 628, Fig 3d (n. var., described, comp. note) Thailand

Platypleura cespiticola Boulard 2007e: 77–80, Figs 50–51, Plate 37 (illustrated, sonogram, comp. note) Thailand

Platypleura (Poecilopsaltria) cespiticola Boulard 2008f: 16, 91 (biogeography, listed, comp. note) Thailand

***P. chalybaea* Villet, 1989**

Platypleura chalybaea Villet 1989b: 56–58, 64, 67, Figs 5–6, Fig 18, Fig 29 (n. sp., described, illustrated, habitat, song, sonogram, frequency spectrum, distribution, comp. note) South Africa

Platypleura chalybaea Villet and Capitao 1995: 140 (habitat, comp. note) South Africa

Platypleura chalybaea Villet and Capitao 1996: 282–283, Tables 1–2 (density, habitat, comp. note) South Africa

Platypleura chalybaea Villet, Barker and Lunt 2005: 590–593, Figs 1–5, Table 1 (phylogeny, listed, comp. note) South Africa

P. ciliaris (Linnaeus, 1758) to *Hamza ciliaris*

***P. coelebs* Stål, 1863** = *Platypleura caelebs* (sic) = *Poecilopsaltria coelebs* = *Platypleura (Poecilopsaltria) coelebs*

Platypleura coelebs Wu 1935: 1 (listed) Equals *Poecilopsaltria coelebs* China, Chusan

Platypleura coelebs Chou, Lei, Li, Lu and Yao 1997: 172 (comp. note) China

Platypleura coelebs Hua 2000: 63 (listed, distribution) China, Guangdong, Guangxi, Zhejiang, Vietnam, India

Platypleura coelebs Lee 2008a: 3 (synonymy, distribution, listed) Equals *Platypleura (Poecilopsaltria) coelebs* Vietnam, China, Guangdong, Guangxi, India

Platypleura coelebs Pham and Yang 2009: 13, Table 2 (listed) Vietnam

***P. crampeli* Boulard, 1975**

Platypleura crampeli Medler 1980: 73 (listed) Nigeria, Central African Republic, Zaire

Platypleura crampeli Villet 1999d: 217 (comp. note) Central African Republic

P. decora (Germar, 1834) to *Capcicada decora*

***P. deusta* (Thunberg, 1822)** = *Tettigonia deusta*

Platypleura deusta Robertson, Villet and Palmer 2004: 463–464, 466, 468, Figs 3e–f, Table, 1 (distribution, comp. note)

Platypleura deusta Villet, Barker and Lunt 2005: 590–593, Figs 1–5, Table 1 (phylogeny, listed, comp. note) South Africa

P. dimidiata Karsch, 1893 to *Sadaka dimidiata*

***P. divisa* (Germar, 1834)** = *Cicada divisa* = *Platypleura divisa* var. *a* Stål, 1866 = *Dasypsaltria maera* Haupt, 1917 = *Platypleura (Platypleura) divisa* = *Platypleura hirtipennis* Villet 1987 (nec Germar) = *Platypleura hirtipennis* Villet 1988 (nec Germar)

Dasypsaltria maera Hayashi 1984: 30 (type species of *Dasypsaltria*)

Platypleura hiritpennis (sic) Villet 1987b: 210 (comp. note) South Africa

Platypleura divisa Villet 1987b: 212 (comp. note) South Africa

Platypleura divisa Villet 1988: 72–73, 75–76, Fig 4, Figs 5c–d, Tables 1–2 (song, sonogram, comp. note) South Africa

Platypleura hirtipennis (sic) Villet 1988: 72–76, Fig 4, Figs 6c–d, Tables 1–2 (song, sonogram, comp. note) South Africa

Platypleura divisa Villet 1989b: 58, 67, Fig 29 (distribution, comp. note) South Africa

Platypleura divisa Villet 1989c: 326 (comp. note) South Africa

Platypleura divisa Villet and Reavell 1989: 333–335, Fig 1c, Fig 3 (distribution, habitat, host plants, key, comp. note) South Africa

Platypleura divisa Villet 1992: 94–95, Table 1 (acoustic responsiveness, comp. note) South Africa

Platypleura divisa Villet 1997a: 321–326, Figs 1–5 (described, illustrated, type material, synonymy, habitat, distribution, comp. note) Equals *Cicada divisa* Equals *Platypleura hirtipennis* Villet 1987:210 Equals *Platypleura hirtipennis* Villet 1988:73 South Africa

Platypleura divisa Villet 1999d: 210, 216–217 (comp. note)

Platypleura divisa Sueur 2001: 37, Table 1 (listed)

Platypleura divisa Phillips, Sanborn and Villet 2002: 31 (thermal responses) South Africa, Eastern Cape

Platypleura divisa Sanborn, Villet and Phillips 2002: 32 (endothermy, comp. note)

Platypleura divisa Sanborn, Phillips and Villet 2003a: 348–349, Table 1 (thermal responses, habitat, comp. note) South Africa

Platypleura divisa Sanborn, Phillips and Villet 2003b: 291 (comp. note) South Africa

Platypleura divisa Sanborn, Villet and Phillips 2003: 305, 307, Table 2 (endothermy, thermoregulation, comp. note) South Africa, Eastern Cape

Platypleura divisa Villet, Sanborn and Phillips 2003a: 2 (endothermy, comp. note) South Africa

Platypleura divisa Malherbe, Burger and Stephen 2004: 86 (comp. note) South Africa

Platypleura divisa Sanborn, Villet and Phillips 2004a: 817, 821, Tables 1–2 (thermoregulation, endothermy, habitat, comp. note) South Africa

Platypleura divisa Villet, Barker and Lunt 2005: 591 (comp. note)

P. durvillei (Boulard, 1991) = *Poecilopsaltria durvillei*

Poecilopsaltria durvillei Boulard 1991c: 121–123, Figs 3–6 (n. sp., described, illustrated, comp. note) New Guinea

P. elizabethae Lee, 2009 = *Platypleura* sp. aff. *nobilis* Endo and Hayashi 1979

Platypleura elizabethae Lee 2009h: 2618–2621, Figs 1–2 (n. sp., described, illustrated, distribution, listed, comp. note) Philippines, Palawan

Platypleura elizabethae Lee 2010a: 15–16, Figs 1A–1B (illustrated, distribution, listed, comp. note) Equals *Platypleura* sp. aff. *nobilis* Endo and Hayashi 1979 Philippines, Mindanao, Palawan

P. *fasuayensis* Boulard, 2006 to *Platypleura badia*

P. fulvigera fulvigera Walker, 1850 = *Poecilopsaltria fulvigera* = *Platypleura (Poecilopsaltria) fulvigera*

Platypleura fulvigera Showalter 1935: 38–39, Fig 12 (coloration, comp. note) Java, Malay Archipelago

Platypleura fulvigera Lee 2009h: 2617 (synonymy, distribution, listed) Equals *Platypleura (Poecilopsaltria) fulvigera* Philippines, Palawan, Luzon, Mindanao, Tara Island, Indonesia, Sulawesi

Platypleura fulvigera Lee 2010c: 2 (synonymy, distribution, listed) Equals *Platypleura (Poecilopsaltria) fulvigera* Philippines, Luzon, Mindoro, Tara Island, Palawan, Indonesia, Sulawesi

P. fulvigera var. *a* Distant, 1889 = *Platypleura (Poecilopsaltria) fulvigera* var. *a*

P. "gamtoosflumensis" Price, Barker & Villet, 2009

Platypleura "gamtoosflumensis" Price, Barker and Villet 2009: 618–619, 621–622, 624, Figs 1–2, Fig 4, Table 1 (illustrated, distribution, phylogeny, comp. note) South Africa

P. **"gariepflumensis" Price, Barker & Villet, 2009**

Platypleura "gariepflumensis" Price, Barker and Villet 2009: 618–619, 621–622, 624, Figs 1–2, Fig 4, Table 1 (illustrated, distribution, phylogeny, comp. note) South Africa

P. girardi Boulard, 1977

P. gowdeyi Distant, 1914

Platypleura gowdeyi Boulard 1985e: 1037, Plate I, Fig 7 (coloration, illustrated, comp. note) Kenya

Platypleura gowdeyi Villet, Barker and Lunt 2005: 589–593, Figs 1–5, Table 1 (phylogeny, listed, comp. note) Kenya

Platypleura gowdeyi Boulard 2006g: 180 (listed)

P. haglundi Stål, 1866

Platypleura haglundi Villet 1988: 72–73, 75, Figs 3a–b, Tables 1–2 (song, sonogram, comp. note) South Africa

Platypleura haglundi Villet 1989c: 326 (comp. note) South Africa

Platypleura haglundi Villet and Reavell 1989: 333–335, Fig 1d, Fig 3 (distribution, habitat, host plants, key, comp. note) South Africa, Zimbabwe

Platypleura haglundi Sueur 2001: 37, Table 1 (listed)

Platypleura haglundi Picker, Griffith and Weaving 2002: 158–159 (described, illustrated, habitat, comp. note) South Africa

Platypleura haglundi Villet, Barker and Lunt 2005: 590–593, Figs 1–5, Table 1 (phylogeny, listed, comp. note) South Africa

P. hampsoni (Distant, 1887) = *Poecilopsaltria hampsoni* = *Platypleura (Poecilopsaltria) hampsoni*

Platypleura hampsoni Srivastava and Bhandari 2004: 1479 (comp. note) India

P. harmandi Distant, 1905 = *Platypleura (Poecilopsaltria) harmandi*

Platypleura harmandi Lee 2008a: 3 (synonymy, distribution, listed) Equals *Platypleura (Poecilopsaltria) harmandi* South Vietnam

Platypleura harmandi Pham and Yang 2009: 13, 16, Table 2 (listed, comp. note) Vietnam

Platypleura harmandi Pham and Yang 2010a: 133 (comp. note) Vietnam

P. hilpa Walker, 1850 = *Poecilopsaltria hilpa* = *Platypleura (Platypleura) hilpa* = *Platypleura fenestrata* Uhler, 1861

Platypleura fenestrata Henshaw 1903: 230 (listed) Japan

Platypleura hilpa Chen 1933: 359 (listed, comp. note)

Platypleura hilpa Wu 1935: 2 (listed) Equals *Poecilopsaltria hilpa* Equals *Platypleura fenestrata* China, Canton, Amoy, Foochow, Tonkin

Platypleura hilpa Zeuner 1941: 88 (comp. note) Indo-China

Platypleura hilpa Hill, Hore and Thornton 1982: 50, 161–162, Plate 105 (listed, comp. note) Hong Kong

Platypleura hilpa Wang and Zhang 1987: 295 (key, comp. note) China

Platyplerua hilpa Duffels 1991a: 128–129 (comp. note) China, Hainan, Hong Kong, Taiwan

Platypleura hilpa Liu 1991: 257 (comp. note)
Platypleura hilpa Liu 1992: 16 (comp. note)
Platypleura hilpa Peng and Lei 1992: 97, Fig 276 (listed, illustrated, comp. note) China, Hunan
Platypleura hilpa Wang and Zhang 1992: 119 (listed, comp. note) China
Platypleura hilpa Lei, Chou and Li 1994: 54 (comp. note)
Platypleura hilpa Villet 1994b: 91 (comp. note)
Platypleura hilpa Chou, Lei, Li, Lu and Yao 1997: 28, 168, 170–171, 308, 316–317, 324, 326, 331, 339, Fig 9–40, Plate VII, Fig 77, Tables 10–12 (key, synonymy, described, illustrated, song, oscillogram, spectrogram, comp. note) Equals *Platypleura fenestrata* China
Platypleura hilpa Easton and Pun 1999: 102 (distribution, comp. note) Macao, Hong Kong, China, Guangdong, Hunan
Platypleura fenestrata Sanborn 1999a: 36 (type material, listed)
Platypleura hilpa Hua 2000: 63 (listed, distribution, hosts) China, Hubei, Jiangsu, Jiangxi, Zhejiang, Fujian, Taiwan, Guangdong, Hong Kong, Hainan, Hunan, Guangxi, Sichuan, Yunnan, Vietnam
Platypleura hilpa Sueur 2001: 37, Table 1 (listed)
Platypleura hilpa Lee and Hayashi 2003a: 153, 156 (key, diagnosis, distribution, song, listed, comp. note) Taiwan, China
Platypleura hilpa Chen 2004: 43, 49, 198 (described, illustrated, listed, comp. note) Taiwan
Platypleura hilpa Pham and Thinh 2005a: 238 (synonymy, distribution, listed) Equals *Platyneura* (sic) *fenestrata* Vietnam
Platypleura hilpa Pham and Thinh 2006: 526–527 (listed, comp. note) Vietnam
Platypleura hilpa Shiyake 2007: 2, 16, 18, Fig 9 (illustrated, distribution, listed, comp. note)
Platypleura (Poecilopsaltria) hilpa Boulard 2008c: 361 (comp. note) Vietnam
Platypleura hilpa Lee 2008a: 3 (synonymy, distribution, listed) Equals *Platypleura (Platypleura) hilpa* North Vietnam, South China, Hainan, Taiwan, Penghu Islands
Platypleura hilpa Lee 2009i: 293 (comp. note) Philippines, Panay
Platypleura hilpa Pham and Yang 2009: 13, Table 2 (listed) Vietnam

***P. hirta* Karsch, 1890**
Platypleura hirta Zeuner 1941: 88 (comp. note) South Africa
Platypleura hirta Villet 1988: 72–73, Figs 3c–d (song, sonogram, comp. note) South Africa
Platypleura hirta Villet 1989c: 325–326, Fig 1A (distribution, comp. note) Equals *Platypleura stridula capensis* Distant 1907 Plate XVII South Africa
Platypleura hirta Sueur 2001: 37, Table 1 (listed)

***P. hirtipennis* (Germar, 1834)** = *Cicada hirtipennis* = *Platypleura (Platypleura) hirtipennis* = *Platypleura capensis* (nec Linnaeus) = *Platypleura ocellata* (nec Degeer) = *Platypleura chloronota* Walker, 1850 = *Platypleura hirtipennis* var. *b* Stål, 1866
Platypleura hirtipennis Gess 1981: 7–8 (described, song, habitat, comp. note) South Africa
Platypleura hirtipennis Villet 1989c: 325–326, Fig 1B (distribution, comp. note) Equals *Platypleura divisa* Distant, 1907 Plate XVII South Africa
Platypleura hirtipennis Villet and Reavell 1989: 333–335, Fig 1b, Fig 3 (distribution, habitat, host plants, key, comp. note) South Africa
Platypleura hirtipennis Villet 1997a: 321–322, 325–329, 332, Figs 6–10 (described, illustrated, type material, synonymy, habitat, comp. note) Equals *Cicada hirtipennis* South Africa
Platypleura hirtipennis Sueur 2001: 37, Table 1 (listed)
Platypleura hirtipennis Phillips, Sanborn and Villet 2002: 31 (thermal responses) South Africa, Northern Province
Platypleura hirtipennis Sanborn, Breitbarth, Heath and Heath 2002: 445 (comp. note)
Platypleura hirtipennis Sanborn, Villet and Phillips 2002: 32, Fig 1 (endothermy, comp. note)
Platypleura hirtipennis Sanborn, Phillips and Villet 2003a: 349, Table 1 (thermal responses, habitat, comp. note) South Africa
Platypleura hirtipennis Sanborn, Phillips and Villet 2003b: 291–294, Figs 1–3, Table 1 (song, sonogram, oscillogram, distribution, comp. note) South Africa
Platypleura hirtipennis Sanborn, Villet and Phillips 2003: 305–307, Tables 1–2 (endothermy, thermoregulation, comp. note) South Africa, Eastern Cape
Platypleura hirtipennis Villet, Sanborn and Phillips 2003a: 2 (endothermy, comp. note) South Africa
Platypleura hirtipennis Sanborn, Villet and Phillips 2004a: 817–818, 821, Fig 2, Tables 1–2 (thermoregulation, endothermy, habitat, comp. note) South Africa
Platypleura hirtipennis Chawanji, Hodson and Villet 2005a: 258–261, 263–265, Fig 4, Fig 7, Fig 10,

Figs 20–33 (sperm morphology, comp. note) South Africa

Platypleura hirtipennis Sanborn 2005c: 114, Table 1 (comp. note)

Platypleura hirtipennis Villet, Barker and Lunt 2005: 590–593, Figs 1–5, Table 1 (phylogeny, listed, comp. note) South Africa

Platypleura hirtipennis Chawanji, Hodson and Villet 2006: 386–387 (comp. note)

Platypleura hirtipennis Price, Barker and Villet 2009: 618–619, 621–624, Figs 1–2, Fig 4, Table 1 (illustrated, distribution, phylogeny, comp. note) South Africa

***P.* cf. *hirtipennis* Villet, Barker & Lunt 2005**

Platypleura cf. *hirtipennis* Villet, Barker and Lunt 2005: 590–593, Figs 1–5, Table 1 (phylogeny, listed, comp. note) South Africa

***P. inglisi* Ollenbach, 1929**

***P. insignis* Distant, 1879** = *Platypleura (Poecilopsaltria) insignis* = *Platypleura (Neoplatypleura) insignis*

Platypleura insignis Boulard 2006c: 131 (comp. note)

Platypleura (Neoplatypleura) insignis Boulard 2008c: 361 (comp. note) Indochina

Platypleura (Neoplatyplerua) insignis Boulard 2008f: 17, 91 (synonymy, biogeography, listed, comp. note) Equals *Platypleura insignis* Equals *Platypleura parvula* Thailand, Tenasserim

***P. instabilis* Boulard, 1977**

***P. intermedia* Liu, 1940**

***P. kabindana* Distant, 1919**

***P. kaempferi* var. ______ Kato, 1938**

***P. kaempferi brevipennis* Naruse, 1983**

Platypleura kaempferi var. *brevipennis* Naruse 1983: 151–152, Fig 1A, Fig 2A (n. ssp., described, illustrated, synonymy, comp. note) Equals *Platypleura kaempferi* (partim) Japan

Platypleura kaempferi brevipennis Hayashi 1984: 31 (comp. note)

***P. kaempferi kaempferi* (Fabricius, 1794)** = *Tettigonia kaempferi* = *Tettigomia* (sic) *kaempferi* = *Tettigonia kaemferi* (sic) = *Cicada kaempferi* = *Cicada kaemferi* (sic) = *Platypleura kaempferi* = *Platypleura kampferi* (sic) = *Platypleura koempferi* (sic) = *Platypleura kaempheri* (sic) = *Platypleuria* (sic) *kaempferi* = *Platypleura kaenpferi* (sic) = *Platyleura* (sic) *kaempferi* = *Platypleure* (sic) *kaempferi* = *Platypleura fuscangulis* Butler, 1874 = *Platypleura kaempferi fuscangulis* = *Platypleura hyalinolimbata* Signoret, 1881 = *Platypleura hyalino-limbata* = *Platypleura repanda* Uhler (nec Linnaeus) = *Platypleura kaempferi* var. *formosana* Matsumura, 1917 = *Platypleura kaempferi annamensis* Moulton, 1923 = *Platyplerua kaempferi dentivitta* Kato, 1925 = *Platypleura kaempferi kyotonis* Kato, 1930 = *Platypleura tsuchidai* Kato, 1936 = *Rlatypleura* (sic) *tsuchidai* = *Platypleura tsuchidai* var. *amamiana* Kato, 1940 = *Platypleura retracta* Liu, 1940 = *Platypleura retracta omeishana* Liu, 1940

Platypleura kaempferi Kishida 1929: 118 (fungal host, comp. note) Japan

Platypleura kaempferi Doi 1932: 42 (listed, comp. note) Korea

Platypleura kaempferi Chen 1933: 358–359 (listed, comp. note)

Platypleura kaempferi Isaki 1933: 380 (listed, comp. note) Japan

Platypleura kaempferi Wu 1935: 2 (listed) Equals *Tettigonia kaempferi* Equals *Cicada kaempferi* Equals *Platypleura hyalinolimbata* Equals *Platypleura fuscangulis* Equals *Platypleura repanda* (nec Linnaeus) Equals *Platypleura kaempferi* var. *formosana* Equals *Platypleura kaempferi dentivitta* China, Peiping, Hangchow, Taichow, Soochow, Chusan, Canton, Tsinan, Nanking, Kashing, Japah, Borneo, Formosa

Platypleura kaempferi Chen 1942: 143 (listed, comp. note) Equals *Tettigonia kaempferi* China

Platypleura kaempferi Ishii 1953: 1757 (parasitism) Japan

Platypleura kaempferi Kurokawa 1953: 1, 5, Table 1 (chromosomes, comp. note) Japan

Platypleura kaempferi Ohgushi 1954: 11 (emergence times, ecology, comp. note) Japan

Platypleura kaempferi Cho 1955: 237 (listed, comp. note) Korea, Borneo

Platypleura kaempferi Kim 1958: 94 (listed) Korea

Platypleura kaempferi Halkka 1959: 37, Table 1 (listed, comp. note)

Platypleura kaempferi Ikeda 1959: 484 (comp. note) Japan

Platypleura kaempferi Katsuki 1960: 57, Table 1 (hearing, comp. note) Japan

Platypleura kaempferi Cho 1965: 173 (listed, comp. note) Korea

Platypleura kaempferi Won, Woo, Ham and Chun 1968: 367, Table 6 (predation, listed, comp. note) Korea

Platypleura kaempferi Itô 1978: 157–158, Fig 3.36 (density, comp. note) Japan

Platypleura kaempferi Soper 1981: 51 (fungal host, comp. note) Japan

Platypleura kaempferi Hayashi 1982b: 187–188, Table 1 (distribution, comp. note) Japan

Platypleura kaempferi Naruse 1983: 151–152, Fig 1B, Fig 2B (illustrated, comp. note) Japan, China, Korea, Borneo, Malay Peninsula

Platypleura kaempferi Strümpel 1983: 71, Table 5 (comp. note)

Platypleura kaempferi Welbourn 1983: 135, Table 5 (mite host, listed) Japan

Platypleura kaempferi Hayashi 1984: 29–31, Figs 4–5, Figs 8–9 (key, described, illustrated, synonymy, comp. note) Equals *Tettigonia kaempferi* Equals *Cicada kaempferi* Equals *Platypleura repanda* (nec Linnaeus) Equals *Platypleura tsuchidai* Japan
Platypleura kaempferi ab. *kyotonis* Hayashi 1984: 31 (comp. note)
Platypleura tsuchidai Hayashi 1984: 31 (comp. note)
Platypleura kaempferi Karban 1986: 224, 231, Table 14.2 (comp. note) Southeast Asia, China
Platypleura kaempferi Kirillova 1986b: 45 (listed) Japan
Platypleura kaempferi Kirillova 1986a: 123 (listed) Japan
Platypleura kampferi (sic) Chou 1987: 21 (comp. note) China
Platypleura kaempferi Wang and Zhang 1987: 295 (key, comp. note) China
Platypleura kaempferi Southcott 1988: 103 (parasitism, comp. note) Japan
Platypleura kaempferi Wang Z.-N. 1988: 22, Fig 2 (fungal host, comp. note) Taiwan
Platypleura kaempferi Liu 1990: 46 (comp. note)
Platykleura (sic) *kaempferi* Liu 1990: 46 (comp. note)
Platypleura kaempferi Liu 1991: 257 (comp. note)
Platypleura retracta Liu 1991: 257 (comp. note)
Platypleura kaempferi Liu 1992: 9, 16, 34, 40, 42 (comp. note)
Platypleura retracta Liu 1992: 16, 40, 42 (comp. note)
Platypleura retracta Peng and Lei 1992: 97, Fig 275 (listed, illustrated, comp. note) China, Hunan
Platypleura kaempferi Wang and Zhang 1992: 119 (listed, comp. note) China
Platypleura kaempferi Wang and Zhang 1993: 188 (listed, comp. note) China
Platypleura kaempferi Lei, Chou and Li 1994: 54, 56–57, Fig 12 (oscillogram, comp. note) China, Shaanxi, Fujian, Shandong, Sichuan
Platypleura kaempferi Anonymous 1994: 809 (fruit pest) China
Platypleura kaempferi Lee 1995: 17, 22, 30–31, 33–39, 41–46, 53–57, 135, 139 (listed, key, synonymy, ecology, distribution, illustrated, comp. note) Equals *Tettigonia kaempferi* Korea
Tettigonia kaempferi Lee 1995: 53 (listed)
Platypleura kaempferi Xu, Ye and Chen 1995: 84 (listed, comp. note) China, Zhejiang
Platypleura kaempferi Jiang, Yang and Liu 1996: 313–318, Figs 1–4, Table 1 (song analysis, oscillogram, timbal structure, comp. note) China
Platypleura kaempferi Zaidi 1996: 100–101, 104, Table 1 (comp. note) Borneo, Sarawak
Platypleura kaempferi Chou, Lei, Li, Lu and Yao 1997: 28, 168–170, 307–308, 315–316, 324, 326, 330, 338–339, Fig 9–39, Plate VII, Fig 79, Tables 10–12 (key, synonymy, described, illustrated, song, oscillogram, spectrogram, comp. note) Equals *Tettigonia kaempferi* Equals *Platypleura tsuchidai* Equals *Platypleura retracta* China
Platypleura kaempferi Lei, Jiang, Li and Chou 1997: 349–357, Figs 1–5, Table 1 (song analysis, frequency spectrum, oscillogram, comp. note) China, Beijing, Shaanxi, Sichuan, Shandong, Fujian
Platypleura kaempferi Zaidi 1997: 112 (comp. note) Borneo, Sarawak
Platyleura kaempferi Itô 1998: 494, Table 1 (life cycle evolution, comp. note)
Platypleura kaempferi Zaidi and Ruslan 1998a: 345 (listed, comp. note) Equals *Tettigonia kaempferi* Equals *Platypleura fuscangulis* Borneo, Sarawak, China, Japan, Peninsular Malaysia
Platypleura kaempferi Feng, Chen, Ye, Wang, Chen, and Wang 1999: 518 (comp. note)
Platypleura kaempferi Lee 1999: 2–3 (distribution, song, listed, comp. note) Equals *Tettigonia kaempferi* Korea, China, Japan, Taiwan, Malaysia, Philippines
Platypleura kaempferi Ohbayashi, Sato and Igawa 1999: 341 (parasitism, comp. note)
Platypleura kaempferi Freytag 2000: 55 (on stamp)
Platypleura hyalinolimbata Hua 2000: 63 (listed, distribution) China
Platypleura kaempferi Hua 2000: 64 (listed, distribution, hosts) Equals *Tettigonia kaempferi* Equals *Platypleura kaempferi* var. *formosana* Equals *Platypleura kaempferi dentivittata* Equals *Platypleura retracta* Equals *Platypleura retracta* var. *omeishana* China, Liaoning, Shaanxi, Gansu, Henan, Shandong, Hebei, Hubei, Anhui, Jiangsu, Jiangxi, Zhejiang, Fujian, Taiwan, Guangdong, Hunan, Guangxi, Sichuan
Platypleura kaempferi Maezono and Miyashita 2000: 52 (nymphal cell host, comp. note) Japan
Platypleura kaempferi Nikoh and Fukatsu 2000: 630 (fungal host) Japan
Platypleura kaempferi Lee 2001a: 51 (comp. note) Korea

Platypleura kaempferi Ohara and Hayashi 2001: 1 (comp. note) Japan
Platypleura kempferi Saisho 2001a: 68 (comp. note) Japan
Platypleura kaempferi Sueur 2001: 37, 46, Tables 1–2 (listed)
Platypleura kaempferi Jeon, Kim, Tripotin and Kim 2002: 241 (parasitism, comp. note) Japan
Platypleura kaempferi Sueur 2002b: 121, 129 (listed, comp. note)
Platyplerua kaempferi Uchiyama and Udagawa 2002: 135–136 (fungal host, comp. note) Japan
Platypleura kaempferi Yaginuma 2002: 2 (comp. note)
Platypleura kaempferi Lee 2003a: 47–48, Fig 1–4 (illustrated, synonymy, distribution, comp. note) Equals Platypleura tsichidai Equals *Platypleura kaempferi brevipennis* Equals *Platypleura kaempferi* ab. *kyotonis* Equals *Platypleura retracta* Korea, Japan, China, Taiwan, Malaysia, Philippines
Platypleura tsuchidai Lee 2003a: 47 (comp. note) Japan
Platypleura kaempferi brevipennis Lee 2003a: 47 (comp. note) Japan
Platypleura kaempferi ab. *kyotonis* Lee 2003a: 47 (comp. note) Japan
Platypleura retracta Lee 2003a: 47 (comp. note) China
Platypleura kaempferi Lee 2003b: 8 (comp. note)
Platypleura kaempferi Lee 2003c: 15 (comp. note) Korea
Platypleura kaempferi Lee and Hayashi 2003a: 153–156, Fig 4 (key, diagnosis, illustrated, synonymy, distribution, song, listed, comp. note) Equals *Tettigonia kaempferi* Equals *Pycna repanda* (nec Linnaeus) Equals *Platypleura kaempferi* var. *formosana* Equals *Platypleura kaempferi* var. *formosan* (sic) Equals *Platypleura kaempferi dentivitta* Taiwan, Japan, Korea, China, Malaysia
Platypleura kaempferi Chen 2004: 43–44, 198 (described, illustrated, listed, comp. note) Taiwan
Platypleura kaempferi Hirai 2004: 376 (predation, comp. note) Japan
Platypleura kaempferi Champman 2005: 327 (comp. note)
Platypleura kaempheri (sic) Itô and Kasuya 2005: 60, Table 14 (life cycle, comp. note)
Platypleura kaempferi Lee 2005: 16, 44–49, 157, 160–161, 168 (described, illustrated, synonymy, song, distribution, key, comp. note) Equals *Tettigonia kaempferi* Korea, China, Japan, Taiwan, Malaysia
Platypleura kaempferi Pham and Thinh 2005a: 238–239 (synonymy, distribution, listed) Equals *Tettigonia kaempfer* (sic) Equals *Cicada kaempfer* (sic) Equals *Platypleura tsuchidai* Equals *Platypleura retracta* Vietnam
Platypleura kaempferi Pham and Thinh 2006: 526–527 (listed, comp. note) Vietnam
Platypleura kaempferi Tian, Yuan and Zhang 2006: 243 (reproductive system, comp. note) China
Platypleura kaempferi Boulard 2007e: 62, Fig 42 (genitalia illustrated, comp. note) Thailand
Platypleura kaempferi Shiyake 2007: 2, 17–18, Fig 13 (illustrated, distribution, listed, comp. note)
Platypleura kaempferi Zhang and Xuan 2007: 404 (fungal host, comp. note) China
Platypleura kaempferi Boulard 2008c: 361 (comp. note) China
Platypleura kaempferi Hill 2008: 585 (loquat pest) China
Platypleura kaempferi Lee 2008a: 3 (synonymy, distribution, listed) Equals *Tettigonia kaempferi* Equals *Platypleura kaempferi annamensis* Vietnam, Japan, Korea, China, Malaysia, Taiwan
Platypleura kaempferi Lee 2008b: 446–447, 462, Fig 1A, Table 1 (key, illustrated, synonymy, distribution, listed, comp. note) Equals *Tettigonia kaempferi* Korea, Japan, China, Taiwan, Vietnam, Malaysia
Platypleura kaempferi Osaka Museum of Natural History 2008: 322 (listed, acoustic behavior, ecology, comp. note) Japan
Platypleura kaempferi Hisamitsu, Sokabe, Terada, Osumi, Terada, et al. 2009: 93, 103, Fig 3b (behavior, illustrated, comp. note) Japan
Platypleura kaempferi Moriyama and Numata 2009: 164, Table 1 (comp. note) Japan
Platypleura kaempferi Pham and Yang 2009: 13, 16, Table 2 (listed, comp. note) Vietnam
Platypleura kaempferi Pham and Yang 2010a: 133 (comp. note) Vietnam

***P. kaempferi ridleyana* Distant, 1905** = *Platypleura ridleyana*
Platypleura kaempferi ridleyana Schouten, Duffels and Zaidi 2004: 373, 376, 379, Table 2 (listed, comp. note) Malayan Peninsula

***P.* "karooensis" Price, Barker & Villet, 2009**
Platypleura "karooensis" Price, Barker and Villet 2009: 618–624, Figs 1–2, Fig 3D, Fig 4, Table 1 (illustrated, distribution, phylogeny, comp. note) South Africa

***P. kenyana* Lallemand, 1929**

***P. kuroiwae* Matsumura, 1917** = *Munza kuroiwae* = *Platypleura kuroiwai* (sic)
Platypleura kuroiwae Itô 1976: 67 (sugarcane pest, comp. note) Japan, Okinawa
Platypleura kuroiwae Nagamine and Itô 1982: 80–83, Figs 1–4, Table 1 (sugarcane pest, density, illustrated, comp. note) Okinawa, Ryukyus
Platypleura kuroiwae Hayashi 1984: 29, 32–34, Figs 16–17, Figs 22–23 (key, described, illustrated, comp. note) Japan
Munza (= Platypleura) kurowiae Karban 1986: 224, Table 14.2 (comp. note) Ryukyu Archipelago
Platypleura kuroiwae Wilson 1987: 489, Table 1 (sugarcane pest, listed, comp. note) Japan
Platypleura kuroiwai (sic) Liu 1990: 46 (comp. note)
Platypleura kuroiwai (sic) Liu 1992: 9 (comp. note)
Platypleura kurowiae Sueur 2001: 46, Table 2 (listed)
Platypleura kuroiwae Perepelov, Burgov and Marymanska-Nadachowska 2002: 218, Fig 2 (karyotype, comp. note) Japan
Platyplerua kuroiwae Uchiyama and Udagawa 2002: 135–136 (fungal host, comp. note) Japan
Platypleura kuroiwae Lee 2003a: 47 (comp. note) Japan
Platypleura kuroiwae Lee and Hayashi 2003a: 156 (comp. note) Japan
Platypleura kuroiwae Shiyake 2007: 2, 17, 19, Fig 14 (illustrated, distribution, listed, comp. note)
Platypleura kuroiwae Osaka Museum of Natural History 2008: 322 (listed, acoustic behavior, ecology, comp. note) Japan

P. liberiana Distant, 1912 to *Canualna liberiana*

***P. lindiana* Distant, 1905** = *Platypleura (Poecilopsaltria) lindiana*
Platypleura lindiana Evans 1966: 111 (parasitism, comp. note) South Africa

***P. lineatella* Distant, 1905** = *Platypleura (Oxypleura) lineatella*

***P. longirostris* Ashton, 1914** = *Platypleura divisa* (nec Germar)
Platypleura longirostris Villet 1997a: 325 (valid species, comp. note)
Platypleura longirostris Villet 1999d: 210, 215–217, Figs 9–12 (status, illustrated, comp. note) Equals *Platypleura divisa* (partim) Uganda

***P. lourensi* Lee, 2009**
Platypleura lourensi Lee 2009i: 293–294, Figs 1–2 (n. sp., described, illustrated, distribution, listed, comp. note) Philippines, Panay

P. lyricen Kirkaldy, 1913 to *Hamza ciliaris* (Linnaeus, 1758)

***P. machadoi* Boulard, 1972**
Platypleura machadoi Boulard 1985e: 1027, 1038, Plate III, Fig 26 (coloration, illustrated, comp. note) Angola
Platypleura machadoi Sanborn 1999a: 35 (type material, listed)

***P. mackinnoni* Distant, 1904** = *Platypleura (Oxypleura) mackinnoni* = *Platypleura mackinoni* (sic)
Platypleura mackinnoni Boulard 2008c: 361 (comp. note) Nepal
Platypleura mackinnoni Ahmed and Sanborn 2010: 27, 38 (distribution) Pakistan, India

***P. makaga* Distant, 1904**
Platypleura makaga Boulard 1985e: 1029, 1046–1047, Plate VIII, Fig 49 (coloration, illustrated, comp. note)
Platypleura makaga Villet 1999d: 217 (comp. note) Congo

P. maritzburgensis Distant, 1913 to *Tugelana butleri* Distant, 1912

P. marsabitensis Boulard, 1980 to *Platypleura argus marsabitensis*

***P. maytenophila* Villet, 1987**
Platypleura maytenophila Villet 1987a: 272, Table 2 (song intensity, listed, comp. note) South Africa
Platypleura maytenophila Villet 1987b: 209–211, 213–214, Figs 1c–d, Fig 2a, Fig 3 (n. sp., described, illustrated, distribution, habitat, comp. note) South Africa
Platypleura maytenophila Villet 1988: 71–73, 75, Fig 4, Figs 6a–b, Table 1 (song, sonogram, comp. note)
Platypleura maytenophila Villet and Reavell 1989: 333–335, Fig 3 (distribution, habitat, host plants, key, comp. note) South Africa
Platypleura maytenophila Villet 1992: 94–95, Table 1 (acoustic responsiveness, comp. note) South Africa
Platypleura maytenophila Villet 1997a: 325 (comp. note) South Africa
Platypleura maytenophila Villet and Noort 1999: 228 (comp. note)
Platypleura maytenophila Sueur 2001: 37, Table 1 (listed)
Platypleura maytenophila Sanborn, Breitbarth, Heath and Heath 2002: 445 (comp. note)
Platypleura maytenophila Sanborn, Phillips and Villet 2003b: 292–293 (comp. note)
Platypleura maytenophila Malherbe, Burger and Stephen 2004: 86 (comp. note) South Africa

***P. mijberghi* Villet, 1989**
Platypleura mijburghi Villet 1989b: 54–56, 64, 67–68, Figs 3–4, Fig 17, Fig 31, Figs 35–36 (n. sp., described, illustrated, habitat, song,

sonogram, frequency spectrum, distribution, comp. note) South Africa
Platypleura mijburghi Villet 1997a: 321 (comp. note) South Africa

P. *mira* Distant, 1904 = *Platypleura (Poecilopsaltria) mira* = *Platypleura (Platypleura) mira* = *Platypleura laotiana* nom. nud.
Platypleura mira Zaidi and Ruslan 1995c: 63 (comp. note) Peninsular Malaysia
Platypleura mira Zaidi and Ruslan 1997: 219 (listed, comp. note) Peninsular Malaysia, Laos
Platypleura mira Boulard 2000c: 56–58, Figs 7–9 (illustrated, song analysis, sonogram, comp. note) Thailand
Platypleura mira Boulard 2001a: 128–129, 131–132, 134, Plate I, Figs C–D, Figs 3–4 (illustrated, song analysis, sonogram, comp. note) Thailand
Platypleura mira Sueur 2001: 37, Table 1 (listed)
Platyplerua mira Boulard 2002a: 46, 63, 65, Plate N, Fig 1 (comp. note)
Platypleura mira Boulard 2002b: 203 (comp. note)
Platypleura mira Boulard 2003c: 172, 178 (comp. note)
Platypleura mira Boulard 2005d: 338, 347 (comp. note) Asia
Platypleura mira Boulard 2005o: 147 (listed)
Platypleura mira Boulard 2006g: 39–40, 113–114, 173, Fig 79, Fig (sonogram, illustrated, comp. note) Asia
Platypleura mira Boulard 2007d: 494, 498, 506–507, Fig A, Fig 1 (predation, illustrated, comp. note) Thailand
Platypleura mira Boulard 2007e: 4, 25, 47, Plate 21 (coloration, illustrated, comp. note) Thailand
Platypleura mira Sanborn, Phillips and Sites 2007: 4–5 (listed, distribution, comp. note) Thailand, Laos, Indochina, Malay Peninsula
Platypleura (Platypleura) mira Boulard 2008f: 15–16, 91 (biogeography, listed, comp. note) Equals *Platypleura mira* Thailand, Laos
Platypleura mira Lee 2008a: 4 (synonymy, distribution, listed) Equals *Platypleura (Poecilopsaltria) mira* Vietnam, Laos, Thailand, Malaysia
Platypleura mira Pham and Yang 2009: 13, Table 2 (listed) Vietnam
Platypleura mira Lee 2010b: 20 (distribution, listed) Cambodia, Vietnam, Laos, Thailand, Malaysia

P. *miyakona* (Matsumura, 1917) = *Pycna miyakona* = *Suisha miyakona*
Platypleura miyakona Hayashi 1984: 29, 31–33, Figs 12–13, Figs 18–19 (key, described, illustrated, synonymy, comp. note) Equals *Pycna miyakona* Equals *Suisha miyakona* Equals *Platypleura ?miyakona* Japan
Platypleura miyakona Sueur 2001: 46, Table 2 (listed)
Platypleura miyakona Shiyake 2007: 2, 17, 19, Fig 15 (illustrated, distribution, listed, comp. note)
Platypleura miyakona Osaka Museum of Natural History 2008: 322 (listed, acoustic behavior, ecology, comp. note) Japan

P. mokensis Boulard, 2003 to *Platypleura watsoni* (Distant, 1897)

P. *murchisoni* Distant, 1905
Platypleura murchisoni Boulard 1985e: 1022, 1042–1043, Plate IV, Fig 38 (coloration, illustrated, comp. note)

P. nigromarginata Ashton, 1914 to *Strumoseura nigromarginata*

P. *nigrosignata* Distant, 1913
Platypleura nigrosignata Lee 2008a: 3 (distribution, listed) Vietnam
Platypleura nigrosignata Pham and Yang 2009: 13, 16, Table 2 (listed, comp. note) Vietnam
Platypleura nigrosignata Pham and Yang 2010a: 133 (comp. note) Vietnam

P. *nobilis nobilis* (Germar, 1830) = *Cicada nobilis* = *Oxypleura nobilis* = *Platypleura (Neoplatypleura) nobilis* = *Cicada hemiptera* Guérin-Méneville, 1843 = *Platypleura gemina* Walker, 1850 = *Platypleura semilucida* Walker, 1850
Cicada nobilis Hayashi 1984: 30 (type species of *Neoplatypleura*)
Platypleura nobilis Wang and Zhang 1987: 295 (key, comp. note) China
Platypleura nobilis Wang and Zhang 1992: 119 (listed, comp. note) China
Platypleura nobilis Zaidi and Ruslan 1994: 426, 428, Table 1 (comp. note) Malaysia
Platypleura nobilis Zaidi and Ruslan 1997: 219 (listed, comp. note) Java, Peninsular Malaysia
Platypleura nobilis Zaidi, Ruslan and Azman 1999: 300 (listed, comp. note) Equals *Cicada nobilis* Borneo, Sabah, Java, Peninsular Malaysia, Sarawak
Platypleura nobilis Zaidi, Noramly and Ruslan 2000a: 320–321 (listed, comp. note) Equals *Cicada nobilis* Borneo, Sabah, Java, Peninsular Malaysia, Sarawak
Platypleura nobilis Zaidi, Noramly and Ruslan 2000b: 338–339, Table 1 (listed, comp. note) Borneo, Sabah
Platypleura nobilis Zaidi, Ruslan and Azman 2000: 198 (listed, comp. note) Equals *Cicada nobilis* Borneo, Sabah, Java, Peninsular Malaysia, Sarawak

Platypleura nobilis Boulard 2003a: 104 (comp. note)
Platypleura nobilis Boulard 2003b: 186–188, 191, Figs 1–2, Fig 5, Plate I, Fig 1 (described, song, sonogram, illustrated, comp. note) Equals *Cicada nobilis* Thailand
Platypleura nobilis Zaidi and Azman 2003: 104, Table 1 (listed, comp. note) Borneo, Sabah
Platypleura nobilis Boulard 2005d: 338, 346 (comp. note) Thailand
Platypleura nobilis Boulard 2005o: 148 (listed)
Platypleura nobilis Boulard 2006c: 133 (comp. note)
Platypleura nobilis Boulard 2006f: 620 (comp. note) Thailand
Platypleura nobilis Boulard 2006g: 39–40, 98–99, Fig 69A (illustrated, comp. note) Thailand
Platypleura nobilis Boulard 2007a: 119 (comp. note)
Platypleura (Neoplatypleura) nobilis Boulard 2007d: 495 (comp. note) Thailand
Platypleura nobilis Boulard 2007e: 48, 77, 87, Plate 26 (illustrated, sonogram, comp. note) Thailand
Platypleura nobilis Sanborn, Phillips and Sites 2007: 5 (listed, distribution, comp. note) Thailand, Burma, India, Malysia, Indochina, Sarawak, Borneo, Philippine Republic
Platypleura nobilis Shiyake 2007: 2, 16, 18, Figs 8a–8b (illustrated, distribution, listed, comp. note)
Platypleura (Neoplatypleura) nobilis Boulard 2008c: 361 (comp. note) Malaysia
Platypleura (Neoplatypleura) nobilis Boulard 2008f: 16–17, 91 (type species of *Neoplatypleura*, biogeography, listed, comp. note) Equals *Cicada nobilis* Equals *Cicada hemiptera* Equals *Platypleura nobilis* Equals *Neoplatypleura nobilis* Thailand, Java, Sumatra, Assam, India, Burma, Siam

P. *nobilis* var. *a* Atkinson, 1884 = *Platypleura (Neoplatypleura) nobilis* var. *a*

P. *nobilis* var. *a* Distant, 1889 = *Platypleura (Neoplatypleura) nobilis* var. *a*

P. n. sp. Duffels, 1990
Platypleura n. sp. Duffels 1990b: 64, Table 2 (listed) Sulawesi
Platypleura n. sp. Boer and Duffels 1996b: 330 (distribution) Sulawesi

P. *octoguttata octoguttata* (Fabricius, 1798) = *Tettigonia 8-guttata* = *Tettigonia octoguttata* = *Tettigonia octotoguttata* (sic) = *Cicada 8-guttata* = *Poecilopsaltria octoguttata* = *Poecilopsaltria 8-guttata* = *Poicilopsaltria* (sic) *octoguttata* = *Poicilopsaltria* (sic) *octopunctata* (sic) = *Platypleura (Poecilopsaltria) octoguttata* = *Platypleura octogullata* (sic) = *Oxypleura sanuiflua* Walker, 1850
Platypleura octoguttata Strümpel 1983: 71, Table 5 (comp. note)
Platypleura octoguttata Ewing 1989: 36 (comp. note)
Poecilopsaltria octoguttata Boulard 1991c: 121 (comp. note)
Platypleura octoguttata Malhi and Parashad 1991: 243 (predation, comp. note) India
Platypleura octoguttata Bennet-Clark and Young 1994: 293, Table 1 (body size and song frequency)
Platypleura octogullata (sic) Chari, Shailaja and Reddy 1995: 25–28 (flight dynamics, comp. note) India
P[latypleura] octoguttata Jiang, Yang, Tang, Xu and Chen 1995b: 229 (song frequency, comp. note)
Platypleura octoguttata Sueur 2001: 37, Table 1 (listed)
Platypleura octoguttata Sueur 2002b: 129 (listed)
Platypleura octoguttata Srivastava and Bhandari 2004: 1479 (comp. note) India
Poecilopsaltria octoguttata Boulard 2007e: 62, Fig 42 (genitalia illustrated, comp. note) Thailand
Platypleura octoguttata Shiyake 2007: 2, 16, 18, Fig 11 (illustrated, listed, comp. note)
Poecilopsaltria octoguttata Boulard 2008f: 16 (type species of *Poecilopsaltria*) Equals *Tettigonia octoguttata*
Tettigonia octoguttata Lee 2009h: 2617 (type species of *Poecilopsaltria*, listed) India
Platypleura octoguttata Ahmed and Sanborn 2010: 27, 38 (synonymy, distribution) Equals *Tettigonia 8-guttata* Equals *Oxypleura sanguiflua* Pakistan, India
Tettigonia octoguttata Lee 2010a: 2 (type species of *Poecilopsaltria*, listed) India

P. *octoguttata* var. *a* (Distant, 1889) = *Poecilopsaltria octoguttata* var. *a* = *Platypleura (Poecilopsaltria) octoguttata* var. *a*

P. *octoguttata* var. *b* (Distant, 1889) = *Poecilopsaltria octoguttata* var. *b* = *Platypleura (Poecilopsaltria) octoguttata* var. *b*

P. "olifantflumensis" Price, Barker & Villet, 2009
Platypleura "olifantsflumensis" Price, Barker and Villet 2009: 618–619, 621–624, Figs 1–2, Fig 3A, Fig 4, Table 1 (illustrated, distribution, phylogeny, comp. note) South Africa

P. *parvula* Boulard, 2006
Platypleura parvula Boulard 2006c: 131–133, 135, Fig A, Fig 1, Fig 3 (n. sp., described,

illustrated, song, sonogram, comp. note) Thailand

P. *pinheyi* Boulard, 1980

P. *plagiata* Karsch, 1890 to *Karscheliana plagiata*

P. *plumosa* (Germar, 1834) = *Cicada plumosa* = *Platypleura hirtipennis* var. *b* Stål, 1866 = *Platypleura hirtipennis plumosa* = *Platypleura (Platypleura) plumosa*

Platypleura plumosa Gess 1981: 6, 8 (described, song, habitat, comp. note) South Africa

Platypleura plumosa Villet 1997a: 321–322, 328–332, Figs 11–15 (described, illustrated, type material, synonymy, habitat, comp. note) Equals *Cicada plumosa* South Africa

Platypleura plumosa Phillips, Sanborn and Villet 2002: 31 (thermal responses) South Africa, Eastern Cape

Platypleura plumosa Sanborn, Villet and Phillips 2002: 32 (endothermy, comp. note)

Platypleura plumosa Sanborn, Phillips and Villet 2003a: 348–349, Table 1 (thermal responses, habitat, comp. note) South Africa

Platypleura plumosa Sanborn, Phillips and Villet 2003b: 291–295, Fig 1, Figs 4–5, Table 1 (song, sonogram, oscillogram, distribution, comp. note) South Africa

Platypleura plumosa Villet, Sanborn and Phillips 2003a: 2 (endothermy, comp. note) South Africa

Platypleura plumosa Villet, Sanborn and Phillips 2003c: 1443 (comp. note)

Platypleura plumosa Sanborn, Villet and Phillips 2004a: 817–819, 821, Fig 4, Tables 1–2 (thermoregulation, endothermy, habitat, comp. note) South Africa

Platypleura plumosa Sanborn 2005c: 114, Table 1 (comp. note)

Platypleura plumosa Villet, Barker and Lunt 2005: 590–593, Figs 1–5, Table 1 (phylogeny, listed, comp. note) South Africa

Platypleura plumosa group Price, Barker and Villet 2009: 618, 622–623 (comp. note)

Platypleura plumosa Price, Barker and Villet 2009: 618–624, Figs 1–2, Fig 3B, Fig 4, Table 1 (illustrated, distribution, phylogeny, comp. note) South Africa

Platypleura plumosa Garrick 2010: 475 (comp. note) SouthAfrica

P. *polita polita* (Walker, 1850) = *Oxypleura polita* = *Platypleura (Oxypleura) polita* = *Poecilopsaltria polita*

Platypleura (Oxypleura) polita Boulard 2008c: 361 (comp. note) India

P. *polita* var. *a* (Distant, 1889) = *Poecilopsaltria polita* var. *a* = *Platypleura (Oxypleura) polita* var. *a*

P. *polydorus* (Walker, 1850) = *Oxypleura polydorus* = *Platypleura (Oxypleura) polydorus* = *Poecilopsaltria polydorus*

P. *rothschildi* Melichar, 1911

Platypleura rothschildi Boulard 1996d: 131 (comp. note)

Platypleura rothschildi Villet 1999d: 217 (comp. note) Kenya

P. *schumacheri vitreomaculata* Schumacher, 1913 to *Severiana severini vitreomaculata*

P. *semusta* (Distant, 1887) = *Poecilopsaltria semusta* = *Platypleura (Poecilopsaltria) semusta*

Platypleura semusta Wu 1935: 3 (listed) Equals *Poecilopsaltria semusta* China, Chusan, Shanghai, Shantung

Platypleura semusta Chou, Lei, Li, Lu and Yao 1997: 172 (comp. note) China

Platypleura semusta Hua 2000: 64 (listed, distribution) Equals *Poecilopsaltria semusta* China, Shandong, Jiangsu, Zhejiang

P. *signifera* Walker, 1850

P. sp. 1 Zaidi & Ruslan, 1995

Platypleura sp. 1 Zaidi and Ruslan 1995c: 64, 68, Table 1 (listed, comp. note) Peninsular Malaysia

Platypleura sp. 1 Zaidi, Ruslan and Mahadir 1996: 60, Table 1 (listed, comp. note) Peninsular Malaysia

P. sp. 1 Villet & Noort, 1999

Platypleura sp. 1 Villet and Noort 1999: 228–229, Table 12.2 (habitat, listed, comp. note) Tanzania

P. sp. 1 Villet, Barker & Lund, 2005

Platypleura sp. 1 Villet, Barker and Lunt 2005: 590–593, Figs 1–5, Table 1 (phylogeny, listed, comp. note) South Africa

P. sp. 1 Pham & Thinh, 2006

Platypleura sp. 1 Pham and Thinh 2006: 526 (listed) Vietnam

P. sp. 2 Zaidi, Ruslan & Mahadir, 1996

Platypleura sp. 2 Zaidi, Ruslan and Mahadir 1996: 60, Table 1 (listed, comp. note) Peninsular Malaysia

P. sp. 2 Villet & Noort, 1999

Platypleura sp. 2 Villet and Noort 1999: 228–229, Table 12.2 (habitat, listed, comp. note) Tanzania

P. sp. 3 Zaidi, Ruslan & Mahadir, 1996

Platypleura sp. 3 Zaidi, Ruslan and Mahadir 1996: 60, Table 1 (listed, comp. note) Peninsular Malaysia

P. sp. 4 Sanborn, Villet & Phillips, 2004

Platypleura sp. 4 Sanborn, Villet and Phillips 2004a: 817, 821, Tables 1–2 (thermoregulation, endothermy, habitat, comp. note) South Africa

Platypleura n. sp. Sanborn, Villet and Phillips 2004a: 820 (comp. note)
Platypleura sp. 4 Villet, Barker and Lunt 2005: 590–593, Figs 1–5, Table 1 (phylogeny, listed, comp. note) South Africa

P. sp. 7 Villet, Barker & Lund, 2005
Platypleura sp. 7 Villet, Barker and Lunt 2005: 590–593, Figs 1–5, Table 1 (phylogeny, listed, comp. note) South Africa

P. sp. 10 Price, Barker & Villet, 2009
Platypleura sp. 10 Price, Barker and Villet 2007: 2575, 2577–2578, 2580, 2582–2583, Table 2 (comp. note) South Africa

***P. sphinx* Walker, 1850** = *Poecilopsaltria sphinx* = *Platypleura (Poecilopsaltria) sphinx*
Platypleura (Poecilopsaltria) sphinx Boulard 2006f: 621 (comp. note) Equals *Poecilopsaltria sphinx*

***P. spicata* Distant, 1905** = *Platypleura (Oxypleura) spicata*

***P. stridula* (Linnaeus, 1758)** = *Cicada stridula* = *Tettigonia stridula* = *Platypleura (Platypleura) stridula* = *Cicada catenata* Drury, 1773 = *Cicada catena* (sic) = *Cicada nigrolinea* Degeer, 1773 = *Cicada nigra* (sic) = *Cicada nigrolineata* (sic) = *Platypleura nigrolinea* = *Platypleura stridula* var. *a* Stål, 1866
Cicada stridula Chen 1942: 143 (type species of *Platypleura*)
Platypleura stridula Smit 1964: 137 (comp. note) South Africa
Cicada stridula Hayashi 1984: 30 (type species of *Platypleura*)
Platypleura stridula Theron 1985: 156–157, Fig 16.120–16.121, Plate II, Fig 2 (illustrated) South Africa
Platypleura stridula Boulard 1988f: 33 (type species of Platypleurini)
Cicada stridula Villet 1989a: 329 (comp. note)
Platypleura stridula Villet 1989a: 329–332, Figs 1A–B, Fig 2A, Fig 3A, Fig 4A, Fig 5A, Table 1 (status, type species of *Platypleura*, type species of Platypleurinae, illustrated, song, sonogram, frequency spectrum, habitat, distribution, key, comp. note) South Africa
Cicada stridula Boulard 1997c: 102, 106 (type species of *Platypleura*)
Platypleura stridula Boulard 1997c: 106, 112–113, 117, Plate II, Fig 5 (type species of *Platypleura*) Equals *Cicada stridula*
Cicada stridula Chou, Lei, Li, Lu and Yao 1997: 168 (type species of *Platypleura*)
Platypleura stridula Malaisse 1997: 238, Table 2.11.1 (human food, listed) Zambezi
Platypleura stridula Boulard 2000i: 21 (listed)
Cicada stridula Boulard 2001b: 22, 26 (type species of *Platypleura*)
Platypleura stridula Boulard 2001b: 26, 32–33, 37, Plate II, Fig 5 (type species of *Platypleura*) Equals *Cicada stridula*
Platypleura stridula Sueur 2001: 37, Table 1 (listed)
Platypleura stridula Huis 2003: 168, Table 1 (predation, comp. note) Zambia
Cicada stridula Lee and Hayashi 2003a: 153 (type species of *Platypleura*) South Africa
Platypleura stridula Sanborn, Phillips and Villet 2003b: 293 (comp. note)
Platypleura stridula Moulds 2005b: 388, 396, 399, 417–420, 422, Figs 56–60, Table 1 (type species of *Platypleura*, higher taxonomy, phylogeny, key, listed, comp. note) Australia
Cicada stridula Moulds 2005b: 392, 433 (type species of *Platypleura*)
Cicada stridula Pham and Thinh 2005a: 238 (type species of *Platypleura*)
Platypleura stridula Villet, Barker and Lunt 2005: 590–593, Figs 1–5, Table 1 (phylogeny, listed, comp. note) South Africa
Platypleura stridula complex Price, Barker and Villet 2007: 2575–2577, 2579, Table 1 (comp. note)
Platypleura stridula Price, Barker and Villet 2007: 2575, 2577, 2579–2580, 2582–2584, Fig 1, Fig 4, Table 2 (comp. note) Equals *Cicada catenata* Equals *Cicada nigrolinea* South Africa
Cicada catenata Price, Barker and Villet 2007: 2575, 2582 (comp. note)
Cicada nigrolinea Price, Barker and Villet 2007: 2575, 2582 (comp. note)
Cicada stridula Lee 2008a: 3 (type species of *Platypleura*) South Africa
Cicada stridula Lee 2009h: 2617 (type species of *Platypleura*, listed) South Africa
Cicada stridula Lee 2009i: 293 (type species of *Platypleura*, listed)
Platypleura stridula Pinto, Quartau, and Bruford 2009: 280 (comp. note)
Platypleura stridula Price, Barker and Villet 2009: 621, Fig 2 (illustrated, phylogeny, comp. note) South Africa
Cicada stridula Ahmed and Sanborn 2010: 38 (type species of *Platypleura*)
Platypleura stridula Garrick 2010: 475 (comp. note) South Africa
Cicada stridula Lee 2010a: 15 (type species of *Platypleura*, listed)
Cicada stridula Lee 2010b: 20 (type species of *Platypleura*, listed)
Cicada stridula Lee 2010c: 2 (type species of *Platypleura*, listed) South Africa

Cicada stridula Mozaffarian and Sanborn 2010: 76 (type species of *Platypleura*)

***P. takasagona* Matsumura, 1917** = *Platypleura kuroiwae takasagona* = *Platypleura repanda* (nec Linnaeus) = *Platypleura nepanda* (sic) (nec Linnaeus) = *Platypleura kaempferi takasagona* = *Platypleura kurowiae takasagon* (sic) = *Platypleura kaempferi* var. *formosana* (nec Matsumura)

Platypleura takasagona Liu 1992: 34 (comp. note)

Platypleura takasagona Anonymous 1994: 809 (fruit pest) China

Platypleura takasagona Chou, Lei, Li, Lu and Yao 1997: 172 (comp. note) China

Platypleura kurowiae takasagon (sic) Hua 2000: 64 (listed, distribution) China, Taiwan

Platypleura takasagona Hua 2000: 64 (listed, distribution, host) China, Taiwan

Platypleura takasagona Lee and Hayashi 2003a: 153, 155–156, Fig 5 (key, diagnosis, illustrated, synonymy, distribution, song, listed, comp. note) Equals *Platypleura kuroiwae* var. *takasagona* Equals *Platypleura kuroiwae* var. *formosana* (sic) Equals *Platypleura kaempferi var. takagasona* Equals *Platypleura kaempferi takasagona* Taiwan

Platypleura takasagona Chen 2004: 43, 47, 198 (described, illustrated, listed, comp. note) Taiwan

Platypleura takasagona Shiyake 2007: 2, 16, 18, Fig 10 (illustrated, distribution, listed, comp. note)

Platypleura takagasona Lee, Lin and Wu 2010: 218–224, Figs 1–6, Tables 1–2 (ecology, comp. note) Taiwan

***P. techowi* Schumacher, 1913** = *Platyplerua divisa techowi* = *Platypleura divisa* var. *techowi* = *Platypleura (Platypleura) divisa techowi* = *Systophlochius palochius* Villet, 1989

Systophlochius palochius Villet 1989b: 52–54, 64, 67–68, Figs 1–2, Fig 16, Fig 29, Figs 32–34 (n. sp., type species of *Systophlochius*, described, illustrated, habitat, song, sonogram, frequency spectrum, distribution, comp. note) South Africa

Systopholchius palochius Villet 1997a: 321, 328 (comp. note) South Africa

Systophlochius palochius Sueur 2001: 37, Table 1 (listed)

Platypleura techowi Villet, Barker and Lunt 2005: 590–591, Table 1 (n. stat., phylogeny, synonymy, listed, comp. note) Equals *Platypleura divisa* var. *techowi* Equals *Systophlochius palochius* n. syn. South Africa

Systophlochius palochius Villet, Barker and Lunt 2005: 590–593, Figs 1–5, Table 1 (synonymy, listed, comp. note) South Africa

Platypleura divisa var. *techowi* Villet, Barker and Lunt 2005: 591 (comp. note)

P. tepperi Goding & Froggatt, 1904 to *Yanga guttulata* (Signoret, 1860)

***P. testacea* (Walker, 1858)** = *Zammara testacea* = *Platypleura (Oxypleura) testacea*

Platypleura testacea Dlabola 1981: 198 (distribution, comp. note) Iran

Platypleura testacea Mozaffarian and Sanborn 2010: 76 (synonymy, listed, distribution, comp. note) Equals *Zammara testacea* Iran, India, Oman

***P. turneri* Boulard, 1975**

Platypleura turneri Villet 1989a: 332 (key, comp. note)

***P. vitreolimbata* Breddin, 1905** = *Platypleura vitrolimbata* (sic)

***P. wahlbergi* Stål, 1855** = *Platypleura walberghi* (sic)

Platypleura wahlbergi Zeuner 1941: 88 (comp. note) South Africa

Platypleura wahlbergi Zeuner 1944: 112, Fig 4 (hind wing illustrated, comp. note) South Africa

Platypleura wahlbergi Phillips, Sanborn and Villet 2002: 31 (thermal responses) South Africa, Eastern Cape

Platypleura wahlbergi Sanborn, Villet and Phillips 2002: 32 (endothermy, comp. note)

Platypleura wahlbergi Sanborn, Phillips and Villet 2003a: 349–350, Table 1 (thermal responses, habitat, comp. note) South Africa

Platypleura wahlbergi Villet, Sanborn and Phillips 2003a: 2–3 (endothermy, comp. note) South Africa

Platypleura wahlbergi Sanborn, Villet and Phillips 2004a: 817–821, Fig 5, Tables 1–2 (thermoregulation, endothermy, habitat, comp. note) South Africa

Platypleura wahlbergi Sanborn, Villet and Phillips 2004b: A1097 (endothermy, comp. note)

Platypleura wahlbergi Sanborn 2005c: 114, Table 1 (comp. note)

Platypleura wahlbergi Villet, Barker and Lunt 2005: 590–593, Figs 1–5, Table 1 (phylogeny, listed, comp. note) South Africa

Platypleura wahlbergi Phillips and Sanborn 2010: 74 (endothermy, comp. note) South Africa

***P. watsoni* (Distant, 1897)** = *Poecilopsaltria watsoni* = *Platypleura (Platypleura) watsoni* = *Platypleura (Poecilopsaltria) watsoni* = *Platypleura mokensis* Boulard, 2003

Platypleura mokensis Boulard 2003c: 172–175, 178, Figs 1–6, Plate A, Figs 1–2 (n. sp., described, illustrated, song, sonogram, comp. note) Thailand

Platypleura watsoni Boulard 2005o: 149 (listed, n. syn.) Equals *Platypleura mokensis*

Platypleura watsoni Boulard 2006f: 620 (comp. note) Thailand
Platypleura watsoni Boulard 2007e: 54, 77, Fig 36 (genitalia illustrated, comp. note) Thailand
Platypleura watsoni Sanborn, Phillips and Sites 2007: 5 (listed, synonymy, distribution, comp. note) Equals *Platyplerua mokensis* Thailand, India, Myanmar
Platypleura mokensis Sanborn, Phillips and Sites 2007: 5 (comp. note) Thailand
Platypleura (Poecilopsaltria) watsoni Boulard 2008f: 16, 91 (biogeography, listed, comp. note) Equals *Poecilopsalta watsoni* Equals *Platypleura watsoni* Equals *Platypleura mokensis* Thailand, Siam

***P. westwoodi* Stål, 1863** = *Poecilopsaltria westwoodi* = *Platypleura (Platypleura) westwoodi* = *Platypleura westwoodii* (sic)
P[latypleura] westwoodi Jiang, Yang, Tang, Xu and Chen 1995b: 229 (song frequency, comp. note)
Platypleura westwoodi Sueur 2001: 37, Table 1 (listed)

***P. witteana* Dlabola, 1960**

***P. yayeyamana* Matsumura, 1917** = *Platypleura kaempferi yayeyamana* = *Platypleura yaeyamana* (sic)
Platypleura yayeyamana Hayashi 1984: 29–31, Figs 6–7, Figs 10–11 (key, described, illustrated, synonymy, comp. note) Equals *Platypleura kaempferi yayeyamana* Japan
Platypleura yayeyamana Sueur 2001: 46, Table 1 (listed)
Platypleura yayeyamana Watanabe and Kobayashi 2006: 224, Table 1 (listed) Japan, Ryukyus
Platypleura yaeyamana (sic) Shiyake 2007: 2, 17, 19, Fig 16 (illustrated, distribution, listed, comp. note)
Platypleura yayeyamana Osaka Museum of Natural History 2008: 322 (listed, acoustic behavior, ecology, comp. note) Japan

***Azanicada* Villet, 1989**
Azanicada Villet 1989b: 54 (n. gen., described, listed, comp. note) South Africa
Azanicada Villet 1994b: 91 (comp. note)

***A. zuluensis* (Villet, 1987)** = *Platypleura zuluensis*
Platypleura zuluensis Villet 1987a: 270, 272, Table 2 (song intensity, listed, comp. note) South Africa
Platypleura zuluensis Villet 1987b: 212–215, Figs 1e–g, Fig 2c, Fig 3 (n. sp., described, illustrated, distribution, habitat, comp. note) South Africa
Platypleura zuluensis Villet 1988: 72–74, Fig 4, Figs 7a–b, Table 1 (song, sonogram, comp. note) South Africa
Platypleura zuluensis Villet 1989b: 54 (type species of *Azanicada*, comp. note) South Africa
Azanicada zuluensis Villet 1989b: 154 (n. comb.)
Azanicada zuluensis Villet and Reavell 1989: 333–335, Fig 3 (distribution, habitat, host plants, key, comp. note) South Africa
Azanicada zuluensis Villet 1992: 94–96, Table 1 (acoustic responsiveness, comp. note) South Africa
Azanicada zuluensis Villet and Noort 1999: 228 (comp. note)
Azanicada zuluensis Villet 1999e: 1 (comp. note) South Africa
Azanicada zuluensis Sueur 2001: 36, Table 1 (listed) Equals *Platypleura zuluensis*
Azanicada zuluensis Phillips, Sanborn and Villet 2002: 31 (thermal responses) South Africa, Eastern Cape
Azanicada zuluensis Sanborn, Phillips and Villet 2003a: 349, Table 1 (thermal responses, habitat, comp. note) South Africa
Azanicada zuluensis Sanborn, Phillips and Villet 2003b: 292 (comp. note) South Africa
Azanicada zuluensis Chawanji, Hodson and Villet 2005a: 258–261, 265, Fig 3, Fig 8, Figs 11–12, Figs 35–36 (sperm morphology, comp. note) South Africa
Azanicada zuluensis Chawanji, Pinchuck, Hodson and Villet 2005: 77, Fig 1 (female reproductive tract)
Azanicada zuluensis Chawanji, Hodson and Villet 2006: 386–387 (comp. note)
Azanicada zuluensis Seabra, Pinto-Juma and Quartau 2006: 844 (comp. note)

***Tugelana* Distant, 1912**
Tugelana Duffels 1991a: 128 (comp. note) Africa
Tugelana Villet 1994b: 87–88, 91 (described, phylogeny, comp. note) South Africa
Tugelana Chou, Lei, Li, Lu and Yao 1997: 298 (comp. note)

***T. butleri* Distant, 1912** = *Platypleura maritzburgensis* Distant, 1913
Tugelana butleri Villet 1994b: 87–91, Figs 1–7 (type species of *Tugelana*, described, illustrated, habitat, comp. note) Equals *Platypleura maritzburgensis* n. syn. South Africa
Platypleura maritzburgensis Villet 1994b: 87 (synonymy, comp. note) South Africa

***Capcicada* Villet, 1989**
Capcicada Villet 1989b: 59 (n. gen., described, listed, comp. note) South Africa
Capcicada Goble, Price, Barker, and Villet 2008a: 673 (phylogeny, comp. note) Africa

C. decora (Germar, 1834) = *Cicada decora* = *Platypleura decora* = *Platypleura deusta* (nec Thunburg) = *Platypleura absimilis* Distant, 1897 = *Platypleura (Platypleura) absimilis*

Cicada decora Villet 1989b: 59 (type species of *Capcicada*) South Africa

Capcicada decora Villet 1989b: 59–60, 65–68, Fig 8, Figs 22–23, Fig 30, Figs 40–41 (n. comb., illustrated, habitat, song, sonogram, frequency spectrum, distribution, comp. note) Equals *Cicada decora* Equals *Platypleura decora* Equals *Platypleura deusta* (partim) Equals *Platyplerua absimilis* South Africa

Capcicada decora Villet, Sanborn and Phillips 2003c: 1443 (comp. note)

Capcicada decora Robertson, Villet and Palmer 2004: 463–464, 466, 468, Figs 3c–d, Table 1 (distribution, comp. note)

Capcicada decora Sanborn, Villet and Phillips 2004a: 817, 821, Tables 1–2 (thermoregulation, endothermy, habitat, comp. note) South Africa

Capcicada decora Villet, Barker and Lunt 2005: 590–593, Figs 1–5, Table 1 (phylogeny, listed, comp. note) South Africa

***Albanycada* Villet, 1989**

Albanycada Villet 1989b: 60 (n. gen., described, listed, comp. note) South Africa

A. albigera (Walker, 1850) = *Platypleura albigera* = *Platypleura membranacea* Karsch, 1890

Platypleura albigera Gess 1981: 6 (described, song, habitat, comp. note) South Africa

Platypleura albigera Villet 1989b: 60 (type species of *Albanycada*) South Africa

Albanycada albigera Villet 1989b: 61–62, 66–67, 69, Figs 9–10, Figs 26–28, Fig 31, Figs 42–43 (n. comb., illustrated, habitat, song, sonogram, frequency spectrum, distribution, comp. note) Equals *Platypleura albigera* Equals *Platypleura membranacea* South Africa

Albanycada albigera Villet and Capitao 1995: 140 (habitat, comp. note) South Africa

Albanycada albigera Villet and Capitao 1996: 282–283, Tables 1–2 (density, habitat, comp. note) South Africa

Albanycada albigera Sueur 2001: 36, Table 1 (listed)

Albanycada albigera Phillips, Sanborn and Villet 2002: 31 (thermal responses) South Africa, Eastern Cape

Albanycada albigera Sanborn, Villet and Phillips 2002: 32 (thermoregulation, comp. note)

Albanycada albigera Sanborn, Phillips and Villet 2003a: 349–350, Table 1 (thermal responses, habitat, comp. note) South Africa

Albanycada albigera Sanborn, Villet and Phillips 2003: 305–307, Tables 1–2 (thermoregulation, comp. note) South Africa, Eastern Cape

Albanycada albigera Villet, Sanborn and Phillips 2003a: 3 (ectothermy, comp. note) South Africa

Albanycada albigera Sanborn, Villet and Phillips 2004a: 817–822, Fig 1, Figs 6–7, Tables 1–2 (thermoregulation, habitat, comp. note) South Africa

Albanycada albigera Sanborn, Villet and Phillips 2004b: A1097 (thermoregulation, comp. note)

Albanycada albigera Chawanji, Hodson and Villet 2005a: 258–262, 265, Fig 1, Fig 5, Figs 14–19, Figs 37–38 (sperm morphology, comp. note) South Africa

Albanycada albigera Sanborn 2005c: 114, Table 1 (comp. note)

Albanycada albigera Chawanji, Hodson and Villet 2006: 386–387 (comp. note)

Albanycada albigera Phillips and Sanborn 2010: 74 (endothermy, comp. note) South Africa

***Karscheliana* Boulard, 1990**

Karscheliana Boulard 1990b: 224 (n. gen., described, comp. note)

Karscheliana Moulds 2005b: 433 (higher taxonomy, listed, comp. note)

K. parva Boulard, 1990

Karscheliana parva Boulard 1990b: 181, 183, 223–226, Plate XXI, Fig 5, Figs 24–28 (n. sp., described, illustrated, comp. note)

Karscheliana parva Boulard 1999a: 78, 89 (listed, sonogram) Kenya

Karscheliana parva Sueur 2001: 36, Table 1 (listed)

Karscheliana parva Boulard 2005d: 338, 344 (comp. note) Africa

Karscheliana parva Boulard 2006g: 36–37, 39–40, 73–74, 83, 131–132, 180, Fig 7d, Fig 41 (illustrated, sonogram, listed, comp. note)

K. plagiata (Karsch, 1890) = *Platypleura (Oxypleura) plagiata* = *Platypleura plagiata*

Platypleura plagiata Boulard 1990b: 223–224 (type species of *Karscheliana*, comp. note)

Karscheliana plagiata Boulard 1990b: 224 (comp. note)

***Strumoseura* Villet, 1999**

Strumoseura Villet 1999d: 210–211 (n. gen, described, comp. note)

S. nigromarginata (Ashton, 1914) = *Platypleura nigromarginata* = *Oxypleura quadraticollis* (nec Butler)

Platypleura nigromarginata Villet 1999d: 209–210 (type species of *Strumoseura*, comp. note)

Strumoseura nigromarginata Villet 1999d: 211–214, Figs 1–8 (n. comb., synonymy, described, distribution, habitat, comp. note) Equals *Platypleura nigromarginata* Equals *Platypleura migromaculata* (sic) Equals *Oxypleura quadraticollis* (nec Butler) Uganda, Tanzania

Platypleura nigromarginata Villet and Noort 1999: 228–229, Table 12.2 (habitat, listed, comp. note) Uganda, Tanzania

***Strumosella* Boulard, 1973**

Strumosella Medler 1980: 73 (listed) Nigeria

Strumosella Villet 1999b: 153 (comp. note)

Strumosella spp. Villet and Noort 1999: 227, 229 (comp. note)

Strumosella sp. Villet and Noort 1999: 228–229, Table 12.2 (habitat, listed, comp. note) Tanzania

Strumosella Boulard 2002b: 203 (comp. note)

Strumosella Moulds 2005b: 392, 433 (higher taxonomy, listed, comp. note)

***S. limpida* (Karsch, 1890)** = *Platypleura limpida* = *Platypleura (Poecilopsaltria) limpida* = *Platypleura ladona* (nec Distant) = *Oxypleura contracta* (nec Walker) = *Oxypleura basalis* (nec Signoret)

***S. strumosa* (Fabricius, 1803)** = *Tettigonia strumosa* = *Cicada strumosa* = *Platypleura strumosa* = *Platypleura (Poecilopsaltria) strumosa* = *Platypleura afzelli* (sic) (nec Stål) = *Oxypleura contracta* Stål, 1866

Tettigonia strumosa Dallas 1870: 482 (synonymy) Equals *Platypleura strumosa*

Strumosella strumosa Medler 1980: 73 (listed) Nigeria, Central African Republic, Upper Volta

Strumosella strumosa Boulard 1999a: 78, 87, 89 (listed, sonogram, comp. note)

Strumosella strumosa Sueur 2001: 37, Table 1 (listed)

Strumosella strumosa Boulard 2002b: 193–194, Fig 45 (illustrated, coloration, comp. note) Ivory Coast

Strumosella strumosa Boulard 2005d: 344 (comp. note)

Strumosella strumosa Boulard 2006g: 73–74, 131–132, 140, 180, Fig 100 (sonogram, listed, comp. note) Ivory Coast

***S. truncaticeps* (Signoret, 1884)** = *Oxypleura truncaticeps* = *Platypleura truncaticeps* = *Platypleura (Oxypleura) truncaticeps*

Strumosella truncaticeps Medler 1980: 73 (listed) Nigeria, Central African Republic, Upper Volta

Strumosella truncaticeps Boulard 1999a: 78, 88 (listed, sonogram) Ivory Coast

Strumosella truncaticeps Sueur 2001: 37, Table 1 (listed)

Strumosella truncaticeps Boulard 2006g: 131–132, 140, 180, Fig 101 (sonogram, listed, comp. note) Ivory Coast

***Oxypleura* Amyot & Audinet-Serville, 1843** = *Cicada (Oxypleura)* = *Platypleura (Oxypleura)* = *Oxyplelure* (sic)

Oxypleura Medler 1980: 73 (listed) Nigeria

Oxypleura Hayashi 1984: 31 (comp. note)

Oxypleura Boulard 1985c: 183 (comp. note)

Oxypleura Boulard 1985e: 1022, 1025 (coloration, comp. note)

Oxypleura Villet 1989a: 329 (comp. note)

Oxypleura Villet 1989b: 54 (comp. note)

Oxypleura Boulard 1990b: 167 (comp. note)

Oxypleura Moulds 1990: 12, 31, 52 (diversity, listed, comp. note) Australia, Christmas Island, Africa

Oxypleura Boulard 1996d: 135 (listed)

Oxypleura Villet and Noort 1999: 229 (comp. note)

Oxypleura Moulds and Carver 2002: 12 (listed) Australia

Oxypleura Moulds 2005b: 387–389, 392, 412, 430, 433, Table 2 (higher taxonomy, listed, comp. note) Australia

Oxypleura Shiyake 2007: 2, 15 (listed, comp. note)

Oxypleura Lee 2008a: 2, 4, Table 1 (listed) Vietnam

Oxypleura Pham and Yang 2009: 13, Table 2 (listed) Vietnam

***O. atkinsoni* (Distant, 1912)** = *Platypleura atkinsoni* = *Platypleura (Oxypleura) atkinsoni*

***O. basalis* Signoret, 1891** = *Platypleura (Oxypleura) basalis*

***O. calypso* Kirby, 1889** = *Poecilopsaltria calypso* = *Platypleura calypso* = *Platypleura (Oxypleura) calypso*

Oxypleura calypso Moulds 1990: 53 (distribution, ecology, listed, comp. note) Australia, Christmas Island

Oxypleura calypso Naumann 1993: 12, 112, 149 (listed) Australia

Platypleura calypso Zaidi and Ruslan 1997: 219 (listed, comp. note) Equals *Oxypleura calyso* Equals *Poecilopsaltria calypso* Christmas Island

Oxypleura calypso Moulds and Carver 2002: 12 (synonymy, distribution, ecology, type data, listed) Australia, Christmas Island

Oxypleura calypso Moulds 2005b: 395–396, 399, 417–420, 422, Figs 56–60, Table 1 (higher taxonomy, phylogeny, key, listed, comp. note) Australia

Oxypleura calypso Lee 2008a: 4 (synonymy, distribution, listed) Equals *Platypleura*

(Oxypleura) calypso Vietnam, Australia, Christmas Island, Sri Lanka, India
Oxypleura calypso Pham and Yang 2009: 13, 16, Table 2 (listed, comp. note) Vietnam
Oxypleura calypso Pham and Yang 2010a: 133 (comp. note) Vietnam

***O. centralis* (Distant, 1897)** = *Platypleura centralis* = *Platypleura (Oxypleura) centralis*
Oxypleura centralis Boulard 1985e: 1019 (comp. note)
Platypleura centralis Sueur 2002b: 122, 129 (listed, comp. note)

***O. clara* Amyot & Audinet-Serville, 1843** = *Platypleura clara* = *Platypleura (Oxypleura) clara* = *Oxypleura passa* Walker, 1850 = *Oxypleura basistigma* Walker, 1850
Platypleura clara Linnavuori 1977: 65 (listed) Somalia
Oxypleura clara Medler 1980: 73 (listed) Nigeria
Oxypleura clara Hayashi 1984: 30 (type species of *Oxypleura*)
Oxypleura clara Boulard 1988d: 150–151, 154, Fig 2 (wing abnormalities, illustrated, comp. note) Zaire
Oxypleura clara Moulds 1990: 52 (type species of *Oxypleura*) Africa
Oxypleura clara Moulds and Carver 2002: 12 (type species of *Oxypleura*) Africa
Oxypleura clara Schedl 2004a: 12–14, Figs 1–2 (illustrated, listed, distribution, comp. note) Yemen, West Africa, South Africa, Zaire, Tanganyika, Mozambique, Kenya, Ethiopia, Sudan, Arabia
Oxypleura clara Lee 2008a: 4 (type species of *Oxypleura*) Senegal, Cape of Good Hope

***O. ethiopiensis* Boulard, 1975**
Oxypleura ethiopensis Boulard 1985c: 182 (comp. note)

***O. lenihani* Boulard, 1985**
Oxypleura lenihani Boulard 1985c: 181–182, 187, Figs 18–22, Fig 42 (n. sp., described, illustrated, comp. note) Mozambique, Zimbabwe, Zaire, Malawi, Zululand
Oxypleura lenihani Villet 1987a: 270, 272, Table 2 (song intensity, listed, comp. note) South Africa
Oxypleura lenihani Villet 1988: 72, 74–75, Figs 8a–b, Tables 1–2 (song, sonogram, comp. note) South Africa
Oxypleura lenihani Villet and Reavell 1989: 334–335, Fig 2b, Fig 3 (distribution, habitat, host plants, key, comp. note) South Africa
Oxypleura lenihani Villet 1992: 95, Table 1 (listed, comp. note) South Africa
Oxypleura lenihani Sueur 2001: 37, Table 1 (listed)
Oxypleura lenihani Picker, Griffith and Weaving 2002: 156–157 (described, illustrated, habitat, comp. note) South Africa
Oxypleura lenihana Sanborn, Phillips and Villet 2003b: 292 (comp. note)

***O. pointeli* Boulard, 1985**
Oxypleura pointeli Boulard 1985c: 182–183, 187, Figs 23–27, Fig 43 (n. sp., described, illustrated, comp. note) Burundi

***O. polydorus* Walker, 1850** = *Platypleura polydorus*
Oxypleura polydorus Boulard 1985c: 181–182 (comp. note)
Oxypleura polydorus Boulard 1985e: 1019 (crypsis, comp. note) Kenya

***O. quadraticollis* (Butler, 1874)** = *Platypleura quadraticollis* = *Platypleura (Oxypleura) quadraticollis* = *Oxyplelure* (sic) *quadraticollis* = *Platypleura nigromarginata* Ashton, 1914 = *Platypleura nigromaculata* (sic) = *Ioba veligera* Boulard, 1965
Platypleura quadraticollis Evans 1966: 111 (parasitism, comp. note) South Africa
Platypleura quadraticollis Fenton, Cummings, Hutton, and Swanepoel 1987: 712 (predation, comp. note) Zimbabwe
Oxypleura quadraticollis Villet 1999d: 209 (comp. note)
Oxypleura quadraticollis Villet and Noort 1999: 229 (comp. note) Botswana, Zimbabwe, South Africa
Platypleura quadraticollis Rice 2000: 6–7 (illustrated, human food) Zimbabwe
Oxypleura quadraticollis Bayefsky-Anand 2005: 95–96 (bat predation) Zimbabwe
Oxypleura quadraticollis Villet, Barker and Lunt 2005: 590–593, Figs 1–5, Table 1 (phylogeny, listed, comp. note) South Africa
Oxyplelura quadraticollis Shiyake 2007: 2, 13, 15 (illustrated, listed, comp. note)
Oxyplelure (sic) *quadraticollis* Shiyake 2007: 13, Fig 7 (illustrated)

***O. spoerryae* Boulard, 1980**
Oxypleura spoerryae Boulard 1985e: 1019, 1022, 1037, Plate I, Fig 5 (coloration, illustrated, comp. note) Kenya
Oxypleura spoerryae Boulard 1990b: 188–189, 198–199, Plate XXIII, Fig 2, Plate XXV, Fig 5 (illustrated, comp. note) Kenya
Oxypleura spoerryae Boulard 1996d: 114, 135–137, Fig S10, Color Plate Fig 3 (song analysis, sonogram, listed, comp. note) Central African Republic
Oxypleura spoerryae Sueur 2001: 37, Table 1 (listed)
Oxypleura spoerryae Boulard 2002b: 193 (coloration, comp. note)
Oxypleura spoerryae Boulard 2005d: 338 (comp. note) Africa

Oxypleura spoerryae Boulard 2006g: 39–40, 72–73, 80, 150, 180, Fig 35, Fig 110D (sonogram, illustrated, listed, comp. note) Kenya
Oxypleura spoerryae Boulard 2007f: 136, Fig 73 (illustrated, comp. note)

***Brevisiana* Boulard, 1973**
Brevisiana spp. O'Toole 1995: 108 (illustrated) Kenya
Brevisiana Moulds 2005b: 392, 433 (higher taxonomy, listed, comp. note)
Brevisiana Moulds 2008a: 208 (comp. note)

***B. brevis* (Walker, 1850)** = *Platyplerua brevis* = *Platypleura (Oxypleura) brevis* = *Platypleura simplex* Walker, 1850 = *Platypleura (Oxypleura) simplex* = *Oxypleura sobrina* Stål, 1855 = *Oxypleura patruelis* Stål, 1855 = *Platypleura (Oxypleura) patruelis* = *Cicada (Oxypleura) neurosticta* Schaum, 1853 = *Cicada (Oxypleura) neurosticta* Schaum, 1862 = *Cicada (Oxypleura) neurostica* (sic) = *Oxypleura neurosticta* = *Platypleura neurosticta*
Platypleura brevis Fenton, Cummings, Hutton, and Swanepoel 1987: 712 (predation, comp. note) Zimbabwe
Brevisiana brevis Villet 1987a: 272, Table 2 (song intensity, listed, comp. note) South Africa
Brevisiana brevis Villet 1988: 72, 75, Figs 2c–d, Fig 11, Table 1 (song, sonogram, comp. note) South Africa
Brevisiana brevis Villet and Reavell 1989: 334–335, Fig 2a, Fig 3 (distribution, habitat, host plants, key, comp. note) South Africa
Brevisiana brevis Sanborn and Phillips 1995a: 482 (comp. note)
Brevisiana brevis Sueur 2001: 36, Table 1 (listed)
Platypleura brevis Morris 2004: 70 (human food) Malawi
Brevisiana brevis Boulard 2005d: 338 (comp. note) Africa
Brevisiana brevis Bayefsky-Anand 2005: 95–96 (bat predation) Zimbabwe
Brevisiana brevis Villet, Barker and Lunt 2005: 590–593, Figs 1–5, Table 1 (phylogeny, listed, comp. note) South Africa
Brevisiana brevis Boulard 2006g: 39–40 (comp. note) Africa
Brevisiana brevis Shiyake 2007: 27 (comp. note)
Brevisiana brevis Matthews and Matthews 2010: 292 (comp. note)

***B. niveonotata* (Butler, 1874)** = *Platypleura (Oxypleura) niveonotata* = *Platypleura niveonotata*
Brevisiana niveonotata Boulard 1999a: 79, 92 (listed, sonogram) Kenya
Brevisiana niveonotata Sueur 2001: 36, Table 1 (listed)
Brevisiana niveonotata Boulard 2006g: 180 (listed)

***B. quartai* Boulard, 1972** = *Brevisiana brevis* Stål, 1866 (nec Walker)

***Yanga* Distant, 1904**
Yanga Boulard 1981d: 103 (comp. note) Comoros Islands
Yanga Itô and Nagamine 1981: 282 (comp. note)
Yanga Boulard 1990b: 88, 208, 211 (comp. note) Madagascar, Comoros
Yanga Novotny and Wilson 1997: 437 (listed)
Yanga Moulds 2005b: 392, 433 (higher taxonomy, listed, comp. note)
Yanga Boulard 2006g: 109, 129 (comp. note) Madagascar
Yanga Boulard 2010b: 334–336 (comp. note) Madagascar
Yanga spp. Moulds 2010: 11 (comp. note) Madagascar

***Y. andriana* (Distant, 1899)** = *Platypleura andriana*
Yanga andriana Boulard 1999a: 79, 93, 96–97, Plate II, Fig 3 (listed, sonogram, illustrated) Madagascar
Yanga andriana Sueur 2001: 37, Table 1 (listed)
Yanga andriana Boulard 2006g: 75–76, 92, 181, Fig 58 (sonogram, listed, comp. note) Madagascar
Yanga andriana Boulard 2010b: 334–336, Fig 3, Figs 5–6 (sound system illustrated, genitalia illustrated, comp. note) Madagascar

***Y. antiopa* (Karsch, 1890)** = *Platypleura antiopa* = *Platypleura (Poecilopsaltria) antiopa*

***Y. bouvieri* Distant, 1905**

***Y. brancsiki* (Distant, 1893)** = *Poecilopsaltria brancsiki*
Poecilopsaltria brancsiki Roth 1949: 81 (parasitism) Madagascar
Poecilopsaltria brancsiki Evans 1966: 111 (parasitism, comp. note) Madagascar

***Y. grandidieri* Distant, 1905**

***Y. guttulata* (Signoret, 1860)** = *Platypleura guttulata* = *Yanga guttularis* (sic) = *Yanga guttalata* (sic) = *Yanga guttatula* (sic) = *Platypleura tepperi* Goding & Froggatt, 1904
Yanga guttulata Itô 1976: 67 (sugarcane pest, comp. note) Madagascar
Yanga guttulata Itô and Nagamine 1981: 281–282, Table 5 (sugarcane pest, comp. note) Madagascar
Yanga guttulata Itô 1982b: 139 (comp. note)
Yanga guttulata Hayashi 1984: 34 (comp. note)
Yanga guttulata Boulard 1986b: 31, 34–35, Fig 1 (sugarcane pest, illustrated, comp. note) Madagascar
Yanga guttalata (sic) Boulard 1986b: 31 (comp. note) Madagascar

Yanga guttulata Wilson 1987: 489, Table 1 (sugarcane pest, listed, comp. note) Madagascar
Yanga guttatula (sic) Wolda 1989: 441 (comp. note) Madagascar
Yanga guttulata group Boulard 1990b: 226 (comp. note)
Yanga guttulata Liu 1990: 43, 46 (comp. note)
Yanga guttulata Liu 1992: 6, 9 (comp. note)
Yanga guttatula (sic) Wolda and Ramos 1992: 278 (comp. note) Madagascar
Yanga guttulata Boulard 2002b: 193, 207 (coloration, comp. note)
Yanga guttulata Boulard 2006g: 147, 181, Fig 108 (sonogram, listed, comp. note) Madagascar
Yanga guttulata Boulard 2007f: 81, 137–138, Fig 96 (comp. note)
Platypleura tepperi Moulds 2010: 11–12, Figs 1–7 (status, comp. note) Madagascar
Yanga guttulata Moulds 2010: 11 (synonymy, comp. note) Equals *Platypleura guttulata* Equals *Yanga guttularis* (sic) Equals *Platyplerua tepperi* n. syn. Madagascar

Y. heathi **(Distant, 1899)** = *Platypleura heathi*
Yanga heathi Boulard 1999b: 184–185, Fig 4 (illustrated, comp. note) Madagascar
Yanga heathi Boulard 2006b: 130 (comp. note) Madagascar

Y. hova **(Distant, 1901)** = *Poecilopsaltria hova*

Y. mayottensis **Boulard, 1990**
Yanga mayottensis Boulard 1990b: 211, 226–230, Figs 29–33 (n. sp., described, illustrated, comp. note) Comoros

Y. pembana **(Distant, 1899)** = *Platypleura pembana*
Yanga pembana Boulard 1981d: 103 (comp. note)

Y. pulvera argyrea **(Melichar, 1896)** = *Platypleura argyrea*

Y. pulverea pulverea **(Distant, 1882)** = *Platypleura pulverea* = *Platypleura pulvera* (sic)
Platypleura pulverea Roth 1949: 81 (parasitism) Madagascar
Platypleura pulverea Evans 1966: 111 (parasitism, comp. note) Madagascar
Yanga pulverea Boulard 1985e: 1022 (coloration, comp. note) Madagascar
Yanga pulverea Boulard 2002b: 193 (coloration, comp. note)
Yanga pulverea Boulard 2005d: 338 (comp. note) Madagascar
Yanga pulverea Boulard 2006g: 36–37, 181, Fig 7a (illustrated, listed, comp. note) Madagascar
Yanga pulverea Boulard 2007f: 92, 141, Fig 146 (comp. note) Madagascar

Y. seychellensis Distant, 1912 to *Sechellalna seychellensis*

Y. viettei anjouaniensis **Boulard, 1990**
Yanga viettei anjouaniensis Boulard 1990b: 211, 229–230, Fig 35 (n. ssp., described, illustrated, comp. note) Comoros

Y. viettei moheliensis **Boulard, 1990**
Yanga viettei moheliensis Boulard 1990b: 211, 229–230, Figs 36–37 (n. ssp., described, illustrated, comp. note) Comoros

Y. viettei viettei **Boulard, 1981** = *Yanga viettei*
Yanga viettei Boulard 1981d: 103–106, Figs 1–8 (n. sp., described, illustrated, comp. note) Comoros Islands
Yanga viettei Boulard 1990b: 211, 226, 230 (comp. note) Grand Comoro
Yanga viettei viettei Boulard 1990b: 229–230, Fig 34 (n. ssp., illustrated, comp. note) Grand Comoro

Sechellalna **Boulard, 2010**
Sechellalna Boulard 2010b: 335–336 (n. gen., description, comp. note) Seychelles

S. seychellensis **(Distant, 1912)** = *Yanga seychellensis* = *Yanga andriana* (nec Distant)
Yanga seychellensis Boulard 1981d: 103 (comp. note)
Yanga seychellensis Freytag 2000: 55 (on stamp)
Yanga seychellensis Boulard 2001f: 136 (comp. note) Seychelles
Yanga seychellensis Boulard 2010b: 334–335 (type species of *Sechellalna*, comp. note) Seychelles
Sechellalna seychellensis Boulard 2010b: 335–337, Figs 1–2, Fig 4, Figs 7–9 (n. comb., synonymy, sound system illustrated, genitalia illustrated, comp. note) Equals *Yanga seychellensis* Equals *Yanga andriana* (nec Distant) Seychelles

Kongota **Distant, 1904**
Kongota Boulard 1985e: 1025 (coloration, comp. note)
Kongota Villet 1989b: 60, 62 (listed, comp. note)
Kongota Boulard 1996c: 96–97 (characteristics, comp. note)
Kongota Moulds 2005b: 392, 433 (higher taxonomy, listed, comp. note)
Kongota Moulds 2008a: 208 (comp. note)

K. handlirschi **(Distant, 1897)** = *Poecilopsaltria handlirschi* = *Yanga handlirschi* = *Kongota malgadessica* Boulard, 1996
Kongota malgadessica Boulard 1996c: 97–100, Figs 1–2 (n. sp., described, illustrated, sonogram, comp. note) Madagascar
Kongota malgadessica Sueur 2001: 36, Table 1 (listed)
Kongota malgadessica Boulard 2006g: 74–75, 89, 150, Fig 52, Fig 110E (sonogram, illustrated, comp. note) Madagascar

K. malgadessica Boulard, 1996 to *Kongota handlirschi* (Distant, 1897)
K. muiri Distant, 1905 to *Kongota punctigera* (Walker, 1850)
K. punctigera **(Walker, 1850)** = *Platypleura punctigera* = *Platypleura subfolia* Walker, 1850 = *Kongota muiri* Distant, 1905
Kongota punctigera Villet 1989b: 62–63, 66–67, 69, Figs 11–14, Fig 24, Fig 31, Figs 44–45 (synonymy, illustrated, habitat, song, sonogram, frequency spectrum, distribution, comp. note) Equals *Platypleura punctigera* Equals *Platypleura subfolia* Equals *Kongota muiri* n. syn. South Africa
Kongota muiri Villet 1989b: 62 (synonymy, comp. note)
Kongota punctigera Villet and Reavell 1989: 333–335, Fig 1a, Fig 3 (distribution, habitat, host plants, key, comp. note) South Africa
Kongota punctigera Boulard 1996c: 97 (type species of *Kongota*, comp. note) Equals *Platypleura punctigera* Equals *Kongota muiri*
Kongota punctigera Sanborn, Villet and Phillips 2004a: 817, 819, 821, Tables 1–2 (thermoregulation, endothermy, habitat, comp. note) South Africa
Kongota punctigera Chawanji, Hodson and Villet 2005b: 66 (spermiogenesis, comp. note)
Kongota punctigera Chawanji, Hodson, Villet, Sanborn and Phillips 2007: 338–345, Fig 1, Figs 2C–D, Figs 3A–D, Fig 4C, Figs 4E–F, Figs 6E–H (spermiogenesis, comp. note) South Africa

Canualna **Boulard, 1985**
Canualna Boulard 1985c: 183 (n. gen., described, comp. note)
Canualna Goemans 2010: 7 (comp. note)
C. liberiana **(Distant, 1912)** = *Platypleura liberiana*
Platypleura liberiana Boulard 1985c: 184 (note on type locality, type species of genus *Canualna*) Pagalu (Annobon) Island
Canualna liberiana Boulard 1985c: 185–188, Figs 28–35, Fig 44 (described, illustrated, comp. note) Liberia?, Pagalu (Annobon) Island
Platypleura liberiana Goemans 2010: 7 (type species of *Canualna*)
C. perspicua **(Distant, 1905)** = *Odopoea perspicua*
Odopoea perspicua Goemans 2010: 2, 5, 7 (comp. note)
Canualna perspicua Goemans 2010: 5 (n. comb., comp. note) Saõ Thomé

Pycna **Amyot & Audinet-Serville, 1843** = *Pycnos* (sic) = *Pyona* (sic) = *Platypleura (Pycna)*
Pycna Wu 1935: 3 (listed) China
Pycna Chen 1942: 143 (listed)
Pycna Hayashi 1982a: 78, 81–82, Fig 11 (listed, distribution)
Pycna Hayashi 1984: 29, 32, 34 (listed, comp. note)
Pycna Boulard 1985c: 184, 187 (comp. note)
Pycna Boulard 1985e: 1023, 1025, 1031 (coloration, comp. note)
Pycna Villet 1989a: 329 (comp. note)
Pycna Boulard 1990b: 88, 208 (comp. note) Madagascar
Pycna Pinchuck and Villet 1995: 112 comp. note)
Pycna Boulard 1996d: 132 (listed)
Pycna Chou, Lei, Li, Lu and Yao 1997: 167, 173, 293, Table 6 (key, described, listed, diversity, biogeography) China
Pyona (sic) sp. Malaisse 1997: 238, Table 2.11.1 (human food, listed) Zambezi
Pycna Boulard 1999a: 90 (comp. note)
Pycna Boulard 2002b: 193 (coloration, comp. note) Africa, Asia
Pycna Lee and Hayashi 2003a: 157 (comp. note)
Pycna Malherbe, Burger and Stephen 2004: 85 (comp. note) South Africa
Pycna Morris 2004: 70 (human food) Malawi
Pycna Boulard 2005b: 112 (comp. note)
Pycna Moulds 2005b: 392, 433 (higher taxonomy, listed, comp. note)
Pycna Sanborn, Phillips and Sites 2007: 3, Table 1 (listed) Thailand
Pycna Shiyake 2007: 2, 15 (listed, comp. note)
Pycna Boulard 2008f: 7, 18, 92 (listed) Thailand
Pycna Lee 2008a: 2, 4, Table 1 (listed) Vietnam
Pycna Moulds 2008a: 208 (comp. note)
Pycna Ahmed and Sanborn 2010: 38 (distribution) Madagascar, southern Africa, China, Thailand
P. antinorii **(Lethierry, 1881)** = *Platypleura antinorii*
Pycna antinorii Boulard 1982b: 57, 59 (distribution, comp. note) Ethiopoa
Pycna antinorii Cloudsley-Thompson 1982: 524 (illustrated)
Pycna antinorii Boulard 1985e: 1022, 1032 (coloration, comp. note)
P. baxteri **Distant, 1914**
Pycna baxteri Boulard 1999a: 79, 90 (listed, sonogram) Kenya
Pycna baxteri Sueur 2001: 37, Table 1 (listed)
Pycna baxteri Boulard 2006g: 180 (listed)
Pycna baxteri Boulard 2007f: 122 (comp. note)
P. beccarii **(Lethierry, 1881)** = *Platypleura beccarii*
P. coelestia **Distant, 1904**
Pycna coelestia Wu 1935: 3 (listed) China
Pycna coelestia Chou, Lei, Li, Lu and Yao 1997: 173, 175–176, Fig 9-43, Plate VII, Fig 80 (key, described, illustrated, comp. note) China

Pycna coelestia Hua 2000: 64 (listed, distribution) China, Hainan, Guangxi, Sichuan
Pycna coelestia Lee 2009h: 2619 (comp. note)

P. concinna **Boulard, 2005**
Pycna concinna Boulard 2005b: 115–119, Figs 5–13 (n. sp., described, illustrated, song, sonogram, comp. note) Thailand
Pycna concinna Boulard 2005o: 151 (listed)
Pycna concinna Boulard 2007d: 495, 498, 506, Figs B-C (comp. note) Thailand
Pycna concinna Sanborn, Phillips and Sites 2007: 5 (listed, distribution, comp. note) Thailand
Pycna concinna Boulard 2008f: 18, 92 (biogeography, listed, comp. note) Thailand

P. dolosa **Boulard, 1975**

P. elliotti Distant, 1907 to *Orapa elliotti*

P. gigas **(Distant, 1881)** = *Platypleura gigas*
Pycna gigas Boulard 1985e: 1025, 1039, 1042–1043, Plate II, Fig 15, Pate IV, Fig 29, Plate VI, Fig 39 (coloration, illustrated, comp. note) Madagascar
Pycna gigas Boulard 1986b: 34–35, Figs 4–5 (illustrated) Madagascar
Pycna gigas Boulard 1997b: 20, Plate II, Fig 4 (morphology, comp. note)
Pycna gigas Boulard 1990b: 198–199, Plate XXV, Fig 15 (coloration, illustrated, comp. note)
Pycna gigas Boulard 2007f: 84, Fig 100.1 (illustrated, comp. note)

P. hecuba **Distant, 1904** = *Platypleura graueri* Melichar, 1908
Pycna hecuba Boulard 1985e: 1019, 1022, 1025–1027, 1031–1034, 1044–1047, Plate II, Figs 10–11, Plate VII, Figs 42–46, Plate IX, Fig 50 (crypsis, coloration, illustrated, comp. note) Kenya, East Africa
Pycna hecuba Boulard 1990b: 198–199, 201, Plate XXV, Fig 10 (coloration, illustrated, comp. note) Kenya
Pycna hecuba Boulard 1997b: 16, 22–23, 29, Plate IV, Figs 1–2 (coloration, comp. note) East Africa, Kenya
Pycna hecuba Boulard 2002b: 193–194, Fig 47 (illustrated, coloration, comp. note) East Africa
Pycna hecuba Boulard 2006g: 33–34, 131–132, 138, 180, Fig 99 (illustrated, sonogram, listed, comp. note) Kenya
Pycna hecuba Boulard 2007f: 92, 141, Figs 147–148 (comp. note) Kenya

P. himalayana **Naruse, 1977**
Pycna himalayana Hayashi 1982a: 79, 81–83, Fig 5, Table 1 (illustrated, distribution, comp. note) Nepal

P. indochinensis **Distant, 1913**
Pycna indochinensis Lee 2008a: 4 (distribution, listed) Vietnam
Pycna indochinensis Pham and Yang 2009: 13, 16, Table 2 (listed, comp. note) Vietnam
Pycna indochinensis Pham and Yang 2010a: 133 (comp. note) Vietnam

P. itremensis **Boulard, 2008**
Pycna itremensis Boulard 2008d: 161–165, Figs 1–5 (n. sp., described, illustrated, song, sonogram, comp. note) Madagascar

P. madagascariensis angusta **(Butler, 1882)** = *Platypleura angusta*

P. madagascariensis madagascariensis **(Distant, 1879)** = *Platypleura madagascariensis*
Pycna madagascariensis Boulard 1990b: 101, 103, Plate X, Figs 1–3 (genitalia, comp. note)
Pycna madagascariensis Birkenshaw 2003: 3 (lemur prey) Madagascar
Pycna madagascariensis Boulard 2005d: 344 (comp. note)
Pycna madagascariensis Boulard 2006g: 75–76, 91, 181, Fig 59 (sonogram, listed, comp. note) Madagascar

P. minor **Liu, 1940** = *Pycna minar* (sic)
Pycna minor Hayashi 1982a: 81–83, Table 1 (distribution, comp. note) India
Pycna minar (sic) Liu 1991: 257 (comp. note)
Pycna minar (sic) Liu 1992: 16 (comp. note)

P. moniquae **Boulard, 2012**

P. montana **Hayashi, 1978**
Pycna montana Hayashi 1982a: 79, 81–83, Fig 4, Table 1 (illustrated, distribution, comp. note) Nepal

P. natalensis **Distant, 1905**
Pycna natalensis Villet and Reavell 1989: 333–335, Fig 2d, Fig 3 (distribution, habitat, host plants, key, comp. note) South Africa

P. neavei **Distant, 1912**

P. nigeriana Distant, 1913 to *Ugada nigeriana*

P. passosdecarvalhoi **Boulard, 1975**
Pycna passosdecarvalhoi Boulard 1996d: 114, 132–135, Fig S9(a-d), Fig S9(e-f), Fig S9(g-j), Color Plate Fig 4 (song analysis, sonogram, listed, comp. note) Malawi, Angola
Pycna passosdecarvavahoi Sueur 2001: 37, Table 1 (listed)
Pycna passosdecarvalhoi Boulard 2006g: 150, Fig 110, Fig (sonogram, illustrated, comp. note) Malawi

P. quanza **(Distant, 1899)** = *Platypleura quanza*

P. repanda repanda **(Linnaeus, 1758)** = *Cicada repanda* = *Cicada rependa* (sic) = *Tettogonia repanda* = *Fidicina ? repanda* = *Oxypleura repanda* = *Platypleura repanda* = *Platypleura nepanda* (sic) = *Platypleura phalaenoides* Walker, 1850 = *Platypleura interna* Walker, 1852 = *Platyplerua congrex* Butler, 1874

Pycna repanda Swinton 1908: 380 (song, comp. note) Cashmere
Pycna repanda Chen 1942: 143 (listed, comp. note) China, India
Pycna repanda Chaudhry, Chaudry and Khan 1970: 105 (listed, comp. note) Pakistan
Pycna repanda Hayashi 1982a: 79, 81–83, Fig 3, Table 1 (illustrated, distribution, comp. note) Nepal
Pycna repanda Emeljanov and Kirillova 1991: 800 (chromosome number, comp. note)
Pycna repanda Liu 1991: 257 (comp. note)
Pycna repanda Emeljanov and Kirillova 1992: 63 (chromosome number, comp. note)
Pycna repanda Liu 1992: 16 (comp. note)
Pycna repanda Peng and Lei 1992: 97–98, Fig 277 (listed, illustrated, comp. note) China, Hunan
Cicada repanda Chou, Lei, Li, Lu and Yao 1997: 3 (synonymy, comp. note) Equals *Pycna repanda*
Pycna repanda Chou, Lei, Li, Lu and Yao 1997: 173–174, Fig 9–42, Plate VII, Fig 82 (key, synonymy, described, illustrated, comp. note) Equals *Cicada repanda* Equals *Tettigonia repanda* Equals *Fidicina? repanda* Equals *Oxypleura repanda* Equals *Platypleura repanda* Equals *Platypleura phalaenoides* Equals *Platypleura congrex* China
Pycna repanda Hua 2000: 64 (listed, distribution, host) Equals *Cicada repanda* China, Shaanxi, Jiangxi, Zhejiang, Taiwan, Hunan, Guizhou, Sichuan, Yunnan, Xizang, Kashmir, Sikkim, Burma, India, Bhutan
Pycna repanda McGavin 2000: 96 (illustrated) India
Pycna repanda Boulard 2005b: 112 (comp. note) India, Nepal, China
Pycna repanda Shiyake 2007: 2, 12, 15, Fig 4 (illustrated, distribution, listed, comp. note) China, India, Tibet, Burma
Pycna repanda Ahmed and Sanborn 2010: 27, 38–39 (synonymy, distribution) Equals *Cicada repanda* Equals *Platyptera phalaenoides* Equals *Platypleura interna* Equals *Platypleura congrex* India, Pakistan, Sri Lanka, Burma, Japan, China, Taiwan, Indonesia, Nepal, Bhutan

***P. repanda* var. *β* (Walker, 1850)** = *Platypleura repanda* var. *β*

***P. repanda* var. *δ* (Walker, 1850)** = *Platypleura repanda* var. *δ*

***P. repanda* var. *ε* (Walker, 1850)** = *Platypleura repanda* var. *ε*

***P. repanda* var. *η* (Walker, 1850)** = *Platypleura repanda* var. *η*

***P. repanda* var. *γ* (Walker, 1850)** = *Platypleura repanda* var. *γ*

***P. repanda* var. *ζ* (Walker, 1850)** = *Platypleura repanda* var. *ζ*

***P. rudis* (Karsch, 1890)** = *Platypleura rudis*

***P. schmitzi* Boulard, 1979**

Pycna schmitzi Boulard 1985c: 184 (comp. note)

***P. semiclara* (Germar, 1834)** = *Cicada semiclara* = *Platypleura semiclara* = *Platypleura basifolis* Walker, 1850

Pycna semiclara Gess 1981: 5–8, Fig 2 (described, illustrated, song, habitat, comp. note) South Africa
Pycna semiclara Villet 1987a: 270, 272, Table 2 (song intensity, listed, comp. note) South Africa
Pycna semiclara Villet 1988: 72, 74–75, Fig 10, Tables 1–2 (song, sonogram, comp. note) South Africa
Pycna semiclara Villet and Reavell 1989: 334–335, Fig 2c, Fig 3 (distribution, habitat, host plants, key, comp. note) South Africa
Pycna semiclara Villet 1992: 94–95, Table 1 (acoustic responsiveness, comp. note) South Africa
Pycna semiclara Pinchuck and Villet 1995: 112 (timbal muscle ultrastructure, comp. note)
Pycna semiclara Sanborn and Phillips 1995a: 482 (comp. note)
Pycna semiclara Sueur 2001: 37, 46, Table 1 (listed, comp. note)
Pycna (Platypleura) semiclara Picker, Griffith and Weaving 2002: 156–157 (described, illustrated, habitat, comp. note) South Africa
Pycna semiclara Phillips, Sanborn and Villet 2002: 31 (thermal responses) South Africa, Eastern Cape
Pycna semiclara Villet, Sanborn and Phillips 2002: 38 (calling behavior, endothermy, comp. note) South Africa
Pycna semiclara Sanborn, Phillips and Villet 2003a: 349, Table 1 (thermal responses, habitat, comp. note) South Africa
Pycna semiclara Sanborn, Phillips and Villet 2003b: 291 (comp. note) South Africa
Pycna semiclara Villet, Sanborn and Phillips 2003a: 2–4 (endothermy, comp. note) South Africa
Pycna semiclara Villet, Sanborn and Phillips 2003b: 79 (calling behavior, endothermy, comp. note) South Africa
Pycna semiclara Villet, Sanborn and Phillips 2003c: 1438–1443, Figs 1–4, Tables 1–2 (calling behavior, endothermy, comp. note) South Africa

Pycna semiclara Malherbe, Burger and Stephen 2004: 85–86 (comp. note) South Africa
Pycna semiclara Sanborn, Villet and Phillips 2004a: 817, 819, 821, Tables 1–2 (thermoregulation, endothermy, habitat, comp. note) South Africa
Pycna semiclara Sanborn 2005c: 114–116, Table 1 (comp. note)
Pycna semiclara Chawanji, Hodson and Villet 2006: 386–387 (comp. note)
Pycna semiclara Seabra, Pinto-Juma and Quartau 2006: 844 (comp. note)

***P. strix* Amyot & Audinet-Serville, 1843** = *Cicada strix* = *Platypleura strix* = *Cicada strix* Brullé, 1849 = *Cicada stryx* (sic) = *Pycna stryx* (sic)
Pycna strix Chen 1942: 143 (type species of *Pycna*)
Pycna strix Boulard 1997b: 22–23, Plate IV, Fig 6 (coloration, comp. note)
Cicada strix Chou, Lei, Li, Lu and Yao 1997: 173 (type species of *Pycna*)
Pycna strix Boulard 1999b: 184–185, Fig 8, Fig 8" (illustrated, comp. note)
Pycna strix Lee 2008a: 4 (type species of *Pycna*) Madagascar
Pycna strix Ahmed and Sanborn 2010: 38 (type species of *Pycna*)

***P. sylvia* (Distant, 1899)** = *Platypleura sylvia*
Pycna sylvia Malherbe, Burger and Stephen 2004: 85–88, Fig 1 (distribution, behavior, host plants, comp. note) South Africa

***P. tangana* (Strand, 1910)** = *Platypleura (Pycna) tangana* = *Pycna tanga* (sic)

***P. umbelinae* Boulard, 1975**

***P. verna* Hayashi, 1982**
Pycna verna Hayashi 1982a: 78–83, Figs 102, Figs 7–10 (n. sp., described, illustrated, distribution, comp. note) India, Nepal

***P. vitrea* Schumacher, 1913**

***P. vitticollis* (Jacobi, 1904)** = *Platypleura vitticollis*

***Suisha* Kato, 1928** = *Dasypsaltria* Kato, 1927
Suisha Wu 1935: 3 (listed) China
Suisha Hayashi 1982a: 82 (comp. note)
Suisha Hayashi 1984: 28–29, 32, 34 (key, described, synonymy, listed, comp. note) Equals *Dasypsaltria*
Suisha Lee 1995: 17, 22, 32, 52, 58–59, 135 (listed, key, described, ecology, distribution, illustrated, comp. note) Korea
Suisha Chou, Lei, Li, Lu and Yao 1997: 167, 176, 293, Table 6 (key, synonymy, described, listed, diversity, biogeography) Equals *Dasypsaltria* China
Suisha Novotny and Wilson 1997: 437 (listed)
Suisha Lee 1999: 2 (comp. note) Korea
Suisha Lee and Hayashi 2003a: 153, 156–157 (key, diagnosis) Taiwan
Suisha Chen 2004: 39–40, 51 (phylogeny, listed)
Suisha Lee 2005: 14, 16, 51, 157, 168 (listed, key, comp. note) Korea
Suisha Moulds 2005b: 392, 433 (higher taxonomy, listed, comp. note)
Suisha Shiyake 2007: 2, 14–15 (listed, comp. note)
Suisha Lee 2008b: 446, Table 1 (listed, comp. note) Korea
Suisha Osaka Museum of Natural History 2008: 322 (listed) Japan

***S. coreana* (Matsumura, 1927)** = *Pycna coreana* = *Dasypsaltria coreana* Kato, 1927
Suisha coreana Doi 1932: 30, 42 (listed, comp. note) Korea
Suisha coreana Wu 1935: 3 (listed) Equals *Pycna coreana* Equals *Dasypsaltria coreana* China, Manchuria, Nanking, Soochow, Hangchow
Suisha coreana Hayashi 1984: 35, Figs 24–27 (described, illustrated, synonymy, comp. note) Equals *Pycna coreana* Equals *Dasypsaltria coreana* Japan
Suisha coreana Wang and Zhang 1987: 295 (key, comp. note) China
Suisha coreana Liu 1991: 257 (comp. note)
Suisha coreana Liu 1992: 17, 34 (comp. note)
Suisha coreana Wang and Zhang 1992: 119 (listed, comp. note) China
Suisha coreana Lee 1995: 17, 22, 32–33, 35–39, 41–46, 135, 139 (listed, key, synonymy, ecology, distribution, illustrated, comp. note) Equals *Pycna coreana* Equals *Dasypsaltria coreana* Korea
Pycna coreana Lee 1995: 32, 59 (comp. note) Equals *Suisha coreana* Korea
Dasypsaltria coreana Lee 1995: 59 (comp. note) Korea
Suisha coreana Chou, Lei, Li, Lu and Yao 1997: 176–179, Fig 9–45, Plate VII, Fig 83 (key, synonymy, described, illustrated, comp. note) Equals *Pycna coreana* Equals *Dasypsaltria coreana* China
Suisha coreana Lee 1999: 3–4, Fig 2 (illustrated, distribution, song, listed, comp. note) Equals *Pycna coreana* Equals *Dasypsaltria coreana* Korea, China, Japan
Suisha coreana Freytag 2000: 55 (on stamp)
Suisha coreana Hua 2000: 64 (listed, distribution, host) Equals *Pycna coreana* China, Shaanxi, Gansu, Hubei, Jiangsu, Zhejiang, Hunan, Guangxi, Korea, Japan
Suisha coreana Lee 2003b: 8 (comp. note)
Suisha coreana Lee and Hayashi 2003a: 157 (comp. note)

Suisha coreana Lee 2005: 16, 50–53, 157, 161, 169 (described, illustrated, synonymy, song, distribution, key, comp. note) Equals *Pycna coreana* Equals *Dasypsaltria coreana* Korea, China, Japan

Suisha coreana Shiyake 2007: 2, 13, 15, Fig 5 (illustrated, distribution, listed, comp. note) China, Korea

Suisha coreana Lee 2008b: 446, 463, Fig 1B, Table 1 (key, illustrated, synonymy, distribution, listed, comp. note) Equals *Pycna coreana* Equals *Dasypsaltria coreana* Korea, Japan, China

Suisha coreana Osaka Museum of Natural History 2008: 322 (listed, acoustic behavior, ecology, comp. note) Japan

***S. formosana* (Kato, 1927)** = *Dasypsaltria formosana* = *Platypleura formosana*

Suisha formosana Hayashi 1982a: 79, Fig 6 (illustrated, comp. note) Taiwan

Dasypsaltria formosana Hayashi 1984: 34 (type species of *Suisha*)

Dasypsaltria formosana Lee 1995: 32, 59 (comp. note)

Suisha formosana Lee 1995: 58–59 (type species of *Suisha*, comp. note) Formosa

Dasypsaltria formosana Chou, Lei, Li, Lu and Yao 1997: 176 (type species of *Suisha*)

Suisha formosana Chou, Lei, Li, Lu and Yao 1997: 176–177, Fig 9–44 (key, synonymy, described, illustrated, comp. note) Equals *Dasypsaltria formosana* China

Dasypsaltria formosana Lee 1999: 3 (type species of *Suisha*) Formosa

Suisha formosana Hua 2000: 64 (listed, distribution) Equals *Dasypsaltrua formosana* China, Zhejiang, Taiwan

Dasypsaltria formosana Lee and Hayashi 2003a: 156 (type species of *Suisha*) Formosa

Suisha formosana Lee and Hayashi 2003a: 157–158, Fig 6 (key, diagnosis, illustrated, synonymy, distribution, song, listed, comp. note) Equals *Dasypsaltria formosana* Taiwan, China

Suisha formosana Chen 2004: 51–52, 198 (described, illustrated, listed, comp. note) Taiwan

Suisha formosana Lee 2005: 51 (type species of *Suisha*)

Dasypsaltria formosana Lee 2005: 168 (type species of *Suisha*) Formosa

Dasypsaltria formosana Lee 2008b: 446 (type species of *Suisha*) Formosa

***Umjaba* Distant, 1904**

Umjaba Boulard 1985e: 1025 (coloration, comp. note)

Umjaba Moulds 2005b: 392, 433 (higher taxonomy, listed, comp. note)

Umjaba Boulard 2006g: 109 (comp. note) Madagascar

***U. alluaudi* Distant, 1905**

***U. evanescens* (Butler, 1882)** = *Platypleura evanescens*

Umjaba evanescens Boulard 2006g: 33, 35–36, 75–76, 90, 181, Fig 5, Fig 55 (illustrated, sonogram, listed, comp. note) China

Umjaba evanescens Boulard 2007f: 140, Fig 120 (illustrated, comp. note) Madagascar

***Ugada* Distant, 1904** = *Platypleura (Ugada)*

Ugada Synave 1966: 1030 (listed) Senegal

Ugada Medler 1980: 73 (listed) Nigeria

Ugada Boulard 1985e: 1025, 1029, 1030 (coloration, comp. note)

Ugada Boulard 1986c: 192 (comp. note)

Ugada Boulard 1986d: 225 (listed)

Ugada Boulard, Deeming and Matile 1989: 143 (predation, comp. note) Ivory Coast

Ugada Boulard 1990b: 74–75, 116, 167, 175 (comp. note)

Ugada Boulard 1991f: 16 (warning sound, comp. note)

Ugada Boulard 1995a: 7 (comp. note)

Ugada Boulard 1996d: 115 (listed)

Ugada Boulard 1999a: 91 (comp. note)

Ugada Boulard 2002b: 193, 203 (coloration, comp. note) Africa

Ugada Moulds 2005b: 392, 433 (higher taxonomy, listed, comp. note)

Ugada Villet, Barker and Lunt 2005: 593 (comp. note) West Africa

Ugada Boulard 2006g: 129 (comp. note)

Ugada Schabel 2006: 258 (human food, comp. note) Africa

Ugada Boulard 2007f: 52 (comp. note)

Ugada Shiyake 2007: 2, 15 (listed, comp. note)

Ugada Moulds 2008a: 208 (comp. note)

***U. atratula* Distant, 1919**

***U. cameroni* (Butler, 1874)** = *Platypleura cameroni* = *Ugada violacea* Boulard, 1972

***U. dargei* Boulard, 2012**

***U. giovanninae* Boulard, 1973** = *Ugada giovannina* (sic)

Ugada giovnninae Boulard 1985e: 1029, 1046–1047, Plate VIII, Fig 49 (coloration, illustrated, comp. note)

Ugada giovannina (sic) Nkouka 1987: 174 (human food) Central Africa

Ugada giovanninae Boulard 1989c: 39, 49, Plate IV, Fig 7 (cacao pest, comp. note) Central African Republic

Ugada giovanninae Boulard 1990b: 75, 171, 175, 185, 187 (sonogram, comp. note)

Ugada giovanninae Boulard 1995a: 8 (comp. note)
Ugada giovanninae Boulard 1996d: 114, 118–121, Fig S2(a-f), Fig S2(g-h), Fig S2(i-l) (song analysis, sonogram, listed, comp. note) Central African Republic
Ugada giovannina (sic) Huis 1996: 7, Table 1 (predation, comp. note)
Ugada giovanninae Sueur 2001: 37, Table 1 (listed)
Ugada giovanninae Huis 2003: 168, Table 1 (predation, comp. note) Congo
Ugada giovanninae Boulard 2005d: 346 (comp. note) Africa
Ugada giovanninae Villet, Barker and Lunt 2005: 590–593, Figs 1–5, Table 1 (phylogeny, listed, comp. note)
Ugada giovanninae Boulard 2006g: 74–75, 104–105, 180, Figs 69–70 (comp. note) Central African Republic
Ugada giovanninae Boulard 2007e: 87 (comp. note) Central African Republic
Ugada giovanninae Boulard 2007f: 131, Fig 21 (illustrated, comp. note)

***U. grandicollis grandicollis* (Germar, 1830)** = *Cicada graneidollis* (sic) = *Cicada grandicollis* = *Ugada grandicollis* = *Platypleura confusa* Karsch, 1890 = *Ugada confusa*
Ugada grandicollis Schouteden 1925: 83 (listed, comp. note) Congo
Ugada grandicollis Boulard 1985e: 1029, 1046–1047, Plate VIII, Fig 49 (coloration, illustrated, comp. note)
Ugada grandicollis Boulard 1986d: 228, 237 (synonymy, comp. note) Equals *Platypleura confusa* Cameroon, Central Africa Republic
Ugada grandicollis grandicollis Boulard 1986d: 226–228, 236–238, Table I, Fig 42 (n. comb., neallotype designation, described, illustrated, comp. note) Central African Republic, Congo
Ugada grandicollis Boulard 1990b: 171, 206 (comp. note)
Ugada grandicollis grandicollis Boulard 1996d: 131 (comp. note)
Ugada grandicollis Freytag 2000: 55 (on stamp)
Ugada grandicollis Sueur 2001: 37, Table 1 (listed)
Ugada grandicollis Boulard 2007f: 131, Fig 21 (illustrated, comp. note)

***U. grandicollis taiensis* Boulard, 1986** = *Ugada grandicollis taïensis* = *Ugada taiensis*
Ugada grandicollis taiensis Boulard 1986d: 226–228, 236–238, Table I, Fig 43 (n. ssp., described, illustrated, comp. note) Ivory Coast, Guinea
Ugada grandicollis taiensis Boulard, Deeming and Matile 1989: 144–145, Fig 1 (predation, illustrated, comp. note) Ivory Coast
Ugada taiensis Boulard 2000d: 95, 106–107, Plate P, Figs 1–3 (song, illustrated, sonogram, comp. note) Ivory Coast
Ugada taiensis Boulard 1995a: 8 (comp. note)
Ugada grandicollis taiensis Boulard 1996d: 114, 116–117, Fig S1, Color Plate Figs 1–2 (song analysis, sonogram, listed, comp. note) Ivory Coast
Ugada taiensis Sueur 2001: 37, Table 1 (listed)
Ugada taiensis Boulard 2002b: 194, 200, Fig 46, sonogram 54a (illustrated, sonogram, comp. note)
Ugada taiensis Boulard 2005d: 345–346, Fig 25.42 (sonogram, illustrated, comp. note)
Ugada taiensis Boulard 2006g: 74–75, 86, 100–101, 180, Fig 47, Fig 66 (sonogram, illustrated, listed, comp. note) Ivory Coast
Ugada taiensis Boulard 2007f: 122, 149, Fig 261 (comp. note)

***U. inquinata* (Distant, 1881)** = *Platypleura inquinata* = *Platypleura (Ugada) inquinata*
Ugada inquinata Boulard 1985e: 1032 (coloration, comp. note)

***U. kageraensis* Boulard, 2012**

***U. lamottei* Boulard, 1977**

***U. limbalis* (Karsch, 1890)** = *Platypleura limbalis* = *Ugada limballis* (sic) = *Ugada limbimacula* Dlabola 1962 (nec Karsch)
Ugada limbalis Medler 1980: 73 (listed) Nigeria
Ugada limbalis Boulard 1985e: 1046–1047, Plate VIII, Fig 49 (coloration, illustrated, comp. note)
Ugada limbalis Boulard 1990b: 74–75, 171, 185, 187, Plate XXII, Fig 5 (sonogram, comp. note) Central African Republic
Ugada limbalis Boulard 1996d: 114, 120–123, Fig S3 (song analysis, sonogram, listed, comp. note) Central African Republic
Ugada limbalis Malaisse 1997: 238, Table 2.11.1 (human food, listed) Zambezi
Ugada limballis (sic) Sueur 2001: 37, Table 1 (listed)
Ugada limbalis Boulard 2002b: 209 (comp. note)
Ugada limbalis Boulard 2005d: 344–345 (comp. note)
Ugada limbalis Boulard 2006g: 73, 75, 86, 96, 180, Fig 46 (sonogram, listed, comp. note) Central African Republic
Ugada limbalis Boulard 2007f: 131, Fig 21 (illustrated, comp. note)
Ugada limbalis Shiyake 2007: 2, 13, 15, Fig 6 (illustrated, distribution, listed, comp. note) Africa

U. limbata limbata (Fabricius, 1775) = *Tettigonia limbata* = *Cicada limbata* = *Oxypleura limbata* = *Platypleura limbata* = *Pycna limbata* = *Ugada limbata* = *Cicada armata* Olivier, 1790 = *Oxypleura armata* = *Cicada africana* Palisot de Beauvois, 1813

Platypleura limbata Distant 1881: ii (comp. note) West Africa

Ugada limbata Schouteden 1925: 83 (listed, comp. note) Congo

Ugada limbata Synave 1966: 1030 (listed, comp. note) Equals *Tettigonia limbata* Senegal

Ugada limbata Medler 1980: 73 (listed) Nigeria

Ugada limbata Boulard 1985e: 1025, 1039–1040, Plate III, Fig 23 (coloration, illustrated, comp. note) Guinea, Congo, Central Africa Republic

Tettigonia limbata Boulard 1986d: 225 (type species of *Ugada*)

Ugada limbata Boulard 1986d: 225, 237 (distribution, comp. note) Equals *Cicada armata* Equals *Cicada africana* Ivory Coast

Ugada limbata limbata Boulard 1986d: 226–227, 236–238, Table I, Figs 5–6, Fig 40 (n. comb., described, illustrated, comp. note) Nigeria, Central African Republic, Cameroon, Gabon, Congo, Ivory Coast

Ugada limbata Nkouka 1987: 174 (human food) Central Africa

Ugada limbata Boulard 1989c: 30 (coffee pest, comp. note) Central African Republic

Ugada limbata Boulard 1990b: 206 (comp. note)

Ugada limbata Huis 1996: 7, Table 1 (predation, comp. note)

Ugada limbata Freytag 2000: 55 (on stamp)

Ugada limbata Huis 2003: 168, Table 1 (predation, comp. note) Congo, Democratic Republic of Congo, Zambia

Ugada limbata Boulard 2007f: 131, Fig 21 (illustrated, comp. note)

U. limbata occidentalis Boulard, 1986 = *Ugada limbata occidentalis* var. *intermedia* Boulard, 1986 = *Ugada limbata occidentalis* var. *marmorea* Boulard, 1986

Ugada limbata occidentalis Boulard 1986d: 226–227, 238, Table I, Figs 7–9 (n. ssp., described, illustrated, comp. note) Ivory Coast, Guinea

Ugada limbata occidentalis var. *intermedia* Boulard 1986d: 227, 236–237, Fig 41 (n. var. of ssp., described, comp. note) Central African Republic, Ivory Coast

Ugada limbata occidentalis var. *marmorea* Boulard 1986d: 227 (n. var. of ssp., described, comp. note) Central African Republic, Ivory Coast

U. limbimacula (Karsch, 1893) = *Ugada limbimaculata* (sic) = *Platyleura limbimacula* = *Ugada hahnei* Schmidt, 1919 = *Ugada nutti* (nec Distant) = *Ugada limbimacula* Dlabola (nec Karsch)

Ugada limbimacula Boulard 1985e: 1029, 1046–1047, Plate VIII, Fig 49 (coloration, illustrated, comp. note)

Ugada limbimaculata (sic) Nkouka 1987: 174 (human food) Central Africa

Ugada limbimacula Boulard 1990b: 74, 185, 187, Plate XXII, Fig 4 (sonogram, comp. note)

Ugada limbimacula Boulard 1996d: 114, 122–125, Fig S4 (song analysis, sonogram, listed, comp. note) Central African Republic

Ugada limbimaculata (sic) Huis 1996: 7, Table 1 (predation, comp. note)

Ugada limbimaculata (sic) Malaisse 1997: 238, Table 2.11.1 (human food, listed) Zambezi

Ugada limbimacula Sueur 2001: 37, Table 1 (listed)

Ugada limbimaculata (sic) Huis 2003: 168, Table 1 (predation, comp. note) Democratic Republic of Congo, Congo

Ugada limbimacula Boulard 2002b: 203, 209 (comp. note)

Ugada limbimacula Boulard 2005d: 347 (comp. note)

Ugada limbimacula Boulard 2007f: 131, Fig 21 (illustrated, comp. note)

U. nigeriana (Distant, 1913) = *Pycna nigeriana*

Ugada nigeriana Medler 1980: 73 (listed) Nigeria

U. nigrofasciata nigrofuscata Distant, 1919 = *Ugada nigrofasciata* = *Ugada oswaldebneri* Schmidt, 1920

Ugada nigrofasciata Boulard 1990b: 206 (comp. note)

Ugada nigrofasciata Duffels and van Mastrigt 1991: 174 (predation, comp. note) Central Africa Republic

Ugada nigrofasciata nigrofasciata Boulard 2000b: 21–22, Fig 13 (song, illustrated, comp. note)

U. nigrofasciata sylvicola Boulard, 1973

Ugada nigrofasciata sylvicola Boulard 1985e: 1029, 1046–1047, Plate VIII, Fig 49 (coloration, illustrated, comp. note)

Ugada nigrofasciata sylvicola Boulard 2000b: 21–22, Fig 14 (song, illustrated, comp. note)

Ugada nigrofasciata sylvicola Boulard 2007f: 131, Fig 21 (illustrated, comp. note)

U. nutti Distant, 1904

U. parva Boulard, 1975

U. poensis Boulard, 1975

U. praecellens (Stål, 1863) = *Platypleura praecellens*

Ugada praecellens Boulard 1985e: 1032 (coloration, comp. note) Guinea

Ugada praecellens Boulard 1986d: 238, Table 1 (comp. note) Ivory Coast

U. sp. 1 Villet, Barker & Lunt, 2005

Ugada sp. 1 Villet, Barker and Lunt 2005: 590–593, Figs 1–5, Table 1 (phylogeny, listed, comp. note) South Africa

U. stalina (Butler, 1874) = *Platypleura stalina*

***Hainanosemia* Kato, 1927**

Hainanosemia Moulds 2005b: 392, 433 (higher taxonomy, listed, comp. note)

***H. stigma* Kato, 1927**

Hainanosemia stigma Hua 2000: 62 (listed, distribution) China, Hainan

***Kalabita* Moulton, 1923**

Kalabita Duffels 1986: 319 (distribution, diversity, comp. note) Greater Sunda Islands, Borneo, Sarawak

Kalabita Boulard 1990b: 144 (comp. note)

Kalabita Zaidi and Ruslan 1995a: 174, Table 2 (listed, comp. note) Borneo, Sarawak

Kalabita Zaidi and Ruslan 1995d: 202, Table 2 (listed, comp. note) Borneo, Sabah

Kalabita Zaidi 1996: 103, Table 2 (comp. note) Borneo, Sarawak

Kalabita Zaidi and Hamid 1996: 54, Table 2 (listed, comp. note) Borneo, Sarawak

Kalabita Zaidi 1997: 115, Table 2 (listed, comp. note) Borneo, Sarawak

Kalabita Zaidi and Ruslan 1999: 4, Table 2 (listed, comp. note) Borneo, Sarawak

Kalabita Moulds 2005b: 392, 433 (higher taxonomy, listed, comp. note)

Kalabita Trilar 2006: 341 (comp. note)

***K. operculata* Moulton, 1923**

Kalabita operculata Boulard 1990b: 152–153, Plate XVII, Fig 7 (sound system, comp. note) Borneo

Kalabita operculata Zaidi and Ruslan 1995a: 171, 176 (comp. note) Borneo, Sarawak

Kalabita operculata Zaidi and Ruslan 1998a: 345–346 (listed, comp. note) Borneo, Sarawak

Kalabita operculata Zaidi and Ruslan 1999: 3, 6 (comp. note) Borneo, Sarawak

Kalabita operculata Trilar 2006: 341–345, Figs 1–6 (illustrated, song, sonogram, oscillogram, distribution, comp. note) Borneo, Sabah, Sarawak

Tribe Hamzini Distant, 1905 = Hamzaria

Hamzini Duffels 1991a: 120, 128 (comp. note)

Hamzini Villet 1994b: 87, 91 (comp. note)

Hamzaria Villet 1994b: 87, 91 (comp. note)

Hamzaria Duffels 1991a: 120, 128 (comp. note)

Hamzini Chou, Lei, Li, Lu and Yao 1997: 106, 128 (key, listed, described) Equals Hamzaria

Hamzaria Chou, Lei, Li, Lu and Hamzini Chou, Lei, Li, Lu and Yao 1997: 298 (comp. note)

Hamzini Novotny and Wilson 1997: 437 (listed)

Hamzaria Moulds 2005b: 386 (history, listed, comp. note)

Hamzini Moulds 2005b: 427 (higher taxonomy, listed)

Hamzini Pham and Thinh 2005a: 243 (listed) Vietnam

Hamzini Pham and Thinh 2005b: 287–289 (key, comp. note) Equals Hamzaria Equals Hamzaini Vietnam

***Hamza* Distant, 1904**

Hamza Hayashi 1984: 29 (listed)

Hamza Duffels 1991a: 119 (comp. note)

Hamza Villet 1994b: 91 (comp. note)

Hamza Boer and Duffels 1996b: 302 (comp. note)

Hamza Chou, Lei, Li, Lu and Yao 1997: 298 (comp. note)

Hamza Novotny and Wilson 1997: 437 (listed)

Hamza Lee 2010a: 15 (listed) Philippines, Mindanao

***H. ciliaris* (Linnaeus, 1758)** = *Cicada ciliaris* = *Poecilopsaltria ciliaris* = *Platypleura ciliaris* = *Platypleura (Platypleura) ciliaris* = *Platypleura cilialis* (sic) = *Cicada ocellata* Degeer, 1773 = *Cicada ocella* (sic) = *Platypleura ocellata* = *Cicada varia* Olivier, 1790 = *Platypleura varia* = *Tettigonia marmorata* Fabricius, 1803 = *Oxypleura marmorata* = *Platypleura marmorata* = *Platypleura arcuata* Walker, 1858 = *Platypleura catocaloides* Walker, 1868 = *Platypleura catocalina* (sic) = *Poecilopsaltria catacaloides* = *Platypleura bouruensis* Distant, 1898 = *Platypleura bouruensis* Distant, 1898 = *Hamza bouruensis* = *Hamza boeruensis* (sic) = *Platypleura lyricen* Kirkaldy, 1913 = *Hamza bouruensis* Kato, 1927 (nec Distant, 1898) = *Hamza uchiyamae* Matsumura, 1927 = *Hamza bouruensis uchiyamae* = *Platyplerua uchiyamae* = *Hamza* sp. Endo and Hayashi 1979

Tettigonia marmorata Dallas 1870: 482 (synonymy) Equals *Platypleura marmorata* Equals *Cicada ciliaris* Equals *Cicada ocellata* Equals *Cicada varia* Equals *Platypleura arcuata*

Platypleura ciliaris Zeuner 1941: 88 (comp. note) Indo-China

Cicada ciliaris Boulard 1988f: 68 (type species of *Cicada*)

Hamza ciliaris Boulard 1991c: 120–121, Figs 1–2 (comp. note) Equals *Cicada ocellata* Equals *Platypleura marmorata* Equals *Platypleura ciliaris* Banda, Ambione

Platypleura ciliaris Boulard 1991c: 120–121, Fig 2 (comp. note)

Platypleura marmorata Boulard 1991c: 120–121, Fig 1 (comp. note)

Hamza bouruensis Duffels 1991a: 119–120l 126 (comp. note) Indonesia, Buru

Platypleura bouruensis Duffels 1991a: 119–120, 122, 124 (type species of *Hamza*, comp. note) Buru Island

Hamza uchiyamae Duffels 1991a: 119, 122, 124, 127 (comp. note) Caroline Islands, Palau Islands

Platypleura lyricen Duffels 1991a: 119, 122, 124 (comp. note) Ambon, Ambiona
Platypleura ciliaris Duffels 1991a: 119, 122, 126 (comp. note) Ambon, Ternate, Seram, Morotai, Ambiona
Cicada ciliaris Duffels 1991a: 119–122, 126, Fig 1 (illustrated, comp. note) Indonesia
Hamza ciliaris Duffels 1991a: 119–129, Figs 2–16 (n. comb., illustrated, described, distribution, comp. note) Equals *Cicada ciliaris* Equals *Cicada ocellata* Equals *Cicada varia* Equals *Tettigonia marmorata* Equals *Platypleura arcuata* Equals *Platypleura catacaloides* Equals *Poecilopsaltria ciliaris* Equals *Platypleura bouruensis* Equals *Hamza bouruensis* Equals *Platypleura lyricen* Equals *Hamza uchiyamae* Equals *Platypleura ciliaris* Seram, Buru, Gebe Island, Ternate, Ambon, Koror Island, Sula Island, Banggai, Palau group, Caroline Islands, Philippine Islands, Timor, Kai Island, Banda Island, Mindanao
Cicada ocellata Duffels 1991a: 122 (comp. note)
Cicada varia Duffels 1991a: 122 (comp. note)
Tettigonia marmorata Duffels 1991a: 122 (lectotype designation, comp. note) Amboina
Platypleura arcuata Duffels 1991a: 122, 126 (comp. note) Ceram, Amboina
Platypleura catacaloides Duffels 1991a: 122, 124 (lectotype designation, comp. note) Amboina, Ceram, Morotai
Poecilopsaltria catacaloides Duffels 1991a: 126 (comp. note)
Platypleura sp. Duffels 1991a: 126 (comp. note)
Poecilopslatria ciliaris Duffels 1991a: 126 (comp. note) Ambiona, Saparau
Poecilopslatria basi-viridis var. ? Duffels 1991a: 126 (comp. note) Buru
Platypleura uchiyamae Duffels 1991a: 127 (comp. note) Anguar Island
Hamza sp. Endo and Hayashi, 1979 Duffels 1991a: 127 (comp. note) Mindanao
Hamza ciliaris Boer 1994c: 133 (comp. note) Molucca Islands, Mindano, Caroline Islands
Hamza ciliaris Villet 1994b: 91 (comp. note)
Hamza ciliaris Boer 1995d: 230 (distribution, comp. note) Maluku, Banda Islands, Timor, Philippines, Caroline Islands
Hamza ciliaris Boer and Duffels 1996b: 310, 316, 330 (distribution, comp. note) Equals *Hamza uchiyamae* Timor, Kai Island, Banda Islands, South Maluku, Sula Islands, North Maluku, Mindanao, Caroline Islands
Hamza uchiyamae Boer and Duffels 1996b: 309 (comp. note) Caroline Islands
Cicada ciliaris Boulard 1997c: 119 (type species of *Cicada*)
Cicada ciliaris Boulard 2001b: 39 (type species of *Cicada*)
Platypleura ciliaris Lee 2008a: 3 (synonymy, distribution, listed) Equals *Cicada ciliaris* Equals *Poecilopsaltria ciliaris* Equals *Platypleura (Platypleura) ciliaris* South Vietnam, Philippines, Malayian Archipelago, India
Hamza ciliaris Lee 2009i: 293 (comp. note) Philippines, Panay
Platypleura ciliaris Pham and Yang 2009: 13, Table 2 (listed) Vietnam
Cicada ciliaris Lee 2010a: 15 (type species of *Hamza*, synonymy, listed) Equals *Platypleura bouruensis* India
Hamza ciliaris Lee 2010a: 15–16, Figs 1C–1D (synonymy, illustrated, distribution, listed, comp. note) Equals *Cicada ciliaris* Equals *Hamza* sp. Endo and Hayashi 1979 Philippines, Mindanao, Palau, Indonesia, Maluku, Timor, Banggai Archipelago

Tribe Orapini Boulard, 1985
Orapini Boulard 1985e: 1032 (n. tribe, characteristics, comp. note)
Orapini Boulard 1996d: 115, 144 (listed)
Orapini Sueur 2001: 39, Table 1 (listed)

***Orapa* Distant, 1905** = *Orapia* (sic)
Orapa Boulard 1985e: 1022–1023, 1025, 1027 (type genus of Orapini, coloration, comp. note)
Orapa Boulard 1990b: 75, 207 (comp. note)
Orapa Boulard 1996b: 4 (comp. note) Africa
Orapa Boulard 1996d: 144 (listed)
Orapa sp. Malaisse 1997: 237, Table 2.11.1 (human food, listed) Zambezi
Orapa Boulard 2002b: 193 (coloration, comp. note) Africa
Orapa sp. Huis 2003: 168, Table 1 (predation, comp. note) Botswana
Orapa Moulds 2005b: 403, 416 (comp. note) East Africa

***O. africana* Kato, 1927**

O. elliotti* (Distant, 1907)** = ***Pycna elliotti
Orapa elliotti Boulard 1985e: 1019, 1031–1033, 1036–1037, 1044–1047, Plate IV, Figs 30, Plate IX, Fig 50 (n. comb., crypsis, coloration, illustrated, comp. note) Equals *Pycna elliotti*
Pycna elliotti Boulard 1985e: 1019, 1032 (crypsis, comp. note)
Orapa elliotti Boulard 1990b: 169, 201 (coloration, comp. note)

Orapa elliotti Boulard 1997b: 15–16, 20, Plate II, Fig 3 (coloration, morphology, comp. note) East Africa
Orapa elliotti Boulard 2002b: 209 (comp. note)
Orapa elliotti Moulds 2005b: 403, 416 (comp. note)
Orapa elliotti Boulard 2007f: 84, Fig 100.1 (illustrated, comp. note)

O. *lateritia* Jacobi, 1910
Orapa lateritia Boulard 1985e: 1032 (comp. note)
Orapa lateritia Moulds 2005b: 403, 416 (comp. note)

O. cf. *lateritia* Villet & Noort, 1999
Orapa cf. *lateritia* Villet and Noort 1999: 228, 230, Table 12.2 (comp. note) Tanzania

O. *numa* (Distant, 1904) = *Pycna numa*
Orapa numa Boulard 1985e: 1019, 1032, 1037–1038, Plate I, Fig 6, Plate III, Fig 25 (synonymy, crypsis, coloration, illustrated, comp. note) Equals *Pycna numa* East Africa, Kenya
Pycna numa Boulard 1985e: 1032 (type species of *Orapa*, comp. note)
Orapa numa Boulard 1996d: 115, 144–146, Fig S14(a-d), Fig S14(e-f) (song analysis, sonogram, listed, comp. note) Malawi
Orapa numa Villet and Noort 1999: 230 (comp. note) Zimbabwe
Orapa numa Sueur 2001: 39, Table 1 (listed)
Orapa numa Boulard 2006g: 49, 51, 54, 180, Fig 12 (sonogram, listed, comp. note) Malawi

O. *uwembaiensis* Boulard, 2012

Tribe Zammarini Distant, 1905 = Zammararia
Zammarini Itô and Nagamine 1981: 281, Table 5 (listed)
Zammarini Ramos 1983: 62 (listed) Dominican Republic
Zammarini Boulard 1988f: 61 (comp. note)
Zammarini Boulard and Mondon 1995: 72 (comp. note)
Zammarini Boulard and Sueur 1996: 105–106 (comp. note) Equals Zammararia
Zammarini Novotny and Wilson 1997: 437 (listed)
Zammarini Boulard 1998: 76 (comp. note)
Zammarini Moulds 1999: 1 (comp. note)
Zammarini Sanborn 1999a: 36 (listed)
Zammarini Villet and Zhao 1999: 322 (comp. note)
Zammarini Moulds 2001: 196 (comp. note)
Zammarini Boulard 2002b: 198, 204 (comp. note)
Zammarini Sueur 2002a: 380, 387 (listed) Mexico, Veracruz
Zammarini Boulard 2005d: 348 (stridulatory apparatus, comp. note)
Zammarini Moulds 2005b: 397, 419, 421–422, 427, 434, Fig 58, Fig 60 (phylogeny, comp. note)
Zammarini Boulard 2006g: 153–154 (stridulation, comp. note)
Zammarini Sanborn 2006a: 76 (listed)
Zammarini Sanborn 2006b: 256 (listed)
Zammarini Sanborn 2007a: 26 (listed) Venezuela
Zammarini Sanborn 2007b: 2, 31 (listed) Mexico
Zammarini Shiyake 2007: 2, 22 (listed, comp. note)
Zammarini Sanborn 2008f: 688 (listed)
Zammarini Sanborn 2009a: 89 (listed) Cuba
Zammarini Goemans 2010: 2–3, 5, 7, Fig 1 (history, diversity, key, listed, comp. note) Equals Zammararia
Zammararia Goemans 2010: 2 (comp. note)
Zammarini Sanborn 2010a: 1579 (listed) Colombia

***Odopoea* Stål, 1861** = *Odopea* (sic) = *Odopaea* (sic) = *Adusella* Haupt, 1918 = *Edhombergia* Delétang, 1919
Odopoea Ramos 1983: 62, 66 (key, comp. note) Dominican Republic, Mexico, Central America, South America, Jamaica
Odopoea Arnett 1985: 217 (distribution) United States (error)
Odopoea Ramos 1988: 62–63, Table 4.1 (listed, distribution, comp. note) Greater Antilles, Hispaniola, Jamaica
Odopoea Heath, Heath and Noriega 1990: 1 (distribution, comp. note) Argentina
Odopoea Boulard and Sueur 1996: 105 (comp. note)
Odopoea Poole, Garrison, McCafferty, Otte and Stark 1997: 251, 332 (listed) Equals *Adusella* Equals *Edholmbergia* Equals *Odopea* (sic) North America
Odopoea Arnett 2000: 298 (distribution) United States (error)
Odopoea Sanborn 2004f: 365 (comp. note)
Odopoea Moulds 2005b: 412–414, 421 (higher taxonomy, listed, comp. note)
Odopoea Sanborn 2007b: 5 (comp. note) Mexico
Odopoea Perez-Gerlabert 2008: 202 (listed) Hispaniola
Odopoea Goemans 2010: 2–3, 5, 8 (diversity, key, characteristics, synonymy, comp. note) Equals *Adusella* Equals *Edholmbergi*
Adusella Goemans 2010: 2, 5 (comp. note)
Edholmbergia Goemans 2010: 2, 5 (comp. note)

O. *azteca* Distant, 1881
Odopoea azteca Sanborn 2007b: 5, 31 (distribution, comp. note) Mexico
Odopoea azteca Goemans 2010: 5 (listed) Mexico

O. *cariboea* Uhler, 1892
Odopoea cariboea Henshaw 1903: 230 (listed) San Domingo
Odopoea cariboea Ramos 1983: 62, 67–68, Fig 1 (lectotype designation, described, illustrated, key, listed, comp. note) Dominican Republic
Odopoea cariboea Sanborn 1999a: 36 (type material, listed)

Odopoea cariboea Perez-Gerlabert 2008: 202 (listed) Dominican Republic
Odopoea cariboea Goemans 2010: 5 (listed) Hispaniola

O. degiacomii **Distant, 1912** = *Odopea* (sic) *degiacomii*
Odopoea degiacomii Ramos 1983: 62, 67–68, Fig 2 (described, illustrated, key, listed, comp. note) Dominican Republic
Odopoea degiacomii Perez-Gerlabert 2008: 202 (listed) Dominican Republic
Odopoea degiacomii Sanborn 2009d: 6 (comp. note) Hispaniola
Odopoea degiacomii Goemans 2010: 5 (listed) Espirito Santo, Brazil

O. dilatata **(Fabricius, 1775)** = *Tettigonia dilatata* = *Cicada dilatata* = *Odopoea dilata* (sic) = *Zammara plena* Walker, 1850 = *Odopoea plena* = *Zammara cuncta* Walker, 1850 = *Odopoea cuncta* = *Zammara praxita* Walker, 1850 = *Zammara erato* Walker, 1850 = *Zammara crato* (sic) = *Odopoea erato* = *Odopoea domingensis* Uhler, 1892
Odopoea domingensis Henshaw 1903: 230 (listed) San Domingo
Odopoea cuncta Wolcott 1927: 187, Fig 53 (illustrated, agricultural pest, comp. note) Haiti
Odopoea dilatata Ramos 1983: 63 (comp. note) Jamaica
Odopoea dilata (sic) Williams 1991: 25, 27–28, Fig 1, Figs 3–5 (coffee pest, illustrated, emergence, comp. note) Jamaica
Odopoea dilatata Moulds 2005b: 396–397, 399, 417–420, 422, Figs 56–60, Table 1 (phylogeny, comp. note)
Odopoea dilatata Sanborn 2009d: 6 (comp. note)
Tettigonia dilatata Goemans 2010: 5 (type species of *Odopoea*) Jamaica
Odopoea dilatata Goemans 2010: 5 (listed, synonymy) Equals *Zammara plena* Equals *Zammara cuncta* Equals *Zammara praxita* Equals *Zammara erato* Equals *Odopoea dominensis* Jamaica
Zammara plena Goemans 2010: 5 (listed) Jamaica
Zammara cuncta Goemans 2010: 6 (listed) Jamaica
Zammara praxita Goemans 2010: 6 (listed)
Zammara erato Goemans 2010: 6 (listed) Jamaica
Odopoea domingensis Goemans 2010: 6 (listed) Hispaniola

O. diriangani **Distant, 1881**
Odopoea diriangani Maes 1998: 143 (listed, distribution) Nicaragua
Odopoea diriangani Sanborn 2006b: 256 (listed, distribution) Mexico, Queretaro, Nicaragua
Odopoea diriangani Sanborn 2007b: 5, 31 (distribution, comp. note) Mexico, Queretaro, Hidalgo
Odopoea diriangani Goemans 2010: 6 (listed) Nicaragua

O. funesta **(Walker, 1858)** = *Odopoea fumesta* (sic) = *Zammara funesta*
Odopoea fumesta (sic) Arnett 1985: 217 (listed) United States (error)
Odopoea funesta Poole, Garrison, McCafferty, Otte and Stark 1997: 332 (listed) Equals *Zammara funesta* North America
Odopoea fumesta (sic) Arnett 2000: 298 (listed) United States (error)
Odopoea funesta Goemans 2010: 6 (listed, synonymy) Equals *Zammara funesta* North America
Zammara funesta Goemans 2010: 6 (listed) North America

O. insignifera **Berg, 1879**
Odopoea insignifera Bolcatto, Medrano and De Santis 2006: 7, 10, Map 2 (distribution) Argentina
Odopoea insignifera De Santis, Medrano, Sanborn and Bolcatto 2007: 11, 13, Fig 3 (distribution, comp. note) Argentina
Odopoea insignifera Goemans 2010: 6 (listed) Argentina, Salta

O. jamaicensis **Distant, 1881**
Odopoea jamaicensis Goemans 2010: 6 (listed) Jamaica

O. *lebruni* (Distant, 1906) to *Odopoea signata* (Haupt, 1918)

O. minuta **Sanborn, 2007**
Odopoea minuta Sanborn 2007b: 2–5, 31, Figs 1–6 (n. sp., described, illustrated, distribution, comp. note) Mexico, Colima
Odopoea minuta Goemans 2010: 6 (listed) Colima, Mexico

O. *perspicua* Distant, 1905 to *Canualna perspicua*

O. signata **(Haupt, 1918)** = *Adusella signata* Haupt, 1918 = *Tettigades lebruni* (partim) = *Edholmbergia lebruni* Delétang, 1919
Odopoea lebruni Goemans 2010: 6 (listed) Equals *Tettigades lebruni* Patagonia
Edholmbergia lebruni Goemans 2010: 6 (listed) Catamarca
Adusella signata Goemans 2010: 6 (listed) Catamarca, Argentina

O. signoreti **Stål, 1864**
Odopoea signoreti Sanborn 2006a: 76, 78, Fig 1 (listed, distribution) Honduras, Mexico
Odopoea signoreti Sanborn 2007b: 5, 31 (distribution, comp. note) Mexico, Veracruz, Honduras
Odopoea signoreti Goemans 2010: 6 (listed) Mexico

O. strigipennis **(Walker, 1858)** = *Zammara strigipennis*
Odopoea strigipennis Ramos 1983: 62, 67–68, Fig 3 (synonymy, described, illustrated, key, listed, comp. note) Equals *Zammara strigipennis* Dominican Republic

Odopoea strigipennis Perez-Gerlabert 2008: 202 (listed) Hispaniola
Odopoea strigipennis Sanborn 2009d: 6 (comp. note) Hispaniola
Odopoea strigipennis Goemans 2010: 6 (listed, synonymy) Equals *Zammara strigipennis* Haiti
Zammara strigipennis Goemans 2010: 6 (listed) Haiti

***O. suffusa* (Walker, 1850)** = *Zammara suffusa*
Odopoea suffusa Ramos 1983: 62–63, 67–68, Fig 4 (synonymy, described, illustrated, key, listed, comp. note) Equals *Zammara suffusa* Dominican Republic
Odopoea suffusa Perez-Gerlabert 2008: 202 (listed) Dominican Republic
Odopoea suffusa Sanborn 2009d: 6 (comp. note) Hispaniola
Odopoea suffusa Goemans 2010: 6 (listed, synonymy) Equals *Zammara suffusa* Dominican Republic
Zammara suffusa Goemans 2010: 6 (listed) Dominican Republic

***O. vacillans* (Walker, 1858)** = *Zammara vacillans*
Odopoea vacillans Ramos 1983: 62–63, 67–69, Figs 7–9 (synonymy, described, illustrated, key, listed, comp. note) Equals *Zammara vacillans* Dominican Republic
Odopoea vacillans Perez-Gerlabert 2008: 202 (listed) Dominican Republic
Odopoea vacillans Goemans 2010: 6 (listed, synonymy) Equals *Zammara vacillans* Dominican Republic
Zammara vacillans Goemans 2010: 6 (listed) Dominican Republic

***O. venturii* Distant, 1906**
Odopoea venturii Bolcatto, Medrano and De Santis 2006: 7, 10, Map 2 (distribution) Argentina
Odopoea venturii De Santis, Medrano, Sanborn and Bolcatto 2007: 11, 13, Fig 3 (comp. note) Argentina
Odopoea venturii Goemans 2010: 6 (listed, synonymy) Equals *Edholmbergia lebruni* Equals *Adusella signata* Argentina

***Miranha* Distant, 1905**
Miranha Boulard and Sueur 1996: 105–106 (comp. note)
Miranha Boulard 2002b: 193 (coloration, comp. note)
Miranha spp. Sueur 2002a: 387 (comp. note)
Miranha Shiyake 2007: 22 (comp. note)
Miranha Goemans 2010: 2–4, 6, 8 (diversity, key, characteristics, comp. note)

***M. imbellis* (Walker, 1858)** = *Zammara imbellis* = *Odopoea imbellis* = *Odopaea* (sic) *imbelis* (sic)
Miranha imbellis Wolda and Ramos 1992: 272, Table 17.1 (distribution, listed) Panama
Miranha imbellis Sueur 2002a: 380–381, 383–384, 387–391, Fig 5, Fig 10, Figs 12–14, Table 1 (acoustic behavior, song, sonogram, oscillogram, comp. note) Mexico, Veracruz
Miranha imbellis Sanborn 2006a: 76, 78, Fig 1 (listed, distribution) Honduras, Guatemala, Central America
Miranha imbellis Sanborn 2007b: 31 (distribution, comp. note) Mexico, Chiapas, Veracruz, Jalisco, Guatemala, Honduras
Zammara imbellis Goemans 2010: 2, 6 (type species of *Miranha*, listed, comp. note)
Miranha imbellis Goemans 2010: 6 (listed, synonymy) Equals *Zammara imbellis* Mexico

***Zammara* Amyot & Audinet-Serville, 1843** = *Zamara* (sic) = *Cicada (Zammara)* = *Cicada (Tibicen)*
Zammara Young 1981a: 139, 141 (comp. note) Costa Rica
Zammara sp. Young 1981b: 180 (comp. note) Costa Rica
Zammara Young 1981b: 190 (comp. note) Costa Rica
Zammara Young 1981c: 13 (comp. note) Costa Rica
Zammara spp. Young 1981c: 14 (comp. note) Costa Rica
Zammara Strümpel 1983: 18 (listed)
Zammara Young 1983b: 24 (comp. note) Costa Rica
Zammara Hayashi 1984: 26 (comp. note)
Zammara Heath 1983: 12 (comp. note) Tropical America
Zammara Arnett 1985: 217 (distribution) United States (error)
Zammara Boulard 1985e: 1025, 1027 (coloration, comp. note)
Zammara Boulard 1988f: 61 (comp. note)
Zammara Boulard 1990b: 146 (comp. note)
Zammara Young 1991: 213, 215–217, 221, 224–226 (comp. note) Costa Rica
Zammara spp. Wolda and Ramos 1992: 272 (comp. note) Panama
Zammara sp. Wolda and Ramos 1992: 272, Table 17.1 (distribution, listed) Panama
Zammara Hogue 1993b: 233 (comp. note)
Zammara Boulard and Mondon 1995: 72 (comp. note)
Zammara Boulard and Sueur 1996: 105–106, 110 (listed, comp. note)
Zammara Moore 1996: 222 (comp. note) Mexico
Zammara Novotny and Wilson 1997: 437 (listed)
Zammara Poole, Garrison, McCafferty, Otte and Stark 1997: 251, 333 (listed) North America
Zammara Moulds 1999: 1 (comp. note)
Zammara Arnett 2000: 298 (distribution) United States (error)

Zammara Boulard 2002b: 193, 198, 204 (coloration, comp. note) Neotropics
Zammara spp. Sueur 2002a: 387 (comp. note)
Zammara Sueur 2002b: 124 (comp. note)
Zammara Campos, Silva, Ferreira and Dias 2004: 125 (comp. note) Brazil
Zammara Sanborn 2004f: 365, 371 (comp. note)
Zammara Moulds 2005b: 386, 412–414, 421–422 (phylogeny, listed, comp. note)
Zammara Salazar Escobar 2005: 194 (comp. note) Colombia
Zammara sp. Salazar Escobar 2005: 195 (comp. note) Colombia
Zammara Shiyake 2007: 2, 22, 95 (listed, comp. note)
Zammara calochroma Shiyake 2007: 2, 20, 22, Fig 17 (illustrated, listed, comp. note)
Zammara Sanborn 2008f: 688 (comp. note)
Zammara Goemans 2010: 2–7 (diversity, key, characteristics, comp. note)
Zammara Sanborn 2010a: 1579, 1581, 1585 (listed) Colombia

***Z. brevis* (Distant, 1905)** = *Orellana brevis* = *Zammara columbia* (nec Distant)
Zammara columbia (sic) Sanborn 2008f: 688 (comp. note) Brazil, Rondonia
Zammara brevis Goemans 2010: 4–6 (listed, synonymy, comp. note) Equals *Orellana brevis* Colombia
Orellana brevis Goemans 2010: 2–4, 6 (listed, comp. note) Colombia
Zammara brevis Sanborn 2010a: 1579, 1581 (synonymy, distribution, comp. note) Equals *Orellana brevis* Colombia, Ecuador

***Z. calochroma* Walker, 1858** = *Zammara callichroma* Stål, 1864
Zammara calochroma Showalter 1935: 38–39, Fig 11 (illustrated, coloration, comp. note) Mexico, South America
Zammara calochroma Wolda 1989: 438–440, Figs 1–3, Tables 1–2 (emergence pattern, comp. note) Panama
Zammara calochroma Wolda and Ramos 1992: 272–273, Fig 17.1, Table 17.1 (distribution, listed, seasonality, comp. note) Panama
Zammara calochroma Sanborn 2006b: 256 (listed, distribution, comp. note) Mexico, Tamaulipas, San Luis Potosi, Venezuela, Colombia, Ecuador
Zammara calochroma Sanborn 2007a: 26 (distribution, comp. note) Colombia, Mexico, Panama, Venezuela, Ecuador, Central America, South America, Neotropical America
Zammara calochroma Sanborn 2007b: 31 (distribution, comp. note) Mexico, Veracruz, Tamaulipas, San Luis Potosí, Ecuador
Zammara calochroma Goemans 2010: 6 (listed) Colombia
Zammara calochroma Sanborn 2010a: 1579 (synonymy, distribution, comp. note) Equals *Zammara callichroma* Colombia, Mexico, Panama, Venezuela, Ecuador, Central America, South America

***Z. erna* Schmidt, 1919**
Zammara erna Goemans 2010: 6 (listed) Peru

***Z. eximia* (Erichson, 1848)** = *Cicada (Zammara) exima* = *Zammara brevis* (nec Distant)
Zammara brevis (sic) Thouvenot 2007a: 13–14, Fig 1 (illustrated, comp. note) French Guiana
Zammara eximia Thouvenot 2007b: 284 (comp. note) Equals *Zammara brevis* Thouvenot 2007a French Guiana
Zammara eximia Goemans 2010: 6 (listed) Equals *Cicada (Zammara) eximia* British Guiana
Cicada (Zammara) eximia Goemans 2010: 6 (listed) British Guiana

***Z. hertha* Schmidt, 1919**
Zammara hertha Goemans 2010: 6 (listed) Peru, *Ecuador*

***Z. intricata* Walker, 1850**
Zammara intricata Moulds 2001: 202, Figs 12–13 (illustrated)
Zammara intricata Moulds 2005b: 396–397, 399, 417–420, 422, Figs 56–60, Table 1 (phylogeny, comp. note)
Zammara intricata Goemans 2010: 6 (listed) Puerto Rico

***Z. lichyi* Boulard & Sueur, 1996**
Zammara lichyi Boulard and Sueur 1996: 106–109, Plate I, Figs 1–5, Plate II, Figs 1–2 (n. sp., described, illustrated, comp. note) Venezuela
Zammara lichyi Sanborn 2007a: 26 (distribution, comp. note) Venezuela
Zammara lichyi Goemans 2010: 6 (listed) Venezuela

***Z. luculenta* Distant, 1883**
Zammara luculenta Sanborn 2004f: 365, 371 (comp. note)
Zammara luculenta Goemans 2010: 6 (listed)

***Z. medialinea* Sanborn, 2004**
Zammara medialinea Sanborn 2004f: 367–370, Figs 3–4 (n. sp., described, illustrated, comp. note) Venezuela
Zammara medialinea Sanborn 2007a: 26 (distribution, comp. note) Venezuela
Zammara medialinea Goemans 2010: 6 (listed) Venezuela

***Z. nigriplaga* Walker, 1858** = *Orellana nigriplaga*
Zammara nigriplaga Goemans 2010: 3–6 (listed, comp. note) South America

Orellana nigriplaga Goemans 2010: 4 (comp. note)
Orellana nigriplaga Sanborn 2010a: 1585 (synonymy, distribution, comp. note) Equals *Zammara nigriplaga* Colombia

***Z. olivacea* Sanborn, 2004**
Zammara olivacea Sanborn 2004f: 365–368, 370–371, Figs 1–2 (n. sp., described, illustrated, comp. note) Colombia
Zammara olivacea Goemans 2010: 6 (listed) Colombia
Zammara olivacea Sanborn 2010a: 1579–1580 (distribution, comp. note) Colombia

***Z. smaragdina* Walker, 1850** = *Zammara zmaragdina* (sic) = *Zammara angulosa* Walker, 1850 = *Zammara angnlosa* (sic)
Zammara smaragdina Young 1981b: 191, 193 (comp. note) Costa Rica
Zammara smaragdina Young 1981c: 14 (comp. note) Costa Rica
Zammara smaragdina Young 1981d: 827 (acoustic behavior, comp. note) Costa Rica
Zammara smaradina Young 1982: 50–52, 103, 161–162, 184, 200, 208–209, 413, Figs 2.15–2.16, Fig 3.16, Fig 5.7, Table 5.6 (illustrated, activity patterns, hosts, comp. note) Costa Rica
Zammara smaragdina Young 1983b: 25 (comp. note) Costa Rica
Zammara smaragdina Young 1984a: 169–170, 173, 176, 178–179, 181, Tables 1–3 (hosts, comp. note) Costa Rica
Zammara smaragdina Arnett 1985: 217 (listed) United States (error)
Zammara smaragdina Young 1991: 190, 212, 214, 217, 248 (illustrated, comp. note) Costa Rica
Zammara smaragdina Wolda and Ramos 1992: 272, 276, Fig 17.5, Table 17.1 (distribution, listed, seasonality, comp. note) Panama
Zammara smaragdina Hogue 1993b: 232–233, 570, Fig 8.6b (illustrated, listed, comp. note)
Zammara smaragdina Poole, Garrison, McCafferty, Otte and Stark 1997: 333 (listed) Equals *Zammara angulosa* Equals *Zammara zmaragdina* (sic) Equals *Zammara angnlosa* (sic) North America
Zammara smaragdina Shelly and Whittier 1997: 279, Table 16–1 (listed, comp. note)
Zammara smaragdina Maes 1998: 143 (listed, synonymy, distribution, host plants) Equals *Zammara angulosa* Nicaragua, USA (error), Mexico, Costa Rica, Panama, Colombia, Venezuela, Ecuador
Zammara smaragdina Arnett 2000: 298 (listed) United States (error)
Zammara smaragdina Humphreys 2005: 71, 74, Table 4 (population density, comp. note) Costa Rica
Zammara smaragdina Salazar Escobar 2005: 195, 199, Fig 1 (illustrated, comp. note) Colombia
Zammara smaragdina Sanborn 2006a: 76, 78, Fig 1 (listed, distribution) Guatemala, Honduras, Central America
Zammara smaragdina Sanborn 2007a: 26 (distribution, comp. note) Colombia, Mexico, Guatemala, Honduras, Nicaragua, Costa Rica, Panama, Ecuador, Venezuela, Central America, Neotropical America
Zammara smaragdina Sanborn 2007b: 31 (distribution, comp. note) Mexico, Tabasco, Ecuador
Zammara smaragdina Salazar and Sanborn 2009: 271 (comp. note) Colombia
Zammara smaragdina Goemans 2010: 6 (listed, synonymy) Equals *Zammara angulosa*
Zammara angulosa Goemans 2010: 6 (listed) Mexico
Zammara smaragdina Sanborn 2010a: 1580 (synonymy, distribution, comp. note) *Zammara angulosa* Equals *Zammara zmaragdina* (sic) Colombia, Mexico, Guatemala, Honduras, Nicaragua, Costa Rica, Panama, Ecuador, Venezuela, Central America, Neotropical Region

***Z. smaragdula* Walker, 1858**
Zammara smaragdula Young 1981a: 128–129, 131–132, 134–139, 141, Fig 4, Fig 7, Fig 9, Tables 1–3 (seasonality, habitat, comp. note) Costa Rica
Zammara smaragdula Young 1981c: 5–8, 10–14, 18, 20–21, 25–26, 28, Fig 6, Fig 11, Tables 4–6 (illustrated, seasonality, habitat, comp. note) Costa Rica
Zammara smaragdula Young 1984a: 170–173, 176–177, 179, Tables 1–3 (hosts, comp. note) Costa Rica
Zammara smaragdula Bartholomew and Barnhart 1984: 135–136, Fig 2, (energetics, morphometry) Panama
Zammara smaragdula Sanborn 2004f: 371 (comp. note)
Zammara smaragdula Sanborn 2006a: 76, 78, Fig 1 (listed, distribution) Guatemala, Central America, South America
Zammara smaragdula Sanborn 2006b: 256 (listed, distribution, comp. note) Mexico, Chiapas, Veracruz
Zammara smaragdula Sanborn 2007b: 31 (distribution, comp. note) Mexico, Veracruz, Chiapas, Guatemala, Costa Rica

Zammara smaragdula Goemans 2010: 6 (listed) South America

Zammara smaragdula Sanborn 2010a: 1580 (distribution, comp. note) Colombia, Mexico, Guatemala, Costa Rica, Panama

Z. strepens **Amyot & Audinet-Serville, 1843** = *Zammara streppens* (sic) = *Cicada tympanum* Palisot de Beauvois, 1813 (nec *Cicada tympanum* Fabricius, 1803)

Zammara strepens Thouvenot 2007a: 13–14, Fig 2 (illustrated, comp. note) French Guiana

Zammara strepens Goemans 2010: 3–6 (listed, synonymy) Brazil

Z. tympanum **(Fabricius, 1803)** = *Tettigonia tympanum* = *Cicada tympanum* = *Tibicen tympanum* = *Cicada (Tibicen) tympanum* = *Zamara* (sic) *tympanum* = *Zammara tympanium* (sic) = *Zammara typanum* (sic) = *Zammara typmanum* (sic)

Zammara tympanum Young 1981b: 181–183, 185–190, Figs 5–6, Tables 1–2 (illustrated, seasonality, habitat, comp. note) Costa Rica

Zammara tympanum Young 1984a: 170, Table 1 (hosts, comp. note) Costa Rica

Zammara tympanum Boulard 1988f: 61 (type species of *Zammara*)

Zammara tympanum Boulard 1990b: 146 (comp. note)

Zammara tympanum Young 1991: 215, 217, 219, 221 (comp. note) Costa Rica

Zammara tympanum Boulard and Mondon 1995: 72 (type species of *Zammara*)

Zammara tympanum Boulard and Sueur 1996: 105–106 (type species of *Zammara*, comp. note)

Zammara tympanum Sanborn 2004f: 365 (type species of *Zammara*, comp. note)

Zammara typanum (sic) Salazar Escobar 2005: 195, 199, Fig 2 (illustrated, comp. note) Colombia

Zammara tympanum Sanborn 2006a: 76, 78, Fig 1 (listed, distribution) Belize, Guatemala, Honduras, South America, Central America

Zammara typmanum (sic) Shiyake 2007: 2 (listed, comp. note)

Zammara tympanum Shiyake 2007: 20, 22, Fig 18 (illustrated, listed, comp. note)

Zammara tympanum Salazar and Sanborn 2009: 270 (comp. note) Equals *Zammara typanum* (sic) Colombia

Zammara tympanum Goemans 2010: 3, 6 (listed, synonymy, comp. note) Equals *Tettigonia tympanum* Brazil

Tettigonia tympanum Goemans 2010: 6 (type species of *Zammara*)

Tettigonia tympanum Sanborn 2010a: 1579 (type species of *Zammara*) Brazil

Zammara tympanum Sanborn 2010a: 1580–1581 (synonymy, distribution, comp. note) Equals *Tettigonia tympanum* Equals *Zammara typanum* (sic) Colombia, Belize, Honduras, Guatemala, Costa Rica, Brazil, Peru, Paraguay, Argentina

Z. **sp. Wolda & Ramos, 1992**

Zammaralna **Boulard & Sueur, 1996**

Zammaralna Boulard and Sueur 1996: 110 (n. gen., described, comp. note)

Zammaralna Sanborn 2004f: 365 (comp. note)

Zammaralna Goemans 2010: 2–4, 6, 8 (diversity, key, characteristics, comp. note) Venezuela

Z. bleuzeni **Boulard & Sueur, 1996**

Zammaralna bleuzeni Boulard and Sueur 1996: 109–112, Plate II, Figs 3–5, Plate III, Figs 1–4 (n. sp., type species of *Zammaralna*, described, illustrated, comp. note) Venezuela

Zammaralna bleuzeni Sanborn 2007a: 26 (distribution, comp. note) Venezuela

Zammaralna bleuzeni Goemans 2010: 6 (type species of *Zammaralna*, listed) Venezuela

Juanaria **Distant, 1920**

Juanaria Ramos 1983: 67 (comp. note)

Juanaria De Zayas 1988: 14 (key) Cuba

Juanaria Ramos 1988: 62–63, Table 4.1 (listed, distribution, comp. note) Greater Antilles, Cuba

Juanaria Sanborn 2004f: 365 (comp. note)

Juanaria Sanborn 2009a: 89 (listed, comp. note) Cuba

Juanaria Goemans 2010: 2–3, 6, 8 (diversity, key, characteristics, comp. note) Cuba

J. poeyi **(Guérin-Méneville, 1856)** = *Cicada (Platypleura) poeyi* = *Cicada poeyi* = *Zammara poeyi* = *Odopoea poeyi* = *Juanaria mimica* Distant, 1920

Juanaria poeyi De Zayas 1988: 12, Fig 2 (described, illustrated, comp. note) Cuba

Juanaria mimica Sanborn 2009a: 89 (type species of *Juanaria*, listed, comp. note) Cuba

Juanaria poeyi Sanborn 2009a: 89–90 (listed, synonymy, key, comp. note) Equals *Cicada (Platypleura) poeyi* Equals *Juanaria mimica* Cuba

Juanaria mimica Goemans 2010: 6 (type species of *Juanaria*, comp. note)

Juanaria poeyi Goemans 2010: 6 (listed, synonymy) Equals *Cicada (Platypleura) poeyi* Cuba

Cicada (Platypleura) poeyi Goemans 2010: 6 (listed) Cuba

Borencona **Davis, 1928**

Borencona Ramos 1983: 67 (comp. note)

Borencona Ramos 1988: 62–63, Table 4.1 (listed, distribution, comp. note) Greater Antilles, Puerto Rico

Borencona Sanborn 2004f: 365 (comp. note)
Borencona Goemans 2010: 2–3, 6, 9 (diversity, key, characteristics, comp. note) Puerto Rica

B. aguadilla **Davis, 1928**
Borencona aguadilla Wolcott 1948: 103 (distribution, predation, habitat, comp. note) Puerto Rico
Borencona aguadilla Sanborn 1999a: 36 (type material, listed)
Borencona aguadilla Goemans 2010: 6 (type species of *Borencona*, listed, comp. note)

Chinaria **Davis, 1934 =** ***Chineria*** **(sic)**
Chinaria Ramos 1983: 66 (comp. note) Mexico, Dominican Republic
Chinaria Boulard 1985e: 1025 (coloration, comp. note)
Chinaria Ramos 1988: 62–63, Table 4.1 (listed, distribution, comp. note) Greater Antilles, Hispaniola
Chinaria Sanborn 2004f: 365 (comp. note)
Chinaria Sanborn 2007b: 9 (comp. note) Mexico
Chinaria Perez-Gerlabert 2008: 202 (listed) Hispaniola
Chinaria Goemans 2010: 2–3, 6, 9 (diversity, key, characteristics, comp. note) Mexico, Dominican Republic

C. mexicana **Davis, 1934** = *Chineria* (sic) *mexicana*
Chinaria mexicana Moore 1996: 222, Fig 17.15a (illustrated, distribution, comp. note) Mexico
Chinaria mexicana Sanborn 1999a: 36 (type material, listed)
Chinaria mexicana Sanborn 2007b: 9, 31 (distribution, comp. note) Mexico, Morales, Guerrero, Sinaloa, Nayarit, Michoacán
Chinaria mexicana Goemans 2010: 6 (type species of *Chinaria*, listed, comp. note) Morelos, Mexico

C. pueblaensis **Sanborn, 2007**
Chinaria pueblaensis Sanborn 2007b: 5–9, 31, Figs 7–12 (n. sp., described, illustrated, distribution, comp. note) Mexico, Puebla
Chinaria pueblaensis Goemans 2010: 6 (listed) Puebla, Mexico

C. similis **Davis, 1942**
Chinaria similis Ramos 1983: 63 (comp. note) Mexico
Chinaria similis Sanborn 1999a: 36 (type material, listed)
Chinaria similis Sanborn 2007b: 9, 31 (comp. note) Mexico, Guerrero
Chinaria similis Goemans 2010: 6 (listed) Guerrero, Mexico

C. vivianae **Ramos, 1983**
Chinaria vivianae Ramos 1983: 62–64, 67, 69, Figs 10–13 (n. sp., described, illustrated, key, listed, comp. note) Dominican Republic
Chinaria vivianae Sanborn 1999a: 36 (type material, listed)
Chinaria vivianae Perez-Gerlabert 2008: 202 (listed) Dominican Republic
Chinaria vivianae Goemans 2010: 7 (listed) Dominican Republic

Orellana **Distant, 1905** = *Oriellana* (sic)
Orellana Boulard 1985e: 1025 (coloration, comp. note)
Orellana Boulard 1990b: 146 (comp. note)
Orellana Boulard and Sueur 1996: 105–106 (comp. note)
Orellana Sanborn 2004f: 365 (comp. note)
Oriellana (sic) Shiyake 2007: 22 (comp. note)
Orellana Sanborn 2008f: 688 (comp. note)
Orellana Goemans 2010: 2–6, 8–9 (diversity, key, characteristics, comp. note)
Orellana Sanborn 2010a: 1581, 1585 (listed, comp. note) Colombia

O. bigibba **Schmidt, 1919** = *Zammara bigibba*
Orellana bigibba Boulard 1990b: 146 (comp. note)
Zammara bigibba Boulard 1990b: 159, 161, Plate XIX, Fig 7 (sound system, comp. note)
Orellana bigibba Salguero 2003: 30 (comp. note) Amazon
Orellana bigibba Goemans 2010: 4–5, 7 (listed, comp. note) Brazil

O. brevis Distant, 1905 to *Zammara brevis*

O. brunneipennis **Goding, 1925** = *Orellana brunnpennis* (sic)
Orellana brunneipennis Goemans 2010: 4–5, 7 (listed) Ecuador

O. castaneamaculata **Sanborn, 2010** = *Zammara* (sic) *castaneamaculata*
Orellana castaneamaculata Goemans 2010: 7 (listed) Colombia
Orellana castaneamaculata Sanborn 2010a: 1581–1585, Fig 1 (n. sp., described, illustrated, distribution, comp. note) Colombia
Zammara (sic) *castaneamaculata* Sanborn 2010a: 1583 (listed) Colombia

O. columbia **(Distant, 1881)** = *Zammara columbia*
Zammara columbia Moulds 2005b: 414 (comp. note)
Zammara columbia Goemans 2010: 2–3, 7 (type species of *Orellana*, comp. note)
Orellana columbia Goemans 2010: 4–5, 7 (listed, comp. note)
Zammara columbia Sanborn 2010a: 1581 (type species of *Orellana*) Colombia
Orellana columbia Sanborn 2010a: 1585 (synonymy, distribution, comp. note) Equals *Zammara columbia* Colombia

O. nigriplaga (Walker, 1858) to *Zammara nigriplaga*

***O. pollyae* Sanborn, 2011**

***O. pulla* Goding, 1925**

Orellana pulla Goemans 2010: 4–5, 7 (listed) Ecuador

***Uhleroides* Distant, 1912** = *Uhlerioides* (sic) = *Uhlerodes* (sic)

Uhleroides Ramos 1983: 61–62, 66 (key, comp. note) Dominican Republic, Cuba

Uhleroides De Zayas 1988: 13, 15 (key, comp. note) Cuba

Uhleroides Ramos 1988: 62–63, Table 4.1 (listed, distribution, comp. note) Greater Antilles, Cuba, Hispaniola

Uhleroides Moulds 2001: 195–196 (tribal status, comp. note) South America

Uhleroides Sanborn 2004f: 365 (comp. note)

Uhleroides Moulds 2005b: 434 (comp. note)

Uhleroides Perez-Gerlabert 2008: 202 (listed) Hispaniola

Uhleroides Sanborn 2009a: 89 (listed, comp. note) Cuba

Uhleroides Goemans 2010: 2–3, 6, 9 (diversity, key, characteristics, comp. note) Cuba, Hispaniola

***U. chariclo* (Walker, 1850)** = *Cicada chariclo* = *Tympanoterpes chariclo* = *Odopoea chariclo* = *Proarna chariclo* = *Proalba* (sic) *chariclo*

Uhleroides chariclo De Zayas 1988: 13–14, Fig 6 (illustrated, comp. note) Cuba

Uhleroides chariclo Sanborn 2009a: 89, 91 (listed, synonymy, key, comp. note) Equals *Cicada chariclo* Cuba

Uhleroides chariclo Goemans 2010: 7 (listed, synonymy) Equals *Cicada chariclo* Cuba

Cicada chariclo Goemans 2010: 7 (listed) Cuba

***U. cubensis* Distant, 1912**

Uhleroides cubensis De Zayas 1988: 14, Fig 7 (illustrated, comp. note) Cuba

Uhleroides cubensis Moulds 2001: 196, 202, Figs 14–16 (illustrated, comp. note)

Uhleroides cubensis Sanborn 2009a: 89–90 (type species of *Uhleroides*, listed, key, comp. note) Cuba

Uhleroides cubensis Goemans 2010: 6 (type species of *Uhleroides*, comp. note)

***U. hispaniolae* Davis, 1939**

Uhleroides hispaniolae Ramos 1983: 62, 64, 67, 69, Fig 14 (described, illustrated, key, listed, comp. note) Dominican Republic

Uhleroides hispaniolae Sanborn 1999a: 36 (type material, listed)

Uhleroides hispaniolae Perez-Gerlabert 2008: 202 (listed) Dominican Republic

Uhleroides hispaniolae Goemans 2010: 7 (listed) Dominican Republic

***U. maestra* Davis, 1939** = *Uhlerodes* (sic) *maestra*

Uhleroides maestra De Zayas 1988: 14 (comp. note) Cuba

Uhleroides maestra Sanborn 1999a: 36 (type material, listed)

Uhleroides maestra Sanborn 2009a: 89–90 (listed, key, comp. note) Cuba

Uhleroides maestra Goemans 2010: 7 (listed) Cuba

***U. sagrae* (Guérin-Méneville, 1856)** = *Cicada sagrae* = *Zammara sagrae* = *Odopoea sagrae* = *Odopaea* (sic) *sagrae*

Uhleroides sagrae De Zayas 1988: 13–14, Fig 6 (illustrated, comp. note) Cuba

Uhleroides sagrae Hidalgo-Gato González 2001: 10 (distribution, predation, comp. note) Cuba

Uhleroides sagrae Sanborn 2009a: 89, 91 (listed, synonymy, key, comp. note) Equals *Cicada sagrae* Cuba

Uhleroides sagrae Goemans 2010: 7 (listed, synonymy) Equals *Cicada sagrae* Cuba

Cicada sagrae Goemans 2010: 7 (listed) Cuba

***U. samanae* Davis, 1939**

Uhleroides samanae Ramos 1983: 62, 64, 67, 70, Fig 15 (described, illustrated, key, listed, comp. note) Dominican Republic

Uhleroides samanae Sanborn 1999a: 36 (type material, listed)

Uhleroides samanae Perez-Gerlabert 2008: 202 (listed) Dominican Republic

Uhleroides samanae Sanborn 2009d: 6 (comp. note) Hispaniola

Uhleroides samanae Goemans 2010: 7 (listed) Dominican Republic

***U. walkerii* (Guérin-Méneville, 1856)** = *Cicada walkerii* = *Zammara walkerii* = *Odopoea walkerii* = *Odopoea walkeri* (sic) = *Uhleroides walkeri* (sic)

Uhleroides walkeri (sic) De Zayas 1988: 13–14, Figs 4–5 (described, illustrated, comp. note) Cuba

Uhleroides walkeri (sic) Genaro and Juarrero de Varona 1998: 323 (predation) Cuba

Uhleroides walkeri (sic) Sanborn 2009a: 89, 91 (listed, synonymy, key, comp. note) Equals *Cicada walkerii* Cuba

Uhleroides walkerii Goemans 2010: 7 (listed, synonymy) Equals *Cicada walkerii* Cuba

Cicada walkerii Goemans 2010: 7 (listed) Cuba

Uhleroides walkerii Holliday, Hastings and Coelho 2009: 3–5, 12–14, Tables 1–2 (predation, comp. note) Cuba

Uhleroides walkerii Hastings, Holliday, Long, Jones and Rodriguez 2010: 419 (predation, comp. note) Cuba

Tribe Plautillini Distant, 1905 = Plautillaria

Plautillaria Boulard 1988f: 52 (comp. note)

Plautillini Boulard 2005d: 348 (stridulatory apparatus, comp. note)

Plautillini Moulds 2005b: 386, 388, 412, 419, 421–422, 427, Fig 58, Fig 60 (n. stat., history, phylogeny, listed, comp. note)
Plautillaria Moulds 2005b: 388 (history, listed, comp. note)
Plautillini Boulard 2006g: 153–154 (stridulation, comp. note)
Plautillini Sanborn 2010a: 1585 (listed) Colombia

***Plautilla* Stål, 1865**
Plautilla Boulard 1982e: 181 (comp. note) Andes
Plautilla Heath 1983: 12 (comp. note) Colombia, Ecuador
Plautilla Strümpel 1983: 18 (listed)
Plautilla Hayashi 1984: 26 (comp. note)
Plautilla Boulard 1990b: 141, 213 (comp. note)
Plautilla Duffels 1993: 1225 (comp. note)
Plautilla Boer 1994a: 2 (comp. note)
Plautilla Boer 1995b: 3 (comp. note) Africa (error)
Plautilla Moulds 1999: 1 (comp. note)
Plautilla Boulard 2002b: 193, 197–198, 204 (coloration, sound system, comp. note) Neotropics
Plautilla Sanborn 2004c: 511 (comp. note)
Plautilla Moulds 2005b: 386, 388, 412–414, 416, 421 (type genus of Plautillidae, type genus of Plautillini, phylogeny, listed, comp. note)
Plautilla Goemans 2010: 7 (comp. note)
Plautilla Sanborn 2010a: 1585–1586 (listed) Colombia

***P. hammondi* Distant, 1914** = *Plautilla hamondi* (sic) = *Odopoea* sp. Salazar Escobar, 2005
Plautilla hammondi Varley 1939: 99 (comp. note)
Plautilla hammondi Strümpel 1983: 17, Fig 28 (illustrated, comp. note)
Plautilla hamondi (sic) Boulard 1990b: 159, 161, Plate XIX, Figs 1–2 (sound system, comp. note)
Plautilla hammondi Duffels 1993: 1225 (comp. note)
Plautilla hammondi Sanborn 2004c: 513, Fig 174 (illustrated)
Odopoea (sic) sp. Salazar Escobar 2005: 196, 204, Fig 12 (illustrated, comp. note) Colombia
Plautilla hammondi Salazar and Sanborn 2009: 271 (comp. note) Equals *Odopoea* sp. Salazar Escobar, 2005 Colombia
Plautilla hammondi Sanborn 2010a: 1585–1586 (synonymy, distribution, comp. note) Equals *Odopoea* sp. Salazar Escobar, 2005 Colombia, Ecuador
Odopoea sp. (sic) Sanborn 2010a: 1586 (comp. note) Colombia

***P. stalagmoptera* Stål, 1865**
Plautilla stalagmoptera Boulard 1988f: 52 (type species of Plautillidae)
Plautilla stalagmoptera Duffels 1993: 1225 (comp. note)
Plautilla stalagmoptera Sanborn 2010a: 1579 (type species of *Plautilla*) Ecuador

***P. venedictoffae* Boulard, 1978**
Plautilla venedictoffae Boulard 1990b: 159, 161, Plate XIX, Fig 3 (sound system, comp. note)
Plautilla venedictoffae Duffels 1993: 1225 (comp. note)
Plautilla venedictoffae Boulard 2000d: 88, Plate D. Fig 3 (stridulatory apparatus)
Plautilla venedictoffae Moulds 2005b: 397, 399, 417–420, 422, Figs 56–60, Table 1 (phylogeny, comp. note)

Tribe Polyneurini Amyot & Audinet-Serville, 1843 = Plyneurini (sic)
Plyneurini (sic) Chen 1942: 143 (listed)
Polyneurini Hayashi 1984: 45 (listed, comp. note)
Polyneurini Chou and Yao 1986: 190 (comp. note)
Polyneurini Kirillova 1986b: 46 (listed)
Polyneurini Kirillova 1986a: 123 (listed)
Polyneurini Boulard 1994a: 61 (comp. note)
Polyneurini Chou, Lei, Li, Lu and Yao 1997: 106, 157–158 (key, listed, synonymy, described) Equals Reticelli Equals Polyneurides Equals Polyneuridae Equals Polyneurinae Equals Polyneuraria Equals Polyneurarini
Polyneurini Novotny and Wilson 1997: 437 (listed)
Polyneurini Boulard 2000c: 59 (comp. note)
Polyneurini Boulard 2001a: 128 (comp. note)
Polyneurini Sueur 2001: 38, Table 1 (listed)
Polyneurini Boulard 2002b: 196–197 (comp. note)
Polyneurini Lee and Hayashi 2003a: 150, 153, 168 (listed, key, diversity, comp. note) Taiwan
Polyneurini Lee and Hayashi 2003b: 359 (listed) Taiwan
Polyneurini Chen 2004: 39–40 (phylogeny, listed)
Polyneurini Lee and Hayashi 2004: 45, 61 (listed) Taiwan
Polyneurini Boulard 2005c: 233, 237 (comp. note)
Polyneurini Boulard 2005j: 80 (comp. note)
Polyneurini Chawanji, Hodson and Villet 2005a: 258 (listed)
Polyneurini Lee 2005: 14, 16, 156, 171 (key, listed, comp. note) Korea
Polyneurini Moulds 2005b: 427 (higher taxonomy, listed)
Polyneurini Pham and Thinh 2005a: 245 (listed) Vietnam
Polyneurini Pham and Thinh 2005b: 287–289 (key, comp. note) Equals Reticelli Equals Polyneurides Equals Polyneuridae Equals Polyneurinae Equals Polyneuraria Vietnam
Polyneurini Chawanji, Hodson and Villet 2006: 373 (listed)

Polyneurini Sanborn, Phillips and Sites 2007: 3, 10, 37, Fig 2, Table 1 (listed, distribution) Thailand
Polyneurini Shiyake 2007: 2, 22 (listed, comp. note)
Polyneurini Boulard 2008f: 7, 18 (listed) Thailand
Polyneurini Boulard 2008c: 362 (listed)
Polyneurini Lee 2008a: 2, 6, Table 1 (listed) Vietnam
Polyneurini Lee 2008b: 446, 448, 463, Table 1 (listed, comp. note) Korea
Polyneurini Osaka Museum of Natural History 2008: 322 (listed) Japan
Polyneurini Pham and Yang 2009: 7, 13, Table 2 (listed) Vietnam
Polyneurini Lee 2010d: 171 (comp. note)
Polyneurini Lee and Hill 2010: 278, 301 (comp. note)

Subtribe Polyneurina Myers, 1929 = Polyneuraria
Polyneuraria Wu 1935: 1 (listed) China
Polyneurina Boulard 2008f: 7, 18 (listed) Thailand

Proretinata Chou & Yao, 1986 to *Angamiana* Distant, 1890
P. fuscula Chou & Yao, 1986 to *Anganiana fuscula*
P. vemacula Chou & Yao, 1986 to *Angamiana vemacula*
P. yunnanensis Chou & Yao, 1986 to *Amganiana yunnanensis*

***Angamiana* Distant, 1890** = *Proretinata* Chou & Yao, 1986 = *Agamiana* (sic) = *Agaminana* (sic)
Angamiana Wu 1935: 3 (listed) China
Angamiana Hayashi 1984: 45 (listed)
Angamiana Chou and Yao 1986: 190, 194 (comp. note)
Proretinata Chou and Yao 1986: 189–190 (key, n. gen., described, comp. note) China
Angamiana Hayashi 1993: 15 (comp. note)
Angamiana Boulard 1994a: 61 (comp. note)
Proretinata Chou, Lei, Li, Lu and Yao 1997: 158, 163–164 (key, described, listed)
Angamiana Chou, Lei, Li, Lu and Yao 1997: 158, 163, 166, 293, Table 6 (key, synonymy, described, listed, diversity, biogeography) Equals *Proretinata* n. syn. China
Angamiana Lee and Hayashi 2003a: 168 (comp. note)
Angamiana Boulard 2005c: 233, 235 (comp. note)
Proretinata Boulard 2005c: 235 (comp. note)
Proretinata Boulard 2005j: 81–82 (comp. note)
Angamiana Boulard 2005j: 80–82 (comp. note)
Angamiana Boulard 2005m: 125 (comp. note)
Proretinata Pham and Thinh 2005a: 245 (listed) Vietnam
Proretinata Pham and Thinh 2005b: 289 (comp. note) Vietnam
Angamiana Boulard 2006g: 129 (comp. note)
Angamiana Boulard 2007e: 4 (venation, comp. note) Thailand
Angamiana Sanborn, Phillips and Sites 2007: 3, 10, Table 1 (listed) Thailand
Angamiana Shiyake 2007: 2, 22 (listed, comp. note)
Angamiana Boulard 2008f: 7, 18, 92 (listed) Thailand
Angamiana Lee 2008a: 2, 6, Table 1 (listed) Vietnam
Angamiana Pham and Yang 2009: 7, 13, Table 2 (listed, synonymy) Equals *Proretinata* Vietnam
Proretinata Pham and Yang 2009: 8 (comp. note)

***A. aetherea* Distant, 1890**
Angamiana aetherea Boulard 1994a: 61 (comp. note) Assam
Angamiana aertherea Chou, Lei, Li, Lu and Yao 1997: 166 (type species of *Angamiana*)
Angamiana aertherea McGavin 2000: 96 (illustrated) India
Angamiana aertherea Boulard 2002b: 196 (venation, comp. note)
Angamiana aertherae Boulard 2005c: 234 (type species of *Angamiana*, comp. note)
Angamiana aertherae Boulard 2005j: 81 (type species of *Angamiana*, comp. note)
Angamiana aertherae Boulard 2005m: 127 (comp. note) India
Angamiana aetherea Boulard 2008f: 18 (type species of *Angamiana*)
Angamiana aetherea Lee 2008a: 6 (type species of *Angamiana*) Continental India
Angamiana aetherea Pham and Yang 2009: 7 (type species of *Angamiana*) India

***A. floridula* Distant, 1904** = *Agaminana* (sic) *floridula* = *Angamiana florida* (sic) = *Angamiana bauflei* Boulard, 1994 = *Angamiana beauflei* (sic) = *Angamiana floridula bauflei* = *Angamiana floridula* var. *beauflei* (sic) = *Agamiana* (sic) *floridula*
Angamiana floridula Wu 1935: 3 (listed) China
Angamiana bauflei Boulard 1994a: 61–63, Figs 1–2 (n. sp., described, illustrated, comp. note) Thailand
Angamiana floridula Boulard 1994a: 61, 63 (comp. note) South China
Angamiana floridula Chou, Lei, Li, Lu and Yao 1997: 167 (described, comp. note) China
Angamiana bauflei Boulard 1999b: 184–185, Fig 2, Fig 2" (illustrated, comp. note) Thailand
Angamiana floridula Boulard 2000c: 59–61, 63, Figs 10–13 (illustrated, song analysis, sonogram, comp. note) Thailand
Angamiana floridula var. *beauflei* (sic) Boulard 2000c: 59 (n. comb, comp. note) Equals *Angamiana beauflei* (sic)

Angamiana floridula Hua 2000: 60 (listed, distribution) China, Yunnan, Vietnam
Angamiana floridula Boulard 2001a: 128–131, Plate I, Figs A-B, Figs 1–2 (illustrated, song analysis, sonogram, comp. note) Equals *Angamiana bauflei* Equals *Angamiana floridula* var. *beauflei* (sic) Thailand
Angamiana floridula Boulard 2001c: 62 (comp. note)
Angamiana floridula Sueur 2001: 38, Table 1 (listed)
Angamiana florida (sic) Boulard 2002b: 196 (venation, comp. note)
Angamiana florida (sic) Boulard 2005c: 234 (comp. note) China, Tonkin
Angamiana bauflei Boulard 2005c: 235 (comp. note)
Angamiana floridula Boulard 2005c: 235–239, Figs 6–11 (comp. note) Thailand
Angamiana floridula Boulard 2005d: 342, 244 (comp. note)
Angamiana florida (sic) Boulard 2005j: 81 (comp. note) China, Tonkin
Angamiana floridula Boulard 2005j: 81–84, 86, 88, Figs 1–6 (synonymy, comp. note) Equals *Angamiana bauflei*
Angamiana bauflei Boulard 2005j: 81–82 (comp. note)
Angamiana floridula var. *bauflei* Boulard 2005j: 83, Figs 5–6 (comp. note)
Angamiana floridula Boulard 2005m: 125, 127 (comp. note)
Angamiana floridula Boulard 2005o: 147, 150 (listed)
Angamiana floridula Lee 2005: 37 (illustrated)
Angamiana floridula Boulard 2006g: 61–62, 180 (llisted, comp. note)
Aganinana (sic) *floridula* Boulard 2006g: 73–74 (comp. note)
Angamiana floridula Trilar 2006: 344 (comp. note)
Angamiana floridula Boulard 2007e: 4, 35, 37, Fig 21, Plate 4 (coloration, illustrated, sonogram, comp. note) Thailand
Angamiana floridula Sanborn, Phillips and Sites 2007: 10 (listed, synonymy, comp. note) Equals *Angamiana bauflei* Equals *Angamaiana floridula* var. *bauflei* Thailand, China, Vietnam
Angamiana bauflei Sanborn, Phillips and Sites 2007: 10 (comp. note)
Angamiana floridula Shiyake 2007: 2, 20, 22, Fig 19 (illustrated, listed, comp. note)
Angamiana floridula Boulard 2008f: 18–19 (biogeography, listed, comp. note) Equals *Angamiana bauflei* Equals *Angamiana floridula* var. *bauflei* Thailand, China, Tonkin
Angamiana florida (sic) Boulard 2008f: 92 (listed) Thailand
Angamiana floridula Lee 2008a: 6 (distribution, listed) North Vietnam, China, Guangxi, Thailand
Angamiana floridula Pham and Yang 2009: 8, 13, Table 2 (comp. note)
Anganiama floridula Pham and Yang 2010a: 133 (comp. note) Vietnam

***A. fuscula* (Chou & Yao, 1986)** = *Proretinata fuscula*
Proretinata fuscula Chou and Yao 1986: 192, 194, Fig 8 (key, n. sp., described, illustrated, comp. note) China
Proretinata fuscula Chou, Lei, Li, Lu and Yao 1997: 164, 166, Fig 9–38 (key, described, illustrated, comp. note) China
Proretinata fuscula Hua 2000: 64 (listed, distribution) China
Proretinata fuscula Boulard 2005c: 235 (comp. note)
Angamiana fuscula Pham and Yang 2009: 8 (n. comb., comp. note)

***A. masamii* Boulard, 2005** = *Angamiana florida* (sic) Hayashi 1982 (nec Distant)
Angamiana masamii Boulard 2005m: 126–127, Figs 1–2 (n. sp., described, illustrated, synonymy, comp. note) Equals *Angamiana florida* (sic) Hayashi 1982 (nec Distant, 1904) Laos
Angamiana masamii Boulard 2005o: 150 (listed)
Angamiana masamii Boulard 2008f: 19, 92 (biogeography, listed, comp. note) Thailand, Laos

***A. melanoptera* Boulard, 2005**
Angamiana melanoptera Boulard 2005c: 234–239, Figs 1–5, Figs 12–14 (n. sp., described, illustrated, sonogram, comp. note) Thailand
Angamiana melanoptera Boulard 2005j: 84–89, Figs 7–13 (n. sp., described, illustrated, sonogram, comp. note) Thailand
Angamiana melanoptera Boulard 2005o: 150–151 (listed)
Angamiana melanoptera Boulard 2006g: 73–74, 84, Fig 42 (comp. note) Thailand
Angamiana melanoptera Trilar 2006: 344 (comp. note)
Angamiana melanoptera Sanborn, Phillips and Sites 2007: 10 (listed, distribution, comp. note) Thailand
Angamiana melanoptera Boulard 2008f: 19, 92 (biogeography, listed, comp. note) Thailand

***A. vemacula* (Chou & Yao, 1986)** = *Proretinata vemacula* = *Proretinata vemaculata* (sic)
Proretinata vemacula Chou and Yao 1986: 190–191, 194, Fig 6 (key, type species of *Proretinata*, n. sp., described, illustrated, comp. note) China
Proretinata vemacula Chou, Lei, Li, Lu and Yao 1997: 163–165, Fig 9–36, Plate VII, Fig 75

(key, type species of *Proretinata*, described, illustrated, comp. note) China
Proretinata vemacula Hua 2000: 64 (listed, distribution) China, Jiangxi
Proretinata vemaculata (sic) Boulard 2005c: 235 (type species of *Proretinata*, comp. note)
Proretina vemaculata (sic) Boulard 2005j: 81 (comp. note)
Proretinata vemacula Pham and Thinh 2005a: 245 (type species of *Proretinata*, distribution, listed) Vietnam
Proretinata vemacula Pham and Yang 2009: 7 (type species of *Proretinata*)
Angamiana vemacula Pham and Yang 2009: 8, 13, 16, Table 2 (n. comb., listed, distribution, listed, comp. note) Vietnam, China
Anganiama vemacula Pham and Yang 2010a: 133 (comp. note) Vietnam
Angamiana vemaculata (sic) Pham and Yang 2009: 8 (n. comb., comp. note)

***A. yunnanensis* (Chou & Yao, 1986)** = *Proretinata yunnanensis*
Proretinata yunnanensis Chou and Yao 1986: 191–192, 194, Fig 7 (key, n. sp., described, illustrated, comp. note) China
Proretinata yunnanensis Chou, Lei, Li, Lu and Yao 1997: 164–166, Fig 9–37 (key, described, illustrated, comp. note) China
Proretinata yunnanensis Hua 2000: 64 (listed, distribution) China, Yunnan
Proretinata yunnanensis Boulard 2005c: 235 (comp. note)
Proretinata yunnanensis Boulard 2005j: 81 (comp. note)
Angamiana yunnanensis Pham and Yang 2009: 8 (n. comb., comp. note)

***Polyneura* Westwood, 1842** = *Polynevra* (sic) = *Cicada (Polyneura)*
Polyneura Wu 1935: 4 (listed) China
Polyneura Chen 1942: 143 (listed)
Polyneura Shcherbakov 1981a: 830 (listed)
Polyneura Shcherbakov 1981b: 66 (listed)
Polyneura Strümpel 1983: 18 (listed)
Polyneura Hayashi 1984: 45 (listed)
Polyneura Chou and Yao 1986: 185, 193 (key, comp. note)
Polyneura Boulard 1994a: 61 (comp. note) Yunnan
Polyneura Chou, Lei, Li, Lu and Yao 1997: 158–159, 293, 307, 315, 324, 326, Table 6, Tables 10–12 (key, described, listed, diversity, biogeography) China
Polyneura Novotny and Wilson 1997: 437 (listed)
Polyneura Lee and Hayashi 2003a: 168 (comp. note)
Polyneura Boulard 2005c: 233 (comp. note)
Polyneura Boulard 2005j: 80–81 (comp. note)
Polyneura Shiyake 2007: 2, 22 (listed, comp. note)
Polyneura Boulard 2008f: 7, 18, 92 (listed) Thailand
Polyneura Lee 2008a: 2, 6, Table 1 (listed) Vietnam
Polyneura Pham and Yang 2009: 13, Table 2 (listed) Vietnam

***P. cheni* Chou & Yao, 1986** = *Polyneura distanti* Boulard, 1994 = *Polyneura hügelii* Graber, 1876 nom. nud.
Polyneura cheni Chou and Yao 1986: 186–187, 193, Fig 1 (key, n. sp., described, illustrated, comp. note) China
Polyneura distanti Boulard 1994a: 63–65, Figs 3–4 (n. sp., described, illustrated, comp. note) Equals *Polyneura Hügelii* Graber, 1876 nom. nud. Yunnan, China
Polyneura cheni Chou, Lei, Li, Lu and Yao 1997: 158–160, Fig 9–31, Plate VII, Fig 78 (key, described, illustrated, comp. note) China
Polyneura cheni Hua 2000: 64 (listed, distribution) China, Sichuan
Polyneura cheni Boulard 2005c: 234 (comp. note)
Polyneura cheni Boulard 2005j: 81 (comp. note)
Polyneura hügelii nom. nud. Boulard 2005c: 233 (comp. note)
Polyneura distanti Boulard 2005c: 234 (comp. note)
olyneura (sic) *hügelii* Boulard 2005j: 80 (comp. note)
Polyneura distanti Boulard 2005j: 81 (comp. note)
Polyneura cheni Lee 2005: 37 (illustrated)
Polyneura cheni Shiyake 2007: 2, 20, 22, Fig 20 (illustrated, distribution, listed, comp. note) China
Polyneura cheni Boulard 2008f: 18, 92 (synonymy, biogeography, listed, comp. note) Equals *Polyneura distanti* Thailand, China, Yunnan

P. distanti Boulard, 1994 to Polyneura cheni Chou & Yao, 1986

***P. ducalis* Westwood, 1840** = *Cicada (Polyneura) ducalis* = *Cicada ducalis* = *Polyneura ducatlis* (sic) = *Polyneura ducaslis* (sic) = *Polyneura linearis* (sic)
Polyneura ducalis Swinton 1908: 380 (song, comp. note) India
Polyneura ducalis Wu 1935: 4 (listed) China, Szechuan, Tibet, Assam, Sikhim
Polyneura ducalis Chen 1942: 144 (listed, comp. note) China
Polyneura ducalis Chou and Yao 1986: 185, 187, 189, 193–194 (type species of *Polyneura*, comp. note)
Polyneura ducalis Kirillova 1986b: 46 (listed) India
Polyneura ducalis Kirillova 1986a: 123 (listed) India
Polyneura ducalis Chou 1987: 22 (comp. note) China
Polyneura ducalis Emeljanov and Kirillova 1991: 800 (chromosome number, comp. note)
Polyneura ducalis Emeljanov and Kirillova 1992: 63 (chromosome number, comp. note)

Polyneura ducaslis (sic) Emeljanov and Kirillova 1992: 63 (chromosome number, comp. note)
Polyneura ducalis Peng and Lei 1992: 100, Fig 284 (listed, illustrated, comp. note) China, Hunan
Polyneura ducalis Wang and Zhang 1992: 120 (listed, comp. note) China
Polyneura ducalis Wang and Zhang 1993: 188 (listed, comp. note) China
Polyneura ducalis Boulard 1994a: 63, 65 (comp. note)
Polyneura ducalis Chou, Lei, Li, Lu and Yao 1997: 158, 160, 307, 315, 324, 326, 330, 337, Tables 10–12 (type species of *Polyneura*, song, oscillogram, spectrogram, comp. note)
Polyneura ducalis Boulard 1999b: 184–185, Fig 10 (illustrated, comp. note)
Polyneura ducalis Hua 2000: 64 (listed, distribution, host) China, Guangxi, Hunan, Sichuan, Yunnan, Xizang, Indonesia, Burma, Nepal, Sikkim, India
Polyneura linearis (sic) Sueur 2001: 38, Table 1 (listed)
Polyneura ducalis Boulard 2002b: 196 (venation, comp. note)
Polyneura ducalis Boulard 2005c: 233–234 (comp. note)
Polyneura ducalis Boulard 2005j: 80–81 (comp. note)
Polyneura ducalis Shiyake 2007: 22 (comp. note)
Polyneura ducalis Boulard 2008f: 18 (type species of *Polyneura*)
Polyneura ducalis Lee 2008a: 6 (type species of *Polyneura*, distribution, listed) South Vietnam, Chinese Tibet, Myanmar, Nepal, India
Polyneura ducalis Pham and Yang 2009: 13, 16, Table 2 (listed, comp. note) Vietnam
Polyneura ducalis Pham and Yang 2010a: 133 (comp. note) Vietnam

P. hügelii Graber, 1876 nom. nud. to *Polyneura cheni* Chou & Yao, 1986

***P. laevigata* Chou & Yao, 1986**
Polyneura laevigata Chou and Yao 1986: 185, 189, 194, Fig 5 (key, n. sp., described, illustrated, comp. note) China
Polyneura laevigata Chou, Lei, Li, Lu and Yao 1997: 158, 162–163, Fig 9–35 (key, synonymy, described, illustrated, comp. note) China
Polyneura laevigata Hua 2000: 64 (listed, distribution) China, Jiangxi
Polyneura laevigata Boulard 2005c: 235 (comp. note)
Polyneura laevigata Boulard 2005j: 81 (comp. note)
Polyneura laevigata Boulard 2008c: 357, 362, Fig 14 (illustrated, comp. note) Nepal

***P.* nr. *laevigata* Boulard, 2008**
Polyneura nr. *laevigata* Boulard 2008c: 362 (comp. note) India

***P. parapuncta* Chou & Yao, 1986**
Polyneura parapuncta Chou and Yao 1986: 186–188, 193, Fig 3 (key, n. sp., described, illustrated, comp. note) China
Polyneura parapuncta Chou, Lei, Li, Lu and Yao 1997: 159–160, Fig 9–33 (key, described, illustrated, comp. note) China
Polyneura parapuncta Hua 2000: 64 (listed, distribution) China, Guizhou
Polyneura parapuncta Boulard 2005c: 235 (comp. note)
Polyneura parapuncta Boulard 2005j: 81 (comp. note)

***P. tibetana* Chou & Yao, 1986**
Polyneura tibetana Chou and Yao 1986: 185, 187–189, 193–194, Fig 4 (key, n. sp., described, illustrated, comp. note) Tibet
Polyneura tibetana Chou, Lei, Li, Lu and Yao 1997: 158, 161–162, Fig 9–34, Plate VII, Fig 81 (key, described, illustrated, comp. note) China
Polyneura tibetana Hua 2000: 64 (listed, distribution) China, Xizang
Polyneura tibetana Boulard 2005c: 235 (comp. note)
Polyneura tibetana Boulard 2005j: 81 (comp. note)
Polyneura tibetana Shiyake 2007: 2, 20, 22, Fig 21 (illustrated, listed, comp. note)

***P. xichangensis* Chou & Yao, 1986** = *Polyneura xichengensis* (sic)
Polyneura xichangensis Chou and Yao 1986: 185, 187, 193, Fig 2 (key, n. sp., described, illustrated, comp. note) China
Polyneura xichangensis Chou, Lei, Li, Lu and Yao 1997: 159–160, Fig 9–32 (key, described, illustrated, comp. note) China
Polyneura xichangensis Hua 2000: 64 (listed, distribution) China, Sichuan
Polyneura xichengensis (sic) Boulard 2005c: 235 (comp. note)
Polyneura xichengensis (sic) Boulard 2005j: 81 (comp. note)

***Graptopsaltria* Stål, 1866** = *Graphopsaltria* (sic) = *Graptosaltria* (sic) = *Graptopsalria* (sic) = *Graptosaltria* (sic) = *Graptosaltia* (sic) = *Grapsaltria* (sic) = *Graptopsaltiria* (sic)
Graptopsaltria Wu 1935: 4 (listed) China
Graptopsaltria Chen 1942: 144 (listed)
Graptopsaltria sp. Fujiyama 1982: 181, 184 (comp. note)
Graptopsaltria Strümpel 1983: 109, Table 8 (comp. note)
Graptopsaltria Hayashi 1984: 28, 45 (key, described, listed, comp. note) Equals *Graptosaltria* (sic)
Graptopsaltria spp. Liu 1992: 30 (comp. note)
Graptopsaltria Lee 1995: 22, 52, 75–76, 136 (listed, key, described, ecology, distribution, illustrated, comp. note) Korea

Graptopsaltria Boer and Duffels 1996b: 315 (comp. note)
Graptopsaltria Chou, Lei, Li, Lu and Yao 1997: 180, 184, 293, 308, 317, 324, 326, Table 6, Tables 10–12 (key, described, listed, diversity, biogeography) China
Graptopsaltria Novotny and Wilson 1997: 437 (listed)
Graptopsaltria Lee 1999: 6 (comp. note) Korea
Graptopsaltria Boulard 2002b: 193 (coloration, comp. note)
Graptopsaltria Lee and Hayashi 2003a: 168 (comp. note)
Graptopsaltria Lee 2005: 14, 16, 71, 171 (listed, key, comp. note) Korea
Graptopsaltria Shiyake 2007: 3, 10, 23 (listed, comp. note)
Graptopsaltria Lee 2008b: 446, 448, Table 1 (listed, comp. note) Korea
Graptopsaltria Osaka Museum of Natural History 2008: 322 (listed) Japan

***G. bimaculata* Kato, 1925** = *Graptopsaltria colorata* Matsumura, 1905 (nec Stål) = *Graptopsaltria tienta* Kato, 1929 (nec Karsch)
Graptopsaltria bimaculata Fujiyama 1982: 181, Fig 2C, Table 1 (comp. note) Japan, Okinawa
Graptopsaltria colorata Welbourn 1983: 135, Table 5 (mite host, listed) Japan
Graptopsaltria bimaculata Hayashi 1984: 29, 46–47, Figs 75–78 (key, described, illustrated, synonymy, comp. note) Equals *Graptopsaltria colorata* (nec Stål) Equals *Graptopsaltria tienta* (nec Karsch) Japan
Graptopsaltria bimaculata Lee 1995: 76 (comp. note)
Graptopsaltria bimaculata Chou, Lei, Li, Lu and Yao 1997: 184, 186–187, 357 (key, synonymy, described, comp. note) Equals *Graptopsaltria colorata* Matsumura, 1905 Equals *Graptopsaltria tienta* Kato, 1929 China
Graptopsaltria colorata Chou, Lei, Li, Lu and Yao 1997: 184 (type species of *Graptopsaltria*) China
Graptopsaltria bimaculata Sueur 2001: 46, Table 2 (listed)
Graptopsaltria bimaculata Lee 2005: 71 (comp. note)
Graptopsaltria bimaculata Shiyake 2007: 3, 21, 23, Fig 24 (illustrated, distribution, listed, comp. note) Japan
Graptopsaltria bimaculata Osaka Museum of Natural History 2008: 322 (listed, acoustic behavior, ecology, comp. note) Japan
Graptropsaltria bimaculata Saisho 2010c: 60 (comp. note) Japan

***G. nigrofuscata badia* Kato, 1925** = *Graptopsaltria nigrofuscata* ab. *badia* = *Graptopsaltria colorata badius* = *Graptopsaltria colorata badia*
Graptopsaltria nigrofuscata ab. *badia* Hayashi 1984: 46 (described, comp. note) Japan

***G. nigrofuscata nigrofuscata* (de Motschulsky, 1866)** = *Fidicina nigrofuscata* = *Fidicina nigrofasciata* (sic) = *Cryptotympana nigrofuscata* = *Goaptopsaltria* (sic) *nigsofuscata* (sic) = *Graptopsalria* (sic) *nigrofuscata* = *Graptosaltria* (sic) *nigrofuscata* = *Graptopsaltria migrofuscata* (sic) = *Graptopsaltria colorata* Stål, 1866 = *Graptosaltia* (sic) *corolata* (sic) = *Grapsaltria* (sic) *colorata* = *Graptopsaltria corolata* (sic) = *Graptosaltria* (sic) *colorata* = *Graptopsaltiria* (sic) *corolata* (sic) = *Grapsaltria* (sic) *colorata* = *Graptopsaltria colotata* (sic) = *Platypleura colorata* = *Graptopsaltria nigrofuscata tsuchidai* Matsumura, 1939 = *Graptopsaltria nigrofuscata* var. *testaceomaculata* Kato, 1937 = *Graptopsaltria nigrofuscata* ab. *testaceomaculata*
Graptopsaltria colorata Doi 1932: 42 (listed, comp. note) Korea
Graptopsaltria nigrofuscata Isaki 1933: 380 (listed, comp. note) Japan
Graptopsaltria colorata Wu 1935: 4 (listed) China, Manchuria, Hangchow, Japan, Korea, New Guinea
Graptopsaltria nigrofuscata Anonymous 1937: 25 (comp. note) Japan
Graptopsaltria colorata Chen 1942: 144 (type species of *Graptopsaltria*)
Graptopsaltria colorata Ishii 1953: 1757 (parasitism) Japan
Graptopsaltria nigrofuscata Ohgushi 1954: 11 (emergence times, ecology, comp. note) Japan
Graptopsaltria nigrofuscata Ikeda 1959: 484–485, 488–491, 494, Figs 1–3, Figs 5–9 (timbal muscle potentials, comp. note) Japan
Graptopsaltria nigrofuscata Katsuki 1960: 57, Table 1 (hearing, comp. note) Japan
Graptopsaltria nigrofuscata Kim 1960: 23 (listed) Korea
Graptopsaltria nigrofuscata Fujiyama 1982: 181–184, Fig 2B, table 1 (comp. note) Equals *Graptopsaltria colorata* Japan, Korea, China
Graptopsaltria nigrofuscata Hayashi 1982b: 187–188, Table 1 (distribution, comp. note) Japan
Graptopsaltria nigrofuscata Nagamine and Itô 1982: 80 (comp. note)
Graptopsasltria nigrofuscata Sato, Yamada and Tani 1982: 197 (control) Japan
Graptopsaltria nigrofuscata Aizu, Sekita and Yamada 1984: 140–143, Tables 1–5, Figs 1–4 (apple pest) Japan
Graptopsaltria colorata Hayashi 1984: 45 (type species of *Graptopsaltria*)
Graptopsaltria nigrofuscata Hayashi 1984: 29, 45–46, Figs 71–74 (key, described, illustrated, synonymy, comp. note) Equals *Fidicina*

nigrofuscata Equals *Graptopsaltria colorata* (partim) Japan
Graptopsaltria nigrofuscata ab. *testaceomaculata* Hayashi 1984: 46 (described, comp. note) Japan
Graptopsaltria nigrofusecata (sic) Huang, Komiya and Maruyama 1984: 512, Table 1 (comp. note)
Graptopsaltria nigrofuscata Lloyd 1984: 80 (comp. note)
Graptopsaltria nigrofuscata Kevan 1985: 22, 68, 76, 79 (illustrated, comp. note) Japan
Graptopsaltria nigrofuscata Karban 1986: 224, Table 14.2 (comp. note) Japan, Manchuria
Graptopsaltria nigrofuscata Seki, Fujishita, Ito, Matsuoka and Tsukida 1987: 102, Table 2 (retinoid composition, listed, comp. note) Japan
Graptopsaltria nigrofuscata Wang and Zhang 1987: 295 (key, comp. note) China
Graptopsaltria nigrofuscata Nakamura, Murayama, Kubota and Shigemori 1988: 17 (predation) Japan
Graptopsaltria colorata Southcott 1988: 103 (parasitism, comp. note) Japan
Graptopsaltria nigrofuscata Shimazu 1989: 430, 432, 434, Fig 1 (illustrated, fungal host, comp. note) Japan
Graptopsaltria nigrofuscata Yamada and Tutumi 1989: 154 (control) Japan
Graptopsaltria nigrofuscata Boulard 1990b: 89 (comp. note)
Graptopsaltria nigrofuscata Hara, Hayashi, Hirano, Zhong, Yasuda and Komae 1990: 1251, Table 1 (comp. note)
Graptopsaltria nigrofuscata Liu 1990: 46 (comp. note)
Graptopsaltria nigrofuscata Liu 1991: 256–257 (comp. note)
Graptopsaltria nigrofuscata Liu 1992: 9, 15–16, 34 (comp. note)
Graptopsaltria colorata Liu 1992: 37 (comp. note)
Graptopsaltria nigrofuscata Tzean, Hsieh, Chen and Wu 1993: 514 (fungal host, comp. note) Japan
Graptopsaltria nigrofuscata Wang and Zhang 1993: 188 (listed, comp. note) China
Graptopsaltria colorata Anonymous 1994: 809 (fruit pest) China
Graptopsaltria nigrofuscata Yoshioka, Narai, Shinkai, Tokuda, Saito et al. 1994: 472 (ion composition, comp. note) Japan
Graptopsaltria nigrofuscata Lee 1995: 18, 22, 38–39, 41–46, 63–68, 76–79, 136, 139 (listed, key, synonymy, ecology, distribution, illustrated, comp. note) Equals *Fidicina nigrofuscata* Equals *Graptopsaltria colorata* Korea
Graptopsaltria corolata (sic) Lee 1995: 29 (comp. note)
Graptopsaltria colorata Lee 1995: 30, 33–36, 42, 75–76 (type species of *Graptopsaltria*, comp. note) Equals *Graptopsaltria nigrofuscata*
Fidicina nigrofuscata Lee 1995: 76 (comp. note)
Graptopsaltria nigrofuscata Pemberton and Yamasaki 1995: 227 (human food, comp. note) Japan
Graptopsaltria nigrofuscata Fujiyama 1996: 546 (comp. note)
Graptopsaltria nigrofuscata Yoshida, Motoyama, Kosaku and Miyamoto 1996: 526 (comp. note)
Graptopsaltria nigrofuscata Chou, Lei, Li, Lu and Yao 1997: 184–185, Plate VIII, Fig 88 (key, synonymy, described, illustrated, comp. note) Equals *Fidicina nigrofuscata* Equals *Graptopsaltria colorata* China
Graptopsaltria nigrofuscata Itô 1998: 494, Table 1 (life cycle evolution, comp. note) Japan
Graptopsaltria colorata Lee 1999: 6 (type species of *Graptopsaltria*) Japan
Graptopsaltria nigrofuscata Lee 1999: 6 (distribution, song, listed, comp. note) Equals *Fidicina nigrofuscata* Equals *Graptopsaltria colorata* Equals *Graptopsaltia* (sic) *corolata* (sic) Korea, China, Japan
Graptosaltria (sic) *nigrofuscata* Ohbayashi, Sato and Igawa 1999: 341 (parasitism, comp. note)
Graptopsaltria nigrofuscata Okane, Nakagiri and Ito 1999: 117 (parasitism) Japan
Graptopsaltria colorata Hua 2000: 62 (listed, distribution, hosts) China, Liaoning, Shaanxi, Henan, Jiangsu, Jiangxi, Zhejiang, Guangdong, Human, Guangxi, Japan, Korea, New Guinea
Graptopsaltria nigrofuscata Hua 2000: 62 (listed, distribution) Equals *Fidicina nigrofuscata* China, Zhejiang, Hunan, Guizhou, Sichuan, Korea, Japan
Graptopsaltria nigrofuscata Lee, Hatakeyama and Oishi 2000: 191, 193–194, Figs 2–3 (sequence, comp. note) Japan
Graptopsaltria nigrofuscata Lee, Nishimori, Hatakeyama and Oishi 2000: 2–6, Figs 1–5 (sequence, comp. note) Japan
Graptopsaltria nigrofuscata Maezono and Miyashita 2000: 52–54, Fig 1 (nymphal cell host, comp. note) Japan
Graptopsaltria nigrofuscata Lee 2001b: 1 (control, comp. note) Korea
Graptopsaltria nigrofuscata Ohara and Hayashi 2001: 1 (comp. note) Japan
Graptopsaltria nigrofuscata Saisho 2001a: 68 (comp. note) Japan

Graptopsaltria nigrofuscata Sueur 2001: 38, 46, Tables 1–2 (listed)
Graptopsaltria nigrofuscata Jeon, Kim, Tripotin and Kim 2002: 241 (parasitism, comp. note) Japan
Graptopsaltria nigrofuscata Kubo-Irie, Irie, Nakazawa and Mohri 2003: 984–990, Figs 1–6 (sperm anatomy, comp. note) Japan
Graptopsaltria nigrofuscata Lee 2003b: 8 (comp. note)
Graptopsaltria nigrofuscata Lee and Hayashi 2003b: 363 (comp. note)
Graptopsaltria nigrofuscata Piulachs, Guiduglli, Barchuk, Cruz, Simões and Bellés 2003: 461–462, Fig 1 (sequnce, phylogeny, comp. note) Japan
Graptopsaltria nigrofuscata Saisho 2003: 51 (comp. note) Japan
Graptopsaltria nigrofuscata Hirai 2004: 376 (predation, comp. note) Japan
Graptopsaltria nigrofuscata Irie, Udagawa, Nagata and Kubo-Irie 2004: 1282 (sperm structure, comp. note) Japan
Graptopsaltria nigrofuscata Nuorteva, Lodenius, Nagasawa, Tulisalo and Nuorteva 2004: 55–57, 59, 62, Tables 1–2, Table 9 (cadmium, metal levels) Japan
Graptopsaltria nigrofuscata Wakakuwa, Ozaki and Aarikawa 2004: 1480, 1484, Fig 7 (protein, comp. note) Japan
Graptopsaltria nigrofuscata Chawanji, Hodson and Villet 2005a: 258, 266 (comp. note)
Graptopsaltria nigrofuscata Itô and Kasuya 2005: 60, Table 14 (life cycle, comp. note)
Graptopsaltria nigrofuscata Lee 2005: 16, 70–73, 156, 163, 171–172 (described, illustrated, synonymy, song, distribution, key, comp. note) Equals *Fidicina nigrofuscata* Equals *Graptopsaltria colorata* Equals *Graptopsaltia* (sic) *corolata* (sic) Equals *Graptosaltria* (sic) *colorata* Korea, China, Japan
Graptopsaltria colorata Lee 2005: 171 (type species of *Graptopsaltria*) Japan
Graptopsaltria nigrofuscata Saisho 2005: 44 (comp. note) Japan
Graptopsaltria nigrofuscata Sanborn 2005c: 113 (comp. note)
Graptopsaltria nigrofuscata Chawanji, Hodson and Villet 2006: 385–387 (comp. note)
Graptopsaltria nigrofuscata Holman and Snook 2006: 1666 (comp. note)
Graptopsaltria nigrofuscata Minoretti and Baur 2006: 277 (comp. note)
Graptopsaltria nigrofuscata Moriyama and Numata 2006: 1220 (comp. note) Japan
Graptopsaltria nigrofuscata Saisho 2006b: 61 (comp. note) Japan
Graptopsaltria nigrofuscata Chawanji, Hodson, Villet, Sanborn and Phillips 2007: 337, 345–346 (comp. note)
Graptopsaltria nigrofuscata Hayakawa 2007: 115 (sperm morphology, comp. note)
Graptopsaltria nigrofuscata Kawahara 2007: 163 (comp. note) Japan
Graptopsaltria nigrofuscata Shiyake 2007: 3, 21, 23, Fig 23 (illustrated, distribution, listed, comp. note) China, Korea
Graptopsaltria nigrofuscata Spatafora, Sung, Sung, Hywell-Jones and White 2007: 1702, Fig 1c (illustrated, fungal host, comp. note) Japan
Graptopsaltria nigrofuscata Takakura and Yamazaki 2007: 730–734, Figs 1–4 (predation, comp. note) Japan
Graptopsaltria nigrofuscata Yamazaki 2007: 347 (comp. note) Japan
Graptopsaltria nigrofuscata Lee 2008b: 446–447, 449, 463, Fig 1E, Table 1 (key, illustrated, synonymy, distribution, listed, comp. note) Equals *Fidicina nigrofuscata* Equals *Graptopsaltria colorata* Equals *Graptopsaltia* (sic) *corolata* (sic) Korea, China, Japan
Graptopsaltria colorata Lee 2008b: 449 (type species of *Graptopsaltria*) Japan
Graptopsaltria nigrofuscata Moriyama and Numata 2008: 1487–1493, Figs 1–7, Table 1 (diapause, comp. note) Japan
Graptopsaltria nigrofuscata Osaka Museum of Natural History 2008: 322 (listed, acoustic behavior, ecology, comp. note) Japan
Graptopsaltria nigrofuscata Tufail and Takeda 2008: 1450–1453, Fig 1–3 (vitellogenin, phylogeny, comp. note) Japan
Graptopsaltria nigrofuscata Hisamitsu, Sokabe, Terada, Osumi, Terada, et al. 2009: 93, 103, Fig 3a (behavior, illustrated, comp. note) Japan
Graptopsaltria nigrofuscata Mitani, Mihara, Ishii and Koike 2009: 901, Table 1 (listed, comp. note) Japan
Graptopsaltria nigrofuscata Moriyama and Numata 2009: 163–168, Figs 1–2, Tables 1–3 (cold tolerance, comp. note) Japan
Graptopsaltria nigrofuscata Ye and Ng 2009: 188 (protein, comp. note)
Graptopsaltria nigrofuscata Yoshida 2009: 949 (comp. note) Japan
Graptopsaltria nicrofuscata Lim Huang, Cai, Zhao, Peng and Wu 2010: 164, 167, Fig 3 (phylogeny, comp. note)
Graptopsaltria nigrofuscata Moriyama and Numata 2010: 69–72, Fig 1, Table 1 (diapause, comp. note) Japan
Graptopsaltria nigrofuscata Oh, Kim and Kim 2010: 134, Table 1 (predation) Korea

Graptropsaltria nigrofuscata Saisho 2010c: 60 (comp. note) Japan
Graptopsaltria nigrofuscata Zhang, Zhang, Jiang, Wang, Luo et al. 2010: 2263 (comp. note)

G. tienta **Karsch, 1894**
Graptopsaltria tienta Wu 1935: 5 (listed) China
Graptopsaltria tienta Chen 1942: 144 (listed, comp. note) China
Graptopsaltria tienta Fujiyama 1982: 181–182, Table 1 (comp. note) China
Graptopsaltria tienta Wang and Zhang 1987: 295 (key, comp. note) China
Graptopsaltria tienta Liu 1991: 256–257 (comp. note)
Graptopsaltria tienta Liu 1992: 15–16 (comp. note)
Graptopsaltria tienta Peng and Lei 1992: 101, Fig 288 (listed, illustrated, comp. note) China, Hunan
Graptopsaltria tienta Wang and Zhang 1993: 188 (listed, comp. note) China
Graptopsaltria tienta Lee 1995: 76 (comp. note)
Graptopsaltria tienta Chou, Lei, Li, Lu and Yao 1997: 28, 184–186, 308, 317–318, 324, 326, 331, 340, Fig 9–49, Plate VIII, Fig 89, Tables 10–12 (key, described, illustrated, song, oscillogram, spectrogram, comp. note) China
Graptopsaltria tienta Hua 2000: 62 (listed, distribution, hosts) China, Liaoning, Hebei, Shandong, Henan, Fujian, Hunan, Guangxi, Sichuan
Graptopsaltria tienta Sueur 2001: 38, Table 1 (listed)
Graptopsaltria tienta Lee 2005: 71 (comp. note)
Graptopsaltria tienta Shiyake 2007: 3, 21, 23, Fig 25 (illustrated, distribution, listed, comp. note) China

Subtribe Formotosenina Boulard, 2008
Formotosenina Boulard 2008f: 7, 19 (n. subtribe, listed) Thailand

Formotosena **Kato, 1925**
Formotosena Hayashi 1984: 45 (listed, comp. note)
Formotosena Chou and Yao 1985: 123 (comp. note) China
Formotosena Duffels 1991a: 128 (comp. note)
Formotosena Chou, Lei, Li, Lu and Yao 1997: 128–129, 293, 297–298, Table 6 (described, key, listed, diversity, biogeography) China
Formotosena Novotny and Wilson 1997: 437 (listed)
Formotosena Lee and Hayashi 2003a: 170–172 (key, diagnosis) Taiwan
Formotosena Chen 2004: 39–40, 54 (phylogeny, listed)
Formotosena Pham and Thinh 2005a: 243 (listed) Vietnam
Formotosena Pham and Thinh 2005b: 289 (comp. note) Vietnam
Formotosena Sanborn, Phillips and Sites 2007: 3, 10, Table 1 (listed, comp. note) Thailand
Formotosena Shiyake 2007: 3, 22 (listed, comp. note)
Formotosena Boulard 2008f: 7, 19, 92 (listed) Thailand
Formotosena Pham and Yang 2009: 7, 13, Table 2 (listed) Vietnam

F. montivaga **(Distant, 1889)** = *Tosena montivaga*
Formotosena montivaga Boulard 2000d: 103, Plate L, Figs 1–2 (illustrated, sonogram, comp. note)
Formotosena montivaga Boulard 2000e: 124 (comp. note)
Formotosena montivaga Boulard 2001a: 128–129, 135–137, Plate I, Fig F, Fig 6 (n. comb., illustrated, song analysis, sonogram, comp. note) Equals *Tosena montivaga* Thailand
Formotosena montivaga Sueur 2001: 39, Table 1 (listed)
Formotosena montivaga Boulard 2002b: 193, 199, 203, Fig 52 (coloration, illustrated, comp. note) Thailand
Formotosena montivaga Boulard 2005o: 147 (listed) Equals *Tosena montivaga*
Formotosena montivaga Lee 2005: 37 (illustrated)
Formotosena montivaga Boulard 2006g: 117–118 (comp. note)
Formotosena montivaga Boulard 2007d: 496, 498, 506, Figs D-E (comp. note) Thailand
Formotosena montivaga Boulard 2007e: 55, 69, 71, Fig 37.2, Plate 19a, Plate 31c (genitalia, illustrated, comp. note) Thailand
Formotosena montivaga Sanborn, Phillips and Sites 2007: 10 (listed, distribution, comp. note) Thailand
Formotosena montivaga Boulard 2008f: 19, 92 (biogeography, listed, comp. note) Equals *Tosena montivaga* Thailand, India, Assam, Fukien
Formotosena montivaga Boulard 2009e: 175–179, Plate A, Figs 1–10, Plate B, Figs 11–13 (eclosion, illustrated, comp. note) Thailand

F. seebohmi **(Distant, 1904)** = *Tosena seebohmi* = *Tosena siebohmi* (sic) = *Formotosena siebohmi* (sic) = *Tosena seebohmi interrupta* Schumacher, 1915 = *Formotosena seebohmi interrupta* = *Formotosena* (sic) *apicalis*
Formotosena seebohmi Chou and Yao 1985: 123 (comp. note) China
Formotosena seebohmi interrupta Chou and Yao 1985: 123 (comp. note) China
Formotosena seebohmi Karban 1986: 222 (comp. note) Malaysia
Tosena seebohmi Chou, Lei, Li, Lu and Yao 1997: 129 (type species of *Formotosena*)

Formotosena seebohmi Chou, Lei, Li, Lu and Yao 1997: 129, Fig 9–12, Plate IV, Fig 50 (key, described, illustrated, synonymy, comp. note) Equals *Tosena seebohmi* Equals *Tosena montivaga (?)* Equals *Tosena seebohmi interrupta* China
Formotosena seebohmi Hua 2000: 61 (listed, distribution) Equals *Tosena seebohmi* China, Jiangxi, Taiwan, Hainan, Japan
Formotosena apicalis Hua 2000: 61 (listed, distribution) Equals *Tosena seebohmi interrupta* China, Jiangxi, Fujian, Taiwan
Formotosena seebohmi Boulard 2001a: 136 (comp. note) Taiwan
Tosena seebohmi Lee and Hayashi 2003a: 149, 168 (type species of *Formotosena*, comp. note) Formosa
Formotosena seebohmi Lee and Hayashi 2003a: 168–170, Fig 13 (key, diagnosis, illustrated, synonymy, song, listed, comp. note) Equals *Tosena seebohmi* Equals *Tosena siebohmi* (sic) Equals *Tosena seebohmi* var. *interrupta* Equals *Formotosena seebohmi* var. *interrupta* Equals *Formotosen seebohmi* ab. *interrupta* Taiwan, China
Formotosena seebohmi Chen 2004: 55, 198, front cover (described, illustrated, listed, comp. note) Taiwan
Tosena seebohmi Pham and Thinh 2005a: 243 (type species of *Formotosena*)
Formotosena seebohmi Pham and Thinh 2005a: 243 (synonymy, distribution, listed) Equals *Formotosena seebohmi* Vietnam
Formotosena seebohmi Shiyake 2007: 3, 21, 23, Fig 22 (illustrated, distribution, listed, comp. note) China
Tosena seebohmi Pham and Yang 2009: 7 (type species of *Formotosena*)
Formotosena seebohmi Pham and Yang 2009: 7, 13, 16, Table 2 (listed, synonymy, distribution, comp. note) Equals *Tosena seebohmi* Vietnam, China, Taiwan, Japan
Formotosena seebohmi Pham and Yang 2010a: 133 (comp. note) Vietnam

Tribe Tacuini Distant, 1904 = Tacuarini
Tacuini Novotny and Wilson 1997: 437 (listed)
Tacuini Moulds 2005b: 427 (higher taxonomy, listed)
Tacuini Shiyake 2007: 3, 23 (listed, comp. note)
Tacuini Lee and Hill 2010: 278, 301 (comp. note)

***Tacua* Amyot & Audinet-Serville, 1843** = *Tama* (sic) = *Cicada (Tacua)*
Tacua sp. Hill, Hore and Thornton 1982: 49, 160 (listed, comp. note) Hong Kong
Tacua Hayashi 1993: 15 (comp. note)
Tacua Zaidi and Ruslan 1995a: 174, Table 2 (listed, comp. note) Borneo, Sarawak
Tacua Zaidi and Ruslan 1995b: 220, Table 2 (listed, comp. note) Borneo, Sabah
Tacua Zaidi and Ruslan 1995c: 70, Table 2 (listed, comp. note) Peninsular Malaysia
Tacua Zaidi 1996: 102–103, Table 2 (comp. note) Borneo, Sarawak
Tacua Zaidi and Hamid 1996: 54–55, Table 2 (listed, comp. note) Borneo, Sarawak
Tacua Novotny and Wilson 1997: 437 (listed)
Tacua Zaidi 1997: 114–115, Table 2 (listed, comp. note) Borneo, Sarawak
Tacua Zaidi and Ruslan 1998b: 4, Table 2 (listed, comp. note) Borneo, Sabah
Tacua Zaidi and Ruslan 1999: 4, Table 2 (listed, comp. note) Borneo, Sarawak
Tacua Shiyake 2007: 3, 23 (listed, comp. note)

***T. speciosa decolorata* Boulard, 1994** = *Tacua speciosa* var. *decolorata* = *Tacua speciosa* var. *a* Distant, 1889
Tacua speciosa var. *decolorata* Boulard 1994b: 66 (n. var., described, comp. note) Malacca, Malaysia
Tacua speciosa decolorata Boulard 1999b: 184–185, Fig 15 (illustrated, comp. note) Malacca

***T. speciosa speciosa* (Illiger, 1800)** = *Tettigonia speciosa* = *Cicada speciosa* = *Cicada indica* Donovan, 1800 = *Tacua indica* = *Tettigonia gigantea* Weber, 1801
Cicada speciosa Bergier 1941: 128, 219 (human food, comp. note) Malaysia
Tacua speciosa Hayashi 1993: 15 (comp. note)
Tacua speciosa Zaidi and Ruslan 1995b: 219, 222, Table 1 (listed, comp. note) Borneo, Sabah
Tacua speciosa Zaidi 1996: 100–101, Table 1 (comp. note) Borneo, Sarawak
Tacua speciosa Zaidi, Ruslan and Mahadir 1996: 60, Table 1 (listed, comp. note) Peninsular Malaysia
Tacua speciosa Zaidi 1997: 112–113, Table 1 (listed, comp. note) Borneo, Sarawak
Tacua speciosa Zaidi and Ruslan 1997: 219–220 (listed, comp. note) Equals *Tettigonia speciosa* Sumatra, Java, Borneo, Sabah, Peninsular Malaysia
Tacua speciosa Zaidi and Ruslan 1998a: 346 (listed, comp. note) Equals *Tettigonia speciosa* Borneo, Sarawak, Sumatra, Java, Peninsular Malaysia, Sabah
Tacua speciosa Zaidi and Ruslan 1998b: 2, Table 1 (listed, comp. note) Borneo, Sabah
Tacua speciosa speciosa Boulard 1999b: 184–185, Fig 13 (illustrated, comp. note) Borneo
Tacua speciosa Zaidi, Ruslan and Azman 1999: 301 (listed, comp. note) Equals *Tettigonia speciosa* Borneo, Sabah, Sumatra, Java, Peninsular Malaysia, Sarawak

Tacua speciosa Zaidi, Noramly and Ruslan 2000a: 321 (listed, comp. note) Equals *Tettigonia speciosa* Borneo, Sabah, Sumatra, Java, Peninsular Malaysia, Sarawak
Tacua speciosa Zaidi, Ruslan and Azman 2000: 199 (listed, comp. note) Equals *Tettigonia speciosa* Borneo, Sabah, Sumatra, Java, Peninsular Malaysia, Sarawak
Tacua speciosa Sueur 2001: 46, Table 2 (listed)
Tacua speciosa Zaidi and Nordin 2001: 183, 185, 189, 193, Table 1 (listed, comp. note) Equals *Tettigonia speciosa* Borneo, Sabah, Sumatra, Java, Peninsular Malaysia, Sarawak
Tacua speciosa Boulard 2002b: 193 (coloration, comp. note)
Tacua speciosa Zaidi, Nordin, Maryati, Wahab, Norashikin et al. 2002: 2, 9–10, Tables 1–2 (listed, comp. note) Equals *Tettigonia speciosa* Borneo, Sabah, Sumatra, Peninsular Malausia, Sarawak,
Tacua speciosa Zaidi and Azman 2003: 104, Table 1 (listed, comp. note) Borneo, Sabah
Tacua speciosa Schouten, Duffels and Zaidi 2004: 373, 378, Table 2 (listed, comp. note) Malayan Peninsula
Tacua speciosa Shiyake 2007: 3, 21, 23, Fig 26 (illustrated, distribution, listed, comp. note) Malaysia, Indonesia

T. speciosa var. *a* Distant to *Tacua speciosa decolorata*

Tribe Talaingini Myers, 1929

Talaingini Chou, Lei, Li, Lu and Yao 1997: 106, 149 (key, listed, described)
Talaingini Boulard 2003b: 198 (listed)
Talaingini Moulds 2005b: 386, 427 (history, listed, comp. note)
Talaingini Pham and Thinh 2005a: 245 (listed) Vietnam
Talaingini Pham and Thinh 2005b: 287–289 (key, comp. note) Vietnam
Talaingini Sanborn, Phillips and Sites 2007: 3, 5, 37, Fig 2, Table 1 (listed, distribution) Thailand
Talaingini Boulard 2008c: 366 (listed)
Talaingini Boulard 2008f: 7, 52 (listed) Thailand
Talaingini Lee 2008a: 2, 8, Table 1 (listed) Vietnam
Talaingini Pham and Yang 2009: 7, 13, Table 2 (listed) Vietnam
Talaingini Lee 2010d: 171 (comp. note)

Subtribe Talaingina Myers, 1929

Talaingaria Boulard 2003b: 198 (listed)
Talaingaria Boulard 2005g: 366 (comp. note)
Talaingaria Sanborn, Phillips and Sites 2007: 3, 5, Table 1 (listed) Thailand

***Talainga* Distant, 1890**

Talainga Wu 1935: 20 (listed) China
Talainga He 1984: 223, 227 (comp. note)
Talainga Chou, Lei and Yao 1992: 171, 176 (comp. note)
Talainga Hayashi 1993: 15 (comp. note)
Talainga Chou, Lei, Li, Lu and Yao 1997: 149, 152, 293, 297, 359, Table 6 (key, described, listed, diversity, biogeography, comp. note) China
Talainga Boulard 2005g: 366 (comp. note)
Talainga Pham and Thinh 2005a: 245 (listed) Vietnam
Talainga Pham and Thinh 2005b: 289 (comp. note) Vietnam
Talainga Boulard 2007e: 4 (coloration, comp. note) Thailand
Talainga Sanborn, Phillips and Sites 2007: 3, 5, Table 1 (listed) Thailand
Talainga Boulard 2008f: 7, 52, 96 (listed) Thailand
Talainga Lee 2008a: 2, 8, Table 1 (listed) Vietnam
Talainga Pham and Yang 2009: 7, 13, Table 2 (type genus of Talaingini, listed)

***T. binghami* Distant, 1890**

Talainga binghami Chou, Lei, Li, Lu and Yao 1997: 149–150, Plate VI, Fig 67 (key, type species of *Talainga*, described, illustrated, comp. note) China
Talainga binghami Boulard 1999b: 184–185, Fig 11 (illustrated, comp. note) Vietnam
Talainga binghami Hua 2000: 64 (listed, distribution) China, Jiangxi, Yunnan, Vietnam, Burma
Talainga binghami Boulard 2003b: 187, 198–199, Fig 10, Plate I, Fig 7 (described, song, sonogram, illustrated, comp. note) Thailand, Indochina, China, Burma
Talainga binghami Boulard 2005d: 346 (comp. note)
Talainga binghami Boulard 2005g: 369 (comp. note)
Talainga binghami Boulard 2005o: 148 (listed)
Talainga binghami Lee 2005: 37 (illustrated)
Talainga binghami Pham and Thinh 2005a: 245 (type species of *Talainga*, distribution, listed) Vietnam
Talainga binghami Boulard 2006g: 102–103, Fig 68 (sonogram, comp. note) Thailand
Talainga binghami Boulard 2007e: 4, 32–33, 66, Fig 18, Plate 2, Fig 4, Plate 19b (illustrated, coloration, sonogram, comp. note) Thailand
Talainga binghami Sanborn, Phillips and Sites 2007: 10–11 (listed, distribution, comp. note) Thailand, Burma, China, Vietnam, Indochina
Talainga binghami Boulard 2008c: 366 (comp. note) Tonkin, Yunnan
Talainga binghami Boulard 2008f: 52, 96 (biogeography, listed, comp. note) Thailand, Burma, Indochina, Tonkin, China

Talainga binghami Lee 2008a: 8 (type species of *Talainga*, distribution, listed) Vietnam, China, Yunnan, Thailand, Myanmar
Talainga binghami Pham and Yang 2009: 13, Table 2 (listed) Vietnam

T. chinensis **Distant, 1900**
Talainga chinensis Wu 1935: 20 (listed) China
Talainga chinensis Peng and Lei 1992: 102, Fig 291 (listed, illustrated, comp. note) China, Hunan
Talainga chinensis Chou, Lei, Li, Lu and Yao 1997: 149–151, Fig 9–26, Plate VI, Fig 66 (key, described, illustrated, comp. note) China
Talainga chinensis Hua 2000: 64 (listed, distribution) China, Jiangxi, Fujian, Hunan, Guangxi, Guizhou, Sichuan, Yunnan
Talainga chinensis Boulard 2008c: 368 (comp. note) China

T. distanti Jacobi, 1902 to *Parataïlinga distanti*

T. japrona **Ollenbach, 1929** = *Talainga japroa* (sic)

T. naga **Ollenbach, 1929**

T. omeishana **Chen, 1957**
Talainga omeishana Chou, Lei, Li, Lu and Yao 1997: 149, 151, Plate VI, Fig 68 (key, described, illustrated, comp. note) China
Talainga omeishana Hua 2000: 64 (listed, distribution) China, Sichuan

Paratalainga **He, 1984** = *Chouia* Yao nom. nud.
Paratalainga He 1984: 223–224, 227–228 (n. gen., described, comp. note) China, Guangxi
Paratalainga Chou, Lei and Yao 1992: 170, 176 (key, described, comp. note) Equals *Chouia* Yao China, Guangxi, Hainan, Yunnan, Guizhou, Sichuan, Hunan
Paratailinga Chou, Lei, Li, Lu and Yao 1997: 149, 152–153, 293, 297, 359, Table 6 (key, described, listed, diversity, biogeography, comp. note) China
Paratalainga Boulard 2005g: 366 (comp. note)
Paratalainga Pham and Thinh 2005a: 245 (listed) Vietnam
Paratalainga Pham and Thinh 2005b: 289 (comp. note) Vietnam
Paratalainga Boulard 2008f: 7, 52, 96 (listed) Thailand
Paratalainga Lee 2008a: 2, Table 1 (listed) Vietnam
Paratalainga Pham and Yang 2009: 7, 13, Table 2 (listed) Vietnam

P. distanti **(Jacobi, 1902)** = *Talainga distanti*
Paratailinga distanti Chou, Lei and Yao 1992: 171, 175–178 (key, n. comb., comp. note) Equals *Talainga distanti* China, Fujian, Tonkin
Paratailinga distanti Chou, Lei, Li, Lu and Yao 1997: 153, 157, 359–360 (key, synonymy, described, comp. note) Equals *Talainga distanti* China
Paratalainga distanti Hua 2000: 63 (listed, distribution) Equals *Talainga distanti* China, Fujian, Vietnam
Paratalainga distanti Pham and Thinh 2005a: 244 (synonymy, distribution, listed) Equals *Talinga* (sic) *distanti* Vietnam
Paratalainga distanti Boulard 2008f: 52–53, 96 (biogeography, listed, comp. note) Thailand, Tonkin, China
Paratalainga distanti Lee 2008a: 7 (synonymy, distribution, listed) Equals *Talainga distanti* Vietnam, China, Fujian
Paratlainga distanti Pham and Yang 2009: 13, Table 2 (listed) Vietnam

P. fucipennis **He, 1984** = *Paratalainga fusipenis* (sic) = *Paratalainga fuscipennis* (sic) = *Chouia liuziensis* Yao nom. nud.
Paratalainga fucipennis He 1984: 224–225, 228, Fig 5, Plate II, Fig 4 (n. sp., described, illustrated, comp. note) China, Guangxi
Paratailinga fucipennis Chou, Lei and Yao 1992: 171, 175–178 (key, comp. note) Equals *Chouia liuziensis* Yao China
Paratailinga fusipenis (sic) Chou, Lei and Yao 1992: 170 (comp. note)
Paratailinga fucipennis Chou, Lei, Li, Lu and Yao 1997: 153–154, 157, 359–360, Plate VI, Fig 72 (key, described, illustrated, comp. note) China
Paratalainga fuscipennis (sic) Hua 2000: 63 (listed, distribution) China, Hainan

P. fumosa **Chou & Lei, 1992**
Paratailinga fumosa Chou, Lei and Yao 1992: 171–174, 177, Fig 2 (key, n. sp. described, illustrated, comp. note) China, Yunnan
Paratailinga fumosa Chou, Lei, Li, Lu and Yao 1997: 152, 155–157, 359–360, Fig 9–29, Plate VI, Fig 69 (key, described, illustrated, comp. note) China

P. guizhouensis **Chou & Lei, 1992**
Paratailinga guizhouensis Chou, Lei and Yao 1992: 171, 173–174, 177–178, Fig 3 (key, n. sp. described, illustrated, comp. note) China, Guizhou
Paratailinga guizhouensis Chou, Lei, Li, Lu and Yao 1997: 152, 156–157, 359–360, Fig 9–30, Plate VI, Fig 74 (key, described, illustrated, comp. note) China
Paratalainga guizhouensis Boulard 2008c: 367–368, Fig 22 (comp. note) China

P. reticulata **He, 1984** = *Chouia pulchra* Yao nom. nud.
Paratalainga reticulata He 1984: 223–224, 227–228, Fig 4, Plate II, Fig 3 (n. sp., described, illustrated, type species of *Paratalainga*, comp. note) China, Guangxi

Parataílinga reticulata Chou, Lei and Yao 1992: 170–173, 176–177, Fig 1 (type species of *Parataílinga*, key, illustrated, comp. note) Equals *Chouia pulchra* Yao China, Guangxi

Paratalainga reticulata Peng and Lei 1992: 102, Fig 292 (listed, illustrated, comp. note) China, Hunan

Parataílinga reticulata Chou, Lei, Li, Lu and Yao 1997: 152–156, 359–360, Fig 9–28, Plate VI, Fig 71 (key, type species of *Parataílinga*, described, illustrated, comp. note) China

Paratalainga reticulata Hua 2000: 63 (listed, distribution) China, Hunan, Guangxi

Paratalainga reticulata Pham and Thinh 2005a: 245 (type species of *Paratalainga*)

Paratalainga reticulata Boulard 2008f: 52 (type species of *Paratalainga*)

Paratalainga reticulata Lee 2008a: 8 (type species of *Paratalainga*) China, Guangxi, Hunan

Paratalainga reticulata Pham and Yang 2009: 7 (type species of *Paratalainga*)

***P. yunnanensis* Chou & Lei, 1992**

Parataílinga yunnanensis Chou, Lei and Yao 1992: 171, 174–175, 177–178, Fig 4 (key, n. sp. described, illustrated, comp. note) China, Yunnan

Parataílinga yunnanensis Chou, Lei, Li, Lu and Yao 1997: 153–154, 359–360, Fig 9–27, Plate VI, Fig 70 (key, described, illustrated, comp. note) China

Paratalainga yunnanensis Pham and Thinh 2005a: 244 (distribution, listed) Vietnam

Paratalainga yunnanensis Pham and Yang 2009: 7, 13, 16, Table 2 (listed, distribution, comp. note) Vietnam, China

Chouia Yao, nom. nud.

C. liuziensis Yao, nom. nud. to *Parataílinga fucipennis* He, 1984

C. pulchra Yao, nom. nud. to *Parataílinga reticulata* He, 1984

Tribe Thophini Distant, 1904 = Tophini (sic)

Thophini Moulds 1990: 31 (listed) Australia

Thophini Novotny and Wilson 1997: 437 (listed)

Thophini Moulds 1999: 1 (comp. note)

Thophini Sanborn 1999a: 36 (listed)

Thophini Moulds 2001: 195–196 (described, composition, comp. note) Australia

Tophini (sic) Sueur 2001: 38, Table 1 (listed)

Thophini Moulds and Carver 2002: 12 (listed) Australia

Thophini Sanborn, Heath, Heath and Phillips 2004: 66 (listed)

Thophini Moulds 2005b: 391, 397, 414, 419, 422–423, 427, 429–433, Fig 58, Fig 60, Table 2 (described, distribution, key, phylogeny, listed, comp. note) Australia

Thophini Shiyake 2007: 3, 26 (listed, comp. note)

Thophini Moulds 2008a: 208–209 (key, comp. note)

Thophini Goemans 2010: 2 (comp. note)

***Thopha* Amyot & Audinet-Serville, 1843** = *Topha* (sic) = *Thophra* (sic) = *Thopa* (sic) = *Tropha* (sic) = *Thorpha* (sic)

Thopha Dallas 1870: 482 (listed)

Thophra (sic) Strümpel 1983: 18 (listed)

Thopha sp. Ewart 1988: 185 (natural history, comp. note) Queensland

Thopha sp. Lithgow 1988: 65 (listed) Australia, Queensland

Thopha Ewart 1989a: 80 (comp. note)

Topha (sic) Boulard 1990b: 143 (comp. note) Australia

Thopha Moulds 1990: 7, 10, 31, 53–54, 58 (listed, diversity, comp. note) Australia

Thopha spp. Moulds 1990: 55 (comp. note)

Thopha Moulds and Carver 1991: 467 (comp. note) Australia

Thopha sp. Ewart 1998a: 72 (listed) Queensland

Thopha spp. Ewart and Popple 2001: 65 (comp. note) Queensland

Thopha Moulds 2001: 195–196 (type species of Thophini, comp. note) Australia, Queensland

Thopha spp. Moulds 2001: 198–201, Fig 2 (comp. note) Australia

Topha (sic) Boulard 2002b: 198 (comp. note) Australia

Thopha Moulds and Carver 2002: 13 (listed) Australia

Thopha Moulds 2005b: 387–389, 393, 412–413, 430, 434, Table 2 (type genus of Thophini, higher taxonomy, listed, comp. note) Australia

Thopha Shiyake 2007: 3, 26 (listed, comp. note)

Thopha Moulds 2008b: 129, 135, 137, 139 (comp. note) Australia, New South Wales, Queensland, Northern Territory

Thopha Sanborn, Moore and Young 2008: 1 (comp. note)

Thopha spp. Ewart 2009b: 140 (comp. note) Queensland

***T. colorata* Distant, 1907** = *Topha* (sic) *colorata*

Thopha colorata Moulds 1990: 54–55, Plate 7, Fig 3 (described, illustrated, distribution, ecology, listed, comp. note) Australia, Northern Territory, Western Australia

Thopha colorata Naumann 1993: 25, 127, 149 (listed) Australia

Thopha colorata Moulds 2001: 195, 197–199, 201, Fig 2, Figs 10–11 (illustrated, key, distribution, comp. note)

Thopha colorata Moulds and Carver 2002: 13 (distribution, ecology, type data, listed) Australia, Northern Territory

Thopha colorata Moulds 2008b: 129, 137 (comp. note) Australia

***T. emmotti* Moulds, 2001**

Thopha emmotti Ewart and Popple 2001: 54–56, 69, Fig 1B, Fig 7A (illustrated, described, habitat, song, oscillogram) Queensland

Thopha emmotti Moulds 2001: 196–200, Figs 1–2, Figs 6–7 (n. sp., described, illustrated, key, distribution, comp. note) Australia, Queensland

Thopha emmotti Moulds and Carver 2002: 13 (distribution, ecology, type data, listed) Australia, Queensland

Thopha emmotti Moulds 2008b: 129, 137, 139 (comp. note) Australia

Thopha emmotti Ewart 2009b: 140 (comp. note) Queensland

***T. hutchinsoni* Moulds, 2008**

Thopha hutchinsoni Moulds 2008b: 129–137, 139–140, Fig 1, Figs 8–11 (n. sp., described, illustrated, key, distribution, comp. note) Australia, Western Australia

***T. saccata* (Fabricius, 1803)** = *Tettigonia saccata* = *Cicada saccata* = *Topha* (sic) *saccata* = *Thopa* (sic) *saccata* = *Tropha* (sic) *saccata*

Thopha saccata Swinton 1880: 224 (illustrated, timbal cover, comp. note)

Thopha saccata Goode 1980: 55 (comp. note) Australia

Thopha saccata Josephson 1981: 30–31, 34, Figs 7–8, Fig 10 (muscle function, comp. note)

Thopha saccata Humphreys and Mitchell 1983: 70, Table 2 (turrets, population density, comp. note) New South Wales

Thopha saccata Moulds 1983: 429, 435 (illustrated, comp. note) Australia, Qeensland, New South Wales

Thopa (sic) *saccata* Strümpel 1983: 108, Fig 144 (oscillogram, comp. note)

Thopha saccata Brunet 1986: 12–13 (emergence, illustrated, comp. note) New South Wales

Thopha saccata Ewart 1986: 51, 56, Fig 4, Table 1 (illustrated, natural history, wasp predation, listed, comp. note) Queensland, New South Wales

Thopha saccata MacNally and Doolan 1986a: 35, 37, 40–42, Figs 1–4, Table 2 (habitat, comp. note) New South Wales

Thopha (sic) MacNally and Doolan 1986a: 37–38, 43 (comp. note) New South Wales

Thopha saccata MacNally and Doolan 1986b: 281, 285–286, 288, Fig 2, Fig 4, Table 2 (habitat, comp. note) New South Wales

Thopha (sic) MacNally and Doolan 1986b: 284–291, Figs 1–3, Table 3, Table 5 (habitat, comp. note) New South Wales

Thopha saccata Ewart 1988: 185, 191, 194, 198–199, Fig 10F, Plate 3A (illustrated, oscillogram, listed, comp. note) Queensland

Thopha saccata (MacNally 1988a: 247 (distribution model, comp. note) New South Wales

Thopha saccata MacNally 1988b: 1976–1977, Tables 1–2 (comp. note) New South Wales

Tettigonia saccata Moulds 1990: 53 (type species of *Thopha*, type data, listed)

Thopha saccata Moulds 1990: 3, 5, 10–11, 18, 21–22, 53, 55–56, 58, Plate 7, Fig 1, Fig 1a (described, illustrated, distribution, ecology, listed, comp. note) Australia, New South Wales, Queensland

Thopha saccata Coombs and Toolson 1991: 100 (distribution, comp. note) Queensland, New South Wales

Thopha saccata Naumann 1993: 19, 127, 149 (listed) Australia

Thopha saccata Ewart 1995: 79 (described, illustrated, natural history, acoustic behavior, oscillogram) Queensland

Thopha saccata Paton, Humphreys and Mitchell 1995: 54, Table 3.5 (burrowing, listed) Australia, New South Wales

Thopha saccata Ewart 1996: 14 (listed)

Thopha saccata Moulds 1996b: 93–94, Plate 69 (illustrated, acoustic behavior, comp. note) Australia, Queensland, New South Wales

Thopha saccata Bennet-Clark 1999: 3354 (comp. note)

Thopha saccata Sanborn 1999a: 36 (type material, listed)

Thopha saccata Moss 2000: 69 (comp. note) Queensland

Thopha saccata Moss and Popple 2000: 58 (listed, comp. note) Australia, New South Wales

Thopha saccata Moulds 2001: 196–200, Figs 2–4 (illustrated, key, distribution, comp. note)

Thopha saccata Sueur 2001: 38, Table 1 (listed)

Topha (sic) *saccata* Boulard 2002b: 198 (comp. note)

Tettigonia saccata Moulds and Carver 2002: 13–14 (type species of *Thopha*, type data, listed)

Thopha saccata Moulds and Carver 2002: 14 (synonymy, distribution, ecology, type data, listed) Equals *Tettigonia saccata* Australia, New South Wales, Queensland

Thopha saccata Popple and Ewart 2002: 116 (illustrated) Australia, Queensland

Thopha saccata Sueur 2002b: 130 (listed)

Thopha saccata Williams 2002: 156–157 (listed) Australia, Queensland, New South Wales

Thopha saccata Lee 2003b: 5–6, Fig 2 (illustrated, comp. note) Australia

Thopha saccata Emery, Emery, Emery and Popple 2005: 102–105, 107, Tables 1–3 (comp. note) New South Wales
Thopha saccata Humphreys 2005: 69, 71–75, Table 2, Table 4 (population density, comp. note) New South Wales
Thopha saccata Lee 2005: 30–31 (illustrated, comp. note)
Thopha saccata Shiyake 2007: 3, 24, 26, Fig 27 (illustrated, distribution, listed, comp. note) Australia
Tropha (sic) *saccata* Costa Neto 2008: 453 (comp. note)
Thopha saccata Moulds 2008b: 129, 137, 139–140 (comp. note) Australia
Thopha saccata Watson, Myhra, Cribb and Watson 2008: 3353 (comp. note)
Thopha saccata Watson, Watson, Hu, Brown, Cribb and Myhra 2010: 117 (wing microstructure, comp. note)

***T. sessiliba* Distant, 1892** = *Thorpha* (sic) *sessiliba* = *Topha* (sic) *sessiliba* = *Thopha stentor* Buckton, 1898 = *Thopha nigricans* Distant, 1910
Thopha sessiliba Ewart 1988: 185, 191, 194, 198–199, Fig 10G, Plate 3B (illustrated, oscillogram, listed, comp. note) Queensland
Topha (sic) *sessibila* Boulard 1990b: 155, 157, Plate XVIII, Figs 2–3 (sound system, comp. note) Australia
Thopha sessiliba Moulds 1990: 27, 53, 56–58, Plate 7, Fig 2, Figs 2a–2b, Plate 24, Fig 6 (synonymy, described, illustrated, distribution, ecology, type data, listed, comp. note) Equals *Thopha stentor* Equals *Thopha nigricans* Australia, New South Wales, Northern Territory, Queensland, Western Australia
Thopha stentor Moulds 1990: 56 (type data, listed)
Thopha nigricans Moulds 1990: 56 (type data, listed) Queensland
Thopha sessiliba Coombs and Toolson 1991: 100 (comp. note) New South Wales
Thopha sessiliba Naumann 1993: 40, 127, 149 (listed) Australia
Thopha sessiliba Ewart 1998b: 135 (listed) Queensland
Thopha sessiliba Ewart and Popple 2001: 54–55, Fig 1A (described, habitat, song, oscillogram) Queensland
Thopha sessibila Moulds 2001: 196–199, 201, Fig 2, Figs 8–9 (illustrated, key, distribution, comp. note) Equals *Thopha nigricans*
Thopha sessiliba Sueur 2001: 38, Table 1 (listed)
Thopha sessiliba Moulds and Carver 2002: 14 (synonymy, distribution, ecology, type data, listed) Equals *Thopha stentor* Equals *Thopha nigricans* Australia, New South Wales, Northern Territory, Queensland, Western Australia
Thopha stentor Moulds and Carver 2002: 14 (type data, listed)
Thopha nigricans Moulds and Carver 2002: 14 (type data, listed) Queensland
Thopha sessiliba Ewart 2005a: 174, 176, Fig 6C (described, song, oscillogram) Queensland
Tettigonia saccata Moulds 2005b: 393, 434 (type species of *Thopha*, listed)
Thopha saccata Moulds 2005b: 395–396, 399, 401, 403, 4087, 417–419, 422, Figs 33–34, Fig 37, Fig 47, Figs 56–60, Table 1 (higher taxonomy, phylogeny, key, listed, comp. note) Australia
Thopha sessibila Moulds 2008b: 129–139, Figs 2–7, Fig 11 (illustrated, key, distribution, comp. note) Australia, Western Australia
Thopha sessiliba Ewart 2009b: 140 (comp. note) Queensland

***Arunta* Distant, 1904**
Arunta Strümpel 1983: 109, Table 8 (comp. note)
Arunta Popov 1985: 43 (acoustic system, comp. note)
Arunta Ewart 1989a: 79 (comp. note)
Arunta Boulard 1990b: 143 (comp. note) Australia
Arunta Moulds 1990: 23, 31, 54, 58 (listed, diversity, comp. note) Australia
Arunta Moulds and Carver 1991: 467 (comp. note) Australia
Arunta Moulds 2001: 195–196 (comp. note) Australia
Arunta Boulard 2002b: 198, 206 (comp. note) Queensland
Arunta Moulds and Carver 2002: 12 (listed) Australia
Arunta Sanborn, Heath, Heath and Phillips 2004: 66, 70 (comp. note) Australia
Arunta Moulds 2005b: 387–389, 393, 413, 430, 434, Table 2 (higher taxonomy, listed, comp. note) Australia
Arunta Shiyake 2007: 3, 26 (listed, comp. note)

***A. interclusa* (Walker, 1858)** = *Thopha interclusa* = *Henicopsaltria interclusa* = *Cicada graminea* Distant, 1904 (nec Fabricius) = *Cicada queenslandica* Kirkaldy, 1909 = *Arunta flava* Ashton, 1912
Arunta interclusa Ewart 1989a: 78–79 (comp. note)
Arunta interclusa Moulds 1990: 59–61, Plate 7, Fig 5 (synonymy, described, illustrated, distribution, ecology, type data, listed) Equals *Thopha interclusa* Equals *Cicada graminea* n. syn. Equals *Cicada queenslandica* n. syn. Equals *Arunta flava* Australia, New South Wales, Queensland

Cicada graminea Moulds 1990: 60 (type data, listed) Australia, Queensland
Cicada queenslandica Moulds 1990: 60 (synonymy, type data, listed)
Arunta interclusa Naumann 1993: 37, 73, 148 (listed) Australia
Arunta interclusa Ewart 1995: 84 (described, illustrated, natural history, acoustic behavior, oscillogram) Queensland
Arunta interclusa Ewart 1996: 14 (listed)
Arunta interclusa Moulds 1996b: 92, Plate 65 (acoustic behavior, comp. note) Australia, Queensland, New South Wales
Arunta interclusa Heath, Sanborn, Heath and Phillips 1999: 1 (thermal tolerances, comp. note) Queensland
Arunta interclusa Moss 2000: 69 (comp. note) Queensland
Arunta interclusa Ewart 2001a: 499–500, Fig 1 (acoustic behavior) Queensland
Arunta interclusa Ewart 2001b: 69, 74 (listed, comp. note) Queensland
Arunta interclusa Sueur 2001: 38, Table 1 (listed)
Arunta interclusa Moulds and Carver 2002: 12–13 (synonymy, distribution, ecology, type data, listed) Equals *Thopha interclusa* Equals *Cicada graminea* Equals *Cicada queenslandica* Equals *Arunta flava* Australia, New South Wales, Queensland
Thopha interclusa Moulds and Carver 2002: 12 (type data, listed)
Cicada graminea Moulds and Carver 2002: 12–13 (type data, listed) Australia, Queensland
Cicada queenslandica Moulds and Carver 2002: 13 (type data, listed)
Arunta interclusa Williams 2002: 156–157 (listed) Australia, Queensland, New South Wales
Arunta interclusa Sanborn, Heath, Heath and Phillips 2004: 66–71, Figs 1–2, Table 1 (thermal responses, comp. note) Australia, Queensland

A. intermedia Ashton, 1921 to *Arunta perulata* (Guérin-Méneville, 1831)

***A. perulata* (Guérin-Méneville, 1831)** = *Cicada perulata* = *Thopha perulata* = *Henicopsaltria perulata* = *Arunta pemlata* (sic) = *Arunta perolata* (sic) = *Arunta intermedia* Ashton, 1921

Henicopsaltria perulata Evans 1966: 114 (parasitism, comp. note) Australia
Arunta perulata Josephson 1981: 30–31, 34, Figs 7–8, Fig 10 (muscle function, comp. note)
Arunta perulata Josephson and Young 1981: 227 (comp. note)
Arunta perulata Moulds 1983: 429, 433 (illustrated, comp. note) Australia, Queensland, New South Wales
Arunta perulata Strümpel 1983: 108, Fig 144 (oscillogram, comp. note)
Arunta perulata Young and Josephson 1983a: 184–186, 192–194, Fig 2, Fig 8, Table 1 (oscillogram, timbal muscle kinetics, comp. note) Australia, New South Wales
Arunta perulata Young and Josephson 1984: 286, Fig 1 (comp. note)
Arunta perulata Claridge 1985: 303, Fig 2 (oscillogram)
Arunta perulata Josephson and Young 1985: 206 (comp. note)
Arunta perulata Popov 1985: 43 (acoustic system, comp. note)
Arunta perulata MacNally and Doolan 1986a: 35, 37, 40–42, Figs 1–4, Table 2 (habitat, comp. note) New South Wales
Arunta (sic) MacNally and Doolan 1986a: 37–40, 43 (comp. note) New South Wales
Arunta perulata MacNally and Doolan 1986b: 281, 285–286, 288, Fig 2, Fig 4, Table 2 (habitat, comp. note) New South Wales
Arunta (sic) MacNally and Doolan 1986b: 284, 286–291, Figs 1–3, Table 3, Table 5 (habitat, comp. note) New South Wales
Arunta perulata Josephson and Young 1987: 995–996, Figs 2–3 (comp. note)
Arunta perulata MacNally 1988a: 247 (distribution model, comp. note) New South Wales
Arunta perulata MacNally 1988b: 1976–1977, Tables 1–2 (comp. note) New South Wales
Arunta perulata Moss 1988: 29 (distribution, song, comp. note) Queensland
Arunta perulata Ewart 1989a: 79 (comp. note)
Cicada perulata Moulds 1990: 58 (type species of *Arunta*, listed) Australia, New South Wales
Arunta perulata Moulds 1990: 25, 58–61, Plate 7, Fig 4, Fig 4a, Plate 24, Fig 10 (synonymy, described illustrated, distribution, ecology, type data, listed) Equals *Cicada perulata* Equals *Thopha perulata* Equals *Henicopsaltria perulata* Equals *Arunta intermedia* n. syn. Australia, New South Wales, Queensland
Arunta intermedia Moulds 1990: 58 (synonymy, type data, listed) Queensland
Arunta perulata Naumann 1993: 62, 73, 148 (listed) Australia
Arunta perulata Brunet 1995: 22 (listed) Australia
Arunta perulata Ewart 1995: 85 (described, illustrated, natural history, acoustic behavior, oscillogram) Queensland
A[runta] perulata Jiang, Yang, Tang, Xu and Chen 1995b: 229 (song frequency, comp. note)
Arunta perolata (sic) Fonseca 1996: 28 (comp. note)

Arunta perulata Bennet-Clark 1999: 3354 (comp. note)
Arunta perulata Gosper 1999: 142 (predation) New South Wales
Arunta perulata Heath, Sanborn, Heath and Phillips 1999: 1 (thermal tolerances, comp. note) Queensland
Arunta perulata Moss 2000: 69 (comp. note) Queensland
Arunta perulata Ewart 2001b: 69–71, 73–75, 79, 82, Fig 6, Fig 10, Table 2 (emergence pattern, population density, nymph illustrated, predation, listed, comp. note) Queensland
Arunta perulata Sueur 2001: 38, Table 1 (listed)
Arunta perulata Boulard 2002b: 1989 (comp. note)
Arunta perulata Nation 2002: 260, Table 9.3 (comp. note)
Cicada perulata Moulds and Carver 2002: 12–13 (type species of *Arunta*, type data, listed) Australia, New South Wales
Arunta perulata Moulds and Carver 2002: 13 (synonymy, distribution, ecology, type data, listed) Equals *Cicada perulata* Equals *Arunta intermedia* Australia, New South Wales, Queensland
Arunta intermedia Moulds and Carver 2002: 13 (type data, listed) Queensland
Arunta perulata Syme and Josephson 2002: 765 (timbal muscle, comp. note)
Arunta perulata Williams 2002: 156–157 (listed) Australia, Queensland, New South Wales
Arunta perulata Sanborn, Heath, Heath and Phillips 2004: 66–71, Fig 1, Table 1 (thermal responses, comp. note) Australia, Queensland
Arunta perulata Emery, Emery, Emery and Popple 2005: 97 (comp. note) New South Wales
Arunta perulata Ewart 2005a: 173–174, 179, Fig 6B (described, song, oscillogram, comp. note) Queensland
Arunta perulata Moulds 2005b: 383–384, 395–396, 398, 410, 417–420, 422, Fig 52, Figs 56–60, Table 1 (illustrated, higher taxonomy, phylogeny, key, listed, comp. note) Australia
Arunta perulata Shiyake 2007: 3, 24, 26, Fig 28 (illustrated, distribution, listed, comp. note) Australia
Arunta perulata Nation 2008: 273, Table 10.1 (comp. note)

Tribe Cyclochilini Distant, 1904 = Cyclochilaria
Cyclochilini Kirillova 1986b: 45 (listed)
Cyclochilini Kirillova 1986a: 122 (listed)
Cyclochilini Moulds 1990: 31 (listed) Australia
Cyclochilini Boer and Duffels 1996b: 309 (diversity, comp. note) Australia
Cyclochilini Novotny and Wilson 1997: 437 (listed)
Cyclochilini Moulds 1999: 1 (comp. note)
Cyclochilini Sanborn 1999a: 36 (listed)
Cyclochilini Moulds 2001: 195 (comp. note)
Cyclochilini Sueur 2001: 38, Table 1 (listed)
Cyclochilini Moulds and Carver 2002: 5 (listed) Australia
Cyclochilini Moulds 2005b: 377, 391, 396, 414, 419, 422–423, 427, 429–430, Fig 58, Fig 60, Table 2 (described, key, history, phylogeny, distribution, key, listed, comp. note) Equals Cyclochilaria Australia
Cyclochilini Shiyake 2007: 3, 26 (listed, comp. note)
Cyclochilini Moulds 2008a: 209 (key, comp. note)

***Cyclochila* Amyot & Audinet-Serville, 1843** = *Cyclochida* (sic) = *Cychlochila* (sic) = *Cyclophila* (sic)
Cyclochila Strümpel 1983: 18 (listed)
Cyclochila Hayashi 1984: 35 (listed)
Cyclochila Smith 1984: 142 (timbal muscle, comp. note)
Cyclochila Wooton 1984: 42 (illustrated) Australia, Queensland
Cyclochila Popov 1985: 43 (acoustic system, comp. note)
Cyclochila Ewart 1986: 50, 57 (natural history, wasp predation, comp. note) Queensland
Cyclochila Moulds 1990: 7, 32, 61 (listed, diversity, comp. note) Australia
Cyclochila Bennet-Clark and Young 1992: 127–129, 131, 135–137, 139, 141, 143–145, 147–150 (comp. note) Australia, Victoria
Cyclochila Bennet-Clark 1994: 170, 172 (song resonating system) Australia
Cyclochila Bennet-Clark 1995: 213, 216 (comp. note)
Cyclochila Young and Bennet-Clark 1995: 1007, 1015, 1018–1019, Table 1 (comp. note)
Cyclochila New 1996: 88–89, 95, Fig 36a (illustrated, comp. note) Australia
Cyclochila Novotny and Wilson 1997: 437 (listed)
Cyclochila Bennet-Clark 1998a: 411–412, 414 (comp. note)
Cyclochila Moulds 2001: 196 (comp. note) Australia
Cyclochila Moulds and Carver 2002: 6 (synonymy, distribution, ecology, listed) Australia
Cyclochila Wooton 2002: 42 (illustrated) Australia, Queensland
Cyclochila Moulds 2005b: 377, 387, 391, 413–414, 423, 430–431, 435, Table 2 (type genus of Cyclochilini, illustrated, stridulatory apparatus, key, phylogeny, listed, comp. note) Australia
Cyclochila Shiyake 2007: 3, 26 (listed, comp. note)

***C. australasiae* (Donovan, 1805)** = *Tettigonia australasiae* = *Cyclophila* (sic) *australasiae* = *Cyclochila austrasiae* (sic) = *Cicada olivacea* Germar, 1830 = *Cyclochila australasiae* var. *spreta* Goding & Froggatt, 1904 = *Cyclochila australasiae* form *spreta*

Cyclochila australasiae Pringle 1957: 25, 34–35, 88, Fig 19C, Fig 22D, Table 1 (illustrated, flight muscles, innervation, comp. note)

Cyclochila australasiae Weis-Fogh 1964: 248 (muscle structure, comp. note)

Cyclochila australasiae Whitten 1965: 80, Table 1 (listed, chromosome number, comp. note) Australia, New South Wales

Cyclochila australasiae Evans 1966: 114 (parasitism, comp. note) Australia

Cyclochila australasiae Cullen 1974: 34, Table 6 (flight muscles structure, comp. note)

Cyclochila australasiae Goode 1980: 55 (comp. note) Australia

Cyclochila australasiae Josephson 1981: 24–25, 30–31, 34, Figs 2A–B, Figs 7–8, Fig 10 (timbal muscle structure, comp. note)

Cyclochila australasiae Josephson and Young 1981: 219–221, 224–235, Figs 1C–1D, Figs 2–6, Figs 7D–F, Tables 1–2 (timbal muscle structure and function, comp. note) Australia

Cyclochila australasiae MacNally and Young 1981: 195 (comp. note)

Cyclochila australasiae Pringle 1981: 6–8, Fig 8 (oscillogram, timbal muscles, comp. note) Australia

Cyclochila australasiae Moulds 1983: 429, 431 (illustrated, comp. note) Australia, Queensland, New South Wales, Victoria, South Australia

Cyclochila australasiae Strümpel 1983: 132–133, Figs 187–188 (sound system, comp. note)

Cyclochila australasiae Young and Josephson 1983a: 192, Table 1 (comp. note)

Cyclochila australasiae Chadwick 1984: 24 (predation) New South Wales

Cyclochila australasiae Marshall 1984: 216–225, Figs 2–10, Tables 1–3 (filter chamber, hemolymph, comp. note) New South Wales

Cyclochila australasiae Smith 1984: 142 (timbal muscle, comp. note) Australia

Cyclochila australasiae Young and Josephson 1984: 286, Fig 1 (comp. note)

Cyclochila australasiae Boulard 1985e: 1021 (coloration, comp. note)

Cyclochila australasiae Claridge 1985: 303, Fig 2 (oscillogram)

Cyclochila australasiae Maitland and Maitland 1985: 331–332 (illustrated, comp. note) Australia

Cyclochila australasiae Ewart 1986: 50–52, 56, Fig 1B-F, Fig 2A, Fig 4, Table 1 (illustrated, natural history, oscillograms, wasp predation, listed, comp. note) Queensland

Cyclochila australasiae Karban 1986: 224, Table 14.2 (comp. note) Australia, Fiji

Cyclochila australasiae Kirillova 1986b: 45 (listed) Japan

Cyclochila australasiae Kirillova 1986a: 122 (listed) Japan

Cyclochila australasiae MacNally and Doolan 1986a: 35, 37, 40–42, Figs 1–4, Table 2 (habitat, comp. note) New South Wales

Cyclochila (sic) MacNally and Doolan 1986a: 37–38, 43 (comp. note) New South Wales

Cyclochila australasiae MacNally and Doolan 1986b: 281, 285–286, 288, Fig 2, Fig 4, Table 2 (habitat, comp. note) New South Wales

Cyclochila (sic) MacNally and Doolan 1986b: 284–291, Figs 1–3, Table 3, Table 5 (habitat, comp. note) New South Wales

Cyclochila australasiae Josephson and Young 1987: 995–996, Figs 2–3 (comp. note)

Cyclochila australasiae MacNally 1988a: 247 (distribution model, comp. note) New South Wales

Cyclochila australasiae MacNally 1988b: 1976–1977, Tables 1–2 (comp. note) New South Wales

Cyclochila australasiae Duffels 1988d: 9 (comp. note) Australia

Cyclochila australasiae Ewart 1989a: 79 (comp. note)

Cyclochila australasiae Humphreys 1989: 99–107, Figs 1–2 (turrets, comp. note) New South Wales

Cyclochila australasiae Carver 1990: 260 (comp. note)

Cyclochila australasiae Faithfull 1990: 97–98 (predation) Victoria

Tettigonia australasiae Moulds 1990: 61 (type species of *Cyclochila*, listed) Australia

Cyclochila australasiae Moulds 1990: 3–8, 10–11, 16, 18–22, 25, 61–66, 194, Figs 6–7, Plate 1, Figs 2–3, Plate 2, Figs 1–4, Plate 5, Fig 2, Plate 8, Fig 2, Figs 2a–2b (illustrated, distribution, ecology, listed) Equals *Cyclochila australasiae* form *spreta* Equals *Cicada olivacea* Australia, New South Wales, Queensland, South Australia, Victoria

Cyclochila australasiae form *spreta* Moulds 1990: 3, 61–63 (listed, comp. note) Australia

Cyclochila australasiae Young 1990: 43–48, 50, 52–55, Figs 1–2, Fig 6, Tables 1–3, volume cover (oscillogram, frequency spectrum, song intensity, sound system, sound radiation, illustrated, comp. note) Australia, Victoria

Cyclochila australasie Carver, Gross and Woodward 1991: 432, Fig 30.4F (wings illustrated)

Cyclochila australasiae Moulds and Carver 1991: 467, Fig 30.28 (sound system, illustrated, comp. note) Australia

Cyclochila australasiae Bennet-Clark and Young 1992: 124, 126, 133–136, 149–150, Fig 1, Table 1 (song production, sound system illustrated) Victoria, Australia
Cyclochila australasiae Parry-Jones and Augee 1992: 9 (predation) Australia, New South Wales
Cyclochila australasiae Naumann 1993: 7, 12, 26, 37, 84, 149 (listed) Australia
Cyclochila australasiae Andersen 1994: 29 (comp. note)
Cyclochila australasiae Bennet-Clark 1994: 165–167, 169–169, 172, 174, Figs 1–2, Tables 1–3 (song intensity, acoustic system illustrated) Australia
Cyclochila australasiae Bennet-Clark and Young 1994: 292–293, Table 1 (body size and song frequency)
Cyclochila australasiae Fonseca and Popov 1994: 350, 360 (comp. note)
Cyclochila australasiae Humphreys 1994: 430 (turrets, comp. note) New South Wales
Cyclochila australasiae Woodall 1994: 44 (predation) Australia, Queensland
C[yclochila] australasiae Yang and Jiang 1994: 258, 260–261 (comp. note)
C[yclochila] austrasiae (sic) Yang, Jiang, Chen and Xu 1994: 217 (comp. note)
Cyclochila australasiae Bennet-Clark 1995: 205–206, 214–215, Fig 4, Fig 8, Table 1, Tables 3–4 (song amplifying system illustrated)
Cyclochila australasiae Brunet 1995: 22 (listed) Australia
Cyclochila australasiae Frith 1995: 44 (predation) Queensland
C[yclochila] australasiae Jiang, Yang, Tang, Xu and Chen 1995b: 229 (song frequency, comp. note)
Cyclochila australasiae Yang and Jiang 1995: 177 (comp. note)
Cyclochila australasiae Young and Bennet-Clark 1995: 1002, 1004, 1008, Fig 1, Figs 4–5 (timbal mechanics, sound production, oscillogram, frequency spectrum, song intensity, illustrated, comp. note) Australia, Victoria, New South Wales
Cyclochila australasiae Coombs 1996: 57, Table 1 (emergence time, comp. note) New South Wales
Cyclochila australasiae Daws and Hennig 1996: 175–187, Figs 1–6 (auditory responses, illustrated, comp. note) Victoria
Cyclochila australasiae Ewart 1996: 14 (listed)
Cyclochila australasiae Moulds 1996b: 94 (acoustic behavior, comp. note) Australia, Queensland, New South Wales, South Australia
Cyclochila australasiae New 1996: 95 (comp. note) Australia
Cyclochila australasiae Young 1996: 96–98, 103–104, Fig 5.1 (oscillogram, acoustic behavior, comp. note) Australia
Cyclochila australasiae Bennet-Clark 1997: 1681–1682, 1684–1693, Fig 2–9, Tables 1–4 (timbal mechanics, song frequency, timbal illustrated) Australia, Victoria
Cyclochila australasiae Daws, Hennig and Young 1997: 175–177, 180, 183, 185–186, Fig 1B (photaxis, comp. note) Victoria
Cyclochila australasiae Bennet-Clark 1998a: 408, 412–413, Table 2 (comp. note)
Cyclochila australasiae Bennet-Clark 1998b: 58–59, Fig (sound production, illustrated) Australia
Cyclochila australasiae Bennet-Clark and Young 1998: 701, 705–706, 711, 714 (comp. note) Australia
Cyclochila australasiae Courts 1998: 189 (predation, comp. note)
Cyclochila australasiae Fonseca and Bennet-Clark 1998: 717, 726, 728–729 (comp. note) Australia
Cyclochila australasiae Monteith 1998: 39 (life history, illustrated)
Cyclochila australasiae Bennet-Clark and Daws 1999: 1803–1804, 1807–1810, 1812–1813, 1815, Fig 1, Figs 3–12, Tables 1–2 (song production energy, timbal illustrated, acoustic system illustrated) Australia, Victoria
Cyclochila australasiae Bennet-Clark 1999: 3349, 3351–3352, 3354–3356, Fig 4A, Fig 5, Table 2 (comp. note, song production)
Cyclochila australasiae Curtis 1999: 5656 (life cycle, comp. note) Australia
Cyclochila australasiae Jongebloed, Rosenzweig, Kalicharan, van der Want and Ishay 1999: 64 (comp. note) Australia
Cyclochila australasiae Hangay and German 2000: 58 (illustrated, behavior) Australia
Cyclochila australasiae Hawkeswood 2000: 419 (predation, comp. note)
Cyclochila australasiae Moss and Popple 2000: 58 (listed, comp. note) Australia, New South Wales
Cyclochila australasiae Lee 2001b: 3 (comp. note)
Cyclochila australasiae Sanborn 2001a: 16, Fig 5 (comp. note)
Cyclochila australasiae Sueur 2001: 35, 38, 46, Table 1 (listed, comp. note)
Cyclochila australasiae Boulard 2002b: 193 (coloration, comp. note)
Cyclochila australasiae Gerhardt and Huber 2002: 22, 33, Fig 2.6B (timbal illustrated, comp. note)

Cyclochila australasiae Nation 2002: 259–260, Table 9.3 (comp. note)
Tettigonia australasiae Moulds and Carver 2002: 6–7 (type species of *Cyclochila*, type data, listed) Australia
Cyclochila australasiae Moulds and Carver 2002: 6–7 (synonymy, distribution, ecology, type data, listed) Equals *Tettigonia australasiae* Equals *Cicada olivacea* Australia, New South Wales, Queensland, South Australia, Victoria
Cicada olivacea Moulds and Carver 2002: 7 (type data, listed) Australiasia
Cyclochila australasiae Sueur 2002b: 119, 129 (listed, comp. note)
Cyclochila australasiae Williams 2002: 156–157 (listed) Australia, Queensland, New South Wales, Victoria
Cyclochila australasiae Lee 2003b: 5–6, Fig 1 (illustrated, comp. note) Australia
Cyclochila australasiae Moulds 2003a: 186, Fig 1 (illustrated, comp. note)
Cyclochila australasiae Sueur and Aubin 2003b: 486 (comp. note)
Cyclochila australasiae Young 2003: 30 (illustrated, sound production, life history, distribution, comp. note) Australia, Queensland, South Australia
Cyclochila australasiae Fonseca and Hennig 2004: 407 (comp. note)
Cyclochila australasiae Stokes and Josephson 2004: 281 (timbal muscle, comp. note)
Cyclochila australasiae Champman 2005: 327 (comp. note)
Cyclochila australasiae Emery, Emery, Emery and Popple 2005: 103–105, 107, Tables 1–3 (comp. note) New South Wales
Cyclochila australasiae Haywood 2005: 9–12, Fig 2, Table 1 (illustrated, distribution, described, habitat, song, listed, comp. note) South Australia, New South Wales, Queensland, Victoria
Cyclochila australasiae Humphreys 2005: 69–74, Tables 3–4 (population density, comp. note) New South Wales
Cyclochila australasiae Lee 2005: 30–31 (illustrated, comp. note)
Tettigonia australasiae Moulds 2005b: 391, 431 (type species of *Cyclochila*, listed) Australia
Cyclochila australasiae Moulds 2005b: 395–396, 398, 403–404, 414, 417–420, 422, 431, Fig 37, Figs 44–45, Figs 56–60, Table 1 (illustrated, stridulatory apparatus, higher taxonomy, phylogeny, key, listed, comp. note) Australia
Cyclochila australasiae Harvey and Thompson 2006: 905–909, Figs 1–3, Tables 1–3 (emergence energetics, comp. note) New South Wales
Cyclochila australasiae Haywood 2006a: 14, 18, 23, Tables 1–2 (illustrated, distribution, described, habitat, song, listed, comp. note) South Australia, Queensland, New South Wales, Victoria
Cyclochila australasiae Seabra, Pinto-Juma and Quartau 2006: 844 (comp. note)
Cyclochila australasiae Stoddart, Cadusch, Boyce, Erasmus and Comins 2006: 681–683, Fig 5b, Table 1 (wing microstructure, comp. note)
Cyclochila australasiae Sueur, Windmill and Robert 2006: 4125 (comp. note)
Cyclochila australasiae Bennet-Clark 2007: Fig 1 (timbal illustrated) Australia
Cyclochila australasiae Oberdörster and Grant 2007b: 19, Table 2 (comp. note)
Cyclochila australasiae Shiyake 2007: 3, 25–27, Fig 30 (illustrated, distribution, listed, comp. note) Australia
Cyclophila (sic) *australasiae* Emery 2008: 35 (color variant, illustrated, comp. note) New South Wales
Cyclochila australasiae Kostovski, White, Mitchell, Mitchell, Austin and Stoddart 2008: 70042H-2, Fig 2 (illustrated, comp. note)
Cyclochila australasiae Medler and Hulme 2008: 409, Table 1 (comp. note)
Cyclochila australasiae Nation 2008: 272–273, Table 10.1 (comp. note)
Cyclochila australasiae Kostovski, White, Mitchell, Austin and Stoddart 2009: 1532, Fig 1 (illustrated, comp. note)
Cyclochila australasiae Moulds 2009b: 163, Fig 1 (illustrated, comp. note)
Cyclochila australasiae Wilson 2009: 40, 48, Appendix (listed, comp. note)

C. *virens* Distant, 1906 = *Cyclochila laticosta* Ashton, 1912
Cyclochila virens Moulds 1990: 9, 19, 63–66, Plate 8, Fig 1, Fig 1a, Plate 24, Fig 4 (illustrated, distribution, ecology, listed) Equals *Cyclochila laticosta* Australia, Queensland
Cyclochila virens Naumann 1993: 149 (listed) Australia
Cyclochila virens Moulds and Carver 2002: 7 (synonymy, distribution, ecology, type data, listed) Equals *Cyclochila laticosta* Australia, Queensland
Cyclochila laticosta Moulds and Carver 2002: 7 (type data, listed) Queensland
Cyclochila virens Zborowski and Storey 2003: front cover (illustrated) Australia
Cyclochila virens Moulds 2005b: 395–396, 398, 404, 417–420, 422, Figs 56–60, Table 1 (higher taxonomy, phylogeny, key, listed, comp. note) Australia

Cyclochila virens Shiyake 2007: 3, 25–26, Fig 29 (illustrated, distribution, listed, comp. note) Australia

Tribe Jassopsaltriini Moulds, 2005

Jassopsaltriini Moulds 2005b: 419, 422–423, 425, 427, 429–430, Fig 60, Table 2 (n. tribe, diagnosis, described, composition, key, phylogeny, comp. note) Australia

Jassopsaltriini Moulds 2008a: 208–209 (key, comp. note)

***Jassopsaltria* Ashton, 1914**

Jassopsaltria Moulds 1990: 8, 32, 111 (listed, diversity, comp. note) Australia

Jassopsaltria Moulds 1996a: 19 (comp. note) Australia

Jassopsaltria Moulds and Carver 2002: 49 (type data, listed) Australia

Jassopsaltria Moulds 2005b: 392, 416, 424–425, 430–433, Table 2 (type genus of Jassopsaltriini, higher taxonomy, listed, comp. note) Australia

***J. rufifacies* Ashton, 1914**

Jassopsaltria rufifacies Moulds 1990: 111, Plate 14, Fig 12 (type species of *Jassopsaltria*, illustrated, distribution, ecology, listed, comp. note) Australia, Western Australia

Jassopsaltria rufifacies Moulds and Carver 2002: 49 (type species of *Jassopsaltria*, distribution, ecology, type data, listed) Australia, Western Australia

Jassopsaltria rufifacies Moulds 2005b: 395–398, 417–420, 422, 433, Table 1, Figs 56–60 (type species of *Jassopsaltria*, phylogeny, comp. note) Australia

***J.* sp. A Moulds, 2005**

Jassopsaltria sp. A Moulds 2005b: 395–398, 417–420, 422, Table 1, Figs 56–60 (type species of *Jassopsaltria*, phylogeny, comp. note) Australia

Tribe Burbungini Moulds, 2005

Burbungini Moulds 2005b: 419, 422–423, 425, 427, 429–430, 433, Fig 58, Fig 60, Table 2 (n. tribe, diagnosis, described, composition, key, phylogeny, comp. note) Australia

Burbungini Moulds 2008a: 208–209 (key, comp. note)

***Burbunga* Distant, 1905** = *Burbanga* (sic)

Burbunga Ewart 1989a: 79 (comp. note)

Burbunga Moulds 1990: 7, 32, 102, 127 (type data, listed) Australia

Burbunga Moulds and Carver 2002: 52 (type data, listed) Australia

Burbunga Moulds 2005b: 393, 416, 425, 429–432, Table 2 (type genus of Burbungini, higher taxonomy, listed, comp. note) Australia

Burbunga Shiyake 2007: 8, 103 (listed, comp. note)

***B. albofasciata* Distant, 1907**

Burbunga albofasciata Moulds 1990: 127 (distribution, listed) Australia, Northern Territory

Burbunga albofasciata Moulds 1994: 97 (comp. note) Australia

Burbunga albofasciata Moulds and Carver 2002: 52 (distribution, ecology, type data, listed) Australia, Northern Territory

Burbunga albofasciata 2005b: 395–398, 417–420, 422, Figs 56–60, Table 1 (higher taxonomy, phylogeny, listed, comp. note) Australia

***B. aterrima* (Distant, 1914)** = *Macrotristria aterrima* = *Macrostristria vulpina* Ashton, 1914

Macrotristria aterrima Moulds 1990: 102 (synonymy, distribution, ecology, listed, comp. note) Equals *Burbunga aterrima* Equals *Macrotristria vulpina* n. syn. Australia, Western Australia

Burbunga aterrima Moulds 1990: 102 (type data, listed) Western Australia

Macrotristria vulpina Moulds 1990: 102 (type data, listed) Western Australia

Macrotristria aterrima Moulds and Carver 2002: 15 (synonymy, distribution, ecology, type data, listed) Equals *Burbunga aterrima* Equals *Macrotristria vulpina* Australia, Western Australia

Burbunga aterrima Moulds and Carver 2002: 15 (type data, listed) Western Australia

Macrotristria vulpina Moulds and Carver 2002: 15 (type data, listed) Western Australia

***B. gilmorei* (Distant, 1882)** = *Tibicen gilmorei* = *Burbanga* (sic) *gilmori* (sic)

Burbunga gilmorei Ewart 1988: 184 (comp. note)

Burbunga gilmorei Lithgow 1988: 65 (listed) Australia, Queensland

Burbunga gilmorei Moulds 1990: 127–128 (distribution, ecology, listed, comp. note) Australia, New South Wales, Queensland

Burbunga gilmorei Naumann 1993: 4, 76, 149 (listed) Australia

Burbunga gilmorei Moulds 1994: 97–103, Fig 1, Figs 6–7, Fig 12 (lectotype designation, synonymy, described, illustrated, distribution, comp. note) Equals *Tibicen gilmorei* Australia, Western Australia, South Australia

Tibicen gilmorei Moulds 1994: 97 (comp. note) Australia, Western Australia

Burbunga gilmorei Moulds and Carver 2002: 52 (synonymy, distribution, ecology, type data, listed) Equals *Tibicen gilmorei* Australia, New South Wales, Queensland

Tibicen gilmorei Moulds and Carver 2002: 52 (type species of *Burbunga*, type data, listed) Western Australia

Burbunga gilmorei 2005b: 395–398, 417–420, 422, Figs 56–60, Table 1 (higher taxonomy, phylogeny, listed, comp. note) Australia

Tibicen gilmorei Moulds 2005b: 429 (type species of *Burbunga*)

Burbunga gilmorei Shiyake 2007: 8, 100, 103, Fig 172 (illustrated, distribution, listed, comp. note) Australia

***B. hillieri* (Distant, 1907)** (nec hillieri Distant, 1906) = *Macrotristria hillieri*

Macrotristria hillieri Moulds 1990: 99–102, Plate 13, Fig 7, Fig 7a, Plate 24, Fig 8 (illustrated, distribution, ecology, listed, comp. note) Australia, New South Wales, Northern Territory, Queensland, South Australia, Victoria, Western Australia

Burbunga hillieri Moulds 1994: 97 (comp. note) Australia

Macrotristria hillieri Ewart and Popple 2001: 55, 62, Fig 1C (described, habitat, song, oscillogram) Queensland

Macrotristria hillieri Moulds and Carver 2002: 16 (distribution, ecology, type data, listed) Australia, New South Wales, Northern Territory, Queensland, South Australia, Victoria, Western Australia

Macrotristria hillieri Moulds 2005b: 395–396, 399, 417–420, 422, 429, Figs 56–60, Table 1 (higher taxonomy, phylogeny, key, listed, comp. note) Australia

Macrotristria hillieri Ewart 2009b: 140 (comp. note) Queensland

***B. inornata* Distant, 1905**

Burbunga inornata Moulds 1990: 127 (distribution, listed) Australia, Western Australia

Burbunga inornata Moulds 1994: 97, 103 (comp. note) Australia

Burbunga inornata Moulds and Carver 2002: 53 (distribution, ecology, type data, listed) Australia, Western Australia

***B. mouldsi* Olive, 2012**

***B. nanda* (Burns, 1964)** = *Macrotristria nanda*

Macrotristria nanda Moulds 1990: 87 (listed, comp. note) Australia, Western Australia

Macrotristria nanda Moulds and Carver 2002: 18 (distribution, ecology, type data, listed) Australia, Western Australia

***B. nigrosignata* (Distant, 1904)** = *Macrotristria nigrosignata*

Macrotristria nigrosignata Moulds 1990: 100–101, Plate 12, Fig 3 (illustrated, distribution, ecology, listed, comp. note) Australia, Western Australia

Macrotristria nigrosignata Moulds and Carver 2002: 18 (distribution, ecology, type data, listed) Australia, Western Australia

***B. occidentalis* (Distant, 1912)** = *Macrotristria occidentalis*

Macrotristria occidentalis Moulds 1990: 101–102 (distribution, ecology, listed, comp. note) Australia, Western Australia

Macrotristria occidentalis Moulds and Carver 2002: 18 (distribution, ecology, type data, listed) Australia, Western Australia

***B. parva* Moulds, 1994**

Burbunga parva Moulds 1994: 97–100, 102–103, Figs 4–5, Figs 10–12 (n. sp., described, illustrated, distribution, comp. note) Australia, Northern Territory

Burbunga parva Moulds and Carver 2002: 53 (distribution, ecology, type data, listed) Australia, Northern Territory

***B. queenslandica* Moulds, 1994** = *Burbunga* sp. Ewart, 1988 = *Burbunga gilmorei* (nec Distant)

Burbunga sp. Ewart 1988: 184, 190–191, 193, 198–199, Fig 8, Fig 10C, Plate 2B (illustrated, oscillogram, natural history, listed, comp. note) Queensland

Burbunga gilmorei (sic) Moulds 1990: Plate 15, Fig 7, Fig 7a, Plate 24, Fig 18 (illustrated, comp. note) Australia, New South Wales, Queensland

Burbunga queenslandica Moulds 1994: 97–103, Figs 2–3, Figs 8–9, Fig 12 (n. sp., synonymy, described, illustrated, distribution, comp. note) Equals *Burbunga gilmorei* (nec Distant) Equals *Burbunga* sp. Ewart, 1988 Australia, Queensland, New South Wales

Burbunga queenslandica Ewart 1998a: 61–62 (described, natural history, song) Equals *Burbunga gilmorei* Moulds 1990 Queensland

Burbunga queenslandica Ewart and Popple 2001: 56, 63, 65, Fig 2A (described, habitat, song, oscillogram) Queensland

Burbunga queenslandica Sueur 2001: 42, Table 1 (listed) Equals *Burbunga* sp.

Burbunga queenslandica Moulds and Carver 2002: 53 (distribution, ecology, type data, listed) Australia, New South Wales, Queensland

Burbunga queenslandica Popple and Strange 2002: 28, Table 1 (listed, habitat, comp. note) Equals *Burbunga gilmorei* Moulds 1990 Australia, Queensland

B. venosa Distant, 1907 to *Parnquila venosa*

Tribe Talcopsaltriini Moulds, 2008

Talcopsaltriini Moulds 2008a: 207–209 (n. tribe, described, key, listed, comp. note) Australia, Queensland

***Talcopsaltria* Moulds, 2008**

Tacopsaltria Moulds 2008a: 208–210, 212 (n. gen., described, type genus of Talcopsaltriini, key, comp. note) Australia

***T. olivei* Moulds, 2008** = *Macrotristria* sp. B Ewart, 1993 = Heathlands sp. B Ewart, 2005

Macrotristria sp. B Ewart 1993: 137–138, 143, Fig 5 (labeled Fig 6 *Pauropsalta basalis*) (natural history, acoustic behavior, song analysis) Queensland

Heathlands sp. B Ewart 2005a: 177, Fig 10 (described, song, oscillogram) Queensland

Tacopsaltria olivei Moulds 2008a: 207–214, Figs 1–10 (n. sp., described, illustrated, type species of *Tacopsaltria*, type species of Talcopsaltriini, song, sonogram, oscillogram, distribution, comp. note) Australia, Queensland

Tribe Cryptotympanini Handlirsch, 1925 = Tibicenini Van Duzee, 1916 = Tibicinae (sic) = Tibiceni (sic) = Lyristini Gomez-Menor, 1957 = Lyristarini

Tibicenini Chen 1942: 144 (listed
Lyristini Boulard 1982f: 184 (listed)
Tibicinae (sic) Boulard 1984c: 169 (status, type genus *Tibicen*, type species *Cicada plebeja*, comp. note)
Lyristarini Boulard 1984c: 175 (comp. note)
Tibicenini Lyristini Tibicinae (sic) Lyristarini Lyristini Boulard 1984c: 177–179 (comp. note)
Tibicenini Hayashi 1984: 28, 35 (key, listed)
Cryptotympanini Duffels 1986: 327–328, 331, Fig 9 (listed, distribution, comp. note)
Tibicenini Kirillova 1986a: 122 (listed)
Tibicenini Kirillova 1986b: 46 (listed)
Lyristini Schedl 1986a: 3 (listed, key) Istria
Tibicenini Hayashi 1987a: 124–125 (comp. note)
Cryptotympanini Hayashi 1987a: 125 (comp. note)
Lyristini Boulard 1988e: 154 (listed)
Lyristini Boulard 1988f: 44–45, 72 (synonymy, comp. note) Equals Cryptotympanini
Lyristarini Boulard 1988f: 72 (comp. note)
Cryptotympanini Boulard 1988f: 72 (n. syn. to Lyristini)
Cryptotympanini Duffels 1988d: 81 (comp. note)
Cryptotympanini Boulard 1990b: 214 (comp. note)
Lyristini Boulard 1990b: 214 (comp. note)
Tibiceni (sic) Liu 1992: 42 (listed)
Cryptotympanini Boulard and Mondon 1995: 65–66 (key, listed) France
Cryptotympanini Boer and Duffels 1996b: 304 (comp. note)
Cryptotympanini Boulard 1996c: 96, 100 (comp. note)
Tibicenini Boer and Duffels 1996b: 304, 330 (distribution, comp. note)
Tibicenini Boulard 1997c: 101–103, 105–106, 112, 121 (comp. note)
Lyristarini Boulard 1997c: 105, 121 (comp. note)
Lyristini Boulard 1997c: 106, 121 (comp. note)
Cryptotympanini Boulard 1997c: 105–106, 112, 121 (comp. note)
Lyristini Chou, Lei, Li, Lu and Yao 1997: 106, 264–265 (key, described, synonymy, listed, described) Equals Tettigonides Equals Cicadaria Equals Tibiceninae Equals Tibicenini
Cryptotympanini Novotny and Wilson 1997: 437 (listed)
Tibicenini Novotny and Wilson 1997: 437 (listed)
Cryptotympanini Moulds 1999: 1 (comp. note)
Tibicenini Sanborn 1999a: 36 (listed)
Cryptotympanini Sanborn 1999a: 42 (listed)
Tibicenini Sanborn 1999b: 1 (listed) North America
Lyristini Villet and Zhao 1999: 322 (comp. note)
Tibicenini Boulard 2001b: 21–25, 41 (comp. note)
Lyristini Boulard 2001b: 25–26, 41 (comp. note)
Cryptotympanini Boulard 2001b: 25–26, 32 (comp. note)
Lyristini Boulard 2001b: 25, 41 (comp. note)
Tibicenini Moulds and Carver 2002: 14 (listed) Australia
Tibicenini Chen 2004: 39, 41 (phylogeny, listed)
Tibicenini Sanborn, Heath, Heath and Phillips 2004: 66 (listed)
Tibicenini Sanborn, Villet and Phillips 2004a: 820 (comp. note)
Lyristini Chawanji, Hodson and Villet 2005a: 258 (listed)
Tibicenini Lee 2005: 14, 16, 156–157, 168 (key, listed, comp. note) Korea
Cryptotympanini Moulds 2005b: 377, 390–393, 396, 404, 410, 414, 419, 422–423, 427–429, 430–432, 434, Fig 58, Fig 60, Table 2 (described, diagnosis, composition, distribution, key, listed, comp. note) Equals Cryptotympanaria Equals Tibicenini Equals Lyristarini Equals Lyristini
Lyristini Pham and Thinh 2005a: 240 (listed) Vietnam
Lyristini Pham and Thinh 2005b: 287–288, 290 (key, comp. note) Equals Tettigonides Equals Tibiceninae Equals Tibicenini Vietnam
Cryptotympanini Yaakop, Duffels and Visser 2005: 248 (comp. note)
Cryptotympanini Chawanji, Hodson and Villet 2006: 373 (listed)

Cryptotympanini Puissant 2006: 19 (listed) France
Tibicenini Sanborn 2006a: 76 (listed)
Tibicenini Sanborn 2006b: 256 (listed)
Cryptotympanini Boulard 2007a: 94 (listed)
Cryptotympanini Boulard 2007d: 493 (listed)
Cryptotympanini Chawanji, Hodson, Villet, Sanborn and Phillips 2007: 338 (listed)
Cryptotympanini Sanborn 2007a: 27 (listed) Venezuela
Cryptotympanini Sanborn 2007b: 9, 31 (listed) Mexico
Cryptotympanini Sanborn, Phillips and Sites 2007: 3, 5, 36, Fig 1, Table 1 (listed, distribution) Thailand
Cryptotympanini Shiyake 2007: 4, 42 (listed, comp. note)
Cryptotympanini Sueur and Puissant 2007b: 62 (listed) France
Cryptotympanini Boulard 2008c: 360 (listed) Equals Lyristini
Cryptotympanini Boulard 2008f: 7, 9 (listed) Thailand
Cryptotympanini Demir 2008: 476 (listed) Turkey
Lyristini Fonseca, Serrão, Pina-Martins, Silva, Mira, et al. 2008: 24 (listed)
Cryptotympanini Lee 2008b: 446–447, 463, Table 1 (listed, comp. note) Korea
Cryptotympanini Lee 2008a: 2, 4, Table 1 (listed) Vietnam
Cryptotympanini Moulds 2008a: 208–209 (key, comp. note)
Tibicenini Osaka Museum of Natural History 2008: 322 (listed) Japan
Cryptotympanini Sanborn, Phillips and Gillis 2008: 5 (listed) Florida
Cryptotympanini Boulard 2009d: 39 (comp. note)
Cryptotympanini Lee 2009h: 2620 (listed) Philippines, Palawan
Cryptotympanini Lee 2009i: 294 (listed) Philippines, Panay
Cryptotympanini Sanborn 2009a: 89 (listed) Cuba
Lyristini Kemal and Koçak 2010: 5 (listed) Turkey
Cryptotympanini Lee 2010a: 16 (listed) Philippines, Mindanao
Cryptotympanini Lee 2010b: 20 (listed) Cambodia
Cryptotympanini Lee 2010c: 2 (listed) Philippines, Luzon
Cryptotympanini Lee 2010d: 171 (comp. note)
Cryptotympanini Lee and Hill 2010: 278 (comp. note)
Cryptotympanini Mozaffarian and Sanborn 2010: 76 (listed)
Cryptotympanini Phillips and Sanborn 2010: 74 (comp. note)
Cryptotympanini Sanborn 2010a: 1586 (listed) Colombia
Cryptotympanini Sanborn 2010b: 67, 74 (listed)
Cryptotympanini Sueur, Janique, Simonis, Windmill and Baylac 2010: 932 (comp. note)
Cryptotympanini Wei, Ahmed and Rizvi 2010: 33 (comp. note)

Subtribe Cryptotympanina Handlirsch, 1925 = Cryptotympanaria = Tibicenaria Van Duzee, 1916

Cryptotympanaria Hayashi 1987a: 125 (comp. note)
Cryptotympanaria Boulard 1997c: 102, 105–106, 112, 117 (comp. note)
Cryptotympanaria Duffels 1988d: 81 (comp. note)
Cryptotympanaria Sanborn 1999a: 42 (listed)
Tibicenaria Sanborn 1999b: 1 (listed) North America
Cryptotympanaria Villet and Zhao 1999: 322 (comp. note)
Cryptotympanaria Boulard 2000i: 21 (listed)
Cryptotympanaria Boulard 2001b: 22, 26, 32, 41 (comp. note)
Cryptotympanaria Boulard 2001c: 51 (listed)
Cryptotympanaria Boulard 2002a: 55 (listed)
Cryptotympanaria Boulard 2003a: 98 (listed)
Cryptotympanaria Boulard 2003c: 172 (listed)
Cryptotympanaria Moulds 2005b: 386, 391 (listed, comp. note)
Cryptotympanaria Boulard 2005h: 6 (listed)
Cryptotympanina Moulds 2005b: 391–392, 394, 396, 431 (listed, comp. note)
Cryptotympanaria Sanborn, Phillips and Sites 2007: 3, 5, Table 1 (listed) Thailand
Cryptotympanaria Mozaffarian and Sanborn 2010: 76 (listed)

***Arenopsaltria* Ashton, 1921**

Arenopsaltria Moulds 1990: 7, 32, 66 (listed, diversity, comp. note) Australia, Western Australia, South Australia, Victoria
Arenopsaltria Moulds 2001: 195–196 (comp. note) Australia
Arenopsaltria Moulds and Carver 2002: 5 (listed) Australia
Arenopsaltria Moulds 2005b: 377, 387, 391, 430–431, Table 2 (higher taxonomy, listed, comp. note) Australia
Arenopsaltria Sanborn, Moore and Young 2008: 1 (comp. note)

***A. fullo* (Walker, 1850)** = *Fidicina fullo* = *Henicopslatria fullo*

Henicopsaltria fullo Goode 1980: 55 (illustrated) Australia
Arenopsaltria fullo Gwynne and Schatral 1988: 32, 36–38, 43, 64–65, Fig 5, Fig 29 (illustrated, oscillogram, emergence time, key, listed, comp. note) Western Australia

Fidicina fullo Moulds 1990: 66 (type species of *Arenopsaltria*, listed)
Arenopsaltria fullo Moulds 1990: 66–67, Plate 8, Fig 5, Plate 24, Fig 9 (illustrated, distribution, ecology, listed, comp. note) Australia, Western Australia
Arenopsaltria fullo Naumann 1993: 50, 72, 148 (listed) Australia
Fidicina fullo Moulds and Carver 2002: 5–6 (type species of *Arenopsaltria*, type data, listed) Australia, Western Australia
Arenopsaltria fullo Moulds and Carver 2002: 6 (synonymy, distribution, ecology, type data, listed) Equals *Fidicina fullo* Australia, Western Australia
Arenopsaltria fullo Moulds 2005b: 395–396, 398, 402, 417–420, 422, Fig 35, Figs 56–60, Table 1 (higher taxonomy, phylogeny, key, listed, comp. note) Australia
Henicopsaltria fullo Robinson 2005: 217 (comp. note) Australia

***A. nubivena* (Walker, 1858)** = *Fidicina nubivena* = *Henicopsaltria nubivena*
Arenopsaltria nubivena Moulds 1990: 67, Plate 8, Fig 3, Fig 3a (illustrated, distribution, ecology, listed, comp. note) Australia, South Australia, Victoria
Arenopsaltria nubivena Naumann 1993: 20, 72, 148 (listed) Australia
Arenopsaltria nubivena Sanborn 1999a: 36 (type material, listed)
Arenopsaltria nubivena Moulds and Carver 2002: 6 (synonymy, distribution, ecology, type data, listed) Equals *Fidicina nubivena* Australia, South Australia, Victoria
Fidicina nubivena Moulds and Carver 2002: 6 (type data, listed) Australia, South Australia

***A. pygmaea* (Distant, 1904)** = *Henicopsaltria pygmaea*
Arenopsaltria pygmaea Moulds 1990: 67–68, Plate 8, Fig 4 (illustrated, distribution, ecology, listed, comp. note) Australia, Western Australia
Arenopsaltria pygmaea Naumann 1993: 47, 72, 148 (listed) Australia
Arenopsaltria pygmaea Moulds and Carver 2002: 6 (synonymy, distribution, ecology, type data, listed) Equals *Henicopsaltria pygmaea* Australia, Western Australia
Henicopsaltria pygmaea Moulds and Carver 2002: 6 (type data, listed) Australia

A. unicolor Ashton, 1921 to *Parnquila unicolor*

***Psaltoda* Stål, 1861** = *Pfaltoda* (sic)
Psaltoda spp. Josephson and Young 1981: 233 (comp. note)
Psaltoda Strümpel 1983: 18, 109, Table 8 (listed, comp. note)
Psaltoda spp. Young and Josephson 1983a: 193 (comp. note)
Psaltoda Young and Josephson 1983a: 194 (comp. note)
Psaltoda Young and Josephson 1983b: 206 (comp. note)
Psaltoda MacNally and Doolan 1986a: 35 (comp. note) New South Wales
Psaltoda Hayashi 1984: 35 (listed)
Psaltoda Moulds 1984: 27, 30–31 (key, comp. note) Australia
Psaltoda Hayashi 1987a: 124 (listed)
Psaltoda Ewart 1989a: 76–78, 80 (comp. note)
Psaltoda Ewing 1989: 36 (comp. note)
Psaltoda Moulds 1990: 7, 23, 32, 71–72, Plate 4 (listed, diversity, comp. note) Australia
Psaltoda spp. Moulds 1990: 78, 87 (comp. note)
Psaltoda Moulds and Carver 1991: 467 (comp. note)
Psaltoda Ewart 1996: 15 (listed)
Psaltoda Moulds 1996b: 92, Plates 63–64, Plate 77 (illustrated, acoustic behavior, comp. note) Australia
Psaltoda Moss and Moulds 2000: 47, 57, 60 (comp. note) Australia
Psaltoda spp. Ewart 2001b: 72, 74–75 (comp. note) Queensland
Psaltoda Boulard 2002b: 206 (comp. note) Queensland
Psaltoda Moulds 2002: 325–326, 328–330, 332–334 (key, comp. note) Australia
Psaltoda Moulds and Carver 2002: 8 (listed) Australia
Psaltoda sp. Gallagher, O'Hare, Stephenson, Waite and Vock 2003: 35, 198 (illustrated, macadamia pest, oviposition damage, comp. note) Australia
Psaltoda sp. Young 2003: 34 (illustrated, comp. note) Australia
Psaltoda Moulds 2005b: 377, 387–389, 390–391, 412–413, 430–431, Table 2 (higher taxonomy, listed, comp. note) Australia
Psaltoda Salmah, Duffels and Zaidi 2005: 15 (comp. note)
Psaltoda Shiyake 2007: 3, 30 (listed, comp. note)
Psaltoda Sanborn, Moore and Young 2008: 1 (comp. note)

***P. adonis* Ashton, 1914**
Psaltoda adonis Moulds 1984: 27, 31 (key, comp. note) Australia, Queensland
Psaltoda adonis Ewart 1989a: 79 (comp. note)
Psaltoda adonis Moulds 1990: 79–80, Plate 10, Fig 3 (illustrated, distribution, ecology, listed, comp. note) Australia, Queensland

Psaltoda adonis Moulds 2002: 326 (key) Australia
Psaltoda adonis Moulds and Carver 2002: 8 (distribution, ecology, type data, listed) Australia, Queensland

***P. antennetta* Moulds, 2002**
Psaltoda antennetta Moulds 2002: 326–331, 333–334, Figs 1–2, Figs 7–9, Fig 15 (n. sp., described, illustrated, key, distribution, comp. note) Australia, Queensland

***P. aurora* Distant, 1881**
Psaltoda aurora Moulds 1984: 27, 30–31 (key, comp. note) Australia, Queensland
Psaltoda aurora Moulds 1990: 78–79 (illustrated, distribution, ecology, listed, comp. note) Australia, Queensland
Psaltoda aurora Naumann 1993: 48, 118, 149 (listed) Australia
Psaltoda aurora Moulds 2002: 326 (key) Australia
Psaltoda aurora Moulds and Carver 2002: 9 (distribution, ecology, type data, listed) Australia, Queensland

***P. brachypennis* Moss & Moulds, 2000** = *Psaltoda* sp. nov. Ewart, 1998
Psaltoda sp. nov. Ewart 1998a: 72 (listed) Queensland
Psaltoda brachypennis Moss and Moulds 2000: 47–60, Fig 1, Fig 4, Figs 7–11, Fig 15, Tables 1–2 (n. sp., described, illustrated, oscillogram, comp. note) Australia, Queensland
Psaltoda brachypennis Moss and Popple 2000: 54, 57 (listed, comp. note) Australia, New South Wales, Queensland
Psaltoda brachypennis Moulds 2002: 326, 328, 334 (key, comp. note) Australia
Psaltoda brachypennis Moulds and Carver 2002: 9 (distribution, ecology, type data, listed) Australia, New South Wales, Queensland
Psaltoda brachypennis Popple and Ewart 2002: 113 (illustrated, distribution, ecology, song, oscillogram, comp. note) Australia, Queensland, New South Wales
Psaltoda brachypennis Williams 2002: 156–157 (listed) Australia, Queensland, New South Wales

***P. claripennis* Ashton, 1921** = *Pfaltoda* (sic) *claripennis*
Psaltoda claripennis Josephson 1981: 30–31, 34, Figs 7–8, Fig 10 (muscle function, comp. note)
Psaltoda claripennis Josephson and Young 1981: 233 (comp. note)
Psaltoda claripennis Young and Josephson 1983a: 184, 189–190, 192–193, Fig 5, Table 1 (oscillogram, timbal muscle kinetics, comp. note) Australia, Queensland
Psaltoda claripennis Josephson 1984: 90 (comp. note)
Psaltoda claripennis Moulds 1984: 27, 31 (key, comp. note) Australia, Queensland, New South Wales
Psaltoda claripennis Smith 1984: 142 (timbal muscle, comp. note)
Psaltoda claripennis Young and Josephson 1984: 286–287, Fig 1 (comp. note)
Psaltoda claripennis Claridge 1985: 303, Fig 2 (oscillogram)
Psaltoda claripennis Ewart 1986: 51, 54, Table 1 (natural history, listed, comp. note) Queensland
Psaltoda claripennis Heller 1986: 106 (comp. note)
Psaltoda claripennis Josephson and Young 1987: 995–996, Figs 2–3 (comp. note)
Psaltoda claripennis Ewart 1988: 184, 191, 195, 200–201, Fig 11A, Plate 4A (illustrated, oscillogram, natural history, listed, comp. note) Queensland
Psaltoda claripennis Ewart 1989a: 77 (comp. note) Queensland
Psaltoda claripennis Ewing 1989: 36 (comp. note)
Psaltoda claripennis Moulds 1990: 24, 73, 82, 84–86, Fig 9, Plate 9, Fig 4, Fig 4a (illustrated, oscillogram, distribution, ecology, listed, comp. note) Australia, Queensland
Psaltoda claripennis Bailey 1991b: 76 (timbal contraction rate comp. note)
Psaltoda claripennis Bennet-Clark and Young 1994: 293, Table 1 (body size and song frequency)
Psaltoda claripennis Ewart 1995: 81 (described, illustrated, natural history, acoustic behavior, oscillogram) Queensland
P[saltoda] claripennis Jiang, Yang, Tang, Xu and Chen 1995b: 229 (song frequency, comp. note)
Psaltoda claripennis Ewart 1996: 14 (listed)
Psaltoda claripennis Fonseca 1996: 29 (comp. note)
Psaltoda claripennis Moulds 1996b: 91–92, Plate 64 (illustrated, comp. note) Australia, Queensland, New South Wales
Psaltoda claripennis Woodall 1996: 44 (predation) Australia, Queensland
Psaltoda claripennis Young 1996: 100 (comp. note) Australia, Queensland
Psaltoda claripennis Ewart 1998a: 72 (listed) Queensland
Psaltoda claripennis Moss and Moulds 2000: 48–49, 53–59, Fig 3, Fig 6, Fig 10, Fig 13, Tables 1–2 (illustrated, oscillogram, comp. note) Australia
Psaltoda claripennis Moss and Popple 2000: 58 (listed, comp. note) Australia, New South Wales
Psaltoda claripennis Ewart 2001a: 500–502, 507–508, Fig 2 (acoustic behavior) Queensland

Psaltoda claripennis Ewart 2001b: 69–75, 77, 82–83, Fig 2, Fig 8, Tables 1–2 (emergence pattern, population density, nymph illustrated, predation, listed, comp. note) Queensland

Psaltoda claripennis Sanborn 2001a: 16, Fig 5 (comp. note)

Psaltoda claripennis Sueur 2001: 38, Table 1 (listed)

Psaltoda claripennis Nation 2002: 260, Table 9.3 (comp. note)

Psaltoda claripennis Moulds 2002: 326 (key) Australia

Psaltoda claripennis Moulds and Carver 2002: 9 (distribution, ecology, type data, listed) Australia, Queensland

Psaltoda claripennis Popple and Ewart 2002: 117 (illustrated) Australia, Queensland

Psaltoda claripennis Williams 2002: 156–157 (listed) Australia, Queensland, New South Wales

Pfaltoda (sic) *claripennis* Watson and Watson 2004: 139 (wing microstructure, comp. note)

Psaltoda claripennis Watson and Watson 2004: 141 (wing microstructure, comp. note)

Psaltoda claripennis Humphreys 2005: 74, Table 4 (population density, comp. note) Queensland

Psaltoda claripennis Watson, Myhra, Cribb and Watson 2006: 9 (wing microstructure, comp. note)

Pfaltoda (sic) *claripennis* Watson, Myhra, Cribb and Watson 2006: 10–11, Fig 3, Fig 5 (wing microstructure, comp. note)

Psaltoda claripennis Nation 2008: 273, Table 10.1 (comp. note)

Psaltoda claripennis Watson, Myhra, Cribb and Watson 2008: 3353 (wing microstructure, comp. note)

Pfaltoda (sic) *claripennis* Watson, Myhra, Cribb and Watson 2008: 3354, Fig 1 (wing microstructure, comp. note)

Psaltoda claripennis Rosenbluth, Szent-Györgyi and Thompson 2010: 2438 (comp. note)

Psaltoda claripennis Watson, Watson, Hu, Brown, Cribb and Myhra 2010: 116, 118–119, 122, 126, Figs 1a–d, Fig 6, Table 1 (wing microstructure, comp. note)

***P. flavescens* Distant, 1892**

Psaltoda flavescens Moulds 1984: 27, 31 (key, comp. note) Australia, Queensland

Psaltoda flavescens Moulds 1990: 74–75, Plate 11, Fig 2, Fig 2a, Plate 24, Fig 2 (illustrated, distribution, ecology, listed, comp. note) Australia, Queensland

Psaltoda flavescens Burwell 1991: 124 (distribution, comp. note) Queensland

Psaltoda flavescens Moulds 2002: 326, 334 (key, comp. note) Australia

Psaltoda flavescens Moulds and Carver 2002: 9 (distribution, ecology, type data, listed) Australia, Queensland

***P. fumipennis* Ashton, 1912**

Psaltoda fumipennis Moulds 1984: 27, 31–32 (key, comp. note) Australia, Queensland

Psaltoda fumipennis Southcott 1988: 115 (parasitism, comp. note) Australia, Queensland

Psaltoda fumipennis Moulds 1990: 80–81, Plate 9, Fig 3, Fig 3a (illustrated, distribution, ecology, type locality, listed, comp. note) Australia, Queensland

Psaltoda fumipennis Moulds 2002: 326 (key) Australia

Psaltoda fumipennis Moulds and Carver 2002: 9 (distribution, ecology, type data, listed) Australia, Queensland

***P. harrisii* (Leach, 1814)** = *Tettigonia harrisii* = *Cicada harrisii* = *Psaltoda harrissii* (sic) = *Neopsaltoda morrisi* (sic) = *Cicada dichroa* Boisduval, 1835 = *Cicada dichroma* (sic) = *Psaltoda dichroa* = *Fidicina subguttata* Walker, 1850 = *Fidicina subgutatta* (sic) = *Psaltoda longirostris* Chisholm, 1932

Psaltoda harrisii Moreira 1928: 153–154 (behavior, comp. note) Australia, New South Wales

Psaltoda harrissii (sic) Josephson 1981: 30–31, 34, Figs 7–8, Fig 10 (muscle function, comp. note)

Psaltoda harrisii Josephson and Young 1981: 220, 233 (comp. note)

Psaltoda harrisi (sic) Moulds 1983: 435 (comp. note) Australia

Psaltoda harrisii Strümpel 1983: 108, Fig 144 (oscillogram, comp. note)

Psaltoda harrisii Young and Josephson 1983a: 184, 190–194, Fig 6, Table 1 (oscillogram, timbal muscle kinetics, comp. note) Australia, New South Wales

Psaltoda harrisii Moulds 1984: 27, 31–32 (key, comp. note) Equals *Psaltoda longirostris* Australia, Queensland, New South Wales

Psaltoda longirostris Moulds 1984: 27 (comp. note) Australia

Psaltoda harrisii Young and Josephson 1984: 286, Fig 1 (comp. note)

Psaltoda harrisii Claridge 1985: 303, Fig 2 (oscillogram)

Psaltoda harrissii (sic) Popov 1985: 45 (acoustic system, comp. note)

Psaltoda harrisii Ewart 1986: 51, 54, Table 1 (natural history, listed, comp. note) Queensland, New South Wales

Psaltoda harrisii MacNally and Doolan 1986a: 35, 37–38, 40–43, Figs 1–4, Table 2 (habitat, comp. note) New South Wales

Psaltoda harrisii MacNally and Doolan 1986b: 281, 283–291, Figs 1–4, Tables 2–3, Table 5 (habitat, comp. note) New South Wales
Psaltoda harrisi Josephson and Young 1987: 995–996, Figs 2–3 (comp. note)
Psaltoda harrisii Ewart 1988: 184 (comp. note) Queensland, New South Wales
Psaltoda harrissii (sic) Ewart 1988: 200–201, Fig 11C (oscillogram, comp. note) Queensland
Psaltoda harrisii MacNally 1988a: 247 (distribution model, comp. note) New South Wales
Psaltoda harrisii MacNally 1988b: 1976–1977, Tables 1–2 (comp. note) New South Wales
Psaltoda harrisii Ewart 1989a: 77 (comp. note) Queensland
Tettigonia harrisii Ewart 1990: 4 (type material, type locality) Australia
Psaltoda harrisii Moulds 1990: 10, 28, 82–85, Plate 10, Fig 2, Fig 2a (synonymy, distribution, ecology, type data, listed) Equals *Tettigonia harrisii* Equals *Cicada dichroa* Equals *Fidicina subguttata* Australia, New South Wales, Queensland
Tettigonia harrisii Moulds 1990: 28 (comp. note) Equals *Cicada dichroa* Equals *Fidicina subguttata*
Cicada dichroa Moulds 1990: 28 (comp. note)
Fidicina subgutata Moulds 1990: 28 (comp. note)
Psaltoda harrisii Bailey 1991a: 48 (oscillogram, comp. note)
Psaltoda harrisii Boulard 1991c: 118 (comp. note) Equals *Cicada dichroa* Australia
Psaltoda harrisii Naumann 1993: 64, 118, 149 (listed) Australia
Psaltoda harrisii Ewart 1995: 81 (described, illustrated, natural history, acoustic behavior, oscillogram) Queensland
P[saltoda] harrisii Jiang, Yang, Tang, Xu and Chen 1995b: 229 (song frequency, comp. note)
Psaltoda harrisii Ewart 1996: 14 (listed)
Psaltoda harrisii Moulds 1996b: 91 (comp. note) Australia, Queensland, New South Wales
Psaltoda harrisii Sanborn 1999a: 36 (type material, listed)
Psaltoda harrisii Moss 2000: 69 (comp. note) Queensland
Psaltoda harrisii Moss and Moulds 2000: 53–59, Fig 10, Fig 14, Tables 1–2 (oscillogram, comp. note) Australia
Psaltoda harrisii Moss and Popple 2000: 57 (listed, comp. note) Australia, New South Wales
Psaltoda harrisii Ewart 2001a: 500, Fig 2 (acoustic behavior) Queensland
Psaltoda harrisii Ewart 2001b: 69–75, 78, 82, Fig 3, Fig 9, Tables 1–2 (emergence pattern, population density, nymph illustrated, predation, listed, comp. note) Queensland
Psaltoda harrisii Sanborn 2001a: 16, Fig 5 (comp. note)
Psaltoda harrisii Sueur 2001: 38, Table 1 (listed)
Psaltoda harrisii Moulds 2002: 326 (key) Australia
Psaltoda harrisii Nation 2002: 260, Table 9.3 (comp. note)
Psaltoda harrisii Moulds and Carver 2002: 10 (synonymy, distribution, ecology, type data, listed) Equals *Tettigonia harrisii* Equals *Cicada dichroa* Equals *Fidicina subguttata* Equals *Psaltoda longirostris* Australia, New South Wales, Queensland
Tettigonia harrisii Moulds and Carver 2002: 10 (type data, listed) New Holland
Cicada dichroa Moulds and Carver 2002: 10 (type data, listed) New South Wales
Fidicina subgutata Moulds and Carver 2002: 10 (type data, listed) New Holland
Psaltoda longirostris Moulds and Carver 2002: 10 (type data, listed) New South Wales
Psaltoda harrisii Popple and Ewart 2002: 117 (illustrated) Australia, Queensland
Psaltoda harrisii Williams 2002: 156–157 (listed) Australia, Queensland, New South Wales
Psaltoda harrisii Emery, Emery, Emery and Popple 2005: 102–105, 107, Table 1–3 (comp. note) New South Wales
Psaltoda harrisii Shiyake 2007: 3, 28, 30, Fig 31 (illustrated, distribution, listed, comp. note) Australia
Psaltoda harrisii Nation 2008: 273, Table 10.1 (comp. note)

***P. insularis* Ashton, 1914**

Psaltoda insularis Moulds 1984: 27, 31–32 (key, comp. note) Australia, Lord Howe Island
Psaltoda insularis Moulds 1990: 86, Plate 9, Fig 5 (illustrated, distribution, ecology, listed, comp. note) Australia, Lord Howe Island
Psaltoda insularis Naumann 1993: 35, 118, 149 (listed) Australia
Psaltoda insularis Moulds 2002: 326 (key) Australia
Psaltoda insularis Moulds and Carver 2002: 10 (distribution, ecology, type data, listed) Australia, Lord Howe Island

***P. macallumi* Moulds, 2002**

Psaltoda maccallumi Moulds 2002: 326–328, 330–331, 333–334, Figs 3–4, Figs 10–11, Fig 14 (n. sp., described, illustrated, key, distribution, comp. note) Australia, Queensland

***P. magnifica* Moulds, 1984**

Psaltoda magnifica Moulds 1984: 27–31, Figs 1–5 (n. sp., described, illustrated, key, distribution, comp. note) Australia, Queensland

Psaltoda magnifica Moulds 1990: 73, 79, 82, Plate 9, Fig 2, Fig 2a (illustrated, distribution, ecology, listed, comp. note) Australia, Queensland

Psaltoda magnifica Naumann 1993: 26, 118, 149 (listed) Australia

Psaltoda magnifica Moulds 2002: 326 (key) Australia

Psaltoda magnifica Moulds and Carver 2002: 10 (distribution, ecology, type data, listed) Australia, Queensland

***P. moerens* (Germar, 1834)** = *Cicada moerens* = *Psaltoda meorens* (sic) = *Psaltoda moerans* (sic)

Psaltoda moerens Greenup 1964: 23 (comp. note) Australia, New South Wales

Psaltoda moerens Humphreys and Mitchell 1983: 70, Table 2 (turrets, population density, comp. note) New South Wales

Psaltoda moerens Moulds 1983: 429, 433 (illustrated, comp. note) Australia, Queensland, New South Wales

Psaltoda moerens Moulds 1984: 27, 30–32 (key, comp. note) Australia, Queensland, New South Wales, Victoria, South Australia

Psaltoda moerens Claridge 1985: 303, Fig 2 (oscillogram)

Psaltoda moerens Brunet 1986: 44 (comp. note) New South Wales

Psaltoda moerens Ewart 1986: 51–53, Fig 2B, Table 1 (illustrated, natural history, listed, comp. note) Queensland, New South Wales, Victoria, South Australia, Tasmania

Psaltoda moerens MacNally and Doolan 1986a: 35, 37–38, 40–43, Figs 1–4, Table 2 (habitat, comp. note) New South Wales

Psaltoda moerens MacNally and Doolan 1986b: 281, 283–291, Figs 1–4, Tables 2–3, Table 5 (habitat, comp. note) New South Wales

Psaltoda moerens MacNally 1988a: 247 (distribution model, comp. note) New South Wales

Psaltoda moerens MacNally 1988b: 1976–1977, Tables 1–2 (comp. note) New South Wales

Psaltoda moerens Humphreys 1989: 106 (comp. note) New South Wales

Psaltoda moerens Moss 1989: 105 (comp. note)

Cicada moerens Ewart 1990: 1, 3 (type material, type locality) Equals *Psaltoda moerens* Australia

Psaltoda moerens Moss 1990: 7 (comp. note)

Cicada moerens Moulds 1990: 74 (type species of *Psaltoda*, listed)

Psaltoda moerens Moulds 1990: 3, 5, 7, 10, 12, 16, 22, 75–79, 82, Plate 3, Fig 4, Plate 10, Fig 5, Fig 5a (illustrated, distribution, ecology, listed, comp. note) Equals *Cicada moerens* Australia, New South Wales, Queensland, South Australia, Tasmania, Victoria

Psaltoda moerens Lepschi 1993: 197 (predation) New South Wales

Psaltoda moerens Naumann 1993: 48, 118, 149 (listed) Australia

Psaltoda moerens Paton, Humphreys and Mitchell 1995: 54, Table 3.5 (burrowing, listed) Australia, New South Wales

Psaltoda moerens Coombs 1996: 57, Table 1 (emergence time, comp. note) New South Wales

Psaltoda moerans (sic) New 1996: 95 (comp. note) Australia, Tasmania

Psaltoda moerens Lepschi 1997: 86 (predation) New South Wales

Psaltoda moerens Steinbauer 1997: 169–171, Figs 1–2 (life history, comp. note) Australia, Tasmania

Psaltoda moerens Sanborn 1999a: 36 (type material, listed)

Psaltoda moerens Moss and Popple 2000: 58 (listed, comp. note) Australia, New South Wales

Psaltoda moerens Sueur 2001: 38, Table 1 (listed)

Psaltoda moerens Moulds 2002: 326 (key) Australia

Cicada moerens Moulds and Carver 2002: 8, 10 (type species of *Psaltoda*, type data, listed) Australiasia

Psaltoda moerens Moulds and Carver 2002: 10–11 (synonymy, distribution, ecology, type data, listed) Equals *Cicada moerens* Australia, New South Wales, Queensland, South Australia, Tasmania, Victoria

Psaltoda moerens Williams 2002: 156–157 (listed) Australia, Queensland, New South Wales, Victoria, South Australia, Tasmania

Psaltoda moerens Emery, Emery, Emery and Popple 2005: 102–105, 107, Tables 1–3 (comp. note) New South Wales

Psaltoda meorens (sic) Haywood 2005: 9, Table 1 (listed) South Australia

Psaltoda moerens Haywood 2005: 13–15, Fig 3 (illustrated, distribution, described, habitat, song, comp. note) South Australia, New South Wales, Queensland, Victoria

Psaltoda moerens Humphreys 2005: 69, 71, 73–74, Table 2, Table 4 (population density, comp. note) New South Wales

Psaltoda moerens Moulds 2005b: 395–396, 399, 409, 417–420, 422, Fig 51, Figs 56–60, Table 1 (higher taxonomy, phylogeny, key, listed, comp. note) Australia

Psaltoda moerens Haywood 2006a: 14–15, 23, Tables 1–2 (illustrated, distribution, described, habitat, song, listed, comp. note) South Australia, New South Wales, Queensland, Victoria

Psaltoda moerens Shiyake 2007: 3, 28, 30, Fig 32 (illustrated, distribution, listed, comp. note) Australia

***P. mossi* Moulds, 2002** = *Psaltoda* sp. A Ewart, 1988 = *Psaltoda* sp. nov. Lithgow, 1988

Psaltoda sp. A Ewart 1988: 184, 191, 195, 200–201, Fig 11B, Plate 4B (illustrated, oscillogram, natural history, listed, comp. note) Queensland

Psaltoda sp. nov. Lithgow 1988: 65 (listed) Australia, Queensland

Psaltoda sp. A Ewart 1998a: 72 (listed) Equals *Psaltoda* sp. A. Ewart 1988 Queensland

Psaltoda mossi Moulds 2002: 326-327, 330-334, Figs 5–6 (n. sp., described, illustrated, key, synonymy, distribution, comp. note) Equals *Psaltoda* sp. A Ewart. 1988 Australia, Queensland

Psaltoda sp. A Popple and Strange 2002: 28, Table 1 (listed, habitat) Australia, Queensland

***P. pictibasis* (Walker, 1858)** = *Cicada pictibasis*

Psaltoda pictibasis Moulds 1984: 27, 31–32 (key, comp. note) Australia, Queensland, New South Wales

Psaltoda pictibasis Ewart 1988: 184, 191–192, 198–199, Fig 10D, Plate 1A (illustrated, oscillogram, natural history, listed, comp. note) Queensland, New South Wales

Psaltoda pictibasis Lithgow 1988: 65 (listed) Australia, Queensland

Psaltoda pictibasis Moulds 1990: 76–78, 82, Plate 10, Fig 4, Fig 4a (illustrated, distribution, ecology, listed, comp. note) Australia, New South Wales, Queensland

Psaltoda pictibasis Naumann 1993: 6, 118, 149 (listed) Australia

Psaltoda pictibasis Ewart 1995: 82 (described, illustrated, natural history, acoustic behavior, oscillogram) Queensland

Psaltoda pictibasis Ewart 1998a: 72 (listed) Queensland

Psaltoda pictibasis Sueur 2001: 38, Table 1 (listed)

Psaltoda pictibasis Moulds 2002: 326 (key) Australia

Psaltoda pictibasis Moulds and Carver 2002: 11 (synonymy, distribution, ecology, type data, listed) Equals *Cicada pictibasis* Australia, New South Wales, Queensland

Psaltoda pictibasis Popple and Ewart 2002: 117 (illustrated) Australia, Queensland

Psaltoda pictibasis Popple and Strange 2002: 28, Table 1 (listed, habitat) Australia, Queensland

Psaltoda pictibasis Williams 2002: 156–157 (listed) Australia, Queensland, New South Wales

Psaltoda pictibasis Shiyake 2007: 3, 28, 30, Fig 33 (illustrated, distribution, listed, comp. note) Australia

***P. plaga* (Walker, 1850)** = *Cicada plaga* = *Cicada argentata* Germar, 1834 = *Psaltoda argentata*

Psaltoda argentata Josephson 1981: 30–31, 34, Figs 7–8, Fig 10 (muscle function, comp. note)

Psaltoda argentata Josephson and Young 1981: 233, Fig 8 (comp. note)

Psaltoda argentata MacNally and Young 1981: 192–193, Table 3 (song energetics, comp. note) New South Wales

Psaltoda plaga Moulds 1983: 429, 435 (illustrated, comp. note) Australia, New South Wales

Psaltoda argentata Young and Josephson 1983a: 184, 187–189, 192–194, Fig 4, Table 1 (oscillogram, timbal muscle kinetics, comp. note) Australia, New South Wales

Psaltoda plaga Moulds 1984: 27, 31–32 (key, comp. note) Australia, Queensland, New South Wales

Psaltoda argentata Young and Josephson 1984: 286, Fig 1 (comp. note)

Psaltoda argentata Claridge 1985: 303, Fig 2 (oscillogram)

Psaltoda argentata Popov 1985: 47 (acoustic system, comp. note)

Psaltoda argentata MacNally and Doolan 1986a: 35, 37–38, 40–43, Figs 1–4, Table 2 (habitat, comp. note) New South Wales

Psaltoda argentata MacNally and Doolan 1986b: 281, 284–288, 291, Figs 1–4, Tables 2–3, Table 5 (habitat, comp. note) New South Wales

Psaltoda argentata Josephson and Young 1987: 995–996, Figs 2–3 (comp. note)

Psaltoda argentata Villet 1987a: 269 (comp. note) Australia

Psaltoda argentata MacNally 1988a: 247, 250 (distribution model, comp. note) New South Wales

Psaltoda argentata MacNally 1988b: 1976–1977, Tables 1–2 (comp. note) New South Wales

Psaltoda plaga Ewart 1989a: 79 (comp. note)

Psaltoda argentata Ewing 1989: 36 (comp. note)

Cicada argentata Ewart 1990: 1, 3 (type material, type locality) Australia

Psaltoda plaga Moulds 1990: 3, 7, 10, 19, 24, 73, 76, 79, 82, 84–85, Fig 8, Plate 4, Fig 3 (oscillogram, illustrated, synonymy, distribution, ecology, listed, comp. note) Equals *Cicada argentata* Equals *Cicada plaga* Equals *Psaltoda argentata* Australia, New South Wales, Queensland

Cicada argentata Moulds 1990: 81 (listed)

Cicada plaga Moulds 1990: 81 (listed)

Psaltoda argentata Moulds 1990: 81 (status)

Psaltoda argentata Young 1990: 55 (comp. note)

Psaltoda plaga Secomb 1992: 29 (predation) Australia, New South Wales
Psaltoda plaga Naumann 1993: 6, 118, 149 (listed) Australia
Psaltoda plaga Brunet 1995: 22 (listed) Australia
P[saltoda] argentata Jiang, Yang, Tang, Xu and Chen 1995b: 229 (song frequency, comp. note)
Psaltoda plaga Coombs 1996: 57, Table 1 (emergence time, comp. note) New South Wales
Psaltoda plaga Moulds 1996b: 91, Plate 63 (illustrated, comp. note) Australia, Queensland, New South Wales
Psaltoda plaga New 1996: 95 (comp. note) Australia
Psaltoda plaga Moss and Moulds 2000: 48–49, 53–59, Fig 2, Fig 5, Fig 10, Fig 12, Tables 1–2 (illustrated, oscillogram, comp. note) Australia
Psaltoda plaga Moss and Popple 2000: 57 (listed, comp. note) Australia, New South Wales
Psaltoda plaga Ewart 2001a: 499–500, Fig 1 (acoustic behavior) Queensland
Psaltoda plaga Ewart 2001b: 69–76, 82–83, Fig 1, Fig 7, Tables 1–2 (emergence pattern, population density, nymph illustrated, predation, listed, comp. note) Equals *Psaltoda argentata* Young and Josephson 1983 Queensland
Psaltoda argentata Sanborn 2001a: 16, Fig 5 (comp. note)
Psaltoda plaga Sueur 2001: 38, Table 1 (listed)
Psaltoda plaga Moulds 2002: 326, 332, 334 (key, comp. note) Australia
Psaltoda argentata Nation 2002: 260, Table 9.3 (comp. note)
Psaltoda plaga Moulds and Carver 2002: 11 (synonymy, distribution, ecology, type data, listed) Equals *Cicada argentata* Equals *Cicada plaga* Australia, New South Wales, Queensland
Cicada argentata Moulds and Carver 2002: 11 (type data, listed) Australiasia
Cicada plaga Moulds and Carver 2002: 11 (type data, listed) New Holland
Psaltoda plaga Popple and Ewart 2002: 113 (illustrated, distribution, ecology, song, oscillogram, comp. note) Australia, Queensland, New South Wales
Psaltoda plaga Williams 2002: 156–157 (listed) Australia, Queensland, New South Wales
Psaltoda argentata Stokes and Josephson 2004: 281 (timbal muscle, comp. note)
Psaltoda plaga Emery, Emery, Emery and Popple 2005: 102–107, Tables 1–3 (comp. note) New South Wales
Psaltoda plaga Humphreys 2005: 74, Table 4 (population density, comp. note) Queensland
Psaltoda plaga Shiyake 2007: 3, 28, 30, Fig 34 (illustrated, distribution, listed, comp. note) Australia
Psaltoda plaga Emery 2008: 34 (life cycle, comp. note) New South Wales
Psaltoda argentata Nation 2008: 273, Table 10.1 (comp. note)

P. plebeia Goding & Froggatt, 1904 to *Tibicen plebejus* (Scopoli, 1763)

***Anapsaltoda* Ashton, 1921** = *Anapsaltodea* (sic)
Anapsaltoda Moulds 1990: 32, 74 (listed, diversity, comp. note) Australia
Anapsaltoda Moulds 2002: 329 (comp. note) Australia
Anapsaltoda Moulds and Carver 2002: 5 (listed) Australia
Anapsaltoda Moulds 2005b: 377, 387–389, 390–391, 412–413, 423, 430–431, Table 2 (higher taxonomy, listed, comp. note) Australia
Anapsaltoda Salmah, Duffels and Zaidi 2005: 15 (comp. note)

***A. pulchra* (Ashton, 1921)** = *Anapsaltodea* (sic) *pulchra* = *Psaltoda pulchra*
Psaltoda pulchra Moulds 1990: 71 (type species of *Anapsaltoda*, listed)
Anapsaltoda pulchra Moulds 1990: 71–72, Plate 11, Fig 1, Fig 1a, Plate 24, Fig 5 (synonymy, distribution, ecology, type data, listed) Equals *Psaltoda pulchra* Australia, Queensland
Anapsaltodea (sic) *pulchra* Naumann 1993: 25, 70, 148 (listed) Australia
Psaltoda pulchra Moulds and Carver 2002: 5 (type species of *Anapsaltoda*, type data, listed) Australia, Queensland
Anapsaltoda pulchra Moulds and Carver 2002: 5 (synonymy, distribution, ecology, type data, listed) Equals *Psaltoda pulchra* Australia, Queensland
Anapsaltoda pulchra Moulds 2005b: 395–396, 398, 405, 408, 417–420, 422, Fig 38, Fig 47, Figs 56–60, Table 1 (illustrated, higher taxonomy, phylogeny, key, listed, comp. note) Australia

***Neopsaltoda* Distant, 1910**
Neopsaltoda Moulds 1990: 32, 72, 74 (listed, diversity, comp. note) Australia
Neopsaltoda Moulds and Carver 2002: 8 (listed) Australia
Neopsaltoda Moulds 2005b: 377, 387–389, 390–391, 412–413, 430–431, Table 2 (higher taxonomy, listed, comp. note) Australia

Neopsaltoda Salmah, Duffels and Zaidi 2005: 15 (comp. note)

N. *crassa* Distant, 1910

Neopsaltoda crassa Moulds 1990: 12, 71–73, 82, 85 (type species of *Neopsaltoda*, distribution, ecology, listed) Australia, Queensland

Neopsaltoda crassa Moulds 2002: 329, 344 (comp. note) Australia

Neopsaltoda crassa Moulds and Carver 2002: 8 (type species of *Neopsaltoda*, distribution, ecology, type data, listed) Australia, Queensland

Neopsaltoda crassa Moulds 2005b: 381, 395–396, 399, 406, 417–420, 422, Figs 14–15, Figs 56–60, Table 1 (illustrated, higher taxonomy, phylogeny, key, listed, comp. note) Australia

***Henicopsaltria* Stål, 1866**

Henicopsaltria Ewart 1989a: 80 (comp. note)

Henicopsaltria Moulds 1990: 7, 32, 68 (listed, diversity, comp. note) Australia

Henicopsaltria Moulds 1993: 23 (diversity, comp. note) Australia

Henicopsaltria spp. Moulds 1993: 24 (comp. note)

Henicopsaltria Novotny and Wilson 1997: 437 (listed)

Henicopsaltria Moulds and Carver 2002: 7 (listed) Australia

Henicopsaltria Moulds 2005b: 377, 387–389, 391, 413, 423, 430–431, Table 2 (higher taxonomy, listed, comp. note) Australia

Henicopsaltria Robinson 2005: 217 (comp. note) Australia

Henicopsaltria Shiyake 2007: 3, 30 (listed, comp. note)

***H. danielsi* Moulds, 1993**

Henicopsaltria danielsi Moulds 1993: 23–26, Figs 1–3 (n. sp., described, illustrated, key, comp. note) Australia, Queensland

Henicopsaltria danielsi Moulds and Carver 2002: 7 (distribution, ecology, type data, listed) Australia, Queensland

***H. eydouxii* (Guérin-Méneville, 1838)** = *Cicada eydouxii* = *Henicopsaltria eydouxi* (sic) = *Henicopsaltria edouxi* (sic) = *Henicopsaltria eudouxi* (sic) = *Psaltoda flavescens* Froggatt (nec Distant)

Henicopsaltria eydouxii Moulds 1983: 430, 433 (illustrated, comp. note) Australia, Queensland, New South Wales

Henicopsaltria eydouxi (sic) Claridge 1985: 303, Fig 2 (oscillogram)

Henicopsaltria eudouxi (sic) Popov 1985: 43 (acoustic system, comp. note)

Henicopsaltria eydouxii Ewart 1986: 51, 54, Table 1 (natural history, listed, comp. note) Queensland, New South Wales

Henicopsaltria eydouxi (sic) MacNally and Doolan 1986a: 35, 37, 40–42, Figs 1–4, Table 2 (habitat, comp. note) New South Wales

Henicopsaltria (sic) MacNally and Doolan 1986a: 37–39, 41, 43 (comp. note) New South Wales

Henicopsaltria eydouxi (sic) MacNally and Doolan 1986b: 281, 285–286, 288, Fig 2, Fig 4, Table 2 (habitat, comp. note) New South Wales

Henicopsaltria (sic) MacNally and Doolan 1986b: 283–291, Figs 1–3, Table 3, Table 5 (habitat, comp. note) New South Wales

Henicopsaltria eydouxi (sic) MacNally 1988a: 247 (distribution model, comp. note) New South Wales

Henicopsaltria eydouxi (sic) MacNally 1988b: 1976–1977, 1980, Tables 1–2 (comp. note) New South Wales

Cicada eydouxii Moulds 1990: 68 (type species of *Henicopsaltria*, listed)

Henicopsaltria eydouxii Moulds 1990: 5, 9, 22, 25, 68–71, Plate 11, Fig 5, Fig 5a, Plate 24, Fig 1 (illustrated, distribution, ecology, listed, comp. note) Australia, New South Wales, Queensland

Henicopsaltria eydouxii Moulds 1993: 23, 26 (key, distribution, comp. note) Australia, Queensland, New South Wales

Henicopsaltria eydouxii Naumann 1993: 47, 95, 149 (listed) Australia

Henicopsaltria eydouxii Ewart 1995: 80 (described, illustrated, natural history, acoustic behavior, oscillogram) Queensland

Henicopsaltria eydouxii O'Toole 1995: 164–165 (illustrated)

Henicopsaltria eydouxii Moulds 1996b: 94 (acoustic behavior, comp.

Henicopsaltria eydouxii Young 1996: 97, 99, 103, Fig 5.2 (oscillogram, acoustic behavior, comp. note)

Henicopsaltria eydouxii Curtis 1999: 5656 (life cycle, comp. note) Australia

Henicopsaltria eydouxii Sanborn 1999a: 36 (type material, listed)

Henicopsaltria eydouxii Hangay and German 2000: 59 (illustrated, behavior) Queensland, New South Wales

Henicopsaltria eydouxi Moss 2000: 69 (comp. note) Queensland

Henicopsaltria eydouxii Moss and Popple 2000: 57 (listed, comp. note) Australia, New South Wales

Henicopsaltria eydouxi (sic) Sueur 2001: 38, Table 1 (listed)

Cicada eydouxii Moulds and Carver 2002: 7 (type species of *Henicopsaltria*, type data, listed)

Henicopsaltria eydouxii Moulds and Carver 2002: 7–8 (synonymy, distribution, ecology, type data, listed) Equals *Cicada eydouxii* Equals *Psaltoda flavescens* Froggatt 1895 Australia, New South Wales, Queensland

Psaltoda flavescens Froggatt 1895 Moulds and Carver 2002: 7 (synonymy, distribution, ecology, type data, listed) Australia

Henicopsaltria eydouxii Popple and Ewart 2002: 116 (illustrated) Australia, Queensland

Henicopsaltria eydouxii Williams 2002: 156–157 (listed) Australia, Queensland, New South Wales

Henicopsaltria eydouyxii Emery, Emery, Emery and Popple 2005: 102-103-105, 107, Tables 1–3 (comp. note) New South Wales

Henicopsaltria eydouxii Moulds 2005b: 395–396, 398, 417–420, 422, Figs 56–60, Table 1 (higher taxonomy, phylogeny, key, listed, comp. note) Australia

Henicopsaltrai eydouxii (sic) Robinson 2005: 217 (comp. note) Australia

Henicopsaltria eydouxii Shiyake 2007: 3, 29–30, Fig 35 (illustrated, distribution, listed, comp. note) Australia

H. fullo Walker, 1850 to *Arenopsaltria fullo*

***H. kelsalli* Distant, 1910**

Henicopsaltria kelsalli Moulds 1990: 70–71, Plate 11, Fig 3, Fig 3a (distribution, ecology, type data, listed) Australia, Queensland

Henicopsaltria kelsalli Moulds 1993: 23, 26 (key, comp. note) Australia

Henicopsaltria kelsalli Moulds and Carver 2002: 8 (distribution, ecology, type data, listed) Australia, Queensland

***H. rufivelum* Moulds, 1978**

Henicopsaltria rufivelum Moulds 1990: 68–70, Plate 11, Fig 4, Fig 4a (illustrated, distribution, ecology, listed, comp. note) Australia, Queensland

Henicopsaltria rufivelum Moulds 1993: 23, 26 (key, comp. note) Australia

Henicopsaltria rufivelum Moulds and Carver 2002: 8 (distribution, ecology, type data, listed) Australia, Queensland

Henicopsaltria rufivelum Moulds 2005b: 395, 398, 417–420, 422, Figs 56–60, Table 1 (higher taxonomy, phylogeny, key, listed, comp. note) Australia

***Illyria* Moulds, 1985**

Illyria Moulds 1985: 25–27, 29, 31–34 (n. gen., described, key, distribution, comp. note) Australia

Illyria Moulds 1990: 8, 32, 107 (listed, diversity, comp. note) Australia

Illyria Moulds and Carver 2002: 3 (listed) Australia

Illyria Moulds 2005b: 387, 390 (higher taxonomy, listed, comp. note) Australia

Illyria Wei, Ahmed and Rizvi 2010: 33 (comp. note) Australia

***I. australensis* (Kirkaldy, 1910)** = *Cicada interrupta* Walker, 1850 (nec *Cicada interrupta* Linnaeus, 1758)

Cicada interrupta Moulds 1985: 25, 34 (comp. note) Equals Australia

Cicada australensis Moulds 1985: 25, 34 (comp. note) Equals *Cicada innterrupta* Australia

Tettigia australensis Moulds 1985: 34 (comp. note) Equals *Cicada interrupta* Walker

Illyria australensis Moulds 1985: 26, 29, 31, 33–35, Fig 7, Figs 11–13, Fig 21 (n. comb., synonymy, described, key, illustrated, distribution, comp. note) Equals *Cicada interrupta* Equals *Tibicen interruptus* Equals *Tettigia interrupta* Equals *Tettigia australensis* Equals *Cicada australensis* Australia, Western Australia, Northern Territory

Illyria australensis group Moulds 1985: 34 (comp. note) Australia

Illyria australensis Moulds 1990: 109–110, Plate 14, Fig 11 (illustrated, distribution, ecology, listed, comp. note) Equals *Cicada interrupta* Equals *Tettigia australensis* Australia, Northern Territory, Western Australia

Illyria australensis Moulds and Carver 2002: 3 (synonymy, distribution, ecology, type data, listed) Equals *Cicada interrupta* Equals *Tettigia australensis* Australia, Northern Territory, Western Australia

Cicada interrupta Moulds and Carver 2002: 3 (type data, listed) Australia

Tettigia australensis Moulds and Carver 2002: 3 (type data, listed) Australia

***I. burkei* (Distant, 1882)** = *Tibicen burkei* = *Cicada burkei* = *Tettigia variegata* Goding & Froggatt, 1904

Tettigia variegata Hahn 1962: 10 (listed) Australia, Queensland

Tibicen burkei Moulds 1985: 25, 27 (type species of *Illyria*, comp. note) Australia

Tettigia burkei Moulds 1985: 25 (comp. note) Australia

Illyria burkei Moulds 1985: 26–29, 31, 33–35, Figs 3–4, Figs 17–20 (n. comb., lectotype designation, synonymy, described, illustrated, key, distribution, comp. note) Equals *Tibicen burkei* Equals *Tettigia variegata* Equals *Tettigia burkei* Equals *Cicada burkei* Australia, Queensland, Western Australia, Northern Territory

Tettigia variegata Moulds 1985: 27 (comp. note) Australia

Illyria burkei species group Moulds 1985: 34–35 (comp. note) Australia
Tettigia variegata Stevens and Carver 1986: 265 (type material) Australia, Queensland
Illyria burkei Ewart 1988: 185, 191–192, 198–199, Fig 10H, Plate 1C (illustrated, oscillogram, listed, comp. note) Queensland, Northern Territory
Illyria burkei Lithgow 1988: 65 (listed) Australia, Queensland
Illyria burkei Carver 1990: 260 (authority)
Tibicen burkei Moulds 1990: 107 (type species of *Illyria*, listed)
Illyria burkei Moulds 1990: 108–110, Plate 4, Fig 4, Plate 14, Fig 9 (illustrated, distribution, ecology, listed, comp. note) Australia, Northern Territory, Queensland, Western Australia
Illyria burkei Burwell 1991: 124 (distribution, comp. note) Queensland
Illyria burkei Ewart 1993: 137, 141, Fig 1 (natural history, acoustic behavior, song analysis) Queensland
Illyria burkei Ewart 1998a: 62–63, Figs 1–2 (described, natural history, song) Queensland
Illyria burkei Ewart 1998b: 135 (listed) Queensland
Illyria burkei Sueur 2001: 40, Table 1 (listed)
Illyria burkei Moulds and Carver 2002: 4 (synonymy, distribution, ecology, type data, listed) Equals *Tibicen burkei* Equals *Tettigia variegata* Australia, Northern Territory, Queensland, Western Australia
Tibicen burkei Moulds and Carver 2002: 4 (type data, listed) Australia, Queensland
Tettigia variegata Moulds and Carver 2002: 4 (type data, listed) Australia, Queensland
Illyria burkei Popple and Strange 2002: 28, Table 1 (listed, habitat) Australia, Queensland
Illyria burkei Sanders 2004: 79–80 (behavior, comp. note) Australia, Queensland
Illyria burkei Ewart 2005a: 174–176, 179, Fig 6A (described, song, oscillogram, comp. note) Queensland
Illyria burkei Moulds 2005b: 395–396, 398, 407, 417–419, 422, Fig 46, Figs 56–60, Table 1 (illustrated, higher taxonomy, phylogeny, key, listed, comp. note) Australia

***I. hilli* (Ashton, 1914)** = *Tettigia hilli* = *Cicada hilli*
Tettigia hilli Moulds 1985: 25 (comp. note) Australia
Illyria hilli Moulds 1985: 26, 29, 31–35, Figs 5–6, Figs 14–16, Fig 21 (n. comb., synonymy, described, illustrated, key, distribution, comp. note) Equals *Tettigia hilli* Equals *Cicada hilli* Australia, Western Australia, Northern Territory
Illyria hilli Moulds 1990: 105, 109–110, Plate 14, Fig 10, Fig 10a, Plate 24, Fig 21 (illustrated, distribution, ecology, listed, comp. note) Australia, Northern Territory, Western Australia
Illyria hilli Moulds and Carver 2002: 4 (synonymy, distribution, ecology type data, listed) Equals *Tettigia hilli* Australia, Northern Territory, Western Australia
Tettigia hilli Moulds and Carver 2002: 4 (type data, listed) Australia, Northern Territory

***I. major* Moulds, 1985**
Illyria major Moulds 1985: 26–35, Figs 1–2, Figs 8–10, Fig 20 (n. sp., described, key, illustrated, distribution, comp. note) Australia, Western Australia, Northern Territory
Illyria major Moulds 1990: 107–109 (distribution, ecology, listed, comp. note) Australia, Northern Territory, Western Australia
Illyria major Moulds and Carver 2002: 4 (distribution, ecology, type data, listed) Australia, Northern Territory, Western Australia

***Macrotristria* Stål, 1870** = *Macrotristia* (sic) = *Macruouistria* (sic) = *Marcrotristria* (sic) = *Cicada (Macrotristria)* = *Cicada (Macrotistria)* (sic)
Macrotristria Hayashi 1984: 35 (listed)
Macrostristria Hayashi 1987a: 124 (listed)
Macrotristria Ewart 1989a: 80 (comp. note)
Macrotristria Moulds 1990: 7–8, 32, 86–87, 90, 102 (listed, diversity, comp. note) Australia
Macrotristria Moulds and Carver 1991: 467 (comp. note) Australia
Macrotristria Bennet-Clark and Young 1992: 127, 141, 147 (comp. note) Australia, Queensland
Macrotristria Moulds 1992: 133–134, 138 (diversity, comp. note) Australia
Macrotristria Boulard 1996c: 96, 100 (characteristics, comp. note)
Macrotristria Ewart 1996: 14 (listed)
Macrotristria sp. Preston-Mafham and Preston-Mafham 2000: 42–43 (illustrated)
Macrotristria spp. Ewart and Popple 2001: 65 (comp. note) Queensland
Macrotristria Moulds 2001: 196 (comp. note) Australia
Macrotristria Moulds and Carver 2002: 14 (listed) Australia
Macrotristria spp. Ewart 2005a: 175, 179 (comp. note)
Macrotristria Moulds 2005b: 387–391, 423, 429–431, Table 2 (higher taxonomy, listed, comp. note) Australia

Macrotristria Robinson 2005: 217 (comp. note) Australia
Macrotristria Shiyake 2007: 3, 31 (listed, comp. note)
Macrotristria Sanborn, Moore and Young 2008: 1 (comp. note)

***M. angularis* (Germar, 1834)** = *Cicada angularis* = *Fidicina angularis* = *Cicada (Macriotristria) angularis* = *Macruouistria* (sic) *angularis* = *Macrotristria augularis* (sic) = *Macrotristra* (sic) *angularis* = *Marcrotristria* (sic) *angularis*

Macrotristria angularis Evans 1966: 114 (parasitism, comp. note) Australia
Macrotristria angularis Moulds 1983: 429, 435 (illustrated, comp. note) Australia, Queensland, New South Wales
Macrotristria angularis Ewart 1986: 51, Table 1 (listed) Queensland
Macrotristria angularis Duffels 1988d: 9 (comp. note) Australia
Macrotristria angularis Ewart 1988: 185, 191, 193, 198–199, Fig 10E, Plate 2A (illustrated, oscillogram, natural history, listed, comp. note) Queensland, New South Wales
Macrotristria angularis Lithgow 1988: 65 (listed) Australia, Queensland
Cicada angularis Ewart 1990: 1 (type locality) Australia
Macrotristria angularis Moulds 1990: 3, 5, 10, 19, 22, 95–97, Plte 5, Figs 3–4, Plate 12, Fig 1, Figs 1a–1b (synonymy, distribution, ecology, type data, listed) Australia, New South Wales, Queensland, South Australia, Victoria
Cicada angularis Moulds 1990: 86 (type species of *Macrotristria*, listed)
Macrotristria angularis Young 1990: 43–44, 46–51, 53, 55, Figs 2–5, Tables 1–4 (oscillogram, frequency spectrum, song intensity, sound system, sound radiation, comp. note) Australia, Queensland
Macrotristria angularis Moulds and Carver 1991: 466, Fig 30.27C (nymph illustrated) Australia
Macrotristria angularis Bennet-Clark and Young 1992: 126, 133–135, 149, Table 1 (song production, comp. note) Queensland, Australia
Macrotristria angularis Naumann 1993: 12, 104, 149 (listed) Australia
Macrotristria angularis Bennet-Clark 1994: 169, Table 1 (comp. note)
Macrotristria angularis Bennet-Clark and Young 1994: 292–293, Table 1 (body size and song frequency)
Macrotristria angularis Carver, Gross and Woodward 1994: 322, Fig 30.8G (nymph illustrated)
Macrotristria angularis Fonseca and Popov 1994: 350, 360 (comp. note)
Macrotristria angularis Humphreys 1994: 430, Fig 4A (turrets, comp. note) New South Wales
Macrotristria angularis Ewart 1995: 80 (described, illustrated, natural history, acoustic behavior, oscillogram) Queensland
Macrotristria angularis Sanborn and Phillips 1995a: 483 (comp. note)
Macrotristria angularis Coombs 1996: 57, Table 1 (emergence time, comp. note) New South Wales
Macrotristria angularis Moulds 1996b: 94, Plate 76 (illustrated, acoustic behavior, comp. note) Australia
Macrotristria angularis Bennet-Clark and Young 1998: 701 (comp. note) Australia
Macrotristria angularis Ewart 1998a: 72 (listed) Queensland
Macrotristria angularis Sanborn 1999a: 40 (type material, listed)
Macrotristria angularis Hangay and German 2000: 59 (illustrated, behavior) Queensland, Victoria, South Australia
Macrotristria angularis Moss and Popple 2000: 58 (listed, comp. note) Australia, New South Wales
Macrotristria angularis Ewart and Popple 2001: 56 (described, habitat, song) Queensland
Macrotristria angularis Sueur 2001: 37, Table 1 (listed)
Cicada angularis Moulds and Carver 2002: 14 (type species of *Macrotristria*, type data, listed)
Macrotristria angularis Moulds and Carver 2002: 14–15 (synonymy, distribution, ecology, type data, listed) Equals *Cicada angularis* Australia, New South Wales, Queensland, South Australia, Victoria
Macrotristria angularis Williams 2002: 156–157 (listed) Australia, Queensland, New South Wales, South Australia
Macrotristria angularis Emery, Emery, Emery and Popple 2005: 102–105, 107, Tables 1–3 (comp. note) New South Wales
Macrotristria angularis Moulds 2005b: 395–396, 399, 405, 417–420, 422, Fig 41, Figs 56–60, Table 1 (illustrated, higher taxonomy, phylogeny, key, listed, comp. note) Australia
Macrotristria angularis Robinson 2005: 217 (comp. note) Australia
Macrotristria angularis Watson, Myhra, Cribb and Watson 2006: 9 (comp. note)
Macrotristria angularis Oberdörster and Grant 2007b: 19, Table 2 (comp. note)
Macrotristria angularis Shiyake 2007: 3, 29, 31, Fig 36 (illustrated, distribution, listed, comp. note) Australia

Macrotristria angularis Watson, Myhra, Cribb and Watson 2008: 3353 (comp. note)
Macrotristria angularis Wilson 2009: 40, 48, Appendix (listed, comp. note)
Macrotristria angularis Watson, Watson, Hu, Brown, Cribb and Myhra 2010: 117 (wing microstructure, comp. note)

M. aterrima (Distant, 1914) to *Burbunga aterrima*

***M. bindalia* Burns, 1964**
Macrotristria bindalia Moulds 1990: 87 (listed, comp. note) Australia, Queensland
Macrotristria bindalia Moulds and Carver 2002: 15 (distribution, ecology, type data, listed) Australia, Queensland

***M. doddi* Ashton, 1912**
Macrotristria doddi Moulds 1990: 9, 98–99, Plate 13, Fig 3 (illustrated, distribution, ecology, listed, comp. note) Australia, Northern Territory
Macrotristria doddi Moulds and Carver 2002: 15 (distribution, ecology, type data, listed) Australia, Northern Territory

***M. dorsalis* Ashton, 1912**
Macrotristria dorsalis Moulds 1990: 87–89, Plate 13, Fig 6 (distribution, ecology, type data, listed) Australia, Queensland
Macrotristria dorsalis Moulds and Carver 2002: 15 (distribution, ecology, type data, listed) Australia, Queensland
Macrotristria dorsalis 2005a: 174 (comp. note) Queensland
Macrotristria dorsalis? Ewart 2005a: 175–176, Fig 7C, Fig 8C (oscillogram) Queensland

***M. douglasi* Burns, 1964**
Macrotristria douglasi Moulds 1990: 87 (listed, comp. note) Australia, Western Australia
Macrotristria douglasi Moulds and Carver 2002: 15 (distribution, ecology, type data, listed) Australia, Western Australia

***M. extrema* (Distant, 1892)** = *Cicada extrema*
Macrotristria extrema Moulds 1990: 90–92 (distribution, ecology, listed, comp. note) Australia, Western Australia
Macrotristria extrema Moulds and Carver 2002: 16 (synonymy, distribution, ecology, type data, listed) Equals *Cicada extrema* Australia, Western Australia
Cicada extrema Moulds and Carver 2002: 16 (type data, listed) Western Australia

***M. frenchi* Ashton, 1914**
Macrotristria frenchi Carver 1990: 260 (authority)
Macrotristria frenchi Moulds 1990: 90–91, 101, Plate 12, Fig 4 (illustrated, distribution, ecology, listed, comp. note) Australia, Western Australia
Macrotristria frenchi Moulds and Carver 2002: 16 (distribution, ecology, type data, listed) Australia, Western Australia

***M. godingi* Distant, 1907**
Macrotristria godingi Moulds 1990: 94–95, Plate 13, Fig 1 (illustrated, distribution, ecology, listed, comp. note) Australia, Queensland
Macrotristria godingi Ewart 1993: 137 (comp. note)
Macrotristria godingi Naumann 1993: 58, 104, 149 (listed) Australia
Macrotristria godingi Moulds and Carver 2002: 16 (distribution, ecology, type data, listed) Australia, Queensland

***M. hieroglyphicalis* (Kirkaldy, 1909)** = *Cicada hieroglyphica* = *Cicada hieroglyphicalis* (sic) = *Cicada hieroglyphica* Goding & Froggatt (nec Say) = *Rihana hieroglyphica* = *Macrotristria hieroglyphica* (sic) = *Macrotristria hieroglyphicus* (sic)
Macrotristria hieroglyphica (sic) Hahn 1962: 8 (listed)
Macrotristria hieroglyphicalis Moulds 1990: 93–94, Plate 13, Fig 4 (illustrated, synonymy, distribution, ecology, type data, listed, comp. note) Equals *Cicada hieroglyphica* Goding and Froggatt Equals *Rihana? hieroglyphica* Equals *Cicada hieroglyphicalis* Equals *Macrotristria hieroglyphica* Australia, Western Australia
Cicada hieroglyphica Moulds 1990: 93 (listed)
Cicada hieroglyphicalis Moulds 1990: 93 (type data, listed) Equals *Cicada hieroglyphica* Goding and Froggatt
Rihana? hieroglyphica Moulds 1990: 93 (listed)
Macrotristria hieroglyphica (sic) Moulds 1990: 93 (listed)
Macrotristria hieroglyphicalis Boulard 1996c: 101 (comp. note) Australia
Macrotristria hieroglyphicus (sic) Sanborn 1999a: 40 (type material, listed)
Macrotristria hieroglyphicalis Moulds and Carver 2002: 16 (synonymy, distribution, ecology, type data, listed) Equals *Cicada hieroglyphica* Equals *Cicada hieroglyphicalis* Australia, Western Australia
Cicada hieroglyphica Moulds and Carver 2002: 16 (type data, listed) Western Australia
Cicada hieroglyphicalis Moulds and Carver 2002: 16 (type data, listed) Equals *Cicada hieroglyphica* Goding and Froggatt Australia

M. hillieri Distant, 1907 to *Burbunga hillieri*

***M. intersecta* (Walker, 1850)** = *Fidicina intersecta* = *Fidicina internata* Walker, 1850 = *Macrotristria internata* = *Fidicina prasina* Walker, 1850 = *Cicada sylvanella* Goding & Froggatt, 1904 = *Macrotristria sylvanella*
Macrotristria intersecta Moulds 1990: 88–91, Plate 13, Fig 2, Fig 2a-2b (illustrated, distribution,

ecology, listed, comp. note) Australia, Northern Territory, Queensland, Western Australia

Macrotristria intersecta Ewart 1993: 137, 142, Fig 3 (natural history, acoustic behavior, song analysis, comp. note) Queensland

Macrotristria intersecta Naumann 1993: 15, 26, 104, 149 (listed) Australia

Macrotristria intersecta Ewart 1998b: 135 (listed) Queensland

Macrotristria internata Sanborn 1999a: 40 (type material, listed)

Macrotristria intersecta Sueur 2001: 37, Table 1 (listed)

Macrotristria intersecta Moulds and Carver 2002: 16–17 (synonymy, distribution, ecology, type data, listed) Equals *Fidicina internata* Equals *Fidicina intersecta* Equals *Fidicina prasina* Equals *Cicada sylvanella* Australia, Northern Territory, Queensland, Western Australia

Fidicina internata Moulds and Carver 2002: 16–17 (type data, listed) Northern Territory

Fidicina intersecta Moulds and Carver 2002: 17 (type data, listed) Northern Territory

Fidicina prasina Moulds and Carver 2002: 17 (type data, listed)

Cicada sylvanella Moulds and Carver 2002: 17 (type data, listed) Queensland

Macrotristria intersecta Ewart 2005a: 175–176, Fig 7A, Fig 8A (described, song, oscillogram) Queensland

Macrotristria intersecta Moulds 2005b: 395, 399, 417–420, 422, Figs 56–60, Table 1 (higher taxonomy, phylogeny, key, listed, comp. note) Australia

Macrotristria intersecta Shiyake 2007: 3, 29, 31, Fig 37 (illustrated, distribution, listed, comp. note) Australia

***M. kabikabia* Burns, 1964**

Macrotristria kabikabia Moulds 1990: 98, Plate 12, Fig 6 (illustrated, distribution, ecology, listed, comp. note) Australia, Queensland

Macrotristria kabikabia Ewart and Popple 2001: 55, 62, 65, Fig 1E (described, habitat, song, oscillogram) Queensland

Macrotristria kabikabia Moulds and Carver 2002: 17 (distribution, ecology, type data, listed) Australia, Queensland

***M. kulungura* Burns, 1964**

Macrotristria kulungura Moss 1988: 29 (distribution, song, comp. note) Queensland

Macrotristria kulungura Moulds 1990: 88–89, Plate 13, Fig 5 (illustrated, distribution, ecology, listed, comp. note) Australia, Queensland, Torres Strait Islands

Macrotristria kulungura Moulds and Carver 2002: 17 (distribution, ecology, type data, listed) Australia, Queensland, Torres Strait Islands

Macrotristria kulungura Ewart 2005a: 174–176, Fig 7D, Fig 8D (oscillogram, comp. note) Queensland

M. sp. nr. *kulungura/dorsalis* Ewart, 2005

Macrotristria sp. nr. *kulungura/dorsalis* Ewart 2005a: 173–175, 178, Fig 7C-D (described, song, oscillogram, comp. note) Queensland

***M. lachlani* Moulds, 1992** = *Macrotristria* sp. A Ewart, 1993

Macrotristria lachlani Moulds 1992: 133–135, 137–138, Fig 1, Figs 3–4, Fig 7 (n. sp., described, illustrated, key, distribution, comp. note) Australia, Queensland

Macrotristria sp. A Ewart 1993: 137, 142, Fig 4 (natural history, acoustic behavior, song analysis, comp. note) Equals *Macrotristria lachlani* Queensland

Macrotristria lachlani Ewart 1993: 137 (comp. note) Equals *Macrotristria* sp. A Queensland

Macrotristria lachlani Sueur 2001: 37, Table 1 (listed)

Macrotristria lachlani Moulds and Carver 2002: 17 (distribution, ecology, type data, listed) Australia, Queensland

Macrotristria lachlani Ewart 2005a: 174–176, Fig 7B, Fig 8B (described, song, oscillogram) Equals *Macrotristria* sp. A Queensland

***M. maculicollis* Ashton, 1914**

Macrotristria maculicollis Moulds 1990: 96–97, Plate 12, Fig 2 (illustrated, distribution, ecology, listed, comp. note) Australia, New South Wales, Queensland

Macrotristria maculicollis Boulard 1996c: 101 (comp. note) Australia

Macrotristria maculicollis Moulds and Carver 2002: 17 (synonymy, distribution, ecology, type data, listed) Australia, New South Wales, Queensland

***M. madegassa* Boulard, 1996**

Macrotristria madegassa Boulard 1996c: 101–103, Figs 3–4 (n. sp., described, illustrated, sonogram, comp. note) Madagascar

Macrotristria madegassa Sueur 2001: 37, Table 1 (listed)

Macrotristria madegassa Boulard 2006g: 181 (listed)

M. nanda Burns, 1964 to *Burbunga nanda*

M. nigronervosa Distant, 1904 to *Macrotristria sylvara* (Distant, 1901)

M. nigrosignata Distant, 1904 to *Burbunga nigrosignata*

M. occidentalis Distant, 1912 to *Burbunga occidentalis*

***M. sylvara* (Distant, 1901)** = *Cicada sylvara* = *Cicada sylvana* (sic) = *Macrotristria nigronervosa* Distant, 1904

Macrotristria sylvara Moulds 1983: 434 (illustrated, comp. note) Australia, Queensland
Macrotristria sylvara Moulds 1990: 92–93, 95, Plate 3, Fig 2, Plate 12, Fig 5, Plate 24, Fig 3 (illustrated, synonymy, distribution, ecology, type data, listed, comp. note) Equals *Cicada sylvara* Equals *Magrotristria nigronervosa* n. syn. Australia, Queensland
Cicada sylvara Moulds 1990: 92 (listed)
Macrotristria nigronervosa Moulds 1990: 92 (listed, comp. note) Queensland
Macrotristria sylvara Naumann 1993: 40, 104, 149 (listed) Australia
Macrotristria sylvara Moulds 2003a: 186, Fig 2 (illustrated, comp. note) Australia
Macrotristria sylvara Moulds and Carver 2002: 18 (synonymy, distribution, ecology, type data, listed) Equals *Cicada sylvara* Equals *Magrotristria nigronervosa* Australia, Queensland
Cicada sylvara Moulds and Carver 2002: 18 (type data, listed) Queensland
Macrotristria nigronervosa Moulds and Carver 2002: 18 (type data, listed) Queensland
Macrotristria sylvara Moulds 2009b: 163, Fig 2 (illustrated, comp. note) Australia

***M. thophoides* Ashton, 1914**

Macrotristria thophoides Moulds 1990: 87 (listed, comp. note) Australia, Western Australia
Macrotristria thophoides Moulds and Carver 2002: 18 (distribution, ecology, type data, listed) Australia, Western Australia

***M. vittata* Moulds, 1992**

Macrotristria vittata Moulds 1992: 134–138, Fig 2, Figs 5–6, Fig 8 (n. sp., described, illustrated, key, distribution, comp. note) Australia, Queensland
Macrotristria vittata Moulds and Carver 2002: 18–19 (distribution, ecology, type data, listed) Australia, Queensland

M. vulpina Ashton, 1914 to *Burbunga aterrima* (Distant, 1914)

***M. worora* Burns, 1964**

Macrotristria worora Moulds 1990: 87 (listed, comp. note) Australia, Northern Territory
Macrotristria worora Moulds and Carver 2002: 19 (distribution, ecology, type data, listed) Australia, Northern Territory

***Chremistica* Stål, 1870** = *Cicada (Chremistica)* = *Chremestica* (sic) = *Cheremistica* (sic) = *Chremistrica* (sic) = *Rihana* Distant, 1904 = *Rhiana* (sic)

Chremistica Wu 1935: 5 (listed) Equals *Rihana* China
Chremistica Duffels 1983a: 492 (comp. note) Lesser Sunda Islands
Rihana Boulard 1984c: 175 (comp. note)
Chremistica Hayashi 1984: 35 (listed)
Chremistica Bregman 1985: 37, 39, 41–43, 46, Fig 15 (history, diversity, biogeography, comp. note) Equals *Rihana* Malaysia, Southeast Asia, India, Sri Lanka, Madagascar, Lombok, Sumbawa, Sumba, Timor, Sulawesi
Chremistica sp. Yukawa and Yamane 1995: 693 (listed) Indonesia, Krakataus
Chremistica Duffels 1986: 319 (diversity, distribution) Sulawesi, Lesser Sunda Islands, Madagascar, Southeast Asia, Greater Sunda Islands
Chremistica Hayashi 1987a: 121, 124–125 (comp. note)
Chremistica Zaidi, Mohammedsaid and Yaakob 1990: 261 (comp. note) Malaysia
Cheremistica (sic) Zaidi, Mohammedsaid and Yaakob 1990: 267, Table 2 (listed, comp. note) Malaysia
Chremistica Carpenter 1992: 222 (comp. note) Equals *Rihana*
Chremistica Hayashi 1993: 14 (comp. note)
Chremistica Zaidi and Ruslan 1994: 427, Table 2 (comp. note) Malaysia
Chremistica Lei, Zhou, Chou and Li 1995: 236, 238 (comp. note)
Chremistica Zaidi and Ruslan 1995a: 174, Table 2 (listed, comp. note) Borneo, Sarawak
Chremistica Zaidi and Ruslan 1995b: 220–221, Table 2 (listed, comp. note) Borneo, Sabah
Chremistica Zaidi and Ruslan 1995c: 70, Table 2 (listed, comp. note) Peninsular Malaysia
Chremistica Zaidi and Ruslan 1995d: 202, Table 2 (listed, comp. note) Borneo, Sabah
Chremistica Boer and Duffels 1996b: 304 (comp. note)
Chremistica Zaidi 1996: 103, Table 2 (comp. note) Borneo, Sarawak
Chremistica Zaidi and Hamid 1996: 54, Table 2 (listed, comp. note) Borneo, Sarawak
Chremistica Chou, Lei, Li, Lu and Yao 1997: 265–266, 293, Table 6 (key, described, listed, diversity, biogeography) China
Chremistica Novotny and Wilson 1997: 437 (listed)
Chremistica Zaidi 1997: 115, Table 2 (listed, comp. note) Borneo, Sarawak
Rihana sp. Chen, Wongsiri, Jamyanya, Rinderer, Vongsamanode, et al. 1998: 27 (predation) Thailand
Chremistica Holloway 1998: 299 (comp. note)
Chremistica Zaidi and Ruslan 1998b: 4–5, Table 2 (listed, comp. note) Borneo, Sabah
Chremistica Zaidi and Ruslan 1999: 4, Table 2 (listed, comp. note) Borneo, Sarawak

Chremistica Boulard 2000a: 255–256 (comp. note)
Rihana Boulard 2000a: 255 (comp. note)
Rihana Boulard 2001e: 117–118 (comp. note)
Chremistica Boulard 2001e: 114, 117–118 (comp. note)
Chremistica Boulard 2001f: 129–130, 136 (diagnosis, comp. note)
Chremistica sp. Zaidi, Azman and Ruslan 2001a: 112, 115, 117, Table 1 (listed, comp. note) Langkawi Island
Chremistica Boulard 2002a: 58 (comp. note)
Chremistica Salmah and Zaidi 2002: 226, 243 (comp. note) Peninsular Malaysia
Rihana Salmah and Zaidi 2002: 226 (comp. note) Peninsular Malaysia
Chremistica Lee and Hayashi 2003a: 158, 160 (key, diagnosis) Equals *Rihana* Taiwan
Chremistica Boulard 2004: 237, 243 (comp. note)
Chremistica Chen 2004: 39, 41, 62 (phylogeny, listed)
Chremistica Salmah, Zaidi and Duffels 2004: 2–3 (comp. note) Malaysia
Chremistica Schouten, Duffels and Zaidi 2004: 379 (comp. note)
Chremistica Boulard 2005d: 346 (comp. note) Seychelles
Chremistica Boulard 2005n: 130 (comp. note)
Chremistica Gogala and Trilar 2005: 68, 70 (comp. note)
Chremistica Moulds 2005b: 390, 431 (higher taxonomy, listed, comp. note) Australia
Chremistica Salmah, Duffels and Zaidi 2005: 15–16, 21–22 (comp. note) Malaysia
Cicada (Chremistica) Salmah, Duffels and Zaidi 2005: 15 (comp. note)
Rihana Salmah, Duffels and Zaidi 2005: 15 (comp. note)
Chremistica Yaakop, Duffels and Visser 2005: 247–251, 253, 270, 283, 285 (described, synonymy, distribution, comp. note) Equals *Cicada (Chremistica)* Sundaland, China, Madagascar, India, Sri Lanka, Southeast Asia, Taiwan, Philippines, Malayan Peninsula, Sumatra, Borneo, Java, Sulawesi, Lesser Sunda Islands
Rihana Yaakop, Duffels and Visser 2005: 248 (comp. note) Equals *Cicada (Chremistica)*
Chremistica spp. Yaakop, Duffels and Visser 2005: 263 (comp. note)
Chremistica Boulard 2006c: 138 (comp. note)
Chremistica Boulard 2006g: 94, 129 (comp. note)
Chremistica Boulard 2007a: 96 (comp. note)
Chremistica Boulard 2007e: 5, 20, 25 (comp. note) Thailand
Chremistica sp. Boulard 2007e: 62, Fig 42 (genitalia illustrated) Thailand
Chremistica Sanborn, Phillips and Sites 2007: 3, 5, 35, Table 1 (listed) Thailand
Rihana Sanborn, Phillips and Sites 2007: 7 (comp. note)
Chremistica Shiyake 2007: 3, 34 (listed, comp. note)
Chremistica Boulard 2008b: 109 (comp. note)
Chremistica Boulard 2008e: 352 (comp. note)
Chremistica sp. Boulard 2008e: 352 (comp. note)
Chremistica Boulard 2008f: 5, 7, 87, 91 (listed, comp. note) Thailand
Chremistica Lee 2008a: 2, 4, Table 1 (listed) Equals *Rihana* Vietnam
Chremistica Sanborn, Moore and Young 2008: 1 (comp. note)
Rihana Sanborn, Moore and Young 2008: 1 (comp. note)
Chremistica Pham and Yang 2009: 13, Table 2 (listed) Vietnam
Chremistica Lee 2010a: 16–17 (listed, comp. note) Philippines, Mindanao
Chremistica Lee 2010b: 20 (listed) Cambodia
Chremistica Lee 2010c: 2 (synonymy, comp. note) Equals *Rihana* Philippines, Luzon

C. *atra* (Distant, 1909) = *Rihana atra*
Chremistica atra Bregman 1985: 39–40, 44 (key, comp. note) Philippines
Chremistica atra Salmah, Duffels and Zaidi 2005: 22 (comp. note)
Chremistica atra Yaakop, Duffels and Visser 2005: 248 (comp. note)
Chremistica atra Lee 2010c: 2 (synonymy, distribution, comp. note) Equals *Rihana atra* Philippines, Luzon

C. *atratula* Boulard, 2007 = *Chremistica atrata* (sic)
Chremistica atratula Boulard 2007a: 94–97, Figs 1–6 (n. sp., described, illustrated, song, sonogram, comp. note) Thailand
Chremistica atratula Boulard 2008f: 9 (biogeography, listed) Thailand
Chremistica atrata (sic) Boulard 2008f: 91 (listed) Thailand

C. *banksi* Liu, 1940
Chremistica banksi Bregman 1985: 39 (comp. note)
Chremistica banks Liu 1992: 40 (comp. note)

C. *biloba* Bregman, 1985
Chremistica biloba Bregman 1985: 38–44, 55–59, Figs 2–3, Fig 13, Fig 16, Figs 35–36, Fig 42 (key, n. sp., described, illustrated, phylogeny, distribution, comp. note) Sarawak, Kalimantan
Chremistica biloba Zaidi and Ruslan 1995d: 197 (comp. note) Borneo, Sarawak
Chremistica biloba Zaidi 1996: 99 (comp. note) Borneo, Sarawak

Chremistica biloba Zaidi and Hamid 1996: 51 (comp. note) Borneo, Sarawak
Chremistica biloba Zaidi 1997: 110 (comp. note) Borneo, Sarawak
Chremistica biloba Zaidi and Ruslan 1999: 1 (comp. note) Borneo, Sarawak
Chremistica biloba Salmah, Zaidi and Duffels 2004: 2–3 (comp. note) Malaysia
Chremistica biloba Salmah, Duffels and Zaidi 2005: 15–17, 20–22, Fig 6b, Table 1 (key, illustrated, distribution, listed, comp. note) Borneo
Chremistica biloba Yaakop, Duffels and Visser 2005: 248–250, 254–255, 285, 287–288, 291, 297, 303, Figs 52–55, Fig 70, Fig 89, Table 1 (key, distribution, listed, comp. note) Borneo, Sarawak, Kalimantan

***C. bimaculata* (Olivier, 1791)** = *Cicada bimaculata* = *Rihana bimaculata* = *Chremistica bimcaulata* (sic) = *Chremistica binaculata* (sic) = *Cicada atro-virens* Guérin-Méneville, 1838 = *Chremistica atrovirens* = *Cicada viridis* (nec Fabricius) = *Rihana viridis* (nec Fabricius) = *Chremistica viridis* (nec Fabricius)

Chremistica atrovirens Bregman 1985: 39 (comp. note)
Chremistica bimaculata Bregman 1985: 50 (comp. note) Java
Chremistica bimaculata Zaidi and Ruslan 1995c: 64 (comp. note) Peninsular Malaysia
Chremistica binaculata (sic) Zaidi and Ruslan 1995c: 68, Table 1 (listed, comp. note) Peninsular Malaysia
Chremistica bimaculata Zaidi 1996: 100–101, 104, Table 1 (comp. note) Borneo, Sarawak
Chremistica bimaculata Zaidi 1997: 112 (comp. note) Borneo, Sarawak
Chremistica bimaculata Zaidi and Ruslan 1997: 220 (listed, comp. note) Java, Malaya, Peninsular Malaysia, Singapore
Chremistica bimaculata Zaidi, Ruslan and Azman 1999: 301 (listed, comp. note) Equals *Cicada bimaculata* Equals *Rihana bimaculata* Borneo, Sabah, Java, Cambodia, Tonkin, Philippines
Tettigonia viridis Boulard 2000a: 255 (comp. note)
Chremistica viridis Boulard 2000a: 255–256 (type species of *Chremistica*, comp. note)
Chremistica bimaculata Zaidi, Noramly and Ruslan 2000a: 320 (listed, comp. note) Equals *Cicada bimaculata* Equals *Rihana bimaculata* Borneo, Sabah, Java, Cambodia, Tonkin, Philippine Islands, Sarawak
Chremistica bimaculata Zaidi, Noramly and Ruslan 2000b: 337–338, Table 1 (listed, comp. note) Borneo, Sabah
Chremistica bimaculata Zaidi, Ruslan and Azman 2000: 199–200 (listed, comp. note) Equals *Cicada bimaculata* Equals *Rihana bimaculata* Borneo, Sabah, Java, Cambodia, Tonkin, Philippine Islands, Sarawak
Tettigonia viridis Boulard 2001e: 114, 116–117 (comp. note)
Cicada bimaculata Boulard 2001e: 114–118, Fig B (illustrated, comp. note) Java
Cicada (Chremistica) viridis Boulard 2001e: 115–116 (comp. note) Equals *Cicada viridis*
Cicada viridis Boulard 2001e: 116–117 (comp. note)
Cicada atro-virens Boulard 2001e: 116 (comp. note)
Chremistica viridis Boulard 2001e: 117–118 (type species of *Chremistica*, comp. note) Equals *Cicada bimaculata*
Rihana bimaculata Boulard 2001e: 117–118 (comp. note) Equals *Cicada bimaculata*
Rihana viridis Boulard 2001e: 117 (comp. note)
Tettigonia viridis Boulard 2001f: 130 (type species of *Chremistica*, comp. note) Equals *Cicada bimaculata*
Chremistica bimaculata Boulard 2001e: 118–119 (synonymy, comp. note) Equals *Cicada bimaculata* Equals *Cicada atro-virens* Equals *Cicada viridis* [non Fabr.] Equals *Rihana viridis* [non Fabr.] Equals *Chremistica bimaculata* Equals *Rihana bimaculata* Equals *Chremistica viridis* [non Fabr.] Equals *Chremistica bimcaulata* (sic) Equals *Chremistica atrovirens* (sic) Equals *Chremistica viridis* Boulard
Rihana viridis Boulard 2002a: 58 (comp. note)
Chremistica bimaculata Zaidi and Azman 2003: 104, Table 1 (listed, comp. note) Borneo, Sabah
Chremistica bimaculata Salmah, Zaidi and Duffels 2004: 2–3, 5–6, Fig 4, Table 1 (type material, illustrated, synonymy, comp. note) Equals *Cicada bimaculata* Equals *Rihana bimaculata* Malaysia
Rihana bimaculata Salmah, Zaidi and Duffels 2004: 2 (comp. note) Malaysia
Chremistica bimaculata Boulard 2005d: 344 (comp. note)
Chremistica bimaculata Boulard 2005o: 148 (listed)
Chremistica bimaculata Salmah, Duffels and Zaidi 2005: 15–16, 18–19, 22, Table 1 (key, synonymy, distribution, listed, comp. note) Equals *Cicada bimaculata* Equals *Rihana bimaculata* Malaysia, Java
Rihana bimaculata Salmah, Duffels and Zaidi 2005: 18–19 (comp. note) Malaysia
Chremistica atrovirens Salmah, Duffels and Zaidi 2005: 22 (comp. note)

Rihana bimaculata Yaakop, Duffels and Visser 2005: 248 (comp. note) Malaysia
Chremistica atrovirens Yaakop, Duffels and Visser 2005: 248 (comp. note)
Chremistica bimaculata group Yaakop, Duffels and Visser 2005: 249, 254–255, 267, 272, 275 (key, comp. note) Sundaland
Chremistica bimaculata Yaakop, Duffels and Visser 2005: 249–250, 254, 267–270, 274, 297, 301, Figs 21–24, Fig 71, Fig 81, Table 1 (described, illustrated, synonymy, key, distribution, listed, comp. note) Equals *Cicada bimaculata* Equals *Cicada atrovirens* Equals *Cicada viridis* (nec Fabricius) Equals *Rihana bimaculata* Equals *Chremistica atrovirens* Java
Cicada bimaculata Yaakop, Duffels and Visser 2005: 251, 253, 269 (type species of *Chremistica*, comp. note) Java
Chremistica bimaculata Boulard 2006c: 138 (comp. note) Java
Chremistica bimaculata Boulard 2006g: 74–75 (comp. note)
Chremistica bimaculata Boulard 2007e: 5, 40, Fig 24, Plate 5b (coloration, illustrated, sonogram, comp. note) Thailand
Chremistica bimaculata Sanborn, Phillips and Sites 2007: 34 (comp. note)
Chremistica bimaculata Boulard 2008f: 9 (type species of *Chremistica*, biogeography, listed) Thailand
Cicada bimaculata Lee 2008a: 4 (type species of *Chremistica*)
Cicada bimaculata Lee 2010b: 20 (type species of *Chremistica*, listed)
Cicada bimaculata Lee 2010c: 2 (type species of *Chremistica*, synonymy, listed) Equals *Cicada (Cicada) bimaculata* Java

C. *bimaculata inthanonensis* Boulard, 2006 to *Chremistica inthanonensis*

C. sp. nr. *bimaculata* Sanborn, Phillips & Sites, 2007
Chremistica sp. nr. *bimaculata* Sanborn, Phillips and Sites 2007: 3 (listed, comp. note) Thailand

C. *borneensis* Yaakop & Duffels, 2005
Chremistica borneensis Yaakop, Duffels and Visser 2005: 249–250, 254–255, 285, 291–293, 299, 304, Figs 60–63, Fig 75, Fig 91, Table 1 (n. sp., described, illustrated, key, distribution, listed, comp. note) Borneo, Sarawak

C. *brooksi* Yaakop & Duffels, 2005
Chremistica brooksi Yaakop, Duffels and Visser 2005: 249–250, 253–254, 267–268, 273–275, Figs 29–31, Fig 71, Fig 83, 297, 301, Table 1 (n. sp., described, illustrated, key, distribution, listed, comp. note) Sumatra

C. *cetacauda* Yaakop & Duffels, 2005
Chremistica cetacauda Yaakop, Duffels and Visser 2005: 249–250, 253–254-255, 265–267, 297, 301, Figs 18–20, Fig 70, Fig 80, Table 1 (n. sp., described, illustrated, key, distribution, listed, comp. note) Sumatra

C. *coronata* (Distant, 1889) = *Cicada coronata* = *Rihana coronata*
Chremistica coronata group Duffels 1986: 319 (comp. note) Lesser Sunda Islands
Chremistica coronata group Bregman 1985: 37–39, 41–42, Fig 15 (comp. note) Lesser Sunda Islands
Chremistica coronata Bregman 1985: 38–39 (comp. note)
Chremistica coronata Salmah, Duffels and Zaidi 2005: 21 (comp. note) Lesser Sunda Islands
Chremistica coronata group Salmah, Duffels and Zaidi 2005: 21 (comp. note) Lesser Sunda Islands
Chremistica coronata group Yaakop, Duffels and Visser 2005: 248–249 (comp. note) Lesser Sunda Islands
Chremistica coronata Yaakop, Duffels and Visser 2005: 248 (comp. note)

C. *echinaria* Yaakop & Duffels, 2005 = *Chremistica echinaaria* (sic)
Chremistica echinaria Yaakop, Duffels and Visser 2005: 249–250, 254, 267–268, 272, 274, 278–280, 298, 302, Figs 36–39, Fig 73, Fig 85, Table 1 (n. sp., described, illustrated, key, distribution, listed, comp. note) Malayan Peninsula
Chremistica echinaaria (sic) Sanborn, Phillips and Sites 2007: 5 (listed, distribution, comp. note) Thailand, Malaysia, Borneo, Sarawak
Chremistica echinaria Boulard 2008c: 360 (comp. note) Malaysia
Chremistica echinaria Boulard 2008f: 10, 91 (biogeography, listed) Thailand

C. *elenae madagascariensis* Boulard, 2001 = *Chremistica helenae* (sic) *madagascariensis*
Chremistica elenae madagascariensis Boulard 2001f: 134–135, 138, Figs 5–6, Carte B (n. sp., n. ssp., described, illustrated, sonogram, comp. note) Madagascar
Chremistica helenae (sic) *madagascariensis* Boulard 2001f: 137, Plate I, Figs 5–6 (illustrated)
Chremistica elenae Yaakop, Duffels and Visser 2005: 248 (comp. note) Madagascar

C. *elenae seychellensis* Boulard, 2001 = *Chremistica helenae* (sic) *seychellensis* = *Chremistica elenae mahiensis* (sic)
Chremistica elenae seychellensis Boulard 2001f: 136, Figs 5–6 (n. sp., n. ssp., described, illustrated, comp. note) Seychelles
Chremistica helenae (sic) *seychellensis* Boulard 2001f: 137, Plate I, Figs 7–8 (illustrated)

Chremistica elenae mahiensis (sic) Boulard 2001f: 136, 139, Carte C (sonogram, comp. note) Seychelles

C. *euterpe* (Walker, 1850) = *Cicada euterpe* = *Rihana euterpe*

Chremistica euterpe Bregman 1985: 39 (comp. note)

C. *germana* (Distant, 1888) = *Cicada germana* = *Rihana germana* = *Chremistrica* (sic) *germana*

Chremistica germana Strümpel 1983: 71, Table 5 (comp. note)

Chremistica germana Bregman 1985: 39 (comp. note)

Chremistrica (sic) *germana* Zaidi, Mohammedsaid and Yaakob 1990: 261 (comp. note) Malaysia

Chremistica germana Zaidi and Ruslan 1994: 426, 428, Table 1 (comp. note) Malaysia

Cheremistica (sic) *germana* Zaidi and Ruslan 1994: 426, 428 (comp. note) Malaysia

Chremistica germana Zaidi, Ruslan and Mahadir 1996: 60, Table 1 (listed, comp. note) Peninsular Malaysia

Chremistica germana Zaidi and Ruslan 1998a: 346–347 (listed, comp. note) Equals *Cicada germana* Equals *Rihana germana* Borneo, Sarawak, Burma, Peninsular Malaysia, Siam, Sumatra, India, Southeast Asia

Chremistica germana Zaidi, Noramly and Ruslan 2000a: 320, 322 (listed, comp. note) Equals *Cicada germana* Equals *Rihana germana* Borneo, Sabah, Burma, Peninsular Malaysia, Siam, Java, Sumatra, India, Sarawak

Chremistica germana Zaidi, Noramly and Ruslan 2000b: 338–339, Table 1 (listed, comp. note) Borneo, Sabah

Chremistica germana Salmah and Zaidi 2002: 227, 230–231 (key, synonymy, comp. note) Equals *Cicada germana* Peninsular Malaysia

Chremistica germana Zaidi and Azman 2003: 104, Table 1 (listed, comp. note) Borneo, Sabah

Chremistica germana Salmah, Zaidi and Duffels 2004: 2–3, 5–6, Fig 3, Table 1 (type material, illustrated, synonymy, comp. note) Equals *Cicada germana* Malaysia

Chremistica germana Salmah, Duffels and Zaidi 2005: 15–16, 19, 22, Table 1 (key, illustrated, synonymy, distribution, listed, comp. note) Equals *Cicada germana* Peninsular Malaysia, Burma, Siam, Java, Sumatra

Rihana germana Yaakop, Duffels and Visser 2005: 248 (comp. note) Malaysia

Chremistica germana Yaakop, Duffels and Visser 2005: 248 (comp. note)

Chremistica germana Sanborn, Phillips and Sites 2007: 6 (listed, synonymy, distribution, comp. note) Equals *Rihana germana* Thailand, India, Burma, Malaysia, Indonesia

Chremistica germana Boulard 2008f: 10, 91 (synonymy, biogeography, listed) Equals *Cicada germana* Equals *Rihana germana* Thailand, Burma, Siam

C. *guamusangensis* Salmah & Zaidi, 2002 = *Chremistica chueatae* Boulard, 2004

Chremistica guamusangensis Salmah and Zaidi 2002: 226, 231–237, Figs 1–4 (n. sp., described, illustrated, key, comp. note) Peninsular Malaysia

Chremistica chueatae Boulard 2004: 237–243, Figs 1–7 (n. sp., described, song, sonogram, illustrated, comp. note) Thailand

Chremistica guamusangensis Salmah, Zaidi and Duffels 2004: 2–3 (comp. note) Malaysia

Chremistica chueatae Boulard 2005o: 149 (listed)

Chremistica guamusangensis Gogala and Trilar 2005: 66, 68–70, 76–77, Figs 4–5 (song, sonogram, oscillogram, acoustic behavior, comp. note) Malaysia

Chremistica guamusangensis Salmah, Duffels and Zaidi 2005: 15–18, 20–22, Fig 2a, Fig 2b, Table 1 (key, illustrated, distribution, listed, comp. note) Peninsular Malaysia, Sarawak

Chremistica guamusangensis Yaakop, Duffels and Visser 2005: 249–250, 252, 254–255, 260–263, 296, 300, Fig 1, Figs 8–13, Fig 69, Fig 78, Table 1 (described, illustrated, key, distribution, listed, comp. note) Malayan Peninsula

Chremistica guamusangensis Boulard 2007d: 494, 506–507, Fig 2 (synonymy, illustrated, comp. note) Equals *Chremistica chueatae* Thailand, Malaysia

Chremistica chueatae Boulard 2007e: 55, Fig 37.6, Plate 20 (genitalia, illustrated, comp. note) Thailand

Chremistica chueatae Sanborn, Phillips and Sites 2007: 5 (listed, distribution, comp. note) Thailand

Chremistica guamusangensis Sanborn, Phillips and Sites 2007: 6, 34 (listed, distribution, comp. note) Thailand, Peninsular Malaysia

Chremistica guamusangensis Boulard 2008f: 10, 91 (synonymy, biogeography, listed) Equals *Chremistica chueatae* Thailand

C. *hollowayi* Yaakop & Duffels, 2005

Chremistica hollowayi Yaakop, Duffels and Visser 2005: 249–250, 254–255, 263–265, 297, 300, Figs 14–17, Fig 70, Fig 79, Table 1 (n. sp., described, illustrated, key, distribution, listed, comp. note) Borneo, Sarawak, Sabah, Brunei

C. *hova* (Distant, 1905) = *Rihana hova*

Chremistica hova Bregman 1985: 38 (comp. note)

Chremistica hova Salmah, Duffels and Zaidi 2005: 21 (comp. note) Madagascar

Chremistica hova Yaakop, Duffels and Visser 2005: 248 (comp. note)

***C. inthanonensis* Boulard, 2006** = *Chremistica bimaculata inthanonensis* = *Cicada bimaculata* (nec Olivier) = *Rihana bimaculata* (nec Olivier) = *Chremistica bimaculata* (nec Olivier) = *Chremistica atrovirens* (nec Guérin-Méneville) = *Chremistica viridis* (nec Fabricius) = *Chremistica* undescribed species A Sanborn, Phillips and Sites 2007

Chremistica bimaculata Boulard 2002a: 55–59, 61, 64–65, Plate I, Figs 1–2, Plate I", Plate J, Figs 1–2, Figs 14–15 (n. sp., described, illustrated, sonogram, comp. note) Equals *Cicada bimaculata* Equals *Cicada atrovirens* Equals *Cicada viridis* Equals *Rihana viridis* Thailand

Chremistica bimaculata inthanonensis Boulard 2006c: 138 (n. ssp., described, comp. note) Equals *Cicada maculata* (nec Olivier) Equals *Rihana maculata* (nec Olivier) Equals *Chremistica atrovirens* (partim) Equals *Chremistica bimaculata* (nec Olivier) Thailand

Chremistica bimaculata inthanonensis Boulard 2006g: 133, Fig, 88 (sonogram, comp. note) Thailand

Chremistica bimaculata inthanonensis Boulard 2007e: 35, 64, Fig 45, Plate 47 (n. comb., illustrated, comp. note) Thailand

Chremistica undescribed species A Sanborn, Phillips and Sites 2007: 7 (listed, synonymy, distribution, comp. note) *Chremistica viridis* Boulard, 2000, 2001 Equals *Chremistica bimaculata* Boulard, 2002 Thailand

Chremistica inthanonensis Boulard 2008c: 360 (comp. note) Equals *Chremistica bimaculata inthanonensis* Equals *Chremistica bimaculata* Boulard, 2002 [55] (nec Olivier) Thailand

Chremistica intanonensis Boulard 2008f: 9–10, 91 (synonymy, biogeography, listed) Equals *Chremistica bimaculata inthanonensis* Equals *Chremistica bimaculata* (nec Olivier) Equals *Cicada bimaculata* (nec Olivier) Equals *Rihana bimaculata* (nec Olivier) Equals *Chremistica viridis* (nec Fabricius) Equals *Chremistica* undescribed species A Sanborn et al., 2007 Equals *Chremistica atrovirens* (partim) Thailand

Chremistica viridis Lee 2008a: 4 (synonymy, distribution, listed) Equals *Tettigonia viridis* Equals *Rihana viridis* Equals *Cicada bimaculata* (nec Olivier) Equals *Rihana bimaculata* (nec Olivier) Equals *Chremistica atrovirens* (nec Guérin-Méneville) Vietnam, Philippines, Cambodia, Thailand, Malysia, Indonesia, Java, Sumatra

Chremistica inthanonensis Lee 2010b: 20 (synonymy, distribution, listed) Equals *Chremistica bimaculata inthanonensis* Equals *Cicada bimaculata* (nec Olivier) Equals *Rihana bimaculata* (nec Olivier) Equals *Chremistica atrovirens* (nec Guérin-Méneville) Equals *Chremistica viridis* (nec Fabricius) Cambodia, Vietnam, Thailand

***C. kecil* Salmah & Zaidi, 2002**

Chremistica kecil Salmah and Zaidi 2002: 227, 237–243, Figs 5–8 (n. sp., described, illustrated, key, comp. note) Peninsular Malaysia

Chremistica kecil Salmah, Zaidi and Duffels 2004: 2–3 (comp. note) Malaysia

Chremistica kecil Salmah, Duffels and Zaidi 2005: 15–22, Fig 4a, Fig 4b, Table 1 (key, illustrated, distribution, listed, comp. note) Malaysia

Chremistica kecil Yaakop, Duffels and Visser 2005: 249–250, 254, 267–268, 275, 283–285, 298, 302, Figs 44–47, Fig 73, Fig 87, Table 1 (key, distribution, listed, comp. note) Malayan Peninsula

***C. kyoungheeae* Lee, 2010**

Chremistica kyoungheeae Lee 2010a: 16–18, Fig 2 (n. sp., described, illustrated, distribution, listed) Philippines, Mindanao

***C. loici* Boulard, 2000**

Chremistica loici Boulard 2000a: 256–259, Figs 1–6 (n. sp., described, illustrated, sonogram, comp. note) Madagascar

Chremistica loici Sueur 2001: 36, Table 1 (listed)

Chremistica loici Yaakop, Duffels and Visser 2005: 248 (comp. note) Madagascar

Chremistica loici Boulard 2006g: 136, 150, 181, Fig 93, Fig 110B (sonogram, illustrated, comp. note) Madagascar

***C. longa* Lei, Chou & Li, 1997**

Chremistica longa Lei, Zhou, Chou and Li 1995: 236–238, Figs 1–6 (n. sp., described, illustrated, key, comp. note) China, Hainan

Chremistica longa Chou, Lei, Li, Lu and Yao 1997: 17, 20, 265, 272–273, Fig 3-6B, Fig 3-9B, Fig 9-91 (key, described, genitalia, abdomen, illustrated, comp. note)

Chremistica longa Yaakop, Duffels and Visser 2005: 248 (comp. note) China

***C. maculata* Chou & Lei, 1997** = *Chremistica maculate* (sic)

Chremistica maculata Chou, Lei, Li, Lu and Yao 1997: 266, 271–272, 357, 369, Figs 9-89-9-90 (n. sp., key, described, illustrated, comp. note) China

Chremistica maculata Yaakop, Duffels and Visser 2005: 248 (comp. note) China

Chremistica maculata Sun, Watson, Zheng, Watson and Liang 2009: 3148–3154, Fig 1A, Fig 2A,

Fig 3A, Fig 4A, Tables 1–2 (illustrated, wing microstructure, comp. note) China, Yunnan

Chremistica maculate (sic) Sun, Watson, Zheng, Watson and Liang 2009: 3154 (comp. note)

***C. malayensis* Yaakop & Duffels, 2005**

Chremistica malayensis Yaakop, Duffels and Visser 2005: 249–250, 254, 267, 270–272, 274, 297, 301, Figs 25–28, Fig 71, Fig 82, Table 1 (n. sp., described, illustrated, key, distribution, listed, comp. note) Malayan Peninsula

Chremistica malayensis Sanborn, Phillips and Sites 2007: 6 (listed, distribution, comp. note) Thailand, Peninsular Malaysia

Chremistica malayensis Boulard 2008f: 10 (biogeography, listed) Thailand

***C. martini* (Distant, 1905)** = *Rihana martini*

Chremistica martini group Bregman 1985: 37–39, 41–42, Fig 15 (comp. note) Madagascar, Seychelles

Chremistica martini Bregman 1985: 38 (comp. note)

Chremistica martini Boulard 2000a: 259 (comp. note)

Chremistica martini group Salmah, Duffels and Zaidi 2005: 21 (comp. note) Madagascar

Chremistica martini Salmah, Duffels and Zaidi 2005: 21 (comp. note) Madagascar

Chremistica martini group Yaakop, Duffels and Visser 2005: 248–249 (comp. note) Madagascar

Chremistica martini Yaakop, Duffels and Visser 2005: 248 (comp. note)

***C. matilei* Boulard, 2000**

Chremistica matilei Boulard 2000a: 259–262, Figs 7–12 (n. sp., described, illustrated, sonogram, comp. note) Madasascar

Chremistica matilei Sueur 2001: 36, Table 1 (listed)

Chremistica matilei Boulard 2004: 237 (comp. note) Madagascar

Chremistica matilei Boulard 2006g: 136, 181, Fig 94 (sonogram, listed, comp. note) Madagascar

***C. minor* Bregman, 1985**

Chremistica minor Bregman 1985: 39–44, 46–48, 59, Fig 3, Fig 9, Fig 16, Figs 20–23, Fig 40 (key, n.sp., described, phylogeny, distribution, comp. note) Kalimantan, Sarawak

Chremistica minor Zaidi and Ruslan 1995d: 197 (comp. note) Borneo, Sarawak

Chremistica minor Zaidi 1996: 99 (comp. note) Borneo, Sarawak

Chremistica minor Zaidi and Hamid 1996: 51 (comp. note) Borneo, Sarawak

Chremistica minor Zaidi and Ruslan 1999: 1 (comp. note) Borneo, Sarawak

Chremistica minor Salmah, Zaidi and Duffels 2004: 2–3 (comp. note) Malaysia

Chremistica minor Salmah, Duffels and Zaidi 2005: 15–17, 20–22, Fig 8a, Fig 8b, Table 1 (key, illustrated, distribution, listed, comp. note) Sarawak

Chremistica minor Yaakop, Duffels and Visser 2005: 248–250, 254–255, 284–285, 293–295, 299, 304, Figs 64–67, Fig 75, Fig 92, Table 1 (described, illustrated, key, distribution, listed, comp. note) Malayan Peninsula, Borneo, Sarawak

Chremistica minor Lee 2010a: 16–17 (distribution, listed, comp. note) Philippines, Mindanao, Peninsular Malysia, Sarawak, Indonesia, Kalimantan

***C. minuta* Liu, 1940 nom. nud.** = *Chemistica minuda* (sic)

***C. mixta* (Kirby, 1891)** = *Dundubia mixta* = *Cicada mixta* = *Rihana mixta*

Chremistica mixta Bregman 1985: 39–44, 50–52, 54, Fig 3, Fig 11, Fig 16, Figs 28–31 (key, described, phylogeny, distribution, comp. note) Equals *Dundubia mixta* Equals *Cicada mixta* Equals *Rihana mixta* Sri Lanka

R[ihana] mixta Jiang, Yang, Tang, Xu and Chen 1995b: 229 (song frequency, comp. note)

Chremistica mixta Sueur 2001: 39, Table 1 (listed)

Chremistica mixta Sueur 2002b: 129 (listed) Equals *Rihana mixta*

Chremistica mixta Salmah, Duffels and Zaidi 2005: 22 (comp. note)

Chremistica mixta Yaakop, Duffels and Visser 2005: 248, 285 (comp. note) Sri Lanka

***C. moultoni* Boulard, 2002**

Chremistica moultoni Boulard 2002a: 55, 57–61, 64, Plate J, Figs 3–4, Plate K, Figs 1–3, Plate K", Figs 16–17 (n. sp., described, illustrated, sonogram, comp. note) Thailand

Chremistica moultoni Boulard 2005d: 345 (comp. note) Thailand

Chremistica moultoni Boulard 2005o: 148 (listed)

Chremistica moultoni Yaakop, Duffels and Visser 2005: 248 (comp. note) Thailand

Chremistica moultoni Boulard 2006g: 74–75, 96 (comp. note)

Chremistica moultoni Boulard 2007a: 94, 96 (comp. note)

Chremistica moultoni Sanborn, Phillips and Sites 2007: 6 (listed, distribution, comp. note) Thailand

Chremistica moultoni Boulard 2008c: 360 (comp. note) Tonkin, Thailand, Laos

Chremistica moultoni Boulard 2008f: 10, 91 (biogeography, listed) Thailand

Chremistica moultoni Lee 2010b: 20 (distribution, listed) Cambodia, Vietnam, Laos, Thailand

C. mussarens Boulard, 2005

Chremistica mussarens Boulard 2005n: 130–134, Figs 1–5, Plate A, Figs 1–2 (n. sp., described, illustrated, song, sonogram, comp. note) Thailand

Chremistica mussarens Boulard 2005o: 150 (listed)

Chremistica mussarens Trilar 2006: 344 (comp. note)

Chremistica mussarens Sanborn, Phillips and Sites 2007: 6 (listed, distribution, comp. note) Thailand

Chremistica mussarens Boulard 2008f: 11, 91 (biogeography, listed) Thailand

C. nana Chen, 1943

Chremistica nana Bregman 1985: 39 (comp. note)

Chremistica nana Lei, Zhou, Chou and Li 1995: 236, 238 (comp. note)

Chremistica nana Chou, Lei, Li, Lu and Yao 1997: 265, 267–268, Fig 9–86, Plate XII, Fig 129 (key, described, illustrated, comp. note) China

Chremistica nana Hua 2000: 61 (listed, distribution) China, Hainan

C. nesiotes Breddin, 1905 = *Chremistica bimaculata* (nec Olivier)

Chremistica nesiotes Bregman 1985: 39 (comp. note)

Chremistica nesiotes Salmah and Zaidi 2002: 227, 230 (key, comp. note) Peninsular Malaysia

Chremistica nesiotes Salmah, Zaidi and Duffels 2004: 2–4, 6, Fig 2, Table 1 (type material, illustrated, comp. note) Malaysia

Chremistica nesiotes Schouten, Duffels and Zaidi 2004: 372–373, Table 2 (listed, comp. note) Malayan Peninsula

Chremistica nesiotes Salmah, Duffels and Zaidi 2005: 15–22, Fig 3a, Fig 3b, Table 1 (key, illustrated, distribution, listed, comp. note) Peninsular Malaysia, Sarawak, Sabah, Banguey Island

Chremistica nesiotes Yaakop, Duffels and Visser 2005: 248–250, 254, 267–268, 270, 275–278, 298, 302, Figs 32–35, Fig 72, Fig 84, Table 1 (described, illustrated, key, distribution, listed, comp. note) Equals *Chremistica bimaculata* (nec Olivier) Malayan Peninsula, Borneo, Sarawak, Sabah, Kalimantan, Brunei, Thailand

Chremistica nesiotes Sanborn, Phillips and Sites 2007: 6 (listed, distribution, comp. note) Thailand, Malaysia, Borneo, Sarawak

Chremistica nesiotes Boulard 2008f: 11, 91 (biogeography, listed) Thailand, Borneo

C. niasica Yaakop & Duffels, 2005

Chremistica niasica Yaakop, Duffels and Visser 2005: 249–250, 255, 259–26, 296, 300, Fig 69, Fig 77, Table 1 (n. sp., described, illustrated, distribution, listed, comp. note) Sumatra, Nias

C. nigra Chen, 1940

Chremistica nigra Bregman 1985: 39 (comp. note)

Chremistica nigra Peng and Lei 1992: 99, Fig 281 (listed, illustrated, comp. note) China, Hunan

Chremistica nigra Chou, Lei, Li, Lu and Yao 1997: 266, 268–269, Fig 9–87, Plate XII, Fig 128 (key, described, illustrated, comp. note) China

Chremistica nigra Hua 2000: 61 (listed, distribution) China, Fujian, Hunan, Sichuan

C. nigrans (Distant, 1904) = *Cicada nigrans* = *Rihana nigrans*

Chremistica nigrans Bregman 1985: 38 (comp. note)

Chremistica nigrans Boulard 2000a: 256 (comp. note)

Chremistica nigrans Salmah, Duffels and Zaidi 2005: 21 (comp. note) Madagascar

Chremistica nigrans Yaakop, Duffels and Visser 2005: 248 (comp. note)

Chremistica nigrans Boulard 2006g: 181 (listed)

C. numida (Distant, 1911) = *Rihana numida*

Chremistica numida Wu 1935: 5 (listed) Equals *Rihana numida* China

Chremistica numida Bregman 1985: 39 (comp. note)

Chremistica numida Chou, Lei, Li, Lu and Yao 1997: 266, 270–271, 369, Fig 9–88, Plate XII, Fig 127 (key, described, illustrated, synonymy, comp. note) Equals *Rihana numida* China

Chremistica numida Hua 2000: 61 (listed, distribution) Equals *Rihana numida* China, Yunnan

Chremistica numida Boulard 2003c: 172, 176–179, Figs 1–6, Plate B, Figs 1–2, Plate C, Figs 1–2 (described, illustrated, song, sonogram, comp. note) Equals *Rihana numida* Thailand, China

Chremistica numida Boulard 2005d: 338, 344 (comp. note)

Chremistica numida Boulard 2005o: 149 (listed)

Chremistica numida Salmah, Duffels and Zaidi 2005: 22 (comp. note)

Chremistica numida Yaakop, Duffels and Visser 2005: 248 (comp. note)

Chremistica numida Boulard 2006c: 137 (comp. note)

Chremistica numida Boulard 2006g: 36, 39, 74–76, 90, Fig 54 (sonogram, comp. note)

Chremistica numida Boulard 2007e: 35, 41, Fig 25 (sonogram, comp. note) Thailand

Chremistica numida Sanborn, Phillips and Sites 2007: 6–7 (listed, distribution, comp. note) Thailand, China

Chremistica numida Boulard 2008f: 11, 91 (synonymy, biogeography, listed) Equals *Rihana numida* Thailand

C. *ochracea gracilis* **(Kato, 1927)** = *Rihana ochracea gracilis*
Chremistica ochracea gracilis Hua 2000: 61 (listed, distribution) Equals *Rihana ochracea gracilis* China, Taiwan

C. *ochracea interrupta* **(Kato, 1927)** = *Rihana ochracea interrupta*
Chremistica ochracea interrupta Hua 2000: 61 (listed, distribution) Equals *Rihana ochracea interrupta* China, Taiwan

C. *ochracea ochracea* **(Walker, 1850)** = *Fidicina ochracea* = *Cicada ochracea* = *Rihana ochracea* = *Rhiana* (sic) *ochracea* = *Rihana ochrea* (sic) = *Tibicen ochracea* = *Chremistica cohraces* (sic) = *Chremistica ochraceus* (sic) = *Cicada ferrifera* Walker, 1850 = *Dundubia fasciceps* Stål, 1854 = *Dundubia fuscipes* (sic) = *Dundubia fusciceps* (sic) = *Cicada fuscipes* (sic)
Rihana ochracea Poulton 1907: 336, 344 (predation) China
Chremistica ochracea Chen 1933: 359 (listed, comp. note)
Chremistica ochracea Wu 1935: 5 (listed) Equals *Fidicina ochracea* Equals *Cicada ochracea* Equals *Rihana ochracea* Equals *Cicada ferrifera* Equals *Dundubia fascipes* Equals *Cicada fuscipes* China
Chremistica ochracea Bregman 1985: 39 (comp. note)
Rihana ochracea Liu 1991: 255 (comp. note)
Rihana ochracea Liu 1992: 14 (comp. note)
Rihana ochracea Wang and Zhang 1993: 188 (listed, comp. note) China
Chremistica ochracea Anonymous 1994: 808 (fruit pest) China
Chremistica ochracea Chou, Lei, Li, Lu and Yao 1997: 28, 265–267, Fig 9–85, Plate XII, Fig 126 (key, described, illustrated, synonymy, comp. note) Equals *Fidicina ochracea* Equals *Cicada ferrifera* Equals *Dundubia fusciceps* (sic) Equals *Cicada fuscipes* (sic) Equals *Cicada ochracea* Equals *Rihana ochracea* Equals *Tibicen ochracea* China
Fidicina ochracea Chou, Lei, Li, Lu and Yao 1997: 265 (type species of *Chremistica*)
Chremistica ochracea Easton and Pun 1999: 101 (distribution, comp. note) Equals *Dundubia ochracea* Macao, Hong Kong, Taiwan, China, Guangdong
Fidicina ochracea Boulard 2000a: 255 (type species of *Rihana*, comp. note)
Chremistica ochraceus (sic) Hua 2000: 61 (listed, distribution, hosts) Equals *Fidicina ochracea* China, Fujian, Taiwan, Guangdong, Guangxi, India, Malaysia
Chremistica ochracea Boulard 2001c: 51–53, Plate A, Figs 1–2 (song, illustrated, sonogram, comp. note) Equals *Fidicina ochracea* Equals *Cicada ferrifera* Equals *Dundubia fasciceps* Equals *Rihana ochracea* China, Taiwan, Japan, India, Malaysia, Thailand
Fidicina ochracea Boulard 2001e: 117 (type species of *Rihana*, comp. note)
Fidicina ochracea Lee and Hayashi 2003a: 160 (type species of *Chremistica* and *Rihana*) China
Chremistica ochracea Lee and Hayashi 2003a: 161–162, Fig 9 (key, diagnosis, illustrated, synonymy, distribution, song, listed, comp. note) Equals *Fidicina ochracea* Equals *Rihana ochracea* Equals *Cicada ochracea* Equals *Rihana ochracea* var. *takesakiana* Equals *Chremistica ochracea* var. *takesakiana* Equals *Rihana ochracea* var. *gracilis* Equals *Chremistica ochracea* var. *gracilis* Equals *Rihana ochracea* var. *interrunpta* Equals *Chremistica ochracea* var. *interrupta* Taiwan, China
Chremistica ochracea Chen 2004: 26–29, 62–63, 198 (described, illustrated, eclosion, listed, comp. note) Taiwan
Chremistica ochracea Salmah, Zaidi and Duffels 2004: 2–3 (comp. note) Malaysia
Chremistica ochracea Boulard 2005o: 148 (listed) Equals *Aola bindusara* sonogram Boulard 2001
Chremistica ochracea Salmah, Duffels and Zaidi 2005: 15–16, 19, 22, Table 1 (key, synonymy, distribution, listed, comp. note) Equals *Fidicina ochracea* Equals *Cicada ferrijera* Equals *Dundubia fascipes* Equals *Rihana ochracea* Peninsular Malaysia
Fidicina ochreacea Yaakop, Duffels and Visser 2005: 248 (type species of *Rihana*)
Chremistica ochracea Yaakop, Duffels and Visser 2005: 248 (comp. note)
Chremistica ochracea Sanborn, Phillips and Sites 2007: 7 (listed, distribution, comp. note) Thailand, India, China, Malaysia, Taiwan
Chremistica ochracea Shiyake 2007: 3, 32, 43, Fig 38 (illustrated, distribution, listed, comp. note) China, Taiwan
Chremistica ochracea Boulard 2008c: 360 (comp. note) Malaysia
Fidicina ochracea Lee 2008a: 4 (type species of *Rihana*)
Chremistica ochracea Lee, Lin and Wu 2010: 218–224, Figs 1–6, Tables 1–2 (ecology, comp. note) Taiwan
Fidicina ochracea Lee 2010c: 15 (type species of *Rihana*, listed) China

C. *ochracea* var. *a* **(Kato, 1925)** = *Rihana ochracea* var. *a*

C. *ochracea* var. *b* **(Kato, 1925)** = *Rihana ochracea* var. *b*

C. *ochracea takesakiana* **(Kato, 1927)** = *Rihana ochracea takesakiana*

Chremistica ochracea takesakiana Hua 2000: 61 (listed, distribution) Equals *Rihana ochracea takesakiana* China, Taiwan

***C. operculissima* (Distant, 1897)** = *Cicada operculissima* = *Rihana operculissima*

Chremistica operculissima Bregman 1985: 38 (comp. note)

Chremistica operculissima Salmah, Duffels and Zaidi 2005: 21 (comp. note) Lesser Sunda Islands

Chremistica operculissima Yaakop, Duffels and Visser 2005: 248 (comp. note)

***C. phamiangensis* Boulard, 2009**

Chremistica phamiangensis Boulard 2009d: 39–41, 49–50, 58, Figs 1–6 (n. sp., described, illustrated, song, sonogram, comp. note) Thailand

***C. polyhymnia* (Walker, 1850)** = *Fidicina polyhymnia* = *Cicada polyhymnia* = *Rihana polyhymnia*

Chremistica polyhymnia Bregman 1985: 39–40, 44 (key, comp. note) Philippines

Chremistica polyhymnia Salmah, Duffels and Zaidi 2005: 22 (comp. note)

Chremistica polyhymnia Yaakop, Duffels and Visser 2005: 248 (comp. note)

***C. pontianaka* (Distant, 1888)** = *Cicada pontianaka* = *Rihana pontianaka* = *Cicada daiaca* Breddin, 1900

Chremistica pontianaka group Bregman 1985: 38–39, 41–42, 50, Fig 15 (comp. note) Southeast Asia, Sundaland, Greater Sunda Islands, Java, Sumatra, Kalimantan

Chremistica pontianaka Bregman 1985: 39 (comp. note)

Chremistica pontianaka Zaidi, Mohammedsaid and Yaakob 1990: 262–265, Fig 1, Table 1 (seasonality, comp. note) Malaysia

Chremistica pontianaka Zaidi 1993: 958, Table 1 (listed, comp. note) Borneo, Sarawak

Chremistica pontianaka Zaidi 1996: 101–102, 104, Table 1 (comp. note) Borneo, Sarawak

Chremistica pontianaka Zaidi and Hamid 1996: 53, 56, Table 1 (listed, comp. note) Borneo, Sarawak

Chremistica pontianaka Zaidi 1997: 113–114, 116, Table 1 (listed, comp. note) Borneo, Sarawak

Chremistica pontianaka Zaidi and Ruslan 1997: 220 (listed, comp. note) Equals *Cicada pontianaka* Equals *Rihana pontianaka* Borneo, Peninsular Malaysia, Sumatra, Java, Sulu, New Guinea, Philippines

Chremistica pontianaka Zaidi and Ruslan 1998a: 347 (listed, comp. note) Equals *Cicada pontianaka* Equals *Rihana pontianaka* Borneo, Sarawak, Peninsular Malaysia, Sumatra, Java, Sulu, Banguey, Philippines, New Guinea, Singapore

Chremistica pontianaka Zaidi, Ruslan and Azman 1999: 302 (listed, comp. note) Equals *Cicada pontianaka* Equals *Rihana pontianaka* Borneo, Sabah, Peninsular Malaysia, Sumatra, Java, Sulu, Banguey, Philippines, New Guinea, Sarawak

Chremistica pontianaka Freytag 2000: 55 (on stamp)

Chremistica pontianaka Zaidi, Ruslan and Azman 2000: 200 (listed, comp. note) Equals *Cicada pontianaka* Equals *Rihana pontianaka* Borneo, Sabah, Peninsular Malaysia, Sumatra, Java, Sulu, Banguey Island, Sarawak, Philippines, New Guinea, Johor, Singapore

Chremistica pontianaka Zaidi, Azman and Ruslan 2001a: 112, 115, 117, Table 1 (listed, comp. note) Equals *Cicada pontianaka* Equals *Rihana pontianaka* Langkawi Island, Borneo, Peninsular Malaysia, Sumatra, Java, Sulu, Banguey Island, Philippines, New Guinea, Johor, Singapore

Chremistica pontianaka Salmah and Zaidi 2002: 227–229, 237 (key, synonymy, comp. note) Equals *Cicada pontanaka* Equals *Rihana pontianaka* Peninsular Malaysia

Chremistica pontianaka Zaidi and Azman 2003: 104, Table 1 (listed, comp. note) Borneo, Sabah

Chremistica pontianaka Salmah, Zaidi and Duffels 2004: 2–4, 6, Fig 1, Table 1 (type material, illustrated, synonymy, comp. note) Malaysia

Cicada pontianaka Salmah, Zaidi and Duffels 2004: 2 (comp. note) Malaysia

Chremistica pontianaka Schouten, Duffels and Zaidi 2004: 372–373, Table 2 (listed, comp. note) Malayan Peninsula

Chremistica pontianaka Gogala and Trilar 2005: 66, 69–70, 78, Fig 6 (song, sonogram, oscillogram, acoustic behavior, comp. note) Malaysia

Chremistica pontianaka Salmah, Duffels and Zaidi 2005: 15–18, 20–22, Fig 1a, Fig 2a, Table 1 (key, synonymy, illustrated, distribution, listed, comp. note) Equals *Cicada pontianaka* Equals *Rihana pontianaka* Malaysia, Java, Sumatra, Sulu Island, Siam, Mollucca, Ambiona, Palawan, Billiton Island, New Guinea

Chremistica pontianaka group Salmah, Duffels and Zaidi 2005: 21–22 (comp. note)

Rihana pontianaka Yaakop, Duffels and Visser 2005: 248 (comp. note) Malaysia

Chremistica pontianaka group Yaakop, Duffels and Visser 2005: 248–249, 255 (listed, comp. note) Southeast Asia, Sundaland

Chremistica pontianaka Yaakop, Duffels and Visser 2005: 248–250, 253–254-255–260, 296, 300, Fig 2, Figs 4–7, Fig 68, Fig 76, Table 1 (lectotype designation, described, illustrated, synonymy, key, distribution, listed, comp. note) Equals *Cicada pontianaka* Equals *Cicada daiaca* Equals *Rihana pontianaka* Malayan Peninsula, Borneo, Sarawak, Sabah, Kalimantan

Cicada pontianaka Yaakop, Duffels and Visser 2005: 255 (type material, comp. note) Borneo

Chremistica pontianaka Sanborn, Phillips and Sites 2007: 7 (listed, distribution, comp. note) Equals *Cicada pontianaka* Equals *Rihana pontianaka* Equals *Chremistica siamensis* Thailand, Malaysia, Indonesia, Sarawak, Borneo, New Guinea

Chremistica pontianaka Boulard 2008f: 11, 90–91 (synonymy, biogeography, listed) Equals *Cicada pontianaka* Equals *Rihana pontianaka* Equals *Chremistica siamensis* (nec Bregman) Thailand, Peninsular Malaysia, Malaysia, Siam

Chremistica pontianaka Lee 2009h: 2637 (comp. note) Sulu Island

Chremistica pontiananka Lee 2010c: 2 (synonymy, distribution, comp. note) Equals *Cicada pontianaka* Philippines, Luzon, Malaysia, Peninsular Malaysia, Sabah, Sarawak, Indonesia, Kallimantan

C. nr. *pontianaka* Zaidi, Ruslan & Mahadir, 1996

Chremistica nr. *pontianaka* Zaidi, Ruslan and Mahadir 1996: 60, Table 1 (listed, comp. note) Peninsular Malaysia

C. *pulverulenta madagascariensis* Boulard, 2001

Chremistica pulverulenta madagascariensis Boulard 2001f: 130–132, 137, Figs 1–2, Plate I, Figs 1–2 (n. ssp., described, illustrated, comp. note) Equals *Cicada pulverulenta* Equals *Tibicen pulverulenta* Madagascar

C. *pulverulenta pulverulenta* (Distant, 1905) = *Cicada pulverulenta* = *Tibicen pulverulenta* = *Antankaria pulverulenta madegassa* Boulard, 1999

Chremistica pulverulenta Bregman 1985: 38 (comp. note)

Antankaria pulverulenta madegassa Boulard 1999a: 95 (n. comb., synonymy, sonogram) Equals *Cicada pulverulenta* Equals *Chremistica pulverulenta* Madagascar

Chremistica pulverulenta Boulard 2001f: 129–130, 138, Carte A (listed, sonogram, comp. note)

Antankaria pulverulenta madegassa Sueur 2001: 36, Table 1 (listed)

Chremistica pulverulenta Salmah, Duffels and Zaidi 2005: 21 (comp. note) Madagascar

Chremistica pulverulenta Yaakop, Duffels and Visser 2005: 248 (comp. note)

C. *pulverulenta seychellensis* (Boulard, 1999) = *Chremistica pulverulenta mahiensis* (sic) = *Antankaria pulverulenta madegassa*

Antankaria pulverulenta seychellensis Boulard 1999a: 79, 95–96 (n. ssp., listed, sonogram, comp. note) Equals *Cicada pulverulenta* Equals *Chremistica pulverulenta* Seychelles

Chremistica pulverulenta seychellensis Boulard 2001f: 133, 136, Figs 3–4 (n. ssp., described, illustrated, comp. note) Seychelles, Aldabra

Chremistica pulverulenta mahiensis (sic) Boulard 2001f: 137, Plate I, Figs 3–4 (illustrated)

Antankaria pulverulenta sechellensis Sueur 2001: 36, Table 1 (listed)

C. *seminiger* (Distant, 1909) = *Rihana seminiger*

Chremistica seminiger Bregman 1985: 39–46, Fig 3, Fig 8, Figs 16–19 (key, described, phylogeny, distribution, comp. note) Equals *Rihana semingira* India

Chremistica seminiger Salmah, Duffels and Zaidi 2005: 22 (comp. note)

Chremistica seminiger Yaakop, Duffels and Visser 2005: 248 (comp. note)

C. *semperi* Stål, 1870 = *Cicada (Chremistica) semperi* = *Cicada semperi* = *Rihana semperi*

Chremistica semperi Bregman 1985: 39–40, 44 (key, comp. note) Philippines

Chremistica semperi Salmah, Duffels and Zaidi 2005: 15, 22 (comp. note)

Cicada (Chremistica) semperi Yaakop, Duffels and Visser 2005: 248, 253 (comp. note)

Chremistica semperi Yaakop, Duffels and Visser 2005: 248 (comp. note)

C. *siamensis* Bregman, 1985

Chremistica siamensis Bregman 1985: 39–44, 48–50, 54, 59, Fig 3, Fig 10, Figs 24–27, Fig 41 (key, n. sp., described, illustrated, phylogeny, distribution, comp. note) Thailand

Chremistica siamensis Boulard 2002a: 58 (comp. note)

Chremistica siamensis Salmah, Zaidi and Duffels 2004: 2–3 (comp. note) Malaysia

Chremistica siamensis Salmah, Duffels and Zaidi 2005: 15–16, 19, 22, Table 1 (key, distribution, listed, comp. note) Thailand

Chremistica siamensis Yaakop, Duffels and Visser 2005: 248, 285 (comp. note) Thailand

Chremistica siamensis Boulard 2008f: 6, 12, 87, 89, 91, Fig 43, Fig 45 (song, sonogram, biogeography, listed, comp. note) Thailand

C. *sibilussima* Boulard, 2006

Chremistica sibilussima Boulard 2006c: 136–138, Fig C, Figs 4–5 (n. sp., described, illustrated, song, sonogram, comp. note) Thailand

Chremistica sibilussima Boulard 2008f: 12, 91 (biogeography, listed) Thailand

Chremistica sibilussima Boulard 2009d: 39 (comp. note)

C. sp. 1 Zaidi, Ruslan & Mahadir, 1996
Chremistica sp. 1 Zaidi, Ruslan and Mahadir 1996: 60, Table 1 (listed, comp. note) Peninsular Malaysia

C. sp. 2 Zaidi, Ruslan & Mahadir, 1996
Chremistica sp. 2 Zaidi, Ruslan and Mahadir 1996: 60, Table 1 (listed, comp. note) Peninsular Malaysia

***C. sumatrana* Yaakop & Duffels, 2005**
Chremistica sumatrana Yaakop, Duffels and Visser 2005: 249–250, 254, 267–268, 278, 280–283, 298, 302, Figs 40–43, Fig 73, Fig 86, Table 1 (n. sp., described, illustrated, key, distribution, listed, comp. note) Sumatra

***C. tagalica* Stål, 1870** = *Cicada (Chremistica) tagalica* = *Cicada tagalica* = *Rihana tagalica*
Chremistica tagalica Bregman 1985: 39–40, 44 (key, comp. note) Philippines
Cicada (Chremistica) tagalica Boulard 2000a: 255 (comp. note)
Cicada (Chremistica) tagalica Boulard 2001e: 115 (comp. note)
Chremistica tagalica Salmah, Duffels and Zaidi 2005: 15 (comp. note)
Cicada (Chremistica) tagalica Yaakop, Duffels and Visser 2005: 248, 253 (comp. note)
Chremistica tagalica Yaakop, Duffels and Visser 2005: 248 (comp. note)

***C. timorensis* (Distant, 1892)** = *Cicada timorensis* = *Rihana timorensis*
Chremistica timorensis Bregman 1985: 38 (comp. note)
Chremistica timorensis Salmah, Duffels and Zaidi 2005: 21 (comp. note) Lesser Sunda Islands
Chremistica timorensis Yaakop, Duffels and Visser 2005: 248 (comp. note)

***C. tondana* (Walker, 1870)** = *Fidicina tondana* = *Cryptotympana tondana*
Chremistica tondana Bregman 1985: 39 (comp. note) Sulawesi
Chremistica tondana Duffels 1990b: 64, Table 2 (listed) Sulawesi
Chremistica tondana Boer and Duffels 1996b: 330 (comp. note) Sulawesi

***C. tridentigera* (Breddin, 1905)** = *Cicada tridentigera*
Chremistica tridentigera group Bregman 1985: 37–43, Fig 1, Fig 3, Figs 15–16 (illustrated, phylogeny, comp. note) Southeast Asia, Malaysian Archipelago, Philippines, Banguey, India, Sri Lanka
Chremistica tridentigera Bregman 1985: 39–44, 53, 55–58, Figs 3–5, Fig 14, Fig 16, Figs 37–39 (key, lectotype designation, phylogeny, comp. note) Equals *Cicada tridentigera* Kalimantan, Brunei, Banguey
Cicada tridentigera Bregman 1985: 57 (comp. note)
Chremistica tridentigera group Duffels 1986: 319 (comp. note) Philippines
Chremistica tridentigera Zaidi and Ruslan 1995d: 197 (comp. note) Borneo, Sabah, Brueni, Sarawak
Chremistica tridentigera Zaidi 1996: 99 (comp. note) Borneo, Sarawak
Chremistica tridentigera Zaidi and Hamid 1996: 51 (comp. note) Borneo, Sabah, Brunei, Sarawak
Chremistica tridentigera Zaidi 1997: 110 (comp. note) Borneo, Sabah, Brunei, Sarawak
Chremistica tridentigera group Holloway 1998: 305, Fig 10 (distribution, phylogeny, comp. note) Philippines
Chremistica tridentigera Zaidi and Ruslan 1999: 1 (comp. note) Borneo, Sarawak, Sabah, Brunei
Chremistica tridentigera Zaidi and Azman 2003: 99, 104, Table 1 (listed, comp. note) Borneo, Sabah
Chremistica tridentigera Salmah, Zaidi and Duffels 2004: 2–3 (comp. note) Malaysia
Chremistica tridentigera group Salmah, Zaidi and Duffels 2004: 2 (comp. note) Malaysia
Chremistica tridentigera Salmah, Duffels and Zaidi 2005: 15–17, 19–22, Fig 5a, Fig 5b, Table 1 (key, illustrated, synonymy, distribution, listed, comp. note) Peninsular Malaysia, Banguey Island, Borneo, Buru, Sumatra
Chremistica tridentigera group Salmah, Duffels and Zaidi 2005: 21–22 (comp. note)
Chremistica tridentigera group Yaakop, Duffels and Visser 2005: 248–250, 255, 285, 291, 293 (comp. note) Southeast Asia, Sri Lanka, India, Sundaland
Chremistica tridentigera Yaakop, Duffels and Visser 2005: 248–250, 253–255, 285–288, 291, 299, 303, Fig 3, Figs 48–51, Fig 74, Fig 88, Table 1 (described, illustrated, synonymy, key, distribution, listed, comp. note) Equals *Cicada tridentigera* Borneo, Sarawak, Sabah, Kalimantan, Brunei

***C. umbrosa* (Distant, 1904)** = *Cicada umbrosa* = *Rihana umbrosa* = *Rihana pisanga* Moulton, 1923 = *Chremistica pisanga*
Chremistica umbrosa Bregman 1985: 39–44, 52–55, Fig 3, Figs 6–7, Fig 12, Fig 16, Figs 32–34 (key, described, phylogeny, comp. note) Equals *Cicada umbrosa* Equals *Rihana umbrosa* Equals *Rihana pisanga* Equals *Chremistica pisanga* Malacca, Palau Tujuh, Palau Pisang, Palau Buru, Banka Island, Malayia, Sumatra
Chremistica pisanga Bregman 1985: 52 (comp. note) Palau Pisang

Rihana pisanga Bregman 1985: 52 (comp. note)
Chremistica viridis Bregman 1985: 39 (comp. note)
Chremistica pisanga Zaidi, Ruslan and Mahadir 1996: 60, Table 1 (listed, comp. note) Peninsular Malaysia
Chremistica umbrosa Zaidi and Ruslan 1998a: 348 (listed, comp. note) Equals *Cicada umbrosa* Equals *Rihana umbrosa* Equals *Rihana pisanga* Equals *Chremistica pisanga* Borneo, Sarawak, Buru, Peninsular Malaysia, Pisang, Sumatra, Singapore
Chremistica umbrosa Salmah and Zaidi 2002: 226, 229–230 (key, synonymy, comp. note) Equals *Cicada umbrosa* Equals *Rihana umbrosa* Equals *Rihana pisanga* Equals *Chremistica pisanga* Peninsular Malaysia
Chremistica umbrosa Salmah, Zaidi and Duffels 2004: 2–3 (comp. note) Malaysia
Chremistica umbrosa Schouten, Duffels and Zaidi 2004: 372–373, Table 2 (listed, comp. note) Malayan Peninsula
Chremistica umbrosa Salmah, Duffels and Zaidi 2005: 15–16, 20–22, Fig 7a, Fig 7b, Table 1 (key, illustrated, synonymy, distribution, listed, comp. note) Equals *Cicada umbrosa* Equals *Rihana umbrosa* Equals *Rihana pisanga* Equals *Chremistica pisanga* Peninsular Malaysia, Borneo, Sumatra
Rihana pisanga Yaakop, Duffels and Visser 2005: 248 (comp. note) Malaysia
Chremistica umbrosa Yaakop, Duffels and Visser 2005: 248–251, 254–255, 285, 289–291, 299, 303, Figs 56–59, Fig 74, Fig 90, Table 1 (described, illustrated, synonymy, key, distribution, listed, comp. note) Equals *Cicada umbrosa* Equals *Rihana umbrosa* Equals *Rihana pisanga* Equals *Chremistica pisanga* Malayan Peninsula, Sumatra

C. undescribed species B Sanborn, Phillips & Sites, 2007
Chremistica undescribed species B Sanborn, Phillips and Sites 2007: 7 (listed, distribution, comp. note) Thailand

C. undescribed species C Sanborn, Phillips & Sites, 2007
Chremistica undescribed species C Sanborn, Phillips and Sites 2007: 7 (listed, distribution, comp. note) Thailand

C. undescribed species Pham & Yang, 2010
Chremistica undescribed species Pham and Yang 2010a: 133 (comp. note) Vietnam

***C. viridis* (Fabricius, 1803)** = *Tettigonia viridis* = *Cicada viridis* = *Cicada (Chremistica) viridis* = *Rihana viridis* = *Cicada bimaculata* Jacobi (nec Olivier)
Tettigonia viridis Dallas 1870: 482 (synonymy) Equals *Cicada viridis* Equals *Cicada bimaculata* Equals *Cicada atro-virens*
Chremistica viridis Zaidi, Mohammedsaid and Yaakob 1990: 262 (comp. note) Malaysia
Chremistica viridis Zaidi and Ruslan 1994: 428 (comp. note) Malaysia
C[hremistica] viridis Jiang, Yang, Tang, Xu and Chen 1995b: 229 (song frequency, comp. note)
Chremistica viridis Zaidi, Ruslan and Mahadir 1996: 60, Table 1 (listed, comp. note) Peninsular Malaysia
Chremistica viridis Salmah and Zaidi 2002: 227, 231 (key, synonymy, comp. note) Equals *Tettigonia viridis* Equals *Cicada viridis* Equals *Rihana viridis* Peninsular Malaysia
Chremistica viridis Salmah, Zaidi and Duffels 2004: 2–3 (comp. note) Malaysia
Chremistica viridis Salmah, Duffels and Zaidi 2005: 15–16, 18–19, 22, Table 1 (type species of *Chremistica*, key, synonymy, distribution, listed, comp. note) Equals *Tettigonia viridis* Equals *Cicada viridis* Equals *Rihana viridis* Malaysia
Cicada viridis Yaakop, Duffels and Visser 2005: 247, 251, 253 (comp. note)
Cicada (Chremistica) viridis Yaakop, Duffels and Visser 2005: 248 (type species of *Chremistica*)
Rihana viridis Yaakop, Duffels and Visser 2005: 248 (comp. note) Malaysia
Chremistica viridis Yaakop, Duffels and Visser 2005: 248 (comp. note) Malaysia
Chremistica viridis Boulard 2007e: 5 (coloration, comp. note)
Chremistica viridis Sanborn, Phillips and Sites 2007: 7 (listed, synonymy, distribution, comp. note) Equals *Rihana viridis* Thailand, Philippine Republic, Indonesia, Malaysia, Indochina, Sri Lanka
Chremistica viridis Shiyake 2007: 3, 32, 43, Fig 39 (illustrated, distribution, listed, comp. note) Southeast Asia
Chremistica viridis Pham and Yang 2009: 13, Table 2 (listed) Vietnam

***Diceroprocta* Stål, 1870** = *Cicada (Diceroprocta)* = *Diceropracta* (sic) = *Diceroprocter* (sic) = *Tibicen (Diceroprocta)* = *Rehana* (sic) = *Tibcien* (sic)
Diceroprocta Tinkham 1941: 172–174, Fig 2 (singing posture, comp. note)
Diceroprocta Heath 1983: 13 (comp. note) North America
Diceroprocta Strümpel 1983: 18 (listed)
Diceroprocta Boulard 1984c: 171 (comp. note)
Diceroprocta Hayashi 1984: 35 (listed)
Diceroprocta Arnett 1985: 217 (diversity) United States

Diceroprocta Bregman 1985: 39 (comp. note)
Diceroprocta Hayashi 1987a: 124 (listed)
Diceroprocta De Zayas 1988: 14 (key) Cuba
Diceroprocta Ramos 1988: 62–63, Table 4.1 (listed, distribution, comp. note) Greater Antilles, Cuba, Jamaica, New World
Diceroprocta Sanborn, Heath and Heath 1992: 756 (comp. note)
Diceroprocta Lamberton 1994: 9 (life cycle, predation, comp. note) New Mexico
Diceroprocta Veenstra and Hagedorn 1995: 393 (comp. note)
Diceroprocta Moore 1996: 221 (comp. note) Mexico
Diceroprocta Novotny and Wilson 1997: 437 (listed)
Diceroprocta Poole, Garrison, McCafferty, Otte and Stark 1997: 251, 331 (listed) Equals *Diceropracta* (sic) North America
Diceroprocta Ohbayashi, Sato and Igawa 1999: 340 (parasitism, comp. note)
Diceroprocta Sanborn 1999b: 1 (listed) North America
Diceroprocta Arnett 2000: 298 (diversity) United States
Diceroprocta Maw, Footit, Hamilton and Scudder 2000: 49 (listed)
Diceroprocta Cioara and Sanborn 2001: 25 (song, comp. note)
Diceroprocta Cooley 2001: 756 (comp. note)
Diceroprocta spp. Sanborn 2001c: 651 (comp. note) West Indies
Diceroprocta Kondratieff, Ellingson and Leatherman 2002: 14, 16–17, 19 (key, listed, diagnosis, diversity, comp. note) Colorado
Diceroprocta Sueur 2002b: 122, 124 (comp. note)
Diceroprocta Sanborn, Heath, Heath and Phillips 2004: 66, 70 (comp. note) Florida
Diceroprocta Moulds 2005b: 390, 431 (higher taxonomy, listed, comp. note)
Cicada (Diceroprocta) Salmah, Duffels and Zaidi 2005: 15 (comp. note)
Diceroprocta Ganchev and Potamitis 2007: 304–305, 319, Fig 11 (listed, comp. note)
Diceroprocta sp. Phillips and Sanborn 2007: 456, Fig 1 (listed) Texas
Diceroprocta Sanborn 2007b: 13, 18, 31, 40 (comp. note) Mexico
Diceroprocta Shiyake 2007: 3, 34 (listed, comp. note)
Diceroprocta Sanborn, Moore and Young 2008: 1 (comp. note)
Diceroprocta Sanborn, Phillips and Gillis 2008: 3, 5, 7 (key, diversity, described, comp. note) Florida
Diceroprocta spp. Sanborn, Phillips and Gillis 2008: 5 (comp. note)
Diceroprocta Holliday, Hastings and Coelho 2009: 3, 13 (predation, diversity, comp. note)
Diceroprocta spp. Holliday, Hastings and Coelho 2009: 12 (predation, comp. note)
Diceroprocta McCutcheon, McDonald and Moran 2009b: 15397 (symbiont, comp. note)
Diceroprocta Sanborn 2009a: 89 (listed, comp. note) Cuba
Tibicen (sic) Strausfeld, Sinakevitch, Brown and Farris 2009: 273 (comp. note)
Diceroprocta Strausfeld, Sinakevitch, Brown and Farris 2009: 277 (comp. note)
Diceroprocta Stucky 2009: 123, 125 (comp. note) Kansas
Diceroprocta Matthews and Matthews 2010: 114 (comp. note)
Diceroprocta Sanborn 2010a: 1586 (listed) Colombia
Diceroprocta Sanborn and Phillips 2010: 863 (comp. note)

***D. alacris alacris* (Stål, 1864)** = *Cicada (Diceroprocta) alacris* = *Cicada transversa* Walker, 1858 = *Pomponia transversa* = *Cicada (Diceroprocta) transversa* = *Rihana transversa* = *Diceroprocta transversa* = *Tibicen transversa*
Diceroprocta alacris Poole, Garrison, McCafferty, Otte and Stark 1997: 331 (listed) Equals *Cicada alacris* Equals *Cicada transversa* Equals *Diceroprocta campechensis* North America
Cicada transversa Kondratieff, Ellingson and Leatherman 2002: 14–15 (type species of *Diceroprocta*)
Diceroprocta alacris Sanborn 2007b: 31 (distribution, comp. note) Mexico, Veracruz, Nuevo León, Tabasco, Yucután, Jalisco
Cicada alacris Sanborn, Phillips and Gillis 2008: 5 (type species of *Diceroprocta*)
Cicada alacris Sanborn 2009a: 89 (type species of *Diceroprocta*)
Cicada alacris Sanborn 2010a: 1586 (type species of *Diceroprocta*) Mexico

***D. alacris campechensis* Davis, 1938** = *Diceroprocta alacris var. campechensis*
Diceroprocta alacris var. *campechensis* Sanborn 1999a: 37 (type material, listed)
Diceroprocta alacris campechensis Sanborn 2007b: 32 (distribution, comp. note) Mexico, Campeche

***D. albomaculata* Davis, 1928**
Diceroprocta albomaculata Sanborn 1999a: 37 (type material, listed)
Diceroprocta albomaculata Sanborn 2007b: 32 (distribution, comp. note) Mexico, San Luis Potosí, Veracruz

***D. apache* (Davis, 1921)** = *Tibicen apache* = *Diceropracta* (sic) *apache* = *Diceroprocter* (sic) *apache* = *Cicada transversa* (nec Walker) = *Cicada vitripennis* (nec Say) = *Diceroprocta apache ochroleuca* Davis, 1942 = *Diceroprocta apache ochraleuca* (sic) = *Tibicen cinctifera* (nec Uhler) = *Diceroprocta cinctifera* (nec Uhler)

Cicada transversa (sic) Snow 1904: 349 (listed) Arizona
Tibicen apache Davis 1922: 73 (comp. note) California
Diceroprocta cinctifera (sic) Glick 1923: 957 (citrus pest) Arizona
Diceroprocta apache Evans 1966: 110 (parasitism, comp. note)
Tibicen cinctifera (sic) Johnson and Lyon 1976: 434 (agricultural pest, comp. note) California, Arizona
Diceroprocta apache Heinrich 1979: 1270 (thermal maximum, comp. note)
Diceroprocta apache Krombein, Hurd, Smith and Burks 1979: 1698 (parasitism)
Diceroprocta apache Powell and Hogue 1979: 113–114, Figs 120–121 (illustrated, described, distribution, comp. note) California
Diceroprocta apache Sorenson and Mills 1981: 1 (life history, oviposition damage, control, comp. note) Nevada
Diceroprocta apache Burk 1982: 96 (comp. note)
Diceroprocta apache Rosenberg, Ohmart and Anderson 1982: 270 (predation, comp. note) Arizona
Diceroprocta apache Glinski and Ohmart 1983: 202, 206 (predation, comp. note) Arizona
Diceroprocta apache Karban 1983b: 328 (comp. note)
Diceroprocta apache Walker 1983: 63 (comp. note)
Diceroprocta apache Glinski and Ohmart 1984: 73–79, Figs 1–2, Table 1 (natural history, population density, host plants, comp. note) Arizona
Diceroprocta apache Claridge 1985: 306 (comp. note) North America
Diceroprocta apache Toolson 1985: 144A (evaporative cooling)
Diceroprocta apache Karban 1986: 222, 224, Table 14.2 (comp. note) Arizona, Mexico
Diceroprocta apache Toolson 1987: 380–385, Fig 1, Tables 1–3 (evaporative cooling, comp. note) Arizona
Diceroprocta apache Toolson and Hadley 1987: 439–443, Figs 1–2, Table 1 (evaporative cooling, illustrated, comp. note) Arizona
Diceroprocta apache Ohmart, Anderson and Hunter 1988: 92 (ecology, predation, density, comp. note) Arizona
Diceroprocta apache Hadley, Toolson and Quinlan 1989: 220–229, Figs 1–3, Table 1 (evaporative cooling, comp. note) Arizona
Diceroprocta apache Heath, Heath, Sanborn and Noriega 1990: 1 (comp. note) Arizona
Diceroprocter (sic) *apache* Martin and Simon 1990a: 1073, Table 3 (comp. note)
Diceroprocta apache Cloudsley-Thompson 1991: 84 (comp. note) North America
Diceroprocta apache Dean and Milton 1991: 117 (comp. note)
Diceroprocta apache Hadley, Quinlan and Kennedy 1991: 269–280, Figs 1–4, Table 1 (evaporative cooling, comp. note) Arizona
Diceroprocta apache Toms 1991: 31 (comp. note)
Diceroprocta apache Toolson and Toolson 1991: 109, 113–114 (comp. note)
Diceroprocta apache Milton and Dean 1992: 290 (comp. note)
Diceroprocta apache Rethwisch, McDaniel and Thiessen 1992: 77 (asparagus pest, control) Arizona
Diceroprocta apache Sanborn, Heath and Heath 1992: 751, 755–756 (comp. note) Arizona
Diceroprocta apache Sanborn and Phillips 1993: 29 (song intensity, comp. note)
Diceroprocta apache Andersen 1994: 26–32 (density, ecosystem function) Arizona
Diceroprocta apache Hadley 1994: 25, 83, 85, 97, 233, 299, 336–341, Figs 9.5–9.8, Table 2.1, Table 3.1 (water balance, listed, comp. note) Arizona
Diceroprocta apache Machin, Smith and Lambert 1994: 84 (comp. note)
Diceroprocta apache Prange and Pinshow 1994: 75 (comp. note)
Diceroprocta apache Simon, Frati, Beckenbach, Crespi, Liu and Flook 1994: 700 (gene sequence, comp. note)
Diceroprocta apache Toolson and Stanley-Samuelson 1994: 279–283, Figs 2–8, Fig 10 (evaporative cooling, comp. note) New Mexico
Diceroprocta apache Sanborn, J.E. Heath, M.S. Heath and Noriega 1995: 326, Table 3 (thermal responses, comp. note)
Diceroprocta apache Sanborn, M.S. Heath, J.E. Heath and Noriega 1995: 456–458, Table 4 (comp. note)
Diceroprocta apache Sanborn and Phillips 1995a: 481–483, Table 1 (song intensity, comp. note)
Diceroprocta apache Veenstra and Hagedorn 1995: 393–395, Figs 1–2 (peptide, comp. note)
Diceroprocta apache Williams and Simon 1995: 273, 278 (comp. note)
Diceroprocta apache Heinrich 1996: 75–77, 141–142 (illustrated, evaporative cooling, thermal tolerances, comp. note) Southwestern United States
Diceroprocta apache Prange 1996: 496 (comp. note)
Diceroprocta apache Roxburgh, Pinshow and Prange 1996: 335 (comp. note)

Diceroprocta apache Sanborn and Phillips 1996: 137–138 (comp. note)
Diceroprocta apache Poole, Garrison, McCafferty, Otte and Stark 1997: 331 (listed) Equals *Tibicen apache* Equals *Diceroprocta ochroleuca* North America
Diceroprocta apache Sanborn 1998: 90–92, 96–100, Table 1 (thermal responses, comp. note)
Diceroprocta apache Toolson 1998: 574 (comp. note)
Diceroprocta apache Miller and Harley 1999: 658 (comp. note)
Diceroprocta apache Sanborn 1999a: 37 (type material, listed)
Diceroprocta apache var. *ochroleuca* Sanborn 1999a: 37 (type material, listed)
Diceroprocta apache Sanborn 1999b: 1 (listed)
Diceroprocta apache Maw, Footit, Hamilton and Scudder 2000: 49 (comp. note)
Diceroprocta apache Sanborn and Maté 2000: 145 (comp. note)
Diceroprocta apache Howard, Moore, Giblin-Davis and Abad 2001: 157 (pest species, comp. note) USA
Diceroprocta apache Callaham, Whiles and Blair 2002: 99 (density, comp. note)
Diceroprocta apache Cook and Holt 2002: 215 (comp. note)
Diceroprocta apache Ellingson and Andersen 2002: 16 (spatial distribution, comp. note) Arizona
Diceroprocta apache Ellingson, Andersen and Kondratieff 2002: 283–288, Figs 1–3, Table 1 (life history, comp. note) Arizona
Diceroprocta apache Kondratieff, Ellingson and Leatherman 2002: 13, 17–18, 67, Fig 12, Table 2 (key, diagnosis, illustrated, listed, distribution, song, comp. note) Equals *Tibicen apache* Colorado, Arizona, California, Nevada, Utah, Mexico
Diceroprocta apache Sanborn 2002b: 458, 462–466, Table 1 (thermal biology, comp. note)
Diceroprocta apache Sanborn, Noriega and Phillips 2002c: 368 (comp. note)
Diceroprocta apache Heinrich 2003: 1122–1123, 1125 (evaporative cooling, thermal tolerances, comp. note) Southwestern United States
Diceroprocta apache Sanborn 2004e: 2225 (comp. note)
Diceroprocta apache Sanborn, Heath, Heath, Noriega and Phillips 2004: 286 (comp. note)
Diceroprocta apache Verdú, Díaz and Galante 2004: 36 (comp. note)
Diceroprocta apache Humphreys 2005: 74, Table 4 (population density, comp. note) Arizona
Diceroprocta apache Sanborn 2005c: 114, Table 1 (comp. note)
Diceroprocta apache Shafroth, Cleverly, Dudley, Taylor, Van Riper et al. 2005: 236 (comp. note)
Diceroprocta apache Wiesenborn 2005: 656 (comp. note) Nevada
Diceroprocta apache Harvey and Thompson 2006: 905 (comp. note)
Diceroprocta apache Smith, Kelly and Finch 2006: 1609 (comp. note)
Diceroprocta apache Krell, Boyd, Nay, Park and Perring 2007: 212 (vector, comp. note) California
Diceroprocta apache Sanborn 2007b: 32 (distribution, comp. note) Mexico, Baja California Norte, Arizona, California, Utah, Colorado, Nevada
Diceroprocta apache Shiyake 2007: 3, 32, 43, Fig 40 (illustrated, distribution, listed, comp. note) United States
Diceroprocta apache Sanborn 2008e: 3759 (comp. note)
Diceroprocta apache Holliday, Hastings and Coelho 2009: 4–5, Table 1 (predation, comp. note) California
Diceroprocta apache Fonseca, Silva, Samuels, DaMatta, Terra, and Silva 2010: 20 (comp. note)

***D. arizona* (Davis, 1916)** = *Cicada arizona* = *Tibicen arizona*
Diceroprocta arizona Poole, Garrison, McCafferty, Otte and Stark 1997: 331 (listed) Equals *Cicada arizona* North America
Diceroprocta arizona Sanborn 1999a: 37 (type material, listed)
Diceroprocta arizona Sanborn 1999b: 1 (listed) North America
Diceroprocta arizona Sanborn 2007b: 32 (distribution, comp. note) Mexico, Chiapas, Durango, Sinaloa, Arizona

***D. aurantiaca* Davis, 1938** = *Diceroprocta delicata* var. *aurantiaca* = *Diceroprocta delicata aurantiaca*
Diceroprocta delicata var. *aurantiaca* Sanborn and Phillips 1995a: 481, Table 1 (song intensity, comp. note)
Diceroprocta delicata var. *aurantiaca* Sanborn 1999a: 38 (type material, listed)
Diceroprocta aurantiaca Sanborn 1999b: 1 (listed) North America
Diceroprocta delicata var. *aurantiaca* Sanborn and Phillips 2001: 159–161 (comp. note) Texas, New Mexico
Diceroprocta aurantiaca Sanborn and Phillips 2001: 161–164, Figs 1–2, Figs 5–6, Tables 1–3 (illustrated, morphology, song, sonogram, oscillogram, thermal responses, distribution, habitat, comp. note) Louisiana, Texas, Mexico, Nuevo Leon, Tamaulipes

Diceroprocta aurantiaca Sanborn 2002b: 465, Table 1 (thermal biology, comp. note)
Diceroprocta aurantiaca Sanborn and Phillips 2010: 863–864 (comp. note)

D. averyi **Davis, 1941**
Diceroprocta averyi Poole, Garrison, McCafferty, Otte and Stark 1997: 331 (listed) North America
Diceroprocta averyi Sanborn 1999a: 37 (type material, listed)
Diceroprocta averyi Sanborn 1999b: 1 (listed) North America

D. azteca **(Kirkaldy, 1909)** = *Cicada pallida* Distant, 1881 (nec *Cicada pallida* Goeze, 1778) = *Tibicen pallida* = *Cicada azteca* = *Tibicen azteca*
Diceroprocta azteca Sanborn and Phillips 1995a: 481, Table 1 (song intensity, comp. note)
Diceroprocta azteca Poole, Garrison, McCafferty, Otte and Stark 1997: 331 (listed) Equals *Cicada azteca* North America
Diceroprocta azteca Sanborn 1999b: 1 (listed) North America
Diceroprocta azteca Kondratieff, Schmidt, Opler and Garhart 2005: 97, 216 (listed, comp. note) Oklahoma
Diceroprocta azteca Sanborn 2007a: 27 (distribution, comp. note) Venezuela, Mexico, United States
Diceroprocta azteca Sanborn 2007b: 32 (distribution, comp. note) Mexico, Nuevo León, Tamaulipas, Michoacán, Texas, Oklahoma, Venezuela
Diceroprocta azteca Stucky 2009: 123–125, Fig 1 (illustrated, distribution, habitat, comp. note) Equals *Cicada pallida* Equals *Cicada azteca* Kansas
Cicada azteca Stucky 2009: 123 (comp. note) Kansas
Cicada pallida Stucky 2009: 123 (comp. note) Kansas

D. bakeri **(Distant, 1911)** = *Rihana bakeri*
Diceroprocta bakeri Sanborn 2007b: 32 (distribution, comp. note) Mexico, Veracruz

D. belizensis **(Distant, 1910)** = *Rihana belizensis*
Diceroprocta belizensis Sanborn 2010b: 74 (comp. note)

D. bequaerti **(Davis, 1917)** = *Tibicen bequaerti* = *Tibicen viridifascia bequaerti* = *Tibicen vitripennis bequaerti* = *Diceroprocta vitripennis bequaerti*
Diceroprocta bequaerti Poole, Garrison, McCafferty, Otte and Stark 1997: 331 (listed) Equals *Tibicen bequaerti* North America
Diceroprocta bequaerti Sanborn 1999a: 37 (type material, listed)
Diceroprocta bequaerti Sanborn 1999b: 1 (listed) North America
Diceroprocta bequaerti Walker and Moore 2004: 1 (listed, comp. note) Florida
Diceroprocta bequaerti Sanborn, Phillips and Gillis 2008: 16–17 (comp. note)

D. bibbyi **Davis, 1928**
Diceroprocta bibbyi Tinkham 1941: 175, 178–179 (distribution, comp. note) Texas
Diceroprocta bibbyi Sanborn and Phillips 1995a: 481, Table 1 (song intensity, comp. note)
Diceroprocta bibbyi Van Pelt 1995: 19 (listed, distribution, comp. note) Texas
Diceroprocta bibbyi Poole, Garrison, McCafferty, Otte and Stark 1997: 331 (listed) North America
Diceroprocta bibbyi Sanborn 1999a: 37 (type material, listed)
Diceroprocta bibbyi Sanborn 1999b: 1 (listed) North America
Diceroprocta bibbyi Phillips and Sanborn 2007: 456–457, 460, Table 1, Fig 1, Fig 4 (illustrated, habitat, distribution, comp. note) Texas
Diceroprocta bibbyi Sanborn 2010b: 74 (listed, distribution) Mexico, Texas

D. bicolora **Davis, 1935** = *Diceroprocta bicolor* (sic)
Diceroprocta bicolor (sic) Sanborn 1999a: 38 (type material, listed)
Diceroprocta bicolora Sanborn 2007b: 32 (distribution, comp. note) Mexico, Morelos, Puebla, Veracruz

D. biconica **(Walker, 1850)** = *Cicada biconica* = *Rihana biconica* = *Tibicen biconica* = *Tibicen (Diceroprocta) biconica* = *Diceroprocta bicornica* (sic) = *Rehana* (sic) *biconica*
Diceroprocta biconica Hennessey 1990: 471 (type material)
Diceroprocta biconica Sanborn and Phillips 1995a: 481, Table 1 (song intensity, comp. note)
Diceroprocta biconica Poole, Garrison, McCafferty, Otte and Stark 1997: 331 (listed) Equals *Cicada biconica* Equals *Diceroprocta bicornica* (sic) Equals *Doceroprocta obscurior* North America
Diceroprocta biconica Heath, Sanborn, Heath and Phillips 1999: 1 (thermal tolerances, comp. note) Florida
Diceroprocta biconica Sanborn 2001c: 651 (comp. note) Florida
Diceroprocta biconica Sanborn, Heath, Heath and Phillips 2004: 66–71, Fig 1, Table 1 (thermal responses, comp. note) Florida
Diceroporcta biconica Walker and Moore 2004: 2 (listed, comp. note) Florida
Diceroprocta biconica Chawanji, Hodson, Villet, Sanborn and Phillips 2007: 338, 340–345, Figs 2A–B, Fig 3E, Fig 4D, Figs 4G–I, Fig 5A, Figs 6A–B (spermiogenesis, comp. note) Florida
Diceroprocta biconica Sanborn 2007b: 39 (comp. note)

Diceroprocta biconica Sanborn, Phillips and Gillis 2008: 4–6, 16, 23–24, Figs 1–10 (key, synonymy, illustrated, distribution, comp. note) Equals *Cicada biconica* Florida, Cuba

Diceroprocta biconica Sanborn 2009a: 89–90 (listed, synonymy, listed, key, comp. note) Cuba, Florida

***D. bicosta* (Walker, 1850)** = *Cicada bicosta* = *Cicada bicostata* (sic) = *Rihana bicosta* = *Tibicen bicosta* = *Diceroprocta transversa* (nec Walker)

Diceroprocta bicosta Young 1981c: 6 (comp. note)

Diceroprocta sp. Maes and Téllez 1988: 66, 82 (listed, agricultural pest) Nicaragua

Diceroprocta bicosta Poole, Garrison, McCafferty, Otte and Stark 1997: 331 (listed) Equals *Cicada bicosta* North America

Diceroprocta bicosta Maes 1998: 143 (listed, distribution, host plants) Equals *Cicada bicosta* Nicaragua, Mexico, Costa Rica

Diceroprocta transversa (sic) (nec Walker) Maes 1998: 144 (listed, synonymy, distribution, host plants) Equals *Cicada transversa* Equals *Pomponia transversa* Equals *Rihania* (sic) *transversa* Equals *Tibicen transversa* Equals *Cicada alacris* Nicaragua, USA (error), Mexico, Costa Rica

Diceroprocta sp. Maes 1998: 144 (listed, distribution, host plants) Nicaragua

Diceroprocta bicosta Sanborn 1999b: 1 (listed) North America

Diceroprocta bicosta Sanborn 2001b: 449 (listed, comp. note) El Salvador

Diceroprocta bicosta Sanborn 2001c: 651 (comp. note) Mexico

Diceroprocta bicosta Sanborn 2006a: 76 (listed, distribution) Honduras, El Salvador

Diceroprocta bicosta Sanborn 2007b: 32, 39 (distribution, comp. note) Mexico, Michoacán, Puebla, Morelos, Jalisco, Sonora, Costa Rica, Honduras, El Salvador

Diceroprocta bicosta Sanborn, Phillips and Gillis 2008: 16 (comp. note)

Diceroporcta bicosta Sanborn 2010a: 1586 (synonymy, distribution, comp. note) Equals *Cicada bicosta* Equals *Cicada bicostata* (sic) Equals *Rihana bicosta* Equals *Tibicen bicosta* Colombia, Mexico, Honduras, El Salvador, Costa Rica

***D.* sp. nr. *bicosta* Young, 1981**

Diceroprocta sp. Young 1981c: 6 (comp. note) near *Diceroprocta bicosta*

Diceroprocta sp. nr. *bicosta* Young 1981c: 5–7, 9, 11–13, 17–20, 24, 26–27, Figs 4–6, Fig 8, Fig 11, Table 3, Tables 5–6 (illustrated, seasonality, habitat, comp. note) Costa Rica

Diceroprocta Young 1981c: 7–10, 13, 21–23, Table 2 (comp. note) Costa Rica

Diceroprocta sp. Young 1982: 118, 209, Fig 3.20, Table 5.6 (illustrated, hosts, comp. note) Costa Rica

Diceroprocta sp. Young 1984a: 170, 177, 179, Tables 1–3 (hosts, comp. note) Costa Rica

***D. biguttula* Billberg, 1820 nom. nud.** = *Tettigonia 2-guttulus*

***D. bonhotei* (Distant, 1901)** = *Cicada bonhotei* = *Rihana bonhotei*

Diceroprocta bonhotei Sanborn 2001c: 651 (listed, comp. note) Bahamas

***D. bulgara* (Distant, 1906)** = *Rihana bulgara* = *Rihana operculissima* Distant, 1906 (nec *Rihana operculissima* Distant, 1897)

Diceroprocta bulgara Sanborn 2006a: 76, 78, Fig 1 (listed, distribution) Guatemala, Mexico

Diceroprocta bulgara Sanborn 2007b: 32 (distribution, comp. note) Mexico, Sinaloa, Jalisco, Sonora, Chihuahua, Colima, Oaxaca, Coahuila, Nayarit, Guatemala

D. canescens Davis, 1935

Diceroprocta canescens Tinkham 1941: 171, 175, 178–179 (distribution, habitat, comp. note) Texas

Diceroprocta canescens Sanborn and Phillips 1994: 45 (distribution, comp. note) Texas

Diceroprocta canescens Sanborn and Phillips 1995a: 481, Table 1 (song intensity, comp. note)

Diceroprocta canescens Van Pelt 1995: 19 (listed, distribution, comp. note) Texas

Diceroprocta canescens Poole, Garrison, McCafferty, Otte and Stark 1997: 331 (listed) North America

Diceroprocta canescens Sanborn and Phillips 1997: 36 (distribution, comp. note) Texas

Diceroprocta canescens Sanborn 1999a: 38 (type material, listed)

Diceroprocta canescens Sanborn 1999b: 1 (listed) North America

Diceroprocta canescens Phillips and Sanborn 2007: 456, 458–459, Table 1, Fig 1, Fig 5 (illustrated, habitat, distribution, comp. note) Texas

***D. caymanensis* Davis, 1939**

Diceroprocta caymanensis Sanborn 2001c: 651 (comp. note) Cayman Islands

***D. chinensis* Billberg, 1820 nom. nud.** = *Tettigonia chinensis*

***D. cinctifera cinctifera* (Uhler, 1892)** = *Cicada cinctifera* = *Cicada acutifera* (sic) = *Tibicen cinctifera* = *Tibicen cinctifer* (sic) = *Diceroprocta cinctifera* = *Diceroprocta cinetifera* (sic) = *Tibicen ochreoptera* Uhler, 1892 = *Tibicen ochrapterus*

Diceroprocta cinctifera Tinkham 1941: 169–174, 178–179, Fig 1 (distribution, habitat, comp. note) Texas, New Mexico

Tibicen cinctifera Johnson and Lyon 1976: 434 (agricultural pest, comp. note) New Mexico
Tibicen cinctifer (sic) Johnson and Lyon 1991: 490 (comp. note)
Diceroprocta cinctifera Sanborn and Phillips 1995a: 481, Table 1 (song intensity, comp. note)
Diceroprocta cinctifera Van Pelt 1995: 19 (listed, distribution, comp. note) Texas
Diceroprocta cinctifera Sanborn and Phillips 1996: 136–138, Fig 1, Tables 1–2 (distribution, thermal responses, comp. note) Texas
Diceroprocta cinctifera species group Sanborn and Phillips 1996: 136 (comp. note)
Diceroprocta cinctifera Poole, Garrison, McCafferty, Otte and Stark 1997: 331 (listed) Equals *Cicada cinctifera* Equals *Diceroprocta viridicosta* Equals *Diceroprocta limpia* North America
Tibicen ochreoptera Poole, Garrison, McCafferty, Otte and Stark 1997: 333 (listed) Equals *Cicada ochreoptera* North America
Diceroprocta cinctifera Sanborn 1998: 98–99, Table 1 (thermal responses, comp. note)
Diceroprocta cinctifera Sanborn 1999a: 38 (type material, listed)
Diceroprocta cinctifera Sanborn 1999b: 1 (listed) North America
Tibicen ochreopterus Sanborn 1999b: 1 (listed) North America
Diceroprocta cinctifera Kondratieff, Ellingson and Leatherman 2002: 18 (comp. note)
Diceroprocta cinctifera Sanborn 2002b: 465, Table 1 (thermal biology, comp. note)
Tibicen cinctifer (sic) Robinson 2005: 217 (comp. note)
Diceroprocta cinctifera Smith, Kelly and Finch 2006: 1612 (comp. note)
Diceroprocta cinctifera Phillips and Sanborn 2007: 456, 458, 460, Table 1, Fig 1, Fig 5 (illustrated, habitat, distribution, comp. note) Texas, New Mexico
Diceroprocta cinctifera Sanborn 2007b: 40 (comp. note)
Diceroprocta cinctifera Shiyake 2007: 3, 32, 43, Fig 41 (illustrated, distribution, listed, comp. note) United States
Diceroprocta cinctifera Hastings, Holliday and Coelho 2008: 658–659, Table 1 (predation, comp. note) Texas
Diceroprocta cinctifera Holliday, Hastings and Coelho 2009: 4–5, 13, Table 1 (predation, comp. note) Texas
Diceroprocta cinctifera cinctifera Sanborn 2010b: 74 (listed, distribution) Mexico, Texas, New Mexico
Diceroprocta cinctifera species group Sanborn and Phillips 2010: 864 (comp. note)

***D. cinctifera limpia* Davis, 1932** = *Diceroprocta cinctifera* var. *limpia* = *Diceroprocta cinctifera limpia*

Diceroprocta cinctifera var. *limpia* Tinkham 1941: 171 (distribution, habitat, comp. note) Texas
Diceroprocta cinctifera limpia Tinkham 1941: 175, 178–179 (distribution, habitat, comp. note) Texas
Diceroprocta cinctifera var. *limpia* Sanborn and Phillips 1995a: 481, Table 1 (song intensity, comp. note)
Diceroprocta cinctifera var. *limpia* Sanborn and Phillips 1996: 136–138, Fig 1, Tables 1–2 (distribution, thermal responses, comp. note) Texas
Diceroprocta cinctifera var. *limpia* Sanborn 1998: 99, Table 1 (thermal responses, comp. note)
Diceroprocta cinctifera var. *limpia* Sanborn 1999a: 38 (type material, listed)
Diceroprocta cinctifera var. *limpia* Sanborn 1999b: 1 (listed) North America
Diceroprocta cinctifera var. *limpia* Sanborn 2002b: 465, Table 1 (thermal biology, comp. note)
Diceroprocta cinctifera limpia Phillips and Sanborn 2007: 460 (comp. note) Texas

***D. cinctifera viridicosta* Davis, 1930** = *Diceroprocta cinctifera* var. *viridicosta* = *Diceroprocta cinctifera viridicosta* = *Diceroprocta cinetifera* (sic) *viridicosta*

Diceroprocta cinctifera var. *viridicosta* Tinkham 1941: 175, 178 (distribution, comp. note) Texas
Diceroprocta cinctifera viridicosta Tinkham 1941: 178 (distribution, habitat, comp. note) Texas
Diceroprocta cinctifera var. *viridicosta* Sanborn and Phillips 1995a: 481, Table 1 (song intensity, comp. note)
Diceroprocta cinctifera var. *viridicosta* Sanborn and Phillips 1996: 136–138, Fig 1, Tables 1–2 (distribution, thermal responses, comp. note) Texas
Diceroprocta cinctifera var. *viridicosta* Sanborn 1998: 99, Table 1 (thermal responses, comp. note)
Diceroprocta cinctifera var. *viridicosta* Sanborn 1999a: 38 (type material, listed)
Diceroprocta cinctifera var. *viridicosta* Sanborn 1999b: 1 (listed) North America
Diceroprocta cinctifera var. *viridicosta* Sanborn 2002b: 465, Table 1 (thermal biology, comp. note)
Diceroprocta cinctifera viridicosta Phillips and Sanborn 2007: 460 (comp. note) Texas
Diceroprocta cinctifera viridicosta Sanborn 2007b: 32, 40 (distribution, comp. note) Mexico, Nuevo León, Texas

D. cleavesi **Davis, 1930**

Diceroprocta cleavesi Hennessey 1990: 471 (type material)

Diceroprocta cleavesi Sanborn 1999a: 38 (type material, listed)

Diceroprocta cleavesi Sanborn 2001c: 651 (comp. note) Cayman Islands

D. crucifera **(Walker, 1850)** = *Cicada crucifera* = *Rihana crucifera*

Diceroprocta crucifera Sanborn 2007b: 32 (distribution, comp. note) Mexico, Distrito Federal

D. delicata **(Osborn, 1906)** = *Cicada delicata* = *Tibicen delicata*

Diceroprocta delicata Tinkham 1941: 178 (habitat, comp. note) Texas

Diceroprocta delicata Meagher, Wilson, Blocker, Eckel and Pfannenstiel 1993: 510–511, Table 1 (sugarcane pest, comp. note) Texas

Diceroprocta delicata Sanborn and Phillips 1995a: 480–481, Table 1 (song intensity, comp. note)

Diceroprocta delicata Sanborn 1996: 69 (predation, comp. note)

Diceroprocta delicata Poole, Garrison, McCafferty, Otte and Stark 1997: 331 (listed) Equals *Cicada delicata* Equals *Diceroprocta aurantiaca* North America

Diceroprocta delicata Sanborn 1999b: 1 (listed) North America

Diceroprocta delicata Sanborn and Phillips 2001: 159–164, Figs 1–4, Tables 1–3 (illustrated, morphology, song, sonogram, oscillogram, thermal responses, distribution, habitat, comp. note) Louisiana, Texas, Mexico

Diceroprocta delicata species complex Sanborn and Phillips 2001: 159 (comp. note)

Diceroprocta delicata species group Sanborn and Phillips 2001: 162 (comp. note)

Diceroprocta delicata Sanborn 2002b: 465, Table 1 (thermal biology, comp. note)

Diceroprocta delicata Sanborn 2007b: 32 (distribution, comp. note) Mexico, Nuevo León, Tamaulipas, Puebla, Texas, Louisiana

Diceroprocta delicata Shiyake 2007: 3, 42–43, Fig 42 (illustrated, distribution, listed, comp. note) United States

Diceroprocta delicata Sanborn and Phillips 2010: 864 (comp. note)

D. digueti **(Distant, 1906)** = *Rihana digueti*

Diceroprocta digueti Poole, Garrison, McCafferty, Otte and Stark 1997: 331 (listed) Equals *Rihana digueti* North America

Diceroprocta digueti Sanborn 2007b: 32 (distribution, comp. note) Mexico, Baja California Norte, Baja California Sur

D. distanti **Metcalf, 1963** = *Cicada intermedia* Distant, 1881 (nec *Cicada intermedia* Signoret, 1847) = *Rihana intermedia*

Diceroprocta distanti Sanborn 2007b: 32 (distribution, comp. note) Mexico

D. eugraphica **(Davis, 1916)** = *Cicada eugraphica* = *Tibicen eugraphica* = *Cicada vitripennis* (nec Say)

Cicada vitripennis (sic) Snow 1904: 349 (listed) Arizona

Diceroprocta eugraphica Tinkham 1941: 169–173, 175, 178–179, Fig 1 (distribution, habitat, comp. note) Texas, New Mexico, Kansa, Oklahoma, Texas, Arizona

Diceroprocta eugraphica Heath 1983: 13 (comp. note) North America

Diceroprocta eugraphica Sanborn and Phillips 1995a: 481, Table 1 (song intensity, comp. note)

Diceroporcta eugraphica Van Pelt 1995: 19 (listed, distribution, comp. note) Texas

Diceroprocta eugraphica Poole, Garrison, McCafferty, Otte and Stark 1997: 331 (listed) Equals *Cicada eugraphica* North America

Diceroprocta eugraphica Sanborn and Phillips 1997: 36 (distribution, comp. note) Texas

Diceroprocta eugraphica Sanborn 1999a: 38 (type material, listed)

Diceroprocta eugraphica Sanborn 1999b: 1 (listed) North America

Diceroprocta eugraphica Kondratieff, Ellingson and Leatherman 2002: 7, 10, 12–13, 17–19, 70, 75, Fig 20, Fig 38, Tables 1–2 (key, diagnosis, illustrated, listed, distribution, song, comp. note) Equals *Cicada eugraphica* Colorado, Arizona, Kansas, New Mexico, Oklahoma, Texas

Diceroprocta eugraphica Sanborn 2002b: 459, Fig 3 (illustrated, thermal biology, comp. note)

Diceroprocta eugraphica Sanborn 2006b: 256 (listed, distribution) Mexico, Chihuahua, Arizona, Texas, Kansas

Diceroprocta eugraphica Phillips and Sanborn 2007: 456–460, Table 1, Fig 1, Fig 4 (illustrated, habitat, distribution, comp. note) Texas, New Mexico

Diceroprocta eugraphica Sanborn 2007b: 32 (distribution, comp. note) Mexico, Chihuahua, United States

Diceroprocta eugraphica Shiyake 2007: 3, 42–43, Fig 43 (illustrated, distribution, listed, comp. note) United States

Diceroprocta eugraphica Holliday, Hastings and Coelho 2009: 4–5, Table 1 (predation, comp. note) New Mexico

Diceroprocta eugraphica Stucky 2009: 123 (comp. note) Kansas

D. *fraterna* Davis, 1935
Diceroprocta fraterna Sanborn 1999a: 38 (type material, listed)
Diceroprocta fraterna Sanborn 2007b: 32 (distribution, comp. note) Mexico, Yucután, Guerrero, Puebla, Nayarit

D. *fuscomaculata* Billberg, 1820 nom. nud. = *Tettigonia fuscomaculata*

D. *fusipennis* (Walker, 1858) = *Fidicina fusipennis* = *Rihana fusipennis* = *Cicada reticularis* Uhler, 1892
Cicada reticularis Henshaw 1903: 229 (listed) Jamaica
Diceroprocta reticularis Sanborn 1999a: 39 (type material, listed)

D. *grossa* (Fabricius, 1775) = *Tettigonia grossa* = *Cicada grossa* = *Tympanoterpes grossa* = *Cicada (Cicada) grossa* = *Rihana grossa*
Cicada grossa Froeschner 1952: 2 (comp. note)
Diceroprocta grossa Poole, Garrison, McCafferty, Otte and Stark 1997: 331 (listed) Equals *Tettigonia grossa* North America
Diceroprocta (Tibicen) grossa Sanborn, Phillips and Gillis 2008: 16–17 (comp. note)

D. *knighti* (Davis, 1917) = *Tibicen knighti*
Diceroporcta knighti Sanborn and Phillips 1995a: 481, Table 1 (song intensity, comp. note)
Diceroprocta knighti Poole, Garrison, McCafferty, Otte and Stark 1997: 331 (listed) Equals *Tibicen knighti* North America
Diceroprocta knighti Sanborn 1999a: 38 (type material, listed)
Diceroprocta knighti Sanborn 1999b: 1 (listed) North America
Diceroprocta knighti Sanborn 2007b: 32–33 (distribution, comp. note) Mexico, Lower California, Sonora, Arizona, California

D. *lata* Davis, 1941 = *Diceroprocta texana* var. *lata* = *Diceroprocta texana lata*
Diceroprocta texana var. *lata* Sanborn and Phillips 1995a: 481, Table 1 (song intensity, comp. note)
Diceroprocta texana var. *lata* Sanborn 1999a: 40 (type material, listed)
Diceroprocta texana var. *lata* Sanborn 1999b: 1 (listed) North America
Diceroprocta texana lata Sanborn 2007b: 33 (distribution, comp. note) Mexico, Nuevo León, Tamaulipas, Coahuila, Texas
Diceroprocta texana var. *lata* Sanborn and Phillips 2010: 860 (comp. note)
Diceroprocta lata Sanborn and Phillips 2010: 860–864, Fig 1, Figs 2C–D, Fig 3, Fig 4B (n. status, illustrated, morphology, song, sonogram, oscillogram, thermal responses, habitat, distribution, comp. note) Mexico, Texas

D. *lucida* Davis, 1934
Diceroprocta lucida Sanborn 1999a: 39 (type material, listed)
Diceroprocta lucida Sanborn 2007b: 33 (distribution, comp. note) Mexico, Morelos

D. *marevagans* Davis, 1928
Diceroprocta marevagans Poole, Garrison, McCafferty, Otte and Stark 1997: 331 (listed) North America
Diceroprocta marevagans Sanborn 1999a: 39 (type material, listed)
Diceroprocta marevagans Sanborn 1999b: 1 (listed) North America
Diceroprocta marevagans Sanborn 2007b: 33 (distribution, comp. note) Mexico, Tamaulipas, Texas

D. *mesochlora* (Walker, 1850) = *Cicada mesochlora* = *Odopoea mesochlora* = *Rihana mesochlora*

D. *oaxacaensis* Sanborn, 2007
Diceroprocta oaxacaensis Sanborn 2007b: 9–14, 33, Figs 13–20 (n. sp., described, illustrated, distribution, comp. note) Mexico, Oaxaca

D. *obscurior* Davis, 1935 = *Diceroprocta biconica* var. *obscurior* = *Diceroprocta biconica obscurior*
Diceroprocta biconica obscurior De Zayas 1988: 13 (described, comp. note) Cuba
Diceroprocta biconica De Zayas 1988: 13, Fig 2 (illustrated) Cuba
Diceroprocta biconica var. *obscurior* Sanborn 1999a: 38 (type material, listed)
Diceroporcta obscurior Sanborn 2009a: 89–90 (n. status, listed, synonymy, key, comp. note) Equals *Diceroprocta biconica* var. *obscurior* Cuba

D. *oculata* Davis, 1935
Diceroprocta oculata Sanborn 1999a: 39 (type material, listed)
Diceroprocta oculata Sanborn 2007b: 33 (distribution, comp. note) Mexico, Nayarit

D. *olympusa* (Walker, 1850) = *Fidicina olympusa* = *Cicada olympusa* = *Rihana olympusa* = *Tibicen olympusa* = *Diceroprocta olympusia* (sic) = *Cicada milvus* Walker, 1858 = *Cicada sordidata* Uhler, 1892 = *Tibicen sordidata* = *Diceroprocta sordida*
Cicada soridada Henshaw 1903: 229 (listed) Florida
Diceroprocta olympusa Wilson 1930: 61 (asparagus pest, host, comp. note) Florida
Diceroprocta olympusa Walker 1986: 683 (comp. note) Florida
Diceroprocta olympusa Sanborn and Phillips 1995a: 480–481, Table 1 (song intensity, comp. note)
Diceroprocta olympusa Poole, Garrison, McCafferty, Otte and Stark 1997: 331 (listed) Equals

Fidicina olympusa Equals *Cicada milvus* Equals *Cicada sordida* North America
Diceroprocta olympusa Sanborn and Maté Nankervis 1997: 9 (thermoregulation, song, comp. note)
Diceroprocta sordida Sanborn 1999a: 39 (type material, listed)
Diceroprocta olympusa Sanborn 1999b: 1 (listed) North America
Diceropocta olympusa Quartau, Seabra and Sanborn 2000: 194 (comp. note) Portugal
Diceroprocta olympusa Sanborn and Maté 2000: 142–146, Figs 1–5 (thermoregulation, song, sonogram, oscillogram, comp. note) Florida
Diceroprocta olympusa Sanborn and Phillips 2001: 164 (comp. note)
Diceroprocta olympusa Sueur 2001: 36, 46, Tables 1–2 (listed)
Diceroprocta olympusa Sanborn 2002b: 465, Table 1 (thermal biology, comp. note)
Diceroprocta olympusa Sanborn, Breitbarth, Heath and Heath 2002: 443 (comp. note)
Diceroprocta olympusa Sanborn, Noriega and Phillips 2002c: 368 (comp. note)
Diceroporcta olympusa Walker and Moore 2004: 1 (listed, comp. note) Florida
Diceroprocta olympusa Sanborn 2005c: 114, Table 1 (comp. note)
Diceroprocta olympusa Samietz, Kroder, Schneider and Dorn 2006: 155 (comp. note)
Diceroprocta olympusa Shiyake 2007: 3, 42–44, Fig 44 (illustrated, distribution, listed, comp. note) United States
Diceroprocta olympusa Hastings, Holliday and Coelho 2008: 659, 662, Table 1 (predation, comp. note) Florida
Diceroprocta olympusia (sic) Hastings, Holliday and Coelho 2008: 662 (predation, comp. note)
Diceroprocta olympusa Sanborn, Phillips and Gillis 2008: 4–6, 16, 24–25, Fig 11, Figs 16–24 (key, synonymy, illustrated, distribution, comp. note) Equals *Fidicina olympusa* Equals *Cicada milvus* Equals *Cicada sordidata* Florida
Diceroprocta olympusa Holliday, Hastings and Coelho 2009: 4–5, Table 1 (predation, comp. note) Florida
Diceroprocta olympusa Sanborn and Maté-Nankervis 2009: 75, 82, Table 1 (listed, comp. note) Florida
Diceroprocta olympusa Hastings, Holliday, Long, Jones and Rodriguez 2010: 414–416, 419, Table 2 (predation, comp. note) Florida

D. operculabrunnea **Davis, 1934**
Diceroprocta operculabrunnea Sanborn 1999a: 39 (type material, listed)
Diceroprocta operculabrunnea Sanborn 2007b: 33 (distribution, comp. note) Mexico, Morelos, Puebla

D. ornea **(Walker, 1850)** = *Cicada ornea* = *Rihana ornea* = *Tibicen ornea*
Diceroprocta ornea Bauman, Rivera-Orduño and Stock 2004: 104 (nematode host) Sonora, Mexico
Diceroprocta ornea Rivera-Orduño and Stock 2004: 98 (asparagus pest, nematode host, comp. note) Mexico, Sonora
Diceroprocta ornea Sanborn 2007b: 33 (distribution, comp. note) Mexico, Lower California, Sonora, Jalisco, Puebla, Chihuahua
Diceroprocta ornea Stock, Rivera-Orduño and Flores-Lara 2009: 175–176 (asparagus pest, parasitism, comp. note) Mexico, Sonora

D. ovata **Davis, 1939**

D. pinosensis **Davis, 1935**
Diceroprocta pinosensis De Zayas 1988: 13 (described, comp. note) Cuba
Diceroprocta pinosensis Sanborn 1999a: 39 (type material, listed)
Diceroprocta pinosensis Sanborn 2009a: 90 (listed, key, comp. note) Cuba, Isle of Youth

D. pronotolinea **Sanborn, 2007**
Diceroprocta pronotolinea Sanborn 2007b: 14–18, 33, Figs 21–29 (n. sp., described, illustrated, distribution, comp. note) Mexico, Guerrero

D. psophis **(Walker, 1850)** = *Cicada psophis* = *Rihana psophis*
Diceroprocta psophis Sanborn 2007b: 33 (distribution, comp. note) Mexico, Veracruz

D. pusilla **Davis, 1942**
Diceroprocta pusilla Sanborn 1999a: 39 (type material, listed)
Diceroprocta pusilla Sanborn 2006a: 76, 78, Fig 1 (listed, distribution) Guatemala, Honduras, Mexico
Diceroprocta pusilla Sanborn 2007b: 33 (distribution, comp. note) Mexico, Michoacán, Chiapas, San Luis Potosí, Queretaro, Guatemal, Honduras
Diceroprocta pusilla Sanborn 2010b: 74 (listed, distribution) Belize, Guatemala, Honduras, Mexico

D. pygmaea **(Fabricius, 1803)** = *Tettigonia pymaea*

D. ruatana **(Distant, 1891)** = *Tympanoterpes ruatana* = *Rihana ruatana*
Diceroprocta ruatana Sanborn 2006a: 76 (listed, distribution) Honduras
Diceroprocta ruatana Sanborn 2010b: 74 (listed, distribution) Guatemala, Honduras

D. semicincta **(Davis, 1925)** = *Tibicen semicincta* = *Tibcien* (sic) *semicincta* = *Diceroprocta semicincta* var. *nigricans* Davis, 1942

Cicada sp. (sic) Martel and Law 1992: 562 (comp. note) Arizona
Diceroprocta semicincta Sanborn and Phillips 1992: 285 (activity, comp. note) Arizona
Diceroprocta semicincta Sanborn and Phillips 1995a: 481, Table 1 (song intensity, comp. note)
Diceroprocta semicincta var. n. Sanborn and Phillips 1995a: 481, Table 1 (song intensity, comp. note)
Diceroprocta semicincta Veenstra and Hagedorn 1995: 392 (peptide, comp. note) Arizona
Diceroprocta semicincta Poole, Garrison, McCafferty, Otte and Stark 1997: 331 (listed) Equals *Tibicen semicincta* Equals *Diceroprocta nigricans* North America
Diceroprocta semicincta Sanborn 1999a: 39 (type material, listed)
Diceroprocta semicincta Sanborn 1999b: 1 (listed) North America
Diceroprocta semicincta var. *nigricans* Sanborn 1999a: 39 (type material, listed)
Diceroprocta semicincta var. *nigricans* Sanborn 1999b: 1 (listed) North America
Diceroprocta apache (sic) Moran, Tran and Gerardo 2005: 8803, 8805–8806, 8808, Figs 1–2, Fig 4, Table 1 (comp. note)
Diceroprocta semicincta Sanborn 2007b: 33 (distribution, comp. note) Mexico, Nayarit, Sonaloa, Sonora, Arizona, New Mexico
Diceroprocta semicincta Holliday, Hastings and Coelho 2009: 4–5, Table 1 (predation, comp. note) Arizona
Diceroprocta semicincta McCutcheon, McDonald and Moran 2009a: 1, 5, 9, Fig 4 (symbiont, comp. note)
Diceroprocta semicincta McCutcheon, McDonald and Moran 2009b: 15394 (symbiont, comp. note) Arizona
Diceroprocta semicincta Stock, Rivera-Orduño and Flores-Lara 2009: 175 (asparagus pest, parasitism, comp. note) Mexico, Sonora
Diceroprocta semicincta Strausfeld, Sinakevitch, Brown and Farris 2009: 267, 276–277, 284, Figs 5F–5G, Fig 12 (mushroom bodies, comp. note) Arizona
Diceroprocta semicincta Delaye and Moya 2010: 282 (symbiont, comp. note)
Diceroprocta semicincta Gil, Latorre, and Moya 2010: 223, 226 (comp. note)
Diceroprocta semicincta Gosalbes, Latorre, Lamelas and Moya 2010: 272 (comp. note)
Diceroprocta semicincta McCutcheon and Moran 2010: 709 (comp. note) Arizona
Diceroprocta semicincta Van Etten, Lane and Dunigan 2010: 84 (comp. note)
Diceroprocta semicincta Woyke, Tighe, Mavromatis, Clum, Copeland et al. 2010: 2 (comp. note)

***D. swalei davisi* Metcalf, 1963** = *Cicada castanea* Davis, 1916 (nec *Cicada castanea* Gmelin, 1789) = *Tibicen castanea* = *Diceroprocta castanea* = *Diceroprocta swalei* var. *castanea*
Diceroprocta swalei var. *davisi* Sanborn 1999a: 39 (type material, listed)
Diceroprocta swalei var. *davisi* Sanborn 1999b: 1 (listed) North America

***D. swalei swalei* (Distant, 1904)** = *Rihana swalei* = *Diceroprocta swalei*
Diceroprocta swalei Sanborn and Phillips 1995a: 480–481, Table 1 (song intensity, comp. note)
Diceroprocta swalei Poole, Garrison, McCafferty, Otte and Stark 1997: 331 (listed) Equals *Rihana swalei* North America
Diceroprocta swalei Sanborn 1999b: 1 (listed) North America
Diceroprocta swalei Sanborn 2007b: 33 (distribution, comp. note) Mexico, Sonora, Arizona
Diceroprocta swalei McCutcheon, McDonald and Moran 2009a: 5, 9, Fig 4 (symbiont, comp. note) Arizona

***D. tepicana* Davis, 1938**
Diceroprocta tepicana Sanborn 1999a: 39–40 (type material, listed)
Diceroprocta tepicana Sanborn 2004c: 512, Fig 172 (illustrated)
Diceroprocta tepicana Sanborn 2007b: 33 (distribution, comp. note) Mexico, Nayarit, Sinaloa, Jalisco, Oaxaca

***D. texana* (Davis, 1916)** = *Cicada texana* = *Tibicen texana*
Diceroprocta texana Tinkham 1941: 175, 178 (distribution, comp. note) Texas, New Mexico
Diceroprocta texana Sanborn and Phillips 1995a: 481, Table 1 (song intensity, comp. note)
Diceroprocta texana Poole, Garrison, McCafferty, Otte and Stark 1997: 331 (listed) Equals *Cicada texana* Equals *Diceroprocta lata* North America
Diceroprocta texana Sanborn 1999a: 40 (type material, listed)
Diceroprocta texana Sanborn 1999b: 1 (listed) North America
Diceroprocta texana Phillips and Sanborn 2007: 460 (comp. note) Texas, New Mexico
Diceroprocta texana Sanborn 2007b: 33 (distribution, comp. note) Mexico, Morelos, Tamaulipas, Texas, New Mexico
Diceroprocta texana Shiyake 2007: 3, 42–44, Fig 45 (illustrated, distribution, listed, comp. note) United States, Mexico
Diceroprocta texana Sanborn and Phillips 2010: 860–864, Fig 1, Figs 2A–B, Fig 3, Fig 4A

(illustrated, morphology, song, sonogram, oscillogram, thermal responses, habitat, distribution, comp. note) Texas, New Mexico
Diceroprocta texana group Sanborn and Phillips 2010: 860–861 (comp. note)
D. texana lata Davis, 1941 to *Diceroprocta lata* Davis
D. tibicen Linnaeus, 1758 to *Tibicen tibicen tibicen* (Linnaeus)
***D. virgulata* (Distant, 1904)** = *Rihana virgulata*
Diceroprocta virgulata Sanborn 2007b: 13–14, 33 (distribution, comp. note) Mexico
***D. viridifascia* (Walker, 1850)** = *Cicada viridifascia* = *Tibicen viridifascia* = *Tibicen viridifasciata* (sic) = *Cicada reperta* Uhler, 1892 = *Tibicen reperta*
Cicada reperta Henshaw 1903: 229 (listed) Florida, North Carolina, Louisiana
Diceroprocta viridifascia Wilson 1930: 61 (host, comp. note) Florida
Diceroprocta viridifascia Poole, Garrison, McCafferty, Otte and Stark 1997: 331 (listed) Equals *Cicada viridifascia* Equals *Cicada reperta* North America
Diceroprocta reperta Sanborn 1999a: 39 (type material, listed)
Diceroprocta viridifascia Sanborn 1999b: 1 (listed) North America
Diceroprocta viridifascia Sueur 2001: 36, 46, Tables 1–2 (listed)
Diceroprocta viridifascia Walker and Moore 2004: 1, 5 (listed, comp. note) Florida
Diceroprocta viridifascia Sanborn, Phillips and Gillis 2008: 3, 6–7, 16–17, 24, 26, Fig 12, Figs 25–33 (key, synonymy, illustrated, distribution, comp. note) Equals *Cicada viridifascia* Equals *Cicada reperta* Florida
Diceroprocta viridifascia Holliday, Hastings and Coelho 2009: 4–5, Table 1 (predation, comp. note) Florida
***D. vitripennis* (Say, 1830)** = *Cicada vitripennis* = *Cicada albipennis* (sic) = *Rihana vitripennis* = *Tibicen vitripennis* = *Cicada erratica* Osborn, 1906 = *Tibicen erratica*
Diceroprocta vitripennis Tinkham 1941: 179 (habitat, comp. note) Texas
Diceroporcta vitripennis McCoy 1965: 41, 43 (key, comp. note) Arkansas
Tibicen vitripennis Evans 1966: 103–104, Table 10 (parasitism, comp. note) Arkansas
Tibicen vitripennis Krombein, Hurd, Smith and Burks 1979: 1698 (parasitism)
Diceroprocta vitripennis Marshall, Cooley, Alexander and Moore 1996: 165–167, Fig 1 (distribution, comp. note) Michigan
Tibicen vitripennis Coelho 1997: 374 (predation, comp. note)
Diceroprocta vitripennis Poole, Garrison, McCafferty, Otte and Stark 1997: 331 (listed) Equals *Cicada vitripennis* Equals *Cicada albipennis* Equals *Cicada erratica* North America
Diceroprocta vitripennis Sanborn 1999b: 1–2 (type material, listed) North America
Diceropocta vitripennis Quartau, Seabra and Sanborn 2000: 196 (comp. note) Portugal
Diceroprocta vitripennis Sueur 2001: 36, 46, Tables 1–2 (listed)
Diceroprocta vitripennis Kondratieff, Ellingson and Leatherman 2002: 10–12, Table 1 (listed, comp. note) Colorado
Diceroprocta vitripennis Walker and Moore 2004: 2 (listed, comp. note) Florida
Diceroprocta vitripennis Kondratieff, Schmidt, Opler and Garhart 2005: 97 (comp. note)
Diceroprocta vitripennis Sanborn, Phillips and Gillis 2008: 16–17 (comp. note) Equals *Diceroprocta erratica*
Diceroprocta erratica Sanborn, Phillips and Gillis 2008: 16–17 (comp. note)
Diceroprocta vitripennis Holliday, Hastings and Coelho 2009: 4–5, Table 1 (predation, comp. note) Arkansas

***Tibicen* Latreille, 1825** = *Tubicen* (sic) = *Tibicens* (sic) = *Tibiceus* (sic) = *Tibicin* (sic) = *Tibicien* (sic) = *Tettigonia* Fabricius, 1775 = *Telligonia* (sic) = *Tetigonia* (sic) = *Cicada* Latreille, 1810 (nec Linnaeus) = *Cicada (Cicada)* (nec Linnaeus) = *Cicada* (= *Tibicen*) = *Cicadina* (sic) = *Lyristes* Horváth, 1926 = *Lyrtistes* (sic) = *Lyriste* (sic)
Tibicen Solier 1837: 202 (comp. note)
Cicada (sic) Carpenter and Dallas 1867: 186 (comp. note)
Tibicen Dallas 1870: 482 (listed)
Cicada (sic) Strobl 1900: 201 (listed) Austria
Cicada (sic) Banks 1904a: 10–16 (comp. note) America
Cicada (sic) Banks 1904b: 597–598 (comp. note) America
Cicada (sic) Davis 1906b: 239 (comp. note)
Tibicen Wilson 1930: 63 (comp. note) Kansas
Lyristes Wu 1935: 6 (listed) China
Tibicen Sugden 1940: 117 (comp. note)
Tibicen Tinkham 1941: 173, Fig 2 (singing posture, comp. note)
Tibicen Chen 1942: 144 (listed
Tibicen Zeuner 1944: 113 (comp. note) Equals *Tibicina*
Lyristes Gómez-Menor 1951: 60–62 (key, comp. note) Equals *Cicada* Equals *Tettigia* Spain
Lyristes Soós 1956: 411 (listed) Equals *Cicada* Romania
Cicada (Tibicen) Halkka 1959: 42 (comp. note)
Tibicen Smith 1960: 448 (neuromuscular junction, comp. note)

Tibicen Smith 1961: 122 (timbal muscle, comp. note)
Tibicen McCoy 1965: 40–41, 44 (key, comp. note) Arkansas
Tibicen Smith 1965: 384 (muscle ultrastructure, comp. note)
Tibicen Evans 1966: 103 (parasitism, comp. note)
Tibicen sp. Evans 1966: 108, Fig 58 (illustrated, parasitism) Massachusetts, New York, Ohio
Tibicen Smith 1966: 115, Table 2 (listed, muscle ultrastructure, comp. note)
Tibicen Smith, Gupta and Smith 1966: 50 (timbal muscle, comp. note)
Tibicen Weber 1968: 236, 344 (comp. note)
Cicada (sic) sp. Anderson, Ostry and Anderson 1979: 478 (ovipostition damage, comp. note) Wisconsin
Tibicen Young 1979: 149 (comp. note)
Tibicen O'Shea 1980: 207 (comp. note)
Tibicen Boulard 1981b: 48 (comp. note)
Lyristes Boulard 1982f: 184 (listed)
Tibicen Fujiyama 1982: 185 (comp. note)
Tibicen Hayashi 1982b: 188–189, 192, Table 1 (distribution, comp. note) Japan
Lyristes Koçak 1982: 150 (listed, comp. note) Turkey
Tettigonia Koçak 1982: 151 (listed, comp. note) Turkey
Tibicen Koçak 1982: 151 (listed, synonymy, comp. note) Equals *Tettigonia* Fabricius (nec *Tettigonia* Linnaeus, 1758) Turkey
Tibicen Wiley and Richards 1982: 162 (comp. note) North America
Lyristes Boulard 1983b: 120 (comp. note) France
Tibicen Strümpel 1983: 18, 134 (listed, comp. note)
Tibicen spp. Walker 1983: 63 (comp. note) United States
Tibicen Boulard 1984c: 167, 169–170, 172–179 (status)
Lyristes Boulard 1984c: 170–171, 175, 178–179 (status, type species *Cicada plebeja*)
Tibicen Hayashi 1984: 28, 35 (key, described, listed, comp. note) Equals *Cicada* (nec Linnaeus) Equals *Lyristes*
Lyristes Hayashi 1984: 36 (comp. note)
Tibicen spp. Lloyd 1984: 79 (comp. note)
Tibicen Lloyd 1984: 80 (comp. note)
Lyristes Melville and Sims 1984: 163–165, 180 (status, history)
Tibicen Melville and Sims 1984: 163–165, 179–181 (status, history)
Tibicen Smith 1984: 142 (timbal muscle, comp. note)
Tibicen Young 1984b: 358 (comp. note)
Tibicen Arnett 1985: 217 (diversity) United States
Tibicen Boulard 1985d: 212–213 (status)
Lyristes Boulard 1985d: 212–213 (status)
Lyristes Boulard 1985e: 1030 (coloration, comp. note)
Tibicen Bregman 1985: 39 (comp. note)
Lyristes Claridge 1985: 301 (song, comp. note)
Tibicen Hamilton 1985: 211 (status, comp. note)
Lyristes Hamilton 1985: 211 (status, comp. note)
Tibicen Lauterer 1985: 214 (status)
Tibicen Moulds 1985: 26 (comp. note)
Tibicen Popov 1985: 38–39, 43–44 (acoustic system, comp. note)
Tibicen sp. Smith 1985: 1 (life cycle, comp. note)
Tibicen Duffels 1986: 328 (comp. note)
Tibicen sp. Lloyd 1986: 363 (comp. note)
Tibicen Lloyd 1986: 363, 367 (comp. note)
Tibicen spp. Lloyd 1986: 364, Fig 2 (illustrated, comp. note)
Tibicen sp. Lockley and Young 1986: 393 (predation) Mississippi
Tibicen Moulds 1986: 40 (comp. note)
Tibicen sp. Smith, Wilkinson, Williams and Steward 1987: 277 (predation, comp. note) Arkansas
Tibicen Hayashi 1987a: 124–125 (comp. note)
Lyristes Hayashi 1987a: 124–125 (comp. note)
Lyrsistes Lodos and Boulard 1987: 644–648 (comp. note) Turkey
Tibicen Anufriev and Emelyanov 1988: 313, 314, 316–317 (key, characteristics) Soviet Union
Lyristes Boulard 1988e: 154 (listed, described, comp. note)
Tibicen Boulard 1988f: 8, 22–25, 28–30, 33–35, 44, 70 (history, comp. note)
Lyristes Boulard 1988f: 29, 33, 44–45 (comp. note)
Tibicen Duffels 1988d: 25 (comp. note)
Tibicen (= Lyristes) Duffels 1988d: 81 (comp. note)
Lyristes Lodos and Kalkandelen 1988: 18 (comp. note) Turkey
Tibicen sp. Dietrich 1989: 146, Table 1 (abdominal sensilla, comp. note)
Lyristes Boulard 1990b: 172, 205, 214 (comp. note)
Tibicen Brown and Brown 1990: 5 (comp. note)
Tibicen spp. Cranshaw and Kondratieff 1991: 10 (life cycle, comp. note) Colorado
Tibicen sp. Jakubczak, Burke and Eickbush 1991: 3295, 3297, Fig 2 (sequence, comp. note)
Tibicen Johnson and Lyon 1991: 490 (comp. note)
Tibicen Popov, Aronov and Sergeeva 1991: 282 (comp. note)
Tibicen spp. Young 1991: 188 (comp. note)
Tibicen Young 1991: 241 (comp. note)
Tibicen Carpenter 1992: 222 (comp. note)
Tibicen Liu 1992: 42 (comp. note)
Tibicen Popov, Arnov and Sergeeva 1992: 151 (comp. note)
Tibicen Sanborn, Heath and Heath 1992: 756 (comp. note)
Tibicen Moore 1993: 271–272, 282, Fig 1 (comp. note)
Lyristes Remane and Wachman 1993: 75 (comp. note)

Tibicen Niehuls and Simon 1994: 258 (comp. note) Germany
Lyristes Boulard 1995a: 33 (listed)
Tibicen Boulard 1995c: 177 (comp. note)
Lyristes Boulard 1995c: 174, 177 (comp. note)
Tibicen spp. Jameson 1995: 3 (sound production, comp. note)
Tibicen Lee 1995: 17, 22, 41, 52, 63, 71, 135 (listed, key, described, ecology, distribution, illustrated, comp. note) Korea
Tibicen Mirzayans 1995: 14 (listed) Iran
Lyristes Riou 1995: 74–76 (comp. note)
Tibicen sp. Sorensen, Campbell, Gill and Steffen-Campbell 1995: 32 (gene sequence, phylogeny, comp. note)
Tibicen spp. Weaver 1995: 3 (parasitism) West Virginia
Tibicen sp. Wheeler, Schuh and Bang 1995: 123, Table 1 (listed)
Tibicen Wheeler, Schuh and Bang 1995: 124–129, 131–134, Figs 3–5, Table 2–3 (sequence, phylogeny, comp. note)
Tibicen Boer and Duffels 1996b: 304, 315 (comp. note) Equals *Lyristes*
Tibicen Holzinger 1996: 505 (listed) Austria
Tibicen Maier 1996: 5 (comp. note) Connecticut
Tibicen Moore 1996: 221 (comp. note) Mexico
Lyristes Vernier 1996: 151 (comp. note) France
Tibicen Boulard 1997c: 83, 92–100, 102–105, 120–121 (history, comp. note)
Tribicen (sic) Boulard 1997c: 101 (comp. note)
Lyristes Boulard 1997c: 99, 102, 104, 112 (comp. note)
Lyristes Chou, Lei, Li, Lu and Yao 1997: 265, 281–282, 293, Table 6 (key, described, synonymy, listed, diversity, biogeography) Equals *Cicada* Equals *Tibicen* China
Tibicen Chou, Lei, Li, Lu and Yao 1997: 281–282 (comp. note)
Tibicen spp. Coelho 1997: 372 (predation, comp. note)
Tibicen Coelho 1997: 372–375 (predation, comp. note)
Lyristes Holzinger, Frölich, Günthart, Pauterer, Nickel et al. 1997: 49 (listed) Europe
Tibicen Hostetler 1997: 27 (comp. note)
Tibicen Novotny and Wilson 1997: 437 (listed)
Tibicen Poole, Garrison, McCafferty, Otte and Stark 1997: 251, 333 (listed) Equals *Tettigonia* Equals *Tubicen* (sic) Equals *Tibicin* (sic) Equals *Tibicens* (sic) Equals *Tibiceus* (sic) Equals *Lyristes* North America
Tibicen Hyche 1998: 12 (comp. note) Alabama
Tibicen Anonymous 1999c: 489 (emergence time)
Tibicen Coelho and Ladage 1999: 481 (predation, comp. note) Illinois
Tibicen spp. Coelho and Wiedman 1999: 6 (predation, comp. note) Illinois
Tibicen Lee 1999: 4 (comp. note) Equals *Lyristes* Korea
Tibicen Sanborn 1999b: 1 (listed) North America
Lyristes Villet and Zhao 1999: 322 (comp. note)
Tibicen Arnett 2000: 298 (diversity) United States
Tibicen Boulard 2000a: 255 (comp. note)
Lyristes Boulard 2000a: 255 (comp. note)
Tibicen Boulard 2000i: 20 (comp. note)
Lyristes Boulard 2000i: 24, 39 (key, comp. note) France
Tibicen Maw, Footit, Hamilton and Scudder 2000: 49 (listed)
Tibicen Milius 2000: 408 (comp. note)
Tibicen spp. Sanborn 2000: 554 (comp. note)
Lyristes Schedl 2000: 262 (listed)
Tibicen Schedl 2000: 262, 264 (listed)
Tibicen Strange 2000: 1–2 (predation, comp. note) Florida
Tibicen Beaupre and Roberts 2001: 45 (comp. note) Arkansas
Tibicen Boulard 2001b: 4, 13–24, 26 (history, comp. note)
Tribicen (sic) Boulard 2001b: 22 (comp. note)
Lyristes Boulard 2001b: 19, 22–23, 25–26, 32 (type species of Lyristini, comp. note)
Lyristes Boulard 2001e: 117 (comp. note)
Tibicen Boulard 2001e: 117 (comp. note)
Tibicen Boulard 2001f: 130 (comp. note)
Tibicen Cioara and Sanborn 2001: 25 (song, comp. note)
Tibicen sp. Coelho 2001: 109 (predation, comp. note)
Tibicen spp. Coelho and Holliday 2001: 346 (predation, comp. note)
Tibicen Cooley 2001: 756 (comp. note) North America
Tibicen Cooley and Marshall 2001: 848 (comp. note)
Cicada (Tibicen) Levinson and Levinson 2001: 11, 22–23, Fig 6 (figurine illustrated)
Tibicen Ohara and Hayashi 2001: 2 (comp. note) Japan
Tibicen spp. Ohara and Hayashi 2001: 6 (listed) Japan
Tibicen spp. Kondratieff, Ellingson and Leatherman 2002: 1 (song, comp. note) North America, Colorado
Tibicen Kondratieff, Ellingson and Leatherman 2002: 14, 16, 19, 23–25 (key, diagnosis, listed, diversity, comp. note) Colorado
Tibicen sp. Rakitov 2002: 119–120, 123, Fig 1, Figs 11–12 (malpighian tubules, comp. note) Illinois
Tibicen spp. Rakitov 2002: 119 (comp. note) Illinois

Tibicen sp. Ratcliff and Hammond 2002: 45, Fig 27 (illustrated, comp. note) Nebraska
Lyristes Sueur 2002b: 122, 124 (comp. note)
Tibicen Boulard 2003f: 371–372 (comp. note)
Lyristes Boulard 2003f: 371–372 (comp. note)
Tibicen spp. Callaham, Blair, Todd, Kitchen and Whiles 2003: 1081, 1086–1087, Figs 5–6, Table 1 (density, comp. note) Kansas
Tibicen Callaham, Blair, Todd, Kitchen and Whiles 2003: 1081, 1084–1085, 1090–1091, Tables 2–3 (density, comp. note) Kansas
Tibicen DeFoliart 2003: 434 (life cycle, comp. note) Illinois
Tibicen sp. Dietrich 2003: 66, Fig 2(9) (illustrated) Illinois
Lyristes Holzinger, Kammerlander and Nickel 2003: 479 (listed) Europe
Tibicen Lee and Hayashi 2003a: 158, 160 (key, diagnosis) Taiwan
Lyristes Lee and Hayashi 2003a: 158 (comp. note)
Tibicen spp. Lee and Hayashi 2003a: 160 (comp. note)
Tibicen Moulds 2003b: 246 (comp. note)
Tibicen spp., Strange 2003: 1 (predation, comp. note) Florida
Tibicen Chen 2004: 39, 41, 58 (phylogeny, listed)
Tibicen Cooley, Marshall and Simon 2004: 201 (comp. note)
Tibicen spp. Kritsky 2004a: 110 (illustrated, comp. note)
Tibicen spp. Ohya 2004: 441–443 (comp. note) Japan
Tibicen Ohya 2004: 441–442 (comp. note) Japan
Tibicen spp. Sanborn 2004b: 97 (comp. note)
Tibicen Sanborn 2004b: 97 (comp. note)
Tibicen sp. Walker and Moore 2004: 4 (illustrated) Florida
Tibicen spp. Walker and Moore 2004: 4 (comp. note) Florida
Tibicen Walker and Moore 2004: 4 (comp. note) Florida
Lyristes Boulard 2005d: 338 (comp. note) Mediterranean
Tibicen spp. Champman 2005: 321 (comp. note)
Tibicen spp. Grant 2005: 173 (comp. note)
Tibicen Lee 2005: 14, 16, 55, 157, 169 (listed, key, comp. note) Equals *Lyristes* Korea
Tibicen Moulds 2005b: 384, 388–389, 392–393 (type genus of Tibiceninae, type genus of Tibicenini, history, phylogeny, listed, comp. note)
Lyristes Moulds 2005b: 387–390, 392–393, 412–413, 431 (type genus of Lyristini, history, phylogeny, listed, comp. note) Equals *Tibicen*
Tibicen Moulds 2005b: 428 (comp. note)
Tibicen sp. Robinson 2005: 208, Fig 8.3f (illustrated, comp. note)
Tibicen spp. Robinson 2005: 217 (comp. note)
Tibicen Salmah, Duffels and Zaidi 2005: 15 (comp. note)
Tibicen Yaakop, Duffels and Visser 2005: 248–249 (comp. note) Equals *Lyristes* Eurasia, North America
Lyristes Boulard 2006g: 39–40, 96, 129 (comp. note) Mediterranean, Turkey
Tibicen Elliott and Hershberger 2006: 184 (listed)
Tibicen spp. Grant 2006: 540 (predation, comp. note) New Jersey
Tibicen Grant 2006: 540–541 (predation, comp. note) New Jersey
Tibicen spp. Marshall 2006: 139 (comp. note) North America
Lyristes Seabra, Pinto-Juma and Quartau 2006: 850 (comp. note)
Tibicen Eaton and Kaufman 2007: 88 (diversity, comp. note) North America
Tibicen Ganchev and Potamitis 2007: 304–305, 319, Fig 11 (listed, comp. note)
Tibicen Oberdörster and Grant 2007a: 2, 10–11 (flight system, flight speeds, comp. note) New Jersey
Tibicen spp. Oberdörster and Grant 2007a: 4, 6, 8–11 (comp. note) New Jersey
Tibicen Oberdörster and Grant 2007b: 15, 18–19, 21 (sound system, predation, comp. note) New Jersey
Tibicen spp, Oberdörster and Grant 2007b: 15–21 (sound system, predation, comp. note) New Jersey
Tibicen sp. Phillips and Sanborn 2007: 457, Fig 2 (listed) Texas
Tibicen Sanborn 2007b: 20–21, 34, 40 (comp. note) Mexico
Tibicen Shiyake 2007: 3, 10–11, 38, 95 (listed, comp. note)
Tibicen sp. Shiyake 2007: 11 (illustrated)
Lyristes Sueur and Puissant 2007b: 55 (comp. note)
Lyristes Boulard 2008c: 363 (comp. note) Anatolia
Tibicen Cole 2008: 815, 819, 821–822 (comp. note) North America
Lyristes Demir 2008: 476 (listed) Turkey
Tibicen Farris, Oshinshy, Forrest, and Hoy 2008: 24 (comp. note)
Lyristes Fonseca, Serrão, Pina-Martins, Silva, Mira, et al. 2008: 25, 27–30 (comp. note)
Tibicen Fonseca, Serrão, Pina-Martins, Silva, Mira, et al. 2008: 29 (comp. note) North America
Tibicen Hastings, Holliday and Coelho 2008: 659, 662 (predation, comp. note)

Tibicen spp. Hastings, Holliday and Coelho 2008: 662 (predation, comp. note)
Lyristes Lee 2008b: 446–447, Table 1 (listed, comp. note) Equals *Tibicen* Korea
Tibicen Osaka Museum of Natural History 2008: 322 (listed) Japan
Lyristes Pinto-Juma, Seabra and Quartau 2008: 2 (comp. note)
Lyristes Sanborn, Moore and Young 2008: 1 (comp. note)
Tibicen Sanborn, Moore and Young 2008: 1 (comp. note)
Tettigonia Sanborn, Moore and Young 2008: 1 (comp. note)
Tibicen Sanborn, Phillips and Gillis 2008: 3, 7–8, 12, 17 (key, diversity, described, comp. note) Florida
Lyristes Sanborn, Phillips and Gillis 2008: 7 (comp. note)
Tibicen sp. Bennett and Hobson 2009: 18 (isotope analysis, comp. note) Saskatchewan
Tibicen Hill and Marshall 2009: 63 (comp. note)
Tibicen Holliday, Hastings and Coelho 2009: 3 (predation, diversity, comp. note)
Tibicen spp. Holliday, Hastings and Coelho 2009: 12 (predation, comp. note)
Tibicen Stucky 2009: 123 (comp. note) Kansas
Tibicen spp. Yasukawa 2009: 404 (comp. note) Wisconsin
Tibicen Bernhardt, Kutschbach-Brohl, Washburn, Chipman and Francoeur 2010: 448, Table 2 (bird predation) Long Island, New York
Tibicen sp. Dmitriev 2010: 597 (listed)
Tibicen Duffels 2010: 11 (comp. note)
Tibicen spp. Hastings, Holliday, Long, Jones and Rodriguez 2010: 415 (predation, comp. note) Florida
Lyristes Kemal and Koçak 2010: 5 (listed, comp. note) Turkey
Tibicen Mozaffarian and Sanborn 2010: 76 (distribution, listed) northern hemisphere
Tibicen Phillips and Sanborn 2010: 74 (comp. note)
Tibicen spp. Saljoqi, Rahimi, Khan and Rehman 2010: 69 (control, comp. note) Afghanistan
Tibicen spp. Saljoqi, Rahimi, Rehman and Khan 2010: 75 (density, control, comp. note) Afghanistan
Tibicen Sanborn 2010b: 70 (comp. note)

***T. altaiensis* (Schmidt, 1932)** = *Lyristes altaiensis*
Lyristes altaiensis Wu 1935: 6 (listed) China, Altai
Lyristes altaiensis Chou, Lei, Li, Lu and Yao 1997: 287 (described, comp. note) China
Lyristes altaiensis Hua 2000: 62 (listed, distribution) China, Xizang

T. annulatus (Brullé, 1832) to *Pagiphora annulata*

***T. armeniacus* (Kolenati, 1857)** = *Cicada plebejus armeniaca* = *Tibicen plebejus armeniacus* = *Lyristes armeniacus*
Lyristes armeniacus Boulard 1988e: 153, 166 (n. comb., status, comp. note) Turkey

***T. atrofasciatus* (Kirkaldy, 1909)** = *Cicada atrofasciatus* = *Cicada sinensis* Distant, 1890 (nec *Cicada sinensis* Gmelin, 1789) = *Tibicen sinensis* = *Lyristes sinensis* = *Talainga sinensis*
Lyristes sinensis Wu 1935: 6 (listed) Equals *Cicada sinensis* China, Chia-kou-ho
Tibicen sinensis Chen 1942: 144 (listed, comp. note) China
Tibicen sinensis Lee 1995: 65 (comp. note)
Lyristes atrofasciatus Chou, Lei, Li, Lu and Yao 1997: 282–283, 286, Plate XIV, Fig 139 (key, described, illustrated, synonymy, comp. note) Equals *Cicada sinensis* Equals *Cicada atrofasciatus* Equals *Tibicen sinensis* Equals *Lyristes sinensis* Equals *Talainga sinensis* Equals *Tibicen atrofasciatus* China
Lyristes atrofasciatus Hua 2000: 62 (listed, distribution) Equals *Cicada atrofasciata* Equals *Cicada sinensis* China, Fujian, Sichuan, Yunnan, Xizang

***T. auletes* (Germar, 1834)** = *Cicada auletes* = *Cicada auletus* (sic) = *Tibicen aurestes* (sic) = *Lyristes auletes* = *Tettigonia grossa* Fabricius, 1775 = *Tympanoterpes grossa* = *Rihana grossa* (nec Fabricius) = *Cicada grossa* (nec Fabricius) = *Tibicen grossa* (nec Fabricius) = *Diceroprocta grossa* = *Fidicina literata* Walker, 1850 = *Cicada literata* = *Tibicen literata* = *Cicada sonora* Walker 1850 = *Tibicen sonora* = *Cicada marginata* (nec Say) = *Cicada emarginata* (sic) (nec Say)
Tibicen auletes McCoy 1965: 42, 44 (key, comp. note) Arkansas
Tibicen auletes Walker 1983: 63 (comp. note) United States
Tibicen auletes Daniel, Knight and Charles 1993: 69–75, Fig 3, Fig 6, Fig 9, Tables 1–2 (song, illustrated, comp. note) North Carolina
Tibicen auletes Maier 1996: 6 (comp. note) Connecticut
Tibicen auletes Marshall, Cooley, Alexander and Moore 1996: 165 (comp. note) Michigan
Tibicen auletes Poole, Garrison, McCafferty, Otte and Stark 1997: 333 (listed) Equals *Cicada auletes* Equals *Fidicina literata* Equals *Cicada sonora* Equals *Cicada auletus* (sic) North America
Tibicen auletes Sanborn 1999b: 1–2 (type material, listed) North America
Tibicen auletes Maw, Footit, Hamilton and Scudder 2000: 49 (comp. note) Canada, Ontario
Tibicen auletes Sanborn 2001a: 17 (comp. note)
Lyristes auletes Sueur 2001: 36, 46, Tables 1–2 (listed) Equals *Tibicen auletes*

Tibicen auletes Sanborn 2002b: 461 (thermal biology, comp. note)
Tibicen auletes Walker and Moore 2004: 2–3, Fig 3 (illustrated, listed, comp. note) Florida
Tibicen auletes Elliott and Hershberger 2006: 32, 192–193 (illustrated, sonogram, listed, distribution, comp. note) New Jersey, Delaware, Maryland, Virginia, North Carolina, South Carolina, Georgia, Florida, Alabama, Mississippi, Louisiana, Texas, Arkansas, Texas, Oklahoma, Kansas, Missouri, Illinois, Tennessee, Kentucky, Indiana, Ohio, Michigan
Tibicen auletes Ganchev and Potamitis 2007: 305 (listed)
Tibicen auletes Sanborn 2007b: 39 (comp. note)
Tibicen auletes Sanborn, Phillips and Gillis 2008: 4, 7–8, 16–17, 24, 27, Fig 13, Figs 34–42 (key, synonymy, illustrated, distribution, comp. note) Florida
Tibicen grossa Sanborn, Phillips and Gillis 2008: 16 (comp. note)
Tibicen auletes Holliday, Hastings and Coelho 2009: 4–5, Table 1 (predation, comp. note) Georgia, Virginia
Tibicen auletes Phillips and Sanborn 2010: 74 (comp. note)

***T. auriferus* (Say, 1825)** = *Cicada aurifera* = *Rihana aurifera* = *Tibicen aurifera* = *Tibicen canicularis auriferei* (sic)
Tibicen aurifera (sic) McCoy 1965: 42, 44 (key, comp. note) Arkansas
Tibicen canicularis auriferei (sic) Young 1984b: 358 (comp. note)
Tibicen aurifera (sic) Poole, Garrison, McCafferty, Otte and Stark 1997: 333 (listed) Equals *Cicada aurifera* North America
Tibicen auriferus Sanborn 1999b: 1–2 (type material, listed) North America
Tibicen aurifera (sic) Callaham 2000: 3957 (density, comp. note) Kansas
Tibicen aurifera (sic) Smith, Kelly, Dean, Bryson, Parker et al. 2000: 125 (illustrated, comp. note) Kansas
Tibicen aurifera (sic) Callaham, Whiles and Blair 2002: 92–100, Fig 1, Fig 3, Tables 1–2 (density, comp. note) Kansas
Tibicen aurifera (sic) Callaham, Whiles, Meyer, Brock and Charlton 2000: 537–542, Figs 1–3, Tables 1–3 (density, comp. note) Kansas
Tibicen auriferus Kondratieff, Ellingson and Leatherman 2002: 10–12, Table 1 (listed, comp. note) Colorado
Tibicen aurifera (sic) Kondratieff, Schmidt, Opler and Garhart 2005: 96–97, 216–217 (listed, comp. note) Oklahoma
Tibicen auriferus Cole 2008: 818–819 (comp. note) Kansas

***T. bermudianus* (Verrill, 1902)** = *Cicada bermudiana* = *Rihana bermudiana* = *Diceroprocta bermudiana* = *Lyristes bermudianus* = *Tibicen bermudiana* (sic)
Tibicen bermudiana (sic) Wilson and Hilburn 1991: 415–416 (listed, comp. note) Bermuda
Tibicen bermudianus Moore 1993: 271–272, 274, Fig 1, Fig 3 (illustrated, distribution, song, sonogram, comp. note) Bermuda
Diceroprocta bermudiana Poole, Garrison, McCafferty, Otte and Stark 1997: 331 (listed) Equals *Cicada bermudiana* North America
Lyristes bermudianus Sueur 2001: 36, Table 1 (listed) Equals *Tibicen bermudianus*

***T. bifidus* (Davis, 1916)** = *Cicada bifida* = *Tibicen bifida*
Tibicen bifida (sic) Matsuda 1960: 724–725, 728, Plate IV, Fig 24 (illustrated, mesothorax, comp. note)
Tibicen bifida (sic) Poole, Garrison, McCafferty, Otte and Stark 1997: 333 (listed) Equals *Cicada bifida* Equals *Tibicen simplex* North America
Tibicen bifidus Toolson 1998: 569–580, Figs 1–5, Table 1 (thermal biology, comp. note) New Mexico
Tibicen bifidus Sanborn 1999a: 40 (type material, listed)
Tibicen bifidus Sanborn 1999b: 1 (listed) North America
Tibicen bifidus Kondratieff, Ellingson and Leatherman 2002: 5, 12–13, 20–21, 71, 75, Fig 24, Fig 39, Tables 1–2 (key, diagnosis, illustrated, listed, distribution, song, comp. note) Equals *Cicada bifida* Colorado, Kansas, New Mexico, Texas

T. bifidus simplex Davis, 1941 to *Tibicen simplex*

***T. bihamatus andrewsi* (Distant, 1904)** = *Cicada andrewsi* = *Tibicen andrewsi* = *Tibicen bihamatus* ab. *andrewsi* = *Lyristes andrewsi*
Tibicen bihamatus ab. *andrewsi* Hayashi 1984: 36–37, Fig 33a (described, illustrated, comp. note) Japan

***T. bihamatus babai* Kato, 1938** = *Tibicen bihamatus* var. *babai* = *Lyristes bihamata babai*
Tibicen bihamatus var. *babai* Hayashi 1984: 36–37, Fig 33d (described, illustrated, comp. note) Japan

***T. bihamatus bihamatus* (de Motschulsky, 1861)** = *Cicada bihamatus* = *Cicada bihammata* (sic) = *Cicadina* (sic) *bihamata* = *Tibicen bihamata* = *Tibicen bihamoata* (sic) = *Lyristes bihamatus* = *Lyristes bihamata* = *Cicada andrewsi* Distant, 1904 = *Cicada nagashimai* Kato, 1925 = *Tibicen nagashimai* = *Tibicen esakii* (nec Kato)
Cicada bihamata Kishida 1929: 109 (comp. note) Korea

Tibicen bihamatus Anonymous 1936: 766 (comp. note) Japan
Tibicen bihamatus Hayashi 1982b: 187–192, Table 1 (synonymy, distribution, comp. note) Equals *Tibicen bihamatus* var. *fujiensis* Japan
Tibicen bihamatus Hayashi 1984: 29, 36–37, Figs 28–33 (key, described, illustrated, synonymy, comp. note) Equals *Cicada bihamata* Equals *Tibicen bihamoata* (sic) Equals *Lyristes bihamata* Equals *Cicada andrewsi* Equals *Cicada nagashimai* Japan
Tibicen bihamatus Wang and Zhang 1987: 295 (key, comp. note) China
Tibicen bihamatus Anufriev and Emelyanov 1988: 314, 316 Figs 236.2, 238.1–238.4 (illustrated, genitalia illustrated, key) Soviet Union
Tibicen bihamatus Wang and Zhang 1992: 121 (listed, comp. note) China
Lyristes bihamatus Boulard 1995c: 174 (comp. note)
Tibicen bihamatus Lee 1995: 18, 22, 28, 37–39, 41–42, 44–46, 63, 119–120, 134, 138 (comp. note) Equals *Cicada bihamata* Equals *Tibicen bihamata*
Cicada bihamata Lee 1995: 32, 33, 119–120 (comp. note)
Tibicen bihamata Lee 1995: 33, 35, 119 (comp. note)
Tibicen bihamatus(?) Lee 1995: 41 (comp. note)
Cicada bihamatus Kurahashi 1996: 109 (predation, comp. note) Japan
Lyristes bihamatus Chou, Lei, Li, Lu and Yao 1997: 282, 284 (key, described, comp. note) Equals *Cicada bihamatus* Equals *Cicada nagashimai* Equals *Tibicen bihamata* Equals *Tibicen bihamatus* Equals *Lyristes bihamata* China
Lyristes bihamatus Hua 2000: 62 (listed, distribution) Equals *Cicada bihamata* China, Zhejiang, Fujian, Japan, Korea, Russia
Tibicen bihamatus Ohara and Hayashi 2001: 1–3, Fig 1 (illustrated, comp. note) Japan
Tibicen bihamatus Perepelov, Burgov and Marymanska-Nadachowska 2002: 218, Fig 1 (karyotype, comp. note) Japan
Tibicen bihamatus Ohya 2004: 441–443, Figs 1–2, Table 1 (song, comp. note) Japan
Tibicen bihamatus Lee, Sato and Sakai 2005: 317 (parasitism, comp. note) North America
Tibicen bihamatus Lee 2005: 55 (comp. note)
Cicada bihamata Lee 2005: 68 (comp. note)
Tibicen bihamatus Shiyake 2007: 3, 36, 38, Fig 46 (illustrated, distribution, listed, comp. note) Japan, China
Tibicen bihamatus Osaka Museum of Natural History 2008: 322 (listed, acoustic behavior, ecology, comp. note) Japan
Tibicen bihamatus Kohira, Okada, Nakanishi and Yamanaka 2009: 14 (predation, comp. note) Japan

***T. bihamatus daisenensis* Kato, 1940** = *Tibicen bihamatus* var. *daisenensis* = *Tibicen bihamatus dasenensis* (sic)
Tibicen bihamatus var. *daisenensis* Hayashi 1984: 37, Fig 33c (described, illustrated, comp. note) Japan

***T. bihamatus fujiensis* Masuda, 1942** = *Tibicen bihamatus* f. *fujisanus* (sic)
Tibicen bihamatus f. *fujisanus* (sic) Hayashi 1984: 37, Fig 33e (described, illustrated, comp. note) Japan

***T. bihamatus harutai* Kato, 1939** = *Tibicen bihamatus* ab. *harutai*
Tibicen bihamatus ab. *harutai* Hayashi 1984: 36 (described, comp. note) Japan

***T. bihamatus nagashimai* (Kato, 1925)** = *Cicada nagashimai* = *Tibicen nagashimai* = *Lyristes nagashimai*
Tibicen bihamatus nagashimai Hayashi 1984: 36 (described, comp. note) Japan

***T. bihamatus nakayamai* Kato, 1939** = *Tibicen bihamatus* var. *nakayamai*
Tibicen bihamatus var. *nakayamai* Hayashi 1984: 36–37 (described, comp. note) Japan

***T. bihamatus saitoi* Kato, 1939** = *Tibicen bihamatus* var. *saitoi* = *Tibicen bihamatus satoi* (sic)
Tibicen bihamatus var. *saitoi* Hayashi 1984: 36 (described, comp. note) Japan

***T. bihamatus takeoi* Kato, 1939** = *Tibicen bihamatus* var. *takeoi*
Tibicen bihamatus var. *takeoi* Hayashi 1984: 36 (described, comp. note) Japan

***T. bihamatus tazawai* Kato, 1939** = *Tibicen bihamatus* ab. *tazawai*
Tibicen bihamatus ab. *tazawai* Hayashi 1984: 36–37, Fig 33f (described, illustrated, comp. note) Japan

***T. bihamatus tsuguonis* Kato, 1940** = *Tibicen bihamatus* var. *tsuguonis*
Tibicen bihamatus var. *tsuguonis* Hayashi 1984: 36 (described, comp. note) Japan

***T. bimaculatus* Sanborn, 2010** = *Tibicen nigriventris* Distant 1881 (nec Walker)
Tibicen bimaculatus Sanborn 2010b: 71–74, Figs 10–18 (n. sp., described, illustrated, listed, distribution, comp. note) Mexico

***T. canicularis* (Harris, 1841)** = *Cicada canicularis* = *Cicada tibicen canicularis* = *Rihana canicularis* = *Tibicen caniculris* (sic) = *Tibicen canulatus* (sic) = *Lyristes canicularis*
Cicada canicularis Peck 1879: 20 (comp. note)
Cicada canicularis Davis 1906b: 237–239, Table (comp. note) New York, Staten Island, New Jersey, Canada, Quebec
Cicada canicularis Davis 1906c: 240 (comp. note)

Tibicen canicularis Davis 1920: 129 (parasitism, comp. note) New York, Staten Island, Long Island
Cicada canicularis Roth 1949: 81 (parasitism) America
Tibicen canicularis McCoy 1965: 42, 44 (key, comp. note) Arkansas
Tibicen canicularis Evans 1966: 104, Table 10 (parasitism, comp. note) Massachusetts, New York, Ohio
Tibicen canicularis Krombein, Hurd, Smith and Burks 1979: 1698 (parasitism)
Tibicen canicularis Young 1979: 149 (predation) Wisconsin
Tibicen canicularis Arnett and Jacques 1981: 168 (life history) Eastern United States, Florida (error), California (error)
Tibicen canicularis Young 1984b: 358 (comp. note) Wisconsin
Tibicen canicularis Arnett 1985: 217 (life history) Northeast North America, Southeast United States, Texas (error), Colorado (error), Kansas, Missouri
Tibicen canicularis Brown and Brown 1990: 5 (distribution, comp. note) Virginia
Tibicen canicularis Bennet-Clark and Young 1994: 293, Table 1 (body size and song frequency)
Tibicen canicularis Sanborn and Phillips 1995a: 481, Table 1 (song intensity, comp. note)
T[ibicen] caniculris (sic) Jiang, Yang, Tang, Xu and Chen 1995b: 229 (song frequency, comp. note)
Tibicen canicularis Marshall, Cooley, Alexander and Moore 1996: 165 (comp. note) Michigan
Tibicen canicularis Hostetler 1997: 29 (comp. note)
Tibicen canicularis Poole, Garrison, McCafferty, Otte and Stark 1997: 333 (listed) Equals *Cicada canicularis* North America
Tibicen canicularis Lakes-Harlan, Stölting and Stumpner 1999: 35 (comp. note) Michigan
Tibicen canicularis Sanborn 1999b: 1 (listed) North America
Tibicen canicularis Arnett 2000: 298 (life history) Northeast North America, Southeast United States, Texas (error), Colorado (error), Kansas, Missouri
Tibicen canicularis Maw, Footit, Hamilton and Scudder 2000: 49 (comp. note) Canada, Saskatchewan, Manitoba, Ontario, New Brunswick, Prince Edward Island, Nova Scotia
Tibicen canicularis Sahli and Ware 2000: 193 (comp. note) Virginia
Tibicen canicularis Beaupre and Roberts 2001: 44–45 (snake predation) Arkansas
Lyristes canicularis Sueur 2001: 33, 36, 46, Tables 1–2 (listed, comp. note) Equals *Tibicen canicularis*
Tibicen canicularis Hutchinson 2002: 6 (life cycle, natural history) Quebec
Tibicen canicularis Kondratieff, Ellingson and Leatherman 2002: 10, 12, 24–25, Table 1 (comp. note) Equals *Cicada canicularis*
Lyristes canicularis Sueur 2002b: 124–125, 129 (listed, comp. note) Equals *Tibicen canicularis*
Tibicen canicularis Walker and Moore 2004: 2 (listed, comp. note) Florida
Tibicen canicularis Elliott and Hershberger 2006: 32, 194–195 (illustrated, sonogram, listed, distribution, comp. note) West Virginia, Ohio, Pennsylvania, New York, Connecticut, Rhode Island, Massachusetts, Maine, New Hampshire, Vermont, Michigan, Illinois, Wisconsin, North Dakota, South Dakota, Montana, Idaho, Washington, Canada
Tibicen canicularis Grant 2006: 542–543, 454, Figs 2–3, Table 2 (predation, comp. note) New Jersey
Tibicen canicularis Marshall 2006: 102, 139 (illustrated, comp. note) North America
Tibicen canicularis Nahirney, Forbes, Morris, Chock and Wang 2006: 2018 (comp. note)
Tibicen canicularis Bunker, Neckermann and Cooley 2007: 357 (comp. note) Connecticut
Tibicen canicularis Ganchev and Potamitis 2007: 288–289, 304–305, Fig 4, Fig 11 (sonogram, listed, comp. note)
Tibicen canicularis Hastings, Holliday and Coelho 2008: 659, Table 1 (predation, comp. note) Massachusetts
Tibicen canicularis Sanborn, Phillips and Gillis 2008: 17 (comp. note)
Tibicen canulatus (sic) Shiyake 2007: 3, 36, 38, Fig 47 (illustrated, distribution, listed, comp. note) United States
Tibicen canicularis Holliday, Hastings and Coelho 2009: 3–5, 12, Table 1 (predation, comp. note) Connecticut, Illinois, Indiana, Massachusetts, Michigan, Minnesota, New Jersey, New York, Ohio, Rhode Island, South Dakota, Wisconsin
Tibicen canicularis Mengesha, Vallance, Barraja and Mittal 2009: 2 (comp. note)
Tibicen canicularis Sims, Palazotto and Norris 2010: 131–135, 138–139 (wing structure, comp. note)

T. chihuahuaensis Sanborn, 2007

Tibicen chihuahuaensis Sanborn 2007b: 18–21, 33, Figs 30–34 (n. sp., described, illustrated, distribution, comp. note) Mexico, Chiricahua

***T. chinensis* (Distant, 1905)** = *Tibicina chinensis* = *Lyristes chinensis*
Tibicina chinensis Varley 1939: 99 (comp. note)
Tibicen chinensis Wu 1935: 25 (listed) Equals *Tibicina chinensis* China, Tatsienlu
Lyristes chinensis Chou, Lei, Li, Lu and Yao 1997: 287 (described, synonymy, comp. note) Equals *Tibicina chinensis* China
Lyristes chinensis Hua 2000: 62 (listed, distribution) Equals *Tibicen chinensis* China, Sichuan

***T. chiricahua* Davis, 1923**
Tibicen chiricahua Hastings and Toolson 1991: 514–519, Figs 1–5, Tables 1–2 (thermoregulation, comp. note) New Mexico
Tibicen chiricahua Sanborn, Heath and Heath 1992: 755 (comp. note)
Tibicen chiricahua Sanborn and Phillips 1995a: 481, Table 1 (song intensity, comp. note)
Tibicen chiricahua Poole, Garrison, McCafferty, Otte and Stark 1997: 333 (listed) North America
Tibicen chiricahua Sanborn 1998: 91–92, 96, 98 (thermal biology, comp. note)
Tibicen chiricahua Toolson 1998: 574 (comp. note)
Tibicen chiricahua Sanborn 1999a: 40 (type material, listed)
Tibicen chiricahua Sanborn 1999b: 1 (listed) North America
Tibicen chiricahua Fonseca and Revez 2002b: 976 (comp. note)
Tibicen chiricahua Sanborn 2002b: 459, 462, 464 (thermal biology, comp. note)

***T. chisosensis* Davis, 1934**
Tibicen chisosensis Tinkham 1941: 171, 174, 178 (distribution, habitat, comp. note) Texas
Tibicen chisosensis Sanborn and Phillips 1994: 45 (distribution, comp. note) Texas
Tibicen chisosensis Sanborn and Phillips 1995a: 481, Table 1 (song intensity, comp. note)
Tibicen chisosensis Van Pelt 1995: 20 (listed, distribution, comp. note) Texas
Tibicen chisosensis Poole, Garrison, McCafferty, Otte and Stark 1997: 333 (listed) North America
Tibicen chisosensis Sanborn and Phillips 1997: 36 (distribution, comp. note) Texas
Tibicen chisosensis Sanborn 1999a: 40 (type material, listed)
Tibicen chisosensis Sanborn 1999b: 1 (listed) North America
Tibicen chisosensis Phillips and Sanborn 2007: 456–461, Table 1, Fig 2, Fig 7 (illustrated, habitat, distribution, comp. note) Texas
Tibicen chisosensis Sanborn 2007b: 33–34 (distribution, comp. note) Mexico, Chihuahua, Texas

***T. chujoi* Esaki, 1935** = *Tibicen atrofasciatus* (nec Kirkaldy)
Tibicen chujoi Lee and Hayashi 2003a: 157–158, Fig 7 (key, diagnosis, illustrated, distribution, song, listed, comp. note) Equals *Tibicen atrofasciatus* (nec Kirkaldy) Taiwan
Tibicen chujoi Chen 2004: 59, 195, 198 (described, illustrated, listed, comp. note) Taiwan

***T. cristobalensis* (Boulard, 1990)** = *Lyristes cristobalensis*
Tibicen (= Lyristes) sp. Duffels 1986: 327, Fig 9 (distribution, comp. note) San Cristobal Island, Solomon Islands
Lyristes cristobalensis Boulard 1990b: 214, 217, 231–233, Plate XXVI, Figs 38–41 (n. sp., described, illustrated, comp. note) Solomon Islands
Lyristes cristobalensis Boer and Duffels 1996b: 302, 330 (distribution, comp. note) San Critobal, Solomon Islands
Lyristes cristobalensis Duffels 1997: 552 (comp. note) Solomon Islands

***T. cultriformis* (Davis, 1915)** = *Cicada cultriformis* = *Tibicen canicularis* (nec Harris)
Tibicen canicularis (sic) Arnett and Jacques 1981: 169 (illustrated)
Tibicen cultriformis Moore 1993: 282 (comp. note)
Tibicen cultiformis Poole, Garrison, McCafferty, Otte and Stark 1997: 333 (listed) Equals *Cicada cultriformis* North America
Tibicen cultriformis Sanborn 1999a: 40 (type material, listed)
Tibicen cultriformis Sanborn 1999b: 1 (listed) North America
Tibicen cultriformis Sanborn 2004a: A1098 (thermoregulation, endothermy, comp. note)
Tibicen cultriformis Sanborn 2004b: 98–100, Fig 1 (thermoregulation, endothermy, comp. note) Arizona
Tibicen cultriformis Sanborn 2005c: 114, Table 1 (comp. note)
Tibicen cultriformis Sanborn 2007b: 34 (distribution, comp. note) Mexico, Sonora, Arizona, New Mexico

***T. davisi davisi* (Smith & Grossbeck, 1907)** = *Cicada davisi* = *Rihana davisi* = *Cicada canicularis* (nec Harris)
Tibicen davisi Wilson 1930: 61–62, 65 (asparagus pest, control, predation, comp. note) Florida
Tibicen davisi McCoy 1965: 44 (comp. note) Arkansas
Tibicen davisi Hennessey 1990: 471 (type material)
Tibicen davisi Sanborn 1999a: 40 (type material, listed)
Tibicen davisi Sanborn 1999b: 1 (listed) North America
Tibicen davisi Walker and Moore 2004: 2, 5 (agricultural pest, listed, comp. note) Florida
Tibicen davisi Elliott and Hershberger 2006: 32, 184–185, 196–197 (illustrated, sonogram, listed, comp. note) New Jersey, Delaware,

Maryland, Virginia, North Carolina, South Carolina, Georgia, Florida, Alabama, Mississippi, Louisiana, Arkansas

Tibicen davisi Hastings, Holliday and Coelho 2008: 659, Table 1 (predation, comp. note) North Carolina

Tibicen davisi Sanborn, Phillips and Gillis 2008: 4, 8, 17, 24, 28, Fig 14, Figs 43–51 (key, synonymy, illustrated, distribution, comp. note) Equals *Cicada davisi* Equals *Cicada canicularis* (partim) Florida, New Jersey

Tibicen davisi Holliday, Hastings and Coelho 2009: 4–5, Table 1 (predation, comp. note) Florida, Georgia, North Carolina, South Carolina, Texas, Virginia

Tibicen davisi Sanborn and Maté-Nankervis 2009: 75, 82, Table 1 (listed, comp. note) Florida

T. davisi hardeni **Davis, 1918** = *Tibicen davisi* var. *hardeni*

Tibicen davisi var. *hardeni* Sanborn 1999a: 40, 55 (type material, listed, comp. note)

Tibicen davisi var. *hardeni* Sanborn 1999b: 1 (listed) North America

T. dealbatus **(Davis, 1915)** = *Cicada marginata dealbata* = *Cicada dealbata* = *Tibicen dealbata*

Tibicen dealbata (sic) Krombein, Hurd, Smith and Burks 1979: 1698 (parasitism)

Tibicen dealbatus Toolson 1984: 551–555, Tables 1–2 (hydrocarbons, water loss, comp. note) New Mexico

Tibicen dealbatus Toolson 1987: 385 (comp. note)

Tibicen dealbatus Toolson and Hadley 1987: 439 (comp. note)

Tibicen dealbatus Stanley-Samuelson, Howard and Toolson 1990: 285–288, Tables 1–2 (fatty acids, metabolism, comp. note) New Mexico

Tibicen dealbatus Hadley, Quinlan and Kennedy 1991: 278 (comp. note)

Tibicen dealbatus Toolson, Ashby, Howard and Stanley-Samuelson 1994: 279–283, Figs 2–8, Fig 10 (evaporative cooling, comp. note) New Mexico

Tibicen dealbatus Cranshaw and Kondratieff 1995: 78 (listed)

Tibicen dealbata (sic) Poole, Garrison, McCafferty, Otte and Stark 1997: 333 (listed) Equals *Cicada dealbata* North America

Tibicen dealbatus Sanborn 1998: 96–97 (thermal biology, comp. note)

Tibicen dealbatus Stanley and Howard 1998: 375 (comp. note)

Tibicen dealbatus Tunaz, Bedick, Miller, Hoback, Rana and Stanley 1998: 925, 930 (comp. note)

Tibicen dealbatus Hoback, Rana and Stanley 1999: 360 (comp. note)

Tibicen dealbatus Sanborn 1999a: 40 (type material, listed)

Tibicen dealbatus Sanborn 1999b: 1 (listed) North America

Tibicen dealbatus Stanley 2000: 53, 176 (comp. note)

Tibicen dealbatus Bashan, Akbas and Yurdakoc 2002: 378–379 (comp. note)

Tibicen dealbatus Kondratieff, Ellingson and Leatherman 2002: 3–4, 6, 12–13, 19, 21–22, 26, 70–71, Figs 21–22, Tables 1–2 (key, diagnosis, illustrated, listed, distribution, song, predation, comp. note) Equals *Cicada marginata dealbata* Colorado, Kansas, Oklahoma, Montana, Nebraska, New Mexico, North Dakota, South Dakota, Texas

Tibicen dealbatus Sanborn 2002b: 462, 464 (thermal biology, comp. note)

Tibicen dealbatus Villet, Sanborn and Phillips 2003c: 1442 (comp. note)

Tibicen dealbatus Sanborn 2004b: 100 (comp. note)

Tibicen dealbatus Bashan and Cakak 2005: 578 (comp. note)

Tibicen dealbata (sic) Kondratieff, Schmidt, Opler and Garhart 2005: 96, 217 (listed, comp. note) Oklahoma

Tibicen dealbatus Smith, Kelly and Finch 2006: 1608–1610, 1612–1613, 1615–1617, Plate 1 (illustrated, emergence, density, comp. note) New Mexico

Tibicen dealbatus Smith, Kelly and Finch 2007: 418 (predation, comp. note) New Mexico

Tibicen dealbatus Cole 2008: 819, 821 (comp. note)

Tibicen dealbatus Hastings, Holliday and Coelho 2008: 659, 661, Table 1 (predation, comp. note) Colorado

Tibicen dealbatus Holliday, Hastings and Coelho 2009: 4–7, Table 1 (predation, comp. note) Colorado, Kansas, New Mexico

T. distanti **Metcalf, 1963** = *Cicada hilaris* Distant, 1881 (nec *Cicada hilaris* Germar, 1834) = *Rihana hilaris* = *Tibicen hilaris*

Tibicen distanti Sanborn 2007b: 34 (distribution, comp. note) Mexico

T. dorsatus **(Say, 1825)** = *Cicada dorsata* = *Rihana dorsata* = *Fidicina dorsata* = *Tibicen dorsata* = *Thopha varia* Walker, 1850 = *Cicada varia* = *Fidicina crassa* Walker, 1858

Tibicen dorsata (sic) Tinkham 1941: 171–173, 178–179 (distribution, habitat, comp. note) Texas

Tibicen dorsata (sic) McCoy 1965: 42 (key, comp. note) Arkansas

Tibicien dorsata (sic) Evans 1966: 103–104, Table 10 (parasitism, comp. note) Kansas

Tibicen dorsata (sic) Krombein, Hurd, Smith and Burks 1979: 1698 (parasitism)

Tibicen dorsata (sic) Cranshaw and Kondratieff 1995: 78 (listed)

Tibicen dorsata (sic) Poole, Garrison, McCafferty, Otte and Stark 1997: 333 (listed) Equals *Cicada dorsata* Equals *Thopha varia* Equals *Fidicina crassa* North America

Tibicen dorsatus Sanborn 1999b: 1–2 (type material, listed) North America

Tibicen dorsata (sic) Callaham, Whiles and Blair 2002: 93 (density, comp. note) Kansas

Tibicen dorsata (sic) Callaham, Whiles, Meyer, Brock and Charlton 2000: 537–542, Figs 1–3, Tables 1–3 (density, comp. note) Kansas

Tibicen dorsata (sic) Smith, Kelly, Dean, Bryson, Parker et al. 2000: 125 (illustrated, comp. note) Kansas

Tibicen dorsatus Kondratieff, Ellingson and Leatherman 2002: Cover, 3–5, 12–13, 19, 22–23, 26, 71, 76, Fig 23, Figs 40–41, Tables 1–2 (key, diagnosis, illustrated, listed, distribution, song, predation, comp. note) Equals *Cicada dorsata* Colorado, Arkansas, Arizona, Iowa, Illinois, Kansas, Missouri, Nebraska, Ohio, Oklahoma, South Dakota, Texas

Tibicen dorsata (sic) Kondratieff, Schmidt, Opler and Garhart 2005: 97, 217, Fig 58 (listed, iluatrated, comp. note) Oklahoma

Tibicen dorsata (sic) Macnamara 2005: 48–49 (illustrated, comp. note) Illinois

Tibicen dorsata (sic) Eaton and Kaufman 2007: 88–89 (illustrated, comp. note) Central and Eastern United States

Tibicen dorsatus Phillips and Sanborn 2007: 460–461 (comp. note) Texas, South Dakota

Tibicen dorsatus Shiyake 2007: 3, 36, 38, Fig 49 (illustrated, listed, comp. note)

Tibicen dorsatus Cole 2008: 815–820, 822, Fig 1A, Fig 2A, Fig 3, Fig 5 (synonymy, neotype designation, described, illustrated, song, sonogram, distribution, comp. note) Equals *Cicada dorsata* Equals *Thopha varia* Equals *Fidicina crassa* Equals *Tibicen dorsata* Kansas

Cicada dorsata Cole 2008: 816 (type status)

Thopha varia Cole 2008: 818 (comp. note)

Fidicina crassa Cole 2008: 818 (comp. note)

Tibicen dorsatus Hastings, Holliday and Coelho 2008: 659, 661, Table 1 (predation, comp. note) Colorado

Tibicen dorsatus Holliday, Hastings and Coelho 2009: 4–5, Table 1 (predation, comp. note) Colorado

***T. duryi* Davis, 1917**

Tibicen duryi Tinkham 1941: 172, 174, 179 (distribution, habitat, comp. note) Texas, Utah, Arizona, Colorado, New Mexico

Tibicen duryi Kaser and Hastings 1981: 1016 (thermal physiology)

Tibicen duryi Toolson 1984: 554 (comp. note)

Tibicen duryi Hastings 1986: 263–264, 265–267, Table 1 (predation, comp. note) Arizona

Tibicen duryi Hastings 1989a: 255 (predation, comp. note)

Tibicen duryi Hastings 1989b: 145–148, Fig 1 (thermoregulation, comp. note) New Mexico

Tibicen duryi Hastings 1990: 4 (predation, comp. note)

Tibicen duryi Hadley, Quinlan and Kennedy 1991: 279 (comp. note)

Tibicen duryi Hastings and Toolson 1991: 514–519, Figs 1–5, Tables 1–2 (thermoregulation, comp. note) New Mexico

Tibicen duryi Sanborn, Heath and Heath 1992: 755–756 (comp. note)

Tibicen duryi Sanborn and Phillips 1995a: 482, Table 1 (song intensity, comp. note)

Tibicen duryi Van Pelt 1995: 20 (listed, distribution, comp. note) Texas

Tibicen duryi Poole, Garrison, McCafferty, Otte and Stark 1997: 333 (listed) North America

Tibicen duryi Sanborn 1998: 91, 96–98 (thermal biology, comp. note)

Tibicen duryi Sanborn 1999a: 41 (type material, listed)

Tibicen duryi Sanborn 1999b: 1 (listed) North America

Tibicen duryi Fonseca and Revez 2002b: 976 (comp. note)

Tibicen duryi Kondratieff, Ellingson and Leatherman 2002: 9, 12–13, 20, 23, 67, 71, Fig 13, Fig 25, Tables 1–2 (key, diagnosis, illustrated, listed, distribution, song, comp. note) Colorado, Arizona, New Mexico, Texas, Utah

Tibicen duryi Sanborn 2002b: 459, 462–464 (thermal biology, comp. note)

Tibicen duryi Phillips and Sanborn 2007: 455–460, Table 1, Fig 2, Fig 6 (illustrated, habitat, distribution, comp. note) Texas

Tibicen duryi Hastings, Holliday and Coelho 2008: 661 (predation, comp. note) Arizona

Tibicen duryi Holliday, Hastings and Coelho 2009: 3–5, 12–14, Tables 1–2 (predation, comp. note) Arizona

***T. esakii* Kato, 1958** = *Tibicen bihamatus* (nec Motschulsky)

Tibicen esakii Hayashi 1984: 29, 39–40, Figs 40–43 (key, described, illustrated, synonymy, comp. note) Equals *Tibicen bihamatus* (nec de Motschulsky) Japan

Tibicen esakii Ohara and Hayashi 2001: 2 (comp. note) Japan

Tibicen esakii Shiyake 2007: 3, 36, 38, Fig 50 (illustrated, distribution, listed, comp. note) Japan

Tibicen esakii Osaka Museum of Natural History 2008: 322 (listed, acoustic behavior, ecology, comp. note) Japan

***T. esfandiarii* Dlabola, 1970** = *Lyristes esfandiarii*

Tibicen esfandiarii Mirzayans, Hashemi, Borumand, Zairi and Rajabi 1976: 110 (distribution, listed) Iran

Tibicen esfandiarii Dlabola 1981: 199 (distribution, comp. note) Iran

Lyristes esfandiarii Boulard 1988e: 153 (n. comb., status, comp. note) Iran

Tibicen esfandiarii Mirzayans 1995: 14 (listed, distribution) Iran

Tibicen esfandiarii Mozaffarian and Sanborn 2010: 76–77 (listed, distribution, comp. note) Iran

***T. figuratus* (Walker, 1858)** = *Fidicina figurata* = *Cicada figurata* = *Tibicen figurata*

Tibicen figuratus McCoy 1965: 44 (comp. note) Arkansas

Tibicen figurata (sic) Poole, Garrison, McCafferty, Otte and Stark 1997: 333 (listed) Equals *Fidicina figurata* North America

Tibicen figuratus Sanborn 1999b: 1 (listed) North America

Tibicen figurata (sic) Walker and Moore 2004: 2 (listed, comp. note) Florida

Tibicen figuratus Hastings, Holliday and Coelho 2008: 659, Table 1 (predation, comp. note) Florida

Tibicen figuratus Sanborn, Phillips and Gillis 2008: 4, 8–9, 24, 29, Fig 15, Figs 52–60 (key, synonymy, illustrated, distribution, comp. note) Equals *Fidicina figurata* Florida

Tibicen figuratus Hill and Marshall 2009: 65 (key)

Tibicen figuratus Holliday, Hastings and Coelho 2009: 6–7, Table 1 (predation, comp. note) Florida, Louisiana

Tibicen figuratus Phillips and Sanborn 2010: 74 (comp. note)

***T. flammatus adonis* Kato, 1933** = *Tibicen flammatus* f. *adonis* = *Lyristes flammata adonis*

Tibicen flammatus f. *adonis* Hayashi 1984: 40–41, Fig 48b (described, illustrated, comp. note) Japan

***T. flammatus concolor* Kato, 1934** = *Tibicen flammatus* ab. *concolor* = *Lyristes flammata concolor*

Tibicen flammatus ab. *concolor* Hayashi 1984: 40–41, Fig 48a (described, illustrated, comp. note) Japan

***T. flammatus flammatus* (Distant, 1892)** = *Cicada flammata* = *Cicada flamma* (sic) = *Tibicen flammata* = *Lyristes flammata* = *Lyristes flammatus* = *Cicada pyropa* Matsumura, 1904

Tibicen flammata (sic) Chen 1933: 359 (listed, comp. note)

Lyristes flammata (sic) Wu 1935: 6 (listed) Equals *Cicada flammata* Equals *Tibicen flammata* Equals *Cicada pyropa* China, Lushan, Korea, Japan

Tibicen flammatus Hayashi 1982b: 187–192, Fig 1, Table 1 (distribution, comp. note) Japan

Tibicen flammatus Hayashi 1984: 29, 40–41, Figs 44–48 (key, described, illustrated, synonymy, comp. note) Equals *Cicada flammata* Equals *Lyristes flammata* Equals *Cicada pyroga* Japan

Tibicen flammatus Kevan 1985: 31, 70 (illustrated, comp. note) Japan

Tibicen flammatus Wang and Zhang 1987: 295 (key, comp. note) China

Cicada flammata Kiho, Miyamoto, Nagai and Ukai 1988: 207 (parasitism)

Cicada flammata Namba, Ma and Inagaki 1988: 253 (medicinal use, comp. note) China

Tibicen flammatus Wang and Zhang 1992: 121 (listed, comp. note) China

Tibicen flammatus Lee 1995: 18, 22, 28, 37–38, 40–46, 63–65, 120–122, 134, 138 (comp. note) Equals *Cicada flammatus* Equals *Tibicen flammata*

Tibicen flammata Lee 1995: 35, 120–121 (listed, comp. note)

Cicada flammata Lee 1995: 69, 120 (comp. note) Equals *Tibicen flammatus*

Lyristes flammatus Chou, Lei, Li, Lu and Yao 1997: 282, 284–285 (key, described, synonymy, comp. note) Equals *Cicada flammata* Equals *Cicada pyrogea* Equals *Tibicen flammata* Equals *Lyristes flammata* Equals *Tibicen flammatus* China

Tibicen flammatus Itô 1998: 494 (life cycle evolution, comp. note) Japan

Cicada flammata Feng, Chen, Ye, Wang, Chen, and Wang 1999: 518 (comp. note)

Tibicen flammatus Lee 1999: 4 (comp. note)

Cicada flammata Mizuno 1999: 251 (fungal host, comp. note)

Lyristes flammata (sic) Hua 2000: 62 (listed, distribution) Equals *Cicada flammata* China, Hubei, Jiangxi, Guangxi, Sichuan, Japan, Korea

Tibicen flammatus Ohara and Hayashi 2001: 1, 5–6, Fig 4 (illustrated, comp. note) Japan

Tibicen flammatus Saisho 2001a: 68 (comp. note) Japan

Lyristes flammatus Sueur 2001: 46, Table 2 (listed) Equals *Tibicen flammatus*

Cicada flammatus Wang and Ng 2002: 7 (anti-fungal protein, comp. note) China

Cicada flammata Kuo, Weng, Chou, Chang and Tsai 2003: 895 (fungal host, comp. note)

Tibicen flammatus Ohya 2004: 441–443, Figs 1–2, Table 1 (song, comp. note) Japan
Cicada flammata Zhao 2004: 477 (medicinal use, comp. note) China
Tibicen flammatus Lee 2005: 55 (comp. note)
Tibicen flammatus Inoue, Goto, Makino, Okabe, Okochi et al. 2007: 251, Table 2 (trap collection, comp. note) Japan
Tibicen flammatus Shiyake 2007: 4, 36, 38, Fig 51 (illustrated, distribution, listed, comp. note) Japan, China
Cicada flammata Zhang and Xuan 2007: 404 (fungal host, comp. note) China
Tibicen flammatus Kanasugi 2008: 83–84, Figs 1–2 (illustrated, distribution, comp. note) Japan
Tibicen flammatus Osaka Museum of Natural History 2008: 322 (listed, acoustic behavior, ecology, comp. note) Japan
Cicada flammata Ye and Ng 2009: 189 (protein, medicinal use, comp. note) China, Fujian

***T. flammatus nakamurai* Kato, 1940** = *Tibicen flammatus* var. *nakamurai*
Tibicen flammatus var. *nakamurai* Hayashi 1984: 41 (described, comp. note) Japan

***T. flammatus viridiflavus* Kato, 1939** = *Tibicen flammatus* ab. *viridiflavus*
Tibicen flammatus ab. *viridiflavus* Hayashi 1984: 41 (described, comp. note) Japan

***T. flavomarginatus* Hayashi, 1977**
Tibicen flavomarginatus Hua 2000: 65 (listed, distribution) China, Taiwan
Tibicen flavomarginatus Lee and Hayashi 2003a: 158–160, Fig 8 (key, diagnosis, illustrated, distribution, song, listed, comp. note) Taiwan
Tibicen flavomarginatus Chen 2004: 195, 198 (listed, comp. note) Taiwan

***T. fuscus* Davis, 1934** = *Tibicen fusca*
Tibicen fuscus Sanborn 1999a: 41 (type material, listed)
Tibicen fuscus Sanborn 2007b: 21, 34 (distribution, comp. note) Mexico, Distrito Federal

***T. gemellus* (Boulard, 1988)** = *Lyristes gemellus*
Lyrsistes gemellus Lodos and Boulard 1987: 645–648, Figs 1–2 (song, oscillogram, frequency spectrum, comp. note) Turkey
Lyristes gemellus Boulard 1988e: 157–165, Figs 3–4, Fig 8, Fig 10, Fig 12, Plate 1, Fig 15, Figs 20–23 (n. sp., described, illustrated, song, comp. note) Turkey
Lyristes gemellus Lodos and Kalkandelen 1988: 18 (comp. note) Turkey
Lyristes gemellus Boulard 1995a: 12, 33–34, 36, Fig 13, Fig $S11_1$, Fig $S11_2$ (illustrated, song analysis, sonogram, listed, comp. note) Turkey
Lyristes gemellus Boulard 2000b: 33–35, Fig 21, Fig 23 (song, illustrated, sonogram, comp. note) Turkey
Lyristes gemellus Sueur 2001: 36, Table 1 (listed)
Lyristes gemellus Boulard 2005d: 339–340 (comp. note) Asia Minor
Lyristes gemellus Boulard 2006g: 50–51, 180 (listed, comp. note) Asia Minor
Lyristes gemellus Kemal and Koçak 2010: 5 (distribution, listed, comp. note) Turkey

***T. heathi* Sanborn, 2010**
Tibicen heathi Sanborn 2010b: 67–71, 73–74, Figs 1–9 (n. sp., described, illustrated, listed, distribution, comp. note) Honduras, Belize, Guatemala, Mexico

***T. hidalgoensis* Davis, 1941**
Tibicen hidalgoensis Sanborn 1999a: 41 (type material, listed)
Tibicen hidalgoensis Sanborn 2007b: 34 (distribution, comp. note) Mexico, Hidalgo

***T. inauditus* Davis, 1917** = *Tibicen inauditis* (sic) = *Lyristes inauditus*
Tibicen inauditus Davis 1919a: 108 (comp. note) Texas
Tibicen inauditus Tinkham 1941: 169, 171–172, 174, 176, 178–179 (distribution, habitat, comp. note) Texas, New Mexico, Oklahoma
Tibicen inauditus Sanborn and Phillips 1995a: 482, Table 1 (song intensity, comp. note)
Tibicen inauditus Van Pelt 1995: 20 (listed, distribution, comp. note) Texas
Tibicen inauditus Poole, Garrison, McCafferty, Otte and Stark 1997: 333 (listed) Equals *Tibicin inauditis* (sic) North America
Tibicen inauditus Sanborn 1999a: 41 (type material, listed)
Tibicen inauditus Sanborn 1999b: 1 (listed) North America
Tibicen inauditus Kondratieff, Ellingson and Leatherman 2002: 7, 13, 20, 24, 72, 77, Figs 26–27, Fig 42, Table 2 (key, diagnosis, illustrated, listed, distribution, song, comp. note) Colorado, New Mexico, Oklahoma, Texas
Lyristes inauditus Sueur 2002b: 124, 129 (listed, comp. note) Equals *Tibicen inauditus*
Tibicen inauditus Phillips and Sanborn 2007: 456–460, Table 1, Fig 2, Fig 6 (illustrated, habitat, distribution, comp. note) Texas, United States

***T. intermedius* Mori, 1931** = *Tibicen intermedia* = *Lyristes intermedius* = *Lyristes horni* Schmidt, 1932 = *Tibicen horni* = *Tibicen japonica* (nec Kato) = *Lyristes flammata* Liu (nec Distant)
Tibicen intermedius Hayashi 1984: 42 (comp. note)

Tibicen intermedius Lee 1995: 17, 22, 35, 37, 39, 41–46, 63–68, 70, 135, 139 (listed, key, synonymy, ecology, distribution, illustrated, comp. note) Equals *Tibicen intermedia* Equals *Tibicen japonica* (nec Kato) Equals *Tibicen japonica* (nec Kato) Equals *Tibicen flammatus* (nec Distant) Korea

Tibicen intermedia Lee 1995: 35–36, 44, 64 (comp. note) Korea

Lyrists horni Lee 1995: 39 (comp. note)

Tibicen intermedius Lee 1999: 3–5, Fig 3 (distribution, song, illustrated, listed, comp. note) Equals *Tibicen japonensis* (nec Kato) Equals *Tibicen intermedia* Equals *Lyristes horni* Equals *Tibicen flammatus* (nec Distant) Korea, China

Tibicen intermedius Lee 2001b: 3 (comp. note) Korea

Tibicen intermedius Lee 2003b: 6, 8 (comp. note)

Tibicen intermedius Lee and Hayashi 2003a: 160 (comp. note)

Tibicen intermedius Lee 2005: 16, 54–58, 157, 161–162, 169–170 (described, illustrated, synonymy, song, distribution, key, comp. note) Equals *Cicada bihamata* (nec Motschulsky) Equals *Tibicen bihamata* (nec Motschulsky) Equals *Tibicen bihamatus* (nec Motschulsky) Equals *Tibicen japonica* (nec Kato) Equals *Tibicen japonicus* (nec Kato) Equals *Tibicen flammata* (nec Distant) Equals *Tibicen flammatus* (nec Distant) Equals *Lyristes horni* Korea, China

Lyristes intermedius Lee 2008b: 446–447, 463, Fig 1C, Table 1 (key, illustrated, synonymy, distribution, listed, comp. note) Equals *Cicada bihamata* (nec Motschulsky) Equals *Tibicen bihamata* (nec Motschulsky) Equals *Tibicen bihamatus* (nec Motschulsky) Equals *Lyristes bihamatus* (nec Motschulsky) Equals *Tibicen japonica* (nec Kato) Equals *Tibicen japonicus* (nec Kato) Equals *Tibicen intermedia* Equals *Tibicen intermedius* Equals *Tibicen flammata* (nec Distant) Equals *Tibicen flammatus* (nec Distant) Equals *Lyristes flammatus* (nec Distant) Equals *Lyristes horni* Korea, China

***T. isodol* (Boulard, 1988)** = *Lyristes isodol*

Lyrsistes isodol Lodos and Boulard 1987: 645–648, Figs 1–2 (song, oscillogram, frequency spectrum, comp. note) Turkey

Lyristes isodol Boulard 1988e: 157–158, 160, 162–164, 166, Figs 5–6, Fig 9, Fig 13, Plate 1, Fig 16 (n. sp., described, illustrated, song, comp. note) Turkey

Lyristes isodol Lodos and Kalkandelen 1988: 18 (comp. note) Turkey

Lyristes isodol Boulard 1995a: 12, 34, 37, Fig $S12_1$, Fig $S12_2$ (song analysis, sonogram, listed, comp. note) Turkey

Lyristes isodol Boulard 2000b: 33–35, Fig 21, Fig 24 (song, illustrated, sonogram, comp. note) Turkey

Lyristes isodol Sueur 2001: 36, Table 1 (listed)

Lyristes isodol Boulard 2005d: 339 (comp. note) Asia Minor

Lyristes isodol Boulard 2006g: 50–51 (comp. note)

Lyristes isodol Kemal and Koçak 2010: 5 (distribution, listed, comp. note) Turkey

***T. jai* Ouchi, 1938** = *Lyristes jai* = *Tibicen orientalis* Ouchi, 1938 = *Tibicen wui* Kato, 1934 = *Lyristes wui* = *Lyristes katoi* Liu, 1939

Lyristes wui Wu 1935: 7 (listed) China, Kwangsi

Lyristes jai Chou, Lei, Li, Lu and Yao 1997: 282–284, 358 (key, described, synonymy, comp. note) Equals *Tibicen jai* Equals *Tibicen orientalis* n. syn. Equals *Tibicen wui* Equals *Lyristes katoi* China

Tibicen orientalis Chou, Lei, Li, Lu and Yao 1997: 358 (synonymy)

Lyristes jai Hua 2000: 62 (listed, distribution) Equals *Tibicen jai* Equals *Tibicen wui* Equals *Tibicen orientalis* Equals *Lyristes katoi* China, Heilongjiang, Hebei, Shaanxi, Jiangxi, Zhejiang

***T. japonicus* var. ______ Kato, 1933**

***T. japonicus echigo* Kato, 1936** = *Tibicen japonicus* f. *echigo*

Tibicen japonicus f. *echigo* Hayashi 1984: 38–39, Fig 39a (described, illustrated, comp. note) Japan

***T. japonicus hooshianus* (Matsumura, 1936)** = *Cicada hooshiana* = *Cicada hoshiana* (sic) = *Lyristes japonica hooshiana*

***T. japonicus immaculatus* Kato, 1933** = *Tibicen japonicus* ab. *immaculatus* = *Tibicen japonicus inmaculatus* (sic)

Tibicen japonicus ab. *immaculatus* Hayashi 1984: 38 (described, comp. note) Japan

***T. japonicus interruptus* Kato, 1943** = *Tibicen japonicus* var. *interruptus*

Tibicen japonicus var. *interruptus* Hayashi 1984: 38–39, Fig 39c (described, illustrated, comp. note) Japan

***T. japonicus itoi* Kato, 1939** = *Tibicen japonicus* ab. *itoi*

Tibicen japonicus ab. *itoi* Hayashi 1984: 39 (described, comp. note) Japan

***T. japonicus iwaoi* Kato, 1939** = *Tibicen japonicus* ab. *iwaoi*

Tibicen japonicus ab. *iwaoi* Hayashi 1984: 38–39, Fig 39b (described, illustrated, comp. note) Japan

***T. japonicus japonicus* (Kato, 1925)** = *Cicada japonica* = *Tibicen japonica* (sic) = *Tibicen japonensis* (sic) = *Lyristes japonica* = *Lyristes japonicus* = *Cicada flammata* Matsumura, 1898 (nec Distant) = *Tibicen dolichoptera* Mori, 1931 = *Tibicen japonicus dolichopterus* = *Tibicen japonica dolichoptera* = *Lyristes japonica* *dolichoptera* = *Tibicen dorichoptera* (sic) = *Lyristes hooshiana* Matsumura, 1936

Tibicen japonica (sic) Isaki 1933: 381 (listed, comp. note) Japan

Tibicen japonicus Anonymous 1936: 766 (comp. note) Japan

Tibicen japonicus Hayashi 1982b: 187–192, Table 1 (distribution, comp. note) Japan

Tibicen japonicus Hayashi 1984: 29, 38–39, Figs 34–39 (key, described, illustrated, synonymy, comp. note) Equals *Cicada flammata* (nec Distant) Equals *Cicada japonica* Equals *Lyristes japonica* Equals *Tibicen dolichoptera* Equals *Lyristes hooshiana* Japan

Lyristes japonicus Boulard 1988d: 150, 152 (anatomical abnormalities, comp. note)

Lyristes japonicus Boulard 1995c: 174 (comp. note)

Tibicen japonicus Lee 1995: 17, 22, 35, 37–40, 43–46, 63–64, 66–71, 135, 139 (listed, key, synonymy, ecology, distribution, illustrated, comp. note) Equals *Cicada japonica* Korea

Tibicen dolichopterus Lee 1995: 27, 63 (comp. note)

Cicada japonica Lee 1995: 31 (comp. note)

Tibicen dolichoptera Lee 1995: 35, 63, 68, 70 (comp. note)

Tibicen japonica Lee 1995: 35, 42, 44, 70 (comp. note) Korea

Tibicen japonicus dolichopterus Lee 1995: 37, 41, 63 (comp. note)

Tibicen japonicus var. *dolichoptera* Lee 1995: 40, 68 (comp. note)

Tibicen japonica var. *dolichoptera* Lee 1995: 42, 69 (comp. note)

Tibicen japonicus var. *dolichopterus* Lee 1995: 68, 70 (comp. note)

Tibicen japonica var. *dodichopterus* (sic) Lee 1995: 69 (comp. note)

Lyristes japonicus Chou, Lei, Li, Lu and Yao 1997: 282, 285, Plate XIV, Fig 140 (key, described, illustrated, synonymy, comp. note) Equals *Cicada flammata* Equals *Cicada japonica* Equals *Tibicen japonica* Equals *Tibicen dolichoptera* Equals *Lyristes hooshiana* Equals *Lyristes japonica* China

Tibicen japonicus Lee 1999: 4–5 (distribution, song, illustrated, listed, comp. note) Equals *Cicada japonica* Equals *Tibicen dolichoptera* Equals *Tibicen japonica* Equals *Tibicen japonica* var. *dolichoptera* Equals *Tibicen japonicus* var. *dolichoptera* Equals *Tibicen japonica* var. *dodichopterus* (sic) Korea, China

Tibicen dolichoptera Lee 1999: 5 (comp. note)

Tibicen japonensis (sic) Freytag 2000: 55 (on stamp)

Lyristes japonicus Hua 2000: 62 (listed, distribution) Equals *Cicada japonica* China, Japan, Korea

Tibicen japonicus Ohara and Hayashi 2001: 1, 3–4, Fig 2 (comp. note) Japan

Lyristes japonicus Sueur 2001: 46, Table 2 (listed) Equals *Tibicen japonicus*

Tibicen japonicus Ohya 2004: 441–443, Figs 1–2, Table 1 (song, comp. note) Japan

Tibicen japonicus Lee 2005: 16, 60–61, 157, 162, 170 (distribution, comp. note) Equals *Cicada japonica* Equals *Tibicen japonica* Equals *Tibicen dolichoptera* Equals *Tibicen japonicus* var. *dolichopterus* Equals *Tibicen japonicus* var. *dolichoptera* Equals *Tibicen japonica* var. *dolichoptera* Equals *Tibicen japonica* var. *dodichoptera* (sic) Korea, China

Tibicen dolichoptera Lee 2005: 60–61 (comp. note)

Tibicen japonicus Inoue, Goto, Makino, Okabe, Okochi et al. 2007: 251, Table 2 (trap collection, comp. note) Japan

Tibicen japonicus Shiyake 2007: 4, 37, 39, Fig 52 (illustrated, distribution, listed, comp. note) Japan

Lyristes japonicus Lee 2008b: 445, 448, 462 (comp. note)

Tibicen japonicus Kanasugi 2008: 83 (comp. note) Japan

Tibicen japonicus Osaka Museum of Natural History 2008: 322 (listed, acoustic behavior, ecology, comp. note) Japan

***T. japonicus kobayashii* Kato, 1939** = *Tibicen japonicus* var. *kobayashii*

Tibicen japonicus var. *kobayashii* Hayashi 1984: 39 (described, comp. note) Japan

***T. japonicus niger* Kato, 1933**

***T. japonicus nigrofasciatus* Kato, 1940** = *Tibicen japonicus* var. *nigrofasciatus* = *Tibicen japonicus ingrofasciatus* (sic)

Tibicen japonicus var. *nigrofasciatus* Hayashi 1984: 39 (described, comp. note) Japan

T. knowlesi (Distant, 1907) to *Raiateana knowlesi*

T. kuruduadua (Distant, 1881) to *Raiateana kuruduadua*

***T. kyushyuensis* (Kato, 1926)** = *Cicada kyushyuensis* = *Tibicen kyushuensis* (sic) = *Lyristes kyushyuensis* = *Lyristes kyushuensis* (sic) = *Tibicen ishiharai* Kato, 1959 = *Tibicen tsukushiensis* Kato, 1959 = *Tibicen shikokuanus* Kato, 1959

Tibicen kyushyuensis Anonymous 1936: 766 (comp. note) Japan

Tibicen kyushyuensis Hayashi 1984: 29, 41–42, Figs 49–54 (key, described, illustrated, synonymy, comp. note) Equals *Cicada kyushyuensis* Equals *Tibicen kyushynensis* (sic) Equals *Lyristes kyushuensis* (sic) Equals *Tibicen tsukushiensis* Equals *Tibicen shikokuamus* Equals *Tibicen ishiharai* Japan

Tibicen kyushyuensis Wang and Zhang 1987: 295 (key, comp. note) China

Tibicen kyushyuensis Wang and Zhang 1993: 189 (listed, comp. note) China

Tibicen kyushyuensis Lee 1995: 65 (comp. note)

Lyristes kyushyuensis Chou, Lei, Li, Lu and Yao 1997: 282, 286–287, 357, Plate XVI, Fig 153 (key, described, illustrated, synonymy, comp. note) Equals *Cicada kyushyuensis* Equals *Tibicen kyushyuensis* Equals *Tibicen tsukushiensis* Equals *Tibicen shikokuanus* Equals *Tibicen ishiharai* China

Tibicen kyushyuensis Hua 2000: 65 (listed, distribution) Equals *Cicada kyushyuensis* China, Sichuan, Yunnan, Xizang, Japan

Tibicen kyushyuensis Ohara and Hayashi 2001: 1, 4–6, Fig 3 (illustrated, comp. note) Japan

Lyristes kyushuensis Sueur 2001: 46, Table 2 (listed) Equals *Tibicen kyushuensis*

Tibicen kyushuensis (sic) Shiyake 2007: 4, 37, 39, Fig 53 (illustrated, distribution, listed, comp. note) Japan

Tibicen kyushyuensis Osaka Museum of Natural History 2008: 322 (listed, acoustic behavior, ecology, comp. note) Japan

***T. latifasciatus* (Davis, 1915)** = *Cicada pruinosa latifasciata* = *Tibicen latifasciata* = *Tibicen pruinosus latifasciata*

Tibicen latifasciata (sic) Davis, W.T. 1933: 29 (comp. note) New Jersey

Tibicen latifasciata (sic) Davis 1936: 44 (comp. note) New Jersey

Tibicen latifasciatus Sanborn 1999a: 41, 55 (type material, listed, comp. note)

Tibicen latifasciatus Sanborn 1999b: 1 (listed) North America

***T. leechi* (Distant, 1890)** = *Cicada leechi* = *Tibicen leachi* (sic) = *Lyristes leechi* = *Lyrtistes* (sic) *leechi*

Lyristes leechi Wu 1935: 6 (listed) Equals *Cicada leechi* China, Wa-shan, Chia-kou-ho

Tibicen leechi Jiang, Lin and Wang 1987: 106, Fig 1B (timbal structure, illustrated, comp. note) China

T[ibicen] leachi (sic) Jiang, Yang, Tang, Xu and Chen 1995b: 228 (song, phonotaxis, flight behavior, comp. note) China

Lyristes leechi Chou, Lei, Li, Lu and Yao 1997: 287 (described, synonymy, comp. note) Equals *Cicada leechi* China

Lyristes leechi Hua 2000: 62 (listed, distribution) Equals *Cicada leechi* China, Wa Shan, Chia-Kuo-Ho

***T. linnei* (Smith & Grossbeck, 1907)** = *Tettigonia tibicen* Fabricius 1775 (nec *Cicada tibicen* Linnaeus, 1758) = *Cicada tibicen* (nec Linnaeus) = *Cicada tubicen* (sic) (nec Linnaeus) = *Fidicina tibicen* (nec Linnaeus) = *Rihana tibicen* (nec Linnaeus) = *Tettigonia tibicen* var. ______ Fabricius 1781 = *Tibicen linnei* var. ______ = *Cicada linnei* Smith & Grossbeck 1907 = *Lyristes linnei*

Tibicen linnei Davis 1920: 129 (parasitism, comp. note) New York, Staten Island, Long Island

Tibicen linnei Steinhaus 1941: 763, 776, 779, 781, Table 1 (listed, bacterial host) Ohio

Tibicen linnei Wright 1951: 10 (larva illustrated, adult illustrated, comp. note)

Tibicen linnei Reger 1961: 114 (neuromuscular junction, comp. note)

Tibicen linnei McCoy 1965: 42, 44 (key, comp. note) Arkansas

Tibicen linnei Evans 1966: 104, Table 10 (parasitism, comp. note) New York, Ohio

Tibicen linnei Johnson and Lyon 1976: 434 (comp. note)

Tibicen linnei Krombein, Hurd, Smith and Burks 1979: 1698 (parasitism)

Cicada linnei Hennessey 1990: 470 (type material)

Tibicen linnei Johnson and Lyon 1991: 490–491, Plate 236D (illustrated, life cycle, comp. note)

Tibicen linnei Hennig, Kleindienst, Weber, Huber, Moore and Popov 1992: 180, Fig A (song, oscillogram, hearing, comp. note)

Tibicen linnei Hennig, Weber, Moore, Kleindienst, Huber and Popov 1992: 261 (song, tensor muscle, comp. note)

Tibicen linnei Hennig, Weber, Moore, Kleindienst, Huber and Popov 1994: 46–53, Figs 1–4 (hearing, comp. note) Michigan

Tibicen linnei Hennig, Weber, Moore, Huber, Kleindienst and Popov 1994: 34–44, Figs 1–7 (song, oscillogram, tensor muscle, comp. note) Michigan

Tibicen linnei Lane 1995: 388 (comp. note)

Tibicen linnei Riede and Kroker 1995: 49–50 (comp. note)

Tibicen linnei Sanborn and Phillips 1995a: 482, Table 1 (song intensity, comp. note)

Tibicen linnei Weaver 1995: 9 (parasitism) West Virginia

Tibicen linnei Young and Bennet-Clark 1995: 1017 (comp. note)

Tibicen linnei Marshall, Cooley, Alexander and Moore 1996: 165 (comp. note) Michigan

Tibicen linnei Hilton 1997: 25, 27 (natural history, comp. note)

Tibicen linnei Poole, Garrison, McCafferty, Otte and Stark 1997: 333 (listed) Equals *Cicada linnei* North America

Tibicen linnei Ostry and Anderson 1998: 2 (oviposition damage, comp. note)

Tibicen linnei Sanborn 1999a: 41, 55 (type material, lectotype designation, listed)

Tibicen linnei Sanborn 1999b: 1 (listed) North America

Tibicen linnei var. Sanborn 1999b: 1–2 (type material, listed) North America

Tibicen linnei Maw, Footit, Hamilton and Scudder 2000: 49 (comp. note) Canada, Ontario
Tibicen linnei Smith, Kelly, Dean, Bryson, Parker et al. 2000: 125 (comp. note) Kansas
Lyristes linnei Sueur 2001: 36, 46, Tables 1–2 (listed) Equals *Tibicen linnei*
Tibicen linnei Kondratieff, Ellingson and Leatherman 2002: 10, 12, 25, Table 1 (comp. note) Equals *Cicada linnei* nom. nov. pro *Cicada tibicen* Fabricius, 1794 nec *Cicada tibicen* Linnaeus, 1758
Lyristes linnei Sueur 2002b: 122, 129 (listed, comp. note) Equals *Cicada tibicen*
Lyristes linnei Sueur and Aubin 2003b: 486 (comp. note)
Tibicen linnei Walker and Moore 2004: 2 (listed, comp. note) Florida
Tibicen linnei Elliott and Hershberger 2006: 32, 190–191 (illustrated, sonogram, listed, distribution, comp. note) Florida, Georgia, Alabama, South Carolina, North Carolina, Tennessee, Virginia, West Virginia, Maryland, Delaware, New Jersey, New York, Connecticut, Pennsylvania, Ohio, Kentucky, Michigan, Indiana, Illinois, Iowa, Minnesota, Missouri, Kansas, Nebraska
Tibicen linnei Grant 2006: 541–545, Figs 1–3, Tables 1–2 (predation, emergence time, comp. note) New Jersey
Tibicen linnei Bunker, Neckermann and Cooley 2007: 357 (comp. note) Connecticut
Tibicen linnei Eaton and Kaufman 2007: 88–89 (illustrated, comp. note) Eastern United States
Tibicen linnei Ganchev and Potamitis 2007: 304–305, Fig 11 (listed, comp. note)
Tibicen linnei Oberdörster and Grant 2007a: 2–4, 6–9, 11, Fig 1, Fig 4, Tables 2–5 (flight system, flight speeds, comp. note) New Jersey
Tibicen linnei Oberdörster and Grant 2007b: 17–20, Tables 1–2 (sound system, predation, comp. note) New Jersey
Tibicen linnei Cole 2008: 818 (comp. note)
Tibicen linnei Hastings, Holliday and Coelho 2008: 659, Table 1 (predation, comp. note) Virginia
Tibicen linnei Sanborn, Phillips and Gillis 2008: 4, 9–10, 12, 30–31, Figs 61–70 (key, synonymy, illustrated, distribution, comp. note) Equals *Cicada linnei* Florida
Tibicen linnei Holliday, Hastings and Coelho 2009: 3, 6–7, 10, 12–13, Tables 1–2 (predation, comp. note) Connecticut, Illinois, Indiana, Kentucky, Maryland, Michigan, Minnesota, New Jersey, New York, North Carolina, Ohio, Pennsylvania, Virginia, Wisconsin, West Virginia

T. linnei var. ______ (Fabricius, 1781) to *Tibicen linnei* (Smith & Grossbeck, 1907)

***T. longioperculus* Davis, 1926** = *Tibicen longiopercula*
Tibicen longiopercula Poole, Garrison, McCafferty, Otte and Stark 1997: 333 (listed) North America
Tibicen longioperculus Sanborn 1999a: 41 (type material, listed)
Tibicen longioperculus Sanborn 1999b: 1 (listed) North America

T. luctuosus Costa, 1883 to *Tibicina nigronervosa* Fieber, 1876

***T. lyricen engelhardti* (Davis, 1910)** = *Cicada engelhardti* = *Rihana engelhardti* = *Tibicen engelhardti* = *Tibicen lyricen* var. *engelhardti* = *Tibicen lyricen englehardti* (sic)
Cicada engelhardti Davis 1911a: 216 (comp. note) Georgia
Tibicen lyricen var. *engelhardti* Moore 1993: 272–273, Figs 1–2 (illustrated, distribution, song, comp. note)
Tibicen lyricen var. *engelhardti* Sanborn 1999a: 41 (type material, lectotype designation, listed)
Tibicen lyricen var. *engelhardti* Sanborn 1999b: 1 (listed) North America
Tibicen lyricen engelhardti Holliday, Hastings and Coelho 2009: 6–7, Table 1 (predation, comp. note) Ohio

***T. lyricen lyricen* (Degeer, 1773)** = *Cicada lyricen* = *Cicada lyricea* (sic) = *Fidicina lyricen* = *Rihana lyricen* = *Tibicen lyricen* = *Lyristes lyricen* = *Tibicen lyricen* var. *lyricen* = *Cicada fulvula* Osborn, 1906 = *Tibicen fulvula* = *Tibicen fulvulus*
Cicada lyricen Hasselt 1881: 192 (comp. note)
Cicada lyricen Davis 1911a: 216 (comp. note)
Tibicen lyricen Davis 1920: 129 (parasitism, comp. note) New York, Staten Island, Long Island
Tibicen lyricen McCoy 1965: 41, 44 (key, comp. note) Arkansas
Tibicen lyricen Evans 1966: 104, Table 10 (parasitism, comp. note) Massachusetts, New York, Ohio
Tibicen lyricen Krombein, Hurd, Smith and Burks 1979: 1698 (parasitism)
Tibicen lyricen Wilson and Hilburn 1991: 416 (comp. note) United States, Canada
Tibicen lyricen Moore 1993: 269, 271, 274, Fig 3 (distribution, song, comp. note) North America
Tibicen lyricen var. *lyricen* Moore 1993: 271–273, Figs 1–2 (illustrated, distribution, song, comp. note)
Tibicen lyricen Lane 1995: 407 (comp. note)
Tibicen lyricen Weaver 1995: 9 (parasitism) West Virginia
Tibicen lyricen Marshall, Cooley, Alexander and Moore 1996: 165 (comp. note) Michigan

Tibicen lyricen Poole, Garrison, McCafferty, Otte and Stark 1997: 333 (listed) Equals *Cicada lyricen* Equals *Cicada lyricea* (sic) Equals *Cicada fulvula* Equals *Cicada engelhardti* Equals *Tibicen virescens* North America
Tibicen lyricen Sanborn 1999b: 1–2 (type material, listed) North America
Tibicen lyricen var. *lyricen* Sanborn 1999b: 1 (listed) North America
Tibicen lyricen Callaham, Whiles, Meyer, Brock and Charlton 2000: 537–540 (comp. note) Kansas
Tibicen lyricen Maw, Footit, Hamilton and Scudder 2000: 49 (comp. note) Canada, Ontario
Tibicen lyricen Sanborn 2001a: 17 (comp. note)
Lyristes lyricen Sueur 2001: 36, 46, Tables 1–2 (listed) Equals *Tibicen lyricen*
Tibicen lyricen Walker and Moore 2004: 2 (listed, comp. note) Florida
Tibicen lyricen Kondratieff, Schmidt, Opler and Garhart 2005: 96, 217 (listed, comp. note) Oklahoma
Lyristes lyricen Sueur 2005: 213 (song variability, comp. note) North America
Tibicen lyricen Elliott and Hershberger 2006: 32, 198–199 (illustrated, sonogram, listed, distribution, comp. note) Maine, New Hampshire, Vermont, Massachusetts, Rhode Island, Connecticut, New York, Pennsylvania, New Jersey, Delaware, Maryland, Virginia, West Virginia, North Carolina, South Carolina, Georgia, Florida, Alabama, Mississippi, Louisiana, Texas, Oklahoma, Arkansas, Kansas, Nebraska, Missouri, Iowa, Illinois, Indiana, Michigan, Ohio, Kentucky, Tennessee, Ontario, New Brunswick
Tibicen lyricen Grant 2006: 541–544, Figs 1–3, Tables 1–2 (predation, emergence time, comp. note) New Jersey
Tibicen lyricen Bunker, Neckermann and Cooley 2007: 357 (comp. note) Connecticut
Tibicen lyricen Ganchev and Potamitis 2007: 305 (listed)
Tibicen lyricen Oberdörster and Grant 2007a: 2–3, 6–8, Tables 2–5 (flight system, flight speeds, comp. note) New Jersey
Tibicen lyricen Oberdörster and Grant 2007b: 17, 19 (sound system, predation, comp. note) New Jersey
Tibicen lyricen Cole 2008: 818 (comp. note)
Tibicen lyricen Hastings, Holliday and Coelho 2008: 659, Table 1 (predation, comp. note) Massachusetts, Illinois
Tibicen lyricen Sanborn 2008a: 227 (comp. note)
Tibicen lyricen lyricen Sanborn, Phillips and Gillis 2008: 5, 9–10, 31–32, Fig 71, Figs 76–84 (key, synonymy, illustrated, distribution, comp. note) Equals *Cicada lyricen* Equals *Cicada fulva* Florida
Tibicen lyricen Holliday, Hastings and Coelho 2009: 3, 6–7, 12, Table 1 (predation, comp. note) Connecticut, Georgia, Illinois, Indiana, Iowa, Louisiana, Massachusetts, Maryland, Michigan, New Jersey, New York, North Carolina, Ohio, Pennsylvania, South Carolina, Virginia, Virginia, Wisconsin
Tibicen lyricen Hughes, Nuttall, Katz and Carter 2009: 961 (comp. note) Connecticut

***T. lyricen virescens* Davis, 1935** = *Tibicen lyricen* var. *virescens* = *Tibicen virescens*
Tibicen lyricen var. *virescens* Moore 1993: 271–273, Figs 1–2 (illustrated, distribution, song, comp. note)
Tibicen lyricen var. *virescens* Sanborn 1999a: 41 (type material, listed)
Tibicen lyricen var. *virescens* Sanborn 1999b: 1 (listed) North America
Tibicen lyricen virescens Sanborn, Phillips and Gillis 2008: 5, 9–10, 31, 33, Fig 71, Figs 85–93 (key, illustrated, distribution, comp. note) Florida
Tibicen virescens Holliday, Hastings and Coelho 2009: 6–7, Table 1 (predation, comp. note) Florida
Tibicen lyricen virescens Holliday, Hastings and Coelho 2009: 6, Table 1 (listed)

***T. maculigena* (Signoret, 1860)** = *Cicada maculigena* = *Cicada stigmosa* Stål, 1866

***T. minor* Davis, 1934**
Tibicen minor Sanborn 1999a: 41 (type material, listed)
Tibicen minor Sanborn 2007b: 21, 34 (distribution, comp. note) Mexico, Distrito Federal, Jalisco

***T. montezuma* (Distant, 1881)** = *Cicada montezuma*
Tibicen montezuma Poole, Garrison, McCafferty, Otte and Stark 1997: 333 (listed) Equals *Cicada montezuma* North America
Tibicen montezuma Sanborn 1999b: 1 (listed) North America
Tibicen montezuma Sanborn 2007b: 34 (distribution, comp. note) Mexico, Morelos

***T. nigriventris* (Walker, 1858)** = *Cicada nigriventris* = *Rihana vitripennis nigriventris* = *Tibicen nigroventris* (sic) = *Diceroprocta nigriventris*
Diceroprocta nigriventris Wolda and Ramos 1992: 273, Table 17.2 (listed) Panama
Tibicen nigriventris Sanborn 2007b: 34 (distribution, comp. note) Mexico, Veracruz, Costa Rica, Panama
Tibicen nigriventris Sanborn 2010b: 70–71, 73–74 (comp. note)

***T. occidentis* (Walker, 1850)** = *Cicada occidentis* = *Cicada occidentalis* (sic) = *Cicada crassimargo* Spinola, 1852 = *Tettigades crassimargo* = *Tibicina crassimargo* = *Tibicen crassimargo*
Tettigades crassimargo Breddin 1897: 30 (listed)
Tibicina crassimargo Varley 1939: 99 (comp. note)
Tibicen crassimargo Peña 1987: 110 (comp. note) Chile

Tibicen ochreoptera Uhler, 1892 nom. nud. to *Diceroprocta cinctifera cinctifera* (Uhler, 1892)

***T. oleacea* (Distant, 1891)** = *Fidicina oleacea* = *Cicada oleacea*
Tibicen oleacea Sanborn 2007b: 34 (distribution, comp. note) Mexico, Oaxaca, Veracruz

T. orientalis Ouchi, 1938 to *Tibicen jai* Ouchi, 1938

***T. parallelus* Davis, 1923** = *Tibicen parallela*
Tibicen parallela (sic) Hastings 1986: 263–267, Table 1 (predation, comp. note) Arizona
Tibicen parallela (sic) Hastings 1989a: 255 (predation, comp. note)
Tibicen parallela (sic) Hastings 1990: 4 (predation, comp. note)
Tibicen parallela (sic) Poole, Garrison, McCafferty, Otte and Stark 1997: 333 (listed) North America
Tibicen parallelus Sanborn 1999a: 41 (type material, listed)
Tibicen parallelus Sanborn 1999b: 1 (listed) North America
Tibicen parallelus Sanborn 2006b: 256 (listed, distribution) Mexico, Chihuahua, Arizona, New Mexico
Tibicen parallelus Sanborn 2007b: 34 (distribution, comp. note) Mexico, Chihuahua, Arizona, New Mexico
Tibicen paralella (sic) Hastings, Holliday and Coelho 2008: 661 (predation, comp. note) Arizona
Tibicen parallelus Holliday, Hastings and Coelho 2009: 3–5, 13–14, Table 1 (predation, comp. note) Arizona

***T. paralleloides* Davis, 1934**
Tibicen paralleloides Hennessey 1990: 471 (type material)
Tibicen paralleloides Sanborn 1999a: 41 (type material, listed)
Tibicen paralleloides Sanborn 2007b: 34 (distribution, comp. note) Mexico, Nayarit, Sinaloa, Jalisco

***T. pekinensis* (Haupt, 1924)** = *Cicada pekinensis* = *Lyristes pekinensis*
Lyristes pekingensis Wu 1935: 6 (listed) Equals *Cicada pekingensis* China, Peiping
Lyristes pekinensis Chou, Lei, Li, Lu and Yao 1997: 287 (described, synonymy, comp. note) Equals *Cicada pekinensis* China
Lyristes pekinensis Hua 2000: 62 (listed, distribution) Equals *Cicada pekinensis* China, Beijing

***T. pieris* Kirkaldy, 1909** = *Lyristes pieris*
Lyristes pieris Chou, Lei, Li, Lu and Yao 1997: 288 (described, synonymy, comp. note) Equals *Tibicen pieris* China
Lyristes pieris Hua 2000: 62 (listed, distribution) Equals *Tibicen peiris* China, Yunnan

T. plebejus armeniaca (Kolenati, 1857) to *Tibicen armeniaca*

***T. plebejus martorellii* Martorell & Pena, 1879, nom. nud.**

***T. plebejus* (Scopoli, 1763)** = *Cicada plebeja* = *Tettigonia plebeja* = *Tettigonia plebeia* (sic) = *Cicada plebeia* (sic) = *Cicada plebia* (sic) = *Cicada plebeya* (sic) = *Cicada pelebeja* (sic) = *Cicada (Cicada) plebeja* = *Cicada plebja* (sic) = *Cicada (= Tibicen) plebia* (sic) = *Rihana plebeja* = *Tibicina plebeja* = *Lyristes plebejus* = *Lyriste* (sic) *plebejus* = *Lyristes plebeja* = *Lyristes plebeius* = *Liristes* (sic) *plebejus* = *Lyristes plebeius* (sic) mph. *plebeius* (sic) = *Tibicen plebeia* (sic) = *Tibicen plebeius* (sic) = *Tibicen pelbeia* (sic) = *Tettigonia orni* Fabricius, 1775 (nec Linnaeus) = *Tibicen orni* (nec Linnaeus) = *Tettigonia fraxini* Fabricius, 1803 = *Cicada fraxini* = *Tettigonia obscura* Fabricius, 1803 = *Cicada plebeja obscura* = *Psaltoda plebeia* Goding & Froggatt, 1904 = *Fidicina africana* Metcalf, 1955 = *Lyristes plebeius* (sic) mph. *castanea* Boulard, 1982 = *Lyristes plebejus* mph. *castanea* = *Lyristes plebejus* var. *castanea*
Tettigonia plebeja Carus 1829: 150, 158, 160, Fig VII, Fig XVI (comp. note)
Cicada plebejae (sic) Sahlberg 1842: 89 (comp. note)
Cicada plebeia (sic) Swinton 1879: 81–82 (hearing, comp. note)
Cicada plebeja Swinton 1880: 23, 222–223, 228, 233, vii, Plate III, Fig 5 (illustrated, sound system, song, hearing, comp. note)
Cicada fraxini Swinton 1880: 23 (comp. note)
Tettigonia plebeja Swinton 1880: 225 (comp. note)
Cicada plebeja Linnaeus Swinton 1880: 23 (comp. note) Africa
Cicada (Tettigonia) plebeja Hasselt 1881: 180 (comp. note)
Cicada plebeja Hasselt 1881: 187, 189, 192, 204, 207 (comp. note)
Cicada plebeja Strobl 1900: 201 (comp. note) Austria
Cicada plebeja Swinton 1908: 378–379 (song, comp. note)
Cicada plebeja Bachmetjew 1907: 251 (body length, comp. note)
Cicada plebeja Hagan 1929: 902 (fig pest, comp. note) Turkey
Cicada plebeja Weber 1931: 111, 113, Fig 110a, Fig 112 (auditory system illustrated, comp. note)

Cicada plebeia (sic) Bergier 1941: 22 (human food, comp. note) Greece
Cicada plebeja Chen 1942: 144 (type species of *Tibicen*)
Cicada plebeja Roth 1949: 81 (comp. note)
Lyristes plebeja Gómez-Menor 1951: 62–63, 65, Fig 13 (illustrated, listed, described, comp. note) Spain
Lyristes plebejus Castellani 1952: 15 (comp. note) Italy
Lyristes plebejus Wagner 1952: 15 (listed, comp. note) Italy
Cicada plebeia (sic) Alexinschi 1954: 221 (comp. note) Romania
Lyristes plebejus Soós 1956: 411 (listed, distribution) Romania
Cicada plebeja Bunescu 1959: 91, 99–101, Fig 3 (distribution, ecology, comp. note) Romania
Lyristes plebejus Cantoreanu 1960: 332 (distribution, ecology, comp. note) Romania, Czechoslovakia, Balkans, Crimea, Caucasus
Psaltoda plebeia Hahn 1962: 10 (listed)
Cicada plebeja Weber 1968: 81, 83–86 (comp. note)
Cicada fraxini Weber 1968: 126, Fig 100 (cerebral ganglia illustrated, comp. note)
Tibicen plebeja (sic) Popov 1971: 302, 316, Fig 11 (hearing, oscillogram, listed, comp. note)
Lyristes plebeja (sic) Villiers 1977: 68, Fig 37 (sound system illustrated, comp. note)
Lyristes plebejus Villiers 1977: 165, Plate XIV, Fig 20 (illustrated, listed, comp. note) France
Tibicen plebejus Drosopoulous 1980: 190 (listed) Greece
Tibicen plebejus Dlabola 1981: 199 (distribution, comp. note) Iran
Lyristes plebejus Lodos and Kalkandelen 1981: 69–70 (synonymy, distribution, comp. note) Equals *Tettigonia fraxini* Equals *Tettigonia obscura* Equals *Cicada plebeja armeniaca* Equals *Fidicina africana* Turkey, Albania, Austria, Cyprus, Czechoslovvakia, France, Germany, Greece, Hungary, Iran, Italy, Poland, Portugal, Romania, Spain, Syria, Switzerland, USSR, Yugoslavia
Lyristes plebeius (sic) mph. *plebeius* (sic) Boulard 1982a: 49 (comp. note) France
Lyristes plebeius (sic) mph. *castanea* Boulard 1982a: 49 (n. morph, described, comp. note) France
Lyristes plebeius (sic) Boulard 1982f: 182, 196 (distribution, comp. note) Portugal
Lyrsistes plebejus Boulard 1982g: 39, 43, 46, Plate II, Fig 4, Plate IV, Fig 16 (described, illustrated, ecolsion illustrated, comp. note) France
Cicada plebeja Koçak 1982: 148, 150 (type species of *Lyristes*, type species of *Tibicen*, comp. note)
Lyrsistes plebejus Boulard 1983b: 119, 121, Figs 1–2 (parasitism, illustrated, comp. note) France
Tibicen plebeja (sic) Panov and Krjutchkova 1983: 141, 143–145, 147, Figs 7–8 (neurosecretory cells, comp. note) Crimea, Azerbaijan
Tibicen plebejus Strümpel 1983: 71, 108, 168, Table 5 (comp. note)
Cicada plebeja Boulard 1984c: 166–171, 174–179 (type species of genus *Cicada*, type species of *Tibicen*, type species of *Lyristes*, comp. note) Equals *Tibicen plebeius*
Tibicen plebeius (sic) Boulard 1984c: 169–170 (status, type genus *Tibicen*, type species *Cicada plebeja*, comp. note)
Lyristes plebejus Boulard 1984c: 170 (status, comp. note)
Tibicen plebeja Boulard 1984c: 171, 173, 175 (status, comp. note)
Cicada plebeia (sic) Boulard 1984c: 171–172 (comp. note)
Lyristes plebejus Boulard 1984d: 5 (comp. note) France
Lyristes (= Tibicen) plebejus Burton and Johnson 1984: 101 (song, comp. note) Europe, Germany, Poland
Cicada plebeja Melville and Sims 1984: 163–164, 180 (type species of *Cicada*, type species of *Tibicen*, history)
Psaltoda plebeia Goding and Froggatt Moulds 1984: 27 (status, comp. note) Australia
Cicada plebeja Arambourg 1985: 18, 19 (olive pest) Turkey
Lyristes plebejus Boulard 1985b: 60, 63 (illustrated, characteristics, comp. note) France
Lyristes plebejus Boulard 1985e: 1024, 1030, 1042–1043, Plate VI, Fig 37 (coloration, illustrated, comp. note)
Lyristes plebeius (sic) Claridge 1985: 301–302, 312 (timbal, illustrated, song, comp. note) Europe, France, USSR
Cicada plebejus Lauterer 1985: 214 (type species of *Tibicen*)
Tibicen plebejus Popov 1985: 40–41, 45, 68–71, 78, 80, 181, 183, 185, Fig 7.1, Fig 23.2, Fig 24 (illustrated, oscillogram, acoustic system, auditory system, comp. note)
Tibicen plebeja (sic) Popov 1985: 45 (acoustic system, comp. note)
Tibicen plebejus Sander 1985: 593 (comp. note) Bulgaria
Tibicen plebejus Chinery 1986: 88–89 (illustrated, comp. note) Europe
Lyristes plebeius (sic) Schedl 1986a: 3–4, 10–14, 22–24, Fig 21, Map 1, Table 1 (listed, key, synonymy, distribution, ecology, comp. note) Equals *Cicada plebeja* Equals *Tibicen plebejus*

Equals *Tettigonia fraxini* Istria, Slovakia, Austria, Italy, Poland, Hungary, Romania, Anatolia, Russia, Ukraine, Transcaucasia, Syria, Iran

Cicada plebeius (sic) Schedl 1986a: 12 (ecology, comp. note)

Psaltoda plebeia Stevens and Carver 1986: 263, 265 (type material, comp. note) Australia, New South Wales

Lyrsistes plebejus Lodos and Boulard 1987: 645–648, Figs 1–2 (song, oscillogram, frequency spectrum, comp. note) Turkey

Lyristes plebejus Jamieson 1987: 155–156 (sperm anatomy, comp. note)

Lyristes plebejus Joermann and Schneider 1987: 284–287, 293–295, Figs 1–3, Fig 4a, Table 1 (song analysis, sonogram, oscillogram, comp. note) Yugoslavia

Lyristes plebejus Nast 1987: 568–569 (distribution, listed) Germany, Poland, Czechoslovakia, France, Switzerland, Austria, Hungary, Ukraine, Romania, Russia, Portugal, Spain, Italy, Yugoslavia, Albania, Bulgaria, Greece, Turkey

Lyrsites plebejus Boulard 1988a: 9, 12, Table (comp. note) France

Lyristes plebejus Boulard 1988d: 152 (wing abnormalities, comp. note) France

Lyristes plebejus Boulard 1988e: 153–164, 166, Figs 1–2, Fig 7, Fig 11, Plate 1, Fig 14 (type species of *Lyristes*, homotype and neallotype designation, described, song, comp. note) Equals *Cicada plebeia* Equals *Tettigonia obscura* Equals *Tettigonia fraxini* Equals *Cicada plebeia* var. *obscura* Equals *Cicada plebeja* Equals *Tibicen plebeja* Equals *Tibicen plebeia* Equals *Lyristes plebeius* Equals *Tibicen plebeius* Equals *Tibicen plebejus* Equals *Tibicina plebeja* Equals *Lyristes plebeja* Equals *Fidicina africana* Turkey

Cicada plebeja Boulard 1988f: 7, 16–17, 20–21, 23–28, 30, 33–34, 68–70, 72 (type species of *Cicada*, *Tibicen*, Tibiceninae, comp. note) Equals *Fidicina africana*

Tettigonia plebeia (sic) Boulard 1988f: 20–21 (comp. note)

Cicada fraxini Boulard 1988f: 20–21 (comp. note)

Tibicen plebeja (sic) Boulard 1988f: 23, 32, 70 (comp. note)

Cicada plebeia (sic) Boulard 1988f: 24 (comp. note)

Tettigonia plebeja Fab Boulard 1988f: 25 (comp. note)

Tibicen plebejus Boulard 1988f: 29, 31 (comp. note)

Lyristes plebejus Boulard 1988f: 29, 31, 44, 49, Fig 2 (illustrated, comp. note)

Lyristes plebeja (sic) Boulard 1988f: 45 (comp. note)

Fidicina africana Boulard 1988f: 68–69 (comp. note)

Lyristes plebejus Boulard 1988g: 3–5, Plate II, Fig 4, Plate III, Fig 11 (illustrated, comp. note) France

Lyristes plebejus Quartau and Fonseca 1988: 367, 371 (synonymy, distribution, comp. note) Equals *Cicada plebeja* Equals *Cicada plebeia* Equals *Tettigonia fraxini* Equals *Tettigonia obscura* Portugal, Albania, Austria, Cyprus, Czechoslovakia, France, Germany, Greece, Hungary, Iran, Italy, Sardinia, Sicily, Poland, Romania, Spain, Syria, Switzerland, Turkey, Anatolia, USSR, Armenia, Azerbaijan, Georgia, Russia, Ukraine, Yugoslavia

Lyristes plebejus Boulard 1990b: 66, 75, 84, 167, 169, 177–180, 205 (comp. note) France

Psaltoda plebeia Moulds 1990: 74 (status)

Lyrists plebejus Moulds 1990: 74 (synonymy) Equals *Tibicen plebejus* Equals *Psaltoda plebeia* Europe

Tibicen plebejus Popov 1990a: 302–303, Fig 2 (comp. note)

Tibicen plebejus Tautz 1990: 105, Fig 12 (song frequency, comp. note)

Lyristes plebejus Boulard 1991d: 262 (oviposition damage, comp. note)

Tibicen plebejus Popov, Aronov and Sergeeva 1991: 282, 284, 288–289, 293–294, Fig 5, Table (auditory system, ecology, comp. note) Russia

Liristes (sic) *plebejus* Bairyamova 1992: 47, Table 1 (listed) Bulgaria

Lyristes plebejus Boulard 1992c: 69 (comp. note)

Tibicen plebejus Brodsky 1992: 42, Fig 4A (mesothoracic terga, illustrated, comp. note)

Lyristes plebejus Moalla, Jardak and Ghorbel 1992: 34–35 (comp. note) France, Italy

Tibicen plebejus Popov, Arnov and Sergeeva 1992: 152, 154, 158–159, 163–164, Fig 5, Table (auditory system, ecology, comp. note) Russia

Lyristes plebejus Villet 1992: 93 (comp. note) France

Lyristes plebejus Boulard 1993c: 30–30, Figs 1–15 (eclosion, illustrated)

Tibicen plebejus Brodskiy 1993: 11, Fig 4A (mesothoracic terga, illustrated, comp. note)

Cicada plebeja Helfert and Sänger 1993: 41 (comp. note) Europe

Tibicen plebejus Moore 1993: 271 (comp. note) Equals *Lyristes plebejus*

Lyristes plebejus Pillet 1993: 99–100 (described, ecology, distribution, comp. note) Switzerland

Tibicen plebejus Brodsky 1994: 142–143, Fig 8.10a, Fig 8.11 (mesothorax, forewing base, illustrated, comp. note)

Lyristes (Tibicen) plebejus Fonseca and Popov 1994: 350 (comp. note)
Lyristes (= Tibicen, Cicada) plebejus Hornig 1994: 65 (comp. note) France
Cicada plebeja Hornig 1994: 65 (comp. note)
Lyristes plebejus Quartau and Rebelo 1994: 139–140, Figs 7–8 (song, comp. note) Portugal, Europe
Lyristes plebejus Boulard 1995a: 6–9, 11–12, 33, 35, Figs 4–6, Fig $S10_{1-2}$ (illustrated, song analysis, sonogram, listed, comp. note) France
Lyristes plebejus Boulard and Mondon 1995: 25, 30, 32–33, 35, 37, 49, 54–56, 64, 66–67, 70–71, 73–74, 84, 159 (larva, illustrated, sonogram, listed, comp. note) Equals *Cicada plebeia* Equals *Cicada plebaja* Equals *Tettigonia obsura* (sic) Equals *Tettigonia fraxini* France
Tibicen plebejus Mirzayans 1995: 14 (listed, distribution) Iran
Lyristes plebejus Quartau 1995: 34–35, 38 (illustrated, comp. note) Portugal
Lyristes plebejus Riou 1995: 75 (comp. note)
Lyristes plebejus Gogala, Popov and Ribaric 1996: 46 (comp. note)
Tibicen plebejus Holzinger 1996: 505 (listed) Austria
Lyristes plebejus Vernier 1996: 147–149, 151, Fig 1 (life history, illustrated, comp. note) France
Cicada plebeja Boulard 1997c: 83–84, 86, 88, 90–92, 94–97, 99, 101–102, 104, 106, 120–121 (type species of *Cicada*, *Tibicen*, Tibiceninae, comp. note) Equals *Cicada fraxini* Equals *Fidicina africana*
Tettigonia plebeia (sic) Fab. Boulard 1997c: 90–91 (type species of *Cicada*)
Tettigonia plebeja Boulard 1997c: 95 (comp. note)
Cicada fraxini Boulard 1997c: 90–91 (comp. note)
Tibicen plebeja Boulard 1997c: 93, 98, 100, 103, 120 (type species of *Tibicen*, comp. note)
Cicada plebeia (sic) Boulard 1997c: 90, 92, 94 (comp. note)
Tibicen plebejus Boulard 1997c: 99 (comp. note)
Lyristes plebejus Boulard 1997c: 99–100, 106, 112–113, Plate II, Fig 2 (illustrated, comp. note)
Cicada plebeja L. Boulard 1997c: 94–97 (comp. note)
Tettigonia plebeja L. Boulard 1997c: 95 (comp. note)
Tibicina (sic) *plebeja* Boulard 1997c: 101 (comp. note)
Fidicina africana Boulard 1997c: 120 (comp. note)
Cicada plebeja Chou, Lei, Li, Lu and Yao 1997: 281 (type species of *Lyristes*)
Lyristes plebejus Holzinger, Frölich, Günthart, Pauterer, Nickel et al. 1997: 49 (listed) Europe
Lyristes plebejus Arias Giralda, Pérez Rodríguez and Gallego Pintor 1998: 262 (olive pest)
Cicada plebejus Dajoz 1998: 322 (comp. note) Mediterranean
Lyristes plebejus Gogala 1998b: 394, 399 (illustrated, oscillogram, comp. note) Slovenia
Lyristes plebejus Ragge and Reynolds 1998: 477–478, Fig 1628 (oscillogram, comp. note) Europe
Tibicen plebejus Okáli and Janský 1998: 38, 42, Fig 1 (distribution, comp. note) Equals *Cicada plebeja* Equals *Lyristes plebejus* Slovakia
Lyristes plebejus Boulard 1999b: 184–185, Fig 9 (illustrated, comp. note) France
Lyristes plebejus Gogala and Gogala 1999: 120–121, 124, Fig 1 (distribution, comp. note) Slovenia, Italy, Mediterranean, Asia
Lyristes plebejus Jamieson, Dallai and Afzelius 1999: 235–236, 303 (sperm anatomy, comp. note)
Cicada plebejus Lee 1999: 4 (type species of *Tibicen*, type species of *Lyristes*) Europe
Lyristes plebeius (sic) Schedl 1999b: 824 (listed, distribution, ecology, comp. note) Israel, western Mediterranean
Cicada plebeja Boulard 2000a: 255 (type species of *Tibicen*, comp. note)
Lyristes plebejus Boulard 2000b: 33–35, Figs 2122 (song, illustrated, sonogram, comp. note) Turkey
Lyristes plebejus Boulard 2000d: 90–91, Plate E, Fig 5 (auditory receptor, comp. note)
Lyrsistes plebejus Boulard 2000h: 108–109, Plates XXI-XXII, Figs 1–15 (larva illustrated, ecolsion illustrated)
Lyristes plebejus Boulard 2000i: 9, 15, 20, 24, 29–33, 101, 119, Fig 10a, Fig 13, Fig 23a, Plate A, Map 1 (synonymy, illustrated, song, sonogram, oscillogram, frequency spectrum, distribution, ecology, comp. note) Equals *Cicada plebeja* Equals *Tettigonia plebeja* Equals *Tettigonia obscura* Equals *Tettigonia fraxini* Equals *Cicada plebeia* (sic) var. *obscura* Equals *Tibicen plebeia* (sic) Equals *Tibicen plebeja* (sic) Equals *Lyristes plebeius* (sic) Equals *Tibicen plebejus* Equals *Tibicen plebeja* (sic) Equals *Lyristes plebeja* (sic) France
Cicada plebeja Boulard 2000i: 20 (comp. note)
Tibicen plebeja (sic) Boulard 2000i: 20 (comp. note)
Cicada (= Lyristes) plebejus Dajoz 2000: 298 (comp. note) Mediterranean
Tibicen plebejus Freytag 2000: 55 (on stamp)
Lyristes plebejus Gogala and Popov 2000: 11 (comp. note)
Lyristes plebejus Guglielmino, d'Urso and Alma 2000: 167 (listed, distribution) Sardinia, Spain, France, Corsica, Italy, Sicily
Lyristes plebejus Puissant and Sueur 2000: 69 (comp. note) France

Lyristes plebeius (sic) Schedl 2000: 258, 262, 266–267, Plate 1, Fig 1, Map 1 (illustrated, key, distribution, listed, comp. note) France, Switzerland, Slovakia, Slovenia, Austria, Italy
Cicada plebeja Boulard 2001b: 4–6, 11–13, 15, 17–18, 20, 23, 26 (type species of *Cicada*, *Tibicen*, Tibiceninae, comp. note) Equals *Cicada fraxini* Equals *Fidicina africana*
Tettigonia plebeia (sic) Fab. Boulard 2001b: 11–12, 15 (type species of *Cicada*)
Tettigonia plebeja Boulard 2001b: 15 (comp. note)
Cicada fraxini Boulard 2001b: 11–12 (comp. note)
Tibicen plebeia (sic) Boulard 2001b: 13, 17–18, 21 (type species of *Tibicen*, comp. note)
Tibicen plebeja (sic) Boulard 2001b: 14, 18–19, 40 (type species of *Tibicen*, comp. note)
Cicada plebeia (sic) Boulard 2001b: 13–14 (comp. note)
Tibicen plebejus Boulard 2001b: 19–20 (comp. note)
Lyristes plebejus Boulard 2001b: 19–20, 26, 32–33, Plate II, Fig 2 (type species of *Lyristes*, illustrated, comp. note) Equals *Cicada plebeja*
Cicada plebeja L. Boulard 2001b: 15–16, 18, 20 (comp. note)
Tettigonia plebeja L. Boulard 2001b: 15 (comp. note)
Tibicina (sic) *plebeja* Boulard 2001b: 22 (comp. note)
Fidicina africana Boulard 2001b: 40 (comp. note)
Cicada plebeja Boulard 2001e: 117 (type species of *Tibicen*, comp. note)
Cicada plebeja Boulard 2001f: 129 (comp. note)
Tibicen plebejus Krpač, Trilar and Gogala 2001: 29 (comp. note) Macedonia
Lyristes plebejus Puissant and Sueur 2001: 430 (comp. note) France
Lyristes plebeius (sic) Schedl 2001b: 27 (comp. note)
Lyristes plebejus Sueur 2001: 36, 46, Tables 1–2 (listed, comp. note)
Lyristes plebejus Sueur and Puissant 2001: 19, 21–22 (illustrated, comp. note) France
Lyristes plebejus Boulard 2002b: 193, 1998, 202, 205–206 (coloration, comp. note)
Cicada plebeja Levinson and Levinson 2001: 24 (comp. note)
Lyristes plebejus Gogala 2002: 243, 245, 248, Fig 3 (song, sonogram, oscillogram, comp. note)
Lyristes plebejus Günther and Mühlenthaler 2002: 331 (listed) Switzerland
Tibicen plebeja (sic) Kondratieff, Ellingson and Leatherman 2002: 19 (type species of *Tibicen*)
Lyristes plebejus Kysela 2002: 22 (on jewelry)
Lyristes plebejus Nickel, Holzinger and Wachmann 2002: 292, Table 4 (listed)
Tibicen plebejus Rakitov 2002: 125 (comp. note)
Lyristes plebeius (sic) Schedl 2002: 232–233, 237, Photo 1, Map 1 (illustrated, distribution, comp. note) Austria
Lyristes plebejus Sueur 2002b: 111, 117–120, 123, 129 (sound system illustrated, listed, comp. note)
Lyristes plebejus Sueur and Aubin 2002b: 128, 131, 134–135, Fig 7 (song, sonogram, oscillogram, comp. note) France
Cicada plebeja Boulard 2003f: 371 (type species of *Tibicen*, *Lyristes*, comp. note)
Cicada plebeia (sic) Boulard 2003f: 372 (comp. note)
Lyristes plebejus Gogala and Seljak 2003: 350 (comp. note) Slovenia
Cicada plebeja Holzinger, Kammerlander and Nickel 2003: 479 (type species of *Lyristes*)
Lyristes plebejus Holzinger, Kammerlander and Nickel 2003: 476, 479, 616–617, 631, 635, Plate 42, Fig 1, Tables 2–3 (key, listed, comp. note) Equals *Cicada plebeja* Equals *Tibicen plebejus* Equals *Tettigonia fraxini* Europe, Mediterranean, Asia Minor, Slovenia
Lyristes plebejus Ibanez 2003: 3, 9, 14, Figs (nymph illustrated, sonogram, oscillogram, comp. note) France
Cicada plebejus Lee and Hayashi 2003a: 158 (type species of *Tibicen*, type species of *Lyristes*) Europe
Tibicen plebejus Tishechkin 2003: 141–142, Figs 143–147 (oscillogram, comp. note) Azerbaijan
Lyristes plebejus Vezinet 2003: 13 (comp. note) France
Lyristes plebejus Öztürk, Ulusoy, Erkiliç and S. (Ö.) Bayhan 2004: 5, 8 (apricot pest, listed) Turkey
Lyristes plebeius (sic) Schedl 2004b: 914–915, Fig 1 (listed, distribution, comp. note) Austria
Lyristes plebejus Sueur, Puissant, Simões, Seabra, Boulard and Quartau 2004: 178–182, Figs 1–2 (distribution, ecology, song, sonogram, oscillogram, comp. note) Portugal
Lyristes plebejus Boulard 2005d: 336, 339–340, 345, 347 (auditory system, comp. note) Europe, Asia Minor
Lyristes plebejus Chawanji, Hodson and Villet 2005a: 258, 266 (comp. note)
Lyristes plebejus Gogala, Trilar and Krpač 2005: 104–107, Fig 1 (distribution, comp. note) Macedonia
Tibicen plebejus Lee 2005: 55, 95 (type species of *Tibicen*, type species of *Lyristes*, comp. note) Europe
Cicada plebeja Lee 2005: 169 (type species of *Tibicen*, type species of *Lyristes*, comp. note) Europe

Cicada plebeja Moulds 2005b: 389, 392–393 (type species of *Lyristes*, type species of *Tibicen*, described, key, history, phylogeny, listed, comp. note)
Cicada plebeja Linnaeus Moulds 2005b: 383 (comp. note)
Lyristes plebejus Moulds 2005b: 396, 399, 406, 410, 417–420, 422, Fig 43, Figs 56–60, Table 1 (illustrated, higher taxonomy, phylogeny, key, listed, comp. note) Equals *Tibicen plebejus* Australia
Lyristes plebejus Puissant and Defaut 2005: 116–118, 121–125, Table 1, Tables 5–9 (listed, habitat association, distribution, comp. note) France
Lyristes plebejus Reynaud 2005: 141–143, Photo 1 (illustrated, comp. note) France
Lyristes plebejus Robinson 2005: 217 (comp. note) Europe
Tibicen plebejus Sanborn 2005c: 116 (comp. note)
Lyristes plebejus Bernier 2006: 4–5, 7, Figs 2–3 (distribution, illustrated, oscillogram) France
Lyristes plebejus Boulard 2006e: 381 (comp. note) Europe
Lyristes plebejus Boulard 2006g: 30, 40, 50–51, 55, 61–62, 115, 127–128, 180, Fig 8, Figs 14–15, Fig 25, Fig 87B (auditory system, sonogram, illustrated, listed, comp. note) Asia Minor, Europe
Lyristes plebejus Chawanji, Hodson and Villet 2006: 387 (comp. note)
Lyristes plebejus Puissant 2006: 19, 21, 25, 30, 32–35, 37–39, 41–44, 48, 64, 70, 86, 89–90, 93–94, 108–115, 134–162, 178–180, 183–187, Figs 8–12, Tables 1–2, Appendix Table 1, Appendix Tables 5–9 (illustrated, synonymy, distribution, habitat association, ecology, listed, comp. note) Equals *Cicada plebeia* Equals *Cicada plebeja* Equals *Tettigonia obscura* Equals *Tettigonia fraxini* France
Lyristes plebejus Sueur, Windmill and Robert 2006: 4126 (comp. note)
Lyristes plebejus Trilar, Gogala and Popa 2006: 177–178 (comp. note) Romania
Lyristes plebejus Màjsky and Janský 2006: 74–75 (illustrated, distribution, comp. note) Equals *Tibicen plebejus* Slovakia
Lyristes plebejus Boulard 2007e: 21, 85 (auditory system, predation, comp. note) Europe
Lyristes plebejus Chawanji, Hodson, Villet, Sanborn and Phillips 2007: 337, 345 (comp. note)
Lyristes plebejus Demir 2007a: 51 (distribution, comp. note) Turkey
Lyristes plebejus Hertach 2007: 51 (comp. note) Switzerland
Tibicen plebeja (sic) Shiyake 2007: 4, 37, 39, Fig 54 (illustrated, distribution, listed, comp. note) Europe
Lyristes plebejus Sueur and Puissant 2007b: 62 (listed) France
Tibicen plebeja (sic) Yamazaki 2007: 349 (comp. note) France
Lyristes plebejus Demir 2008: 476 (distribution) Turkey
Lyristes plebejus Fonseca, Serrão, Pina-Martins, Silva, Mira, et al. 2008: 19, 24–31, Fig 1, Table 1 (song phylogeny, comp. note) Portugal
Lyristes plebejus Hertach 2008b: 12, 14, 19, Fig 11 (illustrated, comp. note) Switzerland
Lyristes plebejus Hugel, Matt, Callot, Feldtrauer and Brua 2008: 5 (comp. note) France
Cicada plebeja Lee 2008b: 447 (type species of *Lyristes*) Europe
Lyristes plebejus Nickel 2008: 207 (distribution, hosts, comp. note) Mediterranean
Tibicen plebejus Sanborn, Phillips and Gillis 2008: 7 (type species of *Tibicen*)
Lyristes plebejus Sueur, Windmill and Robert 2008a: 2380 (comp. note)
Lyristes plebejus Sueur, Windmill and Robert 2008b: 59 (comp. note)
Cicada plebeja Vikberg 2008: 61 (comp. note)
Tibicen plebejus Vikberg 2008: 61 (comp. note)
Lyristes plebejus Trilar and Gogala 2008: 32 (comp. note) Romania
Lyristes plebejus Dourlet 2009: 84 (life history, illustrated)
Lyristes plebejus Gogala, Drosopoulos and Trilar 2009: 19, 23 (comp. note)
Lyristes plebejus Pinto, Quartau, and Bruford 2009: 267, 270, 274, 277–278, Fig 5, Table 4 (phylogeny, comp. note) Portugal
Lyristes plebejus Windmill, Sueur and Robert 2009: 4082 (comp. note)
Lyristes plebejus Kemal and Koçak 2010: 5 (distribution, listed, comp. note) Turkey
Cicada plebeja Mozaffarian and Sanborn 2010: 76 (type species of *Tibicen*)
Tibicen plebejus Mozaffarian and Sanborn 2010: 77 (synonymy, listed, distribution, comp. note) Equals *Cicada plebeja* Equals *Tettigonia orni* (nec Linnaeus) Equals *Tettigonia obscura* Equals *Tettigonia fraxini* Iran, Europe, western Asia, France, Portugal, Georgia, Armenia, Sicily
Lyristes plebejus Sueur, Janique, Simonis, Windmill and Baylac 2010: 923–932, Fig 1, Figs 3–9, Table 1 (tympanum structure, comp. note) France

Lyristes plebejus Sueur, Windmill and Robert 2010: 1681–1687, Figs 1–5, Tables I-III (auditory processing, sonogram, comp. note) France

***T. pronotalis pronotalis* Davis, 1938** = *Tibicen pronotalis* = *Tibicen marginalis pronotalis* = *Tibicen walkeri* var. *pronotalis* = *Tibicen walkeri pronotalis*

Tibicen walkeri var. *pronotalis* Sanborn and Phillips 1995a: 482, Table 1 (song intensity, comp. note)

Tibicen walkeri var. *pronotalis* Sanborn 1999a: 42 (type material, listed)

Tibicen walkeri var. *pronotalis* Sanborn 1999b: 1 (listed) North America

Tibicen pronotalis Cole 2008: 818–819, 821 (comp. note) Equals *Tibicen marginalis* Equals *Tibicen walkeri*

Tibicen pronotalis Hill and Marshall 2009: 64 (comp. note)

Tibicen marginalis var. *pronotalis* Hill and Marshall 2009: 65 (comp. note)

Tibicen pronotalis pronotalis Hill and Marshall 2009: 65 (comp. note)

Tibicen pronotalis walkeri Hill and Marshall 2009: 63–66, Fig 1 (stat. nov., synonymy, illustrated, song, sonogram, oscillogram, key, comp. note) Equals *Tibicen walkeri* Equals *Tibicen marginalis* Florida, Georgia

Tibicen walkeri pronotalis Holliday, Hastings and Coelho 2009: 10–11, Table 1 (predation, comp. note) Iowa

***T. pronotalis walkeri* Metcalf, 1955** = *Tibicen walkeri* = *Cicada marginata* Say, 1825 (nec *Cicada marginata* Olivier, 1790) = *Tibicen marginata* = *Cicada marginalis* Walker, 1852 (nec *Cicada marginalis* Scopoli, 1763) = *Tibicen marginalis* = *Tibicen marginales* (sic) = *Lyristes marginalis*

Cicada marginata Davis 1895: 142 (comp. note) New York, Staten Island

Tibicen marginalis Davis 1922: 73 (comp. note) Kansas

Tibicen marginalis McCoy 1965: 42 (key, comp. note) Arkansas

Tibicen marginalis Evans 1966: 104, Table 10 (parasitism, comp. note) North Carolina

Tibicen marginalis Krombein, Hurd, Smith and Burks 1979: 1698 (parasitism)

Tibicen marginalis Moore 1993: 282 (comp. note)

Tibicen marginalis Wolda 1993: 376 (comp. note) New Mexico

Tibicen walkeri Sanborn and Phillips 1995a: 480, 482, Table 1 (song intensity, comp. note)

Tibicen walkeri Poole, Garrison, McCafferty, Otte and Stark 1997: 333 (listed) Equals *Cicada marginata* Equals *Cicada marginalis* Equals *Tibicen pronotalis* North America

Tibicen walkeri Sanborn 1998: 93, 97 (thermal biology, comp. note)

Tibicen walkeri Sanborn 1999b: 1–2 (type material, listed) North America

Tibicen walkeri Sanborn 2001a: 17 (comp. note)

Tibicen walkeri Kondratieff, Ellingson and Leatherman 2002: 10, 12, 25–26, Table 1 (comp. note) Equals *Cicada marginata* Walker, 1852 nec *Cicada marginalis* Scopoli, 1763

Tibicen marginatus Kondratieff, Ellingson and Leatherman 2002: 21 (comp. note) Equals *Tibicen walkeri*

Lyristes marginalis Sueur 2002a: 388 (comp. note) New Mexico

Tibicen marginalis Walker and Moore 2004: 2 (listed, comp. note) Florida

Tibicen marginalis Taber and Fleenor 2005: 243, Fig 10–2 (illustrated, comp. note) Texas

Tibicen marginalis Elliott and Hershberger 2006: 32, 200–201 (illustrated, sonogram, listed, distribution, comp. note) South Dakota, Nebraska, Iowa, Illinois, Indiana, Kansas, Oklahoma, Texas, Louisiana, Arkansas, Missouri, Ohio, Kentucky, Tennessee, Mississippi, Alabama, Georgia, Florida

Tibicen walkeri Shiyake 2007: 27 (comp. note)

Tibicen walkeri Sanborn, Phillips and Gillis 2008: 17 (comp. note)

Cicada marginata Hill and Marshall 2009: 65 (comp. note)

Cicada marginalis Hill and Marshall 2009: 65 (comp. note)

Tibicen walkeri Hill and Marshall 2009: 65 (comp. note)

Tibicen walkeri Holliday, Hastings and Coelho 2009: 10–11, Table 1 (predation, comp. note) Florida, Illinois, Missouri

***T. pruinosus fulvus* Beamer, 1924** = *Tibicen pruinosa fulva* = *Tibicen pruinosus fulva* = *Tibicen pruinosus* var. *fulvus*

Tibicen pruinosus var. *fulvus* Sanborn 1999a: 41 (type material, listed)

Tibicen pruinosus var. *fulvus* Sanborn 1999b: 1 (listed) North America

***T. pruinosus pruinosus* (Say, 1825)** = *Cicada pruinosa* = *Cicada pruinora* (sic) = *Cicada pruinoso* (sic) = *Tibicen pruinosa* = *Lyristes pruinosus* = *Cicada bruneosa* Wild, 1852 = *Cicada brunneosa* (sic)

Cicada bruneosa Wild 1852: xviii (comp. note) Maryland

Cicada pruinosa Walsh and Riley 1868b: 66 (comp. note)

Cicada pruinosa Swinton 1880: 227 (song, comp. note) United States

Cicada pruinosa Ward 1885: 476 (comp. note) Virginia

Cicada pruinosa Davis 1906b: 237–239, Table (comp. note) New York, Staten Island, New Jersey
Cicada pruinosa Davis 1906c: 240 (comp. note)
Cicada pruinosa Roth 1949: 81 (parasitism) America
Tibicen pruinosa (sic) Fitch 1960: 200, 209 (predation, comp. note) Kansas
Tibicen pruinosa (sic) McCoy 1965: 43–44 (key, comp. note) Arkansas
Tibicen pruinosa (sic) Evans 1966: 103–104, 110, Table 10 (parasitism, comp. note) District of Columbia, Ohio
Cicada pruinosa Weber 1968: 86 (comp. note)
Tibicen pruinosa (sic) Krombein, Hurd, Smith and Burks 1979: 1698 (parasitism)
Tibicen pruinosa (sic) O'Connor and Nash 1981: 299 (comp. note)
Lyristes pruinosa (sic) Boulard 1985d: 213 (comp. note)
Tibicen pruinosa (sic) Hamilton 1985: 211 (status, comp. note)
Tibicen pruinosa (sic) Ettling 1986: 32 (predation, comp. note)
Cicada pruinosa Hennessey 1990: 470 (type material)
Tibicen pruinosus Sanborn and Phillips 1995a: 482, Table 1 (song intensity, comp. note)
Tibicen pruinosa (sic) Marshall, Cooley, Alexander and Moore 1996: 165–169, Fig 1 (distribution, comp. note) Michigan
Tibicen pruinosa (sic) Coelho 1997: 373–374, Table 2 (predation, comp. note) Illinois
Tibicen pruinosa (sic) Poole, Garrison, McCafferty, Otte and Stark 1997: 333 (listed) Equals *Cicada pruinosa* Equals *Cicada bruneosa* Equals *Cicada brunneosa* (sic) Equals *Cicada pruinora* (sic) Equals *Cicada pruinoso* (sic) Equals *Cicada bruenosa* (sic) Equals *Tibicen pruniosa* (sic) Equals *Cicada winnemanna* Equals *Cicada latifasciata* Equals *Tibicen fulva* North America
Tibicen pruinosa (sic) Robert, Miles and Hoy 1999: 1866 (comp. note) Mississippi
Tibicen pruinosus Sanborn 1999b: 1–2 (type material, listed) North America
Tibicen pruinosa (sic) Callaham 2000: 3957 (density, comp. note) Kansas
Tibicen pruinosa (sic) Callaham, Whiles, Meyer, Brock and Charlton 2000: 537–540, Fig 1, Tables 1–2 (density, comp. note) Kansas
Tibicen pruinosa (sic) Punzo 2000: 97, Table 6 (listed, comp. note)
Tibicen pruinosa (sic) Smith, Kelly, Dean, Bryson, Parker et al. 2000: 125–126 (illustrated, comp. note) Kansas
Tibicen pruinosus Sanborn 2001a: 17 (comp. note)
Lyristes pruinosus Sueur 2001: 46, Table 2 (listed) Equals *Tibicen pruinosa*
Lyristes pruinosus Sueur 2002b: 122, 124, 129 (listed, comp. note) Equals *Cicada pruinosa*
Tibicen pruinosa (sic) Walker and Moore 2004: 2 (listed, comp. note) Florida
Tibicen pruinosa (sic) Kondratieff, Schmidt, Opler and Garhart 2005: 96, 217 (listed, comp. note) Oklahoma
Tibicen pruinosa (sic) Robinson 2005: 217 (comp. note)
Tibicen pruinosa (sic) Taber and Fleenor 2005: 243–244, Fig 10–3 (illustrated, comp. note) Texas
Tibicen pruinosa (sic) Elliott and Hershberger 2006: 32, 186–187 (illustrated, sonogram, distribution, listed, comp. note) Eastern United States
Tibicen pruinosa (sic) Elliott and Hershberger 2006: 32, 186–187 (distribution, listed, comp. note) Michigan, Ohio, Indiana, Illinois, Iowa, Missouri, Nebraska, Kansas, Oklahoma, Texas, Arkansas, Louisiana, Mississippi, Tennessee, Kentucky
Tibicen pruinosa (sic) Grant 2006: 542, 544, Table 2 (predation, comp. note) New Jersey, Indiana
Tibicen pruinosa (sic) Eaton and Kaufman 2007: 88–89 (illustrated, comp. note) Eastern United States
Tibicen pruinosa (sic) Ganchev and Potamitis 2007: 305 (listed)
Tibicen pruinosus Cole 2008: 818 (comp. note)
Tibicen pruinosa (sic) Farris, Oshinshy, Forrest, and Hoy 2008: 17–24, Fig 1a, Fig 2b, Fig 3a, Fig 5b, Tables 1–2 (song, spectrogram, power spectrum, parasitism) Mississippi
Tibicen pruinosa (sic) Hastings, Holliday and Coelho 2008: 659, Table 1 (predation, comp. note) North Carolina, Kentucky, Missouri
Tibicen pruinosus Sanborn, Phillips and Gillis 2008: 4, 10, 31, 34, Fig 70, Figs 94–102 (key, synonymy, illustrated, distribution, comp. note) Equals *Cicada pruinosa* Equals *Cicada bruneosa* Florida
Tibicen pruinosus Holliday, Hastings and Coelho 2009: 3, 6–9, 10, 12–14, Tables 1–2 (predation, comp. note) Illinois, Indiana, Iowa, Kansas, Kentucky, Maryland, Minnesota, Missouri, Mississippi, Nebraska, North Carolina, Ohio, South Dakota, Tennessee, Texas, Virginia, Wisconsin, West Virginia
Tibicen pruinosus Hastings, Holliday, Long, Jones and Rodriguez 2010: 415–416, Table 2 (predation, comp. note) Florida

***T. resh* (Haldeman, 1852)** = *Cicada resh* = *Cicada resch* (sic) = *Cicada robertsonii* Fitch, 1855 = *Cicada robertsoni* (sic) = *Tibicen robertsonii* = *Tibicen superba* (sic) (nec Fitch)

Tibicen resh McCoy 1965: 42, 44 (key, comp. note) Arkansas

Tibicen resh Arnett 1985: 217 Fig 21.12 (distribution, illustrated) Southwestern United States, Arkansas, Louisiana, Mississippi, Alabama, Utah, Kansas, Missouri

Cicada robertsonii Barnes 1988: 90 (type locality) Equals *Tibicen resh* Oklahoma

Tibicen resh Sanborn and Phillips 1995a: 482, Table 1 (song intensity, comp. note)

Tibicen superba (sic) Coelho 1997: 373–374, Table 2 (predation, comp. note) Illinois

Tibicen resh Hostetler 1997: 29 (comp. note)

Tibicen resh Poole, Garrison, McCafferty, Otte and Stark 1997: 333 (listed) Equals *Cicada resh* Equals *Cicada robertsonii* Equals *Cicada resch* (sic) North America

Tibicen resh Sanborn 1999b: 1–2 (type material, listed) North America

Tibicen resh Arnett 2000: 298 Fig 21.12 (distribution, illustrated) Southwestern United States, Arkansas, Louisiana, Mississippi, Alabama, Utah, Kansas, Missouri

Tibicen resh Kondratieff, Ellingson and Leatherman 2002: 10–12, Table 1 (listed, comp. note) Colorado

Tibicen resh Taber and Fleenor 2003: 104–105, Fig 90 (illustrated, comp. note) Texas

Tibicen resh Sanborn and Phillips 2004: 647 (comp. note) Equals *Tibicen superbus* Coehlo 1997

Tibicen resh Kondratieff, Schmidt, Opler and Garhart 2005: 97 (comp. note)

Tibicen resh Grant 2006: 543, Fig 3 (predation, comp. note) New Jersey

Tibicen resh Holliday, Hastings and Coelho 2009: 3, 8–9, 14, Table 1 (predation, comp. note) Illinois, Kentucky, Louisiana, Oklahoma, Texas

***T. resonans* (Walker, 1850)** = *Cicada resonans* = *Lyristes resonans*

Tibicen resonans McCoy 1965: 42 (key, comp. note) Arkansas

Tibicen resonans Walker 1983: 63 (comp. note) United States

Tibicen resonans Walker 1986: 683 (comp. note) Florida

Tibicen resonans Hostetler 1997: Plate VII (illustrated)

Tibicen resonans Poole, Garrison, McCafferty, Otte and Stark 1997: 333 (listed) Equals *Cicada resonans* North America

Tibicen resonans Sanborn 1999b: 1 (listed) North America

Tibicen resonans Sanborn 2001a: 17 (comp. note)

Tibicen resonans Sanborn 2002b: 461 (thermal biology, comp. note)

Lyristes resonans Sueur 2002b: 124, 129 (listed, comp. note) Equals *Tibicen resonans*

Tibicen resonans Walker and Moore 2004: 2 (listed, comp. note) Florida

Tibicen resonans Hastings, Holliday and Coelho 2008: 659, 662, Table 1 (predation, comp. note) Florida

Tibicen resonans Sanborn, Phillips and Gillis 2008: 4, 8, 11, 31, 35, Fig 72, figs 103–111 (key, synonymy, illustrated, distribution, comp. note) Equals *Cicada resonans* Florida

Tibicen resonans Holliday, Hastings and Coelho 2009: 3, 8–9, Table 1 (predation, comp. note) Florida, Texas

Tibicen resonans Hastings, Holliday, Long, Jones and Rodriguez 2010: 414–416, 419, Table 2 (predation, comp. note) Florida

Tibicen resonans Phillips and Sanborn 2010: 74 (comp. note)

***T. robinsonianus* Davis, 1922** = *Tibicen robinsoniana* = *Lyristes robinsonianus*

Tibicen robinsoniana (sic) McCoy 1965: 43, 44 (key, comp. note) Arkansas, Missouri

Tibicen robinsoniana (sic) Evans 1966: 104, Table 10 (parasitism, comp. note) Ohio

Tibicen robinsoniana (sic) Krombein, Hurd, Smith and Burks 1979: 1698 (parasitism)

Tibicen robinsoniana (sic) Poole, Garrison, McCafferty, Otte and Stark 1997: 333 (listed) North America

Tibicen robinsonianus Sanborn 1999a: 41 (type material, listed)

Tibicen robinsonianus Sanborn 1999b: 1 (listed) North America

Lyristes robinsonianus Sueur 2002b: 124, 129 (listed, comp. note) Equals *Tibicen robinsoniana*

Tibicen robinsoniana (sic) Elliott and Hershberger 2006: 32, 202–203 (illustrated, sonogram, listed, distribution, comp. note) Virginia, West Virginia, Pennsylvania, Ohio, Kentucky, Alabama, Mississippi, Missouri, Arkansas, Texas

Tibicen robinsonianus Sanborn, Phillips and Gillis 2008: 4, 11, 31, 36, Fig 70, Figs 112–120 (key, synonymy, illustrated, distribution, comp. note) Equals *Tibicen robinsoniana* Florida

Tibicen robinsonianus Holliday, Hastings and Coelho 2009: 8–9, Table 1 (predation, comp. note) Ohio, Tennessee

***T. robustus* (Distant, 1881)** = *Cicada robusta* = *Rihana robusta*

Tibicen robustus Sanborn 2007b: 34 (distribution, comp. note) Mexico, Morelos, Puebla

T. rubicinctus Goding & Froggat, 1904 to *Marteena rubricincta*

***T. similaris* (Smith & Grossbeck, 1907)** = *Cicada similaris* = *Rihana similaris* = *Tibicen simularis* (sic)

Tibicen similaris McCoy 1965: 44 (comp. note) Mississippi

Tibicen similaris Arnett 1985: 217 Fig 21.16 (distribution, illustrated) Southeast United States

Tibicen similaris Weaver 1995: 9 (parasitism) West Virginia

Tibicen similaris Poole, Garrison, McCafferty, Otte and Stark 1997: 333 (listed) Equals *Cicada similaris* Equals *Tibicen simularis* (sic) North America

Tibicen similaris Sanborn 1999a: 42 (type material, listed)

Tibicen similaris Sanborn 1999b: 1 (listed) North America

Tibicen similaris Arnett 2000: 298 Fig 21.16 (distribution, illustrated) Southeast United States

Tibicen similaris Walker and Moore 2004: 2 (listed, comp. note) Florida

Tibicen similaris Hastings, Holliday and Coelho 2008: 659, Table 1 (predation, comp. note) Florida

Tibicen similaris Sanborn, Phillips and Gillis 2008: 4, 11–12, 24, 37, Fig 10, Figs 121–129 (key, synonymy, illustrated, distribution, comp. note) Florida

Tibicen similaris Holliday, Hastings and Coelho 2009: 8–9, Table 1 (predation, comp. note) Florida

Tibicen similaris Hastings, Holliday, Long, Jones and Rodriguez 2010: 415–416, Table 2 (predation, comp. note) Florida

***T. simplex* Davis, 1941** = *Tibicen bifida simplex* = *Tibicen bifidus simplex* = *Tibicen bifidus* var. *simplex*

Tibicen bifidus var. *simplex* Sanborn 1999a: 40 (type material, listed)

Tibicen bifidus var. *simplex* Sanborn 1999b: 1 (listed) North America

Tibicen bifidus simplex Cole 2008: 815 (comp. note) Arizona

***T. slocumi* Chen, 1943** = *Lyristes slocumi*

Lyristes slocumi Chou, Lei, Li, Lu and Yao 1997: 282, 285–286 (key, described, synonymy, comp. note) Equals *Tibicen slocumi* China

Lyristes slocumi Hua 2000: 62 (listed, distribution) Equals *Tibicen slocumi* China, Sichuan

***T. sublaqueatus* (Uhler, 1903)** = *Cicada sublaqueata*

Cicada sublaqueata Henshaw 1903: 229 (listed) Brazil

Tibicen sublaqueata Sanborn 1999a: 42 (type material, listed)

***T. sugdeni* Davis, 1941**

Tibicen sugdeni Sanborn 1999a: 42 (type material, listed)

Tibicen sugdeni Sanborn 2007b: 34 (distribution, comp. note) Mexico, Nuevo León

***T. superbus* (Fitch, 1855)** = *Cicada superba* = *Rihana superba* = *Tibicen superba* = *Tibicen superb* (sic)

Tibicen superba (sic) McCoy 1965: 41, 44 (key, comp. note) Arkansas

Tibicen superbus Arnett 1985: 217 Fig 21.14 (distribution, illustrated) Arkansas, New Mexico, Oklahoma, Texas, Kansas, Missouri

Cicada superba Barnes 1988: 90 (type locality) Equals *Tibicen superbus* Oklahoma

Tibicen superbus Sanborn and Phillips 1995a: 482, Table 1 (song intensity, comp. note)

Tibicen superba (sic) Poole, Garrison, McCafferty, Otte and Stark 1997: 333 (listed) Equals *Cicada superba* North America

Tibicen superbus Sanborn 1999b: 1–2 (type material, listed) North America

Tibicen superbus Arnett 2000: 298 Fig 21.14 (distribution, illustrated) Arkansas, New Mexico, Oklahoma, Texas, Kansas, Missouri

Tibicen superba (sic) Taber and Fleenor 2003: 105 (comp. note) Texas

Tibicen superbus Sanborn and Phillips 2004: 647–651, Figs 1–3 (neotype designation, allotype designation, synonymy, described, illustrated, song, sonogram, oscillogram, distribution, comp. note) Equals *Cicada superba* Equals *Rihana superba* Equals *Tibicen superba* Equals *Tibicen superb* (sic) Oklahoma, Texas, Arkansas, Missouri, Kansas

Cicada superba Sanborn and Phillips 2004: 647 (comp. note) Oklahoma

Tibicen superba Kondratieff, Schmidt, Opler and Garhart 2005: 96, 217 (listed, comp. note) Oklahoma

Tibicen superba (sic) Taber and Fleenor 2005: 242–243, Fig 10–2 (illustrated, comp. note) Texas

Tibicen superba (sic) Grant 2006: 54 (predation, comp. note) Indiana

Tibicen superba (sic) Eaton and Kaufman 2007: 88–89 (illustrated, comp. note) South-central United States, New Mexico

Tibicen superbus Cole 2008: 821 (comp. note)

***T. texanus* Metcalf, 1963** = *Tibicen tigrina* Davis, 1927 (nec *Tibicen tigrinus* Distant, 1888)

Tibicen texanus Sanborn and Phillips 1995a: 482, Table 1 (song intensity, comp. note)

Tibicen texana (sic) Poole, Garrison, McCafferty, Otte and Stark 1997: 333 (listed) Equals *Tibicen tigrina* North America
Tibicen texanus Sanborn 1999a: 42 (type material, listed)
Tibicen texanus Sanborn 1999b: 1 (listed) North America
Tibicen texanus Sanborn 2007b: 34 (comp. note)

***T. tibicen australis* (Davis, 1912)** = *Rihana sayi australis* = *Rihana sayi* var. *australis* = *Cicada sayi australis* = *Tibicen australis* = *Tibicen chloromera australis* = *Tibicen chloromerus* var. *australis*
Tibicen chloromerus var. *australis* Sanborn 1999a: 40 (type material, listed)
Tibicen chloromerus var. *australis* Sanborn 1999b: 1 (listed) North America
Tibicen tibicen australis Sanborn, Phillips and Gillis 2008: 4–5, 12, 31, 39, Fig 73, Figs 139–147 (key, synonymy, illustrated, distribution, comp. note) Equals *Rihana sayi* var. *australis* Florida, Georgia
Tibicen chloromerus australis Sanborn, Phillips and Gillis 2008: 13 (comp. note)
Tibicen tibicen australis Holliday, Hastings and Coelho 2009: 10–11, Table 1 (predation, comp. note)
Tibicen tibicen australis Hastings, Holliday, Long, Jones and Rodriguez 2010: 414–416, 419, Table 2 (predation, comp. note) Florida

***T. tibicen tibicen* (Linnaeus, 1758)** = *Cicada tibicen* = *Tettigonia tibicen* = *Cicada (Tettigonia) tibicen* = *Fidicina tibicen* = *Cicada (Cicada) tibicen* = *Rihana tibicen* = *Diceroprocta tibicen* = *Thopha chloromera* Walker, 1850 = *Tibicen chloromerus* = *Tibicen chloromera* = *Tibicen chloromea* (sic) = *Cicada chloromera* = *Lyristes chloromerus* = *Cicada sayi* Smith & Grossbeck, 1907 = *Rihana sayi* = *Tibicen sayi*
Tettigonia tibicen Dallas 1870: 482 (synonymy) Equals *Cicada tibicen* Equals *Cicada opercularis*
Cicada tibicen Hasselt 1881: 192 (comp. note)
Cicada tibicen Van Duzee 1893: 410 (listed) New York
Cicada tibicen Davis 1906b: 237–239, Table (comp. note) New York, Staten Island, New Jersey
Cicada tibicen Davis 1906c: 240 (comp. note)
Cicada tibicen Surface 1907: 71 (comp. note)
Cicada tibicen Tucker 1907: 192 (distribution, comp. note) Kansas
Cicada sayi Davis 1911a: 216 (comp. note) Georgia
Tibicen sayi Davis 1919b: 342 (comp. note) New York, Staten Island
Tibicen sayi Davis 1920: 129 (parasitism, comp. note) New York, Staten Island, Long Island
Cicada tibicen Malloch 1922: 45 (predation) Maryland
Tibicen sayi Champlain and Knull 1923: 212 (predation, comp. note) Pennsylvania
Tibicen chloromera (sic) Davis 1927: 118 (predation, comp. note) New York, Staten Island
Cicada tibicen Roth 1949: 81 (parasitism) America
Cicada tibicen Halkka 1959: 5–6, 37, Table 1 (chromosome numbers, comp. note)
Tibicen chloromera (sic) McCoy 1965: 41, 44 (key, comp. note) Arkansas
Tibicen chloromera (sic) Evans 1966: 104, Table 10 (parasitism, comp. note) New York, Ohio
Cicada tibicen Weber 1968: 474, Fig 307 (parasitism)
Tibicen chloromera (sic) Krombein, Hurd, Smith and Burks 1979: 1698 (parasitism)
Tibicen sayi Krombein, Hurd, Smith and Burks 1979: 1698 (parasitism)
Cicada tibicen Boulard 1984c: 168 (status, type species of *Cicada*, comp. note)
Cicada tibicen Melville and Sims 1984: 163 (type species of *Cicada*, history)
Diceroprocta tibicen Kirillova 1986b: 46 (listed) Equals *Cicada tibicen* USA
Diceroprocta tibicen Kirillova 1986a: 123 (listed) Equals *Cicada tibicen* USA
Tibicen chloromera (sic) Rypsra and Gregg 1986: 34 (parasitism)
Cicada tibicen Boulard 1988f: 68–69 (type species of *Cicada*, comp. note)
Cicada sayi Hennessey 1990: 471 (type material)
Rihana sayi Hennessey 1990: 471 (type material)
Diceroprocta tibicen Emeljanov and Kirillova 1991: 800 (chromosome number, comp. note)
Tibicen similaris (sic) Sanborn 1991: 5753B (thermoregulation, comp. note)
Tibicen chloromera (sic) Young 1991: 187 (comp. note) New York
Diceroprcta tibicen Emeljanov and Kirillova 1992: 63 (chromosome number, comp. note)
Tibicen chloromera (sic) Daniel, Knight and Charles 1993: 69–75, Fig 4 and Fig 7, Fig 9, Tables 1–2 (song, illustrated, comp. note) North Carolina
Tibicen chloromera (sic) Sweet 1993: 27 (predation) Massachusetts
Diceroprocta tibicen Berenbaum 1995: 132 (historic drawing error)
Tibicen chloromerus Sanborn and Phillips 1995a: 482, Table 1 (song intensity, comp. note)
Tibicen chloromera (sic) Weaver 1995: 9 (parasitism) West Virginia
Tibicen chloromera (sic) Daniel 1996: 14 (poem)
Tibicen chloromera (sic) Marshall, Cooley, Alexander and Moore 1996: 165–169, Fig 1 (distribution, comp. note) Michigan
Tibicen chloromerus Matthews and Hildreth 1997: 37 (comp. note)

Tibicen chloromera (sic) Navarro 1997: 42, 47, Fig 8B (sonogram, comp. note)
Tibicen chloromea (sic) Poole, Garrison, McCafferty, Otte and Stark 1997: 333 (listed) Equals *Thopha chloromera* Equals *Cicada sayi* Equals *Rihana australis* Equals *Tibicen chloromea* (sic) North America
Tibicen chloromera (sic) Bennet-Clark and Young 1998: 705 (comp. note)
Tibicen chloromerus Sanborn 1998: 100 (thermal biology, comp. note)
Tibicen sayi Sanborn 1999a: 42, 55 (type material, listed, comp. note)
Tibicen chloromerus Sanborn 1999a: 55 (comp. note)
Tibicen chloromerus Sanborn 1999b: 1 (listed) North America
Tibicen sayi Maw, Footit, Hamilton and Scudder 2000: 49 (comp. note) Canada, Ontario
Tibicen chloromerus Sanborn 2000: 551–555, Figs 1–2, Tables 1–3 (thermoregulation, thermal responses, comp. note) Tennessee
Cicada tibicen Boulard 2001b: 39 (type species of *Cicada*, comp. note)
Lyristes chloromerus Sueur 2001: 36, 46, Tables 1–2 (listed) Equals *Tibicen chloromerus*
Tibicen chloromerus Sanborn 2002b: 465–466, Table 1 (thermal biology, comp. note)
Tibicen chloromerus Sanborn, Breitbarth, Heath and Heath 2002: 443 (comp. note)
Tibicen chloromerus Sanborn, Noriega and Phillips 2002c: 368 (comp. note)
Diceroprocta tibicen Sueur 2002b: 121, 129 (listed, comp. note)
Tibicen chloromera (sic) Robinson 2005: 217 (comp. note)
Tibicen chloromera (sic) Walker and Moore 2004: 2 (listed, comp. note) Florida
Tibicen chloromerus Sanborn 2005c: 114, Table 1 (comp. note)
Tibicen chloromera (sic) Elliott and Hershberger 2006: 32, 188–189 (illustrated, sonogram, distribution, listed, comp. note) Massachusetts, Connecticut, New York, New Jersey, Delaware, Maryland, Pennsylvania, Ohio, Michigan, West Virginia, Virginia, Indiana, Illinois, Kentucky, Tennessee, Arkansas, Louisiana, Texas, Mississippi, Alabama, Georgia, South Carolina, North Carolina, Florida
Tibicen chloromera (sic) Grant 2006: 541–544, Figs 1–3, Tables 1–2 (predation, emergence time, comp. note) New Jersey
Tibicen chloromera (sic) Bunker, Neckermann and Cooley 2007: 357–360 (distribution, comp. note) Connecticut
Tibicen chloromera (sic) Eaton and Kaufman 2007: 88–89 (illustrated, comp. note) North America
Tibicen chloromera (sic) Ganchev and Potamitis 2007: 305 (listed)
Tibicen chloromera (sic) Oberdörster and Grant 2007a: 2–10, Fig 1, Fig 2B Figs 3–4, Tables 2–5 (flight system, flight speeds, comp. note) New Jersey
Tibicen chloromera (sic) Oberdörster and Grant 2007b: 17–19, Tables 1–2 (sound system, predation, comp. note) New Jersey
Tibicen chloromera (sic) Shiyake 2007: 3, 36, 38, Fig 48 (illustrated, distribution, listed, comp. note) United States
Tibicen chloromera (sic) Farris, Oshinshy, Forrest, and Hoy 2008: 17–20, 22–24, Fig 1b, Fig 2c, Fig 3b, Table 1 (song, spectrogram, parasitism) Mississippi
Tibicen chloromera (sic) Hastings, Holliday and Coelho 2008: 659, Table 1 (predation, comp. note) Pennsylvania, Illinois, Virginia, North Carolina, Kentucky, Florida
Tibicen chlolomera (sic) Hastings, Holliday and Coelho 2008: 659, Table 1 (predation, comp. note) Missouri
Cicada tibicen Sanborn 2008a: 227–230, Fig 1 (lectotype designation, illustrated, synonymy, comp. note) Equals *Tibicen chloromerus* Equals *Cicada sayi* America
Tibicen chloromerus Sanborn 2008a: 227–230, Fig 2 (illustrated, comp. note) Equals *Thopha chloromerus*
Thopha chloromerus Sanborn 2008a: 230 (comp. note)
Tibicen tibicen Sanborn 2008a: 230 (comp. note)
Cicada sayi Sanborn 2008a: 227 (comp. note)
Tibicen tibicen tibicen Sanborn, Phillips and Gillis 2008: 5, 12–13, 31, 38, Fig 73, Figs 130–138 (key, synonymy, illustrated, distribution, comp. note) Equals *Cicada tibicen* Equals *Thopha chloromera* Equals *Cicada sayi* Equals *Cicada pruinosa* (partim) Florida
Tibicen chloromerus Sanborn, Phillips and Gillis 2008: 12 (comp. note)
Tibicen tibicen Sanborn, Phillips and Gillis 2008: 13 (comp. note)
Tibicen tibicen Holliday, Hastings and Coelho 2009: 3, 8–10, 12–14, Tables 1–2 (predation, comp. note) Equals *Tibicen chloromerus* Arkansas, Connecticut, Delaware, Florida, Georgia, Indiana, Kentucky, Louisiana, Maryland, Missouri, New Jersey, New York, North Carolina, Ohio, Pennsylvania, South Carolina, Tennessee, Virginia, West Virginia

Tibicen chloromerus Holliday, Hastings and Coelho 2009: 3 (predation, comp. note)
Tibicen chloromera (sic) Hughes, Nuttall, Katz and Carter 2009: 961 (comp. note) Connecticut

T. toradja Breddin, 1901 to *Brachylobopyga toradja*

***T. townsendii* (Uhler, 1905)** = *Cicada townsendii* = *Cicada townsendi* (sic) = *Rihana townsendi* (sic) = *Tibicen townsendi* (sic)
Tibicen townsendi (sic) Tinkham 1941: 166, 169–174, 178–179, Fig 1 (behavior, distribution, habitat, comp. note) Texas, New Mexico, Arizona
Tibicen townsendii Sanborn and Phillips 1995a: 482, Table 1 (song intensity, comp. note)
Tibicen townsendi (sic) Van Pelt 1995: 20 (listed, distribution, comp. note) Texas
Tibicen townsendii Poole, Garrison, McCafferty, Otte and Stark 1997: 333 (listed) Equals *Cicada townsendii* Equals *Rihana townsendi* (sic) North America
Tibicen townsendi (sic) Sanborn and Phillips 1997: 36 (distribution, comp. note) Texas
Tibicen townsendi (sic) Sanborn 1999a: 42 (type material, listed)
Tibicen townsendii Sanborn 1999b: 1 (listed) North America
Tibicen townsendii Phillips and Sanborn 2007: 456–460, Table 1, Fig 2, Fig 7 (illustrated, habitat, distribution, comp. note) Texas
Tibicen townsendii Lenhart, Mata-Silva and Johnson 2010: 112–113, Table 1 (predation) Texas

***T. tremulus* Cole, 2008**
Tibicen tremulus Cole 2008: 816–822, Fig 1B, Fig 2B, Figs 4–5 (n. sp., described, illustrated, song, sonogram, distribution, comp. note) Kansas, Colorado

***T. tsaopaonensis* Chen, 1943** = *Lyristes tsaopaonensis*
Lyristes tsaopaonensis Chou, Lei, Li, Lu and Yao 1997: 282, 286 (key, described, synonymy, comp. note) Equals *Tibicen tsaopaonensis* China
Lyristes tsaopaonensis Hua 2000: 62 (listed, distribution) Equals *Tibicen tsaopaonensis* China, Sichuan, Yunnan

***T. variegatus* (Fabricius, 1794)** = *Tettigonia variegata* = *Tettigonia variata* (sic) = *Cicada variegata* = *Cicada variegata* var. ______ Germar, 1830 = *Tibicen variegata*
Tibicen variegata Poole, Garrison, McCafferty, Otte and Stark 1997: 333 (listed) Equals *Tettigonia variegata* Equals *Tettigonia variata* (sic) North America

***T. winnemanna* (Davis, 1912)** = *Cicada winnemanna* = *Cicada pruinosa winnemanna* = *Tibicen pruinosus winnemanna* = *Tibicen pruinosa winnemanna* = *Lyristes winnemanna*
Tibicen winnemanna Davis, W.T. 1933: 29 (comp. note) Washington, D.C.
Tibicen winnemanna Davis 1936: 44 (comp. note) Maryland
Cicada winnemanna Hennessey 1990: 471 (type material)
Tibicen winnemanna Sanborn 1991: 5753B (endothermy, comp. note)
Tibicen winnemanna Sanborn and Phillips 1993: 29 (song intensity, comp. note)
Tibicen winnemanna Sanborn and Phillips 1995a: 480, 482–483, Table 1 (song intensity, comp. note)
Tibicen winnemanna Sanborn 1997: 258, 260–262, Fig 1, Table 1 (song, sonogram, oscillogram, song intensity, comp. note)
Tibicen winnemanna Sanborn 1998: 93–95, 100 (thermal biology, comp. note)
Tibicen winnemanna Sanborn 1999a: 42, 55 (type material, listed, comp. note)
Tibicen winnemanna Sanborn 1999b: 1 (listed) North America
Tibicen winnemanna Sanborn 2000: 551–555, Figs 1–3, Tables 1–3 (thermoregulation, endothermy, thermal responses, comp. note) Tennessee
Tibicen winnemanna Sanborn and Maté 2000: 142 (comp. note)
Tibicen winnemanna Lee 2001b: 1 (comp. note)
Tibicen winnemanna Sanborn 2001a: 10–17, Figs 1–5, Tables 1–2 (timbal muscle, song, comp. note) Tennessee
Lyristes winnemanna Sueur 2001: 37, Table 1 (listed)
Tibicen winnemanna Sanborn 2002b: 460–461, 465–466, Table 1 (thermal biology, comp. note)
Tibicen winnemanna Sanborn, Breitbarth, Heath and Heath 2002: 442 (comp. note)
Tibicen winnemanna Sanborn, Noriega and Phillips 2002c: 368 (comp. note)
Tibicen winnemanna Sanborn, Villet and Phillips 2003: 307 (comp. note)
Lyristes winnemanna Sueur and Sanborn 2003: 341 (comp. note)
Tibicen winnemanna Villet, Sanborn and Phillips 2003c: 1442–1443 (comp. note)
Tibicen winnemanna Sanborn 2004b: 100 (comp. note)
Tibicen winnemanna Sanborn 2004d: 2056, Fig 943 (illustrated)
Tibicen winnemanna Sanborn, Villet and Phillips 2004a: 820 (comp. note)
Tibicen winnemanna Sanborn 2005c: 114, 1116, Fig 7.3, Table 1 (comp. note)
Tibicen pruinosa (sic) *winnemanna* Elliott and Hershberger 2006: 186 (distribution, comp.

note) New York, New Jersey, Maryland, West Virginia, Ohio, Kentucky, Virginia, Tennessee, North Carolina, South Carolina, Georgia, Alabama

Tibicen winnemanna Kroder, Samietz, Schneider and Dorn 2006: 110 (comp. note)

Tibicen winnemanna Sanborn 2008d: 3469, Fig 79 (illustrated, comp. note)

Tibicen winnemanna Holliday, Hastings and Coelho 2009: 10–11, Table 1 (predation, comp. note) Maryland, North Carolina, Pennsylvania, Tennessee, Virginia

Tibicen winnemanna McLister 2010: 77 (comp. note)

***Cornuplura* Davis, 1944**

Cornuplura Arnett 1985: 217 (diversity) Southwestern United States

Cornuplura Poole, Garrison, McCafferty, Otte and Stark 1997: 251, 331 (listed) North America

Cornuplura Sanborn 1999b: 1 (listed) North America

Cornuplura Arnett 2000: 298 (diversity) Southwestern United States

Cornuplura Cioara and Sanborn 2001: 25 (song, comp. note)

Cornuplura Moulds 2005b: 390, 431 (higher taxonomy, listed, comp. note)

Cornuplura Sanborn, Moore and Young 2008: 1 (comp. note)

***C. curvispinosa* (Davis, 1936)** = *Tibicen curvispinosa*

Cornuplura curvispinosa Sanborn 1999a: 37 (type material, listed)

Cornuplura curvispinosa Sueur 2000: 217 (comp. note) Mexico, Colima

Cornuplura curvispinosa Sanborn 2007b: 34 (distribution, comp. note) Mexico, Sinaloa, Nayarit, Colima, Jalisco

***C. nigroalbata* (Davis, 1936)** = *Tibicen nigroalbata*

Cornuplura nigroalbata Sanborn and Phillips 1995a: 481, Table 1 (song intensity, comp. note)

Cornuplura nigroalbata Poole, Garrison, McCafferty, Otte and Stark 1997: 331 (listed) Equals *Tibicen nigroalbata* North America

Cornuplura nigroalbata Sanborn 1999a: 37 (type material, listed)

Cornuplura nigroalbata Sanborn 1999b: 1 (listed) North America

Cornuplura nigroalbata Sanborn 2007b: 34 (distribution, comp. note) Mexico, Sonora, Arizona

***C. rudis* (Walker, 1858)** = *Fidicina rudis* = *Cicada rudis* = *Rihana rudis* = *Tibicen rudis*

Cornuplura rudis Poole, Garrison, McCafferty, Otte and Stark 1997: 331 (listed) Equals *Fidicina rudis* North America

Cornuplura rudis Sanborn 2007b: 34 (distribution, comp. note) Mexico, Veracruz, Morelos, Distrito Federal, Chihuahua

***Hea* Distant, 1906** = *Hua* (sic) = *Kinoshitaia* Ouchi, 1938

Hea Lei, Chou, Yao and Lu 1995: 201, 203 (synonymy, comp. note) Equals *Kinoshitaia* China

Kinoshitaia Lei, Chou, Yao and Lu 1995: 201, 203 (comp. note) China

Hea Chou, Lei, Li, Lu and Yao 1997: 80, 293, 297, Table 6 (key, described, synonymy, listed, diversity, biogeography) Equals *Kinoshitaia* China

Hea Moulds 2005b: 393, 437 (higher taxonomy, listed, comp. note)

Hea Pham and Thinh 2005a: 237 (listed) Vietnam

Hea Pham and Thinh 2005b: 287 (comp. note) Vietnam

Hea Pham and Yang 2009: 4, 15, Table 2 (listed) Vietnam

***H. choui* Lei, 1992** = *Hua* (sic) *choui*

Hea choui Peng and Lei 1992: 104–105, Fig 300 (n. sp., described, illustrated, listed, comp. note) China, Hunan

Hea choui Lei, Chou, Yao and Lu 1995: 201–204 (listed, key, distribution, comp. note) China, Hunan, Jiangxi, Fujian, Guangxi

Hea choui Chou, Lei, Li, Lu and Yao 1997: 80, 83–84, Fig 8–25, Plate II, Fig 28 (key, described, illustrated, comp. note) China

Hua (sic) *choui* Chou, Lei, Li, Lu and Yao 1997: 83 (listed) China

***H. fasciata* Distant, 1906** = *Kinoshitaia sinensis* Ouchi, 1938

Hea fasciata Peng and Lei 1992: 104–105, Fig 299 (listed, illustrated, comp. note) China, Hunan

Hea fasciata Wang and Zhang 1993: 190 (listed, comp. note) China

Hea fasciata Lei, Chou, Yao and Lu 1995: 201–204 (type species of *Hea*, synonymy, listed, key, comp. note) Equals *Kinoshitaia sinensis* China, Sichuan, Hunan, Anhui, Zhejiang, Yunnan, Guizhou

Kinoshitaia sinensis Lei, Chou, Yao and Lu 1995: 201, 203–204 (type species of *Kinoshitaia*, comp. note)

Hea fasciata Xu, Ye and Chen 1995: 85 (listed, comp. note) China, Zhejiang

Hea fasciata Chou, Lei, Li, Lu and Yao 1997: 80–82, Fig 8–23, Plate II, Fig 26 (type species of *Hea*, key, described, illustrated, synonymy, comp. note) Equals *Kinoshitaia sinensis* China

Hea fasciata Hua 2000: 62 (listed, distribution) Equals *Kinoshitaia sinensis* China, Hubei, Anhui, Jiangxi, Zhejiang, Hunan, Guizhou, Sichuan, Yunnan, Nepal

Hea fasciata Pham and Thinh 2005a: 237 (type species of *Hea*)
Hea fasciata Pham and Thinh 2006: 525, 527 (listed, comp. note) Vietnam
Hea fasciata Pham and Yang 2009: 4, 15–16, Table 2 (type species of *Hea*, listed, distribution, comp. note) Vietnam, China

***H. yunnanensis* Chou & Yao, 1995**
Hea yunnanensis Lei, Chou, Yao and Lu 1995: 201–204, Fig 1 (n. sp., described, illustrated, key, comp. note) China, Yunnan
Hea yunnanensis Chou, Lei, Li, Lu and Yao 1997: 80, 82–83, Fig 8-24, Plate II, Fig 30 (key, described, illustrated, comp. note) China
Hea yunnanensis Pham and Thinh 2005a: 237 (distribution, listed) Vietnam
Hea yunnanensis Pham and Yang 2009: 4, 15–16, Table 2 (listed, distribution, comp. note) Vietnam, China

***Antankaria* Distant, 1904**
Antankaria Novotny and Wilson 1997: 437 (listed)
Antankaria Moulds 2005b: 390, 431 (higher taxonomy, listed, comp. note)

A. pulverulenta madegassa Boulard, 1999 to *Chremistica pulverulenta pulverulenta* (Distant, 1905)

A. pulverea (sic) *seychellensis* Boulard, 1999 to *Chremistica pulverulenta seychellensis*

***A. signoreti* (Metcalf, 1955)** = *Cicada signoreti* = *Cicada punctipes* Signoret, 1860 (nec *Cicada punctipes* Zetterstedt, 1828) = *Tettigia punctipes* = *Cicada madagascariensis* Distant, 1892 = *Antankaria madagascariensis*
Antankaria signoreti Boulard 1999a: 79, 94, 96–97, Plate II, Fig 4 (n. comb., synonymy, listed, sonogram, comp. note) Equals *Cicada madagascariensis* Equals *Antankaria madagascariensis* Equals *Cicada signoreti* Madagascar
Antankaria signoreti Sueur 2001: 36, Table 1 (listed)
Antankaria signoreti Boulard 2005d: 338, 344 (comp. note) Madagascar
Antankaria signoreti Boulard 2006g: 36–37, 39–40, 74–75, 87, Fig 7f, Fig 49 (illustrated, sonogram, comp. note) Madagascar

***Cacama* Distant, 1904**
Cacama Tinkham 1941: 173, Fig 2 (singing posture, comp. note)
Cacama Hayashi 1984: 35 (listed)
Cacama Arnett 1985: 217 (diveristy) Southwestern United States, Utah, Colorado
Cacama sp. Cranshaw and Kondratieff 1991: 10 (life cycle, comp. note) Colorado
Cacama Veenstra and Hagedorn 1995: 393 (comp. note)
Cacama Novotny and Wilson 1997: 437 (listed)
Cacama Poole, Garrison, McCafferty, Otte and Stark 1997: 251, 331 (listed) North America
Cacama Sanborn 1999b: 1 (listed) North America
Cacama Arnett 2000: 298 (diveristy) Southwestern United States, Utah, Colorado
Cacama Kondratieff, Ellingson and Leatherman 2002: 14–15 (key, listed, diagnosis, comp. note) Colorado
Cacama Sueur 2002b: 122, 124 (comp. note)
Cacama Moulds 2005b: 390, 431 (higher taxonomy, listed, comp. note)
Cacama sp. Phillips and Sanborn 2007: 456, Fig 1 (listed) Texas
Cacama Shiyake 2007: 4, 39 (listed, comp. note)
Cacama sp. Shiyake 2007: 4, 37, 39, Fig 56 (illustrated, distribution, listed, comp. note)
Cacama Sanborn 2007b: 40 (comp. note)

***C. californica* Davis, 1919**
Cacama califonica Poole, Garrison, McCafferty, Otte and Stark 1997: 331 (listed) North America
Cacama californica Sanborn 1999a: 36 (type material, listed)
Cacama californica Sanborn 1999b: 1 (listed) North America
Cacama californica Sanborn 2007b: 35 (distribution, comp. note) Mexico, Baja California Norte, California

***C. carbonaria* Davis, 1919**
Cacama carbonaria Hennessey 1990: 470 (type material)
Cacama carbonaria Sanborn 1999a: 37 (type material, listed)
Cacama carbonaria Sanborn 2007b: 35 (distribution, comp. note) Mexico, Morelos, Oaxaca, Michoacán

***C. collinaplaga* Sanborn & Heath, 2011** = *Cacama* n. sp. 1 Sanborn and Phillips, 1995
Cacama n. sp. 1 Sanborn and Phillips 1995a: 481, Table 1 (song intensity, comp. note)
Cacama collinaplaga Sanborn, Heath, Phillips and Heath 2011 (*Zootaxa* 2897:35–50): 37, 40–44, 48–50, Fig 2, Fig 4 (n. sp., described, illustrated, key, distribution, comp. note) Texas

***C. crepitans* (Van Duzee, 1914)** = *Proarna crepitans*
Cacama crepitans Sugden 1940: 117 (behavior, song, comp. note) California
Cacama crepitans Poole, Garrison, McCafferty, Otte and Stark 1997: 331 (listed) Equals *Proarna crepitans* North America
Cacama crepitans Sanborn 1999a: 37 (type material, listed)

Cacama crepitans Sanborn 1999b: 1 (listed) North America
Cacama crepitans Sueur 2002b: 124, 129 (listed, comp. note)
Cacama crepitans Sanborn 2007b: 35 (distribution, comp. note) Mexico, Baja California Sur, Baja California Norte, California

C. *dissimilis* (Distant, 1881) = *Cicada dissimilis*
Cacama dissimilis Poole, Garrison, McCafferty, Otte and Stark 1997: 331 (listed) Equals *Cicada dissimilis* North America
Cacama dissimilis Sanborn 2007b: 35 (distribution, comp. note) Mexico, Baja California Sur, Hidalgo, Sinaloa

C. *furcata* Davis, 1919
Cacama furcata Sanborn 2007b: 35 (distribution, comp. note) Mexico, Baja California Sur, Tamaulipas

C. *longirostris* (Distant, 1881) = *Proarna longirostris* = *Ploarna* (sic) *longirostris*
Cacama longirostris Sanborn 2007b: 35 (distribution, comp. note) Mexico

C. *maura* (Distant, 1881) = *Proarna maura*
Cacama maura Kondratieff, Ellingson and Leatherman 2002: 15 (type species of *Cacama*)
Cacama maura Sanborn 2006a: 76, 78, Fig 1 (listed, distribution) Honduras, Mexico
Cacama maura Sanborn 2007b: 35 (distribution, comp. note) Mexico, Morelos, Yucután, Oaxaca, Honduras

C. *moorei* Sanborn & Heath, 2011 = *Cacama dissimilis* (nec Distant) = *Cacama* n. sp. 2 Sanborn and Phillips, 1995
Cacama dissimilis von Dohlen and Moran 1995: 213, 215–221, Figs 2–8 (listed, phylogeny, comp. note) Arizona
Cacama n. sp. 2 Sanborn and Phillips 1995a: 481, Table 1 (song intensity, comp. note)
Cacama dissimilis Sanborn 1999b: 1 (listed) North America
Cacama moorei Sanborn, Heath, Phillips and Heath 2011 (*Zootaxa* 2897:35–50): 37, 43–50, Figs 3–4 (n. sp., described, illustrated, key, distribution, comp. note) Texas

C. *pygmaea* Sanborn, 2011

C. *valvata* (Uhler, 1888) = *Proarna valvata* = *Cacama valvada* (sic) = *Cacama valuate* (sic)
Proarna valvata Henshaw 1903: 230 (listed) Texas, Arizona
Cacama valvata Tinkham 1941: 169–171, 173, 175, 178–180, Fig 1 (distribution, habitat, comp. note) Texas, Colorado, Utah, New Mexico, Arizona
Cacama (sic) Tinkham 1941: 172 (comp. note) Texas
Cacama valvada (sic) Cranshaw and Kondratieff 1991: 10 (comp. note) Colorado
Cacama valvata Sanborn and Phillips 1992: 285 (activity, comp. note) Arizona
Cacama valvata Sanborn, J.E. Heath, M.S. Heath and Noriega 1995: 326, Table 3 (thermal responses, comp. note)
Cacama valvata Sanborn, M.S. Heath, J.E. Heath and Noriega 1995: 456–458, Table 4 (comp. note)
Cacama valvata Sanborn and Phillips 1995a: 481, Table 1 (song intensity, comp. note)
Cacama valvata Van Pelt 1995: 19 (listed, distribution, comp. note) Texas
Cacama valvata Veenstra and Hagedorn 1995: 392 (peptide, comp. note) Arizona
Cacama valvata Poole, Garrison, McCafferty, Otte and Stark 1997: 331 (listed) Equals *Proarna valvata* North America
Cacama valvata Sanborn 1998: 91, 96–97, 99, Table 1 (thermal biology, thermal responses, comp. note)
Cacama valvata Toolson 1998: 569–580, Figs 1–5, Table 1 (thermal biology, comp. note) New Mexico
Cacama valvata Sanborn 1999a: 37 (type material, listed)
Cacama valvata Sanborn 1999b: 1 (listed) North America
Cacama valvata Sanborn and Phillips 2001: 164 (comp. note)
Cacama valvata Kondratieff, Ellingson and Leatherman 2002: 3, 5, 8, 12–13, 15–16, 67, 70, Fig 11, Figs 18–19, Tables 1–2 (diagnosis, illustrated, listed, distribution, song, predation, comp. note) Equals *Proarna valvata* Colorado, Arizona, California, New Mexico, Texas, Utah
Cacama valvata Sanborn 2002b: 458, 462, 465, Fig 2, Table 1 (illustrated, thermal biology, comp. note)
Cacama valvata Sanborn, Breitbarth, Heath and Heath 2002: 443 (comp. note)
Cacama valvata Sanborn, Noriega and Phillips 2002c: 368 (comp. note)
Cacama valvata Sanborn 2004b: 100 (comp. note)
Cacama valvata Sanborn, Heath, Heath, Noriega and Phillips 2004: 286 (comp. note)
Cacama valvata Sanborn 2005c: 114, Table 1 (comp. note)
Cacama valvata Eaton and Kaufman 2007: 88–89 (illustrated, comp. note) Western United States
Cacama valvata Heath and Sanborn 2007: 488 (comp. note)
Cacama valvata Phillips and Sanborn 2007: 456–457, 460, Table 1, Fig 1, Fig 3 (illustrated, habitat, distribution, comp. note) Texas

Cacama valvata Sanborn 2007b: 35 (distribution, comp. note) Mexico, Chihuahua, Coahuila, Durango, Nuevo León, Sonora, Veracruz, Arizona, California, Colorado, New Mexico, Texas, Utah, Oklahoma
Cacama valuate (sic) Shiyake 2007: 4, 37, 39, Fig 55 (illustrated, distribution, listed, comp. note) United States
Cacama valvata Lenhart, Mata-Silva and Johnson 2010: 112–113, Table 1 (predation) Texas
Cacama valvata Sanborn and Phillips 2010: 864 (comp. note)

***C. variegata* Davis, 1919**
Cacama variegata Poole, Garrison, McCafferty, Otte and Stark 1997: 331 (listed) North America
Cacama variegata Sanborn 1999a: 37 (type material, listed)
Cacama variegata Sanborn 1999b: 1 (listed) North America
Cacama variegata Sanborn 2007b: 35 (distribution, comp. note) Mexico, Cohuila, Durango, Nuevo León, Tamaulipas, Texas

***Orialella* Metcalf, 1952** = *Oria* Distant, 1904 (nec *Oria* Huebner, 1821)
Orialella Boulard 1986e: 346 (comp. note) Equals *Oria* Distant
Oria Novotny and Wilson 1997: 437 (listed)
Orialella Moulds 2005b: 390, 431 (higher taxonomy, listed, comp. note) Australia

***O. aerizulae* Boulard, 1986**
Orialella aerizulae Boulard 1986e: 345–347, Figs 1–3 (n. sp., described, illustrated, comp. note) French Guiana

***O. boliviana* (Distant, 1904)** = *Cicada boliviana* = *Oria boliviana*
Orialella boliviana Boulard 1986e: 347 (type species of genus *Orialella*, comp. note) French Guiana
Orialella boliviana Pogue 1996: 314 (listed) Peru

***Cryptotympana* Stål, 1861** = *Chryptotympana* (sic) = *Crptotympana* (sic) = *Cryptolympana* (sic) = *Ciyptotympana* (sic) = *Cryptotympanus* (sic) = *Cyrister* (sic) = *Gryptotympana* (sic) = *Cryptotympaua* (sic) = *Criptotympana* (sic) = *Cryplolympana* (sic) = *Crystotympana* (sic) = *Cryptympana* (sic) = *Chryptotympana* (sic) = *Magicicada* (nec Davis)
Cryptotympana Dallas 1870: 482 (listed)
Cryptotympana Wu 1935: 7 (listed) China
Cryptotympana Chen 1942: 144 (listed)
Cryptotympana Duffels 1983a: 492 (comp. note) Lesser Sunda Islands
Cryptotympana Strümpel 1983: 18 (listed)
Cryptotympana Hayashi 1984: 28, 35, 42 (key, described, listed, comp. note)
Cryptotympana Marshall 1984: 225, Table 3 (hemolymph, comp. note)
Cryptotympana Bregman 1985: 39 (comp. note)
Cryptotympana Duffels 1986: 319, 328 (diversity, distribution) Sulawesi, Lesser Sunda Islands, Southeast Asia, Greater Sunda Islands
Cryptotympana Hayashi 1987a: 119–122, 124–125 (history, described, diversity, distribution, comp. note) India, Lesser Sunda Islands, Timor, Japan, Korea, China
Cryptotympana Hayashi 1987b: 96 (listed)
Cryptotympana Duffels 1988d: 81 (comp. note)
Cryptotympana spp. Namba, Ma and Inagaki 1988: 251 (medicinal use, comp. note) China
Cryptotympana Duffels 1990b: 70 (comp. note) Sulawesi
Cryptotympana spp. Liu 1990: 46 (comp. note)
Cryptotympana Zaidi, Mohammedsaid and Yaakob 1990: 261, 267, Table 2 (listed, comp. note) Malaysia
Cryptotympana sp. Liu 1991: 255 (comp. note)
Cryptotympana spp. Liu 1992: 9 (comp. note)
Cryptotympana sp. Liu 1992: 14 (comp. note)
Cryptotympana Peng and Lei 1992: 105 (comp. note) China, Hunan
Cryptotympana Hayashi 1993: 14–16 (distribution, comp. note)
Cryptotympana Lei, Chou and Li 1994: 54 (comp. note)
Cryptotympana Zaidi and Ruslan 1994: 427, Table 2 (comp. note) Malaysia
Cryptotympana Lee 1995: 18, 22, 52, 71, 135 (listed, key, described, ecology, distribution, illustrated, comp. note) Korea
Cryptotympana Zaidi and Ruslan 1995a: 174, Table 2 (listed, comp. note) Borneo, Sarawak
Chryptotympana (sic) Zaidi and Ruslan 1995c: 70, Table 2 (listed, comp. note) Peninsular Malaysia
Cryptotympana Zaidi and Ruslan 1995d: 202, Table 2 (listed, comp. note) Borneo, Sabah
Cryptotympana Boer and Duffels 1996b: 304 (comp. note)
Cryptotympana Zaidi 1996: 103, Table 2 (comp. note) Borneo, Sabah, Brunei, Sarawak
Cryptotympana Zaidi and Hamid 1996: 54–55, Table 2 (listed, comp. note) Borneo, Sarawak
Cryptotympana Boulard 1997c: 106, 112, 117 (type genus of Cryptotympanaria)
Cryptotympana Chou, Lei, Li, Lu and Yao 1997: 265, 293, 309, 321, 325, 327, 347, Table 6, Tables 10–12 (key, described, listed, diversity, biogeography) China
Cryptotympana Holloway 1997: 219 (comp. note) Sulawesi

Cryptotympana spp. Holloway 1997: 219 (comp. note)
Cryptotympana Novotny and Wilson 1997: 437 (listed)
Cryptotympana Zaidi 1997: 114–115, Table 2 (listed, comp. note) Borneo, Sarawak
Cryptotympana Holloway 1998: 297, 304 (comp. note)
Cryptotympana Easton 1999: 28 (comp. note) Macao
Cryptotympana Easton and Pun 1999: 99 (comp. note) Macao
Cryptotympana Lee 1999: 5 (comp. note) Korea
Cryptotympana Zaidi and Ruslan 1999: 4, Table 2 (listed, comp. note) Borneo, Sarawak
Cryptotympana Boulard 2000i: 21 (listed)
Cryptotympana sp. Noda, Kubota, Miyata and Miyahara 2000: 1749 (medicinal use) China
Cryptotympana Boulard 2001b: 26, 32, 37 (type genus of Cryptotympanaria)
Cryptotympana Lee 2003b: 7 (comp. note)
Cryptotympana Lee and Hayashi 2003a: 158, 162 (key, diagnosis) Taiwan
Cryptotympana Rakitov 2002: 123 (comp. note) Illinois
Cryptotympana Chen 2004: 39, 41, 66, 192 (phylogeny, listed)
Magicicada (sic) Song, Lee, Soh, Zhu and Bai 2004: 3036 (wing mechanics, comp. note) Hong Kong
Cryptotympana Boulard 2005h: 8–9 (listed)
Cryptotympana Lee 2005: 14, 16, 63, 170 (listed, key, comp. note) Korea
Cryptotympana Moulds 2005b: 387–389, 390–392, 412–413, 431 (type genus of Cryptotympanini, listed, comp. note)
Cryptotympana Pham 2005b: 232 (key, synonymy, distribution, comp. note) Vietnam
Cryptotympana Pham and Thinh 2005a: 240 (listed) Vietnam
Cryptotympana Pham and Thinh 2005b: 290 (comp. note) Vietnam
Cryptotympana spp. Robinson 2005: 217 (comp. note) Asia
Cryptotympana Salmah, Duffels and Zaidi 2005: 15 (comp. note)
Cryptotympana Yaakop, Duffels and Visser 2005: 248–249 (comp. note) Southeast Asia
Cryptotympana sp. Xu, Lee, Han, Oh, Park, Tian et al. 2006: 7827 (comp. note) Korea
Cryptotympana Boulard 2007e: 3, 20, Fig 2a (comp. note) Thailand
Cryptotympana Sanborn, Phillips and Sites 2007: 3, 8, Table 1 (listed) Thailand
Cryptotympana Shiyake 2007: 4, 42 (listed, comp. note)
Cryptotympana Boulard 2008f: 7, 91 (listed) Thailand
Cryptotympana Lee 2008a: 2, Table 1 (listed) Vietnam
Cryptotympana Lee 2008b: 446, 448, Table 1 (listed, comp. note) Korea
Cryptotympana Osaka Museum of Natural History 2008: 322 (listed) Japan
Cryptotympana Sanborn, Moore and Young 2008: 1 (comp. note)
Cryptotympana Lee 2009h: 2620 (listed) Philippines, Palawan
Cryptotympana Pham and Yang 2009: 13, Table 2 (listed) Vietnam
Cryptotympana Lee 2010a: 19 (listed) Philippines, Mindanao
Cryptotympana Lee 2010b: 21 (listed) Cambodia
Cryptotympana Lee 2010c: 2 (listed) Philippines, Mindanao
Cryptotympana Phillips and Sanborn 2010: 74 (comp. note) Asia

***C. accipiter* (Wallker, 1850)** = *Fidicina accipiter*
Cryptotympana accipiter group Hayashi 1987a: 126–128, Fig 3 (described, distribution, diversity, listed, comp. note) Philippines, Sulawesi
Cryptotympana accipiter group Hayashi 1987b: 1, 3, 24, 27 (listed, described, distribution, comp. note) Philippines, Sulawesi
Cryptotympana accipiter Hayashi 1987b: 1–8, 10, 21–22, 24–25, 28, 102, Fig 1, Figs 11–12, Fig 15, Fig 16A, Figs 17–22, Fig 57, Fig 58A (described, illustrated, synonymy, distribution, key, comp. note) Equals *Fidicina accipiter* Philippines, Samar, Leyte, Luzon
Cryptotympana accipiter Duffels 1990b: 70 (comp. note) Philippines
Cryptotympana accipiter group Hayashi 1993: 16–17, Table 1 (listed, comp. note)
Cryptotympana accipiter Hayashi 1993: 18 (comp. note)
Cryptotympana accipiter group Holloway 1997: 219 (distribution, comp. note) Sulawesi, Philippines
Cryptotympana accipiter group Holloway 1998: 305, Fig 10 (distribution, phylogeny, comp. note) Philippines
Cryptotympana accipiter-group Shiyake 2007: 42 (comp. note)
Cryptotympana accipiter Lee 2010c: 3 (synonymy, distribution, listed) Equals *Fidicina accipiter* Philippines, Luzon, Samar, Leyte

***C. acuta* (Signoret, 1849)** = *Cicada acuta* = *Fidicina acuta* = *Cicada vicinia* Signoret, 1849 = *Fidicina vicinia* = *Fidicina nivifera* Walker, 1850 = *Fidicina vinifera (nivifera)* = *Fidicina bicolor* Walker, 1852 = *Fidicina blicolor* (sic) = *Cryptotympana bicolor*

Cryptotympana acuta Wu 1935: 7 (listed) Equals *Cicada acuta* Equals *Cicada vicina* Equals *Fidicina nivifera* Equals *Fidicina bicolor* Equals *Fidicina timorica* China, Canton, Bhutan, Java, Borneo, Lonbok, Philippines, Timor

Cicada acuta Hayashi 1987a: 119 (comp. note)

Cicada vicina Hayashi 1987a: 119 (comp. note)

Cryptotympana acuta group Hayashi 1987a: 126–127, 144, Fig 3 (described, distribution, diversity, listed, comp. note) Sumatra, Malay Peninsula, Lesser Sunda Islands, Java

Cryptotympana acuta Hayashi 1987a: 144, 153–160, 167, 177, 184, Figs 42–43, Figs 65–67, Figs 68a–c, Figs 69–73 (described, illustrated synonymy, distribution, comp. note) Equals *Cicada acuta* Equals *Fidicina acuta* Equals *Cicada vicina* Equals *Fidicina navifera* Equals *Fidicina blicolor* (sic) Java, Bali

Cryptotympana acuta Hayashi 1987b: 97 (key)

Cryptotympana acuta group Hayashi 1993: 16–17, Table 1 (listed, comp. note)

Cryptotympana acuta Hayashi 1993: 17 (comp. note)

Cryptotympana acuta Chou, Lei, Li, Lu and Yao 1997: 281 (described, synonymy, comp. note) Equals *Cicada acuta* China

Cryptotympana acuta group Holloway 1998: 305, Fig 10 (distribution, phylogeny, comp. note) Banda Islands

Cryptotympana acuta Hua 2000: 61 (listed, distribution) Equals *Cicada acuta* China, Guangzhou, Philippines, Kalimantan, Timor, Indonesia, Malaysia, Bhutan, India

Cryptotympana acuta Shiyake 2007: 4, 40, 42, Fig 57 (illustrated, distribution, listed, comp. note) Indonesia

Cryptotympana acuta-group Shiyake 2007: 42 (comp. note)

***C. albolineata* Hayashi, 1987**

Cryptotympana albolineata Hayashi 1987b: 2–3, 4, 6, 8–9, 24, 103, Fig 2, Figs 13–14, Figs 16C–D, Figs 23–25, Fig 57 (n. sp., described, illustrated, distribution, key, comp. note) Philippines, Luzon

Cryptotympana albolineata Duffels 1990b: 70 (comp. note) Philippines

Cryptotympana albolineata Hayashi 1993: 18 (comp. note)

Cryptotympana albolineata Sanborn 1999a: 42 (type material, listed)

Cryptotympana albolineata Shiyake 2007: 4, 40, 42, Fig 59 (illustrated, distribution, listed, comp. note) Phillipines

Cryptotympana albolineata Lee 2010c: 3 (distribution, listed) Philippines, Luzon

***C. alorensis* Hayashi, 1987**

Cryptotympana alorensis Hayashi 1987a: 144–145, 156, 165, 170–172, Fig 48, Fig 68j, Fig 89, Figs 102–104 (n. sp., described, illustrated, distribution, comp. note) Alor

Cryptotympana alorensis Hayashi 1987b: 98 (key)

***C. aquila* (Walker, 1850)** = *Fidicina aquila* = *Cryptotympana acquila* (sic) = *Crptotympana* (sic) *aquila* = *Cryptolympana* (sic) *aquila* = *Ciyptotympana* (sic) *aquila*

Cryptotympana aquila group Hayashi 1987a: 126–127, 173, Fig 3 (described, distribution, diversity, listed, comp. note) Sumatra, Malay Peninsula, Borneo, Indo-China

Cryptotympana aquila Hayashi 1987a: 173–180, 184–185, Fig 107, Figs 110–111, Figs 114–115, Figs 117–121 (described, illustrated, synonymy, distribution, comp. note) Equals *Fidicina aquila* Sumatra, Borneo, Malay Peninsula, Thailand, Burma, Laos, Vietnam

Cryptotympana aquila Hayashi 1987b: 97 (key)

Cryptotympana aquila Zaidi, Mohammedsaid and Yaakob 1990: 262–265, Fig 1, Table 1 (seasonality, comp. note) Malaysia

Cryptotympana aquila group Hayashi 1993: 16–17, Table 1 (listed, comp. note)

Cryptotympana aquila Hayashi 1993: 17, Fig 7 (illustrated, comp. note)

Cryptotympana aquila Zaidi 1993: 958, 960, Table 1 (listed, comp. note) Borneo, Sarawak

Cryptotympana aquila Zaidi and Ruslan 1994: 426, 428–429, Table 1 (comp. note) Malaysia

Cryptotympana aquila Lee 1995: 18, 22, 28–31, 33–34, 37, 45–46, 71, 122–123, 134, 138 (comp. note)

Fidicina aquila Lee 1995: 122–123 (comp. note)

Crptotympana (sic) *aquila* Zaidi 1996: 101, Table 1 (comp. note) Borneo, Sarawak

Cryptotympana aquila Zaidi 1996: 104 (comp. note) Borneo, Sarawak

Cryptotympana aquila Zaidi, Ruslan and Mahadir 1996: 60, Table 1 (listed, comp. note) Peninsular Malaysia

Cryptotympana aquila Chou, Lei, Li, Lu and Yao 1997: 281 (described, synonymy, comp. note) Equals *Fidicina aquila* China

Crptotympana (sic) *aquila* Zaidi 1997: 113, Table 1 (listed, comp. note) Borneo, Sarawak

Cryptotympana aquila Zaidi 1997: 114 (comp. note) Borneo, Sarawak

Cryptotympana aquila Zaidi and Ruslan 1997: 221 (listed, comp. note) Equals *Fidicina aquila* Borneo, Sarawak, Java, Sumatra, Peninsular Malaysia, Brunei, Korea

Cryptotympana aquila Zaidi and Ruslan 1998a: 348–349 (listed, comp. note) Equals *Fidicina*

aquila Borneo, Sarawak, Korea, Peninsular Malaysia, Sumatra, Java, Brunei

Cryptotympana aquila Zaidi, Ruslan and Azman 1999: 302–303 (listed, comp. note) Equals *Fidicina aquila* Borneo, Sabah, Korea, Peninsular Malaysia, Sumatra, Sarawak, Brunei

Fidicina aquila Duffels and Zaidi 2000: 238 (comp. note)

Cryptotympana aquila Hua 2000: 61 (listed, distribution) Equals *Fidicina aquila* China, Taiwan

Cryptotympana aquila Zaidi, Noramly and Ruslan 2000a: 322 (listed, comp. note) Equals *Fidicina aquila* Borneo, Sabah, Sarawak, Brunei, Korea, Peninsular Malaysia, Sumatra, Java, Brunei

Cryptotympana aquila Zaidi, Ruslan and Azman 2000: 200–201 (listed, comp. note) Equals *Fidicina aquila* Borneo, Sabah, Korea, Peninsular Malaysia, Sumatra, Sarawak, Java, Brunei

Cryptotympana aquila Zaidi, Azman and Ruslan 2001a: 112, 115, 117–118, Table 1 (listed, comp. note) Equals *Fidicina aquila* Langkawi Island, Korea, Peninsular Malaysia, Sumatra, Borneo, Java, Brunei

Cryptotympana aquila Zaidi and Nordin 2001: 183, 185–186, 189, 193–194, Table 1 (listed, comp. note) Equals *Fidicina aquila* Borneo, Sabah, Sarawak, Brunei, Korea, Peninsular Malaysia, Sumatra, Java, Brunei

Cryptolympana (sic) *aquila* Zaidi, Nordin, Maryati, Wahab, Norashikin et al. 2002: 2 (comp. note) Borneo, Sabah

Cryptotympana aquila Zaidi, Nordin, Maryati, Wahab, Norashikin et al. 2002: 3, 10–11, Table 1 (listed, comp. note) Equals *Fidicina aquila* Borneo, Sabah, Peninsular Malaysia, Sumatra, Korea, Java, Sarawak, Brunei

Ciyptotympana (sic) *aquila* Zaidi, Nordin, Maryati, Wahab, Norashikin et al. 2002: 9, Table 2 (comp. note) Borneo, Sabah

Cryptotympana aquila Boulard 2003a: 98–100, Color Fig 1, sonogram 1 (described, song, sonogram, illustrated, comp. note) Equals *Fidicina aquila* Thailand, Korea, Southeast Asia

Cryptotympana aquila Zaidi and Azman 2003: 104, Table 1 (listed, comp. note) Borneo, Sabah

Cryptotympana aquila Chen 2004: 66, 193 (comp. note)

Cryptotympana aquila Boulard 2005d: 344, 346 (comp. note) Thailand

Cryptotympana aquila Boulard 2005o: 149 (listed)

Cryptotympana aquila Gogala and Trilar 2005: 66, 70, 79–80, Figs 7–8 (song, sonogram, oscillogram, acoustic behavior, comp. note) Malaysia, Thailand

Cryptotympana aquila Lee 2005: 63 (comp. note)

Cryptotympana aquila Boulard 2006g: 72–73, 81, 98–99, Fig 37, Fig 64B (sonogram, illustrated, comp. note) Thailand

Cryptotympana aquila Stoddart, Cadusch, Boyce, Erasmus and Comins 2006: 681, 683, Fig 2, Table 1 (wing microstructure, comp. note)

Cryptotympana aquila Boulard 2007d: 493 (comp. note) Thailand

Cryptotympana aquila Boulard 2007e: 4, 35, 38, 48, Fig 22, Plate 3, Fig 1 (coloration, illustrated, sonogram, comp. note) Thailand

Cryptotympana aquila Sanborn, Phillips and Sites 2007: 8 (listed, distribution, comp. note) Thailand, Malay Penisula, Burma, Laos, Sumatra, Borneo

Cryptotympana aquila Shiyake 2007: 4, 40, 42, Fig 58 (illustrated, distribution, listed, comp. note) Southeast Asia, Indonesia

Cryptotympana aquila-group Shiyake 2007: 42 (comp. note)

Cryptotympana aquila Boulard 2008f: 12, 91 (biogeography, listed, comp. note) Equals *Fidicina aquila* Thailand, Korea, Borneo, Sumatra

Cryptotympana aquila Lee 2008a: 5 (synonymy, distribution, listed) Equals *Fidicina aquila* Vietnam, Laos, Thailand, Malay Peninsula, Borneo, Indonesia, Sumatra, Myanmar

Cryptotympana aquila Lee 2008b: 462 (comp. note)

Cryptotympana aquila Pham and Yang 2009: 13, Table 2 (listed) Vietnam

***C. atrata* (Fabricius, 1775)** = *Tettigonia atrata* = *Cicada atrata* = *Cicada atra* (sic) = *Fidicina atrata* = *Cryptotympana atrat* (sic) = *Cryptotympana atrula* (sic) = *Chryptotympana* (sic) *atrata* = *Tettigonia pustulata* Fabricius, 1787 = *Cryptotympana pustulata* = *Cicada pustulata* = *Cryptotympana pustulala* (sic) = *Cryptotympanus* (sic) *pustulatus* = *Cryptotympana pustaulata* (sic) = *Cryptotympana pusstulafa* (sic) = *Cicada nigra* Olivier, 1790 = *Cryptotympana nigra* = *Fidicina bubo* Walker, 1850 = *Cryptotympana bubo* = *Cryptotympana bubo* = *Cryptotympana sinensis* Distant, 1887 = *Cyrister* (sic) *sinensis* = *Cryptotympana dubia* Haupt, 1917 = *Gryptotympana* (sic) *dubia* = *Cryptotympana coreanus* Kato, 1925 = *Cryptotympana coreana* = *Cryptotympana dubia coreana* = *Cryptotympana santoshonis* Matsumura, 1927 = *Cryptotympana santoshoesnis* (sic) = *Cryptotympana pustulata* Kato 1930 (nec Fabricius) = *Cryptotympana wenchewensis* Ouchi, 1938 = *Cryptotypmana pustulata castanea* Liu, 1940 = *Cryptotympana atrata castanea* = *Cryptotympana pustulata fukienensis* Liu,

1940 = *Cryptotympana pustulata fukiensis* (sic) = *Cryptotympana atrata fukienensis*

Tettigonia pustulata Dallas 1870: 482 (synonymy) Equals *Cryptotympana pustulata* Equals *Tettigonia atrata* Equals *Cicada nigra* Equals *Cicada atra*

Cryptotympana coreana Doi 1932: 42–43 (listed, comp. note) Korea

Cryptotympana pustulata Chen 1933: 359 (listed, comp. note)

Cryptotympana dubia Wu 1935: 7 (listed) China, Tsingtau

Cryptotympana pustulata Wu 1935: 9 (listed) Equals *Tettigonia pustulata* Equals *Cicada pustulata* Equals *Tettigonia atrata* Equals *Cicada atrata* Equals *Cicada nigra* Equals *Cicada atra* Equals *Fidicina bubo* China, Peiping, Lushan, Canton, Hongkong, Soochow, Hangchow, Malaysia, Australia, Formosa, Japan

Cryptotympana santoshonis Wu 1935: 9 (listed) China, Santosho

Cryptotympana sinensis Wu 1935: 9 (listed) China, Shantung

Tettigonia pustulata Chen 1942: 144 (type species of *Cryptotympana*)

Cryptotympana coreana Kim 1958: 94 (listed) Korea

Tettigonia pustulata Hayashi 1984: 42 (type species of *Cryptotympana*)

Cryptotympana pustulata He and Chen 1985: 324–327, 329–330, Fig 1B, Fig 3B, Fig 5B, Table 1 (song, sonogram, oscillogram, comp. note) China

Cryptotypmana atrata Jiang 1985: 263 (comp. note) China

Cryptotypmana atrata Jiang, Nie, Liu, Xu, Sun et al. 1986: 175–184, Figs 1–8, Table 1 (timbal function, illustrated, song, oscillogram, comp. note) China

Cryptotympana atrata Wang, Liu, Wang, Chang, Shih and Chu 1986: 217 (medicinal use) China

Cryptotympana atrata Chang and But 1987: 1229 (medicinal use)

Cicada atrata Hayashi 1987a: 119 (comp. note)

Cryptotympana pustulata Hayashi 1987a: 119 (type species of *Cryptotympana*)

Tettigonia pustulata Hayashi 1987a: 124 (type species of *Cryptotympana*)

Cryptotympana atrata group Hayashi 1987a: 126, 128, Fig 3 (described, distribution, diversity, listed, comp. note) China

Cryptotympana atrata group Hayashi 1987b: 33–34 (listed, comp. note)

Cryptotympana atrata Hayashi 1987b: 33–43, 54, 99–100, Fig 80, Figs 85–90, Figs 95–101 (described, illustrated, distribution, synonymy, key, comp. note) Equals *Tettigonia atrata* Equals *Cicada atrata* Equals *Cicada atra* (sic) Equals *Fidicina atrata* Equals *Tettigonia pustulata* Equals *Cicada nigra* Equals *Fidicina bubo* Equals *Cryptotympana sinensis* n. syn. Equals *Cryptotympana dubia* n. syn. Equals *Cryptotympana coreana* n. syn. Equals *Cryptotympana coreanus* (sic) Equals *Cryptotympana santoshonis Equals Cryptotympana wenchewensis* n. syn. *Equals Cryptotympana pustulata castanea* Equals *Cryptotympana atrata* var. *castanea* Equals *Cryptotympana pustulata fukiensis* Equals *Cryptotympana atrata* var. *fukiensis* China, Taiwan, Korea, Indo-China

Cryptotympana pustulata Hayashi 1987b: 40, 43 (comp. note)

Cryptotympana dubia Hayashi 1987b: 40 (comp. note)

Cryptotympana coreana Hayashi 1987b: 40, 42–43 (comp. note)

Cryptotympana wenchewensis Hayashi 1987b: 42–43 (comp. note)

Cryptotympana santoshonis Hayashi 1987b: 54 (comp. note)

Cryptotypmana atrata Jiang, Lin and Wang 1987: 106, Fig 1A (timbal structure, illustrated, song, oscillogram, comp. note) China

Cryptotympana pustulata Wang and Zhang 1987: 295 (key, comp. note) China

Cryptotympana atrata Yuan 1987: 150 (cotton pest, comp. note) China

Cryptotympana atrata Feng 1988: 25–29, Figs 1–4 (song analysis, comp. note) China

Cryptotympana atrata Feng and Jiang 1988: 351–356, Figs 1–4 (sound production system, comp. note) China

Cryptotypmana atrata Jiang, Wang and Lin 1988: 129–139, Figs 1–5, Table 1 (song analysis, timbal structure, illustrated, oscillogram, comp. note) China

Chryptotympana (sic) *atrata* Namba, Ma and Inagaki 1988: 253 (medicinal use, comp. note) China

Cryptotympana atrata Namba, Ma and Inagaki 1988: 253 (medicinal use, comp. note) China

Cryptotympana pustulata Namba, Ma and Inagaki 1988: 253 (medicinal use, comp. note) China

Cryptotympana atrata Shen 1988: 165–169, Figs 1–5 (hearing, comp. note) China

Cryptotympana pustulata Wang L.Y. 1988: 232 (comp. note)

Cryptotympana pustulata Wang Z.-N. 1988: 22, Fig 1 (fungal host, comp. note) Tawian

Cryptotypmana atrata Jiang 1989a: 63–68, Figs 1–5, Table 1 (sound field, oscillogram, comp. note) China

Cryptotypmana atrata Jiang 1989b: 64–77, Figs 1–8, Table 1 (timbal function, illustrated, oscillogram, comp. note) China
Cryptotympana atrata Ma, Qu and Hang 1989: 491 (medicinal use) China
Cryptotympana atrata Feng 1990: 443–448, Figs 1–4 (sound system, song analysis, comp. note) China
Cryptotympana atrata Feng and Shen 1990: 468, Figs 5C (comp. note) China
Cryptotypmana atrata Jiang, Chen and Xu 1990a: 1181–1191, Figs 1–6, Tables 1–3 (sound system functions, oscillogram, comp. note) China
Cryptotypmana atrata Jiang, Chen and Xu 1990b: 159–168, Figs 1–6, Table 1 (timbal function, oscillogram, comp. note) China
Cryptotympana atrata Jiang, Xu, and Chen 1990: 183–185, Figs 1–2 (song production, comp. note) China
Cryptotympana atrata Jiang, Chen, Xu and Zhang 1990: 161–169, Tables 1–2 (sound production, comp. note) China
Cryptotympana atrata Hsieh, Peng, Yeh, Tsai and Chang 1991: 83 (medicinal use)
Cryptotympana atrata Jiang, Wang, Xu, Zhang and Chen 1991: 585 (comp. note)
Cryptotympana pustulata Liu 1991: 256 (comp. note)
Cyrister (sic) *sinensis* Liu 1991: 256 (comp. note)
Cryptotympana atrata Feng and Shen 1992: 8–13, Figs 1–4, Table 1 (song analysis, comp. note) China
Cryptotympana pustulata Liu 1992: 15, 35 (comp. note)
Cyrister (sic) *sinensis* Liu 1992: 16 (comp. note)
Cryptotympana atrata Liu and Jiang 1992: 422–435, Figs 1–2 (timbal, structure and function, comp. note) China
Cryptotympana atrata Peng and Lei 1992: 98, Fig 278 (listed, illustrated, synonymy, comp. note) Equals *Cryptotympana pusstulafa* (sic) China, Hunan
Cryptotympana atrata Shen, Gao and Tang 1992: 383–385, Figs 1–4 (neural response, hearing, comp. note) China
Cryptotympana pustulata Wang and Zhang 1992: 121 (listed, comp. note) China
Cryptotympana atrata Feng and Nie 1993: 415–420, Figs 1–4, Table 1 (song analysis, comp. note) China
Cryptotympana atrata group Hayashi 1993: 16, 18, Table 1 (listed, comp. note)
Cryptotympana atrata Hayashi 1993: 18 (comp. note)
Cryptotympana atrata Moore, Huber, Weber, Klein and Bock 1993: 219 (comp. note) Equals *Cryptotympana pustulata*
Cryptotympana atrata Shen 1993: 363–366, Figs 1–4 (neural response, hearing, comp. note) China
Cryptotympana pustulata Wang and Zhang 1993: 188 (listed, comp. note) China
Cryptotympana atrata Shen 1994: 772–776, Figs 1–4 (neural response, hearing, comp. note) China
Cryptotympana atrata Anonymous 1994: 436–437, 808 (fruit pest) China
Cryptotympana pustulata Anonymous 1994: 809 (fruit pest) China
Cryptotympana atrata Lei, Chou and Li 1994: 53, 56 (comp. note) China
Cryptotympana atrata Chen and Yang 1995: 259, Fig 3(1) (metatarsi illustrated)
C[ryptotympana] atrata Jiang, Yang, Tang, Xu and Chen 1995b: 228 (song, phonotaxis, flight behavior, comp. note) China
Cryptotympana atrata Jiang, Yang, Wang, Xu, Chen and Yang 1995: 676 (comp. note) China
Cryptotympana dubia Lee 1995: 18, 22, 39, 45–46, 71–75, 135, 139 (listed, key, synonymy, ecology, distribution, illustrated, comp. note) Equals *Cryptotympana dubia* var. *coreana* Equals *Cryptotympana dubia* f. *coreana* Equals *Cryptotympana dubia coreana* Equals *Cryptotympana coreanus* Equals *Cryptotympana coreana* Equals *Cryptotympana dubia* f. *coreana* Equals *Cryptotympana dubia* var. *coreana* Equals *Cryptotympana dubia* var. *coreana* Equals *Cryptotympana dubia coreana* Korea
Cryptotympana coreanus Lee 1995: 30–31, 71–72 (comp. note)
Cryptotympana coreana Lee 1995: 31–38, 42–45, 72–73 (comp. note)
Cryptotympana dubia var. *coreana* Lee 1995: 39, 72–73 (comp. note) Korea
Cryptotympana dubia f. *coreana* Lee 1995: 40, 72–73 (comp. note) Korea
Cryptotympana dubia coreana Lee 1995: 41, 72 (comp. note) Korea
Cryptotympana pustulata Lee 1995: 71 (type species of *Cryptotympana*) Formosa, China, Malaysia
Cryptotympana atrata Xu, Ye and Chen 1995: 84 (listed, comp. note) China, Zhejiang
Cryptotympana atrata Yang and Jiang 1995: 177 (comp. note)
Cryptotympana atrata Jiang, Yang and Liu 1996: 314 (comp. note)
Cryptotympana atrata Boulard 1997c: 105–106, 112–113, 117, Plate II, Fig 3 (type species of

Cryptotympanini) Equals *Tettigonia atrata* Equals *Cicada pustulata* Equals *Tettigonia pustulata*

Tettigonia pustulata Boulard 1997c: 105 (comp. note)

Tettigonia atrata Boulard 1997c: 106 (comp. note)

Cryptotympana atrata Chou, Lei, Li, Lu and Yao 1997: 7, 12, 28, 276–277, 309, 321–322, 325, 327, 332, 345–347, Fig 2–1, Fig 3–2, Plate XV, Fig 147, Tables 10–12 (key, described, synonymy, illustrated, anatomy, listed, song, oscillogram, spectrogram, comp. note) Equals *Tettigonia atrata* Equals *Cicada atrata* Equals *Tettigonia pustulata* Equals *Cicada nigra* Equals *Fidicina bubo* Equals *Fidicina atrata* Equals *Cryptotympana sinensis* Equals *Cryptotympana dubia* Equals *Cryptotympana coreana* Equals *Cryptotympana santoshonis* Equals *Cryptotympana wenchewensis* Equals *Cryptotympana pustulata castanea* Equals *Cryptotympnana pustulata fudiensis* China

Tettigonia pustulata Chou, Lei, Li, Lu and Yao 1997: 276 (type species of *Cryptotympana*)

Cryptotympana pustulata Huang, Kang, Liu, Oh, Nam and Kim 1997: 111 (medicinal use)

Cryptotympana atrata Lei, Jiang, Li and Chou 1997: 349 (comp. note) China

Cryptotympana atrata Yang, Chen, Xu and Jiang 1998: 576 (auditory responses, comp. note) China

Cryptotympana atrata Easton and Pun 1999: 101 (distribution, comp. note) Macao, Hong Kong, China, Guangdong, Hebei, Zhejiang, Malaysia, Japan, Taiwan

Cryptotympana atrata Feng, Chen, Ye, Wang, Chen, and Wang 1999: 515–518, Tables 1–7 (nutrient content, comp. note) Yunnan

Cryptotympana atrata Lee 1999: 1, 5–6 (distribution, song, listed, comp. note) Equals *Cryptotympana dubia* n. syn. Equals *Tettigonia atrata* Equals *Cryptotympana coreana* Equals *Cryptotympana coreanus* Equals *Cryptotympana dubia* f. *coreana* Equals *Cryptotympana dubia* var. *coreana* Equals *Cryptotympana dubia coreana* Korea, China, Taiwan, Indo-China

Tettigonia pustulata Lee 1999: 5 (type species of *Cryptotympana*) China

Cryptotympana dubia Lee 1999: 1 (status, comp. note) Korea

Cryptotympana atrata Shin, Park and Kim 1999: 319–324, Figs 1–4, Tables 1–2 (medicinal use, comp. note) Korea

Cryptotympana atrata Stewart, Malik and Redmond 1999: 7, 15, Table 2 (Longan fruit pest, listed, comp. note)

Cryptotympana atrata Boulard 2000i: 21 (listed)

Cryptotympana atrata Chen 2000: 551, 553 (control)

Cryptotympana atrata Hua 2000: 61 (listed, distribution, hosts) Equals *Tettigonia atrata* Equals *Cryptotympana santoshonis* Equals *Cryptotympana pustulata* Equals *Cryptotympana sinensis* Equals *Cryptotympana pustulata* var. *castanea* Equals *Cryptotympana pustulala* (sic) var. *fukienensis* China, Liaoning, Inner Mongolia, Shaanxi, Shandong, Shatosho, Anhui, Hebei, Henan, Jiangsu, Jiangxi, Zhejiang, Fujian, Taiwan, Guangdong, Hainan, Hunan, Guizhou, Hubei, Guangxi, Sichuan, Laos, Korea, Vietnam, Malaysia, Philippines

Cryptotympana dubia Hua 2000: 61 (listed, distribution) China, Shandong

Cryptotympana atrata Boulard 2001b: 25–26, 32–33, 37, Plate II, Fig 3 (type species of Cryptotympanini) Equals *Tettigonia atrata* Equals *Cicada pustulata* Equals *Tettigonia pustulata*

Tettigonia pustulata Boulard 2001b: 25–26 (comp. note)

Tettigonia atrata Boulard 2001b: 26 (comp. note)

Cryptotympana atrata Lee 2001a: 51 (comp. note) Korea

Cryptotympana atrata Lee 2001b: 2 (comp. note) Korea

Cryptotympana atrata Sueur 2001: 36, Table 1 (listed)

Cryptotympana pustulata Yaylayan, Paré, Manti and Bèlanger 2001: 188, 194 (skin extracts, comp. note)

Cryptotympana atrula (sic) Ahn, Hahn, Ryu and Cho 2002: 67, Table 1 (medicinal use, comp. note) China

Cryptotympana pustulata Yaginuma 2002: 2 (comp. note)

Cryptotympana atrata Fu, Ma, Wu, Wei, Yan, and Li 2003: 105, Table 1 (medicinal use) China

Cryptotympana atrata Lee 2003b: 6, 8 (comp. note)

Cryptotympana atrata Lee 2003c: 15 (comp. note) Korea

Tettigonia pustulata Lee and Hayashi 2003a: 162 (type species of *Cryptotympana*) China

Cryptotympana atrata Lee and Hayashi 2003a: 163–164, Fig 10 (key, diagnosis, illustrated, synonymy, song, listed, comp. note) Equals *Tettigonia atrata* Equals *Tettigonia pustulata* Equals *Cryptotympana pustulata* Taiwan, Japan, Korea, China, Indo-China

Cryptotympana atrata Chen 2004: 22, 66–67, 198 (described, illustrated, listed, comp. note) Taiwan

Cryptotympana atrat (sic) Chen 2004: 66 (wing) Taiwan

Cryptotympana pustulata Na, Shin, Kim, Kwon, Park, et al. 2004: 292 (medicinal use) Korea

Cryptotympana pustulata Zhao 2004: 477 (medicinal use, comp. note) China, Zhejiang, Shandong, Jiangsu, Hebei

Cryptotympana atrata Lee 2005: 16, 74–78, 156, 163–164, 172–173 (described, illustrated, synonymy, song, distribution, key, comp. note) Equals *Tettigonia atrata* Equals *Cryptotympana dubia* Equals *Cryptotympana coreanus* Equals *Cryptotympana coreana* Equals *Cryptotympana dubia* f. *coreana* Equals *Cryptotympana dubia* var. *coreana* Equals *Cryptotympana dubia coreana* Korea, China, Taiwan, Japan, Indo-China

Cryptotympana coreanus Lee 2005: 62 (comp. note) Korea

Tettigonia pustulata Lee 2005: 170 (type species of *Cryptotympana*) China

Tettigonia atrata Moulds 2005b: 390 (type species of *Cryptotympana*) Equals *Cryptotympana pustulata*

Cryptotympana atrata Moulds 2005b: 391–392 (type species of *Cryptotympana*, listed, comp. note)

Tettigoina atrata Moulds 2005b: 431 (type species of *Cryptotympana*) Equals *Cryptotympana pustulata*

Cryptotympana atrata Pham 2005b: 232–233 (key, synonymy, distribution, comp. note) Equals *Tettigonia atrata* Equals *Cicada atrata* Equals *Tettigonia pustulata* Equals *Cicada nigra* Equals *Fidicin bubo* Equals *Fidicina atrata* Equals *Cryptotympana sinensis* Equals *Cryptotympana dibia* (sic) Equals *Cryptotympana coreana* Equals *Cryptotympana santoshonis* Equals *Cryptotympana wenchewensis* Equals *Cryptotympana pustulata castanea* Equals *Cryptotympana pustulata fukiensis* Vietnam

Tettigonia pustulata Pham 2005a: 232 (type species of *Cryptotympana*)

Tetttigonia pustulata Pham and Thinh 2005a: 240 (type species of *Cryptotympana*)

Cryptotympana atrata Pham and Thinh 2005a: 240 (distribution, listed) Equals *Tettigonia atrata* Equals *Cicada atrata* Equals *Tettigonia pustulata* Equals *Cicada nigra* Equals *Fidicina bubo* Equals *Fidicina atrata* Equals *Cryptotympana sinensis* Equals *Cryptotympana dibia* (sic) Equals *Cryptotympana coreana* Equals *Cryptotympana santoshonis* Equals *Cryptotympana wenchewensis* Equals *Cryptotympana pustulata castanea* Equals *Cryptotympana pustulata fukiensis* Vietnam

Cryptotympana atrata Saisho 2005: 43 (emergence, comp. note) Japan

Cryptotympana atrata Smolinske 2005: 192 (comp. note)

Cryptotympana pustulata Fattorusso, Frutos, Sun, Sucher, and Pellecchia 2006: 18, Table 1 (medicinal use, listed) China

Cryptotympana atrata Tian, Yuan and Zhang 2006: 243–244, Fig 2a (illustrated, reproductive system, spermatogenesis, comp. note) China

Cryptotympana pustulata Tse, Che, Liu and Lin 2006: 135, Table 1 (medicinal use, listed) China

Cryptotympana pustulata Xu, Lee, Han, Oh, Park, Tian et al. 2006: 7826 (medicinal use, comp. note) Korea

Cryptotympana atrata Zhang, Zhang, Xie, Liu and Shao 2006: 1440, Fig 1 (illustrated, wing microstructure, comp. note) China

Cryptotympana dubia Park, Leem, Suh, Hur, Oh and Park 2007: 205, 211 (medicinal use, comp. note) Korea

Cryptotympana atrata Shiyake 2007: 4, 40, 42, Fig 60 (illustrated, distribution, listed, comp. note) China, Korea, Southeast Asia

Cryptotympana atrata-group Shiyake 2007: 42 (comp. note)

Cryptotympana pustulata Zhang and Xuan 2007: 404 (fungal host, comp. note) China

Cryptotympana atrata Boulard 2008f: 12 (type species of *Cryptotympana*) Thailand

Cryptotympana atrata Júnior, Almeida, Lima, Nunes, Siqueira et al. 2008: 805, Table 1 (listed, comp. note) Taiwan

Tettigonia pustulata Lee 2008a: 5 (type species of *Cryptotympana*) China, Taiwan

Cryptotympana atrata Lee 2008a: 5 (synonymy, distribution, listed) Equals *Tettigonia atrata* Equals *Cryptotympana pustulata* Vietnam, Japan, Korea, China, Taiwan, Indo-China

Cryptotympana atrata Lee 2008b: 446–448, 463, Fig 1D, Table 1 (key, illustrated, synonymy, distribution, listed, comp. note) Equals *Tettigonia atrata* Equals *Cryptotympana dubia* Equals *Cryptotympana coreanus* Equals *Cryptotympana coreana* Equals *Cryptotympana dubia* f. *coreana* Equals *Cryptotympana dubia* var. *coreana* Equals *Cryptotympana dubia coreana* Korea, China, Japan, Taiwan, Indo-China

Tettigonia pustulata Lee 2008b: 448 (type species of *Cryptotympana*) China
Cryptotympana dubia Leem, Park, Suh, Hur, Oh and Park 2008: 106 (erratum for Park et al. 2007)
Cryptotympana atrata Osaka Museum of Natural History 2008: 322 (listed, acoustic behavior, ecology, comp. note) Japan
Cryptotympana dubia Ajeesh and Sreejith 2009: 1001 (medicinal uses) Korea
Cryptotympana pustulata Karalliedde and Kappagoda 2009: H1967, Table 1 (medicinal use) China
Tettigonia atrata Lee 2009h: 2620 (type species of *Cryptotympana*, listed) China
Tettigonia atrata Lee 2009i: 294 (type species of *Cryptotympana*, listed)
Cryptotympana pustulata Li, Wang, Schofield, Lin, Huang and Wang 2009: 115 (medicinal use)
Cryptotympana atrata Pham and Yang 2009: 13, 16, Table 2 (listed, comp. note) Vietnam
Cicada atrata Sanborn, Heath and Heath 2009: 25 (comp. note)
Cryptotympana dubia Ye and Ng 2009: 189 (protein, comp. note) Korea
Tettigonia atrata Lee 2010a: 19 (type species of *Cryptotympana*, listed)
Tettigonia atrata Lee 2010c: 2 (type species of *Cryptotympana*, listed) China
Cryptotympana atrata Lee and Hill 2010: 278, 298–300, Figs 7–8 (type species of Cryptotympanini, comp. note)
Gryptotympana (sic) *dubia* Oh, Kim and Kim 2010: 134, Table 1 (predation) Korea
Cryptotympana atrata Peng, Chien and Hsaio 2010: 145–146, 150–151, Figs 1A–1C (illustrated, fungal host, comp. note) Taiwan
Cryptotympana atrata Pham and Yang 2010a: 133 (comp. note) Vietnam
Cryptotympana atrata Zheng, Zhang and Wang 2010: 752, Table 3 (mercury content, comp. note) China

C. auropilosa **Hayashi, 1987**
Cryptotympana auropilosa Hayashi 1987b: 34, 39, 45, 61–62, 100, Fig 83, Fig 93, Fig 106 (n. sp., described, illustrated, distribution, key, comp. note) Burma
Cryptotympana auropilosa Hayashi 1993: 18 (comp. note)

C. brevicorpus **Hayashi, 1987** = *Cryptotympana epithesia* Moulton, 1923 (partim)
Cryptotympana brevicorpus Hayashi 1987a: 130, 136, 138–139, 146, Fig 7, Figs 27–30 (n. sp., described, illustrated, synonymy distribution, comp. note) Equals *Cryptotympana epithesia* (partim) (nec Distant) Nias Island, Sumatra
Cryptotympana brevicorpus Hayashi 1987b: 102 (key)

C. brunnea **Hayashi, 1987**
Cryptotympana brunnea Hayashi 1987a: 174–175, 180–182, Fig 108, Fig 112, Figs 122–126 (n. sp., described, illustrated, distribution, comp. note) Simeulue
Cryptotympana brunnea Hayashi 1987b: 99 (key)
Cryptotympana brunnea Hayashi 1993: 17 (comp. note)

C. consanguinea **Distant, 1916**
Cryptotympana consanguinea Hayashi 1987b: 2–3, 6, 14–15, 18–22, 102, Fig 6, Figs 16B–C, Figs 40–42, Figs 47–51, Fig 57 (described, illustrated, distribution, key, comp. note) Philippines, Mindanao, Basilan, Dinagat
Cryptotympana consanguinea Duffels 1990b: 70 (comp. note) Philippines
Cryptotympana consanguinea Hayashi 1993: 18 (comp. note)
Cryptotympana consanguinea Lee 2010a: 19 (distribution, listed) Philippines, Mindanao, Dinagat, Basilan

C. coreana Kato, 1925 to *Cryptotympana atrata* (Fabricius, 1775)

C. corvus **(Walker, 1850)** = *Fidicina corvus* = *Cryptotympana corva* = *Fidicina invarians* Walker, 1858 = *Cryptotympana invarians*
Cryptotympana corvus group Hayashi 1987a: 126, 128–129, Fig 3 (described, distribution, diversity, listed, comp. note) China, Indo-China, Himalayas, India
Cryptotympana corvus group Hayashi 1987b: 65, 69 (described, listed, comp. note)
Cryptotympana corvus subgroup Hayashi 1987b: 65, 69–70 (described, listed, comp. note)
Cryptotympana corvus Hayashi 1987b: 66, 69–73, 87, 100, Fig 130, Fig 141, Figs 146–147, Figs 150–154 (described, illustrated, synonymy, distribution, key, comp. note) Equals *Fidicina corvus* Equals *Cryptotympana corva* (sic) Equals *Fidicina invarians* Bangladesh, India, Bhutan, Nepal
Cryptotympana corvus group Hayashi 1993: 16, 18, Table 1 (listed, comp. note)
Cryptotympana corvus Hayashi 1993: 18 (comp. note)
Cryptotympana corvus Pham 2005b: 234 (comp. note)
Cryptotympana corvus-group Shiyake 2007: 43 (comp. note)

C. demissitia **Distant, 1891** = *Cryptotympana demissita* (sic)
Cryptotympana demissitia Hayashi 1987a: 130, 136–139, 143, Fig 6, Fig 22, Fig 39 (described, illustrated, distribution, comp. note) Sumatra, Batu Islands
Cryptotympana demissitia Hayashi 1987b: 102 (key)

***C. diomedea* (Walker, 1858)** = *Fidicina diomedea* = *Cryptotympana dimeda* (sic)
Cryptotympana diomedea group Hayashi 1987a: 125–126, 129, Fig 3 (described, distribution, diversity, listed, comp. note) Sumatra, Malay Peninsula, Borneo
Cryptotympana diomedea Hayashi 1987a: 129–133, 135, 142–143, Fig 4, Figs 9–12, Figs 15–18, Fig 38A, Fig 39 (described, illustrated, synonymy, distribution, comp. note) Equals *Fidicina diomedea* Equals *Cryptotympana diomeda* (sic) Sumatra
Cryptotympana diomedea group Hayashi 1987b: 101 (key)
Cryptotympana diomedea Hayashi 1987b: 101 (key)
Cryptotympana diomedea group Hayashi 1993: 16–17, Table 1 (listed, comp. note)
Cryptotympana diomedea Hayashi 1993: 17 (comp. note)
Cryptotympana diomedea group Holloway 1998: 305, Fig 10 (distribution, phylogeny, comp. note) Banda Islands

***C. distanti* Hayashi, 1987** = *Cryptotympana suluensis* Distant, 1906 (partim)
Cryptotympana distanti Hayashi 1987b: 3, 6, 25–30, 32–33, 103, Fig 8, Figs 16B–D, Fig 58A, Fig 59A, Figs 60–61, Figs 65–71 (n. sp., described, illustrated, distribution, key, comp. note) Sulawesi
Cryptotympana distanti Duffels 1990b: 64, 70, Table 2 (listed, comp. note) Sulawesi
Cryptotympana distanti Hayashi 1993: 18 (comp. note)
Cryptotympana distanti Boer and Duffels 1996b: 330 (distribution) Sulawesi
Cryptotympana distanti Sanborn 1999a: 42 (type material, listed)

***C. dohertyi* Hayashi, 1987**
Cryptotympana dohertyi Hayashi 1987b: 92–93, 95, 98, Fig 193, Fig 197 (n. sp., described, illustrated, distribution, key, comp. note) Enggano, Indonesia
Cryptotympana dohertyi Hayashi 1993: 18 (comp. note)

C. dubia Haupt, 1917 to *Cryptotympana atrata* (Fabricius, 1775)

***C. edwardsi* Kirkaldy, 1902**
Cryptotympana edwardsi Hayashi 1987a: 192–193, 208, 210–212, Fig 155, Figs 204–205, Figs 210–213 (described, illustrated, distribution, comp. note) India
Cryptotympana edwardsi Hayashi 1987b: 98 (key)

***C. epithesia* Distant, 1888**
Cryptotympana epithesia Hayashi 1987a: 130, 135, 139–146, Fig 8, Figs 31–37, Figs 38B–D, Fig 39 (described, illustrated, distribution, comp. note) Borneo, Sumatra, Siberut
Cryptotympana epithesia Hayashi 1987b: 101 (key)
Cryptotympana epithesia Hayashi 1993: 17 (comp. note)

***C. exalbida* Distant, 1891**
Cryptotympana exalbida group Hayashi 1987a: 126, 129, Fig 3 (described, distribution, diversity, listed, comp. note) India, Sri Lanka
Cryptotympana exalbida group Hayashi 1987b: 92 (listed)
Cryptotympana exalbida Hayashi 1987b: 92–95, 101, Fig 192, Figs 194–196, Figs 198–202 (lectotype designation, described, illustrated, distribution, key, comp. note) India, Sri Lanka
Cryptotympana exalbida group Hayashi 1993: 16, 18, Table 1 (listed, comp. note)
Cryptotympana exalbida Hayashi 1993: 18 (comp. note)
C[ryptotympana] exalbida Jiang, Yang, Tang, Xu and Chen 1995b: 229 (song frequency, comp. note)
Cryptotympana exalbida Sueur 2001: 36, Table 1 (listed)

***C. facialis bimaculata* Kato, 1936** = *Cryptotympana facialis* var. *bimaculata*
Cryptotympana facialis var. *bimaculata* Nakao 1987: 19 (comp. note) Japan

***C. facialis facialis* (Walker, 1858)** = *Cicada facialis* = *Cicada fascialis* (sic) = *Cryptotympaua* (sic) *facialis* = *Cryptotympana facialis* = *Cryptotympana fascialis* (sic) = *Cryptotympana pustulata* Uhler, 1896 (nec Fabricius) = *Cryptotympana pnstulata* (sic) (nec Fabricius) = *Cryptotympana intermedia* Matsumura, 1907 (nec Signoret) = *Cryptotympana facialis formosana* Kato, 1925 = *Cryptotympana fascialis* (sic) *formosana* = *Cryptotympana japonensis* Kato, 1925 = *Oryptotympana* (sic) *japonensis* = *Criptotympana* (sic) *japonensis* = *Cryptotympana intermedia* (nec Signoret) = *Cryptotympana iutermedia* (sic) (nec Signoret) = *Cryptotympana japonensis riukiuensis* Kato, 1925 = *Cryptotympana riukiuensis* = *Cryptotympana okinawana* Matsumura, 1927 = *Cryptotympana facialis* var. *okinawana* = *Cryptotympana facialis okinawana* = *Cryptotympana facialis* Matsumura, 1929 (nec Walker) (partim) = *Cryptotympana facialis yonakunina* Ishihara, 1968
Cryptotympana facialis Chen 1933: 359 (listed, comp. note)
Cryptotympana japonensis Chen 1933: 359 (listed, comp. note)
Cryptotympana japonensis Isaki 1933: 381 (listed, comp. note) Japan
Cryptotympana facialis Wu 1935: 7 (listed) Equals *Cicada facialis* Equals *Fidicina nigrofasciata*

Equals *Cryptotympana facialis* var. *formosana* China, Tsinan, Japan, Formosa, Siam

Cryptotympana japonensis Wu 1935: 8 (listed) Equals *Cryptotympana pustulata* Equals *Cryptotympana intermedia* Equals *Cryptotympana japonensis* var. *riukiuensis* Equals *Cryptotympaa facialis* (partim) China, Santosho, Japan, Liu-kiu Island

Cryptotympana japonensis Ohgushi 1954: 11 (emergence times, ecology, comp. note) Japan

Cryptotympana japonensis Ikeda 1959: 484 (comp. note) Japan

Cryptotympana facialis Hayashi 1982b: 187–189, Table 1 (distribution, comp. note) Japan

Cryptotympana facialis Hayashi 1984: 29, 43–45, Figs 55–60, Figs 63–68 (key, described, illustrated, synonymy, comp. note) Equals *Cicada facialis* Equals *Cryptotympana pustulata* (nec Fabricius) Equals *Cryptotympana intermedia* (nec Signoret) Equals *Cryptotympana japonensis* Japan

Cryptotympana facialis facialis Hayashi 1984: 43–45, Figs 55–56, Figs 63–64 (described, illustrated, comp. note) Japan

Cryptotympana facialis okinawana Hayashi 1984: 43–45, Figs 57–58, Figs 65–66 (described, illustrated, comp. note) Japan

Cryptotympana facialis yonakunina Hayashi 1984: 43–45, Figs 59–60, Figs 67–68 (described, illustrated, comp. note) Japan

Cryptotympana facialis group Hayashi 1987a: 126, 128, Fig 3 (described, distribution, diversity, listed, comp. note) Japan, China, Indo-China

Cryptotympana facialis Hayashi 1987b: 34, 39, 43–56, 58–61, 100, Fig 81, Fig 91, Figs 102–103, Figs 109–116 (lectotype designation, described, illustrated, synonymy, distribution, key, comp. note) Equals *Cicada facialis* Equals *Cryptotympana fascialis* (sic) Equals *Cryptotympana facialis* var. *formosana* Equals *Cryptotympana pustulata* (nec Fabricius) Equals *Cryptotympana intermedia* (nec Signoret) Equals *Cryptotympana japonenis* Equals *Cryptotympana japonensis* var. *riukiuensis* Equals *Cryptotympana okinawana* n. syn. Equals *Cryptotympana facialis* var. *okinawana* Equals *Cryptotympana facialis okinawana* Equals *Cryptotympana facialis yonakunina* n. syn. Japan, Tsushima Island, Gotô Islands, Koshiki Islands, Ôsumi Islands, Tokara Islands, Ryukyus, Taiwan, China

Cryptotympana facialis group Hayashi 1987b: 43, 45, 62 (comp. note)

Cryptotympana japonensis Hayashi 1987b: 57 (comp. note)

Cryptotympana facialis Nakao 1987: 19 (acoustic activity, comp. note) Japan

Cryptotympana facialis facialis Nakao 1987: 19 (acoustic activity, comp. note) Japan

Cryptotympana facialis okinawana Nakao 1987: 19 (acoustic activity, comp. note) Japan

Cryptotympana facialis var. *okinawana* Nakao 1987: 19 (acoustic activity, comp. note) Japan

Cryptotympana facialis yonakunina Nakao 1987: 19 (acoustic activity, comp. note) Japan

Cryptotympana japonensis Nakao 1987: 19 (comp. note) Japan

Criptotympana (sic) *japonensis* Seki, Fujishita, Ito, Matsuoka and Tsukida 1987: 100, Fig 6A (retinoid composition, comp. note) Japan

Cryptotympana japonensis Seki, Fujishita, Ito, Matsuoka and Tsukida 1987: 101–102, Table 2 (retinoid composition, listed, comp. note) Japan

Cryptotympana japonensis Hara, Hayashi, Hirano, Zhong, Yasuda and Komae 1990: 1251, Table 1 (comp. note)

Cryptotympana fascialis (sic) Liu 1992: 33, 35, 37 (comp. note)

Cryptotympana japonensis Liu 1992: 34–35 (comp. note)

Cryptotympana okinawana Liu 1992: 34–35 (comp. note)

Cryptotympana facialis Tzean, Hsieh, Chen and Wu 1992: 782–784, Fig 2–3 (fungal host, comp. note) Taiwan

Cryptotympana facialis Hayashi 1993: 16, 18 (comp. note)

Cryptotympana facialis group Hayashi 1993: 16, 18, Table 1 (listed, comp. note)

Cryptotympana facialis Tzean, Hsieh, Chen and Wu 1993: 517 (comp. note) Taiwan

Cryptotympana japonensis Anonymous 1994: 809 (fruit pest) China

Cryptotympana japonensis Seki, Isono, Ito, and Katsuta 1994: 692, 694, Table 1 (retinoid composition, comp. note) Japan

Cryptotympana facialis Lee 1995: 30 (comp. note)

Cryptotympana facialis Chou, Lei, Li, Lu and Yao 1997: 281 (described, synonymy, comp. note) Equals *Cicada facialis* China

Cryptotympana japonensis Kartal and Zeybekoglu 1999a: 62 (comp. note)

Cryptotympana facialis Hua 2000: 61 (listed, distribution) Equals *Cicada facialis* China, Shandong, Taiwan, Japan, Thailand

Cryptotympana facialis Kuboto and Tanase 1999: 64 (comp. note) Japan
Cryptotympana japonensis Hua 2000: 61 (listed, distribution, hosts) China, Liaoning, Hebei, Henan, Shandong, Santosho, Anhui, Jianxi, Zhejiang, Guangxi, Japan
Cryptotympana japonensis Lee, Nishimori, Hatakeyama and Oishi 2000: 2–3 (comp. note) Japan
Cryptotympana facialis Ohara and Hayashi 2001: 1 (comp. note) Japan
Cryptotympana facialis Saisho 2001a: 68 (comp. note) Japan
Cryptotympana facialis Sueur 2001: 46, Table 2 (listed)
Cryptotympana facialis Kuboto and Tanase 2002: 114 (comp. note) Japan
Cryptotympana facialis Sueur 2002b: 121, 129 (listed, comp. note)
Cryptotympana facialis Kubo-Irie, Irie, Nakazawa and Mohri 2003: 988 (comp. note) Japan
Cryptotympana facialis Morikawa, Matsushita, Nishina and Kono 2003: 242–244, Fig 1, Table 2 (comp. note) Japan
Cryptotympana fasialis (sic) Morikawa, Matsushita, Nishina and Kono 2003: 246 (comp. note) Japan
Cryptotympana facialis Lee and Hayashi 2004: 71 (comp. note) Equals *Cicada facialis* Japan
Cryptotympana facialis Boucias, Pendland and Kalkar 2004: 1559 (fungus host) Taiwan
Cryptotympana facialis Chen 2004: 66, 193 (comp. note)
Cryptotympana facialis (= japonensis) Nuorteva, Lodenius, Nagasawa, Tulisalo and Nuorteva 2004: 56–57, 62, Tables 1–2, Table 9 (cadmium, metal levels) Japan
Cryptotympana facialis Moulds 2005b: 396, 398, 417–420, 422, Figs 56–60, Table 1 (higher taxonomy, phylogeny, key, listed, comp. note)
Cryptotympana japonica (sic) Pham 2005b: 232 (comp. note)
Cryptotympana japonensis (sic) Pham 2005b: 232 (comp. note)
Cryptotympana facialis Moriyama and Numata 2006: 1219–1225, Figs 1–5, Table 1 (egg hatching, comp. note) Japan
Cryptotympana facialis Saisho 2006b: 59 (comp. note) Japan
Cryptotympana facialis Watanabe and Kobayashi 2006: 224, Table 1 (listed) Japan, Ryukyus
Cryptotympana facialis Cyranoski 2007: 977 (life cycle, comp. note) Japan
Cryptotympana facialis Holden 2007: 1301 (comp. note) Japan
Cryptotympana facialis Sanborn, Phillips and Sites 2007: 8–9 (listed, synonymy, distribution, comp. note) Equals *Cicada facialis* Equals *Cryptotympana fascialis* (sic) Thailand, Japan, China, Taiwan, Indochina
Cryptotympana facialis Shiyake 2007: 4, 40, 42, Fig 61 (illustrated, distribution, listed, comp. note) Japan
Cryptotympana facialis-group Shiyake 2007: 42–43 (comp. note)
Cryptotympana facialis Takakura and Yamazaki 2007: 730–734, Figs 1–3 (predation, comp. note) Japan
Cryptotympana facialis Yamazaki 2007: 347–348, Fig 1a (illustrated, feeding behavior, comp. note) Japan
Cryptotympana facialis Boucias, Pendland and Kalkar 2008: 2613 (fungus host) Taiwan
Cryptotympana facialis Boulard 2008f: 12–13, 91 (biogeography, listed, comp. note) Equals *Cicada facialis* Equals *Cryptotympana fascialis* (sic) Equals *Cryptotympana facialis yonakunuina* Thailand, Siam, China, Japan, Kyukyus Islands, Taiwan
Cryptotympana facialis Moriyama and Numata 2008: 1487–1493, Figs 1–7, Table 1 (diapause, comp. note) Japan
Cryptotympana facialis Osaka Museum of Natural History 2008: 322 (listed, acoustic behavior, ecology, comp. note) Japan
Cryptotympana japonensis Hisamitsu, Sokabe, Terada, Osumi, Terada, et al. 2009: 93 (behavior, comp. note) Japan
Cryptotympana facialis Moriyama and Numata 2009: 163–168, Figs 1–2, Tables 1–3 (cold tolerance, comp. note) Japan
Cryptotympana facialis Yamazaki 2009: 86 (feeding behavior, comp. note) Japan
Cryptotympana facialis Yoshida 2009: 949 (comp. note) Japan
Cryptotympana facialis Moriyama and Numata 2010: 69–72, Fig 1, Table 1 (diapause, comp. note) Japan
Cryptotympana facialis Peng, Chien and Hsaio 2010: 146 (comp. note)

C. facialis yonakunina Ishihara, 1968 to *Cryptotympana facialis facialis* (Walker, 1858)

***C. fumipennis* (Walker, 1858)** = *Fidicina fumipennis* = *Cryptotympana viridipennis* Distant, 1911 = *Cryptotympana viridipennis infuscata* Moulton & China, 1926 = *Cryptotympana viridipennis* var. *infuscata*
Cryptotympana fumipennis Hayashi 1987a: 144, 147–153, 177, Fig 41, Figs 51–53, Fig 55, Figs 61–64 (lectotype designation, described, illustrated, synonymy, distribution, comp. note) Equals *Fidicina fumipennis*

Equals *Cryptotympana viridipennis* n. syn. Equals *Cryptotympana viridipennis* var. *infuscata* Thailand, Sumatra, Malay Peninsula

Cryptotympana viridipennis Hayashi 1987a: 153 (comp. note)

Cryptotympana fumipennis Hayashi 1987b: 97 (key)

Cryptotympana fumipennis Hayashi 1993: 17 (comp. note)

Cryptotympana fumipennis Sanborn, Phillips and Sites 2007: 9 (listed, synonymy, distribution, comp. note) Equals *Fidicina fumipennis* Thailand, Malay Peninsula, Sumatra

Cryptotympana fumipennis Boulard 2008f: 13, 91 (biogeography, listed, comp. note) Equals *Fidicina fumipennis* Thailand, Siam

C. fusca Kato, 1925 to *Cryptotympana holsti holsti* Distant, 1904

***C. gracilis* Hayashi, 1987** = *Cryptotympana vesta* (nec Distant)

Cryptotympana gracilis Hayashi 1987a: 192–193, 201, 205–207, Fig 153, Figs 180–181, Figs 196–202 (n. sp., described, illustrated, distribution, comp. note) Thailand

Cryptotympana gracilis Hayashi 1987b: 101 (key)

Cryptotympana gracilis Eve 1991: 195, 221 (acoustic behavior, song frequency, comp. note) Thailand

Cryptotympana gracilis Hayashi 1993: 17 (comp. note)

Cryptotympana gracilis Boulard 2003c: 178, 185 (comp. note)

Cryptotympana gracilis Boulard 2005h: 6–9, 52–53, Figs 1–3, Color Plate, Fig A (described, illustrated, song, sonogram, listed, comp. note) Thailand

Cryptotympana gracilis Boulard 2005o: 150 (listed)

Cryptotympana gracilis Boulard 2007d: 494 (comp. note) Thailand

Cryptotympana gracilis Boulard 2007e: 47 (comp. note) Thailand

Cryptotympana gracilis Sanborn, Phillips and Sites 2007: 9, 34 (listed, synonymy, distribution, comp. note) Equals *Cryptotympana vesta* (nec Distant) Thailand

Cryptotympana gracilis Boulard 2008f: 13–14, 91 (biogeography, listed, comp. note) Thailand

C. heuertzi Lallemand & Synave, 1953 to *Cryptotympana timorica* (Walker, 1870)

***C. holsti holsti* Distant, 1904** = *Cryptotympana mandarina* Schumacher, 1915 (nec Distant) = *Cryptotympana vitalasi* Distant, 1917 = *Cryptotympana fusca* Kato, 1925 = *Cryptotympana fusa* (sic) = *Cryptotympana capillata* Kato, 1925 = *Cryptotympana holsti capillata* = *Cryptotympana holsti inornata* Matsumura, 1927 = *Cryptotympana holsti inorrata* (sic) = *Cryptotympana kagiana* Matsumura, 1927 = *Cryptotympana holsti kagiana*

Cryptotympana holsti Wu 1935: 8 (listed) China, Formosa

Cryptotympana holsti Hayashi 1987a: 122, Fig 2A (illustrated, comp. note)

Cryptotympana holsti Hayashi 1987b: 69, 79–87, 91, 97, Fig 143, Figs 161–162, Figs 167–170, Figs 173–178, Fig 179A, Fig 180 (described, illustrated, synonymy, distribution, key, comp. note) Equals *Cryptotympana mandarina* (nec Distant) Equals *Cryptotympana vitalisi* n. syn. Equals *Cryptotympana fusca* n. syn. Equals *Cryptotympana fusa* (sic) Equals *Cryptotympana capillata* Equals *Cryptotympana holsti* var. *capillata* Equals *Cryptotympana holsti* var. *inornata* Equals *Cryptotympana holsti* var. *inorrata* (sic) Equals *Cryptotympana kagiana* Equals *Cryptotympana holsti* var. *kagiana* Taiwan, China, Indo-China, Vietnam, Cambodia, laos

Cryptotympana holsti Peng and Lei 1992: 98–99, Fig 280 (listed, illustrated, comp. note) China, Hunan

Cryptotympana holsti Hayashi 1993: 18 (comp. note)

Cryptotympana holsti Lei, Chou and Li 1994: 55, Fig 7 (song frequency spectrum, comp. note)

Cryptotympana holsti Chou, Lei, Li, Lu and Yao 1997: 276–279, 309, 322–323, 325, 327, 332, 346–347, Fig 9–93, Plate XVI, Fig 148, Plate XVI, Fig 150, Tables 10–12 (key, described, illustrated, synonymy, song, oscillogram, spectrogram, comp. note) Equals *Cryptotympana mandarina* Equals *Cryptotympana vitalisi* Equals *Cryptotympana fusca* Equals *Cryptotympana capillata* Equals *Cryptotympana holsti* var. *inornata* Equals *Cryptotympana kagiana* China

Cryptotympana holsti Hua 2000: 61 (listed, distribution) Equals *Cryptotympana holsti* var. *inornata* Equals *Cryptotympana kagiana* China, Fujian, Taiwan, Guangdong, Hunan, Guangxi, Guizhou, Sichuan, Yunnan, India

Cryptotympana holsti Sueur 2001: 36, Table 1 (listed)

Cryptotympana holsti Lee 2003b: 7 (comp. note)

Cryptotympana holsti Lee and Hayashi 2003a: 163–167, Fig 12 (key, diagnosis, illustrated, synonymy, song, listed, comp. note) Equals *Cryptotympana mandarina* (nec Distant) Equals *Cryptotympana vitalisi* Equals *Cryptotympana fusca* Equals *Cryptotympana*

holsti var. *takahashii* Equals *Cryptotympana capillata* Equals *Cryptotympana holsti* ab. *inornata* Equals *Cryptotympana holsti ab. inorrata* (sic) Equals *Cryptotympana holsti* var. *inornata* Equals *Cryptotympana kagiana* Equals *Cryptotympana kaniana* (sic) Equals *Cryptotympana holsti* var. *capillata* Equals *Cryptotympana holsti* var. *kagiana* Equals *Cryptotympana holsti* ab. *kagiana* Taiwan, China, Indo-China

Cryptotympana holsti Chen 2004: 66, 72, 192, 198 (described, illustrated, listed, comp. note) Taiwan

Cryptotympana fusca Chen 2004: 192 (comp. note)

Cryptotympana holsti Lee 2005: 35, 63 (comp. note)

Cryptotympana holsti Pham 2005b: 232–234, Fig 1a, Fig 2a (key, synonymy, illustrated, distribution, comp. note) Equals *Cryptotympana vitalisi* Equals *Cryptotympana fusca* Equals *Cryptotympana holsti* var. *takahashii* Equals *Cryptotympana capillata* Equals *Cryptotympana holsti* ab. *inornata* Equals *Cryptotympana kagiana* Equals *Cryptotympana holsti* var. *capillata* Equals *Cryptotympana holsti* var. *kagiana* Vietnam

Cryptotympana holsti Pham and Thinh 2005a: 241 (distribution, listed) Equals *Cryptotympana vitalisi* Equals *Cryptotympana fusca* Equals *Cryptotympana holsti* var. *takahashii* Equals *Cryptotympana capillata* Equals *Cryptotympana holsti* ab. *inornata* Equals *Cryptotympana kagiana* Equals *Cryptotympana holsti* var. *capillata* Equals *Cryptotympana holsti* var. *kagiana* Vietnam

Cryptotympana holsti Shiyake 2007: 4, 41, 43, Fig 63 (illustrated, distribution, listed, comp. note) China, Southeast Asia

Cryptotympana holsti Lee 2008a: 5–6 (synonymy, song, listed, comp. note) Equals *Cryptotympana vitalisi* Vietnam, South China, Hainan, Taiwan, Liudau Island, Laos, Cambodia

Cryptotympana holsti Pham and Yang 2009: 13, Table 2 (listed) Vietnam

Cryptpotympana holsti Lee 2010b: 21 (distribution, listed) Cambodia, Vietnam, South China, Hainan, Taiwan, Liudau Island, Laos

***C. holsti takahashii* Kato, 1925**

***C. immaculata* (Olivier, 1790)** = *Cicada immaculata* = *Fidicina immaculata*

Cicada immaculata Hayashi 1987a: 120–121 (comp. note) Java

***C. insularis* Distant, 1887**

Cryptotympana insularis Hayashi 1987a: 192–193, 200–202, Fig 157, Figs 177–178, Figs 182–188 (lectotype designation, described, illustrated, distribution, comp. note) Andaman Islands

Cryptotympana insularis Hayashi 1987b: 98 (key)

Cryptotympana insularis Hayashi 1993: 17 (comp. note)

Cryptotympana insularis Terradas 1999: 48–49, Fig 1 (illustrated, comp. note) Andaman Islands

***C. intermedia* (Signoret, 1849)** = *Cicada intermedia* = *Fidicina immaculata* Walker, 1850 (nec Olivier) = *Cryptotympana immaculata* (nec Olivier) = *Cryptotympana iutermedia* (sic) = *Cryplolympana* (sic) *intermedia* = *Crystotympana* (sic) *inermidia* (sic)

Cicada intermedia Hayashi 1987a: 119 (comp. note)

Cryptotympana intermedia group Hayashi 1987a: 126, 128, Fig 3 (described, distribution, diversity, listed, comp. note) Himalayas

Cryptotympana intermedia group Hayashi 1987b: 34–35 (comp. note) Burma

Cryptotympana intermedia Hayashi 1987b: 34, 39, 45, 60–65, 100, Fig 84, Fig 94, Figs 107–108, Figs 124–128 (n. sp., described, illustrated, synonymy, distribution, key, comp. note) Equals *Cicada intermedia* Equals *Fidicina immaculata* (nec Olivier) India, Nepal

Cryptotympana intermedia group Hayashi 1993: 16, 18, Table 1 (listed, comp. note)

Cryptotympana intermedia Hayashi 1993: 18 (comp. note)

Crystotympana (sic) *inermidia* (sic) Srivastava and Bhandari 2004: 1479 (comp. note) India

***C. izzardi* Lallemand & Synave, 1953**

Cryptotympana izzardi Hayashi 1987a: 144–145, 159, 161–164, Fig 45, Fig 68e, Fig 76, Figs 80–84 (described, illustrated, distribution, comp. note) Sumba

Cryptotympana izzardi Hayashi 1987b: 99 (key)

***C. jacobsoni* China, 1926** = *Cryptotympana epithesia* Moulton, 1925 (nec *Cryptotympana epithesia* Distant, 1888)

Cryptotympana jacobsoni Hayashi 1987a: 130–131, 134, 143, 146, Fig 5, Figs 13–14, Fig 39 (described, illustrated, synonymy, distribution, comp. note) Equals *Cryptotympana epithesia* (nec Distant) Sumatra, Malay Peninsula

Cryptotympana jacobsoni Hayashi 1987b: 102 (key)

Cryptotympana jacobsoni Boer 1994c: 139 (comp. note) Sumatra

***C. karnyi* Moulton, 1923** = *Cryptotympana leopoldi* Lallemand, 1931

Cryptotympana karnyi Hayashi 1987a: 185–189, 191, Fig 132, Figs 134–135, Figs 138–143

(described, synonymy, illustrated, distribution, comp. note) Equals *Cryptotympana leopoldi* n. syn. Sumatra

Cryptotympana karnyi Hayashi 1987b: 99 (key)

Cryptotympana karnyi Hayashi 1993: 17–18 (comp. note)

***C. kotoshoensis* Kato, 1925** = *Cryptotympana shirakii* Matsumura, 1927

Cryptotympana kotoshoensis Hayashi 1987b: 69, 80, 82, 84–87, 90, 97, Fig 144, Fig 163–164, Fig 171, Fig 179B, Figs 181–184 (described, illustrated, synonymy, distribution, key, comp. note) Equals *Cryptotympana shirakii* Taiwan

Cryptotympana kotoshoensis Chou, Lei, Li, Lu and Yao 1997: 281 (described, comp. note) China

Cryptotympana kotoshoensis Hua 2000: 61 (listed, distribution) China, Taiwan

Cryptotympana kotoshsoensis Lee and Hayashi 2003a: 163, 168 (key, diagnosis, synonymy, song, listed, comp. note) Equals *Cryptotympana shirakii* Taiwan

Cryptotympana kotoshoensis Chen 2004: 66, 75, 198 (described, illustrated, listed, comp. note) Taiwan

C. leopoldi Lallemand, 1931 to *Cryptotympana karnyi* Moulton, 1923

***C. limborgi* Distant, 1888** = *Cryptotympana recta* (nec Walker)

Cryptotympana limborgi Hayashi 1987a: 192–193, 201–206, Fig 152, Fig 179, Figs 189–195 (lectotype designation, described, illustrated, distribution, synonymy, comp. note) Equals *Cryptotympana recta* (nec Walker) Burma

Cryptotympana limborgi Hayashi 1987b: 101 (key)

Cryptotympana limborgi Hayashi 1993: 17 (comp. note)

***C. lombokensis* Distant, 1912**

Cryptotympana lombokensis Hayashi 1987a: 144–145, 156, 159–162, Fig 44, Fig 68d, Figs 74–75, Figs 77–79 (described, illustrated, distribution, comp. note) Lombok, Sumbawa

Cryptotympana lombokensis Hayashi 1987b: 98–99 (key)

Cryptotympana lombokensis Hayashi 1993: 17 (comp. note)

***C. mandarina* Distant, 1891** = *Cryptympana* (sic) *mandarina* = *Fidicina operculata* Walker, 1850 nom. nud. = *Cryptotympana corvus* (nec Walker) = *Cryptotympana mimica* Distant, 1917

Cryptotympana mandarina Chen 1933: 359–360 (listed, comp. note)

Cryptotympana mandarina Wu 1935: 8 (listed) Equals *Fidicina operculata* China, Hongkong, Hainan, Tenasserim, Tonkin, Formosa

Cryptotympana mandarina Chen 1942: 144 (listed, comp. note) China

Cryptotympana mimica Hill, Hore and Thornton 1982: 49, 161, Plate 103 (listed, illustrated, comp. note) Hong Kong

Cryptotympana mandarina Hayashi 1987a: 122, 196, Fig 2B (illustrated, comp. note)

Cryptotympana mandarina Wang and Zhang 1987: 295 (key, comp. note) China

Cryptotympana mandarina Liu 1988: 192 (fungal host) China

Cryptotympana mandarina Shimazu 1989: 434 (comp. note)

Cryptotympana mandarina Peng and Lei 1992: 98, Fig 279 (listed, illustrated, comp. note) China, Hunan

Cryptotympana mandarina Wang and Zhang 1992: 121 (listed, comp. note) China

Cryptotympana mandarina Hayashi 1993: 18 (comp. note)

Cryptotympana mimica Andersen 1994: 29 (comp. note)

Cryptotympana mandarina Anonymous 1994: 809 (fruit pest) China

Cryptotympana mandarina Lei, Chou and Li 1994: 55, Fig 6 (song frequency spectrum, comp. note)

Cryptotympana mandarina Chou, Lei, Li, Lu and Yao 1997: 276, 279, 309, 322, 325, 327, 332, 346–347, Plate XVI, Fig 152, Tables 10–12 (key, described, illustrated, synonymy, song, oscillogram, spectrogram, comp. note) Equals *Fidicina operculata* Equals *Cryptotympana corvus* Equals *Cryptotympana mimica* China

Cryptotympana mandarina Easton and Pun 1999: 101 (distribution, comp. note) Macao, Hong Kong, China, Guangdong, Hainan, Taiwan

Cryptotympana mandarina Hua 2000: 61 (listed, distribution, hosts) China, Hubei, Jiangsu, Jiangxi, Zhejiang, Fujian, Taiwan, Guangdong, Hong Kong, Hainan, Hunan, Guangxi, Guizhou, Sichuan, Yunnan, Laos, Vietnam, India

Cryptotympana mimica Hua 2000: 61 (listed, distribution) China, Hong Kong

Cryptotympana mandarina Sueur 2001: 36, Table 1 (listed)

Cryptotympana mandarina Boulard 2003a: 98–101, Color Fig 2, sonogram 2 (described, song, sonogram, illustrated, comp. note) Equals *Cryptotympana corvus* (nec Walker) = *Cryptotympana mimica* (partim) Thailand China, Southeast Asia

Cryptotympana mandarina Boulard 2005o: 149 (listed)
Cryptotympana mandarina Pham 2005b: 232–234, Fig 1b, Fig 2b (key, synonymy, illustrated, distribution, comp. note) Equals *Fidicina operculata* Equals *Cryptotympana corvus* Equals *Cryptotympana mimica* Vietnam
Cryptotympana mandarina Pham and Thinh 2005a: 241 (distribution, listed) Equals *Fidicina operculata* Equals *Cryptotympana corvus* Equals *Cryptotympana mimica* Vietnam
Cryptotympana mandarina Pham and Thinh 2006: 525–527 (listed, comp. note) Vietnam
Cryptotympana mandarina Sanborn, Phillips and Sites 2007: 9 (listed, distribution, comp. note) Thailand, China, India, Taiwan, Indochina
Cryptotympana mandarina Shiyake 2007: 4, 41, 43, Fig 64 (illustrated, distribution, listed, comp. note) China, Southeast Asia
Cryptotympana mandarina Boulard 2008b: 109 (comp. note)
Cryptotympana mandarina Boulard 2008c: 361 (comp. note) South Vietnam
Cryptotympana mandarina Boulard 2008e: 352 (comp. note)
Cryptotympana mandarina Boulard 2008f: 14, 91 (biogeography, listed, comp. note) Thailand, China, Indochina
Cryptotympana mandarina Lee 2008a: 5 (synonymy, distribution, listed) Equals *Cryptotympana mimica* Vietnam, China, Hainan, Laos, Cambodia, Thailand, Myanmar
Cryptotympana mandarina Pham and Yang 2009: 13, Table 2 (listed) Vietnam
Cryptotympana mandarina Lee 2010b: 21 (distribution, listed) Cambodia, China, Hainan, Vietnam, Laos, Thailand, Myanmar

***C. moultoni* Hayashi, 1987** = *Cryptotympana robinsoni* (nec Moulton)
Cryptotympana moultoni Hayashi 1987a: 1192–193, 206–210, Fig 154, Fig 203, figs 206–209 (n. sp., described, illustrated, distribution, comp. note) Equals *Cryptotympana robinsoni* (nec Moulton) Sumatra
Cryptotympana moultoni Hayashi 1987b: 99 (key)
Cryptotympana moultoni Hayashi 1993: 17 (comp. note)
Cryptotympana moultoni Sanborn 1999a: 42 (type material, listed)

***C. niasana* Distant, 1909**
Cryptotympana niasana Hayashi 1987a: 144, 146–149, Fig 40, Fig 50, Fig, 54, Figs 56–60 (lectotype designation, described, illustrated, distribution, comp. note) Nias Island
Cryptotympana niasana Hayashi 1987b: 97 (key)
Cryptotympana niasana Hayashi 1993: 17 (comp. note)

***C. nitidula* Hayashi, 1987**
Cryptotympana nitidula Hayashi 1987a: 192–194, 196–198, Fig 149, Fig 158, Figs 170–171 (n. sp., described, illustrated, distribution, comp. note) Vietnam
Cryptotympana nitidula Hayashi 1987b: 97 (key)
Cryptotympana nitidula Lee 2008a: 5 (distribution, listed) Vietnam, Annam
Cryptotympana nitidula Pham and Yang 2009: 13, 16, Table 2 (listed, comp note) Vietnam
Cryptotympana nitidula Pham and Yang 2010a: 133 (comp. note) Vietnam

***C. ochromelas* Hayashi, 1987** = *Cryptotympana lombokensis* (nec Distant)
Cryptotympana ochromelas Hayashi 1987a: 144–145, 156, 163–168, 170, 172, Fig 46, Figs 68f–g, Figs 85–88, Figs 94–97 (n. sp., described, illustrated, distribution, comp. note) Equals *Cryptotympana lombokensis* (nec Distant) Flores, Solor
Cryptotympana ochromelas Hayashi 1987b: 98 (key)
Cryptotympana ochromelas Hayashi 1993: 17 (comp. note)

C. okinawana Matsumura, 1927 to *Cryptotympana facialis facialis* (Walker, 1858)

***C. pelengensis* Hayashi, 1987**
Cryptotympana pelengensis Hayashi 1987b: 3, 6, 25–27, 30–31, 103, Fig 9, Figs 16C–D, Fig 59B, Figs 62–63, Figs 72–75 (n. sp., described, illustrated, distribution, key, comp. note) Peleng Island, Sulawesi
Cryptotympana pelegensis Duffels 1990b: 64, 70, Table 2 (listed, comp. note) Sulawesi
Cryptotympana pelegensis Boer and Duffels 1996b: 330 (distribution) Sulawesi

***C. praeclara* Hayashi, 1987**
Cryptotympana praeclara Hayashi 1987a: 174–177, 182–185, Fig 109, Figs 113–114, Fig 116, Figs 127–131 (n. sp., described, illustrated, distribution, comp. note) Sarawak, Borneo
Cryptotympana praeclara Hayashi 1987b: 97 (key)
Cryptotympana praeclara Hayashi 1993: 17, Fig 8 (illustrated, comp. note)
Cryptotympana praeclara Sanborn 1999a: 42 (type material, listed)
Cryptotympana praeclara Zaidi and Azman 2003: 104, Table 1 (listed, comp. note) Borneo, Sabah

C. pustulata (Fabricius, 1787) to *Cryptotympana atrata* (Fabricius, 1775)

***C. recta* (Walker, 1850)** = *Fidicina recta*
Cryptotympana recta group Hayashi 1987a: 126–127, 191, Fig 3 (described, distribution, diversity, listed, comp. note) Sumatra, Malay Peninsula, Borneo, India, Indo-China

Cryptotympana recta Hayashi 1987a: 191–196, Fig 148, Figs 156–157, Figs 167–169 (described, illustrated, synonymy, distribution, comp. note) Equals *Fidicina recta* India, Bangladesh, China, Vietnam, Laos, Thailand
Cryptotympana recta Hayashi 1987b: 100 (key)
Cryptotympana recta group Hayashi 1993: 16–17, Table 1 (listed, comp. note)
Cryptotympana recta Hayashi 1993: 17 (comp. note)
Cryptotympana recta Chou, Lei, Li, Lu and Yao 1997: 276, 280, Plate XVI, Fig 151 (key, described, illustrated, synonymy, comp. note) Equals *Fidicina recta* China
Cryptotympana recta group Holloway 1998: 305, Fig 10 (distribution, phylogeny, comp. note) Sunda, Asia
Cryptotympana recta Hua 2000: 61 (listed, distribution) Equals *Fidicina recta* China, Guangxi, India, Laos, Thailand
Cryptotympana recta Pham 2005b: 232–234, Fig 1c, Fig 2c (key, synonymy, illustrated, distribution, comp. note) Equals *Fidicina recta* Vietnam
Cryptotympana recta Pham and Thinh 2005a: 241 (distribution, listed) Equals *Fidicina recta* Vietnam
Cryptotympana recta Sanborn, Phillips and Sites 2007: 9 (listed, distribution, comp. note) Thailand, India, Bangladesh, China, Vietnam, Laos
Cryptotympana recta Boulard 2008f: 14, 91 (biogeography, listed, comp. note) Equals *Fidicina recta* Thailand, India
Cryptotympana recta Lee 2008a: 5 (synonymy, distribution, listed) Equals *Fidicina recta* Vietnam, China, Laos, Thailand, Bangladesh, India
Cryptotympana recta Pham and Yang 2009: 13, Table 2 (listed) Vietnam
Cryptotympana recta Lee 2010b: 21 (synonymy, distribution, listed) Equals *Fidicina recta* Cambodia, China, Vietnam, Laos, Thailand, Bangladesh, India

***C. robinsoni* Moulton, 1923**
Cryptotympana robinsoni Hayashi 1987a: 1192–194, 197–199, 206, Fig 150, Figs 159–160, Figs 172–176 (described, illustrated, distribution, comp. note) Malay Peninsula
Cryptotympana robinsoni Hayashi 1987b: 99 (key)
Cryptotympana robinsoni Hayashi 1993: 17 (comp. note)
Cryptotympana robinsoni Zaidi, Ruslan and Mahadir 1996: 60, Table 1 (listed, comp. note) Peninsular Malaysia

***C. sibuyana* Hayashi, 1987**
Cryptotympana sibuyana Hayashi 1987b: 2–3, 6, 9–14, 24, 103, Fig 3, Fig 16B, Fig 26, Figs 28–29, Figs 31–34, Fig 37A, Fig 57 (described, illustrated, distribution, key, comp. note) Philippines, Sibuyan Island, Romblon, Tablas
Cryptotympana sibuyana Sanborn 1999a: 42 (type material, listed)

C. sinensis Distant, 1887 to *Cryptotympana atrata* (Fabricius, 1775)

***C. socialis* Hayashi, 1987**
Cryptotympana socialis Hayashi 1987b: 2–3, 6, 11, 13–14, 24, 102, Fig 4, Fig 16C, Fig 30, Figs 35–36, Fig 37B, Fig 57 (n. sp., described, illustrated, distribution, key, comp. note) Philippines, Panay, Cebu, Negros
Cryptotympana socialis Sanborn 1999a: 43 (type material, listed)
Cryptotympana socialis Lee 2009i: 293–294 (distribution, listed) Philippines, Panay, Cebu, Negros

***C. suluensis* Distant, 1906**
Cryptotympana suluensi Hayashi 1987b: 3, 6, 21–24, 103, Fig 7, Fig 16D, Figs 52–57 (described, illustrated, distribution, key, comp. note) Philippines, Islo Island, Sula Island
Cryptotympana suluensis Hayashi 1993: 18 (comp. note)

C. sumbawensis Jacobi, 1904 to *Cryptotympana varicolor* Distant, 1904

***C. takasagona* Kato, 1925** = *Cryptotympana intermedia* Schumacher, 1915 (nec *Cicada intermedia* Signoret, 1849) = *Cryptotympana intermedia* Oshanin, 1908 (nec Signoret) = *Cryptotymapana iutermedia* (sic)(nec Signoret) = *Cryptotympana argenteus* Kato, 1925
Cryptotympana takasagona Hayashi 1984: 43 (comp. note) Japan
Cryptotympana takasagona Hayashi 1987a: 122, Figs 2C–D (illustrated, comp. note)
Cryptotympana takasagona Hayashi 1993: 18 (comp. note)
Cryptotympana takasagona Chou, Lei, Li, Lu and Yao 1997: 276, 280–281, Plate XVI, Fig 149 (key, described, illustrated, synonymy, comp. note) Equals *Cryptotympana intermedia* Equals *Cryptotympana argenteus* China
Cryptotympana takasagona Hua 2000: 61 (listed, distribution) China, Fujian, Taiwan
Cryptotympana takasagona Lee 2003b: 7 (comp. note)
Cryptotympana takasagona Lee and Hayashi 2003a: 163–166, Fig 11 (key, diagnosis, illustrated, synonymy, song, listed, comp. note) Equals *Cryptotympana intermedia* (nec Signoret) Equals *Cryptotympana iutermedia* (sic)(nec Signoret) Equals *Cryptotympana facialis* Equals *Cryptotympana fascialis* (sic) Equals

Cryptotympana facialis var. formosana (nec Walker) Equals *Cryptotympana fascialis* (sic) var. *formosana* (nec Walker) Equals *Cryptotympana argenteus* Equals *Cryptotympana okinawana* (nec Matsumura) Taiwan, China

Cryptotympana takasagona Chen 2004: 66, 69, 192–193, 198 (described, illustrated, listed, comp. note) Taiwan

Cryptotympana argenteus Chen 2004: 192 (comp. note)

Cryptotympana takasagona Lee 2005: 35, 63 (comp. note)

Cryptotympana takasagona Shiyake 2007: 4, 41, 43, Fig 62 (illustrated, distribution, listed, comp. note) China, Taiwan

Cryptotympana takagasona Lee, Lin and Wu 2010: 218–224, Figs 1–6, Tables 1–2 (ecology, comp. note) Taiwan

C. *timorica* (Walker, 1870) = *Fidicina timorica* = *Cryptotympana heuertzi* Lallemand & Synave, 1953 = *Cryptotympana acuta* (nec Signoret)

Cryptotympana timorica Hayashi 1987a: 144–145, 156, 166, 168–173, Fig 47, Figs 68h–I, Figs 90–92, Figs 98–100 (described, synonymy, illustrated, distribution, comp. note) Equals *Fidicina timorica* Equals *Cryptotympana acuta* (partim) (nec Signoret) Equals *Cryptotympana heuertzi* n. syn. Timor

Cryptotympana timorica Hayashi 1987b: 98 (key)

Cryptotympana timorica Hayashi 1993: 17 (comp. note)

C. *varicolor* Distant, 1904 = *Cryptotympana sumbawensis* Jacobi, 1941

Cryptotympana varicolor group Hayashi 1987a: 126–127, 185, Fig 3 (described, distribution, diversity, listed, comp. note) Sumatra, Sumbawa, Lesser Sunda Islands

Cryptotympana varicolor Hayashi 1987a: 1185–186, Fig 133, Figs 136–137 (described, illustrated, synonymy, distribution, comp. note) Equals *Cryptotympana sumbawensis* n. syn. Sumbawa

Cryptotympana varicolor Hayashi 1987b: 95, 101 (key, comp. note)

Cryptotympana varicolor group Hayashi 1993: 16–18, Table 1 (listed, comp. note)

C. *ventralis* Hayashi, 1987

Cryptotympana ventralis Hayashi 1987b: 3, 6, 25–27, 102, Fig 10, Fig 16C, Fig 59C, Fig 64, Figs 76–79 (described, illustrated, distribution, key, comp. note) Sulawesi

Cryptotympana ventralis Duffels 1990b: 64, 70, Table 2 (listed, comp. note) Sulawesi

Cryptotympana ventralis Boer and Duffels 1996b: 330 (distribution) Sulawesi

C. *vesta* (Distant, 1904) = *Cicada vesta*

Cryptotympana vesta subgroup Hayashi 1987b: 65 (described, listed, comp. note)

Cryptotympana vesta Hayashi 1987b: 65–68, 101, Fig 129, Figs 131–140 (lectotype designation, described, illustrated, synonymy, distribution, key, comp. note) Equals *Cicada vesta* India

Cryptotympana vesta Hayashi 1993: 18 (comp. note)

Chremistica vesta Sanborn, Phillips and Sites 2007: 34 (comp. note) India

Cryptotympana vesta Boulard 2008f: 13–14, 90 (status, valid species, comp. note) Thailand

C. *viridicostalis* Hayashi, 1987 = *Cryptotympana acuta* (nec Signoret) = *Cryptotympana? varicolor* Banks (nec Distant)

Cryptotympana viridicostalis Hayashi 1987b: 2–3, 6, 10, 14–18, 24, 99, Fig 5, Fig 16E, Fig 27, Figs 38–39, Figs 43–46, Fig 57 (n. sp., described, illustrated, distribution, key, comp. note) Equals *Cryptotympana acuta* (nec Signoret) (partim) Equals *Cryptotympana? varicolor* Banks (nec Distant) Philippines, Palawan

Cryptotympana viridicostalis Hayashi 1993: 18 (comp. note)

Cryptotympana viridicostalis Sanborn 1999a: 43 (type material, listed)

Cryptotympana viridicostalis Lee 2009h: 2620–2621 (synonymy, distribution, listed) Equals *Cryptotympana acuta* (nec Signoret) Equals *Cryptotympana varicolor* (nec Distant) Equals *Cryptotympana varicolor?* (nec Distant) Philippines, Palawan

C. *viridipennis* Distant, 1911 to *Cryptotympana fumipennis* (Walker, 1858)

C. *vitalisi* Distant, 1917 to *Cryptotympana holsti holsti* Distant, 1904

C. *wenchewensis* Ouchi, 1938 to *Cryptotympana atrata* (Fabricius, 1775)

C. *wetarensis* Hayashi, 1987

Cryptotympana wetarensis Hayashi 1987a: 144–145, 156, 166, 170–173, Fig 49, Fig 68k, Fig 93, Fig 101, Figs 105–106 (n. sp., described, illustrated, distribution, comp. note) Wetar

Cryptotympana wetarensis Hayashi 1987b: 98 (key)

C. *yayeyamana* Kato, 1925 = *Cryptotympana yaeyamana* (sic) = *Cryptotympana yayeyama* (sic) = *Cryptotympana ishigakiana* Matsumura, 1927

Cryptotympana yayeyamana Hayashi 1984: 29, 43–45, Figs 61–62, Figs 69–70 (key, described, illustrated, synonymy, comp. note) Equals *Cryptotympana ishigakiana* Japan

Cryptotympana yaeyamana Hayashi 1993: 18 (comp. note)

Cryptotympana yaeyamana Hayashi 1987b: 69, 81–82, 84, 87–92, 99, Fig 145, Figs 165–166, Fig 172, Fig 179C, Figs 185–189 (described, illustrated, synonymy, distribution, key, comp. note) Equals *Cryptotympana yayeyamana* (sic) Equals *Cryptotympana ishigakiana* Ryukyus

Cryptotympana yayeyamana Hua 2000: 61 (listed, distribution) China, Inner Mongolia, Japan

Cryptotympana yayeyamana Sueur 2001: 46, Table 2 (listed)

Cryptotympana yayeyamana Sueur 2002b: 125, 129 (listed, comp. note)

Cryptotympana yaeyamana Saisho 2006d: 62 (eclosion, illustrated, comp. note) Japan

Cryptotympana yaeyamana (sic) Shiyake 2007: 4, 41, 43, Fig 65 (illustrated, distribution, listed, comp. note) Japan

Cryptotympana yaeyamana (sic) Osaka Museum of Natural History 2008: 322 (listed, acoustic behavior, ecology, comp. note) Japan

Pulchrocicada He, 1984 to *Salvazana* Distant, 1913

P. guangxiensis He, 1984 to *Salvazana mirabilis imperialis* Distant, 1918

P. sinensis He, 1984 to *Salvazana mirabilis mirabilis* Distant, 1918

***Salvazana* Distant, 1913** = *Salvasana* (sic) = *Pulchrocicada* He, 1984

Salvazana Hayashi 1984: 31 (comp. note)

Pulchrocicada He 1984: 221, 226 (n. gen., described, comp. note) China

Salvazana Chou, Lei, Li, Lu and Yao 1997: 265, 293, 357, 361, Table 6 (key, described, synonymy, listed, diversity, distribution, comp. note) Equals *Pulchrocicada* n. syn. China

Pulchrocicada Chou, Lei, Li, Lu and Yao 1997: 357, 361 (comp. note)

Salvazana Hayashi 1987a: 125 (comp. note)

Salvazana Hayashi 1993: 15 (comp. note)

Salvazana Novotny and Wilson 1997: 437 (listed)

Salvazana Villet and Zhao 1999: 322 (comp. note) Equals *Pulchrocicada*

Salvazana Moulds 2005b: 390, 431 (higher taxonomy, listed, comp. note) Australia

Salvazana Boulard 2007e: 3, 13 (comp. note) Thailand

Salvazana Sanborn, Phillips and Sites 2007: 3, 5, Table 1 (listed) Thailand

Salvazana Boulard 2008f: 3–4, 7, 14, 91 (listed, comp. note) Thailand

Salvazana Lee 2008a: 2, 5, Table 1 (listed) Vietnam

Salvazana Boulard 2009c: 63 (comp. note)

Salvazana Pham and Yang 2009: 13, Table 2 (listed) Vietnam

Salvazana Phillips and Sanborn 2010: 74 (comp. note) Asia

S. imperialis Distant, 1918 to *Salvazana mirabilis*

***S. mirabilis imperialis* Distant, 1918** = *Salvazana imperialis* = *Salvazana mirabilis var. imperialis* = *Pulchrocicada guangxiensis* He, 1984

Pulchrocicada guangshiensis He 1984: 223, 227, Fig 3, Plate II, Fig 2 (n. sp., described, illustrated, comp. note) China, Guangxi

Salvazana imperialis Chou, Lei, Li, Lu and Yao 1997: 358, 361 (synonymy)

Pulchrocicada guangxiensis Chou, Lei, Li, Lu and Yao 1997: 358, 361 (synonymy)

Salvazana mirabilis imperialis Boulard 2002a: 63–64, Plate M (synonymy, comp. note) Equals *Salvazana mirabilis* Equals *Salvazana imperialis* Equals *Pulchrocicada sinensis* Equals *Pulchrocicada guangxiensis* Thailand, China, Tonkin

Salvazana imperialis Boulard 2005d: 341 (comp. note) Indochina

Salvazana imperialis Lee 2005: 37 (illustrated)

Salvazana imperialis Boulard 2006g: 53 (comp. note)

Salvazana mirabilis imperialis Boulard 2007e: 13, 15, Plate 2, Fig 8, Plates 10–11, Plate 46b, front cover (illustrated, eclosion, comp. note) Thailand

Salvazana imperialis Boulard 2007e: 32 (comp. note) Thailand

Salvazana mirabilis imperialis Boulard 2008f: 15 (biogeography, listed, comp. note) Equals *Salvazana imperialis* Equals *Pulchrocicada guangxiensis* Equals *Salvazana mirabilis var. imperialis* Thailand, Indochina, China

***S. mirabilis mirabilis* Distant, 1913** = *Salvazana mirabilis var. mirabilis* = *Pulchrocicada sinensis* He, 1984

Pulchrocicada sinensis He 1984: 221–223, 226–227, Figs 1–2, Plate II, Fig 1 (n. sp., described, illustrated, type species of *Pulchrocicada*, comp. note) China, Guangxi

Salvazana mirabilis Chou, Lei, Li, Lu and Yao 1997: 274–276, 358, 361, Fig 9-92, Plate XV, Figs 144–146 (type species of *Salvazana*, described, illustrated, synonymy, comp. note) Equals *Salvazana imperialis* n. syn. Equals *Pulchrocicada sinensis* n. syn. Equals *Pulchrocicada guangxiensis* n. syn. China

Pulchrocicada sinensis Chou, Lei, Li, Lu and Yao 1997: 358, 361 (synonymy)

Salvazana mirabilis Boulard 1999b: 184–185, Fig 1 (illustrated, comp. note) Laos

Salvazana mirabilis Hua 2000: 64 (listed, distribution) Equals *Pulchrocicada guangxiensis* Equals *Pulchrocicada sinensis* China, Guangxi

Salvazana mirabilis Boulard 2002a: 61, 63 (synonymy, comp. note) Equals *Salvazana mirabilis* Equals *Salvazana imperialis* Equals

Pulchrocicada sinensis Equals *Pulchrocicada guangxiensis* Thailand, China, Tonkin
Salvazana mirabilis mirabilis Boulard 2002a: 62–64, Plate L, Figs 1–2 (synonymy, comp. note) Equals *Salvazana mirabilis* Equals *Salvazana imperialis* Equals *Pulchrocicada sinensis* Equals *Pulchrocicada guangxiensis* Thailand, China, Tonkin
Salvasana (sic) *mirabilis* Boulard 2002b: 193 (coloration, comp. note)
Salvazana mirabilis Boulard 2005d: 341, 344 (comp. note)
Salvazana mirabilis Boulard 2005o: 148 (listed)
Salvazana mirabilis Boulard 2006g: 53, 75–76, 91, 180–181, Fig 56, Fig (sonogram, illustrated, listed, comp. note) Thailand
Salvazana mirabilis Boulard 2007e: 4, 13, 32, 35, 43, Fig 27, Plate 46 (sonogram, eclosion, illustrated, comp. note) Thailand
Salvazana mirabilis mirabilis Boulard 2007e: 13, 15, Plate 2, Fig 7, Plates 8–11, Plate 46a (illustrated, eclosion, comp. note) Thailand
Salvazana mirabilis Sanborn, Phillips and Sites 2007: 9–10 (listed, distribution, comp. note) Thailand, Indochina
Salvazana mirablis Boulard 2008f: 14, 91 (type species of *Salvazana*)
Salvazana mirabilis Boulard 2008b: 109 (comp. note)
Salvazana mirabilis Boulard 2008e: 352 (comp. note)
Salvazana mirabilis mirabilis Boulard 2008f: 14–15 (biogeography, listed, comp. note) Equals *Salvazana mirabilis* Equals *Pulchrocicada sinensis* Equals *Salvazana mirabilis var. mirabilis* Thailand, Tonkin, Indochina, China
Salvazana mirabilis Lee 2008a: 5 (type species of *Salvazana*, distribution, listed) Vietnam, South China, Thailand
Salvazana mirabilis Boulard 2009c: 63–64, Figs 1–3 (sonogram, illustrated, comp. note) Thailand
Salvazana mirabilis Pham and Yang 2009: 13, 16, Table 2 (listed, comp. note) Vietnam
Salvazana mirabilis Pham and Yang 2010a: 133 (comp. note) Vietnam

***Raiateana* Boulard, 1979**

Raiateana Duffels 1986: 327, 331, Fig 9 (phylogeny, comp. note)
Raiateana Hayashi 1987a: 124–125 (listed)
Raiateana Duffels 1988d: 8, 11–13, 25, 80–81, 91, 93, Fig 4, Figs 7–8, Table 1 (phylogeny, listed, distribution, comp. note) Fiji Islands, Samoa Islands, Society Islands
Raiateana Boulard 1995c: 174 (distribution, comp. note)
Raiateana Boer and Duffels 1996b: 302, 307 (comp. note) Fiji, Samoa, Society Islands
Raiateana group Holloway 1998: 304–305, 307, Fig 10, Fig 12 (distribution, phylogeny, comp. note) Solomon Islands
Raiateana Moulds 2005b: 390, 431 (higher taxonomy, listed, comp. note) Australia
Raiateana Salmah, Duffels and Zaidi 2005: 15 (comp. note)
Raiateana Yaakop, Duffels and Visser 2005: 249 (comp. note)

***R. knowlesi* (Distant, 1907)** = *(?) Tibicen knowlesi* = *Tibicen knowlesi* = *Lyristes knowlesi* = *Cicada knowlesi*

(?) Tibicen knowlesi Duffels 1986: 328 (comp. note) Fiji Islands
Raiateana knowlesi Duffels 1988d: 9, 11–12, 14, 26, 28, 81, Fig 4, Figs 172–177, Table 1 (n. comb., key, lectotype designation, described, illustrated, phylogeny, distribution, comp. note) Equals *Cicada knowlesi* Equals *Lyristes knowlesi* Fiji Islands
Tibicen knowlesi Duffels 1988d: 8, 80, 91 (comp. note) Fiji Islands
Cicada knowlesi Duffels 1988d: 9, 91 (lectotype designation, comp. note) Fiji Islands
Raiateana knowlesi Moore 1991: 214 (comp. note) Fiji
Raiateana knowlesi Boer and Duffels 1996b: 330 (distribution) Fiji Islands
Raiateana knowlesi Wilson 2009: 40, Appendix (listed, distribution, comp. note) Fiji

***R. kuruduadua bifasciata* Duffels, 1988**

Raiateana kuruduadua bifasciata Duffels 1988d: 11–12, 14, 29, 82–83, 85–90, Fig 4, Fig 151, Fig 157, Figs 165–171, Tables 1–2 (n. spp., key, described, illustrated, sonogram, phylogeny, distribution, listed, comp. note) Fiji Islands
Raiateana kuruduadua bifasciata Boulard 1995c: 174 (comp. note) Vanua Levu, Taveuni
Raiateana kuruduadua bifasciata Boer and Duffels 1996b: 299, 330, Plate 1e (illustrated, distribution, comp. note) Fiji Islands
Raiateana kuruduadua bifasciata Wilson 2009: 40, Appendix (listed, distribution, comp. note) Fiji

***R. kuruduadua kuruduadua* (Distant, 1881)** = *Cicada kuruduadua* = *Tibicen kuruduadua* = *Lyristes kuruduaduus* = *Raiateana kuruduadua taveuniensis* (sic)

Raiateana kuruduadua Duffels 1986: 327, 331, Fig 9 (distribution, comp. note) Fiji, Samoa Island
Tibicen kuruduadua Duffels 1986: 327 (comp. note)

Raiateana kuruduadua Duffels 1988d: 9, 10–11, 13, 24, 26, 28, 30, 80–81 (key, listed, comp. note) Equals *Cicada kuruduadua* Fiji, Samoa Island

Tibicen kuruduadua Duffels 1988d: 8, 10, 80 (comp. note) Fiji, Samoa

Cicada kuruduadua Duffels 1988d: 9, 82 (lectotype designation, comp. note) Fiji Islands

Raiateana kuruduadua kuruduadua Duffels 1988d: 11–12, 14, 28, 82–87, Fig 4, Figs 150, Figs 153–156, Figs 158–165, Tables 1–2 (key, described, illustrated, phylogeny, distribution, listed, comp. note) Equals *Cicada kuruduadua* Equals *TIbicen kuruduadua* Equals *Lyristes kuruduaduus* Fiji Islands

Raiateana kuruduadua kuruduadua Boulard 1995c: 174 (comp. note) Viti Levu

Raiateana kuruduadua taveuniensis (sic) Boer and Duffels 1996b: 305, 330, Plate 3 (illustrated, distribution, comp. note) Fiji Islands

Raiateana kuruduadua kuraduadua Wilson 2009: 40, Appendix (listed, distribution, comp. note) Fiji

R. oulietea **Boulard, 1979** = *Raiateana oulietea* morph *poinsignoni* Boulard, 1995 = *Raiateana oulietea* morph *suavis* Boulard, 1995

Raiateana oulietea Duffels 1986: 327, Fig 9 (distribution, comp. note) Raiatea Island, Society Islands

Raiateana oulietea Duffels 1988d: 8, 11, 14, 80–81, Fig 4 (type species of *Raiateana*, phylogeny, comp. note) Society Islands

Raiateana oulietea Boulard 1995c: 163–173, 175–176, Fig 2, Figs 5–12 (described, illustrated, sonogram, comp. note) Society Islands

Raiateana oulietea morph *poinsignoni* Boulard 1995c: 165–166, Fig 4 (n. mph., described, illustrated, comp. note) Society Islands

Raiateana oulietea morph *suavis* Boulard 1995c: 165–166, Fig 3 (n. mph., described, illustrated, comp. note) Society Islands

Raiateana oulietea Anonymous 1996b: 82 (on stamp) French Polynesia

Raiateana oulietea Boer and Duffels 1996b: 310 (comp. note) Society Islands

Raiateana oulietea Anonymous 1998: 135 (on stamp) French Polynesia

Raiateana oulietea Freytag 2000: 55 (on stamp)

Raiateana oulietea Sueur 2001: 37, Table 1 (listed)

Raiateana ouleitea Boulard 2002b: 176, 193, Fig 5 (comp. note) French Polynesia

Raiateana oulietea Boulard 2005d: 338, 346 (comp. note) French Polynesia

Raiateana oulietea Boulard 2006g: 36–37, 39, 94–95, 181, Fig 7c, Fig 63 (illustrated, sonogram, listed, comp. note) Polynesia

R. samoensis **Duffels, 1988** = *Raiateana kuruduadua samoensis*

Raiateana kuruduadua samoensis Duffels 1988d: 11–12, 14, 29, 82, 85, 87, 90, Fig 4, Fig 152, Fig 165, Tables 1–2 (n. ssp., key, described, illustrated, phylogeny, distribution, listed, comp. note) Samoa

Raiateana kuruduadua samoensis Boulard 1995c: 174 (comp. note) Samoa

Raiateana samoensis Boer and Duffels 1996b: 330 (distribution) Samoa Islands

Subtribe Heteropsaltriina Distant, 1905 = Heteropsaltriaria

Heteropsaltriaria Duffels 1988d: 81 (comp. note)

Heteropsaltriaria Moulds 2005b: 386, 391, 431 (history, higher taxonomy, listed, comp. note)

Heteropsaltria **Jacobi, 1902**

Heteropsaltria Duffels 1986: 327, 331, Fig 9 (phylogeny, distribution, comp. note)

Heteropsaltria Hayashi 1987a: 124 (listed)

Heteropsaltria Duffels 1988d: 14, 25, 81 (comp. note) Solomon Islands

Heteropsaltria Boulard 1990b: 214 (comp. note) Bougainville

Heteropsaltria Duffels and Boer 1990: 265 (comp. note) Solomon Islands

Heteropsaltria Boer and Duffels 1996b: 302, 307 (comp. note) Solomon Islands

Heteropsaltria Duffels 1997: 552 (comp. note) Solomon Islands

Heteropsaltria Moulds 2005b: 390, 431 (higher taxonomy, listed, comp. note)

Heteropsaltria Yaakop, Duffels and Visser 2005: 249 (comp. note)

H. aliena **Jacobi, 1902**

Heteropsaltria aliena Duffels 1986: 327, Fig 9 (distribution, comp. note) Bougainville, Solomon Islands

Heteropsaltria aliena Boulard 1990b: 217, Plate XXVI (illustrated, comp. note)

Heteropsaltria aliena Boer and Duffels 1996b: 330 (distribution) Solomon Islands

Heteropsaltria aliena Duffels 1997: 552 (comp. note) Bougainville

Subtribe Nggelianina Boulard, 1979 = Nggelianaria

Nggelianaria Duffels 1988d: 81 (comp. note)

Nggelianaria Moulds 2005b: 386, 431 (history, listed, comp. note)

***Nggeliana* Boulard, 1979**

Nggeliana Duffels 1986: 327, 331, Fig 9 (phylogeny, comp. note)
Nggeliana Hayashi 1987a: 124–125 (listed)
Nggeliana Duffels 1988d: 14, 81 (comp. note) Solomon Islands
Nggeliana Duffels and Boer 1990: 265 (comp. note) Solomon Islands
Nggeliana Boulard 1995c: 174, 177 (comp. note)
Nggeliana Boer and Duffels 1996b: 302, 307 (comp. note) Solomon Islands
Nggeliana Duffels 1997: 552 (comp. note) Solomon Islands
Nggeliana Moulds 2005b: 390, 431 (higher taxonomy, listed, comp. note)
Nggeliana Salmah, Duffels and Zaidi 2005: 15 (comp. note)

***N. leveri* Boulard, 1979**

Nggeliana leveri Duffels 1986: 327, Fig 9 (distribution, comp. note) Florida Island, Solomon Islands
Nggeliana leveri Boulard 1990b: 214, 217, Plate XXVI (illustrated, comp. note) Florida Island
Nggeliana leveri Boer and Duffels 1996b: 330 (distribution) Solomon Islands
Nggeliana leveri Duffels 1997: 552 (comp. note) Florida Island

***N. typica* Boulard, 1979**

Nggeliana typica Duffels 1986: 327, Fig 9 (distribution, comp. note) Guadalcanal Island, Solomon Islands
Nggeliana typica Boulard 1990b: 214, 217, Plate XXVI (illustrated, comp. note) Guadalcanal
Nggeliana typica Boer and Duffels 1996b: 330 (distribution) Solomon Islands
Nggeliana typica Duffels 1997: 552 (comp. note) Guadalcanal

Tribe Fidicinini Distant, 1905

Fidicinini Boulard 1982d: 108 (comp. note)
Fidicinini Ramos 1983: 64 (listed) Dominican Republic
Fidicinini Boulard 1988f: 61 (comp. note)
Fidicinini Boulard 1990b: 213 (comp. note)
Fidicinini Boulard and Mondon 1995: 72 (comp. note)
Fidicinini Boulard 1996d: 114, 139 (listed)
Fidicinini Boulard and Martinelli 1996: 12, 14, 19 (deacribed, comp. note)
Fidicinini Boulard 1997c: 120 (comp. note)
Fidicinini Novotny and Wilson 1997: 437 (listed)
Fidicinini Boulard 1998: 76 (comp. note)
Fidicinini Boulard 1999a: 78, 99, 112 (listed)
Fidicinini Sanborn 1999a: 43 (listed)
Fidicinini Sanborn 1999b: 1 (listed) North America
Fidicinini Sueur 2000: 217, 221 (comp. note)
Fidicinini Boulard 2001b: 40 (comp. note)
Fidicinini Sanborn 2001b: 449 (listed) El Salvador
Fidicinini Sueur 2001: 38, Table 1 (listed)
Fidicinini Kondratieff, Ellingson and Leatherman 2002: 13, Table 2 (listed) Colorado
Fidicinini Sueur 2002a: 380, 387 (listed) Mexico, Veracruz
Fidicinini Sanborn, Villet and Phillips 2003: 307 (listed)
Fidicinini Santos and Martinelli 2004a: 630 (diversity) Neotropics
Fidicinini Sanborn, Villet and Phillips 2004a: 820 (comp. note)
Fidicinini Moulds 2005b: 427 (higher taxonomy, listed)
Fidicinini Sanborn 2006a: 76 (listed)
Fidicinini Sanborn 2006b: 256 (listed)
Fidicinini Sanborn 2007a: 27 (listed) Venezuela
Fidicinini Sanborn 2007b: 35 (listed) Mexico
Fidicinini Shiyake 2007: 4, 46 (listed, comp. note)
Fidicinini Sanborn 2008f: 686, 688 (listed)
Fidicinini Sanborn, Moore and Young 2008: 3 (listed) Costa Rica
Fidicinini Santos and Martinelli 2009b: 638 (diversity)
Fidicinini Sanborn 2009a: 89 (listed) Cuba
Fidicinini Phillips and Sanborn 2010: 74 (comp. note)
Fidicinini Santos, Martinelli, Maccagnan, Sanborn and Ribeiro 2010: 48–50 (diversity, comp. note)
Fidicinini Sanborn 2010a: 1586 (listed) Colombia
Fidicinini Sanborn 2010b: 74 (listed)

Subtribe Fidicinina Distant, 1905

Fidicinina Boulard and Martinelli 1996: 12, 14, 20–21 (n. subtribe, key, described, listed, comp. note)
Fidicinina Boulard 2003e: 260 (comp note)
Fidicinina Sanborn 2006a: 77 (listed)
Fidicinina Sanborn 2007a: 27 (listed) Venezuela
Fidicinina Sanborn 2007b: 36 (listed) Mexico
Fidicinina Sanborn 2008f: 688 (listed)
Fidicinina Sanborn, Moore and Young 2008: 3, 19 (listed, comp. note) Costa Rica
Fidicinina Sanborn 2010a: 1589 (listed) Colombia
Fidicinina Santos, Martinelli, Maccagnan, Sanborn and Ribeiro 2010: 50 (listed)

***Fidicina* Amyot & Audinet-Serville, 1843** = *Cicada (Fidicina)* = *Fidicula* (sic)

Fidicina Dallas 1870: 482 (listed)
Fidicina Pachas 1960: 14 (control, comp. note) Argentina, Misiones
Fidicina Josephson 1973: 796 (comp. note)
Fidicina Young 1981a: 135, 139–141 (comp. note) Costa Rica

Fidicina spp. Young 1981a: 137–138 (comp. note) Costa Rica
Fidicina Young 1981c: 13–14 (comp. note) Costa Rica
Cicada (sic) Carnevali and Valvassori 1981: 2 (muscle structure, comp. note)
Fidicina Souza, Reis and Melles 1983: 16 (comp. note) Brazil, Minas Gerias
Fidicina sp. Souza, Reis and Melles 1983: 16–17, Fig 6 (illustrated, coffee pest, control, comp. note) Brazil, Minas Gerias
Fidicina Strümpel 1983: 109, Table 8 (comp. note)
Fidicina sp. Wolda 1983a: 114 (emergence pattern, comp. note) Panama
Fidicina Ewing 1984: 229 (comp. note)
Fidicina Martinelli and Zucchi 1984: 1 (comp. note) Brazil
Fidicina Reis, Souza and Melles 1984: 8 (comp. note) Brazil
Fidicina Young 1984a: 173 (comp. note) Costa Rica
Fidicina spp. Young 1984a: 189 (comp. note) Costa Rica
Fidicina Boulard 1988d: 158 (comp. note) French Guiana
Fidicina Boulard 1988f: 61, 69 (comp. note)
Fidicina Aidley 1989: 44–45 Fig 9 (timbal muscle contraction properties)
Fidicula (sic) Ewing 1989: 36 (comp. note)
Fidicina Martinelli and Zucchi 1989a: 6, 8 (comp. note) Brazil
Fidicina Martinelli, Zucchi, Silveira Neto and Parra 1989: 168 (coffee pest, comp. note) Brazil, Minas Gerias
Fidicina Martinelli and Zucchi 1989b: 6–7 (comp. note) Brazil
Fidicina Boulard 1990b: 166 (comp. note)
Fidicina Heath, Heath and Noriega 1990: 1 (distribution, comp. note) Argentina
Fidicina Martinelli 1990: 12 (comp. note) Brazil
Fidicina Young 1991: 214–215, 225–226 (comp. note) Costa Rica
Fidicina Boulard and Mondon 1995: 72 (comp. note)
Fidicina Boulard 1996d: 139 (listed)
Fidicina Boulard and Martinelli 1996: 12–13, 19–24 (key, type genus of Fidicinini, comp. note)
Fidicina Cocroft and Pogue 1996: 94 (comp. note) Peru
Fidicina Moore 1996: 221–222 (comp. note) Mexico
Fidicina Boulard 1997c: 120 (comp. note)
Fidicina Martinelli and Zucchi 1997b: 346 (comp. note) Brazil
Fidicina Martinelli and Zucchi 1997c: 134 (comp. note) Brazil
Fidicina Novotny and Wilson 1997: 437 (listed)
Fidicina Poole, Garrison, McCafferty, Otte and Stark 1997: 251, 331 (listed) North America
Fidicina Henríquez 1998: 86 (comp. note)
Fidicina Ohbayashi, Sato and Igawa 1999: 340 (parasitism, comp. note)
Fidicina Arnett 2000: 298 (diveristy) United States (error)
Fidicina Fornazier and Martinelli 2000: 1175 (comp. note) Brazil
Fidicina Boulard 2001b: 40 (comp. note)
Fidicina Sanborn 2001b: 449 (comp. note)
Fidicina Reis, Souza and Venzon 2002: 84 (coffee pest, comp. note) Brazil
Fidicina spp. Sueur 2002a: 387 (comp. note)
Fidicina sp. Almeida 2004: 107 (control) Brazil
Fidicina Campos, Silva, Ferreira and Dias 2004: 125 (comp. note) Brazil
Fidicina Martinelli 2004: 520, 524 (comp. note) Brazil
Fidicina Santos and Martinelli 2004a: 630 (comp. note) Brazil
Fidicina Boulard 2005d: 337 (comp. note) Neotropics
Fidicina Champman 2005: 329 (comp. note)
Fidicina Moulds 2005b: 410 (comp. note)
Fidicina spp. Salazar Escobar 2005: 196 (comp. note) Colombia
Fidicina Sanborn 2005a: 190 (comp. note) Costa Rica
Fidicina Boulard 2006g: 35–36 (comp. note)
Fidicina spp. Bento 2007: 27 (coffee pest) Brazil
Fidicina Sanborn 2007a: 26 (comp. note)
Fidicina Shiyake 2007: 4, 46, 95 (listed, comp. note)
Fidicina spp. Zinn, Lal, Bingham and Resck 2007: 1223 (comp. note) Brazil
Fidicina Costa Neto 2008: 453 (comp. note)
Fidicina Sanborn 2008a: 228–229 (comp. note)
Fidicina Sanborn, Moore and Young 2008: 1–2, 19 (comp. note) Costa Rica
Fidicina Santos and Martinelli 2009a: 559–560 (comp. note) Brazil
Fidicina Santos and Martinelli 2009b: 638 (comp. note)
Fidicina Sanborn 2010a: 1589 (listed) Colombia
Fidicina Santos, Martinelli, Maccagnan, Sanborn and Ribeiro 2010: 49 (comp. note) Brazil

***F. affinis* Haupt, 1918**
***F. aldegondae* Kuhlgatz, 1902**
F. amoena Distant, 1899 to *Dorisiana amoena*
F. amoena Distant, 1899 to *Dorisiana amoena*
F. bogotana Distant, 1892 to *Dorisiana bogotana*
F. brisa (Walker, 1850) to *Dorisiana brisa*
F. cachla Distant, 1899 to *Fidicinoides cachla*
F. chlorogena Walker, 1850 to *Guyalna chlorogena*
***F. christinae* Boulard & Martinelli, 1996** = *Fidicina torresi* (nec Boulard & Martinelli)

Fidicina christinae Boulard and Martinelli 1996: 32–36, Plate VI, Figs 1–5, Plate VII, Figs 5–6 (n. sp., described, illustrated, comp. note) French Guiana
Fidicina christinae Sanborn 2007a: 27 (distribution, comp. note) Venezuela, French Guiana
Fidicina torresi (sic) Sanborn 2008f: 688 (comp. note) Brazil, Minas Gerais, French Guiana, Venezuela

F. "coffea" Young, 1977 to *Fidicinoides coffea* Sanborn, Moore & Young, 2008
F. compostela Davis, 1934 to *Fidicinoides compostela*
F. cuta Walker, 1850 to *Guyalna cuta*
F. determinata Walker, 1858 to *Fidicinoides determinata*
F. distanti Goding, 1925 to *Fidicinoides distanti*
F. drewseni (Stål, 1854) to *Dorisiana drewseni*

F. ethelae **(Goding, 1925)** = *Majeorona ethelae*
Fidicina ethelae Boulard and Martinelli 1996: 36–38, 40, 42–43, 77, Plate VIII, Figs 1–5, Plate XI, Figs 1–2 (n. comb., described, illustrated, comp. note) Equals *Majeorona ethelae* Colombia, Ecuador, Brazil, Venezuela, Panama, Peru, Costa Rica?
Fidicina ethelae Sanborn 2005a: 190 (comp. note) Costa Rica
Fidicina ethelae Sanborn 2007a: 28 (distribution, comp. note) Colombia, Ecuador, Peru, Venezuela, Panama, Costa Rica, Brazil
Fidicina ethelae Sanborn, Moore and Young 2008: 19 (comp. note) Costa Rica
Fidicina ethelae Sanborn 2010a: 1589 (synonymy, distribution, comp. note) Equals *Majeorona ethelae* Colombia, Ecuador, Peru, Venezuela, Panama, Costa Rica, Brazil

F. explanata **Uhler, 1903**
Fidicina explanata Henshaw 1903: 230 (listed) Brazil

F. flavibasalis Distant, 1905 to *Fidicinoides flavibasalis*
F. fumea Distant, 1883 to *Fidicinoides fumea*
F. glauca Goding, 1925 to *Dorisiana glauca*
F. "guayabana" Young, 1977 to *Fidicinoides guayabana* Sanborn, Moore & Young, 2008

F. innotabilis **(Walker, 1858)** = *Cicada innotabilis*
Fidicina innotabilis Boer 1995b: 40 (status) Equals *Cicada innotabilis*

F. lacteipennis Distant, 1905 to *Fidicinoides lacteipennis*

F. mannifera mannifera **(Fabricius, 1803)** = *Tettigonia mannifera* = *Cicada mannifera* = *Tibicen mannifera* = *Cicada (Fidicina) plebeja mannifera* = *Fidicina vinifera* (sic) = *Fidicina manifera* (sic) = *Cicada cantatrix* Germar, 1821 = *Fidicina cantatrix* = *Fidicina rana* Walker, 1850 = *Fidicina excavata* Walker, 1850 = *Fidicina divisa* Walker, 1858
Fidicina mannifera Vayssière 1955: 254 (listed, coffee pest) Brazil
Fidicina mannifera Mariconi 1958: 331–332, Fig 108 (illustrated, described, coffee pest, listed, comp. note) Brazil
Fidicina mannifera Heinrich 1967: 43 (coffee pest, comp. note) Brazil
Fidicina rana Cullen 1974: 38 (comp. note)
Fidicina mannifera Wolda 1977: 244, Table 1 (emergence, comp. note) Panama
Fidicina mannifera Bonaccorso 1979: 386 (predation, comp. note) Panama
Fidicina mannifera Young 1981a: 132–134, 136, 139–140, Fig 8, Tables 1–2 (illustrated, seasonality, habitat, comp. note) Costa Rica
Fidicina mannifera Young 1981b: 181–182, 184–185, 187, 189–192, Figs 4–6, Tables 1–2 (illustrated, seasonality, habitat, comp. note) Costa Rica
Fidicina mannifera Young 1981c: 5–8, 11–14, 18, 20, 26, 28, Fig 6, Fig 11, Tables 5–6 (illustrated, seasonality, habitat, comp. note) Costa Rica
Fidicina mannifera Young 1981d: 827 (acoustic behavior, comp. note) Costa Rica
Fidicina mannifera Young 1982: 102, 200, 290, Fig 3.15, Fig 5.7 (illustrated, activity patterns, comp. note) Costa Rica
Fidicina mannifera Young 1983a: 725–726, Fig 11.31 (illustrated, behavior, habitat, seasonality, comp. note) Costa Rica
Fidicina mannifera Young 1983b: 24–25 (illustrated, comp. note) Costa Rica
Fidicina mannifera Bartholomew and Barnhart 1984: 131, 135–144, Figs 2–6, Tables 1–4 (energetics, endothermy, morphometry) Panama, Central America
Fidicina mannifera Martinelli and Zucchi 1984: 1 (coffee pest, comp. note) Brazil
Fidicina mannifera Young 1984a: 170–171, 173, 176–177, 179, Tables 1–3 (hosts, comp. note) Costa Rica
Fidicina rana Popov 1985: 43–45, Fig 8 (acoustic system, comp. note)
Fidicina mannifera Johnson and Foster 1986: 416–421 (host preferences, comp. note) Panama, Costa Rica
Fidicina mannifera Klein, Bock, Kafka and Moore 1987b: 97 (comp. note)
Fidicina mannifera Martinelli and Zucchi 1987b: 469 (coffee pest, comp. note) Brazil
Fidicina divisa Boulard 1988d: 156–158, Figs 20–22 (head abnormality, illustrated, valid species, comp. note) French Guiana
Fidicina mannifera Boulard 1988d: 158 (comp. note)
Cicada mannifera Boulard 1988f: 25–26 (comp. note)
Fidicina mannifera Boulard 1988f: 61, 70 (type species of *Fidicina*, comp. note) Equals *Fidicina divisa*
Fidicina divisa Boulard 1988f: 69–70 (synonymy, comp. note)

Tettigonia mannifera Boulard 1988f: 69 (comp. note)
Fidicina mannifera Klein, Bock, Kafka and Moore 1988: 166 (comp. note)
Fidicina mannifera Martinelli and Zucchi 1989a: 5–8, 11, Fig 1 (coffee pest, described, illustrated, comp. note) Brazil, São Paulo
Fidicina mannifera Martinelli, Zucchi, Silveira Neto and Parra 1989: 168 (coffee pest, comp. note) Brazil, Minas Gerias
Fidicina mannifera Wolda 1989: 438–440, Figs 1–3, Tables 1–2 (emergence pattern, comp. note) Panama
Fidicina divisa Boulard 1990b: 178, 181, 183, Plate XXI, Figs 1–3 (illustrated, comp. note) Brazil
Fidicina mannifera Boulard 1990b: 85 (comp. note)
Fidicina mannifera Martinelli 1990: 12–13 (illustrated, comp. note) Brazil
Fidicina mannifera Toolson and Toolson 1991: 114 (comp. note)
Fidicina mannifera Young 1991: 191–192, 213, 229 (illustrated, comp. note) Costa Rica
Fidicina mannifera Wolda and Ramos 1992: 271–273, 276–277, Fig 17.1, Fig 17.5, Table 17.1 (distribution, listed, seasonality, comp. note) Panama
Fidicina mannifera Hogue 1993b: 233, 570 (behavior, listed, comp. note)
Fidicina mannifera Martinelli and Zucchi 1993a: 3 (exuvia described, comp. note) Brazil
Fidicina mannifera Martinelli and Zucchi 1993b: 174 (exuvia described, comp. note) Brazil
Fidicina mannifera Moore, Huber, Weber, Klein and Bock 1993: 217 (comp. note) Panama
Fidicina mannifera Wolda 1993: 370–378, Figs 3–4 (seasonality, acoustic behavior, comp. note) Panama, Costa Rica
Fidicina mannifera Braker and Greene 1994: 250 (population abundance, comp. note) Costa Rica
Fidicina divisa Boulard 1995a: 7–8 (comp. note)
Fidicina mannifera Boulard and Mondon 1995: 72 (type species of *Fidicina*)
F[idicina] rana Jiang, Yang, Tang, Xu and Chen 1995b: 229 (song frequency, comp. note)
Fidicina mannifera Sanborn, J.E. Heath, M.S. Heath and Noriega 1995: 325 (comp. note)
Fidicina mannifera Sanborn, M.S. Heath, J.E. Heath and Noriega 1995: 451–452, 455, 457–458, Table 3 (endothermy, comp. note) Central America, South America
Fidicina rana Sanborn and Phillips 1995a: 482 (comp. note)
Fidicina mannifera Sanborn and Phillips 1995a: 482 (comp. note)
Fidicina mannifera Boulard 1996d: 114, 139–140, Fig S11(a-d), Fig S11(e-f) (song analysis, sonogram, listed, comp. note) French Guiana
Tettigonia mannifera Boulard and Martinelli 1996: 12, 22, 24 (type species of *Fidicina*, comp. note)
Fidicina mannifera Boulard and Martinelli 1996: 14–18, 20–26, Plate I, Figs 1–5, Plate II, Figs 1–5, Plate III, Figs 1–2 (type species of *Fidicina* and Fidicinina, described, illustrated, synonymy, comp. note) Equals *Tettigonia mannifera* Equals *Cicada plebeja* Linnaeus Equals *Cicada cantatrix* Equals *Tibicen mannifera* Equals *Cicada (Fidicina) plebeja mannifera* Equals *Fidicina rana* Equals *Fidicina divisa* Equals *Fidicina africana* French Guiana, Brazil, Peru
Cicada plebeja Linnaeus Boulard and Martinelli 1996: 23–24 (comp. note)
Fidicina africana Boulard and Martinelli 1996: 23 (comp. note)
Fidicina divisa Boulard and Martinelli 1996: 23 (comp. note)
Fidicina excavata Boulard and Martinelli 1996: 23 (comp. note)
Fidicina mannifera Cocroft and Pogue 1996: 85–95, Figs 1–4, Table 1–4 (acoustic behavior, sonogram, comp. note) Peru
Fidicina rana Cocroft and Pogue 1996: 93 (comp. note) Brazil
Fidicina mannifera Pogue 1996: 314, 320 (listed, comp. note) Peru, Costa Rica
Cicada mannifera Boulard 1997c: 95 (comp. note)
Fidicina mannifera Boulard 1997c: 120 (comp. note)
Fidicina mannifera Martinelli and Zucchi 1997c: 135–141, Figs 1–2, Figs 6–7, Fig 20, Table 1 (distribution, host plants, key, illustrated, comp. note) Brazil, Argentina, Bolivia, Peru, Ecuador, Colombia, Panama, Guyana, Surinam, French Guiana
Fidicina mannifera Martinelli and Zucchi 1997d: 271–279, Table 1 (host plants, comp. note) Brazil
Fidicina mannifera Boulard 1998: 75, 80–81, 83, 87, Fig 1, Plate I, Fig 1 (illustrated, sonogram, comp. note) French Antilles, Guadeloupe, French Guiana
Fidicina mannifera Sanborn 1998: 94–95 (thermal biology, comp. note)
Fidicina mannifera Kartal and Zeybekoglu 1999a: 62 (comp. note)
Fidicina mannifera Albuquerque, Martinelli, Ros and Stülp 2000: 196 (coffee pest) Brazil
Fidicina mannifera Boulard 2000d: 95 (song, comp. note)
Fidicina mannifera Sanborn 2000: 552, 555 (comp. note)
Cicada mannifera Boulard 2001b: 16 (comp. note)
Fidicina mannifera Boulard 2001b: 40 (comp. note)

Fidicina mannifera Cooley and Marshall 2001: 851 (comp. note)
Fidicina mannifera Sueur 2001: 38, Table 1 (listed) Equals *Fidicin rana*
Fidicina mannifera Sanborn 2002b: 460–461 (thermal biology, comp. note)
Fidicina mannifera Sueur 2002a: 387–388 (comp. note) Panama
Fidicina mannifera Motta 2003: 19–22, Fig 2, Fig 6, Table 1 (exuvia, described, illustrated, key, comp. note) Brazil, Distrito Federal
Fidicina mannifera Salguero 2003: 30 (comp. note) Amazon
Fidicina mannifera Sanborn, Villet and Phillips 2003: 306–307 (comp. note)
Fidicina mannifera Sueur 2003: 2942 (comp. note)
Fidicina mannifera Villet, Sanborn and Phillips 2003c: 1442 (comp. note)
Fidicina mannifera Campos, Silva, Ferreira and Dias 2004: 125 (comp. note) Brazil
Fidicina mannifera Cominctti, Aoki and Souza 2004: 125 (sex ratio) Brazil
Fidicina mannifera Ferrerira, Aoki, and Souza 2004: 125 (host plants) Brazil, Mato Grosso do Sul
Fidicina mannifera Marques, Martinelli, Azevedo, Coutinho and Serra 2004: 121 (coffee pest, distribution, comp. note) Brazil, Bahia
Fidicina mannifera Martinelli 2004: 517, 519–524, 527, 529–531, Figs 18.1–18.2, Figs 18.3c-d, Table 18.1 (synonymy, described, illustrated, coffee pest, host plants, distribution, control, comp. note) Equals *Tettigonia mannifera* Equals *Cicada contratix* Equals *Cicada mannifera* Equals *Fidicina rana* Equals *Fidicina excavata* Equals *Fidicina divisa* Brazil, Argentina, Bolivia, Peru, Ecuador, Colombia, Guyana, Surinam, French Guiana, Panama, Guatemala
Fidicina mannifera Sanborn 2004b: 98 (comp. note)
Fidicina mannifera Sanborn, Villet and Phillips 2004a: 818, 820 (comp. note)
Fidicina mannifera Boulard 2005d: 337, 345 (comp. note)
Fidicina mannifera Cocroft and Pogue 2005: 326, Fig 2e (sonogram, comp. note)
Fidicina rana Salazar Escobar 2005: 193 (comp. note) Colombia
Fidicina mannifera Sanborn 2005a: 190 (comp. note) Costa Rica
Fidicina mannifera Bolcatto, Medrano and De Santis 2006: 7, 10, Map 1 (illustrated, distribution) Argentina
Fidicina mannifera Boulard 2006g: 33–34, 96, 150, Fig 110G (illustrated, comp. note) Neotropics, French Guiana
Fidicina mannifera Maccagnan, Martinelli, Sene and Prado 2006: 1 (sex ratio, comp. note) Brazil
Fidicina mannifera Maccagnan, Prado, Sene and Martinelli 2006a: 1 (song analysis, comp. note) Brazil, São Paulo
Fidicina mannifera Martinelli 2006: 1 (coffee pest, comp. note) Brazil
Fidicina mannifera Seabra, Pinto-Juma and Quartau 2006: 843 (comp. note)
Fidicina mannifera Veiga and Ferrari 2006: 212, Table 1(predation) Brazil, Amazonia
Fidicina mannifera Sanborn 2007a: 30 (comp. note)
Fidicina mannifera Santos and Martinelli 2007: 311 (coffee pest, comp. note) Brazil
Fidicina mannifera Shiyake 2007: 4, 44, 46, Fig 68 (illustrated, distribution, listed, comp. note)
Fidicina mannifera Thouvenot 2007b: 282–283, Fig 5 (illustrated, comp. note) French Guiana
Fidicina mannifera Sanborn 2008a: 227 (comp. note)
Fidicina mannifera Sanborn, Moore and Young 2008: 19 (comp. note) Costa Rica
Fidicina mannifera Aoki, Lopes and de Souza 2010: 162 (distribution) Mato Grosso do Sul
Tettigonia mannifera Sanborn 2010a: 1589 (type species of *Fidicina*) South America
Fidicina mannifera Sanborn 2010a: 1589–1590 (synonymy, distribution, comp. note) Equals *Cicada plebeja* (nec Scopoli) Equals *Tettigonia mannifera* Equals *Cicada cantratix* Equals *Tibicen mannifera* Equals *Cicada (Fidicina) mannifera* Equals *Fidicina rana* Equals *Fidicina africana* Colombia, Antilles, Bolivia, Brazil, Costa Rica, Ecuador, French Guiana, Guyana, Panama, Paraguay, Peru, Surinam, Neotropical Region, South America

***F. mannifera umbrilinea* Walker, 1858** = *Fidicina umbrilinea*
Fidicina umbrilinea Boulard and Martinelli 1996: 23 (comp. note)

***F. muelleri* Distant, 1892** = *Fidicina mülleri*

***F. obscura* Boulard & Martinelli, 1996**
Fidicina obscura Boulard and Martinelli 1996: 40–43, Plate X, Figs 1–5, Plate XI, Figs 5–6 (n. sp., described, illustrated, comp. note) Brazil

F. opalina (Germar, 1821) to *Fidicinoides opalina*

***F. parvula* Jacobi, 1904**

F. passerculus (Walker, 1850) to *Fidicinoides passerculus*

F. picea Walker, 1850 to *Fidicinoides picea*

F. pronoe (Walker, 1850) to *Fididinoides pronoe*

F. pullata Berg, 1879 to *Bergalna pullata*

F. roberti Distant, 1905 to *Fidicinoides roberti*

***F. robini* Boulard & Martinelli, 1996**
Fidicina robini Boulard and Martinelli 1996: 38–40, 42–43, Plate IX, Figs 1–5, Plate XI, Figs 3–4 (n. sp., described, illustrated, comp. note) Brazil

***F. rosacordis* (Walker, 1850)** = *Cicada rosacordis*

***F. rubricata* Distant, 1892**

***F. sawyeri* Distant, 1912**

***F. sciras* (Walker, 1850)** = *Carineta sciras* = *Fidicina mannifera* (nec Fabricius)
Fidicina mannifera (sic) Salazar Escobar 2005: 196, 202, Fig 7 (illustrated, comp. note) Colombia
Fidicina sciras Sanborn 2007a: 28 (distribution, comp. note) Colombia, Venezuela
Fidicina sciras Salazar and Sanborn 2009: 271 (comp. note) Equals *Fidicina mannifera* Salazar Escobar, 2005 (nec Fabricius) Colombia
Fidicina sciras Sanborn 2010a: 1590 (synonymy, distribution, comp. note) Equals *Carineta sciras* Colombia, Venezuela

F. sericans Stål, 1854 to *Fidicinoides sericans*

***F.* sp. 1 Pogue, 1997**
Fidicina sp. 1 Pogue 1996: 314 (listed) Peru

***F.* sp. 2 Pogue, 1997**
Fidicina sp. 2 Pogue 1996: 314 (listed) Peru

***F.* sp. 3 Pogue, 1997**
Fidicina sp. 3 Pogue 1996: 314 (listed) Peru

***F.* sp. 4 Pogue, 1997**
Fidicina sp. 4 Pogue 1996: 314 (listed) Peru

***F.* sp. 5 Pogue, 1997**
Fidicina sp. 5 Pogue 1996: 314 (listed) Peru

***F.* sp. 6 Pogue, 1997**
Fidicina sp. 6 Pogue 1996: 314 (listed) Peru

***F.* sp. 7 Pogue, 1997**
Fidicina sp. 7 Pogue 1996: 314 (listed) Peru

***F.* sp. 8 Pogue, 1997**
Fidicina sp. 8 Pogue 1996: 314 (listed) Peru

***F.* sp. 9 Pogue, 1997**
Fidicina sp. 9 Pogue 1996: 314 (listed) Peru

***F.* sp. 10 Pogue, 1997**
Fidicina sp. 10 Pogue 1996: 314 (listed) Peru

F. spinicosta (Walker, 1850) to *Fidicinoides spinicosta*

F. steindachneri Kuhlgatz & Melichar, 1902 to *Fidicinoides steindachneri*

***F. torresi* Boulard & Martinelli, 1996** = *Fidicina mannifera* (nec Fabricius) = *Fidicina christinae* (nec Boulard & Martinelli)
Fidicina mannifera (sic) Lizer y Trelles 1955: 3 (mate pest, distribution, life cycle, control, comp. note) Argentina, Misiones
Fidicina mannifera (sic) Lizer y Trelles 1957: 81 (mate pest, distribution, life cycle, control, comp. note) Argentina, Misiones
Fidicina mannifera (sic) Pachas 1960: 5, 13, 15 (mate pest, natural history, control, comp. note) Argentina, Misiones
Fidicina mannifera (sic) Sanborn, Noriega, Heath and Heath 1988: 603 (endothermy, comp. note) Argentina
Fidicina mannifera (sic) Sanborn 1991: 5753B (endothermy, comp. note)
Fidicina mannifera (sic) Sanborn, J.E. Heath, M.S. Heath and Noriega 1995: 326, Table 3 (thermal responses, comp. note)
Fidicina mannifera (sic) Sanborn, M.S. Heath, J.E. Heath and Noriega 1995: 451, 455–458, Table 2, Table 4 (endothermy, thermoregulation, comp. note) Argentina
Fidicina torresi Boulard and Martinelli 1996: 27–29, 34–35, Plate IV, Figs 1–5, Plate VII, Figs 1–2 (n. sp., described, illustrated, comp. note) Equals *Fidicina mannifera* Torres (partim) Argentina, Brazil
Fidicina mannifera Sanborn 1998: 95, 98–99, Table 1 (thermal responses, comp. note)
Fidicina torresi Sanborn 2002b: 460–461, 465, Table 1 (thermal biology, comp. note)
Fidicina torresi Sueur 2002a: 388 (comp. note)
Fidicina torresi Sanborn, Phillips and Villet 2003a: 349 (comp. note)
Fidicina mannifera De Santis, Medrano, Sanborn and Bolcatto 2007: 11–12, Fig 2 (distribution, comp. note)
Fidicina torresi De Santis, Medrano, Sanborn and Bolcatto 2007: 11 (comp. note) Equals *Fidicina mannifera* Argentina
Fidicina torresi Heath and Sanborn 2007: 488 (comp. note)
Fidicina mannifera (sic) Costa Neto 2008: 455 (comp. note)
Fidicina christinae (sic) Sanborn 2008f: 688 (comp. note) Brazil, Mato Grosso del Sur, Argentina

***F. toulgoeti* Boulard & Martinelli, 1996** = *Fidicina toulgoëti*
Fidicina toulgoëti Boulard and Martinelli 1996: 29–32, 34–35, Plate V, Figs 1–5, Plate VII, Figs 3–4 (n. sp., described, illustrated, comp. note) Brazil

F. "*variegata*" Young, 1977 to *Fidicinoides varigata* (Sanborn, 2005)

F. variegata Sanborn, 2005 to *Fidicinoides variegata*

F. viridifemur (Walker, 1850) to *Dorisiana viridifemur*

***F. vitellina* (Jacobi, 1904)** = *Cicada vitellina*

***Fidicinoides* Boulard & Martinelli, 1996**
Fidicinoides Boulard 1996d: 141 (listed)
Fidicinoides Boulard and Martinelli 1996: 20, 44 (n. gen., key, described, comp. note)
Fidicinoides Sanborn 2001b: 449 (comp. note)
Fidicinoides spp. Sueur 2002a: 387 (comp. note)
Fidicinoides Martinelli, Maccagnan, Ribeiro, Pereira and Santos 2004: 619 (comp. note) Brazil

Fidicinoides Santos and Martinelli 2004a: 630 (diversity) Brazil
Fidicinoides Boulard 2005d: 337 (comp. note)
Fidicinoides Boulard 2006g: 33, 35–36, 109 (comp. note) French Guiana
Fidicinoides Sanborn 2007b: 36 (comp. note) Mexico
Fidicinoides Santos and Martinelli 2007: 311 (comp. note) Brazil
Fidicinoides Costa Neto 2008: 453 (comp. note)
Fidicinoides Sanborn 2008a: 228 (comp. note)
Fidicinoides Sanborn 2008f: 688–689 (comp. note)
Fidicinoides Sanborn, Moore and Young 2008: 1–2, 17, 19 (key, comp. note) Costa Rica
Fidicinoides Ribeiro, Pereira, Martinelli and Maccagnan 2009: 265 (comp. note) Brazil, São Paulo
Fidicinoides Santos and Martinelli 2009a: 559 (comp. note) Brazil
Fidicinoides Santos and Martinelli 2009b: 638 (diversity, comp. note)
Fidicinoides Sanborn 2010a: 1590, 1593 (listed, comp. note) Colombia
Fidicinoides Sanborn 2010b: 26, 30 (comp. note)
Fidicinoides Santos, Martinelli, Maccagnan, Sanborn and Ribeiro 2010: 49, 52–54 (described, diversity, key, comp. note) Brazil, Costa Rica

F. *besti* Boulard & Martinelli, 1996
Fidicinoides besti Boulard and Martinelli 1996: 59–63, Plate XIX, Figs 1–5, Plate XXI, Fig 3 (n. sp., described, illustrated, comp. note) Brazil, Venezuela
Fidicinoides besti Sanborn 2007a: 26, 28, 30 (distribution, comp. note) Venezuela, Brazil
Fidicinoides besti Santos and Martinelli 2009b: 638 (listed)
Fidicinoides besti Santos, Martinelli, Maccagnan, Sanborn and Ribeiro 2010: 53–55, Fig 12 (illustrated, key, comp. note) Brazil

F. brisa (Walker, 1850) to *Dorisiana brisa*

F. *brunnea* Boulard & Martinelli, 1996
Fidicinoides brunnea Boulard and Martinelli 1996: 61–64, Plate XX, Figs 1–4, Plate XXI, Fig 4 (n. sp., described, illustrated, comp. note) Brazil, Bolivia, Peru
Fidicinoides brunnea Santos and Martinelli 2009b: 638 (listed)
Fidicinoides brunnea Santos, Martinelli, Maccagnan, Sanborn and Ribeiro 2010: 53–55, Fig 11 (illustrated, key, comp. note) Brazil

F. *cachla* (Distant, 1899) = *Fidicina cachla*
Fidicina cachla Sanborn 2005a: 190 (comp. note) Costa Rica
Fidicina cachla Sanborn 2006a: 77–78, Fig 1 (listed, distribution) Honduras, Costa Rica
Fidicinoides cachla Sanborn, Moore and Young 2008: 10, 17–19, Fig 28 (n. comb., illustrated, key, comp. note) Costa Rica

F. nr. *cachla* (Wolda, 1989) = *Dorisiana* nr. *cachla*
Dorisiana nr. *cachla* Wolda 1989: 438–440, Figs 1–3, Tables 1–2 (emergence pattern, comp. note) Panama
Dorisiana sp. nr. *cachla* Wolda and Ramos 1992: 272–273, Fig 17.1, Table 17.1 (distribution, listed, seasonality, comp. note) Panama
Dorisiana sp. nr. *cachla* Wolda 1993: 371, 374–379, Fig 6 (seasonality, acoustic behavior, comp. note) Panama

F. *carmenae* Santos & Martinelli, 2009
Fidicinoides carmenae Santos and Martinelli 2009b: 640–642, Fig 3 (n. sp., described, illustrated, comp. note) Brazil, Mato Grosso
Fidicinoides carmenae Santos, Martinelli, Maccagnan, Sanborn and Ribeiro 2010: 54–55 (comp. note) Brazil

F. *coffea* Sanborn, Moore & Young, 2008 = *Fidicina* "*coffea*" Young, 1977 = *Fidicina* sp. 2, the "coffee cicada" Young, 1984 = *Fidicina coffea* Young, 1984
Fidicina "coffea" Young 1981a: 131, 133–141, Fig 7, Fig 9, Tables 1–4 (illustrated, seasonality, habitat, comp. note) Costa Rica
Fidicina sp. 2 Young 1984a: 170–172, 176–177, Tables 1–2 (hosts, comp. note) Costa Rica
Fidicina coffea Young 1984a: 179, Table 3 (hosts, comp. note) Costa Rica
Fidicina "coffea" Sanborn, Moore and Young 2008: 2 (comp. note)
Fidicinoides coffea Sanborn, Moore and Young 2008: 2–10, 17, 19, Figs 1–13 (n. sp., synonymy, described, illustrated, song, sonogram, oscillogram, key, comp. note) Equals *Fidicina* " *coffea*" Equals *Fidicina* sp. 2, the "coffee cicada" Young, 1984 Equals *Fidicina coffea* Young, 1984 Costa Rica

F. *compostela* (Davis, 1934) = *Fidicina compostela*
Fidicina compostela Sanborn 1999a: 43 (type material, listed)
Fidicinoides compostela Sanborn 2007b: 36 (n. comb., distribution, comp. note) Mexico, Nayarit, Colima

F. *descampsi* Boulard & Martinelli, 1996
Fidicinoides descampsi Boulard and Martinelli 1996: 69–72, Plate XXIV, Figs 5–6, Plate XXV, Figs 1–4 (n. sp., described, illustrated, comp. note) Colombia
Fidicinoides descampsi Santos and Martinelli 2009b: 638 (listed)
Fidicinoides descampsi Sanborn 2010a: 1591 (distribution, comp. note) Colombia

***F. determinata* (Walker, 1858)** = *Fidicina determinata* = *Fidicina pronoe* (nec Walker) = *Fidicinoides picea* (nec Walker)

Fidicinoides determinata Boulard and Martinelli 1996: 46–48, Plate XIII, Figs 1–5, Plate XVI, Fig 1 (n. comb., described, illustrated, comp. note) Equals *Fidicina determinata* Equals *Fidicina picea* Equals *Fidicina pronoe* (partim) Venezuela, Guatemala

Fidicnoides determinata Sanborn 2001b: 449 (listed, comp. note) El Salvador

Fidicinoides determinata Motta 2003: 19–21, Fig 4, Table 1 (exuvia, described, illustrated, key, comp. note) Brazil, Distrito Federal

Fidicinoides determinata Sanborn 2006a: 77–78, Fig 1 (listed, distribution) Guatemala, Honduras, El Salvador

Fidicinoides determinata Sanborn 2007a: 28, 30 (distribution, comp. note) Venezuela, Mexico, Guatemala, Honduras, El Salvador

Fidicinoides determinata Sanborn 2007b: 36 (distribution, comp. note) Mexico, Veracruz, Jalisco, Nayarit,Venezuela, Guatemala, Honduras, El Salvador

Fidicinoides determinata Sanborn 2008f: 688–689 (synonymy, comp. note) Equals *Fidicinoides picea* (partim) Brazil, Rondonia, Venezuela, Guatemala, El Salvador, Honduras, Mexico

Fidicinoides determinata Santos and Martinelli 2009a: 559 (comp. note) Brazil

Fidicinoides determinata Santos and Martinelli 2009b: 638 (listed)

Fidicinoides determinata Santos, Martinelli, Maccagnan, Sanborn and Ribeiro 2010: 54–55 (key, comp. note) Brazil

***F. distanti* (Goding, 1925)** = *Fidicina distanti*

Fidicina distanti Sanborn 2005a: 190 (comp. note) South America

Fidicina distanti Sanborn 2007a: 26 (comp. note)

Fidicinoides distanti Sanborn 2007a: 26, 28, 30 (n. comb., distribution, comp. note) Equals *Fidicina distanti* Ecuador, Venezuela

Fidicinoides distanti Sanborn 2008f: 689 (comp. note) Brazil, Rondonia, Ecuador, Venezuela

Fidicinoides distanti Sanborn 2010a: 1591 (synonymy, distribution, comp. note) Equals *Fidicina distanti* Colombia, Ecuador, Venezuela, Brazil

Fidicinoides distanti Santos, Martinelli, Maccagnan, Sanborn and Ribeiro 2010: 53–55, Fig 8 (illustrated, key, comp. note) Brazil

***F. dolosa* Santos & Martinelli, 2009**

Fidicinoides dolosa Santos and Martinelli 2009b: 639–640, Fig 1 (n. sp., described, illustrated, comp. note) Brazil, Mato Grosso

Fidicinoides dolosa Santos, Martinelli, Maccagnan, Sanborn and Ribeiro 2010: 52, 55 (key, comp. note) Brazil

***F. duckensis* Boulard & Martinelli, 1996**

Fidicinoides duckensis Boulard and Martinelli 1996: 75–77, 80–81, Plate XXVII, Figs 1–5, Plate XXIX, Figs 3–4 (n. sp., described, illustrated, comp. note) Brazil

Fidicinoides duckensis Sanborn 2007a: 28, 30 (distribution, comp. note) Venezuela, Brazil

Fidicinoides duckensis Santos and Martinelli 2009a: 562 (comp. note) Brazil

Fidicinoides duckensis Santos and Martinelli 2009b: 638 (listed)

Fidicinoides duckensis Santos, Martinelli, Maccagnan, Sanborn and Ribeiro 2010: 52, 55 (key, comp. note) Brazil

***F. ferruginosa* Sanborn & Heath, 2011, nom. nud.**

Fidicina sp. Bolcatto, Medrano and De Santis 2006: 7–8, Fig 1F (illustrated, distribution) Argentina, Santa Fe, Córdoba

Fidicina sp. De Santis, Medrano, Sanborn and Bolcatto 2007: 4–5, 11, 14, 17, 19, Fig 1F, Fig 4, Table 1 (listed, distribution) Argentina, Córdoba, Santa Fe

***F. flavibasalis* (Distant, 1905)** = *Fidicina flavibasalis* = *Cicada* (circa) *rudis* Salazar Escobar, 2005

Cicada (circa) *rudis* Salazar Escobar 2005: 196, 203, Fig 10 (illustrated, comp. note) Colombia

Fidicinoides flavibasalis Sanborn, Moore and Young 2008: 17 (n. comb., comp. note) Ecuador

Fidicinoides flavibasalis Salazar and Sanborn 2009: 270 (comp. note) Equals *Cicada* (circa) *rudis* Salazar Escobar 2005 Colombia

Fidicinoides flavibasalis Sanborn 2010a: 1591–1592 (synonymy, distribution, comp. note) Equals *Fidicina flavibasalis* Equals *Cicada* (circa) *rudis* Salazar Escobar, 2005 Colombia, Ecuador

***F. flavipronotum* Sanborn, 2007**

Fidicinoides flavipronotum Sanborn 2007a: 22–26, 28, 30, Figs 1–4 (n. sp., described, illustrated, listed, comp. note) Venezuela

Fidicinoides flavipronotum Sanborn 2010a: 1592, 1603 (distribution, comp. note) Colombia, Venezuela

***F. fumea* (Distant, 1883)** = *Fidicina fumea* = *Dorisiana fumea*

Dorisiana fumea Wolda 1989: 438–440, Figs 1–3, Tables 1–2 (emergence pattern, comp. note) Panama

Dorisiana fumea Wolda and Ramos 1992: 272–273, Fig 17.1, Table 17.1 (distribution, listed, seasonality, comp. note) Panama

Fidicina fumea Sanborn 2005a: 190 (comp. note) Costa Rica

Fidicinoides fumea Sanborn, Moore and Young 2008: 17–19, Fig 26 (n. comb., illustrated, key, comp. note) Panama, Costa Rica

F. glauca (Goding, 1925) to *Dorisiana glauca*

***F. guayabana* Sanborn, Moore & Young, 2008** = *Fidicina "guayabana"* Young, 1977 = *Fidicina* sp. 3, the "guayabana cicada" Young, 1984 = *Fidicina guayabana* Young 1984

Fidicina sp. 3 Young 1984a: 171–172, 177, Tables 1–2 (hosts, comp. note) Costa Rica

Fidicina guayabana Young 1984a: 179, Table 3 (hosts, comp. note) Costa Rica

Fidicina "guayabana" Sanborn, Moore and Young 2008: 2 (comp. note)

Fidicinoides guayabana Sanborn, Moore and Young 2008: 2, 10–17, 19, Figs 14–25 (n. sp., synonymy, described, illustrated, song, sonogram, oscillogram, key, comp. note) Equals *Fidicina "guayabana"* Young Equals *Fidicina* sp. 3, the "guayabana cicada" Young, 1984 Equals *Fidicina guayabana* Young, 1984 Costa Rica

***F. jauffreti* Boulard & Martinelli, 1996**

Fidicinoides jauffreti Boulard and Martinelli 1996: 67–71, Plate XXIII, Figs 1–5, Plate XXIV, Figs 3–4 (n. sp., described, illustrated, comp. note) Brazil

Fidicinoides jauffreti Santos and Martinelli 2009b: 638 (listed)

Fidicinoides jauffreti Santos, Martinelli, Maccagnan, Sanborn and Ribeiro 2010: 53–55, Fig 15 (illustrated, key, comp. note) Brazil

***F. lacteipennis* (Distant, 1905)** = *Fidicina lacteipennis*

Fidicinoides lacteipennis Boulard and Martinelli 1996: 52–54, Plate XVI, Fig 3 (n. comb., described, illustrated, comp. note) Equals *Fidicina lacteipennis* Amazonas

Fidicinoides lacteipennis Santos and Martinelli 2009b: 638 (listed)

Fidicinoides lacteipennis Santos, Martinelli, Maccagnan, Sanborn and Ribeiro 2010: 54–55 (key, comp. note) Brazil

***F. opalina* (Germar, 1821)** = *Cicada opalina* = *Fidicina opalina* = *Fidicina phaeochlora* Walker, 1858 = *Fidicina phoeochlora* (sic) = *Fidicina poeochlora* (sic)

Cicada opalina Boulard 1988f: 25 (comp. note)

Cicada opalina Boulard 1997c: 95–96 (comp. note)

Cicada opalina Boulard 2001b: 16 (comp. note)

Fidicinoides opalina Sanborn, Moore and Young 2008: 17 (n. comb., comp. note) Argentina, Bolivia, Brazil, Peru

Fidicinoides opalina Santos, Martinelli, Maccagnan, Sanborn and Ribeiro 2010: 53–54, Fig 3 (illustrated, key, comp. note) Brazil

***F. passerculus* (Walker, 1850)** = *Cicada passerculus* = *Cicada lacrines* Walker, 1850 = *Fidicina lacrines*

Fidicina passerculus Wolda and Ramos 1992: 273, Table 17.2 (listed) Panama

Fidicinoides passerculus Sanborn, Moore and Young 2008: 17 (n. comb., comp. note) Brazil, Colombia, French Guiana

Fidicinoides passerculus Sanborn 2010a: 1592–1593 (synonymy, distribution, comp. note) Equals *Cicada passerculus* Equals *Cicada lacrines* Equals *Fidicina lacrines* Equals *Fidicina passerculus* Colombia, Brazil, French Guiana

Fidicinoides passerculus Santos, Martinelli, Maccagnan, Sanborn and Ribeiro 2010: 54–55 (key, comp. note) Brazil

***F. pauliensis* Boulard & Martinelli, 1996**

Fidicinoides pauliensis Boulard and Martinelli 1996: 55–57, 62–63, Plate XVII, Figs 1–5, Plate XXI, Fig 1 (n. sp., described, illustrated, comp. note) Brazil

Fidicinoides pauliensis Santos and Martinelli 2004b: 630 (coffee pest) Brazil

Fidicinoides pauliensis Santos and Martinelli 2007: 311–313, Figs 1–2 (coffee pest, described, illustrated, comp. note) Brazil, São Paulo

Fidicinoides pauliensis Santos, Pereira and Martinelli 2009: 627 (coffee pest) Brazil

Fidicinoides pauliensis Santos and Martinelli 2009b: 638 (listed)

Fidicinoides pauliensis Santos, Martinelli, Maccagnan, Sanborn and Ribeiro 2010: 53–55, Fig 5 (illustrated, key, comp. note) Brazil

***F. picea* (Walker, 1850)** = *Fidicina picea* = *Fidicina pertinax* Stål, 1864

Fidicina picea Wolda and Ramos 1992: 273, Table 17.2 (listed) Panama

Fidicinoides picea Boulard and Martinelli 1996: 20–21, Plate III, Figs 3–4, Plate XII, Figs 1–5 (n. comb., described, neallotype designation, illustrated, type species of *Fidicinoides*, comp. note) Equals *Fidicina picea* Equals *Fidicina pertinax* (nec *Fidicina determinata*) Mexico, Venezuela, Colombia, Ecuador

Fidicinoides picea Sueur 2000: 219 (comp. note) Mexico

Fidicinoides picea Sueur 2002a: 380–381, 385, 387–391, Fig 7, Fig 10, Figs 12–14, Table 1 (acoustic behavior, song, sonogram, oscillogram, comp. note) Mexico, Veracruz

Fidicinoides picea Sueur 2005: 215 (song, comp. note) Neotropics

Fidicina picea Boulard 2006g: 180 (listed)

Fidicinoides picea Sanborn 2007a: 28, 30 (distribution, comp. note) Colombia, Mexico, Panama, Venezuela, Ecuador, Brazil, Guyana, Central America

Fidicinoides picea Sanborn 2007b: 36 (distribution, comp. note) Mexico, Veracruz, Colombia,

Panama, Venezuela, Ecuador, Brazil, Guyana, Central America

Fidicinoides pertinax Sanborn 2008f: 689 (comp. note)

Fidicinoides picea Sanborn 2008f: 689 (comp. note)

Fidicinoides picea Santos and Martinelli 2009a: 559–562, Fig 1 (synonymy, illustrated, comp. note) Equals *Fidicina picea* Equals *Fidicina pertinax* Equals *Fidicina determinata* Equals *Fidicina picea* Brazil, Pará

Fidicinoides picea Santos and Martinelli 2009b: 638 (listed)

Fidicina picea Sanborn 2010a: 1590 (type species of *Fidicinoides*) Mexico

Fidicinoides picea Sanborn 2010a: 1592 (synonymy, distribution, comp. note) Equals *Fidicina picea* Equals *Fidicina pertinax* Colombia, Mexico, Panama, Venezuela, Ecuador, Brazil, Guyana

Fidicinoides picea Santos, Martinelli, Maccagnan, Sanborn and Ribeiro 2010: 53–55, Fig 6 (illustrated, key, comp. note) Brazil

***F. poulaini* Boulard & Martinelli, 1996**

Fidicinoides poulaini Boulard and Martinelli 1996: 73–75, 80–81, Plate XXVI, Figs 1–4, Plate, XXIX, Figs 1–2 (n. sp., described, illustrated, comp. note) Peru, Ecuador

Fidicinoides poulaini Santos and Martinelli 2009a: 559–562, Fig 2 (illustrated, comp. note) Brazil, Pará, Amazonas

Fidicinoides poulaini Santos and Martinelli 2009b: 638 (listed)

Fidicinoides poulaini Sanborn 2010a: 1592–1593 (distribution, comp. note) Colombia, Peru, Ecuador, Brazil

Fidicinoides poulaini Santos, Martinelli, Maccagnan, Sanborn and Ribeiro 2010: 52, 55 (key, comp. note) Brazil

***F. pronoe* (Walker, 1850)** = *Cicada pronoe* = *Fidicina pronoe* = *Fidicina pronae* (sic) = *Fidicina prone* (sic) = *Fidicina vinula* Stål, 1854 = *Cicada compacta* Walker, 1858 = *Fidicina compacta*

Fidicina pronoe Escalante 1974: 120 (comp. note) Peru

Fidicina pronoe Young 1981a: 132–136, 139–140, Fig 8, Table 1–2 (illustrated, seasonality, habitat, comp. note) Costa Rica

Fidicina pronoe Young 1981c: 6, 8, 10–12, 18, 20–21, 26, Fig 6, Fig 11, Table 5 (illustrated, seasonality, habitat, comp. note) Costa Rica

Fidicina pronoe Young 1981d: 827 (acoustic behavior, comp. note) Costa Rica

Fidicina pronoe Young 1982: 102–103, 119, 184, 200, 208–209, 417, Fig 3.16, Fig 3.21, Fig 5.7, Table 5.6 (illustrated, activity patterns, hosts, comp. note) Costa Rica

Fidicina pronoe Young 1983b: 24 (illustrated, comp. note) Costa Rica

Fidicina pronoe Martinelli and Zucchi 1984: 1 (coffee pest, comp. note) Brazil

Fidicina sp. Reis, Souza and Melles 1984: 4, 6, 8–9, Fig 7, Fig 9, Table 1 (coffee pest, listed) Brazil

Fidicina pronoe Young 1984a: 170–171, 173, Table 1 (hosts, comp. note) Costa Rica

Fidicina pronoe Martinelli and Zucchi 1987b: 469 (coffee pest, comp. note) Brazil

Fidicina pronoe Martinelli and Zucchi 1987c: 470 (coffee pest, comp. note) Brazil, São Paulo

Fidicina pronoe Martinelli and Zucchi 1989a: 5–6, 8–9, 12, Fig 2 (coffee pest, described, illustrated, comp. note) Brazil, São Paulo

Fidicina pronoe Martinelli, Zucchi, Silveira Neto and Parra 1989: 168 (coffee pest, comp. note) Brazil, Minas Gerias, São Paulo, Mato Grosso, Mexico, Costa Rica, Guatemala, Panama, Colombia, Venezuela, Peru

Fidicina pronoe Martinelli 1990: 12–13 (illustrated, comp. note) Brazil

Fidicina pronoe Young 1991: 189, 212, 228 (illustrated, comp. note) Costa Rica

Fidicina pronoe Villet 1992: 96 (comp. note)

Fidicina pronoe Wolda and Ramos 1992: 272–273, Table 17.1 (distribution, listed, comp. note) Panama

Fidicina pronoe Wolda 1993: 370–377, Figs 3–4 (seasonality, acoustic behavior, comp. note) Panama

Fidicinoides pronoe Boulard and Martinelli 1996: 46, 49–51, Plate XIV, Figs 1–5 (n. comb., described, neallotype designation, illustrated, comp. note) Equals *Cicada pronoe* Equals *Cicada vinula* Equals *Cicada compacta* Equals *Fidicina pronoe* Mexico, Costa Rica, Ecuador, Colombia, Peru

Fidicina pronoe Moore 1996: 222 (comp. note) Mexico

Fidicina pronoe Martinelli and Zucchi 1997c: 135–141, Figs 1–2, Fig 8, Fig 19, Table 1 (distribution, host plants, key, illustrated, comp. note) Brazil, Peru, Colombia, Venezuela, Panama, Costa Rica, Guatemala, Mexico

Fidicina pronoe Poole, Garrison, McCafferty, Otte and Stark 1997: 331 (listed) Equals *Cicada pronoe* Equals *Fidicina vinula* Equals *Cicada compacta* North America

Fidicina pronoe Shelly and Whittier 1997: 279, Table 16–1 (listed, comp. note)

Fidicina pronoe Albuquerque, Martinelli, Ros and Stülp 2000: 196 (coffee pest) Brazil

Fidicina pronoe Arnett 2000: 298 (distribution) United State (error)

Fidicinoides pronoe Sanborn 2001b: 449 (listed, comp. note) El Salvador
Fidicina pronoe Sueur 2001: 38, Table 1 (listed)
Fidicina pronoe Reis, Souza and Venzon 2002: 84, Fig 1 (coffee pest, illustrated, control, comp. note) Brazil
Fidicinoides pronoe Sueur 2002a: 380–381, 385–391, Fig 8, Fig 10, Figs 12–14, Table 1 (acoustic behavior, song, sonogram, oscillogram, comp. note) Mexico, Veracruz, Panama
Fidicinoides pronoe Motta 2003: 19–22, Fig 8, Table 1 (exuvia, described, illustrated, key, comp. note) Brazil, Distrito Federal
Fidicina prone (sic) Almeida 2004: 104–105 (control) Brazil
Fidicina pronoe Almeida 2004: 105, 107 (control) Brazil
Fidicinoides pronoe Martinelli 2004: 518–521, 523–525, 529, Figs 18.1–18.2, Figs 18.3e-f (synonymy, described, illustrated, coffee pest, host plants, distribution, control, comp. note) Equals *Cicada pronoe* Equals *Cicada compacta* Equals *Fidicina pronoe* Equals *Fidicina vinula* Brazil, Peru, Colombia, Venezuela, Costa Rica, Mexico
Fidicinoides pronoe Martinelli, Maccagnan, Ribeiro, Pereira and Santos 2004: 619 (coffee pest, comp. note) Brazil
Fidicinoides pronoe Santos and Martinelli 2004b: 630 (coffee pest) Brazil
Fidicinoides pronae (sic) Scarpellini 2004: 114 (coffee pest, control, comp. note) Brazil, Minas Gerias, São Paulo
Fidicina pronoe Salazar Escobar 2005: 193, 195, 201, Fig 6 (illustrated, comp. note) Colombia
Fidicinoides pronoe Sueur 2005: 215 (song, comp. note) Neotropics
Fidicinoides pronoe Martinelli 2006: 1 (coffee pest, comp. note) Brazil
Fidicina pronoe Santos and Martinelli 2007: 311 (coffee pest, comp. note) Brazil
Fidicinoides pronoe Sanborn 2005b: 77–78, Fig 1 (listed, distribution) Honduras, El Salvador, Guatemala
Fidicinoides pronoe Sanborn 2007a: 28, 30 (distribution, comp. note) Colombia, Mexico, Honduras, El Salvador, Guatemala, Costa Rica, Panama, Venezuela, Ecuador, Peru, Brazil, Trinidad and Tobago, Central America, South America
Fidicinoides pronoe Sanborn 2007b: 36 (distribution, comp. note) Mexico, Oaxaca, Veracruz, Colombia, Honduras, El Salvador, Guatemala, Costa Rica, Panama, Venezuela, Ecuador, Peru, Brazil, Trinidad and Tobago, Central America, South America
Fidicinoides pronoe Sanborn, Moore and Young 2008: 17–19, Fig 27 (illustrated, key, comp. note)
Fidicinoides pronoe Salazar and Sanborn 2009: 270 (comp. note) Equals *Fidicina pronoe* Colombia
Fidicinoides pronoe Santos and Martinelli 2009a: 559–560 (comp. note) Brazil
Fidicinoides pronoe Santos and Martinelli 2009b: 638 (listed)
Fidicinoides pronoe Santos, Pereira and Martinelli 2009: 627 (coffee pest) Brazil
Fidicinoides pronoe Lara, Perioto and de Ffreitas 2010: 116 (coffee pest) Brazil
Fidicinoides pronoe Sanborn 2010a: 1593 (synonymy, distribution, comp. note) Equals *Cicada pronoe* Equals *Cicada compacta* Equals *Fidicina pronoe* Colombia, Mexico, Honduras, El Salvador, Guatemala, Costa Rica, Panama, Venezuela, Ecuador, Peru, Brazil, Trinidad and Tobago
Fidicinoides pronoe Santos, Martinelli, Maccagnan, Sanborn and Ribeiro 2010: 53–55, Fig 7 (illustrated, key, comp. note) Brazil

F. pseudethelae **Boulard & Martinelli, 1996**

Fidicinoides pseudethelae Boulard 1996d: 115, 141–142, Fig S12(a-d), Fig S12(e-f) (song analysis, sonogram, listed, comp. note) French Guiana
Fidicinoides pseudethelae Boulard and Martinelli 1996: 77–81, Plate XXVIII, Figs 1–5, Plate XXIX, Figs 5–6 (n. sp., described, illustrated, comp. note) French Guiana
Fidicinoides pseudethelae Boulard 1998: 83, 85, 87, Fig 3, Plate I, Fig 4 (illustrated, sonogram, comp. note) French Guiana
Fidicinoides pseudethelae Sueur 2001: 38, Table 1 (listed)
Fidicinoides pseudethelae Boulard 2005d: 345 (comp. note) French Guiana
Fidicinoides pseudethelae Boulard 2006g: 74–75, 88, 180, Fig 51 (sonogram, listed, comp. note) French Guiana
Fidicinoides pseudethelae Sanborn 2007a: 28, 30 (distribution, comp. note) Venezuela, French Guiana, Brazil
Fidicinoides pseudethelae Thouvenot 2007b: 282, Fig 4 (illustrated, comp. note) French Guiana
Fidicinoides pseudethelae Santos and Martinelli 2009b: 638 (listed)
Fidicinoides pseudethelae Santos, Martinelli, Maccagnan, Sanborn and Ribeiro 2010: 53–55, Fig 16 (illustrated, key, comp. note) Brazil

F. roberti **(Distant, 1905)** = *Fidicina roberti*

Fidicinoides roberti Boulard and Martinelli 1996: 52–55, Plate XVI, Fig 4 (n. comb., described,

illustrated, comp. note) Equals *Fidicina roberti* Brazil
Fidicinoides roberti Santos and Martinelli 2009b: 638–639 (listed, comp. note)
Fidicinoides roberti Santos, Martinelli, Maccagnan, Sanborn and Ribeiro 2010: 54–55 (key, comp. note) Brazil

***F. rosabasalae* Santos & Martinelli, 2009**
Fidicinoides rosabasalae Santos and Martinelli 2009b: 639–641, Fig 2 (n. sp., described, illustrated, comp. note) Brazil, Rio de Janiero
Fidicinoides rosabasalae Santos, Martinelli, Maccagnan, Sanborn and Ribeiro 2010: 53–55, Fig 10 (illustrated, key, comp. note) Brazil

***F. saccifera* Boulard & Martinelli, 1996**
Fidicinoides saccifera Boulard and Martinelli 1996: 65–67, 70–71, Plate XXII, Figs 1–5, Plate XXIV, Figs 1–2 (n. sp., described, illustrated, comp. note) French Guiana, Brazil
Fidicinoides saccifera Santos and Martinelli 2009b: 638 (listed)
Fidicinoides saccifera Santos, Martinelli, Maccagnan, Sanborn and Ribeiro 2010: 53–55, Fig 14 (illustrated, key, comp. note) Brazil

***F. sarutaiensis* Santos, Martinelli & Maccagnan, 2010**
Fidicinoides sp. Ribeiro, Pereira, Martinelli and Maccagnan 2009: 265–266, Tables 1–2 (coffee pest, comp. note) Brazil, São Paulo
Fidicinoides sarutaiensis Santos, Martinelli, Maccagnan, Sanborn and Ribeiro 2010: 50–52, 54–55, Figs 1–2 (n. sp., described, illustrated, key, comp. note) Brazil, São Paulo

***F. sericans* (Stål, 1854)** = *Fidicina sericans* = *Fidicina sericana* (sic)
Fidicina sericans Young 1981a: 132–136, 139–140, Fig 8, Tables 1–2 (illustrated, seasonality, habitat, comp. note) Costa Rica
Fidicina sericans Young 1981b: 180, 182–183, 185–191, 193, Figs 5–6, Tables 1–2 (illustrated, seasonality, habitat, comp. note) Costa Rica
Fidicina sericans Young 1981c: 14 (comp. note) Costa Rica
Fidicina sericans Young 1981d: 827 (acoustic behavior, comp. note) Costa Rica
Fidicina sericans Young 1982: 102–103, 162, 200, 208, Fig 3.16, Fig 5.7, Table 5.6 (illustrated, activity patterns, hosts, comp. note) Costa Rica
Fidicina sericans Young 1983b: 24–25 (illustrated, comp. note) Costa Rica
Fidicina sericans Young 1984a: 169–171, 173, 175–176, 178–179, 181, 187, 189, Tables 1–3 (hosts, comp. note) Costa Rica
Fidicina sericana (sic) Williams 1991: 28 (comp. note)
Fidicina sericans Young 1991: 191, 212, 214, 248 (illustrated, comp. note) Costa Rica
Fidicina sericans Wolda and Ramos 1992: 278 (comp. note) Costa Rica
Fidicina sericans Braker and Greene 1994: 249 (population abundance, comp. note) Costa Rica
Fidicina sericans Shelly and Whittier 1997: 279, Table 16–1 (listed, comp. note)
Fidicina sericans Humphreys 2005: 74, Table 4 (population density, comp. note) Costa Rica
Fidicina sericans Sanborn 2005a: 190 (comp. note) Costa Rica
Fidicinoides sericans Sanborn, Moore and Young 2008: 17–19, Fig 29 (n. comb., illustrated, key, comp. note) Brazil, Costa Rica
Fidicinoides sericans Santos, Martinelli, Maccagnan, Sanborn and Ribeiro 2010: 53–54, Fig 4 (illustrated, key, comp. note) Brazil

***F. spinicosta* (Walker, 1850)** = *Cicada spinicosta* = *Fidicina spinicosta* = *Fidicina spinocosta* (sic) = *Carinetta* (sic) *spinicosta* = *Carineta* (sic) *spinicosta*
Fidicina spinicosta Young 1981a: 132–134, 136, 139–140, Fig 8, Tables 1–2 (illustrated, seasonality, habitat, comp. note) Costa Rica
Fidicina spinicosta Young 1981d: 827 (acoustic behavior, comp. note) Costa Rica
Fidicina spinicosta Young 1983b: 23 (illustrated, comp. note) Costa Rica
Fidicina spinicosta Young 1984a: 170–171, 173, 175, 189, Table 1 (hosts, comp. note) Costa Rica
Carineta (sic) *spinicosta* Young 1984a: 176, 179, Tables 2–3 (hosts, comp. note) Costa Rica
Carinetta (sic) *spinicosta* Humphreys 2005: 71 (population density, comp. note) Costa Rica
Fidicina spinicosta Humphreys 2005: 74, Table 4 (population density, comp. note) Costa Rica
Fidicina spinicosta Salazar Escobar 2005: 196, 201, Fig 8 (illustrated, comp. note) Colombia
Fidicina spinicosta Sanborn 2005a: 190 (comp. note) Costa Rica
Fidicinoides spinicosta Sanborn, Moore and Young 2008: 18–19, Fig 31 (n. comb., illustrated, key, comp. note) Brazil, Colombia, Costa Rica, Panama
Fidicinoides spinicosta Salazar and Sanborn 2009: 270 (comp. note) Equals *Fidicina spinicosta* Colombia
Fidicinoides spinicosta Sanborn 2010a: 1593 (synonymy, distribution, comp. note) Equals *Cicada spinicosta* Equals *Fidicina spinicosta* Colombia, Brazil, Panama, Costa Rica
Fidicinoides spinicosta Santos, Martinelli, Maccagnan, Sanborn and Ribeiro 2010: 54–55 (key, comp. note) Brazil

***F. steindachneri* (Kuhlgatz & Melichar, 1902)** = *Fidicina steindachneri*

Fidicina steindachneri Sanborn 2007a: 28 (distribution, comp. note) Venezuela, Brazil
Fidicinoides steindachneri Sanborn, Moore and Young 2008: 19 (n. comb., comp. note) Brazil, Venezuela
Fidicinoides steindachneri Santos, Martinelli, Maccagnan, Sanborn and Ribeiro 2010: 54–55 (key, comp. note) Brazil

***F. sucinalae* Boulard & Martinelli, 1996**

Fidicinoides sucinalae Boulard and Martinelli 1996: 57–59, 62–63, Plate XVIII, Figs 1–5, Plate XXI, Fig 2 (n. sp., described, illustrated, comp. note) Brazil
Fidicinoides sucinalae Santos and Martinelli 2009b: 638 (listed)
Fidicinoides sucinalae Santos, Martinelli, Maccagnan, Sanborn and Ribeiro 2010: 53–55, Fig 13 (illustrated, key, comp. note) Brazil

***F. variegata* (Sanborn, 2005)** = *Fidicina variegata* = *Fidicina* "variegata" Young, 1977 = *Fidicina* n. sp. Young, 1981 = *Fidicina* sp. 1 Young, 1984

Fidicina n. sp. Young 1981b: 179, 182–183, 185–187, 189–191, 194, Fig 3, Figs 5–6, Tables 1–2 (illustrated, seasonality, habitat, comp. note) Costa Rica
Fidicina sp. 1 Young 1984a: 170, Table 1 (hosts, comp. note) Costa Rica
Fidicina variegata Sanborn 2005a: 187–190, Figs 1–4 (n. sp., described, illustrated, distribution, comp. note) Equals *Fidicina "variegata"* Young Costa Rica
Fidicinoides variegata Sanborn, Moore and Young 2008: 17–19, Fig 30 (n. comb., illustrated, key, comp. note) Costa Rica

F. viridifemur (Walker, 1850) to *Dorisiana viridifemur*

***F. yavitensis* Boulard & Martinelli, 1996**

Fidicinoides yavitensis Boulard and Martinelli 1996: 51–54, Plate XV, Figs 1–4, Plate XVI, Fig 2 (n. sp., described, illustrated, comp. note) Venezuela
Fidicinoides yavitensis Sanborn 2007a: 28, 30 (distribution, comp. note) Venezuela
Fidicinoides yavitensis Santos and Martinelli 2009b: 638 (listed)

***Bergalna* Boulard & Martinelli, 1996**

Bergalna Boulard and Martinelli 1996: 20 (n. gen., key, described, comp. note)
Bergalna Sanborn 2010a: 1593–1594 (listed) Colombia

***B. pullata* (Berg, 1879)** = *Fidicina pullata* = *Fidicina fullata* (sic) = *Fidicina pulata* (sic)

Fidicina pullata d'Utra 1908: 355–356, Figs 3–4 (illustrated, coffee pest, control, comp. note) Brazil
Fidicina pullata Hempel 1913a: 140 (illustrated, coffee pest, comp. note) Brazil
Fidicina pullata Moreira 1921: 125 (coffee pest, comp. note) Brazil
Fidicina pulata (sic) Costa Lima 1922: 111 (coffee pest, comp. note) Brazil
Fidicina pullata Costa Lima 1922: 81 (coffee pest, comp. note) Brazil, São Paulo
Fidicina pulata (sic) Moreira 1928: 24 (coffee pest, comp. note) Brazil
Fidicina pullata Vayssière 1955: 254 (listed, coffee pest) Brazil
Fidicina pullata Mariconi 1958: 331, 333, Fig 109 (illustrated, described, coffee pest, listed, comp. note) Brazil
Fidicina pullata Heinrich 1967: 43 (coffee pest, comp. note) Brazil
Bergalna pullata Boulard and Martinelli 1996: 20 (n. comb., type species of *Bergalna*, comp. note) Equals *Fidicina pullata*
Fidicina pullata Martinelli and Zucchi 1984: 1 (coffee pest, comp. note) Brazil
Fidicina pullata Martinelli and Zucchi 1987b: 469 (coffee pest, comp. note) Brazil
Fidicina pullata Martinelli and Zucchi 1989a: 5–6, 9 (described, comp. note) Brazil
Fidicina pullata Martinelli, Zucchi, Silveira Neto and Parra 1989: 168 (coffee pest, comp. note) Brazil
Fidicina pullata Martinelli 1990: 13 (comp. note) Brazil
Fidicina pullata Martinelli and Zucchi 1997c: 135–138, 141, Figs 1–2, Table 1 (distribution, host plants, key, comp. note) Brazil, Argentina
Fidicina pullata Albuquerque, Martinelli, Ros and Stülp 2000: 196 (coffee pest) Brazil
Fidicina pullata Martinelli 2004: 518–521, 524, 529, Figs 18.1–18.2, Figs 18.3a-b, Table 18.1 (described, coffee pest, host plants, distribution, control, comp. note) Brazil, Argentina
Fidicina pullata Bolcatto, Medrano and De Santis 2006: 7, 10, Map 1 (distribution) Argentina, Santa Fe
Fidicina pullata Martinelli 2006: 1 (coffee pest, comp. note) Brazil
Fidicina pullata De Santis, Medrano, Sanborn and Bolcatto 2007: 11–12, Fig 2 (distribution, comp. note)
Fidicina pullata Santos and Martinelli 2007: 311 (coffee pest, comp. note) Brazil
Fidicina pullata Sanborn 2010a: 1593 (type species of *Bergalna*) Argentina

***B. xanthospila* (Germar, 1830)** = *Cicada xanthospila*

Bergalna xanthospila Sanborn 2010a: 1593–1594 (n. comb., synonymy, distribution, comp. note) Equals *Cicada xanthospila* Colombia, Brazil

Fidicina xanthospila Sanborn 2010a: 1593 (synonymy, distribution, comp. note) Colombia

Subtribe Guyalnina Boulard & Martinelli, 1996

Guyalnina Boulard and Martinelli 1996: 12, 14, 20 (n. subtribe, key, described, comp. note)
Guyalnina Sanborn 2006a: 77 (listed)
Guyalnina Sanborn 2007a: 28 (listed) Venezuela
Guyalnina Sanborn 2007b: 36 (listed) Mexico
Guyalnina Sanborn 2008f: 688 (listed)
Guyalnina Sanborn 2010a: 1594 (listed) Colombia
Guyalnina Sanborn 2010b: 75 (listed)
Guyalnina Santos, Martinelli, Maccagnan, Sanborn and Ribeiro 2010: 54 (comp. note)

***Proarna* Stål, 1864** = *Proarno* (sic) = *Ploarna* (sic) = *Proarma* (sic) = *Tympanoterpes (Proarua)* (sic) = *Proaria* (sic)

Proarno (sic) sp. Wetmore 1916: 77, 82 (predation) Puerto Rico
Proarna Tinkham 1941: 172 (comp. note)
Proarna sp. Ramos-Elorduy de Conconi and Pino Moreno 1979: 566, 569, 571, Tables I-II (chemical composition, human food, listed) Mexico, Hidalgo
Proarna sp. Ramos Elorduy de Conconi 1981: 566, 569, 571, Tables I-II (nutrient content, human food, listed, comp. note) Mexico, Hidalgo
Proarna Boulard 1982d: 108 (comp. note)
Proarna Ramos 1983: 62 (key) Dominican Republic
Proarna Arnett 1985: 217 (diversity) Southwestern United States (error)
Proarna Boulard 1985e: 1022, 1025 (coloration, comp. note)
Proarna Ramos 1988: 62–63, Table 4.1 (listed, distribution, comp. note) Greater Antilles, Cuba, Hispaniola, Jamaica, Puerto Rico, New World
Proarna Heath, Heath and Noriega 1990: 1 (distribution, comp. note) Argentina
Proarna spp. Sanborn, J.E. Heath, M.S. Heath and Noriega 1995: 321–322, 324, 326–327 (comp. note) Argentina
Proarna Moore 1996: 221 (comp. note) Mexico
Proarna Novotny and Wilson 1997: 437 (listed)
Proarna Poole, Garrison, McCafferty, Otte and Stark 1997: 251 (listed) North America
Proarna spp. Sanborn and Maté 2000: 144 (comp. note)
Proarna Boulard 2002b: 192 (comp. note) America
Proarna Santos and Martinelli 2004a: 630 (comp. note) Brazil
Proarna Sanborn, Villet and Phillips 2004a: 818 (comp. note)
Proarna Salazar Escobar 2005: 194 (comp. note) Colombia
Proarna Shiyake 2007: 4, 46, 95 (listed, comp. note)
Proarna Costa Neto 2008: 453, 456 (comp. note)
Proarna Perez-Gerlabert 2008: 202 (listed) Hispaniola
Proarna Sanborn 2008f: 685, 688 (diversity, comp. note) Neotropics, Mexico, Argentina
Proarna Santos and Martinelli 2009a: 559 (comp. note) Brazil
Proarna Sanborn 2009a: 90 (listed, comp. note) Cuba
Proarna Sanborn 2010a: 1586 (listed) Colombia
Proarna Santos, Martinelli, Maccagnan, Sanborn and Ribeiro 2010: 49 (comp. note) Brazil

***P. alalonga* Sanborn & Heath, 2011, nom. nud.**

***P. bergi* (Distant, 1892)** = *Tympanoterpes bergi* = *Proarma* (sic) *bergi* = *Proarna bergii* (sic)

Proarna bergi Costilla and Pastor 1986: 121 (control, comp. note)
Proarna bergi Wilson 1987: 489, Table 1 (sugarcane pest, listed, comp. note) Argentina
Proarna bergi Wang and Castor 1988: 101–102, 105, Fig 1 (illustrated, fungal host, sugarcane pest, comp. note) Argentina
Proarna bergi Ramallo 1989: 143 (fungal host, sugarcane pest, comp. note) Argentina
Proarna bergii (sic) Ramallo 1989: 144 (nymph) Argentina
Proarna bergi Sanborn, Heath, Noriega and Heath 1989: 154 (endothermy, comp. note) Argentina
Proarna bergi Sanborn 1991: 5753B (endothermy, comp. note)
Proarna bergi Sanborn, J.E. Heath, M.S. Heath and Noriega 1995: 320–327, Fig 1, Tables 1–3 (thermoregulation, endothermy, comp. note) Argentina
Proarna bergi Sanborn and Phillips 1996: 137 (comp. note)
Proarna bergi Sanborn 1998: 95–96, 99, Table 1 (thermal responses, comp. note)
Proarna bergi Sanborn 2000: 555 (comp. note)
Proarna bergi Sanborn and Maté 2000: 145–146 (comp. note)
Proarna bergi Sanborn 2002b: 460–461, 465, Table 1 (thermal biology, comp. note)
Proarna bergi Sanborn, Noriega and Phillips 2002c: 368 (comp. note)
Proarna bergi Sanborn, Villet and Phillips 2003: 307 (comp. note)
Proarna bergi Villet, Sanborn and Phillips 2003c: 1442 (comp. note)
Proarna bergi Sanborn, Villet and Phillips 2004a: 820 (comp. note)

Proarna bergi Sanborn 2005c: 114, Table 1 (comp. note)
Proarna bergi Bolcatto, Medrano and De Santis 2006: 7, 9, Fig 1H (illustrated, distribution) Argentina, Santa Fe
Proarna bergi De Santis, Medrano, Sanborn and Bolcatto 2007: 5, 11, 14, 17, 19, Fig 1H, Fig 4, Table 1 (listed, described, distribution) Argentina, Santa Fe
Proarna bergi Shiyake 2007: 4, 44, 46, Fig 66 (illustrated, listed, comp. note)
Proarna bergi Toledo, Marino de Remes Lenicov and López Lastra 2007: 226 (fungal host, comp. note)
Proarna bergi Phillips and Sanborn 2010: 74 (endothermy, comp. note) Argentina
Proarna bergi Sanborn 2010a: 1601 (synonymy, distribution, comp. note) Equals *Tympanoterpes bergi* Argentina

***P. bufo* Distant, 1905**
Proarna bufo Sanborn, J.E. Heath, M.S. Heath and Noriega 1995: 323–324 (status, comp. note) Argentina
Proarna bufo Bolcatto, Medrano and De Santis 2006: 7, 9–10, Fig 1I, Map 2 (illustrated, distribution) Argentina, Santa Fe
Proarna bufo De Santis, Urteaga and Bolcatto 2006: 1 (comp. note) Argentina, Santa Fe
Proarna bufo De Santis, Medrano, Sanborn and Bolcatto 2007: 6–7, 11, 13–14, 17, 19, Fig 1I, Figs 3–4, Table 1 (listed, described, distribution, comp. note) Argentina, Córdoba, La Rioja, Santa Fe

***P. cocosensis* Davis, 1935**
Proarna cocosensis Sanborn 1999a: 43 (type material, listed)

***P. dactyliophora* Berg, 1879**
Proarna dactyliophora Bolcatto, Medrano and De Santis 2006: 7, 9–10, Fig 1J, Map 2 (illustrated, distribution) Argentina, Santa Fe, Córdoba
Proarna dactyliophora De Santis, Urteaga and Bolcatto 2006: 1 (comp. note) Argentina, Santa Fe
Proarna dactyliophora De Santis, Medrano, Sanborn and Bolcatto 2007: 7–9, 11, 13–14, 17, 19, Fig 1J, Figs 3–4, Table 1 (listed, described, distribution) Argentina, Santa Fe, Córdoba
Proarna dactyliophora Sanborn 2007a: 27 (distribution, comp. note) Venezuela, Bolivia, Argentina
Proarna dactyliophora Sanborn 2010a: 1586–1587 (distribution, comp. note) Colombia, Venezuela, Bolivia, Argentina

***P. germari* Distant, 1905** = *Cicada grisea* Germar, 1821 (nec *Tettigonia grisea* Fabricius, 1775) = *Tympanoterpes grisea* (nec Fabricius)
Proarna germari Wolda and Ramos 1992: 273, Table 17.2 (listed) Panama

***P. gianucai* Sanborn 2008**
Proarna gianucai Sanborn 2008f: 686–688, Figs 1–6 (n. sp., described, illustrated, listed, comp. note) Brazil, Rio Grande de Sul

***P. grisea* (Fabricius, 1775)** = *Tettigonia grisea* = *Cicada grisea*
Tettigonia grisea Dallas 1870: 482 (synonymy) Equals *Proarna grisea*
Proarna grisea Sanborn 2007a: 27 (distribution, comp. note) Venezuela, West Indies, Guyana, Ecuador, Peru, Argentina, South America

***P. guttulosa* (Walker, 1858)** = *Cicada guttulosa*
Proarna guttulosa Pogue 1996: 314 (listed) Peru
Proarna guttulosa Sanborn 2010a: 1587 (synonymy, distribution, comp. note) Equals *Cicada guttulosa* Colombia, Ecuador, Peru, South America

***P. hilaris* (Germar, 1834)** = *Cicada hilaris* = *Tympanoterpes hilaris* = *Poecilopsaltria hilaris* = *Tympanoterpes (Proarua)* (sic) *hilaris* = *Odopoea hilaris* = *Proalba* (sic) *hilaris* = *Cicada subtincta* Walker, 1850 = *Tympanoterpes subtincta* = *Cicada albiflos* Walker, 1850 = *Tympanoterpes albiflos* = *Cicada tomentosa* Walker, 1858 = *Odopoea tomentosa* = *Tympanoterpes tomentosa*
Proarno (sic) *hilaris* Wetmore 1916: 57–59, 62–63, 77, 80–82, 95–98, 106, 114, 116, 118–119 (predation) Puerto Rico
Proarna hilaris Wolcott 1948: 103 (distribution, predation, habitat, comp. note) Puerto Rico
Proarna hilaris Ramos 1983: 62, 64, 66–67, 70, Fig 16 (synonymy, described, illustrated, key, listed, comp. note) Equals *Cicada hilaris* Dominican Republic
Cicada hilaris Ewart 1990: 1 (comp. note)
Proarna hilaris Boulard 1998: 83 (comp. note) French Antilles, Guadeloupe, Martinique
Proarna hilaris Bartlett 2000: 122 (listed) Guana Island, British West Indies
Proarna hilaris Howard, Moore, Giblin-Davis and Abad 2001: 157 (pest species, comp. note) Jamaica
Proarna hilaris Sanborn 2007a: 27 (distribution, comp. note) Venezuela, Caribbean Basin
Proarna hilaris Sanborn 2007b: 35 (distribution, comp. note) Mexico, Caribbean Islands, Venezuela
Proarna hilaris Perez-Gerlabert 2008: 202 (listed) Hispaniola
Proarna hilaris Sanborn 2009a: 90 (type species of *Proarna*, listed, synonymy, key, comp. note) Equals *Cicada hilaris* Equals *Cicada subtincta* Equals *Cicada albiflos* Equals *Cicada tomentosa* Cuba, Mexico, Venezuela, South America, Caribbean Basin

Proarna hilaris Sanborn 2009d: 6 (comp. note) Hispaniola

Cicada hilaris Sanborn 2010a: 1586 (type species of *Proarna*)

P. *insignis* Distant, 1881 = *Proarna alba insignis* = *Proaria* (sic) *insignis*

Proarna insignis Young 1984a: 171, Table 1 (hosts, comp. note) Costa Rica

Proarna insignis Boulard 1985e: 1019, 1032, 1037, Plate I, Fig 3 (crypsis, coloration, illustrated, comp. note) Brazil, Neotropics

Proarna insignis Sanborn, Heath, Noriega and Heath 1989: 154 (endothermy, comp. note) Argentina

Proarna insignis Sanborn 1991: 5753B (endothermy, comp. note)

Proarna insignis Wolda and Ramos 1992: 273, Table 17.2 (listed) Panama

Proarna insignis Sanborn, J.E. Heath, M.S. Heath and Noriega 1995: 320–327, Fig 2, Tables 1–3 (thermoregulation, endothermy, comp. note) Argentina, Nicaragua, Panama

Proarna insignis Maes 1998: 144 (listed, distribution, host plants) Nicaragua, Mexico, Costa Rica, Panama, Colombia, Venezuela, Ecuador, Peru, Brazil

Proarna sp. (sic) Maes 1998: 144 (listed, distribution) Nicaragua

Proarna insignis Sanborn 1998: 94–96, 99, Table 1 (thermal responses, comp. note)

Proarna insignis Sanborn 2000: 555 (comp. note)

Proarna insignis Sanborn 2002b: 460–461, 463, 465, Fig 5, Table 1 (illustrated, thermal biology, comp. note)

Proarna insignis Sanborn, Noriega and Phillips 2002c: 368 (comp. note)

Proarna insignis Sanborn, Villet and Phillips 2003: 307 (comp. note)

Proarna insignis Villet, Sanborn and Phillips 2003c: 1442 (comp. note)

Proarna insignis Sanborn, Villet and Phillips 2004a: 820 (comp. note)

Proarna insignis Salazar Escobar 2005: 193 (comp. note) Colombia

Proarna insignis Sanborn 2005c: 114, Table 1 (comp. note)

Proarna insignis Sanborn 2006a: 76, 78, Fig 1 (listed, distribution) Guatemala, Honduras, Central America

Proarna insignis Heath and Sanborn 2007: 488 (comp. note)

Proarna insignis Sanborn 2007a: 27 (distribution, comp. note) Colombia, Mexico, Guatemala, Honduras, Nicaragua, Panama, Venezuela, French Guiana, Ecuador, Peru, Brazil, Argentina, South America, Central America

Proarna insignis Sanborn 2007b: 35 (distribution, comp. note) Mexico, Colombia, Guatemala, Honduras, Nicaragua, Panama, Venezuela, French Guiana, Ecuador, Peru, Brazil, Argentina, South America, Central America

Proarna insignis Thouvenot 2007b: 283, Fig 6 (illustrated, comp. note) French Guiana

Proarna insignis Sanborn 2010a: 1587 (synonymy, distribution, comp. note) Equals *Proarna albida insignis* Colombia, Mexico, Guatemala, Honduras, Nicaragua, Panama, Venezuela, French Guiana, Ecuador, Peru, Brazil, Argentina, South America

P. *invaria* (Walker, 1850) = *Cicada invaria* = *Tympanoterpes invaria* = *Cicada dexithea* Walker, 1850 = *Tympanoterpes dexithea* = *Cicada ovatipennis* Walker, 1858 = *Cicada fulvoviridis* Walker, 1858 = *Proarna* (circa) *insignis* Salazar Escobar, 2005

Proarna invaria Boulard 1998: 83 (comp. note) French Antilles, St. Vincent

Proarna (circa) *insignis* Salazar Escobar 2005: 196, 204, Fig 12 (illustrated, comp. note) Colombia

Proarna invaria Sanborn 2007a: 27 (distribution, comp. note) Venezuela, Costa Rica, Colombia, Ecuador, Peru, Brazil, Guyana, Surinam, French Guiana, Trinidad and Tobago, Jamaica, Cuba, Central America, South America

Proarna invaria Salazar and Sanborn 2009: 270 (comp. note) Equals *Proarna* (circa) *insignis* Salazar Escobar, 2005 Colombia

Proarna invaria Sanborn 2010a: 1587–1588 (synonymy, distribution, comp. note) Equals *Cicada grisea* (nec Fabricius) Equals *Cicada invaria* Equals *Cicada dexithea* Equals *Cicada ovatipennis* Equals *Cicada fulvoviridis* Equals *Tympanoterpes grisea* (nec Fabricius) Equals *Tympanoterpes invaria* Equals *Tympanoterpes dexithea* Equals *Proarna germari* Equals *Proarna* (circa) *insignis* Salazar Escobar, 2005 Colombia, Costa Rica, Ecuador, Peru, Brazil, Guyana, French Guiana, Honduras, Antilles, Central America, South America

Proarna invaria Sanborn 2010b: 74 (listed, distribution) Honduras, Central America, South America, Antilles, Costa Rica, Colombia, Venezuela, Guyana, French Guiana, Peru, Brazil

P. *montevidensis* Berg, 1882 = *Proarna capistrata* Distant, 1885

Proarna montevidensis Ruffinelli 1970: 4 (distribution, comp. note) Equals *Proarna capistrata* Uruguay

Proarna montevidensis Martinelli and Zucchi 1997d: 272 (host plant, comp. note) Uruguay

Proarna montevidensis Bolcatto, Medrano and De Santis 2006: 7, 9, Fig 1K (illustrated, distribution) Argentina, Santa Fe

Proarna montevidensis De Santis, Medrano, Sanborn and Bolcatto 2007: 9, 11, 14, 17, 19, Fig 1K, Fig 4, Table 1 (listed, distribution) Argentina, Santa Fe

P. *olivieri* Metcalf, 1963 = *Cicada albida* Olivier, 1790 (nec *Cicada albida* Gmelin, 1789) = *Tympanoterpes albida* = *Proarna albida* = *Proarna celbida* (sic)

Proarna celbida (sic) Stal Cullen 1974: 25, Table 2 (flight muscles structure, comp. note) Trinidad

Proarna (sic) Cullen 1974: 30–31, Fig 11 (flight muscles structure illustrated)

Proarna olivieri Young 1981c: 6 (comp. note)

Proarna albida Salazar Escobar 2005: 196 (comp. note) Colombia

Proarna olivieri Sanborn 2006a: 76, 78, Fig 1 (listed, distribution) Guatemala, Central America

Proarna olivieri Sanborn 2007a: 27 (distribution, comp. note) Mexico, Guatemala, Costa Rica, Venezuela, Brazil, Guyana, French Guiana, Surinam, Trinidad and Tobago, Jamaica, Cuba, Central America, South America

Proarna olivieri Sanborn 2007b: 35 (distribution, comp. note) Mexico, Guatemala, Costa Rica, Brazil, Venezuela, Guyana, Surinam, French Guiana, Trinidad and Tobago, Jamaica, Cuba, Central America, South America

Proarna olivieri Salazar and Sanborn 2009: 270 (comp. note) Equals *Proarna albida* Salazar Escobar 2005 Colombia

Proarna olivieri Sanborn 2009a: 90 (listed, synonymy, key, comp. note) Equals *Cicada albida* Cuba, Surinam, Caribbean Basin, Mexico, Brazil

Proarna olivieri Sanborn 2010a: 1588 (synonymy, distribution, comp. note) Equals *Cicada albida* Equals *Tympanoterpes albida* Equals *Proarna albida* Colombia, Mexico, Guatemala, Costa Rica, Venezuela, Brazil, Guyana, French Guiana, Surinam, Trinidad and Tobago, Jamaica, Cuba

P. sp. nr. *olivieri* Young, 1981

Proarna sp. Young 1981c: 6 (comp. note) near *Proarna olivieri*

Proarna sp. nr. *olivieri* Young 1981c: 5–6, 8–9, 11–13, 18, 20, 26–28, Figs 6–7, Fig 11, Tables 5–6 (illustrated, seasonality, habitat, comp. note) Costa Rica

Proarna (sic) Young 1981c: 7–8, 14, 21–23, Table 2 (comp. note) Costa Rica

Proarna sp. Young 1982: 200, Fig 5.7 (illustrated, comp. note) Costa Rica

Proarna sp. Young 1984a: 172–173, 177, 179, Tables 1–3 (hosts, comp. note) Costa Rica

P. *palisoti* (Metcalf, 1963) = *Cicada palisoti* = *Cicada bicolor* Palisot de Beauvois, 1813 (nec *Cicada bicolor* Olivier, 1790)

Proarna palisoti Ramos 1983: 62, 64–65, 67, 70, Fig 17–18 (n. comb., synonymy, described, illustrated, key, listed, comp. note) Equals *Cicada bicolor* Dominican Republic

Proarna palisoti Perez-Gerlabert 2008: 202 (listed) Dominican Republic

P. *parva* Sanborn & Heath, 2011, nom. nud.

P. *praegracilis* Berg, 1881

P. *pulverea* (Olivier, 1790) = *Cicada pulvera* = *Tympanoterpes pulvera*

Proarna pulverea Ruffinelli 1970: 4 (distribution, comp. note) Uruguay

Proarna pulverea Sanborn 2007b: 35–36, 39 (distribution, comp. note) Mexico, Veracruz, Surinam, Brazil, Uruguay, Argentina, South America

P. *sallaei* Stål, 1864 = *Proarna sallei* (sic) = Species A Sueur 2002

Proarna sallei (sic) Young 1982: 200, Fig 5.7 (illustrated, comp. note) Costa Rica

Species A Sueur 2002a: 380–381, 386–391, Figs 9–10, Figs 12–14, Table 1 (acoustic behavior, song, sonogram, oscillogram, comp. note) Mexico, Veracruz

Proarna sallei (sic) Sueur 2002a: 380 (listed) Equals Species A Mexico, Veracruz

Proarna sallei (sic) Salazar Escobar 2005: 196–197 (comp. note) Colombia

Proarna sallaei Sanborn 2006a: 76, 78, Fig 1 (listed, distribution) Honduras, Mexico

Proarna sallaei Sanborn 2007b: 36 (distribution, comp. note) Mexico, Veracruz, Oaxaca, Chiapas, Costa Rica, Honduras

Proarna sallaei Salazar and Sanborn 2009: 270 (comp. note) Equals *Proarna sallei* (sic) Colombia

Proarna sallaei Sanborn 2010a: 1588 (synonymy, distribution, comp. note) Equals *Proarna sallei* (sic) Colombia, Mexico, Costa Rica, Honduras, Nicaragua

Proarna sallaei Sanborn 2010b: 74 (listed, distribution) Nicaragua, Mexico, Honduras, Costa Rica, Panama, Colombia

P. sp. 1 Pogue, 1997

Proarna sp. 1 Pogue 1996: 314 (listed) Peru

P. sp. 2 Pogue, 1997

Proarna sp. 2 Pogue 1996: 314 (listed) Peru

P. sp. 3 Pogue, 1997

Proarna sp. 3 Pogue 1996: 314 (listed) Peru

P. sp. 4 Pogue, 1997
Proarna sp. 4 Pogue 1996: 314 (listed) Peru
P. sp. 5 Pogue, 1997
Proarna sp. 5 Pogue 1996: 314 (listed) Peru
P. sp. 6 Pogue, 1997
Proarna sp. 6 Pogue 1996: 314 (listed) Peru
P. sp. 7 Pogue, 1997
Proarna sp. 7 Pogue 1996: 314 (listed) Peru
P. sp. 8 Pogue, 1997
Proarna sp. 8 Pogue 1996: 314 (listed) Peru
P. sp. 9 Pogue, 1997
Proarna sp. 9 Pogue 1996: 314 (listed) Peru
P. sp. 10 Pogue, 1997
Proarna sp. 10 Pogue 1996: 314 (listed) Peru
***P. squamigera* Uhler, 1895**
Proarna squamigera Henshaw 1903: 230 (listed) St. Vincent
Proarna squamigera Boulard 1998: 83 (comp. note) French Antilles, Guadeloupe, Martinique
Proarna squamigera Sanborn 1999a: 43 (type material, listed)
***P. strigicollis* Jacobi, 1907**
Proarna strigicollis Sanborn 2010a: 1588 (distribution, comp. note) Colombia
***P. uruguayensis* Berg, 1882**
Proarna uruguayensis Ruffinelli 1970: 4 (distribution, comp. note) Uruguay
Proarna uruguayensis Sanborn 2008f: 688 (comp. note)

***Prasinosoma* Torres, 1963**
Prasinosoma Heath, Heath and Noriega 1990: 1 (distribution, comp. note) Argentina
***P. fuembuenai* Torres, 1963**
***P. heidemanni* (Distant, 1905)** = *Proarna heidemanni*
***P. inconspicua* (Distant, 1906)** = *Proarna inconspicua*
***P. medialinea* Sanborn & Heath, 2011, nom. nud.**

***Elassoneura* Torres, 1964**
Elassoneura Heath, Heath and Noriega 1990: 1 (distribution, comp. note) Argentina
***E. carychrous* Torres, 1964**

***Beameria* Davis, 1934**
Beameria Tinkham 1941: 174 (singing posture, comp. note)
Beameria Arnett 1985: 217 (diversity) Southwestern United States, Colorado, Kansas
Beameria Poole, Garrison, McCafferty, Otte and Stark 1997: 251, 331 (listed) North America
Beameria Sanborn 1999b: 1 (listed) North America
Beameria Arnett 2000: 298 (diversity) Southwestern United States, Colorado, Kansas
Beameria Kondratieff, Ellingson and Leatherman 2002: 14, 26 (key, listed, comp. note) Colorado
Beameria Sanborn 2006b: 256 (comp. note) Mexico
Beameria sp. Phillips and Sanborn 2007: 456, Fig 1 (listed) Texas
Beameria Shiyake 2007: 4 (listed, comp. note)
***B. ansercollis* Sanborn & Heath, 2011** = *Beameria* n. sp. Sanborn and Phillips 1995
Beameria n. sp. Sanborn and Phillips 1995a: 481, Table 1 (song intensity, comp. note)
Beameria ansercollis Sanborn, Heath, Phillips and Heath 2011 (*Journal of Natural History* 45:1589–1605): 1593–1603, Figs 1–4, Tables 1–2 (n. sp., described, illustrated, key, sonogram, oscillogram, distribution, comp. note) Utah
***B. venosa* (Uhler, 1888)** = *Prunasis venosa* = *Proarna venosa* = *Beameria vanosa* (sic)
Prunasis venosa Henshaw 1903: 230 (listed) Texas
Beameria venosa Tinkham 1941: 169, 171–172, 176, 179 (distribution, habitat, comp. note) Texas, Nebraska, Kansas, Oklahoma, Colorado, New Mexico, Arizona
Beameria venosa Sanborn and Phillips 1995a: 481, 483, Table 1 (song intensity, comp. note)
Beameria venosa Van Pelt 1995: 19 (listed, distribution, comp. note) Texas
Beameria venosa Poole, Garrison, McCafferty, Otte and Stark 1997: 331 (listed) Equals *Prunasis venosa* North America
Beameria venosa Sanborn 1999a: 43 (type material, listed)
Beameria venosa Sanborn 1999b: 1 (listed) North America
Beameria venosa Smith, Kelly, Dean, Bryson, Parker et al. 2000: 127 (comp. note) Kansas
Beameria venosa Kondratieff, Ellingson and Leatherman 2002: 12–13, 26–27, 68, 72, Fig 15, Fig 28, Tables 1–2 (diagnosis, illustrated, listed, distribution, song, comp. note) Equals *Prunasis venosa* Colorado, Arizona, Kansas, Nebraska, New Mexico, Oklahoma, Texas
Prunasis venosa Kondratieff, Ellingson and Leatherman 2002: 26 (type species of *Beameria*)
Beameria venosa Sanborn 2004d: 2056, Fig 942 (illustrated)
Beameria venosa Kondratieff, Schmidt, Opler and Garhart 2005: 97, 215 (listed, comp. note) Oklahoma, Arizona, Colorado, Kansas, Nebraska, New Mexico, Texas
Beameria venosa Sanborn 2006b: 256 (listed, distribution) Mexico, Chihuahua
Beameria venosa Phillips and Sanborn 2007: 456–457, 460, Table 1, Fig 1, Fig 3 (illustrated, habitat, distribution, comp. note) Texas, United States

Beameria venosa Sanborn 2007b: 36 (distribution, comp. note) Mexico, Chihuahua, United States

Beameria venosa Shiyake 2007: 4, 44, 46, Fig 67 (illustrated, distribution, listed, comp. note) United States

Beameria vanosa (sic) Whitford and Jackson 2007: 550 (predation) New Mexico

Beameria venosa Sanborn 2008b: 34, 36, Fig 18, Fig 20 (sonogram, oscillogram, timbal illustrated, comp. note)

Beameria venosa Sanborn 2008d: 3468, Fig 78 (illustrated, comp. note)

Beameria venosa Sanborn, Heath and Heath 2009: 25–29, Figs 1–2 (sound field, song, sonogram, oscillogram, comp. note) New Mexico

***B. wheeleri* Davis, 1934**

Beameria wheeleri Sanborn and Phillips 1995a: 481, Table 1 (song intensity, comp. note)

Beameria wheeleri Poole, Garrison, McCafferty, Otte and Stark 1997: 331 (listed) North America

Beameria wheeleri Sanborn 1999a: 43 (type material, listed)

Beameria wheeleri Sanborn 1999b: 1 (listed) North America

Beameria wheeleri Sanborn 2006b: 256 (listed, distribution) Mexico, Chihuahua

Beameria wheeleri Sanborn 2007b: 36 (distribution, comp. note) Mexico, Chihuahua, New Mexico

***Dorisiana* Metcalf, 1952** = *Dorisia* Delétang, 1919 (nec *Dorisia* Moeschler, 1883)

Dorisiana Martinelli and Zucchi 1989b: 5–7 (comp. note) Brazil

Dorisiana Martinelli 1990: 12 (comp. note) Brazil

Dorisiana Heath, Heath and Noriega 1990: 1 (distribution, comp. note) Argentina

Dorisiana Boulard 1996d: 142 (listed)

Dorisia Boulard and Martinelli 1996: 12 (comp. note)

Dorisiana Boulard and Martinelli 1996: 20 (key, comp. note)

Dorisiana Martinelli and Zucchi 1997c: 134 (comp. note) Brazil

Dorisiana Boulard 1998: 89 (listed)

Dorisiana Ohbayashi, Sato and Igawa 1999: 340 (parasitism, comp. note)

Dorisiana Fornazier and Martinelli 2000: 1175 (comp. note) Brazil

Dorisiana Sueur 2000: 218, 221 (comp. note)

Dorisiana Reis, Souza and Venzon 2002: 84 (coffee pest, comp. note) Brazil

Dorisiana spp. Reis, Souza and Venzon 2002: 84 (coffee pest, control, comp. note) Brazil

Dorisiana spp. Sueur 2002a: 387 (comp. note)

Dorisiana sp. Motta 2003: 19–22, Fig 9, Table 1 (exuvia, described, illustrated, key, comp. note) Brazil, Distrito Federal

Dorisiana Santos and Martinelli 2004a: 630 (comp. note) Brazil

Dorisiana Boulard 2006g: 109 (comp. note)

Dorisiana spp. Bento 2007: 27 (coffee pest) Brazil

Dorisia De Santis, Medrano, Sanborn and Bolcatto 2007: 11 (comp. note)

Dorisiana De Santis, Medrano, Sanborn and Bolcatto 2007: 11 (comp. note)

Dorisiana Costa Neto 2008: 453 (comp. note)

Dorisiana Sanborn, Moore and Young 2008: 1 (comp. note)

Dorisiana Santos and Martinelli 2009a: 559 (comp. note) Brazil

Dorisiana Sanborn 2010a: 1594 (listed) Colombia

Dorisiana Santos, Martinelli, Maccagnan, Sanborn and Ribeiro 2010: 49, 54 (comp. note) Brazil

***D. amoena* (Distant, 1899)** = *Fidicina amoena*

Fidicina amoena Young 1981a: 128–129, 131, 134–136, 139–140, Fig 7, Tables 1–3 (illustrated, seasonality, habitat, comp. note) Costa Rica

Fidicina amoena Young 1981c: 6, 8, 10–12, 18, 20–21, 26, Fig 6, Fig 11, Table 5 (illustrated, seasonality, habitat, comp. note) Costa Rica

Fidicina amoena Young 1981d: 827 (acoustic behavior, comp. note) Costa Rica

Fidicina amoena Young 1982: 103, 200, Fig 3.16, Fig 5.7 (illustrated, comp. note) Costa Rica

Fidicina amoena Young 1984a: 170–173, 176, 179, Tables 1–3 (hosts, comp. note) Costa Rica

Fidicina amoena Young 1991: 189, 212 (illustrated, comp. note) Costa Rica

Dorisiana prob. *amoena* Wolda and Ramos 1992: 272, Table 17.1 (distribution, listed) Panama

Fidicina amoena Wolda and Ramos 1992: 278 (comp. note) Costa Rica

Fidicina amoena Maes 1994: 6 (listed, coffee pest) Nicaragua

Fidicina amoena Maes 1998: 144 (listed, distribution, host plants) Nicaragua, Costa Rica

Dorisiana amoena Sueur 2000: 218, 221 (n. comb., synonymy, comp. note) Equals *Fidicina amoena* Costa Rica, Panama

Dorisiana amoena Sanborn 2005b: 77–78, Fig 1 (listed, distribution) Guatemala, Costa Rica

***D. beniensis* Boulard & Martinelli, 2011**

***D. bicolor* (Olivier, 1790)** = *Cicada bicolor* = *Fidicina bicolor* = *Fidicina cayennensis* Kirkaldy, 1909

Dorisiana bicolor Boulard 1996d: 115, 142–144, Fig S13(a-d), Fig S13(e-f) (song analysis, sonogram, listed, comp. note) Brazil

Fidicina bicolor Martinelli and Zucchi 1997b: 347 (comp. note) Brazil, Amazonas

Dorisiana bicolor Boulard 1998: 80, 82, 87, Fig 2, Plate I, Fig 2 (illustrated, sonogram, comp. note) French Guiana
Dorisiana bicolor Sueur 2001: 38, Table 1 (listed)
Dorisiana bicolor Boulard 2005d: 345 (comp. note) French Guiana
Dorisiana bicolor Boulard 2006g: 96, 180 (listed, comp. note) French Guiana

***D. bogotana* (Distant, 1892)** = *Fidicina bogotana* = *Fidicina bgotoma* (sic)
Fidicina bogotana Pogue 1996: 314 (listed) Peru
Dorisiana bogotana Sanborn 2010a: 1594 (n. comb., synonymy, distribution, comp. note) Equals *Fidicina bogotana* Colombia, Ecuador, Brazil, Peru

D. bonaerensis (Berg, 1879) to *Guyalna bonaerensis*

***D. brisa* (Walker, 1850)** = *Cicada brisa* = *Cicada briza* (sic) = *Fidicina brisa* = *Fidicinoides brisa* = *Gyualna brisa* = *Guylana briza* (sic) = *Fidicina amazona* Distant, 1892
Fidicina brisa Martinelli and Zucchi 1997b: 347 (comp. note) Brazil, Amazonas
Fidicina brisa Sanborn 2005a: 190 (comp. note) South America
Fidicinoides brisa Sanborn, Moore and Young 2008: 17 (n. comb., comp. note) Brazil, Guyana, Peru
Fidicinoides brisa Sanborn 2010a: 1590–1591 (synonymy, distribution, comp. note) Equals *Cicada brisa* Equals *Fidicina brisa* Equals *Fidicina amazona* Colombia, Brazil, Guyana, Peru
Fidicinoides brisa Santos, Martinelli, Maccagnan, Sanborn and Ribeiro 2010: 53–55, Fig 9 (illustrated, key, comp. note) Brazil

D. nr. *cachla* Wolda, 1989 to *Fidicinoides* nr. *cachla*

***D. christinae* Boulard & Martinelli, 2011**

***D. crassa* Boulard, 1998**
Dorisiana crassa Boulard 1998: 88–90, Plate II, Figs 3–5 (n. sp., described, illustrated, comp. note) French Guiana

***D. drewseni* (Stål, 1854)** = *Cicada drewseni* = *Fidicina drewseni* = *Fidicina drewensi* (sic) = *Dorisia drewseni* = *Fidicina gastracathophora* Berg, 1879
Fidicina drewseni Hempel 1913b: 196 (illustrated, coffee pest, comp. note) Brazil
Fidicina drewseni Vayssière 1955: 254 (listed, coffee pest) Brazil
Fidicina drewseni Mariconi 1958: 331–332 (described, coffee pest, listed, comp. note) Brazil
Fidicina drewseni Heinrich 1967: 43 (coffee pest, comp. note) Brazil
Dorisiana drewseni Ruffinelli 1970: 4 (distribution, comp. note) Equals *Fidicina gastracanthophora* Uruguay, Argentina, Brazil
Dorisiana drewseni Martinelli and Zucchi 1984: 1 (n. comb., coffee pest, comp. note) Brazil
Dorisiana drewseni Martinelli and Zucchi 1987b: 469 (coffee pest, comp. note) Brazil
Dorisiana drewseni Martinelli and Zucchi 1987c: 470 (coffee pest, comp. note) Brazil, Minas Gerias, São Paulo, Paraná
Dorisiana drewseni Martinelli and Zucchi 1989b: 5–8, 10, Fig 1 (coffee pest, described, illustrated, comp. note) Brazil, Minas Gerias, São Paulo, Paraná
Dorisiana drewseni Martinelli 1990: 12 (illustrated, comp. note) Brazil
Dorisiana drewseni Martinelli and Zucchi 1993a: 3 (exuvia described, comp. note) Brazil
Dorisiana drewseni Martinelli and Zucchi 1993b: 174 (exuvia described, comp. note) Brazil
Dorisiana drewseni Martinelli and Zucchi 1997c: 135–141, Figs 1–2, Figs 9–10, Fig 21, Table 1 (distribution, host plants, key, illustrated, comp. note) Brazil, Argentina, Uruguay
Dorisiana drewseni Martinelli and Zucchi 1997d: 271, 273–279, Table 1 (host plants, comp. note) Brazil
Dorisiana drewseni Martinelli and Lusvarghi 1998: 115 (coffee pest, control, comp. note) Brazil
Dorisiana drewseni Martinelli, Matuo, Yamada and Malheiros 1998: 135–139, Tables 1–4 (coffee pest, control, comp. note) Brazil
Dorisiana drewseni Albuquerque, Martinelli, Ros and Stülp 2000: 196 (coffee pest) Brazil
Dorisiana drewseni Martinelli, Matuo and Fiorelli 2000: 103 (coffee pest, control, comp. note) Brazil
Dorisiana drewseni Martinelli, Matuo, Martins and Malheiros 2000: 209 (coffee pest, control, comp. note) Brazil
Dorisiana drewseni Martinelli, Lima and Matuo 2001: 31–33 (coffee pest, control, comp. note) Brazil
Dorisiana drewseni Paião, Menegium, Casagrande and Leite 2002: S67 (vector) Brazil, Paraná
Dorisiana drewseni Reis, Souza and Venzon 2002: 84 (coffee pest, comp. note) Brazil, Minas Gerias, São Paulo, Paraná
Fidicina drewseni Reis, Souza and Venzon 2002: 84 (coffee pest, comp. note) Brazil
Dorisiana drewseni Martinelli, Matuo, Martins and Malheiros 2003: 31–33 (coffee pest, control, comp. note) Brazil
Dorisiana drewseni Motta 2003: 19–22, Fig 10, Table 1 (exuvia, described, illustrated, key, comp. note) Brazil, Distrito Federal
Dorisiana drewseni Almeida 2004: 104–105 (control) Brazil

Dorisiana drewseni Martinelli 2004: 518–521, 525–526, 529–531, Figs 18.1–18.2, Figs 18.4b, Table 18.1 (synonymy, described, illustrated, coffee pest, host plants, distribution, control, comp. note) Equals *Cicada drewseni* Equals *Fidicina gastracanthophora* Equals *Fidicina drewseni* Equals *Dorisia drewseni* Brazil, Argentina, Uruguay

Dorisiana drewseni Martinelli, Maccagnan, Ribeiro, Pereira and Santos 2004: 619 (coffee pest, comp. note) Brazil

Dorisiana drewseni Santos and Martinelli 2004b: 630 (coffee pest) Brazil

Dorisiana drewseni Bolcatto, Medrano and De Santis 2006: 7–8, 10, Fig 1D, Map 1 (illustrated, distribution) Equals *Dorisia drewseni* Argentina, Santa Fe

Dorisiana drewseni Martinelli 2006: 1 (coffee pest, comp. note) Brazil

Dorisiana drewseni De Santis, Urteaga and Bolcatto 2006: 1 (comp. note) Argentina, Santa Fe

Dorisiana drewseni De Santis, Medrano, Sanborn and Bolcatto 2007: 4, 11, 14, 17, 19, Fig 1D, Fig 4, Table 1 (listed, described, distribution) Argentina, Santa Fe

Dorisia (sic) *drewseni* De Santis, Medrano, Sanborn and Bolcatto 2007: 11–12, Fig 2 (distribution, comp. note)

Fidicina drewseni Santos and Martinelli 2007: 311 (coffee pest, comp. note) Brazil

Dorisiana drewseni Santos and Martinelli 2007: 311 (coffee pest, comp. note) Brazil

Dorisiana drewseni Krause, Brown, Bellosi and Genise 2008: 412, 414, 416 (comp. note) Argentina

Dorisiana drewseni Aoki, Lopes and de Souza 2010: 162 (distribution) Mato Grosso do Sul

***D. glauca* (Goding, 1925)** = *Fidicina glauca* = *Fidicinoides glauca*

Fidicinoides glauca Sanborn 2008f: 689 (comp. note) Brazil, Rondonia, Ecuador

Fidicinoides glauca Sanborn, Moore and Young 2008: 17 (n. comb., comp. note) Ecuador

Dorisiana glauca Santos, Martinelli, Maccagnan, Sanborn and Ribeiro 2010: 54 (n. comb., synonymy, comp. note) Equals *Fidicina glauca* Equals *Fidicinoides glauca*

***D. noriegai* Sanborn & Heath, 2011, nom. nud.**

***D. panamensis* (Davis, 1939)** = *Fidicina panamensis* = *Fidicina panamaensis* (sic)

Dorisiana panamensis Wolda 1977: 244, Table 1 (emergence, comp. note) Panama

Dorisiana panamensis Wolda and Ramos 1992: 273, Table 17.2 (listed) Panama

Fidicina panamaensis Sanborn 1999a: 43 (type material, listed)

***D. semilata* (Walker, 1850)** = *Cicada semilata* = *Fidicina semilata* = *Cicada passer* Walker, 1850 = *Cicada brizo* Walker, 1850 = *Fidicina brizo* = *Cicada melisa* Walker, 1850 = *Cicada melina* Walker, 1850 = *Cicada panyases* Walker, 1850 = *Cicada pidytes* Walker, 1850 = *Cicada physcoa* Walker, 1850 = *Cicada braure* Walker, 1850 = *Cicada solennis* Walker, 1850 = *Cicada solemis* (sic)

Fidicina semilata Young 1981a: 126, 131, 133–141, Fig 2, Fig 7, Fig 9, Tables 1–4 (illustrated, seasonality, habitat, comp. note) Costa Rica

Fidicina semilata Young 1982: 411–412, 416, Table 9.7 (host, comp. note) Costa Rica

Dorisiana semilata Boulard and Martinelli 1996: 20 (type species of *Dorisiana*, comp. note) Equals *Fidicina semilata*

Fidicina semilata Salazar Escobar 2005: 196 (comp. note) Colombia

Dorisiana semilata Sanborn 2007a: 28 (distribution, comp. note) Colombia, Uruguay, Paraguay, Brazil, Guyana, Surinam, French Guiana, Venezuela, Peru, Ecuador, Panama, Costa Rica, Trinidad and Tobago, Central America, South America

Cicada semilata Sanborn 2010a: 1594 (type species of *Dorisiana*) French Guiana

Dorisiana semilata Sanborn 2010a: 1594 (synonymy, distribution, comp. note) Equals *Cicada semilata* Equals *Cicada passer* Equals *Cicada brizo* Equals *Cicada melisa* Equals *Cicada melina* Equals *Cicada panyases* Equals *Cicada pidytes* Equals *Cicada physcoa* Equals *Cicada braure* Equals *Cicada solennis* Equals *Fidicina semilata* Colombia, Uruguay, Paraguay, Brazil, Guyana, Surinam, French Guiana, Venezuela, Peru, Ecuador, Panama, Costa Rica, Trinidad and Tobago, Central America, South America

***D.* sp. B Wolda, 1989**

Dorisiana sp. B Wolda 1989: 438–440, Figs 1–3, Tables 1–2 (emergence pattern, comp. note) Panama

Dorisiana sp. B Wolda and Ramos 1992: 272–273, Fig 17.1, Table 17.1 (distribution, listed, seasonality, comp. note) Panama

***D.* sp. 1 Pogue, 1997**

Dorisiana sp. 1 Pogue 1996: 314 (listed) Peru

***D. sutori* Sueur, 2000**

Dorisiana sutori Sueur 2000: 218–221, Figs 1–13 (n. sp., described, illustrated, song, sonogram, oscillogram, ecologycomp. note) Mexico, Veracruz

Dorisiana sutori Sueur 2001: 38, Table 1 (listed)

Dorisiana sutori Sueur 2002a: 380–381, 384–385, 387–391, Fig 6, Fig 10, Figs 12–14, Table 1 (acoustic behavior, song, sonogram, oscillogram, comp. note) Mexico, Veracruz

Dorisiana sutori Sanborn 2007b : 36 (distribution, comp. note) Mexico, Chiapas
Dorisiana sutori Sanborn 2010b: 75 (listed, distribution) Guatemala, Mexico

D. toulgoueti **Boulard & Martinelli, 2011**

D. viridifemur **(Walker, 1850)** = *Cicada viridifemur* = *Fidicina viridifemur* = *Fidicinoides viridifemur*
Fidicina viridifemur Martinelli and Zucchi 1997b: 347 (comp. note) Brazil, Mato Grosso
Fidicinoides viridifemur Sanborn, Moore and Young 2008: 19 (n. comb., comp. note) Brazil
Dorisiana viridifemur Santos, Martinelli, Maccagnan, Sanborn and Ribeiro 2010: 54 (n. comb., synonymy, comp. note) Equals *Cicada viridifemur* Equals *Fidicina viridifemur* Equals *Fidicinoides viridifemur*

D. viridis **(Olivier, 1790)** = *Fidicina viridis* = *Dorisia viridis* = *Dorisiana virides* (sic)
Dorisiana viridis Wolda 1977: 239, 244, Table 1 (emergence, comp. note) Panama
Dorisiana viridis Wolda 1983b: 95, 99, Fig 4 (emergence pattern, comp. note) Panama
Dorisiana viridis Martinelli and Zucchi 1986: 16 (coffee pest, comp. note) Brazil
Dorisiana viridis Martinelli and Zucchi 1987b: 469 (coffee pest, comp. note) Brazil
Dorisiana viridis Martinelli and Zucchi 1987c: 470 (coffee pest, comp. note) Brazil, São Paulo
Dorisiana viridis Martinelli and Zucchi 1989b: 5–6, 8–9, 11, Fig 2 (coffee pest, described, illustrated, comp. note) Brazil, Goiás
Dorisiana viridis Martinelli 1990: 12–13 (illustrated, comp. note) Brazil
Dorisiana viridis Martinelli and Zucchi 1997c: 135–141, Figs 1–2, Fig 11, Fig 17, Table 1 (distribution, host plants, key, illustrated, comp. note) Brazil, Argentina, Paraguay, Bolivia, Peru, Colombia, Venezuela, Guyana, Surinam, French Guiana, Costa Rica
Dorisiana viridis Albuquerque, Martinelli, Ros and Stülp 2000: 196 (coffee pest) Brazil
Dorisiana virides (sic) Paião, Menegium, Casagrande and Leite 2002: S67 (vector) Brazil, Paraná
Dorisiana viridis Motta 2003: 19–22, Fig 7, Table 1 (exuvia, described, illustrated, key, comp. note) Brazil, Distrito Federal
Dorisiana viridis Almeida 2004: 104–105 (control) Brazil
Dorisiana viridis Cominetti, Aoki and Souza 2004: 125 (sex ratio) Brazil
Dorisiana viridis Ferrerira, Aoki, and Souza 2004: 125 (host plants) Brazil, Mato Grosso do Sul
Dorisiana viridis Martinelli 2004: 518–521, 524–526, 529–531, Figs 18.1–18.2, Figs 18.4a, Fig 18.4c, Table 18.1 (synonymy, described, illustrated, coffee pest, host plants, distribution, control, comp. note) Equals *Cicada viridis* Equals *Cicada semilata* Equals *Cicada passer* Equals *Cicada brizo* Equals *Cicada melisa* Equals *Cicada melina* Equals *Cicada panyases* Equals *Cicada pidytes* Equals *Cicada physcoa* Equals *Cicada braure* Equals *Cicada solennis* Equals *Fidicina semilata* Equals *Fidicina brizo* Equals *Fidicina viridis* Equals *Dorisia viridis* Equals *Dorisiana semilata* Brazil, Argentina, Paraguay, Bolivia, Peru, Colombia, Venezuela, Guyana, Surinam, French Guiana, Panama, Costa Rica
Dorisiana viridis Martinelli, Maccagnan, Ribeiro, Pereira and Santos 2004: 619 (coffee pest, comp. note) Brazil
Dorisiana viridis Santos and Martinelli 2004b: 630 (coffee pest) Brazil
Fidicina viridis Salazar Escobar 2005: 193 (comp. note) Colombia
Dorisiana viridis Bolcatto, Medrano and De Santis 2006: 7–8, Fig 1E (illustrated, distribution) Argentina, Santa Fe
Dorisiana viridis De Santis, Urteaga and Bolcatto 2006: 1 (comp. note) Argentina, Santa Fe
Dorisiana viridis Maccagnan, Martinelli, Sene and Prado 2006: 1 (sex ratio, comp. note) Brazil
Dorisiana viridis Maccagnan, Prado, Sene and Martinelli 2006a: 1 (song analysis, comp. note) Brazil, São Paulo
Dorisiana viridis Martinelli 2006: 1 (coffee pest, comp. note) Brazil
Dorisiana viridis De Santis, Medrano, Sanborn and Bolcatto 2007: 4, 11, 14, 17, 19, Fig 1E, Fig 4, Table 1 (listed, described, distribution) Argentina, Santa Fe, Santiago del Estero
Dorisiana viridis Santos and Martinelli 2007: 311 (coffee pest, comp. note) Brazil
Dorisiana viridis Sanborn 2007a: 28 (distribution, comp. note) Venezuela, Colombia, Ecuador, Peru, Brazil, Bolivia, Paraguay, Argentina, Surinam, Central America, South America
Dorisiana viridis Aoki, Lopes and de Souza 2010: 162 (distribution) Mato Grosso do Sul
Dorisiana viridis Sanborn 2010a: 1595 (synonymy, distribution, comp. note) Equals *Cicada viridis* Equals *Fidicina viridis* Equals *Dorisia viridis* Colombia

Tympanoterpes **Stål, 1861** = *Tympanoterpis* (sic) = *Tympanotherpes* (sic)
Tympanoterpes Boulard 1982d: 108 (comp. note)
Tympanoterpes Heath, Heath and Noriega 1990: 1 (distribution, comp. note) Argentina
Tympanoterpes Boulard 2001e: 116 (comp. note)
Tympanoterpes Sanborn, Moore and Young 2008: 1 (comp. note)

T. cordubensis Berg, 1884

T. elegans Berg, 1882

Tympanoterpes elegans Ruffinelli 1970: 5 (distribution, comp. note) Uruguay

Tympanoterpes elegans Sanborn 2002b: 456, Fig 1 (illustrated, thermal biology, comp. note)

Tympanoterpes elegans Bolcatto, Medrano and De Santis 2006: 7, 9, Fig 1M (illustrated, distribution) Argentina, Santa Fe

Tympanoterpes elegans De Santis, Medrano, Sanborn and Bolcatto 2007: 10–11, 14, 17, 19, Fig 1M, Fig 4, Table 1 (listed, distribution) Argentina, Santa Fe

T. perpulchra (Stål, 1854) = *Cicada perpulchra* = *Cicada (Cicada) perpulchra*

T. serricosta (Germar, 1834) = *Cicada serricosta* = *Fidicina pusilla* Berg, 1879 = *Tympanoterpes pusilla*

Fidicina pusilla Hennessey 1990: 471 (type material)

Tympanoterpes pusilla Sanborn 1999a: 43 (type material, listed)

Tympanoterpes serricosta Sanborn 2007a: 27 (distribution, comp. note) Venezuela, Guyana, Brazil, Argentina

T. xanthogramma (Germar, 1834) = *Cicada xanthogramma* = *Cicada fuscovenosa* Stål, 1854

***Pompanonia* Boulard, 1982**

Pompanonia Boulard 1982d: 108 (n. gen., described, comp. note)

Pompanonia Boulard 1990b: 166, 175 (comp. note)

P. buziensis Boulard, 1982

Pomponia buziensis Boulard 1982b: 109–112, Figs 1–7 (n. sp., described, illustrated, sonogram, comp. note) Rio de Janiero, Brazil

Pompanonia buziensis Boulard 1985e: 1021 (coloration, comp. note)

Pompanonia buziensis Boulard 1990b: 170, 185, 187, Plate XXII, Fig 2 (sonogram, comp. note)

Pompanonia buziensis Boulard 1999a: 79, 102, 112–113, Plate III, Fig 1 (listed, sonogram, illustrated) Brazil

Pompanonia buziensis Sanborn 2001a: 17 (comp. note)

Pompanonia buziensis Sueur 2001: 38, Table 1 (listed)

Pompanonia buziensis Boulard 2002b: 203 (comp. note)

Pompanonia buziensis Boulard 2005d: 338, 347 (comp. note) Neotropics

Pompanonia buziensis Boulard 2006g: 36–37, 39, 72–73, 82, 111, 180, Fig 7b, Fig 39 (illustrated, sonogram, listed, comp. note) Brazil

Pompanonia buziensis Boulard 2007f: 92, 141, Fig 145 (comp. note) Brazil

***Ollanta* Distant, 1905**

Ollanta Ramos 1983: 66 (comp. note) Dominican Republic, Mexico, Central America, Caicos Islands

Ollanta Ramos 1988: 62–63, Table 4.1 (listed, distribution, comp. note) Greater Antilles, Mexico, South Caicos Islands

Ollanta Sanborn 2001c: 651 (comp. note) Bahamas, Central America, Hispaniola

Ollanta Perez-Gerlabert 2008: 202 (listed) Hispaniola

O. caicosensis Davis, 1939

Ollanta caicosensis Sanborn 1999a: 43 (type material, listed)

Ollanta caicosensis Sanborn 2001c: 651 (listed, comp. note) Caicos Islands, Bahamas

O. mexicana Distant, 1905

Ollanta mexicana Sanborn 2007b: 36 (distribution, comp. note) Mexico, Veracruz

O. melvini Ramos, 1983

Ollanta melvini Ramos 1983: 62, 65–67, 70, Fig 19–20 (synonymy, described, illustrated, key, listed, comp. note) Equals *Cicada hilaris* Dominican Republic

Ollanta melvini Perez-Gerlabert 2008: 202 (listed) Dominican Republic

O. modesta (Distant, 1881) = *Selymbria modesta* = *Diceroprocta belizensis* (nec Distant) = *Conibosa* sp. (nec Distant)

Conibosa (sic) sp. (partim) Maes 1998: 143 (listed, distribution) Nicaragua

Ollanta modesta Maes 1998: 144 (listed, distribution, host plants) Equals *Selymbria modesta* Nicaragua, Mexico

Diceroprocta belizensis (sic) Sanborn 2001b: 449 (listed, comp. note) El Salvador

Diceroprocta belizensis (sic) Sanborn 2006a: 76, 78, Fig 1 (listed, distribution) Belize, Honduras, El Slavador

Ollanta modesta Sanborn 2007b: 36 (distribution, comp. note) Mexico, Morelos, Nicaragua, Central America

Ollanta modesta Sanborn 2010b: 74 (listed, distribution, comp. note) Equals *Diceroprocta belizensis* (nec Distant) Sanborn, 2001, 2006 Honduras, El Salvador, Guatemala, Central America, Mexico, Nicaragua

***Pacarina* Distant, 1905**

Pacarina Tinkham 1941: 173, Fig 2 (singing posture, comp. note)

Pacarina sp. Young 1981a: 134–136, Tables 1–2 (seasonality, habitat, comp. note) Costa Rica

Pacarina Young 1981c: 2, 7–8, 10 (comp. note) Costa Rica

Pacarina Young 1983b: 23 (illustrated, comp. note) Costa Rica

Pacaraina sp. Young 1984a: 171, Table 1 (hosts, comp. note) Costa Rica

Pacarina sp. Young 1981c: 5–6, 12, 18–20, Fig 6, Figs 9–10 (illustrated, seasonality, habitat, comp. note) Costa Rica
Pacarina Wolda 1989: 441 (comp. note) Panama
Pacarina Moore 1996: 221 (comp. note) Mexico
Pacarina Poole, Garrison, McCafferty, Otte and Stark 1997: 251, 332 (listed) North America
Pacarina Sanborn 1999b: 1 (listed) North America
Pacarina Cioara and Sanborn 2001: 25 (song, comp. note)
Pacarina Kondratieff, Ellingson and Leatherman 2002: 14, 27 (key, diagnosis, listed, comp. note) Colorado
Pacarina Sueur 2002b: 122, 124 (comp. note)
Pacarina Sanborn 2007b: 39 (comp. note)

***P. championi* (Distant, 1881)** = *Proarna championi* = *Conibosa* sp. (nec Distant)
Proarna championi Arnett 1985: 217 (distribution) Southwestern United States (error)
Pacarina championi Wolda 1988: 10 (comp. note) Panama
Pacarina championi Wolda 1989: 438–440, Figs 1–3, Tables 1–2 (emergence pattern, comp. note) Panama
Pacarina championi Wolda and Ramos 1992: 272–273, 276, Fig 17.1, Fig 17.5, Table 17.1 (distribution, listed, seasonality, comp. note) Panama
Pacarina championi Poole, Garrison, McCafferty, Otte and Stark 1997: 332 (listed) Equals *Proarna championi* North America
Conibosa (sic) sp. (partim) Maes 1998: 143 (listed, distribution, host plants) Nicaragua
Pacarina championi Sanborn 2006a: 76–78, Fig 1 (listed, distribution) Belize, Honduras, Guatemala, Central America
Pacarina championi Sanborn 2007b: 36 (distribution, comp. note) Mexico, Guatemala, Costa Rica, Central America

***P. puella* Davis, 1923** = *Cicada signifera* Walker, 1858 (nec *Cicada signifera* Germar, 1830) = *Proarna signifera* = *Pacarina signifera*
Pacarina puella Tinkham 1941: 172, 176, 178–179 (distribution, habitat, comp. note) Texas, Louisiana, Oklahoma, Mexico
Pacarina puella McCoy 1965: 44 (comp. note) Louisiana, Oklahoma, Texas
Pacarina puella Wolda 1977: 239, 244, Table 1 (emergence, comp. note) Panama
Pacarina puella Arnett 1985: 217 Fig 21.15 (distribution, illustrated) Louisiana, Arizona, Texas, Oklahoma
Pacarina signifera Wolda 1989: 438–441, Figs 1–3, Fig 5, Tables 1–2 (emergence pattern, comp. note) Panama
Pacarina puella Wolda and Ramos 1992: 272–273, 275–276, Fig 17.1, Fig 17.5, Table 17.1 (distribution, listed, seasonality, comp. note) Panama
Pacarina puella Meagher, Wilson, Blocker, Eckel and Pfannenstiel 1993: 510–511, Table 1 (sugarcane pest, comp. note) Texas
Pacarina puella Wolda 1993: 371, 374–375, 378–379, Fig 6 (seasonality, acoustic behavior, comp. note) Panama
Pacarina puella Sanborn and Phillips 1995a: 480, 482 (comp. note)
Pacarina puella (mesquite variety) Sanborn and Phillips 1995a: 481, Table 1 (song intensity, comp. note)
Pacarina puella Poole, Garrison, McCafferty, Otte and Stark 1997: 332 (listed) Equals *Cicada signifera* North America
Pacarina puella Sanborn 1999b: 1 (listed) North America
Pacarina puella Arnett 2000: 298 Fig 21.15 (distribution, illustrated) Louisiana, Arizona, Texas, Oklahoma
Pacarina signifera Kondratieff, Ellingson and Leatherman 2002: 27 (type species of *Pacarina*)
Pacarina puella Kondratieff, Ellingson and Leatherman 2002: 14–15 (distribution, comp. note) Arizona, Oklahoma, Texas
Pacarina puella Bukhvalova 2005: 199 (emergence, comp. note) Panama
Pacarina puella Kondratieff, Schmidt, Opler and Garhart 2005: 98, 216 (listed, comp. note) Oklahoma, Texas
Pacarina puella Sanborn 2006a: 77 (listed, distribution) Guatemala, Central America
Pacarina puella Phillips and Sanborn 2007: 460–461 (comp. note) Texas, Arizona
Pacarina puella Sanborn 2007b: 36 (distribution, comp. note) Mexico, Veracruz, Nuevo León, Tamaulipas, San Luis Potosí, Guatemala, texas, Oklahoma, Arizona
Pacarina puella Holliday, Hastings and Coelho 2009: 4–5, Table 1 (predation, comp. note) Texas
Pacarina puella Sanborn 2010b: 75 (listed, distribution) Honduras, Central America, Mexico, Guatemala, Panama, USA

***P. schumanni* Distant, 1905**
Pacarina schumanni Sueur 2000: 219 (comp. note) Mexico
Pacarina schumanni Sanborn 2001b: 449 (listed, comp. note) El Salvador, Mexico
Pacarina schumanni Sueur 2002a: 380–381, 383–385, 387–390, Fig 4, Fig 10, Figs 12–14,

Table 1 (acoustic behavior, song, sonogram, oscillogram, comp. note) Mexico, Veracruz
Pacarina schumanni Sanborn 2005b: 77–78, Fig 1 (listed, distribution) Belize, Honduras, El Salvador
Pacarina schumanni Sanborn 2007b: 36 (distribution, comp. note) Mexico, Veracruz, Belize, Honduras, El Salvador
Pacarina schumanni Sanborn 2010b: 75 (listed, distribution) El Salvador, Guatemala, Mexico, Belize, Honduras

***P. shoemakeri* Sanborn & Heath, 2012** = *Pacarina puella* (nec Davis) = *Pacarina puella* (juniper variety) Sanborn and Phillips, 1995a
Pacarina puella (juniper variety) Sanborn and Phillips 1995a: 481, Table 1 (song intensity, comp. note)
Pacarina puella (sic) Kondratieff, Ellingson and Leatherman 2002: 7, 13, 28, 68, Fig 14, Table 2 (diagnosis, illustrated, listed, distribution, song, comp. note) Colorado
Pacarina shoemakeri Sanborn, Heath, Phillips and Heath 2012 (*Journal of Natural History* 46:923–941): 927–938, Figs 1–4, Tables 1–2 (n. sp., described, illustrated, key, sonogram, oscillogram, distribution, comp. note) Arizona, New Mexico, Colorado, Texas

***P.* sp. 1 Young, 1984**
Pacarina sp. 1 Young 1981c: 6, 26, Table 5 (seasonality, habitat, comp. note) Costa Rica

***Ariasa* Distant, 1905**
Ariasa Heath, Heath and Noriega 1990: 1 (distribution, comp. note) Argentina
Ariasa Boulard 2001e: 118 (comp. note)
Ariasa Boulard 2008f: 10 (comp. note)
Ariasa Sanborn, Moore and Young 2008: 1 (comp. note)
Ariasa Sanborn 2010a: 1589 (listed) Colombia

***A. albiplica* (Walker, 1858)** = *Fidicina albiplica* = *Tympanoterpes albiplica*

***A. alboapicata* (Distant, 1905)** = *Tympanoterpes alboapicata*
Ariasa alboapicata Bolcatto, Medrano and De Santis 2006: 7–8, Fig 1A (illustrated, distribution) Argentina, Santa Fe, Santiago del Estero
Ariasa alboapicata De Santis, Medrano, Sanborn and Bolcatto 2007: 3, 11, 14, 17, 19, Fig 1A, Fig 4, Table 1 (listed, distribution) Argentina, Santa Fe, Santiago del Estero

***A. arechavaletae* (Berg, 1884)** = *Tympanoterpes arechavaletae*
Tympanoterpes arechavaletae Ruffinelli 1970: 4 (distribution, comp. note) Uruguay

***A. bilaqueata* (Uhler, 1903)** = *Cicada bilaqueata* = *Cicada bilagueta* (sic) = *Rihana bilaqueta* (sic) = *Diceroprocta bilaqueata*
Cicada bilaqueata Henshaw 1903: 229 (listed) Brazil
Diceroprocta bilaqueata Sanborn 1999a: 38 (type material, listed)
Ariasa bilaqueata Sanborn 2007a: 26–27 (n. comb., distribution, comp. note) Equals *Diceroprocta bilaqueata* Venezuela, Brazil
Cicada bilaqueata Sanborn 2007a: 27 (comp. note)

A. brasiliorum Kirkaldy, 1909 to *Ariasa marginata* (Olivier, 1790)

***A. colombiae* (Distant, 1892)** = *Tympanoterpes colombiae*
Ariasa colombiae Sanborn 2007a: 27 (distribution, comp. note) Colombia, Venezuela
Tympanoterpes colombiae Sanborn 2010a: 1589 (type species of *Ariasa*) Colombia
Ariasa colombiae Sanborn 2010a: 1589, 1602 (synonymy, distribution, comp. note) Equals *Tympanoterpes colombiae* Colombia, Venezuela

***A. diupsilon* (Walker, 1850)** = *Cicada diupsilon* = *Tympanoterpes diupsilon*

***A. marginata* (Olivier, 1790)** = *Cicada marginata* = *Tympanoterpes marginata* = *Cicada viridis* Stoll, 1788 = *Tettigonia viridis* Fabricius, 1803 (nec *viridis* Stål, 1870, nec *viridis* Distant, 1892, nec *viridis* Moulton, 1923) = *Cicada viridis* Germar, 1830 = *Ariasa brasiliorum* Kirkaldy, 1909
Cicada marginata Boulard 2001e: 114–116, 118, Fig A (illustrated, comp. note) Surinam
Ariasa marginata Boulard 2001e: 118 (n. comb., synonymy) Equals *Cicada marginata* Equals *Tettigonia viridis* Fab. [nec *viridis* Stäl, nec *viridis* Distant, nec *viridis* Moulton] Equals *Cicada viridis* Germar Equals *Tympanoterpes marginata* Equals *Ariasa marginata* Equals *Ariasa brasilianorum* Kirkaldy
Cicada marginata Oliv. Boulard 2002a: 58 (comp. note)
Tettigonia viridis Yaakop, Duffels and Visser 2005: 251, 253 (comp. note) Surinam
Cicada marginata Yaakop, Duffels and Visser 2005: 253 (comp. note) Equals *Tettigonia viridis*
Ariasa marginata Yaakop, Duffels and Visser 2005: 253 (comp. note)
Tettigonia viridis Boulard 2008f: 10 (comp. note)

***A. nigrorufa* (Walker, 1850)** = *Fidicina nigrorufa* = *Tympanoterpes nigrorufa*
Ariasa nigrorufa Boulard 2006g: 180 (listed)
Ariasa nigrorufa Sanborn 2007a: 27 (distribution, comp. note) Colombia, Venezuela
Ariasa nigrorufa Sanborn 2010a: 1589 (synonymy, distribution, comp. note) Equals *Fidicina nigrorufa* Equals *Tympanoterpes nigrorufa* Colombia, Venezuela

***A. nigrovittata* Distant, 1905**

***A.* sp. 1 Pogue, 1997**
Ariasa sp. 1 Pogue 1996: 314 (listed) Peru

***A.* sp. 2 Pogue, 1997**
Ariasa sp. 2 Pogue 1996: 314 (listed) Peru
***A.* sp. 3 Pogue, 1997**
Ariasa sp. 3 Pogue 1996: 314 (listed) Peru
***A.* sp. 4 Pogue, 1997**
Ariasa sp. 4 Pogue 1996: 314 (listed) Peru
***A.* sp. 5 Pogue, 1997**
Ariasa sp. 5 Pogue 1996: 314 (listed) Peru
***A.* sp. 6 Pogue, 1997**
Ariasa sp. 6 Pogue 1996: 314 (listed) Peru
***A. urens* (Walker, 1850)** = *Cicada urens* = *Cicada torrida* Walker, 1850 (nec *Cicada torrida* Erichson, 1842) = *Ariasa quaerenda* Kirkaldy, 1909

***Guyalna* Boulard & Martinelli, 1996**
Guyalna Boulard and Martinelli 1996: 20 (key, described, comp. note)
Guyalna Boulard 1998: 90 (listed)
Guyalna Boulard 1999a: 99 (comp. note)
Guyalna Boulard 2006g: 109 (comp. note)
***G. atalapae* Boulard & Martinelli, 2011**
***G. bleuzeni* Boulard & Martinelli, 2011**
***G. bonaerensis* (Berg, 1879)** = *Fidicina bonaerensis* = *Dorisia bonaerensis* = *Dorisiana bonaerensis* = *Dorisiana bonaërensis*
Dorisiana bonaërensis Ruffinelli 1970: 3 (distribution, host, comp. note) Uruguay
Dorisiana bonaerensis Sanborn, Noriega, Heath and Heath 1988: 603 (endothermy, comp. note) Argentina
Dorisiana bonaerensis Sanborn 1991: 5753B (endothermy, comp. note)
Dorisiana bonaerensis Sanborn, J.E. Heath, M.S. Heath and Noriega 1995: 326, Table 3 (thermal responses, comp. note)
Dorisiana bonaerensis Sanborn, M.S. Heath, J.E. Heath and Noriega 1995: 451–457, Figs 1–4, Tables 1–4 (endothermy, thermoregulation, comp. note) Argentina
Dorisiana bonaerensis Sanborn and Phillips 1996: 137 (comp. note)
Guyalna bonaerensis Boulard and Martinelli 1996: 20 (n. comb., type species of *Guyalna*, comp. note) Equals *Fidicina bonaerensis*
Dorisiana bonaërensis Martinelli and Zucchi 1997d: 272 (host plant, comp. note) Uruguay
Dorisiana bonaerensis Sanborn 1998: 92, 94–95, 99, Table 1 (thermal responses, comp. note)
Guyalna bonaerensis Sanborn 2002b: 460–461, 465, Table 1 (thermal biology, comp. note)
Guyalna bonaerensis Sanborn, Noriega and Phillips 2002c: 368 (comp. note)
Guyalna bonaerensis Sueur 2002a: 388, 390 (comp. note)
Dorisiana bonaerensis Merrick and Smith 2004: 730, Table 2 (comp. note)
Guyalna bonaerensis Sanborn 2005c: 114, Table 1 (comp. note)
Dorisiana bonaerensis Bolcatto, Medrano and De Santis 2006: 7, 10, Map 2 (distribution) Equals *Dorisia bonaerensis* Argentina
Guyalna bonaerensis Bolcatto, Medrano and De Santis 2006: 7, 9, Fig 1G, (illustrated, distribution) Argentina
Guyalna bonaerensis De Santis, Medrano, Sanborn and Bolcatto 2007: 5, 11, 14, 17, 19, Fig 1G, Fig 4, Table 1 (listed, described, distribution) Argentina, Santa Fe, Córdoba
Dorisia bonaerensis De Santis, Medrano, Sanborn and Bolcatto 2007: 11, 13, Fig 3 (distribution, comp. note)
Guyalna bonaerensis Heath and Sanborn 2007: 488 (comp. note)
Dorisiana bonaerensis Costa Neto 2008: 455 (comp. note)
***G. bonaerensis bergi* Delétang, 1919 nom. nud.**
***G. bonaerensis dominiquei* Delétang, 1919 nom. nud.**
G. brisa (Walker, 1850) to *Dorisiana brisa*
***G. chlorogena* (Walker, 1850)** = *Fidicina chlorogena* = *Fidicina basispes* Walker, 1858
Fidicina chlorogena Strümpel 1983: 57 (comp. note)
Fidicina chlorogena Hogue 1993b: 233, 570 (behavior, listed, comp. note)
Fidicina chlorogena Boulard and Martinelli 1996: 12 (comp. note)
Guyalna chlorogena Boulard and Martinelli 1996: 14 (n. comb., type species of *Guyalna* and Guyalnina, comp. note)
Fidicina chlorogena Boulard 2002b: 208 (comp. note)
Fidicina chlorogena Thouvenot 2007b: 281 (comp. note) Brazil
***G. cuta* (Walker, 1850)** = *Cicada cuta* = *Fidicina cuta* = *Cicada lucastia* Walker, 1850
Guyalna cuta Boulard 1999a: 79, 99 (listed, sonogram) French Guiana
Guyalna cuta Sueur 2001: 38, Table 1 (listed)
Guyalna cuta Boulard 2006g: 180 (listed)
***G. densusa* Boulard & Martinelli, 2011**
***G. jauffreti* Boulard & Martinelli, 2011**
***G. nigra* Boulard, 1999**
Guyalna nigra Boulard 1999a: 79, 100–101 (n. sp., described, illustrated, listed, sonogram) Brazil
Guyalna nigra Sueur 2001: 38, Table 1 (listed)
Guyalna nigra Boulard 2005d: 343 (comp. note)
Guyalna nigra Boulard 2006g: 72, 96–97, 180 (listed, comp. note)
***G. platyrhina* Sanborn & Heath, 2011, nom. nud.**
***G. rufapicalis* Boulard, 1998**
Guyalna rufapicalis Boulard 1998: 88, 90–91, Plate II, Figs 6–8 (n. sp., described, illustrated, comp. note) French Guiana

***Hemisciera* Amyot & Audinet-Serville, 1843**
Hemisciera He 1984: 221, 226 (comp. note)
Hemisciera Boulard and Martinelli 1996: 20 (key, comp. note)
Hemisciera Sanborn, Moore and Young 2008: 1 (comp. note)
***H. durhami* Distant, 1905**
Hemisciera durhami Boulard 1985e: 1021, 1039, Plate III, Fig 24 (coloration, illustrated, comp. note) Neotropics
***H. maculipennis* (de Laporte, 1832)** = *Cicada maculipennis* = *Fidicina maculipennis* = *Hemisciera pictipennis* (sic) = *Cicada versicolor* Brullé, 1835 = *Hemisciera versicolor* = *Cicada sumptuosa* Blanchars, 1840 = *Hemisciera sumptuosa* = *Fidicina floslofia* Walker, 1858
Hemisciera maculipennis Wolda and Ramos 1992: 272–273, Table 17.1 (distribution, listed, comp. note) Panama, Brazil, Ecuador, Cayenne
Hemisciera maculipennis Boulard and Martinelli 1996: 20 (type species of *Hemisciera*, comp. note) Equals *Cicada maculipennis*
Hemisciera maculipennis Boulard 2002b: 193 (coloration, comp. note) Amazon
Hemisciera maculipennis Thouvenot 2007b: 283–284, Fig 8 (illustrated, comp. note) French Guiana
***H. taurus* (Walker, 1850)** = *Fidicina taurus*

***Majeorona* Distant, 1905** = *Majerona* (sic) = *Majeroma* (sic)
Majeorona Boulard and Martinelli 1996: 20 (key, comp. note)
Majeorona spp. Zinn, Lal, Bingham and Resck 2007: 1223 (comp. note) Brazil
Majeorona Sanborn, Moore and Young 2008: 1 (comp. note)
Majeorona Sanborn 2010a: 1595 (listed) Colombia
***M. aper* (Walker, 1850)** = *Fidicina aper* = *Majerona* (sic) *aper*
Majeorona aper Wolda and Ramos 1992: 273, Table 17.2 (listed) Panama
Majeorona aper Boulard and Martinelli 1996: 20 (type species of *Majeorona*, comp. note) Equals *Fidicina aper*
Majeorona aper Motta 2003: 19–21, Fig 1, Table 1 (exuvia, described, illustrated, key, comp. note) Brazil, Distrito Federal
Majeorona aper Maccagnan, Martinelli, Prado and Sene 2006: 205 (song analysis, comp. note) Brazil
Fidicina aper Sanborn 2010a: 1595 (type species of *Majeorona*) French Guiana
Majeorona aper Sanborn 2010a: 1595 (synonymy, distribution, comp. note) Equals *Fidicina aper* Colombia, Brazil, Ecuador, French Guiana, Panama
***M. bovilla* Distant, 1905**
Majeorona bovilla Young 1983a: 725 (comp. note) Costa Rica
Majeorona bovilla Pogue 1996: 314 (listed) Peru
Majeorona bovilla Salazar Escobar 2005: 196 (comp. note) Colombia
Majeorona bovilla Sanborn 2010a: 1595 (distribution, comp. note) Colombia, Brazil, Costa Rica, Peru
***M. durantoni* Boulard & Martinelli, 2011**
***M. ecuatoriana* Goding, 1925**
M. ethelae Goding, 1925 to *Fidicina ethelae*
***M. lutea* Distant, 1906** = *Majeroma* (sic) *lutea*
***M. orvoueni* Boulard & Martinelli, 2011**
***M. truncata* Goding, 1925**
Majeorona truncata Sanborn 2005b: 77–78, Fig 1 (listed, distribution) Honduras, Ecuador
Majeorona truncata Sanborn 2008f: 689 (comp. note) Brazil, Matto Grosso, Ecuador, Honduras
Majeorona truncata Sanborn 2010b: 75 (listed, distribution) Guatemala, Honduras, Ecuador, Brazil

Tribe Hyantiini Distant, 1905 = Hyantinii (sic) = Hyantini (sic)
Hyantiini Boulard 1996d: 115, 146 (listed)
Hyantiini Sanborn 1999b: 1 (listed) North America
Hyantiini Sanborn 2001b: 449 (listed) El Salvador
Hyantiini Sueur 2001: 40, Table 1 (listed)
Hyantiini Sueur 2002a: 380 (listed) Mexico, Veracruz
Hyantiini Moulds 2005b: 427 (higher taxonomy, listed)
Hyantinii (sic) Sanborn 2006a: 77 (listed)
Hyantiini Sanborn 2007a: 28 (listed) Venezuela
Hyantiini Sanborn 2007b: 36 (listed) Mexico
Hyantiini Shiyake 2007: 4, 46 (listed, comp. note)
Hyantiini Phillips and Sanborn 2010: 74 (comp. note)
Hyantiini Sanborn 2010a: 1595 (listed) Colombia

***Quesada* Distant, 1905** = *Queseda* (sic) = *Quezada* (sic)
Quesada sp. Wolda 1977: 239 (emergence, comp. note) Panama
Quesada spp. Zambon, Nakayama and Nakano 1980: 255–256, Table 1 (coffee pest, control) Brazil
Quesada Young 1981a: 139 (comp. note) Costa Rica
Quesada Strümpel 1983: 18 (listed)
Quesada Boulard 1984c: 171 (comp. note)
Quesada spp. Reis, Souza and Melles 1984: 9 (comp. note) Brazil
Quesada Martinelli and Zucchi 1987a: 52–53, 57–59, Fig 1 (coffee pest, described, illustrated, comp. note) Brazil
Quesada Heath, Heath and Noriega 1990: 1 (distribution, comp. note) Argentina
Quesada Boulard 1995a: 7 (comp. note)
Quesada Boulard 1996d: 146 (listed)

Quesada Martinelli and Zucchi 1997c: 134 (comp. note) Brazil
Quesada Poole, Garrison, McCafferty, Otte and Stark 1997: 251, 333 (listed) Equals *Queseda* (sic) North America
Quezada (sic) Henríquez 1998: 86 (comp. note)
Quesada Sanborn 1999b: 1 (listed) North America
Quesada Fornazier and Martinelli 2000: 1175 (comp. note) Brazil
Quesada Reis, Souza and Venzon 2002: 84 (coffee pest, comp. note) Brazil
Quesada Salazar Escobar 2005: 194 (comp. note) Colombia
Quesada Boulard 2006g: 129 (comp. note)
Quesada spp. Bento 2007: 27 (coffee pest) Brazil
Quesada Sanborn 2007b: 39 (comp. note)
Quesada Shiyake 2007: 4, 46 (listed, comp. note)
Quesada spp. Zinn, Lal, Bingham and Resck 2007: 1223 (comp. note) Brazil
Quesada Costa Neto 2008: 453 (comp. note)
Quesada Sanborn 2010a: 1595 (listed) Colombia

Q. gigas (Olivier, 1790) = *Cicada gigas* = *Tympanoterpes gigas* = *Queseda* (sic) *gigas* = *Cicada triupsilon* Walker, 1850 = *Cicada triypsilon* (sic) = *Cicada trupsilon* (sic) = *Cicada sonans* Walker, 1850 = *Cicada consonans* Walker, 1850 = *Cicada vibrans* Walker, 1850 = *Tympanoterpes grossa* (nec Fabricius) = *Cicada grossa* (nec Fabricius) = *Cicada (Cicada) grossa* (nec Fabricius) = *Tympanoterpes sibilarix* Berg, 1879 = *Tympanoterpes sibilantis* (sic) = *Tympanoterpis* (sic) *sibilantes* (sic) = *Tympanotherpes* (sic) *gigas*

Cicada gigas Swinton 1880: 224 (song, comp. note) Tobago
Cicada grossa Sörensen 1884: 17 (acoustic behavior, comp. note) Paraguay
Quesada gigas Mariconi 1958: 331–333, Fig 108 (illustrated, described, coffee pest, listed, comp. note) Brazil
Quesada gigas Heinrich 1967: 43–46 (coffee pest, natural history, control, comp. note) Equals *Cicada triupsilon* Equals *Cicada consonans* Equals *Cicada grossa* Equals *Tympanoterpes sibilatrix* Equals *Tympanoterpes gigas* Brazil
Cicada triupsilon Heinrich 1967: 44 (comp. note)
Cicada consonans Heinrich 1967: 44 (comp. note)
Cicada grossa Heinrich 1967: 44 (comp. note)
Tympanoterpes sibilatrix Heinrich 1967: 44 (comp. note)
Tympanoterpes gigas Heinrich 1967: 44 (comp. note)
Quesada gigas Weber 1968: 58, 81–83, 85, 112, Fig 45, Figs 68–69, Fig 94 (wing coupling illustrated, sound system illustrated, thorax illustrated, comp. note)
Quesada gigas Ruffinelli 1970: 4 (distribution, host, comp. note) Uruguay, Argentina, Texas, Brazil
Tympanoterpes gigas Escalante 1974: 120 (comp. note) Peru
Quesada gigas Wolda 1977: 239 (emergence, comp. note) Panama
Quesada gigas Young 1981a: 129, 131, 134, 136, 138–139, Fig 7, Fig 9, Tables 1–2 (illustrated, seasonality, habitat, comp. note) Costa Rica
Quesada gigas Young 1981c: 7–8 (seasonality, habitat, comp. note) Costa Rica
Quesada gigas Young 1981d: 827 (acoustic behavior, comp. note) Costa Rica
Quesada gigas Boulard 1982d: 112 (comp. note)
Quesada gigas Young 1982: 102–103, 119, 184–185, 208–209, Fig 3.16, Fig 3.21, Fig 5.2, Table 5.6 (illustrated, activity patterns, hosts, comp. note) Costa Rica
Quesada gigas Souza, Reis and Melles 1983: 15, 17, Fig 5, Cover (illustrated, coffee pest, control, comp. note) Brazil, Minas Gerias
Quesada gigas Strümpel 1983: 107, Fig 142 (illustrated, comp. note)
Quesada gigas Young 1983a: 725 (comp. note) Costa Rica
Quesada gigas Reis, Souza and Melles 1984: 4, 6, 8–10, Fig 6, Fig 9, Table 1 (coffee pest, illustrated, habitat, control, listed, comp. note) Brazil
Quesada gigas Souza, Reis and Melles 1984: 152 (coffee pest, control, comp. note) Brazil, Minas Gerias
Quesada gigas Wolda 1984: 451 (comp. note) Panama
Quesada gigas Young 1984a: 170, 176, 178–179, Tables 1–3 (hosts, comp. note) Costa Rica
Quesada gigas Arnett 1985: 217 Fig 21.13 (distribution, illustrated) Southwestern United States (error)
Quesada gigas Martinelli, Vieira and Zucchi 1986b: 119 (coffee pest, comp. note) Brazil, São Paulo
Quesada gigas Martinelli, Vieira and Zucchi 1986b: 5–6, Fig 1 (coffee pest, illustrated, comp. note) Brazil, São Paulo
Quesada gigas Martinelli and Zucchi 1987a: 52–55, 57–59, Fig 1 (coffee pest, described, illustrated, comp. note) Brazil, São Paulo, Minas Gerias
Quesada gigas Martinelli and Zucchi 1987b: 469 (coffee pest, comp. note) Brazil
Quesada gigas Martinelli and Zucchi 1987c: 470 (coffee pest, comp. note) Brazil, Minas Gerias, São Paulo
Quesada gigas Boulard 1988d: 154 (wing abnormalities, comp. note)
Quesada gigas Sanborn, Noriega, Heath and Heath 1988: 603 (endothermy, comp. note) Argentina

Quesada gigas Gonçalves and Faria 1989: 96–106 (control) São Paulo, Brazil
Quesada gigas Martinelli and Zucchi 1989a: 8 (comp. note) Brazil
Quesada gigas Martinelli and Zucchi 1989b: 8 (comp. note) Brazil
Quesada gigas Boulard 1990b: 170 (comp. note)
Quesada gigas Wolda 1989: 438–441, Fig 4, Tables 1–2 (emergence pattern, comp. note) Panama
Quesada gigas Carvalho, Martinelli and Lusvarghi 1990: 39 (control) Brazil
Quesada gigas Martinelli 1990: 12–13 (illustrated, comp. note) Brazil
Quesada gigas Sanborn 1991: 5753B (endothermy, comp. note)
Quesada gigas Young 1991: 212–213 (illustrated, comp. note) Costa Rica
Quesada gigas Wolda and Ramos 1992: 271–272, 275–277, Fig 17.5, Fig 17.8, Table 17.1 (distribution, listed, seasonality, comp. note) Panama
Quesada gigas Hogue 1993b: 232–233, 570, Fig 8.6a (illustrated, behavior, listed, comp. note) Venezuela
Quesada gigas Martinelli and Zucchi 1993a: 3 (exuvia described, comp. note) Brazil
Quesada gigas Martinelli and Zucchi 1993b: 174 (exuvia described, comp. note) Brazil
Quesada gigas Moore 1993: 270 (distribution, song, comp. note)
Quesada gigas Wolda 1993: 370–379, Figs 2–4 (seasonality, acoustic behavior, comp. note) Panama
Quesada gigas Maes 1994: 6 (listed, coffee pest) Nicaragua
Quesada gigas Scarpellini, Takematsu, Jocys, Santos and Dias Neto 1994: 55, 57, Table 2 (coffee pest, control, comp. note) Brazil, São Paulo
Quesada gigas Boulard 1995a: 8 (comp. note)
Quesada gigas Sanborn, J.E. Heath, M.S. Heath and Noriega 1995: 326, Table 3 (thermal responses, comp. note)
Quesada gigas Sanborn, M.S. Heath, J.E. Heath and Noriega 1995: 451, 453, 455–458, Tables 2–4 (endothermy, thermoregulation, comp. note) Argentina
Quesada gigas Boulard 1996d: 115, 146–148, Fig S15(a-d), Fig S15(e-f) (song analysis, sonogram, listed, comp. note) Guyana
Quesada gigas Moore 1996: 221–222, Fig 17.15b (illustrated, distribution, comp. note) Mexico
Quesada gigas Pogue 1996: 314, 320 (listed, comp. note) Peru, Costa Rica
Quesada gigas Martinelli and Zucchi 1997c: 135–141, Figs 1–4, Fig 23, Table 1 (distribution, host plants, key, illustrated, comp. note) Brazil, Argentina, Uruguay, Paraguay, Bolivia, Peru, Ecuador, Colombia, Venezuela, Guyana, Surinam, French Guiana, Panama, Guatemala, Mexico
Quesada gigas Martinelli and Zucchi 1997d: 271–279, Table 1 (host plants, comp. note) Brazil
Quesada gigas Poole, Garrison, McCafferty, Otte and Stark 1997: 333 (listed) Equals *Cicada gigas* Equals *Cicada triupsilon* Equals *Cicada sonans* Equals *Cicada consonans* Equals *Cicada vibrans* Equals *Cicada triyupsilon* (sic) Equals *Tympanoterpes sibilatrix* Equals *Cicada trupsilon* (sic) Equals *Tympanoterpes sibilantis* (sic) Equals *Tympanoterpes sibilantes* (sic) North America
Quesada gigas Shelly and Whittier 1997: 279, Table 16–1 (listed, comp. note)
Quesada gigas Boulard 1998: 83, 86–87, Fig 4, Plate I, Fig 3 (illustrated, sonogram, comp. note) French Guiana
Quesada gigas Maes 1998: 144 (listed, distribution, host plants) Equals *Cicada gigas* Equals *Tympanoterpes gigas* Equals *Cicada triupsilon* Equals *Cicada sonans* Equals *Cicada consonans* Equals *Cicada vibrans* Equals *Typmanoterpes grossa* (nec Fabricius) Equals *Tympanoterpes sibilatrix* Nicaragua, USA, Mexico, Guatemala, Belize, Costa Rica, Panama, Colombia, Trinidad, Tobago, Venezuela, Ecuador, Peru, Guyana, Brazil, Bolivia, Paraguay, Argentina
Quesada gigas Martinelli and Lusvarghi 1998: 115 (coffee pest, control, comp. note) Brazil
Quesada gigas Martinelli, Matuo, Yamada and Malheiros 1998: 135–139, Tables 1–4 (coffee pest, control, comp. note) Brazil
Quesada gigas Sanborn 1998: 94–95, 99, Table 1 (thermal responses, comp. note)
Quesada gigas Sanborn 1999b: 1–2 (type material, listed) North America
Quesada gigas Terradas 1999: 50 (comp. note)
Quesada gigas Albuquerque, Martinelli, Ros and Stülp 2000: 196 (coffee pest) Brazil
Quesada gigas Arnett 2000: 217 Fig 21.13 (distribution, illustrated) Southwestern United States (error)
Quesada gigas Fornazier and Martinelli 2000: 1175 (comp. note) Brazil
Quesada gigas Martinelli, Matuo and Fiorelli 2000: 103 (coffee pest, control, comp. note) Brazil
Quesada gigas Martinelli, Matuo, Martins and Malheiros 2000: 209 (coffee pest, control, comp. note) Brazil
Quesada gigas Sueur 2000: 219 (comp. note) Mexico

Quesada gigas Martinelli, Lima and Matuo 2001: 31–33 (coffee pest, control, comp. note) Brazil
Quesada gigas Sanborn 2001b: 449 (listed, comp. note) El Salvador
Quesada gigas Sueur 2001: 40, Table 1 (listed)
Quesada gigas Boulard 2002b: 176, 203, Fig 6 (comp. note) Neotropics
Quesada gigas Maccagnan and Martinelli 2002: 234 (nymphal morphology, comp. note) Brazil, Minas Gerias
Quesada gigas Reis, Souza and Venzon 2002: 84–85, Fig 1 (coffee pest, illustrated, control, comp. anote) Brazil
Quesada gigas Sanborn 2002b: 460–461, 465, Table 1 (thermal biology, comp. note)
Quesada gigas Sueur 2002a: 380–383, 387–391, Fig 2, Fig 10, Figs 12–14, Table 1 (acoustic behavior, song, sonogram, oscillogram, comp. note) Mexico, Veracruz, Panama, French Guiana
Quesada gigas Sueur 2002b: 119, 121, 123, 129 (listed, comp. note)
Quesada gigas Martinelli, Matuo, Martins and Malheiros 2003: 31–33 (coffee pest, control, comp. note) Brazil
Quesada gigas Motta 2003: 19–21, Fig 3, Fig 5, Table 1 (exuvia, described, illustrated, key, comp. note) Brazil, Distrito Federal
Quesada gigas Salguero 2003: 30 (distribution, comp. note) Guatemala, Argentina, Texas
Quesada gigas Almeida 2004: 102–105 (control) Brazil
Quesada gigas Almeida, Fialho, Schneider and Pinheiro 2004: 125 (adult population dynamics) Brazil
Quesada gigas Cominetti, Aoki and Souza 2004: 125 (sex ratio) Brazil
Quesada gigas Ferrerira, Aoki, and Souza 2004: 125 (host plants) Mato Grosso do Sul, Brazil
Quesada gigas Fialho, Almeida, Schneider, and Pinheiro 2004: 126 (predation) Federal District, Brazil
Quesada gigas Maccagnan and Martinelli 2004: 439–446, Figs 1–6, Tables 1–2 (nymph described, illustrated, instar key, comp. note) Brazil, Minas Gerias
Quesada gigas Marques, Martinelli, Azevedo, Coutinho and Serra 2004: 120–121 (coffee pest, distribution, comp. note) Brazil, Bahia
Quesada gigas Martinelli 2004: 517–519, 521–525, 529–531, 533–535, Figs 18.1–18.2, Figs 18.3a-b, Tables 18.1–18.3 (synonymy, described, illustrated, coffee pest, host plants, distribution, control, comp. note) Equals *Cicada gigas* Equals *Cicada triupsilon* Equals *Cicada sonana* Equals *Cicada consonans* Equals *Cicada vibrans* Equals *Tympanoterpes sibilatrix* Equals *Tympanoterpes gigas* Brazil, Argentina, Uruguay, Paraguay, Bolivia, Peru, Ecuador, Colombia, Guyana, Panama, Guatemala, Mexico
Quesada gigas Martinelli, Maccagnan, Ribeiro, Pereira and Santos 2004: 619 (coffee pest, comp. note) Brazil
Quesada gigas Santos and Martinelli 2004b: 630 (coffee pest) Brazil
Quesada gigas Scarpellini 2004: 114 (coffee pest, control, comp. note) Brazil, Minas Gerias, São Paulo
Quesada gigas Zanuncio, Pereira, Zanuncio, Martinelli, Moreira Pinon and Guimarães 2004: 943–945, Figs 1–2 (illustrated, hosts, comp. note) Brazil, Maranão, Pará, North America, Antilles
Quesada gigas Boulard 2005d: 345, 347 (comp. note)
Tympanotherpes (sic) (= *Quesada) gigas* Salazar Escobar 2005: 193 (comp. note) Colombia
Quesada gigas Salazar Escobar 2005: 196, 203, Fig 9 (illustrated, comp. note) Colombia
Quesada gigas Sanborn 2005b: 77 (listed, distribution) Central America, Belize, Guatemala, Honduras, El Salvador
Quesada gigas Taber and Fleenor 2005: 241–242, Fig 10-1 (illustrated, comp. note) Texas
Quesada gigas Bolcatto, Medrano and De Santis 2006: 6–7, 9–10, Fig 1L, Map 1 (illustrated, distribution) Argentina, Santa Fe
Quesada gigas Boulard 2006g: 96, 113–114, 120–121, 148, 180, Fig 78 (sonogram, listed, comp. note) French Guiana
Quesada gigas De Santis, Urteaga and Bolcatto 2006: 1–4, Figs 1–6 (song, comp. note) Argentina, Santa Fe
Quesada gigas Maccagnan, Martinelli, Sene and Prado 2006: 1 (sex ratio, comp. note) Brazil
Quesada gigas Maccagnan, Prado, Sene and Martinelli and Martinelli 2006b: 205 (song analysis, comp. note) Brazil, São Paulo
Quesada gigas Martinelli 2006: 1 (coffee pest, comp. note) Brazil
Quesada gigas Martinelli, Maccagnan, Matuo, Silveira Neto, Prado and Sene 2006: 246 (acoustic response, comp. note) Brazil
Quesada gigas Boulard 2007e: 87 (comp. note) Neotropics
Quesada gigas De Santis, Medrano, Sanborn and Bolcatto 2007: 9–12, 14, 17, 19, Fig 1L, Fig 2, Fig 4, Table 1 (listed, described, distribution) Argentina, Santa Fe, Córdoba
Quesada gigas Heath and Sanborn 2007: 488 (comp. note)

Quesada gigas Maccagnan, Martinelli, Prado and Sene 2007: 230–232, Figs 1–3, Table 1 (emergence pattern, comp. note) Brazil, Minas Gerias, São Paulo

Quesada gigas Maccagnan, Martinelli and Sene 2007: 230–232, Figs 1–3, Table 1 (emergence pattern, comp. note) Brazil, Minas Gerias, São Paulo

Quesada gigas Santos and Martinelli 2007: 311 (coffee pest, comp. note) Brazil

Quesada gigas Sanborn 2007a: 28 (distribution, comp. note) Venezuela, Argentina, United States

Quesada gigas Sanborn 2007b: 36–37 (distribution, comp. note) Mexico, Argentina, United States

Quesada gigas Shilva, Moino, Oliviera, Ferreira and Rosa 2007: 127 (control, comp. note) Brazil

Quesada sp. (sic) Shiyake 2007: 4, 44, 46, Fig 69 (illustrated, listed, comp. note)

Quesada gigas Thouvenot 2007b: 281, Fig 2 (illustrated, comp. note) French Guiana

Quesada gigas Costa Neto 2008: 455 (comp. note)

Quesada gigas Matuo, Raetano, Matuo, Martinelli and Leite 2008: 547–549, Tables 1–2 (coffee pest, control, comp. note) Brazil, Mato Grosso

Quesada gigas Soares, Zaneti, Santos and Leite 2008: 252 (distribution, density, comp. note) Brazil, Pará

Quesada gigas Holliday, Hastings and Coelho 2009: 4–5, 13, Table 1 (predation, comp. note) Texas

Quesada gigas Marshall, S.A. 2009: 8 (illustrated, comp. note) Costa Rica

Quesada gigas Sazima 2009: 260–261 (predation) Brazil, São Paulo

Quesada gigas Aoki, Lopes and de Souza 2010: 162 (distribution) Mato Grosso do Sul

Quesada gigas Fonseca, Silva, Samuels, DaMatta, Terra, and Silva 2010: 20–24, Figs 1–6, Tables 1–2 (gut anatomy, gut physiology, comp. note) Rio de Janiero, Brazil

Quesada gigas Lara, Perioto and de Ffreitas 2010: 116 (coffee pest) Brazil

Quesada gigas Lunz, Azevedo, Júnior, Monteiro, Lechinoski and Zaneti 2010: 632–636, Fig 1–4, Tables 1–2 (control, comp. note) Brazil

Cicada gigas Sanborn 2010a: 1595 (type species of *Quesada*)

Quesada gigas Sanborn 2010a: 1596, 1603 (synonymy, distribution, comp. note) Equals *Cicada gigas* Equals *Cicada triupsilon* Equals *Cicada sonans* Equals *Cicada consonans* Equals *Cicada vibrans* Equals *Cicada triypsilon* (sic) Equals *Cicada grossa* (nec Fabricius) Equals *Tympanoterpes sibilatrix* Equals *Cicada trupsilon* (sic) Equals *Tympanoterpes sibilantis* (sic) Equals *Tympanoperpis* (sic) *sibilantes* (sic) Equals *Quesada gicas* (sic) Equals *Typnaothermpes* (sic) *(Quesada) gigas* Colombia, Argentina, United States

***Q. sodalis* (Walker, 1850)** = *Cicada sodalis* = *Tympanoterpes sodalis* = *Fidicina vultur* Walker, 1858

Quesada sodalis Vayssière 1955: 254 (listed, coffee pest) Brazil

Quesada sodalis Mariconi 1958: 332–333, Fig 108 (illustrated, described, coffee pest, listed, comp. note) Brazil

Quesada sodalis Reis, Souza and Melles 1984: 4, Table 1 (coffee pest, listed) Brazil

Quesada sodalis Martinelli and Zucchi 1987a: 52–53, 55–58, 60, Fig 2 (coffee pest, described, illustrated, comp. note) Brazil, São Paulo, Santa Catarina

Quesada sodalis Martinelli and Zucchi 1987b: 469 (coffee pest, comp. note) Brazil

Quesada sodalis Martinelli 1990: 12 (illustrated, comp. note) Brazil

Quesada sodalis Martinelli and Zucchi 1997c: 135–141, Figs 1–2, Fig 5, Fig 24, Table 1 (distribution, host plants, key, illustrated, comp. note) Brazil, Peru

Quesada sodalis Albuquerque, Martinelli, Ros and Stülp 2000: 196 (coffee pest) Brazil

Quesada sodalis Martinelli 2004: 517–519, 521–522, 529, Figs 18.1–18.2 (synonymy, described, coffee pest, host plants, distribution, control, comp. note) Equals *Cicada sodalis* Equals *Fidicina vultur* Equals *Tympanoterpes sodalis* Brazil, Peru

Quesada sodalis Martinelli 2006: 1 (coffee pest, comp. note) Brazil

Quesada sodalis Santos and Martinelli 2007: 311 (coffee pest, comp. note) Brazil

***Mura* Distant, 1905**

***M. elegantula* Distant, 1905**

Mura elegantula Wolda and Ramos 1992: 272, Table 17.1 (distribution, listed) Panama

***Hyantia* Stål, 1866**

Hemiciera (sic) Boulard 1985e: 1023 (coloration, comp. note)

***H. bahlenhorsti* Sanborn, 2011**

***H. honesta* (Walker, 1850)** = *Cyclochila honesta*

Hyantia honesta Boulard 1985e: Plate III, Fig 24 (coloration, illustrated, comp. note)

Hyantia honesta Boulard 2002b: 193 (coloration, comp. note)

Tribe Tamasini Moulds, 2005

Tamasini Moulds 2005b: 419, 422, 427, 429–430, 434, Fig 58, Fig 60, Table 2 (n. tribe,

diagnosis, described, composition, key, phylogeny, comp. note) Australia
Tamasini Moulds 2008a: 208–209 (key, comp. note)
Tamasini Wei, Ahmed and Rizvi 2010: 33 (comp. note)

***Tamasa* Distant, 1905**
Tamasa Moulds 1985: 26 (comp. note)
Tamasa Ewart 1989a: 79 (comp. note)
Tamasa Moulds 1990: 25, 32, 104, 183 (listed, diversity, comp. note) Australia
Tamasa Moulds 1996b: 92 (comp. note) Australia
Tamasa Moulds and Carver 2002: 4 (listed) Australia
Tamasa Ewart 2005a: 179 (comp. note) Queensland
Tamasa Moulds 2005b: 387, 390, 412, 425, 430, 434, Table 2 (type genus of Tamasini, higher taxonomy, listed, comp. note) Australia
Tamasa Shiyake 2007: 6, 75 (listed, comp. note)
Tamasa Wei, Ahmed and Rizvi 2010: 33 (comp. note) Australia

***T. burgessi* (Distant, 1905)** = *Abricta burgessi*
Abricta burgessi Moulds 1990: 119 (distribution, listed) Australia, Queensland
Abricta burgessi Moulds and Carver 2002: 50–51 (distribution, ecology, type data, listed) Australia, Queensland
Abricta burgessi Moulds 2003b: 246–247 (comp. note) Queensland

***T. doddi* (Goding & Froggatt, 1904)** = *Tibicen doddi* = *Abricta doddi* = *Tamasa tristigma doddi* = *Tamasa tristigma* var. *doddi*
Tibicen doddi Hahn 1962: 10 (listed) Australia, Queensland
Tibicen doddi Stevens and Carver 1986: 264 (type material) Australia, Queensland
Tamasa doddi Southcott 1988: 107, 115 (parasitism, comp. note) Australia, Queensland
Tamasa doddi Moulds 1990: 11, 106–107 (synonymy, illustrated, distribution, ecology, listed, comp. note) Equals *Tibicen doddi* Equals *Abricta doddi* Equals *Tamasa tristigma* var. *doddi* Australia, Queensland
Tibicen doddi Moulds 1990: 106 (listed, comp. note)
Abricta doddi Moulds 1990: 106 (listed, comp. note)
Tamasa tristigma var. *doddi* Moulds 1990: 106 (listed, comp. note)
Tamasa doddi Naumann 1993: 19, 125, 149 (listed) Australia
Tamasa doddi Moulds and Carver 2002: 4 (synonymy, distribution, ecology, type data, listed) Equals *Tibicen doddi* Australia, Queensland
Tibicen doddi Moulds and Carver 2002: 4 (type data, listed) Australia, Queensland

***T. rainbowi* Ashton, 1912**
Tamasa rainbowi Ewart 1989a: 79 (comp. note)
Tamasa rainbowi Moulds 1990: 104–105, Plate 14, Fig 3, Plate 24, Fig 17 (illustrated, distribution, ecology, listed, comp. note) Australia, New South Wales, Queensland
Tamasa rainbowi Naumann 1993: 26, 125, 149 (listed) Australia
Tamasa rainbowi Ewart 1996: 14 (listed)
Tamasa rainbowi Moulds 1996b: 92 (comp. note) Australia, New South Wales
Tamasa rainbowi Moss and Popple 2000: 54, 58 (listed, comp. note) Australia, New South Wales
Tamasa rainbowi Moulds and Carver 2002: 5 (distribution, ecology, type data, listed) Australia, New South Wales, Queensland
Tamasa rainbowi Williams 2002: 156–157 (listed) Australia, Queensland, New South Wales

***T. tristigma* (Germar, 1834)** = *Cicada tristigma* = *Tettigia (Tettigia) tristigma* = *Tettigia tristigma* = *Tibicen kurandae* Goding & Froggatt, 1904 = *Abricta kurundae* = *Tamasa kurandae*
Tettigia tristigma Hahn 1962: 10 (listed) Australia, Northern Territory, South Australia
Tibicen kurudae Hahn 1962: 10 (listed) Australia, Queensland
Tamasa tristigma Josephson 1981: 30–31, 34, Figs 7–8, Fig 10 (muscle function, comp. note)
Tamasa tristigma Young and Josephson 1983a: 184, 191–193, Figs 7–8, Table 1 (oscillogram, timbal muscle kinetics, comp. note) Australia, Queensland
Tamasa tristigma Young and Josephson 1984: 286, Fig 1 (comp. note)
Tamasa tristigma Popov 1985: 45 (acoustic system, comp. note)
Tamasa tristigma Ewart 1986: 51–52, 54, 57, Fig 1A, Table 1 (oscillogram, natural history, listed, comp. note) Queensland, New South Wales, South Australia
Tibicen kurandae Stevens and Carver 1986: 265 (type material) Australia, Queensland
Tettigia tristigma Stevens and Carver 1986: 263–264 (type material, comp. note) Australia, South Australia
Tamasa tristigma Josephson and Young 1987: 995–996, Figs 2–3 (comp. note)
Tamasa tristigma Moulds 1987a: 25 (comp. note)
Tamasa tristigma Ewart 1989a: 78 (comp. note) Queensland
Cicada tristigma Ewart 1990: 1, 3 (type material) Equals *Abroma tristigma* Australia
Cicada tristigma Moulds 1990: 25, 104 (type data, listed)

Tamasa tristigma Moulds 1990: 26, 104–106, 109, Plate 14, Fig 1, Fig 1a, Plate 24, Fig 11 (illustrated, distribution, ecology, listed, comp. note) Australia, New South Wales, Queensland
Tamasa tristigma Ewart 1993: 136–137, 141, Fig 2 (natural history, acoustic behavior, song analysis) Queensland
Tamasa tristigma Naumann 1993: 8, 125, 149 (listed) Australia
Tamasa tristigma Bennet-Clark and Young 1994: 293, Table 1 (body size and song frequency)
Tamasa tristigma Ewart 1995: 85 (described, illustrated, natural history, acoustic behavior, oscillogram) Queensland
T[amasa] tristigma Jiang, Yang, Tang, Xu and Chen 1995b: 229 (song frequency, comp. note)
Tamasa tristigma Fonseca 1996: 28 (comp. note)
Tamasa tristigma Hangay and German 2000: 59–60 (illustrated, behavior) Queensland, New South Wales
Tamasa tristigma Ewart 2001a: 500–508, Fig 6, Fig 8D, Fig 9, Tables 1–2 (acoustic behavior) Queensland
Tamasa tristigma Sanborn 2001a: 16, Fig 5 (comp. note)
Tamasa tristigma Sueur 2001: 40, Table 1 (listed)
Tamasa tristigma Nation 2002: 260, Table 9.3 (comp. note)
Tamasa tristigma Moulds and Carver 2002: 5 (synonymy, distribution, ecology, type data, listed) Equals *Cicada tristigma* Equals *Tibicen kurandae* Australia, New South Wales, Queensland
Cicada tristigma Moulds and Carver 2002: 5 (type data, listed)
Tibicen kurandae Moulds and Carver 2002: 5 (type data, listed) Australia, Queensland
Tamasa tristigma Popple and Ewart 2002: 117 (illustrated) Australia, Queensland
Tamasa tristigma Syme and Josephson 2002: 769, Fig 4B (timbal muscle, comp. note)
Tamasa tristigma Williams 2002: 156–157 (listed) Australia, Queensland, New South Wales
Tamasa tristigma Moulds 2003a: 187, Fig 3 (illustrated, comp. note)
Tamasa tristigma Watson and Watson 2004: 141 (wing microstructure, comp. note)
Tamasa tristigma Emery, Emery, Emery and Popple 2005: 103–107, Tables 1–3 (comp. note) New South Wales
Tamasa tristigma Ewart 2005a: 177–179, Fig 11 (described, song, oscillogram, comp. note) Queensland
Tamasa tristigma Moulds 2005b: 395–396, 399, 403, 410, 417–419, 422, Fig 37, Fig 52, Figs 56–60, Table 1 (illustrated, higher taxonomy, phylogeny, key, listed, comp. note) Australia
Cicada tristigma Moulds 2005b: 434 (type species of *Tamasa*)
Tamasa tristigma Watson, Myhra, Cribb and Watson 2006: 9 (comp. note)
Tamasa tristigma Shiyake 2007: 6, 73, 75, Fig 122 (illustrated, distribution, listed, comp. note) Australia
Tamasa tristigma Nation 2008: 273, Table 10.1 (comp. note)
Tamasa tristigma Watson, Myhra, Cribb and Watson 2008: 3353 (comp. note)
Tamasa tristigma Moulds 2009b: 164, Fig 3 (illustrated, comp. note)
Tamasa tristigma Watson, Watson, Hu, Brown, Cribb and Myhra 2010: 117 (wing microstructure, comp. note)

***T.* sp. nr. *tristigma* Moss, 1988**
Tamasa sp. near *T. tristigma* Moss 1988: 29–30 (distribution, song, comp. note) Queensland
Tamasa sp. near *T. tristigma* Moss 2000: 69 (comp. note) Queensland

***Parnkalla* Distant, 1905**
Parnkalla Moulds 1990: 7, 32, 124 (listed, diversity, comp. note) Australia
Parnkalla Ewart 1999: 2 (listed)
Parnkalla Moulds and Carver 2002: 54 (type data, listed) Australia
Parnkalla Moulds 2005b: 393, 403, 412–413, 416, 425, 430, 434, Table 2 (higher taxonomy, listed, comp. note) Australia
Parnkalla Shiyake 2007: 8, 102 (listed, comp. note)
Parnkalla Ewart 2009b: 144 (comp. note)

P. gregoryi (Distant, 1882) to *Parnkalla muelleri* (Distant, 1882)
P. magna Distant, 1913 to *Parnquila magna*

***P. muelleri* (Distant, 1882)** = *Tibicen muelleri* = *Parnkalla müllerі* = *Parnkalla melleri* (sic) = *Parnkalla mueller* (sic) = *Tibicen gregoryi* Distant, 1882 = *Parnkalla gregoryi*
Parnkalla meulleri Itô 1976: 67 (sugarcane pest, comp. note) Australia, Queensland
Parnkalla muelleri Itô and Nagamine 1981: 281, Table 5 (sugarcane pest, comp. note) Queensland
Parnkalla muelleri Itô 1982b: 139 (comp. note)
Parnkalla muelleri Hitchcock 1983: 9 (control) Queensland
Parnkalla muelleri McKilligan 1984: 137–138, 140, Table 4 (predation) Queensland
Parnkalla muelleri Anonymous 1985: 22 (sugar cane pest, control) Australia
Parnkalla muelleri Wilson 1987: 489, Table 1 (sugarcane pest, listed, comp. note) Australia, Queensland

Parnkalla mueller (sic) Liu 1990: 46 (comp. note)
Parnkalla muelleri Moulds 1990: 3, 5, 8, 124–125Plate 16, Fig 5, Fig 5a, Plate 24, Fig 20 (synonymy, type material, illustrated, distribution, ecology, listed, comp. note) Equals *Tibicen muelleri* Equals *Tibicen gregoryi* n. syn. Equals *Parnkalla gregoryi* Australia, Queensland
Tibicen muelleri Moulds 1990: 124 (type species of *Parnkalla*, listed, comp. note) Queensland
Tibicen gregoryi Moulds 1990: 124 (listed) Queensland
Parnkalla gregoryi Moulds 1990: 124 (listed) Queensland
Parnkalla muelleri Moulds and Carver 1991: 465 (comp. note) Australia
Parnkalla mueller (sic) Liu 1992: 9 (comp. note)
Parnkalla muelleri Naumann 1993: 65, 113, 149 (listed) Australia
Tibicen muelleri Moulds and Carver 2002: 54 (type species of *Parnkalla*, type data, listed) Queensland
Parnkalla muelleri Moulds and Carver 2002: 54 (synonymy, distribution, ecology, type data, listed) Equals *Tibicen gregoryi* Equals *Tibicen muelleri* Australia, Queensland
Tibicen gregoryi Moulds and Carver 2002: 54 (type data, listed) Queensland
Parnkalla muelleri Williams 2002: 156–157 (listed) Australia, Queensland, New South Wales
Parnkalla muelleri Brambila and Hodges 2004: 364 (sugar cane pest, comp. note) Australia
Parnkalla muelleri Moulds 2005b: 395–397, 399, 417–420, 422, Figs 56–60, Table 1 (higher taxonomy, phylogeny, listed, comp. note) Australia
Parnkalla muelleri Shiyake 2007: 8, 100, 103, Fig 171 (illustrated, distribution, listed, comp. note) Australia
Parnkalla muelleri Brambila and Hodges 2008: 604 (sugarcane pest, comp. note) Australia

***Parnquila* Moulds, 2012**

***P. hillieri* (Distant, 1906)** (nec *hillieri* Distant, 1907) = *Burbunga hillieri*
Burbunga hillieri Moulds 1990: 128, Plate 21, Fig 6 (illustrated, distribution, ecology, listed, comp. note) Australia, South Australia
Burbunga hillieri Moulds and Carver 2002: 53 (distribution, ecology, type data, listed) Australia, South Australia

***P. magna* (Distant, 1913)** = *Parnkalla magna*
Parnkalla magna Moulds 1990: 124 (distribution, listed) Australia, Western Australia
Parnkalla magna Moulds and Carver 2002: 54 (distribution, ecology, type data, listed) Australia, Western Australia

***P. unicolor* (Ashton, 1921)** = *Areopsaltria unicolor*
Arenopsaltria unicolor Moulds 1990: 66 (listed, comp. note) Australia, Western Australia
Arenopsaltria unicolor Moulds and Carver 2002: 6 (distribution, ecology, type data, listed) Australia, Western Australia

***P. venosa* (Distant, 1907)** = *Burbunga venosa* = *Burbunga vernosa* (sic)
Burbunga venosa Moulds 1990: 127 (distribution, listed) Australia, Northern Territory
Burbunga vernosa (sic) Moulds 1994: 97 (comp. note) Australia
Burbunga venosa Moulds and Carver 2002: 53 (distribution, ecology, type data, listed) Australia, Northern Territory
Burbunga venosa Ewart 2009b: 137–138, 141–142, 144, 157, Fig 1, Figs 5A–B, Plate 1G (oscillogram, illustrated, distribution, emergence times, habitat, comp. note) Equals Cravens Spinifex Buzzer, no. 30 Queensland, Northern Territory

Tribe Dundubiini Atkinson, 1886 = Dundubini (sic) = Platylomiini Metcalf, 1955
Dundubiini Chen 1942: 145 (listed
Dundubiini Duffels 1983b: 10 (comp. note)
Platylomiini Duffels 1983b: 10 (comp. note)
Dundubiini Hayashi 1984: 28, 56, Fig 3b (illustrated, key, comp. note)
Platylomiini Kirillova 1986a: 122 (listed)
Platylomiini Kirillova 1986b: 45 (listed)
Dundubiini Kirillova 1986b: 45 (listed)
Dundubiini Kirillova 1986a: 123 (listed)
Dundubiini Duffels 1990b: 65 (listed)
Dundubiini Moulds 1990: 31 (listed) Australia
Dundubiini Duffels 1991a: 128 (comp. note)
Dundubini (sic) Gogala 1995: 111 (comp. note)
Dundubiini Boer and Duffels 1996a: 155 (diveristy)
Dundubiini Boer and Duffels 1996b: 298, 301–302, 328, 330 (diversity, comp. note)
Dundubiini Chou, Lei, Li, Lu and Yao 1997: 106, 235 (key, listed, described, synonymy) Equals Dundubiaria Equals Platylomiini
Dundubiini Novotny and Wilson 1997: 437 (listed)
Dundubiini Boer 1999: 115 (comp. note)
Dundubiini Moulds 1999: 1 (comp. note)
Dundubiini Sanborn 1999a: 43 (listed)
Dundubiini Villet and Zhao 1999: 321 (comp. note)
Platylomiini Boulard 2000c: 51 (comp. note)
Dundubiini Duffels and Zaidi 2000: 195, 196, 199, 202 (comp. note)

Dundubiini Kos and Gogala 2000: 2 (comp. note)
Platylomiini Boulard 2001a: 141 (comp. note)
Dundubiini Boulard 2001c: 51, 54, 56 (listed)
Platylomiini Boulard 2001c: 58 (listed)
Dundubiini Sueur 2001: 39, Table 1 (listed)
Platylomiini Sueur 2001: 39, Table 1 (listed)
Platylomiini Boulard 2002b: 197 (comp. note)
Dundubiini Duffels and Turner 2002: 235–236, 244 (comp. note)
Platylomiini Duffels and Turner 2002: 235–236 (comp. note)
Dundubiini Moulds and Carver 2002: 11 (listed) Australia
Dundubiini Schouten and Duffels 2002: 29 (comp. note)
Dundubiini Boulard 2003a: 98 (listed)
Dundubiini Boulard 2003b: 193 (listed)
Dundubiini Boulard 2003c: 172 (listed)
Dundubiini Boulard 2003e: 260, 266, 270 (listed)
Dundubiini Lee and Hayashi 2003a: 150, 153, 170 (key, listed, comp. note) Taiwan
Dundubiini Lee and Hayashi 2003b: 359 (listed) Taiwan
Dundubiini Chen 2004: 39–41 (phylogeny, listed)
Dundubiini Gogala, Trilar, Kozina and Duffels 2004: 2 (listed)
Dundubiini Lee and Hayashi 2004: 45–46 (listed) Taiwan
Dundubiini Boulard 2005e: 118 (listed)
Dundubiini Lee 2005: 14, 17, 156–157, 173 (key, listed, comp. note) Korea
Dundubiini Moulds 2005b: 385–386, 390, 396, 404, 407, 413, 415, 419, 422–423, 427, 429–430, 432, 434, Fig 58, Table 2 (described, diagnosis, composition, higher taxonomy, phylogeny, key, listed, comp. note) Equals Dundubia (sic) Equals Dundubiaria Equals Dundubini (sic) Equals Platylomiini Australia
Platylomiini Moulds 2005b: 387 (history, listed, comp. note)
Dundubiini Pham and Thinh 2005a: 241 (listed) Vietnam
Dundubiini Pham and Thinh 2005b: 287–288, 290 (key, comp. note) Equals Dundubiaria Equals Platylomiini Vietnam
Dundubiini Boulard 2006d: 195 (comp. note)
Dundubiini Boulard 2007a: 99 (listed)
Dundubiini Boulard 2007d: 496 (listed)
Dundubiini Boulard 2007e: 20, 67 (comp. note) Thailand
Dundubiini Sanborn, Phillips and Sites 2007: 3, 15, Table 1 (listed) Thailand
Platylomiini Sanborn, Phillips and Sites 2007: 3, 22, 39, Table 1 (listed) Thailand
Dundubiini Shiyake 2007: 4, 46 (listed, comp. note)
Dundubiini Boulard 2008c: 352, 363 (listed)
Dundubiini Boulard 2008f: 7, 23 (listed) Thailand
Dundubiini Lee 2008a: 1 (status) Vietnam
Dundubiini Moulds 2008a: 209 (key, comp. note)
Dundubiini Osaka Museum of Natural History 2008: 324 (listed) Japan
Dundubiini Duffels 2009: 303 (comp. note)
Dundubiini Lee 2009a: 87 (listed, comp. note)
Dundubiini Pham and Yang 2009: 12 (listed, synonymy) Equals Dundubiaria Equals Platylomiini Vietnam
Dundubiini Lee and Hill 2010: 278 (comp. note)
Platylomiini Lee and Hill 2010: 301 (comp. note)
Dundubiini Wei, Ahmed and Rizvi 2010: 33 (comp. note)

Subtribe Dundubiina Matsumura, 1917 = Dundubiaria = Platylomaria Metcalf, 1955

Dundubiaria Duffels 1983b: 11 (composition, comp. note)
Dundubaria Duffels 1990b: 70 (comp. note)
Dundubiaria Duffels and Zaidi 2000: 195–198 (comp. note) Southeast Asia, Greater Sunda Islands, Lesser Sunda Islands, Philippines
Dundubiaria Boulard 2001c: 51, 56 (listed)
Platylomaria Boulard 2001c: 51, 58 (listed)
Dundubiina Duffels and Turner 2002: 235–236, 238–239, 244 (comp. note)
Dundubiaria Boulard 2003a: 98 (listed)
Dundubiaria Boulard 2003b: 193 (listed)
Dundubiaria Boulard 2003c: 172 (listed)
Dundubiina Lee and Hayashi 2003a: 150, 170 (key, listed) Taiwan
Dundubiina Lee and Hayashi 2003b: 359 (listed) Taiwan
Dundubiina Lee and Hayashi 2004: 45 (listed) Taiwan
Dundubiaria Boulard 2005e: 118 (listed)
Dundubiina Lee 2005: 157, 173 (key, listed, comp. note) Korea
Dundubiina Moulds 2005b: 391–392, 39, 423, 432 (higher taxonomy, listed, comp. note) Equals Dundubiaria
Dundubiaria Sanborn, Phillips and Sites 2007: 3, 15, 38, Fig 3, Table 1 (listed, distribution) Thailand
Platylomiaria Sanborn, Phillips and Sites 2007: 3, 22, 39, Fig 4, Table 1 (listed, distribution) Thailand
Dundubiaria Shiyake 2007: 5, 58 (listed, comp. note)
Dundubiina Boulard 2008f: 7, 23 (listed) Thailand
Dundubiina Lee 2008a: 1–2, 13, Table 1 (listed) Vietnam

Dundubiina Cranston 2009: 93 (distribution, comp. note) Southeast Asia
Dundubiina Lee 2009a: 87 (comp. note)
Dundubiina Lee 2009b: 330 (comp. note)
Dundubiina Lee 2009c: 338 (comp. note)
Dundubiina Lee 2009e: 87 (comp. note)
Dundubiina Lee 2009h: 2629 (listed) Philippines, Palawan
Dundubiina Lee 2009i: 294 (listed) Philippines, Panay
Dundubiina Lee and Sanborn 2009: 31 (listed)
Dundubiina Pham and Yang 2009: 14, Table 2 (listed) Vietnam
Dundubiina Sanborn 2009b: 307 (comp. note) Vietnam
Dundubiina Lee 2010a: 22–23 (listed) Philippines, Mindanao
Dundubiina Lee 2010b: 24 (listed) Cambodia
Dundubiina Lee 2010c: 13 (listed) Philippines, Luzon
Dundubiina Lee 2010d: 167 (comp. note)
Dundubiina Lee and Hill 2010: 278, 302 (comp. note)
Dundubiina Wei, Ahmed and Rizvi 2010: 33 (comp. note)

***Dundubia* Amyot & Audinet-Serville, 1843** = *Dandubia* (sic) = *Dundbia* (sic) = *Dundula* (sic) = *Dendubia* (sic) = *Dubia* (sic)
Dundubia Wu 1935: 11 (listed) China
Dundubia sp. Meer Mohr 1965: 104 (human food, listed) Sumatra
Dundubia sp. Hill, Hore and Thornton 1982: 49, 161, Plate 104 (listed, illustrated, comp. note) Hong Kong
Dundubia Duffels 1983b: 3, 9 (comp. note)
Dundubia Strümpel 1983: 18, 134 (listed, comp. note)
Dundubia Hayashi 1984: 58 (listed, comp. note)
Dundubia Boulard 1985e: 1021 (coloration, comp. note) Orient
Dundubia Duffels 1986: 319 (distribution, diversity, comp. note) Philippines, Greater Sunda Islands, Southeast Asia
Dundubia Zaidi, Mohammedsaid and Yaakob 1990: 261, 267, Table 2 (listed, comp. note) Malaysia
Dundubia Duffels and van Mastrigt 1991: 174 (predation, comp. note) Sumatra
Dundubia Hayashi 1993: 14 (comp. note)
Dundubia Zaidi and Ruslan 1994: 427, Table 2 (comp. note) Malaysia
Dundubia Riede and Kroker 1995: 43 (comp. note) Sabah, Malaysia
Dundubia Zaidi and Ruslan 1995a: 174, Table 2 (listed, comp. note) Borneo, Sarawak
Dundubia Zaidi and Ruslan 1995b: 220–221, Table 2 (listed, comp. note) Borneo, Sabah
Dundubia Zaidi and Ruslan 1995c: 70, Table 2 (listed, comp. note) Peninsular Malaysia
Dundubia Zaidi and Ruslan 1995d: 201–202, Table 2 (listed, comp. note) Borneo, Sabah
Dundubia Beuk 1996: 129–131, 133, 135 (characteristics, history, comp. note)
Dundubia Zaidi 1996: 102–103, Table 2 (comp. note) Borneo, Sarawak
Dundubia Zaidi and Hamid 1996: 54–55, Table 2 (listed, comp. note) Borneo, Sarawak
Dundubia Chou, Lei, Li, Lu and Yao 1997: 236, 247, 293, 308, 320, 325, 327, Table 6, Tables 10–12 (key, described, listed, diversity, biogeography) China
Dundubia Novotny and Wilson 1997: 437 (listed)
Dundubia Zaidi 1997: 114–115, Table 2 (listed, comp. note) Borneo, Sarawak
Dundubia Beuk 1998: 147–148, 150 (comp. note)
Dundula (sic) sp. Chen, Wongsiri, Jamyanya, Rinderer, Vongsamanode, et al. 1998: 27 (predation) Thailand
Dundubia Duffels and Zaidi 1998: 320–321 (comp. note)
Dundubia Holloway 1998: 299 (comp. note)
Dundubia Zaidi and Ruslan 1998b: 4–5, Table 2 (listed, comp. note) Borneo, Sabah
Dundubia Beuk 1999: 1–2, 475, 9 (history, characteristics) Southeast Asia, Greater Sunda Islands, Sulawesi, Philippines
Dundubia Zaidi and Ruslan 1999: 4, Table 2 (listed, comp. note) Borneo, Sarawak
Dundubia Duffels and Zaidi 2000: 195–196, 198 (diversity, comp. note)
Dundubia Boulard 2001d: 82 (comp. note)
Dendubia (sic) Howard, Moore, Giblin-Davis and Abad 2001: 157 (pest species, comp. note) Malaysia
Dundubia Boulard 2002b: 198, 203 (comp. note)
Dundubia Boulard 2003c: 183, 185 (comp. note)
Dundubia Pham 2004: 61 (key, described, distribution, comp. note) Vietnam
Dundubia Prešern, Gogala and Trilar 2004a: 240 (diversity, comp. note)
Dundubia Prešern, Gogala and Trilar 2004b: 62 (comp. note)
Dundubia Schouten, Duffels and Zaidi 2004: 379 (comp. note)
Dundubia Boulard 2005d: 341 (comp. note)
Dundubia Boulard 2005k: 97 (comp. note)
Dundubia Boulard 2005n: 141 (comp. note)
Dundubia Moulds 2005b: 387–389, 391, 412–413, 423, 430, 432, Table 2 (type genus of Dundubiini, higher taxonomy, listed, comp. note)
Dundubia Pham and Thinh 2005a: 241 (listed) Vietnam

Dundubia Pham and Thinh 2005b: 290 (comp. note) Vietnam
Dundubia Boulard 2006g: 113–114 (comp. note)
Dundubia Boulard 2007e: 20 (comp. note) Thailand
Dundubia Sanborn, Phillips and Sites 2007: 3, 15, Table 1 (listed) Thailand
Dundubia Shiyake 2007: 5, 58 (listed, comp. note)
Dundubia sp. Shiyake 2007: 5, 56, 58, Fig 90 (illustrated, listed, comp. note)
Dundubia Boulard 2008f: 6–7, 23, 92 (listed, comp. note) Thailand
Dundubia Lee 2008a: 2, 17, Table 1 (listed) Vietnam
Dundubia Lee 2009h: 2631, 2632 (listed) Philippines, Palawan
Dundubia Lee 2009i: 295 (listed) Philippines, Panay
Dundubia Pham and Yang 2009: 12, 15, Table 2 (type species of Dundubiini, listed) Vietnam
Dundubia Lee 2010a: 24 (listed) Philippines, Mindanao
Dundubia Lee 2010b: 26 (listed) Cambodia
Dundubia Lee 2010c: 13 (listed) Philippines, Luzon
Dundubia Lee and Hill 2010: 302 (comp. note)

D. aerata Distant, 1888 to *Champaka aerata*

***D. andamansidensis* (Boulard, 2001)** = *Platylomia andamansidensis*
Platylomia andamansidensis Boulard 2001c: 63–70, Plate F, Figs 1–4, Plate G, Figs 1–2, Plate H, Figs 1–2 (n. sp., described, song, illustrated, sonogram, comp. note) Thailand
Platylomia andamansidensis Boulard 2002b: 203 (comp. note)
Dundubia andamansidensis Boulard 2005d: 347 (comp. note)
Platylomia andamansidensis Boulard 2005o: 148 (listed)
Platylomia andamansidensis Trilar 2006: 344 (comp. note)
Dundubia andamansidensis Boulard 2007d: 496 (comp. note) Thailand
Dundubia andamansidensis Boulard 2007e: 47 (comp. note) Thailand
Platylomia andamansidensis Sanborn, Phillips and Sites 2007: 22–23 (listed, distribution, comp. note) Thailand
Dundubia andamansidensis Boulard 2008f: 23, 92 (biogeography, listed, comp. note) Equals *Platylomia andamansidensis* Thailand

***D. annandalei* Boulard, 2007** nom. nov. pro *Dundubia intemerata* Boulard, 2003 (nec Walker) = *Dundubia rufivena* (nec Walker) = *Dundubia terpsichore* (nec Walker)
Dundubia intemerata Boulard 2003c: 172, 182–183, Plate F, Figs 1–2 (described, illustrated, song, sonogram, comp. note) Thailand, Malacca, Peninsular Malaysia
Dundubia annandalei Boulard 2007d: 496, 498, Plate, Figs G-H (nom. nov. pro *Dundubia intemerata* Boulard, 2003, illustrated, comp. note) Equals *Dundubia intemerata* (nec Walker) Equals *Dundubia rufivena* (nec Walker) Equals *Dundubia terpsichore* (nec Walker) Thailand
Dundubia annandalei Boulard 2008f: 23, 92 (biogeography, listed, comp. note) Equals *Dundubia intemerata* Boulard 2003 Thailand

***D. ayutthaya* Beuk, 1996**
Dundubia ayutthaya Beuk 1996: 137, 139, 169–172, Fig 4, Figs 65–71 (n. sp., key, described, illustrated, comp. note) southeast Thailand
Dundubia ayutthaya Sanborn, Phillips and Sites 2007: 15 (listed, distribution, comp. note) Thailand
Dundubia ayutthaya Boulard 2008f: 24, 92 (biogeography, listed, comp. note) Thailand

D. bifasciata Liu, 1940 to *Dundubia hainanensis* (Distant, 1901)

***D. crepitans* Boulard, 2005**
Dundubia crepitans Boulard 2005e: 118–121, Figs 1–5, Plate A, Figs 1–2 (n. sp., described, illustrated, song, sonogram, listed, comp. note) Thailand
Dundubia crepitans Boulard 2005o: 151 (listed)
Dundubia crepitans Sanborn, Phillips and Sites 2007: 15 (listed, distribution, comp. note) Thailand
Dundubia crepitans Boulard 2008f: 24, 92 (biogeography, listed, comp. note) Thailand

D. cinctimanus Walker, 1850 to *Pomponia cinctimanus*

***D. cochlearata* Overmeer & Duffels, 1967**
Dundubia cochlearata Lee 2009h: 2632 (comp. note)

***D. dubia* Lee, 2009**
Dundubia dubia Lee 2009h: 2631–2633, Figs 9–10 (n. sp., described, illustrated, distribution, listed) Philippines, Palawan

***D. emanatura* Distant, 1889**
Dundubia emanatura Boulard 2005e: 118 (comp. note)

***D. ensifera* Bloen & Duffels, 1976**
Dundubia ensifera Ahmed and Sanborn 2010: 44 (distribution) Bangladesh

***D. euterpe* Bloem & Duffels, 1976**
Dundubia euterpe Zaidi and Ruslan 1995c: 6 (comp. note) Peninsular Malaysia
Dundubia euterpe Zaidi, Ruslan and Mahadir 1996: 60, Table 1 (listed, comp. note) Peninsular Malaysia
Dundubia euterpe Zaidi, Ruslan and Azman 1999: 307 (listed, comp. note) Borneo, Sabah, Peninsular Malaysia
Dundubia euterpe Zaidi, Ruslan and Azman 2000: 205 (listed, comp. note) Borneo, Sarawak, Peninsular Malaysia, Sabah

Dundubia euterpe Zaidi, Azman and Ruslan 2001b: 124–125 (listed, comp. note) Borneo, Sarawak, Peninsular Malaysia, Sabah
Dundubia euterpe Zaidi and Azman 2003: 104, Table 1 (listed, comp. note) Borneo, Sabah
Dundubia euterpe Schouten, Duffels and Zaidi 2004: 372–373, Table 2 (listed, comp. note) Malayan Peninsula
Dundubia euterpe Gogala and Trilar 2005: 66–67, 70, 75, Fig 1 (song, sonogram, oscillogram, acoustic behavior, comp. note) Malaysia, Thailand

D. feae **(Distant, 1892)** = *Cosmopsaltria feae* = *Orientopsaltria feae* = *Dundubia longina* Distant, 1917
Cosmopsaltria feae Wu 1935: 12 (listed) China, Soochow, Burma, Karennee
Cosmopsaltria feae Duffels 1983b: 4 (comp. note)
Orientopsaltria feae Duffels 1983b: 9 (comp. note)
Dundubia feae Beuk 1996: 131, 137–138, 140–143, 154–155, Fig 2, Figs 6–11 (n. comb., n. syn., key, lectotype designation, described, illustrated, comp. note, distribution) Equals *Cosmopsaltria feae* Equals *Orientopsaltria feae* Equals *Dundubia longina* Burma, Laos, Thailand, Vietnam, China?
Dundubia feae Chou, Lei, Li, Lu and Yao 1997: 247–248, Plate XIII, Fig 134 (key, described, illustrated, synonymy, comp. note) Equals *Cosmopsaltria feae* Equals *Orientopsaltria feae* China
Platylomia feae Beuk 1998: 152 (comp. note)
Cosmopsaltria feae Duffels and Zaidi 2000: 196 (comp. note)
Dundubia feae Hua 2000: 61 (listed, distribution) Equals *Cosmopsaltria feae* China, Jiangsu, Unnan, Sichuan, Burma, Japan, India
Platylomia radha (sic) Boulard 2001a: 142, Fig 11 (sonogram) Thailand
Dundubia feae Boulard 2003c: 172, 184–185, Plate G, Figs 1–2 (described, illustrated, song, sonogram, comp. note) Equals *Cosmopsaltria feae* Equals *Orientopsaltria feae* Equals *Dundubia longina* Thailand, Tenasserim, Laos, Tonkin, China
Dundubia feae Pham 2004: 61 (comp. note)
Dundubia feae Boulard 2005d: 341, 344, 347 (comp. note)
Dundubia feae Boulard 2005o: 149 (listed)
Dundubia longina Pham and Thinh 2005a: 236 (comp. note) Vietnam
Dundubia feae Pham and Thinh 2005a: 236 (comp. note) Vietnam
Dundubia feae Pham and Thinh 2005a: 241 (synonymy, distribution, listed) Equals *Cosmopsaltria feae* Equals *Orentopsaltria feae* Equals *Dundubia longina* Vietnam
Dundubia feae Boulard 2006g: 73–74, 83, 125, Fig 40 (sonogram, comp. note)
Dundubia feae Boulard 2007e: 32, 34, 47, 69, 83, Fig 19 (sonogram, comp. note) Thailand
Dundubia feae Sanborn, Phillips and Sites 2007: 15 (listed, distribution, comp. note) Thailand, Burma, Laos, Vietnam, China
Dundubia feae Boulard 2008f: 24, 92 (biogeography, listed, comp. note) Equals *Cosmopsaltria feae* Equals *Orientopsaltria feae* Thailand, Burma, Laos, Vietnam
Dundubia feae Lee 2008a: 18 (synonymy, distribution, listed) Equals *Cosmopsaltria feae* Equals *Dundubia longina* Vietnam, China, Guangxi, Hainan, Laos, Thailand, Myanmar
Dundubia feae Pham and Yang 2009: 15, Table 2 (listed) Vietnam

D. flava **Lee, 2009**
Dundubia flava Lee 2009h: 2633–2636, Figs 11–12 (n. sp., described, illustrated, distribution, listed) Philippines, Palawan

D. gravesteini **Duffels, 1976** = *Dundubia graveistini* (sic)
Dundubia graveistini (sic) Zaidi 1993: 957 (comp. note) Borneo, Sarawak
Dundubia gravesteini Zaidi 1993: 958, 960, Table 1 (listed, comp. note) Borneo, Sarawak
Dundubia gravesteini Zaidi and Ruslan 1994: 426, 428, Table 1 (comp. note) Malaysia
Dundubia gravesteini Zaidi and Ruslan 1995d: 197 (comp. note) Borneo
Dundubia gravesteini Zaidi 1996: 98, 101, 104, Table 1 (comp. note) Borneo
Dundubia gravesteini Zaidi and Hamid 1996: 51, 53, 56–57, Table 1 (listed, comp. note) Borneo, Sarawak
Dundubia gravesteini Zaidi, Ruslan and Mahadir 1996: 60, Table 1 (listed, comp. note) Peninsular Malaysia
Dundubia gravesteini Zaidi and Ruslan 1999: 1 (comp. note) Borneo
Dundubia gravesteini Schouten, Duffels and Zaidi 2004: 372–373, Table 2 (listed, comp. note) Malayan Peninsula
Dundubia gravesteini Lee 2009h: 2632 (comp. note)

D. hainanensis **(Distant, 1901)** = *Cosmopsaltria hainanensis* = *Platylomia hainanensis* = *Dundubia hainanensi* (sic) = *Dundubia bifasciata* Liu, 1940
Platylomia hainanensis Wu 1935: 13 (listed) Equals *Cosmopsaltria hainanensis* China, Hainan, Tonkin
Dundubia bifasciata Liu 1992: 40 (comp. note)
Dundubia hainanensis Beuk 1996: 131, 137–138, 143–147, Fig 2, Figs 12–18 (n. comb., n. syn., key, described, illustrated, distribution, comp. note) Equals *Cosmopsaltria hainanensis*

Equals *Platylomia hainanensis* Equals *Dundubia bifasciata* n. syn. Vietnam, China

Dundubia hainanensis Chou, Lei, Li, Lu and Yao 1997: 28, 247–249, 308, 320–321, 325, 327, 332, 346, 358, Plate XIII, Fig 133, Tables 10–12 (n. comb., key, described, illustrated, synonymy, song, oscillogram, spectrogram, comp. note) Equals *Cosmopsaltria hainanensis* Equals *Platylomia hainanensis* Equals *Dundubia bifasciata* n. syn. China

Dundubia bifasciata Chou, Lei, Li, Lu and Yao 1997: 358 (synonymy, comp. note)

Dundubia hainanensis Hua 2000: 61 (listed, distribution) Equals *Cosmopsaltria hainanensis* Equals *Platylomia hainanensis* Equals *Dundubia bifasciata* China, Zhejiang, Hainan, Guangxi, Thailand, Vietnam

Dundubia hainanensi (sic) Sueur 2001: 39, Table 1 (listed)

Dundubia hainanensis Pham 2004: 61, 63–64 (key, synonymy, distribution, comp. note) Equals *Cosmopsaltria hainanensis* Equals *Platylomia hainanensis* Vietnam

Dundubia hainanensis Pham and Thinh 2005a: 241 (synonymy, distribution, listed) Equals *Cosmopsaltria hainanensis* Equals *Platylomia hainanensis* Equals *Dundubia bifasciata* Vietnam

Dundubia hainanensis Pham and Thinh 2006: 525, 527 (listed, comp. note) Vietnam

Dundubia bifasciata Sanborn, Phillips and Sites 2007: 16, 34 (comp. note)

Dundubia hainanensis Sanborn, Phillips and Sites 2007: 16, 34 (comp. note)

Dundubia hainanensis Lee 2008a: 18 (synonymy, distribution, listed) Equals *Cosmopsaltria hainanensis* Equals *Platylomia hainanensis* Vietnam, China, Guangxi, Hainan, Thailand

Dundubia hainanensis Pham and Yang 2009: 15, Table 2 (listed) Vietnam

***D. hastata* (Moulton, 1923)** = *Cosmopsaltria hastata* = *Orientopsaltria hastata* = *Dundubia spiculata* (nec Nouhalhier)

Orientopsaltria hastata Duffels 1983b: 9 (n. comb., comp. note)

Dundubia hastata Beuk 1996: 172–173 (comp. note)

Dundubia hastata Boulard 2003b: 195 (comp. note)

Dundubia hastata Boulard 2005n: 138, 140–143, Plate D, Figs 1–2, Fig 6b (synonymy, described, illustrated, song, sonogram, comp. note) Equals *Cosmopsaltria hastata* Equals *Orientopsaltria hastata* Thailand

Dundubia hastata Boulard 2005o: 150 (listed)

Dundubia hastata Sanborn, Phillips and Sites 2007: 15–16, 18 (listed, distribution, comp. note) Thailand, Indochina, India, Malaysia

Cosmopsaltria hastata Sanborn, Phillips and Sites 2007: 15, 18 (comp. note)

Dundubia hastata Boulard 2008c: 363 (comp. note) Vietnam

Dundubia hastata Boulard 2008f: 24, 92 (biogeography, listed, comp. note) Equals *Cosmopsaltria hastata* Equals *Orientopsaltria hastata* Equals *Dundubia spiculata* (nec Nouhalhier) Thailand

Dundubia hastata Lee 2008a: 18 (synonymy, distribution, listed) Equals *Cosmopsaltria hastata* Vietnam, Thailand, Malaysia, India

Dundubia hastata Pham and Yang 2009: 15, Table 2 (listed) Vietnam

D. helena Distant, 1912 to *Dundubia nagarasingna* Distant, 1881

***D. jacoona* (Distant, 1888)** = *Cosmposaltria jacoona* = *Orientopsaltria jacoona*

Cosmopsaltria jacoona Duffels 1983b: 4 (comp. note)

Orientopsaltria jacoona Duffels 1983b: 9 (comp. note)

Orientopsaltria jacoona Zaidi, Mohammedsaid and Yaakob 1990: 262 (comp. note) Malaysia

Orientopsaltria jacoona Zaidi and Ruslan 1994: 428 (comp. note) Malaysia

Orientopsaltria jacoona Zaidi and Ruslan 1995c: 65, 68, Table 1 (listed, comp. note) Peninsular Malaysia

Dundubia jacoona Beuk 1996: 131, 138, 160–164, 169, Fig 3, Figs 51–57 (n. comb., key, described, illustrated, distribution, comp. note) Equals *Cosmopsaltria jacoona* Equals *Orientopsaltria jacoona* Borneo, Sumatra, southern Malay Peninsula

Dundubia jacoona assemblage Beuk 1996: 129–131, 133 (described, comp. note)

Orientopsaltria jacoona Zaidi, Ruslan and Mahadir 1996: 60, Table 1 (listed, comp. note) Peninsular Malaysia

Dundubia jacoona Zaidi and Ruslan 1997: 223–224 (listed, comp. note) Equals *Cosmopsaltria jacoona* Equals *Orientopsaltria jacoona* Peninsular Malaysia, Johore, Borneo, New Guinea, Sarawak, Brunei, Kalimantan, Sumatra, Indonesia, Bantam Island

Orientopsaltria jacoona Duffels and Zaidi 1998: 320 (comp. note)

Dundubia jacoona assemblage Duffels and Zaidi 1998: 320 (comp. note)

Dundubia jacoona Zaidi and Ruslan 1998a: 355 (listed, comp. note) Equals *Cosmopsaltria jacoona* Equals *Orientopsaltria jacoona* Borneo, Sarawak, Johore, New Guinea, Peninsular Malaysia, Sabah, Brunei, Kalimantan, Sumatra, Indonesia, Bantam Island

Cosmopsaltria jacoona Duffels and Zaidi 2000: 196 (comp. note)
Dundubia jacoona assemblage Duffels and Zaidi 2000: 196–197 (comp. note)
Dundubia jacoona Zaidi, Noramly and Ruslan 2000b: 338–339, Table 1 (listed, comp. note) Borneo, Sabah
Dundubia jacoona Zaidi, Ruslan and Azman 2000: 205–206 (listed, comp. note) Equals *Cosmopsaltria jacoona* Equals *Orientopsaltria jacoona* Borneo, Sarawak, Johore, New Guinea, Peninsular Malaysia, Sabah, Brunei, Kalimantan, Sumatra, Indonesia, Bantam Island
Dundubia jacoona assemblage Turner, Hovencamp and van Welzen 2001: 219, 225, Tables 1–2 (comp. note) JAva
Dundubia jacoona Zaidi, Azman and Ruslan 2001b: 125–126 (listed, comp. note) Equals *Cosmopsaltria jacoona* Equals *Orientopsaltria jacoona* Borneo, Sarawak, Johore, New Guinea, Peninsular Malaysia, Sabah, Brunei, Kalimantan, Sumatra, Indonesia, Bantam Island
Dundubia jacoona Boulard 2002b: 202 (comp. note)
Dundubia jacoona Boulard 2003b: 187, Plate I, Fig 5 (described, song, sonogram, illustrated, comp. note) Thailand
Dundubia jacoona Zaidi and Azman 2003: 105, Table 1 (listed, comp. note) Borneo, Sabah
Dundubia jacoona Schouten, Duffels and Zaidi 2004: 373–374, Table 2 (listed, comp. note) Malayan Peninsula

***D. kebuna* Moulton, 1923** = *Dundubia intemerata* (nec Walker)
Dundubia kebuna Zaidi 1996: 100–101, Table 1 (comp. note) Borneo, Sarawak
Dundubia kebuna Zaidi and Hamid 1996: 53, 56–57, Table 1 (listed, comp. note) Borneo, Sarawak
Dundubia kebuna Zaidi, Ruslan and Mahadir 1996: 60, Table 1 (listed, comp. note) Peninsular Malaysia
Dundubia kebuna Zaidi 1997: 112 (comp. note) Borneo, Sarawak
Dundubia kebuna Boulard 2007d: 497, 498, 506, Fig I (illustrated, comp. note) Thailand
Dundubia kebuna Boulard 2007e: Plate 3, Fig 3 (illustrated) Thailand
Dundubia kebuna Sanborn, Phillips and Sites 2007: 16 (listed, distribution, comp. note) Thailand, Malaysia
Dundubia kebuna Boulard 2008f: 24–25, 92 (biogeography, listed, comp. note) Equals *Dundubia intemerata* (partim) Thailand, Siam

***D. laterocurvata* Beuk, 1996** = *Dundubia mannifera* (nec Walker) var. *a* Distant 1889 (partim) = *Cosmopsaltria hastata* Mounton, 1923 (partim)
Dundubia laterocurvata Beuk 1996: 137, 139, 157, 176–179, Fig 5, Figs 79–85 (n. sp., key, described, illustrated, comp. note) Equals *Dundubia mannifera* (nec Walker) var. *a* Distant 1889 (partim) Equals *Cosmopsaltria hastata* Mounton, 1923 (partim) Northern India, Northern Burma

D. longina Distant, 1917 to *Dundubia feae* (Distant, 1892)

***D. myitkyinensis* Beuk, 1996**
Dundubia myitkyinensis Beuk 1996: 137–138, 153–157, Fig 2, Figs 37–44 (n. sp., key, described, distribution, illustrated, comp. note) Burma

***D. nagarasingna* Distant, 1881** = *Cosmopsaltria nagarasingna* = *Platylomia nagarasingna* = *Platylomia magarasingna* (sic) = *Platylomia nagalasingna* (sic) = *Dundubia nagarasingha* (sic) = *Dundubia helena* Distant, 1912 = *Cosmopsaltria fratercula* Distant, 1912 = *Orientopsaltria fratercula* = ?*Dundubia bifasciata* Ishihara, 1961 = *Platylomia* sp. Gogala, 1995
Orientopsaltria fratercula Duffels 1983b: 9 (n. comb., comp. note)
Platylomia nagarasingna Gogala 1994b: 6 (song, comp. note) Peninsular Malaysia
Platylomia nagarasingna Gogala 1995: 101, 104, 106, 111, Fig 3 (song, sonogram, oscillogram, comp. note) Thailand
Platylomia nagarasingna Gogala and Riede 1995: 305 (comp. note) Thailand
Platylomia sp. Gogala 1995: 101–102, 104, 107, Fig 4 (song, sonogram, oscillogram, comp. note) Thailand
Dundubia nagarasingna Beuk 1996: 131, 137–138, 147–153, Fig 2, Figs 19–36 (n. comb., n. syn., key, lectotype designation, described, illustrated, distribution, comp. note) Equals *Cosmopsaltria nagarasingna* Equals *Platylomia nagalasingna* (sic) Equals *Dundubia helena* n. syn. Equals *Cosmopsaltria fratercula* Equals *Orientopsaltria fratercula* n. syn. Equals *Platylomia* sp. Gogala 1995 Equals ?*Dundubia bifasciata* Ishihara, 1961
Dundubia nagarasingna group Beuk 1996: 135 (comp. note)
Dundubia nagarasingna Chou, Lei, Li, Lu and Yao 1997: 251 (key, described, synonymy, comp. note) Equals *Platylomia magarasingna* (sic) China
Platylomia nagarasingna Boulard 2000c: 50–56, Figs 1–6 (illustrated, song analysis, sonogram, comp. note) Thailand
Dundubia nagarasingna Hua 2000: 61 (listed, distribution) China, Yunnan, India, Burma, Thailand, Cambodia, Laos

Platylomia nagarasingna Boulard 2001a: 128, 133, 141–144, Plate II, Figs L-M, Figs 12–13 (illustrated, song analysis, sonogram, comp. note) Equals *Dundubia nagarasingna* Equals *Cosmopsaltria nagarasingna* Thailand
Platylomia nagarasingna Boulard 2001c: 63, 66, 70, Plate F, Fig 5 (illustrated, comp. note)
Platylomia nagarasingna Sueur 2001: 39, Table 1 (listed)
Dundubua nagarasingna Boulard 2003c: 178, 185 (comp. note)
Dundubia nagarasingna Boulard 2005d: 341, 345 (comp. note)
Platylomia nagarasingna Boulard 2005o: 147 (listed) Equals *Dundubia nagarasingna*
Dundubia nagarasingna Pham and Thinh 2005a: 241 (synonymy, distribution, listed) Equals *Cosmopsaltria nagarasingna* Equals *Platylomia nagarasingna* Equals *Cosmopsaltria fratercula* Equals *Dundubia helena* Equals *Orientopsaltria fratercula* Vietnam
Dundubia nagarasingna Boulard 2006g: 96, 125 (comp. note)
Dundubia nagarasingna Trilar 2006: 344 (comp. note)
Platylomia nagarasingna Trilar 2006: 344 (comp. note)
Dundubia nagarasingna Boulard 2007d: 494, 497 (comp. note) Thailand
Dundubia nagarasingna Boulard 2007e: 4, 47, 69, Plate 27c (coloration, illustrated, comp. note) Thailand
Dundubia nagarasingna Boulard 2007f: 122, 149, Fig 260 (comp. note)
Dundubia nagarasingha (sic) Sanborn, Phillips and Sites 2007: 16, 34–35 (listed, distribution, synonymy, comp. note) Equals *Platylomia nagarasingna* Equals *Dundubia bifasciata* (partim) Equals *Platylomia* sp. Gogala 1995 Thailand, Burma, India, Indochina, China
Dundubia nagarasingna Boulard 2008c: 363 (comp. note) Thailand
Dundubia nagarasingna Boulard 2008f: 25, 92 (biogeography, listed, comp. note) Equals *Cosmopsaltria nagarasingna* Equals *Platylomia nagarasingna* Equals *Platylomia nagalasingna* (sic) Equals *Dundubia helena* Equals *Cosmopsaltria fratercula* Equals *Platylomia* sp. Gogala 1995 Equals *Dundubia nagarasingha* (sic) Thailand, Burma, Tenasserim, Conchin China, Siam
Dundubia nagarasingna Lee 2008a: 18 (synonymy, distribution, listed) Equals *Cosmopsaltria nagarasingna* Equals *Platylomia nagarasingna* Equals *Cosmopsaltria fratercula* Vietnam, China, Yunnan, Laos, Cambodia, Thailand, Malay Peninsula, Myanmar
Dundubia nagarasingna Pham and Yang 2009: 15, Table 2 (listed) Vietnam
Dundubia nagarasingna Sun, Watson, Zheng, Watson and Liang 2009: 3148–3149, 3151, Fig 1L, Table 1 (illustrated, wing microstructure, comp. note) China, Yunnan
Dundubia nagarasingna Lee 2010b: 27 (distribution, listed) Cambodia, China, Yunnan, Vietnam, Laos, Thailand, Malay Peninsula, Myanmar
Dundubia nagarasingna Lee and Hill 2010: 299–300, Figs 7–8 (phylogeny, comp. note)

***D. nigripes* (Moulton, 1923)** = *Dundubia mannifera* (nec Walker) var. *a* Distant, 1889 = *Dundubia mannifera* Distant 1906 (partim) = *Cosmopsaltria nigripes* = *Orientopsaltria nigripes* = *Dundubia rafflesii* Overmeer & Duffels, 1967 (partim)
Orientopsaltria nigripes Duffels 1983b: 9 (n. comb., comp. note)
Dundubia nigripes Beuk 1996: 131–132, 135–137, 139, 157–160, Fig 5, Figs 45–50 (n. comb., key, lectotype designation, described, illustrated, distribution, comp. note) Equals *Dundubia mannifera* (nec Walker) var. *a* Distant, 1889 (partim) Equals *Dundubia mannifera* Distant 1906 (partim) Equals *Cosmopsaltria nigripes* Equals *Orientopsaltria nigripes* Equals *Dundubia rafflesii* Overmeer & Duffels, 1967 (partim) Malay Peninsula
Orientopsaltria nigripes Zaidi, Ruslan and Mahadir 1996: 61, Table 1 (listed, comp. note) Peninsular Malaysia
Dundubia nigripes Zaidi and Ruslan 1997: 224 (listed, comp. note) Equals *Cosmopsaltria nigripes* Equals *Orientopsaltria nigripes* Peninsular Malaysia
Dundubia nigripes Zaidi, Azman and Ruslan 2001a: 112, 115, 119, Table 1 (listed, comp. note) Equals *Cosmopsaltria nigripes* Equals *Orientopsaltria nigripes* Langkawi Island, Peninsular Malaysia
Dundubia nigripes Sanborn, Phillips and Sites 2007: 17, 34 (listed, distribution, comp. note) Thailand, Malaysia
Dundubia nigripes Boulard 2008f: 26, 61, 92 (biogeography, listed, comp. note) Thailand, Peninsular Malaysia

***D. nigripesoides* Boulard, 2008**
Dundubia nigripesoides Boulard 2008f: 26, 61–62, 92, Figs 5–8 (n. sp., described, illustrated, biogeography, listed, comp. note) Thailand

***D. oopaga* (Distant, 1881)** = *Cosmopsaltria oopaga* = *Cosmopsaltria oopaqa* (sic) = *Orientopsaltria oopaga*

= *Orientopsaltria oophaga* (sic) = *Cosmopsaltria andersoni* Distant, 1883 = *Terpnosia andersoni* Wu, 1935 (nec Distant, 1892) = *Orientopsaltria andersoni*

Terpnosia andersoni Wu 1935: 17 (listed) Equals *Cosmopsaltria andersoni* China, Yunnan, Nanking, India, Burma

Orientopsaltria andersoni Duffels 1983b: 9 (n. comb., comp. note)

Orientopsaltria oopaga Duffels 1983b: 9 (n. comb., comp. note)

Orientopsaltria andersoni Zaidi and Ruslan 1995c: 64–65, 68, Table 1 (listed, comp. note) Peninsular Malaysia

Orientopsaltria oophaga (sic) Zaidi and Ruslan 1995c: 64, 68, Table 1 (listed, comp. note) Peninsular Malaysia

Orientopsaltria oophaga (sic) Zaidi and Ruslan 1995d: 198, 201 (comp. note) Borneo, Sarawak

Dundubia oopaga Beuk 1996: 131, 135, 137, 139, 164–169, 172, Fig 4, Figs 58–64 (n. comb., n. syn., key, lectotype designation, described, illustrated, distribution, comp. note) Equals *Cosmopsaltria oopaga* Equals *Cosmopsaltria oopaqa* (sic) Equals *Orientopsaltria oopaga* Equals *Cosmopsaltria andersoni* Equals *Terpnosia andersoni* Wu, 1935 (partim) Equals *Orientopsaltria andersoni* n. syn. Burma, Cambodia, Laos, Thailand, Vietnam, Malay Peninsula, Sumatra

Orientopsaltria andersoni Zaidi, Ruslan and Mahadir 1996: 60, Table 1 (listed, comp. note) Peninsular Malaysia

Orientopsaltria oophaga (sic) Zaidi, Ruslan and Mahadir 1996: 61, Table 1 (listed, comp. note) Peninsular Malaysia

Cosmopsaltria andersoni Hua 2000: 61 (listed, distribution) China, Jiansu, Cambodia, Malaysia, Conchin-China, Thailand

Dundubia oopaga Zaidi, Azman and Ruslan 2001a: 112, 115, 119, Table 1 (listed, comp. note) Equals *Cosmopsaltria oopaga* Equals *Orientopsaltria oopaga* Langkawi Island, Burma, Vietnam, Cambodia, Laos, Burma, Thailand, Peninsular Malaysia, Sumatra

Dundubia oopaga Zaidi and Nordin 2001: 185, 189, 194, Table 1 (listed, comp. note) Equals *Cosmopsaltria oopaga* Equals *Orientopsaltria oopaga* Borneo, Sabah, Burma, Vietnam, Cambodia, Laos, Burma, Thailand, Peninsular Malaysia, Sumatra

Dundubia oopaga Boulard 2003b: 193–195, Figs 7–8 (described, song, sonogram, illustrated, comp. note) Equals *Cosmopasaltria oopaga* Equals *Orientopsaltria oopaga* Equals *Cosmopsaltria andersoni* Thailand, Burma, Indochina

Cosmopsaltria andersoni Boulard 2003b: 193 (comp. note) Thailand

Dundubia oopaga Zaidi and Azman 2003: 105, Table 1 (listed, comp. note) Borneo, Sabah

Dundubia oopaga Pham 2004: 61 (comp. note)

Dundubia oopaga Boulard 2005d: 347 (comp. note)

Dundubia oopaga Boulard 2005o: 148 (listed)

Dundubia oopaga Gogala and Trilar 2005: 66–67, 70, 75, Fig 2 (song, sonogram, oscillogram, acoustic behavior, comp. note) Malaysia, Thailand

Dundubia oopaga Pham and Thinh 2005a: 242 (synonymy, distribution, listed) Equals *Cosmopsaltria oopaga* Equals *Cosmopsaltria andersoni* Equals *Orientopsaltria oopaga* Equals *Terpnosia andersoni* Equals *Orientopsaltria andersoni* Vietnam

Dundubia oopaga Boulard 2006g: 113 (comp. note)

Dundubia oopaga Boulard 2007d: 496, 498, 506, Fig F (illustrated, comp. note) Thailand

Dundubia oopaga Boulard 2007e: 47, 83 (comp. note) Thailand

Dundubia oopaga Sanborn, Phillips and Sites 2007: 17 (listed, synonymy, distribution, comp. note) Equals *Cosmopsaltria oopaga* Equals *Cosmopsaltria andersoni* Thailand, Burma, India, Indochina

Dundubia oopaga Boulard 2008f: 26, 92 (biogeography, listed, comp. note) Equals *Cosmopsaltria oopaga* Equals *Cosmopsaltria andersoni* Thailand, Burma, Tenasserim, Siam

Dundubia oopaga Lee 2008a: 17 (synonymy, distribution, listed) Equals *Cosmopsaltria oopaga* Equals *Cosmopsaltria andersoni* Vietnam, Laos, Cambodia, Thailand, Malay Peninsula, Indonesia, Sumatra, Myanmar

Dundubia oopaga Pham and Yang 2009: 15, Table 2 (listed) Vietnam

Dundubia oopaga Lee 2010b: 26 (synonymy, distribution, listed) Equals *Cosmopsaltria oopaga* Equals *Cosmopsaltria andersoni* Cambodia, Vietnam, Laos, Thailand, Malay Peninsula, Indonesia, Sumatra, Myanmar

***D. rafflesii* Distant, 1883**

Dundubia rafflesi Zaidi and Ruslan 1995c: 62 (comp. note) Peninsular Malaysia

Dundubia rafflesii Shiyake 2007: 5, 56, 58, Fig 89 (illustrated, distribution, listed, comp. note) Indonesia

D. ramifera Walker, 1850 to *Pomponia ramifera*

***D. rhamphodes* Bloem & Duffels, 1976**

***D. rufivena rufivena* Walker, 1850** = *Dundubia intemerata* Walker, 1857 = *Dundubia intenerata* (sic) = *Dundubia*

intermerata (sic) = *Pomponia intemerata* = *Fidicina confinis* Walker, 1870 = *Dundubia mellea* Distant, 1889

Dundubia intemerata Bergier 1941: 115, 121, 219 (human food, comp. note) Laos

Dendubia (sic) *intemerata* Lepesme 1947: 94, 168 (listed, on palm) Malaysia

Dendubia (sic) *rufivena* Lepesme 1947: 94, 168 (listed, on palm) Malaysia

Dundubia rufivena Vayssière 1955: 254 (listed, coffee pest, comp. note) Malaysia

Dundubia rufivena Strümpel 1983: 126, Fig 173 (comp. note)

Dundubia intermerata (sic) Zaidi, Mohammedsaid and Yaakob 1990: 262 (comp. note) Malaysia

Dundubia rufivena Zaidi, Mohammedsaid and Yaakob 1990: 262–265, Fig 1, Table 1 (seasonality, comp. note) Malaysia

Dunbudia rufivena Moulds 1990: 53 (comp. note)

Dundubia intemerata Duffels and van Mastrigt 1991: 174 (predation, comp. note) Equals *Dundubia rufivena* Thailand

Dundubia rufivena Kahono 1993: 213, Table 5 (listed) Irian Jaya

Dundubia rufivena Zaidi 1993: 958, 960, Table 1 (listed, comp. note) Borneo, Sarawak

Dundubia rufivena Zaidi and Ruslan 1994: 428–429 (comp. note) Malaysia

Dundubia rufivena Yukawa and Yamane 1995: 693, 696–697 (listed, acoustic activity, comp. note) Indonesia, Krakatau

Dundubia rufivena Zaidi and Ruslan 1995a: 172, 175, Table 1 (listed, comp. note) Borneo, Sarawak

Dundubia rufivena Zaidi and Ruslan 1995b: 219, Table 1 (listed, comp. note) Borneo, Sabah

Dundubia rufivena Zaidi and Ruslan 1995c: 68, Table 1 (listed, comp. note) Peninsular Malaysia

Dundubia rufivena Zaidi and Ruslan 1995d: 200, 203, Table 1 (listed, comp. note) Borneo, Sabah

Dundubia rufivena Zaidi 1996: 101, 104, Table 1 (comp. note) Borneo, Sarawak

Dundubia rufivena Zaidi and Hamid 1996: 53, 56, Table 1 (listed, comp. note) Borneo, Sarawak

Dundubia rufivena Zaidi, Ruslan and Mahadir 1996: 60, Table 1 (listed, comp. note) Peninsular Malaysia

Dundubia rufivena Zaidi 1997: 113–114, 116, Table 1 (listed, comp. note) Borneo, Sarawak

Dundubia rufivena Zaidi and Ruslan 1997: 224–225 (listed, comp. note) Equals *Dundubia intemerata* Peninsular Malaysia, Java, New Guinea, Krakatau, Verlaten, Sebesi, Nias, Mentawei, Amboina, Sumatra, Borneo, Siam, Singapore, Indonesia, Bantam Island

Dundubia rufivena Zaidi and Ruslan 1998a: 355–356 (listed, comp. note) Borneo, Sarawak, Java, New Guinea, Krakatau, Verlaten, Sebesi, Nias, Mentawi, Ambiona, Sumatra, Siam, Peninsular Malaysia, Singapore, Celebes, Sabah, Indonesia, Bantam Island

Dundubia rufivena Zaidi and Ruslan 1998b: 3, Table 1 (listed, comp. note) Borneo, Sabah

Dundubia rufivena Sueur 1999: 57 (human food, comp. note) Laos, Thailand, Vietnam

Dundubia rufivena Zaidi and Ruslan 1999: 3, 5, Table 1 (listed, comp. note) Borneo, Sarawak

Dundubia rufivena Zaidi, Ruslan and Azman 1999: 307–308 (listed, comp. note) Borneo, Sabah, Java, New Guinea, Krakatau, Verlaten, Sebesi, Nias, Mentawei, Ambiona, Sumatra, Siam, Peninsular Malaysia, Celebes, Singapore, Sarawak, Indonesia, Bantam Island

Dundubia rufivena Zaidi, Noramly and Ruslan 2000a: 325 (listed, comp. note) Borneo, Sabah, Sarawak, Java, New Guinea, Krakatau, Verlaten, Sebesi, Nias, Mentawi, Ambiona, Sumatra, Siam, Peninsular Malaysia, Celebes, Siam, Singapore, Indonesia, Bantam Island

Dundubia rufivena Zaidi, Noramly and Ruslan 2000b: 337–338, Table 1 (listed, comp. note) Borneo, Sabah

Dundubia rufivena Zaidi, Ruslan and Azman 2000: 206–207 (comp. note) Borneo, Sabah, Java, New Guinea, Krakatau, Verlaten, Sebesi, Nias, Mentawei, Ambiona, Sumatra, Siam, Peninsular Malaysia, Celebes, Singapore, Sarawak, Indonesia, Bantam Island

Dundubia rufivena Zaidi, Azman and Ruslan 2001b: 126 (listed, comp. note) Borneo, Sarawak, Java, New Guinea, Krakatau, Verlaten, Sebesi, Nias, Mentawi, Ambiona, Sumatra, Siam, Peninsular Malaysia, Celebes, Siam, Singapore, Sabah, Indonesia, Bantam Island

Dundubia rufivena Zaidi and Nordin 2001: 183, 185–186, 189, 195, Table 1 (comp. note) Borneo, Sabah, Java, New Guinea, Krakatau, Verlaten, Sebesi, Nias, Mentawei, Ambiona, Sumatra, Siam, Peninsular Malaysia, Celebes, Singapore, Sarawak, Indonesia, Bantam Island

Dundubia rufivena Sueur 2002b: 111, 129 (listed, comp. note) Thailand

Dundubia rufivena Zaidi, Nordin, Maryati, Wahab, Norashikin et al. 2002: 2–3, 9–11, Tables 1–2 (listed, comp. note) Borneo, Sabah, Java, New Guinea, Krakatau, Verlaten, Sebesi, Nias, Mentawei, Ambiona, Sumatra, Siam, Singapore

Dundubia rufivena Zaidi and Azman 2003: 105, Table 1 (listed, comp. note) Borneo, Sabah

Dundubia rufivena Schouten, Duffels and Zaidi 2004: 373–374, Table 2 (listed, comp. note) Malayan Peninsula
Dundubia rufivena Boulard 2005e: 118 (comp. note)
Dundubia intemerata Boulard 2005o: 149 (listed)
Dundubia intemerata Sanborn, Phillips and Sites 2007: 16 (listed, distribution, comp. note) Thailand, India, Burma, Laos, Malaysia, Indonesia, Borneo
Dundubia rufivena Sanborn, Phillips and Sites 2007: 17, 34 (listed, distribution, comp. note) Thailand, Malaysia, Indonesia, Borneo, Sarawak, New Guinea, Sabah, Philippine Republic
Dundubia rufivena Shiyake 2007: 5, 56, 58, Fig 88 (illustrated, distribution, listed, comp. note) Malaysia, Indonesia
Dundubia rufivena Boulard 2008f: 26–27, 92 (biogeography, listed, comp. note) Equals *Dundubia intemerata* Equals *Dundubia mannifera* var. a (*Cephaloxys terpsichore*) Equals *Dundubia mellea* Equals *Pomponia intermerata* Thailand, Java, Malacca, India, Malaysia, Peninsular Malaysia, Borneo, Elopura, Assam, Siam
Dundubia rufivena Lee 2009h: 2634, 2637 (comp. note) Moluccas, Sumbawa

***D. rufivena* var. *a* Distant, 1889** = *Dundubia mellea* var. *a*

D. siamensis Haupt, 1918 to *Dundubia spiculata* Noualhier, 1896

***D. simalurensis* Overmeer & Duffels, 1967**

***D. sinbyudaw* Beuk, 1996**

Dundubia sinbyudaw Beuk 1996: 137, 139, 176, 179–182, Fig 5, Figs 86–91 (n. sp., key, described, illustrated, comp. note) Burma, Laos, Thailand
Dundubia sinbyudaw Pham and Thinh 2006: 525, 527 (listed, comp. note) Vietnam
Dundubia sinbyudaw Sanborn, Phillips and Sites 2007: 17–18 (listed, distribution, comp. note) Thailand, Burma, Laos, Vietnam
Dundubia sinbyudaw Boulard 2008f: 27, 92 (biogeography, listed, comp. note) Thailand
Dundubia sinbyudaw Lee 2008a: 17 (distribution, listed) Vietnam, Laos, Thailand, Myanmar
Dundubia sinbyudaw Pham and Yang 2009: 15, Table 2 (listed) Vietnam

***D. solokensis* Overmeer & Duffels, 1967**

Dundubia solokensis Chou, Lei, Li, Lu and Yao 1997: 251 (comp. note) China

***D. somraji* Boulard, 2003**

Dundubia somraji Boulard 2003b: 187, 195–197, Fig 9 (n. sp., described, song, sonogram, illustrated, comp. note) Thailand
Dundubia somraji Boulard 2005n: 143 (comp. note)
Dundubia somraji Boulard 2005o: 148 (listed)
Dundubia somraji Boulard 2006g: 109, 111, Fig 75 (sonogram, comp. note)
Dundubia somraji Boulard 2007e: 4, 47, 67, Plate 3, Fig 2 (coloration, illustrated, comp. note) Thailand
Dundubia somraji Sanborn, Phillips and Sites 2007: 18 (listed, distribution, comp. note) Thailand
Dundubia somraji Boulard 2008f: 27, 92 (biogeography, listed, comp. note) Thailand

***D.* sp. 1 Pham & Thinh, 2006**

Dundubia sp. 1 Pham and Thinh 2006: 525 (listed) Vietnam

***D. spiculata* Noualhier, 1896** = *Platylomia spiculata* = *Dundubia speculata* (sic) = *Dundubia siamensis* Haupt, 1918 = *Cosmopsaltria hastata* Moulton, 1923 (partim) = *Orientopsaltria hastata* (partim)

Dundubia spiculata Beuk 1996: 131–132, 137, 139, 172–180, 182, Fig 5, Figs 72–78 (n. comb., n. syn., key, lectotype designation, described, illustrated, distribution, comp. note) Equals *Platylomia spiculata* Equals *Dundubia siamensis* n. syn. Equals *Cosmopsaltria hastata* (partim) Equals *Orientopsaltria hastata* (partim) Burma, Cambodia, Laos, Thailand, Vietnam
Dundubia siamensis Beuk 1996: 173 (comp. note)
Orientopsaltria hastata Zaidi, Ruslan and Mahadir 1996: 60, Table 1 (listed, comp. note) Peninsular Malaysia
Dundubia spiculata Chou, Lei, Li, Lu and Yao 1997: 247, 250–251, Fig 9-79, Plate XIII, Fig 131 (key, described, illustrated, comp. note) China
Dundubia speculata (sic) Hua 2000: 61 (listed, distribution) China, Yunnan, Burma, Thailand, Laos, Vietnam, Cambodia, Malaysia
Dundubia spiculata Boulard 2001a: 128–130, 137–138, Plate I, Fig G, Fig 7 (illustrated, song analysis, sonogram, comp. note) Equals *Platylomia spiculata* Thailand
Dundubia spiculata Zaidi, Azman and Ruslan 2001a: 112, 115, 120, Table 1 (listed, comp. note) Langkawi Island, Thailand, Burma, Laos, Cambodia, Vietnam, Peninsular Malaysia
Dundubia spiculata Pham 2004: 61, 63–64 (key, distribution, comp. note) Vietnam
Dundubia spiculata Boulard 2005n: 137–141, 143, Plate C, Figs 1–2, Fig 6a (synonymy, described, illustrated, song, sonogram, comp. note) Equals *Platylomia spiculata* Thailand
Dundubia spiculata Boulard 2005o: 147, 150 (listed)
Dundubia spiculata Pham and Thinh 2005a: 242 (synonymy, distribution, listed) Equals *Platylomia spiculata* Equals *Dundubia siamensis* Equals *Cosmopsaltria hastata* Equals *Orientopsaltria hastata* Vietnam

Dundubia spiculata Boulard 2007d: 497 (comp. note) Thailand
Dundubia spiculata Boulard 2007e: 21, Fig 10 (illustrated, comp. note) Thailand
Dundubia spiculata Sanborn, Phillips and Sites 2007: 15, 18 (listed, synonymy, distribution, comp. note) Equals *Dundubia siamensis* Equals *Cosmopsaltria hastata* (partim) Thailand, Vietnam, Cambodia, Laos, Burma, Peninsular Malaysia
Dundubia siamensis Sanborn, Phillips and Sites 2007: 18 (comp. note) Thailand
Dundubia spiculata Boulard 2008c: 363 (comp. note) Laos, Thailand
Dundubia spiculata Boulard 2008f: 27–28, 93 (biogeography, listed, comp. note) Equals *Platylomia spiculata* Equals *Dundubia siamensis* Thailand, Cambodia
Dundubia spiculata Lee 2008a: 17 (distribution, listed) Vietnam, Laos, Cambodia, Thailand, Malay Peninsula, Myanmar
Dundubia spiculata Pham and Yang 2009: 15, Table 2 (listed) Vietnam
Dundubia spiculata Lee 2010b: 26–27 (synonymy, distribution, listed) Equals *Platylomia spiculata* Cambodia, Vietnam, Laos, Thailand, Malay Peninsula, Myanmar
Dundubia spiculata Lee and Hill 2010: 299–300, Figs 7–8 (phylogeny, comp. note)

D. terpsichore **(Walker, 1850)** = *Cephaloxys terpsichore* = *Mogannia terpsichore* = *Dundubia intemerata* Distant, 1889 and Moulton, 1923 (nec Walker) = *Dundubia intermerata* (sic) (partim) = *Dundubia mannifera* Breddin (nec Linnaeus, 1754 nom. nud.) = *Dundubia manifera* (sic) = *Dundubia mannifera terpsichore* Distant, 1917 = *Dundubia mannifera* var. *a* Distant, 1892 = *Dundubia vaginata terpischore*

Dundubia manifera (sic) Liu 1990: 39 (comp. note)
Dundubia manifera (sic) Liu 1992: 2, 41 (comp. note)
Dundubia terpsichore Lei, Chou and Li 1994: 53 (comp. note) China
Dundubia intermerata (sic) Zaidi and Ruslan 1994: 428 (comp. note) Malaysia
Dundubia intermerata (sic) Zaidi and Ruslan 1995a: 169 (comp. note) Borneo, Sarawak
Dundubia intermerata (sic) Zaidi and Ruslan 1995d: 197 (comp. note) Borneo, Sarawak
Dundubia terpsichore Zaidi and Ruslan 1995d: 197 (comp. note) Borneo, Sabah
Dundubia intermerata (sic) Zaidi 1996: 98 (comp. note) Borneo, Sarawak
Dundubia terpsichore Zaidi 1996: 98 (comp. note) Borneo, Sarawak
Dundubia intermerata (sic) Zaidi and Hamid 1996: 51 (comp. note) Borneo
Dundubia terpsichore Zaidi and Hamid 1996: 51 (comp. note) Borneo
Dundubia terpsichore Chou, Lei, Li, Lu and Yao 1997: 247, 249–250, 357, Plate XI, Fig 113 (key, described, synonymy, comp. note) Equals *Cephaloxys terpsichore* Equals *Mogannia terpsichore* Equals *Dundubia mannifera* var. *a* Equals *Dundubia mannifera* var. *terpsichore* Equals *Dundubia intermerata* China
Dundubia intermerata (sic) Zaidi 1997: 11 (comp. note) Borneo, Sarawak
Dundubia intermerata (sic) Zaidi and Ruslan 1998a: 357 (comp. note)
Dundubia terpisichore Zaidi and Ruslan 1998a: 356–357 (listed, comp. note) Equals *Cephaloxys terpsichore* Equals *Dundubia intermerata* (partim) Borneo, Sarawak, Myanmar. Indo-China, Thailand, Sumatra, Peninsular Malaysia, Tenesserim, Assam
Dundubia terpsichore group Beuk 1998: 148 (comp. note)
Dundubia terpsichore group Holloway 1998: 305, Fig 10 (distribution, phylogeny, comp. note) Sunda, Asia
Dundubia intermerata (sic) Zaidi and Ruslan 1999: 1 (comp. note) Borneo, Sarawak
Dundubia terpsichore Zaidi and Ruslan 1999: 1 (comp. note) Equals *Dundubia intermerata* (sic) Borneo, Sarawak
Dundubia intemerata Boulard 2001c: 56 (comp. note)
Dundubia terpsichore Boulard 2001c: 56–58, Plate C, Figs 1–2 (song, illustrated, sonogram, comp. note) Equals *Cephaloxys terpsichore* Equals *Dundubia intemerata* India Burma, Indochina, Tenasserim, Sumatra, Thailand
Dundubia terpsichore Boulard 2002b: 202–203 (comp. note)
Dundubua terpsichore Boulard 2003c: 183 (comp. note)
Dundubia terpsichore Pham 2004: 61, 63 (key, synonymy, distribution, comp. note) Equals *Cephaloxys terpsichore* Equals *Mogannia terpsichore* Equals *Dundubia mannifera* var. *a* Equals *Dundubia intermerata* Vietnam
Dundubia terpsichore Boulard 2005d: 346 (comp. note)
Dundubia terpsichore Boulard 2005o: 148 (listed)
Dundubia terpsichore Pham and Thinh 2005a: 242 (synonymy, distribution, listed) Equals *Cephaloxys terpsichore* Equals *Mogannia terpsichore* Equals *Dundubia intermerrata* (sic) Vietnam
Dundubia terpsichore Boulard 2006g: 102–103, 180, Fig 67 (sonogram, listed, comp. note)
Dundubia terpsichore Pham and Thinh 2006: 525, 527 (listed, comp. note) Vietnam

Dundubia terpsichore Trilar 2006: 344 (comp. note)
Dundubia terpsichore Boulard 2007d: 497 (comp. note) Thailand
Dundubia terpsichore Boulard 2007e: 67–68, Fig 46 (sonogram, comp. note) Thailand
Dundubia terpsichore Sanborn, Phillips and Sites 2007: 16, 18 (listed, synonymy, distribution, comp. note) Equals *Dundubia intemerata* (partim) Thailand, India, Indochina, Malaysia
Dundubia terpsichore Boulard 2008f: 27, 93 (biogeography, listed, comp. note) Equals *Cephaloxys terpsichore* Equals *Dundubia intemerata* (nec Walker) Equals *Mogannia intemerata* Thailand, India
Dundubia terpsichore Lee 2008a: 17–18 (synonymy, distribution, listed) Equals *Cephaloxys terpsichore* Equals *Dundubia mannifera* (nec Linnaeus) Equals *Dundubia mannifera* var. *terpsichore Equals Dundubia vaginata* (nec Fabricius) Equals *Dundubia vaginata* var. *terpsichore* Vietnam, China, Yunnan, Laos, North Thailand, Malaysia, Indonesia, Sumatra, Myanmar, India
Dundubia terpsichore Pham and Yang 2009: 15, Table 2 (listed) Vietnam

D. urania Walker, 1850 to *Pomponia urania*

***D. vaginata nigrimacula* Walker, 1950** = *Dundubia vaginata* var. *nigrimacula* = *Dundubia nigrimacula*
Dundubia vaginata var. *nigrimacula* Prešern, Gogala and Trilar 2004a: 240 (song, comp. note) Java
Dundubia vaginata var. *nigrimacula* Prešern, Gogala and Trilar 2004b: 62 (comp. note) Java

***D. vaginata vaginata* (Fabricius, 1787)** = *Tettigonia vaginata* = *Cicada vaginata* = *Dundubia vaginat* (sic) = *Cicada virescens* Olivier, 1790 = *Dundubia immacula* Walker, 1850 = *Dundubia mannifera immacula* = *Dundubia sobria* Walker, 1850 = *Cicada mannifera* Walker, 1850 nom. nud. = *Dundubia mannifera* = *Dundubia manifera* (sic) = *Mogannia mannifera* = *Fidicina mannifera* (nec Fabricius) = *Dundubia mannifera* var. *a* Distant, 1892 = *Dundubia mannifera terpsichore* Distant, 1892 (nec Walker) = *Dundubia mannifera* Banks, 1910 nom. nud. = *Maua* sp. Gogala and Riede 1995
Fidicina mannifera Weber 1931: 111, Fig 110b (auditory system illustrated, comp. note) Java
Dundubia mannifera Wu 1935: 11 (listed) Equals *Cicada mannifera* Equals *Tettigonia vaginata* Equals *Cicada virescens* Equals *Dundubia immacula* Equals *Dundubia nigrimacula* Equals *Dundubia sobria* Equals *Dundubia varians* Equals *Fidicina confinis* Equals *Cephaloxys terpsichore* Equals *Mogannia terpsichore* China, Malaysia, Australia, Sikkim, Assam, Burma, Tenasserim, Sumatra, Java, Borneo, Celebes, Philippines
Dundubia vaginata Zaidi, Mohammedsaid and Yaakob 1990: 262–265, Fig 1, Table 1 (seasonality, comp. note) Malaysia
Dundubia vaginata Helfert and Sänger 1993: 37–41, Fig 1, Table 1 (eclosion, comp. note) Thailand
Dundubia vaginata Kahono 1993: 213, Table 5 (listed) Irian Jaya
Dundubia vaginata Zaidi 1993: 958, 960, Table 1 (listed, comp. note) Borneo, Sarawak
Dundubia vaginata Lei, Chou and Li 1994: 53 (comp. note) China
Dundubia vaginata Zaidi and Ruslan 1994: 426, 428–429, Table 1 (comp. note) Malaysia
Dundubia vaginata Gogala and Riede 1995: 298–299, 302–304, Figs 3–4 (song, sonogram, oscillogram, acoustic behavior, comp. note) Sabah, Borneo
Maua sp. Gogala and Riede 1995: 299, Fig 1d (song, sonogram, oscillogram, acoustic behavior, comp. note) Malaysia
Dundubia vaginata Riede 1995: 45, Fig 1 (calling activity) Borneo
Dundubia vaginata Riede and Kroker 1995: 43–46, 48–50, Figs 1–2, Fig 5 (song, sonogram, oscillogram, calling activity, comp. note) Sabah, Malaysia, Greater Sunda Islands, Philippines, Southeast Asia
Dundubia vaginata Zaidi and Ruslan 1995a: 172, 175, Table 1 (listed, comp. note) Borneo, Sarawak
Dundubia vaginata Zaidi and Ruslan 1995b: 219, 221, Table 1 (listed, comp. note) Borneo, Sabah
Dundubia vaginata Zaidi and Ruslan 1995c: 68, Table 1 (listed, comp. note) Peninsular Malaysia
Dundubia vaginata Zaidi and Ruslan 1995d: 200, 203, Table 1 (listed, comp. note) Borneo, Sabah
Dundubia vaginata Beuk 1996: 130, 132, Fig 1, Table 1 (phylogeny, comp. note, type species of genus *Dundubia*)
Dundubia vaginata Riede 1996: 81, Fig 5 (sonogram, calling activity, comp. note) Sabah, Malaysia
Dundubia vaginata Zaidi 1996: 101, 104, Table 1 (comp. note) Borneo, Sarawak
Dundubia vaginata Zaidi and Hamid 1996: 53, 56, Table 1 (listed, comp. note) Borneo, Sarawak
Dundubia vaginata Zaidi, Ruslan and Mahadir 1996: 60, Table 1 (listed, comp. note) Peninsular Malaysia
Tettigonia vaginata Chou, Lei, Li, Lu and Yao 1997: 247 (type species of *Dundubia*)

Dundubia vaginata Chou, Lei, Li, Lu and Yao 1997: 247–248 (key, described, synonymy, comp. note) Equals *Tettigonia vaginata* China
Dundubia vaginata Riede 1997a: 276, 278, Fig 2 (calling activity, comp. note) Sabah, Malaysia
Dundubia vaginata Riede 1997b: 447, Fig 21.3 (sonogram, calling activity, comp. note) Sabah, Malaysia
Dundubia vaginata Zaidi 1997: 113–114, 116, Table 1 (listed, comp. note) Borneo, Sarawak
Dundubia vaginata Zaidi and Ruslan 1997: 225 (listed, comp. note) Equals *Tettigonia vaginata* Equals *Fidicina confinis* Peninsular Malaysia, Sumatra, Australia, China. India, Java, Borneo, Sarawak, Brunei
Dundubia vaginata Zaidi and Ruslan 1998a: 357 (listed, comp. note) Equals *Tettigonia vaginata* Borneo, Sarawak, Sumatra, Australia, China, India, Java, Peninsular Malaysia, Hong Kong, Tenasserim, Sabah
Dundubia vaginata Zaidi and Ruslan 1998b: 3, 5, Table 1 (listed, comp. note) Borneo, Sabah, Peninsular Malaysia
Dundubia vaginata Beuk 1999: 5–6, 8, Fig 1a, Table 1 (cladogram, comp. note)
Dundubia vaginata Zaidi and Ruslan 1999: 3, 5, Table 1 (listed, comp. note) Borneo, Sarawak
Dundubia vaginata Zaidi, Ruslan and Azman 1999: 308–309 (listed, comp. note) Equals *Tettigonia vaginata* Borneo, Sabah, Sumatra, Australia, China, India, Java, Sarawak, Peninsular Malaysia, Hong Kong, Tenasserim
Dundubia mannifera Hua 2000: 61 (listed, distribution) Equals *Cicada mannifera* Equals *Cicada virescens* Equals *Dundubia immacula* Equals *Fidicina confinis* Equals *Dundubia linearis* China, Taiwan, Hainan, Guangdong, Hong Kong, Hunan, Guizhou, Japan, Philippines, Indonesia, Malaysia, Kalimantan, Sikkim, Burma, India, Celebes, Australia
Dundubia vaginata Zaidi, Noramly and Ruslan 2000a: 325 (listed, comp. note) Equals *Tettigonia vaginata* Borneo, Sabah, Sumatra, Australia, China, India, Java, Peninsular Malaysia, Hong Kong, Tenasserim, Brunei, Sarawak
Dundubia vaginata Zaidi, Noramly and Ruslan 2000b: 337–338, 340, Table 1 (listed, comp. note) Borneo, Sabah
Dundubia vaginata Zaidi, Ruslan and Azman 2000: 207–208 (listed, comp. note) Equals *Tettigonia vaginata* Borneo, Sabah, Brunei, Sarawak, Sumatra, Australia, China, India, Java, Peninsular Malaysia, Hong Kong, Tenasserim, Brunei
Dundubia vaginata Sueur 2001: 39, 46, Tables 1–2 (listed)
Dundubia vaginata Zaidi, Azman and Ruslan 2001a: 112, 115, 120, Table 1 (listed, comp. note) Equals *Tettigonia vaginata* Langkawi Island, Sumatra, Australia, China, India, Java, Borneo, Peninsular Malaysia, Hong Kong, Tenasserim, Brunei
Dundubia vaginata Zaidi, Azman and Ruslan 2001b: 126 (listed, comp. note) Equals *Tettigonia vaginata* Borneo, Sarawak, Sumatra, Australia, China, India, Java, Peninsular Malaysia, Hong Kong, Tenasserim, Sabah, Brunei
Dundubia vaginata Zaidi and Nordin 2001: 183, 185–186, 189, 195–196, Table 1 (listed, comp. note) Equals *Tettigonia vaginata* Borneo, Sabah, Brunei, Sarawak, Sumatra, Australia, China, India, Java, Peninsular Malaysia, Hong Kong, Tenasserim, Brunei
Dundubia vaginata Boulard 2002b: 193 (coloration, comp. note)
Dundubia vaginata Zaidi, Nordin, Maryati, Wahab, Norashikin et al. 2002: 4, 9–11, Tables 1–2 (listed, comp. note) Equals *Tettigonia vaginata* Borneo, Sabah, Sumatra, Australia, China, India, Java, Sarawak, Peninsular Malaysia, Hong Kong, Tenasserim, Brunei
Dundubia vaginat (sic) Zaidi, Nordin, Maryati, Wahab, Norashikin et al. 2002: 12 (comp. note) Borneo, Sabah
Dundubia vaginata Boulard 2003b: 195 (comp. note)
Dundubia vaginata Zaidi and Azman 2003: 100, 105, Table 1 (listed, comp. note) Borneo, Sabah
Dundubia vaginata Gogala, Trilar, Kozina and Duffels 2004: 14 (comp. note) Equals *Maua?* Gogala and Riede 1995
Dundubia vaginata Pham 2004: 61, 63–64 (key, synonymy, distribution, comp. note) Equals *Tettigonia vaginata* Vietnam
Dundubia mannifera Pham 2004: 61 (comp. note)
Tettigonia vaginata Pham 2004: 63 (type species of *Pomponia*)
Dundubia vaginata Prešern, Gogala and Trilar 2004a: 240–247, Fig 2–5, Table 1 (song, oscillogram, sonogram, comp. note) Equals *Maua* sp. Gogala and Riede 1995 Peninsular Malaysia, Borneo, China, India, Philippines, Greater Sunda Islands
Maua sp. Prešern, Gogala and Trilar 2004a: 247 (synonymy)
Dundubia vaginata Prešern, Gogala and Trilar 2004b: 62 (song, comp. note) Malaysia, China, India, Malaysian Peninsula, Greater Sunda Islands, Philippines, Borneo

Dundubia vaginata Schouten, Duffels and Zaidi 2004: 373–374, Table 2 (listed, comp. note) Malayan Peninsula
Tettigonia vaginata Moulds 2005b: 391, 432 (type species of *Dundubia*)
Dundubia vaginata Moulds 2005b: 396, 398, 417–420, 422–423, Figs 56–60, Table 1 (higher taxonomy, phylogeny, key, listed, comp. note) Australia
Dundubia vaginata Pham and Thinh 2005a: 236, 242 (synonymy, distribution, listed) Equals *Tettigonia vaginata* Equals *Cicada mannifera* Equals *Dundubia mannifera* Equals *Cicada virescens* Equals *Dundubia immacula* Equals *Dundubia varians* Equals *Fidicina confinis* Equals *Cephaloxys mannifera* Equals *Mogannia mannifera* Vietnam
Dundubia manifera (sic) Pham and Thinh 2005a: 236 (comp. note)
Tettigonia vaginata Pham and Thinh 2005a: 241 (type species of *Dundubia*)
Dundubia vaginata Pham and Thinh 2006: 525, 527 (listed, comp. note) Vietnam
Dundubia vaginata Trilar 2006: 344 (comp. note)
Dundubia vaginata Sanborn, Phillips and Sites 2007: 18–19 (listed, distribution, comp. note) Thailand, India, Malaysia, Australia, Celebes, Philippine Republic, Japan, China
Dundubia vaginata Shiyake 2007: 5, 57–58, Fig 91 (illustrated, distribution, listed, comp. note) China, Southeast Asia, Indonesia
Dundubia vaginata Boulard 2008c: 363 (comp. note) Malaysia
Dundubia vaginata Boulard 2008f: 23, 90 (type species of *Dundubia*) Equals *Tettigonia vaginata* Thailand
Tettigonia vaginata Lee 2008a: 17 (type species of *Dundubia*) Sumatra
Tettigonia vaginata Lee 2009h: 2631 (type species of *Dundubia*, listed) Sumatra
Dundubia vaginata Lee 2009h: 2631–2632 (synonymy, distribution, listed) Equals *Tettigonia vaginata* Equals *Dundubia mannifera* (Walker, nom. nud.) Philippines, Palawan, Luzon, Sibuyan, Mindanao, southern China, Malay Archipelago, Southern Myanmar, India
Tettigonia vaginata Lee 2009i: 295 (type species of *Dundubia*, listed) Philippines, Panay
Dundubia vaginata Lee 2009i: 295 (synonymy, distribution, listed) Equals *Tettigonia vaginata* Philippines, Panay, southern China, Malay Archipelago, Southern Myanmar, India
Dundubia vaginata Sun, Watson, Zheng, Watson and Liang 2009: 3148–3149, 3151, Fig 1K, Table 1 (illustrated, wing microstructure, comp. note) China, Yunnan
Tettigonia vaginata Lee 2010a: 24 (type species of *Dundubia*, listed)
Dundubia vaginata Lee 2010a: 24 (synonymy, distribution, listed) Equals *Tettigonia vaginata* Equals *Dundubia mannifera* (Walker, nom. nud.) Philippines, Mindanao, Luzon, Sibuyan, Panay, Palawan, Malay Archipelago, South Myanmar, India
Tettigonia vaginata Lee 2010b: 26 (type species of *Dundubia*, listed)
Tettigonia vaginata Lee 2010c: 13 (type species of *Dundubia*, listed) Sumatra
Dundubia vaginata Lee 2010c: 15 (synonymy, distribution, listed) Equals *Tettigonia vaginata* Philippines, Luzon, Sibuyan, Panay, Palawan, Mindanao, South China, Malay Archipelago, South Myanmar, India
Dundubia vaginata Lee and Hill 2010: 299–300, 302, Figs 7–8 (phylogeny, comp. note)
D. vanna Chou, Lei, Li, Lu & Yao 1997 nom. nud.
Dundubia vanna Chou, Lei, Li, Lu and Yao 1997: 247 (key) China

***Acutivalva* Yao, 1985 nom. nud.**
***A. choui* Yao, 1985 nom. nud.**
Acutivalva choui Yao Jiang 1985: 257–259, Fig 1A, Fig 2 (illustrated, song analysis, oscillogram, comp. note) Yunnan
Acutivalva choui Lei, Chou and Li 1994: 53 (comp. note) China
A[cutivalva] choui Jiang, Yang, Tang, Xu and Chen 1995b: 228 (song, phonotaxis, flight behavior, comp. note) China
Acutivalva choui Sanborn, Breitbarth, Heath and Heath 2002: 445 (comp. note)

***Linguvalva* Chou & Yao, 1985 nom. nud.**
***L. sinensis* Chou & Yao, 1985 nom. nud.**
Linguvalva sinensis Chou & Yao, Jiang 1985: 257–258, 261–262, Fig 1C, Fig 4 (illustrated, song analysis, oscillogram, comp. note) Yunnan
Linguvalva sinensis Lei, Chou and Li 1994: 53 (comp. note) China
L[inguvalva] sinensis Jiang, Yang, Tang, Xu and Chen 1995b: 228 (song, phonotaxis, flight behavior, comp. note) China
Linguvalva sinensis Sanborn, Breitbarth, Heath and Heath 2002: 445 (comp. note)

***Spilomistica* Chou & Yao, 1984 nom. nud.**
***S. sinensis* Chou & Yao, 1984 nom. nud.**
Spilomistica sinensis Chou & Yao, Hua 2000: 64 (listed, distribution) China, Guangxi

***Mata* Distant, 1906**

Mata Chou, Lei, Li, Lu and Yao 1997: 187, 293, Table 6 (key, described, listed, diversity, biogeography) China

Mata Moulds 2005b: 391, 432 (higher taxonomy, listed, comp. note)

***M. kama* (Distant, 1881)** = *Pomponia kama*

Pomponia kama Chou, Lei, Li, Lu and Yao 1997: 196 (type species of *Mata*)

Mata kama Boulard 2008c: 366 (comp. note) India

***M. rama* Distant, 1912**

Mata rama Hua 2000: 62 (listed, distribution) China, Xizang, Bhutan

***M.* sp. 1 Pham & Thinh, 2006**

Mata sp. 1 Pham and Thinh 2006: 526 (listed) Vietnam

***Sinosemia* Matsumura, 1927** = *Sinosomia* (sic) = *Senosemia* (sic)

Sinosemia Wu 1935: 12 (listed) China

Sinosemia Chou, Lei, Li, Lu and Yao 1997: 187, 293, 297, Table 6 (key, described, listed, diversity, biogeography) China

Sinosemia Beuk 1999: 2 (comp. note)

Sinosemia Moulds 2005b: 391, 432 (higher taxonomy, listed, comp. note)

Sinosemia Lee 2009b: 330 (comp. note)

Sinosemia Lee 2009e: 87 (comp. note)

Sinosemia Pham and Yang 2009: 12, 14, Table 2 (listed, distribution) Vietnam

***S. shirakii* Matsumura, 1927** = *Senosemia* (sic) *shirakii* = *Cryptotympana shirakii* = *Senosemia* (sic) sp. 1 Pham & Thinh, 2006

Sinosemia shirakii Wu 1935: 12 (listed) China

Sinosemia shirakii Chou, Lei, Li, Lu and Yao 1997: 194–196, Fig 9–54, Plate VIII, Fig 85 (type species of *Sinosemia*, described, illustrated, comp. note) China

Sinosemia shirakii Hua 2000: 64 (listed, distribution) Equals *Cryptotympana shirakii* China, Shanxi, Taiwan, Hainan

Senosemia (sic) *shirakii* Pham and Thinh 2006: 526–527 (listed, comp. note) Vietnam

Senosemia (sic) sp. 1 Pham and Thinh 2006: 526 (listed) Vietnam

Sinosemia shirakii Lee 2009b: 331 (comp. note)

Sinosemia shirakii Pham and Yang 2009: 12, 14, 16, Table 2 (listed, distribution, comp. note) Vietnam, China, Japan

Sinosemia shirakii Pham and Yang 2010a: 133 (comp. note) Vietnam

***Champaka* Distant, 1905**

Champaka Duffels 1990b: 70 (comp. note) Sulawesi, Borneo

Champaka Duffels 1991b: 177–178 (key, described, comp. note) Sulawesi, Borneo

Champaka Zaidi and Ruslan 1995a: 174, Table 2 (listed, comp. note) Borneo, Sarawak

Champaka Zaidi and Ruslan 1995b: 220–221, Table 2 (listed, comp. note) Borneo, Sabah

Champaka Zaidi and Ruslan 1995d: 202, Table 2 (listed, comp. note) Borneo, Sabah

Champaka Boer and Duffels 1996b: 302 (comp. note) Sulawesi

Champaka Zaidi 1996: 102–103, Table 2 (comp. note) Borneo, Sarawak

Champaka Zaidi and Hamid 1996: 54–55, Table 2 (listed, comp. note) Borneo, Sarawak

Champaka Holloway 1997: 219 (diversity, comp. note) Sulawesi, Borneo

Champaka Zaidi 1997: 114–115, Table 2 (listed, comp. note) Borneo, Sarawak

Champaka Zaidi and Ruslan 1998b: 4–5, Table 2 (listed, comp. note) Borneo, Sabah

Champaka Beuk 1999: 1–2, 9 (history, synonymy to *Platylomia*)

Champaka Zaidi and Ruslan 1999: 4, Table 2 (listed, comp. note) Borneo, Sarawak

Champaka Lee 2010a: 24–25 (status, listed, comp. note) Philippines, Mindanao

Champaka Lee 2010c: 15 (listed, comp. note) Philippines, Luzon

Champaka Phillips and Sanborn 2010: 74 (comp. note) Asia

***C. abdulla* (Distant, 1881)** = *Cosmopsaltria abdulla* = *Platylomia abdulla* = *Platylomia distanti* Moulton, 1923 = *Platylomia spinosa distanti*

Platylomia abdulla Beuk 1999: 4, 6, 8, 12–14, 25–32, Fig 1a, Figs 13–20, Tables 1–2 (n. syn., key, lectotype designation, described, illustrated, distribution, cladogram, comp. note) Equals *Cosmopsaltria abdulla* Equals *Platylomia distanti* n. syn. Equals *Platylomia spinosa distanti* Singapore, West Malaysia, East Malaysia, Borneo, Sabah, Sarawak, Brunei, Lalimantan, Sumatra, Java, Philippines?

Platylomia distanti Beuk 1999: 4, 13, 25, 27 (n. syn., history, comp. note)

Platylomia abdulla Zaidi and Azman 2003: 105, Table 1 (listed, comp. note) Borneo, Sabah

Platylomia abdulla Schouten, Duffels and Zaidi 2004: 373, 376, Table 2 (listed, comp. note) Malayan Peninsula

Platylomia abdulla Lee 2009c: 338 (comp. note)

Champaka abdulla Lee 2010a: 25 (n. comb., comp. note)

***C. aerata* (Distant, 1888)** = *Dundubia aerata* = *Platylomia aerata*

Dundubia aerata Zaidi and Ruslan 1995d: 200, 203, Table 1 (listed, comp. note) Borneo, Sabah

Dundubia aerata Zaidi and Ruslan 1998a: 354–355 (listed, comp. note) Borneo, Sarawak, Sumatra, Mentawi, Sabah
Platylomia aerata Beuk 1999: 4, 6, 8–9, 74–80, Fig 1a, Figs 89–103, Table 1 (n. comb., cladogram, comp. note) Equals *Dundubia aerata* West Malaysia, East Malaysia, Sabah, Sarawak, Borneo, Kalimantan, Sumatra, Java
Dundubia aerata Zaidi, Noramly and Ruslan 2000b: 337–338, Table 1 (listed, comp. note) Borneo, Sabah
Dundubia aerata Zaidi, Ruslan and Azman 2000: 205 (listed, comp. note) Borneo, Sabah, Sarawak, Peninsular Malaysia, Sumatra, Mentawi
Platylomia aerata Zaidi and Azman 2003: 105, Table 1 (listed, comp. note) Borneo, Sabah
Dundubia aerata Shiyake 2007: 5, 56, 58, Fig 87 (illustrated, distribution, listed, comp. note) Malaysia, Indonesia
Platylomia aerata Lee 2009c: 338 (comp. note)
Champaka aerata Lee 2010a: 25 (n. comb., comp. note)

***C. celebensis* Distant, 1913** = *Platylomia celebensis* = *Champaka maculipennis* Haupt, 1917
Champaka celebensis Duffels 1990b: 64, Table 2 (listed) Sulawesi
Champaka celebensis Duffels 1991b: 177–179, 181–182, Figs 2–3, Fig 5, Fig 7 (key, lectotype designation, synonymy, described, illustrated, distribution, comp. note) Equals *Champaka maculipennis* Celebes, Sulawesi
Champaka maculipennis Duffels 1991b: 177–178 (lectotype designation, comp. note) Celebes
Champaka celebensis Boer and Duffels 1996b: 330 (distribution) Sulawesi
Platylomia celebensis Beuk 1999: 4, 6, 8, 11, 13, 25, 69–74, Figs 1a–1b, Figs 81–88, Table 1 (n. comb., key, described, illustrated, cladogram, comp. note) Equals *Champaka celebensis* Equals *Champaka maculipennis* Indonesia, Sulawesi
Champaka maculipennis Beuk 1999: 4, 69 (history, comp. note)
Platylomia celebensis Lee 2009c: 338 (comp. note)
Champaka celebensis Lee 2010a: 24–25 (n. comb., comp. note)

***C. constanti* (Lee, 2009)** = *Platylomia constanti*
Platylomia constanti Lee 2009c: 338–342, Fig 1A, Fig 1C, Fig 2 (n. sp., described, illustrated, comp. note) Philippines, Luzon
Champaka constanti Lee 2010a: 25 (n. comb., comp. note)
Champaka constanti Lee 2010a: 15 (distribution. listed) Philippines, Luzon

C. harveyi Distant, 1912 to *Champaka viridimaculata* (Distant, 1889)

C. maculipennis Haupt, 1917 to *Champaka celebensis*

C. majuscula Moulton, 1923 to *Champaka meyeri* (Distant, 1883)

***C. maxima* (Lee, 2009)** = *Platylomia maxima*
Platylomia maxima Lee 2009c: 341–342, Fig 1B, Fig 1D, Fig 3 (n. sp., described, illustrated, comp. note) Philippines, Luzon
Champaka maxima Lee 2010a: 15 (distribution. listed) Philippines, Luzon
Champaka maxima Lee 2010a: 25 (n. comb., comp. note)

***C. meyeri* (Distant, 1883)** = *Cosmopsaltria meyeri* = *Platylomia meyeri* = *Cosmopsaltria majuscula* Distant, 1889 = *Dundubia majuscula* = *Cosmopsaltria maiuscula* (sic) = *Platylomia majuscula*
Platylomia majuscula Duffels 1990b: 64, Table 2 (listed) Sulawesi
Platylomia meyeri Duffels 1990b: 64, Table 2 (listed) Sulawesi
Platylomia meyeri Boer and Duffels 1996b: 330 (distribution) Sulawesi
Platylomia majuscula Boer and Duffels 1996b: 330 (distribution) Sulawesi
Platylomia meyeri Beuk 1999: 4–6, 8, 10, 54-, Figs 1a–1b, Figs 57–69, Table 1 (n. syn., key, lectotype designation, described, illustrated, cladogram, comp. note) Equals *Cosmopsaltria meyeri* Equals *Dundubia majuscula* Equals *Cosmopsaltria majuscula* n. syn. Equals *Cosmopsaltria maiuscula* (sic) Equals *Platylomia majuscula* Indonesia, Sulawesi, Celebes,
Platylomia majuscula Beuk 1999: 4, 54 (n. syn., lectotype designation, history, comp. note)
Platylomia meyeri Lee 2009c: 338–339 (comp. note)
Champaka meyeri Lee 2010a: 25 (n. comb., comp. note)

***C. nigra* (Distant, 1888)** = *Dundubia spinosa* Walker, 1850 (nec Fabricius) = *Cosmopsaltria nigra* = *Platylomia nigra* = *Platylomia albomaculata* Distant, 1905
Platylomia nigra Beuk 1999: 4–6, 8–9, 11, 38-, Figs 1a–1b, Figs 30–42, Table 1 (n. syn., key, lectotype designation, described, illustrated, cladogram, comp. note) Equals *Dundubia spinosa* (nec Fabricius) Equals *Cosmopsalta nigra* Equals *Platylomia albomaculata* Philippines, Luzon, Samar, Sibujan, Mindanao
Platylomia albomaculata Beuk 1999: 4, 38–39 (n. syn., lectotype designation, history, comp. note)
Platylomia nigra Lee 2009c: 338–339 (comp. note)
Champaka nigra Lee 2010a: 23, 25, Figs 6C–6D (n. comb., synonymy, illustrated, distribution,

listed) Equals *Cosmopsaltria nigra* Equals *Platylomia nigra* Philippines, Mindanao, Luzon, Sibuyan, Samar

Champaka nigra Lee 2010a: 14 (synonymy, distribution. listed) Equals *Cosmopsaltria nigra* Equals *Platylomia nigra* Philippines, Luzon, Sibuyan, Samar, Mindanao

***C. spinosa* (Fabricius, 1787)** = *Tettigonia spinosa* = *Tettigonia spinnosa* (sic) = *Tettigonia bispinosa* (sic) = *Cicada spinosa* = *Tettigonia 2spinosa* = *Dundubia spinosa* = *Cosmopsaltria spinosa* = *Cosmopsaltria (Cosmopsaltria) spinosa* = *Platylomia spinosa* = *Platylomia umbrata* Moulton, 1911 (nec Distant) = *Platylomia maculata* Liu, 1940

Platylomia spinosa Distant 1906: 37 (type species of *Platylomia*, comp. note)

Platylomia spinosa Zaidi, Mohammedsaid and Yaakob 1990: 262–265, Fig 1, Table 1 (seasonality, comp. note) Malaysia

Platylomia spinosa Duffels 1991b: 178 (comp. note) Borneo

Platylomia spinosa Zaidi and Ruslan 1994: 426, 428–429, Table 1 (comp. note) Malaysia

Platylomia spinosa Yukawa and Yamane 1995: 693 (listed) Indonesia, Krakataus

Platylomia spinosa Zaidi and Ruslan 1995a: 172, 175, Table 1 (listed, comp. note) Borneo, Sarawak

Platylomia spinosa Zaidi and Ruslan 1995b: 218–219, 221, Table 1 (listed, comp. note) Borneo, Sabah

Platylomia spinosa Zaidi and Ruslan 1995c: 65, 69, Table 1 (listed, comp. note) Peninsular Malaysia

Platylomia spinosa Zaidi and Ruslan 1995d: 198 (comp. note) Borneo, Sabah

Platylomia spinosa Zaidi 1996: 100–101, Table 1 (comp. note) Borneo, Sarawak

Platylomia spinosa Zaidi and Hamid 1996: 53, 56, Table 1 (listed, comp. note) Borneo, Sarawak

Platylomia spinosa Zaidi, Ruslan and Mahadir 1996: 61, Table 1 (listed, comp. note) Peninsular Malaysia

Platylomia spinosa Zaidi 1997: 112–113, 116, Table 1 (listed, comp. note) Borneo, Sarawak

Platylomia spinosa Zaidi and Ruslan 1997: 227–228 (listed, comp. note) Equals *Tettigonia spinosa* Equals *Cosmopsaltria spinosa* Peninsular Malaysia, Sumatra, Borneo, Philippines, Sarawak, Singapore

Platylomia spinosa Zaidi and Ruslan 1998a: 361 (listed, comp. note) Equals *Tettigonia spinosa* Borneo, Sarawak, Sumatra, Philippines, Peninsular Malaysia, New Guinea, Singapore, Baram, Sabah

Platylomia spinosa Zaidi and Ruslan 1998b: 2–3, 5, Table 1 (listed, comp. note) Borneo, Sabah, Peninsular Malaysia

Platylomia spinosa Beuk 1999: 4–6, 8, 12–25, 27–28, 31, 33, 59, 70–71, 73, Fig 1a, Figs 2–12, Tables 1–2 (key, lectotype designation, described, illustrated, distribution, cladogram, comp. note) Equals *Tettigonia spinosa* Equals *Cicada spinosa* Equals *Dundubia spinosa* Equals *Cosmopsaltria spinosa* Equals *Cosmopsaltria (Cosmopsaltria) spinosa* Equals *Platylomia umbrata* Moulton, 1911 (nec Distant) Equals *Platylomia maculata* Sumatra, Thailand, West Malaysia, East Malaysia, Borneo, Sabah, Sarawak, Brunei, Kalimantan, Java, India?

Platylomia spinosa group Beuk 1999: 1–2, 4–5, 8–10, Fig 1 (described, comp. note)

Tettigonia spinosa Beuk 1999: 13 (lectotype designation, history, comp. note)

Platylomia maculata Beuk 1999: 4, 13–15 (n. syn., history, comp. note)

Platylomia spinosa Zaidi and Ruslan 1999: 3, 5, Table 1 (listed, comp. note) Borneo, Sarawak

Platylomia spinosa Zaidi, Ruslan and Azman 1999: 312–313 (listed, comp. note) Equals *Tettigonia spinosa* Borneo, Sabah, Sumatra, Philippines, Sarawak, Peninsular Malaysia, Singapore

Platylomia spinosa Duffels and Zaidi 2000: 197–199, 202, Figs 1–2, Figs 4–5, Tables 1–2 (phylogeny, comp. note) Malayan Peninsula, Sumatra, Borneo

Platylomia spinosa Freytag 2000: 55 (on stamp)

Platylomia spinosa Zaidi, Noramly and Ruslan 2000a: 328–329 (listed, comp. note) Equals *Tettigonia spinosa* Borneo, Sabah, Sumatra, Philippines, Peninsular Malaysia, New Guinea, Sarawak, Singapore

Platylomia spinosa Zaidi, Noramly and Ruslan 2000b: 337, 339, Table 1 (listed, comp. note) Borneo, Sabah

Platylomia spinosa Zaidi, Ruslan and Azman 2000: 211–212 (listed, comp. note) Equals *Tettigonia spinosa* Borneo, Sabah, Sumatra, Philippines, Peninsular Malaysia, New Guinea, Sarawak, Singapore

Platylomia spinosa Zaidi and Nordin 2001: 183, 185, 190, 200, Table 1 (listed, comp. note) Equals *Tettigonia spinosa* Borneo, Sabah, Sumatra, Philippines, Peninsular Malaysia, New Guinea, Sarawak, Singapore

Platylomia spinosa Zaidi, Nordin, Maryati, Wahab, Norashikin et al. 2002: 5–6, 9–10,

Tables 1–2 (listed, comp. note) Equals *Tettigonia spinosa* Borneo, Sabah, Sumatra, Philipines, Peninsular Malaysia, Sarawak, Singapore

Tettigonia spinosa Lee and Hayashi 2003a: 175 (type species of *Platylomia*)

Platylomia spinosa Zaidi and Azman 2003: 105, Table 1 (listed, comp. note) Borneo, Sabah

Tettigonia spinnosa (sic) Pham 2004: 64 (type species of *Pomponia*)

Platylomia spinosa Schouten, Duffels and Zaidi 2004: 373, 376, Table 2 (listed, comp. note) Malayan Peninsula

Platylomia spinosa Boulard 2005i: 62 (comp. note)

Platylomia spinosa Pham and Thinh 2005a: 243 (synonymy, distribution, listed) Equals *Tettigonia spinosa* Equals *Cicada spinosa* Equals *Cosmopsaltria spinosa* Equals *Dundubia spinosa* Equals *Platylomia umbrata* Vietnam

Platylomia spinosa Sanborn, Phillips and Sites 2007: 34 (comp. note) Malaysia, Indonesia, Brunei

Platylomia spinosa Boulard 2008c: 364 (comp. note) Malaysia

Platylomia spinosa Boulard 2008f: 90 (comp. note) Thailand

Platylomia spinosa species group Lee 2009c: 338, 341 (characteristics, diversity, comp. note) Philippines, Luzon

Platylomia spinosa Lee 2009c: 338 (comp. note)

Platylomia spinosa species group Lee 2010a: 24–25 (status, comp. note)

Champaka spinosa Lee 2010a: 25 (n. comb., comp. note)

***C. virescens* (Distant, 1905)** = *Platylomia virescens*

Platylomia virescens Beuk 1998: 149 (comp. note)

Platylomia virescens Beuk 1999: 4–6, 8, 11, 46–53, Figs 1a–1b, Figs 43–56, Table 1 (key, described, lectotype designation, cladogram, comp. note) Philippines, Camigun Island, Luzon, Negros, Mindanao

Platylomia virescens Lee 2009c: 338 (comp. note)

Champaka virescens Lee 2010a: 25 (n. comb., distribution, listed, comp. note) Philippines, Mindanao, Luzon, Camigun Island, Negros, Samar, Basalin, Borneo

Champaka virescens Lee 2010a: 14–15 (synonymy, distribution. listed) Equals *Platylomia virescens* Philippines, Luzon, Camigun Island, Negros, Samar, Mindanao, Basalin, Borneo

***C. viridimaculata* (Distant, 1889)** = *Pomoponia viridimaculata* = *Platylomia viridimaculata* = *Champaka harveyi* Distant, 1912 = *Champaka viridimaculata harveyi*

Pomponia viridimaculata Duffels 1991b: 177, 179 (type species of *Champaka*, comp. note) Borneo

Champaka viridimaculata Duffels 1991b: 177–182, Fig 1, Fig 4, Figs 6–7 (type species of *Champaka*, key, lectotype designation, synonymy, described, illustrated, distribution, comp. note) Equals *Pomponia viridimaculata* Equals *Champaka harveyi* Equals *Champaka viridimaculata harveyi* Borneo

Champaka harveyi Duffels 1991b: 177, 179, 181 (comp. note) Borneo

Champaka viridimaculata Zaidi 1993: 958, 960, Table 1 (listed, comp. note) Borneo, Sarawak

Champaka viridimaculata Zaidi and Ruslan 1995a: 170 (comp. note) Equals *Champaka harveyi* Borneo, Sarawak

Champaka harveyi Zaidi and Ruslan 1995a: 170 (comp. note) Borneo, Kalimantan

Champaka harveryi Zaidi and Ruslan 1995d: 198 (comp. note) Borneo, Kalimantan

Champaka viridimaculata Zaidi and Ruslan 1995d: 198 (comp. note) Equals *Champaka harveyi* Borneo

Champaka harveyi Zaidi 1996: 99 (comp. note) Borneo, Kalimantan

Champaka viridimaculata Zaidi 1996: 99, 101, 104, Table 1 (comp. note) Equals *Champaka harveyi* Borneo, Sarawak

Champaka harveyi Zaidi and Hamid 1996: 51 (comp. note) Borneo, Kalimantan

Champaka viridimaculata Zaidi and Hamid 1996: 51, 53, 56, Table 1 (listed, comp. note) Equals *Champaka harveryi* Borneo, Sarawak

Champaka viridimaculata Zaidi, Ruslan and Mahadir 1996: 60, Table 1 (listed, comp. note) Peninsular Malaysia

Champaka harveyi Zaidi 1997: 110 (comp. note) Borneo, Kalimantan

Champaka viridimaculata Zaidi 1997: 110, 113–114, Table 1 (listed, comp. note) Equals *Champaka harveyi* Borneo, Sarawak

Champaka viridimaculata Zaidi and Ruslan 1998a: 353–354 (listed, comp. note) Equals *Pomponia viridimaculata* Borneo, Sarawak, Peninsular Malaysia

Platylomia viridimaculata Beuk 1999: 4, 6, 8, 10, 13, 25, 32–38, 69, Fig 1a, Figs 21–29, Table 1 (n. comb., key, described, illustrated, cladogram, comp. note) Equals *Pomponia viridimaculata* Equals *Champaka harveyi* Equals *Champaka viridimaculata harveyi* West Malaysia, East Malaysia, Sabah, Sarawak, Brunei, Kalimantan, Philippines?

Champaka harveyi Beuk 1999: 4, 32 (history, comp. note)
Champaka harveyi Zaidi and Ruslan 1999: 1 (comp. note) Borneo, Sarawak
Champaka viridimaculata Zaidi and Ruslan 1999: 1 (comp. note) Equals *Champaka harveyi* Borneo, Sarawak
Champaka viridimaculata Zaidi, Ruslan and Azman 1999: 306 (listed, comp. note) Equals *Pomponia viridimaculata* Borneo, Sabah, Peninsular Malaysia, Sarawak
Champaka viridimaculata Zaidi, Noramly and Ruslan 2000a: 324 (listed, comp. note) Equals *Pomponia viridimaculata* Borneo, Sabah, Peninsular Malaysia, Sarawak, Brunei, Kalimantan
Champaka viridimaculata Zaidi, Noramly and Ruslan 2000b: 337–338, Table 1 (listed, comp. note) Borneo, Sabah
Champaka viridimaculata Zaidi, Ruslan and Azman 2000: 204 (listed, comp. note) Equals *Pomponia viridimaculata* Borneo, Sabah, Peninsular Malaysia, Sarawak, Brunei, Kalimantan
Platylomia viridimaculata Zaidi and Nordin 2001: 185–186, 190, 200–201, Table 1 (listed, comp. note) Equals *Champaka viridimaculata* Borneo, Sabah, Peninsular Malaysia, Sarawak, Brunei, Kalimantan
Platylomia viridimaculata Lee 2009c: 338 (comp. note)
Platylomia viridimaculata Zaidi and Azman 2003: 105, Table 1 (listed, comp. note) Borneo, Sabah
Pomponia viridimaculata Lee 2010a: 24 (type species of *Champaka*, listed, comp. note)
Champaka viridimaculata Lee 2010a: 24–25 (n. comb., comp. note)

C. viridimaculata harveyi Distant, 1912 to *Champaka viridimaculata* (Distant, 1889)

***C. wallacei* (Beuk, 1999)** = *Platylomia wallacei*

Platylomia wallacei Beuk 1999: 4–6, 8, 11, 61–69, Figs 1a–1b, Figs 69–80, Table 1 (n. sp., key, described, illustrated, cladogram, comp. note) Indonesia, Sulawesi
Platylomia wallacei Lee 2009c: 338 (comp. note)
Champaka wallacei Lee 2010a: 25 (n. comb., comp. note)

***Platylomia* Stål, 1870** = *Platlomia* (sic) = *Platytomia* (sic) = *Platylonia* (sic) = *Paltylomia* (sic) = *Cosmopsaltria (Platylomia)*

Platylomia Distant 1906: 36 (comp. note)
Platylomia Wu 1935: 13 (listed) China
Platylomia sp. Meer Mohr 1965: 104 (human food, listed) Sumatra
Platylomia Duffels 1983b: 3, 9 (comp. note)
Platylomia Hayashi 1984: 58 (listed)
Platylomia Duffels 1986: 319 (diversity, distribution) Sulawesi, Greater Sunda Islands, Southeast Asia
Platylomia Zaidi, Mohammedsaid and Yaakob 1990: 261, 267, Table 2 (listed, comp. note) Malaysia
Platylomia Duffels 1991b: 178 (comp. note)
Platylomia Duffels and van Mastrigt 1991: 174 (predation, comp. note) Sumatra
Platylonia (sic) sp. Kahono 1993: 213, Table 5 (listed) Irian Jaya
Platylomia Gogala 1994a: 33–34, Fig 6 (illustrated, comp. note) Thailand
Platylomia Lei, Chou and Li 1994: 54 (comp. note)
Platylomia Zaidi and Ruslan 1994: 427, Table 2 (comp. note) Malaysia
Platylomia spp. Gogala 1995: 111 (comp. note) Thailand
Platylomia Lei and Li 1995: 93 (comp. note) China
Platylomia Zaidi and Ruslan 1995a: 173–174, Table 2 (listed, comp. note) Borneo, Sarawak
Platylomia Zaidi and Ruslan 1995b: 220–221, Table 2 (listed, comp. note) Borneo, Sabah
Platylomia Zaidi and Ruslan 1995c: 70, Table 2 (listed, comp. note) Peninsular Malaysia
Platylomia Zaidi and Ruslan 1995d: 202, Table 2 (listed, comp. note) Borneo, Sabah
Platylomia Beuk 1996: 129–131, 133 (characteristics, history, comp. note)
Platylomia Boer and Duffels 1996b: 302 (comp. note) Sulawesi
Platylomia Zaidi 1996: 102–103, Table 2 (comp. note) Borneo, Sarawak
Platylomia Zaidi and Hamid 1996: 54–55, Table 2 (listed, comp. note) Borneo, Sarawak
Platylomia Chou, Lei, Li, Lu and Yao 1997: 236, 251–252, 293, 308, 319, 325, 327, Table 6, Tables 10–12 (key, described, listed, diversity, biogeography) China
Platylomia Novotny and Wilson 1997: 437 (listed)
Platylomia Zaidi 1997: 114–115, Table 2 (listed, comp. note) Borneo, Sarawak
Platylomia Beuk 1998: 147, 150 (history, characteristics) India, Bhutan, Nepal, Indo-China, Burma, Thailand, Laos, Cambodia, Vietnam, Peninsular Thailand, China, Taiwan
Platylomia Duffels and Zaidi 1998: 320 (comp. note)
Platylomia Zaidi and Ruslan 1998b: 4–5, Table 2 (listed, comp. note) Borneo, Sabah
Platylomia Beuk 1999: 1–2, 5, 9 (history, characteristics) Southeast Asia, Greater Sunda Islands, Sulawesi, Philippines, Malaysian Peninsula, Sumatra, Borneo, Java, Indochina?

Platylomia Zaidi and Ruslan 1999: 4, Table 2 (listed, comp. note) Borneo, Sarawak
Platylomia Duffels and Zaidi 2000: 195, 198–199 (comp. note)
Platylomia sp. Noda, Kubota, Miyata and Miyahara 2000: 1749 (medicinal use) China
Platylomia Boulard 2002b: 198 (comp. note)
Platylomia Duffels and Turner 2002: 235 (comp. note)
Platylomia Lee and Hayashi 2003a: 170, 172–175 (key, diagnosis) Taiwan
Platylomia Chen 2004: 39, 41, 77 (phylogeny, listed)
Platylomia Pham 2004: 61 (key, described, distribution, comp. note) Vietnam
Platylomia Boulard 2005h: 33, 39 (comp. note)
Platylomia Moulds 2005b: 391, 432 (higher taxonomy, listed, comp. note)
Platylomia Pham and Thinh 2005b: 290 (comp. note) Vietnam
Platylomia Pham and Thinh 2005a: 242 (listed) Vietnam
Platylomia Boulard 2007e: 20 (comp. note) Thailand
Platylomia Sanborn, Phillips and Sites 2007: 3, 22, Table 1 (listed) Thailand
Platylomia Shiyake 2007: 5, 58 (listed, comp. note)
Platylomia Boulard 2008b: 108 (comp. note)
Platylomia Boulard 2008c: 368 (comp. note)
Platylomia Boulard 2008f: 7, 28, 93 (listed) Thailand
Platylomia Lee 2008a: 2, 15, 17, Table 1 (listed, comp. note) Vietnam
Platylomia Lee 2009c: 338 (comp. note)
Platylomia Lee and Sanborn 2009: 33 (comp. note)
Platylomia Pham and Yang 2009: 15, Table 2 (listed) Vietnam
Platylomia Ahmed and Sanborn 2010: 39 (distribution) Philippines, Taiwan, Indonesia, China, Indochina, Thailand, Burma, Cambodia, Malaysia, India, Tibet, Nepal, Bhutan, Sri Lanka, Pakistan
Platylomia Lee 2010a: 24–25 (comp. note)
Platylomia Lee 2010b: 26 (listed) Cambodia
Platylomia Lee and Hill 2010: 302 (comp. note)
Platylomia Phillips and Sanborn 2010: 74 (comp. note) Asia

P. abdulla (Distant, 1881) to *Champaka abdulla*

P. albomaculata Distant, 1905 to *Champaka nigra* (Distant, 1888)

P. aerata (Distant, 1888) to *Champaka aerata*

***P. amicta* (Distant, 1889)** = *Dundubia amicta* = *Cosmopsaltria amicta*
Platylomia amicta Beuk 1999: 7 (comp. note)

P. andamansidensis Boulard, 2001 to *Dundubia andamansidensis*

P. assamensis Distant, 1905 to *Macrosemia assamensis*

P. bagueyensis Distant, 1912 to *Orientopsaltria alticola* (Distant, 1905)

***P. bivocalis* (Matsumura, 1907)** = *Cosmopsaltria bivocalis* = *Comopsaltria* (sic) *bivocalis*
Platylomia bivocalis Chou, Lei, Li, Lu and Yao 1997: 259 (described, synonymy, comp. note) Equals *Cosmopsaltria bivocalis* China
Platylomia bivocalis Beuk 1998: 149, 152–153, 168–174, Fig 1, Fig 3, Figs 37–44, Table 1 (phylogeny, key, described, illustrated, distribution, comp. note) Equals *Cosmopsaltria bivocalis* Taiwan
Platylomia bivocalis Hua 2000: 63 (listed, distribution) Equals *Cosmopsaltria bivocalis* China, Taiwan
Platylomia bivocalis Sueur 2002b: 124, 129 (listed, comp. note)
Platylomia bivocalis Lee and Hayashi 2003a: 175–176, Fig 19 (key, diagnosis, illustrated, synonymy, song, listed, comp. note) Equals *Cosmopsaltria bivocalis* Taiwan
Platylomia bivocalis Chen 2004: 77–78, 198 (described, illustrated, listed, comp. note) Taiwan

***P. bocki* (Distant, 1882)** = *Platylomia bbccki* (sic) = *Platylomia bccki* (sic) = *Platylomia bock* (sic) = *Dundubia bocki* = *Cosmopsaltria bocki*
Dundubia bocki Distant 1882: 159 (n. sp., described, comp. note) Sumatra
Platylomia bocki Beuk 1996: 131–133, Fig 1, Table 1 (phylogeny, comp. note)
Platylomia bccki (sic) Chou, Lei, Li, Lu and Yao 1997: 252 (key) China
Platylomia bocki Chou, Lei, Li, Lu and Yao 1997: 256, 357, Plate XII, Fig 130 (described, illustrated, synonymy, comp. note) Equals *Dundubia bocki* Equals *Cosmopsaltria bocki* China
Platylomia bocki Beuk 1998: 149, 152–153, 159–161, 163, 165, Fig 1, Fig 3, Figs 16–22, Table 1 (phylogeny, key, lectotype designation, described, illustrated, distribution, comp. note) Equals *Dundubia bocki* Equals *Cosmopsaltria bocki* Laos, Thailand, Vietnam, Southern China
Platylomia bocki Pham 2004: 61, 64 (key, synonymy, distribution, comp. note) Equals *Dundubia bocki* Equals *Cosmopsaltria bocki* Equals *Platylomia bock* (sic) Vietnam
Platylomia bocki Boulard 2005h: 7, 33–35, Figs 44–46 (described, illustrated, song, sonogram, listed, comp. note) Equals *Dundubia bocki* Equals *Cosmopsaltria bocki* Thailand, China
Platylomia bocki Boulard 2005o: 150 (listed)
Platylomia bocki Pham and Thinh 2005a: 242–243 (synonymy, distribution, listed) Equals

Dundubia bocki Equals *Cosmopsaltria bocki* Vietnam
Platylomia bocki Pham and Thinh 2006: 526–527 (listed, comp. note) Vietnam
Platylomia bocki Trilar 2006: 344 (comp. note)
Platylomia bocki Sanborn, Phillips and Sites 2007: 23 (listed, distribution, comp. note) Thailand, China, Laos, Vietnam
Platylomia bocki Boulard 2008b: 109 (comp. note)
Platylomia bocki Boulard 2008e: 352 (comp. note)
Platylomia bocki Boulard 2008f: 28–29, 93 (biogeography, listed, comp. note) Equals *Dundubia bocki* Equals *Cosmopsaltria bocki* Thailand, Sumatra, Siam
Platylomia bocki Lee 2008a: 17 (synonymy, distribution, listed) Equals *Dundubia bocki* Vietnam, South China, Laos, Thailand
Platylomia bocki Pham and Yang 2009: 15, Table 2 (listed) Vietnam

***P. brevis* Distant, 1912**
Platylomia brevis Lei and Li 1995: 92, 94 (comp. note)
Platylomia brevis Sueur 2002b: 119, 121, 129 (listed, comp. note)

P. celebensis (Distant, 1913) to *Champaka celebensis*
P. constanti (Lee, 2009) to *Champaka constanti*
P. distanti Moulton, 1923 to *Champaka abdulla* (Distant, 1881)

***P. ficulnea* (Distant, 1892)** = *Cosmopsaltria ficulnea*
Platylomia ficulnea Beuk 1998: 149, 152–153, 162–164, 173, Fig 1, Fig 3, Figs 23–29, Table 1 (phylogeny, key, lectotype designation, described, illustrated, distribution, comp. note) Equals *Cosmopsaltria ficulnea* Burma, India
Platylomia ficulnea Boulard 2005h: 36 (comp. note)

***P. flavida* (Guérin-Méneville, 1834)** = *Cicada flavida* = *Cosmopsaltria flavida* = *Cosmopsaltria (Platylomia) flavida* = *Dundubia flavida* = *Dundubia flava* (sic)
Platylomia flavida Distant 1906: 36–37 (type species of *Platylomia*, comp. note)
Platylomia flavida Zaidi and Ruslan 1994: 426, 428, Table 1 (comp. note) Malaysia
Platylomia flavida Zaidi and Ruslan 1995c: 69, Table 1 (listed, comp. note) Peninsular Malaysia
Platylomia flavida Zaidi, Ruslan and Mahadir 1996: 61, Table 1 (listed, comp. note) Peninsular Malaysia
Cicada flavida Chou, Lei, Li, Lu and Yao 1997: 251 (type species of *Platylomia*)
Platylomia flavida Zaidi and Ruslan 1997: 227 (listed, comp. note) Equals *Cicada flavida* Peninsular Malaysia, Java, Sumatra,
Cicada flavida Beuk 1998: 147 (type of genus *Platylomia*)
Platylomia flavida Beuk 1998: 148–149, Fig 1, Table 1 (phylogeny, comp. note)
Cicada flavida Beuk 1999: 1, 5, 8–9, Fig 1a, Table 1 (type of genus *Platylomia*, cladogram)
Platylomia flavida Schouten, Duffels and Zaidi 2004: 373, 376, Table 2 (listed, comp. note) Malayan Peninsula
Cicada flavida Pham and Thinh 2005a: 242 (type species of *Platylomia*)
Platylomia flavida Boulard 2006g: 126–127, Fig 86 (illustrated, comp. note)
Platylomia flavida Duffels and Hayashi 2006: 189 (comp. note)
Platylomia flavida Boulard 2007e: 69–70, Fig 47 (illustrated, comp. note) Thailand
Platylomia flavida Shiyake 2007: 5, 57–58, Fig 92 (illustrated, distribution, listed, comp. note) Malaysia, Indonesia
Platylomia flavida Boulard 2008c: 363 (comp. note) Malaysia
Dundubia flavida Boulard 2008f: 28 (type species of *Platylomia*)
Cicada flavida Lee 2008a: 17 (type species of *Platylomia*) Java, Sumatra
Cicada flavida Lee 2009c: 338 (type species of *Platylomia*, comp. note)
Dundubia flava (sic) Ahmed and Sanborn 2010: 38 (type species of genus)
Platylomia flavida Lee 2010a: 25 (type species of *Platylomia*, comp. note)
Platylomia flavida Lee 2010b: 26 (type species of *Platylomia*, comp. note)
Platylomia flavida Lee and Hill 2010: 299–300, Figs 7–8 (phylogeny, comp. note)

P. fuliginosa (Walker, 1850) to *Orientopsaltria fuliginosa*
P. hainanensis (Distant, 1901) to *Dundubia hainanensis*

***P. insignis* Distant, 1912**

***P. juno* Distant, 1905** = *Cosmopsaltria juno*
Platylomia juno Wu 1935: 13 (listed) China, Szechuan
Platylomia juno Peng and Lei 1992: 99–100, Fig 283 (listed, illustrated, comp. note) China, Hunan
Platylomia juno Chou, Lei, Li, Lu and Yao 1997: 252–254, Fig 9–80 (key, described, illustrated, synonymy, comp. note) Equals *Cosmopsaltria juno* China
Platylomia juno Hua 2000: 63 (listed, distribution) China, Zhejiang, Fujian, Taiwan, Guangdong, Hunan, Guangxi, Sichuan, Xizang, India, Malaysia
Platylomia juno Boulard 2008c: 363, 367, Fig 17 (comp. note) Laos

P. kareisana Matsumura, 1907 to *Macrosemia kareisana*
P. kareisana hopponis Kato, 1925 to *Macrosemia kareisana* (Matsumura, 1907)

P. kareisana karapinensis Kato, 1925 to *Macrosemia kareisana* (Matsumura, 1907)

P. larus **(Walker, 1858)** = *Dundubia larus* = *Cosmopsaltria larus*

Platylomia larus Wu 1935: 13 (listed) China, Soochow, India, Ceylon

Platylomia larus Gogala 1995: 111 (comp. note)

Platylomia larus Beuk 1996: 133 (comp. note)

Platylomia larus Beuk 1998: 148 (comp. note)

Platylomia larus Hua 2000: 63 (listed, distribution) Equals *Dundubia larus* China, Jiangsu, India, Sri Lanka

P. larus **var. ______ (Distant, 1889)** = *Cosmopsaltria larus* var. ______

P. lemoultii **Lallemand, 1924** = *Playlonia* (sic) *lemoultii*

Platylomia lemoultii Wu 1935: 13 (listed) China, Tibet

Platylomia lemoultii Chou, Lei, Li, Lu and Yao 1997: 258 (described, comp. note) China

Platylomia lemoultii Hua 2000: 63 (listed, distribution) China, Xizang

P. maculata Liu, 1940 to *Champaka spinosa* (Fabricius, 1787)

P. majuscula (Distant, 1889) to *Champaka meyeri* (Distant, 1883)

P. malickyi **Beuk, 1998**

Platylomia malickyi Beuk 1998: 149, 152–153, 164–168, Fig 1, Fig 3, Figs 30–36, Table 1 (phylogeny, key, n. sp., described, illustrated, distribution, comp. note) Burma, Laos, Thailand, Vietnam, Southern China, Yunnan

Platylomia malickyi Boulard 2003c: 172, 185–187, Plate H, Figs 1–2 (described, illustrated, song, sonogram, comp. note) Thailand

Platylomia malickyi Boulard 2005o: 149 (listed)

Platylomia malickyi Pham and Thinh 2005a: 243 (distribution, listed) Vietnam

Platylomia malickyi Boulard 2007e: 47 (comp. note) Thailand

Platylomia malickyi Sanborn, Phillips and Sites 2007: 23 (listed, distribution, comp. note) Thailand, Burma, Laos, Vietnam, China

Platylomia malickyi Boulard 2008c: 363 (comp. note) Thailand

Platylomia malickyi Boulard 2008f: 29, 93 (biogeography, listed, comp. note) Thailand

Platylomia malickyi Lee 2008a: 17 (distribution, listed) Vietnam, South China, Yunnan, Laos, Thailand, Myanmar

Platylomia malickyi Pham and Yang 2009: 15, Table 2 (listed) Vietnam

Platylomia malickyi Lee 2010b: 26 (distribution, listed) Cambodia, South China, Yunnan, Vietnam, Laos, Thailand, Myanmar

Platylomia malickyi Lee and Hill 2010: 299–300, Figs 7–8 (phylogeny, comp. note)

P. maxima (Lee, 2009) to *Champaka maxima*

P. meyeri (Distant, 1883) to *Champaka meyeri*

P. nagarasingna (Distant, 1881) to *Dundubia nagarasingna*

P. **n. sp. Duffels, 1990**

Platylomia n. sp. Duffels 1990b: 64, Table 2 (listed) Sulawesi

Platylomia n. sp. Boer and Duffels 1996b: 330 (distribution) Sulawesi

P. nigra (Distant, 1888) to *Champaka nigra*

P. operculata **Distant, 1913** = *Platylomia radha* (nec Distant)

Platylomia operculata Beuk 1998: 152, 157–158 (comp. note) Equals *Platylomia radha* (error)

Platylomia radha Boulard 2001a: 128, 133, 141–142, Plate II, Fig K, Fig 11 (illustrated, song analysis, sonogram, comp. note) Thailand

Platylomia operculata Pham 2004: 61 (comp. note)

Platylomia operculata Boulard 2005h: 6–7, 36–40, Figs 47–54 (status, described, neallotype designation, illustrated, song, sonogram, listed, comp. note) Equals *Cosmopsaltria radha* (nec Distant) Equals *Platylomia radha* (partim Beuk, 1996) Equals *Platylomia radha* Boulard, 2001, 2002, Equals sonogram of *Dundubia feae* Boulard, 2001 Thailand, Burma, Laos, India

Platylomia operculata Boulard 2005o: 148, 150 (listed) Equals *Platylomia radha* Boulard, 2001 sonogram *Dundubia feae*

Platylomia operculata Boulard 2007e: 10, 15, 21, Plates 16–17 (eclosion, illustrated, comp. note) Thailand

Platylomia operculata Sanborn, Phillips and Sites 2007: 23–24, 34 (listed, distribution, comp. note) Equals *Platylomia radha* (partim) Thailand, Burma, Indochina

Platylomia operculata Boulard 2008b: 106–109, 112–114, Figs 1–2, Figs 5–9 (illustrated, comp. note) Thailand

Platylomia operculata Boulard 2008c: 363 (comp. note) Thailand

Platylomia operculata Boulard 2008e: 345–346, 348–354, Figs 1–2, Figs 5–9 (illustrated, comp. note) Thailand

Platylomia operculata Boulard 2008f: 29, 93 (biogeography, listed, comp. note) Equals *Platylomia radha* (partim) Equals *Platylomia radha* (nec Distant) Thailand, Indochina

Platylomia opercularis Lee 2008a: 17 (synonymy, distribution, listed) Equals *Platylomia radha* (nec Distant) Vietnam, China, Yunnan, Guangxi, Jiangxi, Hainan, Laos, Cambodia, Thailand, Myanmar

Platylomia operculata Pham and Yang 2009: 15, Table 2 (listed) Vietnam

Platylomia operculata Lee 2010b: 26 (synonymy, distribution, listed) Equals *Platylomia radha* (nec Distant) China, Yunnan, Guangxi, Jiangxi, Hainan, Cambodia, Vietnam, Laos, Thailand, Myanmar

P. pieli Kato, 1938 to *Macrosemia pieli*

P. pieli elongata (Liu, 1939) to *Macrosemia pieli* (Kato, 1938)

P. pieli trifuscata (Liu, 1939) to *Macrosemia pieli* (Kato, 1938)

***P. pendleburyi* Moulton, 1923**

Platylomia pendleburyi Beuk 1998: 149, 152–153, 173–175, Fig 1, Fig 3, Figs 45–51, Table 1 (phylogeny, key, described, illustrated, distribution, comp. note) Peninsular Thailand

Platylomia pendleburyi Sanborn, Phillips and Sites 2007: 24 (listed, distribution, comp. note) Thailand

Platylomia pendleburyi Boulard 2008f: 30, 93 (biogeography, listed, comp. note) Thailand

***P. plana* Lei & Li, 1994**

Platylomia plana Lei and Li 1995: 91–92, 94, Figs 1–2 (n. sp., described, illustrated, key, comp. note) Tibet

Platylomia plana Chou, Lei, Li, Lu and Yao 1997: 252, 257–258, Fig 9–82, Plate XIV, Fig 143 (key, described, illustrated, comp. note) China

***P. radha* (Distant, 1881)** = *Dundubia radha* = *Cosmopsaltria radha* = *Dundubia similis* Distant, 1882 = *Cosmopsaltria similis* = *Platylomia similis*

Dundubia radha Distant 1882: 160 (comp. note)

Platylomia radha Boulard 1990b: 155, 157, Plate XVIII, Fig 5 (sound system, comp. note) India

Platylomia radha Lei and Li 1995: 92, 94 (synonymy, measurements, distribution, comp. note) Equals *Dundubia radha* Equals *Cosmopsaltria radha* China, Yunnan, Burma, Sikkim, Nepal, India, Thailand

Platylomia radha Beuk 1996: 130, 133–134 (phylogeny, comp. note)

Platylomia radha Chou, Lei, Li, Lu and Yao 1997: 17, 252, 255–256, 357, Fig 3-6C, Fig 9–81, Plate XIII, Fig 132 (key, described, illustrated, synonymy, comp. note) Equals *Dundubia radha* Equals *Cosmopsaltria radha* China

Platylomia similis Zaidi and Ruslan 1997: 227 (listed, comp. note) Equals *Dundubia similis* Peninsular Malaysia, Sikkim, Assam

Platylomia radha Beuk 1998: 149, 152–159, 161, 163–165, 173, Figs 1–2, Figs 4–15, Table 1 (phylogeny, key, lectotype designation, described, illustrated, distribution, comp. note) Equals *Dundubia radha* Equals *Cosmopsaltria radha* Equals *Dundubia similis* Distant, 1882 Equals *Cosmopsaltria similis* Equals *Platylomia similis* Equals *Platylomia operculata* (error) Bhutan, India, Assam, Sikkim, Nepal, Burma, Cambodia, Laos, Thailand, Vietnam, China Hainan, Sichuan, yunnan

Platylomia radha group Beuk 1998: 147–152, 159, 163, 165, 168, 173 (described, comp. note) India, Bhutan, Nepal, Indo-China, Burma, Thailand, Laos, Cambodia, Vietnam, Peninsular Thailand, China, Taiwan

Platylomia similis Beuk 1998: 152, 157–158 (comp. note) Equals *Platylomia radha*

Platylomia radha group Beuk 1999: 60–7 (comp. note)

Platylomia radha Hua 2000: 63 (listed, distribution) Equals *Dundubia radha* China, Guangxi, Jingxi, Hainan, Yunnan, India, Sikkim, Burma, Nepal, Thailand

Platylomia radha Boulard 2002b: 195, 198, Fig 50 (illustrated, comp. note) Southeast Asia

Platylomia radha Pham 2004: 61, 64 (key, synonymy, distribution, comp. note) Equals *Dundubia radha* Equals *Cosmopsaltria radha* Vietnam

Platylomia radha Boulard 2005h: 36 (comp. note)

Platylomia radha Pham and Thinh 2005a: 243 (synonymy, distribution, listed) Equals *Dundubia radha* Equals *Cosmopsaltria radha* Vietnam

Platylomia radha Duffels and Hayashi 2006: 189 (comp. note)

Platylomia radha Pham and Thinh 2006: 526–527 (listed, comp. note) Vietnam

Platylomia radha Sanborn, Phillips and Sites 2007: 24, 34 (comp. note) India

Platylomia similis Sanborn, Phillips and Sites 2007: 34 (comp. note) India

Platylomia radha Sun, Watson, Zheng, Watson and Liang 2009: 3148–3149, 3151, Fig 1J, Table 1 (illustrated, wing microstructure, comp. note) China, Yunnan

P. saturata (Walker, 1858) to *Macrosemia saturata saturata*

P. spiculata (Noualhier, 1896) to *Dundubia spiculata*

P. spinosa (Fabricius, 1787) to *Champaka spinosa*

***P. stasserae* Boulard, 2005**

Platylomia stasserae Boulard 2005i: 62–64, Figs 6–10 (n. sp., described, illustrated, comp. note) Thailand

Platylomia stasserae Boulard 2005o: 150 (listed)

Platylomia stasserae Sanborn, Phillips and Sites 2007: 24 (listed, distribution, comp. note) Thailand

Platylomia stasserae Boulard 2008f: 30, 93 (biogeography, listed, comp. note) Thailand

***P. strongata* Lei, 1997**

Platylomia strongata Chou, Lei, Li, Lu and Yao 1997: 258 (comp. note)

P. tonkiniana Jacobi, 1905 to *Macrosemia tonkiniana*

P. umbrata (Distant, 1888) to *Macrosemia umbrata*

***P.* undescribed species A Sanborn, Phillips & Sites, 2007** = *Platylomia similis* (nec Distant)

Platylomia sp. Sanborn, Phillips and Sites 2007: 3 (listed, comp. note) Thailand

Platylomia undescribed species A Sanborn, Phillips and Sites 2007: 24 (listed, synonymy, distribution, comp. note) Equals *Platylomia similis* (partim) Thailand, Peninsular Malaysia

***P. vibrans* (Walker, 1850)** = *Dundubia vibrans* = *Cosmopsaltria vibrans*

Platylomia vibrans Lee 2009b: 331 (comp. note)

P. virescens Distant, 1905 to *Champaka virescens*

P. viridimaculata Distant, 1889 to *Champaka viridimaculata*

P. wallacei Beuk, 1999 to *Champaka wallacei*

***Sinotympana* Lee, 2009**

Sinotympana Lee 2009e: 470–88 (n. gen., described, comp. note) China

***S. incomparabilis* Lee, 2009**

Sinotympana incomparabilis Lee 2009e: 88–90, Figs 1–2 (n. sp., described, illustrated, type species of *Sinotympana*, comp. note) China

***Ayesha* Distant, 1905**

Ayesha Duffels 1983b: 11 (comp. note)

Ayesha Hayashi 1984: 58 (listed)

Ayesha Lee 1995: 23 (listed)

Ayesha Zaidi and Ruslan 1995a: 174, Table 2 (listed, comp. note) Borneo, Sarawak

Ayesha Zaidi and Ruslan 1995b: 220–221, Table 2 (listed, comp. note) Borneo, Sabah

Ayesha Zaidi and Ruslan 1995d: 202, Table 2 (listed, comp. note) Borneo, Sabah

Ayesha Beuk 1996: 129 (comp. note)

Ayesha Boer and Duffels 1996b: 302 (comp. note) Sulawesi

Ayesha Zaidi 1996: 103, Table 2 (comp. note) Borneo, Sarawak

Ayesha Zaidi and Hamid 1996: 54, Table 2 (listed, comp. note) Borneo, Sarawak

Ayesha Zaidi 1997: 115, Table 2 (listed, comp. note) Borneo, Sarawak

Ayesha Beuk 1998: 147 (comp. note)

Ayesha Zaidi and Ruslan 1998b: 4–5, Table 2 (listed, comp. note) Borneo, Sabah

Ayesha Beuk 1999: 2 (comp. note)

Ayesha Zaidi and Ruslan 1999: 4, Table 2 (listed, comp. note) Borneo, Sarawak

Ayesha Duffels and Turner 2002: 235 (comp. note)

Ayesha Moulds 2005b: 391, 432 (higher taxonomy, listed, comp. note)

Ayesha Lee 2009h: 2630 (listed) Philippines, Palawan

Ayesha sp. Lee and Hill 2010: 299–300, Figs 7–8 (phylogeny, comp. note)

Ayesha Lee and Hill 2010: 302 (comp. note)

A. operculissima (Distant, 1881) to *Ayesha serva* (Walker, 1850)

***A. serva* (Walker, 1850)** = *Dundubia serva* = *Cosmopsaltria serva* = *Cosmopsaltria (Cosmopsaltria) spathulata* Stål, 1870 = *Cosmopsaltria spathulata* = *Ayesha spathulata* = *Cosmopsaltria operculissima* Distant, 1881 = *Ayesha operculissima* = *Cicada elopurina* Distant, 1888 = *Cosmopsaltria vomerigera* Breddin, 1901 = *Dundubia lelita* Kirkaldy, 1904

Ayesha spathulata Duffels 1990b: 64, 70, Table 2 (listed, comp. note) Sulawesi, Philippines, Borneo

Ayesha serva Lee 1995: 19, 23, 28–31, 33–34, 37, 45–46, 130, 135, 138 (comp. note) Equals *Dundubia serva*

Dundubia serva Lee 1995: 122–123 (comp. note)

Ayesha spathulata Boer and Duffels 1996b: 330 (distribution) Sulawesi, Borneo, Philippines

Ayesha spathulata Holloway 1997: 219 (comp. note) Philippines, Lesser Sunda Islands, Sulawesi

Ayesha spathulata Holloway 1998: 308 (distribution, comp. note) Philippines, Sulawesi, Lesser Sunda Islands

Ayesha spathulata Zaidi and Ruslan 1998a: 361–362 (listed, comp. note) Equals *Cosmopsaltria spathulata* Borneo, Sarawak, Philippines, Celebes, Natuna Islands

Ayesha spathulata Zaidi, Ruslan and Azman 1999: 314 (listed, comp. note) Equals *Cosmopsaltria spathulata* Borneo, Sabah, Philippines, Celebes, Sarawak, Natuna Islands

Dundubia serva Duffels and Zaidi 2000: 238 (comp. note)

Ayesha spathulata Zaidi, Ruslan and Azman 2000: 212–213 (listed, comp. note) Borneo, Sarawak

Ayesha spathulata Zaidi and Nordin 2001: 127 (listed, comp. note) Borneo, Sarawak

Ayesha operculissima Zaidi and Azman 2003: 99–100, 105, Table 1 (listed, comp. note) Borneo, Sabah

Ayesha spatulata Zaidi and Azman 2003: 105, Table 1 (listed, comp. note) Borneo, Sabah

Ayesha serva Lee 2008b: 462 (comp. note)

Cosmopsaltria (Cosmopsaltria) spathulata Lee 2009h: 2630 (type species of *Ayesha*, listed) Philippines
Ayesha spathulata Lee 2009h: 2630–2631, Fig 8 (illustrated, synonymy, distribution, listed, comp. note) Philippines, Palawan
Ayesha operculissima Lee 2009h: 2630–2631 (comp. note)
Ayesha serva Lee 2009h: 2630 (comp. note)
A. spathulata (Stål, 1870) to *Ayesha serva* (Walker, 1850)

***Khimbya* Distant, 1905** = *Khimbia* (sic)
Khimbya Beuk 1996: 129 (comp. note)
Khimbya Beuk 1998: 147 (comp. note)
Khimbya Beuk 1999: 2 (comp. note)
Khimbya Duffels and Turner 2002: 236 (comp. note)
Khimbya Moulds 2005b: 391, 432 (higher taxonomy, listed, comp. note)
Khimbya Lee 2008a: 2, 13, Table 1 (listed) Vietnam
Khimbya Lee 2009b: 330 (comp. note)
Khimbya Pham and Yang 2009: 14, Table 2 (listed) Vietnam

***K. cuneata* (Distant, 1897)** = *Pomponia sita*

***K. diminuta* (Walker, 1850)** = *Dundubia dimunita* = *Cosmopsaltria dimunita* = *Khimbia* (sic) *diminuta*

***K. evanescens* (Walker, 1858)** = *Dundubia evanescens* = *Pomponia evanescens* = *Khimbia* (sic) *evanescens*
Pomponia evanescens Boulard 2002a: 46 (comp. note)
Pomponia evanescens Boulard 2005h: 20 (comp. note)
Dundubia evanescens Lee 2008a: 12 (type species of *Khimbya*) Hindustan

***K. immsi* Distant, 1912**

***K. sita* (Distant, 1881)** = *Cosmopsaltria sita* = *Khimbia* (sic) *sita*
Khimbia (sic) *sita* Hua 2000: 62 (listed, distribution, hosts) Equals *Cosmopsaltria sita* China, Guangxi, india
Khimbya sita Lee 2008a: 13–14 (synonymy, distribution, listed) Equals *Cosmopsaltria sita* Vietnam, India
Khimbya sita Pham and Yang 2009: 14, Table 2 (listed) Vietnam

Subtribe Megapomponiina Boulard, 2008
Megapomponiina Boulard 2008f: 7, 36 (n. subtribe, listed) Thailand

***Megapomponia* Boulard, 2005**
Megapomponia Boulard 2005k: 100–102 (n. gen., described, comp. note) Thailand
Pomponia spp. (sic) Gogala and Trilar 2005: 67 (comp. note)
Megapomponia Boulard 2006g: 51, 111, 115, 117–118, 129, 148 (comp. note)
Megapomponia Duffels and Hayashi 2006: 189 (comp. note)
Megapomponia Boulard 2007d: 499 (comp. note) Thailand
Megapomponia Boulard 2007e: 48, 83 (comp. note) Thailand
Megapomponia Sanborn, Phillips and Sites 2007: 3, 30–31, 35, Table 1 (listed) Thailand
Megapomponia Boulard 2008f: 7, 36, 93, 97–98 (listed) Thailand
Megapomponia Lee 2008a: 2, Table 1 (listed) Vietnam
Megapomponia Lee 2009e: 87 (comp. note)
Megapomponia Lee and Sanborn 2009: 31–33, 38 (diagnosis, diversity, distribution, key, listed, comp. note) southeast Asia, Greater Sunda Islands
Megapomponia Pham and Yang 2009: 15, Table 2 (listed) Vietnam
Megapomponia Lee 2010b: 25 (listed) Cambodia
Megapomponia Lee and Hill 2010: 302 (comp. note)
Megapomponia Phillips and Sanborn 2010: 74 (comp. note) Asia

***M. atrotunicata* Lee & Sanborn, 2009**
Megapomponia atrotunicata Lee and Sanborn 2009: 31–37, Fig 9, Figs 14–19 (n. sp., described, illustrated, key, distribution, listed, comp. note) Cambodia
Megapomponia atrotunicata Lee 2010b: 25 (distribution, listed) Cambodia

***M. castanea* Lee & Sanborn, 2009**
Megapomponia castanea Lee and Sanborn 2009: 31–33, 36, 38–39, Figs 24–27 (n. sp., described, illustrated, key, distribution, listed, comp. note) Cambodia
Megapomponia castanea Lee 2010b: 26 (distribution, listed) Cambodia

***M. clamorigravis* Boulard, 2005**
Megapomponia clamorigravis Boulard 2005d: 341 (comp. note) Thailand
Megapomponia clamorigravis Boulard 2005k: 102–106, Figs 11–12 (n. sp., described, illustrated, song, sonogram, comp. note) Thailand
Megapomponia clamorigravis Boulard 2005o: 150 (listed)
Megapomponia clamorigravis Boulard 2006g: 51–52, 58, 60, 180, Fig 24 (sonogram, listed, comp. note)
Megapomponia clamorigravis Boulard 2007d: 499 (comp. note) Thailand
Megapomponia clamorigravis Boulard 2007e: 48, 50, 69, Fig 32, Plate 25 (sonogram, illustrated, comp. note) Thailand
Megapomponia clamorigravis Sanborn, Phillips and Sites 2007: 30 (listed, distribution, comp. note) Thailand
Megapomponia clamorigravis Boulard 2008f: 36, 93 (biogeography, listed, comp. note) Thailand

Megapomponia clamorigravis Lee and Sanborn 2009: 31–32, 35 (key, distribution, comp. note) southern Thailand

M. decem (Walker, 1857) to *Pomponia decem*

M. sp. nr. *decem* Boulard, 2008

Megapomponia sp. nr. *decem* Boulard 2008c: 365 (comp. note) Laos

***M. foksnodi* Boulard, 2010**

Megapomponia foksnodi Boulard 2010a: 15–16, Fig 1 (n. sp., described, illustrated, comp. note) Thailand

***M. imperatoria* (Westwood, 1842)** = *Cicada imperatoria* = *Dundubia imperatoria* = *Pomponia imperatoria* = *Pomponia impteratoria* (sic)

Dundubia imperatoria Distant 1881: iii (comp. note) Malaysia

Pomponia imperatoria Boulard 1988d: 152, 154, Fig 14 (wing abnormalities, illustrated, comp. note) Malaysia

Pomponia impteratoria (sic) Hayashi 1993: 13, Fig 1 (distribution, comp. note)

Pomponia imperatoria Hayashi 1993: 14, Fig 2 (illustrated, key, comp. note)

Pomponia imperatoria complex Hayashi 1993: 14 (comp. note)

Pomponia imperatoria Boulard 1995b: 121 (comp. note)

Pomponia imperatoria Zaidi and Ruslan 1997: 229 (listed, comp. note) Equals *Cicada imperatoria* Equals *Dundubia imperatoria* Peninsular Malaysia, Borneo, Nepal, Indian Islands

Pomponia imperatoria Boulard 1999b: 184–185, Fig 12 (illustrated, comp. note) Malaysia

Pomponia imperatoria Boulard 2001d: 81–86, Plate A, Fig 1 (comp. note) Malaysia

Cicada imperatoria Boulard 2001d: 81 (comp. note)

Pomponia imperatoria Sueur 2001: 40, 46, Tables 1–2 (listed)

Pomponia imperatoria Boulard 2002b: 174–175, Fig 1 (size, comp. note)

Pomponia imperatoria Sueur 2002b: 123, 129 (listed, comp. note)

Pomponia imperatoria Schouten, Duffels and Zaidi 2004: 373, 376, 379, Table 2 (listed, comp. note) Malayan Peninsula

Pomponia imperatoria Trilar and Gogala 2004: 81 (song, comp. note) Peninsular Malaysia

Pomponia imperatoria Boulard 2005k: 93–100, Fig 1 (comp. note)

Cicada imperatoria Boulard 2005k: 94 (comp. note)

Megapomponia imperatoria Boulard 2005k: 100–102, Figs 3–4 (n. comb., type species of *Megapomponia*, illustrated, comp. note) Malaysia

Pomponia imperatoria Gogala and Trilar 2005: 67 (comp. note)

Pomponia imperatoria Lee 2005: 36 (illustrated, comp. note)

Pomponia imperatoria Benko and Perc 2006: 557 (comp. note)

Pomponia imperatoria Duffels and Hayashi 2006: 189 (type species of *Megapomponia*, comp. note)

Megapomponia imperatoria Sanborn, Phillips and Sites 2007: 30 (listed, synonymy, type species of *Megapomponia*, distribution, comp. note) Equals *Dundubia imperatoria* Thailand, India, Nepal, Malaysia, Borneo, Cambodia, Laos, Sarawak, Indonesia

Pomponia imperatoria Shiyake 2007: 6, 84, 86, Fig 140 (illustrated, distribution, listed, comp. note) Malaysia

Megapomponia imperatoria Boulard 2008f: 36, 90 (type species of *Megapomponia*)

Cicada imperatoria Lee 2008a: 16 (type species of *Megapomponia*)

Pomponia imperatoria Moulds 2008b: 140 (comp. note)

Megapomponia imperatoria Lee and Sanborn 2009: 31–33, Figs 1–2, Fig 5 (type species of *Megapomponia*, illustrated, key, distribution, comp. note) Peninsular Malaysia

Cicada imperatoria Lee and Sanborn 2009: 32 (type species of *Megapomponia*, listed)

Cicada imperatoria Lee 2010b: 25 (type species of *Megapomponia*)

Megapomponia imperatoria Lee 2010b: 30 (comp. note) Peninsular Malaysia

***M. intermedia* (Distant, 1905)** = *Pomponia intermedia*

Pomponia intermedia Hayashi 1993: 14, Fig 3 (illustrated, distribution, key, comp. note)

Pomponia intermedia Boulard 2001d: 80–81, 83, 85–86, 90–91, 94–95, Plate B, Fig 4, Plate E, Figs 3–4, Plate H, Plate I (illustrated, song analysis, sonogram, comp. note) Thailand, India

Pomponia intermedia Boulard 2002a: 63 (comp. note)

Pomponia intermedia Boulard 2002b: 203 (comp. note)

Pomponia intermedia Pham 2004: 61–62 (key, distribution, comp. note) Vietnam

Pomponia intermedia Trilar and Gogala 2004: 81 (song, comp. note) Thailand

Megapomponia intermedia Boulard 2005d: 344, 347 (comp. note)

Pomponia intermedia Boulard 2005k: 93, 95, 97, 99–100, Fig 1 (comp. note)

Megapomponia intermedia Boulard 2005k: 101–102, Figs 5–6 (n. comb., illustrated, comp. note) Thailand

Pomponia intermedia Boulard 2005o: 148 (listed)

Megapomponia intermedia Boulard 2005o: 150 (listed)
Pomponia intermedia Gogala and Trilar 2005: 70 (comp. note)
Pomponia intermedia Pham and Thinh 2005a: 240 (distribution, listed) Vietnam
Megapomponia intermedia Boulard 2006g: 50, 52, 58–59, 74–75, 115, Fig 22 (sonogram, comp. note) Thailand, Burma
Megapomponia intermedia Boulard 2007d: 499, 506, 508, Fig 5 (comp. note) Thailand
Megapomponia intermedia Boulard 2007e: 1, 35, 42, 69, Fig 26, Plate 1 (sonogram, illustrated, comp. note) Thailand
Pomponia intermedia Boulard 2007e: 42, Fig 26 (sonogram, comp. note) Thailand
Megapomponia intermedia Sanborn, Phillips and Sites 2007: 30–31 (listed, synonymy, distribution, comp. note) Equals *Pomponia intermedia* Thailand, India, Burma, Indochina
Pomponia intermedia Shiyake 2007: 6, 84, 86, Fig 141 (illustrated, distribution, listed, comp. note) Thailand
Megapomponia intermedia Boulard 2008b: 109 (comp. note)
Megapomponia intermedia Boulard 2008c: 365 (comp. note) Laos, Thailand
Megapomponia intermedia Boulard 2008f: 36, 93 (biogeography, listed, comp. note) Equals *Pomponia intermedia* Thailand
Megapomponia intermedia Lee 2008a: 16 (synonymy, distribution, listed) Equals *Pomponia intermedia* Vietnam, Thailand, Myanmar, India
Pomponia intermedia Ras, Sahramo, Malm, Raula and Karppinen 2008: 11252–11253, S1-S2, Figs S1-S2 (nutrient content, human food, listed, comp. note) Thailand
Pomponia intermedia Senter 2008: 256, Fig 1B (illustrated, comp. note)
Megapomponia intermedia Lee and Sanborn 2009: 31–34, Figs 3–4, Fig 8 (illustrated, key, distribution, comp. note) Myanmar, Thailand, Indochina
Megapomponia intermedia Pham and Yang 2009: 15, Table 2 (listed) Vietnam
Megapomponia intermedia Boulard 2010a: 15 (comp. note)

***M. macilenta* Lee, 2012**

***M. merula* (Distant, 1905)** = *Pomponia merula* = *Pomponia imperatoria* (nec Westwood)
Pomponia merula Hayashi 1993: 13–14, Fig 1, Fig 4 (illustrated, distribution, key, comp. note)
Pomponia merula Zaidi 1993: 958, 960, Table 1 (listed, comp. note) Borneo, Sarawak
Pomponia merula Gogala 1994b: 6–8, Figs 7–8 (illustrated, song, sonogram, comp. note) Peninsular Malaysia
Pomponia imperatoria (sic) Gogala and Riede 1995: 298–299, 302–304, Figs 3–4 (song, sonogram, oscillogram, acoustic behavior, comp. note) Borneo
Pomponia merula Gogala and Riede 1995: 298 (comp. note) Sabah, Borneo
Pomponia imperatoria (sic) Riede 1995: 45, Fig 1 (calling activity) Borneo
Pomponia merula Riede and Kroker 1995: 46 (comp. note) Borneo
Pomponia merula Zaidi and Ruslan 1995a: 172, 175, Table 1 (listed, comp. note) Borneo, Sarawak
Pomponia merula Zaidi and Ruslan 1995d: 200, 203, Table 1 (listed, comp. note) Borneo, Sabah
Pomponia merula Zaidi 1996: 101, 104, Table 1 (comp. note) Borneo, Sarawak
Pomponia merula Zaidi and Hamid 1996: 53, 56, Table 1 (listed, comp. note) Borneo, Sarawak
Pomponia merula Zaidi, Ruslan and Mahadir 1996: 61, Table 1 (listed, comp. note) Peninsular Malaysia
Pomponia imperatoria (sic) Riede 1997a: 278, Fig 2 (calling activity, comp. note) Sabah, Malaysia
Pomponia imperatoria (sic) Riede 1997b: 447, Fig 21.3 (sonogram, calling activity, comp. note) Sabah, Malaysia
Pomponia merula Zaidi 1997: 113–114, Table 1 (listed, comp. note) Borneo, Sarawak
Pomponia merula Zaidi and Ruslan 1998a: 365 (listed, comp. note) Borneo, Sarawak, Sabah
Pomponia merula Zaidi and Ruslan 1999: 2–3, 5, Table 1 (listed, comp. note) Borneo, Sarawak
Pomponia merula Zaidi, Ruslan and Azman 1999: 316 (listed, comp. note) Borneo, Sabah, Sarawak
Pomponia merula Zaidi, Noramly and Ruslan 2000a: 329–330 (listed, comp. note) Borneo, Sabah, Sarawak
Pomponia merula Zaidi, Noramly and Ruslan 2000b: 337, 339, Table 1 (listed, comp. note) Borneo, Sabah
Pomponia merula Zaidi, Ruslan and Azman 2000: 215 (listed, comp. note) Borneo, Sabah, Sarawak
Pomponia merula Boulard 2001d: 81–85, Plate A, Fig 2 (comp. note) Borneo
Pomponia merula Sueur 2001: 40, 46, Tables 1–2 (listed)
Pomponia merula Zaidi and Nordin 2001: 183, 185, 191, 203, Table 1 (listed, comp. note) Borneo, Sabah, Sarawak

Pomponia merula Zaidi, Nordin, Maryati, Wahab, Norashikin et al. 2002: 2, 7, 9–10, Tables 1–2 (listed, comp. note) Borneo, Sabah, Sarawak

Pomponia merula Zaidi and Azman 2003: 106, Table 1 (listed, comp. note) Borneo, Sabah

Pomponia merula Trilar and Gogala 2004: 81 (song, comp. note) Borneo

Pomponia merula Boulard 2005k: 93, 95, 98–100, Fig 1 (comp. note) Borneo

Megapomponia merula Boulard 2005k: 101–102 (n. comb., illustrated, comp. note)

Pomponia merula Gogala and Trilar 2005: 67 (distribution, comp. note) Equals *Pomponia imperatoria* Gogala and Riede, 1995 Borneo

Pomponia merula Shiyake 2007: 7, 85–86, Fig 143 (illustrated, distribution, listed, comp. note) Borneo

Megapomponia merula Lee and Sanborn 2009: 31–33, 35, 37, Fig 7 (illustrated, key, distribution, comp. note) Borneo

Pomponia merula Chung 2010: 148 (predation, comp. note) Borneo

***M. pendleburyi* (Boulard, 2001)** = *Megrapomponia* (sic) *pendylburyi* = *Pomponia pendleburyi* = *Pomponia adusta* (nec Walker) = *Pomponia merula* Gogala & Riede 1995 (nec Distant) = *Pomponia* sp. Gogala & Trilar 2005 = *Pomponia imperatoria* Riede & Kroker 1995 (nec Westwood)

Pomponia merula Gogala and Riede 1995: 298–302, 305, Fig 1g, Fig 2g (song, sonogram, oscillogram, acoustic behavior, comp. note) Malaysia

Pomponia imperatoria (sic) Riede and Kroker 1995: 43, 45–50, Figs 3–4 (song, sonogram, oscillogram, calling activity, comp. note) Sabah, Malaysia

Pomponia pendleburyi Boulard 2001d: 83, 86–93, 96–101, Plate C, Plate D, Figs 1–2, Plate E, Figs 1–2, Plate F, Plate G, Plate J, Figs 1–4, Plate K, Figs 1–5, Plate L (n. sp., described, illustrated, song analysis, sonogram, larval morphology, comp. note) Thailand

Pomponia pendleburyi Boulard 2002b: 199, 203, 208 (illustrated, comp. note) Thailand

Pomponia pendleburyi Trilar and Gogala 2004: 81 (song, comp. note) Thailand

Megapomponia pendleburyi Boulard 2005d: 341, 344, 347 (comp. note) Thailand

Pomponia pendleburyi Boulard 2005k: 93, 95, 99–100 Fig 1 (comp. note) Equals *Pomponia imperatoria* Riede & Kroker 1995 Thailand

Megapomponia pendleburyi Boulard 2005k: 101–102, 105, 107 (n. comb., illustrated, song, sonogram, comp. note) Borneo

Megrapomponia (sic) *pendleburyi* Boulard 2005o: 150 (listed)

Pomponia pendleburyi Boulard 2005o: 148 (listed)

Pomponia pendleburyi Gogala and Trilar 2005: 66–68, 70–71, 76, Fig 3 (song, sonogram, oscillogram, acoustic behavior, comp. note) Equals *Pomponia merula* Gogala and Riede 1995 Equals Malaysian *Pomponia adusta* Malaysia, Thailand

Pomponia sp. Gogala and Trilar 2005: 68 (comp. note) Equals *Pomponia pendleburyi* Malaysia

Megapomponia pendleburyi Boulard 2006g: 51–52, 58–59, 74–75, 105, 180, Fig 23 (sonogram, comp. note) Thailand, Malay Peninsula

Megapomponia pendleburyi Boulard 2007d: 499, 506–507, Figs 3–4 (illustrated, comp. note) Thailand

Megapomponia pendleburyi Boulard 2007e: 9, 48–49, 67, 69, 76, Fig 6, Fig 31, Plate 24, Plates 38–40 (illustrated, nymph, turrets, sonogram, comp. note) Thailand

Megapomponia pendleburyi Sanborn, Phillips and Sites 2007: 31 (listed, synonymy, distribution, comp. note) Equals *Pomponia adusta* (partim) Equals *Pomponia pendleburyi* Thailand

Pomponia pendleburyi Sanborn, Phillips and Sites 2007: 34 (comp. note)

Megapomponia pendleburyi Boulard 2008f: 36, 93 (biogeography, listed, comp. note) Equals *Pomponia pendleburyi* Equals *Pomponia adusta* (nec Walker) Thailand

Megapomponia pendleburyi Lee and Sanborn 2009: 31–33, 35, Fig 6 (illustrated, key, distribution, comp. note) southern Thailand, Peninsular Malaysia

M. rajah (Moulton, 1923) to *Pomponia rajah*

***M. sitesi* Sanborn & Lee, 2009** = *Megapomponia* undescribed species A Sanborn, Phillips & Sites 2007

Megapomponia undescribed species A Sanborn, Phillips and Sites, 2007: 31 (listed, distribution, comp. note) Thailand

Megapomponia sitesi Lee and Sanborn 2009: 31–33, 35, 37–38, Fig 10, Figs 20–23 (n. sp., described, illustrated, synonymy, key, distribution, listed, comp. note) Equals *Megapomponia* undescribed species A Sanborn, Phillips and Sites, 2007 southern Thailand

***M.* undescribed species B Sanborn, Phillips & Sites, 2007**

Megapomponia undescribed species B Sanborn, Phillips and Sites, 2007: 31 (listed, distribution, comp. note) Thailand

Subtribe Macrosemiina Kato, 1925 = Macrosemaria = Macrosemiaria (sic)

Macrosemaria Boulard 2003a: 98 (listed)

Macrosemaria Boulard 2003e: 266 (listed)

Macrosemiaria (sic) Sanborn, Phillips and Sites 2007: 3, 20, 38, Fig 3, Table 1 (listed, distribution) Thailand

***Macrosemia* Kato, 1925**

Macrosemia Beuk 1996: 129 (comp. note)
Macrosemia Chou, Lei, Li, Lu and Yao 1997: 236, 259, 293, 308, 320, 325, 327, Table 6, Tables 10–12 (key, described, listed, diversity, biogeography) China
Macrosemia Beuk 1998: 147 (comp. note)
Macrosemia Beuk 1999: 2 (comp. note)
Macrosemia Boulard 2001a: 128 (comp. note)
Macrosemia Duffels and Turner 2002: 235 (comp. note)
Macrosemia Boulard 2003d: 187, 189 (comp. note)
Macrosemia Lee and Hayashi 2003a: 170–172 (key, diagnosis) Taiwan
Macrosemia Chen 2004: 39, 41, 77, 81 (phylogeny, listed)
Macrosemia Boulard 2005b: 112 (comp. note)
Macrosemia Moulds 2005b: 391, 432 (higher taxonomy, listed, comp. note)
Macrosemia Boulard 2007e: 71 (comp. note) Thailand
Macrosemia Sanborn, Phillips and Sites 2007: 3, 20, Table 1 (listed) Thailand
Macrosemia Shiyake 2007: 5, 59 (listed, comp. note)
Macrosemia Boulard 2008f: 31, 93 (listed) Thailand
Macrosemia Lee 2008a: 2, 15–16, Table 1 (listed, comp. note) Vietnam
Macrosemia Pham and Yang 2009: 14, Table 2 (listed) Vietnam
Macrosemia Lee and Hill 2010: 302 (comp. note)

***M. anhweiensis* Ouchi, 1938**

Macrosemia anhwaiensis Chou, Lei, Li, Lu and Yao 1997: 260 (described, synonymy, comp. note) China
Macrosemia anhweiensis Hua 2000: 62 (listed, distribution) China, Anhui

***M. assamensis* (Distant, 1905)** = *Platylomia assamensis*

Platylomia assamensis Chou, Lei, Li, Lu and Yao 1997: 258 (described, comp. note) China
Platylomia assamensis Hua 2000: 63 (listed, distribution) China, Yunnan, India, Thailand, Vietnam
Platylomia assamensis Pham 2004: 61 (comp. note)
Platylomia assamensis Lee 2008a: 15 (comp. note)
Macrosemia assamensis Lee 2008a: 16 (n. comb., synonymy, distribution, listed, comp. note) Equals *Platylomia assamensis* Vietnam, China, Yunnan, Thailand, India
Macrosemia assamensis Pham and Yang 2009: 14, Table 2 (listed) Vietnam

M. chantrainei Boulard. 2003 to *Macrosemia umbrata* (Distant, 1888)

***M. diana* (Distant, 1905)** = *Platylomia diana* = *Cosmopsaltria diana*

Platylomia diana Wu 1935: 13 (listed) China, Szechuan
Platylomia diana Chou, Lei, Li, Lu and Yao 1997: 258 (described, comp. note) China
Platylomia diana Hua 2000: 63 (listed, distribution) China, Fujian, Sichuan
Platylomia diana Lee 2008a: 15 (comp. note)
Macrosemia diana Pham and Yang 2009: 14, Table 2 (listed) Vietnam

***M. divergens* (Distant, 1917)** = *Cosmopsaltria divergens* = *Platylomia divergens* = *Orientopsaltria divergens*

Orientopsaltria divergens Duffels 1983b: 9 (n. comb., comp. note)
Platylomia divergens Chou, Lei, Li, Lu and Yao 1997: 258 (described, synonymy, comp. note) Equals *Cosmopsaltria divergens* China
Platylomia divergens Hua 2000: 63 (listed, distribution) Equals *Cosmopsaltria divergens* China, Yunnan, Thailand, Laos
Orientopsaltria divergens Duffels and Zaidi 2000: 196 (comp. note)
Cosmopsaltria divergens Duffels and Zaidi 2000: 196 (comp. note)
Macrosemia divergens Lee 2008a: 16 (n. comb., synonymy, distribution, listed, comp. note) Equals *Platylomia diana* Vietnam, China, Sichuan, Fujian
Platylomia divergens Lee 2008a: 15 (comp. note)
Macrosemia divergens Lee 2008a: 16 (n. comb., synonymy, distribution, listed, comp. note) Equals *Cosmopsaltria divergens* Vietnam, Laos
Macrosemia divergens Pham and Yang 2009: 14, 16, Table 2 (listed, comp. note) Vietnam
Macrosemia divergens Pham and Yang 2010a: 133 (comp. note) Vietnam

M. hopponis Kato, 1925 to *Macrosemia kareisana* (Matsumura, 1907)

***M. kareisana* (Matsumura, 1907)** = *Cosmopsaltria kareisana* = *Platylomia karaisana* (sic) = *Platylomia kareisana* = *Platylomia kareipana* (sic) = *Marosemia karaisana* (sic) = *Platylomia hopponis* Kato, 1925 = *Macrosemia hopponis* = *Macrosemia kareisana hopponis* = *Platylomia karapinensis* Kato, 1925 = *Macrosemia kareisana karapinensis* = *Cosmopsaltria montana* Kato, 1927 = *Orientopsaltria montana*

Cosmopsaltria montana Duffels 1983b: 4 (comp. note)
Orientopsaltria montana Duffels 1983b: 9 (comp. note)
Platylomia hopponis Chou, Lei, Li, Lu and Yao 1997: 259 (type species of *Macrosemia*)
Macrosemia kareisana Chou, Lei, Li, Lu and Yao 1997: 259–260, 308, 320, 325, 327, 332, 343, Tables 10–12 (key, described, synonymy, song, oscillogram, spectrogram, comp. note)

Equals *Cosmopsaltria kareisana* Equals *Platylomia kareisana* China
Cosmopsaltria montana Duffels and Zaidi 2000: 196 (comp. note)
Platylomia montana Duffels and Zaidi 2000: 196 (comp. note)
Orientopsaltria montana Duffels and Zaidi 2000: 196 (comp. note)
Cosmopsaltria montana Hua 2000: 61 (listed, distribution) China, Taiwan
Macrosemia kareisana Hua 2000: 62 (listed, distribution) Equals *Cosmopsaltria kareisana* China, Taiwan, Japan
Macrosemia kareisana Sueur 2001: 39, Table 1 (listed)
Platylomia? hopponis Lee and Hayashi 2003a: 170 (type species of *Macrosemia*) Formosa
Macrosemia kareisana Lee and Hayashi 2003a: 171–175, Fig 14, Figs 16–17 (key, diagnosis, illustrated, synonymy, song, listed, comp. note) Equals *Cosmopsaltria karëisana* Equals *Platylomia karëisana* Equals *Platylomia karäisana* (sic) Equals *Platylomia kareisana* Equals *Macrosemia karëisana* Equals *Cosmopsaltria multivocalis* (nec Matsumura) Equals *Platylomia karapinensis* Equals *Platylomia? hopponis Equals Macrosemia hopponis* Equals *Cosmopsaltria montana* n. syn. Equals *Orientopsaltria montana* Equals *Platylomia montana* Equals *Macrosemia karëisana* var. *karapinensis* Equals *Macrosemia kareisana* var. *karapinensis* Equals *Platylomia kareisana* var. *karapinensis* Equals *Macrosemia karëisana* f. *hopponis* Equals *Macrosemia kareisana* f. *hopponis* Equals *Macrosemia kareisana* var. *hopponis* Equals *Platylomia kareisana* var. *hoppinis* Taiwan
Macrosemia kareisana Chen 2004: 81–82, 192, 198 (described, illustrated, listed, comp. note) Taiwan
Orientopsaltria montana Chen 2004: 81, 192 (comp. note)
Macrosemia kareisana Shiyake 2007: 5, 57, 59, Fig 93 (illustrated, distribution, listed, comp. note) Taiwan
Macrosemia kareisana Boulard 2008f: 31 (type species of *Macrosemia*)
Platylomia? hopponis Lee 2008a: 15 (type species of *Macrosemia*) Formosa
Macrosemia kareisana Lee and Hill 2010: 299–300, Figs 7–8 (phylogeny, comp. note)

***M. khuanae* Boulard, 2001** = *Macrosemia longiterebra* Boulard, 2004
Macrosemia khuanae Boulard 2001a: 144–146, Figs 14–17 (n. sp., described, illustrated, song analysis, sonogram, comp. note) Thailand
Macrosemia khuanae Boulard 2003d: 187 (comp. note)
Macrosemia longiterebra Boulard 2003d: 187–191, Figs 1–5 (n. sp., described, illustrated, comp. note) Thailand
Macrosemia longiterebra Boulard 2005b: 111–114, Figs 2–4 (song, sonogram, comp. note) Thailand
Macrosemia khuanae Boulard 2005o: 148 (listed)
Macrosemia longiterebra Boulard 2005o: 149, 151 (listed)
Macrosemia longiterebra Boulard 2006g: 148–149, 180, Fig 109 (sonogram, listed, comp. note) Thailand
Macrosemia khuanae Boulard 2007d: 497, 499 (synonymy, comp. note) Equals *Macrosemia longiterebra* Thailand
Macrosemia longiterebra Boulard 2007e: 55, 71, Fig 37.3, Plate 32b-d (genitalia, illustrated, comp. note) Thailand
Macrosemia khunae Sanborn, Phillips and Sites 2007: 20 (listed, distribution, comp. note) Thailand
Macrosemia longiterebra Sanborn, Phillips and Sites 2007: 20–21 (listed, distribution, comp. note) Thailand
Macrosemia khuanae Boulard 2008c: 365 (comp. note) South Vietnam
Macrosemia khuanae Boulard 2008f: 31, 93 (biogeography, listed, comp. note) Equals *Macrosemia longiterebra* Thailand

***M. kiangsuensis kiangsuensis* Kato, 1938** = *Platylomia kingvosana* Liu, 1940 = *Platylomia kinvosana* (sic)
Platylomia kinvosana (sic) Liu 1992: 41 (comp. note)
Macrosemia kiangsuensis Chou, Lei, Li, Lu and Yao 1997: 259–260, Plate XIII, Fig 138 (key, described, illustrated, comp. note) China
Macrosemia kiangsuensis Hua 2000: 62 (listed, distribution) China, Jiangsu

***M. kiangsuensis viridescens* (Metcalf, 1955)** = *Platylomia kingvosana viridescens* = *Platylomia kingvosana virescens* Liu, 1940

M. kingvosana viridescens (Metcalf, 1955) to *Macrosemia kiangsuensis viridescens* (Metcalf, 1955)

***M. matsumurai* (Kato, 1928)** = *Macrosemia matsumura* (sic) = *Macrosemia matsumarai* (sic) = *Platylomia matsumurai*
Macrosemia matsumurai Wang and Zhang 1993: 189 (listed, comp. note) China
Macrosemia matsumurai Chou, Lei, Li, Lu and Yao 1997: 259–260, Plate XIII, Fig 137 (key, described, illustrated, synonymy, comp. note) Equals *Platylomia matsumurai* China

Macrosemia matsumura (sic) Chou, Lei, Li, Lu and Yao 1997: 308, 320, 325, 327, 332, 343, Tables 10–12 (song, oscillogram, spectrogram, comp. note) China

Macrosemia matsumurai Hua 2000: 62 (listed, distribution) China, Fujian, Taiwan, Hunan, Sichuan, Japan, Malaysia

Macrosemia matsumarai (sic) Sueur 2001: 39, Table 1 (listed)

Macrosemia matsumurai Lee and Hayashi 2003a: 171, 174–175, Fig 15 (key, diagnosis, illustrated, synonymy, song, listed, comp. note) Equals *Platylomia matsumurai* Taiwan, China

Macrosemia matsumurai Chen 2004: 81, 84, 198 (described, illustrated, listed, comp. note) Taiwan

***M. perakana* (Moulton, 1923)** = *Platylomia saturata perakana*

Platylomia saturata perakana Sanborn, Phillips and Sites 2007: 24 (listed, distribution, comp. note) Thailand, Malay Peninsula, Malay Archipelago

Macrosemia perkana Boulard 2008c: 365 (comp. note) Malaysia

Macrosemia perakana Boulard 2008f: 31, 93 (biogeography, listed, comp. note) Equals *Platylomia saturata perakana* Thailand

***M. pieli* (Kato, 1938)** = *Platylomia pieli* = *Platylomia piei* (sic) = *Platylomia pieli elongata* Liu, 1939 = *Platylomia pieli* var. *elongata* = *Platylomia pieli trifuscata* Liu, 1939 = *Platylomia pieli* var. *trifuscata* = *Platylomia chusana* Kato, 1940

Platylomia piei (sic) Chou, Lei, Li, Lu and Yao 1997: 252 (key) China

Platylomia pieli Chou, Lei, Li, Lu and Yao 1997: 252–253, 308, 319, 325, 327, 332, 343, 358, Plate XIII, Fig 136, Tables 10–12 (described, illustrated, synonymy, song, oscillogram, spectrogram, comp. note) Equals *Platylomia pieli* var. *elongata* n. syn. Equals *Platylomia pieli* var. *trifuscata* n. syn. Equals *Platylomia chusana* China

Platylomia pieli var. *elongata* Chou, Lei, Li, Lu and 1997: 358 (synonymy)

Platylomia pieli var. *trifuscata* Chou, Lei, Li, Lu and Yao 1997: 358 (synonymy)

Platylomia chusana Chou, Lei, Li, Lu and Yao 1997: 358 (synonymy)

Platylomia pieli Hua 2000: 63 (listed, distribution) Equals *Platylomia peili elongata* Equals *Platylomia peili trifuscata* Equals *Platylomia* (sic) *chusana* China, Anhui, Jiangsu, Jiangxi, Zhejiang, Fujian, Guangdong, Hunan, guangxi, Sichuan, Vietnam

Platylomia pieli Sueur 2001: 39, Table 1 (listed)

Platylomia peili Xu, Wang, Xu, Hua, and Lin 2001: 396 (life cycle, predation, comp. note) China

Platylomia pieli Yaginuma 2002: 2 (comp. note)

Platylomia pieli Zhang and Xuan 2007: 404 (fungal host, comp. note) China

Platylomia pieli Lee 2008a: 15 (comp. note)

Macrosemia pieli Lee 2008a: 16 (n. comb., synonymy, distribution, listed) Equals *Platylomia pieli* Vietnam, China, Sichuan, Hunan, Jiangxi, Zhejiang, Anhui, Fujian

Macrosemia pieli Pham and Yang 2009: 14, Table 2 (listed) Vietnam

M. pieli elongata (Liu, 1939) to *Macrosemia pieli* (Kato, 1938)

M. pieli trifuscata (Liu, 1939) to *Macrosemia pieli* (Kato, 1938)

***M. saturata saturata* (Walker, 1858)** = *Dundubia saturata* = *Cosmopsaltria saturata* = *Cosmopsaltria saturate* (sic) = *Platylomia saturata* = *Platylomia satura* (sic) = *Paltylomia* (sic) *saturata* = *Dundubia obtecta* Walker, 1850 (nec Fabricius)

Platylomia saturata Swinton 1908: 380 (song, comp. note) India

Platylomia saturata Kirillova 1986a: 122 (listed, chromosome number) India

Platylomia saturata Kirillova 1986b: 45 (listed, chromosome number) India

Platylomia saturata Zaidi, Mohammedsaid and Yaakob 1990: 261 (comp. note) Malaysia

Platylomia saturata Emeljanov and Kirillova 1991: 800 (chromosome number, comp. note)

Paltylomia (sic) *saturata* Emeljanov and Kirillova 1992: 63 (chromosome number, comp. note)

Platylomia saturata Zaidi and Ruslan 1994: 428 (comp. note) Malaysia

Platylomia saturata Lei and Li 1995: 92, 94 (comp. note)

Platylomia saturata Zaidi, Ruslan and Mahadir 1996: 61, Table 1 (listed, comp. note) Peninsular Malaysia

Platylomia saturata Chou, Lei, Li, Lu and Yao 1997: 258 (comp. note)

Platylomia saturata Beuk 1998: 147 (comp. note)

Platylomia saturata Biswas, Ghosh, Bal and Sen 2004: 240, 247–248 (listed, key, described, distribution) Equals *Dundubia saturata* India, Manipur, Assam, Naga Hills, Utta Pradesh, West Bengal, Indochina, Malay Peninsula, Nepal

Cosmopsaltria saturata Ahmed, Siddiqui, Akhter and Rizvi 2005: 843–845 fig 1A-H (redescribed, illustrated) Equals *Dundubia saturata* Walker Equals *Dundubia obtecta* Walker Pakistan

Cosmopsaltria saturate (sic) Ahmed, Siddiqui, Akhter and Rizvi 2005: 845 (comp. note) India

Platylomia saturata Boulard 2005h: 36 (comp. note)
Macrosemia saturata Boulard 2008c: 365 (comp. note) India
Platylomia saturata Lee 2008a: 15 (comp. note)
Macrosemia saturata Lee 2008a: 16 (n. comb., synonymy, distribution, listed) Equals *Dundubia saturata* Equals *Platylomia saturata* Vietnam, Malay Peninsula, Indonesia, Java, Bangladesh, Bhutan, Nepal, India
Macrosemia saturata Pham and Yang 2009: 14, 16, Table 2 (listed, comp. note) Vietnam
Platylomia saturata Ahmed and Sanborn 2010: 39 (synonymy, distribution) Equals *Dundubia obtecta* Equals *Dundubia saturata* Pakistan, India, Nepal, Bhutan, Indochina
Macrosemia saturata Pham and Yang 2010a: 133 (comp. note) Vietnam

***M. saturata* var. *a* (Distant, 1891)** = *Cosmopsaltria saturata* var. *a*

***M. saturata* var. *b* (Distant, 1891)** = *Cosmopsaltria saturata* var. *b*

M. saturata perakana (Moulton, 1923) to *Macrosemia perakana*

***M. suavicolor* Boulard, 2008**
Macrosemia suavicolor Boulard 2008c: 352–354, Figs 1–3 (n. sp., described, illustrated, comp. note) South Vietnam

***M. tonkiniana* (Jacobi, 1905)** = *Cosmopsaltria tonkiniana* = *Cosmoscarta* (sic) *tonkiniana* = *Platylomia tonkiniana* = *Orientopsaltria tonkiniana*
Orientopsaltria tonkiniana Duffels 1983b: 10 (n. comb., comp. note)
Platylomia tonkiniana Chou, Lei, Li, Lu and Yao 1997: 258 (described, synonymy, comp. note) Equals *Cosmopsaltria tonkiniana* China
Platylomia tonkiniana Beuk 1998: 148–149, Fig 1, Table 1 (phylogeny, comp. note)
Platylomia tonkiniana Beuk 1999: 5–6, 8, Fig 1a, Table 1 (cladogram, comp. note)
Orientopsaltria tonkiniana Duffels and Zaidi 2000: 196 (comp. note)
Cosmopsaltria tonkiniana Duffels and Zaidi 2000: 196 (comp. note)
Platylomia tonkiniana Duffels and Zaidi 2000: 196 (comp. note)
Platylomia tonkiniana Hua 2000: 63 (listed, distribution) Equals *Cosmopsaltria tonkiniana* China, Hainan, Yunnan, India, Burma, Thailand, Laos, Vietnam
Macrosemia tonkiniana Boulard 2003a: 98, 104–106, Figs 2–3, sonogram 6 (n. comb., described, allotype female, paratype females, song, sonogram, illustrated, comp. note) Equals *Cosmopsaltria tonkiniana* Equals *Cosmocarta* (sic) *tonkiniana* Equals *Platylomia tonkiniana* Thailand, Tonkin, China
Macrosemia tonkiniana Boulard 2003e: 266 (n. comb., comp. note) Indochina
Platylomia tonkiniana Pham 2004: 61 (comp. note)
Macrosemia tonkiniana Boulard 2005o: 149 (listed, n. comb.) Equals *Cosmopsaltria tonkiniana*
Platylomia tonkiniana Boulard 2006g: 109 (comp. note)
Platylomia tonkiniana Boulard 2007e: 32, 47 (comp. note) Thailand
Macrosemia tonkiniana Ganchev and Potamitis 2007: 286, Fig 1 (sonogram, comp. note)
Macrosemia tonkiniana Sanborn, Phillips and Sites 2007: 22 (listed, synonymy, distribution, comp. note) Equals *Platylomia tonkiniana* Thailand, Vietnam, Indochina, China
Platylomia tonkiniana Boulard 2008c: 364 (comp. note) Myanmar, Laos
Platylomia tonkiniana Boulard 2008f: 30, 93 (biogeography, listed, comp. note) Equals *Cosmopsaltria tonkiniana* Equals *Macrosemia tonkiniana* Thailand, Tonkin
Macrosemia tonkiniana Lee 2008a: 16 (synonymy, distribution, listed) Equals *Cosmopsaltria tonkiniana* Equals *Platylomia tonkiniana* Vietnam, China, Yunnan, Hainan, Laos, Thailand, Myanmar, India
Macrosemia tonkiniana Pham and Yang 2009: 15, Table 2 (listed) Vietnam
Macrosemia tonkiniana Lee and Hill 2010: 299–300, Figs 7–8 (phylogeny, comp. note)

***M. umbrata* (Distant, 1888)** = *Cosmopsaltria umbrata* = *Platylomia umbrata* = *Macrosemia chantrainei* Boulard, 2003
Platylomia umbrata Lei and Li 1995: 92, 94 (synonymy, measurements, distribution, comp. note) Equals *Cosmopsaltria umbrata* China, Yunnan, Jiangxi, Hainan, Burma, Sikkim, Nepal, India, Thailand
Platylomia umbrata Chou, Lei, Li, Lu and Yao 1997: 252, 254–255, 357, Plate XIII, Fig 135 (key, described, illustrated, synonymy, comp. note) Equals *Cosmopsaltria umbrata* China
Platylomia umbrata Hua 2000: 63 (listed, distribution) Equals *Cosmopsaltria umbrata* China, Yunnan, India, Burma, Laos, Thailand, Bhutan, Nepal, Sri Lanka
Platylomia umbrata Boulard 2003c: 187 (comp. note)
Macrosemia umbrata Boulard 2003e: 266 (n. comb., comp. note) Sikkim
Macrosemia chantrainei Boulard 2003e: 266–270, Color Plate, Fig D, Figs 7–10, sonogram 2 (n. sp., described, song, sonogram, illustrated, comp. note) Thailand

Platylomia umbrata Biswas, Ghosh, Bal and Sen 2004: 240, 247 (listed, key, described, distribution) Equals *Cosmopsaltria umbrata* India, Manipur, Assam, Meghalaya, Uttar Pradesh, West Bengal, Myanmar
Cosmopsaltria umbrata Ahmed, Siddiqui, Akhter and Rizvi 2005: 844–845 (comp. note) Pakistan
Macrosemia chantrainei Boulard 2005o: 149 (listed)
Platylomia chantrainei Boulard 2006g: 109 (comp. note)
Macrosemia chantrainei Sanborn, Phillips and Sites 2007: 20 (listed, distribution, comp. note) Thailand
Macrosemia umbrata Sanborn, Phillips and Sites 2007: 22 (listed, synonymy, distribution, comp. note) Equals *Platylomia umbrata* Thailand, India, Burma, Nepal, Himalayas
Macrosemia umbrata Boulard 2008c: 352, 354, 365 (synonymy, comp. note) Equals *Macrosemia chantrainei* India, Thailand, Laos
Macrosemia umbrata Boulard 2008f: 31, 93 (synonymy, biogeography, listed, comp. note) Equals *Cosmopsaltria umbrata* Equals *Platylomia umbrata* Equals *Macrosemia chantrainei* Thailand

Subtribe Aolina Distant, 1905 = Aolaria
Aolaria Boulard 2003c: 172 (listed)
Aolaria Sanborn, Phillips and Sites 2007: 3, 32, 40, Fig 4, Table 1 (listed, distribution) Thailand

Aola Distant, 1905 to *Haphsa* Distant, 1905
A. bindusara (Distant, 1881) to *Haphsa bindusara*
A. scitula (Distant, 1888) to *Haphsa scitula*
A. sulaiyai Boulard, 2005 to *Haphsa sulaiyai*

***Haphsa* Distant, 1905** = *Aola* Distant, 1905 = *Pomponia (Aola)*
Haphsa Wu 1935: 12 (listed) China
Haphsa Beuk 1996: 129 (comp. note)
Aola Chou, Lei, Li, Lu and Yao 1997: 187, 193, 293, Table 6 (key, described, listed, diversity, biogeography) China
Haphsa Chou, Lei, Li, Lu and Yao 1997: 236, 293, Table 6 (key, described, listed, diversity, biogeography) China
Haphsa Beuk 1998: 147 (comp. note)
Aola Beuk 1998: 149 (comp. note)
Haphsa Beuk 1999: 2 (comp. note)
Aola Beuk 1999: 2 (comp. note)
Aola Boulard 2005i: 60 (comp. note)
Aola Moulds 2005b: 391, 432 (higher taxonomy, listed, comp. note)
Haphsa Moulds 2005b: 391, 432 (higher taxonomy, listed, comp. note)
Aola Boulard 2006g: 35–36 (comp. note)
Aola Sanborn, Phillips and Sites 2007: 3, 32, Table 1 (listed) Thailand
Haphsa Boulard 2008c: 368 (comp. note)
Haphsa Boulard 2008f: 7, 32, 93 (listed) Thailand
Aola Boulard 2008f: 7, 40, 94 (listed) Thailand
Haphsa Lee 2008a: 2, 13–15, Table 1 (listed, comp. note) Equals *Aola* n. syn. Vietnam
Aola Lee 2008a: 14 (status, comp. note)
Haphsa Lee 2009b: 330–331 (comp. note) Equals *Aola* India, Thailand
Aola Lee 2009b: 330 (comp. note)
Haphsa Lee 2009e: 87 (comp. note)
Haphsa Pham and Yang 2009: 14, Table 2 (listed) Vietnam
Haphsa Sanborn 2009b: 311 (comp. note) Vietnam
Haphsa Ahmed and Sanborn 2010: 39 (distribution) China, Indochina, Nepal, India, Pakistan
Haphsa Lee 2010b: 24 (synonymy, listed) Equals *Aola* Cambodia
Aola Lee 2010b: 24 (listed)
Haphsa Lee and Hill 2010: 298, 302 (comp. note)
H. aculeus Lee, 2009 to *Kaphsa aculeus*
***H. bicolora* Sanborn, 2009**
Haphsa bicolora Sanborn 2009b: 307, 310–312, Figs 12–22 (n. sp., described, illustrated, comp. note) Vietnam
***H. bindusara* (Distant, 1881)** = *Haphsa bindusura* (sic) = *Pomponia bindusara* = *Aola bindusara*
Aola sp.? *bindusara* Chaudhry, Chaudry and Khan 1970: 104 (listed, comp. note) Pakistan, Burma, India
Aola bindusara Jiang 1985: 257–258, 260–261, Fig 1B, Fig 3 (illustrated, song analysis, oscillogram, comp. note) Yunnan
Aola bindusara Lei, Chou and Li 1994: 53 (comp. note) China
A[ola] bindusara Jiang, Yang, Tang, Xu and Chen 1995b: 228 (song, phonotaxis, flight behavior, comp. note) China
Pomponia bindusara Chou, Lei, Li, Lu and Yao 1997: 193 (type species of *Aola*) China
Aola bindusara Chou, Lei, Li, Lu and Yao 1997: 15, 193–194, Fig 3-5, Fig 9-53, Plate X, Fig 103 (synonymy, described, illustrated, comp. note) Equals *Pomponia bindusara* China
Aola bindusara Hua 2000: 60 (listed, distribution) Equals *Pomponia bindusara* China, Burma, Laos, India
Aola bindusara Sanborn, Breitbarth, Heath and Heath 2002: 445 (comp. note)
Aola bindusara Boulard 2003c: 172, 170, 181, 183, Plate E, Figs 1–3 (described, illustrated, song, sonogram, comp. note) Equals *Pomponia bindusara* Thailand, Tenasserim, Burma, India, Laos

Aola bindusara Boulard 2005d: 336 (comp. note) Asia
Chremistica ochracea (sic) Boulard 2005o: 148 (listed) Equals *Aola bindusara* sonogram Boulard 2001
Aola bindusara Boulard 2005o: 149 (listed)
Aola bindusara Boulard 2006g: 32 (comp. note)
Aola bindusara Boulard 2007e: 23, 47, Plate 18, Plate 45a (illustrated, comp. note) Thailand
Aola bindusara Sanborn, Phillips and Sites 2007: 32 (listed, distribution, comp. note) Thailand, India, Burma, Laos, Indochina, China
Aola bindusara Boulard 2008f: 40, 94 (type species of *Aola*, biogeography, listed, comp. note) Equals *Pomponia bindusara* Thailand, Tenasserim, Burma, India, Indochina
Pomponia bindusara Lee 2008a: 14 (type species of *Aola*)
Aola bindusara Lee 2008a: 14 (comp. note)
Haphsa bindusara Lee 2008a: 15 (n. comb., synonymy, distribution, listed) Equals *Pomponia bindusara* Equals *Aola bindusara* North Vietnam, China, Yunnan, Laos, Thailand, Myanmar, India
Aola bindusara Lee 2009b: 330 (comp. note)
Haphsa bindusara Lee 2009b: 330–331, 335–336, Fig 3C, Fig 4C (illustrated, comp. note)
Haphsa bindusara Pham and Yang 2009: 14, Table 2 (listed) Vietnam
Aola bindusara Sun, Watson, Zheng, Watson and Liang 2009: 3148–3149, 3151, Fig 1G, Table 1 (illustrated, wing microstructure, comp. note) China, Yunnan
Haphsa bindusura (sic) Ahmed and Sanborn 2010: 44 (distribution) Bangladesh, Burma, India, Laos, Indochina, China, Thailand
Pomponia bindusara Lee 2010b: 24 (type species of *Aola*, comp. note)
Haphsa bindusara Lee and Hill 2010: 299–300, Figs 7–8 (phylogeny, comp. note)

H. crassa Distant, 1905 to *Meimuna crassa*

***H. conformis* Distant, 1917**

Haphsa conformis Lee 2008a: 14 (distribution, listed) North Vietnam
Haphsa conformis Lee 2009b: 330–331 (comp. note)
Haphsa conformis Pham and Yang 2009: 14, 16, Table 2 (listed, comp. note) Vietnam
Haphsa conformis Pham and Yang 2010a: 133 (comp. note) Vietnam

***H. dianensis* Chou, Lei, Li, Lu & Yao, 1997**

Haphsa dianensis Chou, Lei, Li, Lu and Yao 1997: 237 (n. sp., described, comp. note) China
Haphsa dianensis Lee 2009b: 330–331 (comp. note)

***H. durga* (Distant, 1881)** = *Cosmopsaltria durga* = *Meimuna durga* = *Dundubia durga*

Meimuna durga Lei 1994: 186, 188 (synonymy, measurements, distribution, comp. note) Equals *Cosmopsaltria durga* China, Yunnan, India, Laos
Meimuna durga Chou, Lei, Li, Lu and Yao 1997: 238, 244, 369, Plate XI, Fig 117 (key, described, illustrated, synonymy, comp. note) Equals *Cosmopsaltria durga* China
Meimuna durga Hua 2000: 62 (listed, distribution) Equals *Cosmopsaltria durga* China, Fujian, Guangdong, Yunnan, India, Thailand, Laos, Vietnam
Meimuna durga Boulard 2003a: 98–99, 101–103, Color Figs 3–4, sonogram 3–4 (described, allotype female, song, sonogram, illustrated, comp. note) Equals *Cosmopsaltria durga* Thailand, India, Laos, Yunnan
Meimuna durga Boulard 2005d: 342, Fig 25.18 (illustrated, sonogram, comp. note)
Meimuna durga Boulard 2005o: 149 (listed)
Meimuna durga Boulard 2006g: 61–63, Fig 25 (sonogram, illustrated, comp. note) Thailand
Meimuna durga Boulard 2007e: 4, 52, Figs 34–35 (sonogram, illustrated, comp. note) Thailand
Meimuna durga Boulard 2007f: 140, Fig 122 (illustrated, comp. note) Southeast Asia
Meimuna durga Sanborn, Phillips and Sites 2007: 19 (listed, distribution, comp. note) Thailand, India, Laos
Meimuna durga Boulard 2008f: 33, 93 (biogeography, listed, comp. note) Equals *Cosmopsaltria durga* Thailand, Assam, India, Laos
Meimuna durga group Lee 2008a: 13 (listed, comp. note) Vietnam
Meimuna durga Lee 2008a: 13 (synonymy, distribution, listed) Equals *Cosmopsaltria durga* Vietnam, China, Yunnan, Guangdong, Fujian, Laos, Thailand, India
Haphsa durga Boulard 2009f: 253 (n. comb., comp. note) Equals *Cosmopsaltria durga* Equals *Meimuna durga*
Meimuna durga Lee 2009b: 331 (comp. note)
Haphsa durga Lee 2009b: 331, 335–336, Fig 3A, Figs 4A (n. comb., illustrated, comp. note)
Meimuna durga Pham and Yang 2009: 14, Table 2 (listed) Vietnam
Meimuna durga Sun, Watson, Zheng, Watson and Liang 2009: 3148–3149, 3151, Fig 1F, Table 1 (illustrated, wing microstructure, comp. note) China, Yunnan
Haphsa durga Lee and Hill 2010: 299–300, Figs 7–8 (phylogeny, comp. note)

H. fratercula **Distant, 1917**
Haphsa fratercula Lee 2008a: 14 (distribution, listed) Vietnam
Haphsa fratercula Lee 2009b: 330–331 (comp. note)
Haphsa fratercula Pham and Yang 2009: 14, 16, Table 2 (listed, comp. note) Vietnam
Haphsa fratercula Pham and Yang 2010a: 133 (comp. note) Vietnam

H. jsguillotsi **(Boulard, 2005)** = *Haphsa jdguillotsi* (sic) = *Meimuna jsguillotsi* = *Meimuna guillotsi* (sic)
Meimuna jsguillotsi Boulard 2005e: 118, 121–124, Plate B, Figs 1–2, Figs 6–11 (n. sp., described, illustrated, song, sonogram, listed, comp. note) Thailand
Meimuna jsguillotsi Boulard 2005o: 151 (listed)
Meimuna jsguillotsi Sanborn, Phillips and Sites 2007: 19 (listed, distribution, comp. note) Thailand
Haphsa jdguillotsi Boulard 2008f: 32, 93 (biogeography, listed, comp. note) Thailand
Meimuna jsguillotsi Lee 2009b: 330 (comp. note)
Haphsa jsguillotsi Lee 2009b: 330–331 (comp. note)

H. karenensis **Ollenbach, 1928** = *Meimuna nauhkae* Boulard, 2005
Meimuna nauhkae Boulard 2005h: 7, 40–44, Figs 55–61, Color Plate, Fig D (n. sp., described, illustrated, song, sonogram, listed, comp. note) Thailand
Meimuna nauhkae Boulard 2005o: 150 (listed)
Meimuna nauhkae Boulard 2006g: 19–20, 74–75, 89, 109, Fig 53 (sonogram, comp. note) Thailand
Meimuna nauhkae Trilar 2006: 344 (comp. note)
Meimuna nauhkae Boulard 2007e: 47, Plate 13a (illustrated, comp. note) Thailand
Meimuna nauhkae Sanborn, Phillips and Sites 2007: 19 (listed, distribution, comp. note) Thailand
Haphsa karenensis Boulard 2008c: 364 (synonymy, comp. note) Equals *Meimuna nauhkae* Thailand
Haphsa karenensis Boulard 2008f: 32, 93 (synonymy, biogeography, listed, comp. note) Equals *Meimuna nauhkae* Thailand
Haphsa karenensis Lee 2009b: 330–331, 335–336, Fig 3B, Figs 4B (illustrated, comp. note)
Haphsa karenensis Lee and Hill 2010: 299–300, Figs 7–8 (phylogeny, comp. note)

H. nana **Distant, 1913**
Haphsa nana Lee 2008a: 14 (distribution, listed) Vietnam
Haphsa nana Lee 2009b: 330–331 (comp. note)
Haphsa nana Pham and Yang 2009: 14, 16, Table 2 (listed, comp. note) Vietnam
Haphsa nana Pham and Yang 2010a: 133 (comp. note) Vietnam

H. nicomache **(Walker, 1850)** = *Dundubia nicomache* = *Cosmopsaltria nicomache* = *Haphsa nicoache* (sic) = *Cicada delineata* Walker, 1858
Dundubia nicomache Chou, Lei, Li, Lu and Yao 1997: 236, Fig 9–76 (type species of *Haphsa*, illustrated, comp. note)
Haphsa nicomache Boulard 2008f: 32 (type species of *Haphsa*) India
Dundubia nicomache Lee 2008a: 14 (type species of *Haphsa*) North India
Haphsa nicomache Lee 2009b: 330–331 (type species of *Haphsa*, comp. note)
Haphsa nicomache Ahmed and Sanborn 2010: 27, 39 (synonymy, distribution, type species of *Haphsa*) Equals *Dundubia nicomache* Pakistan, India
Dundubia nicomache Lee 2010b: 24 (type species of *Haphsa*, comp. note)

H. opercularis **Distant, 1917**
Haphsa opercularis Chou, Lei, Li, Lu and Yao 1997: 237 (described, comp. note) China
Haphsa opercularis Lee 2008a: 14 (distribution, listed) Vietnam, China, Yunnan
Haphsa opercularis Lee 2009b: 330–331 (comp. note)
Haphsa opercularis Pham and Yang 2009: 14, Table 2 (listed) Vietnam

H. scitula **(Distant, 1888)** = *Pomponia scitula* = *Aola scitula* = *Pomponia (Aola) scitula*
Pomponia scitula Wu 1935: 15 (listed) Equals *Aola scitula* China, Yunnan, Burma, Annam, Cambodia
Aola scitula Chaudhry, Chaudry and Khan 1970: 104 (comp. note) India
Pomponia (Aola) scitula Chaudhry, Chaudry and Khan 1970: 104 (listed, comp. note) Pakistan, India
Pomponia scitula Chou, Lei, Li, Lu and Yao 1997: 188, 191–192, Fig 9–51 (key, synonymy, described, illustrated, comp. note) Equals *Aola scitula* China
Pomponia scitula Hua 2000: 64 (listed, distribution) Equals *Aola scitula* China, Xinjiang, Guangxi, Yunnan, Burma, India
Pomponia scitula Boulard 2003c: 172, 179–180, 183, Plate D, Figs 1–3 (described, illustrated, song, sonogram, comp. note) Equals *Aola scitula* Thailand, Assam, Burma, Yunnan, Tonkin
Aola scitula Srivastava and Bhandari 2004: 1479 (comp. note) India
Pomponia scitula Boulard 2005d: 336 (comp. note) Asia
Pomponia scitula Boulard 2005i: 60 (comp. note)
Pomponia scitula Boulard 2005o: 149 (listed)
Pomponia scitula Pham and Thinh 2005a: 240 (synonymy, distribution, listed) Equals *Aola scitula* Vietnam

Pomponia scitula Boulard 2006g: 32, 130 (comp. note)
Pomponia scitula Boulard 2007e: 23, Plate 18, Plate 45b (illustrated, comp. note) Thailand
Pomponia scitula Boulard 2007f: 140, Fig 123 (illustrated, comp. note) Thailand
Pomponia scitula Sanborn, Phillips and Sites 2007: 29–30 (listed, distribution, comp. note) Thailand, Burma, India, Vietnam, Cambodia, China, Indochina
Pomponia scitula Boulard 2008f: 40, 94 (biogeography, listed, comp. note) Equals *Aola scitula* Thailand, Burma, India, Assam, Indochina, yunnan, China
Pomponia scitula Lee 2008a: 14 (comp. note)
Haphsa scitula Lee 2008a: 14 (n. comb., synonymy, distribution, listed) Equals *Pomponia scitula* Equals *Aola scitula* North Vietnam, China, Xinjiang, Yunnan, Guangxi, Cambodia, Thailand, Myanmar, India
Meimuna scitula Lee 2009b: 330 (comp. note)
Haphsa scitula Lee 2009b: 330–331, 335–336, Fig 3F, Fig 4F (illustrated, comp. note)
Haphsa scitula Pham and Yang 2009: 14, Table 2 (listed) Vietnam
Pomponia scitula Sun, Watson, Zheng, Watson and Liang 2009: 3148–3149, 3151, 3154, Fig 1B, Table 1 (illustrated, wing microstructure, comp. note) China, Yunnan
Haphsa scitula Ahmed and Sanborn 2010: 44 (distribution) Bangladesh, Burma, India, Vietnam, Cambodia, Indochina, China, Thailand
Haphsa scitula Lee 2010b: 24–25 (synonymy, distribution, listed) Equals *Pomponia scitula* Equals *Aola scitula* Cambodia, China, Xinjiang, Yunnan, Guangxi, North Vietnam, Thailand, Myanmar, India

H. stellata **Lee, 2009**
Haphsa stellata Lee 2009b: 331–333, 335–336, Fig 1, Fig 3D, Fig 4D (n. sp., described, illustrated, comp. note) India

H. sulaiyai **(Boulard, 2005)** = *Aola sulaiyai*
Aola sulaiyai Boulard 2005i: 59–62, Figs 1–5 (n. sp., described, illustrated, comp. note) Thailand
Aola sulaiyai Boulard 2005o: 150 (listed)
Aola sulaiyai Sanborn, Phillips and Sites 2007: 32 (listed, distribution, comp. note) Thailand
Aola sulaiyai Boulard 2008f: 40, 94 (biogeography, listed, comp. note) Thailand

H. velitaris (Distant, 1897) to *Meimuna velitaris*

Kaphsa **Lee, 2012**

K. aculeus **(Lee, 2009)** = *Haphsa aculeus*
Haphsa aculeus Lee 2009b: 333–336, Fig 2, Fig 3E, Fig 4E (n. sp., described, illustrated, comp. note) Thailand

K. concordia **Lee, 2012**

Subtribe Cosmopsaltriina Kato, 1932
= Cosmopsaltriaria

Cosmopsaltriaria Jong and Duffels 1981: 53 (comp. note) Philippines, Sulawesi, New Guinea
Cosmopsaltriaria Duffels 1982: 156, 160 (comp. note)
Cosmopsaltriaria Jong 1982: 184 (comp. note)
Cosmopsaltriina Duffels 1983a: 492–498, Figs 1–2 (listed) Sulawesi, New Guinea, Queensland, Bismarck Archipelago, Solomon Islands, New Hebrides, Kusaie Island, Fiji and Tonga Islands
Cosmopsatriaria Duffels 1983b: 10, 39 (composition, comp. note)
Cosmopsaltriaria Duffels 1985: 275, 278 (comp. note)
Cosmopsaltriaria Duffels 1986: 320, 325, 330, 333 (listed, distribution, comp. note) Sulawesi, Maluku, New Guinea, Bismarck Archipelago, Solomon Islands, Vanuatu, Kusaie Island, Fiji and Tonga Islands
Cosmopsaltriaria Duffels 1988a: 10 (comp. note)
Cosmopsaltriaria Duffels 1988c: 187 (comp. note)
Cosmopsaltriaria Duffels 1988d: 8 (comp. note)
Cosmopsaltriaria Duffels 1989: 123 (comp. note)
Cosmopsaltriina Duffels 1990b: 64–65, 69–71, Table 1 (listed, comp. note)
Cosmopsaltriaria Moulds 1990: 31 (listed) Australia
Cosmopsaltriaria Duffels 1993: 1226–1227 (comp. note)
Cosmopsaltriaria Duffels and Boer 1990: 261 (comp. note)
Cosmopsaltriaria Boer 1995c: 201 (comp. note)
Cosmopsaltriaria Boer 1995d: 172–174, 200–203, 219, 221–225, 228–229, 233, 237, Fig 28 (distribution, diversity, phylogeny, comp. note)
Cosmopsaltriaria Boer and Duffels 1996a: 155, 157, 163, 167–171, Fig 1, Fig 11 (distribution, diversity, phylogeny, comp. note) Sulawesi, Maluku, Bismarck Archipelago, East Melanesia, Polynesian Archipelago, Solomon Islands, Vanuatu, Fiji, Tonga, Samoa
Cosmopsaltriaria Boer and Duffels 1996b: 298, 301–303, 306–309, 313–315, 320, 328, Fig 1, Fig 3 (distribution, diversity, phylogeny, comp. note)
Cosmopsaltriaria Boer 1997: 93 (comp. note)
Cosmposaltriaria Duffels 1997: 549, 551 (diversity, comp. note)
Cosmopsaltriaria Holloway 1997: 219–223, Figs 5–6 (distribution, phylogeny, comp. note) New Guinea, Bismarck Islands, Fiji, Samoa

Cosmposaltriaria Holloway 1998: 304–305, 307–309, 311, Fig 13 (distribution, phylogeny, comp. note) Melanesia, Bismarck Islands, Fiji, Samoa, New Guinea
Cosmopsaltriaria Holloway and Hall 1998: 9–10, Fig 3 (distribution, comp. note) Melanesia, Fiji, Caroline Islands, New Guinea
Cosmopsaltriaria Boer 1999: 115–116, 120 (comp. note) New Guinea, Maluku, Sulawesi
Cosmopsaltriaria Sanborn 1999a: 43 (listed)
Cosmopsatriaria Duffels and Zaidi 2000: 195, 197 (diversity, comp. note) Sulawesi, New Guinea, Samoa Islands, Moluccas
Cosmopsaltriaria Turner, Hovencamp and van Welzen 2001: 219, 225, Tables 1–2 (phylogeny, comp. note) Vanuatu, Fiji, Tonga
Cosmopsaltriina Duffels and Turner 2002: 235–240, 242–244, 248–253, 256, 258, Figs 13–16, Appendices 1–2 (phylogeny, distribution, comp. note) Equals Cosmopsaltriaria Sulawesi, Papua New Guinea, Samoa
Cosmopsaltriaria Duffels and Turner 2002: 250 (comp. note)
Cosmopsaltriaria Moulds and Carver 2002: 11 (listed) Australia
Cosmposaltriina Lee and Hayashi 2003a: 170 (listed, comp. note) Taiwan
Cosmopsaltriina Chen 2004: 37 (comp. note)
Cosmopsaltriina Moulds 2005b: 387, 391–392, 396, 429–433, Table 2 (described, distribution, listed, comp. note) Equals Cosmopsaltriaria
Cosmopsaltriina Duffels and Boer 2006: 536 (listed) New Guinea
Cosmopsaltriaria Shiyake 2007: 5, 66 (listed, comp. note)
Cosmopsaltriina Boulard 2008f: 7, 32 (listed) Thailand
Cosmposaltriina Lee 2008a: 1–2, 12, Table 1 (listed) Vietnam
Cosmposaltriina Lee 2008b: 446, 450, 452, 463, Table 1 (listed, comp. note) Korea
Cosmopsaltriina Moulds 2008a: 209 (key, comp. note)
Cosmopsaltriina Cranston 2009: 93 (distribution, comp. note) Fiji, Vanuatu, Tonga
Cosmposaltriina Lee 2009a: 87 (listed, comp. note)
Cosmopsaltriina Pham and Yang 2009: 14, Table 2 (listed) Vietnam
Cosmopsaltriina Lee 2010d: 167 (comp. note)
Cosmopsaltriina Lee and Hill 2010: 278, 302 (comp. note)
Cosmopsaltriina Wei, Ahmed and Rizvi 2010: 33 (comp. note)

***Lethama* Distant, 1905**
Lethama Duffels 1983b: 10, 11 (comp. note)
Lethama Duffels and Turner 2002: 235 (comp. note)
Lethama Moulds 2005b: 391, 432 (higher taxonomy, listed, comp. note)

***L. locusta* (Walker, 1850)** = *Cephaloxys locusta* = *Mogannia locusta* = *Dundubia locusta*
Lethama locusta Sueur 2002b: 119, 121–122, 129 (listed, comp. note) India

***Sinapsaltria* Kato, 1940**
Sinapsaltria Duffels and Turner 2002: 235 (comp. note)
Sinapsaltria Moulds 2005b: 391, 432 (higher taxonomy, listed, comp. note)
Sinapsaltria Lee 2008a: 2, 15, Table 1 (listed) Vietnam
Sinapsaltria Lee 2009b: 330 (comp. note)
Sinapsaltria Pham and Yang 2009: 14, Table 2 (listed) Vietnam

***S. annamensis* Kato, 1940**
Sinapsaltria annamensis Lee 2008a: 15 (distribution, listed) Vietnam
Sinapsaltria annamensis Pham and Yang 2009: 14, 16, Table 2 (listed, comp. note) Vietnam
Sinapsaltria annamensis Pham and Yang 2010a: 133 (comp. note) Vietnam

***S. typica* Kato, 1940**
Sinapsaltria typica Lee 2008a: 15 (type species of *Sinapsaltria*)

***Dilobopyga* Duffels, 1977**
Dilobopyga Duffels 1982: 156, 159–160 (biogeography, phylogeny, comp. note) Sulawesi
Dilobopyga Duffels 1983a: 492–495, 497, Figs 1–2, Fig 7 (diversity, distribution, phylogeny, comp. note) Sulawesi, Moluccas
Dilobopyga Duffels 1986: 320–322, 326, 329–330, Figs 1–3, Fig 8, Fig 10 (diversity, distribution, phylogeny, comp. note) Sulawesi, Maluku, Sangihe Island
Dilobopyga Duffels 1989: 123, 127 (comp. note) Sulawesi, Muna, Buton, Sangihe Island, Selayar, Banggai Archipelago, Sula Island, South Maluku
Dilobopyga Duffels 1990a: 323 (comp. note) Sulawesi, Muna, Buton, Sangihe Island, Selayar, Banggai Archipelago, Sula Island, South Maluku
Dilobopyga Duffels 1990b: 64–71, Figs 1–6, Table 1, Table 3 (diversity, distribution, phylogeny, comp. note) (diversity, distribution, phylogeny, comp. note) Sulawesi, Selayar, Banggai, Sula, Sangihe, Maluku
Dilobopyga Duffels and Boer 1990: 256–257, 260, 266, Fig 21.1, Fig 21.8 (distribution,

diversity, phylogeny, comp. note) Sulawesi, Selayar, Banggai, Sula, Sangihe, Maluku

Dilobopyga Duffels 1993: 1227, 1230, 1232, Fig 6 (phylogeny, comp. note)

Dilobopyga Boer 1995d: 172, 219, 221–224, 228–229, Fig 51 (distribution, phylogeny, comp. note) Sulawesi

Dilobopyga Boer and Duffels 1996a: 155, 167, 169, 171, Fig 9, Fig 11 (distribution, phylogeny, comp. note) Sulawesi

Dilobopyga Boer and Duffels 1996b: 329 (distribution)

Dilobopyga Duffels 1997: 549, 552 (comp. note)

Dilobopyga Holloway 1997: 221, 224 (distribution, comp. note) Sulawesi

Dilobopyga Holloway 1998: 305, 307–308, Fig 10, Fig 13 (distribution, phylogeny, comp. note) Sulawesi, Moluccas

Dilobopyga Holloway and Hall 1998: 10, Fig 3 (distribution, comp. note)

Dilobopyga Duffels 1999: 81 (diversity, comp. note) Sulawesi

Dilobopyga Duffels and Zaidi 2000: 198 (comp. note)

Dilobopyga Duffels and Turner 2002: 236, 239, 243–245, 247, 249, 257, Fig 10, Figs 13–15 (phylogeny, distribution, illustrated, listed, comp. note) Sulawesi, Maluku

Dilobopyga Moulds 2005b: 391, 400, 432–433 (higher taxonomy, listed, comp. note)

Dilobopyga Shiyake 2007: 5, 66 (listed, comp. note)

Dilobopyga Lee 2009a: 87 (listed, diversity)

***D. alfura alfura* (Breddin, 1900)** = *Cosmopsaltria alfura* = *Meimuna alfura*

Dilobopyga alfura Duffels 1990b: 65–67, Fig 2, Fig 6 (phylogeny, comp. note) Sulawesi

Dilobppyga alfura Duffels and Boer 1990: 257, Fig 21.1 (phylogeny) Sulawesi

Dilobopyga alfura Boer and Duffels 1996b: 329 (distribution) Sulawesi

Dilobopyga alfura Duffels and Turner 2002: 248, 256, 259, Fig 11, Appendices 1–2 (phylogeny, distribution, listed) Sulawesi

Dilobopyga alfura Lee 2009a: 87 (comp. note)

***D. alfura* var. ______ (Breddin, 1901)** = *Cosmopsaltria alfura* var. ______ = *Meimuna alfura* var. ______.

***D. aprina* Lee, 2009**

Dilobopyga aprina Lee 2009a: 87–89, Figs 1–3 (n. sp., described, illustrated, comp. note) Sulawesi

***D. breddini* (Duffels, 1970)** = *Diceropyga breddini* = *Cosmopsaltria opercularis* Breddin, 1901 (nec Walker)

Dilobopyga breddini Duffels 1990b: 65–66, 69, Fig 2, Fig 6 (phylogeny, comp. note) Sulawesi

Dilobppyga breddini Duffels and Boer 1990: 257, Fig 21.1 (phylogeny) Sulawesi

Dilobopyga breddini Boer and Duffels 1996b: 329 (distribution) Sulawesi

Dilobopyga breddini Duffels 1999: 81–82 (comp. note) Sulawesi

Dilobopyga breddini Duffels and Turner 2002: 240, 248, 256, 259, Fig 3B, Fig 11, Appendices 1–2 (illustrated, phylogeny, distribution, listed, comp. note) Sulawesi

Dilobopyga breddini Lee 2009a: 87 (comp. note)

***D. chlorogaster* (Boisduval, 1835)** = *Cicada chlorogaster* = *Dundubia chlorogaster* = *Cosmopsaltria chlorogaster* = *Cosmopsaltria chlorogastri* (sic) = *Diceropyga chlorogaster* = *Dundubia maculosa* Walker, 1858 = *Diceropyga maculosa* = *Dundubia fuliginosa* (nec Walker)

Dilobopyga chlorogaster Duffels 1990b: 65–66, Fig 2, Fig 6 (phylogeny) Sulawesi

Dilobopyga chlorogaster group Duffels 1990b: 65, 67, Fig 2 (phylogeny, comp. note) Sulawesi

Dilobppyga chlorogaster Duffels and Boer 1990: 257, Fig 21.1 (phylogeny) Sulawesi

Dilobppyga chlorogaster group Duffels and Boer 1990: 257, Fig 21.1 (diversity, distribution, phylogeny) Sulawesi, Maluku, Selayar Island

Dilobopyga chlorogaster Boulard 1991c: 119 (comp. note) Equals *Cicada chlorogaster* Vanikoro

Dilobopyga chlorogaster Boer and Duffels 1996b: 329 (distribution) Sulawesi

Dilobopyga chlorogaster Duffels and Zaidi 2000: 276 (comp. note) Equals *Dundubia fuliginosa* (partim) Equals *Cosmopsaltria fuliginosa* (partim)

Dilobopyga chlorogaster Duffels and Turner 2002: 238, 240, 246–248, 256, 259, Fig 10C, Fig 11, Appendices 1–2 (illustrated, phylogeny, distribution, listed, comp. note) Sulawesi

Dilobopyga chlorogaster Lee 2009a: 87 (type species of *Dilobopyga*, comp. note)

***D. gemina gemina* (Distant, 1888)** = *Cosmopsaltria gemina* = *Diceropyga gemina* = *Diceropyga minahassae* (nec Distant) = *Dundubia vaginata* Walker, 1868 (nec Fabricius) = *Dundubia vibrans* Walker, 1868 (nec Walker, 1850)

Diceropyga gemina Duffels 1983a: 494 (comp. note) Moluccas, Buru, Seram, Gorong, Saparua, Mysol?

Dilobopyga gemina gemina Duffels 1986: 321 (comp. note) Seram, Sarapua, Gorong

Dilobopyga gemina Duffels 1986: 323 (comp. note) Maluku

Dilobopyga gemina Duffels 1990b: 65–67, 69, Fig 2, Fig 6 (phylogeny, comp. note) Maluku, Buru, Seram, Ambon

Dilobppyga gemina Duffels and Boer 1990: 257, 260, Fig 21.1 (phylogeny, comp. note) Maluku

Dilobppyga gemina gemina Duffels and Boer 1990: 260 (distribution) Seram, Saparoa, Goram, Misoöl

Dilobopyga gemina Boer 1995d: 221, 228, 230 (distribution, comp. note) Maluku, Buru, Gorong, Sapuru, Seram

Dilobopyga gemina gemina Boer and Duffels 1996b: 329 (distribution) South Maluku

Dilobopyga gemina Duffels and Turner 2002: 238, 240–241, 247, Fig 2D, Fig 4F (illustrated, comp. note)

Dilobopyga gemina gemina Duffels and Turner 2002: 248, 256, 259, Fig 11, Appendices 1–2 (phylogeny, distribution, listed) Maluku

Dilobopyga gemina Lee 2009a: 87 (comp. note)

***D. gemina toxopei* (Schmidt, 1926)** = *Diceropyga toxopei* = *Dicercopyga* (sic) *roteri* Schmidt, 1926 = *Diceropyga roteri*

Dilobopyga gemina toxopei Duffels 1986: 321 (comp. note) Buru

Dilobppyga gemina toxopei Duffels and Boer 1990: 260 (distribution) Buru

Dilobopyga gemina toxopei Boer and Duffels 1996b: 329 (distribution) South Maluku

Dilobopyga gemina toxopei Duffels and Turner 2002: 247–248, 256, 259, Fig 10D, Fig 11, Appendices 1–2 (illustrated, phylogeny, distribution, listed, comp. note) Maluku

***D. janskocki* Duffels, 1990**

Dilobopyga janstocki Duffels 1982: 323–327, Figs 1–6 (n. sp., described, illustrated, comp. note) Sulawesi

Dilobopyga janstocki Boer and Duffels 1996b: 329 (distribution) Sulawesi

Dilobopyga janstocki Duffels 1999: 81–82 (comp. note) Sulawesi

Dilobopyga janstocki Lee 2009a: 87 (comp. note)

***D. margarethae margarethae* Duffels, 1977** = *Diceropyga gemina* (nec Distant) = *Diceropyga chlorogaster* (nec Boisduval)

Dilobopyga margarethae Duffels 1989: 123 (comp. note) Sulawesi

Dilobopyga margarethae Duffels 1990b: 65–67, 69, Fig 2, Fig 6 (phylogeny, comp. note) Sulawesi

Dilobppyga margarethae Duffels and Boer 1990: 257, Fig 21.1 (phylogeny) Sulawesi

Dilobopyga margarethae margarethae Boer and Duffels 1996b: 329 (distribution) Sulawesi

Dilobopyga margarethae Duffels and Turner 2002: 238, 240, 242, 247, Fig 3C, Fig 5A (illustrated, comp. note)

Dilobopyga margarethae margarethae Duffels and Turner 2002: 248, 256, 259, Fig 11, Appendices 1–2 (phylogeny, distribution, listed) Sulawesi

Dilobopyga margarethae Lee 2009a: 87 (comp. note)

***D. margarethae parvula* Duffels, 1977**

Dilobopyga margarethae parvula Boer and Duffels 1996b: 329 (distribution) Sulawesi

Dilobopyga margarethae parvula Duffels and Turner 2002: 248, 256, 259, Fig 11, Appendices 1–2 (phylogeny, distribution, listed) Sulawesi

***D. minahassae* (Distant, 1888)** = *Cosmopsaltria minahasae* (sic) = *Cosmopsaltria minahassae* = *Diceropyga minabassae* (sic) = *Diceropyga minahassae* = *Diceropyga gemina* auct. (partim) (nec Distant, 1888)

Dilobopyga minahassae Duffels 1983a: 494 (comp. note) Sulawesi, Moena Island

Dilobopyga minahassae Duffels 1989: 123 (comp. note) Sulawesi

Dilobopyga minahassae Duffels 1990b: 64–65, 68, Fig 2 (phylogeny, comp. note) Sulawesi, Sangihe Island

Dilobopyga minahassae group Duffels 1990b: 67–69, Fig 7 (distribution, comp. note)

Dilobppyga minahassae Duffels and Boer 1990: 257, Fig 21.1 (phylogeny) Sulawesi, Sangihe Island

Dilobppyga minahassae group Duffels and Boer 1990: 257, Fig 21.1 (phylogeny) Sulawesi, Sangihe Island

Dilobopyga minahassae Boer and Duffels 1996b: 329 (distribution) Sulawesi

Dilobopyga minahassae Duffels and Turner 2002: 240, 245, 247–248, 256, 259, Fig 5C, Fig 10B, Fig 11, Appendices 1–2 (phylogeny, distribution, listed, comp. note) Sulawesi

Dilobopyga minahassae Lee 2009a: 87 (comp. note)

***D. multisignata* (Breddin, 1901)** = *Cosmopsaltria multisignata* = *Diceropyga* (?) *multisignata*

Dilobopyga multisignata Duffels 1990b: 65–66, Fig 2, Fig 6 (phylogeny) Sulawesi

Dilobppyga multisignata Duffels and Boer 1990: 257, Fig 21.1 (phylogeny) Sulawesi

Dilobopyga multisignata Boer and Duffels 1996b: 329 (distribution) Sulawesi

Dilobopyga multisignata Duffels and Turner 2002: 238, 240, 246–248, 256, 259, Fig 11, Appendices 1–2 (phylogeny, distribution, listed, comp. note) Sulawesi

Dilobopyga multisignata Lee 2009a: 87 (comp. note)

***D.* n. sp. Duffels, 1990**

Dilobopyga n. sp. Duffels 1990b: 65–66, 68, 70, Figs 27 (distribution, phylogeny) Sulawesi, Banggai Island, Sula Island, Sangihe Island, Selyar Island

Dilobppyga n. sp. Duffels and Boer 1990: 257, 262, Fig 21.1, Fig 21.4 (distribution, phylogeny)

Sulawesi, Banggai Island, Sula Island, Selayar Island

***D. opercularis* (Walker, 1858)** = *Dundubia insularis* Waker, 1858 = *Cosmopsaltria insularis* = *Diceropyga insularis*

Dilobopyga opercularis group Duffels 1990a: 323 (comp. note) Sulawesi

Dilobopyga opercularis Duffels 1990b: 65–66, Fig 2, Fig 6 (phylogeny) Sulawesi

Dilobopyga opercularis group Duffels 1990b: 65, 67, 69, Fig 2 (phylogeny, comp. note)

Dilobppyga opercularis Duffels and Boer 1990: 257, Fig 21.1 (phylogeny) Sulawesi

Dilobppyga opercularis group Duffels and Boer 1990: 257, Fig 21.1 (diversity, phylogeny) Sulawesi, Sangihe Island, Banggai Island, Sula Island

Dilobopyga opercularis Boer and Duffels 1996b: 329 (distribution) Sulawesi

Dilobopyga opercularis Duffels 1999: 81–82 (comp. note) Sulawesi

Dilobopyga opercularis Duffels and Turner 2002: 241, 248, 256, 259, Fig 4G, Fig 11, Appendices 1–2 (illustrated, phylogeny, distribution, listed, comp. note) Sulawesi

Dilobopyga opercularis Shiyake 2007: 5, 64, 66, Fig 100 (illustrated, distribution, listed, comp. note)

Dilobopyga opercularis Lee 2009a: 87 (comp. note)

***D. ornaticeps* (Breddin, 1901)** = *Cosmopsaltria ornaticeps* = *Diceropyga ornaticeps*

Dilobopyga ornaticeps Duffels 1990b: 65–66, Fig 2, Fig 6 (phylogeny) Sulawesi

Dilobppyga ornaticeps Duffels and Boer 1990: 257, Fig 21.1 (phylogeny) Sulawesi

Dilobopyga ornaticeps Boer and Duffels 1996b: 329 (distribution) Sulawesi

Dilobopyga ornaticeps Duffels and Turner 2002: 238, 240, 246–248, 256, 259, Fig 11, Appendices 1–2 (phylogeny, distribution, listed, comp. note) Sulawesi

Dilobopyga ornaticeps Lee 2009a: 87 (comp. note)

***D. remanei* Duffels, 1999**

Dilobopyga remanei Duffels 1999: 81–86, Figs 1–6 (n. sp. described, illustrated, comp. note) Sulawesi

Dilobopyga remanei Lee 2009a: 87 (comp. note)

***D. similis* Duffels, 1977** = *Cosmopsaltria opercularis* Breddin (nec Walker) (partim)

Dilobopyga similis Duffels 1990b: 65–67, 69, Fig 2, Fig 6 (phylogeny, comp. note) Sulawesi

Dilobppyga similis Duffels and Boer 1990: 257, Fig 21.1 (phylogeny) Sulawesi

Dilobopyga similis Boer and Duffels 1996b: 329 (distribution) Sulawesi

Dilobopyga similis Duffels 1999: 81–82 (comp. note) Sulawesi

Dilobopyga similis Duffels and Turner 2002: 247–248, 256, 259, Fig 10A, Fig 11, Appendices 1–2 (illustrated, phylogeny, distribution, listed, comp. note) Sulawesi

Dilobopyga similis Lee 2009a: 87 (comp. note)

***Brachylobopyga* Duffels, 1982**

Brachylobopyga Duffels 1982: 156, 159–160 (n. gen., described, biogeography, phylogeny, comp. note) Sulawesi

Brachylobopyga Duffels 1983a: 492–495, 497, Figs 1–2, Fig 7 (diversity, distribution, phylogeny, comp. note) Sulawesi

Brachylobopyga Duffels 1986: 320–322, 326, 329–330, Figs 1–3, Fig 8, Fig 10 (diversity, distribution, phylogeny, comp. note) Sulawesi

Brachylobopyga Duffels 1989: 123, 127 (key, biogeography, comp. note) Sulawesi

Brachylobopyga Duffels 1990b: 64–65, 68, 70–71, Fig 1, Table 1 (diversity, distribution, phylogeny, comp. note) Sulawesi

Brachylobopyga Duffels and Boer 1990: 256, 266, Fig 21.8 (distribution, diversity, phylogeny, comp. note) Sulawesi

Brachylobopyga Duffels 1993: 1227, 1232, Fig 6 (phylogeny, comp. note)

Brachylobopyga Boer 1995d: 219, 222–224, 228–229, Fig 51 (distribution, phylogeny, comp. note) Sulawesi

Brachylobopyga Boer and Duffels 1996a: 155, 167, 169, 171, Fig 9, Fig 11 (distribution, phylogeny, comp. note) Sulawesi

Brachylobopyga Boer and Duffels 1996b: 301, 304, 306, 308, 313–314, 328, Fig 4 (distribution, phylogeny, comp. note) Sulawesi

Brachylobopyga Duffels 1997: 549, 552 (comp. note)

Brachylobopyga Holloway 1997: 221, 224 (distribution, comp. note) Sulawesi

Brachylobopyga Holloway 1998: 307–308, Fig 13 (distribution, phylogeny, comp. note) Sulawesi

Brachylobopyga Holloway and Hall 1998: 10, Fig 3 (distribution, comp. note)

Brachylobopyga Duffels 1999: 81 (comp. note) Sulawesi

Brachylobopyga Duffels and Turner 2002: 236, 238–239, 243–245, 247, 249, 257, Figs 13–15 (phylogeny, distribution, listed, comp. note) Sulawesi

Brachylobopyga Moulds 2005b: 391, 431–433 (higher taxonomy, listed, comp. note)

B. decorata Duffels, 1982 to *Brachylobopyga toradja* (Breddin, 1901)

***B. montana* Duffels, 1988**

Brachylobopyga montana Duffels 1989: 123–12, Figs 1–8 (n. sp., key, described, illustrated, comp. note) Sulawesi

Brachylobopyga montana Boer 1995d: 221 (distribution, comp. note) Sulawesi
Brachylobopyga montana Boer and Duffels 1996b: 328 (distribution) Sulawesi
Brachylopyga montana Duffels and Turner 2002: 248, 257, 258, Fig 11, Appendices 1–2 (phylogeny, distribution, listed) Sulawesi

***B. toradja* (Breddin, 1901)** = *Cicada toradja* = *Cosmopsaltria toradja* = *Tibicen toradja* = *Tibicen toradjus* = *Brachylobopyga decorata* Duffels, 1982
Brachylobopyga decorata Duffels 1982: 156–159, Figs 1–8 (n. sp., type species of *Brachylobopyga*, described, illustrated, comp. note) Sulawesi
Brachylobopyga decorata Duffels 1989: 123–124 (type species of *Brachylobopyga*, comp. note) Equals *Brachylobopyga toradja* Breddin
Brachylobopyga toradja Duffels 1989: 123 (key, n. comb., synonymy, comp. note) Equals *Cicada toradja* Equals *Cosmopsaltria toradja* Equals *Tibicen toradjus* Equals *Brachylobopyga decorata* n.syn. Sulawesi
Cicada toradja Duffels 1989: 124 (comp. note)
Brachylobopyga decorata Duffels 1990b: 68 (comp. note) Sulawesi
Brachylobopyga toradja Boer 1995d: 221 (distribution, comp. note) Sulawesi
Brachylobopyga toradja Boer and Duffels 1996b: 328 (distribution) Sulawesi
Brachylopyga toradja Duffels and Turner 2002: 238, 241, 245, 248, 257, 258, Fig 2E, Fig 4H, Fig 8A, Fig 11, Appendices 1–2 (illustrated, phylogeny, distribution, listed, comp. note) Sulawesi

Fatima Distant, 1905 to *Cosmopsaltria* Stål, 1866
F. capitata (Distant, 1888) to *Cosmopsaltria capitata*
F. loriae (Distant, 1897) to *Cosmopsaltria loriae*

Sawda Distant, 1905 to *Cosmopsaltria* Stål, 1866
S. froggatti Distant, 1911 to *Cosmopsaltria capitata* Distant, 1888
S. gestroei Distant, 1905 to *Cosmopsaltria gestroei*
S. mimica (Distant, 1897) to *Cosmopsaltria mimica*
S. pratti Distant, 1905 to *Cosmopsaltria gigantea gigantea* (Distant, 1897)
S. sharpi Distant, 1905 to *Cosmopsaltria gigantea gigantea* (Distant, 1897)
S. vitiensis Distant, 1906 to *Cosmopsaltria vitiensis*

***Cosmopsaltria* Stål, 1866** = *Cosmopsaltria (Cosmopsaltria)* = *Cosmopsatria* (sic) = *Cosmoscarta* (sic) = *Dundubia (Cosmopsaltria)* = *Fatima* Distant, 1905 = *Sawda* Distant, 1905
Cosmopsaltria Distant 1906: 37 (comp. note)
Cosmopsaltria Wu 1935: 12 (listed) China
Cosmopsaltria Duffels 1982: 159–160 (biogeography, phylogeny, comp. note) New Guinea
Cosmopsaltria Duffels 1983a: 492–494, 496–497, Figs 1–3, Fig 7 (diversity, distribution, phylogeny, comp. note) Moluccas, New Guinea, Bismarck Archipelago
Cosmopsaltria Duffels 1983b: 2–4, 6, 8–24, 27, 32–37, 39–43, 123, Figs 2–8 (described, illustrated, synonymy, phylogeny, distribution, diversity, comp. note) Equals *Fatima* Equals *Sawda* New Guinea, Bismarck Archipelago, Aru Island, Moluccas, Fiji Islands
Fatima Duffels 1983b: 2, 8–10, 43 (synonymy, comp. note)
Sawda Duffels 1983b: 2, 8–10, 43 (synonymy, comp. note)
Cosmopsaltria Strümpel 1983: 18 (listed)
Cosmopsaltria Hayashi 1984: 58 (listed)
Cosmopsaltria Duffels 1986: 320–322, 325, 330–331, Figs 1–3, Fig 8, Fig 10 (distribution, diversity, phylogeny, comp. note) New Guinea, Bismarck Archipelago
Cosmopsaltria Duffels 1988b: 23 (comp. note)
Cosmopsaltria Duffels 1988c: 187 (comp. note) New Guinea, Moluccas, Aru Islands, Bismarck Archipelago, Tjendrawasih
Cosmopsaltria Duffels 1988d: 12, 14, 75, 77, 80, Fig 8, Table 1 (distribution, listed, comp. note) Equals *Fatima* Equals *Sawda* New Guinea, Fiji Islands
Sawda Duffels 1988d: 77 (comp. note)
Cosmopsaltria Duffels 1990b: 65, Fig 1 (phylogeny, comp. note) Maluku, New Guinea
Cosmopsaltria Duffels and Boer 1990: 266, Fig 21.8 (phylogeny) New Guinea, Maluku
Cosmopsaltria Duffels and van Mastrigt 1991: 174, 178–179 (predation, comp. note)
Cosmopsaltria Liu 1991: 258 (comp. note)
Sawda Duffels 1993: 1226 (comp. note) New Guinea
Cosmopsaltria Duffels 1993: 1226–1227, 1230, 1232, Fig 6 (phylogeny, comp. note) Equals *Sawda*
Cosmopsaltria Boer 1994b: 161, 163 (comp. note)
Cosmopsaltria Boer 1995d: 172, 202, 204–206, 215, 222–225, 230, 232–233, Fig 51 (distribution, phylogeny, comp. note) New Guinea
Cosmopsaltria Lee 1995: 23 (listed)
Cosmopsaltria Beuk 1996: 130 (comp. note)
Cosmopsaltria Boer 1996: 358 (comp. note)
Cosmopsaltria Boer and Duffels 1996a: 155, 163–167, 169–172, Fig 6, Fig 9, Fig 11 (distribution, phylogeny, comp. note) New Guinea

Cosmopsaltria Boer and Duffels 1996b: 301, 306, 308, 313–314, 316, 328 (distribution, phylogeny, comp. note) New Guinea
Cosmopsaltria Chou, Lei, Li, Lu and Yao 1997: 358 (comp. note)
Cosmopsaltria Duffels 1997: 549, 552 (comp. note)
Cosmopsaltria Novotny and Wilson 1997: 437 (listed)
Cosmopsatria (sic) sp. Chen, Wongsiri, Jamyanya, Rinderer, Vongsamanode, et al. 1998: 27 (predation) Thailand
Cosmopsaltria Beuk 1998: 147 (comp. note)
Cosmopsaltria Duffels and Zaidi 1998: 320 (comp. note)
Cosmopsaltria Holloway 1998: 304–308, 311, Fig 10, Figs 12–13 (distribution, phylogeny, comp. note) New Guinea
Cosmopsaltria Holloway and Hall 1998: 9–10, Fig 3 (distribution, comp. note) New Guinea
Cosmopsaltria Brown and Toft 1999: 150 (comp. note)
Cosmopsaltria Boer 2000: 2, 7 (comp. note) Papua New Guinea
Cosmopsaltria Duffels and Zaidi 2000: 196–198, 209 (diversity, comp. note) Sumatra, Borneo, Malayan Peninsula
Cosmopsaltria Heads 2001: 902 (distribution, comp. note) New Guinea
Cosmopsaltria Duffels and Turner 2002: 235–237, 239–240, 242–244, 247, 249–253, 257, Figs 14–15 (distribution, listed, comp. note) New Guinea, Moluccas
Sawda Duffels and Turner 2002: 235 (comp. note)
Cosmposaltria Heads 2002b: 286–287, Fig 2 (distribution, comp. note) New Guinea
Cosmopsaltria Clarke, Balagawi, Clifford, Drew, Leblanc, et al. 2004: 151 (comp. note) Papua New Guinea
Cosmopsaltria Ahmed, Siddiqui, Akhter and Rizvi 2005: 843 (distribution) Oriental Region
Cosmopsaltria Moulds 2005b: 387–389, 391, 407, 412–413, 423, 430, 432, Table 2 (type genus of Cosmopsaltriina, higher taxonomy, listed, comp. note)
Cosmopsaltria Duffels and Boer 2006: 537 (distribution, comp. note) New Guinea, Kai Island
Cosmopsaltria Shiyake 2007: 5, 66 (listed, comp. note)
Cosmopsaltria sp. Shiyake 2007: 5, 65–66, Fig 104 (illustrated, listed, comp. note)
Cosmopsaltria sp. Shiyake 2007: 5, 65–66, Fig 105 (illustrated, listed, comp. note)
Cosmopsaltria Lee 2009b: 330 (comp. note)
Cosmopsaltria Lee 2009c: 338 (comp. note)
Cosmopsaltria Lee and Hill 2010: 302 (comp. note)

C. alticola Distant, 1905 to *Orientopsaltria alticola*
C. alticola pontianaka Kirkaldy, 1913 to *Orientopsaltria alticola*
C. andersoni Distant, 1883 to *Dundubia oopaga* (Distant, 1881)

***C. aurata* Duffels, 1983**

Cosmopsaltria aurata Duffels 1983a: 495, Fig 4 (distribution)
Cosmopsaltria aurata Duffels 1983b: 18–19, 21, 23, 34, 38, 42, 45, 73, 75–77, 123, Fig 12, Fig 20, Fig 61, Figs 69–71, Figs 73–78, Plate 3B-C, Table 3, Table 5 (n. sp., described, illustrated, key, distribution, listed, comp. note) New Guinea
*Cosmopsaltria aurata*Duffels 1986: 322, Fig 4 (distribution) New Guinea
Cosmopsaltria aurata Duffels and van Mastrigt 1991: 176, 179 (comp. note) Irian Jaya
Cosmopsaltria aurata Schmid and Sands 1991: 69–71, Fig 3 (illustrated, fungal host, life cycle, comp. note) New Guinea
Cosmopsaltria aurata Boer 1995d: 204, Fig 31 (distribution, comp. note) New Guinea
Cosmopsaltria aurata Boer and Duffels 1996a: 165, Fig 6 (distribution)
Cosmopsaltria aurata Boer and Duffels 1996b: 317, 328, Fig 10 (distribution, comp. note) NewGuinea
Cosmopsaltria aurata Sanborn 1999a: 43 (type material, listed)
Cosmopsaltria aurata Duffels and Turner 2002: 237, 248, 257–258, Fig 11, Appendices 1–2 (phylogeny, distribution, listed, comp. note) New Guinea
Cosmopsaltria aurata Mogia 2002: 18 (human food) New Guinea
Cosmopsaltria aurata Moulds 2009a: 503–504, Fig 14.4(9) (illustrated, comp. note) Papua New Guinea

***C. bloetei* Duffels, 1965**

Cosmopsaltria bloetei Duffels 1983a: 495, Fig 4 (distribution)
Cosmopsaltria bloetei Duffels 1983b: 4, 8, 12, 18–19, 21, 33–34, 44, 52, 92–93, 97, Fig 12, Fig 20, Fig 90, Plate 4E, Table 1, Table 4 (described, illustrated, key, distribution, listed, comp. note) New Guinea
Cosmopsaltria bloetei Duffels 1986: 322, Fig 4 (distribution) New Guinea
Cosmopsaltria bloetei Boer 1995d: 204, Fig 31 (distribution, comp. note)
Cosmopsaltria bloetei Boer 1996: 358 (comp. note)
Cosmopsaltria bloetei Boer and Duffels 1996a: 165, Fig 6 (distribution)
Cosmopsaltria bloetei Boer and Duffels 1996b: 317, 328, Fig 10 (distribution, comp. note) New Guinea

Cosmopsaltria bloetei Duffels and Turner 2002: 237, 239, 247–248, 257–258, Fig 11, Appendices 1–2 (phylogeny, distribution, listed, comp. note) New Guinea

C. brooksi Moulton, 1923 to *Orientopsaltria brooksi*

***C. capitata* Distant, 1888** = *Fatima capitata* = *Sawda froggatti* Distant, 1911 = *Cosmopsaltria froggatti*

Cosmopsaltria capitata Duffels 1983a: 493–495, Fig 4 (distribution, diversity, comp. note) New Guinea, Aru Island

Cosmopsaltria capitata group Duffels 1983a: 494–496, Fig 4 (distribution, comp. note)

Cosmopsaltria capitata Duffels 1983b: 12, 15–16, 18, 21, 35–36, 39, 46, 98–105, 108–109, Figs 9–12, Figs 96–103, Plate 5B-C, Tables 1–3, Table 5 (n. comb., lectotype designation, synonymy, described, illustrated, key, distribution, listed, comp. note) Equals *Fatima capitata* Equals *Sawda capitata* New Guinea, Aru Island

Fatima capitata Duffels 1983b: 12, 99, Table 1 (synonymy, listed, comp. note)

Sawda frogatti Duffels 1983b: 12, 99–100, 102, Table 1 (synonymy, comp. note)

Cosmopsaltria capitata group Duffels 1983b: 12, 17, 19–20, 22–23, 35–37, 40–56, 75, 99, 104, Fig 8, Table 1, Table 5 (described, phylogeny, comp. note)

Cosmopsaltria capitata group Duffels 1986: 322, 324, Fig 4 (distribution, comp. note) New Guinea

Cosmopsaltria capitata Duffels 1986: 322, Fig 4 (distribution, comp. note) New Guinea

Cosmopsaltria capitata group Duffels and Boer 1990: 261 (comp. note) New Guinea

Cosmopsaltria capitata Boer 1995d: 204–205, Fig 31 (distribution, comp. note) New Guinea

Cosmopsaltria capitata group Boer 1995d: 205 (distribution, comp. note) New Guinea

Cosmopsaltria capitata Boer and Duffels 1996a: 165, Fig 6 (distribution)

Cosmopsaltria capitata Boer and Duffels 1996b: 317, 328, Fig 10 (distribution, comp. note) New Guinea

Cosmopsaltria capitata Duffels and Turner 2002: 242, 247–248, 252, 257–258, Fig 11, Appendices 1–2 (phylogeny, distribution, listed, comp. note) New Guinea

C. ceslaui Lallemand & Synave, 1953 to *Orientopsaltria ceslaui*

***C. delfae* Duffels, 1983**

Cosmopsaltria delfae Duffels 1983a: 495–495, Fig 4 (distribution, comp. note) New Guinea, Waigeo

Cosmopsaltria delfae Duffels 1983b: 12, 18, 21, 23, 34–36, 46, 98–100, 104–107, Fig 12, Fig 20, Figs 103–109, Plate 5D, Table 1, Table 3, Table 5 (n. sp., described, illustrated, key, distribution, listed, comp. note) New Guinea, Waigeo, Salawati

Cosmopsaltria delfae Duffels 1986: 322, Fig 4 (distribution) New Guinea

Cosmopsaltria delfae Boer 1995d: 204–205, Fig 31 (distribution, comp. note) New Guinea, Aru Island

Cosmopsaltria delfae Boer and Duffels 1996a: 165, Fig 6 (distribution)

Cosmopsaltria delfae Boer and Duffels 1996b: 317, 328, Fig 10 (distribution, comp. note) New Guinea

Cosmopsaltria delfae Duffels and Turner 2002: 242–243, 247–248, 257–258, Fig 6D, Fig 11, Appendices 1–2 (illustrated, phylogeny, distribution, listed, comp. note) New Guinea

C. divergens Distant, 1917 to *Macrosemia divergens*

***C. doryca* (Boisduval, 1835)** = *Cicada doryca* = *Cicada dorei* (sic) = *Dundubia doryca* = *Dundubia dorei* (sic) = *Dundubia doryca*

Cosmopsaltria doryca Distant 1906: 37 (type species of *Cosmopsaltria*, comp. note)

Cosmopsaltria doryca Duffels 1983a: 493–494, Fig 3 (distribution, diversity) Moluccas, New Guinea, Ternate, Waigeo, Japen Island

Cosmopsaltria doryca group Duffels 1983a: 494, 497, Fig 7 (phylogeny, comp. note) New Guinea

Cosmopsaltria doryca Duffels 1983b: 4, 8, 12, 15–18, 21, 23, 29, 32, 34, 39, 45, 47–52, 55, 123, Figs 9–12, Fig 20, Figs 22–29, Plate 1A-B, Tables 1–3, Table 5 (n. sp., described, illustrated, type species of *Cosmopsaltria*, synonymy, key, distribution, listed, comp. note) Equals *Cicada doryca* Equals *Dundubia doryca* Equals *Cicada dorei* Equals *Dundubia dorei* New Guinea, Moluccas, Waigeo, Japen Island

Cosmopsaltria doryca group Duffels 1983b: 12, 14, 17, 19–20, 23, 32, 39, 47, 56, Fig 8, Table 1, Table 5 (described, phylogeny, comp. note)

Cicada doryca Duffels 1983b: 47 (synonymy, comp. note)

Dundubia doryca Duffels 1983b: 47 (synonymy, comp. note)

Cicada dorei Duffels 1983b: 47 (synonymy, comp. note)

Dundubia dorei Duffels 1983b: 47 (synonymy, comp. note)

Cosmopsaltria doryca group Duffels 1986: 321, 323, 330, Fig 3, Fig 5, Fig 10 (distribution, diversity, phylogeny, comp. note) New Guinea, Maluku, Halmahera

Cosmopsaltria doyca Duffels 1986: 322–323, Fig 5 (distribution) New Guinea, Ternate, Waigeo, Japen Island

Cosmopsaltria doryca group Duffels 1988a: 11 (comp. note) Maluku, Irian Jaya
Cosmopsaltria doryca group Duffels 1988b: 20, 23 (comp. note)
Cosmopsaltria doryca Duffels 1988b: 20, 23–24 (type species of *Cosmopsaltria*, comp. note) Ternate, Waigeo, Tjendrawasih, Japen Island, New Guinea
Cosmopsaltria doryca Duffels 1988d: 75 (type species of *Cosmopsaltria*)
Cosmopsaltria doryca group Duffels and Boer 1990: 258–260, Fig 21.2 (distribution, phylogeny, comp. note)
Cosmopsaltria doryca Duffels and Boer 1990: 258, 260, Fig 21.2 (distribution, phylogeny, comp. note) New Guinea, Waigeo, Ternate, Bacan
Cosmopsaltria doryca Boulard 1991c: 118–119 (comp. note) Equals *Cicada doryca* New Guinea
Cosmopsaltria doryca Boer 1995d: 173, 204–206, 232, Fig 31 (distribution, comp. note) New Guinea, Maluku, Halmahera
Cosmopsaltria doryca group Boer 1995d: 205, 230, 232 (distribution, comp. note)
Cosmopsaltria doryca Boer and Duffels 1996a: 165, Fig 6 (distribution)
Cosmopsaltria doryca Boer and Duffels 1996b: 316–317, 328, Fig 10 (distribution, comp. note) North Maluku, New Guinea
Cosmopsaltria doryca group Boer and Duffels 1996b: 304, 316, 328, Fig 4 (phylogeny, comp. note) Maluku, New Guinea
Cosmopsaltria doryca Duffels and Zaidi 2000: 196 (comp. note)
Cosmopsaltria doryca group Duffels and Turner 2002: 237, 239, 247, 257, Appendix 1 (distribution, listed) New Guinea, Halmahera, Waigeo, Buru, Amboina, Moluccas
Cosmopsaltria doryca Duffels and Turner 2002: 239, 241, 244, 248–249, 252, 257–258, Fig 2B, Fig 11, Fig 13, Appendices 1–2 (phylogeny, distribution, illustrated, listed, comp. note) New Guinea, Halmahera, Waigeo
Cosmopsaltria doryca Lee and Hayashi 2004: 71 (comp. note) Equals *Cicada doryca* New Guinea
Cicada doryca Moulds 2005b: 391 (type species of *Cosmopsaltria*)
Cosmopsaltria doryca Moulds 2005b: 396, 398, 405–406, 417–420, 422, Fig 38, Fig 44, Figs 56–60, Table 1 (higher taxonomy, phylogeny, key, listed, comp. note)
Cosmopsaltria doryca Lee and Hill 2010: 302 (comp. note)

***C. emarginata* Duffels, 1969**
Cosmopsaltria emarginata Duffels 1983a: 494–495, Fig 4 (distribution, comp. note) New Guinea
Cosmopsaltria emarginata Duffels 1983b: 4, 8, 12, 18–19, 21, 33, 35, 37, 44, 92, 97–99, Fig 12, Fig 90, Plate 5A, Table 1, Table 4 (n. sp., described, illustrated, key, distribution, listed, comp. note) New Guinea
Cosmopsaltria emarginata Boer 1995d: 204, Fig 31 (distribution, comp. note)
Cosmopsaltria emarginata Boer and Duffels 1996a: 165, Fig 6 (distribution)
Cosmopsaltria emarginata Boer and Duffels 1996b: 317, 328, Fig 10 (distribution, comp. note) New Guinea
Cosmopsaltria emarginata Duffels and Turner 2002: 237, 239, 247–248, 257–258, Fig 11, Appendices 1–2 (phylogeny, distribution, listed, comp. note) New Guinea

C. feae Distant, 1892 to *Dundubia feae*
C. fratercula Distant, 1912 to *Dundubia naragasingna* Distant, 1881
C. froggatti (Distant, 1911) to *Cosmopsaltria capitata* Distant, 1888
C. fuliginosa (Walker, 1850) to *Orientopsaltria fuliginosa*

***C. gestroei* (Distant, 1905)** = *Sawda gestroei* = *Sawda gestroi* (sic)
Cosmopsaltria gestroei Duffels 1983a: 494 (distribution, comp. note) New Guinea
Cosmopsaltria gestroei Duffels 1983b: 12, 18–19, 21, 32, 34, 38–39, 42, 46, 57–58, 63, 65–68, Fig 12, Fig 20, Figs 47–51, Plate 2C, Table 1, Table 3 (n. comb., described, illustrated, key, distribution, listed, comp. note) New Guinea
Sawda gestroei Duffels 1983b: 12, 65–68, 102, Table 1 (synonymy, comp. note)
Sawda gestroi (sic) Duffels 1983b: 65 (synonymy, comp. note)
Cosmopsaltria gestroei Boer 1995d: 204, Fig 31 (distribution, comp. note) Papuan Peninsula
Cosmopsaltria gestroei Boer and Duffels 1996a: 165, Fig 6 (distribution)
Cosmopsaltria gestroei Boer and Duffels 1996b: 317, 328, Fig 10 (distribution, comp. note) Papuan Peninsula
Cosmopsaltria gestroei Duffels and Turner 2002: 247 (comp. note)
Cosmopsaltria gestroei Duffels and Turner 2002: 248–249, 257–258, Fig 11, Fig 13, Appendices 1–2 (phylogeny, phylogeny, distribution, listed, comp. note) New Guinea

***C. gigantea gigantea* (Distant, 1897)** = *Pomponia gigantea* = *Sawda pratti* Distant, 1905 = *Cosmopsaltria pratti* = *Sawda sharpi* Distant, 1905 = *Cosmopsaltria sharpi* = *Sawda* sp. Brongersma & Venema, 1960
Cosmopsaltria gigantea Duffels 1983a: 493–495, Figs 3–4 (distribution, diverstity, comp. note) New Guinea

Cosmopsaltria gigantea group Duffels 1983a: 494–495, Fig 4 (distribution, phylogeny, comp. note)
Cosmopsaltria gigantea Duffels 1983b: 12, 17–20, 22, 32, 34–35, 38, 42, 68–71, Fig 20, Figs 55–58, Table 1 (n. comb., described, illustrated, synonymy, key, distribution, listed, comp. note) Equals *Pomponia gigantea* Equals *Sawda pratti* Equals *Sawda sharpi* New Guinea
Cosmopsaltria gigantea gigantea Duffels 1983b: 12, Table 1 (listed, comp. note)
Pomponia gigantea Duffels 1983b: 12, 68–71, Table 1 (synonymy, comp. note)
Sawda pratti Duffels 1983b: 12, 68–71, Table 1 (synonymy, comp. note)
Sawda sharpi Duffels 1983b: 12, 69–71, Table 1 (synonymy, comp. note)
Cosmopsaltria gigantea group Duffels 1983b: 12, 14, 38, 40, 68, Fig 8, Table 1 (described, phylogeny, comp. note)
Cosmopsaltria gigantea group Duffels 1986: 322, 324, Fig 4 (distribution, comp. note) New Guinea
Cosmopsaltria gigantea Duffels 1986: 322, 324, Fig 4 (distribution, comp. note) New Guinea
Cosmopsaltria gigantea Duffels and van Mastrigt 1991: 178 (comp. note) Irian Jaya
Cosmopsaltria gigantea Boer 1995b: 66 (comp. note) Equals *Sawda pratti* Equals *Sawda sharpi*
Sawda pratti Boer 1995b: 66 (comp. note)
Sawda sharpi Boer 1995b: 66 (comp. note)
Cosmopsaltria gigantea Boer 1995d: 204, Fig 31 (distribution, comp. note)
Cosmopsaltria gigantea Boer and Duffels 1996a: 165, Fig 6 (distribution)
Cosmopsaltria gigantea Boer and Duffels 1996b: 317, Fig 10 (distribution, comp. note)
Cosmopsaltria gigantea gigantea Boer and Duffels 1996b: 328 (distribution) New Guinea
Cosmopsaltria gigantea Duffels and Turner 2002: 237, 247 (comp. note)
Cosmopsaltria gigantea gigantea Duffels and Turner 2002: 248, 257–258, Fig 11, Appendices 1–2 (phylogeny, distribution, listed) New Guinea
Cosmopsaltria gigantea gigantea Mogia 2002: 18 (human food) New Guinea
Cosmopsaltria gigantea Shiyake 2007: 5, 64, 66, Fig 103 (illustrated, distribution, listed, comp. note)
Cosmopsaltria gigantea Moulds 2009a: 503–504, Fig 14.4(6) (illustrated, comp. note) Papua New Guinea

***C. gigantea occidentalis* Duffels, 1983** = *Cosmopsaltria giganta* (sic) *occidentalis*

Cosmopsaltria gigantea occidentalis Duffels 1983b: 12, 15–17, 21, 38, 45, 58, 68–75, Figs 9–12, Fig 52, Fig 54, Figs 58–59, Plate 2D, Tables 1–2 (n. ssp., described, illustrated, key, distribution, listed, comp. note) New Guinea
Cosmopsaltria gigantea occidentalis Boulard 1988d: 154–155, Fig 15 (wing abnormalities, illustrated, comp. note) New Guinea
Cosmopsaltria gigantea occidentalis Duffels and van Mastrigt 1991: 178 (comp. note) Irian Jaya
Cosmopsaltria gigantea occidentalis Boer and Duffels 1996b: 328 (distribution) New Guinea
Cosmopsaltria giganta (sic) *occidentalis* Sanborn 1999a: 43 (type material, listed)
Cosmopsaltria gigantea occidentalis Duffels and Turner 2002: 248, 257–258, Fig 11, Appendices 1–2 (phylogeny, distribution, listed) New Guinea

***C. gracilis* Duffels, 1983**

Cosmopsaltria gracilis Duffels 1983a: 493, Fig 3 (distribution, diversity) New Guinea
Cosmopsaltria gracilis group Duffels 1983a: 494, 496–498, Fig 7 (phylogeny, distribution, comp. note)
Cosmopsaltria gracilis Duffels 1983b: 12, 18–23, 23, 32, 34, 38, 44, 46, 49, 56–63, 65, Fig 12, Fig 20, Fig 34, Figs 37–43, Fig 47, Plate 1D, Plate 2A, Table 1, Table 3 (n. sp., described, illustrated, key, distribution, listed, comp. note) New Guinea
Cosmopsaltria gracilis group Duffels 1983b: 12, 17, 19–20, 23, 32, 35, 38–42, 56–57, 99, Fig 8, Fig 21, Table 1, Table 5 (described, phylogeny, comp. note)
Cosmopsaltria gracilis group Duffels 1986: 321, 324, 330, Fig 3, Fig 10 (distribution, diversity, phylogeny, comp. note) New Guinea
Cosmopsaltria gracilis group Duffels 1988b: 23 (comp. note)
Cosmopsaltria gracilis group Duffels 1988d: 14, 25, Fig 8 (distribution, comp. note) New Guinea
Cosmopsaltria gracilis group Duffels and Boer 1990: 258, 265, 267, Fig 21.2 (distribution, phylogeny, comp. note) New Guinea, Papuan Peninsula
Cosmopsaltria gracilis Boer 1995d: 204–205, 233, Fig 31 (distribution, comp. note) New Guinea
Cosmopsaltria gracilis group Boer 1995d: 205, 230, 233 (distribution, comp. note)
Cosmopsaltria gracilis Boer and Duffels 1996a: 165, Fig 6 (distribution)
Cosmopsaltria gracilis Boer and Duffels 1996b: 317, 328, Fig 10 (distribution, comp. note) New Guinea
Cosmopsaltria gracilis group Boer and Duffels 1996b: 304, 328, Fig 4 (phylogeny, comp. note) New Guinea, Papuan Peninsula

Cosmopsaltria gracilis Sanborn 1999a: 44 (type material, listed)
Cosmopsaltria gracilis group Duffels and Turner 2002: 244, 252, 257, Appendix 1 (distribution, listed, comp. note) New Guinea
Cosmopsaltria gracilis Duffels and Turner 2002: 244, 248–249, 257–258, Fig 11, Fig 13, Appendices 1–2 (phylogeny, distribution, listed, comp. note) New Guinea

C. guttigera (Walker, 1857) to *Orientopsaltria guttigera*

C. hainanensis Distant, 1901 to *Dundubia hainanensis*

***C. halmaherae* Duffels, 1988**

Cosmopsaltria halmaherae Duffels 1988b: 20–24, Figs 1–4 (n. sp., described, illustrated, comp. note) Halmahera
Cosmopsaltria halmaherae Duffels and Boer 1990: 258, 260, Fig 21.2 (distribution, phylogeny, comp. note) Halmahera
Cosmopsaltria halmaherae Boer 1995d: 204–205, Fig 31 (distribution, comp. note) Halmahera Island
Cosmopsaltria halmaherae Boer and Duffels 1996a: 165, Fig 6 (distribution)
Cosmopsaltria halmaherae Boer and Duffels 1996b: 317, 328, Fig 10 (distribution, comp. note) North Maluku
Cosmopsaltria halmaherae Duffels and Turner 2002: 243, 248–249, 257–258, Fig 6B, Fig 11, Fig 13, Appendices 1–2 (illustrated, phylogeny, distribution, listed, comp. note) Halmahera

C. hastata Moulton, 1923 to *Dundubia hastata*

***C. huonensis* Duffels, 1983**

Cosmopsaltria huonensis Duffels 1983b: 12, 18–22, 32, 38, 44, 46, 56–58, 62–65, Fig 12, Figs 44–47, Plate 2B, Table 1, Table 3 (n. sp., described, illustrated, key, distribution, listed, comp. note) New Guinea
Cosmopsaltria huonensis Boer 1995d: 204–205, Fig 31 (distribution, comp. note) New Guinea, Huon Peninsula
Cosmopsaltria huonensis Boer 1996: 358 (comp. note)
Cosmopsaltria huonensis Boer and Duffels 1996a: 165, Fig 6 (distribution)
Cosmopsaltria huonensis Boer and Duffels 1996b: 317, 328, Fig 10 (distribution, comp. note) New Guinea
Cosmopsaltria huonensis Sanborn 1999a: 44 (type material, listed)
Cosmopsaltria huonensis Duffels and Turner 2002: 243–244, 248–249, 257–258, Fig 6C, Fig 11, Fig 13, Appendices 1–2 (illustrated, phylogeny, distribution, listed, comp. note) New Guinea

C. ida Moulton, 1911 to *Orientopsaltria ida*

***C. inermis* Stål, 1870** = *Cosmopsaltria (Cosmopsaltria) inermis*

C. jacoona Distant, 1888 to *Dundubia jacoona*

***C. kaiensis* Duffels, 1988**

Cosmopsaltria kaiensis Duffels 1988c: 187–188, Figs 1–4 (n. sp., described, illustrated, comp. note) Kai Islands
Cosmopsaltria kaiensis Boer 1995d: 204–205, Fig 31 (distribution, comp. note) Kai Islands
Cosmopsaltria kaiensis Boer and Duffels 1996a: 165, Fig 6 (distribution)
Cosmopsaltria kaiensis Boer and Duffels 1996b: 317, 328, Fig 10 (distribution, comp. note) New Guinea
Cosmopsaltria kaiensis Duffels and Turner 2002: 242, 247–248, 252–253, 257–258, Fig 11, Appendices 1–2 (phylogeny, distribution, listed, comp. note) New Guinea

***C. lata* (Walker, 1868)** = *Dundubia lata*

Cosmopsaltria lata Duffels 1983a: 494 (distribution, comp. note) Moluccas
Cosmopsaltria lata Duffels 1983b: 4, 8, 12, 18, 20–21, 32, 39, 45, 47, 49, 52–56, Fig 12, Figs 29–33, Plate 1C, Table 1, Table 3 (described, illustrated, key, synonymy, distribution, listed, comp. note) Equals *Dundubia lata* Buru Island, Ambon Island
Dundubia lata Duffels 1983b: 53 (synonymy, comp. note) New Guinea, Waigeo, Salawati
Cosmopsaltria lata Duffels 1986: 322–323, Fig 5 (distribution, comp. note) Maluku
Cosmopsaltria lata Duffels 1988b: 20, 23–24 (comp. note) Buru, Ambon, Maluku Selatan
Cosmopsaltria lata Duffels 1990b: 70 (comp. note) Buru, Ambon
Cosmopsaltria lata Duffels and Boer 1990: 258, 260, Fig 21.2 (distribution, phylogeny, comp. note) Ambon, Buru
Cosmopsaltria lata Boer 1995d: 204–205, Fig 31 (distribution, comp. note) Ambon, Buru, Timor?
Cosmopsaltria lata Boer and Duffels 1996a: 165, Fig 6 (distribution)
Cosmopsaltria lata Boer and Duffels 1996b: 317, 328, Fig 10 (distribution, comp. note) South Maluku
Cosmopsaltria lata Duffels and Zaidi 2000: 196 (comp. note)
Cosmopsaltria lata Duffels and Turner 2002: 239, 243, 248–249, 252, 257–258, Fig 6A, Fig 11, Fig 13, Appendices 1–2 (illustrated, phylogeny, distribution, listed, comp. note) Buru, Amboina

C. *loriae* Distant, 1897 = *Fatima loriae*
Cosmopsaltria loriae Duffels 1983a: 494–495, Fig 4 (distribution, comp. note) New Guinea
Cosmopsaltria loriae Duffels 1983b: 12, 18, 22, 35–36, 42, 44–46, 99, 104, 106–108, Fig 12, Fig 103, Figs 110–112, Table 1, Table 3 (n. comb., lectotype designation, described, illustrated, key, synonymy, distribution, listed, comp. note) Equals *Fatima loriae* New Guinea
Fatima loriae Duffels 1983b: 12, 107, Table 1 (synonymy, comp. note)
Cosmopsaltria loriae Duffels 1986: 322, Fig 4 (distribution) New Guinea
Cosmopsaltria loriae Boer 1995d: 204–205, Fig 31 (distribution, comp. note) Papuan Penisula
Cosmopsaltria loriae Boer and Duffels 1996a: 165, Fig 6 (distribution)
Cosmopsaltria loriae Boer and Duffels 1996b: 317, 328, Fig 10 (distribution, comp. note) Papuan Peninsula
Cosmopsaltria loriae Duffels and Turner 2002: 242, 247–248, 257–258, Fig 11, Appendices 1–2 (phylogeny, distribution, listed, comp. note) New Guinea

C. *majuscula* Moulton, 1923 to *Champaka meyeri* (Distant, 1883)

C. *meeki* (Distant, 1906) = *Haphsa meeki*
Cosmopsaltria meeki Duffels 1983a: 493, 495, Figs 3–4 (distribution, diversity, comp. note) New Guinea
Cosmopsaltria meeki group Duffels 1983a: 494–495, Fig 4 (distribution, comp. note)
Cosmopsaltria meeki Duffels 1983b: 4, 8, 12, 15–16, 18–19, 21, 33, 34, 42, 44, 91–94, Figs 9–12, Fig 20, Figs 90–91, Plate 4A, Tables 1–2, table 4 (described, illustrated, key, synonymy, distribution, listed, comp. note) Equals *Haphsa meeki* New Guinea
Cosmopsaltria meeki group Duffels 1983b: 12, 14, 17–19, 22–23, 33, 35, 37–38, 40, 91–92, 97, Fig 8, Table 5 (described, phylogeny, comp. note)
Haphsa meeki Duffels 1983b: 12, 92, Table 1 (synonymy, comp. note)
Cosmopsaltria meeki group Duffels 1986: 322, 324, Fig 4 (distribution, comp. note) New Guinea
Cosmopsaltria meeki Duffels 1986: 322, Fig 4 (distribution) New Guinea
Cosmopsaltria meeki group Duffels and Boer 1990: 261 (comp. note) New Guinea
Cosmopsaltria meeki Boer 1995d: 204, Fig 31 (distribution, comp. note) Papuan Peninsula
Cosmopsaltria meeki Boer and Duffels 1996a: 165, Fig 6 (distribution)
Cosmopsaltria meeki Boer and Duffels 1996b: 317, 328, Fig 10 (distribution, comp. note) Papuan Peninsula
Cosmopsaltria meeki species group Boer 2000: 7 (comp. note)
Cosmopsaltria meeki group Duffels and Turner 2002: 237 (comp. note)
Cosmopsaltria meeki Duffels and Turner 2002: 238, 248, 257–258, Fig 2A, Fig 2F, Fig 11, Appendices 1–2 (illustrated, phylogeny, distribution, listed, comp. note) New Guinea
Haphsa meeki Lee 2009b: 330 (comp. note)
Cosmopsaltria meeki Lee 2009b: 330 (comp. note)

C. *mimica* Distant, 1897 = *Sawda mimica*
Cosmopsaltria mimica Duffels 1983a: 493–495, Figs 3–4 (distribution, diversity, comp. note) New Guinea
Cosmopsaltria mimica complex Duffels 1983a: 493–498, Figs 3–4, Fig 7 (diversity, distribution, phylogeny, comp. note) New Guinea
Cosmopsaltria mimica group Duffels 1983a: 494–495, Fig 4 (distribution, comp. note)
Cosmopsaltria mimica Duffels 1983b: 12, 15–16, 18–19, 21–22, 33–34, 37–38, 42, 45, 73, 75–83, 86, 90, Fig 12, Fig 20, Fig 60, Figs 62–68, Fig 72, Plate 3A, Tables 1–3 (n. comb., described, illustrated, key, synonymy, distribution, listed, comp. note) New Guinea, Bismarck Archipelago
Cosmopsaltria mimica group Duffels 1983b: 12, 14, 17, 19–20, 22–23, 33, 35, 37–38, 40, 75, 99, Fig 8, Table 1, Table 5 (phylogeny, comp. note)
Sawda mimica Duffels 1983b: 12, 75, Table 1 (synonymy, comp. note)
Cosmopsaltria mimica complex Duffels 1983b: 14, 35, 38–40, 42, 75, Fig 8 (phylogeny, comp. note)
Cosmopsaltria mimica complex Duffels 1986: 321–322, 324, 330, Figs 3–4, Fig 10 (distribution, diversity, phylogeny, comp. note) New Guinea
Cosmopsaltria mimica group Duffels 1986: 322, 324, Fig 4 (distribution, comp. note) New Guinea
Cosmopsaltria mimica Duffels 1986: 322, 325, Fig 4 (distribution, comp. note) New Guinea
Cosmopsaltria mimica complex Duffels 1988b: 23 (comp. note)
Cosmopsaltria mimica complex Duffels and Boer 1990: 258, 261, 263, 267, Fig 21.2 (diversity, distribution, phylogeny, comp. note)
Cosmopsaltria mimica group Duffels and Boer 1990: 261 (comp. note) New Guinea
Cosmopsaltria mimica Duffels and van Mastrigt 1991: 179 (comp. note) Irian Jaya

Cosmopsaltria mimica Kahono 1993: 213, Table 5 (listed) Irian Jaya
Cosmopsaltria mimika (sic) Boer 1995d: 204, Fig 31 (distribution, comp. note)
Cosmopsaltria mimica Boer 1995d: 205 (distribution, comp. note) New Guinea, Bismarck Archipelago
Cosmopsaltria mimica complex Boer 1995d: 205, 233 (distribution, comp. note)
Cosmopsaltria mimica group Boer 1995d: 230 (distribution, comp. note)
Cosmopsaltria mimika (sic) Boer and Duffels 1996a: 165, Fig 6 (distribution)
Cosmopsaltria mimika (sic) Boer and Duffels 1996b: 204, Fig 31 (distribution, comp. note)
Cosmopsaltria mimica Boer and Duffels 1996b: 317, 329, Fig 10 (distribution, comp. note) New Guinea, Bismarck Archipelago
Cosmopsaltria mimica complex Boer and Duffels 1996b: 304, 328, Fig 4 (phylogeny, comp. note)
Cosmopsaltria mimica group Boer and Duffels 1996b: 230 (distribution, comp. note)
Cosmopsaltria mimica? Boer 2000: 1, Table 1 (comp. note) Papua New Guinea
Cosmopsaltria mimica Boer 2000: 6–7 (comp. note) New Guinea
Cosmopsaltria mimica species group Boer 2000: 6 (comp. note)
Cosmopsaltria mimica group Duffels and Turner 2002: 237 (comp. note)
Cosmopsaltria mimica Duffels and Turner 2002: 237, 247–248, 253, 257–258, Fig 11, Appendices 1–2 (phylogeny, distribution, listed, comp. note) New Guinea, Bismarck Archipelago
Cosmopsaltria mimica complex Duffels and Turner 2002: 244, 247, 249, 252, 257, Fig 13, Appendix 1 (phylogeny, distribution, listed, comp. note) New Guinea, Bismarck Archipelago
Cosmopsaltria mimica Mogia 2002: 18–19, Fig 3 (human food, illustrated) New Guinea
Cosmopsaltria mimica complex Duffels and Boer 2006: 534 (diversity, distribution, comp. note) New Guinea
Cosmopsaltria mimica Shiyake 2007: 5, 64, 66, Fig 101 (illustrated, distribution, listed, comp. note)
Cosmopsaltria mimica Moulds 2009a: 503–504, Fig 14.4(5) (illustrated, comp. note) Papua New Guinea

C. montana Kato, 1927 to *Macrosemia kareisana* (Matsumura, 1907)
C. moultoni China, 1926 to *Orientopsaltria moultoni*
C. multivocalis Matsumura, 1907 to *Meimuna multivocalis*
C. nigripes Moulton, 1923 to *Dundubia nigripes*
C. oopaga Distant, 1881 to *Dundubia oopaga*
C. padda Distant, 1887 to *Orientopsaltria padda*

***C. papuensis* Duffels, 1983** = *Fatima capitata* Distant, 1912: pl. 6

Cosmopsaltria papuensis Duffels 1983a: 493–495, Figs 3–4 (dstribution, diversity, comp. note) New Guinea
Cosmopsaltria papuensis group Duffels 1983a: 494–495, Fig 4 (distribution, comp. note)
Cosmopsaltria papuensis Duffels 1983b: 12, 15–16, 18, 20–21, 23, 34, 36, 42, 46, 99, 108–117, Figs 9–10, Fig 12, Figs 113–126, Plate 6A-B, Tables 1–3, Table 5 (n. sp., described, illustrated, key, distribution, listed, comp. note) New Guinea
Cosmopsaltria papuensis group Duffels 1983b: 12, 14, 17, 22–23, 35–37, 40, 56, 99, 108, Fig 8,Ttable 1, Table 5 (phylogeny, comp. note)
Cosmopsaltria papuensis group Duffels 1986: 322, 324, Fig 4 (distribution) New Guinea
Cosmopsaltria papuensis Duffels 1986: 322, 324, Fig 4 (distribution, comp. note) New Guinea
Cosmopsaltria papuensis group Duffels 1988c: 189–190 (comp. note) New Guinea
Cosmopsaltria papuensis Duffels 1988c: 189 (comp. note)
Cosmopsaltria papuensis Duffels and van Mastrigt 1991: 176, 178–179 (comp. note) New Guinea
Cosmopsaltria papuaensis Boer 1995d: 204–205, Fig 31 (distribution, comp. note) Papua New Guinea
Cosmopsaltria papuaensis Boer and Duffels 1996a: 165, Fig 6 (distribution)
Cosmopsaltria papuaensis Boer and Duffels 1996b: 299, 317, 329, Fig 10, Plate1a (distribution, illustrated, comp. note) New Guinea
Cosmopsaltria papuensis Boer 2000: 1, 7, Table 1 (comp. note) Papua New Guinea
Cosmopsaltria papuensis Duffels and Boer 2006: 536, Fig 4.4.3 (illustrated) New Guinea
Cosmopsaltria papuensis Duffels and Turner 2002: 242–243, 247–248, 257–258, Fig 6E, Fig 11, Appendices 1–2 (illustrated, phylogeny, distribution, listed, comp. note) New Guinea
Cosmopsaltria papuensis Mogia 2002: 18 (human food) New Guinea
Cosmopsaltria papuensis Shiyake 2007: 5, 64, 66, Fig 102 (illustrated, distribution, listed, comp. note)
Cosmopsaltria papuensis Costa Neto 2008: 453 (comp. note)

***C. personata* Duffels, 1983**

Cosmopsaltria personata Duffels 1983a: 495, Fig 4 (distribution)

Cosmopsaltria personata Duffels 1983b: 12, 18–19, 21, 23, 33–34, 44, 91–93, 95–97, Fig 12, Fig 20, Fig 90, Figs 92–95, Plate 4C-D, Table 1, Tables 4–5 (n. sp., described, illustrated, key, distribution, listed, comp. note) New Guinea

Cosmopsaltria personata Duffels 1986: 322, Fig 4 (distribution) New Guinea

Cosmopsaltria personata Duffels and van Mastrigt 1991: 176, 177 (comp. note) Irian Jaya

Cosmopsaltria personata Boer 1995d: 204, Fig 31 (distribution, comp. note)

Cosmopsaltria personata Boer and Duffels 1996a: 165, Fig 6 (distribution)

Cosmopsaltria personata Boer and Duffels 1996b: 317, 329, Fig 10 (distribution, comp. note) New Guinea

Cosmopsaltria personata Boer 2000: 1, 7, Table 1 (comp. note) Papua New Guinea

Cosmopsaltria personata Duffels and Turner 2002: 248, 257–258, Fig 11, Appendices 1–2 (phylogeny, distribution, listed) New Guinea

Cosmopsaltria personata Costa Neto 2008: 453 (comp. note)

C. phaeophila (Walker, 1850) to *Orientopsaltria phaeophila*

C. pratti Distant, 1905 to *Cosmopsaltria gigantea gigantea* (Distant, 1897)

C. retrorsa **Duffels, 1983**

Cosmopsaltria retrorsa Duffels 1983a: 495, Fig 4 (distribution)

Cosmopsaltria retrorsa Duffels 1983b: 12, 18, 20–21, 34–36, 46, 108, 110, 117–120, Fig 12, Fig 20, Figs 125–130, Plate 6C, Table 1, Table 4 (n. sp., described, illustrated, key, distribution, listed, comp. note) New Guinea

Cosmopsaltria retrorsa Duffels 1986: 322, Fig 4 (distribution) New Guinea

Cosmopsaltria retrorsa Duffels 1988c: 189 (comp. note)

Cosmopsaltria retrorsa Duffels and van Mastrigt 1991: 179 (comp. note) New Guinea

Cosmopsaltria retrorsa Boer 1995d: 204, Fig 31 (distribution, comp. note)

Cosmopsaltria retrorsa Boer 1996: 358 (comp. note)

Cosmopsaltria retrorsa Boer and Duffels 1996a: 165, Fig 6 (distribution)

Cosmopsaltria retrorsa Boer and Duffels 1996b: 317, 329, Fig 10 (distribution, comp. note) New Guinea

Cosmopsaltria retrorsa Boer 2000: 6 (comp. note) New Guinea

Cosmopsaltria retrorsa Duffels and Turner 2002: 242, 247–248, 257–258, Fig 11, Appendices 1–2 (phylogeny, distribution, listed, comp. note) New Guinea

C. satyrus **Duffels, 1969**

Cosmopsaltria satyrus Duffels 1983a: 495, Fig 4 (distribution)

Cosmopsaltria satyrus Duffels 1983b: 4, 8, 12, 15–16, 18–19, 21, 33–34, 44, 91–95, 95–97, Fig 12, Fig 20, Fig 90, Plate 4B, Tables 1–2, Table 4 (described, illustrated, key, distribution, listed, comp. note) New Guinea

Cosmopsaltria satyrus Duffels 1986: 322, Fig 4 (distribution) New Guinea

Cosmopsaltria satyrus Boer 1995d: 204, Fig 31 (distribution, comp. note)

Cosmopsaltria satyrus Boer and Duffels 1996a: 165, Fig 6 (distribution)

Cosmopsaltria satyrus Boer and Duffels 1996b: 317, 329, Fig 10 (distribution, comp. note) New Guinea

Cosmopsaltria satyrus Duffels and Turner 2002: 247–248, 257–258, Fig 11, Appendices 1–2 (phylogeny, distribution, listed, comp. note) New Guinea

C. sharpi Distant, 1905 to *Cosmopsaltria gigantea gigantea* (Distant, 1897)

C. signata **Duffels, 1983**

Cosmopsaltria signata Duffels 1983a: 495, Fig 4 (distribution)

Cosmopsaltria signata Duffels 1983b: 12, 15–16, 18–19, 21, 34–35, 38, 42, 44–45, 75, 77, 86–91, Figs 9–12, Fig 20, Figs 79–89, Plate 3D, Tables 2–3 (n. sp., described, illustrated, key, distribution, listed, comp. note) New Guinea

Cosmopsaltria signata Duffels 1986: 322, Fig 4 (distribution) New Guinea

Cosmopsaltria signata Duffels and Boer 1990: 261 (comp. note) New Guinea

Cosmopsaltria signata Duffels and van Mastrigt 1991: 176 (comp. note) Irian Jaya

Cosmopsaltria signata Boer 1995d: 204, Fig 31 (distribution, comp. note)

Cosmopsaltria signata Boer 1996: 358 (comp. note)

Cosmopsaltria signata Boer and Duffels 1996a: 165, Fig 6 (distribution)

Cosmopsaltria signata Boer and Duffels 1996b: 317, 329, Fig 10 (distribution, comp. note) New Guinea

Cosmopsaltria signata Sanborn 1999a: 44 (type material, listed)

Cosmopsaltria signata Duffels and Turner 2002: 237, 248, 257–258, Fig 11, Appendices 1–2 (phylogeny, distribution, listed, comp. note) New Guinea

C. **sp. a Duffels, 1986**

Cosmopsaltria sp. a Duffels 1986: 323, Fig 5 (distribution)

C. sumatrana Moulton, 1917 to *Orientopsaltria sumatrana*

C. tonkiniana Jacobi, 1905 to *Macrosemia tonkiniana*

C. torjada (Breddin, 1901) to *Brachylobopyga torjada*

***C. toxopeusi* Duffels, 1983**

Cosmopsaltria toxopeusi Duffels 1983a: 495, Fig 4 (distribution)

Cosmopsaltria toxopeusi Duffels 1983b: 12, 15–16, 18, 20–22, 34–36, 46, 108, 110, 117, 120–123, Figs 9–12, Fig 20, Fig 125, Figs 131–135, Plate 6D, Tables 1–3 (n. sp., described, illustrated, key, distribution, listed, comp. note) New Guinea

Cosmopsaltria toxopeusi Duffels 1986: 322, Fig 4 (distribution) New Guinea

Cosmopsaltria toxopeusi Duffels 1988c: 189 (comp. note)

Cosmopsaltria toxopeusi Duffels and van Mastrigt 1991: 179 (comp. note) New Guinea

Cosmopsaltria toxopeusi Boer 1995d: 204, Fig 31 (distribution, comp. note)

Cosmopsaltria toxopeusi Boer 1996: 358 (comp. note)

Cosmopsaltria toxopeusi Boer and Duffels 1996a: 165, Fig 6 (distribution)

Cosmopsaltria toxopeusi Boer and Duffels 1996b: 317, 329, Fig 10 (distribution, comp. note) New Guinea

Cosmopsaltria toxopeusi Boer 2000: 6 (comp. note) New Guinea

Cosmopsaltria toxopeusi Duffels and Turner 2002: 242, 247–248, 257–258, Fig 11, Appendices 1–2 (phylogeny, distribution, listed, comp. note) New Guinea

***C. vitiensis* (Distant, 1906)** = *Sawda ? vitiensis* = *Sawda vitiensis* = *(?) Cosmopsaltria vitiensis* = *Cosmopsaltria (?) vitiensis*

Cosmopsaltria vitiensis Duffels 1983b: 12, 123, Table 1 (n. comb., synonymy, comp. note) Equals *Sawda vitiensis* Fiji Islands

Sawda vitiensis Duffels 1983b: 12, 123, Table 1 (synonymy, comp. note)

(?) Cosmopsaltria vitiensis Duffels 1986: 328 (comp. note) Fiji Islands

Cosmopsaltria vitiensis Duffels 1988c: 187 (comp. note) Fiji

Cosmopsaltria vitiensis Duffels 1988d: 9, 12, 25, 28, 77, Table 1 (key, distribution, comp. note) Equals *Sawda vitiensis* Fiji

Sawda vitiensis Duffels 1988d: 9, 76 (lectotype designation, comp. note) Fiji Islands

Cosmopsaltria (?) vitiensis Duffels 1988d: 75–77, Figs 141–143 (lectotype designation, described, illustrated, comp. note) Equals *Sawda (?) vitiensis* Equals *Sawda vitiensis* Fiji Islands

Sawda (?) vitiensis Duffels 1988d: 76, 103 (comp. note)

Cosmopsaltria vitiensis Boer 1995d: 204 (distribution, comp. note) Fiji Islands

Cosmopsaltria vitiensis Boer and Duffels 1996b: 329 (distribution) Fiji Islands

Cosmopsaltria vitiensis Duffels and Turner 2002: 236, 250, 257–258, Appendices 1–2 (distribution, listed, comp. note) Fiji

Cosmopsaltria vitiensis Wilson 2009: 40, Appendix (listed, distribution, comp. note) Fiji

***C. waine* Duffels, 1991**

Cosmopsaltria waine Duffels 1983a: 175, 177–181, Fig 1–5 (n. sp. described, illustrated, predation, comp. note)

Cosmopsaltria waine Duffels and van Mastrigt 1991: 175, 177–181, Fig 1–5 (n. sp. described, illustrated, predation, comp. note)

Cosmopsaltria waine Boer 1995d: 204, Fig 31 (distribution, comp. note)

Cosmopsaltria waine Boer and Duffels 1996a: 165, Fig 6 (distribution)

Cosmopsaltria waine Boer and Duffels 1996b: 317, 329, Fig 10 (distribution, comp. note) New Guinea

Cosmopsaltria waine Duffels and Turner 2002: 242–243, 247–248, 253, 257–258, Fig 6F, Fig 11, Appendices 1–2 (illustrated, phylogeny, distribution, listed, comp. note) New Guinea

***Diceropyga* Stål, 1870** = *Cosmopsaltria (Diceropyga)* = *Cosmopsaltria (Diceropygis)* (sic)

Diceropyga sp. Meyer-Rochow 1973: 675 (human food, listed) New Guinea

Diceropyga Duffels 1982: 156, 159–160 (biogeography, phylogeny, comp. note) Melanesia

Diceropyga Duffels 1983a: 492–494, 496–498, Figs 1–2, Fig 5, Fig 7 (diversity, distribution, phylogeny, comp. note) Moluccas, New Guinea, Solomon Islands, Bismarck Archipelago

Diceropyga Duffels 1983b: 2–3, 10–11, 19–20, 39–42, Fig 7, Fig 21 (phylogeny, comp. note)

Diceropyga Strümpel 1983: 18 (listed)

Diceropyga Hayashi 1984: 58 (listed, comp. note)

Diceropyga Duffels 1985: 275, 278–279 (comp. note) Maluku, New Guinea, Papuan Peninsula, Bismark Archipelago, Solomon Islands, Batjan

Diceropyga Duffels 1986: 320–322, 324–325, 330–331, Figs 1–3, Fig 10 (distribution, diversity, phylogeny, comp. note) New Guinea, Maluku, Solomon Islands, Bismarck Archipelago, Australia

Diceropyga Duffels 1988a: 8 (comp. note)
Diceropyga Duffels 1988d: 10 (comp. note)
Diceropyga Boer 1989: 7 (comp. note)
Diceropyga Boer 1990: 66 (comp. note)
Diceropyga Duffels 1990b: 65, Fig 1 (phylogeny) New Guinea, Maluku, Solomon Islands, Bismarck Archipelago, Australia
Diceropyga Duffels and Boer 1990: 260–261, 263, 265–266, Fig 21.8 (distribution, phylogeny, comp. note) New Guinea, Maluku, Bismarck Islands, Solomon Islands, Australia
Diceropyga Moulds 1990: 12, 17, 32, 102–103 (listed, diversity, comp. note) Australia, New Guinea
Diceropyga Boer 1992a: 24–25 (distribution, comp. note) Maluku, New Guinea, Papuan Peninsula, Bismarck Archipelago, Northern Queensaland, Solomon Islands
Diceropyga Boer 1992b: 165 (comp. note)
Diceropyga Boer 1993b: 144–145 (phylogeny, comp. note)
Diceropyga Duffels 1993: 1227, 1230, 1232, Fig 6 (phylogeny, comp. note)
Diceropyga Boer 1994b: 161, 163 (comp. note)
Diceropyga Boer 1994c: 134 (comp. note)
Diceropyga Boer 1994d: 90 (comp. note) Papuan Peninsula
Diceropyga Boer 1995a: 7 (comp. note)
Diceropyga Boer 1995b: 15–16, 18 (comp. note) Papua New Guinea, Waigeu Salawati, Maluku, Solomon Islands
Diceropyga Boer 1995d: 172–173, 206–208, 217–219, 222–223, 225–226, 229–231, 233–237, Fig 51 (distribution, phylogeny, comp. note) New Guinea, Maluku, Bismarck Archipelago, East Melanesia, d'Entrecasteaux Islands, Louisiade Archipelago, Woodlark Island, Australia
Diceropyga Boer 1996: 358 (comp. note) New Britain, New Ireland
Diceropyga Boer and Duffels 1996a: 155, 1665–166, 169, 171–173, Fig 9, Fig 11 (distribution, phylogeny, comp. note) New Guinea, Queensland, Solomon Islands, Bismarck Archipelago
Diceropyga Boer and Duffels 1996b: 301, 306–308, 313–314, 316, 318, 320, 329 (distribution, phylogeny, comp. note) Maluku, Buru, Seram, Bismarck Archipelago, Manus Island, Mussau Island
Diceropyga Duffels 1997: 549, 551–552, Fig 2 (phylogeny, comp. note)
Diceropyga Novotny and Wilson 1997: 437 (listed)
Diceropyga Holloway 1998: 305, 307–308, 311, Fig 10, Figs 12–13 (distribution, phylogeny, comp. note) New Guinea
Diceropyga Holloway and Hall 1998: 9–10, Fig 3 (distribution, comp. note)
Diceropyga Boer 1999: 120 (comp. note) New Guinea
Diceropyga Brown and Toft 1999: 150 (comp. note)
Diceropyga Duffels and Turner 2002: 236–237, 239–240, 242, 244–247, 249–252, 256, Fig 9, Figs 13–15 (phylogeny, distribution, illustrated, listed, comp. note) Bismarck Archipelago, Solomon Islands, Moluccas, New Guinea
Diceropyga Moulds and Carver 2002: 11 (listed) Australia
Diceropyga Moulds 2005b: 387–389, 391, 394, 400, 407, 412–413, 423, 430, 432–433, Table 2 (higher taxonomy, listed, comp. note) Australia
Diceropyga Sanborn 2005b: 21, 23 (comp. note)
Diceropyga Duffels and Boer 2006: 537 (distribution, comp. note) Biak, Numfoor
Diceropyga Shiyake 2007: 5, 66 (listed, comp. note)
Diceropyga sp. Shiyake 2007: 5, 65–66, Fig 107 (illustrated, listed, comp. note)

D. acuminata Duffels, 1977 to *Rhadinopyga acuminata*

***D. acutipennis* (Walker, 1858)** = *Cicada acutipennis*

***D. auriculata* Duffels, 1977**

Diceropyga auriculata Duffels 1983a: 495, Fig 6 (distribution)
Diceropyga auriculata Duffels 1986: 323, Fig 6 (distribution)
Diceropyga auriculata Duffels and Boer 1990: 262, Fig 21.4 (distribution)
Diceropyga auriculata Boer 1995d: 207, Fig 33 (distribution, comp. note)
Diceropyga auriculata Boer and Duffels 1996b: 315, 329, Fig 9 (distribution, comp. note) Papuan Peninsula
Diceropyga auriculata Duffels and Turner 2002: 248, 256, 258, Fig 11, Appendices 1–2 (phylogeny, distribution, listed) Goodenough Island, Normanby Island

***D. aurita* Duffels, 1977**

Diceropyga aurita Duffels 1983a: 495, Fig 6 (distribution)
Diceropyga aurita Duffels 1986: 323, Fig 6 (distribution)
Diceropyga aurita Duffels and Boer 1990: 262, Fig 21.4 (distribution)
Diceropyga aurita Boer 1995d: 207, Fig 33 (distribution, comp. note)
Diceropyga aurita Boer and Duffels 1996b: 315, 329, Fig 9 (distribution, comp. note) Solomon Islands
Diceropyga aurita Duffels and Turner 2002: 242, 247–248, 256, 259, Fig 5E, Fig 11, Appendices 1–2 (illustrated, phylogeny,

distribution, listed, comp. note) Solomon Islands

D. bacanensis Duffels, 1988

Diceropyga bacanensis Duffels 1988a: 7, 11, Figs 1–5, Table 1 (n. sp., described, illustrated, comp. note) Bacan

Diceropyga bacanensis Duffels 1988b: 23 (comp. note) Bacan

Diceropyga bacanensis Duffels and Boer 1990: 259, 262, Figs 21.3–21.4 (distribution, phylogeny) Bacan

Diceropyga bacanensis Boer 1995d: 207, Fig 33 (distribution, comp. note)

Diceropyga bacanensis Boer and Duffels 1996b: 315, 329, Fig 9 (distribution, comp. note) North Maluku

Diceropyga bacanensis Duffels and Turner 2002: 246–248, 256, 258, Fig 11, Appendices 1–2 (phylogeny, distribution, listed, comp. note) Maluku

D. bicornis Duffels, 1977

Diceropyga bicornis Duffels 1983a: 495, Fig 6 (distribution)

Diceropyga bicornis Duffels 1986: 323, Fig 6 (distribution)

Diceropyga bicornis Duffels and Boer 1990: 262, Fig 21.4 (distribution)

Diceropyga bicornis Boer 1995d: 207, Fig 33 (distribution, comp. note)

Diceropyga bicornis Boer and Duffels 1996b: 315, 329, Fig 9 (distribution, comp. note) Papuan Peninsula

Diceropyga bicornis Duffels and Turner 2002: 248, 256, 258, Fig 11, Appendices 1–2 (phylogeny, distribution, listed) New Guinea

D. bihamata Duffels, 1977

Diceropyga bihamata Duffels 1983a: 494–495, Fig 6 (distribution, comp. note) New Guinea

Diceropyga bihamata Duffels 1986: 323, Fig 6 (distribution)

Diceropyga bihamata Duffels and Boer 1990: 262, Fig 21.4 (distribution)

Diceropyga bihamata Boer 1992b: 168 (comp. note) New Guinea

Diceropyga bihamata Boer 1993a: 20–21 (comp. note)

Diceropyga bihamata Boer 1995b: 16 (comp. note) New Guinea

Diceropyga bihamata Boer 1995d: 207, Fig 33 (distribution, comp. note) New Guinea

Diceropyga bihamata Boer and Duffels 1996b: 315, 329, Fig 9 (distribution, comp. note) New Guinea

Diceropyga bihamata Duffels and Turner 2002: 244, 246, 248, 256, 258, Fig 7B, Figs 9E–9G, Fig 11, Appendices 1–2 (illustrated, phylogeny, distribution, listed, comp. note) New Guinea

D. bougainvillensis Duffels, 1977

Diceropyga bougainvillensis Duffels 1983a: 495, Fig 6 (distribution)

Diceropyga bougainvillensis Duffels 1986: 323, Fig 6 (distribution)

Diceropyga bougainvillensis Duffels and Boer 1990: 262, Fig 21.4 (distribution)

Diceropyga bougainvillensis Boer 1995d: 207, Fig 33 (distribution, comp. note)

Diceropyga bougainvillensis Boer and Duffels 1996b: 315, 329, Fig 9 (distribution, comp. note) Solomon Islands

Diceropyga bougainvillensis Duffels and Turner 2002: 247–248, 256, 259, Fig 11, Appendices 1–2 (phylogeny, distribution, listed, comp. note) Solomon Islands

D. didyma (Boisduval, 1835) = *Cicada didyma* = *Cosmopsaltria didyma* = *Meimuna didyma*

Diceropyga didyma Duffels 1983a: 495, Fig 6 (distribution)

Diceropyga didyma Duffels 1986: 323, Fig 6 (distribution)

Diceropyga didyma Duffels and Boer 1990: 262, Fig 21.4 (distribution)

Diceropyga didyma Boulard 1991c: 119 (comp. note) Equals *Cicada didyma* Kilinailu Island

Diceropyga didyma Boer 1995d: 207, Fig 33 (distribution, comp. note)

Diceropyga didyma Boer and Duffels 1996b: 315, 329, Fig 9 (distribution, comp. note) Solomon Islands

Diceropyga didyma Duffels and Turner 2002: 236, 256, Appendix 1 (distribution, listed, comp. note) Kinilailau Island, Bougainville Island

D. gravesteini Duffels, 1977 = *Diceropyga pigafettae* Lallemand (nec Distant) (partim)

Diceropyga gravesteini Duffels 1983a: 494–495, Fig 6 (distribution, comp. note) New Guinea, Bismarck Archipelago

Diceropyga gravesteini Duffels 1986: 323, 325, Fig 6 (distribution, comp. note) Bismarck Archipelago, New Guinea

Diceropyga gravesteini Duffels and Boer 1990: 262, Fig 21.4 (distribution)

Diceropyga gravesteini Boer 1995b: 16 (comp. note)

Diceropyga gravesteni Boer 1995d: 207, Fig 33 (distribution, comp. note) New Guinea, Bismarck Archipelago, Admiralty Islands

Diceropyga gravesteni Boer and Duffels 1996a: 173 (comp. note) New Guinea, Bismarck Archipelago

Diceropyga gravesteni Boer and Duffels 1996b: 315, 329, Fig 9 (distribution, comp. note) New Guinea, Bismarck Archipelago

Dundubia gravesteini Zaidi 1997: 110, 112–113, 1116, Table 1 (listed, comp. note) Borneo, Sarawak
Diceropyga gravesteini Sanborn 1999a: 44 (type material, listed)
Diceropyga gravesteini Duffels and Turner 2002: 248, 253, 256, 258, Fig 11, Appendices 1–2 (phylogeny, distribution, listed, comp. note) New Guinea, Bismarck Archipelago

***D. guadalcanalensis* Duffels, 1977** = *Diceropyga pigafettae* (nec Distant) = *Diceropyga guadalcanensis* (sic)
*Diceropyga guadalcanalensis*Duffels 1983a: 495, Fig 6 (distribution)
Diceropyga guadalcanalensis Duffels 1986: 323, Fig 6 (distribution)
Diceropyga guadalcanalensis Duffels and Boer 1990: 262, Fig 21.4 (distribution)
Diceropyga guadalcanensis (sic) Boer 1995d: 207, Fig 33 (distribution, comp. note)
Diceropyga guadalcanalensis Boer and Duffels 1996b: 299, 329, Plate 1c (illustrated, distribution) Solomon Islands
Diceropyga guadalcanensis (sic) Boer and Duffels 1996b: 315, Fig 9 (distribution, comp. note)
Diceropyga guadalcanalensis Duffels and Turner 2002: 240, 246–248, 256, 258, Fig 3A, Fig 9A, Fig 11, Appendices 1–2 (illustrated, phylogeny, distribution, listed, comp. note) Solomon Islands

D. impar Walker, 1868 to *Rhadinopyga impar*

***D. junctivitta* (Walker, 1868)** = *Dundubia junctivitta* = *Cosmopsaltria junctivitta* = *Diceropyga junctivitta* = *Diceropyga obtecta* (nec Fabricius) = *Dundubia subapicalis* Walker, 1870 (partim) = *Cosmopsaltria pigafettae* Distant, 1888 (partim) = *Diceropyga pigafettae* (nec Distant)
Diceropyga junctivitta Duffels 1983a: 495, Fig 6 (distribution)
Diceropyga junctivitta Duffels 1986: 323, 325, Figs 5–6 (comp. note) Maluku
Diceropyga junctivitta Duffels 1988a: 7–8, 10–11, Table 1 (comp. note) Bacan
Diceropyga junctivitta Duffels 1988b: 23 (comp. note) Morotai, Halmahera, Ternate, Kajoa, Bacan, Obi, Talaud
Diceropyga junctivitta Duffels and Boer 1990: 259, 262, Figs 21.3–21.4 (distribution, phylogeny) Moluccas
Diceropyga junctivitta Boer 1995d: 207, Fig 33 (distribution, comp. note)
Diceropyga junctivitta Boer and Duffels 1996b: 315, 329, Fig 9 (distribution, comp. note) North Maluku, New Guinea
Diceropyga junctivitta Duffels and Turner 2002: 237, 246–248, 256, 258, Fig 1A, Fig 11, Appendices 1–2 (illustrated, phylogeny, distribution, listed, comp. note) Maluku, Waigeo
Diceropyga junctivitta Shiyake 2007: 5, 65–66, Fig 106 (illustrated, distribution, listed, comp. note)

***D. major* Duffels, 1977**
Diceropyga major Duffels 1983a: 495, Fig 6 (distribution)
Diceropyga major Duffels 1986: 323, Fig 6 (distribution)
Diceropyga major Duffels and Boer 1990: 262, Fig 21.4 (distribution)
Diceropyga major Boer 1995d: 207, Fig 33 (distribution, comp. note)
Diceropyga major Boer and Duffels 1996b: 315, 329, Fig 9 (distribution, comp. note) Solomon Islands
Diceropyga major Duffels and Turner 2002: 247–248, 256, 259, Fig 11, Appendices 1–2 (phylogeny, distribution, listed, comp. note) Solomon Islands

***D. malaitensis* Duffels, 1977**
Diceropyga malaitensis Duffels 1983a: 495, Fig 6 (distribution)
Diceropyga malaitensis Duffels 1986: 323, Fig 6 (distribution)
Diceropyga malaitensis Duffels and Boer 1990: 262, Fig 21.4 (distribution)
Diceropyga malaitensis Boer 1995d: 207, Fig 33 (distribution, comp. note)
Diceropyga malaitensis Boer and Duffels 1996b: 315, 329, Fig 9 (distribution, comp. note) Solomon Islands
Diceropyga malaitensis Duffels and Turner 2002: 247–248, 256, 259, Fig 11, Appendices 1–2 (phylogeny, distribution, listed, comp. note) Solomon Islands

***D.* n. sp. Biak Boer, 1995** = *Diceropyga* sp. A Duffels & Turner, 2002
Diceropyga n. sp. Biak Boer 1995d: 207, Fig 33 (distribution, comp. note)
Diceropyga n. sp. Biak Boer and Duffels 1996b: 315, 329, Fig 9 (distribution, comp. note) New Guinea
Diceropyga sp. A Duffels and Turner 2002: 236, 247–248, 256, 258, Fig 11, Appendices 1–2 (phylogeny, distribution, listed, comp. note) Biak

***D.* n. sp. Nunfer Boer, 1995** = *Diceropyga* sp. B Duffels & Turner, 2002
Diceropyga n. sp. Nunfer Boer 1995d: 207, Fig 33 (distribution, comp. note)
Diceropyga n. sp. Nunfer Boer and Duffels 1996b: 315, 329, Fig 9 (distribution, comp. note) New Guinea
Diceropyga sp. B Duffels and Turner 2002: 236, 247–248, 256, 259, Fig 11, Appendices 1–2

(phylogeny, distribution, listed, comp. note) Numfur

D. *noonadani* Duffels, 1977

Diceropyga noonadani Duffels 1983a: 495, Fig 6 (distribution)

Diceropyga noonadani Duffels 1986: 323, Fig 6 (distribution)

Diceropyga noodadani Duffels and Boer 1990: 262, Fig 21.4 (distribution)

Diceropyga noonadani Boer 1995d: 207, Fig 33 (distribution, comp. note)

Diceropyga noonadani Boer and Duffels 1996b: 315, 329, Fig 9 (distribution, comp. note) Bismarck Archipelago

Diceropyga noonadani Duffels and Turner 2002: 246, 248, 253, 256, 258, Fig 9D, Fig 11, Appendices 1–2 (illustrated, phylogeny, distribution, listed, comp. note) Bismarck Archipelago

D. *novaebritannicae* Duffels, 1977

Diceropyga novaebritannicae Duffels 1983a: 495, Fig 6 (distribution)

Diceropyga novaebritannicae Duffels 1986: 323, Fig 6 (distribution)

Diceropyga novaebritannicae Duffels and Boer 1990: 262, Fig 21.4 (distribution)

Diceropyga novaebritannicae Boer 1995d: 207, Fig 33 (distribution, comp. note)

Diceropyga novaebritannicae Boer and Duffels 1996b: 315, 329, Fig 9 (distribution, comp. note) Bismarck Archipelago

Diceropyga novaebritannicae Duffels and Turner 2002: 248, 253, 256, 258, Fig 11, Appendices 1–2 (phylogeny, distribution, listed, comp. note) Bismarck Archipelago

D. *novaeguinae* Distant, 1912

Diceropyga novaeguinae Duffels 1983a: 495, Fig 6 (distribution)

Diceropyga novaeguinae Duffels and Boer 1990: 262, Fig 21.4 (distribution)

Diceropyga novaeguinae Duffels and Turner 2002: 245, 248–249, 256, 259, Fig 11, Fig 13, Appendices 1–2 (phylogeny, distribution, listed, comp. note) New Guinea

D. *novaeguinea* Duffels, 1977

Diceropyga novaeguinea Duffels 1986: 323, Fig 6 (distribution)

Diceropyga novaeguinea Boer 1995d: 207, Fig 33 (distribution, comp. note)

Diceropyga novaeguinea Boer and Duffels 1996b: 304, 315, Fig 4, Fig 9 (distribution, phylogeny, comp. note) Papuan Peninsula

D. *obliterans* Duffels, 1977

Diceropyga obliterans Duffels 1983a: 493, 495, Figs 5–6 (diversity, distribution) Bismarck Archipelago

Diceropyga obliterans group Duffels 1983a: 494–497, Figs 6–7 (diversity distribution, phylogeny, comp. note) Bismarck Archipelago

Diceropyga obliterans group Duffels 1983b: 40, 42, Fig 21 (diversity, distribution, comp. note) Bismarck Archipelago

Diceropyga obliterans group Duffels 1986: 321, 323, 330–331, Fig 3, Fig 6, Fig 10 (distribution, diversity, phylogeny, comp. note) Bismarck Archipelago

Diceropyga obliterans Duffels 1986: 323, Fig 6 (distribution)

Diceropyga obliterans Boer 1989: 7 (comp. note)

Diceropyga obliterans group Duffels and Boer 1990: 262, Fig 21.4 (distribution) Bismarck Archipelago

Diceropyga obliterans Duffels and Boer 1990: 262, Fig 21.4 (distribution)

Diceropyga obliterans group Boer 1995d: 207, 232, 235, Fig 33 (distribution, comp. note) Bismarck Archipelago

Diceropyga obliterans Boer 1995d: 207, Fig 33 (distribution, comp. note)

Diceropyga obliterans group Boer and Duffels 1996a: 173 (comp. note) Bismarck Archipelago

Diceropyga obliterans group Boer and Duffels 1996b: 304, 315, 329, Fig 4, Fig 9 (distribution, phylogeny, comp. note) Bismarck Archipelago

Diceropyga obliterans Boer and Duffels 1996b: 315, 329, Fig 9 (distribution, comp. note) Bismarck Archipelago

Diceropyga obliterans group Duffels 1997: 552 (diversity, comp. note) Bismarck Archipelago

Diceropyga obliterans group Duffels and Turner 2002: 245, 247, 249, 256, Fig 13, Appendix 1 (phylogeny, distribution, listed, comp. note) Bismarck Archipelago

Diceropyga obliterans Duffels and Turner 2002: 245, 248, 253, 256, 258, Fig 11, Appendices 1–2 (phylogeny, distribution, listed, comp. note) Bismarck Archipelago

D. *obtecta* (Fabricius, 1803) = *Tettigonia obtecta* = *Cicada obtecta* = *Cosmopsaltria obtecta* = *Cosmopsaltria (Diceropyga) obtecta* = *Cosmopsaltria (Diceropygis)* (sic) *obtecta* = *Dundubia obtecta* = *Dundubia bicaudata* Walker, 1858 = *Cosmopsaltria bicaudata* = *Cosmopsaltria (Diceropyga) bicaudata* = *Dundubia subapicalis* (nec Walker) = *Cosmopsaltria pigafettae* (nec Distant)

Tettigonia obtecta Dallas 1870: 482 (synonymy) Equals *Cosmopsaltria obtecta*

Diceropyga obtecta Duffels 1983a: 493, 495, Figs 5–6 (diversity, distribution) Moluccas

Diceropyga obtecta group Duffels 1983a: 494–495, 497, Figs 6–7 (diversity distribution, phylogeny, comp. note) Moluccas, Waigeo, Gebe Island

Diceropyga obtecta group Duffels 1983b: 40, 42, Fig 21 (diversity, distribution, comp. note) Moluccas, Waigeo, Gebe Island

Diceropyga obtecta group Duffels 1988a: 10–11 (comp. note)

Diceropyga obtecta group Duffels 1986: 321, 323, 330, Fig 3, Figs 5–6, Fig 10 (distribution, diversity, phylogeny, comp. note) Maluku

Diceropyga obtecta Duffels 1986: 323, Fig 6 (comp. note) Maluku

Diceropyga obtecta Duffels 1988a: 10–11 (comp. note) Maluku, Selatan

Diceropyga obtecta group Duffels 1988b: 23 (comp. note)

Diceropyga obtecta Duffels 1988b: 23 (comp. note) Buru, Seram, Ambon, Sula

Diceropyga obtecta Duffels 1990b: 70 (comp. note) Seram, Buru, Ambon, Sula

Diceropyga obtecta group Duffels and Boer 1990: 259–260, 262, Figs 21.3–21.4 (distribution, phylogeny, comp. note)

Diceropyga obtecta Duffels and Boer 1990: 259, 262, Figs 21.3–21.4 (distribution, phylogeny) Moluccas, Sula

Tettigonia obtecta Moulds 1990: 102 (type species of *Diceropyga*)

Diceropyga obtecta group Boer 1994c: 133 (comp. note) Molucca Islands

Diceropyga obtecta group Boer 1995d: 206–207, Fig 33 (distribution, comp. note) Maluku

Diceropyga obtecta Boer 1995d: 206–207, 229, Fig 33 (distribution, comp. note) Sula

Diceropyga obtecta group Boer and Duffels 1996b: 304, 315, 329, Fig 4, Fig 9 (distribution, phylogeny, comp. note) Maluku

Diceropyga obtecta Boer and Duffels 1996b: 315, 329, Fig 9 (distribution, comp. note) South Maluku

Diceropyga obtecta group Duffels 1997: 552 (diversity, comp. note) Maluku

Diceropyga obtecta group Duffels and Turner 2002: 244–247, 249–250, 252, 256, Fig 13, Appendix 1 (phylogeny, distribution, listed, comp. note) Maluku

Diceropyga obtecta Duffels and Turner 2002: 239, 244, 246–248, 256, 258, Fig 7A, Fig 9B, Fig 11, Appendices 1–2 (illustrated, phylogeny, distribution, listed, comp. note) Maluku

Tettigonia obtecta Moulds and Carver 2002: 11 (type species of *Diceropyga*)

***D. ochrothorax* Duffels, 1977**

Diceropyga ochrothorax Duffels 1983a: 495, Fig 6 (distribution)

Diceropyga ochrothorax Duffels 1986: 323, 325, Figs 5–6 (comp. note) Gebe Island

Diceropyga ochrothorax Duffels 1988a: 10 (comp. note)

Diceropyga junctivitta Duffels 1988b: 23 (comp. note) Gebe Island

Diceropyga ochrothorax Duffels and Boer 1990: 259, 262, Figs 21.3–21.4 (distribution, phylogeny) Gebe

Diceropyga ochrothorax Boer 1995d: 207, Fig 33 (distribution, comp. note)

Diceropyga ochrothorax Boer and Duffels 1996b: 315, 329, Fig 9 (distribution, comp. note) North Maluku

Diceropyga ochrothorax Duffels and Turner 2002: 246–248, 256, 258, Fig 9C, Fig 11, Appendices 1–2 (illustrated, phylogeny, distribution, listed, comp. note) Maluku

***D. pigafettae* (Distant, 1888)** = *Cosmopsaltria pigafettae*

D. recedens Walker, 1870 to *Rhadinopyga recedens*

***D. rennellensis* Duffels, 1977**

Diceropyga rennellensis Duffels 1983a: 495, Fig 6 (distribution)

Diceropyga rennellensis Duffels 1986: 323, Fig 6 (distribution)

Diceropyga rennellensis Duffels and Boer 1990: 262, Fig 21.4 (distribution)

Diceropyga rennellensis Boer 1995d: 207, Fig 33 (distribution, comp. note)

Diceropyga rennellensis Boer and Duffels 1996b: 315, 329, Fig 9 (distribution, comp. note) Solomon Islands

Diceropyga rennellensis Duffels and Turner 2002: 241, 247–248, 256, 258, Fig 4A, Fig 11, Appendices 1–2 (illustrated, phylogeny, distribution, listed, comp. note) Solomon Islands

***D. subapicalis* (Walker, 1870)** = *Dundubia subapicalis* (partim) = *Diceropyga apicalis* (sic) = *Cosmopsaltria atra* Distant, 1897 = *Diceropyga atra* = *Diceroprocta obtecta* (nec Fabricius)

Diceropyga subapicalis Duffels 1983a: 493–495, Figs 5–6 (diversity, distribution) New Guinea, Bismarck Archipelago, Solomon Islands, Aru Island, Queensland

Diceropyga subapicalis group Duffels 1983a: 494–497, Figs 6–7 (diversity distribution, phylogeny, comp. note) New Guinea, Solomon Islands

Diceropyga subapicalis Duffels 1983b: 39 (comp. note)

Diceropyga subapicalis group Duffels 1983b: 40, 42, Fig 21 (diversity, distribution, comp. note) New Guinea, Bismarck Archipelago, Solomon Islands

Diceropyga subapicalis group Duffels 1986: 321, 323, 325, 330–331, Fig 3, Fig 6,

Fig 10 (distribution, diversity, phylogeny, comp. note) New Guinea, Australia, Bismarck Archipelago, Solomon Islands, d'Entrecasteaux Islands, Woodlark Island, Louisaide Archipelago

Diceropyga subapicalis Duffels 1986: 323, 325, Fig 6 (distribution, comp. note) New Guinea, Aru Island, Australia

Diceropyga apicalis (sic) Boulard 1988d: 154–155, 158, Fig 19 (operculum abnormalities, illustrated, comp. note) Indonesia

Dundubia subapicalis Duffels 1988a: 8, 10 (listed) Bacan

Diceropyga subapicalis Duffels 1988a: 10 (comp. note)

Diceropyga subapicalis group Duffels and Boer 1990: 259, 262–263, 265, 267, Figs 21.3–21.4 (distribution, phylogeny, comp. note) New Guinea, Queensland, Bismarck Islands, Solomon Islands

Diceropyga subapicalis Duffels and Boer 1990: 262–263, Fig 21.4 (distribution, comp. note)

Diceropyga subapicalis Moulds 1990: 12, 103–104, Plate 14, Fig 4, Plate 24, Fig 31 (illustrated, distribution, ecology, listed, comp. note) Australia, Queensland, Torres Straits Islands, New Guinea

Diceropyga subapicalis Boer 1992a: 25 (comp. note) New Guinea

Diceropyga subapicalis Boer 1995b: 16 (comp. note)

Diceropyga subapicalis group Boer 1995d: 207, 235, Fig 33 (distribution, comp. note) New Guinea, Queensland, Bismarck Archipelago, Admiralty Islands

Diceropyga subapicalis Boer 1995d: 207, 209, 222, 228, 233, Fig 33 (distribution, comp. note) Queensland, Aru Islands

Diceropyga subapicalis group Boer and Duffels 1996b: 304, 315, 329, Fig 4, Fig 9 (distribution, phylogeny, comp. note) Papuan Peninsula, New Guinea, Solomon Islands

Diceropyga subapicalis Boer and Duffels 1996b: 315, 329, Fig 9 (distribution, comp. note) New Guinea

Diceropyga subapicalis Boer 1997: 94 (comp. note) Queensland

Diceropyga subapicalis group Duffels 1997: 552 (diversity, comp. note) Solomon Islands, Biak, Numfer Island, New Guinea, Bismarck Archipelago

Diceropyga subapicalis group Duffels and Turner 2002: 245, 249, 252, 256, Fig 13, Appendix 1 (phylogeny, distribution, listed, comp. note) Goodenough Island, Normanby Island, Solomon Islands, New Guinea, Bismarck Archipelago, Aru Island, Misima Island, Woodlark Island, Biak, Numfer

Diceropyga subapicalis Duffels and Turner 2002: 239, 244–246, 248, 256, 258, Fig 11, Fig 13, Appendices 1–2 (phylogeny, distribution, listed, comp. note) New Guinea, Aru Island

Diceropyga subapicalis Moulds and Carver 2002: 11–12 (synonymy, distribution, ecology, type data, listed) Equals *Dundubia subapicalis* Equals *Cosmopsaltria atra* Equals *Dundubis obtecta* Kirkaldy 1905 Australia, Queensland, Torres Straits Islands

Dundubia subapicalis Moulds and Carver 2002: 11–12 (type data, listed) Aru Island

Cosmopsaltria atra Moulds and Carver 2002: 12 (type data, listed) Australia

Dundubia obtecta Kirkaldy 1905 Moulds and Carver 2002: 12 (type data, listed) Australia

Diceropyga subapicalis species group Arensburger, Simon, and Holsinger 2004: 1778 (comp. note)

Diceropyga subapicalis Moulds 2005b: 395–396, 398, 402–403, 417–420, 422, Figs 36–37, Figs 56–60, Table 1 (higher taxonomy, phylogeny, key, listed, comp. note) Australia

Diceropyga subapicalis Moulds 2009a: 503–504, Fig 14.4(8) (illustrated, comp. note) Papua New Guinea, Australia

Diceropyga subapicalis Lee and Hill 2010: 302 (comp. note)

***D. subjuga* Duffels, 1977**

Diceropyga subjuga Duffels 1983a: 495, Fig 6 (distribution)

Diceropyga subjuga Duffels 1986: 323, Fig 6 (distribution)

Diceropyga subjuga Duffels and Boer 1990: 262, Fig 21.4 (distribution)

Diceropyga subjuga Boer 1995d: 207, Fig 33 (distribution, comp. note)

Diceropyga subjuga Boer and Duffels 1996b: 315, 329, Fig 9 (distribution, comp. note) New Guinea

Diceropyga subjuga Duffels and Turner 2002: 246, 248, 256, 258, Fig 11, Appendices 1–2 (phylogeny, distribution, listed, comp. note) New Guinea

***D. tortifer* Duffels, 1977**

Diceropyga tortifer Duffels 1983a: 495, Fig 6 (distribution)

Diceropyga tortifer Duffels 1986: 323, Fig 6 (distribution)

Diceropyga tortifer Duffels and Boer 1990: 262, Fig 21.4 (distribution)

Diceropyga tortifer Boer 1995d: 207, Fig 33 (distribution, comp. note)

Diceropyga tortifer Boer and Duffels 1996b: 315, 329, Fig 9 (distribution, comp. note) Solomon Islands

Diceropyga tortifer Duffels and Turner 2002: 247, 248, 256, 258, Fig 11, Appendices 1–2 (phylogeny, distribution, listed, comp. note) Solomon Islands

D. triangulata **Duffels, 1977**

Diceropyga triangulata Duffels 1983a: 495, Fig 6 (distribution)

Diceropyga triangulata Duffels 1986: 323, Fig 6 (distribution)

Diceropyga triangulata Duffels and Boer 1990: 262, Fig 21.4 (distribution)

Diceropyga triangulata Boer 1995d: 207, Fig 33 (distribution, comp. note) Admiralty Islands

Diceropyga triangulata Boer and Duffels 1996b: 315, 329, Fig 9 (distribution, comp. note) Bismarck Archipelago

Diceropyga triangulata Duffels and Turner 2002: 248, 256, 258, Fig 11, Appendices 1–2 (phylogeny, distribution, listed) Bismarck Archipelago

D. woodlarkensis inexpectata **Duffels, 1977**

Diceropyga woodlarkensis inexpectata Duffels 1983a: 495, Fig 6 (distribution)

Diceropyga woodlarkensis inexpecta Duffels 1986: 323, Fig 6 (distribution)

Diceropyga woodlarkensis inexpecta Duffels and Boer 1990: 262, Fig 21.4 (distribution)

Diceropyga woodlarkensis inexpectata Boer and Duffels 1996b: 329 (distribution) Papuan Peninsula

Diceropyga woodlarkensis inexpectata Duffels and Turner 2002: 248, 256, 258, Fig 11, Appendices 1–2 (phylogeny, distribution, listed) Woodlark Islands

D. woodlarkensis woodlarkensis **Duffels, 1977**

Diceropyga woodlarkensis Duffels 1983a: 495, Fig 6 (distribution)

Diceropyga woodlarkensis woodlarkensis Duffels 1983a: 495, Fig 6 (distribution)

Diceropyga woodlarkensis Duffels 1986: 323, Fig 6 (distribution)

Diceropyga woodlarkensis woodlarkensis Duffels 1986: 323, Fig 6 (distribution)

Diceropyga woodlarkensis Duffels and Boer 1990: 262, Fig 21.4 (distribution)

Diceropyga woodlarkensis woodlarkensis Duffels and Boer 1990: 262, Fig 21.4 (distribution)

Diceropyga woodlarkensis Boer 1995b: 16 (comp. note)

Diceropyga woodlarkensis Boer 1995d: 207, Fig 33 (distribution, comp. note)

Diceropyga woodlarkensis Boer and Duffels 1996b: 315, Fig 9 (distribution, comp. note)

Diceropyga woodlarkensis woodlarkensis Boer and Duffels 1996b: 329 (distribution) Papuan Peninsula

Diceropyga woodlarkensis Boer 1999: 120 (comp. note) Louisiade Archipelago, Woodlark Island

Diceropyga woodlarkensis Duffels and Turner 2002: 244, Fig 7B (illustrated, comp. note)

Diceropyga woodlarkensis woodlarkensis Duffels and Turner 2002: 248, 256, 258, Fig 11, Appendices 1–2 (phylogeny, distribution, listed) Misima Island

Rhadinopyga **Duffels, 1985**

Rhadinopyga Duffels 1985: 275–276, 278–279 (n. gen, described, comp. note) New Guinea, Batjan

Rhadinopyga Boer 1986: 169 (comp. note) Vogelkop Peninsula, New Guinea

Rhadinopyga Duffels 1986: 320–322, 324, 330–331, Figs 1–3, Fig 7, Fig 10 (distribution, diversity, phylogeny, comp. note) New Guinea, Misöol, Salawati, Waigeo, Bacan

Rhadinopyga Duffels 1988a: 10–11 (comp. note) Bacan, Tjendrawasih, Waigeo, Salawati, Misoöl

Rhadinopyga Boer 1989: 7 (comp. note)

Rhadinopyga Boer 1990: 65–66 (comp. note)

Rhadinopyga Duffels 1990b: 65, Fig 1 (phylogeny, comp. note) New Guinea, Misöol, Salawati, Waigeo, Bacan

Rhadinopyga Duffels and Boer 1990: 261, 263, 266, Fig 21.8 (distribution, diversity, phylogeny, comp. note) New Guinea, Waigeo, Slalwati, Bacan, Misoöl

Rhadinopyga Boer 1993b: 144–145 (phylogeny, comp. note)

Rhadinopyga Duffels 1993: 1227, 1230, 1232, Fig 6 (phylogeny, comp. note)

Rhadinopyga Boer 1994b: 161, 163 (comp. note)

Rhadinopyga Boer 1994d: 90 (comp. note) New Guinea

Rhadinopyga Boer 1995a: 7 (comp. note)

Rhadinopyga Boer 1995d: 172, 209–211, 221–223, 227, 230–231, 235–236, Fig 36, Fig 51 (distribution, phylogeny, comp. note)

Rhadinopyga Boer and Duffels 1996a: 155, 167, 169, 171–173, Fig 89, Fig 11 (distribution, phylogeny, comp. note)

Rhadinopyga Boer and Duffels 1996b: 301, 306, 313–314, 316, 320, 330 (distribution, phylogeny, comp. note) New Guinea

Rhadinopyga Duffels 1997: 549, 551–552, Fig 2 (phylogeny, comp. note)

Rhadinopyga Holloway 1997: 222 (distribution, comp. note) New Guinea

Rhadinopyga Holloway 1998: 308–309, Fig 13 (distribution, phylogeny, comp. note) New Guinea

Rhadinopyga Holloway and Hall 1998: 9–10, 15, Fig 3 (distribution, comp. note) New Guinea
Rhadinopyga Duffels and Turner 2002: 236–237, 239–240, 242–246, 249–251, 253, 257, Figs 13–15 (phylogeny, distribution, listed, comp. note) New Guinea, Salawati, Waigeo, Maluku, Misool
Rhadinopyga Moulds 2005b: 391, 432–433 (higher taxonomy, listed, comp. note)
Rhadinopyga Sanborn 2005b: 21, 23 (comp. note) New Guinea
Rhadinopyga spp. Sanborn 2005b: 21 (comp. note) New Guinea
Rhadinopyga Duffels and Boer 2006: 533–534 (distribution, diversity, comp. note) New Guinea, Bacan, Salawati, Waigeo, Roon

***R. acuminata* (Duffels, 1977)** = *Diceropyga acuminata*
Diceropyga acuminata Duffels 1983a: 495, Fig 6 (distribution)
Diceropyga acuminata Duffels 1985: 275, 278, Fig 10 (distribution, comp. note) Waigeo
Rhadinopyga acuminata Duffels 1985: 275 (n. comb.)
Rhadinopyga acuminata Duffels 1986: 324, Fig 7 (distribution, comp. note) Waigeo
Rhadinopyga acuminata Boer 1995d: 209–210, 231, Fig 36 (distribution, comp. note) Waigeu Island
Rhadinopyga acuminata Boer and Duffels 1996b: 330 (distribution) New Guinea
Rhadinopyga acuminata Duffels and Turner 2002: 248, 256, 259, Fig 11, Appendices 1–2 (phylogeny, distribution, listed) Waigeo
Rhadinopyga acuminata Sanborn 2005b: 23 (comp. note)

***R. duffelsi* Sanborn, 2005**
Rhadinopyga duffelsi Sanborn 2005b: 21–25, Figs 1–10 (n. sp., described, illustrated, comp. note) Indonesia

***R. epiplatys* Duffels, 1985**
Rhadinopyga epiplatys Duffels 1985: 275–278, Figs 1–9 (n. sp., described, illustrated, comp. note) New Guinea, Misool Island, Batjan
Rhadinopyga epiplatys Duffels 1986: 324, Fig 7 (distribution, comp. note) New Guinea, Batjan, Misöol
Rhadinopyga epiplatys Duffels 1988a: 8, 11, Table 1 (comp. note) Bacan, Misoöl, Tjendrawasih
Rhadinopyga epiplatys Boer 1995d: 209–210, Fig 36 (distribution, comp. note) New Guinea, Misool, Bacan
Rhadinopyga epiplatys Boer and Duffels 1996b: 330 (distribution) North Maluku, New Guinea
Rhadinopyga epiplatys Duffels and Turner 2002: 241, 248, 256, 259, fig 4C, Fig 11, Appendices 1–2 (illustrated, phylogeny, distribution, listed, comp. note) Maluku, Misool, New Guinea

***R. impar impar* (Walker, 1868)** = *Dundubia impar* = *Cosmopsaltria impar* = *Diceropyga impar*
Diceropyga impar Duffels 1983a: 495, Fig 6 (distribution)
Diceropyga impar Duffels 1985: 275, 278, Fig 10 (distribution, comp. note) Waigeo
Rhadinopyga impar Duffels 1985: 275 (n. comb.)
Rhadinopyga impar Boer 1995d: 209–210, 231, Fig 36 (distribution, comp. note) Waigeu Island
Rhadinopyga impar Boer and Duffels 1996b: 330 (distribution) New Guinea
Rhadinopyga impar Duffels and Turner 2002: 248, 256, 259, Fig 11, Appendices 1–2 (phylogeny, distribution, listed) Waigeo

***R. impar* var. *β* (Distant, 1891)** = *Cosmopsaltria impar* var. *β* = *Diceropyga impar* var. *β*

***R. recedens* (Walker, 1870)** = *Dundubia recedens* = *Cosmopsaltria recedens* = *Diceropyga recedens* = *Cosmopsaltria lutulenta* Distant, 1888 = *Diceropyga lutulenta*
Diceropyga recedens Duffels 1983a: 493, 495, Figs 5–6 (diversity, distribution) New Guinea, Waigeo
Diceropyga recedens group Duffels 1983a: 494–497, Figs 6–7 (diversity distribution, phylogeny, comp. note) New Guinea, Waigeo, Salawati
Diceropyga recedens group Duffels 1983b: 40, 42, Fig 21 (diversity, distribution, comp. note) Waigeo, Salawati, New Guinea
Diceropyga recedens group Duffels 1985: 275 (comp. note)
Diceropyga recedens Duffels 1985: 275, 278, Fig 10 (distribution, comp. note) Sulawati
Dundubia recedens Duffels 1985: 275 (type species of *Rhadinopyga*)
Rhadinopyga recedens Duffels 1985: 275 (n. comb.)
Diceropyga recedens group Duffels 1986: 320 (comp. note)
Rhadinopyga recedens Duffels 1986: 324, Fig 7 (distribution, comp. note) Salawati
Rhadinopyga recedens Boer 1995d: 209–210, 231, Fig 36 (distribution, comp. note) Sulawati
Rhadinopyga recedens Boer and Duffels 1996b: 330 (distribution) New Guinea
Rhadinopyga recedens Duffels and Turner 2002: 242, 245, 248, 256, 259, Fig 5D, Fig 8H, Fig 11, Appendices 1–2 (illustrated, phylogeny, distribution, listed, comp. note) Salawati
Diceropyga recedens group Sanborn 2005b: 21 (comp. note)
Rhadinopyga recedens group Sanborn 2005b: 23 (comp. note)

***Inflatopyga* Duffels, 1997** = Genus "I" Duffels

Genus "I" Duffels in prep. Boer 1995d: 219, 221–223, 226, 236–237, Fig 51 (distribution, phylogeny, comp. note) Solomon Islands

Genus "I" Duffels in prep. Boer and Duffels 1996a: 155, 168–169, 171, 173, Fig 9, Fig 11 (distribution, phylogeny, comp. note) Solomon Islands, Bougainville, Santa Isabel, Malaita, Guadalcanal, New Georgia

Genus "I" Duffels in prep. Boer and Duffels 1996b: 301, 304, 313, 320, 330, Fig 4 (distribution, comp. note) Solomon Islands

Inflatopyga Duffels 1997: 549–553, Figs 1–2 (n. gen., described, distribution, phylogeny, key, comp. note) Solomon Islands

Inflatopyga sp. Duffels 1997: 567 (comp. note) Bougainville

Inflatopyga Holloway 1998: 305, 307–308, Fig 10, Figs 12–13 (distribution, phylogeny, comp. note) New Guinea

Inflatopyga Holloway and Hall 1998: 9–10, 15, Fig 3 (distribution, comp. note) Solomon Islands

Inflatopyga Duffels and Turner 2002: 236–239, 242–244, 247, 249–252, 257, Figs 13–15 (phylogeny, distribution, listed, comp. note) Solomon Islands

***I. boulardi* Duffels, 1997**

Inflatopyga boulardi Duffels 1997: 550, 553–555, Fig 1, Figs 3–4 (key, n. sp., described, illustrated, distribution, comp. note) Solomon Islands, Bougainville

Inflatopyga boulardi Duffels and Turner 2002: 242, 248, 257, 259, Fig 5B, Fig 11, Appendices 1–2 (illustrated, phylogeny, distribution, listed, comp. note) Solomon Islands

***I. ewarti* Duffels, 1997**

Inflatopyga ewarti Duffels 1997: 550, 553, 555–557, Fig 1, Figs 5–8 (key, n. sp., described, illustrated, distribution, comp. note) Solomon Islands, Bougainville

Inflatopyga ewarti Duffels and Turner 2002: 238, 248, 257, 259, Fig 2C, Fig 11, Appendices 1–2 (illustrated, phylogeny, distribution, listed, comp. note) Solomon Islands

***I. langeraki* Duffels, 1997**

Inflatopyga langeraki Duffels 1997: 550, 553, 555, 559–563, Fig 1, Figs 14–28 (key, n. sp., described, illustrated, distribution, comp. note) Solomon Islands, Guadalcanal, New Georgia Islands

Inflatopyga langeraki Duffels and Turner 2002: 237, 248, 257, 259, Fig 1B, Fig 11, Appendices 1–2 (illustrated, phylogeny, distribution, listed, comp. note) Solomon Islands

***I. mouldsi* Duffels, 1997**

Inflatopyga mouldsi Duffels 1997: 550, 553, 555, 565–567, Fig 1, Figs 35–38 (key, n. sp., described, illustrated, distribution, comp. note) Solomon Islands, Malaita Island

Inflatopyga mouldsi Duffels and Turner 2002: 248, 257, 259, Fig 11, Appendices 1–2 (phylogeny, distribution, listed) Solomon Islands

***I. verlaani* Duffels, 1997**

Inflatopyga verlaani Duffels 1997: 550, 553, 555–559, Fig 1, Figs 9–13 (key, n. sp., described, illustrated, distribution, comp. note) Solomon Islands, Bougainville

Inflatopyga verlaani Duffels and Turner 2002: 248, 257, 259, Fig 11, Appendices 1–2 (phylogeny, distribution, listed) Solomon Islands

***I. webbi* Duffels, 1997**

Inflatopyga webbi Duffels 1997: 550, 553, 555, 563–565, Fig 1, Figs 29–34 (key, n. sp., described, illustrated, distribution, comp. note) Solomon Islands, Santa Isabel

Inflatopyga webbi Duffels and Turner 2002: 241, 248, 257, 259, Fig 4D, Fig 11, Appendices 1–2 (illustrated, phylogeny, distribution, listed, comp. note) Solomon Islands

***Aceropyga* Duffels, 1977**

Aceropyga Duffels 1982: 156, 159–160 (biogeography, phylogeny, comp. note) Melanesia

Aceropyga Duffels 1983a: 492–498, Figs 1–2, Fig 7 (diversity, distribution, phylogeny, comp. note) Bismarck Archipelago, Somolon Islands, Tonga Island, West Pacific

Aceropyga Duffels 1983b: 2, 10–11, 19–20, 37–42, Fig 7, Fig 21 (phylogeny, comp. note) Bismarck Archipelago, Somolon Islands, Vanuatu, Fiji Islands, Tonga Islands

Aceropyga Duffels 1986: 320–322, 325, 330–331, Figs 1–3, Fig 10 (distribution, diversity, phylogeny, comp. note) Bismarck Archipelago, Solomon Islands, Fiji and Tonga Islands, Vanuatu

Aceropyga Duffels 1988d: 8, 10, 12, 14, 24–25, 28, 30–31, 43–44, 75, 78, 80, Fig 8, Table 1 (key, distribution, diversity, phylogeny, listed, comp. note) Bismarck Archipelago, Somolon Islands, Fiji and Tonga Islands, Vanuatu, Kusaie Island

Aceropyga Duffels 1990b: 65, Fig 1 (phylogeny, comp. note) Bismarck Archipelago, Somolon Islands, West Pacific

Aceropyga Duffels and Boer 1990: 262–263, 265–266, Fig 21.5, Fig 21.8 (distribution, diversity, phylogeny, comp. note) Bismarck Archipelago, Somolon Islands, Fiji and Tonga Islands, Vanuatu

Aceropyga Duffels 1993: 1226–1227, 1230, 1232–1233, Fig 6 (phylogeny, listed, comp. note)
Aceropyga Boer 1994b: 161, 163 (comp. note)
Aceropyga Boer 1995d: 172–173, 219–220, 222–223, 227, 236–237, Fig 33, Fig 49, Fig 51 (distribution, phylogeny, comp. note) New Guinea, Maluku, Bismarck Archipelago, East Melanesia, Caroline Islands, Fiji, Tonga
Aceropyga Boer and Duffels 1996a: 155, 168–169, 171–173, Fig 9, Fig 11 (distribution, phylogeny, comp. note) East Melanesia, Vanuatu, Kusaie, Tonga, Fiji
Aceropyga Boer and Duffels 1996b: 301, 304, 307, 313–314, 316, 319–320, 328, Fig 4, Fig 14 (distribution, phylogeny, comp. note) Fiji Islands, Vanuatu, Tonga
Aceropyga Duffels 1997: 549, 552 (comp. note)
Aceropyga Holloway 1998: 304–305, 307–308, Fig 10, Figs 12–13 (distribution, phylogeny, comp. note) Fiji, Bismarck Islands
Aceropyga Holloway and Hall 1998: 10, Fig 3 (distribution, comp. note)
Aceropyga Duffels and Turner 2002: 236, 239, 243–245, 247, 249, 256, Figs 13–15, Appendix 1 (phylogeny, distribution, listed, comp. note) Fiji
Aceropyga Eberhard 2004: 139, Table 2 (listed)
Aceropyga Moulds 2005b: 387, 391, 400, 415, 432–433 (higher taxonomy, listed, comp. note)
Aceropyga Wilson 2009: 35 (comp. note) Fiji
Aceropyga Lee and Hill 2010: 302 (comp. note)

A. acuta **Duffels, 1988**
Aceropyga acuta Duffels 1988d: 12, 30–32, 58, 63–65, 68, Fig 13, Figs 115–118, Table 1 (n. sp., key, described, illustrated, distribution, phylogeny, comp. note) Fiji Islands
Aceropyga acuta Duffels and Boer 1990: 262, Fig 21.5 (distribution)
Aceropyga acuta Duffels 1993: 1233–1234 (listed, comp. note) Fiji Islands
Aceropyga acuta Boer 1995d: 220, Fig 49 (distribution, comp. note)
Aceropyga acuta Boer and Duffels 1996b: 320, 328, Fig 14 (distribution, comp. note) Fiji Islands
Aceropyga acuta Duffels and Turner 2002: 237–238, 248, 256, 258, Fig 11, Appendices 1–2 (phylogeny, distribution, listed, comp. note) Fiji Islands
Aceropyga acuta Wilson 2009: 40, Appendix (listed, distribution, comp. note) Fiji

A. albostriata **(Distant, 1888)** = *Cosmopsaltria albostriata* = *Diceropyga albostriata*
Aceropyga albostriata Duffels 1986: 325, 331 (comp. note) Tonga Islands
Aceropyga albostriata Duffels 1988d: 10, 12, 24–25, 29, 31–32, 40–42, 80, Fig 2, Fig 13, Figs 35–41, Table 1 (key, described, illustrated, phylogeny, distribution, listed, comp. note) Equals *Cosmopsaltria albostriata* Tonga Islands
Cosmopsaltria albostriata Duffels 1988d: 10 (comp. note) Tonga Islands
Aceropyga albostriata Duffels and Boer 1990: 262, Fig 21.5 (distribution)
Aceropyga albostriata Duffels 1993: 1234 (comp. note) Tongo Islands
Aceropyga albostriata Boer 1995d: 220, Fig 49 (distribution, comp. note)
Aceropyga albostriata Boer and Duffels 1996b: 320, 328, Fig 14 (distribution, comp. note) Tonga Islands
Aceropyga albostriata Duffels and Turner 2002: 244, 248, 256, 258, Fig 11, Appendices 1–2 (phylogeny, distribution, listed, comp. note) Tonga Islands

A. corynetus corynetus **Duffels, 1977** = *Aceropyga corynetus*
Aceropyga corynetus group Duffels 1986: 325 (comp. note) Fiji Islands
Aceropyga corynetus Duffels 1986: 325 (comp. note) Fiji Islands
Aceropyga corynetus group Duffels 1988d: 11, 13, 29, 31, 45, Fig 3, Fig 6 (key, phylogeny, comp. note) Fiji Islands
Aceropyga corynetus Duffels 1988d: 11, 13, 24, 26, 29–32, 45, 47, 51, 53, 56, 66, Fig 13 (key, phylogeny, listed, comp. note) Fiji Islands
Aceropyga corynetus cornyetus Duffels 1988d: 11–12, 29, 52, 54–58, Fig 3, Figs 71–74, Figs 81–85, Figs 91–92, Figs 100–101, Table 1 (key, described, illustrated, sonogram, phylogeny, distribution, listed, comp. note) Fiji Islands
Aceropyga corynetus complex Duffels 1988d: 24, 31–32, 42–43, 47, Fig 13 (diversity, phylogeny, listed, comp. note) Fiji Islands
Aceropyga corynetus corynetus Duffels and Boer 1990: 262, Fig 21.5 (distribution)
Aceropyga corynetus complex Duffels 1993: 1227, 1230, 1232–1234, Fig 5 (phylogeny, listed comp. note)
Aceropyga corynetus Duffels 1993: 1233–1234 (comp. note) Fiji Islands
Aceropyga corynetus group Duffels 1993: 1234 (listed)
Aceropyga corynetus corynetus Boer 1995d: 220, Fig 49 (distribution, comp. note)
Aceropyga corynetus corynetus Boer and Duffels 1996b: 299, 320, 328, Fig 14, Plate 1b (distribution, illustrated, comp. note) Fiji Islands

Aceropyga corynetus Duffels and Turner 2002: 237, 239–240, 242, 245, Fig 3E, Fig 5F, Figs 8E–8G (illustrated, comp. note)

Aceropyga corynetus complex Duffels and Turner 2002: 244–245, 256, Appendix 1 (distribution, listed, comp. note)

Aceropyga corynetus corynetus Duffels and Turner 2002: 248, 256, 258, Fig 11, Appendices 1–2 (phylogeny, distribution, listed) Fiji Islands

Aceropyga corynetus corynetus Wilson 2009: 40, Appendix (listed, distribution, comp. note) Fiji

A. *corynetus monacantha* Duffels, 1988

Aceropyga corynetus monacantha Duffels 1988d: 11–12, 29, 51, 54, 58, Fig 3, Figs, 89–90, Fig 101, Table 1 (n. ssp., key, described, illustrated, phylogeny, distribution, listed, comp. note) Fiji Islands

Aceropyga corynetus monacantha Duffels and Boer 1990: 262, Fig 21.5 (distribution)

Aceropyga corynetus monacantha Boer 1995d: 220, Fig 49 (distribution, comp. note)

Aceropyga corynetus monacantha Boer and Duffels 1996b: 320, 328, Fig 14 (distribution, comp. note) Fiji Islands

Aceropyga corynetus monacantha Duffels and Turner 2002: 248, 256, 258, Fig 11, Appendices 1–2 (phylogeny, distribution, listed) Fiji Islands

Aceropyga corynetus monacantha Wilson 2009: 40, Appendix (listed, distribution, comp. note) Fiji

A. *corynetus ungulata* Duffels, 1988 = *Aceropyga corynetus ungulatus* (sic)

Aceropyga corynetus ungulata Duffels 1988d: 11–12, 29, 51, 53–58, Fig 3, Figs 75–80, Figs 86–88, Figs 93–99, Fig 101, Table 1 (n. ssp., key, described, illustrated, phylogeny, distribution, listed, comp. note) Fiji Islands

Aceropyga corynetus ungulata Duffels and Boer 1990: 262, Fig 21.5 (distribution)

Aceropyga corynetus ungulata Boer 1995d: 220, Fig 49 (distribution, comp. note)

Aceropyga corynetus ungulata Boer and Duffels 1996b: 305, 320, 328, Fig 14, Plate 3 (distribution, illustrated, comp. note) Fiji Islands

Aceropyga corynetus ungulatus (sic) Sanborn 1999a: 43 (type material, listed)

Aceropyga corynetus ungulata Duffels and Turner 2002: 248, 256, 258, Fig 11, Appendices 1–2 (phylogeny, distribution, listed) Fiji Islands

Aceropyga corynetus ungulata Wilson 2009: 40, Appendix (listed, distribution, comp. note) Fiji

A. *distans distans* (Walker, 1858) = *Dundubia distans* = *Cosmopsaltria distans* = *Diceropyga distans* = *Dundubia subfascia* Walker, 1858

Aceropyga distans Duffels 1986: 325 (comp. note) Fiji Islands

Aceropyga distans Duffels 1988d: 9, 11, 13, 24–26, 29–32, 37, 40, 80, 104, Figs 12–13 (key, type species of *Aceropyga*, illustrated, distribution, phylogeny, comp. note) Equals *Dundubia distans* Equals *Dundubia subfascia* Equals *Dundubia lineifera* Fiji Islands

Dundubia distans Duffels 1988d: 9 (comp. note) Fiji Islands

Dundubia subfascia Duffels 1988d: 9 (comp. note) Fiji Islands

Aceropyga distans group Duffels 1988d: 10, 13, 24, 26, 29, 31–32, Fig 2, Fig 5, Fig 13 (key, phylogeny, listed, comp. note) Fiji and Tonga Islands

Aceropyga distans distans Duffels 1988d: 10, 12, 29, 33–38, 40, Fig 2, Figs 14–21, Figs 23–33, Table 1 (key, described, illustrated, sonogram, phylogeny, distribution, listed, comp. note) Equals *Dundubia subfascia* Fiji and Tonga Islands

Aceropyga distans distans Duffels and Boer 1990: 262, Fig 21.5 (distribution)

Aceropyga distans group Duffels 1993: 1227, 1230, 1232–1233, Fig 5 (phylogeny, listed, comp. note)

Aceropyga distans Duffels 1993: 1233 (type species of *Aceropyga*, listed, comp. note) Fiji Islands

Aceropyga distans distans Boer 1995d: 220, Fig 49 (distribution, comp. note)

Aceropyga distans distans Boer and Duffels 1996b: 305, 320, 328, Fig 14, Plate 3 (distribution, illustrated, comp. note) Fiji Islands

Aceropyga distans Duffels and Turner 2002: 240–241, 244–245, Fig 3F, Fig 4E, Fig 8D (illustrated, comp. note)

Aceropyga distans group Duffels and Turner 2002: 247, 256, Appendix 1 (distribution, listed, comp. note)

Aceropyga distans distans Duffels and Turner 2002: 248, 256, 258, Fig 11, Appendices 1–2 (phylogeny, distribution, listed) Fiji Islands

Aceropyga distans distans Wilson 2009: 40, Appendix (listed, distribution, comp. note) Fiji

Aceropyga distans distans Lee and Hill 2010: 299–300, Figs 7–8 (phylogeny, comp. note)

A. *distans lineifera* (Walker, 1858) = *Dundubia lineifera* = *Cosmopsaltria lineifera* = *Cosmopsaltria (Diceropyga) lineifera*

Dundubia lineifera Duffels 1988d: 9 (comp. note) Fiji Islands

Aceropyga distans lineifera Duffels 1988d: 10, 12, 29, Fig 2, Fig 34, Table 1 (key, described, phylogeny, distribution, listed, comp. note) Fiji and Tonga Islands

Aceropyga distans lineifera Duffels and Boer 1990: 262, Fig 21.5 (distribution)

Aceropyga distans lineifera Boer 1995d: 220, Fig 49 (distribution, comp. note)

Aceropyga distans lineifera Boer and Duffels 1996b: 320, 328, Fig 14 (distribution, comp. note) Fiji Islands

Aceropyga distans lineifera Duffels and Turner 2002: 248, 256, 258, Fig 11, Appendices 1–2 (phylogeny, distribution, listed) Fiji Islands

Aceropyga distans lineifera Wilson 2009: 40, Appendix (listed, distribution, comp. note) Fiji

***A. distans taveuniensis* Duffels, 1977**

Aceropyga distans taveuniensis Duffels 1988d: 10, 12, 29, 34, 38–40, Fig 2, Fig 22, Fig 34, Table 1 (key, described, illustrated, phylogeny, distribution, listed, comp. note) Fiji and Tonga Islands

Aceropyga distans taveuniensis Duffels and Boer 1990: 262, Fig 21.5 (distribution)

Aceropyga distans taveuniensis Boer 1995d: 220, Fig 49 (distribution, comp. note)

Aceropyga distans taveuniensis Boer and Duffels 1996b: 320, 328, Fig 14 (distribution, comp. note) Fiji Islands

Aceropyga distans taveuniensis Duffels and Turner 2002: 248, 256, 258, Fig 11, Appendices 1–2 (phylogeny, distribution, listed) Fiji Islands

Aceropyga distans taveuniensis Wilson 2009: 40, Appendix (listed, distribution, comp. note) Fiji

***A. egmondae* Duffels, 1988**

Aceropyga egmondae Duffels 1988d: 12, 24, 30–32, 58, 64, 67–70, Fig 13, Fig 118, Fig 124, Figs 126–129, Table 1 (n. sp., key, described, illustrated, distribution, phylogeny, listed, comp. note) Fiji Islands

Aceropyga egmondae Duffels and Boer 1990: 262, Fig 21.5 (distribution)

Aceropyga egmondae Duffels 1993: 1233–1234 (listed, comp. note) Fiji Islands

Aceropyga egmondae Boer 1995d: 220, Fig 49 (distribution, comp. note)

Aceropyga egmondae Boer and Duffels 1996b: 320, 328, Fig 14 (distribution, comp. note) Fiji Islands

Aceropyga egmondae Duffels and Turner 2002: 237–238, 240, 248, 256, 258, Fig 3D, Fig 11, Appendices 1–2 (illustrated, phylogeny, distribution, listed, comp. note) Fiji Islands

Aceropyga egmondae Wilson 2009: 40, Appendix (listed, distribution, comp. note) Fiji

***A. huireka* Duffels, 1988**

Aceropyga huireka Duffels 1988d: 12, 24, 29, 31–32, 42–45, 64–67, 69–71, Fig 13, Figs 42–48, Fig 118, Fig 123, Table 1 (n. sp., key, described, illustrated, distribution, phylogeny, listed, comp. note) Fiji Islands

Aceropyga huireka Duffels and Boer 1990: 262, Fig 21.5 (distribution)

Aceropyga huireka Duffels 1993: 1232, 1234, Fig 5 (phylogeny, listed, comp. note) Fiji Islands

Aceropyga huireka Boer 1995d: 220, Fig 49 (distribution, comp. note)

Aceropyga huireka Boer and Duffels 1996b: 320, 328, Fig 14 (distribution, comp. note) Fiji Islands

Aceropyga huireka Duffels and Turner 2002: 237–238, 248, 256, 258, Fig 11, Appendices 1–2 (phylogeny, distribution, listed, comp. note) Fiji Islands

Aceropyga huireka Wilson 2009: 40, Appendix (listed, distribution, comp. note) Fiji

***A. macracantha* Duffels, 1988**

Aceropyga macracantha Duffels 1988d: 12, 30–32, 58, 63–66, 68, Fig 13, Figs 118–121, Table 1 (n. sp., key, described, illustrated, distribution, phylogeny, listed, comp. note) Fiji Islands

Aceropyga macracantha Duffels and Boer 1990: 262, Fig 21.5 (distribution)

Aceropyga macracantha Duffels 1993: 1233–1234 (listed, comp. note) Fiji Islands

Aceropyga macracantha Boer 1995d: 220, Fig 49 (distribution, comp. note)

Aceropyga macracantha Boer and Duffels 1996b: 320, 328, Fig 14 (distribution, comp. note) Fiji Islands

Aceropyga macracantha Duffels and Turner 2002: 237–238, 248, 256, 258, Fig 11, Appendices 1–2 (phylogeny, distribution, listed, comp. note) Fiji Islands

Aceropyga macracantha Wilson 2009: 40, Appendix (listed, distribution, comp. note) Fiji

***A. philorites* Duffels, 1988** = *Aceropyga philoritis* (sic)

Aceropyga philorites Duffels 1988d: 12, 30–32, 58–64, 67, Fig 13, Figs 102–114, Fig 118, Fig 122, Table 1 (n. sp., key, described, illustrated, distribution, phylogeny, listed, comp. note) Fiji Islands

Aceropyga philorites group Duffels 1988d: 24, 31–32, 58, 74–775, Fig 13 (phylogeny, listed, comp. note) Fiji Islands

Aceropyga philorites Duffels and Boer 1990: 262, Fig 21.5 (distribution)

Aceropyga philorites group Duffels 1993: 1232–1234, Fig 5 (phylogeny, listed, comp. note)
Aceropyga philorites Duffels 1993: 1233–1234 (listed, comp. note) Fiji Islands
Aceropyga philorites Boer 1995d: 220, Fig 49 (distribution, comp. note)
Aceropyga philorites Boer and Duffels 1996b: 320, 328, Fig 14 (distribution, comp. note) Fiji Islands
Aceropyga philorites group Duffels and Turner 2002: 237–238 (comp. note)
Aceropyga philorites Duffels and Turner 2002: 237–238, 248, 256, 258, Fig 2B, Fig 11, Appendices 1–2 (illustrated, phylogeny, distribution, listed, comp. note) Fiji Islands
Aceropyga philoritis (sic) Wilson 2009: 40, Appendix (listed, distribution, comp. note) Fiji

***A. poecilochlora* (Walker, 1858)** = *Dundubia poecilochlora* = *Cosmopsaltria poecilochlora* = *Cosmopsaltria (Diceropyga) poecilochlora* = *Diceropyga poecilochlora* = *Aceropyga poecilochora* (sic) = *Dundubia connata* Walker, 1858 = *Khimbya kusaiensis* Esaki, 1939 = *Diceropyga kusaiensis* = *Khimbya musaiensis* (sic)
Aceropyga poecilochlora Duffels 1983a: 494 (distribution) Vanuatu, Kusaie Island
Aceropyga poecilochlora Duffels 1986: 322, 325 (comp. note) Vanuata, Kusaie Island
Aceropyga poecilochlora Duffels 1988d: 10, 13, 31–32, Fig 2, Fig 5, Fig 13 (phylogeny, comp. note) Vanuata
Aceropyga poecilochlora Boulard 1990b: 214 (comp. note) Vanuatu
Aceropyga poecilochlora Duffels and Boer 1990: 262, Fig 21.5 (distribution)
Aceropyga poecilochora (sic) Duffels 1993: 1227, 1230, 1232–1233, Fig 5 (phylogeny, comp. note)
Aceropyga poecilochlora Duffels 1993: 1234 (listed) New Hebrides, Kusaie Island, Caroline Islands
Aceropyga poecilochlora Boer 1995d: 220, Fig 49 (distribution, comp. note)
Aceropyga poecilochlora Boer and Duffels 1996b: 309, 320, 328, Fig 14 (distribution, comp. note) Equals *Diceropyga kusaiensis* Caroline Islands, Vanuatu
Aceropyga poecilochlora Duffels and Turner 2002: 244–245, 248, 256, 258, Fig 11, Appendices 1–2 (phylogeny, distribution, listed, comp. note) Caroline Islands, Vanuatu

***A. pterophon* Duffels, 1988**
Aceropyga pterophon Duffels 1988d: 12, 26, 30–32, 58, 64, 67–75, Fig 13, Fig 118, Fig 125, Figs 130–140, Table 1 (n. sp., key, described, illustrated, sonogram, distribution, phylogeny, listed, comp. note) Fiji Islands
Aceropyga pterophon Duffels and Boer 1990: 262, Fig 21.5 (distribution)
Aceropyga pterophon Duffels 1993: 1233–1234 (listed, comp. note) Fiji Islands
Aceropyga pterophon Boer 1995d: 220, Fig 49 (distribution, comp. note)
Aceropyga pterophon Boer and Duffels 1996b: 320, 328, Fig 14 (distribution, comp. note) Fiji Islands
Aceropyga pterophon Gogala and Trilar 1998a: 8 (comp. note) Fiji Islands
Aceropyga pterophon Duffels and Turner 2002: 237–238, 248, 256, 258, Fig 11, Appendices 1–2 (phylogeny, distribution, listed, comp. note) Fiji Islands
Aceropyga pterophon Gogala and Trilar 2003: 9 (comp. note) Fiji Islands
Aceropyga pterophon Wilson 2009: 40, Appendix (listed, distribution, comp. note) Fiji

***A. stuarti pallens* Duffels, 1977**
Aceropyga stuarti pallens Duffels 1988d: 11–12, 28, 30, 46, 50–51, Fig 3, Figs 50–51, Figs 56–57, Figs 67–70, Table 1 (key, described, illustrated, phylogeny, listed, comp. note) Fiji Islands
Aceropyga stuarti pallens Duffels and Boer 1990: 262, Fig 21.5 (distribution)
Aceropyga stuarti pallens Boer 1995d: 220, Fig 49 (distribution, comp. note)
Aceropyga stuarti pallens Boer and Duffels 1996b: 320, 328, Fig 14 (distribution, comp. note) Fiji Islands
Aceropyga stuarti pallens Duffels and Turner 2002: 248, 256, 258, Fig 11, Appendices 1–2 (phylogeny, distribution, listed) Fiji Islands
Aceropyga stuarti pallens Wilson 2009: 40, Appendix (listed, distribution, comp. note) Fiji

***A. stuarti stuarti* (Distant, 1882)** = *Cosmopsaltria stuarti* = *Diceropyga stuarti*
Aceropyga stuarti Duffels 1986: 325 (comp. note) Fiji Islands
Aceropyga stuarti Duffels 1988d: 11, 24, 26, 29, 31–32, 45, 50–51, 104, Fig 13, Fig 70 (key, phylogeny, comp. note) Equals *Cosmopsaltria stuarti* Fiji Islands
Cosmopsaltria stuarti Duffels 1988d: 9 (comp. note) Fiji Islands
Aceropyga stuarti stuarti Duffels 1988d: 11–12, 29, 46–50, Fig 3, Fig 49, Figs 52–55, Figs 58–66, Fig 70, Table 1 (key, described, illustrated, sonogram, phylogeny, distribution, listed, comp. note) Fiji Islands
Aceropyga stuarti stuarti Duffels and Boer 1990: 262, Fig 21.5 (distribution)

Aceropyga stuarti Duffels 1993: 1233–1234 (listed, comp. note) Fiji Islands
Aceropyga stuarti stuarti Boer 1995d: 220, Fig 49 (distribution, comp. note)
Aceropyga stuarti stuarti Boer and Duffels 1996b: 320, 328, Fig 14 (distribution, comp. note) Fiji Islands
Aceropyga stuarti Duffels and Turner 2002: 237–238 (comp. note)
Aceropyga stuarti stuarti Duffels and Turner 2002: 248, 256, 258, Fig 11, Appendices 1–2 (phylogeny, distribution, listed) Fiji Islands
Aceropyga stuarti stuarti Wilson 2009: 40, Appendix (listed, distribution, comp. note) Fiji

***Meimuna* Distant, 1905** = *Meinuna* (sic) = *Meinuma* (sic) = *Miemuna* (sic) = *Meimuua* (sic) = *Maimuna* (sic) = *Meimura* (sic)
Meimuna Wu 1935: 13 (listed) China
Meimuna Chen 1942: 145 (listed)
Meimuna Kurokawa 1953: 1 (chromosomes, comp. note) Japan
Meimuna Josephson and Young 1981: 220 (comp. note)
Meimuna Fujiyama 1982: 184–186 (comp. note)
Meimuna Strümpel 1983: 189, Table 8 (listed, comp. note)
Meimura (sic) Ewing 1984: 228 (comp. note)
Meimuna Hayashi 1984: 28, 58, Fig 3b (key, described, listed, comp. note)
Meimuna Popov 1985: 35 (acoustic system, comp. note)
Meimuna Duffels 1990b: 71 (comp. note) Japan, Bonin Island, Ryukyus, Taiwan, Philippines
Meimuna Zhang and Zhang 1990: 337, 346 (listed) China
Meimuna Liu 1991: 258 (comp. note)
Meimuna Liu 1992: 17 (comp. note)
Meimuna sp. Anonymous 1994: 809 (fruit pest) China
Meimuna Peng and Lei 1992: 105 (comp. note) China, Hunan
Meimuna Lei 1994: 184 (listed)
Meimuna Zhang, Sun and Zhang 1994: 25, 56 (listed, comp. note) China
Meimuna Boer 1995d: 174, 201, 223, Fig 28, Fig 54 (distribution, phylogeny, comp. note)
Meimuna Gogala 1995: 112 (comp. note)
Meimuna Lee 1995: 18, 22, 53, 85–86, 136 (listed, key, described, ecology, distribution, illustrated, comp. note) Equals *Chosenosemia* Korea
Meimuna Zaidi and Ruslan 1995c: 70, Table 2 (listed, comp. note) Peninsular Malaysia
Meimuna Beuk 1996: 129, 133–134 (characteristics, comp. note)
Meimuna Boer and Duffels 1996a: 155, 163, 169, 171, Fig 1, Fig 9, Fig 11 (distribution, phylogeny, comp. note) Southeast Asia, Ryukyu
Meimuna Boer and Duffels 1996b: 301–302, 304, 315, Fig 1, Fig 4 (distribution, phylogeny, comp. note) Southeast Asia
Meimuna sp. Zaidi, Ruslan and Mahadir 1996: 61, Table 1 (listed, comp. note) Peninsular Malaysia
Meimuna Chou, Lei, Li, Lu and Yao 1997: 236–238, 293, 308, 318, 325–326, Table 6, Tables 10–12 (key, described, listed, diversity, biogeography) China
Meimuna Holloway 1997: 221, 223, Fig 6 (distribution, comp. note) Asia, Ryukyu Islands
Meimuna Novotny and Wilson 1997: 437 (listed)
Meimuna Beuk 1998: 147 (comp. note)
Meimuna Holloway 1998: 308–309, Fig 13 (distribution, phylogeny, comp. note) Ryukyu Islands, Asia
Meimuna Holloway and Hall 1998: 10, 15, Fig 3 (distribution, comp. note)
Meimuna Beuk 1999: 2, 5 (comp. note)
Meimuna Lee 1999: 8 (comp. note) Korea
Meimuna Ohbayashi, Sato and Igawa 1999: 340 (parasitism, comp. note)
Meimuna Duffels and Turner 2002: 236, 239, 242, 244–247, 249–252, Figs 14–15 (distribution, comp. note)
Meimuna spp. Lee, Choe, Lee and Woo 2002: 5 (comp. note)
Meimuna Boulard 2003a: 103 (comp. note)
Meimuna Lee and Hayashi 2003a: 170, 176–177 (key, diagnosis) Taiwan
Meimuna Matsushima, Yokotani, Lui, Sumida, Honda et al. 2003: 420 (comp. note) Japan
Meimuna Chen 2004: 39, 41, 86, 192 (phylogeny, listed)
Meimuna sp. Chen 2004: 86, 94, 198 (described, illustrated, listed, comp. note) Taiwan
Meimuna Boulard 2005e: 121 (comp. note)
Meimuna Lee 2005: 15, 17, 81, 157–158 (listed, key, comp. note) Korea
Meimuna Moulds 2005b: 391, 432–433 (higher taxonomy, listed, comp. note)
Meimuna Pham and Thinh 2005a: 242 (listed) Vietnam
Meimuna Pham and Thinh 2005b: 290 (comp. note) Vietnam
Meimuna Iwamoto, Inoue and Yagi 2006: 679, 681, Figs 2t–u (flight muscle, comp. note)
Meimuna Wang, Ren, Liang, Liu and Wang 2006: 295, Table 1 (listed) China
Meimuna Boulard 2007e: 81 (comp. note) Thailand
Meimuna Sanborn, Phillips and Sites 2007: 3, 19, 35, Table 1 (listed) Thailand

Meimuna Shiyake 2007: 5, 10, 46, 62 (listed, comp. note)
Meimuna Boulard 2008f: 7, 32, 93 (listed) Thailand
Meimuna Iwamoto 2008: 614 (flight muscle, comp. note)
Meimuna Lee 2008a: 2, 12, 14, Table 1 (listed, comp. note) Vietnam
Meimuna Lee 2008b: 446, 450, Table 1 (listed, comp. note) Korea
Meimuna Osaka Museum of Natural History 2008: 324 (listed) Japan
Meimuna Lee 2009b: 330–331 (comp. note)
Meimuna Pham and Yang 2009: 14, Table 2 (listed) Vietnam
Meimuna Ahmed and Sanborn 2010: 39 (distribution) China, Mongolia, Indochina, Thailand, Burma, Malaysia, Taiwan, Indonesia, India, Pakistan
Meimuna Lee and Hill 2010: 302 (comp. note)

***M. boninensis* (Distant, 1905)** = *Diceropyga boninensis* = *Cosmopsaltria ogasawarensis* Matsumura, 1906 = *Meimuna ogasawarensis*
Meimuna boninensis Hayashi 1984: 29, 61–63, Figs 143–146 (key, described, illustrated, synonymy, comp. note) Equals *Diceropyga boninensis* Equals *Cosmopsaltria ogasawaraensis* (partim) Bonin Islands
Meimuna boninensis Ohbayashi, Sato and Igawa 1999: 339, 341–342, Fig 1, Fig 3 (illustrated, parasitism, comp. note) Ogasawara (Bonin) Islands
Meimuna boninensis Yoshimura and Okochi 2005: 48 (predation, comp. note) Japan, Ogasawara Islands
Meimuna boninensis Shiyake 2007: 5, 60, 62, Fig 94 (illustrated, distribution, listed, comp. note) Japan
Meimuna boninensis Osaka Museum of Natural History 2008: 326 (listed, acoustic behavior, ecology, comp. note) Japan
Meimuna boninensis Yamanaka and Liebhold 2009: 339 (predation) Ogasawara Islands
Meimuna boninensis Toda, Takahashi, Nakagawa and Surigara 2010: 148 (predation) Bonin Islands

***M. bonininsulana* Kato, 1928** = *Meimuna bonin-insulana*
Meimuna bonin-insulana Hayashi 1984: 62 (comp. note)

***M. cassandra* Distant, 1912** = *Meimuna casandra* (sic)

***M. chekianga* Kato, 1940**
Meimuna chekianga Chou, Lei, Li, Lu and Yao 1997: 246 (described, comp. note) China
Meimuna chekianga Hua 2000: 62 (listed, distribution) China, Zhejiang

***M. chekiangensis* Chen, 1940**
Meimuna chekiangensis Chou, Lei, Li, Lu and Yao 1997: 238, 240, Plate XII, Fig 123 (key, described, illustrated, comp. note) China
Meimuna chekiangensis Hua 2000: 62 (listed, distribution) China, Zhejiang, Guangxi

***M. choui* Lei, 1994**
Meimuna choui Lei 1994: 184–185, 187–188, Figs 1–2 (n. sp., described, illustrated, comp. note) China, Shaanxi, Hunnan
Meimuna choui Chou, Lei, Li, Lu and Yao 1997: 238, 242–243, Fig 9–78, Plate XI, Fig 115 (key, described, illustrated, comp. note) China
Meimuna choui Beuk 1999: 5 (comp. note)

***M. crassa* (Distant, 1905)** = *Haphsa crassa* = *Haphsa craasa* (sic)
Haphsa crassa Wu 1935: 12 (listed) China, Yunnan
Haphsa crassa Chou, Lei, Li, Lu and Yao 1997: 237 (described, comp. note) China
Haphsa crassa Hua 2000: 62 (listed, distribution) China, Yunnan
Haphsa crassa Lee 2008a: 15 (distribution, listed) Vietnam, China, Yunnan
Haphsa crassa Lee 2009b: 330–331 (comp. note)
Meimuna crassa Lee 2009b: 331 (n. comb., comp. note)
Haphsa crassa Pham and Yang 2009: 14, Table 2 (listed) Vietnam

***M. duffelsi* Boulard, 2005**
Meimuna duffelsi Boulard 2005e: 118, 124–128, Figs 12–17, Plate C, Figs 1–2 (n. sp., described, illustrated, song, sonogram, listed, comp. note) Thailand
Meimuna duffelsi Boulard 2005o: 151 (listed)
Meimuna duffelsi Boulard 2006a: 3–4 (comp. note) Thailand
Meimuna duffelsi Boulard 2007e: 81–82, Figs 52–53, Plates 42–44, back cover (illustrated, comp. note) Thailand
Meimuna duffelsi Sanborn, Phillips and Sites 2007: 19 (listed, distribution, comp. note) Thailand
Meimuna duffelsi Boulard 2008c: 364, 367, Fig 19 (comp. note)
Meimuna duffelsi Boulard 2008f: 32, 93 (biogeography, listed, comp. note) Thailand

M. durga (Distant, 1881) to *Haphsa durga*

***M. gakokizana* Matsumura, 1917**
Meimuna gakokizana Chou, Lei, Li, Lu and Yao 1997: 246 (described, comp. note) China
Meimuna gakokizana Hua 2000: 62 (listed, distribution) China, Taiwan
Meimuna gakokizana Lee and Hayashi 2003a: 177, 181–182 (key, diagnosis, song, listed, comp. note) Taiwan
Meimuna gakokizana Chen 2004: 86, 195, 198 (comp. note)

Meimuna gakokizana Lee and Hill 2010: 299–300, Figs 7–8 (phylogeny, comp. note)

M. gamameda **(Distant, 1902)** = *Cosmopsaltria gamameda* = *Meimuna ganameda* (sic)

Meimuna gamameda Lei 1994: 186, 188 (synonymy, measurements, distribution, comp. note) Equals *Cosmopsaltria gamameda* China, Yunnan, Sikkim

Meimuna gamameda Chou, Lei, Li, Lu and Yao 1997: 246 (described, synonymy, comp. note) Equals *Cosmopsaltria gamameda* China

Meimuna gamameda Hua 2000: 62 (listed, distribution) Equals *Cosmopsaltria gamameda* China, Yunnan, Sri Lanka

M. goshizana **Matsumura, 1917** = *Meimuna gozhizana* (sic)

Meimuna goshizana Chou, Lei, Li, Lu and Yao 1997: 246 (described, comp. note) China

Meimuna goshizana Hua 2000: 62 (listed, distribution) China, Taiwan

Meimuna goshizana Lee and Hayashi 2003a: 177, 182 (key, diagnosis, song, listed, comp. note) Taiwan

Meimuna goshizana Chen 2004: 86–87, 92, 198 (described, illustrated, listed, comp. note) Taiwan

M. infuscata **Lei & Beuk, 1997** = *Meinuna* (sic) *infuscata*

Meimuna infuscata Chou, Lei, Li, Lu and Yao 1997: 238, 243–244, 357 (n. sp., key, described, synonymy, comp. note) China

Meinuna (sic) *infuscata* Chou, Lei, Li, Lu and Yao 1997: 369 (comp. note)

Meimuna infuscata Pham and Thinh 2006: 526–527 (listed, comp. note) Vietnam

Meimuna infuscata Lee 2008a: 13 (distribution, listed) Vietnam, China, Yunnan

Meimuna infuscata Lee 2009b: 331 (comp. note)

Meimuna infuscata Pham and Yang 2009: 14, Table 2 (listed) Vietnam

M. iwasakii **Matsumura, 1913** = *Meimuna ishigakina* Kato, 1925 = *Meimuna iwasakii ishigakina* = *Meimuna uraina* Kato, 1925 = *Meimuna uraina* var. *a* Kato, 1927 = *Meimuna uraina* var. *b* Kato, 1927 = *Meimuna uraina* var. *c* Kato, 1927 = *Meimuna iriomoteana* Kato, 1936

Meimuna iwasakii Hayashi 1984: 29, 62–63, Figs 147–150 (key, described, illustrated, synonymy, comp. note) Equals *Meimuna ishigakina* Equals *Meimuna uraina* Equals *Meimuna iriomoteana* Japan

Meimuna iwasakii Lee 1995: 30 (comp. note)

Meimuna iwasakii Hua 2000: 62 (listed, distribution, hosts) China, Henan, Taiwan, Japan

Meimuna ishigakia Hua 2000: 62 (listed, distribution) China, Jiangxi, Japan

Meimuna uraina Hua 2000: 63 (listed, distribution) China, Jiangxi

Meimuna iwasakii Turner, Hovencamp and van Welzen 2001: 219, Table 1 (outgroup)

Meimuna iwasakii Sueur 2001: 46, Table 2 (listed)

Meimuna iwasakii Duffels and Turner 2002: 236–237, 239, 241–242, 244–245, 247, 259, Fig 4I, Fig 8B, Appendix 2 (illustrated, listed, comp. note)

Meimuna iwasakii Lee and Hayashi 2003a: 177–183, Figs 22–24 (key, diagnosis, illustrated, synonymy, song, oscillogram, listed, comp. note) Equals *Meimuna uraina* Equals *Meimuna uraina* var. *a* Equals *Meimuna uraina* var. *b* Equals *Meimuna uraina* var. *c* Taiwan, Japan

Meimuna uraina Lee and Hayashi 2003a: 180, 182 (comp. note)

Meimuna iwasakii Chen 2004: 86–87, 90, 198 (described, illustrated, listed, comp. note) Taiwan

Meimuna uraina Chen 2004: 86, 192 (comp. note)

Meimuna iwasakii Watanabe and Kobayashi 2006: 218, 224–227, Tables 2–5 (listed, comp. note) Japan, Ryukyus

Meimuna iwasakii Shiyake 2007: 5, 60, 62, Figs 95a–95c (illustrated, distribution, listed, comp. note) Taiwan

Meimuna iwasakii Osaka Museum of Natural History 2008: 326 (listed, acoustic behavior, ecology, comp. note) Japan

M. khadiga **(Distant, 1904)** = *Cosmopsaltria khadiga*

M. kuroiwae **Matsumura, 1917** = *Meimuna sakaguchii* Matsumura, 1927 = *Meimuna amamioshimana* Kato, 1928 = *Meimuna amami-oshimana* = *Meimuna sakaguchii* (nec Matsumura) = *Meimuna kikaigashimana* Kato, 1937 = *Meimuna tsuchidai* Kato, 1931 = *Meimuna hentonaensis* Kato, 1960 = *Meimuna oshimensis* (nec Matsumura) = *Meimura* (sic) *oshimensis* (nec Matsumura) = *Meimuna nakaoi* Ishihara, 1968 = *Meimuna kuroiwai* (sic)

Meimuna kuroiwae Hayashi 1984: 29, 60–61, Figs 139–142 (key, described, illustrated, synonymy, comp. note) Equals *Meimuna oshimensis* (partim) Equals *Meimuna sakaguchii* (partim) Equals *Meimuna amami-oshimana* Equals *Meimuna tsuchidai* Equals *Meimuna kikaigashimana* Equals *Meimuna nakoi* (partim) Japan

Meimuna kuroiwae Kanmiya and Nakao 1988: 59–80, Figs 2–10, Tables 1–18 (song analysis, comp. note) Japan

Meimuna kuroiwae Nakao 1988: 43 (ecology, distribution, comp. note) Ryukyu Islands

Meimuna kuroiwae Nakao and Kanmiya 1988: 25–45, Figs 1–9, Table 1–10 (song analysis, comp. note) Ryukyu Islands

Meimuna kuroiwae Ohbayashi, Sato and Igawa 1999: 339 (comp. note) Japan, Kyushu Islands, Amami Islands, Okinawa Islands
Meimuna tsuchidai Hua 2000: 63 (listed, distribution) Equals *Meimuna kuroiwae* China, Guangxi
Meimuna kuroiwae Sueur 2001: 46, Table 2 (listed)
Meimuna kuroiwae Sanborn, Breitbarth, Heath and Heath 2002: 445 (comp. note)
Meimuna kuroiwae Lee and Hayashi 2003a: 182 (comp. note)
Meimuna kuroiwae Shiyake 2007: 5, 60, 62, Figs 96–97(sic) (illustrated, distribution, listed, comp. note) Japan
Meimuna kuroiwae Osaka Museum of Natural History 2008: 324 (listed, acoustic behavior, ecology, comp. note) Japan
Meimuna kuroiwae Saisho 2010c: 60 (comp. note) Japan

***M. microdon* (Walker, 1850)** = *Dundubia microdon* = *Cosmopsaltria microdon* = *Meimuna omeinensis* Chen, 1940

Meimuna microdon Lei 1994: 186, 188 (synonymy, measurements, distribution, comp. note) Equals *Dundubia microdon* Equals *Cosmopsaltria microdon* China, Yunnan, Tibet, India, Sikkim
Meimuna microdon Chou, Lei, Li, Lu and Yao 1997: 238–240 (key, described, synonymy, comp. note) Equals *Dundubia microdon* Equals *Cosmopsaltria microdon* Equals *Meimuna omeinensis* n. syn. China
Meimuna omeinensis Chou, Lei, Li, Lu and Yao 1997: 238–239 (key, synonymy) China
Meimuna microdon Pham and Thinh 2005a: 242 (synonymy, distribution, listed) Equals *Dundubia microdon* Equals *Cosmopsaltria microdon* Equals *Meimuna omeinensis* Vietnam
Meimuna microdon Lee 2008a: 13 (synonymy, distribution, listed) Equals *Dundubia microdon* Equals *Cosmopsaltria microdon* Vietnam, South China, India
Meimuna microdon Pham and Yang 2009: 14, Table 2 (listed) Vietnam
Meimuna microdon Sun, Watson, Zheng, Watson and Liang 2009: 3148–3153, Fig 1H, Fig 2C, Fig 3C, Fig 4C, Tables 1–2 (illustrated, wing microstructure, comp. note) China, Hainan

***M. mongolica* (Distant, 1881)** = *Cosmopsaltria mongolica* = *Meimuna suigensis* Matsumura, 1927 = *Meimuna chosensis* Matsumura, 1927 = *Meimuna heijonis* Matsumura, 1927 = *Meimuna santoshonis* Matsumura, 1927 = *Meimuna gallosi* Matsumura, 1927 = *Meimuna galloisi* (sic)

Meimuna chosensis Doi 1932: 43 (listed, comp. note) Korea
Meimuna suigensis Doi 1932: 43 (listed, comp. note) Korea
Meimuna mongolica Chen 1933: 360 (listed, comp. note)
Meimuna mongolica Wu 1935: 13 (listed) Equals *Cosmopsaltria mongolica* Equals *Meimuna suigensis* Equals *Meimuna chosenensis* Equals *Meimuna heijonis* Equals *Meimuna santoshonis* Equals *Meimuna gallosi* China, Mongolia, Peiping, Kiangsu, Shanghai, Chekiang, Haichow, Nanking, Hangchow, Kashing, Korea
Meimuna mongolica Kim 1958: 94 (listed) Korea
Meimuna mongolica Cho 1955: 237, 256, Table 1 (listed, comp. note) Korea
Meimuna mongolica Cho 1965: 173 (listed, comp. note) Korea
Meimuna mongolica Hayashi 1984: 62 (comp. note)
Meimuna mongolica Jiang, Lin and Wang 1987: 106–107, Fig 1D, Fig 2 (timbal structure, illustrated, song, oscillogram, comp. note) China
Meimuna mongolica Wang and Zhang 1987: 295 (key, comp. note) China
Meimuna mongolica Liu 1992: 42 (comp. note)
Meimuna mongolica Peng and Lei 1992: 101, Fig 287 (listed, illustrated, comp. note) China, Hunan
Meimuna mongolica Lei, Chou and Li 1994: 53 (comp. note) China
Meimuna mongolica Jiang, Yang, Tang, Xu and Chen 1995b: 228 (song, phonotaxis, flight behavior, comp. note) China
Meimuna mongolica Lee 1995: 18, 23, 30–32, 37–38, 40–41, 43–46, 85–86, 91–96, 136, 139 (listed, key, synonymy, ecology, distribution, illustrated, comp. note) Equals *Cosmopsaltria mongolica* Equals *Meimuna suigensis* Equals *Meimuna chosensis* Equals *Meimuna heijonis* Equals *Meimuna galloisi* Korea
Meimuna suigensis Lee 1995: 28, 32–33, 35–36, 38, 92–93 (comp. note)
Meimuna chosensis Lee 1995: 28, 32–33, 35–36, 38, 92–93 (comp. note)
Meimuna heijonis Lee 1995: 28, 32–33, 35–36, 38, 92–93 (comp. note)
Meimuna galloisi Lee 1995: 28, 32–33, 35–36, 38, 92–93 (comp. note)
Meimuna santoshonis Lee 1995: 38 (comp. note)
Cosmopsaltria mongolica Lee 1995: 91 (comp. note)
Meimuna mongolica Xu, Ye and Chen 1995: 84 (listed, comp. note) China, Zhejiang
Meimuna mongolica Chou, Lei, Li, Lu and Yao 1997: 20, 238, 240–241, 308, 318–319, 325–326,

331, 342, Fig 3–9C, Fig 9–77, Plate XI, Fig 116, Tables 10–12 (key, described, synonymy, illustrated, song, oscillogram, spectrogram, comp. note) Equals *Cosmopsaltria mongolica* Equals *Meimuna suigensis* Equals *Meimuna chosensis* Equals *Meimuna heijonis* Equals *Meimuna santoshonis* Equals *Meimuna gallosi* China

Meimuna mongolica Beuk 1998: 148–149, Fig 1, Table 1 (phylogeny, comp. note)

Meimuna mongolica Lee 1999: 8–9, Figs 5–6 (distribution, song, illustrated, listed, comp. note) Equals *Cosmopsaltria mongolica* Equals *Meimuna suigensis* Equals *Meimuna heijonis* Equals *Meimuna chosensis* Equals *Meimuna galloisi* Equals *Tana* (sic) *japonensis* (nec Distant) Korea, China

Meimuna mongolica Hua 2000: 62–63 (listed, distribution, hosts) Equals *Cosmopsaltria mongolica* Equals *Meimuna santoshonis* China, Liaoning, Inner Mongolia, Hebei, Henan, Shaanxi, Shandong, Shantosho, Hubei, Anhui, Jiangsu, Zhejiang, Fujian, Jiangxi, Guangdong, Hunana, Guangxi, Korea, Mongolia

Meimuna mongolica Lee 2001a: 51 (comp. note) Korea

Meimuna mongolica Lee 2001b: 2 (comp. note) Korea

Meimuna mongolica Sueur 2001: 39, Table 1 (listed)

Meimuna mongolica Lee, Choe, Lee and Woo 2002: 3–9, Figs 1–3, Tables 2–3 (gene sequence, phylogeny, comp. note) Korea

Meimuna mongolica Lee 2003b: 5, 8 (comp. note)

Meimuna mongolica Lee 2005: 17, 86–89, 158, 165–166, 174–175 (described, illustrated, synonymy, song, distribution, key, comp. note) Equals *Cosmopsaltria mongolica* Equals *Meimuna suigensis* Equals *Meimuna chosensis* Equals *Meimuna heijonis* Equals *Meimuna galloisi* Equals *Tanna japonensis* (nec Distant) Korea, China

Meimuna mongolica Shiyake 2007: 5, 61, 63, Fig 97 (illustrated, distribution, listed, comp. note) China

Meimuna mongolica group Lee 2008a: 13, 15 (listed) Vietnam

Meimuna mongolica Lee 2008a: 15 (comp. note)

Meimuna mongolica Lee 2008b: 446–447, 450–452, 463, Fig 1H, Table 1 (key, illustrated, synonymy, distribution, listed, comp. note) Equals *Cosmopsaltria mongolica* Equals *Meimuna suigensis* Equals *Meimuna chosensis* Equals *Meimuna heijonis* Equals *Meimuna galloisi* Equals *Tanna japonensis* (nec Distant) Korea, China

Meimuna mongolica Lee 2009b: 331 (comp. note)

Meimuna mongolica Sun, Watson, Zheng, Watson and Liang 2009: 3148–3149, 3151, 3154, Fig 1I, Table 1 (illustrated, wing microstructure, comp. note) China, Beijing

***M. multivocalis* (Matsumura, 1917)** = *Cosmpsaltria multivocalis* = *Orientopsaltria multivocalis*

Cosmopsaltria multivocalis Duffels 1983b: 4 (comp. note)

Orientopsaltria multivocalis Duffels 1983b: 9 (comp. note)

Meimuna multivocalis Chou, Lei, Li, Lu and Yao 1997: 246 (described, synonymy, comp. note) Equals *Cosmopsaltria multivocalis* China

Cosmopsaltria multivocalis Duffels and Zaidi 2000: 196 (comp. note)

Meimuna multivocalis Duffels and Zaidi 2000: 196 (comp. note)

Orientopsaltria multivocalis Duffels and Zaidi 2000: 196 (comp. note)

Meimuna multivocalis Hua 2000: 63 (listed, distribution) Equals *Cosmopsaltria multivocalis* China, Taiwan

Meimuna multivocalis Lee and Hayashi 2003a: 177, 182–183 (key, diagnosis, synonymy, song, listed, comp. note) Equals *Cosmopsaltria multivocalis* Equals *Orientopsaltria multivocalis* Taiwan

Meimuna multivocalis Chen 2004: 86, 195, 198 (comp. note)

Orientopsaltria multivocalis Chen 2004: 192 (comp. note)

M. nauhkae Boulard, 2005 to *Haphsa karenensis* Ollenbach, 1928

***M. neomongolica* Liu, 1940**

Meimuna neomongolica Liu 1992: 41–42 (comp. note)

Meimuna neomongolica Chou, Lei, Li, Lu and Yao 1997: 246 (described, comp. note) China

Meimuna neomongolica Hua 2000: 63 (listed, distribution) China, Hubei

M. omeinensis Chen, 1940 to *Meimuna microdon* (Walker, 1850)

***M. opalifera* (Walker, 1850)** = *Dundubia opalifera* = *Cosmopsaltria opalifera* = *Dundubia (Cosmopsaltria) opalifera* = *Meinuma* (sic) *opalifera* = *Miemuna* (sic) *opalifera* = *Meimuua* (sic) *opalifera* = *Maimuna* (sic) *opalifera* = *Meimuna oplifera* (sic) = *Meimuna operlifera* (sic) = *Meimuna opaliferia* (sic) = *Meimura* (sic) *opalifera* = *Meimura* (sic) *oparifera* (sic) = *Meimuna opalifer* (sic) = *Meimuna opalifera formosana* Kato, 1925 = *Meimuna opalifera* var. *formosana* = *Meimuna opalifera formosans* (sic) = *Meimuna opalifera nigroventris* Kato, 1927 = *Meimuna opalifera* var. *nigroventris* = *Meimuna opalifera nigoventria* (sic) = *Meimuna opalifera punctata* Kato, 1932 = *Meimuna* *opalifera* f. *punctata* = *Meimuna opalifera* ab. *punctata* = *Meimuna opalifera* var. *punctata* = *Meimuna longipennis* Kato, 1937 = *Meimura* (sic) *longipennis*

Meimuna opalifera Doi 1932: 43 (listed, comp. note) Korea
Meimuna opalifera Chen 1933: 360 (listed, comp. note)
Meimuna opalifera Isaki 1933: 381 (listed, comp. note) Japan
Meimuna opalifera Seki 1933: 102 (turrets, comp. note)
Meimuna opalifera Wu 1935: 14 (listed) Equals *Dundubia opalifera* Equals *Cosmopsaltria opalifera* China, Peiping, Chekiang, Canton, Japan, Korea, Liu-kiu Island
Meimuna oplifera (sic) Chen 1942: 145 (listed, comp. note) China
Meimura (sic) *opalifera* Nakao 1952: 17 (acoustic behavior, ecology, comp. note) Japan
Meimuna opalifera Nakao 1952: 18, 25 (acoustic behavior, ecology, comp. note) Japan
Meimuna opalifera Ishii 1953: 1757 (parasitism) Japan
Meimuna opalifera Kurokawa 1953: 1, 5, Table 1 (chromosomes, comp. note) Japan
Meimuna opalifera Ohgushi 1954: 11 (emergence times, ecology, comp. note) Japan
Meimuna opalifera Cho 1955: 237, 256, Table 1 (listed, comp. note) Korea
Meimuna opalifera Halkka 1959: 37, Table 1 (listed, comp. note)
Meimuna opalifera Ikeda 1959: 484, 493, Fig 12A (timbal muscle potentials, comp. note) Japan
Meimuna opalifera Katsuki 1960: 57, Table 1 (hearing, comp. note) Japan
Meimura (sic) *opalifera* Kim 1960: 24 (listed) Korea
Meimuna opalifera Cho 1965: 173 (listed, comp. note) Korea
Cosmopsaltria opalifera Hearn 1971: Plate V, Fig 1 (illustrated) Japan
Meimuna opalifer (sic) Asahina 1972: 17 (comp. note) Okinawa
Meimuna opalifera Fujiyama 1982: 184–185, Fig 4B (comp. note)
Meimuna opalifera Hayashi 1982b: 188, Table 1 (distribution, comp. note) Japan
Meimuna opalifera Welbourn 1983: 135, Table 5 (mite host, listed) Japan
Meimuna opalifera formosana Hayashi 1984: 58–59 (described, comp. note) Japan
Meimuna opalifera var. *nigroventris* Hayashi 1984: 58–59 (described, comp. note) Japan
Meimuna opalifera Hayashi 1984: 29, 58–59, Figs 127–130 (key, described, illustrated, synonymy, comp. note) Equals *Dundubia opalifera* Equals *Cosmopsaltria opalifera* Equals *Meimuna longipennis* Japan, Korea
Meimuna opalifera f. *punctata* Hayashi 1984: 58 (described, comp. note) Equals *Meimuna opalifera formosana* Japan
Meimuna opalifera Kevan 1985: 91 (comp. note) Japan
Meimuna opalifera Hu, Kimura and Maruyama 1986: 1487, Table 1 (comp. note)
Meimuna opalifera Kirillova 1986b: 46 (listed) Japan
Meimuna opalifera Kirillova 1986a: 123 (listed) Japan
Meimura (sic) *opalifera* Chou 1987: 21 (comp. note) China
Meimuna opalifera var. *formosana* Jiang, Lin and Wang 1987: 108–109, Figs 4–5 (timbal structure, illustrated, song, oscillogram, comp. note) China
Meimuna opalifera Seki, Fujishita, Ito, Matsuoka and Tsukida 1987: 102, Table 2 (retinoid composition, listed, comp. note) Japan
Meimuna opalifera Wang and Zhang 1987: 295 (key, comp. note) China
Meimuna opalifera Namba, Ma and Inagaki 1988: 259 (comp. note) China
Meimuna opalifera Southcott 1988: 103 (parasitism, comp. note) Japan
Meimuna opalifera Wang 1988: 232 (comp. note)
Meimuna opalifera var. *formosana* Jiang 1989b: 77 (comp. note)
Meimuna longipennis Zhang and Zhang 1990: 339 (comp. note) China
Meimuna opalifera Zhang and Zhang 1990: 337, 339 (comp. note) China
Meimuna opalifera Liu 1991: 258 (comp. note)
Cosmopsaltria opalifera Liu 1991: 258 (comp. note)
Meimuna opalifera var. *formosana* Jiang, Wang, Xu, Zhang and Chen 1992: 158, 161–165, Figs 3–5 (song production, sonogram, oscillogram, comp. note) China
Meimuna opalifera ab. *punctata* Jiang, Wang, Xu, Zhang and Chen 1992: 158–161, Figs 1–2 (song production, sonogram, oscillogram, comp. note) China
Meimuna opalifera Liu 1992: 17 (comp. note)
Cosmopsaltria opalifera Liu 1992: 17, 34 (comp. note)
Meimuna operlifera (sic) Liu 1992: 19 (comp. note)
Meimuna opalifera Peng and Lei 1992: 100–101, Fig 286 (listed, illustrated, comp. note) China, Hunan
Meimuna opalifera Anonymous 1994: 809 (fruit pest) China
Meimuna opalifera Lei 1994: 185, 187 (comp. note)
Meimuna opalifera formosana Lei, Chou and Li 1994: 53 (comp. note) China
Meimuna opalifera var. *punctata* Lei, Chou and Li 1994: 53 (comp. note) China

Meimuna opalifera Seki, Isono, Ito, and Katsuta 1994: 692–694, Table 1 (retinoid composition, comp. note) Japan
Meimuna opalifera Gogala 1995: 112–113, Fig 9 (song, sonogram, oscillogram, comp. note) Japan
Meimuna opalifera var. *formosana* Gogala 1995: 112 (comp. note)
Meimuna opalifera ab *punctata* Gogala 1995: 112 (comp. note)
Meimuna opalifera ab. *punctata* Jiang, Yang, Tang, Xu and Chen 1995b: 228 (song, phonotaxis, flight behavior, comp. note) China
Meimuna opalifera var. *formosana* Jiang, Yang, Tang, Xu and Chen 1995b: 228 (song, phonotaxis, flight behavior, comp. note) China
Meimuna opalifera Lee 1999: 8 (distribution, song, listed, comp. note) Equals *Dundubia opalifera* Equals *Cosmopsaltria opalifera* Equals *Meimuna opalifera* var. *nigroventris* Korea, China, Japan, Taiwan
Meimuna opalifera Lee 1995: 18, 23, 29–35, 37–38, 40–46, 85–91, 136, 139 (listed, key, synonymy, ecology, distribution, illustrated, comp. note) Equals *Dundubia opalifera* Equals *Cosmopsaltria opalifera* Equals *Meimuna opalifera* var. *nigroventris* Equals *Meimuna opalifera* f. *nigroventris* Korea
Dundubia opalifera Lee 1995: 28, 86 (comp. note)
Cosmopsaltria opalifera Lee 1995: 29, 86 (comp. note)
Meimuna opalifera var. *nigroventris* Lee 1995: 37, 40–42, 87 (comp. note)
Meimuna opalifera f. *nigroventris* Lee 1995: 87 (comp. note)
Meimuna opalifera Xu, Ye and Chen 1995: 84 (listed, comp. note) China, Zhejiang
Meimuna opalifera Beuk 1996: 133 (comp. note)
Meimuna opalifera Boer and Duffels 1996b: 315 (comp. note) Japan, Korea, Taiwan, China
Meimuna opalifera Chou, Lei, Li, Lu and Yao 1997: 238, 241–243, 308, 318, 325–326, 331, 341–342, Plate XII, Fig 124, Tables 10–12 (key, described, synonymy, song, oscillogram, spectrogram, comp. note) Equals *Dundubia opalifera* Equals *Cosmopsaltria opalifera* Equals *Meimuna longipennis* China
Meimuna opalifera Beuk 1999: 5–6, 8, Fig 1a, Table 1 (cladogram, comp. note)
Meimuna opalifera Ohbayashi, Sato and Igawa 1999: 341 (parasitism, comp. note)
Meimuna longipennis Hua 2000: 62 (listed, distribution) China, Jiangxi, Zhejiang, Fujian, Taiwan, Guizhou, Japan
Meimuna opalifera Hua 2000: 63 (listed, distribution, hosts) Equals *Dundubia opalifera* Equals *Cosmopsaltria opalifera* China, Hebei, Henan, Shandong, Hubei, Anhui, Zhejiang, Jiangxi, Fujian, Taiwan, Guangdong, Macao, Hunan, Guangxi, Guizhou, Korea, Japan
Meimuna opalifera Lee, Nishimori, Hatakeyama and Oishi 2000: 2–3 (comp. note) Japan
Meimuna opalifera Lee 2001a: 51 (comp. note) Korea
Meimuna opalifera Lee 2001b: 2 (comp. note) Korea
Meimuna opalifera Ohara and Hayashi 2001: 1 (comp. note) Japan
Meimuna opalifera Saisho 2001a: 68 (comp. note) Japan
Meimuna opalifera Sato, Chuman, Matsushima, Shimohigashi, Tominaga and Shimohigashi 2001: 875 (cDNA, protein, comp. note)
Meimuna opalifera Sueur 2001: 39, 46, Tables 1–2 (listed)
Meimuna opalifera var. *formosana* Sueur 2001: 39, Table 1 (listed)
Meimuna opalifera var. *punctata* Sueur 2001: 39, Table 1 (listed)
Meimuna opalifera Jeon, Kim, Tripotin and Kim 2002: 240–241, Figs 3–4 (parasitism, comp. note) Kores, Japan
Meimuna opalifera Lee, Choe, Lee and Woo 2002: 3–9, Figs 1–3, Tables 2–3 (gene sequence, phylogeny, comp. note) Korea
Meimuna opalifera Sato, Chuman, Matsushima, Tominaga, Shimohigashi and Shimohigashi 2002: 822–827, Figs 1–3, Table 1 (cDNA, neuropeptide, comp. note) Japan
Meimuna opalifera Sudo, Tsuyuki and Kanno 2002: 258–259, 261, Fig 4, Figs 7–8, Table 1 (wing movement, comp. note)
Meimura (sic) *oparifera* (sic) Sudo, Tsuyuki and Kanno 2002: 262, Fig 10 (wing movement, comp. note)
Meimuna opalifera Sueur 2002b: 121, 129 (listed, comp. note)
Meimuna opalifera Yaginuma 2002: 227, 229, Fig 3A, Table 1 (fungal host, comp. note) Japan
Meimuna opalifera Kubo-Irie, Irie, Nakazawa and Mohri 2003: 988 (comp. note) Japan
Meimuna opalifera Lee 2003b: 5, 8 (comp. note)
Meimuna opalifera Lee 2003c: 15 (comp. note) Korea
Meimuna opalifera Lee and Hayashi 2003a: 177–180, 182, Figs 20–21 (key, diagnosis, illustrated, synonymy, song, oscillogram, listed, comp. note) Equals *Dundubia opalifera* Equals *Cosmopsaltria opalifera* Equals *Meimuna opalifera* var. *formosana* Equals *Meimuna opalifera formosana* Equals

Meimuna longipennis Taiwan, Japan, Korea, China
Meimuna opalifera var. *formosana* Lee and Hayashi 2003a: 177 (comp. note)
Meimuna oparifera (sic) Matsushima, Yokotani, Lui, Sumida, Honda et al. 2003: 420 (comp. note) Japan
Meimuna opalifera Chen 2004: 86–87, 192, 198 (described, illustrated, listed, comp. note) Taiwan
Meimuna longipennis Chen 2004: 192 (comp. note)
Meimuna opalifera Matsushima, Sato, Chuman, Takeda, Yokotani et al. 2004: 83, 87 (comp. note) Japan
Meimuna opalifera Nuorteva, Lodenius, Nagasawa, Tulisalo and Nuorteva 2004: 56–57, 62, Tables 1–2, Table 9 (cadmium, metal levels) Japan
Meimuna opalifera Hamasaka, Mohrherr, Predel and Wegener 2005: 8, Table 1 (predation, comp. note) Kentucky
Meimuna opalifera Lee 2005: 17, 80–84, 158, 165–166, 173–174 (described, illustrated, synonymy, song, distribution, key, comp. note) Equals *Dundubia opalifera* Equals *Cosmopsaltria opalifera* Equals *Meimuna opalifera* var. *nigroventris* Korea, Japan, China, Taiwan
Meimuna opalifera Owada, Arita, Jinbo, Kishida, Nakajima et al. 2005: 58, 118, Fig 5 (illustrated, parasitism) Japan
Meimuna opalifera Sudo, Tsuyuki and Kanno 2005: 2–4, Fig 4, Figs 7–8, Table 1 (wing movement, comp. note)
Meimura (sic) *oparifera* (sic) Sudo, Tsuyuki and Kanno 2002: 6, Fig 12 (wing movement, comp. note)
Meimuna opalifera Yoshimura and Okochi 2005: 49 (comp. note) Japan
Meimuna opalifera Trilar 2006: 344 (comp. note)
Meimuna opalifera Shiyake 2007: 5, 61, 63, Fig 98 (illustrated, distribution, listed, comp. note) China, Japan, Korea
Meimuna opalifera Takakura and Yamazaki 2007: 732 (comp. note) Japan
Meimuna opalifera Lee 2008b: 446–447, 451–452, 463, Fig 1I, Table 1 (key, illustrated, synonymy, distribution, listed, comp. note) Equals *Dundubia opalifera* Equals *Cosmopsaltria opalifera* Equals *Meimuna opalifera* var. *nigroventris* Korea, China, Japan, Taiwan
Meimuna opalifera Osaka Museum of Natural History 2008: 324 (listed, acoustic behavior, ecology, comp. note) Japan
Meimuna opalifera Mitani, Mihara, Ishii and Koike 2009: 901, Table 1 (listed, comp. note) Japan
Meimuna opalifer (sic) Sun, Watson, Zheng, Watson and Liang 2009: 3148–3149, 3151, Fig 1M, Table 1 (illustrated, wing microstructure, comp. note) China, Shaanxi
Meimuna opalifera Ye and Ng 2009: 188 (protein, comp. note)
Meimuna opalifera Lee, Lin and Wu 2010: 218 (comp. note) Taiwan
Meimuna opalifera Lee and Hill 2010: 299–300, Figs 7–8 (phylogeny, comp. note)
Meimuna opalifera Meelkop, Temmerman, Schoofs and Janssen 2010: 134 (comp. note) Japan
Meimuna opalifera Raksakantong, Meeso, Kubola and Siriamornpun 2010: 350–354, Fig 1, Tables 1–4 (fatty acid composition, human food, listed, comp. note)

***M. oshimensis* (Matsumura, 1906)** = *Cosmopsaltria oshimensis* = *Meimura* (sic) *oshimensis* = *Meimuna oshimensis tokunoshimana* Ishihara, 1968
Meimuna oshimensis Boulard 1988d: 150 (abdominal abnormalities, comp. note)
Meimuna oshimensis Hayashi 1984: 29, 59–60, Figs 131–138 (key, described, illustrated, synonymy, comp. note) Equals *Cosmopsaltria oshimensis* Japan
Meimuna oshimensis Sueur 2001: 46, Table 2 (listed)
Meimuna oshimensis Shiyake 2007: 5, 61, 63, Fig 99 (illustrated, distribution, listed, comp. note) Japan
Meimuna oshimensis Osaka Museum of Natural History 2008: 324 (listed, acoustic behavior, ecology, comp. note) Japan
Meimuna oshimensis Saisho 2010c: 60 (comp. note) Japan

***M. pallida* Ollenbach, 1929**

***M. raxa* Distant, 1913**
Meimuna raxa Lee 2008a: 13 (distribution, listed) Vietnam, Laos
Meimuna raxa Pham and Yang 2009: 14, Table 2 (listed) Vietnam
Meimuna raxa Pham and Yang 2010a: 133 (comp. note) Vietnam

***M. silhetana* (Distant, 1888)** = *Cosmopsaltria silhetana* = *Cosmopsalta* (sic) *silhetana*
Meimuna silhetana Wu 1935: 14 (listed) Equals *Cosmopsaltria silhetana* China, Sylhet
Meimuna silhetana Wang and Zhang 1993: 189 (listed, comp. note) China
Meimuna silhetana Hua 2000: 63 (listed, distribution) Equals *Cosmopsaltria silhetana* China, Fujian, Guangdong, Sichuan, Yunnan, India
Meimuna silhetana Lo and Fiona 2002: 7 (distribution, comp. note) Hong Kong,

India, China, Fujian, Guangdong, Sichuan, Yunnan

***M. subviridissima* Distant, 1913**

Meimuna subviridissima Lei 1994: 186, 188 (measurements, distribution, comp. note) China, Guangxi, Laos

Meimuna subviridissima Chou, Lei, Li, Lu and Yao 1997: 238, 245–246, Plate XI, Fig 118 (key, described, illustrated, synonymy, comp. note) China

Meimuna subviridissima Hua 2000: 63 (listed, distribution) China, Guangxi, Laos, Vietnam, thailand

Meimuna subviridissima Boulard 2008c: 364 (comp. note) China

Meimuna subviridissima Lee 2008a: 13 (distribution, listed) Vietnam, Laos, Thailand

Meimuna subviridissima Pham and Yang 2009: 14, Table 2 (listed) Vietnam

***M. tavoyana* (Distant, 1888)** = *Dundubia tavoyana* = *Cosmopsaltria tavoyana*

Meimuna tavoyana Gogala 1994a: 33, 35, Fig 7 (illustrated, sonogram, comp. note) Thailand

Meimuna tavoyana Gogala 1994b: 9 (song, comp. note) Peninsular Malaysia

Meimuna tavoyana Lei 1994: 186, 188 (synonymy, measurements, distribution, comp. note) Equals *Dundubia tavoyana* Equals *Cosmopsaltria tavoyana* China, Yunnan, Burma, Thailand, Malaysia

Meimuna tavoyana Gogala 1995: 101, 103–105, 111–113, Figs 1–2 (song, sonogram, oscillogram, comp. note) Thailand

Meimuna tavoyana Beuk 1996: 133 (comp. note)

Meimuna tavoyana Chou, Lei, Li, Lu and Yao 1997: 238, 244, Plate XII, Fig 125 (key, described, illustrated, synonymy, comp. note) Equals *Dundubia tavoyana* Equals *Cosmopsaltria tavoyana* China

Meimuna tavoyana Hua 2000: 63 (listed, distribution) Equals *Dundubia tavoyana* China, Yunnan, Burma, Thailand, Malaysia

Meimuna tavoyana Sueur 2001: 39, 46, Tables 1–2 (listed)

Meimuna tavoyana Boulard 2003a: 98–99, 103–105, color Fig 10, Fig 1, sonogram 5 (described, allotype female, song, sonogram, illustrated, comp. note) Thailand, Tenasserim, Burma, Peninsular Malaysia

Meimuna tavoyana Boulard 2005d: 344, 346 (comp. note) Thailand

Meimuna tavoyana Boulard 2005e: 124 (comp. note)

Meimuna tavoyana Boulard 2005l: 111 (comp. note)

Tosena melanoptera (sic) Boulard 2005o: 147 (listed) Equals *Meimuna tavoyana* sonogram Boulard 2000

Meimuna tavoyana Boulard 2005o: 149 (listed)

Miemuna tavoyana Boulard 2006g: 73, 75, 98–99, 180, Fig 65 (sonogram, listed, comp. note) Thailand

Meimuna tavoyana Trilar 2006: 344 (comp. note)

Meimuna tavoyana Boulard 2007d: 497 (comp. note) Thailand

Meimuna tavoyana Boulard 2007e: 48, 51, 67, Fig 33 (sonogram, comp. note) Thailand

Meimuna tavoyana Sanborn, Phillips and Sites 2007: 19–20 (listed, distribution, comp. note) Thailand, Burma, Malay Peninsula

Meimuna tavoyana Boulard 2008e: 352 (comp. note)

Meimuna tavoyana Boulard 2008f: 33, 93 (biogeography, listed, comp. note) Equals *Dundubia tavoyana* Equals *Cosmopsaltria tavoyana* Thailand, Burma, Malaysia

***M. tripurasura* (Distant, 1881)** = *Dundubia tripurasura* = *Cosmopsaltria tripurasura*

Meimuna tripurasura Wu 1935: 15 (listed) Equals *Dundubia tripurasura* Equals *Cosmopsaltria tripurasura* China, Kwangtung, Kwangsi, Sikhim, Assam, Margherita, Khasi Hills

Dundubia tripurasura Chen 1942: 145 (type species of *Meimuna*)

Dundubia tripurasura Hayashi 1984: 58 (type species of *Meimuna*)

Meimuna tripurasura Zhang and Zhang 1990: 338 (type species of *Meimuna*)

Meimuna tripurasura Wang and Zhang 1993: 189 (listed, comp. note) China

Dundubia tripurasura Lei 1994: 184 (type species of *Meimuna*)

Meimuna tripurasura Lee 1995: 85 (type species of *Meimuna*) India

Dundubia tripurasura Chou, Lei, Li, Lu and Yao 1997: 238 (type species of *Meimuna*, comp. note)

Meimuna tripurasura Chou, Lei, Li, Lu and Yao 1997: 238–239 (key, described, synonymy, comp. note) Equals *Dundubia tripurasura* Equals *Cosmopsaltria tripurasura* China

Dundubia tripurasura Lee 1999: 8 (type species of *Meimuna*) India, Indochina, China

Meimuna tripurasura Hua 2000: 63 (listed, distribution) Equals *Dundubia tripurasura* China, Hubei, Guangdong, Guangxi, Sichuan, Yunnan, Sri Lanka, Sikkim, India

Dundubia tripurasura Lee and Hayashi 2003a: 176 (type species of *Meimuna*) India

Meimuna tripurasura Lee and Hayashi 2003a: 177 (comp. note)

Meimuna tripurasura Lee 2005: 81 (type species of *Meimuna*)
Dundubia tripurasura Lee 2005: 173 (type species of *Meimuna*) India, Indochina, China
Dundubia tripurasura Pham and Thinh 2005a: 242 (type species of *Meimuna*)
Meimuna tripurasura Boulard 2008c: 364, 367, Fig 18 (comp. note) Indochina
Meimuna tripurasura Boulard 2008e: 352 (comp. note)
Meimuna tripurasura Boulard 2008f: 32 (type species of *Meimuna*) India
Dundubia tripurasura Lee 2008a: 12 (type species of *Meimuna*)
Meimuna tripurasura Lee 2008a: 12 (synonymy, distribution, listed) Equals *Dundubia tripurasura* Equals *Cosmopsaltria tripurasura* Vietnam, South China, India
Dundubia tripurasura Lee 2008b: 450 (type species of *Meimuna*) Assam
Meimuna tripurasura Lee 2009b: 330–331 (type species of *Meimuna*, comp. note)
Meimuna tripurasura Pham and Yang 2009: 14, Table 2 (listed) Vietnam
Dundubia tripurasura Ahmed and Sanborn 2010: 39 (type species of genus *Meimuna*)

***M.* undescribed species A Sanborn, Phillips & Sites, 2007**
Meimuna undescribed species A Sanborn, Phillips and Sites 2007: 20 (listed, distribution, comp. note) Thailand

***M.* undescribed species B Sanborn, Phillips & Sites, 2007**
Meimuna undescribed species B Sanborn, Phillips and Sites 2007: 20 (listed, distribution, comp. note) Thailand

***M. velitaris* (Distant, 1897)** = *Cosmopsaltria velitaris* = *Haphsa velitaris*
Meimuna velitaris Ahmed and Sanborn 2010: 39 (synonymy, distribution) Equals *Cosmopsaltria velitaris* Pakistan, Burma
Haphsa velitaris Lee 2009b: 330–331 (comp. note)
Meimuna velitaris Lee 2009b: 331 (n. comb., comp. note)

***M. viridissima* Distant, 1912**
Meimuna viridissima Boulard 2005e: 124 (comp. note)

***Moana* Myers, 1928**
Moana Strümpel 1983: 18, 109 (listed, comp. note)
Moana Hayashi 1984: 26 (comp. note)
Moana Duffels 1988d: 12, 25, 77–78, Table 1 (listed, comp. note)
Moana Duffels 1993: 1223, 1226–1227, 1230, 1232–1233, Fig 6 (phylogeny, listed, comp. note) Samoa
Moana Boer 1994b: 161, 163 (comp. note)
Moana Boer 1995d: 219–220, 223, 227, 236–237, Fig 49, Fig 51 (distribution, phylogeny, comp. note) Bismarck Archipelago, Solomon Islands, Samoa, New Britain, New Ireland
Moana Boer and Duffels 1996a: 155, 168–169, 171–173, Fig 9, Fig 11 (distribution, phylogeny, comp. note) East Melanesia, Bismarck Archipelago, Solomon Islands, Samoa
Moana Boer and Duffels 1996b: 301, 304, 307–308, 313–314, 316, 319–320, 329, Fig 4, Fig 14 (distribution, phylogeny, comp. note) Bismarck Archipelago, Solomon Islands, Solomon Islands, Samoa, Southwest Pacific
Moana Duffels 1997: 549, 552 (comp. note)
Moana Holloway 1998: 304–305, 307–308, Fig 10, Figs 12–13 (distribution, phylogeny, comp. note) Bismarck Islands
Moana Holloway and Hall 1998: 10, 12–13, Fig 3 (distribution, comp. note)
Moana Villet and Zhao 1999: 321 (comp. note)
Moana Duffels and Turner 2002: 236, 239, 242, 244–247, 249–252, 256, Figs 13–15 (phylogeny, distribution, listed, comp. note) Solomon Islands, Bismarck Archipelago
Moana Moulds 2005b: 387, 391, 414, 422, 432–433 (higher taxonomy, listed, comp. note)

***M. aluana aluana* (Distant, 1905)** = *Aceropyga aluana* = *Diceropyga aluana* = *Diceropyga aluana* Lallemand (partim) (nec Distant)
Aceropyga aluana Duffels 1983a: 494 (diversity, distribution) Solomon Islands
Aceropyga aluana Duffels 1986: 325, 331 (comp. note) Solomon Islands
Aceropyga aluana group Duffels 1988d: 14, 31–32, Fig 13 (phylogeny, comp. note) Solomon Islands, Bismarck Archipelago
Aceropyga aluana Duffels 1988d: 32, Fig 13 (phylogeny)
Aceropyga aluana aluana Duffels and Boer 1990: 262, Fig 21.5 (distribution)
Aceropyga aluana group Duffels 1993: 1227, 1233 (comp. note)
Moana aluana Duffels 1993: 1230, 1232, Fig 5 (n. comb., phylogeny, listed, comp. note) Solomon Islands
Aceropyga aluana group Boer 1995d: 220 (distribution, comp. note)
Moana aluana aluana Boer 1995d: 220, Fig 49 (distribution, comp. note)
Moana aluana aluana Boer and Duffels 1996b: 320, 329, Fig 14 (distribution, comp. note) Solomon Islands
Moana aluana Duffels 1997: 552 (comp. note) Solomon Islands
Moana aluana Duffels and Turner 2002: 244–245, 252 (comp. note)

Moana aluana aluana Duffels and Turner 2002: 248, 256, 259, Fig 11, Appendices 1–2 (phylogeny, distribution, listed) Solomon Islands

***M. aluana minima* (Duffels, 1977)** = *Aceropyga aluana minima* = *Aceropyga aluana minuta* (sic)

Aceropyga aluana minuta (sic) Duffels and Boer 1990: 262, Fig 21.5 (distribution)

Moana aluana mimima Boer 1995d: 220, Fig 49 (distribution, comp. note)

Moana aluana mimima Boer and Duffels 1996b: 320, 329, Fig 14 (distribution, comp. note) Solomon Islands

Moana aluana minima Duffels and Turner 2002: 248, 256, 259, Fig 11, Appendices 1–2 (phylogeny, distribution, listed) Solomon Islands

***M. aluana torquata* (Duffels, 1977)** = *Aceropyga aluana torquata* = *Diceropyga pigafettae* (nec Distant)

Aceropyga aluana torquata Duffels and Boer 1990: 262, Fig 21.5 (distribution)

Moana aluana torquta Boer 1995d: 220, Fig 49 (distribution, comp. note)

Moana aluana torquta Boer and Duffels 1996b: 320, 329, Fig 14 (distribution, comp. note) Solomon Islands

Moana aluana torquata Duffels and Turner 2002: 248, 256, 259, Fig 11, Appendices 1–2 (phylogeny, distribution, listed) Solomon Islands

***M. expansa* Myers, 1928**

Moana expansa Boulard 1986c: 203 (stridulatory apparatus, comp. note)

Moana expansa Duffels 1986: 331 (comp. note) Samoa

Moana expansa Boulard 1988f: 52 (type species of Moaninae)

Moana expansa Duffels 1988d: 9, 11–12, 25, 28, 30, 77–80, Figs 144–149, Table 1 (key, type species of *Moana*, described, illustrated, distribution, comp. note) Samoa

Moana expansa Boulard 1990b: 148, 159, 161, Plate XX, Figs 4–5 (sound system, comp. note) Samoa

Moana expansa Duffels 1993: 1223–1236, Figs 1–5 (described, illustrated, stridulatory apparatus, phylogeny, listed, comp. note) Samoa

Moana expansa Boer 1995d: 220, Fig 49 (distribution, comp. note)

Moana expansa Boer and Duffels 1996b: 320, 329, Fig 14 (distribution, comp. note) Samoa Islands

Moana expansa Duffels 1997: 551 (comp. note)

Moana expansa Boulard 2000d: 88, Plate D, Figs 4–5 (stridulatory apparatus)

Moana expansa Boulard 2002b: 204 (stridulation, comp. note)

Moana expansa Boulard 2005d: 349 (stridulatory apparatus, comp. note)

Moana expansa Moulds 2005b: 387, 394 (listed, comp. note)

Moana expansa Boulard 2006g: 156, Fig 113 (stridulation, illustrated, comp. note) Samoa

Moana expansa Duffels and Turner 2002: 236, 244–245, 248, 251, 253, 256, 259, Fig 8c, Fig 11, Appendices 1–2 (illustrated, phylogeny, distribution, listed, comp. note) Samoa Islands

***M. novaeirelandicae* (Duffels, 1977)** = *Moana irelandica* (sic) = *Aceropyga novaeirelandicae* = *Diceropyga pigafettae* (nec Distant) = *Diceropyga aluana* (nec Distant)

Aceropyga novaeirelandicae Duffels 1986: 325 (comp. note) New Ireland

Aceropyga novaeirelandicae Duffels 1988d: 32, Fig 13 (phylogeny)

Aceropyga novaeirelandica Duffels and Boer 1990: 262, Fig 21.5 (distribution)

Moana novaeirelandicae Duffels 1993: 1230, 1232–1233, Fig 5 (n. comb., phylogeny, listed, comp. note) New Ireland, Bismarck Archipelago

Moana novaeirelandicae Boer 1995d: 220, Fig 49 (distribution, comp. note)

Moana novaeirelandicae Boer and Duffels 1996b: 320, 330, Fig 14 (distribution, comp. note) Bismarck Archipelago

Moana irelandica (sic) Duffels and Turner 2002: 244 (comp. note)

Moana novaeirelandica Duffels and Turner 2002: 245, 248, 253, 256, 259, Fig 11, Appendices 1–2 (phylogeny, distribution, listed, comp. note) Bismarck Archipelago

***M. obliqua* (Duffels, 1977)** = *Aceropyga obliqua*

Aceropyga obliqua Duffels 1986: 325 (comp. note) New Britain, Umboi Island

Aceropyga obliqua Duffels 1988d: 32, Fig 13 (phylogeny)

Aceropyga obliqua Duffels and Boer 1990: 262, Fig 21.5 (distribution)

Moana obliqua Duffels 1993: 1230, 1232–1233, Fig 5 (n. comb., phylogeny, listed, comp. note) New Britain, Umbor Island, Bismark Archipelago

Moana obliqua Boer 1995d: 220, Fig 49 (distribution, comp. note)

Moana obliqua Boer and Duffels 1996b: 320, 330, Fig 14 (distribution, comp. note) Bismarck Archipelago

Moana obliqua Duffels and Turner 2002: 244–245, 248, 256, 259, Fig 11, Appendices 1–2 (phylogeny, distribution, listed, comp. note) Bismarck Archipelago

Subtribe Orientopsaltriina Boulard, 2004
= Orientopsaltriaria
Orientopsaltriaria Boulard 2003e: 270 (listed)
Oreintopsaltriaria Sanborn, Phillips and Sites 2007: 3, 22, 39, Fig 4, Table 1 (listed, distribution) Thailand
Orientopsaltriina Boulard 2008f: 7, 34 (listed) Thailand

***Orientopsaltria* Kato, 1944**
Orientopsaltria Duffels 1983b: 3–4, 8–11, 19, 39, 41 (diversity, comp. note)
Orientopsaltria Hayashi 1984: 58 (listed)
Orientopsaltria Duffels 1986: 319 (distribution, diversity, comp. note) Greater Sunda Islands, Southeast Asia
Orientopsaltria Zaidi, Mohammedsaid and Yaakob 1990: 261, 267, Table 2 (listed, comp. note) Malaysia
Orientopsaltria Zaidi and Ruslan 1994: 427, Table 2 (comp. note) Malaysia
Orientopsaltria Zaidi and Ruslan 1995a: 174, Table 2 (listed, comp. note) Borneo, Sarawak
Orientopsaltria Zaidi and Ruslan 1995b: 220–221, Table 2 (listed, comp. note) Borneo, Sabah
Orientopsaltria Zaidi and Ruslan 1995c: 70, Table 2 (listed, comp. note) Peninsular Malaysia
Orientopsaltria Zaidi and Ruslan 1995d: 201–202, Table 2 (listed, comp. note) Borneo, Sabah
Orientopsaltria Beuk 1996: 129–131, 134–135 (characteristics, history, comp. note)
Orientopsaltria Zaidi 1996: 103, Table 2 (comp. note) Borneo, Sarawak
Orientopsaltria Zaidi and Hamid 1996: 54–55, Table 2 (listed, comp. note) Borneo, Sarawak
Orientopsaltria Zaidi 1997: 114–115, Table 2 (listed, comp. note) Borneo, Sarawak
Orientopsaltria Beuk 1998: 147 (comp. note)
Orientopsaltria Duffels and Zaidi 1998: 319–321 (distribution, comp. note) Borneo, Sumatra, Peninsular Malaysia, Philippines
Orientopsaltria Zaidi and Ruslan 1998b: 4–5, Table 2 (listed, comp. note) Borneo, Sabah
Orientopsaltria Beuk 1999: 2 (comp. note)
Orientopsaltria Zaidi and Ruslan 1999: 4, Table 2 (listed, comp. note) Borneo, Sarawak
Orientopsaltria sp. Zaidi, Ruslan and Azman 1999: 312 (listed, comp. note) Borneo, Sabah
Orientopsaltria Duffels and Zaidi 2000: 195–199, 219, 249, 259, Figs 1–2, Figs 4–5, Table 1 (described, diversity, phylogeny, comp. note) Malayan Peninsula, Sumatra, Borneo, Palawan, Philippines
Orientopsaltria spp. Duffels and Zaidi 2000: 207, 216, 225, 228, 254, 260, 270, 276 (comp. note)
Orientopsaltria Duffels and Schouten 2002: 49–50 (diversity, distribution, comp. note) Peninsular Malaysia, Sumatra, Borneo, Philippines
Orientopsaltria Duffels and Turner 2002: 235, 239, 244 (comp. note)
Orientopsaltria spp. Duffels and Turner 2002: 239 (comp. note)
Orientopsaltria Schouten, Duffels and Zaidi 2004: 379 (comp. note)
Oreintopsaltria Moulds 2005b: 391, 432 (higher taxonomy, listed, comp. note)
Orientopsaltria Boulard 2007a: 100 (comp. note)
Orientopsaltria Sanborn, Phillips and Sites 2007: 3, 22, Table 1 (listed) Thailand
Orientopsaltria Boulard 2008f: 6–7, 34, 93 (listed, comp. note) Thailand
Orientipsaltria Lee 2009h: 2629 (listed) Philippines, Palawan
Orientopssaltria Lee 2009i: 294 (listed) Philippines, Panay
Orientipsaltria Lee 2010a: 22 (listed) Philippines, Mindanao
Oreintopsaltria Lee 2010b: 25 (listed) Cambodia
Orientopsaltria Lee 2010c: 13 (listed) Philippines, Luzon

***O. agatha* (Moulton, 1911)** = *Cosmopsaltria agatha* = *Cosmopsaltria* sp. aff. *agatha* Endo & Hayashi, 1979 = *Cosmopsaltria montivaga agatha*
Cosmopsaltria agatha Duffels 1983b: 4 (comp. note)
Orientopsaltria agatha Duffels 1983b: 9 (comp. note)
Orientopsaltria agatha Zaidi 1993: 958, 960, Table 1 (listed, comp. note) Borneo, Sarawak
Orientopsaltria agatha Zaidi 1996: 101, 104, Table 1 (comp. note) Borneo, Sarawak
Orientopsaltria agatha Zaidi and Hamid 1996: 53, 56–57, Table 1 (listed, comp. note) Borneo, Sarawak
Orientopsaltria agatha Zaidi 1997: 113–114, 116, Table 1 (listed, comp. note) Borneo, Sarawak
Orientopsaltria agatha Duffels and Zaidi 1998: 321–322 (comp. note)
Oreintopsaltria agatha Zaidi and Ruslan 1998a: 357–358 (listed, comp. note) Equals *Cosmopsaltria agatha* Borneo, Sarawak,
Orientopsaltria agatha Duffels and Zaidi 2000: 196–197, 199–200, 202–205, 215, 217, 219, 225–227, 237, 284, Figs 1–2, Fig 4, Figs 17, Figs 37–41, Fig 63, Plate 2 Fig 3, Tables 1–2 (described, illustrated, distribution, phylogeny, key, comp. note) Equals *Cosmopsaltria agatha* Equals *Cosmopsaltria* sp. aff. *agatha* Endo and Hayashi, 1979 Equals *Cosmopsaltria montivaga* var. *agatha* Borneo, Sarawak, Sabah, Kalimantan, Brunei
Cosmopsaltria agatha Duffels and Zaidi 2000: 196–197, 219 (comp. note)

Orientopsaltria agatha Zaidi, Noramly and Ruslan 2000a: 325–326 (listed, comp. note) Equals *Cosmopsaltria agatha* Borneo, Sabah, Sarawak, Brunei, Kalimantan

Orientopsaltria agatha Boulard 2001c: 58 (comp. note) Sarawak

Orientopsaltrai agatha Zaidi and Nordin 2001: 185, 190, 196–197, Table 1 (listed, comp. note) Equals *Cosmopsaltria agatha* Borneo, Sabah, Sarawak, Brunei, Kalimantan

Orientopsaltria agatha Duffels and Turner 2002: 239 (comp. note)

Orientopsaltria agatha Zaidi and Azman 2003: 105, Table 1 (listed, comp. note) Borneo, Sabah

***O. alticola* (Distant, 1905)** = *Cosmopsaltria alticola* = *Cosmopsaltria alticola pontiaka* Kirkaldy, 1913 = *Cosmopsaltria alticola* var. *pontianka* = *Platylomia bangueyensis* Distant, 1912 = *Cosmopsaltria inermis* (nec Stål)

Orientopsaltria alticola Duffels 1983b: 9 (n. comb., comp. note)

Orientopsaltria alticola Zaidi and Ruslan 1995d: 200, 203, Table 1 (listed, comp. note) Borneo, Sabah

Orientopsaltria alticola Zaidi, Ruslan and Mahadir 1996: 60, Table 1 (listed, comp. note) Peninsular Malaysia

Orientopsaltria alticola Duffels and Zaidi 1998: 321 (comp. note) Peninsular Malaysia, Borneo

Oreintopsaltria alticola Zaidi and Ruslan 1998a: 358 (listed, comp. note) Equals *Cosmopsaltria alticola* Borneo, Sarawak, Sabah

Orientopsaltria alticola Zaidi, Ruslan and Azman 1999: 309 (listed, comp. note) Equals *Cosmopsaltria alticola* Borneo, Sabah, Sarawak, Peninsular Malaysia

Orientopsaltria alticola Duffels and Zaidi 2000: 196–197, 199–200, 202, 204, 264–270, 276, 292, Figs 1–2, Fig 4, Fig 104, Figs 111–114, Fig 124, Plate 5 Fig 3, Tables 1–2 (described, illustrated, synonymy, distribution, phylogeny, key, comp. note) Equals *Cosmopsaltria alticola* Equals *Cosmopsaltria alticola* var. *pontianka* Equals *Cosmopsaltria inermis* (nec Stål)(partim) Equals *Platylomia bangueyensis* Malayan Peninsula, Sarawak, Borneo, Sabah, Kalimantan, Brunei

Cosmopsaltria alticola Duffels and Zaidi 2000: 197, 265 (comp. note)

Platylomia bangueyensis Duffels and Zaidi 2000: 197, 265 (synonymy, comp. note)

Orientopsaltria alticola group Duffels and Zaidi 2000: 201–202, 231, 261, 264–265 (described, diversity, comp. note)

Cosmopsaltria alticola var. *pontianka* Duffels and Zaidi 2000: 265 (synonymy, comp. note)

Orientopsaltria alticola Zaidi, Noramly and Ruslan 2000a: 326 (listed, comp. note) Equals *Cosmopsaltria alticola* Borneo, Sabah, Peninsular Malaysia, Sarawak, Brunei, Kalimantan

Orientopsaltria alticola Zaidi, Noramly and Ruslan 2000b: 337–339, Table 1 (listed, comp. note) Borneo, Sabah

Orientopsaltria alticola Zaidi, Ruslan and Azman 2000: 208 (listed, comp. note) Equals *Cosmopsaltria alticola* Borneo, Sabah, Peninsular Malaysia, Sarawak, Brunei, Kalimantan

Orientopsaltrai alticola Zaidi and Nordin 2001: 183, 185–186, 190, 197, Table 1 (listed, comp. note) Equals *Cosmopsaltria alticola* Borneo, Sabah, Peninsular Malaysia, Sarawak, Brunei, Kalimantan

Orientopsaltria alticola Zaidi, Nordin, Maryati, Wahab, Norashikin et al. 2002: 4, 9–11, Tables 1–2 (listed, comp. note) Equals *Cosmopsaltria alticola* Borneo, Sabah, Sarawak, Peninsular Malaysia, Kalimantan

Platylomia bangueyensis Zaidi and Azman 2003: 99–100, 105, Table 1 (listed, comp. note) Borneo, Sabah

Orientopsaltria alticola Zaidi and Azman 2003: 105, Table 1 (listed, comp. note) Borneo, Sabah

Orientopsaltria alticola Lee 2009h: 2617 (comp. note)

O. andersoni (Distant, 1883) to *Dundubia oopaga* (Distant, 1881)

***O. angustata* Duffels & Zaidi, 2000** = *Cosmopsaltria guttigera* (nec Walker) = *Orientopsaltria guttigera* (nec Walker) = *Orientopsaltria* nr. *guttigera* n. sp. Zaidi and Ruslan 1995

Orientopsaltria guttigera (sic) Duffels 1983b: 9 (n. comb., comp. note)

Orientopsaltria guttigera (sic) Zaidi 1993: 958, 960, Table 1 (listed, comp. note) Borneo, Sarawak

Orientopsaltria guttigera (sic) Zaidi and Ruslan 1995b: 218–219, 221, Table 1 (listed, comp. note) Borneo, Sabah

Orientopsaltria nr. *guttigera* n. sp. Zaidi and Ruslan 1995d: 197–198, 200–201, 203, Table 1 (listed, comp. note) Borneo, Sabah

Orientopsaltria guttigera (sic) Zaidi 1996: 101, 104, Table 1 (comp. note) Borneo, Sarawak

Orientopsaltria guttigera (sic) Zaidi and Hamid 1996: 53, 56, Table 1 (listed, comp. note) Borneo, Sarawak

Orientopsaltria guttigera (sic) Zaidi, Ruslan and Mahadir 1996: 60, Table 1 (listed, comp. note) Peninsular Malaysia

Orientopsaltria guttigera (sic) Zaidi 1997: 113–114, Table 1 (listed, comp. note) Borneo, Sarawak
Orientopsaltria angustata Duffels and Zaidi 2000: 197, 199–200, 202, 205, 228, 230–239, 243, 245, 288, Figs 1–2, Fig 4, Fig 46, Fig 56, Fig 58, Fig 60, Fig 62, Fig 65, Plate 3 Fig 1, Tables 1–2 (n. sp., described, illustrated, synonymy, distribution, phylogeny, key, comp. note) Equals *Cosmopsaltria guttigera* (partim) Equals *Orientopsaltria guttigera* (partim) Sarawak, Kalimantan, Borneo
Orientopsaltria angustata Zaidi, Azman and Ruslan 2001b: 127 (listed, comp. note) Borneo, Sarawak, Kalimantan
Orientopsaltria angustata Duffels and Schouten 2002: 50 (comp. note) Borneo
Orientopsaltria angustata Zaidi, Azman and Nordin 2005: 28–29, Table 1 (listed, comp. note) Borneo, Sabah, Indonesia, Kalimantan

***O. bardi* Boulard, 2008**
Orientopsaltria bardi Boulard 2008f: 34, 63–65, 93, Figs 9–13 (13 labeled 18) (n. sp., described, illustrated, song, sonogram, biogeography, listed, comp. note) Thailand

***O. beaudouini* Boulard, 2003**
Oreintopsaltria beaudouini Boulard 2003e: 270–274, Color Plate, Fig E, sonogram 3 (n. sp., described, song, sonogram, illustrated, comp. note) Thailand
Orientopsaltria beaudouini Boulard 2005o: 149 (listed)
Orientopsaltria beaudouini Sanborn, Phillips and Sites 2007: 22 (listed, distribution, comp. note) Thailand
Orientopsaltria beaudouini Boulard 2008c: 364 (comp. note) Thailand, Myanmar, Laos
Orientopsaltria beaudouini Boulard 2008f: 34, 93 (biogeography, listed, comp. note) Thailand

***O. brooksi* (Moulton, 1923)** = *Cosmopsaltria brooksi*
Orientopsaltria brooksi Duffels 1983b: 9 (n. comb., comp. note)
Orientopsaltria brooksi Zaidi 1993: 958–959, Table 1 (listed, comp. note) Borneo, Sarawak
Orientopsaltria brooksi Zaidi and Ruslan 1995c: 64–65, 68, Table 1 (listed, comp. note) Peninsular Malaysia
Orientopsaltria brooksi Zaidi and Ruslan 1995d: 198, 201 (comp. note)
Orientopsaltria brooksi Zaidi 1996: 101, 104, Table 1 (comp. note) Borneo, Sarawak
Orientopsaltria brooksi Zaidi and Hamid 1996: 53, 56–57, Table 1 (listed, comp. note) Borneo, Sarawak
Orientopsaltria brooksi Zaidi 1997: 113–114, 116, Table 1 (listed, comp. note) Borneo, Sarawak
Orientopsaltria brooksi Duffels and Zaidi 1998: 321 (comp. note) Peninsular Malaysia, Sumatra, Borneo
Orientopsaltria brooksi Duffels and Zaidi 2000: 196–197, 199–200, 202–205, 227–231, 233, 237, 284, Figs 1–2, Fig 4, Figs 51–55, Fig 64, Plate 2 Fig 4, Tables 1–2 (described, illustrated, synonymy, distribution, phylogeny, key, comp. note) Equals *Cosmopsaltria brooksi* Malayan Peninsula, Sumatra
Cosmopsaltria brooksi Duffels and Zaidi 2000: 197 (comp. note)
Orientopsaltria brooksi group Duffels and Zaidi 2000: 200, 227 (described, diversity, comp. note) Malayan Peninsula, Sumatra

***O. cantavis* Boulard, 2003**
Orientopsaltria cantavis Boulard 2003c: 172, 187–191, Figs 7–12, Plate F, Figs 1–3 (n. sp., described, illustrated, song, sonogram, comp. note) Thailand
Orientopsaltria cantavis Boulard 2005d: 344 (comp. note)
Orientopsaltria cantavis Boulard 2005o: 149 (listed)
Orientopsaltria cantavis Boulard 2006g: 73, 75, 84, 180, Fig 43 (sonogram, listed, comp. note) Thailand
Orientopsaltria cantavis Boulard 2007e: 35, 46, 55, 67, Fig 30, Fig 37.5 (sonogram, genitalia illustrated, comp. note) Thailand
Orientopsaltria cantavis Sanborn, Phillips and Sites 2007: 22 (listed, distribution, comp. note) Thailand
Orientopsaltria cantavis Boulard 2008b: 109 (comp. note)
Orientopsaltria cantavis Boulard 2008e: 352 (comp. note)
Orientopsaltria cantavis Boulard 2008f: 34, 93 (biogeography, listed, comp. note) Thailand

***O. ceslaui* (Lallemand & Synave, 1953)** = *Cosmopsaltria ceslaui*
Orientopsaltria ceslaui Duffels 1983b: 9 (n. comb., comp. note)
Orientopsaltria ceslaui Duffels and Zaidi 1998: 321 (comp. note)
Orientopsaltria ceslaui Duffels and Zaidi 2000: 196 (comp. note)
Dundubia ceslaui Duffels and Zaidi 2000: 196 (comp. note)

***O. confluens* Duffels & Zaidi, 2000**
Orientopsaltria confluens Duffels and Zaidi 2000: 197, 199–200, 202–204, 264–265, 270, 272–277, 293, Figs 1–2, Fig 4, Figs 119–123, 126, Plate 6 Fig 1, Tables 1–2 (n. sp. described,

illustrated, distribution, phylogeny, key, comp. note) Basilan Island, Philippines

O. divergens (Distant, 1917) to *Macrosemia divergens*

***O. duarum* (Walker, 1857)** = *Dundubia duarum* = *Cosmopsaltria duarum* = *Cosmopsaltria lauta* Distant, 1888 = *Orientopsaltria lauta*

Cosmposaltria duarum Duffels 1983b: 3 (comp. note)

Orientopsaltria duarum Duffels 1983b: 9 (comp. note)

Orientopsaltria duarum Zaidi and Ruslan 1995b: 218–219, 222, Table 1 (listed, comp. note) Borneo, Sabah

Orientopsaltria duarum Zaidi and Ruslan 1995d: 198, 200 (comp. note) Borneo, Sabah

Orientopsaltria duarum Beuk 1996: 131–132, Fig 1, Table 1 (phylogeny, comp. note)

Orientopsaltria duarum Zaidi, Ruslan and Mahadir 1996: 60, Table 1 (listed, comp. note) Peninsular Malaysia

Orientopsaltria duarum Zaidi and Ruslan 1997: 225 (listed, comp. note) Equals *Dundubia duarum* Equals *Cosmopsaltria nigripes* Peninsular Malaysia, Borneo, Sarawak

Orientopsaltria duarum Duffels and Zaidi 1998: 320–321 (type species of *Orientopsaltria*, comp. note) Malayan Peninsual, Sumatra, Borneo

Oreintopsaltria duarum Zaidi and Ruslan 1998a: 358 (listed, comp. note) Equals *Dundubia duarum* Equals *Cosmopsaltria duarum* Borneo, Sarawak, Peninsular Malaysia

Orientopsaltria duarum Zaidi and Ruslan 1998b: 2–3, 5, Table 1 (listed, comp. note) Borneo, Sabah

Orientopsaltria duarum Zaidi, Ruslan and Azman 1999: 309–310 (listed, comp. note) Equals *Dundubia duarum* Equals *Cosmopsaltria duarum* Borneo, Sabah, Sarawak, Peninsular Malaysia

Orientopsaltria duarum Duffels and Zaidi 2000: 196–197, 199–200, 202–209, 211, 216–217, 281, Figs 1–2, Fig 4, Figs 6–9, Fig 18, Plate 1 Fig 1, Tables 1–2 (described, illustrated, synonymy, distribution, phylogeny, key, comp. note) Equals *Dundubia duarum* Equals *Cosmopsaltria duarum* Equals *Cosmopsaltria lauta* Malayan Peninsula, Sumatra, Sarawak, Borneo, Sabah, Kalimantan

Cosmopsaltria duarum Duffels and Zaidi 2000: 196 (type species of *Orientopsaltria*)

Dundubia duarum Duffels and Zaidi 2000: 196, 203, 207 (type species of *Orientopsaltria*, comp. note) Sarawak

Cosmopsaltria lauta Duffels and Zaidi 2000: 207, 211 (comp. note)

Orientopsaltria duarum Zaidi, Noramly and Ruslan 2000b: 337, 339 (comp. note) Borneo, Sabah

Orientopsaltria duarum Boulard 2001c: 58 (comp. note) Equals *Cosmopsaltria lauta* Borneo

Orientopsaltria lauta Boulard 2003e: 270 (comp. note)

Orientopsaltria duarum Zaidi and Azman 2003: 105, Table 1 (listed, comp. note) Borneo, Sabah

Orientopsaltria lauta Boulard 2005e: 128 (comp. note)

Orientopsaltria duarum Boulard 2008f: 34 (type species of *Orientopsaltria*) Equals *Dundubia duarum*

Dundubia duarum Lee 2009h: 2621–2622 (type species of *Orientopsaltria*, listed) Sarawak, Borneo

Dundubia duarum Lee 2009i: 294 (type species of *Orientopsaltria*, listed)

Dundubia duarum Lee 2010a: 22 (type species of *Orientopsaltria*, listed)

Dundubia duarum Lee 2010b: 25 (type species of *Orientopsaltria*)

Dundubia duarum Lee 2010c: 13 (type species of *Orientopsaltria*, listed) Sarawak, Borneo

***O. endauensis* Duffels & Schouten, 2002**

Orientopsaltria endauensis Duffels and Schouten 2002: 49–54, Figs 1–3 (comp. note) Malaysia

Orientopsaltria endauensis Boulard 2003e: 270 (comp. note)

Orientopsaltria endauensis Schouten, Duffels and Zaidi 2004: 370, 373, 375, Table 2 (listed, comp. note) Malayan Peninsula

***O. fangrayae* Boulard, 2001**

Orientopsaltria fangrayae Boulard 2001c: 58–62, Plate D, Figs 1–3, Plate E, Figs 1–2 (n. sp., described, song, illustrated, sonogram, comp. note) Thailand

Orientopsaltria fangrayae Boulard 2003d: 187 (comp. note)

Orientopsaltria fangrayae Boulard 2005o: 148 (listed)

Orientopsaltria fangrayae Boulard 2006g: 119, 174, Fig (sonogram, comp. note)

Orientopsaltria fangrayae Boulard 2007d: 497 (comp. note) Thailand

Orientopsaltria fangrayae Sanborn, Phillips and Sites 2007: 22–23 (listed, distribution, comp. note) Thailand

Orientopsaltria fangrayae Boulard 2008f: 34, 63, 93 (biogeography, listed, comp. note) Thailand

O. feae (Distant, 1892) to *Dundubia feae*

O. fratercula (Distant, 1912) to *Dundubia nagarasingna* Distant, 1881

***O. fuliginosa* (Walker, 1850)** = *Dundubia fuliginosa* = *Cosmopsaltria fuliginosa* = *Cosmopsaltria (Cosmopsaltria) fuliginosa* = *Platylomia fuliginosa* = *Cosmopsaltria* sp. aff. *fuliginosa* Endo & Hayashi, 1979 = *Dundubia melpomene* Walker, 1850

Orientopsaltria fuliginosa Duffels and Zaidi 2000: 196–197, 199–200, 202, 204, 260, 264–265, 275–280, 282–283, 286, 293, Figs 1–2, Fig 4, Fig 105, Fig 127, Figs 129–135, Plate 6 Fig 2, Tables 1–2 (n. comb., described, illustrated, synonymy, distribution, phylogeny, key, comp. note) Equals *Platylomia fuliginosa* Equals *Dundubia fuliginosa* Equals *Cosmopsaltria fuliginosa* Equals *Cosmopsaltria* sp. aff. *fuliginosa* Endo and Hayashi, 1979 Equals *Dundubia melpomene* Philippines

Platylomia fuliginosa Duffels and Zaidi 2000: 196 (comp. note)

Dundubia fuliginosa Duffels and Zaidi 2000: 196 (comp. note)

Orientopsaltria fuliginosa Lee 2009i: 293–294 (synonymy, distribution, listed) Equals *Dundubia fuliginosa* Philippines, Panay

Orientopsaltria fuliginosa Lee 2010a: 24 (synonymy, distribution, listed) Equals *Dundubia fuliginosa* Equals *Cosmopsaltria* sp. aff. *fuliginosa* Philippines, Mindanao, Luzon, Mindoro, Homonhon, Samar, Leyte, Panay, Negros, Basalin

Orientopsaltria fuliginosa Lee 2010c: 13 (synonymy, distribution, listed) Equals *Dundubia fuliginosa* Philippines, Luzon, Mindoro, Homonhon, Samar, Leyte, Panay, Negros, Mindanao, Basalin

***O. guttigera* (Walker, 1857)** = *Dundubia guttigera* = *Cosmopsaltria guttigera*

Orientopsaltria guttigera Duffels 1983b: 9 (n. comb., comp. note)

Dundubia guttigera Duffels 1983b: 197 (comp. note)

Orientopsaltria guttigera Zaidi and Ruslan 1995d: 198, 200–201 (comp. note)

Orientopsaltria guttigera Zaidi and Ruslan 1997: 225 (listed, comp. note) Equals *Dundubia guttigera* Equals *Cosmopsaltria nigripes* Peninsular Malaysia, Singapore, Borneo, Sarawak

Orientopsaltria guttigera Duffels and Zaidi 1998: 321 (comp. note)

Oreintopsaltria guttigera Zaidi and Ruslan 1998a: 359 (listed, comp. note) Equals *Dundubia guttigera* Equals *Cosmopsaltria guttigera* Borneo, Sarawak, Singapore, Baram, Sabah, Peninsular Malaysia

Orientopsaltria guttigera Zaidi and Ruslan 1998b: 2–3, 5, Table 1 (listed, comp. note) Borneo, Sabah

Orientopsaltria guttigera Duffels and Zaidi 2000: 196–197, 199–200, 202, 205, 228, 230, 232–233, 235–239, 243, 246, 288, Figs 1–2, Fig 4, Fig 47, Fig 57, Fig 59, Figs 61–62, Fig 66, Plate 3 Fig 2, Tables 1–2 (described, illustrated, synonymy, distribution, phylogeny, key, comp. note) Equals *Dundubia guttigera* Equals *Cosmopsaltria guttigera* Malayan Peninsula, Sumatra

Dundubia guttigera Duffels and Zaidi 2000: 197 (comp. note)

Orientopsaltria guttigera group Duffels and Zaidi 2000: 199, 200, 202, 230 (described, diversity, comp. note)

Orientopsaltria guttigera group Duffels and Schouten 2002: 50 (comp. note)

Orientopsaltria guttigera Duffels and Schouten 2002: 50 (comp. note) Malayan Peninsula, Sumatra

Orientopsaltria guttigera Sanborn, Phillips and Sites 2007: 23, 34 (listed, distribution, comp. note) Thailand, Malaysia, Indonesia, Borneo, Sarawak

Orientopsaltria guttigera Boulard 2008f: 90 (comp. note) Thailand

O. n. sp. 1 nr. *guttigera* Zaidi & Ruslan, 1995 to *Orientopsaltria maculosa* Duffels & Zaidi, 2000

O. hastata (Moulton, 1923) to *Dundubia spiculata* Noualhier, 1896

***O. hollowayi* Duffels & Zaidi, 2000**

Orientopsaltria hollowayi Duffels and Zaidi 2000: 197, 199–205, 230, 244–248, 254, 262, 289, Figs 1–2, Fig 4, Figs 76–79, Fig 87, Plate 4 Fig 1, Tables 1–2 (n. sp., described, illustrated, distribution, phylogeny, key, comp. note) Sarawak, Borneo, Sabah, Kalimantan, Brunei

Orientopsaltrai hollowayi Zaidi and Nordin 2001: 183, 185, 190, 197–198, Table 1 (listed, comp. note) Borneo, Sabah, Sarawak

Orientopsaltria hollowayi Duffels and Schouten 2002: 50 (comp. note) Borneo

Orientopsaltria hollowayi Zaidi, Nordin, Maryati, Wahab, Norashikin et al. 2002: 5, 9–10, Tables 1–2 (listed, comp. note) Borneo, Sabah, Sarawak

Orientopsaltria hollowayi Zaidi and Azman 2003: 105, Table 1 (listed, comp. note) Borneo, Sabah

***O. ida* (Moulton, 1911)** = *Cosmopsaltria ida*

Orientopsaltria ida Duffels 1983b: 9 (n. comb., comp. note)

Orientopsaltria ida Zaidi and Ruslan 1995b: 218–219, 221, Table 1 (listed, comp. note) Borneo, Sabah

Orientopsaltria ida Zaidi and Ruslan 1995d: 198, 200, 203, Table 1 (listed, comp. note) Borneo, Sabah

Orientopsaltria ida Zaidi and Hamid 1996: 53, 56, Table 1 (listed, comp. note) Borneo, Sarawak

Orientopsaltria ida Zaidi, Ruslan and Mahadir 1996: 60, Table 1 (listed, comp. note) Peninsular Malaysia

Orientopsaltria ida Duffels and Zaidi 1998: 321 (comp. note) Peninsular Malaysia, Borneo

Oreintopsaltria ida Zaidi and Ruslan 1998a: 359–360 (listed, comp. note) Equals *Cosmopsaltria ida* Borneo, Sarawak, Sabah, Peninsular Malaysia

Orientopsaltria ida Zaidi and Ruslan 1998b: 2–3, 5, Table 1 (listed, comp. note) Borneo, Sabah

Orientopsaltria ida Zaidi, Ruslan and Azman 1999: 310 (listed, comp. note) Equals *Cosmopsaltria ida* Borneo, Sabah, Sarawak, Peninsular Malaysia

Orientopsaltria ida Duffels and Zaidi 2000: 196–197, 199–200, 202, 205, 228, 248–254, 259, 289, Figs 1–2, Fig 4, Fig 50, Figs 80–82, Fig 84, Fig 88, Plate 4 Fig 2, Tables 1–2 (described, illustrated, synonymy, distribution, phylogeny, key, comp. note) Equals *Cosmopsaltria ida* Malayan Peninsula, Sumatra, Sarawak, Borneo, Kalimantan, Brunei, Sabah

Cosmopsaltria ida Duffels and Zaidi 2000: 197, 249 (comp. note)

Orientopsaltria ida group Duffels and Zaidi 2000: 200, 202, 230, 249, 255 (described, diversity, comp. note)

Orientopsaltria ida Zaidi, Noramly and Ruslan 2000a: 326–327 (listed, comp. note) Equals *Cosmopsaltria ida* Borneo, Sabah, Sarawak, Brunei, Kalimantan, Peninsular Malaysia

Orientopsaltria ida Zaidi, Noramly and Ruslan 2000b: 337, 339 (comp. note) Borneo, Sabah

Orientopsaltria ida Zaidi, Ruslan and Azman 2000: 208 (listed, comp. note) Equals *Cosmopsaltria ida* Borneo, Sabah, Sarawak, Brunei, Kalimantan, Peninsular Malaysia

Orientopsaltrai ida Zaidi and Nordin 2001: 183, 185–186, 190, 198, Table 1 (listed, comp. note) Equals *Cosmopsaltria ida* Borneo, Sabah, Sarawak, Brunei, Kalimantan, Peninsular Malaysia

Orientopsaltria ida Zaidi, Nordin, Maryati, Wahab, Norashikin et al. 2002: 5, 9–10, Tables 1–2 (listed, comp. note) Equals *Cosmopsaltria ida* Borneo, Sabah, Sarawak, Brunei, Kalimantan

Orientopsaltria ida Boulard 2003c: 197 (comp. note)

Orientopsaltria ida Zaidi and Azman 2003: 105, Table 1 (listed, comp. note) Borneo, Sabah

Orientopsaltria ida Boulard 2005e: 128 (comp. note)

***O. inermis* (Stål, 1870)** = *Cosmopsaltria inermis*

Orientopsaltria inermis Duffels 1983b: 9 (n. comb., comp. note)

Orientopsaltria inermis Duffels and Zaidi 1998: 321 (comp. note)

Orientopsaltria inermis Duffels and Zaidi 2000: 196–197, 199–200, 202, 204, 264, 269–272, 275–276, 286, 292, Figs 1–2, Fig 4, Figs 115–118, Fig 123, Fig 125, Plate 5 Fig 4, Tables 1–2 (lectotype designation, described, illustrated, synonymy, distribution, phylogeny, key, comp. note) Equals *Cosmopsaltria inermis* Philippines Mindanao

Cosmopsaltria inermis Duffels and Zaidi 2000: 196 (comp. note)

Orientopsaltria inermis Lee 2009h: 2617 (comp. note)

Orientopsaltria inermis Lee 2010a: 22–24, Figs 6A–6B (synonymy, illustrated, distribution. listed) Equals *Cosmopsaltria inermis* Philippines, Mindanao, Samar

O. jacoona (Distant, 1888) to *Dundubia jacoona*

***O. jmalleti* Boulard, 2005**

Orientopsaltria jmalleti Boulard 2005i: 67–68, Figs 13–17 (n. sp., described, illustrated, comp. note) Thailand

Orientopsaltria jmalleti Boulard 2005o: 150 (listed)

***O. kinabaluana* Duffels & Zaidi, 2000**

Orientopsaltria kinabaluana Duffels and Zaidi 2000: 197, 199–202, 205, 231, 246, 260–264, 292, Figs 1–2, Fig 4, Fig 92, Fig 103, Figs 107–110, Plate 5 Fig 2, Tables 1–2 (n. sp., described, illustrated, synonymy, distribution, phylogeny, key, comp. note) Sabah, Borneo

Orientopsaltria kinabaluana group Duffels and Zaidi 2000: 201, 261 (described, diversity, comp. note)

Orientopsaltrai kinabaluana Zaidi and Nordin 2001: 183, 185, 190, 198–199, Table 1 (listed, comp. note) Borneo, Sabah

Orientopsaltria kinabulana Zaidi, Nordin, Maryati, Wahab, Norashikin et al. 2002: 5, 9–10, Tables 1–2 (listed, comp. note) Borneo, Sabah

Orientopsaltria kinabulana Zaidi and Azman 2003: 99, 105, Table 1 (listed, comp. note) Borneo, Sabah

Orientopsaltria kinabuluana Zaidi, Azman and Nordin 2005: 28–29, Table 1 (listed, comp. note) Borneo, Sabah

Orientopsaltria kinabaluana Boulard 2008c: 369 (comp. note) Borneo

***O. latispina* Duffels & Zaidi, 2000**

Orientopsaltria latispina Duffels and Zaidi 2000: 197, 199–200, 202, 204, 264–265, 275–277, 283, 285, 293, Figs 1–2, Fig 4, Fig 123, Fig 128, Figs 136–137, Plate 6 Fig 3, Tables 1–2 (n. sp., described, illustrated, distribution, phylogeny, key, comp. note) Lubang Island, Philippines

***O. lucieni* Boulard, 2005**

Oreintopsaltria lucieni Boulard 2005i: 65–66, Figs 11–12 (n. sp., described, illustrated, comp. note) Thailand

Orientopsaltria lucieni Boulard 2005o: 150 (listed)

Orientopsaltria lucieni Sanborn, Phillips and Sites 2007: 23 (listed, distribution, comp. note) Thailand

Orientopsaltria lucieni Boulard 2008f: 35, 93 (biogeography, listed, comp. note) Thailand

***O. maculosa* Duffels & Zaidi, 2000** = *Orientopsaltria* n. sp. 1 nr. *guttigera* Zaidi & Ruslan, 1995 = *Orientopsaltria* nr. *guttigera* Zaidi, Ruslan & Mahadir, 1996 = *Orientopsaltria* nr. *guttiger* (sic) = *Oreintopsaltria* nr. *guttifer* (sic)

Orientopsaltria n. sp. 1 nr. *guttigera* Zaidi and Ruslan 1995c: 64, 69, Table 1 (listed, comp. note) Peninsular Malaysia

Orientopsaltria nr. *guttigera* Zaidi, Ruslan and Mahadir 1996: 60, Table 1 (listed, comp. note) Peninsular Malaysia

Orientopsaltria maculosa Duffels and Zaidi 2000: 197, 199–200, 202, 205, 228, 230, 237, 241–246, 288, Figs 1–2, Fig 4, Fig 49, Fig 68, Figs 72–75, Plate 3 Fig 4, Tables 1–2 (n. sp. described, illustrated, distribution, phylogeny, key, comp. note) Malayan Peninsula, Sumatra, Sarawak, Borneo, Sabah, Kalimantan, Brunei

Orientopsaltria maculosa Zaidi, Noramly and Ruslan 2000a: 327 (listed, comp. note) Borneo, Sabah, Sarawak, Brunei, Kalimantan

Orientopsaltria nr. *guttiger* (sic) Zaidi, Noramly and Ruslan 2000b: 337 (comp. note) Borneo, Sabah

Orientopsaltria maculosa Zaidi, Noramly and Ruslan 2000b: 337–338, Table 1 (listed, comp. note) Equals *Orientopsaltria* nr. *guttifer* (sic) Borneo, Sabah

Orientopsaltria maculosa Duffels and Schouten 2002: 50 (comp. note) Borneo

Orientopsaltria maculosa Zaidi and Azman 2003: 105, Table 1 (listed, comp. note) Borneo, Sabah

O. montana (Kato, 1927) to *Macrosemia kareisana* (Matsumura, 1907)

***O. montivaga* (Distant, 1889)** = *Cosmopsaltria montivaga*

Cosmopsaltria montivaga Duffels 1983b: 4 (comp. note)

Orientopsaltria montivaga Duffels 1983b: 9 (comp. note)

Orientopsaltria montivaga Zaidi, Mohammedsaid and Yaakob 1990: 262–265, Fig 1, Table 1 (seasonality, comp. note) Malaysia

Orientopsaltria montivaga Zaidi 1993: 958, 960, Table 1 (listed, comp. note) Borneo, Sarawak

Orientopsaltria montivaga Zaidi and Ruslan 1994: 426, 428–429, Table 1 (comp. note) Malaysia

Orientopsaltria montivaga Zaidi and Ruslan 1995d: 200, 203, Table 1 (listed, comp. note) Borneo, Sabah

Orientopsaltria montivaga Zaidi 1996: 101, 104, Table 1 (comp. note) Borneo, Sarawak

Orientopsaltria montivaga Zaidi and Hamid 1996: 53, 56–57, Table 1 (listed, comp. note) Borneo, Sarawak

Orientopsaltria montivaga Zaidi, Ruslan and Mahadir 1996: 60, Table 1 (listed, comp. note) Peninsular Malaysia

Orientopsaltria montivaga Zaidi 1997: 113, 116, Table 1 (listed, comp. note) Borneo, Sarawak

Orientopsaltria montivaga Zaidi and Ruslan 1997: 226 (listed, comp. note) Equals *Cosmopsaltria montivaga* Peninsular Malaysia, Borneo, Sumatra

Orientopsaltria montivaga Duffels and Zaidi 1998: 319–322 (comp. note) Borneo

Oreintopsaltria montivaga Zaidi and Ruslan 1998a: 360 (listed, comp. note) Equals *Cosmopsaltria montivaga* Borneo, Sarawak, Sumatra, Peninsular Malaysia, Sabah

Orientopsaltria montivaga Zaidi, Ruslan and Azman 1999: 310–311 (listed, comp. note) Equals *Cosmopsaltria montivaga* Borneo, Sabah, Sarawak, Peninsular Malaysia

Orientopsaltria montivaga Duffels and Zaidi 2000: 196–197, 199–200, 202, 204, 216, 219–2223, 225–226, 228, 284, Figs 1–2, Fig 4, Fig 22, Figs 27–32, Fig 44, Plate 2 Fig 1, Tables 1–2 (described, illustrated, synonymy, distribution, phylogeny, key, comp. note) Equals *Cosmopsaltria montivaga* Sarawak, Borneo, Sabah, Kalimantan, Brunei

Orientopsaltria montivaga group Duffels and Zaidi 2000: 199, 200, 202, 205, 216, 219, 227 (described, diversity, comp. note)

Cosmopsaltria montivaga Duffels and Zaidi 2000: 196–197, 219 (comp. note) Borneo

Orientopsaltria montivaga Zaidi, Noramly and Ruslan 2000a: 327 (listed, comp. note) Equals *Cosmopsaltria montivaga* Borneo, Sabah, Sumatra, Sarawak, Peninsular Malaysia, Brunei, Kalimantan

Orientopsaltria montivaga Zaidi, Noramly and Ruslan 2000b: 337, 339 (comp. note) Borneo, Sabah

Orientopsaltria montivaga Zaidi, Ruslan and Azman 2000: 209 (listed, comp. note) Equals *Cosmopsaltria montivaga* Borneo, Sabah, Sumatra, Sarawak, Peninsular Malaysia, Brunei, Kalimantan

Orientopsaltria montivaga Boulard 2001c: 58 (comp. note) Equals *Cosmopsaltria montivaga* Sarawak

Orientopsaltrai montivaga Zaidi and Nordin 2001: 185, 190, 199, Table 1 (listed, comp. note) Equals *Cosmopsaltria montivaga* Borneo, Sabah, Sumatra, Sarawak, Peninsular Malaysia, Brunei, Kalimantan

Orientopsaltria montivaga Duffels and Turner 2002: 239 (comp. note)

Orientopsaltria montivaga Zaidi, Nordin, Maryati, Wahab, Norashikin et al. 2002: 12 (comp. note) Borneo, Sabah

Orientopsaltria montivaga Zaidi and Azman 2003: 100, 105, Table 1 (listed, comp. note) Borneo, Sabah

Orientopsaltria montivaga Zaidi, Azman and Nordin 2005: 28–29, Table 1 (listed, comp. note) Borneo, Sabah, Sarawak, Brunei, Kalimantan, Sumatra, Peninsular Malaysia

***O. moultoni* (China, 1926)** = *Cosmopsaltria moultoni*

Orientopsaltria moultoni Duffels 1983b: 9 (n. comb., comp. note)

Orientopsaltria moultoni Duffels and Zaidi 1998: 321 (comp. note)

Orientopsaltria moultoni Duffels and Zaidi 2000: 197, 199–200, 202–205, 209, 215–219, 281, Figs 1–2, Fig 4, Fig 17, Fig 21, Figs 24–26, Plate 1 Fig 4, Tables 1–2 (described, illustrated, synonymy, distribution, phylogeny, key, comp. note) Equals *Cosmopsaltria moultoni* Sumatra

Cosmopsaltria moultoni Duffels and Zaidi 2000: 197 (comp. note)

O. multivocalis (Matsumura, 1917) to *Meimuna multivocalis*

***O. musicus* Boulard, 2005**

Orientopsaltria musicus Boulard 2005e: 118, 128–131, Figs 18–22, Plate D, Figs 1–2, Plate J, Figs 1–2 (n. sp., described, illustrated, listed, song, sonogram, comp. note) Thailand

Orientopsaltria musicus Boulard 2005o: 151 (listed)

Orientopsaltria musicus Sanborn, Phillips and Sites 2007: 23 (listed, distribution, comp. note) Thailand

Orientopsaltria musicus Boulard 2008f: 35, 93 (biogeography, listed, comp. note) Thailand

O. nigripes (Moulton, 1923) to *Dundubia nigripes*

***O. noonadani* Duffels & Zaidi, 2000** = *Cosmopsaltria inermis* (Stål) = *Cosmopsaltria alticola* (nec Distant)

Orientopsaltria noonadani Duffels and Zaidi 2000: 197, 199–200, 202, 204, 260, 264–265, 275, 285–287, 290–191, 293–294, Figs 1–2, Fig 4, Fig 106, Fig 123, Figs 138–147, Plate 6 Fig 4, Tables 1–2 (described, illustrated, distribution, phylogeny, key, comp. note) Palawan, Balabac Island, Busuangu Island

Orientopsaltria noonadani Lee 2009h: 2629–2930 (synonymy, distribution, listed) Equals *Cosmopsaltria inermis* (nec Stål) Equals *Cosmopsaltria alticola* (nec Distant) Philippines, Palawan, Busuanga Island, Balabac Island

O. oopaga (Distant, 1881) to *Dundubia oopaga*

***O. padda* (Distant, 1887)** = *Cosmopsaltria padda* = *Cosmopsaltria latilinea* (nec Walker) = *Cosmopsaltria duarum vera* Moulton, 1911 = *Cosmopsaltria duarum* var. *vera* = *Cosmopsaltria duarum padda* = *Cosmopsaltria duarum* var. *padda*

Orientopsaltria padda Duffels 1983b: 10 (n. comb., comp. note)

Cosmopsaltria padda Naruse 1983: 256 (comp. note)

Cosmopsaltria padda O'Connor, Nash and Anderson 1983: 81 (comp. note)

Orientopsaltria padda Zaidi and Ruslan 1994: 428 (comp. note) Malaysia

Orientopsaltria padda Zaidi and Ruslan 1995b: 218–219, 222, Table 1 (listed, comp. note) Borneo, Sabah

Orientopsaltria padda Zaidi and Ruslan 1995c: 65, 68, Table 1 (listed, comp. note) Peninsular Malaysia

Orientopsaltria padda Zaidi and Ruslan 1995d: 198, 203 (comp. note) Borneo, Sabah

Orientopsaltria padda Zaidi 1996: 100–101, 104, Table 1 (comp. note) Borneo, Sarawak

Orientopsaltria padda Zaidi and Hamid 1996: 53, 56–57, Table 1 (listed, comp. note) Borneo, Sarawak

Orientopsaltria padda Zaidi, Ruslan and Mahadir 1996: 61, Table 1 (listed, comp. note) Peninsular Malaysia

Orientopsaltria padda Zaidi 1997: 112–113, 116, Table 1 (listed, comp. note) Borneo, Sarawak

Orientopsaltria padda Zaidi and Ruslan 1997: 226–227 (listed, comp. note) Equals *Cosmopsaltria padda* Peninsular Malaysia, Java, New Guinea, Sumatra, Borneo, Sarawak, Singapore

Orientopsaltria padda Duffels and Zaidi 1998: 319, 321, 330–331 (comp. note) Peninsular Malaysia, Sumatra, Borneo

Oreintopsaltria padda Zaidi and Ruslan 1998a: 360–361 (listed, comp. note) Equals *Cosmopsaltria padda* Borneo, Sarawak, Peninsular Malaysia, Java, New Guinea, Sabah, Singapore

Orientopsaltria padda Zaidi and Ruslan 1998b: 2–3, 5, Table 1 (listed, comp. note) Borneo, Sabah
Orientopsaltria padda Zaidi, Ruslan and Azman 1999: 311 (listed, comp. note) Equals *Cosmopsaltria padda* Borneo, Sabah, Java, New Guinea, Sumatra, Sarawak, Peninsular Malaysia, Singapore
Orientopsaltria padda Duffels and Zaidi 2000: 196–197, 199–200, 202–206, 209–218, 227–228, 281, Figs 1–2, Fig 4, Figs 10–13, Fig 19, Fig 42, Plate 1 Fig 2, Tables 1–2 (described, illustrated, synonymy, distribution, phylogeny, key, comp. note) Malayan Peninsula, Sumatra, Sarawak, Sabah, Kalimantan, Thailand
Cosmopsaltria padda Duffels and Zaidi 2000: 196, 209 (comp. note)
Orientopsaltria padda group Duffels and Zaidi 2000: 200, 202, 205, 227 (described, diversity, comp. note)
Cosmopsaltria latilinea (nec Walker) Duffels and Zaidi 2000: 209 (comp. note)
Cosmopsaltria duarum var. *vera* Duffels and Zaidi 2000: 209 (comp. note) Sarawak
Cosmopsaltria duarum var. *padda* Duffels and Zaidi 2000: 209 (comp. note)
Orientopsaltria padda Zaidi, Noramly and Ruslan 2000a: 327–328 (listed, comp. note) Equals *Cosmopsaltria padda* Borneo, Sabah, Peninsular Malaysia, Sumatra, Sarawak, Singapore, Kalimantan, Thailand
Orientopsaltria padda Zaidi, Noramly and Ruslan 2000b: 337–338, Table 1 (listed, comp. note) Borneo, Sabah
Orientopsaltria padda Zaidi, Ruslan and Azman 2000: 209–210 (listed, comp. note) Equals *Cosmopsaltria padda* Borneo, Sabah, Peninsular Malaysia, Sumatra, Sarawak, Singapore, Kalimantan, Thailand
Oreintopsaltria padda Turner, Hovencamp and van Welzen 2001: 219, Table 1 (outgroup)
Orientopsaltria padda Duffels and Turner 2002: 236–237, 239, 241–242, 244, 247, 259, Fig 4J, Appendix 2 (illustrated, listed, comp. note)
Orientopsaltria padda Zaidi and Azman 2003: 105, Table 1 (listed, comp. note) Borneo, Sabah
Orientopsaltria padda Schouten, Duffels and Zaidi 2004: 373, 375, Table 2 (listed, comp. note) Malayan Peninsula
Orientopsaltria padda Boulard 2007a: 98–100, Figs 7–10 (described, illustrated, song, sonogram, comp. note) Equals *Cosmopsaltria padda* Equals *Cosmopsaltria latilinea* (nec *Dundubia latilinea*) Thailand
Orientopsaltria padda Sanborn, Phillips and Sites 2007: 23 (listed, synonymy, distribution, comp. note) Equals *Orientopsaltria* cf. *padda* Duffels and Zaidi 2000 Thailand, Peninsular Malaya, Sumatra, Sarawak, Sabah, Kalimantan
Orientopsaltria padda Boulard 2008c: 364 (comp. note) Malaysia
Orientopsaltria padda Boulard 2008f: 35, 93 (biogeography, listed, comp. note) Equals *Cosmopsaltria padda* Equals *Cosmopsaltria latilinea* Thailand
Oreintopsaltria padda Lee 2010b: 25 (distribution, listed) Cambodia, Thailand, Malaysia, Sabah, Sarawak, Peninsular Malaysia, Indonesia, Kalimantan, Sumatra

O. nr. *padda* Zaidi, Ruslan & Mahadir, 1996
Orientopsaltria nr. *padda* Zaidi, Ruslan and Mahadir 1996: 61, Table 1 (listed, comp. note) Peninsular Malaysia

***O. palawana* Duffels & Zaidi, 2000**
Orientopsaltria palawana Duffels and Zaidi 2000: 197, 199–200, 202, 205, 230, 249–255, 258–260, 289, Figs 1–2, Fig 4, Figs 84–86, Fig 89, Fig 100, Plate 4 Fig 3, Tables 1–2 (n. sp., described, illustrated, distribution, phylogeny, key, comp. note) Equals Palawan, Philippines
Orientopsaltria palawana Duffels and Schouten 2002: 50 (comp. note)
Orientopsaltria palawana Lee 2009h: 2629 (distribution, listed) Philippines, Palawan

***O. phaeophila* (Walker, 1850)** = *Dundubia phaeophila* = *Dundubia phoeopila* (sic) = *Cosmopsaltria phaeophila* = *Cosmopsaltria phoeophila* (sic) = *Cosmopsaltria phaephila* (sic)
Orientopsaltria phaeophila Duffels 1983b: 10 (n. comb., comp. note)
Cosmopsaltria phaeophila Lee 1995: 19, 23, 28–29, 31, 33–34, 37, 45–46, 130, 135, 138 (comp. note) Equals *Dundubia phaeophila*
Dundubia phaeophila Lee 1995: 122–123, 130 (comp. note)
Orientopsaltria phaeophila Zaidi and Ruslan 1995c: 64, 68, Table 1 (listed, comp. note) Peninsular Malaysia
Orientopsaltria phaeophila Zaidi and Ruslan 1995d: 198, 200–201, 203, Table 1 (listed, comp. note) Borneo, Sabah
Orientopsaltria phaeophila Zaidi, Ruslan and Mahadir 1996: 61, Table 1 (listed, comp. note) Peninsular Malaysia
Orientopsaltria phaeophila Duffels and Zaidi 1998: 321 (comp. note) Peninsular Malaysia, Borneo

Orientopsaltria phaeophila Zaidi, Ruslan and Azman 1999: 311 (listed, comp. note) Equals *Dundubia phaeophila* Equals *Cosmopsaltria phaeophila* Borneo, Sabah, Sarawak, Peninsular Malaysia

Orientopsaltria phaeophila Duffels and Zaidi 2000: 196–197, 199–200, 202, 205, 228, 230, 232–233, 235–241, 243, 246, 288, Figs 1–2, Fig 4, Fig 48, Fig 62, Fig 67, Figs 69–71, Plate 3 Fig 3, Tables 1–2 (lectotype designation, described, illustrated, synonymy, distribution, phylogeny, key, comp. note) Equals *Dundubia phaeophila* Equals *Dundubia phoeophila* (sic) Equals *Cosmopsaltria phaeophila* Malayan Peninsula, Sarawak, Borneo, Sabah, Kalimantan, Brunei

Dundubia phaeophila Duffels and Zaidi 2000: 196 (comp. note)

Orientopsaltria phaeophila Zaidi, Noramly and Ruslan 2000a: 328 (listed, comp. note) Equals *Dundubia phaeophila* Equals *Cosmopsaltria phaeophila* Borneo, Sabah, Peninsular Malaysia, Brunei, Kalimantan

Orientopsaltria phaeophila Zaidi, Noramly and Ruslan 2000b: 337, 339, Table 1 (listed, comp. note) Borneo, Sabah

Orientopsaltria phaeophila Zaidi, Ruslan and Azman 2000: 210–211 (listed, comp. note) Equals *Dundubia phaeophila* Equals *Cosmopsaltria phaeophila* Borneo, Sabah, Peninsular Malaysia, Brunei, Kalimantan

Orientopsaltrai phaeophila Zaidi and Nordin 2001: 185, 190, 199–200, Table 1 (listed, comp. note) Equals *Dundubia phaeophila* Equals *Cosmopsaltria phaeophila* Borneo, Sabah, Peninsular Malaysia, Brunei, Kalimantan

Orientopsaltria phaeophila Duffels and Schouten 2002: 50 (comp. note) Malayan Peninsula, Borneo

Orientopsaltria phaeophila Zaidi and Azman 2003: 105, Table 1 (listed, comp. note) Borneo, Sabah

Orientopsaltria phaeophila Lee 2008b: 462 (comp. note)

***O. ruslani* Duffels & Zaidi, 1998** = *Cosmopsaltria montivaga* (nec Distant) = *Orientopsaltria montivaga* (nec Distant)

Orientopsaltria ruslani Duffels and Zaidi 1998: 319, 321–331, Figs 1–6 (n. sp., described, illustrated, comp. note) Equals *Cosmopsaltria montivaga* (partim) Equals *Orientopsaltria montivaga* (partim) Peninsular Malaysia

Orientopsaltria ruslani Duffels and Zaidi 1999: 147–148, Fig (illustrated)

Orientopsaltria ruslani Duffels and Zaidi 2000: 196–197, 199–200, 202, 204, 216, 219, 223–225, 228, 284, Figs 1–2, Fig 4, Fig 23, Figs 32–36, Fig 45, Plate 2 Fig 2, Tables 1–2 (described, illustrated, synonymy, distribution, phylogeny, key, comp. note) Equals *Cosmopsaltria montivaga* (partim) Equals *Orientopsaltria montivaga* (partim) Malayan Peninsula, Sumatra

Orientopsaltria ruslani Zaidi, Azman and Ruslan 2001a: 111–112, 115, 120, Table 1 (listed, comp. note) Langkawi Island, Peninsular Malaysia, Sumatra

Orientopsaltria ruslani Duffels and Turner 2002: 239 (comp. note)

Orientopsaltria ruslani Schouten, Duffels and Zaidi 2004: 373, 375, Table 2 (listed, comp. note) Malayan Peninsula

Orientopsaltria ruslani Boulard 2008c: 364 (comp. note) Malaysia

***O. saudarapadda* Duffels & Zaidi, 1998** = *Cosmopsaltria montivaga* (nec Distant)

Orientopsaltria saudarapadda Duffels and Zaidi 1998: 319, 330–339, Figs 7–10 (n. sp., described, illustrated, comp. note) Peninsular Malaysia

Orientopsaltria saudarapadda Duffels and Zaidi 2000: 197, 199–200, 202, 204–205, 209, 213–217, 228, 281, Figs 1–2, Fig 4, Figs 14–17, Fig 20, Fig 43, Plate 1 Fig 3, Tables 1–2 (described, illustrated, synonymy, distribution, phylogeny, key, comp. note) Malayan Peninsula

Orientopsaltria saudarapadda Zaidi, Ruslan and Azman 2000: 211 (comp. note) Borneo, Sabah

Orientopsaltria saudarapadda Zaidi and Nordin 2001: 18 (comp. note) Borneo, Sabah

Orientopsaltria saudarapadda Duffels and Schouten 2002: 50 (comp. note) Malayan Peninsula

Orientopsaltria saudarapadda Zaidi, Nordin, Maryati, Wahab, Norashikin et al. 2002: 12 (comp. note) Borneo, Sabah

Orientopsaltria saudarapadda Zaidi and Azman 2003: 105, Table 1 (listed, comp. note) Borneo, Sabah

Orientopsaltria saudarapadda Schouten, Duffels and Zaidi 2004: 373, 375, Table 2 (listed, comp. note) Malayan Peninsula

***O.* sp. 1 Zaidi, Ruslan & Azman, 2000**

Orientopsaltria sp. 1 Zaidi, Ruslan and Azman 2000: 211 (comp. note) Borneo, Sabah

***O.* sp. 2 Zaidi, Ruslan & Azman, 2000**

Orientopsaltria sp. 2 Zaidi, Ruslan and Azman 2000: 211 (comp. note) Borneo, Sabah

***O. sumatrana* (Moulton, 1917)** = *Cosmopsaltria sumatrana*

Orientopsaltria sumatrana Duffels 1983b: 10 (n. comb., comp. note)
Orientopsaltria sumatrana Duffels and Zaidi 1998: 321 (comp. note)
Orientopsaltria sumatrana Duffels and Zaidi 2000: 196–197, 199–200, 202, 205, 254–260, 289, Figs 1–2, Fig 4, Fig 90, Figs 93–95, Fig 97, Fig 101, Plate 4 Fig 4, Tables 1–2 (described, illustrated, synonymy, distribution, phylogeny, key, comp. note) Equals *Cosmopsaltria sumatrana* Sumatra
Cosmposaltria sumatrana Duffels and Zaidi 2000: 197 (comp. note) Sumatra
Orientopsaltria sumatrana group Duffels and Zaidi 2000: 201–202, 249, 255 (described, diversity, comp. note)

O. tonkiniana (Jacobi, 1905) to *Meimuna tonkiniana*

***O. vanbreei* Duffels & Zaidi, 2000** = *Orientopsaltria sumatrana* Zaidi & Hamid 1996, Zaidi 1997 = *Orientopsaltria* sp. n. 2 (near *sumatrana* (Moulton)) Zaidi & Ruslan, 1995 = *Orientopsaltria* near *sumatrana* Zaidi, Rusland & Mahadir, 1996
Orientopsaltria n. sp. 2 nr. *sumatrana* Zaidi and Ruslan 1995c: 64–65, 69, Table 1 (listed, comp. note) Peninsular Malaysia
Orientopsaltria sumatrana (sic) Zaidi and Hamid 1996: 53, 56–57, Table 1 (listed, comp. note) Borneo, Sarawak
Orientopsaltria nr. *sumatrana* Zaidi, Ruslan and Mahadir 1996: 61, Table 1 (listed, comp. note) Peninsular Malaysia
Orientopsaltria sumatrana (sic) Zaidi 1997: 112–114, 116, Table 1 (listed, comp. note) Borneo, Sarawak
Orientopsaltria vanbreei Duffels and Zaidi 2000: 197, 199–200, 202, 205, 254–261, 292, Figs 1–2, Fig 4, Fig 91, Figs 96–99, Fig 102, Plate 5 Fig 1, Tables 1–2 (n. sp., described, illustrated, synonymy, distribution, phylogeny, key, comp. note) Equals *Orientopsaltria sumatrana* Zaidi and Hamid 1996, Zaidi 1997 Equals *Orientopsaltria* sp. n. 2 (near *sumatrana* (Moulton)) Zaidi and Ruslan Equals *Orientopsaltria* near *sumatrana* Zaidi et al. 1996 Malayan Peninsula, Sumatra, Sarawak, Borneo
Orientopsaltria vanbreei Boulard 2008c: 364 (comp. note) Malaysia

Subtribe Crassopsaltriina Boulard, 2008
Crassopsaltriina Boulard 2008f: 7, 35 (n. subtribe, listed) Thailand

***Crassopsaltria* Boulard, 2008**
Crassopsaltria Boulard 2008f: 7, 35, 59, 93 (n. gen., described, listed) Thailand

***C. giovannae* (Boulard, 2005)** = *Macrosemia giovannae*
Macrosemia giovannae Boulard 2005i: 69–72, Figs 18–23 (n. sp., described, illustrated, comp. note) Thailand
Macrosemia giovannae Boulard 2005o: 150 (listed)
Macrosemia giovannae Sanborn, Phillips and Sites 2007: 20 (listed, distribution, comp. note) Thailand
Crassopsaltria giovannae Boulard 2008f: 35, 58–59, 93, Figs 1–2 (type species of *Crassopsaltria*, biogeography, illustrated, listed, comp. note) Equals *Macrosemia giovannae* Thailand

Tribe Tosenini Amyot & Audinet-Serville, 1843 = Tosenaria = Toseniini (nec Distant) = Tacuaria (nec Distant)
Tacuaria (sic) (nec Distant) Wu 1935: 4 (listed) China
Tosenaria Duffels 1991a: 128 (comp. note)
Tosenini Chou, Lei, Li, Lu and Yao 1997: 106, 180 (key, listed, synonymy, described) Equals Tosenides Equals Tosenidae Equals Toseninae Equals Tacuaria (nec Distant) Equals Tosenaria
Tosenaria Boulard 2001c: 51, 54 (listed)
Tosenini Sueur 2001: 39, Table 1 (listed)
Tosenini Boulard 2002a: 37 (listed)
Tosenini Boulard 2002b: 203 (comp. note)
Tosenini Boulard 2003b: 189 (listed)
Tosenini Boulard 2003e: 260 (listed)
Tosenaria Boulard 2003b: 189 (listed)
Tosenina Boulard 2003e: 260 (listed)
Tosenaria Boulard 2003e: 260 (listed)
Tosenina Lee and Hayashi 2003a: 170 (listed, comp. note) Taiwan
Tosenini Boulard 2005l: 111 (comp. note)
Tosenina Moulds 2005b: 391, 432 (higher taxonomy, listed, comp. note) Equals Tosenaria
Tosenini Pham and Thinh 2005a: 245 (listed) Vietnam
Tosenini Pham and Thinh 2005b: 287–289 (key, comp. note) Equals Tosenides Equals Tosenidae Equals Toseninae Equals Tacuaria Equals Tosenaria Vietnam
Tosenini Boulard 2007d: 495 (listed)
Tosenini Boulard 2007e: 20 (comp. note) Thailand
Tosenini Sanborn, Phillips and Sites 2007: 3, 24, 41, Fig 5, Table 1 (listed, distribution) Thailand
Tosenaria Shiyake 2007: 5, 70 (listed, comp. note)
Tosenini Boulard 2008c: 362 (comp. note)
Toseniini (sic) Boulard 2008f: 7 (listed) Thailand
Tosenini Boulard 2008f: 20 (listed)
Tosenina Lee 2008a: 1–2, 18, Table 1 (listed) Vietnam
Tosenina Lee 2009a: 87 (comp. note)

Tosenina Pham and Yang 2009: 15, Table 2 (listed) Vietnam
Tosenina Lee 2010b: 27 (listed) Cambodia
Tosenina Lee 2010d: 167 (comp. note)
Tosenina Wei, Ahmed and Rizvi 2010: 33 (comp. note)

***Tosena* Amyot & Audinet-Serville, 1843** = *Cicada (Tosena)* = *Tossena* (sic) = *Josena* (sic) = *Tosema* (sic)
Tosena Duffels 1991a: 128 (comp. note)
Tosena Hayashi 1993: 15 (comp. note)
Tossena (sic) Gogala 1994b: 8–9, Fig 9 (illustrated, comp. note) Peninsular Malaysia
Tosena sp. Gogala and Riede 1995: 299–300, Fig 2 (song, sonogram, oscillogram, acoustic behavior, comp. note) Malaysia
Tosena Zaidi and Ruslan 1995a: 173–174, Table 2 (listed, comp. note) Borneo, Sarawak
Tosena Zaidi and Ruslan 1995c: 70, Table 2 (listed, comp. note) Peninsular Malaysia
Tosena Zaidi and Ruslan 1995d: 202, Table 2 (listed, comp. note) Borneo, Sabah
Tosena Zaidi 1996: 103, Table 2 (comp. note) Borneo, Sarawak
Tosena Zaidi and Hamid 1996: 54–55, Table 2 (listed, comp. note) Borneo, Sarawak
Tosena Chou, Lei, Li, Lu and Yao 1997: 180–181, 293, Table 6 (key, described, listed, diversity, biogeography) China
Tosena Novotny and Wilson 1997: 437 (listed)
Tosena Zaidi 1997: 115, Table 2 (listed, comp. note) Borneo, Sarawak
Tosena Zaidi and Ruslan 1999: 4, Table 2 (listed, comp. note) Borneo, Sarawak
Tosena Boulard 2000d: 94 (song, comp. note)
Tosena Boulard 2001c: 54 (comp. note)
Tosena Boulard 2002b: 202 (coloration, comp. note)
Tosena Boulard 2003b: 192 (comp. note)
Tosena Boulard 2003e: 260 (comp. note)
Tosena Boulard 2005d: 345 (comp. note)
Tosena Boulard 2005l: 112 (comp. note)
Tosena Moulds 2005b: 391, 432 (higher taxonomy, listed, comp. note)
Tosena Pham and Thinh 2005a: 245 (listed) Vietnam
Tosena Pham and Thinh 2005b: 289 (comp. note) Vietnam
Tosena Boulard 2006g: 33–34, 39–40, 50, 52, 117–118, 130–131 (comp. note)
Tosena Boulard 2007e: 23, 25, 35, 69 (comp. note) Thailand
Tosena Sanborn, Phillips and Sites 2007: 3, 10, 24, Table 1 (listed, comp. note) Thailand
Tosena Shiyake 2007: 5, 70 (listed, comp. note)
Tosena Sueur and Puissant 2007b: 55 (comp. note)
Tosena Boulard 2008f: 7, 20, 92 (listed) Thailand
Tosena Lee 2008a: 2, 18, Table 1 (listed) Vietnam
Tosena Pinto-Juma, Seabra and Quartau 2008: 2 (comp. note)
Tosena Boulard 2009c: 325 (comp. note)
Tosena Pham and Yang 2009: 15, Table 2 (listed) Vietnam
Tosena Lee 2010b: **27** (listed) Cambodia

***T. albata* Distant, 1878** = *Tosena melanoptera* var. *albata* = *Tosena melanoptera* var. *c* Distant, 1906 = *Tosena albata* var. *c* = *Tosena melanopterix* (sic) var. *albata* = *Tosena melanopterix* (sic) (nec White)
Tosena albata Boulard 1999b: 184–185, Fig 7 (illustrated, comp. note) Thailand
Tosena albata Sueur 2001: 39, Table 1 (listed)
Tosena albata Boulard 2002b: 193, 201–202, sonogram 54c (coloration, sonogram, comp. note)
Tosena albata Boulard 2003e: 266 (comp note)
Tosena albata Boulard 2005d: 333, 340, 344 (comp. note)
Tosena albata Boulard 2005l: 112–114, 116–119, 121, Plate I, Fig 2, Plate II, Fig 4, Plate IV (synonymy, illustrated, song, sonogram, comp. note) Equals *Tosena melanopteryx* Equals *Tosena melanoptera* (error) Equals *Tosena melanopteryx* var. *c albata* Equals *Tosena melanoptera* var. *albata* Equals *Tosena melanopterix* var. *albata* Equals *Tosena melanopteryx* (nec Distant) Thailand
Tosena melanopteryx var. *albata* Boulard 2005l: 113 (comp. note) Nepal
Tosena albata var. *c* Boulard 2005l: 113 (comp. note)
Tosena albata Boulard 2005o: 147, 150 (listed)
Tosena albata Boulard 2006g: 50, 52, 56, 74–75, 117–118, 130–131, 180, Fig 17 (sonogram, listed, comp. note)
Tosena albata Trilar 2006: 344 (comp. note)
Tosena albata Boulard 2007e: 4, 27, 29, 25, 69, 86, Fig 15 (coloration, sonogram, comp. note) Thailand
Tosena albata Sanborn, Phillips and Sites 2007: 24–25 (listed, distribution, comp. note) Thailand, India, Nepal, Himalayas
Tosena albata Boulard 2008b: 109 (comp. note)
Tosena albata Boulard 2008c: 362 (comp. note) Vietnam
Tosena albata Boulard 2008e: 352 (comp. note)
Tosena albata Boulard 2008f: 20, 92 (biogeography, listed, comp. note) Equals *Tosena melanoptera* var. *albata* Equals *Tosena melanoptera* (nec White) Equals *Tosena fasciata* (nec Fabricius) Equals *Tosena melanopterix* (sic) var. *albata* Equals *Tosena melanopterix* (sic) (partim) Thailand, Burma

***T. depicta* Distant, 1888**
Tosena depicta Gogala and Riede 1995: 299–301, 304, Fig 1b, Fig 2b (song, sonogram,

oscillogram, acoustic behavior, comp. note) Malaysia
Tosena depicta Zaidi and Ruslan 1995c: 65, 69, Table 1 (listed, comp. note) Peninsular Malaysia
Tosena depicta Zaidi, Ruslan and Mahadir 1996: 61, Table 1 (listed, comp. note) Peninsular Malaysia
Tosena depicta Zaidi and Ruslan 1998a: 362 (listed, comp. note) Borneo, Sarawak, Peninsular Malaysia
Tosena depicta Zaidi, Ruslan and Azman 1999: 314 (listed, comp. note) Borneo, Sabah, Sarawak, Peninsular Malaysia
Tosena depicta Zaidi, Ruslan and Azman 2000: 213 (comp. note) Borneo, Sabah, Sarawak, Peninsular Malaysia
Tosena depicta Sueur 2001: 39, 46, Tables 1–2 (listed)
Tosena depicta Zaidi and Azman 2003: 105, Table 1 (listed, comp. note) Borneo, Sabah
Tosena depicta Benko and Perc 2006: 556–557, 563–564, Fig 1, Fig 5 (song analysis) Malaysia
Tosena depicta Shiyake 2007: 5, 68, 70, Fig 109 (illustrated, distribution, listed, comp. note) Malaysia, Indonesia
Tosena depicta Boulard 2008c: 362 (comp. note) Malaysia

T. dives **(Westwood, 1842)** = *Cicada dives* = *Gaeana dives* = *Huechys transversa* Walker, 1858
Tosena dives Boulard 2001c: 54 (comp. note)
Tosena dives Boulard 2003b: 192 (comp. note)
Tosena dives Shiyake 2007: 5, 69–70, Fig 112 (illustrated, listed, comp. note)
Tosena dives Boulard 2008c: 357, 362, Fig 15 (illustrated, comp. note) Bhutan

T. fasciata fasciata **(Fabricius, 1787)** = *Tettigonia fasciata fasciata* = *Cicada fasciata* = *Josena* (sic) *fasciata* = *Tosena faciata* (sic)
Tosena fasciata Zaidi and Ruslan 1995a: 171–172, 175, Table 1 (listed, comp. note) Borneo, Sarawak
Tettigonia fasciata Chou, Lei, Li, Lu and Yao 1997: 180 (type species of *Tosena*) China
Tosena fasciata Zaidi and Ruslan 1999: 2–3, 5, Table 1 (listed, comp. note) Borneo, Sarawak
Tosena fasciata Zaidi, Ruslan and Azman 1999: 314–315 (listed, comp. note) Equals *Tettigonia fasciata* Borneo, Sabah, Java, Sumatra, Ambiona
Tosena fasciata Zaidi, Ruslan and Azman 2000: 213 (listed, comp. note) Equals *Tettigonia fasciata* Borneo, Sabah, Java, Sumatra, Ambiona
Tosena fasciata Zaidi and Nordin 2001: 183, 185, 190, 201, Table 1 (listed, comp. note) Equals *Tettigonia fasciata* Borneo, Sabah, Java, Sumatra, Ambiona
Tosena fasciata Zaidi, Nordin, Maryati, Wahab, Norashikin et al. 2002: 6, 9–11, Tables 1–2 (listed, comp. note) Equals *Tettigonia fasciata* Borneo, Sabah, Java, Sumatra, Ambiona
Tettigonia fasciata Boulard 2003e: 260 (type species of *Tosena*) Java
Tosena fasciata Boulard 2003e: 261–262, Figs 1–2 (genitalia illustrated, comp note)
Tosena fasciata fasciata Boulard 2003e: 261 (comp note)
Tosena faciata (sic) Zaidi and Azman 2003: 105, Table 1 (listed, comp. note) Borneo, Sabah
Tettigonia fasciata Boulard 2005l: 112 (type species of *Tosena*, comp. note) Sumatra
Tosena fasciata Boulard 2005l: 115 (comp. note)
Tettigonia fasciata Pham and Thinh 2005a: 245 (type species of *Tosena*, distribution, listed) Vietnam
Tosena fasciata Yen, Robinson and Quicke 2005b: 363, 366, Fig 2, Table 1 (illustrated, listed, mimicry, comp. note)
Tosena fasciata Shiyake 2007: 5, 68, 70, Fig 108 (illustrated, distribution, listed, comp. note) Malaysia, Indonesia, Southeast Asia
Tosena fasciata Boulard 2008f: 20 (type species of *Tosena*) Equals *Tettigonia fasciata*
Tettigonia fasciata Lee 2008a: 18 (type species of *Tosena*) Java
Tosena fasciata Lee 2009h: 2638 (comp. note) Ambiona
Tosena fasciata Chung 2010: 148 (predation, comp. note) Borneo
Tosena fasciata Lee 2010b: 27 (type species of *Tosena*, listed)

T. fasciata **var. *a* Distant, 1889**

T. fasciata **var. *b* Distant, 1889**

T. fasciata **var. *c* Distant, 1889**

T. mearesiana **(Westwood, 1842)** = *Cicada mearesiana* = *Cicada mareaseana* (sic)
Tosena mearesiana Boulard 2007e: 62, Fig 42 (genitalia illustrated, comp. note)

T. melanoptera **(White, 1846)** = *Cicada (Tosena) melanoptera* = *Cicada melanoptera* = *Tosena fasciata* Moulton, 1923 (nec Fabricius) = *Tosena melanopteryx* Kirkaldy, 1909 = *Tosema* (sic) *melanopteryx* = *Tosena fasciata* var. *d* Distant, 1889
Tosena melanopteryx Chou, Lei, Li, Lu and Yao 1997: 181–182, 357, Plate VIII, Fig 84 (key, synonymy, described, illustrated, comp. note) Equals *Cicada melanoptera* Equals *Tosena melanoptera* China

Tosema (sic) *melanopteryx* Freytag 2000: 55 (on stamp)

Tosena melanopteryx Hua 2000: 65 (listed, distribution) Equals *Cicada melanoptera* Equals *Tosena melanoptera* China, Guangxi, Xizang, India, Sikkim, Burma, Bangladesh, Himalaya

Tosena melanoptera Sueur 2001: 39, Table 1 (listed)

Tosena melanoptera Boulard 2003e: 260–266, Color Plate, Figs A-C, Figs 3–6, sonogram 1 (n. sp., described, song, sonogram, illustrated, comp. note) Equals *Cicada (Tosena) melanoptera* Equals *Tosena fasciata* var. *d* Equals *Tosena melanopteryx* Equals *Tosena fasciata melanoptera* Thailand

Tosena fasciata melanoptera Boulard 2003e: 261 (comp note)

Cicada (Tosena) melanoptera Boulard 2003e: 260 (listed)

Tosena melanoptera Boulard 2005d: 340–344 (comp. note)

Tosena melanoptera Boulard 2005l: 111–120, Plate I, Fig 1, Plate II, Fig 3, Plate III (synonymy, illustrated, song, sonogram, comp. note) Equals *Cicada (Tosena) melanoptera* Equals *Tosena melanopteryx* Thailand

Cicada (Tosena) melanoptera Boulard 2005l: 112, 114 (comp. note)

Tosena melanopterix (sic) Boulard 2005l: 112–113 (comp. note)

Tosena melanopteryx Boulard 2005l: 114 (comp. note)

Tosena melanoptera Boulard 2005o: 147, 149–150 (listed) Equals *Meimuna tavoyana* sonogram Boulard, 2000

Tosena melanoptera Lee 2005: 37 (illustrated)

Tosena melanopteryx Pham and Thinh 2005a: 246 (synonymy, distribution, listed) Equals *Cicada melanoptera* Equals *Tosena melanoptera* Vietnam

Tosena melanoptera Boulard 2006g: 50, 52, 74–75, 117–118, 130–131, 180, Fig 82 (illustrated, listed, comp. note)

Tosena melanoptera Trilar 2006: 344 (comp. note)

Tosena melanoptera Boulard 2007d: 495 (comp. note) Thailand

Tosena melanoptera Boulard 2007e: 4, 27–28, 35, 55, 69, Fig 14, Fig 37.1, Plate 2, Fig 1, Plate 31a-b (coloration, genitalia, illustrated, sonogram, comp. note) Thailand

Tosena melanoptera Sanborn, Phillips and Sites 2007: 25 (listed, distribution, comp. note) Thailand, India, Burma, Vietnam, Indochina, Nepal, Himalayas

Tosena melanopteryx Sanborn, Phillips and Sites 2007: 25 (comp. note)

Tosena melanoptera Boulard 2008c: 362 (comp. note) Vietnam

Tosena melanoptera Boulard 2008e: 352 (comp. note)

Tosena melanoptera Boulard 2008f: 20–21, 92 (biogeography, listed, comp. note) Equals *Cicada (Tosena) melanoptera* Equals *Tosena melanopteryx* (sic) Equals *Tosena fasciata* (partim) Thailand, Assan (sic), India, Burma

Tosena melanoptera Lee 2008a: 18–19 (synonymy, distribution, listed) Equals *Cicada (Tosena) melanoptera* Equals *Tosena fasciata* (nec Fabricius) Equals *Tosena albata* var. *melanopteryx* Equals *Tosena melanopteryx* Vietnam, Laos, Thailand, Myanmar, Nepal, India

Tosena melanoptera Pham and Yang 2009: 15, Table 2 (listed) Vietnam

***T. melanoptera* var. *a* Distant, 1889** = *Tosena melanopteryx* var. *a*

***T. melanoptera* var. *b* Distant, 1889** = *Tosena melanopteryx* var. *b*

T. montivaga Distant, 1889 to *Formotosena montivaga*

***T. paviei* (Noualhier, 1896)** = *Gaeana paviei* = *Gaeana pavici* (sic) = *Tosena dives* (nec Westwood) = *Huechys transversa* (nec Walker)

Tosena dives (sic) Chou, Lei, Li, Lu and Yao 1997: 183–184, Fig 9-48, Plate VIII, Fig 86 (illustrated, synonymy, described, distribution, comp. note) Equals *Cicada dives* Equals *Gaeana dives* Equals *Huechys transversa* China

Tosena paviei Boulard 1999b: 184–185, Fig 6 (illustrated, comp. note) Laos

Tosena paviei Boulard 2001c: 54–55, Plate B, Figs 1–2 (song, illustrated, sonogram, comp. note) Equals *Gaeana paviei* Equals *Tosena dives* (nec Westwood) Laos, Thailand

Tosena paviei Boulard 2002b: 196 (comp. note) Southeast Asia, Thailand

Tosena paviei Boulard 2003b: 187, 191–193, Fig 6, Plate I, Fig 4 (described, song, sonogram, illustrated, comp. note) Equals *Gaeana paviei* Equals *Tosena dives* Chou et al. 1997 Thailand, Burma, Tonkin, Cambodia

Tosena paviei Boulard 2005o: 148–149 (listed) Equals *Gaeana paviei* Equals *Gaeana cheni* sonogram Boulard, 2001

Gaeana paviei Yen, Robinson and Quicke 2005b: 363, 365, Fig 1, Table 1 (illustrated, listed, mimicry, comp. note)

Tosena paviei Boulard 2006g: 130–1311, 137, Fig 96 (sonogram, comp. note) Thailand

Tosena paviei Boulard 2007e: 35, Plate 2, Fig 3, Plate 22a (illustrated, comp. note) Thailand

Tosena paviei Sanborn, Phillips and Sites 2007: 25 (listed, synonymy, distribution, comp. note) Equals *Gaeana paviei* Thailand, Laos, Cambodia, Indochina
Tosena paviei Boulard 2008c: 357, 362, Fig 16 (illustrated, comp. note) Thailand
Tosena paviei Boulard 2008f: 21–22, 92 (biogeography, listed, comp. note) Equals *Gaeana paviei* Equals *Tosena dives* Chou et al. 1997 Thailand, Laos, Cambodia
Tosena paviei Lee 2008a: 19 (synonymy, distribution, listed) Equals *Gaeana paviei* Vietnam, Laos, Cambodia, Thailand
Tosena paviei Pham and Yang 2009: 15, Table 2 (listed) Vietnam
Tosena paviei Lee 2010b: 27 (synonymy, distribution, listed) Equals *Gaeana paviei* Cambodia, Vietnam, Laos, Thailand

***T.* sp. 1 Pham & Thinh, 2006**
Tosena sp. 1 Pham and Thinh 2006: 526 (listed) Vietnam

***T.* sp. 2 Pham & Thinh, 2006**
Tosena sp. 2 Pham and Thinh 2006: 526 (listed) Vietnam

T. splendida Distant, 1878 to *Distantalna splendida splendida*
T. splendida var. *a* Distant, 1889 to *Distantalna splendida* var. *a*
T. splendida var. *b* Distant, 1889 to *Distantalna splendida* var. *b*

***Distantalna* Boulard, 2009**
Distantalna Boulard 2009c: 325–326 (n. gen., described, comp. note)

***D. splendida splendida* (Distant, 1878)** = *Tosena splendida* = *Tosena splendidula* (sic)
Tosena splendidula (sic) Boulard 1985e: 1024 (coloration, comp. note)
Tosena splendida Hayashi 1993: 15, Fig 5 (illustrated, comp. note)
Tosena splendida Chou, Lei, Li, Lu and Yao 1997: 181–182, Fig 9–47, Plate VIII, Fig 87 (key, described, illustrated, comp. note) China
Tosena splendida Boulard 1999b: 184–185, Fig 5, Fig 5" (illustrated, comp. note) India
Tosena splendida Hua 2000: 65 (listed, distribution) China, Yunnan, India, Burma
Tosena splendida Boulard 2003a: 104 (comp. note)
Tosena splendida Boulard 2003b: 187, 189–191, Figs 3–5, Plate I, Figs 2–3 (described, song, sonogram, illustrated, comp. note) Equals *Tosena splendidula* (sic) Thailand, Assam, Southeast Asia
Tosena splendida Boulard 2005d: 345–346 (comp. note) Thailand
Tosena splendida Boulard 2005o: 148 (listed)
Tosena splendida Pham and Thinh 2005a: 246 (synonymy, distribution, listed) Vietnam
Tosena splendida Boulard 2006e: 373–374, 376–381, Plate I, Figs 1–12, Fig A (eclosion, illustrated, comp. note) Thailand
Tosena splendida Boulard 2006g: 96, 98–99, 130–131, 137, Fig 64, Fig 95 (sonogram, comp. note) Thailand
Tosena splendida Boulard 2007d: 495 (comp. note) Thailand
Tosena splendida Boulard 2007e: 2, 4–5, 15, 35, 48, 65, Fig 1, Plate 2, Fig 5, Plates 6–7, Plate 26, Plates 28–29 (coloration, eclosion, illustrated, sonogram, comp. note) Thailand
Tosena splendida Sanborn, Phillips and Sites 2007: 25 (listed, distribution, comp. note) Thailand, India, Burma, Indochina
Tosena splendida Shiyake 2007: 5, 69–70, Fig 111 (illustrated, distribution, listed, comp. note) China, India, Southeast Asia
Tosena splendida Boulard 2008b: 109 (comp. note)
Tosena splendida Boulard 2008f: 22, 92 (biogeography, listed, comp. note) Thailand, India, Burma, Assam
Tosena splendida Lee 2008a: 19 (distribution, listed) Vietnam, Thailand, Myanmar, India
Distantalna splendida Boulard 2009c: 325–327, Figs 1–2 (n. comb., type species of *Distantalna*, illustrated, comp. note) Equals *Tosena splendida*
Tosena splendida Boulard 2009c: 325 (comp. note)
Tosena splendida Pham and Yang 2009: 15, Table 2 (listed) Vietnam

***D. splendida* var. *a* (Distant, 1889)** = *Tosena splendida* var. *a*
***D. splendida* var. *b* (Distant, 1889)** = *Tosena splendida* var. *b*

***Trengganua* Moulton, 1923** = *Teregganua* (sic)
Trengganua Hayashi 1993: 15 (comp. note)
Trengganua Zaidi and Ruslan 1995c: 70, Table 2 (listed, comp. note) Peninsular Malaysia
Trengganua Novotny and Wilson 1997: 437 (listed)
Treggannua Moulds 2005b: 391, 432 (higher taxonomy, listed, comp. note)
Trengganua Boulard 2008f: 7, 22, 92 (listed) Thailand

***T. sibylla* (Stål, 1863)** = *Gaeana sibylla* = *Tosena sibylla* = *Teregganua* (sic) *sibylla*
Tosena sibylla Boulard 1985e: 1024 (coloration, comp. note)
Terengganua (sic) *sibylla* Gogala and Riede 1995: 299, 301, 304, Fig 1c, Fig 1c (comp. note) Malaysia

Trengganua sibylla Zaidi, Ruslan and Mahadir 1996: 61, Table 1 (listed, comp. note) Peninsular Malaysia
Trengganua sibylla Zaidi and Ruslan 1997: 228 (listed, comp. note) Equals *Gaeana sibylla* Peninsular Malaysia
Terengganua (sic) *sibylla* Sueur 2001: 39, Table 1 (listed)
Trengganua sibylla Sueur 2001: 46, Table 2 (listed)
Trengganua sibylla Lee 2005: 37 (illustrated)
Tosena sibylla Shiyake 2007: 5, 68, 70, Fig 109 (illustrated, distribution, listed, comp. note) Malaysia
Trengganua sibylla Boulard 2008c: 362 (comp. note) Malaysia
Trengganua sibylla Boulard 2008f: 22, 92 (type species of *Trengganua*, biogeography, listed, comp. note) Equals *Gaeana sibylla* Equals *Tosena sibylla* Equals *Terengganua* (sic) *sibylla* Thailand, Peninsular Malaysia
Tosena sibylla Watson, Watson, Hu, Brown, Cribb and Myhra 2010: 117, 119–120, Figs 2g–h, Table 1 (wing microstructure, listed, comp. note)

***Ayuthia* Distant, 1919**
Ayuthia Hayashi 1993: 15 (comp. note)
Ayuthia Zaidi and Ruslan 1995c: 70, Table 2 (listed, comp. note) Peninsular Malaysia
Ayuthia Novotny and Wilson 1997: 437 (listed)
Ayuthia Moulds 2005b: 391 (higher taxonomy, listed, comp. note)
Ayuthia Sanborn, Phillips and Sites 2007: 3, 25, Table 1 (listed) Thailand
Ayuthia Shiyake 2007: 5, 70 (listed, comp. note)
Ayuthia Boulard 2008f: 7, 23, 92 (listed) Thailand
Ayuthia Lee 2008a: 2, 19, Table 1 (listed) Vietnam
Ayuthia Pham and Yang 2009: 15, Table 2 (listed) Vietnam

***A. spectabile* Distant, 1919**
Ayuthia spectabile Zaidi, Ruslan and Mahadir 1996: 61, Table 1 (listed, comp. note) Peninsular Malaysia
Ayuthia spectabile Zaidi and Ruslan 1997: 228 (listed, comp. note) Peninsular Malaysia, Indo-China
Ayuthia spectabile Boulard 1999b: 184–185, Fig 3, Fig 3" (illustrated, comp. note) Malaysia
Ayuthia spectabile Boulard 2002a: 37–39, 65, Plate B, Figs 1–2, Plate N, Fig 5 (described, illustrated, sonogram, comp. note) Thailand, Indochina, Siam, Malay Peninsula
Ayuthia spectabile Boulard 2002b: 175, 193, Fig 4 (coloration, illustrated, comp. note) Southeast Asia
Ayuthia spectabile Boulard 2005d: 337, 344 (comp. note)
Ayuthia spectabile Boulard 2005o: 148 (listed)
Ayuthia spectabile Lee 2005: 37 (illustrated)
Ayuthia spectabile Boulard 2006g: 33–34, 73–74 (illustrated, comp. note)
Ayuthia spectabile Boulard 2007e: 16–17, 23, 35–36, Figs 8–9, Fig 20, Plate 3, Fig 4 (eye morphology, illustrated, sonogram, comp. note) Thailand
Ayuthia spectabile Boulard 2007f: 92–93, 141–142, Fig 149 (comp. note) Thailand
Ayuthia spectabile Sanborn, Phillips and Sites 2007: 25 (listed, distribution, comp. note) Thailand, Indochina, Malay Peninsula
Ayuthia spectabile Shiyake 2007: 5, 69–70, Fig 113 (illustrated, distribution, listed, comp. note) Malaysia, Thailand
Ayuthia spectabile Boulard 2008b: 109 (comp. note)
Ayuthia spectabile Boulard 2008c: 362 (comp. note) Malaysia
Ayuthia spectabile Boulard 2008e: 352 (comp. note)
Ayuthia spectabile Boulard 2008f: 23, 92 (type species of *Ayuthia*, biogeography, listed, comp. note) Thailand
Ayuthia spectabile Lee 2008a: 19 (type species of *Ayuthia*, distribution, listed) Vietnam, Thailand, Malay Peninsula
Ayuthia spectabile Pham and Yang 2009: 15, Table 2 (listed) Vietnam

Tribe Distantadini Orian, 1963
Distantadini Novotny and Wilson 1997: 437 (listed)
Distantadini Moulds 2005b: 386, 427 (higher taxonomy, history, listed, comp. note) Australia

***Distantada* Orian, 1963**
Distantada Williams and Williams 1988: 4 (listed) Mascarene Islands
Distantada Novotny and Wilson 1997: 437 (listed)

***D. thomasseti* Orian, 1963** = *Distantada thomaseti* (sic)
Distantada thomasseti Williams and Williams 1988: 4 (listed, comp. note) Equals *Cicada thomasseti* Rodriguez Island
Distantada thomasseti Boulard 1990b: 210 (comp. note)

Tribe Sonatini Lee, 2010
Oncotympanini (sic) Shiyake 2007: 6, 71 (listed, comp. note)
Oncotympanini (sic) Osaka Museum of Natural History 2008: 324 (listed) Japan
Sonatini Lee 2010d: 167, 171 (n. tribe, diagnosis, comp. note)

Sonata Lee, 2009 to *Hyalessa* China, 1925
S. ella Lei & Chou, 1997 to *Hyalessa ella*
S. expansa (Walker, 1858) to *Hyalessa expansa*
S. fuscata (Distant, 1905) to *Hyalessa fuscata*
S. maculaticollis (de Motschulsky, 1866) to *Hyalessa maculaticollis maculaticollis*
S. maculaticollis var. *a* Kato, 1932 to *Hyalessa maculaticollis maculaticollis* (de Motschulsky, 1866)
S. maculaticollis var. *b* Distant, 1891 to *Hyalessa maculaticollis maculaticollis* (de Motschulsky, 1866)
S. maculaticollis var. *b* Kato, 1932 to *Hyalessa maculaticollis maculaticollis* (de Motschulsky, 1866)
S. maculaticollis var. *c* Liu, 1940 to *Hyalessa maculaticollis maculaticollis* (de Motschulsky, 1866)
S. maculaticollis var. *d* Liu, 1940 to *Hyalessa maculaticollis* (de Motschulsky, 1866) *maculaticollis*
S. maculaticollis aka Kato, 1934 to *Hyalessa maculaticollis maculaticollis* (de Motschulsky, 1866)
S. maculaticollis mikido Kato, 1925 to *Hyalessa maculaticollis mikado*
S. maculaticollis nigroventris Kato, 1932 to *Hyalessa maculaticollis* (de Motschulsky, 1866)
S. maculaticollis pygmaea Kato, 1932 to *Hyalessa maculaticollis* (de Motschulsky, 1866)
S. mahoni Distant, 1906 to *Hyalessa mahoni*
S. melanoptera (Distant, 1904) to *Hyalessa melanoptera*
S. obnubila (Distant) to *Hyalessa obnubila*
S. stratoria Distant, 1905 to *Hyalessa stratoria*
S. virescens Distant, 1905 to *Hyalessa virescens*

***Hyalessa* China, 1925** = *Sonata* Lee, 2009
Hyalessa Wu 1935: 4 (listed) China
Oncotympana (sic) Wu 1935: 16 (listed) China
Oncotympana (sic) Chen 1942: 145 (listed)
Oncotympana (sic) Fujiyama 1982: 185 (comp. note)
Oncotympana (sic) Peng and Lei 1992: 105 (comp. note) China, Hunan
Onctympana (sic) Zhang, Sun and Zhang 1994: 25 (comp. note) China
Oncotympana (sic) Lee 1995: 18, 22, 53, 79–80, 136 (listed, key, described, ecology, distribution, illustrated, comp. note) Korea
Hyalessa Chou, Lei, Li, Lu and Yao 1997: 187, 293, Table 6 (listed, diversity, biogeography) China
Hyalessa Novotny and Wilson 1997: 437 (listed)
Oncotympana (sic) Lee 1999: 6 (comp. note) Equals *Pomponia (Oncotympana)* Korea
Oncotympana (sic) Chou, Lei, Li, Lu and Yao 1997: 261–262, 293, 308, 321, 325, 327, Table 6, Tables 10–12 (key, described, listed, diversity, biogeography) Equals *Pomponia (Oncotympana)* China
Oncotympana (sic) Lee 2005: 15–16, 75, 172 (listed, key, comp. note) Equals *Pomponia (Oncotympana)* Korea
Hyalessa Moulds 2005b: 391, 432 (higher taxonomy, listed, comp. note)
Oncotympana (sic) Shiyake 2007: 6, 10, 71 (listed, comp. note)
Oncotympana (sic) Lee 2008b: 446, 449, Table 1 (listed, comp. note) Equals *Pomponia (Oncotympana)* Korea
Oncotympana (sic) Osaka Museum of Natural History 2008: 324 (listed) Japan
Sonata Ahmed and Sanborn 2010: 40 (distribution) Japan, Korea, China, Nepal, Bhutan, India, Pakistan
Sonata Lee 2010a: 20 (n. gen., described, diversity, distribution, listed) Asia, Russia, Korea, China
Sonata Lee 2010d: 167, 171 (comp. note)

***H. ella* (Lei & Chou, 1997)** = *Oncotympana ella* = *Sonata ella*
Oncotympana ella Chou, Lei, Li, Lu and Yao 1997: 263–264, 357, 369, Fig 9–84, Plate XIV, Fig 141 (key, described, illustrated, comp. note) China
Oncotympana ella Lee 2010a: 19 (listed, comp. note)
Sonata ella Lee 2010a: 20 (n. comb., illustrated, listed, comp. note)

***H. expansa* (Walker, 1858)** = *Carineta expansa* = *Pomponia expansa* = *Oncotympana expansa* = *Sonata expansa*
Oncotympana expansa Lee 2010a: 19 (listed, comp. note)
Sonata expansa Lee 2010a: 20 (n. comb., illustrated, listed, comp. note)

***H. fuscata* (Distant, 1905)** = *Oncotympana fuscata* = *Sonata fuscata* = *Pomponia maculaticollis* (nec de Motschulsky) = *Oncotympana coreanus* Kato, 1925 = *Oncotympana coreana* = *Oncotympana nigrodorsalis* Mori, 1931 = *Oncotympana nigridorsalis* (sic) = *Oncotympana coreana nigrodorsalis* = *Oncotympana maculaticollis fuscata* = *Oncotympana maculaticollis pruinosa* Kato, 1927 = *Oncotympana coreana pruinosa* = *Oncotympana coreana* var. *pruinosa*
Oncotympana coreana Kishida 1929: 109 (comp. note) Korea
Oncotympana coreana Doi 1932: 43 (listed, comp. note) Korea
Oncotympana coreana Wu 1935: 16 (listed) Equals *Oncotympana coreanus* Equals *Oncotympana maculaticollis* China, Manchuria, Japan, Korea

Oncotympana coreana var. *pruinosa* Wu 1935: 16 (listed) Equals *Oncotympana vaculaticollis* var. *pruinosa* China, Manchuria
Oncotympana fuscata Wu 1935: 16 (listed) China
Oncotympana fuscata Chen 1942: 145 (comp. note) China
Oncotympana coreana Chen 1942: 146 (comp. note) Korea
Oncotympana coreana var. *pruinosa* Chen 1942: 146 (comp. note) China
Oncotympana coreana Kim 1958: 94 (listed) Korea
Oncotympana fuscata Chou 1987: 21 (comp. note) China
Oncotympana coreana Karban 1986: 224, Table 14.2 (comp. note) Japan, Korea, China
Oncotympana coreana Liu 1992: 37 (comp. note)
Oncotympana fuscata Lee 1995: 18, 22, 30, 45–46, 79–85, 136, 139 (listed, key, synonymy, ecology, distribution, illustrated, comp. note) Korea
Oncotympana maculaticollis (nec de Motschulsky) Lee 1995: 18, 22, 28, 30–31, 34–35, 37, 43, 45–46, 79–80, 123–125, 135, 138 (comp. note) Equals *Pomponia maculaticollis* (nec de Motschulsky) Equals *Ancotympana* (sic) *maculaticollis* (nec de Motschulsky) Equals *Cicada maculaticollis* (nec de Motschulsky)
Pomponia maculaticollis (nec de Motschulsky) Lee 1995: 29, 80, 82, 123 (comp. note)
Oncotympana nigrodorsalis Lee 1995: 35, 81–82 (comp. note)
Ancotympana (sic) *maculaticollis* (nec de Motschulsky) Lee 1995: 36 (comp. note)
Oncotympana coreanus Lee 1995: 30–31, 80, 82, 123–124 (comp. note)
Oncotympana coreana Lee 1995: 31, 33, 35–38, 40–45, 80–81 (comp. note)
Oncotympana coreana var. *nigrodorsalis* Lee 1995: 37–43, 81–82 (comp. note)
Cicada maculaticollis (nec de Motschulsky) Lee 1995: 123 (comp. note)
Oncotympana fuscata Lee 1999: 6–7, Fig 4 (distribution, song, illustrated, listed, comp. note) Equals *Pomponia maculaticollis* (nec de Motschulsky) Equals *Oncotympana maculaticollis* (nec de Motschulsky) Equals *Ancotympana* (sic) *maculaticollis* Equals *Oncotympana coreana* Equals *Oncotympana coreanus* Equals *Oncotympana nigrodorsalis* Equals *Oncotympana coreana* var. *nigradorsalis* Equals *Oncotympana coreanus* f. *nigrodorsalis* Equals *Oncotympana maculaticollis fuscata* Korea, China, Russia
Oncotympana maculaticollis fuscata Lee 1999: 7 (comp. note)
Pomponia maculaticollis Hua 2000: 64 (listed, distribution) Equals *Cicada maculaticollis* China, Shandong, Gansu, Sichuan
Oncotympana fuscata Lee 2001a: 51, Fig 1 (illustrated, comp. note) Korea
Oncotympana fuscata Lee 2001b: 2 (comp. note) Korea
Oncotympana fuscata Jeon, Kim, Tripotin and Kim 2002: 240–241, Fig 5 (parasitism, comp. note) Korea
Oncotympana fuscata Lee 2003b: 5, 8 (comp. note)
Oncotympana fuscata Lee 2003c: 15 (comp. note) Korea
Oncotympana fuscata Lee 2005: 16, 74–78, 156, 163–164, 172–173 (described, illustrated, synonymy, song, distribution, key, comp. note) Equals *Pomponia maculaticollis* (nec de Motschulsky) Equals *Oncotympana maculaticollis* (nec de Motschulsky) Equals *Oncotympana coreanus* Equals *Oncotympana coreana* Equals *Oncotympana nigrodorsalis* Equals *Oncotympana coreana* var. *nigrodorsalis* Equals *Oncotympana coreana* var. *nigroventris* Equals *Oncotympana maculaticollis fuscata* Korea, Russia, China
Oncotympana fuscata Lee 2008b: 446–447, 449, 463, Fig 1F, Table 1 (key, illustrated, synonymy, distribution, listed, comp. note) Equals *Pomponia maculaticollis* (nec de Motschulsky) Equals *Oncotympana maculaticollis* (nec de Motschulsky) Equals *Oncotympana coreanus* Equals *Oncotympana coreana* Equals *Oncotympana nigrodorsalis* Equals *Oncotympana coreana* var. *nigrodorsalis* Equals *Onctotympana coreana* f. *nigrodorsalis* Equals *Oncotympana maculaticollis fuscata* Korea, China, Russia
Oncotympana fuscata Lee 2010a: 19–21, Figs 3D–3E (n. comb., illustrated, listed, comp. note)
Sonata fuscata Lee 2010a: 20–21, Fig 4 (n. comb., illustrated, listed, comp. note)

***H. maculaticollis maculaticollis* (de Motschulsky, 1866)**
= *Cicada maculaticollis* = *Pomponia maculaticollis* = *Pomponia maculicollis* (sic) = *Pomponia maculaticollia* (sic) = *Oncotympana maculaticollis* = *Onchotympana* (sic) *maculaticollis* = *Acotympana* (sic) *maculaticollis* = *Oncotympana naculaticollis* (sic) = *Oncotympana maculaticoliis* (sic) = *Oncotympana maculicollis* (sic) = *Sonata maculaticollis* = *Oncotympana maculaticollis aka* Kato, 1934 = *Oncotympana maculaticollis* ab. *aka* = *Oncotympana maculaticollis clara* Kato, 1932 = *Oncotympana maculaticollis* ab. *clara* = *Oncotympana macullaticollis* (sic) *clara* = *Oncotympana maculaticollis pygmaea* Kato, 1932 = *Oncotympana maculaticollis* ab. *pygmaea* = *Oncotympana maculaticollis nigroventris* Kato, 1932 = *Oncotympana maculaticollis* f. *nigroventris*

= *Oncotympana fuscata* (nec Distant) = *Oncotympana maculaticollis fuscata* = *Cicada orni* (nec Linnaeus)

Oncotympana maculaticollis Chen 1933: 360 (listed, comp. note)

Oncotympana maculaticollis Isaki 1933: 381 (listed, comp. note) Japan

Oncotympana maculaticollis Wu 1935: 16 (listed) Equals *Cicada maculaticollis* Equals *Pomponia maculaticollis* China, Tsinan, Kansu, Shantung, Szechuan, Hangchow, Japan, Korea, Formosa

Oncotympana maculaticollis f. *a* Chen 1942: 145–146 (listed, comp. note) China, Chekiang, Anhwei, Kiangsi, Shensi, Szechuan, Sikang

Oncotympana maculaticollis f. *b* Chen 1942: 145–146 (listed, comp. note) China, Nanking

Oncotympana maculaticollis Ohgushi 1954: 11 (emergence times, ecology, comp. note) Japan

Oncotympana maculaticollis Ikeda 1959: 484, 492–493, Figs 10–11, Fig 12B (timbal muscle potentials, comp. note) Japan

Oncotympana maculaticollis Kim 1960: 24 (listed) Korea

Oncotympana maculaticollis Hayashi 1982b: 187–188, 190, 192–193, Table 1 (distribution, comp. note) Japan

Oncotympana maculaticollis Hayashi 1984: 56–58, Figs 121–126 (described, illustrated, synonymy, comp. note) Equals *Cicada maculaticollis* Equals *Pomponia maculaticollis* Equals *Oncotympana coreanus* (sic) Japan

Oncotympana maculaticollis ab. *aka* Hayashi 1984: 57 (described, comp. note) Japan

Oncotympana maculaticollis ab. *clara* Hayashi 1984: 57 (described, comp. note) Japan

Oncotympana maculaticollis fuscata Hayashi 1984: 57 (described, comp. note) Japan

Oncotympana maculaticollis f. *nigroventris* Hayashi 1984: 57 (described, comp. note) Japan

Oncotympana maculaticollis ab. *pygmaea* Hayashi 1984: 57 (described, comp. note) Japan

Oncotympana maculaticollis Lloyd 1984: 80 (comp. note)

Oncotympana maculaticollis He and Chen 1985: 324–327, 329–330, Fig 1C, Fig 3C, Fig 5C, Table 1 (song, sonogram, oscillogram, comp. note) China

Oncotympana maculaticollis Wang and Zhang 1987: 295 (key, comp. note) China

Oncotympana maculaticollis Anufriev and Emelyanov 1988: 312–314, 318–319 Figs 235, 236.4, 240.1–240.4 (illustrated, genitalia illustrated, key) Soviet Union

Onchotympana (sic) *maculaticollis* Boulard 1988d: 150 (pigmentation abnormalities, comp. note)

Oncotympana maculaticollis Nakamura, Murayama, Kubota and Shigemori 1988: 17 (predation) Japan

Oncotympana maculaticollis Wang 1988: 229 (parasitism, comp. note) China

Oncotympana maculaticollis Feng and Shen 1990: 463–469, Figs 1–4, Fig 5A (song analysis, comp. note) China

Oncotympana maculaticollis Jiang, Wang, Xu, Zhang and Chen 1991: 576–586, Figs 1–7 (song production, sonogram, illustrated, song, oscillogram, comp. note) China

Oncotympana maculaticolle (sic) Liu 1991: 256 (comp. note)

Oncotympana maculaticollis Liu 1991: 256 (comp. note)

Oncotympana maculaticollis Jiang, Wang, Xu, Zhang and Chen 1992: 236–244, Figs 1–3, Tables 1–2 (timbal function, illustrated, song, oscillogram, comp. note) China

Oncotympana maculaticolle (sic) Liu 1992: 15 (comp. note)

Oncotympana maculaticollis Liu 1992: 15, 33–34 (comp. note)

Oncotympana maculaticollis Liu and Jiang 1992: 435 (comp. note)

Oncotympana maculaticollis Peng and Lei 1992: 103, Fig 296 (listed, illustrated, comp. note) China, Hunan

Oncotympana maculaticollis Wang and Zhang 1992: 121 (listed, comp. note) China

Oncotympana maculaticollis Wang and Zhang 1993: 189 (listed, comp. note) China

Oncotympana maculaticollis Anonymous 1994: 809 (fruit pest) China

Oncotympana maculaticollis Lei, Chou and Li 1994: 53 (comp. note) China

Oncotympana maculaticollis Yang and Jiang 1994: 258, 262 (timbal muscle structure, comp. note) China

Oncotympana maculaticollis Yang, Jiang, Chen and Xu 1994: 216 (timbal muscle electrical activity, song production, comp. note) China

Oncotympana maculaticollis Gogala 1995: 112 (comp. note)

Oncotympana maculaticollis Jiang, Yang, Tang, Xu and Chen 1995a: 71–74, Figs 1–2 (timbal muscle activation, oscillogram, comp. note) China

Oncotympana maculaticollis Jiang, Yang, Tang, Xu and Chen 1995b: 228 (song, phonotaxis, flight behavior, comp. note) China

Oncotympana maculaticollis Jiang, Yang, Wang, Xu, Chen and Yang 1995: 676–686, Figs 1–6 (timbal muscle activation, oscillogram, comp. note) China
Oncotympana maculaticollis Xu, Ye and Chen 1995: 85 (listed, comp. note) China, Zhejiang
Oncotympana maculaticollis Yang and Jiang 1995: 173, 178 (timbal muscle structure, comp. note) China
Oncotympana maculaticollis Jiang, Yang and Liu 1996: 314 (comp. note)
Oncotympana maculaticollis Yoshida, Motoyama, Kosaku and Miyamoto 1996: 526 (comp. note)
Oncotympana maculaticollis Chou, Lei, Li, Lu and Yao 1997: 11, 13–14, 17–18, 22, 28, 262–263, 308, 321, 325, 327, 332, 344, Fig 3-1, Figs 3-3-3-4, Figs 3-6-3-7, Figs 3-11-3-12, Fig 9-83, Plate XIV, Fig 142, Tables 10–12 (key, described, synonymy, illustrated, anatomy, genitalia, wings, sound system, nymph, egg, song, oscillogram, spectrogram, comp. note) Equals *Cicada maculaticollis* Equals *Pomponia maculaticollis* Equals *Oncotympana coreanus* Equals *Oncotympana fuscata* China
Oncotypmapan maculaticollis Itô 1998: 494 (life cycle evolution, comp. note) Japan
Oncotympana maculaticollis Lee 1999: 7 (comp. note)
Oncotympana maculaticollis Ohbayashi, Sato and Igawa 1999: 341 (parasitism, comp. note)
Oncotympana maculaticollis Usuda, Johkura, Hachiya and Nakazawa 1999: 1120, 1125, fig 5E (wing microstruture, illustrated, comp. note) China
Oncotympana maculaticollis Hua 2000: 63 (listed, distribution, hosts) Equals *Cicada maculaticollis* Equals *Oncotympana fuscata* Equals *Oncotympana coreana* Equals *Oncotympana coreana pruinosa* Equals *Oncotympana maculaticollis* var. *pruinosa* Equals *Oncotympana coreana* f. *nigrodorsalis* China, Liaoning, Hebei, Shanxi, Shaanxi, Gansu, Henan, Shandong, Xinjiang, Hubei, Jiangsu, Jiangxi, Zhejiang, Taiwan, Hunan, Guizhou, Sichuan, Korea, Siberia, Japan
Oncotympana maculaticollis Lee, Nishimori, Hatakeyama and Oishi 2000: 2–3 (comp. note) Japan
Oncotympana maculaticollis Lee 2001a: 51 (comp. note) Japan
Oncotympana maculaticollis f. *aka* Lee 2001a: 51 (comp. note) Japan
Oncotympana maculaticollis Ohara and Hayashi 2001: 1 (comp. note) Japan
Oncotympana maculaticollis Sueur 2001: 39, Table 1 (listed)
Oncotypmana maculaticollis Jeon, Kim, Tripotin and Kim 2002: 241 (parasitism, comp. note) Japan
Oncotympana maculaticollis Kubo-Irie, Irie, Nakazawa and Mohri 2003: 988 (comp. note) Japan
Oncotympana maculaticollis Saisho 2003: 51 (comp. note) Japan
Pomponia maculaticollis Lee 2005: 74 (comp. note)
Oncotympana maculaticollis Lee 2005: 75 (comp. note)
Oncotympana maculaticollis Sanborn 2005c: 113 (comp. note)
Cicada orni (sic) Sun, Feng, Gao and Jiang 2005: 648, Fig 4B (wing microstruture, illustrated, comp. note) China
Oncotympana maculaticollis Tian, Yuan and Zhang 2006: 243, Fig 1a (illustrated, reproductive system, comp. note) China
Oncotympana maculaticollis Shiyake 2007: 6, 69, 71, Fig 114 (illustrated, distribution, listed, comp. note) China, Korea, Japan
Oncotympana maculaticollis Osaka Museum of Natural History 2008: 324 (listed, acoustic behavior, ecology, comp. note) Japan
Oncotympana maculaticollis Hisamitsu, Sokabe, Terada, Osumi, Terada, et al. 2009: 93 (behavior, comp. note) Japan
Oncotympana maculaticollis Mitani, Mihara, Ishii and Koike 2009: 901, Table 1 (listed, comp. note) Japan
Oncotympana maculaticollis Lee 2010a: 19 (n. comb., listed, comp. note)
Sonata maculaticollis Lee 2010a: 20 (n. comb., illustrated, listed, comp. note)
Oncotympana maculaticollis Lee 2010d: 167 (type species of Oncotympanini, comp. note)

H. maculaticollis var. *a* (Kato, 1932) to *Hyalessa maculaticollis maculaticollis* (de Motschulsky, 1866)

H. maculaticollis var. *b* (Distant, 1891) to *Hyalessa maculaticollis maculaticollis* (de Motschulsky, 1866)

H. maculaticollis var. *b* (Kato, 1932) to *Hyalessa maculaticollis maculaticollis* (de Motschulsky, 1866)

H. maculaticollis var. *c* (Liu, 1940) to *Hyalessa maculaticollis maculaticollis* (de Motschulsky, 1866)

H. maculaticollis var. *d* (Liu, 1940) to *Hyalessa maculaticollis maculaticollis* (de Motschulsky, 1866)

***H. maculaticollis mikido* (Kato, 1925)** = *Oncoympana maculaticollis mikado* = *Oncotympana maculaticollis micado* (sic) = *Sonata maculaticollis mikido*
Oncotympana maculaticollis f. *mikado* Hayashi 1984: 56–57, Fig 121 (described, illustrated, comp. note) Japan

***H. mahoni* (Distant, 1906)** = *Oncotympana mahoni* = *Sonata mahoni*
Oncotympana mahoni Lee 2010a: 19 (listed, comp. note)
Sonata mahoni Lee 2010a: 20 (n. comb., illustrated, listed, comp. note)

***H. melanoptera* (Distant, 1904)** = *Pomponia melanoptera* = *Oncotympana melanoptera* = *Sonata melanoptera*
Oncotympana melanoptera Lee 2010a: 19 (listed, comp. note)
Sonata melanoptera Lee 2010a: 20 (n. comb., illustrated, listed, comp. note)

***H. obnubila* (Distant, 1888)** = *Pomponia obnubila* = *Oncotympana obnubila* = *Sonata obnubila*
Oncotympana obnubila Chaudhry, Chaudry and Khan 1966: 94 (listed, comp. note) Pakistan
Sonata obnubila Ahmed and Sanborn 2010: 27, 40 (synonymy, distribution) Equals *Pomponia obnubila* Pakistan, India, Bhutan, Nepal
Oncotympana obnubila Lee 2010a: 19 (listed, comp. note)
Sonata obnubila Lee 2010a: 20 (n. comb., illustrated, listed, comp. note)

***H. ronshana* China, 1925**
Hyalessa ronshana Wu 1935: 4 (listed) China, Yunnan
Hyalessa ronshana Chou, Lei, Li, Lu and Yao 1997: 187 (listed) China
Hyalessa ronshana Hua 2000: 62 (listed, distribution) China, Yunnan

***H. stratoria* (Distant, 1905)** = *Oncotympana stratoria* = *Sonatoa stratoria*
Oncotympana stratoria Wu 1935: 17 (listed) China, Yunnan
Oncotympana stratoria Chou, Lei, Li, Lu and Yao 1997: 264 (described, comp. note) China
Oncotympana stratoria Hua 2000: 63 (listed, distribution) China, Yunnan, Nepal
Oncotympana stratoria Lee 2010a: 19 (listed, comp. note)
Sonata stratoria Lee 2010a: 20 (n. comb., illustrated, listed, comp. note)

***H. virescens* (Distant, 1905)** = *Oncotympana virescens* = *Sonata virescens*
Oncotympana virescens Wu 1935: 17 (listed) China, Peiping, Luchan, Szechuan, Yunnan, Tibet
Oncotympana virescens Chou, Lei, Li, Lu and Yao 1997: 264, 369 (described, comp. note) China
Oncotympana virescens Hua 2000: 63 (listed, distribution) China, Beijing, Jiangxi, Zhejiang, Anhui, Sichuan, Yunnan, Xizang
Oncotympana virescens Lee 2010a: 19 (listed, comp. note)
Sonata virescens Lee 2010a: 20 (n. comb., illustrated, listed, comp. note)

Tribe Gaeanini Distant, 1905
Gaeanini Chou and Yao 1985: 123, 138 (comp. note) China
Gaeanini Duffels 1991a: 120, 128 (comp. note)
Gaeanini Villet 1994b: 87, 91 (comp. note)
Gaeanini Chou, Lei, Li, Lu and Yao 1997: 106 (key, synonymy, listed, described) Equals Gaeanaria Equals Tosenaria
Gaeanini Novotny and Wilson 1997: 437 (listed)
Gaeanini Boulard 2002a: 37 (listed)
Gaeanini Boulard 2003a: 98 (listed)
Gaeanini Boulard 2005e: 118 (listed)
Gaeanini Boulard 2005g: 366 (listed)
Gaeanini Moulds 2005b: 427 (higher taxonomy, listed)
Gaeanini Pham and Thinh 2005a: 239 (listed) Vietnam
Gaeanini Pham and Thinh 2005b: 287–289 (key, comp. note) Equals Gaeanaria Vietnam
Gaeanini Boulard 2007d: 502 (listed)
Gaeanini Sanborn, Phillips and Sites 2007: 3, 25, Table 1 (listed) Thailand
Gaeanini Shiyake 2007: 6, 74 (listed, comp. note)
Gaeanini Boulard 2008c: 366 (listed)
Gaeanini Boulard 2008f: 7, 48 (listed) Thailand
Gaeanini Lee 2008a: 2, 6, Table 1 (listed) Vietnam
Gaeanini Pham and Yang 2009: 5, 13, Table 2 (listed) Equals Gaeanaria Vietnam
Gaeanini Lee 2010d: 171 (comp. note)
Gaeanini Lee and Hill 2010: 278, 301 (comp. note)
Gaeanini Mozaffarian and Sanborn 2010: 77 (listed)

Subtribe Gaeanina Distant, 1905 = Gaeanaria
Gaeanaria Wu 1935: 19 (listed) China
Gaeanaria Chou, Lei, Li, Lu and Yao 1997: 298 (comp. note)
Gaeanaria Boulard 2003a: 98 (comp. note)
Gaeanaria Boulard 2005e: 118 (listed, comp. note)
Gaeanaria Boulard 2005g: 366 (comp. note)
Gaeanaria Sanborn, Phillips and Sites 2007: 25, 41, Fig 5, Table 1 (listed, distribution) Thailand
Gaeanina Lee 2008a: 2, 6, Table 1 (listed) Vietnam
Gaeanina Pham and Yang 2009: 13, Table 2 (listed) Vietnam

***Gaeana* Amyot & Audinet-Serville, 1843** = *Geana* (sic) = *Geaena* (sic) = *Gaena* (sic)
Gaeana Dallas 1870: 482 (listed)
Gaeana Wu 1935: 19 (listed) China

Gaeana sp. Hill, Hore and Thornton 1982: 160, Plate 102 (illustrated, comp. note) Hong Kong
Gaeana Strümpel 1983: 18 (listed)
Gaeana Boulard 1984c: 168 (type genus of Gaeaninae, comp. note)
Gaeana Marshall 1984: 225, Table 3 (hemolymph, comp. note)
Gaeana Boulard 1985e: 1029 (coloration, comp. note)
Gaeana Chou and Yao 1985: 123–127, 130, 138–139 (key, comp. note) China
Gaeana sp. Chou and Yao 1985: 125, Fig 2F (illustrated)
Gaeana Duffels 1991a: 128 (comp. note)
Gaeana Hayashi 1993: 15 (comp. note)
Gaeana Gogala 1994a: 33, Fig 4 (illustrated, comp. note) Thailand
Gaeana Zaidi and Ruslan 1995c: 70, Table 2 (listed, comp. note) Peninsular Malaysia
Gaeana sp. Lau, Reels and Fallows 1996: 35 (listed) Hong Kong
Gaeana Chou, Lei, Li, Lu and Yao 1997: 132, 135–136, 145, 293, 298–299, 301, 358, Fig 12-1, Table 6 (described, listed, key, diversity, biogeography, phylogeny) China
Gaeana Novotny and Wilson 1997: 437 (listed)
Gaeana Rakitov 2002: 123 (comp. note)
Gaeana Boulard 2005g: 365–366 (comp. note)
Gaeana sp. Lee 2005: 37 (illustrated)
Gaeana Pham and Thinh 2005a: 239 (listed) Vietnam
Gaeana Pham and Thinh 2005b: 289 (comp. note) Vietnam
Gaeana Yen, Robinson and Quicke 2005b: 374 (comp. note)
Gaeana Boulard 2007e: 3, 23, 66, Fig 2c (illustrated, comp. note) Thailand
Gaeana Sanborn, Phillips and Sites 2007: 3, 25, Table 1 (listed) Thailand
Gaeana Shiyake 2007: 6, 74–75 (listed, comp. note)
Gaeana Boulard 2008f: 7, 48, 95 (listed) Thailand
Gaeana Lee 2008a: 2, 6, Table 1 (listed) Vietnam
Gaeana Pham and Yang 2009: 5–6, 13, Table 2 (type genus of Gaeanini, listed) Vietnam

G. annamensis annamensis Distant, 1913 to *Callogaena annamensis annamensis*

G. annamensis annamensis var. *b* Distant, 1892 to *Callogaena annamensis* var. *b*

***G. atkinsoni* Distant, 1892**

Gaeana atkinsoni Hayashi 1993: 15 (comp. note)

***G. cheni* Chou & Yao, 1985** = *Gaeana yunnanensis* Chou & Yao, 1985 = *Gaena* (sic) *maculata sheshuangpana* = *Gaeana maculata xishuanbanna*

Gaeana cheni Chou and Yao 1985: 124–125, 129, 135, 139, Fig 1A, Fig 2B, Fig 6, Plate I, Figs 7–8 (n. sp., described, illustrated, comp. note) Equals *Gaena* (sic) *maculata sheshuangpana* Equals *Gaeana maculata xishuanbanna* China
Gaeana yunnanensis Chou and Yao 1985: 125, 129–130, 136, 139, Fig 2A, Fig 7, Plate II, Figs 1–2 (key, n. sp., described, illustrated, comp. note) China
Gaeana maculata sheshuangpana Chou and Yao 1985: 129–130 (comp. note)
Gaeana cheni Chou, Lei, Li, Lu and Yao 1997: 136, 138, 358, Fig 9-16, Plate IV, Fig 53 (key, synonymy, described, illustrated, comp. note) Equals *Gaeana yunnanensis* China
Gaeana yunnanensis Chou, Lei, Li, Lu and Yao 1997: 358 (synonymy)
Gaeana cheni Hua 2000: 61 (listed, distribution) China, Yunnan
Gaeana yunnanensis Hua 2000: 62 (listed, distribution) China, Yunnan
Gaeana cheni Boulard 2003a: 98–99, 116–117, Color Figs 11–12, sonogram 13 (described, song, sonogram, illustrated, comp. note) Thailand, China
Gaeana cheni Boulard 2005g: 369 (comp. note)
Tosena paviei (sic) Boulard 2005o: 148–149 (listed) Equals *Gaeana paviei* Equals *Gaeana cheni* sonogram Boulard 2001
Gaeana cheni Boulard 2005o: 149 (listed)
Gaeana cheni Lee 2005: 37 (illustrated)
Gaeana cheni Pham and Thinh 2005a: 239 (synonymy, distribution, listed) Equals *Gaeana yunnanensis* Vietnam
Gaeana cheni Boulard 2006g: 98–99, Fig 64B (illustrated, comp. note) Thailand
Gaeana cheni Pham and Thinh 2006: 525–527 (listed, comp. note) Vietnam
Gaeana cheni Boulard 2007d: 502 (comp. note) Thailand
Gaeana cheni Boulard 2007e: 4, 32, 66, Plate 2, Fig 2 (illustrated, coloration, comp. note) Thailand
Gaeana cheni Sanborn, Phillips and Sites 2007: 26 (listed, distribution, comp. note) Thailand, China
Gaeana cheni Shiyake 2007: 6, 72, 74, Fig 116 (illustrated, listed, comp. note)
Gaeana cheni Boulard 2008c: 366 (comp. note) Vietnam
Gaeana cheni Boulard 2008f: 48, 95 (biogeography, listed, comp. note) Thailand, China
Gaeana cheni Pham and Yang 2009: 6, 13, 16, Table 2 (listed, distribution, comp. note) Vietnam, China
Gaeana cheni Watson, Watson, Hu, Brown, Cribb and Myhra 2010: 117, 119–120,

124–125, Figs 2c–f, Fig 8, Table 1 (wing microstructure, listed, comp. note)

G. chinensis Kato, 1940

G. consors Atkinson, 1884 = *Gaeana maculata consors* = *Gaeana maculata* var. *a* Distant, 1892

Gaeana consors Chou and Yao 1985: 123, 135, 135, Plate I, Figs 1–2 (comp. note) China

Gaeana consors Peng and Lei 1992: 101–102, Fig 289 (listed, illustrated, comp. note) China, Hunan

Gaeana maculata consors Wang and Zhang 1992: 120 (listed, comp. note) China

Gaeana consors Chou, Lei, Li, Lu and Yao 1997: 136–137 (key, described, synonymy, comp. note) Equals *Gaeana maculata consors* China

Gaeana consors Hua 2000: 61 (listed, distribution, hosts) China, Guangdong, Hunan, Guangxi, Guizhou, Sichuan, Yunnan, Burma, Sikkim, India, Burma

Gaeana consors Sanborn, Phillips and Sites 2007: 26, 34 (listed, distribution, comp. note) Thailand, India, Burma, China

Gaeana consors Boulard 2008f: 48, 95 (biogeography, listed, comp. note) Equals *Gaeana maculata* var. a Equals *Gaeana maculata* var. *consors* Equals *Gaeana maculata consors* Thailand, Assam, India, Sikkim, Burma, Tonkin, China

Gaeana maculata consors Wang, Ding, Wheeler, Purcell and Zhang 2009: 1136 (parasitism, comp. note) China

G. electa Jacobi, 1902 to *Becquartina electa*

G. festiva festiva (Fabricius, 1803) to *Callogaena festiva festiva*

G. festiva var. *a* Distant, 1892 to *Callogaena festiva* var. *a*

G. hageni hageni Distant, 1889 to *Callogaena hageni hageni*

G. hageni var. *a* Distant, 1889 to *Callogaena hageni* var. *a*

G. hainanensis Chou & Yao, 1985

Gaeana hainanensis Chou and Yao 1985: 128, 135, 139, Fig 5, Plate I, Fig 6 (key, n. sp., described, illustrated, comp. note) China

Gaeana hainanensis Chou, Lei, Li, Lu and Yao 1997: 136–138, Fig 9-15, Plate IV, Fig 55 (key, described, illustrated, comp. note) China

Gaeana hainanensis Hua 2000: 61 (listed, distribution) China, Hainan

Gaeana hainanensis Pham and Thinh 2005a: 239 (distribution, listed) Vietnam

Gaeana hainanensis Yen, Robinson and Quicke 2005b: 363, Table 1 (listed, mimicry, comp. note)

Gaeana hainanensis Pham and Yang 2009: 6, 13, 16, Table 2 (listed, distribution, comp. note) Vietnam, China

G. laosensis Distant, 1917 to *Ambragaeana laosensis*

G. maculata barbouri Liu, 1940

Gaeana maculata barbouri Chou and Yao 1985: 123 (comp. note) China

Gaeana maculata babouri Hua 2000: 62 (listed, distribution) China, Guangxi

G. maculata distanti Liu, 1940

Gaeana maculata distanti Chou and Yao 1985: 123 (comp. note) China

Gaeana maculata distanti Hua 2000: 62 (listed, distribution) China, Hong Kong

G. maculata maculata (Drury, 1773) = *Cicada maculata* = *Tettigonia maculata* = *Cicada flavomaculata* (sic) = *Geaena* (sic) *maculata* = *Gaeana maculatus* (sic) = Equals *Gaeana maculata* var. *a*

Gaeana maculata Wu 1935: 19 (listed) Equals *Cicada maculata* Equals *Tettigonia maculata* Equals *Gaeana maculata* var. *a* Equals *Gaeana consors* China, Kwangsi, Sikhim, Assam, Naga Hills, Khasi Hills, Margherita, Samagooting Valley, Dhansiri Valley, Burma, Karenne, Tonkin

Gaeana maculata Hill, Hore and Thornton 1982: 49, 160, 163–164, Plate 101, Plate 109 (listed, illustrated, comp. note) Hong Kong

Gaeana maculata Boulard 1984c: 168 (type species of *Gaeana*, comp. note)

Gaeana maculata Boulard 1985e: 1024, 1046–1047, Plate IX, Fig 52 (coloration, illustrated, comp. note)

Gaeana maculata Chou and Yao 1985: 123, 127–128, 130, 135, 139, Plate I, Figs 3–4 (type species of *Gaeana*, key, comp. note) China

Gaeana maculata Chou 1987: 22 (comp. note) China

Gaeana maculata Jiang, Lin and Wang 1987: 106, 108, Fig 1F, Fig 3(3) (timbal structure, illustrated, song, oscillogram, comp. note) China

Gaeana maculata Wang and Zhang 1987: 296 (key, comp. note) China

Gaeana maculata Wang and Zhang 1992: 119 (listed, comp. note) China

Gaeana maculata Hayashi 1993: 15 (comp. note)

Gaeana maculata Wang and Zhang 1993: 190 (listed, comp. note) China

Gaeana maculata Andersen 1994: 29 (comp. note)

Gaeana maculata Jiang, Yang, Tang, Xu and Chen 1994: 290 (song analysis, comp. note) China

Gaeana maculata Jiang, Yang, Tang, Xu and Chen 1995b: 227 (song, phonotaxis, flight behavior, comp. note) China

Cicada maculata Chou, Lei, Li, Lu and Yao 1997: 136 (type species of *Gaeana*)

Gaeana maculata Chou, Lei, Li, Lu and Yao 1997: 136–138, 145, 366, Plate IV, Fig 56 (key, synonymy, illustrated, described, comp. note) Equals *Cicada maculata* Equals *Tettigonia maculata* China

Gaeana maculata Easton and Pun 1999: 101 (distribution, comp. note) Macao, Hong Kong, Myanmar, India, Assam, Vietnam, China, Guangxi

Gaeana maculata Hua 2000: 61–62 (listed, distribution, hosts) Equals *Cicada maculata* China, Jiangxi, Fujian, Guangdong, Hong Kong, Hainan, Guangxi, Guizhou, Sichuan, Japan, Indonesia, Malaysia, Vietnam, Burma, Bangladesh, Sikkim, India, Australia

Cicada maculata Boulard 2005g: 365 (type species of *Gaeana*, comp. note)

Cicada maculata Pham and Thinh 2005a: 239 (type species of *Gaeana*)

Gaeana maculata Pham and Thinh 2005a: 239 (synonymy, distribution, listed) Equals *Cicada maculata* Equals *Tettigonia maculata* Vietnam

Gaeana maculata Yen, Robinson and Quicke 2005a: 198 (mimicry, comp. note)

Gaeana maculata Yen, Robinson and Quicke 2005b: 363, 365, 375, Fig 1, Table 1 (illustrated, listed, mimicry, comp. note)

Gaeana maculata Sanborn, Phillips and Sites 2007: 26 (comp. note)

Gaeana maculata Shiyake 2007: 6, 72, 74, Fig 118 (illustrated, distribution, listed, comp. note) China, India, Southeast Asia

Gaeana maculata Boulard 2008c: 366 (comp. note) Vietnam

Gaeana maculata Boulard 2008f: 48 (type species of *Gaeana*)

Cicada maculata Lee 2008a: 6 (type species of *Gaeana*)

Gaeana maculata Lee 2008a: 6 (synonymy, distribution, listed) Equals *Cicada maculata* North Vietnam, South China, Myanmar, Sri Lanka, India

Cicada maculata Pham and Yang 2009: 6 (type species of *Gaeana*)

Gaeana maculata Pham and Yang 2009: 13, Table 2 (listed) Vietnam

***G. nigra* Lei & Chou, 1997**

Gaeana nigra Chou, Lei, Li, Lu and Yao 1997: 17, 136, 138–139, 357, 366, Fig 3–6E, Fig 9–17, Plate IV, Fig 54 (genitalia illustrated, key described, comp. note) China

G. stellata stellata (Walker, 1858) to *Ambragaeana stellata stellata*

G. stellata var. *a* Distant, 1892 to *Ambragaeana stellata* var. *a*

G. sticta Chou & Yao, 1985 to *Ambragaeana sticta*

G. sulphurea (Westwood, 1839) to *Sulphogaeana sulphurea*

G. sultana Distant, 1913 to *Callogaena sultana*

***G. variegata* Yen, Robinson & Quicke, 2005 nom. nud.**

Gaeana variegata Yen, Robinson and Quicke 2005b: 363, 365, 375, Fig 1, Table 1 (illustrated, listed, mimicry, comp. note)

G. vestita Distant, 1905 to *Sulphogaeana vestita*

G. vestita var. ____ Distant, 1917 to *Sulphogaeana vestita*

G. vitalisi Distant, 1913 to *Callogaena vitalisi*

***Sulphogaeana* Chou & Yao, 1985**

Sulphogaeana Chou and Yao 1985: 124, 138 (key, n. gen., described, comp. note) China

Sulphogaeana Lei 1997: 72 (listed)

Sulphogaeana Chou, Lei, Li, Lu and Yao 1997: 132, 139–140, 293, 298–299, 301, Fig 12–1, Table 6 (described, listed, key, diversity, biogeography, phylogeny) China

Sulphogaeana Pham and Thinh 2005a: 239 (listed) Vietnam

Sulphogaeana Pham and Thinh 2005b: 289 (comp. note) Vietnam

Sulphogaeana Pham and Yang 2009: 6, 13, Table 2 (listed) Vietnam

***S. dolicha* Lei, 1997**

Sulphogaeana dolicha Lei 1997: 72–74, Figs 1–2 (n. sp., described, illustrated, comp. note) China, Yunnan

Sulphogaeana dolicha Chou, Lei, Li, Lu and Yao 1997: 140–143, Fig 9–20, Plate V, Fig 62 (key, synonymy, described, illustrated, comp. note) China

Sulphogaeana dolicha Pham and Thinh 2005a: 240 (distribution, listed) Vietnam

Sulphogaeana dolicha Pham and Yang 2009: 7, 13, 16, Table 2 (listed, distribution, comp. note) Vietnam, China

***S. sulphurea* (Westwood, 1839)** = *Cicada sulphurea* = *Gaeana sulphurea* = *Sulphogaeana sulphura* (sic) = *Cicada pulchella* Westwood, 1942 = *Cicada pulcilla* (sic)

Gaeana sulphurea Showalter 1935: 38–39, Fig 13 (illustrated, coloration, comp. note) India

Gaeana sulphurea Chou and Yao 1985: 123, 125, 138 (type species of *Sulphogaeana*, comp. note) China

Sulphogaeana sulphurea Chou and Yao 1985: 124–126, 136, Fig 1D, Fig 2E, Fig 3, Plate II, Figs 5–6 (n. comb., illustrated, comp. note) Equals *Cicada sulphurea* Equals *Gaeana sulphurea* China

Gaeana sulphurea Hayashi 1993: 15 (comp. note)

Cicada sulphurea Chou, Lei, Li, Lu and Yao 1997: 140 (type species of *Gaeana*)
Sulphogaeana sulphurea Chou, Lei, Li, Lu and Yao 1997: 140–141, Fig 9–19, Plate V, Fig 64 (key, synonymy, described, illustrated, comp. note) Equals *Cicada sulphurea* Equals *Cicada pulchella* Equals *Gaeana sulphurea* China
Sulphogaeana sulphurea Lei 1997: 72, 74 (comp. note)
Sulphogaeana sulphurea Hua 2000: 64 (listed, distribution) Equals *Cicada sulphurea* Equals *Sulphogaeana sulphura* (sic) Equals *Gaeana sulphurea* China, Yunnan, India, Sikkim, Bangladesh, Nepal
Cicada sulphurea Pham and Thinh 2005a: 239 (type species of *Sulphogaeana*)
Cicada sulphurea Pham and Yang 2009: 6 (type species of *Sulphogaeana*)

***S. vestita* (Distant, 1905)** = *Gaeana vestita* = *Callogaeana vestita* = *Gaeana vestita* var. ______ Distant, 1917 = *Sulphogaeana vestita* var. _______

Gaeana vestita Wu 1935: 20 (listed) China, Yunnan
Gaeana vestita Boulard 1985e: 1024, 1046–1047, Plate IX, Fig 51 (coloration, illustrated, comp. note)
Sulphogaeana vestita Chou and Yao 1985: 126, 138 (n. comb., comp. note) Equals *Gaeana vestita*
Gaeana vestita Boulard 1997b: 15 (coloration, comp. note)
Sulphogaeana vestita Chou, Lei, Li, Lu and Yao 1997: 142 (synonymy, comp. note) Equals *Gaeana vestita* China
Callogaeana vestita Hua 2000: 61 (listed, distribution) Equals *Gaeana vestita* China, Yunnan
Sulphogaeana vestita Hua 2000: 64 (listed, distribution) Equals *Gaeana vestita* China, Yunnan, Laos
Gaeana vestita Lee 2008a: 7 (synonymy, distribution, listed) Equals *Gaeana vestita* var. ______ Vietnam, South China, Laos
Gaeana vestita Pham and Yang 2009: 13, 16, Table 2 (listed, comp. note) Vietnam
Gaeana vestita Pham and Yang 2010a: 133 (comp. note) Vietnam

***Ambragaeana* Chou & Yao, 1985**

Ambragaeana Chou and Yao 1985: 124, 138–139 (key, n. gen., described, comp. note) China
Ambragaeana Chou, Lei, Li, Lu and Yao 1997: 132, 142–143, 293, 298–299, 301, Fig 12–1, Table 6 (described, listed, key, diversity, biogeography, phylogeny) China
Ambragaeana Pham and Thinh 2005a: 239 (listed) Vietnam
Ambragaeana Pham and Thinh 2005b: 289 (comp. note) Vietnam
Ambragaeana Boulard 2007e: 3 (comp. note) Thailand
Ambragaeana Sanborn, Phillips and Sites 2007: 3, 26, Table 1 (listed) Thailand
Ambragaeana Boulard 2008f: 7, 49, 96 (listed) Thailand
Ambragaeana Pham and Yang 2009: 5, 13, Table 2 (listed) Vietnam

***A. ambra* Chou & Yao, 1985**

Ambragaeana ambra Chou and Yao 1985: 124–125, 136, 139, Fig 1B, Fig 2C, Plate II, Figs 3–4 (n. sp., type species of *Ambragaeana*, described, illustrated, comp. note) China
Ambragaeana ambra Chou, Lei, Li, Lu and Yao 1997: 141–145, Fig 9–20, Fig 9–21, Plate V, Fig 58, Plate V, Fig 65 (key, described, illustrated, type species of *Ambragaeana*, comp. note) Equals China
Ambragaeana ambra Boulard 1999b: 184–185, Fig 14 (illustrated, comp. note) China
Ambragaeana ambra Hua 2000: 60 (listed, distribution) China, Yunnan
Ambragaeana ambra Boulard 2003a: 98–99, 117–118, Color Fig 13, sonogram 14 (described, song, sonogram, illustrated, comp. note) Thailand, Yunnan
Ambragaeana ambra Boulard 2005o: 149 (listed)
Ambragaeana ambra Lee 2005: 37 (illustrated)
Ambragaeana ambra Pham and Thinh 2005a: 239 (type species of *Ambragaeana* distribution, listed) Vietnam
Ambragaeana ambra Sanborn, Phillips and Sites 2007: 26 (listed, distribution, comp. note) Thailand, China
Ambragaeana ambra Boulard 2008f: 49, 96 (type species of *Ambragaeana*, biogeography, listed, comp. note) Thailand, Yunnan, China
Ambragaeana ambra Pham and Yang 2009: 5–6, 13, 16, Table 2 (type species of *Ambragaeana*, listed, distribution, listed, comp. note) Vietnam, China

***A. laosensis* (Distant, 1917)** = *Gaeana laosensis*

Ambragaeana laosensis Chou and Yao 1985: 127, 139 (n. comb., comp. note) Equals *Gaeana laosensis*
Gaeana laosensis Shiyake 2007: 6, 72, 74, Fig 115 (illustrated, listed, comp. note)
Gaeana laosensis Lee 2008a: 7 (distribution, listed) Vietnam, Laos

Gaeana laosensis Pham and Yang 2009: 13, Table 2 (listed) Vietnam

Gaeana laosensis Pham and Yang 2010a: 133 (comp. note) Vietnam

***A. stellata stellata* (Walker, 1858)** = *Huechys stellata* = *Gaeana stellata*

Gaeana stellata Chou and Yao 1985: 127 (comp. note)

Ambragaeana stellata Chou and Yao 1985: 127, 139 (n. comb., comp. note) Equals *Huechys stellata* Equals *Gaeana stellata*

Gaeana stellata Wang and Zhang 1992: 120 (listed, comp. note) China

Gaeana stellata Hayashi 1993: 15 (comp. note)

Ambragaeana stellata Sanborn, Phillips and Sites 2007: 26, 34 (listed, distribution, comp. note) Thailand, India

Ambragaeana stellata Boulard 2008f: 50, 90, 96 (biogeography, listed, comp. note) Equals *Huechys stellata* Equals *Gaeana stellata* Thailand, Hindustan, India, China

***A. stellata* var. *a* (Distant, 1892)** = *Gaeana stellata* var. *a*

***A. sticta* (Chou & Yao, 1985)** = *Gaeana sticta*

Gaeana sticta Chou and Yao 1985: 128, 130, 135, 139, Fig 8, Plate I, Fig 5 (key, n. sp., described, illustrated, comp. note) China

Ambragaeana sticta Chou, Lei, Li, Lu and Yao 1997: 143, 145, 358, Plate V, Fig 60 (key, synonymy, described, illustrated, comp. note) Equals *Gaeana sticta* China

Ambragaeana sticta Hua 2000: 60 (listed, distribution) Equals *Gaeana sticta* China, Hainan

***Callogaeana* Chou & Yao, 1985** = *Callagaeana* (sic) = *Callagaena* (sic)

Callogaeana Chou and Yao 1985: 124, 130, 138–139 (key, n. gen., described, comp. note) China

Callogaeana Chou, Lei, Li, Lu and Yao 1997: 132, 145–146, 293, 298–299, 301, Fig 12-1, Table 6 (described, listed, key, diversity, biogeography, phylogeny) China

***C. annamensis annamensis* (Distant, 1913)** = *Callagaena* (sic) *annamensis* = *Gaeana annamensis*

Callagaeana (sic) *annamensis* Chou and Yao 1985: 134 (n. comb., comp. note) Equals *Gaeana annamensis*

Gaeana annamensis Lee 2008a: 7 (distribution, listed) Vietnam, Laos

Gaeana annamensis Pham and Thinh 2005a: 239 (distribution, listed) Vietnam

Gaeana annamensis Pham and Yang 2010a: 133 (comp. note) Vietnam

Gaeana annamensis Pham and Yang 2009: 13, 16, Table 2 (listed, comp. note) Vietnam

***C. annamensis* var. *b* Distant, 1892** = *Gaeana annamensis* var. *b* = *Gaeana festiva* var. *b*

***C. aurantiaca* Chou & Yao, 1985**

Callogaeana aurantica Chou and Yao 1985: 131–132, 137, 140, Fig 10, Plate III, Fig 3 (key, n. sp., described, illustrated, comp. note) China

Callogaeana aurantiaca Chou, Lei, Li, Lu and Yao 1997: 146–147, Fig 9-24, Plate V, Fig 59 (key, described, illustrated, comp. note) China

Callogaeana aurantica Hua 2000: 60 (listed, distribution) China, Yunnan

Callogaeana aurantiaca Lee 2005: 37 (illustrated)

***C. festiva festiva* (Fabricius, 1803)** = *Tettigonia festiva* = *Cicada festiva* = *Gaeana festiva* = *Gaeana fastiva* (sic) = *Callagaeana* (sic) *festiva* = *Cicada thalassina* Perchon, 1838 (nec *Cicada thalassina* Germar, 1830) = *Gaeana thalassina* = *Cicada percheronii* Guérin-Méneville, 1844 = *Gaeana consobrina* Walker, 1850 nom. nud.

Tettigonia festiva Dallas 1870: 482 (synonymy) Equals *Gaeana festiva*

Gaeana festiva Boulard 1985e: 1024, 1046–1047, Plate IX, Fig 53 (coloration, illustrated, comp. note)

Gaeana festiva Chou and Yao 1985: 133 (comp. note)

Callagaeana (sic) *festiva* Chou and Yao 1985: 133 (n. comb., comp. note) Equals *Tettigonia festiva* Equals *Gaeana festiva* Equals *Gaeana chinensis*

Gaeana festiva Wang and Zhang 1992: 119 (listed, comp. note) China

Gaeana festiva Hayashi 1993: 15 (comp. note)

Callogaeana festiva Peng and Lei 1992: 99, Fig 282 (listed, illustrated, comp. note) China, Hunan

Gaeana festiva Boulard 1999b: 184–185, Fig 16 (illustrated, comp. note) Southeast Asia

Callogaeana festiva Hua 2000: 60–61 (listed, distribution) Equals *Tettigonia festiva* Equals *Gaeana festiva* Equals *Gaeana chinensis* China, Hunan, Guangxi, Yunnan, Indonesia, Sikkim, India

Gaeana festiva Boulard and Chueata 2004: 141–143, Figs 1–3 (life cycle, illustrated, comp. note)

Gaeana festiva Boulard 2005a: 31 (coloration, comp. note)

Gaeana festiva Boulard 2005e: 118, 134–135, 137, Plate F, Figs 1–2, Fig 25 (synonymy, described, illustrated, listed, song, sonogram, comp. note) Equals *Tettigonia festiva* Equals *Cicada festiva* Equals *Cicada thalassina* Equals *Cicada percheronii* Thailand, Sumatra

Gaeana festiva Boulard 2005o: 149, 151 (listed)

Gaeana festiva Boulard 2007e: 4, 66–67, 72, Plate 3, Fig 7, Plate 27b, Plate 30 (coloration, illustrated, comp. note) Thailand

Gaeana festiva Sanborn, Phillips and Sites 2007: 26 (listed, distribution, comp. note) Thailand, India, Laos, Malaysia, Indochina, Bhutan

Gaeana festiva Shiyake 2007: 6, 72, 74, Fig 117 (illustrated, distribution, listed, comp. note) Southeast Asia, Malaysia, Indonesia

Gaeana festiva Boulard 2008c: 366 (comp. note) Malaysia

Gaeana festiva Boulard 2008f: 49, 95 (biogeography, listed, comp. note) Equals *Tettigonia festiva* Equals *Cicada thalassina* Percheron (nec Germar) Equals *Cicada percheronii* Thailand, Sumatra, India, Laos

***C. festiva* var. *a* (Distant, 1892)** = *Gaeana festiva* var. *a*

***C. guangxiensis* Chou & Yao, 1985**

Callogaeana guangxiensis Chou and Yao 1985: 131–133, 137, 140, Fig 11, Plate III, Fig 1 (key, n. sp., described, illustrated, comp. note) China

Callogaeana guangxiensis Chou, Lei, Li, Lu and Yao 1997: 146–148, Plate IV, Fig 57 (key, synonymy, described, illustrated, comp. note) China

Callogaeana guangxiensis Hua 2000: 61 (listed, distribution) China, Guangxi

***C. hageni hageni* (Distant, 1889)** = *Callagaeana* (sic) *hageni* = *Gaeana hageni*

Callagaeana (sic) *hageni* Chou and Yao 1985: 134 (n. comb., comp. note) Equals *Gaeana hageni*

Gaeana hageni Hayashi 1993: 15 (comp. note)

***C. hageni* var. *a* (Distant, 1889)** = *Gaeana hageni* var. *a*

***C. jinghongensis* Chou & Yao, 1985**

Callogaeana jinghongensis Chou and Yao 1985: 124–125, 131–133, 137, 140, Fig 1C, Fig 2D, Fig 12, Plate III, Figs 4–5 (n. sp., key, described, illustrated, comp. note) China

Callogaeana jinghongensis Chou, Lei, Li, Lu and Yao 1997: 145–148, Fig 9–25, Plate V, Fig 61 (key, synonymy, described, illustrated, comp. note) China

Callogaeana jinghongensis Hua 2000: 61 (listed, distribution) China, Yunnan

***C. sultana* Distant, 1913** = *Callagaeana* (sic) *sultana* = *Gaeana sultana*

Callagaeana (sic) *sultana* Chou and Yao 1985: 134 (n. comb., comp. note) Equals *Gaeana sultana*

Gaeana sultana Lee 2008a: 7 (distribution, listed) Vietnam, Laos

Gaeana sultana Pham and Yang 2009: 13, 16, Table 2 (listed, comp. note) Vietnam

Gaeana sultana Pham and Yang 2010a: 133 (comp. note) Vietnam

***C. viridula* Chou & Yao, 1985**

Callogaeana viridula Chou and Yao 1985: 130–131, 137, 139, Fig 9, Plate III, Fig 2 (type species of *Callogaeana*, key, n. sp., described, illustrated, comp. note) China

Callogaeana viridula Chou, Lei, Li, Lu and Yao 1997: 145–147, Fig 9–23, Plate V, Fig 63 (key, described, illustrated, type species of *Callogaeana*, comp. note) China

Callogaeana viridula Hua 2000: 61 (listed, distribution) China, Guangxi, Yunnan

***C. vitalisi* (Distant, 1913)** = *Callagaeana* (sic) *vitalisi* = *Gaeana vitalisi* = *Gaeana vitalis* (sic)

Callagaeana (sic) *vitalisi* Chou and Yao 1985: 134 (n. comb., comp. note) Equals *Gaeana vitalis* (sic)

Gaeana vitalisi Lee 2008a: 6 (distribution, listed) Vietnam

Gaeana vitalisi Pham and Yang 2009: 13, 16, Table 2 (listed, comp. note) Vietnam

Gaeana vitalisi Pham and Yang 2010a: 133 (comp. note) Vietnam

***Balinta* Distant, 1905**

Balinta Wu 1935: 20 (listed) China

Balinta Chou and Yao 1985: 123 (comp. note) China

Balinta sp. Chou and Yao 1985: 137, Plate III, Figs 8–9 (illustrated, comp. note)

Balinta Hayashi 1993: 15 (comp. note)

Balinta Chou, Lei, Li, Lu and Yao 1997: 132–134, 293, 298–299, 301, Fig 12–1, Table 6 (described, listed, key, diversity, biogeography, phylogeny) China

Balinta Boulard 2005g: 366 (comp. note)

Balinta Pham and Thinh 2005a: 239 (listed) Vietnam

Balinta Pham and Thinh 2005b: 289 (comp. note) Vietnam

Balinta Sanborn, Phillips and Sites 2007: 3, 26, Table 1 (listed) Thailand

Balinta Boulard 2008f: 6–7, 50, 74, 77, 96 (listed, comp. note) Thailand

Balinta Lee 2008a: 2, Table 1 (listed) Vietnam

Balinta Pham and Yang 2009: 6, 13, Table 2 (listed) Vietnam

***B. auriginea* Distant, 1905**

Balinta auriginea Chou, Lei, Li, Lu and Yao 1997: 134, Plate III, Fig 41 (key, described, illustrated, comp. note) China

Balinta aurigenia Hua 2000: 60 (listed, distribution) China, Hainan, Guangxi, Yunnan

***B. delinenda* (Distant, 1888)** = *Gaeana delinenda* = *Huechys octonotata* Walker, 1850 (nec Westwood)

Balinta delinenda Pham and Thinh 2005a: 239 (synonymy, distribution, listed) Equals *Huechys octonotata* (partim) Vietnam
Balinta delinenda Lee 2008a: 7 (synonymy, distribution, listed) Equals *Gaeana delinenda* Vietnam, India
Balinta delinenda Pham and Yang 2009: 13, Table 2 (listed) Vietnam

***B. flavoterminalia* Boulard, 2008**
Balinta flavoterminalia Boulard 2008f: 50, 74–77, 96, Figs 24–28 (n. sp., described, illustrated, song, sonogram, biogeography, listed, comp. note) Thailand

***B. kershawi* Kirkaldy, 1909** = *Balinta rershawi* (sic)
Balinta kershawi Wu 1935: 20 (listed) China, Macao
Balinta kershawi Chou and Yao 1985: 123 (comp. note) China
Balinta kershawi Easton 1999: 28 (comp. note) Macao
Balinta kershawi Hua 2000: 60 (listed, distribution) China, Guangdong, Macao, Guangxi

***B. minuta* Boulard, 2008**
Balinta minuta Boulard 2008f: 50, 77–80, 96, Figs 29–33 (n. sp., described, illustrated, song, sonogram, biogeography, listed, comp. note) Thailand

***B. octonotata octonotata* (Westwood, 1842)** = *Cicada 8-notata* = *Huechys octonotata* = *Cicada octonotata* = *Gaeana octonotata* = *Huechys picta* Walker, 1858
Balinta octonotata Boulard 2008f: 50 (type species of *Balinta*)
Cicada 8-notata Chou, Lei, Li, Lu and Yao 1997: 133 (type species of *Balinta*)
Cicada octonotata Pham and Thinh 2005a: 239 (type species of *Balinta*)
Cicada octonotata Lee 2008a: 7 (type species of *Balinta*) Assam

***B. octonotata* var. *a* Distant, 1892**

***B. octonotata* var. *b* Distant, 1892**

***B. pulchella* Distant, 1913**
Balinta pulchella Lee 2008a: 7 (distribution, listed) Vietnam
Balinta pulchella Pham and Yang 2009: 13, 16, Table 2 (listed, comp. note) Vietnam
Balinta pulchella Pham and Yang 2010a: 133 (comp. note) Vietnam

***B. sanguiniventris* Ollenbach, 1929**

***B. tenebricosa tenebricosa* (Distant, 1888)** = *Gaena* (sic) *tenebricosa* = *Gaeana tenebricosa* = *Gaeana tenebriscosa* (sic) = *Gaeana tenbricosa* (sic)
Balinta tenebricosa Chou, Lei, Li, Lu and Yao 1997: 134–135, 357, Fig 9–14, Plate IV, Fig 51 (key, described, illustrated, comp. note) China
Balinta tenebricosa Hua 2000: 60 (listed, distribution) Equals *Gaeana tenebricosa* China, Guangxi, Burma
Balinta tenebriscosa (sic) Boulard 2002a: 36–37, 65, Plate A, Figs 1–3, Plate N, Fig 4 (n. sp., described, illustrated, sonogram, comp. note) Equals *Gaeana tenebriscosa* Thailand, Burma
Balinta tenebriscosa (sic) Boulard 2005d: 345 (comp. note)
Balinta tenebriscosa (sic) Boulard 2005o: 148 (listed)
Balinta tenebriscosa (sic) Boulard 2006g: 96–97, 109 (comp. note)
Balinta tenebriscosa (sic) Boulard 2007e: 47, Plate 22b (illustrated, comp. note) Thailand
Balinta tenebriscosa (sic) Sanborn, Phillips and Sites 2007: 26–27 (listed, distribution, comp. note) Thailand, Burma, Laos, China
Balinta tenebriscosa (sic) Boulard 2008b: 109, 116, Fig 14 (illustrated, comp. note)
Balinta tenebriscosa (sic) Boulard 2008e: 352–353, 355, Fig 15 (illustrated, comp. note)
Balinta tenebricosa Boulard 2008f: 50, 74, 76–77, 79, 96 (biogeography, listed, comp. note) Equals *Gaena* (sic) *tenebricosa* Equals *Gaena tenbriscosa* (sic) Equals *Gaeana tenebricosa* Thailand, Burma, Laos
Balinta tenebricosa Pham and Yang 2009: 5, 13, 16, Table 2 (listed, synonymy, distribution, comp. note) Equals *Gaeana tenebriscosa* Vietnam, Laos, Myanmar, China

***B. tenebricosa* var. *a* Distant, 1892**

***Taona* Distant, 1909**
Taona Wu 1935: 20 (listed) China
Taona Chou and Yao 1985: 123 (comp. note) China
Taona Chou, Lei, Li, Lu and Yao 1997: 132–133, 293, 297–299, 301, Fig 12-1, Table 6 (described, listed, key, diversity, biogeography, phylogeny) China

***T. immaculata* Chen, 1940**
Taona immaculata Chou and Yao 1985: 123, 137, Plate III, Figs 6–7 (illustrated, comp. note) China
Taona immaculata Chou, Lei, Li, Lu and Yao 1997: 133, Plate IV, Fig 52 (key, described, illustrated, comp. note) China
Taona immaculata Hua 2000: 64 (listed, distribution) China, Guangxi, Sichuan

***T. versicolor* Distant, 1909**
Taona versicolor Wu 1935: 20 (listed) China, Shensi
Taona versicolor Chou and Yao 1985: 123 (comp. note) China

Taona versicolor Wang and Zhang 1992: 120 (listed, comp. note) China
Taona versicolor Chou, Lei, Li, Lu and Yao 1997: 132–133 (key, described, type species of *Taona*, comp. note) China
Taona versicolor Hua 2000: 64 (listed, distribution) China, Shaanxi

***Chloropsalta* Haupt, 1920** = *Chloropsaltria* (sic) = *Chloropsatia* (sic)
Chloropsalta Koçak 1982: 148 (listed, comp. note) Turkey
Chloropsalta Kartal 1988: 9 (comp. note)
Chloropsalta Mirzayans 1995: 15 (listed) Iran
Chloropsalta Chou, Lei, Li, Lu and Yao 1997: 298 (comp. note)
Chloropsalta Mozaffarian and Sanborn 2010: 77 (distribution, listed) Turkey, Syria, Afghanistan, Iran, USSR, Western Asia

***C. ochreata* (Melichar, 1902)** = *Cicadatra ochreata* = *Chloropsatia* (sic) *ochreata*
Cicadatra ochreata Bâbâi 1968: 56, 58–59, 62 (natural history, distribution, control, comp. note) Iran, Uzbekistan, Turkmenistan
Chloropsalta ochreata Dubovsky 1982: 50 (listed, comp. note) Uzbekistan
Chloropsatia (sic) *ochreata* Emeljanov 1984: 10 (listed) USSR
Chloropsalta ochreata Emeljanov 1984: 56 (key) USSR
Cicadatra ochreata Kozhevnikova 1986: 45 (cotton pest) Uzbekistan
Chloropsalta ochreata Kozhevnikova 1986: 46 (cotton pest) Uzbekistan
Cicadatra ochreata Kholmuminov 1987: 49 (comp. note) Uzbekistan
Cicadatra ochreata Saboori and Lazarboni 2008: 57 (parasitism) Iran
Chloropsalta ochreata Mozaffarian and Sanborn 2010: 77 (listed, distribution) Iran, Persia, Afghanistan, Turkey, Turkestan, USSR

***C. smaragdula* Haupt, 1920**
Chloropsalta smaragdula Mirzayans, Hashemi, Borumand, Zairi and Rajabi 1976: 110 (distribution, listed) Iran
Chloropsalta smaragdula Dlabola 1981: 200 (distribution, comp. note) Iran
Chloropsalta smaragdula Kartal 1988: 9–15, Fig A_2, Fig B_2, Fig C_2, Fig D_2, Fig E_2, Fig F_2 (described, illustrated, comp. note) Turkey, Iran
Chloropsalta smaragdula Mirzayans 1995: 15 (listed, distribution) Iran
Chloropsalta smaragdula Schedl 2003: 426 (comp. note)
Chloropsalta smaragdula Mozaffarian and Sanborn 2010: 77 (listed, distribution) Iran, Mesopotamia, Turkey, Iraq

***C. viridiflava* (Distant, 1914)** = *Psalmocharias viridiflava* = *Cicadetta viridiflava* Horváth nom. nud. = *Cicadetta (Melampsalta) viridiflava* Horváth nom. nud. = *Chloropsaltria* (sic) *viridiflava*
Chloropsalta viridiflava Mirzayans, Hashemi, Borumand, Zairi and Rajabi 1976: 110 (distribution, listed) Iran
Chloropsalta viridiflava Dlabola 1981: 200 (distribution, comp. note) Iran
Psalmocharias viridiflava Koçak 1982: 148 (type species of *Chloropsalta*)
Chloropsalta viridiflava Kartal 1988: 9–13, 15, Fig A_1, Fig B_1, Fig C_1, Fig D_1, Fig E_1, Fig F_1 (described, illustrated, comp. note) Turkey, Afghanistan
Chloropsalta viridiflava Mirzayans 1995: 15 (listed, distribution) Iran
Psalmocharias viridiflava Mozaffarian and Sanborn 2010: 77 (type species of *Chloropsalta*)
Chloropsalta viridiflava Mozaffarian and Sanborn 2010: 77 (synonymy, listed, distribution) Equals *Psalmocharias viridiflava* Iran, Afghanistan, Turkestan, USSR

C. viridissima (Walker, 1858) to *Klapperischicen viridissima*

Subtribe Becquartinina Boulard, 2005 = Becquartinaria
Becquartinaria Boulard 2005e: 118 (n. subtribe, comp. note) Thailand
Becquartinaria Boulard 2005g: 365 (n. subtribe, comp. note)
Becquartinaria Sanborn, Phillips and Sites 2007: 3, 27, Table 1 (listed) Thailand
Becquartinina Boulard 2008f: 7, 51 (listed) Thailand
Becquartinina Lee 2008a: 2, 7, Table 1 (listed) Vietnam
Becquartinina Pham and Yang 2009: 13, Table 2 (listed) Vietnam

Sinopsaltria Chen, 1943 to *Becquartina* Kato, 1940
S. bifasciata Chen, 1943 to *Bequartina bifasciata* (Chen, 1943)

***Becquartina* Kato, 1940** = *Becquardina* (sic) = *Sinopsaltria* Chen, 1943
Becquardina (sic) Chou and Yao 1985: 123 (comp. note) China
Becquartina Duffels 1991a: 128 (comp. note)
Becquartina Chou, Lei, Li, Lu and Yao 1997: 128, 130, 293, 297–298, Table 6 (described, key, synonymy, listed, diversity, biogeoegraphy) Equals *Sinopsaltria* China

Becquartina Boulard 2005g: 365–366, 369 (synonymy, comp. note) Equals *Sinopsaltria* Equals *Becquardina* (sic)
Becquartina Boulard 2007e: 3, Fig 2b (illustrated, comp. note) Thailand
Becquartina Sanborn, Phillips and Sites 2007: 3, 27, Table 1 (listed) Thailand
Becquartina Shiyake 2007: 75 (comp. note)
Becquartina Boulard 2008f: 3–4, 7, 51, 96 (listed, comp. note) Thailand
Becquartina Lee 2008a: 2, 7, Table 1 (listed) Vietnam
Becquartina Pham and Yang 2009: 13, Table 2 (listed) Vietnam

***B. bifasciata* (Chen, 1943)** = *Sinopsaltria bifasciata* Chen, 1943
Sinopsaltria bifasciata Hua 2000: 64 (listed, distribution) China
Becquartina bifasciata Boulard 2005g: 366 (listed, n. comb.) Equals *Sinopsaltria bifasciata*

***B. bleuzeni* Boulard, 2005**
Becquartina bleuzeni Boulard 2005g: 371–372, 374, Fig 3 (n. sp., described illustrated, comp. note) Vietnam, Thailand
Becquartina bleuzeni Boulard 2005o: 149 (listed)
Becquartina bleuzeni Sanborn, Phillips and Sites 2007: 27 (listed, distribution, comp. note) Thailand
Becquartina bleuzeni Boulard 2008f: 51, 96 (biogeography, listed, comp. note) Thailand
Becquartina bleuzeni Lee 2008a: 7–8 (distribution, listed) Vietnam, North Thailand
Becquartina bleuzeni Pham and Yang 2009: 13, Table 2 (listed) Vietnam

***B. decorata* (Kato, 1940)** = *Gaeana ? decorata* = *Gaeana decorata* = *Gaeana decolata* (sic)
Gaeana decorata Chou and Yao 1985: 134 (type species of *Becquartina*, comp. note)
Gaeana decolata (sic) Chou and Yao 1985: 140 (type species of *Becquartina*, comp. note)
Gaeana decorata Hua 2000: 61 (listed, distribution) China, Guizhou

***B. electa* (Jacobi, 1902)** = *Gaeana electa* = *Gaeana electra* (sic)
Becquartina electa Chou and Yao 1985: 123, 134, 136, 140, Plate II, Fig 7 (illustrated, comp. note) China
Becquartina electa Peng and Lei 1992: 102, Fig 290 (listed, illustrated, comp. note) China, Hunan
Gaeana electa Chou, Lei, Li, Lu and Yao 1997: 130 (type species of *Becquartina*)
Becquartina electa Chou, Lei, Li, Lu and Yao 1997: 130–131, Fig 9–13 (key, described, illustrated, synonymy, comp. note) Equals *Gaeana electa* Equals *Gaeana (?) decorata* Equals *Becquartina decorata* Equals *Sinopsaltria bifasciata* China
Bequartina electa Hua 2000: 60 (listed, distribution, hosts) Equals *Gaeana electa* China, Fujian, Hunan, Guizhou, Yunnan
Becquartina electa Boulard 2005e: 118, 136–137, Plate G, Figs 1–2, Plate J, Fig 3 (type species of Becquartinaria, described, illustrated, song, sonogram, listed, comp. note) Equals *Becquartina decorata* Equals *Sinopsalta bifasciata* Tonkin, China, Thailand
Becquartina electa Boulard 2005g: 365 (type species of Bequartinaria, comp. note) Equals *Gaeana electa* Equals *Gaeana ? decorata* Kato Equals *Becquartina decorata* Equals *Sinopsaltria bifasciata*
Becquartina electa Boulard 2005o: 149, 151 (listed, n. comb.) Equals *Gaeana electa*
Becquartina electa Boulard 2007e: Plate 3, Fig 6, Plate 14 (illustrated, comp. note) Thailand
Becquartina electa Sanborn, Phillips and Sites 2007: 27 (listed, distribution, comp. note) Thailand, Vietnam, China
Becquartina electa Boulard 2008f: 51, 96 (type species of *Becquartina*, biogeography, listed, comp. note) Equals *Gaeana electa* Equals *Bequartina decorata* Equals *Sinopsalta bifasciata* Thailand, Tonkin
Gaeana electa Lee 2008a: 7 (type species of *Becquartina*)
Becquartina electa Lee 2008a: 7 (synonymy, distribution, listed) Equals *Gaeana electa* North Vietnam, China, Guizhou, Thailand
Becquartina electa Pham and Yang 2009: 13, Table 2 (listed) Vietnam

***B. ruiliensis* Chou & Yao, 1985** = *Becquartina uiliensis* (sic)
Becquartina ruiliensis Chou and Yao 1985: 134, 136, Fig 13, Plate II, Fig 8 (n. sp., described, illustrated, comp. note) China, Yunnan, Shaanxi
Becquartina uiliensis (sic) Chou and Yao 1985: 134 (illustrated) China
Becquartina ruileinsis Chou, Lei, Li, Lu and Yao 1997: 131 (key, described, comp. note) China
Bequartina ruiliensis Hua 2000: 60 (listed, distribution) China, Yunnan
Becquartina ruiliensis Boulard 2008f: 51, 96 (biogeography, listed, comp. note) Thailand

B. versicolor var. *chantrainei* Boulard, 2005 to *Becquartina versicolor* Boulard, 2005

B. versicolor var. *flammea* Boulard, 2005 to *Becquartina versicolor* Boulard, 2005

B. versicolor var. *fuscalae* Boulard, 2005 to *Becquartina versicolor* Boulard, 2005

B. versicolor var. *obscura* Boulard, 2005 to *Becquartina versicolor* Boulard, 2005

***B. versicolor* Boulard, 2005** = *Becquartina versicolor* var. *chantrainei* Boulard, 2005 = *Becquartina versicolor* var. *flammea* Boulard, 2005 = *Becquartina versicolor* var. *fuscalae* Boulard, 2005 = *Becquartina versicolor* var. *obscura* Boulard, 2005

Becquartina versicolor Boulard 2005a: 31 (coloration, comp. note)

Becquartina versicolor Boulard 2005g: 366–370, 373, Figs 1–2, Plate I, Figs A-B, Plate II (n. sp., described, illustrated, sonogram, coloration, comp. note) Thailand

Becquartina versicolor versicolor Boulard 2005g: 366–367, Plate I, Fig C (n. var., described, illustrated, coloration, comp. note)

Becquartina versicolor var. *chantrainei* Boulard 2005g: 366–367, 371, Plate I, Fig E (n. var., described, illustrated, coloration, comp. note) Thailand

Becquartina versicolor var. *flammea* Boulard 2005g: 366–367, 371, (n. var., described, coloration, comp. note) Thailand

Becquartina versicolor var. *fuscalae* Boulard 2005g: 366–367, 371, Plate I, Fig F (n. var., described, illustrated, coloration, comp. note) Thailand

Becquartina versicolor var. *obscura* Boulard 2005g: 366–367, 371, Plate I, Fig D (n. var., described, illustrated, coloration, comp. note) Thailand

Becquartina versicolor Boulard 2005o: 149 (listed)

Becquartina versicolor var. *chantrainei* Boulard 2005o: 149 (listed)

Becquartina versicolor var. *flammea* Boulard 2005o: 149 (listed)

Becquartina versicolor var. *fuscalae* Boulard 2005o: 149 (listed)

Becquartina versicolor var. *obscura* Boulard 2005o: 149 (listed)

Becquartina versicolor Boulard 2006g: 120 (comp. note)

Becquartina versicolor Boulard 2007e: 4, 66, 72, Plate 3, Fig 5 (illustrated, coloration, comp. note) Thailand

Becquartina versicolor Sanborn, Phillips and Sites 2007: 27 (listed, synonymy, distribution, comp. note) Equals *Becquartina versicolor* var. *chantrainei* Equals *Becquartina versicolor* var. *flammea* Equals *Becquartina versicolor* var. *fuscalae* Equals *Becquartina versicolor* var. *obscura* Thailand

Becquartina versicolor Boulard 2008f: 51, 96 (biogeography, listed, comp. note) Thailand

Becquartina versicolor var. *versicolor* Boulard 2008f: 52 (biogeography, listed, comp. note) Thailand

Becquartina versicolor var. *chantrainei* Boulard 2008f: 51 (biogeography, listed, comp. note) Thailand

Becquartina versicolor var. *flammea* Boulard 2008f: 51 (biogeography, listed, comp. note) Thailand

Becquartina versicolor var. *fuscalae* Boulard 2008f: 51 (biogeography, listed, comp. note) Thailand

Becquartina versicolor var. *obscura* Boulard 2008f: 52 (biogeography, listed, comp. note) Thailand

Tribe Lahugadini Distant, 1905 = Lahudadini (sic)

Lahugadini Moulds 2005b: 386, 427 (history, listed, comp. note)

***Lahugada* Distant, 1905**

***L. dohertyi* (Distant, 1891)** = *Pomponia dohertyi*

Tribe Cicadini Latreille, 1802 = Cicadaria = Psithyristriini Distant, 1905 = Psithyrisrini (sic) = Leptopsaltriini Moulton, 1923 = Leptopsaltrini (sic) = Pomponiini Kato, 1932 = Pomponiaria = Terpnosiina Kato, 1932 = Oncotympanini Ishihara, 1961

Cicadaria Wu 1935: 5 (listed) China

Leptopsaltriini Chen 1942: 146 (listed)

Cicadini Boulard 1982f: 184 (listed)

Cicadini Boulard 1984c: 177, 179 (comp. note)

Oncotympanini Hayashi 1984: 56 (listed)

Cicadini Hayashi 1984: 28, 47, 56, 58, Fig 3a (illustrated, listed, comp. note)

Cicadini Moulds 1985: 25 (listed, comp. note)

Cicadini Schedl 1986a: 3 (listed, key) Istria

Cicadini Boulard 1988f: 36, 44, 72 (comp. note)

Cicadini Boulard 1990b: 116 (comp. note)

Cicadini Moulds 1990: 31 (listed) Australia

Cicadini Chou, Lei and Wang 1993: 82, 85–86 (comp. note) China

Cicadini Boulard and Mondon 1995: 65–66 (key, listed) France

Cicadini Boulard 1997c: 105, 112, 117, 121 (comp. note)

Cicadini Chou, Lei, Li, Lu and Yao 1997: 106, 120 (key, listed, described)

Pomponiini Chou, Lei, Li, Lu and Yao 1997: 106, 187 (key, listed, described, synonymy) Equals Pomponiaria

Leptopsaltriini Chou, Lei, Li, Lu and Yao 1997: 106, 197 (key, listed, described, synonymy) Equals Leptopsaltriaria Equals Semiaria Equals Leptopsaltriini (sic)

Oncotympanini Chou, Lei, Li, Lu and Yao 1997: 106, 261 (key, listed, described)
Cicadini Novotny and Wilson 1997: 437 (listed)
Psithyristriini Novotny and Wilson 1997: 437 (listed)
Oncotympanini Novotny and Wilson 1997: 437 (listed)
Cicadini Moulds 1999: 1 (comp. note)
Cicadini Sanborn 1999a: 44 (listed)
Psithyristriini Sanborn 1999a: 44 (listed)
Cicadini Sanborn 1999b: 1 (listed) North America
Leptopsaltrini (sic) Villet and Zhao 1999: 322 (comp. note)
Cicadini Boulard 2000i: 21, 24, 35 (listed, key)
Cicadini Boulard 2001b: 26, 32, 37, 41 (comp. note)
Cicadini Puissant and Sueur 2001: 430 (listed) Corsica
Oncotympanini Sueur 2001: 39, Table 1 (listed)
Cicadini Sueur 2001: 39, Table 1 (listed)
Psithyristriini Sueur 2001: 40, Table 1 (listed)
Pomponiini Boulard 2002a: 39 (listed)
Psithyristiini Boulard 2002b: 196 (wing morphology, comp. note) Philippines
Cicadini Boulard 2002b: 198 (comp. note)
Leptopsaltriini Boulard 2002a: 49 (listed)
Cicadini Sueur 2002a: 380 (listed) Mexico, Veracruz
Cicadini Moulds and Carver 2002: 3 (listed) Australia
Cicadini Boulard 2003a: 98 (listed)
Pomponiini Boulard 2003a: 98 (listed)
Cicadini Boulard 2003c: 172 (listed)
Pomponiini Boulard 2003c: 172 (listed)
Cicadini Lee and Hayashi 2003a: 170 (status, listed, comp. note) Taiwan
Leptopsaltriini Chen 2004: 39–41 (phylogeny, listed)
Cicadini Sueur, Puissant, Simões, Seabra, Boulard and Quartau 2004: 178 (listed) Portugal
Cicadini Boulard 2005e: 118 (listed)
Cicadini Boulard 2005h: 7 (listed)
Leptopsaltriini Boulard 2005h: 7 (listed)
Cicadini Chawanji, Hodson and Villet 2005a: 258 (listed)
Oncotympanini Lee 2005: 14, 16, 156, 172 (key, listed, comp. note) Korea
Cicadini Moulds 2005b: 386, 390–391, 396, 410, 412–413, 423, 427, 430, 432, Table 2 (higher taxonomy, described, distribution, listed, comp. note) Australia
Oncotympanini Moulds 2005b: 386, 427 (history, listed, comp. note)
Terpnosiina Moulds 2005b: 386, 391, 432 (history, listed, comp. note) Equals Terpnosiaria
Psithyristriini Moulds 2005b: 427 (higher taxonomy, listed)
Pomponiini and Thinh 2005a: 240 (listed) Vietnam
Leptopsaltrini (sic) Pham and Thinh 2005a: 243 (listed) Vietnam
Cicadini Pham and Thinh 2005a: 245 (listed) Vietnam
Cicadini Pham and Thinh 2005b: 287–289 (key, comp. note) Equals Cicadatraria Equals Cicadaria Equals Cicadatrini Vietnam
Pomponiini Pham and Thinh 2005b: 287–289 (key, comp. note) Equals Pomponaria Vietnam
Leptopsaltrini (sic) Pham and Thinh 2005b: 287–290 (key, comp. note) Equals Leptopsaltraria Equals Semiaria Equals Leptopsaltriaria Equals Leptopsaltriini Vietnam
Cicadini Sanborn, Heath, Sueur and Phillips 2005: 191 (comp. note)
Cicadini Boulard 2006e: 129 (comp. note)
Cicadini Chawanji, Hodson and Villet 2006: 373 (listed)
Cicadini Puissant 2006: 19 (listed) France
Cicadini Sanborn 2006a: 77 (listed)
Leptopsaltriini Boulard 2007a: 99 (listed)
Pomponiini Boulard 2007d: 499 (listed)
Leptopsaltriini Boulard 2007d: 500 (listed)
Cicadini Boulard 2007e: 20 (comp. note) Thailand
Leptopsaltriini Duffels, Schouten and Lammertink 2007: 367 (comp. note)
Cicadini Sanborn 2007b: 37 (distribution, comp. note) Mexico
Cicadini Sanborn, Phillips and Sites 2007: 3, 11, Table 1 (listed) Thailand
Pomponiini Sanborn, Phillips and Sites 2007: 3, 27, 40, Table 1 (listed) Thailand
Cicadini Shiyake 2007: 6, 74 (listed, comp. note)
Psithyristrini Shiyake 2007: 6, 86 (listed, comp. note)
Cicadini Sueur and Puissant 2007b: 62 (listed) France
Leptopsaltriini Boulard 2008c: 356, 366 (listed)
Pomponiini Boulard 2008c: 354, 365 (listed)
Cicadini Boulard 2008c: 366 (listed)
Cicadini Boulard 2008f: 7, 46 (listed) Thailand
Pomponiini Boulard 2008f: 7, 36 (listed) Thailand
Leptopsaltriini Boulard 2008f: 7, 41 (listed) Thailand

Cicadini Demir 2008: 475 (listed) Turkey
Cicadini Lee 2008a: 1–2, 8, Table 1 (listed) Equals Dundubiini Vietnam
Oncotympanini Lee 2008b: 446, 449, 463, Table 1 (listed, comp. note) Korea
Cicadini Osaka Museum of Natural History 2008: 324 (listed) Japan
Leptopsaltriini Boulard 2009d: 39 (comp. note)
Cicadini Lee 2008b: 446, 450, 463, Table 1 (listed, comp. note) Korea
Cicadini Sanborn, Phillips and Gillis 2008: 13 (listed) Florida
Cicadini Lee 2009a: 87 (listed, comp. note)
Cicadini Lee 2009b: 330 (comp. note)
Cicadini Lee 2009c: 338 (comp. note)
Cicadini Lee 2009d: 470 (comp. note)
Cicadini Lee 2009e: 87 (comp. note)
Cicadini Lee 2009f: 1487 (comp. note)
Cicadini Lee 2009g: 306 (comp. note)
Cicadini Lee 2009h: 2621 (listed) Philippines, Palawan
Cicadini Lee 2009i: 294 (listed) Philippines, Panay
Cicadini Lee and Sanborn 2009: 31 (listed)
Pomponiini Pham and Yang 2009: 8 (listed) Vietnam
Cicadini Sanborn 2009b: 307 (comp. note) Vietnam
Cicadini Kemal and Koçak 2010: 4 (listed) Turkey
Cicadini Pham and Yang 2009: 10, 13, Table 2 (listed) Vietnam
Oncotympanini Lee 2010a: 2621 (listed) Philippines, Mindanao
Oncotympanini Lee 2010d: 167 (status, comp. note)
Oncotympanini Lee and Hill 2010: 301 (comp. note)
Cicadini Lee 2010a: 21 (listed) Philippines, Mindanao
Cicadini Lee 2010b: 21 (listed) Cambodia
Cicadini Lee 2010c: 3, 13, 15 (listed, comp. note) Philippines, Luzon
Cicadini Lee 2010d: 167, 171 (synonymy, listed, comp. note) Equals Oncotympanini
Cicadini Lee and Hill 2010: 278, 301–303 (synonymy, comp. note) Equals Psithyristriini
Psithyristriini Lee and Hill 2010: 278 (comp. note)
Cicadini Mozaffarian and Sanborn 2010: 77 (listed)
Cicadini Pham, Thinh and Yang 2010: 63 (listed) Vietnam
Cicadini Sueur, Janique, Simonis, Windmill and Baylac 2010: 931 (comp. note)
Cicadini Wei, Ahmed and Rizvi 2010: 28, 33 (comp. note)
Psithyristriini Wei, Ahmed and Rizvi 2010: 33 (comp. note)

Subtribe Cicadina Latreille, 1802 = Cicadaria

Cicadaria Boulard 1997c: 102, 105, 112 (comp. note)
Cicadaria Boulard 2000i: 21 (listed)
Cicadaria Boulard 2001b: 22, 26, 32, 27 (comp. note)
Cicadaria Boulard 2003a: 98 (listed)
Cicadaria Boulard 2003c: 172 (listed)
Cicadina Lee and Hayashi 2003a: 150, 170 (key, listed) Taiwan
Cicadina Lee and Hayashi 2003b: 359–360 (listed) Taiwan
Cicadina Lee and Hayashi 2004: 45–46 (listed) Taiwan
Cicadaria Boulard 2005h: 7 (listed)
Cicadina Lee 2005: 157, 175 (key, listed, comp. note) Korea
Cicadina Moulds 2005b: 386, 391 (higher taxonomy, listed, comp. note)
Cicadaria Sanborn, Phillips and Sites 2007: 3, 14, 37, Fig 2, Table 1 (listed, distribution) Thailand
Cicadina Boulard 2008f: 7, 46 (listed) Thailand
Cicadina Lee 2008a: 1–2, 8, Table 1 (listed) Vietnam
Cicadina Lee 2008b: 446, 450, 463, Table 1 (listed, comp. note) Korea
Cicadina Lee 2009a: 87 (comp. note)
Cicadina Lee 2009d: 470 (comp. note)
Cicadina Lee 2009f: 1487 (comp. note)
Cicadina Lee 2009g: 306 (comp. note)
Cicadina Lee 2009h: 2621 (listed) Philippines, Palawan
Cicadaria Pham and Yang 2009: 12 (listed) Vietnam
Cicadina Pham and Yang 2009: 13, Table 2 (listed) Vietnam
Cicadina Sanborn 2009b: 307 (comp. note) Vietnam
Cicadina Pham, Thinh and Yang 2010: 63 (listed) Vietnam
Cicadina Lee 2010a: 21 (listed) Philippines, Mindanao
Cicadina Lee 2010b: 21 (listed) Cambodia
Cicadina Lee 2010c: 3 (listed, comp. note) Philippines, Luzon
Cicadina Lee and Hill 2010: 278, 301–302 (comp. note)
Cicadina Wei, Ahmed and Rizvi 2010: 33 (comp. note)

***Cicada* Linnaeus, 1758** = *Cicada (Cicada)* = *Cicda* (sic) = *Gicada* (sic) = *Cicida* (sic) = *Cidada* (sic) = *Tettigia* Kolenati, 1857 = *Cicada (Tettigia)* = *Tettigia (Tettigia)* = *Tettigea* (sic) = *Teltigia* (sic) = *Testigia* (sic) = *Macroprotopus* Costa, 1877

Cicada Carus 1829: 143 (comp. note)
Cicada Dallas 1870: 482 (listed)
Cicada Swinton 1877: 78, 81 (sound production, comp. note) France
Cicada Swinton 1879: 81 (hearing, comp. note) Australia
Tettigonia Swinton 1880: 18 (comp. note)
Cicada Putnam 1881: 67–68 (comp. note)
Cicada Claus 1884: 567 (comp. note)
Cicada Hudson 1904: 94 (predation) New Zealand
Cicada Poulton 1907: 343 (predation) Natal
Cicada Anonymous 1933b: 184 (listed)
Cicada sp. Bergier 1941: 219 (human food, comp. note) China, Tonkin, Indochina, Australia
Cicada Gómez-Menor 1951: 61–62, 65 (key, comp. note) Equals *Tettigia* Spain
Cicada Soós 1956: 411 (listed) Equals *Tettigia* Romania
Cicada sp. Lepley, Thevenot, Guillaume, Ponel and Bayle 2004: 162 (predation) France
Cicada Medler 1980: 73 (listed)
Cicada Boulard 1981b: 48 (comp. note)
Cicada Boulard 1982f: 182 (listed)
Cicada Koçak 1982: 148 (listed, comp. note) Turkey
Macroprotopus Koçak 1982: 150 (listed, comp. note) Turkey
Tettigia Koçak 1982: 151 (listed, comp. note) Turkey
Cicada Strümpel 1983: 18 (listed)
Cicada sp. Strümpel 1983: 106, Fig 139.5 (wing illustrated)
Cicada Boulard 1984c: 166–167, 169–170, 172–176, 178–179 (type genus of Cicadinae, comp. note)
Tettigia Boulard 1984c: 176 (comp. note)
Cicada Hayashi 1984: 47 (listed, comp. note)
Cicada Melville and Sims 1984: 163, 180 (history)
Cicada Arnett 1985: 217 (diversity) Texas
Cicada Boulard 1985e: 1022 (coloration, comp. note)
Cicada Bregman 1985: 37 (comp. note)
Cicada Claridge 1985: 301, 310 (song, comp. note) Portugal
Cicada sp. Hussain and Amin 1985: 189–191, Figs 1–3 (electroretinogram, comp. note) India
Tettigia Moulds 1985: 25 (status comp. note)
Cicada Moulds 1985: 25 (comp. note) Equals *Tettigia*
Cicada Popov 1985: 38–39 (acoustic system, comp. note)
Cicada Lodos and Boulard 1987: 644 (comp. note) Turkey
Cicada Quartau 1987: 37 (comp. note) Portugal
Tettigonia Boulard 1988f: 15, 17, 63 (comp. note)
Cicada Boulard 1988f: 7, 15–18, 20–25, 30, 36, 44–45, 67 (comp. note)
Tettigonia Boulard 1988f: 15, 17, 63 (comp. note)
Cicada Boulard 1990b: 205 (comp. note) Syria, Egypt, Maghreb
Cicada Moulds 1990: 4, 25 (comp. note)
Cicada Piggozi 1991: 300 (predation) Italy
Cicada Popov, Aronov and Sergeeva 1991: 282 (comp. note)
Cicada Carpenter 1992: 222 (comp. note)
Cicada Popov, Arnov and Sergeeva 1992: 151 (comp. note)
Cicada Remane and Wachman 1993: 75, 134 (comp. note)
Cicada Quartau and Rebelo 1994: 141 (comp. note)
Cicada Boulard 1995a: 14 (listed)
Cicada Boulard and Mondon 1995: 19, 65 (key, comp. note) France
Cicada Riou 1995: 76 (comp. note)
Cicada Fonseca 1996: 28–29 (comp. note)
Cicada Holzinger 1996: 505 (listed) Austria
Cicada Moulds 1996b: 91 (comp. note)
Cicada Patterson Massei and Genov 1996: 141 (comp. note)
Cicada Boulard 1997c: 83, 85–86, 88–95, 97, 99–100, 102, 104–105, 112, 117, 119–120 (history, type genus of Cicadaria, Cicadini, Cicadinae, Gaeaninae, Cicadidae, comp. note)
Tettigonia Boulard 1997c: 85–86, 88, 105–106 (status, comp. note)
Tettigia Boulard 1997c: 85, 90–93 (status, comp. note)
Cicada Chou, Lei, Li, Lu and Yao 1997: 2 (comp. note)
Cicada Holzinger, Frölich, Günthart, Pauterer, Nickel et al. 1997: 49 (listed) Europe
Cicada sp. Llewellyn, Bell and Moczydlowski 1997: 895, Table 1 (comp. note)
Cicada Novotny and Wilson 1997: 437 (listed)
Cicada Poole, Garrison, McCafferty, Otte and Stark 1997: 251, 331 (listed) Equals *Tettigia* Equals *Macroprotopus* Equals *Tettigea* (sic) Equals *Cicda* (sic) North America
Cicada Quartau, Coble, Viegas-Crespo, Rebelo, Ribeiro, et al. 1997: 138 (comp. note) Portugal
Cicada sp. González, Alvarado, Durán, Serrano and De la Rosa 1998: 803–814, Figs 1–15 (olive pest, natural history, illustrated) Spain
Cicada González, Alvarado, Durán, Serrano and De la Rosa 1998: 803- (distribution, comp. note) France, North Africa, Italy, Portugal, Russia, Greece, Tunisia, Spain
Cicada Pinto, Quartau, Morgan and Hemingway 1998: 59 (comp. note)
Cicada Kritsky 1999: 3, 24 (comp. note)
Cicada Ohbayashi, Sato and Igawa 1999: 340 (parasitism, comp. note)

Cicada spp. Quartau, Rebelo and Simões 1999: 73 (comp. note)
Cicada Boulard 2000a: 255–256 (comp. note)
Cicada Boulard 2000b: 18 (comp. note)
Cicada Boulard 2000i: 20–21, 25, 60 (key, comp. note) France
Cicada Quartau, Ribeiro, Simões and Crespo 2000: 1677 (isozymes, comp. note) Portugal
Cicada Schedl 2000: 262 (listed)
Cicada Boulard 2001b: 4, 6, 8–13, 15, 17, 20, 24–26, 32, 37, 39–40 (history, type genus of Cicadaria, Cicadini, Cicadinae, Gaeaninae, Cicadidae, comp. note)
Tettigonia Boulard 2001b: 6, 8, 25 (status, comp. note)
Tettigia Boulard 2001b: 6, 11–13 (status, comp. note)
Cicada Boulard 2001d: 81 (comp. note)
Cicada Boulard 2001e: 114–115, 117–118 (comp. note)
Cicada Boulard 2001f: 129 (comp. note)
Cicada Kritsky 2001a: 187 (comp. note)
Cicada Boulard 2002b: 192 (comp. note) Europe
Cicada Puissant and Sueur 2002: 197 (comp. note) Slovenia
Cicada Salmah and Zaidi 2002: 226 (comp. note) Peninsular Malaysia
Cicada Seabra, Wilcox, Quartau and Bruford 2002: 173–174 (comp. note)
Cicada Dietrich 2003: 72 (comp. note)
Cicada Holzinger, Kammerlander and Nickel 2003: 476–477 (key, listed, comp. note) Europe
Cicada Sueur and Aubin 2003b: 481 (comp. note)
Cicada sp. Almeida 2004: 105 (control) Brazil
Cicada Boulard 2005k: 96 (comp. note)
Cicada spp. Jami, Kazempour, Elahinia and Khodakaramian 2005: 371 (plum pest, comp. note) Iran
Cicada Quartau and Simões 2005a: 489 (comp. note)
Cicada spp. Quartau and Simões 2005a: 491–492 (comp. note)
Cicada Quartau and Simões 2005b: 227–228, 232, 235, Fig 17.1 (habitat, comp. note)
Cicada spp. Quartau and Simões 2005b: 230, 233, Fig 17.9 (habitat, comp. note)
Cicada Salmah, Duffels and Zaidi 2005: 15 (comp. note)
Cicada Moulds 2005b: 384, 387, 390, 423, 427 (type genus of Cicadidae, type genus of Cicadinae, type genus of Cicadini, history, phylogeny, listed, comp. note)
Cicada (Cicada) Salmah, Duffels and Zaidi 2005: 15 (comp. note)
Cicada Sanborn, Heath, Sueur and Phillips 2005: 191, 193 (comp. note)
Cicada Yaakop, Duffels and Visser 2005: 247 (comp. note)
Cicada Boulard 2006e: 129 (comp. note)
Cicada Seabra, Pinto-Juma and Quartau 2006: 844 (comp. note)
Cicada Simões, Boulard and Quartau 2006: 17 (comp. note)
Cicada Simões and Quartau 2006: 52, 57 (comp. note)
Cicada Shiyake 2007: 6, 10, 82 (listed, comp. note)
Cicada Simões and Quartau 2007: 246 (comp. note)
Cicada Sueur and Puissant 2007b: 55 (comp. note)
Cicada Vignal and Kelly 2007: 487 (comp. note)
Cicada Aldini 2008: 104 (historical reference) Italy
Cicada Demir 2008: 475 (listed) Equals *Macroprotopus* Equals *Tettigia* Equals *Tibicen* Turkey
Cicada Fonseca, Serrão, Pina-Martins, Silva, Mira, et al. 2008: 25, 27, 30–31 (comp. note)
Cicada spp. Fonseca, Serrão, Pina-Martins, Silva, Mira, et al. 2008: 27, 29–30 (comp. note)
Cicada Sanborn, Moore and Young 2008: 1 (comp. note)
Cicada Sanborn, Phillips and Gillis 2008: 17 (comp. note)
Cicada Simões and Quartau 2008: 135, 137, 142, 144 (comp. note)
Cicada spp. Simões and Quartau 2008: 137, 142 (comp. note)
Cicada Pinto, Quartau, and Bruford 2009: 266, 268, 270–271, 273–274, 276, 282–283 (comp. note) Mediterranean
Cicada spp. Pinto, Quartau, and Bruford 2009: 274, 280 (comp. note)
Cicada Seabra, Quartau and Bruford 2009: 249–250 (comp. note)
Cicada spp. Seabra, Quartau and Bruford 2009: 259 (comp. note)
Cicada Simões and Quartau 2009: 393–403 (comp. note)
Cicada spp. Simões and Quartau 2009: 401 (comp. note)
Cicada Kemal and Koçak 2010: 3–4 (diversity, listed, comp. note) Equals *Tettigia* Equals *Macroprotopus* Turkey
Cicada Lee 2010c: 3 (type genus of Cicadina, comp. note)
Cicada Lee 2010d: 167 (type genus of Cicadini, comp. note)
Cicada Lee and Hill 2010: 302 (comp. note)
Cicada Mozaffarian and Sanborn 2010: 76, 80 (distribution, listed, comp. note) Europe, North Africa, Central Asia
Cicada Simões and Quartau 2010: 138–139 (comp. note)
Cicada Wei, Ahmed and Rizvi 2010: 33 (comp. note)

C. *albicans* Walker, 1858

C. *albiceps* Stephens, 1829 nom. nud.

C. *albida* Gmelin, 1789

C. *afona* Costa, 1834 of uncertain position
C. *albella* Billberg, 1820 nom. nud. of uncertain position
C. *alboguttata* Goeze, 1778 of uncertain position
C. *albula* Zetterstedt, 1828 of uncertain position
C. *ambigua* Donovan, 1798 of uncertain position
C. *americana* Gmelin, 1789 of uncertain position
C. *amphibia* Pontoppidan, 1763 of uncertain position
C. *apicalis* Billberg, 1820 nom. nud. of uncertain position
C. *aptera* Linnaeus, 1758 of uncertain position
C. *argentea* De Villers, 1789 of uncertain position
C. *arvensis* Müller, 1776 of uncertain position
C. *asius* Walker, 1850 = *Tympanoterpes asius*
C. *australensis* Kirkaldy, 1909 to *Illyria australensis*
C. *barbara barbara* (Stål, 1866) = *Tettigia barbara*

Cicada barbara Gómez-Menor 1951: 65 (comp. note) Spain
Cicada barbara Boulard 1981a: 37 (comp. note) Algeria
Cicada barbara Boulard 1982f: 182 (comp. note)
Tettigia barbara Boulard 1982f: 183 (comp. note)
Cicada barbara Nast 1987: 568–569 (distribution, listed) Spain, Italy
Cicada barbara Boulard 1990b: 205 (comp. note) Maghreb
Cicada barbara barbara Boulard 1995a: 12, 19–20, 23–24, Fig $S3_1$, Fig $S3_2$, Fig $S3_3$ (song analysis, sonogram, listed, comp. note) Algeria
Cicada barbara Boulard and Mondon 1995: 159 (listed) Algeria
Cicada barbara Arias Giralda, Pérez Rodríguez and Gallego Pintor 1998: 262–269 Figs 1–3 (olive pest, population dynamics, song)
Cicada barbara Guglielmino, d'Urso and Alma 2000: 167 (listed, distribution) Sardinia?, Spain, Sicily
Cicada barbara barbara Sueur 2001: 39, Table 1 (listed)
Cicada barbara Seabra, Wilcox, Quartau and Bruford 2002: 173–175, Table 1 (genetic variability, comp. note) Iberian Peninsula, North Africa
Cicada barbara Sueur, Puissant, Simões, Seabra, Boulard and Quartau 2004: 177 (comp. note) Portugal
Cicada barbara Quartau and Simões 2005a: 489 (comp. note) Algeria, Italy, Libya, Portugal, Sardinia, Sicily, Spain, Tunisia
Cicada barbara Quartau and Simões 2005b: 227–231, 235, 237, Fig 17.2, Fig 17.5, Fig 17.6a, Fig 17.7.5 (illustrated, sonogram, oscillogram, habitat, distribution, comp. note) Portugal, Spain, Morocco, Tunisia, Algeria, Italy, Sardinia, Sicily, Libya
Cicada barbara Boulard 2006g: 131–132 (comp. note)
Cicada barbara Seabra, Pinto-Juma and Quartau 2006: 843–851, Figs 1–5, Tables 1–3 (song variability, sonogram, oscillogram, comp. note) Iberian Peninsula, North Africa
Cicada barbara barbara Seabra, Pinto-Juma and Quartau 2006: 843 (comp. note) North Africa
Cicada barbara Simões, Boulard and Quartau 2006: 17 (comp. note)
Cicada barbara Simões and Quartau 2006: 51–55, 57–58, Fig 1.4, Fig 2, Table 1 (song, sonogram, oscillogram, comp. note)
Cicada barbara Oberdörster and Grant 2007b: 22 (comp. note)
Cicada barbara Pinto, Quartau, and Bruford 2008: 15–23, Figs 1–3, Tables 1–5 (gene sequence, comp. note) Portugal
Cicada barbara barbara Pinto, Quartau, and Bruford 2008: 16, 23 (comp. note) North Africa, Morocco
Cicada barbara Pinto-Juma, Seabra and Quartau 2008: 1–11, Fig 1, Figs 3–5, Tables 1–6 (song variation, comp. note) North Africa, Iberian Peninsula, Italy, Portugal, Spain, Morocco
Cicada barbara barbara Pinto-Juma, Seabra and Quartau 2008: 2–4, 9–11, Fig 2a, Fig 6b, Table 1, Table 7 (song, sonogram, oscillogram, comp. note) North Africa
Cicada barbara Quartau, André, Pinto-Juma, Seabra and Simões 2008: 128 (comp. note)
Cicada barbara Pinto, Quartau, and Bruford 2009: 266–283, 287–288, Figs 1–2, Figs 4–6, Tables 1–6, Appendix (gene sequence, phylogeny, comp. note) North Africa, Portugal, Spain, Morocco
Cicada barbara Seabra, Quartau and Bruford 2009: 250–254, 256–262, Figs 1–2, Figs 4–5, Table 1 (comp. note) North Africa, Iberian Peninsula
Cicada barbara barbara Seabra, Quartau and Bruford 2009: 260–261 (comp. note) Africa, Morocco
Cicada barbara Simões and Quartau 2009: 393–394, 397 (morphometric variation, comp. note)
Cicada barbara barbara Simões and Quartau 2009: 395–402, Figs 4–7, Tables 1–3, Tables 5–6, Table 8 (morphometric variation, comp. note) Algeria, Morocco
Cicada barbara Puissant and Sueur 2010: 556 (comp. note) Spain
Cicada barbara Simões and Quartau 2010: 138 (comp. note)

C. *barbara lusitanica* Boulard, 1982

Cicada barbara lusitanica Boulard 1982f: 182–185, 195–196, Figs 1–12, Fig 46 (n. subspecies, described, illustrated, sonogram, comp. note, listed) Portugal

Cicada barbara lusitanica Boulard 1985a: 33 (distribution, listed) Portugal
Cicada barbara lusitanica Nast 1987: 568–569 (distribution, listed) Portugal
Cicada barbara lusitanica Quartau 1987: 37 (numerical taxonomy, comp. note) Portugal
Cicada barbara lusitanica Quartau 1988: 171–173, 175–181, Figs 10–18, Figs 22–24, Table 1 (numerical taxonomy, comp. note) Portugal
Cicada barbara (sic) Quartau 1988: 171 (comp. note) Portugal
Cicada barbara lusitanica Quartau and Fonseca 1988: 367–368 (synonymy, distribution, comp. note) Equals *Cicada barbara* Portugal
Cicada barbara lusitanica Boulard 1990b: 166, 180, 205 (comp. note) Portugal
Cicada barbara lusitanica Fonseca 1991: 174–179, 189–190, Fig 2 (song analysis, oscillogram, comp. note) Portugal
Cicada barbara lusitanica Quartau and Rebelo 1994: 138–139, Figs 1–2, Figs 5–6 (song, oscillogram, comp. note) Portugal
Cicada barbara lusitanica Boulard 1995a: 12, 20, 24, Fig S4 (song analysis, sonogram, listed, comp. note) Portugal
Cicada barbara lusitanica Quartau 1995: 354–36 (sonogram, comp. note) Portugal
Cicada barbara lusitanica Cocroft and Pogue 1996: 93–94 (comp. note) Brazil
Cicada barbara lusitanica Fonseca 1996: 15, 22–23, 27, Fig 4, Tables 1–2 (timbal muscle physiology, oscillogram, comp. note) Portugal
Cicada barbara (sic) Fonseca 1996: 29 (comp. note)
Cicada barbara lusitanica Quartau, Ribeiro, Almeida and Crespo 1996: 138 (allozymes, comp. note) Portugal
Cicada barbara lusitanica Fonseca and Popov 1997: 418–427, Figs 1–10 (song structure, oscillogram, sound apparatus, comp. note) Portugal
Cicada barbara (sic) Fonseca and Popov 1997: 420, 422, 425–426 (song structure, oscillogram, sound apparatus, comp. note) Portugal
Cicada barbara (sic) Quartau, Coble, Viegas-Crespo, Rebelo, Ribeiro, et al. 1997: 138 (behavior, allozymes, morphometrics, comp. note) Portugal
Cicada barbara lusitanica Arias Giralda, Pérez Rodríguez and Gallego Pintor 1998: 262 (olive pest)
Cicada barbara lusitanica Quartau and Pinto 1998: 106 (mtDNA, comp. note) Portugal
Cicada barbara (sic) Ribeiro, Quartau, Simões, Fernandes, Rebelo et al. 1998: 106–107 (morphometric analysis, isozymes, comp. note) Portugal
Cicada barbara lusitanica Quartau, Rebelo and Simões 1999: 71 (comp. note)
Cicada barbara (sic) Quartau, Simões, Seabra, Drosopoulos, Claridge and Morgan 1999: 1 (song, gene sequences, comp. note) Portugal
Cicada barbara (sic) Quartau, Ribeiro, Simões and Crespo 2000: 1678–1683, Figs 1–3, Tables 1–3 (isozymes, comp. note) Portugal
Cicada barbara lusitanica Cooley and Marshall 2001: 851 (comp. note)
Cicada barbara (sic) Quartau, Ribeiro, Simões and Coehlo 2001: 99–105, Figs 1–2, Tables 1–3, Table 5 (isozymes, comp. note) Portugal
Cicada barbara lusitanica Sueur 2001: 39, Table 1 (listed)
Cicada barbara lusitanica Fonseca and Revez 2002a: 1285–1286 (song structure, oscillogram, sound apparatus, comp. note) Portugal
Cicada barbara (sic) Fonseca and Revez 2002a: 1285–1291, Figs 2–6 (song structure, oscillogram, sound apparatus, comp. note) Portugal
Cicada barbara (sic) Fonseca and Revez 2002b: 975 (comp. note)
Cicada barbara (sic) Quartau, Pinto, Seabra, Simões, Drosopoulos, et al. 2002: 21 (morphometric variation, comp. note) Portugal
Cicada barbara (sic) Quartau and Simões 2003: 118 (comp. note) Portugal
Cicada barbara lusitanica Sueur 2003: 2942 (comp. note)
Cicada barbara lusitanica Sueur and Aubin 2003b: 487 (comp. note)
Cicada barbara lusitanica Sueur, Puissant, Simões, Seabra, Boulard and Quartau 2004: 178–182, Figs 1–2 (distribution, ecology, song, sonogram, oscillogram, comp. note) Portugal
Cicada barabara lusitanica Puissant 2005: 312 (comp. note)
Cicada barbara lusitanica Seabra, Pinto-Juma and Quartau 2006: 843 (comp. note) Iberian Peninsula
Cicada barbara lusitanica Pinto, Quartau, and Bruford 2008: 16, 23 (comp. note) Portugal
Cicada barbara lusitanica Pinto-Juma, Seabra and Quartau 2008: 2–4, 9–10, Fig 2b, Fig 6a, Table 1, Table 7 (song, sonogram, oscillogram, comp. note) Portugal
Cicada barbara (sic) Fonseca, Serrão, Pina-Martins, Silva, Mira, et al. 2008: 19, 24, 26–28, 31, Fig 1, Table 1 (song phylogeny, comp. note) Portugal
Cicada barbara lusitanica Quartau 2008: 267 (habitat, comp. note) Portugal
Cicada barbara (sic) Seabra, André and Quartau 2008: 1–6, 8, Figs 1–2, Fig 4, Tables 1–4 (song

variability, sonogram, oscillogram, listed) Portugal

Cicada barbara lusitanica Sueur, Windmill and Robert 2008a: 2380, 2386 (comp. note)

Cicada barbara lusitanica Seabra, Quartau and Bruford 2009: 250, 260–261 (comp. note) Iberian Peninsula

Cicada barbara lusitanica Simões and Quartau 2009: 395–402, Figs 4–7, Tables 1–3, Tables 5–6, Table 8 (morphometric variation, comp. note) Portugal, Spain

Cicada barbara (sic) Simões and Quartau 2009: 398, Fig 4 (morphometric variation, comp. note) Portugal, Spain

***C. bella* Billberg, 1820 nom. nud.** of uncertain position

***C. biguttulus* Billberg, 1820 nom. nud.** of uncertain position

C. bilaqueata Uhler, 1903 to *Ariasa bilaqueata*

***C. bilineata* De Villers, 1789** of uncertain position

***C. brazilensis* Metcalf, 1963** = *Cicada vitrea* Germar, 1830 (nec *Cicada vitrea* Fabricius, 1803) of uncertain position

***C. bupthalmica* Gravenhorst, 1807** of uncertain position

C. burkei (Distant, 1882) to *Illyria burkei*

***C. cafra* Olivier, 1790** of uncertain position

***C. canaliculata* De Villiers, 1789** of uncertain position

***C. cancellata* Gmelin, 1789** = *Cicada reticulata* Müller, 1776 (nec *Cicicada reticulata* Linnaeus, 1758) of uncertain position

***C. carnaria* Stephens, 1829 nom. nud.** of uncertain position

***C. carniolica* Gmelin, 1789** = *Cicada lineata* Scopoli, 1763 (nec *Cicada lineata* Linnaeus, 1758) of uncertain position

***C. casmatmema* Capanni, 1894**

***C. casta* Gmelin, 1789** of uncertain position

***C. cerisyi* Guérin-Méneville, 1844** = *Cicada* cf. *cerisyi* Schedl, 1993

Cicada cerisyi Boulard 1990b: 205 (comp. note)

Cicada cf. *cerisyi* Schedl 1993: 797–798, 800, Figs 1–4 (illustrated, listed, synonymy, distribution, comp. note) Equals *Tettigia cerisyi* Egypt, Libya

Tettigia cerisyi Schedl 1993: 798 (comp. note) Egypt

Cicada cerisyi Schedl 1993: 798 (comp. note)

Cicada cerisyi Quartau and Simões 2005a: 489, 492 (comp. note) Egypt, Libya

Cicada cerisyi Quartau and Simões 2005b: 227–228, 232 (distribution, comp. note) Egypt, Libya

Cicada cerisyi Pinto, Quartau, and Bruford 2009: 267 (comp. note) Egypt

Cicada cerisyi Simões and Quartau 2009: 393 (comp. note) Egypt, Libya

***C. cinerea* Rossi, 1792**

***C. clarisona* Hancock, 1834** of uncertain position

***C. coeca* de Fourcroy, 1785** of uncertain position

***C. collaris* Degeer, 1773** of uncertain position

***C. complex* Walker, 1850**

***C. confusa* Metcalf, 1963** = *Cicada rufipes* Germar, 1830 (nec *Cicada rufipes* Fabricius, 1803)

***C. conspurcata* (Fabricius, 1777)** = *Tettigonia conspurcata* of uncertain position

***C. costa* Dohrn, 1859 nom. nud.** of uncertain position

***C. costai* Metcalf, 1963** = *Cicada flava* Costa, 1834 (nec *Cicada flava* Linnaeus, 1758) of uncertain position

***C. cretensis* Quartau & Simões, 2005**

Cicada cretensis Quartau and Simões 2005a: 489–493, Figs 1–2, Fig 3-I, Fig 4, Table 1, Tables 3–4 (n. sp., described, illustrated, sonogram, oscillogram, comp. note) Crete

Cicada cretensis Quartau and Simões 2005b: 227–228, 230–232, 234, 236–237, Fig 17.6e, Fig 17.7.3, Fig 17.10, Table 17.1 (illustrated, sonogram, oscillogram, habitat, distribution, comp. note) Portugal, Spain, Morocco, Tunisia, Algeria, Italy, Sardinia, Sicily, Libya

Cicada cretensis Seabra, Pinto-Juma and Quartau 2006: 850 (comp. note)

Cicada cretensis Simões, Boulard and Quartau 2006: 18 (comp. note)

Cicada cretensis Simões and Quartau 2006: 50–52, 54–55, Fig 1.2, Fig 2, Table 1 (song, sonogram, oscillogram, comp. note)

Cicada cretensis Simões and Quartau 2008: 137–140, 142–144, Fig 3, Figs 6–7, Table I, Tables III-IV (song variability, sonogram, oscillogram, distribution, comp. note) Crete

Cicada cretensis Pinto, Quartau, and Bruford 2009: 266–268, 271–278, 280–282, 287–288, Figs 1–2, Figs 4–6, Tables 1–4, Table 6, Appendix (gene sequence, phylogeny, comp. note) Crete

Cicada cretensis Seabra, Quartau and Bruford 2009: 250–253, 256–258, 260, 262, Fig 1, Fig 3, Table 1 (comp. note) Greece

Cicada cretensis Simões and Quartau 2009: 393–402, Figs 3–7, Tables 1–3, Tables 5–6, Table 8 (morphometric variation, comp. note) Greece, Crete

Cicada cretensis Simões and Quartau 2010: 138 (comp. note)

Cicada cretensis Trilar and Gogala 2010: 6 (comp. note) Crete

***C. crocata* Gmelin, 1789** of uncertain position

***C. dinigrata* Gmelin, 1789** of uncertain position

***C. devilliersi* Metcalf, 1963** = *Cicada maculata* De Villiers, 1789 (nec *Cicada maculata* Drury, 1773) of uncertain position

***C. dipalliata* Capanni, 1894** = *Cicada dippalliata* (sic)

***C. dwigubskyi* Metcalf, 1963** = *Cicada nebulosa* Dwigubsky, 1802 (nec *Cicada nebulosa* Fabricius, 1781) of uncertain position

***C. effecta* Walker, 1858** of uncertain position

***C. egregia* Uhler, 1903**

Cicada egregia Henshaw 1903: 229 (listed) Brazil

***C. elegans* Billberg, 1820 nom. nud.** of uncertain position
C. elongella Zetterstedt, 1840 nom. nud.
***C. erythroptera* Gmelin, 1789** of uncertain position
***C. fasciaaurata* De Villiers, 1789** of uncertain position
***C. flava* Linnaeus, 1758** of uncertain position
***C. flavicollis* Stephens, 1829 nom. nud.**
***C. flavifrons* Gmelin, 1789** of uncertain position
***C. flavofasciata* Goeze, 1778** = *Cicada fasciolata* de Fourcroy, 1785 of uncertain position
***C. fuschsii* Eversmann, 1837 nom. nud.** of uncertain position
***C. fuscomaculata* Goeze, 1778**
***C. fuscomaculosa* Metcalf, 1963** = *Cicada maculosa* Gmelin, 1789 of uncertain position
***C. geographica* Goeze, 1778** of uncertain position
***C. gibberula* Gmelin, 1789** of uncertain position
***C. gigantea* Germar, 1830** of uncertain position
***C. glabra* Müller, 1776** of uncertain position
***C. gmelini* Metcalf, 1963** = *Cicada flava* Gmelin, 1789 (nec *Cicada flava* Linnaeus, 1758) of uncertain position
***C. goezei* Metcalf, 1963** = *Cicada maculata* Goeze, 1778 (nec *Cicada maculata* Drury, 1773)
C. graminea Distant, 1904 to *Arunta interclusa* (Walker, 1858)
***C. guineensis* Metcalf, 1963** = *Cicada flavescens* Turton, 1802 (nec *Cicada flavescens* Fabricius, 1794) of uncertain position
***C. guttata* Forster, 1771** of uncertain position
Cicada guttata Arnett 1985: 217 (questionable record) United States
C. hilli (Ashton, 1921) to *Illyria hilli*
***C. inexacta* Walker, 1858** of uncertain position
***C. intacta* Walker, 1852** of uncertain position
***C. javanensis* Metcalf, 1963** = *Cicada undata* Germar, 1830 (nec *Cicada undata* Degeer, 1773) of uncertain position
***C. kirkaldyi* Metcalf, 1963** = *Cicada obtusa* Uhler, 1903 (nec *Cicada obtusa* Fabricius, 1787) = *Cicada brasiliensis* Kirkaldy, 1909 (nec *Cicada brasiliensis* Gmelin, 1789)
Cicada obtusa Henshaw 1903: 229 (listed) Brazil
Cicada obtusa Sanborn 1999a: 44 (type material, listed)
***C. lactea* Gmelin, 1789** of uncertain position
***C. leucophaea* Preissler, 1792** of uncertain position
***C. leucothoe* Walker, 1852** of uncertain position
***C. lineola* von Schrank, 1801**
***C. lineolata* Gmelin, 1789**
***C. lodosi* Boulard, 1979**
Cicada lodosi Lodos and Kalkandelen 1981: 75 (distribution, comp. note) Turkey
Cicada lodosi Nast 1982: 311 (listed) Turkey, Anatolia
Cicada lodosi Boulard 1995a: 7, 12, 21, 25, Fig 8, Fig $S5_1$, Fig $S5_2$ (illustrated, song analysis, sonogram, listed, comp. note) Turkey
Cicada lodosi Sueur 2001: 39, Table 1 (listed)
Cicada lodosi Quartau and Simões 2005a: 489, 492 (comp. note) Turkey
Cicada lodosi Quartau and Simões 2005b: 227–228, 230–232, 234, 237, Fig 17.6d, Fig 17.7.6, Fig 17.11 (illustrated, sonogram, oscillogram, habitat, distribution, comp. note) Turkey, Greece
Cicada lodosi Boulard 2006g: 131–132 (comp. note)
Cicada lodosi Seabra, Pinto-Juma and Quartau 2006: 850 (comp. note)
Cicada lodosi Simões and Quartau 2006: 51–52, 54, Fig 1.5, Fig 2, Table 1 (song, sonogram, oscillogram, comp. note)
Cicada lodosi Boulard 2007f: 140, Fig 121 (illustrated, comp. note) Anatolia
Cicada lodosi Demir 2007a: 52 (distribution, comp. note) Turkey
Cicada lodosi Demir 2008: 475 (distribution) Turkey
Cicada lodosi Simões and Quartau 2008: 137–138, Table I (song variability, sonogram, oscillogram, distribution, comp. note) Turkey
Cicada lodosi Pinto, Quartau, and Bruford 2009: 267–268, 271, 274–278, 280, 287–288, Fig 1, Figs 4–5, Table 1, Table 4, Appendix (gene sequence, phylogeny, comp. note) Turkey
Cicada lodosi Simões and Quartau 2009: 393–402, Figs 4–7, Tables 1–2, Tables 5–6, Table 8 (morphometric variation, comp. note) Turkey
Cicada lodosi Kemal and Koçak 2010: 4 (distribution, listed, comp. note) Turkey
Cicada lodosi Simões and Quartau 2010: 138 (comp. note)
***C. longicornis* Poda, 1761** of uncertain position
***C. lugubris* Gmelin, 1789** of uncertain position
***C. lutea* Gmelin, 1789** of uncertain position
***C. majuscula* Germar, 1818 nom. nud.** of uncertain position
***C. mannifica* Forskal, 1775**
***C. marginata* Sulzer, 1761** of uncertain position
***C. melanaria* Germar, 1830** of uncertain position
***C. melanoptera* Gmelin, 1789** = *Cicada marginata* Degeer, 1773 (nec *Cicada marginata* Sulzer, 1761) of uncertain position
Cicada melanoptera Gmelin Boulard 2000e: 120 (comp. note)
***C. membranacea* de Fourcroy, 1785** of uncertain position
***C. microcephala* La Guillou, 1841** of uncertain position
***C. mordoganensis* Boulard, 1979** = *Cicada oleae* (partim) = *Cicada* sp. nr. *orni* Quartau et al. 1998
Cicada mordoganensis Lodos and Kalkandelen 1981: 75 (distribution, comp. note) Turkey
Cicada mordoganensis Nast 1982: 311 (listed) Turkey, Anatolia

Cicada mordoganensis Boulard 1988e: 161 (comp. note) Turkey
Cicada mordoganensis Boulard 1995a: 7, 12, 19, 22, Fig $S2_1$, Fig $S2_2$ (song analysis, sonogram, listed, comp. note) Turkey
Cicada sp. nr. *orni* Quartau, Simões, Rebelo, Drosopoulos, Claridge and Morgan 1998: 66 (song, comp. note) Samos Island
Cicada mordoganensis Quartau, Simões, Seabra, Drosopoulos, Claridge and Morgan 1999: 1 (song, gene sequences, comp. note) Greece, Samos Island, Ikaria Island
Cicada mordoganensis Boulard 2000b: 32, Fig 20 (song, sonogram, comp. note) Turkey
Cicada mordoganensis Quartau, Seabra and Sanborn 2000: 197 (comp. note)
Cicada mordoganensis Seabra, Simões, Drosopoulos and Quartau 2000: 143–145, Fig 1, Tables 1–3 (genetic variability, comp. note) Greece
Cicada mordoganensis Sueur 2001: 39, Table 1 (listed)
Cicada mordoganensis Quartau, Pinto, Seabra, Simões, Drosopoulos, et al. 2002: 21 (morphometric variation, comp. note) Greece
Cicada mordoganensis Simões, Boulard, Rebelo, Drosopoulos, Claridge, Morgan and Quartau 2002: 437–439, Fig 1, Figs 3–6, Fig 11, Table 1 (song, sonogram, oscillogram, comp. note) Samos Island, Ikaria Island
Cicada mordoganensis Boulard 2005d: 340 (comp. note) Samos, Ikaria
Cicada mordoganensis Gogala, Trilar and Krpač 2005: 124 (comp. note) Greece
Cicada mordoganensis Pinto-Juma, Simões, Seabra and Quartau 2005: 82, 90 (comp. note)
Cicada mordoganensis Quartau and Simões 2005a: 489–493, fig 3-III, Fig 4, Tables 2–4 (sonogram, oscillogram, comp. note) Equals *Cicada oleae* Greece, Samos Island
Cicada mordoganensis Quartau and Simões 2005b: 227–228, 230–232, 234–237, Fig 17.6c, Fig 17.7.4, Fig 17.12, Table 17.1 (illustrated, sonogram, oscillogram, habitat, distribution, comp. note) Greece, Turkey, Samos Island
Cicada mordoganensis Boulard 2006g: 50, 52, 131–132, 180 (listed, comp. note) Samos, Ikaria
Cicada mordoganensis Simões, Boulard and Quartau 2006: 17 (comp. note)
Cicada mordoganensis Simões and Quartau 2006: 50–58, Fig 1.3, Fig 2, Table 1 (song, sonogram, oscillogram, comp. note)
Cicada mordoganensis Demir 2007a: 51 (distribution, comp. note) Turkey
Cicada mordoganensis Demir 2008: 476 (distribution) Turkey
Cicada mordoganensis Quartau, André, Pinto-Juma, Seabra and Simões 2008: 128–129 (comp. note)
Cicada mordoganensis Simões and Quartau 2008: 137–144, Fig 3, Fig 5, Fig 7, Table I, Table IV (song variability, sonogram, oscillogram, distribution, comp. note) Greece, Turkey
Cicada mordoganensis Pinto, Quartau, and Bruford 2009: 266–268, 271–278, 280–282, 287–288, Figs 1–2, Figs 4–6, Tables 1–4, Table 6, Appendix (gene sequence, phylogeny, comp. note) Turkey, Greece
Cicada mordoganensis Seabra, Quartau and Bruford 2009: 250–253, 256–258, 260, 262, Fig 1, Fig 3, Table 1 (comp. note) Greece
Cicada mordoganensis Simões and Quartau 2009: 393–402, Figs 4–7, Tables 1–3, Tables 5–6, Table 8 (morphometric variation, comp. note) Greece
Cicada mordoganensis Kemal and Koçak 2010: 4 (distribution, listed, comp. note) Turkey
Cicada mordoganensis Simões and Quartau 2010: 138 (comp. note)

***C. muscilliformis* Capanni, 1894**

***C. nervoptera* de Fourcroy, 1785** of uncertain position

***C. nigropunctata* Goeze, 1778**

***C. nodosa* (Walker, 1850)** = *Cicada nodosa* = *Diceropyga* (?) *nodosa* = *Diceropyga nodosa*

***C. notata* Scopoli, 1763** of uncertain position

***C. novella* Metcalf, 1963** = *Cicada gracilis* Germar, 1830 (nec *Cicada gracilis* Schellenberg, 1802) of uncertain position

***C. oculata* de Tigny, 1802** of uncertain position

***C. olivierana* Metcalf, 1963** = *Cicada ferruginea* Olivier, 1790 (nec *Cicada ferruginea* Fabricius, 1787) of uncertain position

***C. olivieri* Metcalf, 1963** = *Cicada guttata* Olivier, 1790 (nec *Cicada guttata* Foster, 1771) of uncertain position

***C. opercularis* Olivier, 1790** = *Fidicina opercularis* = *Cicada tibicen opercularis* = *Cicada variegata opercularis*

C. orientalis (Distant, 1912) to *Pomponia orientalis*

***C. orni* Linnaeus, 1758** = *Tettigonia orni* = *Tibicen orni* = *Cicida* (sic) *orni* = *Cidada* (sic) *orni* = *Cicada ornia* (sic) = *Tetigia (Tettigia) orni* = *Cicada (Tettigia) orni* = *Teltigia* (sic) *orni* = *Testigia* (sic) *orni* = *Cicadetta orni* = *Tettigonia punctata* Fabricius, 1798 = *Tettigia orni punctata* = *Tettigonia orni* var. *b* Billberg, 1820 nom. nud. = *Macroprotopus oleae* Costa, 1877 = *Cicada pallida* var. *a* Petagna,1787 = *Cicada orni lesbosiensis* Boulard, 2000

Tettigonia orni Carus 1829: 150, 158, Fig I, XVII (comp. note)
Tibicen orni Solier 1837: 213 (comp. note)
Tibicen orni Swinton 1877: 78–79 (sound production, comp. note) France
Cicada orni Swinton 1879: 81 (hearing, comp. note)

Cicada orni Swinton 1880: 25, 228, x, Plate VII, Fig 1 (illustrated, nervous system, song, comp. note)
Tettigonia orni Swinton 1880: 220, 222–223 (timbal, comp. note)
Cicada (Tettigonia) orni Hasselt 1881: 180 (comp. note)
Cicada orni Hasselt 1881: 192, 209 (comp. note)
Cicada orni Claus 1884: 571, Fig 474 (illustrated, comp. note)
Cicada orni Binet 1894: 580, Plate XV, Figs LXVI-LXIX (nervous system illustrated, comp. note)
Tettigonia orni Bachmetjew 1907: 251 (body length, comp. note)
Cicada orni Swinton 1908: 377, 379 (song, comp. note)
Tettigia orni Swinton 1908: 379–380, 382 (song, comp. note)
Tettigia orni Mader 1922: 179 (comp. note) Austria
Tettigia orni Marcu 1929: 479 (comp. note) Romania
Cicada orni Weber 1931: 93–94, 97–98, Fig 94, Fig 97 (larva illustrated, fungal host, comp. note)
Cicada orni Gómez-Menor 1951: 62, 65, Fig 14 (illustrated, described, listed, comp. note) Equals *Tettigia orni* Spain
Tettigonia orni Bator 1952: 201 (comp. note) Austria
Cicada orni Wagner 1952: 15 (listed, comp. note) Italy
Cicada orni Soós 1956: 411 (listed, distribution) Romania
Cicada orni Cloudsley-Thomopson 1957: 242–243, Fig 2 (illustrated, comp. note) France
Cicada orni Cloudsley-Thomopson and Sankey 1957: 420 (behavior, comp. note) France
Tettigonia orni Bunescu 1959: 89, 91, 99–100, Fig 3 (distribution, ecology, comp. note) Romania
Cicada orni Cantoreanu 1960: 332 (distribution, ecology, comp. note) Romania, Europe, Cyprus, Algeria, Anatolia, Transcaucasia
Cicada orni Cloudsley-Thomopson 1960: 50, 53 (behavior, comp. note) France
Cicada orni Weber 1968: 81, 83–84, 220, 302, Fig 163c (digestive system illustrated, comp. note)
Cicada orni Cloudsley-Thomopson 1970: 20 (comp. note)
Cicada orni Popov 1971: 302 (hearing, listed, comp. note)
Cicada orni Villiers 1977: 165, Plate XIV, Fig 21 (illustrated, listed, comp. note) France
Cicada orni Drosopoulous 1980: 190 (listed) Greece
Cicada orni Boulard 1981b: 47 (comp. note)
Cicada orni Dlabola 1981: 206 (distribution, comp. note) Iran
Cicada orni Lodos and Kalkandelen 1981: 75–76 (synonymy, distribution, comp. note) Equals *Tettigonia punctata* Equals *Macroprotopus oleae* Turkey, Albania, Austria, Cyprus, Czechoslovakia, Egypt, France, Greece, Hungary, Israel, Italy, Jordan, Lebanon, Romania, Switzerland, Spain, Balearic Island, Tunisia, USSR, Yugoslavia
Cicada orni Boulard 1982f: 182, 184, 195–196, Fig 45 (distribution, sonogram, comp. note, listed) Portugal
Cicada orni Boulard 1982g: 40, 42, 46–47, Plate I, Figs 1–3, Plate II, Fig 5, Figs 7–9, Plate III, Figs 1–3 (described, illustrated, sound system illustrated, comp. note) France
Cicada orni Koçak 1982: 148, 152 (type species of *Cicada*, type species of *Tettigia*)
Macroprotopus oleae Koçak 1982: 150 (type species of *Macroprotopus*)
Cicada orni Panov and Krjutchkova 1983: 141–147, Figs 3–6, Fig 10, Figs 12–13, Fig 15 (neurosecretory cells, comp. note) Crimea, Azerbaijan
Cicada orni Strümpel 1983: 4, 57, 70, 77, 88, 103, 118, 132, 168–169, Fig 132, Fig 156, Fig 186 (illustrated, comp. note)
Cicada orni Boulard 1984c: 166–167, 169–170, 176, 178–179 (comp. note)
Tettigia orni Boulard 1984c: 168 (member of Gaeaninae, comp. note)
Tettigonia orni Boulard 1984c: 171 (comp. note)
Cicada orni Boulard 1984d: 5, Plate, Fig 2 (larva illustrated, comp. note) France
Cicada orni Burton and Johnson 1984: 101 (song, comp. note) Europe, Germany, Czechoslovakia
Cicada orni Melville and Sims 1984: 164, 180 (type species of *Cicada*, history)
Cicada orni Arambourg 1985: 18, 19 (olive pest) Greece
Cicada orni Boulard 1985b: 59, 61–63 (illustrated, sound production, sonogram, characteristics, comp. note) France
Cicada orni Boulard 1985e: 1018, 1021, 1030, 1032, 1037, Plate I, Figs 1–2 (crypsis, illustrated, comp. note) France
Cicada orni Claridge 1985: 312 (comp. note) France, USSR
Cicada orni Michelsen and Larsen 1985: 513, Fig 15 (hearing organ, illustrated, comp. note)
Cicada orni Moulds 1985: 25 (type species of *Cicada*, type species of *Tettigia*, comp. note)
Cicada orni Popov 1985: 40–41, 68, 71, 73, 75–77, 80, 181–183, Fig 7.2, Fig 21.2, Fig 23.4, Fig 24, Fig 59.3 (illustrated, oscillogram, acoustic system, auditory system, comp. note)
Cicada orni Sander 1985: 593 (comp. note) Bulgaria

Cicada orni Boulard 1986c: 193 (comp. note)
Cicada orni Chinery 1986: 88–89 (illustrated, comp. note) Europe
Cicada orni Schedl 1986a: 3–4, 11–14, 22–24, Fig 21, Map 2, Table 1 (listed, key, synonymy, distribution, ecology, comp. note) Equals *Tettigonia orni* Equals *Tettigonia punctata* Equals *Macropterotypus oleae* Istria, Switzerland, Austria, Slovakia, Hungary, Caucasus, Transcaucasia, Ukraine, Turkmenistan, Iran, North Africa, Spain, France, Italy
Cicada orni Lodos and Boulard 1987: 644 (comp. note) Turkey
Cicada orni Jamieson 1987: 155–156 (sperm anatomy, comp. note)
Cicada orni Joermann and Schneider 1987: 284, 287–289, 294–295, Fig 4b, Figs 5–6, Table 2 (song analysis, sonogram, oscillogram, comp. note) Yugoslavia
Cicada orni Moschidis 1987: 1023–1026, Figs 1–2, Table III (phospholipids, comp. note) Greece
Cicada orni Nast 1987: 568–569 (distribution, listed) Czechoslovakia, France, Switzerland, Austria, Hungary, Ukraine, Romania, Russia, Portugal, Spain, Italy, Yugoslavia, Albania, Bulgaria, Greece
Cicada orni Quartau 1987: 37 (numerical taxonomy, comp. note) Portugal
Cicada orni Anufriev and Emelyanov 1988: 14 Fig 2.1 (antenna) Soviet Union
Cicada orni Boulard 1988a: 8–13, Table, Figs 3–4, Figs 7–8, Figs 1–3 (illustrated, comp. note) France
Cicada orni Boulard 1988d: 150–154, 156, Fig 1, Fig 5, Fig 7, Fig 13 (wing abnormalities, illustrated, comp. note) France
Cicada orni Boulard 1988f: 7, 16–17, 19, 20–22, 30, 36, 49, 65, 68, Fig 1 (type species of *Cicada*, Cicadinae, Gaeaninae and Cicadidae, comp. note)
Tettigonia orni Boulard 1988f: 23 (comp. note)
Cicada orni Boulard 1988g: 3–5, Plate I, Figs 1–3 (illustrated, larva illustrated, comp. note) France
Cicada orni Quartau 1988: 171–181, Figs 1–21, Table 1 (numerical taxonomy, comp. note) Portugal
Cicada orni Quartau and Fonseca 1988: 367–368 (synonymy, distribution, comp. note) Equals *Tettigonia punctata* Equals *Macroptopus oleae* Portugal, Albania, Austria, Cyprus, Czechoslovakia, Egypt, France, Greece, Crete, Hungary, Israel, Italy, Sardinia, Sicily, Jordan, Lebanon, Romania, Switzerland, Spain, Balearic Island, Tunisia, USSR, Armenia, Azerbaijan, Georgia, Russia, Turkmenia, Ukraine, Yugoslavia
Cicada orni Schedl 1989: 108–111, Fig 1, Fig 2a (illustrated, distribution, ecology, comp. note) Istria, Croatia
Cicada orni Boulard 1990b: 66, 68, 70–71, 75–77, 84, 87, 94–99, 112–115, 118–119, 121, 123, 125, 127, 129, 131–132, 135, 143, 165, 168–169, 177, 179, 198–199, 205, Figs 1–7, Plate V, Figs 1–8, Plate VII, Figs 1–8, Plate VII, Figs 1–6, Plate VIII, Figs 1–6, Plate IX, Figs 1–6, Plate XVI, Figs 1–2, Plate XI, Figs 1–4, Plate XII, Figs 1–6, Plate XIII, Figs 1–3, Plate XIV, Figs 1–5, Plate XXV, Fig 1 (larval anatomy, life cycle, genitalia, sound system, comp. note) France, Europe, Portugal, Spain
Cicada orni Popov 1990a: 302–303, Figs 2–3 (comp. note)
Cicada orni Tautz 1990: 105–106, Fig 12, Fig 13D (song frequency, hearing sensistivity, comp. note)
Cicada orni Boulard 1991d: 262 (oviposition damage, comp. note)
Cicada orni Fonseca 1991: 174, 178–179, 189–190, Fig 3 (song analysis, oscillogram, comp. note) Portugal
Cicada orni Patterson, Cavallini and Rolando 1991: 85–86 (predation, comp. note) Italy
Cicada orni Popov, Aronov and Sergeeva 1991: 282, 284, 289–293, 295, Fig 6, Table (auditory system, ecology, comp. note) Russia
Cicada orni Bairyamova 1992: 47, Table 1 (listed) Bulgaria
Cicada orni Boulard 1992c: 69 (comp. note)
Cicada orni Moalla, Jardak and Ghorbel 1992: 34–35 (comp. note) France, Italy, North Africa
Cicada orni Popov, Arnov and Sergeeva 1992: 152, 154, 159–163, 165, Fig 6, Table (auditory system, ecology, comp. note) Russia
Cicada orni Valier, Lucchi and Lovari 1992: 640 (predation) Italy
Cicada orni Villet 1992: 93 (comp. note) France
Cicada orni Moore 1993: 271 (comp. note)
Cicada orni Pillet 1993: 96–106, 112, Fig 1, Plate XI (described, illustrated, ecology, distribution, comp. note) Switzerland
Cicada orni Remane and Wachman 1993: 12–13, 134–135 (illustrated, comp. note)
Tettigia orni Schedl 1993: 798 (comp. note)
Cicada orni Schedl 1993: 798 (comp. note)
Cicada orni Hornig 1994: 65 (comp. note)
Cicada orni Lovari, Valier and Ricci Lucchi 1994: 323, 326–329, 332, 336, Fig 1, Fig 4, Table 1 (predation) Italy
Cicada orni Quartau and Rebelo 1994: 138–139, Figs 3–4 (song, comp. note) Portugal

Cicada orni Boulard 1995a: 6–8, 11–12, 14–18, Figs 2–3, Fig $S1_1$, Fig $S1_2$, Fig $S1_3$, Fig 7 (illustrated, song analysis, sonogram, listed, comp. note) France
Cicada orni Boulard and Mondon 1995: 7–10, 12–20, 24–33, 35–36, 44, 46–48, 50–53, 62, 64, 66–67, 70–72, 74, 120, 159 (anatomy, larva, illustrated, sonogram, listed, comp. note) Equals *Tettigonia punctata* Equals *Tettigia orni* France, Mediterranean
Cicada orni Holzinger 1995: 1124 (comp. note) Austria
Cicada orni Quartau 1995: 35–38 (sonogram, ecology, comp. note) Portugal
Cicada orni Gogala, Popov and Ribaric 1996: 46 (comp. note)
Cicada orni Holzinger 1996: 505 (listed) Austria
Cicada orni Jardak and Ksantini 1996: 26, Fig 1 (olive pest, comp. note) Tunisia
Cicada orni Patterson and Cavallini 1996: 206, 208–209, Fig 2, Tables 1–2 (song intensity, predation, comp. note) Italy
Cicada orni Patterson, Massei and Genov 1996: 141–145, Fig 2, Tables I-V (density, song intensity, comp. note) Italy
Cicada orni Quartau, Ribeiro, Almeida and Crespo 1996: 138 (allozymes, comp. note) Portugal
Cicada orni Vernier 1996: 147, 151 (comp. note) France
Cicada orni Boulard 1997c: 83, 86, 88–93, 99–100, 104–105, 112–113, 117, 120, Plate II, Fig 1 (type species of *Cicada*, Cicadinae, Gaeaninae and Cicadidae, comp. note)
Tettigonia orni Boulard 1997c: 86, 93 (comp. note)
Cicada orni Fonseca and Popov 1997: 425 (comp. note)
Cicada orni Genov and Massei 1997: 561–563, Fig 1 (larva predation) Itlay
Cicada orni Holzinger, Frölich, Günthart, Pauterer, Nickel et al. 1997: 49 (listed) Europe
Cicada orni Quartau, Coble, Viegas-Crespo, Rebelo, Ribeiro, et al. 1997: 138 (behavior, allozymes, morphometrics, comp. note) Portugal
Cicada orni Arias Giralda, Pérez Rodríguez and Gallego Pintor 1998: 262–264 (olive pest, comp. note) Spain
Cicada orni Gogala 1998b: 393, 398 (illustrated, oscillogram, comp. note) Slovenia
Cicada orni Pinto, Quartau, Morgan and Hemingway 1998: 59–60, 63–64, Fig 2, Table 1 (gene sequence, comp. note) Portugal
Cicada orni Quartau and Pinto 1998: 106 (mtDNA, comp. note) Portugal
Cicada orni Quartau, Simões, Rebelo, Drosopoulos, Claridge and Morgan 1998: 66 (song, comp. note) Greece
Cicada orni Ragge and Reynolds 1998: 476–478, Fig 1627 (oscillogram, comp. note) Europe
Cicada orni Ribeiro, Quartau, Simões, Fernandes, Rebelo et al. 1998: 106–107 (morphometric analysis, isozymes, comp. note) Portugal
Cicada orni Rolando 1998: 253 (predation) Italy
Cicada orni Okáli and Janský 1998: 34, 38, 42, Fig 1 (hosts, distribution, comp. note) Slovakia
Cicada orni Gogala and Gogala 1999: 120–121, 124, Fig 2 (distribution, comp. note) Slovenia, Mediterranean
Cicada orni Holzinger 1999: 431, 441 (listed) Austria
Cicada orni Jamieson, Dallai and Afzelius 1999: 235 (sperm anatomy, comp. note)
Cicada orni Quartau, Rebelo and Simões 1999: 71 (comp. note)
Cicada orni Quartau, Rebelo, Simões, Fernandes, Claridge, Drosopoulos and Morgan 1999: 71–80, Figs 1–17, Table 1 (song variability, oscillogram, sonogram, comp. note) Portugal, Greece, Europe, Iberian Peninsula, Turkey, Crimea, Near East
Cicada orni Quartau, Simões, Seabra, Drosopoulos, Claridge and Morgan 1999: 1 (song, gene sequences, comp. note) Portugal, Greece
Cicada orni Schedl 1999b: 824 (listed, distribution, ecology, comp. note) Israel, Mediterranean, Jordan, Iran
Cicada orni Boulard 2000a: 255 (type species of *Cicada*, comp. note)
Cicada orni Boulard 2000b: 18–19, 32, Fig 9, Fig 19 (song, sonogram, comp. note) Mediterranean, Turkey
Cicada orni lesbosiensis Boulard 2000b: 18–19, Fig 10 (n. ssp., song, sonogram, comp. note) Lesbos Island
Cicada orni Boulard 2000d: 82, 90–91, 95, Plate A, Figs 1–2, Plate E, Fig 6 (sound system, auditory receptor, song, comp. note)
Cicada orni Boulard 2000i: 4–5. 10–13, 20–21, 27, 35–37, 101, 114, Figs 1a–1d, Figs 2–5, Fig 12, Figs 23b–23c, Plate B, Map 2 (synonymy, illustrated, song, sonogram, oscillogram, frequency spectrum, distribution, ecology, predation, comp. note) Equals *Tettigonia punctata* Equals *Tibicen orni* Equals *Tettigia orni* Equals *Cicada (Tettigia) orni* Equals *Macropterus oleae* France
Cicada orni Gogala and Popov 2000: 11 (comp. note)
Cicada orni Guglielmino, d'Urso and Alma 2000: 167 (listed, distribution) Sardinia, Spain, France, Corsica, Italy, Sicily
Cicada orni Puissant and Sueur 2000: 69 (comp. note) France

Cicada orni Quartau, Ribeiro, Simões and Crespo 2000: 1678–1683, Figs 1–3, Tables 1–3 (isozymes, comp. note) Portugal
Cicada orni Quartau, Seabra and Sanborn 2000: 194–197, Fig 1, Tables I-III (song, oscillogram, comp. note) Portugal
Cicada orni Schedl 2000: 258, 262, 266–267, Plate 1, Fig 2, Map 2 (key, distribution, listed, comp. note) Switzerland, Italy, Austria, Slovakia, Hungary, Slovenia
Cicada orni Seabra, Simões, Drosopoulos and Quartau 2000: 143–145, Fig 1, Tables 1–3 (genetic variability, comp. note) Greece
Cicada orni Boulard 2001b: 4, 6, 9–14, 20, 24, 26, 32–33, 37, 39–40, Plate II, Fig 1 (type species of *Cicada*, Cicadinae, Gaeaninae and Cicadidae, comp. note)
Tettigonia orni Boulard 2001b: 8, 14 (comp. note)
Tettigia orni Boulard 2001b: 11 (comp. note)
Cicada orni Boulard 2001e: 117 (type species of *Cicada*, comp. note)
Cicada orni Coffin 2001a: 122 (comp. note)
Cicada orni Coffin 2001b: 128 (comp. note)
Cicada orni Ewart 2001b: 83, Table 3 (population density, comp. note) Italy
Cicada orni Krpač, Trilar and Gogala 2001: 29 (comp. note) Macedonia
Cicada (Tettigia) orni Levinson and Levinson 2001: 24 (comp. note)
Cicada orni Puissant 2001: 148 (comp. note)
Cicada orni Puissant and Sueur 2001: 430–432, Fig 1 (illustrated, distribution, comp. note) Corsica, France
Cicada orni Quartau, Ribeiro, Simões and Coehlo 2001: 99–105, Figs 1–2, Tables 1–5 (isozymes, comp. note) Portugal
Cicada orni Schedl 2001b: 27 (comp. note)
Cicada orni Sueur and Puissant 2001: 19, 22–23 (distribution, illustrated, life cycle, comp. note) France
Cicada orni Sueur 2001: 34, 39–40, 46, Figs 1–3, Table 1, Table 2 (sonogram, oscillogram, frequency spectrum, listed, comp. note)
Cicada orni lesbosiensis Sueur 2001: 40, Table 1 (listed)
Cicadetta (sic) *orni* Achtziger and Nigmann 2002: 4 Fig 2 (illustrated) Mediterranean
Cicada orni Boulard 2002b: 175, 193, 200, 202, 205, Fig 2 (nymph illustrated, coloration, life cycle, comp. note) France
Cicada orni Fonseca and Revez 2002a: 1286–1290, Figs 2–3 (song power spectrum, comp. note)
Cicada orni Gogala 2002: 243, 245, 248, Fig 2 (song, sonogram, oscillogram, comp. note)
Cicada orni Günther and Mühlenthaler 2002: 331 (listed) Switzerland
Cicada orni Kysela 2002: 22 (on jewelry)
Cicada orni Nickel, Holzinger and Wachmann 2002: 292, Table 4 (listed)
Cicada orni Nickel and Remane 2002: 37 (listed, hosts, distribution) Germany, Mediterranean
Cicada orni Quartau, Pinto, Seabra, Simões, Drosopoulos, et al. 2002: 21 (morphometric variation, comp. note) Portugal
Cicada orni Rakitov 2002: 125–126 (comp. note)
Cicada orni Schedl 2002: 233, Photo 2, Map 2 (illustrated, distribution, comp. note) Austria
Cicada orni Seabra, Wilcox, Quartau and Bruford 2002: 173–175, Table 1 (genetic variability, comp. note) Iberian Peninsula, Greece
Cicada orni Simões, Boulard, Rebelo, Drosopoulos, Claridge, Morgan and Quartau 2002: 437–439, Fig 2, Figs 7–11, Table 1 (song, sonogram, oscillogram, comp. note) Greece
Cicada orni Sueur 2002b: 111, 115, 117, 119–120, 123, 129 (sound system, illustrated, listed, comp. note)
Cicada orni Sueur and Aubin 2002b: 127–128, 130–135, Fig 5, Fig 7 (song, sonogram, oscillogram, comp. note) France
Cicada orni Boulard 2003a: 109 (comp. note)
Tettigonia orni Boulard 2003f: 372 (comp. note)
Cicada orni Gogala and Seljak 2003: 349–350 (illustrated, comp. note) Slovenia, Istria, Dalmatia
Cicada orni Holzinger, Kammerlander and Nickel 2003: 476–479, 606–607, 616–617, 628, 634, 636, Plate 37, Fig 8, Plate 42, Fig 2, Tables 2–3 (key, type species of *Cicada*, listed, comp. note) Equals *Tettigia orni* Europe, Mediterranean, Caucasus, Italy
Cicada orni Ibanez 2003: 3, 6, Fig (life cycle, comp. note) France
Cicada orni Nickel 2003: 70 (distribution, hosts, comp. note) Germany
Cicada orni Nickel and Remane 2003: 138 (listed, distribution) Germany
Cicada orni Quartau and Simões 2003: 114–115, 118 (comp. note) Portugal
Cicada orni Sueur 2003: 2942 (comp. note)
Cicada orni Sueur and Aubin 2003a: 322–325, Fig 1B, Fig 2 (song propagation, ecology, comp. note) France
Cicada orni Sueur and Aubin 2003b: 488 (comp. note)
Cicada orni Tishechkin 2003: 141–142, Figs 136–142 (oscillogram, comp. note) Chechnya
Cicada orni Vezinet 2003: 13 (comp. note) France
Cicada orni Villet, Sanborn and Phillips 2003a: 4 (comp. note)
Cicada orni Schedl 2004b: 916 (comp. note)

Cicada orni Sueur, Puissant, Simões, Seabra, Boulard and Quartau 2004: 178–182, Figs 1–2 (distribution, ecology, song, sonogram, oscillogram, comp. note) Portugal

Cicada orni Trilar and Holzinger 2004: 1385 (comp. note)

Cicada orni Boulard 2005d: 336–338, 340, 343, 345, 347, Figs 25.3–25.4, Fig 25.9 (auditory system, song, sonogram, illustrated, comp. note) Europe

Cicada orni Chawanji, Hodson and Villet 2005a: 258, 266 (comp. note)

Cicada orni Gogala, Trilar and Krpač 2005: 104–105, 107–108, Fig 2 (distribution, comp. note) Macedonia

Cicada orni Humphreys 2005: 74, Table 4 (population density, comp. note) Italy

Cicada orni Lee 2005: 95 (comp. note)

Cicada orni Moulds 2005b: 387, 390, 396, 398, 408, 410, 417–420, 422, 427, Fig 47, Figs 56–60, Table 1 (type species of Cicadidae, type species of Cicadinae, type species of *Cicada*, illustrated, higher taxonomy, phylogeny, key, listed, comp. note)

Cicada orni Pinto-Juma, Simões, Seabra and Quartau 2005: 81–92, Figs 1–6, Tables 1–6 (song variability, comp. note) Mediterranean, Iberian Peninsula, Greece, Turkey, Portugal, Spain, Corsica

Cicada orni Puissant and Defaut 2005: 116–118, 123–127, Table 1, Tables 7–10 (listed, habitat association, distribution, comp. note) France

Cicada orni Quartau and Simões 2005a: 489–493, fig 3-II, Fig 4, Tables 2–4 (sonogram, oscillogram, comp. note) Equals *Cicada oleae* Greece, Turkey, Iberian Peninsula

Cicada orni Quartau and Simões 2005b: 227–237, Figs 17.3–17.4, Fig 17.6b, Figs 17.7.1–17.7.2, Fig 17.8, Fig 17.13, Table 17.1 (illustrated, sonogram, oscillogram, habitat, distribution, comp. note) Portugal, Greece, Turkey, Albania, Armenia, Azerbaijan, Cyprus, Egypt, Georgia, Isreal, Jordan, Lebanon, Romania, Russia, Slovakia, Slovenia, Switzerland, Tunisia, Turkmenistan, Ukraine, Yugoslavia

Cicada orni Sanborn, Heath, Sueur and Phillips 2005: 191, 193 (comp. note)

Cicada orni Bernier 2006: 4–5, 7–8, Figs 2–3 (distribution, illustrated, oscillogram) France

Cicada orni Boulard 2006g: 20–21, 31, 33–36, 44–47, 50–51, 72, 96, 102, 115, 127, 131–132, 180, Fig 1, Figs 3–4, Fig 11, Fig 87A (sound system, auditory system, sonogram, illustrated, comp. note) Europe

Cicada orni Chawanji, Hodson and Villet 2006: 387 (comp. note)

Cicada orni Pierre and Francois 2006: 19 (comp. note) France

Cicada orni Puissant 2006: 19, 21, 25, 28–29, 32–35, 38, 40–43, 45, 48, 64, 67, 70, 84, 86, 89–90, 93–94, 108–119, 129, 131–163, 178–180, 185–189, Fig 8, Figs 13–16, Tables 1–2, Appendix Table 1, Appendix Tables 7–10 (illustrated, synonymy, distribution, habitat association, ecology, listed, comp. note) Equals *Tettigonia punctata* Equals *Macroprotopus oleae* France

Cicada orni Seabra, Pinto-Juma and Quartau 2006: 843–851, Figs 1–5, Tables 1–3 (song variability, sonogram, oscillogram, comp. note) Iberian Peninsula, Greece

Cicada orni Simões, Boulard and Quartau 2006: 17–21, Figs 1–3, Tables 1–4 (song variability, distribution, comp. note) Equals *Cicada orni lesbosiensis* Iberian Peninsula, Greece, Turkey, Near East

Cicada orni Simões and Quartau 2006: 48–58, Fig 1.1, Figs 2–5, Tables 1–2 (song, sonogram, oscillogram, comp. note) Portugal

Cicada orni lesbosiensis Simões, Boulard and Quartau 2006: 18 (status, comp. note) Equals *Cicada orni* Lesbos Island

Cicada orni Sueur, Windmill and Robert 2006: 4126 (comp. note)

Cicada orni Trilar, Gogala and Popa 2006: 177–178 (comp. note) Romania

Cicada orni Màjsky and Janský 2006: 74–75 (distribution, comp. note) Slovakia

Cicada orni Boulard 2007e: 85 (life cycle, comp. note) Europe

Cicada orni Chawanji, Hodson, Villet, Sanborn and Phillips 2007: 337, 345 (comp. note)

Cicada orni Hertach 2007: 51 (comp. note) Switzerland

Cicada orni Shiyake 2007: 6, 80, 82, Fig 134 (illustrated, distribution, listed, comp. note) Europe, Middle East, Africa

Cicada orni Simões and Quartau 2007: 246–251, Figs 1–7 (comp. note) Portugal

Cicada orni Sueur and Puissant 2007b: 62 (listed) France

Cicada orni Demir 2008: 476 (distribution) Equals *Cicada oleae* Equals *Cicada punctata* Turkey

Tettigia orni Demir 2008: 475 (comp. note) Turkey

Cicada orni Fonseca, Serrão, Pina-Martins, Silva, Mira, et al. 2008: 19, 24, 26–28, 31, Fig 1, Table 1 (song phylogeny, comp. note) Portugal

Cicada orni Hertach 2008a: 211 (comp. note) Switzerland

Cicada orni Hertach 2008b: 12, 14, 19, Fig 11 (illustrated, comp. note) Switzerland

Cicada orni Holzinger and Komposch 2008: 140 (comp. note) Austria
Cicada orni Nickel 2008: 207 (distribution, hosts, comp. note) Mediterranean
Cicada orni Pinto, Quartau, and Bruford 2008: 15 (comp. note) Portugal
Cicada orni Pinto-Juma, Seabra and Quartau 2008: 10–11 (comp. note)
Cicada orni Quartau, André, Pinto-Juma, Seabra and Simões 2008: 128–129, Fig 1 (song variation, comp. note) Greece, Turkey, Iberian Peninsula, France
Cicada orni Sanborn 2008c: 875, Fig 57 (illustrated, comp. note)
Cicada orni Seabra, André and Quartau 2008: 1–10, Fig 1, Fig 3, Fig 5, Tables 1–4 (song variability, sonogram, oscillogram, comp. note) Portugal
Cicada orni Simões and Quartau 2008: 137–138, 140–141, 143–144, Figs 3–4, Fig 7, Table I, Table IV (song variability, sonogram, oscillogram, distribution, comp. note) Greece, Turkey
Cicada orni Sueur, Pavoine, Hemerlynck and Duvall 2008: 4, Fig 1 (comp. note) France
Cicada orni Sueur, Windmill and Robert 2008a: 2380–2386, Figs 1–5 (tympanum mechanics, comp. note) France
Cicada orni Sueur, Windmill and Robert 2008b: 59 (comp. note)
Cicada orni Trilar and Gogala 2008: 29 (listed, distribution, comp. note) Romania
Cicada orni Vikberg 2008: 61 (comp. note)
Cicada orni Zorba, Stratakis, Barberoglou, Spanakis, Tzanetakis et al. 2008a: 4049 (comp. note)
Cicada orni Zorba, Stratakis, Barberoglou, Spanakis, Tzanetakis et al. 2008b: 819 (comp. note)
Cicada orni Barberoglou, Zorba, Stratakis, Spanakis, Tzanetakis, Anastasiadis and Fotakis 2009: 5425 (comp. note)
Cicada orni Dourlet 2009: 84–85 (life history, illustrated)
Cicada orni Gogala, Drosopoulos and Trilar 2009: 19, 23 (comp. note)
Cicada orni Ibanez 2009: 212 (listed, comp. note) France
Cicada orni Ojie Ardebilie and Nozari 2009: A13, A15-A20, Fig 3, Tables 1–2 (song, comp. note)
Cicada orni Oliver 2009: 146 (predation)
Cicada orni Pinto, Quartau, and Bruford 2009: 266–283, 287–288, Figs 1–2, Figs 4–6, Tables 1–6, Appendix (gene sequence, phylogeny, comp. note) Europe, Portugal, France, Croatia, Slovenia, Macedonia, Greece
Cicada orni Seabra, Quartau and Bruford 2009: 250–254, 256–262, Figs 1–3, Fig 5, Table 1 (genetic variability, comp. note) Southern Europe, West Asia, Middle East, France, Greece, Iberian Peninsula
Cicada orni Simões and Quartau 2009: 393–402, Figs 4–7, Tables 1–3, Tables 5–6, Table 8 (morphometric variation, comp. note) Cyprus, Portugal, France, Greece, Turkey
Cicada orni Windmill, Sueur and Robert 2009: 4082 (comp. note)
Cicada orni Kemal and Koçak 2010: 4 (distribution, listed, comp. note) Turkey
Cicada orni Lee and Hill 2010: 299–300, 302, Figs 7–8 (phylogeny, comp. note)
Cicada orni Mozaffarian and Sanborn 2010: 80 (type species of *Cicada*, synonymy, listed distribution) Equals *Tettigonia punctata* Equals *Macropterus oleae* Iran, Italy, Europe, North Africa, Western Asia
Cicada orni Obrist, Pavan, Sueur, Riede, Liusia and Márquez 2010: 72 (sonogram, oscillogram)
Cicada orni Puissant and Sueur 2010: 556 (comp. note) Spain
Cicada orni Simões and Quartau 2010: 138 (comp. note)
Cicada orni Sueur, Janique, Simonis, Windmill and Baylac 2010: 923–932, Figs 1–9, Table 1 (tympanum structure, comp. note) France
Cicada orni Sueur, Windmill and Robert 2010: 1681–1687, Figs 1–2, Fig 5, Tables I-III (auditory processing, sonogram, comp. note) France

***C. osalbum* De Villiers, 1789** of uncertain position

***C. palearctica* Metcalf, 1963** = *Cicada hyalina* Gmelin, 1789 (nec *Cicada hyalina* Fabricius, 1775)

C. palisoti Metcalf, 1963 to *Proarna palisoti*

***C. pallens* Gmelin, 1789**

***C. pallescens* Müller, 1776** of uncertain position

***C. pallida* Goeze, 1778** = *Cicada pallida* var. *A* de Fourcroy, 1785 = *Cicada pallida* var. *B* de Fourcroy, 1785 = *Cicada pallida* var. *C* de Fourcroy, 1785 of uncertain origin

C. pallida var. *a* Petagna, 1787 to *Cicada orni* Linnaeus, 1758

***C. pennata* (Distant, 1881)** = *Tettigia pennata*
Cicada pennata Sanborn 2005b: 77 (listed, distribution) Guatemala

***C. permagna* (Haupt, 1917)** = *Tettigia permagna*
Cicada permagna Dlabola 1981: 206 (distribution, comp. note) Turkey
Cicada permagna Lodos and Kalkandelen 1981: 76 (distribution, comp. note) Turkey
Cicada permagna Lodos and Boulard 1987: 644 (comp. note) Turkey
Cicada permagna Quartau and Simões 2005a: 489, 492 (comp. note) Turkey
Cicada permagna Quartau and Simões 2005b: 227–228, 232 (distribution, comp. note) Turkey

Cicada permagna Demir 2007a: 52 (distribution, comp. note) Turkey
Cicada permagna Demir 2008: 475 (distribution) Turkey
Cicada permagna Pinto, Quartau, and Bruford 2009: 267 (comp. note) Turkey
Cicada permagna Simões and Quartau 2009: 393 (comp. note) Turkey
Cicada permagna Kemal and Koçak 2010: 4 (distribution, listed, comp. note) Turkey

***C. poae* Capanni, 1849**
***C. prasina* Eversmann, 1837 nom. nud.** of uncertain position
***C. purpurescens* Metcalf, 1963** = *Cicada coerulescens* Germar, 1830 (nec *Cicada coerulescens* Fabricius, 1803)
***C. pusilla* Müller, 1776** of uncertain position
***C. quadrinotata* Gmelin, 1789** = *Cicada quadrimaculata* Müller, 1776 (nec *Cicada quadrimaculata* von Schrank, 1776) of uncertain position
***C. quadristrigata* Gmelin, 1789** = *Cicada 4-striata* of uncertain position
C. queenslandica Kirkaldy, 1909 to *Arunta interclusa* (Walker, 1858)
***C. regalis* de Fourcroy, 1785** of uncertain position
***C. regina* Graber, 1876**
***C. rivulosa* Gmelin, 1789** of uncertain position
***C. roselii* Gravenhorst, 1807** of uncertain position
***C. rubiniae* Capanni, 1894 nom. nud.**
***C. rubroelytrata* Goeze, 1778** of uncertain position
***C. rufgescens* Stephens, 1829 nom. nud.** of uncertain position
***C. sahlbergi* Stål, 1854** = *Cicada (Cicada) sahlbergi* of uncertain position
***C. scutata* Goeze, 1778** of uncertain position
***C. sexpunctata* Gmelin, 1789** of uncertain position
***C. sebai* Metcalf, 1963** = *Cicada fusca* Goeze, 1778 (nec *Cicada fusca* Müller, 1776)
***C. signata* Billberg, 1820 nom. nud.** of uncertain position
C. signoreti Metcalf, 1955 to *Antankaria signoreti*
***C. smaragdina* Capanni, 1894**
***C. squalida* Müller, 1776** of uncertain position
***C. strigosa* Gmelin, 1789** of uncertain position
***C. superba* Billberg, 1820 nom. nud.** of uncertain position
***C. thomasseti* China, 1924**
***C. thalassina* Germar, 1830** of uncertain position
***C. thlaspi* Piller & Mitterpacher, 1783** of uncertain position
***C. torquata* de Fourcroy, 1785** of uncertain position
***C. triangulum* De Villiers, 1789** of uncertain position
***C. tristis* de Fourcroy, 1785** of uncertain position
***C. turtoni* Metcalf, 1963** = *Cicada 3-punctata* Turton, 1802 (nec *Cicada tripunctata* Fabricius, 1781)
***C. variola* Gravenhorst, 1807** of uncertain position
***C. venulosa* Turton, 1802** of uncertain position
***C. vertumnus* Stephens, 1829 nom. nud.**
***C. viridiflava* McAtee, 1934 nom. nud.** of uncertain position
***C. vitrata* De Villiers, 1789** of uncertain position
C. xanthospila Germar, 1830 to *Bergalna xanthospila*

***Cicadalna* Boulard, 2006**

Cicadalna Boulard 2006e: 529 (n. gen., described, comp. note) Thailand
Cicadalna Boulard 2008f: 7, 46, 95 (listed) Thailand

***C. takensis* Boulard, 2006** = *Cicadalna takensis* mph. *suffusca* Boulard, 2006

Cicadalna takensis Boulard 2006b: 529–533, Figs 1–8, Plate, Figs B-C (n. sp., described, illustrated, song, sonogram, comp. note) Thailand
Cicadalna takensis mph. *suffusca* Boulard 2006b: 532–533, Plate, Fig E (sic actually D) (n. sp., n. morph, described, illustrated, comp. note) Thailand
Cicadalna takensis Boulard 2006g: 132, 146, Fig 107 (comp. note) Thailand
Cicadalna takensis Boulard 2007d: 494 (comp. note) Thailand
Cicadalna takensis Boulard 2008f: 46, 95 (type species of *Cicadalna*, biogeography, listed, comp. note) Thailand

***Inthaxara* Distant, 1913**

Inthaxara Lei and Li 1996: 412 (comp. note)
Inthaxara Chou, Lei, Li, Lu and Yao 1997: 197, 233, 293, 297, Table 6 (key, described, listed, diversity, biogeography, comp. note) China
Inthaxara Moulds 2005b: 391, 432 (higher taxonomy, listed, comp. note)
Inthaxara Pham and Thinh 2005a: 243 (listed) Vietnam
Inthaxara Pham and Thinh 2005b: 290 (comp. note) Vietnam
Inthaxara Lee 2008a: 2, Table 1 (listed) Vietnam
Inthraxa Lee 2009f: 1487 (comp. note)
Inthaxara Pham and Yang 2009: 12, 14, Table 2 (listed) Vietnam
Inthaxara Lee and Hill 2010: 302 (comp. note)

***I. flexa* Lei & Li, 1996**

Inthaxara flexa Lei and Li 1996: 410–412, Figs 1–2 (n. sp., described, illustrated, key, comp. note) China, Fujian
Inthaxara flexa Chou, Lei, Li, Lu and Yao 1997: 233–234, Fig 9–74, Plate VI, Fig 73 (n. sp., key, described, illustrated, comp. note) China
Inthaxara flexa Pham and Thinh 2005a: 240 (distribution, listed) Vietnam
Inthaxara flexa Pham and Yang 2009: 12, 14, 16, Table 2 (listed, distribution, comp. note) Vietnam, China

***I. olivacea* Chen, 1940**

Inthaxara olivacea Lei and Li 1996: 411–412 (comp. note)

Inthaxara olivacea Chou, Lei, Li, Lu and Yao 1997: 233–234 (key, described, comp. note) China
Inthaxara olivacea Hua 2000: 62 (listed, distribution) China, Sichuan

I. rex **Distant, 1913**
Inthaxara rex Chou, Lei, Li, Lu and Yao 1997: 233 (type species of *Inthaxara*) China
Inthaxara rex Pham and Thinh 2005a: 243 (type species of *Inthaxara*)
Inthaxara rex Lee 2008a: 12 (type species of *Inthaxara*, distribution, listed) Vietnam, Laos
Inthaxara rex Pham and Yang 2009: 12, 14, Table 2 (type species of *Inthaxara*, listed) Vietnam
Inthaxara rex Pham and Yang 2010a: 133 (comp. note) Vietnam

Rustia **Stål, 1866**
Rustia Moulds 2005b: 391, 432 (higher taxonomy, listed, comp. note)
Rustia Sanborn, Phillips and Sites 2007: 3, 13, Table 1 (listed) Thailand
Rustia Boulard 2008f: 7, 47, 95 (listed) Thailand
Rustia Lee 2008a: 2, 15, Table 1 (listed) Vietnam
Rustia Pham and Yang 2009: 14, Table 2 (listed) Vietnam
Rustia Lee 2010b: 25 (listed) Cambodia

R. dentivitta **(Walker, 1862)** = *Cicada dentivitta* = *Rustia pedunculata* Stål, 1866 = *Tibicen amussitata* Distant, 1892 = *Rustia dentivitta amussitata* = *Rustia dentivitta* var. *amussitata*
Rustia dentivitta Sanborn, Phillips and Sites 2007: 13 (listed, synonymy, distribution, comp. note) Equals *Cicada dentivitta* Equals *Rustia dentivitta amussitata* Thailand, Cambodia, Indonesia, Borneo, Philippine Republic, Korea
Rustia dentivitta Boulard 2008c: 366 (comp. note) Thailand
Rustia dentivitta Boulard 2008f: 47, 95 (type species of *Rustia*, biogeography, listed, comp. note) Equals *Rustia pedonculata* Thailand, Siam, Cambodia, Assam, Burma, India, Himalaya, Indochina, Nepal
Rustia dentivitta Lee 2008a: 15 (synonymy, distribution, listed) Equals *Cicada dentivitta* Vietnam, Cambodia, Thailand, Myanmar, Nepal, India
Rustia dentivitta Pham and Yang 2009: 14, Table 2 (listed) Vietnam
Cicada dentivitta Lee 2010b: 25 (type species of *Rustia*)
Rustia dentivitta Lee 2010b: 25 (synonymy, distribution, listed) Equals *Cicada dentivitta* Equals *Rustia penduncula*ta Equals *Tibicen amuissitatus* Equals *Rustia dentivitta* var. *amussitata* Cambodia, Vietnam, Thailand, Myanmar, Nepal, India

R. tigrina **(Distant, 1888)** = *Tibicen tigrinus*
Rustia tigrina Chou, Lei, Li, Lu and Yao 1997: 218 (comp. note)

Subtribe Oncotympanina Ishihara, 1961
Oncotympanina Lee 2010d: 167 (n. status, comp. note) Philippines

Dokuma Distant, 1905 to *Oncotympana* Stål, 1870
D. nigristigma (Walker, 1850) to *Oncotympana nigristigma*
D. consobrina Distant, 1906 to *Oncotympana viridicincta* (Stål, 1870)

Oncotympana **Stål, 1870** = *Pomponia (Oncotympana)* = *Octotymana* (sic) = *Onctympana* (sic) = *Onchotympana* (sic) = *Acotympana* (sic) = *Dokuma* Distant, 1905
Oncotympana Strümpel 1983: 109, Table 8 (comp. note)
Oncotympana Hayashi 1984: 28, 56–57 (key, listed, comp. note) Equals *Pomponia (Oncotympana)*
Dokuma Duffels 1986: 319 (distribution, diversity, comp. note) Philippines
Oncotympana Boulard 1990b: 143 (comp. note)
Dokuma Hayashi 1993: 18 (comp. note)
Oncotympana Hayashi 1993: 18 (comp. note)
Oncotympana Novotny and Wilson 1997: 437 (listed)
Onchotympana (sic) Boulard 2002b: 198 (comp. note) Philippines
Dokuma Moulds 2005b: 391, 432 (higher taxonomy, listed, comp. note)
Oncotympana Wang, Ren, Liang, Liu and Wang 2006: 295, Table 1 (listed) China
Oncotympana Lee 2010a: 19–21 (synonymy, diversity, listed, comp. note) Equals *Pomponia (Oncotympana)* Equals *Dokuma* n. syn. Philippines, Mindanao
Dokuma Lee 2010a: 19 (status, listed, comp. note)
Oncotympana Lee 2010c: 3 (synonymy, diagnosis, key, listed, comp. note) Equals *Pomponia (Oncotympana)* Philippines, Luzon
Oncotympana Lee 2010d: 167–169 (synonymy, diversity, listed, comp. note) Equals *Pomponia (Oncotympana)* Equals *Dokuma* Philippines
Dokuma Lee 2010d: 168 (listed, comp. note)

O. averta **Lee, 2010**
Oncotympana averta Lee 2010d: 168–169, Fig 1 (n. sp., described, illustrated, listed, comp. note) Philippines

O. brevis **Lee, 2010**
Oncotympana brevis Lee 2010c: 4, 9–11, Fig 3 (n. sp. described, illustrated, key, listed, comp. note) Philippines, Luzon
Oncotympana brevis Lee 2010d: 168 (comp. note) Philippines, Luzon

O. consobrina Distant, 1906 to *Oncotympana viridicincta* (Stål, 1870)
O. coreana Kato, 1925 to *Hyalessa fuscata* (Distant, 1905)
O. coreana nigrodorsalis Mori, 1931 to *Hyalessa fuscata* (Distant, 1905)
O. coreana pruinosa Kato, 1927 to *Hyalessa fuscata* (Distant, 1905)
O. ella Lei & Chou, 1997 to *Hyalessa ella*
O. expansa (Walker, 1858) to *Hyalessa expansa*
O. fuscata (Distant, 1905) to *Hyalessa fuscata*

***O. grandis* Lee, 2010**

Oncotympana grandis Lee 2010c: 4–6, Fig 1 (n. sp. described, illustrated, key, listed, comp. note) Philippines, Luzon
Oncotympana grandis Lee 2010d: 168–169 (comp. note) Philippines, Luzon

O. maculaticollis (de Motschulsky, 1866) to *Hyalessa maculaticollis maculaticollis*
O. maculaticollis var. *a* Kato, 1932 to *Hyalessa maculaticollis maculaticollis* (de Motschulsky, 1866)
O. maculaticollis var. *b* Distant, 1891 to *Hyalessa maculaticollis maculaticollis* (de Motschulsky, 1866)
O. maculaticollis var. *b* Kato, 1932 to *Hyalessa maculaticollis maculaticollis* (de Motschulsky, 1866)
O. maculaticollis var. *c* Liu, 1940 to *Hyalessa maculaticollis maculaticollis* (de Motschulsky, 1866)
O. maculaticollis var. *d* Liu, 1940 to *Hyalessa maculaticollis* (de Motschulsky, 1866) *maculaticollis*
O. maculaticollis aka Kato, 1934 to *Hyalessa maculaticollis maculaticollis* (de Motschulsky, 1866)
O. maculaticollis mikido Kato, 1925 to *Hyalessa maculaticollis mikado*
O. maculaticollis nigroventris Kato, 1932 to *Hyalessa maculaticollis* (de Motschulsky, 1866)
O. maculaticollis pygmaea Kato, 1932 to *Hyalessa maculaticollis* (de Motschulsky, 1866)
O. mahoni Distant, 1906 to *Hyalessa mahoni*
O. melanoptera (Distant, 1904) to *Hyalessa melanoptera*

***O. nigristigma* (Walker, 1850)** = *Dundubia nigrostigma* = *Pomponia nigrostigma* = *Dokuma nigristigma*

Dundubia nigristigma Lee 2010a: 19–20 (type species of *Dokuma*, listed, comp. note)
Dokuma nigristigma Lee 2010a: 19 (status, comp. note)
Oncotympana nigristigma Lee 2010a: 19–21, Fig 3C (n. comb., key, illustrated, listed, comp. note)
Oncotympana nigristigma Lee 2010c: 3–4, 8 (key, listed, comp. note)
Dundubia nigristigma Lee 2010d: 168 (type species of *Dokuma*, comp. note) Philippines
Oncotympana nigristigma Lee 2010d: 168 (comp. note) Philippines

O. obnubila (Distant, 1888) to *Hyalessa obnubila*

***O. pallidiventris* (Stål, 1870)** = *Pomponia (Oncotympana) pallidiventris* = *Pomponia pallidiventris* = *Pomponia (Oncotympana) pallidiveniris* (sic)

Pomponia (Oncotympana) palliventris Hayashi 1984: 57 (type species of *Oncotympana*)
Oncotympana pallidiventris Lee 1995: 79 (type species of *Oncotympana*) Philippines
Oncotympana pallidiventris Hayashi 1993: 18 (comp. note)
Pomponia (Oncotympana) pallidiventris Chou, Lei, Li, Lu and Yao 1997: 261 (type species of *Oncotympana*)
Pomponia (Oncotympana) pallidiventris Lee 1999: 6 (type species of *Oncotympana*) Philippines
Oncotympana pallidiventris Lee 2005: 75 (type species of *Oncotympana*)
Pomponia (Oncotympana) pallidiventris Lee 2005: 172 (type species of *Oncotympana*) Philippines
Pomponia (Oncotympana) pallidiventris Lee 2008b: 449 (type species of *Oncotympana*) Philippines
Pomponia (Oncotympana) pallidiventris Lee 2010a: 19 (type species of *Oncotympana*, listed, comp. note)
Oncotympana pallidiventris Lee 2010a: 19–21, Figs 3A–3B (type species of *Oncotympana*, key, illustrated, listed, comp. note)
Pomponia (Oncotympana) pallidiventris Lee 2010c: 3 (type species of *Oncotympana*, listed, comp. note)
Oncotympana pallidiventris Lee 2010c: 3–4 (key, listed, comp. note)
Oncotympana pallidiventris Lee 2010d: 167–169 (type species of *Oncotympana*, comp. note) Philippines
Pomponia (Oncotympana) pallidiventris Lee 2010d: 168 (type species of *Oncotympana*, listed)
Pomponia (Oncotympana) pallidiventris Lee 2010d: 168 (type species of *Oncotympana*, listed)

***O. simonae* Lee, 2010**

Oncotympana simonae Lee 2010c: 4, 6–9, Fig 2 (n. sp. described, illustrated, key, listed, comp. note) Philippines, Luzon
Oncotympana simonae Lee 2010d: 168 (comp. note) Philippines, Luzon

***O.* sp. Lee, 2010** = *Oncotympana viridicincta* Endo and Hayashi 1979

Oncotympana sp. Lee 2010a: 19–20 (synonymy, key, distribution, listed) Equals *Oncotympana viridicincta* Endo and Hayashi 1979 Philippines, Mindanao
Oncotympana sp. Lee 2010c: 3–4 (synonymy, key, distribution, listed) Equals *Oncotympana viridicincta* Endo and Hayashi 1979 Philippines
Oncotympana sp. Lee 2010d: 168 (comp. note) Philippines, Mindanao

O. stratoria Distant, 1905 to *Hyalessa stratoria*

***O. undata* Lee, 2010**

Oncotympana undata Lee 2010c: 4, 11–13, Fig 4 (n. sp. described, illustrated, key, listed, comp. note) Philippines, Luzon

Oncotympana undata Lee 2010d: 168 (comp. note) Philippines, Luzon

O. virescens Distant, 1905 to *Hyalessa virescens*

***O. viridicincta* (Stål, 1870)** = *Onchotympana* (sic) *viridicincta* = *Pomponia (Oncotympana) viridicincta* = *Pomponia viridicincta* = *Dokuma consobrina* Distant, 1906

Oncotympana viridicincta Boulard 1990b: 155, 157, Plate XVIII, Fig 1 (sound system, comp. note) Philippines

Oncotympana viridicincta Hayashi 1993: 18 (comp. note)

Onchotympana (sic) *viridicincta* Boulard 2002b: 198 (comp. note)

Oncotympana viridicincta Lee 2010a: 19–21, Figs 3D–3E (n. comb., key, illustrated, listed, comp. note) Equals *Doluma consobrina* n. syn.

Dokuma consobrina Lee 2010a: 19 (status, comp. note)

Oncotympana viridicincta Lee 2010c: 3–4, 6 (key, listed, comp. note)

Oncotympana viridicincta Lee 2010d: 168 (comp. note)

***Neoncotympana* Lee, 2010**

Neoncotympana Lee 2010d: 168–169 (n. gen., described, listed, comp. note) Philippines

***N. leeseungmoi* Lee, 2010**

Neoncotympana leeseungmoi Lee 2010d: 169–171, Fig 2 (n. sp., described, illustrated, listed, comp. note) Philippines, Mindanao

Subtribe Psithyristriina Distant, 1905 = Psithyristriaria = Pomponiaria Kato, 1932 = Pomponariaria (sic) = Terpnosiina Kato, 1932 = Terpnosiaria

Pomponariaria (sic) Sanborn 1999a: 44 (listed)

Pomponiaria Boulard 2003a: 98 (listed)

Pomponiaria Boulard 2003c: 172 (listed)

Terpnosiina Lee and Hayashi 2003a: 170 (status, listed, comp. note) Taiwan

Pomponiini Boulard 2005h: 6 (listed)

Pomponiaria Boulard 2005h: 6 (listed)

Pomponiaria Sanborn, Phillips and Sites 2007: 3, 27, 40, Fig 4, Table 1 (listed, distribution) Thailand

Pomponiina Boulard 2008f: 7, 38 (listed) Thailand

Psithyristriina Lee 2010a: 15 (listed, diversity, comp. note) Philippines, Luzon

Psithyristriina Lee and Hill 2010: 278, 301, 304 (n. status, synonymy, comp. note) Equals Psithyristriaria Equals Psithyristriini Equals Psithysristrini (sic) Equals Semiaria (partim) Equals Pomponiaria n. syn.

Psithyristriaria Lee and Hill 2010: 278 (comp. note)

Pomponiaria Lee and Hill 2010: 278, 301 (status, comp. note)

Pomponiina Lee and Hill 2010: 278 (comp. note)

Psithyristriina Wei, Ahmed and Rizvi 2010: 33 (comp. note)

Pomponiina Wei, Ahmed and Rizvi 2010: 33 (comp. note)

Terpnosiina Wei, Ahmed and Rizvi 2010: 33 (comp. note)

***Psithyristria* Stål, 1870**

Psithyristria Strümpel 1983: 18 (listed)

Psithyristria Duffels 1986: 319 (distribution, diversity, comp. note) Philippines

Psithyristria Hayashi 1993: 14, 18 (comp. note)

Psithyristria Chou, Lei, Li, Lu and Yao 1997: 15 (comp. note)

Psithyristria Lee 2010c: 1, 15, 17 (listed, key, comp. note) Philippines, Luzon

Psithyristria Lee and Hill 2010: 278–287, 289–290, 296–301, 303 (described, song, distribution, diversity, listed, comp. note) Philippines

***P. albiterminalis* Lee & Hill, 2010**

Psithyristria albiterminalis Lee 2010c: 16–17 (key, listed, comp. note) Philippines, Luzon

Psithyristria albiterminalis Lee and Hill 2010: 281–287, 297, 299–300, Fig 2B, Fig 3B, Fig 4B, Fig 5B, Figs 6–8 (n. sp., described, illustrated, key, distribution, comp. note) Philippines, Luzon

***P. crassinervis* Stål, 1870**

Psithyristria crassinervis Lee 2010c: 16, 20 (key, listed, comp. note) Philippines, Luzon

Psithyristria crassinervis Lee and Hill 2010: 278, 281–282, 284–285, 290–292, 297, 299–301, Fig 2F, Fig 3F, Fig 4E, Fig 5E, Figs 6–8 (illustrated, key, comp. note) Philippines, Luzon

***P. genesis* Lee & Hill, 2010**

Psithyristria genesis Lee 2010c: 16, 23 (key, listed, comp. note) Philippines, Luzon

Psithyristria genesis Lee and Hill 2010: 281–282, 284–286, 293–301, Fig 2J, Fig 3J, Fig 4H, Fig 5H, Figs 6–8 (n. sp., described, illustrated, key, distribution, comp. note) Philippines, Luzon

***P. grandis* Lee & Hill, 2010**

Psithyristria grandis Lee 2010c: 16 (key, listed, comp. note) Philippines, Luzon

Psithyristria grandis Lee and Hill 2010: 280–286, 297, 299–300, Fig 2A, Fig 3A, Fig 4A, Fig 5A, Figs 6–8 (n. sp., described, illustrated, key, distribution, comp. note) Philippines, Luzon

***P. incredibilis* Lee & Hill, 2010**

Psithyristria incredibilis Lee 2010c: 16–17 (key, listed, comp. note) Philippines, Luzon

Psithyristria incredibilis Lee and Hill 2010: 281–282, 284–288, 297, 299–300, Fig 2C, Fig 3C, Fig 4C, Fig 5C, Figs 6–8 (n. sp., described, illustrated, key, distribution, comp. note) Philippines, Luzon

***P. moderabilis* Lee & Hill, 2010**

Psithyristria moderabilis Lee 2010c: 16, 22 (key, listed, comp. note) Philippines, Luzon

Psithyristria moderalibis Lee and Hill 2010: 279, 281–282, 284–286, 293–295, 297–300, Fig 2I, Fig 3I, Fig 4G, Fig 5G, Figs 6–8 (n. sp., described, illustrated, key, distribution, comp. note) Philippines, Luzon

***P. nodinervis* Stål, 1870**

Psithyristria nodinervis Hayashi 1993: 19 (comp. note)

Psithyristria nodinervis Lee 2010c: 16, 23 (key, listed, comp. note) Philippines, Luzon

Psithyristria nodinervis Lee and Hill 2010: 278, 281–282, 284–286, 290, 296–301, 303, Fig 2K, Fig 3K, Fig 4I, Fig 5I, Figs 6–8 (described, illustrated, key, comp. note) Philippines, Luzon

***P. paracrassis* Lee, 2010**

Psithyristria paracrassis Lee 2010c: 16, 19–22, Figs 6–7 (n. sp., described, illustrated, key, comp. note) Philippines, Luzon (key, comp. note) Philippines

***P. paraspecularis* Lee & Hill, 2010**

Psithyristria paraspecularis Lee 2010c: 16–17 (key, listed, comp. note) Philippines, Luzon

Psithyristria paraspecularis Lee and Hill 2010: 281–282, 284–286, 289–290, 297–300, Fig 2E, Fig 3E, Fig 4D, Fig 5D, Figs 6–8 (n. sp., described, illustrated, key, distribution, comp. note) Philippines, Luzon

***P. paratenuis* Lee, 2010**

Psithyristria paratenuis Lee 2010c: 16–19, Fig 5 (n. sp., described, illustrated, key, comp. note) Philippines, Luzon

***P. peculiaris* Lee & Hill, 2010**

Psithyristria peculiaris Lee 2010c: 16, 22 (key, listed, comp. note) Philippines, Luzon

Psithyristria peculiaris Lee and Hill 2010: 281–282, 284–286, 292–293, 297, 299–300, Fig 2H, Fig 3H, Fig 4F, Fig 5F, Figs 6–8 (n. sp., described, illustrated, key, distribution, comp. note) Philippines, Luzon

***P. simplicinervis* Stål, 1870**

Psithyristria simplicinervis Lee 2010c: 15, 23 (key, listed, comp. note) Philippines, Luzon

Psithyristria simplicinervis Lee and Hill 2010: 278, 281–282, 297, 290, 297, 300–301, Fig 2L, Fig 3L, Fig 6 (described, illustrated, key, comp. note) Philippines, Luzon

***P. specularis* Stål, 1870**

Psithyristria specularis Hayashi 1993: 18–19, Fig 9 (illustrated, comp. note)

Psithyristria specularis Lee 2010c: 15–17 (type species of *Psithyristria*, listed, key, comp. note) Philippines, Luzon

Psithyristria specularis Lee and Hill 2010: 278, 281–282, 288–289, 297, 301, 303, Fig 2D, Fig 3D, Fig 6 (type species of *Psithyristria*, described, illustrated, listed, key, song, comp. note) Philippines, Luzon

***P. tenuinervis* Stål, 1870**

Psithyristria tenuinervis Hayashi 1993: 19 (comp. note)

Psithyristria tenuinervis Lee 2010c: 16–17 (key, listed, comp. note) Philippines, Luzon

Psithyristria tenuinervis Lee and Hill 2010: 278, 281–282, 290, 297, 300–301, Fig 2F, Fig 3F, Fig 6 (described, illustrated, key, comp. note) Philippines, Luzon

***Basa* Distant, 1905**

Basa Moulds 2005b: 391, 432 (higher taxonomy, listed, comp. note)

Basa Lee 2010c: 15 (comp. note)

Basa Lee and Hill 2010: 301, 303 (diversity, comp. note)

B. singularis (Walker, 1858) = *Dundubia singularis* = *Pomponia singularis*

Basa singularis Lee and Hill 2010: 301 (comp. note) India

***Pomponia* Stål, 1866** = *Pomponia (Pomponia)* = *Pomonia* (sic)

Pomponia Wu 1935: 15 (listed) China

Pomponia Chen 1942: 145 (listed)

Pomponia sp. Meer Mohr 1965: 104 (human food, listed) Sumatra

Pomponia sp. O'Connor, Nash and Anderson 1983: 81 (comp. note)

Pomponia Strümpel 1983: 18 (listed)

Pomponia Hayashi 1984: 28, 47, 55 (key, described, listed, comp. note)

Pomponia Boulard 1985e: 1021 (coloration, comp. note) Orient

Pomponia Duffels 1986: 319 (distribution, diversity, comp. note) Greater Sunda Islands, Southeast Asia

Pomponia Zaidi, Mohammedsaid and Yaakob 1990: 261, 267, Table 2 (listed, comp. note) Malaysia

Pomponia Duffels and van Mastrigt 1991: 174 (predation, comp. note) Sumatra
Pomponia Peng and Lei 1992: 105 (comp. note) China, Hunan
Pomponia Zaidi and Ruslan 1994: 427, Table 2 (comp. note) Malaysia
Pomponia Zaidi and Ruslan 1995a: 174, Table 2 (listed, comp. note) Borneo, Sarawak
Pomponia Zaidi and Ruslan 1995b: 220–221, Table 2 (listed, comp. note) Borneo, Sabah
Pomponia Zaidi and Ruslan 1995c: 70, Table 2 (listed, comp. note) Peninsular Malaysia
Pomponia Zaidi and Ruslan 1995d: 202, Table 2 (listed, comp. note) Borneo, Sabah
Pomponia Zaidi 1996: 103, Table 2 (comp. note) Borneo, Sarawak
Pomponia Zaidi and Hamid 1996: 54, Table 2 (listed, comp. note) Borneo, Sarawak
Pomponia Chou, Lei, Li, Lu and Yao 1997: 187–188, 293, 307, 314, 324, 326, Table 6, Tables 10–12 (key, described, listed, diversity, biogeography) China
Pomponia Novotny and Wilson 1997: 437 (listed)
Pomponia Zaidi 1997: 115, Table 2 (listed, comp. note) Borneo, Sarawak
Pomponia Zaidi and Azman 1998: 161 (comp. note) Oriental Region, China, Japan, Malaysia
Pomponia Zaidi and Ruslan 1998b: 4–5, Table 2 (listed, comp. note) Borneo, Sabah
Pomponia Zaidi and Azman 1999: 291 (comp. note) Peninsular Malaysia
Pomponia Zaidi and Azman 2000: 189–190 (comp. note) Sabah
Pomponia spp. Zaidi and Azman 2000: 192 (comp. note)
Pomponia Boulard 2001d: 80–83, 91 (comp. note) Thailand, Borneo
Pomponia Boulard 2002a: 42, 46 (comp. note)
Pomponia Boulard 2002b: 202–203, 207–208 (comp. note)
Pomponia Azman and Zaidi 2002: 2 (comp. note)
Pomponia Shcherbakov 2002: 33 (comp. note)
Pomponia Boulard 2003a: 108 (comp. note)
Pomponia Lee and Hayashi 2003b: 381, 383 (key, diagnosis) Taiwan
Pomponia Chen 2004: 39, 42, 122 (phylogeny, listed)
Pomponia Trilar and Gogala 2004: 81 (song, comp. note)
Pomponia Boulard 2005d: 341, 347–348 (comp. note) Thailand
Pomponia Boulard 2005h: 10, 16 (comp. note)
Pomponia Boulard 2005k: 94, 96–99 (comp. note) Malaysia, Sabah
Pomponia Pham 2004: 61 (key, described, distribution, comp. note) Vietnam
Pomponia and Thinh 2005a: 240 (listed) Vietnam
Pomponia Pham and Thinh 2005b: 289 (comp. note) Vietnam
Pomponia Boulard 2006g: 41, 52, 111–112, 117–118, 130–131 (comp. note)
Pomponia Duffels and Hayashi 2006: 189–190, 197 (comp. note) Sumatra
Pomponia Sanborn 2006c: 829 (comp. note)
Pomponia Boulard 2007e: 20, 25, 47, 69 (comp. note) Thailand
Pomponia Sanborn, Phillips and Sites 2007: 3, 27, 35, Table 1 (listed) Thailand
Pomponia Shiyake 2007: 6, 86 (listed, comp. note)
Pomponia sp. Shiyake 2007: 7, 85–86, Fig 14 (illustrated, listed, comp. note)
Pomponia Boulard 2008c: 368 (comp. note)
Pomponia Boulard 2008f: 7, 38, 94 (listed) Thailand
Pomponia Lee 2008a: 2, 14, Table 1 (listed) Vietnam
Pomponia Osaka Museum of Natural History 2008: 324 (listed) Japan
Pomponia Pinto-Juma, Seabra and Quartau 2008: 2 (comp. note)
Pomponia Lee 2009g: 306 (comp. note)
Pomponia Lee 2009h: 2628 (listed) Philippines, Palawan
Pomponia Lee and Sanborn 2009: 31 (comp. note)
Pomponia Pham and Yang 2009: 2, 8, 14, Table 2 (listed, key, comp. note) Vietnam
Pomponia Sanborn 2009b: 308 (comp. note) Vietnam
Pomponia Lee 2010b: 21 (listed) Cambodia
Pomponia Lee 2010c: 15 (comp. note)
Pomponia Lee and Hill 2010: 278, 298, 301, 303 (comp. note)

***P. adusta* (Walker, 1850)** = *Cicada adusta* = *Cicada buddha* Kirkaldy, 1909 = *Pomponia budda*
Pomponia adusta Zaidi, Mohammedsaid and Yaakob 1990: 261 (comp. note) Malaysia
Pomponia adusta Hayashi 1993: 13, 15, Fig 1 (distribution, key, comp. note)
Pomponia adusta Zaidi and Ruslan 1994: 428 (comp. note) Malaysia
Pomponia adusta Zaidi, Noramly and Ruslan 2000a: 112, 115, 121, Table 1 (listed, comp. note) Equals *Cicada adusta* Langkawi Island, Java, Sumatra, Borneo, Peninsular Malaysia, Siam
Pomponia adusta Boulard 2001d: 80–83 (comp. note) Malaysia
Cicada adusta Boulard 2001d: 81 (comp. note)
Pomponia adusta Zaidi, Azman and Ruslan 2001a: 112, 115, 121, Table 1 (listed, comp. note) Equals *Cicada adusta* Langkawi Island, Java, Sumatra, Borneo, Peninsular Malaysia, Siam
Cicada adusta Boulard 2005k: 96 (comp. note)

Pomponia buddha Boulard 2005k: 96 (comp. note) Equals *Cicada adusta*
Pomponia adusta Boulard 2005k: 97, 99 (comp. note) Java
Pomponia adusta Gogala and Trilar 2005: 70 (distribution, comp. note)
Pomponia adusta Sanborn, Phillips and Sites 2007: 31, 34 (comp. note) Java
Pomponia adusta Lee and Sanborn 2009: 32 (comp. note) Java

***P. backanensis* Pham & Yang, 2009** = Pham, 2004 drawing = *P.* sp. Lee, 2008
Pomponia sp. Pham 2004: 61–63, Fig c (key, described, illustrated, distribution, comp. note) Vietnam
Pomponia sp. Lee 2008a: 10 (distribution, listed) Vietnam
Pomponia sp. Pham and Yang 2009: 2 (comp. note) Vietnam
Pomponia backanensis Pham and Yang 2009: 8–10, 14, 16, Fig 2, Table 2 (n. sp., described, illustrated, listed, key, distribution, comp. note) Vietnam
Pomponia sp. Sanborn 2009b: 308 (comp. note) Vietnam
Pomponia backanensis Pham and Yang 2010a: 133 (comp. note) Vietnam

***P. bullata* Schmidt, 1924**

***P. bulu* Zaidi & Azman, 2000** = *Pomponia* sp. Zaidi, Ruslan & Azman, 1999
Pomponia sp. Zaidi, Ruslan and Azman 1999: 317 (listed, comp. note) Borneo, Sabah
Pomponia bulu Zaidi and Azman 2000: 190–194, Figs 1–6 (n. sp. described, illustrated, comp. note) Equals *Pomponia* sp. 1 Zaidi and Azman 1999 Equals *Pomponia* sp. Zaidi, Ruslan and Azman 1999 Peninsular Malaysia, Langkawi Island
Pomponia bulu Zaidi and Azman 2003: 99–100, 105, Table 1 (listed, comp. note) Borneo, Sabah

***P. cinctimanus* (Walker, 1850)** = *Dundubia cinctimanus* = *Dundubia linearis cinctimanus*
Pomponia cinctimanus Duffels and Hayashi 2006: 197 (n. comb., status, comp. note) Equals *Dundubia cinctimanus* Sylhet
Dundubia cinctimanus Duffels and Hayashi 2006: 197 (comp. note)
Pomponia cinctimanus Lee 2009g: 306, 315 (key, comp. note)
Pomponia cinctimanus Sanborn 2009b: 308 (comp. note) Vietnam

***P. cyanea* Fraser, 1948**
Pomponia cyanea Hayashi 1984: 56 (comp. note)
Pomponia cyanea Boulard 1990b: 139, 152–153, Plate XVII, Fig 4 (sound system illustrated, comp. note) India
Pomponia cyanea Boulard 2002b: 198 (comp. note)
Pomponia cyanea Boulard 2005d: 333 (comp. note)
Pomponia cyanea Boulard 2005h: 16 (comp. note)
Pomponia cyanea Boulard 2006g: 19–20 (comp. note)
Pomponia cyanea Shiyake 2007: 6, 84, 86, Fig 139 (illustrated, listed, comp. note)

***P. daklakensis* Sanborn, 2009**
Pomponia daklakensis Sanborn 2009b: 307–310, Figs 1–11 (n. sp., described, illustrated, comp. note) Vietnam

***P. decem* (Walker, 1857)** = *Dundubia decem* = *Megapomponia decem* = *Pomponia diffusa* Breddin, 1900
Pomponia decem Zaidi and Ruslan 1995a: 172, 175, Table 1 (listed, comp. note) Borneo, Sarawak
Pomponia decem Zaidi and Ruslan 1995b: 219, Table 1 (listed, comp. note) Borneo, Sabah
Pomponia decem Zaidi, Ruslan and Mahadir 1996: 61, Table 1 (listed, comp. note) Peninsular Malaysia
Pomponia decem Zaidi and Ruslan 1997: 228–229 (listed, comp. note) Equals *Dundubia decem* Peninsular Malaysia, Borneo, Sarawak, Banguey Island
Pomponia decem Zaidi and Ruslan 1998a: 362–363 (listed, comp. note) Equals *Dundubia decem* Borneo, Sarawak, Banguey Island, Peninsular Malaysia, Sabah
Pomponia decem Zaidi and Ruslan 1998b: 3, Table 1 (listed, comp. note) Borneo, Sabah
Pomponia decem Zaidi and Ruslan 1999: 3, 5, Table 1 (listed, comp. note) Borneo, Sarawak
Pomponia decem Zaidi, Ruslan and Azman 1999: 315 (listed, comp. note) Equals *Dundubia decem* Borneo, Sabah, Sarawak, Peninsular Malaysia, Banguey ISland
Pomponia decem Zaidi, Noramly and Ruslan 2000a: 329 (listed, comp. note) Equals *Dundubia decem* Borneo, Sabah, Sarawak, Banguey Island, Peninsular Malaysia
Pomponia decem Zaidi, Noramly and Ruslan 2000b: 339, Table 1 (listed, comp. note) Borneo, Sabah
Pomponia decem Zaidi, Ruslan and Azman 2000: 213–214 (listed, comp. note) Equals *Dundubia decem* Borneo, Sabah, Sarawak, Banguey Island, Peninsular Malaysia
Pomponia decem Boulard 2001d: 81–83, 85, Plate B, Fig 3 (comp. note) Borneo
Pomponia diffusa Boulard 2001d: 82 (comp. note)
Pomponia decem Zaidi and Nordin 2001: 183, 185, 191, 201–202, Table 1 (listed, comp. note) Equals *Dundubia decem* Borneo, Sabah, Sarawak, Banguey Island, Peninsular Malaysia

Pomponia decem Zaidi, Nordin, Maryati, Wahab, Norashikin et al. 2002: 6, 9–10, Tables 1–2 (listed, comp. note) Equals *Dundubia decem* Borneo, Sabah, Sarawak, Banguey Island, Peninsular Malaysia
Pomponia decem Zaidi and Azman 2003: 105, Table 1 (listed, comp. note) Borneo, Sabah
Pomponia decem Boulard 2005k: 93, 95, 97, 100, Fig 1 (comp. note) Borneo
Pomponia diffusa Boulard 2005k: 97–98 (comp. note)
Dundubia decem Boulard 2005k: 98 (comp. note)
Megapomponia decem Boulard 2005k: 93, 102 (n. comb., comp. note)
Megapomponia decem Lee and Sanborn 2009: 31 (comp. note)
Pomponia decem Lee and Sanborn 2009: 31, 33, Figs 11–12 (n. comb., illustrated, comp. note)

***P. dolosa* Boulard, 2001** = *Pomponia fusca* (nec Olivier) = *Cicada dolosa*
Pomponia fusca (sic) Boulard 2000c: 62–65, Figs 14–17 (illustrated, song analysis, sonogram, comp. note) Thailand
Pomponia fusca (sic) Boulard 2001a: 128, 133, 139–141, Plate II, Figs I-J, Figs 9–10 (illustrated, song analysis, sonogram, comp. note) Equals *Cicada fusca* Equals *Dundubia linearis* Equals *Dundubia fusca* Thailand
Pomponia dolosa Boulard 2001g: 161–163, Figs 4–5 (n. sp., described, illustrated, comp. note) Equals *Pomponia fusca* (nec Olivier) Thailand
Pomponia fusca (sic) Sueur 2001: 37, Table 1 (listed)
Pomponia dolosa Boulard 2002a: 39, 42, 63–64 (comp. note)
Cicada dolosa Boulard 2002a: 39 (comp. note)
Pomponia dolosa Boulard 2002b: 203 (comp. note)
Pomponia dolosa Boulard 2003a: 108 (comp. note)
Pomponia fusca (sic) Trilar and Gogala 2004: 81 (song, comp. note) Thailand
Pomponia dolosa Boulard 2005d: 341, 342, 347–348 (comp. note) Thailand
Pomponia dolosa Boulard 2005h: 13–14 (comp. note)
Pomponia dolosa Boulard 2005n: 135 (comp. note)
Pomponia dolosa Boulard 2005o: 147–148 (listed) Equals *Pomponia fusca* Boulard, 2001
Pomponia dolosa Boulard 2006g: 50, 52–53, 58, 64, 111–112, 119, 130, 180, Fig 20, Fig 83 (sonogram, illustrated, listed, comp. note) Equals *Pomponia fusca* Boulard, 2001 (nec Olivier) Thailand
Pomponia dolosa Duffels and Hayashi 2006: 192, 197 (comp. note) Equals *Pomponia fusca* Boulard, 2000; Boulard, 2001 Thailand
Pomponia dolosa Boulard 2007d: 499 (comp. note) Thailand
Pomponia dolosa Boulard 2007e: 27, 30, 53, 69, 71, 83, 87, Fig 16, Plate 32a (sonogram, illustrated, comp. note) Thailand
Pomponia dolosa Sanborn, Phillips and Sites 2007: 27–28 (listed, synonymy, distribution, comp. note) Equals *Pomponia fusca* Boulard, 2000 Thailand
Pomponia dolosa Boulard 2008c: 354, 365 (comp. note) Myanmar, Thailand
Pomponia dolosa Boulard 2008f: 37, 94 (biogeography, listed, comp. note) Equals *Pomponia fusca* (nec Olivier) Thailand
Pomponia dolosa Lee 2009g: 306, 315 (key, comp. note)
Pomponia dolosa Sanborn 2009b: 308 (comp. note) Vietnam

P. evanescens (Walker, 1858) to *Khimbya evanescens*

P. fuscacuminis Boulard, 2005 to *Minipomponia fuscacuminis*

***P. fuscoides* Boulard, 2002**
Pomponia fuscoides Boulard 2002a: 39–42, 64, Plate C, Figs 1–2, Figs 1–2 (n. sp., described, illustrated, sonogram, comp. note) Equals *Cicada fusca* (partim) Thailand
Pomponia fuscoides Boulard 2003a: 107–108 (comp. note)
Pomponia fuscoides Trilar and Gogala 2004: 81 (song, comp. note) Thailand
Pomponia fuscoides Boulard 2005d: 347 (comp. note)
Pomponia fuscoides Boulard 2005n: 135 (comp. note)
Pomponia fuscoides Boulard 2005o: 148 (listed)
Pomponia fuscoides Boulard 2006g: 53, 109–110, 130, 180, Figs 73–74 (sonogram, listed, comp. note)
Pomponia fuscoides Duffels and Hayashi 2006: 197 (comp. note) Thailand
Pomponia fuscoides Boulard 2007d: 499 (comp. note) Thailand
Pomponia fuscoides Boulard 2007e: 27, 69 (comp. note) Thailand
Pomponia fuscoides Sanborn, Phillips and Sites 2007: 28 (listed, distribution, comp. note) Thailand
Pomponia fuscoides Boulard 2008b: 108–109 (comp. note)
Pomponia fuscoides Boulard 2008c: 365 (comp. note) Thailand
Pomponia fuscoides Boulard 2008e: 352 (comp. note)
Pomponia fuscoides Boulard 2008f: 37, 94 (biogeography, listed, comp. note) Thailand

Pomponia fuscoides Lee 2009g: 306 (comp. note)
Pomponia fuscoides Sanborn 2009b: 308 (comp. note) Vietnam

***P. gemella* Boulard, 2005**
Pomponia gemella Boulard 2005d: 341, 347 (comp. note) Thailand
Pomponia gemella Boulard 2005h: 6, 13–16,, 52–53, Figs 9–13, Color Plate, Fig E (n. sp., described, illustrated, song, sonogram, listed, comp. note) Thailand
Pomponia gemella Boulard 2005o: 150 (listed)
Pomponia gemella Boulard 2006g: 50, 52–53, 58, 111–112, 130, Fig 21 (sonogram, comp. note) Thailand
Pomponia gemella Boulard 2007e: 27, 31, 83, Fig 17 (sonogram, comp. note) Thailand
Pomponia gemella Sanborn, Phillips and Sites 2007: 28 (listed, distribution, comp. note) Thailand
Pomponia gemella Boulard 2008f: 38, 94 (biogeography, listed, comp. note) Thailand

P. graecina Distant, 1889 to *Terpnosia graecina*

P. hieroglyphica Kato, 1938 to *Pomponia piceata* Distant, 1905

P. imperatoria (Westwood, 1842) to *Megapomponia imperatoria*

P. intermedia (Distant, 1905) to *Megapomponia intermedia*

***P. kiushiuensis* Kato, 1925** = *Pomponia kiusiuensis* (sic) = *Pomponia huishiuensis* (sic)
Pomponia kiushiuensis Hayashi 1984: 27 (comp. note)
Pomponia huishiuensis (sic) Boulard 1985e: 1032 (coloration, comp. note) Japan

P. lactea (Distant, 1887) to *Terpnosia lactea*

***P. langkawiensis* Zaidi & Azman, 1999**
Pomponia langkawiensis Zaidi and Azman 1999: 291–296, Figs 1–6 (n. sp., described, illustrated, comp. note) Peninsular Malaysia, Langkawi Island
Pomponia langkawiensis Zaidi and Azman 2000: 190 (comp. note)
Pomponia langkawiensis Zaidi, Azman and Ruslan 2001a: 111–113, 115, 121, Table 1 (listed, comp. note) Langkawi Island, Pulau Timun, Thailand
Pomponia langkawiensis Duffels and Hayashi 2006: 197 (comp. note) Langkawi Island, Malaysia
Pomponia langkawiensis Lee 2009g: 306, 316 (key, comp. note)
Pomponia langkawiensis Sanborn 2009b: 308 (comp. note) Vietnam

P. lachna Lei & Chou, 1997 to *Semia lachna*

***P. linearis* (Walker, 1850)** = *Dundubia linearis* = *Pomponia fusca* (nec Olivier) = *Pomponia fusca fusca* = *Dundubia linealis* (sic) = *Pomponia linealis* (sic) = *Pomonia* (sic) *fusca*
Cicada fusca Hayashi 1984: 55 (type species of *Pomponia*)
Pomponia linearis Hayashi 1984: 55–56, Figs 117–120 (described, illustrated, synonymy, comp. note) Equals *Cicada fusca* Equals *Dundubia linearis* Equals *Dundubia cinctimanus* Equals *Dundubia ramifera* Equals *Dundubia urania* Equals *Pomponia fusca yayemana* Japan
Pomponia linearis Zaidi, Mohammedsaid and Yaakob 1990: 262–265, Fig 1, Table 1 (seasonality, comp. note) Malaysia
Pomponia linearis Peng and Lei 1992: 100, Fig 285 (listed, illustrated, comp. note) China, Hunan
Pomponia linearis Tzean, Hsieh, Chen and Wu 1993: 514–517, Fig 1a, Fig 2 (fungal host, comp. note) Taiwan
Pomponia linearis Zaidi and Ruslan 1994: 426, 428, Table 1 (comp. note) Malaysia
Pomponia linearis Yukawa and Yamane 1995: 693 (listed) Indonesia, Kratatau
Pomponia linearis Zaidi and Ruslan 1995b: 218–219, 222, Table 1 (listed, comp. note) Borneo, Sabah
Pomponia linearis Zaidi, Ruslan and Mahadir 1996: 61, Table 1 (listed, comp. note) Peninsular Malaysia
Dundubia linearis Chou, Lei, Li, Lu and Yao 1997: 188 (type species of *Pomponia*) China
Pomponia linearis Chou, Lei, Li, Lu and Yao 1997: 188–190, 193, 307, 314–315, 324, 326, 329, 337, 366, Fig 9-50, Plate IX, Fig 91, Tables 10–12 (key, synonymy, described, illustrated, song, oscillogram, spectrogram, comp. note) Equals *Cicada fusca* Equals *Dundubia linearis* Equals *Dundubia cinctimanus* Equals *Dundubia ramifra* Equals *Dundubia urania* Equals *Pomponia fusca* Equals *Pomponia urania* Equals *Pomponia fusca yayeyamana* Equals *Pomponia fusca fusca* China
Pomponia linearis Zaidi and Ruslan 1997: 229–230 (listed, comp. note) Equals *Dundubia linearis* Peninsular Malaysia, Assam, Indo-China, Borneo, New Guinea, Java
Pomponia linearis Zaidi and Ruslan 1998a: 364–365 (listed, comp. note) Equals *Dundubia linearis* Borneo, Sarawak, Assam, Indo-China, New Guinea, Java, Peninsular Malaysia
Pomponia linearis Zaidi and Ruslan 1998b: 2–3, 5, Table 1 (listed, comp. note) Borneo, Sabah
Pomponia linearis Ohbayashi, Sato and Igawa 1999: 341 (parasitism, comp. note)
Pomponia linearis Hua 2000: 64 (listed, distribution, host) Equals *Dundubia linearis* Equals *Cicada fusca* Equals *Pomponia fusca* China,

Shaanxi, Anhui, Jiangxi, Zhejiang, Fujian, Taiwan, Guangdong, Hunan, Guangxi, Sichuan, Yunnan, Xizang, Japan, Philippines, Indonesia, Malaysia, India

Pomponia linearis Sueur 2001: 40, 46, Tables 1–2 (listed)

Pomponia linearis Boulard 2003a: 98–99, 106–110, Color Fig 5, Figs 4–6, sonogram 7 (described, allotype female, paratype females, song, sonogram, illustrated, comp. note) Equals *Dundubia linearis* Equals *Pomponia fusca* (partim) Thailand, Malay Peninsula, Borneo, Java, China

Pomponia linearis Lee and Hayashi 2003b: 383–386, Figs 27–28 (key, diagnosis, illustrated, synonymy, distribution, song, oscillogram, listed, comp. note) Equals *Dundubia linearis* Equals *Cicada fusca* Equals *Pomponia fusca* Equals *Pomponia fusca fusca* Taiwan, Japan, China, Vietnam, Laos, Cambodia, Thailand, Malaysia, Singapore, Indonesia, Myanmar, Bangladesh, Nepal, India

Pomponia linearis Biswas, Ghosh, Bal and Sen 2004: 240, 248 (listed, described, distribution) Equals *Cicada fusca* India, Manipur, Assam, Margherita, Naga Hills, Nilgiri Hills, Sibsagar, Sylhet, Java, Japan, Philippines

Pomponia linearis Boucias, Pendland and Kalkar 2004: 1559 (fungus host) Taiwan

Pomponia linearis Chen 2004: 122–123, 199 (described, illustrated, listed, comp. note) Taiwan

Pomponia linearis Pham 2004: 61–63, Fig a (key, illustrated, distribution, comp. note) Vietnam

Pomponia linearis Boulard 2005d: 344 (comp. note)

Pomponia linearis Boulard 2005h: 16 (comp. note)

Pomponia linearis Boulard 2005n: 134, 137 (comp. note)

Pomponia linearis Boulard 2005o: 149 (listed)

Pomponia linearis Lee 2005: 125 (comp. note)

Dundubia linearis Pham and Thinh 2005a: 240 (type species of *Pomponia*)

Dundubia linearis Pham and Thinh 2005a: 240 (type species of *Pomponia*)

Pomponia linearis Pham and Thinh 2005a: 240 (synonymy, distribution, listed) Equals *Cicada fusca* Equals *Dundubia linearis* Equals *Dundubia cinctimanus* Equals *Dundubia ramifra* (sic) Equals *Dundubia urania* Equals *Pomponia fusca* Equals *Pomponia urania* Vietnam

Pomponia linearis Duffels and Hayashi 2006: 189–192, 196–200, Fig 2, Figs 9–13 (synonymy, status, described, illustrated, distribution, comp. note) Equals *Cicada linearis* Assam, Bhutan, Nepal, Thailand, Vietnam

Dundubia linearis Duffels and Hayashi 2006: 190, 197 (comp. note)

Pomponia linearis group Duffels and Hayashi 2006: 197 (comp. note)

Pomponia linearis species complex Duffels and Hayashi 2006: 197–198 (comp. note)

Pomponia linearis Pham and Thinh 2006: 526–527 (listed, comp. note) Vietnam

Pomponia linearis Saisho 2006a: 57 (eclosion, illustrated, comp. note) Japan

Pomponia linearis Watanabe and Kobayashi 2006: 224–227, Tables 2–5 (listed) Japan, Ryukyus

Pomponia linearis Sanborn, Phillips and Sites 2007: 28–29 (listed, distribution, comp. note) Thailand, Malaysia, Indonesia, India, Indochina, China, Taiwan, Japan, Nepal, Philippine Republic

Pomponia linearis Shiyake 2007: 7, 85–86, Fig 142 (illustrated, distribution, listed, comp. note) China, India, Southeast Asia, Taiwan, Philippines, Malaysia, Indonesia

Pomponia linearis Boucias, Pendland and Kalkar 2008: 2613 (fungus host) Taiwan

Pomponia linearis group Boulard 2007e: 25 (comp. note) Thailand

Pomponia linearis Boulard 2007e: 26, 35, 39, Fig 12, Fig 23 (illustrated, sonogram, comp. note) Thailand

Pomponia linearis Boulard 2008b: 109 (comp. note)

Pomponia linearis Boulard 2008c: 366 (comp. note) Thailand

Pomponia linearis Boulard 2008e: 352 (comp. note)

Pomponia linearis Boulard 2008f: 38–39, 94 (biogeography, listed, comp. note) Thailand

Pomponia linearis Lee 2008a: 9–10 (synonymy, distribution, listed) Equals *Dundubia linearis* Equals *Pomponia fusca* (nec Olivier) Vietnam, Laos, Cambodia, Thailand, Myanmar, Bangladesh, Bhutan, Nepal, India

Pomponia linearis Osaka Museum of Natural History 2008: 324 (listed, acoustic behavior, ecology, comp. note) Japan

Pomponia linearis species group Lee 2009g: 306, 312 (comp. note)

Pomponia linearis Lee 2009g: 306–307, 309–310, 315, Fig 1 (illustrated, key, comp. note)

Dundubia linearis Pham and Yang 2009: 8 (type species of *Pomponia*)

Pomponia linearis group Pham and Yang 2009: 8 (comp. note) Vietnam

Pomponia linearis Pham and Yang 2009: 8–9, 14, Table 2 (listed, key, distribution, comp. note) Vietnam

Pomponia linearis species group Sanborn 2009b: 308 (comp. note) Vietnam

Pomponia linearis Sanborn 2009b: 308 (comp. note) Vietnam

Pomponia linearis Lee, Lin and Wu 2010: 218–224, Figs 1–6, Tables 1–2 (ecology, comp. note) Taiwan

Pomponia linearis Lee 2010b: 21 (synonymy, distribution, listed) Equals *Dundubia linearis* Cambodia, Vietnam, Laos, Thailand, Myanmar, Bangladesh, Bhutan, Nepal, India

P. nr. *linearis* n. sp. Zaidi & Ruslan, 1995

Pomponia nr. *linearis* n. sp. Zaidi and Ruslan 1995d: 198, 201 (comp. note) Borneo, Sabah

P. littldollae Boulard, 2002 to *Minipomponia littldollae*

P. merula (Distant, 1905) to *Megapomponia merula*

P. *mickwanae* Boulard, 2009 = *Pomponia lactea* (nec Distant)

Pomponia lactea (sic) Boulard 2001a: 128, 133, 138–139, Plate II, Fig H, Fig 8 (illustrated, song analysis, sonogram, comp. note) Equals *Leptopsaltria lactea* Thailand

Pomponia mickwanae Boulard 2007d: 499–500 (comp. note) Thailand

Pomponia mickwanae Boulard 2008c: 366–367, Fig 21 (comp. note)

Pomponia mickwanae Boulard 2008f: 39, 94 (biogeography, listed, comp. note) Equals *Pomponia lactea* Boulard 2001 (nec Distant) Thailand

Pomponia mickwanae Boulard 2009a: 39–43, Figs 1–7 (n. sp., synonymy, described, illustrated, song, sonogram, comp. note) Equals *Pomponia lactea* Boulard, 2001 [138] (nec Distant)

P. *namtokola* Boulard, 2005

Pomponia namtokola Boulard 2005h: 6, 16–19, Figs 14–18 (n. sp., described, illustrated, song, sonogram, listed, comp. note) Thailand

Pomponia namtokola Boulard 2005o: 150 (listed)

Pomponia namtokola Boulard 2006g: 130 (comp. note)

Pomponia namtokola Boulard 2007d: 500 (comp. note) Thailand

Pomponia namtokola Boulard 2007e: 27 (comp. note) Thailand

Pomponia namtokola Sanborn, Phillips and Sites 2007: 29 (listed, distribution, comp. note) Thailand

Pomponia namtokola Boulard 2008f: 39, 94 (biogeography, listed, comp. note) Thailand

P. *nasanensis* Boulard, 2005

Pomponia nasanensis Boulard 2005n: 134–137, Plate B, Figs 1–2 (n. sp., described, illustrated, song, sonogram, comp. note) Thailand

Pomponia nasanensis Boulard 2005o: 150 (listed)

Pomponia nasanensis Sanborn, Phillips and Sites 2007: 29 (listed, distribution, comp. note) Thailand

Pomponia nasanensis Boulard 2008f: 39, 94 (biogeography, listed, comp. note) Thailand

P. *noualhieri* Boulard, 2005

Pomponia noualhieri Boulard 2005h: 6, 10–13, Figs 4–8, Color Plate, Fig B (n. sp., described, illustrated, song, sonogram, listed, comp. note) Thailand

Pomponia noualhieri Boulard 2005o: 150 (listed)

Pomponia noualhieri Boulard 2006g: 130 (comp. note)

Pomponia noualhieri Sanborn, Phillips and Sites 2007: 29–30 (listed, distribution, comp. note) Thailand

Pomponia noualhieri Boulard 2008f: 39, 94 (biogeography, listed, comp. note) Thailand

P. *orientalis* (Distant, 1912) = *Tettigia orientalis* = *Cicada orientalis* = *Pomponia occidentalis* (sic)

Pomponia orientalis Lee 2008a: 10–11, Fig 1 (n. comb., illustrated, synonymy, distribution, listed, comp. note) Equals *Tettigia orientalis* South Vietnam, Thailand

Pomponia orientalis Pham and Yang 2009: 8, 14, Table 2 (listed, key, distribution, comp. note) Vietnam, China

Pomponia occidentalis Sanborn 2009b: 308 (comp. note) Vietnam

P. *parafusca* Kato, 1944

P. pendleburyi Boulard, 2001 to *Megapomponia pendylburyi*

P. *piceata* Distant, 1905 = *Pomponia hierogryphica* (sic) Kato, 1938 = *Pomponia hieroglyphica*

Pomponia piceata Chou, Lei, Li, Lu and Yao 1997: 188, 190, Plate IX, Fig 92 (key, synonymy, described, illustrated, comp. note) Equals *Pomponia hieroglyphica* China

Pomponia hieroglyphica Hua 2000: 64 (listed, distribution) China, Guangxi, Vietnam

Pomponia piceata Hua 2000: 64 (listed, distribution) China, Guangxi

Pomponia piceata Pham 2004: 61–63, Fig b (key, illustrated, distribution, comp. note) Vietnam

Pomponia piceata Pham and Thinh 2005a: 240 (synonymy, distribution, listed) Equals *Pomponia hieroglyphica* Vietnam

Pomponia piceata Pham and Thinh 2006: 526–527 (listed, comp. note) Vietnam

Pomponia piceata Lee 2008a: 10 (synonymy, distribution, listed) Equals *Pomponia hieroglyphica* Equals *Pomponia hierogryphica* (sic) North Vietnam, China, Guangxi

Pomponia piceata Pham and Yang 2009: 8–9, 14, Table 2 (listed, key, distribution, comp. note) Vietnam, China
Pomponia piceata Sanborn 2009b: 308 (comp. note) Vietnam

***P. picta* (Walker, 1868)** = *Dundubia picta* = *Cicada fusca* Olivier, 1790 (nec Müller, 1776) = *Dundubia fusca* = *Pomponia (Pomponia) fusca* = *Pomponia fusca fusca* = *Pomponia linearis* (partim) = *Pomponia fusca* Boulard, 2001 (nec Olivier)

Pomponia fusca Chen 1933: 360 (listed, comp. note)
Pomponia fusca Wu 1935: 15 (listed) Equals *Cicada fusca* Equals *Dundubia linearis* Equals *Dundubia cinctimanus* Equals *Dundubia ramifera* Equals *Dundubia urania* Equals *Pomponia linearis* China, Han-lik, Canton, Kwangsi, Sylhet, Assam, Margherita, Naga Hills, Sibsagar, Cachar, Nilgiri Hils, Malay Peninsula, Java, Philippines, Japan
Cicada fusca Chen 1942: 145 (type species of *Pomponia*)
Pomponia fusca Chen 1942: 145 (listed, comp. note) China
Pomponia fusca Wang and Zhang 1987: 295 (key, comp. note) China
Pomponia fusca Wang and Zhang 1992: 120 (listed, comp. note) China
Pomponia fusca Wang and Zhang 1993: 189 (listed, comp. note) China
Pomponia fusca Anonymous 1994: 810 (fruit pest) China
Pomponia picta Zaidi and Ruslan 1994: 426, 428, Table 1 (comp. note) Malaysia
Pomponia fusca Zaidi and Ruslan 1995b: 218–219, 221, Table 1 (listed, comp. note) Borneo, Sabah
Pomponia picta Zaidi and Ruslan 1995b: 218–219, Table 1 (listed, comp. note) Borneo, Sabah
Pomponia fusca Zaidi and Ruslan 1995d: 200 (comp. note)
Pomponia picta Zaidi and Ruslan 1995d: 200 (comp. note)
Pomponia picta Zaidi, Ruslan and Mahadir 1996: 61, Table 1 (listed, comp. note) Peninsular Malaysia
Pomponia fusca Zaidi and Azman 1998: 164, 166–167, Figs 11–13 (illustrated, comp. note)
Pomponia fusca Zaidi and Ruslan 1998a: 363 (listed, comp. note) Equals *Cicada fusca* Borneo, Sarawak, Sumatra, Sabah
Pomponia picta Zaidi and Ruslan 1998a: 365 (listed, comp. note) Equals *Dundubia picta* Borneo, Sarawak, Sumatra, Peninsular Malaysia, Java, New Guinea, Sabah
Pomponia fusca Zaidi and Ruslan 1998b: 2–3, 5, Table 1 (listed, comp. note) Borneo, Sabah
Pomponia picta Zaidi and Ruslan 1998b: 2–3, Table 1 (listed, comp. note) Borneo, Sabah
Pomponia fusca Zaidi and Azman 1999: 291–292, 296 (comp. note) Peninsular Malaysia
Pomponia picta Hua 2000: 64 (listed, distribution) Equals *Dundubia picta* China, Taiwan, Indonesia, Sri Lanka
Pomponia fusca Boulard 2001g: 158–161, Figs 1–3 (homotype designation, described, illustrated, comp. note) Sumatra
Pomponia picta Boulard 2001g: 161 (comp. note)
Pomponia fusca Boulard 2002a: 39, 64 (comp. note)
Pomponia fusca Boulard 2002b: 193 (coloration, comp. note)
Pomponia fusca Boulard 2003a: 106–107 (comp. note)
Cicada fusca Lee and Hayashi 2003b: 383 (type species of *Pomponia*) Sumatra
Pomponia fusca Lee and Hayashi 2003b: 384 (comp. note)
Pomponia picta Zaidi and Azman 2003: 106, Table 1 (listed, comp. note) Borneo, Sabah
Cicada fusca Pham 2004: 61 (type species of *Pomponia*)
Pomponia fusca Boulard 2005d: 341 (comp. note)
Pomponia fusca group Boulard 2005d: 347 (comp. note)
Pomponia fusca clade Boulard 2005h: 10 (comp. note)
Pomponia fusca group Boulard 2005h: 13 (comp. note)
Pomponia fusca Boulard 2005h: 16 (comp. note)
Pomponia fusca Boulard 2005k: 94, 100 (type species of *Pomponia*, comp. note)
Pomponia fusca group Boulard 2005k: 94 (comp. note)
Pomponia fusca Boulard 2005o: 148 (listed)
Pomponia fusca Boulard 2006g: 52–53 (comp. note)
Pomponia picta Duffels and Hayashi 2006: 189–195, 197, Fig 1, Figs 3–8 (synonymy, status, described, illustrated, distribution, comp. note) Equals *Cicada fusca* Olivier Equals *Cicada fusca* Müller Equals *Pomponia fusca* Equals *Dundubia picta* Equals *Pomponia linearis* (partim) Sumatra, Nias Island
Pomponia fusca Duffels and Hayashi 2006: 189–192, 194 (type species of *Pomponia*, synonymy, comp. note)
Cicada fusca Olivier Duffels and Hayashi 2006: 190 (comp. note) Sumatra
Cicada fusca Müller Duffels and Hayashi 2006: 190 (comp. note)
Dundubia picta Duffels and Hayashi 2006: 191 (comp. note)
"*Pomponia fusca*" Boulard 2007e: 27 (comp. note) Thailand

Pomponia fusca Sanborn, Phillips and Sites 2007: 28 (listed, distribution, comp. note) Thailand, Sumatra, India, Java, Philippine Republic, Malay Peninsula, Malay Archipelago, Japan, Malacca, Borneo, Taiwan, Indochina, Sarawak, China

Pomponia picta Boulard 2008c: 365 (comp. note) Equals *Pomponia fusca* Duffels & Hayashi, 2006 Malaysia

Pomponia picta group Boulard 2008f: 7 (biogeography, listed) Equals *Pomponia fusca* Muller

Pomponia picta Boulard 2008f: 37, 90 (type species of *Pomponia*) Equals *Pomponia fusca* Muller

Cicada fusca Lee 2008a: 9 (type species of *Pomponia*) Equals *Dundubia picta* Sumatra

Pomponia picta Lee 2008a: 10 (comp. note) Equals *Terponsia crowfooti* Vietnam, Nepal, India

Pomponia fusca Lee 2009g: 310 (comp. note)

Dundubia picta Lee 2009h: 2628 (type species of *Pomponia*, listed, comp. note) Sumatra

Dundubia picta Lee 2010b: 21 (type species of *Pomponia*, listed, comp. note)

Pomponia picta Lee and Hill 2010: 301 (comp. note)

P. polei Henry, 1931 to *Terpnosia polei*

***P. ponderosa* Lee, 2009** = *Pomponia linearis* (nec Walker) = *Pomponia fusca* (nec Olivier)

Pomponia ponderosa Lee 2009g: 310–313, 315, Figs 4–5 (n. sp., described, illustrated, key, comp. note) Equals *Pomponia linearis* (partim) Equals *Pomponia fusca* (partim) China

***P. promiscua* Distant, 1887**

***P. quadrispinae* Boulard, 2002**

Pomponia quadrispinae Boulard 2002a: 42–46, 65, Plate D, Figs 1–2, Plate E, Figs 1–2, Plate N, Fig 4, Figs 3–4 (n. sp., described, illustrated, sonogram, comp. note) Thailand

Pomponia quadrispinae Trilar and Gogala 2004: 81 (song, comp. note) Thailand

Pomponia quadrispinae Boulard 2005o: 148 (listed)

Pomponia quadrispinae Boulard 2006g: 130 (comp. note)

Pomponia quadrispinae Boulard 2007e: 4, Plate 27a (coloration, illustrated, comp. note) Thailand

Pomponia quadrispinae Sanborn, Phillips and Sites 2007: 29 (listed, distribution, comp. note) Thailand

Pomponia quadrispinae Boulard 2008f: 39, 94 (biogeography, listed, comp. note) Thailand

***P. rajah* Moulton, 1923** = *Megapomponia rajah*

Pomponia rajah Kahono 1993: 213, Table 5 (listed) Irian Jaya

Pomponia rajah Zaidi and Ruslan 1995a: 171, 176 (comp. note) Borneo, Sarawak

Pomponia rajah Zaidi and Ruslan 1998a: 366 (listed, comp. note) Borneo, Sarawak

Pomponia rajah Zaidi and Ruslan 1999: 3, 6 (comp. note) Borneo, Sarawak

Pomponia rajah Zaidi, Ruslan and Azman 1999: 316–317 (listed, comp. note) Borneo, Sabah, Sarawak, Peninsular Malaysia

Pomponia rajah Zaidi, Ruslan and Azman 2000: 215 (comp. note) Borneo, Sabah, Sarawak

Pomponia rajah Zaidi and Nordin 2001: 18 (comp. note) Borneo, Sabah

Pomponia rajah Zaidi, Nordin, Maryati, Wahab, Norashikin et al. 2002: 12 (comp. note) Borneo, Sabah

Pomponia rajah Zaidi and Azman 2003: 106, Table 1 (listed, comp. note) Borneo, Sabah

Pomponia rajah Boulard 2005k: 93, 98, 100 (comp. note) Sarawak

Megapomponia rajah Boulard 2005k: 93, 102 (n. comb., comp. note)

Megapomponia rajah Lee and Sanborn 2009: 31 (comp. note)

Pomponia rajah Lee and Sanborn 2009: 31, 33, Fig 13 (n. comb., illustrated, comp. note)

***P. ramifera* (Walker, 1850)** = *Dundubia ramifera* = *Dundubia linearis ramifera*

Pomponia ramifera Duffels and Hayashi 2006: 197 (n. comb., status, comp. note) Equals *Dundubia ramifera* Sylhet

Dundubia ramifera Duffels and Hayashi 2006: 197 (comp. note)

Pomponia ramifera Lee 2009g: 306, 315 (key, comp. note)

Pomponia ramifera Sanborn 2009b: 308 (comp. note) Vietnam

P. scitula Distant, 1888 to *Haphsa scitula*

***P. secreta* Hayashi, 1978**

***P. siamensis* China, 1925 nom. nud.**

P. similis similis Schmidt, 1924 to *Terpnosia similis similis*

P. similis obsolete China, 1926 to *Terpnosia similis obsolete*

P. simusa Boulard, 2008 to *Terpnosia simusa*

***P. solitaria* Distant, 1888**

***P.* sp. 1 Kahono, 1993**

Pomponia sp. 1 Kahono 1993: 213, Table 5 (listed) Irian Jaya

***P.* sp. 2 Kahono, 1993**

Pomponia sp. 2 Kahono 1993: 213, Table 5 (listed) Irian Jaya

***P. subtilita* Lee, 2009** = *Pomponia linearis* (nec Walker) = *Pomponia fusca* (nec Olivier)

Pomponia linearis (partim) Chou, Lei, Li, Lu and Yao 1997: 188–190, 193, 307, 314–315, 324, 326, 329, 337, 366, Fig 9–50, Plate IX, Fig 91, Tables 10–12 (key, synonymy, described, illustrated, song, oscillogram, spectrogram, comp. note) Equals *Cicada*

fusca Equals *Dundubia linearis* Equals *Dundubia cinctimanus* Equals *Dundubia ramifra* Equals *Dundubia urania* Equals *Pomponia fusca* Equals *Pomponia urania* Equals *Pomponia fusca yayeyamana* Equals *Pomponia fusca fusca* China

Pomponia subtilita Lee 2009g: 313–316, Figs 6–8 (n. sp., described, illustrated, key, comp. note) Equals *Pomponia linearis* (partim) Equals *Pomponia fusca* (partim) Thailand

Pomponia subtilita Lee 2010b: 21 (distribution, listed) Cambodia, Thailand

***P. surya* Distant, 1904**

Pomponia surya Swinton 1908: 380 (song, comp. note) India

***P.* undescribed species A Sanborn, Phillips & Sites, 2007**

Pomponia undescribed species A Sanborn, Phillips and Sites 2007: 30 (listed, distribution, comp. note) Thailand

***P.* undescribed species B Sanborn, Phillips & Sites, 2007**

Pomponia undescribed species B Sanborn, Phillips and Sites 2007: 30 (listed, distribution, comp. note) Thailand

***P.* undescribed species C Sanborn, Phillips & Sites, 2007**

Pomponia undescribed species C Sanborn, Phillips and Sites 2007: 30 (listed, distribution, comp. note) Thailand

P. tuba Lee, 2009 to *P. yayeyamana* Kato, 1933

***P. urania* (Walker, 1850)** = *Dundubia urania*

Pomponia urania Duffels and Hayashi 2006: 197 (n. comb., status, comp. note) Equals *Dundubia urania* East India?

Pomponia urania Lee 2009g: 306, 315 (key, comp. note)

Pomponia urania Sanborn 2009b: 308 (comp. note) Vietnam

***P. yayeyamana* Kato, 1933** = *Pomponia fusca yayeyamana* = *Pomponia tuba* Lee, 2009 = *Pomponia linearis* (nec Walker) = *Pomponia fusca* (nec Olivier)

Pomponia fusca yayeyamana Duffels and Hayashi 2006: 191 (comp. note) Ryukyu Island

Pomponia linearis (partim) Lee and Hayashi 2003b: 383–386, Figs 27–28 (key, diagnosis, illustrated, synonymy, distribution, song, oscillogram, listed, comp. note) Equals *Dundubia linearis* Equals *Cicada fusca* Equals *Pomponia fusca* Equals *Pomponia fusca fusca* Taiwan, Japan, China, Vietnam, Laos, Cambodia, Thailand, Malaysia, Singapore, Indonesia, Myanmar, Bangladesh, Nepal, India

Pomponia tuba Lee 2009g: 307–310, 315, Figs 2–3 (n. sp., described, illustrated, key, comp. note) Equals *Pomponia linearis* (partim) Equals *Pomponia fusca* (partim) Taiwan

Pomponia fusca yayeyamana Lee 2009g: 306 (comp. note)

Pomponia tuba Lee and Hill 2010: 298–301, Figs 7–8 (phylogeny, comp. note)

***P. zakrii* Zaidi & Azman, 1998**

Pomponia zakrii Zaidi and Azman 1998: 161–167, Figs 1–10 (n. sp. described, illustrated, comp. note) Peninsular Malaysia

Pomponia zakrii Zaidi and Azman 2000: 190 (comp. note)

Pomponia zakrii Duffels and Hayashi 2006: 197 (comp. note) Malaysian Peninsula

Pomponia zakrii Lee 2009g: 306, 316 (key, comp. note)

Pomponia zakrii Sanborn 2009b: 308 (comp. note) Vietnam

***P. zebra* Bliven, 1964**

Pomponia zebra Sanborn 1999a: 44 (type material, listed)

***Terpnosia* Distant, 1892** = *Terprosia* (sic) = *Ternopsia* (sic) = *Trepnosia* (sic) = *Ierpnosia* (sic)

Terpnosia Wu 1935: 17 (listed) China

Terpnosia Fujiyama 1982: 185 (comp. note)

Terpnosia Boulard 1985e: 1032 (coloration, comp. note) Japan

Terpnosia Anufriev and Emelyanov 1988: 313, 317–318 (key, characteristics) Soviet Union

Terpnosia Zaidi and Ruslan 1995a: 173–174, Table 2 (listedcomp. note) Borneo, Sarawak

Terpnosia Zaidi and Ruslan 1995b: 220–221, Table 2 (listed, comp. note) Borneo, Sabah

Terpnosia Zaidi and Ruslan 1995c: 70, Table 2 (listed, comp. note) Peninsular Malaysia

Terpnosia Zaidi and Ruslan 1995d: 202, Table 2 (listed, comp. note) Borneo, Sabah

Terpnosia Boer and Duffels 1996b: 315 (comp. note)

Terpnosia Zaidi 1996: 103, Table 2 (comp. note) Borneo, Sarawak

Terpnosia Zaidi and Hamid 1996: 54, Table 2 (listed, comp. note) Borneo, Sarawak

Terprosia (sic) Chou, Lei, Li, Lu and Yao 1997: 4 (comp. note)

Terpnosia Chou, Lei, Li, Lu and Yao 1997: 197, 203, 293, Table 6 (key, described, synonymy, listed, diversity, biogeography) Equals *Yezoterpnosia* China

Terpnosia Zaidi 1997: 115, Table 2 (listed, comp. note) Borneo, Sarawak

Terpnosia Zaidi and Ruslan 1998b: 4–5, Table 2 (listed, comp. note) Borneo, Sabah

Terpnosia Zaidi and Ruslan 1999: 3–4, Table 2 (listed, comp. note) Borneo, Sarawak

Terpnosia sp. Zaidi, Ruslan and Azman 1999: 303 (listed, comp. note) Borneo, Sabah

Terpnosia sp. Zaidi, Ruslan and Azman 2000: 201 (comp. note) Borneo, Sabah
Terpnosia sp. Zaidi and Nordin 2001: 18 (comp. note) Borneo, Sabah
Ternopsia (sic) sp. Zaidi, Nordin, Maryati, Wahab, Norashikin et al. 2002: 12 (comp. note) Borneo, Sabah
Terpnosia Lee and Hayashi 2003b: 363, 381 (comp. note)
Terpnosia sp. Zaidi and Azman 2003: 104, Table 1 (listed, comp. note) Borneo, Sabah
Terpnosia Moulds 2005b: 391, 432 (higher taxonomy, listed, comp. note)
Terpnosia Pham and Thinh 2005a: 244 (listed) Vietnam
Terpnosia Pham and Thinh 2005b: 290 (comp. note) Vietnam
Terpnosia Duffels and Hayashi 2006: 189 (comp. note)
Terpnosia Iwamoto, Inoue and Yagi 2006: 679, 681, Figs 2v–w (flight muscle, comp. note)
Terpnosia Sanborn 2006c: 829, 833 (comp. note)
Terpnosia Sanborn, Phillips and Sites 2007: 3, 14, Table 1 (listed) Thailand
Terpnosia Shiyake 2007: 4, 10, 46 (listed, comp. note)
Terpnosia sp. Shiyake 2007: 11 (illustrated)
Terpnosia Boulard 2008f: 7, 46, 95 (listed) Thailand
Terpnosia Lee 2008a: 2, 8, Table 1 (listed) Vietnam
Terpnosia Pham and Yang 2009: 13, Table 2 (listed) Vietnam
Terpnosia Lee 2010c: 23 (listed)
Terpnosia Lee and Hill 2010: 302 (comp. note)
Terpnosia Pham and Yang 2010b: 206 (comp. note) Vietnam

***T. abdullah* Distant, 1904** = *Terpnosia adbulla* (sic) = *Terpnosia abdullahi* (sic)
Terpnosia abdullah Boulard 2003a: 98, 111–113, Fig 12, sonogram 10 (described, song, sonogram, illustrated, comp. note) Equals Terpnosia *abdullahi* (sic) Thailand, Malay Peninsula, India, Sikkim
Terpnosia abdullah Boulard 2005o: 149 (listed)
Terpnosia abdullahi (sic) Sanborn, Phillips and Sites 2007: 14 (listed, synonymy, distribution, comp. note) Equals *Terpnosia abdullah* Thailand, Malaysia, India, Java
Terpnosia abdullah Boulard 2008f: 46, 95 (biogeography, listed, comp. note) Equals *Terpnosia abdullahi* (sic) Thailand, Peninsular Malaysia, India, Sikhim (sic)

***T. andersoni* Distant, 1892**
Terpnosia andersoni Chou, Lei, Li, Lu and Yao 1997: 204–205, Fig 9–58 (described, illustrated, key, comp. note) China
Terpnosia andersoni Hua 2000: 64 (listed, distribution) China, Jiangsu, Jiangxi, Yunnan, Burma, India, Nepal, Bhutan, Thailand
Terpnosia andersoni Boulard 2003a: 113 (comp. note) Yunnan
Terpnosia andersoni Sanborn, Phillips and Sites 2007: 34 (comp. note) Laos, India, China, Burma
Terpnosia andersoni Shiyake 2007: 4, 44, 46, Fig 70 (illustrated, distribution, listed, comp. note) China, India, Southeast Asia
Terpnosia andersoni Boulard 2008f: 90 (comp. note) Thailand

***T.* nr. *andersoni* Zaidi & Ruslan, 1995**
Terpnosia nr. *andersoni* Zaidi and Ruslan 1995a: 172, Table 1 (listed, comp. note) Borneo, Sarawak
Terpnosia nr. *andersoni* sp. n, Zaidi and Ruslan 1999: 3, 5, Table 1 (listed, comp. note) Borneo, Sarawak

***T. chapana* Distant, 1917**
Terpnosia chapana Pham and Thinh 2005a: 244 (distribution, listed) Vietnam
Terpnosia chapana Lee 2008a: 8 (distribution, listed) North Vietnam, South China
Terpnosia chapana Pham and Yang 2009: 13, 16, Table 2 (listed, comp. note) Vietnam
Terpnosia chapana Pham and Yang 2010a: 133 (comp. note) Vietnam

***T. clio* (Walker, 1850)** = *Dundubia clio*
Terpnosia clio Wu 1935: 18 (listed) Equals *Dundubia clio* China, Yunnan, Himalaya, Sikkim, Burma, Tenasserim
Terpnosia clio Chou, Lei, Li, Lu and Yao 1997: 204, Fig 9–59 (described, illustrated, key, synonymy, comp. note) Equals *Dundubia clio* China
Terpnosia clio Hua 2000: 64–65 (listed, distribution) Equals *Dundubia clio* China, Yunnan, Sikkim, Burma, Himalaya, India, Nepal
Terpnosia clio Shiyake 2007: 4, 45, 47, Fig 71 (illustrated, distribution, listed, comp. note) China, India, Southeast Asia
Terpnosia clio Lee 2008a: 9 (synonymy, distribution, listed) Equals *Dundubia clio* Vietnam, China, Yunnan, Myanmar, Nepal, India
Terpnosia clio Pham and Yang 2009: 14, Table 2 (listed) Vietnam

***T. collina* (Distant, 1888)** = *Pomponia collina*

***T. confusa* Distant, 1905** = *Terpnosia psecas* (nec Walker)

***T. elegans* (Kirby, 1891)** = *Pomponia elegans*

T. fuscoapicalis Kato, 1938 to *Yezoterpnosia fuscoapicalis*

***T. ganesa* Distant, 1904**

***T. graecina* (Distant, 1889)** = *Pomponia graecina* = *Pomponia graecinea* (sic)
Pomponia graecinea (sic) Boulard 1990b: 101, 103, Plate X, Figs 5–6 (genitalia, comp. note)

Pomponia graecina Zaidi and Ruslan 1998a: 363 (listed, comp. note) Borneo, Sarawak
Pomponia graecina Zaidi, Ruslan and Azman 1999: 315 (listed, comp. note) Borneo, Sabah, Sarawak
Pomponia graecina Zaidi, Ruslan and Azman 2000: 214 (listed, comp. note) Borneo, Sabah, Sarawak
Pomponia graecina Zaidi and Nordin 2001: 183, 185, 191, 202, Table 1 (listed, comp. note) Borneo, Sabah, Sarawak
Pomponia graecina Zaidi, Nordin, Maryati, Wahab, Norashikin et al. 2002: 6, 9–10, Tables 1–2 (listed, comp. note) Borneo, Sabah, Sarawak
Pomponia graecina Zaidi and Azman 2003: 105, Table 1 (listed, comp. note) Borneo, Sabah
Pomponia graecina Zaidi, Azman and Nordin 2005: 27–29, Table 1 (listed, comp. note) Borneo, Sabah
Pomponia graecina Boulard 2007e: 62, Fig 42 (genitalia illustrated, comp. note)

T. ichangensis Liu, 1940 to *Yezoterpnosia ichangensis*

***T. jenkinsi* Distant, 1912**

***T. jinpingensis* Lei & Chou, 1997**

Terpnosia jinpingensis Chou, Lei, Li, Lu and Yao 1997: 204, 211, 357, 367, Fig 9–62, Plate X, Fig 109 (n. sp., described, illustrated, key, comp. note) China
Terpnosia jinpingensis Sun, Watson, Zheng, Watson and Liang 2009: 3149–3154, Fig 1O, Fig 2D, Fig 3D, Fig 4D, Tables 1–2 (illustrated, wing microstructure, comp. note) China, Hainan

***T. lactea* (Distant, 1887)** = *Pomponia lactea* = *Pomponia lacteal* (sic) = *Leptopsaltria lactea*

Pomponia lactea Boulard 1985e: 1032 (coloration, comp. note) Malaysia
Pomponia lactea Zaidi and Ruslan 1995a: 172, 175, Table 1 (listed, comp. note) Borneo, Sarawak
Pomponia lactea Zaidi and Ruslan 1995d: 198, 201 (comp. note) Borneo, Sabah
Pomponia lactea Zaidi and Ruslan 1997: 229 (listed, comp. note) Equals *Leptopsaltria lactea* Peninsular Malaysia, Sumatra, Borneo, Brunei, Java
Pomponia lactea Zaidi and Ruslan 1998a: 364 (listed, comp. note) Equals *Leptopsaltria lactea* Borneo, Sarawak, Sumatra, Brueni, Java, Peninsular Malaysia
Pomponia lacteal (sic) Zaidi and Ruslan 1999: 3, Table 1 (listed, comp. note) Borneo, Sarawak
Pomponia lactea Zaidi and Ruslan 1999: 3, 5 (comp. note) Borneo, Sarawak
Pomponia lactea Zaidi, Ruslan and Azman 1999: 315–316 (listed, comp. note) Equals *Leptopsaltria lactea* Borneo, Sabah, Sumatra, Brunei, Java, Peninsular Malaysia
Pomponia lactea Zaidi, Ruslan and Azman 2000: 214 (listed, comp. note) Equals *Leptopsaltria lactea* Borneo, Sabah, Sumatra, Brunei, Java, Peninsular Malaysia, Sarawak
Pomponia lactea Zaidi and Nordin 2001: 183, 185–186, 191, 202–203, Table 1 (listed, comp. note) Equals *Leptopsaltria lactea* Borneo, Sabah, Sumatra, Brunei, Java, Peninsular Malaysia, Sarawak
Pomponia lactea Zaidi, Nordin, Maryati, Wahab, Norashikin et al. 2002: 6–7, 9–10, 12, Tables 1–2 (listed, comp. note) Equals *Leptopsaltria lactea* Borneo, Sabah, Sumatra, Brunei, Java, Peninsular Malaysia, Sarawak
Pomponia lactea Zaidi and Azman 2003: 105, Table 1 (listed, comp. note) Borneo, Sabah
Pomponia lactea Pham 2004: 61 (comp. note)
Pomponia lactea Boulard 2005o: 147 (listed)
Pomponia lactea Boulard 2006g: 130 (comp. note)
Pomponia linearis Boulard 2006g: 73, 75, 87, 130, 180, Fig 48 (listed, comp. note) Thailand
Pomponia lactea Sanborn, Phillips and Sites 2007: 28 (listed, distribution, comp. note) Thailand, Malaysia, Indonesia, Borneo, Indochina, Brunei
Pomponia lactea Boulard 2008c: 366–367, Fig 20 (comp. note) Malaysia
Pomponia lactea Boulard 2008f: 90 (comp. note)
Pomponia lactea Boulard 2009a: 39 (comp. note)
Pomponia lactea Lee 2009h: 2628–2629, Figs 7 (illustrated, synonymy, distribution, listed, comp. note) Equals *Leptopsaltria lactea* Philippines, Palawan, Vietnam, Thailand, Malay Peninsula, Indonesia, Northern India
Pomponia lactea Pham and Yang 2009: 8, 14, Table 2 (listed, key, distribution, comp. note) Vietnam, China
Pomponia lactea Sanborn 2009b: 308 (comp. note) Vietnam

***T. maculipes* (Walker, 1850)** = *Dundubia maculipes*

Terpnosia maculipes Kirillova 1986b: 45 (listed) India
Terpnosia maculipes Kirillova 1986a: 123 (listed) India

T. majuscula Distant, 1917 to *Semia majuscula*

***T. mawi* Distant, 1909** = *Terpnosia nawi* (sic)

Terpnosia mawi Wu 1935: 18 (listed) China, Shensi
Terpnosia nawi Chou, Lei, Li, Lu and Yao 1997: 212 (listed, comp. note) China
Terpnosia mawi Hua 2000: 65 (listed, distribution, host) China, Shaanxi, Henan, Sichuan
Terpnosia mawi Lee 2008a: 9 (distribution, listed) Vietnam, China, Shaanxi, Sichuan
Terpnosia mawi Pham and Yang 2009: 14, Table 2 (listed) Vietnam

T. mega Chou & Lei, 1997
Terpnosia mega Chou, Lei, Li, Lu and Yao 1997: 204, 210–211, 357, 367, Fig 9–61, Plate IX, Fig 100 (n. sp., described, illustrated, key, comp. note) China

T. mesonotalis Distant, 1917 = *Calcagninus salvazanus* Distant, 1917
Terpnosia mesonotalis Lee 2008a: 9 (synonymy, distribution, listed) Equals *Calcagninus salvazanus* North Vietnam
Terpnosia mesonotalis Pham and Yang 2009: 14, 16, Table 2 (listed, comp. note) Vietnam
Terpnosia mesonotalis Pham and Yang 2010a: 133 (comp. note) Vietnam

T. neocollina Liu, 1940
Terpnosia neocollina Hua 2000: 65 (listed, distribution) China
Terpnosia neocollina Sanborn, Phillips and Sites 2007: 14 (listed, distribution, comp. note) Thailand
Terpnosia neocollina Boulard 2008f: 46, 90, 95 (biogeography, listed, comp. note) Thailand, Siam

T. nigella Chou & Lei, 1997
Terpnosia nigella Chou, Lei, Li, Lu and Yao 1997: 204, 211–212, 357, 367–368, Fig 9–63, Plate X, Fig 110 (n. sp., described, illustrated, key, comp. note) China

T. nigricosta (de Motschulsky, 1866) to *Yezoterpnosia nigricosta*

T. nonusaprilis Boulard, 2003
Terpnosia nonusaprilis Boulard 2003a: 98–99, 112–116, Color Figs 8–9, Figs 13–17, sonogram 11–12 (n. sp., described, song, sonogram, illustrated, comp. note) Thailand
Terpnosia nonusaprilis Boulard 2005d: 347 (comp. note)
Terpnosia nonusaprilis Boulard 2005o: 149 (listed)
Terpnosia nonusaprilis Boulard 2006g: 49, 113, 115, Fig 80 (sonogram, comp. note) Asia
Terpnosia nonusaprilis Boulard 2007e: 47, 83 (comp. note) Thailand
Terpnosia nonusaprilis Sanborn, Phillips and Sites 2007: 14 (listed, distribution, comp. note) Thailand
Terpnosia nonusaprilis Boulard 2008f: 46, 95 (biogeography, listed, comp. note) Thailand

T. oberthuri Distant, 1912

T. obscura (Kato, 1938) to *Yezoterpnosia obscura*

T. obscurana Metcalf, 1955 = *Terpnosia obscura* Liu, 1940 (nec *Terpnosia obscura* Kato, 1938)

T. polei (Henry, 1931) = *Pomponia polei*

T. posidonia Jacobi, 1902 = *Terponia* (sic) *posidonia* = *Cicada stipata* Walker, 1850 (nec *Dundubia stipata* Walker, 1850)
Terpnosia posidonia Lee 2008a: 8 (synonymy, distribution, listed) Equals *Terponia* (sic) *posidonia* North Vietnam, South China
Terpnosia posidonia Pham and Yang 2009: 13, Table 2 (listed) Vietnam

T. psecas (Walker, 1850) = *Dundubia psecas* = *Cicada psecas* = *Yezoterpnosia psecas* = *Ternopsia* (sic) *psecas*
Dundubia psecas Hayashi 1984: 47 (type species of *Terpnosia*)
Dundubia psecas Chou, Lei, Li, Lu and Yao 1997: 203 (type species of *Terpnosia*) China
Ternopsia (sic) *psecas* Zaidi and Ruslan 1997: 221 (listed, comp. note) Equals *Dundubia psecas* Java, Peninsular Malaysia, Ceylon
Terpnosia psecas Zaidi and Ruslan 1998a: 349 (listed, comp. note) Equals *Dundubia psecas* Borneo, Sarawak, Java, Ceylon, Siam, Peninsular Malaysia, Sabah
Terpnosia psecas Zaidi and Azman 2003: 10, 104, Table 1 (listed, comp. note) Borneo, Sabah
Dundubia psecas Pham and Thinh 2005a: 244 (type species of *Terpnosia*)
Terpnosia psecas Sanborn, Phillips and Sites 2007: 34 (comp. note)
Terpnosia pescas Boulard 2008f: 46 (type species of *Terpnosia*)
Dundubia psecas Lee 2008a: 8 (type species of *Terpnosia*) Java

T. pumila (Distant, 1891) = *Pomponia pumila*
Terpnosia pumila Zaidi and Azman 2003: 99–100, 104, Table 1 (listed, comp. note) Borneo, Sabah

T. puriticis Lei & Chou, 1997
Terpnosia puriticis Chou, Lei, Li, Lu and Yao 1997: 203, 208–209, 357, 367, Fig 9–60, Plate IX, Fig 99, Plate IX, Fig 101 (n. sp., described, illustrated, key, comp. note) China

T. ransonneti (Distant, 1888) = *Pomponia ransonneti* = *Euterpnosia ransonneti* = *Pomponia ransonetti* (sic) = *Terpnosia ransonetti* (sic) = *Pomponia greeni* Kirby, 1891
Pomponia ransonetti (sic) Boulard 1985e: 1032 (coloration, comp. note) Japan
T[erpnosia] ransonetti (sic) Jiang, Yang, Tang, Xu and Chen 1995b: 229 (song frequency, comp. note)
Terpnosia ransonetti (sic) Sueur 2001: 40, Table 1 (listed)
Terpnosia ransonneti Lee 2008a: 8–9 (synonymy, distribution, listed) Equals *Pomponia ransonneti* Vietnam, Sri Lanka, India
Terpnosia ransonneti Pham and Yang 2009: 13, Table 2 (listed) Vietnam

T. renschi Jacobi, 1941
Terpnosia renschi Schönefeld and Göllner-Scheiding 1984: 71 (type material, listed) Lombok

***T. ridens* Pringle, 1955**
Terpnosia ridens Villet 1988: 71 (comp. note)
T[erpnosia] ridens Jiang, Yang, Tang, Xu and Chen 1995b: 229 (song frequency, comp. note)
Terpnosia ridens Sueur 2001: 40, Table 1 (listed)

***T. rustica* Distant, 1917**
Terpnosia rustica Pham and Thinh 2005a: 244 (distribution, listed) Vietnam
Terpnosia rustica Lee 2008a: 9 (distribution, listed) North Vietnam
Terpnosia rustica Pham and Yang 2009: 14, 16, Table 2 (listed, comp. note) Vietnam
Terpnosia rustica Pham and Yang 2010a: 133 (comp. note) Vietnam

T. shaanxiensis Sanborn, 2006 to *Yezoterpnosia shaanxiensis*

***T. similis obsoleta* (China, 1926)** = *Pomponia similis obsoleta*

***T. similis similis* Schmidt, 1924** = *Pomponia similis* = *Pomponia picta* Distant 1891: pl. vii
Pomponia similis Duffels and Hayashi 2006: 192 (comp. note) Equals *Pomponia picta* Distant, 1891
Pomponia similis Lee 2009h: 2628 (comp. note) Sumatra

***T. simusa* (Boulard, 2008)** = *Pomponia simusa*
Pomponia simusa Boulard 2008c: 353–355, Figs 4–6 (n. sp., described, illustrated, comp. note) Tonkin
Pomponia simusa Sanborn 2009b: 308 (comp. note) Vietnam

***T.* n. sp. Zaidi & Ruslan, 1995**
Terpnosia n. sp. Zaidi and Ruslan 1995a: 175 (comp. note) Borneo, Sarawak

***T.* sp. 1 Zaidi, Ruslan & Mahadir, 1996**
Terpnosia sp. 1 Zaidi, Ruslan and Mahadir 1996: 60, Table 1 (listed, comp. note) Peninsular Malaysia

***T.* sp. 1 Pham & Thinh, 2006**
Terpnosia sp. 1 Pham and Thinh 2006: 526 (listed) Vietnam

***T.* sp. 2 Zaidi, Ruslan & Mahadir, 1996**
Terpnosia sp. 2 Zaidi, Ruslan and Mahadir 1996: 60, Table 1 (listed, comp. note) Peninsular Malaysia

***T. stipata* (Walker, 1950)** = *Dundubia stipata* = *Ternopsia* (sic) *stipata* = *Dundubia clonia* Walker, 1850 = *Dundubia chlonia* (sic)
Terpnosia stipata Boulard 1985e: 1032 (coloration, comp. note) Sri Lanka
Terpnosia stipata Bennet-Clark and Young 1994: 293, Table 1 (body size and song frequency)
T[erpnosia] stipata Jiang, Yang, Tang, Xu and Chen 1995b: 229 (song frequency, comp. note)
Terpnosia stipata Sueur 2001: 40, Table 1 (listed)

***T. translucida* (Distant, 1891)** = *Pomponia translucida*

T. vacua (Olivier, 1790) to *Yezoterpnosia vacua*

T. vacua nigra Kato, 1927 to *Yezoterpnosia vacua* (Olivier, 1790)

***T. versicolor* Distant, 1912**
Terpnosia versicolor Chou, Lei, Li, Lu and Yao 1997: 367 (comp. note)

***T. viridissima* Boulard, 2005**
Terpnosia viridissima Boulard 2005i: 72–75, Figs 24–28 (n. sp., described, illustrated, comp. note) Thailand
Terpnosia viridissima Boulard 2005o: 150 (listed)
Terpnosia viridissima Sanborn, Phillips and Sites 2007: 14 (listed, distribution, comp. note) Thailand
Terpnosia viridissima Boulard 2008f: 47, 95 (biogeography, listed, comp. note) Thailand

***Semia* Matsumura, 1917**
Semia Hayashi 1984: 47 (listed)
Tanna (sic) Lee and Hayashi 2003b: 361 (key) Taiwan
Semia Lee and Hayashi 2003b: 381, 383 (key, diagnosis) Taiwan
Semia Chen 2004: 39, 42, 145 (phylogeny, listed)
Semia Moulds 2005b: 391, 432 (higher taxonomy, listed, comp. note)
Semia Lee 2010c: 15 (comp. note)
Semia Lee and Hill 2010: 298, 301, 303 (diversity, distribution, comp. note) Taiwan, South China

***S. klapperichi* Jacobi, 1944**
Semia klapperichi Hua 2000: 64 (listed, distribution) China, Fujian
Semia klapperichi Lampe, Rohwedder and Rach 2006: 17 (type specimens) China, Fujian

***S. lachna* (Lei & Chou, 1997)** = *Pomponia lachna*
Pomponia lachna Chou, Lei, Li, Lu and Yao 1997: 188, 192–193, 357, 366–367, Fig 9–52, Plate IX, Fig 90 (key, synonymy, described, illustrated, comp. note) Equals *Cicada fusca* Equals *Dundubia linearis* Equals *Dundubia cinctimanus* Equals *Dundubia ramifra* Equals *Dundubia urania* Equals *Pomponia fusca* Equals *Pomponia urania* Equals *Pomponia fusca yayeyamana* Equals *Pomponia fusca fusca* China

***S. majuscula* (Distant, 1917)** = *Terpnosia majuscula* = *Terpnosia majuscule* (sic)
Terpnosia majuscula Lee 2008a: 9 (distribution, listed) Vietnam, Laos
Terpnosia majuscula Pham and Yang 2009: 14, 16, Table 2 (listed, comp. note) Vietnam
Terpnosia majuscule (sic) Pham and Yang 2010a: 133 (comp. note) Vietnam

S. spinosa **Pham, Hayashi & Yang, 2012**

S. watanabei **(Matsumura, 1907)** = *Leptopsaltria watanabei* = *Pomponia watanabei*

Leptopsaltria watanabei Hua 2000: 62 (listed, distribution) China, Taiwan

Semia watanabei Hua 2000: 64 (listed, distribution) Equals *Leptopsaltria watanabei* China, Guangxi

Leptopsaltria watanabei Lee and Hayashi 2003b: 381 (type species of *Semia*) Formosa

Semia watanabei Lee and Hayashi 2003b: 381–384, Figs 25–26 (key, diagnosis, illustrated, synonymy, distribution, song, listed, comp. note) Equals *Leptopsaltria watanabei* Equals *Pomponia watanabei* Taiwan

Semia watanabei Chen 2004: 145–146, 199 (described, illustrated, listed, comp. note) Taiwan

Semia watanabei Lee and Hill 2010: 298–300, Figs 7–8 (phylogeny, comp. note)

Subtribe Minipomponiina Boulard, 2008

Minipomponiina Boulard 2008f: 7, 41 (n. subtribe, listed) Thailand

Minipomponia **Boulard, 2008**

Minipomponia Boulard 2008f: 7, 41, 59, 94 (n. gen., described, listed) Thailand

M. fuscacuminis **(Boulard, 2005)** = *Pomponia fuscacuminis*

Pomponia fuscacuminis Boulard 2005d: 339 (comp. note)

Pomponia fuscacuminis Boulard 2005h: 6, 20–22, Figs 19–24 (n. sp., described, illustrated, song, sonogram, listed, comp. note) Thailand

Pomponia fuscacuminis Boulard 2005o: 150 (listed)

Pomponia fuscacuminis Boulard 2006g: 49, 76, 130, 144, Fig 105 (sonogram, comp. note) Thailand

Pomponia fuscacuminis Sanborn, Phillips and Sites 2007: 28 (listed, distribution, comp. note) Thailand

Minipomonia fuscacuminis Boulard 2008f: 41, 94 (biogeography, listed, comp. note) Equals *Pomponia fuscacuminis* Thailand

M. littldollae **(Boulard, 2002)** = *Pomponia littldollae*

Pomponia littldollae Boulard 2002a: 46–49, 64, Plate F, Figs 1–3, Figs 5–7 (n. sp., described, illustrated, sonogram, comp. note) Thailand

Pomponia littldollae Boulard 2003c: 191 (comp. note)

Pomponia littldollae Trilar and Gogala 2004: 81 (song, comp. note) Thailand

Pomponia littldollae Boulard 2005d: 344 (comp. note) Thailand

Pomponia littldollae Boulard 2005o: 148 (listed)

Pomponia littldollae Boulard 2006g: 76, 92, 95, 109, 130, 180, Fig 59 (sonogram, listed, comp. note) Thailand

Pomponia littldollae Boulard 2007e: 3, 35, 44, 47, Fig 28, Plate 1 (illustrated, sonogram, comp. note) Thailand

Pomponia littldollae Sanborn, Phillips and Sites 2007: 29–30 (listed, distribution, comp. note) Thailand

Minipomonia littldollae Boulard 2008f: 41, 58–59, 94, Figs 3–4 (biogeography, illustrated, listed, comp. note) Equals *Pomponia littldollae* Thailand

Subtribe Leptopsaltriina Moulton, 1923 = Leptopsaltriaria = Semiaria (partim) = Terpnosiaria

Leptopsaltriaria Sanborn 1999a: 43 (listed)

Leptopsaltriaria Kos and Gogala 2000: 2 (comp. note)

Leptopsaltriaria Schouten and Duffels 2002: 29 (comp. note)

Leptopsaltriaria Boulard 2003a: 98 (listed)

Leptopsaltriaria Boulard 2003c: 172 (listed)

Leptopsaltriaria Boulard 2003e: 260, 274 (listed)

Leptopsaltriina Lee and Hayashi 2003a: 170 (status, listed, comp. note) Taiwan

Leptopsaltriina Gogala, Trilar, Kozina and Duffels 2004: 2 (listed)

Leptopsaltriaria Boulard 2005e: 118 (listed)

Leptopsaltriaria Boulard 2005h: 7 (listed)

Leptopsaltriina Boulard 2006d: 195 (comp. note)

Leptopsaltraria Duffels 2004: 463 (comp. note) Equals Leptopsaltriaria

Leptopsaltriina Moulds 2005b: 386, 391, 432 (history, listed, comp. note) Equals Leptopsaltriaria

Leptopsaltraria Duffels, Schouten and Lammertink 2007: 367 (comp. note)

Terpnosiaria Shiyake 2007: 4, 46 (listed, comp. note)

Leptopsaltriaria Shiyake 2007: 5, 54 (listed, comp. note)

Leptopsaltriaria Sanborn, Phillips and Sites 2007: 3, 11, 37, Fig 2, Table 1 (listed, distribution) Thailand

Leptopsaltriina Duffels 2009: 303, 305 (comp. note)

Leptopsaltraria Duffels 2009: 303 (comp. note)

Leptopsaltriina Pham and Yang 2009: 10 (listed, synonymy) Equals Leptopsaltrini Equals Leptopsaltraria Equals Leptopsaltriaria Vietnam

Leptopsaltriina Lee 2010c: 23 (listed, diversity, comp. note) Philippines, Luzon

Leptopsaltriina Lee and Hill 2010: 298, 301–302 (synonymy, comp. note) Equals Leptopsaltriaria Equals Leptopsaltraria (sic) Equals Semiaria (partim) Equals Terpnosiaria n. syn.
Leptopsaltriina Wei, Ahmed and Rizvi 2010: 33 (comp. note)

***Yezoterpnosia* Matsumura, 1917** = *Yzoterpnosia* (sic)
Terpnosia (sic) Hayashi 1984: 28–29, 47, Fig 3a (key, described, listed, comp. note) Equals *Yezoterpnosia*
Terpnosia (sic) Osaka Museum of Natural History 2008: 324 (listed) Japan

***Y. fuscoapicalis* (Kato, 1938)** = *Terpnosia fuscoapicalis*
Terpnosia fuscoapicalis Chou, Lei, Li, Lu and Yao 1997: 203, 205–206, Plate IX, Fig 96 (described, illustrated, key, comp. note) China
Terpnosia fuscoapicalis Hua 2000: 65 (listed, distribution) China, Jiangxi, Zhejiang, Fujian, Hunan, Guangxi

***Y. ichangensis* (Liu, 1940)** = *Terpnosia ichangensis* = *Terpnosia ichangens* (sic)
Terpnosia ichangensis Liu 1992: 41–42 (comp. note)
Terpnosia ichangens (sic) Chou, Lei, Li, Lu and Yao 1997: 212 (listed, comp. note) China
Terpnosia ichangensis Chou, Lei, Li, Lu and Yao 1997: 212 (listed) China
Terpnosia ichangensis Hua 2000: 65 (listed, distribution) China, Hubei

***Y. nigricosta* (de Motschulsky, 1866)** = *Cicada nigricosta* = *Terpnosia nigricosta* = *Terpnosia nigrocosta* (sic) = *Yezoterpnosia nigricosta* = *Yezoterpnosia sapporensis* Kato, 1925 = *Terpnosia sapporensis* = *Terpnosia nigricosta sapporensis* = *Terpnosia obscura* Liu (nec Kato)
Terpnosia nigricosta Hayashi 1982b: 187–188, Table 1 (distribution, comp. note) Japan
Terpnosia nigricosta Hayashi 1984: 29, 48–49, Figs 83–87 (key, described, illustrated, synonymy, comp. note) Equals *Cicada nigricosta* Equals *Yezoterpnosia nigricosta* Equals *Yezoterpnosia sapporensis* Japan
Cicada nigricosta Hayashi 1984: 47 (type species of *Yezoterpnosia*)
Terpnosia sapporensis Hayashi 1984: 49 (comp. note) Japan
Terpnosia nigricosta Anufriev and Emelyanov 1988: 314, 317–318 Figs 236.3, 239.1–239.6 (illustrated, genitalia illustrated, key) Soviet Union
Terpnosia nigricosta Liu 1992: 42 (comp. note)
Terpnosia nigricosta Chou, Lei, Li, Lu and Yao 1997: 203, 206–207, 357 (key, described, synonymy, comp. note) Equals *Cicada nigricosta* Equals *Yezoterpnosia nigricosta* Equals *Yezoterpnosia sapporensis* China
Terpnosia nigricosta Hua 2000: 65 (listed, distribution) Equals *Cicada nigricosta* China, Shaanxi, Hubei, Siberia, Japan
Terpnosia nigricosta Ohara and Hayashi 2001: 1 (comp. note) Japan
Terpnosia nigricosta Sueur 2001: 46, Table 2 (listed)
Terpnosia nigricosta Saisho 2003: 50 (comp. note) Japan
Terpnosia nigricosta Lee and Hayashi 2004: 58 (comp. note) Taiwan
Terpnosia nigricosta Lee, Sato and Sakai 2005: 317 (parasitism, comp. note) North America
Terpnosia nigricosta Sanborn 2006c: 834 (comp. note) China
Terpnosia nigricosta Inoue, Goto, Makino, Okabe, Okochi et al. 2007: 251, Table 2 (trap collection, comp. note) Japan
Terpnosia nigricosta Shiyake 2007: 4, 10, 45, 47, Fig 72 (illustrated, distribution, listed, comp. note) Japan, China
Terpnosia nigricosta Osaka Museum of Natural History 2008: 324 (listed, acoustic behavior, ecology, comp. note) Japan
Terpnosia nigricosta Kohira, Okada, Nakanishi and Yamanaka 2009: 14 (predation, comp. note) Japan

***Y. obscura* Kato, 1938** = *Terpnosia obscura* = *Terpnosia obscula* (sic) = *Terpnosia obscurana* (sic)
Terpnosia obscura Liu 1992: 41–42 (comp. note)
Terpnosia obscurana (sic) Liu 1992: 42 (comp. note)
Terpnosia obscura Chou, Lei, Li, Lu and Yao 1997: 203, 208, 212, 367, Plate IX, Fig 97 (described, illustrated, key, comp. note) China
Terpnosia obscura Hua 2000: 65 (listed, distribution) China, Shaanxi, Jiangxi, Jiangsu, Fujian

***Y. shaanxiensis* (Sanborn, 2006)** = *Terpnosia shaanxiensis*
Terpnosia shaanxiensis Sanborn 2006c: 831–834, Figs 4–6 (n. sp., described, illustrated, comp. note) China, Shaanxi

***Y. vacua* (Olivier, 1790)** = *Cicada vacua* = *Terpnosia vacua* = *Terpnosia vacau* (sic) = *Trepnosia* (sic) *vacuna* (sic) = *Terpnosia vacuna* (sic) = *Terpnosia vacva* (sic) = *Terpnosia nacua* (sic) = *Ierpnosia* (sic) *vacua* = *Cicada clara* de Motschulsky, 1866 = *Terpnosia pryeri* Distant, 1892 = *Terpnosia kawamurae* Matsumura, 1913 = *Terpnosia kuramensis* Kato, 1925 = *Terpnosia knramensis* (sic) = *Terpnosia vacua kuramensis* = *Terpnosia vacua nigra* Kato, 1927 = *Terpnosia vacua* var. *nigra* = *Terpnosia vacuna* (sic) *nigra*
Terpnosia vacua Chen 1933: 360 (listed, comp. note)
Terpnosia vacua Isaki 1933: 380 (listed, comp. note) Japan

Terpnosia vacua Ohgushi 1954: 11 (emergence times, ecology, comp. note) Japan
Terpnosia vacua Ikeda 1959: 484 (comp. note) Japan
Terpnosia vacua Hayashi 1982b: 188, Table 1 (distribution, comp. note) Japan
Terpnosia vacua Hayashi 1984: 29, 47–48, Figs 79–82 (key, described, illustrated, synonymy, comp. note) Equals *Cicada vacua* Equals *Cicada clara* Equals *Terpnosia pryeri* Equals *Terpnosia kawamurae* Equals *Terpnosia kuramensis* Japan
Terpnosia vacua Kevan 1985: 66 (comp. note) Japan
Ierpnosia (sic) *vacua* Liu 1991: 258 (comp. note)
Ierpnosia (sic) *vacua* Liu 1992: 17 (comp. note)
Terpnosia vacua Liu 1992: 33–34, 36 (comp. note)
Terpnosia vacua Chou, Lei, Li, Lu and Yao 1997: 203, 207–208, Plate IX, Fig 94 (described, illustrated, key, synonymy, comp. note) Equals *Cicada vacua* Equals *Cicada clara* Equals *Terpnosia pryeri* Equals *Terpnosia kawamurae* Equals *Terpnosia vacua* f. *kuramensis* Equals *Terpnosia vacua* f. *nigra* China
Terpnosia pryeri Hua 2000: 65 (listed, distribution) China
Terpnosia vacua Hua 2000: 65 (listed, distribution) Equals *Cicada vacua* China, Shandong, Nanjing, Zhejiang, Japan
Terpnosia vacua Ohara and Hayashi 2001: 1 (comp. note) Japan
Terpnosia vacua Sueur 2001: 46, Table 2 (listed)
Terpnosia vacua Saisho 2003: 49 (emergence, comp. note) Japan
Terpnosia vacua Shiyake 2007: 4, 45, 47, Fig 73 (illustrated, distribution, listed, comp. note) Japan, China
Terpnosia vacua Osaka Museum of Natural History 2008: 324 (listed, acoustic behavior, ecology, comp. note) Japan
Terpnosia vacua Sun, Watson, Zheng, Watson and Liang 2009: 3148–3149, 3151, 3154, Fig 1N, Table 1 (illustrated, wing microstructure, comp. note) Japan

***Euterpnosia* Matsumura, 1917** = *Euterponosia* (sic) = *Futerpnosia* (sic) = *Euternosia* (sic) = *Terpnosia (Euterpnosia)*
Euterpnosia Chen 1942: 146 (listed)
Euterpnosia spp. Ohno 1982: 51 (bibliography)
Euterpnosia Hayashi 1984: 28–29, 47 (key, describedlisted, comp. note)
Euterpnosia spp. Ohno 1985: 105 (bibliography)
Euterpnosia Lee 1995: 22, 129 (listed, comp. note)
Euterpnosia Chou, Lei, Li, Lu and Yao 1997: 197–198, 206, 293, Table 6 (key, described, listed, diversity, biogeography, comp. note) China
Euterpnosia Lee and Hayashi 2003b: 360–361, 363, 381 (key, diagnosis) Taiwan
Euterpnosia spp. Lee and Hayashi 2003b: 365 (comp. note) Taiwan
Euterpnosia Chen 2004: 39, 42, 95–96 (phylogeny, listed)
Euterpnosia sp. Chen 2004: 95–96, 117–118, 120, 199 (described, illustrated, listed, comp. note) Taiwan
Euterpnosia Chen 2005: 37–39, 47–49, Fig 1 (illustrated, comp. note) Taiwan
Euterpnosia Moulds 2005b: 391, 432 (higher taxonomy, listed, comp. note)
Euterpnosia Chen 2006: 71, 74 (diversity, comp. note) Taiwan
Euterpnosia Sanborn, Phillips and Sites 2007: 3, 14, Table 1 (listed) Thailand
Euterpnosia Boulard 2008f: 7, 47, 95 (listed) Thailand
Euterpnosia Chen and Shiao 2008: 81, 88, 91 (diversity, comp. note) Japan, China, Korea, Nepal, Taiwan
Euterpnosia Lee 2008a: 2, 9, Table 1 (listed) Vietnam
Euterpnosia Shiyake 2007: 4, 10, 50 (listed, comp. note)
Euterpnosia Osaka Museum of Natural History 2008: 324 (listed) Japan
Euterpnosia Pham and Yang 2009: 12, 14, Table 2 (listed) Vietnam
Euterpnosia Lee, Lin and Wu 2010: 221 (diversity, comp. note) Taiwan
Euterpnosia Lee 2010c: 23 (listed)
Euterpnosia Lee and Hill 2010: 302 (comp. note)
Euterpnosia Pham, Thinh and Yang 2010: 63 (diversity, listed, comp. note) Vietnam, Oriental Region, Taiwan

***E. alipina* Chen, 2005**
Euterpnosia alpina Chen 2005: 38–49, Fig 3, Fig 5C, Figs 6C–D, Fig 7C, Figs 8E–F, Figs 9E–F, Fig 10B, Fig 12, Fig 14, Table 1 (n. sp.. described, illustrated, song, sonogram, comp. note) Taiwan
Euterpnosia alpina Chen 2006: 71–75, Fig 1B, Table 1 (key, comp. note) Taiwan
Euterpnosia alpina Chen and Shiao 2008: 81 (song, sonogram, comp. note) Taiwan

***E. ampla* Chen, 2006**
Euterpnosia ampla Chen 2006: 71–75, Fig 1A, Figs 1D–1F, Tables 1–2 (n. sp., described, illustrated, comp. note) Taiwan

***E. arisana* Kato, 1925** = *Euterpnosia arisanus* (sic)
Euterpnosia arisana Chou, Lei, Li, Lu and Yao 1997: 202 (listed) China
Euterpnosia arisana Hua 2000: 61 (listed, distribution) China, Taiwan

Euterpnosia arisana Lee and Hayashi 2003b: 363, 374 (key, diagnosis, synonymy, distribution, song, listed, comp. note) Equals *Euterpnosia arisanus* (sic) Taiwan

Euterpnosia arisana Chen 2004: 95–96, 101, 198 (described, illustrated, listed, comp. note) Taiwan

Euterpnosia arisana Chen 2005: 48 (comp. note) Taiwan

Euterpnosia arisana Chen and Shiao 2008: 81, 83–90, Figs 1B–B1, Fig 2B, Fig 3B, Fig 4B, Fig 5, Fig 8B, Fig 9B (illustrated, song, oscillogram, sonogram, comp. note) Taiwan

***E. chibensis chibensis* Matsumura, 1917** = *Leptopsaltria tuberosa* Matsumura, 1907 (nec Signoret)

Euterpnosia chibensis Hayashi 1982b: 188, Table 1 (distribution, comp. note) Japan

Euterpnosia chibensis Hayashi 1984: 29, 47–52, Figs 88–93, Figs 96–104 (key, type species of *Euterpnosia*, described, illustrated, synonymy, comp. note) Equals *Leptopsaltria tuberosa* (nec Signoret) Japan

Euterpnosia chibensis chibensis Hayashi 1984: 50–52, Figs 88–89, Fig 96, Figs 99–100 (key, described, illustrated, comp. note) Japan

Euterpnosia chibensis Chou, Lei, Li, Lu and Yao 1997: 198–199, 202, 357, 367, Fig 9–55, Plate IX, Fig 95 (type species of *Euterpnosia*, key, described, illustrated, comp. note) China

Euterpnosia chibensis Freytag 2000: 55 (on stamp)

Euterpnosia chibensis chibensis Ohara and Hayashi 2001: 1 (comp. note) Japan

Euterpnosia chibensis chibensis Saisho 2001a: 67 (phototaxis, comp. note) Japan

Euterpnosia chibensis Sueur 2001: 46, Table 2 (listed)

Euterpnosia chibensis Jeon, Kim, Tripotin and Kim 2002: 241 (parasitism, comp. note) Japan

Euterpnosia chibensis Lee and Hayashi 2003b: 363 (type species of *Euterpnosia*) Japan

Euterpnosia chibensis chibensis Saisho 2003: 51 (comp. note) Japan

Euterpnosia chibensis Boulard 2005h: 45 (comp. note) Honshu, Ryukyu, China

Euterpnosia chibensis chibensis Boulard 2005h: 45 (comp. note)

Euterpnosia chibensis Chen 2005: 48 (comp. note)

Euterpnosia chibensis Saisho 2006b: 60 (comp. note) Japan

Euterpnosia chibensis chibensis Saisho 2006b: 60 (comp. note) Japan

Euterpnosia chibensis Shiyake 2007: 4, 48, 50, Fig 74 (illustrated, distribution, listed, comp. note) Japan

Euterpnosia chibensis Boulard 2008f: 47 (type species of *Euterpnosia*)

Euterpnosia chibensis Lee 2008a: 9 (type species of *Euterpnosia*) Japan

Euterpnosia chibensis Osaka Museum of Natural History 2008: 324 (listed, acoustic behavior, ecology, comp. note) Japan

Euterpnosia chibensis Pham and Yang 2009: 12 (type species of *Euterpnosia*)

Euterpnosia chibensis Pham and Yang 2010b: 206 (type species of *Euterpnosia*) Japan

***E. chibensis daitoensis* Matsumura, 1917** = *Euterpnosia chibensis* var. *daitoensis* = *Leptopsaltria tuberosa* Matsumura (nec Signoret) = *Euterpnosia chibensis gotoi* Kato, 1936 = *Euterpnosia chibensis* var. *gotoi*

Euterpnosia chibensis daitoensis Hayashi 1984: 50–52, Figs 90–91, Fig 97, Figs 101–102 (key, described, illustrated, synonymy, comp. note) Equals *Leptopsaltria tuberosa* (nec Signoret) Equals *Euterpnosia chibensis var. daitoensis* Equals *Euterpnosia chibensis var. gotoi* Equals *Euterpnosia chibensis gotoi* Japan

Euterpnosia chibensis gotoi Hayashi 1984: 51 (comp. note)

Euterpnosia chibensis daitoensis Hayashi, Sasaki and Saisho 2000: 51–54, Figs 1–4 (acoustic behavior, ecolsion, comp. note) Ryukyus

Euterpnosia chibensis daitoensis Saisho 2001b: 69 (comp. note) Japan

Euterpnosia chibensis daitoensis Osaka Museum of Natural History 2008: 324 (listed, acoustic behavior, ecology, comp. note) Japan

***E. chibensis nuaiensis* Boulard, 2005**

Euterpnosia chibensis nuaiensis Boulard 2005h: 7, 45–48, 52–53, Figs 62–67, Color Plate, Fig C (n. ssp., described, illustrated, song, sonogram, listed, comp. note) Thailand

Euterpnosia chibensis nuaiensis Boulard 2005o: 150 (listed)

Euterpnosia chibensis nuaiensis Boulard 2007c: 621 (illustrated, comp. note) Thailand

Euterpnosia chibensis nuaiensis Boulard 2007d: 500 (comp. note) Thailand

Euterpnosia chibensis nauiensis Sanborn, Phillips and Sites 2007: 14–15 (listed, distribution, comp. note) Thailand

Euterpnosia chibensis nuaiensis Boulard 2008f: 47, 95 (biogeography, listed, comp. note) Thailand

***E. chinensis* Kato, 1940**

Euterpnosia chinensis Chou, Lei, Li, Lu and Yao 1997: 198–200 (key, described, comp. note) China

Euterpnosia chinensis Hua 2000: 61 (listed, distribution, hosts) China, Henan, Zhejiang

Euterpnosia chinensis Chen 2005: 38–39, 42, 45, Fig 6G, Fig 14 (illustrated, comp. note) Taiwan

E. chishingsana **Chen & Shiao, 2008**
Euterpnosia chishingsana Chen and Shiao 2008: 83–90, Figs 1D–D1, Fig 2D, Fig 3D, Fig 4D, Fig 7, Fig 8D, Fig 9D (n. sp.. described, illustrated, song, oscillogram, sonogram, comp. note) Taiwan

E. crowfooti **(Distant, 1912)** = *Terpnosia crowfooti*
Euterpnosia crowfooti Chen 2005: 48 (comp. note)
Euterpnosia crowfooti Lee 2008a: 9 (synonymy, distribution, listed) Equals *Terponsia crowfooti* Vietnam, Nepal, India
Euterpnosia crowfooti Pham and Yang 2009: 14, Table 2 (listed) Vietnam
Euterpnosia crowfooti Pham, Thinh and Yang 2010: 63–65, 67, Fig 3A, Fig 3E (illustrated, listed, key, comp. note) Vietnam

E. cucphuongensis **Pham, Ta & Yang, 2010**
Euterpnosia undescribed species Pham and Yang 2010a: 133 (comp. note) Vietnam
Euterpnosia cucphuongensis Pham, Thinh and Yang 2010: 64–67, Figs 1–2, Fig 3D, Fig 3H (n. sp., described, illustrated, distribution, listed, key, comp. note) Vietnam

E. elongata Lee, 2003 to *E. varicolor* Kato, 1926

E. gina **Kato, 1931**
Euterpnosia gina Chou, Lei, Li, Lu and Yao 1997: 202 (listed) China
Euterpnosia gina Hua 2000: 61 (listed, distribution) China, Taiwan
Euterpnosia gina Lee and Hayashi 2003b: 363, 377–379, Figs 19–20 (key, diagnosis, illustrated, distribution, song, listed, comp. note) Taiwan
Euterpnosia gina Chen 2004: 95–96, 112, 199 (described, illustrated, listed, comp. note) Taiwan
Euterpnosia gina Chen 2005: 48–49 (comp. note)
Euterpnosia gina Shiyake 2007: 4, 48, 50, Fig 75 (illustrated, distribution, listed, comp. note) Taiwan
Euterpnosia gina Chen and Shiao 2008: 81 (song, sonogram, comp. note) Taiwan

E. hohogura **Kato, 1933**
Euterpnosia hohoguro Lee and Hayashi 2003b: 371 (diagnosis, distribution, song, listed, comp. note) Taiwan
Euterpnosia hohoguro Chen 2004: 95, 196, 199 (listed, comp. note)
Euterpnosia hohoguro Chen and Shiao 2008: 81 (song, sonogram, comp. note) Taiwan

E. hoppo **Matsumura, 1917** = *Euterpnosia sozanensis* Kato, 1925
Euterpnosia hoppo Chou, Lei, Li, Lu and Yao 1997: 202 (listed) China
Euterpnosia hoppo Hua 2000: 61 (listed, distribution) China, Taiwan
Euterpnosia hoppo Lee and Hayashi 2003b: 363, 371–376, 379, Figs 11–12 (key, diagnosis, illustrated, synonymy, distribution, song, listed, comp. note) Equals *Euterpnosia sozanensis* Taiwan
Euterpnosia hoppo Chen 2004: 95–97, 198 (described, illustrated, listed, comp. note) Taiwan
Euterpnosia hoppo Chen 2005: 48 (comp. note)
Euterpnosia hoppo Chen and Shiao 2008: 81 (song, sonogram, comp. note) Taiwan

E. inanulata Kishida, 1929 to *Leptosemia takanonis* Matsumura, 1917

E. iwasakii **Matsumura, 1913** = *Purana iwasakii* = *Euternosia* (sic) *iwasakii* = *Euterpnosia iwasaku* (sic)
Euterpnosia iwasakii Hayashi 1984: 29, 50, 52–53, Figs 94–95, Figs 105–106 (key, described, illustrated, synonymy, comp. note) Equals *Purana iwasakii* Japan
Euterpnosia iwasakii Boulard 1988d: 150 (wing abnormalities, comp. note)
Euterpnosia iwasakii Saisho 2001a: 69 (comp. note) Japan
Euterpnosia iwasakii Sueur 2001: 46, Table 2 (listed)
Euterpnosia iwasakii Chen 2005: 48 (comp. note)
Euterpnosia iwasakii Shiyake 2007: 4, 48, 50, Fig 76 (illustrated, distribution, listed, comp. note) Japan
Euterpnosia iwasakii Osaka Museum of Natural History 2008: 324 (listed, acoustic behavior, ecology, comp. note) Japan

E. koshunensis **Kato, 1927**
Euterpnosia koshunensis Chou, Lei, Li, Lu and Yao 1997: 202 (listed) China
Euterpnosia koshunensis Hua 2000: 61 (listed, distribution) China, Taiwan
Euterpnosia koshunensis Lee and Hayashi 2003b: 369 (diagnosis, distribution, song, listed, comp. note) Taiwan
Euterpnosia koshunensis Chen 2004: 95, 196, 199 (listed, comp. note)
Euterpnosia koshunensis Chen 2005: 48 (comp. note)
Euterpnosia koshunensis Chen and Shiao 2008: 81 (song, sonogram, comp. note) Taiwan
Euterpnosia koshunensis Lee, Lin and Wu 2010: 218–224, Figs 1–6, Tables 1–2 (ecology, comp. note) Taiwan

E. kotoshoensis **Kato, 1925**
Euterpnosia kotoshoensis Lee and Hayashi 2003b: 363, 374–376, 379, Figs 15–16 (key, diagnosis, illustrated, distribution, song, listed, comp. note) Taiwan
Euterpnosia kotoshoensis Chen 2004: 95–96, 114, 199 (described, illustrated, listed, comp. note) Taiwan

Euterpnosia kotoshoensis Chen 2005: 48 (comp. note)
Euterpnosia kotoshoensis Shiyake 2007: 4, 49–50, Fig 77 (illustrated, distribution, listed, comp. note) Taiwan
Euterpnosia kotoshoensis Chen and Shiao 2008: 81 (song, sonogram, comp. note) Taiwan

E. laii **Lee, 2003** = *Euterpnosia viridifrons* var. *a* Kato, 1927 (partim) = *Euterpnosia viridifrons* f. *a* (partim)
Euterpnosia laii Lee and Hayashi 2003b: 363, 378–381, Figs 21–24 (n. sp., key, diagnosis, illustrated, distribution, song, oscillogram, listed, comp. note) Equals *Euterpnosia viridifrons* var. *a* Equals *Euterpnosia viridifrons* f. *a* Taiwan
Euterpnosia laii Chen 2004: 95–96, 116, 199 (described, illustrated, listed, comp. note) Taiwan
Euterpnosia laii Chen 2005: 48 (comp. note) Taiwan
Euterpnosia laii Shiyake 2007: 4, 49–50, Fig 78 (illustrated, distribution, listed, comp. note) Taiwan
Euterpnosia laii Chen and Shiao 2008: 81 (song, sonogram, comp. note) Taiwan

E. latacauta **Chen & Shiao, 2008**
Euterpnosia latacauta Chen and Shiao 2008: 82–90, Figs 1C–C1, Fig 2C, Fig 3C, Fig 4C, Fig 6, Fig 8C, Fig 9C (n. sp., described, illustrated, key, song, oscillogram, sonogram, comp. note) Taiwan

E. madhava **(Distant, 1881)** = *Pomponia madhava* = *Terpnosia madhava*
Euterpnosia madhava Chou, Lei, Li, Lu and Yao 1997: 17, 198, 200–201, 357, Fig 3–6E, Fig 9–56, Plate IX, Fig 98 (key, illustrated, described, synonymy, comp. note) Equals *Pomponia madhava* Equals *Terpnosia madhava* China
Euterpnosia madhava Chen 2005: 48 (comp. note)
Euterpnosia madhava Lee 2008a: 9 (synonymy, distribution, listed) Equals *Pomponia madhava* Vietnam, China, Xizang, Bhutan, India
Euterpnosia madhava Pham and Yang 2009: 14, Table 2 (listed) Vietnam
Euterpnosia madhava Pham, Thinh and Yang 2010: 63–65, 67, Fig 3B, Fig 3F (illustrated, listed, key, comp. note) Vietnam

E. madawdawensis **Chen, 2005**
Euterpnosia madawdawensis Chen 2005: 41–49, Fig 4, Fig 5D, Figs 6E–F, Fig 7D, Figs 8G–H, Figs 9G–H, Fig 10C, Fig 10E, Fig 13, Fig 14, Table 1 (n. sp., described, illustrated, song, sonogram, comp. note) Taiwan
Euterpnosia madawdawensis Chen 2006: 71–75, Fig 1C, Table 1 (key, comp. note) Taiwan
Euterpnosia madawdawensis Chen and Shiao 2008: 81 (song, sonogram, comp. note) Taiwan

E. okinawana **Ishihara, 1968** = *Euterpnosia chibensis okinawana*
Euterpnosia chibensis okinawana Hayashi 1984: 49–52, Figs 92–93, Fig 98, Figs 103–104 (key, described, illustrated, synonymy, comp. note) Equals *Euterpnosia okinawana* Japan
Euterpnosia chibensis okinawana Saisho 2006b: 60 (comp. note) Japan
Euterpnosia chibensis okinawana Osaka Museum of Natural History 2008: 324 (listed, acoustic behavior, ecology, comp. note) Japan

E. olivacea **Kato, 1927**
Euterpnosia olivacea Chou, Lei, Li, Lu and Yao 1997: 202 (listed) China
Euterpnosia olivacea Hua 2000: 61 (listed, distribution) China, Taiwan
Euterpnosia olivacea Lee and Hayashi 2003b: 363, 375–377, 379, Figs 17–18 (key, diagnosis, illustrated, distribution, song, listed, comp. note) Taiwan
Euterpnosia olivacea Chen 2004: 95–96, 110, 198 (described, illustrated, listed, comp. note) Taiwan
Euterpnosia olivacea Chen 2005: 48 (comp. note)
Euterpnosia olivacea Shiyake 2007: 4, 49, 51, Fig 79 (illustrated, distribution, listed, comp. note) Taiwan
Euterpnosia olivacea Chen and Shiao 2008: 81 (song, sonogram, comp. note) Taiwan

E. ruida **Lei & Chou, 1997** = *Euterpnosia* sp. 1 Pham & Thinh, 2006
Euterpnosia ruida Chou, Lei, Li, Lu and Yao 1997: 198, 201–202, 357, 367, Fig 9–57, Plate IX, Fig 93 (n. sp., described, illustrated, key, comp. note) China
Euterpnosia ruida Chen 2005: 48 (comp. note)
Euterpnosia ruida Pham and Thinh 2006: 525–527 (listed, comp. note) Vietnam
Euterpnosia sp. 1 Pham and Thinh 2006: 525 (listed) Vietnam
Euterpnosia ruida Pham and Yang 2009: 12, 14, 16, Table 2 (listed, distribution, comp. note) Vietnam, China
Euterpnosia ruida Pham, Thinh and Yang 2010: 63–65, 67, Fig 3C, Fig 3G (illustrated, listed, key, comp. note) Vietnam

E. sinensis hwaisana **Chen, 1940** = *Euterpnosia sinensis hwasiana* (sic)
Euterpnosia sinensis hwasiana (sic) Chen 1942: 146 (listed, comp. note) China
Euterpnosia sinensis hwasiana (sic) Hua 2000: 61 (listed, distribution) China, Sichuan

***E. sinensis sinensis* Chen, 1940**
Euterpnosia sinensis Chou, Lei, Li, Lu and Yao 1997: 202 (listed) China
Euterpnosia sinensis Hua 2000: 61 (listed, distribution) China, Zhejiang

***E. suishana* Kato, 1931**
Euterpnosia suishana Chou, Lei, Li, Lu and Yao 1997: 202 (listed) China
Euterpnosia suishana Hua 2000: 61 (listed, distribution) China, Taiwan
Euterpnosia suishana Lee and Hayashi 2003b: 363, 373–376, 379, Figs 13–14 (key, diagnosis, illustrated, distribution, song, listed, comp. note) Taiwan
Euterpnosia suishana Chen 2004: 95–96, 99, 198 (described, illustrated, listed, comp. note) Taiwan
Euterpnosia suishana Chen 2005: 48 (comp. note)
Euterpnosia suishana Shiyake 2007: 4, 49, 51, Fig 80 (illustrated, distribution, listed, comp. note) Taiwan
Euterpnosia suishana Chen and Shiao 2008: 81, 83–86, 88–90, Figs 1A–A1, Fig 2A, Fig 3A, Fig 4A, Fig 5, Fig 8A, Fig 9A (illustrated, song, oscillogram, sonogram, comp. note) Taiwan

***E. varicolor* Kato, 1926** = *Euterpnosia elongata* Lee, 2003
Euterpnosia varicolor Chou, Lei, Li, Lu and Yao 1997: 202 (listed) China
Euterpnosia varicolor Hua 2000: 61 (listed, distribution) China, Taiwan
Euterpnosia varicolor Lee and Hayashi 2003b: 363–366, 371, 378–379, Figs 3–5 (key, diagnosis, illustrated, distribution, song, listed, comp. note) Taiwan
Euterpnosia elongata Lee and Hayashi 2003b: 363, 367–369, 371, 378–379, Figs 6–8 (n. sp., key, diagnosis, illustrated, distribution, song, listed, comp. note) Taiwan
Euterpnosia elongata Chen 2004: 9, 106, 191, 198 (described, illustrated, listed, comp. note) Taiwan
Euterpnosia varicolor Chen 2004: 95–96, 103, 198 (described, illustrated, listed, comp. note) Taiwan
Euterpnosia varicolor Chen 2005: 37–49, Fig 2, Figs 5A–B, Figs 6A–B, Figs 7A–B, Figs 8A–D, Figs 9A–D, Fig 10A, Figs 10D, Fig 11, Fig 14 (synonymy, described, neotype designation, song, sonogram, comp. note) Equals *Euterpnosia elongata* Taiwan
Euterpnosia elongata Chen 2005: 42, 45, 47, Fig 7B, Fig 14 (synonymy, illustrated, comp. note)
Euterpnosia varicolor group Chen 2006: 71, 74 (key, comp. note) Taiwan
Euterpnosia varicolor Chen 2006: 71–72, 75, Table 1 (key, comp. note) Taiwan
Euterpnosia varicolor Chen and Shiao 2008: 81, 83–90, Figs 1E–E1, Fig 2E, Fig 3E, Fig 4E, Fig 7, Fig 8E, Fig 9E (illustrated, song, oscillogram, sonogram, comp. note) Taiwan

***E. viridifrons* Matsumura, 1917** = *Euterpnosia viridifrous* (sic) = *Euternosia* (sic) *viridifrons* = *Euterpnosia viridifrons* var. *a* Kato, 1927 (partim) = *Euterpnosia viridifrons* f. *a* (partim)
Euterpnosia viridifrons Hayashi 1984: 53 (comp. note)
Euterpnosia viridifrons Chou, Lei, Li, Lu and Yao 1997: 202 (listed) China
Euterpnosia viridifrons Hua 2000: 61 (listed, distribution) China, Taiwan
Euterpnosia viridifrons Lee and Hayashi 2003b: 363, 369–371, 376, 378, Figs 9–10 (key, diagnosis, illustrated, synonymy, distribution, song, listed, comp. note) Equals *Euterpnosia viridifrons* var. *a* Equals *Euterpnosia viridifrons* f. *a* Taiwan
Euterpnosia viridifrons Chen 2004: 95–96, 107, 198 (described, illustrated, listed, comp. note) Taiwan
Euterpnosia viridifrons Chen 2005: 47–48 (comp. note) Taiwan
Euterpnosia viridifrons Chen and Shiao 2008: 81 (song, sonogram, comp. note) Taiwan

E. viridifrons var. *a* Kato, 1927 to *Euterpnosia viridifrons* Matsumura, 1917 (partim) and *Euterpnosia laii* Lee, 2003 (partim)

***Leptosemia* Matsumura. 1917** = *Leptosomia* (sic) = *Leptopsalta* (sic) = *Chosenosemia* Doi, 1931
Leptosemia Doi 1932: 33 (comp. note) Equals *Chosenosemia*
Leptosemia Wu 1935: 10 (listed) Equals *Chosenosemia* China
Leptosemia Hayashi 1984: 47 (listed, comp. nte)
Leptosemia Lee 1995: 18, 23, 29, 53, 95–96, 136 (listed, key, described, ecology, distribution, illustrated, comp. note) Equals *Chosenosemia* Korea
Chosenosemia Lee 1995: 96 (comp. note) Korea
Leptosemia Chou, Lei, Li, Lu and Yao 1997: 197, 213, 293, Table 6 (key, described, listed, diversity, biogeography) China
Leptosemia Novotny and Wilson 1997: 437 (listed)
Leptosemia Lee 1999: 9 (comp. note) Equals *Chosenosemia* Korea
Leptosemia Villet and Zhao 1999: 322 (comp. note)
Leptosemia Lee 2003b: 7 (comp. note)
Leptosemia Lee and Hayashi 2003b: 360–361, 363, 381 (key, diagnosis) Taiwan

Leptosemia Chen 2004: 39, 42, 152 (phylogeny, listed)
Leptosemia Lee 2005: 15, 17, 91, 175 (listed, key, comp. note) Equals *Chosenosemia* Korea
Leptosemia Moulds 2005b: 390 (higher taxonomy, listed, comp. note)
Leptosemia Duffels and Hayashi 2006: 189 (comp. note)
Leptosemia Shiyake 2007: 6, 74 (listed, comp. note)
Leptosemia Lee 2008b: 446, 450, Table 1 (listed, comp. note) Equals *Chosenosemia* Korea
Chosenosemia Lee 2008b: 450 (comp. note) Korea
Leptosemia Lee 2010c: 23 (listed)
Leptosemia Lee and Hill 2010: 298, 301–302 (comp. note)
Leptosemia Pham and Yang 2010b: 206 (comp. note) Vietnam
Leptosemia Wei, Ahmed and Rizvi 2010: 33 (comp. note)

L. huasipana **Chen, 1943**
Leptosemia hunsipana Chou, Lei, Li, Lu and Yao 1997: 212 (listed, comp. note) China
Leptosemia huasipana Hua 2000: 62 (listed, distribution) China, Sichuan

L. sakaii **(Matsumura, 1913)** = *Leptopsaltria sakaii* = *Cicada sakaii* = *Terpnosia fuscolimbata* Schumacher, 1915 = *Terpnosia (Euterpnosia) fuscolimbata* == *Euterpnosia fuscolimbata* = *Leptosemia fuscolimbata* = *Leptosemia conica* Kato, 1926
Leptosemia sakaii Doi 1932: 33 (comp. note) Korea
Leptosemia sakaii Kim 1958: 94 (listed) Korea
Terpnosia fuscolimbata Hennessey 1990: 471 (type material)
Leptosemia sakaii Wang and Zhang 1993: 189 (listed, comp. note) China
Leptosemia sakaii Lee 1995: 28, 95–96 (type species of *Leptosemia*, comp. note) Formosa
Leptosemia sakaii Chou, Lei, Li, Lu and Yao 1997: 213–214, Fig 9–64 (key, described, illustrated, synonymy, type species of *Leptosemia*, comp. note) Equals *Terpnosia fuscolimbata* Equals *Leptosemia conica* Equals *Cicada sakaii* China
Leptosemia sakaii Lee 1999: 9 (type species of *Leptosemia*) Formosa
Leptosemia fuscolimbata Sanborn 1999a: 44 (type material, listed)
Leptosemia sakaii Hua 2000: 62 (listed, distribution) Equals *Leptopsaltria sakaii* Equals *Leptosemia conica* China, Hubei, Jiangxi, Zhejiang, Taiwan, Sichuan, Japan
Leptosemia sakaii Lee 2003b: 7–8, Fig 6 (illustrated, comp. note)
Leptosemia sakaii Lee and Hayashi 2003b: 361–363, Figs 1–2 (type species of *Leptosemia*, key, diagnosis, illustrated, synonymy, distribution, song, listed, comp. note) Equals *Leptopsaltria sakaii* Equals *Cicada sakaii* Equals *Terpnosia fuscolimbata* Equals *Terpnosia (Euterpnosia?) fuscolimbata* Equals *Terpnosia? fuscolimbata* Equals *Leptosemia conica* Taiwan, China
Leptosemia sakaii Chen 2004: 152–153, 199 (described, illustrated, listed, comp. note) Taiwan
Leptosemia sakaii Lee 2005: 35, 91, 175 (type species of *Leptosemia*) Formosa
Leptosemia sakaii Shiyake 2007: 6, 73–74, Fig 119 (illustrated, distribution, listed, comp. note) China, Taiwan
Leptosemia sakaii Lee 2008b: 450 (type species of *Leptosemia*) Formosa

L. takanonis **Matsumura, 1917** = *Cicada takanonis* = *Leptosomia* (sic) *takanonis* = *Leptosemia takanois* (sic) = *Euterpnosia inanulata* Kishida, 1929 = *Leptosemia sakaii* (nec Matsumura) = *Chosenosemia souyoensis* Doi, 1931 = *Leptosemia souyoensis* = *Tana* (sic) *japonensis* (nec Distant)
Chosenosemia souyoensis Doi 1932: 33 (comp. note) Korea
Leptosemia souyoensis Doi 1932: 33, 43 (listed, comp. note) Korea
Leptosemia souyoensis Kamijo 1933: 58 (listed, comp. note) Korea
Leptosemia takanonis Wu 1935: 10 (listed) China, Szechuan
Leptosemia takanonis Lee 1995: 18, 23, 29, 40–41, 45–46, 95–99, 136, 139 (listed, key, synonymy, ecology, distribution, illustrated, comp. note) Equals *Leptosemia sakaii* (nec Matsumura) Equals *Chonosemia souyoensis* Equals *Leptosemia souyoensis* Equals *Cicada takanonis* Equals *Tana* (sic) *japonensis* Equals *Leptopsalta souyoensis* Korea
Euterpnosia inanulata Lee 1995: 18, 22, 27–28, 33, 35, 37, 40–42, 44–46, 128, 135, 138 (comp. note) Equals *Euterpnosia? inanulata*
Leptosemia sakaii Lee 1995: 35 (comp. note) Equals *Leptosemia takanonis* (nec Matsumura) Korea
Chosenosemia souyoensis Lee 1995: 35, 96 (comp. note)
Leptosemia souyoensis Lee 1995: 36–38, 42, 44–45, 96 (comp. note)
Euterpnosia? inanulata Lee 1995: 43, 128 (comp. note)
Cicada takanonis Lee 1995: 43, 96 (comp. note)
Tana (sic) *japonensis* Lee 1995: 96 (comp. note)
Leptosemia takanonis Chou, Lei, Li, Lu and Yao 1997: 213, 215, Fig 9–65, Plate X, Fig 105 (key, described, illustrated, synonymy, comp. note) Equals *Chosenosemia souyoensis* Equals *Leptosemia souyoensis* Equals *Cicada takanonis* China

Chosensemia souyoensis Lee 1999: 9 (type species of *Chosenosemia*) Korea
Leptosemia takanonis Lee 1999: 9–10 (distribution, song, listed, comp. note) Equals *Leptosemia sakaii* (nec Matsumura) Equals *Chosenosemia souyoensis* Equals *Leptosemia souyoensis* Equals *Cicada takanonis* Equals *Tana* (sic) *japonensis* (nec Distant) Equals *Leptopsalta souyoensis* Korea, China
Leptosemia takanonis Hua 2000: 62 (listed, distribution) China, Jiangxi, Zhejiang, Fujian, Guangdong, Guangxi, Sichuan, Japan, Korea
Leptosemia takanonis Lee 2001a: 51 (comp. note) Korea
Leptosemia takanonis Lee, Choe, Lee and Woo 2002: 3–9, Figs 1–3, Tables 2–3 (gene sequence, phylogeny, comp. note) Russia, China, Korea
Leptosemia takanonis Lee 2003b: 7–8, Fig 5 (illustrated, comp. note)
Leptosemia takanonis Lee 2005: 17, 90–93, 157, 166, 175 (described, illustrated, synonymy, song, distribution, key, comp. note) Equals *Cicada takanonis* Equals *Euterpnosia inanulata* Equals *Euterpnosia? inanulata* Equals *Leptosemia sakaii* (nec Matsumura) Equals *Chosenosemia souyoensis* Equals *Leptosemia souyoensis* Equals *Tana* (sic) *japonensis* (nec Distant) Korea, China
Chosenosemia souyoensis Lee 2005: 90, 175 (type species of *Chosenosemia*, comp. note) Korea
Leptosemia takanonis Shiyake 2007: 6, 73–74, Fig 120 (illustrated, distribution, listed, comp. note) China, Korea
Leptosemia takanonis Lee 2008b: 445–447, 450, 463, Fig 1G, Table 1 (key, illustrated, synonymy, distribution, listed, comp. note) Equals *Cicada takonensis* Equals *Euterpnosia inanulata* n. syn. Equals *Euterpnosia? inanulata* Equals *Leptosemia sakaii* (nec Motschulsky) Equals *Chosenosemia souyoensis* Equals *Leptosemia souyoensis* Equals *Leptosalta* (sic) *souyoensis* Equals *Tana* (sic) *japonensis* (nec Distant) Korea, China
Euterpnosia inanulata Lee 2008b: 445, 450 (synonymy) Korea
Chosenosemia souyoensis Lee 2008b: 450 (type species of *Chosenosemia*) Korea
Leptosemia takanonis Lee and Hill 2010: 299–300, Figs 7–8 (phylogeny, comp. note)

L. umbrosa **Ku. In Wagner, 1968**

Onomacritus **Distant, 1912**

Onomacritus Duffels 1986: 319 (distribution, comp. note) Sumatra
Onomacritus Moulds 2005b: 390 (higher taxonomy, listed, comp. note)
Onomacritus Wei, Ahmed and Rizvi 2010: 33 (comp. note)

O. sumatranus **Distant, 1912**

Neocicada **Kato, 1932**

Neocicada Arnett 1985: 217 (diversity) Northeastern United States, Southeastern United States, Kansas, Missouri
Neocicada Poole, Garrison, McCafferty, Otte and Stark 1997: 251, 332 (listed) North America
Neocicada Sanborn 1999b: 1 (listed) North America
Neocicada Arnett 2000: 299 (diversity) Northeastern United States, Southeastern United States, Kansas, Missouri
Cicada (sic) Arnett 2000: 299 (diversity) United States
Neocicada Moulds 2005b: 390 (higher taxonomy, listed, comp. note)
Neocicada Sanborn, Heath, Sueur and Phillips 2005: 191–192, 205 (key, described, comp. note) North America, Central America
Neocicada spp. Sanborn, Heath, Sueur and Phillips 2005: 195, 199, 202–203, 205 (comp. note)
Neocicada Elliott and Hershberger 2006: 184 (listed)
Neocicada sp. Phillips and Sanborn 2007: 457, Fig 2 (listed) Texas
Neocicada Phillips and Sanborn 2007: 460 (comp. note) North America, Central America
Neocicada Sanborn 2007b: 39 (comp. note)
Neocicada Shiyake 2007: 6, 74 (listed, comp. note)
Neocicada sp. Shiyake 2007: 6, 73–74, Fig 121 (illustrated, listed, comp. note)
Neocicada Sanborn, Phillips and Gillis 2008: 3, 13 (key, diversity, described, comp. note) Florida
Neocicada Lee 2010c: 23 (listed)
Neocicada Lee and Hill 2010: 298, 301–302 (comp. note)
Neocicada Wei, Ahmed and Rizvi 2010: 33 (comp. note)

N. australamexicana **Sanborn & Sueur, 2005** = *Neocicada* sp. Sueur, 2002 = *Tettigia hieroglyphica* (nec Say)

Neocicada sp. Sueur 2002a: 380–381, 383, 387–391, Fig 3, Fig 10, Figs 12–14, Table 1 (acoustic behavior, song, sonogram, oscillogram, comp. note) Mexico, Veracruz
Neocicada australamexicana Sanborn, Heath, Sueur and Phillips 2005: 191–192, 195, 200–202, 205, Fig 2, Fig 8 (n. sp., key, described, illustrated, distribution, comp. note) Equals *Tettigia hieroglyphica* (nec Say) Distant, 1881 Equals *Neocicada* sp. Sueur, 2002 Mexico, Veracruz, Chiapas
Neocicada australamexicana Sanborn 2007b: 37, 39 (distribution, comp. note) Mexico, Veracruz, Chiapas

***N. centramericana* Sanborn, 2005**

Neocicada centramericana Sanborn, Heath, Sueur and Phillips 2005: 191–192, 194, 200, 202–203, 205–206, Fig 1, Fig 8 (n. sp., key, described, illustrated, distribution, comp. note) Honduras, Guatemala, Belize

Neocicada centramericana Sanborn 2005b: 77 (listed, distribution) Belize, Guatemala, Honduras

***N. chisos* (Davis, 1916)** = *Cicada chisos* = *Tettigia hieroglyphica* (nec Say) = *Neocicada hieroglyphica* (nec Say)

Cicada chisos Tinkham 1941: 171, 176, 179 (distribution, habitat, comp. note) Texas

Cicada chisos Arnett 1985: 217 (diversity) Texas

Neocicada chisos Sanborn and Phillips 1994: 45 (distribution, comp. note) Texas

Neocicada chisos Sanborn and Phillips 1995a: 481, Table 1 (song intensity, comp. note)

Neocicada chisos Van Pelt 1995: 19 (listed, distribution, comp. note) Texas

Cicada chisos Poole, Garrison, McCafferty, Otte and Stark 1997: 331 (listed) North America

Neocicada chisos Sanborn 1999a: 44 (type material, listed)

Neocicada chisos Sanborn 1999b: 1 (listed) North America

Cicada chisos Arnett 2000: 299 (diversity)

Neocicada chisos Sanborn, Heath, Sueur and Phillips 2005: 191–192, 199–201, 205, Fig 6, Fig 8 (key, synonymy, described, illustrated, distribution, comp. note) Equals *Tettigia hieroglyphica* (partim) Equals *Cicada chisos* Equals *Neocicada hieroglyphica* (partim) Texas, Mexico, Nuevo Leon, Tamaulipas

Neocicada chisos Phillips and Sanborn 2007: 456–459, Table 1, Fig 2, Fig 6 (illustrated, habitat, distribution, comp. note) Texas

Neocicada chisos Sanborn 2007b: 37 (distribution, comp. note) Mexico, Nuevo León, Tamaulipas, Texas

Neocicada chisos Lee and Hill 2010: 299–300, Figs 7–8 (phylogeny, comp. note)

***N. hieroglyphica hieroglyphica* (Say, 1830)** = *Cicada hieroglyphica* = *Tettigia hieroglyphica* = *Tettigea* (sic) *hieroglyphica* = *Tibicen hieroglyphica* = *Cicada hieroblyphica* (sic) = *Cicada hierglyphica* (sic) = *Neocicada hieroglyphica* = *Cicada characterea* Germar, 1830 = *Cicada characteria* (sic) = *Tettigia characteria* (sic)

Tettigea (sic) *hieroglyphica* Davis 1906c: 240 (comp. note)

Tettigea (sic) *hieroglyphica* Davis 1911a: 216 (comp. note) Georgia

Cicada hieroglyphica Davis 1919c: 28 (comp. note) New York, Long Island

Cicada hieroglyphica McCoy 1965: 41, 43 (key, comp. note) Arkansas

Neocicada hieroglyphica Walker 1983: 63 (comp. note) Florida

Neocicada hieroglyphica Arnett 1985: 217 (diversity) Northeastern United States, Southeastern United States, Kansas, Missouri

Neocicada hieroglyphica Daniel, Knight and Charles 1993: 69–70, 72–75, Fig 5, Figs 8–9, Tables 1–22 (song, illustrated, comp. note) North Carolina

Neocicada hieroglyphica Sanborn and Phillips 1995a: 481, Table 1 (song intensity, comp. note)

Neocicada hieroglyphica Hostetler 1997: 27 (comp. note)

Neocicada hieroglyphica hieroglyphica Hostetler 1997: 29–30 (comp. note)

Neocicada hieroglyphica Poole, Garrison, McCafferty, Otte and Stark 1997: 332 (listed) Equals *Cicada hieroglyphica* Equals *Cicada characterea* Equals *Cicada johannis* Equals *Cicada sexguttata* Equals *Cicada hieroblyphica* (sic) Equals *Cicada hierglyphica* (sic) North America

Neocicada hieroglyphica Sanborn 1999b: 1–2 (type material, listed) North America

Cicada hieroglyphica Lago and Testa 2000: 188 (listed) Mississippi

Neocicada hieroglyphica Sueur 2001: 40, 46, Tables 1–2 (listed) Equals *Cicada hieroglyphica*

*Neocicada hieroglyphica*Taber and Fleenor 2003: 105–106, Fig 90 (illustrated, comp. note) Texas

Cicada hieroglyphica Maccagnan and Martinelli 2004: 440 (comp. note)

Neocicada hieroglyphica Walker and Moore 2004: 2, Fig 2 (illustrated, listed, comp. note) Florida

Neocicada hieroglyphica Kondratieff, Schmidt, Opler and Garhart 2005: 216 (listed) Oklahoma

Cicada hieroglyphica Sanborn, Heath, Sueur and Phillips 2005: 191 (type species of *Neocicada*)

Neocicada hieroglyphica Sanborn, Heath, Sueur and Phillips 2005: 191–197, 199, 205, Fig 4, Fig 7 (key, type species of *Neocicada*, synonymy, described, illustrated, distribution, comp. note) Equals *Cicada hieroglyphica* Equals *Cicada characterea* Equals *Tettigia hieroglyphica* Equals *Cicada johannis* Equals *Cicada sexguttata* Equals *Tettigea hieroglyphica* Equals *Tibicen hieroglyphica* Equals *Tettigia hieroglyphica* Equals *Cicada hierglyphica* (sic) New York, Florida, Missouri, Illinois, Texas, Kansas, Oklahoma,

Louisiana, Arkansas, Alabama, Mississippi, Georgia, South Carolina, North Carolina, Tennessee, Virginia, Maryland, New Jersey, Pennsylvania

Cicada characterea Sanborn, Heath, Sueur and Phillips 2005: 193 (comp. note)

Neocicada hieroglyphica Elliott and Hershberger 2006: 32, 208–209 (illustrated, sonogram, listed, comp. note) Texas, Oklahoma, Kansas, Missouri, Arkansas, Louisiana, Mississippi, Alabama, Florida, Georgia, South Carolina, North Carolina, Tennessee, Kentucky, Indiana, Ohio, Virginia, Maryland, Delaware, New Jersey, New York, Connecticut

Neocicada hieroglyphica Marshall 2006: 139 (illustrated, comp. note) North Carolina

Neocicada hieroglyphica Eaton and Kaufman 2007: 88–89 (illustrated, comp. note) Eastern United States

Neocicada hieroglyphica Sanborn 2007b: 39 (comp. note)

Neocicada hieroglyphica Farris, Oshinshy, Forrest, and Hoy 2008: 17–20, 22, Fig 2d (song, parasitism) Mississippi

Neocicada hieroglyphica Hastings, Holliday and Coelho 2008: 659, 662, Table 1 (predation, comp. note) Florida

Neocicada hieroglyphica hieroglyphica Sanborn, Phillips and Gillis 2008: 3, 13–14, 31, 40, Fig 74, Figs 148–156 (key, synonymy, illustrated, distribution, comp. note) Equals *Cicada hieroglyphica* Equals *Cicada characterea* Florida, New Jersey, Pennsylvania

Neocicada hieroglyphica Sanborn, Phillips and Gillis 2008: 13–14 (type species of *Neocicada*, comp. note)

Neocicada hieroglyphica Holliday, Hastings and Coelho 2009: 3–5, Table 1 (predation, comp. note) Florida, Georgia, Louisiana, South Carolina

Neocicada hieroglyphica Hastings, Holliday, Long, Jones and Rodriguez 2010: 414–416, 419, Table 2 (predation, comp. note) Florida

***N. hieroglyphica johannis* (Walker, 1850)** = *Cicada johannis* = *Tettigia johannis* = *Cicada hieroglyphica johannis* = *Cicada sexguttata* Walker, 1850 = *Tettigia sexguttata*

Neocicada hieroglyphica var. *johannis* Sanborn 1999b: 1 (listed) North America

Neocicada hieroglyphica var. *johannis* Sanborn, Heath, Sueur and Phillips 2005: 191–192, 195–199, 205, Fig 5, Fig 7 (key, synonymy, described, illustrated, distribution, comp. note) Equals *Cicada johannis* Equals *Cicada sexguttata* Equals *Tettigia johannis* Equals *Neocicada johannis* Equals *Cicada hieroglyphica johannis* Equals *Neocicada hieroglyphica* var. *johannis* Florida, Georgia

Neocicada johannis Sanborn, Heath, Sueur and Phillips 2005: 196 (comp. note)

Neocicada sexguttata Sanborn, Heath, Sueur and Phillips 2005: 196 (comp. note)

Neocicada hieroglyphica johannis Sanborn, Phillips and Gillis 2008: 3, 14, 31, 41, Fig 74, Figs 157–165 (key, synonymy, illustrated, distribution, comp. note) Equals *Cicada johannis* Equals *Cicada sexguttata* Florida

Neocicada hieroglyphica johannis Holliday, Hastings and Coelho 2009: 4–5, Table 1 (predation, comp. note) Florida

Neocicada hieroglyphica johannis Lee and Hill 2010: 299–300, Figs 7–8 (phylogeny, comp. note)

***N. mediamexicana* Sanborn, 2005**

Neocicada mediamexicana Sanborn, Heath, Sueur and Phillips 2005: 191–192, 196, 200, 203–206Fig 3, Fig 8 (n. sp., key, described, illustrated, distribution, comp. note) Mexico, San Luis Potosi, Queretaro

Neocicada mediamexicana Sanborn 2006b: 257 (type locality) Mexico

Neocicada mediamexicana Sanborn 2007b: 37 (distribution, comp. note) Mexico, Queretaro, San Luis Potosí

***Puranoides* Moulton, 1917**

Puranoides Zaidi and Ruslan 1995c: 70, Table 2 (listed, comp. note) Peninsular Malaysia

Puranoides sp. Zaidi, Ruslan and Azman 2000: 201 (comp. note) Borneo, Sabah

Puranoides sp. Zaidi and Nordin 2001: 18 (comp. note) Borneo, Sabah

Puranoides Azman and Zaidi 2002: 2 (type species)

Puranoides sp. Zaidi, Nordin, Maryati, Wahab, Norashikin et al. 2002: 12 (comp. note) Borneo, Sabah

Puranoides sp. Zaidi and Azman 2003: 104, Table 1 (listed, comp. note) Borneo, Sabah

Puranoides Moulds 2005b: 391, 432 (higher taxonomy, listed, comp. note)

Puranoides Sanborn, Phillips and Sites 2007: 3, 31, Table 1 (listed) Thailand

Puranoides Boulard 2008f: 7, 41, 94 (listed) Thailand

Puranoides Lee 2009h: 2621 (listed) Philippines, Palawan

Puranoides Lee 2010a: 21 (listed) Philippines, Mindanao

Puranoides Lee and Hill 2010: 302 (comp. note)

***P. abdullahi* Azman & Zaidi, 2002**

Puranoides abdullahi Azman and Zaidi 2002: 9–13, Figs 8–14 (n. sp., described, comp. notes, illustrated) Malaysia

***P. goemansi* Lee, 2009**
Puranoides goemansi Lee 2009h: 2622–2625, Figs 3–4 (n. sp., described, illustrated, distribution, listed) Philippines, Palawan

***P. jaafari* Azman & Zaidi, 2002**
Puranoides jaafari Azman and Zaidi 2002: 3–8, 11 Figs 1–7 (n. sp., described, comp. notes, illustrated) Malaysia

***P. klossi* Moulton, 1917**
Puranoides klossi Azman and Zaidi 2002: 2, 6, 11 (distribution, type locality, comp. note) Thailand, Peninsular Malaysia, Sumatra
Puranoides klossi Sanborn, Phillips and Sites 2007: 31 (listed, distribution, comp. note) Thailand, Sumatra, Malay Peninsula
Puranoides klossi Boulard 2008f: 41, 94 (type species of *Puranoides*, biogeography, listed, comp. note) Thailand, Sumatra
Puranoides klossi Lee 2009h: 2621–2622 (type species of *Puranoides*, listed, comp. note) Sumatra

***P.* cf. *klossi* Schouten, Duffels & Zaidi, 2004**
Puranoides cf. *klossi* Schouten, Duffels and Zaidi 2004: 373, 378, Table 2 (listed, comp. note) Malayan Peninsula

***Leptopsaltria* Stål, 1866** = *Leptosaltria* (sic)
Leptopsaltria Hayashi 1984: 47, 53 (listed, comp. note)
Leptopsaltria Villet and Zhao 1999: 322 (comp. note)
Leptopsaltria Kos and Gogala 2000: 2 (comp. note)
Leptopsaltria Azman and Zaidi 2002: 2 (comp. note)
Leptopsaltria Schouten and Duffels 2002: 29 (comp. note)
Leptopsaltria Duffels 2004: 463 (comp. note)
Leptopsaltria Gogala, Trilar, Kozina and Duffels 2004: 4 (comp. note)
Leptopsaltria Lee and Hayashi 2004: 50 (comp. note) Taiwan
Leptopsaltria Moulds 2005b: 391, 432 (higher taxonomy, listed, comp. note)
Leptopsaltria Pham and Thinh 2005a: 243 (listed) Vietnam
Leptopsaltria Pham and Thinh 2005b: 290 (comp. note) Vietnam
Leptopsaltria Boulard 2006d: 195 (comp. note)
Leptopsaltria Boulard 2007a: 107 (comp. note)
Leptopsaltria Duffels, Schouten and Lammertink 2007: 367 (comp. note)
Leptopsaltria Sanborn, Phillips and Sites 2007: 3, 11, Table 1 (listed) Thailand
Leptopsaltria Boulard 2008f: 7, 41, 94 (listed) Thailand
Leptopsaltria Lee 2008a: 2, Table 1 (listed) Vietnam
Leptopsaltria Boulard 2009b: 47, 49 (comp. note)
Leptopsaltria Duffels 2009: 303–304 (comp. note)
Leptopsaltria Lee 2009f: 1487 (comp. note)
Leptopsaltria Pham and Yang 2009: 10, 14, Table 2 (type genus of *Leptopsaltria*)
Leptopsaltria Lee 2010c: 23 (listed)
Leptopsaltria Lee and Hill 2010: 302 (comp. note)

***L. andamanensis* Distant, 1888**
Leptopsaltria andamanensis Terradas 1999: 48–49, Fig 1 (illustrated, comp. note) Andaman Islands

***L. draluobi* Boulard, 2003**
Leptopsaltria draluobi Boulard 2003c: 172, 191–195, Figs 13–19, Plate J, Figs 1–2 (n. sp., described, illustrated, song, sonogram, comp. note) Thailand
Leptopsaltria draluobi Boulard 2005d: 344 (comp. note)
Leptopsaltria draluobi Boulard 2005o: 149 (listed)
Leptopsaltria draluobi Boulard 2006g: 74–75, 88, 180, Fig 50 (sonogram, listed, comp. note) Thailand
Leptopsaltria draluobi Sanborn, Phillips and Sites 2007: 11 (listed, distribution, comp. note) Thailand
Leptopsaltria draluobi Boulard 2008f: 41, 94 (biogeography, listed, comp. note) Thailand

***L. jaesornensis* Boulard, 2008** = *Leptopsaltria jaesornensis* Boulard, 2009
Leptopsaltria n. sp. Boulard 2008b: 109, 116, Fig 13 (illustrated, comp. note)
Leptopsaltria jaesornensis Boulard 2008e: 352–353, 355, Fig 14, (n. sp., described, illustrated, comp. note) Thailand
Leptopsaltria jaesornensis Boulard 2009b: 47–50, Figs 1–6 (n. sp., described, illustrated, song, sonogram, comp. note) Thailand

***L. mickwanae* Boulard, 2007**
Leptopsaltria mickwanae Boulard 2007a: 104, 106–107, Figs 17–19 (n. sp., described, illustrated, song, sonogram, comp. note) Thailand
Leptopsaltria mickwanae Boulard 2008f: 42, 94 (biogeography, listed, comp. note) Thailand

***L. phra* Distant, 1913**
Leptopsaltria phra Pham and Thinh 2005a: 240 (distribution, listed) Vietnam
Leptopsaltria phra Lee 2008a: 12 (distribution, listed) North Vietnam, China, Yunnan, India
Leptopsaltria phra Pham and Yang 2009: 14, 16, Table 2 (listed, comp. note) Vietnam
Leptopsaltria phra Pham and Yang 2010a: 133 (comp. note) Vietnam

***L. samia* (Walker, 1850)** = *Dundubia samia* = *Pomponia samia* = *Formosemia samia* = *Purana samia*
Purana samia Chou, Lei, Li, Lu and Yao 1997: 227, 229–230, Fig 9–73, Plate X, Fig 108 (illustrated, key, described, synonymy,

comp. note) Equals *Dundubia samia* Equals *Leptopsaltria samia* Equals *Formotosemia samia* China
Purana samia Hua 2000: 64 (listed, distribution) Equals *Dundubia samia* China, Guangxi, Yunnan, India, Sikkim, Vietnam
Purana samia Boulard 2006d: 197 (listed, comp. note) Equals *Dundubia samia* Equals *Leptopsaltria samia* India, Tonkin, Thailand
Leptopsaltria samia Boulard 2008f: 6, 42, 88–89, 94, Fig 44, Fig 46 (song, sonogram, biogeography, listed, comp. note) Equals *Dundubia samia* Thailand, India, Tonkin, Sikkim, China
Purana samia Lee 2008a: 11 (synonymy, distribution, listed) Equals *Dundubia samia* Equals *Leptopsaltria samia* Equals *Formosemia samia* North Vietnam, China, Yunnan, India
Purana samia Pham and Yang 2009: 14, Table 2 (listed) Vietnam

***L. tuberosa* (Signoret, 1847)** = *Cicada tuberosa* = *Dundubia tuberosa* = *Leptosaltria* (sic) *tuberosa* = *Leptopsaltria toberosa* (sic) = *Leptopsaltria tuberose* (sic)
Leptopsaltria tuberosa Boulard 1990b: 89, 101, 103, Plate X, Fig 9 (genitalia, comp. note)
Leptopsaltria tuberose (sic) Chen 2004: 193 (comp. note)
Leptopsaltria tuberosa Lee and Hayashi 2004: 71 (comp. note) Equals *Cicada tuberosa* Nepal, India
Leptopsaltria tuberosa Boulard 2007e: 62, Fig 42 (genitalia illustrated, comp. note) Thailand
Leptopsaltria tuberosa Boulard 2008f: 41 (type species of *Leptopsaltria*)
Cicada tuberosa Lee 2008a: 12 (type species of *Leptopsaltria*)

***Tanna* Distant, 1905** = *Tann* (sic) = *Tana* (sic)
Tanna Wu 1935: 11 (listed) China
Tanna Chen 1942: 146 (listed)
Tann (sic) sp. Chen 1942: 146 (listed, comp. note) China
Tanna Fujiyama 1982: 183, 186 (comp. note)
Tanna Hayashi 1984: 28, 47, 53, 55 (key, described, listed, comp. note) Equals *Neotanna*
Tanna Kevan 1985: 41, 84 (illustrated, comp. note) Japan
Tanna Peng and Lei 1992: 105 (comp. note) China, Hunan
Tanna sp. Lei, Chou and Li 1994: 55, Fig 5 (song frequency spectrum, comp. note)
Tanna Gogala 1995: 112 (comp. note)
Tanna Lee 1995: 22–31, 71–72 (comp. note)
Tanna Zaidi and Ruslan 1995a: 173, 175, Table 2 (listed, comp. note) Borneo, Sarawak
Tanna Zaidi and Ruslan 1995b: 219–221, Table 2 (listed, comp. note) Borneo, Sabah
Tanna Zaidi and Ruslan 1995d: 202, Table 2 (listed, comp. note) Borneo, Sabah
Tanna Boer and Duffels 1996b: 315 (comp. note)
Tanna Zaidi 1996: 102–103, Table 2 (comp. note) Borneo, Sarawak
Tanna Zaidi and Hamid 1996: 54–55, Table 2 (listed, comp. note) Borneo, Sarawak
Tanna Chou, Lei, Li, Lu and Yao 1997: 197, 218, 293, 307, 313, 324, 326, Table 6, Tables 10–12 (key, described, synonymy, listed, diversity, biogeography) Equals *Neotanna* Hayashi, 1978 China
Tanna sp. Chou, Lei, Li, Lu and Yao 1997: 307, 314, 324, 326, 329, 336, Tables 10–12 (song, oscillogram, spectrogram) China
Tanna Zaidi 1997: 115, Table 2 (listed, comp. note) Borneo, Sarawak
Tanna Zaidi and Ruslan 1998b: 4–5, Table 2 (listed, comp. note) Borneo, Sabah
Tanna Zaidi and Ruslan 1999: 3–4, Table 2 (listed, comp. note) Borneo, Sarawak
Tanna Kos and Gogala 2000: 2 (comp. note)
Tanna spp. Lee 2001b: 3 (comp. note)
Tanna Yoshizawa and Saigusa 2001: 12 (listed)
Tanna Azman and Zaidi 2002: 2 (comp. note)
Tanna Schouten and Duffels 2002: 29 (comp. note)
Tanna Lee 2003b: 7 (comp. note)
Tanna Chen 2004: 39, 42, 129–130, 192 (phylogeny, listed)
Tanna Duffels 2004: 463 (comp. note)
Tanna Lee and Hayashi 2004: 46, 49, 57, 59 (key, diagnosis) Equals *Neotanna* Taiwan, Japan, China, Nepal
Tanna Lee 2005: 35 (comp. note)
Tanna Moulds 2005b: 391, 432 (higher taxonomy, listed, comp. note)
Tanna Boulard 2006g: 41 (comp. note)
Tanna Iwamoto, Inoue and Yagi 2006: 679, 681, Fig 2x (flight muscle, comp. note)
Tanna Özdikmen and Kury 2006: 219 (status)
Tanna Sanborn 2006c: 829 (comp. note)
Tanna Duffels, Schouten and Lammertink 2007: 367 (comp. note)
Tanna Sanborn, Phillips and Sites 2007: 3, 35, Table 1 (listed) Thailand
Tanna Shiyake 2007: 5, 46, 54 (listed, comp. note)
Tanna Boulard 2008f: 6–7, 45, 95 (listed, comp. note) Thailand
Tanna Osaka Museum of Natural History 2008: 324 (listed) Japan
Tanna Duffels 2009: 303 (comp. note)
Tanna Lee 2009f: 1487 (comp. note)

Tanna Pham and Yang 2009: 11 (listed, synonymy, comp. note) Equals *Neotanna* Vietnam, China
Neotanna Pham and Yang 2009: 11 (comp. note)
Tanna Lee, Lin and Wu 2010: 221 (diversity, comp. note) Taiwan
Tanna Lee 2010b: 22–23 (listed, comp. note) Cambodia
Tanna Lee 2010c: 23 (listed)
Tanna Lee and Hill 2010: 298, 301–302 (comp. note)

***T. abdominalis* Kato, 1938** = *Neotanna abdominalis* = *Neotanna sinensis* Ouchi, 1938
Neotanna sinensis Chou, Lei, Li, Lu and Yao 1997: 17, 222, Fig 3-6D, Fig 9-69, Plate XI, Fig 119 (illustrated, key, described, synonymy, comp. note) Equals *Neotanna abdominalis* China
Neotanna sinensis Hua 2000: 63 (listed, distribution) Equals *Neotanna abdominalis* China, Anhui, Zhejiang, Fujian, Hunan, Guangxi

***T. apicalis* Chen, 1940** = *Tana* (sic) *apicalis*
Tanna apicalis Chou, Lei, Li, Lu and Yao 1997: 221, 368 (listed, comp. note) China
Tanna apicalis Hua 2000: 64 (listed, distribution) China, Guangxi

***T. auripennis* Kato, 1930**
Tanna auripennis Hua 2000: 64 (listed, distribution) China, Taiwan
Tanna auripennis Chen 2004: 129, 199 (listed, comp. note) Taiwan
Tanna auripennis Lee and Hayashi 2004: 47, 51–52 (key, diagnosis, song, listed, comp. note) Taiwan

***T. bakeri* Moulton, 1923** = *Tana* (sic) *bakeri*
Tanna bakeri Zaidi and Ruslan 1995a: 172, Table 1 (listed, comp. note) Borneo, Sarawak
Tana (sic) *bakeri* Zaidi and Ruslan 1995b: 219, Table 1 (listed, comp. note) Borneo, Sabah
Tanna bakeri Zaidi 1996: 100–101, Table 1 (comp. note) Borneo, Sarawak
Tanna bakeri Zaidi and Hamid 1996: 53, 56, Table 1 (listed, comp. note) Borneo, Sarawak
Tanna bakeri Zaidi 1997: 112 (comp. note) Borneo, Sarawak
Tana (sic) *bakeri* Zaidi and Ruslan 1998b: 2, Table 1 (listed, comp. note) Borneo, Sabah
Tanna bakeri Zaidi and Ruslan 1999: 3, Table 1 (listed, comp. note) Borneo, Sarawak
Tanna bakeri Zaidi, Ruslan and Azman 1999: 304–305 (listed, comp. note) Borneo, Sabah, Sarawak
Tanna bakeri Zaidi and Nordin 2001: 183, 185, 189, 194, Table 1 (comp. note) Borneo, Sabah, Sarawak
Tanna bakeri Zaidi, Nordin, Maryati, Wahab, Norashikin et al. 2002: 3, 9–10, Tables 1–2 (listed, comp. note) Borneo, Sabah, Sarawak
Tanna bakeri Zaidi and Azman 2003: 104, Table 1 (listed, comp. note) Borneo, Sabah

***T. bhutanensis* Distant, 1912** = *Neotanna bhutanensis*

***T. chekiangensis* Ouchi, 1938** = *Tanna sinensis* Kato, 1938 = *Tanna japonensis* (nec Distant)
Tanna japonensis chekiangensis Hua 2000: 64 (listed, distribution) China, Zhejiang

***T. harpesi* Lallemand & Synave, 1953**

***T. herzbergi* Schmidt, 1932** = *Tanna herrbergi* (sic)
Tanna herzbergi Wu 1935: 11 (listed) China, Kwangsi
Tanna herzbergi Chou, Lei, Li, Lu and Yao 1997: 221 (listed, comp. note) China
Tanna herzbergi Hua 2000: 64 (listed, distribution) China, Fujian, Guangxi

T. horiensis Kato, 1926 to *Tanna taipinensis* (Matsumura, 1907)

T. horishana Kato, 1925 to *Tanna viridis* Kato, 1925

***T. infuscata* Lee & Hayashi, 2004** = *Tanna infuscate* (sic)
Tanna infuscate (sic) Chen 2004: 24, 129–130, 143, 199 (described, illustrated, listed, comp. note) Taiwan
Tanna infuscata Lee and Hayashi 2004: 47, 57–59, Figs 1c–d, Figs 17–21 (n. sp., key, described, illustrated, synonymy, song, listed, comp. note) Equals *Tanna* sp. Okada 2002 Taiwan

***T. insignis* Distant, 1906**
Tanna insignis Zaidi and Ruslan 1995b: 218–219, Table 1 (listed, comp. note) Borneo, Sabah
Tanna insignis Zaidi and Ruslan 1995d: 200 (comp. note)
Tanna insignis Zaidi and Ruslan 1998b: 2–3, Table 1 (listed, comp. note) Borneo, Sabah
Tanna insignis Zaidi, Ruslan and Azman 1999: 305 (listed, comp. note) Borneo, Sabah, Java
Tanna insignis Zaidi and Azman 2003: 104, Table 1 (listed, comp. note) Borneo, Sabah

***T. ishigakiana* Kato, 1960** = *Tanna japonensis ishigakiana*
Tanna japonensis ishigakiana Hayashi 1984: 53–55, Figs 111–112, Figs 115–116 (key, described, illustrated, synonymy, comp. note) Equals *Tanna ishigakiana* Japan
Tanna japonensis ishigakiana Saisho 2006c: 62 (mating, illustrated, comp. note) Japan
Tanna japonensis ishigakiana Osaka Museum of Natural History 2008: 324 (listed, acoustic behavior, ecology, comp. note) Japan

***T. japonensis japonensis* (Distant, 1892)** = *Pomponia japonensis* = *Tanna japonensie* (sic) = *Tanna nipponensis* (sic) Matsumura = *Tanna japonnensis* (sic) = *Tanna japonesis* (sic) = *Tanna joponensis* (sic) = *Leptopsaltria japonica* Horváth, 1892 = *Tanna* (?) *sasaii* Matsumura, 1939 = *Tanna japonensis kimotoi* Kato, 1943 = *Tanna japonensis* f. *kimotoi* = *Tanna* *japonensis nigrofusca* Kato, 1940 = *Tanna japonensis* var. *nigrofusca* = *Tanna obliqua* Liu, 1940

Tanna japonensis Chen 1933: 360 (listed, comp. note)
Tanna japonensis Isaki 1933: 381 (listed, comp. note) Japan
Tanna japonensis Kamijo 1933: 58 (listed, comp. note) Korea
Tanna japonensis Wu 1935: 11 (listed) Equals *Pomponia japonensis* Equals *Leptopsaltria japonica* Equals *Tanna nipponensis* China, Manchuria, Hangchow, Japan, Korea
Pomponia japonensis Chen 1942: 146 (type species of *Tanna*)
Tanna japonensis Ohgushi 1954: 11 (emergence times, ecology, comp. note) Japan
Tanna japonensis Ikeda 1959: 484 (comp. note) Japan
Tanna japonensis Katsuki 1960: 54–55, 57, Fig 1, Table 1 (hearing, comp. note) Japan
Tanna japonensis Hayashi 1982b: 187–188, Table 1 (distribution, comp. note) Japan
Tanna japonensis Hayashi 1984: 53–55, Figs 107–116 (key, described, illustrated, synonymy, comp. note) Equals *Pomponia japonensis* Equals *Leptopsaltria japonica* Japan
Pomponia japonensis Hayashi 1984: 53 (type species of *Tanna*)
Tanna japonensis japonensis Hayashi 1984: 53–54, Figs 107–110, Figs 113–114 (key, described, illustrated, comp. note) Japan
Tanna japonensis f. *kimotoi* Hayashi 1984: 54 (comp. note) Japan
Tanna japonensis var. *nigrofusca* Hayashi 1984: 54 (comp. note) Japan
Tanna japonensis Kevan 1985: 3, 78, 80, Fig 2 (illustrated, comp. note) Japan
Tanna japonensis Ishii 1990: 441 (parasitism, comp. note) Japan
Tanna japonensis Liu 1991: 257 (comp. note)
Tanna japonensis Liu 1992: 16, 34, 40 (comp. note)
Tanna japonensis japonensis Liu 1992: 34 (comp. note)
Tanna obliqua Liu 1992: 40 (comp. note)
Tanna japonensis Metevelis 1992: 190 (comp. note) Japan
Tanna obliqua Peng and Lei 1992: 102–103, Fig 293 (listed, illustrated, comp. note) China, Hunan
Tanna japonensis Lei, Chou and Li 1994: 54–55, Fig 4 (song frequency spectrum, comp. note)
Tanna japonensis Gogala 1995: 112, 115, Fig 11 (song, sonogram, oscillogram, comp. note)
Tanna japonensis Lee 1995: 18, 22, 28, 30–31, 33, 35–38, 40, 42–46, 125–128, 135, 138 (comp. note) Equals *Pomponia japonensis* Equals *Leptopsaltria japonensis* Equals *Tanna japonenis japonensis*
Leptopsaltria japonensis Lee 1995: 29, 125–126 (comp. note) Equals *Tanna japonensis*
Pomponia japonensis Lee 1995: 125 (comp. note)
Tanna japonensis japonensis Lee 1995: 126 (comp. note)
Tanua (sic) *japonensis* Lee 1995: 34, 126–127 (comp. note)
Tanna obliqua Xu, Ye and Chen 1995: 84–85 (listed, comp. note) China, Zhejiang
Tanna japonensis Hayashi, Yasui and Sato 1996: 39–40, Fig 1 (predation) Japan
Tanna japonensis Kurahashi 1996: 109–111, 113 (predation, comp. note) Japan
Pomponia japonensis Chou, Lei, Li, Lu and Yao 1997: 218 (type species of *Tanna*) China
Tanna japonensis Chou, Lei, Li, Lu and Yao 1997: 219–220, 307, 313–314, 324, 326, 329, 336, Fig 9-67, Plate X, Fig 102, Tables 10–12 (key, described, illustrated, synonymy, song, oscillogram, spectrogram, comp. note) Equals *Pomponia japonensis* Equals *Leptopsaltria japonica* Equals *Tanna obliqua* Equals *Tanna japonensis* var. *nigrofusca* Equals *Tanna japonensis* var. *kimotoi* China
Tanna japonensis Freytag 2000: 55 (on stamp)
Tanna japonensis Hua 2000: 64 (listed, distribution) Equals *Pomponia japonensis* Equals *Tanna obliqua* China, Anhui, Hubei, Jiangxi, Zhejiang, Fujian, Hunan, Guangxi, Sichuan, Xizang, Japan, Korea, Querpart Islands, India, Laos
Tanna japonensis Maezono and Miyashita 2000: 52 (nymphal cell host, comp. note) Japan
Tanna japonensis japonensis Ohara and Hayashi 2001: 1 (comp. note) Japan
Tanna japonensis Sueur 2001: 39, 46, Tables 1–2 (listed)
Pomponia japonensis Lee and Hayashi 2004: 46 (type species of *Tanna*) Japan
Tanna japonensis Lee and Hayashi 2004: 55 (comp. note)
Tanna japonensis Boulard 2005h: 10 (comp. note)
Pomponia japonensis Boulard 2005h: 10 (comp. note) Laos
Leptopsaltria japonica Boulard 2005h: 10 (comp. note) Equals *Pomponia japonensis*?
Tanna japonensis Lee 2005: 86 (comp. note)
Leptopsaltria japonica Lee 2005: 85 (comp. note)
Tanna japonensis Sanborn 2005c: 113 (comp. note)
Tanna japonensis Saisho 2006b: 57 (calling activity, comp. note) Japan
Tanna japonensis japonensis Tsuyuki, Sudo and Tani 2006: 745–747, 750–751, Fig 5a, Fig 6a, Fig 7, Fig 10, Fig 12 (wing shape, comp. note) Japan

Tanna japonensis Inoue, Goto, Makino, Okabe, Okochi et al. 2007: 251, Table 2 (trap collection, comp. note) Japan
Tanna japonensis Shiyake 2007: 5, 52, 54, Fig 83 (illustrated, distribution, listed, comp. note) Japan
Tanna japonensis Boulard 2008f: 45 (type species of *Tanna*)
Tanna japonensis Lee 2008b: 462 (comp. note)
Tanna japonensis Osaka Museum of Natural History 2008: 324 (listed, acoustic behavior, ecology, comp. note) Japan
Tanna japonensis Hisamitsu, Sokabe, Terada, Osumi, Terada, et al. 2009: 93, 103, Fig 3c (behavior, illustrated, comp. note) Japan
Pomponia japonensis Pham and Yang 2009: 11 (type species of *Tanna*)
Tanna japonensis Lee and Hill 2010: 299–300, Figs 7–8 (phylogeny, comp. note)
Pomponia japonensis Lee 2010b: 22 (type species of *Tanna*, listed)

***T. japonensis* var. _____ Ishihara, 1939**

T. japonensis chekiangensis Ouchi, 1938 to *Tanna chekiangensis*

***T. karenkonis* Kato, 1939**

Tanna karenkonis Hua 2000: 64 (listed, distribution) China, Taiwan
Tanna karenkonis Chen 2004: 129–130, 137, 199 (described, illustrated, listed, comp. note) Taiwan
Tanna karenkonis Lee and Hayashi 2004: 47, 52 (key, diagnosis, song, listed, comp. note) Taiwan

***T. kimtaewooi* Lee, 2010**

Tanna kimtaewooi Lee 2010b: 22–24, Fig 1 (n. sp., described, illustrated, distribution, listed, comp. note) Cambodia

***T. minor* Hayashi, 1978**

Tanna minor Chou, Lei, Li, Lu and Yao 1997: 224, 368 (comp. note)

T. obliqua Liu, 1940 to *Tanna japonensis japonensis* (Distant, 1892)

***T. ornata* Kato, 1940**

Tanna ornata Chou, Lei, Li, Lu and Yao 1997: 221–222 (listed, comp. note) China
Tanna ornata Hua 2000: 64 (listed, distribution) China, Zhejiang

***T. ornatipennis* Esaki, 1933** = *Tanna (Neotanna) ornatipennis* = *Neotanna ornatipennis*

Tanna ornatipennis Hua 2000: 64 (listed, distribution) China, Taiwan
Tanna ornatipennis Chen 2004: 129–130, 141, 196, 199 (described, illustrated, listed, comp. note) Taiwan
Tanna ornatipennis Lee and Hayashi 2004: 47, 57–59, Figs 22–23 (key, diagnosis, illustrated, synonymy, song, listed, comp. note) Equals *Tanna (Neotanna) ornatipennis* Equals *Neotanna ornatipennis* Taiwan

***T. pallida* Distant, 1906**

Tanna pallida Zaidi and Azman 2003: 99–100, 104, Table 1 (listed, comp. note) Borneo, Sabah
Tanna pallida Lee 2009h: 2637 (comp. note) Sulu Island

***T. pseudocalis* Lei & Chou, 1997**

Tanna pseudocalis Chou, Lei, Li, Lu and Yao 1997: 219–221, 357, 368, Fig 9–68, Plate X, Fig 104 (n. sp., key, described, illustrated, comp. note) China

***T. puranoides* Boulard, 2008**

Tanna puranoides Boulard 2008f: 45, 66–70, 95, Figs 14–18 (n. sp., described, illustrated, song, sonogram, biogeography, listed, comp. note) Thailand

***T. sayurie* Kato, 1926**

Tanna sayurie Hua 2000: 64 (listed, distribution) China, Taiwan
Tanna sayurie Chen 2004: 129–130, 135, 199 (described, illustrated, listed, comp. note) Taiwan
Tanna sayurie Lee and Hayashi 2004: 46–47, 52–53, Fig 1b, Figs 11–13 (key, diagnosis, illustrated, song, listed, comp. note) Taiwan

***T. sinensis* (Ouchi, 1938)** = *Neotanna sinensis*

Tanna sinensis Hua 2000: 64 (listed, distribution) China, Zhejiang
Neotanna sinensis Pham and Yang 2009: 11 (comp. note)
Tanna sinensis Pham and Yang 2009: 11–12, 14, 16, Table 2 (n. comb., synonymy, listed, distribution, comp. note) Equals *Neotanna sinensis* Equals *Neotanna abdominalis* Vietnam, China

***T. sozanensis* Kato, 1926**

Tanna sozanensis Hua 2000: 64 (listed, distribution) China, Taiwan
Tanna sozanensis Chen 2004: 23, 129–131, 199 (described, illustrated, listed, comp. note) Taiwan
Tanna sozanensis Lee and Hayashi 2004: 45–52, 54, 57, Fig 1a, Figs 205, Fig 9 (key, diagnosis, illustrated, song, oscillogram, listed, comp. note) Taiwan
Tanna sozanensis Lee 2005: 123 (comp. note)
Tanna sozanensis Lee and Hill 2010: 299–300, Figs 7–8 (phylogeny, comp. note)

***T.* sp. 1 Pham & Thinh, 2006**

Neotanna sp. 1 Pham and Thinh 2006: 526 (listed) Vietnam

T. taikosana Kato, 1925 to *Tanna taipinensis* (Matsumura, 1907)

***T. taipinensis* (Matsumura, 1907)** = *Leptopsaltria taipinensis* = *Tanna tapinensis* (sic) = *Tanna taikosana* Kato, 1925 = *Tanna horiensis* Kato, 1926

Tanna taikosana Chou, Lei, Li, Lu and Yao 1997: 222 (listed, comp. note) China

Tanna taipinensis Chou, Lei, Li, Lu and Yao 1997: 222 (listed, comp. note) China

Tanna taikosana Hua 2000: 64 (listed, distribution) China, Taiwan

Tanna taipinensis Hua 2000: 64 (listed, distribution) China, Fujian, Taiwan, Japan

Tanna taipinensis Chen 2004: 129–130, 133, 192, 199 (described, illustrated, listed, comp. note) Taiwan

Tanna horiensis Chen 2004: 192 (comp. note)

Tanna taikosana Chen 2004: 192 (comp. note)

Tanna taipinensis Lee and Hayashi 2004: 47, 49–54, Figs 6–10 (key, diagnosis, illustrated, synonymy, song, oscillogram, listed, comp. note) Equals *Leptopsaltria taipinensis* Equals *Tanna taikosana* n. syn. Equals *Tanna horiensis* n. syn. Taiwan

Tanna taikosana Lee and Hayashi 2004: 50 (comp. note)

Tanna horiensis Lee and Hayashi 2004: 50 (comp. note)

Leptopsaltria taipenensis Lee and Hayashi 2004: 50 (lectotype designation)

Tanna taikosana Lee 2005: 139 (comp. note)

Tanna taipinensis Lee 2005: 125 (comp. note)

***T. tairikuana* Kato, 1940**

***T. tigroides* (Walker, 1858)** = *Dundubia tigroides* = *Pomponia tigroides* = *Leptopsaltria tigroides* = *Purana tigroides* = *Neotanna tigroides*

Purana tigroides Shiyake 2007: 5, 52, 54, Fig 82 (illustrated, listed, comp. note)

***T.* undescribed species A Sanborn, Phillips & Sites, 2007**

Tanna undescribed species A Sanborn, Phillips and Sites 2007: 14 (listed, distribution, comp. note) Thailand

***T. ventriroseus* Boulard, 2003**

Tanna ventriroseus Boulard 2003a: 98–99, 109–111, Color Figs 6–7, Figs 7–11, sonograms 8–9 (n. sp., described, song, sonogram, illustrated, comp. note) Thailand

Tanna ventriroseus Boulard 2003c: 197 (comp. note)

Tanna ventriroseus Boulard 2005d: 344 (comp. note) Thailand

Tanna ventriroseus Boulard 2005o: 149 (listed)

Tanna ventriroseus Boulard 2006g: 76, 93, 180, Fig 60 (sonogram, listed, comp. note) Thailand

Tanna ventriroseus Boulard 2007e: 35, 45, Fig 29 (sonogram, comp. note) Thailand

Tanna ventriroseus Sanborn, Phillips and Sites 2007: 13 (listed, distribution, comp. note) Thailand

Tanna ventriroseus Boulard 2008f: 45, 95 (biogeography, listed, comp. note) Thailand

***T. viridis* Kato, 1925** = *Neotanna viridis* = *Neottanna* (sic) *viridis* = *Tanna horishana* Kato, 1925 = *Neotanna horishana* = *Neotanna horishaua* (sic) = *Euterpnosia horishana* = *Tanna viridis niitakaensis* Kato, 1926 = *Neotanna viridis niitakaensis* = *Neotanna tarowanensis* Matsumura, 1927 = *Neotanna rantaizana* Matsumura 1927 = *Neotanna viridis inmaculata* Kato, 1932 = *Tanna viridis immaculata* (sic)

Tanna viridis Hayashi 1984: 53 (type species of *Neotanna*)

Tanna viridis Chou, Lei, Li, Lu and Yao 1997: 222 (type species of *Neotanna*) China

Neotanna viridis Chou, Lei, Li, Lu and Yao 1997: 226 (listed, comp. note) China

Neotanna horishana Chou, Lei, Li, Lu and Yao 1997: 226 (listed, comp. note) China

Neotanna tarowanensis Chou, Lei, Li, Lu and Yao 1997: 226 (listed, comp. note) China

Neotanna horishana Hua 2000: 63 (listed, distribution) Equals *Tanna horishana* China, Taiwan

Neotanna rantaizana Hua 2000: 63 (listed, distribution) China, Taiwan

Neotanna tarowanensis Hua 2000: 63 (listed, distribution) China, Taiwan

Neottanna (sic) *viridis* Hua 2000: 63 (listed, distribution) China, Taiwan

Tanna viridis Hua 2000: 64 (listed, distribution) China, Taiwan

Tanna viridis niitakaensis Hua 2000: 64 (listed, distribution) China, Taiwan

Tanna viridis Chen 2004: 129–130, 138, 199 (described, illustrated, listed, comp. note) Taiwan

Neotanna tarowarensis Chen 2004: 192 (comp. note)

Neotanna horishana Chen 2004: 192 (comp. note)

Tanna viridis Lee and Hayashi 2004: 46, 50, 53–57, Figs 14–16 (type species of *Neotanna*, key, diagnosis, illustrated, synonymy, song, oscillogram, listed, comp. note) Equals *Neotanna viridis* Equals *Tanna horishana* n. syn. Equals *Neotanna horishana* Equals *Tanna viridis* var. *niitakaensis* Equals *Neotanna viridis* var. *niitakaensis* Equals *Neotanna tarowanensis* n. syn. Equals *Tanna tarowanensis* Equals *Neotanna rantaizana* Equals *Neotanna viridis* f. *inmaculata* Equals *Tanna viridis* f. *immaculata* (sic) Equals *Tanna* sp. 2 Hayashi 1979 Taiwan

Neotanna viridis Lee and Hayashi 2004: 54 (comp. note)

Tanna horishana Lee and Hayashi 2004: 54, 56 (comp. note)

Tanna tarowanensis Lee and Hayashi 2004: 54–56 (comp. note)
Tanna viridis Lee 2005: 123 (comp. note)
Tanna viridis Shiyake 2007: 5, 53–54, Fig 84 (illustrated, distribution, listed, comp. note) Taiwan
Tanna viridis Pham and Yang 2009: 11 (type species of *Neotanna*)

***T. yanni* Boulard, 2003**
Tanna yanni Boulard 2003c: 172, 196–198, Figs 20–24, Plate K, Figs 1–2 (n. sp., described, illustrated, song, sonogram, comp. note) Thailand
Tanna yanni Boulard 2005o: 149 (listed)
Tanna yanni Boulard 2006g: 42–43, 48–49, Fig 9d (illustrated, comp. note) Asia, Thailand
Tanna yanni Boulard 2007e: Plate 13b (illustrated) Thailand
Tanna yanni Sanborn, Phillips and Sites 2007: 13 (listed, distribution, comp. note) Thailand
Tanna yanni Boulard 2008b: 109 (comp. note)
Tanna yanni Boulard 2008e: 352 (comp. note)
Tanna yanni Boulard 2008f: 45, 66, 95 (biogeography, listed, comp. note) Thailand

***T. yunnanensis* (Lei & Chou, 1997)** = *Neotanna yunnanensis*
Neotanna yunnanensis Chou, Lei, Li, Lu and Yao 1997: 222, 225–226, 357, 368, Fig 9–71, Plate X, Fig 107 (n. sp., key, described, illustrated, comp. note) China
Neotanna yunnanensis Sanborn 2006c: 833 (comp. note) China
Neotanna yunnanensis Pham and Yang 2009: 11 (comp. note)
Tanna yunnanensis Pham and Yang 2009: 11, 14, 16, Table 2 (n. comb., synonymy, listed, distribution, comp. note) Equals *Neotanna yunnanensis* Vietnam, China

***Paratanna* Lee, 2012**

***P. parata* Lee, 2012**

***Formocicada* Lee & Hayashi, 2004**
Formocicada Chen 2004: 39 (phylogeny, listed)
Formocicada Lee and Hayashi 2004: 59–60 (n. gen., key, diagnosis) Taiwan
Formocicada Lee 2009f: 1487 (comp. note)
Formocicada Lee and Hill 2010: 302 (comp. note)

***F. taiwana* Lee & Hayashi, 2004**
Formocicada taiwana Chen 2004: 156–157, 191, 199 (described, illustrated, listed, comp. note) Taiwan
Formocicada taiwana Lee and Hayashi 2004: 59–61, Figs 24–27 (n. sp., type species of *Formocicada*, described, illustrated, listed, comp. note) Taiwan

***Formosemia* Matsumura, 1917** = *Formosia* (sic)
Formosemia Wu 1935: 11 (listed) China
Formosemia Lee 2009f: 1498–1501, 1503, Table 2 (status, described, key, comp. note)
Formosemia Lee 2010c: 23 (listed)
Formosemia Lee and Hill 2010: 302 (comp. note)

***F. apicalis* (Matsumura, 1907)** = *Leptopsaltria apicalis* = *Purana apicalis* = *Formosemia apicalia* (sic)
Leptopsaltria apicalis Hua 2000: 62 (listed, distribution) China, Taiwan
Leptopsaltria apicalis Lee and Hayashi 2003b: 386 (type species of *Formosemia*) Formosa
Purana apicalis Lee and Hayashi 2003b: 386–388, Figs 29–31 (key, diagnosis, illustrated, synonymy, distribution, song, oscillogram, listed, comp. note) Equals *Leptopsaltria apicalis* Equals *Formosemia apicalis* Taiwan
Purana apicalis Chen 2004: 126–127, 199 (described, illustrated, listed, comp. note) Taiwan
Purana apicalis Shiyake 2007: 5, 52, 54, Fig 81 (illustrated, distribution, listed, comp. note) Taiwan
Leptopsaltria apicalis Lee 2008a: 11 (type species of *Formosemia*) Formosa
Purana apicalis Lee 2009f: 1498 (comp. note)
Formosemia apicalis Lee 2009f: 1498–1501, Fig 10A, Fig 12, Table 2 (n. comb., illustrated, comp. note)
Leptopsaltria apicalis Lee 2009f: 1498 (type species of *Formosemia*, comp. note) Formosa
Purana apicalis Lee, Lin and Wu 2010: 218–224, Figs 1–6, Tables 1–2 (ecology, comp. note) Taiwan
Formosemia apicalis Lee and Hill 2010: 299–300, Figs 7–8 (phylogeny, comp. note)

F. australis Kato, 1944 to *Purana australis*
F. gigas Kato, 1930 to *Purana gigas*
F. samia (Walker, 1850) to *Leptopsaltria samia*
F. taipinensis Kato, 1944 to *Purana taipinensis*

***Maua* Distant, 1905** = *Mau* (sic)
Maua Wu 1935: 10 (listed) China
Maua Hayashi 1984: 47 (listed)
Maua Gogala 1994b: 7, Fig 8 (sonogram, comp. note) Peninsular Malaysia
Maua Lee 1995: 23 (listed)
Maua Zaidi and Ruslan 1995a: 174, Table 2 (listed, comp. note) Borneo, Sarawak
Maua Zaidi and Ruslan 1995b: 220–221, Table 2 (listed, comp. note) Borneo, Sabah
Maua Zaidi and Ruslan 1995c: 70, Table 2 (listed, comp. note) Peninsular Malaysia
Maua Zaidi and Ruslan 1995d: 202, Table 2 (listed, comp. note) Borneo, Sabah
Maua Zaidi 1996: 103, Table 2 (comp. note) Borneo, Sarawak

Maua Zaidi and Hamid 1996: 54, Table 2 (listed, comp. note) Borneo, Sarawak
Maua Chou, Lei, Li, Lu and Yao 1997: 197, 231, 293, Table 6 (key, described, listed, diversity, biogeography) China
Maua Zaidi 1997: 115, Table 2 (listed, comp. note) Borneo, Sarawak
Maua Zaidi and Ruslan 1998b: 4–5, Table 2 (listed, comp. note) Borneo, Sabah
Maua Zaidi and Ruslan 1999: 4, Table 2 (listed, comp. note) Borneo, Sarawak
Maua Duffels and Zaidi 2000: 209 (comp. note)
Maua Kos and Gogala 2000: 2–3 (comp. note)
Maua Azman and Zaidi 2002: 2 (comp. note)
Maua Schouten and Duffels 2002: 29, 31 (comp. note)
Maua Duffels 2004: 463 (comp. note)
Maua Gogala, Trilar, Kozina and Duffels 2004: 4 (comp. note)
Maua Moulds 2005b: 391, 432 (higher taxonomy, listed, comp. note)
Maua Duffels, Schouten and Lammertink 2007: 367–369 (comp. note)
Maua Sanborn, Phillips and Sites 2007: 3, Table 1 (listed) Thailand
Maua Boulard 2008f: 7, 42, 94 (listed) Thailand
Maua Duffels 2009: 303–306, 314, 316, 320, 330 (characteristics, phylogeny, comp. note) Malay Peninsula, Java, Sumatra, Borneo, Palawan, China, Thailand, Philippines
Mau (sic) Duffels 2009: 303 (comp. note)
Maua Lee 2009f: 1487, 1495, 1501, 1503, Table 2 (key, comp. note)
Maua Lee 2009h: 2628 (listed) Philippines, Palawan
Maua Lee 2010a: 22 (listed) Philippines, Mindanao
Maua Lee 2010c: 23 (listed)
Maua Lee and Hill 2010: 302 (comp. note)

M. ackermanni Schmidt, 1924 to *Maua quadrituberculata quadrituberculata* (Signoret, 1847)

***M. affinis* Distant, 1905** = *Leptopsaltria quadrituberculata* (nec Signoret) = *Maua quadrituberculata* (nec Signoret)
Maua affinis Zaidi 1996: 100–101, 104, Table 1 (comp. note) Borneo, Sarawak
Maua affinis Zaidi and Hamid 1996: 53, 56–57, Table 1 (listed, comp. note) Borneo, Sarawak
Maua affinis Zaidi 1997: 112–114, 116, Table 1 (listed, comp. note) Borneo, Sarawak
Maua affinis Zaidi and Ruslan 1998a: 352–353 (listed, comp. note) Borneo, Sarawak, Balabac, Palawan
Maua affinis Zaidi, Ruslan and Azman 1999: 306 (listed, comp. note) Borneo, Sabah, Balabac, Palawan, Sarawak
Maua affinis Kos and Gogala 2000: 3 (comp. note)
Maua affinis Zaidi, Noramly and Ruslan 2000a: 324 (listed, comp. note) Borneo, Sabah, Balabec, Palawan, Sarawak
Maua affinis Zaidi, Ruslan and Azman 2000: 204 (comp. note) Borneo, Sabah, Balabec Palawan, Sarawak
Maua affinis Zaidi and Nordin 2001: 18 (comp. note) Borneo, Sabah
Maua affinis Zaidi, Nordin, Maryati, Wahab, Norashikin et al. 2002: 12 (comp. note) Borneo, Sabah
Maua affinis Zaidi and Azman 2003: 104, Table 1 (listed, comp. note) Borneo, Sabah
Maua affinis Duffels, Schouten and Lammertink 2007: 368 (comp. note)
Maua affinis Duffels 2009: 303–306, 309–317, Figs 6–9 (key, lectotype designation, described, illustrated, distribution, comp. note) Borneo. Sarawak, Sabah
Maua affinis Lee 2009f: 1495, 1497, 1500–1501, 1503, Fig 11, Tables 1–2 (illustrated, comp. note)
Maua affinis Lee 2009h: 2628 (synonymy, distribution, listed) Equals *Leptopsaltria quadrituberculata* (nec Signoret) Equals *Maua quadrituberculata* (nec Signoret) Philippines, Palawan, Balabac Island

***M. albigutta* (Walker, 1857)** = *Dundubia albigutta* = *Dundubia albiguttata* (sic) = *Leptopsaltria albigutta* = *Leptopsaltria albiguttata* (sic) = *Purana albigutta* = *Purana albiguttata* (sic) = *Maua albiguttata* (sic) = *Mau* (sic) *albigutta* = *Naua* (sic) *albigutta*
Purana albigutta Zaidi and Ruslan 1997: 221 (listed, comp. note) Equals *Dundubia albigutta* Equals *Leptopsaltria albigutta* Equals *Maua albigutta* Peninsular Malaysia, Sumatra, Java, Borneo
Purana albigutta Zaidi, Noramly and Ruslan 2000a: 320, 322–323 (listed, comp. note) Equals *Dundubia albigutta* Equals *Leptopsaltria albigutta* Equals *Maua albigutta* Borneo, Sabah, Sumatra, Peninsular Malaysia, Java, Sarawak, Singapore
Purana albigutta Zaidi, Azman and Ruslan 2001a: 112–113, 115, 118, Table 1 (listed, comp. note) Equals *Dundubia albigutta* Equals *Leptopsaltria albigutta* Equals *Maua albigutta* Langkawi Island, Sumatra, Peninsular Malaysia, Java, Borneo, Singapore
Maua albigutta Zaidi and Azman 2003: 104, Table 1 (listed, comp. note) Borneo, Sabah
Maua albigutta Gogala, Trilar, Kozina and Duffels 2004: 2–14, Figs 1–11, Table 1 (illustrated, described, song, sonogram, oscillogram, distribution, comp. note) Malaysia, Borneo, Sumatra, Sabah

Maua albigutta Schouten, Duffels and Zaidi 2004: 373–374, 380, Table 2 (listed, comp. note) Malayan Peninsula

Maua albigutta Boulard 2005d: 338 (comp. note) Southeast Asia

Purana albigutta Boulard 2006d: 196 (listed, comp. note) Equals *Dundubia albigutta* Equals *Leptopsaltria albiguttata* Malaysia, Peninsular Malaysia

Maua albigutta Boulard 2006g: 35–36 (comp. note)

Maua albigutta Trilar 2006: 344 (comp. note)

Maua albigutta Duffels, Schouten and Lammertink 2007: 368 (comp. note)

Maua albigutta Duffels 2009: 303–304, 306, 323–328, 330, Fig 22, Figs 24–26 (key, described, illustrated, distribution, comp. note) Equals *Dundubia albigutta* Equals *Leptopsaltria albiguttata* (sic) Equals *Purana albigutta* Malay Peninsula, Sumatra, Siberut Island

Maua albigutta group Duffels 2009: 305, 323 (comp. note)

Purana albigutta Lee 2009f: 1501, Table 2 (comp. note)

***M. albistigma* (Walker, 1850)** = *Dundubia albistigma* = *Leptopsaltria albistigma*

Maua albistigma Wu 1935: 10 (listed) Equals *Dundubia albistigma* Equals *Leptopsaltria albistigma* China, Hangshow

Maua albistigma Chou, Lei, Li, Lu and Yao 1997: 232 (listed, synonymy, comp. note) Equals *Dundubia albistigma* China

Maua albistigma Hua 2000: 62 (listed, distribution) Equals *Dundubia albistigma* China, Jiangxi, Zhejiang, Guangdong, Java

Maua albistigma Duffels 2009: 303 (comp. note) China

***M. borneensis* Duffels, 2009** = *Maua affinis* (nec Distant)

Maua borneensis Duffels 2009: 303–306, 313–317, Figs 10–12, Fig 14 (key, n. sp., described, illustrated, distribution, comp. note) Equals *Maua affinis* Zaidi, 1996:101; Zaidi 1997:113; Zaidi and Ruslan 1998:352; Zaidi, Ruslan and Azman 1999:306; Zaidi, Ruslan and Azman 2000:204; Zaidi, Noramly and Ruslan 2000:324 Borneo, Sabah, Sarawak, Kalimantan

M. dohrni Schmidt, 1912 to *Maua latilinea* (Walker, 1868)

***M. fukienensis* Liu, 1940** = *Maua quikiensis* (sic) = *Purana fukiensis* (sic)

Maua quikiensis (sic) Liu 1992: 40 (comp. note)

Maua fukiensis Chou, Lei, Li, Lu and Yao 1997: 232 (listed, comp. note) China

Maua fukienensis Hua 2000: 62 (listed, distribution) China, Fujian

Maua fukienensis Duffels 2009: 303 (comp. note) China

***M. latilinea* (Walker, 1868)** = *Dundubia latilinea* = *Cosmopsaltria latilinea* = *Cosmopsatria duarum laticincta* (sic) = *Maua dohrni* Schmidt, 1912

Cosmopsaltria latilinea Duffels and Zaidi 2000: 209 (comp. note)

Maua latilinea Duffels and Zaidi 2000: 209 (comp. note) Equals *Cosmopsaltria latilinea*

Maua latilinea Duffels, Schouten and Lammertink 2007: 368 (comp. note)

Maua latilinea Duffels 2009: 303–306, 317, 319–323, Fig 14, Figs 17–21 (synonymy, key, described, illustrated, distribution, comp. note) Equals *Dundubia latilinea* Equals *Cosmopsaltria latilinea* Equals Equals *Maua dohrni* Borneo Sumatra

Maua dohrni Duffels 2009: 303, 319–320, 322, Figs 20–21 (synonymy, illustrated, comp. note) Sumatra

Dundubia latilinea Duffels 2009: 321 (comp. note) Penang

Maua dohrni Lee 2009f: 1495, 1501, 1503, Table 2 (comp. note)

***M. linggana* Moulton, 1923**

Maua linggana Gogala, Trilar, Kozina and Duffels 2004: 4–5 (comp. note)

Maua linggana Duffels 2009: 303–305, 323–324, 327–330, Fig 23, Fig 26–28 (key, described, illustrated, distribution, comp. note) Borneo, Sarawak, Sabah, Kalimantan, Brunei

***M. palawanensis* Duffels, 2009** = *Maua affinis* (nec Distant)

Maua affinis (sic) Chou, Lei, Li, Lu and Yao 1997: 232, 357, Plate XI, Fig 114 (described, illustrated, comp. note) China

Maua palawanensis Duffels 2009: 303–306, 311, 313, 316–319, Figs 13–16 (key, n. sp., described, illustrated, distribution, comp. note) Equals *Maua affinis* Distant, 1905:61 (partim); Distant, 1912: 41, Plate 5, Fig 33a-c; Moulton, 1923:124–125, 168 (partim);Chou et al. 1997:232, Fig 114 Palawan, Balabac

***M. philippinensis* Schmidt, 1924**

Maua philippinensis Duffels 2009: 303 (comp. note) Philippines

Maua philippinensis Lee 2009f: 1495, 1501, 1503, Table 2 (comp. note)

Maua philippinensis Lee 2010a: 22–23, Fig 5 (illustrated, distribution, listed) Philippines, Mindanao

***M. platygaster* Ashton, 1912**

Maua platygaster Zaidi and Ruslan 1998a: 353 (listed, comp. note) Borneo, Sarawak

Maua platygaster Duffels 2009: 303–305, 326, 329–330, Fig 26, Fig 29 (key, described, illustrated, distribution, comp. note) Borneo, Sarawak

M. *quadrituberculata quadrituberculata* (Signoret, 1847) = *Cicada quadrituberculata* = *Cicada quadrimaculata* (sic) = *Dundubia quadrituberculata* = *Leptopsaltria quadrituberculata* = *Maua 4-tuberculata* = *Maua quadrimaculata* (sic) = *Maua tuberculata* (sic) = *Maua ackermanni* Schmidt, 1924

Maua quadrituberculata Wu 1935: 10 (listed) Equals *Cicada quadrituberculata* Equals *Dundubia quadrituberculata* Equals *Leptopsaltria quadrituberculata* China, Java, Philippines

Maua tuberculata (sic) Gogala 1994b: 9 (song, comp. note) Peninsular Malaysia

Maua quadrituberculata Gogala and Riede 1995: 299, Fig 4 (song, sonogram, oscillogram, acoustic behavior, comp. note) Sabah, Malaysia

Maua quadrituberculata Lee 1995: 19, 23, 45–46, 129, 135, 138 (comp. note) Equals *Cicada quadrituberculata*

Cicada quadrituberculata Lee 1995: 129 (comp. note)

Maua quadrituberculata Zaidi and Ruslan 1995d: 199–200, 203, Table 1 (listed, comp. note) Borneo, Sabah

Maua quadrituberculata Zaidi and Hamid 1996: 53, 56–57, Table 1 (listed, comp. note) Borneo, Sarawak

Maua quadrituberculata Zaidi, Ruslan and Mahadir 1996: 60, Table 1 (listed, comp. note) Peninsular Malaysia

Cicada quadrituberculata Chou, Lei, Li, Lu and Yao 1997: 231 (type species of *Maua*) China

Maua quadrituberculata Chou, Lei, Li, Lu and Yao 1997: 233 (listed, synonymy, comp. note) Equals *Cicada quadrituberculata* China

Maua quadrituberculata Zaidi and Ruslan 1998a: 353 (listed, comp. note) Equals *Cicada quadrituberculata* Borneo, Sarawak, Java, Peninsular Malaysia, Sabah

Maua quadrituberculata Hua 2000: 62 (listed, distribution) Equals *Cicada quadrituberculata* China, Hong Kong, Japan, Philippines, Malaysia, Kalimantan, Java

Maua quadrituberculata Kos and Gogala 2000: 3, 5 (comp. note)

Maua quadrituberculata Zaidi, Noramly and Ruslan 2000b: 337, 399 (comp. note) Borneo, Sabah

Maua quadrituberculata Schouten and Duffels 2002: 29 (comp. note)

Maua quadrituberculata Zaidi and Azman 2003: 104, Table 1 (listed, comp. note) Borneo, Sabah

Maua quadrituberculata Schouten, Duffels and Zaidi 2004: 373–374, Table 2 (listed, comp. note) Malayan Peninsula

Maua quadrituberculata Boulard 2007a: 101, Fig 11 (illustrated, comp. note) Java

Maua quadrituberculata Duffels, Schouten and Lammertink 2007: 368 (comp. note)

Maua quadrituberculata Sanborn, Phillips and Sites 2007: 13 (listed, distribution, comp. note) Thailand, China, Indonesia, Borneo, Philippine Republic, Korea

Maua quadrituberculata Boulard 2008f: 42 (type species of *Maua*)

Maua quadrituberculata Lee 2008b: 462 (comp. note)

Cicada quadrituberculata Duffels 2009: 303 (type species of *Maua*, comp. note)

Maua quadrituberculata Duffels 2009: 303–313, 315, 317, Figs 1–3, Fig 5, Fig 7 (synonymy, type species of *Maua*, key, described, illustrated, distribution, comp. note) Equals *Cicada quadrituberculata* Equals *Leptopsaltria quadrituberculata* Equals *Maua ackermanni* China, Thailand, Malay Peninsula, Java?, Sumatra, Borneo, Sabah, Sarawak, Brunei, Nias Island

Maua ackermanni Duffels 2009: 303, 306, 308, Fig 4 (synonymy, illustrated, comp. note)

Maua quadrituberculata group Duffels 2009: 304–306, 314, 327 (comp. note)

Maua ackermanni Lee 2009f: 1495, 1501, 1503, Table 2 (comp. note)

Maua quadrituberculata Lee 2009f: 1495, 1497, 1501, 1503, Tables 1–2 (comp. note)

Maua quadrituberculata Lee 2009h: 2617 (comp. note)

Cicada quadrituberculata Lee 2009h: 2628 (type species of *Maua*, listed) Java

Cicada quadrituberculata Lee 2010a: 22 (type species of *Maua*, listed)

M. *quadrituberculata tavoyana* (Ollenbach, 1929) = *Maua tavoyana* = *Purana tavoyana* = *Maua quadrituberculata khunpaworensis* Boulard, 2007 = *Maua quadrituberculata kpaworensis* (sic) = *Maua 4-drituberculata* (sic) *kpaworensis*

Purana tavoyana Boulard 2006d: 197 (listed, comp. note) Burma

Maua quadrituberculata khunpaworensis Boulard 2007a: 101–105, Figs 11–16 (n. ssp., described, illustrated, song, sonogram, comp. note) Thailand

Maua quadrituberculata kpaworensis Boulard 2008f: 42 (synonymy, type species of *Maua*, biogeography, listed, comp. note) Equals *Purana tavoyana* Equals *Maua quadrituberculata* (partim) Thailand, Burma

Maua tavoyana Boulard 2008f: 94 (listed) Equals *Maua 4-drituberculata kpaworensis* Thailand

Maua quadrituberculata tavoyana Boulard 2009f: 253 (n. comb., comp. note) Equals *Purana tavoyana* Equals *Maua quadrituberculata kpaworensis*

Maua quadrituberculata khunpaworensis Duffels 2009: 310 (comp. note) Thailand
Purana tavoyana Lee 2009f: 1501, Table 2 (comp. note)

***Purana* Distant, 1905** = *Paruna* (sic)
Purana Wu 1935: 10 (listed) China
Purana Hayashi 1984: 47 (listed)
Purana Duffels 1986: 319 (distribution, diversity, comp. note) Sulawesi, Philippines, Southeast Asia, Greater Sunda Islands
Purana Zaidi, Mohammedsaid and Yaakob 1990: 261–262, 267–268, Table 2 (listed, comp. note) Malaysia
Purana Gogala 1994a: 33 (comp. note) Thailand
Purana Zaidi and Ruslan 1994: 427, Table 2 (comp. note) Malaysia
Purana spp. Gogala 1995: 111 (comp. note) Thailand, Ceylon
Purana Lee 1995: 96 (comp. note)
Purana Zaidi and Ruslan 1995a: 174, Table 2 (listed, comp. note) Borneo, Sarawak
Purana Zaidi and Ruslan 1995b: 220–221, Table 2 (listed, comp. note) Borneo, Sabah
Purana Zaidi and Ruslan 1995c: 70, Table 2 (listed, comp. note) Peninsular Malaysia
Purana Zaidi and Ruslan 1995d: 202, Table 2 (listed, comp. note) Borneo, Sabah
Purana Zaidi 1996: 103, Table 2 (comp. note) Borneo, Sarawak
Purana Zaidi and Hamid 1996: 54, Table 2 (listed, comp. note) Borneo, Sarawak
Purana Chou, Lei, Li, Lu and Yao 1997: 197, 226–227, 293, Table 6 (key, described, listed, diversity, biogeography) China
Purana Novotny and Wilson 1997: 437 (listed)
Purana Zaidi 1997: 114–115, Table 2 (listed, comp. note) Borneo, Sarawak
Purana Zaidi and Ruslan 1998b: 4–5, Table 2 (listed, comp. note) Borneo, Sabah
Purana Gogala and Trilar 1999b: 1 (comp. note)
Purana Zaidi and Ruslan 1999: 4, Table 2 (listed, comp. note) Borneo, Sarawak
Purana Kos and Gogala 2000: 2–3, 5, 8 (comp. note) Greater Sunda Islands, Philippines, Peninsular Malaysia, Thailand, India, Indo-China, China, Japan, Sulawesi
Purana spp. Kos and Gogala 2000: 5 (comp. note)
Purana sp. Zaidi, Ruslan and Azman 2000: 203 (comp. note) Borneo, Sabah
Purana sp. Zaidi and Nordin 2001: 18 (comp. note) Borneo, Sabah
Purana sp. Zaidi and Nordin 2001: 125 (listed, comp. note) Borneo, Sarawak
Purana Boulard 2003e: 275 (comp. note) India
Purana Azman and Zaidi 2002: 2 (comp. note)
Purana Schouten and Duffels 2002: 29–31 (comp. note) Myanmar, Thailand, Cambodia, Vietnam, Malay Peninsula, Philippines, Greater Sunda Islands, India, Sri Lanka, China, Taiwan, Japan
Purana Trilar and Gogala 2002a: 47 (comp. note)
Purana Trilar and Gogala 2002b: 41 (comp. note)
Purana sp. Zaidi, Nordin, Maryati, Wahab, Norashikin et al. 2002: 12 (comp. note) Borneo, Sabah
Purana Lee and Hayashi 2003b: 361, 386, 389 (key, diagnosis) Equals *Formosemia* Taiwan
Purana Chen 2004: 39, 42, 126 (phylogeny, listed)
Purana Duffels 2004: 463 (comp. note)
Purana Gogala, Trilar, Kozina and Duffels 2004: 4, 13 (comp. note)
Purana spp. Gogala, Trilar, Kozina and Duffels 2004: 5, 13 (comp. note) Malaysia, Thailand
Purana Lee and Hayashi 2004: 59 (comp. note) Taiwan
Purana Boulard 2005d: 345 (comp. note)
Purana Boulard 2005e: 131 (comp. note)
Purana Moulds 2005b: 391, 432 (higher taxonomy, listed, comp. note)
Purana Pham and Thinh 2005a: 244 (listed) Vietnam
Purana Pham and Thinh 2005b: 290 (comp. note) Vietnam
Purana Boulard 2006d: 195 (comp. note)
Purana Boulard 2007a: 107–108, 112–113 (comp. note)
Purana Boulard 2007d: 502 (comp. note)
Purana Boulard 2007e: 35 (comp. note) Thailand
Purana Duffels, Schouten and Lammertink 2007: 367–369 (comp. note)
Purana Gogala and Trilar 2007: 398 (comp. note)
Purana spp. Gogala and Trilar 2007: 397 (comp. note)
Purana Sanborn, Phillips and Sites 2007: 3, 11, 35, Table 1 (listed) Thailand
Purana Shiyake 2007: 5, 54 (listed, comp. note)
Purana Boulard 2008b: 109 (comp. note)
Purana Boulard 2008e: 352 (comp. note)
Purana Boulard 2008f: 5–7, 42, 66, 94 (listed, comp. note) Thailand
Purana Lee 2008a: 2, 11, Table 1 (listed) Equals *Formosemia* Vietnam
Purana Boulard 2009b: 47, 50, 54 (comp. note)
Purana Duffels 2009: 303–305 (comp. note)
Purana Lee 2009d: 470–471 (comp. note)
Purana Lee 2009f: 1487, 1495–1496, 1501–1502, Tables 1–2 (key, comp. note) Philippines
Purana Lee 2009h: 2625 (listed) Philippines, Palawan
Purana Pham and Yang 2009: 11, 14, Table 2 (listed) Vietnam

Purana Lee 2010b: 21, 23 (listed, comp. note) Cambodia
Purana Lee 2010c: 23 (listed)
Purana Lee and Hill 2010: 302 (comp. note)

***P. abdominalis* Lee, 2009**
Purana abdominalis species group Lee 2009f: 1487, 1501, 1503, Table 2 (key, comp. note)
Purana abdominalis Lee 2009d: 1488–1491, 1493, 1501, Figs 1–3, Table 2 (n. sp., described, illustrated, comp. note) Philippines, Negros
Purana abdominalis Lee 2010c: 23 (comp. note) Philippines, Negros
Purana abdominalis species group Lee 2010c: 23 (comp. note)

P. albigutta Walker, 1857 to *Maua albigutta*

P. apicalis (Matsumura, 1907) to *Formosemia apicalis*

***P. atroclunes* Boulard, 2002**
Purana atroclunes Boulard 2002a: 52–54, Plate H, Figs 1–3, Figs 11–13 (n. sp., described, illustrated, sonogram, comp. note) Thailand
Purana atroclunes Trilar and Gogala 2002b: 41 (comp. note) Thailand
Purana atroclunes Gogala, Trilar, Kozina and Duffels 2004: 13 (comp. note) Thailand
Purana atroclunes Boulard 2005d: 343 (comp. note)
Purana atroclunes Boulard 2005o: 148 (listed)
Purana atroclunes Boulard 2006d: 196–197, 200–201, Fig 8A (listed, sonogram, comp. note) Thailand
Purana atroclunes Boulard 2006g: 72 (comp. note)
Purana atroclunes Boulard 2007e: 32 (comp. note) Thailand
Purana atroclunes Gogala and Trilar 2007: 398 (comp. note)
Purana atroclunes Sanborn, Phillips and Sites 2007: 11 (listed, distribution, comp. note) Thailand
Purana atroclunes Boulard 2008f: 42–43, 94 (biogeography, listed, comp. note) Thailand
Purana atroclunes Boulard 2009d: 43 (comp. note)
Purana atroclunes Lee 2009f: 1499, 1501–1502, Fig 9B, Fig 13, Table 2 (illustrated, comp. note)

***P. australis* (Kato, 1944)** = *Formosemia australis*
Purana australis Lee 2009f: 1501, Table 2 (comp. note)

***P. barbosae* (Distant, 1889)** = *Leptopsaltria barbosae* = *Purana carmente barbosae*
Purana barbarosae Zaidi, Ruslan and Azman 1999: 303 (listed, comp. note) Equals *Leptopsaltria barbarosae* Borneo, Sabah, Sulu Islands
Purana barbarosae Zaidi, Noramly and Ruslan 2000a: 323 (listed, comp. note) Equals *Leptopsaltria barbarosae* Borneo, Sabah, Sulu, Sarawak, Peninsular Malaysia
Purana barbarosae Zaidi, Ruslan and Azman 2000: 201–202 (listed, comp. note) Equals *Leptopsaltria barbarosae* Borneo, Sabah, Sulu, Sarawak, Peninsular Malaysia, Sabah
Purana barbarosae Zaidi, Azman and Ruslan 2001a: 112, 115, 118–119, Table 1 (listed, comp. note) Equals *Leptopsaltria barbarosae* Langkawi Island, Sulu, Borneo, Sarawak, Peninsular Malaysia, Sabah
Purana barbosae Schouten and Duffels 2002: 29–31, 33–34, 37, 39, Fig 1, Figs 5–7, Fig 11, Table 1 (described, illustrated, synonymy, phylogeny, distribution, key, comp. note) Philippines
Purana carmente barbosae Boulard 2006d: 196 (listed, comp. note) Equals *Leptopsaltria barbosae* Peninsular Malaysia
Purana barbosae Duffels, Schouten and Lammertink 2007: 369 (comp. note)
Purana barbosae Lee 2009f: 1496, 1501, Tables 1–2 (key, comp. note)

***P. campanula* Pringle, 1955**
Purana campanula Gogala 1995: 112, 114, Fig 10 (song, sonogram, oscillogram, comp. note) Sri Lanka
P[urana] campanula Jiang, Yang, Tang, Xu and Chen 1995b: 229 (song frequency, comp. note)
Purana campanula Sueur 2001: 39, Table 1 (listed)
Purana campanula Boulard 2007a: 112 (comp. note)
Purana campanula Gogala and Trilar 2007: 397 (comp. note)
Purana campanula Lee 2009f: 1501, Table 2 (comp. note)

***P. capricornis* Kos & Gogala, 2000**
Purana capricornis Kos and Gogala 2000: 2–3, 5, 9, 13, 15–17, 19, 24, Figs 1–2, Figs 25–27, Fig 31, Fig 40, Tables 1–2 (n. sp., key, described, illustrated, phylogeny, distribution, comp. note) Borneo, Sabah, Sarawak, Brunei, Kalimantan
Purana capricornis Zaidi and Azman 2003: 104, Table 1 (listed, comp. note) Borneo, Sabah
Purana capricornis Lee 2009f: 1497, 1501, Tables 1–2 (comp. note)

***P. carmente* (Walker, 1850)** = *Dundubia carmente* = *Leptopsaltria carmente* = *Chremistica* (sic) *carmente* = *Purana carmente carmente* = *Leptopsaltria nigrescens* Distant, 1889
Purana carmente Zaidi, Mohammedsaid and Yaakob 1990: 262–265, 267, Fig 1, Table 1 (seasonality, comp. note) Malaysia
Purana carmente Zaidi and Ruslan 1994: 428–429 (comp. note) Malaysia
Chremistica (sic) *carmente* Zaidi and Ruslan 1994: 428 (comp. note) Malaysia
Purana carmente Chou, Lei, Li, Lu and Yao 1997: 230, 368 (comp. note) China

Purana carmente Zaidi and Ruslan 1998a: 349 (listed, comp. note) Equals *Dundubia carmente* Borneo, Sarawak, Java, Peninsular Malaysia
Purana carmente Zaidi, Noramly and Ruslan 2000a: 320, 232 (listed, comp. note) Equals *Dundubia carmente* Borneo, Sabah, Java
Purana carmente Boulard 2002a: 49 (comp. note)
Purana carmente group Schouten and Duffels 2002: 29–31, 33, 39, 41, 44 (described, diversity, comp. note) Vietnam, China, Philippines, Malayan Peninsula, Borneo, Sumatra, Java
Purana carmente Schouten and Duffels 2002: 29–33, 35, 37, 39–41, 43, 45, Figs 1–4, Fig 11, Table 1 (described, illustrated, synonymy, phylogeny, distribution, key, comp. note) Equals *Dundubia carmente* Equals Leptopsaltria carmente Equals *Leptopsaltria nigrescens* Equals *Purana carmente carmente* Java
Purana carmente group Trilar and Gogala 2002a: 47 (comp. note)
Purana carmente group Duffels, Schouten and Lammertink 2007: 367–369 (comp. note)
Purana carmente Boulard 2008c: 356 (comp. note)
Purana carmente group Duffels 2009: 303, 305 (comp. note)
Purana carmente species group Lee 2009d: 470 (comp. note)
Purana carmente species group Lee 2009f: 1495, 1497, 1501, 1503, Tables 1–2 (key, comp. note)
Purana carmente Lee 2009f: 1496, 1501, Tables 1–2 (comp. note)
Purana carmente species group Lee 2009h: 2625 (comp. note)

***P. carolettae* Esaki, 1936**
Purana carolettae Duffels 1986: 319 (comp. note) Caroline Islands
Purana carolettae Boer and Duffels 1996b: 309–310 (comp. note) Caroline Islands
Purana carolettae Schouten and Duffels 2002: 30 (comp. note) Caroline Islands
Purana carolettae Lee 2009f: 1501, Table 2 (comp. note)

***P. celebensis* (Breddin, 1901)** = *Leptopsaltria celebensis*
Purana celebensis Duffels 1990b: 64, Table 2 (listed) Sulawesi
Purana celebensis Boer and Duffels 1996b: 330 (distribution) Sulawesi
Purana celebensis Kos and Gogala 2000: 3 (comp. note) Sulawesi
Purana celebensis Schouten and Duffels 2002: 30 (comp. note) Sulawesi
Purana celebensis Lee 2009f: 1497, 1501, Tables 1–2 (comp. note)

***P. chueatae* Boulard, 2007**
Purana chueatae Boulard 2006d: 208–211, Figs 21–26 (n. sp., described, illustrated, song, sonogram, comp. note) Thailand
Purana chueatae Boulard 2008f: 43, 94 (biogeography, listed, comp. note) Thailand
Purana chueatae Lee 2009f: 1501, Table 2 (comp. note)

***P. clavohyalina* Liu, 1939**
Purana clavohyalina Liu 1992: 40 (comp. note)
Purana clavohyalina Lee 2009f: 1501, Table 2 (comp. note)

***P. conifacies* (Walker, 1858)** = *Cicada conifacies* = *Dundubia conifacies*

***P. conspicua* Distant, 1910**
Purana conspicua Zaidi and Hamid 1996: 53, 56, Table 1 (listed, comp. note) Borneo, Sarawak
Purana conspicua Zaidi 1997: 112–113, 116, Table 1 (listed, comp. note) Borneo, Sarawak
Purana conspicua Zaidi and Ruslan 1998a: 349–350 (listed, comp. note) Borneo, Sarawak
Purana conspicua Lee 2009f: 1501, Table 2 (comp. note)

***P. davidi* Distant, 1905** = *Purana davini* (sic) = *Purana dividi* (sic) = *Purana davidii* (sic) = *Leptopsaltria davidi*
Purana davidi Wu 1935: 10 (listed) China, Kiangsi, Hangchow, Canton
Purana davidi Chou, Lei, Li, Lu and Yao 1997: 231 (listed, comp. note) China
Purana davidii (sic) Hua 2000: 64 (listed, distribution) China, Jiangxi, Zhejiang, Guangzhou, Hong Kong, Guangxi
Purana davidi Boulard 2008c: 366 (comp. note) China
Purana davidi Lee 2009f: 1501, Table 2 (comp. note)

***P. dimidia* Chou & Lei, 1997**
Purana dimidia Chou, Lei, Li, Lu and Yao 1997: 227, 230–231, 357, 368–369, Fig 9–74, Plate X, Fig 111 (n. sp., key, described, illustrated, comp. note) China
Purana dimidia Schouten and Duffels 2002: 29–31, 33, 45, Fig 1, Fig 11, Figs 19–20, Table 1 (described, illustrated, synonymy, phylogeny, distribution, key, comp. note) Vietnam, China, Yunnan
Purana dimidia Boulard 2006d: 196 (listed, comp. note) Tonkin
Purana dimidia Pham and Thinh 2006: 526–527 (listed, comp. note) Vietnam
Purana dimidia Lee 2008a: 11 (distribution, listed) North Vietnam, China, Yunnan
Purana dimidia Lee 2009f: 1496, 1501, Tables 1–2 (comp. note)
Purana dimidia Pham and Yang 2009: 14, Table 2 (listed) Vietnam

P. *doiluangensis* Boulard, 2005

Purana doiluangensis Boulard 2005h: 7, 26–29, Figs 31–37 (n. sp., described, illustrated, song, sonogram, listed, comp. note) Thailand

Purana doiluangensis Boulard 2005o: 150 (listed)

Purana doiluangensis Boulard 2006d: 196, 203 (listed, comp. note) Thailand

Purana doiluangensis Gogala and Trilar 2007: 397–398 (comp. note)

Purana doiluangensis Sanborn, Phillips and Sites 2007: 11 (listed, distribution, comp. note) Thailand

Purana doiluangensis Boulard 2008f: 43, 94 (biogeography, listed, comp. note) Thailand

Purana doihuangensis Lee 2009f: 1501, Table 2 (comp. note)

P. *gemella* Boulard, 2007

Purana gemella Boulard 2006d: 206–208, Figs 18–20 (n. sp., described, illustrated, song, sonogram, comp. note) Thailand

Purana gemella Boulard 2008f: 43, 94 (biogeography, listed, comp. note) Thailand

Purana gemella Lee 2009f: 1501, Table 2 (comp. note)

P. *gigas* (Kato, 1930) = *Purana clavohyalina* Liu, 1940 = *Formosemia gigas* = *Formosia* (sic) *gigas*

Formosemia gigas Wu 1935: 11 (listed) China, Cheking

Formosemia gigas Jiang, Lin and Wang 1987: 108, Fig 3(1) (illustrated, song, oscillogram, comp. note) China

Purana gigas Chou, Lei, Li, Lu and Yao 1997: 227–228, Fig 9–72, Plate X, Fig 106 (illustrated, key, described, synonymy, comp. note) Equals *Formotosemia gigas* Equals *Purana clavohyalina* Equals *Maua fukienensis* n. syn. China

Purana gigas Hua 2000: 64 (listed, distribution, host) Equals *Formotosemia gigas* Equals *Purana clavohyalina* China, Jiangxi, Zhejiang, Fujian, Hunan, Guangxi

Purana gigas Lee 2009f: 1501, Table 2 (comp. note)

P. *guttularis* (Walker, 1858) = *Cicada guttularis* = *Leptopsaltria guttularis* = *Purana gutturalis* (sic) = *Purana* sp. 1 Pham & Thinh, 2006

Purana guttularis Zaidi, Mohammedsaid and Yaakob 1990: 261 (comp. note) Malaysia

Purana guttularis Zaidi and Ruslan 1994: 428 (comp. note) Malaysia

Purana guttularis Zaidi, Ruslan and Mahadir 1996: 60, Table 1 (listed, comp. note) Peninsular Malaysia

Purana guttularis Chou, Lei, Li, Lu and Yao 1997: 227–229 (key, described, synonymy, comp. note) Equals *Cicada guttularis* Equals *Leptopsaltria guttularis* China

Purana guttularis Zaidi and Ruslan 1997: 222 (listed, comp. note) Equals *Cicada guttularis* Equals *Leptopsaltria guttularis* Peninsular Malaysia, Burma, Philippines, Nias Island, Borneo, Sarawak

Purana guttularis Zaidi and Ruslan 1998a: 350 (listed, comp. note) Equals *Cicada guttularis* Borneo, Sarawak, Burma, Nias Island, Philippines, Peninsular Malaysia, Java, Singapore

Purana guttularis Zaidi, Ruslan and Azman 1999: 304 (listed, comp. note) Equals *Cicada guttularis* Borneo, Sabah, Burma, Nias Island, Philippines, Sarawak, Peninsular Malaysia, Java, Singapore

Purana guttularis Hua 2000: 64 (listed, distribution) Equals *Cicada guttularis* China, Guangdong, Philippines, Kalimantan, Malaysia, Burma, India

Purana guttularis Kos and Gogala 2000: 3 (comp. note)

Purana guttularis Zaidi, Ruslan and Azman 2000: 202 (comp. note) Equals *Cicada guttularis* Borneo, Sabah, Burma, Nias Island, Philippines, Sarawak, Peninsular Malaysia, Java, Singapore

Purana guttularis Zaidi and Nordin 2001: 18 (comp. note) Borneo, Sabah

Purana guttularis Zaidi, Nordin, Maryati, Wahab, Norashikin et al. 2002: 12 (comp. note) Borneo, Sabah

Purana guttularis Zaidi and Azman 2003: 104, Table 1 (listed, comp. note) Borneo, Sabah

Purana guttularis Schouten, Duffels and Zaidi 2004: 373, 377, Table 2 (listed, comp. note) Malayan Peninsula

Purana guttularis Pham and Thinh 2005a: 244 (synonymy, distribution, listed) Equals *Cicada guttularis* Equals *Leptopsaltria guttularis* Vietnam

Purana guttularis Boulard 2006d: 196 (listed, comp. note) Equals *Cicada guttularis* Equals *Leptopsaltria guttularis* Burma, Siam

Purana sp. 1 Pham and Thinh 2006: 526 (listed) Vietnam

Purana guttularis Duffels, Schouten and Lammertink 2007: 368 (comp. note)

Purana guttularis Sanborn, Phillips and Sites 2007: 11 (listed, distribution, comp. note) Thailand, Burma, India, Malaysia, Indonesia, Borneo, Sarawak, Philippine Republic

Purana guttularis Lee 2009f: 1497, 1501, Tables 1–2 (comp. note)

Purana guttularis Pham and Yang 2009: 11, 14, 16, Table 2 (listed, synonymy, distribution, comp. note) Equals *Cicada guttularis*

Equals *Leptopsaltria guttularis* Vietnam, China, India, Siam, Borneo, Sarawak, Java, Myanmar, Philippine, Brunei
Purana guttularis Lee 2010a: 22 (synonymy, distribution, listed) Equals *Cicada guttularis* Equals *Purana gutturalis* (sic) Philippines, Mindanao, South China, Indonesia, Borneo, Java, Sumatra, Nias, Malaysia, Thailand, Myanmar, India

P. nr. *guttularis* Zaidi, Ruslan & Mahadir, 1996
Purana nr. *guttularis* Zaidi, Ruslan and Mahadir 1996: 60, Table 1 (listed, comp. note) Peninsular Malaysia

P. *hermes* Schouten & Duffels, 2002
Purana hermes Schouten and Duffels 2002: 29–31, 33, 38–39, 41–42, 44, Fig 1, Fig 12, Fig 15, Fig 18, Table 1 (n. sp., described, illustrated, phylogeny, distribution, key, comp. note) Borneo, Sabah, Sarawak
Purana hermes Zaidi and Azman 2003: 104, Table 1 (listed, comp. note) Borneo, Sabah
Purana hermes Lee 2009f: 1496, 1501, Tables 1–2 (comp. note)

P. *hirundo* (Walker, 1850) = *Cicada hirundo*

P. *infuscata* Schouten & Duffels, 2002
Purana infuscata Schouten and Duffels 2002: 29–31, 33, 38–39, 42–44, Fig 1, Fig 13, Fig 16, Fig 18, Table 1 (n. sp., described, illustrated, phylogeny, distribution, key, comp. note) Borneo, Malaysia, Sabah, Indonesia, Kalimantan
Purana infuscata Lee 2009f: 1496, 1501, Tables 1–2 (comp. note)

P. *jacobsoni* Distant, 1912
Purana jacobsoni Duffels 2009: 324–325 (valid species, comp. note)

P. *jdmoorei* Boulard, 2005 = *Purana jdmoorrei* (sic)
Purana jdmoorei Boulard 2005d: 338 (comp. note) Asia
Purana jdmoorei Boulard 2005h: 7, 30–33, Figs 38–43 (n. sp., described, illustrated, song, sonogram, listed, comp. note) Thailand
Purana jdmoorei Boulard 2005o: 150 (listed)
Purana jdmoorei Boulard 2006d: 196 (listed, comp. note) Thailand
Purana jdmoorei Boulard 2006g: 39–40 (comp. note) Asia
Purana jdmoorei Gogala and Trilar 2007: 397 (comp. note)
Purana jdmoorei Sanborn, Phillips and Sites 2007: 11 (listed, distribution, comp. note) Thailand
Purana jdmoorei Boulard 2008f: 43, 94 (biogeography, listed, comp. note) Thailand
Purana jdmoorrei (sic) Lee 2009f: 1501, Table 2 (comp. note)

P. *johanae* Boulard, 2005
Purana johanae Boulard 2005h: 7, 23–25, Figs 25–30 (n. sp., described, illustrated, song, sonogram, listed, comp. note) Thailand
Purana johanae Boulard 2005o: 150 (listed)
Purana johanae Boulard 2006d: 196 (listed, comp. note) Thailand
Purana johanae Trilar 2006: 344 (comp. note)
Purana johanae Gogala and Trilar 2007: 397 (comp. note)
Purana johanae Sanborn, Phillips and Sites 2007: 12 (listed, distribution, comp. note) Thailand
Purana johanae Boulard 2008f: 43, 94 (biogeography, listed, comp. note) Thailand
Purana johanae Lee 2009f: 1501, Table 2 (comp. note)

P. *karimunjawa* Duffels & Schouten, 2007
Purana karimunjawa Duffels, Schouten and Lammertink 2007: 367, 369, 383–386, Fig 25, Figs 27–30 (n. sp., described, illustrated, distribution, comp. note) Karimunjawa Island
Purana karimunjawa Lee 2009f: 1496, 1501, Tables 1–2 (comp. note)

P. *khaosokensis* Boulard, 2007
Purana khaosokensis Boulard 2007a: 113 (comp. note)
Purana khaosokensis Boulard 2007d: 500–502, 506, 509–510, Figs 6–8 (n. sp., described, illustrated, song, sonogram, comp. note) Thailand
Purana khaoskensis Boulard 2008f: 43, 94 (biogeography, listed, comp. note) Thailand
Purana khaosokensis Lee 2009f: 1501, Table 2 (comp. note)

P. *khuanae* Boulard, 2002 = *Purana khunpaworensis* (sic)
Purana khuanae Boulard 2002a: 49–52, Plate G, Figs 1–3, Figs 8–10 (n. sp., described, illustrated, sonogram, comp. note) Thailand
Purana khuanae Trilar and Gogala 2002b: 41 (comp. note) Thailand
Purana khuanae Gogala, Trilar, Kozina and Duffels 2004: 13 (comp. note) Thailand, Malaysia
Purana khuanae Boulard 2005d: 339 (comp. note)
Purana khuanae Boulard 2005o: 148 (listed)
Purana khuanae Boulard 2006d: 196–197 (listed, comp. note) Thailand
Purana khuanae Boulard 2006g: 48–49 (comp. note)
Purana khunpaworensis (sic) Boulard 2007d: 500 (comp. note) Thailand
Purana khuanae Sanborn, Phillips and Sites 2007: 12 (listed, distribution, comp. note) Thailand
Purana khuanae Boulard 2008f: 43, 70, 95 (biogeography, listed, comp. note) Thailand
Purana khuanae Boulard 2009d: 41 (comp. note)
Purana khuanae Lee 2009f: 1501, Table 2 (comp. note)

P. khuniensis Boulard, 2005

Purana khuniensis Boulard 2005e: 118, 131–133, Plate E, Figs 1–2, Figs 23–24 (n. sp., described, illustrated, song, sonogram, listed, comp. note) Thailand
Purana khuniensis Boulard 2005o: 151 (listed)
Purana khuniensis Boulard 2006d: 196 (listed, comp. note) Thailand
Purana khuniensis Gogala and Trilar 2007: 397 (comp. note)
Purana khuniensis Sanborn, Phillips and Sites 2007: 12 (listed, distribution, comp. note) Thailand
Purana khuniensis Boulard 2008f: 44, 95 (biogeography, listed, comp. note) Thailand
Purana khuniensis Lee 2009f: 1501, Table 2 (comp. note)

P. kpaworensis Boulard, 2007

Purana kpaworensis Boulard 2006d: 203–206, Figs 13–17 (n. sp., described, illustrated, song, sonogram, comp. note) Thailand
Purana kpaworensis Boulard 2008f: 44, 95 (biogeography, listed, comp. note) Thailand
Purana kpaworensis Lee 2009f: 1498–1499, 1501, Fig 7, Fig 9C, Table 2 (illustrated, comp. note) Thailand

P. latifascia Duffels & Schouten, 2007

Purana latifascia Duffels, Schouten and Lammertink 2007: 367, 369, 381–384, Figs 22–26 (n. sp., described, illustrated, distribution, comp. note) Borneo, Sabah
Purana latifascia Gogala and Trilar 2007: 389–390, 394–399, Figs 6–10 (song, sonogram, oscillogram, comp. note) Sabah, Borneo
Purana latifascia Lee 2009f: 1496, 1501, Tables 1–2 (comp. note)

P. metallica Duffels & Schouten, 2007 = *Purana* aff. *tigrina* Gogala, 1995 = *Purana langkawi* (sic)

Purana aff. *tigrina* Gogala 1995: 101–102, 104, 108–112, 114–115, Figs 5–8 (song, sonogram, oscillogram, comp. note) Thailand
Purana aff. *tigrina* Kos and Gogala 2000: 8 (comp. note)
Purana aff. *tigrina* Gerhardt and Huber 2002: 13, Fig 2.1A (oscillogram, comp. note)
Purana aff. *tigrina* Trilar and Gogala 2002a: 47 (comp. note) Thailand
Purana aff. *tigrina* Trilar and Gogala 2002b: 41 (comp. note) Thailand
Purana aff. *tigrina* Gogala, Trilar, Kozina and Duffels 2004: 13 (comp. note) Thailand
Purana aff. *tigrina* Trilar 2006: 344 (comp. note)
Purana metallica Duffels, Schouten and Lammertink 2007: 367, 369, 373–377, Figs 7–12 (n. sp., described, illustrated, distribution, comp. note) Equals *Purana* aff. *tigrina* Gogala, 1995 Thailand, Langkawi Island, Tarutao Island, Malaysia
Purana langkawi Duffels, Schouten and Lammertink 2007: 368 (comp. note)
Purana metallica Gogala and Trilar 2007: 389–394, 398, Figs 1–5 (song, sonogram, oscillogram, synonymy, comp. note) Equals *Purana* aff. *tigrina* Gogala 1995 Thailand
Purana aff. *tigrina* Sanborn, Phillips and Sites 2007: 3, 12, 34 (listed, distribution, comp. note) Thailand
Purana metallica Lee 2009f: 1496, 1501, Tables 1–2 (comp. note)

P. mickhuanae Boulard, 2009

Purana mickhuanae Boulard 2009d: 41–43, 51–52, 58, Figs 7–10 (n. sp., described, illustrated, song, sonogram, comp. note) Thailand

P. montana Kos & Gogala, 2000

Purana montana Kos and Gogala 2000: 2–3, 5, 9, 13, 15–19, 24, Figs 1–2, Figs 28–31, Fig 42, Tables 1–2 (key, described, illustrated, phylogeny, distribution, comp. note) Borneo, Sabah, Sarawak, Brunei
Purana montana Lee 2009f: 1497, 1501, Tables 1–2 (comp. note)

P. morrisi (Distant, 1892) = *Leptopsaltria morrisi*

Purana morrisi Boulard 2002a: 49, 52 (comp. note)
Purana morrisi Boulard 2006d: 196 (listed, comp. note) Equals *Leptopsaltria morrisi* India
Purana morrisi Lee 2009f: 1501, Table 2 (comp. note)

P. mulu Duffels & Schouten, 2007

Purana mulu Duffels, Schouten and Lammertink 2007: 367, 369, 379–381, 383, Figs 19–21, Fig 25 (n. sp., described, illustrated, distribution, comp. note) Sarawak, Brunei
Purana mulu Lee 2009f: 1496, 1501, Tables 1–2 (comp. note)

P. nana Lee, 2009

Purana nana Lee 2009d: 1491–1495, 1501, Figs 4–6, Table 2 (n. sp., described, illustrated, comp. note) Philippines, Mindanao
Purana nana Lee 2010a: 22 (distribution, listed) Philippines, Mindanao
Purana nana Lee 2010c: 23 (comp. note) Philippines, Mindanao

P. natae Boulard, 2007

Purana natae Boulard 2006d: 201–203, Figs 9–12 (n. sp., described, illustrated, song, sonogram, comp. note) Thailand
Purana natae Boulard 2008c: 356 (comp. note)
Purana natae Boulard 2008f: 44, 95 (biogeography, listed, comp. note) Thailand
Purana natae Lee 2009f: 1501, Table 2 (comp. note)

P. *nebulilinea* (Walker, 1868) = *Dundubia nebulilinea* = *Dundubia nebulinea* (sic) = *Leptopsaltria nebulilinea* = *Leptopsaltria nebulinea* (sic) = *Purana tigrina* (nec Walker) = *Purana pryeri* (nec Distant)

Purana nebulilinea Zaidi, Mohammedsaid and Yaakob 1990: 262–265, Fig 1, Table 1 (seasonality, comp. note) Malaysia

Purana nebulilinea Zaidi and Ruslan 1994: 426, 428, Table 1 (comp. note) Malaysia

Purana pryeri (sic) Zaidi and Ruslan 1995c: 64, 68, Table 1 (listed, comp. note) Peninsular Malaysia

Purana nebulilinea Zaidi 1997: 112–113, Table 1 (listed, comp. note) Borneo, Sarawak

Purana nebulilinea Zaidi and Ruslan 1997: 222 (listed, comp. note) Equals *Dundubia nebulilinea* Peninsular Malaysia, Sumatra, Borneo, Singapore

Purana pryeri (sic) Zaidi and Ruslan 1997: 222–223 (listed, comp. note) Equals *Leptopsaltria pryeri* Peninsular Malaysia, Borneo, Singapore

Purana nebulilinea Zaidi and Ruslan 1998a: 350 (listed, comp. note) Equals *Dundubia nebulilinea* Borneo, Sarawak, Sumatra, Singapore

Purana nebulilinea group Kos and Gogala 2000: 2–3, 5, 8–9, 13, 18–19, 22–23 (described, phylogeny, distribution, comp. note) Laos, Peninsular Malaysia, Sumatra, Borneo

Purana nebulilinea Kos and Gogala 2000: 2–5, 9–15, 18–22, 24, Figs 1–15, Figs 18–20, Fig 24, Fig 39, Tables 1–2 (key, described, illustrated, phylogeny, distribution, comp. note) Equals *Dundubia nebulilinea* Equals *Leptopsaltria nebulinea* (sic) Borneo, Sabah, Sarawak, Kalimantan, Sumatra, Paninsular Malaysia

Purana nebulilinea group Schouten and Duffels 2002: 29 (comp. note)

Purana nebulilinea Trilar and Gogala 2002a: 47 (comp. note) Peninsular Malaysia

Purana nebulilinea Trilar and Gogala 2002b: 41 (comp. note) Peninsular Malaysia

Purana nebulilinea Zaidi and Azman 2003: 104, Table 1 (listed, comp. note) Borneo, Sabah

Purana nebulilinea Schouten, Duffels and Zaidi 2004: 373, 377, Table 2 (listed, comp. note) Malayan Peninsula

Purana nebulilinea Gogala, Trilar, Kozina and Duffels 2004: 13 (comp. note) Malaysia

Purana nebulilinea Boulard 2006d: 196 (listed, comp. note) Equals *Dundubia nebulilinea* Equals *Leptopsaltria pryeri* Equals *Leptopsaltria nebulinea* (sic) Sumatra, Bornea, Peninsular Malaysia

Purana nebulilinea Trilar 2006: 344 (comp. note)

Purana nebulilinea group Duffels, Schouten and Lammertink 2007: 367 (comp. note)

Purana n. sp. Duffels, Schouten and Lammertink 2007: 370 (comp. note) Equals *Purana tigrina* Boulard 2006

Purana nebulilinea Gogala and Trilar 2007: 397, 399 (comp. note)

Purana nebulilinea group Duffels 2009: 303, 305 (comp. note)

Purana nebulilina species group Lee 2009f: 1495, 1497, 1501–1502, Fig 9A, Tables 1–2 (key, illustrated, comp. note)

Purana nebulilinea Lee 2009f: 1497, 1501, Tables 1–2 (comp. note)

P. nr. *nebulilinea* n. sp. Zaidi & Ruslan, 1995

Purana sp. nr. *nebulilinea* Zaidi and Ruslan 1995c: 64, 68, Table 1 (listed, comp. note) Peninsular Malaysia

P. *niasica* Kos & Gogala, 2000

Purana niasica Kos and Gogala 2000: 2–3, 5, 9, 14, 19–22, 24, Figs 1–2, Fig 24, Figs 32–34, Fig 43, Table 1 (key, described, illustrated, phylogeny, distribution, comp. note) Nias Island

Purana niasica Lee 2009f: 1497, 1501, Tables 1–2 (comp. note)

P. *notatissima* Jacobi, 1944

Purana notatissima Hua 2000: 64 (listed, distribution) China, Fujian

Purana notatissima Boulard 2006d: 197 (listed, comp. note) Fukien

Purana notatissima Lampe, Rohwedder and Rach 2006: 23–24 (type specimen) China, Fujian

Purana notatissima Lee 2009f: 1501, Table 2 (comp. note)

P. *obducta* Schouten & Duffels, 2002

Purana obducta Schouten and Duffels 2002: 29–31, 33, 36, 39–41, 43–45, Fig 1, Figs 8–11, Table 1 (n. sp., described, illustrated, phylogeny, distribution, key, comp. note) Malayan Peninsula, Borneo, Sabah

Purana obducta Zaidi and Azman 2003: 104, Table 1 (listed, comp. note) Borneo, Sabah

Purana obducta Duffels, Schouten and Lammertink 2007: 369 (comp. note)

Purana obducta Lee 2009f: 1496, 1501, Tables 1–2 (comp. note)

Purana obducta Lee 2009h: 2626 (comp. note)

P. *opaca* Lee, 2009

Purana opaca Lee 2009h: 2625–2628, Figs 5–6 (n. sp., described, illustrated, distribution, listed) Philippines, Palawan

P. *parvituberculata* Kos & Gogala, 2000

Purana parvituberculata Kos and Gogala 2000: 2–3, 5, 9, 22–24, Figs 1–2, Figs 35–38, Fig 44, Table 1 (key, described, illustrated, phylogeny, distribution, comp. note) Laos

Purana parvituberculata Lee 2009f: 1497, 1501–1502, Fig 9A, Fig 14, Tables 1–2 (illustrated, comp. note)
Purana parvituberculata Lee 2010b: 22 (distribution, listed) Cambodia, Laos

***P. phetchabuna* Boulard, 2008** = *Purana phechabuna* (sic)
Purana phechabuna (sic) Boulard 2008f: 44, 95 (biogeography, listed, comp. note) Thailand
Purana phetchabuna Boulard 2008f: 70–73, Figs 19–23 (n. sp., described, illustrated, song, sonogram, biogeography, listed, comp. note) Thailand
Purana phetchabuna Boulard 2009d: 41 (comp. note)
Purana phetchabuna Lee 2009f: 1501, Table 2 (comp. note)

***P. pigmentata* Distant, 1905**
Purana pigmentata Boulard 2006d: 197 (listed, comp. note) Siam
Purana pigmentata Sanborn, Phillips and Sites 2007: 12 (listed, distribution, comp. note) Thailand, China, Cambodia
Purana pigmentata Boulard 2008f: 44, 95 (biogeography, listed, comp. note) Thailand
Purana pigmentata Lee 2008a: 11 (distribution, listed) South Vietnam, Cambodia, Thailand
Purana pigmentata Lee 2009f: 1501, Table 2 (comp. note)
Purana pigmentata Pham and Yang 2009: 14, Table 2 (listed) Vietnam
Purana pigmentata Lee 2010b: 22 (distribution, listed) Cambodia, South Vietnam, Thailand

***P. pryeri* (Distant, 1881)** = *Leptopsaltria pryeri* = *Paruna* (sic) *pryeri*
Purana pryeri Zaidi and Ruslan 1998a: 351 (listed, comp. note) Equals *Leptopsaltria pryeri* Borneo, Sarawak
Purana pryeri Zaidi, Ruslan and Azman 1999: 304 (listed, comp. note) Equals *Leptopsaltria pryeri* Borneo, Sabah, Peninsular Malaysia, Singapore
Purana pryeri Kos and Gogala 2000: 2–3, 5, 9–10, 12–17, 19, 24, Figs 1–2, Figs 21–24, Fig 41, Tables 1–2 (rev. status, key, described, illustrated, phylogeny, distribution, comp. note) Equals *Leptopsaltria pryeri* Equals *Paruna* (sic) *pryeri* Borneo, Sabah, Sarawak, Kalimantan
Purana pryeri Zaidi, Noramly and Ruslan 2000a: 323–324 (listed, comp. note) Equals *Leptopsaltria pryeri* Borneo, Sabah, Peninsular Malaysia, Singapore, Sarawak
Purana pryeri Zaidi, Ruslan and Azman 2000: 202 (listed, comp. note) Equals *Leptopsaltria pryeri* Borneo, Sabah, Sumatra, Peninsular Malaysia, Sarawak, Java, Singapore
Purana pryeri Zaidi and Azman 2003: 104, Table 1 (listed, comp. note) Borneo, Sabah
Purana pryeri Lee 2009f: 1497, 1501, Tables 1–2 (comp. note)

***P. ptorti* Boulard, 2007**
Purana ptorti Boulard 2007a: 108–112, Figs 20–23 (n. sp., described, illustrated, song, sonogram, comp. note) Thailand
Purana ptorti Boulard 2008f: 44, 95 (biogeography, listed, comp. note) Thailand
Purana ptorti Lee 2009f: 1501, Table 2 (comp. note)

P. samia (Walker, 1850) to *Leptopsaltria samia*

***P. sagittata* Schouten & Duffels, 2002**
Purana sagittata Schouten and Duffels 2002: 29–31, 33, 38, 42–45, Fig 1, Fig 14, Figs 17–18, Table 1 (n. sp., described, illustrated, phylogeny, distribution, key, comp. note) Malayan Peninsula
Purana sagittata Trilar and Gogala 2002a: 47–55, Figs 1–6, Tables 1–2 (song, sonogram, oscillogram, comp. note) Peninsular Malaysia
Purana sagittata Trilar and Gogala 2002b: 41 (comp. note) Peninsular Malaysia
Purana sagittata Zaidi and Azman 2003: 104, Table 1 (listed, comp. note) Borneo, Sabah
Purana sagittata Gogala, Trilar, Kozina and Duffels 2004: 13 (comp. note) Malaysia
Purana sagittata Schouten, Duffels and Zaidi 2004: 370, 373, 377, Table 2 (listed, comp. note) Malayan Peninsula
Purana sagittata Gogala and Trilar 2005: 71 (comp. note) Malaysia
Purana sagittata Trilar 2006: 344 (comp. note)
Purana sagittata Duffels, Schouten and Lammertink 2007: 369 (comp. note)
Purana sagittata Gogala and Trilar 2007: 397 (comp. note)
Purana sagittata Lee 2009f: 1496, 1501, Tables 1–2 (comp. note)

P. samia (Walker, 1850) to *Leptopsaltria samia*

***P.* sp. Lee, 2010**
Purana sp. Lee 2010c: 23 (distribution, listed, comp. note) Philippines, Lúzon

***P. taipinensis* (Kato, 1944)** = *Formosemia taipinensis*
Purana taipinensis Lee 2009f: 1501, Table 2 (comp. note)

***P. tanae* Boulard, 2007**
Purana tanae Boulard 2006d: 197–201, Figs 1–7, Fig 8B (n. sp., described, illustrated, song, sonogram, comp. note) Thailand
Purana tanae Boulard 2008f: 44, 95 (biogeography, listed, comp. note) Thailand
Purana tanae Lee 2009f: 1501, Table 2 (comp. note)

P. tavoyana Ollenbach, 1929 to *Maua quadrituberculata tavoyana*

***P. tigrina* (Walker, 1850)** = *Dundubia tigrina* = *Leptopsaltria tigrina* = *Purana tigrina mjöbergi* Moulton, 1923

Purana tigrina Chaudhry, Chaudry and Khan 1970: 104 (listed, comp. note) Pakistan, India, Tibet, Peninsular Malaysia

Purana tigrina Zaidi, Mohammedsaid and Yaakob 1990: 262–265, Fig 1, Table 1 (seasonality, comp. note) Malaysia

Purana tigrina Zaidi and Ruslan 1994: 428 (comp. note) Malaysia

Purana tigrina Zaidi and Ruslan 1995c: 68, Table 1 (listed, comp. note) Peninsular Malaysia

Purana tigrina Zaidi, Ruslan and Mahadir 1996: 60, Table 1 (listed, comp. note) Peninsular Malaysia

Dundubia tigrina Chou, Lei, Li, Lu and Yao 1997: 227 (type species of *Purana*) China

Purana tigrina Zaidi 1997: 112–113, Table 1 (listed, comp. note) Borneo, Sarawak

Purana tigrina Zaidi and Ruslan 1997: 223 (listed, comp. note) Equals *Dundubia tigrina* Equals *Leptopsaltria tigrina* Peninsular Malaysia, Malabar, India, Tibet, Borneo, Philippines, Singapore

Purana tigrina Zaidi and Ruslan 1998a: 351–352 (listed, comp. note) Equals *Dundubia tigrina* Borneo, Sarawak, Malabar, India, Tibet, Peninsular Malaysia, Philippines

Purana tigrina Kos and Gogala 2000: 5 (comp. note)

Purana tigrina Zaidi, Noramly and Ruslan 2000b: 338–339, Table 1 (listed, comp. note) Borneo, Sabah

Purana tigrina Zaidi, Ruslan and Azman 2000: 203 (listed, comp. note) Equals *Dundubia tigrina* Borneo, Sabah, Malabar, India, Tibet, Peninsular Malaysia, Sarawak, Philippines, Java, Singapore

Purana tigrina Schouten and Duffels 2002: 30–31, Fig 1, Table 1 (phylogeny, comp. note)

Purana tigrina group Schouten and Duffels 2002: 30 (comp. note)

Dundubia tigrina Lee and Hayashi 2003b: 386 (type species of *Purana*) Malabar

Purana tigrina Zaidi and Azman 2003: 104, Table 1 (listed, comp. note) Borneo, Sabah

Purana tigrina group Schouten, Duffels and Zaidi 2004: 377(comp. note) Malayan Peninsula

Purana tigrina Boulard 2005d: 338–339, Fig 25.8 (sonogram, comp. note) Asia

Purana tigrina Boulard 2005h: 26 (comp. note)

Purana tigrina Boulard 2005o: 149 (listed)

Dundubia tigrina Pham and Thinh 2005a: 244 (type species of *Purana*)

Purana tigrina group Boulard 2006d: 196 (comp. note)

Purana tigrina Boulard 2006g: 42, 44, Fig 10 (sonogram, comp. note)

Purana tigrina Boulard 2007a: 93, 112 (comp. note) Madras, India, Tibet, Philippines, Borneo, Sumatra, Burma, Malaysia, Thailand

Dundubia tigrina Boulard 2007a: 112 (type species of *Purana*, comp. note)

Purana tigrina Boulard 2007e: 4, Plate 2, Fig 6 (illustrated, comp. note) Thailand

Purana tigrina group Duffels, Schouten and Lammertink 2007: 367–369 (diagnosis, phylogeny, distribution, key, comp. note) Sundaland, Nias Island, Sumatra, Malayan Peninsula, Bunguran Island, Borneo, Kalimantan Timur

Purana tigrina Duffels, Schouten and Lammertink 2007: 367, 369–375, 377, 380, Figs 1–6 (synonymy, described, illustrated, comp. note) Equals *Dundubia tigrina* Equals *Leptopsaltria tigrina* Equals *Purana tigrina mjöbergi* Malayan Peninsula, Bunguran Island, Borneo, Kalimantan Timur, Sumatra, Nias Island

Purana tigrina species group Gogala and Trilar 2007: 389 (comp. note) Sundaland

Purana tigrina Gogala and Trilar 2007: 389, 397 (comp. note)

Purana tigrina Sanborn, Phillips and Sites 2007: 12 (listed, distribution, comp. note) Thailand, India, Tibet, Burma, Malaysia, Indonesia, Philippine Republic

Purana tigrina Boulard 2008f: 42, 44–45, 95 (type species of *Purana*, biogeography, listed, comp. note) Equals *Dundubia tigrina* Equals *Leptopsaltria tigrina* Equals *Purana* aff. *tigrina* Gogala Thailand, Madras, India, Malaysia, Tibet, Peninsula Malaysia

Dundubia tigrina Lee 2008a: 11 (type species of *Purana*) Malabar

Purana tigrina group Duffels 2009: 303 (comp. note)

Purana tigrina Lee 2009f: 1487, 1496, 1501, Tables 1–2 (type species of *Purana*, comp. note)

Purana tigrina species group Lee 2009f: 1495, 1497, 1501, 1503, Tables 1–2 (key, comp. note)

Dundubia tigrina Pham and Yang 2009: 11 (type species of *Purana*)

Purana tigrina Ahmed and Sanborn 2010: 44 (distribution) Bangladesh, India, Tibet, Burma, Thailand, Malaysia, Indonesia, Philippines

Dundubia tigrina Lee 2009h: 2625 (type species of *Purana*, listed) Malabar

Dundubia tigrina Lee 2010a: 21 (type species of *Purana*, listed)

Dundubia tigrina Lee 2010b: 22 (type species of *Purana*, listed)

P. cf. *tigrina* Schouten, Duffels & Zaidi, 2004
Purana cf. *tigrina* Schouten, Duffels and Zaidi 2004: 373, 377, Table 2 (listed, comp. note) Malayan Peninsula

P. *tigrinaformis* Boulard, 2007 = *Purana tigrina* (nec Walker)
Purana tigrina (sic) Boulard 2003e: 274–277, Color Plate, Figs F-G, sonogram 4 (n. sp., described, song, sonogram, illustrated, comp. note) Thailand
Purana tigrina (sic) Boulard 2006d: 195 (type species of *Purana*, comp. note) Equals *Dundubia tigrina* Equals *Leptopsaltria tigrina* Madras, India, Peninsular Malaysia, Burma, Thailand
Purana tigrinaformis Boulard 2007a: 93, 112–115, Figs 24–27 (n. sp., described, illustrated, song, sonogram, comp. note) Equals *Purana tigrina* Boulard, 2003 [274] Equals *Purana tigrina* Boulard, 2006 [197] Thailand
Purana tigrinaformis Boulard 2008f: 45, 95 (biogeography, listed, comp. note) Thailand
Purana tigrinaformis Lee 2009f: 1501, Table 2 (comp. note)

P. *tripunctata* Moulton, 1923
Purana tripunctata Zaidi and Hamid 1996: 53, 56, Table 1 (listed, comp. note) Borneo, Sarawak
Purana tripunctata Boulard 2006d: 197 (listed, comp. note) Siam
Purana tripunctata Duffels, Schouten and Lammertink 2007: 368 (comp. note)
Purana tripunctata Lee 2009f: 1501, Table 2 (comp. note)

P. *trui* Pham, Schouten & Yang, 2012

P. *ubina* Moulton, 1923 = *Purana tigrina* (nec Walker)
Purana ubina Zaidi and Ruslan 1998a: 352 (listed, comp. note) Borneo, Sarawak, Sabah, Peninsular Malaysia, Sumatra
Purana ubina Kos and Gogala 2000: 3, 5 (comp. note)
Purana ubina Schouten and Duffels 2002: 29 (comp. note)
Purana ubina Zaidi and Azman 2003: 104, Table 1 (listed, comp. note) Borneo, Sabah
Purana ubina Gogala, Trilar, Kozina and Duffels 2004: 13 (comp. note)
Purana ubina Schouten, Duffels and Zaidi 2004: 373, 377, Table 2 (listed, comp. note) Malayan Peninsula
Purana ubina Boulard 2006d: 197 (listed, comp. note) Siam
Purana ubina Boulard 2007a: 108 (comp. note)
Purana ubina Duffels, Schouten and Lammertink 2007: 368 (comp. note)
Purana ubina Sanborn, Phillips and Sites 2007: 12–13 (listed, distribution, comp. note) Thailand, Malaysia, Indonesia
Purana ubina Boulard 2008f: 45, 95 (biogeography, listed, comp. note) Equals *Purana tigrina* Boulard 2004, 2007 (nec Walker) Thailand
Purana ubina Lee 2009f: 1495, 1497–1499, 1501, 1503, Fig 8, Fig 10B, Tables 1–2 (illustrated, comp. note)
Purana ubina species group Lee 2009f: 1495, 1497, 1501, 1503, Tables 1–2 (key, comp. note)

P. nr. *ubina* Zaidi, Ruslan & Mahadir, 1996
Purana nr. *ubina* Zaidi, Ruslan and Mahadir 1996: 60, Table 1 (listed, comp. note) Peninsular Malaysia

P. undescribed species A Sanborn, Phillips & Sites, 2007
Purana undescribed species A Sanborn, Phillips and Sites 2007: 13 (listed, distribution, comp. note) Thailand

P. *usnani* Duffels & Schouten, 2007
Purana usnani Duffels, Schouten and Lammertink 2007: 367, 369, 375, 377–381, 383–384, Figs 12–18 (n. sp., described, illustrated, distribution, comp. note) Singapore, Lingga Island, Bunguran, Borneo, Sarawak, Brunei
Purana usnani Lee 2009f: 1496, 1501, Tables 1–2 (key, comp. note)

P. *vesperalba* Boulard, 2009
Purana vesperalba Boulard 2009b: 50–54, Figs 7–10 (n. sp., described, illustrated, song, sonogram, comp. note) Thailand

P. *vindevogheli* Boulard, 2008
Purana vindevogheli Boulard 2008c: 353, 356–358, Figs 9–10 (n. sp., described, illustrated, comp. note) Malaysia

***Kamalata* Distant, 1889**
Kamalata Moulds 2005b: 391, 432 (higher taxonomy, listed, comp. note)

K. *javanensis* Distant, 1905
Kalamata javanensis Sanborn, Phillips and Sites 2007: 34 (comp. note) Java

K. *pantherina* Distant, 1889 = *Kamalata pantheriana* (sic)
Kamalata pantherina Zaidi and Ruslan 1998a: 352 (listed, comp. note) Borneo, Sarawak, Sumatra

***Calcagninus* Distant, 1892**
Calcagninus Moulds 2005b: 391, 432 (higher taxonomy, listed, comp. note)
Calcagninus Shiyake 2007: 5, 54 (listed, comp. note)
Calcagninus Lee 2009d: 470 (comp. note)
Calcagninus Lee 2009f: 1487 (comp. note)
Calcagninus Lee 2010c: 23 (listed)
Calcagninus Lee and Hill 2010: 302 (comp. note)

C. divaricatus **Bliven, 1964**
Calcaginus divaricatus Sanborn 1999a: 43 (type material, listed)

C. nilgirensis **(Distant, 1887)** = *Leptopsaltria nilgirensis* = *Calcagninus nilgiriensis* (sic)
Calcagninus nilgiriensis Shiyake 2007: 5, 53–54, Fig 86 (illustrated, listed, comp. note)

C. picturatus **(Distant, 1888)** = *Leptopsaltria picturata* = *Calcagninus pictulatus* (sic)

Qurana **Lee, 2009**
Qurana Lee 2009d: 470 (n. gen., described, comp. note)
Qurana Lee 2010b: 22 (listed) Cambodia

Q. ggoma **Lee, 2009**
Qurana ggoma Lee 2009d: 470–472, Fig 1 (n. sp., described, illustrated, type species of *Qurana*, comp. note) Cambodia
Qurana ggoma Lee 2010b: 22 (type species of *Qurana*, distribution, listed) Cambodia

Nabalua **Moulton, 1923** = *Nabula* (sic)
Nabalua Hayashi 1984: 47 (listed)
Nabalua Duffels 1986: 319 (distribution, diversity, comp. note) Greater Sunda Islands, Borneo, Sarawak
Nabalua Zaidi and Ruslan 1995a: 173–174, Table 2 (listed, comp. note) Borneo, Sarawak
Nabalua Zaidi and Ruslan 1995b: 220–221, Table 2 (listed, comp. note) Borneo, Sabah
Nabalua Zaidi and Ruslan 1995c: 70, Table 2 (listed, comp. note) Peninsular Malaysia
Nabalua Zaidi and Ruslan 1995d: 202, Table 2 (listed, comp. note) Borneo, Sabah
Nabalua Zaidi 1996: 103, Table 2 (comp. note) Borneo, Sarawak
Nabalua Zaidi and Hamid 1996: 54, Table 2 (listed, comp. note) Borneo, Sarawak
Nabalua Zaidi and Ruslan 1998b: 4–5, Table 2 (listed, comp. note) Borneo, Sabah
Nabalua Zaidi and Ruslan 1999: 3–4, Table 2 (listed, comp. note) Borneo, Sarawak
Nabalua sp. Zaidi, Ruslan and Azman 1999: 306 (listed, comp. note) Borneo, Sabah
Nabalua Kos and Gogala 2000: 2 (comp. note)
Nabalua Azman and Zaidi 2002: 2 (comp. note)
Nabalua Schouten and Duffels 2002: 29 (comp. note)
Nabalua sp. Zaidi and Azman 2003: 104, Table 1 (listed, comp. note) Borneo, Sabah
Nabalua Duffels 2004: 463–465 (described, key, comp. note)
Nabalua Moulds 2005b: 391, 432 (higher taxonomy, listed, comp. note)
Nabalua Duffels, Schouten and Lammertink 2007: 367 (comp. note)
Nabalua Duffels 2009: 303 (comp. note)
Nabulua Lee 2009f: 1487 (comp. note)
Nabalua Lee 2010c: 23 (listed)
Nabalua Lee and Hill 2010: 302 (comp. note)

N. borneensis **Duffels, 2004** = *Nabula* (sic) *mascula* Zaidi & Ruslan, 1995
Nabula (sic) *mascula* (sic) Zaidi and Ruslan 1995a: 172, 175, Table 1 (listed, comp. note) Borneo, Sarawak
Nabalua borneensis Duffels 2004: 463, 465, 467–472, 478–480, 488–489, Figs 4–6, Figs 9–10, Figs 13–14, Fig 43, Fig 48 (key, n. sp., described, illustrated, distribution, comp. note) Equals *Nabula (sic) mascula* Zaidi and Ruslan 1998:172 Borneo, Sabah, Sarawak

N. maculata **Duffels, 2004**
Nabalua maculata Duffels 2004: 463–464, 474–476, 485–489, Figs 29–31, Figs 36–37, Figs 40–41, Figs 47–48 (key, n. sp., described, illustrated, distribution, comp. note) Borneo, Sarawak

N. mascula **(Distant, 1889)** = *Leptopsaltria mascula* = *Nabula* (sic) *mascula*
Nabula (sic) *mascula* Zaidi and Ruslan 1999: 3, 5, Table 1 (listed, comp. note) Borneo, Sarawak
Nabalua mascula Zaidi, Ruslan and Azman 1999: 305 (listed, comp. note) Borneo, Sabah Johore
Nabalua mascula Zaidi, Ruslan and Azman 2000: 203 (listed, comp. note) Equals *Leptopsaltria mascula* Borneo, Sabah, Johore
Nabalua mascula Zaidi, Nordin, Maryati, Wahab, Norashikin et al. 2002: 12 (comp. note) Borneo, Sabah
Nabalua mascula Zaidi and Azman 2003: 104, Table 1 (listed, comp. note) Borneo, Sabah
Nabalua mascula Duffels 2004: 463–467, 475, 477, 479–480, 488–489, Figs 1–3, Figs 7–8, Figs 11–12, Fig 42, Fig 48 (key, type species of *Nabalua*, lectotype designation, described, illustrated, distribution, comp. note) Equals *Leptopsaltria mascula* Equals *Nabula* (sic) *mascula* Borneo, Sabah

N. **nr.** ***mascula*** **sp. n. Zaidi & Ruslan, 1995**
Nabalua nr. *mascula* sp. n. Zaidi and Ruslan 1995d: 198, 200 (comp. note) Borneo, Sabah

N. neglecta **Moulton, 1923** = *Leptopsaltria neglecta*
Nabalua neglecta Zaidi and Ruslan 1998a: 354 (listed, comp. note) Borneo, Sabah
Nabalua neglecta Zaidi, Ruslan and Azman 1999: 305 (listed, comp. note) Borneo, Sabah
Nabalua neglecta Zaidi and Azman 2003: 99–100, 104, Table 1 (listed, comp. note) Borneo, Sabah
Leptopsaltria neglecta Duffels 2004: 465 (comp. note)

Nabalua neglecta Duffels 2004: 463, 465, 472–474, 484, 486–489, Figs 26–28, Figs 34–35, Figs 38–39, Fig 45, Fig 48 (key, described, illustrated, distribution, comp. note) Borneo, Sabah

***N. sumatrana* Duffels, 2004**

Nabalua sumatrana Duffels 2004: 463, 465, 467, 469–471, 481, 483, 488–489, Figs 15–18, Figs 22–23, Fig 44, Fig 48 (key, n. sp., described, illustrated, distribution, comp. note) Sumatra

***N. zaidii* Duffels, 2004** = *Nabalua mascula* Zaidi, Ruslan & Mahadir, 1996

Nabalua mascula (sic) Zaidi, Ruslan and Mahadir 1996: 60, Table 1 (listed, comp. note) Peninsular Malaysia

Nabalua zaidii Duffels 2004: 463, 465, 467, 469, 471–472, 482–483, 486, 488–489, Figs 19–21, Figs 24–25, Figs 32–33, Fig 46, Fig 48 (key, n. sp., described, illustrated, distribution, comp. note) Equals *Nabalua mascula* Zaidi, Ruslan and Mahadir 1996:60 Malayan Peninsula

***Neotanna* Kato, 1927 = Tanna (Neotanna)**

Neotanna Chou, Lei, Li, Lu and Yao 1997: 197, 222, 293, Table 6 (key, described, listed, diversity, biogeography) China

Neotanna Chen 2004: 129 (comp. note)

Neotanna Sanborn 2006c: 829, 833 (comp. note)

***N. conyla* Chou & Lei, 1997**

Neotanna condyla Chou, Lei, Li, Lu and Yao 1997: 222–225, 357, 368, Fig 9–70, Plate X, Fig 112 (n. sp., key, described, illustrated, comp. note) China

Neotanna condyla Sanborn 2006c: 833 (comp. note) China

***N. shensiensis* Sanborn, 2006**

Neotanna shensiensis Sanborn 2006c: 829–831, 833, Figs 1–3 (n. sp., described, illustrated, comp. note) China, Shaanxi

***N. simultaneous* Chen, 1940**

Neotanna simultaneous Chou, Lei, Li, Lu and Yao 1997: 226 (listed, comp. note) China

Neotanna simultaneous Hua 2000: 63 (listed, distribution) China, Zhejiang

N. sinensis Ouchi, 1938 to *Tanna sinensis*

N. tarowanensis Matsumura, 1927 to *Tanna viridis* Kato, 1925

***N. thalia* (Walker, 1850)** = *Dundubia thalia* = *Pomponia thalia* = *Neotanna thalia* = *Tanna thalia* = *Dubia* (sic) *thalia* = *Pomponia thaila* (sic) = *Cicada sphinx* Walker, 1850 = *Puranoides sphinx* = *Pomponia horsfieldi* Distant, 1893

Pomponia thalia Wu 1935: 16 (listed) Equals *Dundubia thalia* Equals *Cicada sphinx* Equals *Pomponia horsfieldi* China, Yunna, Tibet, India, Sikkim, Darjiling

Pomponia thalia Wang and Zhang 1993: 189 (listed, comp. note) China

Neotanna thalia Chou, Lei, Li, Lu and Yao 1997: 226, 368 (listed, comp. note) China

Pomponia thalia Hua 2000: 64 (listed, distribution) Equals *Dundubia thalia* China, Yunnan, Sichuan, Xizang, Indonesia, Sikkim, India

Pomponia thalia Boulard 2002a: 42 (comp. note)

Puranoides sphinx Azman and Zaidi 2002: 2, 6, 11 (distribution, comp. note) Java

Tanna thalia Shiyake 2007: 5, 53–54, Fig 85 (illustrated, listed, comp. note)

N. yunnanensis Lei & Chou, 1997 to *Tanna yunnanensis*

***Taiwanosemia* Matsumura, 1917** = *Higurasia* Kato, 1925

Taiwanosemia Hayashi 1984: 47 (listed)

Taiwanosemia Lee and Hayashi 2003b: 361, 389 (key, diagnosis, synonymy) Equals *Higurasia* Taiwan

Taiwanosemia Chen 2004: 39, 42, 148 (phylogeny, listed)

Taiwanosemia Lee and Hayashi 2004: 46 (comp. note) Taiwan

Taiwanosemia Moulds 2005b: 391, 432 (higher taxonomy, listed, comp. note)

Taiwanosemia Lee 2009f: 1487 (comp. note)

Taiwanosemia Lee 2010c: 23 (listed)

Taiwanosemia Lee and Hill 2010: 298, 301–302 (comp. note)

***T. hoppoensis* (Matsumura, 1907)** = *Leptopsaltria hoppoensis* = *Tanna hoppoensis* = *Purana hoppoensis* = *Terpnosia bicolor* Kato, 1925 = *Higurasia bicolor*

Taiwanosemia hoppoensis Liu 1992: 34 (comp. note)

Leptopsaltria hoppoensis Hua 2000: 62 (listed, distribution) China, Taiwan

Taiwanosemia hoppoensis Hua 2000: 64 (listed, distribution) China, Taiwan

Leptopsaltria hoppoensis Lee and Hayashi 2003b: 389 (type species of *Taiwanosemia*) Formosa

Terpnosia? bicolor Lee and Hayashi 2003b: 389 (type species of *Higurasia*) Formosa

Taiwanosemia hoppoensis Lee and Hayashi 2003b: 389–391, Figs 32–34 (key, diagnosis, illustrated, synonymy, distribution, song, listed, comp. note) Equals *Leptopsaltria hoppoensis* Equals *Purana hoppoensis* Equals *Terpnosia hoppoensis* Equals *Terpnosia? bicolor* Equals *Huigasia bicolor* Taiwan

Taiwanosemia hoppoensis Chen 2004: 148–149, 199 (described, illustrated, listed, comp. note) Taiwan

Taiwanosemia hoppoensis Lee and Hill 2010: 299–300, Figs 7–8 (phylogeny, comp. note)

Subtribe Gudabina Boulard, 2008
Gudabina Boulard 2008f: 7, 48 (n. subtribe, listed) Thailand

***Gudaba* Distant, 1906**
Gudaba Lei 1996: 210, 212 (listed)
Gudaba Chou, Lei, Li, Lu and Yao 1997: 197, 216, 293, 357, Table 6 (key, described, listed, diversity, biogeography, comp. note) China
Gudaba Moulds 2005b: 391, 432 (higher taxonomy, listed, comp. note)
Gudaba Boulard 2008f: 7, 48, 95, 97–98 (listed) Thailand
Gudaba Lee 2008a: 2, 12, Table 1 (listed) Vietnam
Gudaba sp. Ewart 2009b: 140 (comp. note) Queensland
Gudaba Lee 2009d: 470 (comp. note)
Gudaba Lee 2009f: 1487 (comp. note)
Gudaba Pham and Yang 2009: 14, Table 2 (listed) Vietnam
Gudaba Lee 2010c: 23 (listed)
Gudaba Lee and Hill 2010: 302 (comp. note)

***G. apicata* Distant, 1906**
Gudaba apicata Lee 2008a: 12 (distribution, listed) South Vietnam
Gudaba apicata Pham and Yang 2009: 14, 16, Table 2 (listed, comp. note) Vietnam
Gudaba apicata Pham and Yang 2010a: 133 (comp. note) Vietnam

***G. longicauda* Lei, 1996**
Gudaba longicauda Lei 1996: 210–21, Figs 1–6 (n. sp., described, illustrated, comp. note) China, Yunnan
Gudaba longicauda Chou, Lei, Li, Lu and Yao 1997: 216–218, Fig 9–66, Plate III, Fig 44 (key, described, illustrated, comp. note) China

***G. maculata* Distant, 1912**
Gudaba maculata Lei 1996: 210, 212 (comp. note) China, Yunnan, Hainan
Gudaba maculata Chou, Lei, Li, Lu and Yao 1997: 216–217, Plate III, Fig 45 (key, described, illustrated, comp. note) China
Gudaba maculata Boulard 2006c: 138–140, Figs 6–8, Fig D (described, illustrated, song, sonogram, comp. note) Thailand, Sikkim, India, China
Gudaba maculata Boulard 2008f: 48, 95 (type species of *Gudaba*, biogeography, listed, comp. note) Thailand, Sikkim, India, China

***G. marginatus* (Distant, 1897)** = *Calcagninus marginatus*
Calcagninus marginatus Chou, Lei, Li, Lu and Yao 1997: 216 (type species of *Gudaba*) China
Calcagninus marginatus Lee 2008a: 12 (type species of *Gudaba*) Burma

Tribe Cicadatrini Distant, 1905 = Cicadatraria = Moganniini Distant, 1905 = Moganniaria = Moganini (sic) = Mogannini (sic) = Moganiini (sic)
Cicadatraria Wu 1935: 17 (listed) China
Moganniaria Wu 1935: 20 (listed) China
Moganini (sic) Chen 1942: 146 (listed)
Cicadatrini Chen 1942: 146 (listed)
Mogannini (sic) Itô and Nagamine 1981: 281, Table 5 (listed) Taiwan, Okinawa
Moganniini Hayashi 1984: 63 (listed)
Moganniini Chou, Lei and Wang 1993: 82, 85–86 (comp. note) China
Moganniini Chou, Lei, Li, Lu and Yao 1997: 106–107 (key, synonymy, listed, described) Equals Moganniaria
Moganniini Sanborn 1999a: 44 (listed)
Moganiini (sic) Sueur 2001: 41, Table 1 (listed)
Moganniini Lee and Hayashi 2003a: 150, 152 (key, listed) Taiwan
Moganniini Lee and Hayashi 2003b: 359 (listed) Taiwan
Moganniini Chen 2004: 39, 41 (phylogeny, listed)
Moganniini Lee and Hayashi 2004: 45, 61 (listed) Taiwan
Moganniini Boulard 2005e: 118 (listed)
Cicadatraria Moulds 2005b: 386 (history, listed, comp. note)
Moganniini Moulds 2005b: 386, 427 (higher taxonomy, listed, comp. note)
Moganniini Pham and Thinh 2005a: 244 (listed) Vietnam
Moganniini Pham and Thinh 2005b: 287–289 (key, comp. note) Vietnam
Moganniini Boulard 2007e: 20 (comp. note) Thailand
Moganniini Sanborn 2007a: 29 (listed) Venezuela
Moganniini Sanborn, Phillips and Sites 2007: 3, 32, 41, Fig 5, Table 1 (listed, distribution) Thailand
Mogannini (sic) Shiyake 2007: 6, 82 (listed, comp. note)
Moganniini Boulard 2008f: 7, 53 (listed) Thailand
Moganniini Lee 2008a: 2, 19, Table 1 (listed) Vietnam
Moganniini Osaka Museum of Natural History 2008: 326 (listed) Japan
Moganniini Lee 2009h: 2636 (listed) Philippines, Palawan
Moganniini Pham and Yang 2009: 5, 15, Table 2 (listed) Vietnam
Moganniini Lee 2010b: 27 (listed) Cambodia
Cicadatrini Lee 2010c: 24 (listed, comp. note) Philippines, Luzon
Cicadatrini Lee 2010d: 171 (comp. note)

Cicadatrini Lee and Hill 2010: 303 (status, synonymy, comp. note) Equals Cicadatraria Equals Moganniaria Equals Moganniini
Moganniini Lee and Hill 2010: 302–303 (comp. note)
Moganniini Sanborn 2010a: 1596 (listed) Colombia
Cicadatrini Wei, Ahmed and Rizvi 2010: 28, 33 (comp. note)
Moganniini Wei, Ahmed and Rizvi 2010: 28 (comp. note)

***Emathia* Stål, 1866**
Emathia Lee and Hill 2010: 303 (comp. note)
Emathia Moulds 2005b: 390 (higher taxonomy, listed, comp. note)
Emathia Wei, Ahmed and Rizvi 2010: 33 (comp. note)
***E. aegrota* Stål, 1866** = *Tibicen aurengzebe* Distant, 1881

***Vagitanus* Distant, 1918**
Vagitanus Moulds 2005b: 390 (higher taxonomy, listed, comp. note)
Vagitanus Wei, Ahmed and Rizvi 2010: 33 (comp. note)
***V. luangensis* Distant, 1918**
***V. vientianensis* Distant, 1918**

***Triglena* Fieber, 1875**
Triglena Koçak 1982: 152 (listed, comp. note) Turkey
Triglena Moulds 2005b: 390 (higher taxonomy, listed, comp. note)
Triglena Kemal and Koçak 2010: 3–4 (diversity, listed, comp. note) Turkey
Triglena Wei, Ahmed and Rizvi 2010: 33 (comp. note)
***T. virescens* Fieber, 1876**
Triglena virescens Lodos and Kalkandelen 1981: 70 (distribution, comp. note) Turkey, Syria
Triglena virescens Koçak 1982: 152 (type species of *Triglena*)
Triglena virescens Kemal and Koçak 2010: 4 (distribution, listed, comp. note) Turkey

***Taungia* Ollenbach, 1929**
Taungia Moulds 2005b: 390 (higher taxonomy, listed, comp. note)
Taungia Boulard 2009d: 43 (comp. note)
Taungia Wei, Ahmed and Rizvi 2010: 33 (comp. note)
***T. abnormis* Ollenbach, 1929**
Taungia abnormis Boulard 2009d: 43 (comp. note)
***T. albantennae* Boulard, 2009** = *Purana* (sic) *albantennae*
Taungia albantennae Boulard 2009d: 43–45, 53–54, 58, Figs 11–16 (n. sp., described, illustrated, song, sonogram, comp. note) Equals *Purana* (sic) *albantennae* Thailand
Purana (sic) *albantennae* Boulard 2009d: 58 (comp. note)

***Cicadatra* Kolenati, 1857** = *Cicada (Cicadatra)* = *Tettigia (Cicadatra)* = *Cicadrata* (sic) = *Cicadatra (Rustavelia)* Horváth, 1912 = *Rustavelia*
Cicadatra Wu 1935: 18 (listed) China
Cicadatra Chen 1942: 146 (listed)
Cicadatra sp. Chen 1942: 146 (listed, comp. note) China
Cicadatra Gómez-Menor 1951: 61 (key) Spain
Cicadatra Soós 1956: 412 (listed) Romania
Cicadatra spp. Gardenhire 1959: 58 (grape pest, comp. note) Iran
Cicadatra sp. Chaudhry, Chaudry and Khan 1966: 94 (listed, comp. note) Pakistan
Cicadatra sp. Chaudhry, Chaudry and Khan 1970: 104 (listed, comp. note) Pakistan
Cicadatra Boulard 1981b: 48 (comp. note)
Cicadatra Dlabola 1981: 202 (comp. note)
Cicadatra Koçak 1982: 148–149 (listed, comp. note) Turkey
Rustavelia Koçak 1982: 151 (listed, comp. note) Turkey
Cicadatra Strümpel 1983: 18 (listed)
Cicadatra Hayashi 1984: 47 (listed)
Cicadatra Claridge 1985: 301 (song, comp. note)
Cicadatra Popov 1985: 22–23, 38, 70, 183 (acoustic system, comp. note)
Cicadatra Boulard 1986c: 192 (comp. note)
Cicadatra Popov and Sergeeva 1987: 684, 689 (comp. note)
Cicadatra Boulard 1988f: 41 (comp. note)
Cicadatra Popov 1989a: 305 (comp. note)
Cicadatra Popov 1989b: 74 (comp. note)
Cicadatra Boulard 1990b: 116, 205 (comp. note)
Cicadatra Popov, Aronov and Sergeeva 1991: 282, 294 (comp. note)
Cicadatra Boulard 1992c: 56 (comp. note)
Cicadatra Popov, Arnov and Sergeeva 1992: 151, 164 (comp. note)
Cicadatra spp. Fonseca 1993: 767 (comp. note)
Cicadatra Boulard 1995a: 26, 28 (listed, comp. note)
Cicadatra Boulard and Mondon 1995: 19, 65 (key, comp. note) France
Cicadatra Mirzayans 1995: 16 (listed) Iran
Cicadatra Riou 1995: 76 (comp. note)
Cicadatra Boulard 1997c: 110 (comp. note)
Cicadatra Chou, Lei, Li, Lu and Yao 1997: 88, 120, 293, Table 6 (described, key, listed, diversity, biogeography, comp. note) China
Cicadatra Holzinger, Frölich, Günthart, Pauterer, Nickel et al. 1997: 49 (listed) Europe

Cicadatra Gogala and Trilar 1998a: 6, 8 (comp. note)
Cicadatra sp. Gogala and Trilar 1998a: 7 (comp. note)
Cicadatra spp. Gogala and Trilar 1998a: 8, 13 (comp. note)
Cicadatra Trilar and Gogala 1998: 69 (comp. note) Macedonia
Cicadatra spp. Gogala and Gogala 1999: 127 (comp. note) Macedonia
Cicadatra Gogala and Trilar 1999b: 1 (comp. note)
Cicadatra spp. Schedl 1999b: 827 (listed, distribution, ecology, comp. note) Turkey, Lebanon, Israel
Cicadatra Schedl 1999b: 830 (comp. note)
Cicadatra Boulard 2000i: 25 (key) France
Cicadatra Schedl 2000: 262 (listed)
Cicadatra Boulard 2001b: 30 (comp. note)
Cicadatra Gogala and Trilar 2001a: 28 (comp. note)
Cicadatra Gogala and Trilar 2001b: 14 (comp. note)
Cicadatra Schedl 2001a: 1287 (comp. note)
Cicadatra Gogala and Trilar 2003: 8 (comp. note)
Cicadatra Holzinger, Kammerlander and Nickel 2003: 477, 479 (key, listed, comp. note) Europe
Cicadatra Schedl 2003: 423 (comp. note)
Cicadatra Boulard 2005d: 338 (comp. note) Mediterranean
Cicadatra spp. Gogala, Trilar and Krpač 2005: 112 (comp. note) Macedonia
Cicadatra Moulds 2005b: 390, 410, 415 (higher taxonomy, listed, comp. note)
Cicadatra Boulard 2006e: 129 (comp. note)
Cicadatra Boulard 2006g: 39 (comp. note) Mediterranean
Cicadatra Shiyake 2007: 6, 78 (listed, comp. note)
Cicadatra sp. Shiyake 2007: 6, 78, 80, Fig 124 (illustrated, listed, comp. note)
Cicadatra sp. Shiyake 2007: 6, 78, 80, Fig 125 (illustrated, listed, comp. note)
Cicadatra sp. Shiyake 2007: 6, 78, 80, Fig 126 (illustrated, listed, comp. note)
Cicadatra Sueur and Puissant 2007b: 55 (comp. note)
Cicadatra Demir 2008: 475 (listed) Equals *Psalmocharias* Equals *Rustavelia* Turkey
Cicadatra sp. Ahmed and Rizvi 2010: 25 (generic diversity) Pakistan
Cicadatra Ahmed and Sanborn 2010: 40 (distribution) Europe, North Africa, Central Asia
Cicadatra sp. Ahmed, Sanborn and Hill 2010: 75 (generic diversity) Europe, North Africa, Middle East, Central Asia, Pakistan, India, Israel, Iran, Afghanistan
Cicadatra Kemal and Koçak 2010: 3–4 (diversity, listed, comp. note) Turkey
Cicadatra Lee and Hill 2010: 298 (comp. note)
Cicadatra Mozaffarian and Sanborn 2010: 77 (distribution, listed) Europe, North Africa, central Asia
Cicadatra Mozaffarian, Sanborn and Phillips 2010: 33 (comp. note)
Cicadatra Wei, Ahmed and Rizvi 2010: 28, 33 (comp. note)

***C. abdominalis* Schumacher, 1923**

***C. acberi* (Distant, 1888)** = *Tibicen acberi* = *Sena acberi* = *Psalmocharias acberi*
Cicadatra acberi Shiyake 2007: 6, 78, 80, Fig 123 (illustrated, listed, comp. note)
Cicadatra acberi Ahmed and Sanborn 2010: 40 (synonymy, distribution) Equals *Tibicen acberi* Pakistan, India
Cicadatra acerbi Ahmed, Sanborn and Hill 2010: 78, 80 (comp. note)

***C. adanai* Kartal, 1980** = *Cicadatra adanae* (sic)
Cicadatra adanae (sic) Lodos and Kalkandelen 1981: 71 (distribution, comp. note) Turkey
Cicadatra adanai Nast 1982: 311 (listed) Turkey, Anatolia
Cicadatra adanai Demir 2007a: 51 (distribution, comp. note) Turkey
Cicadatra adanai Demir 2008: 475 (distribution) Turkey
Cicadatra adanai Kemal and Koçak 2010: 4 (distribution, comp. note) Turkey

***C. alhageos* (Kolenati, 1857)** = *Cicada (Cicadatra) atra alhageos* = *Cicada alhageos* = *Cicadatra atra alhageos* = *Cicadatra alhageos alhageos* = *Cicadatra algheos* (sic) = *Cicadatra olivacea* Melichar, 1913 = *Cicadatra alhageos olivacea* = *Cicadatra stenoptera* Haupt, 1917 = *Cicadatra alhageos stenoptera*
Psalmocharias alhageos Mirzayans, Hashemi, Borumand, Zairi and Rajabi 1976: 110 (distribution, listed) Iran
Cicadatra alhageos Bâbâi 1968: 55 (comp. note)
Cicadatra alhageos Drosopoulous 1980: 190 (listed) Greece
Cicadatra alhageos Dlabola 1981: 201 (distribution, comp. note) Iran
Cicadatra alhageos Nast 1987: 568–569 (distribution, listed) Italy?
Cicadatra alhageos Mirzayans 1995: 16 (listed, distribution) Iran

Psalmocharias alhageos Shekarian and Rezwani 2001: 113–118, Tables 1–5 (grape pest, comp. note) Iran
Psalmocharias alhageos Zamanian, Mehdipour and Ghaemi 2008: 2064–2071, Figs 1–13, Table 1 (acoustic behavior, song, oscillogram, comp. note) Iran, Afghanistan, Afghanistan, Russia, Turkey, Iraq
Psalmocharias alhageos Ahmed and Sanborn 2010: 44 (distribution)
Cicadatra alhageos Ahmed, Sanborn and Hill 2010: 78 (comp. note)
Cicadatra alhageos Mozaffarian and Sanborn 2010: 77–78 (synonymy, listed, distribution, comp. note) Equals *Cicada (Cicadatra) atra alhageos* Equals *Cicada (Cicadatra) atra glycyrrhizae* Equals *Cicadatra olivacea* Equals *Cicadatra alhageos stenoptera* Iran, Transcaucasia, Italy, Israel, Caucasus, Georgia, Afghanistan, USSR

***C. anoea* (Walker, 1850)** = *Cicada anoea* = *Cicada anae* (sic)
Cicadatra anoea Ahmed, Sanborn and Hill 2010: 78 (comp. note)

***C. appendiculata* Linnavuori, 1954**
Cicadatra atra appendiculata Sanborn 1999a: 44 (type material, listed)

***C. atra atra* (Olivier, 1790)** = *Cicada atra* = *Cicadatra atra* = *Tettigia (Cicadatra) atra* = *Tettigia atra* = *Cicadrata* (sic) *atra* = *Cicada arfa* (sic) = *Cicada arta* (sic) (nec Germar) = *Cicadatra concinna* (nec Germar) = *Cicada transversa* Germar, 1830 = *Cicadatra transversa* = *Cicadatra atra concinna* = *Cicadatra atra aquila* Fieber, 1876 = *Cicadatra atra pallipes* Fieber, 1876 = *Cicadatra atra tau* Fieber, 1876
Cicadatra atra Bachmetjew 1907: 251 (body length, comp. note)
Cicadatra atra Swinton 1908: 382 (song, comp. note) Israel
Cicada atra Chen 1942: 146 (type species of *Cicadatra*)
Cicadatra atra Castellani 1952: 15 (comp. note) Italy
Cicadatra atra Wagner 1952: 15 (listed, comp. note) Italy
Cicadatra atra Soós 1956: 412 (listed, distribution) Romania
Cicadatra atra Villiers 1977: 165, Plate XIV, Fig 22 (illustrated, listed, comp. note) France
Cicadatra atra Drosopoulous 1980: 190 (listed) Greece
Cicada atra Boulard 1981b: 41 (comp. note)
Cicadatra atra Boulard 1981b: 50 (comp. note)
Cicadatra atra Dlabola 1981: 205 (distribution, comp. note) Iran
Cicadatra atra Lodos and Kalkandelen 1981: 72 (synonymy, distribution, comp. note) Equals *Cicada concinna* Equals *Cicada transversa* Equals *Tibicen vitreus* Equals *Tibicen hyalinatus* Equals *Cicada helianthemi* Equals *Cicadatra atra tau* Equals *Cicadatra appendiculata* Turkey, Albania, Bulgaria, Cyprus, Czechoslovakia, France, Greece, Iran, Italy, Lebanon, Palestine, Rhodes, Romania, Spain, Switzerland, Syria, Turkey, USSR, Yugoslavia
Cicada atra Koçak 1982: 148 (type species of *Cicadatra*)
Cicadatra atra Boulard 1983b: 119 (comp. note) France
Cicada atra Arambourg 1985: 18, 19 (olive pest) Turkey
Cicadatra atra Boulard 1985b: 60, 63 (illustrated, characteristics, comp. note) France
Cicadatra atra Boulard 1985e: 1033 (coloration, comp. note)
Cicadatra atra Claridge 1985: 301–302 (timbal, illustrated, song, comp. note)
Cicadatra atra Popov 1985: 71, 73, 75, 77, 80, 181–183, 185, Fig 21.4, Fig 23.1, Fig 24, Fig 59.2, Fig 59.4 (oscillogram, acoustic system, auditory system, comp. note)
Cicadatra atra Sander 1985: 593 (comp. note) Bulgaria
Cicadatra atra Chinery 1986: 88–89 (illustrated, comp. note) Europe
Cicadatra atra Schedl 1986a: 3–4, 7, 11, 14–15, 22–24, Fig 9, Fig 21, Map 3, Table 1 (illustrated, listed, key, synonymy, distribution, ecology, comp. note) Equals *Cicada atra* Equals *Cicadetta atra* Equals *Cicada transversa* Equals *Tibicen vitreus* Equals *Tibicen hyalinatus* Equals *Cicada helianthemi* Equals *Cicadatra appendiculata* Istria, Germany, Hungary, Switzerland, Spain, Italy, Balkans, Ukraine, Turkey, Syria, Lebanon, Itan, Azerbaijan, North Africa
Cicadatra atra Nast 1987: 568–569 (distribution, listed) Czechoslovakia, France, Switzerland, Ukraine, Romania, Spain, Italy, Yugoslavia, Albania, Bulgaria, Greece, Turkey
Cicadatra atra Boulard 1988a: 12, Table (comp. note) France
Cicadatra atra Boulard 1988e: 161 (comp. note) Turkey
Cicada atra Boulard 1988f: 16 (comp. note)
Cicadatra atra Boulard 1988g: 4–5, Plate II, Fig 5 (illustrated, comp. note) France
Cicadatra atra Duffels 1988d: 75 (comp. note) Europe
Cicadatra atra Boulard 1990b: 132–133, 143, 167, 175–176, 185, 187, Plate XV, Fig 3, Plate XXII, Fig 11–14 (genitalia, sonogram, comp. note)

Cicadatra atra Popov 1990a: 302–303, Fig 2, Fig 4 (comp. note)
Cicadatra atra Tautz 1990: 105–106, Fig 12, Fig 13A (song frequency, hearing sensitivity, comp. note)
Cicadatra atra Popov, Aronov and Sergeeva 1991: 282–284, 291–292, Fig 1, Table (auditory system, ecology, comp. note) Russia
Cicadatra atra atra Boulard 1992c: 56–59, 61–85, Plate I, Figs 1–7, Fig 9, Plate II, Figs 1–4, Plate III, Figs 1–3, Plate IV, Figs 1–3, Plate V, Figs 4–8, Plate VI, Figs 1–4, Plate VII, Figs 1–6, Plate VIII, Graphs 1–2 (n. sp., neotype designation, neallotype designation, described, illustrated, sonogram, comp. note) Equals *Cicada concinna* (nec Germar) Equals *Cicada transversa* Equals *Cicada (Cicadatra) atra* Equals *Cicadatra atra atra* Equals *Cicadatra atra (concinna)* France, Spain, Italy, Greece, Yugoslavia, Corsica
Cicada atra Boulard 1992c: 56–57 (comp. note)
Cicadatra atra Popov, Arnov and Sergeeva 1992: 152–154, 161–162, Fig 1, Table (auditory system, ecology, comp. note) Russia
Cicadatra atra d'Urso and Ippolitto 1994: 217, Figs 19–20 (wing couling apparatus, illustrated, comp. note)
Cicadatra atra Boulard 1995a: 7–8, 10, 12, 26–28, Fig $S6_1$, Fig $S6_2$ (song analysis, sonogram, listed, comp. note)
Cicadatra atra Boulard and Mondon 1995: 30, 32–34, 43, 53, 57, 60, 66–67, 70–71, 73–74, 159 (illustrated, listed, comp. note) Equals *Cicada transversa* Equals *Cicada concinna* Equals *Cicadatra atra concinna* France
Cicada atra Boulard 1997c: 86 (comp. note)
Cicada atra Chou, Lei, Li, Lu and Yao 1997: 120 (type species of *Cicadatra*)
Cicadatra atra Holzinger, Frölich, Günthart, Pauterer, Nickel et al. 1997: 49 (listed) Europe
Tettigetta atra Arias Giralda, Pérez Rodríguez and Gallego Pintor 1998: 262 (olive pest)
Cicadatra atra Gogala 1998a: 156 (comp. note) Macedonia
Cicadatra atra Gogala 1998b: 394, 396 (oscillogram, comp. note) Slovenia
Cicadatra atra Gogala and Trilar 1998a: 7–9 (comp. note) Macedonia
Cicadatra atra Trilar and Gogala 1998: 69 (comp. note) Macedonia
Cicadatra atra Gogala and Gogala 1999: 120–121, 124, 127, Fig 3 (distribution, comp. note) Slovenia, Mediterranean
Cicadatra atra Gogala and Trilar 1999b: 1 (comp. note)
Cicadatra atra Schedl 1999b: 826, 830 (comp. note)
Cicadatra atra Boulard 2000d: 96 (song, comp. note)
Cicadatra atra Boulard 2000i: 15, 17, 24, 39–44, 99, 120, Fig 7, Figs 10b–10c, Fig 14, Fig 22b, Plates C-C', Map 3 (synonymy, illustrated, song, sonogram, oscillogram, frequency spectrum, distribution, ecology, comp. note) Equals *Cicada atra* Equals *Cicada concinna* (nec Germar) Equals *Cicada transversa* Equals *Cicada (Cicadatra) atra* Equals *Cicadatra atra atra* Equals *Cicadatra atra concinna* France
Cicadatra atra atra Boulard 2000i: 39–44 (comp. note)
Cicadatra atra Puissant and Sueur 2000: 70, 72 (comp. note) France
Cicadatra atra Schedl 2000: 259–260, 262, 264, 266–267, Fig 1, Plate 1, Fig 3, Map 3 (key, illustrated, distribution, listed, comp. note) Switzerland, Czech Republic, Hungary, Slovenia
Cicada atra Boulard 2001b: 6 (comp. note)
Cicadatra atra Gogala and Trilar 2001a: 28 (comp. note)
Cicadatra atra Krpač, Trilar and Gogala 2001: 29 (comp. note) Macedonia
Cicadatra atra Puissant 2001: 152 (comp. note)
Cicadatra atra Schedl 2001b: 27 (comp. note)
Cicadatra atra Sueur 2001: 40, Table 1 (listed)
Cicadatra atra Sueur and Puissant 2001: 19, 22–23 (comp. note) France
Cicadatra atra Boulard 2002b: 193, 205–206 (coloration, comp. note) France
Cicadatra atra Gogala 2002: 244, 248, Fig 5 (song, sonogram, oscillogram, comp. note)
Cicadatra atra Günther and Mühlenthaler 2002: 331 (listed) Switzerland
Cicadatra atra Nickel, Holzinger and Wachmann 2002: 292, Table 4 (listed)
Cicadatra atra Sueur and Aubin 2002b: 128, 131, 134–135, Fig 7 (song, sonogram, oscillogram, comp. note) France
Cicadatra atra Sueur 2002b: 115, 120, 129 (listed, comp. note)
Cicadatra atra Gogala and Seljak 2003: 350 (comp. note) Slovenia
Cicadetta atra Gogala and Trilar 2003: 6, 8–9, 12–13, Figs 3–4 (sound production, comp. note) Slovenia
Cicadatra atra Holzinger, Kammerlander and Nickel 2003: 478–479, 628, 637, Fig 259C, Plate 42, Fig 13, Tables 2–3 (listed, illustrated, comp. note) Europe, Czech Republic, Mediterranean, Asia Minor, France
Cicadatra atra Sueur 2003: 2942 (comp. note)

Cicadatra atra Sueur and Aubin 2003b: 488 (comp. note)
Cicadatra atra Tishechkin 2003: 141, 145, Figs 161–165 (oscillogram, comp. note) Chechnya
Cicadatra atra Boulard 2005d: 342–343 (comp. note)
Cicadatra atra Gogala, Trilar and Krpač 2005: 104, 108–112, 124, Fig 3, Fig 5 (distribution, sonogram, oscillogram, comp. note) Macedonia
Cicadatra atra Moulds 2005b: 410 (comp. note)
Cicadatra atra Puissant and Defaut 2005: 116–125, Tables 1–7 (listed, habitat association, distribution, comp. note) France
Cicadatra atra Boulard 2006g: 28–29, 61–62, 64–67, 180, Figs 2b–2c, Figs 26–27 (illuatrated, sonogram, listed, comp. note)
Cicadatra atra Didier 2006: 28 (comp. note) France
Cicada atra Duffels and Hayashi 2006: 190 (comp. note) France
Cicadatra atra Puissant 2006: 19, 21, 25, 28, 32, 34–35, 42–45, 49–50, 61, 70, 76, 84, 86, 89–90, 92, 95, 108–115, 134–140, 145–1149, 152–154, 157–159, 161, 178–187, Fig 8, Figs 17–19, Tables 1–2, Appendix Tables 1–7 (illustrated, synonymy, distribution, habitat association, ecology, listed, comp. note) France
Cicadatra atra Sueur, Windmill and Robert 2006: 4116–4127, Figs 1–15 (illustrated, tympanum mechanics, comp. note) France
Cicadatra atra Trilar, Gogala and Popa 2006: 178 (comp. note) Romania
Cicadatra atra Shiyake 2007: 6, 79, 81, Fig 128 (illustrated, listed, comp. note)
Cicadatra atra Sueur and Puissant 2007b: 62 (listed) France
Cicadatra atra Demir 2008: 475 (distribution) Equals *Cicadatra appendiculata* Equals *Cicadatra transversa* Turkey
Cicadatra atra Nation 2008: 304 (comp. note)
Cicadatra atra Nickel 2008: 207 (distribution, hosts, comp. note) Mediterranean
Cicadatra atra Sueur, Windmill and Robert 2008a: 2380, 2385–2386, Fig 6 (tympanum mechanics, comp. note)
Cicadatra atra Sueur, Windmill and Robert 2008b: 59 (comp. note)
Cicadatra atra Trilar and Gogala 2008: 29 (listed, distribution, comp. note) Romania
Cicadatra atra Gogala, Drosopoulos and Trilar 2009: 19 (comp. note)
Cicadatra atra Windmill, Robert and Sueur 2009: S130 (tympanal vibrations, comp. note)
Cicadatra atra Windmill, Sueur and Robert 2009: 4079–4082, Figs 1–5 (tympanal vibrations, comp. note) France
Cicada atra Ahmed and Sanborn 2010: 40 (type species of genus *Cicadatra*)
Cicadatra atra atra Ahmed, Sanborn and Hill 2010: 78 (comp. note)
Cicadatra atra Kemal and Koçak 2010: 4 (distribution, comp. note) Turkey
Cicadatra atra Lee and Hill 2010: 299–300, Figs 7–8 (phylogeny, comp. note)
Cicada atra Mozaffarian and Sanborn 2010: 77 (type species of *Cicadatra*)
Cicadatra atra Mozaffarian and Sanborn 2010: 78 (synonymy, listed, distribution, comp. note) Equals *Cicada atra* Equals *Cicada concinna* Equals *Cicada transversa* Equals *Tibicen vitreus* Equals *Tibicen hyalinatus* Equals *Cicada helianthemi* Iran, Spain, France, southern Europe, Central Europe, Georgia
Cicadatra atra Mozaffarian, Sanborn and Phillips 2010: 33 (comp. note)
Cicadatra atra Sueur, Janique, Simonis, Windmill and Baylac 2010: 923–932, Fig 1, Figs 3–9, Table 1 (tympanum structure, comp. note) France
Cicadatra atra Sueur, Windmill and Robert 2010: 1681–1687, Figs 1–2, Fig 5, Tables I-III (auditory processing, sonogram, comp. note) France

***C. atra hyalinata* (Brullé, 1832)** = *Cicadatra hyalinata* = *Tibicen hyalinatus* = *Cicadatra atra* var. *hyalinata*
Cicadatra atra var. *hyalinata* Boulard 1992c: 60, 78–79, Plate V, Figs 2–3 (n. comb., neotype designation, described, illustrated, comp. note) Equals *Tibicen hyalinatus* Greece, France
Cicadatra atra var. *hyalinata* Boulard and Mondon 1995: 66 (listed) France
Cicadatra hyalinata Gogala and Trilar 1998a: 7 (comp. note) Macedonia
Cicadatra hyalinata Gogala and Gogala 1999: 127 (comp. note)
Cicadatra hyalinata Gogala and Trilar 1999b: 1 (comp. note)
Cicadatra atra var. *hyalinata* Boulard 2000i: 44 (described, distribution, comp. note) France, Greece

***C. atra putoni* Boulard, 1992** = *Cicadatra atra* var. *putoni*
Cicadatra atra var. *putoni* Boulard 1992c: 60, 78–79, Plate V, Fig 1 (n. var., described, illustrated, comp. note) France
Cicadatra atra var. *putoni* Boulard and Mondon 1995: 66 (listed) France

***C. atra sufflava* Boulard, 1992** = *Cicadatra atra* var. *sufflava*
Cicadatra atra var. *sufflava* Boulard 1992c: 60, 70–71, Fig 8 (n. var., described, illustrated, comp. note) France

Cicadatra atra var. *sufflava* Boulard and Mondon 1995: 66 (listed) France
Cicadatra atra var. *sufflava* Boulard 2000i: 44 (described, distribution, comp. note) Turkey

C. *atra vitrea* (Brullé, 1832) = *Tibicen vitreus* = *Cicada vitreus* = *Cicadatra atra* var. *vitrea*
Cicadatra atra var. *vitrea* Boulard 1992c: 60 (n. comb., neotype designation, described, comp. note) Equals *Tibicen vitreus* Greece, France
Cicadatra atra var. *vitrea* Boulard and Mondon 1995: 66 (listed) France
Cicadatra atra var. *vitrea* Boulard 2000i: 44 (described, distribution, comp. note) France, Greece

C. *bistunensis* Mozaffarian & Sanborn, 2010
Cicadatra bistunensis Mozaffarian and Sanborn 2010: 69–73, 78, Figs 1–7 (n. sp., described, illustrated, comp. note) Iran

C. *cataphractica* Popov, 1989 to *Psalmocharias cataphractica*

C. *erkowitensis* Linnavuori, 1973
Cicadatra erkowitensis Sanborn 1999a: 44 (type material, listed)

C. *flavicollis* Horváth, 1911 = *Psalmocharias flavicollis* = *Cicadatra foveicollis* (sic)
Cicadatra foveicollis (sic) Bey 1913: 141–142 (comp. note) Egypt
Psalmocharias flavicollis Walker and Pittaway 1987: 40–41 (illustrated, natural history, distribution, comp. note) Saudi Arabia
Cicadatra flavicollis Schedl 1993: 797 (listed, distribution, comp. note) Egypt, Israel, Arabia
Cicadatra flavicollis Sanborn 1999a: 44 (type material, listed)
Cicadatra flavicollis Schedl 1999b: 825 (listed, distribution, ecology, comp. note) Israel, Egypt, Arabia

C. *genoina* Dlabola, 1979
Cicadatra genoina Nast 1982: 311 (listed) Iran
Cicadatra genoina Mozaffarian and Sanborn 2010: 78 (listed, distribution) Iran

C. *gingat* China, 1926
Cicadatra gingat Ahmed and Sanborn 2010: 40 (distribution) Pakistan, India
Cicadatra gingat Ahmed, Sanborn and Hill 2010: 78 (comp. note)

C. *glycyrrhizae* (Kolenati, 1857) = *Cicada (Cicadatra) glycyrrhizae* = *Cicada glycyrrhizae* = *Cicadatra glycyrrhizae* = *Cicadatra atra glycyrrhizae* = *Cicadatra glacyrrhiza* (sic) = *Cicadatra glycyrrhiza* (sic) = *Cicadatra glycirrhizae* (sic) = *Cicadatra glycyrrhizicae* (sic) = *Cicadatra atra glycirrhizae* (sic)
Cicadatra glycyrrhizae Castellani 1952: 15 (comp. note) Italy
Cicadatra glycyrrhizae Wagner 1952: 15 (listed, comp. note) Italy
Cicadatra glycirrhizae Boulard 1995a: 12, 28, 30–31, Fig S9$_1$, Fig S9$_2$, Figs 11–12 (illustrated, song analysis, sonogram, listed, comp. note) Turkey
Cicadatra glycirrhizae Gogala and Trilar 1998a: 8–9 (comp. note)
Cicadatra glycirrhizae Boulard 2000g: 144 (sonogram correction)
Cicadatra glycyrrhizae Sueur 2001: 40, Table 1 (listed)
Cicadatra glycyrrhizae Boulard 2006g: 180 (listed)

C. *gregoryi* China, 1925
Cicadatra gregoryi Wu 1935: 18 (listed) China, Yunnan
Cicadatra gregoryi Chou, Lei, Li, Lu and Yao 1997: 120–121 (key, described, comp. note) China
Cicadatra gregoryi Hua 2000: 61 (listed, distribution) China, Yunnan
Cicadatra gregoryi Ahmed, Sanborn and Hill 2010: 78 (comp. note)

C. *hagenica* Dlabola, 1987 = nom. nud. pro *Cicada virens* Hagen, 1856 (nec Herrich Schäffer, 1835) = *Cicadatra hyalina virens* = *Cicadatra virens*
Cicadatra hagenica Dlabola 1987: 306–308, Figs 55–60 (nom. nud., illustrated, comp. note) Equals *Cicada virens* Hagen Equals *Cicadatra hyalina virens* Equals *Cicadatra virens* Turkey
Cicadatra "virens" Dlabola 1987: 306 (comp. note)
Cicadatra hagenica Lodos and Kalkandelen 1988: 18 (comp. note) Turkey
Cicadatra hagenica Ahmed, Sanborn and Hill 2010: 78 (comp. note)
Cicadatra hagenica Kemal and Koçak 2010: 4 (distribution, comp. note) Turkey
Cicadatra hagenica Mozaffarian and Sanborn 2010: 78 (synonymy, listed, distribution) Equals *Cicada virens* (nec Herrich-Schäffer) Equals *Cicadatra hyalina virens* Iran, Turkey

C. *hyalina* (Fabricius, 1798) = *Tettigonia hyalina* = *Cicada hyalina* = *Tibicen hyalina* = *Cicada (Cicadatra) hyalina* = *Tettigia (Cicadatra) hyalina* = *Tettigonia hyalina* = *Cicadetta hyalina* = *Cicadatra hyaline* (sic) = *Cicadatra (Rustavelia) burriana* Horváth, 1912 = *Cicadatra burriana* = *Rustavelia burriana* = *Cicada (Cicadatra) geodesma* Kolenati, 1857 = *Cicada geodesma* = *Cicadatra geodesma* = *Cicadatra goedesma* (sic) = *Cicadatra hyalina geodesma* = *Cicada (Cicadrata)* (sic) *hyalina geodesma* = *Cicadatra geodesma geodesma* = *Cicadatra geodesma discrepans* Schumacher, 1923 =

Cicadatra hyalina discrepens = *Cicadatra geodesma rossica* Schumacher, 1923 = *Cicadatra hyalina rossica* = *Cicadatra taurica* Fieber, 1876 = *Cicadatra hyalina taurica* = *Cicadatra geodesma taurica* = *Cicadatra hyalina virens* Fieber, 1876 = *Cicada virens* = *Cicadatra hyalina vireus* (sic) = *Cicadatra viridis* Haupt, 1917 = *Cicadetta coriaria* (nec Puton)

Cicadatra hyalina Mirzayans, Hashemi, Borumand, Zairi and Rajabi 1976: 111 (distribution, listed) Iran

Cicadatra hyalina Drosopoulous 1980: 190 (listed) Greece

Cicadatra hyalina Boulard 1981b: 48, 51 (synonymy, comp. note) Balkans, East Mediterranean

Cicadatra hyalina Dlabola 1981: 205 (distribution, comp. note) Iran

Cicadetta hyalina Jankovic and Papovic 1981: 131 (comp. note) Yugoslavia

Cicadatra hyalina Lodos and Kalkandelen 1981: 72–73 (synonymy, distribution, comp. note) Equals *Cicada (Cicadatra) hyalina geodesma* Equals *Cicadatra hyalina taurica* Equals *Cicadatra hyalina virens* Equals *Cicadatra (Rustavelia) burriana* Equals *Cicadatra viridis* Equals *Cicadatra geodesma discrepans* Equals *Cicadatra geodesma rossica* Turkey, Balkan Peninsula, Iran, Israel, Jordan, Syria, USSR

Cicadatra hyalina Popov 1981: 279 (comp. note)

Cicadatra (Rustavelia) burriana Koçak 1982: 151 (type species of *Rustavelia*)

Tettigonia hyalina Boulard 1984c: 171, 176 (comp. note)

Cicadatra hyalina Popov 1985: 48, 68–76, 80, 181–183, 185, Figs 17–19, Fig 24, Fig 59.1, Fig 59.4 (oscillogram, acoustic system, auditory system, comp. note)

Cicadatra hyalina Popov, Aronov and Sergeeva 1985a: 461 (comp. note)

Cicadatra hyalina Popov, Aronov and Sergeeva 1985b: 297 (comp. note)

Cicadatra hyalina Dlabola 1987: 305–308, Figs 61–64 (listed, illustrated, comp. note) Equals *Tettigonia hyalina* Turkey, Greece, USSR, Iran

Cicadatra hyalina discrepans Dlabola 1987: 306 (comp. note)

Cicadatra viridis Dlabola 1987: 306, 308 (comp. note)

Cicadatra hyalina Nast 1987: 568–569 (distribution, listed) Ukraine, Russia, Yugoslavia

Tettigonia hyalina Boulard 1988f: 23 (comp. note)

Cicadatra hyalina Popov 1990a: 302, Fig 2 (comp. note)

Cicadatra hyalina Tautz 1990: 105, Fig 12 (song frequency, comp. note)

Cicadatra hyalina Popov, Aronov and Sergeeva 1991: 282, 284–286, 289, 292, 295, Fig 2–3, Table (auditory system, ecology, comp. note) Russia

Cicadatra hyalina Popov, Arnov and Sergeeva 1992: 152, 154–156, 159, 162, 165, Fig 2–3, Table (auditory system, ecology, comp. note) Russia

Cicadatra hyalina Boulard 1995a: 12, 28–29, Fig $S8_1$, Figs 9–10 (illustrated, song analysis, sonogram, listed, comp. note) Turkey

Cicadatra hyalina Mirzayans 1995: 16–17 (listed, distribution) Iran

Tettigonia hyalina Boulard 1997c: 93 (comp. note)

Cicadatra hyalina Gogala 1998a: 156 (valid species, comp. note) Macedonia

Cicadatra hyalina Gogala and Trilar 1998a: 7, 9 (comp. note) Macedonia

Cicadatra hyalina Trilar and Gogala 1998: 69 (comp. note) Macedonia

Cicadatra hyalina Gogala and Gogala 1999: 127 (comp. note)

Cicadatra hyalina Kartal and Zeybekoglu 1999b: 63–65, Figs 1–3 (described, illustrated, comp. note) Anatolia, Turkey

Cicadatra hyalina discrepans Kartal and Zeybekoglu 1999b: 65 (comp. note)

Cicadatra hyalina Schedl 1999b: 825, 834, Fig 3 (illustrated, listed, distribution, ecology, comp. note) Israel, Palestine

Tettigonia hyalina Boulard 2001b: 14 (comp. note)

Cicadatra hyalina Krpač, Trilar and Gogala 2001: 29 (comp. note) Macedonia

Cicadatra hyalina Gogala 2002: 244 (comp. note)

Tettigonia hyalina Boulard 2003f: 372 (comp. note)

Cicadatra hyalina Sueur 2003: 2942 (comp. note)

Cicadatra hyalina Gogala, Trilar and Krpač 2005: 104, 110–112, 124, Fig 4 (distribution, comp. note) Macedonia

Cicadatra hyalina Moulds 2005b: 410 (comp. note)

Cicadatra hyalina Boulard 2006g: 36–37, Fig 7h (illustrated, comp. note)

Cicadatra hyalina Trilar, Gogala and Popa 2006: 177 (listed, distribution, comp. note) Romania

Cicadatra hyalina Demir 2007a: 51 (distribution, comp. note) Turkey

Cicadatra hyalina Demir 2007b: 484 (distribution, comp. note) Turkey

Cicadatra hyaline (sic) Shiyake 2007: 6, 79, 81, Fig 127 (illustrated, distribution, listed, comp. note) Middle East

Cicadatra hyalina Demir 2008: 475 (distribution) Equals *Cicadatra burriana* Equals *Cicadatra goedesma* Equals *Cicadatra taurica* Equals *Cicadatra virens* Equals *Cicadatra viridis* Turkey

Cicadatra hyalina Trilar and Gogala 2008: 30 (comp. note) Romania

Cicadatra hyalina Ahmed, Sanborn and Hill 2010: 78 (comp. note)

Cicadatra hyalina Kemal and Koçak 2010: 2, 4, 16, 18–19, Fig 19, Figs 23–25 (illustrated, predation, distribution, listed, comp. note) Turkey

Cicadatra hyalina Koçak and Kemal 2010: 135 (listed, distribution, comp. note) Turkey

Cicadatra hyalina Mozaffarian and Sanborn 2010: 78 (synonymy, listed, distribution) Equals *Tettigonia hyalina* Equals *Cicada geodesma* Equals *Cicada hyalina geodesma* Equals *Cicadatra taurica* Equals *Cicadatra (Rustavelia) burriana* Equals *Cicadatra viridis* Equals *Cicadatra geodesma discrepens* Equals *Cicadatra geodesma rossica* Iran, Southern Russia, Balkan Peninsula, Turkey, Syria, Jordan, Georgia, USSR Armenia, Transcaucasia, Caucasus

Cicadatra hyalina Mozaffarian, Sanborn and Phillips 2010: 32–33 (comp. note)

C. hyalinata (Brullé, 1832) to *Cicadatra atra* var. *hyalinata*

***C. icari* Simões, Sanborn & Quartau, 2012**

***C. inconspicua* Distant, 1912**

Cicadatra inconspicua Ahmed, Sanborn and Hill 2010: 78 (comp. note)

***C. karachiensis* Ahmed, Sanborn & Hill, 2010**

Cicadatra karachiensis Ahmed, Sanborn and Hill 2010: 76–80 Figs 1–7 (n. sp., comp. notes, illustrated) Pakistan

***C. karpathosensis* Simões, Sanborn & Quartau, 2012**

***C. kermanica* Dlabola, 1970**

Cicadatra kermanica Mirzayans, Hashemi, Borumand, Zairi and Rajabi 1976: 111 (distribution, listed) Iran

Cicadatra kermanica Dlabola 1981: 205 (distribution, comp. note) Iran

Cicadatra kermanica Mirzayans 1995: 17 (listed, distribution) Iran

Cicadatra kermanica Mozaffarian and Sanborn 2010: 78 (listed, distribution) Iran

***C. longipennis* Schumacher, 1923**

Cicadatra longipennis Dlabola 1981: 201–203, Figs 130–131 (comp. note)

Cicadatra longipennis Schedl 1999b: 826, 834, Fig 4 (illustrated, listed, distribution, ecology, comp. note) Israel, Jordan, Egypt

***C. loristanica* Mozaffarian & Sanborn, 2010**

Cicadatra lorestanica Mozaffarian, Sanborn and Phillips 2010: 28–37 Figs 1–11 (n. sp., described, illustrated, song, sonogram, oscillogram, distribution, comp. note) Iran

***C. mirzayansi* Dlabola, 1981** = *Cicadatra mirzyansi* (sic)

Cicadatra mirzayansi Dlabola 1981: 203–205, Figs 132–133 (n. sp., described, illustrated, distribution, comp. note) Iran

Cicadatra mirzyansi (sic) Mirzayans 1995: 17 (listed, distribution) Iran

Cicadatra mirzayansi Mozaffarian and Sanborn 2010: 79 (listed, distribution) Iran

***C. naja* Dlabola, 1979** = *Cicadetta* (sic) *naja*

Cicadatra naja Nast 1982: 311 (listed) Iran

Cicadetta (sic) *naja* Mirzayans 1995: 18 (listed, distribution) Iran

Cicadatra naja Sanborn 1999a: 44 (type material, listed)

Cicadatra naja Mozaffarian and Sanborn 2010: 79 (listed, distribution) Iran

***C. pallisi* Schumacher, 1923**

***C. persica* Kirkaldy, 1909** = *Cicadatra lineola* Hagen, 1856 (nec *Cicada lineola* von Schrank, 1801) = *Tettigia (Cicadatra) lineola* = *Cicadatra lineola*

Cicadatra persica Dlabola 1981: 205 (distribution, comp. note) Iran, Turkey

Cicadatra persica Lodos and Kalkandelen 1981: 73 (synonymy, distribution, comp. note) Equals *Cicada lineola* Turkey, Iran, Israel, Italy, Syria, USSR

Cicadatra persica Nast 1987: 568–569 (distribution, listed) Italy?

Cicadatra persica Mirzayans 1995: 17–18 (listed, distribution) Iran

Cicadatra persica Gogala 1998a: 156 (comp. note) Macedonia

Cicadatra persica Gogala and Trilar 1998a: 6–13, Figs 1–5 (illustrated, song, sonogram, oscillogram, comp. note) Macedonia, Iran, Caucasus, Asia Minor, Syria, Israel, Sicily

Cicadatra persica Trilar and Gogala 1998: 69 (comp. note) Macedonia

Cicadatra persica Gogala and Gogala 1999: 127 (comp. note)

Cicadatra persica Kartal and Zeybekoglu 1999a: 59–62, Figs 1–3 (described, illustrated, fecundity, comp. note) Anatolia, Turkey

Cicadatra persica Gogala and Trilar 2001a: 28 (comp. note)

Cicadatra persica Gogala and Trilar 2001b: 14 (comp. note)

Cicadatra persica Krpač, Trilar and Gogala 2001: 29 (comp. note) Macedonia

Cicadatra persica Sueur 2001: 40, Table 1 (listed)

Cicadatra persica Gogala 2002: 244 (comp. note)

Cicadetta persica Gogala and Trilar 2003: 6–9, 11, Fig 2 (sound production, comp. note) Macedonia

Cicadatra persica Tishechkin 2003: 141, 143–144, Figs 148–160 (oscillogram, comp. note) Azerbaijan
Cicadatra persica Boulard 2005d: 343 (comp. note)
Cicadatra persica Gogala, Trilar and Krpač 2005: 105, 113, 123–124, Fig 7 (distribution, comp. note) Macedonia
Cicadatra persica Demir 2007b: 484 (distribution, comp. note) Turkey
Cicadatra persica Demir 2008: 475 (distribution) Equals *Cicadatra lineola* Turkey
Cicadatra persica Ahmed and Sanborn 2010: 40 (synonymy, distribution) Equals *Cicada lineola* (nec von Schank) Pakistan, Italy, Sicily, Turkey, Israel, Syria, Georgia, USSR, Transcaucasia, Iran
Cicadatra persica Ahmed, Sanborn and Hill 2010: 78 (comp. note)
Cicadatra persica Kemal and Koçak 2010: 2, 5, 12, 14, Figs 11–12, Fig 16 (illustrated, distribution, listed, comp. note) Turkey
Cicadatra persica Koçak and Kemal 2010: 135 (listed, distribution, comp. note) Turkey
Cicadatra persica Mozaffarian and Sanborn 2010: 73, 79 (synonymy, distribution, listed, comp. note) Equals *Cicada lineola* (nec von Schrank) Iran, Syria, Persia, Italy, Sicily, Turkey, Israel, Syria, Goergia, USSR, Transcaucasia
Cicadatra persica Mozaffarian, Sanborn and Phillips 2010: 33 (comp. note)

***C. platyptera* Feiber, 1876** = *Cicadatra atra platyptera* = *Cicadatra decumana* Schumacher, 1923 = *Cicadatra hyalina decumana* = *Cicadatra platypleura livida* Schumacher, 1923 = *Cicadatra livida* = *Cicadatra platyptera livens* (sic) = *Cicadatra platyptera melanaria* Schumacher, 1923
Cicadatra platyptera Dlabola 1981: 206 (distribution, comp. note) Turkey
Cicadatra platyptera Lodos and Kalkandelen 1981: 73–74 (synonymy, distribution, comp. note) Equals *Cicadatra platyptera livens* Equals *Cicadatra decumana* Equals *Cicadatra platyptera melanaria* Turkey, Iran, Israel, Italy, Syria, USSR, Lebanon
Cicadatra hyalina decumana Dlabola 1987: 306 (comp. note) Iran
Cicadatra platyptera Boulard 1995a: 6, 12, 26, 28–29, Fig S7$_1$ (song analysis, sonogram, listed, comp. note) Turkey
Cicadatra livida Schedl 1999b: 830 (comp. note)
Cicadatra platyptera Schedl 1999b: 826, 835, Fig 5 (illustrated, listed, distribution, ecology, comp. note) Israel, Syria, Turkey, Jordan
Cicadatra platyptera Boulard 2000g: 144 (sonogram correction)
Cicadatra platyptera Gogala and Trilar 2001a: 28 (comp. note)
Cicadatra platyptera Krpač, Trilar and Gogala 2001: 29 (comp. note) Macedonia
Cicadatra platyptera Sueur 2001: 40, Table 1 (listed)
Cicadatra platyptera Gogala 2002: 244 (comp. note)
Cicadatra platyptera Gogala and Trilar 2003: 8 (comp. note)
Cicadatra platyptera Gogala, Trilar and Krpač 2005: 105, 111–112, 124, Figs 5–6 (distribution, sonogram, oscillogram, comp. note) Macedonia
Cicadatra platyptera Ahmed, Sanborn and Hill 2010: 78 (comp. note)
Cicadatra platyptera Kemal and Koçak 2010: 5 (distribution, listed, comp. note) Turkey
Cicadatra platyptera Mozaffarian and Sanborn 2010: 79 (synonymy, listed, distribution) Equals *Cicadatra platyptera livida* Equals *Cicadatra decumana* Equals *Cicadatra livida* Iran, Turkey, Italy, Greece, Syria, Israel, Jordan, Palestine, Asia Minor
Cicadatra platyptera Mozaffarian, Sanborn and Phillips 2010: 33 (comp. note)

***C. raja* Distant, 1906**
Cicadatra raja Ahmed and Sanborn 2010: 40–41 (distribution) Pakistan, India
Cicadatra raja Ahmed, Sanborn and Hill 2010: 78 (comp. note)

***C. ramanensis* Linnavouri, 1962**
Cicadatra ramanensis Dlabola 1981: 203, 205–206 (comp. note)
Cicadatra ramanensis Mirzayans 1995: 18 (listed, distribution) Iran
Cicadatra ramanensis Sanborn 1999a: 44 (type material, listed)
Cicadatra ramanensis Schedl 1999b: 826, 835, Fig 6 (illustrated, listed, distribution, ecology, comp. note) Israel, Syria, Iran
Cicadatra ramanensis Ahmed, Sanborn and Hill 2010: 78 (comp. note)
Cicadatra ramanensis Mozaffarian and Sanborn 2010: 79 (listed, distribution) Iran, Israel
Cicadatra ramanensis Mozaffarian, Sanborn and Phillips 2010: 32 (comp. note)

***C. reinhardi* Kartal, 1980**
Cicadatra reinhardti Nast 1982: 311 (listed) Iraq

***C. sankara* (Distant, 1904)** = *Tibicen sankara* = *Cicadatra sankana* (sic)
Cicadatra sankara Ahmed and Sanborn 2010: 41 (synonymy, distribution) Equals *Tibicen sankara* Pakistan, India
Cicadatra sankara Ahmed, Sanborn and Hill 2010: 78 (comp. note)

C. *shaluensis* China, 1925
Cicadatra shaluensis Wu 1935: 18 (listed) China, Yunnan
Cicadatra shaluensis Chou, Lei, Li, Lu and Yao 1997: 121 (key, described, comp. note) China
Cicadatra shaluensis Hua 2000: 61 (listed, distribution) China, Yunnan
Cicadatra shaluensis Ahmed, Sanborn and Hill 2010: 78 (comp. note)

C. *shapur* Dlabola, 1981
Cicadatra shapur Dlabola 1981: 201–203, Figs 123–129 (n. sp., described, illustrated, distribution, comp. note) Iran
Cicadatra shapur Mozaffarian and Sanborn 2010: 79 (listed, distribution) Iran

C. *tenebrosa* Fieber, 1876
Cicadatra tenebrosa Lodos and Kalkandelen 1981: 74–75 (distribution, comp. note) Turkey
Cicadatra tenebrosa Kemal and Koçak 2010: 5 (lsited, comp. note) Turkey

C. *vitreus* (Brullé, 1832) to *Cicadatra atra* var. *vitrea*

C. *vulcania* Dlabola & Heller, 1962
Cicadatra vulcania Mirzayans, Hashemi, Borumand, Zairi and Rajabi 1976: 111 (distribution, listed) Iran
Cicadatra vulcana Dlabola 1981: 205–206 (distribution, comp. note)
Cicadatra vulcania Mirzayans 1995: 18 (listed, distribution) Iran
Cicadatra vulcania Mozaffarian and Sanborn 2010: 79 (listed, distribution) Iran
Cicadatra vulcania Mozaffarian, Sanborn and Phillips 2010: 32–33 (comp. note)

C. *walkeri* Metcalf, 1963 = *Cicada striata* Walker, 1850 (nec *Cicada striata* Linnaeus, 1758) = *Cicadatra striata*
Cicadatra walkeri Ahmed and Sanborn 2010: 41 (synonymy, distribution) Equals *Cicada striata* (nec Linnaeus) Pakistan, India
Cicadatra walkeri Ahmed, Sanborn and Hill 2010: 78 (comp. note)

C. *xantes* (Walker, 1850) = *Cicada xantes* = *Cicada subvenosa* Walker, 1858 = *Cicadatra xanthes* (sic)
Cicadatra xantes Ahmed and Sanborn 2010: 41 (distribution) Pakistan, India
Cicadatra xanthes (sic) Ahmed, Sanborn and Hill 2010: 78 (comp. note)

C. *zahedanica* Dlabola, 1970
Cicadatra zahedanica Mirzayans, Hashemi, Borumand, Zairi and Rajabi 1976: 111 (distribution, listed) Iran
Cicadatra zahedanica Dlabola 1981: 206 (distribution, comp. note) Iran
Cicadatra zahedanica Mirzayans 1995: 18 (listed, distribution) Iran
Cicadatra zahedanica Ahmed, Sanborn and Hill 2010: 78 (comp. note)
Cicadatra zahedanica Mozaffarian and Sanborn 2010: 79 (listed, distribution) Iran, Afghanistan

C. *ziaratica* Ahmed, Sanborn & Akhter, 2012

***Klapperichicen* Dlabola, 1957** = *Klapperischicen* (sic) = *Klapparichicen* (sic)
Klapperischicen Mirzayans 1995: 18 (listed) Iran
Klapparichicen (sic) Schedl 2001a: 1287 (comp. note)
Klapperichicen Schedl 2003: 426 (comp. note)
Klapperischicen (sic) Kemal and Koçak 2010: 3, 5 (diversity, listed, comp. note) Turkey
Klapperichicen Mozaffarian and Sanborn 2010: 81 (distribution, listed) Europe, Central Asia

K. *acoloratus* Dlabola, 1960 = *Klapperichicen accoloratus* (sic) = *Klapperichicen* (sic) *accoloratus* (sic)
Klapperichicen accoloratus (sic) Dlabola 1981: 207 (distribution, comp. note) Iran
Klapperichicen (sic) *accoloratus* (sic) Mirzayans 1995: 18 (listed, distribution) Iran
Klapperichicen acoloratus Schedl 2003: 424–427, 431, Photo 4 (listed, key, comp. note) Iran
Klapperichicen acoloratus Mozaffarian and Sanborn 2010: 81 (listed, distribution, comp. note) Iran

K. *dubius* (Jacobi, 1927) = *Cicadatra dubius*
Cicadatra (?) dubia Schedl 2003: 423, 426 (comp. note) Afghanistan
Klapperichicen dubius Schedl 2003: 424, 426–428, 431, Fig 1, Photo 1 (illustrated, listed, synonymy, key, illustrated, comp. note) Equals *Cicadatra (?) dubia* Afghanistan
Klapperichicen dubius Mozaffarian and Sanborn 2010: 81 (type species of *Klapperichicen*)

K. *turbatus* (Melichar, 1902) = *Klapperichicen* (sic) *turbatus* = *Tibicen turbatus* = *Tibicina turbata*
Klapperichicen turbatus Dlabola 1981: 207 (distribution, comp. note) Iran
Klapperichicen (sic) *turbatus* Mirzayans 1995: 18 (listed, distribution) Iran
Tibicen turbatus Schedl 2003: 423 (comp. note)
Klapperichicen turbatus Schedl 2003: 424–427, 429, 431, Fig 2, Photo 2 (illustrated, listed, synonymy, key, comp. note) Equals *Tibicen turbatus* Iran
Klapperichicen turbatus Mozaffarian and Sanborn 2010: 81 (synonymy, listed, distribution, comp. note) Equals *Tibicen turbatus* Iran

K. *viridissimus* (Walker, 1858) = *Cicada viridissima* = *Cicadatra viridissima* = *Chloropsalta viridissima* = *Klapperischen viridissima* (sic) = *Klapperichien* (sic) *viridissima*
Klapperichien (sic) *viridissima* (sic) Lodos and Kalkandelen 1981: 71 (synonymy, distribution, comp. note) Equals *Cicadatra*

atra var. *alhageos* Equals *Cicadatra atra* var. *glycyrrhizae* Equals *Cicada ochreata* Equals *Cicada olivacea* Equals *Cicadatra stenoptera* Equals *Chloropsalta smaragdula* Turkey, Afghanistan, Greece, Iran, Iraq, Israel, Italy, Jordan, USSR

Klapperischen viridissima (sic) Kartal 1988: 9 (comp. note)

Chloropsalta viridissima (sic) Kartal and Zeybekoglu 1999a: 62 (comp. note)

Klapperichicen viridissima (sic) Freytag 2000: 55 (on stamp)

Klapperichicen viridissimus Schedl 2003: 425–427, 430–431, Fig 3, Photo 3 (listed, synonymy, key, comp. note) Equals *Cicada viridissimus* Iraq, Iran, Afghanistan, Turkey, Israel, Jordan, USSR

Cicada viridissimus Schedl 2003: 427 (comp. note)

Klapperischicen viridissima (sic) Kemal and Koçak 2010: 5 (distribution, listed, comp. note) Turkey

Klapperichicen viridissimus Mozaffarian and Sanborn 2010: 81 (synonymy, listed, distribution, comp. note) Equals *Cicada viridissimus* Iran, Iraq, Italy, Greece, Afghanistan

***Psalmocharias* Kirkaldy, 1908** = *Cicadatra (Psalmocharias)* = *Psalmochavias* (sic) = *Sena* Distant, 1905

Psalmocharias Boulard 1981b: 44 (comp. note) North Africa

Psalmocharias Koçak 1982: 150–151 (listed, synonymy, comp. note) Equals *Sena* Distant (nec *Sena* Walker, 1862) Turkey

Sena Koçak 1982: 151 (listed, comp. note) Turkey

Psalmocharias Hayashi 1984: 47 (listed)

Psalmocharias Mirzayans 1995: 15 (listed) Equals *Cicadatra* Iran

Psalmocharias Chou, Lei, Li, Lu and Yao 1997: 120, 289, 293, 297, 357, Table 6 (described, key, synonymy, listed, diversity, biogeography, comp. note) Equals *Sena* China

Psalmocharias Moulds 2005b: 390 (higher taxonomy, listed, comp. note)

Psalmocharias Shiyake 2007: 6 (listed, comp. note)

Psalmocharias sp. Shiyake 2007: 6 (illustrated, distribution, listed, comp. note)

Psalmocharias Lee and Hill 2010: 298 (comp. note)

Psalmocharias Mozaffarian and Sanborn 2010: 79 (distribution, listed) Mongolia, Tunisia

Psalmocharias Wei, Ahmed and Rizvi 2010: 28, 33 (comp. note)

***P. balochii* Ahmed & Sanborn, 2010**

Psalmocharias balochii Ahmed and Sanborn 2010: 28–30, 41, Figs 1–5 (n. sp., key, described, comp. note, distribution) Pakistan

***P. cataphractica* (Popov, 1989)** = *Cicadatra cataphractica* = *Cicadatra* ex. gr. *querula* Dlabola, 1961

Cicadatra ex. gr. *querula* (sp. nov.) Popov 1985: 68, 80, Fig 24 (acoustic system, comp. note)

Cicadatra cataphractica Popov, Aronov and Sergeeva 1985a: 451–452, 453, 455–458, 460–461, Figs 4.4–4.7, Fig 7 (n. sp., song, oscillogram, comp. note) Tadzhikistan

Cicadatra cataphractica Popov, Aronov and Sergeeva 1985b: 289–295, 297, Figs 4.4–4.7, Fig 7 (n. sp., song, oscillogram, comp. note) Tadzhikistan

Cicadatra sp. ex. gr. *querula* Popov 1989a: 291 (comp. note)

Cicadatra sp. ex. gr. *querula* Popov 1989b: 62 (comp. note)

Cicadatra cataphractica Popov 1989a: 291–306, Figs 1.1–1.2, Fig 2, Fig 3.2, Figs 4.3–4.5, Figs 5.4–5.5, Figs 6.3–6.4, Fig 7, Figs 8.3–8.4, Fig 9, Fig 10B, Fig 11B, Fig 12, Table 1 (illustrated, song, oscillogram, hearing, comp. note) Equals *Cicadatra* sp. ex. gr. *querula* Dlabola 1961 Tajikistan

Cicadatra cataphractica Popov 1989b: 62–77, Figs 1.1–1.2, Fig 2, Fig 3.2, Figs 4.3–4.5, Figs 5.4–5.5, Figs 6.3–6.4, Fig 7, Figs 8.3–8.4, Fig 9, Fig 10B, Fig 11B, Fig 12, Table 1 (illustrated, song, oscillogram, hearing, comp. note) Equals *Cicadatra* sp. ex. gr. *querula* Dlabola 1961 Tajikistan

Cicadatra cataphractica Popov 1990a: 302–303, Figs 2–3 (comp. note)

Cicadatra cataphractica Tautz 1990: 105–106, Fig 12, Fig 13C (song frequency, hearing sensitivity, comp. note)

Cicadatra cataphractica Popov, Aronov and Sergeeva 1991: 294–295 (comp. note)

Cicadatra cataphractica Popov, Arnov and Sergeeva 1992: 164–165 (comp. note)

Cicadatra cataphractica Bennet-Clark and Young 1994: 293, Table 1 (body size and song frequency)

Cicadatra cataphractica Sueur 2001: 40, Table 1 (listed)

***P. chitralensis* Ahmed & Sanborn, 2010**

Psalmocharias chitralensis Ahmed and Sanborn 2010: 28, 31–33, 35, 37, 41, Figs 6–10 (n. sp., key, described, distribution) Pakistan

***P. flava* Dlabola, 1970**

Psalmocharias flava Mirzayans, Hashemi, Borumand, Zairi and Rajabi 1976: 111 (distribution, listed) Iran

Pslamocharias flava Dlabola 1981: 201 (distribution, comp. note) Iran

Psalmocharias flava Mirzayans 1995: 15 (listed, distribution) Iran

Psalmocharias flava Mozaffarian and Sanborn 2010: 79–80 (listed, distribution, comp. note) Iran

***P. gizarensis* Ahmed & Sanborn, 2010**

Psalmocharias gizarensis Ahmed and Sanborn 2010: 33–35, 37, 41, Figs 11–15 (n. sp., key, described, comp. note, distribution) Pakistan

***P. japokensis* Ahmed & Sanborn, 2010**

Psalmocharias japokensis Ahmed and Sanborn 2010: 33, 35–37, 42, Figs 16–20 (n. sp., key, described, comp. note, distribution) Pakistan

***P. paliur* (Kolenati, 1857)** = *Cicada (Cicadatra) querula paliuri* = *Cicada paliuri* = *Cicadatra querula paliuri* = *Cicada (Cicadatra) querula paliuri* = *Cicadatra querula palinri* (sic) = *Cicadatra paliuri* = *Psalmocharias querula paliuri* = *Psalmocharias paiiri* (sic)

Psalmocharias paliuri Ahmed and Sanborn 2010: 35 (comp. note)

Psalmocharias paiiuri (sic) Ahmed and Sanborn 2010: 37 (comp. note) Pakistan

***P. plagifera* (Schumacher, 1922)** = *Cicadatra (Psalmocharias) plagifera* = *Cicadatra plagifera*

Psalmocharias plagifera Villiers 1977: 165, Plate XIV, Fig 23 (illustrated, listed, comp. note) North Africa

Psalmocharias plagifera Boulard 1981b: 43–45, 50, Figs 3–4, Fig 6 (neallotype designation, illustrated, comp. note)

Psalmocharias plagifera Nast 1987: 570–571 (distribution, listed) France

Psalmocharias plagifera Moalla, Jardak and Ghorbel 1992: 34–42, Figs 1–5, Plates I-III, Table 4 (olive pest, described, illustrated, larva, distribution, comp. note) Tunisia

Psalmocharias plagifera Jardak and Ksantini 1996: 26, Fig 1 (olive pest, comp. note) Tunisia

Psalmocharias plagifera Schedl 1999b: 824–825, 833, Fig 1 (illustrated, listed, distribution, comp. note) Israel, Northern Africa

Pslamocharias plagifera Lee and Hill 2010: 299–300, Figs 7–8 (phylogeny, comp. note)

***P. querula* (Pallas, 1773)** = *Cicada querula* = *Tettigonia querula* = *Cicada concinna querula* Germar, 1830 = *Cicada quaerula* (sic) = *Cicada quercula* (sic) = *Cicada (Cicadatra) querula* = *Tettigia (Cicadatra) querula* = *Cicadatra querula* = *Cicadatra guerula* (sic) = *Cicadatra gurula* (sic) = *Sena querula* = *Sena quaerula* (sic) = *Cicadatra (Psalmocharias) querula* = *Psalmocharias quaerula* (sic) = *Cephaloxys quadrimacula* Walker, 1850 = *Cephaloxys quadrimaculata* (sic) = *Cicadatra quadrimacula* = *Cicadatra quadrimaculata* (sic) = *Mogannia quadrimacula* = *Cicada steveni* Stål, 1854 = *Cicadatra steveni*

Cicadatra querula Gómez-Menor 1951: 56 (comp. note)

Cicadatra querula Grechkin 1956: 1489 (comp. note) Tajikistan

Psalmocharias querula Dlabola 1971: 382 (listed, comp. note) Iran, Afghanistan

Psalmocharias querula Boulard 1981b: 43–45, 50, Fig 2, Fig 5 (illustrated, comp. note) Europe, West Asia, Equals *Cicada querula* Equals *Cicadatra querula*

Pslamocharias querula Dlabola 1981: 201 (distribution, comp. note) Iran

Cicadatra querula Lodos and Kalkandelen 1981: 74 (synonymy, distribution, comp. note) Equals *Cicada steveni* Equals *Cicada querula* var. *paliuri* Turkey, Afghanistan, Algeria, France, Iran, Israel, Mongolia, Tunisia, USSR

Cicadatra querula Dubovsky 1982: 50 (listed, comp. note) Uzbekistan

Cicada querula Koçak 1982: 150–151 (type species of *Psalmocharias*, type species of *Sena*)

Cicadatra querula Emeljanov 1984: 10, 56 (listed, key)

Cicadatra guerula (sic) Popova and Dubovskaya 1984: 3 (agricultural pest, listed) Uzbekistan

Cicadatra querula Popov 1985: 40–41, 68, 71, 73, 75–76, 80, Fig 7.5, Fig 24 (illustrated, acoustic system, comp. note)

Cicadatra querula Popov, Aronov and Sergeeva 1985a: 451, 455–461, Figs 4.1–4.3, Fig 6, Figs 8c–d (song, oscillogram, hearing, comp. note) Tadzhikistan

Cicadatra querula Popov, Aronov and Sergeeva 1985b: 289, 291–297, Figs 4.1–4.3, Fig 6, Figs 8c–d (song, oscillogram, hearing, comp. note) Tadzhikistan

Cicadatra querula Kozhevnikova 1986: 46–47 (cotton pest) Uzbekistan

Cicadatra guerula (sic) Kholmuminov 1987: 49 (comp. note) Uzbekistan

Psalmocharias querula Nast 1987: 570–571 (distribution, listed) France, Ukraine, Russia, Spain

Cicadatra querula Popov 1989a: 291–306, Figs 1.3–1.4, Fig 2, Fig 3.1, Figs 4.1–4.2, Figs 5.1–5.3, Figs 6.1–6.2, Figs 8.1–8.2, Fig 10A, Fig 11A, Table 1 (illustrated, song, oscillogram, hearing, comp. note) Tajikistan, Kirghizia, Transcaucasia, Crimea

Cicadatra querula Popov 1989b: 62–75, 77, Figs 1.3–1.4, Fig 2, Fig 3.1, Figs 4.1–4.2, Figs 5.1–5.3, Figs 6.1–6.2, Figs 8.1–8.2, Fig 10A, Fig 11A, Table 1 (illustrated, song, oscillogram, hearing, comp. note) Tajikistan, Kirghizia, Transcaucasia, Crimea

Cicadatra querula Popov 1990a: 302–303, Fig 2 (comp. note)
Cicadatra querula Tautz 1990: 105, Fig 12 (song frequency, comp. note)
Cicadatra querula Popov, Aronov and Sergeeva 1991: 294–295 (comp. note)
Cicadatra guerula (sic) Luk"yanova 1992: 31 (listed, comp. note) Tajikistan
Cicadatra querula Popov, Arnov and Sergeeva 1992: 164–165 (comp. note)
Psalmocharias querula Schedl 1993: 797 (listed, synonymy, distribution, comp. note) Equals *Cicadatra querula* Egypt, Algeria, Tunisia, Israel, Turkey, Iran, Afghanistan, Pakistan, India, USSR, Mongolia
Cicadatra querula Bennet-Clark and Young 1994: 293, Table 1 (body size and song frequency)
Psalmocharias querula Mirzayans 1995: 15–16 (listed, distribution) Equals *Cicada querula* Iran
Cicadatra querula Kozhevnikova 1996: 97–98 (cotton pest) Uzbekistan
Cicada querula Chou, Lei, Li, Lu and Yao 1997: 120 (type species of *Psalmocharias*)
Psalmocharias querula Chou, Lei, Li, Lu and Yao 1997: 122–123, 357, Plate III, Fig 49 (described, illustrated, synonymy, comp. note) Equals *Cicada querula* Equals *Cicada concinna querula* Equals *Cephaloxys quadrimacula* Equals *Cicada steveni* Equals *Cicadatra querula* Equals *Sena querula* China
Psalmocharias querula Schedl 1999b: 825, 833, Fig 2 (illustrated, listed, distribution, ecology, comp. note) Israel, Turkey, Egypt, Afghanistan, Pakistan, India, Central Asia, Mongolia, Algeria, Tunisia
Psalmocharias quaerula (sic) Hua 2000: 64 (listed, distribution, hosts) Equals *Cicada querula* China, Xinjiang, India, Afghanistan, Turkestan, Iran
Psalmocharias querula Sueur 2001: 40, Table 1 (listed) Equals *Cicadatra querula*
Cicadatra querula Kadamshoev, Miralibekov and Abdulnazarov 2002: 45 (comp. note)
Psalmocharias querula Moulds 2005b: 410 (comp. note)
Psalmocharias quaerula (sic) Shiyake 2007: 6, 79, 81, Fig 129 (illustrated, distribution, listed, comp. note) Europe
Psalmocharias querula Ojie Ardebilie and Nozari 2009: A15 (comp. note)
Psalmocharias querula Ahmed and Sanborn 2010: 28, 30, 41–42 (key, comp. note, type species of genus *Psalmocharias*, synonymy, distribution) Equals *Cicada querula* Equals *Cephaloxys quadrimacula* Equals *Cicada steveni* Russia, Egypt, Mongolia, Central Asia, Pakistan
Cicada querula Mozaffarian and Sanborn 2010: 79 (type species of *Psalmocharias*)
Psalmocharias querula Mozaffarian and Sanborn 2010: 80 (synonymy, listed, distribution, comp. note) Equals *Cicada querula* Equals *Cephaloxys quadrimacula* Equals *Cicada steveni* Iran, Mongoia, Egypt

***P. rugipennis* (Walker, 1858)** = *Cicada rugipennis* = *Cicadatra rugipennis* = *Sena rugipennis* = *Sena querula* var. ______ Distant, 1906 = *Cicadatra quaerula* (sic) var. ______ = *Psalmochavias* (sic) *rugipennis*
Psalmocharias (Sena) rugipennis Chaudhry, Chaudry and Khan 1970: 105 (listed, comp. note) Pakistan, India
Psalmocharias rugipennis Shiyake 2007: 6, 79, 81, Fig 130 (illustrated, distribution, listed, comp. note) Afghanistan
Psalmocharias rugipennis Ahmed and Sanborn 2010: 28, 42 (key, synonymy, distribution) Equals *Cicada rugipennis* Afghanistan, Pakistan, India

***Nipponosemia* Kato, 1925** = *Niponosemia* (sic)
Nipponosemia Wu 1935: 26 (listed) China
Nipponosemia Chen 1942: 144 (listed)
Nipponosemia Hayashi 1984: 28, 63 (key, described, listed, comp. note)
Nipponosemia Chou, Lei and Wang 1993: 82, 85–86 (key, comp. note)
Nipponosemia Chou, Lei, Li, Lu and Yao 1997: 120, 293, Table 6 (described, key, listed, diversity, biogeography) China
Nipponosemia Chen 2004: 39, 41, 158, 192 (phylogeny, listed)
Nipponosemia Lee and Hayashi 2004: 61, 63 (key, diagnosis) Taiwan, Japan, China
Nipponosemia Moulds 2005b: 390 (higher taxonomy, listed, comp. note)
Nipponosemia Pham and Thinh 2005a: 245 (listed) Vietnam
Nipponosemia Pham and Thinh 2005b: 289 (comp. note) Vietnam
Nipponosemia Shiyake 2007: 6, 82 (listed, comp. note)
Nipponosemia Osaka Museum of Natural History 2008: 326 (listed) Japan
Nipponosemia Pham and Yang 2009: 5, 15, Table 2 (listed) Vietnam
Nipponosemia Lee and Hill 2010: 303 (comp. note)
Nipponosemia Wei, Ahmed and Rizvi 2010: 28, 33 (comp. note)

***N. guangxiensis* Chou & Wang, 1993**

Nipponosemia guangxiensis Chou, Lei and Wang 1993: 83–84, 86 (key, n. sp., described, illustrated, comp. note) China

Nipponosemia guangsiensis Chou, Lei, Li, Lu and Yao 1997: 123, 127–128, Fig 9–11, Plate III, Fig 47 (key, described, illustrated, comp. note) China

Nipponosemia guangxiensis Pham and Thinh 2005a: 244 (distribution, listed) Vietnam

Nipponosemia guangxiensis Pham and Yang 2009: 5, 15–16, Table 2 (listed, distribution, comp. note) Vietnam, China

***N. metulata* Chou & Lei, 1993**

Nipponosemia metulata Chou, Lei and Wang 1993: 83–84, 86 (key, n. sp., described, illustrated, comp. note) China

Nipponosemia metulata Chou, Lei, Li, Lu and Yao 1997: 123, 125–127, Fig 9–10, Plate III, Fig 46 (key, described, illustrated, comp. note) China

***N. terminalis* (Matsumura, 1913)** = *Abroma terminalis* = *Lemuriana terminalis* = *Cicada terminalis* = *Gicada* (sic) *terminalis* = *Niponosemia* (sic) *terminalis* = *Cicada fuscoplaga* Schumacher, 1915 = *Gicada* (sic) *fuscoplaga* = *Nipponosemia fuscoplaga* = *Cicada terminalis formosana* Kato, 1925 = *Cicada terminalis* var. *formosana* = *Nipponosemia terminalis formosana* = *Nipponosemia terminalis* var. *formosana*

Nipponosemia terminalis Wu 1935: 26 (listed) Equals *Abroma terminalis* Equals *Lemuriana terminalis* Equals *Cicada terminalis* China, Szechuan, Loochoo Isalnd

Abroma terminalis Chen 1942: 144 (type species of *Nipponosemia*)

Nipponosemia terminalis Chen 1942: 144 (listed, comp. note) China

Abroma terminalis Hayashi 1984: 63 (type species of *Nipponosemia*)

Nipponosemia terminalis Hayashi 1984: 63–65, Figs 151–155 (described, illustrated, synonymy, comp. note) Equals *Abroma terminalis* Equals *Lemuriana terminalis* Equals *Cicada terminalis* Equals *Cicada fuscoplaga* (partim) Japan

Abroma terminalis Chou, Lei and Wang 1993: 82, 85 (type species of *Nipponosemia*, comp. note)

Nipponosemia terminalis Chou, Lei and Wang 1993: 82, 84–86 (key, comp. note)

Abroma terminalis Chou, Lei, Li, Lu and Yao 1997: 123 (type species of *Nipponosemia*)

Nipponosemia terminalis Chou, Lei, Li, Lu and Yao 1997: 123–125, Fig 9–9, Plate III, Fig 48 (key, described, illustrated, synonymy, comp. note) Equals *Abroma terminalis* Equals *Gicada* (sic) *fuscoplaga* Equals *Lemuriana terminalis* Equals *Gicada* (sic) *terminalis* China

Cicada fuscoplaga Hua 2000: 61 (listed, distribution) China, Taiwan

Nipponosemia terminalis Hua 2000: 63 (listed, distribution) Equals *Abroma terminalis* Equals *Cicada terminalis* var. *formosana* China, Fujian, Taiwan, Sichuan, Japan

Nipponosemia terminalis Sueur 2001: 46, Table 2 (listed)

Nipponosemia terminalis Chen 2004: 158–159, 199 (described, illustrated, listed, comp. note) Taiwan

Abroma terminalis Lee and Hayashi 2004: 61 (type species of *Nipponosemia*) Japan, China

Nipponosemia terminalis Lee and Hayashi 2004: 61–63, Fig 28 (key, described, illustrated, synonymy, song, listed, comp. note) Equals *Abroma terminalis* Equals *Cicada fuscoplaga* Equals *Nipponosemia fuscoplaga* Equals *Cicada terminalis* var. *formosana* Equals *Nipponosemia terminalis* var. *formosana* Taiwan, Japan, China

Abroma terminalis Pham and Thinh 2005a: 245 (type species of *Nipponosemia*)

Nipponosemia terminalis Shiyake 2007: 6, 80, 82, Fig 132 (illustrated, distribution, listed, comp. note) China, Taiwan

Nipponosemia terminalis Osaka Museum of Natural History 2008: 326 (listed, acoustic behavior, ecology, comp. note) Japan

Abroma terminalis Pham and Yang 2009: 5 (type species of *Nipponosemia*)

***N. virescens* Kato, 1926** = *Nipponosemia virecens* (sic)

Nipponosemia virescens Chou, Lei and Wang 1993: 82, 86 (key, comp. note)

Nipponosemia virescens Chou, Lei, Li, Lu and Yao 1997: 123, 125 (key, described, comp. note) China

Nipponosemia virescens Hua 2000: 63 (listed, distribution) China, Taiwan, Japan

Nipponosemia virescens Chen 2004: 158, 162, 199 (described, illustrated, listed, comp. note) Taiwan

Nipponosemia virescens Lee and Hayashi 2004: 61–63, Fig 29 (key, described, illustrated, synonymy, song, listed, comp. note) Taiwan

Nipponosemia virecens (sic) Shiyake 2007: 6, 80, 82, Fig 133 (illustrated, distribution, listed, comp. note) Taiwan

Nipponosemia virescens Lee, Lin and Wu 2010: 218–224, Figs 1–6, Tables 1–2 (ecology, comp. note) Taiwan

***Shaoshia* Wei, Ahmed & Rizvi, 2010**

Shaoshia Ahmed and Sanborn 2010: 41 (distribution) Pakistan

Shaoshia Wei, Ahmed and Rizvi 2010: 29, 33 (n. gen., described, comp. note) Pakistan

S. *zhangi* Wei, Ahmed & Rizvi, 2010

Shaoshia zhangi Ahmed and Sanborn 2010: 41 (distribution, type species of genus *Shaoshia*) Pakistan

Shaoshia zhangi Wei, Ahmed and Rizvi 2010: 29–32, Fig 1–3 (n. sp., type species of *Shaoshia*, described, illustrated, host plant, comp. note) Pakistan

***Mogannia* Amyot & Audinet-Serville, 1843** = *Cephaloxys* Signoret, 1847 = *Mogonnia* (sic) = *Mongania* (sic) = *Moggania* (sic) = *Mognnia* (sic) = *Mogana* (sic)

Mogannia Wu 1935: 20 (listed) Equals *Cephaloxys* China

Mogannia Chen 1942: 147 (listed)

Mogannia Itô and Nagamine 1981: 282 (comp. note)

Mogannia Shcherbakov 1981a: 830 (listed)

Moggania (sic) Shcherbakov 1981b: 66 (listed)

Mogannia sp. Hill, Hore and Thornton 1982: 43, 46, 161–162, Plate 107 (listed, comp. note) Hong Kong

Mogannia Strümpel 1983: 18 (listed)

Mogannia Hayashi 1984: 28, 63–64 (key, described, listed, comp. note) Equals *Cephaloxys*

Mogannia Boulard 1985e: 1029, 1032 (coloration, comp. note)

Mogannia Duffels 1986: 319 (distribution, diversity, comp. note) Greater Sunda Islands, Southeast Asia

Mogannia spp. Namba, Ma and Inagaki 1988: 278 (medicinal use, comp. note) China

Mogannia Ramallo 1989: 144 (comp. note)

Mogannia Chou, Lei and Wang 1993: 82, 85 (comp. note) China

Mogannia Hayashi 1993: 15 (comp. note)

Mogannia sp. Kahono 1993: 213, Table 5 (listed) Irian Jaya

Mogannia Zaidi and Ruslan 1994: 427, Table 2 (comp. note) Malaysia

Mogannia Lee 1995: 23 (listed)

Mogannia Zaidi and Ruslan 1995a: 174, Table 2 (listed, comp. note) Borneo, Sarawak

Mogannia Zaidi and Ruslan 1995b: 220–221, Table 2 (listed, comp. note) Borneo, Sabah

Mogannia Zaidi and Ruslan 1995c: 70, Table 2 (listed, comp. note) Peninsular Malaysia

Mogannia Zaidi and Ruslan 1995d: 202, Table 2 (listed, comp. note) Borneo, Sabah

Mogannia Zaidi 1996: 103, Table 2 (comp. note) Borneo, Sarawak

Mogannia Zaidi and Hamid 1996: 54–55, Table 2 (listed, comp. note) Borneo, Sarawak

Mogannia Chou, Lei, Li, Lu and Yao 1997: 88, 107, 293, 299, 301, 307, 312, 324, 326, Fig 12-1, Table 6, Tables 10–12 (described, listed, synonymy, diversity, biogeography, phylogeny, comp. note) Equals *Cephaloxys* China

Mogannia Zaidi 1997: 115, Table 2 (listed, comp. note) Borneo, Sarawak

Mogannia Zaidi and Ruslan 1998b: 4–5, Table 2 (listed, comp. note) Borneo, Sabah

Mogannia Zaidi and Ruslan 1999: 4, Table 2 (listed, comp. note) Borneo, Sarawak

Mogannia Yoshizawa and Saigusa 2001: 12 (listed)

Mogannia Chen 2004: 39, 41, 164 (phylogeny, listed)

Mogannia Lee and Hayashi 2004: 61, 63 (key, diagnosis) Equals *Cephaloxys* Taiwan, Japan, China

Mogannia Schouten, Duffels and Zaidi 2004: 379 (comp. note)

Mogannia Boulard 2005e: 117, 141 (comp. note) Thailand

Mogannia Pham and Thinh 2005a: 244 (listed) Vietnam

Mogannia Pham and Thinh 2005b: 289 (comp. note) Vietnam

Mogannia Shen, Lei and Yang 2006: 175, 177 (comp. note)

Mogannia Boulard 2007e: 3, 5, 72–74, Fig 2d, Fig 48 (illustrated, coloration, comp. note) Thailand

Mogannia Sanborn, Phillips and Sites 2007: 3, 32, Table 1 (listed) Thailand

Mogannia Shiyake 2007: 6, 82 (listed, comp. note)

Mogannia Boulard 2008f: 6–7, 53, 96 (listed, comp. note) Thailand

Mogannia Lee 2008a: 2, 19, Table 1 (listed) Equals *Cephaloxys* Vietnam

Mogannia Osaka Museum of Natural History 2008: 326 (listed) Japan

Mogannia Boulard 2009f: 250 (comp. note)

Mogannia Lee 2009h: 2636 (listed) Philippines, Palawan

Mogannia Pham and Yang 2009: 5, 15, Table 2 (listed, synonymy) Equals *Cephaloxys* Vietnam

Mogannia Lee 2010b: 27 (listed) Cambodia

Mogannia Lee 2010c: 24 (listed) Philippines, Luzon

Mogannia Lee and Hill 2010: 298 (comp. note)

Mogannia Pham and Yang 2010a: 133 (comp. note) Vietnam

Mogannia Wei, Ahmed and Rizvi 2010: 33 (comp. note)

***M. aliena* Distant, 1920**

Mogannia aliena Lee 2008a: 20 (distribution, listed) North Vietnam

Mogannia aliena Pham and Yang 2009: 15–16, Table 2 (listed, comp. note) Vietnam

Mogannia aliena Pham and Yang 2010a: 133 (comp. note) Vietnam

***M. aurea* Fraser, 1942**

***M. basalis* Matsumura, 1913** = *Mogannia flavocapitata* Kato, 1925 = *Mogannia basalis tienmushanensis* Ouchi, 1938 = *Mogannia basalis* var. *tienmushanensis*

Mogannia basalis Chou, Lei, Li, Lu and Yao 1997: 119 (comp. note) Equals *Mogannia basalis* var. *tienmushanensis* China

Mogannia basalis Hua 2000: 63 (listed, distribution) China, Zhejiang, Taiwan, Guangxi

Mogannia basalis tienmushanensis Hua 2000: 63 (listed, distribution) China, Zhejiang

Mogannia basalis Chen 2004: 164, 197, 199 (listed, comp. note)

Mogannia basalis Lee and Hayashi 2004: 63, 68 (key, described, illustrated, synonymy, listed, comp. note) Equals *Mogannia flavocapitata* Taiwan, China

***M. binotata* Distant, 1906** = *Prasia elegans* Kirkaldy, 1913

Mogannia binotata Zaidi and Azman 2003: 99–100, 106, Table 1 (listed, comp. note) Borneo, Sabah

***M. caesar* Jacobi, 1902** = *Mogannia cyanea* (nec Walker)

Mogannia cyanea (sic) Boulard 2005a: 32, Figs 2–4 (illustrated, coloration, comp. note) Thailand

Mogannia caesar Boulard 2005e: 142 (comp. note) Thailand

Mogannia caesar Boulard 2005o: 151 (listed)

Mogannia caesar Boulard 2006c: 140–143, Figs 9–12, Fig E (described, illustrated, song, sonogram, comp. note) Equals *Mogannia cyanea* Boulard 2005 Thailand, Tonkin, Indochina

Mogannia caesar Boulard 2007e: 72, Plate 5a, Plate 34a-b (coloration, illustrated, comp. note) Thailand

Mogannia caesar Sanborn, Phillips and Sites 2007: 32 (listed, distribution, comp. note) Thailand, China, Indochina, Vietnam

Mogannia caesar Boulard 2008f: 53, 96 (biogeography, listed, comp. note) Equals *Mogannia cyanea* Boulard, 2005 (nec Walker) Thailand, Tonkin

Mogannia caesar Lee 2008a: 19 (distribution, listed) North Vietnam, China, Thailand

Mogannia caesar Pham and Yang 2009: 15, Table 2 (listed) Vietnam

***M. conica conica* (Germar, 1830)** = *Cicada conica* = *Mogannia illustrata* Amyot & Audinet-Serville, 1843 = *Mogannia conica illustrata* = *Cephaloxys hemelytra* Signoret, 1847 = *Mogannia indicans* Walker, 1850 = *Mogannia conica indicans* = *Mogannia incidans* var. *β* Walker, 1850 = *Mogannia ignifera* Walker, 1850 = *Mogannia conica ignifera* = *Mogannia avicula* Walker, 1850 = *Mogannia avicula* (sic) = *Mogannia recta* Walker, 1858 = *Mogannia histrionica* Uhler, 1861

Mogannia histrionica Henshaw 1903: 230 (listed) China, Hong Kong

Mogannia conica Wu 1935: 21 (listed) Equals *Cicada conica* Equals *Mogannia illustrata* Equals *Cephaloxys hemelytra* Equals *Mogannia indicans* Equals *Mogannis ignifera* Equals *Mogannia avicula* Equals *Mogannia recta* Equals *Mogannia histrionica* Equals *Mogannia venutissima* China, Assam, Khasi Hills, Margherita, Tenasserim, Thaga, Java, Sumatra, Philippines, Tonkin

Mogannia conica Chen 1942: 147 (type species of *Mogannia*)

Mogannia illustrata Hayashi 1984: 64 (type species of *Mogannia*)

Mogannia ignifera Zaidi and Ruslan 1994: 426, 428, Table 1 (comp. note) Malaysia

Mogannia ignifera Zaidi, Ruslan and Mahadir 1996: 61, Table 1 (listed, comp. note) Peninsular Malaysia

Cicada conica Chou, Lei, Li, Lu and Yao 1997: 107 (type species of *Mogannia*)

Mogannia conica Chou, Lei, Li, Lu and Yao 1997: 108–109, Fig 9–1, Plate III, Fig 36 (key, described, illustrated, synonymy, comp. note) Equals *Cicada conica* Equals *Mogannia illustrata* Equals *Cephaloxys hemelytra* Equals *Mogannia indicans* Equals *Mogannia ignifera* Equals *Mogannia avicula* Equals *Mogannia recta* Equals *Mogannia histrionica* China

Mogannia histrionica Sanborn 1999a: 44 (type material, listed)

Mogannia conica Hua 2000: 63 (listed, distribution) Equals *Cicada conica* China, Guangdong, Hong Kong, Yunnan, Xizang, Philippines, Indonesia, Vietnam, India, Nepal, Malaysia

Cicada conica Lee and Hayashi 2004: 63 (type species of *Cephaloxys*)

Mogannia illustrata Lee and Hayashi 2004: 63 (type species of *Mogannia*) Southeast Asia

Cicada conica Pham and Thinh 2005a: 244 (type species of *Mogannia*)

Mogannia conica Pham and Thinh 2005a: 244 (synonymy, distribution, listed) Equals *Cicada conica* Equals *Mogannia illustrata* Equals *Cephaloxys hemelytra* Equals *Mogannia indicans* Equals *Mogannia avicula* Equals *Mogannia recta* Equals *Mogannia histrionica* Vietnam

Mogannia conica Shen, Lei and Yang 2006: 177–178 (comp. note)

Mogannia conica Sanborn, Phillips and Sites 2007: 32, 34 (listed, distribution, comp. note) Thailand, Nepal, China, India, Vietnam, Malaysia, Sumatra, Java, Philippine Republic

Mogannia conica Boulard 2008f: 53, 96 (type species of *Mogannia*, biogeography, listed, comp. note) Equals *Cicada conica* Thailand, India, Malayan Archipelago, Sumatra, Philippines

Mogannia illustrata Lee 2008a: 19 (type species of *Mogannia*) Java

Cicada conica Lee 2008a: 19 (type species of *Cephaloxys*)

Mogannia conica Lee 2008a: 20 (synonymy, distribution, listed) Equals *Cicada conica* North Vietnam, China, Xizang, Yunnan, Guangdong, Philippines, Thailand, Malysia, Indonesia, Java, Sumatra, Myanmar, Nepal, India

Cicada conica Lee 2009h: 2636 (type species of *Mogannia*, listed) Java

Mogannia conica Lee 2009h: 2636 (synonymy, distribution, listed) Equals *Cicada conica* Philippines, Palawan, China, Xizang, Yunnan, Guangdong, Northern Vietnam, Malaysia, Indonesia, Java, Sumatra, Myanmar, Nepal, India

Cicada conica Pham and Yang 2009: 5 (type species of *Mogannia*)

Mogannia conica Pham and Yang 2009: 15–16, Table 2 (listed, comp. note) Vietnam

Mogannia conica Sun, Watson, Zheng, Watson and Liang 2009: 3148–3152, 3154, Fig 1E, Fig 2B, Fig 3B, Table 1 (illustrated, wing microstructure, comp. note) China, Yunnan

Cicada conica Lee 2010b: 27 (type species of *Mogannia*, listed)

Cicada conica Lee 2010c: 24 (type species of *Mogannia*, listed) Java

***M. conica* var. *d* Distant, 1892**

***M. cyanea* Walker, 1858** = *Mogannia nigrocyanea* Matsumura, 1913 = *Mogannia nigocyanea* (sic) = *Mogannia nigrocyanea flavofascia* Kato, 1925 = *Mogannia cyanea flavofascia* = *Mogannia cyanea flavofescia* (sic) = *Mogannia bella* Kato, 1927 = *Mogannia chekingensis* Ouchi, 1938 = *Mogannia chinensis* Ouchi, 1938 (nec *Mogannia chinensis* Stål, 1865) = *Mogannia tienmushana* Chen, 1957 = *Mogannia ouchii* Metcalf, 1963

Mogannia cyanea Chen 1933: 361 (listed, comp. note)

Mogannia cyanea Wu 1935: 21 (listed) China, Assam, Margherita, Naga Hills, Burma, Ruby Mines, India, Formosa, Arisan

Mogannia cyanea Wang and Zhang 1987: 296 (key, comp. note) China

Mogannia cyanea Liu 1991: 257–258 (comp. note)

Mogannia cyanea Liu 1992: 16, 28, 31 (comp. note)

Mogannia cyanea Peng and Lei 1992: 103, Fig 294 (listed, illustrated, comp. note) China, Hunan

Mogannia cyanea Wang and Zhang 1992: 121 (listed, comp. note) China

Mogannia cyanea Wang and Zhang 1993: 190 (listed, comp. note) China

Mogannia cyanea Xu, Ye and Chen 1995: 85 (listed, comp. note) China, Zhejiang

Mogannia cyanea Chou, Lei, Li, Lu and Yao 1997: 107, 110–111, Fig 9-3, Plate III, Fig 40 (key, described, illustrated, synonymy, comp. note) Equals *Mogannia nigrocyanea* Equals *Mogannia bella* Equals *Mogannia chekingensis* Equals *Mogannia chinensis* Equals *Mogannia tiennulshana* China

Mogannia cyanea Hua 2000: 63 (listed, distribution) Equals *Mogannia nigrocyanea* Equals *Mogannia bella* Equals *Mogannia chekiangensis* China, Liaoning, Shanxi, Gansu, Hubei, Jiangsu, Jiangxi, Zhejiang, Fujian, Taiwan, Guangdong, Hong Kong, Hunan, Guangxi, Sichuan, Yunnan, Korea, Burma, India

Mogannia cyanea flavofascia Hua 2000: 63 (listed, distribution) China, Taiwan, India

Mogannia cyanea Chen 2004: 164, 168, 192, 199 (described, illustrated, listed, comp. note) Taiwan

Mogannia bella Chen 2004: 192 (comp. note)

Mogannia nigrocyanea Chen 2004: 192 (comp. note)

Mogannia cyanea Lee and Hayashi 2004: 63–65, Fig 30 (key, described, illustrated, synonymy, song, listed, comp. note) Equals *Mogannia nigrocyanea* Equals *Mogannia nigrocyanea* var. *flavifascia* Equals *Mogannia bella* Equals *Mogannia cyanea* var. *flavifascia* Equals *Mogannia cyanea* var. *flavofescia* (sic) Taiwan, China, Myanmar, India

Mogannia cyanea Boulard 2005e: 142 (comp. note) Thailand

Mogannia cyanea Boulard 2005o: 151 (listed)

Mogannia cyanea Boulard 2006c: 143 (comp. note) Thailand

Mogannia cyanea Sanborn, Phillips and Sites 2007: 32 (listed, distribution, comp. note) Thailand, India, Burma, China, Taiwan, Indochina

Mogannia cyanea Shiyake 2007: 6, 81–82, Fig 135 (illustrated, distribution, listed, comp. note) China, India, Southeast Asia

Mogannia cyanea Boulard 2008f: 53, 96 (biogeography, listed, comp. note) Thailand, India, China

Mogannia cyanea Lee 2008a: 19 (distribution, listed) Vietnam, South China, Taiwan, Thailand, Myanmar, India

Mogannia cyanea Pham and Yang 2009: 15, Table 2 (listed) Vietnam

Mogannia cyanea Lee and Hill 2010: 299–300, Figs 7–8 (phylogeny, comp. note)

M. cyanea yungshienensis Liu, 1940 to *Mogannia saucia* Noualhier, 1896

***M. doriae* Distant, 1888**

***M. effecta effecta* Distant, 1892** = *Mogannia efecta* (sic) = *Mogannia conica* (nec Germar)

Mogannia effecta Chou, Lei, Li, Lu and Yao 1997: 107, 114, Fig 9–5, Plate III, Fig 35 (key, described, illustrated, synonymy, comp. note) Equals *Mogannia conica* Naruse, 1973 China

Mogannia efecta (sic) Chou, Lei, Li, Lu and Yao 1997: 357 (comp. note)

Mogannia effecta Pham and Thinh 2005a: 244 (synonymy, distribution, listed) Equals *Mogannia conica* (partim) Vietnam

Mogannia effecta Pham and Yang 2009: 5, 15–16, Table 2 (listed, distribution, comp. note) Vietnam, China, Nepal, India, Indonesia

***M. effecta* var. *a* Distant, 1892**

***M. effecta* var. *b* Distant, 1892**

***M. flammifera* Moulton, 1923**

Mogannia flammifera Zaidi and Ruslan 1994: 426 (comp. note) Malaysia

***M.* nr. *flammifera* Zaidi, Ruslan & Mahadir, 1996**

Mogannia nr. *flammifera* Zaidi, Ruslan and Mahadir 1996: 61, Table 1 (listed, comp. note) Peninsular Malaysia

***M. formosana* Matsumura, 1907** = *Mogannia nasalis* (nec White, 1844) = *Mogannia rubricosta* Matsumura, 1917 = *Mogannia formosana rubricosta* = *Mogannia rubrlcosta* (sic) = *Mogannia pallipes* Matsumura, 1917 = *Mogannia formosana pallipes* = *Mogannia fasciata pallipes* = *Mogannia formosana taikosana* Kato, 1925 = *Mogannia formosana* var. *b* Kato, 1925 = *Mogannia fasciata* Kato, 1925 = *Mogannia formosana fasciata* = *Mogannia fasciata nigroventris* Kato, 1925 = *Mogannia formosana fasciata nigroventris* = *Mogannia fuscomaculata* Kato, 1925 = *Mogannia formosana fuscomaculata* = *Mogannia fusculata* (sic) = *Mogannia kashotoensis* Kato, 1925 = *Mogannia kwashotoensis* (sic) = *Mogannia kashotoensis flavoguttata* Kato, 1925 = *Mogannia kwashotoensis* (sic) *flavoguttata* = *Mogannia kasotoensis* (sic) *flavoguttata* = *Mogannia kanoi* Kato, 1925 = *Mogannia formosana elegans* Kato, 1926 = *Mogannia formosana interrupta* Kato, 1926 = *Mogannia formosana rubricosta* Kato, 1930 = *Mogannia formosana uraina* Kato, 1933 = *Mogannia sinensis* Ouchi, 1938

Mogannia formosana Chou, Lei, Li, Lu and Yao 1997: 108, 116–119, Figs 9-7-9-8, Plate III, Fig 43 (key, described, illustrated, synonymy, comp. note) Equals *Mogannis nasalis* (partim) Equals *Mogannia rubricosta* Equals *Mogannia pallipes* Equals *Mogannia fasciata* Equals *Mogannia fuscomaculata* Equals *Mogannia kashotoensis* Equals *Mogannia kanoi* Equals *Mogannia formosana elegans* Equals *Mogannia formosana uraina* Equals *Mogannia sinensis* China

Mogannia formosana Hua 2000: 63 (listed, distribution) Equals *Mogannia pallipes* Equals *Mogannia fomosana elegans* Equals *Mogannia formosana interrupta* Equals *Mogannia sinensis* China, Zhejiang, Fujian, Taiwan, Hainan, Japan

Mogannia kanoi Hua 2000: 63 (listed, distribution) China, Taiwan

Mogannia kashotoensis Hua 2000: 63 (listed, distribution) China, Jiangxi

Mogannia rubricosta Hua 2000: 63 (listed, distribution) China, Taiwan

Mogannia formosana Chen 2004: 164, 170, 192, 199 (described, illustrated, listed, comp. note) Taiwan

Mogannia kashotoensis Chen 2004: 192 (comp. note)

Mogannia kanoi Chen 2004: 192 (comp. note)

Mogannia formosana Lee and Hayashi 2004: 63–66, Fig 31 (key, described, illustrated, synonymy, song, listed, comp. note) Equals *Mogannia nasalis* (nec White) Equals *Mogannia rubricosta* Equals *Mogannia pallipes* Equals *Mogannia formosana* var. *taikosana* Equals *Mogannia formosana* var. *b* Equals *Mogannia fasciata* Equals *Mogannia fasciata* var. *nigroventris* Equals *Mogannia fuscomaculata* Equals *Mogannia kashotoensis* Equals *Mogannia kashotoensis* var. *flavoguttata* Equals *Mogannia kashotoensis* var. *flavofascia* (sic) Equals *Mogannia kanoi* Equals *Mogannia formosana elegans* Equals *Mogannia formosana* var. *elegans* Equals *Mogannia formosana* ab. *interrupta* Equals *Mogannia formosana* var. *rubricosta* Equals *Mogannia formosana* var. *pallipes* Equals *Mogannia formosana* var. *fasciata* Equals *Mogannia formosana* var. *fasciata* f. *nigroventris* Equals *Mogannia formosana* var. *fuscomaculata* Equals *Mogannia formosana uraina* Taiwan, China, Myanmar, India

Mogannia formosana Shiyake 2007: 6, 81, 83, Fig 136 (illustrated, distribution, listed, comp. note) Taiwan, China

Mogannia formosana Lee, Lin and Wu 2010: 218, 221 (comp. note) Taiwan

***M. funebris* Stål, 1865**

Mogannia funebris Boulard 2005e: 141 (comp. note)

Mogannia funebris Lee 2008a: 20 (distribution, listed) Vietnam, Myanmar, Bangladesh, India

Mogannia funebris Pham and Yang 2009: 15, Table 2 (listed) Vietnam

M. *funebris* var. *a* Distant, 1892

M. *funettae* Boulard, 2008

Mogannia funettae Boulard 2008f: 54, 80–85, 96, Figs 34–42 (n. sp., described, illustrated, song, sonogram, biogeography, listed, comp. note) Thailand

M. *grelaka* Moulton, 1923

M. *guangdongensis* Chen, Yang & Wei, 2012

M. *hainana* Shen, Lei & Yang, 2006

Mogannia hainana Shen, Lei and Yang 2006: 175–178, Fig 1 (n. sp. described, illustrated, comp. note) China, Hainan

M. *hebes* (Walker, 1858) = *Cephaloxys hebes* = *Mogonnia* (sic) *hebes* = *Mogannia hebets* (sic) = *Mogannia spurcata* Walker, 1858 = *Mogannia hebes* var. *a* Kato, 1925 = *Mogannia hebes* var. *b* Kato, 1925 = *Mogannia hebes* var. *c* Kato, 1925 = *Mogannia flavescens* Kato, 1925 = *Mogannia hebes flavescens* = *Mogannia delta* Kato, 1925 = *Mogannia hebes* var. *d* Kato, 1925 = *Mogannia ritozana* Matsumura, 1927 = *Mogannia hebes ritozana* = *Mogannia ritozana dorsovitta* Matsumura, 1927 = *Mogannia ritozana dorsovittata* (sic) = *Mogannia ritozana dorsovitta* (sic) = *Mogannia hebes dorsovitta* = *Mogannia hebes dorsivitta* (sic) = *Mogannia katonis* Matsumura, 1927 = *Mogannia hebes katonis* = *Mogannia subfusca* Kato, 1928 = *Mogannia hebes concolor* Kato, 1932 = *Mogannia janea* Hua, 2000

Mogannia hebes Chen 1933: 361 (listed, comp. note)

Mogannia hebes Wu 1935: 22 (listed) Equals *Cephaloxys hebes* Equals *Mogannia spurcata* China, Manchuria, Nanking, Soochow, Chekiang, Foochow, Amoy, Canton, Hongkong, Hangchow, Korea, Formosa

Mogannia hebes Chen 1942: 147 (listed, comp. note) China

Mogannia hebets (sic) Itô 1976: 67 (sugarcane pest, comp. note) Taiwan

Mogannia hebes Itô and Nagamine 1981: 281–282, Table 5 (sugarcane pest, comp. note) Taiwan

Mogannia hebes Costilla and Pastor 1986: 122 (comp. note)

Mogannia hebes Chou 1987: 21 (comp. note) China

Mogannia hebes Jiang, Lin and Wang 1987: 106, 108, Fig 1E, Fig 3(2) (timbal structure, illustrated, song, oscillogram, comp. note) China

Mogannia hebes Wang and Zhang 1987: 296 (key, comp. note) China

Mogannia hebes Wilson 1987: 489, Table 1 (sugarcane pest, listed, comp. note) Taiwan

Mogannia hebes Wang Z.-N. 1988: 25 (fungal host, comp. note) Taiwan

Mogannia hebes Jiang 1989b: 77 (comp. note)

Mogannia hebes Liu 1990: 42 (comp. note)

Mogannia hebes Liu 1991: 257 (comp. note)

Mogannia hebes Liu 1992: 5, 16 (comp. note)

Mogannia hebes Peng and Lei 1992: 103, Fig 295 (listed, illustrated, comp. note) China, Hunan

Mogannia hebes Wang and Zhang 1992: 121 (listed, comp. note) China

Mogannia hebes Wang and Zhang 1993: 190 (listed, comp. note) China

Mogannia hebes Lei, Chou and Li 1994: 55–56, Fig 2, Fig 8 (song frequency spectrum, oscillogram, comp. note) China

Mongania (sic) *hebes* Anonymous 1994: 809 (fruit pest) China

Mogannia hebes Chen and Yang 1995: 259, Fig 5(6) (metatarsi illustrated)

Mogannia hebes Jiang, Yang, Tang, Xu and Chen 1995b: 227 (song, phonotaxis, flight behavior, comp. note) China

Mogannia hebes Lee 1995: 19, 23, 28–29, 31, 33, 45–36, 131, 135, 138 (comp. note) Equals *Cephaloxys hebes*

Cephaloxys hebes Lee 1995: 131 (comp. note)

Mogannia hebes Xu, Ye and Chen 1995: 85 (listed, comp. note) China, Zhejiang

Mogannia hebes Chou, Lei, Li, Lu and Yao 1997: 28, 107, 112–113, 307, 312–313, 324, 326, 328, 335, Fig 9–4, Plate III, Fig 39, Tables 10–12 (key, described, illustrated, synonymy, song, oscillogram, spectrogram, comp. note) Equals *Cephaloxys hebes* Equals *Mogannia spurcata* Equals *Mogannia flavescens* Equals *Mogannia delta* Equals *Mogannia ritozana* Equals *Mogannia katonis* Equals *Mogannia subfusca* China

Mogannia hebes Easton and Pun 1999: 102 (distribution, comp. note) Macao, Korea, India, Myanmar, China, Chejiang, Fujian, Guangdong, Guangxi, Jiangsi, Hunan, Sichuan, Yunnan

Mogannia hebes Hua 2000: 63 (listed, distribution, hosts) Equals *Cephaloxys hebes* Equals *Mogannia katonis* Equals *Mogannia ritozana* Equals *Mogannia ritozana dorsivitta* Equals *Mogannia sufusca* China, Inner Mongolia, Henan, Hubei, Anhui, Jiangsu, Jiangxi, Zhejiang, Fujian, Taiwan, Guangdong, Hong Kong, Guangxi, Sichuan, Korea, Japan, India

Mogannia janea Hua 2000: 63 (listed, distribution) China, Guangxi

Mogannia hebes Sueur 2001: 41, Table 1 (listed)

Mogannia hebes Chen 2004: 164–165, 167, 192, 197, 199, front cover (described, illustrated, listed, comp. note) Taiwan

Mogannia delta Chen 2004: 192 (comp. note)

Mogannia subfusca Chen 2004: 192 (comp. note)

Mogannia hebes Lee and Hayashi 2004: 63, 66–68, Figs 32–33 (key, described, illustrated, synonymy, song, listed, comp. note) Equals

Cephaloxys hebes Equals *Mogannia hebes* var. *a* Equals *Mogannia hebes* var. *b* Equals *Mogannia hebes* var. *c* Equals *Mogannia flavescens* Equals *Mogannia delta* Equals *Mogannia hebes* var. *d* Equals *Mogannia ritozana* Equals *Mogannia ritozana* var. *dorsivitta* Equals *Mogannia katonis* Equals *Mogannia subfusca* Equals *Mogannia hebes* var. *ritozana* Equals *Mogannia hebes* var. *flavescens* Equals *Mogannia hebes* var. *dorsivitta* Equals *Mogannia hebes* var. *concolor* Equals *Mogannia hebes* var. *katonis* Taiwan, China

Mogannia hebes Pham and Thinh 2005a: 244 (synonymy, distribution, listed) Equals *Cephaloxys hebes* Equals *Mogannia spurcata* Equals *Mogannia flavescens* Equals *Mogannia ritozana* Equals *Mogannia delta* Equals *Mogannia katonis* Equals *Mogannia subfusca* Vietnam

Mogannia hebes Shiyake 2007: 6, 81, 83, Fig 137 (illustrated, distribution, listed, comp. note) Taiwan, China, Southeast Asia

Mogannia hebes Lee 2008a: 19 (synonymy, distribution, listed) Equals *Cephaloxys hebes* Vietnam, China, Taiwan, Liudau Island, Lanya Island

Mogannia hebes Pham and Yang 2009: 15, Table 2 (listed) Vietnam

Mogannia hebes Sun, Watson, Zheng, Watson and Liang 2009: 3148–3149, 3151–3154, Fig 1C, Fig 4B, Tables 1–2 (illustrated, wing microstructure, comp. note) China, Fujian

Mogannia hebes Lee and Hill 2010: 299–300, Figs 7–8 (phylogeny, comp. note)

***M. horsfieldi* Distant, 1905**

***M. indigotea* Distant, 1917**

Mogannia indigotea Lei, Chou and Li 1994: 54–56, Fig 3, Fig 9 (song frequency spectrum, oscillogram, comp. note)

Mogannia indigotea Chou, Lei, Li, Lu and Yao 1997: 107, 109–110, 307, 313, 324, 326, 328, 335, Fig 9-2, Plate III, Fig 37, Tables 10–12 (key, described, illustrated, synonymy, song, oscillogram, spectrogram, comp. note) Equals *Mogannia cyanea yungshienensis* China

Mogannia indigotea Hua 2000: 63 (listed, distribution, hosts) China, Fujian, Guangxi, Sichuan, Yunnan, Philippines

Mogannia indigotea Sueur 2001: 41, Table 1 (listed)

Mogannia indicotea Boulard 2005e: 140–141 (synonymy, comp. note)

Mogannia indigotea Pham and Thinh 2005a: 244–245 (distribution, listed) Vietnam

Mogannia indigotea Pham and Thinh 2006: 526–527 (listed, comp. note) Vietnam

Mogannia indicotea Boulard 2007e: 73 (comp. note) Thailand

Mogannia indigotea Lee 2010c: 24 (distribution, listed) Philippines, Luzon, Southern China, North Vietnam

***M. malayana* Moulton, 1923**

Mogannia malayana Zaidi, Ruslan and Mahadir 1996: 61, Table 1 (listed, comp. note) Peninsular Malaysia

***M.* nr. *malayana* Zaidi, Ruslan & Mahadir, 1996**

Mogannia nr. *malayana* Zaidi, Ruslan and Mahadir 1996: 61, Table 1 (listed, comp. note) Peninsular Malaysia

***M. mandarina* Distant, 1905** = *Mogannia mandarinea* (sic)

Mogannia mandarina Poulton 1907: 336 (predation) China

Mogannia mandarina Wu 1935: 22 (listed) China, Hongkong

Mogannia mandarina Chou, Lei, Li, Lu and Yao 1997: 119 (comp. note) China

Mogannia mandarina Hua 2000: 63 (listed, distribution) China, Hong Kong

***M. minuta* Matsumura, 1907** = *Mogannia iwasakii* Matsumura, 1913 = *Mogannia formosana* (nec Matusumura) = *Mogannia iwasakii nigroventris* Kato, 1937

Mogannia minuta Nagamine and Itô 1980: 132 (sugarcane pest, predation, comp. note) Okinawa

Mogannia iwasakii Itô 1976: 65–66, Plate 1, Figs E-F (sugarcane pest, population density, comp. note) Japan, Okinawa

Mogannia minuta Itô 1978: 84, 156–157, Fig 2.35, Fig 3.35 (sugarcane pest, survivorship curve, density, comp. note) Equals *Mogannia iwasakii* Okinawa

Mogannia minuta Itô and Nagamine 1981: 273–282, Figs 1–2, Fig 4, Tables 1–5 (sugarcane pest, life cycle, comp. note) Okinawa

Mogannia minuta Itô 1982b: 137–140, Fig 3, Photos 1–2, Table 1 (life cycle evolution, illustrated, comp. note) Japan

Mogannia iwasakii Hayashi 1984: 65 (comp. note)

Mogannia minuta Hayashi 1984: 64–66 (described, synonymy, comp. note) Equals *Mogannia formosana* (nec Matsumura) Equals *Mogannia iwasakii* (partim) Japan

Mogannia minuta (= iwasakii) Karban 1984a: 1660 (mortality, comp. note)

Mogannia iwasakii Karban 1984b: 355 (development time, comp. note)

Mogannia minuta White and Lloyd 1985: 606 (comp. note) Okinawa

Mogannia minuta Karban 1986: 222–224, 230–231, Fig 14.6, Tables 14.1–14.2 (survival, comp. note) Okinawa, Ryukyu Archipelago
Mogannia minuta Wilson 1987: 489, Table 1 (sugarcane pest, listed, comp. note) Japan
Mogannia minuta Liu 1990: 46 (comp. note)
Mogannia minuta Martin and Simon 1990b: 361, Fig 3 (comp. note)
Mogannia minuta Liu 1992: 9 (comp. note)
Mogannia iwasakii White and Sedcole 1993a: 47, 50 (comp. note)
Mogannia iwasakii Wolda and Ramos 1992: 278 (comp. note) Okinawa
Mogannia minuta Chou, Lei, Li, Lu and Yao 1997: 108, 115–117, Fig 9–6 (key, described, illustrated, synonymy, comp. note) Equals *Mogannia iwasakii* Equals *Mogannia formosana* Kato, 1933 Equals *Mogannia iwasakii* var. *nigroventris* China
Mogannia minuta Karban 1997: 158 (comp. note)
Mogannia minuta Itô 1998: 494–495, Table 1 (life cycle evolution, comp. note) Japan
Mogannia minuta Itô and Nagamine 1998: 533–535, Figs 1–2 (sugarcane pest, distribution, comp. note) Okinawa
Mogannia minuta Hua 2000: 63 (listed, distribution) China, Taiwan, Japan
Mogannia minuta Sueur 2001: 46, Table 2 (listed)
Mogannia minuta Chen 2004: 164, 172, 199 (described, illustrated, listed, comp. note) Taiwan
Mogannia minuta Lee and Hayashi 2004: 63, 68 (key, described, song, listed, comp. note) Taiwan, Japan
Mogannia minuta Maccagnan and Martinelli 2004: 440 (comp. note)
Mogannia minuta Itô and Kasuya 2005: 60, Table 14 (life cycle, comp. note)
Mogannia minuta Moriyama and Numata 2006: 1219 (comp. note) Japan, Okinawa
Mogannia iwasakii Oberdörster and Grant 2007a: 10 (comp. note)
Mogannia minuta Shiyake 2007: 6, 81, 83, Fig 138 (illustrated, distribution, listed, comp. note) Taiwan
Mogannia minuta Moriyama and Numata 2008: 1487 (comp. note)
Mogannia minuta Osaka Museum of Natural History 2008: 326 (listed, acoustic behavior, ecology, comp. note) Japan

***M. montana* Moulton, 1923**
Mogannia montana Zaidi and Ruslan 1997: 230 (listed, comp. note) Peninsular Malaysia, Sumatra

***M. moultoni* Distant, 1916**
Mogannia moultoni Zaidi and Ruslan 1998a: 366 (listed, comp. note) Borneo, Sarawak, Sabah

***M. nasalis fusca* Liu, 1939** = *Mogannia nasalis fuscous* (sic)
Mogannia nasalis fuscous (sic) Hua 2000: 63 (listed, distribution) China

***M. nasalis nasalis* (White, 1844)** = *Cicada (Mogannia) nasalis* = *Mogannia chinensis* Stål, 1865 = *Mogannia kikowensis* Ouchi, 1938 = *Mogannia kikowensis* var. *a* Ouchi, 1938 = *Mogannia kikowensis a* = *Mogannia kikowensis* var. *b* Ouchi, 1938 = *Mogannia kikowensis b* = *Mogannia kikowensis* var. *c* Ouchi, 1938 = *Mogannia kikowensis c* = *Mogannia nasalis kikowensis*
Mogannia nasalis Poulton 1907: 344 (predation) China
Mogannia nasalis Wu 1935: 22 (listed) Equals *Cicada (Mogannia) nasalis* Equals *Mogannia chinensis* China, Macao, Hongkong, Assam, Formosa
Mogannia nasalis Chou, Lei, Li, Lu and Yao 1997: 108, 115, 358, Plate III, Fig 42 (key, described, synonymy, comp. note) Equals *Cicada nasalis* Equals *Mogannia chinensis* Equals *Mogannia kikowensis* Equals *Mogannia kikowensis a* Equals *Mogannia kikowensis b* Equals *Mogannia kikowensis c* China
Mogannia kikowensis Chou, Lei, Li, Lu and Yao 1997: 358 (synonymy)
Mogannia kikowensis a Chou, Lei, Li, Lu and Yao 1997: 358 (synonymy)
Mogannia kikowensis b Chou, Lei, Li, Lu and Yao 1997: 358 (synonymy)
Mogannia kikowensis c Chou, Lei, Li, Lu and Yao 1997: 358 (synonymy)
Mogannia nasalis Easton and Pun 1999: 102 (distribution, comp. note) Macao, Hong Kong, India, Assam
Mogannia nasalis Hua 2000: 63 (listed, distribution) Equals *Cicada nasalis* Equals *Mogannia kikowensis* China, Anhui, Zhejiang, Fujian, Taiwan, Hong Kong, Macao, Hunan, Guangxi, India

***M. obliqua* Walker, 1858** = *Mogannia obligua* (sic)
Mogannia obliqua Chou, Lei, Li, Lu and Yao 1997: 107, 113–114, Plate III, Fig 38 (key, described, illustrated, comp. note) China
Mogannia obligua (sic) Chou, Lei, Li, Lu and Yao 1997: 357 (comp. note)
Mogannia obliqua Zaidi and Azman 2003: 106, Table 1 (listed, comp. note) Borneo, Sabah
Mogannia obliqua Boulard 2008f: 54, 96 (biogeography, listed, comp. note) Thailand, India, Malaysian Archipelago, Java, Bengal, Burma, Malaysia

Mogannia obliqua Lee 2008a: 20 (distribution, listed) Vietnam, Malaysia, Indonesia, Java, Myanmar, India
Mogannia obliqua Pham and Yang 2009: 15, Table 2 (listed) Vietnam

***M. paae* Boulard, 2009** = *Mogannia* sp. Boulard, 2005, 2007
Mogannia sp. Boulard 2005a: 31–32, Fig 5, Fig 7 (illustrated, coloration, comp. note)
Mogannia sp. Boulard 2007e: Plate 34c (illustrated) Thailand
Mogannia paae Boulard 2009f: 250–253, Figs 1–3 (n. sp., described, illustrated, song, sonogram, comp. note) Equals *Mogannina* sp. Boulard, 2005 [32] Equals *Mogannia* sp. Boulard, 2007 [Plate 34, Fig c] Thailand

***M. rosea* Moulton, 1923**

***M. ruiliensis* Chen, Yang & Wei, 2012**

***M. saucia* Noualhier, 1896** = *Mogannia sauciei* (sic) = *Mogannia cyanea yungshienensis* Liu, 1940
Mogannia saucia Boulard 2005a: 31–32, Fig 1, Fig 6, Figs 8–10 (illustrated, sonogram, coloration, comp. note) Thailand
Mogannia saucia Boulard 2005e: 118, 138–142, Plate H, Figs 1–2, Plate I, Figs 1–2, Figs 26–28, Plate J, Fig 4 (synonymy, described, illustrated, song, sonogram, listed, comp. note) Equals *Mogannia sauciei* (sic) Equals *Mogannia indigotea* Laos, Cambodia, Tonkin, Thailand
Mogannia saucia Boulard 2005o: 151 (listed)
Mogannia saucia Boulard 2006g: 132, 145, Fig 106 (sonogram, comp. note) Thailand
Mogannia saucia Boulard 2007e: 72–75, Figs 48–49, Plates 35–36 (coloration, illustrated, sonogram, comp. note) Thailand
Mogannia saucia Sanborn, Phillips and Sites 2007: 32 (listed, distribution, comp. note) Thailand. Laos, Cambodia, Vietnam
Mogannia saucia Boulard 2008f: 54, 80, 83, 96 (biogeography, listed, comp. note) Thailand, Laos
Mogannia saucia Lee 2008a: 20 (synonymy, distribution, listed) Equals *Mogannia indigotea* North Vietnam, South China, Philippines, Laos, Cambodia, Thailand
Mogannia saucia Pham and Yang 2009: 15, Table 2 (listed) Vietnam
Mogannia saucia Lee 2010b: 27 (distribution, listed) Cambodia, South China, Philippines, North Vietnam, Laos, Thailand

***M. sesioides* Walker, 1870**
Mogannia sesioides Zaidi and Ruslan 1994: 426, 428, Table 1 (comp. note) Malaysia
Mogannia sesioides Zaidi, Ruslan and Mahadir 1996: 61, Table 1 (listed, comp. note) Peninsular Malaysia
Mogannia sesioides Zaidi and Ruslan 1997: 230 (listed, comp. note) Peninsular Malaysia, Sumatra, Java, Mentawei Island
Mogannia sesioides Schouten, Duffels and Zaidi 2004: 373, 375, 379, Table 2 (listed, comp. note) Malayan Peninsula

***M.* sp. 1 Pham & Thinh, 2006**
Mogannia sp. 1 Pham and Thinh 2006: 526 (listed) Vietnam

***M. sumatrana* Moulton, 1923**

***M. tienmushana* Chen, 1957**
Mogannia tienmushana Chou, Lei, Li, Lu and Yao 1997: 119 (comp. note) China
Mogannia tienmushana Hua 2000: 63 (listed, distribution) Equals *Mogannia chinensis* China, Zhejiang

***M. tumdactylina* Chen, Yang & Wei, 2012**

***M. unicolor* (Walker, 1852)** = *Cephaloxys unicolor*

***M. venustissima venustissima* Stål, 1865** = *Mogannia venutissima* (sic) = *Mognnia* (sic) *venutissima* (sic)
Mogannia venustissima Yukawa and Yamane 1995: 697 (comp. note) Indonesia, Krakatau

***M. venustissima* var. *a* Stål, 1865**

***M. venustissima* var. *b* Stål, 1865**

***M. viridis* (Signoret, 1847)** = *Cephaloxys viridis* = *Cephaloxys rostrata* Walker, 1850 = *Mogannia distinguenda* Distant, 1920
Cephaloxys viridis Hayashi 1984: 64 (type species of *Cephaloxys*)
Mogannia viridis Zaidi and Ruslan 1998a: 366–367 (listed, comp. note) Equals *Cephaloxys viridis* Borneo, Sarawak, Java, India, Burma, Peninsular Malaysia, Banguey Island, Philippines
Mogannia viridis Zaidi, Ruslan and Azman 1999: 317 (listed, comp. note) Equals *Cephaloxys viridis* Borneo, Sabah, Java, India, Burma, Banguey Island, Philippines, Peninsular Malaysia
Mogannia viridis Zaidi and Azman 2003: 106, Table 1 (listed, comp. note) Borneo, Sabah
Mogannia viridis Schouten, Duffels and Zaidi 2004: 373, 375, Table 2 (listed, comp. note) Malayan Peninsula
Mogannia viridis Lee 2008a: 20 (synonymy, distribution, listed) Equals *Cephaloxys viridis* Equals *Mogannia distinguenda* Vietnam, Philippines, Malaysia, Indonesia, Java, Sumatra, Myanmar, India
Mogannia viridis Pham and Yang 2009: 15, Table 2 (listed) Vietnam

***Pachypsaltria* Stål, 1863** = *Pachysaltria* (sic)
Pachypsaltria Hayashi 1984: 63 (listed)
Pachypsaltria Heath, Heath and Noriega 1990: 1 (distribution, comp. note) Argentina

Pachypsaltria Pham and Thinh 2005b: 289 (comp. note)
Pachypsaltria Sanborn 2010a: 1596 (listed) Colombia

***P. camposi* Goding, 1925**

***P. cinctomaculata* (Stål, 1854)** = *Cicada cinctomaculata* = *Carineta ciliaris* Walker, 1858
Pachypsaltria cintomaculata Sanborn 2007a: 29 (distribution, comp. note) Colombia, Venezuela, Ecuador, Brazil, Bolivia, Argentina
Cicada cinctomaculata Sanborn 2010a: 1596 (type species of *Pachypsaltria*) Venezuela
Pachypsaltria cintomaculata Sanborn 2010a: 1596 (synonymy, distribution, comp. note) Equals *Cicada cinctomaculata* Equals *Cephaloxys cintomaculata* Colombia, Venezuela, Ecuador, Brazil, Bolivia, Argentina

***P. cirrha* Torres, 1960**

***P. haematodes* Torres, 1960**

***P. monrosi* Torres, 1960**

***P. phaedima* Torres, 1960**

***P. peristicta* Torres, 1960**

Species of uncertain position in Cicadidae

Genus A Pogue, 1996
Genus A sp. Pogue 1996: 314 (listed) Peru

Genus B Pogue, 1996
Genus B sp. Pogue 1996: 314 (listed) Peru

Subfamily Cicadettinae Buckton, 1889
Tibicininae (sic) Medler 1980: 73 (listed) Nigeria
Tibicinidae (sic) Popov 1981: 271 (comp. note)
Tibicinidae (sic) Ramos 1983: 66 (listed) Dominican Republic
Tibicinidae (sic) Ramos and Wolda 1985: 177 (listed) Panama
Tibicininae (sic) Ramos and Wolda 1985: 177 (listed) Panama
Tibicininae (sic) Moulds 1986: 39 (listed) Australia
Tibicinidae (sic) Ramos 1988: 62, Table 4.1 (listed) Greater Antilles
Tibicinidae (sic) Boulard, Deeming and Matile 1989: 143 (listed) Ivory Coast
Tibicininae (sic) Moulds 1990: 30–32, 52, 110 (listed, key, comp. note) Australia
Tibicininae (sic) Lee 1995: 18, 21, 23, 99, 135–136, 138–139 (listed, key, comp. note) Korea
Tibicinidae (sic) Pogue 1996: 313, 317 (listed, comp. note) Peru
Tibicininae (sic) Lee 1999: 10 (comp. note) Korea
Tibicininae (sic) Villet and Noort 1999: 226, 228–229, Table 12.1 (listed, comp. note)
Tibicininae (sic) Sueur and Puissant 2000: 262 (comp. note) France
Tibicininae (sic) Sanborn 2001b: 449 (listed) El Salvador
Tibicininae (sic) Moulds and Carver 2002: 19, 23, 46, 49–50 (listed) Australia
Tibicininae (sic) Williams 2002: 156–157 (listed) Australia, Queensland, New South Wales, Victoria, South Australia, Tasmania
Cicadettinae Moulds 2003a: 164 (comp. note)
Tibicininae (sic) Vincent 2006: 64 (listed) France
Tibicninae (sic) Lee 2005: 14, 16, 156 (key, listed, comp. note) Korea
Tibicininae (sic) Ewart 2005b 441 (listed)
Cicadettinae Moulds 2005b: 377, 381–382, 418, 426–427, 430, 435, Fig 57, Table 2 (n. status, described, diagnosis, composition, key, history, phylogeny, listed, comp. note) Equals Tibicininae *auct.* Australia
Tibicininae (sic) Pham and Thinh 2005a: 237 (listed) Vietnam
Tibicininae (sic) Pham and Thinh 2005b: 287 (key, comp. note) Equals Tibicinae Equals Tibicinos Equals Tibicines Equals Tibicinidae Vietnam
Cicadettinae Chawanji, Hodson and Villet 2006: 373–374 (listed, comp. note)
Cicadettinae Puissant 2006: 20 (listed) France
Tibicininae (sic) Sanborn 2006a: 77 (listed)
Cicadettinae Buckley and Simon 2007: 419 (comp. note)
Cicadettinae Chawanji, Hodson, Villet, Sanborn and Phillips 2007: 337 (listed, comp. note)
Cicadettinae Sanborn 2007a: 29 (listed) Venezuela
Cicadettinae Sanborn 2007b: 21, 37 (listed) Mexico
Cicadettinae Sanborn, Phillips and Sites 2007: 3, 33, Table 1 (listed) Thailand
Cicadettinae Sueur and Puissant 2007b: 62 (listed) France
Cicadettinae Boulard 2008c: 351, 358, 368 (listed, comp. note)
Cicadettinae Boulard 2008f: 7, 54, 97 (listed) Thailand
Cicadettinae Ewart and Marques 2008: 153 (listed)
Cicadettinae Lee 2008b: 446, 452, 462, Table 1 (listed, comp. note) Korea
Cicadettinae Lee 2008a: 2, 20, Table 1 (listed) Vietnam
Cicadettini (sic) Sanborn 2008f: 689 (listed)
Cicadettinae Sanborn, Phillips and Gillis 2008: 15 (listed) Florida
Cicadettinae Sanborn 2009d: 1 (listed)
Cicadettinae Boulard 2009d: 39 (comp. note)
Cicadettinae Lee 2009h: 2636 (listed) Philippines, Palawan

Cicadettinae Lee 2009i: 295 (listed) Philippines, Panay
Cicadettinae Pham and Yang 2009: 1, 3, 15–16, Table 2 (listed, comp. note) Vietnam
Cicadettinae Sanborn 2009a: 89 (listed) Cuba
Cicadetttinae Wei, Tang, He and Zhang 2009: 337–338 (listed, comp. note)
Cicadettinae Kemal and Koçak 2010: 5 (listed) Turkey
Cicadettinae Lee 2010a: 25 (listed) Philippines, Mindanao
Cicadettinae Lee 2010b: 27 (listed) Cambodia
Cicadettinae Lee and Hill 2010: 278 (comp. note)
Cicadettinae Mozaffarian and Sanborn 2010: 81 (listed)
Cicadettinae Phillips and Sanborn 2010: 75 (comp. note)
Cicadettinae Pham and Yang 2010a: 133 (diversity, comp. note) Vietnam
Cicadettinae Popple and Emery 2010: 148, 154 (listed) Australia
Cicadetttinae Puissant and Sueur 2010: 556 (listed)
Cicadettinae Sanborn 2010a: 1597 (listed) Colombia
Cicadettini (sic) Sanborn 2010b: 75 (listed)
Cicadettinae Wei, Ahmed and Rizvi 2010: 33 (comp. note)
Cicadettinae Wei and Zhang 2010: 37–38 (comp. note)

Tribe Dazini Kato, 1932
Dazini Boulard 1996b: 3–4 (comp. note)
Dazini Sanborn 1999a: 44 (listed)
Dazini Villet and Noort 1999: 226, Table 12.1 (diversity, listed)
Dazini Villet and Zhao 1999: 322 (comp. note)
Dazini Champanhet, Boulard and Gaiani 2000: 42 (comp. note)
Dazini Sueur 2002a: 380 (listed) Mexico, Veracruz
Dazini Moulds 2005b: 386, 427 (history, listed, comp. note)
Dazini Sanborn 2006a: 77 (listed)
Dazini Sanborn 2007a: 29 (listed) Venezuela
Dazini Sanborn 2007b: 21, 37 (listed) Mexico
Dazini Sanborn 2010b: 75 (listed)

***Daza* Distant, 1905** = *Dazu* (sic)
Daza Strümpel 1983: 17 (listed)
Daza Boulard 1996b: 3 (key, comp. note)
Daza Goemans 2010: 7 (comp. note)
D. *montezuma* (Walker, 1850) = *Zammara montezuma* = *Odopoea montezuma* = *Odopaea* (sic) *montezuma*
Daza montezuma Sueur 2002a: 380–382, 387–391, Fig 1, Fig 10, Figs 12–14, Table 1 (acoustic behavior, song, sonogram, oscillogram, comp. note) Mexico, Veracruz
Daza montezuma Sanborn 2005b: 77–78, Fig 1 (listed, distribution) Guatemala, Mexico
Daza montezuma Sanborn 2007b: 37 (distribution, comp. note) Mexico, Guatemala
Daza montezuma Sanborn 2010b: 75 (listed, distribution) Honduras, Mexico, Guatemala
***D. nayaritensis* Davis, 1934**
Daza nayaritensis Sanborn 1999a: 45 (type material, listed)
Daza nayaritensis Sanborn 2007b: 37 (distribution, comp. note) Mexico, Nayarit

***Procollina* Metcalf, 1963** = *Collina* Distant 1905 (nec *Collina* Bonarelli, 1893)
Procollina Young 1991: 221–223 (comp. note) Costa Rica
Collina Boulard 1996b: 3–4 (comp. note)
Procollina Boulard 1996b: 4 (key, comp. note) Equals *Collina*
Procollina Champanhet, Boulard and Gaiani 2000: 42 (comp. note) Equals *Collina*
Procollina Sanborn, Heath, Sueur and Phillips 2005: 193 (comp. note) Central America
Procollina Sanborn 2007b: 26 (comp. note) Mexico
Procollina Goemans 2010: 7 (comp. note)
***P. biolleyi* (Distant, 1903)** = *Odopoea biolleyi* = *Collina biolleyi*
Procollina biolleyi Young 1981d: 827 (acoustic behavior, comp. note) Costa Rica
Procollina biolleyi Young 1982: 414–415, Figs 9.3b-c (illustrated, comp. note) Costa Rica
Procollina biolleyi Young 1984a: 170, 176, 179, Tables 1–3 (hosts, comp. note) Costa Rica
Procollina biolleyi Young 1991: 221 (comp. note) Costa Rica
Procollina biolleyi Wolda and Ramos 1992: 272, Table 17.1 (distribution, listed) Panama
Procollina biolleyi Humphreys 2005: 70–71, 73–74, Table 4 (population density, comp. note) Costa Rica
***P. medea* (Stål, 1864)** = *Odopoea medea* = *Collina medea*
Procollina medea Sanborn 2007b: 26, 37 (distribution, comp. note) Mexico, Veracruz, Oaxaca, Costa Rica
***P. obesa* (Distant, 1906)** = *Collina obesa*
***P. queretaroensis* Sanborn, 2007**
Procollina queretaroensis Sanborn 2007b: 21–26, 37, Figs 35–43 (n. sp., described, illustrated, distribution, comp. note) Mexico, Queretaro

***Aragualna* Champanhet, Boulard & Gaiani, 2000**
Aragualna Champanhet, Boulard and Gaiani 2000: 42 (n. gen., described, comp. note)
Aragualna Goemans 2010: 7 (comp. note)

***A. plenalinea* Champanhet, Boulard & Gaiani, 2000**
Aragualna plenalinea Champanhet, Boulard and Gaiani 2000: 42–48, Plates I-V (type species of *Aragualna*, n. sp., described, illustrated, comp. note) Venezuela
Aragualna plenalinea Sanborn 2007a: 29 (distribution, comp. note) Venezuela

***Onoralna* Boulard, 1996**
Onoralna Boulard 1996b: 4 (n. gen., key, described, comp. note)
Onoralna Moulds 2005b: 414 (comp. note)
Onoralna Goemans 2010: 7 (comp. note)
***O. falcata* Boulard, 1996**
Onoralna falcata Boulard 1996b: 4–8, Figs 1–8 (n. sp., described, illustrated, song, comp. note) Ecuador

Tribe Huechysini Distant, 1905
Huechysini Boer and Duffels 1996b: 328 (distribution)
Huechysini Chou, Lei, Li, Lu and Yao 1997: 46, 97 (key, synonymy, listed, described) Equals Huechysaria
Huechysini Moulds 1996a: 20 (comp. note)
Huechysini Novotny and Wilson 1997: 437 (listed)
Huechysini Sueur 2001: 41, Table 1 (listed)
Huechysini Lee and Hayashi 2003a: 150 (listed) Taiwan
Huechysini Chen 2004: 39–40 (phylogeny, listed)
Huechysini Lee and Hayashi 2004: 68 (listed) Taiwan
Huechysini Boulard 2005h: 7 (listed)
Huechysini Moulds 2005b: 427 (higher taxonomy, listed)
Huechysini Pham 2005a: 216, 218 (key, comp. note) Vietnam
Huechysini Pham and Thinh 2005a: 237 (listed) Vietnam
Huechysini Pham and Thinh 2005b: 287–288 (key, comp. note) Equals Huechysaria Vietnam
Huechysini Boulard 2007d: 502 (listed)
Huechysini Boulard 2007e: 20 (comp. note) Thailand
Huechysini Sanborn, Phillips and Sites 2007: 3, 33, 41, Fig 5, Table 1 (listed, distribution) Thailand
Huechysini Shiyake 2007: 7, 90 (listed, comp. note)
Huechysini Boulard 2008c: 368 (comp. note)
Huechysini Boulard 2008f: 7, 54 (listed) Thailand
Huechysini Lee 2008a: 2, Table 1 (listed) Vietnam
Huechysini Ibrahim and Leong 2009: 317 (comp. note)
Huechysini Lee 2009h: 2636 (listed) Philippines, Palawan
Huechysini Pham and Yang 2009: 15, Table 2 (listed) Vietnam
Huechysini Lee 2010a: 25 (listed) Philippines, Mindanao
Huechysini Lee 2010c: 24 (listed, comp. note) Philippines, Luzon

Subtribe Huechysina Distant, 1905 = Huechysaria
Huechysaria Wu 1935: 22 (listed) China
Huechysaria Boulard 2005h: 7 (listed)

***Graptotettix* Stål, 1866**
Graptotettix Boer and Duffels 1996b: 302 (comp. note)
Graptotettix Chou, Lei, Li, Lu and Yao 1997: 97–98, 293, Table 6 (key, listed, described, diversity, biogeography) China
Graptotettix Pham 2005a: 216 (comp. note)
Graptotettix Shiyake 2007: 7, 90 (listed, comp. note)
***G. guttatus* Stål, 1866**
Graptotettix guttatus Chou, Lei, Li, Lu and Yao 1997: 19, 97–99, Fig 3–8A, Fig 8–31, Plate XI, Fig 120 (type species of *Graptotettix*, described, illustrated, comp. note) China
***G. rubromaculatus* Distant, 1905**
Graptotettix rubromaculatus Duffels 1990b: 64, Table 2 (listed) Sulawesi
Graptotettix rubromaculatus Boer and Duffels 1996b: 328 (distribution) Sulawesi
***G. thoracicus* Distant, 1892**

***Parvittya* Distant, 1905** = *Paravittya* (sic) = *Tarvittya* (sic)
Parvittya Duffels 1986: 319 (distribution, diversity, comp. note) Philippines
***P. samara* Distant, 1905**

***Huechys* Amyot & Audinet-Serville, 1843** = *Cicada (Huechys)* = *Huechys (Huechys)* = *Huechys (Amplihuechys)* = *Hyechis* (sic) = *Huechis* (sic) = *Heuchys* (sic) = *Hyechys* (sic) = *Huachas* (sic) = *Peuchys* (sic).
Huechys Wu 1935: 23 (listed) China
Huechys (sic) Shcherbakov 1981a: 830 (listed)
Huechis (sic) Shcherbakov 1981b: 66 (listed)
Huechys Strümpel 1983: 17 (listed)
Huechys Boulard 1985e: 1024, 1029 (coloration, comp. note)
Huechys Duffels 1986: 319 (distribution, diversity, comp. note) Philippines
Huechys spp. Namba, Ma and Inagaki 1988: 261 (medicinal use, comp. note) China
Huechys Boer 1990: 64 (comp. note)
Huechys Boulard 1990b: 201 (coloration, comp. note)
Huechys Duffels 1990b: 64 (comp. note)

Huechys Zaidi, Mohammedsaid and Yaakob 1990: 261, 267, Table 2 (listed, comp. note) Malaysia
Huechys Liu 1991: 254, 258 (comp. note)
Huechy (sic) Liu 1992: 2 (comp. note)
Huechys Liu 1992: 13, 17 (comp. note)
Huechys Peng and Lei 1992: 105 (comp. note) China, Hunan
Huechys Hayashi 1993: 15 (comp. note)
Huechys Zaidi and Ruslan 1994: 427, Table 2 (comp. note) Malaysia
Huechys Zaidi and Ruslan 1995a: 174, Table 2 (listed, comp. note) Borneo, Sarawak
Huechys Zaidi and Ruslan 1995b: 220–221, Table 2 (listed, comp. note) Borneo, Sabah
Huechys Zaidi and Ruslan 1995c: 70, Table 2 (listed, comp. note) Peninsular Malaysia
Huechys Zaidi and Ruslan 1995d: 202, Table 2 (listed, comp. note) Borneo, Sabah
Huechys Boer and Duffels 1996b: 302 (comp. note)
Huechys sp. Boer and Duffels 1996b: 299, Plate 1f (illustrated)
Huechys Zaidi 1996: 103, Table 2 (comp. note) Borneo, Sarawak
Huechys Zaidi and Hamid 1996: 54–55, Table 2 (listed, comp. note) Borneo, Sarawak
Huechys Chou, Lei, Li, Lu and Yao 1997: 3, 99, 293, 307, 312, 324, 326, Table 6, Tables 10–12 (described, listed, diversity, biogeography, comp. note) China
Huechis (sic) Novotny and Wilson 1997: 437 (listed)
Huechys Zaidi 1997: 115, Table 2 (listed, comp. note) Borneo, Sarawak
Huechys Zaidi and Ruslan 1998b: 4–5, Table 2 (listed, comp. note) Borneo, Sabah
Huechys Easton and Pun 1999: 99 (comp. note) Macao
Huechys Zaidi and Ruslan 1999: 5, Table 2 (listed, comp. note) Borneo, Sarawak
Huechys Ahn, Hahn, Ryu and Cho 2002: 68, Table 1 (medicinal use, listed)
Huechys Chen 2004: 40, 173 (phylogeny, listed)
Huechys Lee and Hayashi 2004: 68, 70 (key, diagnosis) Taiwan
Huechys Schouten, Duffels and Zaidi 2004: 378 (comp. note) Malayan Peninsula
Huechys Pham 2005a: 216 (key, described, comp. note) Vietnam
Huechys Pham and Thinh 2005a: 237 (listed) Vietnam
Huechys Pham and Thinh 2005b: 287 (comp. note) Vietnam
Huechys Boulard 2007e: 23 (comp. note) Thailand
Huechys Sanborn, Phillips and Sites 2007: 3, 33, Table 1 (listed) Thailand
Huechys Shiyake 2007: 7, 90 (listed, comp. note)
Huechys Xia, Bai, Yi, Liu, Chu, Liang et al. 2007: 960, 962–963, 966, Fig 1, Table 1, Table 3 (illustrated, medicinal use, comp. note) China, Sichuan, Shaanxi, Hubei, Guangdong, Liaoning, Henan, Hong Kong
Huechys Boulard 2008c: 368 (comp. note)
Huechys Boulard 2008f: 7, 54, 96 (listed) Thailand
Huechys Lee 2008a: 2, Table 1 (listed) Vietnam
Huechys Lee 2009h: 2636 (listed) Philippines, Palawan
Huechys Pham and Yang 2009: 15, Table 2 (listed) Vietnam
Huechys Lee 2010a: 25 (listed) Philippines, Mindanao
Huechys Lee 2010c: 24 (listed, comp. note) Philippines, Luzon

H. aenea **Distant, 1913**

H. beata **Distant, 1892** = *Huechys sanguinea* var. *b* Distant, 1892 = *Huechys philaemata* var. *b* = *Huechys sanguinea* (nec Degeer)

Huechys beata Chou, Lei, Li, Lu and Yao 1997: 99, 101, Plate III, Fig 32 (key, described, illustrated, synonymy, comp. note) Equals *Huechys sanguinea* var. *b* China
Huechys beata Hua 2000: 62 (listed, distribution) China, Guangxi, Yunnan, India, Malaysia
Huechys (sic) *sanguinea* Boulard 2005h: 7, 48–53, Figs 68–72, Color Plate, Fig F (described, illustrated, song, sonogram, listed, comp. note) Equals *Cicada sanguinea* Equals *Tettigonia sanguinolenta* Equals *Cicada sanguinolenta* Equals *Huechys vesicatoria* Equals *Huechys (Huechys) sanguinea* Thailand
Huechys beata Pham 2005a: 216–218 (key, synonymy, distribution, comp. note) Equals *Huechys sanguinea* (partim) Vietnam
Huechys beata Pham and Thinh 2005a: 237 (synonymy, distribution, listed) Equals *Huechys sanguinea* var. *b* Vietnam
Huechys beata Boulard 2007d: 502 (comp. note) Equals *Huechys sanguinea* var. b Equals *Huechys sanguinea* Boulard, 2005 [48] Equals *Huechys sanguinea* Boulard, 2006 (nec Degeer) [1] Equals *Huechys sanguinea* Boulard, 2007 (nec Degeer) [4, 12, 15, 20, 47–48, 86] Thailand
Huechys sanguinea (sic) Boulard 2007e: 4, 12, 15, 20, 47–48, 86 (coloration, eclosion, comp. note) Thailand
Huechys beata Sanborn, Phillips and Sites 2007: 33, 34 (listed, distribution, comp. note) Thailand, China, India, Malaysia, Vietnam, Indonesia
Huechys beata Boulard 2008c: 368 (comp. note) Equals *Huechys sanguinea* var. b Equals *Huechys sanguinea* Boulard, 2005 [48] Equals *Huechys sanguinea* Boulard, 2007 [4, 12, 15, 20, 47–48, 86] (nec Degeer) Thailand

Huechys beata Boulard 2008f: 54–55, 96 (biogeography, listed, comp. note) Equals *Huechys sanguinea* var. b Equals *Huechys sanguinea* Boulard 2005, 2006, 2006 (nec Degeer) Thailand, Assam, Malaysia, Sumatra
Huechys beata Lee 2008a: 21 (synonymy, distribution, listed) Equals *Huechys sanguinea* var. *b* Vietnam, South China, Thailand, Malaysia, Indonesia, India
Huechys beata Pham and Yang 2009: 15, Table 2 (listed) Vietnam

***H. celebensis celebensis* Distant, 1888** = *Huechys (Amplihuechys) celebensis*
Huechys celebensis Boer and Duffels 1996b: 328 (distribution) Sulawesi
Huechys celebensis Duffels 1990b: 64, Table 2 (listed) Sulawesi

***H. celebensis* var. *a* Breddin, 1901** = *Huechys (Amplihuechys) celebensis* var. *a*

***H. celebensis* var. *b* Breddin, 1901** = *Huechys (Amplihuechys) celebensis* var. *b*

***H. chryselectra* Distant, 1888** = *Huechys (Amplihuechys) chryselecrta* = *Huechys chtyselectra* (sic)
Huechys chtyselectra (sic) Distant Haitlinger 1990: 53 (parasitic host) Java
Huechys chryselectra Zaidi and Ruslan 1998a: 367 (listed, comp. note) Borneo, Sarawak

***H. curvata* Haupt, 1924** = *Huechys (Amplihuechys) curvata*

***H. dohertyi dohertyi* Distant, 1892**

***H. dohertyi* var. *a* Distant, 1892**

***H. eos* Breddin, 1901**
Huechys eos Duffels 1990b: 64, Table 2 (listed) Sulawesi
Huechys eos Boer and Duffels 1996b: 328 (distribution) Sulawesi

***H. facialis facialis* Walker, 1857** = *Huechys fascialis* (sic) = *Huechys (Amplihuechys) facialis*

***H. facialis obscura* Haupt, 1924** = *Huechys (Amplihuechys) facialis obscura*

***H. fretensis* Haupt, 1924** = *Huechys (Huechys) fretensis*

***H. funebris* Lee, 2011**

***H. fusca fusca* Distant, 1892** = *Huechys (Huechys) fusca*
Huechys sp. nr. *fusca* Boulard 1985e: 1032, Plate IX, Fig 54 (coloration, illustrated, comp. note)
Huechys fusca Duffels 1990b: 64, Table 2 (listed) Sulawesi
Huechys fusca Zaidi, Mohammedsaid and Yaakob 1990: 262 (comp. note) Malaysia
Huechys fusca Zaidi and Ruslan 1994: 428 (comp. note) Malaysia
Huechys fusca Boer and Duffels 1996b: 328 (distribution) Sulawesi, Asia
Huechys fusca Zaidi and Ruslan 1997: 230–231 (listed, comp. note) Peninsular Malaysia, Sumatra, Borneo, Singapore
Huechys fusca Zaidi and Ruslan 1998a: 367–368 (listed, comp. note) Borneo, Sarawak, Sumatra, Peninsular Malaysia, Singapore, Sabah
Huechys fusca Zaidi, Ruslan and Azman 1999: 318 (listed, comp. note) Borneo, Sabah, Sumatra, Sarawak, Peninsular Malaysia, Singapore
Huechys fusca Zaidi, Ruslan and Azman 2000: 216 (listed, comp. note) Borneo, Sabah, Sumatra, Peninsular Malaysia, Singapore, Sarawak
Huechys fusca Zaidi and Azman 2003: 106, Table 1 (listed, comp. note) Borneo, Sabah
Huechys fusca Boulard 2007d: 502, 506, 509, Fig 9 (illustrated, comp. note) Thailand
Huechys fusca Boulard 2008f: 55, 96 (biogeography, listed, comp. note) Equals *Huechys (Huechys) fusca* Thailand, Borneo, Peninsular Malaysia, Sumatra, Sulu, Philippines
Huechys fusca Lee 2009h: 2636 (synonymy, distribution, listed, comp. note) Equals *Huechys (Huechys) fusca* Philippines, Palawan, Sulu, Malay Peninsula, Sabah, Sarawak, Indonesia, Sulawesi, Sumbawa, Kalimantan, Sumatra

***H. fusca* var. *a* Distant, 1892** = *Huechys (Huechys) fusca* var. *a*

***H. fusca nigra* Lallemand, 1931** = *Huechys (Huechys) fusca nigra*

***H. haematica* Distant, 1888** = *Huechys fusca haematica* = *Huechys (Huechys) fusca haematica*
Huechys haematica Liu 1991: 255 (comp. note)
Huechys haematica Liu 1992: 13 (comp. note)
Huechys haematica Chou, Lei, Li, Lu and Yao 1997: 99, 101 (key, described, synonymy, comp. note) Equals *Huechys fusca haematica* China
Huechys haematica Hua 2000: 62 (listed, distribution) China, Guangxi, India, Philippines

***H. incarnata aterrima* Haupt, 1924** = *Huechys (Huechys) incarnata aterrima*

***H. incarnata* var. *b* Distant, 1897** = *Huechys (Huechys) incarnata* var. *b*

***H. incarnata brunnea* Haupt, 1924** = *Huechys (Huechys) incarnata brunnea*

***H. incarnata incarnata* (Germar, 1834)** = *Cicada incarnata* = *Huechis* (sic) *incarnata* = *Huechys sanguinea incarnata* = *Huechys (Huechys) incarnata* = *Cicada sanguinolenta* Brullé, 1835 (nec Fabricius) = *Cicada germari* Guérin-Méneville, 1838 = *Huechys germari*
Huechys incarnata Boulard 1985e: 1024 (coloration, comp. note)
Huechys incarnata Duffels 1990b: 64, Table 2 (listed) Sulawesi
Huechys incarnata Boer and Duffels 1996b: 328 (distribution) Sulawesi, West Melanesia
Huechys incarnata Boulard 2002b: 193 (coloration, comp. note)
Huechys incarnata Shiyake 2007: 7, 88, 90, Fig 146 (illustrated, listed, comp. note)

***H. incarnata novaeguineensis* Haupt, 1924** = *Huechys (Huechys) incarnata novaeguineensis*

***H. incarnata nox* Schmidt, 1919** = *Huechys incarnata* var. *a* Distant, 1897 = *Huechys (Huechys) incarnata nox*

***H. incarnata testacea* (Fabricius, 1787)** = *Tettigonia testacea* = *Cicada testacea* = *Cicada sanguinolenta* Germar (partim) = *Huechys (Huechys) incarnata testacea*

***H. incarnata venosa* Haupt, 1924** = *Huechys (Huechys) incarnata venosa*

***H. insularis borneensis* Haupt, 1924** = *Huechys (Huechys) insularis borneensis*

***H. insularis insularis* Haupt, 1924** = *Huechys (Huechys) insularis*

***H. insularis lombokensis* Haupt, 1924** = *Huechys (Huechys) insularis lombokensis*

***H. insularis niasensis* Haupt, 1924** = *Huechys (Huechys) insularis niasensis*

***H. insularis sumatrana* Haupt, 1924** = *Huechys (Huechys) insularis sumatrana*

***H. lutulenta* Distant, 1892** = *Huechys (Amplihuechys) lutulenta*

Huechys lutulenta Zaidi and Ruslan 1998a: 368 (listed, comp. note) Borneo, Sarawak

Huechys lutulenta Zaidi and Azman 2003: 99–100, 106, Table 1 (listed, comp. note) Borneo, Sabah

***H. nigripennis* Moulton, 1923**

***H. parvula* Haupt, 1924** = *Huechys (Huechys) parvula*

Huechys parvula Lee 2010a: 25 (synonymy, distribution, listed, comp. note) Equals *Huechys (Huechys) parvula* Philippines, Mindanao, Samar

H. phaenicura balabakensis Haupt, 1924 to *Huechys phaenicura* (Germar, 1834)

***H. phaenicura* (Germar, 1834)** = *Cicada phaenicura* = *Cicada phoenicura* (sic) = *Huechys phoenicura* (sic) = *Huechys (Huechys) phaenicura* = *Huechys phaenicura balabakensis* Haupt, 1924 = *Huechys phoenicura* (sic) *balabakensis* = *Huechys (Huechys) phaenicura balabakensis* = *Huechys (Huechys) phoenicura* (sic) f. *balabakensis* = *Huechys phaenicura palawanensis* Haupt, 1924 = *Huechys phoenicura* (sic) *palawanensis* = *Huechys (Huechys) phaenicura palawanensis* = *Huechys (Huechys) phoenicura* (sic) form. *palawanensis*

Huechys phoenicura (sic) Showalter 1935: 38–39, Fig 1 (illustrated, coloration, comp. note) Brazil

Huechys phaenicura Boulard 2008c: 367–368, Fig 23 (comp. note) Laos

Huechys phaenicura Lee 2009h: 2636–2637 (synonymy, distribution, listed, comp. note) Equals *Huechys (Huechys) phoenicura* (sic) f. *palawanensis* Equals *Huechys (Huechys) phoenicura* (sic) f. *balabakensis* Philippines, Palawan, Samar, Luzon, Balabac Island, Malaysia, Indonesia, Java, India

Huechys phaenicura Lee 2010a: 25 (synonymy, distribution, listed, comp. note) Equals *Cicada phaenicura* Equals *Huechys (Huechys) phaenicura* Philippines, Mindanao, Samar

H. phaenicura palawanensis Haupt, 1924 to *Huechys phaenicura* (Germar, 1834)

***H. pingenda breddini* Haupt, 1924** = *Huechys (Amplihuechys) pingenda breddini*

***H. pingenda pingenda* Distant, 1888** = *Huechys (Amplihuechys) pingenda* = *Graptotettix pingenda*

Huechys pingenda Duffels 1990b: 64, Table 2 (listed) Sulawesi

Huechys pingenda Boer and Duffels 1996b: 328 (distribution) Sulawesi, Sumatra

Graptotettix pingenda Shiyake 2007: 7, 88, 90, Fig 145 (illustrated, listed, comp. note)

***H. sanguinea* var. ______ Breddin, 1899** = *Huechys (Huechys) sanguinea* var. ______

***H. sanguinea* var. ______ Walker, 1870** = *Huechys (Huechys) sanguinea* var. ______

***H. sanguinea hainanensis* Kato, 1931** = *Huechys philaemata hainanensis* = *Huechys sanguinea ocher* Liu, 1940 = *Huechys sanguinea ochra* = *Huechys (Huechys) sanguinea hainanensis*

Huechys sanguinea hainanensis Hua 2000: 62 (listed, distribution) China, Hainan

H. sanguinea philaemata (Fabricius, 1803) = *Tettigonia philaemata* = *Cicada philaemata* = *Cicada incarnata philaematis* = *Cicada philoemata* (sic) = *Huechys philaemata* = *Huechys (Huechys) sanguinea philaemata* = *Huechys sanguinea* var. *philaemata* = *Huechys sanguinea philoemata* (sic) = *Huechys sanguinea* var. *a* Distant, 1892 = *Huechys sanguinea* (nec Degeer)

Huechys sanguinea var. *philaemata* Chen 1933: 361 (listed, comp. note)

Huechys sanguinea var. *philaemata* Wu 1935: 23 (listed) Equals *Tettigonia sanguinea* var. *philaemata* Equals *Cicada sanguinea* var. *philaemata* China, Haichos, Nanking, Soochow, Hangchow, Burma, Formosa

Huechys philaemata Namba, Ma and Inagaki 1988: 260–261, Fig 7B (medicinal use, illustrated, comp. note) China

Huechys sanguinea var. *philaemata* Liu 1992: 34 (comp. note)

Huechys sanguinea philaemata Anonymous 1994: 809 (fruit pest) China

Huechys philamata Hua 2000: 62 (listed, distribution, hosts) Equals *Tettigonia philamata* Equals *Huechys sanguinea* var. *philaemata* (sic) China, Anhui, Jiangsu, Jiangxi, Zhejiang, Taiwan, Hainan, Guangxi, Yunnan, India

Huechys philaemata Boulard 2008c: 368 (comp. note) China, Vietnam
Huechys philaemata Boulard 2009f: 253 (comp. note) Equals *Huechys sanguinea* Boulard, 2008 [55] (nec Degeer)

***H. sanguinea sanguinea* (Degeer, 1773)** = *Cicada sanguinea* = *Heuchys* (sic) *sanguinea* = *Hyechys* (sic) *sanguinea* = *Huachas* (sic) *sanguinea* = *Huechys sanquiena* (sic) = *Huechys sanquinea* (sic) = *Huechys (Huechys) sanguinea* = *Scieroptera sanguinea* = *Tettigonia sanguinolenta* Fabricius, 1775 = *Cicada sanguineolenta* = *Tettigonia sangvinolenta* (sic) = *Tettigonia philaemata* Fabricius, 1803 = *Huechys vesicatoria* Smith, 1871 = *Peuchys* (sic) *vesicatoria* = *Huechys quadrispinosa* Haupt, 1924 = *Huechys sanguinea philaemata* = *Huechys sanguinea philaemata albifascia* Kato, 1927 = *Huechys philaemata albifascia*

Tettigonia sanguinea Carus 1829: 150, 158, Fig XV (comp. note)
Huechys sanguinea Wu 1935: 23 (listed) Equals *Cicada sanguinea* Equals *Tettigonia sanguinolenta* Equals *Cicada sanguinolenta* Equals *Tettigonia testacea* Equals *Cicada testacea* China, Soochow, Kwangsi, Canton, Macao, Han-lik, Hong Kong, Hangchow, Sikhim, Assam, Brahmaputra, Calcutta, Burma, Rangoon, Kakhein Hills, Tenasserim, Thagata, Myitta, Malay Peninsula, Sumatra, Borneo, Timor Laut, Tonkin, Than-Moi, Formosa
Huechys sanguinea Weber 1968: 257 (comp. note)
Huechys sanguinea Chaudhry, Chaudry and Khan 1970: 104 (listed, comp. note) Pakistan, India, Burma, Peninsular Malaysia, Sumatra, Borneo, Timor, China
Huechys sanguinea Das and Roy 1980: 33 (tea pest, comp. note) India
Huechys sanguinea Riegel 1981: 15 (medicinal use, comp. note) China
Scieroptera sanguinea Hill, Hore and Thornton 1982: 50, 161–163, Plate 106, Plate 108 (listed, comp. note) Hong Kong
Huechys sanguinea Boulard 1985e: 1024, 1033–1034 (coloration, comp. note) China
Huechys sanguinea Chou 1987: 22 (comp. note) China
Huechys sanguinea Jiang, Lin and Wang 1987: 106, Fig 1C (timbal structure, illustrated, comp. note) China
Huechys sanguinea Kritsky 1987: 140 (medicinal use) Hong Kong
Huechys sanguinea Wang and Zhang 1987: 296 (key, comp. note) China
Huechys sanguinea Namba, Ma and Inagaki 1988: 260–261, Fig 7A (medicinal use, illustrated, comp. note) China
Huechys sanguinea Liu 1990: 39 (comp. note)
Huechys sanguinea Zaidi, Mohammedsaid and Yaakob 1990: 262–265, 267, Fig 1, Table 1 (seasonality, comp. note) Malaysia
Huechys sanguinea Liu 1991: 254 (comp. note)
Huechys sanguinea Liu 1992: 13, 36 (comp. note)
Huechys sanguinea Peng and Lei 1992: 104, Fig 298 (listed, illustrated, comp. note) China, Hunan
Huechys sanguinea Wang and Zhang 1992: 120 (listed, comp. note) China
Huechys sanguinea Wang and Zhang 1993: 190 (listed, comp. note) China
Huechys sanguinea Zaidi 1993: 958, Table 1 (listed, comp. note) Borneo, Sarawak
Huechys sanguinea Anonymous 1994: 809 (fruit pest) China
Huechys sanguinea Zaidi and Ruslan 1994: 428 (comp. note) Malaysia
Huechys sanguinea Jiang, Yang, Tang, Xu and Chen 1995b: 227 (song, phonotaxis, flight behavior, comp. note) China
Huechys sanguinea Tan, Zhang, Wang, Deng and Zhu 1995: 325, 328 (medicinal use, comp. note) China
Huechys sanguinea Zaidi 1996: 102 (comp. note) Borneo, Sarawak
Huechys sanguinea Zaidi, Ruslan and Mahadir 1996: 61, Table 1 (listed, comp. note) Peninsular Malaysia
Cicada sanguinea Chou, Lei, Li, Lu and Yao 1997: 3, 99 (type species of *Huechys*, comp. note)
Huechys sanguinea Chou, Lei, Li, Lu and Yao 1997: 8, 20, 28, 99–101, 307, 312, 324, 326, 328, 335, Fig 3–9A, Fig 8–32, Plate 4–2, Figs 3–4, Plate II, Fig 24, Plate XI, Fig 121, Tables 10–12 (key, described, illustrated, synonymy, song, oscillogram, spectrogram, comp. note) Equals *Cicada sanguinea* Equals *Tettigonia sanguinolenta* Equals *Tettigonia philaemata* Equals *Huechys quadrispinosa* China
Huechys sanguinea Zaidi 1997: 112 (comp. note) Borneo, Sarawak
Huechys sanguinea Zaidi and Ruslan 1997: 231 (listed, comp. note) Equals *Cicada sanguinea* Peninsular Malaysia, China, India, Burma, Siam, Sumatra, Borneo, Singapore
Huechys sanguinea Easton 1999: 28 (comp. note) Macao
Huechys sanguinea Easton and Pun 1999: 101 (distribution, comp. note) Macao, Hong Kong

Huechys sanguinea Liu, Jin and Xu 1999: 74–75, 124, Fig 1K (medicinal use)
Huechys sanguinea Stewart, Malik and Redmond 1999: 7, 15, Table 2 (Longan fruit pest, listed, comp. note)
Huechys sanguinea Sueur 1999: 57 (medicinal use, comp. note) Vietnam, China
Huechys sanguinea Ahn, Ryu, Lee and Kim 2000: 477–480 (medicinal uses) China
Huechys sanguinea Hua 2000: 62 (listed, distribution, hosts) Equals *Cicada sanguinea* China, Shaanxi, Henan, Hubei, Jiangsu, Zhejiang, Jiangxi, Fujian, Taiwan, Guangdong, Hong Kong, Macao, Hainan, Guangxi, Hunan, Sichuan, Yunnan, Philippines, Indonesia, Kalimantan, Malaysia, Vietnam, Sikkim, Maymer, India, Burma
Huechys sanguinea albifascia Hua 2000: 62 (listed, distribution) China, Taiwan
Scieroptera sanguinea Hua 2000: 64 (listed, distribution) China, Hong Kong
Huechys sanguinea Sueur 2001: 41, Table 1 (listed)
Huechys sanguinea Achtziger and Nigmann 2002: 13 (medicinal uses) Orient
Huechys sanguinea Boulard 2002b: 193 (coloration, comp. note)
Huechys sanguinea Ahn, Hahn, Ryu and Cho 2002: 66, 68, Table 1 (medicinal use, listed) China
Huechys sanguinea Chen 2004: 173–174, 199 (described, illustrated, listed, comp. note) Taiwan
Cicada sanguinea Lee and Hayashi 2004: 68 (type species of *Huechys*)
Huechys sanguinea Lee and Hayashi 2004: 69–70, Fig 34 (key, described, illustrated, synonymy, song, listed, comp. note) Equals *Cicada sanguinea* Equals *Tettigonia philaemata* Equals *Huechys philaemata* Equals *Huechys sanguinea* var. *philaemata* Equals *Huechys sanguinea* var. *philaemata* ab. *albifascia* Equals *Huechys sanguinea* ab. *albifascia* Equals *Huechys sanguinea* var. *albifascia* Equals *Huechys philaemata* ab. *albifascia* Taiwan, China, Philippines, Andaman Island, Nicobar Island
Huechys sanguinea Schouten, Duffels and Zaidi 2004: 373, 378–379, Table 2 (listed, comp. note) Malayan Peninsula
Huechys sanguinea Zhao 2004: 471 (medicinal use, comp. note) China, Guangdong, Guangxi, Sichuan, Fujian, Taiwan, Zhejiang, Jiangsu
Huechys sanguinea Boulard 2005d: 337, 345 (coloration, comp. note)
Huechys sanguinea Boulard 2005o: 150 (listed)
Huechys sanguinea Costa-Neto 2005: 98, 104, Table 1 (medicinal use, comp. note) Queensland, New South Wales
Huechys sanguinea Gogala and Trilar 2005: 66, 70–71, 81, Fig 9 (song, sonogram, oscillogram, acoustic behavior, comp. note) Malaysia
Huechys sanguinea Pham 2005a: 216–218 (key, synonymy, distribution, comp. note) Equals *Cicada sanguinea* Equals *Tettigonia sanguinolenta* Equals *Tettigonia philaemata* Vietnam, Myanmar, Malaysia, Philippines, Laos
Cicada sanguinea Pham 2005a: 216 (type species of *Huechys*)
Cicada sanguinea Pham and Thinh 2005a: 237 (type species of *Huechys*)
Huechys sanguinea Pham and Thinh 2005a: 237 (synonymy, distribution, listed) Equals *Cicada sanguinea* Equals *Tettigonia sanguinolenta* Equals *Tettigonia philaemata* Equals *Huechys quadrispinosa* Vietnam
Huechys sanguinea Boulard 2006e: 373, 375, 377–381, Plate II, Figs 1–14, Fig B (eclosion, illustrated, comp. note) Thailand
Huechys sanguinea Boulard 2006g: 76–77, 94, Fig 62 (sonogram, comp. note) Thailand
Huechys sanguinea Pham and Thinh 2006: 525, 527 (listed, comp. note) Vietnam
Huechys sanguinea Boulard 2007e: Plate 15, Plate 23 (illustrated, eclosion, comp. note) Thailand
Huechys sanguinea Li, Chen, Wang, and Hou 2007: 750–752 (control) China
Huechys sanguinea Sanborn, Phillips and Sites 2007: 33 (listed, distribution, comp. note) Thailand, India, Philippine Republic, Southeast Asia, Indonesia
Huechys sanguinea Shiyake 2007: 7, 88, 90, Fig 147 (illustrated, distribution, listed, comp. note) China, Southeast Asia, Taiwan, Malaysia, Indonesia, Borneo, Philippines
Huechys sanguinea Xia, Bai, Yi, Liu, Chu, Liang et al. 2007: 960–962, 964–965, Fig 2, Table 1 (illustrated, medicinal use, comp. note) China, Sichuan, Shaanxi, Hubei, Guangdong, Liaoning, Henan, Hong Kong
Huechys sanguinea Boulard 2008c: 368 (comp. note) Vietnam
Huechys sanguinea Boulard 2008f: 3–4, 7, 54–56, 96 (synonymy, type species of *Huechys*, biogeography, listed, comp. note) Equals *Cicada sanguinea* Equals *Huechys philaemata* Equals *Huechys sanguinolenta* Equals *Huechys sanguinea* var. a *philaemata* Equals *Huechys (Huechys) sanguinea* Equals

Huechys quadrispinosa Thailand, China, Japan, India, Sikkim, Formosa, Taiwan
Huechys sanguinea Costa Neto 2008: 453 (comp. note)
Cicada sanguinea Lee 2008a: 21 (type species of *Huechys*)
Huechys sanguinea Lee 2008a: 21 (synonymy, distribution, listed) Equals *Cicada sanguinea* Equals *Huechys (Huechys) sanguinea* Equals *Huechyys* (sic) *aurantiaca* (nom. nud.) Equals *Huechys sanguinea* var. *aurantiaca* (nom. nud.) Equals *Huechys (Huechys) quadrispinosa* Vietnam, South China, Hainan, Taiwan, Thailand, Malay Peninsula, Timor, Borneo, Indonesia, Sumatra, Myanmar, India
Huachas (sic) *sanguinea* Galacgac and Balisacan 2009: 2050, Table 4 (comp. note)
Huechys sanguinea Ibrahim and Leong 2009: 317–321, Figs 1–6 (illustrated, eclosion, distribution, comp. note) Indonesia, China, Taiwan, Vietnam, Thailand, Malay Peninsula, Timor, Borneo, Sumatra, Myanmar, India
Cicada sanguinea Lee 2009h: 2636 (type species of *Huechys*, listed) China
Huechys sanguinea Lee 2009h: 2637 (synonymy, distribution, listed, comp. note) Equals *Cicada sanguinea* Equals *Huechys (Huechys) sanguinea* Philippines, Palawan, Leyte, southern China, Hainan, Taiwan, Vietnam, Thailand, Malay Peninsula, Timor, Borneo, Indonesia, Sumatra, Myanmar, India
Huechys sanguinea Pham and Yang 2009: 15–16, Table 2 (listed, comp. note) Vietnam
Huechys sanguinea Ahmed and Sanborn 2010: 44 (distribution) Bangladesh, Asia, India, Philippine Republic Southeast Asia, Indonesia
Huechys sanguinea Lee, Lin and Wu 2010: 218 (comp. note) Taiwan
Cicada sanguinea Lee 2010a: 25 (type species of *Huechys*, listed)
Cicada sanguinea Lee 2010c: 24 (type species of *Huechys*, listed) China
Huechys sanguinea Pham and Yang 2010a: 133 (comp. note) Vietnam
Huechys tonkinensis Pham and Thinh 2005a: 237 (distribution, listed) Vietnam

***H. sanguinea suffusa* Distant, 1888** = *Huechys (Huechys) sanguinea suffusa*

***H. sanguinea wuchangensis* Liu, 1940** = *Huechys (Huechys) sanguinea wuchangensis*

***H. sumbana* Jacobi, 1941**

***H. sumbana* var. ______ Jacobi, 1941**

***H. thoracica* Distant, 1879** = *Huechys (Huechys) thoracica*

Huechys thoracica Wang and Zhang 1993: 190 (listed, comp. note) China
Huechys thoracica Chou, Lei, Li, Lu and Yao 1997: 99, 102, 357, Fig 8–33, Plate XI, Fig 122 (key, described, illustrated, comp. note) China
Huechys thoracica Hua 2000: 62 (listed, distribution) China, Yunnan, Burma
Huechys thoracica Shiyake 2007: 7, 88, 90, Fig 148 (illustrated, distribution, listed, comp. note) Malaysia, Southheast Asia
Huechys thoracica Boulard 2008c: 368 (comp. note) Vietnam

***H. tonkinensis* Distant, 1917**

Huechys tonkinensis Lee 2008a: 21 (distribution, listed) North Vietnam
Huechys tonkinensis Pham and Yang 2009: 15, Table 2 (listed) Vietnam
Huechys tonkinensis Pham and Yang 2010a: 133 (comp. note) Vietnam

***H. vidua vidua* White, (1846)** = *Cicada (Huechys) vidua* = *Cicada vidua* = *Huechys (Huechys) vidua*

***H. vidua* var. *a* Distant, 1892** = *Huechys (Huechys) vidua* var. *a*

***Scieroptera* Stål, 1866** = *Schieroptera* (sic) = *Sleroptera* (sic) = *Scieretopra* (sic) = *Scieoptera* (sic) = *Scleroptera* (sic)

Scieroptera Wu 1935: 24 (listed) China
Scieroptera Shcherbakov 1981a: 830 (listed)
Scieroptera Shcherbakov 1981b: 66 (listed)
Scieroptera Boer 1990: 64 (comp. note)
Scieroptera Hayashi 1993: 15 (comp. note)
Scieroptera Zaidi and Ruslan 1995a: 174, Table 2 (listed, comp. note) Borneo, Sarawak
Scieroptera Zaidi and Ruslan 1995c: 70, Table 2 (listed, comp. note) Peninsular Malaysia
Scieroptera Zaidi and Ruslan 1995d: 202, Table 2 (listed, comp. note) Borneo, Sabah
Scieroptera Boer and Duffels 1996b: 302 (comp. note)
Scieroptera Moulds 1996a: 20 (comp. note)
Scieroptera Zaidi 1996: 103, Table 2 (comp. note) Borneo, Sarawak
Scieroptera Zaidi and Hamid 1996: 54, Table 2 (listed, comp. note) Borneo, Sarawak
Scieroptera Chou, Lei, Li, Lu and Yao 1997: 97, 103, 293, Table 6 (described, key, listed, diversity, biogeography) China
Scieroptera Novotny and Wilson 1997: 437 (listed)
Scieroptera Zaidi 1997: 115, Table 2 (listed, comp. note) Borneo, Sarawak
Scieroptera Zaidi and Ruslan 1999: 5, Table 2 (listed, comp. note) Borneo, Sarawak
Scieroptera Biswas, Ghosh, Bal and Sen 2004: 248 (key) India
Scieroptera Chen 2004: 40, 173, 176 (phylogeny, listed)

Scieroptera Lee and Hayashi 2004: 68, 70 (key, diagnosis) Taiwan
Scieroptera Pham 2005a: 216 (key, described, comp. note) Vietnam
Scieroptera Pham and Thinh 2005a: 237 (listed) Vietnam
Scieroptera Pham and Thinh 2005b: 287 (comp. note) Vietnam
Scieroptera Sanborn, Phillips and Sites 2007: 3, 33, Table 1 (listed) Thailand
Scieroptera Shiyake 2007: 7, 90 (listed, comp. note)
Scieroptera Boulard 2008f: 7, 56, 96 (listed) Thailand
Scieroptera Lee 2008a: 2, Table 1 (listed) Vietnam
Scieroptera Pham and Yang 2009: 15, Table 2 (listed) Vietnam
Scieroptera Lee 2010a: 26 (listed) Philippines, Mindanao

***S. borneensis* Schmidt, 1918**

***S. crocea crocea* (Guérin-Méneville, 1838)** = *Cicada crocea*
Scieroptera crocea Hua 2000: 64 (listed, distribution) Equals *Cicada crocea* China, Hainan, Guangxi, India, Indonesia, Kalimantan
Scieroptera crocea Lee 2008a: 22 (synonymy, distribution, listed) Equals *Cicada crocea* North Vietnam, Malaysia, Indonesia, Java, Sumatra, Myanmar, India
Scieroptera crocea Pham and Yang 2009: 15, Table 2 (listed) Vietnam

***S. crocea* var. *a* Distant, 1892**

***S.* nr. *crocea* Zaidi & Ruslan, 1995**
Scieroptera nr. *crocea* Zaidi and Ruslan 1995a: 172, Table 1 (listed, comp. note) Borneo, Sarawak
Scieroptera nr. *crocea* sp. n. Zaidi and Ruslan 1999: 3, 5, Table 1 (listed, comp. note) Borneo, Sarawak

***S. delineata* Distant, 1917** = *Scieroptera delineate* (sic)
Scieroptera delineata Lee 2008a: 22 (distribution, listed) Vietnam, Laos
Scieroptera delineata Pham and Yang 2009: 15, Table 2 (listed) Vietnam
Scieroptera delineate (sic) Pham and Yang 2009: 16 (comp. note) Vietnam
Scieroptera delineate (sic) Pham and Yang 2010a: 133 (comp. note) Vietnam

***S. distanti* Schmidt, 1920** = *Schieroptera* (sic) *distanti* = *Scieroptera distani* (sic)
Scieroptera distani (sic) Wu 1935: 24 (listed) China, Canton
Scieroptera distanti Chou, Lei, Li, Lu and Yao 1997: 105 (comp. note)
Scieroptera distanti Hua 2000: 64 (listed, distribution) China, Guangzhou

***S. flavipes* Schmidt, 1918**

***S. formosana* Schmidt, 1918** = *Scieroptera splendidula* (nec Fabricius) = *Sleroptera* (sic) *splendidula* (nec Fabricius) = *Scieroptera splendidula cuprea* (nec Walker) = *Scieroptera formosana ater* Kato, 1925 = *Scieroptera formosana atra* = *Scieroptera formosana ater* = *Scieroptera formosana trigutta* Kato, 1926 = *Scieroptera formosana albifascia* Kato, 1926 = *Scieretopra* (sic) *formosana* = *Scieoptera* (sic) *formosana* = *Scieroptera formasana* (sic)
Scieroptera formosana Chen 1933: 361 (listed, comp. note)
Scieroptera formosana Wang and Zhang 1992: 120 (listed, comp. note) China
Scieroptera formosana Chou, Lei, Li, Lu and Yao 1997: 103–105, Plate III, Fig 34 (key, described, illustrated, synonymy, comp. note) Equals *Scieroptera splendidula* Matsumura Equals *Scieroptera splendidula cuprea* Schumacher China
Scieroptera formosana Hua 2000: 64 (listed, distribution) China, Jiangxi, Zhejiang, Fujian, Taiwan, Guangdong, Hubei, Hunan, Guangxi, Sichuan, India, Burma, Philippines, Malaysia
Scieroptera formosana albifasciata Hua 2000: 64 (listed, distribution) China, Taiwan
Scieroptera formosana ater Hua 2000: 64 (listed, distribution) China, Taiwan
Scieroptera formosana trigutta Hua 2000: 64 (listed, distribution) China, Taiwan
Scieroptera formosana Chen 2004: 32–33, 173, 177–179, 199, back cover (described, illustrated, listed, comp. note) Taiwan
Scieroptera formosana Lee and Hayashi 2004: 70 (key, described, illustrated, synonymy, song, listed, comp. note) Equals *Scieroptera splendidula* (nec Fabricius) Equals *Scieroptera splendidula* var. *cuprea* Equals *Scieroptera formosana* ab. *ater* Equals *Scieroptera formosana* ab. *trigutta* Equals *Scieroptera formosana* var. *trigutta* Equals *Scieroptera formosana* ab. *albifascia* Equals *Scieroptera formosana* var. *albifascia* Taiwan, China
Scieroptera formasana (sic) Pham 2005a: 216–218 (key, synonymy, distribution, comp. note) Equals *Scieroptera splendidula* (partim) Equals *Scieroptera splendidula cuprea* (partim) Vietnam
Scieroptera formosana Pham 2005a: 216 (key) Vietnam
Scieroptera formasana (sic) Pham and Thinh 2005a: 238 (synonymy, distribution, comp. note) Equals *Scieroptera splendidula* (partim) Equals *Scieroptera splendidula cuprea* (partim) Vietnam
Scieroptera formosana Shiyake 2007: 7, 89, 91, Figs 149a–149b (illustrated, distribution, listed, comp. note) China, Taiwan, Southeast Asia

Scieroptera formosana Lee 2008a: 22 (distribution, listed) Vietnam, China, Taiwan

Scieroptera formosana Pham and Yang 2009: 15, Table 2 (listed) Vietnam

***S. fulgens* Schmidt, 1918**

***S. fumigata* (Stål, 1854)** = *Huechys fumigata*

***S. fuscolimbata* Schmidt, 1918**

***S. hyalinipennis* Schmidt, 1912**

***S. limpida* Schmidt, 1918**

***S. montana* Schmidt, 1918** = *Scieroptera carmente* Schu.

***S. niasana* Schmidt, 1918**

***S. orientalis* Schmidt, 1918**

Scieroptera orientalis Lee 2008a: 22 (distribution, listed) North Vietnam, South China

Scieroptera orientalis Pham and Yang 2009: 15, Table 2 (listed) Vietnam

***S. sanaoensis* Schmidt, 1924**

Scieroptera sanaoensis Lee 2010a: 26 (distribution, listed, comp. note) Philippines, Mindanao

***S. sarasinorum* Breddin, 1901**

Sceiroptera sarasinorum Duffels 1990b: 64, Table 2 (listed) Sulawesi

Scieroptera sarasinorum Boer and Duffels 1996b: 328 (distribution) Sulawesi

S. n. sp. Zaidi & Ruslan, 1995

Scieroptera n. sp. Zaidi and Ruslan 1995a: 175 (comp. note) Borneo, Sarawak

***S. splendidula cuprea* (Walker, 1870)** = *Huechys cuprea* = *Scieroptera splendidula* var. *a* Distant, 1982 = *Huechys cuprae* (sic) = *Scieroptera splendidula cuprae* (sic)

Scieroptera splendidula cuprea Sanborn, Phillips and Sites 2007: 33 (listed, distribution, comp. note) Thailand, India, Southeast Asia, Indonesia, China, Japan, Philippine Republic, Taiwan

Scieroptera splendidula cuprea Boulard 2008f: 56, 96 (biogeography, listed, comp. note) Equals *Huechys cuprea* Equals *Scieroptera splendidula* var. a *cuprea*Thailand, Celebes, Siam

***S. splendidula splendidula* (Fabricius, 1775)** = *Tettigonia splendidula* = *Telligonia* (sic) *splendidula* = *Cicada splendidula* = *Huechys splendidula* = *Scleroptera* (sic) *splendidula* = *Scieroptera splendidula cuprea* (partim) = *Scieroptera splendidula vittata* Kato, 1940

Scieroptera splendidula Wu 1935: 24 (listed) Equals *Tettigonia splendidula* Equals *Cicada splendidula* Equals *Huechys cuprea* Equals *Cicada trabeata* China, Kwangsi, Tien-mu-shan, Sikhim, Assam, Margherita, Khasi Hills Burma, Momeit, Tenasserim, Java, Borneo, Celebes, Formosa

Scieroptera splendidula Wang and Zhang 1987: 296 (key, comp. note) China

Scieroptera splendidula Liu 1991: 258 (comp. note)

Scieroptera splendidula Liu 1992: 17 (comp. note)

Scieroptera splendidula Peng and Lei 1992: 104, Fig 297 (listed, illustrated, comp. note) China, Hunan

Scieroptera splendidula Zaidi, Ruslan and Mahadir 1996: 61, Table 1 (listed, comp. note) Peninsular Malaysia

Tettigonia splendidula Chou, Lei, Li, Lu and Yao 1997: 103 (type species of *Scieroptera*)

Scieroptera splendidula Chou, Lei, Li, Lu and Yao 1997: 103–104, Fig 8–34, Plate 4–2, Figs 1–2, Plate III, Fig 33 (key, described, illustrated, synonymy, comp. note) Equals *Tettigonia splendidula* Equals *Cicada splendidula* Equals *Huechys splendidula* China

Scieroptera splendidula Zaidi and Ruslan 1997: 231–232 (listed, comp. note) Equals *Tettigonia splendidula* Peninsular Malaysia, China, Sumatra, India, Java, Borneo, Sarawak, Langkawi Island

Scieroptera splendidula Zaidi and Ruslan 1998a: 368 (listed, comp. note) Equals *Tettigonia splendidula* Borneo, Sarawak, China, Sumatra, India, Java, Peninsular Malaysia

Scieroptera splendidula Zaidi, Ruslan and Azman 1999: 318 (listed, comp. note) Equals *Tettigonia splendidula* Borneo, Sabah, Sarawak, Peninsular Malaysia

Scieroptera splendidula Hua 2000: 64 (listed, distribution) Equals *Tettigonia splendidula* China, Zhejiang, Fujian, Taiwan, Guangdong, Hunan, Guizhou, Guangxi, Sichuan, Indonesia, Kalimantan, Celebes, Sikkim, Burma, India, Malaysia

Scieroptera splendidula Zaidi, Noramly and Ruslan 2000a: 330 (listed, comp. note) Equals *Tettigonia splendidula* Borneo, Sabah, China, Sumatra, Java, India, Sarawak, Peninsular Malaysia

Scieroptera splendidula Zaidi, Ruslan and Azman 2000: 216 (listed, comp. note) Equals *Tettigonia splendidula* Borneo, Sabah, China, Sumatra, Java, India, Borneo, Peninsular Malaysia

Scieroptera splendidula Zaidi, Azman and Ruslan 2001a: 112–113, 116, 121–122, Table 1 (listed, comp. note) Equals *Tettigonia splendidula* Langkawi Island, China, Sumatra, Java, India, Borneo, Peninsular Malaysia

Scieroptera splendidula Zaidi and Nordin 2001: 183, 185, 191, 203–204, Table 1 (listed, comp. note) Equals *Tettigonia splendidula* Borneo, Sabah, China, Sumatra, Java, India, Peninsular Malaysia

Scieroptera splendidula Zaidi, Nordin, Maryati, Wahab, Norashikin et al. 2002: 7, 10, Tables 1–2 (listed, comp. note) Equals *Tettigonia*

splendidula Borneo, Sabah, China, Sumatra, India, Java, Peninsular Malaysia, Sarawak
Scieroptera splendidula Zaidi and Azman 2003: 106, Table 1 (listed, comp. note) Borneo, Sabah
Scieroptera splendidula Biswas, Ghosh, Bal and Sen 2004: 240, 248–249 (listed, key, described, distribution) Equals *Telligonia* (sic) *splendidula* India, Manipur, Assam, Meghalaya, Sikkim, Myanmar, China
Tettigonia splendidula Lee and Hayashi 2004: 70 (type species of *Scieroptera*) China
Scieroptera splendidula Pham 2005a: 216–218 (key, synonymy, distribution, comp. note) Equals *Tettigonia splendidula* Equals *Cicada splendidula* Equals *Huechys splendidula* Vietnam
Tettigonia splendidula Pham 2005a: 217 (type species of *Scieroptera*)
Tettigonia splendidula Pham and Thinh 2005a: 237 (type species of *Scieroptera*)
Scieroptera splendidula Pham and Thinh 2005a: 237–238 (synonymy, distribution, listed) Equals *Tettigonia splendidula* Equals *Cicada splendidula* Equals *Huechys splendidula* Vietnam
Scieroptera splendidula Boulard 2008c: 368 (comp. note) Malaysia
Scieroptera splendidula Boulard 2008f: 56, 96 (type species of *Scieroptera*, biogeography, listed, comp. note) Equals *Tettigonia splendidula* Equals *Cicada splendidula* Equals *Scieroptera splendidula* var. *cuprea* Thailand, China, India, Sumatra, Tenasserim, Sikkim, Burma, Borneo, Celebes, Kwangsi, Hainan, Kwangtung, Nepal
Tettigonia splendidula Lee 2008a: 21 (type species of *Scieroptera*) China
Scieroptera splendidula Lee 2008a: 21–22 (synonymy, distribution, listed) Equals *Tettigonia splendidula* Equals *Scieroptera splendidula* var. *vittata* Vietnam, China, Philippines, Laos, Thailand, Malaysia, Indonesia, Myanmar, Nepal, India
Scieroptera splendidula Pham and Yang 2009: 15, Table 2 (listed) Vietnam
Tettigonia splendidula Lee 2010a: 26 (type species of *Scieroptera*, listed)

***S. splendidula* var. *c* Distant, 1892**

***S. splendidula* var. *d* Distant, 1892**

***S. splendidula trabeata* (Germar, 1830)** = *Huechys trabeata* = *Scieroptera splendidula* var. *b* Distant, 1892

***S. sumatrana nigripennis* Schmidt, 1918** = *Huechys nigripennis*

***S. sumatrana sumatrana* Schmidt, 1918** = *Scieroptera splendidula* (nec Fabricius) = *Huechys splendidula* (nec Fabricius)
Scieroptera sumatrana Lee 2008a: 22 (distribution, listed) Vietnam, Malaysia, Indonesia, Java, Sumatra, Myanmar, India
Scieroptera sumatrana Pham and Yang 2009: 15, Table 2 (listed) Vietnam

Tribe Carinetini Distant, 1905 = Carinettini (sic) = Sinosenini Boulard, 1975
Carinetini Boulard 1982e: 179 (comp. note)
Sinosenini Boulard 1988f: 51 (key)
Carinetini Boulard 1988f: 64 (comp. note)
Sinosenini Chou, Lei, Li, Lu and Yao 1997: 37 (listed, described)
Carinetini Novotny and Wilson 1997: 437 (listed)
Carinetini Boulard 1998: 76 (comp. note)
Carinetini Sanborn 1999a: 45 (listed)
Carinetini Sanborn 2001b: 449 (listed) El Salvador
Carinettini (sic) Sueur 2002a: 380 (listed) Mexico, Veracruz
Carinetini Jong and Boer 2004: 267 (comp. note)
Carinetini Moulds 2005b: 427 (higher taxonomy, listed, comp. note)
Sinosenini Moulds 2005b: 427 (higher taxonomy, listed)
Carinetini Sanborn 2006a: 77 (listed)
Carinetini Sanborn 2007a: 29 (listed) Venezuela
Carinetini Sanborn 2007b: 26, 37 (listed) Mexico
Carinetini Shiyake 2007: 7, 91 (listed, comp. note)
Carinetini Boulard 2008f: 7, 57 (listed) Thailand
Carinetini Sanborn 2008f: 689 (listed)
Carinetini Sanborn 2009a: 89 (listed) Cuba
Sinosenini Wei, Tang, He and Zhang 2009: 337–338 (listed, comp. note)
Carinetini Sanborn 2010a: 1597 (listed) Colombia
Carinetini Sanborn 2010b: 75 (listed)

Subtribe Carinetiina Distant, 1905 = Carinetaria
Carinetaria Wu 1935: 24 (listed) China
Carinetaria Wei, Tang, He and Zhang 2009: 337 (comp. note)

***Carineta* Amyot & Audinet-Serville, 1843** = *Cerinetha* (sic) = *Carinata* (sic) = *Carinetta* (sic) = *Carinita* (sic) = *Carinenta* (sic) = *Clarineta* (sic)
Carineta Dallas 1870: 482 (listed)
Carineta Heinrich 1967: 43 (coffee pest, comp. note) Brazil
Carineta spp. Young 1981b: 180 (comp. note) Costa Rica
Carineta Young 1981b: 181, 190, Fig 4 (habitat, comp. note) Costa Rica
Carineta Boulard 1982e: 182 (comp. note)
Carineta Souza, Reis and Melles 1983: 9 (comp. note) Brazil, Minas Gerias

Carineta sp. Souza, Reis and Melles 1983: 17, Fig 7 (illustrated, coffee pest, control, comp. note) Brazil, Minas Gerias
Carineta Strümpel 1983: 17 (listed)
Carineta spp. Reis, Souza and Melles 1984: 4, 6, 9, Figs 8–9, Table 1 (coffee pest, listed) Brazil
Carineta Reis, Souza and Melles 1984: 6, 9 (comp. note) Brazil
Carineta sp. Young 1984a: 170, Table 1 (hosts, comp. note) Costa Rica
Carineta spp. Young 1984a: 173 (hosts, comp. note) Costa Rica
Carineta Boulard 1985e: 1021, 1027 (coloration, comp. note)
Carineta Boulard 1986a: 417 (comp. note)
Carineta Boulard 1986c: 191–194, 202–203 (comp. note) Neotropics
Carineta Martinelli and Zucchi 1989c: 14 (comp. note) Brazil
Carineta Boulard 1990b: 89, 116, 137, 144, 149 (comp. note)
Carineta Heath, Heath and Noriega 1990: 1 (distribution, comp. note) Argentina
Carineta Young 1991: 214–215, 222–223 (comp. note) Costa Rica
Carineta Duffels 1993: 1225 (stridulatory apparatus, comp. note)
Carineta Moore 1996: 221 (comp. note) Mexico
Carineta Novotny and Wilson 1997: 437 (listed)
Carineta Champanhet 1999: 66 (comp. note)
Carineta Ohbayashi, Sato and Igawa 1999: 340 (parasitism, comp. note)
Carineta Boulard 2000d: 86, 89 (stridulation, comp. note) Neotropics
Carineta Fornazier and Martinelli 2000: 1175 (comp. note) Brazil
Carineta Champanhet 2001: 105–106 (comp. note)
Carineta Boulard 2002b: 197, 204, 206 (sound system, comp. note)
Carineta sp. Paião, Menegium, Casagrande and Leite 2002: S67 (vector) Brazil, Paraná
Carineta sp. Reis, Souza and Venzon 2002: 84, Fig 1 (coffee pest, illustrated, control, comp. note) Brazil
Carineta Reis, Souza and Venzon 2002: 84 (coffee pest, comp. note) Brazil
Carineta sp. Almeida 2004: 105 (control) Brazil
Carineta Campos, Silva, Ferreira and Dias 2004: 125 (comp. note) Brazil
Carineta Martinelli 2004: 520 (comp. note) Brazil
Carineta Moulds 2005b: 410, 414 (comp. note)
Carineta Salazar Escobar 2005: 194 (comp. note) Colombia
Carineta Boulard 2006g: 157–158 (stridulation, comp. note)
Carineta spp. Bento 2007: 27 (coffee pest) Brazil
Carineta Shiyake 2007: 7, 91 (listed, comp. note)
Carineta sp. Shiyake 2007: 7, 89, 91, Fig 150 (illustrated, listed, comp. note)
Carineta Thouvenot 2007b: 284 (comp. note) French Guiana
Carineta Sanborn, Moore and Young 2008: 1 (comp. note)
Carineta Sanborn 2010a: 1597 (listed) Colombia

***C. aestiva* Distant, 1883**
Carineta aestiva Boulard 1986c: 199 (comp. note)
Carineta aestiva Wolda and Ramos 1992: 273, Table 17.2 (listed) Panama

***C. apicalis* Distant, 1883**
Carineta apicalis Boulard 1986a: 415, 423 (comp. note)
Carineta apicalis group Boulard 1990b: 213 (comp. note)

***C. apicoinfuscata* Sanborn, 2011**

***C. aratayensis* Boulard, 1986**
Carineta aratayensis Boulard 1986a: 416, 419–422, Plate I, Fig B, Figs 11–15 (n. sp., described, illustrated, comp. note) French Guiana

***C. argentea* Walker, 1852** = *Carinita* (sic) *argentea* = *Carineta argentata* (sic)
Carineta argentea Salazar and Sanborn 2009: 271 (comp. note) Colombia
Carineta argentea Sanborn 2010a: 1597 (synonymy, distribution, comp. note) Equals *Carinita* (sic) *argentea* Equals *Carineta argentata* (sic) Colombia

***C. basalis* Walker, 1850**
Carineta basalis Boulard 1986c: 194 (comp. note)
Carineta basalis Sanborn 2007a: 29 (distribution, comp. note) Colombia, Venezuela, Ecuador, Peru
Carineta basalis Sanborn 2010a: 1597 (distribution, comp. note) Colombia, Venezuela, Ecuador, Peru

***C. boliviana* Distant, 1905**

***C. boulardi* Champanhet, 1999**
Carineta boulardi Champanhet 1999: 69–72, Plate II, Plate III, Figs 1–7 (n. sp., described, illustrated, comp. note) Bolivia

***C. calida* Walker, 1858**

***C. carayoni* Boulard, 1986**
Carineta carayoni Boulard 1986c: 194–195, 198, 200–202, Figs 9–11, Fig 19, Fig 26, Fig 30 (n. sp., described, illustrated, comp. note) Ecuador
Carineta carayoni Boulard 2006g: 158, Fig 114A (illustrated, comp. note)

***C. cearana* Distant, 1906**
Carineta cearana Boulard 1986c: 194 (comp. note)
Carineta cearana Champanhet 1999: 69 (comp. note)

C. centralis **Distant, 1892**
Carineta centralis Champanhet 1999: 69 (comp. note)
Carineta centralis Champanhet 2001: 106 (comp. note)
C. cinara **Distant, 1883**
Carineta cinara Wolda and Ramos 1992: 272, 276, Fig 17.6, Table 17.1 (distribution, listed, seasonality, comp. note) Panama
Carineta cinara Sanborn 2010b: 75 (listed, distribution) Guatemala, Costa Rica, Panama
C. cingenda **Distant, 1883**
Carineta cingenda Boulard 1986a: 427 (comp. note)
Carineta cingenda Pogue 1996: 317 (listed) Peru
C. congrua **Walker, 1858**
Carineta congrua Sanborn 2010a: 1597 (distribution, comp. note) Colombia, Ecuador
C. crassicauda **Torres, 1948** = *Carinenta* (sic) *crassicauda*
C. criqualicae **Boulard, 1986**
Carineta criqualicae Boulard 1986a: 416, 423–425, Plate I, Fig D, Figs 20–23 (n. sp., described, illustrated, comp. note) French Guiana
Carineta criqualicae Boulard 1986c: 194 (comp. note)
C. cristalinea **Champanhet, 2001**
Carineta cristalinea Champanhet 2001: 105–108, 110, Plate I, Figs 1–7 (n. sp., described, illustrated, comp. note) Bolivia
C. crocea **Distant, 1883**
Carineta crocea Sanborn 2007a: 29 (distribution, comp. note) Colombia, Venezuela
Carineta crocea Sanborn 2010a: 1597 (distribution, comp. note) Colombia, Venezuela
C. crumena **Goding, 1925**
C. cyrili **Champanhet, 1999**
Carineta cyrili Champanhet 1999: 66–69, Plate I, Fig 1, Figs 3–7 (n. sp., described, illustrated, comp. note) Ecuador
C. detoulgoueti **Champanhet, 2001** = *Carineta detoulgouëti*
Carineta detoulgouëti Champanhet 2001: 110–111, Plate III, Figs 1–5 (n. sp., described, illustrated, comp. note) Ecuador
C. diardi **(Guérin-Méneville, 1829)** = *Cicada formosa* Germar, 1830 = *Carineta formosa* = *Carinata* (sic) *formosa* = *Cicada polychroa* Perty, 1833 = *Cicada duvaucelii* Guérin-Méneville, 1838 = *Cicada duvancellii* (sic)
Carineta formosa Weber 1931: 105 (flight muscle, comp. note)
Carineta formosa Showalter 1935: 38–39, Fig 9 (illustrated, coloration, comp. note) Brazil
Carineta formosa Weber 1968: 295 (circulation, comp. note)
Carineta formosa Otero 1971: 71, 116, 144, 175 (illustrated, natural history, comp. note) Brazil
Carineta diardi Boulard 1986c: 191–195, Figs 1–4 (type species of *Carineta*, stridulatory apparatus, genitalia illustrated, comp. note) Equals *Carineta formosa* (Germar)
Carineta diardi Boulard 1988d: 152 (ectopic structure, comp. note) Equals *Carineta formosa*
Carineta diardi Boulard 1990b: 101, 103, Plate X, Fig 4 (genitalia, comp. note)
Carineta diardi Boulard 2002b: 176, Fig 6 (illustrated, comp. note) Brazil
Carineta formosa Moulds 2005b: 410 (comp. note) Equals *Carineta diardi*
Carineta diardi Bolcatto, Medrano and De Santis 2006: 9, Fig 1B, (illustrated, distribution) Argentina, Misiones
Carineta diardi De Santis, Medrano, Sanborn and Bolcatto 2007: 3, 14, 17, 19, Fig 1B, Fig 4, Table 1 (listed, distribution) Argentina, Misiones
Carineta diardi Shiyake 2007: 7, 89, 91, Fig 151 (illustrated, listed, comp. note)
Cicada formosa Sanborn 2010a: 1597 (type species of *Carineta*) Brazil
C. dolosa **Boulard, 1986**
Carineta dolosa Boulard 1986a: 418–419, Figs 9–10 (n. sp., described, illustrated, comp. note)
Carineta dolosa Champanhet 1999: 66, 68, Plate I, Fig 2 (illustrated, comp. note)
C. doxiptera **Walker, 1858**
C. durantoni **Boulard, 1986**
Carineta durantoni Boulard 1986a: 416, 423–425, 427, Plate I, Fig E, Figs 24–27 (n. sp., described, illustrated, comp. note) French Guiana
Carineta durantoni Sanborn 2008f: 689 (comp. note) Brazil, Rondonia, French Guiana
C. ecuatoriana **Goding, 1925**
C. fasciculata **(Germar, 1821)** = *Cicada fasciculata* = *Carinenta* (sic) *fasciculata* = *Carinetta* (sic) *fasciculata* = *Fidicina fasciculata* = *Cicada bilineosa* Walker, 1858 = *Carineta bilineosa* = *Cicada obtusa* Walker, 1858 = *Carineta obtusa* = *Cicada tenuistriga* Walker, 1858 = *Carineta trenuistriga* = *Carineta diplographa* Berg, 1879
Carineta fasciculata? d'Utra 1908: 354–355, Fig 2 (illustrated, coffee pest, control, comp. note) Brazil
Carineta fasciculata Hempel 1913a: 139 (illustrated, coffee pest, comp. note) Brazil
Carineta fasciculata Hempel 1913b: 197 (illustrated, coffee pest, comp. note) Brazil
Carinetta (sic) *fasciculata* Moreira 1921: 125 (coffee pest, comp. note) Brazil

Carineta fasciculata Costa Lima 1922: 111 (coffee pest, comp. note) Brazil
Carineta fasciculata Costa Lima 1922: 81 (coffee pest, comp. note) Brazil, São Paulo
Carineta fasciculata Moreira 1928: 24 (coffee pest, comp. note) Brazil
Carineta fasciculata Vayssière 1955: 254 (listed, coffee pest) Brazil
Fidicina fasciculata Mariconi 1958: 331–332 (coffee pest, described, listed, comp. note) Brazil
Carineta fasciculata Heinrich 1967: 43 (coffee pest, comp. note) Brazil
Carineta fasciculata Martinelli and Zucchi 1987b: 469 (coffee pest, comp. note) Brazil
Carineta fasciculata Martinelli and Zucchi 1987c: 470 (coffee pest, comp. note) Brazil, São Paulo
Carineta fasciculata Santos 1988: 27 (coffee pest) Brazil, Paraná
Carineta fasciculata Martinelli and Zucchi 1989c: 13–16, 18, 20, Fig 1 (coffee pest, described, illustrated, comp. note) Brazil, São Paulo
Carineta bilineosa Martinelli and Zucchi 1989c: 16 (comp. note) Brazil
Carineta fasciculata Martinelli 1990: 13 (illustrated, comp. note) Brazil
Carineta fasciculata Wolda and Ramos 1992: 272, Table 17.1 (distribution, listed) Panama
Carineta fasciculata Martinelli and Zucchi 1997a: 16 (distribution, comp. note) Brazil, Minas Gerias, São Paulo, Paraná, Argentina, Bolivia, Paraguay
Carineta fasciculata Martinelli and Zucchi 1997c: 135–141, Figs 1–2, Fig 14, Fig 18, Table 1 (distribution, host plants, key, illustrated, comp. note) Brazil, Argentina, Paraguay, Bolivia
Carineta fasciculata Albuquerque, Martinelli, Ros and Stülp 2000: 196 (coffee pest) Brazil
Carineta fasciculata Fornazier and Martinelli 2000: 1176 (coffee pest, distribution, comp. note) Brazil, Espírito Santo, Minas Gerais, Paraíba, Rio de Janeiro, São Paulo, Bolivia, Argentina, Paraguay
Carineta fasciculata Fornazier and Rocha 2000: 1167, 1169 (coffee pest, control, comp. note) Espírito Santo, Brazil
Carineta fasciculata Reis, Souza and Venzon 2002: 84 (coffee pest, comp. note) Brazil
Carineta fasciculata Almeida 2004: 104–105 (control) Brazil
Carineta fasciculata Martinelli 2004: 518–521, 525–526, 529, Figs 18.1–18.2, Fig 18.4d, Fig 18.4g (synonymy, described, illustrated, coffee pest, host plants, distribution, control, comp. note) Equals *Cicada fasciculata* Equals *Cicada bilineosa* Equals *Cicada obtusa* Equals *Cicada tenuistriga* Equals *Carineta ternuistriga* (sic) Equals *Carineta diplographa* Equals *Carineta obtusa* Equals *Carineta bilineosa* Brazil, Argentina, Paraguay, Bolivia
Carineta fasciculata Martinelli, Maccagnan, Ribeiro, Pereira and Santos 2004: 619 (coffee pest, comp. note) Brazil
Carineta fasciculata Santos and Martinelli 2004b: 630 (coffee pest) Brazil
Carineta fasciculata Salazar Escobar 2005: 193 (comp. note) Colombia
Carineta fasciculata Bolcatto, Medrano and De Santis 2006: 7, 10, Map 1 (distribution) Argentina
Carineta fasciculata Martinelli 2006: 1 (coffee pest, comp. note) Brazil
Carineta fasciculata De Santis, Medrano, Sanborn and Bolcatto 2007: 11–12, Fig 2 (distribution, comp. note)
Carineta fasciculata Santos and Martinelli 2007: 311 (coffee pest, comp. note) Brazil
Carineta fasciculata Lara, Perioto and de Ffreitas 2010: 116 (coffee pest) Brazil

***C. fimbriata* Distant, 1891**
Carineta fimbriata Sanborn 2007a: 29 (distribution, comp. note) Venezuela, Ecuador

***C. garleppi* Jacobi, 1907**

***C. gemella* Boulard, 1986**
Carineta gemella Boulard 1986a: 417–419, Figs 5–6 (n. sp., described, illustrated, comp. note) Venezuela
Carineta gemella Sanborn 2007a: 29 (distribution, comp. note) Venezuela, Argentina, Brazil, Bolivia, Ecuador
Carineta gemella Sanborn 2008f: 689 (comp. note) Brazil, Rondonia, Amazonas, Minas Gerais, Roraima, Hyutandaha, Venezuela

***C. genitalostridens* Boulard, 1986**
Carineta genitalostridens Boulard 1986c: 198–200, Fig 16, Figs 21–22, Fig 25, Fig 29 (n. sp., described, illustrated, comp. note) Ecuador
Carineta genitalostridens Boulard 1990b: 152–153, 159, 161, 191, 193, Plate XVII, Fig 5, Plate XIX, Fig 6, Plate, XXIV, Fig 1 (sound system, comp. note) Andes
Carineta genitalostridens Boulard 2000d: 89, Plate D", Figs 1–2 (stridulatory apparatus)
Carineta genitalostridulens Boulard 2006g: 158, Figs 114B–C (illustrated, comp. note)

***C. guianaensis* Sanborn, 2011**

***C. illustris* Distant, 1905**
Carineta illustris Champanhet 1999: 69–70 (comp. note)
Carineta illustris Pogue 1996: 317 (listed) Peru

Carineta illustris Champanhet 2001: 106 (comp. note)
Carineta illustris Sanborn 2010a: 1597–1598 (distribution, comp. note) Colombia, Bolivia, Brazil, Ecuador, Peru

***C. imperfecta* Boulard, 1986**
Carineta imperfecta Boulard 1986c: 194–196, 198, 200, Figs 5–6, Fig 20, Fig 23, Fig 27 (n. sp., described, illustrated, comp. note) Ecuador

***C. indecora* (Walker, 1858)** = *Cicada indecora*
Carineta indecora Wolda and Ramos 1992: 272, 276, Fig 17.7, Table 17.1 (distribution, listed, seasonality, comp. note) Panama
Carineta indecora Sanborn 2010a: 1598 (distribution, comp. note) Colombia, Costa Rica, Panama, Guatemala
Carineta indecora Sanborn 2010b: 75 (listed, distribution) Guatemala, Costa Rica, Panama, Colombia

***C. lichiana* Boulard, 1986**
Carineta lichiana Boulard 1986a: 415–419, Plate I, Fig A, Figs 1–4 (n. sp., described, illustrated, comp. note) Venezuela
Carineta lichiana Sanborn 2007a: 29 (distribution, comp. note) Venezuela, French Guiana

***C. limpida* Torres, 1948** = *Clarineta* (sic) *limpida*
Clarineta (sic) *limpida* Ruffinelli 1970: 3 (distribution, comp. note) Uruguay

***C. liturata* Torres, 1948**

***C. lydiae* Champanhet, 1999**
Carineta lydiae Champanhet 1999: 72–75, Plate IV, Plate V, Figs 1–6 (n. sp., described, illustrated, comp. note) Bolivia

***C. maculosa* Torres, 1948** = *Carinenta* (sic) *maculosa* = *Carineta fasciculata* (nec Germar)
Carineta maculosa Wolda and Ramos 1992: 272–273, 276, Fig 17.7, Table 17.1 (distribution, listed, seasonality, comp. note) Panama
Carineta maculosa Sanborn 2007a: 29 (distribution, comp. note) Colombia, Panama, Venezuela, Ecuador, Brazil, Paraguay, Argentina
Carineta maculosa Sanborn 2010a: 1598 (synonymy, distribution, comp. note) Equals *Carineta fasciculata* (nec Germar) Colombia, Panama, Venezuela, Ecuador, Brazil, Paraguay, Argentina

***C. martiniquensis* Davis, 1934**
Carineta martiniquensis Sanborn 1999a: 45 (type material, listed)

***C. matura* Distant, 1892**
Carineta matura Martinelli and Zucchi 1986: 16 (coffee pest, comp. note) Brazil
Carineta matura Martinelli and Zucchi 1987b: 469 (coffee pest, comp. note) Brazil
Carineta matura Martinelli and Zucchi 1987c: 470 (coffee pest, comp. note) Brazil, Minas Gerias
Carineta matura Martinelli and Zucchi 1989c: 14, 16–18, 21, Fig 2 (coffee pest, described, illustrated, comp. note) Brazil, Minas Gerias
Carineta matura Martinelli 1990: 13 (illustrated, comp. note) Brazil
Carineta matura Martinelli and Zucchi 1997c: 135–141, Figs 1–2, Figs 12–13, Fig 16, Table 1 (distribution, host plants, key, illustrated, comp. note) Brazil, Venezuela
Carineta matura Albuquerque, Martinelli, Ros and Stülp 2000: 196 (coffee pest) Brazil
Carineta matura Fornazier and Martinelli 2000: 1176 (coffee pest, distribution, comp. note) Espírito Santo, Minas Gerais, Brazil, Venezuela
Carineta matura Fornazier and Rocha 2000: 1167, 1169 (coffee pest, control, comp. note) Espírito Santo, Brazil
Carineta matura Paião, Menegium, Casagrande and Leite 2002: S67 (vector) Brazil, Paraná
Carineta matura Reis, Souza and Venzon 2002: 84 (coffee pest, comp. note) Brazil
Carineta matura Almeida 2004: 104–105 (control) Brazil
Carineta matura Martinelli 2004: 518–521, 525, 529, Figs 18.1–18.2, Figs 18.4e (described, illustrated, coffee pest, host plants, distribution, control, comp. note) Brazil, Venezuela
Carineta matura Martinelli, Maccagnan, Ribeiro, Pereira and Santos 2004: 619 (coffee pest, comp. note) Brazil
Carineta matura Santos and Martinelli 2004b: 630 (coffee pest) Brazil
Carineta matura Martinelli 2006: 1 (coffee pest, comp. note) Brazil
Carineta matura Santos and Martinelli 2007: 311 (coffee pest, comp. note) Brazil
Carineta matura Sanborn 2007a: 29 (distribution, comp. note) Venezuela
Carineta matura Lara, Perioto and de Freitas 2010: 116 (coffee pest) Brazil

***C. modesta* Sanborn, 2011**

***C. naponore* Boulard, 1986**
Carineta naponore Boulard 1986a: 416, 426, 428, Plate I, Fig G, Figs 32–35 (n. sp., described, illustrated, comp. note) Ecuador
Carineta naponore Boulard 1986c: 194 (comp. note)

***C. peruviana* Distant, 1905**
Carineta peruviana Boulard 1986a: 428 (comp. note)
Carineta peruviana Boulard 1986c: 194 (comp. note)

***C. picadae* Jacobi, 1907**

***C. pilifera* Walker, 1858**
Carineta pilifera Boulard 1986c: 194 (comp. note)

Carineta pilifera Salazar Escobar 2005: 193 (comp. note) Colombia
Carineta pilifera Sanborn 2007a: 29 (distribution, comp. note) Colombia, Ecuador, Venezuela
Carineta pilifera Sanborn 2008f: 689 (comp. note) Brazil, Santa Catarina, Colombia, Ecuador
Carineta pilifera Sanborn 2010a: 1598 (distribution, comp. note) Colombia, Ecuador, Venezuela

C. *pilosa* Walker, 1850
Carineta pilosa Boulard 1986a: 415 (comp. note)
Carineta pilosa Sanborn 2007a: 29 (distribution, comp. note) Colombia, Venezuela
Carineta pilosa Sanborn 2010a: 1598 (distribution, comp. note) Colombia, Venezuela

C. *platensis* Berg, 1882

C. *porioni* Champanhet, 2001
Carineta porioni Champanhet 2001: 108–110, Plate II, Figs 1–6 (n. sp., described, illustrated, comp. note) Ecuador

C. *postica* Walker, 1858 = *Carineta viridicata* (nec Distant)
Carineta postica Young 1981b: 181–182, 184–187, 189, Figs 5–6, Tables 1–2 (illustrated, seasonality, habitat, comp. note) Costa Rica
Carineta postica Young 1982: 414–415, Fig 9.3a, Fig 9c (illustrated, comp. note) Costa Rica
Carineta postica Young 1984a: 170, Table 1 (hosts, comp. note) Costa Rica
Carineta postica Young 1991: 222 (comp. note) Costa Rica
Carineta postica Wolda and Ramos 1992: 273, Table 17.2 (listed) Panama
Carineta postica Salazar and Sanborn 2009: 271 (comp. note) Equals *Carineta* sp. Salazar Escobar 2005 Colombia
Carineta postica Sanborn 2010a: 1598–1599 (synonymy, distribution, comp. note) Equals *Carineta viridicata* (nec Walker) Salazar Escobar 2005 Colombia, Panama, Peru, Ecuador, Costa Rica, Guatemala
Carineta postica Sanborn 2010b: 75 (listed, distribution) Guatemala, Costa Rica, Panama, Colombia, Ecuador, Peru

C. *producta* Walker, 1858
Carineta producta Sanborn 2007a: 29 (distribution, comp. note) Colombia, Venezuela, Brazil, Peru
Carineta producta Sanborn 2010a: 1599 (distribution, comp. note) Colombia, Venezuela, Brazil, Peru

C. *propinqua* Torres, 1948 = *Carinenta* (sic) *propinqua*

C. *quinimaculata* Sanborn, 2011

C. *rubricata* Distant, 1883

C. *rufescens* (Fabricius, 1803) = *Tettigonia rufescens* = *Cicada rufescens* = *Carineta oberthuri* Distant, 1881 = *Carineta oberthuerii* (sic)
Tettigonia rufescens Dallas 1870: 482 (synonymy) Equals *Carineta rufescens*
Carineta rufescens Boulard 1986c: 194 (comp. note)
Carineta rufescens Boulard 1990b: 117, Figs 8–9 (genitalia, comp. note)
Carineta rufescens Pogue 1996: 317 (listed) Peru
Carineta rufescens Sanborn 2004c: 512, Fig 173 (illustrated)
Carineta rufescens Thouvenot 2007b: 283–284, Fig 7 (illustrated, comp. note) French Guiana
Carineta rufescens Sanborn 2010a: 1599 (synonymy, distribution, comp. note) Equals *Tettigonia rufescens* Equals *Cicada rufescens* Equals *Carineta oberthuri* Equals *Carineta oberthuerii* (sic) Colombia, Brazil, French Guiana, Peru

C. *rustica* Goding, 1925

C. *scripta* Torres, 1948 = *Carinenta* (sic) *scripta*

C. *socia* Uhler, 1875 = *Carineta socio* (sic) = *Carineta* sp. 3 Salazar Escobar, 2005
Carineta socia Henshaw 1903: 229 (listed) Amazons
Carineta socia Boulard 1986a: 415, 418–419, 422, Figs 7–8 (comp. note)
Carineta socia Champanhet 1999: 66 (comp. note)
Carineta socia Sanborn 1999a: 45 (type material, listed)

C. sp. 3 Salazar Escobar, 2005
Carineta sp. 3 Salazar Escobar 2005: 195 (comp. note) Colombia
Carineta socia Salazar and Sanborn 2009: 271 (comp. note) Colombia
Carineta socia Sanborn 2010a: 1599 (synonymy, distribution, comp. note) Equals *Carineta* sp. 3 Salazar Escobar, 2005 Colombia, Bolivia, Brazil, Ecuador, French Guiana, Peru

C. sp. Young, 1981
Carineta sp. Young 1981b: 181–182, 184–187, 189, 194, Figs 5–6, Tables 1–2 (illustrated, seasonality, habitat, comp. note) Costa Rica
Carineta spp. Young 1981d: 827 (acoustic behavior, comp. note) Costa Rica

C. sp. M Wolda & Ramos, 1992
Carineta sp. M Wolda and Ramos 1992: 272, 276, Fig 17.5, Table 17.1 (distribution, listed, seasonality, comp. note) Panama

C. sp. 1 Pogue, 1997
Carineta sp. 1 Pogue 1996: 317 (listed) Peru

C. sp. 2 Pogue, 1997
Carineta sp. 2 Pogue 1996: 317 (listed) Peru

C. sp. 3 Pogue, 1997
Carineta sp. 3 Pogue 1996: 317 (listed) Peru

C. sp. 4 Pogue, 1997
Carineta sp. 4 Pogue 1996: 317 (listed) Peru

C. sp. 5 Pogue, 1997
Carineta sp. 5 Pogue 1996: 317 (listed) Peru

C. sp. 6 Pogue, 1997
Carineta sp. 6 Pogue 1996: 317 (listed) Peru
C. sp. 7 Pogue, 1997
Carineta sp. 7 Pogue 1996: 317 (listed) Peru
C. sp. 1 Salazar Escobar, 2005
Carineta sp. 1 Salazar Escobar 2005: 195 (comp. note) Colombia
C. sp. 2 Salazar Escobar, 2005
Carineta sp. 2 Salazar Escobar 2005: 195, 201, Fig 5 (illustrated, comp. note) Colombia
***C. spoliata* (Walker, 1858)** = *Cicada spoliata*
Carineta spoliata Boulard 1986a: 423 (comp. note)
Carineta spoliata Martinelli and Zucchi 1986: 16 (coffee pest, comp. note) Brazil
Carineta spoliata Martinelli and Zucchi 1987b: 469 (coffee pest, comp. note) Brazil
Carineta spoliata Martinelli and Zucchi 1987c: 470 (coffee pest, comp. note) Brazil, Minas Gerias
Carineta spoliata Martinelli and Zucchi 1989c: 14, 17–18, 22, Fig 3 (coffee pest, described, illustrated, comp. note) Brazil, Minas Gerias, São Paulo
Carineta spoliata Martinelli 1990: 13 (illustrated, comp. note) Brazil
Carineta spoliata Martinelli and Zucchi 1997c: 135–141, Figs 1–2, Fig 15, Fig 22, Table 1 (distribution, host plants, key, illustrated, comp. note) Brazil, Bolivia, Peru, Colombia, Venezuela
Carineta spoliata Albuquerque, Martinelli, Ros and Stülp 2000: 196 (coffee pest) Brazil
Carineta spoliata Reis, Souza and Venzon 2002: 84 (coffee pest, comp. note) Brazil
Carineta spoliata Almeida 2004: 104–105 (control) Brazil
Carineta spoliata Martinelli 2004: 518–521, 525–526, 529, Figs 18.1–18.2, Fig 18.4d, Fig 18.4g (synonymy, described, illustrated, coffee pest, host plants, distribution, control, comp. note) Equals *Cicada spoliata* Brazil, Bolivia, Peru, Colombia, Venezuela
Carineta spoliata Martinelli, Maccagnan, Ribeiro, Pereira and Santos 2004: 619 (coffee pest, comp. note) Brazil
Carineta spoliata Santos and Martinelli 2004b: 630 (coffee pest) Brazil
Carineta spoliata Martinelli 2006: 1 (coffee pest, comp. note) Brazil
Carineta spoliata Martinelli and Maccagnan 2006: 246 (emergence, comp. note) Brazil
Carineta spoliata Santos and Martinelli 2007: 311 (coffee pest, comp. note) Brazil
Carineta spoliata Sanborn 2007a: 29 (distribution, comp. note) Colombia, Venezuela Brazil, Peru, Bolivia
Carineta spoliata Lara, Perioto and de Freitas 2010: 116 (coffee pest) Brazil
Carineta spoliata Sanborn 2010a: 1599 (distribution, comp. note) Colombia, Venezuela Brazil, Peru, Bolivia
***C. strigilifera* Boulard, 1986**
Carineta strigilifera Boulard 1986c: 194–200, Figs 7–8, Fig 18, Fig 31 (n. sp., described, illustrated, comp. note) Ecuador
***C. submarginata* Walker, 1850**
Carineta submarginata Champanhet 1999: 72 (comp. note)
***C. tetraspila* Jacobi, 1907**
***C. tigrina* Boulard, 1986**
Carineta tigrina Boulard 1986a: 416, 426–428, Plate I, Fig F, Figs 28–31 (n. sp., described, illustrated, comp. note) French Guiana
***C. titschacki* Jacobi, 1951** = *Carineta titchaki* (sic) = *Carineta titschachi* (sic)
Carineta titchaki (sic) Boulard 1986c: 194 (comp. note)
Carineta titschachi (sic) Boulard 1986c: 196 (comp. note)
***C. tracta* Distant, 1892**
Carineta tracta Boulard 1986c: 201 (comp. note)
***C. trivittata* Walker, 1858** = *Carineta strigimargo* Walker, 1858
Carineta trivittata Young 1981b: 190 (comp. note) Costa Rica
Carineta trivitatta Wolda and Ramos 1992: 272, 276–277, Fig 17.6, Table 17.1 (distribution, listed, seasonality, comp. note) Panama
Carineta trivittata Champanhet 2001: 106 (comp. note)
Carineta trivittata Salazar Escobar 2005: 195, 200, Fig 4 (illustrated, comp. note) Colombia
Carineta trivittata Sanborn 2005b: 77–78, Fig 1 (listed, distribution) Honduras, Guatemala, Central America
Carineta trivittata Sanborn 2007a: 30 (comp. note)
Carineta trivittata Sanborn 2007b: 37 (distribution, comp. note) Mexico, Chiapas, Colombia, Guatemala, Honduras, Costa Rica, Panama, Brazil, Guiana, Ecuador, Peru, Bolivia
Carineta trivittata Salazar and Sanborn 2009: 271 (comp. note) Colombia
Carineta trivittata Sanborn 2010a: 1600, 1603 (synonymy, distribution, comp. note) Equals *Carineta strigimargo* Colombia, Mexico, Guatemala, Honduras, Costa Rica, Panama, Ecuador, Peru, Brazil, Bolivia, Guyana
***C. turbida* Jacobi, 1907**
***C. urostridulens* Boulard, 1986**
Carineta urostridulens Boulard 1986c: 198, 200–201, Fig 17, Fig 24, Fig 28 (n. sp., described, illustrated, comp. note) Ecuador

C. ventralis **Jacobi, 1907**

C. ventrilloni **Boulard, 1986**

Carineta ventrilloni Boulard 1986a: 416, 420–423, Plate I, Fig C, Figs 16–19 (n. sp., described, illustrated, comp. note) Venezuela

Carineta ventrilloni Boulard 1986c: 194 (comp. note)

Carineta ventrilloni Sanborn 2007a: 29 (distribution, comp. note) Venezuela

C. verna **Distant, 1883**

Carineta verna Wolda and Ramos 1992: 273, Table 17.2 (listed) Panama

C. viridicata **Distant, 1883**

Carineta viridicata Bartholomew and Barnhart 1984: 135 (energetics) Panama

Carineta viridicata Boulard 1986a: 419 (comp. note)

Carineta viridicata Wolda 1989: 438–439, Fig 1 (emergence pattern, comp. note) Panama

Carineta viridicata Salazar Escobar 2005: 193, 195, 200, Fig 3 (illustrated, comp. note) Colombia

Carineta viridicata Sanborn 2010a: 1600 (distribution, comp. note) Colombia, Peru, Panama, Bolivia

C. viridicollis viridicollis **(Germar, 1830)** = *Cicada viridicollis* = *Carineta viricollis* (sic) = *Carineta viridis* (sic)

Carineta viridicollis Boulard 1985e: 1021 (coloration, comp. note)

Carineta viridicollis Boulard 1986a: 415 (comp. note)

Carineta viricollis (sic) Boulard 1986a: 419 (comp. note)

Carineta viridis (sic) Boulard 2002b: 193 (coloration, comp. note)

Carineta viridicollis Sanborn 2007a: 29 (distribution, comp. note) Colombia, Venezuela, Brazil, French Guiana, Paraguay

Carineta viridicollis Sanborn 2010a: 1600 (synonymy, distribution, comp. note) Equals *Cicada viridicollis* Colombia, Venezuela, Brazil, French Guiana, Paraguay

C. viridicollis **var. *b* Stål, 1862**

Toulgoetalna **Boulard, 1982**

Toulgoetalna Boulard 1982e: 179–180 (n. gen., described) French Guiana

Toulgoetalna Boulard 1990b: 212 (comp. note)

T. tavakiliani **Boulard, 1982**

Toulgoetalna tavakiliani Boulard 1982e: 180–182, Figs 1–7 (n. sp., described, illustrated, comp. note) French Guiana

Toulgoetalna tavakiliani Boulard 1990b: 138, 150–151, Plate XVI, Fig 6 (sound system illustrated, comp. note) French Guiana

Toulgoetalna tavakaliani Boulard 2002b: 197 (sound system, comp. note)

Novemcella **Goding, 1925**

N. ecuatoriana **Goding, 1925**

Herrera **Distant, 1905**

Herrera Arnett 1985: 217 (diversity) Southeastern United States (error)

Herrera Poole, Garrison, McCafferty, Otte and Stark 1997: 251, 331 (listed) North America

Herrera Arnett 2000: 299 (diversity) Southeastern United States (error)

H. ancilla **(Stål, 1864)** = *Carineta ancilla* = *Cicada marginella* Walker, 1858 = *Carineta marginella* = *Herrera marginella*

Herrera ancilla Arnett 1985: 217 (distribution) Southeastern United States (error)

Herrera ancilla Wolda and Ramos 1992: 273, Table 17.2 (listed) Panama

Herrera ancilla Poole, Garrison, McCafferty, Otte and Stark 1997: 331 (listed) Equals *Carineta ancilla* Equals *Cicada marginella* North America

Herrera ancilla Arnett 2000: 299 (distribution) Southeastern United States (error)

Herrera ancilla Sanborn 2001b: 449 (listed, comp. note) El Salvador, Guatemala, Honduras, Nicaragua

Herrera ancilla Sueur 2002a: 380, 387 (comp. note) Mexico, Veracruz

Herrera ancilla Sanborn 2005b: 77–78, Fig 1 (listed, distribution) Honduras, Central America, Belize, Guatemala, El Salvador

Herrera ancilla Sanborn 2007b: 28, 37 (distribution, comp. note) Mexico, Oaxaca, Veracruz, Hidalgo, Tamaulipas, Chiapas, Central America, Belize, Guatemala, El Salvador, Honduras, Panama, Costa Rica

H. coyamensis **Sanborn, 2007**

Herrera coyamensis Sanborn 2007b: 26–30, 37, Figs 44–53 (n. sp., described, illustrated, distribution, comp. note) Mexico, Veracruz

Herrera coyamensis Sanborn 2009a: 87 (comp. note) Mexico

H. humilistrata **Sanborn & Heath, 2011, nom. nud.**

H. infuscata **Sanborn 2009**

Herrera infuscata Sanborn 2009a: 86–90, Figs 1–9 (n. sp., described, illustrated, listed, synonymy, key, comp. note) Cuba, Isle of Youth

H. laticapitata **Davis, 1938**

Herrera laticapitata Sanborn 1999a: 45 (type material, listed)

Herrera laticapitata Sanborn 2007b: 28, 37 (comp. note) Mexico, Chiapas

Herrera laticapitata Sanborn 2009a: 88 (comp. note) Mexico

***H. lugubrina compostelensis* Davis, 1938** = *Herrera lugubrina* var. *compostelensis* = *Herrera tugubrina* (sic) *compostelensis*

Herrera lugubrina var. *compostelensis* Sanborn 1999a: 45 (type material, listed)

Herrera lugubrina compostelensis Sanborn 2007b: 28, 37 (comp. note) Mexico, Nayarit, Sinoloa, Puebla

***H. lugubrina lugubrina* (Stål, 1864)** = *Carineta lugubrina*

Herrera lugubrina Sanborn 2007b: 28, 37 (comp. note) Mexico

Herrera lugubrina lugubrina Sanborn 2009a: 88 (comp. note) Mexico

***H.* nr. *lugubrina* Wolda & Ramos, 1992**

Herrera nr. *lugubrina* Wolda and Ramos 1992: 272, 276, Figs 17.5–17.6, Table 17.1 (distribution, listed, seasonality, comp. note) Panama

***H.* sp. F3 Wolda & Ramos, 1992**

Herrera sp. F3 Wolda and Ramos 1992: 272, Table 17.1 (distribution, listed) Panama

***H.* sp. H1 Wolda, 1989**

Herrera sp. H1 Wolda 1989: 438–440, Figs 1–3, Tables 1–2 (emergence pattern, comp. note) Panama

Herrera sp. H1 Wolda and Ramos 1992: 272–273, Fig 17.1, Table 17.1 (distribution, listed, seasonality, comp. note) Panama

***H.* sp. Q5 Wolda & Ramos, 1992**

Herrera sp. Q5 Wolda and Ramos 1992: 272, Table 17.1 (distribution, listed) Panama

***H. umbraphila* Sanborn & Heath, 2011, nom. nud.**

***Paranistria* Metcalf, 1952** = *Tympanistria* Stål, 1861 (nec *Tympanistria* Reichenbach, 1852)

P. fornicata (Linnaeus, 1758) to *Zouga fornicata*

***P. trichiosoma* (Walker, 1850)** = *Carineta trichiosoma* = *Tympanistria trichiosoma*

***P. villosa villosa* (Fabricius, 1781)** = *Tettigonia villosa* = *Cicada villosa* = *Carineta villosa* = *Tympanistria villosa*

***P. villosa* var. *a* (Stål, 1866)** = *Tympanistria villosa* var. *a*

***P. villosa* var. *b* (Stål, 1866)** = *Tympanistria villosa* var. *b*

***P. villosa* var. *c* (Stål, 1866)** = *Tympanistria villosa* var. *c*

***Guaranisaria* Distant, 1905**

***G. bicolor* Torres, 1958**

***G. dissimilis* Distant, 1905**

***G. llanoi* Torres, 1964**

Subtribe Kareniina Boulard, 2008

Kareniina Boulard 2008f: 7, 57 (n. subtribe, listed) Thailand

***Karenia* Distant, 1888** = *Sinosenia* Chen, 1943

Karenia Wu 1935: 25 (listed) China

Karenia Boulard 1988f: 64 (comp. note) Equals *Sinosena*

Karenia Duffels 1993: 1225 (comp. note) Equals *Sinosena*

Karenia Chou, Lei, Li, Lu and Yao 1997: 37, 293, 297, Table 6 (synonymy, described, listed, diversity, biogeography) Equals *Sinosena* China

Karenia Villet and Zhao 1999: 321 (comp. note) China

Karenia Moulds 2005b: 385, 388 (higher taxonomy, listed, comp. note)

Karenia Boulard 2007e: 71 (comp. note) Thailand

Karenia Boulard 2008f: 7, 57, 97 (listed) Thailand

Karenia Wei, Tang, He and Zhang 2009: 337–338, 340, 342 (synonymy, described,, listed, comp. note) Equals *Sinosena* Chna, Burma

Karenia spp. Wei, Tang, He and Zhang 2009: 338 (comp. note) Burma, China, Vietnam

Sinosena Wei, Tang, He and Zhang 2009: 337–338 (type genus of Sinosenini, listed, comp. note)

Karenia Pham and Yang 2010a: 133 (comp. note) Vietnam

Karenia Wei and Zhang 2010: 37–38 (comp. note)

Karenia Boulard 2010b: 336 (comp. note)

***K. caelatata* Distant, 1890** = *Karenia coelata* (sic) = *Karenia coelatata* (sic) = *Sinosena caelatata* = *Sinosena caelatta* (sic)

Karenia caelatata Wu 1935: 25 (listed) China, Chia Kou Ho

Karenia caelatata Boulard 1988f: 53 (type species of Sinosenini) Equals *Sinosenia caelatata*

Karenia caelatata Chou, Lei, Li, Lu and Yao 1997: 38–40, Fig 6–2, Plate II, Fig 29 (key, described, illustrated, comp. note) China

Karenia caelatata Hua 2000: 62 (listed, distribution) China, Shaanxii, Hubei, Fujian, Hunan, Guangxi, Sichuan

Karenia caelatata Boulard 2005d: 349 (comp. note) China

Karenia caelatata Boulard 2006g: 160–161 (comp. note) China

Karenia caelatata Boulard 2007e: 71, Plate 33 (comp. note) Thailand, Burma, China

Karenia caelatata Wei, Tang, He and Zhang 2009: 337–338, 342, Fig 1b, Fig 5 (type species of *Sinosena*, illustrated, key, distribution, comp. note) Equals *Sinosena caelatata* China

Sinosena caelatata Wei, Tang, He and Zhang 2009: 337 (synonymy, comp. note)

***K. chama* Wei & Zhang, 2009**

Karenia chama Wei, Tang, He and Zhang 2009: 337–342, Figs 2–3, Fig 5 (n. sp., described, illustrated, fungal host, habitat, distribution, comp. note) China, Yunnan

***K. hoanglienensis* Pham & Yang, 2012** = *Karenia* undescribed species Pham & Yang, 2010

Karenia undescribed species Pham and Yang 2010a: 133 (comp. note) Vietnam
Karenia hoanglienensis Pham and Yang 2012 (*Zootaxa* 3153: 32–38): 32–37, Figs 1–3, Fig 5E (n. sp., described, illustrated, key, distribution, comp. note) Vietnam

K. ravida **Distant, 1888** = *Sinosena ravida*
Karenia ravida Boulard 1990b: 178, 191, 193, Plate XXIV, Figs 5–6 (sound system, comp. note) China
Karenia ravida Chou, Lei, Li, Lu and Yao 1997: 37–39, 365, Fig 6-1, Plate II, Fig 22 (type species of *Karenia*, key, described, illustrated, comp. note) China
Karenia ravida Hua 2000: 62 (listed, distribution) China, Yunnan, Burma
Karenia ravida Boulard 2002b: 197, 205 (sound system, comp. note) China
Karenia ravida Boulard 2005d: 349 (comp. note) China
Karenia ravida Boulard 2006g: 160–161 (comp. note) China
Karenia ravida Boulard 2007e: 71 (comp. note) Thailand
Karenia ravida Boulard 2008f: 57, 97 (type species of *Karenia*, biogeography, listed, comp. note) Thailand, Burma
Karenia ravida Wei, Tang, He and Zhang 2009: 337–338, 342, Fig 1a, Fig 5 (type species of *Karenia*, illustrated, key, distribution, comp. note) Burma, China, Yunnan

K. sulcata **Lei & Chou, 1997**
Karenia sulcata Chou, Lei, Li, Lu and Yao 1997: 38, 40–41, 357, 365, Fig 6-3 (key, described, illustrated, comp. note) China
Karenia sulcata Wei, Tang, He and Zhang 2009: 337–338, 342, Fig 1c, Fig 5 (illustrated, key, distribution, comp. note) China, Yunnan

K. undescribed species Pham & Yang, 2010 to *Karenia hoanglienensis* Pham & Yang, 2012

Tribe Parnisini Distant, 1905 = Parnisiini (sic) = Parnesini (sic)
Parnisini Heath 1983: 14 (distribution, comp. note) South America, Africa, India, Australia
Parnisini Boulard 1990a: 237 (characteristics)
Parnisini Moulds 1990: 31 (listed)
Parnisiini (sic) Boer and Duffels 1996b: 302 (comp. note)
Parnisini Moulds 1996a: 19–20 (listed)
Parnisini Chou, Lei, Li, Lu and Yao 1997: 46, 86 (key, synonymy, listed, described) Equals Parnisaria
Parnisini Novotny and Wilson 1997: 437 (listed)
Parnisini Boulard 1999a: 80, 110, 112 (listed)
Parnisini Moulds 1999: 1 (comp. note)
Parnisini Sanborn 1999a: 51 (listed)
Parnisini Villet and Noort 1999: 226, Table 12.1 (diversity, listed) South Africa, Zimbabwe, Namibia
Parnisini Sueur 2001: 42, Table 1 (listed)
Parnisini Moulds and Carver 2002: 46 (listed)
Parnisini Jong and Boer 2004: 267 (comp. note)
Parnisini Sanborn, Heath, Heath, Noriega and Phillips 2004: 282 (comp. note)
Parnisini Moulds 2005b: 377, 392, 396, 412, 419, 421, 423–424, 427, 430, 437, Tables 1–2, Fig 58, Fig 61 (described, phylogeny, distribution, listed, comp. note)
Parnisini Chawanji, Hodson and Villet 2006: 374, 376, Table 1 (listed, comp. note)
Parnisini Sanborn 2007a: 1600 (listed) Venezuela
Parnesini (sic) Schedl 2007: 155 (listed)
Parnisini Shiyake 2007: 7, 102 (listed, comp. note)
Parnisini Mozaffarian and Sanborn 2010: 81 (listed)
Parnisini Sanborn 2010a: 1600 (listed) Colombia
Parnisini Sanborn 2010b: 75 (listed)
Parnisini Wei, Ahmed and Rizvi 2010: 33 (comp. note)

Quintilia **Stål, 1866** = *Tibicen (Quintilia)* = *Quintillia* (sic)
Quintilia Boulard 1990a: 238 (comp. note)
Quintillia (sic) Moulds 1990: 7 (comp. note)
Quintilia Moulds 1990: 12, 32, 111 (listed, comp. note) Australia, Africa
Quintilia spp. Dean and Milton 1991: 118 (comp. note) South Africa
Quintilia Moulds 1996a: 19 (comp. note) Australia
Quintilia Novotny and Wilson 1997: 437 (listed)
Quintilia sp. Dean, Milton, du Plessis and Siegfried 2001: 179–180, Fig 6 (density, comp. note) South Africa
Quintilia spp. Dean and Milton 2001: 191 (density, comp. note) South Africa
Quintilia Moulds and Carver 2002: 49 (type data, listed) Australia
Quintilia Picker, Griffith and Weaving 2002: 156–157 (described, illustrated, habitat, comp. note) South Africa
Quintilia Malherbe, Burger and Stephen 2004: 87 (comp. note) South Africa
Quintilia Moulds 2005b: 392, 424, 430, Table 2 (higher taxonomy, listed, comp. note) Australia
Quintilia Goble, Price, Barker, and Villet 2008b: 674 (phylogeny, comp. note) Africa

Q. annulivena **(Walker, 1850)** = *Cicada annulivena* = *Tibicen annulivena*

Q. aurora **(Walker, 1850)** = *Cicada aurora* = *Tibicen aurora* = *Tibicen (Quintilia) sanguinarius* Stål, 1866 = *Quintilia sanguinaria*

Q. carinata **(Thunberg, 1822)** = *Quintilia carinatus* (sic) = *Tettigonia carinata* = *Tibicen carinata* = *Tibicen*

(Quintilia) carinatus = *Cicada tristis* Germar, 1834 = *Cicada nigroviridis* Walker, 1852 = *Tibicen nigroviridis*
Quintilia carinatus (sic) Boulard 1990a: 238 (comp. note)
Quintilia carinata Goble, Price, Barker, and Villet 2008b: 674 (phylogeny, comp. note) Africa

***Q. catena* (Fabricius, 1775)** = *Tettigonia catena* = *Cicada catena* = *Tibicen catenus* = *Cicada catenata* (sic) = *Cicada nebulosa* Olivier, 1790 = *Cicada hyalina* Olivier, 1790 = *Tibicen hyalina* = *Cicada garrula* Olivier, 1790 = *Tibicen garrulus* = *Tettigonia crenata* Thunberg, 1822

***Q. diversa* (Walker, 1850)** = *Cicada diversa*

***Q. dorsalis* (Thunberg, 1822)** = *Tettigonia dorsalis* = *Tibicen dorsalis* = *Cicada maculipennis serricosta* Walker, 1850 nom. nud. = *Cicada serricosta* Germar nom. nud. = *Cicada serricostata* (sic)

Q. infans (Walker, 1850) to *Terepsalta infans*

***Q. maculiventris* Distant, 1905**

***Q. monilifera* (Walker, 1850)** = *Cicada monilifera* = *Tibicen monilifera* = *Tibicen (Quintilia) monolifera* = *Tibicen (Quintilia) maculinervis*

***Q. musca* (Olivier, 1790)** = *Cicada musca* = *Tettigomyia musca* = *Tibicen musca* = *Cicada tachinaria* Germar, 1830

***Q. obliqua* (Walker, 1850)** = *Cicada obliqua* = *Tibicen obliquus*

***Q. pallidiventris* (Stål, 1866)** = *Tibicen (Quintilia) pallidiventris*

***Q. peregrina peregrina* (Linnaeus, 1764)** = *Cicada peregrina* = *Tibicen peregrina* = *Tibicen (Quintilia) peregrinus* (sic) = *Cicada maculipennis* Germar, 1834

***Q. peregrina* var. *a* (Stål, 1866)** = *Tibicen (Quintilia) peregrinus* var. *a*

***Q.* cf. *peregrina* Dean, 1992**
Quintilia cf. *peregrina* Dean 1992: 179 (comp. note) South Africa
Quintilia cf. *peregrina* Dean 1993: 42 (predation) South Africa

***Q. pomponia* Distant, 1912** = *Lycurgus pomponia* = *Lycurgus pomponius*

***Q. primitiva* (Walker, 1850)** = *Cicada primitiva* = *Tibicen primitiva* = *Tibicen (Quintilia) primitiva* = *Tibicen (Quintilia) haematinus* Stål, 1866 = *Quintilia haematinus*

***Q. rufiventris* (Walker, 1850)** = *Cicada rufiventris* = *Cicada holmgreni* Stål, 1856 = *Tibicen holmgreni* = *Tibicen (Quintilia) holmgreni*
Quintilia rufiventris Villet 1999c: 158–159 (comp. note)
Cicada rufiventris Moulds 1990: 111 (type species of *Quintilia*, listed)
Cicada rufiventris Moulds and Carver 2002: 49 (type species of *Quintilia*, listed) Equals *Quintilia holmgreni*
Quintillia holmgreni Moulds and Carver 2002: 49 (listed)
Quintilia rufiventris Goble, Price, Barker, and Villet 2008b: 674 (phylogeny, comp. note) Africa

***Q. semipunctata* Karsch, 1890** = *Cicada semipunctata* Walker, 1850 nom. nud. = *Cicada semipunctata* Dohrn, 1859 nom. nud. = *Melampsalta semipunctata*

***Q.* sp. A Phillips, Sanborn & Villet, 2002**
Quintilia sp. A Phillips, Sanborn and Villet 2002: 31 (thermal responses) South Africa, Eastern Cape

***Q.* sp. B Phillips, Sanborn & Villet, 2002**
Quintilia sp. B Phillips, Sanborn and Villet 2002: 31 (thermal responses) South Africa, Eastern Cape

***Q. umbrosa* (Stål, 1866)** = *Tibicen (Quintilia) umbrosus* = *Tibicen umbrosus* = *Cicada diaphana* Germar, 1830 = *Cicada maculipennis diaphana* = *Tibicen diaphanus* = *Tibicen (Quintilia) peregrinus* var. *b* Stål, 1866

***Q. vitripennis* Karsch, 1890** = *Tibicen (Quintilia) vitripennis* = *Lycurgus vitripennis*

***Q.* cf. *vitripennis* Dean, 1992**
Quintilia cf. *vitripennis* Dean 1992: 179 (comp. note) South Africa
Quintilia cf. *vitripennis* Dean and Milton 1992: 71–74, Fig 1, Table 1 (emergence, density, comp. note) South Africa
Quintilia cf. *vitripennis* Milton and Dean 1992: 288–290 (density, comp. note) South Africa

***Q. vittativentris* (Stål, 1866)** = *Tibicen (Quintilia) vittativentris*

***Q. wallkeri* China, 1925**
Quintilia walkeri Goble, Price, Barker, and Villet 2008b: 674 (phylogeny, comp. note) Africa

***Q. wealei* (Distant, 1892)** = *Tibicen (Quintilia) wealei*
Quintilia wealei Gess 1981: 4, 8–10, Fig 3 (described, song, sonogram, habitat, comp. note) South Africa
Quintilia wealei Phillips, Sanborn and Villet 2002: 31 (thermal responses) South Africa, Eastern Cape
Quintilia wealei Sanborn, Phillips and Villet 2003a: 349, Table 1 (thermal responses, habitat, comp. note) South Africa
Quintilia wealei Chawanji, Hodson and Villet 2006: 374–380, 383, 386, Fig 2, Fig 8, Fig 11, Figs 15–16, Figs 18–38, Tables 1–6 (listed, comp. note) South Africa

***Rhinopsaltria* Melichar, 1908**

***R. sicardi* Melichar, 1908**
Rhinopsalta sicardi Boulard 1999a: 80, 111 (listed, sonogram) Madagascar
Rhinopsalta sicardi Sueur 2001: 42, Table 1 (listed)
Rhinopsaltria sicardi Boulard 2006g: 181 (listed)

***Jafuna* Distant, 1912**

***J. melichari* Distant, 1912**

Malgotilia **Boulard, 1980**

Malgotilia Moulds 2005b: 392 (higher taxonomy, listed, comp. note)

M. sogai **Boulard, 1980**

Lycurgus **China, 1925**

Lycurgus Wu 1935: 18 (listed) China

?*Lycurgus* sp. Chaudhry, Chaudry and Khan 1966: 94 (listed, comp. note) Pakistan

Lycurgus Chou, Lei, Li, Lu and Yao 1997: 87, 293, Table 6 (key, described, listed, diversity, biogeography) China

Lycurgus Moulds 2005b: 392 (higher taxonomy, listed, comp. note)

L. conspersa **(Karsch, 1890)** = *Quintilia conspersa*

L. **cf.** ***conspersa*** **Dean & Milton, 1991**

Quintilia cf. *conspersa* Dean and Milton 1991: 112–117, Table 1 (emergence, density, comp. note) South Africa

Quintilia cf. *conspersa* Dean 1992: 179 (comp. note) South Africa

Quintilia cf. *conspersa* Dean and Milton 1992: 74 (comp. note) South Africa

Quintilia cf. *conspersa* Milton and Dean 1992: 288–291 (density, comp. note) South Africa

Quintilia cf. *conspersa* Dean 1993: 42 (predation) South Africa

Quintilia cf. *conspersa* Ewart 2001b: 83, Table 3 (population density, comp. note) South Africa

Quintilia cf. *conspersa* Humphreys 2005: 73–74, Table 4 (population density, comp. note) South Africa

L. frontalis **(Karsch, 1890)** = *Quintilia frontalis*

L. sinensis **Jacobi, 1944**

Lycurgus sinensis Chou, Lei, Li, Lu and Yao 1997: 87–88 (key, described, comp. note) China

Lycurgus sinensis Hua 2000: 62 (listed, distribution) China, Fujian, Zhejiang

Lycurgus sinensis Lampe, Rohwedder and Rach 2006: 28 (type specimens) China, Fujian

L. subvittus **(Walker, 1850)** = *Cicada subvitta* = *Tibicen subvitta* = *Tibicen subvittatus* (sic) = *Quintilia subvittatus* (sic) = *Quintilia subvitta* = *Lycurgus subvitta* (sic) = *Cicada strigosa* Walker, 1858

Lycurgus subvitta (sic) Wu 1935: 19 (listed) Equals *Cicada subvitta* Equals *Quintilia subvitta* Equals *Tibicen subvitta* Equals *Cicada strigosa* Equals *Tibicen subvittatus* China, Mussooree, Himalya, Sikhim

Cicada subvitta Chou, Lei, Li, Lu and Yao 1997: 87 (type species of *Lycurgus*)

Lycurgus subvittus Chou, Lei, Li, Lu and Yao 1997: 87–88 (key, described, synonymy, comp. note) Equals *Cicada subvitta* Equals *Quintilia subvitta* China

Lycurgus subvitta (sic) Hua 2000: 62 (listed, distribution) Equals *Cicada subvitta* China, Himalaya, India, Sikkim

Psilotympana **Stål, 1861**

Psilotympana Moulds 2005b: 392 (higher taxonomy, listed, comp. note)

P. ruficollis **(Thunberg, 1822)** = *Tettigonia ruficollis* = *Cicada ruficollis*

P. signifera **(Germar, 1830)** = *Cicada signifera* = *Cicada elana* Walker, 1850 = *Cicada brevipennis* Schaum nom. nud.

P. varicolor **Distant, 1920**

P. walkeri **Metcalf, 1963** = *Cicada gracilis* Walker, 1850 (nec *Cicada gracilis* Schellenberg, 1802) = *Psilotympana gracilis*

Abagazara **Distant, 1905**

Abagazara Moulds 2005b: 392 (higher taxonomy, listed, comp. note)

A. bicolorata **(Distant, 1892)** = *Callipsaltria bicolorata*

A. omaruruensis **Hesse, 1925**

Henicotettix **Stål, 1858**

Henicotettix Moulds 2005b: 392 (higher taxonomy, listed, comp. note)

H. hageni **Stål, 1858**

Henicotettix hageni Dean 1993: 42 (comp. note) South Africa

Koranna **Distant, 1905** = *Korania* (sic)

Koranna Goble, Price, Barker, and Villet 2008b: 674 (phylogeny, comp. note) Africa

K. analis **Distant, 1905**

Masupha **Distant, 1892**

Masupha Moulds 2005b: 392 (higher taxonomy, listed, comp. note)

Masupha Goble, Price, Barker, and Villet 2008b: 674 (phylogeny, comp. note) Africa

M. ampliata **Distant, 1892**

M. delicata **Distant, 1892**

M. dregei **Distant, 1892**

Derotettix **Berg, 1882**

Derotettix Heath 1983: 13–14 (comp. note) South America

Derotettix Heath, Heath, Sanborn and Noriega 1988: 304 (thermal responses, comp. note) Argentina

Derotettix Heath, Heath, Sanborn and Noriega 1999: 1 (thermal responses, comp. note) Argentina

Derotettix Sanborn, Heath, Heath, Noriega and Phillips 2004: 282, 287 (comp. note)

Derotettix spp. Sanborn, Heath, Heath, Noriega and Phillips 2004: 286–287 (comp. note)

Derotettix Moulds 2005b: 392 (higher taxonomy, listed, comp. note)
Derotettix Wei, Ahmed and Rizvi 2010: 33 (comp. note)

***D. mendosensis* Berg, 1882** = *Derotettix mendocensis* (sic)
Derotettix mendocensis (sic) Heath, Heath, Sanborn and Noriega 1999: 1 (thermal responses, comp. note) Argentina
Derotettix mendosensis Sanborn, Heath, Heath, Noriega and Phillips 2004: 282–286, Figs 1–2, Tables 1–4 (thermal responses, habitat, distribution, comp. note) Argentina, La Rioja, Rio Negro San Juan

D. proseni Torres, 1945 to *Derotettix wagneri* Distant, 1905

***D. wagneri* Distant, 1905** = *Derotettix proseni* Torres, 1945
Derotettix wagneri Sanborn, Heath, Heath, Noriega and Phillips 2004: 282–285, Fig 1, Tables 1–4 (thermal responses, habitat, distribution, comp. note) Argentina, Santiago del Estero

***Acyroneura* Torres, 1958**
Acyroneura Moulds 2005b: 392 (higher taxonomy, listed, comp. note)

***A. singularis* Torres, 1958**
Acyroneura singularis Ruffinelli 1970: 3 (distribution, comp. note) Uruguay, Argentina, Corrientes, Entre Rios

***Taipinga* Distant, 1905**
Taipinga Wu 1935: 28 (listed) China
Taipinga sp. Theron 1985: 156–157, Fig 16.120–16.121 (illustrated) South Africa
Taipinga Chou, Lei, Li, Lu and Yao 1997: 87–89, 293, Table 6 (key, described, listed, diversity, biogeography) China
Taipinga Moulds 2005b: 392 (higher taxonomy, listed, comp. note)

***T. albivenosa* (Walker, 1858)** = *Cicada albivenosa* = *Psilotympana albivenosa*

***T. consobrina* Distant, 1906**

***T. fuscata* Distant, 1906**

***T. fusiformis* (Walker, 1850)** = *Cicada fusiformis* = *Psilotympana fusiformis*

***T. infuscata* (Distant, 1892)** = *Psilotympana infuscata*

***T. luctuosa* (Stål, 1855)** = *Cicada luctuosa* = *Tibicen luctuosus* = *Tibicen (Quintilia) luctuosus*

***T. nana* (Walker, 1850)** = *Cicada nana* = *Tibicen nanus* = *Tibicen nana*
Taipinga nana Wu 1935: 28 (listed) Equals *Cicada nana* Equals *Tibicen nana* China
Cicada nana Chou, Lei, Li, Lu and Yao 1997: 88 (type species of *Taipinga*) China
Taipinga nana Chou, Lei, Li, Lu and Yao 1997: 89 (described, synonymy, comp. note) Equals *Cicada nana* China
Taipinga nana Hua 2000: 64 (listed, distribution) Equals *Cicada nana* China, Guanxi

***T. nigricans* (Stål, 1855)** = *Cicada nigricans* = *Tibicen nigricans* = *Tibicen (Quintilia) nigricans*
Taipinga nigricans Sueur 2002b: 122–123, 129 (listed, comp. note) Equals *Tibicen nigricans*

***T. rhodesi* Distant, 1920**

***T. undulata* (Thunberg, 1822)** = *Tettigonia undulata* = *Tibicen (Quintilia) undulates* = *Quintilia undulata* = *Tibicen undulates* = *Cicada scripta* Germar, 1834 = *Tibicen scriptus*

***Zouga* Distant, 1906**
Zouga Boulard 1990b: 116, 141 (comp. note)
Zouga Boulard 1994d: 457, 466 (comp. note) North Africa
Zouga Boulard 2002b: 198 (comp. note) North Africa, South Africa
Zouga Moulds 2005b: 392, 421 (higher taxonomy, listed, comp. note) Morocco
Zouga Goble, Price, Barker, and Villet 2008b: 674 (phylogeny, comp. note) Africa

***Z. apiana* Hesse, 1925**
Zouga apiana Boulard 2002b: 198 (comp. note)

***Z. beaumonti* Boulard, 1994**
Zouga beaumonti Boulard 1994d: 458–462, 466, Figs 1–11 (n. sp., described, illustrated, comp. note) Morocco
Zouga beaumonti Boulard 2002b: 198–199, 202, Fig 55 (sound system, illustrated, comp. note)

***Z. delalandei* Distant, 1914**

***Z. festiva* Distant, 1920**

***Z. fornicata* (Linnaeus, 1758)** = *Cicada fornicata* = *Tettigonia fornicata* = *Tympanistria fornicata* = *Paranistria fornicata* = *Cicada fornicata* Olivier, 1790 = *Cicada dimidiata* Olivier, 1790 = *Tettigonia dimidiata* = *Carineta leuconeura* Walker, 1850 = *Tympanistria leuconeura*
Tettigonia dimidiata Boulard 1981b: 42 (comp. note)
Zouga fornicata Boulard 1990b: 146, 150–151, Plate XVI, Fig 3(n. comb., sound system, comp. note) South Africa
Zouga fornicata Boulard 2002b: 199 (comp. note)

***Z. heterochroma* Boulard, 1994**
Zouga heterochroma Boulard 1994d: 462–465, Figs 12–20 (n. sp., described, illustrated, comp. note) Morocco

***Z. hottentota* Distant, 1914**

***Z. peringueyi* (Distant, 1920)** = *Neomuda peringueyi*
Zouga peringueyi Boulard 1990b: 141 (n. comb., comp. note) Equals *Neomuda peringeyi*
Neomuda peringueyi Boulard 1990b: 155, 157, Plate XVIII, Fig 6 (sound system, comp. note) South Africa
Zouga peringeyi Boulard 2002b: 198 (comp. note)

***Z. typica* Distant, 1906**

Luangwana **Distant, 1914**
L. capitata **Distant, 1914**

Calopsaltria **Stål, 1861** = *Callipsaltria* Stål, 1866
Calopsaltria Moulds 2005b: 392 (higher taxonomy, listed, comp. note)
C. convexa **Haupt, 1918** = *Callipsaltria* (sic) *convexa*
C. hottentota **Kirkaldy, 1909** = *Cicada elongata* Stål, 1855 (nec *Cicada elongata* Fabricius, 1781) = *Calopsaltria elongata* = *Callipsaltria elongata*
C. longula **(Stål, 1855)** = *Cicada longula* = *Callipsaltria longula*
Callipsaltria longula Poulton 1907: 409 (predation) Transvaal
C. nigra **(Karsch, 1890)** = *Callipsaltria nigra*

Kageralna **Boulard, 2012**
K. dargei **Boulard, 2012**

Calyria **Stål, 1862** = *Cicada (Calyria)*
Calyria Dallas 1870: 482 (listed)
Calyria Moulds 2005b: 392 (higher taxonomy, listed, comp. note)
Calyria Sanborn 2010a: 1600 (listed) Colombia
C. cuna **(Walker, 1850)** = *Cicada cuna* = *Cicada (Calyria) blanda* Stål, 1862
Cicada (Calyria) blanda Sanborn 2010a: 1600 (type species of *Calyria*)
C. fenestrata **(Fabricius, 1803)** = *Tettigonia fenestrata* = *Cicada fenestrata*
Tettigonia fenestrata Dallas 1870: 482 (synonymy) Equals *Calyria fenestrata*
C. jacobii **Bergroth, 1914**
Calyria jacobii Sanborn 2007a: 30 (distribution, comp. note) Venezuela
C. mogannioides **Jacobi, 1907** = *Calyria mogannioides* (sic)
Calyria mogannioides (sic) Sanborn 2010a: 1600 (listed) Colombia
Calyria mogannoides Sanborn 2010a: 1600–1601 (synonymy, distribution, comp. note) Colombia
C. stigma **(Walker, 1850)** = *Cicada stigma*
Cicada stigma Boer 1981c: 129 (comp. note) Brazil
C. telifera **(Walker, 1858)** = *Cicada telifera* = *Calyria cuna* (nec Walker) = *Calyira* (sic) *telifera*
Calyria telifera Pogue 1996: 317 (listed) Peru
Calyria telifera Sanborn 2007a: 30 (listed, distribution) Costa Rica, Venezuela, Brazil, Peru, Central America
Calyria telifera Sanborn 2010b: 75–76 (listed, distribution) Honduras, Costa Rica, Venezuela, Brazil, Peru, Central America

Parnisa **Stål, 1862** = *Cicada (Parnisa)* = *Pamisa* (sic) = *Parnissa* (sic)
Parnisa Boulard 1981b: 48 (comp. note)
Parnisa Strümpel 1983: 17 (listed)
Parnisa Moulds 2005b: 392, 422, 424 (type genus of Parnisini, higher taxonomy, listed, comp. note)
Parnisa Salazar Escobar 2005: 194 (comp. note) Colombia
Parnisa Sanborn 2010a: 1601 (listed) Colombia
P. angulata **Uhler, 1903** = *Pamisa* (sic) *angulata*
Pamisa (sic) *angulata* Henshaw 1903: 230 (listed) Brazil
P. demittens **(Walker, 1858)** = *Cicada demittens*
P. designata **(Walker, 1858)** = *Cicada designata* = *Cicada casta* Stål, 1854 = *Cicada (Parnisa) casta* = *Parnisa casta*
Parnisa designata Moulds 2005b: 396, 399, 417–419, 421, 424, Table 1, Figs 56–59, Fig 61 (phylogeny, comp. note)
P. fraudulenta **(Stål, 1862)** = *Cicada (Parnisa) fraudulenta* = *Cicada fraudulenta*
P. haemorrhagica **Jacobi, 1904**
P. lineaviridia **Sanborn & Heath, 2011, nom. nud.**
P. moneta **(Germar, 1834)** = *Cicada moneta* = *Melampsalta moneta* = *Cicada tricolor* Walker, 1850 = *Parnisa tricolor*
Parnisa moneta Sanborn 2007a: 30 (distribution, comp. note) Colombia, Venezuela, Brazil
Parnisa moneta Sanborn 2010a: 1601 (synonymy, distribution, comp. note) Equals *Cicada moneta* Equals *Cicada tricolor* Equals *Parnisa tricolor* Equals *Melampsalta moneta* Colombia, Venezuela, Brazil
P. proponens proponens **(Walker, 1858)** = *Cicada proponens* = *Cicada (Parnisa) biplagiata* Stål, 1862
Cicada propones (sic) Moulds 2005b: 392 (type species of *Parnisa*)
Cicada proponens Sanborn 2010a: 1601 (type species of *Parnisa*) Brazil
P. proponens **var. *β*** **(Walker, 1858)** = *Cicada proponens* var. *β*
P. protracta **Uhler, 1903** = *Pamisa* (sic) *protracta*
Pamisa (sic) *protracta* Henshaw 1903: 230 (listed) Brazil
P. viridis **Sanborn & Heath, 2011, nom. nud.**

Mapondera **Distant, 1905**
Mapondera Moulds 2005b: 392 (higher taxonomy, listed, comp. note)
M. capicola **Kirkaldy, 1909** = *Cicada pulchella* Stål, 1855 (nec *Cicada pulchella* Westwood, 1845) = *Psilotympana pulchella* = *Mapondera pulchella*
M. hottentota **Kirkaldy, 1909** = *Cicada abdominalis* Stål, 1855 (nec *Cicada abdominalis* Donovan, 1798) = *Tibicen abdominalis* = *Tibicen (Quintilia) abdominalis* = *Mapondera abdominalis*

***Bijaurana* Distant, 1912**
Bijaurana Moulds 2005b: 392 (higher taxonomy, listed, comp. note)
***B. sita* Distant, 1912**
***B. typica* Distant, 1912**

***Adeniana* Distant, 1905** = *Adenia* Distant, 1905 (nec *Adenia* Robineau-Desvoidy, 1863) = *Hymenogaster* Horváth, 1911
Adenia (sic) Koçak 1982: 148 (synonymy, listed, comp. note) Turkey
Adeniana Koçak 1982: 148 (synonymy, listed, comp. note) Equals *Adenia* Distant (nec *Adenia* Robinwau-Desvoidy, 1863) Turkey
Hymenogaster Koçak 1982: 149 (listed, comp. note) Turkey
Adeniana Boulard 1990b: 139, 141 (comp. note)
Adeniana Boulard 1994d: 457, 466 (comp. note)
Adeniana spp. Schedl 1999b: 830 (comp. note)
Adeniana Boulard 2002b: 198 (comp. note)
Adeniana Moulds 2005b: 392 (higher taxonomy, listed, comp. note)
Adeniana spp. Schedl 2007: 157 (comp. note)
Adeniana Mozaffarian and Sanborn 2010: 81 (distribution, listed) North Africa, Russia, Middle East
***A. decolorata* Korsakoff, 1947**
***A. korsakoffi* Villiers, 1943**
***A. kovacsi* (Horváth, 1911)** = *Hymenogaster kovacsi* = *Zouga kovasci* (sic)
Adeniana kovácsi Schedl 1999b: 830 (comp. note)
***A. leouffrei* Korsakoff, 1947**
***A. licenti* Villiers, 1943**
***A. longiceps* (Puton, 1887)** = *Cicadatra longiceps* = *Hymenogaster longiceps* = *Zouga longiceps* = *Hymenogaster planiceps* Horváth, 1917 = *Adeniana planiceps* = *Hymenogaster tabia* Horváth, 1911 = *Zouga tabia* = *Adeniana tabia*
Cicada longiceps Koçak 1982: 149 (type species of *Hymenogaster*)
Adeniana longiceps Schedl 1993: 799–800, Figs 5–6 (illustrated, listed, synonymy, distribution, comp. note) Equals *Cicadatra longiceps* Equals *Hymenogaster tabida* Equals *Hymenogaster planiceps* Egypt
Adeniana planiceps Sanborn 1999a: 51 (type material, listed)
Adeniana longiceps Schedl 1999b: 828, 830 (listed, distribution, ecology, comp. note) Israel, Algeria, Tunisia, Egypt, Azerbaijan
Adeniana longiceps Schedl 2004a: 12 (comp. note) North Africa, Israel
Adeniana longiceps Mozaffarian 2008: 9 (distribution, comp. note) Iran, Algeria, Egypt, Tunisia, Israel, Azerbaijan
Adeniana longiceps Mozaffarian and Sanborn 2010: 81 (synonymy, listed, distribution) Equals *Cicadatra longiceps* Equals *Hymenogaster planiceps* Iran, Tunisia, North Africa, Russian Armenia, Transcaucasia, Caucasus, Israel, Russia
***A. mairei* (de Bergevin, 1917)** = *Hymenogaster mairei* = *Adeniana bourlieri* de Bergevin, 1931
***A. maroccana* Villiers, 1943**
Adeniana maroccana Boulard 1994d: 462 (comp. note)
***A. minuta* Villiers, 1943**
***A. nigerrima* Villiers, 1943**
***A. nigricans* Distant, 1914**
***A. obokensis* Distant, 1914**
Adeniana obokensis Schedl 1999b: 830 (comp. note)
Adeniana obokensis Schedl 2004a: 12 (comp. note)
***A. robusta* Villiers, 1943**
Adeniana robusta Boulard 1994d: 458 (comp. note)
***A. safadi* Schedl, 2007**
Adeniana safadi Schedl 2007: 155–157, Figs 3–7, Plate 3 (n. sp., described, illustrated, ecology, distribution, listed, comp. note) United Arab Emirates
***A. seitzi* Distant, 1914**
***A. yerburyi* Distant, 1905**
Adenia yerburyi Koçak 1982: 148 (type species of *Adenia* Distant, type species of *Adeniana*)
Adeniana yerburgi Schedl 1999b: 830 (comp. note)
Adeniana yerburyi Schedl 2004a: 12–14, Fig 4 (illustrated, listed, distribution, comp. note) Yemen, Arabia, North Africa
Adenia yerburyi Mozaffarian and Sanborn 2010: 81 (type species of *Adeniana*)
***A. yotvataensis* Schedl, 1999**
Adeniana yotvataensis Schedl 1999b: 828–830, 837, Figs 1–2, Fig 8 (n. sp., described, illustrated, listed, distribution, ecology, comp. note) Israel
Adeniana yotvataensis Schedl 2004a: 12 (comp. note) Israel, Saudi Arabia, Oman

***Arcystasia* Distant, 1882** = *Arcystacea* (sic)
Arcystasia Boer and Duffels 1996b: 302 (comp. note)
Arcystasia Moulds 2005b: 392 (higher taxonomy, listed, comp. note)
***A. godeffroyi* Distant, 1882** = *Arcystasia doddefroyi* (sic)
Arcystasia godeffroyi Boer and Duffels 1996b: 310 (comp. note) Caroline Islands
Arcystasia godeffroyi Boulard 1990b: 138 (comp. note)

***Prunasis* Stål, 1862** = *Cicada (Prunasis)*
Prunasis Boulard 1990b: 144 (comp. note)
Prunasis Boer 1995c: 205 (comp. note) South America

Prunasis Boulard 2002b: 198 (comp. note) South America
Prunasis Moulds 2005b: 392 (higher taxonomy, listed, comp. note)

***P. pulcherrima pulcherrima* (Stål, 1854)** = *Cicada pulcherrima* = *Cicada (Prunasis) pulcherrima* = *Cicada viridula* Walker, 1850 = *Prunasis viridula*
Prunasis pulcherrima Boulard 1990b: 139 (comp. note)

***P. pulcherrima* var. *a* (Stål, 1862)** = *Cicada (Prunasis) pulcherrima* var. *a* Stål, 1862

***P. pulcherrima* var. *b* (Stål, 1862)** = *Cicada (Prunasis) pulcherrima* var. *b* Stål, 1862

***P. pulcherrima* var. *c* (Stål, 1862)** = *Cicada (Prunasis) pulcherrima* var. *c* Stål, 1862

***Crassisternalna* Boulard, 1980**

***C. pauliani* Boulard, 1980**
Crassisternalna pauliani Boulard 1990b: 155, 157, Plate XVIII, Fig 4 (sound system, comp. note) Madagascar

Tribe Taphurini Distant, 1905 = Taphusinini (sic) = Eaphurarini (sic)
Taphurini Medler 1980: 73 (listed) Nigeria
Taphurini Ramos 1983: 66 (listed) Dominican Republic
Taphurini Ramos and Wolda 1985: 177 (listed) Panama
Taphurini Moulds 1986: 39 (comp. note) Australia
Taphurini Boulard 1988c: 62 (comp. note)
Taphurini Boulard 1988f: 64 (comp. note)
Taphurini Boulard 1990b: 209, Fig 12 (phylogeny, comp. note)
Taphurini Moulds 1990: 31, 180 (listed, comp. note) Australia
Taphurini Boulard 1991e: 259 (comp. note)
Taphurini Boulard 1991g: 74 (comp. note)
Taphurini Boulard 1992a: 119 (listed)
Taphurini Boulard 1993d: 91 (comp. note)
Taphurini Boer 1995c: 202, 204–205 (comp. note)
Taphurini Boer and Duffels 1996b: 308, 328 (distribution, comp. note)
Taphurini Boulard 1996d: 115, 148 (listed)
Taphurini Boer 1997: 202, 204–205 (comp. note)
Taphurini Boulard 1997a: 180 (comp. note)
Taphurini Chou, Lei, Li, Lu and Yao 1997: 46, 80 (key, listed, synonymy, described) Equals Taphuraria Equals Lemuriaria
Taphurini Novotny and Wilson 1997: 437 (listed)
Taphusinini (sic) Novotny and Wilson 1997: 437 (listed)
Taphurini Boulard 1998: 83 (comp. note)
Taphurini Boer 1999: 128 (comp. note)
Taphurini Boulard 1999a: 79, 103, 112 (listed)
Taphurini Moulds 1999: 1 (comp. note)
Taphurini Sanborn 1999a: 51 (listed)
Taphurini Villet 1999c: 159 (comp. note)
Taphurini Villet and Noort 1999: 226, Table 12.1 (diversity, listed) Tanzania, Kenya, Uganda, South Africa, Zimbabwe, Namibia, Angola
Taphurini Sueur 2001: 42, Table 1 (listed)
Taphurini Boulard 2002b: 205 (comp. note)
Taphurini Moulds and Carver 2002: 50 (listed) Australia
Taphurini Moulds 2003b: 245 (comp. note)
Taphurini Jong and Boer 2004: 267 (comp. note)
Taphurini Chawanji, Hodson and Villet 2005b: 66 (listed)
Taphurini Moulds 2005b: 377, 393, 396, 404, 413–414, 417, 419, 422–425, 427, 430, 435, 437, Fig 58, Fig 612, Table 2 (described, diagnosis, composition, key, history, phylogeny, distribution, listed, comp. note) Equals Taphuriaria Equals Lemuriaria Equals Eaphurarini (sic)
Taphurini Pham and Thinh 2005a: 237 (listed) Vietnam
Taphurini Pham and Thinh 2005b: 287–288 (key, comp. note) Equals Taphuraria Vietnam
Taphurini Chawanji, Hodson and Villet 2006: 374, 383, 386, Table 1 (listed, comp. note)
Taphurini Sanborn 2006a: 77 (listed)
Taphurini Boulard 2007a: 93, 116 (listed, comp. note)
Taphurini Chawanji, Hodson, Villet, Sanborn and Phillips 2007: 338 (listed)
Taphurini Sanborn 2007a: 30 (listed) Venezuela
Taphurini Shiyake 2007: 7, 102 (listed, comp. note)
Taphurini Boulard 2008c: 351, 358 (listed, comp. note)
Taphurini Boulard 2008d: 168, 173 (comp. note)
Taphurini Boulard 2008f: 7, 57 (listed) Thailand
Taphurini Lee 2008a: 2, 20, Table 1 (listed) Vietnam
Taphurini Marshall 2008: 2786 (comp. note)
Taphurini Osaka Museum of Natural History 2008: 326 (listed) Japan
Taphurini Boulard 2009d: 39 (comp. note)
Taphurini Lee 2009h: 2637 (listed) Philippines, Palawan
Taphurini Pham and Yang 2009: 4, 15, Table 2 (listed) Vietnam
Taphurini Duffels 2010: 12–13 (comp. note)
Taphurini Lee 2010a: 26 (comp. note)
Taphurini Lee 2010b: 27 (listed) Cambodia
Taphurini Pham and Yang 2010b: 205 (comp. note) Vietnam
Taphurini Sanborn 2010a: 1601 (listed) Colombia
Taphurini Sanborn 2010b: 76 (listed)
Taphurini Wei, Ahmed and Rizvi 2010: 33 (comp. note)

Subtribe Taphurina Distant, 1905 = Taphuraria

Taphuraria Wu 1935: 26 (listed) China

Taphuraria Moulds 2003b: 246 (comp. note)

Taphurina Moulds 2005b: 430, 437, Table 2 (described, composition, listed, comp. note)

***Chalumalna* Boulard, 1998**

Chalumalna Boulard 1998: 83–84 (n. gen., described, comp. note)

***C. martinensis* Boulard, 1998**

Chalumalna martinensis Boulard 1998: 84, 88, Plate II, Figs 1–2 (n. sp., type species of *Chalumalna*, described, illustrated, comp. note) St. Martin

***Psallodia* Uhler, 1903**

Psallodia Ramos 1983: 66–67 (comp. note) Dominican Republic, Hispaniola

Psallodia Ramos 1988: 62–63, Table 4.1 (listed, distribution, comp. note) Greater Antilles, Hispaniola

Psallodia Moulds 2005b: 393, 437 (higher taxonomy, listed, comp. note)

Psallodia Perez-Gerlabert 2008: 202 (listed) Hispaniola

***P. espinii* Uhler, 1903**

Psallodia espinii Ramos 1983: 62, 64, 67, 70, Fig 15 (described, illustrated, key, listed, comp. note) Dominican Republic

Psallodia espinii Sanborn 1999a: 51 (type material, listed)

Psallodia espinii Perez-Gerlabert 2008: 202 (listed) Hispaniola, Dominican Republic, Haiti

Psallodia espinii Sanborn 2009d: 6 (comp. note) Hispaniola

***Chrysolasia* Moulds, 2003**

Chrysolasia Moulds 2003b: 245–246, 254–255, 257–258, 260, 262, 265 (n. gen., described, key, diversity, distribution, comp. note) Guatemala

***C. guatemalena* (Distant, 1883)** = *Tibicen guatemalenus* = *Tibicen guatemalanus* (sic) = *Abricta guatemalena*

Chrysolasia guatemalena Moulds 2003b: 250–254, 258, 262–263, 265, Figs 15–17, Fig 21, Figs 35–36, Fig 51, Table 1 (n. comb., described, illustrated, phylogeny, distribution, comp. note) Equals *Tibicen guatemalena* Equals *Tibicen guatemalanus* (sic) Equals *Abricta guatemalena* Guatemala

Abricta guatemalena Moulds 2003b: 254 (comp. note)

Tibicen guatemalena Moulds 2003b: 260 (type species of *Chrysolasia*)

Chrysolasia guatemalena Sanborn 2005b: 77 (listed, distribution) Guatemala

Crysolasia guatemalena Marshall 2008: 2786 (comp. note) Guatemala

***Dorachosa* Distant, 1892**

Dorachosa Moulds 2005b: 393, 437 (higher taxonomy, listed, comp. note)

***D. explicta* Distant, 1892**

Dorachosa explicata Wolda and Ramos 1992: 273, Table 17.2 (listed) Panama

Dorachosa explicata Sanborn 2005b: 77–78, Fig 1 (listed, distribution) El Salvador, Panama

***Dulderana* Distant, 1905**

Dulderana Moulds 2005b: 437 (higher taxonomy, listed, comp. note)

***D. truncatella* (Walker, 1850)**

***Nosola* Stål, 1866**

Nosola Heath, Heath and Noriega 1990: 1 (distribution, comp. note) Argentina

Nosola Moulds 2005b: 393, 437 (higher taxonomy, listed, comp. note)

***N. paradoxa* Stål, 1866**

***Selymbria* Stål, 1861** = *Selymbra* (sic) = *Selymbia* (sic)

Selymbria Ramos and Wolda 1985: 177, 179–180 (listed, comp. note) Panama

Selymbria Boer 1995c: 205 (comp. note) South America

Selymbria Moulds 2005b: 393, 437 (higher taxonomy, listed, comp. note)

***S. ahyetios* Ramos & Wolda, 1985** = *Selymbria achyetios* (sic) = *Selymbria* sp. H2 Wolda, 1977

Selymbria sp. H2 Wolda 1977: 239 (emergence, comp. note) Panama

Selymbria ahyetios Ramos and Wolda 1985: 178–180, 182–183, Plate II, Figs 7–12, Fig 13 (n. sp., described, illustrated, synonymy, listed, comp. note) Equals *Selymbria stigmatica* Wolda 1984 Panama

Selymbria ahyetios Wolda 1988: 10 (comp. note) Panama

Selymbria ahyetios Wolda 1989: 438–440, Figs 1–3, Tables 1–2 (emergence pattern, comp. note) Panama

Selymbria ahyetios Wolda and Ramos 1992: 272–273, Fig 17.1, Table 17.1 (distribution, listed, seasonality, comp. note) Panama

Selymbria ahyetios Wolda 1993: 371, 375, 378, Fig 6 (seasonality, acoustic behavior, comp. note) Panama

Selymbria ahyetios Sanborn 1999a: 51 (type material, listed)

Selymbria achyetios (sic) Bukhvalova 2005: 199 (emergence, comp. note) Panama

S. *danieleae* Sanborn, 2011

S. *pandora* Distant, 1911

S. *pluvialis* Ramos & Wolda, 1985 = *Selymbria stigmatica* (nec Germar) = *Selymbria* sp. 1 Wolda, 1977

Selymbria sp. 1 Wolda 1977: 244, Table 1 (emergence, comp. note) Panama

Selymbria stigmatica (sic) Wolda 1984: 451 (comp. note) Panama

Selymbria pluvialis Ramos and Wolda 1985: 177–181, 183–185, Plate I, Figs 1–6, Figs 13–15 (n. sp., described, illustrated, synonymy, listed, comp. note) Equals *Selymbria stigmatica* Wolda, 1984 Panama

Selymbria pluvialis Wolda 1988: 10 (comp. note) Panama

Selymbria pluvialis Wolda 1989: 438–440, Figs 1–3, Tables 1–2 (emergence pattern, comp. note) Panama

Selymbria pluvialis Wolda and Ramos 1992: 272–273, Fig 17.1, Table 17.1 (distribution, listed, seasonality, comp. note) Panama

Selymbria pluvialis Sanborn 1999a: 51 (type material, listed)

Selymbria pluvialis Sanborn 2010b: 76 (listed, distribution) Guatemala, Panama

S. sp. 1 Pogue, 1996

Selymbra (sic) sp. 1 Pogue 1996: 317 (listed) Peru

S. *stigmatica* (Germar, 1834) = *Cicada signifera* Germar, 1830 (nec *Cicada signifera* Germar, 1830) = *Chicada* (sic) *signifera* = *Cicada stigmatica* = *Cicada macrophthalma* Stål, 1854 = *Cicada macropthalma* (sic)

Cicada stigmatica Ramos and Wolda 1985: 177 (type species of *Selymbria*, listed)

Selymbria stigmatica Ramos and Wolda 1985: 177 (comp. note) Brazil

S. *subolivacea* (Stål, 1862) = *Cicada (Cicada) subolivacea* = *Cicada subolivacea*

***Prosotettix* Jacobi, 1907**

Prosotettix Boer 1995c: 205 (comp. note) South America

Prosotettix Moulds 2005b: 393, 437 (higher taxonomy, listed, comp. note)

P. *sphecoidea* Jacobi, 1907

***Taphura* Stål, 1862** = *Cicada (Taphura)*

Taphura sp. Wolda 1983a: 114, Fig 1C (emergence pattern, comp. note) Panama

Taphura Ramos 1988: 63 (comp. note) Panama, Brazil

Taphura Wolda 1988: 10 (comp. note) Panama

Taphura Boulard 1990b: 115, 177–178 (comp. note)

Taphura Campos, Silva, Ferreira and Dias 2004: 125 (comp. note) Brazil

Taphura Boulard 2005d: 349 (comp. note)

Taphura Moulds 2005b: 393, 407, 412–413, 425, 437 (type genus of Taphurini, higher taxonomy, comp. note)

Taphura Boulard 2006g: 159–160 (comp. note)

Taphura Sanborn 2010a: 1601 (listed) Colombia

T. *boulardi* Sanborn, 2011

T. *charpentierae* Boulard, 1971

T. *debruni* Boulard, 1990

Taphura debruni Boulard 1990b: 177, 191, 193, 219–221, Plate XXIV, Fig 3, Figs 14–18 (n. sp., described, illustrated, comp. note) Brazil

Taphura debruni Boulard 2000d: 89 (comp. note)

Taphura debruni Boulard 2002b: 205 (comp. note)

Taphura debruni Boulard 2005d: 349 (comp. note) Brazil

Taphura debruni Boulard 2006g: 159–161, Fig 116 (illustrated, comp. note) Brazil

T. *egeri* Sanborn, 2011

T. *hastifera* (Walker, 1858) = *Cicada hastifera* = *Cicada frontalis* Walker, 1858 = *Taphura frontalis*

Taphura hastifera Wolda 1988: 10 (comp. note) Panama

Taphura hastifera Wolda 1989: 438–440, Figs 1–3, Tables 1–2 (emergence pattern, comp. note) Panama

Taphura hastifera Wolda and Ramos 1992: 272–273, Fig 17.1, Table 17.1 (distribution, listed, seasonality, comp. note) Panama

Taphura hastifera Boulard 1990b: 219 (comp. note)

Taphura hastifera Pogue 1996: 317 (listed) Peru

Taphura hastifera Sanborn 2007a: 30 (distribution, comp. note) Venezuela, Brazil, Peru, Panama

T. *maculata* Sanborn, 2011

T. *minusculus* Thouvenot, 2012

T. *misella* (Stål, 1854) = *Cicada misella* = *Cicada (Taphura) misella* = *Thaphura* (sic) *misella*

Taphura misella Moulds 2005b: 393, 396, 399, 417–420, 424, 437, Figs 56–59, Fig 61, Table 1 (type genus of *Taphura*, higher taxonomy, phylogeny, listed, comp. note)

Cicada misella Sanborn 2010a: 1601 (type species of *Taphura*) Brazil

T. *nitida* (Degeer, 1773) = *Cicada nitida* = *Cicada (Taphura) nitida*

Taphura nitida Pogue 1996: 317 (listed) Peru

T. *sauliensis* Boulard, 1971

Taphura sauliensis Boulard 1990b: 219 (comp. note)

Taphura sauliensis Pogue 1996: 317 (listed) Peru

Taphura sauliensis Sanborn 2010a: 1601 (distribution, comp. note) Colombia, French Guiana, Peru

T. sp. C Wolda, 1977

Taphura sp. C Wolda 1977: 244, Table 1 (emergence, comp. note) Panama

Taphura n. sp. Wolda 1978: 375 (emergence, comp. note) Panama

***Elachysoma* Torres, 1964**

Elachysoma Heath, Heath and Noriega 1990: 1 (distribution, comp. note) Argentina

Elachysoma Moulds 2005b: 393, 437 (higher taxonomy, listed, comp. note)

***E. quadrivittata* Torres, 1964** = *Elachysoma quadrimarginatus* (sic)

Elachysoma quadrimarginatus (sic) Hua 2000: 61 (listed, distribution) China, Taiwan

***Imbabura* Distant, 1911**

Imbabura Moulds 2005b: 393, 437 (higher taxonomy, listed, comp. note)

***I. typica* Distant, 1911**

Subtribe Tryellina Moulds, 2005

Tryellina Moulds 2005b: 430, 437, Table 2 (n. subtribe, described, disagnosis, composition, listed, comp. note), 412–413, 430, 434, Table 2

Tryellina Marshall 2008: 2786 (comp. note)

***Abricta* Stål, 1866** = *Tibicen (Abricta)*

Abricta Strümpel 1983: 17, 108, 109, Table 8 (listed, comp. note)

Abricta Popov 1985: 38–39 (acoustic system, comp. note)

Abricta Moulds 1986: 40 (comp. note)

Abricta Williams and Williams 1988: 3 (listed) Mascarene Islands

Abricta Boulard 1990b: 209, Fig 12 (phylogeny, comp. note)

Abricta Moulds 1990: 7, 12, 16, 32, 118, 121 (listed, diversity, comp. note) Australia, Queensland, New South Wales, Reunion Island, Mauritius, India, Moluccas, New Caledonia, Guatemala

Abricta spp. Moulds 1990: Plate 5, Fig 1 (illustrated, comp. note)

Abricta spp. Moulds and Carver 1991: 467 (comp. note) Australia

Abricta spp. Ewart 1993: 140 (comp. note)

Abricta Boer 1995c: 203–204, Fig 1 (phylogeny, comp. note) Australia

Abricta Boer and Duffels 1996b: 308–309 (distribution, comp. note) Australia, Africa, Orient

Abricta Ewart 1996: 14 (listed)

Abricta Novotny and Wilson 1997: 437 (listed)

Abricta Boulard 1998: 83 (comp. note)

Abricta Ewart 2001b: 74 (comp. note) Queensland

Abricta Moulds and Carver 2002: 50 (type data, listed) Australia

Abricta Moulds 2003b: 245–251, 254–255, 257–259, 261–262, 279 (described, key, diversity, comp. note) Mauritius

Abricta complex Moulds 2003b: 246–247, 249–255, 260, 262–263 (comp. note) Equals *Tibicen (Abricta)*

Abricta spp. Moulds 2003b: 251, 253 (comp. note)

Abricta Jong and Boer 2004: 267 (comp. note)

Abricta Moulds 2005a: 133 (comp. note)

Abricta Moulds 2005b: 393, 437 (higher taxonomy, listed, comp. note) Australia

Abricta Shiyake 2007: 7, 102 (listed, comp. note)

A. borealis (Goding & Froggatt, 1904) to *Adelia borealis*

***A. brunnea* (Fabricius, 1798)** = *Tettigonia brunnea* = *Cicada brunnea* = *Tibicen (Abricta) brunneus* = *Tibicen brunneus*

Tettigonia brunnea Dallas 1870: 482 (synonymy) Equals *Tibicen brunnea*

Abricta brunnea Williams and Williams 1988: 3 (listed, comp. note) Equals *Tettigonia brunnea* Mauritius, Reuinion Island

Abricta brunnea Boulard 1989a: 129 (comp. note)

Abricta brunnea Boulard 1990b: 210 (comp. note)

Abricta brunnea Bonfils, Quilici and Reynaud 1994: 236 (listed) Réunion Island

Tettigonia brunnea Moulds 2003b: 246, 257 (type species of *Abricta*, comp. note)

Abricta brunnea Moulds 2003b: 246, 250–254, 257–261, Figs 15–17, Fig 19, Figs 37–38, Fig 40, Table 1 (type species of *Abricta*, lectotype designation, described, illustrated, key, phylogeny, distribution, comp. note) Equals *Tettigonia brunnea* Equals *Cicada brunnea* Equals *Tibicen brunnea* Equals *Tibicen (Abricta) brunnea* Equals *Tibicen brunneus* Mauritius, Reunion Island

A. burgessi Distant, 1905 to *Tamasa burgessi*

A. castanea Distant, 1905 to *Tryella castanea*

A. cincta (Fabricius, 1803) to *Diemeniana cincta*

A. curvicosta (Germar, 1834) to *Aleeta curvicosta*

A. elseyi Distant, 1905 to *Tryella rubra* (Goding & Froggatt, 1904)

***A. ferruginosa* (Stål, 1866)** = *Tibicen (Abricta) ferruginosa*

Abricta ferruginosa Schouteden 1907: 287 (listed, comp. note) Maurice Island

Abricta ferruginosa Williams and Williams 1988: 3 (listed, comp. note) Equals *Tibicen (Abricta) ferruginosus* Mauritius

Abricta ferruginosa Moulds 2003b: 245–246, 250–253, 257, 259–261, Figs 15–17, Fig 41, Table 1 (described, illustrated, key, phylogeny, distribution, comp. note) Equals *Tibicen (Abricta) ferruginosus* Mauritius

Abricta ferruginosus Moulds 2003b: 246 (comp. note)

A. *flavoannulata* Distant, 1920 to *Kanakia flavoannulata*
A. *frenchi* Distant, 1907 to *Diemeniana frenchi*
A. *guatemalena* (Distant, 1883) to *Chysolasia guatemalena*
A *occidentalis* (Goding & Froggatt, 1904) to *Pictila occicentalis*
A. *pusilla* (Fabricius, 1803) = *Tettigonia pusilla* = *Cicada pusilla* = *Tibicen pusillus*
Tettigonia pusilla Moulds 2003b: 247 (nomen dubia, comp. note)
A. *rubra* (Goding & Froggatt, 1904) to *Tryella rubra*
A. *willsi* (Distant, 1882) to *Tryella willsi*

***Aleeta* Moulds, 2003**
Abricta (sic) Josephson and Young 1981: 231 (comp. note)
Aleeta Moulds 2003b: 245, 251, 254–255, 257–258, 265 (n. gen., described, key, diversity, comp. note) Australia, Queensland, New South Wales
Aleeta Moulds 2005a: 133–134, 137 (comp. note) Australia
Aleeta Moulds 2005b: 393, 402, 413, 425, 430, 437, Table 2 (higher taxonomy, listed, comp. note)
A. *curvicosta* (Germar, 1834) = *Cicada curvicosta* = *Tibicen curvicosta* = *Abricta curvicosta* = *Abricta curyvicosii* (sic) = *Cicada tephrogaster* Boisduval, 1835 = *Tibicen (Abricta) tephrogaster* = *Abricta tephrogaster*
Abricta curvicosta Josephson 1981: 30–31, 34, Figs 7–8, Fig 10 (muscle function, comp. note)
Abricta curvicosta Moulds 1983: 429, 434 (illustrated, comp. note) Australia, Queensland, New South Wales
Abricta curvicosta Strümpel 1983: 108, Fig 144 (oscillogram, comp. note)
Abricta curvicosta Young and Josephson 1983a: 184–186, 192, 194, Fig 1, Table 1 (oscillogram, timbal muscle kinetics, comp. note) Australia, New South Wales
Abricta curvicosta Young and Josephson 1983b: 198–199, 201, 204, 206, Fig 1D (timbal illustrated, comp. note) Australia
Abricta curvicosta Young and Josephson 1984: 286, Fig 1 (comp. note)
Abricta curvicosta Claridge 1985: 303, Fig 2 (oscillogram)
Abricta MacNally and Doolan 1986a: 37–38, 40, 43 (comp. note) New South Wales
Abricta curvicosta Maitland and Maitland 1985: 331 (illustrated, comp. note) Australia
Abricta curvicosta MacNally and Doolan 1986a: 35, 37, 40–42, Figs 1–4, Table 2 (habitat, comp. note) New South Wales
Abricta curvicosta MacNally and Doolan 1986b: 281, 285–286, 288, Fig 2, Fig 4, Table 2 (habitat, comp. note) New South Wales
Abricta (sic) MacNally and Doolan 1986b: 284–291, Figs 1–3, Table 3, Table 5 (habitat, comp. note) New South Wales
Abricta curvicosta Josephson and Young 1987: 995–996, Figs 2–3 (comp. note)
Abricta curvicosta MacNally 1988a: 247 (distribution model, comp. note) New South Wales
Abricta curvicosta MacNally 1988b: 1976–1977, Tables 1–2 (comp. note) New South Wales
Cicada curvicosta Ewart 1990: 1, 3 (type material, type locality) Equals *Tympanoterpes curvicosta* Australia
Abricta curvicosta Moulds 1990: 3–6, 21, 23, 118–121, Plate 15, Fig 4, Fig 4a, Plate 24, Fig 28 (illustrated, distribution, ecology, listed, comp. note) Australia, New South Wales, Queensland
Abricta curvicosta Boulard 1991c: 118 (comp. note) Equals *Cicada tephrogaster* Australia
Abricta curvicosta Moulds and Carver 1991: 467 (comp. note) Australia
Abricta curvicosta Naumann 1993: 23, 67, 148 (listed) Australia
Abricta curvicosta Bennet-Clark and Young 1994: 293, Table 1 (body size and song frequency)
Abricta curvicosta Brunet 1995: 22 (listed) Australia
Abricta curvicosta Ewart 1995: 82 (described, illustrated, natural history, acoustic behavior, oscillogram) Queensland
A[bricta] curyvicosii (sic) Jiang, Yang, Tang, Xu and Chen 1995b: 229 (song frequency, comp. note)
Abricta curvicosta Moulds 1996b: 94, Plate 78 (comp. note) Australia
Abricta curvicosta Hangay and German 2000: 58 (illustrated, behavior) Queensland, New South Wales
Abricta curvicosta Ewart 2001a: 501–509, Fig 5, Fig 7C, Fig 9, Tables 1–2 (acoustic behavior) Queensland
Abricta curvicosta Ewart 2001b: 69–75, 81–83, Fig 4, Fig 12, Table 2 (emergence pattern, population density, nymph illustrated, predation, listed, comp. note) Queensland
Abricta curvicosta Sanborn 2001a: 16, Fig 5 (comp. note)
Abricta curvicosta Sueur 2001: 42, Table 1 (listed)
Abricta curvicosta Nation 2002: 260, Table 9.3 (comp. note)
Abricta curvicosta Moulds and Carver 2002: 51 (synonymy, distribution, ecology, type data,

listed) Equals *Cicada curvicosta* Equals *Cicada tephrogaster* Australia, New South Wales, Queensland

Cicada curvicosta Moulds and Carver 2002: 51 (type data, listed) Australasia

Cicada tephrogaster Moulds and Carver 2002: 51 (type data, listed) New South Wales

Abricta curvicosta Popple and Ewart 2002: 118 (illustrated) Australia, Queensland

Abricta curvicosta Williams 2002: 156–157 (listed) Australia, Queensland, New South Wales

Abricta curvicosta Lee 2003b: 7, Fig 4 (illustrated, comp. note) Australia

Aleeta curvicosta Moulds 2003b: 245, 247, 250–255, 263–267, 270–271, 297, Figs 1–2, Fig 8, Figs 15–17, Fig 20, Fig 50, Table 1 (n. comb., illustrated, phylogeny, comp. note) Equals *Cicada curvicosta* Equals *Cicada tephrogaster* Equals *Tibicen curvicostus* Equals *Tibicen (Abricta) tephrogaster* Equals *Tibicina curvicosta* Equals *Abricta curvicosta* Australia, Queensland

Cicada tephrogaster Moulds 2003b: 246, 264 (lectotype designation, comp. note)

Abricta tephrogaster Moulds 2003b: 257 (comp. note)

Cicada curvicosta Moulds 2003b: 263–264 (type species of *Aleeta*, lectotype designation, comp. note)

Abricta curvicosta Sueur and Aubin 2003b: 486 (comp. note)

Aleeta curvicosta Emery, Emery, Emery and Popple 2005: 102–105, 107, Tables 1–3 (comp. note) New South Wales

Abricta curvicosta Moulds 2005a: 133 (comp. note) Australia

Aleeta curvicosta Moulds 2005a: 133–136, 138, 140, Fig 6, Fig 19, Table 1 (song analysis, oscillogram, sonogram, comp. note) Australia, Queensland

Aleeta curvicosta 2005b: 395–398, 408, 417–420, 424, Figs 56–59, Fig 61, Table 1 (higher taxonomy, phylogeny, listed, comp. note) Australia

Abricta curvicosta Harvey and Thompson 2006: 905–909, Figs 1–3, Tables 1–3 (emergence energetics, comp. note) New South Wales

Aleeta curvicosta Watson, Myhra, Cribb and Watson 2006: 9, 16, 18, Figs 13–14, Fig 17 (wing microstructure, comp. note)

Abricta curvicosta Shiyake 2007: 7, 100, 102, Fig 169 (illustrated, distribution, listed, comp. note) Australia

Abricta curvicosta Nation 2008: 273, Table 10.1 (comp. note)

Aleeta curvicosta Watson, Myhra, Cribb and Watson 2008: 3353, 3355, 3358, Fig 10 (comp. note)

***Tryella* Moulds, 2003**

Tryella Moulds 2003b: 245–246, 254–258, 263, 271–272, 290 (n. gen., described, key, diversity, distribution, comp. note) Australia, Papua New Guinea

Tryella spp. Moulds 2003b: 250–251, 255–256, 268–269, 272–273, 276–277, 283–286, 290, 293, 296 (comp. note)

Tryella Moulds 2005a: 133–134, 137 (comp. note) Australia

Tryella Moulds 2005b: 393, 400, 413, 425, 430, 437, Table 2 (type genus of Tryellina, higher taxonomy, listed, comp. note)

***T. adela* Moulds, 2003**

Tryella adela Moulds 2003b: 251–254, 256, 268, 271, 275, 277–278, 282, 285, 288–289, Figs 15–17, Fig 30, Fig 53, Fig 75, Figs 85–86, Table 1 (n. sp., described, illustrated, key, phylogeny, distribution, comp. note) Australia, Northern Territory

Tryella adela species group Moulds 2003b: 254, 256 (comp. note) Australia

***T. burnsi* Moulds, 2003**

Tryella burnsi Moulds 2003b: 251–256, 268, 271, 274, 278–280, 282, 289–290, 293, 295, 300, Fig 9, Figs 15–17, Fig 22, Fig 30, Fig 58, Fig 76, Figs 98–99, Table 1 (n. sp., described, illustrated, key, phylogeny, distribution, comp. note) Australia, Queensland

***T. castanea* (Distant, 1905)** = *Abricta castanea*

Abricta castanea Moulds 1990: 121–123, Plate 15, Fig 3 (illustrated, distribution, ecology, listed, comp. note) Australia, Northern Territory, Queensland, Western Australia

Abricta castanea Ewart 1993: 139 (comp. note)

Abricta castanea Ewart 1998b: 135 (listed) Queensland

Abricta castanea Moulds and Carver 2002: 51 (distribution, ecology, type data, listed) Australia, Northern Territory, Queensland, Western Australia

Tryella castanea Moulds 2003b: 245, 250–254, 256, 268, 271–272, 274–275, 278, 281–283, 285–286, 288–289, 295, Figs 1–2, Fig 10, Figs 15–17, Fig 30, Fig 54, Fig 77, Figs 83–84, Table 1 (n. comb., described, illustrated, key, phylogeny, distribution, comp. note) Equals *Abricta castanea* Australia, Western Australia, Northern Territory, Queensland

Tryella castanea Moulds 2005a: 133–136, 138, 141, Fig 3, Figs 21–23, Table 1 (song analysis, oscillogram, sonogram, comp. note) Australia, Northern Territory

***T. crassa* Moulds, 2003**

Tryella crassa Moulds 2003b: 249, 251–256, 271–273, 275, 283–285, 298, Figs 15–17, Fig 23, Fig 30, Fig 61, Fig 78, Figs 81–82, Table 1 (n. sp., described, illustrated, key, phylogeny, distribution, comp. note) Australia, Queensland, Northern Territory

Tryella crassa Moulds 2005a: 133–136, 138–139, 141, Fig 1, Fig 4, Figs 9–14, Fig 20, Table 1 (illustrated, song analysis, oscillogram, sonogram, comp. note) Australia, Queensland, Northern Territory

***T. graminea* Moulds, 2003**

Tryella graminea Moulds 2003b: 249, 251–256, 268, 271–272, 285–286, 293, 300, Figs 15–17, Fig 24, Fig 30, Fig 57, Fig 87, Figs 92–93, Table 1 (n. sp., described, illustrated, key, phylogeny, distribution, comp. note) Australia, Queensland, Northern Territory, South Australia

Tryella graminea Ewart 2009b: 137–138, 140–142, 150, 167, Fig 17, Plate 1H (oscillogram, illustrated, distribution, emergence times, habitat, comp. note) Equals Grass Buzzing Bullet, no. 53 Queensland, South Australia, Northern Territory

***T. infuscata* Moulds, 2003**

Tryella infuscata Moulds 2003b: 250, 252–254, 256, 271–272, 274, 276–277, 287, 297, Figs 15–17, Fig 30, Fig 65, Figs 71–72, Fig 88, Table 1 (n. comb., described, illustrated, key, phylogeny, distribution, comp. note) Australia, Northern Territory, Queensland

***T. kauma* Moulds, 2003**

Tryella kauma Moulds 2003b: 250–256, 268, 271, 274–275, 280, 288–289, 293, 295, Fig 11, Figs 15–17, Fig 25, Fig 30, Fig 52, Fig 89, Figs 96–97, Table 1 (n. sp., described, illustrated, key, phylogeny, distribution, comp. note) Australia, Queensland

Tyrella kauma Moulds 2005a: 135–136, 138, 142, Fig 2, Fig 24, Table 1 (song analysis, oscillogram, sonogram, comp. note) Australia, Queensland

***T. lachlani* Moulds, 2003** = *Abricta* sp. "G" Ewart, 1993 = *Abricta* sp. Zborowski & Storey 1995 = Heathlands sp. G Ewart

Abricta sp. G Ewart 1993: 139–140, 147, Fig 14 (labeled Fig 13 Species F) (natural history, acoustic behavior, song analysis) Queensland

Abricta sp. Zborowski and Storey 1995: 87 (illustrated, comp. note) Australia

Tryella lachlani Moulds 2003b: 250–256, 268, 271, 274, 277, 289–290, 295, Fig 12, Figs 15–17, Fig 30, Fig 56, Figs 73–74, Fig 90, Table 1 (n. sp., described, illustrated, key, synonymy, phylogeny, distribution, comp. note) Equals Species G Ewart 1993 Equals *Abricta* sp. Zborowski and Storey 1995 Australia, Queensland, Torres Straits Islands, Papua New Guinea

Tryella lachlani Ewart 2005a: 178–179, Fig 12 (described, song, oscillogram, comp. note) Equals Heathlands sp. G Queensland

Abricta sp. Zborowski and Storey 2003: 87 (illustrated, comp. note) Australia

***T. noctua* (Distant, 1913)** = *Abricta noctua* = *Abricta rufonigra* Ashton, 1914

Abricta noctua Moulds 1990: 89–90, 121–122, Plate 15, Fig 2, Fig 2a, Plate 24, Fig 27 (illustrated, distribution, ecology, listed, comp. note) Australia, Northern Territory, South Australia, Western Australia

Abricta noctua Moulds and Carver 2002: 51 (synonymy, distribution, ecology, type data, listed) Equals *Abricta rufonigra* Australia, Northern Territory, South Australia, Western Australia

Abricta rufonigra Moulds and Carver 2002: 51 (type data, listed) Western Australia

Tryella noctua Moulds 2003b: 250–256, 271–272, 275–276, 285, 287, 290–293, 298, Figs 15–17, Fig 26, Fig 30, Fig 64, Fig 91, Table 1 (n. comb., described, illustrated, key, synonymy, phylogeny, distribution, comp. note) Equals *Abricta rufonigra* Australia, Western Australia, Northern Territory, South Australia

Abricta noctua Moulds 2003b: 290 (lectotype designation, comp. note)

Abricta rufonigra Moulds 2003b: 290–291 (lectotype designation, comp. note)

***T. occidens* Moulds, 2003**

Tryella occidens Moulds 2003b: 250–254, 256, 271–272, 275–276, 285, 287, 291–294, 298, Figs 15–17, Fig 30, Fig 66, Figs 100–102, Table 1 (n. sp., described, illustrated, key, phylogeny, distribution, comp. note) Australia, Western Australia

Tryella occidens species group Moulds 2003b: 254, 256 (comp. note) Australia

***T. ochra* Moulds, 2003**

Tryella ochra Moulds 2003b: 249, 252–256, 268, 271, 274–275, 277, 280, 286, 294–295, Figs 3–4, Figs 15–17, Fig 27, Fig 30, Fig 55, Figs 69–70, Fig 103, Table 1 (n. sp., described, illustrated, key, phylogeny, distribution, comp. note) Australia, Queensland

Tryella ochra Moulds 2005b: 381, 395, 397, 399, 405, 408, 417–419, 421, 424, 437, Figs 19–21, Figs 56–59, Fig 61, Table 1 (illustrated, higher taxonomy, phylogeny, listed, comp. note) Australia

T. rubra (Goding & Frogatt, 1904) = *Tibicen ruber* = *Abricta ruber* = *Abricta rubra* = *Abricta elseyi* Distant, 1905 = *Abricta kadulua* Burns, nom. nud.

Tibicen ruber Hahn 1962: 11 (listed) Australia, Western Australia, Queensland

Tibicen ruber Stevens and Carver 1986: 264–265 (type material) Equals *Abricta kadulua* Burns nom. nud. Australia, Western Australia, Queensland

Abricta rubra Moulds 1990: 121, 123, Plate 15, Fig 1, Fig 1a (synonymy, illustrated, distribution, ecology, listed, comp. note) Equals *Tibcen rubra* Equals *Abricta elseyi* n. syn. Australia, Northern Territory, Western Australia

Tibicen rubra Moulds 1990: 121 (lectotype designation, listed) Western Australia

Abricta elseyi Moulds 1990: 121 (listed, synonymy) Northern Territory

Abricta rubra Moulds and Carver 2002: 51–52 (synonymy, distribution, ecology, type data, listed) Equals *Tibcen rubra* Equals *Abricta elseyi* Australia, Northern Territory, Western Australia

Tibicen rubra Moulds and Carver 2002: 51–52 (type data, listed) Western Australia

Abricta elseyi Moulds and Carver 2002: 52 (type data, listed) Northern Territory

Abricta elseyi Moulds 2003b: 246, 281, 295 (comp. note)

Abricta rubra Moulds 2003b: 246 (comp. note) Equals *Abricta elseyi*

Tryella rubra Moulds 2003b: 250–256, 271, 273–275, 285, 295–297, Fig 13, Figs 15–17, Fig 28, Fig 30, Fig 62, Figs 79–80, Fig 104, Table 1 (n. comb., described, illustrated, key, synonymy, phylogeny, distribution, comp. note) Equals *Tibicen ruber* Equals *Abricta ruber* (sic) Equals *Abricta elseyi* Equals *Abricta rubra* Australia, Western Australia, Northern Territory

Tibicen ruber Moulds 2003b: 295 (comp. note)

Tyrella rubra Moulds 2005a: 135–136, 138, 140, 142, Fig 5, Figs 15–18, Figs 25–27, Table 1 (song analysis, oscillogram, sonogram, comp. note) Australia, Western Australia, Northern Territory

T. stalkeri (Distant, 1907) = *Abricta stalkeri*

Abricta stalkeri Moulds 1990: 122, Plate 15, Fig 5, Fig 5a (illustrated, distribution, ecology, listed, comp. note) Australia, Western Australia

Abricta stalkeri Moulds and Carver 2002: 52 (distribution, ecology, type data, listed) Australia, Western Australia

Tryella stalkeri Moulds 2003b: 250–256, 271–272, 275–277, 285, 287, 291, 293, 297–298, Fig 14, Figs 15–17, Figs 29–30, Fig 63, Figs 67–68, Fig 105, Table 1 (n. comb., lectotype designation, described, illustrated, key, phylogeny, distribution, comp. note) Australia, Western Australia

T. willsi (Distant, 1882) = *Tibicen willsi* = *Abricta willsi*

Abricta willsi Moulds 1990: 123–124, Plate 15, Fig 6, Fig 6a (illustrated, distribution, ecology, listed, comp. note) Australia, New South Wales, Queensland

Abricta willsi Burwell 1991: 124 (distribution, comp. note) Queensland

Abricta willsi Ewart 1993: 139 (comp. note)

Abricta willsi Ewart and Popple 2001: 63, 65 (described, habitat, song) Queensland

Abricta willsi Moulds and Carver 2002: 52 (synonymy, distribution, ecology, type data, listed) Equals *Tibicen willsi* Australia, New South Wales, Queensland

Tibicen willsi Moulds and Carver 2002: 52 (type data, listed) Queensland, New South Wales

Abricta willsi Popple and Strange 2002: 30, Table 1 (listed, habitat) Australia, Queensland

Tryella willsi Moulds 2003b: 250–254, 256, 268, 271–272, 280, 286, 293, 299–300, Figs 15–17, Fig 30, Fig 59, Figs 94–95, Fig 106, Table 1 (n. comb., described, illustrated, key, synonymy, phylogeny, distribution, comp. note) Equals *Tibicen willsi* Equals *Abricta willsi* Australia

Tibicen willsi Moulds 2003b: 298–299 (lectotype designation, comp. note)

Abricta willsi Shiyake 2007: 8, 100, 102, Fig 170 (illustrated, distribution, listed, comp. note) Australia

Tryella willsi Ewart 2009b: 140 (comp. note) Queensland

***Chrysocicada* Boulard, 1989**

Chrysocicada Boulard 1989d: 65–66, 73 (n. gen., described, comp. note) Western Australia

Chrysocicada Boulard 1990a: 238–239 (n. gen., described, comp. note) Western Australia

Chrysocicada Moulds 2005a: 133 (comp. note) Australia, Queensland

Chrysocicada Moulds 2005b: 377, 392, 430, 437, Table 2 (higher taxonomy, listed, comp. note)

C. franceaustralae Boulard, 1989 = *Chrysocicada franceaustralasae* Boulard, 1990

Chrysocicada franceaustralae Boulard 1989d: 65–71, 76–77, Figs 2–10 (n. sp., type species of *Chrysocicada*, described, illustrated, sonogram, comp. note) Western Australia

Chrysocicada franceaustralae Boulard 1990a: 237–243, Figs 1–10 (n. sp., type species of *Chrysocicada*, described, illustrated, sonogram, comp. note) Western Australia
Chrysocicada franceaustralae Boulard 1999a: 80, 110, 112–113, Plate III, Fig 4 (listed, sonogram) Western Australia
Chrysocicada franceaustralae Sueur 2001: 42, Table 1 (listed)
Chrysocicada franceaustralae Boulard 2005d: 338, 344 (comp. note) Australia
Chrysocicada francoaustralae Moulds 2005b: 395, 397–398, 403, 417–420, 422, 424, Fig 37, Figs 56–61, Table 1 (higher taxonomy, phylogeny, key, listed, comp. note) Australia
Chrysocicada franceaustralae Boulard 2006g: 36–37, 39, 72, 80, 181, Fig 7g, Fig 34 (sonogram, illustrated, listed, comp. note) Western Australia

***Pictila* Moulds, 2012**

***P. occidentalis* (Goding & Froggatt, 1904)** = *Tibicen occidentalis* = *Abricta occidentalis*
Tibicen occidentalis Hahn 1962: 10 (listed) Australia, Western Australia
Tibicen occidentalis Stevens and Carver 1986: 265 (type material) Australia, Western Australia
Abricta occidentalis Moulds 1990: 119 (distribution, listed) Australia, Western Australia
Abricta occidentalis Moulds and Carver 2002: 51 (synonymy, distribution, ecology, type data, listed) Equals *Tibicen occidentalis* Australia, Western Australia
Tibicen occidentalis Moulds and Carver 2002: 51 (type data, listed) Western Australia
Abricta occidentalis Moulds 2003b: 246 (comp. note)
Tibicen occidentalis Moulds 2003b: 247 (comp. note) Western Australia
Abricta occidentalis 2005b: 395–398, 417–419, 421, 424, Figs 56–59, Fig 61, Table 1 (higher taxonomy, phylogeny, listed, comp. note) Australia

***Magicicada* Davis, 1925** = *Magicicadas* (sic) = Magicicadidae (sic) = *Migicicada* (sic) = *Megacicada* (sic) = *Magicicadae* (sic) = *Cicada* (nec Linnaeus) = *Mogannia* (nec Amyot & Audinet-Serville)
Cicada (sic) Walsh 1868: 7 (sting, comp. note) Missouri
Cicada (sic) Forbes 1888: 9–10 (fungal host, comp. note)
Cicada (sic) Davis 1894: 21 (comp. note) New York
Cicada (sic) Goldstein 1929: 394 (fungal host, comp. note)
Magicicada Snodgrass 1935: 230, Fig 128B (illustrated, morphology, comp. note)
Magicicada McCoy 1965: 40–41, 43 (key, comp. note) Arkansas
Magicicada Evans 1966: 103 (parasitism, comp. note)
Magicicada Borror and White 1970: 129 (life cycle, comp. note)
Magicicada spp. Lloyd 1973: 270 (comp. note)
Magicicada sp. Emigh 1979: 327 (periodicity)
Magicicada Skaife 1979: 89 (comp. note) America
Magicicada sp. Arnett and Jacques 1981: 170 (life history) Eastern United States, Iowa
Magicicada Hamilton 1981: 959, Fig 3 (exuvia head illustrated, comp. note)
Mogannia (sic) Itô and Nagamine 1981: 273 (comp. note)
Magicicada spp. Itô and Nagamine 1981: 280, (life cycle, comp. note)
Magicicada Josephson and Young 1981: 231 (comp. note)
Magicicada spp. Karban 1981b: 385 (comp. note) Iowa, New York
Magicicada Popov 1981: 276–279 (comp. note)
Magicicada spp. Simon, Karban and Lloyd 1981: 1525 (comp. note)
Magicicada Simon, Karban and Lloyd 1981: 1527 (comp. note)
Magicicada spp. Soper 1981: 53 (fungal host, comp. note)
Magicicada spp. Strehl 1981: 2216B (predation, comp. note) Illinois
Magicicada spp. Young 1981d: 828 (comp. note)
Magicicada Young 1981d: 828 (comp. note)
Magicicada Cahoon and Donoho 1982: 16 (control, comp. note)
Magicicada spp. Itô 1982b: Fig 5 (life cycle evolution, comp. note)
Magicicada spp. Karban 1982a: 74 (host growth, comp. note) New York
Magicicada spp. Karban 1982b: 321 (comp. note) New York, Iowa
Magicicada spp. Maier 1982a: 14 (comp. note)
Magicicada Maier 1982a: 14, 17 (comp. note)
Magicicada Maier 1982b: 432 (comp. note)
Magicicada Simon and Lloyd 1982: 275 (brood formation, comp. note)
Magicicada spp. White, Lloyd and Karban 1982: 475 (comp. note)
Magicicada Claridge 1983: 113 (song, comp. note)
Magicicada Heath 1983: 13 (comp. note) North America
Magicicadidae (sic) Huber 1983a: 108, 116 (comp. note)
Magicicadidae (sic) Huber 1983b: 3 (comp. note)
Magicicada spp. Karban 1983a: 226 (host growth, comp. note) Pennsylvania

Magicicada spp. Lloyd and Karban 1983: 299, 303 (comp. note)
Magicicada spp. Lloyd, Kritsky and Simon 1983: 1162 (life cycle, comp. note)
Magicicada Lloyd, Kritsky and Simon 1983: 1170, 1178 (life cycle, comp. note)
Magicicada Simon 1983a: 104–105 (wing variation, comp. note) Eastern United States
Magicicada Simon 1983b: 378 (comp. note) Eastern United States
Magicicada Strümpel 1983: 51, 58, 108–110, 117, 168, Tables 8–9 (comp. note)
Magicicada spp. Walker 1983: 66, Fig 7 (comp. note)
Magicicada spp. White, Ganter, McFarland, Stanton and Lloyd 1983: 281 (comp. note)
Magicicada White and Lloyd 1983: 1245 (fungal infection, comp. note)
Magicicada spp. White and Lloyd 1983: 1248, Table 1 (fungal infection, comp. note)
Magicicada Young and Josephson 1983a: 194 (comp. note)
Magicicada Young and Josephson 1983b: 198 (comp. note)
Magicicada spp. Young and Josephson 1983b: 204 (comp. note)
Magicicada Chilcote and Stehr 1984: 53 (distribution, comp. note) Michigan, Ohio, West Virginia, Pennsylvania
Magicicada Ewing 1984: 228 (comp. note)
Magicicada Fleming 1984: 204 (comp. note)
Magicicada spp. Fleming 1984: 204 (comp. note)
Magicicada spp. Karban 1984a: 1656 (comp. note) New York
Magicicada spp. Karban 1984b: 355 (development time, comp. note)
Magicicada Lloyd 1984: 80–82, 85, 90 (periodicity, life history, comp. note)
Magicicada Young 1984a: 178 (comp. note)
Magicicada spp. Young 1984a: 185 (comp. note)
Magicicada Archie, Simon and Wartenberg 1985: 1262 (life history) United States
Magicicada Arnett 1985: 218 (diversity) Eastern United States
Magicicada Cade 1985: 594 (comp. note)
Magicicada Claridge 1985: 302, 307 (song, comp. note) North America
Magicicada James, Williams and Smith 1985: 37 (distribution, comp. note) Arkansas
Magicicada spp. Karban 1985: 269 (comp. note) North America
Magicicada spp. Maier 1985: 1019, 1024 (comp. note)
Magicicada Maier 1985: 1019 (comp. note) Kentucky
Magicicada Popov 1985: 38–39 (acoustic system, comp. note)
Magicicada spp. Schuder 1985: 2 (life history, control, comp. note) Indiana
Magicicada White and Lloyd 1985: 609 (comp. note)
Migicicada (sic) spp. Karban 1986: 224, Table 14.2 (comp. note) USA
Magicicada spp. Karban 1986: 224, Table 14.2 (comp. note) USA
Magicicada spp. Lloyd 1986: 362 (rearing, distribution, comp. note)
Magicicada Lloyd 1986: 363, 366–377 (comp. note)
Magicicada spp. Murphy 1986: 1486 (predation) Kansas
Magicicada spp. Strehl and White 1986: 178 (predation, comp. note) Illinois
Magicicada spp. Williams and Smith 1986: 353 (emergence, predation) Arkansas
Magicicada Andersen 1987: 274 (density) Indiana
Magicicada spp. Davidson and Lyon 1987: 373 (natural history, comp. note)
Magicicada Gwynne 1987: 575 (comp. note)
Magicicada spp. Gwynne 1987: 575 (comp. note)
Magicicada spp. Klein, Bock, Kafka and Moore 1987b: 93–94, 97 (comp. note) Missouri, Ohio, Pennsylvania, West Virginia
Magicicada spp. Lloyd and White 1987: 50 (comp. note)
Magicicada spp. Schuder 1987: 1 (life history, control, comp. note) Indiana
Magicicada spp. Smith, Wilkinson, Williams and Steward 1987: 277 (predation, comp. note) Arkansas
Magicicada spp. Villet 1987a: 271 (comp. note)
Magicicada spp. Weber, Moore, Huber and Klein 1987: 329 (comp. note)
Magicicada spp. Williams 1987: 424 (oviposition damage)
Magicicada Young and Kritsky 1987: 323–325, 327, Fig 1, Table 2 (comp. note) Indiana
Magicicada spp. Cox and Carleton 1988: 183–184, Fig 1 (comp. note)
Magicicada Cox and Carleton 1988: 185–191, Fig 4 (life cycle evolution, comp. note)
Magicicada Feng 1988: 25 (comp. note)
Magicicada Hewitt, Nichols and Ritchie 1988: 206 (brood evolution, comp. note) Illinois, Missouri
Magicicada spp. Klein, Bock, Kafka and Moore 1988: 153–154, 166 (antennal sensilla, antennal responses, comp. note) Missouri, Ohio, Pennsylvania, West Virginia
Magicicada MacNally 1988b: 1979, Table 4 (comp. note)
Magicicada spp. Martin and Simon 1988: 237 (gene sequence, comp. note) North America

Magicicada Samson, Evans and Latgé 1988: 40, Plate 22, Fig E (fungal host, illustrated)
Magicicada Simon 1988: 163, 166, 169, 171, 173–174 (life history, distribution, evolution, comp. note) Eastern United States
Magicicada spp. Simon 1988: 172, Table 2 (comp. note)
Magicicada spp. Steward, Smith and Stephen 1988: 348 (predation, comp. note) Arkansas
Magicicada species Ewing 1989: 36, 38, 158, 160 (comp. note)
Magicicada spp. Luken and Kalisz 1989: 310 (soil disturbance, comp. note) Kentucky
Magicicada spp. Fullard and Heller 1990: 64 (comp. note)
Magicicada sp. Hahus and Smith 1990: 249 (comp. note)
Magicicada spp. Kellner, Smith, Wilkinson and James 1990: 375 (predation, comp. note) Arkansas
Migicicada (sic) Liu 1990: 45 (comp. note) China
Magicicada Martin and Simon 1990a: 1066, 1069, 1073, Fig 1, Table 3 (genetic diversity, comp. note) North America
Magicicada Martin and Simon 1990b: 360–362, 366, Fig 3, Table 2 (life cycle, comp. note) North America
Magicicada Moore 1990: 58 (comp. note)
Magicicada Nakao 1990: 127 (listed)
Magicicada Simon 1990: 309 (wing variation, comp. note) Eastern United States
Magicicada Simon, Pääbo, Kocher and Wilson 1990: 240 (comp. note)
Magicicada Sismondo 1990: 55 (comp. note)
Magicicada Stephen, Wallis and Smith 1990: 369 (predation, comp. note) Arkansas
Magicicada Bailey 1991b: 96 (sound processing comp. note)
Magicicada spp. Cranshaw and Kondratieff 1991: 11 (life cycle)
Magicicada spp. Cox and Carleton 1991: 63 (comp. note)
Magicicada Cox and Carleton 1991: 63–64, 68–69, 71 (life cycle evolution, comp. note) Illinois, Missouri, Indiana, Arkansas, Iowa
Magicicada spp. Dean and Milton 1991: 116, 118 (comp. note)
Magicicada Eaton 1991: 9 (life cycle, comp. note) Ohio
Magicicada Fonseca 1991: 190 (comp. note)
Magicicada Gogala 1991a: 3, 6–7, 9, Fig 8 (life cycle, distribution, comp. note) Ohio, Kentucky, Indiana
Magicicada spp. Johnson and Lyon 1991: 490 (life cycle, comp. note)
Magicicada spp. Karlin, Williams, Smith and Sugg 1991: 213 (comp. note) Arkansas
Magiciada spp. Krohne, Couillard and Riddle 1991: 317 (predation) Indiana
Magicicada spp. Reding and Guttman 1991: 323–324 (comp. note)
Magicicada Simon, Franke and Martin 1991: 53 (comp. note)
Magicicada Toolson and Toolson 1991: 113 (comp. note)
Magicicada spp. Young 1991: 188 (comp. note)
Magicicada Bennet-Clark and Young 1992: 127, 147 (comp. note) Kansas, North America
Magicicada Clark 1992: 47 (contamination, predation, comp. note) Maryland
Magicicada Cox 1992: 845 (life cycle evolution, comp. note)
Magicicada Danks 1992: 176 (predation, comp. note)
Magicicada spp. Dean and Milton 1992: 74 (comp. note)
Magicicada spp. Hennig, Kleindienst, Weber, Huber, Moore and Popov 1992: 180 (comp. note)
Migicicada (sic) Liu 1992: 8 (comp. note) China
Magicicada spp. Sadof 1992: 1 (life history, illustrated, oviposition damage, control) Indiana
Magicicada Wheeler, Williams and Smith 1992: 340 (comp. note)
Magicicada Andersen and Folk 1993: 662 (predation) Indiana
Magicicada spp. Helfert and Sänger 1993: 37 (comp. note)
Magicicada spp. Lane 1993: 56–57 (comp. note)
Magicicada spp. Long 1993: 209 (life cycle, comp. note)
Magicicada Moore 1993: 269, 271, 275–277, 279, 282, Figs 4–6 (distribution, song, comp. note)
Magicicada Moore, Huber, Weber, Klein and Bock 1993: 199, 217, 219 (listed, comp. note)
Magicicada spp. White and Sedcole 1993b: 60 (comp. note)
Magicicada White and Sedcole 1993b: 60–61 (comp. note)
Magicicada Williams, Smith and Stephen 1993: 1143 (comp. note)
Magicicada Branson 1994: 10–11 (life cycle, comp. note) North America
Magicicada spp. Fonseca and Popov 1994: 350 (comp. note)
Magicicada spp. Heliövaara, Väisänen and Simon 1994: 477 (illustrated, comp. note) North America
Magicicada Heliövaara, Väisänen and Simon 1994: 479, Fig 3 (distribution, comp. note) North America
Magicicada spp. Kalisz 1994: 121, Fig 1 (emergence patterns) Kentucky

Magicicada spp. Karlin, Stout, Adams, Duke and English 1994: 89 (comp. note) Arkansas
Magicicada Lei, Chou and Li 1994: 53 (comp. note)
Magicicada spp. Simon, Frati, Beckenbach, Crespi, Liu and Flook 1994: 689, 691–692 (comp. note)
Magicicada Woodall 1994: 20 (comp. note)
Magicicada Boer 1995d: 172 (comp. note) North America
Magicicada spp. Cranshaw and Kondratieff 1995: 79 (life history, comp. note)
Magicicada spp. Jameson 1995: 3 (sound production, comp. note)
Magicicada Lane 1995: 378–379, 388 (comp. note) North America
Magicicada spp. Lane 1995: 388 (comp. note) North America
Magicicada Raina, Pannell, Kochansky and Jaffe 1995: 929 (comp. note) Maryland
Magicicada species complex Raina, Pannell, Kochansky and Jaffe 1995: 929 (neuropeptide, comp. note) United States
Magicicada spp. Raina, Pannell, Kochansky and Jaffe 1995: 930–931, Fig 1 (neuropeptide, comp. note) Maryland
Magicicada Sanborn, M.S. Heath, J.E. Heath and Noriega 1995: 456 (comp. note)
Magicicada Williams and Simon 1995: 269, 282–283, Table 1 (comp. note)
Magicicada spp. Williams and Simon 1995: 270, 273, 275, 279–280, 282–283, 290, Table 1 (natural history, comp. note)
Magicicada sp. Anonymous 1996a: 20 (life history) eastern North America
Magicicada Boer and Duffels 1996a: 157 (life cycle comp. note)
Magicicada Boer and Duffels 1996b: 298 (comp. note) North America
Magicicada spp. Daws and Hennig 1996: 185 (comp. note)
Magicicada spp. Fonseca 1996: 29 (comp. note)
Magicicada spp. Johnsen, Hendeveld, Blank and Yasak 1996: 356 (predation, comp. note) Indiana
Magicicada Patterson Massei and Genov 1996: 141 (comp. note)
Magicicada spp. Simon, Nigro, Sullivan, Holsinger, Martin, et al. 1996: 923 (comp. note)
Magicicada Alexander, Marshall and Cooley 1997: 10 (lek)
Magicicada Chou, Lei, Li, Lu and Yao 1997: 25 (comp. note)
Magicicada Hilton 1997: 25–26 (natural history, comp. note)
Magicicada spp. Hilton 1997: 26 (natural history, comp. note)
Magicicada spp. Karban 1997: 456 (comp. note)
Magicicada Karban 1997: 457 (comp. note)
Magicicada Novotny and Wilson 1997: 437 (listed)
Magicicada Poole, Garrison, McCafferty, Otte and Stark 1997: 251, 331 (listed) North America
Magicicada spp. Rodenhouse, Bohlen and Barrett 1997: 124–127, 129–133, Fig 1, Table 1, Tables 3–5 (distribution, density, comp. note) Ohio
Magicicada spp. Yoshimura 1997: 112 (life cycle evolution, comp. note)
Magicicada spp. Cox and Carleton 1998: 162–163 (life cycle evolution, comp. note)
Magicicada Henríquez 1998: 86 (comp. note)
Magicicada Hyche 1998: 6, Photo 2 (natural history, distribution, comp. note) Alabama
Magicicada spp. Itô 1998: 493, 495 (life cycle evolution, comp. note)
Magicicada Kocher, Thomas, Meyer, Edwards, Pääbo et al. 1998: 6197 (sequence, comp. note)
Magicicada spp. Sanborn 1998: 90 (thermal biology, comp. note)
Magicicada Sanborn 1998: 90, 92, 96 (thermal biology, comp. note)
Magicicada spp. Toolson 1998: 573 (comp. note)
Magicicada Toolson 1998: 574 (comp. note)
Magicicada Anonymous 1999c: 489 (emergence time)
Magicicada Coelho and Wiedman 1999: 11 (comp. note)
Magicicada Kritsky 1999: 24 (comp. note)
Magicicada Sanborn 1999b: 2 (listed) North America
Magicicada sp. Schmidt and Whelan 1999: 66 (predation) Illinois
Magicicada Arnett 2000: 299 (diversity) Eastern United States
Magicicada Bauernfeind 2000: 238, 240 (distribution) Kansas
Magicicadas (sic) Behncke 2000: 414–415 (natural history, life cycle evolution) central USA, northeast USA, south central USA
Magicicada Boulard 2000h: 249 (life cycle, comp. note) America
Magicicada sp. Freytag 2000: 55 (on stamp)
Magicicada Goles, Schulz and Markus 2000: 208 (life cycle model)
Magicicada spp. Irwin and Coelho 2000: 82–89, Figs 1–3 (distribution, comp. note) Illinois
Magicicada spp. Karban, Black, Weinbaum 2000: 253 (development time, comp. note) Pennsylvania
Magicicada Marshall and Cooley 2000: 1313, 1316, 1323–1324, Figs 9–11 (speciation, comp. note)

Magicicada Maw, Footit, Hamilton and Scudder 2000: 49 (listed)
Magicicada Milius 2000: 408 (life cycle, comp. note) North America
Magicicada Nickens 2000: 27 (life cycle) North America
Magicicada spp. Sahli and Ware 2000: 187, 193 (oviposition sites, comp. note) Virginia
Magicicada Sahli and Ware 2000: 187–188 (oviposition sites, comp. note) Virginia
Magicicada Sanborn 2000: 555 (comp. note)
Magicicada Seibt and Wickler 2000: 805 (comp. note)
Magicicada Simon, Tang, Dalwadi, Staley, Deniega and Unnasch 2000: 1333 (speciation, comp. note)
Magicicada spp. Sugden 2000: 435 (comp. note)
Magicicada spp. Vokoun 2000: 261 (predation, comp. note) Missouri
Magicicada Barrett 2001: 1 (natural history) Missouri, Eastern United States, Kansas, Oklahoma, Texas, Arkansas, Louisiana
Magicicada Beaupre and Roberts 2001: 45 (comp. note) Arkansas
Magicicada Cook, Holt and Yao 2001: 51 (comp. note)
Magicicada Cooley 2001: 756, 758 (comp. note) North America
Magicicada Cooley and Marshall 2001: 828–832, 834, 836, 838, 841, 846–851, Table 2 (communication, comp. note) Virginia, Illinois, Arkansas, Ohio, North Carolina
Magicicada Goles, Schulz and Markus 2001: 33 (life cycle model)
Magicicada spp. Marshall 2001: 386 (comp. note) United States
Magicicada Marshall 2001: 386–399, Figs 1–5, Tables 2–3 (life cycle, broods, distribution, comp. note) United States, Illinois, Missouri
Magicicada Ritchie 2001: 59 (life cycle, comp. note) North America
Magicicada spp. Sanborn, Perez, Valdes and Seepersaud 2001: A1106 (wing morphology, flight, comp. note)
Magicicada spp. Valdes and Sanborn 2001: 25 (comp. note)
Magicicada spp. Webb 2001: 387–388 (life cycle model) United States
Magicicada spp. Whiles, Callaham, Meyer, Brock and Charleton 2001: 177 (comp. note) Illinois
Magicicada Achtziger and Nigmann 2002: 3 (life cycle) North America
Magicicada sp. Achtziger and Nigmann 2002: 13 (on stamp) North America
Magicicada Boulard 2002b: 206 (comp. note) North America
Magicicada sp. Day, Pfeiffer and Lewis 2002: 1 (natural history. distribution, comp. note) Virginia
Magicicada Dietrich 2002: 159 (life cycle, comp. note)
Magicicada spp. Ellingson, Anderson and Kondratieff 2002: 283 (comp. note)
Magicicada Gerhardt and Huber 2002: 390 (comp. note) North America
Magicicada Gogala 2002: 242 (comp. note)
Magicicada Hickey and McGhee 2002: 12 (comp. note)
Magicicada spp. Kondratieff, Ellingson and Leatherman 2002: 2 (life cycle, comp. note) North America
Magicicada Kondratieff, Ellingson and Leatherman 2002: 9 (comp. note) New York
Magicicada Marchand 2002: 36–37 (life history, comp. note) Louisiana, Mississippi, Arkansas, Tennessee, Kentucky, Missouri, Illinois
Magicicadae (sic) Markus, Schulz and Goles 2002: 202 (life cycle, comp. note)
Magicicada McGhee 2002: 12 (life cycle) North America
Magicicada spp. Rakitov 2002: 119 (comp. note) Illinois
Magicicada sp. Rakitov 2002: 119, 121–12, Figs 2–10 (malpighian tubules, comp. note) Illinois
Magicicada Sanborn 2002a: 228 (comp. note)
Magicicada spp. Sanborn 2002a: 226–227, Fig 2 (natural history, distribution, comp. note)
Magicicada spp. Valdes, Sanborn, Perez and Gebaide 2002: 21 (wing morphology, flight, comp. note)
Magicicada Woodall 2002: 20 (comp. note)
Magicicada spp. Cooley, Simon and Marshall 2003: 152–153, 155–157, Figs 2–4 (speciation, illustrated, comp. note)
Magicicada spp. Sueur 2002b: 119, 124–125 (comp. note) North America
Magicicada spp. Sueur and Aubin 2002b: 134–135 (comp. note)
Magicicada spp. Cox and Carleton 2003: 428 (comp. note) North America
Magicicada Cox and Carleton 2003: 428–431, Figs 1–2 (evolution, comp. note)
Magicicada DeFoliart 2003: 434 (life cycle, comp. note) Illinois
Magicicada Ibanez 2003: 10 (life cycle, comp. note) North America
Magicicada spp. Ibanez 2003: 10 (life cycle, comp. note) North America

Magicicada spp. Koenig and Liebhold 2003: 1084–1085, 1087, Figs 1–2 (distribution, host growth, comp. note) North America
Magicicada spp. Lee 2003b: 5–6 (comp. note)
Magicicada Marshall, Cooley and Simon 2003: 433–436, Fig 1 (speciation, distribution, comp. note)
Magicicada Moulds 2003a: 187 (comp. note)
Magicicada Tishechkin 2003: 144 (comp. note)
Magicicada spp. Villet, Sanborn and Phillips 2003c: 1441 (comp. note)
Magicicada Boulard and Chueata 2004: 141 (life cycle, comp. note) North America
Magicicada Brambila and Hodges 2004: 364 (life cycle, comp. note) United States
Magicicadas (sic) Campos, Oliveira, Giro and Galvão 2004a: 1–2 (life cycle model)
Magicicadas (sic) Campos, Oliveira, Giro and Galvão 2004b: 1 (life cycle model)
Magicicada Capinera 2004: 787 (fungal host)
Magicicada Chen 2004: 20 (life cycle, comp. note)
Magicicada Cooley 2004: 746–747, 754, 756 (comp. note) North America
Magicicada spp. Cooley and Marshall 2004: 649 (comp. note)
Magicicada Cooley and Marshall 2004: 649–650, 652, 655, 666–670 (mate choice, comp. note)
Magicicada spp. Cooley, Marshall and Simon 2004: 198 (comp. note)
Magicicada Cooley, Marshall and Simon 2004: 198–202, Fig 1, Table 1 (brood VII distribution, comp. note) New York
Magicicada spp. Harwood and Obrycki 2004: 235–238, Figs 1–3 (illustrated, natural history, comp. note) United States
Magicicada Hayes 2004: 401 (life cycle, comp. note)
Magicicada Hopkin 2004: 1 (comp. note)
Magicicada Hutchins, Evans, Garrison and Schlager 2004: 68 (calling behavior, comp. note)
Magicicada Jolivet 2004: 4–5 (life cycle, illustrated, distribution, comp. note) Virginia
Magicicada Khamasi 2004: 12 (comp. note) Tennessee, Kentucky
Magicicada Kritsky 2004a: 87 (comp. note)
Magicicada Kritsky 2004c: 3 (natural history, comp. note)
Magicicada Nelson 2004: 233 (life cycle) eastern United States
Magicicada spp. Ostfeld and Keesing 2004: 1488 (nutrient pulse, comp. note)
Magicicada Powledge 2004: 21 (life cycle, comp. note)
Magicicada Stölting, Moore and Lakes-Harlan 2004: 254–255 (comp. note)
Magicicada spp. Sueur and Aubin 2004: 223 (comp. note)
Magicicada spp. Walker and Moore 2004: 1, 4 (comp. note)
Magicicada Walker and Moore 2004: 4 (comp. note)
Magicicada spp. Yang 2004: 1565 (resource pulse, comp. note) North America
Magicicada spp. Zyla 2004a: 485–486 (distribution, comp. note) Maryland
Magicicada spp. Zyla 2004b: 488 (distribution, comp. note) Maryland, Virginia, District of Colombia
Magicicada Ahern, Frank and Raupp 2005: 2133 (life history) North America
Magicicada Baker 2005: 229–230 (life cycle model) North America
Magicicada Biedermann, Achtziger, Nickel and Stewart 2005b 234 (comp. note)
Magicicada Boulard 2005d: 339, 343 (comp. note) North America
Magicicadas (sic) Campos and Oliveira 2005: 1 (comp. note) New England
Magicicada Chen, Bao and Chen 2005: 1, 5 (life cycle model) North America
Magicicada spp. Champman 2005: 321 (comp. note)
Magicicada Champman 2005: 329 (comp. note)
Magicicada Claridge 2005: I-4 (comp. note)
Magicicada spp. Edwards, Faivre, Crist, Sitvarin and Zyla 2005: 273 (distribution, life cycle, comp. note) Pennsylvania
Magicicada Emery, Emery, Emery and Popple 2005: 106 (comp. note) United States
Magicicada Grant 2005: 169, 171–173 (diversity, life cycle, phylogeny, comp. note)
Magicicada spp. Grant 2005: 169, 171–173, Table 1 (life cycle, comp. note)
Magicicada Greenfield 2005: 3 (comp. note)
Magicicada Grimaldi and Engel 2005: 307–308 (comp. note)
Magicicada Heckel, Keener, Khang and Lee 2005: 2 (mercury bioaccumulation, comp. note) Ohio
Magicicada spp. Hill, Marshall and Cooley 2005: 74–76 (comp. note)
Magicicada spp. Humphreys 2005: 73–74, Table 4 (population density, comp. note) Costa Rica
Magicicada Khamasi 2005: 25 (comp. note) Tennessee, Kentucky
Magicicada spp. Koenig and Liebhold 2005: 1873, 1876, Fig 1 (predation, distribution, comp. note) North America
Magicicada Koenig and Liebhold 2005: 1873 (life cycle, comp. note)
Magicicada spp. Kondratieff, Schmidt, Opler and Garhart 2005: 96 (comp. note)
Magicicada Kondratieff, Schmidt, Opler and Garhart 2005: 96 (life cycle, comp. note) Oklahoma

Magicicada spp. Lee, Sato and Sakai 2005: 317 (parasitism, comp. note) North America
Magicicada spp. Lehmann-Ziebarth, Heideman, Shapiro, Stoddart, Hsaio et al. 2005: 3200 (evolution of periodicity model)
Magicicada Mjølhus, Wikan and Solberg 2005: 2 (comp. note)
Magicicada Moulds 2005b: 377, 415, 417, 425, 437 (tribal placement, phylogeny, listed, comp. note)
Magicicada spp. Pierce, Gibb and Waltz 2005: 455 (comp. note) Indiana
Magicicada spp. Pinto-Juma, Simões, Seabra and Quartau 2005: 81, 92 (comp. note)
Magicicada Quartau and Simões 2005b: 227 (comp. note)
Magicicada Remondino and Cappellini 2005: 4 (life cycle)
Magicicada spp. Robinson 2005: 217 (comp. note) USA
Magicicada spp. Robinson, Sibrell, Broughton, Yang and Hancock 2005: B31A-06 (heavy metal concentrations, comp. note) Virginia, West Virginia
Magicicada spp. Sanborn 2005c: 114, Table 1 (comp. note)
Magicicada Sueur 2005: 209 (song, comp. note)
Magicicada spp. West-Eberhard 2005: 6546 (speciation, comp. note)
Magicicada spp. Yang 2005: 523 (comp. note) North America
Magicicada Benko and Perc 2006: 557 (comp. note)
Magicicada Boulard 2006g: 50–51, 96 (comp. note)
Magicicada Chawanji, Hodson and Villet 2006: 386 (comp. note)
Magicicada spp. Cooley, Marshall, Hill and Simon 2006: 856–859, 864–866, Fig 2 (eggs illustrated, comp. note) North America
Magicicada Elliott and Hershberger 2006: 184, 210 (listed, comp. note)
Magicicada spp. Gilbert and Klass 2006: 78–79, 81–83, Table 1 (distribution, comp. note) New York
Magicicada spp. Grant 2006: 541 (comp. note) New Jersey
Magicicada Marshall 2006: 102, 139 (illustrated, life history, comp. note) North America, United States, Canada
Magicicada sp. Miranda 2006: 384 (comp. note)
Magicicada spp. Moriyama and Numata 2006: 1220, 1224 (comp. note)
Magicicada Nahirney, Forbes, Morris, Chock and Wang 2006: 2018–2019, 2024 (timbal muscles, illustrated, comp. note)
Magicicada spp. Oberdörster and Grant 2006: 409, 412 (ecology, density) New Jersey
Magicicada Oberdörster and Grant 2006: 410, 412 (ecology, density) New Jersey
Magicicada Remondino and Cappellini 2006a: 7 (life cycle)
Magicicada Remondino and Cappellini 2006b: 491 (life cycle)
Magicicada spp. Seabra, Pinto-Juma and Quartau 2006: 844 (comp. note)
Magicicada spp. Simões and Quartau 2006: 250 (comp. note)
Magicicada Simões and Quartau 2006: 251 (comp. note)
Magicicada spp. Smith, Kelly and Finch 2006: 1615 (comp. note)
Magicicada spp. Yang 2006: 2993 (comp. note) North America
Magicicada Boulard 2007e: 65, 85 (life cycle, comp. note) North America
Magicicada spp. Bunker, Neckermann and Cooley 2007: 359–360 (comp. note) Michigan
Magicicada spp. Cardozo, Silvestre and Colato 2007a: 1 (life cycle model)
Magicicada Cardozo, Silvestre and Colato 2007a: 2 (life cycle model)
Magicicada spp. Cardozo, Silvestre and Colato 2007b: 339 (life cycle model)
Magicicada Cardozo, Silvestre and Colato 2007b: 440 (life cycle model)
Magicicada spp. Cooley 2007: 90 (comp. note)
Magicicada spp. Cooley, Simon, Marshall, Slon and Ehrhardt 2007: 661 (comp. note)
Magicicada Cooley, Simon, Marshall, Slon and Ehrhardt 2007: 661–664, 669 (speciation, comp. note)
Magicicada Eaton and Kaufman 2007: 90–91 (illustrated, life cycle, comp. note) North America
Magicicada Fontaine, Cooley, and Simon 2007: 2–4, 6, Fig 6 (illustrated, haplotype, comp. note)
Magicicada spp. Fontaine, Cooley, and Simon 2007: 1, 4, 6 (comp. note) North America
Magicicada Ganchev and Potamitis 2007: 304–305, 319, Fig 11 (listed, comp. note)
Magicicada Heckel and Keener 2007: 79–80 (mercury bioaccumulation, comp. note) Ohio
Magicicada Joy and Crespi 2007: 791 (comp. note)
Magicicada Lin, Cast, Wood and Chen 2007: 750 (comp. note)
Magicicada Nowlin, González, Vanni, Stevens, Fields and Valente 2007: 2174 (resource pulse) North America
Magicicada Oberdörster and Grant 2007a: 1–2, 10 (flight system, flight speeds, comp. note) New Jersey

Magicicada spp. Oberdörster and Grant 2007a: 4–11 (comp. note) New Jersey
Magicicada Oberdörster and Grant 2007b: 15–16, 18–22 (sound system, predation, comp. note) New Jersey
Magicicada spp. Oberdörster and Grant 2007b: 16–17, 20, 22 (sound system, predation, comp. note) New Jersey
Magicicada spp. Phillips and Sanborn 2007: 455 (comp. note)
Magicicada spp. Robinson, Sibrell, Broughton and Yang 2007: 369 (heavy metal concentrations, comp. note) Virginia, West Virginia
Magicicada Shiyake 2007: 7, 98 (listed, comp. note)
Magicicada Sueur and Puissant 2007b: 55 (comp. note)
Magicicada Bergh and Short 2008: 784 (comp. note)
Magicicada Brambila and Hodges 2008: 603 (life cycle, comp. note) United States
Magicicada Capinera 2008: 1344 (fungal host)
Magicicada Cole 2008: 815 (comp. note) North America
Magicicada Costa Neto 2008: 455 (comp. note)
Magicicada Emery 2008: 34 (comp. note)
Magicicada Flory and Mattingly 2008: 650 (oviposition impact) Indiana
Magicicada Fonseca, Serrão, Pina-Martins, Silva, Mira, et al. 2008: 29 (comp. note) America
Magicicada spp. Greenfield and Schul 2008: 289 (comp. note)
Magicicada spp. Marcello, Wilder and Meikle 2008: 41 (predation, comp. note) Ohio, Indiana
Magicicada spp. Marshall 2008: 2785–2794, Figs 25–28, Tables 7–8 (natural history, illustrated, song, comp. note) United States
Magicicada Menninger, Palmer, Craig and Richardson 2008: 1307, 1310 (ecosystem impact, comp. note)
Magicicada spp. Moriyama and Numata 2008: 1487 (comp. note)
Magicicada spp. Nowlin, Vanni and Yang 2008: 647 (resource pulse) North America
Magicicada spp. Pinto-Juma, Seabra and Quartau 2008: 1 (comp. note)
Magicicada spp. Sanborn 2008c: 875 (comp. note)
Magicicada spp. Sanborn, Phillips and Gillis 2008: 2 (comp. note)
Magicicada spp. Smith and Hasiotis 2008: 504 (comp. note)
Magicicada spp. Storm and Whitaker 2008: 353, Table 1 (listed, predation, comp. note) Indiana
Magicicada Yoshimura, Hayashi, Tanaka, Tainaka and Simon 2008: 292 (life cycle evolution, comp. note)
Magicicada Brittain, Meretsky, Gwinn, Hammond and Riegel 2009: 205 (predation, comp. note) Indiana
Magicicada Brown and Zuefle 2009: 346 (comp. note)
Magicicada sp. Buschbeck and Mauser 2009: 367, 371–372, Fig 6c (eye structure, comp. note) Ohio
Magicicada spp. Clay, Shelton and Winkle 2009a: 1689 (oviposition damage, comp. note) Indiana, Kentucky, Tennessee
Magicicada spp. Clay, Shelton and Winkle 2009b: 277 (life history, oviposition damage, comp. note) Indiana
Magicicada Cooley, Kritsky, Edwards, Zyla, Marshall et al. 2009: 106, 109–147, 149 (brood X distribution, comp. note) Long Island, New England, Illinois, Ohio, Maryland, Pennsylvania, Indiana, Delaware, Virginia, West Virginia, Michigan, Tennessee, North Carolina, Georgia, Washington, DC, New York, New Jersey
Magicicada spp. Holliday, Hastings and Coelho 2009: 13 (predation, comp. note)
Magicicada spp. Logan and Maher 2009: 268 (comp. note)
Magicicada McCutcheon, McDonald and Moran 2009b: 15398 (symbiont, comp. note)
Magicicada Méléard and Tran 2009: 882 (comp. note)
Magicicada Moulds 2009b: 164 (comp. note)
Magicicada Pickover 2009: 354 (life cycle) North America
Magicicada Pinto, Quartau, and Bruford 2009: 272, Fig 2 (gene sequence, phylogeny, comp. note) Portugal
Magicicada Pray, Nowlin, and Vanni 2009: 182 (ecosystem nutrients, comp. note)
Magicicada Sanborn and Heath 2009: 14 (comp. note)
Magicicada spp. Smith 2009: 4 (predation, comp. note)
Magicicada Strauβ and Lakes-Harlan 2009: 313 (comp. note)
Magicicada spp. Tanaka, Yoshimura, Simon, Cooley and Tainaka 2009: 8975 (life cycle evolution, comp. note)
Magicicada Tanaka, Yoshimura, Simon, Cooley and Tainaka 2009: 8975 (life cycle evolution, comp. note)
Magicicada spp. Yang and Karban 2009: 105 (tree growth, comp. note) Pennsylvania
Magicicada spp. Yasukawa 2009: 404 (comp. note) Wisconsin
Magicicada spp. Zoratto, Santucci and Alleva 2009: 166 (dilution effect, comp. note)

Magicicada Cristensen and Fogel 2010: 211, 219 (comp. note)
Magicicada spp. Logan, Hill, Connolly, Maher and Dobson 2010: 267 (comp. note)
Magicicada Matthews and Matthews 2010: 1084 (oviposition behavior, comp. note) Indiana
Magicicada Rizza 2010: 103–105, 107–113 (life cycle)
Magicicada spp. Saljoqi, Rahimi, Rehman and Khan 2010: 75 (comp. note)
Magicicada Saulich 2010: 1132 (life cycle, comp. note)
Magicicada Smith, Cooley and Westerman 2010: 73 (comp. note)
Magicicada spp. Smith, Cooley and Westerman 2010: 73 (comp. note)
Magicicada spp. Speer, Clay, Bishop and Creech 2010: 174 (tree growth, comp. note) Eastern United States, Indiana

***M. cassinii* (Fisher, 1852)** = *Cicada cassinii* = *Cicada septendecim cassinii* = *Cicada septendecim cassini* = *Tibicen cassinii* = *Tibicen cassini* (sic) = *Tibicen septendecim cassina* (sic) = *Tibicina septendecim cassini* (sic) = *Tibicina septendecim cassinii* = *Cicada cassinii* = *Cicada septendecim cassinii* = *Tibicena* (sic) *cassinii* = *Magicicada septendecim cassinii* = *Magicicada septendecim* var. *cassinii* = *Magicicada septendecim cassini* (sic) = *Magicicada cassini* (sic) = *Magicicada cassani* (sic) = *Migicicada* (sic) *cassini* (sic) = *Magiciada ccassini* (sic) = *Magicicada casini* (sic) = *Magicicada cassi* (sic)

Cicada cassinii Walsh and Riley 1868b: 63–64 (species, comp. note)
Cicada cassinii Walsh 1870: 335 (speciation, comp. note)
Cicada cassinii Riley 1872: 33 (comp. note)
Cicada cassinii Strecker 1879: 256 (comp. note) Texas
Cicada cassinii Davis 1911b: 2 (comp. note) New York, Staten Island, New Jersey
Cicada cassinii Davis 1912: 2 (comp. note) New York, Staten Island, New Jersey
Tibicina cassinii Davis 1919b: 341–342 (comp. note) Missouri
Tibicina cassini (sic) Snodgrass 1919: 392 (illustrated, natural history, comp. note) Equals *Tibicina septendecim cassini*
Tibicina septendecim cassini (sic) Snodgrass 1919: 392 (comp. note)
Magicicada septendecim var. *cassinii* Hyslop 1935: 1, 3 (comp. note)
Magicicada septendecim var. *cassinii* Hyslop 1940: 2–3 (comp. note)
Magicicada septendecim var. *cassinii* Strandine 1940: 177 (soil impact, habitat, comp. note) Illinois
Tibicen cassinii Zeuner 1944: 113 (comp. note)
Magicicada cassinii Craighead 1949: 124 (oviposition damage, comp. note) United States
Magicicada cassinii Deay 1953: 203 (comp. note) Indiana
Magicicada cassini (sic) McCoy 1965: 41, 43 (key, comp. note) Arkansas
Magicicada cassinii Buck and Buck 1968: 1326 (behavior, comp. note)
Magicicada cassini (sic) Walker 1969: 894 (chorus, comp. note)
Magicicada cassini (sic) Cloudsley-Thomopson 1970: 20 (comp. note)
Magicicada cassini (sic) Itô 1978: 158 (comp. note) America
Magicicada cassi (sic) Mattson 1980: 133 (life cycle, comp. note)
Magicicada cassini (sic) Karban 1981a: 261 (comp. note) New York
Magicicada cassini (sic) Karban 1981b: 385–389, Fig 2 (dispersal, comp. note) Iowa, New York
Magicicada cassini (sic) White 1981: 727–730, 732–737, Fig 1, Fig 3, Fig 5, Table I, Tables IV-V (oviposition damage, comp. note) Kentucky
Magicicada cassini (sic) White and Lloyd 1981: 219–220 (egg mortality, comp. note) Illinois
Magicicada cassini (sic) Itô 1982b: 110–111, Table 1 (life cycle evolution, comp. note)
Magicicada cassini (sic) Karban 1982a: 75 (host growth, comp. note) New York
Magicicada cassini (sic) Karban 1982b: 322, 324 (reproductive success, density, predation, comp. note) New York, Iowa
Magicicada cassini (sic) Lloyd, White and Stanton 1982: 852, 855–856, Table 3 (comp. note) Kansas
Magicicada cassini (sic) Simon and Lloyd 1982: 275, 279 (comp. note) Illinois
Magicicada cassini (sic) White, Lloyd and Karban 1982: 477–478, 480, Fig 7, Table 2 (hatching success, comp. note) Ohio
Magicicada cassini (sic) Huber 1983a: 116–117, 119–121, Fig 8C, Fig 10B, Fig 11B (illustrated, neural responses, oscillogram, comp. note)
Magicicada cassini (sic) Huber 1983b: 5–7, 9–11, 13, 19, 21, 34, Fig 4–6, Figs 8–9, Fig 16, Fig 27 (illustrated, neural responses, phonotaxis, sonogram, oscillogram, comp. note)
Magicicada cassini (sic) Karban 1983a: 227 (host growth, comp. note) Pennsylvania
Magicicada cassini (sic) Karban 1983b: 325–328, Figs 1–3 (reproductive success, predation, comp. note) Kansas

Magicicada cassini (sic) Lloyd and Karban 1983: 299–301, Table 1 (chorusing centers, comp. note) Kansas
Magicicada cassini (sic) Lloyd, Kritsky and Simon 1983: 1162–1163, 1166, 1168, Fig 3 (life cycle, comp. note)
Magicicada cassini (sic) Lloyd and White 1983: 294–301, Fig 1, Tables 1–2, Tables 4–6 (population density, comp. note) Kentucky
Magicicada cassini (sic) Simon 1983b: 105, 114 (comp. note) Eastern United States
Magicicada cassini (sic) Strümpel 1983: 51, 58, 168, Table 9 (comp. note)
Magicicada cassini (sic) White, Ganter, McFarland, Stanton and Lloyd 1983: 282 (comp. note)
Magicicada cassini (sic) White and Lloyd 1983: 1246–1248, Figs 1.2–1.4 (fungal infection, comp. note) Kansas, Ohio
Magicicada cassini (sic) Young and Josephson 1983a: 192, Table 1 (comp. note)
Magicicada cassini (sic) Young and Josephson 1983b: 198–206, Fig 1A, Fig 1C, Figs 2A–2C, Fig 3, Figs 4A–4D, Figs 5A–B, Fig 6, Fig 7A, Tables 1–2 (illustrated, song production, oscillogram, frequency spectrum, timbal muscle mechanics, comp. note) Kansas
Magicicada cassini (sic) Huber 1984: 25–27, 30, Fig 2B, Fig 3, Fig 11 (illustrated, neural responses, sonogram, oscillogram, comp. note)
Magicada cassini (sic) Lloyd 1984: 79–80, 87–88, 90, Fig 1, Fig 6 (illustrated, life history, comp. note)
Magicicada cassini (sic) Young and Josephson 1984: 286, Fig 1 (comp. note)
Magicicada cassini (sic) Claridge 1985: 302, 307, 309 (song, comp. note)
Magicicada cassini (sic) Heliövaara and Väisänen 1985: 88–89, 93 (natural history, life cycle evolution, comp. note) North America
Magicicada cassini (sic) Huber 1985: 56, Fig 2 (neural responses, comp. note)
Magicicada cassini (sic) Karban 1985: 270 (comp. note) Iowa
Magicicada cassini (sic) Michelsen and Larsen 1985: 549 (comp. note)
Magicicada cassinii Popov 1985: 78 (comp. note)
Magicicada cassini (sic) Smith 1985: 4 (life cycle, comp. note) Arkansas
Magicicada cassini (sic) White and Lloyd 1985: 605–609, Tables 1–3 (nymphal growth, comp. note) Ohio
Magicicada cassini (sic) Karban 1986: 227 (comp. note)
Magicicada cassini (sic) Lloyd 1986: 363–365, Fig 2, Table 1 (illustrated, comp. note)
Magicicada cassini (sic) Schildberger, Kleindienst, Moore and Huber 1986: 126 (hearing, comp. note) North America
Magicicada cassini (sic) Strehl and White 1986: 179, 181 (predation, comp. note) Illinois
Magicicada cassini (sic) Davidson and Lyon 1987: 374 (natural history, comp. note)
Magicicada cassini (sic) Klein, Bock, Kafka and Moore 1987a: 211 (antennal sensilla, antennal responses, comp. note)
Magicicada cassini (sic) Klein, Bock, Kafka and Moore 1987b: 93–97, Figs 1–2, Tables 1–2 (antennal sensilla, antennal responses, comp. note) Missouri, Ohio, Pennsylvania, West Virginia
Magicicada cassini (sic) Lloyd and White 1987: 50 (comp. note)
Magicicada cassini (sic) Smith, Wilkinson, Williams and Steward 1987: 277–278 (predation, comp. note) Arkansas
Magicicada cassini (sic) Villet 1987a: 269 (comp. note) North America
Magicicada cassini (sic) Weber, Moore, Huber and Klein 1987: 330–335, Figs 1–2, Figs 5–6 (song, sonogram, oscillogram, comp. note)
Magicicada cassini (sic) Yaginuma 1987: 15–16, Tables 1–2 (illustrated, life history, comp. note)
Magicicada cassini (sic) Young and Kritsky 1987: 323–324 (distribution, comp. note) Indiana
Magicicada cassini (sic) Klein, Bock, Kafka and Moore 1988: 154–165, Figs 1–23, Table 1 (antennal sensilla, antennal responses, comp. note) Missouri, Ohio, Pennsylvania, West Virginia
Magicicada cassini (sic) Kritsky 1988: 168 (distribution, comp. note) Ohio
Magicicada cassini (sic) Simon 1988: 166–167 (comp. note) Eastern United States
Magicicada cassini (sic) Steward, Smith and Stephen 1988: 348 (predation, comp. note)
Magicicada casini (sic) Ewing 1989: 37, Fig 2.10a, c (oscillogram, song frequency, comp. note)
Magicicada cassini (sic) Ewing 1989: 38, 103–104, Fig 4.8 (auditory response, comp. note)
Magicicada cassini (sic) Luken and Kalisz 1989: 310 (soil disturbance, comp. note) Kentucky
Magicicada cassini (sic) Hahus and Smith 1990: 249–251, Table 1 (predation, comp. note) Kentucky
Magicicada cassini (sic) Huber, Kleindienst, Moore, Schildberger and Weber 1990: 218–222, 224–225, 227, Figs 1–4, Fig 7B, Figs 8–9, Tables 1–2 (neural responses, oscillogram, sonogram, comp. note)

Migicicada (sic) *cassini* (sic) Liu 1990: 45 (comp. note)
Magicicada cassini (sic) Martin and Simon 1990a: 1066 (comp. note) North America
Magicicada cassini (sic) Martin and Simon 1990b: 360 (comp. note) North America
Magicicada cassini (sic) Moore, Huber, Kleindienst, Schildberger and Weber 1990: 1 (neuronal responses, comp. note)
Magicicada cassini (sic) Nakao 1990: 127 (life cycle, comp. note)
Magicicada cassini (sic) Simon 1990: 310, 321, Fig 8 (wing variation, comp. note) Eastern United States
Magicicada cassini (sic) Young 1990: 55 (comp. note)
Magicicada cassini (sic) Eaton 1991: 9 (life cycle, comp. note) Ohio
Magicicada cassini (sic) Flowers 1991: 24 (natural history, comp. note) Illinois
Magicicada cassini (sic) Gogala 1991a: 3, 5–7, 9–10, Figs 6–7, Fig 8D, Fig 9, Figs 11–12 (illustrated, sonogram, life cycle, fungal host, distribution, comp. note) Ohio, Kentucky, Indiana
Magicicada cassini (sic) Gogala 1991b: 76 (comp. note)
Magiciada cassini (sic) Kritsky and Young 1991: 45 (distribution, comp. note) Indiana
Magiciada cassini (sic) Krohne, Couillard and Riddle 1991: 317 (predation) Indiana
Magicicada cassini (sic) Reding and Guttman 1991: 323–324, 326–335, Tables 1–6 (morphometric variability, genetic variability, comp. note) Ohio, Indiana
Magicicada cassinii Russell and Stoetzel 1991: 480 (egg nests, comp. note) Maryland
Magicicada cassini (sic) Stoetzel and Russell 1991: 471, 477 (emergence, hosts, oviposition, predation, comp. note) Maryland
Magicicada cassini (sic) Williams and Smith 1991: 276, 286 (comp. note) Arkansas
Magicicada cassini (sic) Bennet-Clark and Young 1992: 126, 148 (song parameters, comp. note) Kansas, North America
Magicicada cassini (sic) Cox and Carleton 1992: 63 (comp. note)
Magicicada cassini (sic) Hennig, Kleindienst, Weber, Huber, Moore and Popov 1992: 180 (comp. note)
Magicicada cassini (sic) Kritsky 1992: 38 (distribution, comp. note) Ohio
Magiciada cassini (sic) Kritsky and Young 1992: 59 (distribution, comp. note) Indiana
Migicicada (sic) *cassini* (sic) Liu 1992: 8 (comp. note)
Magicicada cassini (sic) Sanborn and Davis 1992: 18 (wing morphology, flight, comp. note)
Magicicada cassini (sic) Villet 1992: 93 (comp. note)
Magicicada cassini (sic) Wheeler, Williams and Smith 1992: 340–341 (ecosystem nutrients, comp. note) Arkansas
Magicicada cassini (sic) Lane 1993: 56–57 (comp. note) United States
Magicicada cassini (sic) Long 1993: 214 (comp. note)
Magicicada cassini (sic) Moore 1993: 273, 275–277, Fig 4, Fig 6 (distribution, song, comp. note)
Magicicada cassini (sic) Moore, Huber, Weber, Klein and Bock 1993: 200, 202, 205–208, 210, 212–219, Fig 1, Figs 3–6, Figs 8–12 (head illustrated, vision, phonotaxis, comp. note) Missouri, Oklahoma
Magicicada cassini (sic) Simon, McIntosh and Deniega 1993: 229 (comp. note) Illinois
Magicicada cassini (sic) White and Sedcole 1993a: 47, 49–50 (comp. note)
Magicicada cassini (sic) Williams, Smith and Stephen 1993: 1149 (comp. note)
Magicicada cassini (sic) Bennet-Clark 1994: 169, Table 1 (comp. note)
Magicicada cassini (sic) Bennet-Clark and Young 1994: 292–293, Table 1 (body size and song frequency)
Magicicada cassini (sic) Greenfield 1994: 109 (chorus, comp. note)
Magicicada cassini (sic) Heliövaara, Väisänen and Simon 1994: 476–477, Table 1 (life cycle, comp. note) North America
Magicicada cassini (sic) Hennig, Weber, Moore, Kleindienst, Huber and Popov 1994: 54 (comp. note)
Magicicada cassini (sic) Kalisz 1994: 118–122, Fig 1 (emergence patterns) Kentucky
Magicicada cassini (sic) von Dohlen and Moran 1995: 213, 215–221, Figs 2–8 (listed, phylogeny, comp. note)
M[agicicada] cassini (sic) Jiang, Yang, Tang, Xu and Chen 1995b: 229, Fig 3c (song frequency, comp. note)
Magicicada cassini (sic) Jiang, Yang, Wang, Xu, Chen and Yang 1995: 676 (comp. note) China
Magicicada cassini (sic) Raina, Pannell, Kochansky and Jaffe 1995: 929–931 (neuropeptide, comp. note) Maryland
Magicicada cassini (sic) Sanborn, J.E. Heath, M.S. Heath and Noriega 1995: 326, Table 3 (thermal responses, comp. note)
Magicicada cassini (sic) Sanborn, M.S. Heath, J.E. Heath and Noriega 1995: 456, Table 4 (comp. note)

Magicicada cassini (sic) Williams and Simon 1995: 271, 287 (natural history, comp. note)
Magicicada cassini (sic) Johnsen, Hendeveld, Blank and Yasak 1996: 358 (predation, comp. note) Indiana
Magiciada cassini (sic) Kritsky and Simon 1996: 27 (comp. note) Ohio
Magicicada cassini (sic) Marshall, Cooley, Alexander and Moore 1996: 165 (comp. note) Michigan
Magicicada cassini (sic) Simon, Nigro, Sullivan, Holsinger, Martin, et al. 1996: 924, Fig 1 (comp. note)
Magicicada cassini (sic) Alexander, Marshall and Cooley 1997: 25 (comp. note)
Magicicada cassini (sic) Daws, Hennig and Young 1997: 184 (comp. note)
Magicicada cassini (sic) Karban 1997: 447, 450, 452–456, Tables 2–4 (development time, comp. note) New York
Magicicada cassini (sic) Miller 1997: 225–226 (oviposition damage, control, comp. note) Illinois
Magicicada cassinii Poole, Garrison, McCafferty, Otte and Stark 1997: 331 (listed) Equals *Cicada cassinii* Equals *Cicada cassini* (sic) Equals *Magicicada cassanaii* (sic) North America
Magicicada cassini (sic) Rodenhouse, Bohlen and Barrett 1997: 127, 131–133 (distribution, density, comp. note) Ohio
Magicicada cassini (sic) Shelly and Whittier 1997: 279, Table 16–1 (listed, comp. note)
Magicicada cassini (sic) Blair 1998: 3 (life history) Kansas, Missouri, Arkansas, Oklahoma
Magicicada cassini (sic) Gordon 1998: 70 (recipe) North America
Magicicada cassini (sic) Hyche 1998: 3–4 (comp. note) Alabama
Magicicada cassini (sic) Miller and Crowley 1998: 248 (oviposition damage, control, comp. note) Illinois
Magicicada cassini (sic) Sanborn 1998: 91, 99, Table 1 (thermal biology, thermal responses, comp. note)
Magicicada cassini (sic) Tunaz, Bedick, Miller, Hoback, Rana and Stanley 1998: 925–926, 929 (response to bacterial infection, comp. note) Nebraska
Magicicada cassini (sic) Anonymous 1999a: 1 (natural history) Ohio, Maryland, Pennsylvania, Virginia, West Virginia
Cicada cassini (sic) Kritsky 1999: 23 (comp. note)
Magicicada cassini (sic) Kritsky 1999: 24–25 (illustrated, comp. note)
Magiciada cassini (sic) Kritsky, Smith and Gallagher 1999: 43 (distribution, comp. note) Ohio
Magicicada cassini (sic) Sanborn 1999b: 2 (listed) North America
Magicicada cassini (sic) Bauernfeind 2000: 239–240 (distribution) Kansas
Magicicada cassini (sic) Callaham, Whiles, Meyer, Brock and Charlton 2000: 538–539, 541, Fig 3 (density, comp. note) Kansas
Magicicada cassini (sic) Irwin and Coelho 2000: 82 (comp. note) Illinois
Magicicada cassini (sic) Marshall and Cooley 2000: 1313–1315, Fig 1, Table 1 (distribution, comp. note)
Magicicada cassini (sic) Sahli and Ware 2000: 187–188 (comp. note)
Magicicada cassini (sic) Sanborn 2000: 555 (comp. note)
Magicicada cassini (sic) Sanborn and Maté 2000: 144 (comp. note)
Magicicada cassini (sic) Seibt and Wickler 2000: 805 (comp. note)
Magicicada cassini (sic) Simon, Tang, Dalwadi, Staley, Deniega and Unnasch 2000: 1326, 1333 (speciation, comp. note)
Magicicada cassini (sic) Smith, Kelly, Dean, Bryson, Parker et al. 2000: 127 (comp. note) Kansas
Magicicada cassini (sic) Waldbauer 2000: 220 (life history, comp. note)
Magicicada cassini (sic) Beaupre and Roberts 2001: 45 (comp. note) Arkansas
Magicicada cassini (sic) Bednick, Tunaz, Nor Aliza, Putnam, Ellis and Stanley 2001: 108 (comp. note)
Magicicada cassini (sic) Cook, Holt and Yao 2001: 54, 58 (oviposition damage, comp. note) Kansas
Magicicada cassini (sic) Cooley and Marshall 2001: 828, 833–834, 836, 839–840, 843, 850, Table 1, Table 6 (communication, comp. note)
Magicicada cassini (sic) Ewart 2001b: 83, Table 3 (population density, comp. note) America
Magicicada casini (sic) Kritsky 2001b: 118 (comp. note)
Magicicada cassini (sic) Marshall 2001: 387, 398–399, Table 1 (listed, comp. note) Illinois
Magicicada cassini (sic) Ritchie 2001: 59–60, Fig 2 (life cycle, speciation, distribution, comp. note) North America
Magicicada cassini (sic) Sanborn 2001a: 16, Fig 5 (comp. note)
Magicicada cassini (sic) Sueur 2001: 35, 41, 46, Tables 1–2 (listed, comp. note)
Magicicada cassini (sic) Tunaz, Jurenka and Stanley 2001: 435 (comp. note)

Magicicada cassini (sic) Whiles, Callaham, Meyer, Brock and Charleton 2001: 178–182, Fig 2 (density, nutrients, comp. note) Kansas
Magicicada cassini (sic) Cook and Holt 2002: 215 (oviposition damage, comp. note) Kansas
Magicicada cassini (sic) Dietrich 2002: 159, Fig 3 (illustrated) North America
Magicicada cassini (sic) Fonseca and Revez 2002a: 1291 (comp. note)
Magicicada cassini (sic) Gerhardt and Huber 2002: 33, 35, 45, 153, 156, 186, 260, 271, 390–392, Figs 2.13B-C, Fig 5.14, Fig 11.8 (illustrated, song, oscillogram, hearing, distribution, comp. note)
Magicicada cassini (sic) Nation 2002: 260, Table 9.3 (comp. note)
Magicicada cassini (sic) Page, Cruickchank and Johnson 2002: 367 (sequence, comp. note)
Magicicada cassinii Sanborn 2002a: 226 (comp. note)
Magicicada cassinii Sanborn 2002b: 458, 465, Table 1 (thermal biology, comp. note)
Magicicada cassinii Sanborn, Breitbarth, Heath and Heath 2002: 443 (comp. note)
Magicicada cassinii Sanborn, Noriega and Phillips 2002c: 368 (comp. note)
Magicicada cassinii Sueur 2002a: 391 (comp. note)
Magicicada cassinii Sueur 2002b: 124 (comp. note)
Magicicada cassini (sic) Sueur 2002b: 129 (listed)
Magicicada cassini (sic) Sueur and Aubin 2002b: 133 (comp. note)
Magicicada cassini (sic) Cooley, Simon and Marshall 2003: 153–155, 155–157, Fig 3, Table 1 (speciation, comp. note) Virginia, West Virginia, Connecticut, Maryland, North Carolina, New Jersey, New York, Pennsylvania, Iowa, Illinois, Missouri, Kansas, Nebraska, Oklahoma, Texas, Ohio, Georgia, South Carolina, Delaware, Indiana, Kentucky, Michigan, Tennessee, Wisconsin, Indiana, Kentucky
Magicicada cassini (sic) Dietrich 2003: 66, Fig 2(8) (illustrated) Illinois
Magicicada cassini (sic) Marshall, Cooley and Simon 2003: 433–434, Fig 1 (distribution, comp. note)
Magicicada cassini (sic) Sueur and Aubin 2003b: 486 (comp. note)
Magicicada cassini (sic) Chen 2004: 20 (life cycle, comp. note)
Magicicada cassini (sic) Cooley 2004: 754 (comp. note)
Magicicada cassini (sic) Cooley and Marshall 2004: 666 (comp. note)
Cicada cassinii Kritsky 2004a: 32, 69–70, 76 (described, illustrated, comp. note)
Magicicada cassini (sic) Kritsky 2004a: 87–88, 111–113 (illustrated, key, comp. note)
Magicicada cassinii Zyla 2004a: 486 (comp. note) Maryland
Magicicada cassinii Zyla 2004b: 488 (comp. note)
Magicicada cassini (sic) Ahern, Frank and Raupp 2005: 2134–2135 (control) North America
Magicicada cassini (sic) Boulard 2005d: 345 (comp. note)
Magicicada cassini (sic) Dietrich 2005: 503, Fig 1B (illustrated) Illinois
Magicicada cassini (sic) Edwards, Faivre, Crist, Sitvarin and Zyla 2005: 273 (comp. note) Pennsylvania
Magicicada cassini (sic) Grant 2005: 169–171, Figs 1–2 (life cycle, phylogeny, comp. note) North America
Magicicada cassini (sic) Greenfield 2005: 3, 5 (comp. note)
Magicicada cassini (sic) Heckel, Keener, Khang and Lee 2005: 2, 9 (mercury bioaccumulation, comp. note) Ohio
Magicicada cassini (sic) Kondratieff, Schmidt, Opler and Garhart 2005: 96–97, Fig 57 (illustrated, comp. note) Oklahoma
Magicicada cassinii Kondratieff, Schmidt, Opler and Garhart 2005: 216 (listed) Oklahoma
Magicicada cassinii Moulds 2005a: 133 (comp. note)
Magiciada cassini (sic) Kritsky, Webb, Folsom and Pfiester 2005: 65, 69, Fig 3 (distribution, comp. note) Indiana, Ohio
Magicicada cassini (sic) Pierce, Gibb and Waltz 2005: 455 (comp. note) Indiana
Magicicada cassini (sic) Robinson 2005: 217 (comp. note)
Magicicada cassinii Sanborn 2005c: 114, Table 1 (comp. note)
Magicicada cassini (sic) Yang 2005: 523–525 (resource pulse, comp. note) West Virginia
Magicicada cassinii Boulard 2006g: 97 (comp. note)
Magicicada cassinii Cook and Holt 2006: 809, 811 (dispersal, comp. note) Kansas
Magicicada cassini (sic) English, English, Dukes, Smith 2006: 246 (emergence, comp. note) Arkansas
Magicicada cassini (sic) Gilbert and Klass 2006: 79, 81, Table 1 (distribution, comp. note) New York
Magicicada cassinii Gilbert and Klass 2006: 81 (comp. note) New York
Magicicada cassini (sic) Nahirney, Forbes, Morris, Chock and Wang 2006: 2017–2018, 2020–2021, Figs 1–2 (timbal muscle structure, illustrated, comp. note)

Magicicada cassini (sic) Oberdörster and Grant 2006: 410, 412, 415–416 (ecology, density) New Jersey
Magicicada cassini (sic) Strauβ and Lakes-Harlan 2006: 2–5, Fig 1 (embryonic development, comp. note) Ohio
Magicicada cassini (sic) Sueur, Windmill and Robert 2006: 4116 (comp. note)
Magicicada cassini (sic) Yang 2006: 2995 (oviposition sites, comp. note) West Virginia
Magicicada cassini (sic) Cooley, Simon, Marshall, Slon and Ehrhardt 2007: 662, Fig 1 (speciation, distribution, comp. note)
Magicicada cassini (sic) Eaton and Kaufman 2007: 90 (comp. note) Eastern and Northern United States
Magicicada cassini (sic) Fontaine, Cooley, and Simon 2007: 1–2, 4–5, Fig 3B, Fig 4C, Fig 5, Table 3 (illustrated, mtDNA transmittance, comp. note) Virginia, New Jersey
Magicicada cassini (sic) Ganchev and Potamitis 2007: 289–290, Fig 6a (sonogram, comp. note)
Magicicada cassinii Heath and Sanborn 2007: 488 (comp. note)
Magicicada cassini (sic) Heckel and Keener 2007: 79–80, Fig 1 (mercury bioaccumulation, comp. note) Ohio
Magicicada cassini (sic) Keener, Khang, Lee and Heckel 2007: 435 (mercury content) Ohio
Magicicada cassini (sic) Muehleisen 2007: 2947 (song, comp. note)
Magicicada cassini (sic) Nowlin, González, Vanni, Stevens, Fields and Valente 2007: 2176 (resource pulse) Ohio
Magicicada cassini (sic) Oberdörster and Grant 2007a: 2–11, Figs 1–4, Tables 2–5 (flight system, flight speeds, comp. note) New Jersey
Magicicada cassini (sic) Oberdörster and Grant 2007b: 17–22, Tables 1–2 (sound system, predation, comp. note) New Jersey
Magicicada cassini (sic) Robinson, Sibrell, Broughton and Yang 2007: 372–374, Table 2 (heavy metal concentrations, comp. note) Virginia, West Virginia
Magicicada cassini (sic) Shiyake 2007: 7, 96, 98, Fig 161 (illustrated, distribution, listed, comp. note) United States
Magicicada cassini (sic) Storm and Whitaker 2007: 196 (predation, comp. note) Indiana
Magicicada cassini (sic) Hartbauer 2008: 105 (comp. note)
Magicicada cassini (sic) Marshall 2008: 2786–2787, 2792, Fig 26, Table 7 (distribution, comp. note) United States
Magicicada cassini (sic) Nation 2008: 273, Table 10.1 (comp. note)
Magicicada cassini (sic) Storm and Whitaker 2008: 350, 354 (predation, comp. note) Indiana
Magicicada cassini (sic) Brown and Zuefle 2009: 346 (comp. note)
Magicicada cassini (sic) Clay, Shelton and Winkle 2009a: 1689 (listed)
Magicicada cassini (sic) Clay, Shelton and Winkle 2009b: 278, 285 (listed, comp. note)
Magicicada cassini (sic) Gourley and Kuang 2009: 4653 (population model) USA
Magicicada cassinii Holliday, Hastings and Coelho 2009: 4–5, Table 1 (predation, comp. note)
Magicicada cassini (sic) Hughes, Nuttall, Katz and Carter 2009: 961 (comp. note) Illinois
Magiciada cassini (sic) Kritsky, Hoelmer and Noble 2009: 83 (distribution, comp. note) Indiana, Ohio
Magiciada cassini (sic) Kritsky and Smith 2009: 81 (distribution, comp. note) Indiana, Illinois, Wisconsin, Iowa
Magicicada cassini (sic) McCutcheon, McDonald and Moran 2009a: 5, 9, Fig 4 (symbiont, comp. note) Illinois
Magicicada cassini (sic) Pinto, Quartau, and Bruford 2009: 280 (comp. note)
Magicicada cassini (sic) Pray, Nowlin, and Vanni 2009: 183 (ecosystem nutrients, comp. note)
Magicicada cassinii Sanborn and Heath 2009: 14–19, Figs 1–3, Table 1 (comp. note) Tennessee, Oklahoma, Illinois
Magicicada cassini (sic) Vandegrift and Hudson 2009: 203 (predation) Pennsylvania
Magicicada cassini (sic) Cristensen and Fogel 2010: 211–212, 215 (biochemistry, comp. note) Pennsylvania
Magicicada cassini (sic) Demichelis, Manino, Sartor, Cifuentes and Patetta 2010: 528 (comp. note)
Magicicada cassini (sic) Johnson, Hagen and Martinko 2010: 151 (comp. note) Kansas
Magicicada cassini (sic) Langley, Anderson, Stanley, and Frey 2010: 189–190, Table 1 (predation, comp. note) Kansas
Magicicada cassinii Sanborn, Heath, Phillips, Heath and Noriega 2010: 71 (thermoregulation, comp. note)
Magicicada cassini (sic) Saulich 2010: 1132 (life cycle, comp. note)
Magicicada cassini (sic) Speer, Clay, Bishop and Creech 2010: 175 (comp. note)
Magicicada cassini (sic) Zheng, Zhang and Wang 2010: 749, 752, Table 3 (mercury content, comp. note) China

***M. neotredecim* Marshall & Cooley, 2000**

Magicicada neotredecim Brown 1999: 34–35 (life cycle, distribution, comp. note) United States, Arkansas

Magicicada neotredecim Marshall and Cooley 2000: 1314–1325, Fig 1, Figs 4–8, Table 1, Tables 3–5 (n. sp., described, song, distribution, comp. note) Missouri, Illinois

Magicicada neotredecim Simon, Tang, Dalwadi, Staley, Deniega and Unnasch 2000: 1331–1334, Fig 3 (speciation, comp. note)

Magicicada neotredecim Cooley and Marshall 2001: 828, 834, 839–840, 851, Table 1 (communication, comp. note)

Magicicada neotredecim Marshall 2001: 387, 394, Table 1 (listed, comp. note) Illinois, Indiana

Magicicada neotredecim Ritchie 2001: 59–60, Figs 1–2 (life cycle, speciation, distribution, comp. note) North America

Magicicada neotredecim Sueur 2001: 41, Table 1 (listed)

Magicicada neotredecim Gerhardt and Huber 2002: 390–391, Fig 11.8 (distribution, comp. note)

Magicicada neotredecim Lee, Choe, Lee and Woo 2002: 10 (comp. note)

Magicicada neotredecim Sanborn 2002a: 226–227, Fig 2 (distribution, comp. note)

Magicicada neotredecim Cooley, Simon and Marshall 2003: 153–157, Figs 3–4, Table 1 (speciation, sonogram, comp. note) Arkansas, Indiana, Illinois, Kentucky, Missouri, Oklahoma

Magicicada neotredecim Cox and Carleton 2003: 428–429 (evolution, comp. note) Missouri, Illinois

Magicicada decim spp. (sic) Cox and Carleton 2003: 428 (evolution, comp. note)

Magicicada neotredecim Marshall, Cooley and Simon 2003: 433–436, Fig 1 (distribution, comp. note)

Magicicada neotredecim Izzo and Gray 2004: 832 (comp. note)

Magicicada neotredecim Kritsky 2004a: 88–89, 112–113 (illustrated, key, comp. note)

Magicicada neotredecim Grant 2005: 169–172, Figs 1–2 (life cycle, phylogeny, comp. note) North America

Magicicada neotredecim Sueur 2005: 216 (song, comp. note)

Magicicada neotredecim West-Eberhard 2005: 6546 (speciation, comp. note)

Magicicada neotredecim Cooley, Marshall, Hill and Simon 2006: 856–866, Fig 1, Fig 3, Table 1, Table 4, Table 6 (character displacement, sonogram, comp. note) Illinois, Arkansas, Indiana

Magicicada neotredecim Jang and Gerhardt 2006: 156 (female preference, comp. note)

Magicicada neotredecim Seabra, Pinto-Juma and Quartau 2006: 844 (comp. note)

Magicicada neotredecim Cooley 2007: 90 (comp. note)

Magicicada neotredecim Cooley, Simon, Marshall, Slon and Ehrhardt 2007: 662–670, Figs 1–5, Tables 1–2 (speciation, distribution, sonogram, comp. note)

Magicicada neotredecim Eaton and Kaufman 2007: 90 (comp. note) Southern and Central United States

Magicicada neotredecim Fontaine, Cooley, and Simon 2007: 1 (illustrated, mtDNA transmittance, comp. note)

Magicicada neotredecim Oberdörster and Grant 2007b: 21 (comp. note)

Magicicada neotredecim Shiyake 2007: 98 (comp. note)

Magicicada neotredecim Marshall 2008: 2786–2787, 2793–2794, Fig 26, Table 7 (distribution, comp. note) United States

Magicicada neotredecim Clay, Shelton and Winkle 2009b: 278, 285 (listed, comp. note)

Magicicada neotredecim Cooley, Kritsky, Edwards, Zyla, Marshall et al. 2009: 107, 110 (distribution, comp. note)

Magicicada neotredecim Yoshimura, Hayashi, Tanaka, Tainaka and Simon 2008: 292 (evolution, comp. note)

Magicicada neotredecim Sanborn and Heath 2009: 14–19, Figs 1–3, Table 1 (comp. note) Illinois

Magicicada neotredecim Saulich 2010: 1132 (life cycle, comp. note)

Magicicada neotredecim Speer, Clay, Bishop and Creech 2010: 175 (comp. note)

***M. septendecim* (Linnaeus, 1758)** = *Cicada septendecim* = *Tettigonia septendecim* = *Tettigonia septendedecim* (sic) = *Cicada septemdecim* (sic) = *Tettigonia septemdecim* (sic) = *Tettigonia septendecem* (sic) = *Tettigonia septendecim* = *Tibicen septendecim* = *Cicada 17-decim* = *Tibicen septemdecim* (sic) = *Tibicina septemdecim* = *Tibicim* (sic) *septendecim* = *Cicada septedecim* (sic) = *Tibicena* (sic) *septendecim* = *Ciccada* (sic) *septemdecim* (sic) = *Magicicada septendcim* (sic) = *Megacicada* (sic) *septemdecim* (sic) = *Magicada* (sic) *septendecim* = *Tibicina septendceim* (sic) = *Cicada septednecim* (sic) = *Ciccada* (sic) *septemdecim* (sic) = *Magicicada septendecem* (sic) = *Magicicada septemdecim* (sic) = *Magicicada sependecim* (sic) = *Migicicada* (sic) *septendecim* = *Magicicada septemdecem* (sic) = *Magicicada septendecin* (sic) = *Magicicada septendecium* (sic) = *Magicicada sepiendecim* (sic) = *Magicicada septimdecim* (sic) = *Magicicada septidecim* (sic) = *Magicicada septindecim* (sic) = *Magicicada* *septemdicim* (sic) = *Magicicada septendecimthe* (sic) = *Magicicada septendedim* (sic) = *Magicicada septemdecima* (sic) = *Tettigonia costalis* Fabricius, 1798 = *Cicada costalis*

Cicada septendecim Barton 1804: 56 (comp. note) Equals *Tettigonia septendecim*

Cicada septendecim Van Rensselaer 1828: 224 (life cycle, predation, behavior) New York

Cicada septendecim Carus 1829: 151 (comp. note)

Cicada septendecim Thomas 1832: 188 (distribution, hosts, comp. note) New York

Cicada septemdecim (sic) Spence 1851: 103 (ovipostition damage, behavior) Maryland

Cicada septemdecim (sic) Wild 1852: xviii-xix (life cycle, comp. note) Maryland

Cicada septendecim Walsh 1866: 33 (comp. note) Ohio

Cicada septendecim Walsh 1867: 56 (song, coloration, comp. note) Ohio

Cicada septendecim Chambers 1868: 69 (range, life cycle, comp. note) Massachusetts, Louisiana

Cicada septendecim Walsh and Riley 1868a: 36 (sting, comp. note)

Cicada septemdecim (sic) Walsh and Riley 1868b: 63–64, 68–72 (life cycle, predation, broods, species, comp. note) Connecticut, Pennsylvania, Michigan, Wisconsin, Illinois, Ohio, Massachusetts, Long Island, New York, Virginia, Oklahoma, Arkansas, Missouri, New Jersey, Iowa, Kansas, Indiana, North Carolina, Maryland, South Carolina, Georgia

Cicada septemdecim (sic) Walsh and Riley 1869a: 202 (broods, comp. note) Connecticut, Massachusetts

Cicada septemdecim (sic) Wardner 1869: 117 (oviposition damage, fungal host, comp. note) Ohio

Cicada septemdecim (sic) Walsh 1870: 335 (distribution, life cycle, comp. note)

Cicada septendecim Riley 1872: 31–32 (distribution, comp. note) Massachusetts, Pennsylvania, Ohio, Illinois, Michigan, Iowa, Wisconsin

Cicada septendecim Peck 1879: 19 (fungal host, comp. note) New Jersey, New York

Cicada septendecim Strecker 1879: 256 (comp. note) Texas

Cicada septemdecim (sic) Swinton 1880: 227 (song, comp. note) United States

Cicada septemdecim (sic) Hasselt 1881: 192 (comp. note)

Cicada septendecim Anonymous 1884: 567 (comp. note) New York, Massachusetts, Vermont, Pennsylvania, Delaware, Maryland, Virginia, District of Columbia, Ohio, Michigan, Indiana, Kentucky

Cicada septemdecim (sic) Claus 1884: 571 (comp. note)

Cicada septendecim Ward 1885: 476 (asynchronous emergence, comp. note) Virginia

Cicada septemdecim (sic) Wheeler 1889: 644 (appendage embryology, comp. note)

Cicada (sic) Wheeler 1889: 644–645 (appendage embryology, comp. note)

Cicada septendecim Webster 1892: 87 (distribution, comp. note) Indiana

Tibicen septendecim Van Duzee 1893: 410 (listed) New York

Cicada septendecim Osborn 1896: 125 (comp. note) Iowa

Cicada septendecim Webster 1898: 225 (distribution, comp. note) Indiana

Cicada septendecim Britcher 1899: 737 (comp. note) New York

Cicada septendecim Davis 1906a: 91 (comp. note) New York, Staten Island, Long Island

Tibicen septendecim Johnson 1906: 159 (comp. note) Massachusetts

Tibicen (Cicada) septendecim Surface 1907: 71–72, Plate VI (illustrated, life cycle, hosts, control, comp. note)

Cicada septendecim Davis 1908: 1 (comp. note) New York, Staten Island, New Jersey

Cicada septemdecim (sic) d'Utra 1908: 352 (comp. note) United States

Cicada septendecim Davis 1911b: 1 (comp. note) New York, Staten Island, New Jersey

Tibicen septendecim Davis 1911c: 120 (comp. note) New York, Staten Island, North Carolina, New Jersey

Tibicen septendecim Gortner 1911a: 119 (comp. note)

Tibicen septendecim Gortner 1911b: 153 (coloration, comp. note) New Jersey

Cicada septendecim Davis 1912: 1 (comp. note) New York, Staten Island, New Jersey

Tibicen septendecim Webster 1912: 471 (distribution, comp. note) Iowa

Cicada septendecim Davis 1913: 99 (comp. note) New York, Staten Island, New Jersey

Tibicina septendecim Davis 1919b: 341 (comp. note) Missouri

Tibicena (sic) *septendecim* Snodgrass 1919: 392 (illustrated, natural history, comp. note)

Cicada septendecim Speare 1919: 116 (fungal host) District of Columbia

Tibicena (sic) *septendecim* Snodgrass 1921: 1 (illustrated, mouthpart structure, comp. note)

Tibicina septendecim Speare 1919: 72, 81, Plate 5, Figs 1–2 (fungal host) District of Columbia

Ciccada (sic) *septemdecim* (sic) Moreira 1921: 126 (comp. note)
Tibicina septendecim Rau 1922: 5 (predation) Missouri
Tibicina septendecim Snodgrass 1923: 445, Fig 30 (illustrated, acoustic behavior, comp. note)
Tibicina septendecim Forbush 1924: 468 (predation, comp. note) Massachusetts
Tibicina septendecim Snodgrass 1927: 1 (illustrated, head structure, mouthpart structure, comp. note)
Magicicada septendecim Goldstein 1929: 394 (fungal host, comp. note) New Jersey, New York, Staten Island
Tibicina septendecim Kishida 1929: 117 (comp. note) Equals *Magicicada septendecim*
Magicicada septendecim Snodgrass 1930: Plate 7 (illustrated, natural history, comp. note)
Tibicen septendecim Weber 1931: 74, 81–82, 103, 117, 122, 129–131, 133, Fig 76, Figs 82–83, Fig 102, Fig 115, Fig 119, Fig 127, Fig 128d, Fig 129, Fig 131 (head illustrated, digestive system illustrated, larval foreleg illustrated, mesothoracic muscles illustrated, oviposition, larva illustrated, eclosion illustrated, comp. note)
Magicicada septendecim Anonymous 1932: 424 (comp. note) Delaware, Georgia, Illinois, Indiana, Kentucky, Maryland, Missouri, North Carolina, Ohio, Oklahoma, Pennsylvania, South Carolina, Virginia, West Virginia, Wisconsin
Cicada septendecim Davis, J.J. 1933: 220 (comp. note) Indiana
Cicada septendecim Davis 1935: 264 (comp. note) Indiana
Magicicada septendecim Hyslop 1935: 1, map (illustrated, natural history, distribution, comp. note) Alabama, Delaware, Georgia, Illinois, Indiana, Kentucky, Maryland, District of Columbia, Massachusetts, Michigan, Missouri, New Jersey, New York, North Carolina, Ohio, Pennsylvania, Tennessee, Vermont, Virginia, West Virginia, Wisconsin
Tibicina septendecim Showalter 1935: 38–39, Fig 4, Fig 6 (illustrated, coloration, comp. note)
Magicicada septendecim Snodgrass 1935: 171, 200, 228, 254, 330, 332–333, 335, 337–338, 386, 449, 598–599, 616, Fig 95A, Fig 111C, Fig 127A, Fig 140L, Fig 177B, Figs 178–181, Fig 182A, Fig 210B, Fig 237, Fig 304D, Fig 305J, Fig 316 (illustrated, morphology, comp. note)
Magicicada septendecim Snodgrass 1938: 234–235, Plate 22, Fig F (illustrated, lorum and hypophyarynx morphology, comp. note)
Tibicina septemdecim (sic) Varley 1939: 99 (comp. note)
Magicicada septendecim Hyslop 1940: 1, map (illustrated, natural history, distribution, comp. note) Georgia, Illinois, Indiana, Kentucky, Maryland, Massachusetts, New Jersey, New York, North Carolina, Ohio, Pennsylvania, Tennessee, Virginia, West Virginia, Wisconsin
Magicicada septendecim Strandine 1940: 177, 182–183 (soil impact, habitat, comp. note) Illinois
Cicada septemdecim (sic) Bergier 1941: 169–170, 219 (human food, comp. note) United States
Magicicada septendecim Thoennes 1941: 830 (apple pest, peach pest, oviposition damage, comp. note) Missouri
Tibicen septemdecim (sic) Zeuner 1944: 113 (comp. note)
Magicicada septendecim Craighead 1949: 124–125, Fig 22 (oviposition damage, comp. note) United States
Magicicada septemdecima (sic) Gómez-Menor 1951: 57 (comp. note)
Magicicada septendecim Wright 1951: 11 (adult illustrated, comp. note)
Magicicada septendecim Deay 1953: 203–204 (comp. note) Indiana, Pennsylvania, Maryland, Virginia, West Virginia, South Carolina, Georgia, New York, Wisconsin, Massachusetts, Ohio, Illinois, Iowa, Michigan, North Carolina
Cicada septemdecim (sic) Pierantoni 1953: 36 (symbiont, comp. note)
Magicicada septendecim Halkka 1959: 37, Table 1 (listed, comp. note)
Magicicada septendecim Fitch 1960: 209 (predation, comp. note) Pennsylvania
Tibicina septemdecim (sic) Pachas 1960: 5 (comp. note)
Magicicada septendecim Hendricks 1963: 26 (periodicity, comp. note)
Magicicada septendecim McCoy 1965: 40–41, 43 (key, comp. note) Arkansas
Magicicada septendecim Heinrich 1967: 44 (comp. note)
Tibicen septendecim Weber 1968: 50–52, 69, 84–86, 136, 230, 235, 243, 248, 277, 342–343, 357, 368, 449, 490, Figs 39–40, Fig 172e, Fig 178, Fig 211, Fig 253, Fig 260, Fig 266 (larval foreleg illustrated, turret, mesothoracic muscles illustrated, phototaxis, digestive system illustrated, epithelium illustrated,

oviposition, larva illustrated, eclosion illustrated, comp. note)

Magicicada septendecim Borror and White 1970: Plate 4 (illustrated, comp. note)

Magicicada septendecim Soper 1974: 2801-B (fungal host, comp. note)

Magicicada septendecim Princen 1975: 28 (predation) Illinois

Magicicada septendecim Johnson and Lyon 1976: 343–435, Plate 209, Fig E (illustrated, natural history, comp. note) North America

Magicicada septendecim Kang, Ferguson-Miller and Margoliash 1977: 919 (comp. note)

Tibicen septemdecim (sic) Villiers 1977: 58–59, Fig 31 (nymph illustrated, comp. note) America

Magicicada septendecim Itô 1978: 158 (comp. note) America

Magicicada septendecim Mattson 1980: 133 (life cycle, comp. note)

Magicicada septendecim Karban 1981a: 260–263, Figs 1–3, Tables 1–4 (mating, fecundity, comp. note) New York

Magicicada septendecim Karban 1981b: 385–389, Fig 2 (dispersal, comp. note) Iowa, New York

Magicicada septendecim Popov 1981: 276 (comp. note)

Magicicada septendecim Simon, Karban and Lloyd 1981: 1526 (density, distribution, comp. note) Long Island

Magicicada septendecim White 1981: 727–730, 732–737, Fig 1, Fig 3, Fig 5, Table I, Tables IV-V (oviposition damage, comp. note) Kentucky

Magicicada septendecim White and Lloyd 1981: 219–220, 225–226, Fig 4 (egg mortality, comp. note) Illinois

Magicicada septendecim Itô 1982a: 108–111, Fig 1, Photos 1–3, Table 3 (life cycle evolution, illustrated, comp. note)

Magicicada septendecim Karban 1982a: 75 (host growth, comp. note) New York

Magicicada septendecim Karban 1982b: 321–327, Figs 3–8 (reproductive success, density, predation, comp. note) New York, Iowa

Magicicada septendecim Lloyd, White and Stanton 1982: 852–855, 857, Tables 1–2 (dispersal, comp. note) Ohio, Kansas

Magicicada septendecim Maier 1982a: 14–22, Figs 1–5, Tables 1–2 (emergence pattern, predation, comp. note) Connecticut

Magicicada septendecim Maier 1982b: 430–438, Figs 1–2, Tables 1–2 (density, distribution, comp. note) Connecticut

Magicicada septendecim Simon and Lloyd 1982: 275, 279, 286–288, Table 1 (comp. note) Illinois

Magicicada septendecim White, Lloyd and Karban 1982: 476–480, Fig 1, Figs 4–7, Table 2 (hatching success, comp. note) Ohio

Magicicada septendecim Huber 1983a: 116–117, 119–121, Fig 8, Fig 10A, Fig 11A (illustrated, neural responses, oscillogram, comp. note)

Magicicada septendecim Huber 1983b: 5–6, 9, Fig 4–5, Fig 8, Fig 21, Fig 27 (illustrated, neural responses, sonogram, oscillogram, comp. note)

Magicicada septendecim Karban 1983a: 227 (host growth, comp. note) Pennsylvania

Magicicada septendecim Lloyd and Karban 1983: 299–302, Table 1 (chorusing centers, comp. note) Ohio

Magicicada septendecim Lloyd, Kritsky and Simon 1983: 1162–1163, 1166, 1168, Fig 3 (life cycle, comp. note)

Magicicada septendecim Lloyd and White 1983: 294–300, Fig 1, Tables 1–2, Tables 4–6 (population density, comp. note) Kentucky

Magicicada septendecim Ostry and Anderson 1983: 1092 (oviposition damage, comp. note) Wisconsin

Magicicada septendecim Simon 1983b: 105, 108, 110–111, 114, Figs 2–3, Table 1 (comp. note) Eastern United States

Magicicada septendecim Strümpel 1983: 58, 107, Fig 141.1 (illustrated, comp. note)

Magicicada septemdecim (sic) Strümpel 1983: 71, 112, 117, 168, Fig 147B, Fig 155, Table 5, Table 9 (illustrated, comp. note)

Tibicen septemdecim (sic) Strümpel 1983: 108, Fig 143 (illustrated, comp. note)

Magicicada septendecim White, Ganter, McFarland, Stanton and Lloyd 1983: 282, 284, Fig 2, Tables 1–2 (flight capabilities, comp. note) Pennsylvania

Magicicada septendecim White and Lloyd 1983: 1246–1248, 1250, Fig 1.1, Tables 2–3 (fungal infection, comp. note) Kansas, Ohio

Magicicada septendecim Young and Josephson 1983a: 192, Table 1 (comp. note)

Magicicada septendecim Young and Josephson 1983b: 198–207, Figs 1A–1B, Figs 2D–2F, Fig 3, Figs 4E–4J, Figs 5C–5D, Fig 6, Figs 7B–DTables 1–2 (illustrated, song production, oscillogram, frequency spectrum, timbal muscle mechanics, comp. note) Kansas

Magicicada septendecim Chilcote and Stehr 1984: 53 (distribution, comp. note) Michigan, Ohio, West Virginia, Pennsylvania

Magicicada septendecim Huber 1984: 25–26, 30, Fig 2A, Fig 3, Fig 11 (illustrated, neural responses, sonogram, oscillogram, comp. note)

Magicicada septendecim Karban 1984a: 1656–1660, Figs 1–5, Tables 2–3 (density, mortality, comp. note) New York
Magicada septendecim Lloyd 1984: 79, 87–88, 90, Fig 6 (illustrated, life history, comp. note)
Tibicen septendecim Reichhoff-Riehm 1984: 90 (comp. note)
Magicicada septendecim Young and Josephson 1984: 286, Fig 1 (comp. note)
Magicicada septendecim Archie, Simon and Wartenberg 1985: 1263–1269 Figs 1–4 (brood evolution) United States
Magicicada septendecim Arnett 1985: 218 Fig 21.17 (life history, illustrated) Eastern United States
Magicicada septendecim Beyer, Patte, Sileo, Hoffman and Mulhern 1985: 72, Table 4 (metal contamination) Pennsylvania
Magicicada septendecim Claridge 1985: 302–303, 307, 309 (song, comp. note)
Magicicada septendecim Heliövaara and Väisänen 1985: 88–89, Fig 1 (natural history, life cycle evolution, comp. note) North America
Magicicada septendecim Huber 1985: 55 (neural responses, comp. note)
Magicicada septendecim Karban 1985: 270–271, Table 1 (density, host growth, comp. note) New York, Iowa
Magicicada septendecim Maier 1985: 1020–1021, 1024, Fig 1 (emergence, density, distribution, comp. note) Connecticut, North Carolina
Magicicada septemdecim (sic) Michelsen and Larsen 1985: 549 (comp. note)
Magicicada septendecim Pechuman 1985: 59–60 (distribution, predation, comp. note) New York
Magicicada septendecim Popov 1985: 42 (acoustic system, comp. note)
Magicicada septendecim Smith 1985: 4 (life cycle, comp. note) Arkansas
Magicicada septendecim White and Lloyd 1985: 605–609, Tables 1–3 (nymphal growth, comp. note) Ohio
Magicicada septendecim Karban 1986: 230–231, Fig 14.7 (survival, comp. note)
Magicicada septendecim Kirillova 1986b: 46 (listed) Equals *Cicada (Tibicen) septendecim* USA
Magicicada septendecim Kirillova 1986a: 123 (listed) Equals *Cicada (Tibicen) septendecim* USA
Magicicada septendecim Lloyd 1986: 363–365, Fig 2, Table 1 (illustrated, comp. note)
Magicicada septendecim Schildberger, Kleindienst, Moore and Huber 1986: 126 (hearing, comp. note) North America
Magicicada septendecim Strehl and White 1986: 179 (predation, comp. note) Illinois
Magicicada septendecim Benson 1987: 5 (life history)
Magicicada septendecim Davidson and Lyon 1987: 374–375, Fig 18.9 (natural history, illustrated, comp. note)
Magicicada septendecim Klein, Bock, Kafka and Moore 1987b: 94, 97 (comp. note)
Magicicada septendecim Lloyd and White 1987: 50–52, Fig 2 (illustrated, xylem feeding, comp. note) Virginia
Magicicada septendecim Parker and Wright 1987: 7 (life cycle, comp. note) United States
Magicicada septendecim Villet 1987a: 269 (comp. note) North America
Magicicada septendecim Weber, Moore, Huber and Klein 1987: 330–334, Fig 1, Fig 4 (song, sonogram, oscillogram, comp. note)
Magicicada septendecim Yaginuma 1987: 15–16, Fig 4 (illustrated, life history, comp. note)
Magicicada decim (sic) Yaginuma 1987: 15, Tables 1–2 (comp. note)
Magicicada septendecim Young and Kritsky 1987: 323–324 (distribution, comp. note) Indiana
Magicicada septendecim Hubbell 1988: 9 (illustrated, natural history, comp. note)
Magicicada septendecim Klein, Bock, Kafka and Moore 1988: 154, 166 (comp. note)
Magicicada septendecim Kritsky 1988: 168 (distribution, comp. note) Ohio
Magicicada septendecim Martin and Simon 1988: 237 (comp. note) North America
Magicicada septendecim Simon 1988: 166–167, Fig 2 (illustrated, comp. note) Eastern United States
Magicicada septendecim Southcott 1988: 103 (parasitism, comp. note) United States
Magicicada septendecim Steward, Smith and Stephen 1988: 348 (predation, comp. note)
Magicicada septendecim Barbosa and Wagner 1989: 211 (oviposition damage)
Magicicada septendecim Ewing 1989: 37–38, 103–104, Fig 2.10b, c, Fig 4.8 (oscillogram, song frequency, auditory response,comp. note)
Magicicada septendecim Humphreys 1989: 99, 104–107 (turrets, comp. note) USA
Magicicada septendecim Luken and Kalisz 1989: 310 (soil disturbance, comp. note) Kentucky
Magicicada septendecim Boulard 1990b: 85 (comp. note)
Magicicada septendecim Hogmire, Baugher, Crim and Walter 1990: 2401 (control, comp. note) West Virginia
Magicicada septendecim Huber, Kleindienst, Moore, Schildberger and Weber 1990: 218–219, 221–227, Figs 1–6, Fig 7A, Figs 10–11

(neural responses, oscillogram, sonogram, comp. note)
Cicada septendecim Liu 1990: 45 (comp. note) China
Migicicada (sic) *septendecim* Liu 1990: 45 (comp. note)
Magicicada septendecim Martin and Simon 1990a: 1066–1067, 1071 (comp. note) North America
Magicicada septendecim Martin and Simon 1990b: 359–360, Fig 1 (illustrated, comp. note) North America
Magicicada septendecim Moore, Huber, Kleindienst, Schildberger and Weber 1990: 1 (neuronal responses, comp. note)
Magicicada septendecim Nakao 1990: 127 (life cycle, comp. note)
Magicicada septendecim Simon 1990: 310, 314, 321, Fig 8, Table 1 (wing variation, comp. note) Eastern United States
Magicicada septendecim Young 1990: 55 (comp. note)
Magicicada septendecim Eaton 1991: 9 (life cycle, comp. note) Ohio
Magicicada septendecim Flowers 1991: 24 (natural history, comp. note) Illinois
Magicicada septendecim Gogala 1991a: 3–4, 6–7, 9–10, Fig 1, Fig 8B, Fig 9, Figs 11–12 (illustrated, sonogram, life cycle, distribution, comp. note) Ohio, Kentucky, Indiana
Magicicada decim (sic) Gogala 1991b: 75–76 (genotype, distribution, comp. note)
Magicicada septendecim Johnson and Lyon 1991: 490–491, Plate 236G (illustrated, life cycle, comp. note)
Magiciada septendecim Kritsky and Young 1991: 45 (distribution, comp. note) Indiana
Magiciada septendecim Krohne, Couillard and Riddle 1991: 317 (predation) Indiana
Magicicada septendecim Reding and Guttman 1991: 323 (comp. note)
Magicicada septendecim Russell and Stoetzel 1991: 480 (egg nests, comp. note) Maryland
Magicicada septendecim Stoetzel and Russell 1991: 471, 477 (emergence, hosts, oviposition, predation, comp. note) Maryland
Magicicada septemdecem (sic) Toolson and Toolson 1991: 113 (comp. note)
Magicicada septendecim Williams and Smith 1991: 276, 286 (comp. note) Arkansas
Magicicada septendecim Bennet-Clark and Young 1992: 126, 148–149, 151 (song parameters, comp. note) Kansas, North America
Magicicada septendecim Bessin and Townsend 1992: 3 (control)
Magicicada septendecim Cox and Carleton 1992: 63 (comp. note)
Magicicada septendecim Hennig, Kleindienst, Weber, Huber, Moore and Popov 1992: 180 (comp. note)
Magicicada septendecim Hidvegi and Baugnée 1992: 122 (comp. note)
Magicicada septendecim Kritsky 1992: 38 (distribution, comp. note) Ohio
Magiciada septendecim Kritsky and Young 1992: 59 (distribution, comp. note) Indiana
Cicada septendecim Liu 1992: 8 (comp. note) China
Migicicada (sic) *septendecim* Liu 1992: 8 (comp. note)
Tibicen septendecim Moalla, Jardak and Ghorbel 1992: 35 (comp. note) America
Magicicada septendecim Sanborn and Davis 1992: 18 (wing morphology, flight, comp. note)
Magicicada septendecim Savinov 1992: 267 (comp. note)
Magicicada septendecim Villet 1992: 93 (comp. note)
Magicicada decim (sic) Wheeler, Williams and Smith 1992: 340 (comp. note)
Magicicada septendecim Elven 1993: 5 (comp. note) America
Magicicada septendecim Lane 1993: 56–57 (comp. note) United States
Magicicada septemdecim (sic) Long 1993: 214 (comp. note)
Magicicada septendecim Moore 1993: 273, 275–277, 279, 281, Fig 4, Fig 6 (distribution, song, comp. note)
Magicicada septendecim Moore, Huber, Weber, Klein and Bock 1993: 200, 206, 209, 217 (vision, phonotaxis, comp. note) Missouri
Magicicada septendecim Press and Whittaker 1993: 106 (comp. note)
Magicicada septendecim Savinov 1993: 1 (comp. note)
Magicicada septendecim Simon, McIntosh and Deniega 1993: 229–230, 232, 234, Fig 4, Tables 1–2 (density, distribution, comp. note) Nebraska, Missouri, Kansas
Magicicada septendecim White and Sedcole 1993a: 47, 49–50 (comp. note)
Magicicada septendecim Williams, Smith and Stephen 1993: 1149 (comp. note)
Magicicada septendecim Bennet-Clark 1994: 174, Table 3 (comp. note)
Magicicadada septendecim Bennet-Clark and Young 1994: 292 (comp. note)
Magicicada septemdecim (sic) Brown and Schmitt 1994: 1187 (comp. note) West Virginia
Magicicada septendecim Gäde and Janssens 1994: 807 (comp. note)

Magicicada septendecim Heliövaara, Väisänen and Simon 1994: 476–477, Table 1 (life cycle, comp. note) North America
Magicicada septendecim Woodall 1994: 89, 94 (life span, comp. note) North America
M[agicicada] septendecim Jiang, Yang, Tang, Xu and Chen 1995b: 229, Fig 3b (song frequency, comp. note)
Magicicada septendecim Raina, Pannell, Kochansky and Jaffe 1995: 929–930 (neuropeptide, comp. note) Maryland
Magicicada septendecim Sanborn, M.S. Heath, J.E. Heath and Noriega 1995: 457 (comp. note)
Magicicada septendecim Williams and Simon 1995: 271, 284, 287–288 (natural history, comp. note)
Magicicada septendecim Johnsen, Hendeveld, Blank and Yasak 1996: 358 (predation, comp. note) Indiana
Magiciada septendecim Kritsky and Simon 1996: 27 (comp. note) Ohio
Magicicada septendecim Maier 1996: 5 (broods, comp. note) Connecticut
Magicicada septendecim Marshall, Cooley, Alexander and Moore 1996: 165, 168–169 (distribution, comp. note) Michigan
Magicicada septendecim Schneider 1996: 28 (life cycle, comp. note)
Magicicada septendecim Simon, Nigro, Sullivan, Holsinger, Martin, et al. 1996: 924, Fig 1 (comp. note)
Magicicada septendecim Anonymous 1997a: 4 (life cycle) Ohio
Magicicada septendecim Alexander, Marshall and Cooley 1997: 25 (comp. note)
Magicicada septendecim Bosik 1997: 49, 112, 165 (approved common name)
Magicicada septendecim Chou, Lei, Li, Lu and Yao 1997: 23 (comp. note)
Magicicada septendecim Daws, Hennig and Young 1997: 184 (comp. note)
Magicicada decim (sic) Karban 1997: 447, 449–456, 458, Fig 2, Tables 2–4 (development time, comp. note) New York
Magicicada septendecim Matthews and Hildreth 1997: 37 (comp. note)
Magicicada septendecim Miller 1997: 225–226 (oviposition damage, control, comp. note) Illinois
Magicicada septendecim Poole, Garrison, McCafferty, Otte and Stark 1997: 331 (listed) Equals *Cicada septendecim* Equals *Cicada septemdecim* (sic) Equals *Tettigonia costalis* Equals *Tettigonia septendecem* (sic) North America
Magicicada decim (sic) Shelly and Whittier 1997: 279, Table 16–1 (listed, comp. note)
Magicicada septendecim Rodenhouse, Bohlen and Barrett 1997: 127, 131–133 (distribution, density, comp. note) Ohio
Magicicada septemdecem (sic) van der Zwet, Brown and Estabrook 1997: 36 (pear pest, comp. note) Virginia
Magicicada septendecim Blair 1998: 3 (life history) Kansas, Missouri, Arkansas, Oklahoma
Magicicada septendecim Cooley, Hammond and Marshall 1998: 191–162, 164, 166, Tables 1–2 (marking survivorship, comp. note) Virginia
Magicicada septendecim Hyche 1998: 3–4 (comp. note) Alabama
Magicicada septendecim Itô 1998: 495 (life cycle evolution, comp. note)
Magicicada decim (sic) Itô 1998: 495 (life cycle evolution, comp. note)
Magicicada septendecim Miller and Crowley 1998: 248 (oviposition damage, control, comp. note) Illinois
Magicicada septendecim Ostry and Anderson 1998: 2 (oviposition damage, comp. note)
Magicicada septendecim Tunaz, Bedick, Miller, Hoback, Rana and Stanley 1998: 925–926, 928–930, Tables 1–2 (response to bacterial infection, comp. note) Nebraska
Magicicada septidecim (sic) Tunaz, Bedick, Miller, Hoback, Rana and Stanley 1998: 927 (comp. note)
Magicicada septendecim Anonymous 1999a: 1 (natural history) Ohio, Maryland, Pennsylvania, Virginia, West Virginia
Magicicada septendecim Anonymous 1999b: 2 (natural history) West Virginia
Magicicada septendecim Brown 1999: 32, 34 (life cycle, distribution, comp. note) United States, Massachusetts
Magicicada septendecim Hoback, Rana and Stanley 1999: 356–360, Tables 1–4 (fatty acid metabolism and composition, comp. note) Nebraska
Magicicada septemdecim (sic) Hopper 1999: 551 (comp. note)
Magicicada septemdecim (sic) Kartal and Zeybekoglu 1999a: 62 (comp. note)
Cicada septemdecim (sic) Kritsky 1999: 23 (comp. note)
Magicicada septendecim Kritsky 1999: 24–25 (illustrated, comp. note)
Magiciada septendecim Kritsky, Smith and Gallagher 1999: 43 (distribution, comp. note) Ohio

Magicicada septendecim Sanborn 1999b: 2 (listed) North America
Magicicada septimdecim (sic) Stokes and Josephson 1999: 72A (predation, comp. note) Illinois
Magicicada septemdecim (sic) Villani, Allee, Díaz and Robbins 1999: 245, Fig 3D (illustrated)
Magicicada septendecim Arnett 2000: 299 Fig 21.17 (life history, illustrated) Eastern United States
Magicicada septendecim Bauernfeind 2000: 239–240 (distribution) Kansas
Magicicada septendecim Furlow 2000: 24 (life cycle)
Magicicada septendecim Irwin and Coelho 2000: 82 (comp. note) Illinois
Magicicada septendecim Karban and English-Loeb 2000: 251–254 (density, predation, comp. note) New York
Magicicada septendecim Marshall and Cooley 2000: 1314–1315, 1320, 1322–1324, Fig 1, Table 1 (distribution, comp. note)
Magicicada septendecim Maw, Footit, Hamilton and Scudder 2000: 49 (comp. note) Canada, Ontario
Magicicada septendecim Sahli and Ware 2000: 187–188 (comp. note)
Magicicada septendecim Sanborn 2000: 555 (comp. note)
Magicicada septendecim Simon, Tang, Dalwadi, Staley, Deniega and Unnasch 2000: 1326, 1329, 1331–1334, Fig 3 (speciation, comp. note)
Magicicada septendecim Smith, Kelly, Dean, Bryson, Parker et al. 2000: 126–127 (illustrated, comp. note) Kansas
Magicicada septendecim Sugden 2000: 435 (illustrated)
Magicicada septendecim Waldbauer 2000: 220 (life history, comp. note)
Magicicada septendecim Bednick, Tunaz, Nor Aliza, Putnam, Ellis and Stanley 2001: 108, 115 (comp. note)
Magicicada septemdecim (sic) Coffin 2001a: 122 (comp. note)
Magicicada septendecim Cook, Holt and Yao 2001: 54 (comp. note) Kansas
Magicicada septendecim Cooley and Marshall 2001: 828, 830, 834–845, 850, Fig 3, Fig 6, Table 1, Tables 4–6, Table 8 (communication, sonogram, comp. note)
Magicicada septendecim Ewart 2001b: 83, Table 3 (population density, comp. note) America
Magicicada septendecim Kritsky 2001a: 186 (illustrated)
Cicada septendecim Kritsky 2001a: 187 (comp. note)
Magicicada septendecim Marshall 2001: 387, 394, Table 1 (listed, comp. note)
Magicicada septendecim Ritchie 2001: 59–60, Figs 1–2 (illustrated, life cycle, speciation, distribution, comp. note) North America
Magicicada septendecim Sanborn 2001a: 16, Fig 5 (comp. note)
Magicicada septendecim Stumpner and von Helverson 2001: 164, Fig 2 (interneurons, comp. note)
Magicicada septendecim Sueur 2001: 35, 41, 46, Tables 1–2 (listed, comp. note)
Magicicada septendecim Sueur and Puissant 2001: 20 (comp. note)
Magicicada septendecim Tunaz, Jurenka and Stanley 2001: 435 (comp. note)
Magicicada septendecim Achtziger and Nigmann 2002: 3 (life cycle) North America
Magicicada septendecim Bashan, Akbas and Yurdakoc 2002: 379 (comp. note)
Magicicada septendecim Boulard 2002b: 175 (life cycle, comp. note)
Magicicada septendecim Büyükgüzel, Tunaz, Putnam and Stanley 2002: 436 (comp. note)
Magicicada septendecim Gerhardt and Huber 2002: 34–35, 45, 185–186, 390–391, Fig 2.13A, Fig 2.13C, Fig 11.8 (illustrated, song, oscillogram, hearing, distribution, comp. note)
Magicicada septendecim Nation 2002: 260, Table 9.3 (comp. note)
Magicicada septendecim Sanborn 2002a: 226, Fig 1 (illustrated, comp. note)
Magicicada (Cicada) septendecim Sanborn 2002a: 226 (comp. note)
Magicicada septendecim Sueur 2002b: 119, 124, 129 (listed, comp. note) Equals *Tibicina septendecim*
Magicicada septendecim Sueur and Aubin 2002b: 133 (comp. note)
Magicicada septendecim Woodall 2002: 89, 94 (life span, comp. note) North America
Magicicada septendecim Cooley, Simon and Marshall 2003: 153–157, Figs 2–4, Table 1 (speciation, illustrated, sonogram, comp. note) Virginia, West Virginia, Connecticut, Maryland, North Carolina, New Jersey, New York, Pennsylvania, Iowa, Illinois, Missouri, Kansas, Nebraska, Oklahoma, Texas, Ohio, Georgia, South Carolina, Delaware, Indiana, Kentucky, Michigan, Tennessee, Wisconsin, Indiana, Kentucky
Magicicada decim spp. Cox and Carleton 2003: 428 (evolution, comp. note)
Magicicada septendecim Cox and Carleton 2003: 428 (evolution, comp. note)

Magicicada septendecim Gogala and Trilar 2003: 9 (comp. note)
Magicicada septendecim Marshall, Cooley and Simon 2003: 433–436, Fig 1 (distribution, comp. note)
Magicicada septendecim Sueur and Aubin 2003b: 487 (comp. note)
Magicicada septendecim Agnello 2004: 168 (apple pest) North America
Magicicada septendecim Anonymous 2004: 367 (emergence) United States
Magicicada septendecim Chen 2004: 20 (life cycle, comp. note)
Magicicada septendecim Cooley 2004: 747, 753 (mating behavior, comp. note) Virginia
Magicicada septendecim Cooley and Marshall 2004: 650–651, 653–656, 658, 660, 663–664, 666–669, Figs 1–3, Tables 2–3 (mate choice, comp. note) Virginia, Illinois, Ohio, North Carolina
Magicicada septendecim Cooley, Marshall and Simon 2004: 198 (comp. note)
Magicicada septendecim Gogala, Trilar, Kozina and Duffels 2004: 14 (comp. note)
Magicicada septendecim Harwood and Obrycki 2004: 235 (comp. note)
Magicicada septendecim Hutchins, Evans, Garrison and Schlager 2004: 265, 269–270, Fig 11 (natural history, comp. note) Equals *Cicada septendecim*
Cicada septendecim Hutchins, Evans, Garrison and Schlager 2004: 269 (listed)
Magicicada septemdecim (sic) Jolivet 2004: 3–4 (life cycle, illustrated, distribution, comp. note) Virginia
Cicada septendecim Kritsky 2004a: 32, 69–72, 74, 76, 79 (described, comp. note)
Magicicada septendecim Kritsky 2004a: 87–89, 110–111, 113 (illustrated, key, comp. note)
Cicada septendecim Kritsky 2004b: 43 (comp. note)
Magicicada septendecim Nuorteva, Lodenius, Nagasawa, Tulisalo and Nuorteva 2004: 55 (comp. note) Pennsylvania
Magicicada septendecim Spichiger 2004a: 3 (comp. note) Pennsylvania
Magicicada septendecim Spichiger 2004b: 3 (comp. note) Pennsylvania
Magicicada septendecim Stokes and Josephson 2004: 280–281, 287–289, Fig 6 (timbal muscle, comp. note)
Magicicada septendecim Zyla 2004a: 485–486 (comp. note) Maryland
Magicicada septendecim Zyla 2004b: 488 (comp. note)
Magicicada septendecim Ahern, Frank and Raupp 2005: 2134–2135 (control) North America
Magicicada septendecim Boulard 2005d: 345 (comp. note)
Magicicada septendecim Champman 2005: 321–330, Figs 1–9 (timbal muscle ultrastructure, comp. note)
Magicicada septendedim (sic) Champman 2005: 322 (comp. note)
Magicicada septendecim Edwards, Faivre, Crist, Sitvarin and Zyla 2005: 273 (comp. note) Pennsylvania
Magicicada septendecim Grant 2005: 169–172, Figs 1–2 (life cycle, phylogeny, comp. note) North America
Magicicada septendecium (sic) Heckel, Keener, Khang and Lee 2005: 2, 8–9 (mercury bioaccumulation, comp. note) Ohio
Magicicada septemdecim (sic) Jolivet and Verma 2005: 96–97 (comp. note) USA
Magiciada septendecim Kritsky, Webb, Folsom and Pfiester 2005: 65 (distribution, comp. note) Indiana, Ohio
Magicicada septindecim (sic) Leskey and Bergh 2005: 2122 (comp. note) Virginia
Magicicada septendecim Moran, Tran and Gerardo 2005: 8803, 8805–8806, 8808, Figs 1–2, Fig 4, Table 1 (comp. note)
Magicicada septendecim 2005b: 396–397, 399, 417–419, 421, 424, Figs 56–59, Fig 61, Table 1 (higher taxonomy, phylogeny, listed, comp. note)
Magicicada septendecim Pierce, Gibb and Waltz 2005: 455 (comp. note) Indiana
Magicicada septendecim Remondino and Cappellini 2005: 1, 4, 6, Fig 9 (life cycle, illustrated)
Magicicada septendecim Robinson 2005: 217 (comp. note)
Magicicada septendecim Wang and St. Leger 2005: 938 (fungal host, comp. note)
Magicicada septendecim West-Eberhard 2005: 6546 (speciation, comp. note)
Magicicada septendecim Yang 2005: 523–524 (resource pulse, comp. note) West Virginia
Magicicada septendecim Boulard 2006g: 97 (comp. note)
Magicicada septendecim Chawanji, Hodson and Villet 2006: 386, 388 (comp. note)
Magicicada septendecim Cooley, Marshall, Hill and Simon 2006: 856–865, Fig 1, Figs 3–4, Table 1, Tables 3–5 (character displacement, sonogram, comp. note) New York, West Virginia, North Carolina
Magicicada septendecim Elliott and Hershberger 2006: 8–9, 21–22, 32, 184, 210–213, 216 (illustrated, sonogram, listed, comp. note) New York, Pennsylvania, New Jersey, Delaware, Maryland, West Virginia, Virginia,

North Carolina, South Carolina, Georgia, Connecticut, Illinois, Indiana, Kentucky, Michigan, Tennessee, Iowa, Wisconsin, Missouri, Kansas, Nebraska, Oklahoma, Texas, Massachusetts

Magicicada septendecim Gilbert and Klass 2006: 78–82, Fig 1 (distribution, comp. note) New York

Magicicada septendecim Grant 2006: 541 (predation, comp. note) New Jersey

Magicicada septendecim Marshall 2006: 102, 139 (illustrated, comp. note) North America

Magicicada septendecim Nahirney, Forbes, Morris, Chock and Wang 2006: 2018 (comp. note)

Magicicada septendecim Oberdörster and Grant 2006: 410–412, 416, Plate 1 (illustrated, ecology, density) New Jersey

Magicicada septendecimthe (sic) Pons 2006: 8 (fungal host, comp. note) United States

Magicicada septendecim Remondino and Cappellini 2006a: 1, 7–9, Fig 9 (life cycle, illustrated)

Magicicada septendecim Remondino and Cappellini 2006b: 485, 491, 496–497, Fig 10 (life cycle, illustrated)

Cicada septendecim Strauβ and Lakes-Harlan 2006: 3 (comp. note)

Magicicada septendecim Wang and St. Leger 2006: 6649 (fungal host, comp. note)

Magicicada septendecim Wheeler and Miller 2006: 38 (comp. note)

Magicicada septendecim Yang 2006: 2995 (oviposition sites, comp. note) West Virginia

Magicicada septendecim Anonymous 2007: 3 (emergence) West Virginia

Magicicada septendecim Cooley, Simon, Marshall, Slon and Ehrhardt 2007: 662–663, 665–667, 669–670, Fig 1, Fig 4 (speciation, distribution, comp. note)

Magicicada septendecim Eaton and Kaufman 2007: 90 (comp. note) Eastern and Northern United States

Magicicada septendecim Fontaine, Cooley, and Simon 2007: 1–2, 4–5, Fig 3A, Fig 4, Table 3 (illustrated, mtDNA transmittance, comp. note) Virginia, New Jersey

Magicicada septendecim Ganchev and Potamitis 2007: 289–290, Fig 5, Fig 6b (sonogram, comp. note)

Magicicada septendecim Heckel and Keener 2007: 79 (mercury bioaccumulation, comp. note) Ohio

Magicicada septendecim Heckel, Lee and Keener 2007: 16, Fig 1 (mercury bioaccumulation, illustrated, comp. note) Ohio

Magicicada septendecim Keener, Khang, Lee and Heckel 2007: 435 (mercury content) Ohio

Magicicada septendecim Muehleisen 2007: 2947 (song, comp. note)

Magicicada septendecim Oberdörster and Grant 2007a: 2–10, Figs 1–4, Tables 2–5 (flight system, flight speeds, comp. note) New Jersey

Magicicada septendecim Oberdörster and Grant 2007b: 16–22, Tables 1–2 (sound system, predation, comp. note) New Jersey

Magicicada septendecim Potamitis, Ganchev and Fakotakis 2007: 2 (spectrogram, comp. note)

Magicicada septendecim Robinson, Sibrell, Broughton and Yang 2007: 372–374, Table 2 (heavy metal concentrations, comp. note) Virginia, West Virginia

Magicicada septendecim Shiyake 2007: 7, 96, 98, Fig 162 (illustrated, distribution, listed, comp. note) United States

Magicicada septendecim Storm and Whitaker 2007: 196 (predation, comp. note) Indiana

Magicicada septendecim Agnello 2008: 249 (apple pest) North America

Magicicada septendecim Anonymous 2008: 3 (emergence) West Virginia

Magicicada septendecim Marshall 2008: 2786–2787, 2792–2793, Fig 26, Fig 29, Table 7 (distribution, illustrated, comp. note) United States

Magicicada septendecim Nation 2008: 273, Table 10.1 (comp. note)

Magicicada septendecim Sanborn, Phillips and Gillis 2008: 17 (comp. note)

Magicicada septendecim Storm and Whitaker 2008: 350 (predation, comp. note) Indiana

Magicicada septendecim Visser and Ellers 2008: 1316, Table 1 (listed, comp. note)

Magicicada septendecim Way 2008: 271 (predation, comp. note) Massachusetts

Magicicada septendecim Yoshimura, Hayashi, Tanaka, Tainaka and Simon 2008: 292 (evolution, comp. note)

Magicicada septendecim Yang 2008: 1498–1499 (resource pulse, comp. note) West Virginia

Magicicada septendecim Brown and Zuefle 2009: 346–347, 351–353 (oviposition, comp. note) Delaware

Magicicada septendecim Clay, Shelton and Winkle 2009a: 1689 (listed)

Magicicada septendecim Clay, Shelton and Winkle 2009b: 278 (listed)

Magicicada septendecim Cooley, Kritsky, Edwards, Zyla, Marshall et al. 2009: 110–111, Fig 5 (distribution, comp. note)

Magicicada septendecim Dearborn, MacDade, Robinson, Dowling Fink and Fink 2009: 520 (density, bird predation) Missouri, Illinois, Pennsylvania

Magicicada septendecim Gourley and Kuang 2009: 4653 (population model) USA
Magicicada septendecim Holliday, Hastings and Coelho 2009: 4–5, Table 1 (predation, comp. note) Illinois
Magicicada septendecim Hughes, Nuttall, Katz and Carter 2009: 961 (comp. note) Illinois
Magiciada septendecim Kritsky, Hoelmer and Noble 2009: 83 (distribution, comp. note) Indiana, Ohio
Magiciada septendecim Kritsky and Smith 2009: 81 (distribution, comp. note) Indiana, Illinois, Wisconsin, Iowa
Magicicada septendecim Marshall and Hill 2009: 7 (comp. note)
Magicicada septendecim Ostry and Anderson 2009: 395 (oviposition damage, comp. note) Wisconsin
Magicicada septendecim Pinto, Quartau, and Bruford 2009: 280 (comp. note)
Magicicada septendecim Pray, Nowlin, and Vanni 2009: 183 (ecosystem nutrients, comp. note)
Magicicada septendecim Sanborn and Heath 2009: 14–18, Figs 1–3, Table 1 (comp. note) Illinois, Tennessee
Magicicada septendecim Vandegrift and Hudson 2009: 203 (predation) Pennsylvania
Magicicada septendecim Aspöck and Aspöck 2010: 60 (listed)
Magicicada septendecim Cristensen and Fogel 2010: 211–212, 214, 216 (biochemistry, comp. note) Washington, DC, Pennsylvania
Magicicada septendecim Saulich 2010: 1132 (life cycle, comp. note)
Magicicada septendecim Smith, Cooley and Westerman 2010: 74 (oviposition, nymph density, comp. note) New York
Magicicada septendecim Speer, Clay, Bishop and Creech 2010: 175 (comp. note)

***M. septendecula* Alexander & Moore, 1962** = *Magicicada septendicula* (sic) = *Migicicada* (sic) *septendecula*
Magicicada septendecula McCoy 1965: 41, 43 (key, comp. note) Arkansas
Magicicada septendecula Itô 1978: 158 (comp. note) America
Magicicada septendecula Mattson 1980: 133 (life cycle, comp. note)
Magicicada septendecula White 1981: 727–730, 732–737, Fig 1, Figs 3–6, Table I, Tables IV-V (oviposition damage, comp. note) Kentucky
Magicicada septendecula White and Lloyd 1981: 219–220 (egg mortality, comp. note) Illinois
Magicicada septendecula Itô 1982b: 110, Table 1 (life cycle evolution, comp. note)
Magicicada septendecula Karban 1982b: 327 (comp. note)
Magicicada septendecula Lloyd, White and Stanton 1982: 852, 855 (comp. note)
Magicicada septendecula Simon and Lloyd 1982: 275 (comp. note)
Magicicada septendecula White, Lloyd and Karban 1982: 477–480, Fig 5, Fig 7, Table 2 (hatching success, comp. note) Ohio
Magicicada septendecula Huber 1983b: 5–6, 9, 27, 34, Fig 4–5, Fig 8, Fig 21 (illustrated, neural responses, sonogram, oscillogram, comp. note)
Magicicada septendecula Karban 1983a: 227 (host growth, comp. note) Pennsylvania
Magicicada septendecula Lloyd, Kritsky and Simon 1983: 1162 (life cycle, comp. note)
Magicicada septendecula Lloyd and White 1983: 294–302, Fig 1, Tables 1–2, Tables 4–6 (population density, comp. note) Kentucky
Magicicada septendecula Simon 1983b: 105 (comp. note) Eastern United States
Magicicada septendecula Strümpel 1983: 58, 168, Table 9 (comp. note)
Magicicada septendecula White, Ganter, McFarland, Stanton and Lloyd 1983: 282 (comp. note)
Magicicada septendecula White and Lloyd 1983: 1246–1247, Figs 1.5–1.6 (fungal infection, comp. note) Kansas, Ohio
Magicicada septendecula Huber 1984: 25–26, Fig 2C, Fig 3, Fig 11 (illustrated, neural responses, sonogram, oscillogram, comp. note)
Magicada septendecula Lloyd 1984: 79, 87–90, Fig 6 (illustrated, life history, comp. note)
Magicicada septendecula Heliövaara and Väisänen 1985: 88–89 (natural history, life cycle evolution, comp. note) North America
Magicicada septendecula Maier 1985: 1024–1025 (comp. note) North Carolina
Magicicada septendecula Smith 1985: 4 (life cycle, comp. note) Arkansas
Magicicada septendecula White and Lloyd 1985: 605–606 (nymphal growth, comp. note) Ohio
Magicicada septendecula Lloyd 1986: 363–366, Fig 2, Table 1 (illustrated, comp. note)
Magicicada septendecula Schildberger, Kleindienst, Moore and Huber 1986: 126 (hearing, comp. note) North America
Magicicada septendecula Strehl and White 1986: 179 (predation, comp. note) Illinois
Magicicada septendecula Lloyd and White 1987: 50 (comp. note)
Magicicada septendecula Davidson and Lyon 1987: 374 (natural history, comp. note)
Magicicada septendecula Villet 1987a: 269 (comp. note) North America

Magicicada septendecula Yaginuma 1987: 15–16, Fig 4 (illustrated, life history, comp. note)
*Magicicada decula (*sic) Yaginuma 1987: 15, Tables 1–2 (comp. note)
Magicicada septendecula Young and Kritsky 1987: 323 (distribution, comp. note) Indiana
Magicicada septendecula Kritsky 1988: 168 (distribution, comp. note) Ohio
Magicicada septendecula Simon 1988: 166 (comp. note) Eastern United States
Magicicada septendecula Steward, Smith and Stephen 1988: 348 (predation, comp. note)
Magicicada septendecula Villet 1988: 71 (comp. note)
Magicicada septendecula Luken and Kalisz 1989: 310 (soil disturbance, comp. note) Kentucky
Migicicada (sic) *septendecula* Liu 1990: 45 (comp. note)
Magicicada septendecula Martin and Simon 1990a: 1066 (comp. note) North America
Magicicada septendecula Martin and Simon 1990b: 360 (comp. note) North America
Magicicada septendecula Nakao 1990: 127 (life cycle, comp. note)
Magicicada septendecula Simon 1990: 310, 321, Fig 8 (wing variation, comp. note) Eastern United States
Magicicada septendecula Eaton 1991: 9 (life cycle, comp. note) Ohio
Magicicada septendecula Gogala 1991a: 6–7, 9–10, Fig 8F, Fig 9, Figs 11–12 (sonogram, life cycle, comp. note) Ohio, Kentucky, Indiana
Magicicada septemdecula (sic) Gogala 1991a: 8 (life cycle, distribution, comp. note)
Magicicada decula (sic) Gogala 1991b: 76 (comp. note)
Magiciada septendecula Kritsky and Young 1991: 45 (distribution, comp. note) Indiana
Magiciada septendecula Kritsky and Young 1991: 59 (distribution, comp. note) Indiana
Magiciada septendecula Krohne, Couillard and Riddle 1991: 317 (predation) Indiana
Magicicada septendecula Reding and Guttman 1991: 323 (comp. note)
Magicicada septendecula Russell and Stoetzel 1991: 480 (egg nests, comp. note) Maryland
Magicicada septendecula Stoetzel and Russell 1991: 471, 477 (emergence, hosts, oviposition, predation, comp. note) Maryland
Magicicada septendecula Williams and Smith 1991: 276 (comp. note) Arkansas
Magicicada septendecula Cox and Carleton 1992: 63 (comp. note)
Magicicada septendecula Kritsky 1992: 38 (distribution, comp. note) Ohio
Migicicada (sic) *septendecula* Liu 1992: 8 (comp. note)
Magicicada decula (sic) Wheeler, Williams and Smith 1992: 340 (comp. note)
Magicicada septendecula Moore 1993: 273, 275–277, Fig 4, Fig 6 (distribution, song, comp. note)
Magicicada septendecula Moore, Huber, Weber, Klein and Bock 1993: 217 (comp. note)
Magicicada septendecula Simon, McIntosh and Deniega 1993: 230 (comp. note)
Magicicada septendecula Heliövaara, Väisänen and Simon 1994: 476–477, Table 1 (life cycle, comp. note) North America
Magicicada septendecula Kalisz 1994: 118–122, Fig 1 (emergence patterns) Kentucky
Magicicada septendecula Raina, Pannell, Kochansky and Jaffe 1995: 929–930 (neuropeptide, comp. note) Maryland
Magicicada septendecula Williams and Simon 1995: 271 (natural history, comp. note)
Magicicada septendecula Johnsen, Hendeveld, Blank and Yasak 1996: 358 (predation, comp. note) Indiana
Magicicada septendecula Simon, Nigro, Sullivan, Holsinger, Martin, et al. 1996: 924, Fig 1 (comp. note)
Magicicada decula (sic) Karban 1997: 447 (development time, comp. note) New York
Magicicada septendecula Poole, Garrison, McCafferty, Otte and Stark 1997: 331 (listed) North America
Magicicada septendecula Rodenhouse, Bohlen and Barrett 1997: 127, 131–133 (distribution, density, comp. note) Ohio
Magicicada decula (sic) Shelly and Whittier 1997: 279, Table 16–1 (listed, comp. note)
Magicicada septendecula Hyche 1998: 3–4 (comp. note) Alabama
Magicicada septendecula Anonymous 1999a: 1 (natural history) Ohio, Maryland, Pennsylvania, Virginia, West Virginia
Magicicada septendecula Kritsky 1999: 24–25 (illustrated, comp. note)
Magiciada septendecula Kritsky, Smith and Gallagher 1999: 43 (distribution, comp. note) Ohio
Magicicada septendecula Sanborn 1999b: 2 (listed) North America
Magicicada septendecula Bauernfeind 2000: 239–240 (distribution) Kansas
Magicicada septendecula Irwin and Coelho 2000: 82 (comp. note) Illinois
Magicicada septendecula Marshall and Cooley 2000: 1314–1315, Fig 1, Table 1 (distribution, comp. note)
Magicicada septendecula Smith, Kelly, Dean, Bryson, Parker et al. 2000: 127 (comp. note) Kansas

Magicicada septendecula Waldbauer 2000: 220 (life history, comp. note)
Magicicada septendecula Cook, Holt and Yao 2001: 54 (comp. note)
Magicicada septendecula Cooley and Marshall 2001: 828, 840, Table 1 (communication, comp. note)
Magicicada septendecula Sahli and Ware 2000: 187 (comp. note)
Magicicada septendecula Simon, Tang, Dalwadi, Staley, Deniega and Unnasch 2000: 1326 (speciation, comp. note)
Magicicada septendecula Marshall 2001: 387, Table 1 (listed, comp. note)
Magicicada septendecula Ritchie 2001: 59–60, Fig 2 (life cycle, speciation, distribution, comp. note) North America
Magicicada septendecula Sueur 2001: 41, 46, Tables 1–2 (listed)
Magicicada septendecula Gerhardt and Huber 2002: 390–391, Fig 11.8 (distribution, comp. note)
Magicicada septendecula Sanborn 2002a: 226 (comp. note)
Magicicada septendecula Cooley, Simon and Marshall 2003: 153–155, Fig 3, Table 1 (speciation, comp. note) Virginia, West Virginia, Connecticut, Maryland, North Carolina, New Jersey, New York, Pennsylvania, Iowa, Illinois, Missouri, Kansas, Nebraska, Oklahoma, Texas, Ohio, Georgia, South Carolina, Delaware, Indiana, Kentucky, Michigan, Tennessee, Wisconsin, Indiana, Kentucky
Magicicada septendecula Marshall, Cooley and Simon 2003: 433–434, Fig 1 (distribution, comp. note)
Magicicada septendecula Chen 2004: 20 (life cycle, comp. note)
Magicicada septendecula Kritsky 2004a: 87–88, 112–113 (illustrated, key, comp. note)
Magicicada septendecula Zyla 2004a: 486 (comp. note) Maryland
Magicicada septendecula Zyla 2004b: 488 (comp. note)
Magicicada septendecula Edwards, Faivre, Crist, Sitvarin and Zyla 2005: 273 (comp. note) Pennsylvania
Magicicada septendecula Grant 2005: 169–171, Figs 1–2 (life cycle, phylogeny, comp. note) North America
Magicicada septendecula Heckel, Keener, Khang and Lee 2005: 2, 9 (mercury bioaccumulation, comp. note) Ohio
Magiciada septendecula Kritsky, Webb, Folsom and Pfiester 2005: 65 (distribution, comp. note) Indiana, Ohio
Magicicada septendecula Pierce, Gibb and Waltz 2005: 455 (comp. note) Indiana
Magicicada septendecula Robinson 2005: 217 (comp. note)
Magicicada septendecula Nahirney, Forbes, Morris, Chock and Wang 2006: 2018 (comp. note)
Magicicada septendecula Oberdörster and Grant 2006: 410, 412 (ecology, density) New Jersey
Magicicada septendecula Cooley, Simon, Marshall, Slon and Ehrhardt 2007: 662, Fig 1 (speciation, distribution, comp. note)
Magicicada septendecula Eaton and Kaufman 2007: 90 (comp. note) Eastern and Northern United States
Magicicada septendecula Fontaine, Cooley, and Simon 2007: 1–2, 4–5, Fig 3C, Table 3 (mtDNA transmittance, comp. note) Virginia
Magicicada septendecula Ganchev and Potamitis 2007: 289–290, Fig 6c (sonogram, comp. note)
Magicicada septendecula Heckel and Keener 2007: 79 (mercury bioaccumulation, comp. note) Ohio
Magicicada septendecula Keener, Khang, Lee and Heckel 2007: 435 (mercury content) Ohio
Magicicada septendecula Muehleisen 2007: 2947 (song, comp. note)
Magicicada septendecula Oberdörster and Grant 2007b: 17, 19, 21–22 (comp. note)
Magicicada septendecula Robinson, Sibrell, Broughton and Yang 2007: 372–374, Table 2 (heavy metal concentrations, comp. note) Virginia, West Virginia
Magicicada septendecula Shiyake 2007: 7, 96, 98, Fig 163 (illustrated, distribution, listed, comp. note) United States
Magicicada septendecula Storm and Whitaker 2007: 196 (predation, comp. note) Indiana
Magicicada septendecula Marshall 2008: 2786–2787, 2792, Fig 26, Table 7 (distribution, comp. note) United States
Magicicada septendecula Storm and Whitaker 2008: 350 (predation, comp. note) Indiana
Magicicada septendecula Brown and Zuefle 2009: 346 (comp. note)
Magicicada septendecula Clay, Shelton and Winkle 2009a: 1689 (listed)
Magicicada septendecula Clay, Shelton and Winkle 2009b: 278, 285 (listed, comp. note)
Magicicada septendecula Gourley and Kuang 2009: 4653 (population model) USA
Magiciada septendecula Kritsky, Hoelmer and Noble 2009: 83 (distribution, comp. note) Indiana, Ohio

Magicicada septendecula Kritsky and Smith 2009: 81 (distribution, comp. note) Indiana, Illinois, Wisconsin, Iowa

Magicicada septendecula Pray, Nowlin, and Vanni 2009: 183 (ecosystem nutrients, comp. note)

Magicicada septendecula Sanborn and Heath 2009: 14–19, Figs 1–3, Table 1 (comp. note) Illinois

Magicicada septendecula Vandegrift and Hudson 2009: 203 (predation) Pennsylvania

Magicicada septendecula Cristensen and Fogel 2010: 211 (comp. note)

Magicicada septendecula Saulich 2010: 1132 (life cycle, comp. note)

Magicicada septendecula Speer, Clay, Bishop and Creech 2010: 175 (comp. note)

***M. tredecassini* Alexander & Moore, 1962** = *Magicicada trecassini* (sic) = *Magicicada tredecessini* (sic) = *Magicicada tredicassini* (sic)

Magicicada tredecassini McCoy 1965: 41, 43 (key, comp. note) Arkansas

Magicicada tredecassini Itô 1978: 158 (comp. note) America

Magicicada tredecassini Itô 1982b: 110–111, Table 1 (life cycle evolution, comp. note)

Magicicada tredecassini Simon and Lloyd 1982: 275, 292 (comp. note)

Magicicada tredecassini Huber 1983b: 10 (comp. note)

Magicicada tredecassini Simon 1983b: 105–108, 110, Figs 2–3, Table 1 (comp. note) Eastern United States

Magicicada tredecassini Strümpel 1983: 58, 168, Table 9 (comp. note)

Magicicada tredecassini James, Williams and Smith 1985: 37 (distribution, comp. note) Arkansas

Magicicada tredecassini Smith 1985: 4 (life cycle, comp. note) Arkansas

Magicicada tredecassini Lloyd 1986: 363 (comp. note)

Magicicada tredecassini Davidson and Lyon 1987: 374 (natural history, comp. note)

Magicicada tredecassini Simon 1988: 167 (comp. note) Eastern United States

Magicicada tredecassini Steward, Smith and Stephen 1988: 348–349 (predation, comp. note) Arkansas

Magicicada tredecassini Martin and Simon 1990a: 1066 (comp. note) North America

Magicicada tredecassini Martin and Simon 1990b: 360 (comp. note) North America

Magicicada tredecassini Nakao 1990: 127 (life cycle, comp. note)

Magicicada tredecassini Simon 1990: 310, 314, 321, Fig 8, Table 1 (wing variation, comp. note) Eastern United States

Magicicada tredecassini Stephen, Wallis and Smith 1990: 369 (predation, comp. note) Arkansas, Missouri

Magicicada tredecassini Young and Kritsky 1990: 25 (distribution, comp. note) Indiana, Kentucky, Illinois

Magicicada tredecassini Anonymous 1991: 227 (emergence dynamics)

Magicicada tredecassini Gogala 1991a: 6, Fig 8C (life cycle, distribution, comp. note)

Magicicada cassini (sic) Gogala 1991b: 76 (comp. note)

Magicicada tredecassini Karlin, Williams, Smith and Sugg 1991: 213–214, 216–219, Tables 2–3 (genetic variability, comp. note) Arkansas

Magicicada tredecassini Reding and Guttman 1991: 323 (comp. note)

Magicicada tredecassini Smith, Williams, and Stephen 1991: 253 (emergence, mortality, predation, comp. note) Arkansas

Magicicada tredecassini Williams and Smith 1991: 276–277 (chorusing sites, comp. note) Arkansas

Cassini (sic) Williams and Smith 1991: 277, 279–281, 284–290, Figs 4–5, Tables 1–2 (chorusing sites, comp. note) Arkansas

Magicicada tredecassini Cox and Carleton 1992: 63 (comp. note)

Magicicada cassini (sic) Wheeler, Williams and Smith 1992: 340–341 (comp. note) Arkansas

Magicicada tredecassini Williams and Smith 1992: 353 (oviposition sites, chorusing sites) Arkansas

Magicicada tredecassini Long 1993: 214 (comp. note)

Magicicada tredecassini Moore 1993: 273, 275–277, 279–280, Fig 4, Fig 6 (distribution, song, comp. note)

Magicicada tredecassini Williams and Smith 1993: 489 (oviposition sites) Arkansas

Magicicada tredecassini Williams, Smith and Stephen 1993: 1147 (emergence, mortality, predation, comp. note) Arkansas

Magicicada tredecassini Heliövaara, Väisänen and Simon 1994: 476–477, Table 1 (life cycle, comp. note) North America

Magicicada tredecassini Hennig, Weber, Moore, Kleindienst, Huber and Popov 1994: 54 (comp. note)

Magicicada tredecassini Karlin, Stout, Adams, Duke and English 1994: 89–92, Tables 1–3 (genetic variability, comp. note) Arkansas

Magicicada tredecassini Williams and Simon 1995: 271, 276, 287–288 (natural history, comp. note)

Magicicada tredecassini Simon, Nigro, Sullivan, Holsinger, Martin, et al. 1996: 924, Fig 1 (comp. note)
Magicicada tredecassini Alexander, Marshall and Cooley 1997: 25 (comp. note)
Magicicada tredecassini Poole, Garrison, McCafferty, Otte and Stark 1997: 331 (listed) North America
Magicicada tredecassini Hyche 1998: 3–4 (comp. note) Alabama
Magicicada tredecassini Hestir and Cain 1999: 243 (predation, comp. note) Arkansas
Cicada tredecassini Kritsky 1999: 24 (comp. note)
Magicicada tredecassini Sanborn 1999b: 2 (listed) North America
Magicicada tredecassini Irwin and Coelho 2000: 82 (comp. note) Illinois
Magicicada tredecassini Marshall and Cooley 2000: 1313–1315, 1317, Fig 1, Table 1 (distribution, comp. note)
Magicicada tredecassini Sahli and Ware 2000: 187 (comp. note) Virginia
Magicicada tredecassini Simon, Tang, Dalwadi, Staley, Deniega and Unnasch 2000: 1326, 1333 (speciation, comp. note)
Magicicada tredecassini Smith, Kelly, Dean, Bryson, Parker et al. 2000: 127 (comp. note) Kansas
Magicicada tredecassini Waldbauer 2000: 220 (life history, comp. note)
Magicicada tredecassini Cooley and Marshall 2001: 828, 840, 843, Table 1 (communication, comp. note)
Magicicada tredecassini Ewart 2001b: 83, Table 3 (population density, comp. note) America
Magicicada tredecassini Marshall 2001: 387, 398–399, Table 1 (listed, comp. note) Illinois
Magicicada tredecassini Ritchie 2001: 59–60, Fig 2 (life cycle, speciation, distribution, comp. note) North America
Magicicada tredecassini Sueur 2001: 41, Table 1 (listed)
Magicicada tredecassini Gerhardt and Huber 2002: 390–391, Fig 11.8 (distribution, comp. note)
Magicicada tredecassini Sanborn 2002a: 226 (comp. note)
Magicicada tredecassini Cooley, Simon and Marshall 2003: 153–155, Fig 3, Table 1 (speciation, illustrated, comp. note) Alabama, Arkansas, Georgia, Indiana, Illinois, Kentucky, Louisiana, Maryland, Missouri, Mississippi, North Carolina, Oklahoma, South Carolina, Tennessee, Virginia
Magicicada tredecassini Marshall, Cooley and Simon 2003: 433–434, Fig 1 (distribution, comp. note)
Magicicada tredecassini Kritsky 2004a: 87, 89, 112–113 (illustrated, key, comp. note)
Magicicada tredecassini Zyla 2004a: 486 (comp. note)
Magicicada tredecassini Grant 2005: 169–171, Figs 1–2 (life cycle, phylogeny, comp. note) North America
Magicicada tredecassini Robinson 2005: 217 (comp. note)
Magicicada tredecassini Cooley, Simon, Marshall, Slon and Ehrhardt 2007: 662, Fig 1 (speciation, distribution, comp. note)
Magicicada tredecassini Eaton and Kaufman 2007: 90 (comp. note) Southern and Central United States
Magicicada tredecassini Fontaine, Cooley, and Simon 2007: 1, 4 (mtDNA transmittance, comp. note)
Magicicada tredecassini Shiyake 2007: 7, 97–98, Fig 164 (illustrated, distribution, listed, comp. note) United States
Magicicada tredecassini Marshall 2008: 2786–2787, Fig 26, Table 7 (distribution, comp. note) United States
Magicicada tredicassini (sic) Way 2008: 271 (predation, comp. note)
Magicicada tredecassini Clay, Shelton and Winkle 2009b: 278 (listed)
Magicicada tredecassini Sanborn and Heath 2009: 14–19, Figs 1–3, Table 1 (comp. note)
Magicicada tredecassini Matthews and Matthews 2010: 310, Fig 8.7 (activity, comp. note)
Magicicada tredecassini Saulich 2010: 1132 (life cycle, comp. note)
Magicicada tredecassini Speer, Clay, Bishop and Creech 2010: 175 (comp. note)

***M. tredecim* (Walsh & Riley, 1868)** = *Cicada tredecim* = *Cicada septendecim tredecim* = *Cicada tredecem* (sic) = *Tibicen tredecim* = *Cicada trydecim* (sic) = *Tibicen septendecim tredecim* = *Tibicina septendecim tredecim* = *Magicicada septendecim tredecim* = *Magicicada septendecim tridecim* (sic) = *Megacicada* (sic) *tredecim* = *Magicicada tridecim* (sic) = *Magicicada tredecem* (sic) = *Magicicada tredicim* (sic)
Cicada tredecim Walsh and Riley 1868b: 63–64, 68, 70–71 (life cycle, predation, broods, species, comp. note) Illinois, Mississippi, Louisiana, Missouri, Oklahoma, Arkansas, Kentucky, Tennessee, Alabama, Georgia, Kansas
Cicada tredecim Walsh and Riley 1869b: 217 (asynchronous emergence, comp. note) Missouri
Cicada tredecim Walsh and Riley 1869c: 251 (asynchronous emergence, comp. note) Mississippi

Cicada tredecim Walsh 1870: 335 (distribution, life cycle, comp. note) Ohio
Cicada tredecim Riley 1872: 31–33 (distribution, comp. note) Georgia, Florida, Alabama, Mississippi, Tennessee, Illinois, Louisiana
Cicada tredecim Anonymous 1884: 567 (comp. note)
Magicicada septendecim tredecim Anonymous 1932: 336–337 (comp. note) Alabama, Arkansas, Georgia, Illinois, Iowa, Kansas, Kentucky, Mississippi, Missouri, North Carolina, Oklahoms, South Carolina, Tennessee, Virginia
Magicicada septendecim tredecim Anonymous 1933a: 336–337 (comp. note) Alabama, Arkansas, Georgia, Illinois, Iowa, Kansas, Kentucky, Mississippi, Missouri, North Carolina, Oklahoms, South Carolina, Tennessee, Virginia
Magicicada tredicim (sic) Deay 1953: 203 (comp. note) Indiana
Magicicada tredecim Deay 1953: 205 (comp. note) Indiana, Ohio, Illinois
Magicicada tredecim McCoy 1965: 40–41, 43 (key, comp. note) Arkansas
Magicicada tredecim Itô 1978: 158 (comp. note) America
Magicicada tredecim Itô 1982b: 110–111, Table 1 (life cycle evolution, comp. note)
Magicicada tredecim Simon and Lloyd 1982: 275, 286 (comp. note)
Magicicada tredecim Simon 1983b: 105, 108, 110–111, 114, Figs 2–3, Table 1 (comp. note) Eastern United States
Magicicada tredecim Strümpel 1983: 58, 168, Table 9 (comp. note)
Magicicada tredecim Archie, Simon and Wartenberg 1985: 1263–1269 Figs 1–4 (brood evolution) United States
Magicicada tredecim James, Williams and Smith 1985: 37 (distribution, comp. note) Arkansas
Magicicada tredecim Maier 1985: 1024 (comp. note) South Carolina
Magicicada tredecim Smith 1985: 4 (life cycle, comp. note) Arkansas
Magicicada tredecim Ettling 1986: 32 (predation, comp. note)
Magicicada tredecim Lloyd 1986: 363 (comp. note)
Magicicada tredecim Davidson and Lyon 1987: 374 (natural history, comp. note)
Magicicada tredecim Martin and Simon 1988: 237 (comp. note) North America
Magicicada tredecim Simon 1988: 167 (comp. note) Eastern United States
Magicicada tredecim Steward, Smith and Stephen 1988: 348 (predation, comp. note) Arkansas, Missouri
Magicicada tredecim Martin and Simon 1990a: 1066–1067, 1071 (comp. note) North America
Magicicada tredecim Martin and Simon 1990b: 360 (comp. note) North America
Magicicada tredecim Nakao 1990: 127 (life cycle, comp. note)
Magicicada tredecim Simon 1990: 310, 314, 321, Fig 8, Table 1 (wing variation, comp. note) Eastern United States
Magicicada tredecim Simon, Pääbo, Kocher and Wilson 1990: 235–238, 241–242, Figs 2–4 (gene sequence, comp. note)
Magicicada tredecim Stephen, Wallis and Smith 1990: 369 (predation, comp. note) Arkansas
Magicicada tredecim Young and Kritsky 1990: 25 (distribution, comp. note) Indiana, Kentucky, Illinois
Magicicada tredecim Gogala 1991a: 6, Fig 8A (life cycle, distribution, comp. note)
Magicicada decim (sic) Gogala 1991b: 75–76 (genotype, distribution, comp. note)
Magicicada tredecim Reding and Guttman 1991: 323 (comp. note)
Magicicada tredecim Simon 1991: 41, Fig 1 (RNA sequence, comp. note)
Magicicada tredecem (sic) Toolson and Toolson 1991: 110–114, Figs 1–3, Tables 1–5 (evaporative cooling, endothermy, comp. note) Mississippi
Magicicada tredecim Williams and Smith 1991: 276–277 (chorusing sites, comp. note) Arkansas
Decim (sic) Williams and Smith 1991: 277, 279–280, 284–290, Figs 4–5, Tables 1–2 (chorusing sites, comp. note) Arkansas
Magicicada tredecim Cox and Carleton 1992: 63 (comp. note)
Magicicada decim (sic) Wheeler, Williams and Smith 1992: 340 (comp. note)
Magicicada tredecim Long 1993: 214 (comp. note)
Magicicada tredecim Moore 1993: 273, 275–277, 279, Fig 4, Fig 6 (distribution, song, comp. note)
Magicicada tredecim Simon, McIntosh and Deniega 1993: 229, 232, 234, Fig 4, Table 4 (comp. note)
Magicicada tredecim Williams, Smith and Stephen 1993: 1147 (comp. note) Arkansas
Magicicada tredecim Heliövaara, Väisänen and Simon 1994: 476–477, Table 1 (life cycle, comp. note) North America

Magicicada tredecim Simon, Frati, Beckenbach, Crespi, Liu and Flook 1994: 655, 699–700, Fig 2 (gene sequence, comp. note)
Magicicada tredecim Williams and Simon 1995: 271, 284, 287–288 (natural history, comp. note)
Magicicada tredecim Simon, Nigro, Sullivan, Holsinger, Martin, et al. 1996: 924, Fig 1 (comp. note)
Magicicada tredecim Alexander, Marshall and Cooley 1997: 25 (comp. note)
Magicicada tredecim Poole, Garrison, McCafferty, Otte and Stark 1997: 331 (listed) Equals *Cicada tredecem* (sic) Equals *Magicicada tridecim* (sic) North America
Magicicada tredecim Cohen, Stark, Gawthrop, Burke and Thayer 1998: 476, Table 1 (sequence, comp. note)
Magicicada tredecim Goruch 1998: 1 (comp. note) South Carolina
Magicicada tredecim Hyche 1998: 3–4 (comp. note) Alabama
Magicicada tredecim Sanborn 1998: 96–97 (thermal biology, comp. note)
Magicicada tredecim Brown 1999: 32, 34–35 (life cycle, distribution, comp. note) United States, Louisiana
Magicicada tredecim Hestir and Cain 1999: 243 (predation, comp. note) Arkansas
Cicada tredecim Kritsky 1999: 24 (comp. note)
Magicicada tredecim Sanborn 1999b: 2 (listed) North America
Magicicada tredecim Hickson, Simon and Perry 2000: 531–532, 536, Fig 1, Fig 5 (sequence, comp. note)
Magicicada tredecim Irwin and Coelho 2000: 82 (comp. note) Illinois
Magicicada tredecim Marshall and Cooley 2000: 1314–1322, Fig 1, Table 1, Figs 4–8, Tables 3–5 (distribution, song, comp. note) Missouri, Illinois
Magicicada tredecim Sahli and Ware 2000: 187–188 (oviposition sites, comp. note) Virginia
Magicicada tredecim Simon, Tang, Dalwadi, Staley, Deniega and Unnasch 2000: 1326, 1331–1332, 1334, Fig 3 (speciation, comp. note)
Magicicada tredecim Smith, Kelly, Dean, Bryson, Parker et al. 2000: 127 (comp. note) Kansas
Magicicada tredecim Waldbauer 2000: 220 (life history, comp. note)
Magicicada tredecim Cooley and Marshall 2001: 828, 834, 840, 844, 851, Table 1, Table 7 (communication, comp. note)
Magicicada tredecim Marshall 2001: 387, 394, Table 1 (listed, comp. note) Illinois, Indiana
Magicicada tredecim Ritchie 2001: 59–60, Fig 2 (life cycle, speciation, distribution, comp. note) North America
Magicicada tredecim Sueur 2001: 41, Table 1 (listed)
Magicicada tredecim Boulard 2002b: 175 (life cycle, comp. note)
Magicicada tredecim Gerhardt and Huber 2002: 390–391, Fig 11.8 (distribution, comp. note)
Magicicada tredecim Lee, Choe, Lee and Woo 2002: 10 (comp. note)
Magicicada tredecim Sanborn 2002a: 226–227, Fig 2 (distribution, comp. note)
Magicicada tredecim Sanborn 2002b: 463 (thermal biology, comp. note)
Magicicada tredecim Sanborn, Noriega and Phillips 2002c: 368 (comp. note)
Magicicada tredecim Cooley, Simon and Marshall 2003: 153–157, Figs 3–4, Table 1 (speciation, sonogram, comp. note) Alabama, Arkansas, Georgia, Indiana, Illinois, Kentucky, Louisiana, Maryland, Missouri, Mississippi, North Carolina, Oklahoma, South Carolina, Tennessee, Virginia
Magicicada decim spp. Cox and Carleton 2003: 428–429 (evolution, comp. note)
Magicicada tredecim Cox and Carleton 2003: 428–429 (evolution, comp. note)
Magicicada tredecim Marshall, Cooley and Simon 2003: 433–436, Fig 1 (distribution, comp. note)
Magicicada tredecim Izzo and Gray 2004: 832 (comp. note)
Cicada tredecim Kritsky 2004a: 76 (comp. note)
Magicicada tredecim Kritsky 2004a: 87, 89, 112–113 (illustrated, key, comp. note)
Magicicada tredecim Zyla 2004a: 485–486, Fig 1 (distribution, comp. note) Maryland
Magicicada tredecim Zyla 2004b: 488 (comp. note)
Magicicada tredecim Grant 2005: 169–172, Figs 1–2 (life cycle, phylogeny, comp. note) North America
Magicicada tredecim Macnamara 2005: 42 (illustrated, comp. note) Illinois
Magicicada tredecim Remondino and Cappellini 2005: 1, 4 (life cycle)
Magicicada tredecim Robinson 2005: 217 (comp. note)
Magicicada tredecim West-Eberhard 2005: 6546 (speciation, comp. note)
Magicicada tredecim Cooley, Marshall, Hill and Simon 2006: 856–866, Fig 1, Figs 3–4, Table 1, Tables 3–6 (character displacement, sonogram, comp. note) Mississippi, Arkansas, Illinois, Indiana
Magicicada tredecim Jang and Gerhardt 2006: 156 (female preference, comp. note)

Magicicada tredecim Remondino and Cappellini 2006a: 1, 7 (life cycle)
Magicicada tredecim Remondino and Cappellini 2006b: 485, 491 (life cycle)
Magicicada tredecim Seabra, Pinto-Juma and Quartau 2006: 844 (comp. note)
Magicicada tredecim Cooley 2007: 90 (comp. note)
Magicicada tredecim Cooley, Simon, Marshall, Slon and Ehrhardt 2007: 662–669, Figs 1–4, Tables 1–2 (speciation, distribution, sonogram, comp. note)
Magicicada tredecim Eaton and Kaufman 2007: 90 (comp. note) Southern and Central United States
Magicicada tredecim Fontaine, Cooley, and Simon 2007: 1, 4 (mtDNA transmittance, comp. note)
Magicicada tredecim Oberdörster and Grant 2007b: 21 (comp. note)
Magicicada tredecim Shiyake 2007: 7, 97–98, Fig 165 (illustrated, distribution, listed, comp. note) United States
Magicicada tredecim Marshall 2008: 2786–2787, 2794, Fig 26, Table 7 (distribution, comp. note) United States
Magicicada tredecim Sanborn, Phillips and Gillis 2008: 17 (comp. note)
Magicicada tredecim Yoshimura, Hayashi, Tanaka, Tainaka and Simon 2008: 292 (evolution, comp. note)
Magicicada tredecim Clay, Shelton and Winkle 2009b: 278 (listed)
Magicicada tredecim Cooley, Kritsky, Edwards, Zyla, Marshall et al. 2009: 107, 110 (distribution, comp. note)
Magicicada tredecim Sanborn and Heath 2009: 14 (comp. note)
Magicicada tredecim Matthews and Matthews 2010: 310, Fig 8.7 (activity, comp. note)
Magicicada tredecim Saulich 2010: 1132 (life cycle, comp. note)
Magicicada tredecim Speer, Clay, Bishop and Creech 2010: 175 (comp. note)

***M. tredecula* Alexander & Moore, 1962**

Magicicada tredecula McCoy 1965: 41, 43 (key, comp. note) Arkansas
Magicicada tredecula Itô 1978: 158 (comp. note) America
Magicicada tredecula Itô 1982b: 110, Table 1 (life cycle evolution, comp. note)
Magicicada tredecula Simon and Lloyd 1982: 275 (comp. note)
Magicicada tredecula Simon 1983b: 105 (comp. note) Eastern United States
Magicicada tredecula Strümpel 1983: 58, 168, Table 9 (comp. note)
Magicicada tredecula James, Williams and Smith 1985: 37 (distribution, comp. note) Arkansas
Magicicada tredecula Smith 1985: 4 (life cycle, comp. note) Arkansas
Magicicada tredecula Lloyd 1986: 363 (comp. note)
Magicicada tredecula Davidson and Lyon 1987: 374 (natural history, comp. note)
Magicicada tredecula Steward, Smith and Stephen 1988: 348 (predation, comp. note) Arkansas, Missouri
Magicicada tredecula Martin and Simon 1990a: 1066 (comp. note) North America
Magicicada tredecula Martin and Simon 1990b: 360 (comp. note) North America
Magicicada tredecula Nakao 1990: 127 (life cycle, comp. note)
Magicicada tredecula Simon 1990: 310, 321, Fig 8 (wing variation, comp. note) Eastern United States
Magicicada tredecula Stephen, Wallis and Smith 1990: 369 (predation, comp. note) Arkansas
Magicicada tredecula Young and Kritsky 1990: 25 (distribution, comp. note) Indiana, Kentucky, Illinois
Magicicada tredecula Gogala 1991a: 6, Fig 8E (life cycle, distribution, comp. note)
Magicicada decula (sic) Gogala 1991b: 76 (comp. note)
Magicicada tredecula Reding and Guttman 1991: 323 (comp. note)
Magicicada tredecula Williams and Smith 1991: 276–277 (chorusing sites, comp. note) Arkansas
Decula (sic) Williams and Smith 1991: 277, 279–281, 284–290, Figs 4–5, Tables 1–2 (chorusing sites, comp. note) Arkansas
Magicicada tredecula Cox and Carleton 1992: 63 (comp. note)
Magicicada decula (sic) Wheeler, Williams and Smith 1992: 340 (comp. note)
Magicicada tredecula Moore 1993: 273, 275–277, 279, Fig 4, Fig 6 (distribution, song, comp. note)
Magicicada tredecula Williams, Smith and Stephen 1993: 1147 (comp. note) Arkansas
Magicicada tredecula Heliövaara, Väisänen and Simon 1994: 476–477, Table 1 (life cycle, comp. note) North America
Magicicada tredecula Williams and Simon 1995: 271 (natural history, comp. note)
Magicicada tredecula Simon, Nigro, Sullivan, Holsinger, Martin, et al. 1996: 924, Fig 1 (comp. note)
Magicicada tredecula Poole, Garrison, McCafferty, Otte and Stark 1997: 331 (listed) North America

Magicicada tredecula Hyche 1998: 3–4 (comp. note) Alabama
Magicicada tredecula Hestir and Cain 1999: 243 (predation, comp. note) Arkansas
Magicicada tredecula Kritsky 1999: 24 (comp. note)
Magicicada tredecula Sanborn 1999b: 2 (listed) North America
Magicicada tredecula Irwin and Coelho 2000: 82 (comp. note) Illinois
Magicicada tredecula Marshall and Cooley 2000: 1314–1315, 1317, Fig 1, Table 1 (distribution, comp. note)
Magicicada tredecula Sahli and Ware 2000: 187 (comp. note) Virginia
Magicicada tredecula Simon, Tang, Dalwadi, Staley, Deniega and Unnasch 2000: 1326 (speciation, comp. note)
Magicicada tredecula Smith, Kelly, Dean, Bryson, Parker et al. 2000: 127 (comp. note) Kansas
Magicicada tredecula Waldbauer 2000: 220 (life history, comp. note)
Magicicada tredecula Cooley and Marshall 2001: 828, Table 1 (communication, comp. note)
Magicicada tredecula Marshall 2001: 387, Table 1 (listed)
Magicicada tredecula Ritchie 2001: 59–60, Fig 2 (life cycle, speciation, distribution, comp. note) North America
Magicicada tredecula Sueur 2001: 41, Table 1 (listed)
Magicicada tredecula Gerhardt and Huber 2002: 390–391, Fig 11.8 (distribution, comp. note)
Magicicada tredecula Sanborn 2002a: 226 (comp. note)
Magicicada tredecula Cooley, Simon and Marshall 2003: 153–155, Fig 3, Table 1 (speciation, illustrated, comp. note) Alabama, Arkansas, Georgia, Indiana, Illinois, Kentucky, Louisiana, Maryland, Missouri, Mississippi, North Carolina, Oklahoma, South Carolina, Tennessee, Virginia
Magicicada tredecula Marshall, Cooley and Simon 2003: 433–434, Fig 1 (distribution, comp. note)
Magicicada tredecula Kritsky 2004a: 87, 89, 112–113 (illustrated, key, comp. note)
Magicicada tredecula Zyla 2004a: 486 (comp. note)
Magicicada tredecula Grant 2005: 169–171, Figs 1–2 (life cycle, phylogeny, comp. note) North America
Magicicada tredecula Robinson 2005: 217 (comp. note)
Magicicada tredecula Cooley, Simon, Marshall, Slon and Ehrhardt 2007: 662, Fig 1 (speciation, distribution, comp. note)
Magicicada tredecula Eaton and Kaufman 2007: 90 (comp. note) Southern and Central United States
Magicicada tredecula Fontaine, Cooley, and Simon 2007: 1 (mtDNA transmittance, comp. note)
Magicicada tredecula Shiyake 2007: 7, 97–98, Fig 166 (illustrated, distribution, listed, comp. note) United States
Magicicada tredecula Marshall 2008: 2786–2787, Fig 26, Table 7 (distribution, comp. note) United States
Tredecula sp. Zamanian, Mehdipour and Ghaemi 2008: 2069 (comp. note)
Magicicada tredecula Clay, Shelton and Winkle 2009b: 278 (listed)
Magicicada tredecula Sanborn and Heath 2009: 14 (comp. note) Illinois, Tennessee
Magicicada tredecula Matthews and Matthews 2010: 310, Fig 8.7 (activity, comp. note)
Magicicada tredecula Saulich 2010: 1132 (life cycle, comp. note)
Magicicada tredecula Speer, Clay, Bishop and Creech 2010: 175 (comp. note)

***Ueana* Distant, 1905** = *Ueana (Tibicen)* = *Uena* (sic) = *Neana* (sic)

Ueana Duffels 1988d: 100 (comp. note) New Caledonia
Ueana Ewart 1989b: 292 (comp. note) New Caledonia
Ueana Boer 1990: 64 (comp. note)
Ueana Boulard 1991e: 262 (listed)
Ueana Boulard 1992a: 122 (listed)
Ueana Boulard 1993b: 262 (comp. note) New Caledonia
Ueana Boulard 1993e: 109 (comp. note)
Ueana Boulard 1994c: 147 (comp. note)
Ueana Boulard 1995a: 7 (comp. note)
Ueana Boulard and Mondon 1995: 71 (comp. note) New Caledonia
Ueana Boer and Duffels 1996b: 306, 308–309 (comp. note) Vanuatu, New Caledonia
Ueana Boulard 1997a: 179–180, 182 (listed, comp. note)
Ueana Boer 1999: 118, 128 (comp. note)
Ueana Boulard 1999a: 105–106 (comp. note) New Caledonia
Ueana Boulard and Moulds 2001: 74 (comp. note) New Caledonia
Ueana Boulard 2002b: 206 (comp. note)
Ueana Moulds 2005b: 393, 422, 437 (higher taxonomy, listed, comp. note)
Ueana Boulard 2006g: 131–132 (comp. note) New Caledonia

U. bergerae Boulard, 1993
Ueana bergerae Boulard 1993e: 118–121, Figs 14–20 (n. sp., described, illustrated, sonogram, comp. note) New Caledonia

U. boudinoti Boulard, 1992
Ueana boudinoti Boulard 1992a: 129–131, Figs 24–29 (n. sp., described, illustrated, comp. note) New Caledonia

U. bouleti Boulard, 1993 = *Uena* (sic) *bouleti*
Ueana bouleti Boulard 1993e: 110–113, Figs 1–7 (n. sp., described, illustrated, sonogram, comp. note) New Caledonia
Uena (sic) *bouleti* Sueur 2001: 42, Table 1 (listed)

U. crepitans Boulard, 1992 = *Uena* (sic) *crepitans*
Ueana crepitans Boulard 1992a: 122–125, Figs 7–14 (n. sp., described, illustrated, sonogram, comp. note) New Caledonia
Uena (sic) *crepitans* Sueur 2001: 42, Table 1 (listed)
Ueana crepitans Boulard 2006g: 49, 51, 54, 181, Fig 13 (sonogram, listed, comp. note) New Caledonia

U. dahli (Kuhlgatz, 1905) to *Gymnotympana dahli*

U. desserti Boulard, 1997 = *Uena* (sic) *desserti*
Ueana desserti Boulard 1997a: 180–182, Figs 1–4 (n. sp., described, illustrated, sonogram, comp. note) New Caledonia
Uena (sic) *desserti* Sueur 2001: 42, Table 1 (listed)
Ueana desserti Boulard 2006g: 181 (listed)

U. fungigera Boulard, 1991
Ueana fungifera (sic) Boulard 1991d: 263, Fig 2 (plant damage, illustrated, comp. note) New Caledonia
Ueana fungigera Boulard 1991e: 262–265, Figs 7–12 (n. sp., described, illustrated, comp. note) New Caledonia
Ueana fungifera (sic) Boulard 1992a: 122 (comp. note)
Ueana fungifera (sic) Boulard 1993e: 110 (comp. note)

U. harmonia Kirkaldy, 1905 = *Ueana polymnia* Kirkaldy, 1905
Ueana harmonia Bigot 1985: 327 (distribution) New Caledonia
Ueana harmonia Boulard 1991d: 264, Fig 3 (fungus host, illustrated, comp. note) New Caledonia
Ueana polymnia Boulard 1991e: 262 (comp. note)
Ueana polymnia Boulard 1992a: 129 (comp. note)
Ueana harmonia Boulard 1994c: 147 (comp. note)
Ueana harmonia Boulard 2006g: 121 (comp. note) New Caledonia

U. hyalinata Boulard, 1994
Ueana hyalinata Boulard 1993e: 118 (comp. note)
Ueana hyalinata Boulard 1994c: 143–145, Figs 1–3 (n. sp., described, illustrated, comp. note) New Caledonia

U. latreillae Boulard, 1993 = *Uena* (sic) *latreillae*
Ueana latreillae Boulard 1993e: 114–117, Figs 8–13 (n. sp., described, illustrated, sonogram, comp. note) New Caledonia
Uena (sic) *latreillae* Sueur 2001: 42, Table 1 (listed)
Ueana latreillae Boulard 2006g: 111 (comp. note) New Caledonia

U. letocarti Boulard, 1992
Ueana letocarti Boulard 1992a: 128–129, Figs 21–23 (n. sp., described, illustrated, comp. note) New Caledonia
Ueana letocarti Boulard 1994c: 143 (comp. note)

U. lifuana (Montrouzier, 1861) = *Cicada lifuana* = *Tibicen lifuana* = *Neana* (sic) *lifuana*
Neana (sic) *lifuana* Vayssière 1955: 254 (listed, coffee pest) New Caledonia
Ueana lifuana Bigot 1985: 327 (distribution, habitat notes) New Caledonia
Ueana lifuana Boulard 1991d: 263, Fig 1 (plant damage, illustrated, comp. note) New Caledonia
Baeturia lifuana Boulard 1994c: 147 (comp. note) Equals *Neana* (sic) *lifuana*
Ueana lifuana Boulard 1995a: 7 (comp. note)
Ueana lifuana Boulard 1995c: 177 (comp. note) New Caledonia
Ueana lifuana Boulard and Moulds 2001: 74, 76 (comp. note) New Caledonia

U. lullae Boulard, 1994
Ueana lullae Boulard 1994c: 146–147, Figs 7–9 (n. sp., described, illustrated, comp. note) New Caledonia

U. maculata Distant, 1906
Ueana maculata Bigot 1985: 327 (distribution) New Caledonia
Ueana maculata Boulard 1991e: 262 (comp. note)
Ueana maculata Boulard 1992a: 122, 129 (comp. note)

U. montaguei Distant, 1920
Ueana montaguei Boulard 1991d: 263 (comp. note) New Caledonia
Ueana montaguei Boulard 1994c: 143 (comp. note)

U. procera Boulard, 1994
Ueana procera Boulard 1994c: 145–146, Figs 4–6 (n. sp., described, illustrated, comp. note) New Caledonia
Ueana procera Boulard 1997a: 180 (comp. note)

U. quirosi Boulard, 1994
Ueana quirosi Boulard 1994c: 147–148, Figs 10–12 (n. sp., described, illustrated, comp. note) New Caledonia
Ueana quirosi Boer and Duffels 1996b: 308, 328 (comp. note) Vanuatu

U. rosacea (Distant, 1892) = *Melampsalta rosacea* = *Uena* (sic) *rosacea* = *Parasemia rosacea*
Ueana rosacea Bigot 1985: 327 (distribution, notes) New Caledonia

Ueana rosacea Boulard 1994c: 146 (comp. note)
Ueana rosacea Boulard 1999a: 79, 106–107 (listed, sonogram) New Caledonia
Melampsalta rosacea Hua 2000: 63 (listed, distribution) China, Guangxi
Uena (sic) *rosacea* Sueur 2001: 42, Table 1 (listed)
Ueana rosacea Boulard 2002b: 203 (comp. note) New Caledonia
Ueana rosacea Boulard 2005d: 348 (comp. note) New Caledonia
Ueana rosacea Boulard 2006g: 121–122, 181, Fig 84 (sonogram, listed, comp. note) New Caledonia

U. rufilinea Ashton, 1914 to *Urabunana dameli* (Distant, 1905)

***U. simonae* Boulard, 1993**
Ueana simonae Boulard and Moulds 2001: 74–77, Figs 1–5 (n. sp., described, illustrated, comp. note) New Caledonia

***U. speciosa* Boulard, 1992**
Ueana speciosa Boulard 1992a: 125 128, Figs 15–20 (n. sp., described, illustrated, comp. note) New Caledonia
Ueana speciosa Boulard 1997a: 182 (comp. note)

***U. spectible* Boulard, 1997**
Ueana spectabile Boulard 1997a: 182–184, Figs 5–10 (n. sp., described, illustrated, sonogram, comp. note) New Caledonia

***U. stasserae* Boulard, 1994**
Ueana stasserae Boulard 1991e: 265–268, Figs 13–18 (n. sp., described, illustrated, comp. note) New Caledonia
Ueana stasserae Boulard 1994c: 145 (comp. note)

***U. tintinnabula* Boulard, 1991** = *Uena* (sic) *tintinabula*
Ueana tintinnabula Boulard 1991g: 74–81, Figs 1–11 (n. sp., described, illustrated, sonogram, comp. note) New Caledonia
Ueana tintinnabula Boulard 1993b: 262 (comp. note)
Ueana tintinnabula Boulard 1999a: 79, 105–106 (listed, sonogram) New Caledonia
Uena (sic) *tintinabula* Sueur 2001: 42, Table 1 (listed)
Ueana tintinnabula Boulard 2005d: 344 (comp. note) New Caledonia
Ueana tintinnabula Boulard 2006g: 72–73, 81, 181, Fig 37 (sonogram, listed, comp. note) New Caledonia

***U. tiyanni* Boulard, 1992**
Ueana tiyanni Boulard 1992a: 130, 132, Figs 30–32 (n. sp., described, illustrated, comp. note) New Caledonia

***U. variegata* Boulard, 1993**
Ueana variegata Boulard 1993b: 261–264, Figs 1–6 (n. sp., described, illustrated, comp. note) New Caledonia

***Nelcyndana* Distant, 1906** = *Nelcynda* Stål, 1870 (nec *Nelcynda* Walker, 1862) = *Tibicen (Nelcynda)* = *Necyndana* (sic)
Nelcyndana Wu 1935: 26 (listed) Equals *Nelcynda* China
Nelcyndana Zaidi and Ruslan 1995a: 174, Table 2 (listed, comp. note) Borneo, Sarawak
Nelcyndana Zaidi and Ruslan 1995b: 220–221, Table 2 (listed, comp. note) Borneo, Sabah
Nelcyndana Zaidi and Ruslan 1995d: 202, Table 2 (listed, comp. note) Borneo, Sabah
Nelcyndana Zaidi 1996: 103, Table 2 (comp. note) Borneo, Sarawak
Nelcyndana Zaidi and Hamid 1996: 54–55, Table 2 (listed, comp. note) Borneo, Sarawak
Nelcyndana Chou, Lei, Li, Lu and Yao 1997: 80, 86, 293, Table 6 (key, described, synonymy, listed, diversity, biogeogrpahy) Equals *Tibicen (Nelcyndana)* China
Nelcyndana Zaidi 1997: 115, Table 2 (listed, comp. note) Borneo, Sarawak
Nelcyndana Zaidi and Ruslan 1998b: 4–5, Table 2 (listed, comp. note) Borneo, Sabah
Nelcyndana Zaidi and Ruslan 1999: 5, Table 2 (listed, comp. note) Borneo, Sarawak
Nelcyndana Moulds 2005b: 437 (higher taxonomy, listed, comp. note)
Nelcyndana Duffels 2010: 11–13, 18 (key, described, listed, comp. note) Equals *Tibicen (Nelcynda)* Equals *Nelcynda* Peninsular Malaysia, Borneo, Philippines
Nelcynda Duffels 2010: 11, 13 (listed, comp. note)
Tibicen (Nelcynda) Duffels 2010: 13 (listed)
Tibicen (Nelcynda) tener Duffels 2010: 14 (listed)
Nelcyndana Lee 2010a: 26 (listed, comp. note) Philippines, Mindanao
Nelcyndana Wei, Ahmed and Rizvi 2010: 33 (comp. note)

***N. borneensis* Duffels, 2010**
Nelcyndana borneensis Duffels 2010: 14, 18–23, 26–27, Fig 3, Figs 5–7 (key, n. sp., described, illustrated, comp. note) Borneo, Sabah, Sarawak, Brunei, Kalimantan

***N. madagascariensis* (Distant, 1905)** = *Nelcynda madagascariensis*
Nelcyndana madagascariensis Duffels 2010: 11 (comp. note)

***N. mulu* Duffels, 2010**
Nelcyndana mulu Duffels 2010: 14, 26–30, Figs 11–14 (key, n. sp., described, illustrated, comp. note) Borneo, Sarawak

***N. tener* (Stål, 1870)** = *Tibicen (Nelcynda) tener* = *Nelcynda tener* = *Tibicen tener* = *Necyndana* (sic) *tener*
Nelcyndana tener Wu 1935: 26 (listed) Equals *Tibicen (Nelcynda) tener* Equals *Tibicen tener* China, Foochow, Philippines

Tibicen (Nelcyndana) tener Chou, Lei, Li, Lu and Yao 1997: 86 (type species of *Nelcyndana*)
Nelcyndana tener Chou, Lei, Li, Lu and Yao 1997: 86 (described, synonymy, comp. note) Equals *Tibicen (Nelcyndana) tener* China
Nelcyndana tener Zaidi and Ruslan 1997: 232 (listed, comp. note) Equals *Tibicen tener* Peninsular Malaysia
Nelcyndana tener Zaidi and Ruslan 1998a: 369–370 (listed, comp. note) Equals *Tibicen tener* Borneo, Sarawak, Philippines
Nelcyndana tener Hua 2000: 63 (listed, distribution) Equals *Tibicen tener* China, Fujian, Philippines, Malaysia, Kalimantan
Nelcyndana tener Zaidi, Noramly and Ruslan 2000a: 330–331 (listed, comp. note) Equals *Tibicen tener* Borneo, Sabah, Philippines, Sarawak, Peninsular Malaysia
Nelcyndana tener Zaidi, Ruslan and Azman 2000: 217–218 (listed, comp. note) Equals *Tibicen tener* Borneo, Sabah, Philippines, Sarawak, Peninsular Malaysia
Nelcyndana tener Zaidi, Nordin, Maryati, Wahab, Norashikin et al. 2002: 106, Table 1 (listed, comp. note) Borneo, Sabah
Nelcyndana tener Zaidi and Azman 2003: 106, Table 1 (listed, comp. note) Borneo, Sabah
Nelcynda (sic) *tener* Duffels 2010: 11 (type species of *Nelcynda*, comp. note) Philippines
Nelcyndana tener Duffels 2010: 12, 14–18, Figs 1–2 (type species of *Nelcyndana*, key, lectotype designation, described, illustrated, comp. note) Equals *Tibicen (Nelcynda) tener* Equals *Tibicen tener* Philippines
Tibicen (Nelcynda) tener Lee 2010a: 26 (type species of *Nelcyndana*, listed) Philippines
Nelcyndana tener Lee 2010a: 25 (synonymy, distribution, listed, comp. note) Equals *Tibicen (Nelcynda) tener* Philippines, Mindanao, Samar

***N.* cf. *tener* Schouten, Duffels & Zaidi, 2004**
Nelcyndana cf. *tener* Schouten, Duffels and Zaidi 2004: 373, 379, Table 2 (listed, comp. note) Malayan Peninsula

***N. vantoli* Duffels, 2010**
Nelcyndana vantoli Duffels 2010: 14, 19, 23–26, Fig 4, Figs 8–10 (key, n. sp., described, illustrated, comp. note) Borneo, Sabah, Kalimantan

***Trismarcha* Karsch, 1891** = *Trimarcha* (sic)
Trismarcha Medler 1980: 73 (listed) Nigeria
Trismarcha Boulard 1985e: 1021 (crypsis, comp. note) Africa
Trismarcha Boulard 1986d: 228 (listed)
Trismarcha Boulard, Deeming and Matile 1989: 145 (comp. note) Ivory Coast
Trismarcha Boulard 1990b: 138, 176 (comp. note)
Trismarcha Boer 1995c: 203–204, Fig 1 (phylogeny, comp. note) Africa
Trismarcha Boulard 1996d: 148 (listed)
Trismarcha Boulard 2002b: 197 (comp. note) Africa, Madagascar
Trismarcha Moulds 2003b: 248, 254, 258, 261 (comp. note)
Trismarcha Jong and Boer 2004: 267 (comp. note)
Trismarcha Moulds 2005b: 393, 437 (higher taxonomy, listed, comp. note)
Trismarcha Boulard 2006g: 109, 130–131 (comp. note) Africa

***T. angolensis* Distant, 1905**
Trismarcha angolensis Boulard 1986d: 229–230, Fig 17 (illustrated, comp. note)

***T. atrata* Distant, 1905** = *Trismarcha atra* (sic)
Trismarcha atrata Medler 1980: 73 (listed) Nigeria
Trismarcha atrata Boulard 1990b: 138, 152–153 Plate XVII, Fig 1 (sound system illustrated, comp. note)
Trismarcha atrata Boulard 2002b: 197 (sound system, comp. note)

***T. consobrina* (Distant, 1920)** = *Lemuriana consobrina*
Lemuriana consobrina Schedl 1993: 799 (comp. note)
Lemuriana consobrina Pham and Yang 2010b: 205–207 (key, comp. note) Ethiopian Region

***T. decolorata* Dlabola, 1960**

***T. excludens* (Walker, 1858)** = *Cicada excludens* = *Tibicen excludens* = *Trismarcha sericosta* Karsch, 1891 = *Trismarcha sanguinea*
Trismarcha excludens Boulard 1986d: 229–230, 237–238, Table I, Fig 16 (illustrated, comp. note) Ivory Coast
Trismarcha excludens Boulard 1990b: 138, 152–153, 176, 206, Plate XVII, Fig 1 (sound system illustrated, comp. note) Africa
Trismarcha excludens Boulard 1996d: 115, 148–150, Fig S16(a-d), Fig S16(e-f) (song analysis, sonogram, listed, comp. note) Ivory Coast
Trismarcha excludens Boulard 2000d: 95, 106, Plate O, Fig 2 (illustrated, song, comp. note)
Trismarcha excludens Sueur 2001: 42, Table 1 (listed)
Trismarcha excludens Boulard 2002b: 197 (sound system, comp. note)
Trismarcha excludens Boulard 2005d: 345–346, Fig 25.42 (sonogram, illustrated, comp. note)
Trismarcha excludens Boulard 2006g: 96, 100–101130-131, 180, Fig 66 (sonogram, illustrated, listed, comp. note) Ivory Coast

***T. exsul* Jacobi, 1904**

***T. ferruginosa* Karsch, 1891** = *Trimarcha* (sic) *ferruginosa* = *Trismarcha fuliginosa* Karsch, 1991

Trismarcha fuliginosa Schouteden 1925: 84 (listed, comp. note) Congo
Trismarcha ferruginosa Medler 1980: 73 (listed) Equals *Trismarcha feluginosa* Nigeria
Trismarcha ferruginosa Boulard 1986d: 228–229, 237–238, Table I, (distribution, comp. note) Ivory Coast
Trismarcha ferruginosa Boulard, Deeming and Matile 1989: 143 (behavior, comp. note) Ivory Coast
Trismarcha ferruginosa Boulard 1990b: 206 (comp. note)
Trismarcha ferruginosa Boulard 2006g: 130–131 (comp. note)

***T. flavocostata* (Distant, 1905)** = *Lemuriana flavocostata*
Lemuriana flavocostata Pham and Yang 2010b: 205–207 (key, comp. note) Ethiopian Region

***T. guillaumeti* Boulard, 1986**
Trismarcha guillaumeti Boulard 1986d: 229–230, 237–238, Table I, Figs 18–23 (n. sp., described, illustrated, comp. note) Ivory Coast

***T. guineica* Dlabola, 1971**
Trismarcha guineica Boulard 1986d: 229, 237–238, Table I (distribution, comp. note) Ivory Coast

***T. heimi* Boulard, 1975**

***T. lobayensis* Boulard, 1973**

***T. nana* Dlabola, 1960**

***T. sikorae* (Distant, 1905)** = *Lemuriana sikorae*

***T. sirius* (Distant, 1899)** = *Tibicen sirius* = *Lemuriana sirius* = *Limuriana* (sic) *sirius*
Lemuriana sirius Pham and Yang 2010b: 205–207 (key, comp. note) Ethioipian Region

***T. umbrosa retusa* Boulard, 1986** = *Trismarcha umbrosa* var. *retusa*
Trismarcha umbrosa var. *retusa* Boulard 1986d: 230–231, Fig 25 (n. var., described, illustrated, comp. note) Ivory Coast

***T. umbrosa umbrosa* Karsch, 1891** = *Trismarcha umbrosa*
Trismarcha umbrosa Boulard 1986d: 228, 230–231, 237–238, Table I, Fig 24 (type species of *Trismarcha*, illustrated, comp. note)
Trismarcha umbrosa Boulard 1990b: 167–168, 175, 181, 193, 206, Plate XXI, Fig 4 (illustrated, comp. note) Ivory Coast
Trismarcha umbrosa Boulard 1995a: 8 (comp. note)
Trismarcha umbrosa Boulard 1996d: 115, 150–153, Fig S17(a-d), Fig S17(e-f) (song analysis, sonogram, listed, comp. note) Ivory Coast
Trismarcha umbrosa Sueur 2001: 42, Table 1 (listed)
Trismarcha umbrosa Moulds 2003b: 248, 250–254, 258, 261, Fig 7, Figs 15–18, Figs 31–32, Fig 39, Table 1 (type species of *Trismarcha*, illustrated, phylogeny, comp. note) Republic of Central Africa
Trismarcha umbrosa Boulard 2006g: 130–131, 180 (listed, comp. note)

***T. voeltzkowi* Jacobi, 1917**

***Monomotapa* Distant, 1879** = *Monomatapa* (sic)
Monomotapa Moulds 2003b: 246, 248, 254, 257–258, 262 (key, comp. note)
Monomotapa Moulds 2005b: 393, 437 (higher taxonomy, listed, comp. note)
Monomotapa Shiyake 2007: 8, 103 (listed, comp. note)

***M. insignis* Distant, 1897** = *Mogana* (sic) *insignis* = *Monomatapa* (sic) *insignis* = *Mogannia* (sic) *insignis* = *Abricta offirmata* Dlabola, 1962
Monomatapa (sic) *insignis* Boulard 1985e: 1021 (crypsis, comp. note) Africa
Monomatapa (sic) *insignis* Malaisse 1997: 237, Table 2.11.1 (human food, listed) Zambezi
Mogannia (sic) *insignis* Howard, Moore, Giblin-Davis and Abad 2001: 157 (pest species, comp. note) West Africa
Monomatapa (sic) *insignis* Sueur 2002b: 122, 129 (listed, comp. note)
Monomatapa (sic) *insignis* Huis 2003: 168, Table 1 (predation, comp. note) Botswana
Monomatapa (sic) *insignis* Moulds 2003b: 246, 248, 250–254, Figs 15–17, Figs 33–34, Table 1 (illustrated, phylogeny, comp. note) Zimbabwe
Monomotapa insignis Shiyake 2007: 8, 100, 103, Fig 173 (illustrated, distribution, listed, comp. note)

***M. matoposa* Boulard, 1980**
Monomotapa matoposa Chawanji, Hodson and Villet 2005b: 66 (spermiogenesis, comp. note)
Monomotapa matoposa Chawanji, Hodson and Villet 2006: 374–375, 383, 386–388, Fig 4, Fig 9, Fig 14, Figs 87–90, Tables 1–6 (sperm morphology, listed, comp. note) Zimbabwe
Monomotapa matoposa Chawanji, Hodson, Villet, Sanborn and Phillips 2007: 338, 340–345, Fig 2E, Figs 4A–B, Figs 5B–D, Fig 6D (spermiogenesis, comp. note) Zimbabwe

***M. socotrana* Distant, 1905**

***Kanakia* Distant, 1892**
Kanakia Boulard 1988c: 62 (described, comp. note) New Caledonia
Kanakia Boulard 1991e: 259 (listed)
Kanakia Boer and Duffels 1996b: 308 (comp. note)
Kanakia Moulds 2003b: 246 (comp. note)
Kanakia Moulds 2005b: 393, 437 (higher taxonomy, listed, comp. note)

***K. congrua* Goding & Froggatt, 1904** = *Kanakia congruens* (sic)

Kanakia congrua Hahn 1962: 8 (listed) Australia, Queensland
Kanakia congruens (sic) Hahn 1962: 8 (listed) Australia, Queensland
Kanakia congrua Stevens and Carver 1986: 264 (type material) Australia, Queensland

***K. flavoannulata* (Distant, 1920)** = *Abricta flavoannulata* = *Kanakia annulata* (sic)
Kanakia flavoannulata Boulard 1988c: 62, 64 (n. comb., comp. note)
Kanakia flavoannulata Boer and Duffels 1996b: 308 (comp. note) New Caledonia
Kanakia annulata (sic) Boulard 2002b: 203 (comp. note)
Abricta flavoannulata Moulds 2003b: 246 (comp. note)
Kanakia annulata (sic) Boulard 2005d: 347 (comp. note)
Kanakia flavoannulata Boulard 2006g: 111, 181 (listed, comp. note) New Caledonia

***K. gigas* Boulard, 1988**
Kanakia gigas Boulard 1988c: 62–64, Figs 2–5 (n. sp., described, illustrated, comp. note) New Caledonia
Kanakia gigas Boer and Duffels 1996b: 308 (comp. note) New Caledonia
Kanakia gigas Freytag 2000: 55 (on stamp)
Kanakia gigas Boulard 2002b: 19, 203 (comp. note) New Caledonia
Kanakia gigas Boulard 2005d: 347 (comp. note)
Kanakia gigas Boulard 2006g: 115–116, Fig 81 (sonogram, comp. note) New Caledonia

***K. parva* Boulard, 1991**
Kanakia parva Boulard 1991e: 259–262, Figs 1–6 (n. sp., described, illustrated, comp. note) New Caledonia
Kanakia parva Boer and Duffels 1996b: 308 (comp. note) New Caledonia

***K. typica* Distant, 1892**
Kanakia typica Boulard 1985e: 1032 (coloration, comp. note) New Caledonia
Kanakia typica Boulard 1988c: 62–64, Fig 1 (type species of genus *Kanakia*, illustrated, comp. note) New Caledonia
Kanakia typica Boer and Duffels 1996b: 308 (comp. note) New Caledonia
Kanakia typica Boulard 2002b: 203 (comp. note)
Kanakia typica Boulard 2005d: 347 (comp. note) New Caledonia
Kanakia typica Boulard 2006g: 72, 115, 181 (listed, comp. note) New Caledonia

***Abroma* Stål, 1866** = *Tibicen (Abroma)* = *Abraoma* (sic)
Abroma Wu 1935: 26 (listed) China
Abroma Williams and Williams 1988: 3 (listed) Mascarene Islands
Abroma Boulard 1990b: 209, Fig 12 (phylogeny, comp. note)
Abroma Zaidi, Mohammedsaid and Yaakob 1990: 261, 267, Table 2 (listed, comp. note) Malaysia
Abroma Zaidi and Ruslan 1994: 427, Table 2 (comp. note) Malaysia
Abroma Boer 1995c: 203–204, Fig 1 (phylogeny, comp. note) Mascarene Isles
Abroma Zaidi and Ruslan 1995a: 173–174, Table 2 (listed, comp. note) Borneo, Sarawak
Abroma Zaidi and Ruslan 1995b: 220–221, Table 2 (listed, comp. note) Borneo, Sabah
Abroma Zaidi and Ruslan 1995c: 70, Table 2 (listed, comp. note) Peninsular Malaysia
Abroma Zaidi and Ruslan 1995d: 202, Table 2 (listed, comp. note) Borneo, Sabah
Abroma Boer and Duffels 1996b: 308–309 (comp. note)
Abroma Zaidi 1996: 103, Table 2 (comp. note) Borneo, Sarawak
Abroma Zaidi and Hamid 1996: 54–55, Table 2 (listed, comp. note) Borneo, Sarawak
Abroma Chou, Lei, Li, Lu and Yao 1997: 80, 85, 293, Table 6 (key, described, synonymy, listed, diversity, biogeography) Equals *Tibicen (Abroma)* China
Abroma Zaidi 1997: 115, Table 2 (listed, comp. note) Borneo, Sarawak
Abroma Zaidi and Ruslan 1998b: 4–5, Table 2 (listed, comp. note) Borneo, Sabah
Abroma Zaidi and Ruslan 1999: 4–5, Table 2 (listed, comp. note) Borneo, Sarawak
Abroma Moulds 2003b: 246, 248–249, 254–255, 257, 262 (described, key, diversity, comp. note)
Abroma spp. Moulds 2003b: 253 (comp. note)
Abroma Biswas, Ghosh, Bal and Sen 2004: 248 (key) India
Abroma Jong and Boer 2004: 267 (comp. note)
Abroma Schouten, Duffels and Zaidi 2004: 378 (comp. note) Malayan Peninsula
Abroma Moulds 2005b: 393, 437 (higher taxonomy, listed, comp. note)
Abroma Boulard 2007a: 116, 118 (comp. note)
Abroma Boulard 2008f: 7, 57, 96 (listed) Thailand
Abroma Lee 2008a: 2, 20, Table 1 (listed) Equals *Tibicen (Abroma)* Vietnam
Abroma Lee 2009h: 2637 (listed) Equals *Tibicen (Abroma)* Philippines, Palawan
Abroma Pham and Yang 2009: 15, Table 2 (listed) Vietnam
Abroma Lee 2010b: 28 (comp. note)

***A. antandroyae* Boulard, 2008**
Abroma antandroyae Boulard 2008d: 165–168, Figs 6–10 (n. sp., described, illustrated, song, sonogram, comp. note) Madagascar

A. apicalis **Ollenbach, 1929**

A. apicifera **(Walker, 1850)** = *Cicada terminus* Walker, 1850

A. bengalensis **Distant, 1906**

A. bowringi **Distant, 1905** = *Abroma buringi* (sic)

Abroma bowringi Wu 1935: 26 (listed) China, Hongkong

Abroma bouringi (sic) Chou, Lei, Li, Lu and Yao 1997: 85 (key, described, comp. note) China

Abroma bouringi (sic) Hua 2000: 60 (listed, distribution) China, Hong Kong

A. canopea **Boulard, 2007**

Abroma canopea Boulard 2007a: 116–119, Figs 28–32 (n. sp., described, illustrated, song, sonogram, comp. note) Thailand

Abroma canopea Boulard 2008c: 358 (comp. note)

Abroma canopea Boulard 2008f: 57, 96 (biogeography, listed, comp. note) Thailand

Abroma canopea Boulard 2009d: 45 (comp. note)

A. cincturae **Boulard, 2009**

Abroma cincturae Boulard 2009d: 45–47, 55–58, Figs 17–23 (n. sp., described, illustrated, song, sonogram, comp. note) Thailand

A. egae **(Distant, 1892)**

A. ferraria **(Stål, 1870)** = *Tibicen (Abroma) ferrarius* = *Tibicen ferrarius* = *Abroma ferrarius*

A. guerinii **(Signoret, 1847)** = *Cicada guerinii* = *Tibicen (Abroma) guerinii* = *Abroma guerini* = *Lemuriana sikoriae* Distant

Abroma guerinii Williams and Williams 1988: 3 (comp. note)

Cicada guerinii Chou, Lei, Li, Lu and Yao 1997: 85 (type species of *Abroma*)

Abroma guerinii Moulds 2003b: 248–254, Figs 5–6, Figs 15–17, Table 1 (type species of *Abroma*, illustrated, phylogeny, comp. note)

Abroma guerinii Boulard 2008f: 57 (type species of *Abroma*) Equals *Cicada guerinii* Thailand

Cicada guerinii Lee 2008a: 20 (type species of *Abroma*)

Cicada guerinii Lee 2009h: 2637 (type species of *Abroma*, listed) Madagascar

Cicada guerinii Lee 2010b: 28 (type species of *Abroma*, listed) Madagascar

A. inaudibilis **Boulard, 1999**

Abroma inaudibilis Boulard 1999a: 79, 103–105, 112–113, Plate III, Fig 2 (n. sp., described, illustrated, listed, sonogram) Madagascar

Abroma inaudibilis Boulard 2000d: 94 (song, comp. note) Madagascar

Abroma inaudibilis Sueur 2001: 42, Table 1 (listed)

Abroma inaudibilis Boulard 2005d: 338, 345 (comp. note) Madagascar

Abroma inaudibilis Boulard 2006g: 39, 76–77, 93, 181, Fig 61 (sonogram, listed, comp. note) Madagascar

A. maculicollis **(Guérin-Méneville, 1838)** = *Cicada maculicollis* = *Tibicen (Abroma) maculicollis* = *Abraoma* (sic) *maculicollis*

Abroma maculicollis Zaidi, Mohammedsaid and Yaakob 1990: 262–265, 267, Fig 1, Table 1 (seasonality, comp. note) Malaysia

Abroma maculicollis Zaidi and Ruslan 1994: 428 (comp. note) Malaysia

Abroma maculicollis Zaidi and Ruslan 1995a: 172, 175, Table 1 (listed, comp. note) Borneo, Sarawak

Abroma maculicollis Zaidi and Ruslan 1995b: 218–219, Table 1 (listed, comp. note) Borneo, Sabah

Abroma maculicollis Zaidi and Ruslan 1995c: 69, Table 1 (listed, comp. note) Peninsular Malaysia

Abroma maculicollis Zaidi and Ruslan 1995d: 198 (comp. note) Borneo, Sabah

Abroma maculicollis Zaidi, Ruslan and Mahadir 1996: 61, Table 1 (listed, comp. note) Peninsular Malaysia

Abroma maculicollis Zaidi and Ruslan 1998a: 368–369 (listed, comp. note) Equals *Cicada maculicollis* Borneo, Sarawak, Bengal, India, Ceylon, Peninsular Malaysia, Sabah

Abroma maculicollis Zaidi and Ruslan 1998b: 2–3, Table 1 (listed, comp. note) Borneo, Sabah

Abraoma (sic) *maculicollis* Zaidi and Ruslan 1999: 3, Table 1 (listed, comp. note) Borneo, Sarawak

Abroma maculicollis Zaidi and Ruslan 1999: 5 (comp. note) Borneo, Sarawak

Abroma maculicollis Zaidi, Ruslan and Azman 1999: 318–319 (listed, comp. note) Equals *Cicada maculicollis* Borneo, Sabah, Bengal, India, Ceylon, Sarawak, Peninsular Malaysia

Abroma maculicollis Zaidi, Noramly and Ruslan 2000a: 330–331 (listed, comp. note) Equals *Cicada maculicollis* Borneo, Sabah, Bengal, China, Sumatra, Java, India, Peninsular Malaysia, Sarawak

Abroma maculicollis Zaidi, Ruslan and Azman 2000: 216–217 (listed, comp. note) Equals *Cicada maculicollis* Borneo, Sabah, Bengal, China, Sumatra, Java, India, Peninsular Malaysia, Sarawak

Abroma maculicollis Zaidi and Azman 2003: 106, Table 1 (listed, comp. note) Borneo, Sabah

Abroma maculicollis Biswas, Ghosh, Bal and Sen 2004: 240, 248 (listed, key, described, distribution) Equals *Cicada maculicollis* India, Manipur, West Bengal, Sri Lanka, Malay Peninsula, Borneo

Abroma maculicollis Schouten, Duffels and Zaidi 2004: 373, 378–379, Table 2 (listed, comp. note) Malayan Peninsula

Abroma maculicollis Boulard 2007a: 116 (comp. note)
Cicada maculicollis Boulard 2007a: 116 (comp. note)

***A. mameti* Boulard, 1979**
Abroma mameti Williams and Williams 1988: 3 (listed, comp. note) Mauritius

***A. minor* Jacobi, 1917**

***A. nubifurca* (Walker, 1858)** = *Cicada nubifurca* = *Tibicen nubifurca* = *Cicada nubifera* (sic) = *Cicada apicalis* Kirkaldy, 1891 = *Tibicen apicalis* = *Tibicen nubifurca apicalis*
Abroma nubifurca Poulton 1907: 336 (predation) Equals *Tibicen nubifurca* Ceylon

***A. orhanti* Boulard, 2008**
Abroma orhanti Boulard 2008c: 357–360, Figs 11–13 (n. sp., described, illustrated, comp. note) Laos
Abroma orhanti Boulard 2009d: 45 (comp. note)

***A. philippinensis* Distant, 1905**
Abroma philippinensis Lee 2009h: 2637 (distribution, listed) Philippines, Palawan

***A.* probably n. sp. Schouten, Duffels & Zaidi, 2004**
Abroma probably n. sp. Schouten, Duffels and Zaidi 2004: 373, 378, Table 2 (listed, comp. note) Malayan Peninsula

***A. pumila* (Distant, 1892)** = *Tibicen pumila*
Abroma pumila Bigot 1985: 327 (distribution) New Caledonia
Abroma pumila Boulard 1991d: 263 (plant damage, comp. note) New Caledonia
Abroma pumila Boer and Duffels 1996b: 308 (comp. note) New Caledonia
Abroma pumila Waterhouse 1997: 12, Table 1 (agricultural pest, listed)
Abroma pumila Boulard and Chueata 2004: 141 (life cycle, comp. note) New Caledonia
Abroma pumila Boulard 2006g: 131–132 (comp. note)
Abroma pumila Boulard 2007e: 65 (comp. note) New Caledonia

***A. reducta* (Jacobi, 1902)** = *Tibicen reductus*
Abroma reducta Chou, Lei, Li, Lu and Yao 1997: 85–86 (key, described, synonymy, comp. note) Equals *Tibicen reductus* China
Abroma reducta Hua 2000: 60 (listed, distribution) Equals *Tibicen reducta* China
Tibicen reductus Boulard 2007a: 118 (comp. note)
Abroma reducta Lee 2008a: 20 (synonymy, distribution, listed) Equals *Tibicen reductus* North Vietnam, Hong Kong
Abroma reducta Pham and Yang 2009: 15, Table 2 (listed) Vietnam

***A.* sp. 1 Zaidi, Ruslan & Mahadir, 1996**
Abroma sp. 1 Zaidi, Ruslan and Mahadir 1996: 61, Table 1 (listed, comp. note) Peninsular Malaysia

***A. tahanensis* Moulton, 1923**
Abroma tahanensis Zaidi, Mohammedsaid and Yaakob 1990: 261 (comp. note) Malaysia
Abroma tahanensis Zaidi and Ruslan 1994: 428 (comp. note) Malaysia
Abroma tahanensis Zaidi and Ruslan 1997: 232 (listed, comp. note) Peninsular Malaysia

***A. temperata* (Walker, 1858)** = *Cicada temperata* = *Cicada blandula* Walker, 1858

***A. vinsoni* Boulard, 1979**
Abroma vinsoni Williams and Williams 1988: 3 (listed, comp. note) Mauritius

***Allobroma* Duffels, 2011**

***A. kedenburgi* Breddin, 1905** = *Lemuriana chandaea* Moulton, 1912 = *Abroma maculicollis* (nec Guérin-Méneville, 1838)
Lemuriana chandaea Lee 2010b: 28 (key, listed, comp. note)
Lemuriana chandaea Pham and Yang 2010b: 205–207 (key, comp. note) Oriental Region

***A. portegiesi* Duffels, 2011**

***Sundabroma* Duffels, 2011**

***S. bimaculata* Duffels, 2011**

***S. suffusca* Duffels, 2011**

***Malagasia* Distant, 1882** = *Tibicen (Epora)* Stål, 1866 = *Epora*
Malagasia Boulard 1988c: 62 (comp. note) Madagascar
Malagasia Boulard 1990b: 208 (comp. note) Madagascar
Malagasia Boer and Duffels 1996b: 308 (comp. note) Madagascar
Malagasia Moulds 2005b: 393, 437 (higher taxonomy, listed, comp. note)
Malagasia Boulard 2006g: 109 (comp. note) Madagascar
Malagasia Boulard 2008d: 169 (comp. note)

***M. aperta* (Signoret, 1860)** = *Cicada aperta* = *Tibicen (Epora) apertus*

M. distanti Karsch, 1890 to *Anopercalna distanti*

***M. inflata* Distant, 1882**
Malagasia inflata Boulard 1985e: 1032 (coloration, comp. note) Madagascar
Malagasia inflata Boulard 1990b: 145 (comp. note)
Malagasia inflata Boulard 2005d: 337 (comp. note) Madagascar
Malagasia inflata Boulard 2006g: 35, 72, 29, 181, Fig 6, Fig 33 (sonogram, illustrated, listed, comp. note) Madagascar

***M. mariae* Boulard, 1980**
Malagasia mariae Boulard 1990b: 145 (comp. note)

M. vadoni Boulard, 1980 to *Anopercalna vadoni*

M. virescens Distant, 1905 to *Anopercalna virescens*

Malgachialna **Boulard, 1980**

Malagachialna Moulds 2005b: 393, 437 (higher taxonomy, listed, comp. note)

M. noualhieri **Boulard, 1980**

Malgachialna noualhieri Boulard 1986b: 34–35, Fig 3 (illustrated) Madagascar

Malagachialna nouahlieri Boulard 1990b: 145 (comp. note)

Musimoia **China, 1929**

Musimoia Moulds 2005b: 393, 437 (higher taxonomy, listed, comp. note)

M. burri **China, 1929**

Hylora **Boulard, 1971**

Hylora Boulard 1984a: 52 (comp. note)

Hylora Boulard 1986d: 231 (listed, comp. note)

Hylora Boulard 1988f: 64 (comp. note)

Hylora Boulard 2002b: 205 (comp. note)

Hylora Moulds 2005b: 393, 437 (higher taxonomy, listed, comp. note)

H. bonneti **Boulard, 1975**

Hylora bonneti Boulard 1984a: 52 (comp. note)

H. differata **(Dlabola, 1960)** = *Panka differata* = *Hylora centralensis* Boulard, 1971

Hylora differata Boulard 1984a: 52 (comp. note)

Hylora centralensis Boulard 1986d: 231 (type species of *Hylora*) Central African Republic

Hylora differata Boulard 1988d: 152 (gynandromoph, comp. note)

H. mondziana **Boulard, 1973**

H. villiers **Boulard, 1984**

Hylora villiersi Boulard 1984a: 52–54, Figs 1–6 (n. sp., described, illustrated, comp. note) Ivory Coast

Hylora villiersi Boulard 1986d: 231, 237–238, Table I (distribution, comp. note) Ivory Coast

Ligymolpa **Karsch, 1890** = *Ligymorpa* (sic)

Ligymolpa Moulds 2005b: 393, 437 (higher taxonomy, listed, comp. note)

L. madegassa **Karsch, 1890**

Lemuriana **Distant, 1905** = *Limuriana* (sic)

Lemuriana Boulard 1984c: 171 (comp. note)

Lemuriana Zaidi and Ruslan 1995a: 174, Table 2 (listed, comp. note) Borneo, Sarawak

Lemuriana Zaidi and Ruslan 1995d: 202, Table 2 (listed, comp. note) Borneo, Sabah

Lemuriana Zaidi 1996: 103, Table 2 (comp. note) Borneo, Sarawak

Lemuriana Zaidi and Hamid 1996: 54, Table 2 (listed, comp. note) Borneo, Sarawak

Lemuriana Zaidi 1997: 115, Table 2 (listed, comp. note) Borneo, Sarawak

Lemuriana Zaidi and Ruslan 1999: 5, Table 2 (listed, comp. note) Borneo, Sarawak

Lemuriana Moulds 2005b: 393, 437 (higher taxonomy, listed, comp. note)

Lemuriana Lee 2008a: 2, 21, Table 1 (listed) Vietnam

Lemuriana Pham and Yang 2009: 15, Table 2 (listed) Vietnam

Lemuriana Lee 2010b: 28 (key, diversity, listed, comp. note) Cambodia

Lemuriana Pham and Yang 2010b: 205 (composition, distribution, comp. note) Vietnam

L. apicalis **(Germar, 1830)** = *Cicada apicalis* = *Tibicen apicalis* = *Tibicen (Abroma) apicalis* = *Limuriana* (sic) *apicalis* = *Limuriana* (sic) *apicatus* (sic) = *Cicada semicinta* Walker, 1850 = *Tibicen semicinctus*

Limuriana (sic) *apicatus* (sic) Spichiger 2004a: 1479 (comp. note) India

Cicada apicalis Lee 2008a: 21 (type species of *Lemuriana*)

Lemuriana apicalis Lee 2008a: 21 (synonymy, distribution, listed) Equals *Cicada apicalis* Vietnam, Nepal, India

Lemuriana apicalis Pham and Yang 2009: 15, Table 2 (listed) Vietnam

Cicada apicalis Lee 2010b: 28 (type species of *Lemuriana*)

Lemuriana apicalis Lee 2010b: 28 (key, listed, comp. note) India

Cicada apicalis Pham and Yang 2010b: 205 (type species of *Lemuriana*)

Lemuriana apicalis Pham and Yang 2010b: 205–207 (key, comp. note) Oriental Region

L. cambodiana **Lee, 2010**

Lemuriana cambodiana Lee 2010b: 28–30, Fig 2 (n. sp., described, illustrated, key, distribution, listed, comp. note) Cambodia

L. chandaea Moulton, 1912 to *Allobroma kedenburgi* Breddin, 1905

L. connexa Distant, 1910 to *Unduncus connexa*

L. undescribed species Pham & Yang, 2010 to *Lemuriana vinhcuuensis* Pham & Yang, 2010

L. vinhcuuensis **Pham & Yang, 2010** = *Lemuriana* undescribed species Pham & Yang, 2010

Lemuriana undescribed species Pham and Yang 2010a: 133 (comp. note) Vietnam

Lemuriana vinhcuuensis Pham and Yang 2010b: 205–207, 209–210, Figs 1–2 (n. sp., described, illustrated, key, comp. note) Vietnam

Unduncus **Duffels, 2011**

U. connexa **(Distant, 1910)** = *Lemuriana connexa*

Lemuriana connexa Zaidi and Ruslan 1998a: 369 (listed, comp. note) Borneo, Sarawak

Lemiurana connexa Zaidi, Ruslan and Azman 2000: 217 (listed, comp. note) Borneo, Sabah, Sarawak
Lemuriana connexa Zaidi and Azman 2003: 106, Table 1 (listed, comp. note) Borneo, Sabah
Lemuriana connexa Lee 2010b: 28 (key, listed, comp. note)
Lemuriana connexa Pham and Yang 2010b: 205–207 (key, comp. note) Oriental Region

***Panka* Distant, 1905**
Panka Boulard 1986d: 231 (listed)
Panka Boulard 1990b: 115 (comp. note)
Panka Novotny and Wilson 1997: 437 (listed)
Panka Moulds 2005b: 393, 437 (higher taxonomy, listed, comp. note)
***P. africana* Distant, 1905**
***P. duartei* Boulard, 1975**
***P. lunguncus* Boulard, 1970**
Panka lunguncus Boulard 1986d: 231, 237–238, Table I, (distribution, comp. note) Ivory Coast
Panka lunguncus Boulard 1990b: 132–133, 206, Plate XV, Fig 4 (genitalia, comp. note)
***P. minimuncus* Boulard, 1970**
***P. parvula* Boulard, 1973**
Panka parvula Boulard 1995b: 122–123 (comp. note)
***P. parvulina* Boulard, 1995**
Panka parvulina Boulard 1995b: 121–124, Figs 1–8 (n. sp., described, illustrated, comp. note) Ivory Coast
***P. silvestris* Jacobi, 1912**
***P. simulata* Distant, 1905** = *Tibicen nubifurca* Distant, 1892 (nec *Tibicen nubifurca* Walker, 1858)
Panka simulata Boulard 1986d: 231 (type species of *Panka*) Ceylon
***P. umbrosa* Distant, 1920**

***Nyara* Villet, 1999**
Nyara Villet 1999c: 158, 160–161 (n. gen., described, comp. note) South Africa
***N. thanatotica* Villet, 1999**
Nyara thanatotica Villet 1999c: 158–162, Figs 1–9 (n. sp., type species of *Nyara*, described, illustrated, distribution, habitat, behavior, comp. note) South Africa
Nyara thanatotica Boulard 2002b: 193 (coloration, comp. note)

***Oudeboschia* Distant, 1920**
*Oudeboschia*Villet 1999c: 159 (comp. note)
Oudeboschia Moulds 2005b: 393, 437 (higher taxonomy, listed, comp. note)
Oudeboschia Goble, Price, Barker, and Villet 2008b: 674 (phylogeny, comp. note) Africa
***O. festiva* Distant, 1920**

***Neomuda* Distant, 1920**
Neomuda Villet 1999c: 159 (comp. note)
Neomuda Picker, Griffith and Weaving 2002: 156–157 (described, illustrated, habitat, comp. note) South Africa, Western Cape
Neomuda Moulds 2005b: 393, 437 (higher taxonomy, listed, comp. note)
Neomuda Goble, Price, Barker, and Villet 2008b: 674 (phylogeny, comp. note) Africa
***N. abdominalis* Distant, 1920**
N. peringueyi Distant, 1920 to *Zouga peringueyi*
***N. trimeni* Distant, 1920**

***Viettealna* Boulard, 1980**
Viettealna Moulds 2005b: 393, 437 (higher taxonomy, listed, comp. note)
***V. griveaudi* Boulard, 1980**
Viettealna griveaudi Boulard 1986b: 34–35, Fig 2 (illustrated) Madagascar

Subtribe Anopercalnina Boulard, 2008
Anopercalnina Boulard 2008d: 173 (n. subribe, comp. note)

***Anopercalna* Boulard, 2008**
Anopercalna Boulard 2008d: 169 (n. gen., described, comp. note)
***A. distanti* (Karsch, 1890)** = *Malagasia distanti*
Anopercalna distanti Boulard 2008d: 169, 172–175, Figs 15–18 (n. comb., described, illustrated, song, sonogram, comp. note) Equals *Malagasia distanti* Madagascar
***A. vadoni* (Boulard, 1980)** = *Malagasia vadoni*
Malagasia vadoni Boulard 1986b: 34–35, Fig 7 (illustrated) Madagascar
Malagasia vadoni Boulard 1990b: 145, 150–151, Plate XVI, Fig 4 (sound system, comp. note)
Anopercalna vadoni Boulard 2008d: 169 (n. comb., comp. note) Equals *Malagasia vadoni*
***A. viettei* Boulard, 2008**
Anopercalna viettei Boulard 2008d: 169–173, Figs 11–14 (n. sp., type species of *Anopercalna*, Anopercalnina, described, illustrated, song, sonogram, comp. note) Madagascar
***A. virescens* (Distant, 1905)** = *Malagasia virescens*
Anopercalna virescens Boulard 2008d: 169 (n. comb., comp. note) Equals *Malagasia virescens*

Tribe Cicadettini Buckton, 1889 = Melampsaltini Distant, 1905 = Melampsaltaria
Cicadettini Boulard 1981a: 37, 40, 42 (listed, comp. note)

Melampsaltini Itô and Nagamine 1981: 281, Table 5 (listed) Germany
Cicadettini Boulard 1982f: 184 (listed)
Cicadettini Ramos 1983: 66 (listed) Dominican Republic
Cicadettini Boulard 1984c: 212 (status)
Cicadettini Hayashi 1984: 66 (listed)
Cicadettini Boulard 1985d: 212 (status)
Cicadettini Moulds 1986: 42 (comp. note) Australia
Cicadettini Schedl 1986a: 2, 7 (listed, key, comp. note) Istria
Cicadettini Boulard 1988b: 43 (comp. note)
Cicadettini Boulard 1988c: 65 (comp. note)
Cicadettini Boulard 1988f: 8, 37, 42–45, 72 (comp. note)
Melampsaltini Boulard 1988f: 8, 37, 41 (comp. note)
Cicadettini Duffels 1988d: 100 (comp. note)
Cicadettini Moulds 1988: 39–41 (status, comp. note) Equals Melampsaltini
Melampsaltini Moulds 1988: 39–41 (status, comp. note)
Cicadettini spp. Ewart 1989b: 289 (comp. note)
Cicadettini Boulard 1990b: 115 (comp. note)
Cicadettini Boulard and Nel 1990: 38 (comp. note)
Cicadettini Moulds 1990: 31, 126 (listed, comp. note) Australia
Cicadettini Boulard 1991b: 26 (comp. note)
Cicadettini Boulard 1992a: 119 (listed)
Melampsaltini Boulard 1991b: 26 (comp. note)
Cicadettini Boulard 1992b: 365 (comp. note)
Cicadettini Boer 1995c: 205 (comp. note)
Cicadettini Boer 1995d: 200–202 (comp. note)
Cicadettini Boulard and Mondon 1995: 65–66 (key, listed) France
Cicadettini Boer and Duffels 1996b: 305, 309, 327 (comp. note) New Caledonia
Melampsaltini Moulds 1996a: 20 (comp. note)
Cicadettini Moulds 1996a: 20 (comp. note)
Cicadettini Boulard 1997a: 180 (comp. note)
Cicadettini Boulard 1997c: 84, 106, 111, 117, 122 (comp. note)
Melampsaltini Boulard 1997c: 84, 106 (comp. note)
Cicadettini Chou, Lei, Li, Lu and Yao 1997: 46, 66, 71, 78, 365 (key, listed, synonymy, described, comp. note) Equals Melampsaltaria Equals Melampsaltini
Cicadettini Novotny and Wilson 1997: 437 (listed)
Cicadettini Popov, Beganović and Gogala 1998a: 93 (comp. note)
Cicadettini Boer 1999: 115–117, 120, 137 (comp. note) New Guinea, Australia, New Zealand
Cicadettini Boulard 1999a: 80, 111 (listed)
Cicadettini Moulds 1999: 1 (comp. note)
Cicadettini Sanborn 1999a: 52 (listed)
Cicadettini Sanborn 1999b: 2 (listed) North America
Cicadettini Villet and Noort 1999: 226, Table 12.1 (diversity, listed) Kenya, South Africa
Cicadettini Villet and Zhao 1999: 322 (comp. note)
Cicadettini Puissant and Sueur 2000: 70 (comp. note) France
Cicadettini Schedl 2000: 264 (listed)
Cicadettini Sueur and Puissant 2000: 262 (comp. note) France
Cicadettini Boulard 2001b: 5, 27, 31–32, 37, 41 (comp. note)
Melampsaltini Boulard 2001b: 5, 27 (comp. note)
Cicadettini Boulard 2000i: 20–21, 24–25, 73 (listed, key, comp. note) Equals Melampsaltrini
Melampsaltini Boulard 2000i: 20 (comp. note)
Cicadettini Boulard and Moulds 2001: 74 (comp. note) New Caledonia
Cicadettini Buckley, Simon, Shimodaira and Chambers 2001: 223 (listed)
Cicadettini Puissant and Sueur 2001: 433 (listed) Corsica
Cicadettini Sueur 2001: 42, Table 1 (listed)
Cicadettini Buckley, Arensburger, Simon and Chambers 2002: 5 (comp. note)
Cicadettini Kondratieff, Ellingson and Leatherman 2002: 13, Table 2 (listed) Colorado
Cicadettini Lee, Choe, Lee and Woo 2002: 3 (diversity)
Cicadettini Moulds and Carver 2002: 23 (listed) Australia
Cicadettini Moulds 2003b: 247 (comp. note)
Cicadettini Sueur, Puissant, Simões, Seabra, Boulard and Quartau 2004: 182 (listed) Portugal
Cicadettini Boulard 2005d: 338 (comp. note)
Cicadettini Ewart 2005b 441 (listed)
Cicadettini Hill, Marshall and Cooley 2005: 71 (diversity, comp. note) New Zealand
Cicadettini Lee 2005: 14, 17, 176 (listed, comp. note) Korea
Cicadettini Moulds 2005b: 377, 390, 395, 404, 407, 412–413, 419, 421–423, 425, 430, 435–437, Fig 58, Fig 62, Table 1 (higher taxonomy, described, distribution, phylogeny, key, listed, comp. note) Equals Melampsaltaria Equals Cicadettaria Equals Melampsaltini Australia
Melampsaltini Moulds 2005b: 390 (comp. note)
Cicadettini Pham and Thinh 2005a: 238 (listed) Vietnam
Cicadettini Pham and Thinh 2005b: 287–288 (key, comp. note) Equals Melampsaltriaria Equals Melampsaltini Vietnam

Cicadettini Boulard 2006g: 131–132, 143 (comp. note)
Cicadettini Chawanji, Hodson and Villet 2006: 374, 380, Table 1 (listed, comp. note)
Cicadettini Puissant 2006: 20 (listed) France
Cicadettini Vincent 2006: 64 (listed) France
Cicadettini Boulard 2007a: 93 (comp. note)
Cicadettini Buckley and Simon 2007: 419 (comp. note)
Cicadettini Shiyake 2007: 8, 106 (listed, comp. note)
Cicadettini Sueur and Puissant 2007b: 62 (listed) France
Cicadettini Boulard 2008b: 109 (comp. note)
Cicadettini Demir 2008: 477 (listed) Turkey
Cicadettini Ewart and Marques 2008: 153 (listed)
Cicadettini Fonseca, Serrão, Pina-Martins, Silva, Mira, et al. 2008: 24 (listed)
Cicadettini Lee 2008b: 446, 451–452, 462, Fig 2, Table 1 (listed, illustrated, comp. note) Korea
Cicadettini Osaka Museum of Natural History 2008: 326 (listed) Japan
Cicadettini Sanborn, Phillips and Gillis 2008: 15 (listed) Florida
Cicadettini Ibrahim and Leong 2009: 317 (comp. note)
Cicadettini Lee 2009i: 295 (listed) Philippines, Panay
Cicadettini Marshall and Hill 2009: 1–2, 5–7 (comp. note) Australia
Cicadettini Marshall, Hill, Fontaine, Buckley and Simon 2009: 1–2, 5–7 (comp. note) Australia
Cicadettini Pain 2009: 44–45 (listed) Australia
Cicadettini Pham and Yang 2009: 3, 15, Table 2 (listed) Vietnam
Cicadettini Sanborn 2009d: 1 (listed)
Cicadettini Simon 2009: 104 (comp. note)
Cicadettini Duffels 2010: 13 (comp. note)
Cicadettini Kemal and Koçak 2010: 3, 5 (diversity, listed, comp. note) Turkey
Cicadettini Lee 2010a: 26 (listed, comp. note) Philippines, Mindanao
Cicadettini Lee 2010c: 24 (listed, comp. note) Philippines, Luzon
Cicadettini Mozaffarian and Sanborn 2010: 81 (listed)
Cicadettini Pham and Yang 2010b: 205 (comp. note) Vietnam
Cicadettini Popple and Emery 2010: 148, 150 (listed) Australia
Cicadettini Puissant and Sueur 2010: 556 (listed)
Cicadettini Wei, Ahmed and Rizvi 2010: 33 (comp. note)

Subtribe Cicadettina Buckton, 1889

Melampsaltaria Wu 1935: 26 (listed) China
Melampsaltraria Boulard 1988f: 37 (comp. note)
Melampsaltraria Boulard 1997c: 106 (comp. note)
Cicadettaria Boulard 1997c: 112, 117 (comp. note)
Cicadettaria Boulard 2000i: 21 (listed)
Melampsaltraria Boulard 2001b: 27 (comp. note)
Cicadettaria Boulard 2001b: 32, 37 (comp. note)

***Auta* Distant, 1897**

Auta Boer 1990: 64 (comp. note)
Auta Boer 1995d: 201, 225, Fig 54 (distribution, phylogeny, comp. note) New Guinea
Auta Boer and Duffels 1996b: 306, 327 (comp. note)
Auta Boer 1999: 117–119, 121–122, 128 (described, comp. note) New Guinea
Auta Moulds 2005b: 393, 436 (higher taxonomy, listed, comp. note)

***A. daymanensis* Boer, 1999**

Auta daymanensis Boer 1999: 117–118, 121, 124, 144, Fig 1, Figs 15–21 (n. sp., key, described, illustrated, distribution, comp. note) Papuan Peninsula

***A. insignis* Distant, 1897**

Auta insignis Boer and Duffels 1996b: 327 (distribution) Papuan Peninsula
Auta insignis Boer 1999: 115, 117, 119–126, 144, Fig 1, Figs 1–15 (key, type species of genus *Auta*, described, illustrated, distribution, phylogeny, comp. note) New Guinea

***A. tapinensis* Boer, 1999**

Auta tapinensis Boer 1999: 117–118, 121–122, 124, 126–128, 144, Fig 1, Fig 15, Figs 22–28 (n. sp., key, described, illustrated, distribution, comp. note) Papuan Peninsula

***Nigripsaltria* Boer, 1999** = *Nigripsalta* (sic)

Nigripsaltria Boer 1999: 115, 117–120, 128–129 (n. gen., described, comp. note) New Guinea, Louisiade Archipelago
Nigripsaltria Moulds 2005b: 436 (higher taxonomy, listed)
Nigripsalta (sic) Popple and Emery 2010: 147, 150 (comp. note) New Guinea

***N. carinata* Boer, 1999** = *Nigripsalta* (sic) *carinata*

Nigripsaltria carinata Boer 1999: 115, 117–118, 121, 128–129, 132–136, 144, Fig 1, Figs 44–50 (n. sp., type species of genus *Nigripsaltria*, key, described, illustrated, distribution, phylogeny, comp. note) Papua New Guinea
Nigripsalta (sic) *carinata* Popple and Emery 2010: 150 (comp. note) Australia

N. dali Boer, 1999

Nigripsaltria dali Boer 1999: 115, 117–118, 121, 128–129, 132–136, 144, Fig 1, Fig 44, Figs 51–54 (n. sp., key, described, illustrated, distribution, phylogeny, comp. note) Papua New Guinea

N. mouldsi Boer, 1999

Nigripsaltria mouldsi Boer 1999: 115, 117, 121, 128, 130–132, 144, Fig 1, Figs 37–44 (n. sp., key, described, illustrated, distribution, phylogeny, comp. note) Papua New Guinea

N. tagulaensis Boer, 1999

Nigripsaltria tagulaensis Boer 1999: 115, 117, 121, 129–132, 144, Fig 1, Figs 29–36, Fig 44 (n. sp., key, described, illustrated, distribution, phylogeny, comp. note) Louisiade Archipelago, Sudest Island

Toxopeusella Schmidt, 1926

Toxopeusella Boulard 1981c: 129, 137 (generic described, distribution, comp. note) Moluccas, New Guinea, Solomon Islands

Toxopeusella Duffels 1986: 328 (distribution, diversity, comp. note) Maluku, New Guinea, Solomon Islands

Toxopeusella Boer 1990: 64 (comp. note)

Toxopeusella Boulard 1990b: 214 (comp. note)

Toxopeusella Duffels and Boer 1990: 260–261 (comp. note) New Guinea, Maluku

Toxopeusella Boer 1995c: 204 (comp. note) Buru Island, New Guinea, Bismark Archipelago, Solomon Islands

Toxopeusella Boer 1995d: 201 (distribution, comp. note) New Guinea

Toxopeusella Boer and Duffels 1996b: 306 (comp. note)

Toxopeusella Boer 1999: 115, 117–120, 128, 136–138, 142 (described, key, comp. note) New Guinea, Melanesia

Toxopeusella Moulds 2005b: 436 (higher taxonomy, listed)

Toxopeusella Duffels and Boer 2006: 534 (distribution, comp. note) New Guinea

T. fakfakensis Boer, 1999

Toxopeusella fakfakensis Boer 1999: 115, 117–118, 120–121, 137, 140–141, 144, Fig 1, Figs 67–72 (n. sp., key, described, illustrated, distribution, comp. note) New Guinea

T. moluccana (Kirkaldy, 1909) = *Baeturia moluccana* = *Melampsalta moluccana* = *Cicada stigma* Walker, 1870 (nec *Cicada stigma* Walker, 1850) = *Baeturia stigma* = *Toxopeusella stigma* = *Melampsalta stigma* = *Toxopeusella browni* Boulard, 1981 = *Toxopeusella cheesmanae* Boulard, 1981 = *Toxopeusella smarti* Boulard, 1981

Cicada stigma Boulard 1981c: 129 (comp. note)

Baeturia moluccana Boulard 1981c: 129 (comp. note) Equals *Cicada stigma* Walker, 1870

Toxopeusella moluccana Boulard 1981c: 130, 135 (type species of genus *Toxopeusella*, comp. note) Equals *Cicada stigma* Walker, 1870 Moluccas

Toxopeusella smarti Boulard 1981c: 130–133, Figs 1–8 (n. sp., described, illustrated, comp. note) New Guinea

Toxopeusella cheesmanae Boulard 1981c: 133–135, Figs 9–16 (n. sp., described, illustrated, comp. note) Papua New Guinea

Toxopeusella browni Boulard 1981c: 135–133, Figs 17–25 (n. sp., described, illustrated, comp. note) Solomon Islands, Guadalcanal

Toxopeusella smarti Boulard 1990b: 150–151, Plate XVI, Fig 5 (sound system, comp. note)

Toxopeusella browni Boer and Duffels 1996b: 328 (distribution) Solomon Islands

Toxopeusella cheesmanae Boer and Duffels 1996b: 328 (distribution) Papuan Peninsula

Toxopeusella moluccana Boer and Duffels 1996b: 328 (distribution) North Maluku, South Maluku

Toxopeusella smarti Boer and Duffels 1996b: 328 (distribution) New Guinea

Toxopeusella browni Duffels 1997: 552 (comp. note) Solomon Islands

Toxopeusella moluccana Boer 1999: 115, 117–118, 120, 137, 140–145, Fig 1, Figs 56–67, Fig 77 (key, synonymy, described, illustrated, distribution, comp. note) Equals *Cicada stigma* Equals *Baeturia stigma* Equals *Baeturia moluccana* Equals *Melampsalta stigma* Equals *Toxopeusella moluccana* Equals *Toxopeusella stigma* Equals *Toxopeusella browni* Equals *Toxopeusella cheesmanae* Equals *Toxopeusella smarti* Maluku, New Guinea, Bismark Archipelago, Solomon Islands

Toxopeusella browni Boer 1999: 115, 138–139 (synonymy, comp. note) Solomon Islands

Toxopeusella cheesmanae Boer 1999: 115, 138–139, 142 (synonymy, comp. note) New Guinea

Toxopeusella smarti Boer 1999: 115, 138–139 (synonymy, comp. note) New Guinea

Cicada stigma Boer 1999: 136–137 (type species of genus *Toxopeusella*, synonymy, comp. note) New Guinea, Melanesia

Baeturia moluccana Boer 1999: 137 (synonymy, comp. note)

T. montana Boer, 1999

Toxopeusella montana Boer 1999: 115, 117–118, 120, 137, 140–145, Fig 1, Figs 73–76, Figs 78–82 (n. sp., key, described, illustrated, distribution, comp. note) New Guinea

***Pagiphora* Horváth, 1912** = *Enneaglena* Haupt, 1917
Pagiphora Soós 1956: 413 (listed) Equals *Cicadetta* Romania
Pagiphora Boulard 1981a: 37 (listed) Algeria
Enneaglena Koçak 1982: 149 (listed, comp. note) Turkey
Pagiphora Koçak 1982: 150 (listed, comp. note) Turkey
Pagiphora Boulard 1992b: 365 (comp. note)
Pagiphora Boulard 1993a: 1 (comp. note)
Pagiphora Mirzayans 1995: 19 (listed) Iran
Pagiphora Holzinger, Frölich, Günthart, Pauterer, Nickel et al. 1997: 49 (listed) Europe
Pagiphora Holzinger, Kammerlander and Nickel 2003: 477, 486 (key, listed, comp. note) Europe
Pagiphora Gogala and Trilar 2000: 23 (comp. note)
Pagiphora Gogala and Trilar 2001a: 28 (comp. note)
Pagiphora Gogala and Trilar 2001b: 14 (comp. note)
Pagiphora Moulds 2005b: 390, 415, 436 (higher taxonomy, listed, comp. note)
Pagiphora Demir 2008: 477 (listed) Turkey
Pagiphora Trilar and Gogala 2009: 74–75 (comp. note)
Pagiphora Kemal and Koçak 2010: 3, 5 (diversity, listed, comp. note) Turkey
Pagiphora Mozaffarian and Sanborn 2010: 81 (distribution, listed) Africa, Europe, Asia Minor

***P. annulata* (Brullé, 1832)** = *Pagiphora anullata* (sic) = *Tibicen annulatus* = *Cicada annulata* = *Cicadetta annulata* = *Cicadetta anulata* (sic) = *Parnisa anulata* (sic) = *Melampsalta annulata* = *Melampsalta (Melampsalta) annulata* = *Pagiphora annullata* (sic)
Cicadetta annulata Henshaw 1903: 191 (comp. note) Romania
Pagiphora annulata Soós 1956: 413 (listed, distribution) Romania
Pagiphora annulata Boulard 1981a: 40 (type species of genus *Pagiphora*, comp. note) Greece
Pagiphora annulata Boulard 1981b: 48–51, Fig 11, Fig 13 (n. comb., neotype designation, illustrated, comp. note) Equals *Tibicen annulatus* Greece, Asia Minor
Tibicen annulatus Boulard 1981b: 48–49 (comp. note)
Pagiphora annulata Dlabola 1981: 209 (distribution, comp. note) Turkey
Pagiphora annulata Lodos and Kalkandelen 1981: 77 (distribution, comp. note) Turkey, Albania, Algeria, Bulgaria, Czechoslovakia, Greece, Tunisia, Yugoslavia
Enneaglena annulata Koçak 1982: 149 (type species of *Enneaglena*)
Tibicen annulatus Koçak 1982: 150 (type species of *Pagiphora*)
Pagiphora annulata Nast 1987: 568–569 (distribution, listed) Czechoslovakia, Yugoslavia, Albania, Bulgaria, Greece
Pagiphora annulata Boulard 1992b: 366 (comp. note)
Pagiphora annulata Boulard 1993a: 1 (comp. note)
Pagiphora annulata Mirzayans 1995: 19 (listed, distribution) Iran
Pagiphora annulata Holzinger, Frölich, Günthart, Pauterer, Nickel et al. 1997: 49 (listed) Europe
Pagiphora annulata Gogala 1998b: 399 (comp. note) Slovenia
Pagiphora annulata Trilar and Gogala 1998: 69 (comp. note) Macedonia
Pagiphora annulata Okáli and Janský 1998: 37–38, 42, Fig 1 (distribution, comp. note) Slovakia
Pagiphora annulata Gogala and Gogala 1999: 127 (comp. note) Macedonia
Pagiphora annulata Gogala and Trilar 2000: 22–26, Figs 1–2 (song, sonogram, oscillogram, comp. note) Macedonia
Pagiphora annulata Schedl 2000: 260–261, 264, 266–267, Fig 5, Plate 1, Fig 12, Map 3 (key, illustrated, distribution, listed, comp. note) Slovakia, Hungary
Pagiphora anullata (sic) Gogala and Trilar 2001a: 28 (comp. note)
Pagiphora anullata (sic) Gogala and Trilar 2001b: 14 (comp. note)
Pagiphora annulata Krpač, Trilar and Gogala 2001: 29–30 (comp. note) Macedonia
Pagiphora annulata Schedl 2001b: 27 (comp. note)
Pagiphora annulata Sueur 2001: 43, Table 1 (listed)
Pagiphora annulata Gogala 2002: 247–248, Fig 12 (song, sonogram, oscillogram, comp. note)
Pagiphora annulata Gogala and Trilar 2003: 6, 8–9, 14, Fig 5 (sound production, comp. note) Macedonia
Pagiphora annulata Holzinger, Kammerlander and Nickel 2003: 486, 616–617, 632, Plate 42, Fig 12, Table 2 (key, listed, illustrated, comp. note) Europe
Pagiphora annulata Gogala, Trilar and Krpač 2005: 105, 122–124, Figs 16–17 (distribution, comp. note) Macedonia
Pagiphora annulata Trilar, Gogala and Popa 2006: 178 (comp. note) Romania
Pagiphora annulata Demir 2008: 477 (distribution) Turkey
Pagiphora annulata Trilar and Gogala 2008: 32 (comp. note) Romania
Pagiphora annulata Trilar and Gogala 2009: 74–75 (song, comp. note) Europe, Asia Minor
Pagiphora annulata Kemal and Koçak 2010: 6 (distribution, listed, comp. note) Turkey

Tibicen annulata Mozaffarian and Sanborn 2010: 81 (type species of *Pagiphora*)
Pagiphora annulata Mozaffarian and Sanborn 2010: 81 (synonymy, listed, distribution, comp. note) Equals *Tibicen annulata* Iran, Spain, Turkey, Southern Europe, Western North Africa, Syria

***P. aschei* Kartal, 1978**
Pagiphora aschei Boulard 1981a: 40 (comp. note) Crete
Pagiphora aschei Nast 1982: 311 (listed) Greece, Crete
Pagiphora aschei Kartal 1983: 240 (comp. note) Turkey
Pagiphora aschei Nast 1987: 570–571 (distribution, listed) Greece
Pagiphora aschei Boulard 1992b: 365 (comp. note) Crete
Pagiphora aschei Boulard 1993a: 1 (comp. note) Crete
Pagiphora aschei Demir 2007a: 52 (distribution, comp. note) Turkey
Pagiphora aschei Demir 2008: 477 (distribution) Turkey
Pagiphora aschei Trilar and Gogala 2009: 74–75 (song, comp. note) Crete
Pagiphora aschei Kemal and Koçak 2010: 6 (distribution, listed, comp. note) Turkey
Pagiphora aschei Trilar and Gogala 2010: 6 (comp. note) Crete

***P. hauptosa* Boulard, 1981** = *Enneaglena annulata* Haupt, 1917 = *Pagiphora annulata* (nec Horvath)
Pagiphora hauptosa Boulard 1981b: 49, 51, Fig 12, Fig 14 (illustrated, synonymy, comp. note) Equals *Enneaglena annulata* Equals *Pagiphora annulata* (nec Horváth) Asia Minor
Enneaglena annulata Boulard 1981b: 48–50 (comp. note)
Pagiphora hauptosa Boulard 1992b: 366 (comp. note)
Pagiphora hauptosa Boulard 1993a: 1 (comp. note)

***P. maghrebensis* Boulard, 1981**
Pagiphora maghrebensis Boulard 1981a: 38, 40–1, Figs C, Figs 6–7 (n. sp., described, illustrated, comp. note) Algeria
Pagiphora maghrebensis Boulard 1992b: 365 (comp. note) North Africa
Pagiphora maghrebensis Boulard 1993a: 1 (comp. note) North Africa

***P. yanni* Boulard, 1992**
Pagiphora yanni Boulard 1992b: 365–373, Plate I, Figs 1–7, Plate II, Figs 1–4, Plate III, Figs 1–4, Plate IV, Figs A1-A4, B1-B4 (n. sp., described, illustrated, song, comp. note) Turkey
Pagiphora yanni Boulard 1993a: 1–9, Plate I, Figs 1–5, Plate II, Figs 1–2, Plate III, Figs 1–4, Plate IV, Figs A1-A4, B1-B4, Figs 6–7 (n. sp., described, illustrated, song, comp. note) Turkey
Pagiphora yanni Gogala and Trilar 2000: 22–23 (comp. note)
Pagiphora yanni Krpač, Trilar and Gogala 2001: 30 (comp. note) Anatolia
Pagiphora yanni Puissant 2001: 152 (comp. note)
Pagiphora yanni Sueur 2001: 43, Table 1 (listed)
Pagiphora yanni Boulard 2005d: 342 (comp. note)
Pagiphora yanni Gogala, Trilar and Krpač 2005: 122 (comp. note)
Pagiphora yanni Boulard 2006g: 61–62, 180 (listed, comp. note)
Pagiphora yanni Demir 2008: 477 (distribution) Turkey
Pagiphora yanni Trilar and Gogala 2009: 74–75 (song, comp. note) Anatolia
Pagiphora yanni Kemal and Koçak 2010: 6 (distribution, listed, comp. note) Turkey

***Tympanistalna* Boulard, 1982**
Tympanistalna Boulard 1982f: 190, 192 (n. gen., described, comp. note)
Tympanistalna Fonseca 1996: 17, 28 (comp. note)
Tympanistalna Boulard 1995a: 67 (listed)
Tympanistalna Fonseca and Hennig 1996: 1536 (comp. note)
Tympanistalna Bennet-Clark 1998a: 411–414, Table 2 (comp. note)
Tympanistalna Sueur, Puissant, Simões, Seabra, Boulard and Quartau 2004: 177 (comp. note) Portugal
Tympanistalna Fonseca, Serrão, Pina-Martins, Silva, Mira, et al. 2008: 25, 27, 29–30 (comp. note)

***T. distincta* (Rambur, 1840)** = *Cicada distincta* = *Cicadetta distincta* = *Melampsalta distincta* = *Melampsalta (Melampsalta) distincta*
Tettigetta distincta Boulard 1981a: 41 (comp. note)
Cicada distincta Boulard 1982f: 190 (type species of genus *Tympanistalna*, comp. note)
Tympanistalna distincta Boulard 1982f: 190, 197 (n. comb., synonymy, comp. note, listed) Equals *Cicada distincta*
Tympanistalna distincta Nast 1987: 570–571 (distribution, listed) Spain
Cicadetta distincta Kartal 1999: 100–103, Fig A_2, Fig B_2, Fig C_2, Fig D_2, Fig E_2, Fig F_2, Fig G_2 (described, illustrated, comp. note) Spain

***T. gastrica* (Stål, 1854)** = *Cicada gastrica* = *Melampsalta gastrica* = *Cicadetta gastrica* = *Melampsalta (Melampsalta) gastrica*
Cicadetta gastrica Drosopoulous 1980: 190 (listed) Greece

Tettigetta gastrica Boulard 1981a: 41 (comp. note)
Cicada gastrica Boulard 1982f: 182, 190, 197 (comp. note)
Tympanistalna gastrica Boulard 1982f: 182, 191, 196, Figs 31–35, Figs 37–40, Fig 50 (n. comb., synonymy, described, illustrated, sonogram, comp. note) Equals *Cicada gastrica* Equals *Melampsalta gastrica* Equals *Cicadetta gastrica* Portugal
Tympanistalna gastrica Nast 1987: 570–571 (distribution, listed) Portugal, Italy, Bulgaria, Greece
Tympanistalna gastrica Quartau and Fonseca 1988: 367, 369, 374 (synonymy, distribution, comp. note) Equals *Cicada gastrica* Equals *Cicadetta gastrica* Portugal
Tympanistalna gastrica Boulard 1990b: 166, 205 (comp. note)
Tympanistalna gastrica Fonseca 1991: 174, 187–189-190, Fig 9 (song analysis, comp. note) Portugal
Tympanistalna gastrica Fonseca 1993: 767–773, Figs 1–10 (hearing physiology, sound field, comp. note) Portugal
Tympanistalna gastrica Boulard 1995a: 7–8, 13, 67–69, Figs 21–22, Fig $S28_1$, Fig $S28_2$ (illustrated, song analysis, sonogram, listed, comp. note) Portugal
Tympanistalna gastrica Quartau 1995: 37–38 (ecology, comp. note) Portugal
Tympanistalna gastrica Young and Bennet-Clark 1995: 1018 (comp. note)
Tympanistalna gastrica Daws and Hennig 1996: 186 (comp. note)
Tympanistalna gastrica Fonseca 1996: 15, 21–22, 27, Fig 3, Table 2 (timbal muscle physiology, oscillogram, comp. note) Portugal
Tympanistalna gastrica Fonseca and Hennig 1996: 1535–1538, 1541–1544, Fig 2B, Figs 9–10, Fig 11B, (tensor muscle physiology, comp. note) Portugal
Tympanistalna gastrica Bennet-Clark 1997: 1688 (comp. note)
Tympanistalna gastrica Fonseca and Popov 1997: 418, 425–426 (comp. note)
Tympanistalna gastrica Bennet-Clark 1998a: 412 (comp. note)
Tympanistalna gastrica Bennet-Clark and Young 1998: 702, 714 (comp. note) Portugal
Tympanistalna gastrica Fonseca and Bennet-Clark 1998: 717–730, Figs 1–10, Tables 1–2 (song analysis, oscillogram, comp. note) Portugal
Tympanistalna gastrica Michelsen and Elsner 1999: 1577 (comp. note)
Tympanistalna gastrica Michelsen and Fonseca 2000: 163–168, Figs 1–3, Fig 4B (sound field, comp. note) Portugal
Tympanistalna gastrica Sueur 2001: 45, Table 1 (listed)
Tympanistalna gastrica Fonseca and Revez 2002b: 971–976, Fig 1, Figs 2B–C, Fig 3, Tables 1–2 (song analysis, oscillogram, comp. note) Portugal
Tympanistalna gastrica Gerhardt and Huber 2002: 41 (comp. note)
Tympanistalna gastrica Sueur and Sanborn 2003: 340 (comp. note)
Tympanistalna gastrica Fonseca and Hennig 2004: 401–406, Fig 1, Figs 3A–C, Figs 4A–4B, Figs 5A–B (hearing physiology, comp. note) Portugal
Tympanistalna gastrica Sueur, Puissant, Simões, Seabra, Boulard and Quartau 2004: 178–179, 182, 185, Fig 1, Fig 4 (distribution, ecology, song, sonogram, oscillogram, comp. note) Portugal
Tympanistalna gastrica Boulard 2005d: 338 (comp. note)
Tympanistalna gastrica Ewart 2005b 459 (comp. note)
Tympanistalna gastrica Gogala, Trilar, Kozina and Duffels 2004: 14 (comp. note)
Tympanistalna gastrica Sanborn 2005c: 115 (comp. note)
Tympanistalna gastrica Boulard 2006g: 41–43, 181, Fig 9f (illustrated, listed, comp. note) Portugal
Tympanistalna gastrica Puissant 2006: 87 (comp. note)
Tympanistalna gastrica Seabra, Pinto-Juma and Quartau 2006: 850 (comp. note)
Tympanistalna gastrica Ewart and Marques 2008: 199 (comp. note)
Tympanistalna gastrica Fonseca, Serrão, Pina-Martins, Silva, Mira, et al. 2008: 19, 23–29, Fig 1, Table 1 (song phylogeny, comp. note) Portugal
Tympanistalna gastrica Pinto-Juma, Seabra and Quartau 2008: 9 (comp. note)
Tympanistalna gastrica Sueur, Windmill and Robert 2008a: 2380 (comp. note)
Tympanistalna gastrica Pinto, Quartau, and Bruford 2009: 267, 270, 274, 277–278, Fig 5, Table 4 (phylogeny, comp. note) Portugal
Tympanistalna gastrica Sanborn, Heath and Heath 2009: 25 (comp. note)
Tympanistalna gastrica Mozaffarian and Sanborn 2010: 76 (comp. note) Portugal
Cicadetta gastrica Mozaffarian and Sanborn 2010: 76 (comp. note)

***Cicadetta* Kolenati, 1857** = *Cicada (Cicadetta)* = *Cacadetta* (sic) = *Cicadette* (sic) = *Sicadetta* (sic) = *Cicadelta* (sic) = *Heptaglena* Horváth, 1911 (nec *Heptaglena* Schmarda, 1849) = *Oligoglena* Horváth, 1912 = *Heptaglena (Oligglena)*

Cicadetta Swinton 1880: 18 (comp. note)
Cicadetta Strobl 1900: 201 (listed) Austria
Cicadetta Soós 1956: 413 (listed) Romania
Cicadetta Hamann 1960: 165 (comp. note)
Cicadetta Le Quesne 1965: 4–5 (listed, key) Equals *Melampsalta* Great Britain
Cicadetta Boulard 1981a: 41 (comp. note)
Cicadetta Boulard 1981b: 46, 48 (comp. note)
Cicadetta Osssiannilsson 1981: 223 (described, synonymy) Equals *Melampsalta*
Cicadetta Popov 1981: 271, 274 (comp. note) Equals *Melampsalta*
Cicadetta Koçak 1982: 149 (listed, comp. note) Turkey
Heptaglena Koçak 1982: 149 (listed, comp. note) Turkey
Oligoglena Koçak 1982: 149–150 (comp. note) Turkey
Cicadetta Strümpel 1983: 17, 134 (listed, comp. note)
Cicadetta Boulard 1984c: 171 (comp. note)
Cicadetta Fleming 1984: 201 (comp. note)
Cicadetta Hayashi 1984: 28, 66 (key, described, synonymy, comp. note) Equals *Cicada (Cicadetta)* Equals *Cicada (Tettigetta)* Equals *Cicada (Melampsalta)* Equals *Kosemia* Equals *Parasemia* Equals *Leptopsalta* Equals *Karapsalta*
Cicadetta Arnett 1985: 218 (diversity) Southeastern United States, Missouri
Cicadetta Boulard 1985d: 212–213 (status)
Cicadetta Schedl 1986a: 7 (key) Istria
Cicadetta Schedl 1986b: 580 (comp. note)
Cicadetta Anufriev and Emelyanov 1988: 313–316 (key, characteristics) Soviet Union
Cicadetta Boulard 1988f: 37–39, 40–45, 71–72 (comp. note)
Cicadetta Ewart 1988: 183 (comp. note)
Cicadetta spp. Gwynne and Schatral 1988: 37 (comp. note) Western Australia
Cicadetta Moulds 1988: 39–40 (status, comp. note)
Cicadetta Ramos 1988: 62, Table 4.1 (listed, distribution, comp. note) Greater Antilles, Hispaniola, New World
Cicadetta Ewart 1989b: 289, 292 (comp. note) Australia
Cicadetta Boer 1990: 64 (comp. note)
Cicadetta Boulard 1990b: 172, 174 (comp. note)
Cicadetta Boulard and Nel 1990: 37–38 (comp. note)
Cicadetta Carver 1990: 260 (comp. note)
Cicadetta Heath and Heath 1990: 1 (comp. note) Australia
Cicadetta Moulds 1990: 7–8, 12, 32, 136–138, 143–144, 149 (listed, diversity, comp. note) Australia
Cicadetta sp. Moulds 1990: 11 (comp. note)
Cicadetta Boulard 1991b: 25–26 (comp. note)
Cicadetta sp. Carver, Gross and Woodward 1991: 438, Figs 30.10C-D (female reproductive system illustrated)
Cicadetta Dolling 1991: 7–8, 10, 151 (life history, comp. note)
Cicadetta Moulds and Carver 1991: 467 (comp. note) Australia
Cicadetta Popov and Sergeeva 1987: 681–682, 689 (comp. note)
Cicadetta Popov, Aronov and Sergeeva 1991: 281, 294 (comp. note)
Cicadetta Popov, Arnov and Sergeeva 1992: 151, 164 (comp. note)
Cicadetta spp. Fonseca 1993: 767 (comp. note)
Cicadetta Remane and Wachman 1993: 75, 134 (comp. note)
Cicadetta Schedl 1993: 799 (comp. note)
Cicadetta Lei, Chou and Li 1994: 53 (comp. note)
Cicadetta Boulard 1995a: 55 (listed)
Cicadetta Boulard and Mondon 1995: 19, 65 (key, comp. note) France
Cicadetta Lee 1995: 18, 23, 100, 104, 111–112 (listed, key, described, ecology, distribution, illustrated, comp. note) Korea
Cicadetta Mirzayans 1995: 18 (listed) Iran
Cicadetta Boer and Duffels 1996b: 309 (distribution, comp. note) New Caledonia
Cicadetta Fonseca and Hennig 1996: 1536 (comp. note)
Cicadetta Gogala, Popov and Ribaric 1996: 46 (comp. note)
Cicadetta Holzinger 1996: 505 (listed) Austria
Cicadetta Moulds 1996b: 92 (comp. note) Australia
Cicadetta Boulard 1997a: 185 (comp. note)
Cicadetta Boulard 1997c: 107–113, 117, 121–122 (comp. note)
Cicadetta Chou, Lei, Li, Lu and Yao 1997: 25, 47, 58, 289, 293, 297, 306, 312, 324, 326, 358, Table 6, Tables 10–12 (key, described, synonymy, listed, diversity, biogeography, comp. note) Equals *Heptaglena* Equals *Oligoglena* Equals *Kosemia* Equals *Parasemia* Equals *Karapsalta* China
Cicadetta Feltwell 1997: 102 (comp. note) France
Cicadetta Gogala and Popov 1997a: 12 (comp. note)
Cicadetta Holzinger, Frölich, Günthart, Pauterer, Nickel et al. 1997: 49 (listed) Europe
Cicadetta Poole, Garrison, McCafferty, Otte and Stark 1997: 251, 331 (listed) Equals *Tettigetta*

Equals *Heptaglena* Equals *Oligoglena* Equals *Kosemia* Equals *Parasemiodes* Equals *Melampalta* (sic) Equals *Karapsalta* North America
Cicadetta Popov 1997a: 3, 15 (comp. note)
Cicadella (sic) Popov 1997b: 1 (comp. note)
Cicadetta Popov 1997b: 11 (comp. note)
Cicadetta spp. Bennet-Clark 1998a: 411 (comp. note)
Cicadetta Gogala 1998a: 156 (comp. note)
Cicadetta Gogala 1998b: 396 (comp. note) Slovenia
Cicadetta Gogala and Trilar 1998b: 67–68 (comp. note) Slovenia
Cicadetta spp. Lee 1998: 21 (comp. note) Korea
Cicadetta Popov 1998a: 315–316, 319 (comp. note)
Cicadetta Popov 1998b: 309, 311 (comp. note)
Cicadetta Trilar and Gogala 1998: 69 (comp. note) Macedonia
Cicadetta Boer 1999: 118 (comp. note)
Cicadetta Ewart 1999: 2 (listed)
Cicadetta Gogala and Gogala 1999: 120 (comp. note) Slovenia
Cicadetta Gogala and Trilar 1999b: 1 (comp. note)
Cicadetta Kartal 1999: 99, 103 (listed, comp. note) Spain
Cicadetta Lee 1999: 1, 10 (comp. note) Equals *Melampsalta* Equals *Kosemia* Equals *Leptopsalta* Korea
Cicadetta Ohbayashi, Sato and Igawa 1999: 340 (parasitism, comp. note)
Cicadetta Sanborn 1999b: 2 (listed) North America
Melampsalta Arnett 2000: 299 (diversity) United States
Cicadetta Arnett 2000: 299 (diversity) United States
Cicadetta Boulard 2000i: 21, 26, 80 (listed, key, comp. note) France
Cicadetta Gogala and Popov 2000: 11 (comp. note)
Cicadetta Lee 2000: 4 (comp. note) Korea
Cicadetta Schedl 2000: 264 (listed)
Cicadetta Boulard 2001b: 27–32, 37, 41 (comp. note)
Cicadetta Chambers, Boon, Buckley and Hitchmough 2001: 380 (comp. note) Australia
Cicadetta Cooley 2001: 758 (comp. note) Australia, Asia
Cicadetta Cooley and Marshall 2001: 847–848 (comp. note) Asia, Australia
Cicadetta sp. Krpač, Trilar and Gogala 2001: 29 (comp. note) Macedonia
Cicadetta Puissant 2001: 141 (comp. note)
Cicadetta Puissant and Sueur 2001: 435 (comp. note)
Cicadetta Schedl 2001a: 1287 (comp. note)
Cicadetta Wilson 2001: 14 (comp. note)
Cicadetta Buckley, Arensburger, Simon and Chambers 2002: 6 (comp. note) Australia
Cicadetta Kondratieff, Ellingson and Leatherman 2002: 14, 45 (key, listed, diagnosis, comp. note) Colorado
Cicadetta Lee, Choe, Lee and Woo 2002: 3, 6–7, 9, Fig 3, Table 1 (phylogeny, comp. note) Europe, Africa, Asia, Oceana, Russia, China, Japan, Korea
Cicadetta spp. Lee, Choe, Lee and Woo 2002: 9–10 (comp. note)
Cicadetta Moulds and Carver 2002: 24 (listed) Australia
Cicadetta spp. Callaham, Blair, Todd, Kitchen and Whiles 2003: 1081, 1086–1087, Figs 5–6, Table 1 (density, comp. note) Kansas
Cicadetta Callaham, Blair, Todd, Kitchen and Whiles 2003: 1081, 1084–1085, 1090–1091, Tables 2–3 (density, comp. note) Kansas
Cicadetta Dietrich 2003: 72 (comp. note)
Cicadetta Holzinger, Kammerlander and Nickel 2003: 477 (key, listed, comp. note) Europe
Cicadetta Moulds 2003a: 187 (comp. note)
Cicadetta Popple 2003: 107, 111 (comp. note) Australia, Queensland
Cicadetta Sueur and Aubin 2003b: 481 (comp. note)
Cicadetta Sueur and Puissant 2003: 121 (comp. note) France
Cicadetta spp. Sueur, Puissant and Pillet 2003: 105 (comp. note)
Cicadetta Sueur 2005: 209 (song, comp. note)
Cicadetta Arensburger, Buckley, Simon, Moulds, and Holsinger 2004: 562 (molecular phylogeny) Australia
Cicadetta Arensburger, Simon, and Holsinger 2004: 1770 (comp. note) New Zealand
Cicadetta Boulard 2005d: 341 (comp. note)
Cicadetta Dietrich, Dmitriev, Rakitov, Takiya and Zahniser 2005: S-14, Fig 1 (phylogeny outgroup)
Cicadetta spp. Emery, Emery, Emery and Popple 2005: 106 (comp. note) New South Wales
Cicadetta spp. Gogala, Trilar and Krpač 2005: 124 (comp. note) Greece
Cicadetta Lee 2005: 15, 17, 97, 156, 158, 176 (listed, key, comp. note) Equals *Melampsalta* Equals *Kosemia* Equals *Leptopsalta* Korea
Cicadetta Moulds 2005b: 390, 410, 415, 422–423, 430, 436, Table 2 (type genus of Cicadettini, higher taxonomy, listed, comp. note) Australia
Cicadetta spp. Moulds 2005b: 410 (comp. note)
Cicadetta Bernier 2006: 4 (comp. note) France
Cicadetta Gogala 2006a: 158 (comp. note)
Cicadetta Gogala 2006b: 370 (comp. note)

Cicadetta Haywood 2006a: 13 (comp. note) South Australia
Cicadetta Kartal 2006: 280 (comp. note)
Cicadetta sp. Novikov, Novikova, Anufriev and Dietrich 2006: 308 (distribution, listed) Kyrgyzstan
Cicadetta Puissant 2006: 76, 80–81, 86–87 (comp. note) France
Cicadetta Vincent 2006: 64 (listed) France
Cicadetta Boulard 2007b: 97–98, Figs 3–4 (illustrated, comp. note)
Cicadetta Shiyake 2007: 8, 10, 106 (listed, comp. note)
Cicadetta Söderman 2007: 26 (listed) Finland
Cicadetta spp. Sueur and Puissant 2007a: 127 (comp. note)
Cicadetta spp. Sueur and Puissant 2007b: 55–57, 61–63, 67 (distribution, comp. note) France
Cicadetta Demir 2008: 477 (listed) Turkey
Cicadetta sp. Fonseca, Serrão, Pina-Martins, Silva, Mira, et al. 2008: 28 (comp. note)
Cicadetta spp. Gogala, Drosopoulos and Trilar 2008b: 123 (comp. note)
Cicadetta Gogala and Schedl 2008: 404 (comp. note)
Cicadetta sp. Gogala and Trilar 2008: 31 (comp. note)
Cicadetta Hertach 2008b: 15, 19 (comp. note) Switzerland
Cicadetta Hugel, Matt, Callot, Feldtrauer and Brua 2008: 5–7 (comp. note) France
Cicadetta Lee 2008b: 446, 452, 457–458, Tables 1–2 (listed, comp. note) Equals *Takapsalta* Korea
Cicadetta Perez-Gerlabert 2008: 202 (listed) Hispaniola
Cicadetta Pinto-Juma, Seabra and Quartau 2008: 2 (comp. note)
Cicadetta Sanborn, Phillips and Gillis 2008: 3, 15 (key, diversity, described, comp. note) Florida
Cicadetta Zamanian, Mehdipour and Ghaemi 2008: 2071 (comp. note)
Cicadetta spp. Gogala, Drosopoulos and Trilar 2009: 24 (comp. note)
Cicadetta Moulds 2009b: 164 (comp. note)
Cicadetta Sanborn 2009d: 1 (comp. note) Africa, Asia, Australia, Europe
Cicadetta Duffels 2010: 13 (type genus of Cicadettini, comp. note)
Cicadetta spp. Kemal and Koçak 2010: 2 (diversity, comp. note) Turkey
Cicadetta Kemal and Koçak 2010: 3, 6 (diversity, listed, comp. note) Equals *Melampsalta* Equals *Tettigetta* Equals *Pauropsalta* Equals *Heptaglena* Equals *Oligoglena* Equals *Kosemia* Equals *Karapsalta* Turkey
Cicadetta Lee 2010a: 26 (comp. note)
Cicadetta Mozaffarian and Sanborn 2010: 76 (comp. note)
Cicadetta Puissant and Sueur 2010: 570 (comp. note)

C. aaede (Walker, 1850) to *Yoyetta aaeda*
C. abdominalis (Distant, 1892) to *Yoyetta abdominalis*
C. sp. nr. *abdominalis* Emery, Emery, Emery & Popple, 2005 to *Yoyetta* sp. nr. *abdominalis*
C. sp. aff. *abdominalis* Haywood, 2005 to *Yoyetta* sp. aff. *abdominalis*
C. sp. nr. *abdominalis* Emery, 2008 to *Yoyetta* sp. nr. *abdominalis*

***C. abscondita* Lee, 2008** = *Takapsalta ichinosawana* (nec Matsumura) = *Takapsalta* sp. Kato, 1956 = *Cicadetta montana* (nec Scopoli) = *Cicadetta* sp. Gogala & Trilar 2004
Takapsalta ichinosawana Lee 1995: 18, 23, 32, 42–43, 45–46, 100–104, 136, 139 (type species of *Takapsalta*, listed, key, synonymy, ecology, distribution, illustrated, comp. note) Equals *Takapsalta* sp. Kato 1956 Korea
Takapsalta sp. Kato 1956 Lee 1995: 43, 101–102 (comp. note) Korea
Cicadetta montana Lee 1998: 19–21, Figs 1–2 (illustrated, habitat, behavior, song, comp. note) Equals *Takapsalta ichinosawana* Korea
Cicadetta montana Lee 1999: 1 (synonymy, comp. note) Equals *Takapsalta ichinosawana* Korea
Takapsalta ichinosawana Lee 1999: 1, 11 (comp. note) Korea
Cicadetta sp. Gogala and Trilar 2004a: 322 (comp. note)
Cicadetta abscondita Lee 2008b: 446, 451–457, 463, Fig 2A, Figs 3–5, Table 1 (n. sp., described, illustrated, key, synonymy, song, oscillogram, sonogram, distribution, listed, comp. note) Equals *Takapsalta ichinosawana* (nec Matsumura) Equals *Takapsalta* sp. Kato, 1956 Equals *Cicadetta montana* (nec Scopoli) Equals *Cicadetta* sp. Gogala & Trilar 2004 Korea
Takapsalta sp. Kato, 1956 Lee 2008b: 456 (comp. note)

C. adelaida (Ashton, 1914) to *Clinopsalta adelaida*
C. sp. nr. *adelaida* Emery et al. 2005 to *Clinopsalta* sp. nr. *adelaida*

***C. afghanistanica* Dlabola, 1964**

***C. albipennis* Fieber, 1876** = *Melampsalta albipennis* = *Melampsalta (Melampsalta) albipennis*
Cicadetta albipennis Drosopoulous 1980: 190 (listed) Greece
Cicadetta albipennis Nast 1987: 568–569 (distribution, listed) Italy, Greece

Cicadetta albipennis Kartal 2006: 279–280, Figs 8–17 (described, illustrated, comp. note) Italy, Sicily

C. *anapaistica* Hertach, 2011

C. apicata (Ashton, 1914) to *Kobonga apicata*

C. sp. nr. *apicata* Callitris Clicker Popple & Strange, 2002 to *Kogonga* sp. nr. *apicata*

C. arenaria (Distant, 1907) to *Sylphoides arenaria*

C. binotata (Goding & Froggatt, 1904) to *Myopsalta binotata*

Species near *Cicadetta binotata* Ewart & Popple, 2001 to Species near *Myopsalta binotata*

Cicadetta sp. nr. *binotata* Mitchell Smokey Buzzer Popple & Strange, 2002 to *Myopsalta* sp. nr. *binotata*

Cicadetta sp. nr. *binotata* Moree Smokey Buzzer Popple & Strange, 2002 to *Myopsalta* sp. nr. *binotata*

C. *brevipennis* Fieber, 1872 = *Cicadetta brevipenis* (sic) = *Cicada montana* var. *brevipennis* = *Cicadetta montana brevipennis* = *Melampsalta montana brevipennis* = *Melampsalta (Melampsalta) montana brevipennis* = *Cicadetta montana* var. *brevipennis* = *Cicadetta montana* (nec Scopoli)

Cicadetta montana (sic) Boulard 1995a: 8, 13, 55–56, Fig $S20_1$, Fig $S20_2$ (song analysis, sonogram, listed, comp. note) Turkey

Cicadetta montana brevipennis Schedl 2000: 264 (comp. note)

Cicadetta montana (sic) Puissant 2001: 141–150, Figs 1–7 (illustrated, life history, synonymy, distribution, song, oscillogram, comp. note) Equals *Cicadetta petryi* n. syn. France, Slovenia, Belgium, Europe, China, Korea

Cicadetta montana var. *brevipennis* Gogala and Trilar 2004a: 317 (status, comp. note)

Cicadetta brevipennis Gogala and Trilar 2004a: 317–320, 322–323, Fig 1, Table 1 (n. status, valid species, sonogram, oscillogram, comp. note) Slovenia, Europe

Cicadetta brevipennis Gogala and Trilar 2004b: 35 (comp. note) France, Italy, Slovenia

Cicadetta brevipennis Trilar and Holzinger 2004: 1384–1385, Fig 1b (song, sonogram, oscillogram, comp. note) Austria, France, Germany, Italy, Slovenia, Romania

Cicadetta brevipennis Boulard 2005d: 341 (comp. note) France, Europe, Slovenia

Cicadetta brevipennis Gogala, Trilar and Krpač 2005: 124 (comp. note) Romania

Cicadetta brevipennis Puissant and Defaut 2005: 117–125, 128, Table 1, Table 3, Tables 5–8, Tables 14–15 (listed, habitat association, distribution, comp. note) France

Cicadetta brevipennis Sueur 2005: 215 (song, comp. note) Equals *Cicadetta petryi* France

Cicadetta brevipennis Bernier 2006: 2–5, 7, Figs 2–3 (distribution, illustrated, oscillogram) France

Cicadetta brevipennis Boulard 2006g: 41–43, 50, 52, 180, Fig 9b (illustrated, listed, comp. note) France

Cicadetta brevipennis Gogala 2006b: 371–373, 375, Fig 3 (song, sonogram, oscillogram, comp. note) France, Italy, Slovenia, Switzerland, Austria, Romania

Cicadetta montana brevipennis Gogala 2006b: 372 (comp. note)

Cicadetta brevipennis Puissant 2006: 19–20, 23, 25–27, 32–35, 46, 54, 63, 71–78, 80, 82–83, 90–93, 97, 107–110, 113–114, 133–137, 140–142, 144–145, 148–152, 156, 158–159, 161–163, 179–187, 190, Fig 8, Figs 42–44, Tables 1–2, Appendix Table 1, Appendix Table 3, Appendix Tables 5–8, Appendix Tables 14–15 (illustrated, distribution, habitat association, ecology, listed, comp. note) Equals *Cicadetta montana* Boulard, 1995 Equals *Cicadetta montana* Boulard, 2000 Equals *Cicadetta montana* Puissant, 2001 Equals *Cicadetta petryi* (partim) France

Cicadetta brevipennis Trilar, Gogala and Popa 2006: 177–178, Fig 1 (sonogram, oscillogram, listed, distribution, comp. note) Romania

Cicadetta brevipennis Trilar, Gogala and Szwedo 2006: 313, 319 (comp. note) Equals *Cicadetta montana* (partim)

Cicadetta montana var. *brevipennis* Vincent 2006: 64 (comp. note) France

Cicadetta brevipennis Vincent 2006: 64 (comp. note) France

Cicadetta brevipennis Hertach 2007: 37, 41, 43–49, 51–59, Fig 4C, Fig 5C, Fig 6C, Figs 7E–F, Figs 8C–D, Fig 8F, Fig 9, Fig 10, Figs 13–14, Tables 2–4 (natural history, illustrated, song, oscillogram, habitat, distribution, comp. note) Switzerland

Cicadetta montana var. *brevipennis* Hertach 2007: 37 (comp. note) Switzerland

Cicadetta brevipennis Sueur and Puissant 2007a: 126–128, Figs 6–7 (oscillogram, comp. note) France

Cicadetta brevipennis Sueur and Puissant 2007b: 56–57, 60, 62, 64–66, Fig 8 (listed, distribution, comp. note) France

Cicadetta brevipennis Trilar and Gogala 2007: 6, 9–10, 19, Fig 5f (oscillogram, comp. note) Romania

Cicadetta brevipennis Brua and Hugel 2008: 49 (comp. note)

Cicadetta brevipennis Gogala 2008: 55–56 (song, oscillogram, comp. note) Switzerland, Greece

Cicadetta brevipennis Gogala, Drosopoulos and Trilar 2008b: 123 (comp. note) France, Romania

Cicadetta brevipennis Gogala and Trilar 2008: 31 (comp. note)
Cicadetta brevipennis Hertach 2008b: 9, 13 (comp. note)
Cicadetta brevipennis Hugel, Matt, Callot, Feldtrauer and Brua 2008: 5–9, Figs 1–2, Table 2 (distribution, oscillogram, comp. note) France
Cicadetta brevipennis Lee 2008b: 456 (comp. note)
Cicadetta brevipennis Nickel 2008: 207 (distribution, hosts, comp. note) Europe
Cicadetta brevipennis Trilar and Gogala 2008: 30, 32 (comp. note) Romania
Cicadetta brevipennis Trilar and Hertach 2008: 185–186, Fig 1a (oscillogram, comp. note) Italy
Cicadetta brevipennis Gogala, Drosopoulos and Trilar 2009: 15, 24 (comp. note)
Cicadetta brevipennis Gogala and Trilar 2009a: 32–33 (comp. note)
Cicadetta brevipennis Ibanez 2009: 208, 212 (listed, comp. note) France

C. brevis (Ashton, 1912) to *Ewartia brevis*

***C. calliope calliope* (Walker, 1850)** = *Cicada calliope* = *Melampsalta calliope* = *Cicada parvula* Say, 1825 = *Melampsalta parvula* = *Carineta parvula* = *Cicada pallescens* Germar, 1830 = *Melampsalta pallescens*
Melampsalta calliope Davis 1927: 452 (life cycle, comp. note) Kansas
Melampsalta calliope McCoy 1965: 40, 43 (key, comp. note) Arkansas
Melampsalta calliope Lin 1971: 280 (predation) Oklahoma
Cicadetta calliope Karban 1986: 224, Table 14.2 (comp. note) USA
Melampsalta calliope Poole, Garrison, McCafferty, Otte and Stark 1997: 331 (listed) Equals *Cicada calliope* North America
Cicadetta calliope Sanborn 1999b: 2 (listed) North America
Melampsalta calliope Arnett 2000: 299 (listed) United States
Cicadetta calliope Callaham 2000: 3957 (density, comp. note) Kansas
Cicadetta calliope Callaham, Whiles, Meyer, Brock and Charlton 2000: 537–542, Figs 1–3, Tables 1–3 (density, comp. note) Kansas
Melampsalta calliope Smith, Kelly, Dean, Bryson, Parker et al. 2000: 128 (illustrated, comp. note) Kansas
Cicadetta calliope Dietrich, Rakitov, Holmes and Black 2001: 296, Table 1 (comp. note) Illinois
Cicadetta calliope Callaham, Whiles and Blair 2002: 92–100, Fig 1, Fig 3, Tables 1–2 (density, comp. note) Kansas
Cicadetta calliope Kondratieff, Ellingson and Leatherman 2002: 5, 12–13, 45–47, 66, 73, Fig 10, Fig 33, Tables 1–2 (key, diagnosis, illustrated, listed, distribution, song, comp. note) Equals *Cicada calliope* Colorado, Alabama, Arkansas, Florida, Georgia, Kansas, Illinois, Iowa, Louisiana, Michigan, Mississippi, Nebraska, New Jersey, North Carolina, Ohio, Oklahoma, Tennessee, Texas, Virginia
Cicada parvula Kondratieff, Ellingson and Leatherman 2002: 47 (comp. note) Equals *Melampsalta calliope*
Melampsalta calliope Dietrich 2003: 66, Fig 2(7) (illustrated) Illinois
Melampsalta calliope Walker and Moore 2004: 1 (listed, comp. note) Florida
Cicadetta calliope Kondratieff, Schmidt, Opler and Garhart 2005: 97, 215–216 (listed, comp. note) Oklahoma, Florida, Michigan, Texas
Cicadetta calliope Buckley, Cordeiro, Marshall and Simon 2006: 414 (comp. note) North America
Cicadetta calliope Howard and Hill 2007: 60 (comp. note)
Cicadetta calliope calliope Sanborn, Phillips and Gillis 2008: 3, 15–16, 31, 42, Fig 75, Figs 166–174 (key, synonymy, illustrated, distribution, comp. note) Equals *Cicada parvula* Equals *Cicada pallescens* Equals *Cicada calliope* Florida, North Carolina
Cicadetta calliope Smith and Hasiotis 2008: 503–512, Fig 3, Figs 5–8 (illustrated, burrows, comp. note) Equals *Melampsalta calliope* Kansas
Cicadetta calliope Counts and Hasiotis 2009: 85 (burrow, comp. note)
Cicadetta calliope Marshall and Hill 2009: 4–5, Fig 6C, Table 2 (sonogram, comp. note) USA
Cicadetta calliope Pain 2009: 47 (sonogram) United States
Cicadetta calliope Sanborn 2009d: 1, 6 (comp. note) North America

***C. calliope floridensis* (Davis, 1920)** = *Melampsalta calliope floridensis* = *Cicadetta calliope* var. *floridensis* = *Melampsalta floridensis*
Cicadetta callipe floridensis Frost 1969: 93 (listed) Florida
Cicadetta calliope var. *floridensis* Sanborn 1999a: 52 (type material, listed)
Cicadetta calliope var. *floridensis* Sanborn 1999b: 2 (listed) North America
Cicadetta calliope var. *floridensis* Sueur 2001: 46, Table 2 (listed) Equals *Melampsalta floridensis*

Melampsalta floridensis Walker and Moore 2004: 1–2, Fig 1 (illustrated, listed, comp. note) Florida
Cicadetta calliope floridensis Sanborn, Phillips and Gillis 2008: 3, 15–16, 31, 43, Fig 75, Figs 175–183 (key, synonymy, illustrated, distribution, comp. note) Florida
Cicadetta calliope floridensis Sanborn 2009d: 1 (comp. note) North America

***C. camerona* (Davis, 1920)** = *Melampsalta camerona*
Cicadetta camerona Poole, Garrison, McCafferty, Otte and Stark 1997: 331 (listed) Equals *Melampsalta camerona* North America
Cicadetta camerona Sanborn 1999a: 52 (type material, listed)
Cicadetta camerona Sanborn 1999b: 2 (listed) North America
Cicadetta camerona Sanborn 2009d: 1 (comp. note) North America

***C. cantilatrix* Sueur & Puissant, 2007** = *Cicadetta cerdaniensis* (nec Puissant & Boulard)
Cicadetta sp. Hertach 2004: 59–64, Figs 1–3 (song, oscillogram, illustrated, habitat, comp. note) Switzerland
Cicadetta cantilatrix Hertach 2007: 37–38, 48, 59 (song, oscillogram, illustrated, habitat, comp. note) Switzerland
Cicadetta cantilatrix Sueur and Puissant 2007b: 57–67, Figs 1–5, Fig 8, Table 1 (n. sp., described, illustrated, song, oscillogram, distribution, listed, comp. note) France
Cicadetta cantilatrix Trilar and Gogala 2007: 10, 19, Fig 5c (oscillogram, comp. note) Poland
Cicadetta cantilatrix Brua and Hugel 2008: 49–51, Fig 2 (song, oscillogram, comp. note) Equals *Cicadetta cerdaniensis* Hertach 2007 (partim) France
Cicadetta cantilatrix Chrétien 2008: 3 (oscillogram, comp. note) Switzerland
Cicadetta cantilatrix Gogala 2008: 55–56 (song, oscillogram, comp. note) France, Switzerland, Italy
Cicadetta cantilatrix Gogala, Drosopoulos and Trilar 2008a: 99 (comp. note)
Cicadetta cantilatrix Gogala, Drosopoulos and Trilar 2008b: 123 (comp. note)
Cicadetta cantilatrix Gogala and Trilar 2008: 31 (comp. note)
Cicadetta cantilatrix Hertach 2008b: 2, 7–11, 13, 15–21, 23, 26, Figs 4–8, Tables 5–6, Appendix B (natural history, illustrated, song, oscillogram, comp. note) Switzerland
Cicadetta cantilatrix Hugel, Matt, Callot, Feldtrauer and Brua 2008: 7 (comp. note) France
Cicadetta cantilatrix Lee 2008b: 456 (comp. note)
Cicadetta cantilatrix Schneemann 2008: 2 (comp. note) Austria
Cicadetta cantilatrix Trilar and Gogala 2008: 30–32, Fig 2 (sonogram, oscillogram, listed, distribution, comp. note) Romania
Cicadetta cantilatrix Trilar and Hertach 2008: 186 (comp. note)
Cicadetta cantilatrix Gogala, Drosopoulos and Trilar 2009: 24 (comp. note) Macedonia
Cicadetta cantilatrix Fuhrmann 2010: 136 (comp. note) France

C. capistrata (Ashton, 1912) to *Taurella forresti*

***C. caucasica* Kolenati, 1857** = *Cicada (Tettigetta) tibialis caucasica* = *Tettigetta tibialis caucasica* = *Cicadetta prasina caucasica* = *Cicadetta tibialis caucasica* = *Melampsalta (Pauropsalta) caucasica* = *Pauropsalta tibialis caucasica* = *Melampsalta tibialis caucasica* = *Cicada tibialis imbecilis* Eversman, 1859 = *Cicadetta sareptana* Fieber, 1876 = *Melampsalta sareptana* = *Melampsalta (Melampsalta) sareptana*
Cicadetta caucasica Popov 1985: 75–77, 80, Fig 21.3, Fig 24 (acoustic system, comp. note)
Cicadetta caucasica Joermann and Schneider 1987: 295 (comp. note)
Cicadetta caucasica Nast 1987: 568–569 (distribution, listed) Russia
Cicadetta caucasica Popov 1990a: 302, Fig 2 (comp. note)
Cicadetta caucasica Tautz 1990: 105, Fig 12 (song frequency, comp. note)
Cicadetta caucasica Gogala, Popov and Ribaric 1996: 47, 51 (comp. note)
Cicadetta caucasica Popov 1997a: 3 (comp. note)
Cicadetta caucasica Popov 1997b: 1 (comp. note)
Cicadetta caucasica Sueur and Puissant 2000: 266–267 (comp. note)

C. celis Moulds, 1988 to *Yoyetta celis*

C. sp. nr. *celis* Emery et al., 2005 to *Yoyetta* sp. nr. *celis*

***C. cerdaniensis* Puissant & Boulard, 2000** = *Cicadetta montana* (nec Scopoli)
Cicadetta cerdaniensis Boulard 2000i: 83–86, 125, Plate M, Map 13 (synonymy, illustrated, song, sonogram, oscillogram, frequency spectrum, distribution, ecology, comp. note) France
Cicadetta cerdaniensis Puissant and Boulard 2000: 112–117, Figs 1–3 (n. sp., described, illustrated, song, oscillogram, sonogram, comp. note) France
Cicadetta cerdaniensis Puissant 2001: 145, 148 (comp. note)
Cicadetta cerdaniensis Sueur and Puissant 2001: 22–23 (comp. note) France
Cicadetta cerdaniensis Kapla, Trilar, Gogala and Gogala 2001: 35 (song, comp. note) Pyrenees

Cicadetta cerdaniensis Puissant and Sueur 2001: 435 (comp. note)
Cicadetta cerdaniensis Sueur 2001: 42, Table 1 (listed)
Cicadetta cerdaniensis Gogala 2002: 247 (comp. note)
Cicadetta cerdaniensis Gogala and Trilar 2004a: 317, 319–320, 322–323, Fig 3 (comp. note) France, Poland, Slovenia, Macedonia, Switzerland
Cicadetta cerdaniensis Gogala and Trilar 2004b: 35 (song, comp. note) Pyrenees, Switzerland, Poland, Slovenia, Macedonia
Cicadetta cerdaniensis Hertach 2004: 59, 63–66 (comp. note) Switzerland
Cicadetta cerdaniensis Trilar and Holzinger 2004: 1384–1385, Fig 1c (song, sonogram, oscillogram, comp. note) Austria, France, Germany, Switzerland, Poland, Slovenia, Macedonia
Cicadetta cerdaniensis Chrétien 2005: 5 (comp. note) Switzerland
Cicadetta cerdaniensis Boulard 2005d: 341 (comp. note) France, Slovenia
Cicadetta cerdaniensis Gogala, Trilar and Krpač 2005: 105, 119, Fig 13 (distribution, comp. note) Macedonia
Cicadetta cerdaniensis Puissant 2005: 311 (comp. note)
Cicadetta cerdaniensis Puissant and Defaut 2005: 117–118, 127, Table 1, Table 13 (listed, habitat association, distribution, comp. note) France
Cicadetta cerdaniensis Bernier 2006: 2 (distribution) France
Cicadetta cerdaniensis Boulard 2006g: 50, 52, 57, Fig 19 (sonogram, comp. note) France
Cicadetta cerdaniensis? Boulard 2006g: 50, 52 (comp. note)
Cicadetta cerdaniensis Gogala 2006b: 371–373, 375, Fig 4 (song, sonogram, oscillogram, comp. note) France, Switzerland, Austria, Germany, Poland, Slovenia, Macedonia
Cicadetta cerdaniensis Puissant 2006: 19–20, 23, 25–29, 32, 35, 53, 71, 77–80, 89–93, 97, 146–147, 179–180, 189, Fig 8, Figs 49–50, Tables 1–2, Appendix Table 1, Appendix Table 13 (illustrated, distribution, habitat association, ecology, listed, comp. note) France
Cicadetta cerdaniensis Trilar, Gogala and Szwedo 2006: 313–319, Figs 1–4 (illustrated, oscillogram, distribution, habitat, comp. note) Equals *Cicadetta montana* (partim) Poland, France, Switzerland, Germany, Austria, Slovenia, Montenegro, Macedonia
Cicadetta cerdaniensis Hertach 2007: 37–38, 41–48, 50–51, 53–59, Fig 4B, Fig 5B, Fig 6B, Figs 7C–D, Figs 8E–F, Figs 11–14, Table 2, Table 4 (natural history, illustrated, song, oscillogram, habitat, distribution, comp. note) Switzerland
Cicadetta cerdaniensis Sueur and Puissant 2007a: 126 (comp. note)
Cicadetta cerdaniensis Sueur and Puissant 2007b: 55–56, 59–66, Figs 6–8, Table 1 (song, oscillogram, distribution, listed, comp. note) France, Slovenia, Switzerland, Austria, Germany, Macedonia, Montenegro, Poland
Cicadetta cerdaniensis Trilar and Gogala 2007: 6, 9–10, 19, Fig 5d (oscillogram, comp. note) France, Switzerland
Cicadetta cerdaniensis Brua and Hugel 2008: 51 (comp. note)
Cicadetta cerdaniensis Chrétien 2008: 3 (comp. note) Pyrenees
Cicadetta cerdaniensis Gogala 2008: 55–56 (song, oscillogram, comp. note) France, Austria, Poland, Macedonia, Greece
Cicadetta cf. *cerdaniensis* Hertach 2008a: 211 (comp. note) Switzerland
Cicadetta cerdaniensis Hertach 2008b: 8 (comp. note) France
Cicadetta cerdaniensis Gogala, Drosopoulos and Trilar 2008a: 99 (comp. note)
Cicadetta cerdaniensis Gogala, Drosopoulos and Trilar 2008b: 123 (comp. note)
Cicadetta cerdaniensis group Gogala, Drosopoulos and Trilar 2008b: 123 (comp. note)
Cicadetta cerdaniensis Gogala and Trilar 2008: 31 (comp. note)
Cicadetta cerdaniensis Hugel, Matt, Callot, Feldtrauer and Brua 2008: 7 (comp. note) France
Cicadetta cerdaniensis Lee 2008b: 456 (comp. note)
Cicadetta cerdaniensis Nickel 2008: 207 (distribution, hosts, comp. note) Europe
Cicadetta cerdaniensis Trilar and Hertach 2008: 185–186, Fig 1c (oscillogram, comp. note) Italy, France, Switzerland
Cicadetta cerdaniensis Gogala 2010: 82 (song, comp. note)

***C. chaharensis* (Kato, 1938)** = *Melampsalta chaharensis*
Cicadetta chaharensis Chou, Lei, Li, Lu and Yao 1997: 58, 63–64, Fig 8-12, Plate I, Fig 13 (key, described, illustrated, synonymy, comp. note) Equals *Melampsalta chaharensis* China
Cicadetta chaharensis Hua 2000: 61 (listed, distribution) Equals *Melampsalta chaharensis* China, Hebei, Xizang

***C. chinensis* Distant, 1905** = *Tibicina chinensis* = *Tibicen chinensis*

C. concinna (Germar, 1821) = *Cicada concinna* = *Cicada coccina* (sic) = *Cicadetta concina* (sic) = *Cicadatra concinna* = *Cicada adusta concinna* = *Melampsalta (Melampsalta) anglica concinna* = *Cicadatra concinna concinna* = *"Cicadatra concinna"* = *Cicadetta concina* (sic) = *Cicada aestivalis* Eversmann, 1837 nom. nud. = *Cicada subapicalis* Walker, 1858 = *Cicadetta subapicalis* = *Melampsalta (Melampsalta) subapicalis* = *Cicada montana* var. *adusta* Hagen, 1856 = *Cicada adusta* = *Cicadetta adusta* = *Melampsalta adusta* = *Cicada (Cicadetta) montana adusta* = *Cicada montana* var. *adusta* = *Melampsalta (Melampsalta) adusta* = *Cicadetta montana adusta* = *Cicada concinna* var. Fieber, 1876

Cicadetta adusta Bachmetjew 1907: 251 (body length, comp. note)

Cicadetta adusta Soós 1956: 413 (listed, distribution) Romania

Cicada concinna Weber 1968: 124, Fig 107 (ocellus illustrated, comp. note)

Cicadetta concinna Boulard 1981b: 50 (n. comb., synonymy) Equals *Cicada concinna* (nec *Cicada atra*) Equals *Cicada podolica* n. syn. Equals *Cicada montana* var. *adusta* n. syn. Equals *Cicada concinna* var. Fieber, 1876 Europe

Cicada concinna Boulard 1981b: 41–42 (valid species, comp. note)

Cicada adusta Boulard 1981b: 41–42 (comp. note)

Cicada montana var. *adusta* Boulard 1981b: 41 (comp. note)

Cicadetta montana adusta Schedl 1986b: 579 (comp. note)

Cicadetta adusta Schedl 1986b: 579 (comp. note)

Cicadetta aestivalis Schedl 1986b: 580 (comp. note)

Cicada subapicalis Schedl 1986b: 580 (comp. note)

Melampsalta adusta Schedl 1986b: 584 (comp. note)

Cicadatra concinna Nast 1987: 568–569 (distribution, listed) Yugoslavia

Melampsalta adusta Chou, Lei, Li, Lu and Yao 1997: 49 (comp. note)

"Cicadatra concinna" Gogala, Trilar and Krpač 2005: 124 (comp. note) Albania

Cicadetta concinna Gogala, Trilar and Krpač 2005: 124 (comp. note)

Cicadetta concina (sic) Gogala 2006b: 370 (comp. note)

Cicadetta adusta Gogala 2006b: 373 (comp. note)

Cicadetta concinna Sueur and Puissant 2007a: 126 (comp. note) Equals *Cicadetta podolica*

Cicadetta concinna Sueur and Puissant 2007b: 65 (comp. note) Equals *Cicadetta podolica*

Cicadetta concinna Gogala 2008: 55–56 (song, oscillogram, comp. note)

Cicadetta concinna Gogala and Trilar 2008: 31 (comp. note) Equals *Cicadetta podolica*

C. continua Distant, 1888 to *Linguacicada continua*

C. convergens (Walker, 1850) to *Physeema convergens*

C. crucifera (Ashton, 1912) to *Myopsalta crucifera*

C. sp. nr. *crucifera* Ewart 1998 to *Myopsalta* sp. nr. *crucifera*

C. sp. nr. *crucifera* Fishing Reel Buzzer Popple & Strange, 2002 to *Myopsalta* sp. nr. *crucifera*

C. cuensis (Distant, 1913) to *Ewartia cuensis*

C. denisoni (Distant, 1893) to *Yoyetta denisoni*

C. sp. nr. *denisoni* Brown Firetail Moss & Popple, 2000 to *Yoyetta* sp. nr. *denisoni*

C. sp. nr. *denisoni* Emery et al., 2005 to *Yoyetta* sp. nr. *denisoni*

C. *diminuta* Horváth, 1907 = *Melampsalta diminuta* = *Melampsalta (Melampsalta) diminuta*

C. *dirfica* Gogala, Trilar & Drosopoulos, 2011

C. distincta (Rambur, 1840) to *Tympanistalna distincta*

C. eyrei (Distant, 1882) to *Palapsalta eyrei*

C. *fangoana fangoana* Boulard, 1976 = *Cicadetta (Cicadetta) fangoana* = *Cicada haematodes* (nec Scopoli, nec Linnaeus) = *Cicadetta montana* var. *dimidiata* (nec Fabricius) = *Cicadetta montana* (nec Scopoli, partim)

Cicadetta fangoana Nast 1982: 311 (listed) France, Corsica

Cicadetta fangoana Boulard 1984d: 5 (comp. note) France

Cicadetta (Cicadetta) fangoana Boulard 1985b: 59, 63 (illustrated, characteristics, comp. note) France

Cicadetta fangoana Nast 1987: 568–569 (distribution, listed) France

Cicadetta fangoana Boulard 1988a: 12, Table (comp. note) France

Cicadetta fangoana Boulard 1990b: 132–133, 174, 185, 187, Plate XV, Fig 7, Plate XXII, Fig 10 (genitalia, sonogram, comp. note) Corsica

Cicadetta fangoana Boulard 1995a: 8, 13, 56–57, Fig $S21_1$, Fig $S21_2$ (song analysis, sonogram, listed, comp. note) Corsica

Cicadetta fangoana Boulard and Mondon 1995: 30, 32, 66–67, 80, 159 (illustrated, sonogram, listed, comp. note) France

Cicadetta fangoana Chou, Lei, Li, Lu and Yao 1997: 59 (comp. note)

Cicadetta fangoana Boulard 2000i: 79–82, 125, Fig 18, Plate L, Map 12 (synonymy, illustrated, song, sonogram, oscillogram, frequency spectrum, distribution, ecology, comp. note) Equals *Cicada haematodes* (nec Scopoli, nec Linnaeus) Equals *Cicadetta montana* var. *dimidiata* (nec Fabricius, partim) Equals *Cicadetta montana* (nec Scopoli, partim) France

Cicadetta fangoana Puissant and Boulard 2000: 112 (comp. note)

Cicadetta fangoana Puissant 2001: 145, 148 (comp. note)
Cicadetta fangoana Puissant and Sueur 2001: 430–435, Fig 4, Fig 8, Fig 9 (illustrated, distribution, comp. note) Corsica
Cicadetta fangoana Sueur 2001: 42, Table 1 (listed)
Cicadetta fangoana Sueur and Puissant 2001: 19, 22 (comp. note) France
Cicadetta fangoana Ibanez 2003: 12 (comp. note) France
Cicadetta fangoana Boulard 2005d: 338 (comp. note)
Cicadetta fangoana Puissant and Defaut 2005: 117–118, 125–127, Table 1, Tables 10–12 (listed, habitat association, distribution, comp. note) France
Cicadetta fangoana Bernier 2006: 2 (distribution) Corsica
Cicadetta fangoana Boulard 2006g: 41, 180 (listed, comp. note)
Cicadetta fangoana Puissant 2006: 20, 23, 25–30, 32, 35, 44, 56–57, 67, 71–72, 74–77, 80, 90, 92, 97, 116–133, 179–118, 187–189, Fig 8, Figs 45–48, Tables 1–2, Appendix Table 1, Appendix Tables 10–12 (illustrated, distribution, habitat association, ecology, listed, comp. note) France, Corsica
Cicadetta fangoana Sueur and Puissant 2007a: 126 (comp. note)
Cicadetta fangoana Sueur and Puissant 2007b: 55, 62, 65–66, Fig 8 (listed, distribution, comp. note) France
Cicadetta fangoana Trilar and Gogala 2007: 6, 9, 19, Fig 5e (oscillogram, comp. note) Corsica
Cicadetta fangoana Gogala 2008: 55 (song, comp. note) Corsica
Cicadetta fangoana Lee 2008b: 456 (comp. note)

C. fangoana rufinervis **Boulard, 1980** = *Cicadetta fangoana* var. *rufinervis*
Cicadetta fangoana var. *rufinervis* Boulard and Mondon 1995: 66 (listed) France
Cicadetta fangoana rufinervis Boulard 2000i: 82 (described, comp. note)

C. flaveola (Brullé, 1832) to *Cicadivetta flaveola*
C. forresti (Distant, 1882) to *Taurella forresti*
C. fraseri (China, 1938) to *Melampsalta fraseri*
C. froggatti (Distant, 1907) to *Taurella froggatti*
C. fulva (Goding & Froggatt, 1904) to *Birrima castanea* (Goding & Froggatt, 1904)

C. fuscoclavalis **(Chen, 1943)** = *Melampsalta fuscoclavalis* = *Melampsalta fuscoclavalis chungnanshana* Chen, 1943 = *Cicadetta fuscoclavalis chungnanshana*

C. gastrica (Stål, 1854) to *Tympanistalna gastrica*
C. graminis (Goding & Froggatt, 1904) to *Neopunia graminis*
C. hackeri (Distant, 1915) to *Telmapsalta hackeri*

C. haematophleps **Fieber, 1876**

C. hageni **Fieber, 1872** = *Cicada hageni* = *Melampsalta hageni* = *Melampsalta (Melampsalta) hageni* = *Cicada annulata* Hagen, 1856 (nec *Tibicen annulata* Brullé, 1832) = *Cicadetta annulata* = *Cicada angulata* (sic) = *Tibicen annulatus* (nec Brullé) = *Melampsalta (Melampsalta) annulata*
Cicadetta hageni Drosopoulous 1980: 190 (listed) Greece

C. hannekeae **Gogala, Drosopoulos & Trilar, 2008**
Cicadetta hannekeae Gogala 2008: 55–56 (song, oscillogram, comp. note) Greece
Cicadetta hannekeae Gogala, Drosopoulos and Trilar 2008a: 92–98, Figs 1–10, Figs 14–15, Table 1 (n. sp., described, illustrated, song, sonogram, oscillogram, distribution, comp. note) Greece
Cicadetta hannekeae Gogala, Drosopoulos and Trilar 2008b: 123 (comp. note)
Cicadetta hannekeae Gogala, Drosopoulos and Trilar 2009: 14, 23–24, 27 (comp. note) Greece
Cicadetta hannekeae Gogala and Trilar 2008: 31 (comp. note)
Cicadetta hannekeae Lee 2008b: 456 (comp. note)
Cicadetta hannekeae Gogala and Trilar 2009a: 32–33 (comp. note)
Cicadetta hannekeae Trilar and Gogala 2010: 6 (comp. note) Greece

C. hermansburgensis (Distant, 1907) to *Erempsalta hermansburgensis*

C. hodoharai **(Kato, 1940)** = *Melampsalta hodoharai*
Cicadetta hodoharai Chou, Lei, Li, Lu and Yao 1997: 66 (listed, comp. note) China
Cicadetta hodoharai Hua 2000: 61 (listed, distribution) China
Cicadetta hodoharai Lee, Choe, Lee and Woo 2002: 3, 10 (comp. note) China

C. hunterorum Moulds, 1988 to *Yoyetta hunterorum*
C. incepta (Walker, 1850) to *Yoyetta incepta*
Species complex near *Cicadetta incepta* Ewart & Popple, 2001 to species complex near *Yoyetta incepta*
Species complex near *Cicadetta incepta* (Adavale) Ewart & Popple 2001 to species complex near *Yoyetta incepta* (Advale)
Species complex near *Cicadetta incepta* (Blackall) Ewart & Popple 2001 to species complex near *Yoyetta incepta* (Blackall)
Species complex near *Cicadetta incepta* (Charleville) Ewart & Popple 2001 to species complex near *Yoyetta incepta* (Charleville)
C. sp. nr. *incepta* Ewart, 2009 to species complex near *Yoyetta* sp. nr. *incepta*

C. incipiens (Walker, 1850) to *Plerapsalta incipiens*
C. infuscata (Goding & Froggatt, 1904) to *Pauropsalta infuscata*
***C. inglisi* (Ollenbach, 1929)** = *Melampsalta inglisi*
Melampsalta inglisi Wu 1935: 27 (listed) China, Tibet
Cicadetta inglisi Chou, Lei, Li, Lu and Yao 1997: 66 (listed, synonymy, comp. note) Equals *Melampsalta inglisi* China
Cicadetta inglisi Hua 2000: 61 (listed, distribution) Equals *Melampsalta inglisi* China, Xizang
***C. inserta* Horváth, 1911** = *Melampsalta inserta* = *Melampsalta (Melampsalta) inserta*
Cicadetta inserta Popov 1985: 76, 80, Fig 24 (acoustic system, comp. note)
Cicadetta inserta Popov, Aronov and Sergeeva 1985: 451, 454, 456, 458–461, Figs 1.3–1.4, Fig 5d, Figs 8a–b (song, oscillogram, hearing, comp. note) Tadzhikistan
Cicadetta inserta Popov, Aronov and Sergeeva 1985: 289, 291, 293–297, Figs 1.3–1.4, Fig 5d, Figs 8a–b (song, oscillogram, hearing, comp. note) Tadzhikistan
Cicadetta inserta Popov and Sergeeva 1987: 684, 688–689 (comp. note)
Cicadetta inserta Popov 1990a: 302–303, Fig 2 (comp. note)
Cicadetta inserta Popov 1990b: 579–586 Fig 1A, Figs 2.1–2.2, Fig 3.1, Fig 4, Figs 6.1–6.2, Table (illustrated, song, oscillogram, ecology, comp. note) Tadzhikistan
Cicadetta inserta Tautz 1990: 105, Fig 12 (song frequency, comp. note)
Cicadetta inserta Popov 1991: 17–20, 22–23, Fig 1A, Figs 2.1–2.2, Fig 3.1, Fig 4, Figs 6.1–6.2, Table (illustrated, song, oscillogram, ecology, comp. note) Tadzhikistan
Cicadetta inserta Popov, Aronov and Sergeeva 1991: 281 (comp. note)
Cicadetta inserta Popov, Arnov and Sergeeva 1992: 151 (comp. note)
Cicadetta inserta Bennet-Clark and Young 1994: 293, Table 1 (body size and song frequency)
Cicadetta inserta Sueur 2001: 42, Table 1 (listed)
***C. intermedia* (Ollenbach, 1929)** = *Melampsalta intermedia*
***C. iphigenia* Emeljanov, 1996** = *Cicadetta tibialis* (nec Panzer)
Cicadetta iphigenia Emeljanov 1996a: 1897, 1899 Fig 2 (illustrated, comp. note) Crimea
Cicadetta iphigenia Emeljanov 1996b: 1054–1056, Fig 2 (illustrated, comp. note) Crimea
Cicadetta iphigenia Gogala and Popov 1997b: 34 (comp. note)
Cicadetta iphigenica Popov 1997a: 3, 10, 12–15, Figs 7–9 (song, oscillogram, sonogram, comp. note) Equals *Cicadetta tibialis* Popov, 1975 Crimea
Cicadetta iphigenica Popov 1997b: 1, 8–10, Figs 7–9 (song, oscillogram, sonogram, comp. note) Equals *Cicadetta tibialis* Popov, 1975 Crimea
Cicadetta iphigenia Sueur 2001: 43, Table 1 (listed) Equals *Cicadivetta tibialis*
Cicadetta iphigenia Trilar, Gogala and Popa 2006: 177, 179, Fig 2 (oscillogram, listed, distribution, comp. note) Romania, Crimean Peninsula
Cicadetta iphigenia Trilar and Gogala 2008: 30 (comp. note) Romania
C. isshikii (Kato, 1926) to *Tettigetta isshikii*
C. issoides (Distant, 1905) to *Noongara issoides*
***C. juncta* (Walker, 1850)** = *Cicada juncta* = *Melampsalta juncta*
Cicadetta juncta Moulds 1990: 144 (distribution, listed) Equals *Cicada juncta* Australia
Cicadetta juncta Moulds and Carver 2002: 28–29 (synonymy, distribution, ecology, type data, listed) Equals *Cicada juncta* Australia
Cicada juncta Moulds and Carver 2002: 28 (type data, listed)
Cicadetta juncta Popple and Emery 2010: 150 (comp. note) Australia
***C. kansa* (Davis, 1919)** = *Melampsalta kansa* = *Melampsalta parvula* (nec Say)
Cicadetta kansa Poole, Garrison, McCafferty, Otte and Stark 1997: 331 (listed) Equals *Melampsalta kansa* North America
Cicadetta kansa Sanborn 1999a: 52 (type material, listed)
Cicadetta kansa Sanborn 1999b: 2 (listed) North America
Cicadetta kansa Kondratieff, Ellingson and Leatherman 2002: 5, 12–13, 45–47, Tables 1–2 (key, diagnosis, listed, distribution, song, comp. note) Equals *Melampsalta kansa* Colorado, Kansas, Oklahoma, Texas
Cicadetta kansa Kondratieff, Schmidt, Opler and Garhart 2005: 97, 216 (listed, comp. note) Oklahoma, Colorado, Kansas, Texas
Cicadetta kansa Sanborn 2009d: 1 (comp. note) North America
***C. kissavi* Gogala, Drosopoulos & Trilar, 2009**
Cicadetta kissavi Gogala, Drosopoulos and Trilar 2009: 14, 19–27, Figs 5–10, Table 2 (n. sp., described, illustrated, song, sonogram, oscillogram, comp. note)
Cicadetta kissavi Trilar and Gogala 2010: 6 (comp. note) Greece
***C. kollari* Fieber, 1876** = *Melampsalta kollari* = *Melampsalta (Melampsalta) kollari* = *Cicadetta collari*

(sic) = *Cicada brachyptera* Hagen, 1856 nom. nud. = *Cicada brachyptera* Kolenati, 1857 nom. nud.

***C. konoi* (Kato, 1937)** = *Melampsalta konoi* = *Melampsalta yezoensis* Kato, 1932 (nec *Melampsalta yezoensis* Matsumura, 1898)

C. labeculata (Distant, 1892) to *Galanga labeculata*

C. labyrinthica (Walker, 1850) to *Physeema labyrinthica*

C. lactea (Distant, 1905) to *Myopsalta lactea*

***C. laevifrons* (Stål, 1870)** = *Melampsalta laevifrons*

C. landsboroughi (Distant, 1882) to *Yoyetta landsboroughi*

Cicadetta landsboroughi var. *ashtoni* Metcalf, 1963 to *Physeema convergens* (Walker, 1850)

C. sp. nr. *landsboroughi* Ewart, 1986 to *Yoyetta* sp. nr. *landsboroughi*

C. latorea (Walker, 1850) to *Physeema latorea*

C. n. sp. #75M nr. *latoria* (sic) Gwynne & Schatral 1988 to *Physeema* n. sp. #75M nr. *latorea*

***C. limitata* (Walker, 1852)** = *Cicada limitata* = *Melampsalta limitata* = *Cicada signifera* Waker, 1850 (nec *Cicada signifera* Germar, 1830)

***C. macedonica* Schedl, 1999** = *Cicadetta montana macedonica*

Cicadetta montana macedonica Gogala and Trilar 1999a: 91–96, Figs 1–3, Fig 4a (n. ssp., illustrated, described, song, sonogram, oscillogram, distribution, comp. note) Macedonia

Cicadetta montana macedonica Schedl 1999a: 87–89, Figs 1–4 (n. ssp., described, illustrated, comp. note) Macedonia

Cicadetta montana macedonica Kapla, Trilar, Gogala and Gogala 2001: 35 (song, comp. note) Macedonia

Cicadetta montana macedonica Krpač, Trilar and Gogala 2001: 29–30 (comp. note) Macedonia

Cicadetta montana macedonica Sueur 2001: 43, 46, Tables 1–2 (listed)

Cicadetta montana macedonica Gogala 2002: 247 (comp. note)

Cicadetta montana macedonica Gogala and Trilar 2002: 39 (comp. note) Slovenia

Cicadetta montana macedonica Gogala and Trilar 2004a: 317 (comp. note) Macedonia

Cicadetta macedonica Gogala and Trilar 2004a: 317, 320, 322 (n. status, valid species, comp. note)

Cicadetta macedonica Gogala and Trilar 2004b: 35 (song, comp. note) Macedonia

Cicadetta macedonica Boulard 2005d: 341 (comp. note) Macedonia

Cicadetta macedonica Gogala, Trilar and Krpač 2005: 105, 117–119, 123–124, Fig 11 (distribution, comp. note) Macedonia

Cicadetta montana macedonica Gogala, Trilar and Krpač 2005: 115 (comp. note)

Cicadetta montana macedonica Gogala and Trilar 2005: 71 (comp. note) Europe

Cicadetta montana macedonica Gogala 2006b: 371 (comp. note)

Cicadetta macedonica Gogala 2006b: 372–375, Fig 5 (song, sonogram, oscillogram, comp. note) Macedonia

Cicadetta macedonica Trilar, Gogala and Szwedo 2006: 313 (comp. note)

Cicadetta macedonica Sueur and Puissant 2007a: 126 (comp. note)

Cicadetta montana ssp. *macedonica* Sueur and Puissant 2007b: 55 (comp. note)

Cicadetta macedonica Sueur and Puissant 2007b: 56, 65 (comp. note)

Cicadetta macedonica Trilar and Gogala 2007: 6, 9–10, 14, 19, Fig 5b, Table 2 (oscillogram, comp. note) Macedonia

Cicadetta montana macedonica Gogala 2008: 55 (comp. note) Macedonia

Cicadetta macedonica Gogala 2008: 55–56 (song, oscillogram, comp. note)

Cicadetta macedonica Gogala, Drosopoulos and Trilar 2008a: 92–94, 96–99, Fig 1, Figs 12–15, Table 2 (song, sonogram, oscillogram, distribution, comp. note) Greece

Cicadetta macedonica Gogala, Drosopoulos and Trilar 2008b: 123 (comp. note) Greece

Cicadetta macedonica group Gogala, Drosopoulos and Trilar 2008b: 123 (comp. note)

Cicadetta macedonica Gogala and Trilar 2008: 31 (comp. note)

Cicadetta macedonica Lee 2008b: 456 (comp. note)

Cicadetta macedonica Trilar and Gogala 2008: 30–32, Fig 3 (oscillogram, listed, distribution, comp. note) Romania

Cicadetta macedonica Gogala, Drosopoulos and Trilar 2009: 14, 23, 27 (comp. note) Greece

Cicadetta macedonica Gogala 2010: 82 (song, comp. note)

C. mackinlayi (Distant, 1882) to *Myopsalta mackinlayi*

***C. mediterranea* Fieber, 1872** = *Melampsalta mediterranea* = *Melampsalta (Melampsalta) mediterranea* = *Euryphara mediterranea*

Cicadetta mediterranea Castellani 1952: 15–16 (comp. note) Italy

Cicadetta mediterranea Wagner 1952: 15 (listed, comp. note) Italy

Cicadetta mediterranea Schedl 1986a: 3–4, 6, 8–10, 17, 19, 22–24, Fig 1, Fig 8, Fig 14, Fig 18, Fig 23, Map 8, Table 1 (illustrated, listed, key, synonymy, distribution, ecology, comp. note) Equals *Euryphara mediterranea* Istria, Italy, Yugoslavia, Bulgaria, Jordan, Tunisia

Cicadetta mediterranea Nast 1987: 568–569 (distribution, listed) Italy, Yugoslavia, Bulgaria, Germany?
Cicadetta mediterranea Gogala and Popov 1997a: 12–23, Figs 1–7 (illustrated, song, sonogram, oscillogram, comp. note) Croatia
Cicadetta mediterranea Gogala and Popov 1997b: 34–35 (song, comp. note) Slovenia, Mediterranean
Cicadetta mediterranea Gogala 1998b: 396–397 (oscillogram, comp. note) Slovenia
Cicadetta mediterranea Popov, Beganović and Gogala 1998a: 93 (comp. note)
Cicadetta mediterranea Gogala and Gogala 1999: 127 (comp. note) Croatia
Cicadetta mediterranea Gogala and Popov 2000: 11 (comp. note)
Cicadetta mediterranea Schedl 2000: 260–261, Fig 4 (key, illustrated, distribution, listed, comp. note)
Cicadetta mediterranea Sueur 2001: 43, 46, Tables 1–2 (listed)
Cicadetta mediterranea Gogala 2002: 246–247, Fig 10 (song, sonogram, oscillogram, comp. note)
Cicadetta mediterranea Holzinger, Kammerlander and Nickel 2003: 481–482, 616–617, 628, 636, Plate 42, Fig 9, Tables 2–3 (key, listed, illustrated, comp. note) Europe, Mediterranean, Italy
Cicadetta mediterranea Gogala and Drosopoulos 2006: 278 (comp. note)
Cicadetta mediterranea Trilar and Gogala 2010: 16 (comp. note)

C. melete (Walker, 1850) to *Pyropsalta melete*
C. minima (Goding & Froggatt, 1904) to *Punia minima*
***C. minuta* (Ollenbach, 1929)** = *Melampsalta minuta*
C. mixta (Distant, 1914) to *Platypsalta mixta*
C. sp. 1 nr. *mixta* Black Scrub-buzzer (Popple & Strange, 2002) to *Platypsalta* sp. 1 nr. *mixta*
C. sp. 2 nr. *mixta* Surat Scrub-buzzer (Popple & Strange, 2002) to *Platypsalta* sp. 1 nr. *mixta*
C. sp. 3 nr. *mixta* Little Grass Buzzer (Popple & Strange, 2002) to *Platypsalta* sp. 1 nr. *mixta*
***C. montana montana* (Scolpoli, 1772)** = *Cicada montana* = *Cicada montata* (sic) = *Tettigonia montana* = *Cicada (Cicadetta) montana* = *Melampsalta montana* = *Melampsalta (Melampsalta) montana* = *Tettigia montana* = *Cicadetta (Melampsalta) montana* = *Cicadetta montana montana* = *Cacadetta* (sic) *montana* = *Cicadette* (sic) *montana* = *Cicadetta montana* form *typica* = *Cicada haematodes* Linnaeus, 1767 (nec Scopoli) = *Tettigonia haematodes* (nec Scopoli) = *Cicada hoematodes* (sic) (nec Scopoli) = *Cicada hematodes* (sic) (nec Scopoli) = *Cicada haematodis* (sic) (nec Scolopli) = *Cicada haematodes* (nec Scopoli, nec Linnaeus) = *Cicada haematoda* (sic) (nec Scopoli) = *Cicada haemalodes* (sic) (nec Scopoli) = *Cicadetta haematodes* (nec Scopoli) = *Cicada haematoides* (sic) (nec Scolopli) = *Melampsalta haematodes* (nec Scolopli) = *Melampsalta (Melampsalta) haematodes* (nec Scopoli) = *Tetigonia* (sic) *secunda* Schaeffer, 1767 = *Tetigonia* (sic) *prima* Schaeffer, 1767 = *Cicada flavofenestrata* Goeze, 1778 = *Cicadetta montana flavofenestrata* = *Cicadetta montana* var. *flavofenestrata* = *Cicadetta montana* form *flavofenestrata* = *Cicada schaefferi* Gmelin, 1789 = *Tettigonia schaefferi* = *Melampsalta (Melampsalta) schaefferi* = *Tettigonia dimidiata* Fabricius, 1803 = *Cicadetta montana* var. *dimidiata* = *Cicadetta montana dimidiata* = *Cicadetta montana* form *dimidiata* = *Cicada anglica* Samouelle, 1824 = *Melampsalta (Melampsalta) anglica* = *Cicada parvula* Walker, 1850 = *Cicada saxonica* Hartwig, 1857 = *Melampsalta (Melampsalta) saxonica* = *Cicadetta megerlei* Fieber, 1872 = *Cicadetta montana megerlei* = *Cicadetta montana* var. *megerlei* = *Melampsalta megerlei* = *Melampsalta (Melampsalta) megerlei* = *Cicadetta montana longipennis* Fieber, 1876 = *Cicada orni* (nec Linnaeus) = *Cicada tibialis* (nec Panzer) = *Cicadetta tibialis* (nec Panzer) = *Cicadetta montana* var. *longipennis* (partim) = *Cicadetta montana longipenis* (sic) = *Cicadetta pygmea* (nec Olivier) = *Cicadetta pygmaea* (sic) (nec Olivier) = *Tettigonia sanguinea* (nec Fabricius) = *Cicadetta montana* var. *brevipennis* (partim) = *Cicadetta* cf. *montana* Hugel et al. 2008 = *Takapsalta ichinosawana* (nec Matsumura)

Cicada haematodes (sic) Weaver 1832: 668–669 (behavior) England
Cicada haematodes (sic) Wallis 1841: 203 (acoustic behavior) England
Cicadetta montana Szilády 1870: 119 (listed) Romania
Cicada anglica Swinton 1880: 226 (comp. note) Britain
Cicada montana Swinton 1880: 226 (comp. note) Equals *Cicada anglica*
Cicada montana Hasselt 1881: 192 (comp. note)
Cicadetta montana Swinton 1908: 376 (song, comp. note) Britain
Cicadetta megerlei Strobl 1900: 201 (comp. note) Austria
Cicadetta montana Anonymous 1911: 102 (comp. note) Great Britain
Cicadetta montana Mader 1922: 179 (comp. note) Austria
Cicadetta montana Weber 1931: 109 (compound eye, comp. note)
Melampsalta megerlei Wu 1935: 27 (listed) Equals *Cicadetta megerlei* China, Szechuan, Austria, Tyrolis, Croatia, Dalmatia

Cicadetta montana Vogel 1938: 117 (comp. note)
Cicadetta montana Kühnelt 1949: 116 (listed, comp. note)
Cicadetta montana Soós 1956: 413 (listed, distribution) Romania
Cicadetta montana var. *megerlei* Soós 1956: 414 (listed, distribution) Romania
Cicadetta montana Cloudsley-Thomopson 1957: 241, 243 (comp. note) Great Britain
Cicadetta montana Hamann 1960: 166 (comp. note) Germany
Cicadetta montana var. *megerlei* Hamann 1960: 166 (comp. note) Germany
Cicadetta montana Le Quesne 1965: 4–5, Figs 7–8 (illustrated, described, listed, key) Great Britain
Cicadetta montana Popov 1971: 302 (hearing, listed, comp. note)
Cicadetta montana Cantoreanu 1972b: 210 (distribution, comp. note) Romania
Cicadetta montana Kaltenbach, Steiner and Aschenbrenner 1972: 490 (comp. note) Germany
Cicadetta montana Villiers 1977: 167, Plate XIV, Fig 28 (illustrated, listed, comp. note) France, England
Cicadetta montana Boulard 1981a: 41 (type species of genus *Cicadetta*)
Cicada montana Boulard 1981b: 42, 44 (comp. note)
Cicadetta montana Donat 1981: 145 (comp. note)
Cicadetta montana Lodos and Kalkandelen 1981: 78 (synonymy, distribution, comp. note) Equals *Cicada flavofenestrata* Equals *Tettigonia schaefferi* Equals *Cicada pygmea* Equals *Tettigonia dimidiata* Equals *Cicada anglica* Equals *Cicada parvula* Equals *Cicada saxonica* Equals *Cicadetta megerlei* Equals *Cicadetta montana brevipennis* Equals *Cicadetta montana longipennis* Equals *Cicadetta montana petryi* Turkey
Cicada montana Osssiannilsson 1981: 223 (type species of *Melampsalta*)
Cicadetta montana Osssiannilsson 1981: 223–225, Fig 734 (described, illustrated, distribution, comp. note) Norway, Sweden, Finland, Austria, Belgium, Bulgaria, Bohemia, Moravia, Slovakia, France, Germany, England, Greece, Hungary, Italy, Poland, Romania, Yugoslavia, Spain, Switzerland, Rusia, Anatolia, Palestine, Azerbaijan, Georgia, Kazakhstan, Moldavia, Ukraine, Tadzhikistan, Siberia, China
Cicada montana Koçak 1982: 149 (type species of *Cicadetta*)
Cicada montana Strümpel 1983: 70 (comp. note)
Cicadetta montana Strümpel 1983: 170, Table 10 (listed, comp. note)
Cicadetta montana Whalley 1983: 139 (comp. note) Britain
Cicadetta montana Emmrich 1984: 109, 111, 113–116, Figs 1–2 (distribution, comp. note) Equals *Cicadetta montana* var. A. *longipennis* East Germany
Cicadetta megerlei Emmrich 1984: 109 (comp. note) East Germany
Cicadetta saxonica Emmrich 1984: 109 (comp. note) East Germany
Cicadetta sanguinea haematodes Emmrich 1984: 114 (comp. note)
Cicadetta montana var. *flavofenestrata* Emmrich 1984: 116 (comp. note) Equals *Cicadetta montana* var. B. *brevipennis* East Germany
Cicada montana Hayashi 1984: 66 (type species of *Cicadetta*)
Cicadetta (Cicadetta) montana Boulard 1985b: 63 (characteristics) France
Cicadetta montana Kevan 1985: 81 (illustrated, comp. note) Britain
Cicadetta montana Mey and Steuer 1985: 110 (comp. note) Germany
Cicadetta montana Popov 1985: 76, 80, Fig 24 (acoustic system, comp. note)
Cicadetta montana Chinery 1986: 88–89 (illustrated, comp. note) Europe
Cicadetta montana Schedl 1986a: 3, 6–8, 10, 19–20, 22–24, Fig 6, Fig 13, Fig 21, Table 1 (illustrated, listed, key, synonymy, distribution, ecology, comp. note) Equals *Cicada montana* Equals *Cicada flavofenestrata* Equals *Tettigonia schaefferi* Equals *Tettigonia dimidiata* Equals *Cicadetta megerli* Istria, Europe, England, Sweden, Finland, Belgium, Germany, Poland, Czechoslovakia, Hungary, Romania, USSR, Balkan States, Italy, Spain, France, Alp States, Anatolia, Siberia, China
Cicadetta montana Schedl 1986b: 579–580, 582–583, Map 1 (distribution, comp. note)
Cicadetta montana Artmann 1987: 190, 192 Fig 5 (note, illustrated) Switzerland
Cicadetta montana Gebicki 1987: 94, 96, Table 1 (habitat, comp. note) Poland
Cicadetta montana Joermann and Schneider 1987: 284, 289–291, 294, Figs 7–8, Fig 9a, Table 3 (song analysis, sonogram, oscillogram, comp. note) Yugoslavia
Cicadetta montana Nast 1987: 568–569 (distribution, listed) Norway, Sweden, Finland, Russia, Great Britain, Belgium, Germany, Poland, Czechoslovakia, France, Switzerland, Austria, Hungary, Ukraine,

Moldavia, Romania, Spain, Italy, Yugoslavia, Bulgaria, Greece

Cicadetta montana Remane 1987: 288 (listed, comp. note) Germany

Cicadetta montana Anufriev and Emelyanov 1988: 315 Fig 237.1–237.4 (genitalia illustrated, key) Soviet Union

Cicadetta montana Boulard 1988a: 12, Table (comp. note) France

Cicada montana Boulard 1988f: 7–8, 38, 41–43 (type species of *Cicadetta* and Cicadettini, comp. note) Europe

Cicada (Cicadetta) montana Boulard 1988f: 41 (comp. note)

Cicadetta montana Boulard 1988g: Plate III, Fig 8 (illustrated, comp. note) France

Cicadetta montana Boulard 1988f: 43–45, 49, Fig 4 (type species of *Cicadetta*, illustrated, comp. note)

Cicadetta montana Duffels 1988d: 75 (comp. note) Europe

Cicadetta montana Moulds 1988: 40 (type species of *Cicadetta*, comp. note)

Cicadetta montana Tishechkin 1988: 7 (comp. note) Russia

Cicadetta montana Boulard 1990b: 115, 166, 174, 177, 180, 185, 187, Plate XXII, Fig 9 (sonogram, comp. note) Europe

Cicadetta montana Boulard and Nel 1990: 38 (comp. note)

Cicadetta montana Popov 1990a: 302, Fig 2 (comp. note)

Cicada montana Moulds 1990: 143 (type species of *Cicadetta*, listed)

Cicadetta montana Boulard 1991b: 25 (type species of *Cicadetta*, comp. note) Equals *Cicadetta haematodes* F. Equals *Cicada montana*

Cicadetta montana Dolling 1991: 15, 152, Figs 81–83 (illustrated, comp. note) Great Britain

Cicadetta montana Pigott 1991: 1189 (hosts) Soviet Union, Britain

Cicadetta montana Boulard 1992e: 126 (comp. note) Belgium

Cicadetta montana Fain, Baugnée, and Hidvegi 1992: 336 (mite parasite) Belgium

Cicadetta montana Grant and Ward 1992: 2–38, 42–58, Figs 1–6, Tables 1–8, Appendices 1–2 (natural history, distribution, management, comp. note) England, Europe, Asia, Russia, Georgia, Germany, Austria

Cicadetta montana Hidvegi and Baugnée 1992: 121–124, Figs 1–2 (turrets, phenology, comp. note) Belgium

Cicadetta montana Messner and Adis 1992: 714–715, 719, Figs 7–8 (cuticle, illustrated, comp. note) Switzerland

Cacadetta (sic) *montana* Moalla, Jardak and Ghorbel 1992: 35 (comp. note) Britain, Finland

Cicadetta montana Savinov 1992: 267–282, Figs 1–20 (illustrated, thoracic skeleton, comp. note)

Cicadetta montana Dobosz 1993: 15 (distribution, comp. note) Poland

Cicadetta montana Elven 1993: 5 (illustrated, comp. note) Norway

Sicadetta (sic) *montana* Elven 1993: 6, Table 1 (distribution) Norway

Cicadetta montana Pillet 1993: 96–97, 99, 108, 110–111, Fig 4, Plate XII (described, illustrated, ecology, distribution, comp. note) Switzerland

Cicadetta montana Remane and Wachman 1993: 134–135 (illustrated, comp. note)

Cicadetta montana Sander 1993: 71 (listed) Germany

Cicadetta montana Savinov 1993: 1–15, Figs 1–20 (illustrated, thoracic skeleton, comp. note)

Cicadetta montana Brodsky 1994: 145–146, Fig 8.14 (wing movements, illustrated, comp. note)

Cicadetta montana Nickel 1994: 537–538 (listed, distribution, comp. note) Germany

Cicadetta montana Niehuls and Simon 1994: 253–254, 259–262, Figs 3–4 (distribution, comp. note) Germany

Cicadetta montana Woodall 1994: 105–106 (illustrated, song, comp. note) Britain

Cicadetta montana Boulard 1995a: 8, 13, 55–56, Fig $S20_1$, Fig $S20_2$ (song analysis, sonogram, listed, comp. note) Turkey

Cicadetta montana Boulard and Mondon 1995: 25, 30, 33, 40, 43, 66–67, 84, 159 (illustrated, sonogram, listed, comp. note) Equals *Cicada saxonica* France

Cicadetta montana var. *dimidiata* Boulard and Mondon 1995: 66 (listed) Equals *Cicada megerlei* France

Cicadetta montana var. *flavofenestrata* Boulard and Mondon 1995: 66 (listed) Equals *Tettigonia schaefferi* Equals *Tettigonia sanguinea* Panzer (nec Fabricius) France

Cicadetta montana Ferenca 1996: 147 (natural history, comp. note) Lithuania

Cicadetta montana Holzinger 1996: 505 (listed) Austria

Cicadetta montana Nickel and Sander 1996: 159 (listed) Germany

Cicadetta montana Niehuls and Weitzel 1996: 439–445, Figs 1–2 (distribution, comp. note) Germany

Cicadetta montana Anonymous 1997c: 82 (listed) Germany

Cicada montana Boulard 1997c: 83–84, 108, 110–112 (type species of *Cicadetta* and Cicadettini, comp. note) Equals *Tettigonia haematodes* Fab.

Cicada (Cicadetta) montana Boulard 1997c: 110 (comp. note)

Cicadetta montana Boulard 1997c: 110, 112–113, 117, Plate II, Fig 6 (type species of *Cicadetta*, illustrated, comp. note)

Cicadetta montana Cox 1997: 15, 32 (listed, comp. note) United Kingdom

Cicada montana Chou, Lei, Li, Lu and Yao 1997: 58 (type species of *Cicadetta*)

Cicadetta montana Chou, Lei, Li, Lu and Yao 1997: 58–60, Fig 8–9, Plate I, Fig 10 (key, described, illustrated, synonymy, comp. note) Equals *Cicada montana* Equals *Cicada haematodes* L. Equals *Cicadetta montana* var. *dimidiata* Equals *Cicada montana* var. *flavofenestrata* Equals *Cicada pygmea* (nec Olivier) Equals *Cicada tibialis* Latreille Equals *Cicadetta montana* var. *longipennis* Equals *Cicadetta megerlei* Equals *Melampsalta montana* Equals *Cicadetta montana petryi* China

Cicadetta montana Holzinger, Frölich, Günthart, Pauterer, Nickel et al. 1997: 49 (listed) Europe

Cicadetta montana Remane, Fröhlich, Nickel, Witsack and Achtziger 1997: 64, 66 (listed, comp. note) Germany

Cicadetta montana Schiemenz 1997: 43 (listed) Germany

Cicadetta montana Gogala 1998a: 156 (comp. note) Slovenia

Cicadetta montana Gogala 1998b: 397–398 (illustrated, oscillogram, comp. note) Slovenia

Cicadetta montana Gogala and Trilar 1998b: 67–68 (comp. note) Slovenia

Cicadetta montana McDouall 1998: 71 (listed) England

Cicadetta montana Ragge and Reynolds 1998: 477–478, Fig 1630 (oscillogram, comp. note) Europe, Britain

Cicadetta montana Remane, Achtziger, Fröhlich, Nickel and Witsack 1998: 244, 246 (listed, comp. note) Germany

Cicadetta montana Schiemenz 1998: 37–38 (listed, distributin, comp. note) Germany

Cicadetta montana Schönitzer and Oesterling 1998: 34 (listed, comp. note) Germany

Cicadetta montana Trilar and Gogala 1998: 69 (comp. note) Macedonia

Cicadetta montana Okáli and Janský 1998: 34–36, 43, Fig 2 (distribution, comp. note) Equals *Cicadetta montana* var. *megerlei* Slovakia

Cicadetta montana Dey 1999: 161–172, Figs 1–14 (compound eye morphology, comp. note)

Cicadetta montana Gogala and Gogala 1999: 120, 123, 126–127, Fig 8 (distribution, comp. note) Slovenia, Eurasia

Cicadetta montana Gogala and Trilar 1999a: 94–96, Fig 4b, Fig 4d (song, oscillogram, comp. note) Macedonia

Cicadetta montana complex Gogala and Trilar 1999a: 96 (comp. note)

Cicadetta montana complex Gogala and Trilar 1999b: 1 (comp. note)

Cicadetta montana Holzinger 1999: 431, 441 (listed, comp. note) Equals *Melampsalta montana* Equals *Cicadetta megerlei* Austria

Cicadetta megerlei Holzinger 1999: 441 (comp. note) Austria

Cicadetta montana Lee 1999: 10–11, Fig 7 (distribution, song, illustrated, listed, comp. note) Equals *Cicada montana* Equals *Takapsalta ichinosawana* (nec Matsumura) Korea, China, Russia, Palaearctic Region, Europe

Cicada montana Lee 1999: 10 (type species of *Cicadetta*) Europe

Cicadetta montana Schedl 1999a: 87 (comp. note)

Cicadetta montana montana Schedl 1999a: 87, 89 (comp. note)

Cicadetta montana Witsack 1999: 425 (listed) Germany

Cicadetta montana dimidiata Boulard 2000i: 77 (synonymy, described, distribution, comp. note) Equals *Tettigonia dimidiata* Equals *Cicada haematodes* (nec Scopoli, nec Linnaeus) Equals *Cicadetta megerlei* Equals *Cicadetta montana* var. *longipennis* (partim) Equals *Cicadetta montana* form *dimidiata* France

Cicadetta montana flavofenestrata Boulard 2000i: 76, 79 (synonymy, described, distribution, ecology, comp. note) Equals *Cicada flavo-fenestrata* Equals *Tetigonia* (sic) *secunda* Equals *Tettigonia schaefferi* Equals *Tettigonia sanguinea* Equals *Cicadetta montana* var. *brevipennis* (partim) Equals *Cicadetta montana* form *flavofenestrata* France

Cicada montana Boulard 2000i: 20 (comp. note)

Cicadetta montana Boulard 2000i: 72–80, 83–84, 92, 124, Fig 16, Plate K, Map 11 (synonymy, illustrated, song, sonogram, oscillogram, frequency spectrum, distribution, ecology, comp. note) Equals *Cicada montana* Equals *Tetigonia* (sic) *prima* Equals *Tettigonia montana* Equals *Cicada anglica* Equals *Cicada saxonica* Equals

Cicada (Cicadetta) montana Equals *Cicadetta montana longipenis* (sic) Equals *Cicadetta montana longipennis* Equals *Cicadetta pygmaea* (sic) (nec Olivier) Equals *Cicadetta montana* form *typica* Equals *Tettigia montana* Equals *Cicada haematodes* (nec Scopili, nec Linnaeus) Equals *Cicadetta (Melampsalta) montana* France

Melampsalta megerlei Hua 2000: 63 (listed, distribution) Eqals *Cicadella* (sic) *megerlei* China, Sichuan, Austria, Typolis, Croatia, Dalmatia

Cicadetta montana Hua 2000: 61 (listed, distribution) Equals *Cicada montana* China, Heilongjiang, Sichuan, Europe

Cicadetta montana Martin and Webb 2000: 59 (comp. note) Britain

Cicadetta montana Puissant and Boulard 2000: 112, 114 (comp. note)

Cicadetta montana Schedl 2000: 257, 260–261, 264, 266–267, Fig 6, Plate 1, Fig 6, Map 5 (key, illustrated, distribution, listed, comp. note) Belgium, France, Germany, Poland, Czech Republic, Slovakia, Austria, Italy, Hungary, Slovenia

Cicadetta megerlei Schedl 2000: 264 (comp. note)

Cicadetta montana dimidiata Schedl 2000: 264 (comp. note)

Cicadetta montana Beenen 2001: 201 (mite host) Netherlands

Cicada montana Boulard 2001b: 4–5, 28, 30–32 (type species of *Cicadetta* and Cicadettini, comp. note) Equals *Tettigonia haematodes* Fab.

Tettigonia haematodes Fab Boulard 2001b: 28, 32 (comp. note)

Cicada (Cicadetta) montana Boulard 2001b: 31 (comp. note)

Cicadetta montana Boulard 2001b: 31–33, 37, Plate II, Fig 6 (type species of *Cicadetta*, illustrated, comp. note)

Cicadetta montana Dmitriev 2001a: 58 (comp. note) Russia

Cicadetta montana Dmitriev 2001b: 10 (comp. note) Russia

Cicadetta montana Högmander 2001: 73, Appendix 1 (listed) Russia

Cicadetta montana Kapla, Trilar, Gogala and Gogala 2001: 35 (song, comp. note) Macedonia, Europe, Slovenia, Korea, Soviet Union

Cicadetta montana Krpač, Trilar and Gogala 2001: 29 (comp. note) Macedonia

Cicadetta montana montana Krpač, Trilar and Gogala 2001: 29–30 (comp. note) Macedonia

Cicadetta montana Puissant 2001: 141–150, Figs 1–7 (illustrated, life history, synonymy, distribution, song, oscillogram, comp. note) Equals *Cicadetta petryi* n. syn. France, Slovenia, Belgium, Europe, China, Korea

Cicadetta montana Puissant and Sueur 2001: 435 (comp. note)

Cicadetta montata (sic) Schedl 2001b: 27 (comp. note)

Cicadetta montana Schedl 2001b: 27 (comp. note)

Cicadetta montana Sueur 2001: 43, Table 1 (listed)

Cicadetta montana Sueur and Puissant 2001: 22–23 (comp. note) France

Cicadetta montana Achtziger and Nigmann 2002: 3 Fig 1 (illustrated)

Cicadetta montana Boulard 2002b: 205 (comp. note)

Cicadetta montana Gogala 2002: 246–247, Fig 11 (song, sonogram, oscillogram, comp. note)

Cicadetta montana Gogala and Trilar 2002: 39 (comp. note) France, Great Britain, Europe, Asia, USSR

Cicadetta montana complex Gogala and Trilar 2002: 39 (comp. note)

Cicadetta montana f. *dimidiata* Gogala and Trilar 2002: 39 (comp. note)

Cicadetta montana var. *adusta* Gogala and Trilar 2002: 39 (comp. note)

Cicadetta montana Günther and Mühlenthaler 2002: 331 (listed) Switzerland

Cicadetta montana Kondratieff, Ellingson and Leatherman 2002: 45 (type species of *Cicadetta*)

Cicadetta montana Kysela 2002: 22 (on jewelry)

Cicadetta montana Lee, Choe, Lee and Woo 2002: 4–10, Figs 1–3, Tables 1–3 (gene sequence, phylogeny, comp. note) Palaearctic, Russia, China, Korea

Cicadetta montana Moulds and Carver 2002: 24 (type species of *Cicadetta*, listed)

Cicadetta montana Nickel, Holzinger and Wachmann 2002: 292, Table 4 (listed)

Cicadetta montana Nickel and Remane 2002: 37 (listed, hosts, distribution) Germany, Transpalaearctic

Cicadetta montana Pinchen and Ward 2002: 258–259, 263 (illustrated, ecology, distribution, comp. note) Britain

Cicadetta montana Rakitov 2002: 125–126 (comp. note)

Cicadetta montana Schedl 2002: 234–237, Photo 4, Map 4 (illustrated, distribution, comp. note) Austria, Germany, Italy

Cicadetta montana Sueur 2002b: 122, 129 (listed, comp. note)

Cicadetta montana complex Trilar and Gogala 2002a: 49 (comp. note)
Cicadetta montana Woodall 2002: 105–106 (illustrated, song, comp. note) Britain
Cicadetta montana Gogala and Seljak 2003: 351 (illustrated, comp. note) Slovenia, Europe, Great Britain
Cicadetta montana Holzinger, Kammerlander and Nickel 2003: 475–476, 478, 480–482, 606–607, 616–617, 628, 634, Figs 259A–B, Fig 260, Plate 37, Fig 7, Plate 42, Fig 8, Tables 2–3 (key, listed, illustrated, comp. note) Equals *Cicada montana* Equals *Cicada flavofenestrata* Equals *Tettigonia schaefferi* Equals *Tettigonia dimidiata* Equals *Cicada anglica* Equals *Cicada saxonica* Equals *Cicadetta megerlei* Equals *Cicadetta montana brevipennis* Equals *Cicadetta montana longipennis* Equals *Cicadetta montana petryi* Equals *Cicadetta montana macedonica* Europe, Austria, Palaearctic, Slovenia
Cicadetta montana Ibanez 2003: 3, 12 (comp. note) France
Cicadetta montana Lee 2003b: 8 (comp. note)
Cicadetta montana Nickel 2003: 71 (distribution, hosts, comp. note) Germany
Cicadetta montana Nickel and Remane 2003: 138 (listed, distribution) Germany
Cicadetta montana Oakley 2003: 525, Table 1 (listed)
Cicadetta montana Smith and Smith 2003: 153 (listed) Macedonia
Cicadetta montana Vezinet 2003: 14 (comp. note) France
Cicadetta montana Walter and Bemment 2003: 69 (comp. note) Britain, Belgium
Cicadetta montana Walter, Emmrich and Nickel 2003: 6, 17 (listed) Germany
Cicadetta montana Carl, Huemer, Zanetti and Salvadori 2004: 183 (listed) Italy
Cicadetta montana species group Gogala and Trilar 2004a: 316–317, 319–320, 322 (diversity, comp. note) Europe
Cicadetta montana species complex Gogala and Trilar 2004a: 317, 322–323 (comp. note)
Cicadetta montana Gogala and Trilar 2004a: 317–319, 322, Fig 2, Table 1 (sonogram, oscillogram, comp. note) France, Slovenia, Great Britain, Macedonia, Korea, Germany, Switzerland, Austria, Italy, Croatia, Greece
Cicadetta montana var. *longipennis* Gogala and Trilar 2004a: 320 (comp. note)
Cicadetta montana Gogala and Trilar 2004b: 35 (song, comp. note) Slovenia, Great Britain, Macedonia
Cicadetta montana species complex Gogala and Trilar 2004b: 35 (song, comp. note)
Cicadetta montana Handke 2004a: 150 (listed) Croatia
Cicadetta montana Handke 2004b: 158 (listed) Croatia
Cicadetta montana Hertach 2004: 58–59, 62–64 (comp. note) Switzerland
Cicadetta montana Jäger and Frank 2004: 295 (listed)
Cicadetta montana Nickel 2004: 59, 64 (listed, comp. note) Germany
Cicadetta montana Pilarczyk and Swierczewski 2004: 146 (comp. note) Poland
Cicadetta montana Schedl 2004b: 914–915, Fig 1 (listed, distribution, comp. note) Austria
Cicadetta montana group Schedl 2004b: 914 (comp. note)
Cicadetta montana Snipp and Bemment 2004: 73, 76 (listed, comp. note) Britain
Cicadetta montana complex Trilar and Holzinger 2004: 1383–1384 (comp. note)
Cicadetta montana Trilar and Holzinger 2004: 1384–1385, Fig 1a (song, sonogram, oscillogram, comp. note) Austria, Germany, Slovenia, Macedonia, Britain
Cicadetta montana Witsack and Nickel 2004: 232 (listed) Germany
Cicadetta montana Boulard 2005d: 341 (comp. note) Slovenia
Cicadetta montana Delescaille 2005: 120 (comp. note) Belgium
Cicadetta montana Gogala and Trilar 2005: 71 (comp. note) Europe
Cicadetta montana Gogala, Trilar and Krpač 2005: 105, 116–117, Fig 10 (distribution, comp. note) Macedonia
Cicadetta montana complex Gogala, Trilar and Krpač 2005: 119, 124 (comp. note)
Cicadetta montana Meineke and Krügener 2005: 144 (comp. note) Germany
Cicadetta montana Lee 2005: 17, 96–101, 158, 166, 176 (described, illustrated, synonymy, song, distribution, key, type species of *Cicadetta*, comp. note) Equals *Cicada montana* Equals *Takapsalta ichinosawana* (nec Matsumura) Equals *Cicadetta* sp. Gogala and Trilar 2004 Korea, China, Russia, Palaearctic Region
Cicada montana Moulds 2005b: 390, 436 (type species of *Cicadetta*)
Cicadetta montana Moulds 2005b: 396–398, 410, 412, 417–419, 421–422, 425, Figs 56–59, Fig 62, Table 1 (higher taxonomy, phylogeny, key, listed, comp. note)
Cicadetta montana Puissant 2005: 311–312 (comp. note)

Cicadetta montana Reynaud 2005: 143–144, Photo 3 (illustrated, comp. note) France
Cicadette (sic) *montana* Robinson 2005: 217 (comp. note) Britain
Cicadetta montana Schnappauf and Müller 2005: 28 (listed) Germany
Cicadetta montana Sueur 2005: 215 (song, comp. note) France
Cicadetta montana Janský 2005: 83, Table 1 (listed, comp. note) Slovakia
Cicadetta montana Anufriev 2006: 250 (listed) Equals *Melampsalta montana* Bashkir Republic
Cicadetta montana Bernier 2006: 2 (related species) France
Cicadetta gr. *montana* Bernier 2006: 2–4 (distribution) France, Corsica
Cicadetta montana Boulard 2006g: 50, 52 (comp. note)
Cicadetta megerlei Gogala 2006b: 370 (comp. note)
Cicadetta montana Gogala 2006b: 371–372, 375, Fig 2 (song, sonogram, oscillogram, comp. note) Europe
Cicadetta montana Gogala and Trilar 2006: 16–17 (comp. note)
Cicadetta montana Novikov, Novikova, Anufriev and Dietrich 2006: 308 (distribution, listed) Kyrgyzstan
Cicadetta montana Puissant 2006: 30, 71–72, 78, 91 (comp. note)
Cicadetta montana Trilar, Gogala and Popa 2006: 178 (comp. note) Romania
Cicadetta montana complex Trilar, Gogala and Popa 2006: 178 (comp. note) Romania
Cicadetta montana complex Trilar, Gogala and Szwedo 2006: 313, 315, 319 (comp. note)
Cicadetta montana group Trilar, Gogala and Szwedo 2006: 319 (comp. note)
Cicadetta montana Trilar, Gogala and Szwedo 2006: 313, 315 (comp. note) Britain
Cicadetta montana Vincent 2006: 64 (listed, comp. note) France
Cicadetta montana Dmitriev 2007a: 817 (comp. note) Russia
Cicadetta montana Dmitriev 2007b: 1209(comp. note) Russia
Cicadetta montana Hertach 2007: 37–38, 41–48, 51–59, Figs 2–3, Fig 4A, Fig 5A, Fig 6A, Figs 7A–B, Figs 8A–B, Fig 8D, Figs 13–14, Tables 2–4 (natural history, illustrated, song, oscillogram, habitat, distribution, comp. note) Switzerland
Cicadetta montana complex Hertach 2007: 40, 51 (natural history, illustrated, song, oscillogram, habitat, distribution, comp. note) Switzerland
Cicadetta montana Shiyake 2007: 8, 104, 106–107, Fig 181 (illustrated, distribution, listed, comp. note) Europe, Russia, Asia
Cicadetta montana Söderman 2007: 10, 12, 26, 77, Cover (illustrated, listed, comp. note) Finland
Cicadetta montana Sueur and Puissant 2007a: 126–128, Figs 1–5 (habitat, song, oscillogram, comp. note) France, Slovenia, Great Britain, Germany, Switzerland, Macedonia, Austria
Cicadetta montana Sueur and Puissant 2007b: 55–56, 62, 66, Fig 8 (listed, distribution, comp. note) France
Cicadetta montana species complex Sueur and Puissant 2007b: 55 (comp. note)
Cicadetta montana complex Sueur and Puissant 2007b: 56, 64–65 (comp. note)
Cicadetta montana complex Trilar and Gogala 2007: 6, 9 (comp. note)
Cicadetta montana group Trilar and Gogala 2007: 10 (comp. note)
Cicadetta montana Trilar and Gogala 2007: 6, 9, 19, Fig 5g (oscillogram, comp. note) Britain, Slovenia
Cicadetta montana group Brua and Hugel 2008: 49, 51 (comp. note)
Cicadetta montana Brua and Hugel 2008: 49–50, Fig 1 (song, oscillogram, comp. note) Equals *Cicadetta* cf. *montana* Hugel et al. 2006 France
Cicadetta montana Chrétien 2008: 3 (oscillogram, comp. note) Switzerland
Cicadetta montana Demir 2008: 477 (distribution) Turkey
Melampsalta montana Demir 2008: 477 (listed) Turkey
Cicada montana Gogala 2008: 54, 56 (type location, comp. note) Slovenia
Cicadetta montana Gogala 2008: 54–56 (illustrated, song, oscillogram, comp. note) Slovenia, Europe, Greece
Cicadetta montana complex Gogala, Drosopoulos and Trilar 2008a: 91, 93, 99–100 (diversity, comp. note) Greece
Cicadetta montana Gogala, Drosopoulos and Trilar 2008a: 91–94, 99, Fig 1 (distribution, comp. note) Greece
Cicadetta montana Gogala, Drosopoulos and Trilar 2008b: 123 (comp. note) Great Britain, Europe, Iran
Cicadetta montana Gogala and Trilar 2008: 31 (comp. note)
Cicadetta montana species complex Gogala and Trilar 2008: 31 (comp. note)
Cicadetta montana Hertach 2008b: 2, 5–6, 9–11, 15–18, 20–24, 26, Figs 2–3, Figs 6–8, Tables

5–7, Table 9, Appendix B (natural history, illustrated, song, oscillogram, comp. note) Switzerland
Cicadetta montana Hugel, Matt, Callot, Feldtrauer and Brua 2008: 5, 7 (comp. note) France
Cicadetta cf. *montana* Hugel, Matt, Callot, Feldtrauer and Brua 2008: 5–7, 9, Fig 2, Table 1 (distribution, comp. note) France
Cicadetta montana group Hugel, Matt, Callot, Feldtrauer and Brua 2008: 6–7, Table 1 (comp. note) France
Cicada montana Lee 2008b: 445, 452, 455–456 (type species of *Cicadetta*, comp. note) Carnioliae circa Idriam
Cicadetta montana species complex Lee 2008b: 455–457 (comp. note)
Cicadetta montana Nickel 2008: 207 (distribution, hosts, comp. note) Trans-Palaearctic
Cicadetta montana Sanborn, Phillips and Gillis 2008: 15 (type species of *Cicadetta*)
Cicadetta montana Vikberg 2008: 60 (parasitism) Finland
Cicadetta montana Trilar and Hertach 2008: 185–186, Fig 1b (oscillogram, comp. note) Italy
Cicadetta montana complex Trilar and Hertach 2008: 185 (comp. note) Italy
Cicadetta montana Trilar and Gogala 2008: 30–32, Fig 1 (sonogram, oscillogram, listed, distribution, comp. note) Romania
Cicadetta montana complex Trilar and Gogala 2008: 32 (comp. note)
Cicadetta montana Witsack 2008: 263–264, Table 4.27 (listed, comp. note) Germany
Cicadetta montana Anufriev and Smirnova 2009a: 530 (listed) Chuvashia Republic
Cicadetta montana Anufriev and Smirnova 2009b: 785 (listed) Chuvashia Republic
Cicadetta montana species complex Gogala, Drosopoulos and Trilar 2009: 14, 20 (comp. note) Greece
Cicadetta montana Gogala, Drosopoulos and Trilar 2009: 14–15, 20, 23–24, 27 (comp. note) Greece
Cicadetta montana Gogala and Trilar 2009a: 32–33 (comp. note)
Cicadetta montana species complex Gogala and Trilar 2009b: 21 (distribution, comp. note) Slovenia, France, Switzerland, Austria, Poland, Montenegro, Romania, Macedonia, Bulgaria, Greece
Cicadetta group *montana* Ibanez 2009: 208–209, 212 (listed, comp. note) France
Cicadetta montana Ibanez 2009: 209 (comp. note) France
Cicadetta montana Muraoka 2009: 8 (predation) Austria
Cicadetta montana Simões and Quartau 2009: 402 (comp. note)
Cicadetta montana Aspöck and Aspöck 2010: 60 (listed)
Cicadetta montana Fuhrmann 2010: 135–136, Fig 1 (comp. note) Germany
Cicadetta cf. *montana* Fuhrmann 2010: 136 (comp. note) Germany
Cicadetta montana Gogala 2010: 82 (song, comp. note)
Cicadetta montana Kemal and Koçak 2010: 6 (distribution, listed, comp. note) Turkey
Cicadetta montana Lang and Walentowsski 2010: 75 (listed) Germany
Cicadetta montana Natural England 2010: 47 (status) England
Cicadetta montana complex Puissant and Sueur 2010: 571 (comp. note)
Cicadetta montana Ruckstuhl 2010: 2 (comp. note) Switzerland
Cicadetta montana Saulich 2010: 1132 (life cycle, comp. note)
Cicadetta montana complex Sanborn and Phillips 2010: 860 (comp. note)

C. *montana peregrina* Gogala, 1998
Cicadetta montana peregrina Gogala 1998b: 398 (comp. note) Slovenia
Cicadetta montana peregrina Gogala and Trilar 1999a: 95 (song, oscillogram, comp. note)

C. cf. *montana* Gogala & Trilar, 2004
Cicadetta cf. *montana* Gogala and Trilar 2004a: 323, Fig 5 (sonogram, oscillogram, comp. note)

C. cf. *montana* Vincent 2006
Cicadetta cf. *montana* Vincent 2006: 64–71, Map 1, Table 1 (illustrated, ecology, distribution, listed, comp. note) France

C. cf. *montana* Sueur & Puissant 2007
Cicadetta cf. *montana* Sueur and Puissant 2007b: 66, Fig 8 (distribution) France

C. cf. *montana* Hugel, Matt, Callot, Feldtrauer & Brua 2008 to *Cicadetta montana montana* (Scopoli, 1772)

C. cf. *montana* Niedringhaus, Biedermann & Nickel 2010
Cicadetta cf. *montana* Niedringhaus, Biedermann and Nickel 2010: 27, 40–41, 47, 68, Table 2, Table 8, Appendix 2 (distribution, listed, comp. note) Luxemburg

C. *multifascia* (Walker, 1850) to *Plerapsalta multifasciata*
C. *murrayensis* (Distant, 1907) to *Plerapsalta incipiens* (Walker, 1850)
C. *musiva* (Germar, 1930) to *Tettigetta musiva*

C. nebulosa (Goding & Froggatt, 1904) to *Pauropsalta rubea* (Goding & Froggatt, 1904)

***C. nigropilosa* Logvinenko, 1976**

Cicadetta nigropilosa Nast 1982: 311 (listed) USSR, Azerbaijan

C. oldfieldi (Distant, 1883) to *Ewartia oldfieldi*

C. sp. nr. *oldfieldi* Green Heath-Buzzer Moss & Popple, 2000 to *Ewartia* sp. nr. *oldfieldi*

C. sp. nr. *oldfieldi* Western Wattle Cicada Popple & Strange, 2002 to *Ewartia* sp. nr. *oldfieldi*

***C. olympica* Gogala, Drosopoulos & Trilar, 2009**

Cicadetta olympica Gogala, Drosopoulos and Trilar 2009: 15–19, 24, 27, Figs 1–4, Table 1 (n. sp., described, illustrated, song, sonogram, oscillogram, comp. note) Greece

Cicadetta olympica Trilar and Gogala 2010: 6 (comp. note) Greece

C. oxleyi (Distant, 1892) to *Kobonga oxleyi*

C. sp. nr. *Cicadetta oxleyi* Dunn, 1991 to *Kobonga* sp. nr. *oxleyi*

C. parvula (Fieber, 1876) to *Cicadivetta parvula*

***C. pellosoma* (Uhler, 1861)** = *Cicada pellosoma* = *Melampsalta pellosoma* = *Melampalta* (sic) *pellosoma* = *Melampsalta (Melampsalta) pellosoma* = *Melampsalta pellosoma concolor* Kato, 1932 = *Melampsalta pellosoma* var. *concolor* = *Melampsalta pellosoma flava* Kato, 1927 = *Melampsalta pellosoma* var. *flava*

Cicada pellosoma Henshaw 1903: 229 (listed) China, Hong Kong

Melampsalta pellosoma Doi 1932: 43 (listed, comp. note) Korea

Melampsalta pellosoma Chen 1933: 361 (listed, comp. note)

Melampsalta pellosoma Wu 1935: 28 (listed) Equals *Cicada pellosoma* Equals *Cicadetta pellosoma* China, Shantung, Hong Kong, Manchuria, Korea, Sahalien, Ussuri

Melampsalta pellosoma var. *concolor* Wu 1935: 28 (listed) China, Manchuria

Melampsalta pellosoma var. *flava* Wu 1935: 28 (listed) China, Manchuria

Melampsalta pellosoma He and Chen 1985: 324–327, 329–330, Fig 1A, Fig 3A, Fig 5A, Table 1 (song, sonogram, oscillogram, comp. note) China

Melampsalta pellosoma Wang and Zhang 1987: 296 (key, comp. note) China

Cicadetta pellosoma Anufriev and Emelyanov 1988: 315–316 Fig 237.5–237.8 (genitalia illustrated, key) Soviet Union

Melampalta (sic) *pellosoma* Feng and Shen 1990: 468, Fig 5B (comp. note) China

Cicadetta pellosoma Popov 1990a: 302, Fig 2 (comp. note)

Cicadetta pellosoma Tautz 1990: 105, Fig 12 (song frequency, comp. note)

M[elampsalta] pellosoma Jiang, Yang, Tang, Xu and Chen 1995b: 228 (song, phonotaxis, flight behavior, comp. note) China

Cicadetta pellosoma Lee 1995: 18, 23, 45–46, 104, 112–114, 136, 139 (listed, key, synonymy, ecology, distribution, illustrated, comp. note) Equals *Cicada pellosoma* Equals *Melampsalta pellosoma* Korea

Melampsalta pellosoma Lee 1995: 30–31, 33, 35–44, 131, 139 (comp. note) Korea

Cicadetta pellosoma Chou, Lei, Li, Lu and Yao 1997: 58, 60–61, Fig 8–10, Plate I, Fig 17 (key, described, illustrated, synonymy, comp. note) Equals *Cicada pellosoma* Equals *Melampsalta pellosoma* Equals *Melampsalta pellosoma concolor* Equals *Melampsalta pellosoma flava* China

Cicadetta pellosoma Lee 1998: 21 (comp. note) Korea

Cicadetta pellosoma Popov 1998a: 316, 319–325, Figs 3–6, Figs 7b–c (song, oscillogram, comp. note) Russia

Cicadetta pellosoma Popov 1998b: 309–310, 312–317, Figs 3–6, Figs 7b–c (song, oscillogram, comp. note) Russia

Cicadetta pellosoma Lee 1999: 10, 12–13, Figs 8–9 (distribution, song, illustrated, listed, comp. note) Equals *Cicada pellosoma* Equals *Melampsalta pellosoma* Korea, China, Russia

Cicadetta pellosoma Sanborn 1999a: 52 (type material, listed)

Cicadetta pellosoma Hua 2000: 61 (listed, distribution) Equals *Cicada pellosoma* Equals *Melampsalta pellosoma* Equals *Melampsalta pellosoma* var. *concolor* Equals *Melampsalta pellosoma* var. *flava* China, Liaoning, Shandong, Hebei, Shaanxi, Gansu, Jiangsu, Hong Kong, Sichuan, Korea, Russia

Cicadetta pellosoma Lee 2001b: 3 (comp. note) Korea

Cicadetta pellosoma Puissant 2001: 148, 150 (comp. note)

Cicadetta pellosoma Sueur 2001: 43, Table 1 (listed)

Cicadetta pellosoma Lee, Choe, Lee and Woo 2002: 3–10, Figs 1–3, Tables 1–3 (gene sequence, phylogeny, comp. note) Russia, China, Korea

Melampsalta (= Cicadetta) pellosoma Lee, Choe, Lee and Woo 2002: 10 (comp. note)

Cicadetta pellosoma Lee 2003b: 7 (comp. note)

Cicadetta pellosoma Sueur 2003: 29442 (comp. note)

Cicadetta pellosoma Lee, Oh and Park 2004: 127–131, Fig 2, Figs 3a–c (habitat, illustrated, comp. note) Korea

Cicadetta pellosoma var. *concolor* Lee, Oh and Park 2004: 129 (comp. note) Korea

Cicadetta pellosoma var. *flava* Lee, Oh and Park 2004: 129 (comp. note) Korea
Cicadetta pellosoma Lee 2005: 17, 112–115, 159, 167, 177 (described, illustrated, synonymy, song, distribution, key, comp. note) Equals *Cicada pellosoma* Equals *Melampsalta pellosoma* Equals *Mogannia hebes* (nec Walker) Korea, China, Russia
Melampsalta pellosoma Lee 2005: 112 (comp. note)
Cicadetta pellosoma Puissant 2005: 312 (comp. note)
Cicadetta pellosoma Shiyake 2007: 8, 105, 107, Fig 183 (illustrated, distribution, listed, comp. note) China, Korea, Russia
Cicadetta pellosoma Lee 2008b: 446, 461–462 (comp. note) China, Hong Kong
Cicada pellosoma Lee 2008b: 460–461, Fig 7 (lectotype designation, illustrated, comp. note) Hong Kong
Cicadetta pellosoma Zamanian, Mehdipour and Ghaemi 2008: 2070 (comp. note)

C. *petrophila* Popov, Aronov & Sergeyeva, 1985
Cicadetta petrophila Popov, Aronov and Sergeeva 1985: 451–454, 456, 458–461, Figs 1.1–1.2, Fig 5c, Fig 8f (n. sp., song, hearing, comp. note) Tadzhikistan
Cicadetta petrophila Popov, Aronov and Sergeeva 1985: 289, 291, 293–297, Figs 1.1–1.2, Fig 5c, Fig 8f (n. sp., song, hearing, comp. note) Tadzhikistan
Cicadetta petrophila Popov 1990a: 302–303, Fig 2 (comp. note)
Cicadetta petrophila Popov and Sergeeva 1987: 684, 689 (comp. note)
Cicadetta petrophila Popov 1990b: 579–586, Fig 1B, Figs 2.3–2.4, Fig 3.2, Fig 5, Figs 6.3–6.4, Table (illustrated, song, oscillogram, ecology, comp. note) Tadzhikistan
Cicadetta petrophila Popov 1991: 17–23, Fig 1B, Figs 2.3–2.4, Fig 3.2, Fig 5, Figs 6.3–6.4, Table (illustrated, song, oscillogram, ecology, comp. note) Tadzhikistan
Cicadetta petrophila Popov, Aronov and Sergeeva 1991: 281 (comp. note)
Cicadetta petrophila Popov, Arnov and Sergeeva 1992: 151 (comp. note)
Cicadetta petrophila Bennet-Clark and Young 1994: 293, Table 1 (body size and song frequency)
Cicadetta petrophila Sueur 2001: 43, Table 1 (listed)

C. *petryi* Schumacher, 1924 = *Cicadetta montana petryi* = *Cicadetta montana* f. *petryi*
Cicadetta petryi Boulard 1995a: 13, 58–59, Fig S22$_1$, Fig S22$_2$ (song analysis, sonogram, listed, comp. note) France
Cicadetta petryi Boulard and Mondon 1995: 66–67, 159 (n. comb., listed, comp. note) Equals *Cicadetta montana petryi* France
Cicadetta montana petryi Gogala and Trilar 1999a: 94, 96, Fig 4c (song, oscillogram, comp. note)
Cicadetta montana petryi Boulard 2000i: 77–78, Fig 17 (synonymy, illustrated, song, sonogram, oscillogram, frequency spectrum, distribution, ecology, comp. note) Equals *Cicadetta montana* form *petryi* Equals *Cicadetta petryi* France
Cicadetta montana petryi Schedl 2000: 264 (comp. note)
Cicadetta petryi Puissant 2001: 149 (synonymy, comp. note)
Cicadetta petryi Sueur 2001: 43, Table 1 (listed)
Cicadetta petryi Nickel 2003: 71 (comp. note) France, Germany
Cicadetta petryi Sueur 2005: 215 (status, song, comp. note) France
Cicadetta petryi Puissant 2006: 19, 91 (synonymy, comp. note) France

C. *pieli* (Kato, 1940) = *Melampsalta pieli* = *Melampsalta wulsini* Liu, 1940
Melampsalta wulsini Chou, Lei, Li, Lu and Yao 1997: 53 (comp. note)
Melampsalta wulsini Liu 1992: 41, 43 (comp. note)
Melampsalta pieli Liu 1992: 43 (comp. note)
Cicadetta pieli Liu 1992: 43 (comp. note)
Cicadetta pieli Chou, Lei, Li, Lu and Yao 1997: 58, 65 (key, described, illustrated, synonymy, comp. note) Equals *Melampsalta pieli* Equals *Melampsalta wulsini* China
Cicadetta pieli Hua 2000: 61 (listed, distribution) Equals *Melampsalta pieli* China, Hebei, Shanxi, Gansu, Xinjiang, Jiangsu
Melampsalta wulsini Hua 2000: 63 (listed, distribution) China

C. *pilosa* Horváth, 1901 = *Pauropsalta pilosa*
Cicadetta pilosa Kemal and Koçak 2010: 6 (listed, comp. note) Turkey

C. *podolica* Eichwald, 1830 = *Cicada podolica* = *Melampsalta (Melampsalta) podolica* = *Cicadetta adusta* (nec Hagen) = *Cicadetta concinna* (nec Germar)
Cicada podolica Boulard 1981b: 41–42 (comp. note)
Cicadetta podolica Schedl 1986a: 20 (comp. note)
Cicadetta podolica Schedl 1986b: 579–584, Figs 1–3, Map 1 (synonymy, described, illustrated, distribution, ecology, comp. note) Equals *Cicada aestivalis* Equals *Cicada subapicalis* Poland, Hungary, Romania, Yugoslavia, Soviet Union, Turkey
Cicadetta podolica Gebicki 1987: 96, Table 1 (habitat, comp. note) Poland
Cicadetta podolica Nast 1987: 568–569 (distribution, listed) Poland, Ukraine, Romania, Russia, Spain

Cicadetta podolica Holzinger, Frölich, Günthart, Pauterer, Nickel et al. 1997: 49 (listed) Europe
Cicadetta podolica Gogala 1998b: 399 (comp. note) Slovenia
Cicadetta podolica Gogala and Trilar 1998b: 67–68 (comp. note) Slovenia
Cicadetta podolica Gogala and Gogala 1999: 127 (comp. note)
Cicadetta podolica Schedl 2000: 260–261, 266–268, Fig 7, Plate 1, Fig 7, Map 6 (key, illustrated, distribution, synonymy, listed, comp. note) Equals *Cicadetta adusta* Poland
Cicadetta podolica Dmitriev 2001a: 58 (comp. note) Russia
Cicadetta podolica Dmitriev 2001b: 10 (comp. note) Russia
Cicadetta podolica Schedl 2001b: 27 (comp. note)
Cicadetta podolica Holzinger, Kammerlander and Nickel 2003: 481–483, 616–617, 628, Plate 42, Fig 11, Table 2 (key, listed, illustrated, comp. note) Equals *Cicadetta adusta* Europe, Ukraine
Cicadetta podolica species group Gogala and Trilar 2004a: 316, 319 (comp. note)
Cicadetta podolica Gogala and Trilar 2004a: 317, 319, 321–322, Fig 4, Table 2 (comp. note) Poland, Macedonia
Cicadetta podolica Gogala and Trilar 2004b: 35 (song, comp. note)
Cicadetta podolica Pilarczyk and Swierczewski 2004: 146 (comp. note) Poland
Cicadetta podolica Gogala, Trilar and Krpač 2005: 105, 118, 124 (distribution, comp. note) Macedonia, Poland
Cicadetta podolica Moulds 2005b: 410 (comp. note)
Cicadetta podolica Gogala 2006b: 370, 373–375, Fig 6 (song, sonogram, oscillogram, comp. note) Poland
Cicadetta podolica Trilar, Gogala and Popa 2006: 178 (comp. note) Romania
Cicadetta podolica Dmitriev 2007a: 818 (comp. note) Russia
Cicadetta podolica Dmitriev 2007b: 1210 (comp. note) Russia
Cicadetta podolica Trilar and Gogala 2007: 6–11, Figs 1–4, Fig 5a, Fig 6, Tables 1–2 (illustrated, habitat, song, sonogram, oscillogram, comp. note) Equals *Cicadetta adusta* Poland, Ukraine
Cicadetta adusta Trilar and Gogala 2007: 6 (comp. note)
Cicadetta podolica Gogala 2008: 55 (song, comp. note)
Cicadetta podolica Gogala, Drosopoulos and Trilar 2008b: 123 (comp. note)
Cicadetta podolica Lee 2008b: 456 (comp. note)
Cicadetta podolica Nickel 2008: 207 (distribution, hosts, comp. note) Europe
Cicadetta podolica Trilar and Gogala 2008: 32 (comp. note) Equals *Cicadetta concinna* (partim) Romania

C. cf. *podolica* Gogala, Drosopoulos & Trilar, 2008
Cicadetta cf. *podolica* Gogala, Trilar and Krpač 2005: 118–119, 124, Fig 12 (distribution, comp. note) Macedonia
Cicadetta cf. *podolica* Gogala, Drosopoulos and Trilar 2008a: 99 (comp. note)

C. polita Popple, 2003 to *Heliopsalta polita*

***C. popovi* Emeljanov, 1996** = *Cicadetta* ex. gr. *tibialis* Popov, 1985
Cicadetta ex. gr. *tibialis* (sp. nov.) Popov 1985: 80, Fig 24 (acoustic system, comp. note)
Cicadetta ex. gr. *tibialis* Popov, Aronov and Sergeeva 1985a: 451–454, 456, 458, 460–461, Fig 2, Fig 5e (song, oscillogram, comp. note) Tadzhikistan
Cicadetta ex. gr. *tibialis* Popov, Aronov and Sergeeva 1985b: 289–290, 292–294, Fig 2, Fig 5e (song, oscillogram, comp. note) Tadzhikistan
Cicadetta popovi Emeljanov 1996a: 1897–1900, Fig 3 (illustrated, comp. note) Tadzhikistan
Cicadetta popovi Emeljanov 1996b: 1054–1057, Fig 3 (illustrated, comp. note) Tadzhikistan
Cicadetta popovi Gogala and Popov 1997b: 34 (comp. note)
Cicadetta ex. gr. *tibialis* Popov 1997a: 3 (comp. note) Tadzhikistan
Cicadetta popovi Popov 1997a: 3, 8–11, 14–15, Figs 4–6 (song, oscillogram, sonogram, comp. note) Equals *Cicadetta* ex. gr. *tibialis* Popov, Aronov and Sergeeva, 1985 Tadzhikistan
Cicadetta ex. gr. *tibialis* Popov 1997b: 1 (comp. note) Tadzhikistan
Cicadetta popovi Popov 1997b: 1, 5–8, 10–11, Figs 4–6 (song, oscillogram, sonogram, comp. note) Equals *Cicadetta* ex. gr. *tibialis* Popov, Aronov and Sergeeva, 1985 Tadzhikistan
Cicadetta popovi Popov, Beganović and Gogala 1998a: 93 (comp. note)
Cicadetta popovi Sueur 2001: 43, Table 1 (listed)

C. puer (Walker, 1850) to *Chelapsalta puer*

C. quadricincta (Walker, 1850) to *Physeema quadricincta*

***C. ramosi* Sanborn, 2009** = *Cicadetta calliope* (nec Walker)
Cicadetta calliopesic (sic) Ramos 1983: 62, 64, 66–67, 70, Fig 22 (synonymy, described, illustrated, key, listed, comp. note) Equals *Cicada calliope* Dominican Republic
Cicadetta calliopesic (sic) Perez-Gerlabert 2008: 202 (listed) Hispaniola

Cicadetta ramosi Sanborn 2009d: 1–6, Figs 1–9 (n. sp., synonymy, described, illustrated, comp. note) Equals *Cicadetta calliope* Ramos, 1983 Dominican Republic, Haiti

C. rouxi (Distant, 1914) to *Rouxalna rouxi*

C. rubea (Goding & Froggatt, 1904) to *Pauropsalta rubea*

C. rubricincta (Goding & Froggatt, 1904) to *Pyropsalta melete* (Walker, 1850)

C. rubristrigata (Goding & Froggatt, 1904) to *Pauropsalta rubristrigata*

C. sachalinensis Matsumura, 1817 to *Kosemia yezoensis*

C. sancta (Distant, 1913) to *Simona sancta*

***C. sarasini* (Distant, 1914)** = *Melampsalta sarasini*

C. shansiensis (Esaki & Ishihara, 1950) = *Melampsalta shansiensis*

Cicadetta shansiensis Chou, Lei, Li, Lu and Yao 1997: 58, 76, 62–63, 307, 312, 324, 326, 328, 335, 362, 364, Fig 8–11, Plate I, Fig 14, Tables 10–12 (key, described, illustrated, synonymy, song, oscillogram, spectrogram, comp. note) Equals *Melampsalta shansiensis* China

Cicadetta shansiensis Hua 2000: 61 (listed, distribution) Equals *Melampsalta shansiensis* China, Hebei, Shanxi, Shaanxi, Gansu, Shandong, Hubei, Sichuan

Cicadetta shansiensis Sueur 2001: 43, Table 1 (listed)

***C. sibilatrix* Horváth, 1901** = *Pauropsalta sibilatrix* = *Melampsalta (Pauropsalta) sibilatrix* = *Melampsalta (Cicadetta) sibilatrix* = *Melampsalta sibilathrix* (sic)

Cicadetta sibilatrix Lodos and Kalkandelen 1981: 79 (distribution, comp. note) Turkey, Israel, Syria

Pauropsalta sibilatrix Schedl 1993: 799 (comp. note)

Cicadetta sibilatrix Demir 2008: 477 (distribution) Turkey

*Cicadetta sibilatrix*Kemal and Koçak 2010: 6 (distribution, listed, comp. note) Turkey

C. singula (Walker, 1850) to *Plerapsalta multifascia* (Walker, 1850)

C. sinuatipennis Oshanin, 1906 to *Melampsalta sinuatipennis*

***C.* sp. n. Drosopoulos, Asche & Hoch, 1986**

Cicadetta sp. n. Drosopoulos, Asche and Hoch 1986: 9 (listed) Greece

C. sp. n. Ewart, 1986 of uncertain position in Cicadettini

C. sp. nov. 1 Lithgow, 1988 of uncertain position in Cicadettini

C. sp. nov. 2 Lithgow, 1988 of uncertain position in Cicadettini

C. sp. nov. 3 Lithgow, 1988 of uncertain position in Cicadettini

C. sp. nov. 4 Lithgow, 1988 of uncertain position in Cicadettini

C. sp. nov. 5 Lithgow, 1988 of uncertain position in Cicadettini

C. sp. A, Gidyea Cicada Ewart & Popple, 2001 of uncertain position in Cicadettini

C. sp. A, Brown Sandplain Cicada Ewart, 2009 of uncertain position in Cicadettini

C. sp. B Ewart, 2009 of uncertain position in Cicadettini

C. sp. F Ewart, 2009 of uncertain position in Cicadettini

C. sp. G Enamel Cicada Popple & Strange, 2002 to *Heliopsalta polita* (Popple, 2003)

C. sp. H Capparis Cicada Popple & Strange, 2002 = *Cicadetta* sp. H = *Cicadetta* sp. K Ewart, 2009 of uncertain position in Cicadettini

C. sp. I Pink-wing Cicada Popple & Strange, 2002 of uncertain position in Cicadettini

C. sp. I Small Acacia Cicada Ewart, 2009 of uncertain position in Cicadettini

C. sp. J Ewart, 2009 of uncertain position in Cicadettini

C. sp. K Ewart, 2009 of uncertain position in Cicadettini

C. sp. M Ewart, 2009 of uncertain position in Cicadettini

C. sp. nov. Small Robust Ambertail Popple & Strange, 2002 of uncertain position in Cicadettini

C. sp. nov. Small Treetop Ticker Popple & Strange, 2002 of uncertain position in Cicadettini

C. spinosa (Goding & Froggatt, 1904) to *Auscala spinosa*

C. spreta (Goding & Froggatt, 1904) to *Gelidea torrida* (Erichson, 1842)

C. stradbrokensis (Distant, 1915) to *Limnopsalta stradbrokensis*

C. subgulosa (Ashton, 1914) to *Simona sancta* (Distant, 1913)

C. sulcata (Distant, 1907) to *Taurella froggatti* (Distant, 1907)

***C. surinamensis* Kirkaldy 1909** = *Cicada marginella* Olivier, 1790 (nec *Cicada marginella* Fabricius, 1787) = *Melampsalta marginella*

Cicadetta surinamensis Sanborn 2009d: 1 (comp. note) South America

***C. texana* (Davis, 1936)** = *Melampsalta texana*

Melampsalta texana Tinkham 1941: 179 (habitat, comp. note) Texas

Cicadetta texana Poole, Garrison, McCafferty, Otte and Stark 1997: 331 (listed) Equals *Melampsalta texana* North America

Cicadetta texana Sanborn 1999a: 52 (type material, listed)

Cicadetta texana Sanborn 1999b: 2 (listed) North America

Cicadetta texana Sanborn 2009d: 1 (comp. note) North America

C. tigris (Ashton, 1914) to *Clinopsalta tigris*

C. sp. nr. *tigris* Ewart, 1988 = Species C Ewart & Popple, 2001 to *Clinopsalta* sp. nr. *tigris*
C. sp. nr. *tigris* Ewart, 2009 = *Notopsalta* sp. G Ewart 1998, Popple and Strange 2002 to *Clinopsalta* sp. nr. *tigris*
C. *toowoombae* (Distant, 1915) to *Yoyetta toowoombae*
C. *torrida torrida* (Erichson, 1842) to *Gelidea torrida torrida*
C. *torrida* var. *β* Walker, 1850 to *Gelidea torrida* var. *β*
C. *torrida* var. *δ* Walker, 1850 to *Gelidea torrida* var. *δ*
C. *torrida* var. *γ* Walker, 1850 to *Gelidea torrida* var. *γ*
C. *transylvanica hyalinervis* Boulard, 1981 = *Cicadetta transylvanica* var. *hyalinervis*
Cicadetta transylvanica var. *hyalinervis* Boulard 1981b: 48 (n. var., diagnosis, comp. note)
C. *transylvanica transylvanica* Fieber, 1876 = *Cicadetta transsylvanica* (sic) = *Cicadetta cissylvanica* (sic) = *Melampsalta transylvanica* = *Melampsalta (Melampsalta) transsylvanica* (sic) = *Cicadivetta* (sic) *transylvanica*
Cicadetta transylvanica Soós 1956: 415 (listed, distribution) Equals *Cicadetta cissylvanica* (sic) Romania
Cicadetta transylvanica Boulard 1981b: 48 (lectotype designation, comp. note)
Cicadetta transsylvanica (sic) Boulard 1981b: 48 (comp. note)
Cicadetta transylvanica Nast 1987: 568–569 (distribution, listed) Germany, Poland, Czechoslovakia, Romania, Italy
Cicadivetta transylvanica Schedl 2000: 258 (distribution, comp. note)
Cicadetta transsylvanica (sic) Trilar, Gogala and Popa 2006: 178 (comp. note)
C. *tristrigata* (Goding & Froggatt, 1904) to *Yoyetta tristrigata*
C. sp. nr. *tristrigata* sp. 5 Fiery Ambertail M43 Moss & Popple, 2000 to *Yoyetta* sp. nr. *tristrigata* sp. 5 Fiery Ambertail M43
C. sp. nr. *tristrigata* sp. 16 Robust Ambertail Moss & Popple, 2000 to *Yoyetta* sp. nr. *tristrigata* sp. 16 Robust Ambertail
C. sp. nr. *tristrigata* Ewart & Popple, 2001 to *Yoyetta* sp. nr. *tristigata*
C. sp. nr. *tristrigata* Emery, Emery, Emery and Popple, 2005 to *Yoyetta* sp. nr. *tristigata*
C. sp. nr. *tristrigata* Haywood, 2005 to *Yoyetta* sp. nr. *tristigata*
C. *tumidifrons* Horváth, 1911 = *Cicadetta (Pauropsalta) tumidifrons* = *Melampsalta (Pauropsalta) tumidifrons* = *Melampsalta tumidifrons*
C. *tunisiaca* (Karsch, 1890) = *Melampsalta tunisiaca* = *Melampsalta (Melampsalta) tunisiaca*
C. *turcica* (Schedl, 2001)
Cicadetta turcica Kemal and Koçak 2010: 6 (distribution, listed, comp. note) Turkey
C. *tympanistria* Kirkaldy, 1907 to *Fijipsaltria tympanistria*
C. *undulata* (Waltl, 1837) to *Euryphara undulata*
C. *ventricosa* (Stål, 1866) = *Melampsalta ventricosa*
C. *virens* (Herrich-Schäffer, 1835) to *Euryphara virens*
C. *viridicincta* (Ashton, 1912) to *Physeema quadricincta* (Walker, 1850)
C. *viridis* (Ashton, 1912) to *Taurella viridis*
C. *walkerella* Kirkaldy, 1909 = *Cicada connexa* Walker, 1850: 177 (nec *Cicada connexa* Walker, 1850: 173) = *Melampsalta connexa*
C. *warburtoni* (Distant, 1882) to *Taurella forresti* (Distant, 1882)
C. *waterhousei* (Distant, 1906) to *Myopsalta waterhousi*
C. sp. aff. *waterhousei* Haywood, 2005 to *Myopsalta waterhousei*
C. *yamashitai* (Esaki & Ishihara, 1950) = *Melampsalta yamashitai*
C. *yezoensis* (Matsumura, 1898) to *Kosemia yezoensis*

***Euboeana* Gogala, Drosopoulos & Trilar, 2011**
E. *castaneivaga* Gogala, Drosopoulos & Trilar, 2011

***Cicadivetta* Boulard, 1982** = *Cicadetta (Cicadivetta)*
Cicadivetta Boulard 1982a: 50 (n. gen., described)
Cicadivetta Boulard 1995a: 71 (listed)
Cicadivetta Boulard and Mondon 1995: 65, 68 (illustrated, key) France
Cicadivetta Gogala, Popov and Ribaric 1996: 46 (comp. note)
Cicadivetta Holzinger 1996: 505 (listed) Austria
Cicadivetta Gogala and Popov 1997a: 18 (comp. note)
Cicadivetta Holzinger, Frölich, Günthart, Pauterer, Nickel et al. 1997: 49 (listed) Europe
Cicadivetta Gogala 1998b: 396 (comp. note)
Cicadivetta Gogala and Gogala 1999: 120 (comp. note) Slovenia
Cicadivetta Boulard 2000i: 26 (key) France
Cicadivetta Schedl 2000: 258, 260, 268 (key, listed, comp. note)
Cicadivetta Lee, Choe, Lee and Woo 2002: 3 (comp. note)
Cicadivetta Gogala, Trilar and Krpač 2005: 115 (comp. note)
Cicadivetta Moulds 2005b: 390, 436 (higher taxonomy, listed, comp. note)
Cicadivetta Gogala and Schedl 2008: 404 (comp. note)
Cicadivetta Gogala 2006a: 158 (comp. note)
Cicadivetta Gogala and Drosopoulos 2006: 278 (comp. note)
Cicadivetta Puissant 2006: 87 (comp. note) France

Cicadivetta Kemal and Koçak 2010: 3, 6 (diversity, listed, comp. note) Turkey
Cicadivetta Mozaffarian and Sanborn 2010: 76, 82 (distribution, listed, comp. note) Europe, North Africa, Middle East, Eastern Asia

C. carayoni **(Boulard, 1982)** = *Tettigetta carayoni*
Tettigetta carayoni Boulard 1982c: 103–105, Figs 6–10 (n. sp., described, illustrated) Crete
Tettigetta carayoni Nast 1987: 570–571 (distribution, listed) Greece
Cicadetta carayoni Gogala, Drosopoulos and Trilar 2008a: 99 (comp. note)
Tettigetta carayoni Puissant and Sueur 2010: 558 (comp. note)
Cicadivetta carayoni Puissant and Sueur 2010: 558 (n. comb., comp. note)
Tettigetta carayoni Trilar and Gogala 2010: 6–17, Figs 1–8, Table 1 (illustrated, song, sonogram, oscillogram, comp. note) Crete

C. flaveola **(Brullé, 1832)** = *Tettigetta flaveola* = *Tibicen flaveolus* = *Cicada flaveola* = *Cicadetta flaveola* = *Cicadetta (Cicadivetta) flaveola* = *Melampsalta flaveola* = *Melampsalta (Melampsalta) flaveola*
Cicadetta flaveola Drosopoulous 1980: 190 (listed) Greece
(Cicadetta) flaveola Boulard 1982f: 197 (comp. note) Balkans, Anatolia
Cicadetta flaveola Nast 1987: 568–569 (distribution, listed) Portugal, Spain, Italy, Greece
Cicadetta flaveola Lodos and Kalkandelen 1988: 18 (comp. note) Turkey
Cicadetta flaveola Quartau and Fonseca 1988: 367, 369 (synonymy, distribution, comp. note) Equals *Tibicen flaveola* Portugal, Algeria, Greece, Italy, Sicily, Morocco, Spain, USSR, Georgia, Kazakhstan, Tadzhikistan, Turkmenia
Tettigetta flaveola Schedl 2001a: 1288 (comp. note)
Cicadetta flaveola Sueur, Puissant, Simões, Seabra, Boulard and Quartau 2004: 182 (comp. note)
Cicadetta flaveola Gogala 2006a: 163–164, Figs 3–4 (illustrated, sonogram, oscillogram, comp. note)
Cicadetta flaveola Gogala, Trilar and Krpač 2005: 124 (comp. note)
Cicadetta flaveola Gogala and Drosopoulos 2006: 275–278, Figs 1–7 (illustrated, oscillogram, comp. note) Greece, Algeria, Italy, Sicily, Morocco, Portugal, Spain, Georgia, Kazakhstan, Tadzhikistan, Turkmenia
Cicadetta flaveola Gogala, Drosopoulos and Trilar 2008a: 99 (comp. note)
Cicadetta (Cicadivetta) flaveola Gogala and Schedl 2008: 397, 399–401, 403 (comp. note)
Cicadetta flaveola Kemal and Koçak 2010: 6 (distribution, listed, comp. note) Turkey
Cicadivetta flaveola Puissant and Sueur 2010: Appendiz 2 (n. comb., comp. note)
Cicadetta flaveola Trilar and Gogala 2010: 16–17 (comp. note)

C. goumenissa **Gogala, Drosopoulos & Trilar, 2012**

C. parvula **(Fieber, 1876)** = *Tettigetta parvula* = *Cicadetta parvula* = *Melampsalta parvula* = *Cicadetta fieberi* Oshanin, 1908 = *Cicadetta fieberi* Kirkaldy, 1909 = *Melampsalta (Melampsalta) fieberi* = *Melampsalta fieberi* = *Heptaglena (Oligglena) libanotica* Horváth, 1911 = *Oligglena libanotica*
Cicadetta parvula Lodos and Kalkandelen 1981: 79 (synonymy, distribution, comp. note) Equals *Cicadetta fieberi* Turkey, Morocco
Heptaglena libanotica Koçak 1982: 149–150 (type species of *Heptoglena*, type species of *Oligoglena*)
Tettigetta parvula Schedl 1999b: 827–828, 836, Figs 7a–c (illustrated, listed, synonymy, distribution, ecology, comp. note) Equals *Heptaglena (Oligglena) libanotica* Israel, Turkey, Lebanon
Oligglena libanotica Schedl 1999b: 836, Fig 7d (illustrated)
Cicadetta parvula Kemal and Koçak 2010: 6 (distribution, listed, comp. note) Turkey
Cicadetta parvula Puissant and Sueur 2010: 558 (comp. note)
Cicadivetta parvula Puissant and Sueur 2010: 558 (n. comb., comp. note)

C. tibialis **(Panzer, 1798)** = *Tettigonia tibialis* = *Cicada tibialis* = *Cicadetta tibialis* = *Cicada (Tettigetta) tibialis* = *Melampsalta tibialis* = *Pauropsalta tibialis* = *Melampsalta (Pauropsalta) tibialis* = *Melampsalta (Melampsalta) tibialis* = *Melampsalta tibalis* (sic) = *Cicadetta cissylvanica* Haupt, 1835 = *Cicada minor* Eversmann, 1837 = *Cicada tibialis minor* = *Cicadetta tibialis minor* = *Cicadivetta tibialis* var. *minor* = *Heptaglena libanotica* Horváth, 1911 = *Oligoglena libanotica* = *Cicadetta libanotica* = *Cicadetta tibialis acuta* Dlabola, 1961 = *Cicadivetta tibialis acuta* = *Cicadetta transylvanica* (nec Fieber)
Cicadetta tibialis Szilády 1870: 119 (listed) Romania
Cicadetta tibialis Mader 1922: 179 (comp. note) Austria
Pauropsalta tibialis Gómez-Menor 1951: 70 (comp. note)
Cicadetta tibialis Castellani 1952: 16 (comp. note) Italy
Cicadetta tibialis Wagner 1952: 16 (listed, comp. note) Italy
Cicadetta tibialis Soós 1956: 414 (listed, distribution) Romania

Cicadetta tibialis Cantoreanu 1968a: 343, Table 1 (listed, distribution, comp. note) Romania
Cicadetta tibialis Cantoreanu 1968b: 128 (distribution, comp. note) Romania
Cicadetta tibialis Cantoreanu 1969: 261, Table 1 (listed, distribution, comp. note) Romania
Cicadetta tibialis Cantoreanu 1972a: 111, Table 1 (listed, distribution, comp. note) Romania
Cicadetta tibialis Kaltenbach, Steiner and Aschenbrenner 1972: 490 (comp. note) Germany
Cicadetta tibialis Cantoreanu 1975: 92 (distribution, comp. note) Romania
Cicadetta tibialis Drosopoulous 1980: 190 (listed) Greece
Cicadetta tibialis Dlabola 1981: 208 (distribution, comp. note) Iran
Pauropsalta tibialis Lodos and Kalkandelen 1981: 79–80 (synonymy, distribution, comp. note) Equals *Cicada tibialis minor* Equals *Heptaglena libanotica* Equals *Cicadetta cissilvanica* Equals *Cicadetta tibialis acuta* Turkey, Albania, Austria, Bulgaria, Czechoslovakia, France, Germany, Greece, Hungary, Israel, Italy, Morocco, Portugal, Romania, Spain, Tunisia, USRR
Tettigonia tibialis Boulard 1982a: 50 (type species of genus *Cicadivetta*)
Cicadivetta tibialis Boulard 1982a: 50 (synonymy) Equals *Tettigonia tibialis* Equals *Cicada tibialis* Equals *Melampsalta tibialis* Equals *Cicadetta tibialis* Equals *Pauropsalta tibialis* Equals *Tettigetta tibialis*
Melampsalta tibialis Boulard 1982f: 197 (comp. note)
Cicadivetta tibialis Boulard 1984d: 5 (comp. note) France
Cicadetta tibialis Nosko 1984: 58 (comp. note) Slovakia
Cicadetta tibialis acuta Popova and Dubovskaya 1984: 3 (agricultural pest, listed) Uzbekistan
Cicadetta tibialis Reichhoff-Riehm 1984: 90–91 (natural history, illustrated, comp. note)
Cicadivetta tibialis Boulard 1985b: 63 (characteristics) France
Cicadetta tibialis Popov 1985: 71, 75, 78, 80, Fig 24 (acoustic system, comp. note)
Cicadetta tibialis Popov, Aronov and Sergeeva 1985a: 454, 460–461 (song, comp. note) Crimea
Cicadetta tibialis Popov, Aronov and Sergeeva 1985b: 292, 295–297 (song, comp. note) Crimea
Cicadetta tibialis Chinery 1986: 88 (comp. note) Europe
Cicadivetta tibialis Schedl 1986a: 3, 6, 8–10, 20–24, Fig 7, Fig 15, Fig 19, Fig 21, Map 9, Table 1 (illustrated, listed, key, synonymy, distribution, ecology, comp. note) Equals *Tettigonia tibialis* Equals *Cicadetta tibialis* Equals *Heptaglena libanotica* Equals *Cicadetta cissylvanica* Equals *Pauropsalta tibialis* Istria, Austria, Czechoslovakia, Hungary, Portugal, Spain, France, Italy, Yugoslavia, Albania, Greece, Romania, Russia, Ukraine, Caucasus, Transcaucasia, Lebanon, Isreal, Iran, Morocco, Tunisia
Cicadivetta tibialis Joermann and Schneider 1987: 284, 291–295, Fig 9b, Figs 10–12 (song analysis, sonogram, oscillogram, comp. note) Yugoslavia
Cicadivetta tibialis Nast 1987: 568–569 (distribution, listed) Germany, Poland, Czechoslovakia, France, Austria, Hungary, Ukraine, Moldavia, Romania, Russia, Portugal, Spain, Italy, Yugoslavia, Albania, Bulgaria, Greece
Cicadivetta tibialis Boulard 1988a: 9, 12, Table (comp. note) France
Cicadetta tibialis Quartau and Fonseca 1988: 368–369 (synonymy, distribution, comp. note) Equals *Tettigonia tibialis* Equals *Cicada tibialis minor* Equals *Heptaglena libanotica* Equals *Cicadetta cissilvanica* Equals *Cicadetta tibialis acuta* Portugal, Albania, Austria, Bulgaria, Czechoslovakia, France, Germany, Greece, Crete, Hungary, Italy, Sicily, Morocco, Romania, Spain, Tunisia, USSR, Kirghizia, Moldavia, Russia, Transcaucasia, Ukraine, Uzbekistan
Cicadivetta tibialis Schedl 1989: 112 (listed)
Cicadetta tibialis Popov 1990a: 302, Fig 2 (comp. note)
Cicadetta tibialis Tautz 1990: 105, Fig 12 (song frequency, comp. note)
Cicadetta tibialis Bairyamova 1992: 45, 47, Table 1 (listed) Bulgaria
Cicadivetta tibialis Boulard 1995a: 13, 71–72, Fig S30$_1$, Fig S30$_2$ (illustrated, song analysis, sonogram, listed, comp. note) Italy
Cicadivetta tibialis Boulard and Mondon 1995: 30, 66–67, 159 (listed, comp. note) Equals *Cicadetta transylvanica* France
Cicadetta tibialis Emeljanov 1996a: 1897–1900, Fig 1 (illustrated, comp. note) Western Europe, Moldavia, Ukraine, Causasus, Transcaucasia, Georgia
Cicadetta tibialis Emeljanov 1996b: 1054–1056, Fig 1 (illustrated, comp. note) Western Europe, Moldavia, Ukraine, Causasus, Transcaucasia, Georgia

Cicadetta tibialis Gogala, Popov and Ribaric 1996: 47–51, 54–62, Figs 1–9 (song, sonogram, oscillogram, comp. note) Equals *Cicadivetta tibialis* Slovenia, Georgia, Azerbaijan, Chechnya

Cicadivetta tibialis Holzinger 1996: 505 (listed) Austria

Cicadetta tibialis Gogala and Popov 1997a: 12, 15–19 (comp. note)

Cicadivetta tibialis Gogala and Popov 1997b: 34–35 (song, comp. note) Slovenia, Croatia, Macedonia, South Caucasus, Chechnya

Cicadivetta tibialis Holzinger, Frölich, Günthart, Pauterer, Nickel et al. 1997: 49 (listed) Europe

Cicadetta tibialis Popov 1997a: 3–8, 13–15, Figs 1–3 (song, oscillogram, sonogram, comp. note) Equals *Cicadetta caucasica* Popov 1975 Equals *Cicadivetta tibialis* Crimea, Transcaucasia, Azerbaidzhan, Georgia, Slovenia, Bosnia, Macedonia

Cicadetta tibialis Popov 1997b: 1–7, 9–11, Figs 1–3 (song, oscillogram, sonogram, comp. note) Equals *Cicadetta caucasica* Popov 1975 Equals *Cicadivetta tibialis* Crimea, Transcaucasia, Azerbaidzhan, Georgia, Slovenia, Bosnia, Macedonia

Cicadetta tibialis Gogala 1998b: 392, 396 (illustrated, oscillogram, comp. note) Slovenia

Cicadetta tibialis Popov, Beganović and Gogala 1998a: 91, 93–94 (comp. note)

Cicadivetta tibialis Okáli and Janský 1998: 36–37, 43, Fig 3 (distribution, comp. note) Equals *Cicadetta tibialis* Equals *Pauropsalta tibiais* Slovakia

Cicadetta tibialis Gogala and Gogala 1999: 120, 123, 126, Fig 9 (distribution, comp. note) Slovenia, Mediterranean

Cicadivetta tibialis Boulard 2000i: 95–98, 127, Fig 21, Plate P, Map 16 (synonymy, illustrated, song, sonogram, oscillogram, frequency spectrum, distribution, ecology, comp. note) Equals *Tettigonia tibialis* Equals *Cicada tibialis* Equals *Melampsalta tibialis* Equals *Cicadetta tibialis* Equals *Cicadetta transylvanica* (nec Fieber) Equals *Pauropsalta tibialis* Equals *Heptoglena libanotica* Equals *Oligoglena libanotica* Equals *Melampsalta (Pauropsalta) tibialis* Equals *Cicadetta (Melampsalta) tibialis* France, Europe, Asia Minor

Heptoglena libanotica Boulard 2000i: 96 (comp. note)

Cicadetta tibialis Gogala and Popov 2000: 11 (comp. note)

Cicadetta tibialis Gogala and Trilar 2000: 23 (comp. note)

Cicadivetta tibialis Puissant and Sueur 2000: 68–73, Figs 1–4 (illustrated, metamorphosis, song, behavior, hosts, comp. note) France

Cicadetta tibialis Schedl 2000: 258 (distribution, comp. note)

Cicadivetta tibialis Schedl 2000: 260, 263, 265–266, 268–269, Fig 9, Fig 12b, Fig 13b, Plate 1, Fig 8, Map 8 (key, illustrated, distribution, listed, comp. note) Austria, Czech Republic, Slovakia, Hungary, Italy

Cicadivetta tibialis Sueur and Puissant 2000: 262–267, Figs 1–6, Tables 1–2 (song analysis, sonogram, oscillogram, comp. note) Equals *Cicadivetta tibialis* var. *minor* Equals *Cicadivetta tibilais acuta* France

Cicadivetta tibialis acuta Sueur and Puissant 2000: 267 (status, comp. note) Uzbekistan, Russia

Cicadivetta tibialis var. *minor* Sueur and Puissant 2000: 267 (status, comp. note) Russia

Cicadetta tibialis Krpač, Trilar and Gogala 2001: 29 (comp. note) Macedonia

Cicadivetta tibialis Puissant and Sueur 2001: 435 (comp. note)

Cicadivetta tibialis Puissant 2001: 147 (comp. note)

Cicadivetta tibialis Schedl 2001b: 27 (comp. note)

Cicadivetta tibialis Sueur and Puissant 2001: 19, 21–22 (illustrated, comp. note) France

Cicadivetta tibialis Sueur 2001: 43, 46, Tables 1–2 (listed) Equals *Cicadetta caucasica* Equals *Cicadetta tibialis*

Cicadivetta tibialis Gogala 2002: 246, Fig 9 (song, sonogram, oscillogram, comp. note)

Cicadetta tibialis Gogala 2002: 247 (comp. note)

Cicadivetta tibialis Nickel, Holzinger and Wachmann 2002: 292, Table 4 (listed)

Cicadivetta tibialis Schedl 2002: 236–237, Photo 5, Map 5 (illustrated, distribution, comp. note) Austria, Italy

Cicadetta tibialis Gogala and Seljak 2003: 351 (comp. note) Slovenia

Cicadetta tibialis Holzinger, Kammerlander and Nickel 2003: 481, 483, 616–617, 628, 635, Plate 42, Fig 10, Tables 2–3 (key, listed, illustrated, comp. note) Europe, Mediterranean, Asia, Austria

Cicadetta tibialis Tishechkin 2003: 141, 145, Figs 166–167 (oscillogram, comp. note) Chechnya

Cicadivetta (Cicadetta) tibialis Schedl 2004b: 916 (comp. note)

Cicadivetta tibialis Sueur, Puissant, Simões, Seabra, Boulard and Quartau 2004: 182 (comp. note) France, Italy

Cicadetta tibialis Gogala, Trilar and Krpač 2005: 105, 115–116, Fig 9 (distribution, comp. note) Macedonia
Pauropsalta tibialis Janský 2005: 84, Table 1 (listed, comp. note) Slovakia
Cicadetta tibialis Lee 2005: 95 (comp. note)
Cicadivetta tibialis Puissant and Defaut 2005: 116–119, 124, Tables 1–2, Table 8 (listed, habitat association, distribution, comp. note) France
Cicadetta tibialis Sanborn 2005c: 116 (comp. note)
Cicadivetta tibialis Boulard 2006g: 180 (listed)
Cicadetta tibialis Demir 2006: 102 (distribution, comp. note) Turkey
Cicadetta tibialis Gogala 2006a: 158 (comp. note)
Cicadivetta tibialis Gogala 2006a: 158 (comp. note)
Cicadivetta tibialis Puissant 2006: 20, 23, 25, 27–29, 32, 52, 76, 87–91, 98, 108–109, 178–181, 186, Fig 8, Figs 57–58, Tables 1–2, Appendix Tables 1–2, Appendix Table 8 (illustrated, synonymy, distribution, habitat association, ecology, listed, comp. note) Equals *Cicadetta translyvanica* France
Tettigonia tibialis Puissant 2006: 87 (type species of *Cicadivetta*)
Cicadetta tibialis Trilar, Gogala and Popa 2006: 177, 179 (listed, distribution, comp. note) Romania
Cicadivetta tibialis Demir 2007b: 484 (distribution, comp. note) Turkey
Cicadetta tibialis Shiyake 2007: 8, 105, 107, Fig 185 (illustrated, distribution, listed, comp. note) Europe, Middle East
Cicadivetta tibialis Sueur and Puissant 2007b: 62 (listed) France
Cicadetta tibialis Holzinger and Komposch 2008: 140 (comp. note) Austria
Cicadivetta tibialis Nickel 2008: 207 (distribution, hosts, comp. note) Western Palaearctic
Cicadetta tibialis Trilar and Gogala 2008: 29–30 (listed, distribution, comp. note) Romania
Cicadivetta tibialis Kemal and Koçak 2010: 6 (distribution, listed, comp. note) Turkey
Cicadivetta tibialis Mozaffarian and Sanborn 2010: 76, 82 (synonymy, listed, distribution, comp. note) Equals *Tettigonia tibialis* Iran, Austria, North Africa, Europe, Middle East, Asia Minor
Tettigonia tibialis Mozaffarian and Sanborn 2010: 82 (type species of *Cicadivetta*)

***Dimissalna* Boulard, 2007**

Dimisslana Boulard 2007b: 97–98, Figs 1–2 (n. gen., described, illustrated, comp. note)

***D. dimissa* (Hagen, 1856)** = *Cicada dimissa* = *Cicadetta dimissa* = *Melampsalta dimissa* = *Melampsalta (Melampsalta) dimissa* = *Melampsalta dismissa* (sic) = *Tettigetta dimissa* = *Melampsalta hodoharai* (nec Kato)
Melampsalta dimissa Wu 1935: 27 (listed) Equals *Cicada dimissa* Equals *Cicadetta dimissa* China, Szechuan, Dalmatia, Syria, Caucasus, Siberia
Cicadetta dimissa Soós 1956: 413 (listed, distribution) Romania
Cicadetta dimissa Drosopoulous 1980: 190 (listed) Greece
Cicadetta dimissa Lodos and Kalkandelen 1981: 77–78 (distribution, comp. note) Turkey, Albania, China, Greece, Italy, USSR, Yugoslavia
Cicadetta dimissa Popov 1985: 40–41, 71, 76, 78–80, Fig 7.3, Fig 23.5, Fig 24 (illustrated, oscillogram, acoustic system, comp. note)
Tettigetta dimissa Schedl 1986a: 3–5, 8–10, 17–19, 22–24, Fig 2, Fig 5, Fig 11, Fig 16, Fig 21, Map 7, Table 1 (illustrated, listed, key, synonymy, distribution, ecology, comp. note) Equals *Cicadetta dimissa* Equals *Melampsalta dimissa* Equals *Cicadetta dimissa* Istria, Italy, Yugoslavia, Albania, Greece, Anatolia, USSR, Armenia, Azerbaijan, Kazahkstan, Siberia, China
Cicadetta dimissa Nast 1987: 568–569 (distribution, listed) Italy, Yugoslavia, Albania, Bulgaria, Greece
Tettigetta dimissa Schedl 1989: 109–112, Fig 1, Figs 2b–d (illustrated, distribution, oviposition, ecology, comp. note) Istria, Croatia
Cicadetta dimissa Popov 1990a: 302–303, Fig 2 (comp. note)
Cicadetta dimissa Tautz 1990: 105, Fig 12 (song frequency, comp. note)
Cicadetta dimissa Popov, Aronov and Sergeeva 1991: 281 (comp. note)
Cicadetta dimissa Popov, Arnov and Sergeeva 1992: 151 (comp. note)
Cicadetta dimissa Chou, Lei, Li, Lu and Yao 1997: 66 (listed, synonymy, comp. note) Equals *Cicada dimissa* Equals *Melampsalta hodoharai* (nec Kato) China
Tettigetta dimissa Gogala 1998b: 395 (oscillogram, comp. note) Slovenia
Tettigetta dimissa Gogala and Gogala 1999: 120, 122, 125, 127, Fig 6 (distribution, comp. note) Slovenia, Mediterranean, Asia
Tettigetta dimissa Schedl 1999b: 827 (listed, distribution, ecology, comp. note) Israel, Greece, Turkey, Armenia, Azerbaijan, Kazakhstan, Siberia, China
Cicadetta dimissa Hua 2000: 61 (listed, distribution) Equals *Cicada dimissa* China, Xinjing, Sichuan, USSR, Syria, Dalmatia

Tettigetta dimissa Gogala and Popov 2000: 8–19, Figs 1–8 (illustrated, song, sonogram, oscillogram, comp. note) Slovenia, Croatia, Macedonia
Tettigetta dimissa Schedl 2000: 260, 263, 265–266, 268–269, Fig 8, Fig 12a, Fig 13a, Plate 1, Fig 11, Map 7 (key, illustrated, distribution, synonymy, listed, comp. note) Equals *Cicadetta dimissa* Slovenia
Tettigetta dimissa Krpač, Trilar and Gogala 2001: 29 (comp. note) Macedonia
Tettigetta dimissa Schedl 2001b: 27 (comp. note)
Tettigetta dimissa Sueur 2001: 44, 46, Tables 1–2 (listed) Equals *Cicadetta dimissa*
Tettigetta dimissa Gogala 2002: 244–245, Fig 7 (song, sonogram, oscillogram, comp. note)
Tettigetta dimissa Gogala and Seljak 2003: 350 (comp. note) Slovenia
Tettigetta dimissa Holzinger, Kammerlander and Nickel 2003: 483–484, 616–617, 633, 638, Plate 42, Fig 5, Tables 2–3 (key, listed, illustrated, comp. note) Europe, Bulgaria, Siberia, Oriental Region
Cicadetta dimissa Tishechkin 2003: 141, 145, Figs 168–171 (oscillogram, comp. note) Chechnya
Tettigetta dimissa Gogala, Trilar and Krpač 2005: 105, 120–121, fig 15 (distribution, comp. note) Macedonia
Tettigetta dimissa Gogala 2006a: 158 (comp. note)
Cicadetta dimissa Boulard 2007b: 97 (comp. note)
Cicada dimissa Boulard 2007b: 98 (type species of *Dimissalna*, comp. note)
Cicadetta dimissa Demir 2007a: 52 (distribution, comp. note) Turkey
Tettigetta dimissa Demir 2008: 477 (distribution) Turkey
Melampsalta dimissa Demir 2008: 477 (listed) Turkey
Cicadetta dimissa Demir 2008: 477 (listed) Turkey
Dimissana dimissa Trilar and Gogala 2008: 30–32, Fig 4 (sonogram, oscillogram, listed, distribution, comp. note) Equals *Tettigetta dimissa* Romania
Dimissalna (= Tettigetta) dimissa Gogala, Drosopoulos and Trilar 2009: 19 (comp. note)
Cicadetta dimissa Kemal and Koçak 2010: 6 (distribution, listed, comp. note) Turkey

***Curvicicada* Chou & Lu, 1997** = *Oedicida* (sic)
Curvicicada Chou, Lei, Li, Lu and Yao 1997: 46, 78, 289, 293, 357, Table 6 (n. gen., described, key, listed, diversity, biogeography) China
Oedicida (sic) Chou, Lei, Li, Lu and Yao 1997: 46, 289, 293, Table 6 (key, diversity, biogeography) China
Curvicicada Villet and Zhao 1999: 322 (comp. note)
Oecidada Villet and Zhao 1999: 322 (comp. note)
Curvicicada Moulds 2005b: 390, 436 (higher taxonomy, listed, comp. note)

C. continuata (Distant, 1888) to *Linguacicada continuata*

C. verna (Distant, 1912) to *Toxala verna*

***C. xinjiangensis* Chou & Lu, 1997** = *Curvicicada buerensis* (sic)
Curvicicada xinjiangensis Chou, Lei, Li, Lu and Yao 1997: 77–79, 364, 366, Fig 8-22, Plate I, Fig 15 (n. sp., type species of *Curvicicada*, described, illustrated, comp. note) China
Curvicicada buerensis (sic) Chou, Lei, Li, Lu and Yao 1997: 357 (comp. note)

***Pinheya* Dlabola, 1963**
Pinheya Moulds 2005b: 390, 436 (higher taxonomy, listed, comp. note)

***P. violacea violacea* (Linnaeus, 1758)** = *Cicada violacea* = *Tettigonia violacea* = *Tettigonia violecea* (sic) = *Melampsalta violacea* = *Cicada hottentota* Olivier, 1790 = *Cicada hottentotta* (sic) = *Cicadetta hottentota*
Pinheya violacea Boulard 1981b: 50–51 (synonymy) Equals *Cicada violacea* Equals *Melampsalta violacea* Equals *Cicada hottentota*
Cicada violacea Boulard 1981b: 46–7 (comp. note) Equals *Cicada hottentota*
Cicada hottentota Boulard 1981b: 46–47 (status, comp. note)

***P. violacea* var. *β* Walker, 1850** = *Cicada hottentota* var. *β* = *Cicadetta violacea* var. *β*

***Hilaphura* Webb, 1979** = *Hylaphura* (sic)
Hilaphura Koçak 1982: 149 (listed, comp. note) Turkey
Hilaphura Nast 1982: 310 (listed)
Hilaphura Boulard 1988f: 41, 43 (synonymy, comp. note)
Hilaphura Boulard 1997c: 110 (synonymy, comp. note)
Hilaphura Kartal 1999: 99 (described, comp. note) Spain
Hilaphura Boulard 2001b: 30, 32 (synonymy, comp. note)
Hilaphura Moulds 2005b: 390, 436 (higher taxonomy, listed, comp. note)
Hilaphura Moulds 2005b: 390, 436 (higher taxonomy, listed, comp. note)
Hilaphura Puissant 2005: 302 (comp. note)

H. segetum (Rambur, 1840) to *Hilaphura varipes* (Waltl, 1837)

***H. varipes* (Waltl, 1837)** = *Cicada varipes* = *Cicadetta varipes* = *Melampsalta varipes* = *Melampsalta (Melampsalta) varipes* = *Melapmpsalta varipes* = *Cicada segetum* Rambur, 1840 = *Cicadatra segetum* =

Cicadetta segetum = *Melampsalta segetum* = *Hilaphura segetum* = *Hylaphura* (sic) *segetum* = *Cicada picta* Germar, 1830 (nec *Cicada picta* Fabricius, 1794) = *Melampsalta picta* (nec Fabricius) = *Cicadetta picta* (nec Fabricius) = *Melampsalta (Melampsalta) picta* (nec Fabricius) = *Cicadetta decorata* Kirkaldy, 1909 = *Melampsalta (Melampsalta) decorata* = *Cicada musiva* var. *caspica* (nec Kolenati)

Hilaphura varipes Boulard 1982f: 197 (n. comb., synonymy, comp. note) Equals *Cicada varipes*
Cicada picta Boulard 1982f: 197 (synonymy, comp. note) Equals *Cicada segetum*
Cicada segetum Boulard 1982f: 197 (synonymy, comp. note) Equals *Cicada picta*
Hilaphura segetum Boulard 1982f: 197 (n. comb., synonymy, comp. note) Equals *Cicada segetum* Equals *Cicada picta*
Cicada segetum Koçak 1982: 149 (type species of *Hilaphura*)
Cicada segetum Nast 1982: 310 (type species of *Hilaphura*)
Hilaphura varipes Boulard 1985a: 33 (distribution, listed) Portugal
Hilaphura varipes Nast 1987: 568–569 (distribution, listed) Portugal, Spain
Cicada varipes Boulard 1988f: 38, 40–41, 43 (type species of *Melampsalta*, comp. note) Equals *Cicada segetum*
Cicada segetum Boulard 1988f: 38, 40 (comp. note)
Hilaphura segetum Boulard 1988f: 40 (comp. note)
Cicadetta decorata Boulard 1988f: 40 (comp. note)
Cicadetta segetum Boulard 1988f: 41 (comp. note)
Cicadatra segetum Boulard 1988f: 41 (comp. note)
Melampsalta varipes Boulard 1988f: 41, 43, 45, 49, Fig 6 (n. comb., comp. note)
Cicada varipes Moulds 1988: 40 (comp. note)
Melampsalta varipes Quartau and Fonseca 1988: 367, 370–371 (synonymy, distribution, comp. note) Equals *Cicada varipes* Equals *Cicada segetum* Equals *Hylaphyra* (sic) *segetum* Portugal, Algeria, France, Italy, Sicily, Spain
Cicada varipes Boulard 1991b: 26–27 (type species of *Melampsalta*) Equals *Melampsalta melampsalta*
Cicada varipes Boulard 1997c: 107, 110, 112 (type species of *Melampsalta*, comp. note) Equals *Cicada segetum*
Cicadetta decorata Boulard 1997c: 110 (comp. note)
Melampsalta varipes Boulard 1997c: 110, 112–113, 117, Plate II, Fig 7 (n. comb., comp. note)
Tettigonia picta Boulard 1997c: 86, 93 (comp. note)
Cicada picta Boulard 1997c: 110 (type species of *Melampsalta*, comp. note) Spain
Cicadetta picta Boulard 1997c: 110 (comp. note)
Cicada segetum Boulard 1997c: 110 (comp. note)
Hilaphura segetum Boulard 1997c: 110 (comp. note)
Cicadetta segetum Boulard 1997c: 110 (comp. note)
Cicadatra segetum Boulard 1997c: 110 (comp. note)
Hilaphura segetum Kartal 1999: 99–102, Fig A_1, Fig B_1, Fig C_1, Fig D_1, Fig E_1, Fig F_1, Fig G_1, Fig H_1, Fig I_1 (described, illustrated, comp. note) Spain
Melampsalta varipes Boulard 2000i: 21 (listed)
Cicada varipes Boulard 2000i: 63 (comp. note)
Cicada segetum Boulard 2000i: 63 (comp. note)
Cicada varipes Boulard 2001b: 28, 30, 32 (type species of *Melampsalta*, comp. note) Equals *Cicada segetum*
Melampsalta varipes Boulard 2001b: 30, 32–33, 37, Plate II, Fig 7 (n. comb., comp. note)
Cicadetta decorata Boulard 2001b: 30 (comp. note)
Cicada segetum Boulard 2001b: 30 (comp. note)
Hilaphura segetum Boulard 2001b: 30 (comp. note)
Cicadetta segetum Boulard 2001b: 30 (comp. note)
Cicadatra segetum Boulard 2001b: 30 (comp. note)
Melampsalta varipes Sueur, Puissant, Simões, Seabra, Boulard and Quartau 2004: 182 (comp. note) Portugal
Melampsalta varipes Puissant 2005: 301–312, Figs 2–5, Figs 8–11, Tables 1–3 (neotype designation, taxonomy, synonymy, illustrated, distribution, habitat, song, sonogram, oscillogram, comp. note) Equals *Cicada segetum* Equals *Cicadetta decorata* Equals *Cicada picta* (nec Fabricius) Equals *Tettigetta argentata* (nec Olivier) Equals *Hilaphura varipes* Equals *Hilaphura segetum* Spain, Portugal
Cicada segetum Puissant 2005: 301 (comp. note)
Cicadetta decorata Puissant 2005: 301 (comp. note)
Cicada varipes Puissant 2005: 305 (type species of *Melampsalta*, comp. note) Equals *Cicada musiva* var. *caspica* (partim)
Melampsalta varipes Puissant and Defaut 2005: 127 (comp. note)
Melampsalta varipes Boulard 2006g: 106, 108 (comp. note)
Melampsalta varipes Puissant 2006: 189 (comp. note)
Hilaphura varipes Puissant and Sueur 2010: 555 (comp. note) Equals *Melampsalta varipes*

***Euryphara* Horváth, 1912**

Euryphara Boulard 1981b: 44 (comp. note)
Euryphara Boulard 1982c: 101 (listed)
Euryphara Boulard 1982f: 193 (listed)
Euryphara Koçak 1982: 149 (listed, comp. note) Turkey
Euryphara Moulds 2005b: 390, 436 (higher taxonomy, listed, comp. note)

Euryphara spp. Puissant and Defaut 2005: 127 (comp. note)
Euryphara spp. Puissant 2006: 189 (comp. note)
Euryphara Kartal 2006: 280 (comp. note)

E. cantans **(Fabricius, 1794)** = *Tettigonia cantans* = *Cicada cantans* = *Melampsalta cantans* = *Melampsalta (Melampsalta) cantans* = *Cicadetta cantans* = *Cicada haematodes* Linnaeus, 1767 (nec Scopoli)
Tettigonia cantans Dallas 1870: 482 (synonymy) Equals *Melampsalta cantans*
Euryphara cantans Drosopoulous 1980: 190 (listed) Greece
Euryphara cantans Boulard 1981a: 37 (comp. note) Algeria
Tettigonia cantans Boulard 1981b: 44 (comp. note) North Africa
Euryphara cantans Boulard 1981b: 50 (synonymy) Equals *Cicada haematodes* Linnaeus (nec *Cicada haematodes* Scopoli) Equals *Tettigonia cantans* North Africa
Tettigonia cantans Koçak 1982: 149 (type species of *Euryphara*)
Euryphara cantans Nast 1987: 568–569 (distribution, listed) Portugal, Spain, France?
Tettigonia cantans Boulard 1988f: 71 (comp. note)
Eurphara cantans Quartau and Fonseca 1988: 367, 370 (synonymy, distribution, comp. note) Equals *Tettigonia cantans* Portugal, Algeria, France, Greece, Morocco, Spain, Tunisia
Euryphara cantans Boulard 1997c: 121 (comp. note) Equals *Cicada haematodes* L. Equals *Tettigonia cantans*
Cicada haematodes L. Boulard 1997c: 121 (comp. note)
Tettigonia cantans Boulard 1997c: 121 (comp. note)
Euryphara cantans Boulard 2001b: 41 (comp. note) Equals *Cicada haematodes* L. Equals *Tettigonia cantans*
Cicada haematodes L. Boulard 2001b: 41 (comp. note)
Tettigonia cantans Boulard 2001b: 41 (comp. note)
Euryphara cantans Sueur 2002b: 124, 129 (listed, comp. note)
Euryphara cantans Kartal 2006: 279–280, Figs 1–7 (described, illustrated, comp. note) Equals *Tettigonia cantans* Tunisia
Tettigonia cantans Kartal 2006: 280 (comp. note)

E. contentei **Boulard, 1982**
Euryphara contentei Boulard 1982f: 191, 193–195, 197, Fig 36, Figs 41–44 (n. sp., described, illustrated, comp. note, listed) Portugal
Euryphara contentei Nast 1987: 568–569 (distribution, listed) Portugal
Euryphara contentei Quartau and Fonseca 1988: 367, 369–370 (distribution, comp. note) Portugal
Euryphara contentei Boulard 1995a: 7–8, 13, 69–71, Fig $S29_1$, Figs 23–24, Fig $S29_2$ (illustrated, song analysis, sonogram, listed, comp. note) Spain
Euryphara contentei Quartau 1995: 37 (ecology, comp. note) Portugal
Euryphara contentei Esteban and Sanchiz 1997: 143 (listed) Iberian Peninsula
Euryphara contentei Sueur 2001: 43, Table 1 (listed)
Euryphara contentei Quartau and Simões 2004: 564–565, 568–571, Figs 1–7 (illustrated, song, sonogram, oscillogram, comp. note) Portugal
Euryphara contentei Puissant 2005: 311 (comp. note)
Euryphara contentei Boulard 2006g: 180 (listed)
Euryphara contentei Gogala and Drosopoulos 2006: 278 (comp. note)
Euryphara contentei Quartau 2008: 267 (habitat, comp. note) Portugal
Euryphara contentei Trilar and Gogala 2010: 16 (comp. note)

E. dubia **(Rambur, 1840)** = *Cicada dubia* = *Melampsalta dubia* = *Melampsalta (Melampsalta) dubia* = *Cicadetta euphorbiae* Fieber, 1876 = *Melampsalta (Melampsalta) euphorbiae* = *Cicadetta dubia*
Cicada dubia Koçak 1982: 149 (comp. note)
Cicadetta dubia Nast 1987: 568–569 (distribution, listed) Spain
Euryphara dubia Puissant and Sueur 2010: Appendiz 2 (n. comb., comp. note) Equals *Cicadetta euphorbiae*

E. ribauti **Boulard, 1982**
Euryphara ribauti Boulard 1982c: 101–103, Figs 1–5 (n. sp., described, illustrated) Morocco

E. undulata **(Waltl, 1837)** = *Cicada undulata* = *Cicadetta undulata* = *Melampsalta (Melampsalta) undulata*
Euryphara undulata Boulard 1982f: 193–194, 197 (n. comb., comp. note, listed)
Cicada undulata Koçak 1982: 149 (comp. note) Equals *Cicada dubia*
Euryphara undulata Nast 1987: 568–569 (distribution, listed) Spain
Euryphara undulata Quartau and Fonseca 1988: 367, 370 (synonymy, distribution, comp. note) Equals *Cicada undulata* Equals *Cicada dubia* Equals *Cicadetta euphorbiae* Portugal, Spain

E. virens **(Herrich-Schäffer, 1835)** = *Cicada virens* = *Cicadetta virens* = *Melampsalta (Melampsalta) virens* = *Tettigetta virens*
Euryphara virens Puissant and Sueur 2010: Appendiz 2 (n. comb., comp. note)

***Saticula* Stål, 1866** = *Sacula* (sic)
Saticula Gómez-Menor 1951: 61 (key) Spain
Saticula Boulard 2002b: 196 (wing morphology, comp. note) Maghreb
Saticula Moulds 2005b: 390, 436 (higher taxonomy, listed, comp. note)

***S. coriaria* Stål, 1866** = *Cicadetta coriaria* = *Cicadetta coriacea* (sic) = *Cicadetta coriascea* (sic) = *Melampsalta coriaria* = *Sacula* (sic) *coriaria* = *Cicadetta violacea* Hagen, 1855 (nec Linnaeus) = *Cicadetta auratiaca* Puton, 1883 = *Cicadetta aurantica* (sic) = *Melampsalta aurantiaca* = *Melampsalta (Melampsalta) aurantiaca* = *Saticula aurantiaca*
Cicadetta coriaria Weber 1931: 114, Fig 113 (scolopodium illustrated, comp. note)
Cicadetta coriaria Weber 1968: 113–114, Figs 95–96 (scolopodium illustrated, comp. note)
Cicada violacea Boulard 1981b: 46 (comp. note) Hagen
Cicadetta coriacea (sic) Boulard 1981b: 46 (comp. note)
Cicadetta violacea Boulard 1981b: 46 (comp. note)
Saticula coriaria Boulard 1981b: 46–47, 51, Fig 9 (synonymy, illustrated, comp. note) Equals *Cicada violacea* Hagen (nec Linnaeus) Equals *Cicadetta coriaria* Equals *Cicadetta aurantiaca* Equals *Saticula aurantiaca* North Africa
Cicadetta aurantiaca Boulard 1981b: 47–48, Fig 10 (illustrated, comp. note)
Saticula coriaria Boulard 1988b: 41 (type species of *Saticula*, comp. note, part of Cicadettini not Tettigomyini) Equals *Cicada violacea* Equals *Pinheya violacea*
Sacula (sic) *coriaria* Sueur 2002b: 128–129 (listed, comp. note)

***S. vayssieresi* Boulard, 1988**
Saticula vayssieresi Boulard 1988b: 41–43, Figs 1–3 (n. sp., described, illustrated, comp. note) Morocco

***Xossarella* Boulard, 1980**
Xossarella Nast 1982: 310 (listed)
Xossarella Moulds 2005b: 390, 436 (higher taxonomy, listed, comp. note)

***X. giovanninae* Boulard, 1980**
Xossarella giovanninae Nast 1982: 310–311 (type species of *Xossarella*, listed) Algeria

***X. marmottani* Boulard, 1980**
Xossarella marmottani Nast 1982: 311 (listed) Algeria

***X. soussensis* Boulard, 1980**
Xossarella soussensis Nast 1982: 311 (listed) Tunisia

***Takapsalta* Matsumura, 1927**
Takapsalta Hayashi 1984: 66 (listed)
Takapsalta Lee 1995: 18, 23, 32, 99–100, 104, 136 (listed, key, described, ecology, distribution, illustrated, comp. note) Korea
Takapsalta Lee, Choe, Lee and Woo 2002: 3 (comp. note)
Takapsalta Lee 2005: 15 (comp. note) Korea
Takapsalta Moulds 2005b: 390, 436 (higher taxonomy, listed, comp. note)
Takapsalta Lee 2008b: 450, 457, Table 2 (listed, comp. note) Korea

***T. ichinosawana* Matsumura, 1927** = *Cicadetta ichinosawana*
Takapsalta ichinosawana Lee 2005: 96 (comp. note)
Takapsalta ichinosawana Lee 2008b: 452, 456, 457 (type species of *Takapsalta*) Ichinosawa
Cicadetta ichinosawana Lee 2008b: 456 (comp. note)

***T. komaensis* Kato, 1944**

Leptopsalta Kato, 1928 to *Kosemia* Matsumura, 1927
L. admirabilis (Kato, 1927) to *Kosemia admirabilis*
L. bifuscata (Liu, 1940) to *Kosemia bifuscata*
L. fuscoclavalis (Chen, 1942) to *Kosemia fuscoclavalis*
L. kishidai Kato, 1932 to *Kosemia admirabilis* (Kato, 1927)
L. phaea Chou & Lei, 1997 to *Kosemia phaea*
L. prominentis Lei & Chou, 1997 to *Kosemia prominentis*
L. radiator (Uhler, 1896) to *Kosemia radiator*
L. rubicosta Chou & Lei, 1997 to *Kosemia rubicosta*
L. yamashitai (Esaki & Ishihara, 1950) to *Kosemia yamashitai*
L. yezoensis (Matsumura, 1898) to *Kosemia yezoensis*

***Kosemia* Matsumura, 1927** = *Leptopsalta* Kato, 1928 = *Leptosalta* (sic) = *Loptopsalta* (sic) = *Karapsalta* Matsumura, 1931
Kosemia Koçak 1982: 150 (listed, comp. note) Turkey
Kosemia Lee 1995: 18, 23, 32, 99, 104–105, 136 (listed, key, described, ecology, distribution, illustrated, comp. note) Equals *Leptopsalta* Equals *Karapsalta* Korea
Leptopsalta Lee 1995: 32, 107–108 (comp. note) Korea
Leptopsalta Chou, Lei, Li, Lu and Yao 1997: 46–47, 293, 297, 306, 311, 324, 326, 347, Table 6, Tables 10–12 (key, described, listed, diversity, biogeography) China
Kosemia Lee, Choe, Lee and Woo 2002: 10 (comp. note)
Leptopsalta Lee, Choe, Lee and Woo 2002: 10 (comp. note)
Leptopsalta Moulds 2005b: 390, 436 (higher taxonomy, listed, comp. note)

Kosemia Lee 2008b: 445–446, 457–458, 463, Tables 1–2 (status, listed, comp. note) Equals *Leptopsalta* n. syn. Korea

Leptopsalta Lee 2008b: 457, Table 2 (status, listed, comp. note) Korea

Cicadetta (sic) Osaka Museum of Natural History 2008: 326 (listed) Japan

***K. admirabilis* (Kato, 1927)** = *Melampsalta admirabilis* = *Leptopsalta admirabilis* = *Leptopsalta apmirabilis* (sic) = *Leptopsalta abmirabilis* (sic) = *Leptosalta* (sic) *admirabilis* = *Cicadetta admirabilis* = *Leptopsalta admirabilis kishidae* Kato, 1932 = *Leptopsalta admirabilis* var. *kishidai* = *Leptopsalta kishidae* = *Melampsalta radiator* Liu (nec Uhler)

Leptopsalta admirabilis Doi 1932: 43 (listed, comp. note) Korea

Cicadetta admirabilis Hayashi 1984: 67 (comp. note)

Leptopsalta admirabilis Lei, Chou and Li 1994: 56, Fig 11 (oscillogram, comp. note)

Kosemia kishidai Lee 1995: 18, 23, 105, 110–111, 136, 139 (listed, key, synonymy, ecology, distribution, illustrated, comp. note) Equals *Leptopsalta admirabilis* var. *kishidai* Equals *Leptopsalta kishidai* Korea

Leptopsalta admirabilis Lee 1995: 32–33, 35–46, 105, 107–108, 110, 134, 137–138 (listed, key, synonymy, ecology, distribution, illustrated, comp. note) Equals *Leptopsalta kishidai* Korea

Leptopsalta admirabilis var. *kishidai* Lee 1995: 37, 39–40, 42, 110 (comp. note) Korea

Leptopsalta kishidai Lee 1995: 42–43, 45–46, 110 (comp. note) Korea

Leptopsalta admirabilis Chou, Lei, Li, Lu and Yao 1997: 47–49, 51–52, 306, 311–312, 324, 326, 328, 334, 347, Fig 8–1, Plate I, Fig 4, Tables 10–12 (key, described, illustrated, synonymy, song, oscillogram, spectrogram, comp. note) Equals *Melampsalta admirabilis* China

Cicadetta admirabilis Lee 1999: 1, 10–11 (rev. status, distribution, song, illustrated, listed, comp. note) Equals *Melampsalta? admirabilis* Equals *Leptopsalta admirabilis* Equals *Leptopsalta admirabilis var. kishidai* Equals *Leptopsalta kishidai* Equals *Kosemia kishidai* n. syn. Korea, China

Kosemia kishidai Lee 1999: 1 (status)

Leptopsalta admirabilis Lee 1999: 1 (comp. note)

Kosemia kishidai Lee 1999: 1 (comp. note)

Leptopsalta admirabilis Hua 2000: 62 (listed, distribution) Equals *Melampsalta admirabilis* China, Liaoning, Shaanxi, Ningxia, Gansu, Korea

Cicadetta admirabilis Lee 2000: 4–5, Figs 1–6 (illustrated, comp. note) Korea

Leptopsalta admirabilis var. *kishidai* Lee 2000: 4 (comp. note) Korea

Leptosalta (sic) *admirabilis* Sueur 2001: 43, Table 1 (listed)

Cicadetta admirabilis Lee, Choe, Lee and Woo 2002: 4–10, Figs 1–3, Tables 1–3 (gene sequence, phylogeny, comp. note) China, Korea

Cicadetta admirabilis Lee 2005: 17, 106–107, 158, 166–167, 176–177 (described, illustrated, synonymy, song, distribution, key, comp. note) Equals *Melampsalta? admirabilis* Equals *Leptopsalta admirabilis* Equals *Leptopsalta admirabilis* var. *kishidai* Equals *Cicadetta kishidai* Korea, China

Cicadetta admirabilis var. *kishidai* Lee 2005: 106 (comp. note)

Leptopsalta admirabilis Tian, Yuan and Zhang 2006: 243 (reproductive system, comp. note) China

Cicadetta admirabilis Lee 2008b: 445 (status) Korea

Kosemia admirabilis Lee 2008b: 446, 457–458, 463, Tables 1–2 (n. comb., key, synonymy, distribution, listed, comp. note) Equals *Melampsalta? admirabilis* Equals *Leptopsalta admirabilis* Equals *Cicadetta admirabilis* Equals *Leptopsalta admirabilis* var. *kishidai* Equals *Leptopsalta kishidai* Equals *Kosemia kishidai* Korea, China

Melampsalta admirabilis Lee 2008b: 457 (comp. note)

Leptopsalta admirabilis Lee 2008b: 457 (comp. note)

***K. bifuscata* (Liu, 1940)** = *Melampsalta bifuscata* = *Leptopsalta bifuscata*

Leptopsalta bifuscata Chou, Lei, Li, Lu and Yao 1997: 47, 49–50, Fig 8–2, Plate I, Fig 1 (key, n. comb., described, illustrated, synonymy, comp. note) Equals *Melampsalta bifuscata* China

Leptopsalta bifuscata Hua 2000: 62 (listed, distribution) Equals *Melampsalta bifuscata* China, Qinghai, Hubei, Sichuan

Leptopsalta bifuscata Sun, Watson, Zheng, Watson and Liang 2009: 3148–3149, 3151–3152, Fig 1D, Table 1 (illustrated, wing microstructure, comp. note) China, Hebei

***K. fuscoclavalis* (Chen, 1942)** = *Melampsalta fuscoclavalis* = *Cicadetta fuscoclavalis* = *Leptopsalta fuscoclavalis* = *Leptopsalta fuscoflavelis* (sic) = *Leptopsalta juscoclavalis* (sic) = *Melampsalta fuscoclavalis chungnanshana* Chen, 1943

Leptopsalta fuscoclavalis Chou, Lei, Li, Lu and Yao 1997: 47, 52–53, Fig 8–5 (key, n. comb., described, illustrated, synonymy, comp. note) Equals *Melampsalta fuscoclavalis* Equals *Cicadetta fuscoclavalis* Equals *Melampsalta fuscoclavalis changnanshana* China

Leptopsalta fuscoflavelis (sic) Chou, Lei, Li, Lu and Yao 1997: 57 (comp. note) China

Leptopsalta juscoclavalis (sic) Chou, Lei, Li, Lu and Yao 1997: 358 (comp. note)

Leptopsalta fuscoclavalis Hua 2000: 62 (listed, distribution) Equals *Melampsalta fuscoclavalis* Equals *Melampsalta fuscoclavalis chungnanshana* China, Shaanxi

K. kishidai (Kato, 1932) to *Kosemia admirabilis* (Kato, 1927)

***K. phaea* (Chou & Lei, 1997)** = *Leptopsalta phaea*

Leptopsalta phaea Chou, Lei, Li, Lu and Yao 1997: 47, 55–57, 357, 365, Fig 8-8, Plate I, Fig 5 (key, n. sp., described, illustrated, comp. note) China

***K. prominentis* (Lei & Chou, 1997)** = *Leptopsalta prominentis*

Leptopsalta prominentis Chou, Lei, Li, Lu and Yao 1997: 47, 55–57, 357, 365, Fig 8-7, Plate I, Fig 9 (key, n. sp., described, illustrated, comp. note) China

***K. radiator* (Uhler, 1896)** = *Melampsalta radiator* = *Melampsaltria* (sic) *radiator* = *Cicadetta radiator* = *Melampsalta (Melampsalta) radiator* = *Tibicen radiator* = *Leptopsalta radiator* = *Loptopsalta* (sic) *radiator*

Melampsalta radiator Henshaw 1903: 230 (listed) Japan

Leptopsaltra radiator Isaki 1933: 381 (listed, comp. note) Japan

Melampsalta radiator Ohgushi 1954: 11 (emergence times, ecology, comp. note) Japan

Cicadetta radiator Hayashi 1982b: 188–190, Table 1 (distribution, comp. note) Japan

Cicadetta radiator Hayashi 1984: 29, 66–67, 69, Figs 161–164 (key, described, illustrated, synonymy, comp. note) Equals *Melampsalta radiator* Equals *Kosemia radiator* Equals *Leptopsalta radiator* Japan

Melapsalta radiator Hayashi 1984: 66 (type species of *Leptopsalta*)

Leptopsalta radiator Wang and Zhang 1987: 296 (key, comp. note) China

Leptopsalta radiator Wang and Zhang 1992: 120 (listed, comp. note) China

Kosemia radiator Lee 1995: 19, 23, 28, 131–133, 135, 138 (comp. note) Equals *Leptopsalta radiator* Equals *Melampsalta radiator*

Leptopsalta radiator Lee 1995: 35, 37, 40, 45–46, 107, 132 (comp. note)

Melampsalta radiator Lee 1995: 132 (comp. note)

Melampsalta radiator Chou, Lei, Li, Lu and Yao 1997: 47 (type species of *Leptopsalta*)

Leptopsalta radiator Chou, Lei, Li, Lu and Yao 1997: 53, 57 (described, synonymy, comp. note) Equals *Melampsalta radiator* Equals *Tibicen radiator* China

Melampsalta radiator Lee 1999: 10 (type species of *Leptopsalta*) Japan

Leptopsalta radiator Sanborn 1999a: 52 (type material, listed)

Leptopsalta radiator Hua 2000: 62 (listed, distribution, hosts) Equals *Melampsalta radiator* China, Jilin, Liaoning, Shanxi, Qinghai, Hubei, Korea, Japan

Cicadetta radiator Ohara and Hayashi 2001: 1 (comp. note) Japan

Cicadetta radiator Sueur 2001: 46, Table 2 (listed)

Cicadetta radiator Lee, Choe, Lee and Woo 2002: 4–10, Figs 1–3, Tables 1–3 (gene sequence, phylogeny, comp. note) Japan

Cicadetta radiator Saisho 2004: 1 (emergence, comp. note) Japan

Melampsalta radiator Lee 2005: 176 (type species of *Leptopsalta*)

Cicadetta radiator Inoue, Goto, Makino, Okabe, Okochi et al. 2007: 251, Table 2 (trap collection, comp. note) Japan

Cicadetta radiator Shiyake 2007: 8, 105, 107, Fig 184 (illustrated, distribution, listed, comp. note) Japan

Melampsalta radiator Lee 2008b: 457 (type species of *Leptosalta*, comp. note) Japan

Kosemia radiator Lee 2008b: 457, Table 2 (comp. note)

Leptopsalta radiator Lee 2008b: 457 (comp. note)

Cicadetta radiator Osaka Museum of Natural History 2008: 326 (listed, acoustic behavior, ecology, comp. note) Japan

Kosemia radiator Saisho 2010a: 57 (ecolsion, comp. note) Japan

Kosemia radiator Saisho 2010b: 59 (ecolsion, comp. note) Japan

***K. rubicosta* (Chou & Lei, 1997)** = *Leptopsalta rubicosta*

Leptopsalta rubicosta Chou, Lei, Li, Lu and Yao 1997: 47, 53–55, 357, 365, Fig 8-6, Plate I, Fig 6 (key, n. sp., described, illustrated, comp. note) China

***K. yamashitai* (Esaki & Ishihara, 1950)** = *Melampsalta yamashitai* = *Leptopsalta yamashitai* = *Leptosalta* (sic) *yamashitai*

Leptopsalta yamashitai Lei, Chou and Li 1994: 56, Fig 10 (oscillogram, comp. note)

Leptopsalta yamashitai Chou, Lei, Li, Lu and Yao 1997: 47, 51–52, 306, 311, 324, 326, 328, 334, 347, 358, Fig 8-4, Plate I, Fig 3, Tables 10–12 (key, n. comb., described, illustrated, synonymy, song, oscillogram, spectrogram, comp. note) Equals *Melampsalta yamashitai* China

Leptopsalta yamashitai Hua 2000: 62 (listed, distribution) Equals *Melampsalta yamashitai*

China, Shanxi, Gansu, Ningxia, Shaanxi, Zhejiang

Leptosalta (sic) *yamashitai* Sueur 2001: 43, Table 1 (listed)

***K. yezoensis* (Matsumura, 1898)** = *Melampsaltria* (sic) *yezoensis* = *Melampsalta yezoensis* = *Melampsalta (Melampsalta) yezoensis* = *Melampsalta yezoensis gigas* = *Cicadetta yezoensis* = *Cicadetta jezoensis* (sic) = *Leptopsalta yezoensis* = *Cicadetta sachalinensis* Matsumura, 1917 = *Melampsalta sachalinensis* = *Kosemia sachalinensis* = *Karapsalta sachalinensis* = *Cicada orni* (nec Linnaeus) = *Melampsalta konoi* (nec Kato)

Cicadetta sachaliensis Koçak 1982: 150 (type species of *Kosemia*)

Cicadetta yezoensis Hayashi 1984: 29, 67–69, Figs 165–169 (key, described, illustrated, synonymy, comp. note) Equals *Melampsaltria* (sic) *yezoensis* Equals *Cicadetta jezoensis* (sic) Equals *Kosemia yezoensis* Equals *Cicadetta sachaliensis* (partim) Equals *Melampsalta yezoensis gigas* Equals *Melampsalta konoi* Japan

Cicadetta sachalinensis Hayashi 1984: 66 (type species of *Kosemia*)

Cicadetta yezoensis Popov and Sergeeva 1987: 681–689, Figs 1–5, Table (song, oscillogram, illustrated, hearing, comp. note) USSR

Cicadetta yezoensis Anufriev and Emelyanov 1988: 314–316, Fig 236.1, Figs 237.9–237.13 (illustrated, genitalia illustrated, key) Soviet Union

Cicadetta yezoensis Popov 1990a: 302–303, Fig 2, Fig 4 (comp. note)

Cicadetta yezoensis Tautz 1990: 105–107, Fig 12, Fig 13B, Fig 14 (song frequency, hearing sensitivity, comp. note)

Cicadetta yezoensis Popov, Aronov and Sergeeva 1991: 281 (comp. note)

Cicadetta yezoensis Popov, Arnov and Sergeeva 1992: 151 (comp. note)

Kosemia yezoensis Lee 1995: 18, 23, 32, 105–110, 134, 136, 139 (listed, key, synonymy, ecology, distribution, illustrated, comp. note) Equals *Cicadetta yezoensis* Equals *Melampsalta yezoensis* Equals *Melampsalta? admirabilis* Equals *Melampsaltria* (sic) *yezoensis* Equals *Leptopsalta admirabilis* n. syn. Korea

Melampsalta? admirabilis Lee 1995: 32, 105, 107 (comp. note)

Melampsalta sachalinensis Lee 1995: 34, 37–40, 42, 44, 106 (comp. note)

Melampsalta yezoensis Lee 1995: 41, 43, 106 (comp. note)

Melampsalta saghalinensis (sic) Lee 1995: 45 (comp. note)

Cicadetta yezoensis Lee 1995: 45–46, 106 (comp. note)

Kosemia sachalinensis Lee 1995: 104 (type species of *Kosemia*) Equals *Kosemia yezoensis*

Melampsaltria (sic) *yezoensis* Lee 1995: 45–46, 106 (comp. note)

Cicadetta sachalinensis Lee 1995: 106–107 (comp. note)

Leptopsalta yezoensis Chou, Lei, Li, Lu and Yao 1997: 47, 50–52, 358, Fig 8–3 (key, n. comb., described, illustrated, synonymy, comp. note) Equals *Melamsaltria* (sic) *yezoensis* Equals *Melampsalta yezoensis* Equals *Cicadetta yezoensis* Equals *Cicadetta sachalinensis* Equals *Melampsalta yezoensis gigas* Equals *Melampsalta konoi* China

Cicadetta yezoensis Lee 1999: 1, 10–12 (n. comb., distribution, song, illustrated, listed, comp. note) Equals *Melampsalta yezoensis* Equals *Melampsaltria* (sic) *yezoensis* Equals *Cicadetta sachalinensis* Equals *Melampsalta sachalinensis* Equals *Kosemia yezoensis* Korea, China, Japan, Russia

Kosemia yezoensis Lee 1999: 1 (comp. note)

Melampsalta sachalinensis Lee 1999: 10 (type species of *Kosemia*) Saghalien

Leptopsalta yezoensis Hua 2000: 62 (listed, distribution) Equals *Melampsalta yezoensis* Equals *Melampsalta yezoensis gigas* China, Heilongjiang, Loaoning, Korea, Mongolia, USSR, Japan

Cicadetta yezoensis Lee 2000: 4 (comp. note) Korea

Cicadetta yezoensis Sueur 2001: 43, Table 1 (listed)

Cicadetta yezoensis Lee, Choe, Lee and Woo 2002: 4–10, Figs 1–3, Tables 1–3 (gene sequence, phylogeny, comp. note) Russia, China, Korea

Cicadetta yezoensis Lee 2003b: 8 (comp. note)

Cicada orni (sic) Lee, Jin, Yoo and Lee 2004: 7668, Fig 4 (wing surface, illustrated) Korea

Cicadetta yezoensis Saisho 2004: 3 (comp. note) Japan

Cicadetta yezoensis Lee 2005: 17, 108–111, 159, 167, 177 (described, illustrated, synonymy, song, distribution, key, comp. note) Equals *Melampsalta yezoensis* Equals *Melamsaltria* (sic) *yezoensis* Equals *Kosemia yezoensis* Equals *Cicadetta sachalinensis* Equals *Melampsalta sachalinensis* Korea, China, Russia, Japan

Melampsalta sachalinensis Lee 2005: 108, 176 (type species of *Kosemia*, comp. note)

Cicadetta yezoensis Shiyake 2007: 8, 105, 107, Fig 187 (illustrated, distribution, listed, comp. note) Japan, Korea, China, Mongolia, Russia

Cicadetta yezoensis Lee 2008b: 445, 457 (status, comp. note) Korea

Kosemia yezoensis Lee 2008b: 446, 451, 457–458, 464, Fig 2B, Tables 1–2 (n. comb., illustrated, key, synonymy, distribution, listed, comp. note) Equals *Melampsalta yezoensis* Equals *Melampsaltria* (sic) *yezoensis* Equals *Cicadetta yezoensis* Equals *Cicadetta sachalinensis* Equals *Melampsalta sachalinensis* Korea, China, Russia, Japan

Melampsalta sachalinensis Lee 2008b: 457 (comp. note)

Melampsalta yezoensis Lee 2008b: 457 (comp. note)

Cicadetta sachalinensis Lee 2008b: 458 (type species of *Kosemia*) Saghalien

Cicadetta yezoensis Osaka Museum of Natural History 2008: 326 (listed, acoustic behavior, ecology, comp. note) Japan

Cicada orni (sic) Liu, Chu and Ding 2010: 277, Fig 2 (illustrated, wing anatomy, comp. note)

***Melampsalta* Kolenati, 1857** = *Cicada (Melampsalta)* = *Melampsalta (Melampsalta)* = *Cicadetta (Melampsalta)* = *Cicadetta (Melampsarta)* (sic) = *Melampsaltria* (sic) = *Melampalta* (sic) = *Melampsaltera* (sic) = *Melamosalta* (sic) = *Malampsalta* (sic) = *Melampsata* (sic) = *Parasemia* Matsumura, 1927 (nec *Parasemia* Hübner, 1820) = *Parasemoides* Strand, 1928

Melampsalta Dallas 1870: 482 (listed)

Melampsalta Swinton 1880: 18 (comp. note)

Melampsalta Hudson 1904: 98 (predation) New Zealand

Melampsalta Walker 1920: cxxxi (diversity, comp. note) New Zealand

Melampsalta Doi 1932: 65 (listed, comp. note) Equals *Melamthalta* (sic)

Melampsalta Dumbleton 1966: 977 (comp. note) New Zealand

Melampsalta Wu 1935: 27 (listed) Equals *Cicadetta* Equals *Tettigetta* Equals *Pauropsalta* Equals *Urabunana* Equals *Helptoglena* Equals *Kosemia* Equals *Parasemia* Equals *Karapsalta* China

Melampsalta Varley 1939: 98–99 (comp. note)

Melampsalta Gómez-Menor 1951: 61, 67, 69 (key, comp. note) Spain

Melampsalta Miller 1952: 55 (listed) New Zealand

Melampsalta Evans 1956: 87 (comp. note) New Zealand

Melampsalta sp. Greenup 1964: 23 (comp. note) Australia, New South Wales

Melampsalta Fleming 1966: 1 (comp. note) New Zealand

Melampsalta Grant-Taylor 1966: 347 (comp. note) New Zealand

Melampsalta sp. Chaudhry, Chaudry and Khan 1970: 104 (listed, comp. note) Pakistan

Melampsalta spp. Crowhurst 1970: 123 (listed, predation, comp. note) New Zealand

Melampsalta sp. Douglas 1970: 95 (predation) New Zealand

Melampsalta sp. Coleman 1977: 109, Table 2 (listed, predation, comp. note) New Zealand

Melampsalta Boulard 1981a: 41 (comp. note)

Melampsalta Boulard 1981b: 46, 48 (comp. note)

Melampsalta Hamilton 1981: 962, Fig 14 (head illustrated, comp. note)

Melampsalta Osssiannilsson 1981: 223 (described, synonymy)

Melampsalta Koçak 1982: 150 (listed, comp. note) Turkey

Melampsalta Fleming 1984: 201 (comp. note)

Melampsalta Hayashi 1984: 66 (comp. note)

Melampsalta sp. Goodwin 1986: 8 (grape pest) New South Wales

Melampsalta Boulard 1988f: 37–43, 45, 71 (history, synonymy, comp. note) Equals *Hilaphura*

Melampsalta Moulds 1988: 39–40 (status, comp. note)

Melampsalta Ewart 1989b: 289 (comp. note)

Melampsalta Boulard 1991b: 25–27 (comp. note)

Melampsalta Lee 1995: 100 (comp. note)

Melampsalta Mirzayans 1995: 19 (listed) Iran

Melampsalta Boer and Duffels 1996b: 309 (comp. note)

Melampsalta Boulard 1997a: 185 (comp. note)

Melampsalta Boulard 1997c: 107–113, 117, 121 (history, synonymy, comp. note) Equals *Hilaphura*

Melampsalta Chou, Lei, Li, Lu and Yao 1997: 76, 358, 364 (comp. note)

Melampsalta Poole, Garrison, McCafferty, Otte and Stark 1997: 251, 331 (listed) Equals *Toxopeusella* Equals *Parasemia* North America

Melampsalta Boer 1999: 137 (comp. note)

Melampsalta Villet and Zhao 1999: 322 (comp. note)

Melampsalta Boulard 2000i: 21 (listed)

Melampsalta Boulard 2001b: 27–32, 37, 41 (history, synonymy, comp. note) Equals *Hilaphura*

Melampsalta Cooley 2001: 758 (comp. note) New Zealand

Melampsalta Cooley and Marshall 2001: 847 (comp. note) New Zealand

Melampsalta Kondratieff, Ellingson and Leatherman 2002: 45 (comp. note)

Melampsalta Moulds 2005b: 390, 436 (higher taxonomy, listed, comp. note)

Melampsalta Puissant 2005: 302, 305–306, 309 (comp. note) Equals *Cicadetta* (partim) Equals *Cicadatra* (partim) Equals *Hilaphura*

Melampsalta Londt 2006: 325 (predation) South Africa
Melampsalta Smith and Hasiotis 2008: 504 (comp. note)
Melampsalta Lee 2008b: 457, Table 2 (listed, comp. note) Korea
Melampsalta Ahmed and Sanborn 2010: 43 (distribution) Europe, Asia, Africa, Australia
Melampsalta Mozaffarian and Sanborn 2010: 76, 82 (distribution, listed, comp. note) Europe, Asia, Africa, Australia
Melampsalta Puissant and Sueur 2010: 555 (comp. note)

***M. aethiopica* Distant, 1905** = *Cicadetta aethiopica*

***M. albeola* (Eversmann, 1859)** = *Cicada albeola* = *Cicadetta albeola* = *Melampsalta (Melampsalta) albeola* = *Cicadetta (Melampsalta) albeola*
Melampsalta albeola Nast 1987: 568–569 (distribution, listed) Russia
Cicadetta albeola Chelpakova 1991: 68 (comp. note) Kazakhstan, Kyrghyzstan

***M. artensis* (Montrouzier, 1861)** = *Cicada artensis*

***M. bifuscata* Liu, 1940** = *Melampsalta bifasciata* (sic) = *Cicadetta bifasciata* (sic)
Melampsalta bifasciata (sic) Liu 1992: 41–42 (comp. note)
Cicadetta bifasciata (sic) Liu 1992: 43 (comp. note)

***M. cadisia* (Walker, 1850)** = *Cicada cadisia* = *Cicada variegata* Olivier, 1790 = *Melampsalta variegata* = *Cicada pellucida* Germar, 1830 = *Cicada ruficollis* (nec Thunberg) = *Melampsalta ruficollis* (nec Thunberg)

M. capistrata Ashton, 1912 to *Taurella forresti*

***M. caspica* (Kolenati, 1857)** = *Cicada (Melampsalta) musiva caspica* = *Cicadetta musiva caspica* = *Cicada musiva caspica* = *Melampsalta musiva caspica* = *Melampsalta (Melampsalta) musiva caspica* = *Melampsalta musiva* var. *caspica* = *Cicadetta caspica* = *Melampsalta caspia* (sic) = *Cicadetta musiva* = *Melampsalta (Cicadetta) caspica*
Melampsalta (Cicadetta) caspica Mirzayans, Hashemi, Borumand, Zairi and Rajabi 1976: 111 (distribution, listed) Iran
Melampsalta caspica Boulard 1981a: 41 (type species of genus *Melampsalta*)
Melampsalta caspica Dlabola 1981: 207 (distribution, comp. note) Iran
Melampsalta caspia (sic) Emeljanov 1981a: 743, Figs 29–30 (endopleural structures of mesothorax, illustrated, comp. note)
Melampsalta caspia (sic) Emeljanov 1981b: 25, Figs 29–30 (endopleural structures of mesothorax, illustrated, comp. note)
Melampsalta musiva caspica Koçak 1982: 150 (type species of *Melampsalta*)
Melampsalta caspica Anufriev and Emelyanov 1988: 15 Fig 3.1 (leg) Soviet Union
Cicada caspica Boulard 1988f: 39, 41 (type species of *Melampsalta*, comp. note)
Melampsalta musiva var. *caspica* Moulds 1988: 40 (type species of *Melampsalta*, comp. note)
Melampsalta caspica Moulds 1988: 40 (type species of *Melampsalta*, comp. note)
Melampsalta musiva var. *caspica* Boulard 1991b: 25 (comp. note)
Cicada caspica Boulard 1991b: 26 (comp. note)
Melampsalta caspica Mirzayans 1995: 19 (listed, distribution) Iran
Cicada caspica Boulard 1997c: 108 (type species of *Melampsalta*, comp. note)
Cicada caspica Boulard 2001b: 28, 30 (type species of *Melampsalta*, comp. note)
Melampsalta caspica Puissant 2005: 305 (type species of *Melampsalta*, comp. note)
Cicada musiva var. *caspica* Puissant 2005: 305 (comp. note)
Cicada (Melampsalta) musiva caspica Ahmed and Sanborn 2010: 43 (type species of genus *Melampsalta*)
Melampsalta caspica Mozaffarian and Sanborn 2010: 82 (synonymy, listed, distribution) Equals *Cicada (Melampsalta) musiva caspica* Iran, Turkmenia, Turkey, Mongolia

***M. charharensis* Kato, 1938**

M. depicta Distant, 1920 to *Myersalna depicta*

M. dubia (Rambur, 1840) to *Euryphara dubia*

M. dumbeana Distant, 1920 to *Myersalna dumbeana*

M. expansa Haupt, 1918 = *Cicadetta expansa*

M. eyrei Distant, 1882 to *Palapsalta eyrei*

***M. fraseri* China, 1938** = *Cicadetta fraseri*
Melampsalta fraseri Mirzayans 1995: 19 (listed, distribution) Iran
Melampsalta fraseri Mozaffarian and Sanborn 2010: 82 (listed, distribution) Iran, Iraq

***M. germaini* Distant, 1906** = *Cicadetta germaini* = *Germalna* (sic) *germaini*
Germalna (sic) *germaini* Boulard 2006g: 131–132, 143, 181, Fig 104 (sonogram, listed, comp. note) New Caledonia

M. sp. nr. *incepta* Greenup, 1964 to *Yoyetta* sp. nr. *incepta*

M. infuscata Goding & Froggatt, 1904 to *Pauropsalta infuscata*

M. isshikii Kato, 1926 to *Tettigetta isshikii*

***M. kewelensis* Distant, 1907** = *Melampsalta kewellensis* (sic) = *Cicadetta kewelensis* = *Cicadetta kewellensis* (sic)

***M. kulingana* Kato, 1938**

M. sp. nr. *labeculata* Greenup, 1964 to *Galanga* sp. nr. *labeculata*

***M. leucoptera leucoptera* (Germar, 1830)** = *Cicada leucoptera* = *Melampsalta leucoptera* = *Melampsaltera* (sic) *leucoptera* = *Cicada fusconervosa* Stål, 1855 = *Cicadetta fusconervosa*
Melampsalta leucoptera Chawanji, Hodson and Villet 2006: 374–382, 385–386, Fig 1, Fig 6,

Fig 10, Fig 17, Figs 39–56, Tables 1–6 (sperm morphology, listed, comp. note) South Africa

***M. leucoptera* var. *a* Stål, 1866**

***M. leucoptera* var. *b* Stål, 1866**

***M. literata* (Distant, 1888)** = *Cicadetta literata* = *Cicadetta (Melampsalta) literata*

Melampsalta literata Ahmed and Sanborn 2010: 43 (synonymy, distribution) Equals *Cicadetta literata* Pakistan, India

***M. lobulata* Fieber, 1876** = *Cicadetta lobulata* = *Melampsalta (Melampsalta) lobulata*

Cicadetta lobulata Nast 1987: 568–569 (distribution, listed) Ukraine

***M. mogannia* (Distant, 1905)** = *Quintilia mogannia* = *Lycurgus mogannia* = *Lycurgus monganius* (sic) = *Cicadetta mogannia* = *Quintilia kozanensis* Ouci, 1938

Melampsalta mogannia Wu 1935: 28 (listed) Equals *Quintilia mogannia* China, Soochow

Cicadetta mogannia Chou, Lei, Li, Lu and Yao 1997: 66 (listed, synonymy, comp. note) Equals *Quintilia mogannia* China

Cicadetta mogannia Hua 2000: 61 (listed, distribution) Equals *Quintilia mogannia* Equals *Lycurgus mongannius* (sic) China, Anhui, Zhejiang, Jiangsu, Hong Kong, Yunnan, Xizang

Quintilia kozanensis Hua 2000: 64 (listed, distribution) China, Anhui, Zhejiang

***M. mokanshanensis* Ouchi, 1938** = *Cicadetta mokanshanensis*

M. moluccana (Kirkaldy, 1909) to *Toxopeusella moluccana*

M. musiva (Germar, 1930) to *Tettigetta musiva*

***M. neocruentata* Liu, 1940** = *Melamosalta* (sic) *neocruentata* = *Cicadetta neocruentata*

Melamosalta (sic) *neocruentata* Liu 1992: 41 (comp. note)

Melampsalta neocruentata Liu 1992: 42 (comp. note)

Cicadetta neocruentata Liu 1992: 42 (comp. note)

Melampsalta neocruentata Hua 2000: 63 (listed, distribution) China

***M. occidentalis* (Schumacher, 1922)** = *Cicadetta (Melampsalta) occidentalis* = *Cicadetta occidentalis*

Melampsalta occidentalis Boulard 1981a: 37 (comp. note) Algeria

M. rosacea Distant, 1892 to *Ueana rosacea*

M. rubea Goding & Froggatt, 1904 to *Pauropsalta rubea*

M. sachalinensis (Matsumura, 1817) to *Kosemia yezoensis*

M. sarasini Distant, 1912 to *Poviliana sarasini*

M. segetum (Rambur, 1840) to *Hilaphura varipes* (Waltl, 1837)

***M. sinuatipennis* (Oshanin, 1906)** = *Cicadetta sinuatipennis* = *Cicadetta sinautipennis* (sic) = *Melampsalta (Melampsalta) sinuatipennis* = *Cicadetta (Melampsalta) sinuatipennis*

Melampsalta sinuatipennis Dlabola 1981: 207 (distribution, comp. note) Iran

Cicadetta sinuatipennis Popov 1981: 271–279, Figs 1–7, Table 1 (illustrated, sound production, oscillogram, hearing, comp. note) Turkmenistan

Cicadetta sinuatipennis Hutchins and Lewis 1983: 195 (hearing organ, comp. note)

Cicadetta sinuatipennis Michelsen and Larsen 1985: 549 (comp. note)

Cicadetta situatipennis Popov 1985: 71, 80, 82, 181, 183, Fig 24 (acoustic system, auditory system, comp. note)

Cicadetta sinuatipennis Popov, Aronov and Sergeeva 1985a: 461 (comp. note)

Cicadetta sinuatipennis Popov, Aronov and Sergeeva 1985b: 297 (comp. note)

Cicadetta sinuatipennis Villet 1987a: 269 (comp. note) Russia

Cicadetta sinnuatipennis Ewing 1989: 38–39, Fig 2.11 (stridulatory apparatus illustrated, comp. note)

Cicadetta sinuatipennis Huber, Kleindienst, Moore, Schildberger and Weber 1990: 219 (comp. note)

Cicadetta sinuatipennis Popov 1990a: 302, 304, Fig 2 (comp. note)

Cicadetta sinuatipennis Tautz 1990: 105, Fig 12 (song frequency, comp. note)

Cicadetta sinuatipennis Popov, Aronov and Sergeeva 1991: 281, 294 (comp. note)

[Cicadetta] sinuatipennis Boulard 1992c: 64 (comp. note)

Cicadetta sinuatipennis Popov, Arnov and Sergeeva 1992: 151, 165 (comp. note)

Cicadetta sinuatipennis Fonseca and Popov 1994: 350 (comp. note)

Cicadetta sinuatipennis Gogala and Trilar 1998a: 8 (comp. note) Asia

Cicadetta sinuatipennis Sueur 2001: 43, Table 1 (listed)

Cicadetta sinuatipennis Gogala and Trilar 2003: 9 (comp. note) Asia

Cicadetta sinuatipennis Boulard 2005d: 343 (comp. note)

Melampsalta sinuatipennis Mozaffarian and Sanborn 2010: 82 (synonymy, listed, distribution, comp. note) Equals *Cicadetta sinuatipennis* Iran, Transcaspia, Turkey, Kazakhstan, USSR

***M. soulii soulii* (Distant, 1905)** = *Quintilia soulii* = *Melampsalta soulii* = *Cicadetta soulii* = *Lycurgus soulii*

Melampsalta soulii Wu 1935: 28 (listed) Equals *Quintilia soulii* China, Yunnan

Melampsalta soulii Hua 2000: 63 (listed, distribution) Equals *Quintilia soulii* China, Yunnan

Quintilia soulii Hua 2000: 64 (listed, distribution) China, Yunnan

M. soulii var. a (Distant, 1905) = *Quintilia soulii* var. *a* = *Cicadetta soulii* var. *a*

M. soulii var. b (Distant, 1905) = *Quintilia soulii* var. *b* = *Cicadetta soulii* var. *b*

M. sp. a Phillips, Sanborn & Villet, 2002

Melampsalta sp. a Phillips, Sanborn and Villet 2002: 31 (thermal responses) South Africa, Eastern Cape

Melampsalta sp. a Sanborn, Phillips and Villet 2003a: 348–349, Table 1 (thermal responses, habitat, comp. note) South Africa

M. sp. b Phillips, Sanborn & Villet, 2002

Melampsalta sp. b Phillips, Sanborn and Villet 2002: 31 (thermal responses) South Africa, Eastern Cape

Melampsalta sp. b Sanborn, Phillips and Villet 2003a: 349, Table 1 (thermal responses, habitat, comp. note) South Africa

M. rubristrigata Goding & Froggatt, 1904 to *Pauropsalta rubristrigata*

***M. telxiope* (Walker, 1850)** = *Cicada telxiope* = *Cicadetta telxiope* = *Cicada arche* Walker, 1850 = *Cicadetta arche* = *Cicada duplex* Walker, 1850 = *Melampsalta duplex*

Melampsalta telxiope Buchanan 1879: 214 (comp. note) Equals *Cicada telxiope* Equals *Cicada duplex* Equals *Cicada arche* New Zealand, Australia

***M. uchiyamae* (Matsumura, 1927)** = *Parasemia uchiyamae* = *Melampsalta rosacea* (nec Distant)

Parasemia uchiyamae Hayashi 1984: 66 (type species of *Cicadetta*)

Melampsalta uchiyamae Boer and Duffels 1996b: 310 (comp. note) Caroline Islands

M. viridicincta Ashton, 1912 to *Physeema quadricincta* (Walker, 1850)

***M. yapensis* Esaki, 1947**

Melampsalta yapensis Boer and Duffels 1996b: 310 (comp. note) Caroline Islands

***M. yasumatsui* Esaki, 1947**

Melampsalta yasumatsui Boer and Duffels 1996b: 310 (comp. note) Caroline Islands

M. zenobia Distant, 1912 to *Tibeta zenobia*

***Linguacicada* Chou & Lu, 1997**

Linguacicada Chou, Lei, Li, Lu and Yao 1997: 76, 357 (n. gen., described, listed, comp. note) China

Linguacicada Villet and Zhao 1999: 322 (comp. note)

Linguacicada Moulds 2005b: 390, 436 (higher taxonomy, listed, comp. note)

Linguacicada Ahmed and Sanborn 2010: 43 (distribution) China, India, Pakistan

***L. continuata* (Distant, 1888)** = *Cicadetta continuata* = *Melampsalta continuata* = *Melampsalta (Melampsalta) continuata* = *Cicadetta (Melampsalta) continuata* = *Curvicicada continuata*

Cicadetta continuata Chou, Lei, Li, Lu and Yao 1997: 76, 364 (type species of *Linguacicada*)

Linguacicada continuata Chou, Lei, Li, Lu and Yao 1997: 76–77, 357–358, 364, 366, Fig 8–21, Plate I, Fig 18 (n. comb., described, illustrated, synonymy, comp. note) Equals *Cicadetta continuata* Equals *Melampsalta continuata* China

Curvicicada continuata Chou, Lei, Li, Lu and Yao 1997: 79 (comp. note)

Linguacicada continuata Hua 2000: 62 (listed, distribution) Equals *Cicadetta continuata* China, Xinjiang, Pakistan, India

Linguacicada continuata Ahmed and Sanborn 2010: 43 (synonymy, distribution) Equals *Cicadetta continuata* Pakistan, India, China

***Amphipsalta* Fleming, 1969**

Melampsalta (sic, partim) Buchanan 1879: 213 (comp. note) New Zealand

Amphipsalta sp. Harvie 1973: 51, Table 3 (predation) New Zealand

Amphipsalta sp. Dingley 1974: 535 (fungal host) New Zealand

Amphipsalta Fleming 1975a: 1570 (comp. note) New Zealand

Amphipsalta Chambers 1976: 83 (comp. note) New Zealand

Amphipsalta Fleming 1977: 258 (comp. note) New Zealand

Amphipsalta Chambers 1978: 69 (comp. note) New Zealand

Amphipsalta Kay 1980: 2–6, 8, Figs 6–7 (oviposition, comp. note) New Zealand

Amphipsalta sp. Moeed 1980: 250, Table 1 (predation, listed) New Zealand

Amphipsalta Popov 1981: 271, 275, 279 (comp. note)

Amphipsalta Reid, Ordish and Harrison 1982: 78, 80, 83, Table 1, Appendix 1 (predation, listed, comp. note) New Zealand

Amphipsalta Fleming 1984: 191–192 (comp. note) New Zealand

Amphipsalta Moeed and Meads 1987a: 57, Table 5 (density) New Zealand

Amphipsalta Duffels 1988d: 75 (comp. note) New Zealand

Amphipsalta Andrews and Gibbs 1989: 105 (note) New Zealand

Amphiplata spp. Colbourne, Baird, and Jolly 1990: 538 (predation, comp. note) New Zealand

Amphipsalta Moulds 1990: 176 (comp. note) New Zealand

Amphipsalta spp. Glare, O'Callaghan, and Wigley 1993: 98 (fungal parasite) New Zealand

Amphipsalta Lane 1995: 392, 409 (comp. note) New Zealand

Amphipsalta Boer and Duffels 1996b: 309 (distribution, comp. note) New Zealand

Amphipsalta Novotny and Wilson 1997: 437 (listed)

Amphipsalta Gogala and Trilar 1998a: 8 (comp. note) New Zealand

Amphipsalta Brown 2001: 111–112 (predation, comp. note) New Zealand

Amphipsalta Chambers, Boon, Buckley and Hitchmough 2001: 380, 385 (comp. note) New Zealand

Amphipsalta Cooley 2001: 758 (comp. note) New Zealand, Australia

Amphipsalta Cooley and Marshall 2001: 847 (comp. note) New Zealand

Amphipsalta Arensburger 2002: 5055B (phylogeny) New Zealand

Amphipslata Buckley, Arensburger, Simon and Chambers 2002: 5–6, 15 (phylogeny, diversity, comp. note) New Zealand

Amphipsalta Gogala and Trilar 2003: 9 (comp. note) New Zealand

Amphipsalta Arensburger, Buckley, Simon, Moulds, and Holsinger 2004: 557–558, 562–563, 565–566 (molecular phylogeny, diversity, comp. note) New Zealand

Amphipsalta Arensburger, Simon, and Holsinger 2004: 1769, 1775 (molecular phylogeny, comp. note) New Zealand

Amphipsalta Sueur and Aubin 2004: 223 (comp. note)

Amphipsalta Moulds 2005b: 414–415, 436 (higher taxonomy, listed, comp. note)

Amphipsalta Shiyake 2007: 8, 110 (listed, comp. note)

Amphipsalta Hill and Marshall 2008: 523 (diversity, comp. note) New Zealand

Amphipsalta Simon 2009: 104 (diversity, comp. note) New Zealand

***A. cingulata cingulata* (Fabricius, 1775)** = *Tettigonia cingulata* = *Cicada cingulata* = *Melampsalta cingulata* = *Cicadetta cingulata* = *Cicada mendosa* Walker, 1858 = *Cicada indivulsa* Walker, 1858 = *Cicada indivisula* (sic) = *Cicadetta indivulsa* = *Cicada flexicosta* Stål, 1859 = *Melampsalta flexicosta* = *Cicada flavicosta* (sic)

Melampsalta cingulata Buchanan 1879: 213 (comp. note) Equals *Tettigonia cingulata* Equals *Cicada flexicosta* Equals *Cicada zealandica* (sic) Equals *Cicada indivulsa* New Zealand

Cicadetta cingulata Alfken 1903: 582 (comp. note) New Zealand

Melampsalta cingulata Miller 1952: 16–18, 43, 57 (song, comp. note) New Zealand

Melampsalta cingulata Harford 1958: 4–5, 7–8, 10–11, Fig 1 (illustrated, oviposition, emergence, hosts, parasitism, comp. note) New Zealand

Melampsalta cingulata Fleming 1966: 1–2, Fig 1 (illustrated, comp. note) New Zealand

Melampsalta cingulata Grant-Taylor 1966: 348 (illustrated, comp. note) New Zealand

Amphipsalta cingulata Daniel 1972: 333 (predation, comp. note) New Zealand

Amphipsalta cingulata Daniel 1973: 26, 28, Table 3, Table 5 (predation, comp. note) New Zealand

Amphipsalta cingulata Fleming 1975a: 1569–1570, Fig 2 (illustrated, comp. note) New Zealand

Amphipsalta cingulata Kay 1980: 3–4, 6–7, Fig 5 (comp. note) New Zealand

Amphipsalta cingulata Moeed and Meads 1984: 224, Table 4 (listed) Hen Island

Melampsalta cingulata Claridge 1985: 310 (comp. note) New Zealand

Amphipsalta cingulata Moeed and Meads 1987a: 57–58, Table 6 (listed, comp. note) New Zealand

Amphipsalta cingulata Duffels 1988d: 75 (comp. note) New Zealand

Amphipsalta cingulata Andrews and Gibbs 1989: 105 (note) New Zealand

Amphipsalta? cingulata Noyes and Valentine 1989a: 36 (parasitism) New Zealand

Amphipsalta cingulata Noyes and Valentine 1989b: 33 (parasitism) New Zealand

Amphipsaltria cingulata Boulard 1990b: 177 (comp. note)

Amphipsalta cingulata Colbourne, Baird, and Jolly 1990: 535 (predation, comp. note) New Zealand

Amphipsalta cingulata Boulard 1991c: 118 (comp. note) Equals *Cicada zelandica* New Zealand

Melampsalta cingulata Glare, O'Callaghan, and Wigley 1993: 98, 109 (fungal parasite) New Zealand

Amphipsalta cingulata Glare, O'Callaghan, and Wigley 1993: 109 (fungal parasite) New Zealand

A[mphipsalta] cingulata Jiang, Yang, Tang, Xu and Chen 1995b: 229 (song frequency, comp. note)

Amphipsalta cingulata Lane 1995: 386–387, 392, Figs 17.7c-d (oscillogram, comp. note) New Zealand

Amphipsalta cingulata Chambers, Boon, Buckley and Hitchmough 2001: 380, Fig 3 (phylogeny, comp. note) New Zealand

Amphipslata cingulata Cooley and Marshall 2001: 847, 850 (comp. note)

Amphipsalta cingulata Boulard 2002b: 205 (comp. note)
Amphipsalta cingulata Buckley, Arensburger, Simon and Chambers 2002: 6, 11, Fig 1 (phylogeny, comp. note) New Zealand
Melampsalta cingulata Merrett, Clarkson and Bathgate 2002: 56 (oviposition damage) New Zealand
Amphipsalta cingulata Sueur 2002b: 112, 125, 129 (listed, comp. note) Equals *Cicada cingulata* New Zealand
Amphipsalta cingulata Towns 2002: 595 (comp. note) New Zealand
Amphipsalta cingulata Arensburger, Buckley, Simon, Moulds, and Holsinger 2004: 559, 562, Table 1, Figs 1–2 (molecular phylogeny, comp. note) New Zealand
Amphipsalta cingulata Boulard 2005d: 343 (comp. note) New Zealand
Amphipsalta cingulata Logan and Connolly 2005: 36–42, Figs 2–5, Tables 1–3 (agricultural pest, exuvia key, comp. note) New Zealand
Amphipsalta cingulata Boulard 2006g: 66 (comp. note)
Amphipsalta cingulata Logan 2006: 20 (comp. note) New Zealand
Amphipsalta cingulata Esson 2007: 20–21 (sex ratio) New Zealand
Amphipsalta cingulata Simon 2009: 104, Fig 2 (phylogeny, comp. note) New Zealand
Amphipsalta cingulata Logan, Hill, Connolly, Maher and Dobson 2010: 260, 262–268, 270, Figs 3–6, Table 4 (agricultural pest, density, comp. note) New Zealand

***A. cingulata* var. _____ (Walker, 1852)** = *Cicada zealandica* (sic) var. _____.

***A. strepitans* (Kirkaldy, 1891)** = *Cicadetta strepitans* = *Melampsalta strepitans* = *Cicada cingulara obscura* Hudson, 1891 = *Melampsalta cingulata obscura* = *Melampsalta obscura* = *Cicadetta obscura* = *Melampsalta cingulata* Kirby (nec Fabricius)
Melampsalta strepitans Miller 1952: 16 (comp. note) New Zealand
Melampsalta strepitans Fleming 1966: 2, Fig 2 (illustrated, comp. note) New Zealand
Melampsalta strepitans Grant-Taylor 1966: 348 (comp. note) New Zealand
Amphipsalta strepitans Fleming 1975a: 1570 (comp. note) New Zealand
Amphipsalta strepitans Kay 1980: 3–4, 6–7, Fig 5 (comp. note) New Zealand
Amphipsalta strepitans Duffels 1988d: 75 (comp. note) New Zealand
Amphipsalta strepitans Palma, Lovis and Tither 1989: 38 (status) Equals *Cicada cingulata obscura* New Zealand
Cicada cingulata obscura Palma, Lovis and Tither 1989: 38 (type material, type data) Equals *Amphipsalta strepitans* New Zealand
Cicada strepitans Palma, Lovis and Tither 1989: 38 (type material, type data) Equals *Amphipsalta strepitans* New Zealand
Amphipsalta strepitans Palma, Lovis and Tither 1989: 38 (status) Equals *Cicada strepitans* New Zealand
Amphipsaltra strepitans Boulard 1990b: 177 (comp. note)
Amphipsalta sterpitans Chambers, Boon, Buckley and Hitchmough 2001: 380 (comp. note) New Zealand
Amphipsalta strepitans Boulard 2002b: 205 (comp. note)
Amphipsalta strepitans Sueur 2002b: 129 (listed) Equals *Melampsalta strepitans*
Amphipsalta strepitans Boulard 2005d: 343 (comp. note) New Zealand
Amphipsalta strepitans Boulard 2006g: 66 (comp. note)
Amphipsalta strepitans Hill and Marshall 2008: 524 (comp. note) New Zealand
Amphipsalta strepitans Marshall, Hill, Fontaine, Buckley and Simon 2009: 2006 (comp. note) New Zealand

***A. zelandica* (Boisduval, 1835)** = *Cicada zelandica* = *Cicada zealandica* (sic) = *Cicada zeylandica* (sic) = *Melampsalta zelandica* = *Melampsalta zealandica* (sic) = *Cicadetta zealandica* (sic) = *Amphipsalta zealandica* (sic)
Amphipsalta zelandica Daniel 1972: 333 (predation, comp. note) New Zealand
Amphipsalta zelandica Daniel 1973: 26, 28, Table 3, Table 5 (predation, comp. note) New Zealand
Amphipsalta zelandica Helson 1973: 29 (illustrated, life cycle, agricultural pest, comp. note) New Zealand
Amphipsalta zealandica (sic) Fleming 1975a: 1569–1570, Fig 1 (illustrated, comp. note) New Zealand
Amphipsalta zelandica Fleming 1977: 258 (comp. note) New Zealand
Amphipsalta zelandica Chambers 1978: 69 (comp. note) New Zealand
Amphipsalta zelandica Daniel 1979: 362, Table 2 (predation, comp. note) New Zealand
Amphipsalta zelandica Fitzgerald and Karl 1979: 116, Table 2 (predation, listed, comp. note) New Zealand
Amphipsalta zelandica Kay 1980: 1, 3–4, 6–7, Fig 1, Figs 4–5, Fig 9 (illustrated, comp. note) New Zealand

Amphipsalta zelandica Gill, Powlesland, and Powlesland 1983: 83 (predation) New Zealand
Amphipsalta zealandica (sic) Moeed and Meads 1983: 44, Fig 4D (hosts, comp. note) New Zealand
Amphipsalta zelandica Fleming 1984: 196 (comp. note) New Zealand
Amphipsalta zelandica Fitzgerald, Meads, and Whitaker 1986: 25–26, Table 1 (predation) New Zealand
Amphipsalta zealandica (sic) Cowan 1987: 177 (predation, comp. note) New Zealand
Amphipsalta zelandica Moeed and Meads 1987a: 57–58, Table 6 (listed, comp. note) New Zealand
Amphipsalta zelandica Moeed and Meads 1987b: 200, Table 1 (listed) New Zealand
Amphipsalta zelandica Thomas 1987: 60 (predation) New Zealand
Amphipsalta zelandica Duffels 1988d: 75 (comp. note) New Zealand
Amphipsalta zelandica Moeed and Meads 1988: 488 (listed) New Zealand
Amphipsalta zealandica (sic) Andrews and Gibbs 1989: 105 (note) New Zealand
Amphipsalta zelandica Langham 1990: 252 (predation) New Zealand
Amphipsalta zelandica Glare, O'Callaghan, and Wigley 1993: 109 (fungal parasite) New Zealand
A[mphipsalta] zelandica Jiang, Yang, Tang, Xu and Chen 1995b: 229 (song frequency, comp. note)
Amphipsalta zelandica Lane 1995: 386–387, 392, Figs 17.7a-b (oscillogram, comp. note) New Zealand
Amphipsalta zealandica (sic) Freytag 2000: 55 (on stamp)
Amphipsalta zealandica (sic) Chambers, Boon, Buckley and Hitchmough 2001: 380 (comp. note) New Zealand
Amphipsalta zealandica (sic) Bowie, Marris, Emberson, Andrew, Berry, et al. 2003: 105 (distribution, comp. note) Quail Island
Amphipsalta zelandica Hill, Marshall and Cooley 2005: 73, Fig 2, Fig 4 (sonogram, comp. note) New Zealand
Amphipsalta zelandica Logan and Connolly 2005: 36–42, Fig 2, Figs 4–5, Tables 1–3 (agricultural pest, exuvia key, comp. note) New Zealand
Amphipsalta zelandica Logan 2006: 20 (comp. note) New Zealand
Amphipsalta zealandica (sic) Esson 2007: 20 (sex ratio) New Zealand
Amphipsalta zelandica Logan and Alspach 2007: 235–239, Figs 1–3, Table 1 (agricultural pest, comp. note) New Zealand
Amphipsalta zealandica (sic) Shiyake 2007: 8, 108, 110, Fig 188 (illustrated, listed, comp. note)
Amphipsalta zelandica Hill and Marshall 2008: 523 (comp. note) New Zealand
Amphipsalta zelandica Logan and Maher 2009: 268–271, Table 3 (control, comp. note)
Amphipsalta zelandica Marshall, Hill, Fontaine, Buckley and Simon 2009: 2006 (comp. note) New Zealand
Amphipsalta zelandica Simon 2009: 103, Fig 1a (illustrated, comp. note) New Zealand
Amphipsalta zelandica Logan, Hill, Connolly, Maher and Dobson 2010: 260, 262–265, 267, 269–270, Figs 3–5, Fig 7, Table 5 (agricultural pest, density, comp. note) New Zealand

***Tibeta* Lei & Chou, 1997**
Tibeta Chou, Lei, Li, Lu and Yao 1997: 47, 71, 293, 357, Table 6 (key, n. gen., described, listed, diversity, biogeography) China
Tibeta Villet and Zhao 1999: 322 (comp. note)
Tibeta Moulds 2005b: 390, 436 (higher taxonomy, listed, comp. note)

***T. minensis* Lei & Chou, 1997**
Tibeta minensis Chou, Lei, Li, Lu and Yao 1997: 71, 73–75, 357, 363, Fig 8-19, Plate I, Fig 8 (key, described, illustrated, comp. note) China

***T. nanna* Lei & Chou, 1997**
Tibeta nanna Chou, Lei, Li, Lu and Yao 1997: 71, 75–76, 357, 363–364, Fig 8-20, Plate I, Fig 19 (key, described, illustrated, comp. note) China

***T. planarius* Lei & Chou, 1997**
Tibeta planarius Chou, Lei, Li, Lu and Yao 1997: 71–73, 357, 363, Fig 8-18, Plate I, Fig 11 (key, n. sp., described, illustrated, comp. note) China

***T. zenobia* (Distant, 1912)** = *Melampsalta zenobia* = *Cicadetta zenobia* = *Melampsalta kulingana* Kato, 1938
Melampsalta zenobia Chou, Lei, Li, Lu and Yao 1997: 71, 363 (type species of *Tibeta*, comp. note)
Tibeta zenobia Chou, Lei, Li, Lu and Yao 1997: 71–73, 75, 357–358, 363, Fig 8-17, Plate I, Fig 16 (n. comb., key, described, illustrated, synonymy, comp. note) Equals *Melampsalta zenobia* Equals *Cicadetta zenobia* Equals *Melampsalta kulingana* China
Melampsalta kulingana Chou, Lei, Li, Lu and Yao 1997: 358 (comp. note)
Tibeta zenobia Hua 2000: 65 (listed, distribution) Equals *Melampsalta kulingana* China, Jiangxi, Zhejiang, Yunnan, Xizang

Scolopita Chou & Lei, 1997

Scolopita Chou, Lei, Li, Lu and Yao 1997: 47, 66, 293, 357, 361–362, Table 6 (key, n. gen., described, listed, diversity, biogeography) China

Scolopita Villet and Zhao 1999: 322 (comp. note)

Scolopita Moulds 2005b: 390, 436 (higher taxonomy, listed, comp. note)

Scolopita Pham and Thinh 2005a: 238 (listed) Vietnam

Scolopita Pham and Thinh 2005b: 287 (comp. note) Vietnam

Scolopita Pham and Yang 2009: 3, 15, Table 2 (listed, comp. note) Vietnam

S. cupis Chou & Lu, 1997

Scolopita cupis Chou, Lei, Li, Lu and Yao 1997: 66–69, 357, 362, Fig 8–14, Plate I, Fig 20 (key, n. sp., described, illustrated, comp. note) China

S. fusca Chou & Lei, 1997

Scolopita fusca Chou, Lei, Li, Lu and Yao 1997: 66, 68–69, 357, 362, Fig 8–15, Plate I, Fig 2 (key, n. sp., described, illustrated, comp. note) China

S. lusiplex Chou & Lei, 1997 = *Scolopita* sp. 1 Pham & Thinh, 2006

Scolopita lusiplex Chou, Lei, Li, Lu and Yao 1997: 66–67, 69–70, 357, 362, Fig 8–16, Plate I, Fig 7 (key, n. sp., described, illustrated, comp. note) China

Scolopita lusiplex Pham and Thinh 2005a: 238 (distribution, listed) Vietnam

Scolopita sp. 1 Pham and Thinh 2006: 526 (listed) Vietnam

Scolopita lusiplex Pham and Yang 2009: 3, 15–16, Table 2 (distribution, listed, comp. note) Vietnam, China

S. mokanshanensis (Ouchi, 1938) = *Melampsalta mokanshanensis*

Melampsalta mokanshanensis Chou, Lei, Li, Lu and Yao 1997: 66, 361 (type species of *Scolopita*)

Scolopita mokanshanensis Chou, Lei, Li, Lu and Yao 1997: 66–67, 70, 358, 362, Plate I, Fig 15 (key, n. comb., described, illustrated, synonymy, comp. note) Equals *Melampsalta mokanshanensis* China

Scolopita mokanshanensis Hua 2000: 64 (listed, distribution) Equals *Melampsalta mokanshanensis* China, Zhejiang, Hunan

Melampsalta mokanshanensis Pham and Thinh 2005a: 238 (type species of *Scolopita*)

Melampsalta mokanshanesis Pham and Yang 2009: 3 (type species of *Scolopita*)

Samaecicada Popple & Emery, 2010

Samaecicada Popple and Emery 2010: 148, 150 (n. gen., described, synonymy, distribution, hosts, song, comp. note) Australia, New South Wales, Queensland

S. subolivacea (Ashton, 1912) = *Pauropsalta subolivacea* = *Melampsalta subolivacea*

Pauropsalta subolivacea Ewart 1989b: 293 (listed) New South Wales

Pauropsalta subolivacea Moulds 1990: 131 (distribution, listed) Australia, New South Wales

Pauropsalta subolivacea Moulds and Carver 2002: 44 (distribution, ecology, type data, listed) Australia, New South Wales

Pauropsalta subolivacea Popple and Emery 2010: 147–148, 151 (type species of *Samaecicada*, comp. note) Australia, New South Wales

Samaecicada subolivacea Popple and Emery 2010: 148–155, Figs 1–4 (n. comb., described, illustrated, synonymy, distribution, habitat, behavior, comp. note) Equals *Pauropsalta subolivacea* Australia, New South Wales

Paragudanga Distant, 1913 to *Gudanga* Distant, 1905

P. browni Distant, 1913 to *Gudanga browni*

Gudanga Distant, 1905 = *Paragudanga* Distant, 1913

Gudanga spp, Moulds 1990: 19 (comp. note)

Gudanga Moulds 1990: 8, 32, 116–117 (synonymy, listed, diversity, comp. note) Equals *Paragudanga* n. syn. Australia

Paragudanga Moulds 1990: 116–117 (listed, comp. note)

Gudanga spp. Moulds 1996a: 19–23, 26–27, 29–30 (comp. note) Australia

Gudanga Moulds 1996a: 16, 30 (described, key, distribution, comp. note) Equals *Paragudanga* Australia, Western Australia, Queensland

Paragudanga Moulds 1996a: 19, 30 (comp. note) Australia

Gudanga Moulds and Carver 2002: 47 (synonymy, type data, listed) Equals *Paragudanga* Australia

Paragudanga Moulds and Carver 2002: 47 (type data, listed)

Gudanga Moulds 2005b: 377, 392, 423–424, 430, 436, Table 2 (higher taxonomy, listed, comp. note) Australia

Gudanga Olive 2007: 1 (comp. note) Australia

Gudanga spp. Olive 2007: 3–5 (comp. note) Australia

Gudanga sp. Watson, Watson, Hu, Brown, Cribb and Myhra 2010: 123 (wing microstructure, comp. note)

G. adamsi Moulds, 1996 = *Gudanga* sp. nov. Lithgow, 1988 = *Gudanga* sp. Ewart, 1988 = *Gudanga browni* (nec Distant)

Gudanga sp. nov. Lithgow 1988: 65 (illustrated, listed, comp. note) Australia, Queensland
Gudanga adamsi Moulds 1996a: 20–22, 24–26, 29–30, Figs 5–6, Figs 17–18, Fig 25 (n. sp., synonymy, described, key, illustrated, distribution, comp. note) Equals *Gudanga* sp. nov. Lithgow, 1988 Equals *Gudanga* sp. Ewart, 1988 Equals *Gudanga browni* Moulds, 1990 (partim) Australia, Queensland
Gudanga adamsi Sueur 2001: 42, Table 1 (listed) Equals *Gudanga* sp.
Gudanga adamsi Moulds and Carver 2002: 47–48 (distribution, ecology, type data, listed) Australia, Queensland
Gudanga adamsi Olive 2007: 1, 3–6, Fig 3, Figs 5–6 (key, illustrated, comp. note) Queensland

G. sp. nr. *adamsi* Southern Brigalow Blackwing Popple & Strange 2002
Gudanga sp. nr. *adamsi* Southern Brigalow Blackwing Popple and Strange 2002: 28, Table 1 (listed, habitat) Australia, Queensland

G. sp. nr. *adamsi* Watson, Watson, Hu, Brown, Cribb & Myhra, 2010
Gudanga sp. nr. *adamsi* Watson, Watson, Hu, Brown, Cribb and Myhra 2010: 117, 119–120, 122, 124, Figs 2a–b, Fig 6, Fig 8, Table 1 (wing microstructure, listed, comp. note)

***G. aurea* Moulds, 1996**
Gudanga aurea Moulds 1996a: 20, 22–23, 27–30, Figs 7–8, Figs 21–22, Fig 27 (n. sp., described, key, illustrated, distribution, comp. note) Australia, Western Australia
Gudanga aurea Moulds and Carver 2002: 48 (distribution, ecology, type data, listed) Australia, Western Australia
Gudanga aurea Olive 2007: 1 (key) Western Australia

***G. boulayi* Distant, 1905**
Gudanga boulayi Moulds 1990: 116–118 (type species of *Gudanga*, distribution, ecology, listed, comp. note) Australia, Western Australia
Gudanga boulayi Naumann 1993: 48, 93, 149 (listed) Australia
Gudanga boulayi Moulds 1996a: 19–25, 29–30, Figs 1–2, Figs 13–14, Fig 26 (type species of *Gudanga*, described, key, illustrated, distribution, comp. note) Australia, Western Australia
Gudanga boulayi group Moulds 1996a: 30 (comp. note)
Gudanga boulayi Moulds and Carver 2002: 47–48 (type species of *Gudanga*, distribution, ecology, type data, listed) Australia, Western Australia
Gudanga boulayi 2005b: 395–398, 417–419, 425, Table 1, Figs 56–59, Fig 62 (phylogeny, comp. note) Australia
Gudanga boulayi Olive 2007: 2 (key) Western Australia

***G. browni* (Distant, 1913)** = *Paragudanga browni*
Paragudanga browni Moulds 1990: 116 (type species of *Paragudanga*, listed) Western Australia
Gudanga browni Moulds 1990: 117–118, Plate 21, Fig 2, Fig 2a (type species of *Paragudanga*, synonymy, illustrated, distribution, ecology, listed, comp. note) Equals *Paragudanga browni* Australia, Western Australia
Gudanga browni Naumann 1993: 41, 93, 149 (listed) Australia
Gudanga browni Moulds 1996a: 19–20, 23, 26–30, Figs 9–10, Figs 19–20, Fig 28 (type species of *Paragudanga*, synonymy, described, key, illustrated, distribution, comp. note) Equals *Paragudanga browni* Australia, Western Australia
Gudanga browni group Moulds 1996a: 30 (comp. note)
Gudanga browni Moulds and Carver 2002: 48 (type species of *Paragudanga*, synonymy, distribution, ecology, type data, listed) Equals *Paragudanga browni* Australia, Western Australia
Paragudanga browni Moulds and Carver 2002: 48 (type data, listed) Western Australia
Gudanga browni Olive 2007: 2 (key) Western Australia

***G. kalgoorliensis* Moulds, 1996**
Gudanga kalgoorliensis Moulds 1996a: 20–22, 24–25, 29–30, Figs 3–4, Figs 15–16, Fig 26 (n. sp., described, key, illustrated, distribution, comp. note) Australia, Western Australia
Gudanga kalgoorliensis Moulds and Carver 2002: 48 (distribution, ecology, type data, listed) Australia, Western Australia
Gudanga kalgoorliensis Olive 2007: 1 (key) Western Australia

***G. lithgowae* Ewart & Popple, 2013** = *Gudanga* sp. Ewart, 1988 = *Gudanga* n. sp. Ewart 1989 = *Gudanga* sp. A Ewart and Popple, 2001 = *Gudanga adamsi* (nec Moulds)
Gudanga sp. Ewart 1988: 185, 191, 193, 198–199, Fig 10I, Plate 2E (illustrated, oscillogram, listed, comp. note) Queensland
Gudanga n. sp. Ewart 1989a: 79 (comp. note)
Gudanga adamsi (sic) Ewart 1998a: 62–63 (described, natural history, song) Queensland
Gudanga sp. A Ewart and Popple 2001: 56, 65, Fig 2C (described, habitat, song, oscillogram) Queensland
Gudanga lithgowae Ewart and Popple 2013 (Memoires of the Queensland

Museum – Nature 56: 355–406): 358–364, 366, 368–369, 372, 375–376, 378–379, 381–382, 385, 388–389, 391–393, 398, 402–404, Figs 1A–6A, Fig 7, Fig 8B, Fig 10C, Fig 11A, Figs 15A–E, Fig 16, Figs 18–20, Plate 1, Plate 4A, Table 1 (n.sp., described, illustrated, key, song, oscillogram, frequency spectrum, distribution, comp. note) Equals Gudanga *adamsi* Ewart, 1998 Equals Gudanga sp. A Australia, Queensland

***G. nowlandi* Ewart & Popple, 2013** = *Gudanga* sp. B Ewart and Popple, 2001

Gudanga sp. B near *Gudanga kalgoorliensis* Ewart and Popple 2001: 62, 70, Fig 8C (illustrated, described, habitat, song) Queensland

Gudanga nowlandi Ewart and Popple 2013 (Memoires of the Queensland Museum – Nature 56: 355–406): 358–359, 361, 363–370, 372, 374–375, 377–379, 381–385, 389, 391–393, 397, 399, 401–404, Figs 1B–6B, Fig 7, Fig 9A, Fig 10A, Figs 11–13, Figs 16A–C, Figs 18–20, Plate 2, Plate 4B, Table 1, Tables 3–4 (n.sp., described, illustrated, key, song, oscillogram, frequency spectrum, distribution, comp. note) Australia, Queensland

***G. pterolongata* Olive, 2007**

Gudanga pterolongata Olive 2007: 1–6, Figs 1–2, Fig 4, Figs 7–8 (n. sp., described, illustrated, key, comp. note) Queensland

***G. solata* Moulds, 1996**

Gudanga solata Moulds 1996a: 20, 23, 27–30, Figs 11–12, Figs 23–24, Fig 28 (n. sp., described, key, illustrated, distribution, comp. note) Australia, Western Australia

Gudanga solata Moulds and Carver 2002: 48 (distribution, ecology, type data, listed) Australia, Western Australia

Gudanga solata Olive 2007: 2 (key) Western Australia

***Adelia* Moulds, 2012**

***A. borealis* (Goding & Froggatt, 1904)** = *Tibicen borealis* = *Abricta borealis*

Tibicen borealis Hahn 1962: 10 (listed) Australia, Western Australia

Tibicen borealis Stevens and Carver 1986: 264 (type material) Australia, Western Australia

Abricta borealis Moulds 1990: 118 (distribution, listed) Australia, Western Australia

Abricta borealis Moulds and Carver 2002: 50 (synonymy, distribution, ecology, type data, listed) Equals *Tibicen borealis* Australia, Western Australia

Abricta borealis Moulds 2003b: 246 (comp. note)

Tibicen borealis Moulds 2003b: 247 (status, comp. note) Western Australia

Abricta borealis Moulds 2005b: 377, 395–398, 403, 415, 417–420, 421, 423, 425, 436, Fig 37, Figs 56–59, Fig 62, Table 1 (illustrated, higher taxonomy, phylogeny, listed, comp. note) Australia

***Diemeniana* Distant, 1906** = *Diemenia* Distant, 1905 (nec *Diemenia* Spinola, 1850)

Diemeniana Moss 1989: 106 (comp. note) Australia

Diemeniana Boer 1990: 64 (comp. note)

Diemeniana Moss 1990: 8 (comp. note) Australia

Diemeniana Moulds 1990: 7, 32, 112–113, 149 (listed, diversity, comp. note) Australia

Diemeniana spp. Moulds 1990: 16, 19 (comp. note)

Diemeniana Moulds 1996a: 19 (comp. note) Australia

Diemeniana Chambers, Boon, Buckley and Hitchmough 2001: 380 (comp. note) Australia

Diemeniana Buckley, Arensburger, Simon and Chambers 2002: 6 (phylogeny, comp. note) Australia

Diemeniana Moulds and Carver 2002: 46 (synonymy, type data, listed) Equals *Diemenia* Australia

Diemenia Moulds and Carver 2002: 46 (type data, listed)

Diemeniana Moulds 2003b: 246 (comp. note)

Diemeniana Moulds 2005b: 377, 392, 423–424, 430, 436, Table 2 (higher taxonomy, listed, comp. note) Australia

Diemeniana Shiyake 2007: 7, 102 (listed, comp. note)

***D. cincta* (Fabricius, 1803)** = *Tettigonia cincta* = *Cicada cincta* = *Tibicen cinctus* = *Abricta cincta* = *Diemeniana tillyardi* Hardy, 1918

Tettigonia cincta Dallas 1870: 482 (synonymy) Equals *Tibicen cincta*

Diemeniana tillyardi Moss 1989: 105 (comp. note) Tasmania

Diemeniana tillyardi Moss 1990: 7 (comp. note) Tasmania

Diemeniana tillyardi Moulds 1990: 113–115, Plate 14, Fig 7 (illustrated, distribution, ecology, listed, comp. note) Australia, Tasmania

Abricta cincta Boer and Duffels 1996b: 308 (comp. note) New Caledonia

Diemeniana tillyardi Chambers, Boon, Buckley and Hitchmough 2001: 380, Fig 3 (phylogeny, comp. note) Australia

Diemeniana tillyardi Buckley, Arensburger, Simon and Chambers 2002: 6, 11, Fig 1 (phylogeny, comp. note) Australia

Diemeniana tillyardi Moulds and Carver 2002: 47 (distribution, ecology, type data, listed) Australia, Tasmania

Tettigonia cincta Moulds 2003b: 246 (comp. note) Australia

Diemeniana tillyardi Arensburger, Buckley, Simon, Moulds, and Holsinger 2004: 558–559, 562–563, Table 1, Figs 1–2 (molecular phylogeny, comp. note) New Caledonia
Diemeniana tillyardi Simon 2009: 104, Fig 2 (phylogeny, comp. note) Australia

***D. euronotiana* (Kirkaldy, 1909)** = *Abricta euronotiana* = *Cicada aurata* Walker, 1850 (nec *Cicada aurata* Linnaeus, 1758) = *Tibicen auratus* = *Tibicen(?) auratus* = *Abricta aurata* = *Diemeniana richesi* Distant, 1913
Diemeniana euronotiana Greenup 1964: 22–23 (comp. note) Australia, New South Wales
Diemeniana euronotiana Moss 1989: 103–104 (distribution, song, comp. note) Tasmania
Diemeniana euronotiana Moss 1990: 6–7 (distribution, song, comp. note) Tasmania, New South Wales, Australian Capital Territory, Victoria
Diemeniana euronotiana Moulds 1990: 112–114, 146, Plate 14, Fig 6, Figs 6a–6b, Plate 24, Fig 25 (illustrated, distribution, ecology, listed, comp. note) Australia, Australian Capital Territory, New South Wales, Tasmania, Victoria
Diemeniana richesi Moulds 1990: 112 (distribution, listed) Australia, New South Wales
Diemeniana euronotiana Boulard 1991c: 120 (comp. note) Equals *Abricta euronotiana* Tasmania
Diemeniana euronotiana New 1996: 95 (comp. note) Australia
Diemeniana euronotiana Moss and Popple 2000: 57 (listed, comp. note) Australia, New South Wales
Diemeniana euronotiana Moulds and Carver 2002: 46 (synonymy, distribution, ecology, type data, listed) Equals *Cicada aurata* Equals *Abricta euronotiana* Australia, Australian Capital Territory, New South Wales, Tasmania, Victoria
Cicada aurata Moulds and Carver 2002: 46 (type data, listed)
Abricta euronotiana Moulds and Carver 2002: 46 (type data, listed)
Diemeniana richesi Moulds and Carver 2002: 47 (distribution, ecology, type data, listed) Australia, New South Wales
Diemeniana euronotiana Williams 2002: 156–157 (listed) Australia, New South Wales, Victoria, Tasmania
Abricta aurata Moulds 2003b: 246 (comp. note)
Abricta euronotiana Moulds 2003b: 246 (comp. note)
Diemeniana euronotiana Moulds 2003b: 246 (comp. note)
Diemeniana euronotiana Haywood 2005: 9, 25–27, Fig 7, Table 1 (illustrated, distribution, described, habitat, song, oscillogram, listed, comp. note) South Australia, New South Wales, Victoria, Tasmania
Diemeniana euronotiana Haywood 2006a: 14, 16–17, 23, Tables 1–2 (illustrated, distribution, described, habitat, song, oscillogram, listed, comp. note) South Australia, New South Wales, Victoria, Tasmania
Diemeniana euronotiana Shiyake 2007: 7, 100, 102, Fig 167 (illustrated, distribution, listed, comp. note) Australia

***D. frenchi* (Distant, 1907)** = *Abricta frenchi* = *Cicada coleoptrata* Walker, 1858 (nec *Cicada coleoptrata* Linnaeus, 1758) = *Tibicen coleoptratus* = *Diemeniana coleoptrata* = *Diemeniana tasmani* Kirkaldy, 1909
Diemeniana coleoptrata Moss 1989: 105 (comp. note) Tasmania
Diemeniana tasmani Moss 1989: 105 (comp. note) Equals *Diemeniana coleoptrata* Tasmania
Diemeniana coleoptrata Moss 1990: 7 (comp. note) Tasmania
Diemeniana tasmani Moss 1990: 7 (comp. note) Equals *Diemeniana coleoptrata* Tasmania
Diemeniana frenchi Moulds 1990: 113, 115–116, Plate 14, Fig 5, Plate 24, Fig 24 (n. comb., synonymy, illustrated, distribution, ecology, listed, comp. note) Equals *Cicada coleoptrata* Walker Equals *Abricta frenchi* Equals *Diemeniana tasmani* n. syn. Australia, New South Wales, Victoria
Cicada coleoptrata Moulds 1990: 112, 115 (type species of *Diemeniana*, listed) Equals *Diemeniana frenchi*
Tibicen coleoptratus Moulds 1990: 115 (type species of *Diemeniana* type data, listed)
Diemeniana coleoptrata Moulds 1990: 115 (listed)
Abricta frenchi Moulds 1990: 115 (listed, comp. note) Victoria
Diemeniana tasmani Moulds 1990: 115 (listed, comp. note) Equals *Diemeniana coleoptrata*
Diemeniana frenchi Chambers, Boon, Buckley and Hitchmough 2001: 380, Fig 3 (phylogeny, comp. note) Australia
Diemeniana frenchi Buckley, Arensburger, Simon and Chambers 2002: 6, 11, Fig 1 (phylogeny, comp. note) Australia
Cicada coleoptrata Moulds and Carver 2002: 46 (type species of *Diemenia* type data, listed) Equals *Diemeniana frenchi*
Diemeniana frenchi Moulds and Carver 2002: 46 (synonymy, distribution, ecology, type data, listed) Equals *Cicada coleoptrata* Equals *Abricta frenchi* Equals *Diemeniana tasmani* Australia, New South Wales, Victoria
Abricta frenchi Moulds and Carver 2002: 46 (type data, listed) Victoria
Diemeniana tasmani Moulds and Carver 2002: 46 (type data, listed)

Diemeniana frenchi Moulds 2003b: 246 (comp. note)
Diemeniana frenchi Arensburger, Buckley, Simon, Moulds, and Holsinger 2004: 558–559, 562–563, Table 1, Figs 1–2 (molecular phylogeny, comp. note) Australia
Diemeniana frenchi Moulds 2005b: 395–398, 405, 409, 417–419, 421, 425, Table 1, Fig 40, Fig 49, Figs 56–59, Fig 62 (illustrated, phylogeny, comp. note) Australia
Diemeniana frenchi Simon 2009: 104, Fig 2 (phylogeny, comp. note) Australia

***D. hirsuta* (Goding & Froggatt, 1904)** = *Tibicen hirsutus* = *Tibicen hirsuta* = *Abricta hirsuta* = *Diemeniana hirsutus* = *Diemeniana turneri* Distant, 1914
Tibicen hirsutus Hahn 1962: 10 (listed) Australia, South Australia
Tibicen hirsutus Stevens and Carver 1986: 264 (type material) Australia, South Australia
Diemeniana hirsuta Moss 1989: 104, 106 (distribution, song, comp. note) Tasmania
Diemeniana hirsuta Moss 1990: 6–8 (distribution, song, comp. note) Tasmania
Diemeniana hirsuta Moulds 1990: 113–114, Plate 14, Fig 8 (synonymy, illustrated, distribution, ecology, listed, comp. note) Equals *Tibicen hirsutus* Equals *Abricta? hirsuta* Equals *Diemeniana turneri* Equals *Diemeniana hirsutus* Australia, Tasmania
Tibicen hirsutus Moulds 1990: 113 (listed)
Tibicen hirsuta (sic) Moulds 1990: 113 (comp. note)
Diemeniana turneri Moulds 1990: 113 (listed, comp. note)
Diemeniana hirsuta Moulds and Carver 2002: 47 (synonymy, distribution, ecology, type data, listed) Equals *Tibicen hirsutus* Equals *Diemeniana turneri* Australia, Tasmania
Tibicen hirsutus Moulds and Carver 2002: 47 (type data, listed) South Australia
Diemeniana turneri Moulds and Carver 2002: 47 (type data, listed) Tasmania
Diemeniana hirsuta Moulds 2003b: 246 (comp. note)

***D. neboissi* Burns, 1958**
Diemeniana neboissi Greenup 1964: 23 (comp. note) Australia, New South Wales
Diemeniana neboissi Moulds 1990: 113, 115–116 (distribution, ecology, listed, comp. note) Australia, New South Wales, Victoria
Diemeniana neboissi Moulds and Carver 2002: 47 (distribution, ecology, type data, listed) Australia, New South Wales, Victoria

D. richesi Distant, 1913 to *D. euronotiana* (Kirkaldy, 1909)
D. tasmani Kirkaldy, 1909 to *D. frenchi* (Distant, 1907)
D. tillyardi Hardy, 1918 to *D. cincta* (Fabricius, 1803)

***Uradolichos* Moulds, 2012**

***U. longipennis* (Ashton, 1914)** = *Urabunana longipennis*
Urabunana longipennis Moulds 1990: 173 (distribution, listed) Australia, Northern Territory
Urabunana segmentaria Moulds 1990: 173 (distribution, listed) Australia, Queensland
Urabunana longipennis Moulds and Carver 2002: 45 (distribution, ecology, type data, listed) Australia, Northern Territory

***Pauropsalta* Goding & Froggatt, 1904** = *Cicadetta (Pauropsalta)* = *Melampsalta (Pauropsalta)*
Pauropsalta Gómez-Menor 1951: 61 (key) Spain
Pauropsalta sp. Hill, Hore and Thornton 1982: 43, 161 (listed, comp. note) Hong Kong
Pauropsalta Koçak 1982: 150 (listed, comp. note) Turkey
Pauropsalta Hayashi 1984: 66 (listed)
Pauropsalta Boulard 1988f: 40 (comp. note) Australia
Pauropsalta Duffels 1988d: 100–101 (comp. note) Australia
Pauropsalta Ewart 1988: 182 (comp. note)
Pauropsalta Ewart 1989a: 76–78, 80 (comp. note) Queensland
Pauropsalta Ewart 1989b: 289–290, 293, 308, 366 (described, diversity, distribution, comp. note) Queensland
Pauropsalta spp. Ewart 1989b: 343, 368 (comp. note)
Pauropsalta Boer 1990: 64 (comp. note)
Pauropsalta Carver 1990: 260 (comp. note)
Pauropsalta Moulds 1990: 7–8, 12, 32, 130–131, 136–137 (listed, diversity, comp. note) Australia, Tasmania
Pauropsalta Boulard 1993e: 109 (comp. note)
Pauropsalta Lei, Chou and Li 1994: 53 (comp. note)
Pauropsalta Boer 1995d: 201–202 (distribution, comp. note) New Guinea
Pauropsalta Lane 1995: 376 (comp. note) Australia
Pauropsalta Boer and Duffels 1996b: 306, 308–309 (distribution, comp. note) Australia, New Caledonia
Pauropsalta spp. Coombs 1996: 60 (comp. note) Queensland
Pauropsalta Moulds 1996b: 92 (comp. note) Australia
Pauropsalta Boulard 1997a: 179, 185, 191, 195 (listed, comp. note)
Pauropsalta Boulard 1997c: 109 (comp. note) Australia
Pauropsalta Chou, Lei, Li, Lu and Yao 1997: 25 (comp. note)
Pauropsalta Novotny and Wilson 1997: 437 (listed)

Pauropsalta Boer 1999: 116, 118, 128, 136 (described, comp. note) New Guinea, Australia, New South Wales, Queensland
Pauropsalta Ewart 1999: 2 (listed)
Pauropsalta Boulard 2001b: 29 (comp. note) Australia
Pauropsalta spp. Ewart 2001a: 501, 506 (acoustic behavior) Queensland
Pauropsalta Lee, Choe, Lee and Woo 2002: 3 (comp. note)
Pauropsalta Moulds and Carver 2002: 37 (type data, listed) Australia
Pauropsalta Popple and Strange 2002: 27 (comp. note) Australia, Queensland
Pauropsalta spp. Emery, Emery, Emery and Popple 2005: 106 (comp. note) New South Wales
Pauropsaltas (sic) Ewart 2005a: 179 (comp. note)
Pauropsalta Moulds 2005b: 390, 412, 430, 436, Table 2 (higher taxonomy, listed, comp. note) Australia
Pauropsalta Haywood 2006a: 13 (comp. note) South Australia
Pauropsalta Shiyake 2007: 8, 111 (listed, comp. note)
Pauropsalta Ewart and Marques 2008: 150–151, Table 1 (described, comp. note)
Pauropsalta Popple, Walter and Raghu 2008: 204 (comp. note) Australia, New Guinea
Pauropsalta Ewart 2009b: 141 (comp. note)
Pauropsalta Lee 2009i: 295 (comp. note)
Pauropsalta sp. Pain 2009: 47 (sonogram) Australia
Pauropsalta Popple and Emery 2010: 147, 150 (comp. note) Australia
Pauropsalta Popple and Walter 2010: 554 (comp. note) Australia, New Guinea

P. aktites **Ewart, 1989** = *Psaltoda* (sic) *aktites*

Pauropsalta aktites Ewart 1989b: 291–294, 329–333, 335, 337, 370, Fig 1, Figs 28–29, Figs 32E–F, Fig 57d, Table 2 (n. sp., described, illustrated, key, distribution, song, oscillogram, comp. note) Queensland, New South Wales
Pauropsalta aktites Ewart 1996: 14 (listed)
Psaltoda (sic) *aktites* Moulds 1996b: 92 (comp. note) Australia, Queensland, New South Wales
Pauropsalta aktites Moss 2000: 69 (comp. note) Queensland
Pauropsalta aktites Ewart 2001b: 69 (listed) Queensland
Pauropsalta aktites Sueur 2001: 44, Table 1 (listed)
Pauropsalta aktites Moulds and Carver 2002: 37 (distribution, ecology, type data, listed) Australia, Queensland
Pauropsalta aktites Arensburger, Buckley, Simon, Moulds, and Holsinger 2004: 558 (molecular phylogeny, comp. note) Australia
Pauropsalta akites Emery, Emery, Emery and Popple 2005: 97 (comp. note) New South Wales
Pauropsalta aktites Shiyake 2007: 8, 109, 111, Fig 195 (illustrated, distribution, listed, comp. note) Australia

P. annulata **Goding & Froggatt, 1904**

Pauropsalta annulata Hahn 1962: 9 (listed) Australia, Queensland
Pauropsalta annulata Ewart 1986: 51, Table 1 (listed) Queensland
Pauropsalta annulata Stevens and Carver 1986: 264 (type material, listed) Australia, Queensland
Pauropsalta annulata Ewart 1988: 181, 191, 195–197, Fig 9A, Plate 4C (illustrated, oscillogram, natural history, listed, comp. note) Queensland, New South Wales
Pauropsalta annulata Lithgow 1988: 65 (listed) Australia, Queensland
Pauropsalta annulata Ewart 1989a: 80 (comp. note)
Pauropsalta annulata Ewart 1989b: 291–294, 327, 349–356, 358, 372, Fig 1, Figs 42–44, Figs 45A–B, Fig 59a, Table 5 (lectotype designation, described, illustrated, key, distribution, song, oscillogram, comp. note) Queensland, New South Wales
Pauropsalta annulata group Ewart 1989b: 291, 293, 361, Fig 1 (described, comp. note)
Pauropsalta annulata Moulds 1990: 136, 139–140 (distribution, ecology, listed, comp. note) Australia, New South Wales, Queensland
Pauropsalta annulata Ewart 1995: 87 (described, illustrated, natural history, acoustic behavior, oscillogram) Queensland
Pauropsalta annulata Ewart 1998a: 61 (natural history) Queensland
Pauropsalta annulata Ewart 2001a: 501, Fig 3 (acoustic behavior) Queensland
Pauropsalta annulata Sueur 2001: 44, Table 1 (listed)
Pauropsalta annulata Moulds and Carver 2002: 37 (distribution, ecology, type data, listed) Australia, New South Wales, Queensland
Pauropsalta annulata Popple and Ewart 2002: 117 (illustrated) Australia, Queensland
Pauropsalta annulata Popple and Strange 2002: 27 (comp. note) Australia, Queensland
Pauropsalta annulata Williams 2002: 156–157 (listed) Australia, Queensland, New South Wales
Pauropsalta annulata Shiyake 2007: 8, 109, 111, Fig 196 (illustrated, distribution, listed, comp. note) Australia
Pauropsalta annulata Popple, Walter and Raghu 2008: 204–207, 209–212 (song diversity, comp. note) Australia, Queensland, New South Wales
Pauropsalta annulata Marshall and Hill 2009: 4, Table 2 (comp. note)

Pauropsalta annulata Popple and Walter 2010: 554–556, 559–563, Fig 1 (song diversity, illustrated, comp. note) Australia, Queensland, New South Wales

P. *annulata* song type *A* Popple, Walter & Raghu, 2008

Pauropsalta annulata song type A Popple, Walter and Raghu 2008: 206–207, 209–212, Figs 1–5 (song, oscillogram, habitat, distribution, comp. note) Australia, Queensland

Pauropsalta annulata song type A Popple and Walter 2010: 554–563, Figs 2–5, Tables 1–2 (song, oscillogram, habitat, distribution, comp. note) Australia, Queensland

P. *annulata* song type *B* Popple, Walter & Raghu, 2008

Pauropsalta annulata song type B Popple, Walter and Raghu 2008: 206–207, 209–212, Figs 2–5 (song, habitat, distribution, comp. note) Australia, Queensland

Pauropsalta annulata song type B Popple and Walter 2010: 554–563, Figs 2–5, Tables 1–2 (song, habitat, distribution, comp. note) Australia, Queensland

P. *annulata* song type *C* Popple, Walter & Raghu, 2008

Pauropsalta annulata song type C Popple, Walter and Raghu 2008: 206–207, 209–212, Figs 2–5 (song, habitat, distribution, comp. note) Australia, Queensland, New South Wales

P. *annulata* song type C_N Popple, Walter & Raghu, 2008

Pauropsalta annulata song type C_{North} Popple, Walter and Raghu 2008: 210–212 (song, habitat, distribution, comp. note) Australia, Queensland, New South Wales

Pauropsalta annulata song type C_{North} Popple and Walter 2010: 554–563, Figs 2–5, Tables 1–2 (song, habitat, distribution, comp. note) Australia, Queensland

P. *annulata* song type C_S Popple, Walter & Raghu, 2008

Pauropsalta annulata song type C_{South} Popple, Walter and Raghu 2008: 210–212 (song, habitat, distribution, comp. note) Australia, Queensland, New South Wales

Pauropsalta annulata song type C_{South} Popple and Walter 2010: 554–563, Figs 2–5, Tables 1–2 (song, habitat, distribution, comp. note) Australia, Queensland, New South Wales

P. *annulata* song type *X* Popple & Walter, 2010

Pauropsalta annulata song type X Popple and Walter 2010: 555–563, Figs 2–5, Tables 1–2 (song, habitat, distribution, comp. note) Australia, Queensland

P. sp. nr. *annulata* Southern Red-eyed Squeaker Moss & Popple, 2000

Pauropsalta sp. nr. *annulata* Southern Red-eyed Squeaker Moss and Popple 2000: 58 (listed, comp. note) Australia, New South Wales

P. sp. nr. *annulata* Inland Sprinkler Squeaker Popple & Strange, 2002

Pauropsalta sp. nr. *annulata* Inland Sprinkler Squeaker Popple and Strange 2002: 30, Table 1 (listed, habitat) Australia, Queensland

P. sp. nr. *annulata* Maraca Squeaker Popple & Strange, 2002

Pauropsalta sp. nr. *annulata* Maraca Squeaker Popple and Strange 2002: 23, 26–27, 30, Fig 4B, Fig 7C, Table 1 (described, illustrated, habitat, oscillogram, listed, comp. note) Australia, Queensland

P. sp. nr. *annulata* Static Squeaker Popple & Strange, 2002

Pauropsalta sp. nr. *annulata* Static Squeaker Popple and Strange 2002: 23, 26–27, 30, Fig 4B, Fig 7D, Table 1 (described, illustrated, habitat, oscillogram, listed, comp. note) Australia, Queensland

P. sp. nr. *annulata* Emery, Emery, Emery & Popple, 2005

Pauropsalta sp. nr. *annulata* Emery, Emery, Emery and Popple 2005: 100–102, 104–105, 107, Fig 9, Tables 1–3 (life cycle, illustrated, comp. note) New South Wales

P. *aquilus* Ewart, 1989 = *Pauropsalta* sp. D Ewart, 1988

Pauropsalta sp. D Ewart 1988: 181–182, 187, 191–192, 196–197, Fig 2, Fig 9D, Plate 1F (illustrated, oscillogram, natural history, listed, comp. note) Queensland

Pauropsalta aquilus Ewart 1989b: 291, 293–294, 361–366, 372, Fig 1, Figs 50–51, Figs 52E–F, Fig 59d, Table 6 (n. sp., described, illustrated, key, distribution, song, oscillogram, comp. note) Equals *Pauropsalta* sp. D Ewart, 1988 Queensland

Pauropsalta aquilus group Ewart 1989b: 291, 293, 361, Fig 1 (described, comp. note)

Pauropsalta aquilus Ewart 1998a: 61 (natural history) Queensland

Pauropsalta aquilus Sueur 2001: 44, Table 1 (listed) Equals *Pauropsalta* sp. D

Pauropsalta aquilus Moulds and Carver 2002: 37–38 (distribution, ecology, type data, listed) Australia

P. *ayrensis* Ewart, 1989

Pauropsalta ayrensis Ewart 1989b: 291, 293–294, 349, 353, 355–359, 372, Fig 1, Figs 45C–D, Figs 46–47, Fig 59b (n. sp., described, illustrated, key, distribution, song, oscillogram, comp. note) Queensland

Pauropsalta ayrensis Sueur 2001: 44, Table 1 (listed)

Pauropsalta ayrensis Moulds and Carver 2002: 38 (distribution, ecology, type data, listed) Australia, Queensland

P. *basalis* Goding & Froggatt, 1904 to *Nanopsalta basalis*

P. *bellatrix* Ashton, 1914 to *Physeema bellatrix*

P. *borealis* Goding & Froggatt, 1904 = *Melampsalta borealis*

Pauropsalta borealis Hahn 1962: 9 (listed) Australia, Northern Territory
Pauropsalta borealis Stevens and Carver 1986: 264 (type material) Australia, Northern Territory
Pauropsalta borealis Ewart 1989b: 293 (listed) Northern Territory, South Australia
Pauropsalta borealis Moulds 1990: 131 (distribution, listed) Australia, Northern Territory
Pauropsalta borealis Moulds and Carver 2002: 38 (distribution, ecology, type data, listed) Australia, Northern Territory

P. circumdata (Walker, 1852) to *Palapsalta circumdata*

***P. collina* Ewart, 1989** = *Pauropsalta* sp. C Ewart, 1988
Pauropsalta sp. C Ewart 1988: 181, 187, 191–192, 196–197, Fig 1, Figs 9B-C, Plate 1B (illustrated, oscillogram, natural history, listed, comp. note) Queensland
Pauropsalta collina Ewart 1989b: 291–293, 316, 322–327, 370, Fig 1, Figs 18E–F, Figs 24–25, Fig 57c (n. sp., described, illustrated, key, distribution, song, oscillogram, comp. note) Equals *Pauropsalta* sp. C Ewart, 1988 Queensland
Pauropsalta collina Coombs 1996: 57, 60, Table 1 (emergence time, comp. note) New South Wales
Pauropsalta collina Ewart 1996: 13 (listed)
Pauropsalta collina Sueur 2001: 44, Table 1 (listed) Equals *Pauropsalta* sp. C
Pauropsalta collina Moulds and Carver 2002: 39 (distribution, ecology, type data, listed) Australia, New South Wales, Queensland

***P.* sp. nr. *collina* Emery, Emery, Emery & Popple, 2005**
Pauropsalta sp. nr. *collina* (TNS 422) Emery, Emery, Emery and Popple 2005: 100, 102–107, Fig 8, Tables 1–3 (life cycle, illustrated, comp. note) New South Wales

***P.* sp. nr. *collina/encaustica* Haywood, 2005**
Pauropsalta sp. nr. *collina/encaustica* Haywood 2005: 31, 56–58, Fig 17, Table 2 (illustrated, distribution, described, habitat, song, oscillogram, listed, comp. note) South Australia
Pauropsalta sp. nr. *collina/encaustica* Haywood 2006a: 14, Table 1 (listed) South Australia
Pauropsalta sp. nr. *collina/encaustica* Haywood 2006b: 52–53, Table 1 (illustrated, distribution, described, habitat, song, oscillogram, listed, comp. note) South Australia

***P. corticinus* Ewart, 1989** = *Pauropsaltra* sp. A Ewart 1986 = *Pauropsalta corticina*
Pauropsaltra sp. A Ewart 1986: 51–55, Fig 3A, Table 1 (illustrated, natural history, listed, comp. note) Queensland, New South Wales
Pauropsalta corticinus Ewart 1989b: 291–294, 311–317, 322, 327, 353, 358, 364, 370, Fig 1, Figs 16–17, Figs 18A–B, Fig 57a, Table 1 (n. sp., described, illustrated, key, distribution, song, oscillogram, comp. note) Equals *Pauropsalta* sp. A Ewart, 1988 Queensland, New South Wales, Victoria
Pauropsalta corticinus Ewart 1995: 86 (described, illustrated, natural history, acoustic behavior, oscillogram) Queensland
Pauropsalta corticinus Coombs 1996: 56, Table 1 (emergence time, comp. note) New South Wales
Paurapralta corticinus Ewart 1996: 14 (listed)
Pauropsalta corticinus Moulds 1996b: 92 (comp. note) Australia, Queensland, New South Wales
Pauropsalta corticinus Ewart 2001a: 501, Fig 3 (acoustic behavior) Queensland
Pauropsalta corticinus Sueur 2001: 44, Table 1 (listed)
Pauropsalta corticinus Moulds and Carver 2002: 39 (distribution, ecology, type data, listed) Australia, New South Wales, Queensland, Victoria
Pauropsalta corticinus Popple and Ewart 2002: 115, 117 (illustrated, comp. note) Australia, Queensland
Pauropsalta corticinus Cryan 2005: 565, 568, Fig 2, Table 1 (phylogeny, comp. note)
Pauropsalta corticinus Ceotto, Kergoat, Rasplus and Bourgoin 2008: 668 (outgroup)

***P.* sp. nr. *corticinus* Montane Bark Squeaker M13 Moss & Popple, 2000**
Pauropsalta sp. nr. *corticinus* Montane Bark Squeaker M13 Moss and Popple 2000: 57 (listed, comp. note) Australia, New South Wales
Pauropsalta sp. nr. *corticinus* Williams 2002: 156–157 (listed) Australia

***P.* sp. nr. *corticinus* Emery, Emery, Emery & Popple, 2005**
Pauropsalta sp. nr. *corticinus* (TNS 422) Emery, Emery, Emery and Popple 2005: 100–102, 104–107, Fig 7, Tables 1–3 (life cycle, illustrated, comp. note) New South Wales

P. dameli Distant, 1905 to *Drymopsalta dameli*

***P. dolens* (Walker, 1850)** = *Cicada dolens*
Pauropsalta dolens Ewart 1989b: 293, 309, Figs 13F–H (status, valid species, illustrated, comp. note) Australia
Cicada dolens Ewart 1989b: 310 (comp. note)
Pauropsalta dolens Moulds and Carver 2002: 39 (synonymy, distribution, ecology, type data, listed) Equals *Cicada dolens* Australia
Cicada dolens Moulds and Carver 2002: 39 (type data, listed) New Holland

P. dubia Goding & Froggatt, 1904 to *Platypsalta dubia*

P. sp. nr. *dubia* Broad-wing Scrub-buzzer Popple & Strange, 2002 to *Pauropsalta* sp. nr. *dubia*

***P. elgneri* Ashton, 1912** = *Melampsalta elgneri*

Pauropsalta elgneri Ewart 1989b: 291–293, 300–304, Fig 7, Figs 8A–B, Fig 56 (synonymy, described, illustrated, key, distribution, comp. note) Equals *Melampsalta elgneri* Queensland

Pauropsalta elgneri Moulds 1990: 133–134, Plate 20, Fig 7 (illustrated, distribution, ecology, listed, comp. note) Australia, Queensland

Pauropsalta elgneri Ewart 1993: 138, 144, Fig 7 (natural history, acoustic behavior, song analysis) Queensland

Pauropsalta elgneri Sueur 2001: 44, Table 1 (listed)

Pauropsalta elgneri Moulds and Carver 2002: 40 (distribution, ecology, type data, listed) Australia, Queensland

Pauropsalta elgneri Ewart 2005a: 176–177, Fig 9A (described, song, oscillogram) Queensland

P. emma Goding & Froggatt, 1904 to *Mugadina emma*

***P. encaustica* (Germar, 1834)** = *Cicada encaustica* = *Melampsalta encaustica* = *Cicada arclus* Walker, 1850 = *Melampsalta arclus* = *Pauropsalta arclus* = *Cicada juvenis* Walker, 1850

Melampsalta encaustica Greenup 1964: 22–23 (comp. note) Australia, New South Wales

Pauropsalta encaustica Strümpel 1983: 108, Fig 144 (oscillogram, comp. note)

Pauropsaltria encaustica Ewart 1986: 53 (comp. note) Australia

Pauropsalta encaustica Ewart 1989b: 289, 293, 308–311, 324, 353, Figs 12–15, (lectotype designation, synonymy, described, illustrated, key, distribution, song, oscillogram, comp. note) Equals *Cicada encaustica* Equals *Cicada arclus* Equals *Cicada juvenis* Equals *Melampsalta encaustica* Equals *Melampsalta arclus* Equals *Pauropsalta arclus* New South Wales

Cicada encaustica Ewart 1989b: 308, 310 (comp. note)

Cicada arclus Ewart 1989b: 308–310, Figs 13A–B (comp. note)

Cicada juvenis Ewart 1989b: 308–310, Figs 13C–E (comp. note)

Pauropsalta encaustica Ewart 1990: 1 (type material) Australia

Cicada encaustica Ewart 1990: 1–2 (type material, type locality) Equals *Melampsalta encaustica* Australia

Pauropsalta encaustica Moulds 1990: 135–136, 140, Plate 20, Fig 9, Fig 9a (illustrated, distribution, ecology, listed, comp. note) Australia, New South Wales, Queensland, South Australia, Tasmania, Victoria, Western Australia

Pauropsalta encaustica Naumann 1993: 6, 113, 149 (listed) Australia

Pauropsalta encaustica New 1996: 95 (comp. note) Australia

Pauropsalta encaustica Moss and Popple 2000: 58 (listed, comp. note) Australia, New South Wales

Pauropsalta encaustica Sueur 2001: 44, Table 1 (listed)

Pauropsalta encaustica Moulds and Carver 2002: 40 (synonymy, distribution, ecology, type data, listed) Equals *Cicada encaustica* Equals *Cicada arclus* Equals *Cicada juvenis* Australia, New South Wales, Queensland, South Australia, Tasmania, Victoria, Western Australia

Cicada encaustica Moulds and Carver 2002: 40 (type data, listed) New Holland

Cicada arclus Moulds and Carver 2002: 40 (type data, listed) New Holland

Cicada juvenis Moulds and Carver 2002: 40 (type data, listed) New Holland

Pauropsalta encaustica Williams 2002: 156–157 (listed) Australia, Queensland, New South Wales

Pauropsalta encaustica Moulds 2005b: 395, 399, 405, 417–419, 421, Fig 39, Figs 56–59, Fig 62, Table 1 (illustrataed, higher taxonomy, phylogeny, key, listed, comp. note) Australia

Pauropsalta encaustica Shiyake 2007: 8, 109, 111, Fig 197 (illustrated, distribution, listed, comp. note) Australia

***P.* n. sp. #16M nr. *encaustica* Gwynne & Schatral 1988**

Pauropsalta n. sp. #16M nr. *encaustica* Gwynne and Schatral 1988: 32, 36–38, 43, 66–67, Fig 5, Fig 33 (illustrated, oscillogram, emergence time, key, listed, comp. note) Western Australia

***P.* sp. nr. *P. encaustica* Moss, 1989**

Pauropsalta sp. near *P. encaustica* Moss 1989: 103 (distribution, song, comp. note) Tasmania

Pauropsalta sp. near *P. encaustica* Moss 1990: 6 (distribution, song, comp. note) Tasmania

***P.* sp. nr. *encaustica* Sandstone Squeaker M26 Moss & Popple, 2000**

Pauropsalta sp. nr. *encaustica* Sandstone Squeaker M26 Moss and Popple 2000: 53, 57 (comp. note) Australia, New South Wales

Pauropsalta sp. 2 nr. *encaustica* Williams 2002: 156–157 (listed) Australia

***P.* sp. nr. *encaustica* Montane Squeaker M14 Moss & Popple, 2000**

Pauropsalta sp. nr. *encaustica* Montane Squeaker M14 Moss and Popple 2000: 57 (listed, comp. note) Australia, New South Wales

Pauropsalta sp. 1 nr. *encaustica* Williams 2002: 156–157 (listed) Australia

***P. exaequata* (Distant, 1892)** = *Melampsalta exaequata* = *Melampsalta exequata* (sic) = *Pauropsalta exequata* (sic)

Pauropsalta exaequata Ewart 1989b: 293 (listed) India

***P. extensa* Goding & Froggatt, 1904**

Pauropsalta extensa Ewart 1989b: 293 (listed) South Australia

Pauropsalta extensa Moulds 1990: 131 (distribution,) Australia, South Australia

Pauropsalta extensa Moulds and Carver 2002: 40 (distribution, ecology, type data, listed) Australia, South Australia

***P. extrema* (Distant, 1892)** = *Melampsalta extrema*

Pauropsalta extrema Ewart 1989b: 291–293, 300–301, 304, 368–369, Fig 5, Fig 56 (revised status, synonymy, described, illustrated, key, distribution, comp. note) Equals *Melampsalta extrema* Western Australia

Melampsalta extrema Ewart 1989b: 301 (lectotype designation, comp. note)

Pauropsalta extrema Moulds 1990: 132–134, Plate 20, Fig 4, Fig 4a (rev. stat., synonymy, illustrated, distribution, ecology, listed, comp. note) Equals *Melampsalta extrema* Australia, Western Australia

Melampsalta extrema Moulds 1990: 132 (comp. note)

Pauropsalta extrema Naumann 1993: 60, 113, 149 (listed) Australia

Pauropsalta extrema Moulds and Carver 2002: 40–41 (synonymy, distribution, ecology, type data, listed) Equals *Melampsalta extrema* Australia, Western Australia

Melampsalta extrema Moulds and Carver 2002: 40–41 (type data, listed) Western Australia

Pauropsalta extrema Moulds 2005b: 423 (comp. note) Australia

***P.* near *extrema* Marshall & Hill, 2009**

Pauropsalta "near extrema" Marshall and Hill 2009: 4, Table 2 (comp. note)

P. eyrei (Distant, 1882) to *Palapsalta eyrei*

***P. fuscata* Ewart, 1989** = *Pauorpsalta furcata* (sic) = *Pauropsalta* sp. B Ewart, 1986 = *Melampsalta (Pauropsalta) encaustica* Young (1972)

Pauropsaltra sp. B Ewart 1986: 51–55, Fig 3B, Table 1 (illustrated, natural history, listed, comp. note) Queensland, New South Wales

Pauropsalta fuscata Ewart 1989b: 291–294, 311, 315–322, 327, 353, 358, 364, 370, Fig 1, Figs 18C–D, Figs 19–23, Fig 57b (n. sp., described, illustrated, key, distribution, song, oscillogram, comp. note) Equals *Melampsalta (Pauropsalta) encaustica* Young 1972 Equals *Pauropsalta* sp. B Ewart, 1986 Queensland, New South Wales

Pauropsalta furcata Ewart 1996: 13 (listed)

Pauropsalta fuscata Moss 2000: 69 (comp. note) Queensland

Pauropsalta fuscata Moss and Popple 2000: 57 (listed, comp. note) Australia, New South Wales

Pauropsalta fuscata Ewart 2001a: 501, Fig 3 (acoustic behavior) Queensland

Pauropsalta fuscata Sueur 2001: 44, Table 1 (listed)

Pauropsalta fuscata Moulds and Carver 2002: 41 (distribution, ecology, type data, listed) Australia, New South Wales, Queensland

Pauropsalta fuscata Popple and Ewart 2002: 115 (illustrated, distribution, ecology, song, oscillogram, comp. note) Australia, Queensland, New South Wales

***P.* sp. nr. *fuscata* Montane Grass Squeaker Moss & Popple, 2000**

Pauropsalta sp. nr. *fuscata* Montane Grass Squeaker Moss & Popple 2000: 57 (listed, comp. note) Australia, New South Wales

***P.* sp. nr. *fuscata* Moss and Popple, 2000**

Pauropsalta sp. nr. *fuscata* Moss & Popple 2000: 54, 57 (comp. note) Australia, New South Wales

***P.* sp. nr. *fuscata* Haywood, 2005**

Pauropsalta sp. nr. *fuscata* Haywood 2005: 31–34, Fig 9, Table 2 (illustrated, distribution, described, habitat, song, oscillogram, listed, comp. note) South Australia

Pauropsalta sp. nr. *fuscata* Haywood 2006a: 14, Table 1 (listed) South Australia

Pauropsalta sp. nr. *fuscata* Haywood 2006b: 48–49, 53, Table 1 (illustrated, distribution, described, habitat, song, oscillogram, listed, comp. note) South Australia

P. fuscomarginata Distant, 1914 to *Kobonga fuscomarginata*

***P. infrasila* Moulds, 1987**

Pauropsalta infrasila Moulds 1987b: 17–22, Fig 2, Figs 5–6 (n. sp., described, illustrated, distribution, comp. note) Australia, Queensland

Pauropsalta infrasila Ewart 1989b: 289, 291–294, 298–299, 304–306, 308, 368–369, Fig 1, Figs 4C–D, Fig 9, Fig 56 (n. sp., described, illustrated, key, distribution, comp. note) Queensland

Pauropsalta infrasila Moulds 1990: 134–135, 141, Plate 20, Fig 2, Fig 2a (illustrated, distribution, ecology, listed, comp. note) Australia, Queensland

Pauropsalta infrasila Ewart 1993: 138, 145, Fig 9 (labeled Fig 10 Species C) (natural history, acoustic behavior, song analysis) Queensland

Pauropsalta infrasila Sueur 2001: 44, Table 1 (listed) Equals Species C

Pauropsalta infrasila Moulds and Carver 2002: 41 (distribution, ecology, type data, listed) Australia, Queensland

Pauropsalta infrasila Ewart 2005a: 177–178, Fig 9B (described, song, oscillogram) Queensland

***P. infuscata* (Goding & Froggatt, 1904)** = *Melampsalta infuscata* = *Cicadetta infuscata*

Melampsalta infuscata Hahn 1962: 9 (listed) Australia, South Australia

Melampsalta infuscata Stevens and Carver 1986: 265 (type material) Australia, South Australia

Pauropsalta infuscata Ewart 1989b: 293, 361 (n. comb., comp. note)

Cicadetta infuscata Moulds 1990: 144 (distribution, listed) Australia, South Australia

Cicadetta infuscata Moulds and Carver 2002: 28 (synonymy, distribution, ecology, type data, listed) Equals *Melampsalta infuscata* Australia, South Australia

Melampsalta infuscata Moulds and Carver 2002: 28 (type data, listed) South Australia

***P. johanae* Boulard, 1993**

Pauropsalta johanae Boulard 1993e: 121–125, Figs 21–27 (n. sp., described, illustrated, sonogram, comp. note) New Caledonia

Pauropsalta johanae Boer and Duffels 1996b: 309 (distribution, comp. note) New Caledonia

*Pauropsalta johanae*Boulard 1997a: 180, 192, 195, Fig 36C (oscillogram, comp. note) New Caledonia

Pauropsalta johanae Sueur 2001: 44, Table 1 (listed)

Pauropsalta johanae Buckley, Arensburger, Simon and Chambers 2002: 6, 11, Fig 1 (phylogeny, comp. note) New Caledonia

Pauropsalta johanae Arensburger, Buckley, Simon, Moulds, and Holsinger 2004: 558–559, 561–563, Table 1, Figs 1–2 (molecular phylogeny, comp. note) New Caledonia

Pauropsalta johanae Boulard 2006g: 49, 131–132, 142, Fig 103 (sonogram, comp. note) New Caledonia

Pauropsalta johanae Buckley and Simon 2007: 423 (comp. note) New Caledonia

Pauropsalta johanae Simon 2009: 104, Fig 2 (phylogeny, comp. note) New Caledonia

***P. judithae* Boulard, 1997** = *Pauropsalta juditae* (sic)

Pauropsalta judithae Boulard 1997a: 191–195, Figs 28–35, Fig 36D (n. sp., described, illustrated, sonogram, comp. note) New Caledonia

Pauropsalta juditae (sic) Boulard 1997a: 194 (comp. note)

Pauropsalta judithae Sueur 2001: 44, Table 1 (listed)

Pauropsalta judithae Boulard 2006g: 48–49 (comp. note) New Caledonia

P. sp. aff. *labyrinthica* Haywood, 2005 to *Physeema* sp. aff. *labyrinthica*

***P. lineola* Ashton, 1914** = *Melampsalta lineola*

Pauropsalta lineola Ewart 1989b: 293 (listed) Western Australia

Pauropsalta lineola Moulds 1990: 131 (distribution, listed) Australia, Western Australia

Pauropsalta lineola Moulds and Carver 2002: 41 (distribution, ecology, type data, listed) Australia, Western Australia

P. marginata (Leach, 1814) to *Palapsalta circumdata* (Walker, 1852)

***P. melanopygia* (Germar, 1834)** = *Cicada melanopygia* = *Tibicen melanopygius* = *Melampsalta melanopygia* = *Pauropsalta leurensis* Goding & Froggatt, 1904

Pauropsalta leurensis Hahn 1962: 10 (listed) Australia, New South Wales

Pauropsalta leurensis Koçak 1982: 150 (type species of *Pauropsalta*)

Pauropsalta leurensis Stevens and Carver 1986: 265 (type material) Australia, New South Wales

Pauropsalta melanopygia Moulds 1987b: 17 (status, comp. note) Australia

Pauropsalta melanopygia Ewart 1989b: 291–293, 297–301, 304, 368–369, Fig 1, Fig 5, Fig 56 (synonymy, described, illustrated, key, distribution, comp. note) Equals *Cicada melanopygia* Equals *Tibicen melanopygius* Equals *Melampsalta melanopygia* Equals *Melampsalta mnemae* (sic) Equals *Pauropsalta leurensis* New South Wales, Australia Capital Territory, Victoria, South Australia

Cicada melanopygia Ewart 1989b: 297 (lectotype designation, comp. note) Northern Territory

Pauropsalta leurensis Ewart 1989b: 290, 295 (type species of *Pauropsalta*, comp. note)

Pauropsalta melanopygia Ewart 1990: 1 (type material) Australia

Cicada melanopygia Ewart 1990: 1–2 (type material, type locality) Equals *Melampsalta melanopygia* Australia

Pauropsalta melanopygia Moulds 1990: 132–134 (synonymy, distribution, ecology, listed, comp. note) Equals *Cicada melanopygia* Equals *Tibicen melanopygius* Equals *Melampsalta melanopygia* Australia, Northern Territory

Cicada melanopygia Moulds 1990: 133 (listed)

Tibicen melanopygius Moulds 1990: 133 (listed)

Melampsalta melanopygia Moulds 1990: 133 (listed)

Pauropsalta melanopygia Moulds and Carver 2002: 41 (synonymy, distribution, ecology, type data, listed) Equals *Cicada melanopygia* Australia, Northern Territory

Cicada melanopygia Moulds and Carver 2002: 41 (type data, listed) Northern Territory

Pauropsalta melanopygia Marshall and Hill 2009: 3–4, 7, Table 2 (comp. note)

***P.* sp. nr. *melanopygia* Moss, 1988**

Pauropsalta sp. near *melanopygia* Moss 1988: 30 (distribution, song, comp. note) Queensland

***P. mimica* Distant, 1907**
Pauropsalta mimica Ewart 1989b: 293 (listed) South Australia

P. minima (Goding & Froggatt, 1904) to *Punia minima*

***P. mneme* (Walker, 1850)** = *Cicada mneme* = *Melampsalta mneme* = *Melampsalta mnemae* (sic) = *Cicadetta (Pauropsalta) mneme* = *Cicada antica* Walker, 1850 = *Pauropsalta leurensis* Goding & Froggatt, 1904
Melampsalta mneme Greenup 1964: 23 (comp. note) Australia, New South Wales
Pauropsalta mneme Strümpel 1983: 108, Fig 144 (oscillogram, comp. note)
Pauropsalta prolongata Stevens and Carver 1986: 265 (type material) Australia, South Australia
Pauropsalta mmeme (sic) Ewart 1989b: 290 (comp. note) Equals *Pauropsalta leurensis*
Pauropsalta mneme Ewart 1989b: 291, 293, 295–299, Figs 1–3, Figs 4A–B (synonymy, described, illustrated, key, distribution, song, oscillogram, comp. note) Equals *Cicada mneme* Equals *Cicada antica* Equals *Melampsalta mneme* Equals *Melampsalta mnemae* (sic) Equals *Pauropsalta leurensis* New South Wales, Australia Capital Territory, Victoria, South Australia
Pauropsalta mneme group Ewart 1989b: 291–292, Fig 1 (described, comp. note)
Cicada mneme Ewart 1989b: 295 (lectotype designation, comp. note)
Cicada antica Ewart 1989b: 295 (comp. note)
Pauropsalta mneme Moss 1989: 105 (comp. note)
Pauropsalta mneme Moss 1990: 7 (comp. note)
Cicada mneme Moulds 1990: 130 (type species of *Pauropsalta*, listed)
Pauropsalta mneme Moulds 1990: 131–132, Plate 20, Fig 6, Fig 6a (illustrated, distribution, ecology, listed, comp. note) Australia
Pauropsalta mneme Naumann 1993: 58, 113, 149 (listed) Australia
Pauropsalta mneme New 1996: 95 (comp. note) Australia
Cicada mneme Boer 1999: 136 (type species of *Pauropsalta*)
Pauropsalta mneme Sanborn 1999a: 52 (type material, listed)
Pauropsalta mneme Moss and Popple 2000: 57 (listed, comp. note) Australia, New South Wales
Pauropsalta mneme Sueur 2001: 44, Table 1 (listed)
Pauropsalta leurensis Moulds and Carver 2002: 37, 42 (type species of *Pauropsalta*, type data, listed) Equals *Pauropsalta mneme*
Pauropsalta mneme Moulds and Carver 2002: 37 (type data, listed)
Pauropsalta mneme Moulds and Carver 2002: 37, 42 (synonymy, distribution, ecology, type data, listed) Equals *Cicada antica* Equals *Cicada mneme* Equals *Pauropsalta leurensis* Australia
Cicada antica Moulds and Carver 2002: 42 (type data, listed)
Cicada mneme Moulds and Carver 2002: 42 (type data, listed)
Pauropsalta mneme Williams 2002: 156–157 (listed) Australia, Queensland, New South Wales, Victoria, South Australia
Pauropsalta mneme Emery, Emery, Emery and Popple 2005: 102–107, Tables 1–3 (comp. note) New South Wales
Pauropsalta mneme Moulds 2005b: 395–397, 399, 405, 410, 417–419, 421, 425, Fig 40, Figs 56–59, Fig 62, Table 1 (illustrated, higher taxonomy, phylogeny, key, listed, comp. note) Australia
Pauropsalta mneme Shiyake 2007: 8, 109, 111, Fig 198 (illustrated, distribution, listed, comp. note) Australia
Pauropsalta mneme Ewart 2009b: 141 (comp. note)

***P. nigristriga* Goding & Froggatt, 1904** = *Pauropsalta nigristrigata* (sic)
Pauropsalta nigristriga Hahn 1962: 10 (listed) Australia, Queensland
Pauropsalta nigristriga Stevens and Carver 1986: 265 (type material) Australia, Queensland
Pauropsalta nigristriga Moulds 1987b: 17–19, Fig 1, Figs 3–4 (rev. status, described, illustrated, distribution, comp. note) Australia, Queensland
Pauropsalta nigristriga Ewart 1989b: 291, 293–294, 300, 303, 305–306, 342, 346–348, 371, Fig 1, Figs 8E–F, Figs 40–41, Fig 58d (described, illustrated, key, distribution, song, oscillogram, comp. note) Queensland
Pauropsalta nigristriga Moulds 1990: 133, 135, 141, Plate 20, Fig 3, Fig 3a (illustrated, distribution, ecology, listed, comp. note) Australia, Queensland
Pauropsalta nigristriga Sueur 2001: 44, Table 1 (listed)
Pauropsalta nigristriga Moulds and Carver 2002: 42 (distribution, ecology, type data, listed) Australia, Queensland

P. nodicosta Goding & Froggatt, 1904 to *Clinata nodicosta*

P. sp. aff. *nodicosta* (Little Shrub Cicada) Haywood, 2005 to *Clinata* sp. aff. *nodicosta* (Little Shrub Cicada)

P. sp. aff. *nodicosta* (Little Buloke Cicada) Haywood, 2005 to *Clinata* sp. aff. *nodicosta* (Little Buloke Cicada)

***P. opacus* Ewart, 1989** = *Pauropsalta opaca*

Pauropsalta opacus Ewart 1989b: 291, 293–294, 298–299, 306–308, 368–369, Fig 1, Figs 4E–F, Figs 10–11, Fig 56 (n. sp., described, illustrated, key, distribution, song, oscillogram, comp. note) Queensland
Pauropsalta opacus Sueur 2001: 44, Table 1 (listed)
Pauropsalta opacus Moulds and Carver 2002: 42 (distribution, ecology, type data, listed) Australia, Queensland
Pauropsalta opacus Moulds 2005b: 401, Fig 33 (illustrated, comp. note) Australia

***P. prolongata* Goding & Froggatt, 1904**

Pauropsalta prolongata Hahn 1962: 10 (listed) Australia, South Australia
Pauropsalta prolongata Ewart 1989b: 293, 297 (revised status, listed, comp. note) Australia
Pauropsalta prolongata Moulds and Carver 2002: 42–43 (distribution, ecology, type data, listed) Australia

***P. rubea* (Goding & Froggatt, 1904)** = *Melampsalta rubea* = *Cicadetta rubea* = *Melampsalta nebulosa* Goding & Froggatt, 1904 = *Cicadetta nebulosa* = *Melampsalta geisha* Distant, 1915 = *Cicadetta geisha* = *Pauropsalta geisha*

Melampsalta rubea Hahn 1962: 9 (listed) Australia, Queensland
Pauropsaltria geisha Ewart 1986: 51, Table 1 (listed) Queensland
Melampsalta rubea Stevens and Carver 1986: 265 (type material) Australia, Queensland
Pauropsalta rubea Ewart 1989b: 289, 291, 293–294, 331–337, 371, Fig 1, Figs 30–31, Figs 32C–D, Fig 58a (n. comb., synonymy, described, illustrated, key, distribution, song, oscillogram, comp. note) Equals *Melampsalta rubea* Equals *Cicadetta rubea* Equals *Melampsalta geisha* Equals *Cicadetta geisha* Queensland, New South Wales
Pauropsalta geisha Ewart 1989b: 289 (synonymy, comp. note)
Melampsalta rubea Ewart 1989b: 333, 335 (lectotype designation, comp. note)
Melampsalta geisha Ewart 1989b: 333, 335 (comp. note)
Pauropsalta rubea Moulds 1990: 136, 138, 140, Plate 20, Fig 8 (n. comb., synonymy, illustrated, distribution, ecology, listed, comp. note) Equals *Melampsalta rubea* Equals *Cicadetta rubea* Australia, New South Wales, Queensland
Melampsalta rubea Moulds 1990: 138 (listed) Queensland
Cicadetta rubea Moulds 1990: 138 (listed) Queensland
Cicadetta nebulosa Moulds 1990: 144 (distribution, listed) Australia, New South Wales, Queensland
Pauropsalta rubea Ewart 1998a: 61 (natural history) Queensland
Cicadetta rubea Sanborn 1999a: 52 (type material, listed)
Pauropsalta rubea Moss 2000: 69 (comp. note) Queensland
Pauropsalta rubea Moss and Popple 2000: 57 (listed, comp. note) Australia, New South Wales
Pauropsalta rubea Ewart 2001b: 69 (listed) Queensland
Pauropsalta rubea Sueur 2001: 44, Table 1 (listed)
Cicadetta nebulosa Moulds and Carver 2002: 31 (synonymy, distribution, ecology, type data, listed) Equals *Melampsalta nebulosa* Australia, New South Wales, Queensland
Melampsalta nebulosa Moulds and Carver 2002: 31 (type data, listed) Queensland, New South Wales
Pauropsalta rubea Moulds and Carver 2002: 43 (synonymy, distribution, ecology, type data, listed) Equals *Melampsalta rubea* Equals *Melampsalta geisha* Australia, New South Wales, Queensland
Melampsalta rubea Moulds and Carver 2002: 43 (type data, listed) Queensland
Melampsalta geisha Moulds and Carver 2002: 43 (type data, listed) Queensland
Pauropsalta rubea Popple and Ewart 2002: 114 (illustrated, distribution, ecology, song, oscillogram, comp. note) Australia, Queensland
Pauropsalta rubea Williams 2002: 156–157 (listed) Australia, Queensland, New South Wales

***P. rubra* Goding & Froggatt, 1904** = *Melampsalta rubra*

Pauropsalta rubra Hahn 1962: 10 (listed) Australia, Victoria
Pauropsalta rubra Ewart 1989b: 293 (listed) Victoria
Pauropsalta rubra Moulds 1990: 131 (distribution, listed) Australia, Victoria
Pauropsalta rubra Moulds and Carver 2002: 43 (distribution, ecology, type data, listed) Australia, Victoria

***P. rubristrigata* (Goding & Froggatt, 1904)** = *Melampsalta rubristrigata* = *Melampsalta rubrostrigata* (sic) = *Cicadetta rubristrigata*

Melampsalta rubrostrigata (sic) Hahn 1962: 9 (listed) Australia, South Australia
Melampsalta rubristrigata Greenup 1964: 23 (comp. note) Australia, New South Wales
Melampsalta rubristrigata Stevens and Carver 1986: 265 (type material) Australia, South Australia
Pauropsalta rubristrigata Ewart 1989b: 293, 349 (comp. note)
Pauropsalta rubristrigata Moulds 1990: 136–138 (n. comb., synonymy, distribution, ecology,

type data, listed) Equals *Melampsalta rubristrigata* Equals *Cicadetta rubristrigata* Australia, Australian Capital Territory, New South Wales, South Australia, Victoria

Melampsalta rubristrigata Moulds 1990: 136 (listed)

Cicadetta rubristrigata Moulds 1990: 136 (listed)

Pauropsalta rubristrigata Moss and Popple 2000: 57 (listed, comp. note) Australia, New South Wales

Pauropsalta rubristrigata Moulds and Carver 2002: 43 (synonymy, distribution, ecology, type data, listed) Equals *Melampsalta rubristrigata* Australia, Australian Capital Territory, New South Wales, South Australia, Victoria

Melampsalta rubristrigata Moulds and Carver 2002: 43 (type data, listed) South Australia

Pauropsalta rubristrigata Williams 2002: 156–157 (listed) Australia, Queensland, New South Wales, Victoria, South Australia

Pauropsalta rubristrigata Haywood 2005: 9, 30, Fig 8, Table 1 (distribution, listed) South Australia

Cicadetta rubristrigata Haywood 2005: 28–29 (illustrated, distribution, described, habitat, song, oscillogram, comp. note) South Australia, New South Wales, Australia Capital Territory, Victoria

Pauropsalta rubristigata Haywood 2006a: 14, 23, Tables 1–2 (illustrated, distribution, described, habitat, song, comp. note) South Australia

Cicadetta rubristrigata Haywood 2006a: 21–22 (illustrated, distribution, described, habitat, song, oscillogram, comp. note) South Australia, New South Wales, Australia Capital Territory, Victoria

P. rufifascia **(Walker, 1850)** = *Cicada rufifascia* = *Melampsalta rufifascia*

P. Sandstone Marshall & Hill, 2009

Pauropsalta "Sandstone" Marshall and Hill 2009: 4–5, 7–8, Fig 6H, Table 2 (sonogram, comp. note) Australia

P. siccanus **Ewart, 1989** = *Pauropsalta* sp. F Ewart, 1988

Pauropsalta sp. F Ewart 1988: 182, 188, 191, 195, 200–201, Fig 4, Fig 11G-H, Plate 4G (illustrated, oscillogram, natural history, listed, comp. note) Queensland, New South Wales

Pauropsalta siccanus Ewart 1989b: 291–294, 308, 324, 326–329, 353, 358, 364–365, 370, Fig 1, Figs 26–27, Figs 52C–D, Fig 57c (n. sp., described, illustrated, key, distribution, song, oscillogram, comp. note) Equals *Pauropsalta* sp. F Ewart, 1988 Queensland

Pauropsalta siccanus Ewart 1998a: 61–62, Fig 7 (described, natural history, song, oscillogram) Queensland

Pauropsalta siccanus Sueur 2001: 44, Table 1 (listed) Equals Pauropsalta sp. F

Pauropsalta siccanus Moulds and Carver 2002: 43 (distribution, ecology, type data, listed) Australia, New South Wales, Queensland

Pauropsalta siccanus Callitris Squeaker Popple and Strange 2002: 30, Table 1 (listed, habitat) Australia, Queensland

P. sp. nr. *siccanus* Callistris Squeaker Popple & Strange, 2002

Pauropsalta sp. nr. *siccanus* Callitris Squeaker Popple and Strange 2002: 30, Table 1 (listed, habitat) Australia, Queensland

P. signata Distant, 1914 to *Dipsopsalta signata*

P. sp. nov. 1 Lithgow, 1988

Pauropsalta sp. nov. 1 Lithgow 1988: 65 (listed) Australia, Queensland

P. sp. nov. 2 Lithgow, 1988

Pauropsalta sp. nov. 2 Lithgow 1988: 65 (listed) Australia, Queensland

P. sp. A Ewart and Popple, 2001

Pauropsalta sp. A Ewart and Popple 2001: 59–60, 65, Fig 5A (described, habitat, song, oscillogram) Queensland

Pauropsalta sp. A Ewart 2009b: 140–143, 154, Fig 2, Plate 1A (oscillogram, illustrated, distribution, habitat, comp. note) Equals Black Spinifex Squeaker, no. 692 Queensland

Black Spinifex Squeaker Ewart 2009b: 140–141, Fig 1 (distribution, emergence times, habitat, comp. note) Queensland

P. sp. C Ewart, 1993

Pauropsalta sp. C Ewart 1993: 138–139, 145, Fig 10 (labeled Fig 9 *Pauropsalta infrasila*) (natural history, acoustic behavior, song analysis) Queensland

Pauropsalta Heathlands sp. C Ewart 2005a: 177–178, Fig 9C (described, song, oscillogram) Queensland

P. sp. M Moulds, 2005

Pauropsalta sp. M Moulds 2005b: 395–397, 399, 412, 417–419, 421, 425, Figs 56–59, Fig 62, Table 1 (higher taxonomy, phylogeny, key, listed, comp. note) Australia

P. stigmatica **Distant, 1905** = *Melampsalta stigmatica*

Pauropsalta stigmata Ewart 1989b: 293 (listed) South Australia

Pauropsalta stigmatica Moulds 1990: 131 (distribution, listed) Australia, South Australia

Pauropsalta stigmatica Moulds and Carver 2002: 44 (distribution, ecology, type data, listed) Australia, South Australia

Pauropsalta stigmatica Ewart 2005b: 445 (comp. note)

P. subolivacea Ashton, 1912 to *Samaecicada subolivacea*

P. vernalis Distant, 1916 to *Ggomapsalta vernalis*

P. virgulatus Ewart, 1989 to *Palapsalta virgulatus*

P. vitellinus Ewart, 1989 to *Palapsalta vitellinus*

P. walkeri **Moulds & Owen, 2011**

P. **near walkeri Marshall & Hill, 2009**

Pauropsalta "near walkeri" Marshall and Hill 2009: 4, 7, Table 2 (comp. note)

Graminitigrina **Ewart & Marques, 2008**

Graminitigrina Ewart and Marques 2008: 150–156, 199, Table 1 (n. gen., described, diversity, comp. note)

Graminitigrina spp. Ewart and Marques 2008: 176–177, 197–198, 200 (comp. note)

Graminitigrina Popple and Emery 2010: 150 (comp. note) Australia

G. bolloni **Ewart & Marques, 2008**

Graminitigrina bolloni Ewart and Marques 2008: 152–153, 156–157, 164–169, 176–177, 180, 187–194, 198–199, Fig 5, Figs 9–10, Figs 29–32, Plate 2A-B, Table 4 (n. sp., described, illustrated, key, song, oscillogram, distribution, habitat, listed, comp. note) Queensland

Graminitigrina bolloni Marshall and Hill 2009: 3 (comp. note)

G. bowensis **Ewart & Marques, 2008**

Graminitigrina bowensis Ewart and Marques 2008: 153–161, 175–187, 189–190, 199, 201–202, Figs 1–5, Figs 15–24, Fig A1, Plate 1A-E, Table 2 (n. sp., described, illustrated, key, song, oscillogram, distribution, habitat, listed, comp. note) Queensland

G. carnarvonensis **Ewart & Marques, 2008**

Graminitigrina carnarvonensis Ewart and Marques 2008: 152–153, 156–157, 167, 169–172, 176–177, 180, 191, 193–199, Fig 5, Figs 11–12, Figs 33–35, Plate 2C-D, Table 5 (n. sp., described, illustrated, key, song, oscillogram, distribution, habitat, listed, comp. note) Queensland

G. karumbae **Ewart & Marques, 2008** = *Pauropsalta* sp. D Ewart, 1993 = Heathlands sp. D Ewart, 2005

Pauropsalta sp. D Ewart 1993: 139, 146, Fig 11 (natural history, acoustic behavior, song analysis) Queensland

Heathlands sp. D Ewart 2005a: 172–173, Fig 3A, Fig 5 (described, song, oscillogram, distribution, comp. note) Queensland

Graminitigrina karumbae Ewart and Marques 2008: 153, 156–157, 160–165, 176–178, 184–186, 189–190, 199, Figs 5–8, Figs 25–28, Plate 1F-H, Table 3 (n. sp., described, illustrated, key, song, oscillogram, distribution, habitat, listed, comp. note) Queensland

G. triodiae **Ewart & Marques, 2008** = *Urabunana* sp. Dunn, 2002

Graminitigrina triodiae Ewart and Marques 2008: 153, 156–157, 172–177, 180, 194–199, Fig 5, Figs 13–14, Figs 36–40, Plate 2E-F, Table 6 (n. sp., described, illustrated, key, song, oscillogram, distribution, habitat, listed, comp. note) Queensland

Palapsalta **Moulds, 2012**

P. circumdata **(Walker, 1852)** = *Cicada circumdata* (nom. nud. pro *Cicada marginata* Leach, 1814 nec *Cicada marginata* Olivier, 1790) = *Pauropsalta circumdata* = *Tettigonia marginata* = *Cicada marginata* = *Melampsalta marginata* = *Cicadetta marginata* = *Pauropsalta marginata* = *Cicada themiscura* Walker, 1850 = *Melampsalta themiscura* = *Melampsalta fletcheri* Goding & Froggatt, 1904 = *Cicadetta fletcheri*

Melampsalta fletcheri Hahn 1962: 9 (listed) Australia, New South Wales

Cicadetta marginata Moulds 1983: 432 (comp. note) Australia

Cicadetta marginata Ewart 1986: 51, 54–55, Fig 3C, Table 1 (illustrated, taxonomic position, natural history, listed, comp. note) Queensland, Tasmania

Melampsalta fletcheri Stevens and Carver 1986: 264 (type material) Australia, New South Wales

Cicadetta marginata Ewart 1988: 182, 191–192, 196–197, Fig 9F-G, Plate 1E (illustrated, oscillogram, natural history, listed, comp. note) Queensland, New South Wales

Pauropsalta marginata Ewart 1989b: 289, 291, 293–294, 335–339, 342, Fig 1, Figs 32A–B, Figs 33–34 (n. comb., synonymy, described, illustrated, key, distribution, song, oscillogram, comp. note) Equals *Tettigonia marginata* Equals *Cicada marginata* Equals *Cicada themiscura* Equals *Melampsalta marginata* Equals *Melampsalta themiscura* Equals *Melampsalta fletheri* Equals *Cicadetta marginata* Equals *Cicadetta fletcheri* Queensland, New South Wales

Pauropsalta marginata group Ewart 1989b: 291, 293, 308, Fig 1 (described, comp. note)

Tettigonia marginata Ewart 1989b: 336 (comp. note)

Cicada themiscura Ewart 1989b: 336 (comp. note)

Melampsalta fletcheri Ewart 1989b: 336, 338 (comp. note)

Pauropsalta marginata Moss 1989: 105 (comp. note)

Pauropsalta marginata Ewart 1990: 1 (type material) Australia
Tettigonia marginata Ewart 1990: 4 (type material, type locality) Equals *Cicada themiscura* Australia
Pauropsalta marginata Moss 1990: 7 (comp. note)
Pauropsalta circumdata Moulds 1990: 6, 137–139, Plate 20, fig 5, Fig 5a (n. comb., synonymy, illustrated, distribution, ecology, listed, comp. note) Equals *Tettigonia marginata* Equals *Cicada marginata* Equals *Cicada themiscura* Equals *Cicada circumdata* Equals *Melampsalta marginata* Equals *Melampsalta themiscura* Equals *Melampsalta fletcheri* Equals *Cicadetta marginata* Equals *Cicadetta fletcheri* Australia, New South Wales, Queensland
Tettigonia marginata Moulds 1990: 137 (listed)
Cicada marginata Moulds 1990: 137 (listed)
Cicada themiscura Moulds 1990: 137 (listed)
Cicada circumdata Moulds 1990: 137 (listed)
Melampsalta marginata Moulds 1990: 137 (listed)
Melampsalta themiscura Moulds 1990: 137 (listed)
Melampsalta fletcheri Moulds 1990: 137 (listed)
Cicadetta marginata Moulds 1990: 137 (listed)
Cicadetta fletcheri Moulds 1990: 137 (listed)
Pauropsalta circumdata Ewart 1998a: 61 (natural history) Equals *Pauropsalta marginata* Ewart 1989 Queensland
Pauropsalta circumdata Moss 2000: 69 (comp. note) Queensland
Pauropsalta circumdata Moss and Popple 2000: 58 (listed, comp. note) Australia, New South Wales
Pauropsalta circumdata Ewart 2001a: 501, Fig 3 (acoustic behavior) Queensland
Pauropsalta circumdata Sueur 2001: 44, Table 1 (listed) Equals *Cicadetta marginata* Equals *Pauropsalta marginata*
Pauropsalta circumdata Moulds and Carver 2002: 38–39 (synonymy, distribution, ecology, type data, listed) Equals *Tettigonia marginata* Equals *Cicada marginata* Equals *Cicada themiscura* Equals *Cicada circumdata* Equals *Melampsalta fletcheri* Australia, New South Wales, Queensland
Tettigonia marginata Moulds and Carver 2002: 38 (type data, listed) New Holland
Cicada marginata Moulds and Carver 2002: 39 (type data, listed)
Cicada themiscura Moulds and Carver 2002: 39 (type data, listed)
Cicada circumdata Moulds and Carver 2002: 39 (type data, listed)
Melampsalta fletcheri Moulds and Carver 2002: 39 (type data, listed)
Pauropsalta circumdata Popple and Ewart 2002: 114 (illustrated, distribution, ecology, song, oscillogram, comp. note) Australia, Queensland, New South Wales
Pauropsalta circumdata Williams 2002: 156–157 (listed) Australia, Queensland, New South Wales
Cicadetta circumdata Arensburger, Buckley, Simon, Moulds, and Holsinger 2004: 566 (comp. note) Australia
Pauropsalta circumdata Emery, Emery, Emery and Popple 2005: 102, 104–107, Tables 1–3 (comp. note) New South Wales
Pauropsalta circumdata Moulds 2005b: 395, 399, 417–419, 421, 425, Figs 56–59, Fig 62, Table 1 (higher taxonomy, phylogeny, key, listed, comp. note) Australia

***P. eyrei* (Distant, 1882)** = *Pauropsalta eyreyi* = *Pauropsalta eyrey* (sic) = *Melampalta eyrei* = *Cicadetta eyrei*
Pauropsalta eyrei Ewart 1989b: 289, 291–294, 349, 356, 372, Fig 1, Figs 45E–F, Figs 48–49, Fig 59c (n. comb., synonymy, described, illustrated, key, distribution, song, oscillogram, comp. note) Equals *Melampsalta eyrei* Equals *Cicadetta eyrei* Queensland
Melampsalta eyrei Ewart 1989b: 359 (comp. note)
Pauropsalta eyrei Moulds 1990: 12, 141–142, Plate 18, Fig 6, Fig 6a, Plate 24, Fig 23 (synonymy, illustrated, distribution, ecology, listed, comp. note) Equals *Melampsalta eyrei* Equals *Cicadetta eyrei* Australia, Queensland
Melampsalta eyrei Moulds 1990: 141 (listed)
Cicadetta eyrei Moulds 1990: 141 (listed)
Pauropsalta eyrei Ewart 1993: 138, 143, Fig 8 (natural history, acoustic behavior, song analysis) Queensland
Pauropsalta eyrey (sic) Boer 1995d: 201 (distribution, comp. note) Papua New Guinea
Pauropsalta eyrei Boer 1995d: 234 (distribution, comp. note) Papua New Guinea
Pauropsalta eyrei Boer and Duffels 1996b: 305, 328 (distribution, comp. note) New Guinea, Australia, Papuan Peninsula
Pauropsalta eyrei Ewart 1998b: 135 (natural history, acoustic behavior, listed, comp. note) Queensland
Pauropsaltria eyrei Boer 1999: 115–117, 119–121, 128, 136, 140, 144, Fig 1, Fig 55, Fig 67 (key, described, illustrated, distribution, phylogeny, comp. note) New Guinea, Australia, Queensland
Pauropsalta eyrei Sueur 2001: 44, Table 1 (listed)
Pauropsalta eyrei Moulds and Carver 2002: 41 (synonymy, distribution, ecology, type data,

listed) Equals *Melampsalta eyrei* Australia, Queensland

Melampsalta eyrei Moulds and Carver 2002: 41 (type data, listed) Queensland

Pauropsalta eyrei Moulds 2005b: 395–396, 399, 405, 409, 412–413, 417–419, 421, 425, Fig 41, Fig 50, Figs 56–59, Fig 62, Table 1 (illustrated, higher taxonomy, phylogeny, key, listed, comp. note) Australia

Pauropsalta eyrei Ewart and Marques 2008: 150–151, Table 1 (type species of new genus, described, comp. note)

***P. virgulatus* (Ewart, 1989)** = *Pauropsalta virgulatus*

Pauropsalta virgulatus Ewart 1989b: 291–294, 300, 308, 342–346, 364–365, 371, Fig 1, Figs 38–39, Figs 52A–B, Fig 58c, Table 4 (n. sp., described, illustrated, key, distribution, song, oscillogram, comp. note) Queensland

Pauropsalta virgulatus Ewart 1993: 139 (comp. note)

Pauropsalta virgulatus Sueur 2001: 44, Table 1 (listed)

Pauropsalta virgulatus Moulds and Carver 2002: 44 (distribution, ecology, type data, listed) Australia, Queensland

***P. vitellinus* (Ewart, 1989)** = *Pauropsalta vitellinus* = *Pauropsalta* sp. E Ewart, 1988 = *Pauropsalta* sp. near *melanopygia* St Leger Moss, 1988

Pauropsalta sp. E Ewart 1988: 182, 188, 191–192, 196–197, Fig 3, Fig 9E, Plate 1D (illustrated, oscillogram, natural history, listed, comp. note) Queensland

Pauropsalta vitellinus Ewart 1989b: 291, 293–294, 300–301, 303, 339–343, 345, 371, Fig 1, Figs 8C–D, Figs 35–37, Fig 58b, Table 3 (n. sp., described, illustrated, key, distribution, song, oscillogram, comp. note) Equals *Pauropsalta* sp. E Ewart, 1988 Equals *Pauropsalta* sp. nr. *Pauropsalta melanopygia* Moss, 1988 Queensland

Pauropsalta vitellinus Sueur 2001: 44, Table 1 (listed) Equals *Pauropsalta* sp. E

Pauropsalta vitellinus Moulds and Carver 2002: 44 (distribution, ecology, type data, listed) Australia, Queensland

Pauropsalta vitellinus Popple and Strange 2002: 30, Table 1 (listed, habitat) Australia, Queensland

***Nanopsalta* Moulds, 2012**

***N. basalis* (Goding & Froggatt, 1904)** = *Pauropsalta basalis* = *Melampsalta basalis* = *Pauropsalta endeavourensis* Distant, 1907 = *Melampsalta endeavourensis* = *Cicadetta endeavourensis*

Pauropsalta basalis Hahn 1962: 9 (listed) Australia, Queensland

Pauropsalta basalis Stevens and Carver 1986: 264 (type material) Australia, Queensland

Pauropsalta basalis Moss 1988: 30 (distribution, song, comp. note) Queensland

Pauropsalta basalis Ewart 1989b: 289, 291, 293, 364–365, Fig 1, Figs 52G–H, Figs 53–55 (synonymy, described, illustrated, key, distribution, song, oscillogram, comp. note) Equals *Melampsalta basalis* Equals *Pauropsalta endeavourensis* Equals *Melampsalta endeavourensis* Equals *Cicadetta endeavourensis* Queensland

Pauropsalta endeavourensis Ewart 1989b: 289, 261, 368, Fig 1 (synonymy, comp. note)

Pauropsalta basalis group Ewart 1989b: 291, 293, 366, Fig 1 (described, comp. note)

Pauropsalta basalis Moulds 1990: 142–143, Plate 19, Fig 4, Fig 4a (illustrated, distribution, ecology, listed, comp. note) Australia, Queensland

Pauropsalta endeavourensis Moulds 1990: 131 (type data, listed) Queensland

Pauropsalta basalis Ewart 1993: 138, 143, Fig 6 (labeled Fig 5 Species B) (natural history, acoustic behavior, song analysis) Queensland

Pauropsalta basalis Sueur 2001: 44, Table 1 (listed) Equals Species B

Pauropsalta basalis Moulds and Carver 2002: 38 (synonymy, distribution, ecology, type data, listed) Equals *Pauropsalta endeavourensis* Australia, Queensland

Pauropsalta endeavourensis Moulds and Carver 2002: 38 (type data, listed) Queensland

Pauropsalta basalis Moulds 2005b: 395–397, 399, 405, 412–413, 417–419, 421, 425, Fig 40, Figs 56–59, Fig 62, Table 1 (illustrated, higher taxonomy, phylogeny, key, listed, comp. note) Australia

Pauropsalta basalis Ewart and Marques 2008: 150–151, Table 1 (type species of new genus, described, comp. note)

***Punia* Moulds, 2012**

***P. minima* (Goding & Froggatt, 1904)** = *Cicadetta minima* = *Pauropsalta minima* = *Melampsalta minima*

Pauropsalta minima Ewart 1989b: 293 (listed) Northern Territory, South Australia

Cicadetta minima Moulds 1990: 18, 162–163, Plate 16, Fig 3, Fig 3a (n. comb., synonymy, illustrated, distribution, ecology, listed, comp. note) Equals *Pauropsalta minima* Equals *Melampsalta minima* Australia, Northern Territory

Pauropsalta minima Moulds 1990: 162 (listed)

Melampsalta minima Moulds 1990: 162 (listed)

Cicadetta minima Ewart 1998b: 135 (listed) Queensland
Cicadetta minima Moulds and Carver 2002: 30 (synonymy, distribution, ecology, type data, listed) Equals *Pauropsalta minima* Australia, Northern Territory
Pauropsalta minima Moulds and Carver 2002: 30 (type data, listed) Northern Territory
Cicadetta minima Moulds 2005b: 395–398, 408, 412–413, 417–419, 421, 425, 436, Fig 47, Figs 56–59, Fig 62, Table 1 (illustrated, higher taxonomy, phylogeny, key, listed, comp. note) Australia

***Neopunia* Moulds, 2012**

***N. graminis* (Goding & Froggatt, 1904)** = *Cicadetta graminis* = *Melampsalta graminis*
Cicadetta graminis Moulds 1990: 144 (distribution, listed) Australia, South Australia
Cicadetta graminis Moulds and Carver 2002: 27 (synonymy, distribution, ecology, type data, listed) Equals *Melampsalta graminia* Australia, South Australia
Melampsalta graminis Moulds and Carver 2002: 27 (type data, listed) South Australia
Cicadetta graminis Moulds 2005b: 395, 397–398, 412, 417–419, 421, 425, Figs 56–59, Fig 62, Table 1 (higher taxonomy, phylogeny, key, listed, comp. note) Australia

***Marteena* Moulds, 1986**

Marteena Moulds 1986: 39–40 (n. gen., described, comp. note) Australia
Marteena Moulds 1990: 32, 126, 147 (listed, diversity, comp. note) Australia
Marteena Moulds and Carver 2002: 53 (type data, listed) Australia
Marteena Moulds 2005b: 377, 393, 423, 425, 430, 436, Table 2 (higher taxonomy, listed, comp. note) Australia

***M. rubricincta* (Goding & Froggatt, 1904)** = *Tibicen rubricinctus* = *Tibicen rubricincta* = *Tibicina rubricinctus* = *Tibicina rubricincta*
Tibicen rubricincta Hahn 1962: 11 (listed) Australia, South Australia
Tibicen rubricinctus Moulds 1986: 39–40 (type species of *Marteena*, comp. note) Australia
Marteena rubricincta Moulds 1986: 40–42, Figs 1–4 (n. comb., synonymy, lectotype designation, described, key, illustrated, distribution, comp. note) Equals *Tibicen rubricinctus* Equals *Tibicina rubricincta* Australia, Western Australia, South Australia, New South Wales
Tibicen rubricinctus Stevens and Carver 1986: 265 (type material) Australia, South Australia
Marteena rubricincta Moulds 1990: 126, 147, Plate 18, Fig 3 (illustrated, distribution, ecology, listed, comp. note) Equals *Tibicen rubricinctus* Australia, New South Wales, South Australia, Western Australia
Tibicen rubricinctus Moulds 1990: 126 (type species of *Marteena*, listed) South Australia
Marteena rubricincta Moulds and Carver 2002: 54 (synonymy, distribution, ecology, type data, listed) Equals *Tibicen rubricinctus* Australia, New South Wales, South Australia, Western Australia
Tibicen rubricinctus Moulds and Carver 2002: 54 (type species of *Marteena*, type data, listed) South Australia
Marteena rubricincta 2005b: 395–397, 399, 401, 417–420, 425, Fig 33, Figs 56–59, Fig 62, Table 1 (illustrated, higher taxonomy, phylogeny, listed, comp. note) Australia

***Auscala* Moulds, 2012**

***A. spinosa* (Goding & Froggatt, 1904)** = *Cicadetta spinosa* = *Melampsalta spinosa* = *Kobonga clara* Distant, 1916
Melampsalta spinosa Hahn 1962: 9 (listed) New Holland
Melampsalta spinosa Stevens and Carver 1986: 265 (type material) New Holland
Kobonga clara Ewart 1988: 184, 190–191, 193, 198–199, Fig 7, Fig 10J, Plate 2F (illustrated, oscillogram, natural history, listed, comp. note) Queensland, New South Wales
Kobonga clara Lithgow 1988: 65 (listed) Australia, Queensland
Cicadetta spinosa Moulds 1990: 144–145, Plate 18, Fig 1 (synonymy, illustrated, distribution, ecology, listed, comp. note) Equals *Melampsalta spinosa* Equals *Kobonga clara* Australia, New South Wales, Queensland, Victoria
Melampsalta spinosa Moulds 1990: 144 (lectotype designation, listed, comp. note)
Kobonga clara Moulds 1990: 144 (listed, comp. note)
Cicadetta spinosa Ewart 1998a: 60 (described, natural history, song) Queensland
Kobonga clara Moulds and Kopestonsky 2001: 141 (comp. note)
Cicadetta spinosa Moulds and Kopestonsky 2001: 141 (comp. note) Equals *Kobonga clara*
Cicadetta spinosa Sueur 2001: 43, Table 1 (listed) Equals *Kobonga clara*
Cicadetta spinosa Emery and Emery 2002: 137–139, Fig 2 (illustrated, comp. note) Queensland, New South Wales, Victoria

Cicadetta spinosa Moulds and Carver 2002: 32 (synonymy, distribution, ecology, type data, listed) Equals *Melampsalta spinosa* Equals *Kobonga clara* Australia, New South Wales, Queensland, Victoria
Melampsalta spinosa Moulds and Carver 2002: 32 (type data, listed) New Holland
Kobonga clara Moulds and Carver 2002: 32 (type data, listed) New South Wales
Cicadetta spinosa Popple and Strange 2002: 28, Table 1 (listed, habitat) Australia, Queensland
Cicadetta spinosa Emery, Emery, Emery and Popple 2005: 102–107, Tables 1–3 (comp. note) New South Wales
Cicadetta spinosa Haywood 2005: 9, 19–21, Fig 5, Table 1 (illustrated, distribution, described, habitat, song, oscillogram, listed, comp. note) South Australia, New South Wales, Queensland, Victoria
Cicadetta spinosa Haywood 2006a: 14–16, 23, Tables 1–2 (illustrated, distribution, described, habitat, song, oscillogram, comp. note) South Australia, Queensland, New South Wales, Victoria

***Birrima* Distant, 1906**
Birrima Moulds 1990: 17, 32, 171 (listed, diversity, comp. note) Australia, Queensland, New South Wales
Birrima Moulds and Carver 2002: 23 (listed) Australia
Birrima Moulds 2005b: 390, 402, 413, 430, 436, Table 2 (higher taxonomy, listed, comp. note) Australia
Birrima Shiyake 2007: 9, 114 (listed, comp. note)

***B. castanea* (Goding & Froggatt, 1904)** = *Pauropsalta castanea* = *Melampsalta castanea* = *Cicadetta castanea* = *Melampsalta fulva* Goding & Froggatt, 1904 = *Cicadetta fulva* = *Birrima montrouzieri* Distant, 1906 = *Cicadetta montrouzieri*
Melampsalta fulva Hahn 1962: 9 (listed) Australia, New South Wales
Pauropsalta castanea Hahn 1962: 10 (listed) Australia, South Australia
Pauropsalta castanea Stevens and Carver 1986: 264 (type material) Australia, South Australia
Pauropsalta castanea Ewart 1989b: 293 (listed) South Australia
Cicadetta fulva Moulds 1990: 143 (distribution, listed) Australia, New South Wales
Birrima montrouzieri Moulds 1990: 171 (type species of *Birrima*, listed, comp. note) Equals *Birrima castanea*
Birrima castanea Moulds 1990: 171–172, Plate 21, Fig 4, Figs 4a–4b (synonymy, illustrated, distribution, ecology, listed, comp. note) Equals *Melampsalta castanea* Equals *Birrima montrouzieri* n. syn. Equals *Cicadetta montrouzieri* Equals *Cicadetta castanea* Australia, New South Wales, Queensland
Melampsalta castanea Moulds 1990: 171 (listed) New South Wales
Cicadetta montrouzieri Moulds 1990: 171 (listed) New South Wales
Cicadetta castanea Moulds 1990: 171 (listed) New South Wales
Birrima castanea Naumann 1993: 48, 75, 148 (listed) Australia
Birrima castanea Moss and Popple 2000: 57 (listed, comp. note) Australia, New South Wales
Birrima montrouzieri Moulds and Carver 2002: 23 (type species of *Birrima*, type data, listed) Australia, New South Wales, Queensland
Birrima castanea Moulds and Carver 2002: 23 (synonymy, distribution, ecology, type data, listed) Equals *Melampsalta castanea* Equals *Birrima montrouzieri* Australia, New South Wales, Queensland
Melampsalta castanea Moulds and Carver 2002: 23 (type data, listed) New South Wales
Cicadetta fulva Moulds and Carver 2002: 27 (synonymy, distribution, ecology, type data, listed) Equals *Melampsalta fulva* Australia, New South Wales
Melampsalta fulva Moulds and Carver 2002: 27 (type data, listed) New South Wales
Birrima castanea Williams 2002: 156–157 (listed) Australia, Queensland, New South Wales
Birrima castanea Moulds 2005b: 395–398, 405, 407–409, 411, 417–419, 421, 425, Fig 38, Fig 46, Fig 53, Figs 56–59, Fig 62, Table 1 (illustrated, higher taxonomy, phylogeny, key, listed, comp. note) Australia

B. montrouzieri Distant, 1906 to *B. castanea* (Goding & Froggatt, 1904)

***B. varians* (Germar, 1834)** = *Cicada varians* = *Dundubia varians* = *Melampsalta varians*
Birrima varians Ewart 1986: 51, 56, Table 1 (natural history, listed, comp. note) Queensland
Birrima varians Ewart 1989a: 80 (comp. note)
Birrima varians Moulds 1990: 172–173, Plate 21, Fig 3, Fig 3a-3b, Plate 24, Fig 16 (synonymy, distribution, ecology, type data, listed) Equals *Cicada varians* Australia, New South Wales, Queensland
Birrima varians Naumann 1993: 6, 75, 149 (listed) Australia
Birrima varians Coombs 1996: 56, 60, Table 1 (emergence time, comp. note) New South Wales

Cicada varians Ewart 1990: 1–2 (type material, type locality) Equals *Melampsalta varians* Australia
Birrima varians Ewart 1995: 86 (described, illustrated, natural history, acoustic behavior, oscillogram) Queensland
Birrima varians Ewart 1996: 14 (listed)
Birrima varians Moulds 1996b: 92–93 (acoustic behavior, comp. note) Australia, Queensland, New South Wales
Birrima varians Moss and Popple 2000: 57 (listed, comp. note) Australia, New South Wales
Birrima varians Ewart 2001b: 69 (listed) Queensland
Birrima varians Sueur 2001: 42, Table 1 (listed)
Birrima varians Moulds and Carver 2002: 24 (synonymy, distribution, ecology, type data, listed) Equals *Cicada varians* Australia, New South Wales, Queensland
Cicada varians Moulds and Carver 2002: 24 (type data, listed) Australasia
Birrima varians Popple and Ewart 2002: 116, 118 (illustrated, comp. note) Australia, Queensland
Birrima varians Williams 2002: 156–157 (listed) Australia, Queensland, New South Wales
Birrima varians Arensburger, Buckley, Simon, Moulds, and Holsinger 2004: 558 (molecular phylogeny, comp. note) Australia
Birrima varians Shiyake 2007: 9, 113–114, Fig 203 (illustrated, distribution, listed, comp. note) Australia

***Yoyetta* Moulds, 2012**

***Y. aaede* (Walker, 1850)** = *Melampsalta aaede* = *Cicadetta aaede*
Cicadetta aaede Moulds 1990: 143 (distribution, listed) Australia, South Australia
Cicadetta aaede Moulds and Carver 2002: 24 (synonymy, distribution, ecology, type data, listed) Equals *Cicada aaede* Australia, South Australia
Cicada aaede Moulds and Carver 2002: 24 (type data, listed) South Australia

***Y. abdominalis* (Distant, 1892)** = *Melampsalta abdominalis* = *Cicadetta abdominalis* = *Pauropsalta castanea* Goding & Froggatt
Cicadetta abdominalis Moss 1989: 102–103 (distribution, song, comp. note) Tasmania
Cicadetta abdominalis Moss 1990: 6 (distribution, song, comp. note) Tasmania
Cicadetta abdominalis Moulds 1990: 155–156, Plate 3, Fig 1, Plate 17, Fig 7, fig 7a (illustrated, distribution, ecology, listed, comp. note) Australia, New South Wales, South Australia, Tasmania, Victoria
Cicadetta abdominalis Boulard 1991c: 119 (comp. note) Tasmania
Cicadetta abdominalis Brandstetter 1999: 40 (density, comp. note) Victoria
Cicadetta abdominalis Moulds and Carver 2002: 24 (synonymy, distribution, ecology, type data, listed) Equals *Melampsalta abdominalis* Equals *Pauropsalta castanea* Australia, New South Wales, South Australia, Tasmania, Victoria
Melampsalta abdominalis Moulds and Carver 2002: 24 (type data, listed) Victoria
Pauropsalta castanea Moulds and Carver 2002: 24 (type data, listed)
Cicadetta abdominalis complex Emery, Emery, Emery and Popple 2005: 109 (comp. note) New South Wales
Cicadetta abdominalis Shiyake 2007: 8, 104, 106, Fig 178 (illustrated, distribution, listed, comp. note) Australia

***Y.* sp. nr. *abdominalis* sp. 4 Katydid Cicada (Moss & Popple, 2000)** = Cicadetta sp. nr. *abdominalis* sp. 4 Katydid Cicada
Cicadetta sp. nr. *abdominalis* sp. 4 Katydid Cicada Moss and Popple 2000: 57 (listed, comp. note) Australia, New South Wales
Cicadetta sp. 4 nr. *abdominalis* Williams 2002: 156–157 (listed) Australia

***Y.* sp. nr. *abdominalis* (Emery, Emery, Emery & Popple, 2005)** = *Cicadetta* sp. nr. *abdominalis*
Cicadetta sp. nr. *abdominalis* (TNS 512) Emery, Emery, Emery and Popple 2005: 99, 103–107, 109, Fig 6, Tables 1–3 (life cycle, illustrated, comp. note) New South Wales
Cicadetta nr. *abdominalis* (sp. TNS 512) Emery 2008: 35 (comp. note) New South Wales

***Y.* sp. aff. *abdominalis* (Haywood, 2005)** = *Cicadetta* sp. aff. *abdominalis*
Cicadetta sp. aff. *abdominalis* Haywood 2005: 31, 47–49, Fig 14, Table 2 (illustrated, distribution, described, habitat, song, oscillogram, listed, comp. note) South Australia
Cicadetta sp. aff. *abdominalis* Haywood 2006a: 14–15, Table 1 (illustrated, distribution, described, habitat, song, comp. note) South Australia
Cicadetta sp. aff. *abdominalis* Haywood 2007: 14–15, 18, Appendix 1 (illustrated, distribution, described, habitat, song, oscillogram, comp. note) South Australia

***Y.* sp. nr. *abdominalis* (Emery, 2008)** = *Cicadetta* sp. nr. *abdominalis*
Cicadetta nr. *abdominalis* (sp. TNS 513) Emery 2008: 35 (comp. note) New South Wales

***Y. celis* (Moulds, 1988)** = *Cicadetta celis*

Cicadetta celis Moulds 1988: 39, 43–47, Figs 4–5, Figs 8–9 (n. sp., described, illustrated, distribution, comp. note) Australia, Victoria, Queensland, New South Wales

Cicadetta celis Moulds 1990: 151–152, Plate 17, Fig 1, Fig 1a (illustrated, distribution, ecology, listed, comp. note) Australia, Victoria, New South Wales, Queensland

Cicadetta celis Buckley, Simon, Flook and Misol 2000: 570–572, Figs 2–3 (sequence, comp. note) New Zealand

Cicadetta celis Moss and Popple 2000: 58 (listed, comp. note) Australia, New South Wales

Cicadetta celis Chambers, Boon, Buckley and Hitchmough 2001: 380, Fig 3 (phylogeny, comp. note) Australia

Cicadetta celis Buckley, Arensburger, Simon and Chambers 2002: 6, 11, Fig 1 (phylogeny, comp. note) Australia

Cicadetta celis Emery and Emery 2002: 137, 139 (comp. note) New South Wales

Cicadetta celis Moulds and Carver 2002: 25–26 (distribution, ecology, type data, listed) Australia, Victoria, New South Wales, Queensland

Cicadetta celis Williams 2002: 156–157 (listed) Australia, Queensland, Victoria

Cicadetta celis Arensburger, Buckley, Simon, Moulds, and Holsinger 2004: 558–559, 562–563, Table 1, Figs 1–2 (molecular phylogeny, comp. note) Australia

Cicadetta celis White, Boyce and Stoddart 2004: 292, Fig 1 (wing microstructure, comp. note)

Cicadetta celis Emery, Emery, Emery and Popple 2005: 101–108, Tables 1–3 (life cycle, comp. note) New South Wales

Cicadetta celis Moulds 2005b: 395–398, 400–402, 417–419, 421, 425, Figs 32–33, Fig 36, Figs 56–59, Fig 62, Table 1 (illustrated, higher taxonomy, phylogeny, key, listed, comp. note) Australia

Cicadetta celis Stoddart, Cadusch, Boyce, Erasmus and Comins 2006: 681–685, Fig 1, Fig 5a, Figs 6–8, Table 1 (wing microstructure, comp. note)

Cicadetta celis Shiyake 2007: 8, 104, 106, Fig 179 (illustrated, distribution, listed, comp. note) Australia

Cicadetta celis Emery 2008: 34 (life cycle, comp. note) New South Wales

Cicadetta celis Ewart and Marques 2008: 150–151, Table 1 (type species of new genus, described, comp. note)

Cicadetta celis Kostovski, White, Mitchell, Mitchell, Austin and Stoddart 2008: 70042H-2 (comp. note)

Cicadetta celis Simon 2009: 104, Fig 2 (phylogeny, comp. note) Australia

***Y.* sp. nr. *celis* (Emery, Emery, Emery & Popple, 2005)** = *Cicadetta* sp. nr. *celis*

Cicadetta sp. nr. *celis* (TNS 480) Emery, Emery, Emery and Popple 2005: 98, 101–102, 104–109, Fig 3, Tables 1–3 (life cycle, illustrated, comp. note) New South Wales

***Y. denisoni* (Distant, 1893)** = *Melampsalta denisoni* = *Cicadetta denisoni* = *Melampsalta kershawi* Goding & Froggatt, 1904 = *Cicadetta kershawi*

Melampsalta denisoni Greenup 1964: 23 (comp. note) Australia, New South Wales

Cicadetta denisoni Ewart 1986: 51–52, Table 1 (natural history, listed, comp. note) Queensland, New South Wales, Victoria

Cicadetta denisoni Moulds 1990: 156–157, Plate 17, Fig 8, Fig 8a, Plate 24, Fig 14 (illustrated, distribution, ecology, listed, comp. note) Australia, New South Wales, Victoria

Cicadetta denisoni Naumann 1993: 22, 80, 149 (listed) Australia

Cicadetta denisoni Moss and Popple 2000: 57 (listed, comp. note) Australia, New South Wales

Cicadetta denisoni Moulds and Carver 2002: 26 (synonymy, distribution, ecology, type data, listed) Equals *Melampsalta denisoni* Equals *Melampsalta kershawi* Australia, New South Wales, Victoria

Melampsalta denisoni Moulds and Carver 2002: 26 (type data, listed)

Melampsalta kershawi Moulds and Carver 2002: 26 (type data, listed) Victoria

Cicadetta denisoni Williams 2002: 156–157 (listed) Australia, New South Wales, Victoria

Cicadetta denisoni Arensburger, Buckley, Simon, Moulds, and Holsinger 2004: 558, 566 (molecular phylogeny, comp. note) Australia

Cicadetta denisoni complex Emery, Emery, Emery and Popple 2005: 109 (comp. note) New South Wales

Cicadetta denisoni complex Emery 2008: 35 (comp. note) New South Wales

***Y.* sp. nr. *denisoni* Brown Firetail (Moss & Popple, 2000)** = *Cicadetta* sp. nr. *denisoni*

Cicadetta sp. nr. *denisoni* Brown Firetail Moss and Popple 2000: 57 (listed, comp. note) Australia, New South Wales

Cicadetta sp. nr. *denisoni* Williams 2002: 156–157 (listed) Australia

***Y.* sp. nr. *denisoni* (Emery,Emery, Emery & Popple, 2005)** = *Cicadetta* sp. nr. *denisoni*

Cicadetta sp. nr. *denisoni* (TNS 509) Emery, Emery, Emery and Popple 2005: 99, 103–107, 109, Fig 5, Tables 1–3 (life cycle, illustrated, comp. note) New South Wales

***Y. hunterorum* (Moulds, 1988)** = *Cicadetta hunterorum*
Cicadetta hunterorum Moulds 1988: 39, 41–45, Figs 1–3, Figs 6–7 (n. sp., described, illustrated, distribution, comp. note) Australia, New South Wales
Cicadetta hunterorum Moulds 1990: 152–153 (distribution, ecology, listed, comp. note) Australia, New South Wales
Cicadetta hunterorum Moulds and Carver 2002: 28 (distribution, ecology, type data, listed) Australia, New South Wales
Cicadetta hunterorum Emery, Emery, Emery and Popple 2005: 97 (comp. note) New South Wales

***Y. incepta* (Walker, 1850)** = *Cicada incepta* = *Melampsalta incepta* = *Melempsalta* (sic) *incepta* = *Cicadetta incepta*
Melampsalta incepta Gómez-Menor 1951: 69 (comp. note) Australia
Cicadetta incepta Moulds 1990: 154, Plate 17, Fig 6, Fig 6a (illustrated, distribution, ecology, listed, comp. note) Australia, South Australia
Cicadetta incepta Moulds and Carver 2002: 28 (synonymy, distribution, ecology, type data, listed) Equals *Cicada incepta* Australia, South Australia
Cicada incepta Moulds and Carver 2002: 28 (type data, listed) South Australia
Cicadetta incepta Arensburger, Buckley, Simon, Moulds, and Holsinger 2004: 566 (comp. note) Australia

***Y.* sp. nr. *incepta* (Greenup, 1964)** = *Melampsalta* sp. nr. *incepta*
Melampsalta sp. nr. *incepta* Greenup 1964: 23 (comp. note) Australia, New South Wales

Species complex near *Y. incepta* (Ewart & Popple, 2001) = species complex near *Cicadetta incepta*
Species complex near *Cicadetta incepta* Ewart and Popple 2001: 58–60, 65, Figs 4A–C (described, habitat, song, oscillogram) Queensland

Species complex near *Y. incepta* (Adavale) (Ewart & Popple, 2001) = species complex near *Cicadetta incepta* (Advale)
Species complex near *Cicadetta incepta* (Adavale) Ewart and Popple 2001: 58–60, 69, Fig 4A, Fig 7B (illustrated, described, habitat, song, oscillogram) Queensland

Species complex near *Y. incepta* (Blackall) (Ewart & Popple, 2001) = species complex near *Cicadetta incepta* (Blackall)
Species complex near *Cicadetta incepta* (Blackall) Ewart and Popple 2001: 58–60, Fig 4C (illustrated, described, habitat, song) Queensland

Species complex near *Y. incepta* (Charleville) (Ewart & Popple, 2001) = species complex near *Cicadetta incepta* (Charleville)
Species complex near *Cicadetta incepta* (Charleville) Ewart and Popple 2001: 58–60, 69, Fig 4B, Fig 7C (illustrated, described, habitat, song, oscillogram) Queensland

***Y.* sp. nr. *incepta* (Ewart, 2009)** = *Cicadetta* sp. nr. *incepta*
Cicadetta sp. nr. *Cicadetta incepta* Ewart 2009b: 140 (comp. note) Queensland

***Y. landsboroughi* (Distant, 1882)** = *Melampsalta landsboroughi* = *Melampsalta telxiope* (nec Walker) = *Cicadetta landsboroughi*
Melampsalta landsboroughi Greenup 1964: 23 (comp. note) Australia, New South Wales
Cicadetta landsboroughi Moulds 1990: 11, 151, 153–154, 165, Plate 17, Fig 4, Fig 4a (illustrated, distribution, ecology, listed, comp. note) Australia, New South Wales, Queensland, South Australia, Victoria
Cicadetta landsboroughi Coombs 1996: 56–58, Table 1 (emergence time, comp. note) New South Wales
Cicadetta landsboroughi Ewart 1998a: 61, 71, Fig 15B (illustrated, described, natural history, song) Queensland
Cicadetta landsboroughi Ewart and Popple 2001: 57 (comp. note) Queensland
Cicadetta landsboroughi Moulds and Carver 2002: 29 (synonymy, distribution, ecology, type data, listed) Equals *Melampsalta landsboroughi* Australia, New South Wales, Queensland, South Australia, Victoria
Melampsalta landsboroughi Moulds and Carver 2002: 29 (type data, listed) New South Wales
Cicadetta lansboroughi Callitris Squeaker Popple and Strange 2002: 30, Table 1 (listed, habitat) Australia, Queensland
Cicadetta landsboroughi Williams 2002: 156–157 (listed) Australia, Queensland, New South Wales, Victoria, South Australia
Cicadetta landsboroughi Arensburger, Buckley, Simon, Moulds, and Holsinger 2004: 566 (comp. note) Australia
Cicadetta landsboroughi Emery, Emery, Emery and Popple 2005: 101–102, 104–108, Tables 1–3 (life cycle, comp. note) New South Wales
Cicadetta landsboroughi Haywood 2007: 18 (comp. note) South Australia
Cicadetta landsboroughi Shiyake 2007: 8, 104, 106, Fig 182 (illustrated, distribution, listed, comp. note) Australia
Cicadetta landsboroughi species complex Ewart 2009b: 140 (comp. note) Queensland

***Y.* sp. nr. *landsboroughi* (Ewart, 1986)** = *Cicadetta* sp. nr. *landsboroughi*

Cicadetta sp. nr. *landsboroughi* Ewart 1986: 51, Table 1 (listed) Queensland
Cicadetta sp. nr. *landsboroughi* Ewart and Popple 2001: 57–58, 65, Fig 4D (described, habitat, song, oscillogram) Queensland

***Y. toowoombae* (Distant, 1915)** = *Melampsalta toowoombae* = *Cicadetta toowoombae*
Cicadetta toowoombae Moulds 1990: 144 (distribution, listed) Australia, Queensland
Cicadetta toowoombae Sanborn 1999a: 52 (type material, listed)
Cicadetta toowoombae Moulds and Carver 2002: 33 (synonymy, distribution, ecology, type data, listed) Equals *Melampsalta toowoombae* Australia, Queensland
Melampsalta toowoombae Moulds and Carver 2002: 33 (type data, listed) Queensland

***Y. tristrigata* (Goding & Froggatt, 1904)** = *Melampsalta tristrigata* = *Cicadetta tristrigata*
Melampsalta tristrigata Hahn 1962: 9 (listed) Australia, Queensland
Cicadetta tristrigata Ewart 1986: 51, Table 1 (listed) Queensland
Melampsalta tristrigata Stevens and Carver 1986: 265 (type material) Australia, Queensland, New South Wales
Cicadetta tristrigata Moulds 1990: 151–153, 156, Plate 17, Fig 1, Fig 2a (synonymy, illustrated, distribution, ecology, listed, comp. note) Australia, New South Wales, Queensland, South Australia, Victoria
Cicadetta tristrigata Coombs 1993b: 143–144 (behavior, comp. note) New South Wales
Cicadetta tristrigata Coombs 1996: 56, Table 1 (emergence time, comp. note) New South Wales
Cicadetta tristrigata Moulds and Carver 2002: 34 (synonymy, distribution, ecology, type data, listed) Equals *Melampsalta tristigata* Australia, New South Wales, Queensland, South Australia, Victoria
Melampsalta tristrigata Moulds and Carver 2002: 34 (type data, listed) Queensland, New South Wales
Cicadetta tristrigata Williams 2002: 156–157 (listed) Australia, Queensland, New South Wales, Victoria, South Australia
Cicadetta tristrigata Arensburger, Buckley, Simon, Moulds, and Holsinger 2004: 558, 566 (molecular phylogeny, comp. note) Australia
Cicadetta tristrigata Emery, Emery, Emery and Popple 2005: 101–102, 104–105, 107–108, Tables 1–3 (life cycle, comp. note) New South Wales
Cicadetta tristrigata Moulds 2005b: 410 (comp. note)
Cicadetta tristrigata Shiyake 2007: 8, 105, 107, Fig 186 (illustrated, distribution, listed, comp. note) Australia

Y. sp. nr. *tristrigata* sp. 5 Fiery Ambertail M43 (Moss & Popple, 2000) = *Cicadetta* sp. nr. *tristrigata* sp. 5 Fiery Ambertail M43
Cicadetta sp. nr. *tristrigata* sp. 5 Fiery Ambertail M43 Moss and Popple 2000: 57 (listed, comp. note) Australia, New South Wales
Cicadetta sp. 5 nr. *tristrigata* Williams 2002: 156–157 (listed) Australia

Y. sp. nr. *tristrigata* sp. 16 Robust Ambertail (Moss & Popple, 2000) = *Cicadetta* sp. nr. *tristrigata* sp. 16 Robust Ambertail
Cicadetta sp. nr. *tristrigata* sp. 16 Robust Ambertail Moss and Popple 2000: 57 (listed, comp. note) Australia, New South Wales
Cicadetta sp. 16 nr. *tristrigata* Williams 2002: 156–157 (listed) Australia

Y. sp. nr. *tristrigata* (Ewart & Popple, 2001) = *Cicadetta* sp. nr. *tristrigata*
Species near *Cicadetta tristrigata* Ewart and Popple 2001: 57 (described, habitat, song, oscillogram) Queensland

Y. sp. nr. *tristrigata* (Popple & Ewart, 2002) = *Cicadetta* sp. nr. *tristrigata*
Cicadetta sp. nr. *tristrigata* Popple and Ewart 2002: 116 (illustrated, distribution, ecology, song, oscillogram, comp. note) Australia, Queensland

Y. sp. nr. *tristrigata* (Emery, Emery, Emery & Popple, 2005) = *Cicadetta* sp. nr. *tristrigata*
Cicadetta sp. nr. *tristrigata* (TNS 486) Emery, Emery, Emery and Popple 2005: 99, 102–105, 107, 109, Fig 4, Tables 1–3 (life cycle, illustrated, comp. note) New South Wales

Y. sp. nr. *tristrigata* (Haywood, 2005) = *Cicadetta* sp. nr. *tristrigata*
Cicadetta sp. nr. *tristrigata* Haywood 2005: 31, 53–55, Fig 16, Table 2 (illustrated, distribution, described, habitat, song, oscillogram, listed, comp. note) South Australia
Cicadetta sp. nr. *tristrigata* Haywood 2006a: 14–15, Table 1 (illustrated, distribution, described, habitat, song, comp. note) South Australia
Cicadetta sp. nr. *tristrigata* Haywood 2007: 16–18, Appendix 1 (illustrated, distribution, described, habitat, song, oscillogram, comp. note) South Australia

***Taurella* Moulds, 2012**

***T. forresti* (Distant, 1882)** = *Melampsalta forresti* = *Cicadetta forresti* = *Melampsalta warburtoni* Distant, 1882 = *Cicadetta warburtoni* = *Melampsalta capistrata* Ashton, 1912 = *Cicadetta capistrata*

Cicadetta capistrata Moulds 1990: 161–162, Plate 16, Fig 4, Fig 4a (illustrated, distribution, ecology, listed, comp. note) Australia, Queensland

Cicadetta forresti Moulds 1990: 161–162, Plate 18, Fig 5, Fig 5a (illustrated, distribution, ecology, listed, comp. note) Australia, New South Wales, Queensland

Cicadetta warburtoni Moulds 1990: 144 (distribution, listed) Australia, Queensland

Cicadetta capistrata Moulds and Carver 2002: 25 (synonymy, distribution, ecology, type data, listed) Equals *Melampsalta capistrata* Australia, Queensland

Melampsalta capistrata Moulds and Carver 2002: 25 (type data, listed) Queensland

Cicadetta warburtoni Moulds and Carver 2002: 34 (synonymy, distribution, ecology, type data, listed) Equals *Melampsalta warburtoni* Australia, Queensland

Melampsalta warburtoni Moulds and Carver 2002: 34 (type data, listed) Queensland

Cicadetta forresti Popple and Ewart 2002: 115 (illustrated, distribution, ecology, song, oscillogram, comp. note) Australia, Queensland

Cicadetta forresti Williams 2002: 156–157 (listed) Australia, Queensland, New South Wales

Cicadetta forresti Moulds 2005b: 395–398, 407, 413–414, 417–419, 421, 425, 436, Figs 56–59, Fig 62, Table 1 (higher taxonomy, phylogeny, key, listed, comp. note) Australia

Cicadetta forresti group Popple and Emery 2010: 150 (comp. note) Australia

Cicadetta capistrata Popple and Emery 2010: 150 (comp. note) Australia

***T. froggatti* (Distant, 1907)** = *Melampsalta froggatti* = *Cicadetta froggatti* = *Melampsalta sulcata* Distant, 1907 = *Cicadetta sulcata*

Cicadetta froggatti Moulds 1990: 160, Plate 16, fig 7, Fig 7a (illustrated, distribution, ecology, listed, comp. note) Australia, Queensland

Cicadetta sulcata Moulds 1990: 160–161, Plate 16, Fig 6, Fig 6a (synonymy, distribution, ecology, type data, listed) Australia, Queensland

Cicadetta sulcata Naumann 1993: 41, 80, 149 (listed) Australia

Cicadetta froggatti Naumann 1993: 48, 80, 149 (listed) Australia

Cicadetta forresti Moulds and Carver 2002: 27 (synonymy, distribution, ecology, type data, listed) Equals *Melampsalta forresti* Australia, New South Wales, Queensland

Melampsalta forresti Moulds and Carver 2002: 27 (type data, listed) Queensland

Cicadetta froggatti Moulds and Carver 2002: 27 (synonymy, distribution, ecology, type data, listed) Equals *Melampsalta froggatti* Australia, Queensland

Melampsalta froggatti Moulds and Carver 2002: 27 (type data, listed) Queensland

Cicadetta sulcata Moulds and Carver 2002: 33 (synonymy, distribution, ecology, type data, listed) Equals *Melampsalta sulcata* Australia, Queensland

Melampsalta sulcata Moulds and Carver 2002: 33 (type data, listed) Queensland

Cicadetta sulcata Ewart 2005a: 170–171, 179, Fig 1C (described, song, oscillogram) Queensland

Cicadetta froggatti Moulds 2005b: 403, Fig 37 (illustrated, comp. note) Australia

Cicadetta froggatti Popple and Emery 2010: 150 (comp. note) Australia

Cicadetta sulcata Popple and Emery 2010: 150 (comp. note) Australia

***T. viridis* (Ashton, 1912)** = *Melampsalta viridis* = *Cicadetta viridis* = *Cicadetta* sp. nr. *viridis* Ewart, 1998

Cicadetta viridis Moulds 1990: 144 (distribution, listed) Australia, Queensland

Cicadetta viridis Ewart 1998b: 136 (comp. note) Queensland

Cicadetta sp. nr. *viridis* Ewart 1998b: 136–138, Fig 3 (illustrated, sonogram, natural history, acoustic behavior, listed, comp. note) Queensland

Cicadetta viridis Moulds and Carver 2002: 34 (synonymy, distribution, ecology, type data, listed) Equals *Melampsalta viridis* Australia, Queensland

Melampsalta viridis Moulds and Carver 2002: 34 (type data, listed) Queensland?

Cicadetta viridis Marshall and Hill 2009: 4–5, 7, Fig 6G, Table 2 (sonogram, comp. note) Australia

Cicadetta viridis Pain 2009: 47 (sonogram) Australia

Cicadetta viridis Popple and Emery 2010: 150 (comp. note) Australia

***Urabunana* Distant, 1905**

Urabunana Moulds 1990: 7, 32, 173 (listed, diversity, comp. note) Australia

Urabunana Moulds and Carver 1991: 467 (comp. note) Australia

Urabunana Coombs 1995: 13 (comp. note) Australia

Urabunana Chou, Lei, Li, Lu and Yao 1997: 358 (comp. note)

Urabunana Ewart 1999: 2 (listed)

Urabunana Moulds and Carver 2002: 44 (type data, listed)

Urabunana Arensburger, Buckley, Simon, Moulds, and Holsinger 2004: 566 (comp. note) Australia
Urabunana Moulds 2005b: 390, 430, 436, Table 2 (higher taxonomy, listed, comp. note) Australia
Urabunana Shiyake 2007: 8, 111 (listed, comp. note)
Urabunana sp. Ewart 2009b: 148 (comp. note) Queensland

U. daemeli (Distant, 1905) to *Drymopsalta daemeli*
U. festiva Distant, 1907 to *Mugadina festiva*
U. leichardti (Distant, 1882) to *Paradina leichardti*
U. longipennis Ashton, 1914 to *Uradolichos longipennis*
U. marshalli Distant, 1911 to *Mugadina marshalli*
U. sp. nr. *marshalli* Lime grass-ticker Popple & Strange, 2002 to *Mugadina* sp. nr. *marshalli*
U. rufilinea Ashton, 1914 to *Drymopsalta dameli* (Distant, 1905)

U. sp. nov. Lithgow, 1988
Urabunana sp. nov. Lithgow 1988: 65 (listed) Australia, Queensland

U. sp. nov. Little Green Grass-Ticker Popple & Strange, 2002 to *Mugadina* sp. nov. Little Green Grass-Ticker

***U. segmentaria* Distant, 1905**
Urabunana segmentaria Moulds and Carver 2002: 45 (distribution, ecology, type data, listed) Australia, Queensland
Urabunana segmentaria Ewart 2005b: 454 (comp. note)

***U. sericeivitta* (Walker, 1862)** = *Cicada seiceivitta* = *Melampsalta seicevitta* (sic) = *Pauropsaltria sericeivitta*
Cicada sericeivitta Moulds 1990: 173 (type species of *Urabunana*, type data, listed)
Urabunana sericeivitta Moulds 1990: 173–174, Plate 16, Fig 9, Fig 9a (synonymy, illustrated, distribution, ecology, listed, comp. note) Australia, New South Wales, Queensland
Urabunana sericeivitta Moss and Popple 2000: 58 (listed, comp. note) Australia, New South Wales
Cicada sericeivitta Moulds and Carver 2002: 44–45 (type species of *Urabunana*, type data, listed) New South Wales
Urabunana sericeivitta Moulds and Carver 2002: 45 (synonymy, distribution, ecology, type data, listed) Equals *Cicada sericeivitta* Australia, New South Wales, Queensland
Urabunana sericeivitta Popple and Ewart 2002: 114 (illustrated, distribution, ecology, song, oscillogram, comp. note) Australia, Queensland, New South Wales
Urabunana sericeivitta Williams 2002: 156–157 (listed) Australia, Queensland, New South Wales
Urabunana sericeivitta Moulds 2005b: 395–397, 399, 405–406, 409, 417–419, 421, 425, Fig 41, Fig 43, Fig 50, Figs 56–59, Fig 62, Table 1 (illustrated, higher taxonomy, phylogeny, key, listed, comp. note) Australia

U. sp. A Ewart, 1998
Urabunana sp. A Ewart 1998a: 67–68, Fig 12B (described, natural history, song, oscillogram) Queensland
Urabunana sp. A Ewart 2009b: 148, 164, Figs 12A–B (oscillogram, distribution, habitat, comp. note) Equals Cravens Grass Chirper, no. 634 Queensland
Cravens Grass Chirper Ewart 2009b: 140–141, 147, Fig 1, Plate 2K-L (illustrated, distribution, emergence times, habitat, comp. note) Queensland

U. verna Distant, 1912 to *Toxala verna*
U. wollomombii Coombs, 1995 to *Myopsalta wollomombii*

***Sylphoides* Moulds, 2012**

***S. arenaria* (Distant, 1907)** = *Melampsalta arenaria* = *Melampsalta orenaria* (sic) = *Cicadetta arenaria* = *Cicadetta areuarai* (sic)
Melampsalta orenaria (sic) Hahn 1962: 9 (listed) Australia, New South Wales
Melampsalta arenaria Stevens and Carver 1986: 264 (type material) Australia, New South Wales
Cicadetta arenaria Moulds 1990: 158–160, Plate 19, Fig 5, Fig 5a (illustrated, distribution, ecology, listed, comp. note) Australia, New South Wales
Cicadetta arenaria Naumann 1993: 50, 80, 149 (listed) Australia
Cicadetta areuarai (sic) Gosper 1999: 142 (predation) New South Wales
Cicadetta arenaria Moulds and Carver 2002: 25 (synonymy, distribution, ecology, type data, listed) Equals *Melampsalta arenaria* Australia, New South Wales
Melampsalta arenaria Moulds and Carver 2002: 25 (type data, listed) New South Wales
Cicadetta arenaria Arensburger, Buckley, Simon, Moulds, and Holsinger 2004: 558, 566 (molecular phylogeny comp. note) Australia
Cicadetta arenaria Emery, Emery, Emery and Popple 2005: 97 (comp. note) New South Wales

***Pyropsalta* Moulds, 2012**

***P. melete* (Walker, 1850)** = *Cicada melete* = *Melampsalta melete* = *Cicadetta melete* = *Melampsalta rubricincta* Goding & Froggatt, 1904 = *Cicadetta rubricincta*
Melampsalta rubricincta Hahn 1962: 9 (listed) Australia, Western Australia

Cicadetta melete Gwynne and Schatral 1988: 32, 36–38, 43, 64–65, Fig 5, Fig 31 (illustrated, oscillogram, emergence time, key, listed, comp. note) Western Australia

Cicadetta melete Moulds 1990: 157, Plate 17, Fig 3, Fig 3a (synonymy, illustrated, distribution, ecology, listed, comp. note) Equals *Cicada melete* Equals *Melampsalta melete* Equals *Melampsalta rubricincta* Equals *Cicadetta rubricincta* Australia, Western Australia

Cicada melete Moulds 1990: 157 (listed, comp. note)

Melampsalta melete Moulds 1990: 157 (listed, comp. note)

Melampsalta rubricincta Moulds 1990: 157 (listed, comp. note)

Cicadetta rubricincta Moulds 1990: 157 (listed, comp. note)

Cicadetta melete Nanninga 1991: Plate 3, Fig G (illustrated) Australia

Cicadetta melete Naumann 1993: 48, 80, 149 (listed) Australia

Cicadetta melete Moulds and Carver 2002: 30 (synonymy, distribution, ecology, type data, listed) Equals *Cicada melete* Equals *Melampsalta rubricincta* Australia, Western Australia

Cicada melete Moulds and Carver 2002: 30 (type data, listed) Western Australia

Melampsalta rubricincta Moulds and Carver 2002: 30 (type data, listed) Western Australia

Cicadetta melete Moulds 2005b: 395–397 (comp. note) Australia

***Physeema* Moulds, 2012**

***P. bellatrix* (Ashton, 1914)** = *Pauropsalta bellatrix* = *Melampsalta bellatrix*

Pauropsalta bellatrix Ewart 1989b: 293 (listed) Western Australia

Pauropsalta bellatrix Moulds 1990: 131 (distribution, listed) Australia, Western Australia

Pauropsalta bellatrix Moulds and Carver 2002: 38 (distribution, ecology, type data, listed) Australia, Western Australia

***P. convergens* (Walker, 1850)** = *Cicada convergens* = *Melampsalta convergens* = *Cicadetta convergens* = *Macrotristria intersecta* (nec Walker) = *Melampsalta landsboroughi convergens* = *Melampsalta cylindrica* Ashton, 1912 = *Cicadetta landsboroughi* var. *ashtoni* Metcalf, 1963 = *Melampsalta landsboroughi* var. *ashtoni*

Cicadetta convergens Moulds 1990: 26, 145–146, Plate 18, Fig 4 (n. comb., synonymy, illustrated, distribution, ecology, listed, comp. note) Equals *Cicada convergens* Equals *Melampsalta convergens* Equals *Macrotristria intersecta* (partim) Equals *Melampsalta landsboroughi* var. *convergens* Equals *Melampsalta cylindricata* Equals *Cicadetta landsboroughi* var. *ashtoni* n. syn. Australia, Western Australia

Cicada convergens Moulds 1990: 145 (listed)

Melampsalta convergens Moulds 1990: 145–146 (listed, comp. note)

Macrotristria intersecta Moulds 1990: 145–146 (listed, comp. note)

Melampsalta landsboroughi var. *convergens* Moulds 1990: 145 (listed)

Melampsalta cylindricata Moulds 1990: 145 (listed)

Cicadetta landsboroughi var. *ashtoni* Moulds 1990: 145 (listed)

Cicadetta convergens Moulds and Carver 2002: 26 (synonymy, distribution, ecology, type data, listed) Equals *Cicada convergens* Equals *Melampsalta cylindricata* Australia, Western Australia

Cicada convergens Moulds and Carver 2002: 26 (type data, listed) New Holland

Melampsalta cylindricata Moulds and Carver 2002: 26 (type data, listed) Victoria

***P. labyrinthica* (Walker, 1850)** = *Dundubia labyrinthica* = *Melampsalta labyrinthica* = *Cicadetta labyrinthica* = *Cicada interstans* Walker, 1858 = *Melampsalta interstans*

Melampsalta labyrinthica Greenup 1964: 23 (comp. note) Australia, New South Wales

Cicadetta labyrinthica Moulds 1990: 144 (distribution, listed) Australia, South Australia, Victoria

Cicadetta labyrinthica Moulds and Carver 2002: 29 (synonymy, distribution, ecology, type data, listed) Equals *Dundubia labyrinthica* Equals *Cicada interstans* Australia, South Australia, Victoria

Dundubia labyrinthica Moulds and Carver 2002: 29 (type data, listed) New Holland

Cicada interstans Moulds and Carver 2002: 29 (type data, listed) South Australia

***P.* sp. aff. *labyrinthica* (Haywood, 2005)** = *Pauropsalta* sp. aff. *labyrinthica*

Pauropsalta sp. aff. *labyrinthica* Haywood 2005: 31, 44–46, Fig 13, Table 2 (illustrated, distribution, described, habitat, song, oscillogram, listed, comp. note) South Australia

Pauropsalta sp. aff. *labyrinthica* Haywood 2006a: 14, Table 1 (listed) South Australia

Pauropsalta sp. aff. *labyrinthica* Haywood 2006b: 51–53, Table 1 (illustrated, distribution, described, habitat, song, oscillogram, listed, comp. note) South Australia

P. latorea (Walker, 1850) = *Cicada latorea* = *Melampsalta latorea* = *Cicadetta latorea*

Cicadetta latorea Moulds 1990: 144 (distribution, listed) Australia, Western Australia

Cicadetta latorea Moulds and Carver 2002: 29 (synonymy, distribution, ecology, type data, listed) Equals *Cicada latorea* Australia, Western Australia

Cicada latorea Moulds and Carver 2002: 29 (type data, listed) New Holland

P. n. sp. #75M nr. *latorea* (Gwynne & Schatral, 1988) = *Cicadetta* n. sp. #75M nr. *latoria* (sic)

Cicadetta n. sp. #75M nr. *latoria* (sic) Gwynne and Schatral 1988: 32, 36–38, 43, 66–67, Fig 5, Fig 32 (illustrated, oscillogram, emergence time, key, listed, comp. note) Western Australia

***P. quadricincta* (Walker, 1850)** = *Cicada quadricincta* = *Melampsalta quadricincta* = *Cicadetta quadricincta*

Cicadetta quadricincta Wei, Tang, He and Zhang 2009: 340 (comp. note)

Cicadetta quadricincta Gwynne 1987: 571–575, Figs 1–3, Table 1 (illustrated, acoustic behavior, oscillogram, predation) Western Australia

Cicadetta quadricincta Gwynne and Schatral 1988: 27, 32, 36, 38, 43, 64–65, Fig 5, Fig 30 (illustrated, oscillogram, sound production, emergence time, key, listed, comp. note) Western Australia

Cicadetta quadricincta Moulds 1990: 8, 10, 23, 165–166, Plate 19, Fig 1, Figs 1a–b, Plate 24, Fig 12 (synonymy, illustrated, distribution, ecology, listed, comp. note) Equals *Cicada quadricincta* Equals *Cicada telxiope* Equals *Cicada duplex* Equals *Melampsalta quadricincta* Equals *Melampslata telxiope* Equals *Melampsalta duplex* Equals *Melampsalta viridicincta* Equals *Cicadetta telxiope* Equals *Cicadetta viridicincta* Australia, Western Australia

Cicada quadricincta Moulds 1990: 165 (listed, comp. note)

Cicada duplex Moulds 1990: 165 (listed, comp. note) New Holland

Cicada telxiope Moulds 1990: 165 (listed, comp. note) New Holland

Melampsalta quadricincta Moulds 1990: 165 (listed)

Melampsalta telxiope Moulds 1990: 165 (listed)

Melampsalta duplex Moulds 1990: 165 (listed)

Melampsalta viridicincta Moulds 1990: 165 (listed, comp. note)

Cicadetta telxiope Moulds 1990: 165 (listed)

Cicadetta viridicincta Moulds 1990: 165 (listed)

Cicadetta quadricincta Heady and Nault 1991: 14 (comp. note)

Cicadetta quadricincta Williams and Smith 1991: 289 (comp. note)

Cicadetta quadricincta Naumann 1993: 58, 80, 149 (listed) Australia

Cicadetta quadricincta Bennet-Clark and Young 1994: 292–293, Table 1 (body size and song frequency)

Cicadetta quadricincta Cocroft and Pogue 1996: 94 (comp. note)

Cicadetta quadricincta Jennions and Petrie 1997: 297 (comp. note)

Cicadetta quadricincta Cooley and Marshall 2001: 847–848 (comp. note)

Melampsalta quadricincta Sueur 2001: 43, Table 1 (listed) Equals *Cicadetta melampsalta*

Cicadetta quadricincta Gerhardt and Huber 2002: 26 (comp. note)

Cicadetta quadricincta Moulds and Carver 2002: 31–32 (synonymy, distribution, ecology, type data, listed) Equals *Cicada duplex* Equals *Cicada quadricincta* Equals *Cicada telxiope* Equals *Melampsalta viridicincta* Australia, Western Australia

Cicada duplex Moulds and Carver 2002: 31 (type data, listed) New Holland

Cicada quadricincta Moulds and Carver 2002: 31 (type data, listed) New Holland

Cicada telxiope Moulds and Carver 2002: 31 (type data, listed) New Holland

Melampsalta viridicincta Moulds and Carver 2002: 31–32 (type data, listed) Western Australia

Cicadetta quadricincta Bailey 2003: 159–160, 169, Fig 1, Table 1 (duetting oscillogram, listed, comp. note) Australia

Cicadetta quadricincta Sueur and Aubin 2004: 223 (comp. note)

Cicadetta quadricincta Narhardiyati and Bailey 2005: 104 (predation, comp. note) Australia

***Gelidea* Moulds, 2012**

***G. torrida torrida* (Erichson, 1842)** = *Cicada torrida* = *Melampsalta torrida* = *Cicadetta torrida* = *Cicada basiflamma* Walker, 1850 = *Cicada connexa* Walker, 1850 = *Cicada damater* Walker, 1850 = *Melampsalta damater* = *Cicadetta (Tettigeta) torrida* = *Melampsalta spreta* Goding & Froggatt, 1904 = *Cicadetta spreta*

Cicadetta torrida Moss 1989: 102, 104–105 (distribution, song, comp. note) Equals *Cicadetta spreta* Tasmania

Cicadetta spreta Moss 1989: 105 (comp. note)

Cicadetta torrida Moss 1990: 6–7 (distribution, song, comp. note) Equals *Cicadetta spreta* Tasmania

Cicadetta spreta Moss 1990: 7 (comp. note)

Cicadetta spreta Moulds 1990: 144 (distribution, listed) Australia, Tasmania

Cicadetta torrida Moulds 1990: 113–115, 149, Plate 18, Fig 2, Fig 2a (illustrated, distribution, ecology, listed, comp. note) Australia, Tasmania, Victoria

Cicadetta (Tettigetta) torrida Boulard 1991c: 119 (n. comb., comp. note) Tasmania

Cicadetta torrida Moulds and Carver 2002: 33–34 (synonymy, distribution, ecology, type data, listed) Equals *Cicada torrida* Equals *Cicada basiflamma* Equals *Cicada connexa* Equals *Cicada damater* Australia, Tasmania, Victoria

Cicada torrida Moulds and Carver 2002: 33 (type data, listed)

Cicadetta spreta Moulds and Carver 2002: 32 (synonymy, distribution, ecology, type data, listed) Equals *Melampsalta spreta* Australia, Tasmania

Melampsalta spreta Moulds and Carver 2002: 32 (type data, listed) Tasmania

Cicada basiflamma Moulds and Carver 2002: 33 (type data, listed) New Holland

Cicada connexa Moulds and Carver 2002: 33–34 (type data, listed) New Holland

Cicada damater Moulds and Carver 2002: 33 (type data, listed) New Holland

Cicadetta torrida Arensburger, Buckley, Simon, Moulds, and Holsinger 2004: 558 (molecular phylogeny, comp. note) Australia

Cicadetta torrida Moulds 2005b: 423 (comp. note) Australia

***G. torrida* var. β Walker, 1850** = *Cicada torrida* var. β = *Cicadetta torrida* var. β = *Cicadetta (Tettigeta) torrida* var. β

***G. torrida* var. δ Walker, 1850** = *Cicada torrida* var. δ = *Cicadetta torrida* var. δ = *Cicadetta (Tettigetta) torrida* var. δ

***G. torrida* var. γ Walker, 1850** = *Cicada torrida* var. γ = *Cicadetta torrida* var. γ = *Cicadetta (Tettigetta) torrida* var. γ

***Clinopsalta* Moulds, 2012**

***C. adelaida* (Ashton, 1914)** = *Melampsalta adelaida* = *Cicadetta adelaida*

Cicadetta adelaida Moulds 1986: 42 (comp. note) Australia

Cicadetta adelaida Moulds 1990: 126, 146–147 (distribution, ecology, listed, comp. note) Australia, South Australia, Victoria

Cicadetta adelaida Moulds and Carver 2002: 24 (synonymy, distribution, ecology, type data, listed) Equals *Melampsalta adelaida* Australia, South Australia, Victoria

Melampsalta adelaida Moulds and Carver 2002: 25 (type data, listed) South Australia

***C.* sp. nr. *adelaida* (Emery, Emery, Emery & Popple, 2005)** = *Cicadetta* sp. nr. *adelaida*

Cicadetta sp. nr. *adelaida* (TNS 214) Emery, Emery, Emery and Popple 2005: 98, 102–105, 107, Fig 1, Tables 1–3 (life cycle, illustrated, comp. note) New South Wales

Cicadetta nr. *adelaida* Emery 2008: 35 (comp. note) New South Wales

***C. tigris* (Ashton, 1914)** = *Melampsalta tigris* = *Cicadetta tigris*

Cicadetta tigris Lithgow 1988: 65 (listed) Australia, Queensland

Cicadetta tigris Moulds 1990: 144 (distribution, listed) Australia, South Australia

Cicadetta tigris Moulds and Carver 2002: 33 (synonymy, distribution, ecology, type data, listed) Equals *Melampsalta tigris* Australia, South Australia

Melampsalta tigris Moulds and Carver 2002: 33 (type data, listed) South Australia

***C.* sp. nr. *tigris* (Ewart, 1988)** = *Cicadetta* sp. nr. *tigris* = Species C Ewart & Popple, 2001

Cicadetta sp. nr. *tigris* Ewart 1988: 183, 191, 194, 200–201, Fig 11J, Plate 3E (illustrated, oscillogram, natural history, listed, comp. note) Queensland, New South Wales

Cicadetta sp. nr. *tigris* Ewart 1989a: 80 (comp. note)

Species C Ewart and Popple 2001: 61, 65 (described, habitat, song) Equals *Cicadetta* sp. nr. *tigris* Ewart, 1988 Queensland

***C.* sp. nr. *tigris* (Ewart, 2009)** = *Cicadetta* sp. nr. *tigris* = *Notopsalta* sp. G Ewart 1998, Popple and Strange, 2002

Cicadetta sp. nr. *tigris* Brigalow Tiger Popple and Strange 2002: 28, Table 1 (listed, habitat) Australia, Queensland

Cicadetta sp. nr. *tigris* Ewart 2009b: 149, 166, Figs 14A–C (oscillogram, distribution, habitat, comp. note) Equals Yellow-Spotted Brigalow Cicada, no. 209 Equals *Notopsalta* sp. G Ewart 1998, Popple and Strange, 2002 Queensland

Yellow-Spotted Brigalow Cicada Ewart 2009b: 141, 147, Fig 1, Plate 2O-P (illustrated, distribution, emergence times, habitat, comp. note) Queensland

Notopsalta sp. G Ewart 2009b: 166, Fig 14D (oscillogram, comp. note) Queensland

***Galanga* Moulds, 2012**

***G. labeculata* (Distant, 1892)** = *Melampsalta labeculata* = *Cicadetta labeculata*

Cicadetta labeculata Ewart 1986: 51, 56–57, Table 1 (natural history, listed, comp. note) Queensland, New South Wales

Cicadetta labeculata Moulds 1990: 21, 147–149, Plate 3, Fig 3, Plate, 18, Fig 7, Fig 7a, Plate 24, Fig 13 (illustrated, distribution, ecology, listed, comp. note) Equals *Melampsalta labeculata* Australia, New South Wales, Queensland

Cicadetta laberculata Coombs 1993b: 144 (comp. note) New South Wales

Cicadetta labeculata Naumann 1993: 19, 80, 149 (listed) Australia

Cicadetta laberculata Coombs 1996: 56, Table 1 (emergence time, comp. note) New South Wales

Cicadetta labeculata Moss 2000: 69 (comp. note) Queensland

Cicadetta labeculata Moss and Popple 2000: 57 (listed, comp. note) Australia, New South Wales

Cicadetta labeculata Moulds and Carver 2002: 29 (synonymy, distribution, ecology, type data, listed) Equals *Melampsalta labeculata* Australia, New South Wales, Queensland

Melampsalta labeculata Moulds and Carver 2002: 29 (type data, listed) Australia

Cicadetta labeculata Williams 2002: 156–157 (listed) Australia, Queensland, Victoria

Cicadetta labeculata Lee 2003b: 7 (comp. note)

Cicadetta labeculata Arensburger, Buckley, Simon, Moulds, and Holsinger 2004: 558 (molecular phylogeny, comp. note) Australia

Cicadetta labeculata Emery, Emery, Emery and Popple 2005: 102–105, 107, Tables 1–3 (comp. note) New South Wales

Cicadetta labeculata Moulds 2005b: 398, 410–411, 417–419, 421, 425, Fig 53, Figs 56–59, Fig 62, Table 1 (illustrated, higher taxonomy, phylogeny, key, listed, comp. note) Australia

G. sp. nr. *labeculata* (Greenup, 1964) = *Melampsalta* sp. nr. *labeculata*

Melampsalta sp. nr. *labeculata* Greenup 1964: 23 (comp. note) Australia, New South Wales

***Kobonga* Distant, 1906**

Kobonga Moulds 1990: 7, 32, 129 (listed, diversity, comp. note) Australia

Kobonga Moulds and Kopestonsky 2001: 141–143 (described, diversity, key, distribution, comp. note) Australia, Western Australia, Northern Territory, Queensland

Kobonga spp. Moulds and Kopestonsky 2001: 144, 146, 156 (comp. note)

Kobonga Moulds and Carver 2002: 35 (type data, listed) Australia

Kobonga Arensburger, Buckley, Simon, Moulds, and Holsinger 2004: 566 (comp. note) Australia

Kobonga Moulds 2005b: 390, 430, 436, Table 2 (higher taxonomy, listed, comp. note) Australia

Kobonga sp. Marshall and Hill 2009: 3 (comp. note)

***K. apicans* Moulds & Kopestonsky, 2001**

Kobonga apicans Moulds and Kopestonsky 2001: 142, 144, 146–147, 154–156, Fig 6, Fig 10, Fig 12 (n. sp., described, illustrated, key, distribution, comp. note) Australia, Northern Territory, Western Australia, South Australia

Kobonga apicans Moulds and Carver 2002: 35–36 (distribution, ecology, type data, listed) Australia, Northern Territory, South Australia, Western Australia

Kobonga apicans Ewart 2009b: 137–138, 140–141, 147, 151, 170, Fig 1, Fig 19, Plate 2T (oscillogram, illustrated, distribution, emergence times, habitat, comp. note) Equals Northern Robust Clicker, no. 161 Queensland, Northern Territory, Western Australia, South Australia

Kobonga apicans Marshall and Hill 2009: 4, 7, Table 2 (comp. note)

***K. apicata* (Ashton, 1914)** = *Cicadetta apicata* = *Melampsalta apicata* = *Kobonga* sp. nr. *Cicadetta apicata* Ewart, 2009

Cicadetta apicata Ewart 1988: 183, 191, 195, 200–201, Fig 11I, Plate 4I (illustrated, oscillogram, natural history, listed, comp. note) Queensland

Cicadetta apicata Moulds 1990: 163–164, Plate 18, Fig 9, Fig 9a (illustrated, distribution, ecology, listed, comp. note) Australia, Northern Territory, Queensland, Western Australia

Cicadetta apicata Ewart 1998a: 61 (described, natural history, song) Queensland

Cicadetta apicata Ewart and Popple 2001: 61, 65 (described, habitat, song) Queensland

Cicadetta apicata Sueur 2001: 42, Table 1 (listed)

Cicadetta apicata Moulds and Carver 2002: 25 (synonymy, distribution, ecology, type data, listed) Equals *Melampsalta apicata* Australia, Northern Territory, Queensland, Western Australia

Melampsalta apicata Moulds and Carver 2002: 25 (type data, listed) Western Australia

Kobonga sp. nr. *Cicadetta apicata* Ewart 2009b: 140 (comp. note) Queensland

K. sp. nr. *apicata* Callitris Clicker (Popple & Strange, 2002) = *Cicadetta* sp. nr. *apicata* Callitris Clicker

Cicadetta sp. nr. *apicata* Callitris Clicker Popple and Strange 2002: 28, Table 1 (listed, habitat) Australia, Queensland

***K. froggatti* Distant, 1913** = *Kobonga castanea* Ashton, 1914

Kobonga froggatti Moulds 1990: 129, Plate 21, Fig 5 (illustrated, distribution, ecology, listed, comp. note) Australia, South Australia, Western Australia

Kobonga froggatti Moulds and Kopestonsky 2001: 141–142, 144, 147, 153–156, Fig 4, Fig 12 (described, illustrated, key, synonymy, distribution, comp. note) Equals *Kobonga castanea* Australia, Western Australia

Kobonga castanea Moulds and Kopestonsky 2001: 153 (type material, comp. note)

Kobonga froggatti Moulds and Carver 2002: 36 (synonymy, distribution, ecology, type data, listed) Equals *Kobonga castanea* Australia, South Australia, Western Australia

Kobonga castanea Moulds and Carver 2002: 36 (type data, listed) Western Ausatralia

Kobonga froggatti Moulds 2005b: 423 (comp. note) Australia

***K. fuscomarginata* Distant, 1914** = *Pauropsalta fuscomarginata* = *Pauropsalta fuscomarginatus* (sic) = *Melampsalta fuscomarginata*

Pauropsalta fuscomarginatus (sic) Ewart 1989b: 293 (listed) New South Wales

Pauropsalta fuscomarginata Moulds 1990: 131 (distribution, listed) Australia, New South Wales, Queensland, Victoria

Pauropsalta fuscomarginata Coombs 1993a: 133 (distribution, comp. note) New South Wales

Pauropsalta fuscomarginata Moulds and Kopestonsky 2001: 141 (comp. note)

Kobonga fuscomarginata Moulds and Kopestonsky 2001: 141–143, 146–147, 151-, Fig 9, Fig 11 (n. comb., described, illustrated, key, synonymy, distribution, comp. note) Equals *Pauropsalta fuscomarginatus* (sic) Equals *Melampsalta fuscomarginata* Australia, Queensland, New South Wales, Victoria

Kobonga fuscomarginata Moulds and Carver 2002: 36 (synonymy, distribution, ecology, type data, listed) Equals *Pauropsalta fuscomarginata* Australia, New South Wales, Queensland, Victoria

Pauropsalta fuscomarginata Moulds and Carver 2002: 36 (type data, listed) New South Wales

***K. godingi* Distant, 1905** = *Melampsalta godingi* = *Melampsalta umbrimargo* Goding & Froggatt (nec *Cicada umbrimargo* Walker, 1858)

Kobonga godingi Moulds 1990: 129 (distribution, listed) Australia, New South Wales, South Australia, Victoria

Melampsalta godingi Moulds and Kopestonsky 2001: 141 (comp. note)

Kobonga godingi Moulds and Kopestonsky 2001: 141–144, 146–150, 156, Fig 5, Fig 7, Fig 12 (described, illustrated, key, synonymy, distribution, comp. note) Equals *Melampsalta godingi* Equals *Melampsalta umbrimargo* (partim) Australia, Victoria

Kobonga godingi Moulds and Carver 2002: 36 (synonymy, distribution, ecology, type data, listed) Equals *Melampsalta umbrimargo* (partim) Equals *Melampsalta godingi* Australia, New South Wales, South Australia, Victoria

Melampsalta godingi Moulds and Carver 2002: 36 (type data, listed) South Australia

Melampsalta umbrimargo Moulds and Carver 2002: 36 (type data, listed)

***K. oxleyi* (Distant, 1882)** = *Melampsalta oxleyi* = *Cicadetta oxleyi*

Cicadetta oxleyi Ewart 1988: 182, 191, 193, 196–197, Fig 9H-I, Plate 2D (illustrated, oscillogram, natural history, listed, comp. note) Queensland

Cicadetta oxleyi Lithgow 1988: 65 (listed) Australia, Queensland

Cicadetta oxleyi Ewart 1989a: 80 (comp. note)

Cicadetta oxleyi Moulds 1990: 147, Plate 21, Fig 1 (illustrated, distribution, ecology, listed, comp. note) Australia, New South Wales, Queensland

Cicadetta oxleyi Coombs 1993a: 133 (distribution, comp. note) New South Wales, Queensland, Victoria

Cicadetta oxleyi Ewart 1996: 14 (listed)

Cicadetta oxleyi Moulds 1996b: 92 (comp. note) Australia, Queensland

Cicadetta oxleyi Ewart 1998a: 60, Fig 6 (described, natural history, song, oscillogram) Queensland

Cicadetta oxleyi Ewart and Popple 2001: 61, 66 (described, habitat, song) Queensland

Cicadetta oxleyi Moulds and Kopestonsky 2001: 141 (comp. note)

Kobonga oxleyi Moulds and Kopestonsky 2001: 141–144, 146–147, 150–152, Fig 6, Fig 8, Fig 12 (n. comb., described, illustrated, key, synonymy, distribution, comp. note) Equals *Melampsalta oxleyi* Equals *Cicadetta oxleyi* Australia, Queensland

Cicadetta oxleyi Sueur 2001: 43, Table 1 (listed)

Kobonga oxleyi Moulds and Carver 2002: 36 (synonymy, distribution, ecology, type data, listed) Equals *Melampsalta oxleyi* Australia, New South Wales, Queensland

Melampsalta oxleyi Moulds and Carver 2002: 36 (type data, listed) Queensland

Kobonga oxleyi Popple and Strange 2002: 28, Table 1 (listed, habitat) Australia, Queensland

Kobonga oxleyi Haywood 2005: 9, 16–18, Fig 4, Table 1 (illustrated, distribution, described, habitat, song, oscillogram, listed, comp. note) South Australia, New South Wales, Queensland, Victoria

Kobonga oxleyi Haywood 2006a: 14, 19–20, 23, Tables 1–2 (illustrated, distribution, described, habitat, song, oscillogram, comp. note) South Australia, Queensland, New South Wales, Victoria

Kobonga oxleyi Marshall and Hill 2009: 2–4, 6, 8, Fig 5, Table 2 (predation, sonogram, comp. note) Queensland

Kobonga oxleyi Pain 2009: 44 (predation) Australia

***K.* sp. nr. *Cicadetta oxleyi* (Dunn, 1991)**

Cicadetta sp. nr. *oxleyi* Dunn 1991: 53 (distribution) Victoria

***K.* sp. C Ewart, 2009**

Kobonga sp. C Ewart 2009b: 149, 167, Fig 15 (oscillogram, distribution, habitat, comp. note) Equals Kettledrum Cicada, no. 199 Equals Species C Ewart and Popple, 2001 Queensland

Kettledrum Cicada Ewart 2009b: 140–141, 147, Plate 2S (illustrated, distribution, emergence times, habitat, comp. note) Queensland

***K. umbrimargo* (Walker, 1858)** = *Cicada umbrimargo* = *Melampsalta umbrimargo* = *Kobongo umbrimago* (sic)

Kobonga umbrimargo Ewart 1989a: 80 (comp. note)

Kobonga umbrimargo Moulds 1990: 129–130, 164, Plate 21, Fig 7, Fig 7a, Plate 24, Fig 26 (illustrated, distribution, ecology, listed, comp. note) Australia, Western Australia, South Australia, Victoria, New South Wales

Cicada umbrimargo Moulds and Kopestonsky 2001: 141–142 (type species of *Kobonga*, comp. note)

Kobonga umbrimargo Moulds and Kopestonsky 2001: 141–150, 152, 155, Figs 1–2, Fig 11 (described, illustrated, key, synonymy, distribution, comp. note) Equals *Cicada umbrimargo* Equals *Kobonga umbrimago* (sic) Australia, Western Australia, South Australia, Victoria

Cicada umbrimargo Moulds and Carver 2002: 35, 37 (type species of *Kobonga*, listed)

Kobonga umbrimargo Moulds and Carver 2002: 37 (synonymy, distribution, ecology, type data, listed) Equals *Cicada umbrimargo* Australia

Kobonga umbrimargo Moulds 2005b: 381, 395–397, 399, 406–407, 409, 413–414, 417–419, 421, 423, 425, 436, Figs 16–18, Fig 42, Fig 46, Fig 50, Figs 56–59, Fig 62, Table 1 (illustrated, higher taxonomy, phylogeny, key, listed, comp. note) Australia

Kobonga umbrimago (sic) Haywood 2007: 18 (comp. note) South Australia

Kobonga umbrimargo Marshall and Hill 2009: 4, Table 2 (comp. note)

***Notopsalta* Dugdale, 1972**

Notopsalta Andrews and Gibbs 1989: 106 (note) New Zealand

Notopsalta Ewart 1989b: 292 (comp. note) Australia

Notopsalta Moulds 1990: 7, 12, 32, 169 (listed, diversity, comp. note) Australia, New Zealand

Notopsalta Boer and Duffels 1996b: 309 (distribution, comp. note) New Zealand, Australia

Notopsalta sp. Coombs 1996: 57, 60, Table 1 (emergence time, comp. note) New South Wales

Notopsalta Chou, Lei, Li, Lu and Yao 1997: 71, 363 (comp. note)

Notopsalta Ewart 1999: 2 (listed)

Notopsalta Simon, Arensburger and Chambers 1999: 1 (comp. note) New Zealand

Notopsalta Chambers, Boon, Buckley and Hitchmough 2001: 380, 385 (comp. note) Australia, New Zealand

Notopsalta Cooley 2001: 758 (comp. note) New Zealand

Notopsalta Cooley and Marshall 2001: 847 (comp. note) New Zealand

Notopsalta Arensburger 2002: 5055B (phylogeny) New Zealand

Notopsalta Buckley, Arensburger, Simon and Chambers 2002: 5–6, 15 (phylogeny, diversity, comp. note) New Zealand, Norfolk Island

Notopsalta Moulds and Carver 2002: 37 (type data, listed) Australia

Notopsalta Williams 2002: 38 (diversity, comp. note) Australia, New Zealand

Notopsalta Arensburger, Buckley, Simon, Moulds, and Holsinger 2004: 557–558, 562, 565–566 (molecular phylogeny, diversity, comp. note) New Zealand

Notopsalta Arensburger, Simon, and Holsinger 2004: 1769 (molecular phylogeny, comp. note) New Zealand

Notopsalta spp. Emery, Emery, Emery and Popple 2005: 106 (comp. note) New South Wales

Notopsalta Moulds 2005b: 390, 415, 430, 436, Table 2 (higher taxonomy, listed, comp. note) Australia

Notopsalta Haywood 2006a: 13 (comp. note) South Australia

Notopsalta Shiyake 2007: 8, 110 (listed, comp. note)

Notopsalta Hill and Marshall 2008: 523 (diversity, comp. note) New Zealand

Notopsalta Simon 2009: 104 (diversity, comp. note) New Zealand

N. atrata (Goding & Froggatt, 1904) to *Myopsalta atrata*

N. sp. nr. *atrata* Montane Grass Buzzer Moss & Popple, 2000 to *Myopsalta* sp. nr. *atrata*
N. sp. nr. *atrata* Black Acacia Buzzer Popple & Strange, 2002 to *Myopsalta* sp. nr. *atrata*
N. sp. nr. *atrata* Black Brigalow Buzzer Popple & Strange, 2002 to *Myopsalta* sp. nr. *atrata*
N. sp. nr. *atrata* Emery et al. 2005 to *Myopsalta* sp. nr. *atrata*
***N. melanesiana* (Myers, 1926)** = *Cicadetta melanesiana*
***N. sericea sericea* (Walker, 1850)** = *Cicada sericea* = *Melampsalta sericea* = *Melampsalta cruentata sericea* = *Cicadetta sericea* = *Melamsalta* (sic) *sericea* = *Cicada nervosa* Walker, 1850 = *Melampsalta nervosa* = *Cicadetta nervosa* = *Melampsalta scutellaris* (nec Walker) = *Melampsalta cruentata* (nec Fabricius) = *Melampsalta indistincta* Myers, 1921 = *Melampsalta nervosa* (nec Walker)
Melampsalta? nervosa Buchanan 1879: 214 (comp. note) Equals Cicada nervosaNew Zealand
Cicadetta nervosa Alfken 1903: 582 (comp. note) New Zealand
Melampsalta sericea Miller 1952: 16 (comp. note) New Zealand
Melampsalta sericea Eyles 1960: 995 (listed, distribution, habitat, comp. note) New Zealand
Melampsalta sericea Fleming 1966: 5, Fig 11 (illustrated, comp. note) New Zealand
Melampsalta sericea Grant-Taylor 1966: 348 (comp. note) New Zealand
Notopsalta sericea Fleming 1975a: 1570 (comp. note) New Zealand
Notopsalta sericea Fleming 1984: 204 (comp. note) New Zealand
Notopsalta sericea Moeed and Meads 1987a: 58, Table 6 (listed) New Zealand
Notopsalta sericea Andrews and Gibbs 1989: 105–106 (note) New Zealand
Cicada sericea Moulds 1990: 169 (type species of *Notopsalta*, listed, comp. note) New Zealand
Notopsalta sericea Chambers, Boon, Buckley and Hitchmough 2001: 380, Fig 3 (phylogeny, comp. note) New Zealand
Notopsalta sericea Buckley, Arensburger, Simon and Chambers 2002: 6, 11, Fig 1 (phylogeny, comp. note) New Zealand
Cicada sericea Moulds and Carver 2002: 37 (type species of *Notopsalta*, listed)
Notopsalta sericea Sueur 2002b: 129 (listed) Equals *Melampsalta indistincta*
Notopsalta sericea Arensburger, Buckley, Simon, Moulds, and Holsinger 2004: 558–559, 563 Table 1, Figs 1–2 (molecular phylogeny, comp. note) New Zealand
Notopsalta sericea Logan and Connolly 2005: 36–42, Fig 2, Figs 4–5, Tables 1–3 (agricultural pest, exuvia key, comp. note) New Zealand
Notopsalta sericea Shiyake 2007: 8, 108, 110, Fig 190 (illustrated, distribution, listed, comp. note) New Zealand
Notopsalta sericea Simon 2009: 103–104, Fig 1b, Fig 2 (illustrated, phylogeny, comp. note) New Zealand
***N. sericea* var. β (Walker, 1850)** = *Cicada nervosa* var. β
N. sp. A Ewart, 1988 to *Myopsalta* sp. A
N. sp. B Ewart, 1988 to *Myopsalta* sp. B
N. sp. C Ewart, 1988 to *Myopsalta* sp. C
N. sp. D Popple & Strange, 2002 to *Myopsalta* sp. D
N. sp. E Ewart, 1998 to *Myopsalta* sp. E
N. sp. F Ewart, 1998 to *Myopsalta* sp. F
N. sp. G Ewart, 1998, 2009 to *Myopsalta* sp. G

***Rhodopsalta* Dugdale, 1972**
Melampsalta (sic, partim) Buchanan 1879: 213–214 (comp. note) New Zealand
Rhodopsalta sp. Moeed 1975: 293, Table 2 (predation, listed) New Zealand
Rhodopsalta Andrews and Gibbs 1989: 106 (note) New Zealand
Melampsalta spp. Glare, O'Callaghan, and Wigley 1993: 98 (fungal parasite) New Zealand
Rhodopsalta Boer and Duffels 1996b: 309 (distribution, comp. note) New Zealand
Rhodopsalta Chou, Lei, Li, Lu and Yao 1997: 66, 361 (comp. note)
Rhodopsalta Simon, Arensburger and Chambers 1999: 1 (comp. note) New Zealand
Rhodopsalta Chambers, Boon, Buckley and Hitchmough 2001: 380, 385 (comp. note) New Zealand
Rhodopsalta Cooley 2001: 758 (comp. note) New Zealand
Rhodopsalta Cooley and Marshall 2001: 847 (comp. note) New Zealand
Rhodopsalta Arensburger 2002: 5055B (phylogeny) New Zealand
Rhodopsalta Buckley, Arensburger, Simon and Chambers 2002: 5–6, 12–13, 15 (phylogeny, diversity, comp. note) New Zealand, Australia
Rhodopsalta Arensburger, Buckley, Simon, Moulds, and Holsinger 2004: 557, 561–563, 565–566 (molecular phylogeny, diversity, comp. note) New Zealand
Rhodopsalta Arensburger, Simon, and Holsinger 2004: 1769, 1773, 1776, Fig 2, Table 3 (molecular phylogeny, diversity, comp. note) New Zealand
Rhodopsalta spp. Logan and Connolly 2005: 36 (comp. note) New Zealand
Rhodopsalta Moulds 2005b: 390, 415, 436 (higher taxonomy, listed, comp. note)
Rhodopsalta Buckley and Simon 2007: 421, 426, 431 (comp. note)

Rhodopsalta Shiyake 2007: 8, 110 (listed, comp. note)
Rhodopsalta Hill and Marshall 2008: 523–524 (comp. note) New Zealand
Rhodopsalta Cranston 2009: 92 (distribution, comp. note)
Rhodopsalta Simon 2009: 104 (diversity, comp. note) New Zealand

R. cruentata **(Fabricius, 1775)** = *Tettigonia cruentata* = *Cicada cruentata* = *Melampsalta cruentata* = *Melampsalta muta cruentata* = *Cicadetta cruentata* = *Cicadetta cincta* Walker, 1850 = *Melampsalta cincta* = *Melampsalta muta cincta* = *Cicadetta cincta* = *Cicada muta minor* Hudson, 1891 = *Cicadetta muta minor* = *Cicadetta minor*

Melampsalta cruentata Buchanan 1879: 214 (comp. note) Equals *Tettigonia cruentata* New Zealand
Melampsalta cincta Buchanan 1879: 214 (comp. note) *Cicada cincta* New Zealand
Cicadetta (= Melampsalta) cruentata Alfken 1903: 582 (comp. note) New Zealand
Melampsalta cruentata Miller 1952: 43, 57 (comp. note) New Zealand
Melampsalta cruentata Fleming 1966: 3, Fig 7 (illustrated, comp. note) New Zealand
Melampsalta cruentata Grant-Taylor 1966: 347 (comp. note) New Zealand
Rhodopsalta cruentata Fleming 1975a: 1570 (comp. note) New Zealand
Rhodopsalta cruentata Johns 1977: 322 (comp. note) New Zealand
Cicadetta cruentata Karban 1986: 224, Table 14.2 (comp. note) Australia, New Zealand
Rhodopsalta cruentata White 1987: 149, Fig 2 (abundance, comp. note) New Zealand
Rhodopsalta cruentata Andrews and Gibbs 1989: 105–106 (note) New Zealand
Rhodopsalta cruentata Noyes and Valentine 1989a: 36 (parasitism) New Zealand
Rhodopsalta cruentata Palma, Lovis and Tither 1989: 37 (status) Equals *Melampsalta microdora* New Zealand
Melampsalta cruentata Glare, O'Callaghan, and Wigley 1993: 98, 109 (fungal parasite) New Zealand
Rhodopsalta cruentata Lane 1993: 53–54 (comp. note) New Zealand
Rhodopsalta cruentata White and Sedcole 1993a: 39, 43, 46 (host, comp. note) New Zealand
Rhodopsalta cruentata Simon, Arensburger and Chambers 1999: 1 (comp. note) New Zealand
Rhodopsalta cruentata Chambers, Boon, Buckley and Hitchmough 2001: 380, Fig 3 (phylogeny, comp. note) New Zealand
Rhodopsalta cruentata Buckley, Arensburger, Simon and Chambers 2002: 6, 11, Fig 1 (phylogeny, comp. note) New Zealand
Rhodopsalta cruentata Sueur 2002b: 129 (listed) Equals *Melampsalta cruentata*
Rhodopsalta cruentata Bowie, Marris, Emberson, Andrew, Berry, et al. 2003: 105 (distribution, comp. note) Quail Island
Rhodopsalta cruentata Arensburger, Buckley, Simon, Moulds, and Holsinger 2004: 559, 562–563 Table 1, Figs 1–2 (molecular phylogeny, comp. note) New Zealand
Rhodopsalta cruentata Logan and Connolly 2005: 35, 37–42, Figs 4–5, Tables 1–3 (agricultural pest, exuvia key, comp. note) New Zealand
Rhodopsalta cruentata Buckley, Cordeiro, Marshall and Simon 2006: 412, 417–418, 420, Figs 3–5, Table 1 (phylogeny, sonogram, comp. note) New Zealand
Rhodopsalta cruentata Buckley and Simon 2007: 422, 425–428, Figs 1–4, Table 1 (phylogeny, comp. note) New Zealand
Rhodopsalta cruentata Shiyake 2007: 8, 108, 110, Fig 191 (illustrated, distribution, listed, comp. note) New Zealand
Rhodopsalta cruentata Hill and Marshall 2008: 524 (comp. note) New Zealand
Rhodopsalta cruentata Marshall, Hill, Fontaine, Buckley and Simon 2009: 2006 (comp. note) New Zealand
Rhodopsalta cruentata Simon 2009: 103–104, Fig 1c, Fig 2 (illustrated, phylogeny, comp. note) New Zealand

R. leptomera **(Myers, 1921)** = *Melampsalta leptomera* = *Malampsalta* (sic) *leptomera* = *Cicadetta leptomera*

Melampsalta leptomera Miller 1952: 17, 57 (comp. note) New Zealand
Melampsalta leptomera Fleming 1966: 4, Fig 8 (illustrated, comp. note) New Zealand
Melampsalta leptomera Grant-Taylor 1966: 347 (comp. note) New Zealand
Rhodopsalta leptomera Fleming 1975a: 1570 (comp. note) New Zealand
Rhodopsalta leptomera Buckley, Simon and Chambers 2001a: 77–78, Figs 6–7 (phylogeny, comp. note) New Zealand
Rhodopsalta leptomera Buckley, Simon, Shimodaira and Chambers 2001: 225, 228–229, Figs 2–3, Fig 5, Table 1 (phylogeny, comp. note) New Zealand
Rhodopsalta leptomera Chambers, Boon, Buckley and Hitchmough 2001: 380, Fig 3 (phylogeny, comp. note) New Zealand

Rhodopsalta leptomera Buckley, Arensburger, Simon and Chambers 2002: 6, 11, Fig 1 (phylogeny, comp. note) New Zealand

Rhopdopsalta leptomera Guindon and Gascuel 2002: 542 (phylogeny, comp. note)

Rhodopsalta leptomera Sueur 2002b: 129 (listed) Equals *Melampsalta leptomera*

Rhodopsalta leptomera Buckley, Cordeiro, Marshall and Simon 2006: 412, 417–418, 420, Figs 3–5, Table 1 (phylogeny, sonogram, comp. note) New Zealand

Rhodopsalta leptomera Arensburger, Buckley, Simon, Moulds, and Holsinger 2004: 559, 562–563 Table 1, Figs 1–2 (molecular phylogeny, comp. note) New Zealand

Rhodopsalta leptomera Arensburger, Simon, and Holsinger 2004: 1770–1776, Figs 2–3, Table 1, Table 3 (molecular phylogeny, diversity, comp. note) New Zealand

Rhodpsalta leptomera Buckley, Simon and Chambers 2006: 77–78, Figs 6–7 (phylogeny, comp. note) New Zealand

Rhodopsalta leptomera Buckley and Simon 2007: 422, 425–428, Figs 1–4, Table 1 (phylogeny, comp. note) New Zealand

Rhodopsalta leptomera Esson 2007: 20–21 (sex ratio) New Zealand

Rhodopsalta leptomera Simon 2009: 104, Fig 2 (phylogeny, comp. note) New Zealand

***R. microdora* (Hudson, 1936)** = *Melampsalta microdora* = *Cicadetta microdora*

Melampsalta microdora Palma, Lovis and Tither 1989: 37 (type material, type data) Equals *Rhodopsalta cruentata* New Zealand

Rhodopsalta microdora Hill and Marshall 2008: 524 (comp. note) New Zealand

Rhodopsalta microdora Marshall, Simon and Buckley 2006: 998–999, Figs 4–5 (phylogeny, comp. note) New Zealand

Rhodopsalta microdora Marshall, Slon, Cooley, Hill and Simon 2008: 1056–1058, 1060, 1063, Figs 1–3, Fig 5, Table 1 (phylogeny, oscillogram, comp. note) New Zealand

Rhodopsalta microdora Marshall, D.C. 2009: 109, Fig 1 (phylogeny, comp. note) New Zealand

***Plerapsalta* Moulds, 2012**

***P. incipiens* (Walker, 1850)** = *Cicada incipiens* = *Melampsalta incipiens* = *Cicadetta incipiens* = *Melampsalta murrayensis* Distant, 1907 = *Melampsalta mureensis* (sic) = *Cicadetta murrayensis* = *Cicadetta murrayiensis* (sic) = *Melampsalta abbreviata* (nec Walker)

Cicadetta murrayiensis (sic) Soper 1981: 51, 53 (fungal host, comp. note) Australia, Queensland

Cicadetta murrayensis Moulds 1990: 11, 154, Plate 17, Fig 5, Fig 5a (illustrated, distribution, ecology, listed, comp. note) Australia, New South Wales

Cicadetta incipiens Moulds 1990: 144 (distribution, listed) Australia, South Australia

Cicadetta murrayensis Naumann 1993: 47, 80, 149 (listed) Australia

Cicadetta murrayensis Ewart and Popple 2001: 59, 65, Fig 5B (described, habitat, song, oscillogram) Queensland

Cicadetta incipiens Moulds and Carver 2002: 28 (synonymy, distribution, ecology, type data, listed) Equals *Cicada incipiens* Australia, South Australia

Cicada incipiens Moulds and Carver 2002: 28 (type data, listed) South Australia

Cicadetta murrayensis Moulds and Carver 2002: 31 (synonymy, distribution, ecology, type data, listed) Equals *Melampsalta murrayensis* Australia, New South Wales

Melampsalta murrayensis Moulds and Carver 2002: 31 (type data, listed) Victoria

Cicadetta murrayensis Arensburger, Buckley, Simon, Moulds, and Holsinger 2004: 566 (comp. note) Australia

Cicadetta murrayensis Ewart 2005b 454 (comp. note)

***P. multifascia* (Walker, 1850)** = *Cicada multifasciata* = *Melampsalta multifascia* = *Pauropsalta multifascia* = *Cicadetta multifascia* = *Cicada obscurior* Walker, 1850 = *Cicada singula* Walker, 1850 = *Pauropsalta singula* = *Melampsalta singula* = *Cicadetta singula*

Cicadetta multifascia Moulds 1990: 144 (distribution, listed) Australia, South Australia

Cicadetta singula Moulds 1990: 144 (distribution, listed) Australia, South Australia

Cicadetta multifascia Moulds and Carver 2002: 30 (synonymy, distribution, ecology, type data, listed) Equals *Cicada multifascia* Equals *Cicada obscurior* Australia, South Australia

Cicada multifascia Moulds and Carver 2002: 30 (type data, listed) South Australia, New Holland

Cicada obscurior Moulds and Carver 2002: 30 (type data, listed)

Cicadetta singula Moulds and Carver 2002: 32 (synonymy, distribution, ecology, type data, listed) Equals *Cicada singula* Australia, South Australia

Cicada singula Moulds and Carver 2002: 32 (type data, listed)

Cicadetta multifascia Emery, Emery, Emery and Popple 2005: 101–102, 104–108, Tables 1–3 (life cycle, comp. note) New South Wales

***Caliginopsalta* Ewart, 2005**

Caliginopsalta Ewart 2005b: 440, 470–471, Table 1 (n. gen., described, diversity, comp. note)

***C. percola* Ewart, 2005** = *Notopsalta* sp. D Ewart, 1988

Notopsalta sp. D Ewart 1988: 183–184, 189, 191, 195, Fig 6, Plate 4F (illustrated, natural history, listed, comp. note) Queensland

Notopsalta sp. D Ewart 1998a: 64, 69, Fig 9B, Fig 13A (illustrated, described, natural history, song, oscillogram) Equals *Notopsalta* sp. D Ewart, 1988 Queensland

Notopsalta sp. D Royal Brigalow Ticker Popple and Strange 2002: 30, Table 1 (listed, habitat) Australia, Queensland

Caliginopsalta percola Ewart 2005b: 462, 470–476, 498, Figs 21–24, Fig 42, Table 4 (n. sp., described, type species of *Caliginopsalta*, illustrated, key, song, oscillogram, distribution, habitat, listed, comp. note) Queensland

***Terepsalta* Moulds, 2012**

***T. infans* (Walker, 1850)** = *Cicada infans* = *Tibicen infans* = *Quintilia infans* = *Cicada abbreviata* Walker, 1862 = *Melampsalta abbreviata*

Quintilia infans Moulds 1990: 111 (distribution, comp. note) South Australia

Quintilia infans Moulds and Carver 2002: 49 (synonymy, distribution, ecology, type data, listed) Equals *Cicada infans* Equals *Cicada abbreviata* Australia, South Australia

Cicada infans Moulds and Carver 2002: 49 (type data, listed) South Australia

Cicada abbreviata Moulds and Carver 2002: 49 (type data, listed) South Australia

Quintilia infans Moulds 2005b: 377, 395–396, 399, 417–419, 423–425, 436, Table 1, Figs 56–59, Fig 62 (higher taxonomy, phylogeny, comp. note) Australia

***Telmapsalta* Moulds, 2012**

***T. hackeri* (Distant, 1915)** = *Melampsalta hackeri* = *Cicadetta hackeri*

Cicadetta hackeri Moulds 1990: 150–151, 153, Plate 20, Fig 1, Fig 1a (illustrated, distribution, ecology, listed, comp. note) Australia, New South Wales, Queensland

Cicadetta hackeri Naumann 1993: 42, 80, 149 (listed) Australia

Cicadetta hackeri Ewart 1995: 87 (described, illustrated, natural history, acoustic behavior, oscillogram) Queensland

Cicadetta hackeri Sanborn 1999a: 52 (type material, listed)

Cicadetta hackeri Moss 2000: 69 (comp. note) Queensland

Cicadetta hackeri Ewart 2001b: 69–75, 80, 82–83, Fig 5, Fig 11, Table 2 (emergence pattern, population density, nymph illustrated, predation, listed, comp. note) Queensland

Cicadetta hackeri Sueur 2001: 42, Table 1 (listed)

Cicadetta hackeri Emery and Emery 2002: 137–139, Fig 1 (illustrated, comp. note) Queensland, New South Wales

Cicadetta hackeri Moulds and Carver 2002: 27 (synonymy, distribution, ecology, type data, listed) Equals *Melampsalta hackeri* Australia, New South Wales, Queensland

Melampsalta hackeri Moulds and Carver 2002: 27 (type data, listed) Queensland

Cicadetta hackeri Popple and Ewart 2002: 117 (illustrated) Australia, Queensland

Cicadetta hackeri Williams 2002: 156–157 (listed) Australia, Queensland, New South Wales

Cicadetta hackeri Emery, Emery, Emery and Popple 2005: 97 (comp. note) New South Wales

Cicadetta hackeri Marshall and Hill 2009: 4, Table 2 (comp. note)

***Limnopsalta* Moulds, 2012**

***L. stradbrokensis* (Distant, 1915)** = *Melampsalta stradbrokensis* = *Cicadetta stradbrokensis*

Cicadetta stradbrokensis Ewart 1989a: 77 (comp. note)

Cicadetta stradbrokensis Moulds 1990: 23, 164–165, Plate 19, Fig 2 (illustrated, distribution, ecology, listed, comp. note) Australia, Queensland

Cicadetta stradbrokensis Naumann 1993: 61, 80, 149 (listed) Australia

Cicadetta stradbrokensis Ewart 1995: 88 (described, illustrated, natural history, acoustic behavior, oscillogram) Queensland

Cicadetta stradbrokensis Ewart 1996: 14 (listed)

Cicadetta stradbrokensis Moulds 1996b: 92 (acoustic behavior, comp. note) Australia, Queensland

Cicadetta stradbrokensis Sanborn 1999a: 52 (type material, listed)

Cicadetta stradbrokensis Moss 2000: 69 (comp. note) Queensland

Cicadetta stradbrokensis Ewart 2001b: 69 (listed) Queensland

Cicadetta stradbrokensis Sueur 2001: 43, Table 1 (listed)

Cicadetta stradbrokensis Moulds and Carver 2002: 32–33 (synonymy, distribution, ecology, type data, listed) Equals *Melampsalta stradbrokensis* Australia, Queensland

Melampsalta stradbrokensis Moulds and Carver 2002: 32 (type data, listed) Queensland

***Heliopsalta* Moulds, 2012**

***H. polita* (Popple, 2003)** = *Cicadetta polita* = *Cicadetta* sp. G Popple and Strange, 2002

Cicadetta sp. G Popple and Strange 2002: 18–19, 24, 29, Fig 1B, Figs 5A–B, Table 1 (described, illustrated, habitat, oscillogram, comp. note) Australia, Queensland

Cicadetta polita Popple 2003: 107–114, Figs 1–4 (n. sp., described, illustrated, synonymy, distribution, hosts, song, comp. note) Equals *Cicadetta* sp. G Popple and Strange, 2002 Australia, Queensland

***Tettigetta* Kolenati, 1857** = *Cicada (Tettigetta)* = *Tettigella* (sic) = *Tettigretta* (sic)

Tettigetta Boulard 1981a: 37 (comp. note)

Tettigetta Boulard 1982c: 103 (listed)

Tettigetta Boulard 1982f: 182 (listed)

Tettigetta Koçak 1982: 151 (listed, comp. note) Turkey

Tettigetta Hayashi 1984: 66 (comp. note)

Tettigetta Schedl 1986a: 10 (key) Istria

Tettigetta Boulard 1988c: 65 (comp. note)

Tettigetta Boulard 1988f: 37–39, 42–43, 45 (comp. note)

Tettigetta Quartau and Rebelo 1990: 1 (comp. note) Portugal

Tettigetta Quartau and Rebelo 1994: 141 (comp. note)

Tettigetta Boulard 1995a: 59 (listed)

Tettigetta Boulard and Mondon 1995: 19, 65 (key, comp. note) France

Tettigetta Quartau 1995: 34–35 (comp. note) Portugal

Tettigetta Gogala, Popov and Ribaric 1996: 46 (comp. note)

Tettigetta Boulard 1997c: 107–108, 111–113, 117 (comp. note)

Tettigetta Chou, Lei, Li, Lu and Yao 1997: 76, 364 (comp. note)

Tettigetta Holzinger, Frölich, Günthart, Pauterer, Nickel et al. 1997: 49 (listed) Europe

Tettigella (sic) Quartau, Coble, Viegas-Crespo, Rebelo, Ribeiro, et al. 1997: 138 (comp. note) Portugal

Tettigetta Gogala 1998a: 156 (comp. note)

Tettigetta Popov, Beganović and Gogala 1998a: 90 (comp. note)

Tettigetta Quartau 1998: 381 (comp. note)

Tettigetta Gogala and Trilar 1999b: 1 (comp. note)

Tettigetta Boulard 2000f: 133 (comp. note)

Tettigetta Boulard 2000i: 21, 26, 92 (listed, key, comp. note) France

Tettigetta Gogala and Popov 2000: 11 (comp. note)

Tettigetta Quartau, Ribeiro, Simões and Crespo 2000: 1677 (comp. note) Portugal

Tettigetta Schedl 2000: 260, 268 (key, listed)

Tettigetta Boulard 2001b: 27–29, 31–32, 37 (comp. note)

Tettigetta Schedl 2001a: 1287, 1289 (comp. note)

Tettigetta Lee, Choe, Lee and Woo 2002: 3, 10 (comp. note)

Tettigetta Holzinger, Kammerlander and Nickel 2003: 477, 483 (key, listed, diversity, comp. note) Europe

Tettigetta Quartau and Simões 2003: 118 (comp. note) Portugal

Tettigetta spp. Gogala, Trilar and Krpač 2005: 124 (comp. note) Greece

Tettigetta Moulds 2005b: 390, 436 (higher taxonomy, listed, comp. note)

Tettigetta Puissant 2005: 309, 312 (comp. note)

Tettigetta Puissant 2006: 34, 80, 87 (comp. note) France

Tettigetta Boulard 2007b: 97–98 (comp. note)

Tettigetta Demir 2008: 477 (listed) Turkey

Tettigetta Gogala and Schedl 2008: 395–396, 404 (comp. note)

Tettigetta spp. Gogala and Schedl 2008: 400 (comp. note)

Tettigetta Lee 2008b: 446, 457–458, Tables 1–2 (listed, comp. note) Korea

Tettigetta Pinto-Juma, Seabra and Quartau 2008: 2 (comp. note)

Tettigetta Mozaffarian and Sanborn 2010: 76, 81 (distribution, listed, comp. note) Europe, North Africa, Middle East, Asia, Mongolia

Tettigetta Puissant and Sueur 2010: 555–558, 566–567, 569–570 (listed, comp. note) Spain, Portugal, North Africa

T. aneabi Boulard, 2000 to *Tettigettalna aneabi*

T. argentata argentata (Olivier, 1790) to *Tettigettalna argentata argentata*

T. argentata flavescens Boulard, 1982 to *Tettigettalna argentata flavescens*

T. argentata mauresqua Boulard, 1982 to *Tettigettalna argentata mauresqua*

T. argentata nana (Boulard, 2000) to *Tettigetta argentata* var. *nana*

T. atra (Gomez-Menor Ortega, 1957) to *Tettigettalna argentata argentata* Olivier, 1790)

T. baenai Boulard, 2000 to *Tettigettacula baenai*

T. bergevini Boulard, 1980 to *Tettigettacula bergevini*

T. brullei (Fieber, 1872) to *Tettigettula pygmea* (Olivier, 1790)

T. carayoni Boulard, 1982 to *Cicadivetta carayoni*

***T. dlabolai* Mozaffarian & Sanborn, 2010** = *Cicadetta gastrica* (nec Stål)

Cicadetta gastrica (sic) Dlabola 1981: 207–208, Figs 134–137 (distribution, comp. note) Iran

Cicadetta gastrica (sic) Mirzayans 1995: 18 (listed, distribution) Iran

Tettigetta dlabolai Mozaffarian and Sanborn 2010: 73–76, 81, Figs 8–14 (n. sp., described, illustrated, synonymy, comp. note) Equals *Cicadetta gastrica* Dlabola, 1981 Equals *Cicadetta gastrica* Mirzayans, 1995 Iran

T. estrellae Boulard, 1982 to *Tettigettalna estrellae*

T. floreae Boulard, 1987 to *Tettigettacula floreae*

***T. golestani* Gogala & Schedl, 2008**

Tettigetta golestani Gogala and Schedl 2008: 396–403, Figs 1–8 (n. sp., described, illustrated, song, sonogram, oscillogram, comp. note) Iran

Tettigetta golestani Mozaffarian and Sanborn 2010: 81–82 (listed, distribution, comp. note) Iran

Tettigetta golestani Mozaffarian, Sanborn and Phillips 2010: 27 (comp. note) Iran

Tettigettta golestani Puissant and Sueur 2010: 557 (comp. note)

***T. hoggarensis* Boulard, 1980**

Tettigetta hoggarensis Nast 1982: 311 (listed) Algeria

***T. isshikii* (Kato, 1926)** = *Melampsalta isshikii* = *Cicadetta isshikii* = *Melampsalta isshikii* var. *flavicosta* Kato, 1927 = *Cicadetta isshikii flavicosta* = *Leptopsalta radiator* (nec Uhler) = *Leptopsalta radiater* (sic) (nec Uhler)

Melampsalta isshikii Doi 1932: 43 (listed, comp. note) Korea

Melampsalta isshikii Wu 1935: 27 (listed) China, Manchuria

Cicadetta isshikii Lee 1995: 18, 23, 45–46, 104, 112, 114–118, 136 (listed, key, described, ecology, distribution, illustrated, comp. note) Equals *Melampsalta isshikii* Equals *Melampsalta isshikii* var. *flavicosta* Equals *Leptopsalta radiator* (nec Uhler) Korea

Melampsalta isshikii Lee 1995: 31, 33, 35–45, 115 (comp. note) Korea

Melampsalta isshikii var. *flavicosta* Lee 1995: 35, 37, 39–43, 115–116, 132 (listed, key, described, ecology, distribution, illustrated, comp. note) Korea

Cicadetta isshikii Chou, Lei, Li, Lu and Yao 1997: 58, 63–65, Fig 8–13 (key, described, illustrated, synonymy, comp. note) Equals *Melampsalta isshikii* Equals *Melampsalta isshikii flavicosta* China

Cicadetta isshikii Lee 1998: 21 (comp. note) Korea

Cicadetta isshikii Lee 1999: 10, 12 (distribution, song, listed, comp. note) Equals *Melampsalta isshikii* Equals *Melampsalta isshikii* var. *flavicosta* Equals *Leptopsalta radiator* (nec Uhler) Korea, China, Russia

Cicadetta isshikii Hua 2000: 61 (listed, distribution) Equals *Melampsalta isshikii* China, Liaoning, Shandong, Korea, USSR

Cicadetta isshikii Lee, Choe, Lee and Woo 2002: 3–10, Figs 1–3, Tables 1–3 (gene sequence, phylogeny, comp. note) Russia, China, Korea

Cicadetta isshikii Lee 2003b: 8 (comp. note)

Cicadetta isshikii Lee, Oh and Park 2004: 127–131, Figs 1–2, Figs 3e–f (habitat, illustrated, comp. note) Korea

Cicadetta isshikii var. *flavicosta* Lee, Oh and Park 2004: 129 (comp. note) Korea

Cicadetta isshikii Lee 2005: 17, 116–119, 159, 167–168, 177–178 (described, illustrated, synonymy, song, distribution, key, comp. note) Equals *Melampsalta isshikii* Equals *Melampsalta isshikii* var. *flavicosta* Equals *Leptopsalta radiator* (nec Uhler) Korea, China

Cicadetta isshikii Lee 2008b: 446 (status) Korea

Tettigetta isshikii Lee 2008b: 446, 451, 457–462, Fig 2C, Fig 6, Tables 1–2 (n. comb., key, illustrated, synonymy, distribution, listed, comp. note) Equals *Mogannia hebes* (nec Walker) Equals *Melampsalta pellosoma* (nec Uhler) Equals *Cicadetta pellosoma* (nec Uhler) Equals *Melampsalta isshikii* Equals *Cicadetta isshikii* Equals *Melampsalta isshikii* var. *flavicosta* Equals *Cicadetta isshikii* var. *flavicosta* Equals *Leptopsalta radiator* (nec Uhler) Equals *Melampsalta radiator* (nec Uhler) Equals *Cicadetta pellosoma*-like species Korea, Russia, China

Cicadetta pellosoma-like species Lee 2008b: 461–462 (comp. note) Korea

Tettigettta isshilii Puissant and Sueur 2010: 557 (comp. note)

T. josei Boulard, 1982 to *Tettigettalna josei*

T. leunami Boulard, 2000 to *Pseudotettigetta melanophrys leunami*

T. linaresae Boulard, 1987 to *Tettigettacula linaresae*

T. manueli Boulard, 2000 to *Tettigettalna helianthemi helianthemi* (Rambur, 1840)

T. mariae Quartau & Boulard, 1995 to *Tettigettalna mariae*

***T. maroccana* Boulard, 1980**
Tettigetta maroccana Nast 1982: 311 (listed) Morocco

***T. megalopercula* Mozaffarian & Sanborn, 2012**

***T. merkli* Boulard, 1980**
Tettigetta merkli Nast 1982: 311 (listed) Algeria

***T. musiva* (Germar, 1830)** = *Cicada musiva* = *Melampsalta musiva* = *Cicadetta musiva* = *Melampsalta (Melampsalta) musiva* = *Melampsalta (Cicadetta) musiva* = *Cicadetta (Melampsalta) musiva* = *Cicada tamarisci* Walker, 1870 = *Cicada tamaricis* (sic) = *Cicada tamarisca* (sic) = *Cicadetta musiva suoedicola* de Bergevin, 1913 = *Cicadetta musiva suaedicola* (sic)
Melampsalta musiva Seki 1933: 112 (illustrated, comp. note)
Melampsalta (Cicadetta) musiva Mirzayans, Hashemi, Borumand, Zairi and Rajabi 1976: 111 (distribution, listed) Iran
Melampsalta musiva Dlabola 1981: 207 (distribution, comp. note) Iran
Cicadetta musiva Lodos and Kalkandelen 1981: 78–79 (synonymy, distribution, comp. note) Equals *Cicada (Melampsalta) musiva caspica* Equals *Cicada tamarisci* Equals *Cicadetta musiva sueodicola* Turkey, Afghanistan, Algeria, Cyprus, Egypt, Iran, Iraq, Israel, Jordan, Morocco, Tunisia, USSR
Cicada musiva Osssiannilsson 1981: 223 (type species of *Melampsalta*)
Cicadetta musiva Dadamirzaev and Kholmuminov 1982: 66 (predation) Uzbekistan
Melampsalta musiva Panov and Krjutchkova 1983: 141–148, Figs 1–2, Fig 9, Fig 11, Fig 14, Fig 16 (neurosecretory cells, comp. note) Crimea, Azerbaijan
Cicada musiva Hayashi 1984: 66 (type species of *Cicada (Melampsalta)*)
Melampsalta musiva Kholmuminov 1987: 49 (comp. note) Uzbekistan
Melampsalta musiva Walker and Pittaway 1987: 40–41 (illustrataed, natural history, distribution, comp. note) Saudi Arabia
Cicadetta musiva Schedl 1993: 798–799 (listed, synonymy, distribution, comp. note) Equals *Cicada musiva* Egypt, Israel, Jordan, Morocca, Tunisia, Algeria, Nubia, Saudi Arabia, Syria, Turkey, Persia, Beluchistan, Iraq, Azerbaijan, Kazakhstan, Turkmenia, Uzbekistan
Melampsalta musiva Mirzayans 1995: 19 (listed, distribution) Iran
Cicada musiva Lee 1999: 10 (type species of *Melampsalta*) Europe
Tettigetta musiva Schedl 1999b: 827 (n. comb., synonymy, listed, distribution, ecology, comp. note) Equals *Cicadetta musiva* Israel, Iran, Egypt, North Africa, Jordan, Syria, Cyprus, Turkey, Iraq, Azerbaijan, Kazakhstan, Turkmenistan, Uzbekistan
Tettigetta musiva Schedl 2004a: 17 (comp. note)
Cicada musiva Lee 2005: 176 (type species of *Melampsalta*)
Cicada musiva Puissant 2005: 305 (comp. note)
Melampsalta musiva Ahmed and Sanborn 2010: 43 (synonymy, distribution) Equals *Cicada musiva* Equals *Cicada melampsalta* Equals *Cicada tamarisci* Pakistan, North Africa
Cicadetta musiva Kemal and Koçak 2010: 6 (distribution, listed, comp. note) Turkey
Cicada (Melampsalta) musiva Mozaffarian and Sanborn 2010: 82 (type species of *Melampsalta*)
Melampsalta musiva Mozaffarian and Sanborn 2010: 82 (synonymy, listed, distribution, comp. note) Equals *Cicada musiva* Equals *Cicada melampsalta* Equals *Cicada tamarisci* Iran, Nubia, North Africa, Afghanistan

***T. omar* (Kirkaldy, 1899)** = *Melampsalta omar* = *Cicadetta omar*
Melampsalta omar Poulton 1907: 343 (predation) Yemen
Tettigetta omar Schedl 2004a: 17 (comp. note)

T. parvula (Fieber, 1876) to *Cicadivetta parvula*

***T. prasina* (Pallas, 1773)** = *Cicada prasina* = *Cicada brasina* (sic) = *Cicada prosina* (sic) = *Cicada porasina* (sic) = *Tettigonia prasina* = *Tettigetta pallas* (sic) = *Cicada (Tettigetta) prasina* = *Cicadetta prasina* = *Melampsalta prasina* = *Melampsalta (Melampsalta) prasina* = *Cicada lutescens* Olivier, 1790 = *Melampsalta (Melampsalta) lutescens*
Cicada prasina Boulard 1981a: 41 (type species of genus *Tettigetta*)
Melampsalta prasina Boulard 1982f: 197 (comp. note)
Cicada prasina Koçak 1982: 151 (type species of *Tettigetta*)
Cicada prasina Hayashi 1984: 66 (type species of *Cicada (Tettigetta)*)
Cicadetta prasina Popov 1985: 76, 80, Fig 24 (acoustic system, comp. note)
Tettigetta prasina Nast 1987: 570–571 (distribution, listed) Russia
Cicada prasina Boulard 1988f: 38, 43 (type species of *Tettigetta*, comp. note)
Tettigetta prasina Boulard 1988f: 43, 49, Fig 5 (type species of *Tettigetta*, illustrated, comp. note)

Tettigetta pallas (sic) Boulard 1988f: 45 (type species of *Tettigetta*, comp. note)
Cicadetta prasina Quartau and Fonseca 1988: 368, 369 (synonymy, distribution, comp. note) Equals *Cicada prasina* Equals *Cicada lutescens* Portugal, Morocco, USSR, Armenia, Kazakhstan, Siberia
Cicadetta prasina Popov 1990a: 302, Fig 2 (comp. note)
Cicadetta prasina Tautz 1990: 105, Fig 12 (song frequency, comp. note)
Cicadetta prasina Chelpakova 1991: 67 (comp. note) Kazakhstan, Kyrghyzstan
Cicada prasina Boulard 1997c: 108, 110–112 (type species of *Tettigetta*, comp. note)
Tettigetta prasina Boulard 1997c: 112–113, 117, Plate II, Fig 8 (type species of *Tettigetta*, illustrated, comp. note)
Cicadetta prasina Chou, Lei, Li, Lu and Yao 1997: 66 (listed, synonymy, comp. note) Equals *Cicada prasina* China
Cicadetta prasina Popov 1998a: 316–319, 322–325, Figs 1–2, Fig 7a (song, oscillogram, comp. note) Kyrgystan
Cicadetta prasina Popov 1998b: 309–312, 314, 316–317, Figs 1–2, Fig 7a (song, oscillogram, comp. note) Kyrgystan
Tettigetta prasina Boulard 2000i: 21 (listed)
Cicadetta prasina Hua 2000: 61 (listed, distribution) Equals *Cicada prasina* China, Xinjiang
Cicada prasina Boulard 2001b: 28, 31–32 (type species of *Tettigetta*, comp. note)
Tettigetta prasina Boulard 2001b: 32–33, 37, Plate II, Fig 8 (type species of *Tettigetta*, illustrated, comp. note)
Cicadetta prasina Puissant 2001: 148 (comp. note)
Cicadetta prasina Sueur 2001: 43, Table 1 (listed)
Cicada prasina Holzinger, Kammerlander and Nickel 2003: 483 (type species of *Tettigetta*)
Cicada prasina Puissant 2006: 80 (type species of *Tettigetta*)
Cicadetta prasina Vsevolodova-Perel and Sizemskaya 2007: 633 (habitat, comp. note) Russia
Cicada prasina Lee 2008b: 458 (type species of *Tettigetta*) Russia
Cicada prasina Mozaffarian and Sanborn 2010: 81 (type species of *Tettigetta*)
Cicada prasina Puissant and Sueur 2010: 556–557 (type species of *Tettigetta*, comp. note)
Tettigettta prasina Puissant and Sueur 2010: 557 (comp. note)

T. pygmea (Olivier, 1790) to *Tettigettula pygmea*

***T. safavii* Mozaffarian & Sanborn, 2012**

T. sedlami Boulard, 1981 to *Tettigettacula sedlami*

T. septempulsata Boulard & Quartau, 1991 to *Tettigettalna estrellae* (Boulard, 1982)

***T. turcica* Schedl, 2001**

Tettigetta turcica Schedl 2001a: 1287–1289, Figs 1–4, Photo 1 (n. sp., described, illustrated, comp. note) Turkey
Tettigetta turcica Gogala and Schedl 2008: 399 (comp. note)

T. virens (Herrich-Schäffer, 1835) to *Euryphara virens*

***T. yemenensis* Schedl, 2004**

Tettigetta yemenensis Schedl 2004a: 12, 14–17, Fig 3, Figs 5–6 (n. sp., described, illustrated, illustrated, listed, distribution, comp. note) Yemen

***Tettigettalna* Puissant, 2010**

Tettigetta Fonseca 1996: 17, 28–29 (comp. note)
Tettigetta spp. Fonseca 1996: 22 (comp. note)
Tettigetta Fonseca and Hennig 1996: 1536 (comp. note)
Tettigetta spp. Fonseca and Hennig 1996: 1541, 1544 (comp. note)
Tettigetta Fonseca, Serrão, Pina-Martins, Silva, Mira, et al. 2008: 25, 27, 29–30 (comp. note)
Tettigetta spp. Fonseca, Serrão, Pina-Martins, Silva, Mira, et al. 2008: 25, 27–28, 31 (comp. note)
Tettigettalna Puissant and Sueur 2010: 557–558, 562, 570 (n. gen., described, comp. note) Portugal

***T. aneabi* (Boulard, 2000)** = *Tettigetta aneabi*

Tettigetta aneabi Boulard 2000f: 135–137, 140–143, Plate I, Figs 3–4, Plate II, Fig 2 (n. sp., described, illustrated, comp. note) Spain
Tettigettalna aneabi Puissant and Sueur 2010: 557–558, 561–562, 567, 570–571, Fig 8, Fig 20, Figs 24–27 (n. comb., illustrated, synonymy, habitat, distribution, acoustic behavior, sonogram, oscillogram, comp. note) Equals *Tettigetta aneabi* Spain

***T. argentata argentata* (Olivier, 1790)** = *Cicada argentata* = *Cicadea argentea* (sic) = *Cicadetta argentata* = *Melampsalta argentata* = *Melampsalta (Melampsalta) argentata* = *Tettigia argentata* = *Tettigretta* (sic) *argentata* = *Tettigia* (sic) *argentata* = *Cicadetta (Melampsalta) argentata* = *Cicada sericans* Herrich-Schäffer, 1835 = *Cicadetta sericans* = *Melampsalta (Melampsalta) sericans* = *Melampsalta argentata atra* Gomez-Menor Ortega, 1957 = *Melampsalta argentata* var. *atra* = *Cicadetta argentata atra* = *Tettigetta atra* = *Cicada helianthemi* (nec Rambur)

Cicada argentata Swinton 1880: 223 (comp. note)

Cicadetta argentata Castellani 1952: 15 (comp. note) Italy

Cicadetta argentata Wagner 1952: 15 (listed, comp. note) Italy

Cicadetta argentata Villiers 1977: 167 (listed, comp. note) Mediterranean

Tettigetta argentata mph. *argentata* Boulard 1982a: 50 (comp. note) France

Tettigetta argentata Boulard 1982f: 184, 186–188, 197 (distribution, comp. note, listed) Equals *Cicada argentata* Equals *Cicadetta argentata* Portugal

Tettigetta atra Boulard 1982f: 185–189, 195, 197, Fig 13, Figs 16–19, Fig 47 (n. comb., synonymy, distribution, comp. note, listed) Equals *Melampsalta argentata* var. *atra* Portugal, Spain, France

Melampsalta argentata var. *atra* Boulard 1982f: 182 (comp. note) Portugal

Tettigetta atra Boulard 1985a: 33 (distribution, listed) Portugal

Cicadetta (Tettigetta) argentata Boulard 1985b: 63 (characteristics) France

Cicadetta argentata Chinery 1986: 88–89 (illustrated, comp. note) Europe

Tettigetta argentata Schedl 1986a: 3, 5, 8–10, 16–17, 22–24, Fig 3, Fig 10, Figs 20–21, Map 5, Table 1 (illustrated, listed, key, synonymy, distribution, ecology, comp. note) Equals *Cicada argentata* Equals *Melampsalta argentata* Equals *Cicadetta argentata* Equals *Cicada distincta* Istria, Portugal, Spain, France, Sicily

Tettigetta argentata Nast 1987: 570–571 (distribution, listed) France, Portugal, Spain, Italy

Tettigetta argentata Boulard 1988a: 12, Table (comp. note) France

Cicada argentata Boulard 1988f: 18, 42 (comp. note)

Tettigetta argentata Quartau and Fonseca 1988: 367, 371 (synonymy, distribution, comp. note) Equals *Cicada argentata* Equals *Cicada helianthemi* Equals *Cicadetta argentata* Portugal, France, Italy, Sicily, Spain

Tettigetta atra Quartau and Fonseca 1988: 367, 371–372 (synonymy, distribution, comp. note) Equals *Melampsalta argentata* var. *atra* Portugal, France, Spain

Tettigetta argentata Boulard 1990b: 166, 172, 180, 185, 187, Plate XXII, Fig 7 (sonogram, comp. note)

Cicadetta argentata Moulds 1990: 81 (comp. note) Europe

Cicada argentata Moulds 1990: 81 (comp. note) Europe

Tettigetta argentata Quartau and Rebelo 1990: 1 (oscillogram, comp. note) Portugal

Tettigetta argentata Boulard and Quartau 1991: 50, 54 (comp. note)

Tettigetta argentata Fonseca 1991: 174, 180–181, 189, Fig 4 (song analysis, oscillogram, comp. note) Portugal

Tettigetta sp. Fonseca 1991: 174, 185–186, 189, Fig 7 (song analysis, oscillogram, comp. note) Portugal

Tettigetta atra Fonseca 1991: 174, 180–181, 189–190 (song analysis, oscillogram, comp. note) Portugal

Tettigetta argentata Hennig, Weber, Moore, Kleindienst, Huber and Popov 1994: 54 (comp. note)

Tettigetta atra Hennig, Weber, Moore, Kleindienst, Huber and Popov 1994: 54 (comp. note)

Tettigetta argentata Boulard 1995a: 6–7, 13, 59–61, Fig $S23_1$, Fig $S23_2$ (song analysis, sonogram, listed, comp. note) France

Tettigetta atra Boulard 1995a: 13, 61–62, Fig $S24_1$, Fig $S24_2$ (song analysis, sonogram, listed, comp. note) Portugal

Tettigetta argentata Boulard and Mondon 1995: 30, 32, 42, 66–67, 86, 159 (illustrated, sonogram, listed, comp. note) Equals *Cicada sericans* France

Tettigetta atra Boulard and Mondon 1995: 159 (listed) Spain

Tettigetta argentata Quartau and Boulard 1995: 106–108, 110 (comp. note)

Tettigetta atra Quartau and Boulard 1995: 106–107 (comp. note)

Tettigetta argentata Quartau 1995: 34, 36–38 (illustrated, ecology, comp. note) Portugal

Tettigetta argentata Fonseca 1996: 15, 20, 22, 27–29, Fig 2, Table 2 (timbal muscle physiology, oscillogram, comp. note) Portugal

Tettigetta argentata Fonseca and Hennig 1996: 1535–1536, 1540–1541, 1543–1544, Figs 3–6, Fig 10, Fig 11A, (tensor muscle physiology, comp. note) Portugal

Cicada argentata Boulard 1997c: 89, 111 (comp. note) France

Tettigetta argentata Holzinger, Frölich, Günthart, Pauterer, Nickel et al. 1997: 49 (listed) Europe

Tettigetta argentata Arias Giralda, Pérez Rodríguez and Gallego Pintor 1998: 262, 265, 268 (olive pest, comp. note) Spain

Tettigetta argentata Gogala 1998b: 395 (oscillogram, comp. note) Slovenia

Tettigetta argentata Ragge and Reynolds 1998: 477–478, Fig 1629 (oscillogram, comp. note) France, Iberian Peninsula, Italy, Sicily
Tettigetta argentata Gogala and Gogala 1999: 120, 122, 125, Fig 4 (distribution, comp. note) Slovenia, Mediterranean
Cicadetta argentata Kartal 1999: 100–103, Fig A_3, Fig B_3, Fig C_3, Fig D_3, Fig E_3, Fig F_3, Fig G_3 (described, illustrated, comp. note) Spain
Tettigetta argentata Quartau, Rebelo and Simões 1999: 70–73, Figs 2–5 (ecology, emergence, song, sonogram, oscillogram, comp. note)
Tettigretta (sic) *argentata* Quartau, Rebelo and Simões 1999: 70, Fig 1 (illustrated)
Tettigetta argentata Boulard 2000f: 135, 137, 139 (comp. note)
Tettigetta argentata Boulard 2000i: 84, 87–90, 99, 126, Fig 22c, Plate N, Map 14 (synonymy, illustrated, song, sonogram, oscillogram, frequency spectrum, distribution, ecology, comp. note) Equals *Cicada argentata* Equals *Cicada sericans* Equals *Cicada helianthemi* Equals *Cicadetta argentata* Equals *Melampsalta (Melampsalta) argentata* Equals *Tettigia* (sic) *argentata* Equals *Cicadetta (Melampsalta) argentata* France
Tettigetta argentata Gogala and Popov 2000: 11 (comp. note)
Tettigetta argentata Puissant and Boulard 2000: 116 (comp. note)
Tettigetta argentata Puissant and Sueur 2000: 69, 72 (comp. note) France
Tettigetta argentata Schedl 2000: 262–263, 265–266, 268–269, Fig 10, Plate 1, Fig 9, Fig 12c, Fig 13c, Map 7 (key, illustrated, distribution, synonymy, listed, comp. note) Equals *Cicadetta argentata* Italy, Slovenia
Cicada argentata Boulard 2001b: 9, 31 (comp. note) France
Tettigetta argentata Puissant 2001: 147 (comp. note) France
Tettigetta argentata Schedl 2001b: 27 (comp. note)
Tettigetta argentata Sueur 2001: 35, 44, Tables 1–2 (listed, comp. note) Equals *Tettigetta atra*
Tettigetta atra Sueur 2001: 44, Table 1 (listed) Equals *Tettigetta argentata*
Tettigetta argentata Sueur and Puissant 2001: 22 (comp. note) France
Tettigetta argentata Fonseca and Revez 2002a: 1286–1288, 1291, Fig 2 (song power spectrum, comp. note)
Tettigetta argentata Fonseca and Revez 2002b: 971–976, Fig 1, Fig 2A, Fig 2C, Fig 3, Tables 1–2 (song analysis, oscillogram, comp. note) Portugal
Tettigetta argentata Gogala 2002: 244–245, Fig 6 (song, sonogram, oscillogram, comp. note)
Tettigetta argentata Gogala and Seljak 2003: 350 (comp. note) Slovenia
Tettigetta argentata Holzinger, Kammerlander and Nickel 2003: 483–484, 616–617, 633, 637, Plate 42, Fig 6, Tables 2–3 (key, listed, illustrated, comp. note) Europe, Mediterranean, France
Tettigetta argentata Sueur, Puissant, Simões, Seabra, Boulard and Quartau 2004: 178–180, 182–186, Fig 1, Fig 4 (synonymy, distribution, ecology, song, sonogram, oscillogram, comp. note) Equals *Tettigetta atra* n. syn. Equals *Tettigetta* sp. Fonseca 1991 Portugal
Tettigetta atra Sueur, Puissant, Simões, Seabra, Boulard and Quartau 2004: 182–184 (comp. note)
Tettigetta sp. Sueur, Puissant, Simões, Seabra, Boulard and Quartau 2004: 186 (comp. note)
Tettigetta argentata Boulard 2005d: 337 (comp. note) Europe
Tettigetta argentata Puissant 2005: 309 (comp. note)
Tettigetta argentata Puissant and Defaut 2005: 116, 118–125, 127–128, Tables 1–9, Table 13, Table 15 (listed, habitat association, distribution, comp. note) France
Tettigetta argentata Sanborn 2005c: 115 (comp. note)
Tettigetta argentata Sueur 2005: 214 (song, comp. note) Equals *Tettigetta atra* France
Tettigetta atra Sueur 2005: 214 (song, comp. note) Iberia
Tettigetta argentata Bernier 2006: 4–5, 7, Figs 2–3 (distribution, illustrated, oscillogram) France
Tettigetta argentata Boulard 2006g: 35–36, 72, 76, 180, Fig 30 (sonogram, listed, comp. note) France
Cicada argentata Duffels and Hayashi 2006: 190 (comp. note) France
Tettigetta argentata Gogala 2006a: 158 (comp. note)
Tettigetta argentata Puissant 2006: 20, 23, 25–27, 32–35, 45, 48, 50, 61, 64, 70, 76, 80–85, 89–90, 92–93, 98, 109–115, 134–136, 138–155, 159–163, 178, 180–187, 189–190, Fig 8, Figs 51–53, Tables 1–2, Appendix Tables 1–9, Appendix Table 13, Appendix Table 15 (illustrated, synonymy, distribution, habitat association, ecology, listed, comp. note) Equals *Cicada servicans* Equals *Cicada helianthemi* France
Tettigetta argentata Seabra, Pinto-Juma and Quartau 2006: 850 (comp. note)

Tettigetta argentata Simões and Quartau 2006: 51, 54, Fig 2 (comp. note)
Tettigetta argentata Sueur, Windmill and Robert 2006: 4126 (comp. note)
Cicadetta argentata Shiyake 2007: 8, 104, 106, Fig 180 (illustrated, distribution, listed, comp. note) Europe
Tettigetta argentata Sueur and Puissant 2007b: 62 (listed) France
Tettigetta argentata Fonseca, Serrão, Pina-Martins, Silva, Mira, et al. 2008: 19, 24, 26–27, Fig 1, Table 1 (song phylogeny, comp. note) Portugal
Tettigetta argentata Hertach 2008a: 209–214, Figs 1–4, Table 1 (song, sonogram, oscillogram, distribution, habitat, comp. note) Spain, Portugal, Italy, Slovenia, Switzerland, France
Tettigetta argentata Pinto-Juma, Seabra and Quartau 2008: 9 (comp. note)
Tettigetta argentata Ibanez 2009: 209, 212 (listed, comp. note) France
Cicadetta argentata Ibanez 2009: 209 (comp. note) France
Cicada argentata Puissant and Sueur 2010: 558 (type species of *Tettigettalna*) France
Tettigettalna argentata Puissant and Sueur 2010: 557–559, 561, 570, Fig 1, Figs 8–11 (n. comb., illustrated, habitat, distribution, acoustic behavior, sonogram, oscillogram, comp. note) Equals *Cicada argentata* Equals *Tettigetta argentata* France

***T. argentata flavescens* (Boulard, 1982)** = *Tettigetta argentata* mph. *flavescens* = *Tettigetta argentata* var. *flavescens*
Tettigetta argentata mph. *flavescens* Boulard 1982a: 50 (n. morph, described, comp. note) France
Tettigetta argentata var. *flavescens* Boulard and Mondon 1995: 66 (listed) Equals *Cicada megerlei* France
Tettigetta argentata var. *flavescens* Boulard 2000i: 88 (described, comp. note) France

***T. argentata mauresqua* (Boulard, 1982)** = *Tettigetta argentata* mph. *mauresqua* = *Tettigetta argentata* var. *mauresqua*
Tettigetta argentata mph. *mauresqua* Boulard 1982a: 50 (n. morph, described, comp. note) France
Tettigetta argentata var. *mauresqua* Boulard and Mondon 1995: 66 (listed) Equals *Cicada megerlei* France
Tettigetta argentata var. *mauresqua* Boulard 2000i: 88 (described, comp. note) France

***T. argentata nana* (Boulard, 2000)** = *Tettigetta argentata* var. *nana*
Tettigetta argentata var. *nana* Boulard 2000i: 88 (n. var., described, ecology, distribution, comp. note) France, Iberian Peninsula

***T. armandi* Puissant, 2010**
Tettigettalna armandi Puissant and Sueur 2010: 557–558, 560, 562, 564–565, 570–571, Fig 8, Fig 13, Fig 22, Figs 31–33 (n. sp., described, illustrated, habitat, distribution, acoustic behavior, sonogram, oscillogram, comp. note) Spain

***T. boulardi* Puissant, 2010**
Tettigettalna boulardi Puissant and Sueur 2010: 557–558, 560, 562–564, 570, Fig 8, Fig 13, Fig 21, Figs 28–30 (n. sp., described, illustrated, habitat, distribution, acoustic behavior, sonogram, oscillogram, comp. note) Spain

***T. defauti* Puissant, 2010**
Tettigettalna defauti Puissant and Sueur 2010: 557–558, 560, 565, 570–571, Fig 8, Fig 13, Fig 23, Figs 34–36 (n. sp., described, illustrated, habitat, distribution, acoustic behavior, sonogram, oscillogram, comp. note) Spain

***T. estrellae* (Boulard, 1982)** = *Tettigetta estrellae* = *Tettigetta septempulsata* Boulard & Quartau, 1991
Tettigetta estrellae Boulard 1982f: 185–189, 196–197, Fig 14, Figs 20–25, Fig 48 (n. sp., described, illustrated sonogram, comp. note, listed) Portugal
Tettigetta estrellae Nast 1987: 570–571 (distribution, listed) Portugal
Tettigetta estrellae Quartau and Fonseca 1988: 367, 372 (distribution, comp. note) Portugal
Tettigetta estrellae Boulard 1990b: 172, 185, 187, Plate XXII, Fig 8 (sonogram, comp. note)
Tettigetta estrellae Boulard and Quartau 1991: 50, 52, 54–56, Fig 12 (song, sonogram, comp. note)
Tettigetta septempulsata Boulard and Quartau 1991: 50–56, Figs 1–11 (n. sp., described, illustrated, comp. note)
Tettigetta estrellae Fonseca 1991: 174, 176, 181–183, 189–190, Fig 2, Fig 5 (song analysis, oscillogram, comp. note) Portugal
Tettigetta septempulsata Quartau and Rebelo 1994: 140, Figs 9–10 (song, comp. note) Portugal
Tettigetta estrellae Boulard 1995a: 13, 62–63, Fig $S25_1$, Fig $S25_2$ (song analysis, sonogram, listed, comp. note) Portugal
Tettigetta estrellae Boulard and Mondon 1995: 159 (listed) Spain
Tettigetta estrellae Quartau 1995: 37 (ecology, comp. note) Portugal
Tettigetta septempulsata Quartau 1995: 37–38 (ecology, comp. note) Portugal
Tettigetta estrellae Esteban and Sanchiz 1997: 143 (listed) Iberian Peninsula
Tettigetta septempulsata Esteban and Sanchiz 1997: 143 (listed) Iberian Peninsula
Tettigetta estrellae Sueur 2001: 44, Table 1 (listed)

Tettigetta septempulsata Sueur 2001: 45, Table 1 (listed)
Tettigetta estrellae Sueur, Puissant, Simões, Seabra, Boulard and Quartau 2004: 177–179, 184–185, Fig 1, Fig 4 (synonymy, distribution, ecology, song, sonogram, oscillogram, comp. note) Equals *Tettigetta septempulsata* n. syn. Portugal
Tettigetta septempulsata Sueur, Puissant, Simões, Seabra, Boulard and Quartau 2004: 177, 184 (comp. note) Portugal
Tettigetta estrellae Boulard 2006g: 180 (listed)
Tettigetta estrellae Fonseca, Serrão, Pina-Martins, Silva, Mira, et al. 2008: 19, 24, 26–27, 31, Fig 1, Table 1 (song phylogeny, comp. note) Portugal
Tettigetta estrellae Quartau 2008: 267 (habitat, comp. note) Portugal
Tettigettalna estrellae Puissant and Sueur 2010: 565 (n. comb., comp. note) Portugal

***T. helianthemi galantei* Puissant, 2010** = *Tettigettalna argentata* (nec Olivier)
Tettigettalna helianthemi galantei Puissant and Sueur 2010: 557–558, 560–561, 570, Fig 4, Fig 8, Fig 13, Figs 17–19 (n. ssp., described, illustrated, habitat, distribution, acoustic behavior, sonogram, oscillogram, comp. note) Spain

***T. helianthemi helianthemi* (Rambur, 1840)** = *Cicada helianthemi* = *Melampsalta (Melampsalta) helianthemi* = *Cicadetta helianthemi* = *Cicadatra atra* (nec Olivier) = *Cicadatra concinna* (nec Germar) = *Tettigetta argentata* (nec Olivier) = *Tettigetta argentata?* = *Tettigetta manueli* Boulard, 2000
Tettigetta manueli Boulard 2000f: 137–135, 140–143, Plate I, Figs 5–6, Plate II, Fig 3 (n. sp., described, illustrated, comp. note) Spain
Tettigetta argentata (partim) Puissant 2006: 20, 23, 25–27, 32–35, 45, 48, 50, 61, 64, 70, 76, 80–85, 89–90, 92–93, 98, 109–115, 134–136, 138–155, 159–163, 178, 180–187, 189–190, Fig 8, Figs 51–53, Tables 1–2, Appendix Tables 1–9, Appendix Table 13, Appendix Table 15 (illustrated, synonymy, distribution, habitat association, ecology, listed, comp. note) Equals *Cicada servicans* Equals *Cicada helianthemi* France
Tettigettalna helianthemi Puissant and Sueur 2010: 559, 561, 571 (n. comb., comp. note) Spain
Tettigettalna helianthemi helianthemi Puissant and Sueur 2010: 557–560, 563–565, 570, Figs 2–3, Fig 8, Figs 12–16 (synonymy, illustrated, habitat, distribution, acoustic behavior, sonogram, oscillogram, comp. note) Equals *Cicadatra atra* (nec Olivier) Equals *Cicadatra concinna* (nec Germar) Equals *Tettigetta argentata* (nec Olivier) Equals *Tettigetta argetnata?* Equals *Tettigetta manueli* Spain

***T. josei* (Boulard, 1982)** = *Tettigetta josei*
Tettigetta josei Boulard 1982f: 185, 189–190, 196–197, Fig 15, Figs 26–29, Fig 49 (n. sp., described, illustrated, sonogram, comp. note, listed) Portugal
Tettigetta josei Boulard 1985a: 34 (distribution, listed) Portugal
Tettigetta josei Nast 1987: 570–571 (distribution, listed) Portugal
Tettigetta josie Quartau and Fonseca 1988: 367, 369, 372 (distribution, comp. note) Portugal
Tettigetta josei Fonseca 1991: 174, 183–185, 189–190, Fig 6 (song analysis, oscillogram, comp. note) Portugal
Tettigetta josei Fonseca and Popov 1994: 349–360, Figs 1–10 (song structure, oscillogram, sound apparatus, comp. note) Portugal
Tettigetta josei Boulard 1995a: 13, 65–67, Fig $S27_1$, Fig $S27_2$ (song analysis, sonogram, listed, comp. note) France
Tettigetta josei Boulard and Mondon 1995: 159 (listed) Spain
Tettigetta josei Quartau and Boulard 1995: 110 (comp. note)
Tettigetta josei Quartau 1995: 37–38 (illustrated, ecology, comp. note) Portugal
Tettigetta josei Fonseca 1996: 15, 19, 22, 27–28, Fig 1, Table 2 (timbal muscle physiology, oscillogram, comp. note) Portugal
Tettigetta josei Fonseca and Hennig 1996: 1535–1544, Figs 2A–B, Figs 3–6, Fig 11A (tensor muscle physiology, auditory responses, comp. note) Portugal
Tettigetta josei Esteban and Sanchiz 1997: 143 (listed) Iberian Peninsula
Tettigetta josei Boulard 2000f: 134 (comp. note)
Tettigetta josei Fonseca, Münch, and Hennig 2000: 297–298, Fig 1, Table 1 (hearing physiology, song frequency spectrum, comp. note)
Tettigetta josei Stumpner and von Helverson 2001: 164, Fig 2 (interneurons, comp. note)
Tettigetta josei Sueur 2001: 44, Table 1 (listed)
Tettigetta josei Fonseca and Revez 2002b: 971–976, Fig 1, Fig 2D, Fig 3, Tables 1–2 (song analysis, oscillogram, comp. note) Portugal
Tettigetta josei Gerhardt and Huber 2002: 184–185, Fig 6.7 (hearing, comp. note)
Tettigetta josei Sueur and Aubin 2002b: 135 (comp. note)
Tettigetta josei Fonseca and Hennig 2004: 401–406, Fig 2, Figs 3D–F, Fig 4A, Fig 4C, Figs 5C–D (hearing physiology, comp. note) Portugal
Tettigetta josei Boulard 2005d: 344 (comp. note)

Tettigetta josei Sanborn 2005c: 115 (comp. note)
Tettigetta josei Boulard 2006g: 72–73, 180 (listed, comp. note)
Tettigetta josei Seabra, Pinto-Juma and Quartau 2006: 850 (comp. note)
Tettigetta josei Sueur, Windmill and Robert 2006: 4116 (comp. note)
Tettigetta josei Fonseca and Correia 2007: 1834–1844, Figs 2–8, Tables 1–3 (hearing physiology, sound field, comp. note) Portugal
Tettigetta josei Correia, Moura Santos and Fonseca 2008: 74 (auditory currents, comp. note)
Tettigetta josei Fonseca and Michelsen 2008: 30 (hearing physiology, comp. note)
Tettigetta josei Fonseca, Serrão, Pina-Martins, Silva, Mira, et al. 2008: 19, 24, 26–27, 31, Fig 1, Table 1 (song phylogeny, comp. note) Portugal
Tettigetta josei Pinto-Juma, Seabra and Quartau 2008: 9 (comp. note)
Tettigetta josei Sueur, Windmill and Robert 2008a: 2386 (comp. note)
Tettigetta josei Pinto, Quartau, and Bruford 2009: 267, 270, 274, 277–278, Fig 5, Table 4 (phylogeny, comp. note) Portugal
Tettigettalna josei Puissant and Sueur 2010: 565, 568 (n. comb., comp. note) Portugal
Tettigetta josie Puissant and Sueur 2010: 568 (comp. note)

T. mariae **(Quartau & Boulard, 1995)** = *Tettigetta mariae*
Tettigetta mariae Quartau and Boulard 1995: 106–110, Figs 1–9 (n. sp., described, illustrated, song, sonogram, oscillogram, habitat, comp. note) Portugal
Tettigetta mariae Sueur 2001: 45, Table 1 (listed)
Tettigetta mariae Fonseca, Serrão, Pina-Martins, Silva, Mira, et al. 2008: 19, 24, 26–27, 31, Fig 1, Table 1 (song phylogeny, comp. note) Portugal
Tettigetta mariae Quartau 2008: 267 (habitat, comp. note) Portugal
Tettigettalna mariae Puissant and Sueur 2010: 565 (n. comb., comp. note) Portugal

Tettigettula **Puissant, 2010**
Tettigettula Puissant and Sueur 2010: 566 (n. gen., described, comp. note) Spain

T. pygmea **(Olivier, 1790)** = *Cicada pygmea* = *Cicada pygmaea* (sic) = *Cicada pygmacea* (sic) = *Cicada pygmoea* (sic) = *Melampsalta (Melampsalta) pygmaea* (sic) = *Cicadetta pygmea* = *Tettigetta pygmea* = *Tettigetta pygmaea* (sic) = *Cicadetta brullei* Fieber, 1872 = *Tettigetta brullei* = *Melampsalta brullei* = *Tettigetta brellei* (sic) = *Melampsalta (Melampsalta) brullei* = *Cicadetta transsylvanica* (sic) (nec Fieber) = *Cicadetta tibialis* (nec Panzer) = *Cicadetta montana* (nec Scopoli) = *Melampsalta (Melampsalta) montana* (nec Scopoli)
Melampsalta brulllei Gómez-Menor 1951: 67 (described, listed, comp. note) Spain
Cicadetta brullei Castellani 1952: 15 (comp. note) Italy
Cicadetta brullei Wagner 1952: 15 (listed, comp. note) Italy
Cicadetta brullei Soós 1956: 413 (listed, distribution) Romania
Cicadetta brullei Drosopoulous 1980: 190 (listed) Greece
Cicadetta pygmea Boulard 1981b: 48, 51 (synonymy, comp. note) France, Italy, Spain? Equals *Cicadetta transylvanica* (nec Fieber)
Tettigetta pygmea Boulard 1982f: 189–190 (comp. note) France
Tettigetta pygmea Boulard 1982g: 40, 45, 48, Plate IV, Fig 21 (described, illustrated, comp. note) France
Tettigetta pygmea Boulard 1984d: 5 (comp. note) France
Cicadetta (Tettigetta) pygmea Boulard 1985b: 63 (characteristics) France
Tettigetta pygmaea (sic) Boulard 1985e: 1033 (coloration, comp. note)
Cicadetta pygmea Chinery 1986: 88–89 (illustrated, comp. note) Europe
Tettigetta brullei Schedl 1986a: 3, 5, 8–10, 16–18, 22–24, Fig 4, Fig 12, Fig 17, Fig 21, Map 6, Table 1 (illustrated, listed, key, synonymy, distribution, ecology, comp. note) Equals *Cicadetta brullei* Istria, Spain, France, Italy, Yugoslavia, Albania, Greece
Cicadetta brullei Nast 1987: 568–569 (distribution, listed) France, Spain, Italy, Yugoslavia, Albania, Greece
Cicadetta pygmea Nast 1987: 568–569 (distribution, listed) France, Spain, Italy
Tettigetta pygmea Boulard 1988a: 9 (comp. note) France
Tettigetta pygmaea (sic) Boulard 1988a: 12, Table (comp. note) France
Cicada pygmea Boulard 1988f: 18, 42 (comp. note)
Tettigetta pygmea Boulard 1988f: 18 (type species of *Tettigetta*, comp. note)
Cicada pygmaea (sic) Boulard 1988f: 42 (comp. note)
Tettigetta pygmea Boulard 1988g: 3 (comp. note) France
Cicadetta pygmea Boulard 1988g: 4–5, Plate II, Figs 6–7, Plate III, Fig 9 (illustrated, predation, comp. note) France
Tettigetta brullei Schedl 1989: 112 (listed)
Tettigetta pygmea Boulard 1990b: 166, 179, 180 (comp. note)

Cicadetta (Tettigetta) pygmea Boulard 1992c: 62 (comp. note)
Tettigetta pygmea Boulard 1995a: 6–7, 13, 64–65, Fig $S26_1$, Fig $S26_2$ (song analysis, sonogram, listed, comp. note) France
Tettigetta pygmea Boulard and Mondon 1995: 25, 30, 32, 53, 60, 66–67, 70, 87–88, 159 (illustrated, listed, comp. note) Equals *Cicada parvula* Equals *Cicadetta brullei* France
Tettigetta brullei Gogala, Popov and Ribaric 1996: 51 (comp. note)
Cicada pygmea Boulard 1997c: 89, 111 (comp. note) France
Cicada pygmaea (sic) Boulard 1997c: 111 (comp. note)
Tettigetta pygmaea (sic) Boulard 1997c: 108 (type species of *Tettigetta*, comp. note)
Tettigetta pygmea Gogala and Popov 1997b: 35 (comp. note)
Tettigetta brullei Gogala and Popov 1997b: 34–36 (song, comp. note) Slovenia, Croatia, Macedonia
Tettigetta brullei Holzinger, Frölich, Günthart, Pauterer, Nickel et al. 1997: 49 (listed) Europe
Tettigetta pygmea Gogala 1998b: 395 (comp. note) Slovenia
Tettigetta brullei Gogala 1998b: 395–396 (oscillogram, comp. note) Slovenia
Tettigetta pygmea Popov, Beganović and Gogala 1998a: 90–91, 93 (comp. note) Equals *Tettigetta brullei* France
Tettigetta brullei Popov, Beganović and Gogala 1998a: 90–100, Fig 1, Figs 3–10 (illustrated, song, sonogram, oscillogram, comp. note) Equals *Tettigetta pygmea* Slovenia, Mediterranean, Spain, Greece
Tettigetta brullei Gogala and Gogala 1999: 120, 122, 125, Fig 5 (distribution, comp. note) Slovenia, Mediterranean
Tettigetta pygmea Boulard 2000i: 4–5, 91–95, 98–99, 101, 126, Fig 1e, Fig 20, Fig 22f, Fig 23d, Plate O, Map 15 (synonymy, illustrated, song, sonogram, oscillogram, frequency spectrum, distribution, ecology, predation, comp. note) Equals *Cicada pygmea* Equals *Cicada pygmaea* (sic) Equals *Cicada parvula* (partim) Equals *Cicadetta brullei* Equals *Cicadetta tibialis* (nec Panzer) Equals *Cicadetta montana* (nec Scopoli) Equals *Melampsalta (Melampsalta) montana* (non Scopoli) Equals *Tettigetta brullei* France
Cicadetta brullei Boulard 2000i: 91, 94 (comp. note)
Cicada pygmaea (sic) Boulard 2000i: 92 (comp. note)
Cicada pygmoea (sic) Boulard 2000i: 92 (comp. note)
Tettigetta brullei Gogala and Popov 2000: 11 (comp. note)
Tettigetta brullei Gogala and Trilar 2000: 23 (comp. note)
Tettigetta pygmea Puissant and Sueur 2000: 68–70, 72 (comp. note) France
Tettigetta brullei Schedl 2000: 262–263, 265–266, 268–269, Fig 11, Fig 12d, Fig 13d, Plate 1, Fig 10, Map 6 (key, illustrated, distribution, synonymy, listed, comp. note) Equals *Cicadetta brullei* Slovenia
Cicada pygmea Boulard 2001b: 9, 31 (comp. note) France
Cicada pygmaea (sic) Boulard 2001b: 31 (comp. note)
Tettigetta pygmaea (sic) Boulard 2001b: 28 (type species of *Tettigetta*, comp. note)
Tettigetta brullei Krpač, Trilar and Gogala 2001: 29 (comp. note) Macedonia
Tettigetta brellei (sic) Schedl 2001b: 27 (comp. note)
Tettigetta pygmea Sueur and Puissant 2001: 19, 22–23 (comp. note) France
Tettigetta pygmea Sueur 2001: 45–46, Tables 1–2 (listed, comp. note) Equals *Tettigetta brullei*
Tettigetta pygmea Boulard 2002b: 193, 206 (coloration, comp. note)
Tettigetta brullei Gogala 2002: 243, 245–246, Fig 8 (song, sonogram, oscillogram, comp. note)
Tettigetta pygmea Sueur 2002b: 120, 130 (listed, comp. note)
Tettigetta brullei Holzinger, Kammerlander and Nickel 2003: 478, 483–484, 633, 638, Fig 259D, Plate 42, Fig 7, Tables 2–3 (listed, illustrated, comp. note) Europe, Mediterranean, France, Italy, Greece, Corfu
Tettigetta pygmea Vezinet 2003: 14 (comp. note) France
Tettigetta brullei Gogala and Seljak 2003: 351 (comp. note) Slovenia, France, Macedonia
Tettigetta brullei Schedl 2004b: 916 (comp. note)
Tettigetta pygmea Boulard 2005d: 337, 344 (comp. note) Europe
Tettigetta brullei Gogala, Trilar and Krpač 2005: 105, 120, Fig 14 (distribution, comp. note) Macedonia
Tettigetta brullei Lee 2005: 33 (comp. note)
Tettigetta pygmea Puissant and Defaut 2005: 116–118, 123–125, Table 1, Tables 7–9 (listed, habitat association, distribution, comp. note) France
Tettigetta pygmea Reynaud 2005: 143–144, Photo 4 (illustrated, comp. note) France
Tettigetta brullei Sanborn 2005c: 115 (comp. note)

Tettigetta pygmaea (sic) Bernier 2006: 4–5, 7, Figs 2–3 (distribution, illustrated, oscillogram) France

Tettigetta pygmea Boulard 2006g: 35–36, 72, 180 (listed, comp. note)

Tettigetta pygmea Gogala 2006a: 158 (comp. note)

Tettigetta brullei Gogala 2006a: 158 (comp. note)

Tettigetta pygmea Puissant 2006: 20, 23, 25, 30, 32–34, 42, 44–45, 48, 70, 82–86, 89–90, 92–93, 98, 108–110, 114–115, 133–135, 137–139, 151–162, 178–180, 185–187, Fig 8, Figs 54–56, Tables 1–2, Appendix Table 1, Appendix Tables 7–9 (illustrated, distribution, habitat association, ecology, listed, comp. note) France

Tettigetta brullei Trilar, Gogala and Popa 2006: 177–179 (listed, distribution, comp. note) Equals *Cicadetta transsylvanica* (sic) (partim) Romania, Slovenia

Tettigetta pygmea Trilar, Gogala and Popa 2006: 179 (comp. note) Equals *Cicadetta transsylvanica* (sic)

Tettigetta pygmea Sueur and Puissant 2007b: 62 (listed) France

Tettigetta brullei Gogala, Drosopoulos and Trilar 2009: 19 (comp. note)

Tettigetta pygmaea (sic) Ibanez 2009: 212 (listed, comp. note) France

Cicada pygmea Puissant and Sueur 2010: 566 (type species of *Tettigettula*, comp. note) France

Tettigettula pygmea Puissant and Sueur 2010: 558, 566–568, 571, Fig 6, Fig 8, Fig 37, Figs 41–43 (n. comb., synonymy, illustrated, habitat, distribution, acoustic behavior, sonogram, oscillogram, comp. note) Equals *Cicada pygmea* Equals *Cicadetta brullei* France, Italy, Greece, Istria, Slovenia, Austria, Macedonia, Romania

Tettigetta brullei Puissant and Sueur 2010: 566–568 (synonymy, comp. note)

Tettigetta pygmea Puissant and Sueur 2010: 567–568 (comp. note)

Tettigettacula **Puissant, 2010**

Tettigettacula Puissant and Sueur 2010: 562, 567 (n. gen., described, comp. note) Spain

T. baenai **(Boulard, 2000)** = *Tettigetta baenai* = *?Tettigetta brullei* (nec Fieber) = *?Tettigetta pygmea* (nec Olivier)

Tettigetta baenai Boulard 2000f: 134–135, 140–143, Plate I, Figs 1–2, Plate II, Fig 1 (n. sp., described, illustrated, comp. note) Spain

Tettigetta baenai Puissant and Sueur 2010: 567 (type species of *Tettigettacula*, comp. note) Spain

Tettigettacula baenai Puissant and Sueur 2010: 558, 562, 566–568, 570–571, Fig 8, Fig 24, Fig 38, Fig 40, Figs 44–48 (n. comb., synonymy, illustrated, habitat, distribution, acoustic behavior, sonogram, oscillogram, comp. note) Equals *Tettigetta baenai* Equals *?Tettigetta brullei* (nec Fieber) Equals *?Tettigetta pygmea* (nec Olivier) Spain

T. bergevini **(Boulard, 1980)** = *Tettigetta bergevini*

Tettigetta bergevini Nast 1982: 311 (listed) Algeria

Tettigettacula bergevini Puissant and Sueur 2010: 565, 568 (n. comb., comp. note) Algeria

T. floreae **(Boulard, 1987)** = *Tettigetta floreae*

Tettigetta floreae Boulard 1987: 215–216, Figs 1–4 (n. sp., described, illustrated, comp. note) Morocco

Tettigetta floreae Boulard 1988d: 150–151, 154, 156, Fig 9 (wing abnormalities, illustrated, comp. note) Algeria

Tettigettacula floreae Puissant and Sueur 2010: 565, 568 (n. comb., comp. note) Morocco

T. linaresae **(Boulard, 1987)** = *Tettigetta linaresae*

Tettigetta linaresae Boulard 1987: 217–218, Figs 5–7 (n. sp., described, illustrated, comp. note) Algeria

Tettigettacula linaresae Puissant and Sueur 2010: 565, 568 (n. comb., comp. note) Algeria

T. sedlami **(Boulard, 1981)** = *Tettigetta sedlami*

Tettigetta sedlami Boulard 1981a: 38, 41–2, Fig D, Figs 8–11 (n. sp., described, illustrated, comp. note) Algeria

Tettigetta sedlami Boulard 1982f: 189 (comp. note) Algeria

Tettigettacula sedlami Puissant and Sueur 2010: 565, 568 (n. comb., comp. note) Algeria

Pseudotettigetta **Puissant, 2010**

Pseudotettigetta Puissant and Sueur 2010: 562, 566, 568–569 (n. gen., described, comp. note) Spain

P. melanophyrs leunami **(Boulard, 2000)** = *Tettigetta leunami* = *Cicadetta helianthemi* (nec Rambur) = *Tettigettalna helianthemi* (nec Rambur)

Cicadetta helianthemi (sic) Kartal 1999: 100–103, Fig A_4, Fig B_4, Fig C_4, Fig D_4, Fig E_4, Fig F_4, Fig G_4 (described, illustrated, comp. note) Spain

Tettigetta leumani Boulard 2000f: 139–143, Plate I, Figs 7–8, Plate II, Fig 4 (n. sp., described, illustrated, comp. note) Spain

Pseudotettigetta melanophrys leunami Puissant and Sueur 2010: 558, 562, 566, 569–570, Figs 7–8, Fig 24, Fig 39, Figs 49–52 (n. comb, stat. nov., synonymy, illustrated, habitat, distribution, acoustic behavior, sonogram, oscillogram, comp. note) Equals *Tettigettalna helianthemi* (nec Rambur) Equals *Tettigetta*

leunami Equals *Cicadetta helianthemi* (nec Rambur) Spain
Tettigetta leunami Puissant and Sueur 2010: 569 (comp. note)
***P. melanophrys melanophrys* (Horváth, 1907)** = *Melampsalta (Melampsalta) melanophrys* = *Cicadetta melanophrys* = *Melampsalta melanophrys*
Melampsalta melanophrys Villiers 1977: 167, Plate XIV, Fig 29 (illustrated, listed, comp. note)
Melampsalta melanophrys Boulard 1981a: 37 (comp. note) Algeria
Cicadetta melanophrys Puissant and Sueur 2010: 569 (type species of *Pseudotettigetta*, comp. note)
Pseudotettigetta melanophrys Puissant and Sueur 2010: 569 (n. comb., synonymy, comp. note) Equals *Cicadetta melanophrys* Algeria
Pseudotettigetta melanophrys melanophrys Puissant and Sueur 2010: 569 (distribution, comp. note) Algeria, Tunisia

***Gagatopsalta* Ewart, 2005**
Gagatopsalta Ewart 2005b: 440, 460, Table 1 (n. gen., described, diversity, comp. note)
***G. auranti* Ewart, 2005** = *Notopsalta* sp. H Ewart, 1998
Notopsalta sp. H Ewart 1998a: 64–65, 70, Fig 9A, Fig 14C (illustrated, described, natural history, song, oscillogram) Queensland
Notopsalta sp. nov. Ewart 1998a: 72 (listed) Queensland
Notopsalta sp. H Popple and Strange 2002: 19 (comp. note) Australia, Queensland
Gagatopsalta auranti Ewart 2005b: 460–463, 465, 467–470, 496, Figs 12–13, Fig 15, Fig 17, Figs 19–20, Fig 40, Table 3 (n. sp., described, type species of *Gagatopsalta*, synonymy, illustrated, key, song, oscillogram, distribution, habitat, listed, comp. note) Equals *Notopsalta* sp. H Ewart, 1998 Queensland
***G. obscurus* Ewart, 2005** = species B Ewart & Popple, 2001 = *Gagatopsalta obscura*
Species B (Clipclop cicada) Ewart and Popple 2001: 57, 60–61, 66, 70, Fig 3B, Fig 8B (illustrated, described, habitat, song, oscillogram) Queensland
Gagatopsalta obscurus Ewart 2005b: 460, 461–470, 497, Figs 13–14, Fig 16, Figs 18–20, Fig 41, Table 3 (n. sp., described, illustrated, key, song, oscillogram, distribution, habitat, listed, comp. note) Queensland

***Simona* Moulds, 2012**
***S. sancta* (Distant, 1913)** = *Melampsalta sancta* = *Cicadetta sancta* = *Melampsalta subgulosa* Ashton, 1914 = *Melampsalta subglusa* (sic) = *Cicadetta subgulosa*
Cicadetta sancta Moulds 1990: 144 (distribution, listed) Australia, Western Australia
Cicadetta subgulosa Moulds 1990: 144 (distribution, listed) Australia, Western Australia
Cicadetta sancta Moulds and Carver 2002: 32 (synonymy, distribution, ecology, type data, listed) Equals *Melampsalta sancta* Australia, Western Australia
Melampsalta sancta Moulds and Carver 2002: 32 (type data, listed) Western Australia
Cicadetta subgulosa Moulds and Carver 2002: 33 (synonymy, distribution, ecology, type data, listed) Equals *Melampsalta subglusa* (sic) Equals *Melampsalta subgulosa* Australia, Western Australia
Melampsalta subglusa (sic) Moulds and Carver 2002: 33 (type data, listed) Western Australia
Melampsalta subgulosa Moulds and Carver 2002: 33 (type data, listed)

***Chelapsalta* Moulds, 2012**
***C. puer* (Walker, 1850)** = *Cicada puer* = *Melampsalta puer* = *Melampsata* (sic) *puer* = *Melampsalta pueri* (sic) = *Pauropsalta puer* = *Cicadetta puer* = *Cicadetta (Melampsarta)* (sic) *puer*
Cicadetta (Melampsarta) (sic) *puer* Itô 1976: 67 (sugarcane pest, comp. note) Australia, Queensland
Cicadetta puer Itô and Nagamine 1981: 281, Table 5 (sugarcane pest, comp. note) Queensland
Cicadetta puer Soper 1981: 51, 53 (fungal host, comp. note) Australia, Queensland
Cicadetta puer Hitchcock 1983: 9 (control) Queensland
Cicadetta puer Anonymous 1985: 22 (sugarcane pest, control) Australia
Melampsalta puer Wilson 1987: 489, Table 1 (sugarcane pest, listed, comp. note) Australia, Queensland
Melampsalta puer Liu 1990: 42 (comp. note)
Melampsalta pueri (sic) Liu 1990: 46 (comp. note)
Cicadetta puer Moulds 1990: 3, 159–160, 167 (synonymy, distribution, ecology, listed, comp. note) Australia, New South Wales, Queensland, Victoria
Melampsalta puer Liu 1992: 5 (comp. note)
Melampsalta pueri (sic) Liu 1992: 9 (comp. note)
Cicadetta puer Coombs 1996: 56–57, Table 1 (emergence time, comp. note) New South Wales
Cicadetta puer Moss and Popple 2000: 58 (listed, comp. note) Australia, New South Wales
Cicadetta puer Chambers, Boon, Buckley and Hitchmough 2001: 380, Fig 3 (phylogeny, comp. note) Australia

Cicadetta puer Buckley, Arensburger, Simon and Chambers 2002: 6, 11, Fig 1 (phylogeny, comp. note) Australia

Cicadetta puer Moulds and Carver 2002: 31 (synonymy, distribution, ecology, type data, listed) Equals *Cicada puer* Australia, New South Wales, Queensland, Victoria

Cicada puer Moulds and Carver 2002: 31 (type data, listed) Souteh Australia

Cicadetta puer Popple and Strange 2002: 28, Table 1 (listed, habitat) Australia, Queensland

Cicadetta puer Arensburger, Buckley, Simon, Moulds, and Holsinger 2004: 558–559, 562–563, Table 1, Figs 1–2 (molecular phylogeny, comp. note) Australia

Cicadetta puer Krasnoff, Reátegui, Wagenaar, Gloer and Gibson 2005: 50 (fungal host) Australia

Cicadetta puer Pons 2006: 8 (fungal host, comp. note) Australia

Cicadetta puer Simon 2009: 104, Fig 2 (phylogeny, comp. note) Australia

***Erempsalta* Moulds, 2012**

***E. hermansburgensis* (Distant, 1907)** = *Melampsalta hermansburgensis* = *Melampsalta hermannsburgensis* (sic) = *Cicadetta hermansburgensis* = *Cicadetta hermannsburgensis* (sic) = Species D Ewart & Popple, 2001

Cicadetta hermannsburgensis (sic) Moulds 1990: 144 (distribution, listed) Australia, Northern Territory

Species D near *Cicadetta hermannsburgensis* (sic) Ewart and Popple 2001: 63, 65, 67, 71, Fig 6C, Fig 9B (illustrated, described, habitat, song, oscillogram) Queensland

Cicadetta hermannsburgensis (sic) Moulds and Carver 2002: 27–28 (synonymy, distribution, ecology, type data, listed) Equals *Melampsalta hermannsburgensis* (sic) Australia, Northern Territory

Melampsalta hermannsburgensis (sic) Moulds and Carver 2002: 27–28 (type data, listed) Northern Territory

Cicadetta hermannsbergensis (sic) Popple and Strange 2002: 19 (comp. note) Australia, Queensland

***Pipilopsalta* Ewart, 2005**

Pipilopsalta Ewart 2005b: 440, 476, Table 1 (n. gen., described, diversity, comp. note)

***P. ceuthoviridis* Ewart, 2005** = species E Ewart & Popple, 2001

Species E Ewart and Popple 2001: 64, 66–67, 71, Fig 6A, Fig 9A (illustrated, described, habitat, song, oscillogram) Queensland

Pipilopsalta ceuthoviridis Ewart 2005b 441, 462, 476–481, 499, Figs 25–28, Fig 43, Table 5 (n. sp., described, type species of *Pipilopsalta*, synonymy, illustrated, key, song, oscillogram, distribution, habitat, listed, comp. note) Equals Species E Ewart and Popple 2001 Queensland, South Australia

Pipilopsalta ceuthoviridis Ewart 2009b: 140 (comp. note) Queensland

***Dipsopsalta* Moulds, 2012**

***D. signata* (Distant, 1914)** = *Pauropsalta signata* = *Melampsalta signata*

Pauropsalta signata Ewart 1989b: 293 (listed) Australia

Pauropsalta signata Moulds 1990: 131 (distribution, listed) Australia, Western Australia

Pauropsalta signata Moulds and Carver 2002: 43 (distribution, ecology, type data, listed) Australia, Western Australia

***Paradina* Moulds, 2012**

***P. leichardti* (Distant, 1882)** = *Melampsalta leichardti* = *Pauropsalta leichardti* = *Urabunana leichardti*

Urabunana leichardti Moulds 1990: 173 (distribution, listed) Australia, Queensland

Urabunana leichardti Moulds and Carver 2002: 45 (synonymy, distribution, ecology, type data, listed) Equals *Melampsalta leichardti* Australia, Queensland

Melampsalta leichardti Moulds and Carver 2002: 45 (type data, listed) Queensland

***Mugadina* Moulds, 2012**

***M. emma* (Goding & Froggatt, 1904)** = *Pauropsalta emma* = *Melampsalta emma*

Pauropsalta emma Hahn 1962: 10 (listed) Australia, Queensland

Pauropsalta emma Stevens and Carver 1986: 264 (type material) Australia, Queensland

Pauropsalta emma Ewart 1989b: 293 (listed) Queensland

Pauropsalta emma Moulds 1990: 131 (distribution, listed) Australia, Queensland

Pauropsalta emma Moulds and Carver 2002: 40 (distribution, ecology, type data, listed) Australia, Queensland

Pauropsalta emma Ewart 2005b: 454 (comp. note)

***M. festiva* (Distant, 1907)** = *Urabunana festiva*

Urabunana festiva Moulds 1990: 173 (distribution, listed) Australia, Victoria

Urabunana festiva Coombs 1993a: 133 (distribution, comp. note) New South Wales, Victoria

Urabunana festiva Moulds and Carver 2002: 45 (distribution, ecology, type data, listed) Australia, Victoria

Urabunana festiva Haywood 2005: 9, 22–24, Fig 6, Table 1 (illustrated, distribution, described, habitat, song, oscillogram, listed, comp. note) South Australia, New South Wales, Australia Capital Territory, Victoria

Urabunana festiva Haywood 2006a: 14, 20–21, 23, Tables 1–2 (illustrated, distribution, described, habitat, song, oscillogram, comp. note) South Australia, Victoria, Australia Capital Territory, New South Wales

***M. marshalli* (Distant, 1911)** = *Urabunana marshalli*

Urabunana marshalli Ewart 1988: 184, 191, 193198–199, Fig 10B, Plate 2G (illustrated, oscillogram, natural history, listed, comp. note) Queensland, New South Wales

Urabunana marshalli Lithgow 1988: 65 (listed) Australia, Queensland

Urabunana marshalli Ewart 1989a: 80 (comp. note)

Urabunana marshalli Moulds 1990: 175, Plate 16, Fig 8, Fig 8a (illustrated, distribution, ecology, listed, comp. note) Australia, New South Wales, Queensland

Urabunana marshalli Coombs 1995: 13, 15, Fig 5 (comp. note) New South Wales

Urabunana marshalli Coombs 1996: 56, Table 1 (emergence time, comp. note) New South Wales

Urabunana marshalli Ewart 1998a: 66–68, Fig 12A (described, natural history, song, oscillogram) Queensland

Urabunana marshalli Sueur 2001: 45, Table 1 (listed)

Urabunana marshalli Moulds and Carver 2002: 45 (distribution, ecology, type data, listed) Australia, New South Wales, Queensland

Urabunana marshalli Popple and Strange 2002: 29, Table 1 (listed, habitat) Australia, Queensland

Urabunana marshallli Arensburger, Buckley, Simon, Moulds, and Holsinger 2004: 566 (molecular phylogeny, comp. note) Australia

Urabunana marsahlli Ewart 2005b: 454 (comp. note)

Urabunana marshalli Moulds 2005b: 395–397, 399, 406, 408, 417–419, 421, 425, Fig 44, Fig 47, Figs 56–59, Fig 62, Table 1 (illustrated, higher taxonomy, phylogeny, key, listed, comp. note) Australia

Urabunana marshalli Shiyake 2007: 8, 109–111, Fig 193 (illustrated, distribution, listed, comp. note) Australia

Urabunana marshalli Ewart and Marques 2008: 150–151, Table 1 (type species of new genus, described, comp. note)

Urabunana marshalli Marshall and Hill 2009: 4–5, 7, Fig 6A, Table 2 (sonogram, comp. note) Australia

***M.* sp. nr. *marshalli* Lime Grass-ticker (Popple & Strange, 2002)** = *Urabunana* sp. nr. *marshalli* Lime Grass-ticker Popple & Strange

Urabunana sp. nr. *marshalli* Lime Grass-ticker Popple and Strange 2002: 29, Table 1 (listed, habitat) Australia, Queensland

***M.* sp. nov. Little Green grass-ticker (Popple & Strange, 2002)** = *Urabunana* sp. nov. Little Green grass-ticker Popple & Strange

Urabunana sp. nov. Little Green Grass-ticker Popple and Strange 2002: 29, Table 1 (listed, habitat) Australia, Queensland

***Myopsalta* Moulds, 2012**

***M. atrata* (Goding & Froggatt, 1904)** = *Melampsalta atrata* = *Cicadetta atrata* = *Notopsalta atrata*

Melampsalta atrata Hahn 1962: 8 (listed) Australia

Melampsalta atrata Stevens and Carver 1986: 264 (type material) Australia

Notopsalta atrata Ewart 1988: 183, 191, 194, 198–199, Fig 10A, Plate 3F (illustrated, oscillogram, natural history, listed, comp. note) Queensland

Notopsalta atrata Lithgow 1988: 65 (listed) Australia, Queensland

Notopsalta atrata Moulds 1990: 160, 169, Plate 19, Fig 7, Fig 7a (synonymy, distribution, ecology, type data, listed) Australia, New South Wales, Queensland

Melampsalta atrata Moulds 1990: 169 (comp. note) Australia

Notopsalta atrata Ewart 1998a: 54–57, Figs 1–2 (described, natural history, song, oscillogram) Queensland

Notopsalta atrata Chambers, Boon, Buckley and Hitchmough 2001: 380 (comp. note) Australia

Notopsalta atrata Sueur 2001: 43, Table 1 (listed)

Notopsalta atrata Moulds and Carver 2002: 37 (synonymy, distribution, ecology, type data, listed) Australia, New South Wales, Queensland

Notopsalta atrata Emery, Emery, Emery and Popple 2005: 102–107, Tables 1–3 (comp. note) New South Wales

Notopsalta atrata Moulds 2005b: 395–397, 399, 417–419, 421, 425, Figs 56–59, Fig 62, Table 1 (higher taxonomy, phylogeny, key, listed, comp. note) Australia

Notopsalta atrata Shiyake 2007: 8, 108, 110, Fig 189 (illustrated, distribution, listed, comp. note) Australia

***M.* sp. nr. *atrata* Montane Grass Buzzer (Moss & Popple, 2000)** = *Notopsalta* sp. nr. *atrata*

Notopsalta sp. nr. *atrata* Montane Grass Buzzer Moss and Popple 2000: 54, 57 (listed, comp. note) Australia, New South Wales

Notopsalta sp. nr. *atrata* Williams 2002: 156–157 (listed) Australia, New South Wales

Notopsalta atrata Popple and Strange 2002: 22, 29, Table 1 (listed, habitat, comp. note) Australia, Queensland

Notopsalta sp. nr. *atrata* Williams 2002: 156–157 (listed) Australia, New South Wales

***M.* sp. nr. atrata Black Acacia Buzzer (Popple & Strange, 2002)** = *Notopsalta* sp. nr. *atrata*

Notopsalta sp. nr. *atrata* Black Acacia Buzzer Popple and Strange 2002: 21–22, 25, 29, Fig 3A, Fig 6D, Table 1 (described, illustrated, habitat, oscillogram, listed, comp. note) Australia, Queensland

M. **sp. nr. *atrata* Black Brigalow Buzzer (Popple & Strange, 2002)** = *Notopsalta* sp. nr. *atrata*

Notopsalta sp. nr. *atrata* Black Brigalow Buzzer Popple and Strange 2002: 21–22, 25, 29, Fig 3B, Fig 6E, Table 1 (described, illustrated, habitat, oscillogram, listed, comp. note) Australia, Queensland

***M.* sp. nr. *atrata* (Emery et al., 2005)** = *Notopsalta* sp. nr. *atrata*

Notopsalta sp. nr. *atrata* Emery, Emery, Emery and Popple 2005: 98, 105, 107, Fig 2, Tables 1–3 (life cycle, illustrated, comp. note) New South Wales

***M.* sp. nr. *atrata* (Haywood, 2005)** = *Notopsalta* sp. nr. *atrata*

Notopsalta sp. nr. *atrata* Haywood 2005: 31, 50–52, Fig 15, Table 2 (illustrated, distribution, described, habitat, song, oscillogram, listed, comp. note) South Australia

Notopsalta sp. nr. *atrata* Haywood 2006a: 14–15, Table 1 (illustrated, distribution, described, habitat, song, comp. note) South Australia

Notopsalta sp. nr. *atrata* Haywood 2007: 17–18, Appendix 1 (illustrated, distribution, described, habitat, song, oscillogram, comp. note) South Australia

***M. binotata* (Goding & Froggatt, 1904)** = *Melampsalta binotata* = *Cicadetta binotata*

Melampsalta binotata Hahn 1962: 9 (listed) Australia, South Australia

Melampsalta binotata Greenup 1964: 23 (comp. note) Australia, New South Wales

Melampsalta binotata Stevens and Carver 1986: 264 (type material) Australia, South Australia

Cicadetta binotata Moulds 1990: 143 (distribution, listed) Australia, South Australia

Cicadetta binotata Moulds and Carver 2002: 25 (synonymy, distribution, ecology, type data, listed) Equals *Melampsalta binotata* Australia, South Australia

Melampsalta binotata Moulds and Carver 2002: 25 (type data, listed) South Australia

Cicadetta binotata Popple and Strange 2002: 27 (comp. note) Australia, Queensland

Species near *M. binotata* (Ewart & Popple, 2001) = Species near *Cicadetta binotata*

Species near *Cicadetta binotata* Ewart and Popple 2001: 64, 66–67, Fig 6B (described, habitat, song, oscillogram) Queensland

***M.* sp. nr. *binotata* Mitchell Smokey Buzzer (Popple & Strange, 2002)** = *Cicadetta* sp. nr. *binotata*

Cicadetta sp. nr. *binotata* Mitchell Smokey Buzzer Popple and Strange 2002: 29, Table 1 (listed, habitat) Australia, Queensland

***M.* sp. nr. *binotata* Moree Smokey Buzzer (Popple & Strange, 2002)** = *Cicadetta* sp. nr. *binotata*

Cicadetta sp. nr. *binotata* Moree Smokey Buzzer Popple and Strange 2002: 23, 26–27, 29, Fig 4A, Fig 7B, Table 1 (described, illustrated, habitat, oscillogram, listed, comp. note) Australia, Queensland

***M. crucifera* (Ashton, 1912)** = *Melampsalta crucifera* = *Cicadetta crucifera*

Cicadetta crucifera Lithgow 1988: 65 (listed) Australia, Queensland

Cicadetta crucifera Moulds 1990: 3, 5, 8, 11, 159–160, 167–169, Plate 1, Fig 4, Plate 19, Fig 8, Fig 8a (synonymy, illustrated, distribution, ecology, listed, comp. note) Australia, New South Wales, Queensland

Cicadetta crucifera Moulds and Carver 1991: 465 (comp. note) Australia

Cicadetta crucifera Naumann 1993: 9, 80, 149 (listed) Australia

Cicadetta crucifera Ewart 1998b: 135 (comp. note) Queensland

Cicadetta crucifera Moulds and Carver 2002: 26 (synonymy, distribution, ecology, type data, listed) Equals *Melampsalta crucifera* Australia, New South Wales, Queensland

Melampsalta crucifera Moulds and Carver 2002: 26 (type data, listed)

Cicadetta crucifera Popple and Strange 2002: 22 (comp. note) Australia, Queensland

Cicadetta crucifera Williams 2002: 156–157 (listed) Australia, Queensland, New South Wales, Victoria

Cicadetta crucifera Brambila and Hodges 2004: 364 (sugar cane pest, comp. note) Australia

Cicadetta crucifera Brambila and Hodges 2008: 603 (sugar cane pest, comp. note) Australia

Cicadetta crucifera Ewart 2009b: 150 (comp. note) Queensland

***M.* sp. nr. *crucifera* (Ewart, 1998)** = *Cicadetta* sp. nr. *crucifera*

Cicadetta sp. nr. *crucifera* Ewart 1998b: 135–137, Figs 1–2 (illustrated, sonogram, oscillogram, natural history, acoustic behavior, listed, comp. note) Queensland

***M.* sp. nr. *crucifera* Fishing Reel Buzzer (Popple & Strange, 2002)** = *Cicadetta* sp. nr. *Cicadetta crucifera* = *Cicadetta* sp. H Ewart, 2009

Cicadetta sp. nr. *crucifera* Fishing Reel Buzzer Popple and Strange 2002: 21–22, 26, 29, Fig 3C, Fig 7A, Table 1 (described, illustrated, habitat, oscillogram, listed, comp. note) Australia, Queensland

Fishing Reel Buzzer Ewart 2009b: 140–142, Fig 1, Plate 1J (illustrated, distribution, emergence times, habitat, comp. note) Queensland, Northern Territory

Cicadetta sp. H, sp. nr. *Cicadetta crucifera* Ewart 2009b: 150, 168, Fig 16 (oscillogram, distribution, habitat, comp. note) Equals Fishing Reel Buzzer, no. 291 Equals *Cicadetta* sp. H Ewart, 1998 Equals *Cicadetta* sp. H Popple and Strange, 2002 Queensland, Northern Territory

***M. lactea* (Distant, 1905)** = *Melampsalta lactea* = *Cicadetta lactea*

Cicadetta lactea Moulds 1990: 144 (distribution, listed) Australia, Victoria

Cicadetta lactea Moulds and Carver 2002: 29 (synonymy, distribution, ecology, type data, listed) Equals *Melampsalta lactea* Australia, Victoria

Melampsalta lactea Moulds and Carver 2002: 29 (type data, listed) Victoria

***M. mackinlayi* (Distant, 1882)** = *Melampsalta mackinlayi* = *Cicadetta mackinlayi*

Cicadetta mackinlayi Moulds 1990: 144 (distribution, listed) Australia, Queensland

Cicadetta mackinlayi Moulds and Carver 2002: 30 (synonymy, distribution, ecology, type data, listed) Equals *Melampsalta mackinlayi* Australia, Queensland

Melampsalta mackinlayi Moulds and Carver 2002: 30 (type data, listed) Queensland

***M.* sp. A (Ewart, 1988)** = *Notopsalta* sp. A

Notopsalta sp. A Ewart 1988: 183, 189, 191, 194, Fig 5, Plate 3D (illustrated, natural history, listed, comp. note) Queensland

Notopsalta sp. A Kunzea Cicada Popple and Strange 2002: 30, Table 1 (listed, habitat) Australia, Queensland

***M.* sp. B (Ewart, 1988)** = *Notopsalta* sp. B

Notopsalta sp. B Ewart 1988: 183, 191, 195, 200–201, Fig 11L, Plate 4D (illustrated, oscillogram, natural history, listed, comp. note) Queensland

Notopsalta sp. B Ewart 1998a: 57, 71, Fig 3, Fig 15A (illustrated, described, natural history, song, oscillogram) Queensland

Notopsalta sp. B Copper Shrub-buzzer Popple and Strange 2002: 19, 29, Table 1 (listed, habitat, comp. note) Australia, Queensland

Notopsalta sp. B Popple 2003: 113 (comp. note) Australia

***M.* sp. C (Ewart, 1988)** = *Notopsalta* sp. C

Notopsalta sp. C Ewart 1988: 183, 191, 195, 200–201, Fig 11K, Plate 4E (illustrated, oscillogram, natural history, listed, comp. note) Queensland

Notopsalta sp. C Ewart 1998a: 58 (described) Queensland

Notopsalta sp. C Clear-wing Brigalow Cicada Popple and Strange 2002: 28, Table 1 (listed, habitat) Australia, Queensland

***M.* sp. E (Ewart, 1998)** = *Notopsalta* sp. E

Notopsalta sp. E Ewart 1998a: 58, 70, Fig 4, Fig 14A (illustrated, described, natural history, song, oscillogram) Queensland

Notopsalta sp. E Ewart and Popple 2001: 61 (described, habitat, song) Queensland

Notopsalta sp. E Eramophile Cicada Popple and Strange 2002: 19, 29, Table 1 (listed, habitat, comp. note) Australia, Queensland

***M.* sp. F (Ewart, 1998)** = *Notopsalta* sp. F

Notopsalta sp. F Ewart 1998a: 58–59, 70, Fig 5A, Fig 14B (illustrated, described, natural history, song, oscillogram) Queensland

Notopsalta sp. F Semilunata Cicada Popple and Strange 2002: 28, Table 1 (listed, habitat) Australia, Queensland

***M.* sp. G (Ewart, 1998, 2009)** = *Notopsalta* sp. G

Notopsalta sp. G Ewart 1998a: 59, Fig 5B (described, song, oscillogram) Queensland

Notopsalta sp. G Yellow-spotted Brigalow Cicada Popple and Strange 2002: 28, Table 1 (listed, habitat) Australia, Queensland

Notopsalta sp. G Ewart 2009b: 145, 160, Fig 8 (oscillogram, distribution, habitat, comp. note) Equals Black Brigalow Cicada, no. 282 Equals *Notopsalta* sp. G Popple and Strange, 2002 Queensland

Black Brigalow Cicada Ewart 2009b: 141–142, Fig 1, Plate 1B (illustrated, distribution, emergence times, habitat, comp. note) Queensland

***M. waterhousei* (Distant, 1906)** = *Melampsalta waterhousi* = *Cicadetta waterhousei*

Cicadetta waterhousei Ewart 1988: 182, 191, 195, 200–201, Fig 11E-F, Plate 4H (illustrated, oscillogram, natural history, listed, comp.

note) Queensland, New South Wales, Victoria, South Australia

Cicadetta waterhousei Moulds 1990: 166–167, Plate 19, Fig 6, Fig 6a (synonymy, illustrated, distribution, ecology, listed, comp. note) Equals *Melampsalta waterhousei* Equals *Melampsalta kewelensis* Australia, New South Wales, Queensland, South Australia, Victoria

Melampsalta waterhousei Moulds 1990: 166 (listed, comp. note) South Australia

Melampsalta kewelensis Moulds 1990: 166 (listed, comp. note) Victoria

Cicadetta waterhousei Naumann 1993: 53, 80, 149 (listed) Australia

Cicadetta waterhousi Coombs 1996: 56, Table 1 (emergence time, comp. note) New South Wales

Cicadetta waterhousi Sueur 2001: 43, Table 1 (listed)

Cicadetta waterhousei Moulds and Carver 2002: 34 (synonymy, distribution, ecology, type data, listed) Equals *Melampsalta waterhousei* Equals *Melampsalta kewelensis* Australia, New South Wales, Queensland, South Australia, Victoria

Melampsalta waterhousei Moulds and Carver 2002: 34 (type data, listed) South Australia

Melampsalta kewelensis Moulds and Carver 2002: 34 (type data, listed) Victoria

Cicadetta waterhousei Akers 2007: 5 (note) 6 (illustrated) Australia

***M.* sp. aff. *waterhousei* (Haywood, 2005)** = *Cicadetta* sp. aff. *waterhousei*

Cicadetta sp. aff. *waterhousei* Haywood 2005: 31, 35–37, Fig 10, Table 2 (illustrated, distribution, described, habitat, song, oscillogram, listed, comp. note) South Australia

Cicadetta sp. aff. *waterhousei* Haywood 2006a: 14–15, Table 1 (illustrated, distribution, described, habitat, song, comp. note) South Australia

Cicadetta sp. aff. *waterhousei* Haywood 2007: 13–14, 18, Appendix 1 (illustrated, distribution, described, habitat, song, oscillogram, comp. note) South Australia

***M. wollomombii* (Coombs, 1995)** = *Urabunana wollomombii*

Urabunana wollomombii Coombs 1995: 13–15, Figs 1–5 (n. sp., described, illustrated, comp. note) New South Wales

Urabunana wollomombii Coombs 1996: 56, 60, Table 1 (emergence time, comp. note) New South Wales

***Noongara* Moulds, 2012**

***N. issoides* (Distant, 1905)** = *Melampsalta issoides* = *Cicadetta issoides*

Cicadetta issoides Moulds 1990: 144 (distribution, listed) Australia, Western Australia

Cicadetta issoides Moulds and Carver 2002: 28 (synonymy, distribution, ecology, type data, listed) Equals *Melampsalta issoides* Australia, Western Australia

Melampsalta issoides Moulds and Carver 2002: 28 (type data, listed) Western Australia

Larrakeeya Ashton, 1912 to *Froggattoides* Distant, 1910

L. pallida Ashton, 1912 to *Froggattoides pallida*

***Froggattoides* Distant, 1910** = *Frogattoides* (sic) = *Froggatoides* (sic) = *Larrakeeya* Ashton, 1912

Frogattoides (sic) Shcherbakov 1981a: 830 (listed)

Frogattoides (sic) Shcherbakov 1981b: 66 (listed)

Froggattoides Moulds 1990: 7, 32, 176–177 (listed, diversity, comp. note) Equals *Larrakeeya* Australia

Froggattoides spp. Moulds 1990: 17–18, 23 (comp. note)

Larrakeeya Moulds 1990: 177 (comp. note) Australia

Froggattoides Moulds and Carver 2002: 34–35 (synonymy, type data, listed) Equals *Larrakeeya* Australia

Larrakeeya Moulds and Carver 2002: 35 (type data, listed)

Froggattoides Gogala and Trilar 2003: 8 (comp. note)

Froggattoides spp. Gogala and Trilar 2003: 9 (comp. note)

Frogattoides Arensburger, Buckley, Simon, Moulds, and Holsinger 2004: 566 (comp. note) Australia

Froggattoides Moulds 2005b: 390, 414–415, 430, 436, Table 2 (higher taxonomy, listed, comp. note) Australia

***F. pallida* (Ashton, 1912)** = *Larrakeeya pallida*

Froggattoides pallida Moulds 1990: 177 (rev. stat., n. comb., synonymy, distribution, ecology, listed, comp. note) Equals *Larrakeeya pallida* Australia, Western Australia

Larrakeeya pallida Moulds 1990: 177 (listed, comp. note) Western Australia

Froggattoides pallida Naumann 1993: 62, 92, 149 (listed) Australia

Larrakeeya pallida Moulds and Carver 2002: 35 (type species of *Larrakeeya*, type data, listed) Western Australia

Froggattoides pallida Moulds and Carver 2002: 35 (synonymy, distribution, ecology, type data, listed) Equals *Larrakeeya pallida* Australia, Western Australia

***F. typicus* Distant, 1910** = *Frogattoides* (sic) *typicus*

Froggattoides typicus Evans 1956: 230–231, Figs 24D–E (comp. note)

Froggattoides typicus Boulard 1982e: 181 (comp. note) Australia

Froggatoides typicus Ewart 1988: 185, 191, 194, Plate 3C (illustrated, listed, comp. note) Queensland

Froggattoides typicus Lithgow 1988: 65 (listed) Australia, Queensland

Froggattoides typicus Boulard 1990b: 138, 152–153, Plate XVII, Fig 6 (sound system illustrated, comp. note) Australia

Froggattoides typicus Moulds 1990: 34, 176–177, Plate 16, Fig 12 (type species of *Froggattoides*, illustrated, distribution, ecology, listed, comp. note) Australia, New South Wales, Queensland

Froggattoides typicus Moulds and Carver 1991: 466–467, Fig 30.27B (illustrated, comp. note)

Froggattoides typicus Naumann 1993: 20, 92, 149 (listed) Australia

Froggatoides typicus Ewart 1998a: 63 (described, natural history, song, comp. note) Queensland

Froggattoides typicus Ewart and Popple 2001: 61 (described, habitat, song) Queensland

Froggattoides typicus Boulard 2002b: 197 (sound system, comp. note) Australia

Froggattoides typicus Moulds and Carver 2002: 34 (distribution, ecology, type data, listed) Australia, New South Wales, Queensland

Froggattoides typicus Moulds 2005b: 395–398, 417–419, 421, 425, Figs 56–59, Fig 62, Table 1 (higher taxonomy, phylogeny, key, listed, comp. note) Australia

Froggattoides typicus Boulard 2006c: 142 (comp. note) Australia

Froggattoides typicus Cryan 2005: 565, 568, Fig 2, Table 1 (phylogeny, comp. note) Australia

Froggattoides typicus Ewart and Popple 2007: 127–139, Figs 1–6 (acoustic behavior, oscillogram, sound system illustrated, comp. note) Queensland

Froggattoides typicus Ceotto, Kergoat, Rasplus and Bourgoin 2008: 668 (outgroup)

Froggattoides typicus Cryan and Svenson 2010: 399, 402, Fig 1, Table 1 (phylogeny, comp. note) Australia

***Clinata* Moulds, 2012**

***C. nodicosta* (Goding & Froggatt, 1904)** = *Pauropsalta nodicosta* = *Melampsalta nodicosta*

Pauropsalta nodicosta Hahn 1962: 10 (listed) Australia, Western Australia

Pauropsalta nodicosta Stevens and Carver 1986: 265 (type material) Australia

Pauropsalta nodicosta Ewart 1989b: 293 (listed) Western Australia

Pauropsalta nodicosta Moulds 1990: 131 (distribution, listed) Australia, Western Australia

Pauropsalta nodicosta Moulds and Carver 2002: 42 (distribution, ecology, type data, listed) Australia, Western Australia

Pauropsalta nodicosta Moulds 2005b: 395–396, 399, 408, 413–415, 417–419, 421, 425, Fig 47, Figs 56–59, Fig 62, Table 1 (illustrated, higher taxonomy, phylogeny, key, listed, comp. note) Australia

***C.* sp. aff. nodicosta (Little Bukoke Cicada) (Haywood, 2005)** = *Pauropsalta* sp. aff. *nodicosta* (Little Buloke Cicada) with C.

Pauropsalta sp. aff. *nodicosta* (Little Buloke Cicada) Haywood 2005: 31, 41–43, Fig 12, Table 2 (illustrated, distribution, described, habitat, song, listed, comp. note) South Australia

Pauropsalta sp. aff. *nodicosta* (Little Buloke Cicada) Haywood 2006a: 14, Table 1 (listed) South Australia

Pauropsalta sp. aff. *nodicosta* (Little Buloke Cicada) Haywood 2006b: 50–51, 53, Table 1 (illustrated, distribution, described, habitat, song, listed, comp. note) South Australia

***C.* sp. aff. nodicosta (Little Shrub Cicada) (Haywood, 2005)** = *Pauropsalta* sp. aff. *nodicosta* (Little Shrub Cicada) with C.

Pauropsalta sp. aff. *nodicosta* (Little Shrub Cicada) Haywood 2005: 31, 38–40, Fig 11, Table 2 (illustrated, distribution, described, habitat, song, listed, comp. note) South Australia

Pauropsalta sp. aff. *nodicosta* (Little Shrub Cicada) Haywood 2006a: 14, Table 1 (listed) South Australia

Pauropsalta sp. aff. *nodicosta* (Little Shrub Cicada) Haywood 2006b: 49–50, 53, Table 1 (illustrated, distribution, described, habitat, song, listed, comp. note) South Australia

***Toxala* Moulds, 2012**

***T. verna* (Distant, 1912)** = *Urabunana verna* = *Curvicicada verna* = *Oecicicada* (sic) *verna*

Urabunana verna Lithgow 1988: 65 (listed) Australia, Queensland

Urabunana verna Moulds 1990: 173, 175–176, Plate 16, Fig 1 (illustrated, distribution, ecology, listed, comp. note) Australia, New South Wales

Curvicicada verna Chou, Lei, Li, Lu and Yao 1997: 79 (n. comb., synonymy, comp. note) Equals *Urabunana verna* China
Oecicicada (sic) *verna* Chou, Lei, Li, Lu and Yao 1997: 358 (comp. note)
Curvicicada verna Hua 2000: 61 (listed, distribution) Equals *Urabunana verna* China, Xinjiang, Australia
Urabunana verna Emery and Emery 2002: 137–139, Fig 4 (illustrated, comp. note) New South Wales
Urabunana verna Moulds and Carver 2002: 46 (distribution, ecology, type data, listed) Australia, New South Wales
Urabunana verna Williams 2002: 156–157 (listed) Australia, Queensland, New South Wales
Urabunana verna Gogala and Trilar 2003: 9 (comp. note) Australia
Urabunana verna Emery, Emery, Emery and Popple 2005: 102, 104–108, Tables 1–3 (comp. note) New South Wales
Urabunana verna Ewart 2005b: 454 (comp. note)
Urabunana verna Lee 2005: 36 (comp. note)
Urabunana verna Shiyake 2007: 8, 109, 111, Fig 194 (illustrated, distribution, listed, comp. note) Australia

***Platypsalta* Moulds, 2012**

***P. dubia* (Goding & Froggatt, 1904)** = *Pauropsalta dubia* = *Melampsalta dubia*
Pauropsalta dubia Hahn 1962: 10 (listed) Australia, Victoria, South Australia
Pauropsalta dubia Stevens and Carver 1986: 264 (type material) Australia, Victoria, South Australia
Pauropsalta dubia Moulds 1990: 131 (distribution, listed) Australia, Queensland, South Australia, Victoria
Pauropsalta dubia Ewart 1998a: 65–66, 71, Fig 11, Fig 15C (illustrated, described, natural history, song, oscillogram) Queensland
Pauropsalta dubia Sueur 2001: 44, Table 1 (listed)
Pauropsalta dubia Moulds and Carver 2002: 39–40 (distribution, ecology, type data, listed) Australia, Queensland, South Australia, Victoria
Pauropsalta dubia Ewart 2005b: 454 (comp. note)

***P.* sp. nr. *dubia* Broad-wing Scrub-buzzer (Popple & Strange, 2002)** = *Pauropsalta* sp. nr. *dubia*
Pauropsalta sp. nr. *dubia* Broad-wing Scrub-buzzer Popple and Strange 2002: 28, Table 1 (listed, habitat, comp. note) Equals *Pauropsalta dubia* Fig Ewart, 1998 Australia, Queensland

***P. mixta* (Distant, 1914)** = *Pauropsalta mixta* = *Melampsalta mixta* = *Cicadetta mixta*
Pauropsalta mixta Ewart 1988: 182–183, 191, 195, 200–201, Fig 11D, Plate 4J (illustrated, oscillogram, natural history, listed, comp. note) Queensland
Pauropsalta mixta Ewart 1989b: 293 (listed) New South Wales
Cicadetta mixta Moulds 1990: 157–158, 160, Plate 24, Fig 22 (n. comb., illustrated, distribution, ecology, listed, comp. note) Equals *Pauropsalta mixta* Equals *Melampsalta mixta* Australia, New South Wales, Queensland
Pauropsalta mixta Moulds 1990: 157 (listed)
Melampsalta mixta Moulds 1990: 157 (listed)
Cicadetta mixta Ewart 1998a: 72 (listed) Queensland
Cicadetta mixta Sueur 2001: 43, Table 1 (listed) Equals *Pauropsalta mixta*
Cicadetta mixta Moulds and Carver 2002: 30 (synonymy, distribution, ecology, type data, listed) Equals *Pauropsalta mixta* Australia, New South Wales, Queensland
Pauropsalta mixta Moulds and Carver 2002: 30 (type data, listed) New South Wales
Cicadetta mixta Popple and Strange 2002: 19 (comp. note) Australia, Queensland

***P.* sp. 1 nr. *mixta* Black Scrub-buzzer (Popple & Strange, 2002)** = *Cicadetta* sp. 1 nr. *mixta*
Cicadetta sp. 1 nr. *mixta* Black Scrub-buzzer Popple and Strange 2002: 19–20, 22, 25, 29, Fig 2B, Fig 6B, Table 1 (described, illustrated, habitat, oscillogram, listed, comp. note) Australia, Queensland

***P.* sp. 2 nr. *mixta* Surat Scrub-buzzer (Popple & Strange, 2002)** = *Cicadetta* sp. 2 nr. *mixta*
Cicadetta sp. 2 nr. *mixta* Surat Scrub-buzzer Popple and Strange 2002: 20, 22, 25, 29, Fig 2C, Fig 6C, Table 1 (described, illustrated, habitat, oscillogram, listed, comp. note) Australia, Queensland

***P.* sp. 3 nr. *mixta* Little Grass Buzzer (Popple & Strange, 2002)** = *Cicadetta* sp. 3 nr. *mixta* = *Cicadetta mixta* (partim)
Cicadetta mixta Moulds 1990: Plate 19, Fig 2, Fig 3a (illustrated)
Cicadetta sp. 3 nr. *mixta* Little Grass Buzzer Popple and Strange 2002: 28, Table 1 (listed, habitat, comp. note) Equals *Cicadetta mixta* Fig Moulds 1990 Australia, Queensland

***Drymopsalta* Ewart, 2005**
Drymopsalta Ewart 2005b: 440, 480–483, Table 1 (n. gen., described, diversity, comp. note)
Drymopsalta Ewart 2009a: 146 (comp. note)

***D. crepitum* Ewart, 2005** = species F Ewart, 1993 = Healthlands sp. F Ewart 2005

Heathlands sp. F Ewart 2005a: 170, 178–179, Fig 1A (illustrated, described, natural history, song, oscillogram) Queensland

Drymopsalta crepitum Ewart 2005b: 462, 481–489, 500, Figs 29–35, Fig 44, Table 6 (n. sp., described, type species of *Drymopsalta*, illustrated, key, song, oscillogram, distribution, habitat, listed, comp. note) Queensland

D. dameli **(Distant, 1905)** = *Pauropsalta dameli* (sic) = *Melampsalta daemeli* = *Urabunana daemeli* (sic) = *Urabunana dameli* = *Drymopsaltra daemeli* (sic) = *Urabunana rufilinea* Ashton, 1914 = *Pauropsalta* sp. F Ewart, 1993

Pauropsalta dameli Ewart 1989b: 293 (listed) Australia

Urabunana daemeli (sic) Moulds 1990: 173–174, Plate 16, Fig 2 (n. comb., synonymy, illustrated, distribution, ecology, listed, comp. note) Equals *Pauropsalta daemeli* Equals *Urabunana rufilinea* n. syn. Equals *?Urabunana rufilinea* Australia, New South Wales

Pauropsalta dameli Moulds 1990: 173 (listed, comp. note) Australia

Urabunana rufilinea Moulds 1990: 173–174 (listed, comp. note) New South Wales

Pauropsalta sp. F Ewart 1993: 139, 147, Fig 13 (labeled Fig 14 Species G) (natural history, acoustic behavior, song analysis) Queensland

Urabunana daemeli (sic) Moss and Popple 2000: 58 (listed, comp. note) Australia, New South Wales

Urabunana daemeli (sic) Emery and Emery 2002: 137–139, Fig 3 (illustrated, comp. note) New South Wales

Urabunana daemeli (sic) Moulds and Carver 2002: 44–45 (synonymy, distribution, ecology, type data, listed) Equals *Pauropsalta daemeli* Equals *Urabunana rufilinea* Australia, New South Wales

Pauropsalta daemeli (sic) Moulds and Carver 2002: 44 (type data, listed) Australia

Urabunana rufilinea Moulds and Carver 2002: 44 (type data, listed) New South Wales

Urabunana daemeli (sic) Emery, Emery, Emery and Popple 2005: 102–108, Tables 1–3 (comp. note) New South Wales

Urabunana daemeli (sic) Lee 2005: 33 (comp. note)

Drymopsalta daemeli Ewart 2005b: 481 (n. comb., listed)

Urabunana daemeli (sic) Shiyake 2007: 8, 109–111, Fig 192 (illustrated, distribution, listed, comp. note) Australia

Crotopsalta **Ewart, 2005**

Crotopsalta Ewart 2005b: 440–442, 451, 454, 458, 460, Table 1 (n. gen., described, diversity, comp. note)

Crotopsalta spp. Ewart 2005b: 452, 454, 459, 468 (comp. note)

Crotopsalta Ewart and Marques 2008: 152 (comp. note)

Crotopsalta Ewart 2009a: 139, 147, 149 (comp. note) Queensland

Crotopsalta Ewart 2009b: 148, 163 (comp. note) Queensland

C. fronsecetes **Ewart, 2005** = *Notopsalta* sp. J Ewart, 1998

Notopsalta sp. J Ewart 1998a: 65–66, 69, Fig 10B, Fig 13C (illustrated, described, natural history, song, oscillogram) Queensland

Crotopsalta fronsecetes Ewart 2005b: 441–442, 445–449, 451, 455–459, 493, Figs 2–4, Figs 7–11, Fig 37, Table 2 (n. sp., described, synonymy, illustrated, key, song, oscillogram, distribution, habitat, listed, comp. note) Equals *Notopsalta* sp. J Ewart, 1998 Queensland

Crotopsalta fronsecetes Ewart 2009a: 139, 146, 149, Fig 3, Fig 5 (distribution, comp. note) Queensland

Crotopsalta fronsecetes Ewart 2009b: 163, Fig 11 (comp. note) Queensland

C. leptotigris **Ewart, 2009** = *Crotopsalta* sp. A Ewart, 2009

Crotopsalta leptotigris Ewart 2009a: 140–150, Figs 1–3, Fig 4A-D, Figs 5–6 (n. sp., described, illustrated, key, oscillogram, frequency spectra, natural history, distribution, comp. note) Queensland

Crotopsalta leptotigris Ewart 2009b: 146 (synonymy) Equals *Crotopsalta* sp. A = Cravens Ticker, no. 358 Queensland

Crotopsalta sp. A Ewart 2009b: 141, 146, 162–183, Figs 10A–B, Fig 11 (oscillogram, distribution, habitat, comp. note) Equals Cravens Ticker, no. 358 Queensland

Cravens Ticker Ewart 2009b: 141–142, 147, Fig 1, Plate 1I (illustrated, distribution, emergence times, habitat, comp. note) Queensland

C. plexis **Ewart, 2005** = *Notopsalta* sp. I sp. nr. *Pauropsalta stigmata* Ewart, 1998 = *Notopsalta* sp. I Wilga Ticker

Notopsalta sp. I sp. nr. *Pauropsalta stigmata* Ewart 1998a: 65–66, 69, Fig 10A, Fig 13B (illustrated, described, natural history, song, oscillogram) Queensland

Notopsalta sp. I Wilga Ticker Popple and Strange 2002: 29, Table 1 (listed, habitat) Australia, Queensland

Crotopsalta plexis Ewart 2005b: 441–445, 451, 455–459, 492, Figs 1–2, Figs 7–11, Fig 36, Table 2 (n. sp., described, type species of *Crotopsalta*, illustrated, key, song, oscillogram, distribution, habitat, listed, comp. note) Queensland, New South Wales

Crotopsalta plexis Ewart 2009a: 139, 146, 149, Fig 3, Fig 5 (distribution, comp. note) Queensland

Crotopsalta plexis Ewart 2009b: 163, Fig 11 (comp. note) Queensland

C. *poaecetes* Ewart, 2005

Crotopsalta poaecetes Ewart 2005b: 441–442, 445, 451–459, 495, Fig 2, Figs 6–11, Fig 39, Table 2 (n. sp., described, illustrated, key, song, oscillogram, distribution, habitat, listed, comp. note) Queensland

Crotopsalta poaecetes Ewart 2009a: 139, 146–149, Fig 3, Fig 4E, Fig 5 (key, distribution, comp. note) Queensland

Crotopsalta poaecetes Ewart 2009b: 141, 148, 162–163, Fig 10C, Fig 11 (oscillogram, comp. note) Queensland

C. *strenulum* Ewart, 2005

Crotopsalta strenulum Ewart 2005b: 441–442, 445, 449–452, 455–459, 494, Fig 2, Fig 5, Figs 7–11, Fig 38, Table 2 (n. sp., described, illustrated, key, song, oscillogram, distribution, habitat, listed, comp. note) Queensland

Crotopsalta strenulum Ewart 2009a: 139, 146–147, 149, Fig 3, Fig 5 (key, distribution, comp. note) Queensland

Crotopsalta strenulum Ewart 2009b: 163, Fig 11 (comp. note) Queensland

***Ewartia* Moulds, 2012**

E. *brevis* (Ashton, 1912) = *Melampsalta brevis* = *Cicadetta brevis* = Heathlands sp. E Ewart, 1993 = *Pauropsalta* sp. E Ewart, 1993

Cicadetta brevis Moulds 1990: 150, Plate 18, Fig 8, Fig 8a (illustrated, distribution, ecology, listed, comp. note) Australia, Queensland

Pauropsalta sp. E Ewart 1993: 139, 146, Fig 12 (natural history, acoustic behavior, song analysis) Queensland

Cicadetta brevis Moulds and Carver 2002: 25 (synonymy, distribution, ecology, type data, listed) Equals *Melampsalta brevis* Australia, Queensland

Melampsalta brevis Moulds and Carver 2002: 25 (type data, listed) Queensland

Cicadetta brevis Ewart 2005a: 170, 177, 179, Fig 1B (described, song, oscillogram) Equals Heathlands sp. E Moulds 1990 Queensland

E. *cuensis* (Distant, 1913) = *Melampsalta cuensis* = *Cicadetta cuensis* = *Melampsalta hermannsburgensis* (sic) (nec *Melampsalta hermansburgensis* Distant, 1907)

Cicadetta cuensis Moulds 1990: 143 (distribution, listed) Australia, Western Australia

Cicadetta cuensis Moulds and Carver 2002: 26 (synonymy, distribution, ecology, type data, listed) Equals *Melampsalta cuensis* Australia, Western Australia

Melampsalta cuensis Moulds and Carver 2002: 26 (type data, listed) Western Australia

E. *oldfieldi* (Distant, 1883) = *Melampsalta oldfieldi* = *Cicadetta oldfieldi* = *Cicaderna* (sic) *oldfieldi*

Cicadetta oldfieldi Ewart 1986: 51, Table 1 (listed) Queensland

Cicadetta oldfieldi Ewart 1988: 182–183, 191, 193, 196–197, Fig 9J-M, Plate 2C (illustrated, natural history, listed, comp. note) Queensland

Cicadetta oldfieldi Lithgow 1988: 65 (listed) Australia, Queensland

Cicadetta oldfieldi Ewart 1989a: 80 (comp. note)

Cicadetta oldfieldi Moulds 1990: 163, Plate 16, Fig 10, Fig 10a (illustrated, distribution, ecology, listed, comp. note) Australia, New South Wales, Queensland

Cicadetta oldfieldi Naumann 1993: 61, 80, 149 (listed) Australia

Cicadetta oldfieldi Ewart 1995: 88 (described, illustrated, natural history, acoustic behavior, oscillogram) Queensland

Cicaderna (sic) *oldfieldi* Tomley 1995: 13 (comp. note) Australia, Queensland

Cicadetta oldfieldi Ewart 1998a: 61 (described, natural history, song) Queensland

Cicadetta oldfieldi Ewart 2001b: 69 (listed) Queensland

Cicadetta oldfieldi Ewart and Popple 2001: 62–63 (described, habitat, song) Queensland

Cicadetta oldfieldi Moulds and Carver 2002: 31 (synonymy, distribution, ecology, type data, listed) Equals *Melampsalta oldfieldi* Australia, New South Wales, Queensland

Cicadetta oldfieldi Sueur 2001: 43, Table 1 (listed)

Melampsalta oldfieldi Moulds and Carver 2002: 31 (type data, listed) New Holland

Cicadetta oldfieldi Popple and Ewart 2002: 117 (illustrated) Australia, Queensland

Cicadetta oldfieldi Popple and Strange 2002: 17, 29, Table 1 (listed, habitat, comp. note) Australia, Queensland

Cicadetta oldfieldi Williams 2002: 156–157 (listed) Australia, Queensland, New South Wales

Cicadetta oldfieldi Watson and Watson 2004: 141 (wing microstructure, comp. note)

Cicadetta oldfieldi Palmer, Vitelli and Donnelly 2005: 176–177, 179, Table 1 (host, density, parasitism, listed, comp. note) Australia, Queensland

Cicadetta oldfieldi Watson, Myhra, Cribb and Watson 2006: 9 (comp. note)

Cicadetta oldfieldi Watson, Myhra, Cribb and Watson 2008: 3353 (comp. note)

Cicadetta oldfieldi Watson, Watson, Hu, Brown, Cribb and Myhra 2010: 116, 118–119, 122, 124–126, Figs 1e–f, Fig 6, Fig 8, Table 1 (wing microstructure, listed, comp. note)

E. sp. nr. *oldfieldi* Green Heath-Buzzer (Moss & Popple, 2000) = *Cicadetta* sp. nr. *oldfieldi*
Cicadetta sp. nr. *oldfieldi* Green Heath-Buzzer Moss and Popple 2000: 54, 57 (listed, comp. note) Australia, New South Wales
Cicadetta sp. nr. *oldfieldi* Williams 2002: 156–157 (listed) Australia

E. sp. nr. *oldfieldi* Western Wattle Cicada (Popple & Strange, 2002) = *Cicadetta* sp. nr. *oldfieldi*
Cicadetta sp. nr. *oldfieldi* Western Wattle Cicada Popple and Strange 2002: 17–18, 24, 29, Fig 1A, Fig 5D, Table 1 (described, illustrated, habitat, oscillogram, listed, comp. note) Australia, Queensland

***Aestuansella* Boulard, 1981**
Aestuansella Boulard 1981a: 42 (n. gen., described, comp. note)
Aestuansella Puissant 2005: 309, 311–312 (comp. note)

***A. aestuans* (Fabricius, 1794)** = *Tettigonia aestuans* = *Cicada aestuans* = *Melampsalta aestuans* = *Cicadetta aestuans* = *Melampsalta (Pauropsalta) aestuans* = *Pauropsalta aestuans* = *Tettigonia algira* Fabricius, 1803 = *Tibicen algira* = *Cicada algira* = *Cicadetta algira* = *Tettigetta algira* = *Cicada severa* Stål, 1854 = *Melampsalta severa* = *Cicadetta severa*
Tettigonia aestuans Dallas 1870: 482 (synonymy) Equals *Melampsalta aestuans* Equals *Tettigonia algira* Equals *Tettigonia severa*
Cicadetta aestuans Drosopoulous 1980: 190 (listed) Greece
Aestuansella aestuans Boulard 1981a: 38, 43–44, Fig E, Figs 12–14, (type species of genus *Aestuansella*, illustrated, comp. note) Algeria
Tettigonia aestuans Boulard 1981a: 43 (type species of genus *Aestuansella*)
Cicada aestuans Boulard 1981b: 46 (comp. note)
Tettigonia algira Boulard 1984c: 171 (comp. note)
Aestuansella aestuans Nast 1987: 568–569 (distribution, listed) France, Greece
Tettigonia algica Boulard 1988f: 23 (comp. note)
Cicada aestuans Boulard 1988f: 42 (comp. note)
Tettigonia algica Boulard 1997c: 93 (comp. note)
Cicada aestuans Boulard 1997c: 111 (comp. note)
Tettigetta algira Boulard 2000i: 92 (comp. note)
Tettigonia algica Boulard 2001b: 14 (comp. note)
Cicada aestuans Boulard 2001b: 31 (comp. note)
Tettigonia algira Boulard 2003f: 372 (comp. note)

***A. dogueti* Boulard, 1981**
Aestuansella dogueti Boulard 1981a: 38, 43–44, Figs F, Figs 15–17 (n. sp., described, illustrated, comp. note) Algeria

***Ggomapsalta* Lee, 2009**
Ggomapsalta Lee 2009i: 295 (n. gen., described, illustrated, distribution, listed, comp. note) Philippines, Panay
Ggomapsalta Lee 2010c: 24 (listed, comp. note) Philippines, Luzon

***G. vernalis* (Distant, 1916)** = *Pauropsalta vernalis*
Pauropsalta vernalis Ewart 1989b: 293 (listed) Philippines, Luzon
Pauropsalta vernalis Lee 2009i: 295 (type species of *Ggomapsalta*)
Ggomapsalta vernalis Lee 2009i: 295, Fig 3 (n. comb., illustrated, distribution, listed, comp. note) Philippines, Panay, Luzon
Pauropsalta vernalis Lee 2010c: 24 (type species of *Ggomapsalta*, listed) Philippines, Luzon
Ggomapsalta vernalis Lee 2010c: 25 (synonymy, distribution, listed, comp. note) Equals *Pauropsalta vernalis* Philippines, Luzon, Panay

***Kikihia* Dugdale, 1972** = *Melampslata* (nec Kolenati)
Melampsalta (sic, partim) Buchanan 1879: 213–214 (comp. note) New Zealand
Melampsalta (sic) sp. Southcott 1972: 36 (parasitism) New Zealand
Kikihia Chambers 1975: 95 (comp. note) New Zealand
Kikihia Fleming 1975a: 1570 (comp. note) New Zealand
Kikihia Fleming 1975b: 1592 (comp. note) New Zealand
Melampsalta (sic) sp. King and Moody 1982: 70, 79, Appendix 3, Table 2 (predation, listed, comp. note) New Zealand
Kikihia Fleming 1984: 191–193, 197, 202, 204 (diversity, distribution comp. note) New Zealand, Norfolk Island, Kemadec Island, Chatham Island
Kikihia spp. Fleming 1984: 193–195, 197, 201, 203–205 (comp. note) New Zealand
Kikihia sp. Fitzgerald, Meads, and Whitaker 1986: 26, Table 1 (predation) New Zealand
Kikihia sp. Cowan 1987: 177 (predation, comp. note) New Zealand
Kikihia sp. Moeed and Meads 1987a: 57–58, Table 6 (listed, comp. note) New Zealand
Kikihia Lane 1990a: 1 (comp. note) New Zealand
Kikihia Lane 1990b: 1 (comp. note) New Zealand
Kikihia Moulds 1990: 12, 32, 170 (listed, diversity, comp. note) Australia, Norfolk Island, New Zealand
Melampsalta spp. (sic) Glare, O'Callaghan, and Wigley 1993: 98 (fungal parasite) New Zealand
Kikihia Lane 1993: 53 (comp. note) New Zealand

Kikihia Lei, Chou and Li 1994: 55 (comp. note)
Kikihia Lane 1995: 367, 373, 374, 378–379, 382–383, 388–389, 392, 396, 403, 405, 407–409, Fig 17.5 (illustrated, comp. note) New Zealand
Kikihia Boer and Duffels 1996b: 309 (distribution, comp. note) New Zealand
Kikihia Chou, Lei, Li, Lu and Yao 1997: 25 (comp. note)
Kikihia Jowett, Hayes, Deans and Elson 1998: 319 (comp. note) New Zealand
Kikihia Simon, Arensburger and Chambers 1999: 1 (comp. note) New Zealand
Kikihia Chambers, Boon, Buckley and Hitchmough 2001: 380, 385–386 (comp. note) New Zealand
Kikihia Cooley 2001: 758 (comp. note) New Zealand
Kikihia Cooley and Marshall 2001: 847 (comp. note)
Kikihia spp. Cooley and Marshall 2001: 847 (comp. note)
Kikihia Arensburger 2002: 5055B (molecular phylogeny) New Zealand
Kikihia Buckley, Arensburger, Simon and Chambers 2002: 5–6, 12–13, 15 (phylogeny, diversity, comp. note) New Zealand
Kikihia Lee, Choe, Lee and Woo 2002: 3 (comp. note)
Kikihia Moulds and Carver 2002: 35 (type data, listed) Australia
Kikihia Bowie, Marris, Emberson, Andrew, Berry, et al. 2003: 81 (distribution, comp. note) Quail Island
Kikihia Arensburger, Buckley, Simon, Moulds, and Holsinger 2004: 557–558, 561–563, 565–566 (molecular phylogeny, diversity, comp. note) New Zealand
Kikihia Arensburger, Simon, and Holsinger 2004: 1769–1776, 1779–1780 (molecular phylogeny, diversity, comp. note) New Zealand
Kikihia Sueur and Aubin 2004: 223 (comp. note)
Kikihia Marshall, Cooley, Hill and Simon 2005: S-21-S-22, Fig 1 (diversity, phylogeny, comp. note) New Zealand
Kikihia Moulds 2005b: 390, 415, 430, 436, Table 2 (higher taxonomy, listed, comp. note) Australia
Kikihia spp. Moulds 2005b: 410 (comp. note)
Kikihia Buckley, Cordeiro, Marshall and Simon 2006: 422 (comp. note)
Kikihia Marshall, Simon and Buckley 2006: 993–994, 997–998, 1001, Fig 4, Table 2 (diversity, phylogeny, comp. note) New Zealand
Kikihia Buckley and Simon 2007: 423, 431 (phylogeny, comp. note) New Zealand
Kikihia Hill, Simon, Marshall and Chambers 2007: 680 (comp. note) New Zealand
Kikihia Shiyake 2007: 8, 110, 114 (listed, comp. note)
Kikihia Hill and Marshall 2008: 523 (diversity, comp. note) New Zealand
Kikihia Sueur, Vanderpool, Simon, Ouvrard and Bourgoin 2007: 613 (comp. note)
Kikihia sp. Hill and Marshall 2008: 523 (comp. note) New Zealand
Kikihia Marshall, Slon, Cooley, Hill and Simon 2008: 1055, 1059, 1061–1062, 1064, Fig 4 (diversity, phylogeny, comp. note) New Zealand
Kikihia Cranston 2009: 91–92 (distribution, comp. note) New Zealand
Kikihia Marshall, D.C. 2009: 109–110, 114, 116, Fig 4 (diversity, phylogeny, comp. note) New Zealand
Kikihia sp. Marshall and Hill 2009: 7 (comp. note)
Kikihia Marshall, Hill, Fontaine, Buckley and Simon 2009: 1997, 1999, 2006 (comp. note) New Zealand
Kikihia Simon 2009: 104–106 (diversity, comp. note) New Zealand
Kikihia spp. Simon 2009: 106, Fig 3 (phylogeny, comp. note) New Zealand
Kikihia Fan, Wu, Chen, Kuo, and Lewis 2010: 528, 530 (comp. note) New Zealand

***K.* "acoustica" Marshall, Cooley, Hill & Simon, 2005**
Kikihia "acoustica" Marshall, Cooley, Hill and Simon 2005: S-22, Fig 1 (phylogeny, comp. note) New Zealand
Kikihia "acoustica" Marshall, Simon and Buckley 2006: 998–999, Figs 4–5 (phylogeny, comp. note) New Zealand
Kikihia "acoustica" Marshall, Slon, Cooley, Hill and Simon 2008: 1056–1057, 1060, 1063, Figs 1–3, Fig 5, Table 1 (phylogeny, oscillogram, comp. note) New Zealand
Kikihia "acoustica" Marshall, D.C. 2009: 109, Fig 1 (phylogeny, comp. note) New Zealand
Kikihia "acoustica" Simon 2009: 110, Fig 7 (phylogeny, comp. note) New Zealand

***K. angusta* (Walker, 1850)** = *Cicada angusta* = *Melampsalta angusta* = *Melampsalta muta angusta* = *Cicadetta angusta* = *Cicadetta angustata* (sic) = *Cicada muta cinerascens* (sic) = *Melampsalta muta* (nec Fabricius) = *Melampsalta cruentata* (nec Fabricius) = *Melampsalta cruentata angusta* = *Cicadetta muta cinerascens* (sic)
Melampsalta angusta Buchanan 1879: 214 (comp. note) Equals *Cicada angusta* Equals *Cicada rosea* Equals *Cicada bilinea* New Zealand
Kikihia angusta Fleming 1975a: 1570 (comp. note) New Zealand
Kikihia angusta Fleming 1975b: 1591–1592 (illustrated, comp. note) New Zealand
Kikihia angusta Johns 1977: 322 (comp. note) New Zealand

Kikihia angusta Barratt 1983: 88 (comp. note) New Zealand
Cicada angusta Fleming 1984: 204 (comp. note)
Kikihia angusta Barratt and Patrick 1987: 74, 79 (comp. note, listed, distribution) New Zealand
Kikihia angusta White 1987: 149, Fig 2 (abundance, comp. note) New Zealand
Kikihia angusta Lane 1993: 54 (comp. note) New Zealand
Kikihia angusta White and Sedcole 1993a: 39, 43, 46, 50 (host, comp. note) New Zealand
Kikihia angusta Simon, Arensburger and Chambers 1999: 1 (comp. note) New Zealand
Kikihia angusta Arensburger, Simon, and Holsinger 2004: 1770–1779, Figs 1–3, Table 1, Table 3 (molecular phylogeny, comp. note) New Zealand
Kikihia angusta Hill, Marshall and Cooley 2005: 75 (comp. note) New Zealand
Kikihia angusta Marshall, Cooley, Hill and Simon 2005: S-22, Fig 1 (phylogeny, comp. note) New Zealand
Kikihia angusta Marshall, Simon and Buckley 2006: 998–999, Figs 4–5 (phylogeny, comp. note) New Zealand
Kikihia angusta Hill and Marshall 2008: 523 (comp. note) New Zealand
Kikihia angusta Marshall, Slon, Cooley, Hill and Simon 2008: 1056–1057, 1060–1061, 1063, Figs 1–3, Fig 5, Table 1 (phylogeny, oscillogram, comp. note) New Zealand
Kikihia angusta Marshall, D.C. 2009: 109, Fig 1 (phylogeny, comp. note) New Zealand
Kikihia angusta Marshall and Hill 2009: 4, Table 2 (comp. note)
Kikihia angusta Simon 2009: 110, Fig 7 (phylogeny, comp. note) New Zealand

***K.* "aotea" Marshall & Hill, 2009**
Kikihia "aotea" Marshall and Hill 2009: 4, Table 2 (comp. note)
Kikihia aotea Simon 2009: 105 (comp. note) New Zealand

***K.* "aotea east" Marshall, Cooley, Hill & Simon, 2005**
Kikihia "aotea east" Marshall, Cooley, Hill and Simon 2005: S-22, Fig 1 (phylogeny, comp. note) New Zealand
Kikihia "aotea east" Marshall, Simon and Buckley 2006: 998–999, Figs 4–5 (phylogeny, comp. note) New Zealand
Kikihia "aotea east" Marshall, Slon, Cooley, Hill and Simon 2008: 1056–1057, 1059–1060, 1063, Figs 1–3, Fig 5, Table 1 (phylogeny, oscillogram, comp. note) New Zealand
Kikihia "aotea east" Marshall, D.C. 2009: 109, Fig 1 (phylogeny, comp. note) New Zealand
Kikihia "aotea east" Simon 2009: 106, 110, Fig 3, Fig 7 (oscillogram, phylogeny, comp. note) New Zealand

***K.* "aotea west" Marshall, Cooley, Hill & Simon, 2005**
Kikihia "aotea west" Marshall, Cooley, Hill and Simon 2005: S-22, Fig 1 (phylogeny, comp. note) New Zealand
Kikihia "aotea west" Marshall, Simon and Buckley 2006: 998–999, Figs 4–5 (phylogeny, comp. note) New Zealand
Kikihia "aotea west" Marshall, Slon, Cooley, Hill and Simon 2008: 1056–1057, 1060, 1063, Figs 1–3, Fig 5, Table 1 (phylogeny, oscillogram, comp. note) New Zealand
Kikihia "aotea west" Marshall, D.C. 2009: 109, Fig 1 (phylogeny, comp. note) New Zealand
Kikihia "aotea west" Simon 2009: 106, 110, Fig 3, Fig 7 (oscillogram, phylogeny, comp. note) New Zealand

***K.* "astragali" Marshall, Cooley, Hill & Simon, 2005**
Kikihia "astragali" Marshall, Cooley, Hill and Simon 2005: S-22, Fig 1 (phylogeny, comp. note) New Zealand
Kikihia "astragali" Marshall, Simon and Buckley 2006: 998–999, Figs 4–5 (phylogeny, comp. note) New Zealand
Kikihia "astragali" Marshall, Slon, Cooley, Hill and Simon 2008: 1056–1057, 1059–1060, 1063, Figs 1–3, Fig 5, Table 1 (phylogeny, oscillogram, comp. note) New Zealand
Kikihia "astragali" Marshall, D.C. 2009: 109, Fig 1 (phylogeny, comp. note) New Zealand
Kikihia "astragali" Simon 2009: 110, Fig 7 (phylogeny, comp. note) New Zealand

***K.* "balaena" Fleming, 1984**
Kikihia "balaena" Arensburger, Simon, and Holsinger 2004: 1770, 1778–1779 (molecular phylogeny, comp. note) New Zealand
Kikihia "balaena" Marshall, Cooley, Hill and Simon 2005: S-22, Fig 1 (phylogeny, comp. note) New Zealand
Kikihia "balaena" Marshall, Simon and Buckley 2006: 998–999, Figs 4–5 (phylogeny, comp. note) New Zealand
Kikihia "balaena" Marshall, Slon, Cooley, Hill and Simon 2008: 1056–1057, 1060, 1063, Figs 1–3, Fig 5, Table 1 (phylogeny, oscillogram, comp. note) New Zealand
Kikihia "balaena" Marshall, D.C. 2009: 109, Fig 1 (phylogeny, comp. note) New Zealand
Kikihia "balaena" Simon 2009: 110, Fig 7 (phylogeny, comp. note) New Zealand

***K. cauta* (Myers, 1921)** = *Melampsalta cauta* = *Cicadetta cauta*
Melampsalta cauta Fleming 1966: 4, Fig 10 (illustrated, comp. note) New Zealand

Melampsalta cauta Grant-Taylor 1966: 348 (comp. note) New Zealand
Kikihia cauta Fleming 1975a: 1570 (comp. note) New Zealand
Kikihia cauta Fleming 1977: 258 (comp. note) New Zealand
Kikihia cauta Fleming 1984: 192 (comp. note) New Zealand
Kikihia cauta Moeed and Meads 1984: 224, Table 4 (listed) Hen Island
Kikihia cauta Moeed and Meads 1987a: 58, Table 6 (listed) New Zealand
Kikihia cauta Simon, Arensburger and Chambers 1999: 1 (comp. note) New Zealand
Kikihia cauta Chambers, Boon, Buckley and Hitchmough 2001: 380, Fig 3 (phylogeny, comp. note) New Zealand
Kikihia cauta Buckley, Arensburger, Simon and Chambers 2002: 6, 11, Fig 1 (phylogeny, comp. note) New Zealand
Kikihia cauta Sueur 2002b: 129 (listed) Equals *Melampsalta cuata*
Kikihia cauta Arensburger, Buckley, Simon, Moulds, and Holsinger 2004: 559, 562–563, 565 Table 1, Figs 1–2 (molecular phylogeny, comp. note) New Zealand
Kikihia cauta Arensburger, Simon, and Holsinger 2004: 1770–1776, 1779–1780, Figs 1–3, Table 1, Table 3 (molecular phylogeny, comp. note) New Zealand
Kikihia cauta Marshall, Cooley, Hill and Simon 2005: S-22, Fig 1 (phylogeny, comp. note) New Zealand
Kikihia cauta Marshall, Simon and Buckley 2006: 998–999, Figs 4–5 (phylogeny, comp. note) New Zealand
Kikihia cauta Buckley and Simon 2007: 428, Fig 4 (phylogeny, comp. note) New Zealand
Kikihia cauta Esson 2007: 20–21 (sex ratio) New Zealand
Kikihia cauta Marshall, Slon, Cooley, Hill and Simon 2008: 1056–1057, 1059–1060, 1063, Figs 1–3, Fig 5, Table 1 (phylogeny, oscillogram, comp. note) New Zealand
Kikihia cauta Marshall, D.C. 2009: 109, Fig 1 (phylogeny, comp. note) New Zealand
Kikihia cauta Marshall and Hill 2009: 4, Table 2 (comp. note)
Kikihia cauta Simon 2009: 104, 110, Fig 2, Fig 7 (phylogeny, comp. note) New Zealand

***K. convicta* (Distant, 1892)** = *Melampsalta convicta* = *Cicadetta convicta*

Kikihia convicta Moulds 1990: 170, Plate 19, Fig 9 (illustrated, distribution, ecology, listed, comp. note) Australia, Norfolk Island
Kikihia convicta Naumann 1993: 40, 100, 149 (listed) Australia
Kikihia convicta Simon, Arensburger and Chambers 1999: 1 (comp. note) Norfolk Island
Kikihia convicta Arensburger 2002: 5055B (phylogeny)
Kikihia convicta Moulds and Carver 2002: 35 (synonymy, distribution, ecology, type data, listed) Equals *Melampsalta convicta* Australia, Norfolk Island
Melampsalta convicta Moulds and Carver 2002: 35 (type data, listed) Norfolk Island
Kikihia convicta Arensburger, Buckley, Simon, Moulds, and Holsinger 2004: 557 (diversity) Norfolk Island, Australia
Kikihia convicta Arensburger, Simon, and Holsinger 2004: 1770–1780, Figs 1–3, Table 1, Table 3 (molecular phylogeny, comp. note) Norfolk Island, Australia
Kikihia convicta Marshall, Cooley, Hill and Simon 2005: S-21-S-22, Fig 1 (phylogeny, comp. note) Norfolk Island
Kikihia convicta Moulds 2005b: 397 (comp. note)
Kikihia convicta Marshall, Simon and Buckley 2006: 998–999, Figs 4–5 (phylogeny, comp. note) Norfolk Island
Kikihia convicta Buckley and Simon 2007: 423, 428, Fig 4 (phylogeny, comp. note) Norfolk Island
Kikihia convicta Shiyake 2007: 110 (comp. note)
Kikihia convicta Marshall, Slon, Cooley, Hill and Simon 2008: 1056–1057, 1060, 1063, Figs 1–3, Fig 5, Table 1 (phylogeny, oscillogram, comp. note) Norfolk Island
Kikihia convicta Cranston 2009: 91 (distribution, comp. note) Norfolk Island
Kikihia convicta Marshall, D.C. 2009: 109, Fig 1 (phylogeny, comp. note) Norfolk Island
Kikihia convicta Simon 2009: 110, Fig 7 (phylogeny, comp. note) New Zealand

***K. cutora cumberi* Fleming, 1973**

Kikihia cutora cumberi Chambers 1975: 95 (comp. note) New Zealand
Kikihia cutora cumberi Fleming 1984: 193, 195, 197 (distribution, comp. note) New Zealand
Kikihia cutora cumberi Moeed and Meads 1987a: 58, Table 6 (listed) New Zealand
Kikihia cutora cumberi Palma, Lovis and Tither 1989: 37 (type material, type data) New Zealand
Kikihia cutora cumberi Lane 1990a: 1 (comp. note) New Zealand
Kikihia cutora cumberi Lane 1990b: 1 (comp. note) New Zealand
Kikihia cutora cumberi Lane 1995: 367–371, 374, 377, 384–385, 394–404, 406, 409–410, Fig 17.1, Fig 17.3, Fig 17.5, Fig 17.6d, Fig 17.9a-

b, Fig 17.10–17.17 (distribution, timbal illustrated, sonogram, comp. note) New Zealand

Kikihia cutora cumberi Sueur 2001: 43, Table 1 (listed)

Kikihia cutora cumberi Arensburger, Simon, and Holsinger 2004: 1770–1777, Figs 1–3, Table 1, Table 3 (molecular phylogeny, comp. note) New Zealand

Kikihia cutora cumberi Marshall, Cooley, Hill and Simon 2005: S-22, Fig 1 (phylogeny, comp. note) New Zealand

Kikihia cutora cumberi Marshall, Simon and Buckley 2006: 998–999, Figs 4–5 (phylogeny, comp. note) New Zealand

Kikihia cutora cumberi Buckley and Simon 2007: 428, Fig 4 (phylogeny, comp. note) New Zealand

Kikihia cutora cumberi Marshall, Slon, Cooley, Hill and Simon 2008: 1056–1057, 1060, 1063, Figs 1–3, Fig 5, Table 1 (phylogeny, oscillogram, comp. note) New Zealand

Kikihia cutora cumberi Marshall, D.C. 2009: 109, Fig 1 (phylogeny, comp. note) New Zealand

Kikihia cutora cumberi Simon 2009: 110, Fig 7 (phylogeny, comp. note) New Zealand

***K. cutora cutora* (Walker, 1850)** = *Cicada cutora* = *Melampsalta cuterae* (sic) = *Melampsalta cutora* = *Cicadetta cutora* = *Kikihia cutara* (sic) = *Melampsalta cruentata subalpina* (nec Hudson) = *Melampsalta subalpine* (sic) (nec Hudson) = *Melampsalta muta cutora* = *Cicada muta* var. *cutora* = *Melampsalta muta* var. *c cutora* (nec Walker) = *Melampsalta exulis* = *Melampsalta muta* (nec Fabricius) = *Melampsalta subalpina* (nec Hudson) = *Melampsalta ochrina* (nec Walker) = *Melampsalta cutora* (nec Walker) = *Melampsalta cuterae* Kirby, 1896 (nec Walker)

Melampsalta cutora Fleming 1966: 3, Fig 5 (illustrated, comp. note) New Zealand

Melampsalta cutora Grant-Taylor 1966: 348 (comp. note) New Zealand

Kikihia cutara (sic) Green 1977: 15 (comp. note) New Zealand

Kikihia cutora Fleming 1984: 191, 193, 195, 197, 201–202, 205 (comp. note) Equals *Cicada cutora* New Zealand

Kikihia cutora cutora Fleming 1984: 197 (comp. note) New Zealand

Cicada muta var. *cutora* Fleming 1984: 204 (comp. note)

Kikihia cutora Lane 1995: 367–368, 371, 373, 375, 392, 396, 405–406, Fig 17.1 (distribution, comp. note) New Zealand

Kikihia cutora cutora Lane 1995: 367–368, 371, 374, 397–404, Fig 17.1, Fig 17.3, Fig 17.5, Figs 17.10–17.17 (distribution, timbal illustrated, comp. note) New Zealand

Kikihia cutora group Simon, Arensburger and Chambers 1999: 1 (comp. note) New Zealand

Kikihia cutora cutora Sueur 2001: 43, Table 1 (listed)

Kikihia cutora Sueur 2002b: 129 (listed) Equals *Melampsalta muta* var. *cutora*

Kikihia cutora cutora Arensburger, Simon, and Holsinger 2004: 1770–1778, Figs 1–3, Table 1, Table 3 (molecular phylogeny, comp. note) New Zealand

Kikihia cutora Arensburger, Simon, and Holsinger 2004: 1778 (molecular phylogeny, comp. note) New Zealand

Kikihia cutora cutora Hill, Marshall and Cooley 2005: 76 (comp. note) New Zealand

Kikihia cutora cutora Marshall, Cooley, Hill and Simon 2005: S-22, Fig 1 (phylogeny, comp. note) New Zealand

Kikihia cutora cutora Marshall, Simon and Buckley 2006: 998–999, Figs 4–5 (phylogeny, comp. note) New Zealand

Kikihia cutora Buckley and Simon 2007: 423 (comp. note) New Zealand

Kikihia cutora cutora Buckley and Simon 2007: 423, 428, Fig 4 (phylogeny, comp. note) New Zealand

Kikihia cutora cutora Marshall, Slon, Cooley, Hill and Simon 2008: 1056–1057, 1060, 1063, Figs 1–3, Fig 5, Table 1 (phylogeny, oscillogram, comp. note) New Zealand

Kikihia cutora cutora Marshall, D.C. 2009: 109, Fig 1 (phylogeny, comp. note) New Zealand

Kikihia cutora cutora Simon 2009: 110, Fig 7 (phylogeny, comp. note) New Zealand

***K. cutora exulis* (Hudson, 1950)** = *Melampsalta exulis* = *Melampsalta cruentata subalpina* Hudson = *Melampsalta muta* var. *c. cutora* (nec Walker) = *Cicadetta cutora exulis* = *Cicadetta exulis* = *Melampsalta muta cutora* = *Melampsalta muta subalpina* (nec Hudson)

Kikihia cutora exulis Fleming 1984: 197 (synonymy, distribution, comp. note) Equals *Melampsalta exulis* New Zealand

Melampsalta exulis Palma, Lovis and Tither 1989: 37 (type material, type data) Equals *Kikihia cutora exulis* New Zealand

Kikihia cutora exulis Palma, Lovis and Tither 1989: 37 (status) Equals *Melampsalta exulis* New Zealand

Kikihia cutora exulis Simon, Arensburger and Chambers 1999: 1 (comp. note) New Zealand

Kikihia cutora exulis Sueur 2001: 43, Table 1 (listed)

Kikihia cutora exulis Arensburger 2002: 5055B (phylogeny) Kermadec Islands

Kikihia cutora exulis Arensburger, Simon, and Holsinger 2004: 1771–1777, 1779–1780, Figs 1–3, Table 1, Table 3 (molecular phylogeny, comp. note) Kermadec Islands

Kikihia cutora exulis Marshall, Cooley, Hill and Simon 2005: S-22, Fig 1 (phylogeny, comp. note) New Zealand

Kikihia cutora exulis Marshall, Simon and Buckley 2006: 998–999, Figs 4–5 (phylogeny, comp. note) New Zealand

Kikihia cutora exulis Buckley and Simon 2007: 423 (comp. note) New Zealand

Kikihia cutora exulis Marshall, Slon, Cooley, Hill and Simon 2008: 1056–1057, 1060, 1063, Figs 1–3, Fig 5, Table 1 (phylogeny, oscillogram, comp. note) New Zealand

Kikihia cutora exulis Marshall, D.C. 2009: 109, Fig 1 (phylogeny, comp. note) New Zealand

Kikihia cutora exulis Simon 2009: 110, Fig 7 (phylogeny, comp. note) New Zealand

***K. dugdalei* Fleming, 1984** = *Kikihia* sp. V Fleming, 1975

Kikihia dugdalei Fleming 1984: 195, 198–201, 203–204, Fig 2C, Fig 3B, Figs 5G–H (n. sp., described, illustrated, synonymy, song, oscillogram, distribution, emergence time, habitat, comp. note) Equals *Kikihia* sp. V Fleming, 1975 New Zealand

Kikihia dugdalei Simon, Arensburger and Chambers 1999: 1 (comp. note) New Zealand

Kikihia dugdalei Arensburger, Simon, and Holsinger 2004: 1770, 1778 (molecular phylogeny, comp. note) New Zealand

Kikihia dugdalei Marshall, Cooley, Hill and Simon 2005: S-22, Fig 1 (phylogeny, comp. note) New Zealand

Kikihia dugdalei Marshall, Simon and Buckley 2006: 998–999, Figs 4–5 (phylogeny, comp. note) New Zealand

Kikihia dugdalei Marshall, Slon, Cooley, Hill and Simon 2008: 1056–1057, 1060, 1063, Figs 1–3, Fig 5, Table 1 (phylogeny, oscillogram, comp. note) New Zealand

Kikihia dugdalei Marshall, D.C. 2009: 109, Fig 1 (phylogeny, comp. note) New Zealand

Kikihia dugdalei Simon 2009: 110, Fig 7 (phylogeny, comp. note) New Zealand

***K.* "flemingi" Marshall, Cooley, Hill & Simon, 2005**

Kikihia "flemingi" Marshall, Cooley, Hill and Simon 2005: S-22, Fig 1 (phylogeny, comp. note) New Zealand

Kikihia "flemingi" Marshall, Simon and Buckley 2006: 998–999, Figs 4–5 (phylogeny, comp. note) New Zealand

Kikihia "flemingi" Marshall, Slon, Cooley, Hill and Simon 2008: 1056–1060, 1063, Figs 1–3, Fig 5, Table 1 (phylogeny, oscillogram, comp. note) New Zealand

Kikihia "flemingi" Marshall, D.C. 2009: 109, Fig 1 (phylogeny, comp. note) New Zealand

Kikihia "flemingi" Marshall and Hill 2009: 2, Fig 4 (sonogram, comp. note)

Kikihia "flemingi" Marshall, Hill, Fontaine, Buckley and Simon 2009: 2006 (comp. note) New Zealand

Kikihia "flemingi" Simon 2009: 110, Fig 7 (phylogeny, comp. note) New Zealand

***K. horologium* Fleming, 1984** = *Cicada muta flavescens* Hudson, 1891 (nec *C. flavescens* Fabricius, 1794 (Cicadellidae), nec C. Rossi, 1790, nec Turton, 1802 (Cicadidae) *nomen dubium*) = *Melampsalta muta* var. a *muta* (nec Fabricius) = *Melampsalta muta* var. *d flavescens* (nec Fabricius, etc.) = *Melampsalta subalpina* (nec Hudson) = *Melampsalta muta flavescens* = *Melampsalta cruentata flavescens* = *Cicadetta muta flavescens* = *Kikihia* sp. 7 Fleming, 1975

Kikihia horologium Fleming 1984: 194, 199–201, 203, Fig 1C, Fig 3C, Fig 4B, Figs 5C–D (n. sp., described, illustrated, synonymy, song, oscillogram, distribution, emergence time, habitat, comp. note) Equals *Cicada muta* var. *flavescens* Equals *Melampsalta muta* var. a *muta* (nec Fabricius) Equals *Melampsalta muta* var. d *flavescens* Equals *Melampsalta subalpina* (nec Hudson) Equals *Kikihia* sp. 7 Fleming, 1975 New Zealand

Cicada muta var. *flavescens* Fleming 1984: 201 (comp. note)

Kikihia sp. 7 Fleming 1984: 201 (comp. note)

Kikihia horologium White 1987: 149, Fig 2 (abundance, comp. note) New Zealand

Cicada muta flavescens Palma, Lovis and Tither 1989: 37 (type material, type data) Equals ?*Kikihia horologium* New Zealand

?*Kikihia horologium* Palma, Lovis and Tither 1989: 37 (status) Equals *Cicada muta flavescens* New Zealand

Kikihia horologium Palma, Lovis and Tither 1989: 37 (type material, type data) New Zealand

Kikihia horologium White and Sedcole 1993a: 39, 43, 50 (host, comp. note) New Zealand

Kikihia horologium Arensburger, Simon, and Holsinger 2004: 1770–1776, 1778, 1780, Figs 1–3, Table 1, Table 3 (molecular phylogeny, comp. note) New Zealand

Kikihia horologium Marshall, Cooley, Hill and Simon 2005: S-22, Fig 1 (phylogeny, comp. note) New Zealand

Kikihia horologium Marshall, Simon and Buckley 2006: 998–999, Figs 4–5 (phylogeny, comp. note) New Zealand

Kikihia horologium Buckley and Simon 2007: 430, Table 4 (comp. note) New Zealand
Kikihia horologium Hill and Marshall 2008: 523 (illustrated, comp. note) New Zealand
Kikihia horologium Marshall, Slon, Cooley, Hill and Simon 2008: 1056–1057, 1059–1060, 1063, Figs 1–3, Fig 5, Table 1 (phylogeny, oscillogram, comp. note) New Zealand
Kikihia horologium Marshall, D.C. 2009: 109, Fig 1 (phylogeny, comp. note) New Zealand
Kikihia horologium Simon 2009: 103, 110, Fig 1e, Fig 7 (illustrated, phylogeny, comp. note) New Zealand

***K. laneorum* Fleming, 1984** = *Kikihia* sp. L Fleming, 1975
Kikihia laneorum Chambers 1975: 95 (comp. note) New Zealand
Kikihia sp. H (sic) Bennett 1984: 89 (color variation) New Zealand
Kikihia laneorum Fleming 1984: 193–197, 200, Fig 1B, Fig 2B, Fig 3A (n. sp., described, illustrated, synonymy, song, distribution, emergence time, habitat, comp. note) New Zealand
Kikihia laneorum Palma, Lovis and Tither 1989: 37 (type material, type data) New Zealand
Kikihia laneorum Lane 1990a: 1 (comp. note) New Zealand
Kikihia laneorum Lane 1990b: 1 (comp. note) New Zealand
Kikihia laneorum Lane 1995: 367, 369, 371, 374–377, 384–385, 392–393, 396–404, 406, 409–410, Fig 17.2, Figs 17.5, Fig 17.6c, Fig 17.8, Figs 17.10–17.17 (distribution, timbal illustrated, sonogram, comp. note) New Zealand
Kikihia laneorum Simon, Arensburger and Chambers 1999: 1 (comp. note) New Zealand
Kikihia laneorum Arensburger, Simon, and Holsinger 2004: 1770, 1777–1778 (molecular phylogeny, comp. note) New Zealand
Kikihia laneorum Marshall, Cooley, Hill and Simon 2005: S-22, Fig 1 (phylogeny, comp. note) New Zealand
Kikihia laneorum Marshall, Simon and Buckley 2006: 998–999, Figs 4–5 (phylogeny, comp. note) New Zealand
Kikihia laneorum Marshall, Slon, Cooley, Hill and Simon 2008: 1056–1057, 1060, 1063, Figs 1–3, Fig 5, Table 1 (phylogeny, oscillogram, comp. note) New Zealand
Kikihia laneorum Marshall, D.C. 2009: 109, Fig 1 (phylogeny, comp. note) New Zealand
Kikihia laneorum Simon 2009: 110, Fig 7 (phylogeny, comp. note) New Zealand

***K. longula* (Hudson, 1950)** = *Melampsalta muta longula* = *Cicadetta longula* = *Kikihia muta longula* = *Melampsalta cruentata* (nec Fabricius) = *Cicadetta cruentata muta* (nec Fabricius) = *Cicadetta muta subalpina* (nec Hudson) = *Melampsalta muta muta* (nec Fabricius)
Kikihia longula Macfarlane 1979: 65–66, Table 1 (listed, habitat, comp. note) New Zealand, Chatham Islands
Kikihia longula Simon, Arensburger and Chambers 1999: 1 (comp. note) New Zealand
Kikihia longula Arensburger, Simon, and Holsinger 2004: 1770–1780, Figs 1–3, Table 1, Table 3 (molecular phylogeny, comp. note) Chatham Island, New Zealand
Kikihia muta longula Hill, Marshall and Cooley 2005: 71 (comp. note) Chatham Islands
Kikihia longula Marshall, Cooley, Hill and Simon 2005: S-22, Fig 1 (phylogeny, comp. note) New Zealand
Kikihia longula Marshall, Simon and Buckley 2006: 998–999, Figs 4–5 (phylogeny, comp. note) New Zealand
Kikihia longula Marshall, Slon, Cooley, Hill and Simon 2008: 1056–1057, 1060, 1063, Figs 1–3, Fig 5, Table 1 (phylogeny, oscillogram, comp. note) New Zealand
Kikihia longula Marshall, D.C. 2009: 109, Fig 1 (phylogeny, comp. note) New Zealand
Kikihia longula Simon 2009: 110, Fig 7 (phylogeny, comp. note) New Zealand
Kikihia longula Heenan, Mitchell, de Lange, Keeling and Paterson 2010: 105 (comp. note) Chatham Island

***K.* "murihikua" Fleming, 1984**
Kikihia "murihikua" Arensburger, Simon, and Holsinger 2004: 1770–1776, 1778–1779, Figs 1–3, Table 1, Table 3 (molecular phylogeny, comp. note) New Zealand
Kikihia "murihikua" Marshall, Cooley, Hill and Simon 2005: S-22, Fig 1 (phylogeny, comp. note) New Zealand
Kikihia "murihikua" Marshall, Simon and Buckley 2006: 998–999, Figs 4–5 (phylogeny, comp. note) New Zealand
Kikihia "murihikua" Marshall, Slon, Cooley, Hill and Simon 2008: 1056–1057, 1060–1061, 1063, Figs 1–3, Fig 5, Table 1 (phylogeny, oscillogram, comp. note) New Zealand
Kikihia "murihikua" Marshall, D.C. 2009: 109, Fig 1 (phylogeny, comp. note) New Zealand
Kikihia "murihikua" Simon 2009: 110, Fig 7 (phylogeny, comp. note) New Zealand

***K.* "muta east" Marshall, Cooley, Hill & Simon, 2005**
Kikihia "muta east" Marshall, Cooley, Hill and Simon 2005: S-22, Fig 1 (phylogeny, comp. note) New Zealand
Kikihia "muta east" Marshall, Simon and Buckley 2006: 998–999, Figs 4–5 (phylogeny, comp. note) New Zealand

Kikihia "muta east" Marshall, Slon, Cooley, Hill and Simon 2008: 1056–1057, 1059–1060, 1063, Figs 1–3, Fig 5, Table 1 (phylogeny, oscillogram, comp. note) New Zealand
Kikihia "muta east" Marshall, D.C. 2009: 109, Fig 1 (phylogeny, comp. note) New Zealand
Kikihia "muta east" Simon 2009: 106, 110, Fig 3, Fig 7 (oscillogram, phylogeny, comp. note) New Zealand

***K. muta muta* (Fabricius, 1775)** = *Tettigonia muta* = *Cicada muta* = *Melampsalta muta* = *Melampsalta mudta* (sic) = *Melampsalta muta muta* = *Cicadetta muta* = *Cicada bilinea* Walker, 1858 = *Melampsalta bilinea* = *Cicadetta bilinea* = *Cicada muta cinerescens* Hudson, 1891 = *Cicada muta cinerascens* (sic) = *Cicadetta muta cinerescens* = *Cicadetta muta cinerascens* (sic) = *Cicada cruentata* (nec Fabricius) = *Melampsalta creuentata* (nec Fabricius) = *Cicadetta cruentata* var. *muta* = *Melampsalta fuliginosa* Myers, 1921 = *Cicadetta fuliginosa*
Melampsalta muta Buchanan 1879: 213 (comp. note) Equals *Tettigonia muta* Equals *Cicada cutora* Equals *Cicada ochrina* New Zealand
Cicadetta cruentata var. *muta* Alfken 1903: 582 (comp. note) New Zealand
Melampsalta muta Miller 1952: 16–17, 43, 57 (human food, comp. note) New Zealand
Melampsalta muta Anonymous 1957: 15 (listed) New Zealand
Melampsalta muta Eyles 1960: 994 (listed, distribution, habitat, comp. note) New Zealand
Melampsalta muta Fleming 1966: 2–3, Fig 3 (illustrated, comp. note) New Zealand
Melampsalta mudta (sic) Grant-Taylor 1966: 348 (illustrated, comp. note) New Zealand
Melampsalta muta Grant-Taylor 1966: 348 (comp. note) New Zealand
Kikihia muta Fleming 1975a: 1570, 1572, Fig 1 (illustrated, comp. note) New Zealand
Kikihia muta Barratt 1983: 88 (comp. note) New Zealand
Kikihia muta Fleming 1984: 193–195, 202, 204 (comp. note) New Zealand
Kikihia ("Cicada") muta Fleming 1984: 204 (comp. note) Equals *Cicada cruentata* (partim) New Zealand
Cicada bilinea Fleming 1984: 204 (comp. note)
Kikihia muta Moeed and Meads 1987a: 58, Table 6 (listed) New Zealand
Kikihia muta Moeed and Meads 1988: 488 (listed) New Zealand
Kikihia muta Andrews and Gibbs 1989: 105 (note) New Zealand
Kikihia muta Noyes and Valentine 1989a: 36 (parasitism) New Zealand
Kikihia muta Lane 1993: 53 (comp. note) Equals *Melampsalta cruentata* Cumber, 1952 New Zealand
Kikihia muta White and Sedcole 1993a: 46, 48 (comp. note) New Zealand
Kikihia muta Lane 1995: 378 (comp. note) New Zealand
Kikihia muta muta Lane 1995: 384–385, 392, Fig 17.6b (oscillogram, comp. note) New Zealand
Kikihia muta complex Lane 1995: 409 (comp. note) New Zealand
Kikihia muta Scheele 1997: 21 (flax pest, comp. note) New Zealand
Kikihia muta group Simon, Arensburger and Chambers 1999: 1 (comp. note) New Zealand
Kikihia muta Simon, Arensburger and Chambers 1999: 1 (comp. note) New Zealand
Kikihia muta Sueur 2002b: 124, 129 (listed, comp. note) Equals *Melampsalta muta*
Kikihia muta Arensburger, Simon, and Holsinger 2004: 1770, 1775, 1777–1778 (molecular phylogeny, comp. note) New Zealand
Kikihia muta Hill, Marshall and Cooley 2005: 74–76 (comp. note) New Zealand
Kikihia muta complex Marshall, Cooley, Hill and Simon 2005: S-21 (phylogeny, comp. note) New Zealand
Kikihia muta Marshall, Cooley, Hill and Simon 2005: S-21-S-22, Fig 1 (phylogeny, comp. note) New Zealand
Kikihia muta group Marshall, Cooley, Hill and Simon 2005: S-22, Fig 1 (phylogeny, comp. note) New Zealand
Kikihia muta Buckley, Cordeiro, Marshall and Simon 2006: 414 (comp. note) New Zealand
Kikihia muta Logan 2006: 19–20 (life cycle, comp. note) New Zealand
Kikihia muta Marshall, Simon and Buckley 2006: 998–999, Figs 4–5 (phylogeny, comp. note) New Zealand
Kikihia muta Esson 2007: 20–21 (sex ratio) New Zealand
Kikihia muta Hill and Marshall 2008: 523 (comp. note) New Zealand
Kikihia muta Marshall, Slon, Cooley, Hill and Simon 2008: 1056–1057, 1060, 1063, Figs 1–3, Fig 5, Table 1 (phylogeny, oscillogram, comp. note) New Zealand
Kikihia muta Marshall, D.C. 2009: 109, Fig 1 (phylogeny, comp. note) New Zealand
Kikihia muta grass cicada complex Marshall, Hill, Fontaine, Buckley and Simon 2009: 2005 (comp. note) New Zealand

Kikihia muta Marshall, Hill, Fontaine, Buckley and Simon 2009: 2006 (comp. note) New Zealand
Kikihia muta Simon 2009: 105, 110, Fig 7 (phylogeny, comp. note) New Zealand

***K. muta pallida* (Hudson, 1950)** = *Kikihia muta* var. *pallida* = *Melampsalta muta pallida* = *Melampsalta muta* var. *pallida*
Melampsalta muta var. *pallida* Fleming 1966: 2 (comp. note) New Zealand
Kikihia muta var. *pallida* Boulard 1988d: 152 (gynandromoph, comp. note)
Melampsalta muta pallida Palma, Lovis and Tither 1989: 38 (type material, type data) Equals *Kikihia muta pallida* New Zealand
Kikihia muta pallida Palma, Lovis and Tither 1989: 38 (status) Equals *Melampsalta muta pallida* New Zealand

***K.* "muta west" Simon, 2009**
Kikihia "muta west" Simon 2009: 106, 110, Fig 3, Fig 7 (oscillogram, phylogeny, comp. note) New Zealand

***K.* "nelsonensis" Fleming, 1984**
Kikihia "nelsonensis" Arensburger, Simon, and Holsinger 2004: 1770–1778, Figs 1–3, Table 1, Table 3 (molecular phylogeny, comp. note) New Zealand
Kikihia "nelsonensis" Marshall, Cooley, Hill and Simon 2005: S-22, Fig 1 (phylogeny, comp. note) New Zealand
Kikihia "nelsonensis" Marshall, Simon and Buckley 2006: 998–999, Figs 4–5 (phylogeny, comp. note) New Zealand
Kikihia "nelsonensis" Marshall, Slon, Cooley, Hill and Simon 2008: 1056–1057, 1060, 1063, Figs 1–3, Fig 5, Table 1 (phylogeny, oscillogram, comp. note) New Zealand
Kikihia "nelsonensis" Marshall, D.C. 2009: 109, Fig 1 (phylogeny, comp. note) New Zealand
Kikihia "nelsonensis" Marshall and Hill 2009: 4–5, Fig 6J, Table 2 (sonogram, comp. note) New Zealand
Kikihia "nelsonensis" Simon 2009: 106, 110, Fig 3, Fig 7 (oscillogram, phylogeny, comp. note) New Zealand

***K.* "NWCM" Fleming & Dugdale in Arensburger, Simon and Holsinger, 2004**
Kikihia "NWCM" Arensburger, Simon, and Holsinger 2004: 1770, 1772–1776, 1779, Figs 2–3, Table 1, Table 3 (molecular phylogeny, comp. note) New Zealand

***K. ochrina* (Walker, 1858)** = *Cicada ochrina* = *Melampsalta ochrina* = *Cicadetta ochrina* = *Cicada orbrina* (sic) = *Cicada aprilina* Hudson, 1891 = *Cicadetta aprilina* = *Kikihia aprilina* = *Melampsalta muta* (nec Fabricius) = *Melampsalta cuterae* (nec Walker) = *Melampsalta cutora* (nec Walker)
Melampsalta ochrina Miller 1952: 18, 43, 57 (comp. note) New Zealand
Melampsalta ochrina Anonymous 1957: 7 (listed) New Zealand
Melampsalta ochrina Fleming 1966: 3, Fig 6 (illustrated, comp. note) New Zealand
Melampsalta ochrina Grant-Taylor 1966: 348 (illustrated, comp. note) New Zealand
Kikihia ochrina Fleming 1975a: 1569, Figs 3–4 (illustrated, comp. note) New Zealand
Kikihia ochrina Bennett 1984: 89 (color variation) New Zealand
Kikihia ochrina Fleming 1984: 195, 197–201, 203–204, Fig 2D, Figs 5E–F (described, illustrated, synonymy, song, oscillogram, distribution, emergence time, habitat, comp. note) Equals *Cicada ochrina* Equals *Cicada aprilina* Equals *Melampsalta muta* (nec Fabricius) Equals *Melampsalta ochrina* New Zealand
Cicada aprilina Palma, Lovis and Tither 1989: 37 (type material, type data) Equals *Kikihia aprilina* New Zealand
Kikihia aprilina Palma, Lovis and Tither 1989: 37 (status) Equals *Cicada aprilina* New Zealand
Kikihia ochrina Sueur 2002b: 129 (listed) Equals *Melampsalta ochrina*
Kikihia ochrina Arensburger, Simon, and Holsinger 2004: 1770–1772, 1774–1776, 1778, Figs 1–3, Table 1, Table 3 (molecular phylogeny, comp. note) New Zealand
Kikihia ochrina Hill, Marshall and Cooley 2005: 72–77, Figs 3–4 (distribution, illustrated, song, sonogram, comp. note) New Zealand
Kikihia ochrina Logan and Connolly 2005: 36–42, Fig 2, Figs 4–5, Tables 1–3 (agricultural pest, exuvia key, comp. note) New Zealand
Kikihia ochrina Marshall, Cooley, Hill and Simon 2005: S-22, Fig 1 (phylogeny, comp. note) New Zealand
Kikihia ochrina Logan 2006: 19–21 (life cycle, comp. note) New Zealand
Kikihia ochrina Marshall, Simon and Buckley 2006: 998–999, Figs 4–5 (phylogeny, comp. note) New Zealand
Kikihia ochrina Buckley and Simon 2007: 428, Fig 4 (phylogeny, comp. note) New Zealand
Kikihia ochrina Esson 2007: 20–21 (sex ratio) New Zealand
Kikihia ochrina Shiyake 2007: 8, 112, 114, Fig 199 (illustrated, distribution, listed, comp. note) New Zealand
Kikihia ochrina Hill and Marshall 2008: 523 (comp. note) New Zealand
Kikihia ochrina Marshall, Slon, Cooley, Hill and Simon 2008: 1056–1057, 1060, 1063, Figs

1–3, Fig 5, Table 1 (phylogeny, oscillogram, comp. note) New Zealand

Kikihia ochrina Marshall, D.C. 2009: 109, Fig 1 (phylogeny, comp. note) New Zealand

Kikihia ochrina Simon 2009: 110, Fig 7 (phylogeny, comp. note) New Zealand

***K. paxillulae* Fleming, 1984**

Kikihia paxillulae Fleming 1984: 194, 201–202, Fig 1D, Fig 3D, (n. sp., described, illustrated, song, distribution, emergence time, habitat, comp. note) New Zealand

Kikihia paxillulae Palma, Lovis and Tither 1989: 38 (type material, type data) New Zealand

Kikihia paxillulae Arensburger, Simon, and Holsinger 2004: 1770–1772, 1774–1776, 1778, Figs 1–3, Table 1, Table 3 (molecular phylogeny, comp. note) New Zealand

Kikihia pauxillae Marshall, Cooley, Hill and Simon 2005: S-22, Fig 1 (phylogeny, comp. note) New Zealand

Kikihia pauxillae Marshall, Simon and Buckley 2006: 998–999, Figs 4–5 (phylogeny, comp. note) New Zealand

Kikihia pauxillae Marshall, Slon, Cooley, Hill and Simon 2008: 1056–1057, 1060, 1063, Figs 1–3, Fig 5, Table 1 (phylogeny, oscillogram, comp. note) New Zealand

Kikihia pauxillae Marshall, D.C. 2009: 109, Fig 1 (phylogeny, comp. note) New Zealand

Kikihia pauxillae Simon 2009: 110, Fig 7 (phylogeny, comp. note) New Zealand

***K.* "peninsularis" Fleming, 1984**

Kikihia "peninsularis" Bowie, Marris, Emberson, Andrew, Berry, et al. 2003: 105 (distribution, comp. note) Quail Island

Kikihia "peninsularis" Arensburger, Simon, and Holsinger 2004: 1770–1776, 1778–1779, Figs 1–3, Table 1, Table 3 (molecular phylogeny, comp. note) New Zealand

Kikihia "peninsularis" Marshall, Cooley, Hill and Simon 2005: S-22, Fig 1 (phylogeny, comp. note) New Zealand

Kikihia "peninsularis" Marshall, Simon and Buckley 2006: 998–999, Figs 4–5 (phylogeny, comp. note) New Zealand

Kikihia "peninsularis" Marshall, Slon, Cooley, Hill and Simon 2008: 1056–1057, 1060, 1063, Figs 1–3, Fig 5, Table 1 (phylogeny, oscillogram, comp. note) New Zealand

Kikihia "peninsularis" Marshall, D.C. 2009: 109, Fig 1 (phylogeny, comp. note) New Zealand

Kikihia "peninsularis" Simon 2009: 110, Fig 7 (phylogeny, comp. note) New Zealand

***K. rosea* (Walker, 1850)** = *Cicada rosea* = *Cicada rosa* (sic) = *Melampsalta rosea* = *Cicadetta rosea* = *Melampsalta angusta* (nec Walker) = *Melampsalta muta cruentata* (nec Fabricius) = *Melampsalta cruentata* (nec Fabricius) = *Melampsalta muta muta* (nec Fabricius) = *Melampsalta muta* (nec Fabricius)

Kikihia rosea Fleming 1975a: 1570, 1572, Fig 3 (illustrated, comp. note) New Zealand

Kikihia rosea Fleming 1984: 200–201, 204 (comp. note) New Zealand

Kikihia rosea group Fleming 1984: 201 (comp. note) New Zealand

Kikihia rosea Lane 1995: 378 (comp. note) New Zealand

Kikihia rosea group Simon, Arensburger and Chambers 1999: 1 (comp. note) New Zealand

Kikihia rosea Arensburger, Simon, and Holsinger 2004: 1770–1776, 1778, Figs 1–3, Table 1, Table 3 (molecular phylogeny, comp. note) New Zealand

Kikihia rosea Marshall, Cooley, Hill and Simon 2005: S-22, Fig 1 (phylogeny, comp. note) New Zealand

Kikihia rosea Marshall, Simon and Buckley 2006: 998–999, Figs 4–5 (phylogeny, comp. note) New Zealand

Kikihia rosea Hill and Marshall 2008: 523 (comp. note) New Zealand

Kikihia rosea Marshall, Slon, Cooley, Hill and Simon 2008: 1056–1057, 1059–1060, 1063, Figs 1–3, Fig 5, Table 1 (phylogeny, oscillogram, comp. note) New Zealand

Kikihia rosea Marshall, D.C. 2009: 109, Fig 1 (phylogeny, comp. note) New Zealand

Kikihia rosea Marshall and Hill 2009: 4, Table 2 (comp. note)

Kikihia rosea Simon 2009: 110, Fig 7 (phylogeny, comp. note) New Zealand

***K. scutellaris* (Walker, 1850)** = *Melampsalta scutellaris* = *Cicadetta scutellaris* = *Cicada tristis* Hudson, 1891

Melampsalta scutellaris Buchanan 1879: 213 (comp. note) Equals *Cicada scutellaris* Equals *Cicada sericea* New Zealand

Melampsalta scutellaris Miller 1952: 17, 57 (comp. note) New Zealand

Cicadetta scutellaris Alfken 1903: 582 (comp. note) New Zealand

Melampsalta scutellaris Fleming 1966: 4, Fig 9 (illustrated, comp. note) New Zealand

Melampsalta scutellaris Grant-Taylor 1966: 348 (comp. note) New Zealand

Kikihia scutellaris Fleming 1975a: 1570 (comp. note) New Zealand

Kikihia scutellaris Fleming 1977: 258 (comp. note) New Zealand

Kikihia scutellaris Moeed and Meads 1983: 44, Fig 5A (hosts, comp. note) New Zealand

Kikihia scutellaris Fleming 1984: 192, 199 (comp. note) New Zealand

Kikihia scutellaris Moeed and Meads 1984: 224, Table 4 (listed) Hen Island
Kikihia scutellaris Fitzgerald, Meads, and Whitaker 1986: 26, 30, Table 1 (predation) New Zealand
Kikihia scutellaris Moeed and Meads 1987a: 58, Table 6 (listed) New Zealand
Kikihia scutellaris Moeed and Meads 1988: 488 (listed) New Zealand
Kikihia scutellaris Lane 1995: 383–385, 388, 392, Fig 17.6a (oscillogram, comp. note) New Zealand
Kikihia scutellaris Simon, Arensburger and Chambers 1999: 1 (comp. note) New Zealand
Kikihia scutellaris Buckley, Simon and Chambers 2001a: 77–78, Figs 6–7 (phylogeny, comp. note) New Zealand
Kikihia scutellaris Buckley, Simon, Shimodaira and Chambers 2001: 225, 228–229, Figs 2–3, Fig 5, Table 1 (phylogeny, comp. note) New Zealand
Kikihia scutellaris Chambers, Boon, Buckley and Hitchmough 2001: 380, Fig 3 (phylogeny, comp. note) New Zealand
Kikihia scutellaris Buckley, Arensburger, Simon and Chambers 2002: 6, 11, Fig 1 (phylogeny, comp. note) New Zealand
Kikihia scutellaris Guindon and Gascuel 2002: 542 (phylogeny, comp. note)
Kikihia scutellaris Sueur 2002b: 129 (listed) Equals *Melampsalta scutellaris*
Kikihia scutellaris Arensburger, Buckley, Simon, Moulds, and Holsinger 2004: 559, 562–563, 565 Table 1, Figs 1–2 (molecular phylogeny, comp. note) New Zealand
Kikihia scutellaris Arensburger, Simon, and Holsinger 2004: 1770–1776, 1779–1780, Figs 1–3, Table 1, Table 3 (molecular phylogeny, comp. note) New Zealand
Kikihia scutellaris Hill, Marshall and Cooley 2005: 72–77, Figs 1–2, Fig 5 (distribution, illustrated, song, sonogram, comp. note) New Zealand
Kikihia scutellaris Logan and Connolly 2005: 36–42, Fig 2, Figs 4–5, Tables 1–3 (agricultural pest, exuvia key, comp. note) New Zealand
Kikihia scutellaris Marshall, Cooley, Hill and Simon 2005: S-22, Fig 1 (phylogeny, comp. note) New Zealand
Melampsalta scutellaris Moulds 2005b: 410 (comp. note) Equals *Kikihia scutellaris*
Kikihia scutellaris Buckley, Simon and Chambers 2006: 77–78, Figs 6–7 (phylogeny, comp. note) New Zealand
Kikihia scutellaris Marshall, Simon and Buckley 2006: 998–999, Figs 4–5 (phylogeny, comp. note) New Zealand
Kikihia scutellaris Buckley and Simon 2007: 428, Fig 4 (phylogeny, comp. note) New Zealand
Kikihia scutellaris Esson 2007: 20–21 (sex ratio) New Zealand
Kikihia scutellaris Shiyake 2007: 8, 112, 114, Fig 200 (illustrated, distribution, listed, comp. note) New Zealand
Kikihia scutellaris Marshall, Slon, Cooley, Hill and Simon 2008: 1056–1057, 1059–1060, 1063, Figs 1–3, Fig 5, Table 1 (phylogeny, oscillogram, comp. note) New Zealand
Kikihia scutellaris Marshall, D.C. 2009: 109, Fig 1 (phylogeny, comp. note) New Zealand
Kikihia scutellaris Marshall and Hill 2009: 3–6, Fig 6M, Table 2 (comp. note) New Zealand
Kikihia scutellaris Simon 2009: 104, 110, Fig 2, Fig 7 (phylogeny, comp. note) New Zealand

***K. subalpina* (Hudson, 1891)** = *Cicada muta* var. *sub-alpina* = *Cicada muta subalpina* = *Cicada muta subalpina* = *Cicadetta subalpina* = *Cicadelta* (sic) *muta subalpina* = *Melampsalta cruentata* var. *sub-alpina* = *Melampsalta muta* var. *subalpina* = *Melampsalta muta subalpina* = *Melampsalta cruentata sub-alpina* = *Melampsalta subalpina* = *Kikihia subalpine* (sic) = *Cicada muta rufescens* Hudson, 1891 = *Melampsalta muta rufescens* = *Cicadetta muta rufescens* = *Cicada muta flavescens* Hudson, 1891 = *Cicadetta muta flavescens* = *Melampsalta muta callista* Hudson, 1950 = *Cicadetta muta callista* = *Melampsalta muta muta* Kirby (nec Fabricius) = *Melampsalta muta* Hutton (nec Fabricius) = *Melampsalta cruentata flavescens* Hudson = *Melampsalta muta* var. b *subalpina*

Melampsalta muta var. *subalpina* Miller 1952: 18, 57 (comp. note) New Zealand
Melampsalta subalpina Fleming 1966: 2–3, Fig 4 (illustrated, comp. note) New Zealand
Melampsalta subalpina Grant-Taylor 1966: 348 (comp. note) New Zealand
Kikihia subalpina Fleming 1975a: 1570 (comp. note) New Zealand
Kikihia subalpina Fleming 1975b: 1592 (comp. note) New Zealand
Kikihia subalpine (sic) Green 1977: 15 (comp. note) New Zealand
Kikihia subalpina Johns 1977: 322 (comp. note) New Zealand
Kikihia subalpina Bennett 1984: 89 (color variation) New Zealand
Cicadetta subalpina Fleming 1984: 192 (type species of *Kikihia*) Equals *Cicada muta* var. *subalpina* New Zealand
Kikihia subalpina Fleming 1984: 193–197, 199–201, 203, 205, Fig 1A, Fig 2A, Figs 5A–B (described, illustrated, synonymy, song, oscillogram, distribution, habitat, comp. note) Equals *Cicada muta* var. *sub-alpina*

Equals *Melampsalta cruentata* var. *subalpina* Equals *Melampsalta subalpina* Equals *Melampsalta muta* var b *subalpina* Equals *Melampsalta muta subalpina* Equals *Melampsalta subalpina* Equals *Cicadetta subalpina* New Zealand

Cicada muta var. *subalpina* Fleming 1984: 204 (comp. note) New Zealand

Melampsalta muta var. *subalpina* Fleming 1984: 201 (comp. note)

Melampsalta muta callista Palma, Lovis and Tither 1989: 37 (type material, type data) New Zealand

Cicada muta rufescens Palma, Lovis and Tither 1989: 38 (type material, type data) New Zealand

Cicada muta subalpina Palma, Lovis and Tither 1989: 38 (type material, type data) Equals *Kikihia subalpina* New Zealand

Kikihia subalpina Palma, Lovis and Tither 1989: 38 (status) Equals *Cicada muta subalpina* New Zealand

Kikihia subalpina Lane 1990a: 1 (comp. note) New Zealand

Kikihia subalpina Lane 1990b: 1 (comp. note) New Zealand

Cicada subalpina Moulds 1990: 170 (type species of *Kikihia*, listed)

Kikihia subalpina Harris 1993: 8–9 (emergence time, comp. note) New Zealand

Kikihia subalpina Lane 1995: 367–368, 371–373, 374–377, 392, 394–407, 409–410, Figs 17.4–17.5, Figs 17.9c-e, Figs 17.10–17.17 (distribution, timbal illustrated, sonogram, comp. note) New Zealand

Cicada muta subalpina Moulds and Carver 2002: 35 (type species of *Kikihia*, listed) Equals *Cicada muta* var. *subalpina*

Kikihia subalpina Sueur 2002b: 129 (listed) Equals *Melampsalta muta* var. *subalpina*

Kikihia subalpina Arensburger, Simon, and Holsinger 2004: 1770–1772, 1774–1778, Figs 1–3, Table 1, Table 3 (molecular phylogeny, comp. note) New Zealand

Kikihia subalpina Hill, Marshall and Cooley 2005: 77 (comp. note) New Zealand

Kikihia subalpina Marshall, Cooley, Hill and Simon 2005: S-22, Fig 1 (phylogeny, comp. note) New Zealand

Kikihia subalpina Moulds 2005b: 395–398, 417–419, 421, 425, Figs 56–59, Fig 62, Table 1 (higher taxonomy, phylogeny, key, listed, comp. note)

Kikihia subalpina Marshall, Simon and Buckley 2006: 998–999, Figs 4–5 (phylogeny, comp. note) New Zealand

Kikihia subalpina Hill and Marshall 2008: 523 (comp. note) New Zealand

Kikihia subalpina Marshall, Slon, Cooley, Hill and Simon 2008: 1056–1063, Figs 1–3, Figs 4–5, Table 1, Tables 3–4 (phylogeny, oscillogram, comp. note) New Zealand

Kikihia subalpina Marshall, D.C. 2009: 109, Fig 1 (phylogeny, comp. note) New Zealand

Kikihia subalpina Marshall and Hill 2009: 4–5, Fig 6K, Table 2 (sonogram, comp. note) New Zealand

Kikihia subalpina Marshall, Hill, Fontaine, Buckley and Simon 2009: 1996–2006, Fig 1, Figs 3–6, Tables 1–3 (phylogeny, distribution, oscillogram, comp. note) New Zealand

Kikihia subapina O'Neill, Buckley, Jewell and Ritchie 2009: 534 (comp. note) New Zealand

Kikihia subalpina Pain 2009: 47 (sonogram) Australia

Kikihia subalpina Simon 2009: 109–111, Figs 7–8 (illustrated, phylogeny, comp. note) New Zealand

K. "tasmani" Fleming, 1984

Kikihia "tasmani" Arensburger, Simon, and Holsinger 2004: 1770–1776, 1778–1779, Figs 1–3, Table 1, Table 3 (molecular phylogeny, comp. note) New Zealand

Kikihia "tasmani" Marshall, Cooley, Hill and Simon 2005: S-22, Fig 1 (phylogeny, comp. note) New Zealand

Kikihia "tasmani" Marshall, Simon and Buckley 2006: 998–999, Figs 4–5 (phylogeny, comp. note) New Zealand

Kikihia "tasmani" Marshall, Slon, Cooley, Hill and Simon 2008: 1056–1057, 1060–1061, 1063, Figs 1–3, Fig 5, Table 1 (phylogeny, oscillogram, comp. note) New Zealand

Kikihia "tasmani" Marshall, D.C. 2009: 109, Fig 1 (phylogeny, comp. note) New Zealand

Kikihia "tasmani" Simon 2009: 110, Fig 7 (phylogeny, comp. note) New Zealand

K. "tuta" Marshall, Cooley, Hill & Simon, 2005

Kikihia "tuta" Marshall, Cooley, Hill and Simon 2005: S-22, Fig 1 (phylogeny, comp. note) New Zealand

Kikihia "tuta" Marshall, Simon and Buckley 2006: 998–999, Figs 4–5 (phylogeny, comp. note) New Zealand

Kikihia "tuta" Marshall, Slon, Cooley, Hill and Simon 2008: 1056–1057, 1060, 1063, Figs 1–3, Fig 5, Table 1 (phylogeny, oscillogram, comp. note) New Zealand

Kikihia "tuta" Marshall, D.C. 2009: 109, Fig 1 (phylogeny, comp. note) New Zealand

Kikihia "tuta" Marshall and Hill 2009: 4–5, Fig 6I, Table 2 (sonogram, comp. note) New Zealand
Kikihia "tuta" Simon, Arensburger and Chambers 1999: 1 (comp. note) New Zealand
Kikihia "tuta" Simon 2009: 106, 110, Fig 3, Fig 7 (oscillogram, phylogeny, comp. note) New Zealand

K. "westlandica" Simon, 2009
Kikihia "westlandica" Simon 2009: 105 (comp. note) New Zealand

K. "westlandica North" Marshall, Cooley, Hill & Simon, 2005
Kikihia "westlandica N" Marshall, Cooley, Hill and Simon 2005: S-22, Fig 1 (phylogeny, comp. note) New Zealand
Kikihia "westlandica n" Marshall, Simon and Buckley 2006: 998–999, Figs 4–5 (phylogeny, comp. note) New Zealand
Kikihia "westlandica n" Marshall, Slon, Cooley, Hill and Simon 2008: 1056–1057, 1060–1061, 1063, Figs 1–3, Fig 5, Table 1 (phylogeny, oscillogram, comp. note) New Zealand
Kikihia "westlandica north" Marshall, D.C. 2009: 109, Fig 1 (phylogeny, comp. note) New Zealand
Kikihia "westlandica north" Simon 2009: 106, 110, Fig 3, Fig 7 (oscillogram, phylogeny, comp. note) New Zealand

K. "westlandica South" Marshall, Cooley, Hill & Simon, 2005
Kikihia "westlandica S" Marshall, Cooley, Hill and Simon 2005: S-22, Fig 1 (phylogeny, comp. note) New Zealand
Kikihia "westlandica s" Marshall, Simon and Buckley 2006: 998–999, Figs 4–5 (phylogeny, comp. note) New Zealand
Kikihia "westlandica s" Marshall, Slon, Cooley, Hill and Simon 2008: 1056–1057, 1060, 1063, Figs 1–3, Fig 5, Table 1 (phylogeny, oscillogram, comp. note) New Zealand
Kikihia "westlandica south" Marshall, D.C. 2009: 109, Fig 1 (phylogeny, comp. note) New Zealand
Kikihia "westlandica south" Simon 2009: 106, 110, Fig 3, Fig 7 (oscillogram, phylogeny, comp. note) New Zealand

***Maoricicada* Dugdale, 1972** = *Maoripsalta* (sic) = *Maoricicad* (sic) = *Maoripsalta* (sic) = *Melampsalta* (nec Kolenati)
Melampsalta (sic, partim) Buchanan 1879: 213–214 (comp. note) New Zealand
Melampsalta (sic) Salmon 1950: 1 (listed)
Maoripsalta (sic) Fleming 1975a: 1570 (comp. note) New Zealand
Maoricicada Fleming 1975b: 1592 (comp. note) New Zealand
Maoriciada Henderson 1983: 173 (comp. note) New Zealand
Melampsalta sp. Welbourn 1983: 138, Table 5 (mite host, listed) New Zealand
Maoricicada Fleming 1984: 191 (comp. note) New Zealand
Melampsalta (sic) sp. Southcott 1988: 103 (parasitism, comp. note) New Zealand
Maoricicada spp. Villet 1988: 71 (comp. note)
Melampsalta spp. (sic) Glare, O'Callaghan, and Wigley 1993: 98 (fungal parasite) New Zealand
Maoricicada Lei, Chou and Li 1994: 53 (comp. note)
Maoricicada Boer and Duffels 1996b: 309 (distribution, comp. note) New Zealand
Maoricicad (sic) Chou, Lei, Li, Lu and Yao 1997: 25 (comp. note)
Maoricicada Simon, Arensburger and Chambers 1999: 1 (comp. note) New Zealand
Maoricicada spp. Simon, Arensburger and Chambers 1999: 1 (comp. note) New Zealand
Maoricicada Buckley, Simon and Chambers 2001a: 68 (comp. note) New Zealand
Maoricicada spp. Buckley, Simon and Chambers 2001a: 76, 81 (comp. note) New Zealand
Maoricicada Buckley, Simon and Chambers 2001b: 1395, 1397–1398, 1402–1403 (comp. note) New Zealand
Maoricicada Buckley, Simon, Shimodaira and Chambers 2001: 223–225, 228, 230–232 (phylogeny, comp. note) New Zealand
Maoricicada Chambers, Boon, Buckley and Hitchmough 2001: 380–381, 385–386 (comp. note) New Zealand
Maoricicada Cooley 2001: 758 (comp. note) New Zealand
Maoricicada Cooley and Marshall 2001: 847 (comp. note) New Zealand
Maoricicada Arensburger 2002: 5055B (phylogeny) New Zealand
Maoricicada Buckley, Arensburger, Simon and Chambers 2002: 5–6, 12–13, 15 (phylogeny, diversity, comp. note) New Zealand
Maoricicada Dietrich 2002: 158 (speciation, comp. note) New Zealand
Maoricicada spp. Guindon and Gascuel 2002: 541–542 (phylogeny, comp. note)
Maoricicada Lee, Choe, Lee and Woo 2002: 3 (comp. note)
Maoricicada Arensburger, Buckley, Simon, Moulds, and Holsinger 2004: 557–558, 561, 562–563, 565–566 (molecular phylogeny, diversity, comp. note) New Zealand

Maoricicada Arensburger, Simon, and Holsinger 2004: 1769, 1773–1775 (molecular phylogeny, diversity, comp. note) New Zealand
Maoricicada Quek, Davies, Itino and Pierce 2004: 565, Table 4 (comp. note) New Zealand
Maoricicada spp. Heads 2005: 70 (comp. note) New Zealand
Maoricicada Hill, Simon and Chambers 2005: S-23 (phylogeny, comp. note) New Zealand
Maoricicada Moulds 2005b: 390, 415, 436 (higher taxonomy, listed, comp. note)
Maoricicada Buckley, Cordeiro, Marshall and Simon 2006: 411, 414–416, 419, 421–422 (phylogeny, diversity, comp. note) New Zealand
Maoricicada Buckley, Simon and Chambers 2006: 68, 76 (phylogeny, comp. note) New Zealand
Maoricicada Buckley and Simon 2007: 419–421, 423–424, 426, 429–431, Fig 5 (comp. note) New Zealand
Maoricicada spp. Hill, Simon, Marshall and Chambers 2007: 677 (comp. note) New Zealand
Maoricicada Shiyake 2007: 8, 114 (listed, comp. note)
Maoricicada Sueur, Vanderpool, Simon, Ouvrard and Bourgoin 2007: 613, 623 (comp. note) New Zealand
Maoricicada spp. Fonseca, Serrão, Pina-Martins, Silva, Mira, et al. 2008: 30 (comp. note) New Zealand
Maoricicada Goldberg, Trewick and Paterson 2008: 3326 (comp. note) New Zealand
Maoriciada Hill and Marshall 2008: 523–524 (diversity, comp. note) New Zealand
Maoricicada sp. Hill and Marshall 2008: 524 (comp. note) New Zealand
Maoricicada Marshall, Slon, Cooley, Hill and Simon 2008: 1060, 1064 (phylogeny, comp. note) New Zealand
Maoricicada Cranston 2009: 92 (distribution, comp. note)
Maoricicada Marshall, Hill, Fontaine, Buckley and Simon 2009: 2006 (comp. note) New Zealand
Maoricicada Meng and Kubatko 2009: 42 (comp. note) New Zealand
Maoricicada Simon 2009: 104, 106 (diversity, comp. note) New Zealand

***M. alticola* Dugdale & Fleming, 1978**

Maoricicada alticola Deitz 1979: 25 (listed) New Zealand
Maoricicada alticola Buckley, Simon and Chambers 2001a: 77–78, Figs 6–7 (phylogeny, comp. note) New Zealand
Maoricicada alticola Buckley, Simon, Shimodaira and Chambers 2001: 225, 228–230, Figs 2–3, Figs 5–6, Table 1 (phylogeny, comp. note) New Zealand
Maoricicada alticola Chambers, Boon, Buckley and Hitchmough 2001: 381, Fig 4 (phylogeny, comp. note) New Zealand
Maoricicada alticola Guindon and Gascuel 2002: 542 (phylogeny, comp. note)
Maoricicada alticola Buckley, Cordeiro, Marshall and Simon 2006: 412, 417–422, Figs 3–5, Table 1, Table 7 (phylogeny, sonogram, comp. note) New Zealand
Maoricicada alticola Buckley, Simon and Chambers 2006: 77–78, Figs 6–7 (phylogeny, comp. note) New Zealand
Maoricicada alticola Buckley and Simon 2007: 422–428, Figs 1–4, Tables 1–2 (phylogeny, comp. note) New Zealand

***M. campbelli* (Myers, 1923)** = *Melampsalta campbelli* = *Cicadetta campbelli* = *Cicadetta cambellii* (sic) = *Maoripsalta* (sic) *campbelli* = *Pauropsaltra maorica* Myers, 1923 = *Melampsalta maorica* = *Cicadetta maorica*

Melampsalta campbelli Salmon 1950: 1–2 (synonymy, comp. note) Equals *Pauropsalta campbelli* Equals *Pauropsalta maorica* Equals *Melampsalta maorica* New Zealand
Melampsalta maorica Salmon 1950: 2 (comp. note) New Zealand
Melampsalta campbelli Fleming 1966: 5, Fig 12 (illustrated, comp. note) Equals *Melampsalta maorica* New Zealand
Maoricicada campbelli Johns 1977: 320, 324, Table 21D (listed, comp. note) New Zealand
Maoripsalta (sic) *campbelli* Fox 1982: 287 (comp. note) New Zealand
Maoricicada campbelli White 1987: 149, Fig 2 (abundance, comp. note) New Zealand
Maoricicada cambpelli Patrick 1991: 4 (listed, comp. note) New Zealand
Maoricicada campbelli Lane 1993: 53 (comp. note) New Zealand
Maoricicada campbelli White and Sedcole 1993a: 39, 43 (host, comp. note) New Zealand
C[icadetta] cambellii (sic) Jiang, Yang, Tang, Xu and Chen 1995b: 229 (song frequency, comp. note)
Maoricicada campbelli Simon, Arensburger and Chambers 1999: 1 (comp. note) New Zealand
Maoricicada campbelli Buckley, Simon and Chambers 2001a: 71–72, 76–78, Fig 2, Figs 6–7 (phylogeny, sequence, comp. note) New Zealand

Maoricicada campbelli Buckley, Simon and Chambers 2001b: 1395–1405, Figs 1–4, Tables 1–2 (phylogeny, illustrated, comp. note) Equals *Pauropsalta maorica* New Zealand

Maoricicada campbelli Buckley, Simon, Shimodaira and Chambers 2001: 223–225, 227–232, Figs 2–6, Table 1, Tables 4–5 (phylogeny, comp. note) New Zealand

Maoricicada campbelli Chambers, Boon, Buckley and Hitchmough 2001: 381, Fig 4 (phylogeny, comp. note) New Zealand

Cicadetta campbelli Sueur 2001: 42, Table 1 (listed)

Maoricicada campbelli Guindon and Gascuel 2002: 541–542 (phylogeny, comp. note)

Maoricicada campbelli Heads 2002b: 70 (comp. note) New Zealand

Maoricicada campbelli Lee, Choe, Lee and Woo 2002: 10 (comp. note) New Zealand

Maoricicada campbelli Sueur 2002b: 129 (listed) Equals *Melampsalta campbelli*

Maoricicada campbelli Hill, Simon and Chambers 2005: S-23 (phylogeny, comp. note) New Zealand

Maoricicada campbelli Buckley, Cordeiro, Marshall and Simon 2006: 412, 415, 417–422, Figs 2–5, Table 1, Table 7 (phylogeny, sonogram, comp. note) New Zealand

Maoricicada campbelli Buckley, Simon and Chambers 2006: 71–72, 77–78, Fig 2, Figs 6–7 (sequence, phylogeny, comp. note) New Zealand

Maoricicada campbelli Buckley and Simon 2007: 422–429, 431, Figs 1–4, Tables 1–2 (phylogeny, comp. note) New Zealand

Maoricicada campbelli Greaves, Chapple, Gleeson, Daugherty and Ritchie 2007: 737 (comp. note) New Zealand

Maoricicada campbelli Hill, Simon, Marshall and Chambers 2007: 675, 677–689, Figs 2–5, Tables 1–5 (phylogeny, song, sonogram, oscillogram, comp. note) New Zealand

Maoricicada campbelli Asunce, Ernst, Clark and Nigg 2008: 1449 (comp. note)

Maoricicada campbelli Goldberg, Trewick and Paterson 2008: 3326, Fig 5 (phylogeny, comp. note) New Zealand

Maoricicada campbelli Hill and Marshall 2008: 524 (comp. note) New Zealand

Maoricicada campbelli Marshall and Hill 2009: 2, 4–5, Fig 4, Fig 6D, Table 2 (sonogram, comp. note) New Zealand

Maoricicada campbelli Marshall, Hill, Fontaine, Buckley and Simon 2009: 2005 (comp. note) New Zealand

Maoricicada campbelli Meng and Kubatko 2009: 42 (comp. note) New Zealand

Maoricicada campbelli O'Neill, Chapple, Daugherty and Ritchie 2008: 1174, 1176 (comp. note) New Zealand

Maoricicada campbelli O'Neill, Buckley, Jewell and Ritchie 2009: 533–534 (comp. note) New Zealand

Maoricicada campbelli Simon 2009: 105, 107, 109, Figs 4–6 (sonogram, oscillogram, phylogeny, comp. note) New Zealand

***M. cassiope* (Hudson, 1891)** = *Cicada cassiope* = *Melampsalta cassiope* = *Cicadetta cassiope* = *Maoripsalta* (sic) *cassiope* = *Melampsalta quadricincta* (nec Walker) = *Melampsalta nervosa* (nec Walker)

Melampsalta cassiope Miller 1952: 44, 57 (comp. note) New Zealand

Melampsalta cassiope Fleming 1966: 6, Fig 15 (illustrated, comp. note) New Zealand

Melampsalta cassiope Grant-Taylor 1966: 348 (comp. note) New Zealand

Maoricicada cassiope Fleming 1975b: 1593, Fig 3 (illustrated, comp. note) New Zealand

Maoricicada cassiope Johns 1977: 324–325, Table 21E (comp. note) New Zealand

Maoripsalta (sic) *cassiope* Fox 1982: 287 (comp. note) New Zealand

Maoricicada cassiope White 1987: 149, Fig 2 (abundance, comp. note) New Zealand

Cicada cassiope Palma, Lovis and Tither 1989: 37 (type material, type data) Equals *Maoricicada cassiope* New Zealand

Maoricicada cassiope Palma, Lovis and Tither 1989: 37 (status) Equals *Cicada cassiope* New Zealand

Maoricicada cassiope White and Sedcole 1993a: 39, 43 (host, comp. note) New Zealand

Maoricicada cassiope Simon, Arensburger and Chambers 1999: 1 (comp. note) New Zealand

Maoricicada cassiope Buckley, Simon and Chambers 2001a: 71, 76–78, Figs 6–7 (phylogeny, comp. note) New Zealand

Maoricicada cassiope Buckley, Simon, Shimodaira and Chambers 2001: 225, 228–230, Figs 2–3, Figs 5–6, Table 1 (phylogeny, comp. note) New Zealand

Maoricicada cassiope Chambers, Boon, Buckley and Hitchmough 2001: 380–381, Figs 3–4 (phylogeny, comp. note) New Zealand

Maoricicada cassiope Buckley, Arensburger, Simon and Chambers 2002: 6, 11, Fig 1 (phylogeny, comp. note) New Zealand

Maoricicada cassiope Guindon and Gascuel 2002: 541–542 (phylogeny, comp. note)

Maoricicada cassiope Sueur 2002b: 129 (listed) Equals *Melampsalta cassiope*
Maoricicada cassiope Arensburger, Buckley, Simon, Moulds, and Holsinger 2004: 559, 562–563 Table 1, Figs 1–2 (molecular phylogeny, comp. note) New Zealand
Maoricicada cassiope Arensburger, Simon, and Holsinger 2004: 1770–1772, 1774, 1776, Fig 2, Table 1, Table 3 (molecular phylogeny, diversity, comp. note) New Zealand
Maoricicada cassiope Buckley, Cordeiro, Marshall and Simon 2006: 412, 417–418, 420–422, Figs 3–5, Table 1, Table 7 (phylogeny, sonogram, comp. note) New Zealand
Maoricicada cassiope Buckley, Simon and Chambers 2006: 71, 76–78, Figs 6–7 (phylogeny, comp. note) New Zealand
Maoricicada cassiope Marshall, Simon and Buckley 2006: 998–999, Figs 4–5 (phylogeny, comp. note) New Zealand
Maoricicada cassiope Buckley and Simon 2007: 419, 422–429, Figs 1–4, Tables 1–2 (phylogeny, comp. note) New Zealand
Maoricicada cassiope Shiyake 2007: 8, 112, 114, Fig 201 (illustrated, distribution, listed, comp. note) New Zealand
Maoricicada cassiope Hill and Marshall 2008: 524 (comp. note) New Zealand
Maoricicada cassiope Marshall, Slon, Cooley, Hill and Simon 2008: 1056–1057, 1060, 1063, Figs 1–3, Fig 5, Table 1 (phylogeny, oscillogram, comp. note) New Zealand
Maoricicada cassiope Marshall, D.C. 2009: 109, Fig 1 (phylogeny, comp. note) New Zealand
Maoricicada cassiope Simon 2009: 104, 110, Fig 2, Fig 7 (phylogeny, comp. note) New Zealand

***M. clamitans* Dugdale & Fleming, 1978** = *Maoricicada* sp. C Fleming, 1975
Maoricicada clamitans Fleming 1975b: 1592–1593, Fig 1 (illustrated, comp. note) New Zealand
Maoricicada clamitans Deitz 1979: 25 (listed) New Zealand
Maoricicada clamitans Barratt and Patrick 1987: 74 (comp. note, distribution) New Zealand
Maoricicada clamitans Palma, Lovis and Tither 1989: 37 (type material, type data) New Zealand
Maoricicada clamitans Buckley, Simon and Chambers 2001a: 76–78, Figs 6–7 (phylogeny, comp. note) New Zealand
Maoricicada clamitans Buckley, Simon and Chambers 2001b: 1397–1398, 1401, Fig 4, Table 1 (phylogeny, comp. note) New Zealand
Maoricicada clamitans Buckley, Simon, Shimodaira and Chambers 2001: 225, 228–230, Figs 2–3, Figs 5–6, Table 1 (phylogeny, comp. note) New Zealand
Maoricicada clamitans Chambers, Boon, Buckley and Hitchmough 2001: 381, Fig 4 (phylogeny, comp. note) New Zealand
Maoricicada clamitans Guindon and Gascuel 2002: 542 (phylogeny, comp. note)
Maoricicada clamitans Buckley, Cordeiro, Marshall and Simon 2006: 412, 415, 417–422, Figs 3–5, Table 1, Table 7 (phylogeny, sonogram, comp. note) New Zealand
Maoricicada clamitans Buckley, Simon and Chambers 2006: 76–78, Figs 6–7 (phylogeny, comp. note) New Zealand
Maoricicada clamitans Buckley and Simon 2007: 422–428, Figs 1–4, Tables 1–2 (phylogeny, comp. note) New Zealand
Maoricicada clamitans Hill, Simon, Marshall and Chambers 2007: 678–679 (comp. note) New Zealand
Maoricicada clamitans Hill and Marshall 2008: 524 (comp. note) New Zealand

***M. hamiltoni* (Myers, 1926)** = *Cicadetta hamiltoni*
Melampsalta hamiltoni Fleming 1966: 5, Fig 13 (illustrated, comp. note) New Zealand
Maoricicada hamiltoni Fleming 1975a: 1572 (comp. note) New Zealand
Maoricicada hamiltoni Johns 1977: 319–320, Table 21D (listed, comp. note) New Zealand
Maoricicada hamiltoni Buckley, Simon and Chambers 2001a: 76–78, 80–81, Figs 6–7 (phylogeny, comp. note) New Zealand
Maoricicada hamiltoni Buckley, Simon, Shimodaira and Chambers 2001: 223, 225, 228–231, Figs 2–3, Figs 5–6, Table 1 (phylogeny, comp. note) New Zealand
Maoricicada hamiltoni Chambers, Boon, Buckley and Hitchmough 2001: 380–381, Figs 3–4 (phylogeny, comp. note) New Zealand
Maoricicada hamiltoni Buckley, Arensburger, Simon and Chambers 2002: 6, 11, Fig 1 (phylogeny, comp. note) New Zealand
Maoricicada hamiltoni Guindon and Gascuel 2002: 542 (phylogeny, comp. note)
Maoricicada hamiltoni Sueur 2002b: 129 (listed) Equals *Melampsalta hamiltoni*
Maoricicada hamiltoni Arensburger, Buckley, Simon, Moulds, and Holsinger 2004: 559, 562–563 Table 1, Figs 1–2 (molecular phylogeny, comp. note) New Zealand
Maoricicada hamiltoni Buckley, Cordeiro, Marshall and Simon 2006: 412, 416–420, 422, Figs 3–5, Table 1, Table 7 (phylogeny, sonogram, comp. note) New Zealand

Maoricicada hamiltoni Buckley, Simon and Chambers 2006: 76–78, 80, Figs 6–7 (phylogeny, comp. note) New Zealand
Maoricicada hamiltoni Buckley and Simon 2007: 422–428, 431, Figs 1–4, Tables 1–2 (phylogeny, comp. note) New Zealand
Maoricicada hamiltoni Hill and Marshall 2008: 524 (comp. note) New Zealand
Maoricicada hamiltoni Simon 2009: 104, Fig 2 (phylogeny, comp. note) New Zealand

***M. iolanthe* (Hudson, 1891)** = *Cicada iolanthe* = *Melampsalta iolanthe* = *Cicadetta iolanthe* = *Cicadetta planthe* (sic)
Cicadetta planthe (sic) Alfken 1903: 582 (comp. note) New Zealand
Melampsalta iolanthe Fleming 1966: 6, Fig 14 (illustrated, comp. note) New Zealand
Maoricicada iolanthe Fleming 1975a: 1570 (comp. note) New Zealand
C[icadetta] iolanthe Jiang, Yang, Tang, Xu and Chen 1995b: 229 (song frequency, comp. note)
Maoricicada iolanthe Simon, Arensburger and Chambers 1999: 1 (comp. note) New Zealand
Maoricicada iolanthe Buckley, Simon and Chambers 2001a: 69, 76–78, Figs 6–7 (phylogeny, comp. note) New Zealand
Maoricicada iolanthe Buckley, Simon, Shimodaira and Chambers 2001: 223–232, Figs 1–6, Table 1, Tables 4–5 (phylogeny, illustrated, comp. note) New Zealand
Maoricicada iolanthe Chambers, Boon, Buckley and Hitchmough 2001: 380–381, Fig 4 (phylogeny, comp. note) New Zealand
Cicadetta iolanthe Sueur 2001: 43, Table 1 (listed)
Maoricicada iolanthe Guindon and Gascuel 2002: 542 (phylogeny, comp. note)
Maoricicada iolanthe Buckley, Cordeiro, Marshall and Simon 2006: 411–412, 414–415, 417–422, 425, Figs 3–5, Table 1, Table 7 (phylogeny, sonogram, illustrated, comp. note) New Zealand
Maoricicada iolanthe Buckley, Simon and Chambers 2006: 69, 76–78, Figs 6–7 (phylogeny, comp. note) New Zealand
Maoricicada iolanthe Buckley and Simon 2007: 422–429, Figs 1–4, Tables 1–2 (phylogeny, comp. note) New Zealand
Maoricicada iolanthe Hill, Simon, Marshall and Chambers 2007: 680 (comp. note) New Zealand
Maoricicada iolanthe Meng and Kubatko 2009: 42 (comp. note) New Zealand

***M. lindsayi* (Myers, 1923)** = *Pauropsalta lindsayi* = *Melampsalta linsayi* = *Cicadetta lindsayi* = *Maoricicada mangu lindsayi*
C[icadetta] lindsayi Jiang, Yang, Tang, Xu and Chen 1995b: 229 (song frequency, comp. note)
Maoricicada lindsayi Buckley, Simon and Chambers 2001a: 76–78, 80–81, Figs 6–7 (phylogeny, comp. note) New Zealand
Maoricicada lindsayi Buckley, Simon, Shimodaira and Chambers 2001: 223, 225, 228–231, Figs 2–3, Figs 5–6, Table 1 (phylogeny, comp. note) New Zealand
Maoricicada lindsayi Chambers, Boon, Buckley and Hitchmough 2001: 381, Fig 4 (phylogeny, comp. note) New Zealand
Cicadetta lindsayi Sueur 2001: 43, Table 1 (listed)
Maoricicada mangu lindsayi Guindon and Gascuel 2002: 542 (phylogeny, comp. note)
Maoricicada lindsayi Buckley, Cordeiro, Marshall and Simon 2006: 412, 417–422, Figs 3–5, Table 1, Table 7 (phylogeny, sonogram, comp. note) New Zealand
Maoricicada lindsayi Buckley, Simon and Chambers 2006: 76–78, 80, Figs 6–7 (phylogeny, comp. note) New Zealand
Maoricicada lindsayi Buckley and Simon 2007: 422–429, 431, Figs 1–4, Tables 1–2 (phylogeny, comp. note) New Zealand
Maoricicada lindsayi Hill and Marshall 2008: 524 (comp. note) New Zealand
Maoricicada lindsayi Marshall, Hill, Fontaine, Buckley and Simon 2009: 2006 (comp. note) New Zealand
Maoricicada lindsayi Meng and Kubatko 2009: 42 (comp. note) New Zealand

***M. mangu celer* Dugdale & Fleming, 1978**
Maoricicada mangu celer Deitz 1979: 25 (listed) New Zealand
Maoricicada mangu celer Buckley, Simon and Chambers 2001a: 77–78, Figs 6–7 (phylogeny, comp. note) New Zealand
Maoricicada mangu celer Buckley, Simon, Shimodaira and Chambers 2001: 225, 228–230, Figs 2–3, Figs 5–6, Table 1 (phylogeny, comp. note) New Zealand
Maoricicada mangu celer Guindon and Gascuel 2002: 542 (phylogeny, comp. note)
Maoricicada mangu celer Buckley, Cordeiro, Marshall and Simon 2006: 412, 417–418, Figs 3–4, Table 1 (phylogeny, comp. note) New Zealand
Maoricicada mangu celer Buckley, Simon and Chambers 2006: 77–78, Figs 6–7 (phylogeny, comp. note) New Zealand
Maoricicada mangu celer Buckley and Simon 2007: 422–423, 425, 428, Fig 1, Fig 4, Tables 1–2 (phylogeny, comp. note) New Zealand

M. mangu gourlayi **Dugdale & Fleming, 1978**

Maoricicada mangu gourlayi Deitz 1979: 25 (listed) New Zealand
Maoricicada mangu gourlayi Palma, Lovis and Tither 1989: 37 (type material, type data) New Zealand
Maoricicada mangu gourlayi Buckley, Simon and Chambers 2001a: 76–78, Figs 6–7 (phylogeny, comp. note) New Zealand
Maoricicada mangu gourlayi Buckley, Simon, Shimodaira and Chambers 2001: 225, 228–230, Figs 2–3, Figs 5–6, Table 1 (phylogeny, comp. note) New Zealand
Maoricicada mangu gourlayi Guindon and Gascuel 2002: 542 (phylogeny, comp. note)
Maoricicada mangu gourlayi Buckley, Cordeiro, Marshall and Simon 2006: 412, 417–418, Figs 3–4, Table 1 (phylogeny, comp. note) New Zealand
Maoricicada mangu gourlayi Buckley, Simon and Chambers 2006: 76–78, Figs 6–7 (phylogeny, comp. note) New Zealand
Maoricicada mangu gourlayi Buckley and Simon 2007: 422–423, 425, 428, Fig 1, Fig 4, Tables 1–2 (phylogeny, comp. note) New Zealand

M. mangu mangu **(White, 1879)** = *Melampsalta mangu* = *Cicadetta mangu* = *Melampsalta quadricincta* Myers (nec Walker)

Melampsalta mangu Buchanan 1879: 214 (n. sp., described, comp. note) New Zealand
Melampsalta mangu Miller 1952: 43, 57 (comp. note) New Zealand
Maoricicada mangu Fleming 1975b: 1594–1595, Fig 1 (illustrated, comp. note) New Zealand
Maoricicada mangu Johns 1977: 325, Table 21E (comp. note) New Zealand
Maoricicada mangu White 1987: 149, Fig 2 (abundance, comp. note) New Zealand
Maoricicada mangu mangu Lane 1993: 53 (comp. note) New Zealand
Maoricicada mangu mangu White and Sedcole 1993a: 39, 43, 50 (host, comp. note) New Zealand
Maoricicada mangu Simon, Arensburger and Chambers 1999: 1 (comp. note) New Zealand
Maoricicada mangu Buckley, Simon and Chambers 2001a: 71, 76, 78 (phylogeny, comp. note) New Zealand
Maoricicada mangu mangu Buckley, Simon and Chambers 2001a: 77–78, Figs 6–7 (phylogeny, comp. note) New Zealand
Maoricicada mangu mangu Buckley, Simon, Shimodaira and Chambers 2001: 225, 228–230, Figs 2–3, Figs 5–6, Table 1 (phylogeny, comp. note) New Zealand
Maoricicada mangu Buckley, Simon, Shimodaira and Chambers 2001: 229 (comp. note)
Maoricicada mangu Chambers, Boon, Buckley and Hitchmough 2001: 381, Fig 4 (phylogeny, comp. note) New Zealand
Maoricicada mangu mangu Guindon and Gascuel 2002: 542 (phylogeny, comp. note)
Maoricicada mangu Sueur 2002b: 129 (listed) Equals *Melampsalta mangu*
Maoricicada mangu mangu Buckley, Cordeiro, Marshall and Simon 2006: 412, 415, 417–419, Figs 3–4, Table 1 (phylogeny, comp. note) New Zealand
Maoricicada mangu Buckley, Cordeiro, Marshall and Simon 2006: 414, 419–422, Fig 5, Table 7 (sonogram, comp. note) New Zealand
Maoricicada mangu Buckley, Simon and Chambers 2006: 71, 76, 79 (phylogeny, comp. note) New Zealand
Maoricicada mangu mangu Buckley, Simon and Chambers 2006: 77–78, Figs 6–7 (phylogeny, comp. note) New Zealand
Maoricicada mangu Buckley and Simon 2007: 419, 422, 426–429, Figs 2–4, Table 1 (phylogeny, comp. note) New Zealand
Maoricicada mangu mangu Buckley and Simon 2007: 422–423, 425–428, Figs 1–4, Tables 1–2 (phylogeny, comp. note) New Zealand
Maoricicada mangu Hill and Marshall 2008: 524 (illustrated, comp. note) New Zealand
Maoricicada mangu Marshall and Hill 2009: 4, Table 2 (comp. note)
Maoricicada mangu Simon 2009: 105 (comp. note) New Zealand

M. mangu multicostata **Dugdale & Fleming, 1978**

Maoricicada mangu multicostata Deitz 1979: 25 (listed) New Zealand
Maoricicada mangu multicostata Palma, Lovis and Tither 1989: 37 (type material, type data) New Zealand
Maoricicada mangu multicostata Buckley, Simon and Chambers 2001a: 77–78, Figs 6–7 (phylogeny, comp. note) New Zealand
Maoricicada mangu multicostata Buckley, Simon, Shimodaira and Chambers 2001: 225, 228–230, Figs 2–3, Figs 5–6, Table 1 (phylogeny, comp. note) New Zealand
Maoricicada mangu multicostata Guindon and Gascuel 2002: 542 (phylogeny, comp. note)
Maoricicada mangu multicostata Buckley, Cordeiro, Marshall and Simon 2006: 412, 417–419, Figs 3–4, Table 1 (phylogeny, comp. note) New Zealand
Maoricicada mangu multicostata Buckley, Simon and Chambers 2006: 77–78, Figs 6–7 (phylogeny, comp. note) New Zealand

Maoricicada mangu multicostata Buckley and Simon 2007: 422–423, 425, 428, Fig 1, Fig 4, Tables 1–2 (phylogeny, comp. note) New Zealand

M. myersi **(Fleming, 1971)** = *Cicadetta myersi* = *Melampsalta iolanthe* Myers (nec Hudson)

Cicadetta myersi Palma, Lovis and Tither 1989: 38 (type material, type data) Equals *Maoricicada myersi* New Zealand

Maoricicada myersi Palma, Lovis and Tither 1989: 38 (status) Equals *Cicadetta myersi* New Zealand

C[icadetta] myersi Jiang, Yang, Tang, Xu and Chen 1995b: 229 (song frequency, comp. note)

Maoricicada myersi Buckley, Simon and Chambers 2001a: 76–78, 80–81, Figs 6–7 (phylogeny, comp. note) New Zealand

Maoricicada myersi Buckley, Simon, Shimodaira and Chambers 2001: 223, 225, 228–231, Figs 2–3, Figs 5–6, Table 1 (phylogeny, comp. note) New Zealand

Maoricicada myersi Chambers, Boon, Buckley and Hitchmough 2001: 381, Fig 4 (phylogeny, comp. note) New Zealand

Cicadetta myersi Sueur 2001: 43, Table 1 (listed)

Maoricicada myersi Guindon and Gascuel 2002: 542 (phylogeny, comp. note)

Maoricicada myersi Buckley, Cordeiro, Marshall and Simon 2006: 412, 417–422, Figs 3–5, Table 1, Table 7 (phylogeny, sonogram, comp. note) New Zealand

Maoricicada myersi Buckley, Simon and Chambers 2006: 76–78, 80, Figs 6–7 (phylogeny, comp. note) New Zealand

Maoricicada myersi Buckley and Simon 2007: 422–429, 431, Figs 1–4, Tables 1–2 (phylogeny, comp. note) New Zealand

Maoricicada myersi Marshall, Hill, Fontaine, Buckley and Simon 2009: 2006 (comp. note) New Zealand

Maoricicada myersi Meng and Kubatko 2009: 42 (comp. note) New Zealand

M. nigra frigida **Dugdale & Fleming, 1978** = *Maoricicada nigra frigide* (sic) = *Maoricicada* sp. F Fleming, 1975

Maoricicada nigra frigida Deitz 1979: 25 (listed) New Zealand

Maoricicada nigra frigida Buckley, Simon and Chambers 2001a: 77–78, Figs 6–7 (phylogeny, comp. note) New Zealand

Maoricicada nigra frigida Buckley, Simon, Shimodaira and Chambers 2001: 225, 228–230, Figs 2–3, Figs 5–6, Table 1 (phylogeny, comp. note) New Zealand

Maoricicada nigra frigida Guindon and Gascuel 2002: 542 (phylogeny, comp. note)

Maoricicada nigra frigida Buckley, Cordeiro, Marshall and Simon 2006: 412, 417–420, 422, Figs 3–5, Table 1, Table 7 (phylogeny, sonogram, comp. note) New Zealand

Maoricicada nigra frigida Buckley, Simon and Chambers 2006: 77–78, Figs 6–7 (phylogeny, comp. note) New Zealand

Maoricicada nigra frigida Buckley and Simon 2007: 422–423, 425–428, Figs 1–4, Tables 1–2 (phylogeny, comp. note) New Zealand

Maoricicada nigra frigide (sic) Shiyake 2007: 9 , 112, 114, Fig 202 (illustrated, distribution, listed, comp. note) New Zealand

M. nigra nigra **(Myers, 1921)** = *Melampsalta nigra* = *Cicadetta nigra*

Melampsalta nigra Fleming 1966: 6, Fig 17 (illustrated, comp. note) New Zealand

Maoricicada nigra nigra Helmore 1982: 30–31 (illustrated, comp. note) New Zealand

Maoricicada nigra nigra Buckley, Simon and Chambers 2001a: 77–78, Figs 6–7 (phylogeny, comp. note) New Zealand

Maoricicada nigra nigra Buckley, Simon, Shimodaira and Chambers 2001: 225, 228–230, Figs 2–3, Figs 5–6, Table 1 (phylogeny, comp. note) New Zealand

Maoricicada nigra Chambers, Boon, Buckley and Hitchmough 2001: 381, Fig 4 (phylogeny, comp. note) New Zealand

Maoricicada nigra nigra Guindon and Gascuel 2002: 542 (phylogeny, comp. note)

Maoricicada nigra nigra Buckley, Cordeiro, Marshall and Simon 2006: 412, 417–420, 422, Figs 3–5, Table 1, Table 7 (phylogeny, sonogram, comp. note) New Zealand

Maoricicada nigra Buckley, Simon, Shimodaira and Chambers 2001: 230 (comp. note)

Maoricicada nigra nigra Buckley, Simon and Chambers 2006: 77–78, Figs 6–7 (phylogeny, comp. note) New Zealand

Maoricicada nigra nigra Buckley and Simon 2007: 422–423, 425–428, Figs 1–4, Tables 1–2 (phylogeny, comp. note) New Zealand

Maoricicada nigra Buckley, Cordeiro, Marshall and Simon 2006: 414, 416, 419 (comp. note) New Zealand

Maoricicada nigra Buckley and Simon 2007: 419, 424 (comp. note) New Zealand

Maoricicada nigra nigra Hill and Marshall 2008: 524 (comp. note) New Zealand

M. oromelaena **(Myers, 1926)** = *Melampsalta oromelaena* = *Cicadetta oromelaena*

Melampsalta oromeleana Fleming 1966: 6, Fig 16 (illustrated, comp. note) New Zealand

Melampsalta oromelaena Southcott 1972: 34, 36 (parasitism) New Zealand

Maoricicada oromeleana Fleming 1975b: 1594–1595, Fig 2 (illustrated, comp. note) New Zealand
Maoricicada oromelaena Johns 1977: 325, Table 21E (comp. note) New Zealand
Melampsalta oromelaena Welbourn 1983: 138, Table 5 (mite host, listed) New Zealand
Melampsalta oromelaena Southcott 1988: 103 (parasitism, comp. note) New Zealand
Maoricicada oromelaena Buckley, Simon and Chambers 2001a: 76–78, Figs 6–7 (phylogeny, comp. note) New Zealand
Maoricicada oromelaena Buckley, Simon, Shimodaira and Chambers 2001: 225, 228–231, Figs 2–3, Figs 5–6, Table 1 (phylogeny, comp. note) New Zealand
Maoricicada oromelaena Chambers, Boon, Buckley and Hitchmough 2001: 381, Fig 4 (phylogeny, comp. note) New Zealand
Maoricicada oromelaena Guindon and Gascuel 2002: 542 (phylogeny, comp. note)
Maoricicada oromelaena Buckley, Cordeiro, Marshall and Simon 2006: 412, 417–422, Figs 3–5, Table 1, Table 7 (phylogeny, sonogram, comp. note) New Zealand
Maoricicada oromelaena Buckley, Simon and Chambers 2006: 76–78, Figs 6–7 (phylogeny, comp. note) New Zealand
Maoricicada oromelaena Buckley and Simon 2007: 422–429, Figs 1–4, Tables 1–2 (phylogeny, comp. note) New Zealand
Maoricicada oromelaena Hill and Marshall 2008: 524 (comp. note) New Zealand

***M. otagoensis maceweni* Dugdale & fleming, 1978**
Maoricicada otagoensis macewani Deitz 1979: 25 (listed) New Zealand
Maoricicada otagoensis macewani Palma, Lovis and Tither 1989: 37 (type material, type data) Equals *Kikihia aprilina* New Zealand
Maoricicada otagoensis maceweni Simon, Arensburger and Chambers 1999: 1 (comp. note) New Zealand
Maoricicada otagoensis macewani Buckley, Simon and Chambers 2001a: 68–69 (comp. note) New Zealand
Maoricicada otagoensis maceweni Buckley, Cordeiro, Marshall and Simon 2006: 412, 417–418, 422, Figs 3–4, Table 1, Table 7 (phylogeny, comp. note) New Zealand
Maoricicada otagoensis maceweni Buckley, Simon and Chambers 2006: 68–69 (phylogeny, comp. note) New Zealand
Maoricicada otagoensis maceweni Buckley and Simon 2007: 422–423, 425, 428, Figs 1, Fig 4, Tables 1–2 (phylogeny, comp. note) New Zealand

***M. otagoensis otagoensis* Dugdale & Fleming, 1978** = *Maoricicada* sp. O Fleming, 1975
Maoricicada otagoensis otagoensis Deitz 1979: 25 (listed) New Zealand
Maoricicada otagoensis otagoensis Palma, Lovis and Tither 1989: 38 (type material, type data) New Zealand
Maoricicada otagoensis Villet 1995: 425 (acoustic behavior, comp. note) New Zealand
Maoricicada otagoensis otagoensis Buckley, Simon and Chambers 2001a: 77–78, Figs 6–7 (phylogeny, comp. note) New Zealand
Maoricicada otagoensis otagoensis Buckley, Simon, Shimodaira and Chambers 2001: 225, 228–229, Figs 2–3, Fig 5, Table 1 (phylogeny, comp. note) New Zealand
Maoricicada otagoensis Buckley, Simon, Shimodaira and Chambers 2001: 230, Fig 6 (phylogeny, comp. note) New Zealand
Maoricicada otagoensis Chambers, Boon, Buckley and Hitchmough 2001: 381, Fig 4 (phylogeny, comp. note) New Zealand
Maoricicada otagoensis otagoensis Guindon and Gascuel 2002: 542 (phylogeny, comp. note)
Maoricicada otagoensis otagoensis Buckley, Cordeiro, Marshall and Simon 2006: 412, 417–418, 422, Figs 3–4, Table 1, Table 7 (phylogeny, comp. note) New Zealand
Maoricicada otagoensis Buckley, Cordeiro, Marshall and Simon 2006: 419–420, Fig 5 (phylogeny, sonogram, comp. note) New Zealand
Maoricicada otagoensis otagoensis Buckley, Simon and Chambers 2006: 77–78, Figs 6–7 (phylogeny, comp. note) New Zealand
Maoricicada otagoensis otagoensis Buckley and Simon 2007: 422–423, 425, 428, Fig 1, Fig 4, Tables 1–2 (phylogeny, comp. note) New Zealand
Maoricicada otagoensis Buckley and Simon 2007: 426–427, 429, Figs 2–3, Tables 1–2 (phylogeny, comp. note) New Zealand

***M. phaeoptera* Dugdale & Fleming, 1978** = *Maoricicada* sp. P Fleming, 1975
Maoricicada phaeoptera Deitz 1979: 25 (listed) New Zealand
Maoricicada phaeoptera Palma, Lovis and Tither 1989: 38 (type material, type data) New Zealand
Maoricicada phaeoptera Buckley, Simon and Chambers 2001a: 76–78, Figs 6–7 (phylogeny, comp. note) New Zealand
Maoricicada phaeoptera Buckley, Simon and Chambers 2001b: 1397–1398, 1401, Fig 4, Table 1 (phylogeny, comp. note) New Zealand

Maoricicada phaeoptera Buckley, Simon, Shimodaira and Chambers 2001: 225, 228–230, Figs 2–3, Figs 5–6, Table 1 (phylogeny, comp. note) New Zealand
Maoricicada phaeoptera Chambers, Boon, Buckley and Hitchmough 2001: 381, Fig 4 (phylogeny, comp. note) New Zealand
Maoricicada phaeoptera Guindon and Gascuel 2002: 541–542 (phylogeny, comp. note)
Maoricicada phaeoptera Buckley, Cordeiro, Marshall and Simon 2006: 412, 417–420, 422, Figs 3–5, Table 1, Table 7 (phylogeny, sonogram, comp. note) New Zealand
Maoricicada phaeoptera Buckley, Simon and Chambers 2006: 76–78, Figs 6–7 (phylogeny, comp. note) New Zealand
Maoricicada phaeoptera Buckley and Simon 2007: 422–428, Figs 1–4, Tables 1–2 (phylogeny, comp. note) New Zealand
Maoricicada phaeoptera Hill, Simon, Marshall and Chambers 2007: 678 (comp. note) New Zealand

***M. tenuis* Dugdale & Fleming, 1978** = *Maoriciada* sp. T Fleming, 1975
Maoricicada tenuis Deitz 1979: 25 (listed) New Zealand
Maoricicada tenuis Buckley, Simon and Chambers 2001a: 71, 76–78, Figs 6–7 (phylogeny, comp. note) New Zealand
Maoricicada tenuis Buckley, Simon, Shimodaira and Chambers 2001: 224–225, 228–230, Figs 2–3, Figs 5–6, Table 1 (phylogeny, comp. note) New Zealand
Maoricicada tenuis Chambers, Boon, Buckley and Hitchmough 2001: 381, Fig 4 (phylogeny, comp. note) New Zealand
Maoricicada tenuis Guindon and Gascuel 2002: 541–542 (phylogeny, comp. note)
Maoricicada tenuis Buckley, Cordeiro, Marshall and Simon 2006: 412, 417–422, Figs 3–5, Table 1, Table 7 (phylogeny, sonogram, comp. note) New Zealand
Maoricicada tenuis Buckley, Simon and Chambers 2006: 71, 76–78, Figs 6–7 (phylogeny, comp. note) New Zealand
Maoricicada tenuis Buckley and Simon 2007: 422–428, Figs 1–4, Tables 1–2 (phylogeny, comp. note) New Zealand
Maoricicada tenuis Simon 2009: 103, Fig 1d (illustrated, comp. note) New Zealand

***Mouia* Distant, 1920**
Mouia Boer and Duffels 1996b: 309 (distribution, comp. note) New Caledonia
Mouia Moulds 2005b: 390, 436 (higher taxonomy, listed, comp. note)

***M. variabilis* variabilis Distant, 1920**
Mouia variabilis Boulard 2002b: 193, 206 (coloration, comp. note) New Caledonia

***M. variabilis* var. *a* Distant, 1920**

***M. variabilis* var. *b* Distant, 1920**

***M. variabilis* var. *c* Distant, 1920**

***Buyisa* Distant, 1907**
Buyisa Moulds 2005b: 390, 436 (higher taxonomy, listed, comp. note)

***B. umtatae* Distant, 1907**
Buyisa umtatae Gess 1981: 4, 8 (described, song, habitat, comp. note) South Africa

***Fijipsalta* Duffels, 1988**
Fijipsalta Duffels 1988d: 7, 9, 12, 25, 100, Table 1 (n. gen., described, illustrated, distribution, listed, comp. note) Fiji
Fijipsalta Boer and Duffels 1996b: 306 (distribution, comp. note)
Fijipsalta Moulds 2005b: 390, 436 (higher taxonomy, listed, comp. note)
Fijipsalta Cranston 2009: 93 (distribution, comp. note) Fiji
Fijipsalta Popple and Emery 2010: 147, 150 (comp. note) Fiji

***F. tympanistria* (Kirkaldy, 1907)** = *Cicadetta tympanistria* = *(?) Cicadetta tympanistria*
(?) Cicadetta tympanistria Duffels 1986: 328 (comp. note) Fiji Islands
Cicadetta tympanistria Duffels 1988d: 9, 102–103 (type species of *Fijipsaltria*, comp. note) Fiji Islands
Fijipsalta tympanistria Duffels 1988d: 9, 12, 26, 28, 100–105, Figs 197–207, Table 1 (n. comb., type species of *Fijipsaltria*, key, described, illustrated, distribution, listed, comp. note) Equals *Cicadetta tympanistria* Fiji Islands
Fijipsalta tympanistria Boer and Duffels 1996b: 305, 327, Plate 3 (distribution, illustrated, comp. note) Fiji Islands
Cicadetta tympanistria Sanborn 1999a: 52 (type material, listed)
Fijipsalta tympanistria Wilson 2009: 40, Appendix (listed, distribution, comp. note) Fiji

***Myersalna* Boulard, 1988**
Myersalna Boulard 1988c: 64–65 (n. gen., described, comp. note)
Myersalna Boulard 1992a: 119 (listed)
Myersalna Boer and Duffels 1996b: 309 (distribution, comp. note) New Caledonia
Myersalna Boulard 1997a: 185 (comp. note)
Myersalna Boulard 1999a: 112 (comp. note)
Myersalna Moulds 2005b: 390, 436 (higher taxonomy, listed, comp. note)

M. bigoti Boulard, 1988 to *Myersalna depicta* (Distant, 1920)

***M. depicta* (Distant, 1920)** = *Melampsalta depicta* = *Cicadetta depicta* = *Myersalna bigoti* Boulard, 1988

Myersalna bigoti nom. nud. Bigot 1985: 327 (distribution) New Caledonia

Myersalna bigoti Boulard 1988c: 65–66, Figs 6–11 (n. sp., type species of genus *Myersalna*, described, illustrated, comp. note) New Caledonia

Myersalna depicta Boulard 1992a: 120 (n. comb., synonymy, comp. note) Equals *Melampsalta depicta* Equals *Myersalna bigoti* New Caledonia

Myersalna depicta Boer and Duffels 1996b: 309 (distribution, comp. note) New Caledonia

Myersalna depicta Buckley, Arensburger, Simon and Chambers 2002: 6, 11–12, Fig 1 (phylogeny, comp. note) New Caledonia

Myersalna depicta Arensburger, Buckley, Simon, Moulds, and Holsinger 2004: 558–559, 561–563, Table 1, Figs 1–2 (molecular phylogeny, comp. note) New Caledonia

Myersalna depicta Buckley and Simon 2007: 423 (comp. note) New Caledonia

Myersalna depicta Simon 2009: 104, Fig 2 (phylogeny, comp. note) New Caledonia

***M. dumbeana* Distant, 1920** = *Melampsalta dumbeana* = *Cicadetta dumbeana*

Myersalna dumbeana Boulard 1992a: 120 (n. comb.) Equals *Melampsalta dumbeana* New Caledonia

***M. flavocaerulea* Boulard, 1992**

Myersalna flavocaerulea Boulard 1992a: 120–122, Figs 1–6 (n. sp., described, illustrated, comp. note) New Caledonia

Myersalna flavocaerulea Boer and Duffels 1996b: 309 (distribution, comp. note) New Caledonia

***Rouxalna* Boulard, 1999**

Rouxalna Boulard 1999a: 111–112 (n. gen., described) New Caledonia

***R. rouxi* (Distant, 1914)** = *Cicadetta rouxi* = *Melampsalta rouxi*

(Melampsalta) rouxi Bigot 1985: 327 (distribution) New Caledonia

Rouxalna rouxi Boulard 1999a: 80, 111–112 (n. comb., listed, sonogram) Equals *Melampsalta rouxi* Equals *Cicadetta rouxi* New Caledonia

Rouxalna rouxi Sueur 2001: 44, Table 1 (listed)

Rouxalna rouxi Boulard 2006g: 181 (listed)

***Poviliana* Boulard, 1997**

Poviliana Boulard 1997a: 185, 195 (n. gen., described, comp. note) New Caledonia

Poviliana Moulds 2005b: 390, 436 (higher taxonomy, listed, comp. note)

***P. sarasini* (Distant, 1914)**

Melampsalta sarasini Boulard 1997a: 179 (comp. note)

Poviliana sarasini Boulard 1997a: 185–188, 195, Figs 11–19, Fig 36A (n. comb., type species of *Poviliana*, described, illustrated, sonogram, comp. note) Equals *Melamspalta sarasini* New Caledonia

Poviliana sarasani Sueur 2001: 44, Table 1 (listed)

Poviliana sarasini Boulard 2005d: 338, 341 (comp. note) New Caledonia

Poviliana sarasini Boulard 2006g: 36–37, 39–40, 72–73, 82, 181, Fig 7e, Fig 38 (illustrated, sonogram, listed, comp. note) New Caledonia

***P. vincentiensis* Boulard, 1997**

Poviliana vincentiensis Boulard 1997a: 188–191, 195, Figs 20–27, Fig 36B (n. sp., described, illustrated, sonogram, comp. note) New Caledonia

Poviliana vincentiensis Sueur 2001: 44, Table 1 (listed)

Poviliana vincentiensis Boulard 2006g: 48–49 (comp. note) New Caledonia

***Stellenboschia* Distant, 1920**

Stellenboschia Moulds 2005b: 390, 436 (higher taxonomy, listed, comp. note)

***S. rotundata* (Distant, 1892)** = *Melampsalta rotundata* = *Pauropsalta rotundata*

***Katoa* Ouchi, 1938** = *Katao* (sic) = *Lisu* Liu, 1940

Lisa Liu 1992: 42 (comp. note)

Lisu Moulds 1996a: 20 (comp. note)

Katoa Chou, Lei, Li, Lu and Yao 1997: 89–90, 293, 357, Table 6 (listed, described, synonymy, diversity, biogeography) Equals *Lisu* China

Lisu Chou, Lei, Li, Lu and Yao 1997: 357 (comp. note)

Lisu Hayashi 1998: 62 (comp. note)

Katoa Hayashi 1998: 62 (comp. note)

Katoa Villet and Zhao 1999: 322 (comp. note) Equals *Lisu*

Lisu Pham 2005a: 216 (comp. note)

Katoa Pham 2005a: 216 (comp. note)

Katoa Pham and Yang 2009: 4, 15, Table 2 (listed, synonymy, comp. note) Equals *Lisu* Vietnam

***K. chlorotica* Chou & Lu, 1997** = *Katoa chlorotiea* (sic) = *Katoa* sp. 1 Pham & Thinh, 2006 = *Katoa* sp. 2 Pham & Thinh, 2006

Katoa chlorotica Chou, Lei, Li, Lu and Yao 1997: 90, 94–96, 357, 366, Fig 8-29, Plate II, Fig 31 (key, described, illustrated, comp. note) China

Katoa chlorotiea (sic) Pham and Thinh 2006: 526–527 (listed, comp. note) Vietnam
Katoa sp. 1 Pham and Thinh 2006: 526 (listed) Vietnam
Katoa sp. 2 Pham and Thinh 2006: 526 (listed) Vietnam
Katoa chlorotica Pham and Yang 2009: 4–5 (listed, distribution) Vietnam, China
Katoa chlorotiea (sic) Pham and Yang 2009: 15–16, Table 2 (listed, comp. note) Vietnam

***K. neokanagana* (Liu, 1940)** = *Lisu neokanagana*
Lisa neokanagana Liu 1992: 42 (type species of *Lisu*, comp. note)
Katoa neokanagana Chou, Lei, Li, Lu and Yao 1997: 19, 90–92, 96, 366, Fig 3-8B, Fig 8-26, Plate II, Fig 25 (key, described, illustrated, synonymy, comp. note) Equals *Lisu neokanagana* China
Lisu neokanagana Hayashi 1998: 62 (comp. note)
Katao neokanagana Hua 2000: 62 (listed, distribution) Equals *Lisu neokanagana* China, Sichuan

***K. paucispina* Lei & Chou, 1995**
Katoa paucispina Lei and Chou 1995: 99–100, 102, Fig 1 (n. sp., described, illustrated, comp. note) China, Yunnan
Katoa paucispina Chou, Lei, Li, Lu and Yao 1997: 90, 93–94, Fig 8-28, Plate II, Fig 21 (key, described, illustrated, comp. note) China

***K. paura* Chou & Lu, 1997**
Katoa paura Chou, Lei, Li, Lu and Yao 1997: 90, 96–97, 357, 366, Fig 8-30 (n. sp., key, described, illustrated, comp. note) China

***K. taibaiensis* Chou & Lei, 1995**
Katoa taibaiensis Lei and Chou 1995: 100–102, Fig 2 (n. sp., described, illustrated, comp. note) China, Yunnan, Hainan
Katoa taibaiensis Chou, Lei, Li, Lu and Yao 1997: 90, 92–93, Fig 8-27, Plate II, Fig 23 (key, described, illustrated, comp. note) China

***K. tenmokuensis* Ouchi, 1938** = *Katao* (sic) *tenmokuensis*
Katoa tenmokuensis Chou, Lei, Li, Lu and Yao 1997: 89–91 (type species of *Katoa*, key, described, comp. note) China
Katoa tenmokuensis Hayashi 1998: 62 (comp. note)
Katao tenmokuensis Hua 2000: 62 (listed, distribution) China, Zhejiang
Katoa tenmokuensis Pham and Yang 2009: 4 (type species of *Katoa*)

Species of uncertain position in Cicadettini

***Cicadetta* sp. Soper, 1981**
Cicadetta sp. Soper 1981: 53 (fungal host, comp. note) Australia

***Cicadetta* sp. n. Ewart, 1986**
Cicadetta sp. nov. Ewart 1986: 51, Table 1 (listed) Queensland

***Cicadetta* sp. nov. 1 Lithgow, 1988**
Cicadetta sp. nov. 1 Lithgow 1988: 65 (listed) Australia, Queensland

***Cicadetta* sp. nov. 2 Lithgow, 1988**
Cicadetta sp. nov. 2 Lithgow 1988: 65 (listed) Australia, Queensland

***Cicadetta* sp. nov. 3 Lithgow, 1988**
Cicadetta sp. nov. 3 Lithgow 1988: 65 (listed) Australia, Queensland

***Cicadetta* sp. nov. 4 Lithgow, 1988**
Cicadetta sp. nov. 4 Lithgow 1988: 65 (listed) Australia, Queensland

***Cicadetta* sp. nov. 5 Lithgow, 1988**
Cicadetta sp. nov. 5 Lithgow 1988: 65 (listed) Australia, Queensland

***Cicadetta* sp. A, Gidyea Cicada Ewart & Popple, 2001**
Species A (Gidyea cicada) Ewart and Popple 2001: 57, 60, 66, 70, Fig 3A, Fig 8A (illustrated, described, habitat, song, oscillogram) Queensland
Cicadetta sp. A Gidyea Cicada Popple and Strange 2002: 28, Table 1 (listed, habitat) Australia, Queensland
Cicadetta sp. A Ewart 2009b: 138, 148, 165, Fig 13 (oscillogram, distribution, habitat, comp. note) Equals Gidyea Cicada, no. 207 Equals Species A Ewart and Popple, 2001 Queensland
Gidyea Cicada Ewart 2009b: 141, 147, Fig 1, Plate 2Q (illustrated, distribution, emergence times, habitat, comp. note) Queensland

***Cicadetta* sp. A, Brown Sandplain Cicada Ewart, 2009**
Cicadetta sp. A Ewart 2009b: 143, 156, 158, Fig 4 (oscillogram, distribution, habitat, comp. note) Equals Brown Sandplain Cicada, no. 253 Queensland, Northern Territory
Brown Sandplain Cicada Ewart 2009b: 141–142, Fig 1, Plate 1C (illustrated, distribution, emergence times, habitat, comp. note) Queensland, Northern Territory

***Cicadetta* sp. B Ewart, 2009** = Cravens Buzzing Cicada Ewart, 2009
Cicadetta sp. B Ewart 2009b: 145, 159, Fig 7 (oscillogram, distribution, habitat, comp. note) Equals Cravens Buzzing Cicada, no. 204 Queensland
Cravens Buzzing Cicada Ewart 2009b: 141–142, 145, Fig 1, Plate 1F (illustrated, distribution, emergence times, habitat, comp. note) Queensland

***Cicadetta* sp. F Ewart, 2009** = Species F (Spinifex rattler) Ewart and Popple 2001

Species F (Spinifex rattler) Ewart and Popple 2001: 64, 66–67, 71, Fig 6D, Fig 9C (illustrated, described, habitat, song) Queensland

Cicadetta sp. F Ewart 2009b: 144, 158, Fig 6 (oscillogram, distribution, habitat, comp. note) Equals Spinifex Rattler, no. 259 Equals Species F Ewart and Popple, 2001 Queensland, Northern Territory

Spinifex Rattler Ewart 2009b: 141–142, Fig 1, Plate 1D (illustrated, distribution, emergence times, habitat, comp. note) Queensland, Northern Territory

***Cicadetta* sp. H Capparis Cicada (Popple & Strange, 2002)** = *Cicadetta* sp. H Popple, 2003 = *Cicadetta* sp. K Ewart, 2009

Cicadetta sp. H Capparis Cicada Popple and Strange 2002: 18–19, 24, 29, Fig 1C, Fig 5C, Table 1 (described, illustrated, habitat, oscillogram, listed, comp. note) Australia, Queensland

Cicadetta sp. H Popple 2003: 107, 114 (comp. note) Australia

Cicadetta sp. K Ewart 2009b: 143, 155, Fig 3D (oscillogram, comp. note) Equals Capparis Cicada, no. 219 Queensland

Capparis Cicada Ewart 2009b: 141 (comp. note) Queensland

***Cicadetta* sp. I Pink-wing Cicada Popple & Strange, 2002**

Cicadetta sp. I Pink-wing Cicada Popple and Strange, 2002: 19–20, 25, 29, Fig 2A, Fig 6A, Table 1 (described, illustrated, habitat, listed, comp. note) Australia, Queensland

***Cicadetta* sp. I Ewart, 2009** = Small Acacia Cicada

Cicadetta sp. I Ewart 2009b: 146, 161, Fig 9 (oscillogram, distribution, habitat, comp. note) Equals Small Acacia Cicada, no. 206 Queensland, Northern Territory, Western Australia, South Australia

Small Acacia Cicada Ewart 2009b: 140–141, 147, Fig 1, Plate 2R (illustrated, distribution, emergence times, habitat, comp. note) Queensland, Northern Territory, Western Australia, South Australia

***Cicadetta* sp. J Ewart, 2009** = Roaring Senna Cicada Ewart, 2009

Cicadetta sp. J Ewart 2009b: 143, 155, Figs 3A–C (oscillogram, distribution, habitat, comp. note) Equals Roaring Senna Cicada, no. 210 Queensland

Roaring Senna Cicada Ewart 2009b: 141–143, 145, Fig 1, Plate 1E (illustrated, distribution, emergence times, habitat, comp. note) Queensland

***Cicadetta* sp. K Ewart, 2009** = Capparis Cicada Ewart, 2009

Cicadetta sp. K Ewart 2009b: 143, 155, Fig 3D (oscillogram, comp. note) Equals Capparis Cicada, no. 219 Queensland

Capparis Cicada Ewart 2009b: 141 (comp. note) Queensland

***Cicadetta* sp. M Ewart, 2009** = *Notopsalta* sp. B Ewart 1988 = Copper Shrub Buzzer Ewart, 2009

Notopsalta sp. B Ewart and Popple 2001: 56, 61, 65, Fig 2B (described, habitat, song, oscillogram) Equals *Notopsalta* sp. B Ewart, 1988, 1998 Queensland

Cicadetta sp. M Ewart 2009b: 151, 169, Fig 18 (oscillogram, distribution, habitat, comp. note) Equals Copper Shrub Buzzer, no. 278 Equals *Notopsalta* sp. B Ewart and Popple, 2001 Queensland

Copper Shrub Buzzer Ewart 2009b: 140–141, 147, Fig 1, Plate 2N (illustrated, distribution, emergence times, habitat, comp. note) Queensland

***Cicadetta* sp. nov. Small Robust Ambertail Popple & Strange, 2002**

Cicadetta sp. nov. Small Robust Ambertail Popple and Strange 2002: 30, Table 1 (listed, habitat) Australia, Queensland

***Cicadetta* sp. nov. Small Treetop Ticker Popple & Strange, 2002**

Cicadetta sp. nov. Small Treetop Ticker Popple and Strange 2002: 30, Table 1 (listed, habitat) Australia, Queensland

***Melampsalta alanubilis* Goding & Froggatt nom. nud. in Stevens and Carver 1986**

Melampsalta alanubilis Hahn 1962: 8 (listed) Australia

Melampsalta alanubilis Goding & Froggatt nom. nud. Stevens and Carver 1986: 264 (type material) Australia

Cicadettini sp. a Phillips, Sanborn & Villet, 2002

Cicadettini sp. a Phillips, Sanborn and Villet 2002: 31 (thermal responses) South Africa, Eastern Cape

Cicadettini sp. a Sanborn, Phillips and Villet 2003a: 348–349, Table 1 (thermal responses, habitat, comp. note) South Africa

Cicadettini sp. b Phillips, Sanborn & Villet, 2002

Cicadettini sp. b Phillips, Sanborn and Villet 2002: 31 (thermal responses) South Africa, Eastern Cape

Cicadettini sp. b Sanborn, Phillips and Villet 2003a: 348–349, Table 1 (thermal responses, habitat, comp. note) South Africa

Cunnamulla Smoky Buzzer Ewart, 2009

Cunnamulla Smoky Buzzer, no. 295 Ewart 2009b: 144, 157, Figs 5C–D (oscillogram, comp. note) Queensland, New South Wales

Jundah Grass Ticker Ewart, 2009

Jundah Grass Ticker, no. 344 Ewart 2009b: 141, 147, 148, 164, Figs 12C–D, Plate 2M (illustrated, oscillogram, comp. note) Queensland

Melaleuca sp. A Ewart, 2005
Melaleuca sp. A Ewart 2005a: 171–172, 178, Fig 3B (illustrated, described, natural history, song, oscillogram) Queensland

N. gen. n. sp. Phillips, Sanborn & Villet, 2002
n. gen., n. sp. Phillips, Sanborn and Villet 2002: 31 (thermal responses) South Africa, Eastern Cape
n. gen. n. sp. Sanborn, Phillips and Villet 2003a: 349, Table 1 (thermal responses, habitat, comp. note) South Africa

N. Gen. "Kynuna" Marshall & Hill, 2009
N. Gen. "Kynuna" Marshall and Hill 2009: 4–5, Fig 6E, Table 2 (sonogram, comp. note)

N. Gen. "northwestern" Marshall & Hill, 2009
N. Gen. "northwestern" Marshall and Hill 2009: 4, Table 2 (comp. note)

N. Gen. "Nullarbor wingbanger" Marshall & Hill, 2009
N. Gen. "Nullarbor wingbanger" Marshall and Hill 2009: 4–5, Fig 6B, Table 2 (sonogram, comp. note)

N. Gen. "pale grass cicada" Marshall & Hill, 2009
N. Gen. "pale grass cicada" Marshall and Hill 2009: 4–5, 7–8, Fig 6F, Table 2 (sonogram, comp. note) Australia
Pale grass cicada Pain 2009: 47 (sonogram) Australia

N. Gen. "near pale grass cicada" Marshall & Hill, 2009
N. Gen. "near pale grass cicada" Marshall and Hill 2009: 4, Table 2 (comp. note) Australia

N. Gen. "revving tigris" Marshall & Hill, 2009
N. Gen. "revving tigris" Marshall and Hill 2009: 4, Table 2 (comp. note)

N. Gen. "swinging tigris" Marshall & Hill, 2009
N. Gen. "swinging tigris" Marshall and Hill 2009: 4–5, Fig 6L, Table 2 (comp. note) Australia

N. Gen. "tigris H2" Marshall & Hill, 2009
N. Gen. "tigris H2" Marshall and Hill 2009: 4, Table 2 (comp. note)

N. Gen. "troublesome tigris" Marshall & Hill, 2009
N. Gen. "troublesome tigris" Marshall and Hill 2009: 4–6, Fig 6N, Table 2 (comp. note) Australia

Tribe Prasiini Matsumura, 1917 = Prasini (sic) = Prassini (sic)

Prasiini Medler 1980: 73 (listed) Nigeria
Prasiini Jong and Duffels 1981: 53, 55 (diversity, comp. note) Philippines, Sulawesi, ana
Prasiini Jong 1982: 175, 182, 184 (comp. note)
Prasiini Jong 1985: 165–166 (diversity, comp. note) Philippines, Sanghir, Sulawesi, Java, Lesser Sunda Islands, Misoöl, New Guinea, Queensland, Australia
Prasiini Duffels 1986: 325, 333 (listed, comp. note)
Prasiini Jong 1986: 141–143, 173–174 (comp. note)
Prasiini Jong 1987: 177 (comp. note)
Prasiini Duffels 1990b: 64, 68–69, Table 1 (listed, comp. note)
Prasiini Moulds 1990: 31 (listed) Australia
Prasiini Boulard 1993d: 91 (comp. note)
Prasiini Boer 1995a: 2, Fig 1 (phylogeny, comp. note)
Prasiini Boer 1995b: 3 (comp. note)
Prasiini Boer 1995c: 203–207, 210, Fig 1 (phylogeny, comp. note)
Prasiini Boer 1995d: 172–174, 202–203, 222–225, 228–229, 232, 237, Fig 28, Figs 52–53 (distribution, diversity, phylogeny, comp. note) Sulawesi, Maluku, New Guinea
Prasiini Boer 1996: 351–354, Fig 1 (comp. note)
Prasiini Boer and Duffels 1996a: 155–157, 163, 167, 169–172, Fig 2, Figs 10–11 (distribution, diversity, phylogeny, comp. note)
Prasiini Boer and Duffels 1996b: 298, 301–304, 306–308, 313–315, Figs 2–3, Fig 5 (distribution, diversity, phylogeny, comp. note) Sulawesi, Java, Sumatra, Borneo, Lesser Sunda Islands, New Guinea, Queensland
Prasiini Boulard 1996d: 115 (listed)
Prasiini Boer 1997: 92–93, 113, Fig 1 (comp. note) Sulawesi
Prasiini Holloway 1997: 219–220, 222–223, Figs 5–6 (distribution, phylogeny, diversity, comp. note) Sulawesi, Banda Islands
Prasiini Hayashi 1998: 62 (comp. note)
Prasiini Holloway 1998: 304, 307–309, Fig 13 (distribution, phylogeny, comp. note) Sulawesi
Prasiini Holloway and Hall 1998: 9–12, Fig 3 (distribution, phylogeny, comp. note)
Prasiini Boer 1999: 115–116, 118 (comp. note) Sulawesi, New Guinea, Maluku
Prasiini Boulard and Riou 1999: 136 (listed)
Prasiini Moulds 1999: 1 (comp. note)
Prasiini Villet and Noort 1999: 226, Table 12.1 (diversity, listed) Tanzania, Kenya, South Africa, Botswana, Zimbabwe, Angola
Prasiini Boulard 2002b: 205–206 (comp. note)
Prasiini Moulds and Carver 2002: 49 (listed) Australia
Prasiini Jong and Boer 2004: 265–267, 269, Fig 1 (phylogeny, comp. note)
Prasiini Moulds 2005b: 386, 389, 393–394, 396, 412, 414, 419, 422, 424, 427, 430, 435–437, Fig 58, Fig 61, Table 2 (described, phylogeny, composition, key, distribution, listed, comp. note) Equals Prasinaria Equals Prasiinae (sic) Equals Prasini (sic) Australia
Prasiini Shiyake 2007: 9 (listed, comp. note)
Prassini (sic) Shiyake 2007: 114 (listed)

Subtribe Prasiina Matsumura, 1917

***Arfaka* Distant, 1905**

Arfaka Jong and Duffels 1981: 55 (diversity, comp. note) New Guinea
Arfaka Jong 1985: 165 (diversity, comp. note)
Arfaka Duffels 1986: 325–326, 330 (distribution, diversity, comp. note) New Guinea, Misöol
Arfaka Boer 1989: 7 (comp. note)
Arfaka Boer 1990: 65 (comp. note)
Arfaka Duffels 1990b: 68 (comp. note) New Guinea, Misöol
Arfaka Duffels and Boer 1990: 261 (distribution, diversity, comp. note) New Guinea, Misoöl
Arfaka Boer 1993b: 144–146 (phylogeny, comp. note)
Arfaka Boer 1995c: 203, 206–207, Fig 2 (phylogeny, comp. note) New Guinea, Waigeu, Salawati
Arfaka Boer 1995d: 200–201, 221–222, 225, 228–229, 231–232, Fig 29, Fig 54 (distribution, phylogeny, comp. note) New Guinea
Arfaka Boer and Duffels 1996a: 155, 167 (distribution, comp. note) New Guinea
Arfaka Boer and Duffels 1996b: 301, 306, 327 (distribution, comp. note) New Guinea
Arfaka Boer 1997: 113 (comp. note)
Arfaka Holloway 1997: 219–220, 222, Fig 5 (distribution, phylogeny, comp. note) New Guinea
Arfaka Holloway 1998: 308–309, Fig 13 (distribution, comp. note) New Guinea
Arfaka Jong and Boer 2004: 266–267, 269, Fig 1 (phylogeny, comp. note)
Arfaka Moulds 2005b: 393, 436 (higher taxonomy, listed, comp. note)
Arfaka Duffels and Boer 2006: 533–534 (distribution, diversity, comp. note) New Guinea, Misool

***A. fulva* (Walker, 1870)** = *Cephaloxys fulva* = *Mogannia fulva*

Arfaka fulva Jong 1985: 166 (comp. note)
Arfaka fulva Boer 1995d: 221 (distribution, comp. note) New Guinea
Arfaka fulva Boer and Duffels 1996b: 327 (distribution) New Guinea

***A. hariola* (Stål, 1863)** = *Prasia hariola*

Prasia hariola Jong 1985: 165–166 (comp. note) Misoöl
Arfaka hariola Jong 1985: 166 (n. comb., comp. note) Misoöl, New Guinea
Arfaka hariola Boer 1995d: 221, 231 (distribution, comp. note) New Guinea, Misool
Arfaka hariola Boer and Duffels 1996b: 327 (distribution) New Guinea

***Sapatanga* Distant, 1905** = *Spatanga* (sic)

Sapantanga Jong and Duffels 1981: 55 (diversity, comp. note)
Sapantanga Jong 1985: 165 (diversity, comp. note) South America
Sapantanga Boer 1995c: 205 (comp. note) South America
Sapantanga Boer 1997: 113 (comp. note)
Sapantanga Moulds 2005b: 393, 436 (higher taxonomy, listed, comp. note)

***S. nutans* (Walker, 1850)** = *Cephaloxys nutans* = *Selymbria nutans* = *Selymbia* (sic) *nutans*

***Jacatra* Distant, 1905**

Jacatra Jong and Duffels 1981: 55 (diversity, comp. note) Java
Jacatra Jong 1985: 165 (comp. note)
Jacatra Duffels 1986: 319, 325 (distribution, diversity, comp. note) Java
Jacatra Boulard 1988d: 154–156, Figs 16–17 (wing abnormalities, illustrated, comp. note) Sumatra
Jacatra Duffels 1990b: 68 (comp. note)
Jacatra Boer 1993b: 145 (comp. note)
Jacatra Boer 1995c: 203, 206, Fig 2 (phylogeny, comp. note) Java, Sumatra
Jacatra Boer 1995d: 201, 221–222, 225, 228, 231, Fig 29, Fig 54 (distribution, phylogeny, comp. note) Java, Sumatra
Jacatra Boer and Duffels 1996a: 155, 167 (distribution, comp. note) Java, Sumatra
Jacatra Boer and Duffels 1996b: 301, 327 (distribution, comp. note) Java, Sumatra
Jacatra Boer 1997: 113 (comp. note)
Jacatra Holloway 1997: 219–220, Fig 5 (distribution, phylogeny, comp. note) Java, Sumatra
Jacatra Holloway 1998: 308, Fig 13 (distribution, comp. note) Java, Sumatra
Jacatra Jong and Boer 2004: 266–267, 269, Fig 1 (phylogeny, comp. note)
Jacatra Moulds 2005b: 393, 436 (higher taxonomy, listed, comp. note)

***J. typica* Distant, 1905**

Jacatra typica Jong 1986: 143 (comp. note)
Jacatra typica Duffels 1990b: 68 (comp. note) Java
Jacatra typica Boer 1995d: 221 (distribution, comp. note) Java, Sumatra
Jacatra typica Boer and Duffels 1996b: 301, 327 (distribution, comp. note) Java, Sumatra

***Mariekea* Jong & Boer, 2004**

Mariekea Jong and Boer 2004: 266–269, 271, Fig 1 (n. gen., described, phylogeny, key, comp. note) Sumba

***M. acuta* Jong, 2004**
Mariekea acuta Jong and Boer 2004: 267, 269, 275–277, Fig 2, Figs 24–28 (n. sp., described, illustrated, key, distribution, comp. note) Flores, Sumba

***M. euharderi* Jong, 2004**
Mariekea euharderi Jong and Boer 2004: 267–269, 271, 274–275, Figs 2–4, Fig 6, Figs 19–23 (n. sp., described, illustrated, key, distribution, comp. note) Sumba

***M. floresiensis* Jong, 2004**
Mariekea floresiensis Jong and Boer 2004: 267–269, 271, 277–279, 281, Fig 2, Figs 7–8, Figs 29–33 (n. sp., described, illustrated, key, distribution, comp. note) Flores

***M. groenendaeli* Boer, 2004**
Mariekea groenendaeli Jong and Boer 2004: 267, 269, 271, 279–281, Fig 2, Figs 34–37 (n. sp., described, illustrated, key, distribution, comp. note) Flores

***M. harderi* (Schmidt, 1925)** = *Lembeja harderi*
Lembeja harderi Jong 1982: 183 (comp. note) Lesser Sunda Islands
Lembeja harderi Jong 1986: 141, 149 (comp. note)
Lembeja harderi group Jong and Boer 2004: 265 (comp. note) Lesser Sunda Islands
Lembeja harderi group Boer 1995c: 206–207, Fig 2 (phylogeny, comp. note) Lesser Sunda Islands
Lembeja harderi Boer 1995c: 207 (comp. note) Lesser Sunda Islands
Lembeja harderi group Boer 1995d: 201, 221, 225, 228, 231, Fig 29, Fig 54 (distribution, phylogeny, comp. note)
Lembeja harderi Boer 1995d: 222 (distribution, comp. note) Lesser Sunda Islands
Lembeja harderi group Boer and Duffels 1996a: 167 (comp. note) Lesser Sunda Islands
Lembeja harderi Boer and Duffels 1996b: 327 (distribution) Lesser Sunda Islands
Lembeja harderi group Holloway 1997: 220, Fig 5 (distribution, phylogeny, comp. note) Lesser Sunda Islands
Lembeja harderi group Holloway 1998: 308, Fig 13 (distribution, comp. note) Lesser Sunda Islands
Lembeja harderi Jong and Boer 2004: 266 (type species of *Lembeja*)
Mariekea harderi Jong and Boer 2004: 266–271, 273, Fig 2, Fig 5, Figs 9–13 (n. comb., described, illustrated, key, distribution, comp. note) Equals *Lembeja harderi* Lombok, Sumba, Sumbawa

***M. major* Jong, 2004**
Mariekea major Jong and Boer 2004: 267, 269, 272–273, 275, 277, Fig 2, Figs 14–18 (n. sp., described, illustrated, key, distribution, comp. note) Sumba

***Prasia* Stål, 1863** = *Drepanopsaltria* Breddin, 1901 = *Drepanopsatra* (sic) *(Lembeja)* = *Parasia* (sic)
Prasia Jong and Duffels 1981: 53, 61 (diversity, comp. note) Sulawesi, New Guinea, Mysol
Prasia Jong 1982: 182–183 (comp. note)
Drepanopsaltria Jong 1982: 182 (comp. note)
Prasia Duffels 1983a: 492 (comp. note)
Prasia Strümpel 1983: 17 (listed)
Prasia Hayashi 1984: 69 (listed)
Prasia Jong 1985: 165–168, 178 (described, distribution, diversity, synonymy, key, comp. note) Equals *Drepanopsaltria* Sulawesi, Muna Island
Drepanopsaltria Jong 1985: 166 (comp. note)
Prasia Duffels 1986: 325–326, 329, Fig 8 (distribution, diversity, comp. note) Sulawesi, Muna Island
Prasia Jong 1986: 141–142 (comp. note)
Prasia Jong 1987: 177 (comp. note)
Prasia Duffels 1990b: 64, 68, Table 1 (diversity, distribution, comp. note) Sulawesi
Prasia Duffels and Boer 1990: 256 (distribution, diversity, comp. note) Sulawesi
Prasia Moulds 1990: 178 (comp. note)
Prasia Boer 1993b: 145 (comp. note)
Prasia Boer 1994a: 2 (comp. note)
Prasia Boer 1995b: 3, 8 (comp. note)
Prasia Boer 1995c: 203, 206–207, 210–211, 213, 219, 224–226, 228, 230–231, Figs 2–4 (comp. note) Sulawesi, Muna
Prasia Boer 1995d: 201, 221–222, 225, 235, Fig 29, Fig 54 (distribution, phylogeny, comp. note) New Guinea
Prasia Boer 1996: 353 (comp. note)
Prasia Boer and Duffels 1996a: 155, 167 (distribution, comp. note) Sulawesi
Prasia Boer and Duffels 1996b: 301, 306, 308, 327 (distribution, comp. note) Sulawesi
Prasia Boer 1997: 113 (comp. note)
Prasia Holloway 1997: 219–220, 224, Fig 5 (distribution, phylogeny, diversity, comp. note) Sulawesi
Prasia Holloway 1998: 305, 308, Fig 13 (distribution, comp. note) Sulawesi
Prasia Boulard and Riou 1999: 136–137 (comp. note)
Prasia Duffels 1999: 81 (comp. note) Sulawesi
Prasia Jong and Boer 2004: 266–267, 269, Fig 1 (phylogeny, comp. note)
Prasia Moulds 2005b: 393, 412, 436 (type genus of Prasiini, higher taxonomy, listed, comp. note)

***P. breddini* Jong, 1985**
Prasia breddini Jong 1985: 166, 169–170, 174–176, 178, Fig 7, Figs 21–30 (n. sp., described, illustrated, key, distribution, comp. note) Sulawesi

Prasia breddini Boer and Duffels 1996b: 327 (distribution) Sulawesi

P. culta Distant, 1898 to *Prasia faticina* Stål, 1863

P. faticina **Stål, 1863** = *Prasia culta* Distant, 1898 = *Drepanopsaltria culta* = *Drepanopsatra* (sic) *(Lembeja) culta*

Prasia culta Jong 1982: 182 (comp. note)

Prasia faticina Jong 1985: 165–167, 169–174, 176, 178, 180–181, 189, Figs 7–20, Figs 86–87 (type species of *Prasia*, synonymy, described, illustrated, key, distribution, comp. note) Equals *Prasia culta* n. syn. Equals *Drepanopsaltria (Lembeja) culta* Equals *Drepanopsaltria culta* Sulawesi

Prasia culta Jong 1985: 166, 171, 189, Fig 86 (comp. note) Equals *Drepanopsaltria culta*

Prasia faticina Jong 1986: 141 (type species of *Prasia*, comp. note)

Prasia faticina Boer and Duffels 1996b: 327 (distribution) Sulawesi

Prasia faticina Moulds 2005b: 393, 396, 399, 417–419, 421, 424, 436, Figs 56–59, Fig 61, Table 1 (type species of *Prasia*, phylogeny, comp. note)

P. hariola Stål, 1863 to *Arfaka hariola*

P. nigropercula **Jong, 1985**

Prasia nigropercula Jong 1985: 166, 169–170, 180–182, 190, Figs 6–7, Figs 51–59, Fig 89 (n. sp., described, illustrated, key, distribution, comp. note) Muna Island

Prasia nigropercula Boer and Duffels 1996b: 327 (distribution) Sulawesi

P. princeps **Distant, 1888** = *Drepanopsaltria princeps* = *Prasia culta* (nec Distant)

Prasia princeps Jong 1982: 182 (comp. note)

Prasia princeps Jong 1985: 166, 168–171, 182–186, 190, Figs 1–5, Fig 7, Figs 60–75, Figs 90–91 (synonymy, described, illustrated, key, distribution, comp. note) Equals *Prasia culta* Lallemand, 1931 Equals *Drepanopsaltria (?) princeps* Equals *Drepanopsaltria princepsz* Sulawesi

Prasia princeps Boer and Duffels 1996b: 327 (distribution) Sulawesi

P. sarasinorum **Jong, 1985**

Prasia sarasinorum Jong 1985: 166, 169–170, 178, 182, 186–188, Fig 7, Figs 31–40, Fig 88 (n. sp., described, illustrated, key, distribution, comp. note) Sulawesi

Prasia sarasinorum Boer and Duffels 1996b: 327 (distribution) Sulawesi

P. senilirata **Jong, 1985**

Prasia senilirata Jong 1985: 166, 169–170, 174, 178–180, Fig 7, Figs 76–85 (n. sp., described, illustrated, key, distribution, comp. note) Sulawesi

Prasia senilirata Boer and Duffels 1996b: 327 (distribution) Sulawesi

P. tincta Distant, 1909 to *Lembeja tincta*

P. tuberculata **Jong, 1985**

Prasia tuberculata Jong 1985: 166, 169–170, 174, 178–180, Fig 7, Figs 41–50 (n. sp., described, illustrated, key, distribution, comp. note) Sulawesi

Prasia tuberculata Boer and Duffels 1996b: 327 (distribution) Sulawesi

P. vitticollis Ashton, 1912 to *Lembeja vitticollis*

Lacetas **Karsch, 1890** = *Lacetas (Lacetasiastes)* = *Lacetas (Lacetas)*

Lacetas Medler 1980: 73 (listed) Nigeria

Lacetas Jong and Duffels 1981: 55 (diversity, comp. note) Africa

Lacetas Boulard 1985e: 1026, 1028, 1032 (coloration, comp. note) Africa

Lacetas Jong 1985: 165 (diversity, comp. note) Africa

Lacetas Boer 1995c: 205 (comp. note) Africa

Lacetas Boer 1997: 113 (comp. note) Africa

Lacetas Boulard 1997b: 51 (morphology, comp. note) Africa

Lacetas Moulds 2005b: 393, 436 (higher taxonomy, listed, comp. note)

L. annulicornis **Karsch, 1890** = *Lacetas (Lacetas) annulicornis* = *Lacetas anulicornis* (sic)

Lacetas annulicornis Medler 1980: 73 (listed) Nigeria, Zaire

Lacetas annulicornis Boulard 1985e: 1019, 1027, 1040–1043, Plate II, Fig 16, Plate V, Fig 33, Plate VII, Fig 48 (crypsis, coloration, illustrated, comp. note) Sudan

Lacetas annulicornis Boulard 1990b: 198–199, Plate XXV, Fig 16 (coloration, illustrated, comp. note) Central African Republic

Lacetas annulicornis Boulard 1997b: 21–23, Plate III, Fig 3, Plate IV, Fig 7 (morphology, coloration, comp. note)

Lacetas anulicornis (sic) Boulard 2000b: 9 (parallel evolution)

Lacetas annulicornis Boulard 2002b: 193 (coloration, comp. note)

Lacetas annulicornis Boulard 2007f: 83, 143, Fig 99.3, Fig 170 (illustrated, comp. note)

L. breviceps **Schumacher, 1912** = *Lacetas (Lacetasiastes) breviceps*

L. jacobii **Schumacher, 1912** = *Lacetas (Lacetas) jacobii*

L. longicollis longicollis **Schumacher, 1912** = *Lacetas (Lacetas) longicollis*

Lacetas longicollis Boulard 1985e: 1027 (coloration, comp. note)

L. longicollis tendaguruensis **Schumacher, 1912** = *Lacetas (Lacetas) tendaguruensis*

***Lembeja* Distant, 1892** = *Lembejam* (sic) = *Perissoneura* Distant, 1883 (nec *Perissoneura* Mac-Lachlan, 1871)
Lembeja Jong and Duffels 1981: 53, 61 (diversity, comp. note) Philippines, Sanghir Islands, Sulawesi, Sumba, Flores, New Guinea, Thursday Island, Australia
Lembeja Shcherbakov 1981a: 830 (listed)
Lembeja Shcherbakov 1981b: 66 (listed)
Lembeja Jong 1982: 175, 182–184 (distribution, comp. note) Mindanao, Sulawesi, Lesser Sunda Islands, New Guinea, Queensland, Australia
Lembeja Duffels 1983a: 492 (comp. note) Papua New Guinea
Lembeja Jong 1985: 165–166 (diversity, comp. note)
Lembeja Duffels 1986: 319, 325–326 (diversity, distribution, comp. note) New Guinea, Sulawesi, Philippines, Lesser Sunda Islands, Sangihc Island, Torres Strait Island, Queensland
Lembeja Jong 1986: 141–142 (distribution, comp. note) Equals *Perissoneura* Philippines, Mindanao, Sangihe Islands, Sulawesi, Lesser Sunda Islands, New Guinea, Torres Strait Islands, Queensland
Perissoneura Jong 1986: 141 (comp. note)
Lembeja Jong 1987: 177 (comp. note)
Lembeja Boulard 1990b: 138–139, 142, 144 (comp. note)
Lembeja Duffels 1990b: 68–69 (comp. note) New Guinea, Lesser Sunda Islands
Lembeja Duffels and Boer 1990: 257 (comp. note)
Lembeja Moulds 1990: 12, 16, 19, 23, 32 (synonymy, listed, diversity, comp. note) Australia, New Guinea
Lembeja spp. Moulds 1990: 21 (comp. note)
Lembeja Boer 1993b: 145 (phylogeny, comp. note)
Lembeja Boer 1995c: 203, 206–207 (comp. note)
Lembeja Boer 1995d: 200–201, 221, 224, 229, 234, Fig 29, Fig 54 (distribution, phylogeny, comp. note) Sulawesi, Borneo, Mindanao
Lembeja Boer and Duffels 1996a: 155, 167 (distribution, comp. note) Sulawesi, Borneo, Philippines, New Guinea, Australia
Lembeja Boer and Duffels 1996b: 301, 308, 327 (distribution, phylogeny, comp. note) Sulawesi, New Guinea, Lesser Sunda Islands
Lembeja Boer 1997: 113 (comp. note)
Lembeja Holloway 1997: 219, 221 (distribution, comp. note) New Guinea, Borneo, Mindanao, Lesser Sunda Islands, Queensland, Obi Island, Sulawesi
Lembeja Holloway 1998: 306, 308 (distribution, diversity, comp. note) Lesser Sunda Islands, Sulawesi, Borneo, Mindanao, Queensland, Obi Island
Lembeja Boulard and Riou 1999: 136–137 (comp. note)
Lembeja Boulard 2002b: 197–199 (comp. note) Australia, Southeast Asia
Lembeja Moulds and Carver 2002: 49 (synonymy, type data, listed) Australia
Perissoneura Moulds and Carver 2002: 49 (type data, listed)
Lembeja spp. Lee 2003b: 7 (comp. note)
Lembeja Jong and Boer 2004: 265, 267, 269 (comp. note)
Lembeja Moulds 2005b: 393, 412–413, 430, 436, Table 2 (higher taxonomy, listed, comp. note) Australia
Lembeja Duffels and Boer 2006: 534, 537 (distribution, diversity, comp. note) New Guinea
Lembeja Boulard 2007e: 21 (comp. note) Southeast Asia
Lembeja Shiyake 2007: 9, 114 (listed, comp. note)

L. acutipennis (Karsch, 1890) to *Lembeja paradoxa* (Karsch, 1890)

***L. brendelli* Jong, 1986**
Lembeja brendelli Boer and Duffels 1996b: 327 (distribution) Sulawesi
Lembeja brendelli Jong 1986: 142, 146–147, 178–180, Figs 129–138, Figs 145–146, Map 1 (n. sp., described, illustrated, key, distribution, comp. note) Sulawesi

L. brunneosa Distant, 1910 to *Lembeja paradoxa* (Karsch, 1890)

***L. consanguinea* Jong, 1987**
Lembeja consanguinea Jong 1987: 177–180, 182, 191, 201–203, Figs 49–56, Map 3 (n. sp., described, illustrated, key, distribution, comp. note) Sulawesi
Lembeja consanguinea Boer and Duffels 1996b: 327 (distribution) Sulawesi

L. crassa Distant, 1909 to *Lembeja papuensis* Distant, 1897

***L. dekkeri* Jong, 1986**
Lembeja dekkeri Jong 1986: 142, 146–147, 150, 155–158, 173, Figs 36–45, Map 1 (n. sp., described, illustrated, key, distribution, comp. note) Sulawesi
Lembeja dekkeri Boer and Duffels 1996b: 327 (distribution) Sulawesi

***L. distanti* Jong, 1986**
Lembeja distanti subgroup Jong 1986: 142, 145–146, 170, 173–174 (described, diversity, key, distribution, comp. note) Sulawesi
Lembeja distanti Jong 1986: 142, 146–147, 175–176, 178, 180, Figs 118–128, Map 1 (n. sp., described, illustrated, key, distribution, comp. note) Sulawesi
Lembeja distanti Jong 1987: 188, 194 (comp. note)

Lembeja distanti Boer and Duffels 1996b: 327 (distribution) Sulawesi

***L. elongata* Jong, 1986**

Lembeja elongata Jong 1986: 142, 144, 146–147, 150, 153–155, 157–158, 173, Figs 1–3, Figs 26–35, Map 1 (n. sp., described, illustrated, key, distribution, comp. note) Sulawesi

Lembeja elongata Boer and Duffels 1996b: 327 (distribution) Sulawesi

***L. fatiloqua* (Stål, 1870)** = *Prasia fatiloqua*

Lembeja fatiloqua Jong 1982: 182 (comp. note)

Lembeja fatiloqua group Jong 1985: 166 (comp. note)

Prasia fatiloqua Jong 1985: 166 (comp. note)

Lembeja fatiloqua group Duffels 1986: 326 (diversity, distribution, comp. note) Sulawesi, Philippines, Lesser Sunda Islands, New Guinea, Australia

Prasia fatiloqua Jong 1986: 141 (comp. note)

Prasia fatiloqua Jong 1986: 141 (comp. note)

Lembeja fatiloqua group Jong 1986: 141–142 (comp. note)

Lembeja fatiloqua Jong 1986: 142 (comp. note)

Lembeja fatiloqua group Jong 1987: 177–179, 180–182, 194, Figs 1–5 (described, key, diveristy, comp. note) Mindanao, Philippines, Borneo, Sulawesi, Lesser Sunda Islands, New Quinea, Torres Strait Island, Queensland

Lembeja fatiloqua Jong 1987: 177–178, 180–184, 186–188, 190, 195, 206, Figs 3–11, Figs 74–75 Map 1 (described, illustrated, key, synonymy, distribution, comp. note) Equals *Prasia fatiloqua* Mindanao, Philippines, Borneo

Lembeja fatiloqua group Duffels 1990b: 64, 68–69, Table 1 (diversity, distribution, comp. note) Sulawesi, Sumbawa, Mindanao, Borneo, New Guinea, Australia

Lembeja fatiloqua group Duffels and Boer 1990: 256–257 (distribution, diversity, comp. note) Sulawsesi, Sumba, Sumbawa, Mindanao, Borneo New Guinea, Australia

Lembeja fatiloqua group Boer 1995c: 206–207, Fig 2 (phylogeny, comp. note) Sulawesi, Mindanao, Borneo, Simba, New Guinea, Queensland

Lembeja fatiloqua group Boer 1995d: 221–222, 225, 228–229, 234, Fig 54 (distribution, phylogeny, comp. note) Sulawesi, Mindanao, Borneo, Simba

Lembeja fatiloqua Boer 1995d: 221, 228–229 (distribution, phylogeny, comp. note) Mindanao

Lembeja fatiloqua Boer and Duffels 1996b: 301 (distribution, comp. note) Borneo, Philippines

Lembeja fatiloqua group Holloway 1997: 219–220, Fig 5 (distribution, phylogeny, comp. note) Sulawesi, Borneo, Mindanao, Sunda Islands, Sumbawa, New Guinea, Queensland

Lembeja fatiloqua group Holloway 1998: 308, Fig 13 (distribution, comp. note) Lesser Sunda Islands

Lembeja fatiloqua group Jong and Boer 2004: 265–266, 269, Fig 1 (phylogeny, comp. note)

Lembeja fatiloqua group Schmidt 2005: 276 (distribution, comp. note)

***L. foliata* (Walker, 1858)** = *Cephaloxys foliata* = *Prasia foliata* = *Lembeja foliate* (sic)

Cephaloxys foliata Jong 1985: 165 (type species of *Prasia*, comp. note)

Prasia foliata Jong 1985: 166 (comp. note)

Lembeja foliata group Duffels 1986: 326, 329, Fig 8 (diversity, distribution, comp. note) Sulawesi, Sangihe Island

Lembeja foliata group Jong 1986: 141–146, 158, 160, 162, 166, 170, 174, Figs 1–6 (described, illustrated, diversity, distribution, key, comp. note) Sulawesi, Sangihe Island

Lembeja foliata subgroup Jong 1986: 142, 145–146, 170–171 (described, key, diveristy, comp. note) Sulawesi

Lembeja foliata Jong 1986: 142, 145, 147, 149–151, 153, 158, 160, 173, Figs 7–14, Map 1 (described, illustrated, key, synonymy, distribution, comp. note) Equals *Cephaloxys foliata* Equals *Prasia foliata* Sulawesi

Cephaloxys foliata Jong 1986: 141 (comp. note)

Prasia foliata Jong 1986: 141 (comp. note)

Lembeja foliata group Jong 1987: 177 (comp. note)

Lembeja foliata group Duffels 1990b: 64, 68–69, Table 1 (diversity, distribution, comp. note) Sulawesi, Sangihe Island

Lembeja foliata group Duffels and Boer 1990: 256–257 (distribution, diversity, comp. note) Sulawesi, Sangihe

Lembeja foliate (sic) group Moulds 1990: 178 (comp. note)

Lembeja foliata group Boer 1995c: 206–207, Fig 2 (phylogeny, comp. note) Sulawesi, Sangihe Island

Lembeja foliata group Boer 1995d: 221, 225, Fig 54 (distribution, phylogeny, comp. note)

Lembeja foliata Boer and Duffels 1996b: 327 (distribution) Sulawesi

Lembeja foliata group Holloway 1997: 219–221, 224, Fig 5 (distribution, phylogeny, comp. note) Sulawesi, Sangihe Island

Lembeja foliata group Holloway 1998: 305, 307–308, Fig 10, Fig 13 (distribution, comp. note) Sulawesi

Lembeja foliata group Jong and Boer 2004: 265–266, Fig 1 (phylogeny, comp. note) New Guinea

***L. fruhstorferi* Distant, 1897** = *Prasia fruhstorferi*
Lembeja fruhstorferi Jong 1986: 141–142 (comp. note)
Lembeja fruhstorferi Jong 1987: 177–178, 180, 191, 194–196, Map 3 (described, key, synonymy, distribution, comp. note) Equals *Prasia fruhstorferi* Sulawesi
Lembeja fruhstorferi Boer and Duffels 1996b: 327 (distribution) Sulawesi

L. harderi Schmidt, 1925 to *Mariekea harderi*

***L. hollowayi* Jong, 1986**
Lembeja hollowayi Jong 1986: 142, 146, 148, 162–164, 166, 173, 177, Figs 66–76, Figs 141–142, Map 2 (n. sp., described, illustrated, key, distribution, comp. note) Sulawesi
Lembeja hollowayi Boer and Duffels 1996b: 327 (distribution) Sulawesi

***L. incisa* De Jong, 1986**
Lembeja incisa Jong 1986: 142, 146, 148,, 158, 166, 171, Figs 108–117, Map 2 (n. sp., described, illustrated, key, distribution, comp. note) Sulawesi
Lembeja incisa Boer and Duffels 1996b: 327 (distribution) Sulawesi

***L. lieftincki* Jong, 1987**
Lembeja lieftincki Jong 1987: 177–178, 180–181, 191, 196, 199–200, 202, 207, Fig 2, Figs 42–48, Figs 78–79, Map 3 (n. sp., described, illustrated, key, distribution, comp. note) Sulawesi
Lembeja lieftincki Boer and Duffels 1996b: 299, 327, Plate 1e (illustrated, comp. note) Sulawesi, Borneo, Philippines

***L. maculosa* (Distant, 1883)** = *Perissoneura maculosa* = *Prasia maculosa* = *Prasia faticina* (nec Distant)
Lembeja maculosa Jong 1985: 171 (comp. note) Equals *Prasia faticina* Distant 1982 (partim)
Lembeja maculosa Jong 1986: 141–142 (comp. note)
Lembeja maculosa Jong 1987: 177–178, 180, 188–193, Figs 23–28, Map 3 (described, illustrated, key, synonymy, distribution, comp. note) Equals *Perissoneura maculosa* Equals *Prasia maculosa* Sulawesi
Perissoneura maculosa Moulds 1990: 178 (type species of *Perissoneura*, listed) Sulawesi
Lembeja maculosa Boer and Duffels 1996b: 327 (distribution) Sulawesi
Perissoneura maculosa Moulds and Carver 2002: 49 (type species of *Perissoneura*, listed)

***L. majuscula* Jong, 1986**
Lembeja majuscula Jong 1986: 142, 144, 146, 148, 170–171, 173, 179, Fig 4, Figs 97–107, Figs 143–144, Map 2 (n. sp., described, illustrated, key, distribution, comp. note) Sulawesi
Lembeja majuscula Boer and Duffels 1996b: 327 (distribution) Sulawesi

***L. minahassae* Jong, 1986**
Lembeja minahassae Jong 1986: 142, 145–146, 147, 150–153, 158, 177, Fig 5, Figs 15–25, Figs 139–140, Map 1 (n. sp., described, illustrated, key, distribution, comp. note) Sulawesi
Lembeja minahassae Boer and Duffels 1996b: 327 (distribution) Sulawesi

***L. mirandae* Jong, 1986**
Lembeja mirandae Jong 1986: 142, 145–146, 148, 166–173, Fig 6, Figs 87–96, Map 2 (n. sp., described, illustrated, key, distribution, comp. note) Sulawesi
Lembeja mirandae Boer and Duffels 1996b: 327 (distribution) Sulawesi

***L. oligorhanta* Jong, 1986**
Lembeja oligorhanta Jong 1986: 142, 146, 148, 158, 164–166, 173, Figs 77–86, Map 2 (n. sp., described, illustrated, key, distribution, comp. note) Sulawesi
Lembeja oligorhanta Boer and Duffels 1996b: 327 (distribution) Sulawesi

***L. papuensis* Distant, 1897** = *Prasia papuensis* = *Drepanopsaltria russula* Jacobi, 1903 = *Lembeja crassa* Distant, 1909
Lembeja papuensis Jong and Duffels 1981: 54–55, Fig 1, Figs 3–12 (lectotype designation, synonymy, described, illustrated, distribution, comp. note) Equals *Prasia papuensis* Equals *Drepanopsaltria russula* Equals *Lembeja crassa* n. syn. New Guinea
Drepanopsaltria russula Jong and Duffels 1981: 54–57, Figs 2–3 (lectotype designation, illustrated, comp. note)
Lembeja crassa Jong and Duffels 1981: 55, 57 (lectotype designation, synonymy, comp. note)
Lembeja papuensis Jong 1982: 175, 182–183 (comp. note)
Lembeja papuensis Jong 1986: 141 (comp. note)
Lembeja papuensis Moulds 1990: 178 (comp. note)
Lembeja papuensis Boer 1995d: 222 (distribution, phylogeny, comp. note) New Guinea
Lembeja papuensis Boer and Duffels 1996b: 327 (distribution) New Guinea

***L. paradoxa* (Karsch, 1890)** = *Perissoneura paradoxa* = *Prasia paradoxa* = *Perissoneura acutipennis* Karsch, 1890 = *Prasia acutipennis* = *Lembeja acutipennis* = *Lembeja brunneosa* Distant, 1910 = *Lembeja australis* Ashton, 1912
Lembeja paradoxa Jong 1982: 175–180, 182–184, Figs 1–8, Figs 17–19, Figs 22–23 (n. comb., lectotype designation, synonymy, described, illustrated, comp. note) Equals *Perissoneura paradoxa* Equals *Prasia paradoxa* Equals *Perissoneura acutipennis* Equals *Prasia acutipennis* Equals *Lembeja acutipennis*

n. syn. Equals *Lembeja brunneosa* n. syn. Equals *Lembeja australis* New Guinea, Australia, Thursday Island, Moa Island, Prince of Wales Island, Queensland

Lembeja acutipennis Jong 1982: 175–178, 182, Fig 1, Fig 18 (lectotype designation, synonymy, comp. note) Thursday Island

Lembeja brunneosa Jong 1982: 175–177 (lectotype designation, synonymy, comp. note) Queensland

Lembeja australis Jong 1982: 175–177 (lectotype designation, synonymy, comp. note)

Lembeja brunneosa Young and Josephson 1983b: 197, 207 (comp. note) Australia, Queensland

Lembeja brunneosa Popov 1985: 38–39 (acoustic system, comp. note)

Lembeja paradoxa Jong 1986: 141–142 (comp. note)

Lembeja paradoxa Jong 1987: 178–179, 180, 194, Figs 35–37 (described, illustrated, key, distribution, comp. note) Equals *Perissoneura paradoxa* Equals *Prasia paradoxa* Equals *Perissoneura acutipennis* Equals *Prasia acutipennis* Equals *Lembeja acutipennis* Equals *Lembeja brunneosa* Equals *Lembeja australis* New Guinea, Torres Strait Islands, Queensland

Lembeja brunneosa Ewart 1989a: 77, 79 (comp. note)

Lembeja paradoxa Boulard 1990b: 152, 153, Plate XVII, Fig 3 (sound system illustrated, comp. note) Australia

Lembeja parvula group Duffels and Boer 1990: 256 (distribution, diversity, comp. note) Sulawesi

Lembeja paradoxa Moulds 1990: 12, 18–19, 21, 23–25, 110, 178–179, Figs 12–13, Plate 6, Fig 3, Fig 3a, Plate 24, Fig 19 (illustrated, oscillogram, distribution, ecology, listed, comp. note) Australia, Queensland, Torres Straits Islands, Papua New Guinea

Lembeja paradoxa Boer 1992a: 25 (comp. note) New Guinea, northern Queensland

Lembeja paradoxa Ewart 1993: 140 (comp. note) Queensland

Lembeja paradoxa Naumann 1993: 3, 101, 149 (listed) Australia

Lembeja paradoxa Boer 1995b: 16 (comp. note)

Lembeja paradoxa Boer 1995d: 222, 228–229, 233–234 (distribution, phylogeny, comp. note) Queensland

Lembeja paradoxa Boer and Duffels 1996b: 301, 327 (distribution, comp. note) Queensland, Papuan Peninsula

Lembeja paradoxa Boer 1997: 94 (comp. note) Queensland

Lembeja paradoxa Boulard 2002b: 197 (sound system, comp. note) Australia

Lembeja brunneosa Sueur 2001: 45–46, Table 1 (listed, comp. note)

Lembeja paradoxa Sueur 2001: 45, Table 1 (listed)

Lembeja paradoxa Moulds and Carver 2002: 49–50 (synonymy, distribution, ecology, type data, listed) Equals *Perissoneura acutipennis* Equals *Perissoneura paradoxa* Equals *Lembeja brunneosa* Equals *Lembeja australis* Australia, Queensland, Papua New Guinea

Perissoneura acutipennis Moulds and Carver 2002: 49 (type data, listed) Thursday Island, Torres Straits

Perissoneura paradoxa Moulds and Carver 2002: 50 (type data, listed) New Guinea

Lembeja brunneosa Moulds and Carver 2002: 50 (type data, listed) Queensland

Lembeja australis Moulds and Carver 2002: 50 (type data, listed) Queensland

Lembeja paradoxa Lee 2003b: 7 (comp. note)

Lembeja brunneosa Tishechkin 2003: 144 (comp. note)

Lembeja paradoxa Shiyake 2007: 9, 113, 115, Fig 204 (illustrated, distribution, listed, comp. note) Australia

***L. parvula* Jong, 1987**

Lembeja parvula group Jong 1987: 177, 178–179, 202–204, Figs 57–60 (described, key, diveristy, comp. note) Sulawesi

Lembeja parvula Jong 1987: 178, 180, 191, 202–205, 208, Fig 58, Figs 60–67, Map 3 (n. sp., described, illustrated, key, distribution, comp. note) Sulawesi

Lembeja parvula group Duffels 1990b: 64, 68–69, Table 1 (diversity, distribution, comp. note) Sulawesi

Lembeja distanti (sic) group Duffels 1990b: 69 (comp. note)

Lembeja parvula group Boer 1995c: 206–207, Fig 2 (phylogeny, comp. note) Sulawesi

Lembeja parvula group Boer 1995d: 221, 225, Fig 54 (distribution, phylogeny, comp. note)

Lembeja parvula Boer and Duffels 1996b: 327 (distribution) Sulawesi

Lembeja parvula group Holloway 1997: 219–220, Fig 5 (distribution, phylogeny, comp. note) Sulawesi

Lembeja parvula group Holloway 1998: 308, Fig 13 (distribution, comp. note)

Lembeja parvula group Jong and Boer 2004: 265–266, 269, Fig 1 (phylogeny, comp. note)

***L. pectinulata* Jong, 1986**

Lembeja pectinulata Jong 1986: 142, 144, 146, 148, 158–160, 162, 173, Figs 46–55, Map 2 (n. sp., described, illustrated, key, distribution, comp. note) Sulawesi

Lembeja pectinulata Boer and Duffels 1996b: 327 (distribution) Sulawesi

L. robusta **Distant, 1909**

Lembeja robusta Jong and Duffels 1981: 53, 61 (comp. note)
Lembeja robusta Jong 1982: 175, 182 (comp. note)
Lembeja robusta group Duffels 1986: 326 (diversity, distribution, comp. note) New Guinea
Lembeja robusta Jong 1986: 141–142 (comp. note)
Lembeja robusta group Boer 1995c: 206–207, Fig 2 (phylogeny, comp. note) New Guinea, Obi Island, Queensland
Lembeja robusta group Boer 1995d: 221–222, 225, 228–229, 234, Fig 54 (distribution, phylogeny, comp. note)
Lembeja robusta Boer 1995d: 222 (distribution, phylogeny, comp. note) New Guinea
Lembeja robusta Boer and Duffels 1996b: 327 (distribution) Papuan Peninsula
Lembeja robusta group Holloway 1997: 220, Fig 5 (distribution, phylogeny, comp. note) New Guinea
Lembeja robusta group Holloway 1998: 308, Fig 13 (distribution, comp. note)
Lembeja robusta group Jong and Boer 2004: 265–266, Fig 1 (phylogeny, comp. note) New Guinea

L. roehli **Schmidt, 1925** = *Prasia fatiloqua* (nec Stål)

Lembeja roehli Jong 1982: 182–183 (comp. note) Lesser Sunda Islands
Lembeja roehli Jong 1986: 141–142 (comp. note)
Lembeja roehli Jong 1987: 177–178, 180, 184–186, 188, Figs 12–17, Map 2 (described, illustrated, key, synonymy, distribution, comp. note) Equals *Prasia fatiloqua* Lallemand, 1935 (partim) Sumba
Lembeja roehli Boer 1995d: 221 (distribution, comp. note) Lesser Sunda Islands
Lembeja roehli Boer and Duffels 1996b: 327 (distribution) Lesser Sunda Islands

L. sangihensis **Jong, 1986** = *Prasia foliata* (nec Walker)

Lembeja sangihensis Jong 1986: 142, 146, 148–149, 160–162, Figs 55–56, Map 2 (n. sp., described, illustrated, key, distribution, comp. note) Equals *Prasia foliata* (partim) Sulawesi
Lembeja sangihensis Boer and Duffels 1996b: 327 (distribution) Sulawesi

L. sanguinolenta **Distant, 1909**

Lembeja sanguineolenta Jong 1982: 182 (comp. note)
Lembeja sanguineolenta Jong 1986: 141–142 (comp. note)
Lembeja sanguineolenta Jong 1987: 177–178, 180, 191, 195–198, Figs 38–41, Map 3 (described, illustrated, key, distribution, comp. note) Sulawesi
Lembeja sanguinolenta Boer and Duffels 1996b: 327 (distribution) Sulawesi

L. sumbawensis **Jong, 1987**

Lembeja sumbawensis Jong 1987: 177–178, 180, 186–188, Figs 18–22, Map 2 (n. sp., described, illustrated, key, distribution, comp. note) Sumbawa
Lembeja sumbawensis Boer 1995d: 221 (distribution, comp. note) Lesser Sunda Islands
Lembeja sumbawensis Boer and Duffels 1996b: 327 (distribution) Lesser Sunda Islands

L. tincta **(Distant, 1909)** = *Prasia tincta*

Lembeja tincta Jong 1982: 182 (comp. note)
Prasia tincta Jong 1985: 166 (status, comp. note)
Lembeja tincta Jong 1985: 166 (n. comb., comp. note) Sulawesi
Prasia tincta Jong 1986: 141 (comp. note)
Lembeja tincta Jong 1986: 142 (comp. note)
Lembeja tincta Jong 1987: 177–178, 180–181, 186, 190–193, 196, 206, Fig 1, Figs 29–34, Figs 76–77, Map 3 (described, illustrated, key, synonymy, distribution, comp. note) Equals *Prasia tincta* Sulawesi
Lembeja tincta Boer and Duffels 1996b: 327 (distribution) Sulawesi

L. vitticollis **(Ashton, 1912)** = *Prasia vitticollis* = *Prasia viticollis* (sic)

Lembeja vitticollis Jong 1982: 175, 179–184, Figs 9–16, Figs 20–21, Fig 23 (n. comb., described, illustrated, comp. note) Australia, Queensland
Prasia vitticollis Jong 1982: 175 (comp. note)
Prasia vitticollis Jong 1986: 141 (comp. note)
Lembeja vitticollis Jong 1986: 142 (comp. note)
Lembeja viticollis Boulard 1990b: 138 (comp. note)
Lembeja vitticollis Moulds 1990: 178–180, Plate 6, Fig 4, Figs 4a–b (illustrated, oscillogram, distribution, ecology, listed, comp. note) Australia, Queensland
Lembeja vitticollis Naumann 1993: 9, 101, 149 (listed) Australia
Lembeja vitticollis Boer 1997: 94 (comp. note) Queensland
Lembeja vitticollis Boer 1995d: 222, 233–234 (distribution, phylogeny, comp. note) Queensland
Lembeja vitticollis Boer and Duffels 1996b: 300–301, 327, plate 2 (timbal illustrated, comp. note) Borneo, Philippines, Australia
Lembeja vitticollis Boulard 2002b: 197 (sound system, comp. note) Australia
Lembeja vitticollis Moulds and Carver 2002: 50 (synonymy, distribution, ecology, type data, listed) Equals *Prasia vitticollis* Australia

Prasia vitticollis Moulds and Carver 2002: 50 (type data, listed) Queensland?
Lembeja vitticollis Moulds 2005b: 395–397, 399, 417–419, 421, 424, Figs 56–59, Fig 61, Table 1 (phylogeny, comp. note) Australia

***L. wallacei* Jong, 1987**
Lembeja wallacei Jong 1987: 178, 180, 191, 204, 207–208, Fig 57, Fig 59, Figs 68–73, Figs 80–81, Map 3 (n. sp., described, illustrated, key, distribution, comp. note) Sulawesi
Lembeja wallacei Boer and Duffels 1996b: 327 (distribution) Sulawesi

***L.* 1 new species Boer & Duffels, 1996**
Lembeja 1 new species Boer and Duffels 1996b: 327 (distribution) South New Guinea

***L.* 2 new species Boer & Duffels, 1996**
Lembeja 2 new species Boer and Duffels 1996b: 327 (distribution) Birds Head New Guinea

***L.* 3 new species Boer & Duffels, 1996**
Lembeja 3 new species Boer and Duffels 1996b: 327 (distribution) North Maluku

***L.* 4 new species Boer & Duffels, 1996**
Lembeja 4 new species Boer and Duffels 1996b: 327 (distribution) New Guinea

***L.* 5 new species Boer & Duffels, 1996**
Lembeja 5 new species Boer and Duffels 1996b: 327 (distribution) Lesser Sunda Islands

Subtribe Iruanina Boulard, 1993
= Iruanaria

Iruanaria Boulard 1993d: 92 (n. subtribe, comp. note)
Iruanaria Boulard 2002b: 206 (comp. note)

***Iruana* Distant, 1905**
Iruana Jong and Duffels 1981: 55 (diversity, comp. note) Africa
Iruana Boulard 1982b: 58–59 (comp. note) Ethiopia, Kenya
Iruana Boulard 1982d: 112 (comp. note)
Iruana Boulard 1985e: 1026 (coloration, comp. note)
Iruana sp. Boulard 1985e: Plate II, Fig 12 (coloration, illustrated, comp. note) Kenya
Iruana Jong 1985: 165 (diversity, comp. note) Africa
Iruana Jong 1986: 142 (comp. note) Africa
Iruana Jong 1987: 177 (comp. note) Africa
Iruana Boulard 1990b: 141, 175, 207 (comp. note)
Iruana Boer 1995c: 205 (comp. note) Africa
Iruana Boulard 1993d: 87–88 (comp. note) West Africa
Iruana Boulard 1996d: 157 (listed)
Iruana Boer 1997: 113 (comp. note)
Iruana Boulard and Riou 1999: 136 (comp. note)
Iruana Boulard 2002b: 199, 206 (comp. note) East Africa
Iruana Moulds 2005b: 393, 436 (higher taxonomy, listed, comp. note)

***I. brignolii* Boulard, 1982**
Iruana brignolii Boulard 1982b: 58–59, Plate I Figs 1–3, Plate II Figs 5–7 (n. sp., described, illustrated, comp. note) Ethiopia
Iruana brignollii Boulard 1990b: 155, 157, Plate XVIII, Fig 7 (sound system, comp. note) Ethiopia
Iruana brignolii Boulard 1993d: 92 (comp. note)
Iruana brignolii Boulard 2002b: 206 (comp. note)

***I. meruana* Boulard, 1990**
Iruana meruana Boulard 1990b: 185, 187–189, 221–223, Plate XXII, Fig 1, Plate XXV, Fig 12, Figs 19–23 (n. sp., described, illustrated, sonogram, comp. note) Kenya
Iruana meruana Boulard 1993d: 88, 92 (comp. note)
Iruana meruana Boulard 1996d: 115, 157–158, Fig S20(a-d), Fig S20(e-f) (song analysis, sonogram, listed, comp. note) Kenya
Iruana meruana Boulard 1997b: 22–23, Plate IV, Fig 3 (coloration, comp. note) Kenya
Iruana meruana Sueur 2001: 45, Table 1 (listed)
Iruana meruana Boulard 2002b: 206 (comp. note)
Iruana meruana Boulard 2006g: 180 (listed)
Iruana meruana Boulard 2007f: 134, Fig 58 (illustrated, comp. note) East Africa

***I. rougeoti* Boulard, 1975**
Iruana rougeoti Boulard 1982b: 58–59, Plate I Fig 4, Plate II Figs 8–10 (comp. note) Ethiopia
Iruana rougeoti Boulard 1993d: 92 (comp. note)
Iruana rougeoti Boulard 2002b: 206 (comp. note)

***I. sulcata* Distant, 1905**
Iruana sulcata Boulard 1985e: 1027, 1040–1041, Plate V, Fig 32 (coloration, illustrated, comp. note)
Iruana sulcata Boulard 1993d: 92 (comp. note)
Iruana sulcata Boulard 1997b: 21, Plate III, Fig 2 (morphology, comp. note)
Iruana sulcata Boulard 2002b: 206 (type species of *Iruana*, comp. note)
Iruana sulcata Boulard 2007f: 83, Fig 99.2 (illustrated, comp. note)

Subtribe Bafutalnina Boulard, 1993
= Bafutalnaria

Bafutalnaria Boulard 1993d: 92 (n. subtribe, comp. note)
Bafutalnaria Boulard 2002b: 206 (comp. note)

***Bafutalna* Boulard, 1993**
Bafutalna Boulard 1993d: 88, 91 (n. gen., described, comp. note)
Bafutalna Boulard 2002b: 197, 205 (sound system, comp. note) Africa

Bafutalna Boulard 2006g: 16–17 (comp. note) Cameroon
Bafutalna Boulard 2010b: 336 (comp. note)
***B. mirei* Boulard, 1993**
Bafutalna mirei Boulard 1993d: 88–92, Figs 1–7 (n. sp., type species of *Bafutalna*, described, illustrated, comp. note) Cameroon
Bafutalna mirei Boulard 2002b: 205 (comp. note) Cameroon
Bafutalna mirei Boulard 2005d: 349 (comp. note) Cameroon
Bafutalna mirei Boulard 2006g: 160–161 (comp. note) Cameroon

***Murphyalna* Boulard, 2012**
***M. mughessensis* Boulard, 2012**

Tribe Chlorocystini Distant, 1905 = Chlorocystinae (sic) = Chlorocystaria = Gymnotympanini Boulard, 1979
Chlorocystini Hayashi 1984: 69 (listed, diversity)
Chlorocystini Boulard 1986d: 233 (listed)
Chlorocystini Villet 1987a: 272, Table 2 (listed) South Africa
Gymnotympanini Boulard 1988f: 64 (comp. note)
Chlorocystini Villet 1997b: 348, 355 (comp. note)
Gymnotympanini Villet 1997b: 348, 356 (comp. note)
Chlorocystini Boulard 1990b: 141, 209, Fig 12 (phylogeny, comp. note)
Gymnotympanini Boulard 1990b: 141, 209, Fig 12 (phylogeny, comp. note)
Chlorocystinae (sic) Boulard 1990b: 209, Fig 12 (phylogeny, comp. note)
Chlorocystini Moulds 1990: 31, 180, 197 (listed) Equals Chlorocystaria Australia
Chlorocystaria Moulds 1990: 197 (comp. note)
Chlorocystini Liu 1992: 42 (listed)
Chlorocystini Boulard 1993d: 91 (comp. note)
Gymnotympanini Boulard 1993d: 91–92 (comp. note)
Chlorocystini Villet 1993: 435, 439 (comp. note)
Gymnotympanini Villet 1993: 435, 439 (comp. note)
Chlorocystini Boer 1995c: 201–202, 204–206, 210, 212–214, 219, Figs 1–2, Fig 4 (phylogeny, comp. note)
Gymnotympanini Boer 1995c: 202, 204 (comp. note)
Chlorocystini Boer 1995d: 173–174, 200–203, 207–208, 210, 216–219, 222–225, 228–230, 233, 237, Figs 28–29, Fig 54 (distribution, diversity, phylogeny, comp. note) East Melanesia, Samoa, Tonga
Chlorocystini Boer and Duffels 1996a: 155–157, 163, 167–172, Fig 2, Fig 11 (distribution, diversity, comp. note)
Chlorocystini Boer and Duffels 1996b: 298, 301–303, 306–308, 313–315, Figs 2–3 (distribution, diversity, phylogeny, comp. note) New Guinea
Chlorocystini Boulard 1996a: 157 (comp. note)
Chlorocystini Boulard 1996d: 115, 153 (listed)
Chlorocystini Boer 1997: 91–94, 98, 113–114, 117 (phylogeny, comp. note)
Gymnotympanini Boer 1997: 202, 204 (comp. note)
Chlorocystini Chou, Lei, Li, Lu and Yao 1997: 90 (comp. note)
Chlorocystini Holloway 1997: 219–223, Figs 5–6 (distribution, phylogeny, comp. note) New Guinea, Sula Island, Samoa, Moluccas, Australia
Gymnotympanini Novotny and Wilson 1997: 437 (listed)
Cholorcystini Hayashi 1998: 62 (comp. note)
Chlorocystini Holloway 1998: 304, 307–310, Fig 13 (distribution, phylogeny, comp. note) New Guinea, Queensland, New South Wales, Australia, Sula Islands, Samoa
Chlorocystini Holloway and Hall 1998: 9, 12 (distribution, comp. note)
Chlorocystini Boer 1999: 115–116, 118–120 (comp. note) New Guinea, Maluku
Chlorocystini Moulds 1999: 1 (comp. note)
Chlorocystini Sanborn 1999a: 51 (listed)
Gymnotympanini Sanborn 1999a: 51 (listed)
Chlorocystini Turner, Hovencamp and van Welzen 2001: 219, 225, Tables 1–2 (phylogeny, comp. note) Australia, Solomon Islands, Santa Cruz, Vanuatu, Samoa, Tonga
Chlorocystini Sueur 2001: 42, Table 1 (listed)
Gymnotympanini Sueur 2001: 42, Table 1 (listed)
Chlorocystini Boulard 2002b: 198 (comp. note)
Gymnotympanini Boulard 2002b: 198, 205 (comp. note)
Chlorocystini Dietrich 2002: 158 (comp. note)
Chlorocystini Moulds and Carver 2002: 19 (listed) Australia
Chlorocystini Jong and Boer 2004: 266–267, 269, Fig 1 (phylogeny, comp. note)
Chlorocystini Moulds 2005b: 386, 389–390, 392, 394, 396, 412–415, 419, 422, 424, 427, 435, 437, Tables 1–2, Fig 58, Fig 61 (described, history, phylogeny, higher taxonomy, key, distribution, listed, comp. note) Equals Gymnotympanini Equals Hemidictyini (partim) Australia
Gymnotympanini Moulds 2005b: 386, 389–390, 392, 394, 396, 435 (history, higher taxonomy, listed, comp. note) Australia
Chlorocystini Boulard 2006g: 130 (listed)

Chlorocystini Duffels and Boer 2006: 535 (listed) New Guinea
Chlorocystini Boulard 2007f: 78, 137 (comp. note)
Chlorocystini Sanborn 2007b: 37 (listed) Mexico
Chlorocystini Sanborn, Phillips and Sites 2007: 3, 33, 41, Fig 5, Table 1 (listed, distribution) Thailand
Chlorocystini Boulard 2008f: 7, 57, 90 (listed) Thailand
Gymnotymnpanini Lee and Hill 2010: 278 (comp. note) Equals Chlorocystini
Chlorocystini Sanborn 2010a: 1601 (listed) Colombia
Chlorocystini Sanborn 2010b: 76 (listed)

***Dinarobia* Mamet, 1957** = *Mauricia* Orian, 1954 (nec Mauricia Harris, 1919
Mauricia Shcherbakov 1981a: 830 (listed)
Mauricia Shcherbakov 1981b: 66 (listed)
Dinarobia Williams and Williams 1988: 3 (listed) Mascarene Islands
Dinarobia Boulard 1990b: 144, 209–210, Fig 12 (phylogeny, comp. note)
Dinorobia Boer 1995c: 203–205, Fig 1 (phylogeny, comp. note) Mascarene Isles
Dinarobia Boulard 2002b: 198 (comp. note) Mascarene Islands
Dinarobia Jong and Boer 2004: 267 (comp. note)

***D. claudeae* (Orian, 1954)** = *Mauricia claudeae*
Dinarobia claudea Boulard 2002b: 196 (venation, comp. note) Mauritius
Dinarobia claudeae Williams and Williams 1988: 3 (listed, comp. note) Equals *Mauricia claudeae* Mauritius

***Owra* Ashton, 1912**
Owra Moulds 1990: 32, 183–185 (listed, diversity, comp. note) Australia, Queensland
Owra Boer 1992a: 18–20, 22 (phylogeny, comp. note)
Owra Boer 1993a: 16–17 (comp. note)
Owra Boer 1993b: 142 (phylogeny, comp. note)
Owra Boer 1995a: 2, 6, Fig 1 (phylogeny, comp. note)
Owra Boer 1995b: 8 (comp. note)
Owra Boer 1995c: 202, 204, 210–211, 214–215, 218, Fig 3, Fig 11, Fig 18 (phylogeny, comp. note) Australia
Owra Boer 1995d: 218–219, 222, 224–225, 233, Fig 47, Figs 52–53 (distribution, phylogeny, comp. note) Australia
Owra Boer 1996: 352, 354, Fig 1 (comp. note)
Owra Boer and Duffels 1996a: 155, 168, 170–171, Figs 10–11 (distribution, phylogeny, comp. note) Australia
Owra Boer and Duffels 1996b: 301, 304, 326, Fig 5 (distribution, phylogeny, comp. note) Australia
Owra Boer 1997: 91–93, 96–98, 106–107, 109, 109, Figs 1–2 (phylogeny, described, comp. note) Australia
Owra Holloway 1997: 220, Fig 5 (distribution, phylogeny, comp. note) Australia
Owra Holloway 1998: 308, Fig 13 (distribution, comp. note) Australia
Owra Boer 2000: 4, 12, 15, Fig 1 (phylogeny, comp. note) Australia
Owra Moulds and Carver 2002: 22 (listed) Australia
Owra Moulds 2005b: 390, 413, 430, 435, Table 2 (higher taxonomy, listed, comp. note) Australia

***O. insignis* Ashton, 1912**
Owra insignis Moulds 1990: 183, 185, Plate 22, Fig 6, Fig 6a (illustrated, distribution, ecology, listed, comp. note) Australia, Queensland
Owra insignis Boer 1995c: 213, 224–225, 228, 230–231, Fig 4 (phylogeny, comp. note)
Owra insignis Boer 1995d: 219, Fig 47 (distribution, comp. note) Australia
Owra insignis Boer and Duffels 1996b: 325 (distribution) Australia
Owra insignis Boer 1997: 95, 97–98, 104, 106–107, Fig 2, Figs 30–37 (phylogeny, described, type species of genus *Owra*, illustrated, comp. note) Queensland
Owra insignis Moulds and Carver 2002: 22 (distribution, ecology, type data, listed) Australia, Queensland
Owra insignis Moulds 2005b: 395–397, 399–402, 417–419, 421, 424, Fig 32, Fig 34, Figs 56–59, Fig 61, Table 1 (illustrated, phylogeny, comp. note) Australia

Mardalana Distant, 1905 to *Chlorocysta* Westwood, 1851
M. fumea Ashton, 1914 to *Chlorocysta fumea*
M. suffusa Distant, 1907 to *Chlorocysta suffusa*

***Chlorocysta* Westwood, 1851** = *Cystosoma (Chlorocysta)* = *Chlorocysta (Chlorocysta)* = *Mardalana* Distant, 1905 = *Mardarana* (sic) = *Mardalena* (sic)
Chlorocysta MacNally and Young 1981: 194 (comp. note)
Chlorocysta Strümpel 1983: 17 (listed)
Chlorocysta Hayashi 1984: 69 (listed)
Chlorocysta Ewart 1989a: 79 (comp. note)
Chlorocysta Boer 1990: 64 (comp. note)
Chlorocysta Moulds 1990: 17, 32, 185–186, 190 (synonymy, listed) Equals *Mardelana* n. syn Australia
Mardalana Moulds 1990: 185–187 (listed, comp. note) Australia
Chlorocysta Boer 1991: 2 (comp. note)

Chlorocysta Boer 1992a: 18–20, 22 (phylogeny, comp. note)
Chlorocysta Boer 1992b: 164 (comp. note)
Chlorocysta Boer 1993a: 16–18 (comp. note)
Chlorocysta Boer 1993b: 142 (phylogeny, comp. note)
Chlorocysta Villet 1993: 439 (type genus of Chlorocystini, comp. note)
Chlorocysta Boer 1994a: 3 (comp. note)
Chlorocysta Boer 1995a: 2, 6, Fig 1 (phylogeny, comp. note)
Chlorocysta Boer 1995b: 4, 8, 24 (comp. note) Australia
Chlorocysta Boer 1995c: 202, 204, 210–211, 214–215, 218, Fig 3, Fig 11, Fig 18 (phylogeny, comp. note) Australia
Chlorocysta Boer 1995d: 218–219, 222, 224–225, 233–234, Fig 48, Figs 52–53 (distribution, phylogeny, comp. note) Australia
Chlorocysta Boer 1996: 352, 354, Fig 1 (comp. note)
Chlorocysta Boer and Duffels 1996a: 155, 171, Figs 10–11 (distribution, phylogeny, comp. note) Australia
Chlorocysta Boer and Duffels 1996b: 301, 304, 314, 325, Fig 5 (distribution, phylogeny, comp. note) Australia
Chlorocysta Boer 1997: 91–93, 96–98, 107, 109, Figs 1–2 (phylogeny, described, synonymy, comp. note) Equals *Cystosoma (Chlorocysta)* Equals *Glaucocysta* Equals *Mardalana* Equals *Mardarana* (sic) Equals *Mardalena* (sic) Australia
Chlorocysta Holloway 1997: 220, Fig 5 (distribution, phylogeny, comp. note) Australia
Chlorocysta Holloway 1998: 308, Fig 13 (distribution, comp. note) Australia
Chlorocysta Boer 2000: 4, 12, 15, Fig 1 (phylogeny, comp. note) Australia
Chlorocysta/Mardalana sp. Moss and Popple 2000: 57 (listed, comp. note) Australia, New South Wales
Chlorocysta Moulds and Carver 2002: 19 (synonymy, listed) Equals *Mardelana* Australia
Mardalana Moulds and Carver 2002: 19 (listed) Australia
Chlorocysta Williams 2002: 38 (diversity, comp. note) Australia
Chlorocysta Moulds 2005b: 390, 392, 413, 430, 435, Table 2 (type genus of Chlorocystini, higher taxonomy, listed, comp. note) Australia

C. congrua (Walker, 1862) to *Chloropsalta vitripennis* (Westwood, 1851)

***C. fumea* (Ashton, 1914)** = *Mardalana fumea*
Chlorocysta fumea Moulds 1990: 187 (n. comb., distribution, ecology, type data, listed) Equals *Mardalana fumea* Australia, Queensland
Chlorocysta fumea Boer 1995c: 213, 228, 230–231, Fig 4 (phylogeny, comp. note)
Chlorocysta fumea Boer 1995d: 219, Fig 47 (distribution, comp. note) Australia
Chlorocysta fumea Boer and Duffels 1996b: 325 (distribution) Australia
Chlorocysta fumea Boer 1997: 95, 97–98, 102–106, Fig 2, Figs 14–19 (phylogeny, synonymy, described, illustrated, comp. note) Equals *Mardalana fumea* Queensland
Mardalana fumea Boer 1997: 101 (synonymy, comp. note)
Chlorocysta fumea Moulds and Carver 2002: 19 (synonymy, distribution, ecology, type data, listed) Equals *Mardalana fumea* Australia, Queensland

C. macrula Stål, 1863 to *Chloropsalta vitripennis* (Westwood, 1851)

***C.* sp. n. Ewart, 1998**
Chlorocysta n. sp. Ewart 1998a: 63 (natural history, song, comp. note) Queensland

***C. suffusa* (Distant, 1907)** = *Mardalana suffusa*
Mardalana suffusa Southcott 1988: 115 (parasitism, comp. note) Australia, Queensland
Chlorocysta suffusa Moulds 1990: 11, 186–187, Plate 22, Fig 7, Fig 7a, Plate 24, Fig 30 (n. comb., illustrated, distribution, ecology, listed, comp. note) Equals *Mardalana suffusa* Australia, Queensland
Chlorocysta suffusa Boer 1995c: 213, 228, 230–231, Fig 4 (phylogeny, comp. note)
Chlorocysta suffusa Boer 1995d: 219, Fig 47 (distribution, comp. note) Australia
Chlorocysta suffusa Boer and Duffels 1996b: 325 (distribution) Australia
Chlorocysta suffusa Boer 1997: 95, 97–98, 101–103, 105, Fig 2, Figs 20–29 (phylogeny, synonymy, described, illustrated, comp. note) Equals *Mardalana suffusa* Queensland
Mardalana suffusa Boer 1997: 103 (synonymy, comp. note)
Chlorocysta suffusa Moulds and Carver 2002: 19 (synonymy, distribution, ecology, type data, listed) Equals *Mardalana suffusa* Australia, Queensland

***C. vitripennis* (Westwood, 1851)** = *Cystosoma (Chlorocysta) vitripennis* = *Chlorocysta* (sic) *(Chlorocysta) vitripennis* = *Cystosoma vitripennis* = *Cicada congrua* Walker, 1862 = *Mardalana congrua* = *Mardarana* (sic) *congrua* = *Kanakia congrua* = *Chlorocysta macrula* Stål, 1863
Chlorocysta vitripennis Moulds 1990: 9, 20, 186, 189–190 (synonymy, distribution, ecology, type data, listed) Equals *Cystosoma (Chlorocysta) vitripennis* Equals *Cystosoma vitripennis* Equals *Cicada congrua* n. syn.

Equals *Chlorocysta macrula* n. syn. Equals *Kanakia congrua* n. syn. Equals *Mardalana congrua* Australia, Queensland, New South Wales, New Hollalnd

Cystosoma (Chlorocysta) vitripennis Moulds 1990: 188 (type species of *Chlorocysta*, listed, comp. note)

Cystosoma vitripennis Moulds 1990: 188 (listed)

Cicada congrua Moulds 1990: 188 (type species of *Mardalana*, listed, comp. note) Queensland

Chlorocysta macrula Moulds 1990: 188 (listed, comp. note) Queensland

Kanakia congrua Moulds 1990: 188 (listed, comp. note) Queensland

Mardalana congrua Moulds 1990: 188 (listed)

Chlorocysta vitripennis Naumann 1993: 52, 80, 149 (listed) Australia

Chlorocysta vitripennis Boer 1995a: 5 (comp. note)

Chlorocysta vitripennis Boer 1995b: 16, 77 (comp. note) Equals *Kanakia congrua* (partim)

Kanakia congrua Boer 1995b: 77 (synonymy, comp. note)

Chlorocysta vitripennis Boer 1995c: 213, 228, 230–231, Fig 4 (phylogeny, comp. note)

Chlorocysta vitripennis Boer 1995d: 218–219, Fig 47 (distribution, comp. note) Australia, New South Wales

Chlorocysta vitripennis Ewart 1995: 84 (described, illustrated, natural history, acoustic behavior, oscillogram) Queensland

Chlorocysta vitripennis Boer and Duffels 1996b: 325 (distribution) Australia

Chlorocysta vitripennis Boer 1997: 94, 97–101, Figs 2–13 (phylogeny, synonymy, described, type species of genus *Chlorocysta*, illustrated, comp. note) Equals *Cystosoma (Chlorocysta) vitripennis* Equals *Cicada congrua* Equals *Chlorocysta macrula* Equals *Mardalana congrua* Equals *Mardarana* (sic) *congrua* Equals *Kanakia congrua* Queensland, New South Wales

Chlorocysta macrula Boer 1997: 98, 101 (synonymy, comp. note)

Cicada congrua Boer 1997: 98, 101 (synonymy, comp. note)

Kanakia congrua Boer 1997: 100–101 (synonymy, comp. note)

Mardalana macrula Boer 1997: 100–101 (synonymy, comp. note)

Chlorocysta vitripennis Ewart 1998a: 63 (comp. note) Queensland

Chlorocysta vitripennis Sueur 2001: 42, Table 1 (listed)

Chlorocysta (sic) *(Chlorocysta) vitripennis* Moulds and Carver 2002: 19 (type species of *Chlorocysta*)

Cicada congrua Moulds and Carver 2002: 19 (type species of *Mardalana*) Queensland

Chlorocysta vitripennis Moulds and Carver 2002: 19–20 (synonymy, distribution, ecology, type data, listed) Equals *Cystosoma (Chlorocysta) vitripennis* Equals *Cicada congrua* Equals *Chlorocysta macrula* Equals *Kanakia congrua* Australia, Queensland, New South Wales, New Hollalnd

Chlorocysta macrula Moulds and Carver 2002: 20 (type data, listed) Queensland

Kanakia congrua Moulds and Carver 2002: 20 (type data, listed) Queensland

Chlorocysta vitripennis Popple and Ewart 2002: 118 (illustrated) Australia, Queensland

Chlorocysta vitripennis Williams 2002: 38, 156–157 (listed, comp. note) Australia, Queensland, New South Wales

Chlorocysta vitripennis Moulds 2005b: 390, 395–398, 403, 408, 417–419, 421, 424, Fig 37, Fig 47, Figs 56–59, Fig 61, Table 1 (type species of *Chlorocysta*, illustrated, phylogeny, comp. note) Australia

Cystosoma vitripennis Moulds 2005b: 435 (type species of *Chlorocysta*)

***Glaucopsaltria* Goding & Froggatt, 1904** = *Glaucocysta* (sic)

Glaucopsaltria Moulds 1990: 32, 186, 189 (rev. status, listed, diversity, comp. note) Australia

Glaucopsaltria Boer 1992a: 18–20, 22 (phylogeny, comp. note)

Glaucopsaltria Boer 1993a: 16–17 (comp. note)

Glaucopsaltria Boer 1993b: 142 (phylogeny, comp. note)

Glaucopsaltria Boer 1995a: 2–3, 5–6, Fig 1 (phylogeny, comp. note)

Glaucopsaltria Boer 1995b: 8 (comp. note)

Glaucopsaltria viridis Boer 1995b: 16 (comp. note)

Glaucopsaltria Boer 1995c: 202, 204, 210–211, 214–215, 218, Fig 3, Fig 11, Fig 18 (phylogeny, comp. note) Australia

Glaucopsaltria Boer 1995d: 218–219, 222, 224–225, 233, Fig 47, Figs 52–53 (distribution, phylogeny, comp. note) Australia

Glaucopsaltria Boer 1996: 352, 354–355, Fig 1 (comp. note)

Glaucopsaltria Boer and Duffels 1996a: 155, 168, 171, Figs 10–11 (distribution, phylogeny, comp. note) Australia

Glaucopsaltria Boer and Duffels 1996b: 301, 304, 325, Fig 5 (distribution, phylogeny, comp. note) Australia

Glaucopsaltria Boer 1997: 91–93, 96–98, 107, 109, Figs 1–2 (phylogeny, described, comp. note) Australia

Glaucopsaltria Holloway 1997: 220, Fig 5 (distribution, phylogeny, comp. note) Australia
Glaucopsaltria Holloway 1998: 308, Fig 13 (distribution, comp. note) Australia
Glaucopsaltria Boer 2000: 4, 15, Fig 1 (phylogeny, comp. note)
Glaucopsaltria Moulds and Carver 2002: 21 (listed) Australia
Glaucopsaltria Williams 2002: 38 (diversity, comp. note) Australia
Glaucopsaltria Moulds 2005b: 390, 413, 430, 435, Table 2 (higher taxonomy, listed, comp. note) Australia

G. viridis Goding & Froggatt, 1904 = *Chlorocysta viridis*
Glaucopsaltria viridis Hahn 1962: 8 (listed) Australia, Queensland
Chlorocysta viridis Josephson 1981: 30–31, 34, Figs 7–8, Fig 10 (muscle function, comp. note)
Chlorocysta viridis Young and Josephson 1983a: 184, 186–187, 192, 194, Fig 3, Table 1 (oscillogram, timbal muscle kinetics, comp. note) Australia, Queensland
Chlorocysta viridis Young and Josephson 1983b: 197–199, 204–207, Fig 1E (timbal illustrated, comp. note) Australia
Chlorocysta viridis Young and Josephson 1984: 286, Fig 1 (comp. note)
Chlorocysta viridis Popov 1985: 42 (acoustic system, comp. note)
Glaucopsaltria viridis Stevens and Carver 1986: 265 (type material) Australia, Queensland
Chlorocysta viridis Josephson and Young 1987: 995–996, Figs 2–3 (comp. note)
Chlorocysta viridis Moss 1988: 30 (distribution, song, comp. note) Queensland
Chlorocysta viridis Ewart 1989a: 77–79 (comp. note)
Glaucopsaltria viridis Moulds 1990: 18, 23, 188–190, Plate 4, Fig 2, Plate 22, Fig 3 (type species of *Glaucopsaltria*, illustrated, distribution, ecology, listed, comp. note) Australia, Queensland, New South Wales
Chlorocysta viridis Bennet-Clark and Young 1992: 151 (comp. note)
Glaucopsaltria viridis Naumann 1993: 8, 93, 149 (listed) Australia
Glaucopsaltria viridis Boer 1995c: 213, 228, 230–231, Fig 4 (phylogeny, comp. note)
Glaucopsaltria viridis Boer 1995d: 218–219, Fig 47 (distribution, comp. note) Australia, New South Wales
Glaucopsaltria viridis Ewart 1995: 83 (described, illustrated, natural history, acoustic behavior, oscillogram) Queensland
Glaucopsaltria viridis Boer and Duffels 1996b: 325 (distribution) Australia
Glaucopsaltria viridis Ewart 1996: 14 (listed)
Glaucopsaltria viridis Moulds 1996b: 93, Plate 67 (illustrated, comp. note) Australia, Queensland, New South Wales
Glaucopsaltria viridis Young 1996: 100 (comp. note) Australia, Queensland
Glaucopsaltria viridis Boer 1997: 95, 97, 107–109, Fig 2, Figs 38–44 (phylogeny, described, type species of genus *Glaucopsaltria*, illustrated, comp. note) Queensland
Chlorocysta viridis Bennet-Clark and Young 1998: 702 (comp. note) Australia
Glaucopsaltria viridis Ewart 2001a: 501–509, Fig 4, Fig 7B, Fig 7D, Fig 8A, Fig 9, Tables 1–2 (acoustic behavior) Queensland
Chlorocysta viridis Sanborn 2001a: 16, Fig 5 (comp. note)
Chlorocysta viridis Sueur 2001: 42, Table 1 (listed)
Glaucopsaltria viridis Boulard 2002b: 199 (comp. note)
Chlorocysta viridis Nation 2002: 260, Table 9.3 (comp. note)
Glaucopsaltria viridis Moulds and Carver 2002: 21 (type species of *Glaucopsaltria*, distribution, ecology, type data, listed) Australia, Queensland, New South Wales
Glaucopsaltria viridis Popple and Ewart 2002: 118 (illustrated) Australia, Queensland
Glaucopsaltria viridis Williams 2002: 156–157 (listed) Australia, Queensland, New South Wales
Glaucopsaltria viridis Moulds 2005b: 395–398, 405, 409, 417–419, 421, 424, Fig 41, Fig 49, Figs 56–59, Fig 61, Table 1 (illustrated, phylogeny, comp. note) Australia
Chlorocysta viridis Nation 2008: 273, Table 10.1 (comp. note)

Thaumastopsaltria Kirkaldy, 1900 = *Thaumatopsaltria* (sic) = *Thoumastopsaltria* (sic) = *Taumatopsaltria* (sic) = *Thaumastropsaltria* (sic) = *Acrilla* Stål, 1863 (nec *Acrilla* Adams, 1860)
Thaumastopsaltria Duffels 1986: 328 (distribution, diversity, comp. note) New Guinea, Queensland, Misöol
Thaumastopsaltria Boer 1990: 64 (comp. note)
Thaumastopsaltria Moulds 1990: 12, 32, 191 (listed, diversity, comp. note) Australia, New Guinea
Thaumastopsaltria Boer 1991: 2 (comp. note)
Thaumastopsaltria Boer 1992a: 17–20, 22–25, 28, 30, 37–38 (phylogeny, described, comp. note) Equals *Acrilla* New Guinea, New Britain, Bougainville, Fergusson Island, Normanby Island, northern Queensland, Misoöl

Thaumastopsaltria Boer 1992b: 164 (comp. note)
Acrilla Boer 1992a: 17, 25 (comp. note) to *Thaumastopsaltria*
Thaumastopsaltria Boer 1993a: 16–19 (comp. note)
Thaumastopsaltria Boer 1993b: 141–143, 145, 154, 162 (phylogeny, comp. note)
Thaumastopsaltria Boer 1994c: 131, 134 (comp. note)
Thaumastopsaltria Boer 1994d: 90, 97 (comp. note) Papuan Peninsula
Thaumastopsaltria Boer 1995a: 2–3, 5–7, Fig 1 (phylogeny, comp. note)
Thaumastopsaltria Boer 1995b: 8–9, 15–17 (comp. note) Papua New Guinea, Waigeu, Solomon Islands
Thaumastopsaltria Boer 1995c: 202, 204, 207, 211, 215, 217–219, Fig 3, Fig 11, Fig 18 (phylogeny, comp. note)
Thaumastopsaltria Boer 1995d: 207–208, 217–219, 222, 224–226, 229, 231, 233–235, Fig 34, Figs 52–53 (distribution, phylogeny, comp. note) New Guinea, Papuan Peninsula, Queensland, Maluku
Thaumastopsaltria Boer 1996: 350–354, 356, 358, 360, Fig 1 (comp. note)
Thaumastopsaltria Boer and Duffels 1996a: 156, 165–166, 170–173, Figs 10–11 (distribution, phylogeny, comp. note) New Guinea, Queensland
Thaumastopsaltria Boer and Duffels 1996b: 301, 304, 313–314, 316–317, 326, Fig 5 (distribution, phylogeny, comp. note) Papua New Guinea
Thaumastopsaltria Boer 1997: 92–93, 98, 114, Fig 1 (phylogeny, comp. note)
Thaumastopsaltria Holloway 1997: 220, Fig 5 (distribution, phylogeny, comp. note) New Guinea
Thaumastopsaltria Holloway 1998: 304–308, Fig 10, Figs 12–13 (distribution, phylogeny, comp. note) New Guinea
Thaumastopsaltria Moulds and Carver 2002: 22–23 (listed) Australia
Thaumastopsaltria Moulds 2005b: 390, 412, 430, 435, Table 2 (higher taxonomy, listed, comp. note) Australia
Thaumastopsaltria Duffels and Boer 2006: 534 (distribution, comp. note) Misool

***T. adipata* (Stål, 1863)** = *Acrilla adipata* = *Thoumastopsaltria* (sic) *adipata*
Acrilla adipata Moulds 1990: 191 (type species of *Thaumastopsaltria*, listed)
Acrilla adipata Boer 1992a: 17, 20–23, 25, 27–31, 36, Fig 10, Figs 12–18 (type species of *Acrilla*, key, described, illustrated, comp. note) Equals *Acrilla adipata* Equals *Thoumastopsaltria* (sic) *adipata* Irian Jaya, Misoöl Island
Thaumastopsaltria adipata Boer 1995c: 213, 218, 226, 228, 230–231, Fig 4, Fig 13 (phylogeny, comp. note)
Thaumastopsaltria adipata Boer 1995d: 208, 231, 235, Fig 34 (distribution, comp. note) Misool?, Morotai?
Thaumastopsaltria adipata Boer 1996: 355 (comp. note)
Thaumastopsaltria adipata Boer and Duffels 1996a: 167 (comp. note)
Thaumastopsaltria adipata Boer and Duffels 1996b: 318, 326, Fig 12 (distribution, comp. note) New Guinea
Acrilla adipata Moulds and Carver 2002: 23 (type species of *Thaumastopsaltria*, listed) Equals *Mardalana fumea* Australia

T. glauca Ashton, 1912 to *Thaumastopsaltria globosa* (Distant, 1897)

***T. globosa* (Distant, 1897)** = *Acrilla globosa* = *Thaumastopsaltria glauca* Ashton, 1912 = *Thaumatopsaltria* (sic) *glauca*
Thaumastopsaltria glauca Moulds 1990: 191, Plaste 22, Fig 1, Fig 1a (illustrated, distribution, ecology, listed, comp. note) Australia, Queensland
Thaumastopsaltria globosa Boer 1992a: 19, 21–25, 27, 34–38, 40–42, Fig 3, Figs 7–8, Fig 11, Figs 36–49 (phylogeny, key, lectotype designation, described, illustrated, comp. note) Equals *Acrilla globosa* Equals *Thaumastopsaltria glauca* Equals *Thaumnatopsaltria* (sic) *glauca* New Guinea, Australia, northern Queensland, Northern Terrytory (sic)
Thaumastopsaltria glauca Boer 1992a: 34–35 (synonymy, lectotype designation, comp. note) to *Thaumastopsaltria globosa* n. syn.
Thaumastopsaltria globosa Boer 1995b: 16 (comp. note)
Thaumastopsaltria globosa Boer 1995c: 213, 218, 228, 230–231, Fig 4, Fig 13 (phylogeny, comp. note)
Thaumastopsaltria globosa Boer 1995d: 208–209, 222, 228, 233, Fig 34 (distribution, comp. note) New Guinea, Queensland, Grootte Island, Queensland
Thaumastopsaltria globosa Boer and Duffels 1996b: 318, 326, Fig 12 (distribution, comp. note) New Guinea
Thaumastopsaltria globosa Boer 1997: 91, 94–95 (comp. note) New Guinea, Australia, Queensland, Northern Territory

Thaumastopsaltria glauca Moulds and Carver 2002: 23 (distribution, ecology, type data, listed) Australia, Queensland
Thaumastopsaltria globosa Moulds 2005b: 395–397, 399, 417–419, 421, 424, Figs 56–59, Fig 61, Table 1 (phylogeny, comp. note) Australia
Thaumastopsaltria globosa Moulds 2009a: 503 (comp. note) Equals *Thaumastopsaltria glauca* Papua New Guinea, Australia

***T. lanceola* Boer, 1992**

Thaumastopsaltria lanceola Boer 1992a: 21–23, 25, 27, 38–41, Fig 10, Figs 60–68 (n. sp., phylogeny, key, described, illustrated, comp. note) New Guinea, D"Entrecasteau Islands, Fergusson Island, Normanby Island
Thaumastopsaltria lanceola Boer 1993b: 143, 162 (comp. note)
Thaumastopsaltria lanceola Boer 1995b: 16 (comp. note) New Guinea
Thaumastopsaltria lanceola Boer 1995c: 213, 218, 226, 228, 230–231, Fig 4, Fig 13 (phylogeny, comp. note)
Thaumastopsaltria lanceola Boer 1995d: 208, Fig 34 (distribution, comp. note) New Guinea
Thaumastopsaltria lanceola Boer and Duffels 1996b: 318, 327, Fig 12 (distribution, comp. note) New Guinea
Thaumastopsaltria lanceola Sanborn 1999a: 51 (type material, listed)

T. nana (Jacobi, 1903) to *Papuapsaltria nana*

***T. pneumatica* Boer, 1992**

Thaumastopsaltria pneumatica Boer 1992a: 19–26, 29–31, 33, Figs 5–6, Fig 11, Figs 19–27 (n. sp., phylogeny, key, described, illustrated, comp. note) Papuan Peninsula
Thaumastopsaltria pneumatica Boer 1995c: 213, 218, 228, 230–231, Fig 4, Fig 13 (phylogeny, comp. note)
Thaumastopsaltria pneumatica Boer 1995d: 208, Fig 34 (distribution, comp. note) Papuan Peninsula
Thaumastopsaltria pneumatica Boer 1996: 355 (comp. note)
Thaumastopsaltria pneumatica Boer and Duffels 1996b: 318, 327, Fig 12 (distribution, comp. note) Papuan Peninsula
Thaumastopsaltria pneumatica Sanborn 1999a: 51 (type material, listed)

***T. sarissa* Boer, 1992**

Thaumastopsaltria sarissa Boer 1992a: 19, 21–22, 24, 27, 41–43, Fis 2, Fig 11, Figs 69–76 (n. sp., phylogeny, key, described, illustrated, comp. note) New Guinea
Thaumastopsaltria sarissa Boer 1993b: 162 (comp. note)
Thaumastopsaltria sarissa Boer 1995c: 213, 218, 225, 228, 230–231, Fig 4, Fig 13 (phylogeny, comp. note)
Thaumastopsaltria sarissa Boer 1995d: 208, Fig 34 (distribution, comp. note) New Guinea
Thaumastopsaltria sarissa Boer and Duffels 1996b: 318, 327, Fig 12 (distribution, comp. note) New Guinea

***T. sicula* Boer, 1992**

Thaumastopsaltria sicula Boer 1992a: 20–23, 25–26, 36–38, Figs 9–10, Figs 50–59 (n. sp., phylogeny, key, described, illustrated, comp. note) north Irian Jaya, Waigeu Island
Thaumastopsaltria sicula Boer 1993b: 143 (comp. note)
Thaumastopsaltria sicula Boer 1995c: 213, 218, 225–226, 228, 230–231, Fig 4, Fig 13 (phylogeny, comp. note)
Thaumastopsaltria sicula Boer 1995d: 208, Fig 34 (distribution, comp. note) New Guinea, Waigeu Island
Thaumastopsaltria sicula Boer and Duffels 1996b: 318, 327, Fig 12 (distribution, comp. note) New Guinea

***T. smithersi* Moulds, 2012**

***T. spelunca* Boer, 1992**

Thaumastopsaltria spelunca Boer 1992a: 19–25, 27, 31–38, Fig 1, Fig 4, Fig 11, Figs 28–35 (n. sp., phylogeny, key, described, illustrated, comp. note)
Thaumastopsaltria spelunca Boer 1995b: 16 (comp. note) New Guinea, Umboi Island, New Britain
Thaumastopsaltria spelunca Boer 1995c: 213, 218, 226, 228, 230–231, Fig 4, Fig 13 (phylogeny, comp. note)
Thaumastopsaltria spelunca Boer 1995d: 208, Fig 34 (distribution, comp. note) New Guinea, New Britain, Buka, Bougainville
Thaumastopsaltria spelunca Boer 1996: 358 (comp. note) Papuan Peninsula
Thaumastopsaltria spelunca Boer and Duffels 1996a: 173 (distribution, comp. note) New Guinea, Bismarck Archipelago
Thaumastopsaltria spelunca Boer and Duffels 1996b: 318, 327, Fig 12 (distribution, comp. note) Papuan Peninsula, Bismarck Archipelago, Solomon Islands
Thaumastopsaltria spelunca Duffels 1997: 552 (comp. note) Solomon Islands, Papuan Peninsula, Bismarck Archipelago

***Cystosoma* Westwood, 1842** = *Cicada (Cystosoma)* = *Cystisoma* (sic) = *Cyrtosoma* (sic) = *Cytosoma* (sic)

Cystosoma MacNally and Young 1981: 195 (comp. note)

Cystosoma Jong 1982: 182 (comp. note) Australia
Cystosoma Strümpel 1983: 17, 109, Table 8 (listed, comp. note)
Cystosoma Ewing 1984: 228–229 (comp. note)
Cystosoma Boulard 1985e: 1026, 1032 (coloration, comp. note) Australia
Cystosoma Popov 1985: 43 (acoustic system, comp. note)
Cystosoma Feng 1988: 27 (comp. note)
Cystosoma Ewart 1989a: 79 (comp. note)
Cystosoma Ewing 1989: 35–36, Fig 2.9 (sound apparatus illustrated, comp. note)
Cystosoma Boer 1990: 64 (comp. note)
Cystosoma Boulard 1990b: 142 (comp. note)
Cystosoma Moulds 1990: 16, 18–19, 23, 32 (listed, diversity, comp. note) Australia, Queensland, New South Wales
Cystosoma spp, Moulds 1990: 16, 21 (listed, diversity, comp. note) Australia
Cystosoma Boer 1991: 2 (comp. note)
Cystosoma Boer 1992a: 18–19, 21–22 (phylogeny, comp. note)
Cystosoma Boer 1992b: 164 (comp. note)
Cystosoma Boer 1993a: 16–17 (comp. note)
Cystosoma Boer 1993b: 141–142, 154 (phylogeny, comp. note)
Cystosoma Villet 1993: 439 (comp. note)
Cystosoma Forrest and Raspert 1994: 300 (comp. note)
Cystosoma Prestwich 1994: 640 (comp. note)
Cystosoma Boer 1995a: 2–3, Fig 1 (phylogeny, comp. note)
Cystosoma Boer 1995b: 8, 12, 16 (comp. note)
Cystosoma Boer 1995c: 202, 204, 210–211, 214–215, 218–219, Fig 3, Fig 11, Fig 18 (phylogeny, comp. note) Australia
Cystosoma Boer 1995d: 218–219, 222, 224–225, 233–234, Fig 48, Figs 52–53 (distribution, phylogeny, comp. note) Australia
Cystosoma Boer 1996: 350–355, Fig 1 (comp. note)
Cystosoma Boer and Duffels 1996a: 155, 168, 170–171, Figs 10–11 (distribution, phylogeny, comp. note) Australia
Cystosoma Boer and Duffels 1996b: 301, 304, 325, Fig 5 (distribution, phylogeny, comp. note) Australia
Cystosoma Boer 1997: 91–93, 112–113, 119, 121, Fig 1 (phylogeny, described, comp. note) Australia
Cystosoma Boulard 1997b: 51 (morphology, comp. note) Australia
Cystosoma Holloway 1997: 220, Fig 5 (distribution, phylogeny, comp. note) Australia
Cystosoma Bennet-Clark 1998a: 411 (comp. note)
Cystosoma Holloway 1998: 308, Fig 13 (distribution, comp. note) Australia
Cystosoma Reinhold 1999: 219, Table 1 (listed, comp. note)
Cystosoma Marshall and Cooley 2000: 1320 (comp. note) Australia
Cystosoma Cooley 2001: 758 (comp. note) Australia
Cystosoma Cooley and Marshall 2001: 847 (comp. note) Australia
Cystosoma Moulds and Carver 2002: 20 (listed)
Cystosoma Williams 2002: 38 (diversity, comp. note) Australia
Cystosoma spp. Lee 2003b: 7 (comp. note)
Cystosoma Boulard 2005c: 233 (comp. note)
Cystosoma Boulard 2005j: 80 (comp. note)
Cystosoma Moulds 2005b: 390, 392, 412–413, 415, 422, 430, 435, Table 2 (higher taxonomy, listed, comp. note) Australia
Cystosoma Shiyake 2007: 9, 115 (listed, comp. note)

***C. saundersii* (Westwood, 1842)** = *Cicada (Cystosoma) saundersii* = *Cystosoma saundersi* (sic) = *Cystosoma laudersii* (sic) = *Cytosoma* (sic) *saundersii*

Cystosoma saundersii Swinton 1880: 224 (abdomen, comp. note)
Cystosoma saundersi (sic) Evans 1956: 231, Fig 24C (comp. note) Australia
Cystosoma saundersii Hill, Lewis, Hutchins and Coles 1980: 149 (listed)
Cystosoma saundersii Josephson 1981: 26, 30–31, 34, 39–40, Fig 3, Figs 7–8, Fig 10 (muscle function, comp. note)
Cystosoma saundersii Josephson and Young 1981: 225, 227 (comp. note)
Cystosoma saundersii MacNally and Young 1981: 185–187, 189–195, Fig 1, Fig 4, Tables 1–3 (song energetics, comp. note) New South Wales
Cystosoma saundersii Popov 1981: 276–277 (comp. note)
Cystosoma saundersii MacNally and Doolan 1982: 180–186, Fig 1, Tables 1–5 (reproductive energetics, comp. note) New South Wales
Cystosoma saundersii Huber 1983a: 108, 116 (comp. note)
Cystosoma saundersii Hutchins and Lewis 1983: 194–195 (hearing organ, comp. note)
Cystosoma saundersii Moulds 1983: 429–430 (illustrated, comp. note) Australia, Queensland, New South Wales
Cystosoma saundersii Strümpel 1983: 17, 108–109, 133, Fig 30, Fig 144 (illustrated, oscillogram, comp. note)
Cystosoma saundersii Young and Josephson 1983a: 192, Table 1 (comp. note)
Cystosoma saundersii Young and Josephson 1983b: 204, 206 (comp. note)
Cystosoma saundersii Young and Josephson 1984: 286, Fig 1 (comp. note)

Cystosoma saundersii Boulard 1985e: 1027, 1040–1041, Plate V, Fig 34 (coloration, illustrated, comp. note)
Cystosoma saundersii Claridge 1985: 302–303, 307–309, Fig 2 (song, oscillogram, comp. note) Australia
Cystosoma saundersii Michelsen and Larsen 1985: 505, 549–550 (comp. note)
Cystosoma saundersii Popov 1985: 37–38, 44, 46–47, 68, 105, 130–133, 181, 184, Fig 6, Fig 8 (illustrated, acoustic system, comp. note)
Cystosoma saundersii Ryan 1985: 51 (comp. note)
Cystosoma saundersii Ewart 1986: 51, 54, Table 1 (natural history, listed, comp. note) Queensland, New South Wales
Cystosoma saundersii Heller 1986: 97 (comp. note)
Cystosoma saundersii Johnson and Foster 1986: 421 (comp. note)
Cystosoma saundersii MacNally and Doolan 1986a: 35, 37, 40–42, Figs 1–4, Table 2 (habitat, comp. note) New South Wales
Cystosoma (sic) MacNally and Doolan 1986a: 37–39, 41, 43 (comp. note) New South Wales
Cystosoma saundersii MacNally and Doolan 1986b: 281, 285–286, 288, Fig 2, Fig 4, Table 2 (habitat, comp. note) New South Wales
Cystosoma (sic) MacNally and Doolan 1986b: 282, 284–291, Figs 1–3, Table 3, Table 5 (habitat, comp. note) New South Wales
Cystosoma saundersii Brown 1987: 6 (comp. note) New South Wales
Cystosoma saundersii Gwynne 1987: 575 (comp. note)
Cystosoma saundersii Josephson and Young 1987: 994–996, Figs 1–3 (comp. note)
Cystosoma saundersii Kavanagh 1987: 115–118, Tables 3–4 (song energetics, comp. note)
Cystosoma saundersii Villet 1987a: 269 (comp. note) Australia
Cystosoma saundersii Burk 1988: 402–404, Tables 1–3 (energetics of calling, comp. note)
Cystosoma saundersii Ewart 1988: 181 (comp. note) Queensland
Cystosoma saundersii MacNally 1988a: 247 (distribution model, comp. note) New South Wales
Cystosoma saundersii MacNally 1988b: 1976–1977, Tables 1–2 (comp. note) New South Wales
Cystoma (sic) *saundersii* Ryan 1988: 887, Table 1 (comp. note)
Cystosoma saundersii Ewart 1989a: 76–78 (comp. note)
Cystosoma saundersii Ewing 1989: 36, 66, Fig 3.5 (oscillogram, auditory system, comp. note)
Cystosoma saundersii Popov 1989a: 304 (comp. note)
Cystosoma saundersii Popov 1989b: 74 (comp. note)
Cystosoma saundersii Boulard 1990b: 155, 157, Plate XVIII, Fig 8 (comp. note)
Cystosoma saundersii Feng 1990: 447 (comp. note)
Cystosoma saundersii Fullard and Heller 1990: 63–64 (comp. note)
Cystosoma saundersii Moulds 1990: 3–5, 7–9, 12, 19, 21, 23, 25, 110, 190, 193–196, Plate 4, Fig 1, Plate 23, Fig 1, Figs 1a–1b, Plate 24, Fig 29 (type species of *Cystosoma*, illustrated, distribution, ecology, listed, comp. note) Australia, New South Wales, Queensland, New Holland
Cystosoma saundersii Young 1990: 42, 46, 53–55, Table 2 (comp. note)
Cystosoma saundersii Bailey 1991a: 54 (energetics, comp. note)
Cystosoma saundersii Bailey 1991b: 2, 76–77, 106, Figs 1.1–1.2, Fig 5.10 (illustrated, sound production, timbal anatomy, comp. note)
Cystosoma saundersii Carver, Gross and Woodward 1991: 431, 434, 437, Fig 30.3B, Fig 30.6D, Fig 30.9A (thorax illustrated, ovipositor illustrated, salivary glands illustrated)
Cystosoma saundersii Forrest 1991: 336, Table 2 (comp. note)
Cystosoma saundersii Moulds and Carver 1991: 466–467, Fig 30.27A (illustrated, comp. note) Australia
Cystosoma saundersii Fletcher 1992: 243 (comp. note)
Cystosoma saundersii Jiang, Wang, Xu, Zhang and Chen 1992: 236 (comp. note)
Cystosoma saundersii Villet 1992: 93, 96 (comp. note)
Cystosoma saundersii Daniel, Knight and Charles 1993: 67 (comp. note) Australia
Cystosoma saundersii Fonseca 1993: 767 (comp. note)
Cystosoma saundersii Moore, Huber, Weber, Klein and Bock 1993: 217 (comp. note) Australia
Cystosoma saundersii Naumann 1993: 7, 84, 149 (listed) Australia
Cystosoma saundersii Shen 1993: 363 (comp. note)
Cystosoma saundersii White and Sedcole 1993a: 48 (comp. note) Australia
Cystosoma saundersii Bennet-Clark and Young 1994: 292 (comp. note)
Cystosoma saundersii Carver, Gross and Woodward 1994: 318, Fig 30.3B (thorax illustrated)
Cystosoma saundersii Fonseca and Popov 1994: 350 (comp. note)
Cystosoma saundersii Forrest and Raspert 1994: 300 (comp. note)
Cystosoma saundersii Prestwich 1994: 632, 637, 639, Tables 1–3 (energetics, listed, comp. note)

Cystosoma saundersii Simmons 1994: 190 (comp. note)
Cystosoma saundersii Boer 1995a: 5 (comp. note)
Cystosoma saundersii Boer 1995c: 213, 228, 230–231, Fig 4 (phylogeny, comp. note)
Cystosoma saundersii Boer 1995d: 219, Fig 48 (distribution, comp. note) Australia
Cystosoma saundersii Ewart 1995: 83 (described, illustrated, natural history, acoustic behavior, oscillogram) Queensland
Cystosoma saundersii Frith 1995: 44 (predation) Australia
Cystosoma saundersii Jiang, Yang, Tang, Xu and Chen 1994: 291 (comp. note)
Cystosoma saundersii Jiang, Yang, Wang, Xu, Chen and Yang 1995: 676 (comp. note) China
Cystosoma saundersii Lane 1995: 405, 407 (comp. note)
Cystosoma saundersii Riede and Kroker 1995: 49 (comp. note)
Cystosoma saundersii Sanborn, M.S. Heath, J.E. Heath and Noriega 1995: 457–458 (comp. note)
Cystosoma saundersii Sanborn and Phillips 1995a: 480, 483 (comp. note)
Cystosoma saundersii Boer and Duffels 1996b: 325 (distribution) Australia
Cystosoma saundersii Cocroft and Pogue 1996: 95 (comp. note)
Cystosoma saundersii Coombs 1996: 57, 59, Table 1 (emergence time, comp. note) New South Wales
Cystosoma saundersii Daws and Hennig 1996: 175, 185 (auditory responses, comp. note)
Cystosoma saundersii Ewart 1996: 14 (listed)
Cystosoma saundersii Hoy and Robert 1996: 436, Table 1 (comp. note)
Cystosoma saundersii Jiang, Yang and Liu 1996: 314 (comp. note)
Cystosoma saundersii Moulds 1996b: 93, Plate 66 (illustrated, comp. note) Australia, Queensland, New South Wales
Cystosoma saundersii Young 1996: 98, 102–104, Fig 5.4, Plate 66 (acoustic behavior, sound system illustrated, comp. note)
Cystosoma saundersii Alexander, Marshall and Cooley 1997: 12 (comp. note)
Cystosoma saundersii Boer 1997: 93, 96, 114, 117–121, Fig 56, Figs 69–71, Figs 73–75, Fig 81 (phylogeny, described, illustrated, comp. note) Queensland, New South Wales
Cystosoma saundersii Boulard 1997b: 21, Plate III, Fig 4 (morphology, comp. note)
Cystosoma saundersii Daws, Hennig and Young 1997: 174–177, 179–186, Fig 1A, Figs 3–4 (photaxis, comp. note) New South Wales
Cystosoma saundersii Fonseca and Popov 1997: 425 (comp. note)
Cystosoma saundersii Shelly and Whittier 1997: 279, 286, Table 16-1 (listed, comp. note)
Cystosoma saundersii Bailey 1998: 109 (female acoustic response comp. note)
Cystosoma saundersii Bennet-Clark and Young 1998: 702–712, 714, Figs 1–11, Table 1 (sound production, acoustic system illustrated) New South Wales, Australia
Cystosoma saundersii Moulds and Hangay 1998: 75–76 (distribution, comp. note) Australia, Lord Howe Island, New South Wales
Cystosoma saundersii Römer 1998: 82, 87 (comp. note)
Cystosoma saundersii Sanborn 1998: 93–95 (thermal biology, comp. note)
Cystosoma saundersii Bennet-Clark 1999: 3351, 3354–3356, Fig 4C, Table 2 (comp. note, song production)
Cystosoma saundersii Jamieson, Dallai and Afzelius 1999: 233, Fig 12.1B (illustrated, comp. note)
Cystosoma saundersii Sanborn and Phillips 1999: 454 (comp. note)
Cystosoma saundersii Boulard 2000b: 9 (parallel evolution)
Cystosoma saundersii Sanborn and Maté 2000: 142 (comp. note)
Cystosoma saundersii Cooley and Marshall 2001: 847 (comp. note)
Cystosoma saundersii Ewart 2001a: 499, 501–508, Fig 4, Fig 7A, Fig 8B, Fig 9, Tables 1–2 (acoustic behavior, comp. note) Queensland
Cystosoma saundersii Lee 2001b: 3 (comp. note)
Cystosoma saundersii Sanborn 2001a: 16, Fig 5 (comp. note)
Cystosoma saundersii Sueur 2001: 45–46, Table 2 (listed, comp. note)
Cystosoma saundersii Anonymous 2002: 30 (color morph) Queensland, New South Wales
Cystosoma saundersii Boulard 2002b: 199 (comp. note)
Cystosoma saundersii Fonseca and Revez 2002a: 1290–1291 (comp. note)
Cystosoma saundersii Fonseca and Revez 2002b: 971, 974–975 (comp. note)
Cystosoma saundersii Gerhardt and Huber 2002: 22, 33, 125, 127, 255–256, 345, Fig 2.6A, Fig 8.1 (illustrated, song, hearing, mate selection, comp. note)
Cystosoma saundersii Nation 2002: 260, Table 9.3 (comp. note)
Cystosoma saundersii Moulds and Carver 2002: 20 (type species of *Cystosoma*, distribution,

ecology, type data, listed) Australia, New South Wales, Queensland, New Holland
Cystosoma saundersii Sanborn 2002b: 461 (thermal biology, comp. note)
Cystosoma saundersii Sanborn, Breitbarth, Heath and Heath 2002: 442, 446 (comp. note)
Cystosoma saundersii Sueur 2002b: 119, 123, 129 (listed, comp. note) Australia
Cystosoma saundersii Sueur and Aubin 2002b: 135 (comp. note)
Cystosoma saundersii Williams 2002: 156–157 (listed) Australia, Queensland, New South Wales
Cystosoma saundersii Lee 2003b: 6–7, Fig 3 (illustrated, comp. note) Australia
Cystosoma saundersii Sueur 2003: 2940, 2942–2943 (comp. note)
Cystosoma saundersii Sueur and Aubin 2003b: 486–487 (comp. note)
Cystosoma saundersii Villet, Sanborn and Phillips 2003c: 1441–1442 (comp. note)
Cystosoma saundersii Fletcher 2004: 2337 (song power, comp. note) Australia
Cystosoma saundersii Fonseca and Hennig 2004: 407 (comp. note)
Cystosoma saundersii Gogala, Trilar, Kozina and Duffels 2004: 14 (comp. note)
Cystosoma saundersii Sueur and Aubin 2004: 223 (comp. note)
Cystosoma saundersii Champman 2005: 321, 327 (comp. note)
Cystosoma saundersii Lee 2005: 32–33 (illustrated, comp. note)
Cystosoma saundersii Moulds 2005b: 395–398, 401, 417–419, 421, 424, Fig 33, Figs 56–59, Fig 61, Table 1 (illustrated, phylogeny, comp. note) Australia
Cystosoma saundersii Puissant 2005: 312 (comp. note)
Cystosoma saundersii Sanborn 2005c: 115, Figs 7.1–7.2 (comp. note)
Cystosoma saundersii Seabra, Pinto-Juma and Quartau 2006: 844 (comp. note)
Cystosoma saundersii Boulard 2007f: 83, Fig 99.4 (illustrated, comp. note)
Cystosoma saundersii Shiyake 2007: 9, 113, 115, Fig 205 (illustrated, distribution, listed, comp. note) Australia
Cystosoma saundersii Nation 2008: 273, 412, Table 10.1 (comp. note)
Cystosoma saundersii Sueur, Windmill and Robert 2008a: 2380 (comp. note)
Cystosoma saundersii Zamanian, Mehdipour and Ghaemi 2008: 2070 (comp. note)
Cystosoma saundersii Sanborn, Heath and Heath 2009: 25 (comp. note)
Cystosoma saundersii Strauβ and Lakes-Harlan 2009: 313 (comp. note)

***C. schmeltzi* Distant, 1882** = *Cystosoma schmelzi* (sic)
Cystosoma schmeltzi MacNally and Young 1981: 195 (comp. note)
Cystosoma schmeltzi Boulard 1985e: 1027 (coloration, comp. note)
Cystosoma schmeltzi Lithgow 1988: 65 (listed) Australia, Queensland
Cystosoma schmeltzi Ewart 1989a: 79 (comp. note) Queensland
Cystosoma schmeltzi Moulds 1990: 192–193, Plate 23, Fig 2, Fig 2a (illustrated, distribution, ecology, listed, comp. note) Australia, Queensland, New South Wales
Cystosoma schmeltzi Naumann 1993: 34, 84, 149 (listed) Australia
Cystosoma schmeltzi Boer 1995c: 213, 228, 230–231, Fig 4 (phylogeny, comp. note)
Cystosoma schmeltzi Boer 1995d: 219, Fig 48 (distribution, comp. note) Australia
Cystosoma schmeltzi Boer and Duffels 1996b: 325 (distribution) Australia
Cystosoma schmeltzi Anonymous 1997b: 102–103, Fig 10.1, 10.2 (natural history, control, distribution map, illustrated) Australia
Cystosoma schmeltzi Boer 1997: 93, 96, 114–117, 119, 121, Figs 56–68 (phylogeny, described, type species of genus *Cystosoma*, illustrated, comp. note) Queensland, New South Wales
Cystosoma schmeltzi Moulds and Hangay 1998: 75 (citrus pest, comp. note) Australia
Cystosoma schmeltzi Boulard 2002b: 199 (comp. note)
Cystosoma schmeltzi Moulds and Carver 2002: 21 (distribution, ecology, type data, listed) Australia, Queensland, New South Wales

***Cystopsaltria* Goding & Froggatt, 1904**
Cystopsaltria Moulds 1990: 16, 32, 192, 196–197 (listed, diversity, comp. note) Australia
Cystopsaltria Boer 1992a: 18–19 (phylogeny, comp. note)
Cystopsaltria Boer 1993a: 16–17 (comp. note)
Cystopsaltria Boer 1993b: 141–142, 154 (phylogeny, comp. note)
Cystopsaltria Boer 1995a: 2–3, Fig 1 (phylogeny, comp. note)
Cystopsaltria Boer 1995b: 8 (comp. note)
Cystopsaltria Boer 1995c: 202–204, 206–207, 3211, 215, 218–219, Fig 3, Fig 11, Fig 18 (phylogeny, comp. note) Australia
Cystopsaltria Boer 1995d: 218–219, 222, 224–225, 233, Fig 47, Figs 52–53 (distribution, phylogeny, comp. note) Australia

Cystopsaltria Boer 1996: 350–355, Fig 1 (comp. note)
Cystopsaltria Boer and Duffels 1996a: 155, 168, 170–171, Figs 10–11 (distribution, phylogeny, comp. note) Australia
Cystopsaltria Boer and Duffels 1996b: 301, 304, 314, 325, Fig 5 (distribution, phylogeny, comp. note) Australia
Cystopsaltria Boer 1997: 91–93, 112–113, Fig 1 (phylogeny, described, comp. note) Australia
Cystopsaltria Holloway 1997: 220, Fig 5 (distribution, phylogeny, comp. note) Australia
Cystopsaltria Holloway 1998: 308, Fig 13 (distribution, comp. note) Australia
Cystopsaltria Moulds and Carver 2002: 20 (listed) Australia
Cystopsaltria Moulds 2005b: 390, 413, 430, 435, Table 2 (higher taxonomy, listed, comp. note) Australia

***C. immaculata* Goding & Froggatt, 1904**
Cystopsaltria immaculata Moulds 1990: 19, 193–194, 197–198, Plate 23, Fig 3, Fig 3a (type species of *Cystopsaltria*, illustrated, distribution, ecology, listed, comp. note) Australia, Queensland
Cystopsaltria immaculata Naumann 1993: 47, 84, 149 (listed) Australia
Cystopsaltria immaculata Boer 1995c: 213, 225–226, 228, 230–231, Fig 4 (phylogeny, comp. note)
Cystopsaltria immaculata Boer 1995d: 219, Fig 48 (distribution, comp. note) Australia
Cystopsaltria immaculata Boer and Duffels 1996b: 325 (distribution) Australia
Cystopsaltria immaculata Boer 1997: 93, 95, 114, 118–121, Fig 56, Fig 72, Figs 78–80, Figs 82–87 (phylogeny, described, type species of genus *Cystopsaltria*, illustrated, comp. note) Queensland
Cystopsaltria immaculata Moulds and Carver 2002: 20 (type species of *Cystopsaltria*, distribution, ecology, type data, listed) Australia, Queensland
Cystopsaltria immaculata Moulds 2005b: 395–398, 417–419, 421, 424, Figs 56–59, Fig 61, Table 1 (phylogeny, comp. note) Australia

***Venustria* Goding & Froggatt, 1904** = *Venustra* (sic)
Venustria Boer 1990: 64 (comp. note)
Venustria Moulds 1990: 32, 180 (listed, diversity, comp. note) Australia
Venustria Boer 1991: 2–3 (comp. note)
Venustria Boer 1992a: 19 (phylogeny, comp. note)
Venustria Boer 1992b: 164 (comp. note)
Venustria Boer 1993a: 16–17 (comp. note)
Venustria Boer 1993b: 142 (phylogeny, comp. note)
Venustria Boer 1994a: 3 (comp. note)
Venustria Boer 1994c: 129–130, Fig 1a (phylogeny, comp. note)
Venustria Boer 1995a: 2, 5, Fig 1 (phylogeny, comp. note)
Venustria Boer 1995b: 3–5, 8, 11, Fig 1 (comp. note)
Venustria Boer 1995c: 202, 204, 210–211, 214–215, 218–219, Fig 3, Fig 11, Fig 18 (phylogeny, comp. note) Australia
Venustria Boer 1995d: 218, 222, 224–225, 233, Figs 52–53 (distribution, phylogeny, comp. note) Australia, Queensland
Venustria Boer 1996: 352, Fig 1 (comp. note)
Venustria Boer and Duffels 1996a: 156, 168, 171–173, Figs 10–11 (distribution, phylogeny, comp. note) Australia
Venustria Boer and Duffels 1996b: 301, 304, 327, Fig 5 (distribution, phylogeny, comp. note) Australia
Venustria Boer 1997: 91–94, 109, Fig 1 (phylogeny, described, comp. note) Equals *Venustra* (sic) Australia
Venustria Holloway 1997: 220–221, Fig 5 (distribution, phylogeny, comp. note) Australia
Venustria Holloway 1998: 308, Fig 13 (distribution, comp. note) Australia
Venustria Boer 2000: 4, 15, Fig 1 (phylogeny, comp. note)
Venustria Moulds and Carver 2002: 23 (listed) Australia
Venustria Moulds 2005b: 390, 407, 412, 422, 430, 435, Table 2 (higher taxonomy, listed, comp. note) Australia
Venustria Shiyake 2007: 7, 102 (listed, comp. note)

***V. superba* Goding & Froggatt, 1904**
Venustria superba Hahn 1962: 11 (listed) Australia, Queensland
Venustria superba Stevens and Carver 1986: 265 (type material) Australia, Queensland
Venustria superba Southcott 1988: 103–104, 107, 115, Fig 1 (illustrated, parasitism, comp. note) Australia, Queensland
Venustria superba Moulds 1990: 11, 19, 180–181, Plate 16, Fig 11, Plate 24, Fig 15 (type species of *Venustria*, distribution, ecology, type data, listed) Australia, Queensland
Venustria superba McGavin 1993: 160 (emergence, illustrated) Queensland
Venustria superba Naumann 1993: 23, 130, 149 (listed) Australia
Venustria superba Boer 1995b: 3–4, 9, 13, 15, 74, 77 (comp. note)
Venustria superba Boer 1995c: 213–215, 225, 228, 230–231, Fig 4, Figs 9–10 (phylogeny, comp. note)

Venustria superba Boer 1995d: 206, 218, 233 (distribution, comp. note) Queensland
Venustria superba Boer and Duffels 1996b: 327 (distribution) Australia
Venustria superba Boer 1997: 92–93, 96, 98, 108–112, Figs 45–55 (phylogeny, described, type species of genus *Venustria*, illustrated, comp. note) Queensland
Venustria superba Freytag 2000: 55 (on stamp)
Venustria superba Moulds and Carver 2002: 23 (type species of *Venustria*, distribution, ecology, type data, listed) Australia, Queensland
Venustria superba Moulds 2005b: 395–397, 399, 417–419, 421, 424, Figs 56–59, Fig 61, Table 1 (phylogeny, comp. note) Australia
Venustria superba Shiyake 2007: 7, 100, 102, Fig 168 (illustrated, distribution, listed, comp. note) Australia

***Mirabilopsaltria* Boer, 1996**

Mirabilopsaltria Boer 1995a: 2–3, 6, Fig 1 (phylogeny, comp. note)
Mirabilopsaltria Boer 1995c: 202, 204, 206, 211, 215, 217–219, Fig 3, Fig 11, Fig 18 (phylogeny, comp. note)
Mirabilopsaltria Boer 1995d: 217–218, 222, 224–226, 231–232, 235, Fig 46, Figs 52–53 (distribution, phylogeny, comp. note)
Mirabilopsaltria Boer 1996: 350–357, 376, Fig 1 (n. gen., described, phylogeny, comp. note) New Guinea, New Britain
Mirabilopsaltria Boer and Duffels 1996a: 155, 167, 170–173, Figs 10–11 (distribution, phylogeny, comp. note) New Guinea, Bismarck Archipelago
Mirabilopsaltria Boer and Duffels 1996b: 301, 304, 306, 313–314, 326, Fig 5 (distribution, phylogeny, comp. note) Papua New Guinea
Mirabilopsaltria Boer 1997: 92–93, 98, Fig 1 (phylogeny, comp. note)
Mirabilopsaltria Holloway 1997: 220, Fig 5 (distribution, phylogeny, comp. note) New Guinea
Mirabilopsaltria Holloway 1998: 304–308, Fig 10, Figs 12–13 (distribution, phylogeny, comp. note) New Guinea
Mirabilopsaltria Boer 2000: 12 (comp. note)
Mirabilopsaltria Moulds 2005b: 390, 435 (higher taxonomy, listed, comp. note)
Mirabilopsaltria Duffels and Boer 2006: 537 (distribution, comp. note) New Guinea

***M. globulata* Boer, 1996**

Mirabilopsaltria globulata Boer 1995c: 213, 218, 225–226, 228, 230–231, Fig 4, Fig 13 (phylogeny, comp. note)
Mirabilopsaltria globulata Boer 1995d: 217–218, Fig 46 (distribution, comp. note) New Guinea
Mirabilopsaltria globulata Boer 1996: 351, 353–358, 360–361, 367–371, 373, 375, Fig 2, Figs 30–42 (n. sp., key, described, illustrated, phylogeny, distribution, comp. note) New Guinea
Mirabilopsaltria globulata Boer and Duffels 1996b: 325 (distribution) Papuan Peninsula, Bismarck Archipelago

***M. humilis* (Blöte, 1960)** = *Baeturia humilis*

Baeturia humilis Boer 1993b: 141–142 (comp. note)
Baeturia humilis Boer 1995b: 8 (comp. note)
Mirabilopsaltria humilis Boer 1995c: 213, 218, 225–226, 228, 230–231, Fig 4, Fig 13 (phylogeny, comp. note)
Mirabilopsaltria humilis Boer 1995d: 217–218, Fig 46 (distribution, comp. note) New Guinea, Biak Island
Mirabilopsaltria humilis Boer 1996: 351, 353–359, 361–368, 373, 376, Figs 2–18 (n. comb., type species of genus *Mirabilopsaltria*, key, described, illustrated, phylogeny, distribution, comp. note) Equals *Baeturia humilis* New Guinea, Biak Island
Mirabilopsaltria humilis Boer and Duffels 1996b: 325 (distribution) New Guinea
Mirabilopsaltria humilis Boer 1997: 113 (comp. note)
Mirabilopsaltria humilis Boer 2000: 12 (comp. note)

***M. inconspicua* Boer, 1996**

Mirabilopsaltria inconspicua Boer 1995c: 213, 218, 228, 230–231, Fig 4, Fig 13 (phylogeny, comp. note)
Mirabilopsaltria inconspicua Boer 1995d: 217–218, Fig 46 (distribution, comp. note) New Guinea
Mirabilopsaltria inconspicua Boer 1996: 351, 354–357, 361, 364–367, Fig 2, Figs 19–29 (n. sp., key, described, illustrated, phylogeny, distribution, comp. note) New Guinea
Mirabilopsaltria inconspicua Boer and Duffels 1996b: 325 (distribution) New Guinea

***M. inflata* Boer, 1996**

Mirabilopsaltria inflata Boer 1995c: 213, 218, 225–226, 228, 230–231, Fig 4, Fig 13 (phylogeny, comp. note)
Mirabilopsaltria inflata Boer 1995d: 217–218, Fig 46 (distribution, comp. note) New Guinea, Bismarck Archipelago
Mirabilopsaltria inflata Boer 1996: 351, 354–361, 370–373, Fig 2, Figs 43–51 (n. sp., key, described, illustrated, phylogeny, distribution, comp. note) Bismarck Archipelago, New Britain, New Ireland

Mirabilopsaltria inflata Boer and Duffels 1996b: 325 (distribution) New Guinea, Bismarck Archipelago

***M. toxopeusi* (Blöte, 1960)** = *Baeturia toxopeusi*

Mirabilopsaltria toxopeusi Boer 1995c: 213, 218, 225–226, 228, 230–231, Fig 4, Fig 13 (phylogeny, comp. note)

Mirabilopsaltria toxopeusi Boer 1995d: 217–218, Fig 46 (distribution, comp. note) New Guinea

Mirabilopsaltria toxopeusi Boer 1996: 351, 354–361, 376–378, Fig 2, Figs 64–72 (n. comb., key, described, illustrated, phylogeny, distribution, comp. note) Equals *Baeturia toxopeusi* Irian Jaya

Baeturia toxopeusi Boer 1996: 376 (type series)

Mirabilopsaltria toxopeusi Boer and Duffels 1996b: 325 (distribution) New Guinea

Baeturia toxopeusi Sanborn 1999a: 52 (type material, listed)

***M. viridicata* (Distant, 1897)** = *Baeturia viridicata*

Baeturia viridicata Boer 1992a: 21 (comp. note)

Baeturia viridicata Boer 1993b: 141–142 (comp. note)

Baeturia viridicata Boer 1995b: 8 (comp. note)

Mirabilopsaltria viridicata Boer 1995c: 213, 218, 225–226, 228, 230–231, Fig 4, Fig 13 (phylogeny, comp. note)

Mirabilopsaltria viridicata Boer 1995d: 217–218, Fig 46 (distribution, comp. note) New Guinea

Mirabilopsaltria viridicata Boer 1996: 351, 353–358, 360–361, 373–376, Fig 2, Figs 52–63 (n. comb., key, described, illustrated, phylogeny, distribution, comp. note) Equals *Baeturia viridicata* Papuan Peninsula

Mirabilopsaltria viridicata Boer and Duffels 1996b: 325 (distribution) New Guinea

Mirabilopsaltria viridicata Boer 1997: 113 (comp. note)

***Guineapsaltria* Boer, 1993**

Guineapsaltria Boer 1993a: 15–19, 21–23, 33 (n. gen., described, distribution, comp. note)

Guineapsaltria Boer 1994d: 88 (comp. note)

Guineapsaltria Boer 1995a: 2, 5–6, 41, Fig 1 (phylogeny, comp. note)

Guineapsaltria Boer 1995c: 202, 204, 210–211, 214–219, Fig 3, Fig 11, Fig 15 (phylogeny, comp. note)

Guineapsaltria Boer 1995d: 208–209, 217–219, 222, 224–225, Fig 35, Figs 52–53 (distribution, phylogeny, comp. note)

Guineapsaltria Boer 1996: 351–353, 356, 358, Fig 1 (comp. note)

Guineapsaltria Boer and Duffels 1996a: 155, 167–168, 170–172, Figs 10–11 (distribution, phylogeny, comp. note) New Guinea, Australia

Guineapsaltria Boer and Duffels 1996b: 301, 304, 306, 313–314, 325, Fig 5 (distribution, phylogeny, comp. note) Papua New Guinea

Guineapsaltria Boer 1997: 92–93, 98, Fig 1 (phylogeny, comp. note)

Guineapsaltria Holloway 1997: 220, Fig 5 (distribution, phylogeny, comp. note) New Guinea

Guineapsaltria Holloway 1998: 304–308, Fig 10, Figs 12–13 (distribution, phylogeny, comp. note) New Guinea

Guineapsaltria Boer 2000: 4, 7, 15, Fig 1 (phylogeny, comp. note)

Guineapsaltria Moulds and Carver 2002: 21 (listed) Australia

Guineapsaltria Moulds 2005b: 390, 412–413, 430, 435, Table 2 (higher taxonomy, listed, comp. note) Australia

Guineapsaltria Duffels and Boer 2006: 537 (distribution, comp. note) New Guinea

***G. chinai* (Blöte, 1960)** = *Baeturia chinai*

Guineapsaltria chinai Boer 1993a: 17–19, 21–23, 33–35, 40, Fig 3, Figs (n. comb., key, described, illustrated, distribution, comp. note) Equals *Baeturia chinai* Papuan Peninsula

Guineapsaltria chinai Boer 1995a: 5, 38 (comp. note)

Guineapslatria chinai Boer 1995c: 213, 215–217, 225, 228, 230–231, Fig 4, Fig 12 (phylogeny, comp. note)

Guineapslatria chinai Boer 1995d: 209, Fig 35 (distribution, comp. note) Papuan Peninsula

Guineapsaltria chinai Boer 1996: 358 (comp. note)

Guineapslatria chinai Boer and Duffels 1996b: 325 (distribution) Papuan Peninsula

***G. flava* (Goding & Froggatt, 1904)** = *Tibicen flavus* = *Abricta flava* = *Abricta flavus* = *Baeturia flava* = *Melampsalta flava* = *Cicadetta flava* = *Baeturia modesta* Distant, 1907 = *Baeturia exhausta* (nec Guérin-Méneville) = *Baeturia minuta* Blöte, 1960 = *Baeturia stylata* (nec Blöte)

Melampsalta flava Hahn 1962: 9 (listed) Australia

Baeturia flava Moulds 1990: 12 (comp. note) Australia, New Guinea

Baeturia flava Moulds 1990: 12, 182–183, Plate 22, Fig 4 (synonymy, illustrated, distribution, ecology, listed, comp. note) Equals *Baeturia exhausta* (partim) Equals *Tibicen flavus* Equals *Baeturia modesta* n. syn. Equals *Baeturia exhausta* (partim) Equals *Baeturia stylata* (partim) Australia, Queensland, Papua New Guinea

Tibicen flavus Moulds 1990: 182 (listed, comp. note)

Abricta? flava Moulds 1990: 182 (listed, comp. note)

Abricta flavus Moulds 1990: 182 (listed, comp. note)
Baeturia modesta Moulds 1990: 182 (type data, listed) Queensland
Baeturia exhausta (sic) Moulds 1990: 182 (listed)
Baeturia stylata (sic) Moulds 1990: 182 (listed)
Guineapsaltria flava Boer 1993a: 17–28, 32, 41, Fig 1, Figs 5–24 (n. comb., type species of genus *Guineapsaltria*, key, synonymy, described, illustrated, distribution, comp. note) Equals *Tibicen flavus* Equals *Abricta flava* Equals *Abricta flavus* Equals *Baeturia flava* Equals *Baeturia modesta* Distant, 1907 Equals *Baeturia exhausta* (Guérin-Méneville) (partim) Equals *Baeturia minuta* Blöte, 1960 Equals *Baeturia stylata* Blöte, 1960 (partim) Equals *Melampsalta flava* Northern Queensland, New Guinea
Baeturia flavá Boer 1993b: 144 (comp. note)
Baeturia flava Naumann 1993: 26, 74, 148 (listed) Australia
Guineapsaltria flava Boer 1994c: 133–134 (comp. note) New Guinea, Queensland, Aru Islands
Guineapslatria flava Boer 1995c: 213, 215–217, 228, 230–231, Fig 4, Fig 12 (phylogeny, comp. note)
Guineapslatria flava Boer 1995d: 209, 233, 234, Fig 35 (distribution, comp. note) New Guinea, Aru Island, Queensland
Guineapslatria flava Boer and Duffels 1996a: 173 (comp. note) New Guinea, Queensland
Guineapslatria flava Boer and Duffels 1996b: 325 (distribution) New Guinea
Guineapslatria flava Boer 1997: 213, 215–217, 228, 230–231, Fig 4, Fig 12 (phylogeny, comp. note)
Guineapslatria flava Boer 1999: 119 (comp. note) New Guinea, Queensland
Tibicen flava Moulds and Carver 2002: 21 (type species of *Guineapsaltria*, listed) Queensland
Guineapsaltria flava Moulds and Carver 2002: 21–22 (synonymy, distribution, ecology, type data, listed) Equals *Baeturia exhausta* (partim) Equals *Tibicen flava* Equals *Baeturia modesta* Equals *Baeturia minuta* Equals *Baeturia stylata* (partim) Australia, Queensland, Papua New Guinea
Baeturia exhausta (sic) Moulds and Carver 2002: 21 (listed)
Baeturia modesta Moulds and Carver 2002: 21 (type data, listed) Queensland
Baeturia minuta Moulds and Carver 2002: 21 (type data, listed) Hollandia
Baeturia stylata (sic) Moulds and Carver 2002: 22 (listed)
Cicadetta flava Moulds and Carver 2002: 26–27 (synonymy, distribution, ecology, type data, listed) Equals *Melampsalta flava* Australia
Melampsalta flava Moulds and Carver 2002: 26–27 (type data, listed) South Australia
Abricta flava Moulds 2003b: 246 (comp. note)
Guineapsaltria flava Moulds 2005b: 395–398, 417–419, 421, 424, Figs 56–59, Fig 61, Table 1 (phylogeny, comp. note) Australia
Guineapaltria flava Moulds 2009a: 503 (comp. note) Papua New Guinea

G. *flaveola* Boer, 1993
Guineapsaltria flaveola Boer 1993a: 17–19, 21, 23, 26–27, 41, Fig 1, Figs 25–30 (n. sp., key, described, illustrated, distribution, comp. note) Papuan Peninsula, Sideia Island
Guineapslatria flaveola Boer 1995c: 213, 215–217, 228, 230–231, Fig 4, Fig 12 (phylogeny, comp. note)
Guineapslatria flaveola Boer 1995d: 209, 234, Fig 35 (distribution, comp. note) Papuan Peninsula, Sideia Island
Guineapslatria flaveola Boer and Duffels 1996b: 325 (distribution) Papuan Peninsula

G. *pallida* (Blöte, 1960) = *Baeturia pallida*
Guineapsaltria pallida Boer 1993a: 17–23, 35–38, 41, Fig 3, Figs 73–84 (n. comb., key, described, illustrated, distribution, comp. note) Equals *Baeturia pallida* New Guinea
Guineapslatria pallida Boer 1995c: 213, 215–217, 228, 230–231, Fig 4, Fig 12 (phylogeny, comp. note)
Guineapslatria pallida Boer 1995d: 209, Fig 35 (distribution, comp. note) New Guinea
Guineapslatria pallida Boer and Duffels 1996b: 325 (distribution) New Guinea

G. *pallidula* Boer, 1993
Guineapsaltria pallidula Boer 1993a: 17–21, 23, 37–38, 41, Fig 3, Figs 85–94 (n. sp., key, described, illustrated, distribution, comp. note) Irian Jaya, Papua New Guinea
Guineapslatria pallidula Boer 1995c: 213, 215–217, 228, 230–231, Fig 4, Fig 12 (phylogeny, comp. note)
Guineapslatria pallidula Boer 1995d: 209, Fig 35 (distribution, comp. note) New Guinea
Guineapslatria pallidula Boer and Duffels 1996b: 325 (distribution) New Guinea

G. *pennyi* Boer, 1993
Guineapsaltria pennyi Boer 1993a: 17–23, 39–41, Fig 2, Figs 95–110 (n. sp., key, described, illustrated, distribution, comp. note) Papua New Guinea
Guineapsaltria pennyi Boer 1995a: 38 (comp. note)

Guineapslatria pennyi Boer 1995c: 213, 215–217, 225, 228, 230–231, Fig 4, Fig 12 (phylogeny, comp. note)
Guineapslatria pennyi Boer 1995d: 209, Fig 35 (distribution, comp. note) New Guinea
Guineapsaltria pennyi Boer 1996: 358 (comp. note)
Guineapslatria pennyi Boer and Duffels 1996b: 325 (distribution) New Guinea

G. stylata **(Blöte, 1960)** = *Baeturia stylata*
Guineapsaltria stylata Boer 1993a: 17–20, 22–23, 25, 27–28, 32, 36, 41, Fig 2, Fig 4, Figs 31–50 (n. comb., key, described, illustrated, distribution, comp. note) Equals *Baeturia stylata* Papua New Guinea
Guineapslatria stylata Boer 1995c: 213, 215–217, 225, 228, 230–231, Fig 4, Fig 12 (phylogeny, comp. note)
Guineapslatria stylata Boer 1995d: 209, Fig 35 (distribution, comp. note) New Guinea
Guineapslatria stylata Boer and Duffels 1996b: 325 (distribution) New Guinea

G. viridula **(Blöte, 1960)** = *Baeturia viridula*
Guineapsaltria viridula Boer 1993a: 17–21, 23, 30–33, 35–37, Fig 3, Figs 51–62 (n. comb., key, described, illustrated, distribution, comp. note) Equals *Baeturia viridula* New Guinea, Umboi Island, Admiralty Islands, New Britain, Manus Island
Guineapslatria viridula Boer 1995c: 213, 215–217, 228, 230–231, Fig 4, Fig 12 (phylogeny, comp. note)
Guineapslatria viridula Boer 1995d: 209, Fig 35 (distribution, comp. note) New Guinea, New Britain, Manus Island
Guineapsaltria viridula Boer 1996: 358 (comp. note) Papuan Peninsula
Guineapslatria viridula Boer and Duffels 1996b: 325 (distribution) New Guinea, Solomon Islands

Scottotympana **Boer, 1991**
Scottotympana Boer 1991: 1–4 (n. gen., described, distribution, comp. note) New Guinea
Scottotympana Boer 1992a: 19 (phylogeny, comp. note)
Scottotympana Boer 1993a: 16–17 (comp. note)
Scottotympana Boer 1993b: 142–143 (phylogeny, comp. note)
Scottotympana Boer 1994a: 2–3 (comp. note)
Scottotympana Boer 1994c: 129–130, Fig 1a (phylogeny, comp. note)
Scottotympana Boer 1995a: 2, 5, Fig 1 (phylogeny, comp. note)
Scottotympana Boer 1995b: 3, 5, 8, 11, Fig 1 (comp. note)
Scottotympana Boer 1995c: 202, 204, 210–211, 213–215, 218, Fig 3, Fig 11, Fig 15 (phylogeny, comp. note)
Scottotympana Boer 1995d: 215–216, 222, 224–225, 232, Fig 44, Figs 52–53 (distribution, phylogeny, comp. note)
Scottotympana Boer 1996: 352, 368, Fig 1 (comp. note)
Scottotympana Boer and Duffels 1996a: 156, 168, 170–172, Figs 10–11 (distribution, phylogeny, comp. note) New Guinea
Scottotympana Boer and Duffels 1996b: 301, 304, 306, 313, 326, Fig 5 (distribution, phylogeny, comp. note) Papua New Guinea
Scottotympana Boer 1997: 92–93, Fig 1 (phylogeny, comp. note)
Scottotympana Holloway 1997: 220, Fig 5 (distribution, phylogeny, comp. note) New Guinea
Scottotympana Holloway 1998: 308, Fig 13 (distribution, comp. note) New Guinea
Scottotympana Boer 2000: 4, 15, Fig 1 (phylogeny, comp. note)
Scottotympana Moulds 2005b: 390, 435 (higher taxonomy, listed, comp. note)
Scottotympana Duffels and Boer 2006: 537 (distribution, comp. note) New Guinea
Scottotympana Liebherr 2008: 153 (comp. note) Papua New Guinea

S. biardae **Boer, 1991**
Scottotympana biardae Boer 1991: 2–9, Figs 6–18 (n. sp., type species of genus *Scottotympana*, key, described, illustrated, distribution, comp. note) Papua New Guinea
Scottotympana biardae Boer 1995a: 4 (comp. note)
Scottotympana biardae Boer 1995c: 212, 227, 229, 231, Fig 4 (phylogeny, comp. note)
Scottotympana biardae Boer 1995d: 215–216, Fig 44 (distribution, comp. note) New Guinea
Scottotympana biardae Boer and Duffels 1996b: 325 (distribution) New Guinea

S. huibregtsae **Boer, 1991**
Scottotympana huibregtsae Boer 1991: 2–5, 7–11, Fig 5, Fig 10, Figs 19–24 (n. sp., key, described, illustrated, distribution, comp. note) Papua New Guinea, New Britain?
Scottotympana huibregtsae Boer 1995a: 4 (comp. note)
Scottotympana huibregtsae Boer 1995c: 212, 225, 227, 229, 231, Fig 4 (phylogeny, comp. note)
Scottotympana huibregtsae Boer 1995d: 215–216, Fig 44 (distribution, comp. note) New Guinea
Scottotympana huibregtsae Boer and Duffels 1996b: 325 (distribution) New Guinea

S. sahebdivanii Boer, 1991

Scottotympana sahebdivanii Boer 1991: 2–4, 9–11, Fig 1, Fig 3, Fig 10, Figs 25–33 (n. sp., key, described, illustrated, distribution, comp. note) Irian Jaya

Scottotympana sahebdivanii Boer 1995c: 212, 227, 229, 231, Fig 4 (phylogeny, comp. note)

Scottotympana sahebdivanii Boer 1995d: 215–216, Fig 44 (distribution, comp. note) New Guinea

Scottotympana sahebdivanii Boer and Duffels 1996b: 325 (distribution) New Guinea

***Gymnotympana* Stål, 1861**

Gymnotympana Boulard 1982e: 181 (comp. note) New Guinea

Gymnotympana Hayashi 1984: 69 (listed)

Gymnotympana Duffels 1986: 328 (distribution, diversity, comp. note) Malukku, New Guinea, Woodlark Island

Gymnotympana Duffels 1988a: 7 (comp. note) Maluku, New Guinea, d'Entrecasteaux Islands, Woodlark Island

Gymnotympana Boer 1989: 2 (comp. note)

Gymnotympana Boer 1990: 64 (comp. note)

Gymnotympana Duffels and Boer 1990: 260–261 (comp. note) New Guinea, Maluku

Gymnotympana Boer 1991: 2–3 (comp. note)

Gymnotympana Boer 1992a: 18–20, 22–23 (phylogeny, comp. note)

Gymnotympana Boer 1992b: 164 (comp. note)

Gymnotympana sp. Kuniata and Nagaraja 1992: 65 (sugarcane pest, comp. note) Papua New Guinea

Gymnotympana Boer 1993a: 16–17 (comp. note)

Gymnotympana Boer 1993b: 141–143 (phylogeny, comp. note)

Gymnotympana Villet 1993: 435, 439 (type genus of Gymnotympanini comp. note)

Gymnotympana Boer 1994a: 1–3 (phylogeny, comp. note)

Gymnotympana Boer 1994b: 162 (comp. note)

Gymnotympana Boer 1994c: 129–131, 134, Fig 1a (phylogeny, comp. note)

Gymnotympana Boer 1994d: 88, 90, 97 (comp. note) Papuan Peninsula

Gymnotympana Boer 1995a: 2, 5–7, Fig 1 (phylogeny, comp. note)

Gymnotympana Boer 1995b: 1–5, 8–9, 11–19, 22, 25, 29, 32, 40, 41, 48, 50, 68, 74–75, 77–78, Fig 1 (described, phylogeny, comp. note) New Guinea, Queensland, Maluku

Gymnotympana Boer 1995c: 202, 204, 206, 210–211, 213–215, 218, Fig 3, Fig 11, Fig 18 (phylogeny, comp. note)

Gymnotympana Boer 1995d: 202, 205–208, 214, 217–219, 222, 224–225, 229–230, 233–237, Fig 32, Figs 52–53 (distribution, phylogeny, comp. note) Papua New Guinea

Gymnotympana Boer 1996: 352–356, 358, Fig 1 (comp. note)

Gymnotympana Boer and Duffels 1996a: 155, 159, 163–167, 171–173, Fig 7, Figs 10–11 (distribution, phylogeny, comp. note) Maluku, New Guinea, Australia

Gymnotympana Boer and Duffels 1996b: 301, 304, 306, 313, 316–317, 325, Fig 5 (distribution, phylogeny, comp. note) Papua New Guinea, Maluku

Gymnotympana Boer 1997: 91–94, 98, Fig 1 (phylogeny, comp. note)

Gymnotympana Holloway 1997: 220–221, Fig 5 (distribution, phylogeny, comp. note) New Guinea

Gymnotympana Novotny and Wilson 1997: 437 (listed)

Gymnotympana Villet 1997b: 356 (comp. note)

Gymnotympana Holloway 1998: 304–308, Fig 10, Figs 12–13 (distribution, phylogeny, comp. note) New Guinea

Gymnotympana Boer 1999: 120 (comp. note) Louisiade Archipelago, New Guinea

Gymnotympana Boer 2000: 4, 15, Fig 1 (phylogeny, comp. note)

Gymnotympana Heads 2001: 902–903, Fig 23 (distribution, comp. note) New Guinea, Moluccas

Gymnotympana Boulard 2002b: 197–198 (comp. note)

Gymnotympana Heads 2002b: 287, Fig 3 (distribution, comp. note) New Guinea

Gymnotympana Moulds and Carver 2002: 22 (listed) Australia

Gymnotympana Clarke, Balagawi, Clifford, Drew, Leblanc, et al. 2004: 151 (comp. note) Papua New Guinea

Gymnotympana Moulds 2005b: 390, 392, 412–413, 430, 435, Table 2 (type genus of Gymnotympanini, higher taxonomy, listed, comp. note)

***G. dahli* (Kuhlgatz, 1905)** = *Tibicen dahli* = *Ueana dahli* = *Ueana (Tibicen) dahli* = *Ueana dahlii* (sic) = *Baeturia valida* Blöte, 1960

Gymnotympana dahli Boer 1993a: 21 (comp. note)

Gymnotympana dahli Boer 1995b: 1, 4–6, 8–9, 11, 13–16, 19–20, 25–31, Fig 1, Fig 2a, Fig 12, Figs 18–19, Figs 50–58 (lectotype designation, synonymy, key, described, illustrated, phylogeny, distribution, comp.

note) Equals *Tibicen dahli* Equals *Ueana dahli* Equals *Uena dahli* Equals *Baeturia valida* New Guinea, Papuan Peninsula, Biak Island, Goodenough Island, New Britain, Manus Island

Baeturia valida Boer 1995b: 27–28 (synonymy)

Gymnotympana dahli Boer 1995c: 213–215, 225, 227, 229, 231, Fig 4, Figs 9–10 (phylogeny, comp. note)

Gymnotympana dahli Boer 1995d: 205–206, Fig 32 (distribution, comp. note) New Guinea

Gymnotympana dahli Boer 1996: 358 (comp. note) Papuan Peninsula

Gymnotympana dahli Boer and Duffels 1996a: 1996a: 166, 173, Fig 7 (distribution, comp. note) New Guinea, Bismarck Archipelago

Gymnotympana dahli Boer and Duffels 1996b: 312, 326, Fig 8 (distribution, comp. note) New Guinea, Bismarck Archipelago

***G. hirsuta* Boer, 1995**

Gymnotympana hirsuta Boer 1995b: 4–5, 8–9, 11–12, 14–15, 18, 20, 25, 63–66, 68, 70, 74, Fig 1, Figs 158–170, Fig 182 (n. sp., key, described, illustrated, phylogeny, distribution, comp. note) Papua New Guinea

Gymnotympana hirsuta Boer 1995c: 213–215, 224, 228, 230–231, Fig 4, Figs 9–10 (phylogeny, comp. note)

Gymnotympana hirsuta Boer 1995d: 205–206, Fig 32 (distribution, comp. note) Papua New Guinea

Gymnotympana hirsuta Boer and Duffels 1996a: 166, Fig 7 (distribution)

Gymnotympana hirsuta Boer and Duffels 1996b: 312, 326, Fig 8 (distribution, comp. note) New Guinea

***G. langeraki* De Boer, 1995**

Gymnotympana langerski Boer 1995b: 5, 8–9, 11–15, 20, 52–55, 59, Fig 1, Fig 13, Figs 125–132, Fig 144 (n. sp., key, described, illustrated, phylogeny, distribution, comp. note) d'Entrecasteaux Islands, Goodenough Island, Normanby

Gymnotympana langerski Boer 1995c: 213–215, 228, 230–231, Fig 4, Figs 9–10 (phylogeny, comp. note)

Gymnotympana langerski Boer 1995d: 205–206, Fig 32 (distribution, comp. note) d'Entrecasteaux Islands

Gymnotympana langerski Boer and Duffels 1996a: 166, Fig 7 (distribution)

Gymnotympana langerski Boer and Duffels 1996b: 299, 304, 312, 3226, Fig 5, Fig 8, Plate 1h (distribution, phylogeny, illustrated, comp. note) Papuan Peninsula

G. loriae (Distant, 1897) to *Baeturia loriae*

***G. membrana* Boer, 1995**

Gymnotympana membrana Boer 1995b: 5, 8–16, 19–20, 43–51, Fig 1, Fig 9, Fig 16, Fig 21, Figs 98–108 (n. sp., key, described, illustrated, phylogeny, distribution, comp. note) New Guinea

Gymnotympana membrana Boer 1995c: 213–215, 228, 230–231, Fig 4, Figs 9–10 (phylogeny, comp. note)

Gymnotympana membrana Boer 1995d: 205–206, Fig 32 (distribution, comp. note) New Guinea

Gymnotympana membrana Boer and Duffels 1996a: 166, Fig 7 (distribution)

Gymnotympana membrana group Boer and Duffels 1996b: 304, 325, Fig 5 (distribution, phylogeny, comp. note) Papuan Peninsula, New Guinea

Gymnotympana membrana Boer and Duffels 1996b: 312, 325, Fig 8 (distribution, comp. note) New Guinea

***G. minoramembrana* De Boer, 1995**

Gymnotympana minoramembrana Boer 1995b: 4–5, 8–9, 11–12, 14–15, 20, 46–49, Fig 1, Figs 108–116 (n. sp., key, described, illustrated, phylogeny, distribution, comp. note) New Guinea

Gymnotympana minoramembrana Boer 1995c: 213–215, 228, 230–231, Fig 4, Figs 9–10 (phylogeny, comp. note)

Gymnotympana minoramembrana Boer 1995d: 205, Fig 32 (distribution, comp. note)

Gymnotympana minoramembrana Boer and Duffels 1996a: 166, Fig 7 (distribution)

Gymnotympana minoramembrana Boer and Duffels 1996b: 312, 326, Fig 8 (distribution, comp. note) New Guinea

***G. montana* De Boer, 1995**

Gymnotympana montana Boer 1995b: 4–5, 8–9, 11–15, 20, 25, 70, 72–74, Fig 1, Fig 11, Fig 182, Figs 194–197 (n. sp., key, described, illustrated, phylogeny, distribution, comp. note) Papua New Guinea

Gymnotympana montana Boer 1995c: 213–215, 228, 230–231, Fig 4, Figs 9–10 (phylogeny, comp. note)

Gymnotympana montana Boer 1995d: 205, Fig 32 (distribution, comp. note)

Gymnotympana montana Boer and Duffels 1996a: 166, Fig 7 (distribution)

Gymnotympana montana Boer and Duffels 1996b: 312, 326, Fig 8 (distribution, comp. note) Papuan Peninsula

Gymnotympana montana Boer 1997: 113 (comp. note)

G. *nigravirgula* Boer, 1995

Gymnotympana nigravirgula Boer 1995b: 5, 9, 11–15, 21, 55–59, Fig 1, Fig 14, Figs 133–144 (n. sp., key, described, illustrated, phylogeny, distribution, comp. note) Papuan Peninsula

Gymnotympana nigravirgula Boer 1995c: 213–215, 227, 229, 231, Fig 4, Figs 9–10 (phylogeny, comp. note)

Gymnotympana nigravirgula Boer 1995d: 205, Fig 32 (distribution, comp. note)

Gymnotympana nigravirgula Boer and Duffels 1996a: 166, Fig 7 (distribution)

Gymnotympana nigravirgula Boer and Duffels 1996b: 312, 326, Fig 8 (distribution, comp. note) Papuan Peninsula

G. *obiensis* Boer, 1995

Gymnotympana obiensis Boer 1995b: 4–9, 11–15, 21, 38–42 (n. sp., key, described, illustrated, phylogeny, distribution, comp. note) Obi Island

Gymnotympana obiensis Boer 1995c: 213–215, 225, 227, 229, 231, Fig 4, Figs 9–10 (phylogeny, comp. note)

Gymnotympana obiensis Boer 1995d: 205, Fig 32 (distribution, comp. note)

Gymnotympana obiensis Boer and Duffels 1996a: 166, Fig 7 (distribution)

Gymnotympana obiensis Boer and Duffels 1996b: 312, 326, Fig 8 (distribution, comp. note) North Maluku

G. *olivacea* Distant, 1905

Gymnotympana olivacea Boer 1992a: 20 (comp. note)

Gymnotympana olivacea Boer 1994a: 3 (comp. note)

Gymnotympana olivacea Boer 1995b: 4–5, 8–9, 11–15, 20, 25, 66–72, 74, Fig 1, Figs 171–182 (lectotype designation, key, described, illustrated, phylogeny, distribution, comp. note) Papuan Peninsula

Gymnotympana olivacea Boer 1995c: 213–215, 224–225, 228, 230–231, Fig 4, Figs 9–10 (phylogeny, comp. note)

Gymnotympana olivacea Boer 1995d: 205, Fig 32 (distribution, comp. note)

Gymnotympana olivacea Boer and Duffels 1996a: 166, Fig 7 (distribution)

Gymnotympana olivacea group Boer and Duffels 1996b: 304, 326, Fig 5 (distribution, phylogeny, comp. note) Papuan Peninsula, New Guinea

Gymnotympana olivacea Boer and Duffels 1996b: 312, 326, Fig 8 (distribution, comp. note) Papuan Peninsula

Gymnotympana olivacea Boer 1997: 113 (comp. note)

G. *parvula* Boer, 1995

Gymnotympana parvula Boer 1995b: 5, 8–9, 11–12, 14–15, 20, 59, 61–63, Fig 1, Fig 144, Figs 152–157 (n. sp., key, described, illustrated, phylogeny, distribution, comp. note) Papuan Peninsula

Gymnotympana parvula Boer 1995c: 213–215, 228, 230–231, Fig 4, Figs 9–10 (phylogeny, comp. note)

Gymnotympana parvula Boer 1995d: 205, Fig 32 (distribution, comp. note)

Gymnotympana parvula Boer and Duffels 1996a: 166, Fig 7 (distribution)

Gymnotympana parvula group Boer and Duffels 1996b: 304, 326, Fig 5 (distribution, phylogeny, comp. note) Papuan Peninsula

Gymnotympana parvula Boer and Duffels 1996b: 312, 326, Fig 8 (distribution, comp. note) Papuan Peninsula

G. *phyloglycera* Boer, 1995

Gymnotympana phyloglycera Boer 1995b: 5, 8–9, 11–12, 14–15, 20, 46, 49–52, 76 (n. sp., key, described, illustrated, phylogeny, distribution, comp. note) Papua New Guinea

Gymnotympana phyloglycera Boer 1995c: 213–215, 228, 230–231, Fig 4, Figs 9–10 (phylogeny, comp. note)

Gymnotympana phyloglycera Boer 1995d: 205, Fig 32 (distribution, comp. note)

Gymnotympana phyloglycera Boer and Duffels 1996a: 166, Fig 7 (distribution)

Gymnotympana phyloglycera Boer and Duffels 1996b: 298, 312, 326, Fig 8 (sugar cane pest, distribution, comp. note) Papua New Guinea

G. *rubricata* (Distant, 1897) = *Baeturia rubicata* = *Gymnotympana nenians* Jacobi, 1903 = *Gymnotympana neniens* (sic)

Gymnotympana nenians Hennessey 1990: 471 (type material)

Gymnotympana rubricata Boer 1993a: 21 (comp. note)

Gymnotympana rubricata Boer 1995b: 1, 4–7, 9–16, 18–20, 31–35, Fig 1, Fig 2b, Fig 6, Fig 20, Fig 22, Figs 59–71 (lectotype designation, key, described, illustrated, phylogeny, distribution, comp. note) Equals *Baeturia rubricata* Equals *Gymnotympana nenians* Papua New Guinea, d'Entrecasteaux Islands, Louisiade Archipelago

Gymnotympana nenians Boer 1995b: 1, 31 (synonymy)

Gymnotympana rubricata Boer 1995c: 1995c: 213–215, 225–227, 229, 231, Fig 4, Figs 9–10 (phylogeny, comp. note)

Gymnotympana rubricata Boer 1995d: 205–206, Fig 32 (distribution, comp. note) New Guinea

Gymnotympana rubricata Boer and Duffels 1996a: 166, Fig 7 (distribution)
Gymnotympana rubricata Boer and Duffels 1996b: 300, 312, 326, Fig 8, Plate 2 (distribution, timbal illustrated, comp. note) New Guinea
Gymnotympana rubricata group Boer and Duffels 1996b: 304, 326, Fig 5 (distribution, phylogeny, comp. note) Papuan Peninsula, New Guinea
Gymnotympana nenians Sanborn 1999a: 52 (type material, listed)

***G. rufa* (Ashton, 1914)** = *Baeturia rufa*
Baeturia rufa Moulds 1990: 182–184 (distribution, ecology, listed, comp. note) Australia, Queensland
Gymnotympana rufa Boer 1995b: 1, 4–6, 9, 11–12, 14–15, 18, 20, 74–78, Fig 1, Fig 2b, Figs 198–206 (n. comb., key, described, illustrated, phylogeny, distribution, comp. note) Queensland
Gymnotympana rufa Boer 1995c: 213–215, 228, 230–231, Fig 4, Figs 9–10 (phylogeny, comp. note)
Gymnotympana rufa Boer 1995d: 205, 209, 218, 233–234, Fig 32 (distribution, comp. note) Queensland
Gymnotympana rufa Boer and Duffels 1996a: 166, Fig 7 (distribution)
Gymnotympana rufa group Boer and Duffels 1996b: 304, 326, Fig 5 (distribution, phylogeny, comp. note) Australia
Gymnotympana rufa Boer and Duffels 1996b: 312, 326, Fig 8 (distribution, comp. note) Australia
Gymnotympana rufa Boer 1997: 91–92, 94, 96, 98, 107 (comp. note) Australia, Queensland
Gymnotympana rufa Moulds and Carver 2002: 22 (distribution, ecology, type data, listed) Australia, Queensland

***G. stenocephalis* Boer, 1995**
Gymnotympana stenocephalis Boer 1995b: 4–5, 9, 11–15, 18–19, 21, 58–61, Fig 1, Fig 23, Figs 144–151 (n. sp., key, described, illustrated, phylogeny, distribution, comp. note) Papuan Peninsula
Gymnotympana stenocephalis Boer 1995c: 213–215, 226, 228, 230–231, Fig 4, Figs 9–10 (phylogeny, comp. note)
Gymnotympana stenocephalis Boer 1995d: 205, Fig 32 (distribution, comp. note)
Gymnotympana stenocephalis Boer and Duffels 1996a: 166, Fig 7 (distribution)
Gymnotympana stenocephalis Boer and Duffels 1996b: 312, 326, Fig 8 (distribution, comp. note) Papuan Peninsula
Gymnotympana stenocephalis Sanborn 1999a: 52 (type material, listed)

***G. strepitans* (Stål, 1861)** = *Cicada strepitans*
Gymnotympana strepitans Boer 1995b: 5, 8–16, 18, 21–25, 29–30, 32, 34, Fig 1, Figs 3–5, Figs 25–37 (type species of genus *Gymnotympana*, key, described, illustrated, phylogeny, distribution, comp. note) Equals *Cicada strepitans* Papua New Guinea, d'Entrecasteaux Islands, Louisiade Archipelago
Gymnotympana strepitans Boer 1995c: 213–215, 227, 229, 231, Fig 4, Figs 9–10 (phylogeny, comp. note)
Gymnotympana strepitans Boer 1995d: 205–206, Fig 32 (distribution, comp. note) New Guinea
Gymnotympana strepitans Boer and Duffels 1996a: 166, Fig 7 (distribution)
Gymnotympana strepitans Boer and Duffels 1996b: 312, 326, Fig 8 (distribution, comp. note) New Guinea
Gymnotympana strepitans Boer 1997: 105 (comp. note)
Gymnotympana strepitans Heads 2002a: 267 (distribution, comp. note) New Guinea, Aru Islands
Cicada strepitans Moulds and Carver 2002: 22 (type species of *Gymnotympana*, listed)
Gymnotympana strepitans Moulds 2005b: 395–398, 403, 412, 417–419, 421, 424, Fig 37, Figs 56–59, Fig 61, Table 1 (illustrated, phylogeny, comp. note) Australia

***G. stridens* (Stål, 1861)** = *Cicada stridens* = *Dundubia significata* Walker, 1870
Gymnotympana stridens Duffels 1988a: 7–8, 11, Table 1 (comp. note) Bacan, Morotai, Obi
Gymnotympana stridens Boer 1994a: 3 (comp. note)
Gymnotympana stridens Boer 1995b: 4–15, 21, 35–39, 41–42, Fig 1, Fig 2b, Fig 7, Fig 15, Figs 72–82 (key, described, illustrated, phylogeny, distribution, comp. note) Equals *Cicada stridens* Equals *Dundubia significata* Maluku, Bacan, Morotai, Obi, Ternate
Gymnotympana stridens Boer 1995c: 213–215, 227, 229, 231, Fig 4, Figs 9–10 (phylogeny, comp. note)
Gymnotympana stridens Boer 1995d: 205, Fig 32 (distribution, comp. note)
Gymnotympana stridens Boer and Duffels 1996a: 166, Fig 7 (distribution)
Gymnotympana stridens Boer and Duffels 1996b: 312, 326, Fig 8 (distribution, comp. note) North Maluku

***G. subnotata* (Walker, 1868)** = *Cicada subnotata* = *Baeturia subnotata* = *Cicada innotabilis* Walker, 1868 (nec Walker, 1858) = *Fidicina innotabilis* Kuhlgatz &

Melichar, 1902 (partim: Maluku) = *Gymnotympana innotabilis*

Gymnotympana innotabilis Duffels 1988a: 7–8, 11, Table 1 (comp. note) Equals *Cicada innotabilis* Bacan, Morotai

Gymnotympana subnotata Boer 1995b: 1–2, 4–15, 19, 21, 40–43, Fig 1, Fig 2b, Fig 8, Fig 17, Fig 24, Figs 87–97 (n. sp., key, described, illustrated, phylogeny, distribution, comp. note) Equals *Cicada subnotata* Equals *Baeturia subnotata* Equals *Cicada innobilis* Walker, 1868 Equals *Baeturia innotabilis* Equals *Fidicina innotabilis* (partim) Maluku, Bacan, Halmahera, Morotai

Cicada subnotata Boer 1995b: 1–2, 40 (synonymy)

Cicada innotabilis Boer 1995b: 40 (synonymy)

Gymnotympana subnotata Boer 1995c: 213–215, 227, 229, 231, Fig 4, Figs 9–10 (phylogeny, comp. note)

Gymnotympana subnotata Boer 1995d: 205, Fig 32 (distribution, comp. note)

Gymnotympana subnotata Boer and Duffels 1996a: 166, Fig 7 (distribution)

Gymnotympana subnotata Boer and Duffels 1996b: 300, 312, 326, Fig 8, Plate 2 (distribution, timbal illustrated, comp. note) North Maluku

Gymnotympana subnotata group Boer and Duffels 1996b: 304, 326, Fig 5 (distribution, phylogeny, comp. note) Maluku

***G. varicolor* (Distant, 1907)** = *Baeturia varicolor* = *Kanakia congrua* (nec Walker)

Baeturia varicolor Moulds 1990: 20–22, 182, 184, 188–190, Plate 22, Fig 5, Figs 5a–d (illustrated, distribution, ecology, listed, comp. note) Australia, Queensland

Baeturia varicolor Naumann 1993: 48, 74, 149 (listed) Australia

Gymnotympana varicolor Boer 1995b: 1, 4–6, 8–9, 11–12, 14–15, 19–20, 76–79, Fig 1, Fig 2a, Figs 207–213 (n. comb., key, described, illustrated, phylogeny, distribution, comp. note) Equals *Baeturia varicolor* Equals *Kanakia congrua* (partim) Queensland

Baeturia varicolor Boer 1995b: 1 (synonymy) Australia

Gymnotympana varicolor Boer 1995c: 213–215, 225, 228, 230–231, Fig 4, Figs 9–10 (phylogeny, comp. note)

Gymnotympana varicolor Boer 1995d: 205, 209, 218, 233–234, Fig 32 (distribution, comp. note) Queensland

Gymnotympana varicolor Boer and Duffels 1996a: 166, Fig 7 (distribution)

Gymnotympana varicolor Boer and Duffels 1996b: 312, 326, Fig 8 (distribution, comp. note) Australia

Gymnotympana varicolor Boer 1997: 91, 93–94, 96, 98, 101 (comp. note) Australia, Queensland

Gymnotympana varicolor Moulds and Carver 2002: 22 (synonymy, distribution, ecology, type data, listed) Equals *Baeturia varicolor* Australia

Baeturia varicolor Moulds and Carver 2002: 22 (type data, listed) Australia, Queensland

Baeturia varicolor Lee 2003b: 7 (comp. note)

Gymnotympana varicolor Moulds 2005b: 395–398, 417–419, 421, 424, Figs 56–59, Fig 61, Table 1 (phylogeny, comp. note) Australia

***G. varicolor* var. *a* (Distant, 1907)** = *Baeturia varicolor* var. *a*

***G. varicolor* var. *b* (Distant, 1907)** = *Baeturia varicolor* var. *b*

***G. verlaani* Boer, 1995**

Gymnotympana verlaani Boer 1995b: 4–5, 8–9, 11–12, 14–15, 20, 25, 69–74, Fig 1, Figs 182–193 (n. sp., key, described, illustrated, phylogeny, distribution, comp. note) Papua New Guinea

Gymnotympana verlaani Boer 1995c: 213–215, 228, 230–231, Fig 4, Figs 9–10 (phylogeny, comp. note)

Gymnotympana verlaani Boer 1995d: 205–206, Fig 32 (distribution, comp. note) Papua New Guinea

Gymnotympana verlaani Boer and Duffels 1996a: 166, Fig 7 (distribution)

Gymnotympana verlaani Boer and Duffels 1996b: 312, 326, Fig 8 (distribution, comp. note) New Guinea

Gymnotympana verlaani Boer 1997: 113 (comp. note)

***G. viridis* Boer, 1995**

Gymnotympana viridis Boer 1995b: 4–6, 9–15, 18–20, 25–27, 29–30, Fig 1, Fig 2a, Fig 10, Fig 20, Figs 38–49 (n. sp., key, described, illustrated, phylogeny, distribution, comp. note) Papuan Peninsula

Gymnotympana viridis Boer 1995c: 213–215, 227, 229, 231, Fig 4, Figs 9–10 (phylogeny, comp. note)

Gymnotympana viridis Boer 1995d: 205, Fig 32 (distribution, comp. note)

Gymnotympana viridis Boer and Duffels 1996a: 166, Fig 7 (distribution)

Gymnotympana viridis Boer and Duffels 1996b: 312, 326, Fig 8 (distribution, comp. note) Papuan Peninsula

***Fractuosella* Boulard, 1979**

Fractuosella Williams and Williams 1988: 4 (listed) Mascarene Islands

Fractuosella Boulard 1989a: 129 (comp. note)

Fractuosella Boulard 1990b: 144, 209–210, Fig 12 (phylogeny, comp. note)
Fractuosella Villet 1993: 435, 439 (comp. note)
Fractuosella Boer 1995c: 203–205, Fig 1 (phylogeny, comp. note) Mascarene Isles
Fractuosella Boulard 2002b: 198 (comp. note) Mascareigne Island
Fractuosella Jong and Boer 2004: 267 (comp. note)

***F. breoni* Boulard, 1989**
Fractuosella breoni Boulard 1989a: 130–132, Figs 1–5 (n. sp., described, illustrated, comp. note) Réunion Island
Fractuosella breoni Bonfils, Quilici and Reynaud 1994: 236 (listed) Réunion Island

***F. darwini* (Distant, 1905)** = *Stagira darwini*
Stagira darwini Schouteden 1907: 287 (listed, comp. note) Maurice Island
Fractuosella darwini Williams and Williams 1988: 4 (listed, comp. note) Equals *Stagira darwini* Mauritius
Fractuosella darwini Villet 1997b: 348, 355 (comp. note) Mauritius

***F. vinsoni* Boulard, 1979**
Fractuosella vinsoni Williams and Williams 1988: 4 (listed, comp. note) Mauritius

***F. virginiae* Boulard, 1979**
Fractuosella virginiae Williams and Williams 1988: 4 (listed, comp. note) Mauritius

***Papuapsaltria* Boer, 1995**
Papuapsaltria Boer 1995a: 1–9, 12–13, 38, 41, Fig 1 (n. gen., phylogeny, described, distribution, comp. note) New Guinea
Papuapsaltria Boer 1995c: 202, 204, 210–211, 214–219, 225, Fig 3, Fig 11, Fig 18 (phylogeny, comp. note)
Papuapsaltria Boer 1995d: 216–218, 222, 224–225, 232–236, Fig 45, Figs 52–53 (distribution, phylogeny, comp. note) New Guinea, Normanby Island, Roon Island, Waigeu, Woodlark Island, Japen
Papuapsaltria Boer and Duffels 1996a: 155, 167, 170–172, Figs 10–11 (distribution, phylogeny, comp. note) New Guinea
Papuapsaltria Boer and Duffels 1996b: 301, 306, 313–314, 326, Fig 5 (distribution, phylogeny, comp. note) Papua New Guinea
Papuapsaltria Boer 1997: 92–93, 98, Fig 1 (phylogeny, comp. note)
Papuapsaltria Holloway 1997: 220, Fig 5 (distribution, phylogeny, comp. note) New Guinea
Papuapsaltria Holloway 1998: 304–308, Fig 10, Figs 12–13 (distribution, phylogeny, comp. note) New Guinea
Papuapsaltria Boer 2000: 1, 4, 7, 11, 15, Fig 1 (phylogeny, comp. note) Papua New Guinea
Papuapsaltria Heads 2002b: 287 (distribution, comp. note) New Guinea
Papuapsaltria Clarke, Balagawi, Clifford, Drew, Leblanc, et al. 2004: 151 (comp. note) Papua New Guinea
Papuapsaltria Moulds 2005b: 390, 435 (higher taxonomy, listed, comp. note)

***P. angulata* Boer, 1995**
Papuapsaltria angulata Boer 1995a: 2, 4–5, 7, 9–14, 32, 38, Fig 2, Figs 12–20, Table 1 (n. sp., type species of genus *Papuapsaltria*, key, described, illustrated, distribution, comp. note) Papua New Guinea
Papuapsaltria angulata Boer 1995c: 213, 216–217, 228, 230–231, Fig 4, Fig 12 (phylogeny, comp. note)
Papuapsaltria angulata Boer 1995d: 216, Fig 45 (distribution, comp. note) New Guinea
Papuapsaltria angulata group Boer and Duffels 1996b: 304, Fig 5 (distribution, phylogeny, comp. note) Papuan Peninsula
Papuapsaltria angulata Boer and Duffels 1996b: 325 (distribution) Papuan Peninsula

***P. baasi* Boer, 1995**
Papuapsaltria baasi Boer 1995a: 2–3, 6–7, 9, 16, 30–33, Fig 2, Figs 119–129, Table 1 (n. sp., key, described, illustrated, distribution, comp. note) New Guinea
Papuapsaltria baasi Boer 1995c: 213, 216–217, 228, 230–231, Fig 4, Fig 12 (phylogeny, comp. note)
Papuapsaltria baasi Boer 1995d: 216–217, Fig 45 (distribution, comp. note) New Guinea
Papuapsaltria baasi Boer and Duffels 1996b: 325 (distribution) New Guinea
Papuapsaltria baasi Boer 2000: 4, 15, Fig 1 (phylogeny, comp. note)

***P. bidigitula* Boer, 1995**
Papuapsaltria bidigitula Boer 1995a: 2–4, 6–7, 9, 24–28, Figs 2–3, Figs 92–101, Table 1 (n. sp., key, described, illustrated, distribution, comp. note) Irian Jaya, Yapen Island
Papuapsaltria bidigitula Boer 1995c: 213, 216–217, 225, 228, 230–231, Fig 4, Fig 12 (phylogeny, comp. note)
Papuapsaltria bidigitula Boer 1995d: 216, 230, Fig 45 (distribution, comp. note) New Guinea, Weta Island
Papuapsaltria bidigitula Boer and Duffels 1996b: 325 (distribution) New Guinea
Papuapsaltria bidigitula Boer 2000: 4, 7, 15, Fig 1 (phylogeny, comp. note)

***P. bosaviensis* Boer, 2000**

Papuapsaltria bosaviensis Boer 2000: 1–4, 7, 11–12, 15, 18, Fig 2, Figs 25–32, Tables 1–2 (n. sp., described, illustrated, phylogeny, comp. note) Papua New Guinea

***P. brassi* Boer, 1995**

Papuapsaltria brassi Boer 1995a: 2–3, 5, 7, 9, 13–15, Fig 2, Fig 32, Figs 43–52, Table 1 (n. sp., key, described, illustrated, distribution, comp. note) Papua New Guinea

Papuapsaltria brassi Boer 1995c: 213, 216–217, 228, 230–231, Fig 4, Fig 12 (phylogeny, comp. note)

Papuapsaltria brassi Boer 1995d: 216, Fig 45 (distribution, comp. note) New Guinea

Papuapsaltria brassi Boer and Duffels 1996b: 325 (distribution) New Guinea

Papuapsaltria brassi Sanborn 1999a: 51 (type material, listed)

***P. dioedes* Boer, 1995**

Papuapsaltria dioedes Boer 1995a: 2, 4–7, 9, 34–35, 37–39, Fig 2, Fig 4, Figs 156–164, Table 1 (n. sp., key, described, illustrated, distribution, comp. note) Papua New Guinea

Papuapsaltria diodes Boer 1995c: 211, 213, 216–217, 225, 228, 230–231, Fig 4, Fig 12 (phylogeny, comp. note)

Papuapsaltria diodes Boer 1995d: 216, Fig 45 (distribution, comp. note) New Guinea

Papuapsaltria diodes Boer and Duffels 1996b: 325 (distribution) New Guinea

Papuapsaltria diodes Sanborn 1999a: 51 (type material, listed)

Papuapsaltria dioedes Boer 2000: 6, 15, Fig 1 (phylogeny, comp. note)

***P. dolabrata* Boer, 1995**

Papuapsaltria dolabrata Boer 1995a: 2–3, 5–7, 9, 30–34, 36, Fig 2, Fig 126, Figs 137–146, Table 1 (n. sp., key, described, illustrated, distribution, comp. note) New Guinea, Roon Island, Waigeu Island

Papuapsaltria dolabrata Boer 1995c: 213, 216–217, 228, 230–231, Fig 4, Fig 12 (phylogeny, comp. note)

Papuapsaltria dolabrata Boer 1995d: 216–217, Fig 45 (distribution, comp. note) New Guinea

Papuapsaltria dolabrata Boer and Duffels 1996b: 325 (distribution) New Guinea

Papuapsaltria dolabrata Boer 2000: 4, 12, 15, Fig 1 (phylogeny, comp. note)

***P. goniodes* Boer, 1995**

Papuapsaltria gonoides Boer 1995a: 2, 5–7, 10–11, 13–14, 16, 38, Fig 2, Fig 5, Figs 7–9, Fig 11, Figs 21–32, Table 1 (n. sp., key, described, illustrated, distribution, comp. note) Papua New Guinea

Papuapsaltria goniodes Boer 1995c: 213, 216–217, 225, 228, 230–231, Fig 4, Fig 12 (phylogeny, comp. note)

Papuapsaltria goniodes Boer 1995d: 216, Fig 45 (distribution, comp. note) New Guinea

Papuapsaltria goniodes Boer and Duffels 1996b: 325 (distribution) New Guinea

***P. lachlani* Boer, 1995**

Papuapsaltria lachlani Boer 1995a: 2, 5, 7, 9, 13–17, Fig 2, Figs 32–42, Table 1 (n. sp., key, described, illustrated, distribution, comp. note) New Guinea

Papuapsaltria lachlani Boer 1995c: 213, 216–217, 228, 230–231, Fig 4, Fig 12 (phylogeny, comp. note)

Papuapsaltria lachlani Boer 1995d: 216–217, Fig 45 (distribution, comp. note) New Guinea

Papuapsaltria lachlani Boer and Duffels 1996b: 325 (distribution) New Guinea

***P. nana* (Jacobi, 1903)** = *Acrilla nana* = *Thaumastopsaltria nana* = *Thaumastropsaltria* (sic) *nana* = *Baeturia famulus* Distant, 1906

Acrilla nana Hennessey 1990: 470 (type material)

Papuapsaltria nana Boer 1995a: 2–5, 7, 9, 22–26, Figs 2–3, Fig 6, Figs 83–91, Table 1 (n. comb., synonymy, lectotype designation, key, described, illustrated, distribution, comp. note) Equals *Baeturia beccarii* Distant 1892 (nec Distant, 1888) Equals *Acrilla nana* Equals *Baeturia famulus* Equals *Thaumastopsaltria nana* New Guinea, d'Entrecasteaux Islands, Normanby Island

Papuapsaltria nana Boer 1995c: 213, 216–217, 225, 228, 230–231, Fig 4, Fig 12 (phylogeny, comp. note)

Papuapsaltria nana Boer 1995d: 216, Fig 45 (distribution, comp. note) New Guinea

Papuapsaltria nana Boer and Duffels 1996b: 325 (distribution) New Guinea

Baeturia nana Sanborn 1999a: 51–52 (type material, listed)

***P. novariae* De Boer, 1995**

Papuapsaltria novariae Boer 1995a: 2, 5–7, 9, 30–34, Fig 2, Figs 129–136, Table 1 (n. sp., key, described, illustrated, distribution, comp. note) Papua New Guinea

Papuapsaltria novariae Boer 1995c: 213, 216–217, 228, 230–231, Fig 4, Fig 12 (phylogeny, comp. note)

Papuapsaltria novariae Boer 1995d: 216–217, Fig 45 (distribution, comp. note) New Guinea

Papuapsaltria novariae Boer and Duffels 1996b: 325 (distribution) New Guinea

Papuapsaltria novariae Boer 2000: 4, 15, Fig 1 (phylogeny, comp. note)

P. *phyllophora* (Blöte, 1960) = *Baeturia phyllophora*
Baeturia phyllophora Boer 1993a: 16 (comp. note)
Papuapsaltria phyllophora Boer 1995a: 2–9, 40–43, Fig 2, Fig 4, Figs 176–190, Table 1 (n. comb., key, described, illustrated, distribution, comp. note) Equals *Baeturia phyllophora* New Guinea, Waigeu Island
Papuapsaltria phyllophora Boer 1995c: 213, 216–217, 228, 230–231, Fig 4, Fig 12 (phylogeny, comp. note)
Papuapsaltria phyllophora Boer 1995d: 216–217, Fig 45 (distribution, comp. note) New Guinea
Papuapsaltria phyllophora Boer and Duffels 1996b: 325 (distribution) New Guinea
Papuapsaltria phyllophora Boer 2000: 1, 7, 15, Fig 1, Table 1 (phylogeny, comp. note) Papua New Guinea

P. *plicata* Boer, 1995
Papuapsaltria plicata Boer 1995a: 2–8, 18–21, 26, Figs 2–3, Figs 65–75, Table 1 (n. sp., key, described, illustrated, distribution, comp. note) Papuan Peninsula
Papuapsaltria plicata Boer 1995c: 213, 216–217, 225, 228, 230–231, Fig 4, Fig 12 (phylogeny, comp. note)
Papuapsaltria plicata Boer 1995d: 216, Fig 45 (distribution, comp. note) New Guinea
Papuapsaltria plicata Boer and Duffels 1996b: 325 (distribution) Papuan Peninsula

P. *pusilla* Boer, 2000
Papuapsaltria pusilla Boer 2000: 1–4, 7, 12–14, 19, Figs 33–41, Tables 1–2 (n. sp., comp. note) Papua New Guinea

P. *spinigera* Boer, 1995
Papuapsaltria spinigera Boer 1995a: 2, 4–5, 7, 9, 27–29, 40, Figs 2–3, Figs 102–109, Fig 175, Figs 65–75, Table 1 (n. sp., key, described, illustrated, distribution, comp. note) Papua New Guinea
Papuapsaltria spinigera Boer 1995c: 213, 216–217, 228, 230–231, Fig 4, Fig 12 (phylogeny, comp. note)
Papuapsaltria spinigera Boer 1995d: 216, Fig 45 (distribution, comp. note) New Guinea
Papuapsaltria spinigera Boer and Duffels 1996b: 325 (distribution) New Guinea

P. *stoliodes* Boer, 1995
Papuapsaltria stoliodes Boer 1995a: 2–9, 18–21, 26, Figs 2–3, Figs 53–64, Table 1 (n. sp., key, described, illustrated, distribution, comp. note) Papuan Peninsula
Papuapsaltria stoliodes Boer 1995c: 213, 216–217, 225, 228, 230–231, Fig 4, Fig 12 (phylogeny, comp. note)
Papuapsaltria stoliodes Boer 1995d: 216, Fig 45 (distribution, comp. note) New Guinea
Papuapsaltria stoliodes Boer and Duffels 1996b: 325 (distribution) Papuan Peninsula

P. *toxopei* Boer, 1995
Papuapsaltria toxopei Boer 1995a: 2, 4, 6–7, 9, 34–37, 18–21, 26, Fig 2, Fig 4, Figs 147–155, Figs 65–75, Table 1 (n. sp., key, described, illustrated, distribution, comp. note) Irian Jaya
Papuapsaltria toxopei Boer 1995c: 213, 216–217, 228, 230–231, Fig 4, Fig 12 (phylogeny, comp. note)
Papuapsaltria toxopei Boer 1995d: 216–217, Fig 45 (distribution, comp. note) New Guinea
Papuapsaltria toxopei Boer and Duffels 1996b: 325 (distribution) New Guinea
Papuapsaltria toxopei Boer 2000: 4, 15, Fig 1 (phylogeny, comp. note)

P. *ungula* Boer, 1995
Papuapsaltria ungula Boer 1995a: 2, 4, 7, 9, 21–23, 25–26, Figs 2–3, Figs 76–82, Table 1 (n. sp., key, described, illustrated, distribution, comp. note) Papuan Peninsula
Papuapsaltria ungula Boer 1995c: 213, 216–217, 228, 230–231, Fig 4, Fig 12 (phylogeny, comp. note)
Papuapsaltria ungula Boer 1995d: 216, Fig 45 (distribution, comp. note) New Guinea
Papuapsaltria ungula Boer and Duffels 1996b: 325 (distribution) Papuan Peninsula

P. *ustulata* (Blöte, 1960) = *Baeturia ustulata*
Baeturia ustulata Boer 1993a: 16 (comp. note)
Papuapsaltria ustulata Boer 1995a: 2–3, 6–7, 9, 27, 29–33, Fig 2, Figs 110–118, Fig 129, Table 1 (n. comb., key, described, illustrated, distribution, comp. note) Equals *Baeturia ustulata* New Guinea
Papuapsaltria ustulata Boer 1995c: 213, 216–217, 228, 230–231, Fig 4, Fig 12 (phylogeny, comp. note)
Papuapsaltria ustulata Boer 1995d: 216–217, Fig 45 (distribution, comp. note) New Guinea
Baeturia ustulata Boer 1996: 358 (comp. note)
Papuapsaltria ustulata Boer and Duffels 1996b: 325 (distribution) New Guinea
Papuapsaltria ustulata group Boer and Duffels 1996b: 304, Fig 5 (distribution, phylogeny, comp. note) New Guinea
Papuapsaltria ustulata Boer 2000: 4, 15, Fig 1 (phylogeny, comp. note)

P. *woodlarkensis* Boer, 1995
Papuapsaltria woodlarkensis Boer 1995a: 2–7, 9, 37–41, Fig 2, Fig 4, Fig 10, Figs 165–174, Table 1 (n. sp., key, described, illustrated, distribution, comp. note) Woodlark Island

Papuapsaltria woodlarkensis Boer 1995c: 213, 216–217, 225, 228, 230–231, Fig 4, Fig 12 (phylogeny, comp. note)
Papuapsaltria woodlarkensis Boer 1995d: 216, Fig 45 (distribution, comp. note) Woodlark Island
Papuapsaltria woodlarkensis Boer and Duffels 1996b: 325 (distribution) Papuan Peninsula
Papuapsaltria woodlarkensis Sanborn 1999a: 51 (type material, listed)

***Baeturia* Stål, 1866**

Baeturia sp. Meyer-Rochow 1973: 675 (human food, listed) New Guinea
Baeturia Boulard 1981c: 129 (comp. note)
Baeturia Boer 1982: 57–58, 60 (distribution, diversity, comp. note) Japan, Ryukus islands, Moluccas, New Guinea, Queensland, Solomon Islands, Samoa, New Hebrides
Baeturia Holloway 1983: 524 (comp. note)
Baeturia Hayashi 1984: 28, 69 (key, described, listed, comp. note)
Baeturia sp. Pond 1984: 99 (comp. note) Polynesia
Baeturia Boer 1986: 167, 169–170, 173, 177 (distribution, comp. note) Moluccas, Vogelkop Peninsula, New Guinea, Malaki, Samoa, Japan, Ryukyu, Queensland, Vanuatu
Baeturia Duffels 1986: 336–327, 331, 333 (listed, distribution, diversity, comp. note) Japan, Ryukyu Islands, Maluku, New Guinea, Queensland, Bismarck Archipelago, Solomon Islands, Vanuatu, Rotuma Island, Tonga, Samoa Island
Baeturia Duffels 1988a: 7 (comp. note)
Baeturia Duffels 1988d: 12, 14, 25, 28, 30, 94, 98–99, Fig 8, Table 1 (key, distribution, listed, comp. note) Japan, Ryukyu Islands, Maluku, New Guinea, Queensland, Solomon Islands, Bismarck Archipelago, Samoa, Vanuatu, Rotuma Island, Tonga Islands
Baeturia Boer 1989: 1–4, 6–7, 12, 14, 18–19, 36 (distribution, phylogeny, comp. note) Maluku, New Guinea, Southwest Pacific
Baeturia Boer 1990: 63–64 (comp. note)
Baeturia Boulard 1990b: 214 (comp. note)
Baeturia Duffels and Boer 1990: 261, 265 (comp. note) New Guinea, Papuan Peninsula
Baeturia Moulds 1990: 12, 32, 182 (listed, diversity, comp. note) Australia, New Guinea, Japan, Ryukyu Island, Maluku, Bismark Archipelago
Baeturia spp. Moulds 1990: 189–190 (comp. note)
Baeturia Boer 1991: 1–2 (comp. note)
Baeturia Boer 1992a: 17–19, 21 (phylogeny, comp. note)
Baeturia Boer 1992b: 163–165, 169, 171, 182 (comp. note) New Guinea
Baeturia Boer 1993a: 16–17 (comp. note)
Baeturia Boer 1993b: 141–143, 157 (phylogeny, comp. note)
Baeutria Villet 1993: 435, 439 (comp. note)
Baeturia Boer 1994a: 1–3, 13, 15 (phylogeny, comp. note)
Baeturia Boer 1994b: 161–162, 164 (comp. note)
Baeturia Boer 1994c: 127–128, 130–131, 133–134, 137–138, 142, 151, 154 (comp. note)
Baeturia Boer 1994d: 87–90, 94, 97 (comp. note) New Guinea, Timor, Samoa, Tonga
Baeturia Boer 1995a: 1–3, 5, 18, 41, Fig 1 (phylogeny, comp. note)
Baeturia Boer 1995b: 1–5, 8–9, 11–12, 14, 16–17, 29, 32, 34, 74, 77, Fig 1 (comp. note)
Baeturia Boer 1995c: 201–204, 207, 210–215, 218, 224, Fig 11, Fig 15 (phylogeny, comp. note)
Baeturia Boer 1995d: 172–173, 206, 211–212, 214, 216, 222, 224–225, 229–230, 232, 234, 236–237, Figs 52–53 (distribution, phylogeny, comp. note) New Guinea, Timor, Samoa, Tonga, Bismarck Archipelago
Baeturia Boer 1996: 350–354, 356, 368–369, Fig 1 (comp. note)
Baeturia Boer and Duffels 1996a: 155, 159, 168, 171–172, Figs 10–11 (distribution, phylogeny, comp. note) Maluku, New Guinea, East Melanesia
Baeturia Boer and Duffels 1996b: 301, 306–307, 313–314, 316, 320, 324, Fig 5 (distribution, phylogeny, comp. note) Papua New Guinea, Manus Island, Mussau Island, Solomon Islands
Baeturia Boulard 1996a: 152, 157 (comp. note) Vanuatu, New Caledonia
Baeturia Boer 1997: 91–93, 98, Fig 1 (phylogeny, comp. note)
Baeturia Chou, Lei, Li, Lu and Yao 1997: 90 (comp. note)
Baeturia Holloway 1997: 221 (distribution, comp. note) Moluccas, Samoa, New Guinea
Baeturia Novotny and Wilson 1997: 437 (listed)
Baeturia Hayashi 1998: 62 (history, comp. note)
Baeturia Holloway 1998: 304–305, 309 (distribution, comp. note) New Guinea, Moluccas, Samoa
Baeturia Brown and Toft 1999: 150 (comp. note)
Baeturia Boer 1999: 137 (comp. note)
Baeturia Boer 2000: 1, 4, 8, 15, Fig 1 (phylogeny, comp. note) Papua New Guinea
Baeturia sp. Boer 2000: 1, Table 1 (comp. note) Papua New Guinea
Baeturia Heads 2001: 902 (distribution, comp. note) New Guinea
Baeturia Eberhard 2004: 139, Table 2 (listed)

Baeturia Boulard 2005f: 189, 192 (comp. note) New Caledonia
Baeturia Moulds 2005b: 390, 435 (higher taxonomy, listed, comp. note)
Baeturia Duffels and Boer 2006: 534, 537 (distribution, comp. note) Roon
Baeturia Shiyake 2007: 8, 103 (listed, comp. note)
Baeturia sp. Shiyake 2007: 8, 101, 103, Fig 175 (illustrated, listed, comp. note)
Baeturia sp. Shiyake 2007: 8, 101, 103, Fig 176 (illustrated, listed, comp. note)
Baeturia sp. Shiyake 2007: 8, 101, 103, Fig 177 (illustrated, listed, comp. note)

***B. arabuensis* Blöte, 1960**
Baeturia arabuensis Boer 1982: 58, 60, 64–65, 69, 71–72, 74, 76, Figs 13–18, Fig 44, Fig 48 (key, described, illustrated, distribution, comp. note)
Baeturia arabuensis Boer 1994b: 162–164, Fig 1b (comp. note)
Baeturia arabuensis Boer 1995c: 211–213, 224, 227, 229, 231, Figs 4–5 (phylogeny, comp. note)
Baeturia arabuensis Boer 1995d: 215–216, Fig 44 (distribution, comp. note) New Guinea
Baeturia arabuensis Boer and Duffels 1996b: 325 (distribution) New Guinea
Baeturia arabuensis Duffels and van Mastrigt 1991: 176, 178 (comp. note) Irian Jaya
Baeturia arabuensis Costa Neto 2008: 453 (comp. note)

B. aubertae Boulard, 1979 to *Baeturia edauberti* Boulard, 1979

***B. bemmeleni* Boer, 1994**
Baeturia bemmeleni Boer 1994a: 2–5, 8, 10–13, Fig 1, Figs 21–29 (n. sp., key, described, illustrated, distribution, comp. note) Iriayan Jaya
Baeturia bemmeleni Boer 1995c: 212, 214, 227, 229, 231, Fig 4, Fig 7 (phylogeny, comp. note)
Baeturia bemmeleni Boer 1995d: 214, Fig 42 (distribution, comp. note) New Guinea
Baeturia bemmeleni Boer and Duffels 1996b: 325 (distribution) New Guinea

***B. bicolorata* Distant, 1892**
Baeturia bicolorata Boer 1989: 2, 6, 10 (comp. note)
Baeturia bicolorata Boer 1994a: 23 (comp. note)
Baeturia bicolorata Boer 1994c: 129, 132–135, 137, 139, 141–147, Fig 1b, Fig 2a, Figs 18–32 (key, described, illustrated, phylogeny, comp. note) New Guinea, Aru Islands
Baeturia bicolorata Boer 1995c: 212, 227, 229, 231, Fig 4 (phylogeny, comp. note)
Baeturia bicolorata Boer 1995d: 212–213, 229, Fig 40 (distribution, comp. note) New Guinea
Baeturia bicolorata Boer and Duffels 1996b: 316, 324 (distribution, comp. note) New Guinea
Baeturia bicolorata Boer 2000: 6 (comp. note) New Guinea
Baeturia bicolorata Moulds 2009a: 503–504, Fig 14.4(3) (illustrated, comp. note) Papua New Guinea

***B. bilebanarai* Boer, 1989**
Baeturia bilebanarai Boer 1989: 5, 10, 22, 28–33, Figs 23–24, Figs 113–119, Fig 124 (n. sp., key, described, illustrated, distribution, comp. note)
Baeturia bilebanarai Duffels and Boer 1990: 264, Fig 21.6 (distribution)
Baeturia bilebanarai Boer 1995c: 212–213, 227, 229, 231, Fig 4, Fig 6 (phylogeny, comp. note)
Baeturia bilebanarai Boer 1995d: 212, Fig 38 (distribution, comp. note)
Baeturia bilebanarai Boer and Duffels 1996a: 1996a: 160, Fig 4 (distribution)
Baeturia bilebanarai Boer and Duffels 1996b: 319, 324, Fig 13 (distribution, comp. note) Solomon Islands

***B. bipunctata* Blöte, 1960** = *Baeturia bipuncta* (sic)
Baeturia bipuncta (sic) Boer 1982: 58 (comp. note)
Baeturia bipunctata Boer 1982: 58–60, 68–71, 74, 77, Figs 29–33, Fig 44, Fig 51 (key, described, illustrated, distribution, comp. note) Irian New Guinea
Baeturia bipunctata Boer 1994b: 162–164, Fig 1b (comp. note)
Baeturia bipunctata Boer 1995c: 211–213, 224, 227, 229, 231, Figs 4–5 (phylogeny, comp. note)
Baeturia bipunctata Boer 1995d: 215, Fig 43 (distribution, comp. note) New Guinea
Baeturia bipunctata Boer 1996: 358 (comp. note)
Baeturia bipunctata Boer and Duffels 1996b: 325 (distribution) New Guinea

***B. biroi* Boer, 1994**
Baeturia biroi Boer 1994d: 89–92, 95–98, Fig 1, Figs 24–32 (n. sp., key, described, illustrated, distribution, comp. note) Irian Jaya, Papua New Guinea
Baeturia biroi Boer 1995c: 212, 227, 229, 231, Fig 4 (phylogeny, comp. note)
Baeturia biroi Boer 1995d: 213, Fig 39 (distribution, comp. note)
Baeturia biroi Boer and Duffels 1996b: 325 (distribution) New Guinea

***B. bismarkensis* Boer, 1989**
Baeturia bismarkensis Boer 1989: 3–4, 8–9, 13, 15–19, Figs 9–10, Figs 37–38, Fig 52, Figs 59–65 (n. sp., key, described, illustrated, distribution, comp. note) Bismarck Archipelago
Baeturia bismarkensis Duffels and Boer 1990: 264–265, Fig 21.6 (distribution, comp. note) Bismarck Archipelago

Baeturia bismarkensis Boer 1995c: 212–213, 227, 229, 231, Fig 4, Fig 6 (phylogeny, comp. note)
Baeturia bismarkensis Boer 1995d: 212, 216, Fig 38 (distribution, comp. note) Bismarck Archipelago
Baeturia bismarkensis Boer 1996: 358 (comp. note)
Baeturia bismarkensis Boer and Duffels 1996a: 160, Fig 4 (distribution)
Baeturia bismarkensis Boer and Duffels 1996b: 319, 324, Fig 13 (distribution, comp. note) Bismarck Archipelago

***B. bloetei* Boer, 1989** = *Baeturia bicolorata* Blöte, 1960
Baeturia bloetei Boer 1989: 3–6, 8, 10–14, 16, 26, 41, Figs 1–2, Figs 5–6, Figs, 39–48, Fig 52, Figs 160–161 (n. sp., key, described, illustrated, distribution, comp. note) Biak Island, Japen Island, New Guinea
Baeturia bloetei group Boer 1989: 1–7, 10, 18 (described, comp. note)
Baeturia bloetei group Boer 1990: 66 (comp. note)
Baeturia bloetei group Duffels and Boer 1990: 260–265, 267, Fig 21.6–21.7 (diversity, distribution, comp. note) Admiralty Islands, Bismarck Archipelago, Solomon Islands, Vanuatu, Samoa, Tonga, Rotuma Island
Baeturia bloetei Duffels and Boer 1990: 264, Fig 21.6 (distribution)
Baeturia bloetei Boer 1992b: 167, 170, 174 (comp. note)
Baeturia bloetei group Boer 1992b: 163, 165–167 (described, comp. note)
Baeturia bloetei group Boer 1994a: 1 (comp. note)
Baeturia bloetei group Boer 1994b: 162, Fig 1a (comp. note)
Baeturia bloetei group Boer 1994c: 128–131, 133, 139, 141, Fig 1a (phylogeny, comp. note)
Baeturia bloetei Boer 1994c: 133 (comp. note)
Baeturia bloetei group Boer 1994d: 87–88 (comp. note)
Baeturia bloetei group Boer 1995b: 16–17 (comp. note)
Baeturia bloetei group Boer 1995c: 202, 211, 213–214, Fig 3, Fig 8 (phylogeny, comp. note)
Baeturia bloetei Boer 1995c: 212–213, 227, 229, 231, Fig 4, Fig 6 (phylogeny, comp. note)
Baeturia bloetei group Boer 1995d: 173, 211–212, 221, 232, 236–237, Fig 38 (distribution, comp. note) New Guinea, Maluku, Bismarck Archipelago, East Melanesia, Solomon Islands, Vanuatu, Santa Cruz, Rotuma Island, Samoa, Tonga
Baeturia bloetei Boer 1995d: 212, 229, Fig 38 (distribution, comp. note) New Guinea
Baeturia bloetei group Boer and Duffels 1996a: 159–160, 168, 172–173, Fig 4 (distribution, phylogeny, comp. note) Maluku, New Guinea, Bismarck Archipelago, East Melanesia, Samoa, Tonga
Baeturia bloetei Boer and Duffels 1996a: 160, Fig 4 (distribution)
Baeturia bloetei group Boer and Duffels 1996b: 304, 319–320, 324, Fig 5, Fig 13 (distribution, phylogeny, comp. note) New Guinea, Maluku, Bismarck Archipelago, Southwest Pacific
Baeturia bloetei Boer and Duffels 1996b: 299, 316, 319, 324, Fig 13, Plate 1g (distribution, illustrated, comp. note) New Guinea
Baeturia bloeti group Boulard 1996a: 154, 157 (comp. note)
Baeturia bloetei group Duffels 1997: 552 (diversity, comp. note) Solomon Islands
Baeturia bloetei group Holloway 1997: 220, Fig 5 (distribution, phylogeny, comp. note) Maluku, Bismarck Islands, Solomon Islands, Samoa
Baeturia bloetei group Holloway 1998: 305, 307–308, 310–311, Fig 10, Figs 12–13 (distribution, phylogeny, comp. note) Melanesia
Baeturia bloetei group Holloway and Hall 1998: 13 (distribution, comp. note)
Baeturia bloetei group Boer 1999: 120 (comp. note) Maluku, Solomon Islands, Vanuatu, Samoa, Tonga
Baeturia bloetei group Boer 2000: 15, Fig 1 (phylogeny, comp. note)
Baeturia bloetei species group Arensburger, Simon, and Holsinger 2004: 1778 (comp. note) New Guinea
Baeturia bloetei group Boulard 2005f: 190, 192 (comp. note)

***B. boulardi* Boer, 1989**
Baeturia boulardi Boer 1989: 5, 10, 34–35, 37–39, Figs 29–30, Figs 131–137, Fig 148 (n. sp., key, described, illustrated, distribution, comp. note) Vanuatu
Baeturia boulardi Duffels and Boer 1990: 264, Fig 21.6 (distribution)
Baeturia boulardi Boer 1995c: 212–213, 227, 229, 231, Fig 4, Fig 6 (phylogeny, comp. note)
Baeturia boulardi Boer 1995d: 212, Fig 38 (distribution, comp. note)
Baeturia boulardi Boer and Duffels 1996a: 160, Fig 4 (distribution)
Baeturia boulardi Boer and Duffels 1996b: 319, 324, Fig 13 (distribution, comp. note) Vanuatu

***B. brandti* Boer, 1989**
Baeturia brandti Boer 1989: 4, 6, 9, 20–25, 27–28, 36, 41–42, Figs 15–16, Figs 80–84, Fig 99, Figs

162–163 (n. sp., key, described, illustrated, distribution, comp. note) Solomon Islands

Baeturia brandti Duffels and Boer 1990: 264, Fig 21.6 (distribution)

Baeturia brandti Boer 1994c: 139 (comp. note)

Baeturia brandti Boer 1995c: 212–213, 227, 229, 231, Fig 4, Fig 6 (phylogeny, comp. note)

Baeturia brandti Boer 1995d: 212, Fig 38 (distribution, comp. note)

Baeturia brandti Boer and Duffels 1996a: 160, Fig 4 (distribution)

Baeturia brandti Boer and Duffels 1996b: 319, 324, Fig 13 (distribution, comp. note) Solomon Islands

***B. brongersmai* Blöte, 1960**

Baeturia brongersmai Boer 1989: 10 (comp. note)

Baeturia brongersmai Boer 1992a: 24–25 (comp. note) New Guinea

Baeturia brongersmai Boer 1992b: 164–167, 169, 172–174, 180, Fig 1, Figs 15–23 (phylogeny, key, described, distribution, comp. note) New Guinea, Irian Jaya

Baeturia brongersmai Boer 1994c: 130, 142 (comp. note)

Baeturia brongersmai Boer 1995c: 212, 227, 229, 231, Fig 4 (phylogeny, comp. note)

Baeturia brongersmai Boer 1995d: 212, 214, Fig 41 (distribution, comp. note) New Guinea

Baeturia brongersmai Boer and Duffels 1996b: 325 (distribution) New Guinea

Baeturia brongersmai Boer 2000: 4, 6, 8 (comp. note) New Guinea

B. chinai Blöte, 1960 to *Guineapsaltria chinai*

***B. colossea* Boer, 1994** = *Baeturia gigantea* (sic)

Baeturia colossea Boer 1994c: 129, 132, 134–135, 144–146, Fig 1b, Fig 2a, Figs 33–40 (n. sp., key, described, illustrated, phylogeny, comp. note) Papua New Guinea

Baeturia colossea Boer 1995c: 212, 227, 229, 231, Fig 4 (phylogeny, comp. note)

Baeturia colossea Boer 1995d: 213, Fig 40 (distribution, comp. note)

Baeturia gigantea (sic) Boer 1995d: 213 (comp. note)

Baeturia colossea Boer and Duffels 1996b: 324 (distribution) New Guinea

Baeturia colossea Sanborn 1999a: 51 (type material, listed)

Baeturia colossea Boer 2000: 1, 6–7, Table 1 (comp. note) Papua New Guinea

***B. conviva* (Stål, 1861)** = *Cicada conviva* = *Tibicen convivus* = *Cicada parallela* Walker, 1870

Cicada conviva Hayashi 1984: 69 (type species of *Baeturia*)

Baeturia conviva Boer 1986: 168–169, 171–175, 178, 180, Figs 1–9, Fig 28 (type species of genus *Baeturia*, key, described, illustrated, distribution, comp. note) Equals *Cicada conviva* Maluku Islands, Bacan, Obi

Baeturia conviva group Boer 1986: 168–170 (described, comp. note)

Baeturia conviva Duffels 1988a: 7, 10–11, Table 1 (comp. note) Bacan, Obi

Baeturia conviva group Duffels 1988a: 11 (comp. note) Maluku, Utara, Buru, Tjendrawasih, Aru Islands, Roon Island

Baeturia conviva group Duffels 1988b: 23 (comp. note)

Baeturia conviva Duffels 1988b: 23 (comp. note) Bacan, Obi

Baeturia conviva Duffels 1988d: 94 (type species of *Baeturia*)

Baeturia conviva Boer 1989: 1–3 (illustrated, comp. note) Maluku Islands, Bacan, Obi

Baeturia conviva group Boer 1989: 1 (comp. note)

Baeturia conviva group Duffels and Boer 1990: 258–260, Fig 21.2 (distribution, phylogeny, comp. note)

Baeturia conviva Duffels and Boer 1990: 258, Fig 21.2 (distribution, phylogeny) Obi, Bacan, Ternate

Cicada conviva Moulds 1990: 182 (type species of *Baeturia*, listed)

Baeturia conviva group Boer 1992b: 163–164, 166 (described, comp. note)

Baeturia conviva group Boer 1994a: 1–2 (comp. note)

Baeturia conviva group Boer 1994b: 162, Fig 1a (comp. note)

Baeturia conviva group Boer 1994c: 128–131, 133, 137, 139, Fig 1a (phylogeny, comp. note)

Baeturia conviva Boer 1994c: 137 (comp. note) Buru Island

Baeturia conviva group Boer 1994d: 87–90 (comp. note) New Guinea

Baeturia conviva group Boer 1995c: 202, 211, 21, Fig 3, Fig 8 (phylogeny, comp. note)

Baeturia conviva Boer 1995c: 211–212, 227, 229, 231, Figs 4–5 (phylogeny, comp. note)

Baeturia conviva group Boer 1995d: 173, 206, 211, 213, 230, 232, Fig 39 (distribution, comp. note) New Guinea, Maluku, Bacan, Ori, Halmahera, Buru, Seram, Sula

Baeturia conviva Boer 1995d: 211, 213, 230, Fig 39 (distribution, comp. note) Bacan, Obi Island

Baeturia conviva group Boer and Duffels 1996b: 304, 316, Fig 5 (distribution, phylogeny, comp. note) New Guinea, Maluku

Baeturia conviva Boer and Duffels 1996b: 324 (distribution) North Maluku

Baeturia conviva group Holloway 1998: 305–307, Fig 10, Fig 12 (distribution, phylogeny, comp. note) New Guinea, Maluku
Baeturia conviva group Boer 2000: 4, 15, Fig 1 (phylogeny, comp. note)

***B. cristovalensis* Boer, 1989**
Baeturia cristovalensis Boer 1989: 4, 6, 9, 23, 25–27, Figs 19–20, Figs, 99–105 (n. sp., key, described, illustrated, distribution, comp. note) Solomon Islands
Baeturia cristovalensis Duffels and Boer 1990: 264, Fig 21.6 (distribution)
Baeturia cristovalensis Boer 1995c: 212–213, 227, 229, 231, Fig 4, Fig 6 (phylogeny, comp. note)
Baeturia cristovalensis Boer 1995d: 212, Fig 38 (distribution, comp. note)
Baeturia cristovalensis Boer and Duffels 1996a: 160, Fig 4 (distribution)
Baeturia cristovalensis Boer and Duffels 1996b: 319, 324, Fig 13 (distribution, comp. note) Solomon Islands

***B. daviesi* Boer, 1994**
Baeturia daviesi Boer 1994a: 2–5, 8, 17–19, Fig 2, Figs 60–66 (n. sp., key, described, illustrated, distribution, comp. note) Papuan Peninsula
Baeturia daviesi Boer 1995b: 16 (comp. note)
Baeturia daviesi Boer 1995c: 212, 214, 224, 227, 229, 231, Fig 4 (phylogeny, comp. note)
Baeturia daviesi Boer 1995d: 214, Fig 42 (distribution, comp. note) Papuan Penisula
Baeturia daviesi Boer and Duffels 1996b: 325 (distribution) Papuan Peninsula

B. digitata Blöte, 1960 to *Aedeastria digitata*

***B. edauberti* Boulard, 1979** = *Baeturia aubertae* Boulard, 1979 = *Baeturia efatensis* Boulard, 1979 = *Baeturia epiensis* Boulard, 1979
Baeturia edauberti Duffels 1988d: 99 (comp. note)
Baeturia edauberti Boer 1989: 4–6, 9, 21, 23, 33–38, 40, 42, Figs 31–32, Figs 138–144, Fig 148 (illustrated, comp. note) Equals *Baeturia aubertae* Equals *Baeturia efatensis* Equals *Baeturia epiensis* Vanuatu
Baeturia aubertae Boer 1989: 33, 35 (synonymy, comp. note) to *Baeturia edauberti* n. syn. Vanuatu
Baeturia efatensis Boer 1989: 33, 35 (synonymy, comp. note) to *Baeturia edauberti* n. syn. Vanuatu
Baeturia epiensis Boer 1989: 33, 35 (synonymy, comp. note) to *Baeturia edauberti* n. syn. Vanuatu
Baeturia edauberti Duffels and Boer 1990: 264, Fig 21.6 (distribution)
Baeturia edauberti Boer 1994c: 139 (comp. note)
Baeturia edauberti Boer 1995c: 212–213, 227, 229, 231, Fig 4, Fig 6 (phylogeny, comp. note)
Baeturia edauberti Boer 1995d: 212, Fig 38 (distribution, comp. note)
Baeturia edauberti Boer and Duffels 1996a: 160, Fig 4 (distribution)
Baeturia edauberti Boer and Duffels 1996b: 319, 324, Fig 13 (distribution, comp. note) Vanuatu
Baeturia edauberti Boulard 1996a: 154 (comp. note)
Baeturia edauberti Boulard 2005f: 190 (comp. note) New Hebrides, Vanuatu

B. efatensis Boulard, 1979 to *Baeturia edauberti* Boulard, 1979

B. epiensis Boulard, 1979 to *Baeturia edauberti* Boulard, 1979

***B. exhausta* (Guérin-Méneville, 1831)** = *Cicada exhausta* = *Cicada hastipennis* Walker, 1858 = *Baeturia hastipennis* = *Cicada parallella* Walker, 1868 = *Baeturia parallela* = *Dundubia parabola* Walker, 1858 = *Baeturia parabola*
Baeturia exhausta Boer 1986: 173 (comp. note)
Baeturia exhausta Duffels 1988d: 10, 94 (comp. note) Buru
Baeturia exhausta Boer 1989: 1–3 (comp. note) Maluku, New Guinea
Baeturia exhausta Boer 1992b: 166 (comp. note)
Baeturia exhausta group Boer 1994b: 162, Fig 1a (comp. note)
Baeturia exhausta group Boer 1994c: 127–131, 133–134, 144, 154, Fig 1a (phylogeny, described, comp. note) Timor, Maluku, New Guinea
Baeturia exhausta Boer 1994c: 129, 131–143, 145, 147, Fig 1b, Fig 2a, Figs 3–17 (synonymy, key, neotype designation, described, illustrated, phylogeny, comp. note) Equals *Cicada exhausta* Equals *Cicada hastipennis* Equals *Baeturia hastipennis* Equals *Cicada parallela* n. syn. Equals *Baeturia parallela* Equals *Dundubia parabola* Equals *Baeturia parabola* Equals *Pomponia jacobsoni* China Molucca Islands, Buru Island, Seram, Irian Jaya
Cicada exhausta Boer 1994c: 136–137 (comp. note) Buru Island
Cicada hastipennis Boer 1994c: 136, 138–139 (comp. note) Seram
Baeturia hastipennis Boer 1994c: 136, 140 (comp. note)
Dundubia parabola Boer 1994c: 136, 139 (comp. note) Seram?
Cicada parallela Boer 1994c: 136, 139 (comp. note) Seram
Baeturia exhausta group Boer 1994d: 87 (comp. note)
Baeturia exhausta group Boer 1995c: 202, 211, 214, Fig 3, Fig 8 (phylogeny, comp. note)
Baeturia exhausta Boer 1995c: 212, 227, 229, 231, Fig 4 (phylogeny, comp. note)
Baeturia exhausta group Boer 1995d: 211, 213, Fig 40 (distribution, comp. note) New Guinea

Baeturia exhausta Boer 1995d: 211, 213, 229, Fig 40 (distribution, comp. note) Maluku, Banda Islands, Timor

Baeturia exhausta group Boer 1996: 358 (comp. note)

Baeturia exhausta group Boer and Duffels 1996b: 304, Fig 5 (distribution, phylogeny, comp. note) New Guinea, Maluku

Baeturia exhausta Boer and Duffels 1996b: 300, 306, 316, 324, Fig 13, Plate 2 (distribution, timbal illustrated, comp. note) North Maluku, South Maluku, Banda Islands, New Guinea

Baeturia exhausta group Holloway 1998: 305–307, Fig 10, Fig 12 (distribution, phylogeny, comp. note) New Guinea

Baeturia exhausta species group Boer 2000: 2, 6, 9 (comp. note)

Baeturia exhausta group Boer 2000: 2, 4, 15, Fig 1 (phylogeny, comp. note)

B. famulus Myers, 1928 to *Aedeastria latrifrons* (Blöte, 1860)

B. flava (Goding & Froggatt, 1904) to *Guineapsaltria flava*

***B. fortuini* Boer, 1994**

Baeturia fortuini Boer 1994a: 2–5, 8, 15, 20–24, Fig 2, Figs 76–87 (n. sp., key, described, illustrated, distribution, comp. note) Papua New Guinea

Baeturia fortuini Boer 1995c: 212, 214, 227, 229, 231, Fig 4, Fig 7 (phylogeny, comp. note)

Baeturia fortuini Boer 1995d: 213–214, Fig 42 (distribution, comp. note) New Guinea

Baeturia fortuini Boer 1996: 358 (comp. note)

Baeturia fortuini Boer and Duffels 1996b: 325 (distribution) New Guinea

Baeturia fortuini Boer 2000: 6 (comp. note) New Guinea

***B. furcillata* Boer, 1992**

Baeturia furcillata Boer 1992b: 164–169, 177, 179–181, Fig 1, Figs 50–57 (n. sp., phylogeny, key, described, distribution, comp. note) Irian Jaya, Papua New Guinea

Baeturia furcillata Boer 1994d: 89 (comp. note)

Baeturia furcillata Boer 1995c: 212, 224, 227, 229, 231, Fig 4 (phylogeny, comp. note)

Baeturia furcillata Boer 1995d: 214, Fig 41 (distribution, comp. note) New Guinea

Baeturia furcillata Boer and Duffels 1996b: 325 (distribution) New Guinea

***B. galeata* Boer, 2000**

Baeturia galeata Boer 2000: 1–7, 9–11, 15, 17–18, Fig 1, Figs 16–24, Tables 1–2 (n. sp., described, illustrated, phylogeny, comp. note) Papua New Guinea

***B. gibberosa* Boer, 1994**

Baeturia gibberosa Boer 1994b: 162–164, 168–170, Fig 1b, Fig 2, Figs 23–31 (n. sp., described, illustrated, distribution, comp. note) Papuan Peninsula, Normanby Island

Baeturia gibberosa Boer 1995c: 211–213, 224, 227, 229, 231, Figs 4–5 (phylogeny, comp. note)

Baeturia gibberosa Boer 1995d: 214–216, Fig 43 (distribution, comp. note) Papuan Peninsula

Baeturia gibberosa Boer and Duffels 1996b: 325 (distribution) Papuan Peninsula

***B. gressitti* Boer, 1989**

Baeturia gressitti Boer 1989: 5–6, 10, 26–29, 32, Figs 21–22, Figs 106–112, Fig 124 (n. sp., key, described, illustrated, distribution, comp. note) Solomon Islands

Baeturia gressitti Duffels and Boer 1990: 264, Fig 21.6 (distribution)

Baeturia gressitti Boer 1995c: 212–213, 224, 227, 229, 231, Fig 4, Fig 6 (phylogeny, comp. note)

Baeturia gressitti Boer 1995d: 212, Fig 38 (distribution, comp. note)

Baeturia gressitti Boer and Duffels 1996a: 160, Fig 4 (distribution)

Baeturia gressitti Boer and Duffels 1996b: 319, 324, Fig 13 (distribution, comp. note) Solomon Islands

Baeturia gressitti Sanborn 1999a: 51 (type material, listed)

***B. guttulinervis* Blöte, 1960**

Baeturia guttulinervis Boer 1982: 69 (comp. note)

Baeturia guttulinervis group Duffels and Boer 1990: 258, Fig 21.2 (distribution, phylogeny) New Guinea, Roon

Baeturia guttulinervis Duffels and Boer 1990: 260 (comp. note)

Baeturia guttulinervis Boer 1992b: 164–165 (comp. note)

Baeturia guttulinervis group Boer 1992b: 163–164, 166 (comp. note)

Baeturia guttulinervis Boer 1994a: 2, 10 (comp. note)

Baeturia guttulinervis group Boer 1994b: 162, 164–165, Fig 1a (comp. note)

Baeturia guttulinervis group Boer 1994c: 128–131, Fig 1a (phylogeny, comp. note)

Baeturia guttulinervis group Boer 1994d: 87–91, Fig 1 (described, comp. note) New Guinea

Baeturia guttulinervis Boer 1994d: 89–96, Figs 1–15 (key, described, illustrated, distribution, comp. note) Irian Jaya

Baeturia guttulinervis group Boer 1995c: 202, 211, 214, Fig 3, Fig 8 (phylogeny, comp. note)

Baeturia guttulinervis Boer 1995c: 212, 224, 227, 229, 231, Fig 4 (phylogeny, comp. note)

Baeturia guttulinervis group Boer 1995d: 211, 213, 230, Fig 39 (distribution, comp. note) New Guinea
Baeturia guttulinervis Boer 1995d: 213, Fig 39 (distribution, comp. note)
Baeturia guttulinervis group Boer 1996: 358 (comp. note)
Baeturia guttulinervis Boer 1996: 358 (comp. note)
Baeturia guttulinervis group Boer and Duffels 1996b: 304, 317, 325, Fig 5, Fig 11 (distribution, comp. note) New Guinea
Baeturia guttulinervis Boer and Duffels 1996b: 325 (distribution) New Guinea
Baeturia guttulinervis group Holloway 1998: 305–307, Fig 10, Fig 12 (distribution, phylogeny, comp. note) New Guinea, Maluku
Baeturia guttulinervis Sanborn 1999a: 51 (type material, listed)
Baeturia guttulinervis group Boer 2000: 4, 15, Fig 1 (phylogeny, comp. note)
Baeturia guttulinervis Boer 2000: 5 (comp. note)

B. guttulipennis **Blöte, 1960**
Baeturia guttulipennis Boer 1994a: 2 (comp. note)
Baeturia guttulipennis Boer 1994b: 162–164, Fig 1b, Figs 2–13 (n. sp., key, described, illustrated, distribution, comp. note) Irian Jaya
Baeturia guttulipennis Boer 1994d: 89 (comp. note)
Baeturia guttulipennis Boer 1995c: 211–213, 224, 227, 229, 231, Figs 4–5 (phylogeny, comp. note)
Baeturia guttulipennis Boer 1995d: 215, Fig 44 (distribution, comp. note) New Guinea
Baeturia guttulipennis Boer and Duffels 1996b: 325 (distribution) New Guinea

B. hamiltoni **De Boer, 1994**
Baeturia hamiltoni Boer 1994a: 2–13, Fig 1, Figs 3–20 (n. sp., key, described, illustrated, distribution, comp. note) Irian Jaya, Papua New Guinea
Baeturia hamiltoni Boer 1995c: 212, 214, 227, 229, 231, Fig 4, Fig 7 (phylogeny, comp. note)
Baeturia hamiltoni Boer 1995d: 214, Fig 42 (distribution, comp. note) New Guinea
Baeturia hamiltoni Boer 1996: 358 (comp. note)
Baeturia hamiltoni Boer and Duffels 1996b: 325 (distribution) New Guinea
Baeturia hamiltoni Duffels and Boer 2006: 535, Fig 4.4.2 (illustrated) New Guinea

B. hardyi **De Boer, 1986**
Baeturia hardyi Boer 1986: 169–171, 175, 179–181, Figs 24–28 (key, described, illustrated, distribution, comp. note) Roon Island, New Guinea
Baeturia hardyi Boer 1989: 25 (comp. note)
Baeturia hardyi Duffels and Boer 1990: 258, Fig 21.2 (distribution, phylogeny) Roon Island
Baeturia hardyi Boer 1992b: 165 (comp. note)
Baeturia hardyi Boer 1995c: 212, 227, 229, 231, Fig 4 (phylogeny, comp. note)
Baeturia hardyi Boer 1995d: 212, 214, Fig 42 (distribution, comp. note) New Guinea
Baeturia hardyi Boer and Duffels 1996b: 324 (distribution) New Guinea

B. hartonoi **Boer, 1994**
Baeturia hartonoi Boer 1994a: 19–20, Fig 1, Figs 67–75 (key, described, illustrated, distribution, comp. note) Papua New Guinea
Baeturia hartonoi Boer 1995c: 212, 214, 227, 229, 231, Fig 4, Fig 7 (phylogeny, comp. note)
Baeturia hartonoi Boer 1995d: 213, Fig 39 (distribution, comp. note)
Baeturia hartonoi Boer and Duffels 1996b: 325 (distribution) New Guinea
Baeturia hartonoi Boer 2000: 1, 6–7, Table 1 (comp. note) Papua New Guinea

B. hirsuta Blöte, 1960 to *Baeturia quadrifida* (Walker, 1868)

B. humilis Blöte, 1960 to *Mirabilopsaltria humilis*

B. inconstans **Boer, 1994**
Baeturia inconstans Boer 1994d: 89–92, 97–99, Fig 1, Figs 33–40 (n. sp., key, described, illustrated, distribution, comp. note)
Baeturia inconstans Boer 1995c: 212, 224–227, 229, 231, Fig 4 (phylogeny, comp. note)
Baeturia inconstans Boer 1995d: 213, Fig 39 (distribution, comp. note)
Baeturia inconstans Boer 1997: 98 (comp. note)
Baeturia inconstans Boer and Duffels 1996b: 325 (distribution) New Guinea

B. innotabilis (Walker, 1968) to *Gymnotympana subnotata* (Walker, 1868)

B. intermedia **Boer, 1982**
Baeturia intermedia Boer 1982: 58, 61, 71–73, 77, Figs 39–41, Fig 43, Fig 53 (key, described, illustrated, distribution, comp. note)
Baeturia intermedia Boer 1994b: 162–164, Fig 1b (comp. note)
Baeturia intermedia Boer 1995c: 212–213, 227, 229, 231, Figs 4–5 (phylogeny, comp. note)
Baeturia intermedia Boer 1995d: 215–216, Fig 44 (distribution, comp. note) New Guinea
Baeturia intermedia Boer and Duffels 1996b: 325 (distribution) New Guinea

B. karkarensis **De Boer, 1992**
Baeturia karkarensis Boer 1992b: 164–169, 181–182, Fig 1, Figs 58–65 (n. sp., phylogeny, key, described, distribution, comp. note) Papua New Guinea, Karkar Island
Baeturia karkarensis Boer 1994c: 131 (comp. note)
Baeturia karkarensis Boer 1994d: 90 (comp. note)
Baeturia karkarensis Boer 1995c: 212, 224, 227, 229, 231, Fig 4 (phylogeny, comp. note)

Baeturia karkarensis Boer 1995d: 214, Fig 41 (distribution, comp. note) New Guinea
Baeturia karkarensis Boer and Duffels 1996b: 325 (distribution) New Guinea

B. kuroiwae (Matsumura, 1913) to *Muda kuroiwae*

***B. laminifer* Blöte, 1960**

Baeturia laminifer Boer 1982: 58–60, 70–71, 74, 77, Figs 34–38, Fig 44, Fig 52 (key, described, illustrated, distribution, comp. note)
Baeturia laminifera Boer 1994b: 162–164, 170, Fig 1b (comp. note)
Baeturia laminifer Boer 1995c: 211–213, 224, 227, 229, 231, Figs 4–5 (phylogeny, comp. note)
Baeturia laminifer Boer 1995d: 213, 215–216, Fig 43 (distribution, comp. note) New Guinea, Bismarck Archipelago, Admiralty Islands
Baeturia laminifer Boer 1996: 358 (comp. note) Papuan Peninsula
Baeturia laminifer Boer and Duffels 1996b: 325 (distribution) New Guinea, Bismarck Archipelago

***B. laperousei* Boulard, 2005**

Baeturia laperousei Boulard 2005f: 189–192, Figs 1–9 (n. sp. described, illustrated, comp. note) Solomon Islands, Vanikoro
Ueana laperousei Boulard 2008a: 67–70, Fig 2, Fig 4 (status, song, sonogram, comp. note) Vanikoro

B. latifrons Blöte, 1960 to *Aedeastria latifrons*

***B. laureli* Boer, 1986**

Baeturia laureli Boer 1986: 169, 171, 173–175, 180–181, Figs 10–15, Fig 28 (n. sp., key, described, illustrated, distribution, comp. note) Maluku Islands, Halmahera Island
Baeturia laureli Duffels 1988a: 11 (comp. note) Halmahera
Baeturia laureli Duffels 1988b: 23 (comp. note) Halmahera
Baeturia laureli Boer 1989: 25 (comp. note)
Baeturia laureli Duffels and Boer 1990: 258, Fig 21.2 (distribution, phylogeny) Halmahera, Morotai
Baeturia laureli Boer 1995c: 212, 227, 229, 231, Fig 4 (phylogeny, comp. note)
Baeturia laureli Boer 1995d: 211, 213, Fig 39 (distribution, comp. note) Halmahera
Baeturia laureli Boer and Duffels 1996b: 324 (distribution) North Maluku

***B. lorentzi* Boer, 1992**

Baeturia lorentzi Boer 1992a: 25 (comp. note) New Guinea
Baeturia lorentzi Boer 1992b: 164–167, 169–171, 174–175, 180, Fig 1, Figs 24–32 (n. sp., phylogeny, key, described, distribution, comp. note) New Guinea, Irian Jaya
Baeturia lorentzi Boer 1994c: 130, 142 (comp. note)
Baeturia lorentzi Boer 1995c: 212, 227, 229, 231, Fig 4 (phylogeny, comp. note)
Baeturia lorentzi Boer 1995d: 212, 214, Fig 41 (distribution, comp. note) New Guinea
Baeturia lorentzi Boer and Duffels 1996b: 325 (distribution) New Guinea
Baeturia lorentzi Boer 2000: 4, 6, 8 (comp. note) New Guinea

***B. loriae* Distant, 1897** = *Gymnotympana loriae*

Baeturia loriae Boer 1989: 2 (comp. note)
Baeturia loriae Boer 1991: 3 (comp. note)
Baeturia loriae group Boer 1994a: 1–5, 10, 13–14, 20, 25, Table 1 (described, phylogeny, comp. note)
Baeturia loriae Boer 1994a: 2–5, 8, 15, 21, 23–24, Fig 2, Figs 88–93 (n. comb., key, described, illustrated, distribution, comp. note) Papuan Peninsula
Baeturia loriae Boer 1994b: 162 (comp. note)
Baeturia loriae group Boer 1994b: 162, Fig 1a (comp. note)
Baeturia loriae group Boer 1994c: 128–131, Fig 1a (phylogeny, comp. note)
Baeturia loriae group Boer 1994d: 87–89 (comp. note)
Baeturia loriae group Boer 1995b: 2–3, 5, 11, Fig 1 (comp. note)
Baeturia loriae Boer 1995b: 16 (comp. note)
Gymnotympana loriae Boer 1995b: 1 (synonymy comp. note)
Baeturia loriae group Boer 1995c: 202, 206, 211, 214, Fig 3, Fig 8 (phylogeny, comp. note)
Baeturia loriae Boer 1995c: 212, 214, 227, 229, 231, Fig 4, Fig 7 (phylogeny, comp. note)
Baeturia loriae group Boer 1995d: 214–216, Fig 42 (distribution, comp. note) New Guinea
Baeturia loriae Boer 1995d: 214, Fig 42 (distribution, comp. note) Papuan Peninsula
Baeturia loriae Boer 1996: 358 (comp. note)
Baeturia loriae group Boer and Duffels 1996b: 304, 325, Fig 5 (distribution, phylogeny, comp. note) Papuan Peninsula, New Guinea
Baeturia loriae Boer and Duffels 1996b: 325 (distribution) Papuan Peninsula
Baeturia loriae group Holloway 1998: 305–307, Fig 10, Fig 12 (distribution, phylogeny, comp. note) New Guinea
Baeturia loriae group Boer 2000: 2, 4, 6, 15, Fig 1 (phylogeny, comp. note)
Baeturia loriae species group Boer 2000: 6, 9 (comp. note)
Baeturia loriae Boer 2000: 6 (comp. note) New Guinea

***B. maai* Boer, 1994**

Baeturia maai Boer 1994c: 129, 131–133, 135, 153–154, Fig 1b, Fig 2b, Figs 72–78 (n. sp., key, described, illustrated, phylogeny, comp. note) Irian Jaya

Baeturia maai Boer 1995c: 212, 227, 229, 231, Fig 4 (phylogeny, comp. note)

Baeturia maai Boer 1995d: 213, Fig 40 (distribution, comp. note)

Baeturia maai Boer and Duffels 1996b: 324 (distribution) New Guinea

***B. macgillavryi* Boer, 1989** = *Baeturia famulus* (nec Myers) = *Cephaloxys dilectus* nom. nud. = *Baeturia dilectus* nom. nud. = *Cephaloxys cillogata* nom. nud.

Baeturia macgillavryi Boer 1989: 3, 5–6, 10, 12–16, Figs 49–52 (n. sp., key, described, illustrated, distribution, comp. note) Moluccas, Timor

Baeturia macgillavry Duffels and Boer 1990: 264, Fig 21.6 (distribution)

Baeturia macgillavryi Duffels 1991a: 122, 127 (comp. note) Moluccas, Timor, Talaud Islands

Baeturia macgillavryi Boer 1994c: 133, 137–139 (comp. note) Equals *Cephaloxys dilectus* Equals *Baeturia dilectus* Equals *Cephaloxys cillogata* Molucca Islands, Buru Island, Seram

Baeturia macgillavry Boer 1995a: 23 (comp. note) Equals *Baeturia famulus* Lallemand

Baeturia macgillavry Boer 1995c: 212–213, 227, 229, 231, Fig 4, Fig 6 (phylogeny, comp. note)

Baeturia macgillavry Boer 1995d: 211–212, 229, Fig 38 (distribution, comp. note) Maluku, Halmahera, Moratai, Talaud, Banda Islands, Timor

Baeturia macgillavry Boer and Duffels 1996a: 160, Fig 4 (distribution)

Baeturia macgillavry Boer and Duffels 1996b: 306, 316, 319, 324, Fig 13 (distribution, comp. note) Maluku, Banda Islands

***B. maddisoni* Duffels, 1988** = *Baeturia exhausta* (nec Guérin-Méneville)

Baeturia maddisoni Duffels 1988d: 10–12, 14, 25–26, 28, 30, 94–100, Figs 178–190, Table 1 (n. sp., key, described, illustrated, distribution, listed, comp. note) Equals *Baeturia exhausta* (partim) Samoa Islands, Tonga Islands

Baeturia maddisoni Boer 1989: 3–5, 9, 36, 38–40, 42, Figs 35–36, Figs 148, Figs 153–159 (illustrated, comp. note) Samoa, Tonga

Baeturia maddisoni Duffels and Boer 1990: 264, Fig 21.6 (distribution)

Baeturia maddisoni Boer 1994c: 139 (comp. note)

Baeturia maddisoni Boer 1995c: 212–213, 227, 229, 231, Fig 4, Fig 6 (phylogeny, comp. note)

Baeturia maddisoni Boer 1995d: 212, Fig 38 (distribution, comp. note)

Baeturia maddisoni Boer and Duffels 1996a: 160, Fig 4 (distribution)

Baeturia maddisoni Boer and Duffels 1996b: 319, 324, Fig 13 (distribution, comp. note) Samoa Islands, Tonga Islands

Baeturia maddisoni Boulard 1996a: 154 (comp. note) Samoa

***B. mamillata* Blöte, 1960**

Baeturia mamillata Boer 1982: 58, 60, 62–71, 73, 75, Figs 7–12, Fig 43, Fig 47 (key, described, illustrated, distribution, comp. note) Irian New Guinea, Papua New Guinea

Baeturia mamillata Boer 1994b: 162–164, Fig 1b (comp. note)

Baeturia mamillata Boer 1995c: 211–213, 224, 227, 229, 231, Figs 4–5 (phylogeny, comp. note)

Baeturia mamillata Boer 1995d: 203, 214–215, Fig 43 (distribution, comp. note) New Guinea

Baeturia mamillata Boer and Duffels 1996b: 325 (distribution) New Guinea

Baeturia mamillata Boer 2000: 1, 5, 7, Table 1 (comp. note) Papua New Guinea

***B. manusensis* Boer, 1989**

Baeturia manusensis Boer 1989: 4, 10, 13, 18–20, Figs 11–12, Fig 52, Figs 66–72 (n. sp., key, described, illustrated, distribution, comp. note) Admiralty Islands, Manus Island

Baeturia manuensis Duffels and Boer 1990: 264, Fig 21.6 (distribution)

Baeturia manusensis Boer 1995c: 212–213, 227, 229, 231, Fig 4, Fig 6 (phylogeny, comp. note)

Baeturia manusensis Boer 1995d: 212, Fig 38 (distribution, comp. note) Manus Island

Baeturia manusensis Boer and Duffels 1996a: 160, Fig 4 (distribution)

Baeturia manusensis Boer and Duffels 1996b: 319, 324, Fig 13 (distribution, comp. note) Bismarck Archipelago

***B. marginata* Boer, 1989**

Baeturia marginata Boer 1989: 5, 10, 32–35, Figs 27–28, Figs 124–128 (n. sp., key, described, illustrated, distribution, comp. note) Solomon Islands

Baeturia marginata Duffels and Boer 1990: 264, Fig 21.6 (distribution)

Baeturia marginata Boer 1995c: 212–213, 227, 229, 231, Fig 4, Fig 6 (phylogeny, comp. note)

Baeturia marginata Boer 1995d: 212, Fig 38 (distribution, comp. note)

Baeturia marginata Boer and Duffels 1996a: 160, Fig 4 (distribution)

Baeturia marginata Boer and Duffels 1996b: 319, 324, Fig 13 (distribution, comp. note) Solomon Islands

***B. marmorata* Blöte, 1960**

Baeturia marmorata Boer 1982: 58–60, 67–69, 73, 76, Figs 24–28, Fig 50 (key, described, illustrated, distribution, comp. note) Irian New Guinea

Baeturia marmorata Boer 1994b: 162–164, Fig 1b (comp. note)

Baeturia marmorata Boer 1995c: 211–213, 224, 227, 229, 231, Figs 4–5 (phylogeny, comp. note)

Baeturia marmorata Boer 1995d: 215, Fig 44 (distribution, comp. note) New Guinea

Baeturia marmorata Boer 1996: 358 (comp. note)

Baeturia marmorata Boer and Duffels 1996b: 325 (distribution) New Guinea

Baeturia marmorata Sanborn 1999a: 51 (type material, listed)

Baeturia marmorata Boer 2000: 5 (comp. note)

***B. mendanai* Boer, 1989**

Baeturia mendanai Boer 1989: 5, 10, 29–33, Figs 25–26, Figs 120–124 (n. sp., key, described, illustrated, distribution, comp. note) Solomon Islands

Baeturia mendanai Duffels and Boer 1990: 264, Fig 21.6 (distribution)

Baeturia mendanai Boer 1995c: 212–213, 227, 229, 231, Fig 4, Fig 6 (phylogeny, comp. note)

Baeturia mendanai Boer 1995d: 212, Fig 38 (distribution, comp. note)

Baeturia mendanai Boer and Duffels 1996a: 160, Fig 4 (distribution)

Baeturia mendanai Boer and Duffels 1996b: 319, 324, Fig 13 (distribution, comp. note) Solomon Islands

Baeturia mendanai Boulard 1996a: 154 (comp. note) Guadalcanal

Baeturia mendanai Boulard 2005f: 190 (comp. note) Guadalcanal

B. minuta Blöte, 1960 to *Guineapsaltria flava*

B. modesta Distant, 1907 to *Guineapsaltria flava* (Goding & Froggatt, 1904)

***B. mussauensis* Boer, 1989**

Baeturia mussauensis Boer 1989: 4, 10, 13, 19–21, Figs 13–14, Fig 52, Figs 73–77 (n. sp., key, described, illustrated, distribution, comp. note) Admiralty Islands, Mussau Island

Baeturia mussauensis Duffels and Boer 1990: 264, Fig 21.6 (distribution)

Baeturia mussauensis Boer 1995c: 212–213, 227, 229, 231, Fig 4, Fig 6 (phylogeny, comp. note)

Baeturia mussauensis Boer 1995d: 212, Fig 38 (distribution, comp. note) Mussa Island

Baeturia mussauensis Boer and Duffels 1996a: 160, Fig 4 (distribution)

Baeturia mussauensis Boer and Duffels 1996b: 319, 324, Fig 13 (distribution, comp. note) Bismarck Archipelago

B. nana (Jacobi, 1903) to *Papuapsaltria nana*

***B. nasuta* Blöte, 1960**

Baeturia nasuta Boer 1982: 58, 61–64, 66, 69–73, 75, Figs 1–6, Fig, 42, Figs, 45–46 (key, described, illustrated, distribution, comp. note) Irian New Guinea, Papua New Guinea

Baeturia nasuta group Boer 1986: 168–169 (comp. note)

Baeturia nasuta group Duffels 1986: 327, 330, 331(comp. note) New Guinea

Baeturia nasuta group Boer 1989: 1, 36 (comp. note)

Baeturia nasuta group Duffels and Boer 1990: 261, 263, 267 (diversity, distribution) New Guinea

Baeturia nasuta Boer 1992b: 165 (comp. note)

Baeturia nasuta group Boer 1992b: 163–166, 168 (comp. note)

Baeturia nasuta group Boer 1994a: 1 (comp. note)

Baeturia nasuta Boer 1994a: 2 (comp. note)

Baeturia nasuta group Boer 1994b: 161–165, 167, Fig 1a (described, distribution, comp. note)

Baeturia nasuta Boer 1994b: 162–165, Fig 1b (comp. note)

Baeturia nasuta group Boer 1994c: 128–131, Fig 1a (phylogeny, comp. note)

Baeturia nasuta group Boer 1994d: 87–89 (comp. note)

Baeturia nasuta group Boer 1995a: 5 (comp. note)

Baeturia nasuta group Boer 1995b: 4 (comp. note)

Baeturia nasuta group Boer 1995c: 202, 211, 214, Fig 3, Fig 8 (phylogeny, comp. note)

Baeturia nasuta Boer 1995c: 212–213, 227, 229, 231, Figs 4–5 (phylogeny, comp. note)

Baeturia nasuta Boer 1995d: 213, 215–216, Fig 43 (distribution, comp. note) New Guinea

Baeturia nasuta group Boer 1995d: 173, 213, 215–216, 222, 232–233, Fig 43 (distribution, comp. note) New Guinea

Baeturia nasuta Boer and Duffels 1996b: 325 (distribution) New Guinea

Baeturia nasuta group Boer and Duffels 1996b: 304, 306, 325, Fig 5 (distribution, phylogeny, comp. note) New Guinea

Baeturia nasuta group Holloway 1998: 305–307, Fig 10, Fig 12 (distribution, phylogeny, comp. note) New Guinea

Baeturia nasuta Sanborn 1999a: 52 (type material, listed)

Baeturia nasuta? Boer 2000: 1, Table 1 (comp. note) Papua New Guinea

Baeturia nasuta group Boer 2000: 4–5, 15, Fig 1 (phylogeny, comp. note)

Baeturia nasuta species group Boer 2000: 5 (comp. note)

Baeturia nasuta Boer 2000: 5, 7 (comp. note)

Baeturia nasuta group Duffels and Boer 2006: 534 (diversity, distribution, comp. note) New Guinea

B. pallida Blöte, 1960 to *Guineapsaltria pallida*

***B. papuensis* Boer, 1989**

Baeturia papuensis Boer 1989: 3–4, 6–7, 9, 13–16, Figs 7–8., Figs 52–58 (n. sp., key, described, illustrated, distribution, comp. note) Papua New Guinea

Baeturia papuensis Duffels and Boer 1990: 264–265, Fig 21.6 (distribution, comp. note) Papua New Guinea

Baeturia papuensis Kuniata and Nagaraja 1992: 65–71, Figs 1–3, Tables 1–3 (sugarcane pest, natural history, predation, comp. note) Papua New Guinea

Baeturia papuensis Kuniata 1994: 47, 49 (fungal host) Papua New Guinea

Baeturia papuensis Boer 1995c: 212–213, 224, 227, 229, 231, Fig 4, Fig 6 (phylogeny, comp. note)

Baeturia papuensis Boer 1995d: 212–213, 236, Fig 38 (distribution, comp. note) New Guinea

Baeturia papuensis Boer 1996: 358 (comp. note)

Baeturia papuensis Boer and Duffels 1996a: 160, Fig 4 (distribution)

Baeturia papuensis Boer and Duffels 1996b: 298, 319, 324, Fig 13 (sugar cane pest, distribution, comp. note) Papua New Guinea

Baeturia papuensis Hartemink and Kuniata 1996: 229, 231, Tables 1–2 (comp. note) Papua New Guinea

Baeturia papuensis Waterhouse 1997: 12, Table 1 (agricultural pest, listed)

Baeturia papuensis Kuniata 2001: 339, Table 3 (sugarcane pest) Papua New Guinea

***B. parva* Blöte, 1960**

Baeturia parva Boer 1982: 58–59, 61, 66–67, 73, 76, Figs 19–23, Fig 42, Fig 49 (key, described, illustrated, distribution, comp. note) Irian New Guinea, Papua New Guinea, Bismarck Archipelago, New Britain

Baeturia parva Duffels 1986: 327 (comp. note) New Guinea, New Britain

Baeturia parva Duffels and Boer 1990: 261 (diversity, distribution) New Guinea, New Britain, New Ireland, Admiralty Islands, Biak, Misoöl

Baeturia parva Boer 1992b: 168 (comp. note) Obi Island, Bismarck Archipelago, Admiralty Islands

Baeturia parva Boer 1993a: 21 (comp. note)

Baeturia parva Boer 1994b: 162–163, Fig 1b (comp. note)

Baeturia parva Boer 1995b: 16 (comp. note)

Baeturia parva Boer 1995c: 212–213, 227, 229, 231, Figs 4–5 (phylogeny, comp. note)

Baeturia parva Boer 1995d: 212, 215–216, Fig 43 (distribution, comp. note) New Guinea, Misool, Bismarck Archipelago, Admiralty Islands

Baeturia parva Boer 1996: 358 (comp. note)

Baeturia parva Boer and Duffels 1996b: 325 (distribution) New Guinea, Bismarck Archipelago

Baeturia parva Boer 1999: 120 (comp. note) Misool, Bismark Archipelago, New Guinea

B. phyllophora Blöte, 1960 to *Papuapsaltria phyllophora*

***B. pigrami* Boer, 1994**

Baeturia pigrami Boer 1994a: 2–5, 8, 15–17, Fig 2, Figs 49–59 (n. sp., key, described, illustrated, distribution, comp. note) New Guinea

Baeturia pigrami Boer 1994b: 162, 165 (comp. note)

Baeturia pigrami Boer 1995c: 212, 214, 224, 227, 229, 231, Fig 4, Fig 7 (phylogeny, comp. note)

Baeturia pigrami Boer 1995d: 214, Fig 42 (distribution, comp. note) New Guinea

Baeturia pigrami Boer and Duffels 1996b: 325 (distribution) New Guinea

Baeturia pigrami Moulds 2009a: 504, Fig 14.4(2) (illustrated) Papua New Guinea

***B. polhemi* Boer, 2000**

Baeturia polhemi Boer 2000: 1–4, 6–9, 15–17, Fig 1, Figs 3–15, Tables 1–2 (n. sp., described, illustrated, phylogeny, comp. note) Papua New Guinea

***B. quadrifida* (Walker, 1868)** = *Cicada quadrifida* = *Baeturia hirsuta* Blöte, 1960

Baeturia quadrifida Boer 1986: 169–171, 177–181, Figs 19–23, Fig 28 (key, n. syn., described, illustrated, distribution, comp. note) Equals *Cicada quadrifida* Equals *Baeturia hirsuta* n.syn. Irian New Guinea, Aru Islands, Vogelkop Peninsula

Baeturia hirsuta Boer 1986: 177 (comp. note)

Baeturia quadrifida Duffels and Boer 1990: 258, Fig 21.2 (distribution, phylogeny) New Guinea, Aru

Baeturia quadrifida Boer 1992b: 165 (comp. note)

Baeturia quadrifida Boer 1994c: 137 (comp. note)

Baeturia quadrifida Boer 1995c: 212, 227, 229, 231, Fig 4 (phylogeny, comp. note)

Baeturia quadrifida Boer 1995d: 212–213, Fig 39 (distribution, comp. note) New Guinea

Baeturia quadrifida Boer and Duffels 1996b: 316, 324 (distribution, comp. note) New Guinea

***B. reijnhoudti* Boer, 1989** = *Baeturia reynhoudti* (sic) = *Baeturia reinhoudti* (sic)

Baeturia reijnhoudti Boer 1989: 5, 9, 23–25, 36 (n. sp., key, described, illustrated, distribution, comp. note) Solomon Islands

Baeturia reijnhoudti Duffels and Boer 1990: 264, Fig 21.6 (distribution)
Baeturia reynhoudti (sic) Boer 1995c: 212–213, 227, 229, 231, Fig 4, Fig 6 (phylogeny, comp. note)
Baeturia reinhoudti (sic) Boer 1995d: 212, Fig 38 (distribution, comp. note)
Baeturia reinhoudti (sic) Boer and Duffels 1996a: 160, Fig 4 (distribution)
Baeturia reinhoudti (sic) Boer and Duffels 1996b: 319, Fig 13 (distribution, comp. note)
Baeturia reynhoudti (sic) Boer and Duffels 1996b: 324 (distribution) Solomon Islands

***B. retracta* Boer, 1994**
Baeturia retracta Boer 1994b: 162–164, 168–171, Fig 1b, Fig 2, Figs (n. sp., described, illustrated, distribution, comp. note) Papua New Guinea, Admiralty Islands, Manus Island
Baeturia retracta Boer 1995c: 211–213, 224, 227, 229, 231, Figs 4–5 (phylogeny, comp. note)
Baeturia retracta Boer 1995d: 215–216, Fig 44 (distribution, comp. note) New Guinea, Bismarck Archipelago, Admiralty Islands
Baeturia retracta Boer and Duffels 1996b: 325 (distribution) New Guinea, Bismarck Archipelago
Baeturia retracta Sanborn 1999a: 52 (type material, listed)
Baeturia retracta Boer 2000: 6 (comp. note)

***B. roonensis* Boer, 1994**
Baeturia roonensis Boer 1994d: 89–92, 94–96, 98, Fig 1, Figs 16–23 (n. sp., key, described, illustrated, distribution, comp. note) Roon Island
Baeturia roonensis Boer 1995c: 212, 227, 229, 231, Fig 4 (phylogeny, comp. note)
Baeturia roonensis Boer 1995d: 213, Fig 39 (distribution, comp. note)
Baeturia roonensis Boer and Duffels 1996b: 325 (distribution) New Guinea

***B. rossi* Boer, 1994**
Baeturia rossi Boer 1994c: 129, 131–133, 135, 154–156, Fig 1b, Fig 2b, Figs 79–86 (n. sp., key, described, illustrated, phylogeny, comp. note) Irian Jaya, Yapen Island
Baeturia rossi Boer 1995c: 212, 224, 227, 229, 231, Fig 4 (phylogeny, comp. note)
Baeturia rossi Boer 1995d: 213, Fig 40 (distribution, comp. note)
Baeturia rossi Boer and Duffels 1996b: 324 (distribution) New Guinea
Baeturia rossi Lampe, Rohwedder and Rach 2006: 27 (type specimens) New Guinea

***B. rotumae* Duffels, 1988**
Baeturia rotumae Duffels 1988d: 11–12, 28, 30, 99–100, Figs 191–196, Table 1 (n. sp., key, described, illustrated, distribution, listed, comp. note) Rotuma Island
Baeturia rotumae Boer 1989: 5, 10, 37–40, Figs 33–34, Figs 145–152 (illustrated, comp. note) Rotuma
Baeturia rotumae Duffels and Boer 1990: 264, Fig 21.6 (distribution)
Baeturia rotumae Boer 1995c: 212–213, 227, 229, 231, Fig 4, Fig 6 (phylogeny, comp. note)
Baeturia rotumae Boer 1995d: 212, Fig 38 (distribution, comp. note)
Baeturia rotumae Boer and Duffels 1996a: 160, Fig 4 (distribution)
Baeturia rotumae Boer and Duffels 1996b: 319, 324, Fig 13 (distribution, comp. note) Rotuma Island
Baeturia rotumae Wilson 2009: 40, Appendix (listed, distribution, comp. note) Fiji

B. rubricata (Distant, 1897) to *Gymnotympana rubicata*
B. rufa Ashton, 1914 to *Gymnotympana rufa*

***B. rufula* Blöte, 1960** = *Baeturia rufa* (sic)
Baeturia rufula Boer 1982: 69 (comp. note)
Baeturia rufula Boer 1992b: 164–169, 175–179, Fig 1, Figs 33–39 (phylogeny, key, described, distribution, comp. note) New Guinea, Irian Jaya
Baeturia rufula Boer 1994d: 90 (comp. note)
Baeturia rufula Boer 1995c: 212, 227, 229, 231, Fig 4 (phylogeny, comp. note)
Baeturia rufa (sic) Boer 1995c: 224 (comp. note)
Baeturia rufula Boer 1995d: 214, Fig 41 (distribution, comp. note) New Guinea
Baeturia rufula Boer 1996: 358 (comp. note)
Baeturia rufula Boer and Duffels 1996b: 325 (distribution) New Guinea

***B. schulzi* Schmidt, 1926**
Baeturia schulzi Boer 1986: 169, 171, 175–177, 180, Figs 16–18, Fig 28 (key, lectotype designation, described, illustrated, distribution, comp. note) Maluku Islands, Buru Island
Baeturia schulzi Duffels 1988a: 11 (comp. note) Malluku Selatan
Baeturia schulzi Duffels 1988b: 24 (comp. note) Buru
Baeturia schulzi Duffels and Boer 1990: 258, Fig 21.2 (distribution, phylogeny) Moluccas, Sula
Baeturia schulzi Boer 1992b: 165 (comp. note)
Baeturia schulzi Boer 1994c: 137–139 (comp. note) Buru Island, Seram
Baeturia schulzi Boer 1995c: 212, 227, 229, 231, Fig 4 (phylogeny, comp. note)
Baeturia schulzi Boer 1995d: 211, 213, Fig 39 (distribution, comp. note) Buru, Seram, Sula

Baeturia schulzi Boer and Duffels 1996b: 324 (distribution) South Maluku

***B. sedlacekorum* Boer, 1989**

Baeturia sedlacekorum Boer 1989: 4, 6, 9, 22–25, 32, 36, Figs 17–18, Figs 85–91, Fig 99 (n. sp., key, described, illustrated, distribution, comp. note) Solomon Islands

Baeturia sedlacekorum Duffels and Boer 1990: 264, Fig 21.6 (distribution)

Baeturia sedlacekorum Boer 1995c: 212–213, 227, 229, 231, Fig 4, Fig 6 (phylogeny, comp. note)

Baeturia sedlacekorum Boer 1995d: 212, Fig 38 (distribution, comp. note)

Baeturia sedlacekorum Boer and Duffels 1996a: 160, Fig 4 (distribution)

Baeturia sedlacekorum Boer and Duffels 1996b: 319, 324, Fig 13 (distribution, comp. note) Solomon Islands

***B. silveri* Boer, 1994**

Baeturia silveri Boer 1994a: 2–5, 8, 13–15, 17, Fig 2, Figs 39–48 (n. sp., key, described, illustrated, distribution, comp. note) Papua New Guinea

Baeturia silveri Boer 1994b: 165 (comp. note)

Baeturia silveri Boer 1995c: 212, 214, 227, 229, 231, Fig 4, Fig 7 (phylogeny, comp. note)

Baeturia silveri Boer 1995d: 214, Fig 42 (distribution, comp. note) New Guinea

Baeturia silveri Boer and Duffels 1996b: 325 (distribution) New Guinea

***B.* sp. 1 Kahono, 1993**

Baeturia sp. 1 Kahono 1993: 213, Table 5 (listed) Irian Jaya

***B.* sp. 2 Kahono, 1993**

Baeturia sp. 2 Kahono 1993: 213, Table 5 (listed) Irian Jaya

***B.* sp. 3 Kahono, 1993**

Baeturia sp. 3 Kahono 1993: 213, Table 5 (listed) Irian Jaya

***B. splendida* Boer, 1994**

Baeturia splendida Boer 1994b: 162–164, 166–167, 169, Fig 1b, Fig 2, Figs 14–22 (n. sp., described, illustrated, distribution, comp. note) Papua New Guinea

Baeturia splendida Boer 1995c: 212–213, 227, 229, 231, Figs 4–5 (phylogeny, comp. note)

Baeturia splendida Boer 1995d: 215, Fig 44 (distribution, comp. note) New Guinea

Baeturia splendida Boer and Duffels 1996b: 325 (distribution) New Guinea

Baeturia splendida Boer 2000: 5 (comp. note)

B. stylata Blöte, 1960 to *Guineapsaltria stylata*

B. subnotata (Walker, 1968) to *Gymnotympana subnotata* (Walker, 1868)

***B. tenuispina* Blöte, 1960**

Baeturia tenuispina Boer 1994a: 1 (comp. note)

Baeturia tenuispina Boer 1994a: 1, 24–26, Fig 1, Figs 94–102 (key, described, illustrated, distribution, comp. note) Papua New Guinea, d'Entrecasteaux Islands, Goodenough Island

Baeturia tenuispina Boer 1995c: 211–212, 224, 227, 229, 231, Figs 3–4 (phylogeny, comp. note)

Baeturia tenuispina Boer 1995d: 214 (distribution, comp. note) New Guinea, Normanby Island

Baeturia tenuispina Boer and Duffels 1996b: 325 (distribution) New Guinea

B. toxopeusi Blöte, 1860 to *Mirabilopsaltria toxopeusi*

***B. turgida* Boer, 1992**

Baeturia turgida Boer 1992b: 164–169, 177–182, Fig 1, Figs 40–49 (n. sp., phylogeny, key, described, distribution, comp. note) Papua New Guinea

Baeturia turgida Boer 1994c: 145 (comp. note)

Baeturia turgida Boer 1994d: 90 (comp. note)

Baeturia turgida Boer 1995c: 212, 224, 227, 229, 231, Fig 4 (phylogeny, comp. note)

Baeturia turgida Boer 1995d: 214, Fig 41 (distribution, comp. note) New Guinea

Baeturia turgida Boer and Duffels 1996b: 325 (distribution) New Guinea

B. ustulata Blöte, 1860 to *Paupuapsaltria ustulata*

***B. uveiensis* Boulard, 1996 =** *Baeturia uveaiensis* (sic)

Baeturia uveiensis Boulard 1996a: 152–157, Figs 2–12 (n. sp., described, illustrated, sonogram, comp. note) Wallis Island

Baeturia uveiensis Sueur 2001: 42, Table 1 (listed)

Baeturia uveiensis Boulard 2005d: 344, 347 (comp. note)

Baeturia uveiensis Boulard 2005f: 190, 192 (comp. note) Wallis Island

Baeturia uveaiensis (sic) Boulard 2006g: 72, 79, 111, 181, Fig 32 (sonogram, listed, comp. note) Wallis Island

Baeturia uveaiensis (sic) Boulard 2008a: 67–70, Fig 1, Fig 3 (song, sonogram, comp. note) Wallis Island

***B. valida* Blöte, 1960**

Baeturia valida Kuniata and Nagaraja 1992: 65 (sugarcane pest, comp. note) Papua New Guinea

Baeturia valida Kuniata 1994: 47, 49 (fungal host) Papua New Guinea

Baeturia valida Kuniata 2001: 339 (sugarcane pest) Papua New Guinea

***B. vanderhammeni* Blöte, 1960**

Baeturia vanderhammeni Boer 1994c: 129, 131–135, 146–149, 153–154, Fig 1b, Fig 2a, Figs 41–51 (key, described, illustrated, phylogeny, comp. note) New Guinea

Baeturia vanderhammeni Boer 1995c: 212, 227, 229, 231, Fig 4 (phylogeny, comp. note)

Baeturia vanderhammeni Boer 1995d: 213, Fig 40 (distribution, comp. note) New Guinea
Baeturia vanderhammeni Boer and Duffels 1996b: 324 (distribution) New Guinea

B. varicolor Distant, 1907 to *Gymnotympana varicolor*

***B. versicolor* Boer, 1994**

Baeturia versicolor Boer 1994c: 129, 131–135, 149–152, Fig 1b, Fig 2b, Figs 61–71 (n. sp., key, described, illustrated, phylogeny, comp. note) Papuan Peninsula
Baeturia versicolor Boer 1995c: 212, 227, 229, 231, Fig 4 (phylogeny, comp. note)
Baeturia versicolor Boer 1995d: 213, Fig 40 (distribution, comp. note) Papuan Penisula
Baeturia versicolor Boer and Duffels 1996b: 325 (distribution) Papuan Peninsula
Baeturia versicolor Sanborn 1999a: 52 (type material, listed)

B. viridicata Distant, 1897 to *Mirabilopsaltria viridicata*

***B. viridis* Blöte, 1960**

Baeturia viridis Boer 1989: 6 (comp. note)
Baeturia viridis Boer 1992b: 163–167, 169–175, 180, Figs 1–14 (phylogeny, key, described, distribution, comp. note) New Guinea
Baeturia viridis group Boer 1992b: 163–168 (described, comp. note)
Baeturia viridis group Boer 1993b: 154 (phylogeny, comp. note)
Baeturia viridis group Boer 1994a: 1 (comp. note)
Baeturia viridis group Boer 1994b: 162, 164, Fig 1a (comp. note)
Baeturia viridis group Boer 1994c: 128–131, 145, Fig 1a (phylogeny, comp. note)
Baeturia viridis Boer 1994c: 130, 142 (comp. note)
Baeturia viridis group Boer 1994d: 87–89, 97–98 (comp. note)
Baeturia viridis group Boer 1995a: 5 (comp. note)
Baeturia viridis group Boer 1995c: 202, 211, 214, Fig 3, Fig 8 (phylogeny, comp. note)
Baeturia viridis Boer 1995c: 211–212, 227, 229, 231, Figs 4–5 (phylogeny, comp. note)
Baeturia viridis group Boer 1995d: 211 (distribution, comp. note) New Guinea
Baeturia viridis Boer 1995d: 212, 214, Fig 41 (distribution, comp. note)
Baeturia viridis group Boer 1996: 358 (comp. note)
Baeturia viridis group Boer and Duffels 1996b: 304, 325, Fig 5 (distribution, phylogeny, comp. note) New Guinea
Baeturia viridis Boer and Duffels 1996b: 325 (distribution) New Guinea
Baeturia viridis group Holloway 1998: 305–307, Fig 10, Fig 12 (distribution, phylogeny, comp. note) New Guinea
Baeturia viridis Boer 2000: 4, 6, 8 (comp. note) New Guinea
Baeturia viridis group Boer 2000: 4, 8, 15, Fig 1, (phylogeny, comp. note)
Baeturia viridis species group Boer 2000: 6 (comp. note)
Baeturia viridis species group Arensburger, Simon, and Holsinger 2004: 1778 (comp. note)

B. viridula Blöte, 1960 to *Guineapsaltria viridula*

***B. wauensis* Boer, 1994**

Baeturia wauensis Boer 1994c: 129, 131–135, 148–152, Fig 1b, Fig 2b, Figs 52–60 (n. sp., key, described, illustrated, phylogeny, comp. note) Papua New Guinea
Baeturia wauensis Boer 1995c: 212, 224, 227, 229, 231, Fig 4 (phylogeny, comp. note)
Baeturia wauensis Boer 1995d: 213, Fig 40 (distribution, comp. note)
Baeturia wauensis Boer and Duffels 1996b: 325 (distribution) New Guinea

***B. wegeneri* Boer, 1994**

Baeturia wegeneri Boer 1994a: 2–5, 8, 12–13, Fig 1, Figs 30-38 (n. sp., key, described, illustrated, distribution, comp. note) Papua New Guinea
Baeturia wegeneri Boer 1995c: 212, 214, 227, 229, 231, Fig 4, Fig 7 (phylogeny, comp. note)
Baeturia wegeneri Boer 1995d: 214–215, Fig 42 (distribution, comp. note) New Guinea
Baeturia wegeneri Boer and Duffels 1996b: 324 (distribution) New Guinea

***Aedeastria* Boer, 1990**

Aedeastria Boer 1990: 63–66 (n. gen., described, comp. note)
Aedeastria Boer 1991: 2 (comp. note)
Aedeastria Boer 1992a: 18–19 (phylogeny, comp. note)
Aedeastria Boer 1993a: 16–17, 19 (comp. note)
Aedeastria Boer 1993b: 140–142, 153 (phylogeny, genus described, comp. note)
Aedeastria Boer 1994d: 88, 90 (comp. note) New Guinea
Aedeastria Boer 1995a: 2–3, 7, Fig 1 (phylogeny, comp. note)
Aedeastria Boer 1995c: 202, 204, 211, 215, 217–219, 225, Fig 3, Fig 11, Fig 15 (phylogeny, comp. note)
Aedeastria Boer 1995d: 202, 210–211, 222, 224–227, 230–235, Figs 52–53 (distribution, phylogeny, comp. note) New Guinea, Moluccas
Aedeastria Boer 1996: 351–353, 356, Fig 1 (comp. note)
Aedeastria Boer and Duffels 1996a: 155, 163, 167, 170–171, 173, Fig 5, Figs 10–11 (distribution, phylogeny, comp. note) New Guinea

Aedeastria Boer and Duffels 1996b: 301, 304, 306, 313–314, 316, 324, Fig 5 (distribution, phylogeny, comp. note) Papua New Guinea
Aedeastria Boer 1997: 92–93, 98, Fig 1 (phylogeny, comp. note)
Aedeastria Holloway 1997: 220–222, Fig 5 (distribution, phylogeny, comp. note) New Guinea
Aedeastria Holloway 1998: 304–307, 309, Fig 10, Figs 12–13 (distribution, phylogeny, comp. note) New Guinea
Aedeastria Holloway and Hall 1998: 9 (distribution, comp. note)
Aedeastria Boer 2000: 5 (comp. note) New Guinea
Aedeastria Heads 2001: 902–903, Fig 22 (distribution, comp. note) New Guinea, Moluccas
Aedeastria Moulds 2005b: 390, 435 (higher taxonomy, listed, comp. note)
Aedeastria Duffels and Boer 2006: 533–534 (distribution, diversity, comp. note) New Guinea, Maluku, Waigeo, Misool, Roon, Kai Island

A. bullata **Boer, 1993**
Aedeastria bullata Boer 1993b: 142–144, 146–147, 163–165, Fig 2, Figs 87–94 (n. sp., key, described, illustrated, distribution, comp. note) Papua New Guinea
Aedeastria bullata Boer 1995c: 213, 218, 226, 228, 230–231, Fig 4, Figs 13–14 (phylogeny, comp. note)
Aedeastria bullata Boer 1995d: 210–211, Fig 37 (distribution, comp. note) New Guinea
Aedeastria bullata Boer and Duffels 1996a: 164, Fig 5 (distribution)
Aedeastria bullata Boer and Duffels 1996b: 324 (distribution) New Guinea

A. cheesmanae **Boer, 1993**
Aedeastria cheesmanae Boer 1993b: 142–143, 148, 152, 155–157, Fig 1, Figs 44–53 (n. sp., key, described, illustrated, distribution, comp. note) Waigeu and Misool Islands
Aedeastria cheesmanae Boer 1995c: 213, 217–218, 225, 228, 230–231, Fig 4, Figs 13–14 (phylogeny, comp. note)
Aedeastria cheesmanae Boer 1995d: 210–211, Fig 37 (distribution, comp. note) Waigeu, Misool
Aedeastria cheesmanae Boer and Duffels 1996a: 164, Fig 5 (distribution)
Aedeastria cheesmanae Boer and Duffels 1996b: 324 (distribution) New Guinea

A. cobrops **Boer, 1990**
Aedeastria cobrops Boer 1986: 66–68, Figs 1–13 (n. sp., type species of genus *Aedeastria*, described, illustrated, distribution, comp. note) Irian New Guinea, Vogelkop Peninsula
Aedeastria cobrops Boer 1993a: 19 (comp. note)
Aedeastria cobrops Boer 1993b: 140–143, 147–148 (key, comp. note)
Aedeastria cobrops Boer 1995c: 213, 217–218, 228, 230–231, Fig 4, Figs 13–14 (phylogeny, comp. note)
Aedeastria cobrops Boer 1995d: 210–211, Fig 37 (distribution, comp. note) New Guinea
Aedeastria cobrops Boer and Duffels 1996a: 164, Fig 5 (distribution)
Aedeastria cobrops Boer and Duffels 1996b: 324 (distribution) New Guinea

A. digitata **(Blöte, 1960)** = *Baeturia digitata*
Aedeastria digitata Boer 1993b: 142–144, 146–148, 153–155, Fig 2, Figs 37–43 (n. comb., key, described, illustrated, distribution, comp. note) Equals *Baeturia digitata* Irian Jaya
Aedeastria digitata Boer 1995c: 213, 217–218, 224, 226, 228, 230–231, Fig 4, Figs 13–14 (phylogeny, comp. note)
Aedeastria digitata Boer 1995d: 210–211, Fig 37 (distribution, comp. note) New Guinea
Aedeastria digitata Boer 1996: 358 (comp. note)
Aedeastria digitata Boer and Duffels 1996a: 164, Fig 5 (distribution)
Aedeastria digitata Boer and Duffels 1996b: 324 (distribution) New Guinea
Baeturia digitata Sanborn 1999a: 51 (type material, listed)

A. dilobata **Boer, 1993**
Aedeastria dilobata Boer 1993b: 142–143, 146–147, 164–166, Fig 2, Figs 95–104 (n. sp., key, described, illustrated, distribution, comp. note) Papuan Peninsula
Aedeastria dilobata Boer 1995c: 213, 218, 225–226, 228, 230–231, Fig 4, Figs 13–14 (phylogeny, comp. note)
Aedeastria dilobata Boer 1995d: 210–211, Fig 37 (distribution, comp. note) Papuan Peninsula
Aedeastria dilobata Boer and Duffels 1996a: 164, Fig 5 (distribution)
Aedeastria dilobata Boer and Duffels 1996b: 324 (distribution) Papuan Peninsula
Aedeastria dilobata Boer 2000: 5 (comp. note) Papuan Peninsula

A. hastulata **Boer, 1993**
Aedeastria hastulata Boer 1993b: 142–143, 146–148, 153, 161–163, Fig 1, Figs 75–86 (n. sp., key, described, illustrated, distribution, comp. note) Molucca Islands, Bacan, Halmahera, Morotai, Ternate
Aedeastria hastulata Boer 1995a: 5 (comp. note)
Aedeastria hastulata Boer 1995c: 213, 217–218, 225, 228, 230–231, Fig 4, Figs 13–14 (phylogeny, comp. note)

Aedeastria hastulata Boer 1995d: 210–211, Fig 37 (distribution, comp. note) Moluccas
Aedeastria hastulata Boer and Duffels 1996a: 164, Fig 5 (distribution)
Aedeastria hastulata Boer and Duffels 1996b: 324 (distribution) North Maluku

***A. kaiensis* Boer, 1993**
Aedeastria kaiensis Boer 1993b: 142–143, 146, 148, 150–152, Fig 2, Figs 18–28 (n. sp., key, described, illustrated, distribution, comp. note) New Guinea, Kai Islands
Aedeastria kaiensis Boer 1995c: 213, 217–218, 228, 230–231, Fig 4, Figs 13–14 (phylogeny, comp. note)
Aedeastria kaiensis Boer 1995d: 210–211, Fig 37 (distribution, comp. note) Kai Islands
Aedeastria kaiensis Boer and Duffels 1996a: 164, Fig 5 (distribution)
Aedeastria kaiensis Boer and Duffels 1996b: 324 (distribution) New Guinea

***A. latifrons* (Blöte, 1960)** = *Baeturia latifrons* = *Baeturia famulus* (nec Distant) = *Baeturia famulus* Myers, 1928 = *Baeturia beccarii* (nec Distant)
Baeturia famulus Boer 1993a: 16 (comp. note)
Aedeastria latifrons Boer 1993a: 18, 20 (comp. note)
Aedeastria latifrons Boer 1993b: 142–144, 148–150, 152, Figs 2–17 (n. comb., key, synonymy, described, illustrated, distribution, comp. note) Equals *Baeturia latifrons* Equals *Baeturia famulus* (nec Distant) New Guinea
Aedeastrria latifrons Boer 1994c: 133 (comp. note) Papua New Guinea, Aru Islands
Aedeastria latifrons Boer 1995a: 23 (comp. note) Equals *Baeturia famulus* Myers
Aedeastria latifrons Boer 1995c: 213, 217–218, 225, 228, 230–231, Fig 4, Figs 13–14 (phylogeny, comp. note)
Aedeastria latifrons Boer 1995d: 210–211, Fig 37 (distribution, comp. note) Irian Jaya, Papua New Guinea, Aru Islands
Aedeastria latifrons Boer and Duffels 1996a: 164, Fig 5 (distribution)
Aedeastria latifrons Boer and Duffels 1996b: 324 (distribution) New Guinea
Aedeastria latifrons Boer 1999: 119 (comp. note) New Guinea
Aedeastria latifrons Boer 2000: 1, 5, 7, Table 1 (comp. note) Papua New Guinea

***A. moluccensis* Boer, 1993**
Aedeastria moluccensis Boer 1993b: 142–143, 146–148, 157–159, Fig 2, Figs 54–64 (n. sp., key, described, illustrated, distribution, comp. note) Obi Island
Aedeastria moluccensis Boer 1995c: 213, 218, 228, 230–231, Fig 4, Figs 13–14 (phylogeny, comp. note)
Aedeastria moluccensis Boer 1995d: 210–211, Fig 37 (distribution, comp. note) Moluccas
Aedeastria moluccensis Boer and Duffels 1996a: 164, Fig 5 (distribution)
Aedeastria moluccensis Boer and Duffels 1996b: 324 (distribution) North Maluku

***A. obiensis* Boer, 1993**
Aedeastria obiensis Boer 1993b: 142–143, 148, 159–161, Fig 1, Figs 65–74 (n. sp., key, described, illustrated, distribution, comp. note) Obi Island
Aedeastria obiensis Boer 1995c: 213, 218, 225, 228, 230–231, Fig 4, Figs 13–14 (phylogeny, comp. note)
Aedeastria obiensis Boer 1995d: 210–211, Fig 37 (distribution, comp. note) Obi island
Aedeastria obiensis Boer and Duffels 1996a: 164, Fig 5 (distribution)
Aedeastria obiensis Boer and Duffels 1996b: 324 (distribution) North Maluku

***A. sepia* Boer, 1990**
Aedeastria sepia Boer 1986: 68–71, Fig 1, Figs 14–23 (n. sp., described, illustrated, distribution, comp. note) Irian New Guinea, Vogelkop Peninsula, Roon Island
Aedeastria sepia Boer 1993b: 142–143, 147–148 (key, described, illustrated, distribution, comp. note)
Aedeastria sepia Boer 1995c: 213, 217–218, 228, 230–231, Fig 4, Figs 13–14 (phylogeny, comp. note)
Aedeastria sepia Boer 1995d: 210–211, Fig 37 (distribution, comp. note) New Guinea, Roon Island
Aedeastria sepia Boer and Duffels 1996a: 164, Fig 5 (distribution)
Aedeastria sepia Boer and Duffels 1996b: 324 (distribution) New Guinea

***A. waigeuensis* Boer, 1993**
Aedeastria waigeuensis Boer 1993b: 142–143, 148, 152–153, 155–157, Fig 2, Figs 28–36 (n. sp., key, described, illustrated, distribution, comp. note) Waigeu ISland
Aedeastria waigeuensis Boer 1995c: 213, 217–218, 225, 228, 230–231, Fig 4, Figs 13–14 (phylogeny, comp. note)
Aedeastria waigeuensis Boer 1995d: 210–211, Fig 37 (distribution, comp. note) Waigeu Island
Aedeastria waigeuensis Boer and Duffels 1996a: 164, Fig 5 (distribution)
Aedeastria waigeuensis Boer and Duffels 1996b: 324 (distribution) New Guinea

***Decebalus* Distant, 1920**
Decebalus Boer 1995c: 204–205 (comp. note) Africa

***D. ugandanus* Distant, 1920**

***Muda* Distant, 1897** = *Iwasemia* Matsumura, 1927 = *Nahasemia* Matsumura, 1930
Iwasemia Hayashi 1984: 69 (comp. note)
Muda Duffels 1986: 319 (distribution, comp. note) Java, Sumatra, Borneo, Malayan Peninsula
Muda Villet 1993: 435, 439 (comp. note)
Muda Boer 1995a: 23 (comp. note)
Muda Boer 1995c: 203–207, 210–211, 213, 225–226 228, 230–231, Figs 1–4 (phylogeny, comp. note) Southeast Asia, Greater Suda Islands, Malay Peninsula
Muda Boer 1995d: 174, 201, 224–225, Fig 29, Figs 52–54 (distribution, phylogeny, comp. note)
Muda Zaidi and Ruslan 1995a: 174, Table 2 (listed, comp. note) Borneo, Sarawak
Muda Zaidi and Ruslan 1995b: 220–221, Table 2 (listed, comp. note) Borneo, Sabah
Muda Zaidi and Ruslan 1995c: 70, Table 2 (listed, comp. note) Peninsular Malaysia
Muda Zaidi and Ruslan 1995d: 201–202, Table 2 (listed, comp. note) Borneo, Sabah
Muda Boer and Duffels 1996a: 156, 163, 169–171, Fig 2, Figs 10–11 (distribution, phylogeny, comp. note)
Muda Boer and Duffels 1996b: 301–302, 304, Fig 2, Fig 5 (distribution, phylogeny, comp. note) Malay Peninsula, Greater Sunda Islands, Ryukyu Islands
Muda Zaidi 1996: 103, Table 2 (comp. note) Borneo, Sarawak
Muda Zaidi and Hamid 1996: 54–55, Table 2 (listed, comp. note) Borneo, Sarawak
Muda Boer 1997: 92–93 (comp. note) Southeast Asia, Greater Sunda Islands
Muda Holloway 1997: 219–220, 223, Figs 5–6 (distribution, phylogeny, comp. note) Sundaland, Ryukyu
Muda Zaidi 1997: 115, Table 2 (listed, comp. note) Borneo, Sarawak
Iwasemia Hayashi 1998: 62 (comp. note)
Muda Holloway 1998: 308, Fig 13 (distribution, comp. note) Sundaland, Ryukyu
Muda Hayashi 1998: 62 (comp. note)
Muda Zaidi and Ruslan 1998b: 4–5, Table 2 (listed, comp. note) Borneo, Sabah
Muda Zaidi and Ruslan 1999: 5, Table 2 (listed, comp. note) Borneo, Sarawak
Muda Jong and Boer 2004: 267, 269 (comp. note) Sooutheast Asia
Muda Schouten, Duffels and Zaidi 2004: 380, 383 (comp. note)
Muda Sanborn, Phillips and Sites 2007: 3, 33–35, Table 1 (listed, distribution) Thailand, Peninsular Malaysia, Indonesia, Papua, Australia
Muda Shiyake 2007: 8, 103 (listed, comp. note) Africa
Muda Boulard 2008f: 7, 57, 90 (listed) Thailand
Muda Osaka Museum of Natural History 2008: 326 (listed) Japan
Muda Duffels 2010: 12 (comp. note)

***M. kuroiwae* (Matsumura, 1913)** = *Prasia kuroiwae* = *Parasia* (sic) *kuroiwae* = *Iwasemia kuroiwae* = *Nahasemia kuroiwae* = *Nahasemia knroiwae* (sic) = *Baeturia kuroiwae* = *"Baeturia" kuroiwae*
"Baeturia" kuroiwae Hayashi 1984: 68–69, Figs 170–173 (described, illustrated, synonymy, comp. note) Equals *Prasia kuroiwae* Equals *Iwasemia kuroiwae* Equals *Nahasemis* (nom. nud.) *kuroiwae* Japan
Muda kuroiwae Boer 1995c: 207 (n. comb., comp. note) Equals *Baeturia kuroiwae* Ryukyu Islands, Loochoo Islands, Japan
Muda kuroiwae Hayashi 1998: 62 (n. comb., comp. note) Equals *"Baeturia" kurowiae*
Baeturia kuroiwae Sueur 2001: 46, Table 2 (listed)
Muda kuroiwae Schouten, Duffels and Zaidi 2004: 380 (comp. note) Ryukyu Islands, Loochoo Islands
Muda kuroiwae Shiyake 2007: 8, 101, 103, Fig 174 (illustrated, distribution, listed, comp. note) Japan
Muda kuroiwae Osaka Museum of Natural History 2008: 326 (listed, acoustic behavior, ecology, comp. note) Japan

***M. obtusa* (Walker, 1858)** = *Cephaloxys obtusa* = *Baeturia obtusa* = *Muda concolor* Distant, 1897
Muda obtusa Zaidi and Ruslan 1995b: 219, Table 1 (listed, comp. note) Borneo, Sabah
Muda obtusa Zaidi and Ruslan 1995d: 200, 203, Table 1 (listed, comp. note) Borneo, Sabah
Muda obtusa Zaidi, Ruslan and Mahadir 1996: 61, Table 1 (listed, comp. note) Peninsular Malaysia
Muda obtusa Zaidi and Ruslan 1998a: 370 (listed, comp. note) Equals *Cephaloxys obtusa* Borneo, Sarawak, Java, Peninsular Malaysia, Banguey Island, Sumatra, Sabah
Muda obtusa Zaidi and Ruslan 1998b: 3, Table 1 (listed, comp. note) Borneo, Sabah
Muda obtusa Zaidi, Ruslan and Azman 1999: 319 (listed, comp. note) Equals *Cephaloxys obtusa* Borneo, Sabah, Java, Banguey Island, Sumatra, Peninsular Malaysia
Muda obtusa Zaidi, Noramly and Ruslan 2000a: 331 (listed, comp. note) Equals *Cephaloxys obtusa* Langkawi Island, Borneo, Java, Banguey Island, Sumatra, Peninsular Malaysia
Muda obtusa Zaidi, Noramly and Ruslan 2000b: 337, 399 (comp. note) Borneo, Sabah

Muda obtusa Zaidi, Ruslan and Azman 2000: 218 (listed, comp. note) Equals *Cephaloxys obtusa* Borneo, Sabah, Java, Banguey Island, Sumatra, Peninsular Malaysia, Sarawak

Muda obtusa Zaidi, Azman and Ruslan 2001a: 112, 116, 122, Table 1 (listed, comp. note) Equals *Cephaloxys obtusa* Langkawi Island, Borneo, Java, Banguey Island, Sumatra, Peninsular Malaysia

Muda obtusa Zaidi and Nordin 2001: 185, 191, 204, Table 1 (listed, comp. note) Equals *Cephaloxys obtusa* Borneo, Sabah, Java, Banguey Island, Sumatra, Peninsular Malaysia, Sarawak

Muda obtusa Zaidi and Azman 2003: 106, Table 1 (listed, comp. note) Borneo, Sabah

Muda concolor Boulard 2008f: 57 (type species of *Muda*) Equals *Cephaloxys obtusa* Thailand

Muda obtusa Schouten, Duffels and Zaidi 2004: 380 (comp. note) Java, Sumatra, Borneo

Cephaloxys obtusa Schouten, Duffels and Zaidi 2004: 383 (comp. note)

M. sp. 3 Zaidi, Azman & Ruslan, 2001

Muda sp. 3 Zaidi, Azman and Ruslan 2001a: 112, 116, 122, Table 1 (listed, comp. note) Langkawi Island

M. sp. 4 Zaidi, Azman & Ruslan, 2001

Muda sp. 4 Zaidi, Azman and Ruslan 2001a: 112, 116, 122, Table 1 (listed, comp. note) Langkawi Island

***M. tua* Duffels, 2004**

Muda tua Schouten, Duffels and Zaidi 2004: 373, 379, Figs 1–4, Table 2 (n. sp., described, illustrated, listed, comp. note) Malayan Peninsula

M. undescribed species A Sanborn, Phillips & Sites, 2007

Muda undescribed species A Sanborn, Phillips and Sites 2007: 33–34 (listed, distribution, comp. note) Thailand

M. undescribed species B Sanborn, Phillips & Sites, 2007

Muda undescribed species B Sanborn, Phillips and Sites 2007: 34 (listed, distribution, comp. note) Thailand

***M. virguncula* (Walker, 1857)** = *Cicada virguncula* = *Baeturia beccarii* (nec Distant) = *Muda beccarii* (nec Distant) = *Melampsalta flava* (nec Goding & Froggatt)

Muda virguncula Boer 1995a: 23 (comp. note) Equals *Baeturia famulus* Stål in synonymy of *Baeturia beccarii* Distant

Muda beccarii Boer 1995a: 23 (comp. note synonymy)

Muda virguncula Schouten, Duffels and Zaidi 2004: 380 (comp. note) Java, Sumatra, Borneo

***Musoda* Karsch, 1890**

Musoda Medler 1980: 73 (listed) Nigeria

Musoda Boulard 1985e: 1026 (coloration, comp. note)

Musoda Boulard 1990b: 92–93, 175 (comp. note)

Musoda Boulard 1993d: 91 (comp. note)

Musoda Villet 1993: 435, 439 (comp. note) Africa

Musoda Boer 1995c: 204–205 (comp. note) Africa

Musoda Boulard 1996d: 153 (listed)

Musoda Boulard 2002b: 205 (comp. note)

***M. flavida* Karsch, 1890**

Musoda flavida Medler 1980: 73 (listed) Nigeria

Musoda flavida Boulard 1982d: 112 (comp. note)

Musoda flavida Boulard 1985e: 1021, 1034, 1037, Plate I, Fig 8 (crypsis, coloration, illustrated, comp. note) Central Africa Republic

Musoda flavida Boulard 1990b: 93, 101, 103, 169, 185, 187, 196, 198–199, Plate X, Fig 7, Plate XXII, Fig 3, Plate XXV, Fig 8 (type species of *Musoda*, genitalia, sonogram, coloration, comp. note) Central African Republic

Musoda flavida Boulard 1996d: 115, 153–155, Fig S18 (song analysis, sonogram, listed, comp. note) Central African Republic

Musoda flavida Boulard 1997b: 25 (coloration, comp. note)

Musoda flavida Boulard 2000b: 6 (coloration, comp. note)

Musoda flavida Boulard 2000d: 104, Plate M, Figs 1–2 (illustrated, sonogram, comp. note)

Musoda flavida Sueur 2001: 42, Table 1 (listed)

Musoda flavida Boulard 2002b: 193, 195, Fig 49 (coloration, illustrated, comp. note) Africa

Musoda flavida Boulard 2005d: 347 (comp. note)

Musoda flavida Boulard 2006g: 111–112, 129–130, 180, Fig 76 (sonogram, listed, comp. note) Central African Republic

Musoda flavida Boulard 2007f: 79, 137, Fig 88 (comp. note)

***M. gigantea* Distant, 1914**

Musoda gigantea Boulard 1990b: 93 (comp. note)

***M. occidentalis* Boulard, 1974**

Musoda occidentalis Medler 1980: 73 (listed) Nigeria

Musoda occidentalis Boulard 1985e: 1027, 1040–1041, Plate V, Fig 34 (crypsis, coloration, illustrated, comp. note) Central Africa Republic

Musoda occidentalis Boulard 1990b: 93 (comp. note)

Musoda occidentalis Boulard 1997b: 21, Plate III, Fig 1 (morphology, comp. note)

Musoda occidentalis Boulard 2007f: 83, Fig 99.1 (illustrated, comp. note)

***M. orientalis* Boulard, 1974**

Musoda orientalis Boulard 1990b: 93 (comp. note)

***Cephalalna* Boulard, 2006**
Cephalalna Boulard 2006b: 129 (n. gen., described, comp. note) Madagascar
***C. francimontanum* Boulard, 2006**
Cephalalna francimontanum Boulard 2006b: 129–132, Figs 1–6 (n. sp., described, illustrated, song, sonogram, comp. note) Madagascar
Cephalalna francimontanum Boulard 2006g: 71–72, 77, 181, Fig 29 (sonogram, listed, comp. note) Madagascar

***Nablistes* Karsch, 1891** = *Nabristes* (sic)
Nablistes Boulard 1986d: 233 (listed)
Nablistes Boer 1995c: 204–205 (comp. note) Africa
Nablistes Boulard 1996d: 155 (listed)
***N. heterochroma* Boulard, 1986**
Nablistes heterochroma Boulard 1986d: 233–235, 237–238, Table I, Figs 29–33 (n. sp., described, illustrated, comp. note) Ivory Coast
Nablistes heterochroma Boulard 1996d: 115, 155–156, Fig S19 (song analysis, sonogram, listed, comp. note) Ivory Coast
Nablistes heterochroma Boulard 2000b: 6 (coloration, comp. note)
Nablistes heterochroma Sueur 2001: 42, Table 1 (listed)
Nablistes heterochroma Boulard 2002b: 193 (coloration, comp. note)
Nablistes heterochroma Boulard 2005d: 347 (comp. note) Ivory Coast
Nablistes heterochroma Boulard 2006g: 100–101, 111–112, 129–130, 180, Fig 77 (sonogram, listed, comp. note) Ivory Coast
Nablistes heterochroma Boulard 2007f: 78, 137, Fig 87 (comp. note)
***N. terebrata* Karsch, 1891** = *Nabristes* (sic) *terebrata*
Nablistes terebrata Boulard 1986d: 233 (type species of *Nablistes*, comp. note) Ivory Coast

***Akamba* Distant, 1905** = *Aramba* (sic)
Akamba Boer 1995c: 204–205 (comp. note) Africa
***A. aethiopica* Distant, 1905**

***Kumanga* Distant, 1905**
Kumanga Boer 1995c: 203–205, Fig 1 (phylogeny, comp. note) Southeast Asia
Kumanga Jong and Boer 2004: 267 (comp. note) Southeast Asia
***K. sandaracata* (Distant, 1888)** = *Baeturia sandaracata*

***Conibosa* Distant, 1905**
Conibosa Boulard 1990b: 144 (comp. note)
Conibosa Boer 1995c: 205 (comp. note) Middle America
Conibosa Boulard 2002b: 198 (comp. note) South America
***C. occidentis* (Walker, 1858)** = *Cephaloxys occidentis* = *Calyria occidentis* = *Conibosa occidentris* (sic) = *Conibosa occidentalis* (sic) = *Conibosa occidentes* (sic) = *Calyria virginea* Stål, 1864
Conibosa occidentes (sic) Young 1984a: 172 (comp. note) Costa Rica
Conibosa occidentis Wolda and Ramos 1992: 272, Table 17.1 (distribution, listed) Panama
Conibosa occidentis Sanborn 2007b: 37 (distribution, comp. note) Mexico, Veracruz, Panama, Costa Rica
Conibosa occidentalis (sic) Sanborn 2010b: 76 (listed, distribution) Honduras, Mexico, Panama
C. sp. nr. *occidentis* Young, 1980
Conibosa sp. Young 1981a: 130–131, 134, 136–139, Fig 9, Tables 1–3 (seasonality, habitat, comp. note) Costa Rica
Conibosa sp. Young 1984a: 170–172, 176, 179, Tables 1–3 (hosts, comp. note) near *Conibosa occidentes* (sic) Costa Rica
***C.* sp. D Wolda & Ramos, 1992**
Conibosa sp. D Wolda and Ramos 1992: 272, 276, Fig 17.7, Table 17.1 (distribution, listed, seasonality, comp. note) Panama
***C.* sp. G Wolda & Ramos, 1992**
Conibosa sp. G Wolda and Ramos 1992: 273, Table 17.2 (listed) Panama

***Durangona* Distant, 1911**
Durangona Boer 1995c: 205 (comp. note) South America
Durangona Sanborn 2010a: 1601 (listed) Colombia
***D. tigrina* Distant, 1911**
Durangona tigrina Salazar and Sanborn 2009: 271 (comp. note) Equals Fig 1 Salazar Escobar, 2005 Colombia
Durangona tigrina Sanborn 2010a: 1601 (type species of *Durangona*, distribuion, comp. note) Ecuador, Colombia

Tribe Tettigomyiini Distant, 1905 = Tettigomyiarini (sic) = Tettigomyini (sic)
Tettigomyiarini (sic) Boulard 1981b: 47 (comp. note)
Tettigomyini (sic) Boulard 1981b: 47 (comp. note)
Tettigomyini (sic) Boulard 1988b: 43 (comp. note)
Tettigomyini (sic) Boulard 1990b: 141 (comp. note)
Tettigomyiini Villet 1993: 439 (comp. note) South Africa
Tettigomyiini Villet 1997b: 348, 356 (comp. note)
Tettigomyini (sic) Boulard 1999a: 80, 108, 112 (listed)
Tettigomyiini Villet 1999a: 153 (comp. note)

Tettigomyiini Villet and Noort 1999: 226, Table 12.1 (diversity, listed) Tanzania, Kenya, Uganda, South Africa, Zimbabwe
Tettigomyini (sic) Sueur 2001: 45, Table 1 (listed)
Tettigomyini (sic) Boulard 2002b: 198 (comp. note)
Tettigomyiini Dietrich 2002: 158 (comp. note)
Tettigomyiini Moulds 2005b: 427 (higher taxonomy, listed)
Tettigomyiini Chawanji, Hodson and Villet 2006: 374, 383, Table 1 (listed, comp. note)
Tettigomyiini Price, Bellingan, Villet and Barker 2008: 332 (comp. note)

***Gazuma* Distant, 1905**
***G. barrettae* Distant, 1905** = *Gazuma barettae* (sic) = *Gazuma balettae* (sic)
***G. delalandei* Distant, 1905**
Gazuma delalandei Boulard 2002b: 196 (wing morphology, comp. note) South Africa
***G. pretoriae* Distant, 1905**

***Paectira* Karsch, 1890** = *Inyamana* Distant, 1905 = *Inymana* (sic) = *Inyanama* (sic)
Paectira Boulard 1985e: 1022, 1033 (coloration, comp. note)
Paectira Boulard 1990b: 92, 207 (comp. note) Kenya, Tanzania
Paectira Boulard 1992d: 105–106 (comp. note) Equals *Inyamana*
Paectira Novotny and Wilson 1997: 437 (listed)
Paectira Villet 1997b: 356 (comp. note)
Paectira Villet 1999a: 151 (comp. note) Equals *Inyamana* East Africa
Inyamana Villet 1999a: 152 (comp. note)
Paectira spp. Villet 1999a: 152 (comp. note)
Paectira Villet and Noort 1999: 228, 231 (comp. note) Kenya, Tanzania, Uganda
Paectira spp. Villet and Noort 1999: 228, 230–231 (comp. note)
Paectira Boulard 2002b: 193 (coloration, comp. note)
Paectira Price, Bellingan, Villet and Barker 2008: 332 (comp. note) South Africa, Eastern Cape
***P. boulardi* Heller, 1980**
***P. bouvieri* (Melichar, 1911)** = *Inymana* (sic) *bouvieri* = *Inyamana bouvieri*
***P. capeneri* Boulard, 1992**
Paectira capeneri Boulard 1992d: 109–111, Plate II, Figs 1–7 (n. sp., described, illustrated, comp. note) South Africa
***P. dispar* Heller, 1980**
Paectira dispar Boulard 1992d: 109 (comp. note)
***P. cf. dispar* Villet & Noort, 1999**
Paectira cf. *dispar* Villet and Noort 1999: 228, 230, Table 12.2 (habitat, listed, comp. note) Tanzania
***P. dulcis* Karsch, 1890** = *Paectira ducalis* (sic)
Paectira dulcis group Boulard 1992d: 106, 112 (comp. note)
Paectira dulcis Villet 1999a: 153 (comp. note)
***P. feminavirens* Boulard, 1977** = *Paectira femiravirens* (sic)
Paectira feminavirens Boulard 1985e: 1021, 1037, Plate I, Fig 9 (coloration, illustrated, comp. note) Kenya, Tanzania
Paectira feminavirens Boulard 1990b: 178, 180–181, 183, Plate XXI, Fig 6 (illustrated, comp. note) Kenya
Paectira feminavirens Boulard 1995a: 7 (comp. note)
Paectira feminavirens Boulard 1999a: 80, 108 (listed, sonogram) Kenya
Paectira feminavirens Sueur 2001: 45, Table 1 (listed)
Paectira feminavirens Boulard 2005d: 339 (comp. note)
Paectira feminavirens Boulard 2006g: 49, 131–132, 141, 180, Fig 102 (sonogram, listed, comp. note) Kenya
***P. forchhammeri* Boulard, 1992**
Paectira forchhammeri Boulard 1992d: 111–114, Plate III, Figs 1–7 (n. sp., described, illustrated, comp. note) Tanzania
***P. fuliginosa* (Distant, 1905)** = *Xosopsaltria fuliginosa*
Xosopsaltria fuliginosa Villet 1999a: 151 (comp. note)
Paectira fuliginosa Villet 1999a: 151–153, Figs 1–9 (n. comb., described, illustrated, comp. note) Equals *Xosopslatria fuliginosa* Kenya
***P. hemaris* (Distant, 1905)** = *Inyamana hemaris*
Inyamana hemeris group Boulard 1992d: 106, 111 (comp. note)
Paectira hemaris Villet and Noort 1999: 231 (comp. note)
***P. hyndi* Boulard, 1992**
Paectira hyndi Boulard 1992d: 106–108, Plate I, Figs 1–8 (n. sp., described, illustrated, comp. note) Tanzania
***P. jeanuaudi* Boulard, 1977**
Paectira jeanuaudi Boulard 1992d: 108 (comp. note)
Paectira jeanuaudi Villet 1999a: 153 (comp. note) Kenya
***P. lindneri* Heller, 1980**
***P. modesta* Heller, 1980**
***P. ochracea ochracea* (Distant, 1905)** = *Inyamana ochracea*
Paectira ochracea Villet 1999a: 152 (comp. note)
***P. ochracea viridans* Heller, 1980**
***P. cf. ochracea* Villet & Noort, 1999**
Paectira cf. *ochracea* Villet and Noort 1999: 228, 230, Table 12.2 (habitat, listed, comp. note) Tanzania
***P. oreas* (Jacobi, 1905)** = *Inyamana oreas*
Paectira oreas Villet 1999a: 152–153 (comp. note)

P. scheveni Heller, 1980
Paectira scheveni Boulard 1992d: 111 (comp. note)
P. ventricosa (Melichar, 1914) = *Inyanama* (sic) *ventricosa* = *Inyamana ventricosa*
Paectira venticosa Boulard 1992d: 109 (comp. note)
Paectira ventricosa Villet 1999a: 153 (comp. note)
Paectira ventricosa Villet and Noort 1999: 228, 230, Table 12.2 (habitat, listed, comp. note) Tanzania

***Xosopsaltria* Kirkaldy, 1904** = *Pydna* Stål, 1861 (nec *Pydna* Walker, 1856)
Xosopsaltria Gess 1981: 7 (comp. note) South Africa
Xosopsaltria Shcherbakov 1981a: 830 (listed)
Xosopsaltria Shcherbakov 1981b: 66 (listed)
Xosopsaltria Boulard 1990b: 139, 142 (comp. note)
Xosopsaltria Villet 1993: 439 (comp. note) South Africa
Xosopsaltria Villet 1994a: 295–296 (comp. note) South Africa
Xosopsaltria Boer 1995c: 205 (comp. note) South Africa
Xosopsaltria Villet 1997b: 356 (comp. note)
Xosopsaltria Villet 1999a: 151 (comp. note) South Africa
Xosopslatria Villet 1999e: 2 (comp. note)
Xosopsaltria Boulard 2002b: 199 (comp. note) South Africa
Xosopsaltria Price, Bellingan, Villet and Barker 2008: 332 (comp. note) South Africa, Eastern Cape
X. annulata (Germar, 1830) = *Cicada annulata* = *Tettigomyia annulata* = *Pydna annulata* = *Xosopsaltria capicola* Kairkaldy, 1909
Xosopsaltria annulata Phillips, Sanborn and Villet 2002: 31 (thermal responses) South Africa, Eastern Cape
Xosopsaltria annulata Sanborn, Phillips and Villet 2003a: 349, Table 1 (thermal responses, habitat, comp. note) South Africa
X. fuliginosa Distant, 1905 to *Paectira fuliginosa*
X. n. sp. Phillips, Sanborn & Villet, 2002
Xosopsaltria n. sp. Phillips, Sanborn and Villet 2002: 31 (thermal responses) South Africa, Eastern Cape
Xosopsaltria n. sp. Sanborn, Phillips and Villet 2003a: 349, Table 1 (thermal responses, habitat, comp. note) South Africa
X. scholandi Distant, 1907
X. scurra (Germar, 1830) = *Cicada scurra* = *Tettigomyia scurra* = *Pydna scurra* = *Cicada lutea* Olivier, 1790 = *Tettigomyia lutea* = *Pydna lutea* = *Xosopsaltria lutea*
X. thunbergi Metcalf, 1955 = *Tettigonia punctata* Thunberg, 1822 (nec *Tettigonia punctata* Fabricius, 1798) = *Pydna punctata* = *Xosopsaltria punctata* = *Cicada tabaniformis* Germar nom. nud. = *Tettygomyia* (sic) *tabaniformis* Walker, 1850 nom. nud. = *Tettigomyia tabaniformis* = *Xosopsaltria tabaniformis*
Xosopsaltria punctata Gess 1981: 4, 8–10, Fig 4 (described, song, sonogram, habitat, comp. note) South Africa
Xosopsaltria thunbergi Villet and Capitao 1995: 140 (habitat, comp. note) South Africa
Xosopsaltria thunbergi Villet and Capitao 1996: 282–283, Tables 1–2 (density, habitat, comp. note) South Africa
Xosopsaltria punctata Picker, Griffith and Weaving 2002: 158–159 (described, illustrated, habitat, comp. note) South Africa, Eastern Cape
Xosopsaltria thunbergi Phillips, Sanborn and Villet 2002: 31 (thermal responses) South Africa, Eastern Cape
Xosopsaltria thunbergi Sanborn, Phillips and Villet 2003a: 349, Table 1 (thermal responses, habitat, comp. note) South Africa
Xosopsaltria thunbergi Chawanji, Hodson and Villet 2006: 374–379, 383–386, Fig 3, Fig 7, Fig 13, Fig 58, Fig 64, Figs 68–69, Tables 1–6 (sperm morphology, listed, comp. note) South Africa

***Tettigomyia* Amyot & Audinet-Serville, 1843** = *Cicada (Tettigomyia)* = *Tettigomya* (sic) = *Tettigomia* (sic) = *Tettygomyia* (sic)
Tettigomyia Strümpel 1983: 17 (listed)
Tettigomyia Boulard 1990b: 141 (comp. note)
Tettigomyia Villet 1993: 439 (comp. note) South Africa
Tettigomyia Villet 1994a: 295–297 (comp. note) South Africa
Tettigomia (sic) Boer 1995c: 205 (comp. note) South Africa
Tettigomyia Villet 1997b: 348, 356–357 (comp. note)
Tettigomyia Villet 1999e: 2 (comp. note)
Tettigomyia Villet and Noort 1999: 225 (comp. note)
Tettigomyia Boulard 2002b: 199 (comp. note) South Africa
Tettigomyia Price, Bellingan, Villet and Barker 2008: 332 (comp. note) South Africa, Eastern Cape
T. vespiformis Amyot & Audinet-Serville, 1843
Tettigomyia vespiformis Boulard 2002b: 196 (wing morphology, comp. note) South Africa
Tettigomyia vespiformis Phillips, Sanborn and Villet 2002: 31 (thermal responses) South Africa, Eastern Cape
Tettigomyia vespiformis Sanborn, Phillips and Villet 2003a: 349, Table 1 (thermal responses, habitat, comp. note) South Africa

***Bavea* Distant, 1905** = *Baeva* (sic)
Bavea Villet 1993: 436, 439 (described, phylogeny, comp. note) South Africa
Bavea Villet 1994a: 295–296 (comp. note) South Africa
Bavea Boer 1995c: 204–205 (comp. note) Africa
Bavea Villet 1997b: 356 (comp. note)
Bavea Price, Bellingan, Villet and Barker 2008: 332 (comp. note) South Africa, Eastern Cape

***B. concolor* (Walker, 1850)** = *Cephaloxys concolor*
Xosopsaltria (sic) sp. Gess 1981: 5, 7, Fig 1 (described, illustrated, song, habitat, comp. note) South Africa
Bavea concolor Villet 1993: 435–439, Figs 1–11 (type species of *Bavea*, described, phylogeny, habitat, comp. note) Equals *Cephaloxys concolor* South Africa
Bavea concolor Phillips, Sanborn and Villet 2002: 31 (thermal responses) South Africa, Eastern Cape
Bavea concolor Sanborn, Phillips and Villet 2003a: 349, Table 1 (thermal responses, habitat, comp. note) South Africa

***Stagira* Stål, 1861**
Stagira Boulard 1985e: 1021 (coloration, comp. note) South Africa
Stagira Villet 1993: 435, 439 (comp. note) South Africa, Zimbabwe
Stagira Villet 1994a: 295–296 (comp. note) South Africa, Zimbabwe
Stagira Boer 1995c: 204–205 (comp. note) Africa
Stagira Villet 1997b: 348–349, 353, 355–357, 391 (described, comp. note) South Africa, Zimbabwe, Mozambique
Stagira Villet 1999e: 1–2 (comp. note) South Africa, Zimbabwe
Stagira Villet and Noort 1999: 225 (comp. note)
Stagira Picker, Griffith and Weaving 2002: 158–159 (described, illustrated, habitat, comp. note) South Africa
Stagira Sueur 2003: 2942 (comp. note) South Africa
Stagira Londt 2006: 325 (predation) South Africa
Stagira Price, Bellingan, Villet and Barker 2008: 332 (comp. note) South Africa, Eastern Cape

***S. abagnata* Villet, 1997**
Stagira abagnata Villet 1997b: 352–353, 357, 360–361, 386–387, Fig 3O, Fig 4O, Fig 7K, Fig 40 (n. sp., described, illustrated, key, distribution, habitat, comp. note) South Africa

***S. acrida* (Walker, 1850)** = *Cicada acridia*
Stagira acrida Villet 1997b: 348, 391 (comp. note) Equals *Cicada acrida* South Africa

***S. aethlius* (Walker, 1850)** = *Cicada aethlius*
Stagira aethlius Villet 1997b: 348, 351, 353–354, 356, 358, 363, 365–366, 391, Fig 2E, Fig 4B, Fig 5E, Fig 11 (n. stat., described, illustrated, key, distribution, habitat, comp. note) South Africa

***S. caffrariensis* Villet, 1997**
Stagira caffrariensis Villet 1997b: 351, 354, 356, 358, 363, 367–368, Fig 2I, Fig 5I, Fig 15 (n. sp., described, illustrated, key, distribution, habitat, comp. note) South Africa
Stagira caffrariensis Phillips, Sanborn and Villet 2002: 31 (thermal responses) South Africa, Eastern Cape
Stagira caffrariensis Sanborn, Phillips and Villet 2003a: 349, Table 1 (thermal responses, habitat, comp. note) South Africa

***S. celsus* Villet, 1997**
Stagira celsus Villet 1997b: 352, 356, 359, 361, 380, 384, Fig 3K, Fig 7G, Fig 34 (n. sp., described, illustrated, key, distribution, habitat, comp. note) South Africa

***S. chloana* Villet, 1997**
Stagira chloana Villet 1997b: 352, 356, 359, 361, 380, 384–385, Fig 3L, Fig 7H, Fig 35 (n. sp., described, illustrated, key, distribution, habitat, comp. note) South Africa

***S. consobrina* Distant, 1920**
Stagira consobrina Villet 1997b: 351, 353–354, 357–358, 371–374, Fig 2O, Fig 4H, Fig 5O, Fig 23 (described, illustrated, key, distribution, habitat, comp. note) South Africa

***S. consobrinoides* Villet, 1997**
Stagira consobrinoides Villet 1997b: 351, 354, 357–358, 371, 373–374, Fig 2P, Fig 5P, Fig 22 (n. sp., described, illustrated, key, distribution, habitat, comp. note) South Africa

***S. dendrophila* Villet, 1997**
Stagira dendrophila Villet 1997b: 351, 353–355, 357–358, 370–372, Fig 2M, Fig 4G, Fig 5M, Fig 20 (n. sp., described, illustrated, key, distribution, habitat, comp. note) South Africa

***S. dracomontana* Villet, 1997**
Stagira dracomontana Villet 1997b: 352, 357, 360–361, 386, 390–391, Fig 3T, Fig 7P, Fig 45 (n. sp., described, illustrated, key, distribution, habitat, comp. note) South Africa

***S. dracomontanoides* Villet, 1997**
Stagira dracomontanoides Villet 1997b: 352, 357, 360–361, 386, 390, Fig 3S, Fig 7O, Fig 44 (n. sp., described, illustrated, key, distribution, habitat, comp. note) South Africa

***S. elegans* Villet, 1997**
Stagira elegans Villet 1997b: 350–351, 354, 356–357, 362–364, Fig 1C, Fig 2C, Fig 5C, Fig 9 (n. sp., described, illustrated, key, distribution, habitat, comp. note) South Africa

***S. empangeniensis* Villet, 1997**
Stagira empangeniensis Villet 1997b: 352, 357, 360–361, 386, 388, Fig 3P, Fig 7L, Fig 41 (n. sp., described, illustrated, key, distribution, habitat, comp. note) South Africa

***S. enigmatica* Villet, 1997**
Stagira enigmatica Villet 1997b: 351, 354, 356–357, 363–364, Fig 2D, Fig 5D, Fig 10 (n. sp., described, illustrated, key, comp. note) South Africa

***S. eshowiensis* Villet, 1997**
Stagira eshowiensis Villet 1997b: 350, 352, 357, 359, 361, 380–381, 386, Fig 1E, Fig 3F, Fig 7B, Fig 39 (n. sp., described, illustrated, key, distribution, habitat, comp. note) South Africa

***S. furculata* Villet, 1997**
Stagira furculata Villet 1997b: 351, 354, 356–357, 360, 362, 371, Fig 2A, Fig 5A, Fig 16 (described, illustrated, key, distribution, habitat, comp. note) South Africa

***S. fuscula* Villet, 1997**
Stagira fuscula Villet 1997b: 352, 356, 359, 361, 380, 383, Fig 3J, Fig 7F, Fig 33 (n. sp., described, illustrated, key, distribution, habitat, comp. note) South Africa

***S. microcephala* (Walker, 1850)** = *Cephaloxys microcephala* = *Cicada cereris* Stål, 1855 = *Stagira cereris* = *Cicada ceresis* (sic)
Stagira microcephala Villet 1987a: 270, 272, Table 2 (song intensity, listed, comp. note) South Africa
Stagira microcephala Villet 1992: 95–96, Table 1 (acoustic responsiveness, comp. note) South Africa
Stagira cereris Villet 1997b: 348, 379 (comp. note)
Stagira microcephala Villet 1997b: 352–353, 355, 357, 359, 375, 378–379, Fig 3D, Fig 4L, Fig 6F, Fig 29 (described, illustrated, key, distribution, habitat, comp. note) Equals *Cephaloxys microcephala* Equals ? = *Cicada cereis* South Africa

***S. mystica* Villet, 1997**
Stagira mystica Villet 1997b: 351, 354–355, 357–358, 371–372, Fig 2N, Fig 5N, Fig 21 (described, illustrated, key, distribution, habitat, comp. note) South Africa

***S. nasuta* Villet, 1997**
Stagira nasuta Villet 1997b: 352–353, 357, 359, 361, 379–380, 386, Fig 3E, Fig 4M, Fig 7A, Fig 38 (n. sp., described, illustrated, key, distribution, habitat, comp. note) South Africa

***S. natalensis* Villet, 1997**
Stagira natalensis Villet 1997b: 350, 352–353, 357, 360–361, 386, 388–389, Fig 1D, Fig 3Q, Fig 4P, Fig 7M, Fig 42 (n. sp., described, illustrated, key, distribution, habitat, comp. note) South Africa

***S. ngomiensis* Villet, 1997**
Stagira ngomiensis Villet 1997b: 352, 355, 357, 359, 375, 378, Fig 3C, Fig 6E, Fig 28 (n. sp., described, illustrated, key, distribution, habitat, comp. note) South Africa

***S. nkandlhaensis* Villet, 1997**
Stagira nkandlhaensis Villet 1997b: 352–353, 355, 357, 359, 375, 377–378, Fig 3B, Fig 4K, Fig 6D, Fig 27 (n. sp., described, illustrated, key, distribution, habitat, comp. note) South Africa

***S. oxyclypea* Villet, 1997**
Stagira oxyclypea Villet 1997b: 350–351, 353–354, 357–358, 369–371, Fig 1B, Fig 2L, Fig 4F, Fig 5L, Fig 19 (n. sp., described, illustrated, key, distribution, habitat, comp. note) Zimbabwe

***S. pondoensis* Villet, 1997**
Stagira pondoensis Villet 1997b: 350–351, 353–354, 356–357, 362–363, Fig 1G, Fig 2B, Fig 4A, Fig 5B, Fig 8 (n. sp., described, illustrated, key, distribution, habitat, comp. note) South Africa
Stagira pondoensis group Villet 1997b: 356 (comp. note)

***S. pseudaethlius* Villet, 1997**
Stagira pseudaethlius Villet 1997b: 351, 354, 356, 358, 363, 365–366, Fig 2F, Fig 5F, Fig 12 (n. sp., described, illustrated, key, distribution, habitat, comp. note) South Africa

***S. purpurea* Villet, 1997**
Stagira purpurea Villet 1997b: 352, 355, 357, 359, 375–377, Fig 3A, Fig 6C, Fig 26 (n. sp., described, illustrated, key, distribution, habitat, comp. note) South Africa

S. ruficosta Distant, 1920 to *Stagira segmentaria* Karsch, 1890

S. sanguinea Distant, 1920 to *Stagira segmentaria* Karsch, 1890

***S. segmentaria* Karsch, 1890** = *Stagira sanguinea* Distant, 1920 = *Stagira ruficostata* Distant, 1920
Stagira segmentaria Villet 1997b: 351, 353, 355, 357, 359, 374–376, Fig 2R, Fig 4J, Fig 6B, Fig 25 (described, illustrated, key, synonymy, distribution, habitat, comp. note) Equals *Stagira sanguinea* n. syn. Equals *Stagira ruficostata* n. syn. South Africa
Stagira sanguinea Villet 1997b: 351, 353, 376, Fig 2S (described, illustrated,, distribution, habitat, comp. note) South Africa

***S. sexcostata* Villet, 1997**
Stagira sexcostata Villet 1997b: 352, 356, 359, 361, 380, 382–382, Fig 3I, Fig 7E, Fig 32 (n. sp.,

described, illustrated, key, distribution, habitat, comp. note) South Africa

S. *selindensis* Villet, 1997

Stagira selindensis Villet 1997b: 351, 354, 357–358, 369, 371, Fig 2K, Fig 5K, Fig 18 (n. sp., described, illustrated, key, distribution, habitat, comp. note) Zimbabwe

S. *simplex* (Germar, 1934) = *Cicada simplex* = *Cicada simplicis* (sic)

Stagira simplex Villet 1997b: 349–351, 353–354, 356, 358, 363, 367–368, 391, Fig 1D, Fig 1F, Fig 1H, Fig 2H, Fig 4D, Fig 5H, Fig 14 (type species of *Stagira*, described, illustrated, key, distribution, habitat, comp. note) Equals *Cicada simplex* South Africa

Stagira simplex group Villet 1997b: 356 (comp. note)

Stagira simplex Phillips, Sanborn and Villet 2002: 31 (thermal responses) South Africa, Eastern Cape

Stagira simplex Sanborn, Phillips and Villet 2003a: 349, Table 1 (thermal responses, habitat, comp. note) South Africa

Stagira simplex Chawanji, Hodson and Villet 2006: 374–379, 383–386, Fig 5, Fig 12, Fig 57, Figs 59–63, Figs 65–67, Figs 70–77, Tables 1–6 (sperm morphology, listed, comp. note) South Africa

S. *stygia* Villet, 1997

Stagira stygia Villet 1997b: 352, 356, 359, 361, 380–382, Fig 3G, Fig 7C, Fig 30 (n. sp., described, illustrated, key, distribution, habitat, comp. note) South Africa

S. *sylvia* Villet, 1997

Stagira sylvia Villet 1997b: 351, 353–354, 356, 358, 368–369, 371, Fig 2J, Fig 4E, Fig 5J, Fig 17 (n. sp., described, illustrated, key, distribution, habitat, comp. note) South Africa

S. *virescens* Kirkaldy, 1909 = *Cicada viridula* Walker, 1858 (nec *Cicada viridula* Walker, 1850) = *Stagira viridula* = *Stagira simplex* (nec Germar)

Stagira virescens Villet 1997b: 351, 353–354, 356–357, 363, 366, Fig 2G, Fig 4C, Fig 5G, Fig 13 (described, illustrated, key, distribution, habitat, comp. note) Equals *Cicada viridula* South Africa

S. cf. *viridula* Villet, 1992

Stagira cf. *viridula* Villet 1992: 95–96, Table 1 (acoustic responsiveness, comp. note) South Africa

S. *viridoptera* Villet, 1997

Stagira viridoptera Villet 1997b: 352, 357, 359, 361, 380, 385–387, Fig 3N, Fig 7J, Fig 37 (n. sp., described, illustrated, key, distribution, habitat, comp. note) South Africa

S. *vulgata* Villet, 1997

Stagira vulgata Villet 1997b: 352–353, 356, 359, 361, 380, 385, Fig 3M, Fig 4N, Fig 7I, Fig 36 (n. sp., described, illustrated, key, distribution, habitat, comp. note) South Africa

S. *xenomorpha* Villet, 1997

Stagira xenomorpha Villet 1997b: 351, 353, 355, 357–358, 374–375, Fig 2Q, Fig 4I, Fig 6A, Fig 24 (n. sp., described, illustrated, key, distribution, habitat, comp. note) South Africa

S. *zebrata* Villet, 1997

Stagira zebrata Villet 1997b: 352, 356, 359, 361, 380, 382, Fig 3H, Fig 7D, Fig 31 (n. sp., described, illustrated, key, distribution, habitat, comp. note) South Africa

S. *zuluensis* Villet, 1997

Stagira zuluensis Villet 1997b: 352–353, 357, 361, 386, 389–390, Fig 3R, Fig 4Q, Fig 7N, Fig 43 (n. sp., described, illustrated, key, distribution, habitat, comp. note) South Africa

***Stagea* Villet, 1994**

Stagea Villet 1994a: 294, 296–297 (n. gen., described, phylogeny, comp. note) South Africa

Stagea Villet 1997b: 356–357 (comp. note)

Stagea Price, Bellingan, Villet and Barker 2008: 332 (comp. note) South Africa, Eastern Cape

S. *platyptera* Villet, 1994

Stagea platyptera Villet 1994a: 294–297, Figs 1–11 (n. sp., type species of *Stagea*, described, illustrated, habitat, comp. note) South Africa

***Spoerryana* Boulard, 1974** = *Spoeryana* (sic)

Spoerryana Boulard 1990b: 141, 207 (comp. note) Kenya, Tanzania

Spoeryana (sic) Villet 1999e: 2 (comp. note)

Spoeryana (sic) Villet and Noort 1999: 225 (comp. note)

Spoerryana Boulard 2002b: 199, 202 (comp. note) Est Africa

Spoerryana Boulard 2006g: 41 (comp. note)

S. *llewelyni* Boulard, 1974 = *Spoerryana llewelini* (sic) = *Spoerryana llyewellini* (sic)

Spoerryana llewelyni Boulard 1984d: 6, Plate Fig 3 (illustrated, comp. note) Kenya

Spoerryana llewelyni Boulard 1985e: 1033 (coloration, comp. note)

Spoerryana llewelyni Boulard 1990b: 92, 101, 103, 168, 176, 180, Plate X, Fig 8, Plate XXIII, Figs 4–5 (genitalia, illustrated, comp. note) Kenya

Spoerryana llewelyni Villet 1992: 94, 96 (comp. note)

Spoerryana llyewellini (sic) Boulard 1995a: 7 (comp. note)

Spoerryana llewelini (sic) Boulard and Mondon 1995: 71 (comp. note) Kenya
Spoerryana llewelyni Sueur 2001: 45, Table 1 (listed)
Spoerryana llewelyni Boulard 2002b: 193, 196, 203, 205–206 (coloration, wing morphology, comp. note) East Africa
Spoerryana llewelyni Boulard 2005d: 339, 343–344, 347 (comp. note) Kenya
Spoerryana llewelyni Boulard 2006g: 42–43, 73, 75, 85, 123–125, 180, Fig 9c, Fig 44, Fig 85 (illustrated, sonogram, listed, comp. note) Kenya
Spoerryana llevelyni (sic) Boulard 2006g: 49, 66 (comp. note) Kenya

New Genus in Tettigomyiini Villet & Noort, 1999
New Genus Villet and Noort 1999: 228, Table 12.2 (comp. note)

Tribe Hemidictyini Distant, 1905 = Hemidictyaria = Hemidyctyini (sic)
Hemidictyini Moulds 1990: 31, 197 (comp. note)
Hemidictyini Villet 1993: 439 (comp. note)
Hemidictyini Boer 1995c: 202, 204–205 (comp. note)
Hemidictyini Boer 1997: 113 (comp. note)
Hemidyctyini (sic) Sueur 2001: 45, Table 1 (listed)
Hemidictyini Moulds 2005b: 386, 390, 392, 421, 427 (described, distribution, listed, comp. note) Australia
Hemidictyini Shiyake 2007: 9, 115 (listed, comp. note)

***Hemidictya* Burmeister, 1905** = *Hemidyctya* (sic) = *Hemididicta* (sic) = *Hemidyctia* (sic)
Hemidictya Strümpel 1983: 17 (listed)
Hemidictya Boulard 1985e: 1026, 1032, 1034 (coloration, comp. note) Brazil
Hemidictya Boer 1995c: 205 (comp. note)
Hemidictya Boer 1997: 113 (comp. note)
Hemidictya Boulard 1997b: 16, 51 (coloration, comp. note) Neotropics, Brazil
Hemidictya Boulard 2005c: 233 (comp. note)
Hemidictya Boulard 2005j: 80 (comp. note)
Hemidictya Moulds 2005b: 392 (type genus of Hemidictyini, higher taxonomy, listed, comp. note) Australia

***H. frondosa* Burmeister, 1835** = *Hemidyctya* (sic) *frondosa* = *Hemidyctia* (sic) *frondosa*
Hemidictya frondosa Boulard 1985e: 1027, 1040–1041, Plate V, Fig 35 (coloration, illustrated, comp. note)
Hemidictya frondosa Boer 1997: 113–114 (comp. note) Brazil
Hemidictya frondosa Boulard 1997b: 21, Plate III, Fig 5 (morphology, comp. note)
Hemidictya frondosa Boulard 2000b: 9 (parallel evolution)
Hemidictya frondosa Boulard 2002b: 193 (coloration, comp. note)
Hemidictya frondosa Moulds 2005b: 392 (type species of *Hemidictya*, listed)
Hemidictya frondosa Boulard 2007f: 83, Fig 99.5 (illustrated, comp. note)

***Hovana* Distant, 1905**
Hovana Boulard 1985e: 1026, 1032, 1034 (coloration, comp. note) Madagascar
Hovana Boer 1995c: 205 (comp. note)
Hovana Boer 1997: 113 (comp. note)
Hovana Boulard 1997b: 16, 51 (coloration, comp. note) Madagascar

***H. distanti* (Brancsik, 1893)** = *Hemidictya distanti*
Hovana distanti Boulard 1985e: 1019, 1027, 1040–1043, Plate II, Fig 17, Plate V, Fig 36, Plate VI, Figs 40–41 (crypsis, coloration, illustrated, comp. note) Madagascar
Hovana distanti Boulard 1986b: 34–35, Fig 6 (illustrated) Madagascar
Hovana distanti Boulard 1990b: 198–199, Plate XXV, Fig 17 (coloration, illustrated, comp. note) Madagascar
Hovana distanti Boer 1997: 113 (comp. note) Madagascar
Hovana distanti Boulard 1997b: 21–23, Plate III, Fig 6, Plate IV, Fig 8 (morphology, coloration, comp. note)
Hovana distanti Boulard 2000b: 9 (parallel evolution)
Hovana distanti Boulard 2002b: 193 (coloration, comp. note)
Hovana distanti Boulard 2007f: 83, 95, 143, Fig 99.6, Figs 171–172 (illustrated, comp. note) Madagascar

Subfamily Tibicininae Distant, 1905 = Tibicinae (sic) = Tibiciniinae (sic) = Tibiciinae (sic) = Tettigadinae Distant, 1905 = Tettigadesaria = Platypediinae Kato, 1932 = Platypedinae (sic) = Ydiellinae Boulard, 1973
Tibicininae Wu 1935: 22 (listed) China
Tibicinae (sic) Villiers 1977: 16 (listed, comp. note)
Tibicininae Heinrich 1967: 44 (comp. note)
Platypediinae Medler 1980: 73 (listed) Nigeria
Tettigadinae Heath 1983: 12 (comp. note) South America, Argentina, Chile
Tibicininae Strümpel 1983: 17 (comp. note)
Tettigadinae Strümpel 1983: 17 (comp. note)
Ydiellinae Strümpel 1983: 18 (comp. note)
Platypediinae Strümpel 1983: 18 (comp. note)
Tibicininae Boulard 1984c: 167–168, 174, 176, 178–179 (status, type genus *Tibicina*)
Tibiciniinae (sic) Boulard 1984c: 178 (comp. note)

Tibiciinae (sic) Boulard 1984c: 169 (status, type genus *Tibicina*, type species *Cicada haematodes*, comp. note)
Tibicininae Hayashi 1984: 26–28, 66, Fig 1d-e, Fig 2, Table 1 (key, comp. note)
Platypediinae Hayashi 1984: 26–27, Fig 1g, Table 1 (comp. note)
Ydiellinae Hayashi 1984: 26–27, Fig 1h, Table 1 (comp. note)
Tettigadinae Hayashi 1984: 26–27, Fig 1f, Table 1 (comp. note)
Tibicininae Melville and Sims 1984: 164–165, 180 (status, history)
Tibicininae Boulard 1985d: 212 (status)
Tibicininae Hamilton 1985: 211 (status, comp. note)
Tibicininae Boulard 1986c: 202 (stridulatory apparatus, comp. note)
Tettigadinae Boulard 1986c: 202 (stridulatory apparatus, comp. note)
Ydiellinae Boulard 1986d: 235 (n. comb., comp. note)
Tibicinae (sic) Schedl 1986a: 3 (listed, key) Istria
Tibicininae Martinelli and Zucchi 1987b: 469 (listed) Brazil
Ydiellinae Boulard 1988f: 51, 64 (key, comp. note)
Platypediinae Boulard 1988f: 52, 64 (comp. note)
Tibicininae Boulard 1988f: 8, 30–33, 35–36, 44–45, 51, 64, 71 (key, comp. note)
Tettigadinae Boulard 1988f: 51 (key)
Tettigadesaria Boulard 1988f: 53 (comp. note)
Tettigadinae Boulard 1990b: 148 (comp. note)
Tibicininae Boulard 1990b: 92, 116, 148, 209, Fig 12 (phylogeny, comp. note)
Tettigadinae Heath, Heath, Sanborn and Noriega 1990: 1 (comp. note)
Tettigadinae Heath, Heath and Noriega 1990: 1 (distribution, comp. note) Argentina
Tibicininae Liu 1990: 46 (comp. note)
Platypediinae Moore 1990: 58 (comp. note)
Tibicinae (sic) Moore 1990: 36 (comp. note)
Tibicininae Moulds and Carver 1991: 465, 467 (key, comp. note) Australia
Tibicininae Liu 1992: 9 (comp. note)
Tibicinae (sic) Moalla, Jardak and Ghorbel 1992: 34 (comp. note)
Tibicininae Boulard 1993d: 91–92 (comp. note)
Ydiellinae Boulard 1993d: 92 (comp. note)
Ydiellinae Duffels 1993: 1225 (comp. note)
Tibicininae Moore, Huber, Weber, Klein and Bock 1993: 199 (listed)
Platypediinae Lei, Chou and Li 1994: 52, 54 (comp. note)
Ydellinae Lei, Chou and Li 1994: 52, 54 (comp. note)
Tettigadinae Lei, Chou and Li 1994: 52, 54 (comp. note)
Tibicininae Lei, Chou and Li 1994: 54 (comp. note)
Tibicininae Boulard and Mondon 1995: 65–66 (key, listed) France
Tibicininae Boulard 1996a: 157 (comp. note)
Tibicininae Boulard 1996b: 3 (comp. note)
Tibicininae Moulds 1996a: 19–20 (listed)
Tibicininae Boulard 1997c: 84, 100–101, 104, 106, 112, 117, 121 (comp. note)
Tibicinae (sic) Boulard 1997c: 106 (comp. note)
Tibicininae Chou, Lei, Li, Lu and Yao 1997: 4–5, 30–36, 45, Figs 5-1-5-3 (phylogeny, key, synonymy, listed) Equals Tibicinae Equals Tibicinos Equals Tibicines
Tettigadinae Chou, Lei, Li, Lu and Yao 1997: 5, 30–36, 42, Figs 5-1-5-3 (phylogeny, key, synonymy, listed, described) Equals Tettigadesaria Equals Tettigadesini Equals Tettigadidae
Platypediinae Chou, Lei, Li, Lu and Yao 1997: 5, 24, 30–37, Figs 5-1-5-3 (phylogeny, key, synonymy, listed, comp. note) Equals Platypedidae
Ydiellinae Chou, Lei, Li, Lu and Yao 1997: 24, 30–36, Figs 5-1-5-3 (phylogeny, key, listed, comp. note)
Platypediinae Poole, Garrison, McCafferty, Otte and Stark 1997: 251 (listed) North America
Tettigadinae Poole, Garrison, McCafferty, Otte and Stark 1997: 251 (listed) North America
Tibicininae Poole, Garrison, McCafferty, Otte and Stark 1997: 251 (listed) North America
Tettigadinae Moulds 1999: 1 (comp. note)
Tibicininae Moulds 1999: 1 (comp. note)
Tibicininae Sanborn 1999a: 44 (listed)
Tibicininae Sanborn 1999b: 2 (listed) North America
Tibicininae Villet and Zhao 1999: 321 (comp. note) China
Tettigadinae Villet and Zhao 1999: 321 (comp. note) China
Platypediinae Villet and Zhao 1999: 321–322 (comp. note) China
Tibicininae Boulard 2000i: 21, 23–24, 45, 73 (listed, key, comp. note)
Tibicininae Champanhet, Boulard and Gaiani 2000: 42 (comp. note)
Platypedinae (sic) Maw, Footit, Hamilton and Scudder 2000: 49 (listed)
Tibicininae Maw, Footit, Hamilton and Scudder 2000: 49 (listed)
Tibicininae Boulard 2001b: 5, 20–22, 24–25, 32, 37, 41 (comp. note)
Platypediinae Cooley 2001: 758 (comp. note) North America
Platypediinae Cooley and Marshall 2001: 847 (comp. note) North America

Tibicininae Puissant and Sueur 2001: 432–433 (listed) Corsica
Tibicininae Sueur 2001: 41, 46–47, Tables 1–3 (listed, diversity, distribution, comp. note)
Tettigadinae Sueur 2001: 45–47, Table 1, Table 3 (listed, diversity, distribution, comp. note)
Platypediinae Sueur 2001: 45–47, Table 1, Table 3 (listed, diversity, distribution, comp. note)
Ydiellinae Boulard 2002b: 205 (comp. note)
Tibicininae Boulard 2002b: 197, 205–207 (comp. note)
Tibicinae (sic) Boulard 2002b: 195 (comp. note)
Platypediinae Kondratieff, Ellingson and Leatherman 2002: 13, 47, Table 2 (listed) Colorado
Tibicininae Lee, Choe, Lee and Woo 2002: 3 (comp. note)
Tibicininae Moulds and Carver 2002: 1 (comp. note)
Tettigadinae Moulds and Carver 2002: 1 (comp. note)
Platypediinae Moulds and Carver 2002: 1 (comp. note)
Ydiellinae Moulds and Carver 2002: 1 (comp. note)
Tibicininae Stölting, Moore and Lakes-Harlan 2002: 2 (listed)
Tibicininae Holzinger, Kammerlander and Nickel 2003: 475, 480 (listed, comp. note) Europe
Tibicinae (sic) Moulds 2003a: 187 (comp. note)
Tibicininae Sueur 2003: 2932 (listed)
Tettigadinae Chen 2004: 38 (listed)
Platypediinae Chen 2004: 38 (listed)
Ydiellinae Chen 2004: 38 (listed)
Tibicininae Chen 2004: 38–40 (phylogeny, listed)
Tibicininae Lee and Hayashi 2004: 68 (listed) Taiwan
Tibicininae Sanborn, Heath, Heath, Noriega and Phillips 2004: 282 (comp. note)
Tettigadinae Sanborn, Heath, Heath, Noriega and Phillips 2004: 282 (comp. note)
Tibicininae Sueur, Puissant, Simões, Seabra, Boulard and Quartau 2004: 180, 182 (listed) Portugal
Tibicininae Boulard 2005d: 343, 348 (comp. note)
Tibicininae Chawanji, Hodson and Villet 2005b: 66 (listed)
Tibicininae Moulds 2005b: 384–389, 404, 416–418, 424, 428, 430, Fig 31, Table 2 (described, key, history, phylogeny, listed, comp. note)
Tettigadinae Moulds 2005b: 377, 385–387, 397, 416–418, 427–428, Fig 31, Fig 57 (synonymy, described, key, history, phylogeny, listed, comp. note) Equals Tibicinae n. syn. Equals Platypediinae n. syn.
Tettigadesaria Moulds 2005b: 386, 388 (history, listed, comp. note)
Platypediinae Moulds 2005b: 384–389, 396, 416–418, Fig 31, Fig 57 (described, key, history, phylogeny, listed, comp. note)
Ydiellinae Moulds 2005b: 377, 385–389, 392, Fig 31 (described, key, history, phylogeny, listed, comp. note)
Tibicinae (sic) Reynaud 2005: 141 (listed)
Tibicininae Boulard 2006g: 16–17, 66 (listed)
Tettigadinae Chawanji, Hodson and Villet 2006: 374 (comp. note)
Tettigadinae Puissant 2006: 19 (listed) France
Tibicininae Boulard 2007e: 86, Plate 15 (illustrated, comp. note) Thailand
Tibicininae Chawanji, Hodson, Villet, Sanborn and Phillips 2007: 337 (comp. note)
Tibicininae Sanborn 2007b: 37 (listed) Equals Tettigadinae Mexico
Tettigadinae Sanborn 2007b: 37 (listed) Mexico
Tibicininae Shiyake 2007: 7, 90 (listed, comp. note)
Platypediinae Shiyake 2007: 9, 117 (listed, comp. note)
Tettigadinae Shiyake 2007: 9, 117 (listed, comp. note)
Tettigadinae Sueur and Puissant 2007b: 62 (listed) France
Tettigadinae Sueur, Vanderpool, Simon, Ouvrard and Bourgoin 2007: 612 (comp. note)
Tibicininae Demir 2008: 476 (listed) Turkey
Tettigadinae Pham and Yang 2009: 1, 3, 15, Table 2 (listed, comp. note) Vietnam
Tibicininae Wei, Tang, He and Zhang 2009: 337 (comp. note)
Tettigadinae Wei, Tang, He and Zhang 2009: 337 (comp. note)
Tettigadinae Kemal and Koçak 2010: 3, 5 (diversity, listed, comp. note) Equals Tibicinincae Turkey
Tibicininae Mozaffarian and Sanborn 2010: 80 (listed)
Tibicininae Phillips and Sanborn 2010: 75 (comp. note)
Tettigadinae Pham and Yang 2010a: 133 (diversity, comp. note) Vietnam
Tibicininae Sanborn 2010a: 1602 (listed) Colombia
Tettigadinae Wei and Zhang 2010: 37–38 (comp. note)

Tribe Tettigadini Distant, 1905 = Tettigadesini
Tettigadini Sueur 2001: 45, Table 1 (listed)
Tettigadesini Boulard 2005d: 348 (stridulatory apparatus, comp. note)
Tettigadini Moulds 2005b: 419, 428, Fig 58 (phylogeny, comp. note)
Tettigadesini Boulard 2006g: 153 (comp. note)
Tettigadini Sanborn 2007b: 38 (listed) Mexico
Tettigadini Sanborn 2010a: 1602 (listed) Colombia

***Tettigades* Amyot & Audinet-Serville, 1843** = *Tettigodes* (sic) = *Tettigates* (sic)

Tettigades Varley 1939: 98–99 (comp. note)
Tettigades Weber 1968: 80 (stridulatory apparatus, comp. note)
Tettigades Shcherbakov 1981a: 830 (listed)
Tettigades Shcherbakov 1981b: 66 (listed)
Tettigades Heath 1983: 12–13 (comp. note) Argentina
Tettigades Strümpel 1983: 17, 109, 117 (listed, comp. note)
Tettigades Hayashi 1984: 26 (comp. note)
Tettigades Arnett 1985: 218 (diversity) Arizona (error)
Tettigades Boulard 1990b: 88 (comp. note)
Tettigades Heath, Heath, Sanborn and Noriega 1990: 1 (comp. note) Argentina
Tettigades Heath, Heath and Noriega 1990: 1 (distribution, comp. note) Argentina
Tettigades Duffels 1993: 1224–1225 (stridulatory apparatus, comp. note)
Tettigades Navarro 1997: 42 (comp. note)
Tettigades Poole, Garrison, McCafferty, Otte and Stark 1997: 251, 333 (listed) Equals *Tettigodes* (sic) North America (error)
Tettigades Moulds 1999: 1 (comp. note)
Tettigates (sic) Ohbayashi, Sato and Igawa 1999: 340 (parasitism, comp. note)
Tettigades Arnett 2000: 299 (diversity) Arizona (error)
Tettigades Moulds 2002: 329 (comp. note) Australia
Tettigades Sanborn, Heath, Heath, Noriega and Phillips 2004: 282 (comp. note)
Tettigades Boulard 2005d: 348 (stridulatory apparatus, comp. note)
Tettigades Moulds 2005b: 386, 388, 410, 414, 421, 428 (type genus of Tettigadidae, type genus of Tettigadinae, history, phylogeny, listed, comp. note)
Tettigades Boulard 2006g: 153 (stridulation, comp. note)
Tettigades Shiyake 2007: 9, 117 (listed, comp. note)
Tettigades Marshall, S.A. 2009: 8 (illustrated) Chile

***T. angularis* Torres, 1958**

***T. auropilosa* Torres, 1944**

***T. bosqi* Torres, 1942**

***T. brevicauda* Torres, 1958**

***T. brevivenosa* Torres, 1942**

***T. chilensis* Amyot & Audinet-Serville, 1843** = *Tettigades chiliensis* (sic) = *Cicada rubrolineata* Spinola, 1852 = *Tettigades rubrolineata* = *Fidicina crassivena* Walker, 1858 = *Cicada eremophila* Philippi, 1866

Tettigades chilensis Breddin 1897: 30–31 (synonymy, comp. note) Equals *Cicada rubolineata* Equals *Cicada crassimargo* Equals *Fidicina crassivena* Equals *Cicada cremophila* Equals *Tettigades crassimargo* Chile, Argentina, Chubut
Cicada rubolineata Breddin 1897: 30 (listed)
Fidicina crassivena Breddin 1897: 30 (listed)
Cicada cremophila Breddin 1897: 30 (listed)
Tettigades chilensis Evans 1966: 110 (parasitism, comp. note) Argentina. Mendoza
Tettigades chilensis Weber 1968: 80–81, Fig 67a (stridulatory apparatus illustrated, comp. note)
Tettigades chilensis Peña 1987: 110 (comp. note) Chile
Tettigades chilensis Boulard 1988d: 154 (wing abnormalities, comp. note)
Tettigades chilensis Boulard 1988f: 53 (type species of Tettigadinae)
Tettigades chilensis González 1989: 78–80, 274, Fig 42 (described, illustrated, hosts, pest, listed) Equals *Cicada rubrolineata* Chile
Tettigades chilensis Quiroga, Arretz and Araya 1991: 470–471, Table 1 (agricultural pest, comp. note) Chile
Tettigades chilensis Navarro 1997: 41–44, 46–48, Figs 1-3, Figs 6–7, Fig 8A, Table 1 (acoustic analysis, sonogram, comp. note) Chile
Tettigades chilensis Sueur 2001: 45, Table 1 (listed)
Tettigades chilensis Holliday, Hastings and Coelho 2009: 3, 10–11, Table 1 (predation, comp. note) Argentina, Mendoza
Tettigades chilensis Sueur 2002b: 111, 122, 129–130 (listed, comp. note) Equals *Cicada rubolineata* Chile
Tettigades chilensis Moulds 2005b: 388, 428 (type species of *Tettigades*, comp. note)
Tettigades chilensis Shiyake 2007: 9 116–117, Fig 208 (illustrated, listed, comp. note)

***T. compacta* Walker, 1850**

Tettigades compacta Peña 1987: 110 (comp. note) Chile

***T. crassa* Torres, 1958**

***T. curvicosta* Torres, 1958**

***T. distanti* Torres, 1958**

***T. dumfriesi* Distant, 1920** = *Tettigades dumfriesii* (sic)

Tettigades dumfriesii (sic) Heath, Heath, Sanborn and Noriega 1990: 1 (thermal tolerances, comp. note) Argentina

***T. lacertosa* Torres, 1944**

***T. lebruni* Distant, 1906** = *Odopoea lebruni* (partim)

***T. limbata* Torres, 1958**

***T. lizeri* Torres, 1942**

***T. lizeriana* Delétang, 1919**

***T. major* Torres, 1944**

Tettigades major Heath, Heath, Sanborn and Noriega 1990: 1 (thermal tolerances, comp. note) Argentina

Tettigades major Sanborn, Heath, Heath, Noriega and Phillips 2004: 282–285, Tables 1–4 (thermal responses, habitat, distribution, comp. note) Argentina, Mendoza

***T. mexicana* Distant, 1881**

Tettigades mexicana Arnett 1985: 218 (listed) Arizona (error)

Tettigades mexicana Poole, Garrison, McCafferty, Otte and Stark 1997: 333 (listed) North America (error)

Tettigades mexicana Arnett 2000: 299 (listed) Arizona (error)

Tettigades mexicana Sanborn 2007b: 38–39 (distribution, comp. note) Mexico

***T. opaca* Jacobi, 1907** = *Tettigades porteri* Bréthes, 1920

***T. parva* Distant, 1892** = *Tettigades parvus* = *Tettigades brevivenosa* Torres, 1942

***T. pauxilla* Torres, 1958**

***T. porteri* Bréthes, 1920**

Tettigades porteri Peña 1987: 110 (comp. note) Chile

***T. procera* Torres, 1958**

***T. sarcinatrix* Torres, 1944** = *Tettigades sarsinatrix* (sic)

***T. sordida* Torres, 1944**

***T. ulnaria* Distant, 1906** = *Tettigades urnaria* (sic)

Tettigades ulnaria Peña 1987: 110 (comp. note) Chile

Tettigades ulnaria Moulds 2005b: 396–397, 399, 403, 406, 414, 417–420, Fig 37, Fig 45, Fig 56–59, Table 1 (phylogeny, comp. note)

***T. undata* Torres, 1958**

Tettigades undata Sanborn 2008b: 35, Fig 19 (stridulatory apparatus, illustrated, comp. note)

Tettigades undata Sanborn 2008d: 3470, Fig 81 (stridulatory apparatus, illustrated, comp. note)

***T. unipuncta* Torres, 1958**

***T. varivenosa* Distant, 1906**

***Torresia* Sanborn & Heath, 2011, nom. nud.**

***T. lariojaensis* Sanborn & Heath, 2011, nom. nud.**

***T. sanjuanensis* Sanborn & Heath, 2011, nom. nud.**

***Alarcta* Torres, 1958**

Alarcta Heath, Heath and Noriega 1990: 1 (distribution, comp. note) Argentina

Alarcta Moulds 2005b: 388 (higher taxonomy, listed, comp. note)

***A. albicerata* (Torres, 1949)** = *Tettigades albicerata*

***A. bahiensis* (Torres, 1942)** = *Tettigades bahiensis*

***A. blanchardi* (Torres, 1948)** = *Tettigades blanchardi*

***A. macrogina* (Torres, 1949)** = *Tettigades macrogina*

***A. micromaculata* Sanborn & Heath, 2011, nom. nud.**

***A. minuta* (Torres, 1949)** = *Tettigades minuta*

***A. quadrimacula* Torres, 1958**

***A. terrosa* Torres, 1958**

***Tettigotoma* Torres, 1942**

***T. maculata* Torres, 1942**

***Psephenotettix* Torres, 1958**

Psephenotettix Heath, Heath, Sanborn and Noriega 1990: 1 (comp. note) Argentina

Psephenotettix Heath, Heath and Noriega 1990: 1 (distribution, comp. note) Argentina

Psephenotettix Moulds 2005b: 388 (listed, comp. note)

***P. grandis* Torres, 1958**

***P. minor* Torres, 1958**

Psephenotettix minor Heath, Heath, Sanborn and Noriega 1990: 1 (thermal tolerances, comp. note) Argentina

***Babras* Jacobi, 1907**

Babras Weber 1968: 80 (stridulatory apparatus, comp. note)

Babras Hayashi 1984: 26 (comp. note)

Babras Heath, Heath, Sanborn and Noriega 1988: 304 (thermal responses, comp. note) Argentina

Babras Heath, Heath and Noriega 1990: 1 (distribution, comp. note) Argentina

Babras Heath, Heath, Sanborn and Noriega 1999: 1 (thermal responses, comp. note) Argentina

Babras Sanborn, Heath, Heath, Noriega and Phillips 2004: 282 (comp. note)

Babras Moulds 2005b: 388, 414 (higher taxonomy, listed, comp. note)

***B. sonorivox* Jacobi, 1907**

Babras sonorivox Weber 1968: 80, Fig 67b (stridulatory apparatus illustrated, comp. note)

Babras sonorivox Heath, Heath, Sanborn and Noriega 1988: 304 (thermal responses, comp. note) Argentina

Babras sonorivox Heath, Heath, Sanborn and Noriega 1999: 1 (thermal responses, comp. note) Argentina

Babras sonorivox Sueur 2002b: 122, 124, 129 (listed, comp. note)

Babras sonorivox Sanborn, Heath, Heath, Noriega and Phillips 2004: 282–287, Figs 1–2, Tables 1–4 (thermal responses, habitat, distribution, comp. note) Argentina, La Rioja

***Calliopsida* Torres, 1958**

Calliopsida Moulds 2005b: 388, 414 (higher taxonomy, listed, comp. note)

***C. cinnabarina* (Berg, 1879)** = *Tettigades cinnabarina* = *Chonosia cinnabarina*

Chonosia cinnabarina Peña 1987: 110 (comp. note) Chile

Chonosia cinnabarina Bolcatto, Medrano and De Santis 2006: 7, 10, Map 1 (distribution) Argentina
Chonosia cinnabarina De Santis, Medrano, Sanborn and Bolcatto 2007: 11–12, Fig 2 (distribution, comp. note) Argentina

Coata **Distant, 1906**
Coata Boulard 1990b: 144 (comp. note)
Coata Boer 1995c: 205 (comp. note) South America
Coata Boulard 2002b: 198 (comp. note) South America
Coata Moulds 2005b: 388 (higher taxonomy, listed, comp. note)
Coata Sanborn 2010a: 1602 (listed) Colombia
C. facialis facialis **Distant, 1906** = *Coata fascialis* (sic)
Coata facialis Sanborn 2010a: 1602 (type species of *Coata*) Ecuador
C. facialis **var. ______ Jacobi, 1907**
C. fragilis **Goding, 1925**
C. humeralis **(Walker, 1858)** = *Cicada humeralis* = *Plautilla humeralis*
Coata humeralis Sanborn 2010a: 1602 (synonymy, distribution, comp. note) Equals *Cicada humeralis* Equals *Plautilla humeralis* Colombia

Chonosia **Distant, 1905**
Chonosia Weber 1968: 80 (stridulatory apparatus, comp. note)
Chonosia sp. Bohart and Strange 1976: 313 (comp. note) Argentina, Catamarca
Chonosia Strümpel 1983: 17, 109 (listed, comp. note)
Chonosia Hayashi 1984: 26 (comp. note)
Chonosia Heath, Heath, Sanborn and Noriega 1990: 1 (comp. note) Argentina
Chonosia Moulds 2005b: 388 (higher taxonomy, listed, comp. note)
C. atrodorsalis **Torres, 1945**
Chonosia atrodorsalis Heath, Heath, Sanborn and Noriega 1990: 1 (thermal tolerances, comp. note) Argentina
C. crassipennis **(Walker, 1858)** = *Fidicina crassipennis*
Chonosia crassipennis Varley 1939: 99 (comp. note)
Chonosia crassipennis Wooton 1968: 319, Fig d, Fig l (wing venation, comp. note)
Chonosia crassipennis Heath, Heath, Sanborn and Noriega 1990: 1 (thermal tolerances, comp. note) Argentina
Chonosia crassipennis Bolcatto, Medrano and De Santis 2006: 7–8, 10, Fig 1C, Map 2 (illustrated, distribution) Argentina
Chonosia crassipennis De Santis, Medrano, Sanborn and Bolcatto 2007: 3, 11, 13–14, 17, 19, Fig 1C, Figs 3–4, Table 1 (listed, distribution) Argentina, Córdoba
C. crassipennis metequei **Delétang, 1919 nom. nud.**
C. longiopercula **Sanborn & Heath, 2011, nom. nud.**
C. papa **(Berg, 1882)** = *Tettigades papa*
Tettigades papa Breddin 1897: 31 (comp. note) Argentina, Mendoza, Chubut
Chonosia papa Heath, Heath, Sanborn and Noriega 1990: 1 (thermal tolerances, comp. note) Argentina
C. septentrionala **Sanborn & Heath, 2011, nom. nud.**
C. trigonocelis **Torres, 1945**

Acuticephala **Torres, 1958**
Acuticephala Heath, Heath and Noriega 1990: 1 (distribution, comp. note) Argentina
Acuticephala Moulds 2005b: 388 (higher taxonomy, listed, comp. note)
A. alipuncta **Torres, 1958**
Acuticephala alipuncta Ruffinelli 1970: 3 (distribution, comp. note) Uruguay, Argentina, Entre Rios

Mendozana **Distant, 1906** = *Mendozana (Tettigades)* = *Semaiophora* Haupt, 1918 = *Fadylia* Delétang, 1919
Mendozana Heath, Heath and Noriega 1990: 1 (distribution, comp. note) Argentina
Mendozana spp. Moulds 2002: 329 (comp. note)
Mendozana Moulds 2005b: 388 (higher taxonomy, listed, comp. note)
M. antennaria **(Jacobi, 1907)** = *Tettigades antennaria* = *Mendozana (Tettigades) antennaria* = *Semaiophora dietherti* Schmidt, 1926
Mendozana (Tettigades) antennaria Boulard 1988d: 156 (wing abnormalities, comp. note)
Mendozana antennaria Moulds 2002: 329 (comp. note) South America
M. platypleura **Distant, 1906** = *Semaiophora basimaculata* Haupt, 1918 = *Fadylia bruchi* Delétang, 1919
Mendozana platypleura Varley 1939: 99 (comp. note)
Mendozana platypleura Heath 1983: 13 (comp. note) North America
Mendozana platypleura Moulds 2002: 329 (comp. note) South America

Tribe Lamotialnini Boulard, 1976 = Lamotialnii (sic)
Lamotialnini Boulard 1986d: 231 (n. comb., comp. note) Central African Republic, Ivory Coast
Lamotialnini Boulard 1988f: 51 (key)
Lamotialnii (sic) Jong and Boer 2004: 267 (comp. note)
Lamotialnini Moulds 2005b: 427 (higher taxonomy, listed)

Subtribe Lamotialnina Boulard, 1976 = Lamotialnaria
Lamotialnaria Boulard 1993d: 92 (n. subtribe, comp. note)
Lamotialnaria Boulard 2002b: 205 (comp. note)

***Lamotialna* Boulard, 1976**
Lamotialna Medler 1980: 73 (listed) Nigeria
Lamotialna Boulard 1986d: 231 (listed, comp. note)
Lamotialna Boulard 1993d: 91–92 (comp. note) West Africa
Lamotialna Duffels 1993: 1225 (comp. note)
Lamotialna Boer 1995c: 203–204, Fig 1 (phylogeny, comp. note) Africa
Lamotialna Boulard 2002b: 197, 205 (sound system, comp. note) Africa
Lamotialna Jong and Boer 2004: 267 (comp. note)
Lamotialna Moulds 2005b: 385 (higher taxonomy, listed, comp. note)
Lamotialna Boulard 2006g: 16–17 (comp. note)
Lamotialna Wei, Tang, He and Zhang 2009: 337 (comp. note)
Lamotialna Boulard 2010b: 336 (comp. note)

***L. condamini* Boulard, 1976**
Lamotialna condamini Medler 1980: 73 (listed) Nigeria, Ivory Coast
Lamotialna condamini Boulard 1984d: 7 (comp. note)
Lamotialna condamini Boulard 1986d: 231–232, 237–238, Table I (type species of *Lamotialna*, comp. note) Ivory Coast
Lamotialna condamini Boulard 1988f: 52 (type species of Lamnotialnini)
Lamotialna condamini Boulard 1990b: 132–133, 178, 191, 193, Plate XV, Fig 6, Plate XXIV, Fig 4 (genitalia, comp. note)
Lamotialna condamini Boulard 1993d: 88 (comp. note)
Lamotialna condamini Boulard 2002b: 205 (comp. note) Ivory Coast
Lamotialna condamini Boulard 2005d: 349 (comp. note) Ivory Coast
Lamotialna condamini Boulard 2006g: 160–161 (comp. note) Ivory Coast

***L. couturieri* Boulard, 1986**
Lamotialna couturieri Boulard 1986d: 232–233, 237–238, Table I, Figs 26–28 (n. sp., described, illustrated, comp. note) Ivory Coast
Lamotialna couturieri Boulard 1990b: 178 (comp. note)
Lamotialna couturieri Boulard 1993d: 88 (comp. note)
Lamotialna couturieri Boulard 2002b: 205 (comp. note) Ivory Coast
Lamotialna couturieri Boulard 2005d: 349 (comp. note) Ivory Coast
Lamotialna couturieri Boulard 2006g: 160–161 (comp. note) Ivory Coast

Tribe Platypediini Kato, 1932 = Platypedini (sic)
Platypediini Boulard 1988f: 51 (key)
Platypedini (sic) Moulds 1999: 1 (comp. note)
Platypediini Sanborn 1999a: 53 (listed)
Platypediini Sueur 2001: 45, Table 1 (listed)
Platypediini Kondratieff, Ellingson and Leatherman 2002: 47 (comp. note)
Platypediini Moulds 2005b: 416, 419, 428, Fig 58 (phylogeny, comp. note)
Platypediini Sanborn 2007b: 37 (listed) Mexico
Platypediini Sueur, Vanderpool, Simon, Ouvrard and Bourgoin 2007: 612 (comp. note)

***Platypedia* Uhler, 1888**
Platypedia Powell and Hogue 1979: 113 (described, distribution, comp. note) California
Platypedia Shcherbakov 1981a: 830, 834, Fig 32 (wing structure, comp. note)
Platypedia Shcherbakov 1981b: 66, 70, Fig 32 (wing structure, comp. note)
Platypedia Strümpel 1983: 18 (listed)
Platypedia Boulard 1984c: 171 (comp. note)
Platypedia Hayashi 1984: 26 (comp. note)
Platypedia Arnett 1985: 218 (diversity) Southwestern United States, Pacific Northwest, Montana, Wyoming, Colorado, Nebraska, South Dakota
Platypedia Boulard 1988f: 64 (comp. note)
Platypedia Duffels 1988d: 75 (comp. note) North America
Platypedia Boulard 1990b: 178 (comp. note)
Platypedia spp. Cranshaw and Kondratieff 1991: 10 (life cycle, comp. note) Colorado
Platypedia Carpenter 1992: 222 (comp. note)
Platypedia Duffels 1993: 1225 (diversity, comp. note)
Platypedia Cranshaw and Kondratieff 1995: 79 (life history)
Platypedia Miller and Harley 1999: 537, Fig 32.11 (eclosion, illustrated)
Platypedia Moore 1996: 221 (comp. note) Mexico
Platypediinae (sic) Poole, Garrison, McCafferty, Otte and Stark 1997: 251 (listed) North America
Platypedia Poole, Garrison, McCafferty, Otte and Stark 1997: 332 (listed) North America
Platypedia Moulds 1999: 1 (comp. note)
Platypedia Ohbayashi, Sato and Igawa 1999: 340 (parasitism, comp. note)
Platypedia Sanborn 1999b: 2 (listed) North America

Platypedia Sanborn and Phillips 1999: 451, 454 (comp. note)
Platypedia Villet and Zhao 1999: 322 (comp. note)
Platypedia Arnett 2000: 299 (diversity) Southwestern United States, Pacific Northwest, Montana, Wyoming, Colorado, Nebraska, South Dakota
Platypedia Maw, Footit, Hamilton and Scudder 2000: 49 (listed)
Platypedia Boulard 2002b: 197, 205 (sound system, comp. note) North America
Platypedia Kondratieff, Ellingson and Leatherman 2002: 14, 47–49 (key, listed, diagnosis, diversity, song, comp. note) Colorado, California
Platypedia Sueur 2002b: 122, 124 (comp. note)
Platypedia Lee 2003b: 6 (comp. note)
Platypedia Sanborn 2004d: 2057 (comp. note)
Platypedia Boulard 2005d: 349 (comp. note) North America
Platypedia Moulds 2005b: 385, 387, 414–416, 428 (type genus of Platypediidae, type genus Platypediinae, higher taxonomy, phylogeny, listed, comp. note)
Platypedia Boulard 2006g: 16–17, 160–161 (comp. note) North America
Platypedia Shiyake 2007: 9–10, 95, 117 (listed, comp. note)
Platypedia Sueur, Vanderpool, Simon, Ouvrard and Bourgoin 2007: 612–613 (comp. note)
Platypedia sp. Sueur, Vanderpool, Simon, Ouvrard and Bourgoin 2007: 612–614, 618–619, 621, Fig 1, Figs 3–6, Table 1 (comp. note) Oregon
Platypedia Sanborn 2008d: 3469 (comp. note)
Platypedia Holliday, Hastings and Coelho 2009: 12–13 (comp. note)
Platypedia Wei, Tang, He and Zhang 2009: 337, 340 (comp. note)
Platypedia Boulard 2010b: 336 (comp. note)

P. *affinis* Davis, 1939
Platypedia affinis Poole, Garrison, McCafferty, Otte and Stark 1997: 332 (listed) North America
Platypedia affinis Sanborn 1999a: 53 (type material, listed)
Platypedia affinis Sanborn 1999b: 2 (listed) North America

P. *aperta* Van Duzee, 1915
Platypedia aperta Poole, Garrison, McCafferty, Otte and Stark 1997: 332 (listed) North America
Platypedia aperta Sanborn 1999a: 53 (type material, listed)
Platypedia aperta Sanborn 1999b: 2 (listed) North America
Platypedia aperta Sanborn 2007b: 37 (distribution, comp. note) Mexico, Baja California Norte, California

P. *areolata* (Uhler, 1861) = *Cicada areolata* = *Platypedia areoloata* (sic) = *Platypedia aerolata* (sic)
Platypedia areolata Henshaw 1903: 230 (listed) Equals *Cicada areolata* Washington
Platypedia areolata Arnett 1985: 218 (distribution) California, Pacific Northwest, Montana, Colorado
Platypedia aerolata (sic) Boulard 1988f: 52 (type species of Platypediinae)
Platypedia areolata Poole, Garrison, McCafferty, Otte and Stark 1997: 332 (listed) Equals *Cicada areolata* North America
Platypedia areolata Sanborn 1999a: 53 (type material, listed)
Platypedia areolata Sanborn 1999b: 2 (listed) North America
Platypedia areolata Arnett 2000: 299 (distribution) California, Pacific Northwest, Montana, Colorado
Platypedia areolata Maw, Footit, Hamilton and Scudder 2000: 49 (comp. note) Canada, British Columbia
Platypedia areolata Kondratieff, Ellingson and Leatherman 2002: 48, 52–53 (type species of *Platypedia*, comp. note) California, Pacific Northwest
Platypedia areolata Moulds 2005b: 387 (type species of *Platypedia*)
Platypedia areolata Shiyake 2007: 9, 116–117, Fig 207 (illustrated, listed, comp. note)

P. *australis* Davis, 1941
Platypedia australis Sanborn 1999a: 53 (type material, listed)
Platypedia australis Sanborn 2007b: 38 (distribution, comp. note) Mexico, Nuevo León

P. *balli* Davis, 1936
Platypedia balli Poole, Garrison, McCafferty, Otte and Stark 1997: 332 (listed) North America
Platypedia balli Sanborn 1999a: 53 (type material, listed)
Platypedia balli Sanborn 1999b: 2 (listed) North America

P. *barbata* Davis, 1920
Platypedia barbata Hennessey 1990: 471 (type material)
Platypedia barbata Poole, Garrison, McCafferty, Otte and Stark 1997: 332 (listed) North America
Platypedia barbata Sanborn 1999a: 53 (type material, listed)
Platypedia barbata Sanborn 1999b: 2 (listed) North America

***P. bernardinoensis* Davis, 1932** = *Platypedia rufipes bernardinoensis* = *Platypedia rufipes* var. *bernardoensis* (sic) = *Platypedia bernardoensis* (sic) = *Platypedia rufipes angustipennis* Davis, 1932

Platypedia bernardinoensis Poole, Garrison, McCafferty, Otte and Stark 1997: 332 (listed) Equals *Platypedia angustipennis* Equals *Platypedia bernardoensis* (sic) North America

Platypedia bernardinoensis Sanborn 1999a: 53 (type material, listed)

Platypedia bernardinoensis Sanborn 1999b: 2 (listed) North America

Platypedia rufipes var. *angustipennis* Sanborn 1999a: 54 (type material, listed)

***P. falcata* Davis, 1920** = *Platypedid* (sic) *falcata*

Platypedia falcata Tinkham 1941: 176, 178 (distribution, comp. note) Texas

Platypedia falcata Hennessey 1990: 471 (type material)

Platypedia falcata Poole, Garrison, McCafferty, Otte and Stark 1997: 332 (listed) North America

Platypedia falcata Sanborn 1999a: 53 (type material, listed)

Platypedia falcata Sanborn 1999b: 2 (listed) North America

Platypedia falcata Phillips and Sanborn 2007: 461 (comp. note) Texas, California

***P. intermedia* Van Duzee, 1915** = *Platypedia areolata intermedia*

Platypedia intermedia Poole, Garrison, McCafferty, Otte and Stark 1997: 333 (listed) North America

Platypedia intermedia Sanborn 1999a: 53 (type material, listed)

Platypedia intermedia Sanborn 1999b: 2 (listed) North America

Platypedia intermedia Kondratieff, Ellingson and Leatherman 2002: 49 (comp. note)

***P. laticapitata* Davis, 1921**

Platypedia laticapitata Hennessey 1990: 471 (type material)

Platypedia laticapitata Hogue 1993a: 123–124, Fig 94 (described, illustrated, comp. note) California

Platypedia laticapitata Poole, Garrison, McCafferty, Otte and Stark 1997: 333 (listed) North America

Platypedia laticapitata Sanborn 1999a: 53 (type material, listed)

Platypedia laticapitata Sanborn 1999b: 2 (listed) North America

P. latipennis Davis, 1921 to *Platypedia mohavensis mohavensis*

***P. mariposa* Davis, 1935** = *Platypedia plumbea* Van Duzee nom. nud. (partim)

Platypedia mariposa Poole, Garrison, McCafferty, Otte and Stark 1997: 333 (listed) North America

Platypedia mariposa Sanborn 1999a: 53 (type material, listed)

Platypedia plumbea Sanborn 1999a: 56 (comp. note)

Platypedia mariposa Sanborn 1999b: 2 (listed) North America

***P. middlekauffi* Simons, 1953**

Platypedia middlekauffi Poole, Garrison, McCafferty, Otte and Stark 1997: 333 (listed) Equals *Cicada* North America

Platypedia middlekauffi Sanborn 1999a: 53 (type material, listed)

Platypedia middlekauffi Sanborn 1999b: 2 (listed) North America

***P. minor* Uhler, 1888** = *Platypedia minsi* Johnson & Ledig, 1918 nom. nud. = *Platypedia gressitti* Kato, 1932

Platypedia minor Henshaw 1903: 230 (listed)

Platypedia minor Phillips 1992: 251 (grape pest, control) California

Platypedia gressitti Poole, Garrison, McCafferty, Otte and Stark 1997: 333 (listed) North America

Platypedia minor Poole, Garrison, McCafferty, Otte and Stark 1997: 333 (listed) North America

Platypedia minsi Poole, Garrison, McCafferty, Otte and Stark 1997: 333 (listed) North America

Platypedia minor Sanborn 1999a: 54 (type material, listed)

Platypedia gressitti Sanborn 1999b: 2 (type material, listed) North America

Platypedia minor Sanborn 1999b: 2 (listed) North America

Platypedia minsi Sanborn 1999b: 2 (listed) North America

Platypedia minor Kondratieff, Ellingson and Leatherman 2002: 12–13, 48–50, 80–81, Figs 53–54, Tables 1–2 (key, diagnosis, illustrated, comp. note) Equals *Platypedia intermedia* Colorado, California, Nevada, Mexico

Platypedia minor Krell, Boyd, Nay, Park and Perring 2007: 215 (grapevine pest, comp. note) California

Platypedia minor Sanborn 2007b: 38 (distribution, comp. note) Mexico, Sinaloa, Mazatlan, California, Nevada, Colorado

***P. mohavensis mohavensis* Davis, 1920** = *Platypedia mohavensis* = *Platypedia latipennis* Davis, 1921

Platypedia mohavensis Sugden 1940: 125 (comp. note) Utah

Platypedia latipennis Poole, Garrison, McCafferty, Otte and Stark 1997: 333 (listed) North America

Platypedia mohavensis Poole, Garrison, McCafferty, Otte and Stark 1997: 333 (listed) North America
Platypedia latipennis Sanborn 1999a: 53 (type material, listed)
Platypedia mohavensis Sanborn 1999a: 54 (type material, listed)
Platypedia latipennis Sanborn 1999b: 2 (listed) North America
Platypedia mohavensis Sanborn 1999b: 2 (listed) North America
Platypedia mohavensis Kondratieff, Ellingson and Leatherman 2002: 9, 11–13, 48–51, 55, 79–80, Figs 49–50, Tables 1–2 (key, diagnosis, illustrated, listed, distribution, song, comp. note) Equals *Platypedia latipennis* n. syn. Equals *Platypedia mohavensis rufescens* Colorado, Arizona, California, New Mexico, Utah
Platypedia latipennis Kondratieff, Ellingson and Leatherman 2002: 11–12, 51, Table 1 (listed, comp. note) Colorado

***P. mohavensis rufescens* Davis, 1932** = *Platypedia rufescens* = *Platypedia mohavensis* var. *rufescens*
Platypedia rufescens Poole, Garrison, McCafferty, Otte and Stark 1997: 333 (listed) North America
Platypedia mohavensis var. *rufescens* Sanborn 1999a: 54 (type material, listed)
Platypedia mohavensis var. *rufescens* Sanborn 1999b: 2 (listed) North America
Platypedia mohavensis rufescens Kondratieff, Ellingson and Leatherman 2002: 51 (comp. note)

P. plumbea (manuscript name) Van Duzee to *Platypedia mariposa* Davis, 1935 (partim) and *Platypedia similis* Davis, 1920 (partim)

***P. putnami keddiensis* Davis, 1920** = *Platypedia putnami* var. *keddiensis* = *Platypedia keddiensis*
Platypedia putnami var. *keddiensis* Sanborn 1999a: 54 (type material, listed)
Platypedia putnami var. *keddiensis* Sanborn 1999b: 2 (listed) North America

***P. putnami lutea* Davis, 1920** = *Platypedia putnami* var. *lutea* = *Platypedia lutea*
Platypedia putnami lutea Sugden 1940: 125 (song, habitat, comp. note) Utah
Platypedia putnami var. *lutea* Sanborn 1999a: 54–55 (type material, listed, comp. note)
Platypedia putnami var. *lutea* Sanborn 1999b: 2 (listed) North America
Platypedia putnami var. *lutea* Sanborn and Phillips 1999: 451–454, Figs 1–3, Table 1 (song, sonogram, oscillogram, comp. note) Arizona
Platypedia putnami var. *lutea* Sueur 2001: 45, Table 1 (listed)
Platypedia putnami var. *lutea* Sanborn, Noriega and Phillips 2002a: 21 (thermal biology, comp. note)
Platypedia putnami var. *lutea* Sanborn, Noriega and Phillips 2002b: A38 (thermal biology, comp. note) Arizona
Platypedia putnami var. *lutea* Sanborn, Noriega and Phillips 2002c: 366–368, Figs 1–3 (thermoregulation, thermal responses, comp. note) Arizona
Platypedia putnami var. *lutea* Sanborn 2005c: 114, Table 1 (comp. note)
Platypedia putnami lutea Heath and Sanborn 2007: 488 (comp. note) Arizona

***P. putnami occidentalis* Davis, 1920** = *Platypedia putnami* var. *occidentalis* = *Platypedia occidentalis*
Platypedia putnami var. *occidentalis* Sanborn 1999a: 54 (type material, listed)
Platypedia putnami var. *occidentalis* Sanborn 1999b: 2 (listed) North America
Platypedia putnami occidentalis Kondratieff, Ellingson and Leatherman 2002: 52 (comp. note)

***P. putnami putnami* (Uhler, 1877)** = *Cicada putnami* = *Platypedia putnami* = *Plptypedia* (sic) *putnami*
Cicada putnami Putnam 1881: 67 (comp. note) Colorado
Platypedia putnami Henshaw 1903: 230 (listed) Equals *Cicada putnami* Colorado, Utah
Platypedia putnami Cranshaw and Kondratieff 1991: 10–11 (comp. note) Colorado
Platypedia putnami Cranshaw and Kondratieff 1995: 80 (oviposition damage, illustrated)
Platypedia putnami Poole, Garrison, McCafferty, Otte and Stark 1997: 333 (listed) Equals *Cicada putnami* Equals *Platypedia keddiensis* Equals *Platypedia lutea* Equals *Platypedia occidentalis* North America
Platypedia putnami Sanborn 1999a: 54 (type material, listed)
Platypedia putnami Sanborn 1999b: 2 (listed) North America
Platypedia putnami Maw, Footit, Hamilton and Scudder 2000: 49 (comp. note) Canada, British Columbia
Platypedia putnami Kondratieff, Ellingson and Leatherman 2002: 4, 6–7, 9, 12–13, 48–49, 51–53, 55, 69, 74, 79–80, Fig 17, Figs 34–36, Fig 48, Figs 51–52, Tables 1–2 (key, diagnosis, illustrated, listed, distribution, song, comp. note) Equals *Cicada putnami* Equals *Platypedia putnami lutea* Colorado, Arizona, California, Montana, Nebraska, Nevada, New Mexico, Oregon, Utah, Washington
Platypedia putnami Lee 2003b: 6–7 (comp. note)

Platypedia putnami Moulds 2005b: 396, 399, 414, 417–420, Figs 56–59, Table 1 (phylogeny, comp. note)
Platypedia putnami Sueur, Vanderpool, Simon, Ouvrard and Bourgoin 2007: 612–614, 618–619, 621, Fig 1, Figs 3–6, Table 1 (comp. note) Arizona

***P. rufipes* Davis, 1920**
Platypedia rufipes Poole, Garrison, McCafferty, Otte and Stark 1997: 333 (listed) North America
Platypedia rufipes Sanborn 1999a: 54 (type material, listed)
Platypedia rufipes Sanborn 1999b: 2 (listed) North America

***P. scotti* Davis, 1935** = *Platypedia sierra* Wymore, 1935
Platypedia scotti Hennessey 1990: 471 (type material)
Platypedia scotti Poole, Garrison, McCafferty, Otte and Stark 1997: 333 (listed) Equals *Platypedia sierra* North America
Platypedia scotti Sanborn 1999a: 54–55 (type material, listed, comp. note)
Platypedia sierra Sanborn 1999a: 54 (type material, listed)
Platypedia scotti Sanborn 1999b: 2 (listed) North America

***P. similis* Davis, 1920** = *Platypedia areolata similis* = *Platypedia plumbea* Van Duzee nom. nud. (partim)
Platypedia similis Poole, Garrison, McCafferty, Otte and Stark 1997: 333 (listed) North America
Platypedia plumbea Sanborn 1999a: 56 (comp. note)
Platypedia similis Sanborn 1999a: 54–55 (type material, listed)
Platypedia similis Sanborn 1999b: 2 (listed) North America

***P. sylvesteri* Simons, 1953**
Platypedia sylvesteri Poole, Garrison, McCafferty, Otte and Stark 1997: 333 (listed) North America
Platypedia sylvesteri Sanborn 1999a: 55 (type material, listed)
Platypedia sylvesteri Sanborn 1999b: 2 (listed) North America

***P. tomentosa* Davis, 1942**
Platypedia tomentosa Poole, Garrison, McCafferty, Otte and Stark 1997: 333 (listed) North America
Platypedia tomentosa Sanborn 1999a: 55 (type material, listed)
Platypedia tomentosa Sanborn 1999b: 2 (listed) North America

***P. usingeri* Simons, 1953**
Platypedia usingeri Poole, Garrison, McCafferty, Otte and Stark 1997: 333 (listed) North America
Platypedia usingeri Sanborn 1999a: 55 (type material, listed)
Platypedia usingeri Sanborn 1999b: 2 (listed) North America

***P. vanduzeei* Davis, 1920**
Platypedia vanduzeei Poole, Garrison, McCafferty, Otte and Stark 1997: 333 (listed) North America
Platypedia vanduzeei Sanborn 1999a: 55 (type material, listed)
Platypedia vanduzeei Sanborn 1999b: 2 (listed) North America

***Neoplatypedia* Davis, 1920** = *Neoplaytpedia* (sic)
Neoplatypedia Strümpel 1983: 18 (listed)
Neoplatypedia Hayashi 1984: 26 (comp. note)
Neoplatypedia Arnett 1985: 218 (diversity) California, Arizona, Oregon, Utah, Idaho
Neoplatypedia Boulard 1988f: 64 (comp. note)
Neoplatypedia Duffels 1988d: 75 (comp. note) North America
Neoplatypedia Duffels 1993: 1225 (diversity, comp. note)
Neoplatypedia Poole, Garrison, McCafferty, Otte and Stark 1997: 251 (listed) North America
Neoplatypedia Sanborn 1999b: 2 (listed) North America
Neoplatypedia Arnett 2000: 299 (diversity) California, Arizona, Oregon, Utah, Idaho
Neoplatypedia Boulard 2002b: 197 (sound system, comp. note) North America
Neoplatypedia Kondratieff, Ellingson and Leatherman 2002: 14, 47, 54 (key, listed, diagnosis, comp. note) Colorado
Neoplatypedia Sueur 2002b: 122, 124 (comp. note)
Neoplatypedia Sanborn 2004d: 2057 (comp. note)
Neoplatypedia Moulds 2005b: 385, 387, 416 (higher taxonomy, listed, comp. note)
Neoplatypedia Boulard 2006g: 16–17, 161 (comp. note) North America
Neoplatypedia Shiyake 2007: 9, 117 (listed, comp. note)
Neoplatypedia Sanborn 2008d: 3469 (comp. note)
Neoplatypedia Wei, Tang, He and Zhang 2009: 337 (comp. note)
Neoplatypedia Boulard 2010b: 336 (comp. note)

***N. ampliata* (Van Duzee, 1915)** = *Platypedia ampliata*
Neoplatypedia ampliata Poole, Garrison, McCafferty, Otte and Stark 1997: 332 (listed) Equals *Platypedia ampliata* North America
Neoplatypedia ampliata Sanborn 1999a: 53 (type material, listed)
Neoplatypedia ampliata Sanborn 1999b: 2 (listed) North America

Platypedia ampliata Kondratieff, Ellingson and Leatherman 2002: 45 (type species of *Neoplatypedia*)
Neoplatypedia ampliata Shiyake 2007: 9 116–117, Fig 206 (illustrated, listed, comp. note)

***N. constricta* Davis, 1920**
Neoplatypedia constricta Sugden 1940: 125 (song, habitat, comp. note) Utah
Neoplatypedia constricta Boulard 1990b: 191, 193, Plate XXIV, Fig 2 (sound system, illustrated, comp. note)
Neoplatypedia constricta Hennessey 1990: 471 (type material)
Neoplatypedia constricta Poole, Garrison, McCafferty, Otte and Stark 1997: 332 (listed) North America
Neoplatypedia constricta Gogala and Trilar 1998a: 8 (comp. note) North America
Neoplatypedia constricta Sanborn 1999a: 53, 55 (type material, listed, comp. note)
Neoplatypedia constricta Sanborn 1999b: 2 (listed) North America
Neoplatypedia constricta Kondratieff, Ellingson and Leatherman 2002: 8, 12–13, 54–55, 69, 74, 81, Fig 17, Fig 37, Figs 55–56, Tables 1–2 (diagnosis, illustrated, listed, distribution, song, comp. note) Colorado
Neoplatypedia constricta Boulard 2006g: 160, Fig 115 (illustrated, comp. note) North America

Tribe Ydiellini Boulard, 1973
Ydiellini Boulard 1986d: 235 (n. comb., comp. note)
Ydiellini Moulds 2005b: 389, 427 (n. status, listed, comp. note)

Subtribe Ydiellina Boulard, 1993
= Ydiellaria
Ydiellaria Boulard 1993d: 92 (n. subtribe, comp. note)
Ydiellaria Boulard 2002b: 205 (comp. note)
Ydiellaria Moulds 2005b: 389 (history, listed, comp. note)
Ydiellaria Lee and Hill 2010: 278 (comp. note) Equals Ydiellina

Ydiella Boulard, 1973 to *Maroboduus* Distant, 1920
Ydiella gilloni Boulard, 1973 to *Maroboduus fractus* Distant, 1920
Ydiella maxicapsulifera Boulard, 1986 to *Maroboduus maxicapsulifera*

***Maroboduus* Distant, 1920** = *Ydiella* Boulard, 1973
Ydiella Medler 1980: 73 (listed) Nigeria
Ydiella Strümpel 1983: 18, 109 (listed, comp. note)
Ydiella Hayashi 1984: 26 (comp. note)
Ydiella Boulard 1986c: 203 (stridulatory apparatus, comp. note)
Ydiella Boulard 1986d: 235 (comp. note) Ivory Coast
Maroboduus Boulard 1988f: 64 (comp. note) Equals *Ydiella*
Maroboduus Boulard 1990b: 149 (comp. note) Equals *Ydiella*
Maroboduus Boulard 1993d: 91 (comp. note) Equals *Ydiella*
Ydiella Duffels 1993: 1224 (comp. note)
Maroboduus Duffels 1993: 1224–1225 (comp. note) Equals *Ydiella*
Ydiella Chou, Lei, Li, Lu and Yao 1997: 21 (comp. note)
Ydiella Boulard 2002b: 197 (sound system, comp. note) Africa
Maroboduus Boulard 2002b: 204 (stridulation, comp. note) Equals *Ydiella*
Maroboduus Boulard 2005d: 349 (stridulatory apparatus, comp. note) Equals *Ydiella*
Ydiella Moulds 2005b: 385, 389 (higher taxonomy, listed, comp. note) Equals *Maroboduus*
Maroboduus Moulds 2005b: 389, 414 (higher taxonomy, listed, comp. note) Equals *Ydiella*
Maroboduus Boulard 2006g: 16–17, 154 (stridulation, comp. note) Equals *Ydiella*
Maroboduus Wei, Tang, He and Zhang 2009: 337–338 (comp. note) Equals *Ydiella*
Maroboduus Boulard 2010b: 336 (comp. note) Equals *Ydiella*
Maroboduus Lee and Hill 2010: 278 (synonymy, comp. note) Equals *Ydiella* Africa
Ydiella Lee and Hill 2010: 278 (comp. note)

***M. fractus* Distant, 1920** = *Ydiella gilloni* Boulard, 1973 = *Maroboduus gilloni*
Ydiella gilloni Hayashi 1984: 26 (comp. note)
Ydiella gilloni Medler 1980: 73 (listed) Nigeria, Ivory Coast
Ydiella gilloni Boulard 1986d: 235 (type species of *Ydiella*, comp. note) Ivory Coast
Ydiella gilloni Boulard 1988f: 53 (type species of Ydiellinae)
Maroboduus gilloni Boulard 1990b: 163, Plate XX, Figs 1–4 (sound system, comp. note) Equals *Ydiella gilloni*
Maroboduus fractus Boulard 1993d: 91 (comp. note) Equals *Ydiella gilloni*
Maroboduus fractus Duffels 1993: 1224 (type species of *Maroboduus*, comp. note)
Ydiella gillioni Duffels 1993: 1224 (stridulatory apparatus, comp. note)
Maroboduus gillioni Duffels 1993: 1224 (comp. note)
Ydiella gilloni Chou, Lei, Li, Lu and Yao 1997: 21, Figs 3–10 (stidulatory apparatus illustrated, comp. note)
Maroboduus fractus Boulard 2002b: 205 (comp. note) Equals *Ydiella gilloni*

Maroboduus fractus Moulds 2005b: 389 (type species of *Maroboduus*, history, phylogeny, listed, comp. note)
Maroboduus fractus Boulard 2006g: 154–155, Fig 112 (stridulation, illustrated, comp. note) Equals *Ydiella gilloni*

M. gilloni (Boulard, 1973) to *Maroboduus fractus* Distant, 1920

***M. maxicapsulifera* (Boulard, 1986)** = *Ydiella maxicapsulifera*
Ydiella maxicapsulifera Boulard 1986d: 234–235, 237–238, Table I, Figs 34–36 (n. sp., described, illustrated, comp. note) Ivory Coast
Ydiella maxicapsulifera Boulard 1990b: 132–133, Plate XV, Fig 5 (genitalia, comp. note)
Maroboduus maxicasulifera Boulard 1993d: 91 (comp. note)
Maroboduus maxicapsulifera Boulard 2002b: 205 (comp. note)
Maroboduus maxicapsulifera Boulard 2006g: 154–155 (stridulation, comp. note)

Tribe Tibicinini Distant, 1905 = Tibicenini (sic) = Tibicini (sic) = Tibicinae (sic)
Tibicinaria Wu 1935: 25 (listed) China
Tibicenini (sic)Varley 1939: 99 (comp. note)
Tibicinini Boulard 1981a: 37 (listed)
Tibicinini Schedl 1986a: 3 (listed, key) Istria
Tibicinini Boulard 1988f: 35–36, 44–45 (comp. note)
Tibicinini Boulard 1992c: 63 (comp. note)
Tibicinini Simon, Frati, Beckenbach, Crespi, Liu and Flook 1994: 655 (comp. note)
Tibicinini Boulard and Mondon 1995: 65–66 (key, listed, comp. note) France
Tibicini (sic) Moulds 1996a: 20 (comp. note) Equals Tibicinini
Tibicinini Boulard 1997c: 101–102, 106, 112, 117 (comp. note)
Tibicinini Chou, Lei, Li, Lu and Yao 1997: 46, 89 (key, synonymy, listed, described) Equals Tibicinaria
Tibicinini Novotny and Wilson 1997: 437 (listed)
Tibicinini Moulds 1999: 1 (comp. note)
Tibicinini Sanborn 1999a: 45 (listed)
Tibicinini Sanborn 1999b: 2 (listed) North America
Tibicinini Boulard 2000i: 21, 24, 45 (listed, key)
Tibicinini Schedl 2000: 264 (listed)
Tibicinae (sic) Boulard 2001b: 26 (comp. note)
Tibicinini Boulard 2001b: 21, 23, 32, 37 (comp. note)
Tibicinini Puissant and Sueur 2001: 432 (listed) Corsica
Tibicinini Sueur 2001: 41, Table 1 (listed)
Tibicinini Kondratieff, Ellingson and Leatherman 2002: 13, Table 2 (listed) Colorado
Tibicinini Schedl 2003: 426 (comp. note)
Tibicinini Sanborn, Heath, Heath, Noriega and Phillips 2004: 282 (comp. note)
Tibicinini Sueur, Puissant, Simões, Seabra, Boulard and Quartau 2004: 180 (listed) Portugal
Tibicinini Moulds 2005b: 377, 386, 388, 397, 414–417, 419, 425, 437, Fig 58 (history, phylogeny, listed, comp. note)
Tibicinini Puissant 2006: 19 (listed) France
Tibicinini Sanborn 2007b: 38 (listed) Mexico
Tibicinini Shiyake 2007: 7, 91 (listed, comp. note)
Tibicinini Sueur and Puissant 2007b: 62 (listed) France
Tibicinini Sueur, Vanderpool, Simon, Ouvrard and Bourgoin 2007: 612–613 (comp. note)
Tibicinini Demir 2008: 476 (listed) Turkey
Tibicinini Fonseca, Serrão, Pina-Martins, Silva, Mira, et al. 2008: 24 (listed)
Tibicinini Pham and Yang 2009: 4, 15, Table 2 (listed) Vietnam
Tibicinini Simon, Frati, Beckenbach, Crespi, Liu and Flook 1994: 655 (comp. note)
Tibicinini Mozaffarian and Sanborn 2010: 80 (listed)
Tibicinini Wei and Zhang 2010: 38 (comp. note)

Subtribe Tibicinina Distant, 1905 = Tibicinaria
Tibicinaria Boulard 1997c: 106, 112, 117 (comp. note)
Tibicinaria Boulard 2000i: 21 (listed)
Tibicinaria Boulard 2001b: 106, 112, 117 (comp. note)

***Tibicina* Kolenati, 1857** = *Cicada (Tibicina)* = *Tibicicina* (sic) = *Tebicina* (sic) = *Tibicea* (sic) = *Tibicena* (sic) = *Tybicina* (sic) = *Tibucina* (sic)
Cicada (sic) Kölliker 1888: 700 (comp. note)
Tibicen (sic) Wu 1935: 25 (listed) Equals *Tibicina* China
Tibicina Gómez-Menor 1951: 61 (key) Spain
Tibicen (sic) Soós 1956: 412 (listed) Romania
Tibicina Boulard 1981a: 37 (listed) Algeria
Paharia (sic) Soper 1981: 55 (fungal host, comp. note)
Tibicina Koçak 1982: 151 (listed, comp. note) Turkey
Tibicina Boulard 1983a: 309 (comp. note)
Tibicina Heath 1983: 12 (comp. note) Palearctic
Tibicina Strümpel 1983: 17 (listed)
Tibicina Boulard 1984c: 168, 175–176, 178–179 (status, type genus of Tibicininae, comp. note)
Tibicina Hayashi 1984: 26 (comp. note)

Tibicina Melville and Sims 1984: 163–16, 180 (status, history)
Tibicina Boulard 1985d: 212–213 (status, comp. note)
Tibicina Hamilton 1985: 211 (status, comp. note)
Tibicina Boulard 1985e: 1021(coloration, comp. note)
Tibicina Claridge 1985: 301 (song, comp. note)
Tibicina Lauterer 1985: 214 (status)
Tibicina Popov 1985: 38–39 (acoustic system, comp. note)
Tibicina Moulds 1986: 40 (comp. note)
Tibicina Boulard 1988f: 8, 30, 36–37, 44–45, 71 (type genus of Tibicinini, Tibicininae and Tibicinidae, comp. note)
Tibicina Boulard and Riou 1988: 349, 351 (comp. note)
Tibicina Duffels 1988d: 75 (comp. note) Spain
Tibicina Boulard 1990b: 172 (comp. note)
Tibicina Quartau and Rebelo 1990: 1 (comp. note) Portugal
Tibicina Popov, Aronov and Sergeeva 1991: 282 (comp. note)
Tibicina Popov, Arnov and Sergeeva 1992: 151 (comp. note)
Tibicina Quartau 1993: 93 (comp. note) Portugal
Tibicina Remane and Wachman 1993: 75 (comp. note)
Tibicina Niehuls and Simon 1994: 256 (comp. note) Germany
Tibicina Boulard 1995a: 38 (listed)
Tibicina Boulard and Mondon 1995: 19, 65 (key, comp. note) France
Tibicina Niehuls 1995: 39 (comp. note) Germany
Tibicina sp. Quartau 1995: 33 (illustrated) Portugal
Tibicina Quartau 1995: 38 (comp. note) Portugal
Tibicina Fonseca 1996: 28–29 (comp. note)
Tibicina Holzinger 1996: 505 (listed) Austria
Tibicina Boulard 1997c: 84, 99–101, 104, 106–107, 117, 121 (type genus of Tibicinini, Tibicininae and Tibicinidae, comp. note)
Tibicina Chou, Lei, Li, Lu and Yao 1997: 281 (comp. note)
Tibicina Holzinger, Frölich, Günthart, Pauterer, Nickel et al. 1997: 49 (listed) Europe
Tibicina Quartau, Coble, Viegas-Crespo, Rebelo, Ribeiro, et al. 1997: 138 (comp. note) Portugal
Tibicina Moulds 1999: 1 (comp. note)
Tibicina Quartau, Rebelo and Simões 1999: 71 (comp. note)
Tibicina Boulard 2000b: 18 (comp. note)
Tibicina Boulard 2000i: 21, 25 (key, listed, comp. note) France
Tibicina Prokop and Boulard 2000: 128, 130 (comp. note)
Tibicina Puissant and Sueur 2000: 70 (comp. note) France
Tibicina Schedl 2000: 260, 264 (key, listed)
Tibicina Boulard 2001b: 5, 20, 22, 25–27, 32, 37, 41 (type genus of Tibicinini, Tibicininae and Tibicinidae, comp. note)
Tibicina Cooley 2001: 758 (comp. note) Europe
Tibicina Cooley and Marshall 2001: 847 (comp. note) Europe
Tibicina Puissant 2001: 147 (comp. note)
Tibicina Quartau, Simões, Rebelo and André 2001: 401 (comp. note) Portugal
Tibicina Schedl 2001a: 1287 (comp. note)
Tibicina Sueur and Puissant 2001: 22–23 (comp. note) France
Tibicina Boulard 2002b: 193 (coloration, comp. note)
Tibicina Puissant and Sueur 2002: 197–199 (comp. note) Slovenia
Tibicina spp. Puissant and Sueur 2002: 197–200 (comp. note) Turkey
Tibicina Sueur 2002b: 124 (comp. note)
Tibicina spp. Sueur and Aubin 2002b: 135 (comp. note)
Tibicina Sueur and Puissant 2002: 477–478, 480, 483 (comp. note) Palaearctic, France
Tibicina spp. Sueur and Puissant 2002: 483 (comp. note)
Tibicina Sueur 2003: 2932, 2940, 2942 (comp. note) Palaearctic, France
Tibicina spp. Sueur 2003: 2934, 2940, 2943, 2945 (comp. note) France
Tibicina Sueur and Puissant 2003: 121–123 (diversity, comp. note)
Tibicina Sueur, Puissant and Pillet 2003: 105 (comp. note)
Tibicina Sueur and Sanborn 2003: 340–342 (comp. note) France
Tibicina spp. Sueur and Sanborn 2003: 340–342 (comp. note)
Tibicina Holzinger, Kammerlander and Nickel 2003: 476, 485 (key, listed, comp. note) Europe
Tibicina Lee and Hayashi 2003a: 158 (comp. note)
Tibicina Quartau and Simões 2003: 113, 117–118 (comp. note) Portugal
Tibicina spp. Quartau and Simões 2003: 118 (comp. note) Portugal
Tibicina Schedl 2003: 426 (comp. note)
Tibicina Sueur and Aubin 2003b: 481–483, 485–488 (comp. note) Palaearctic, France, Spain, Portugal, Switzerland
Tibicina spp. Sueur and Aubin 2003b: 483–486, 488 (comp. note)
Tibicina Sueur and Aubin 2004: 218–223 (comp. note) Palaearctic, France, Spain, Portugal, Switzerland

Tibicina spp. Sueur and Aubin 2004: 218 (comp. note)
Tibicina Boulard 2005d: 338, 341 (comp. note) Mediterranean
Tibicina Moulds 2005b: 384, 389, 410, 414–417, 428 (type species of Tibicinidae, type species of Tibicininae, history, phylogeny, listed, comp. note)
Tibicina Puissant 2005: 309, 312 (comp. note)
Tibicina spp. Puissant 2005: 311–312 (comp. note)
Tibicina Reynaud 2005: 143 (comp. note) France
Tibicina Sanborn 2005c: 117 (comp. note)
Tibicina Sueur 2005: 209 (song, comp. note)
Tibicina Boulard 2006g: 106, 108, 129 (comp. note)
Tibicina Puissant 2006: 35, 46–50, 53–54, 56, 69 (comp. note) France
Tibicina spp. Seabra, Pinto-Juma and Quartau 2006: 843 (comp. note)
Tibicina Shiyake 2007: 7, 91 (listed, comp. note)
Tibicina spp. Sueur and Puissant 2007a: 127 (comp. note)
Tibicina spp. Sueur and Puissant 2007b: 64 (comp. note)
Tibicina Sueur, Vanderpool, Simon, Ouvrard and Bourgoin 2007: 612–613, 616–617, 619–620, 623–624 (comp. note) Palaearctic, France, Spain, Portugal, Switzerland
Tibicina spp. Sueur, Vanderpool, Simon, Ouvrard and Bourgoin 2007: 612, 620, 623–624 (comp. note)
Tibicina Demir 2008: 476 (listed) Equals *Euryphara* Turkey
Tibicina Fonseca, Serrão, Pina-Martins, Silva, Mira, et al. 2008: 25, 27–30 (comp. note)
Tibicina spp. Fonseca, Serrão, Pina-Martins, Silva, Mira, et al. 2008: 30 (comp. note)
Tibicina sp. Zamanian, Mehdipour and Ghaemi 2008: 2070 (comp. note)
Tibicina Ahmed and Sanborn 2010: 42 (distribution) central Asia, Europe, Russia
Tibicina Kemal and Koçak 2010: 1, 3, 5 (diversity, listed, comp. note) Turkey
Tibicina Mozaffarian and Sanborn 2010: 80 (distribution, listed) Europe, Russia, Central Asia

T. baetica (Rambur, 1840) to *Tibicina tomentosa* (Olivier, 1790)

***T. casyapae* (Distant, 1888)** = *Tibicen casyapae* = *Paharia casyapae*
Paharia casyapae Varley 1939: 99 (comp. note)
Paharia casyapae Soper 1981: 55 (fungal host, comp. note) Afghanistan
Paharia casyapae Srivastava and Bhandari 2004: 1479 (comp. note) India
Tibicina casyapae Ahmed and Sanborn 2010: 42 (synonymy, distribution) Equals *Tibicen casyapae* Pakistan, Afghanistan, India

***T. cisticola* (Hagen, 1855)** = *Cicada cisticola* = *Tibicen cisticola* = *Tettigia cisticola*
Cicada cisticola Hagen Boulard 1984b: 46 (history of species, comp. note) France
Tibicina cisticola Boulard 1984c: 171 (status, comp. note)
Tibicina cisticola Nast 1987: 570–571 (distribution, listed) France, Spain, Italy
Tibicina cisticola Boulard 1988f: 31 (comp. note)
Tibicina cisticola Boulard 1997c: 100 (comp. note)
Tibicina cisticola Guglielmino, d'Urso and Alma 2000: 167 (listed, distribution) Sardinia, Spain, France, Corsica
Tibicina cisticola Boulard 2000i: 68 (comp. note)
Tibicina cisticola Boulard 2001b: 21 (comp. note)
Cicada cisticola Puissant 2006: 58 (comp. note)

T. cisticola leferstifi Boulard, 1980 to *Tibicina corsica corsica* var. *lerestifi*

***T. corsica corsica* (Rambur, 1840)** = *Cicada corsica* = *Cicadetta corsica* = *Melampsalta corsica* = *Melampsalta (Melampsalta) corsica* = *Tibicen corsicus* = *Tibicina corsica* = *Tibicen tomentosus* (nec Olivier)
Tibicina corsica Boulard 1985b: 59, 63 (illustrated, characteristics, comp. note) France
Tibicina corsica Boulard 1985e: 1021 (coloration, comp. note) France
Tibicina corsica Nast 1987: 570–571 (distribution, listed) France, Spain, Italy
Tibicina corsica Boulard 1988d: 152, 154, Fig 11 (wing abnormalities, illustrated, comp. note) Corsica
Tibicina corsica Boulard 1990b: 172–173, Fig 10 (sonogram, comp. note)
Tibicina corsica Boulard 1995a: 6, 13, 51–52, Fig $S18_1$, Fig $S18_2$ (song analysis, sonogram, listed, comp. note) Corsica
Tibicina corsica Boulard and Mondon 1995: 30, 32, 66–67, 79, 159 (illustrated, sonogram, listed, comp. note) Equals *Cicada cisticola* (nec Fairmaire) France
Tibicina corsica corsica Boulard 2000b: 20–21 (n. comb., sonogram, comp. note) Corsica, Sardinia
Tibicina corsica corsica Boulard 2000i: 64–67, 70, 123, Plate I, Plate I-J, Map 9 (synonymy, illustrated, song, sonogram, oscillogram, frequency spectrum, distribution, ecology, comp. note) Equals *Cicada corsica* Equals *Cicada cisticola* Equals *Tibicina cisticola* Equals *Tibicen cisticola* Equals *Cicadetta corsica* Equals *Melampsalta? corsica* Equals *Melampsalta (Melampsalta) corsica* Equals

Tibicen corsicus Equals *Tibicen tomentosus* (sic) (partim) France, Corsica
Tibicina corsica Boulard 2000i: 66, 68 (comp. note) France, Corsica
Tibicina corsica Guglielmino, d'Urso and Alma 2000: 167 (listed, distribution) Sardinia, Spain, France, Corsica
Tibicina corsica corsica Puissant and Sueur 2001: 430–432, 434–435, Fig 2, Fig 6, Fig 9 (illustrated, distribution, comp. note) Corsica
Tibicina corsica Quartau, Simões, Rebelo and André 2001: 402 (comp. note)
Tibicina corsica Sueur and Puissant 2001: 18, 22 (illustrated, comp. note) France
Tibicina corsica corsica Sueur 2001: 41, Table 1 (listed)
Tibicina corsica corsica Sueur and Puissant 2002: 477–483, Tables 2–5 (ecology, distribution, comp. note) France
Tibicina corsica corsica Sueur 2003: 2932, 2938, 2940, 2942, 2944, Fig 7 (acoustic behavior, aggression, comp. note) France, Corsica
Tibicina corsica Sueur 2003: 2942 (comp. note) France
Tibicina corsica corsica Sueur and Aubin 2003b: 482, 484, 486–490, Figs 5–9, Fig 11, Tables 1–4, Table 6 (song, comp. note) France, Corsica
Tibicina corsica corsica Sueur, Puissant and Pillet 2003: 105 (comp. note)
Tibicina corsica corsica Sueur and Aubin 2004: 218–222, Tables 1–3 (courtship song, comp. note) France, Corsica
Tibicina corsica corsica Boulard 2005d: 341 (comp. note) France
Tibicina corsica corsica Puissant and Defaut 2005: 117–118, 126, Table 1, Table 11 (listed, habitat association, distribution, comp. note) France, Corsica
Tibicina corsica corsica Boulard 2006g: 53, 135, Fig 91 (sonogram, comp. note) France
Tibicina corsica corsica Puissant 2006: 19, 21, 25, 27–28, 32, 50, 55–58, 60–61, 67, 77, 90, 96, 115–118, 120, 122, 125–127, 129, 131–132, 179–180, 188, Fig 8, Figs 27–30, Tables 1–2, Appendix Table 1, Appendix Table 11 (illustrated, synonymy, distribution, habitat association, ecology, listed, comp. note) Equals *Cicada cisticola* France, Corsica
Tibicina corsica Puissant 2006: 21–22, 40 (illustrated, comp. note) France
Tibicina corsica corsica Sueur and Puissant 2007b: 62 (listed) France, Corsica
Tibicina corsica corsica Sueur, Vanderpool, Simon, Ouvrard and Bourgoin 2007: 612–615, 617–624, Figs 1–6, Tables 1–2 (phylogeny, comp. note) France, Corsica
Tibicina corsica Zamanian, Mehdipour and Ghaemi 2008: 2070 (comp. note)

***T. corsica corsica* var. *lerestifi* Boulard, 1980** = *Tibicina cisticola* var. *lerefstifi* = *Tibicina corsica* var. *lerestifi* = *Tibicina corsica larestifi* (sic)
Tibicina cisticola larestifi (sic) Nast 1982: 311 (listed) France, Corsica
Tibicina cisticola larestifi (sic) Nast 1987: 570–571 (distribution, listed) France
Tibicina corsica var. *lerestifi* Boulard and Mondon 1995: 66 (listed) France
Tibicina corsica corsica var. *lerestifi* Boulard 2000i: 67 (described, comp. note) Equals *Tibicina cisticola* var. *lerestifi* France, Corsica

***T. corsica fairmairei* Boulard, 1984** = *Cicada cisticola* Fairmaire, 1855 = *Tettigia cisticola* = *Tibicen cisticola* = *Tibicina cisticola* = *Tibicina fairmairei* = *Tibicina faimairei* (sic) = *Cicada cisticola* (nec Hagen)
Tibicina cisticola Villiers 1977: 166, Plate XIV, Fig 27 (illustrated, listed, comp. note) France
Cicada cisticola Fairmaire Boulard 1984b: 45–46 (history of species, comp. note) France
Tibicina fairmairei Boulard 1984b: 46, Figs 1–3 (nom. nov. pro *Cicada cisticola* Fairmaire, 1884 nec *C. cisticola* Hagen, 1855, distribution, illustrated, host plants) France
Tibicina fairmairei Boulard 1985b: 62, 63 (characteristics, comp. note) France
Tibicina fairmairei Boulard 1985e: 1021 (coloration, comp. note) France
Tibicina fairmairei Nast 1987: 570–571 (distribution, listed) France
Tibicina fairmairei Boulard 1988a: 9–10, 12, Table (illustrated, comp. note) France
Tibicina fairmairei Boulard 1988d: 152 (wing abnormalities, comp. note) France
Tibicina fairmairei Boulard 1988f: 31 (comp. note) Equals *Tibicina cristicola*
Tibicina fairmairei Quartau and Fonseca 1988: 367, 372–373 (synonymy, distribution, comp. note) Equals *Cicada cisticola* Portugal, France, Corsica, Italy, Sardinia, Spain
Tibicina fairmairei Boulard 1990b: 72, 172, 174, 177, 205, Fig 10 (sonogram, comp. note)
Tibicina fairmairei Boulard 1992c: 65 (comp. note)
Tibicina fairmairei Hidvegi and Baugnée 1992: 123 (comp. note)
Tibicina fairmairei Boulard 1995a: 6, 13, 53–54, Fig S19$_1$, Fig S19$_2$, Fig 20 (illustrated, song analysis, sonogram, listed, comp. note) France
Tibicina faimairei (sic) Boulard and Mondon 1995: 25 (comp. note) France
Tibicina fairmairei Boulard and Mondon 1995: 30, 32, 53, 59, 66–67, 159 (illustrated, listed, comp. note) Equals *Cicada cisticola* (nec Hagen) France

Tibicina fairmairei Boulard 1997c: 100 (comp. note) Equals *Tibicina cristicola*
Tibicina corsica fairmairei Boulard 2000b: 20–21, Fig 12 (n. comb., sonogram, comp. note)
Cicada corsica fairmairei Boulard 2000i: 63 (comp. note)
Tibicina corsica fairmairei Boulard 2000i: 64, 67–72, 124, Fig 15, Plate J, Plate I-J, Map 10 (synonymy, illustrated, song, sonogram, oscillogram, frequency spectrum, distribution, ecology, comp. note) Equals *Cicada cisticola* Equals *Tibicina fairmairei* Equals *Tibicina cisticola* Equals *Tibicen cisticola* Equals *Tettigia cisticola* France, Corsica
Tibicina fairmairei Boulard 2000i: 68 (comp. note)
Tibicina faimairei (sic) Boulard 2001b: 21 (comp. note) Equals *Tibicina cristicola*
Tibicina corsica fairmairei Puissant and Sueur 2001: 433 (comp. note)
Tibicina fairmairei Quartau, Simões, Rebelo and André 2001: 401 (comp. note)
Tibicina fairmairei Sueur 2001: 35 (comp. note)
Tibicina corsica fairmairei Sueur 2001: 41, Table 1 (listed, comp. note)
Tibicina faimarei (sic) Boulard 2002b: 205 (comp. note)
Tibicina corsica fairmairei Sueur and Puissant 2002: 477–483, Fig 3, Tables 2–5 (ecology, distribution, comp. note) France
Tibicina fairmairei Quartau and Simões 2003: 117 (comp. note)
Tibicina corsica fairmairei Sueur 2003: 2932–2938, 2940, 2942–2944, Fig 3, Figs 6–7, Tables 1–2 (sonogram, oscillogram, acoustic behavior, aggression, comp. note) France
Tibicina corsica fairmairei Sueur and Aubin 2003b: 482, 484–490, Figs 5–9, Fig 11, Tables 1–4, Table 6 (song, comp. note) France
Tibicina corsica fairmairei Sueur, Puissant and Pillet 2003: 105 (comp. note)
Tibicina corsica fairmairei Sueur and Aubin 2004: 218–223, Fig 2, Tables 1–3 (courtship song, sonogram, oscillogram, comp. note) France
Tibicina corsica fairmairei Sueur, Puissant, Simões, Seabra, Boulard and Quartau 2004: 180 (comp. note) Corsica
Tibicina corsica fairmairei Boulard 2005d: 341 (comp. note) France
Tibicina corsica fairmairei Puissant and Defaut 2005: 116, 118, 120–121, Table 1, Table 4 (listed, habitat association, distribution, comp. note) France
Tibicina corsica fairmairei Boulard 2006g: 53, 66 (comp. note) France
Tibicina corsica fairmairei Puissant 2006: 20, 25–26, 28–29, 32, 50–51, 54–55, 58–63, 66, 70, 90, 96, 111–112, 135, 138, 140, 178, 180, 182–183, Fig 8, Figs 31–33, Tables 1–2, Appendix Table 1, Appendix Table 4 (illustrated, distribution, habitat association, ecology, listed, comp. note) France
Tibicina corsica fairmairei Sueur, Vanderpool, Simon, Ouvrard and Bourgoin 2007: 612–615, 617–624, Figs 1–6, Tables 1–2 (phylogeny, comp. note) France
Tibicina fairmairei Thouvenot 2007b: 281 (comp. note) Corsica

***T. fieberi* Puissant & Sueur, 2002**

Tibicina fieberi Puissant and Sueur 2002: 148–155, Figs 1–5 (n. sp., described, illustrated, comp. note) Turkey, Asia Minor
Tibicina fieberi Sueur, Vanderpool, Simon, Ouvrard and Bourgoin 2007: 612–613 (comp. note) Turkey

***T. garricola* Boulard, 1983**

Tibicina garricola Boulard 1983a: 309–312, Figs 1–5 (n. sp., described, illustrated, sonogram, comp. note) France
Tibicina garricola Boulard 1985b: 60, 63 (illustrated, characteristics, comp. note) France
Tibicina garricola Nast 1987: 570–571 (distribution, listed) France
Tibicina garricola Boulard 1988a: 11–12, Table (illustrated, comp. note) France
Tibicina garricola Boulard 1988d: 150–151, 154, 156, Fig 3, Fig 6 (wing abnormalities, illustrated, comp. note) France
Tibicina garricola Boulard 1988g: Plate III, Fig 10 (illustrated, comp. note) France
Tibicina garricola Boulard 1990b: 172–173, 177, 179, 205, Fig 10 (sonogram, comp. note)
Tibicina garricola Boulard 1992c: 65 (comp. note)
Tibicina garricola Boulard 1995a: 13, 61–62, Fig 16, Fig S16$_1$, Fig S16$_2$, Fig S16$_3$ (illustrated, song analysis, sonogram, listed, comp. note) France
Tibicina garricola Boulard and Mondon 1995: 25, 30, 32, 39, 43, 58, 66–67, 71, 76, 159 (illustrated, sonogram, listed, comp. note) France
Tibicina garricola Boulard 2000i: 56–59, 99, 122, Fig 22e, Plates G-G', Map 7 (synonymy, illustrated, song, sonogram, oscillogram, frequency spectrum, distribution, ecology, comp. note) France, Spain, Portugal
Tibicina garricola Quartau, Simões, Rebelo and André 2001: 401–407, 409, Figs 1–2, Figs, 4–15 (described, illustrated, ecology, comp. note) Portugal

Tibicina garricola Sueur 2001: 41, Table 1 (listed)
Tibicina garricola Sueur and Puissant 2001: 22 (comp. note) France
Tibicina garricola Boulard 2002b: 175, 198, 206, Fig 3 (illustrated, comp. note) France
Tibicina garricola Sueur and Puissant 2002: 477–483, Fig 4, Tables 2–5 (ecology, distribution, comp. note) France
Tibicina garricola Quartau and Simões 2003: 114–118, Fig 1, Figs 3–4, Table 1 (song, sonogram, oscillogram, comp. note) Portugal
Tibicina garricola Sueur 2003: 2932, 2934, 2938, 2942, Table 1 (acoustic behavior, aggression, comp. note) France
Tibicina garricola Sueur and Aubin 2003b: 482–490, Figs 1–9, Fig 11, Tables 1–4, Table 6 (song, sonogram, oscillogram, comp. note) France, Portugal
Tibicina garricola Sueur, Puissant and Pillet 2003: 105 (comp. note)
Tibicina garricola Sueur and Aubin 2004: 217–218, 221–222, Fig 2, Tables 1–3 (courtship song, sonogram, oscillogram, comp. note) France
Tibicina garricola Sueur, Puissant, Simões, Seabra, Boulard and Quartau 2004: 177–182, Figs 1–2 (distribution, ecology, song, sonogram, oscillogram, comp. note) Portugal, France
Tibicina garricola Puissant and Defaut 2005: 116–120, 122–125, 127, Table 1, Table 3, Tables 6–7 (listed, habitat association, distribution, comp. note) France
Tibicina garricola Boulard 2006g: 41, 61–62, 66, 70–71, 180, Fig 28 (illuatrated, sonogram, listed, comp. note)
Tibicina garricola Puissant 2006: 20, 23, 25, 28, 32, 34–35, 47–49, 56, 63–64, 68–70, 72, 75, 90, 97, 110–112, 114, 136–140, 144, 146, 148, 155, 160–162, 178–182, 184–187, 189, Fig 8, Figs 40–41, Tables 1–2, Appendix Table 1, Appendix Table 3, Appendix Tables 6–7 (illustrated, distribution, habitat association, ecology, listed, comp. note) France
Tibicina garricola Sueur, Vanderpool, Simon, Ouvrard and Bourgoin 2007: 612–615, 617–623, Figs 1–6, Tables 1–2 (phylogeny, comp. note) France, Portugal
Tibicina garricola Fonseca, Serrão, Pina-Martins, Silva, Mira, et al. 2008: 19, 24–27, 29–30, Fig 1, Table 1 (song phylogeny, comp. note) Portugal

***T. haematodes flammea* Boulard, 1982** = *Tibicina haematodes* mph. *flammea* = *Tibicina haematodes* var. *flammea*

Tibicina haematodes mph. *flammea* Boulard 1982a: 50 (n. morph, described, comp. note) France
Tibicina haematodes var. *flammea* Boulard and Mondon 1995: 66, 79 (listed, illustrated) France
Tibicina haematodes flammea Boulard 2000i: 49 (described, distribution, comp. note) Equals *Tibicina haematodes* morph *flammea* France

***T. haematodes haematodes* (Scopoli, 1763)** = *Cicada haematodes* = *Cicada haematoides* (sic) = *Cicada (Tettigonia) haematodes* = *Tettigonia haematodes* = *Tibicen haematodes* = *Tibicina haematodes* = *Tebicina* (sic) *haematodes* = *Tibicea* (sic) *haematodes* = *Tibicena* (sic) *haematodes* = *Tybicina* (sic) *haematodes* = *Tibicen haematordes* (sic) = *Tibicina haematedes* (sic) = *Tettigonia haematodes* Fabricius, 1794 = *Tettigonia sanguinea* Fabricius, 1803 = *Cicada sanguinea* = *Cicada (Tettigonia) sanguinea* = *Tibicen sanguinea* = *Cicada plebeja* Germar, 1830 (nec Scopoli) = *Tibicen plebeja* Germar = *Tibicina hematodes* (sic) = *Tibicina haematotes* (sic)

Cicada haematodes Swinton 1877: 78, 80 (sound production, comp. note) France
Cicada haematodes Swinton 1879: 82 (hearing, comp. note)
Cicada haematodes Swinton 1880: 18, 20, 226, 228 (song, comp. note)
Tibicen haematodes Swinton 1880: 220–223 (timbal, comp. note)
Cicada (Tettigonia) sanguinea Hasselt 1881: 180 (comp. note)
Cicada (Tettigonia) haematodes Hasselt 1881: 180 (comp. note)
Cicada haematodes Hasselt 1881: 190 (comp. note)
Cicada haematodes Claus 1884: 571 (comp. note)
Cicada haematodes Kölliker 1888: 691, 694, 709, Plate XLIV, Fig 4, Fig 8 (muscle structure, comp. note)
Tibicina haematodes Bachmetjew 1907: 251 (body length, comp. note)
Tibicina haematodes Swinton 1908: 379, 383 (song, comp. note) France
Tibicina haematodes Mader 1922: 179 (comp. note) Austria
Tettigonia haematodes Stadler 1922: 160 (comp. note) Germany
Tibicina haematodes Faber 1928: 231 (song, comp. note)
Tibicina haematodes Marcu 1929: 479 (comp. note) Romania
Tibicina haematodes Weber 1931: 71–72, 109, 112, Fig 74, Fig 111 (wing anatomy, sound system illustrated, comp. note) Germany
Tibicen haematodes Vogel 1938: 116–117, 121 (larva, distribution, habitat, comp. note) Germany
Tibicen haematodes Zeuner 1944: 113 (type species of *Tibicen*, comp. note)

Tibicina haematodes Castellani 1952: 15 (comp. note) Italy
Tibicina haematodes Wagner 1952: 15 (listed, comp. note) Italy
Tibicen haematodes Soós 1956: 412 (listed, distribution) Romania
Tibicina haematodes Bunescu 1959: 91, 99–101, Fig 3 (distribution, ecology, comp. note) Romania
Tibicen haematodes Cantoreanu 1960: 332 (distribution, ecology, comp. note) Romania, Europe, Asia Minor
Tibicen haematodes Cantoreanu 1968a: 343, Table 1 (listed, distribution, comp. note) Romania
Tibicina haematodes Weber 1968: 86 (comp. note)
Tibicina haematodes Cantoreanu 1969: 261, Table 1 (listed, distribution, comp. note) Romania
Tibicina haematodes Cantoreanu 1972a: 111, Table 1 (listed, distribution, comp. note) Romania
Tibicina haematodes Kaltenbach, Steiner and Aschenbrenner 1972: 490–491, Fig 143.4 (illustrated, comp. note) Germany
Tibicina haematodes Cantoreanu 1975: 92 (distribution, comp. note) Romania
Tibicina haematodes Villiers 1977: 166, Plate XIV, Fig 24 (illustrated, listed, comp. note) France
Tibicina haematodes Drosopoulous 1980: 190 (listed) Greece
Tettigonia haematodes Boulard 1981b: 42 (comp. note)
Cicada haematodes Boulard 1981b: 42 (comp. note)
Tibicina haematodes Boulard 1981b: 46 (comp. note)
Tibicina haematodes Lodos and Kalkandelen 1981: 76–77 (synonymy, distribution, comp. note) Equals *Tettigonia sanguinea* Equals *Cicada helvola* Equals *Cicada steveni* Equals *Tibicina haematodes viridinervis* Turkey, Albania, Austria, Bulgaria, Czechoslovakia, France, Germany, Greece, Hungary, Italy, Palestine, Portugal, Romania, Spain, Switzerland, USSR, Yugoslavia
Tibicina haematodes mph. *haematodes* Boulard 1982a: 50 (comp. note) France
Cicada haematodes Boulard 1982f: 197 (distribution, comp. note) Equals *Tibicina haematodes* Spain
Cicada haematodes Koçak 1982: 151 (type species of *Tibicina*)
Tibicina haematodes Strümpel 1983: 4, 70, 108, 135, Fig 191 (comp. note)
Cicada haematodes Boulard 1984c: 166, 170, 176, 178 (type species of genus *Tibicina* comp. note) Equals *Tibicen haematodes*
Tibicen haematodes Boulard 1984c: 166–167, 170–171, 173, 175, 177 (type species of genus *Tibicen*, comp. note)
Tibicina haematodes Boulard 1984c: 170–171 (status, comp. note)
Cicada haematode (sic) Boulard 1984c: 171 (comp. note)
Tibicina haematodes Boulard 1984d: 5, Plate, Fig 1 (illustrated, comp. note) France
Tibicina haematodes Emeljanov 1984: 10, 56 (listed, key) USSR
Tibicina haematodes Emmrich 1984: 114 (comp. note)
Tettigonia haematodes Emmrich 1984: 114 (comp. note)
Cicada haematodes Melville and Sims 1984: 163–165, 180 (type species of *Tibicen*, type species of *Tibicina*, history)
Tibicina haematodes Nosko 1984: 58 (comp. note) Slovakia
Tibicina haematodes Boulard 1985b: 60, 63 (illustrated, characteristics, comp. note) France
Cicada haematodes Boulard 1985d: 212–213 (type species of genus *Tibicina* comp. note)
Cicada haematodes Lauterer 1985: 214 (type species of *Tibicina*)
Tibicina haematodes Boulard 1986c: 193 (comp. note)
Tibicina haematodes Chinery 1986: 88–89 (illustrated, comp. note) Europe
Tibicina haematodes Schedl 1986a: 3, 7, 11, 15–16, 22–24, Fig 21, Map 4, Table 1 (listed, key, synonymy, distribution, ecology, comp. note) Equals *Cicada haematodes* Equals *Tettigonia sanguinea* Equals *Cicada helvola* Istria, Slovakia, Austria, Hungary, Romania, Bulgaria, Albania, Italy, Switzerland, France, Spain, Portugal, Ukraine, Caucasus, Transcaucasia, Russia
Tibicina haematodes Heller 1987: 93 (comp. note) Germany
Tibicina haematodes Nast 1987: 570–571 (distribution, listed) Germany, Czechoslovakia, France, Switzerland, Austria, Hungary, Ukraine, Moldavia, Romania, Russia, Portugal, Spain, Italy, Yugoslavia, Albania, Bulgaria, Greece
Tibicina haematodes Remane 1987: 288 (listed, comp. note) Germany
Tibicina haematodes Boulard 1988a: 8–9, 11–12, Table (illustrated, comp. note) France
Tibicina haematodes Boulard 1988d: 150–152, 154, Fig 4 (wing abnormalities, illustrated, comp. note) France

Cicada haematodes Boulard 1988f: 7–8, 18, 20–21, 23, 29–30, 34–35 (type species of *Tibicen*, type species of *Tibicina*, comp. note)
Tibicen haematodes Boulard 1988f: 26, 28, 30–32, 70 (comp. note)
Tettigonia sanguinea Boulard 1988f: 29 (comp. note)
Tibicina haematodes Boulard 1988f: 36, 44–45, 49, Fig 3 (illustrated, comp. note)
Tettigonia haematodes Boulard 1988f: 38 (comp. note)
Tibicina haematodes Boulard 1988g: 3 (comp. note) France
Tibicina haematodes Boulard and Riou 1988: 350–351 (illustrated, comp. note)
Cicadetta haematodes Moulds 1988: 40 (comp. note)
Tibicina haematoes Quartau and Fonseca 1988: 368, 373 (synonymy, distribution, comp. note) Equals *Cicada haematodes* Equals *Tettigonia sanguinea* Equals *Cicada helvola* Portugal, Albania, Austria, Bulgaria, Czechoslovakia, France, Corsica, Germany, Greece, Hungary, Italy, Sicily, Palestine, Romania, Spain, Switzerland, Turkey, USSR, Armenia, Azerbaijan, Georgia, Moldavia, Russia, Ukraine, Yugoslavia
Tibicina haematodes Boulard 1989b: 205–207, Figs 1–4 (no antennal regeneration)
Tibicina haematodes Schedl 1989: 108–110, Fig 1 (distribution, ecology, comp. note) Istria, Croatia
Tibicina haematodes Boulard 1990b: 66, 113–114, 132–133, 172–173, 176, 180, Plate XV, Fig 2, Fig 10 (genitalia, sonogram, comp. note) France
Tibicina haematodes Boulard 1992c: 65 (comp. note)
Tibicina haematodes Brackenbury 1992: 42–43, 103, Fig 21, Fig 53 (wing movement, illustrated, comp. note)
Tibicen haematordes (sic) Moalla, Jardak and Ghorbel 1992: 34 (comp. note) France
Tibicina haematodes Pillet 1993: 98–99, 104–109, Fig 3, Plate XI (described, illustrated, ecology, distribution, comp. note) Switzerland
Tibicina haematodes Remane and Wachman 1993: 19, 134 (illustrated, comp. note)
Tibicina haematodes Gogala, Aljancic, Gogala and Sivec 1994: 10, Fig 11 (illustrated, comp. note)
Tibicina haematodes Niehuls and Simon 1994: 253–259, Figs 1–2 (distribution, comp. note) Germany
Cicada haematodes Niehuls and Simon 1994: 255 (comp. note) Germany
Tibicina haematodes Boulard 1995a: 7–8, 12, 38–41, Fig S13$_1$, Fig S13$_2$, Fig 14, Fig S13$_3$ (illustrated, song analysis, sonogram, listed, comp. note) France
Tibicina haematodes Boulard and Mondon 1995: 25, 30, 32–33, 35, 38, 44, 53, 64, 66–67, 71–72, 77–79, 81–83, 159 (illustrated, sonogram, listed, comp. note) Equals *Tettigonia sanguinea* France
Tibicina haematodes Lee 1995: 63 (type species of *Tibicen*)
Tibicina haematodes Niehuls 1995: 37, 39–40, Fig 3 (distribution, comp. note) Germany
Tibicina haematodes Gogala, Popov and Ribaric 1996: 46 (comp. note)
Tibicina haematodes Holzinger 1996: 505 (listed) Austria
Tibicina haematodes Niehuls and Weitzel 1996: 441 (comp. note) Germany
Tibicina haematodes Vernier 1996: 147 (comp. note) France
Cicada haematodes Boulard 1997c: 83–84, 89–91, 93, 98–99, 103–104, 106, 121 (type species of *Tibicen*, *Tibicina*, Tibicininae, Tibicinidae, comp. note) Equals *Tettigonia sanguinea*
Tettigonia sanguinea Boulard 1997c: 98, 103 (comp. note)
Tibicina haematodes Boulard 1997c: 99–100, 106, 112–113, 117, Plate II, Fig 4 (illustrated, comp. note) Equals *Cicada haematodes*
Tettigonia haematodes Boulard 1997c: 86, 108 (comp. note)
Tibicen haematoides (sic) Boulard 1997c: 101 (comp. note)
Tibicina haematodes Holzinger, Frölich, Günthart, Pauterer, Nickel et al. 1997: 49 (listed) Europe
Tibicina haematodes Remane, Fröhlich, Nickel, Witsack and Achtziger 1997: 64, 68 (listed, comp. note) Germany
Tibicina haematodes Gogala 1998b: 394, 399 (illustrated, oscillogram, comp. note) Slovenia
Tibicina haematodes Remane, Achtziger, Fröhlich, Nickel and Witsack 1998: 246 (listed) Germany
Tibicina haematodes Schönitzer and Oesterling 1998: 34 (listed, comp. note) Germany
Tibicina haematodes Okáli and Janský 1998: 38–40, 43, Fig 4 (distribution, comp. note) Equals *Cicada haematodes* Equals *Tibicen haematodes* Equals *Tibicen haematodus* (sic) Slovakia
Tibicina haematodes Gogala and Gogala 1999: 120–121, 126, Fig 7 (distribution, comp. note) Slovenia, Italy, Mediterranean

Tibicina haematodes Abaii 2000: 142 (pest species) Iran
Tibicina haematodes haematodes Boulard 2000b: 6 (coloration, comp. note)
Cicada haematodes Boulard 2000i: 20 (comp. note) France
Tibicina haematodes Boulard 2000i: 20–21, 24, 45–50, 99, 121, Fig 10d, Fig 22d, Plates D-D', Map 4 (synonymy, illustrated, song, sonogram, oscillogram, frequency spectrum, distribution, ecology, comp. note) Equals *Cicada haematodes* Equals *Tettigonia haematodes* Equals *Tettigonia sanguinea* Equals *Tibicina sanguinea* Equals *Cicada (Tibicen) haematodes* Equals *Tibicina (Tibicina) haematodes* France
Tibicina haematodes haematodes Boulard 2000i: 46 (comp. note) France
Tibicina haematodes Freytag 2000: 55 (on stamp)
Tibicina haematodes Schedl 2000: 259–260, 266–267, Fig 2, Plate 1, Fig 4, Map 4 (key, illustrated, distribution, listed, comp. note) Switzerland, Italy, Czech Republic, Austria, Slovakia, Hungary, Slovenia, Germany
Cicada haematodes Boulard 2001b: 4–5, 9–11, 14, 19–21, 25–26, 40 (type species of *Tibicen, Tibicina*, Tibicininae, Tibicinidae, comp. note) Equals *Tettigonia sanguinea*
Tibicen haematodes Boulard 2001b: 16, 20, 40 (comp. note)
Tettigonia sanguinea Boulard 2001b: 19, 24 (comp. note)
Tibicina haematodes Boulard 2001b: 20, 26, 32–33, 37, Plate II, Fig 4 (illustrated, comp. note) Equals *Cicada haematodes*
Tettigonia haematodes Boulard 2001b: 8, 28 (comp. note)
Tibicen haematoides (sic) Boulard 2001b: 22 (comp. note)
Tibicina haematodes Dmitriev 2001a: 58 (comp. note) Russia
Tibicina haematodes Dmitriev 2001b: 10 (comp. note) Russia
Tibicina haematodes Krpač, Trilar and Gogala 2001: 29 (comp. note) Macedonia
Tibicina haematodes Levinson and Levinson 2001: 24 (comp. note)
Tibicina haematodes Schedl 2001b: 27 (comp. note)
Tibicina haematodes Sueur 2001: 41–42, 46, Tables 1–2 (listed)
Tibicina haematodes Sueur and Puissant 2001: 20, 22 (illustrated, comp. note) France
Tibicina haematodes haematodes Boulard 2002b: 193, Fig 43 (coloration, illustrated, comp. note) France
Tibicina haematodes Boulard 2002b: 193, 196, 198, 205–206 (comp. note)
Tibicina haematodes viridis Boulard 2002b: 193, Fig 43 (coloration, illustrated, comp. note) France
Tibicina haematodes Gogala 2002: 244, 248, Fig 4 (song, sonogram, oscillogram, comp. note)
Tibicina haematodes Günther and Mühlenthaler 2002: 331 (listed) Switzerland
Tibicina haematodes Nickel, Holzinger and Wachmann 2002: 292, Table 4 (listed)
Tibicina haematodes Nickel and Remane 2002: 37 (listed, hosts, distribution) Germany, Mediterranean
Tibicina haematodes Puissant and Sueur 2002: 197 (type species of *Tibicina*, comp. note) Slovenia
Tibicina haematodes Schedl 2002: 233–234, 237, Photo 3, Map 3 (illustrated, distribution, comp. note) Austria
Tibicina haematodes Sueur 2002b: 116, 118–120, 128–130 (listed, comp. note)
Tibicina haematodes Sueur and Puissant 2002: 477–483, Fig 2, Tables 2–5 (ecology, distribution, comp. note) France
Cicada haematodes Boulard 2003f: 372 (comp. note)
Tibicina haematodes Gogala and Seljak 2003: 350 (illustrated, comp. note) Slovenia
Tibicina haematodes Holzinger, Kammerlander and Nickel 2003: 485, 616–617, 633–634, 636, 638, Plate 42, Fig 3, Tables 2–3 (key, listed, illustrated, comp. note) Equals *Cicada haematodes* Europe, Slovenia, Mediterranean, Asia Minor, Caucasus
Tybicina (sic) *haematodes* Ibanez 2003: 3 (comp. note) France
Tibicina haematodes Ibanez 2003: 12 (comp. note) France
Tibicina haematodes Nickel 2003: 70 (distribution, hosts, comp. note) Germany
Tibicina haematodes Nickel and Remane 2003: 138 (listed, distribution) Germany
Tibicina haematodes Sueur 2003: 2932–2934, 2936, 2938, 2940, 2943, Table 1 (acoustic behavior, aggression, comp. note) France
Tibicina haematodes Sueur and Aubin 2003b: 482–490, Figs 5–9, Fig 11, Tables 1–4, Table 6 (song, comp. note) France
Tibicina haematodes Sueur, Puissant and Pillet 2003: 105–107, Fig 2 (illustrated, comp. note)
Tibicina haematodes Nickel 2004: 59, 64 (listed, comp. note) Germany
Tibicina haematodes Schedl 2004b: 914–915, Fig 1 (listed, distribution, comp. note) Austria

Tibicina haematodes Sueur and Aubin 2004: 218–219, 221–223, Figs 1–2, Tables 1–3 (courtship song, sonogram, oscillogram, comp. note) France

Tibicina haematodes Sueur, Puissant, Simões, Seabra, Boulard and Quartau 2004: 180 (comp. note) France

Tibicina haematodes Boulard 2005d: 338, 343, 345 (comp. note)

Tibicina haematodes Gogala, Trilar and Krpač 2005: 104–105, 113–114, Fig 8 (distribution, comp. note) Macedonia, Bulgaria

Tibicina haematodes Puissant and Defaut 2005: 116–118, 123, 125, 128, Table 1, Table 7, Table 9, Table 15 (listed, habitat association, distribution, comp. note) France

Tibicina haematodes Reynaud 2005: 142–143, Photo 2 (illustrated, comp. note) France

Tibicina haematodes Janský 2005: 84, Table 1 (listed, comp. note) Slovakia

Cicada haematodes Moulds 2005b: 389, 392–393 (type species of *Tibicen*, type species of *Tibicina*, history, listed, comp. note) Equals *Tettigonia sanguinea*

Tibicina haematodes Moulds 2005b: 396–397, 399, 407, 409, 413–414, 417–420, Fig 46, Fig 51, Figs 56–59, Table 1 (higher taxonomy, phylogeny, listed, comp. note)

Tibicina haematodes Bernier 2006: 3–5, 7, Figs 2–3 (distribution, illustrated, oscillogram) France

Tibicina haematodes Boulard 2006g: 41–43, 61–62, 66, 96, 107, 180, Figs 71–72 (comp. note) France

Tibicina haematodes Puissant 2006: 19, 21, 25–27, 30, 32, 34–35, 45–49, 52–54, 62–64, 69–71, 90, 93, 95, 108–109, 112, 114–115, 133–135, 137–141, 150–152, 154, 156–157, 159–1560, 163, 178–180, 185, 187, 190, Fig 8, Figs 20–22, Tables 1–2, Appendix Table 1, Appendix Table 7, Appendix Table 9, Appendix Table 15 (illustrated, synonymy, distribution, habitat association, ecology, listed, comp. note) Equals *Tettigonia sanguinea* Equals *Cicada helveola* Equals *Tibicen sanguinea* France

Tibicina haematodes Seabra, Pinto-Juma and Quartau 2006: 850 (comp. note)

Tibicina haematodes Trilar, Gogala and Popa 2006: 177–178 (listed, distribution, comp. note) Romania

Tibicina haematodes Demir 2007a: 52 (distribution, comp. note) Turkey

Tibicina haematodes Dmitriev 2007a: 811, 817 (comp. note) Russia

Tibicina haematodes Dmitriev 2007b: 1204, 1209 (comp. note) Russia

Tibicina haematodes Shiyake 2007: 7, 89, 91, Fig 152 (illustrated, distribution, listed, comp. note) Europe

Tibicina haematodes Sueur, Vanderpool, Simon, Ouvrard and Bourgoin 2007: 612–615, 617–623, Figs 1–6, Tables 1–2 (phylogeny, comp. note) France

Tibicina haematodes Demir 2008: 476–477 (distribution) Equals *Tibicina cantans* Equals *Tibicina helveola* Equals *Tibicina sanguinea* Turkey

Tibicina haematodes Hertach 2008b: 14, 19 (comp. note) Switzerland

Tibicina haematodes Hugel, Matt, Callot, Feldtrauer and Brua 2008: 5–6 (comp. note) France

Tibicina haematodes Nickel 2008: 207 (distribution, hosts, comp. note) Mediterranean

Tibicina haematodes Trilar and Gogala 2008: 30 (listed, distribution, comp. note) Romania

Tibicina haematedes (sic) Zamanian, Mehdipour and Ghaemi 2008: 2071 (comp. note)

Tibicina haematodes Zamanian, Mehdipour and Ghaemi 2008: 2071 (comp. note)

Tibicina haematodes Ibanez 2009: 208–209, 212 (listed, comp. note) France

Cicada haematodes Ahmed and Sanborn 2010: 42 (type species of genus *Tibicina*)

Tibicina haematodes Kemal and Koçak 2010: 5 (distribution, listed, comp. note) Turkey

Cicada haematodes Mozaffarian and Sanborn 2010: 80 (type species of *Tibicina*)

Tibicina haematodes Mozaffarian and Sanborn 2010: 80 (synonymy, listed, distribution, comp. note) Equals *Tettigonia sanguinea* Equals *Cicada helvola* Iran, Europe, Russia, North Africa, Asiatic Turkey

***T. haematodes helveola* (Germar, 1821)** = *Cicada helvola* = *Cicada helveola* (sic) = *Cicada helveolus* (sic) = *Tibicen helvola* = *Tibicen helvolus*

Tibicina haematodes var. *helveola* Boulard and Mondon 1995: 66 (listed) France

Tibicina haematodes helveola Boulard 2000i: 49 (described, distribution, comp. note) Equals *Cicada helveola* France

T. haematodes steveni (Krynicki, 1837) to *Tibicina steveni*

***T. haematodes viridinervis* Fieber, 1876** = *Tibicina haematodes* var. *viridinervis*

Tibicina haematodes var. *viridinervis* Boulard and Mondon 1995: 66, 79 (listed, illustrated) France

Tibicina haematodes viridinervis Boulard 2000b: 6 (coloration, comp. note)

Tibicina haematodes viridinervis Boulard 2000i: 49 (described, distribution, comp. note) Equals *Tibicina haematodes* var. *viridinervis* France, Austria

Tibicen (= Tibicina) haematodes f. *viridinervis* Gogala, Trilar and Krpač 2005: 124 (comp. note) Albania

Tibicina haematodes f. *viridinervis* Gogala, Trilar and Krpač 2005: 124 (comp. note)

***T. insidiosa* Boulard, 1977**

Tibicina insidiosa Nast 1982: 311 (listed) Afghanistan

***T. intermedia* Fieber, 1876** = *Tibicen intermedia* = *Tibicen intermedius* (sic) = *Tibicina intermedius* (sic)

Tibicina intermedia Popov 1971: 302 (hearing, listed, comp. note)

Tibicina intermedia Donat 1981: 145 (eclosion, comp. note)

Tibicina intermedia Popov 1985: 40–41, 68, 71, 73, 76–77, 80, 181, 183, Fig 7.4, Fig 21.1, Fig 23.3 (illustrated, oscillogram, acoustic system, auditory system, comp. note)

Tibicina intermedia Anufriev and Emelyanov 1988: 19 Fig 7 (wings) Soviet Union

Tibicinca intermedia Popov 1990a: 302–303, Fig 2 (comp. note)

Tibicina intermedia Tautz 1990: 105, Fig 12 (song frequency, comp. note)

Tibicina intermedia Popov, Aronov and Sergeeva 1991: 282, 284, 287, 292–294, Fig 4, Table (auditory system, ecology, comp. note) Russia

Tibicina intermedia Sueur 2001: 42, Table 1 (listed)

Tibicina intermedia Puissant and Sueur 2002: 197, 199 (comp. note) Turkey

Tibicina intermedia Sueur, Puissant and Pillet 2003: 105–108, Fig 3, Figs 7–8 (comp. note)

Tibicina intermedia Puissant 2006: 52 (comp. note)

T. luctuosus (Costa, 1883) to *Tibicina nigronervosa* Fieber, 1876

***T. maldesi* Boulard, 1981**

Tibicina maldesi Boulard 1981a: 37–40, Figs A-B, Figs 1–5 (n. sp., described, illustrated, comp. note) Algeria

Tibicina maldesi Boulard 1988d: 153, 156, Fig 10 (wing abnormalities, pigmentation abnormalities, illustrated, comp. note) Algeria

Tibicina maldesi Sueur, Vanderpool, Simon, Ouvrard and Bourgoin 2007: 612–613 (comp. note) Algeria

***T. nigronervosa* Fieber, 1876** = *Tibicen nigronervosa* = *Tibicen nigronervosus* = *Tibicina nigronervossa* (sic) = *Tibicina nigrovenosa* (sic) = *Tibicen luctuosa* Costa, 1888 = *Tibicina luctuosa* = *Tibicen luctuosa* (sic) = *Tibicen luctuosus* = *Tibicina tomentosa* (nec Olivier)

Tibicen nigronervosa Zeuner 1944: 113 (comp. note)

Tibicina nigronervosa Villiers 1977: 166, Plate XIV, Fig 25 (illustrated, listed, comp. note) Corsica

Tibicina nigronervosa Boulard 1981b: 45–46, 50, Fig 7 (comp. note) Corsica, Sardinia

Tibicina nigronervosa Dlabola 1981: 209 (distribution, comp. note) Iran

Tibicina nigronervosa Lodos and Kalkandelen 1988: 18 (comp. note) Turkey

Tibicina nigronervosa Boulard 1982f: 197 (distribution, comp. note) Corsica, Sardinia

Tibicen nigronervosa Boulard 1984c: 171 (status, comp. note)

Tibicina nigronervosa Boulard 1984d: 5 (comp. note) France, Corsica

Tibicina nigronervosa Boulard 1985b: 63 (characteristics) France

Tibicina nigronervosa Nast 1987: 570–571 (distribution, listed) France, Spain, Italy, Portugal?

Tibicina luctuosa (sic) Nast 1987: 570–571 (distribution, listed) Italy

Tibicina nigronervosa Boulard 1988a: 12, Table (comp. note) France

Tibicen nigronervosa Boulard 1988f: 31 (comp. note)

Tibicina nigronervosa Boulard 1988g: 3 (comp. note) France, Corsica

Tibicina nigronervosa Quartau and Fonseca 1988: 368, 373 (distribution, comp. note) Portugal, France, Corsica, Spain, Switzerland, USSR, Transcaucasia

Tibicina nigronervosa Boulard 1995a: 6, 13, 44–45, 51, Fig $S15_1$, Fig $S15_2$, Fig 17 (illustrated, song analysis, sonogram, listed, comp. note) France

Tibicina nigronervosa Boulard and Mondon 1995: 30, 32, 66–67, 80, 159 (illustrated, listed, comp. note) Equals *Tibicina luctuosa* France

Tibicen nigronervosa Boulard 1997c: 100 (comp. note)

Tibicina luctuosa (sic) Guglielmino, d'Urso and Alma 2000: 167 (listed, distribution) Sardinia

Tibicina nigronervosa Boulard 2000i: 53–55, 122, Plate F, Map 6 (synonymy, illustrated, song, sonogram, oscillogram, frequency spectrum, distribution, ecology, comp. note) Equals *Tibicina luctuosa* Equals *Tibicen nigronervosus* Equals *Tibicina tomentosa* (nec Olivier) Equals *Tibicen nigronervosa* (sic) France

Tibicen nigronervosa Boulard 2001b: 21 (comp. note)

Tibicina nigronervosa Puissant and Sueur 2001: 429–435, Fig 3, Fig 7, Fig 9 (illustrated, distribution, comp. note) Corsica, France

Tibicina nigronervosa Quartau, Simões, Rebelo and André 2001: 401 (comp. note)
Tibicina nigronervosa Sueur 2001: 42, Table 1 (listed)
Tibicina nigronervosa Sueur and Puissant 2001: 19, 22 (illustrated, comp. note) France
Tibicina nigronervosa Sueur and Puissant 2002: 477–483, Tables 2–5 (ecology, distribution, comp. note) France
Tibicina nigronervosa Gogala and Trilar 2003: 9 (comp. note)
Tibicina nigronervosa Quartau and Simões 2003: 117 (comp. note)
Tibicina nigronervosa Sueur 2003: 2932, 2938, 2942 (acoustic behavior, aggression, comp. note) France, Corsica
Tibicina nigronervosa Sueur and Aubin 2003b: 482, 484, 486–490, Figs 5–9, Fig 11, Tables 1–4, Table 6 (song, comp. note) France, Corsica
Tibicina nigronervosa Sueur, Puissant and Pillet 2003: 105 (comp. note)
Tibicina nigronervosa Sueur and Aubin 2004: 218–223, Fig 2, Tables 1–3 (courtship song, sonogram, oscillogram, comp. note) France, Corsica
Tibicina nigronervosa Sueur, Puissant, Simões, Seabra, Boulard and Quartau 2004: 180 (comp. note) Corsica, Sardinia
Tibicina nigronervosa Moulds 2005b: 414 (comp. note)
Tibicina nigronervosa Puissant and Defaut 2005: 117–118, 125–127, Table 1, Table 10, Table 12 (listed, habitat association, distribution, comp. note) France
Tibicina nigronervosa Boulard 2006g: 41, 135, Fig 92 (sonogram, comp. note)
Tibicina nigronervosa Puissant 2006: 20, 23, 25, 27–28, 32, 48, 55–56, 63–68, 70, 90, 96, 115–120, 122, 125–126, 132, 179–180, 184–189, Fig 8, Figs 37–39, Tables 1–2, Appendix Table 1, Appendix Table 10, Appendix Table 12 (illustrated, synonymy, distribution, habitat association, ecology, listed, comp. note) Equals *Tibicina luctuosa* France, Corsica
Tibicina luctuosa Puissant 2006: 66 (comp. note)
Tibicina nigronervosa Sueur, Vanderpool, Simon, Ouvrard and Bourgoin 2007: 612–615, 617–624, Figs 1–6, Tables 1–2 (phylogeny, comp. note) France, Corsica
Tibicina nigronervosa Kemal and Koçak 2010: 5 (distribution, listed, comp. note) Turkey

***T. picta* (Fabricius, 1794)** = *Tettigonia picta* = *Cicada picta* = *Tibicen picta* = *Tibicen pictus* = *Tibicien* (sic) *pictus* = *Cicadetta picta*
Tettigonia picta Boulard 1984c: 171 (comp. note)
Tettigonia picta Dallas 1870: 482 (synonymy) Equals *Tibicen picta*
Tibicina picta Nast 1987: 570–571 (distribution, listed) France, Spain, Italy
Tettigonia picta Boulard 1988f: 17, 23 (comp. note)
Cicada picta Boulard 1988f: 40 (comp. note)
Cicadetta picta Boulard 1988f: 41 (comp. note)
Tettigonia picta Boulard 1997c: 86, 93 (comp. note)
Cicada picta Boulard 1997c: 110 (type species of *Melampsalta*, comp. note) Spain
Cicadetta picta Boulard 1997c: 110 (comp. note)
Cicada picta Boulard 2000i: 60 (comp. note)
Tettigonia picta Boulard 2000i: 63 (comp. note)
Tibicina picta Guglielmino, d'Urso and Alma 2000: 167 (listed, distribution) Sardinia, Spain, France, Italy
Tettigonia picta Boulard 2001b: 8, 14 (comp. note)
Cicada picta Boulard 2001b: 30 (type species of *Melampsalta*, comp. note) Spain
Cicadetta picta Boulard 2001b: 30 (comp. note)
Tettigonia picta Boulard 2003f: 372 (comp. note)
Cicadetta picta Puissant 2005: 309 (comp. note)

***T. pieris* (Kirkaldy, 1909)** = *Tibicen pieris*
Tibicen pieris Wu 1935: 25 (listed) China, Yunnan

***T. quadrisignata* (Hagen, 1855)** = *Cicada haematodes 4-signata* = *Cicada 4-signata* = *Tibicen quadrisignata* = *Tibicen quadrisignatus* = *Tibicina 4-signata* = *Tibicen quadrosignata* (sic)
Tibicina quadrisignata Varley 1939: 99 (comp. note)
Tibicen quadrosignata (sic) Zeuner 1944: 113 (comp. note)
Tibicina quadrisignata Villiers 1977: 166, Plate XIV, Fig 20 (illustrated, listed, comp. note) Mediterranean
Tibicina quadrisignata Boulard 1981a: 37 (comp. note)
Tibicina quadrisignata Boulard 1981b: 45–46, 50, Fig 8 (comp. note) Equals *Cicada haematoes* var. *quadrisignata* Equals *Tibicina nigronervosa* (nec Fieber) Western Europe
Cicada 4-signata Boulard 1982f: 197 (distribution, comp. note) Equals *Tibicina quadrisignata* Spain
Tibicina quadrisignata Boulard 1983a: 309, 312 (comp. note) France
Tibicina quadrisignata Boulard 1985a: 33 (distribution, listed) Portugal
Tibicina quadrisignata Boulard 1985b: 63 (characteristics) France
Tibicina quadrisignata Nast 1987: 570–571 (distribution, listed) France, Portugal, Spain
Tibicina quadrisignata Boulard 1988a: 9, 12, Table (comp. note) France

Tibicina quadrisignata Quartau and Fonseca 1988: 367, 373 (synonymy, distribution, comp. note) Equals *Cicada haematodes quadrisignata* Portugal, Algeria, Spain, USSR
Tibicina quadrisignata Boulard 1990b: 172–173, Fig 10 (sonogram, comp. note)
Tibicina quadrisignata Quartau and Rebelo 1990: 1 (comp. note) Portugal
Tibicina quadrisignata Fonseca 1991: 174, 186–187, 189, Fig 8 (song analysis, oscillogram, comp. note) Portugal
Tibicina quadrisignata Pillet 1993: 99, 104, 106–109, Fig 3 (described, ecology, distribution, comp. note) Switzerland
Tibicina quadrisignata Fonseca and Popov 1994: 350 (comp. note)
Tibicina quadrisignata Boulard 1995a: 7–8, 12, 42–43, Fig 15, Fig $S14_1$, Fig $S14_2$ (illustrated, song analysis, sonogram, listed, comp. note) Portugal
Tibicina quadrisignata Boulard and Mondon 1995: 30, 32, 66–67, 79, 159 (illustrated, listed, comp. note) France
Tibicina quadrisignata Fonseca 1996: 15, 25–27, Fig 5, Table 2 (timbal muscle physiology, oscillogram, comp. note) Portugal
Tibicina quadrisignata Holzinger, Frölich, Günthart, Pauterer, Nickel et al. 1997: 49 (listed) Europe
Tibicina quadrisignata Fonseca and Bennet-Clark 1998: 729 (comp. note)
Tibicina quadrisignata Boulard 2000i: 50–53, 55, 99, 121, Fig 22a, Plate E, Map 3 (synonymy, illustrated, song, sonogram, oscillogram, frequency spectrum, distribution, ecology, comp. note) Equals *Cicada haematodes* var. *4-signata* Equals *Tibicen quadrisignata* Equals *Tibicina nigronervosa* (nec Fieber) Equals *Tibicina (Tibicina) quadrisignata* France, Portugal, Spain
Tibicina 4-signata Boulard 2000i: 55 (comp. note)
Tibicina quadrisignata Puissant and Sueur 2000: 73 (comp. note) France
Tibicina quadrisignata Schedl 2000: 259–260, 266–267, Fig 3, Plate 1, Fig 5, Map 3 (key, illustrated, distribution, listed, comp. note) Switzerland, Italy
Tibicina quadrisignata Coffin 2001a: 123, 125–127, Color Plate, Figs 1–6 (oviposition, illustrated, comp. note) Europe
Tibicina quadrisignata Puissant 2001: 147 (comp. note) France
Tibicina quadrisignata Quartau, Simões, Rebelo and André 2001: 401, 403, 407–410, Fig 3, Figs 16–21 (described, illustrated, ecology, comp. note) Portugal
Tibicina quadrisignata Schedl 2001b: 27 (comp. note)
Tibicina quadrisignata Sueur 2001: 42, Table 1 (listed)
Tibicina quadrisignata Sueur and Puissant 2001: 22 (comp. note) France
Tibicina quadrisignata Fonseca and Revez 2002a: 1286–1290, Fig 2 (song power spectrum, comp. note)
Tibicina quadrisignata Günther and Mühlenthaler 2002: 331 (listed) Switzerland
Tibicina quadrisignata Sueur 2002b: 122, 130 (listed, comp. note)
Tibicina quadrisignata Sueur and Puissant 2002: 477–483, Fig 2, Tables 2–5 (ecology, distribution, comp. note) France
Tibicina quadricincta Holzinger, Kammerlander and Nickel 2003: 485–486, 616–617, 633, 638, Plate 42, Fig 4, Tables 2–3 (key, listed, illustrated, comp. note) Europe, France, Spain, Portugal, Mediterranean
Tibicina quadrisignata Quartau and Simões 2003: 114–118, Figs 2–4, Table 1 (song, sonogram, oscillogram, comp. note) Portugal
Tibicina quadrisignata Sueur 2003: 2932–2933, 2935, 2938, 2940–2942, Fig 5, Table 2 (sonogram, oscillogram, acoustic behavior, aggression, comp. note) France
Tibicina quadrisignata Sueur and Aubin 2003b: 482, 484–490, Figs 5–9, Fig 11, Tables 1–4, Table 6 (song, comp. note) France
Tibicina quadrisignata Sueur, Puissant and Pillet 2003: 105 (comp. note)
Tibicina quadrisignata Sueur and Aubin 2004: 217–218, 221–222, Fig 2, Tables 1–3 (courtship song, sonogram, oscillogram, comp. note) France
Tibicina quadrisignata Sueur, Puissant, Simões, Seabra, Boulard and Quartau 2004: 177–182, Figs 1–2 (distribution, ecology, song, sonogram, oscillogram, comp. note) Portugal, France
Tibicina quadrisignata Moulds 2005b: 414 (comp. note)
Tibicina quadrisignata Puissant and Defaut 2005: 116–118, 122–125, Table 1, Tables 6–7, Table 9 (listed, habitat association, distribution, comp. note) France
Tibicina quadrisignata Boulard 2006g: 133, 150, Fig 89, Fig 110A (sonogram, illustrated, comp. note) Portugal, France
Tibicina 4-signata Boulard 2006g: 181 (listed)
Tibicina quadrisignata Puissant 2006: 20, 22, 25, 32, 24–25, 4, 48–49, 54, 62–66, 69–70, 90, 96, 08, 110, 135–136, 138, 141–143, 145, 147, 149–150, 152–154, 178–180, 184–187, Fig

8, Figs 34–36, Tables 1–2, Appendix Table 1, Appendix Tables 6–7, Appendix Table 9 (illustrated, distribution, habitat association, ecology, listed, comp. note) France

Tibicina quadrisignata Hertach 2007: 51 (comp. note) Switzerland

Tibicina quadrisignata Sueur, Vanderpool, Simon, Ouvrard and Bourgoin 2007: 612–615, 617–623, Figs 1–6, Tables 1–2 (phylogeny, comp. note) France

***T. reticulata* Distant, 1888** = *Tibicen reticulatus* = *Paharia reticulata*

Tibicina reticulata Ahmed and Sanborn 2010: 42 (synonymy, distribution) Equals *Tibicen reticulatus* Pakistan, India

***T. serhadensis* Koçak & Kemal, 2010**

Tibicina serhadensis Kemal and Koçak 2010: 1–3, 5, 8–17, 19, Figs 2–15, Figs 17–22, Fig 26, Map 1 (illustrated, behavior, distribution, comp. note) Turkey

Tibicina serhadensis Koçak and Kemal 2010: 148 (n. sp., described, listed, distribution, comp. note) Turkey

***T. steveni* (Krynicki, 1837)** = *Cicada steveni* = *Cicada stevenii* (sic) = *Cicada stevensi* (sic) = *Tibicina steveni* = *Cicada (Tibicina) haematodes stevenii* (sic) = *Tibicina haematodes* var. *steveni* = *Tibicen haematodes steveni* = *Tibicina intermedia* (nec Fieber)

Tibicina intermedia (sic) Popov, Arnov and Sergeeva 1992: 152, 154, 157, 162–164, Fig 4, Table (auditory system, ecology, comp. note) Russia

Tibicina haematodes var. *steveni* Pillet 1993: 105 (comp. note)

Tibicina haematodes steveni Trilar and Gogala 1998: 69 (comp. note) Macedonia

Tibicina steveni Krpač, Trilar and Gogala 2001: 29 (comp. note) Macedonia

Tibicen haematodes steveni Krpač, Trilar and Gogala 2001: 29 (comp. note) Macedonia

Tibicina steveni Puissant and Sueur 2002: 197, 199 (comp. note) Turkey

Tibicina steveni Sueur and Aubin 2003b: 482, 484–490, Figs 5–9, Fig 11, Tables 1–4, Table 6 (song, comp. note) France, Switzerland, Georgia

Tibicina steveni Sueur, Puissant and Pillet 2003: 105–110, Fig 1, Figs 4–6, Figs 9–12 (illustrated, described, habitat, song, sonogram, oscillogram, comp. note) Turkey, Ukraine, Armenia, Azerbaijan, Georgia, Macedonia, France, Switzerland

Tibicina steveni Sueur and Aubin 2004: 218, 221–222, Tables 1–3 (courtship song, comp. note) France

Tibicina steveni Gogala, Trilar and Krpač 2005: 104–105, 114–115 (distribution, comp. note) Macedonia, Bulgaria

Tibicina steveni Puissant and Defaut 2005: 117–118, 128, Table 1, Table 14 (listed, habitat association, distribution, comp. note) France, Turkey, Ukraine, Armenia, Azerbaijan, Georgia, Macedonia, Switzerland

Tibicina steveni Puissant 2006: 19–20, 22, 25–27, 30, 32, 34, 48, 52–55, 63–64, 66, 70, 79, 90–91, 95, 150, 179–180, 190, Fig 8, Figs 25–26, Tables 1–2, Appendix Table 1, Appendix Table 14 (illustrated, distribution, habitat association, ecology, listed, comp. note) Equals *Tibicina intermedia* Popov et al. 1992 France, Turkey, Ukraine, Armenia, Azerbaijan, Georgia, Macedonia, Switzerland

Tibicina steveni Hertach 2007: 51 (comp. note) Switzerland

Tibicina steveni Sueur, Vanderpool, Simon, Ouvrard and Bourgoin 2007: 612–615, 617–619, 621–622, Figs 1–6, Tables 1–2 (phylogeny, comp. note) Switzerland

Tibicina steveni Nickel 2008: 207 (distribution, hosts, comp. note) Mediterranean

Tibicina steveni Kemal and Koçak 2010: 5 (distribution, listed, comp. note) Turkey

Tibicina steveni Koçak and Kemal 2010: 148 (comp. note) Turkey

***T. tomentosa* (Olivier, 1790)** = *Cicada tomentosa* = *Cicado* (sic) *tomentosus* (sic) = *Tibicen tomentosa* (sic) = *Tibicen tomentosus* = *Tibicina (Tibicina) tomentosa* = *Tibicin* (sic) *tomentosa* = *Tettigia tomentosa* = *Cicada baetica* (Rambur, 1840) = *Tibicina baetica* = *Tibicina boetica* (sic) = *Tibicen beaticus* = *Tibicen baetica* = *Cicada picta* (nec Germar) = *Tettigonia picta* (nec Fabricius) = *Tibicen pictus* (nec Fabricius) = *Tibicien* (sic) *pictus* (nec Fabricius) = *Tibicina picta* (nec Fabricius) = *Tibicina cisticola* (nec Hagen)

Tibicina tomentosa Varley 1939: 99 (comp. note)

Tibicen tomentosa Zeuner 1944: 113 (comp. note)

Tibicina tomentosa Villiers 1977: 166, Plate XIV, Fig 26 (illustrated, listed, comp. note) France

Tibicina tomentosa Boulard 1982g: 40, 45, 47, Plate IV, Fig 15 (described, illustrated, comp. note) France

Tibicina tomentosa Boulard 1983a: 309 (comp. note) France

Cicada tomentosa Boulard 1984b: 46 (comp. note)

Cicada baetica Boulard 1984b: 46 (comp. note)

Tibicina tomentosa Boulard 1984d: 5 (comp. note) France

Tibicen baeticus Reichhoff-Riehm 1984: 90–91 (natural history, illustrated, comp. note)

Tibicina baetica Boulard 1985a: 33 (distribution, listed) Portugal
Tibicina tomentosa Boulard 1985b: 63 (characteristics) France
Tibicina tomentosa Boulard 1985e: 1021 (coloration, comp. note) France
Tibicina baetica Nast 1987: 570–571 (distribution, listed) Germany, France, Spain, Italy
Tibicina tomentosa Boulard 1988a: 9, 12, Table (illustrated, comp. note) France
Cicada tomentosa Boulard 1988f: 16–17 (comp. note) Equals *Tettigonia picta*
Tibicina baetica Quartau and Fonseca 1988: 367, 372 (synonymy, distribution, comp. note) Equals *Cicada baetica* Portugal, France, Germany, Italy, Sardinia, Spain
Tibicina tomentosa Boulard 1995a: 6, 8, 13, 49–51, Fig $S17_1$, Fig $S17_2$, Fig 18 (illustrated, song analysis, sonogram, listed, comp. note) France
Tibicina tomentosa Boulard and Mondon 1995: 30, 32, 66–67, 70, 80, 159 (illustrated, listed, comp. note) Equals *Tettigonia picta* Equals *Cicada picta* France
Cicada baetica Boulard 1997c: 93 (comp. note)
Cicada tomentosa Boulard 1997c: 86 (comp. note) Equals *Tettigonia picta*
Tibicina tomentosa Boulard 2000i: 60–64, 123, Plate H, Map 8 (synonymy, illustrated, song, sonogram, oscillogram, frequency spectrum, distribution, ecology, comp. note) Equals *Cicada tomentosa* Equals *Tettigonia picta* Equals *Cicada picta* (nec Germar) Equals *Tibicen picta* Equals *Cicado* (sic) *tomentosus* (sic) Equals *Tibicen pictus* Equals *Tibicina cisticola* (partim) Equals *Tibicen tomentosa* (sic) Equals *Tibicen tomentosus* Equals *Tibicien* (sic) *pictus* Equals *Tibicina picta* (partim) Equals *Tibicina (Tibicina) tomentosa* France
Cicada tomentosa Boulard 2000i: 60 (comp. note)
Cicada baetica Boulard 2000i: 63 (comp. note)
Tibicina baetica Boulard 2000i: 64 (comp. note)
Tibicina baetica Guglielmino, d'Urso and Alma 2000: 167 (listed, distribution) Sardinia, Spain, France
Tibicina tomentosa Puissant and Sueur 2000: 69 (comp. note) France
Cicada baetica Boulard 2001b: 14 (comp. note)
Cicada tomentosa Boulard 2001b: 6 (comp. note) Equals *Tettigonia picta*
Tibicina baetica Quartau, Simões, Rebelo and André 2001: 402 (comp. note)
Tibicina tomentosa Sueur 2001: 42, Table 1 (listed)
Tibicina tomentosa Sueur and Puissant 2001: 22 (comp. note) France
Tibicina tomentosa Boulard 2002b: 205 (comp. note)
Tibicina tomentosa Sueur 2002b: 122, 130 (listed, comp. note)
Tibicina tomentosa Sueur and Puissant 2002: 477–483, Fig 3, Tables 2–5 (ecology, distribution, comp. note) France
Tibicina tomentosa Sueur 2003: 2932–2936, 2938–2940, 2942–2944, Figs 1–2, Fig 4, Figs 6–7, Tables 1–2 (sonogram, oscillogram, acoustic behavior, aggression, comp. note) France
Tibicina tomentosa Sueur and Aubin 2003b: 482, 484–490, Figs 5–9, Fig 11, Tables 1–4, Table 6 (song, comp. note) France, Portugal, Spain
Tibicina tomentosa Sueur, Puissant and Pillet 2003: 105 (comp. note)
Tibicina tomentosa Sueur and Aubin 2004: 218–223, Figs 2–3, Tables 1–3 (courtship song, sonogram, oscillogram, comp. note) France
Tibicina baetica Sueur, Puissant, Simões, Seabra, Boulard and Quartau 2004: 177, 180 (comp. note) Portugal
Tibicina tomentosa Sueur, Puissant, Simões, Seabra, Boulard and Quartau 2004: 177, 179–181, Figs 1–2 (distribution, ecology, song, sonogram, oscillogram, comp. note) Portugal, France, Spain
Tibicina tomentosa Boulard 2005d: 343 (comp. note)
Tibicina tomentosa Puissant 2005: 309, 311 (comp. note) Equals *Tettigonia picta*
Tibicina tomentosa Puissant and Defaut 2005: 116, 118, 120–121, Table 1, Tables 4–5 (listed, habitat association, distribution, comp. note) France
Tibicina tomentosa Sueur 2005: 214 (song, comp. note) Equals *Tibicen baetica* France
Tibicina baetica Sueur 2005: 214 (song, comp. note) Iberian Peninsula
Tibicina tomentosa Boulard 2006g: 41–43, 134, 181, Fig 9e, Fig 90 (illustrated, sonogram, listed, comp. note)
Tibicen baetica Boulard 2006g: 53 (comp. note) Equals *Tibicina tomentosa*
Cicada tomentosa Duffels and Hayashi 2006: 190 (comp. note) France
Tibicina tomentosa Puissant 2006: 19–20, 22, 25–28, 32, 41, 49–52, 54, 58, 60–61, 66, 89–90, 93, 95, 138, 140, 154, 178, 180, 182–183, Fig 8, Figs 23–24, Tables 1–2, Appendix Table 1, Appendix Tables 4–5 (illustrated, synonymy, distribution, habitat association, ecology, listed, comp. note) Equals *Tettigonia picta* Equals *Tibicina baetica* France
Tibicina baetica Puissant 2006: 19, 49 (comp. note)
Tibicina tomentosa Sueur, Vanderpool, Simon, Ouvrard and Bourgoin 2007: 612–615,

617–624, Figs 1–6, Tables 1–2 (phylogeny, comp. note) France, Portugal, Spain
Tibicina tomentosa Zamanian, Mehdipour and Ghaemi 2008: 2070 (comp. note)
Tibicina tomentosa Koçak and Kemal 2010: 148 (comp. note)
Tibicina baetica Puissant and Sueur 2010: 555 (comp. note)
Tibicina tomentosa Puissant and Sueur 2010: 555 (comp. note) Equals *Tibicina baetica*

***Subtibicina* Lee, 2012**
***S. tigris* Lee, 2012**

***Paharia* Distant, 1905**
Paharia Koçak 1982: 150 (listed, comp. note) Turkey
Paharia Popov 1985: 38–39 (acoustic system, comp. note)
Paharia Boulard 1988d: 152 (comp. note)
Paharia Mirzayans 1995: 18 (listed) Iran
Paharia Chou, Lei, Li, Lu and Yao 1997: 43, 90 (comp. note)
Paharia Ohbayashi, Sato and Igawa 1999: 340 (parasitism, comp. note)
Paharia Schedl 2001a: 1287 (comp. note)
Paharia Schedl 2003: 426 (comp. note)
Paharia Boulard 2006g: 153 (comp. note)
Paharia Ahmed and Sanborn 2010: 42 (distribution) Pakistan, Iran, central Asia, USSR
Paharia Mozaffarian and Sanborn 2010: 80 (distribution, listed) Central Asia, USSR, Pakistan, Iran
Paharia Wei and Zhang 2010: 37–38 (comp. note) Equals *Subpsultria* (sic)? India, Afghanistan, Turkmenistan

***P. lacteipennis* (Walker, 1850)** = *Cephaloxys lacteipennis* = *Mogannia lacteipennis* = *Tibicen lacteipennis* = *Tibicina lacteipennis*
Cephaloxys lacteipennis Koçak 1982: 150 (type species of *Paharia*)
Paharia lacteipennis Boulard 1988d: 152 (wing abnormalities, comp. note)
Cephaloxys lacteipennis Ahmed and Sanborn 2010: 42 (type species of genus *Paharia*)
Paharia lacteipennis Ahmed and Sanborn 2010: 43 (synonymy, distribution) Equals *Cephaloxys lacteipennis* Pakistan, Kazakhstan, Afghanistan, India
Cephaloxys lacteipennis Mozaffarian and Sanborn 2010: 80 (type species of *Paharia*)

***P. putoni putoni* (Distant, 1892)** = *Tibicen putoni* = *Tibicina putoni* = *Tibicina lacteipennis* Puton, 1883 (nec *Cephaloxys lacteipennis* Walker, 1850) = *Sena lacteipennis* = *Tibicen lacteipennis* = *Tibicina lacteipennis* = *Psalmocharias lacteipennis*
Paharia putoni Mirzayans, Hashemi, Borumand, Zairi and Rajabi 1976: 111 (distribution, listed) Iran
Paharia putoni Dlabola 1981: 207 (distribution, comp. note) Iran
Tibicina (Paharia) putoni Boulard 1988d: 150–151, Fig 8 (wing abnormalities, illustrated, comp. note) Persia
Paharia putoni Boulard 1988d: 152 (wing abnormalities, comp. note)
Paharia putoni Mirzayans 1995: 18 (listed, distribution) Iran
Paharia putoni Ahmed and Sanborn 2010: 43 (synonymy, distribution) Equals *Tibicina lacteipennis* (nec Walker) Pakistan, Iran, Afghanistan, USSR, central Asia
Paharia putoni Mozaffarian and Sanborn 2010: 80–81 (synonymy, listed, distribution) Equals *Tibicina lacteipennis* (nec Walker) Equals *Tibicen lacteipennis* Iran, Persia, Afghanistan, USSR, central Asia

***P. putoni* var. ______ Jacobi, 1928**

***P. semenovi* Oshanin, 1906** = *Tibicen semenovi* = *Tibicina semenovi* = *Psalmocharias semenovi*

***P. zevara* (Kusnezov, 1931)** = *Tibicina zevara* = *Tibucina* (sic) *zevara* = *Paharia zevara*
Tibucina (sic) = *Paharia zevara* Grechkin 1956: 1489 (comp. note) Tajikistan
Paharia zevara Popov 1985: 36, 40–41, 71, 80, Fig 4, Fig 7.6, Fig 23.6, Fig 24 (illustrated, oscillogram, acoustic system, comp. note)
Paharia zevara Popov, Aronov and Sergeeva 1985a: 451, 453–456, 458–461, Fig 3, Figs 5a–b, Fig 8e (song, oscillogram, hearing, comp. note) Tadzhikistan
Paharia zevara Popov, Aronov and Sergeeva 1985b: 289–290, 292–294, 296–297, Fig 3, Figs 5a–b, Fig 8e (song, oscillogram, hearing, comp. note) Tadzhikistan
Paharia zevara Popov 1990a: 302–303, Fig 2 (comp. note)
Paharia zevara Tautz 1990: 105, Fig 12 (song frequency, comp. note)
Paharia zevara Popov, Aronov and Sergeeva 1991: 281 (comp. note)
Paharia zevara Luk"yanova 1992: 31 (listed, comp. note) Tajikistan
Paharia zevara Popov, Arnov and Sergeeva 1992: 164 (comp. note)
Paharia zevara Sueur 2001: 41, Table 1 (listed)

***Tibicinoides* Distant, 1914** = *Tibicenoides* (sic)
Tibicinoides Heath 1983: 12 (comp. note) North America

Tibicinoides Arnett 1985: 217 (diversity) Southwestern United States, Utah, Montana, Colorado, Kansas
Tibicinoides Poole, Garrison, McCafferty, Otte and Stark 1997: 251, 333 (listed) Equals *Tibicenoides* (sic) North America
Tibicinoides Sanborn 1999b: 2 (listed) North America
Tibicinoides Arnett 2000: 299 (diversity) Southwestern United States, Utah, Montana, Colorado, Kansas
Tibicinoides Kondratieff, Ellingson and Leatherman 2002: 33 (comp. note)
Tibicinoides Sueur 2002b: 124 (comp. note)
Tibicinoides Moulds 2005b: 416 (comp. note)

***T. cupreosparsa* (Uhler, 1889)** = *Tibicen cupreosparsa* = *Tibicen cupreosparsus* = *Tibicina cupreosparsa* = *Tibicinoides cupreosparsus* = *Tibicen cupressparsa* (sic)
Tibicen cupreosparsa Henshaw 1903: 230 (listed) California
Tibicinoides cupreosparsa Varley 1939: 99 (comp. note)
Tibicinoides cupreosparsa Sugden 1940: 125 (song, comp. note) California
Tibicinoides cupreosparsa Powell and Hogue 1979: 114 (described, distribution, comp. note) California
Tibicinoides cupreosparsus Hogue 1993a: 36, 123, Fig 123 (described, illustrated, comp. note) California
Tibicinoides cupreosparsa Poole, Garrison, McCafferty, Otte and Stark 1997: 333 (listed) Equals *Tibicen cupreossparsa* (sic) Equals *Tibicinoides cupreosparsus* (sic) North America
Tibicinoides cupreosparsa Sanborn 1999a: 50 (type material, listed)
Tibicinoides cupreosparsa Sanborn 1999b: 2 (listed) North America
Tibicinoides cupreosparsa Sanborn 2007b: 38 (distribution, comp. note) Mexico, Baja California Norte, California

T. hesperia (Uhler, 1872) to *Okanagana hesperia*

***T. mercedita* (Davis, 1915)** = *Okanagana mercedita*
Tibicinoides mercedita Sugden 1940: 125 (song, comp. note) California
Tibicinoides mercedita Poole, Garrison, McCafferty, Otte and Stark 1997: 333 (listed) Equals *Okanagana mercedita* North America
Tibicinoides mercedita Sanborn 1999a: 50–51 (type material, listed)
Tibicinoides mercedita Sanborn 1999b: 2 (listed) North America

***T. minuta* (Davis, 1915)** = *Okanagana minuta*
Okanagana minuta Poole, Garrison, McCafferty, Otte and Stark 1997: 332 (listed) Equals *Cicada* North America
Tibicinoides minuta Sanborn 1999a: 51 (type material, listed)
Tibicinoides minuta Sanborn 1999b: 2 (listed) North America

***Okanagana* Distant, 1905** = *Okanagama* (sic) = *Okanaga* (sic) = *Okanogana* (sic) = *Ocanagana* (sic) = *Kanagana* (sic)
Okanagana Varley 1939: 99 (comp. note)
Okanagana Sugden 1940: 117–118, 120, 123–124 (comp. note)
Okanagana Zeuner 1944: 113 (comp. note)
Okanagana Evans 1966: 103 (parasitism, comp. note)
Okanagana sp. Clark 1983: 30–31, Fig 1 (ant association, nymph illustrated, comp. note) Idaho
Okanagana Heath 1983: 12–13 (comp. note) North America
Okanagana Strümpel 1983: 147, Fig 209E (wing illustrated, comp. note)
Okanagana Arnett 1985: 217 (diversity) United States
Okanagana spp. Maier 1985: 1024 (comp. note) Michigan, Wisconsin
Okanagana Brown and Brown 1990: 5–6 (comp. note)
Okanagana Martin and Simon 1990b: 362 (comp. note)
Okanagana Moore 1990: 58 (comp. note)
Okanagana spp. Cranshaw and Kondratieff 1991: 10 (life cycle, comp. note) Colorado
Okanagana Johnson and Lyon 1991: 490 (comp. note)
Kanagana (sic) Liu 1992: 42 (comp. note)
Okanagana spp. Moore 1993: 279, 282 (comp. note)
Okanagana Heath and Heath 1994: 17 (comp. note) Utah
Okanagana Cranshaw and Kondratieff 1995: 79 (life history)
Okanagana Moore 1996: 221 (comp. note) Mexico
Okanagana Chou, Lei, Li, Lu and Yao 1997: 90 (comp. note)
Okanagana Poole, Garrison, McCafferty, Otte and Stark 1997: 251, 332 (listed) Equals *Okanagama* (sic) Equals *Okanaga* (sic) North America
Okanagana Yoshimura 1997: 120 (comp. note) North America
Okanagana Flook and Rowell 1998: 167–170, Figs 3–6 (phylogeny)
Okanagana Kocher, Thomas, Meyer, Edwards, Pääbo et al. 1998: 6197 (sequence, comp. note)
Okanagana sp. Hamilton 1999: 227, Fig 68 (illustrated, comp. note)
Okanagana Ohbayashi, Sato and Igawa 1999: 340 (parasitism, comp. note)

Okanagana Sanborn 1999b: 2 (listed) North America
Okanagana Arnett 2000: 299 (diversity) United States
Okanagana Maw, Footit, Hamilton and Scudder 2000: 49 (listed)
Okanagana Milius 2000: 408 (comp. note) Arizona
Okanagana Cioara and Sanborn 2001: 25 (song, comp. note)
Okanagana Cooley 2001: 756, 758 (comp. note) North America
Okanagana Cooley and Marshall 2001: 847 (comp. note) North America
Okanagana Sanborn and Webb 2001: 451 (comp. note)
Okanagana Kondratieff, Ellingson and Leatherman 2002: 14, 28, 33, 43–44 (key, diagnosis, listed, diversity, comp. note) Colorado
Okanagana spp. Kondratieff, Ellingson and Leatherman 2002: 34 (comp. note)
Okanagana O'Geen, McDaniel and Busacca 2002: 1585 (burrowing activity, comp. note) Washington, Oregon, Idaho
Okanagana Sanborn, Breitbarth, Heath and Heath 2002: 446 (comp. note)
Okanagana Sueur 2002b: 124 (comp. note)
Okanagana spp. Villet, Sanborn and Phillips 2003c: 1442 (comp. note)
Okanagana Cooley, Marshall and Simon 2004: 201 (comp. note)
Okanagana Sanborn, Heath, Heath and Phillips 2004: 70 (listed)
Okanagana Sanborn, Heath, Heath, Noriega and Phillips 2004: 282 (comp. note)
Okanagana Stölting, Moore and Lakes-Harlan 2004: 251 (comp. note)
Okanagana spp. Stölting, Moore and Lakes-Harlan 2004: 254 (comp. note)
Okanagana Sueur and Aubin 2004: 223 (comp. note)
Okanagana Moulds 2005b: 415–416 (comp. note)
Okanagana Elliott and Hershberger 2006: 184 (listed)
Okanagana spp. Marshall 2006: 139 (comp. note) North America
Okanagana Eaton and Kaufman 2007: 88–89 (diversity, illustrated, comp. note) United States, Canada
Okanagana Ganchev and Potamitis 2007: 304–305, 319, Fig 11 (listed, comp. note)
Okanagana Heath and Sanborn 2007: 483–484 (key, comp. note) Arizona
Okanagana spp. Heath and Sanborn 2007: 487–488 (comp. note) Arizona
Okanagana Shiyake 2007: 7, 94–95 (listed, comp. note)
Okanagana Sueur, Vanderpool, Simon, Ouvrard and Bourgoin 2007: 612, 624 (comp. note)
Okanagana spp. Sueur, Vanderpool, Simon, Ouvrard and Bourgoin 2007: 624 (comp. note)
Okanagana spp. Fonseca, Serrão, Pina-Martins, Silva, Mira, et al. 2008: 29 (comp. note) America
Okanagana Fonseca, Serrão, Pina-Martins, Silva, Mira, et al. 2008: 30 (comp. note) America
Okanagana Cooley, Kritsky, Edwards, Zyla, Marshall et al. 2009: 107 (distribution, comp. note) Maine, Canada
Okanagana Holliday, Hastings and Coelho 2009: 12–13 (comp. note)
Okanagana Sanborn 2009c: 869 (comp. note)

O. *annulata* Davis, 1935
Okanagana annulata Sanborn and Phillips 1995a: 481, Table 1 (song intensity, comp. note)
Okanagana annulata Poole, Garrison, McCafferty, Otte and Stark 1997: 332 (listed) North America
Okanagana annulata Sanborn 1999a: 46 (type material, listed)
Okanagana annulata Sanborn 1999b: 2 (listed) North America

O. *arboraria* Wymore, 1934 = *Okanagana arboraria crocea* Wymore, 1934
Okanagana arboraria Poole, Garrison, McCafferty, Otte and Stark 1997: 332 (listed) Equals *Okanagana crocea* North America
Okanagana arboraria Sanborn 1999a: 46 (type material, listed)
Okanagana arboraria crocea Sanborn 1999a: 46 (type material, listed)
Okanagana arboraria Sanborn 1999b: 2 (listed) North America

O. *arctostaphylae* Van Duzee, 1915
Okanagana arctostaphylae Lloyd 1984: 82 (comp. note)
Okanagana arctostaphylae Poole, Garrison, McCafferty, Otte and Stark 1997: 332 (listed) North America
Okanagana arctostaphylae Sanborn 1999a: 46 (type material, listed)
Okanagana arctostaphylae Sanborn 1999b: 2 (listed) North America

O. *aurantiaca* Davis, 1917
Okanagana aurantiaca Sanborn 1999a: 46 (type material, listed)
Okanagana aurantiaca Sanborn 2007b: 38 (distribution, comp. note) Mexico, Baja California Sur

O. *aurora* Davis, 1936
Okanagana aurora Poole, Garrison, McCafferty, Otte and Stark 1997: 332 (listed) North America

Okanagana aurora Sanborn 1999a: 46 (type material, listed)
Okanagana aurora Sanborn 1999b: 2 (listed) North America

***O. balli* Davis, 1919**
Okanagana balli Poole, Garrison, McCafferty, Otte and Stark 1997: 332 (listed) North America
Okanagana balli Sanborn 1999a: 46 (type material, listed)
Okanagana balli Sanborn 1999b: 2 (listed) North America

***O. bella* Davis, 1919** = *Okanagana belli* (sic) = *Okanagana rubrocaudata* Davis, 1925 = *Okanagana bella* var. *rubrocaudata* = *Okanagana bella rubrocaudata* = *Okanagana rubracaudata* (sic)
Okanagana bella Davis 1922: 73 (comp. note) Oregon
Okanagana bella Sugden 1940: 117, 120–121 (habitat, song, comp. note) Utah
Okanagana bella Tinkham 1942: 155–156 (distribution, hosts, comp. note) Cananda, Alberta, Utah, Colorado, Kansas, New Mexico, Wyoming, Montana, Idaho, Washington, Oregon, California
Okanagana belli (sic) Heath, Heath, Sanborn and Noriega 1990: 1 (comp. note) Arizona
Okanagana bella Cranshaw and Kondratieff 1991: 10 (comp. note) Colorado
Okanagana bella Sanborn and Phillips 1995a: 481, Table 1 (song intensity, comp. note)
Okanagana bella Poole, Garrison, McCafferty, Otte and Stark 1997: 332 (listed) North America
Okanagana rubrocaudata Poole, Garrison, McCafferty, Otte and Stark 1997: 332 (listed) Equals *Okanagana rubracaudata* (sic) North America
Okanagana bella Sanborn 1999a: 46 (type material, listed)
Okanagana rubrocaudata Sanborn 1999a: 48 (type material, listed)
Okanagana bella Sanborn 1999b: 2 (listed) North America
Okanagana rubrocaudata Sanborn 1999b: 2 (listed) North America
Okanagana bella Maw, Footit, Hamilton and Scudder 2000: 49 (comp. note) Canada, British Columbia, Alberta
Okanagana bella Sanborn and Webb 2001: 451–452 (comp. note)
Okanagana bella Kondratieff, Ellingson and Leatherman 2002: 7, 12–13, 31, 35–39, 45, 78, Fig 44, Tables 1–2 (key, diagnosis, illustrated, listed, distribution, song, comp. note) Equals *Okanagana bella* var. *rubrocaudata* Equals *Okanagana rubrocaudata* Colorado, Arizona, Alberta, British Colombia, California, Idaho, Kansas, Montana, Nevada, New Mexico, Oregon, South Dakota, Utah, Wyoming, Washington
Okanagana bella bella Kondratieff, Ellingson and Leatherman 2002: 37 (comp. note) Colorado
Okanagana bella rubrocaudata Kondratieff, Ellingson and Leatherman 2002: 10, 37 (comp. note) Colorado
Okanagana bella Sanborn, Heath, Heath, Noriega and Phillips 2004: 282–286, Fig 3, Tables 1–4 (illustrated, thermal responses, habitat, distribution, comp. note) Arizona, New Mexico, Colorado, Utah, Idaho, Wyoming, Montana, South Dakota
Okanagana bella Heath and Sanborn 2007: 484 (key) Arizona
Okanagana rubrocaudata Heath and Sanborn 2007: 484 (key) Arizona
Okanagana bella Shiyake 2007: 7, 92, 94, Fig 153 (illustrated, distribution, listed, comp. note) United States, Canada
Okanagana bella Sanborn 2009c: 867, 872 (comp. note)

O. californica Distant, 1914 to *Okanagana vanduzeei* Distant, 1914

***O. canadensis* (Provancher, 1889)** = *Cicada canadensis* = *Tibicen canadensis*
Okanagana canadensis Tinkham 1942: 155 (distribution, hosts, comp. note) Canada, Alberta, New Brunswick, New Hampshire, Pennsylvania, Michigan, South Dakota, Colorado
Okanagana canadensis Soper 1974: 2801-B (fungal host, comp. note)
Okanagana canadensis Brown and Brown 1990: 5 (distribution, comp. note) Canada, United States
Okanagana canadensis Sanborn and Phillips 1995a: 481, Table 1 (song intensity, comp. note)
Okanagana canadensis Marshall, Cooley, Alexander and Moore 1996: 165 (comp. note) Michigan
Okanagana canadensis Poole, Garrison, McCafferty, Otte and Stark 1997: 332 (listed) Equals *Cicada canadensis* North America
Okanagana canadensis Yoshimura 1997: 120 (comp. note) North America
Okanagana canadensis Lakes-Harlan, Stölting and Stumpner 1999: 35 (comp. note) Michigan
Okanagana canadensis Sanborn 1999b: 2 (type material, listed) North America
Okanagana canadensis Maw, Footit, Hamilton and Scudder 2000: 49 (comp. note) Canada, Northwest Territory, British Columbia, Alberta, Saskatchewan, Manitoba, Ontario, Quebec, New Brunswick

Okanagana canadensis Cooley 2001: 756–758, Fig 1 (pair formation, song, sonogram, comp. note) Michigan
Okanagana canadensis Cooley and Marshall 2001: 847–848 (comp. note)
Okanagana canadensis Köhler and Lakes-Harlan 2001: 586 (comp. note) Michigan
Okanagana canadensis Sueur 2001: 41, 46, Tables 1–2 (listed)
Okanagana canadensis Kondratieff, Ellingson and Leatherman 2002: 10–12-13, 31, 35, 38, 45, Tables 1–2 (key, diagnosis, listed, distribution, song, comp. note) Equals *Cicada canadensis* Colorado, Alberta, Maine, Manitoba, New Brunswick, New Hampshire, New York, Michigan, Ontario, Pennsylvania, Quebec, South Dakota, Utah
Okanagana canadensis Schneiderkötter and Lakes-Harlan 2004: 2, 6 (comp. note) Michigan
Okanagana canadensis Stölting, Moore and Lakes-Harlan 2004: 244–245, 248, 250, 252–253, 256, Fig 5 (song, oscillogram, comp. note) Michigan
Okanagana canadensis Sueur and Aubin 2004: 223 (comp. note)
Okanagana canadensis Vries and Lakes-Harlan 2005: 240, 244 (comp. note) Michigan
Okanagana canadensis Elliott and Hershberger 2006: 32, 204–205 (illustrated, sonogram, listed, distribution, comp. note) North Dakota, South Dakota, Minnesota, Wisconsin, Michigan, New York, Pennsylvania, Vermont, New Hampshire, Maine, New Brunswick, Quebec, Ontario, Manitoba
Okanagana canadensis Marshall 2006: 139 (illustrated, comp. note) North America
Okanagana canadensis Shiyake 2007: 7, 92, 94–95, Fig 154 (illustrated, distribution, listed, comp. note) United States, Canada
Okanagana canadensis Sueur, Vanderpool, Simon, Ouvrard and Bourgoin 2007: 613–614, 616–619, 621, 624, Fig 1, Figs 3–5, Tables 1–2 (comp. note) Michigan
Okanagana canadensis Sanborn 2009c: 872 (comp. note)
Okanagana canadensis Wei, Tang, He and Zhang 2009: 340 (comp. note)

***O. canescens* Van Duzee, 1915** = *Okanagana albibasalis* Wymore, 1934 = *Okanagana alibasalis* (sic)
Okanagana canescens Poole, Garrison, McCafferty, Otte and Stark 1997: 332 (listed) Equals *Okanagana albibasalis* North America
Okanagana albibasalis Sanborn 1999a: 46 (type material, listed)
Okanagana canescens Sanborn 1999a: 46 (type material, listed)
Okanagana canescens Sanborn 1999b: 2 (listed) North America

***O. catalina* Davis, 1936** = *Okanagana hirsuta* var. *catalina*
Okanagana catalina Miller 1985: 18 (rev. status, listed) California, Channel Islands
Okanagana hirsuta var. *catalina* Sanborn 1999a: 47 (type material, listed)
Okanagana hirsuta var. *catalina* Sanborn 1999b: 2 (listed) North America

O. consobrina Distant, 1914 to *Okanagana vanduzeei* Distant, 1914

***O. cruentifera* (Uhler, 1892)** = *Tibicen cruentifera* = *Tibicina cruentifera* = *Tibicens* (sic) *cruentifera* = *Tibicen cruentiferus* = *Cicada cruentifera*
Tibicen cruentifera Henshaw 1903: 230 (listed) nevada
Okanagana cruentifera Powell and Hogue 1979: 113 (host, distribution, comp. note) California
Okanagana cruentifera Smiley and Giuliani 1991: 258 (host, comp. note) California
Okanagana cruentifera Sanborn and Phillips 1995a: 481, Table 1 (song intensity, comp. note)
Okanagana cruentifera Poole, Garrison, McCafferty, Otte and Stark 1997: 332 (listed) Equals *Tibicen cruentifera* North America
Okanagana cruentifera Sanborn 1999a: 46 (type material, listed)
Okanagana cruentifera Sanborn 1999b: 2 (listed) North America
Okanagana cruentifera Kondratieff, Ellingson and Leatherman 2002: 12, 34, 44, Table 1 (listed, diagnosis, comp. note) Equals *Tibicen cruentifera* California, Montana, Nevada, Utah, Washington
Okanagana cruentifera Shiyake 2007: 7, 92, 94, Fig 155 (illustrated, distribution, listed, comp. note) United States

***O. ferrugomaculata* Davis, 1936**
Okanagana ferrugomaculata Poole, Garrison, McCafferty, Otte and Stark 1997: 332 (listed) North America
Okanagana ferrugomaculata Sanborn 1999a: 46 (type material, listed)
Okanagana ferrugomaculata Sanborn 1999b: 2 (listed) North America

***O. formosa* Davis, 1926**
Okanagana formosa Poole, Garrison, McCafferty, Otte and Stark 1997: 332 (listed) North America
Okanagana formosa Sanborn 1999a: 47 (type material, listed)
Okanagana formosa Sanborn 1999b: 2 (listed) North America

***O. fratercula* Davis, 1915**
Okanagana fratercula Tinkham 1942: 156 (comp. note)

Okanagana fratercula Heath and Heath 1994: 17 (type locality, comp. note) Utah
Okanagana fratercula Poole, Garrison, McCafferty, Otte and Stark 1997: 332 (listed) North America
Okanagana fratercula Sanborn 1999a: 47 (type material, listed)
Okanagana fratercula Sanborn 1999b: 2 (listed) North America
Okanagana fratercula Maw, Footit, Hamilton and Scudder 2000: 49 (comp. note) Canada, Alberta

O. *fumipennis* Davis1932
Okanagana fumipennis Sugden 1940: 124 (behavior, song, comp. note) Utah
Okanagana fumipennis Poole, Garrison, McCafferty, Otte and Stark 1997: 332 (listed) North America
Okanagana fumipennis Sanborn 1999a: 47 (type material, listed)
Okanagana fumipennis Sanborn 1999b: 2 (listed) North America
Okanagana fumipennis Kondratieff, Ellingson and Leatherman 2002: 8, 13, 30–32, Table 2 (key, diagnosis, listed, distribution, song, comp. note) Colorado, Arizona, New Mexico, Utah
Okanagana fumipennis Heath and Sanborn 2007: 484 (key) Arizona

O. *georgi* Heath & Sanborn, 2007
Okanagana georgi Sanborn 1999b: 2 (listed) North America
Okanagana georgi Heath and Sanborn 2007: 484–488, Figs 1–5 (n. sp., described, illustrated, key, natural history, song, comp. note) Arizona

O. *gibbera* Davis, 1927
Okanagana gibbera Poole, Garrison, McCafferty, Otte and Stark 1997: 332 (listed) North America
Okanagana gibbera Sanborn 1999a: 47 (type material, listed)
Okanagana gibbera Sanborn 1999b: 2 (listed) North America
Okanagana gibbera Kondratieff, Ellingson and Leatherman 2002: 10, 12–13, 30, 32, 40, 42, Tables 1–2 (key, diagnosis, listed, distribution, song, comp. note) Colorado, California, Nevada, Oregon, Utah, Washington, Wyoming

O. *hesperia* (Uhler, 1872) = *Cicada hesperia* = *Tibicen hesperina* (sic) = *Tibicen hesperia* = *Tibicina hesperia* = *Tibicinoides hesperia* = *Tibicinoides hesperius* (sic) = *Okanagana striatipes* var. *beameri* Davis, 1930 = *Okanagana striatipes beameri*
Tibicen hesperia Henshaw 1903: 230 (listed) Colorado
Tibicinoides hesperius (sic) Varley 1939: 99 (comp. note)
Okanagana hesperia Sanborn, J.E. Heath, M.S. Heath and Noriega 1995: 326, Table 3 (thermal responses, comp. note)
Okanagana hesperia Sanborn, M.S. Heath, J.E. Heath and Noriega 1995: 456, Table 4 (comp. note)
Okanagana hesperia Sanborn and Phillips 1995a: 480–481, Table 1 (song intensity, comp. note)
Tibicinoides hesperia Poole, Garrison, McCafferty, Otte and Stark 1997: 333 (listed) Equals *Cicada hesperia* North America
Okanagana hesperia Sanborn 1998: 98–99, Table 1 (thermal responses, comp. note)
Okanagana hesperia Sanborn 1999a: 47 (type material, listed)
Okanagana striatipes var. *beameri* Sanborn 1999a: 49 (type material, listed)
Okanagana hesperia Sanborn 1999b: 2 (listed) North America
Okanagana hesperia Kondratieff, Ellingson and Leatherman 2002: 5, 13, 29, 32–33, 42, 73, Fig 30, Table 2 (key, diagnosis, illustrated, listed, distribution, song, comp. note) Equals *Cicada hesperia* Equals *Tibicinoides hesperia* Equals *Okanagana striatipes beameri* Colorado, Arizona, Kansas, Oklahoma, Montana, Nevada, Texas, Utah
Tibicinoides hesperia Kondratieff, Ellingson and Leatherman 2002: 12, Table 1 (listed) Colorado
Okanagana hesperia Sanborn 2002b: 465, Table 1 (thermal biology, comp. note)
Okanagana hesperia Sanborn, Breitbarth, Heath and Heath 2002: 443 (comp. note)
Okanagana hesperia Sanborn, Noriega and Phillips 2002c: 368 (comp. note)
Okanagana hesperia Sanborn 2005c: 114, Table 1 (comp. note)
Okanagana striatipes Sanborn 2005c: 114, Table 1 (comp. note)
Okanagana hesperia Heath and Sanborn 2007: 484, 488 (key, comp. note) Arizona

O. *hirsuta* Davis, 1915
Okanagana hirsuta Miller 1985: 18–19 (listed, comp. note) California, Channel Islands
Okanagana hirsuta Hennessey 1990: 471 (type material)
Okanagana hirsuta Poole, Garrison, McCafferty, Otte and Stark 1997: 332 (listed) Equals *Okanagana catalina* North America
Okanagana hirsuta Sanborn 1999a: 47 (type material, listed)
Okanagana hirsuta Sanborn 1999b: 2 (listed) North America

O. *hirsuta* var. *catalina* Davis, 1936 to *Okanagana catalina*

O. *lurida* Davis, 1919

Okanagana lurida Poole, Garrison, McCafferty, Otte and Stark 1997: 332 (listed) North America

Okanagana lurida Sanborn 1999a: 47 (type material, listed)

Okanagana lurida Sanborn 1999b: 2 (listed) North America

Okanagana lurida Maw, Footit, Hamilton and Scudder 2000: 49 (comp. note) Canada, British Columbia

O. *luteobasalis* Davis, 1935 = *Okanagana fratercula* (nec Davis)

Okanagana luteobasalis Tinkham 1942: 156 (distribution, hosts, comp. note) Equals *Okanagana fratercula* (partim) Canada, Alberta, North Dakota, Montana, Oregon, Idaho, Utah

Okanagana luteobasalis Sugden 1940: 122 (habitat, comp. note) Utah

Okanagana luteobasalis Orians and Horn 1969: 935 (predation) Washington

Okanagana luteobasalis Poole, Garrison, McCafferty, Otte and Stark 1997: 332 (listed) North America

Okanagana luteobasalis Sanborn 1999a: 47 (type material, listed)

Okanagana luteobasalis Sanborn 1999b: 2 (listed) North America

Okanagana luteobasalis Maw, Footit, Hamilton and Scudder 2000: 49 (comp. note) Canada, Alberta

O. *magnifica* Davis, 1919 = *Okanagana magnica* (sic)

Okanagana magnifica Cranshaw and Kondratieff 1995: 79 (comp. note) Colorado, New Mexico, Utah

Okanagana magnifica Sanborn and Phillips 1995a: 481, Table 1 (song intensity, comp. note)

Okanagana magnifica Poole, Garrison, McCafferty, Otte and Stark 1997: 332 (listed) North America

Okanagana magnifica Sanborn 1999a: 47 (type material, listed)

Okanagana magnifica Sanborn 1999b: 2 (listed) North America

Okanagana magnifica Kondratieff, Ellingson and Leatherman 2002: 9, 12–13, 30, 34–35, 44, 66, Fig 9, Tables 1–2 (key, diagnosis, illustrated, listed, distribution, song, comp. note) Colorado, Arizona, California, New Mexico

Okanagana magnifica Heath and Sanborn 2007: 484, 487 (key, comp. note) Arizona

Okanagana magnica (sic) Shiyake 2007: 7, 92, 94, Fig 156 (illustrated, distribution, listed, comp. note) United States

Okanagana magnifica Phillips and Sanborn 2010: 75 (comp. note)

O. *mariposa mariposa* Davis, 1915 = *Okanagana mairposa*

Okanagana mariposa Poole, Garrison, McCafferty, Otte and Stark 1997: 332 (listed) Equals *Okanagana oregonensis* North America

Okanagana mariposa Sanborn 1999a: 47 (type material, listed)

Okanagana mariposa Sanborn 1999b: 2 (listed) North America

Okanagana mariposa Heath and Sanborn 2007: 484, 487 (key, comp. note) Arizona

O. *mariposa oregonensis* Davis, 1939 = *Okanagana mariposa* var. *oregonensis*

Okanagana mariposa var. *oregonensis* Sanborn 1999a: 47 (type material, listed)

Okanagana mariposa var. *oregonensis* Sanborn 1999b: 2 (listed) North America

O. *napa* Davis, 1919

Okanagana napa Poole, Garrison, McCafferty, Otte and Stark 1997: 332 (listed) North America

Okanagana napa Sanborn 1999a: 47 (type material, listed)

Okanagana napa Sanborn 1999b: 2 (listed) North America

O. *nigriviridis* Davis, 1921 = *Okanagana nigrinervis* (sic)

Okanagana nigriviridis Watts and Williams 1992: 381 (coloration, host plants, comp. note) California

Okanagana nigriviridis Poole, Garrison, McCafferty, Otte and Stark 1997: 332 (listed) Equals *Okanagana niginervis* (sic) North America

Okanagana migriviridis Sanborn 1999a: 47–48 (type material, listed)

Okanagana nigriviridis Sanborn 1999b: 2 (listed) North America

O. *nigrodorsata* Davis, 1923

Okanagana nigrodorsata Poole, Garrison, McCafferty, Otte and Stark 1997: 332 (listed) North America

Okanagana nigrodorsata Sanborn 1999a: 48 (type material, listed)

Okanagana nigrodorsata Sanborn 1999b: 2 (listed) North America

O. *noveboracensis* (Emmons, 1854) = *Cicada noveboracensis* = *Cicada novaeboracensis* (sic) = *Tibicen noveboracensis* = *Tibicen rimosa noveboracensis* = *Tibicen novaboracensis* (sic) = *Okanagana novaeboracensis* (sic) = *Okanogana* (sic) *noveboracensis*

Okanagana noveboracensis Poole, Garrison, McCafferty, Otte and Stark 1997: 332 (listed) Equals *Cicada noveboracensis* Equals *Cicada novaeboracensis* (sic) Equals *Tibicen novaboracensis* (sic) North America

Okanagana noveboracensis Sanborn 1999b: 2 (type material, listed) North America

Okanagana noveboracensis Maw, Footit, Hamilton and Scudder 2000: 49 (comp. note) Canada, Manitoba, Ontario
Okanagana noveboracensis Sanborn 2009c: 867–872, Figs 1–4 (neotype designation, synonymy, described, illustrated, comp. note) Equals *Cicada noveboracensis* Equals *Cicada novaeboracensis* (sic) Equals *Tibicen noveboracensis* Equals *Tibicen novaboacensis* (sic) Equals *Okanagana* (sic) *noveeboracensis* New York, Ontario

O. *occidentalis* (Walker, 1866) = *Cicada occidentalis* = *Tibicen occidentalis*
Okanagana occidentalis Davis 1922: 73 (comp. note) Oregon
Okanagana occidentalis Varley 1939: 99 (comp. note)
Okanagana occidentalis Poole, Garrison, McCafferty, Otte and Stark 1997: 332 (listed) Equals *Cicada occidentalis* North America
Okanagana occidentalis Sanborn 1999b: 2 (type material, listed) North America
Okanagana occidentalis Maw, Footit, Hamilton and Scudder 2000: 49 (comp. note) Canada, British Columbia, Alberta, Manitoba
Okanagana occidentalis Sanborn and Webb 2001: 451–452 (lectotype designation, type locality, comp. note) Canada, British Columbia
Cicada occidentalis Sanborn and Webb 2001: 451 (comp. note)
Okanagana occidentalis Kondratieff, Ellingson and Leatherman 2002: 8–9, 13, 30, 35–36, 38–39, 41, 45, 55, Table 2 (key, diagnosis, listed, distribution, song, comp. note) Equals *Cicada occidentalis* Colorado, Alberta, California, British Colombia, Idaho, Montana, Manitoba, Nevada, Oregon, Utah, Washington
Okanagana occidentalis Sanborn 2009c: 867, 872 (comp. note)

O. *opacipennis* Davis, 1926 = *Okanagana arctostaphylae opacipennis*
Okanagana opacipennis Lloyd 1984: 82 (comp. note)
Okanagana opacipennis Watts and Williams 1992: 380 (coloration, host plants, comp. note) California
Okanagana opacipennis Poole, Garrison, McCafferty, Otte and Stark 1997: 332 (listed) North America
Okanagana opacipennis Sanborn 1999a: 48 (type material, listed)
Okanagana opacipennis Sanborn 1999b: 2 (listed) North America

O. *oregona* Davis, 1916
Okanagana oregona Davis 1922: 73 (comp. note) Oregon
Okanagana oregona Poole, Garrison, McCafferty, Otte and Stark 1997: 332 (listed) North America
Okanagana oregona Sanborn 1999a: 48 (type material, listed)
Okanagana oregona Sanborn 1999b: 2 (listed) North America
Okanagana oregona Maw, Footit, Hamilton and Scudder 2000: 49 (comp. note) Canada, British Columbia
Okanagana oregona Shiyake 2007: 7, 93–94, Fig 157 (illustrated, distribution, listed, comp. note) United States

O. *orithyia* Bliven, 1964
Okanagana orithyia Poole, Garrison, McCafferty, Otte and Stark 1997: 332 (listed) North America
Okanagana orithyia Sanborn 1999a: 48 (type material, listed)
Okanagana orithyia Sanborn 1999b: 2 (listed) North America

O. *ornata* Van Duzee, 1915
Okanagana ornata Hennessey 1990: 471 (type material)
Okanagana ornata Poole, Garrison, McCafferty, Otte and Stark 1997: 332 (listed) North America
Okanagana ornata Sanborn 1999a: 48 (type material, listed)
Okanagana ornata Sanborn 1999b: 2 (listed) North America
Okanagana ornata Maw, Footit, Hamilton and Scudder 2000: 49 (comp. note) Canada, British Columbia
Okanagana ornata Shiyake 2007: 7, 93–94, Fig 158 (illustrated, distribution, listed, comp. note) United States

O. *pallidula* Davis, 1917 = *Okanagana pallidula nigra* Davis, 1938 = *Okanagana pallidula* var. *nigra* = *Okanagana nigra*
Okanagana pallidula Hennessey 1990: 471 (type material)
Okanagana pallidula Sanborn and Phillips 1995a: 481, Table 1 (song intensity, comp. note)
Okanagana pallidula Sanborn and Phillips 1995b: 374–375, Table 1 (song intensity, comp. note) California
Okanagana pallidula Poole, Garrison, McCafferty, Otte and Stark 1997: 332 (listed) Equals *Okanagana nigra* North America
Okanagana pallidula Sanborn 1999a: 48 (type material, listed)
Okanagana pallidula var. *nigra* Sanborn 1999a: 48 (type material, listed)
Okanagana pallidula Sanborn 1999b: 2 (listed) North America

O. *pernix* Bliven, 1964

Okanagana pernix Poole, Garrison, McCafferty, Otte and Stark 1997: 332 (listed) North America

Okanagana pernix Sanborn 1999a: 48 (type material, listed)

Okanagana pernix Sanborn 1999b: 2 (listed) North America

O. *rhadine* Bliven, 1964

Okanagana rhadine Poole, Garrison, McCafferty, Otte and Stark 1997: 332 (listed) North America

Okanagana rhadine Sanborn 1999a: 48 (type material, listed)

Okanagana rhadine Sanborn 1999b: 2 (listed) North America

O. *rimosa ohioensis* Davis, 1942 = *Okanagana rimosa* var. *ohioensis*

Okanagana rimosa var. *ohioensis* Sanborn 1999b: 2 (listed) North America

Okanagana rimosa ohioensis Sanborn 2009c: 872 (comp. note)

O. *rimosa rimosa* (Say, 1830) = *Cicada rimosa* = *Tibicen rimosa* = *Okanagana rinosa* (sic) = *Okanaga* (sic) *rimosa* = *Okanagana rimrosa* (sic) = *Ocanagana* (sic) *rimosa* = *Okanagana rumorsa* (sic)

Cicada rimosa Putnam 1881: 67 (host, comp. note) Colorado, Wyoming, Nevada, Utah

Tibicen rimosa Van Duzee 1893: 410 (listed) New York

Okanagana rimosa Sugden 1940: 119 (comp. note)

Okanagana rimosa Tinkham 1942: 156 (comp. note) Canada, Manitoba, North Dakota

Okanagana rimosa Luthy 1974: 791–794 (bacterial host, comp. note)

Okanagana rimosa Soper 1974: 2801-B (fungal host, comp. note)

Okanagana rimosa Mangold 1978: 57 (parasitism, comp. note)

Okanagana rimosa Soper 1981: 53 (fungal host, comp. note)

Okanagana rimosa Hutchins and Lewis 1983: 195 (hearing organ, comp. note)

Okanagana rimosa Tyrell and Weston 1983: 1 (fungal host, comp. note) Ontario

Okanagana rimosa Huber 1985: 57, Fig 3 (neural responses, comp. note)

Okanagana rimosa Michelsen and Larsen 1985: 548, Fig 42 (auditory neurons, illustrated, comp. note)

Okanagana rimosa Popov 1985: 24 (acoustic system, comp. note)

Okanagana rimosa Karban 1986: 224, Table 14.2 (comp. note) USA

Okanagana rimosa Gwynne 1987: 571 (predation, comp. note)

Okanagana rimosa Brown and Brown 1990: 5–7 (distribution, comp. note) North Carolina, Tennessee, Pennsylvania, Ohio, Virginia, Michigan, Canada, United States

Okanagana rimosa Martin and Simon 1990b: 362 (comp. note)

Okanagana rumorsa (sic) Sutherland and De Jong 1991: 16 (comp. note)

Okanagana rimosa Hennig, Kleindienst, Weber, Huber, Moore and Popov 1992: 180, Fig B (hearing, comp. note)

Okanagana rimosa Villet 1992: 96 (comp. note)

Okanagana rimosa Hennig, Weber, Moore, Kleindienst, Huber and Popov 1993: 324–326, Figs 1–2 (song, oscillogram, tensor muscle, comp. note)

Okanagana rimosa Moore 1993: 278–279 (comp. note)

Okanagana rimosa Bennet-Clark and Young 1994: 293, Table 1 (body size and song frequency)

Okanagana rimosa Hennig, Weber, Moore, Kleindienst, Huber and Popov 1994: 46–47, 51, 53 (hearing, comp. note) Michigan

O[kanagana] rimosa Jiang, Yang, Tang, Xu and Chen 1995b: 229 (song frequency, comp. note)

Okanagana rimosa Williams and Simon 1995: 280 (comp. note)

Okanagana rimosa Marshall, Cooley, Alexander and Moore 1996: 165 (comp. note) Michigan

Okanagana rimosa Poole, Garrison, McCafferty, Otte and Stark 1997: 332 (listed) Equals *Cicada rimosa* Equals *Okanagana rinosa* (sic) Equals *Okanagana ohioensis* Equals *Okanagana rimrosa* (sic) North America

Okanagana rimosa Yoshimura 1997: 119–120 (comp. note) North America

Okanagana rimosa Gogala and Trilar 1998a: 8 (comp. note) North America

Okanagana rimosa Itô 1998: 494 (life cycle evolution, comp. note) Canada

Okanagana rimosa Haynes and Yeargan 1999: 965 (predation, comp. note)

Okanagana rimosa Lakes-Harlan, Stölting and Stumpner 1999: 32–33, 35–38, Fig 1, Table 2 (parasitism, oscillogram, spectrogram, comp. note) Michigan

Okanagana rimosa Sanborn 1998: 90 (thermal biology, comp. note)

Okanagana rimosa Sanborn 1999b: 2 (type material, listed) North America

Okanagana rimosa Lakes-Harlan, Stölting and Moore 2000: 1161, 1166 (parasitism, comp. note) Michigan

Okanagana rimosa Maw, Footit, Hamilton and Scudder 2000: 49 (comp. note) Canada, Alberta, Saskatchewan, Manitoba, Ontario, Quebec, New Brunswick, Nova Scotia

Okanagana rimosa Cooley 2001: 756–758, Fig 1, Table 1 (pair formation, song, sonogram, comp. note) Michigan

Okanagana rimosa Cooley and Marshall 2001: 847–848 (comp. note)

Okanagana rimosa Köhler and Lakes-Harlan 2001: 581–582, 584–585, Fig 1 (predation, oscillogram, comp. note) Michigan

Okanagana rimosa Sanborn and Webb 2001: 451–452 (comp. note)

Okanagana rimosa Sueur 2001: 33, 41, 46, Tables 1–2 (listed, comp. note)

Okanagana rimosa Gerhardt and Huber 2002: 380 (parasitism, comp. note)

Okanagana rimosa group Kondratieff, Ellingson and Leatherman 2002: 12 (comp. note)

Okanagana rimosa Kondratieff, Ellingson and Leatherman 2002: 28–29, 35, 40, 44–45 (type species of *Okanagana*, distribution, comp. note) Equals *Cicada rimosa* California, Idaho, Montana, Nevada, New Mexico, North Dakota, Oregon, South Dakota, Utah, Washington, Wyoming

Okanagana rimosa complex Kondratieff, Ellingson and Leatherman 2002: 30, 35–36 (key) Colorado

Okanagana rimosa Schneiderkötter and Lakes-Harlan 2002: 8 (parasitism, comp. note)

Okanagana rimosa Stölting, Moore and Lakes-Harlan 2002: 1–5, Fig 2 (vibrational communication, song, oscillogram, comp. note) Michigan

Okanagana rimosa Sueur 2002b: 125, 129 (listed, comp. note)

Okanagana rimosa Gogala and Trilar 2003: 9 (comp. note) North America

Okanagana rimosa Lakes-Harlan and Köhler 2003: 758–759 (parasitism) Michigan

Okanagana rimosa Schneiderkötter and Lakes-Harlan 2004: 1–6, Figs 1–5 (parasitism, illustrated, comp. note) Michigan

Okanagana rimosa Stölting, Moore and Lakes-Harlan 2004: 244–256, Figs 1–4, Figs 6–10 (song, oscillogram, comp. note) Michigan

Okanagana rimosa Sueur and Aubin 2004: 223 (comp. note)

Okanagana rimosa Grant 2005: 172–173, Table 1 (life cycle, comp. note) Michigan, Ontario

Okanagana rimosa Vries and Lakes-Harlan 2005: 239–240, 242, 244, Fig 1 (oscillogram, parasitism, comp. note) Michigan

Okanagana rimosa Elliott and Hershberger 2006: 32, 206–207 (illustrated, sonogram, listed, comp. note) North Dakota, South Dakota, Minnesota, Wisconsin, Illinois, Michigan, Ohio, New York, Pennsylvania, Virginia, West Virginia, Maryland, Connecticut, Rhode Island, Massachusetts, Vermont, New Hampshire, Maine, New Brunswick, Quebec, Ontario, Manitoba

Okanagana rimosa Bunker, Neckermann and Cooley 2007: 359 (comp. note)

Okanagana rimosa Oberdörster and Grant 2007b: 22 (comp. note)

Okanagana rimosa Shiyake 2007: 7, 93–94, Fig 159 (illustrated, distribution, listed, comp. note) United States, Canada

Okanagana rimosa Sueur, Vanderpool, Simon, Ouvrard and Bourgoin 2007: 624 (comp. note)

Okanagana rimosa Vries and Lakes-Harlan 2007: 477–478 (parasitism, comp. note) Michigan

Okanagana rimosa Farris, Oshinshy, Forrest, and Hoy 2008: 22, 24 (comp. note)

Okanagana rimosa Hill 2008: 116–117 (comp. note) Australia

Okanagana rimosa Ojie Ardebilie and Nozari 2009: A13, A15, A17-A20, Fig 3, Tables 1–2 (song, comp. note)

Okanagana rimosa Sanborn 2009c: 867–868, 870–872, Figs 5–7 (illustrated, comp. note)

Okanagana rimosa rimosa Sanborn 2009c: 872 (comp. note)

Okanagana rimosa Strauβ and Lakes-Harlan 2009: 306–310, 312–314, Figs 1–3, Fig 5, Tables 1–2 (auditory system development, comp. note) Michigan

Okanagana rimosa Wei, Tang, He and Zhang 2009: 340 (comp. note)

Okanaga (sic) *rimosa* Sueur, Janique, Simonis, Windmill and Baylac 2010: 931 (comp. note)

O. rotundifrons Davis, 1916 to *Clidophleps* rotundifrons

O. rubrocaudata Davis, 1925 to *Okanagana bella* Davis, 1919

***O. rubrovenosa* Davis, 1915** = *Okanagana rubrovenosa rubrovenosa* = *Okanagana rubrovenosa rubida* Davis, 1936 = *Okanagana rubrovenosa* var. *rubida* = *Okanagana rubida*

Okanagana rubrovenosa Sugden 1940: 120–121 (habitat, song, comp. note) California, Utah

Okanagana rubrovenosa Lloyd 1984: 82 (comp. note)

Okanagana rubrovenosa Watts and Williams 1992: 380 (coloration, host plants, comp. note) California

Okanagana rubrovenosa Poole, Garrison, McCafferty, Otte and Stark 1997: 332 (listed) Equals *Okanagana rubida* North America
Okanagana rubrovenosa Sanborn 1999a: 48 (type material, listed)
Okanagana rubrovenosa var. *rubida* Sanborn 1999a: 48, 55 (type material, listed, comp. note)
Okanagana rubrovenosa Sanborn 1999b: 2 (listed) North America
Okanagana rubrovenosa Heath and Sanborn 2007: 484, 487 (key, comp. note) Arizona
Okanagana rubrovenosa rubrovenosa Sueur, Vanderpool, Simon, Ouvrard and Bourgoin 2007: 613–614, 618–619, Figs 3–4, Table 1 (comp. note) California

***O. salicicola* Bliven, 1964**
Okanagana salicicola Poole, Garrison, McCafferty, Otte and Stark 1997: 332 (listed) North America
Okanagana salicicola Sanborn 1999a: 48–49 (type material, listed)
Okanagana salicicola Sanborn 1999b: 2 (listed) North America

***O. schaefferi* Davis, 1915** = *Okanagana shaefferi* (sic)
Okanagana schaefferi Sugden 1940: 119, 124–125 (habitat, song, comp. note) Utah
Okanagana shaefferi (sic) Heath and Heath 1994: 17 (type locality, comp. note) Utah
Okanagana schaefferi Poole, Garrison, McCafferty, Otte and Stark 1997: 332 (listed) North America
Okanagana schaefferi Sanborn 1999a: 49 (type material, listed)
Okanagana schaefferi Sanborn 1999b: 2 (listed) North America
Okanagana schaefferi Kondratieff, Ellingson and Leatherman 2002: 12–13, 30–32, 39–40, 42, Tables 1–2 (key, diagnosis, distribution, listed, song, comp. note) Colorado, New Mexico, utah

***O. sequoiae* Bliven, 1964**
Okanagana sequoiae Poole, Garrison, McCafferty, Otte and Stark 1997: 332 (listed) North America
Okanagana sequoiae Sanborn 1999a: 49 (type material, listed)
Okanagana sequoiae Sanborn 1999b: 2 (listed) North America

***O. simulata* Davis, 1921**
Okanagana simulata Poole, Garrison, McCafferty, Otte and Stark 1997: 332 (listed) North America
Okanagana simulata Sanborn 1999a: 49, 55 (type material, listed, comp. note)
Okanagana simulata Sanborn 1999b: 2 (listed) North America

***O. sperata* Van Duzee, 1935**
Okanagana sperata Poole, Garrison, McCafferty, Otte and Stark 1997: 332 (listed) North America
Okanagana sperata Sanborn 1999a: 49 (type material, listed)
Okanagana sperata Sanborn 1999b: 2 (listed) North America

***O. striatipes* (Haldeman, 1852)** = *Cicada striatipes* = *Tibicen striatipes*
Okanagana striatipes Sugden 1940: 120–121, 123 (habitat, song, comp. note) Utah
Okanagana striatipes Heath and Heath 1994: 17 (comp. note) Utah
Okanagana striatipes Sanborn and Phillips 1995a: 480–481, Table 1 (song intensity, comp. note)
Okanagana striatipes Poole, Garrison, McCafferty, Otte and Stark 1997: 332 (listed) Equals *Cicada striatipes* Equals *Okanagana beameri* North America
Okanagana striatipes Sanborn 1999a: 56 (comp. note)
Okanagana striatipes Sanborn 1999b: 2 (type material, listed) North America
Okanagana striatipes Kondratieff, Ellingson and Leatherman 2002: 8–10, 12–13, 29, 33, 40–41, 43, 55, Tables 1–2 (key, diagnosis, listed, distribution, song, comp. note) Colorado, Arizona, Utah, Oregon, California
Okanagana striatipes Sanborn, Breitbarth, Heath and Heath 2002: 437–438, 440–446, Figs 1–2, Fig 5, Tables 1–2 (thermoregulation, song, sonogram, oscillogram, comp. note)
Okanagana striatipes Heath and Sanborn 2007: 484 (key) Arizona

***O. sugdeni* Davis, 1938**
Okanagana sugdeni Sanborn and Phillips 1995a: 481, Table 1 (song intensity, comp. note)
Okanagana sugdeni Poole, Garrison, McCafferty, Otte and Stark 1997: 332 (listed) North America
Okanagana sugdeni Sanborn 1999a: 49 (type material, listed)
Okanagana sugdeni Sanborn 1999b: 2 (listed) North America

***O. synodica nigra* Davis, 1944** = *Okanagana synodica* var. *nigra*
Okanagana synodica (sic) Sanborn and Phillips 1995a: 481, Table 1 (song intensity, comp. note)
Okanagana synodica var. *nigra* Sanborn 1999a: 49 (type material, listed)
Okanagana synodica var. *nigra* Sanborn 1999b: 2 (listed) North America

O. synodica synodica (Say, 1825) = *Cicada synodica* = *Tibicen synodica*

Cicada synodica Putnam 1881: 67 (comp. note) Colorado

Okanagana synodica Sugden 1940: 124 (song, comp. note) Utah

Okanagana synodica Tinkham 1942: 155–156 (distribution, hosts, comp. note) Canada, Alberta, Montana, Colorado, New Mexico, Nebraska, Kansas, Texas

Okanagana synodica Poole, Garrison, McCafferty, Otte and Stark 1997: 332 (listed) Equals *Okanagana nigra* North America

Okanagana synodica Sanborn 1999b: 2 (type material, listed) North America

Okanagana synodica Maw, Footit, Hamilton and Scudder 2000: 49 (comp. note) Canada, Alberta

Okanagana synodica Kondratieff, Ellingson and Leatherman 2002: 5, 12–13, 30, 33, 73, 78, Figs 31–32, Fig 45, Tables 1–2 (key, diagnosis, illustrated, listed, distribution, song, comp. note) Equals *Cicada synodica* Colorado, Arizona, Alberta, Kansas, Montana, Nebraska, New Mexico, North Dakota, South Dakota, Texas, Wyoming, Utah

Okanagana synodica Heath and Sanborn 2007: 484 (key) Arizona

Okanagana synodica Shiyake 2007: 7, 93, 95, Fig 160 (illustrated, distribution, listed, comp. note) United States

O. tanneri Davis, 1930 = *Okanagana schaefferi tanneri*

Okanagana tanneri Poole, Garrison, McCafferty, Otte and Stark 1997: 332 (listed) Equals *Cicada* North America

Okanagana tanneri Sanborn 1999a: 49, 56 (type material, listed, comp. note)

Okanagana tanneri Sanborn 1999b: 2 (listed) North America

Okanagana tanneri Kondratieff, Ellingson and Leatherman 2002: 30, 40, 42, 78, Fig 46 (key, diagnosis, listed, illustrated, comp. note) Equals *Okanagana schaefferi tanneri* Colorado, Utah

Okanagana tanneri Heath and Sanborn 2007: 484 (key) Arizona

O. triangulata Davis, 1915 = *Okanagana triangulata crocea* Wymore, 1934 = *Okanagana triangulata croncina* (sic)

Okanagana triangulata Poole, Garrison, McCafferty, Otte and Stark 1997: 332 (listed) North America

Okanagana triangulata Sanborn 1999a: 49, 56 (type material, listed, comp. note)

Okanagana triangulata croncina (sic) Sanborn 1999a: 49 (type material, listed)

Okanagana triangulata Sanborn 1999b: 2 (listed) North America

Okanagana triangulata Sanborn 2007b: 38 (distribution, comp. note) Mexico, Baja California, California

O. tristis rubrobasalis Davis, 1926 = *Okanagana tristis* var. *rubrobasalis* = *Okanagana tristis rubrobasilis* (sic)

Okanagana tristis var. *rubrobasalis* Sanborn 1999a: 49 (type material, listed)

Okanagana tristis var. *rubrobasalis* Sanborn 1999b: 2 (listed) North America

O. tristis tristis Van Duzee, 1915 = *Okanagana davisi* Simons, 1953

Okanagana tristis Davis 1922: 73 (comp. note) Oregon

Okanagana tristis Hennessey 1990: 471 (type material)

Okanagana tristis Sanborn and Phillips 1995a: 481, Table 1 (song intensity, comp. note)

Okanagana tristis Poole, Garrison, McCafferty, Otte and Stark 1997: 332 (listed) Equals *Okanagana rubrobasalis* Equals *Okanagana davisi* North America

Okanagana davisi Sanborn 1999a: 46 (type material, listed)

Okanagana tristis Sanborn 1999a: 49 (type material, listed)

Okanagana tristis Sanborn 1999b: 2 (listed) North America

Okanagana tristis Heath and Sanborn 2007: 484 (key) Arizona

O. uncinata Van Duzee, 1915

Okanagana uncinata Poole, Garrison, McCafferty, Otte and Stark 1997: 332 (listed) North America

Okanagana uncinata Sanborn 1999a: 49 (type material, listed)

Okanagana uncinata Sanborn 1999b: 2 (listed) North America

O. utahensis Davis, 1919

Okanagana utahensis Sugden 1940: 123 (habitat, song, comp. note) Utah

Okanagana utahensis Campbell, Steffen-Campbell and Gill 1994: 76, 85 (comp. note) California

Okanagana utahensis Heath and Heath 1994: 16–17 (type locality, comp. note) Utah

Okanagana utahensis Campbell, Steffen-Campbell, Sorensen and Gill 1995: 177, 189–194 (sequence, comp. note) California

Okanagana utahensis Sanborn and Phillips 1995a: 480–481, Table 1 (song intensity, comp. note)

Okanagana utahensis Sorensen, Campbell, Gill and Steffen-Campbell 1995: 34 (gene sequence, phylogeny, comp. note) California

Okanagana utahensis Poole, Garrison, McCafferty, Otte and Stark 1997: 332 (listed) North America

Okanagana (sic) Flook and Rowell 1998: 167–170, Figs 3–6 (phylogeny)

Okanagana utahensis Flook and Rowell 1998: 175–176, Table 1 (sequence, listed, comp. note)

Okanagana utahensis Sanborn 1999a: 49 (type material, listed)

Okanagana utahensis Sanborn 1999b: 2 (listed) North America

Okanagana utahensis Ouvrard, Campbell, Bourgoin and Chan 2000: 406, 412–414, Figs 7–9 (sequence, comp. note)

Okanagana utahensis Kondratieff, Ellingson and Leatherman 2002: 8–9, 12–13, 29, 41, 43, 72, 77, 79, Fig 29, Fig 43, Fig 47, Tables 1–2 (key, diagnosis, illustrated, listed, distribution, song, comp. note) Colorado

Okanagana utahensis Sanborn, Breitbarth, Heath and Heath 2002: 437–438, 440–446, Figs 3- 5, Tables 1–2 (thermoregulation, song, sonogram, oscillogram, comp. note)

Okanagana utahensis Sanborn 2005c: 114, Table 1 (comp. note)

Okanagana utahensis Nichols, Nelson and Garey 2006: 55, Table 2 (listed)

Okanagana utahensis Heath and Sanborn 2007: 484 (key) Arizona

Okanagana utahensis Sueur, Vanderpool, Simon, Ouvrard and Bourgoin 2007: 613–614, 617–619, Figs 3–4, Table 1 (comp. note)

***O. vanduzeei* Distant, 1914** = *Okanagana vanduzei* (sic) = *Okanaga* (sic) *vanduziae* (sic) = *Okagana* (sic) *vanduzeei* = *Okanagana californica* Distant, 1914 = *Okanagana californicus* (sic) = *Okanagana vanduzeei californica* = *Okanagana vanduzeei consobrina* Distant, 1914 = *Okanagana vanduzei* (sic) *consobrina* = *Okanagana consobrina*

Okanagana californica Varley 1939: 99 (comp. note)

Okanagana vanduzeei Varley 1939: 99 (comp. note)

Okanagana vanduzeei Sugden 1940: 121–122, 125 (song, behavior, comp. note) California

Okanagana californica Sugden 1940: 122 (comp. note)

Okanagana vanduzeei Josephson and Young 1985: 185–207, Figs 1–11, Table 1 (illustrated, oscillogram, timbal muscle structure and function, comp. note) California

Okanagana vanduzeei Anonymous 1986: 23 (timbal muscle physiology) California

Okanagana vanduzeei Heller 1986: 106 (comp. note)

Okanagana vanduzeei Josephson and Young 1987: 993–995, Fig 1 (comp. note)

Okanagana vanduzeei Simon, Pääbo, Kocher and Wilson 1990: 241 (gene sequence, comp. note)

Okanagana vanduzeei Hogue 1993a: 124 (comp. note) California

Okanagana vanduzeei Bennet-Clark and Young 1994: 293, Table 1 (body size and song frequency)

Okanagana vanduzeei Simon, Frati, Beckenbach, Crespi, Liu and Flook 1994: 655, 700 (gene sequence, comp. note)

Okanagana vanduzeei Sanborn and Phillips 1995b: 375 (comp. note) California

Okanagana vanduzeei Fonseca 1996: 29 (comp. note)

Okanagana vanduzeei Young 1996: 101, Fig 5.3 (timbal illustrated, comp. note)

Okanagana vanduzeei Girgenrath and Marsh 1997: 3101, Fig 10 (comp. note)

Okanagana consobrina Poole, Garrison, McCafferty, Otte and Stark 1997: 332 (listed) North America

Okanagana vanduzeei Poole, Garrison, McCafferty, Otte and Stark 1997: 332 (listed) Equals *Okanagana californica* Equals *Okanagana vanduzei* (sic) North America

Okanagana vanduzeei Sanborn 1998: 94 (thermal biology, comp. note)

Okanagana vanduzeei Girgenrath and Marsh 1999: 3233, Fig 10 (comp. note)

Okanagana consobrina Sanborn 1999b: 2 (listed) North America

Okanagana vanduzeei Sanborn 1999b: 2 (listed) North America

Okanagana vanduzeei Bennet-Clark and Daws 1999: 1813 (comp. note)

Okanagana vanduzeei Marden 2000: 168 (comp. note)

Okanagana vanduzeei Maw, Footit, Hamilton and Scudder 2000: 49 (comp. note) Canada, British Columbia

Okanagana vanduzeei Sanborn 2001a: 16, Fig 5 (comp. note)

Okanagana vanduzeei Sueur 2001: 41, Table 1 (listed)

Okanagana vanduzeei Nation 2002: 260, Table 9.3 (comp. note)

Okanagana vanduzeei Syme and Josephson 2002: 763, 765, Fig 1 (illustrated, timbal muscle, comp. note)

Okanaga (sic) *vanduziae* (sic) Retallack 2004: 225 (burrows, comp. note) Oregon

Okanagana vanduzeei Champman 2005: 321 (comp. note)

Okanagana vanduzeei Sanborn 2007b: 38 (distribution, comp. note) Mexico, Baja California Norte, California, Nevada, Utah, Idaho, Oregon, Washington
Okanagana vanduzeei Nation 2008: 273, Table 10.1 (comp. note)
Okanagana vanduzeei Gillooly and Ophir 2010: 1327 (comp. note)
Okanagana vanduzeei Morris and Askew 2010: Fig S2 (comp. note)
Okagana (sic) *vanduzeei* Wu, Wright, Whitaker and Ahn 2010: 2551 (comp. note)

***O. vandykei* Van Duzee, 1915**
Okanagana vandykei Poole, Garrison, McCafferty, Otte and Stark 1997: 332 (listed) North America
Okanagana vandykei Sanborn 1999a: 50 (type material, listed)
Okanagana vandykei Sanborn 1999b: 2 (listed) North America

***O. venusta* Davis, 1935**
Okanagana venusta Poole, Garrison, McCafferty, Otte and Stark 1997: 332 (listed) North America
Okanagana venusta Sanborn 1999a: 50 (type material, listed)
Okanagana venusta Sanborn 1999b: 2 (listed) North America

***O. villosa* Davis, 1941**
Okanagana villosa Poole, Garrison, McCafferty, Otte and Stark 1997: 332 (listed) North America
Okanagana villosa Sanborn 1999a: 50 (type material, listed)
Okanagana villosa Sanborn 1999b: 2 (listed) North America

***O. viridis* Davis, 1918**
Okanagana viridis McCoy 1965: 41, 43 (key, comp. note) Arkansas
Okanagana viridis Poole, Garrison, McCafferty, Otte and Stark 1997: 332 (listed) North America
Okanagana viridis Sanborn 1999a: 50, 56 (type material, listed, comp. note)
Okanagana viridis Sanborn 1999b: 2 (listed) North America

***O. vocalis* Bliven, 1964**
Okanagana vocalis Poole, Garrison, McCafferty, Otte and Stark 1997: 332 (listed) North America
Okanagana vocalis Sanborn 1999a: 50 (type material, listed)
Okanagana vocalis Sanborn 1999b: 2 (listed) North America

***O. wymorei* Davis, 1935** = *Okanagana wymori* (sic)
Okanagana wymorei Poole, Garrison, McCafferty, Otte and Stark 1997: 332 (listed) Equals *Okanagana wymori* (sic) North America
Okanagana wymorei Sanborn 1999a: 50 (type material, listed)
Okanagana wymorei Sanborn 1999b: 2 (listed) North America

***O. yakimaensis* Davis, 1939**
Okanagana yakimaensis Poole, Garrison, McCafferty, Otte and Stark 1997: 332 (listed) North America
Okanagana yakimaensis Sanborn 1999a: 50 (type material, listed)
Okanagana yakimaensis Sanborn 1999b: 2 (listed) North America

***Okanagodes* Davis, 1919**
Okanagodes Heath 1983: 12–13 (comp. note) North America
Okanagodes Arnett 1985: 217 (diversity) Southeastern United States (error), Utah
Okanagodes Heath, Heath, Sanborn and Noriega 1988: 304 (thermal responses, comp. note)
Okanagodes Sanborn, Heath and Heath 1992: 756 (comp. note)
Okanagodes Poole, Garrison, McCafferty, Otte and Stark 1997: 251, 332 (listed) North America
Okanagodes Heath, Heath, Sanborn and Noriega 1999: 1 (thermal responses, comp. note)
Okanagodes Sanborn 1999b: 2 (listed) North America
Okanagodes Arnett 2000: 299 (diversity) Southeastern United States (error), Utah
Okanagodes Cioara and Sanborn 2001: 25 (song, comp. note)
Okanagodes Sueur 2002b: 122, 124 (comp. note)
Okanagodes Sanborn, Heath, Heath, Noriega and Phillips 2004: 282, 287 (comp. note)
Okanagodes spp. Sanborn, Heath, Heath, Noriega and Phillips 2004: 287 (comp. note)
Okanagodes sp. Phillips and Sanborn 2007: 457, Fig 2 (listed) Texas
Okanagodes Sueur, Vanderpool, Simon, Ouvrard and Bourgoin 2007: 613 (comp. note)

***O. gracilis gracilis* Davis, 1919** = *Okanagodes gracilis pallida* Davis, 1932 = *Okanagodes gracilis* var. *pallida*
Okanagodes gracilis Davis 1922: 73 (comp. note) Arizona
Okanagodes gracilis Varley 1939: 99 (comp. note)
Okanagodes gracilis Sugden 1940: 119 (habitat, song, comp. note) Nevada
Okanagodes gracilis Heath, Heath, Sanborn and Noriega 1988: 304 (thermal responses, comp. note)
Okanagodes gracilis Sanborn, Heath, Heath and Phillips 1990: A106 (evaporative cooling, comp. note) Sonoran Desert

Okanagodes gracilis Sanborn, Heath and Heath 1992: 751–756, Figs 2–7, Table 1 (thermoregulation, evaporative cooling, comp. note) Arizona
Okanagodes gracilis Sanborn and Phillips 1992: 285 (activity, comp. note) Arizona
Okanagodes gracilis Sanborn, J.E. Heath, M.S. Heath and Noriega 1995: 326, Table 3 (thermal responses, comp. note)
Okanagodes gracilis Sanborn, M.S. Heath, J.E. Heath and Noriega 1995: 456–457, Table 4 (comp. note)
Okanagodes gracilis Sanborn and Phillips 1995a: 481, Table 1 (song intensity, comp. note)
Okanagodes gracilis Prange 1996: 496 (comp. note)
Okanagodes gracilis Roxburgh, Pinshow and Prange 1996: 335 (comp. note)
Okanagodes gracilis Poole, Garrison, McCafferty, Otte and Stark 1997: 332 (listed) Equals *Okanagodes pallida* Equals *Okanagodes viridis* North America
Okanagodes gracilis Sanborn 1998: 91–92, 96–100, Table 1 (thermal biology, thermal responses, comp. note)
Okanagodes gracilis Heath, Heath, Sanborn and Noriega 1999: 1 (thermal responses, comp. note)
Okanagodes gracilis Sanborn 1999a: 50 (type material, listed)
Okanagodes gracilis var. *pallida* Sanborn 1999a: 50 (type material, listed)
Okanagodes gracilis Sanborn 1999b: 2 (listed) North America
Okanagodes gracilis Sanborn 2002b: 458–459, 462–466, Fig 4, Fig 6, Table 1 (illustrated, thermal biology, comp. note)
Okanagodes gracilis Sanborn, Breitbarth, Heath and Heath 2002: 443 (comp. note)
Okanagodes gracilis Sanborn, Noriega and Phillips 2002c: 368 (comp. note)
Okanagodes gracilis Sanborn 2004e: 2225 (comp. note)
Okanagodes gracilis Sanborn, Heath, Heath, Noriega and Phillips 2004: 282–287, Figs 1–3, Tables 1–4 (illustrated, thermal responses, habitat, distribution, comp. note) Arizona, Utah, Nevada
Okanagodes gracilis Sanborn 2005c: 114, Table 1 (comp. note)
Okanagodes gracilis Phillips and Sanborn 2007: 455, 459–460 (comp. note)
Okanagodes gracilis Sanborn 2007b: 38 (distribution, comp. note) Mexico, Baja California Norte, Sonora, Arizona, California, Nevada, Utah
Okanagodes gracilis Sueur, Vanderpool, Simon, Ouvrard and Bourgoin 2007: 613–614, 617–619, Figs 3–4, Table 1 (comp. note) California
Okanagodes gracilis Sanborn 2008e: 3759 (comp. note)

***O. gracilis viridis* Davis, 1934** = *Okanagodes gracilis* var. *viridis*
Okanagodes gracilis var. *viridis* Sanborn 1999a: 50 (type material, listed)
Okanagodes gracilis var. *viridis* Sanborn 1999b: 2 (listed) North America
Okanagodes gracilis var. *viridis* Sanborn, Heath, Heath, Noriega and Phillips 2004: 282–286, Figs 1–2, Tables 1–4 (thermal responses, habitat, distribution, comp. note) Arizona

***O. terlingua* Davis, 1932**
Okanagodes terlingua Tinkham 1941: 171, 176, 178–179 (distribution, habitat, comp. note) Texas
Okanagodes terlingua Sanborn and Phillips 1994: 45 (distribution, comp. note) Texas
Okanagodes terlingua Sanborn and Phillips 1995a: 480–481, Table 1 (song intensity, comp. note)
Okanagodes terlingua Van Pelt 1995: 20 (listed, distribution, comp. note) Texas
Okanagodes terlingua Poole, Garrison, McCafferty, Otte and Stark 1997: 332 (listed) North America
Okanagodes terlingua Sanborn and Phillips 1997: 36 (distribution, comp. note) Texas
Okanagodes terlingua Heath, Heath, Sanborn and Noriega 1999: 1 (thermal responses, comp. note)
Okanagodes terlingua Sanborn 1999a: 50 (type material, listed)
Okanagodes terlingua Sanborn 1999b: 2 (listed) North America
Okanagodes terlingua Sanborn, Heath, Heath, Noriega and Phillips 2004: 282–286, Figs 1–2, Tables 1–4 (thermal responses, habitat, distribution, comp. note) Texas
Okanagodes terlingua Phillips and Sanborn 2007: 456–459, Table 1, Fig 2, Fig 5 (illustrated, habitat, distribution, comp. note) Texas

***Clidophleps* Van Duzee, 1915**
Clidophleps Varley 1939: 98–99 (comp. note)
Clidophleps Sugden 1940: 119 (comp. note) California
Clidophleps Heath 1983: 12 (comp. note) North America
Clidophleps Boulard 1984c: 171 (comp. note)
Clidophleps Hayashi 1984: 26 (comp. note)
Clidophleps Arnett 1985: 217 (diversity) California

Clidophleps Poole, Garrison, McCafferty, Otte and Stark 1997: 251, 331 (listed) North America
Clidophleps Sanborn 1999b: 2 (listed) North America
Clidophleps Arnett 2000: 299 (diversity) California
Clidophleps Sueur 2002b: 122, 124 (comp. note)
Clidophleps Moulds 2005b: 414, 416, 421 (comp. note)
Clidophleps Sueur, Vanderpool, Simon, Ouvrard and Bourgoin 2007: 613 (comp. note)

C. *astigma* Davis, 1917
Clidophleps astigma Sanborn 1999a: 45 (type material, listed)
Clidophleps astigma Sanborn 2007b: 38 (distribution, comp. note) Mexico, Baja California Norte

C. *beameri* Davis, 1936
Clidophleps beameri Duffels 1988d: 75 (comp. note) California
Clidophleps beameri Boulard 1990b: 177 (comp. note)
Clidophleps beameri Poole, Garrison, McCafferty, Otte and Stark 1997: 331 (listed) North America
Clidophleps beameri Sanborn 1999a: 45 (type material, listed)
Clidophleps beameri Sanborn 1999b: 2 (listed) North America
Clidophleps beameri Boulard 2002b: 205 (comp. note)
Clidophleps beameri Boulard 2005d: 343 (comp. note) California
Clidophleps beameri Boulard 2006g: 66 (comp. note)

C. *blaisdellii* (Uhler, 1892) = *Tibicen blaisdellii* = *Tibicen blaisdelli* (sic) = *Okanagana blaisdelli* (sic) = *Clidophleps blaisdelli* (sic) =
Tibicen blaisdellii Henshaw 1903: 230 (listed) California
Clidophleps blaisdellii Varley 1939: 98–99 (comp. note)
Clidophleps blaisdellii Heath 1983: 14 (stridulatory apparatus, illustrated)
Clidophleps blaisdellii Watts and Williams 1992: 381 (coloration, host plants, comp. note) California
Clidophleps blaisdellii Poole, Garrison, McCafferty, Otte and Stark 1997: 331 (listed) Equals *Tibicen blaisdellii* Equals *Tibicen blaisdellii* (sic) North America
Clidophleps blaisdellii Sanborn 1999a: 45 (type material, listed)
Clidophleps blaisdellii Sanborn 1999b: 2 (listed) North America

C. *distanti distanti* (Van Duzee, 1914) = *Okanagana dstanti* (sic) = *Okanagana distanti* = *Okanagana distanti pallida* Van Duzee, 1914 = *Clidophleps distanti pallida* = *Clidophleps pallida* = *Okanagana distanti pallidus* (sic) = *Clidophleps distanti* var. *pallidus*
Clidophleps distanti Varley 1939: 97–99, Fig 1 (illustrated, stridulation, comp. note) California
Clidophleps pallida Sugden 1940: 118 (song, behavior, comp. note) California
Clidophleps distanti Sanborn and Phillips 1995a: 481, Table 1 (song intensity, comp. note)
Clidophleps distanti Poole, Garrison, McCafferty, Otte and Stark 1997: 331 (listed) Equals *Okanagana distanti* Equals *Okanagana dstanti* (sic) Equals *Okanagana pallidus* Equals *Okanagana truncata* North America
Clidophleps distanti Sanborn 1999a: 45 (type material, listed)
Clidophleps distanti var. *pallidus* Sanborn 1999a: 45 (type material, listed)
Clidophleps distanti Sanborn 1999b: 2 (listed) North America
Clidophleps distanti Sueur 2001: 47 (comp. note)
Clidophleps distanti Sueur 2002b: 122, 129 (listed, comp. note)
Clidophleps distanti Sanborn 2007b: 38 (distribution, comp. note) Mexico, Baja California Norte, California

C. *distanti truncata* (Van Duzee, 1914) = *Okanagana distanti truncatus* (sic) = *Okanagana distanti truncata* = *Clidophleps distanti* var. *truncatus* (sic) = *Clidophleps truncata* = *Okanagana truncata*
Clidophleps distanti var. *truncatus* (sic) Sanborn 1999a: 45 (type material, listed)
Clidophleps distanti var. *truncatus* (sic) Sanborn 1999b: 2 (listed) North America
Clidophleps distanti truncatus (sic) Sanborn 2007b: 38 (distribution, comp. note) Mexico, Baja California Norte, California

C. *rotundifrons* (Davis, 1916) = *Okanagana rotundifrons*
Okanagana rotundifrons Poole, Garrison, McCafferty, Otte and Stark 1997: 332 (listed) North America
Clidophleps rotundifrons Sanborn 1999a: 45, 55 (type material, listed, comp. note)
Clidophleps rotundifrons Sanborn 1999b: 2 (listed) North America

C. *tenuis* Davis, 1927
Clidophleps tenuis Poole, Garrison, McCafferty, Otte and Stark 1997: 331 (listed) North America
Clidophleps tenuis Sanborn 1999a: 45 (type material, listed)
Clidophleps tenuis Sanborn 1999b: 2 (listed) North America
Clidophleps tenuis Sanborn 2007b: 38 (distribution, comp. note) Mexico, Baja California Norte, California

***C. vagans* Davis, 1925**

Clidophleps vagans Poole, Garrison, McCafferty, Otte and Stark 1997: 331 (listed) North America

Clidophleps vagans Sanborn 1999a: 45 (type material, listed)

Clidophleps vagans Sanborn 1999b: 2 (listed) North America

Clidophleps vagans Sueur, Vanderpool, Simon, Ouvrard and Bourgoin 2007: 613–614, 617–619, Figs 3–4, Table 1 (comp. note) California

***C. wrighti* Davis, 1926**

Clidophleps wrighti Watts and Williams 1992: 381 (coloration, host plants, comp. note) California

Clidophleps wrighti Poole, Garrison, McCafferty, Otte and Stark 1997: 331 (listed) North America

Clidophleps wrighti Sanborn 1999a: 45 (type material, listed)

Clidophleps wrighti Sanborn 1999b: 2 (listed) North America

Clidophleps wrighti Sanborn 2007b: 38 (distribution, comp. note) Mexico, Baja California Norte, California

***Subpsaltria* Chen, 1943** = *Subpsultria* (sic) = *Subpsalta* (sic) = *Subpsaltra* (sic)

Subpsaltria Heath 1983: 12 (comp. note) China

Subpsaltria Hayashi 1984: 26 (comp. note)

Subpsaltria Chou, Lei, Li, Lu and Yao 1997: 42, 293, 297, Table 6 (described, listed, diversity, biogeography) China

Subpsaltria Villet and Zhao 1999: 322 (comp. note)

Subpsaltria Moulds 2005b: 388, 414 (listed, comp. note)

Subpsultria (sic) Wei and Zhang 2010: 38 (comp. note)

Subpsaltra (sic) Wei and Zhang 2010: 38 (comp. note) China, Mongolia

***S. yangi* Chen, 1943** = *S. sienyangensis* Chen, 1943

Subpsaltria sienyangensis Wang and Zhang 1993: 190 (listed, comp. note) China

Subpsaltria yangi Chou, Lei, Li, Lu and Yao 1997: 42–44, 358, Fig 7–1, Plate II, Fig 27 (type species of *Subpsaltria*, synonymy, key, described, illustrated, synonymy, comp. note) Equals *Subpsaltria sienyangensis* China

Supsaltria sienyangensis Chou, Lei, Li, Lu and Yao 1997: 358 (synonymy)

Subpsaltria yangi Hua 2000: 65 (listed, distribution) Equals *Subpsaltria sienyangensis* China, Shaanxi, Sichuan

Subpsaltria yangi Boulard 2006g: 153–154, Fig 111 (illustrated, stridulation, comp. note)

***Ahomana* Distant, 1905**

***A. chilensis* Distant, 1905**

Ahomana chilensis Peña 1987: 110 (comp. note) Chile

***A. neotropicalis* Distant, 1905**

SUMMARY SPECIES LIST

SUPERFAMILY CICADOIDEA Westwood, 1840

Family Tettigarctidae Distant, 1905

Subfamily Tettigarctinae Distant, 1905

Tribe Tettigarctini Distant, 1905

Subtribe Tettigarctina Distant, 1905 = Tettigarctaria

Tettigarcta White, 1845 = *Tettigareta* (sic) = *Tettigarota* (sic) = *Tettigarta* (sic)
T. crinita Distant, 1883 = *Tettigarcta criniti* (sic)
T. tomentosa White, 1845 = *Tettigarcta tormentosa* (sic)

Family Cicadidae Latreille, 1802 = Cicadae (sic) = Cicadiae (sic) = Cicididae (sic) = Cicadide (sic) = Plautillidae Distant, 1905 = Plautilidae (sic) = Tettigadidae Distant, 1905 = Tibicinidae Distant, 1905 = Tibiciinidae (sic) = Platypediidae Kato, 1932 = Tibicenidae Van Duzee, 1916 = Ydiellidae Boulard, 1973

Subfamily Cicadinae Latreille, 1802 = Gaeaninae Distant, 1905 = Daeaninae (sic) = Plautillinae Distant, 1905 = Tibicininae Distant, 1905 = Zammarinae Distant, 1905 = Tibiceninae Van Duzee, 1916 = Tibicenina (sic) = Platypleurinae Schmidt, 1918 = Platypleurnae (sic) = Lyristinae Gomez-Monor, 1957 = Moaninae Boulard, 1976

Tribe Platypleurini Schmidt, 1918 = Platypleuraria = Platypleuriti (sic) = Tibicenini Van Duzee, 1916 = Lyristini Gomez-Menor, 1957 = Hainanosemiina Kato, 1927 = Hainanosemiaria = Hainanosemiiti (sic)

Munza Distant, 1904 = *Numza* (sic)
M. basimacula (Walker, 1850) = *Platypleura basimacula* = *Numza basimacula* (sic) = *Platypleura reducta* Walker, 1850 = *Platypleura basimacula reducta* = *Poecilopsaltria reducta* = *Poecilopsaltera* (sic) *reducta* = *Munza pygmaea* Jacobi, 1910 = *Munza pygmoea* (sic) = *Platypleura sikumba* Ashton, 1914 = *Munza parva* Villet, 1989 = *Cicada decora* (nec Germar) = *Platypleura decora* (nec Germar) = *Platypleura deusta* (nec Thunburg) = *Platypleura absimilis* (nec Distant)
M. furva (Distant, 1897) = *Poecilopsalta furva* = *Munza oculata* Jacobi, 1910
M. laticlavia laticlavia (Stål, 1858) = *Platypleura laticlavia* = *Munza laticlava* (sic)
M. laticlavia lubberti Schumacher, 1913 = *Munza laticlavia lübberti*
M. laticlavia semitransparens Schumacher, 1913 = *Munza laticlava* (sic) *semitransparens*
M. otjosonduensis Schumacher, 1913
M. pallescens Schumacher, 1913
M. popovi Boulard, 1975
M. revoili Distant, 1905
M. signata Distant, 1914
M. straeleni Dlabola, 1960
M. sudanensis Distant, 1913
M. trimeni (Distant, 1892) = *Poecilopsaltria trimeni*
M. venusta Hesse, 1925

Soudaniella Boulard, 1973 = *Soudanielle* (sic)
S. cortustusa Boulard, 1974 = *Soudanielle* (sic) *cortustusa*
S. dlabolai Boulard, 1972 = *Platypleura schoutedeni* Dlabola, 1962
S. laticeps (Karsch, 1890) = *Platypleura laticeps* = *Platypleura (Poecilopsaltria) laticeps*
S. marshalli (Distant, 1897) = *Poecilopsaltria marshalli* = *Platypleura marshalli* = *Platypleura (Poecilopsaltria) marshalli*

S. melania fuscala Boulard, 1975
S. melania melania (Distant, 1904) = *Soudaniella melania* = *Platypleura (Poecilopsaltria) melania* = *Platypleura melania* = *Platypleura melanaria* (sic) = *Poecilopsaltria melania*
S. schoutedeni (Distant, 1913) = *Platypleura schoutedeni* = *Soudaniella schoutedemi* (sic)
S. seraphina (Distant, 1905) = *Platypleura seraphina*
S. sudanensis (Distant, 1913) = *Munza sudanensis*

Esada Boulard, 1973
E. esa (Distant, 1905) = *Platypleura esa* = *Platypleura (Oxypleura) esa*

Ioba Distant, 1904 = *Iobao* (sic) = *Lioba* (sic)
I. bequaerti Distant, 1913
I. horizontalis (Karsch, 1890) = *Platypleura horizontalis* = *Platypleura (Poecilopsaltria) horizontalis*
I. leopardina (Distant, 1881) = *Poecilopsaltria leopardina* = *Platypleura (Joba)* (sic) *leopardina* = *Lioba* (sic) *leopardina*
I. limbaticollis (Stål, 1863) = *Platypleura limbaticollis* = *Oxypleura limbata* Walker, 1858 (nec *Platypleura limbata* Fabricius, 1775) = *Ioba leopardina* (nec Distant)
I. stormsi (Distant, 1893) = *Poecilopsaltria stormsi*
I. veligera (Jacobi, 1904) = *Platypleura veligera* = *Platypleura laticollis* Melichar, 1904

Muansa Distant, 1904
M. clypealis (Karsch, 1890) = *Platypleura clypealis* = *Platypleura (Poecilopsaltria) clypealis*

Sadaka Distant, 1904
S. aurovirens Dlabola, 1960
S. dimidiata (Karsch, 1893) = *Platypleura dimidiata*
S. morini Boulard, 1985
S. radiata centralensis Boulard, 1986
S. radiata radiata (Karsch, 1890) = *Platypleura (Oxypleura) radiata* = *Platypleura radiata* = *Sadaka radiata*
S. sagittifera Boulard, 1985
S. virescens (Karsch, 1890) = *Platypleura (Oxypleura) virescens* = *Platypleura virescens*

Afzeliada Boulard, 1973
A. afzelii (Stål, 1854) = *Platypleura afzelii* = *Platypleura afzelli* (sic) = *Platypleura (Poecilopsaltria) afzelii* = *Platypleura strumosa* (nec Fabricius) = *Platyplerua aerea* Distant, 1881 = *Platypleura area* (sic) = *Platypleura monodi* Lallemand, 1927
A. balachowskyi Boulard, 1975
A. bernardii (Boulard, 1971) = *Platypleura bernardii*
A. caluangana Boulard, 1979
A. christinetta Boulard, 1973 = *Platypleura rutherfordina* Boulard, 1971 nom. nud.
A. circumscripta (Jacobi, 1910) = *Platypleura circumscripta* = *Afzeliada circunscripta* (sic)
A. contracta (Walker, 1850) = *Oxypleura contracta* = *Platypleura contracta* = *Platypleura (Poecilopsaltria) contracta*
A. deheegheri Boulard, 1975
A. donskoffi Boulard, 1979 = *A. crassa* Boulard, 1975 nom. nud.
A. duplex (Dlabola, 1961) = *Platypleura duplex*
A. ericina Boulard, 1973
A. grillotti Boulard, 1985
A. helenae Boulard, 1975
A. hyalina (Distant, 1905) = *Sadaka hyalina*
A. hyaloptera (Stål, 1866) = *Platypleura hyaloptera* = *Platypleura (Oxypleura) hyaloptera*
A. iringana Boulard, 2012
A. izzardi (Dlabola, 1960) = *Platypleura izzardi*
A. ladona (Distant, 1919) = *Platypleura ladona* = *Platypleura (Oxypleura) ladona*
A. longula (Distant, 1904) = *Platypleura (Oxypleura) longula* = *Platypleura longula*
A. lusingana Boulard, 1979

A. *mikessensis* Boulard, 2012
A. *passosi* Boulard, 1972
A. *rutherfordi* (Distant, 1883) = *Platypleura rutherfordi* = *Platypleura (Poecilopsaltria) rutherfordi*

Attenuella Boulard, 1973
A. *tigrina* (Palisot de Beauvois, 1813) = *Cicada tigrina* = *Platypleura tigrina* = *Ugada tigrina* = *Platypleura attenuata* Distant, 1905 = *Platypleura (Oxypleura) attenuata*

Severiana Boulard, 1973
S. *magna* Villet, 1999
S. *severini severini* (Distant, 1893) = *Poecilopsaltria severini* = *Poecsilopsaltria* (sic) *severini* = *Platypleura severini* = *Platypleura (Poecilopsaltria) severini* = *Platypleura fenestrata* Schumacher, 1913 = *Platypleura (Platypleura) schumacheri* Metcalf, 1955 = *Platypleura schumacheri*
S. *severini vitreomaculata* (Schumacher, 1913) = *Platypleura fenestrata vitreomaculata* Schumacher, 1913 = *Platypleura fenestrata* var. *vitreomaculata* = *Platypleura (Platypleura) vitreomaculata*
S. *similis* (Schumacher, 1913) = *Platypleura similis*

Koma Distant, 1904 = *Kuma* (sic)
K. *bombifrons* (Karsch, 1890) = *Platypleura bombifrons* = *Poecilopsaltria bombifrons* = *Kuma* (sic) *bombifrons*
K. *intermedia* Boulard, 1980
K. *semivitrea* (Distant, 1914) = *Platypleura semivitrea* = *Koma basilewskyi* Dlabola, 1958
K. *umbrosa camusa* Boulard, 1980
K. *umbrosa umbrosa* Boulard, 1980

Platypleura Amyot & Audinet-Serville, 1843 = *Cicada (Platypleura)* = *Platypleura (Platypleura)* = *Patypleura* (sic) = *Plathypleura* (sic) = *Platypeura* (sic) = *Platypleure* (sic) = *Rlatypleura* (sic) = *Poecilopsaltria* Stål, 1866 = *Poicilopsaltria* (sic) = *Paecilopsaltria* (sic) = *Poecilopsaltera* (sic) = *Platypleura (Poecilopsaltria)* = *Systophlochius* Villet, 1989
P. *adouma* Distant, 1904 = P. *adoreus* (sic)
P. *affinis affinis* (Fabricius, 1803) = *Tettigonia affinis* = *Cicada affinis* = *Poecilopsaltria affinis* = *Platypleura (Poecilopsaltria) affinis*
P. *affinis distincta* Atkinson, 1884 = *Platypleura nicobarica* Atkinson, 1884 (nec *Platypleura nicobarica* Butler, 1874) = *Poecilopsaltria nicobarica* = *Poecilopsaltria nicobarica* var. *a* Distant, 1889 = *Platypleura (Poecilopsaltria) affinis distincta*
P. *albivannata* Hayashi, 1974
P. *andamana* Distant, 1878 = *Platypleura (Platypleura) andamana* = *Platypleura (Poecilopsaltria) andamana* = *Poecilopsaltria andamana* = *Platypleura roepstorffii* Atkinson, 1884
P. *arabica* Myers, 1928
P. *argentata* Villet, 1987
P. *argus argus* Melichar, 1911
P. *argus marsabitensis* Boulard, 1980 = *Platypleura marsabitensis*
P. *arminops* Noualhier, 1896 = *Platypleura (Poecilopsaltria) arminops*
P. *assamensis* Atkinson, 1884 = *Platypleura repanda* var. *a* Distant, 1889 = *Platypleura repanda assamensis* = *Platypleura (Poecilopsaltria) assamensis*
P. *auropilosa* Kato, 1940 = *Platypleura auripilosa* (sic)
P. *badia* Distant, 1888 = *Platypleura (Poecilopsaltria) badia* = *Afzeliada badia* = *Platypeura (Neoplatypleura) badia* = *Platypleura fasuayensis* Boulard, 2006
P. *basialba* (Walker, 1850) = *Oxypleura basialba* = *Platypleura (Oxypleura) basialba* = *Platypleura (Poecilopsaltria) basialba*
P. *basiviridis* Walker, 1850 = *Poecilopsaltria basiviridis* = *Platypleura (Poecilopsaltria) basiviridis*
P. *bettoni* Distant, 1904 = *Platypleura (Poecilopsaltria) bettoni*
P. "breedeflumensis" Price, Barker & Villet, 2009
P. *brunea* Villet, 1989 = *Platypleura brunnea* (sic)
P. cf. *brunnea* Phillips, Sanborn & Villet, 2002
P. *bufo* (Walker, 1850) = *Oxypleura bufo* = *Platypleura (Oxypleura) bufo* = *Poecilopsaltria bufo* = *Platypleura (Platypleura) bufo* = *Platypleura (Poecilopsaltria) bufo*

P. canescens (Walker, 1870) = *Oxypleura canescens* = *Platypleura (Oxypleura) canescens* = *Poecilopsaltria canescens*
P. capensis (Linnaeus, 1764) = *Cicada capensis* = *Tettigonia capensis* = *Platypleura stridula* var. *capensis* = *Platypleura stridula capensis* = *Platypleura (Platypleura) stridula capensis* = *Platypleura stridula* var. *b* Stål, 1866
P. capitata (Olivier, 1790) = *Cicada capitata* = *Oxypleura capitata* = *Poecilopsaltria capitata* = *Platypleura (Poecilopsaltria) capitata* = *Oxypleura subrufa* Walker, 1850 = *Platypleura subrufa* = *Poecilopsaltria subrufa* = *Paecilopsaltria* (sic) *subrufa*
P. carlinii Distant, 1906 = *Platyplerua testacea* de Carlini, 1892 (nec *Zammara testacea* Walker, 1858)
P. cervina Walker, 1850 = *Platypleura (Oxypleura) cervina* = *Poecilopsaltria cervina* = *Platypleura straminea* Walker, 1850
P. cespiticola Boulard, 2006 = *Platypleura (Poecilopsaltria) cespiticola* = *Platypleura (Poecilopsaltria) cespiticola* var. *fuscalae* Boulard, 2006
P. chalybaea Villet, 1989
P. coelebs Stål, 1863 = *Platypleura caelebs* (sic) = *Poecilopsaltria coelebs* = *Platypleura (Poecilopsaltria) coelebs*
P. crampeli Boulard, 1975
P. deusta (Thunberg, 1822) = *Tettigonia deusta*
P. divisa (Germar, 1834) = *Cicada divisa* = *Platypleura divisa* var. *a* Stål, 1866 = *Dasypsaltria maera* Haupt, 1917 = *Platypleura (Platypleura) divisa* = *Platypleura hirtipennis* Villet 1987 (nec Germar) = *Platypleura hirtipennis* Villet 1988 (nec Germar)
P. durvillei (Boulard, 1991) = *Poecilopsaltria durvillei*
P. elizabethae Lee, 2009 = *Platypleura* sp. aff. *nobilis* Endo & Hayashi, 1979
P. fulvigera fulvigera Walker, 1850 = *Poecilopsaltria fulvigera* = *Platypleura (Poecilopsaltria) fulvigera*
P. fulvigera var. *a* Distant, 1889 = *Platypleura (Poecilopsaltria) fulvigera* var. *a*
P. "gamtoosflumensis" Price, Barker & Villet, 2009
P. "gariepflumensis" Price, Barker & Villet, 2009
P. girardi Boulard, 1977
P. gowdeyi Distant, 1914
P. haglundi Stål, 1866
P. hampsoni (Distant, 1887) = *Poecilopsaltria hampsoni* = *Platypleura (Poecilopsaltria) hampsoni*
P. harmandi Distant, 1905 = *Platypleura (Poecilopsaltria) harmandi*
P. hilpa Walker, 1850 = *Poecilopsaltria hilpa* = *Platypleura (Platypleura) hilpa* = *Platypleura fenestrata* Uhler, 1861
P. hirta Karsch, 1890
P. hirtipennis (Germar, 1834) = *Cicada hirtipennis* = *Platypleura (Platypleura) hirtipennis* = *Platypleura capensis* (nec Linnaeus) = *Platypleura ocellata* (nec Degeer) = *Platypleura chloronota* Walker, 1850 = *Platypleura hirtipennis* var. *b* Stål, 1866
P. cf. *hirtipennis* Villet, Barker & Lunt 2005
P. inglisi Ollenbach, 1929
P. insignis Distant, 1879 = *Platypleura (Poecilopsaltria) insignis* = *Platypleura (Neoplatypleura) insignis*
P. instabilis Boulard, 1977
P. intermedia Liu, 1940
P. kabindana Distant, 1919
P. kaempferi var. _______ Kato, 1938
P. kaempferi brevipennis Naruse, 1983
P. kaempferi kaempferi (Fabricius, 1794) = *Tettigonia kaempferi* = *Tettigomia* (sic) *kaempferi* = *Tettigonia kaemferi* (sic) = *Cicada kaempferi* = *Cicada kaemferi* (sic) = *Platypleura kaempferi* = *Platypleura kampferi* (sic) = *Platypleura koempferi* (sic) = *Platypleura kaempheri* (sic) = *Platypleuria* (sic) *kaempferi* = *Platypleura kaenpferi* (sic) = *Platyleura* (sic) *kaempferi* = *Platypleure* (sic) *kaempferi* = *Platypleura fuscangulis* Butler, 1874 = *Platypleura kaempferi fuscangulis* = *Platypleura hyalinolimbata* Signoret, 1881 = *Platypleura hyalino-limbata* = *Platypleura repanda* (nec Linnaeus) = *Platypleura kaempferi* var. *formosana* Matsumura, 1917 = *Platypleura kaempferi annamensis* Moulton, 1923 = *Platyplerua kaempferi dentivitta* Kato, 1925 = *Platypleura kaempferi kyotonis* Kato, 1930 = *Platypleura tsuchidai* Kato, 1936 = *Rlatypleura* (sic) *tsuchidai* = *Platypleura tsuchidai* var. *amamiana* Kato, 1940 = *Platypleura retracta* Liu, 1940 = *Platypleura retracta omeishana* Liu, 1940

P. kaempferi ridleyana Distant, 1905 = *Platypleura ridleyana*
P. "karooensis" Price, Barker & Villet, 2009
P. kenyana Lallemand, 1929
P. kuroiwae Matsumura, 1917 = *Munza kuroiwae* = *Platypleura kuroiwai* (sic)
P. lindiana Distant, 1905 = *Platypleura (Poecilopsaltria) lindiana*
P. lineatella Distant, 1905 = *Platypleura (Oxypleura) lineatella*
P. longirostris Ashton, 1914 = *Platypleura divisa* (nec Germar)
P. lourensi Lee, 2009
P. machadoi Boulard, 1972
P. mackinnoni Distant, 1904 = *Platypleura (Oxypleura) mackinnoni* = *Platypleura mackinoni* (sic)
P. makaga Distant, 1904
P. maytenophila Villet, 1987
P. mijberghi Villet, 1989
P. mira Distant, 1904 = *Platypleura (Poecilopsaltria) mira* = *Platypleura (Platypleura) mira* = *Platypleura laotiana* nom. nud.
P. miyakona (Matsumura, 1917) = *Pycna miyakona* = *Suisha miyakona*
P. murchisoni Distant, 1905
P. nigrosignata Distant, 1913
P. nobilis nobilis (Germar, 1830) = *Cicada nobilis* = *Oxypleura nobilis* = *Platypleura (Neoplatypleura) nobilis* = *Cicada hemiptera* Guérin-Méneville, 1843 = *Platypleura gemina* Walker, 1850 = *Platypleura semilucida* Walker, 1850
P. nobilis var. *a* Atkinson, 1884 = *Platypleura (Neoplatypleura) nobilis* var. *a*
P. nobilis var. *a* Distant, 1889 = *Platypleura (Neoplatypleura) nobilis* var. *a*
P. n. sp. Duffels, 1990
P. octoguttata octoguttata (Fabricius, 1798) = *Tettigonia 8-guttata* = *Tettigonia octoguttata* = *Tettigonia octotoguttata* (sic) = *Cicada 8-guttata* = *Poecilopsaltria octoguttata* = *Poecilopsaltria 8-guttata* = *Poicilopsaltria* (sic) *octoguttata* = *Poicilopsaltria* (sic) *octopunctata* (sic) = *Platypleura (Poecilopsaltria) octoguttata* = *Platypleura octogullata* (sic) = *Oxypleura sanuiflua* Walker, 1850
P. octoguttata var. *a* (Distant, 1889) = *Poecilopsaltria octoguttata* var. *a* = *Platypleura (Poecilopsaltria) octoguttata* var. *a*
P. octoguttata var. *b* (Distant, 1889) = *Poecilopsaltria octoguttata* var. *b* = *Platypleura (Poecilopsaltria) octoguttata* var. *b*
P. "olifantflumensis" Price, Barker & Villet, 2009
P. parvula Boulard, 2006
P. pinheyi Boulard, 1980
P. plumosa (Germar, 1834) = *Cicada plumosa* = *Platypleura hirtipennis* var. *b* Stål, 1866 = *Platypleura hirtipennis plumosa* = *Platypleura (Platypleura) plumosa*
P. polita polita (Walker, 1850) = *Oxypleura polita* = *Platypleura (Oxypleura) polita* = *Poecilopsaltria polita*
P. polita var. *a* (Distant, 1889) = *Poecilopsaltria polita* var. *a* = *Platypleura (Oxypleura) polita* var. *a*
P. polydorus (Walker, 1850) = *Oxypleura polydorus* = *Platypleura (Oxypleura) polydorus* = *Poecilopsaltria polydorus*
P. rothschildi Melichar, 1911
P. semusta (Distant, 1887) = *Poecilopsaltria semusta* = *Platypleura (Poecilopsaltria) semusta*
P. signifera Walker, 1850
P. sp. 1 Zaidi & Ruslan, 1995
P. sp. 1 Villet & Noort, 1999
P. sp. 1 Villet, Barker & Lund, 2005
P. sp. 1 Pham & Thinh, 2006
P. sp. 2 Zaidi, Ruslan & Mahadir, 1996
P. sp. 2 Villet & Noort, 1999
P. sp. 3 Zaidi, Ruslan & Mahadir, 1996
P. sp. 4 Sanborn, Villet & Phillips, 2004
P. sp. 7 Villet, Barker & Lund, 2005
P. sp. 10 Price, Barker & Villet, 2009
P. sphinx Walker, 1850 = *Poecilopsaltria sphinx* = *Platypleura (Poecilopsaltria) sphinx*
P. spicata Distant, 1905 = *Platypleura (Oxypleura) spicata*

P. stridula (Linnaeus, 1758) = *Cicada stridula* = *Tettigonia stridula* = *Platypleura (Platypleura) stridula* = *Cicada catenata* Drury, 1773 = *Cicada catena* (sic) = *Cicada nigrolinea* Degeer, 1773 = *Cicada nigra* (sic) = *Cicada nigrolineata* (sic) = *Platypleura nigrolinea* = *Platypleura stridula* var. *a* Stål, 1866

P. takasagona Matsumura, 1917 = *Platypleura kuroiwae takasagona* = *Platypleura repanda* (nec Linnaeus) = *Platypleura nepanda* (sic) (nec Linnaeus) = *Platypleura kaempferi takasagona* = *Platypleura kurowiae takasagon* (sic) = *Platypleura kaempferi* var. *formosana* (nec Matsumura)

P. techowi Schumacher, 1913 = *Platyplerua divisa techowi* = *Platypleura divisa* var. *techowi* = *Platypleura (Platypleura) divisa techowi* = *Systophlochius palochius* Villet, 1989

P. testacea (Walker, 1858) = *Zammara testacea* = *Platypleura (Oxypleura) testacea*

P. turneri Boulard, 1975

P. vitreolimbata Breddin, 1905 = *Platypleura vitrolimbata* (sic)

P. wahlbergi Stål, 1855 = *Platypleura walberghi* (sic)

P. watsoni (Distant, 1897) = *Poecilopsaltria watsoni* = *Platypleura (Platypleura) watsoni* = *Platypleura (Poecilopsaltria) watsoni* = *Platypleura mokensis* Boulard, 2003

P. westwoodi Stål, 1863 = *Poecilopsaltria westwoodi* = *Platypleura (Platypleura) westwoodi* = *Platypleura westwoodii* (sic)

P. witteana Dlabola, 1960

P. yayeyamana Matsumura, 1917 = *Platypleura kaempferi yayeyamana* = *Platypleura yaeyamana* (sic)

Azanicada Villet, 1989

A. zuluensis (Villet, 1987) = *Platypleura zuluensis*

Tugelana Distant, 1912

T. butleri Distant, 1912 = *Platypleura maritzburgensis* Distant, 1913

Capcicada Villet, 1989

C. decora (Germar, 1834) = *Cicada decora* = *Platypleura decora* = *Platypleura deusta* (nec Thunburg) = *Platypleura absimilis* Distant, 1897 = *Platypleura (Platypleura) absimilis*

Albanycada Villet, 1989

A. albigera (Walker, 1850) = *Platypleura albigera* = *Platypleura membranacea* Karsch, 1890

Karscheliana Boulard, 1990

K. parva Boulard, 1990

K. plagiata (Karsch, 1890) = *Platypleura (Oxypleura) plagiata* = *Platypleura plagiata*

Strumoseura Villet, 1999

S. nigromarginata (Ashton, 1914) = *Platypleura nigromarginata* = *Oxypleura quadraticollis* (nec Butler)

Strumosella Boulard, 1973

S. limpida (Karsch, 1890) = *Platypleura limpida* = *Platypleura (Poecilopsaltria) limpida* = *Platypleura ladona* (nec Distant) = *Oxypleura contracta* (nec Walker) = *Oxypleura basalis* (nec Signoret)

S. strumosa (Fabricius, 1803) = *Tettigonia strumosa* = *Cicada strumosa* = *Platypleura strumosa* = *Platypleura (Poecilopsaltria) strumosa* = *Platypleura afzelli* (sic) (nec Stål) = *Oxypleura contracta* (nec Walker)

S. truncaticeps (Signoret, 1884) = *Oxypleura truncaticeps* = *Platypleura truncaticeps* = *Platypleura (Oxypleura) truncaticeps*

Oxypleura Amyot & Audinet-Serville, 1843 = *Cicada (Oxypleura)* = *Platypleura (Oxypleura)* = *Oxyplelure* (sic)

O. atkinsoni (Distant, 1912) = *Platypleura atkinsoni* = *Platypleura (Oxypleura) atkinsoni*

O. basalis Signoret, 1891 = *Platypleura (Oxypleura) basalis*

O. calypso Kirby, 1889 = *Poecilopsaltria calypso* = *Platypleura calypso* = *Platypleura (Oxypleura) calypso*

O. centralis (Distant, 1897) = *Platypleura centralis* = *Platypleura (Oxypleura) centralis*

O. clara Amyot & Audinet-Serville, 1843 = *Platypleura clara* = *Platypleura (Oxypleura) clara* = *Oxypleura passa* Walker, 1850 = *Oxypleura basistigma* Walker, 1850

O. ethiopiensis Boulard, 1975

O. lenihani Boulard, 1985

O. pointeli Boulard, 1985
O. polydorus Walker, 1850 = *Platypleura polydorus*
O. quadraticollis (Butler, 1874) = *Platypleura quadraticollis* = *Platypleura (Oxypleura) quadraticollis* = *Oxyplelure* (sic) *quadraticollis* = *Platypleura nigromarginata* Ashton, 1914 = *Platypleura nigromaculata* (sic) = *Ioba veligera* Boulard, 1965
O. spoerryae Boulard, 1980

Brevisiana Boulard, 1973
B. brevis (Walker, 1850) = *Platyplerua brevis* = *Platypleura (Oxypleura) brevis* = *Platypleura simplex* Walker, 1850 = *Platypleura (Oxypleura) simplex* = *Oxypleura sobrina* Stål, 1855 = *Oxypleura patruelis* Stål, 1855 = *Platypleura (Oxypleura) patruelis* = *Cicada (Oxypleura) neurosticta* Schaum, 1853 = *Cicada (Oxypleura) neurosticta* Schaum, 1862 = *Cicada (Oxypleura) neurostica* (sic) = *Oxypleura neurosticta* = *Platypleura neurosticta*
B. niveonotata (Butler, 1874) = *Platypleura (Oxypleura) niveonotata* = *Platypleura niveonotata*
B. quartai Boulard, 1972 = *Brevisiana brevis* (nec Walker)

Yanga Distant, 1904
Y. andriana (Distant, 1899) = *Platypleura andriana*
Y. antiopa (Karsch, 1890) = *Platypleura antiopa* = *Platypleura (Poecilopsaltria) antiopa*
Y. bouvieri Distant, 1905
Y. brancsiki (Distant, 1893) = *Poecilopsaltria brancsiki*
Y. grandidieri Distant, 1905
Y. guttulata (Signoret, 1860) = *Platypleura guttulata* = *Yanga guttularis* (sic) = *Yanga guttalata* (sic) = *Yanga guttatula* (sic) = *Platypleura tepperi* Goding & Froggatt, 1904
Y. heathi (Distant, 1899) = *Platypleura heathi*
Y. hova (Distant, 1901) = *Poecilopsaltria hova*
Y. mayottensis Boulard, 1990
Y. pembana (Distant, 1899) = *Platypleura pembana*
Y. pulvera argyrea (Melichar, 1896) = *Platypleura argyrea*
Y. pulverea pulverea (Distant, 1882) = *Platypleura pulverea* = *Platypleura pulvera* (sic)
Y. viettei anjouaniensis Boulard, 1990
Y. viettei moheliensis Boulard, 1990
Y. viettei viettei Boulard, 1981 = *Yanga viettei*

Sechellalna Boulard, 2010
S. seychellensis (Distant, 1912) = *Yanga seychellensis* = *Yanga andriana* (nec Distant)

Kongota Distant, 1904
K. handlirschi (Distant, 1897) = *Poecilopsaltria handlirschi* = *Yanga handlirschi* = *Kongota malgadessica* Boulard, 1996
K. punctigera (Walker, 1850) = *Platypleura punctigera* = *Platypleura subfolia* Walker, 1850 = *Kongota muiri* Distant, 1905

Canualna Boulard, 1985
C. liberiana (Distant, 1912) = *Platypleura liberiana*
C. perspicua (Distant, 1905) = *Odopoea perspicua*

Pycna Amyot & Audinet-Serville, 1843 = *Pycnos* (sic) = *Pyona* (sic) = *Platypleura (Pycna)*
P. antinorii (Lethierry, 1881) = *Platypleura antinorii*
P. baxteri Distant, 1914
P. beccarii (Lethierry, 1881) = *Platypleura beccarii*
P. coelestia Distant, 1904
P. concinna Boulard, 2005
P. dolosa Boulard, 1975
P. gigas (Distant, 1881) = *Platypleura gigas*
P. hecuba Distant, 1904 = *Platypleura graueri* Melichar, 1908

P. himalayana Naruse, 1977
P. indochinensis Distant, 1913
P. itremensis Boulard, 2008
P. madagascariensis angusta (Butler, 1882) = *Platypleura angusta*
P. madagascariensis madagascariensis (Distant, 1879) = *Platypleura madagascariensis*
P. minor Liu, 1940 = *Pycna minar* (sic)
P. moniquae Boulard, 2012
P. montana Hayashi, 1978
P. natalensis Distant, 1905
P. neavei Distant, 1912
P. passosdecarvalhoi Boulard, 1975
P. quanza (Distant, 1899) = *Platypleura quanza*
P. repanda repanda (Linnaeus, 1758) = *Cicada repanda* = *Cicada rependa* (sic) = *Tettogonia repanda* = *Fidicina ? repanda* = *Oxypleura repanda* = *Platypleura repanda* = *Platypleura nepanda* (sic) = *Platypleura phalaenoides* Walker, 1850 = *Platypleura interna* Walker, 1852 = *Platyplerua congrex* Butler, 1874
P. repanda var. *β* (Walker, 1850) = *Platypleura repanda* var. *β*
P. repanda var. *δ* (Walker, 1850) = *Platypleura repanda* var. *δ*
P. repanda var. *ε* (Walker, 1850) = *Platypleura repanda* var. *ε*
P. repanda var. *η* (Walker, 1850) = *Platypleura repanda* var. *η*
P. repanda var. *γ* (Walker, 1850) = *Platypleura repanda* var. *γ*
P. repanda var. *ζ* (Walker, 1850) = *Platypleura repanda* var. *ζ*
P. rudis (Karsch, 1890) = *Platypleura rudis*
P. schmitzi Boulard, 1979
P. semiclara (Germar, 1834) = *Cicada semiclara* = *Platypleura semiclara* = *Platypleura basifolis* Walker, 1850
P. strix Amyot & Audinet-Serville, 1843 = *Cicada strix* = *Platypleura strix* = *Cicada strix* Brullé, 1849 = *Cicada stryx* (sic) = *Pycna stryx* (sic)
P. sylvia (Distant, 1899) = *Platypleura sylvia*
P. tangana (Strand, 1910) = *Platypleura (Pycna) tangana* = *Pycna tanga* (sic)
P. umbelinae Boulard, 1975
P. verna Hayashi, 1982
P. vitrea Schumacher, 1913
P. vitticollis (Jacobi, 1904) = *Platypleura vitticollis*

Suisha Kato, 1928 = *Dasypsaltria* Kato, 1927
S. coreana (Matsumura, 1927) = *Pycna coreana* = *Dasypsaltria coreana* Kato, 1927
S. formosana (Kato, 1927) = *Dasypsaltria formosana* = *Platypleura formosana*

Umjaba Distant, 1904
U. alluaudi Distant, 1905
U. evanescens (Butler, 1882) = *Platypleura evanescens*

Ugada Distant, 1904 = *Platypleura (Ugada)*
U. atratula Distant, 1919
U. cameroni (Butler, 1874) = *Platypleura cameroni* = *Ugada violacea* Boulard, 1972
U. dargei Boulard, 2012
U. giovanninae Boulard, 1973 = *Ugada giovannina* (sic)
U. grandicollis grandicollis (Germar, 1830) = *Cicada graneidollis* (sic) = *Cicada grandicollis* = *Ugada grandicollis* = *Platypleura confusa* Karsch, 1890 = *Ugada confusa*
U. grandicollis taiensis Boulard, 1986 = *Ugada grandicollis taïensis* = *Ugada taiensis*
U. inquinata (Distant, 1881) = *Platypleura inquinata* = *Platypleura (Ugada) inquinata*
U. kageraensis Boulard, 2012
U. lamottei Boulard, 1977
U. limbalis (Karsch, 1890) = *Platypleura limbalis* = *Ugada limballis* (sic) = *Ugada limbimacula* (nec Karsch)
U. limbata limbata (Fabricius, 1775) = *Tettigonia limbata* = *Cicada limbata* = *Oxypleura limbata* = *Platypleura limbata* = *Pycna limbata* = *Ugada limbata* = *Cicada armata* Olivier, 1790 = *Oxypleura armata* = *Cicada africana* Palisot de Beauvois, 1813

U. limbata occidentalis Boulard, 1986 = *Ugada limbata occidentalis* var. *intermedia* Boulard, 1986 = *Ugada limbata occidentalis* var. *marmorea* Boulard, 1986
U. limbimacula (Karsch, 1893) = *Ugada limbimaculata* (sic) = *Platyleura limbimacula* = *Ugada hahnei* Schmidt, 1919 = *Ugada nutti* (nec Distant)
U. nigeriana (Distant, 1913) = *Pycna nigeriana*
U. nigrofasciata nigrofasciata Distant, 1919 = *Ugada nigrofasciata* = *Ugada oswaldebneri* Schmidt, 1920
U. nigrofasciata sylvicola Boulard, 1973
U. nutti Distant, 1904
U. parva Boulard, 1975
U. poensis Boulard, 1975
U. praecellens (Stål, 1863) = *Platypleura praecellens*
U. sp. 1 Villet, Barker & Lunt, 2005
U. stalina (Butler, 1874) = *Platypleura stalina*

Hainanosemia Kato, 1927
H. stigma Kato, 1927

Kalabita Moulton, 1923
K. operculata Moulton, 1923

Tribe Hamzini Distant, 1905 = Hamzaria

Hamza Distant, 1904
H. ciliaris (Linnaeus, 1758) = *Cicada ciliaris* = *Poecilopsaltria ciliaris* = *Platypleura ciliaris* = *Platypleura (Platypleura) ciliaris* = *Platypleura cilialis* (sic) = *Cicada ocellata* Degeer, 1773 = *Cicada ocella* (sic) = *Platypleura ocellata* = *Cicada varia* Olivier, 1790 = *Platypleura varia* = *Tettigonia marmorata* Fabricius, 1803 = *Oxypleura marmorata* = *Platypleura marmorata* = *Platypleura arcuata* Walker, 1858 = *Platypleura catocaloides* Walker, 1868 = *Platypleura catocalina* (sic) = *Poecilopsaltria catacaloides* = *Platypleura bouruensis* Distant, 1898 = *Platypleura bouruensis* Distant, 1898 = *Hamza bouruensis* = *Hamza boeruensis* (sic) = *Platypleura lyricen* Kirkaldy, 1913 = *Hamza bouruensis* (nec Distant, 1898) = *Hamza uchiyamae* Matsumura, 1927 = *Hamza bouruensis uchiyamae* = *Platyplerua uchiyamae* = *Hamza* sp. Endo & Hayashi, 1979

Tribe Orapini Boulard, 1985

Orapa Distant, 1905 = *Orapia* (sic)
O. africana Kato, 1927
O. elliotti (Distant, 1907) = *Pycna elliotti*
O. lateritia Jacobi, 1910
O. cf. *lateritia* Villet & Noort, 1999
O. numa (Distant, 1904) = *Pycna numa*
O. uwembaiensis Boulard, 2012

Tribe Zammarini Distant, 1905 = Zammararia

Odopoea Stål, 1861 = *Odopea* (sic) = *Odopaea* (sic) = *Adusella* Haupt, 1918 = *Edhombergia* Delétang, 1919
O. azteca Distant, 1881
O. cariboea Uhler, 1892
O. degiacomii Distant, 1912 = *Odopea* (sic) *degiacomii*
O. dilatata (Fabricius, 1775) = *Tettigonia dilatata* = *Cicada dilatata* = *Odopoea dilata* (sic) = *Zammara plena* Walker, 1850 = *Odopoea plena* = *Zammara cuncta* Walker, 1850 = *Odopoea cuncta* = *Zammara praxita* Walker, 1850 = *Zammara erato* Walker, 1850 = *Zammara crato* (sic) = *Odopoea erato* = *Odopoea domingensis* Uhler, 1892
O. diriangani Distant, 1881
O. funesta (Walker, 1858) = *Odopoea fumesta* (sic) = *Zammara funesta*
O. insignifera Berg, 1879

O. jamaicensis Distant, 1881
O. minuta Sanborn, 2007
O. signata (Haupt, 1918) = *Adusella signata* Haupt, 1918 = *Tettigades lebruni* (partim) = *Edholmbergia lebruni* Delétang, 1919
O. signoreti Stål, 1864
O. strigipennis (Walker, 1858) = *Zammara strigipennis*
O. suffusa (Walker, 1850) = *Zammara suffusa*
O. vacillans (Walker, 1858) = *Zammara vacillans*
O. venturii Distant, 1906

Miranha Distant, 1905
M. imbellis (Walker, 1858) = *Zammara imbellis* = *Odopoea imbellis* = *Odopaea* (sic) *imbelis* (sic)

Zammara Amyot & Audinet-Serville, 1843 = *Zamara* (sic) = *Cicada (Zammara)* = *Cicada (Tibicen)*
Z. brevis (Distant, 1905) = *Orellana brevis* = *Zammara columbia* (nec Distant)
Z. calochroma Walker, 1858 = *Zammara callichroma* Stål, 1864
Z. erna Schmidt, 1919
Z. eximia (Erichson, 1848) = *Cicada (Zammara) exima* = *Zammara brevis* (nec Distant)
Z. hertha Schmidt, 1919
Z. intricata Walker, 1850
Z. lichyi Boulard & Sueur, 1996
Z. luculenta Distant, 1883
Z. medialinea Sanborn, 2004
Z. nigriplaga Walker, 1858 = *Orellana nigriplaga*
Z. olivacea Sanborn, 2004
Z. smaragdina Walker, 1850 = *Zammara zmaragdina* (sic) = *Zammara angulosa* Walker, 1850 = *Zammara angnlosa* (sic)
Z. smaragdula Walker, 1858
Z. strepens Amyot & Audinet-Serville, 1843 = *Zammara streppens* (sic) = *Cicada tympanum* Palisot de Beauvois, 1813 (nec *Cicada tympanum* Fabricius, 1803)
Z. tympanum (Fabricius, 1803) = *Tettigonia tympanum* = *Cicada tympanum* = *Tibicen tympanum* = *Cicada (Tibicen) tympanum* = *Zamara* (sic) *tympanum* = *Zammara tympanium* (sic) = *Zammara typanum* (sic) = *Zammara typmanum* (sic)
Z. sp. Wolda & Ramos, 1992

Zammaralna Boulard & Sueur, 1996
Z. bleuzeni Boulard & Sueur, 1996

Juanaria Distant, 1920
J. poeyi (Guérin-Méneville, 1856) = *Cicada (Platypleura) poeyi* = *Cicada poeyi* = *Zammara poeyi* = *Odopoea poeyi* = *Juanaria mimica* Distant, 1920

Borencona Davis, 1928
B. aguadilla Davis, 1928

Chinaria Davis, 1934 = *Chineria* (sic)
C. mexicana Davis, 1934 = *Chineria* (sic) *mexicana*
C. pueblaensis Sanborn, 2007
C. similis Davis, 1942
C. vivianae Ramos, 1983

Orellana Distant, 1905 = *Oriellana* (sic)
O. bigibba Schmidt, 1919 = *Zammara bigibba*
O. brunneipennis Goding, 1925 = *Orellana brunnpennis* (sic)
O. castaneamaculata Sanborn, 2010 = *Zammara castaneamaculata*
O. columbia (Distant, 1881) = *Zammara columbia*
O. pollyae Sanborn, 2011
O. pulla Goding, 1925

Uhleroides Distant, 1912 = *Uhlerioides* (sic) = *Uhlerodes* (sic)
U. chariclo (Walker, 1850) = *Cicada chariclo* = *Tympanoterpes chariclo* = *Odopoea chariclo* = *Proarna chariclo* = *Proalba* (sic) *chariclo*
U. cubensis Distant, 1912
U. hispaniolae Davis, 1939
U. maestra Davis, 1939 = *Uhlerodes* (sic) *maestra*
U. sagrae (Guérin-Méneville, 1856) = *Cicada sagrae* = *Zammara sagrae* = *Odopoea sagrae* = *Odopaea* (sic) *sagrae*
U. samanae Davis, 1939
U. walkerii (Guérin-Méneville, 1856) = *Cicada walkerii* = *Zammara walkerii* = *Odopoea walkerii* = *Odopoea walkeri* (sic) = *Uhleroides walkeri* (sic)

Tribe Plautillini Distant, 1905 = Plautillaria

Plautilla Stål, 1865
P. hammondi Distant, 1914 = *Plautilla hamondi* (sic) = *Odopoea* sp. Salazar Escobar, 2005
P. stalagmoptera Stål, 1865
P. venedictoffae Boulard, 1978

Tribe Polyneurini Amyot & Audinet-Serville, 1843 = Plyneurini (sic)

Subtribe Polyneurina Myers, 1929 = Polyneuraria

Angamiana Distant, 1890 = *Agamiana* (sic) = *Agaminana* (sic) = *Proretinata* Chou & Yao, 1986
A. aetherea Distant, 1890
A. floridula Distant, 1904 = *Agaminana* (sic) *floridula* = *Angamiana florida* (sic) = *Angamiana bauflei* Boulard, 1994 = *Angamiana beauflei* (sic) = *Angamiana floridula bauflei* = *Angamiana floridula* var. *beauflei* (sic) = *Agamiana* (sic) *floridula*
A. fuscula (Chou & Yao, 1986) = *Proretinata fuscula*
A. masamii Boulard, 2005 = *Angamiana florida* (sic) (nec Distant)
A. melanoptera Boulard, 2005
A. vemacula (Chou & Yao, 1986) = *Proretinata vemacula* = *Proretinata vemaculata* (sic)
A. yunnanensis (Chou & Yao, 1986) = *Proretinata yunnanensis*

Polyneura Westwood, 1842 = *Polynevra* (sic) = *Cicada (Polyneura)*
P. cheni Chou & Yao, 1986 = *Polyneura distanti* Boulard, 1994 = *Polyneura hügelii* Graber, 1876 nom. nud.
P. ducalis Westwood, 1840 = *Cicada (Polyneura) ducalis* = *Cicada ducalis* = *Polyneura ducatlis* (sic) = *Polyneura ducaslis* (sic) = *Polyneura linearis* (sic)
P. laevigata Chou & Yao, 1986
P. nr *laevigata* Boulard, 2008
P. parapuncta Chou & Yao, 1986
P. tibetana Chou & Yao, 1986
P. xichangensis Chou & Yao, 1986 = *Polyneura xichengensis* (sic)

Graptopsaltria Stål, 1866 = *Graphopsaltria* (sic) = *Graptosaltria* (sic) = *Graptopsalria* (sic) = *Graptosaltria* (sic) = *Graptosaltia* (sic) = *Grapsaltria* (sic) = *Graptopsaltiria* (sic)
G. bimaculata Kato, 1925 = *Graptopsaltria colorata* (nec Stål) = *Graptopsaltria tienta* (nec Karsch)
G. nigrofuscata badia Kato, 1925 = *Graptopsaltria nigrofuscata* ab. *badia* = *Graptopsaltria colorata badius* = *Graptopsaltria colorata badia*
G. nigrofuscata nigrofuscata (de Motschulsky, 1866) = *Fidicina nigrofuscata* = *Fidicina nigrofasciata* (sic) = *Cryptotympana nigrofuscata* = *Goaptopsaltria* (sic) *nigsofuscata* (sic) = *Graptopsalria* (sic) *nigrofuscata* = *Graptosaltria* (sic) *nigrofuscata* = *Graptopsaltria migrofuscata* (sic) = *Graptopsaltria colorata* Stål, 1866 = *Graptosaltia* (sic) *corolata* (sic) = *Grapsaltria* (sic) *colorata* = *Graptopsaltria corolata* (sic) = *Graptosaltria* (sic) *colorata* = *Graptopsaltiria* (sic) *corolata* (sic) = *Grapsaltria* (sic) *colorata* = *Graptopsaltria colotata* (sic) = *Platypleura colorata* = *Graptopsaltria nigrofuscata tsuchidai* Matsumura, 1939 = *Graptopsaltria nigrofuscata* var. *testaceomaculata* Kato, 1937 = *Graptopsaltria nigrofuscata* ab. *testaceomaculata*
G. tienta Karsch, 1894

Subtribe Formotosenina Boulard, 2008

Formotosena Kato, 1925

F. montivaga (Distant, 1889) = *Tosena montivaga*

F. seebohmi (Distant, 1904) = *Tosena seebohmi* = *Tosena siebohmi* (sic) = *Formotosena siebohmi* (sic) = *Tosena seebohmi interrupta* Schumacher, 1915 = *Formotosena seebohmi interrupta* = *Formotosena apicalis* (sic)

Tribe Tacuini Distant, 1904 = Tacuarini

Tacua Amyot & Audinet-Serville, 1843 = *Tama* (sic) = *Cicada (Tacua)*

T. speciosa decolorata Boulard, 1994 = *Tacua speciosa* var. *decolorata* = *Tacua speciosa* var. *a* Distant, 1889

T. speciosa speciosa (Illiger, 1800) = *Tettigonia speciosa* = *Cicada speciosa* = *Cicada indica* Donovan, 1800 = *Tacua indica* = *Tettigonia gigantean* Weber, 1801

Tribe Talaingini Myers, 1929

Subtribe Talaingina Myers, 1929

Talainga Distant, 1890

T. binghami Distant, 1890

T. chinensis Distant, 1900

T. japrona Ollenbach, 1929 = *Talainga japroa* (sic)

T. naga Ollenbach, 1929

T. omeishana Chen, 1957

Paratalainga He, 1984 = *Chouia* Yao nom. nud.

P. distanti (Jacobi, 1902) = *Talainga distanti*

P. fucipennis He, 1984 = *Paratalainga fusipenis* (sic) = *Paratalainga fuscipennis* (sic) = *Chouia liuziensis* Yao, nom. nud.

P. fumosa Chou & Lei, 1992

P. guizhouensis Chou & Lei, 1992

P. reticulata He, 1984 = *Chouia pulchra* Yao, nom. nud.

P. yunnanensis Chou & Lei, 1992

Tribe Thophini Distant, 1904 = Tophini (sic)

Thopha Amyot & Audinet-Serville, 1843 = *Topha* (sic) = *Thophra* (sic) = *Thopa* (sic) = *Tropha* (sic) = *Thorpha* (sic)

T. colorata Distant, 1907 = *Topha* (sic) *colorata*

T. emmotti Moulds, 2001

T. hutchinsoni Moulds, 2008

T. saccata (Fabricius, 1803) = *Tettigonia saccata* = *Cicada saccata* = *Topha* (sic) *saccata* = *Thopa* (sic) *saccata* = *Tropha* (sic) *saccata*

T. sessiliba Distant, 1892 = *Thorpha* (sic) *sessiliba* = *Topha* (sic) *sessiliba* = *Thopha stentor* Buckton, 1898 = *Thopha nigricans* Distant, 1910

Arunta Distant, 1904

A. interclusa (Walker, 1858) = *Thopha interclusa* = *Henicopsaltria interclusa* = *Cicada graminea* Distant, 1904 (nec *Cicada graminea* Fabricius, 1798) = *Cicada queenslandica* Kirkaldy, 1909 = *Arunta flava* Ashton, 1912

A. perulata (Guérin-Méneville, 1831) = *Cicada perulata* = *Thopha perulata* = *Henicopsaltria perulata* = *Arunta pemlata* (sic) = *Arunta perolata* (sic) = *Arunta intermedia* Ashton, 1921

Tribe Cyclochilini Distant, 1904

Cyclochila Amyot & Audinet-Serville, 1843 = *Cyclochida* (sic) = *Cychlochila* (sic) = *Cyclophila* (sic)
C. australasiae (Donovan, 1805) = *Tettigonia australasiae* = *Cyclophila* (sic) *australasiae* = *Cyclochila austrasiae* (sic) = *Cicada olivacea* Germar, 1830 = *Cyclochila australasiae* var. *spreta* Goding & Froggatt, 1904 = *Cyclochila australasiae* form *spreta*
C. virens Distant, 1906 = *Cyclochila laticosta* Ashton, 1912

Tribe Jassopsaltriini Moulds, 2005

Jassopsaltria Ashton, 1914
J. rufifacies Ashton, 1914
J. sp. A Moulds, 2005

Tribe Burbungini Moulds, 2005

Burbunga Distant, 1905 = *Burbanga* (sic)
B. albofasciata Distant, 1907
B. aterrima Distant, 1914 = *Macrotristria aterrima* = *Macrotristria vulpina* Ashton, 1914
B. gilmorei (Distant, 1882) = *Tibicen gilmorei* = *Burbanga* (sic) *gilmori* (sic)
B. hillieri (Distant, 1907) (nec *hillieri* Distant, 1906) = *Macrotristria hillieri*
B. inornata Distant, 1905
B. mouldsi Olive, 2012
B. nanda (Burns, 1964) = *Macrotristria nanda*
B. nigrosignata (Distant, 1904) = *Macrotristria nigrosignata*
B. occidentalis (Distant, 1912) = *Macrotristria occidentalis*
B. parva Moulds, 1994
B. queenslandica Moulds, 1994 = *Burbunga* sp. Ewart, 1988 = *Burbunga gilmorei* (nec Distant)

Tribe Talcopsaltriini Moulds, 2008

Talcopsaltria Moulds, 2008
T. olivei Moulds, 2008 = *Macrotristria* sp. B Ewart, 1993 = Heathlands sp. B Ewart, 2005

Tribe Cryptotympanini Handlirsch, 1925 = Tibicenini Van Duzee, 1916 = Tibicinae (sic) = Tibiceni (sic) = Lyristini Gomez-Menor, 1957 = Lyristarini

Subtribe Cryptotympanina Handlirsch, 1925 = Cryptotympanaria = Tibicenaria Van Duzee, 1916

Arenopsaltria Ashton, 1921
A. fullo (Walker, 1850) = *Fidicina fullo* = *Henicopslatria fullo*
A. nubivena (Walker, 1858) = *Fidicina nubivena* = *Henicopsaltria nubivena*
A. pygmaea (Distant, 1904) = *Henicopsaltria pygmaea*

Psaltoda Stål, 1861 = *Pfaltoda* (sic)
P. adonis Ashton, 1914
P. antennetta Moulds, 2002
P. aurora Distant, 1881
P. brachypennis Moss & Moulds, 2000 = *Psaltoda* sp. nov. Ewart, 1998
P. claripennis Ashton, 1921 = *Pfaltoda* (sic) *claripennis*
P. flavescens Distant, 1892
P. fumipennis Ashton, 1912
P. harrisii (Leach, 1814) = *Tettigonia harrisii* = *Cicada harrisii* = *Psaltoda harrissii* (sic) = *Neopsaltoda morrisi* (sic) = *Cicada dichroa* Boisduval, 1835 = *Cicada dichroma* (sic) = *Psaltoda dichroa* = *Fidicina subguttata* Walker, 1850 = *Fidicina subgutatta* (sic) = *Psaltoda longirostris* Chisholm, 1932
P. insularis Ashton, 1914
P. macallumi Moulds, 2002

P. magnifica Moulds, 1984
P. moerens (Germar, 1834) = *Cicada moerens* = *Psaltoda meorens* (sic) = *Psaltoda moerans* (sic)
P. mossi Moulds, 2002 = *Psaltoda* sp. A Ewart, 1988 = *Psaltoda* sp. nov. Lithgow, 1988
P. pictibasis (Walker, 1858) = *Cicada pictibasis*
P. plaga (Walker, 1850) = *Cicada plaga* = *Cicada argentata* Germar, 1834 (nec *Cicada argentata* Olivier, 1790) = *Psaltoda argentata*

Anapsaltoda Ashton, 1921 = *Anapsaltodea* (sic)
A. pulchra (Ashton, 1921) = *Anapsaltodea* (sic) *pulchra* = *Psaltoda pulchra*

Neopsaltoda Distant, 1910
N. crassa Distant, 1910

Henicopsaltria Stål, 1866
H. danielsi Moulds, 1993
H. eydouxii (Guérin-Méneville, 1838) = *Cicada eydouxii* = *Henicopsaltria eydouxi* (sic) = *Henicopsaltria edouxi* (sic) = *Henicopsaltria eudouxi* (sic) = *Psaltoda flavescens* (nec Distant)
H. kelsalli Distant, 1910
H. rufivelum Moulds, 1978

Illyria Moulds, 1985
I. australensis (Kirkaldy, 1909) = *Cicada interrupta* Walker, 1850 (nec *Cicada interrupta* Linnaeus, 1758)
I. burkei (Distant, 1882) = *Tibicen burkei* = *Cicada burkei* = *Tettigia variegata* Goding & Froggatt, 1904
I. hilli (Ashton, 1914) = *Tettigia hilli* = *Cicada hilli*
I. major Moulds, 1985

Macrotristria Stål, 1870 = *Macrotristia* (sic) = *Macruouistria* (sic) = *Marcrotristria* (sic) = *Cicada (Macrotristria)* = *Cicada (Macrotistria)* (sic)
M. angularis (Germar, 1834) = *Cicada angularis* = *Fidicina angularis* = *Cicada (Macriotristria) angularis* = *Macruouistria* (sic) *angularis* = *Macrotristria augularis* (sic) = *Macrotristra* (sic) *angularis* = *Marcrotristria* (sic) *angularis*
M. bindalia Burns, 1964
M. doddi Ashton, 1912
M. dorsalis Ashton, 1912
M. douglasi Burns, 1964
M. extrema (Distant, 1892) = *Cicada extrema*
M. frenchi Ashton, 1914
M. godingi Distant, 1907
M. hieroglyphicalis (Kirkaldy, 1909) = *Cicada hieroglyphica* = *Cicada hieroglyphicalis* (sic) = *Cicada hieroglyphica* (nec Say) = *Rihana hieroglyphica* = *Macrotristria hieroglyphica* (sic) = *Macrotristria hieroglyphicus* (sic)
M. intersecta (Walker, 1850) = *Fidicina intersecta* = *Fidicina internata* Walker, 1850 = *Macrotristria internata* = *Fidicina prasina* Walker, 1850 = *Cicada sylvanella* Goding & Froggatt, 1904 = *Macrotristria sylvanella*
M. kabikabia Burns, 1964
M. kulungura Burns, 1964
M. sp. nr. *kulungura/dorsalis* Ewart, 2005
M. lachlani Moulds, 1992 = *Macrotristria* sp. A Ewart, 1993
M. maculicollis Ashton, 1914
M. madegassa Boulard, 1996
M. sylvara (Distant, 1901) = *Cicada sylvara* = *Cicada sylvana* (sic) = *Macrotristria nigronervosa* Distant, 1904
M. thophoides Ashton, 1914
M. vittata Moulds, 1992
M. worora Burns, 1964

Chremistica Stål, 1870 = *Cicada (Chremistica)* = *Chremestica* (sic) = *Cheremistica* (sic) = *Chremistrica* (sic) = *Rihana* Distant, 1904 = *Rhiana* (sic)
C. atra (Distant, 1909) = *Rihana atra*

C. atratula Boulard, 2007 = *Chremistica atrata* (sic)
C. banksi Liu, 1940
C. biloba Bregman, 1985
C. bimaculata (Olivier, 1791) = *Cicada bimaculata* = *Rihana bimaculata* = *Chremistica bimcaulata* (sic) = *Chremistica binaculata* (sic) = *Cicada atro-virens* Guérin-Méneville, 1838 = *Chremistica atrovirens* = *Cicada viridis* (nec Fabricius) = *Rihana viridis* (nec Fabricius) = *Chremistica viridis* (nec Fabricius)
C. sp. nr. *bimaculata* Sanborn, Phillips & Sites, 2007
C. borneensis Yaakop & Duffels, 2005
C. brooksi Yaakop & Duffels, 2005
C. cetacauda Yaakop & Duffels, 2005
C. coronata (Distant, 1889) = *Cicada coronata* = *Rihana coronata*
C. echinaria Yaakop & Duffels, 2005 = *Chremistica echinaaria* (sic)
C. elenae madagascariensis Boulard, 2001 = *Chremistica helenae* (sic) *madagascariensis*
C. elenae seychellensis Boulard, 2001 = *Chremistica helenae* (sic) *seychellensis* = *Chremistica elenae mahiensis* (sic)
C. euterpe (Walker, 1850) = *Cicada euterpe* = *Rihana euterpe*
C. germana (Distant, 1888) = *Cicada germana* = *Rihana germana* = *Chremistrica* (sic) *germana*
C. guamusangensis Salmah & Zaidi, 2002 = *Chremistica chueatae* Boulard, 2004
C. hollowayi Yaakop & Duffels, 2005
C. hova (Distant, 1905) = *Rihana hova*
C. inthanonensis Boulard, 2006 = *Chremistica bimaculata inthanonensis* = *Cicada bimaculata* (nec Olivier) = *Rihana bimaculata* (nec Olivier) = *Chremistica bimaculata* (nec Olivier) = *Chremistica atrovirens* (nec Guérin-Méneville) = *Chremistica viridis* (nec Fabricius) = *Chremistica* undescribed species A Sanborn, Phillips & Sites 2007
C. kecil Salmah & Zaidi, 2002
C. kyoungheeae Lee, 2010
C. loici Boulard, 2000
C. longa Lei, Chou & Li, 1997
C. maculata Chou & Lei, 1997 = *Chremistica maculate* (sic)
C. malayensis Yaakop & Duffels, 2005
C. matilei Boulard, 2000
C. martini (Distant, 1905) = *Rihana martini*
C. minor Bregman, 1985
C. *minuta* Liu, 1940 nom. nud. = *Chemistica minuda* (sic)
C. mixta (Kirby, 1891) = *Dundubia mixta* = *Cicada mixta* = *Rihana mixta*
C. moultoni Boulard, 2002
C. mussarens Boulard, 2005
C. nana Chen, 1943
C. nesiotes Breddin, 1905 = *Chremistica bimaculata* (nec Olivier)
C. niasica Yaakop & Duffels, 2005
C. nigra Chen, 1940
C. nigrans (Distant, 1904) = *Cicada nigrans* = *Rihana nigrans*
C. numida (Distant, 1911) = *Rihana numida*
C. ochracea gracilis (Kato, 1927) = *Rihana ochracea gracilis*
C. ochracea interrupta (Kato, 1927) = *Rihana ochracea interrupta*
C. ochracea ochracea (Walker, 1850) = *Fidicina ochracea* = *Cicada ochracea* = *Rihana ochracea* = *Rhiana* (sic) *ochracea* = *Rihana ochrea* (sic) = *Tibicen ochracea* = *Chremistica cohraces* (sic) = *Chremistica ochraceus* (sic) = *Cicada ferrifera* Walker, 1850 = *Dundubia fasciceps* Stål, 1854 = *Dundubia fuscipes* (sic) = *Dundubia fusciceps* (sic) = *Cicada fuscipes* (sic)
C. ochracea var. *a* (Kato, 1925) = *Rihana ochracea* var. *a*
C. ochracea var. *b* (Kato, 1925) = *Rihana ochracea* var. *b*
C. ochracea takesakiana (Kato, 1927) = *Rihana ochracea takesakiana*
C. operculissima (Distant, 1897) = *Cicada operculissima* = *Rihana operculissima*
C. phamiangensis Boulard, 2009
C. polyhymnia (Walker, 1850) = *Fidicina polyhymnia* = *Cicada polyhymnia* = *Rihana polyhymnia*
C. pontianaka (Distant, 1888) = *Cicada pontianaka* = *Rihana pontianaka* = *Cicada daiaca* Breddin, 1900
C. nr. *pontianaka* Zaidi, Ruslan & Mahadir, 1996

C. pulverulenta pulverulenta (Distant, 1905) = *Cicada pulverulenta* = *Tibicen pulverulenta* = *Antankaria pulverulenta madegassa* Boulard, 1999
C. pulverulenta madagascariensis Boulard, 2001
C. pulverulenta seychellensis (Boulard, 1999) = *Chremistica pulverulenta mahiensis* (sic) = *Antankaria pulverea* (sic) *madegassa*
C. seminiger (Distant, 1909) = *Rihana seminiger*
C. semperi Stål, 1870 = *Cicada (Chremistica) semperi* = *Cicada semperi* = *Rihana semperi*
C. siamensis Bregman, 1985
C. sibilussima Boulard, 2006
C. sp. 1 Zaidi, Ruslan & Mahadir, 1996
C. sp. 2 Zaidi, Ruslan & Mahadir, 1996
C. sumatrana Yaakop & Duffels, 2005
C. tagalica Stål, 1870 = *Cicada (Chremistica) tagalica* = *Cicada tagalica* = *Rihana tagalica*
C. timorensis (Distant, 1892) = *Cicada timorensis* = *Rihana timorensis*
C. tondana (Walker, 1870) = *Fidicina tondana* = *Cryptotympana tondana*
C. tridentigera (Breddin, 1905) = *Cicada tridentigera*
C. umbrosa (Distant, 1904) = *Cicada umbrosa* = *Rihana umbrosa* = *Rihana pisanga* Moulton, 1923 = *Chremistica pisanga*
C. undescribed species B Sanborn, Phillips & Sites, 2007
C. undescribed species C Sanborn, Phillips & Sites, 2007
C. undescribed species Pham & Yang, 2010
C. viridis (Fabricius, 1803) = *Tettigonia viridis* = *Cicada viridis* = *Cicada (Chremistica) viridis* = *Rihana viridis* = *Cicada bimaculata* (nec Olivier)

Diceroprocta Stål, 1870 = *Cicada (Diceroprocta)* = *Diceropracta* (sic) = *Diceroprocter* (sic) = *Tibicen (Diceroprocta)* = *Rehana* (sic) = *Tibcien* (sic)
D. alacris alacris (Stål, 1864) = *Cicada (Diceroprocta) alacris* = *Cicada transversa* Walker, 1858 = *Pomponia transversa* = *Cicada (Diceroprocta) transversa* = *Rihana transversa* = *Diceroprocta transversa* = *Tibicen transversa*
D. alacris campechensis Davis, 1938 = *Diceroprocta alacris* var. *campechensis*
D. albomaculata Davis, 1928
D. apache (Davis, 1921) = *Tibicen apache* = *Diceropracta* (sic) *apache* = *Diceroprocter* (sic) *apache* = *Cicada transversa* (nec Walker) = *Cicada vitripennis* (nec Say) = *Diceroprocta apache ochroleuca* Davis, 1942 = *Diceroprocta apache* var. *ochroleuca* = *Diceroprocta apache ochraleuca* (sic) = *Tibicen cinctifera* (nec Uhler) = *Diceroprocta cinctifera* (nec Uhler)
D. arizona (Davis, 1916) = *Cicada arizona* = *Tibicen arizona*
D. aurantiaca Davis, 1938 = *Diceroprocta delicata* var. *aurantiaca* = *Diceroprocta delicata aurantiaca*
D. averyi Davis, 1941
D. azteca (Kirkaldy, 1909) = *Cicada pallida* Distant, 1881 (nec *Cicada pallida* Goeze, 1778) = *Tibicen pallida* = *Cicada azteca* = *Tibicen azteca*
D. bakeri (Distant, 1911) = *Rihana bakeri*
D. belizensis (Distant, 1910) = *Rihana belizensis*
D. bequaerti (Davis, 1917) = *Tibicen viridifascia* var. *bequaerti* = *Tibicen viridifascia bequaerti* = *Tibicen vitripennis bequaerti* = *Diceroprocta vitripennis bequaerti*
D. bibbyi Davis, 1928
D. bicolora Davis, 1935 = *Diceroprocta bicolor* (sic)
D. biconica (Walker, 1850) = *Cicada biconica* = *Rihana biconica* = *Tibicen biconica* = *Tibicen (Diceroprocta) biconica* = *Diceroprocta bicornica* (sic) = *Rehana* (sic) *biconica*
D. bicosta (Walker, 1850) = *Cicada bicosta* = *Cicada bicostata* (sic) = *Rihana bicosta* = *Tibicen bicosta* = *Diceroprocta transversa* (nec Walker)
D. sp. nr. *bicosta* Young, 1981
D. biguttula Billberg, 1820 nom. nud. = *Tettigonia 2-guttulus*
D. bonhotei (Distant, 1901) = *Cicada bonhotei* = *Rihana bonhotei*
D. bulgara (Distant, 1906) = *Rihana bulgara* = *Rihana operculissima* Distant, 1906 (nec *Rihana operculissima* Distant, 1897)
D. canescens Davis, 1935

D. caymanensis Davis, 1939
D. chinensis Billberg, 1820 nom. nud. = *Tettigonia chinensis*
D. cinctifera cinctifera (Uhler, 1892) = *Cicada cinctifera* = *Cicada acutifera* (sic) = *Tibicen cinctifera* = *Tibicen cinctifer* (sic) = *Diceroprocta cinctifera* = *Diceroprocta cinetifera* (sic) = *Tibicen ochreoptera* Uhler, 1892 = *Tibicen ochrapterus*
D. cinctifera limpia Davis, 1932 = *Diceroprocta cinctifera* var. *limpia* = *Diceroprocta cinctifera limpia*
D. cinctifera viridicosta Davis, 1930 = *Diceroprocta cinctifera* var. *viridicosta* = *Diceroprocta cinctifera viridicosta* = *Diceroprocta cinetifera* (sic) *viridicosta*
D. cleavesi Davis, 1930
D. crucifera (Walker, 1850) = *Cicada crucifera* = *Rihana crucifera*
D. delicata (Osborn, 1906) = *Cicada delicata* = *Tibicen delicata*
D. digueti (Distant, 1906) = *Rihana digueti*
D. distanti Metcalf, 1963 = *Cicada intermedia* Distant, 1881 (nec *Cicada intermedia* Signoret, 1847) = *Rihana intermedia*
D. eugraphica (Davis, 1916) = *Cicada eugraphica* = *Tibicen eugraphica* = *Cicada vitripennis* (nec Say)
D. fraterna Davis, 1935
D. fuscomaculata Billberg, 1820 nom. nud. = *Tettigonia fuscomaculata*
D. fusipennis (Walker, 1858) = *Fidicina fusipennis* = *Rihana fusipennis* = *Cicada reticularis* Uhler, 1892
D. grossa (Fabricius, 1775) = *Tettigonia grossa* = *Cicada grossa* = *Tympanoterpes grossa* = *Cicada (Cicada) grossa* = *Rihana grossa*
D. knighti (Davis, 1917) = *Tibicen knighti*
D. lata Davis, 1941 = *Diceroprocta texana* var. *lata* = *Diceroprocta texana lata*
D. lucida Davis, 1934
D. marevagans Davis, 1928
D. mesochlora (Walker, 1850) = *Cicada mesochlora* = *Odopoea mesochlora* = *Rihana mesochlora*
D. oaxacaensis Sanborn, 2007
D. obscurior Davis, 1935 = *Diceroprocta biconica* var. *obscurior* = *Diceroprocta biconica obscurior*
D. oculata Davis, 1935
D. olympusa (Walker, 1850) = *Fidicina olympusa* = *Cicada olympusa* = *Rihana olympusa* = *Tibicen olympusa* = *Diceroprocta olympusia* (sic) = *Cicada milvus* Walker, 1858 = *Cicada sordidata* Uhler, 1892 = *Tibicen sordidata* = *Diceroprocta sordida*
D. operculabrunnea Davis, 1934
D. ornea (Walker, 1850) = *Cicada ornea* = *Rihana ornea* = *Tibicen ornea*
D. ovata Davis, 1939
D. pinosensis Davis, 1935
D. pronotolinea Sanborn, 2007
D. psophis (Walker, 1850) = *Cicada psophis* = *Rihana psophis*
D. pusilla Davis, 1942
D. pygmaea (Fabricius, 1803) = *Tettigonia pymaea*
D. ruatana (Distant, 1891) = *Tympanoterpes ruatana* = *Rihana ruatana*
D. semicincta (Davis, 1925) = *Tibicen semicincta* = *Tibcien* (sic) *semicincta* = *Diceroprocta semicincta* var. *nigricans* Davis, 1942
D. swalei swalei (Distant, 1904) = *Rihana swalei* = *Diceroprocta swalei*
D. swalei davisi Metcalf, 1963 = *Cicada castanea* Davis, 1916 (nec *Cicada castanea* Gmelin, 1789) = *Tibicen castanea* = *Diceroprocta castanea* = *Diceroprocta swalei* var. *castanea*
D. tepicana Davis, 1938
D. texana (Davis, 1916) = *Cicada texana* = *Tibicen texana*
D. virgulata (Distant, 1904) = *Rihana virgulata*
D. viridifascia (Walker, 1850) = *Cicada viridifascia* = *Tibicen viridifascia* = *Tibicen viridifasciata* (sic) = *Cicada reperta* Uhler, 1892 = *Tibicen reperta*
D. vitripennis (Say, 1830) = *Cicada vitripennis* = *Cicada albipennis* (sic) = *Rihana vitripennis* = *Tibicen vitripennis* = *Cicada erratica* Osborn, 1906 = *Tibicen erratica*

Tibicen Latreille, 1825 = *Tubicen* (sic) = *Tibicens* (sic) = *Tibiceus* (sic) = *Tibicin* (sic) = *Tibicien* (sic) = *Tettigonia* Fabricius, 1775 = *Telligonia* (sic) = *Tetigonia* (sic) = *Cicada* Latreille, 1810 (nec *Cicada* Linnaeus, 1758) = *Cicada (Cicada)* (nec Linnaeus) = *Cicada (= Tibicen)* = *Cicadina* (sic) = *Lyristes* Horváth, 1926 = *Lyrtistes* (sic) = *Lyriste* (sic)
T. altaiensis (Schmidt, 1932) = *Lyristes altaiensis*

T. armeniacus (Kolenati, 1857) = *Cicada plebejus armeniaca* = *Tibicen plebejus armeniacus* = *Lyristes armeniacus*
T. atrofasciatus (Kirkaldy, 1909) = *Cicada atrofasciatus* = *Cicada sinensis* Distant, 1890 (nec *Cicada sinensis* Gmelin, 1789) = *Tibicen sinensis* = *Lyristes sinensis* = *Talainga sinensis*
T. auletes (Germar, 1834) = *Cicada auletes* = *Cicada auletus* (sic) = *Tibicen aurestes* (sic) = *Lyristes auletes* = *Tettigonia grossa* Fabricius, 1775 = *Tympanoterpes grossa* = *Rihana grossa* (nec Fabricius) = *Cicada grossa* (nec Fabricius) = *Tibicen grossa* (nec Fabricius) = *Diceroprocta grossa* = *Fidicina literata* Walker, 1850 = *Cicada literata* = *Tibicen literata* = *Cicada sonora* Walker 1850 = *Tibicen sonora* = *Cicada marginata* (nec Say) = *Cicada emarginata* (sic) (nec Say)
T. auriferus (Say, 1825) = *Cicada aurifera* = *Rihana aurifera* = *Tibicen aurifera* = *Tibicen canicularis auriferei* (sic)
T. bermudianus (Verrill, 1902) = *Cicada bermudiana* = *Rihana bermudiana* = *Diceroprocta bermudiana* = *Lyristes bermudianus* = *Tibicen bermudiana* (sic)
T. bifidus (Davis, 1916) = *Cicada bifida* = *Tibicen bifida*
T. bihamatus andrewsi (Distant, 1904) = *Cicada andrewsi* = *Tibicen andrewsi* = *Tibicen bihamatus* ab. *andrewsi* = *Lyristes andrewsi*
T. bihamatus babai Kato, 1938 = *Tibicen bihamatus* var. *babai* = *Lyristes bihamata babai*
T. bihamatus bihamatus (de Motschulsky, 1861) = *Cicada bihamatus* = *Cicada bihammata* (sic) = *Cicadina* (sic) *bihamata* = *Tibicen bihamata* = *Tibicen bihamoata* (sic) = *Lyristes bihamatus* = *Lyristes bihamata* = *Cicada andrewsi* Distant, 1904 = *Cicada nagashimai* Kato, 1925 = *Tibicen nagashimai* = *Tibicen esakii* (nec Kato)
T. bihamatus daisenensis Kato, 1940 = *Tibicen bihamatus* var. *daisenensis* = *Tibicen bihamatus dasenensis* (sic)
T. bihamatus fujiensis Masuda, 1942 = *Tibicen bihamatus* f. *fujisanus* (sic)
T. bihamatus harutai Kato, 1939 = *Tibicen bihamatus* ab. *harutai*
T. bihamatus nagashimai (Kato, 1925) = *Cicada nagashimai* = *Tibicen nagashimai* = *Lyristes nagashimai*
T. bihamatus nakayamai Kato, 1939 = *Tibicen bihamatus* var. *nakayamai*
T. bihamatus saitoi Kato, 1939 = *Tibicen bihamatus* var. *saitoi* = *Tibicen bihamatus satoi* (sic)
T. bihamatus takeoi Kato, 1939 = *Tibicen bihamatus* var. *takeoi*
T. bihamatus tazawai Kato, 1939 = *Tibicen bihamatus* ab. *tazawai*
T. bihamatus tsuguonis Kato, 1940 = *Tibicen bihamatus* var. *tsuguonis*
T. bimaculatus Sanborn, 2010 = *Tibicen nigriventris* (nec Walker)
T. canicularis (Harris, 1841) = *Cicada canicularis* = *Cicada tibicen canicularis* = *Rihana canicularis* = *Tibicen caniculris* (sic) = *Tibicen canulatus* (sic) = *Lyristes canicularis*
T. chihuahuaensis Sanborn, 2007
T. chinensis (Distant, 1905) = *Tibicina chinensis* = *Lyristes chinensis*
T. chiricahua Davis, 1923
T. chisosensis Davis, 1934
T. chujoi Esaki, 1935 = *Tibicen atrofasciatus* (nec Kirkaldy)
T. cristobalensis (Boulard, 1990) = *Lyristes cristobalensis*
T. cultriformis (Davis, 1915) = *Cicada cultriformis* = *Tibicen canicularis* (nec Harris)
T. davisi davisi (Smith & Grossbeck, 1907) = *Cicada davisi* = *Rihana davisi* = *Cicada canicularis* (nec Harris)
T. davisi hardeni Davis, 1918 = *Tibicen davisi* var. *hardeni*
T. dealbatus (Davis, 1915) = *Cicada marginata dealbata* = *Cicada dealbata* = *Tibicen dealbata*
T. distanti Metcalf, 1963 = *Cicada hilaris* Distant, 1881 (nec *Cicada hilaris* Germar, 1834) = *Rihana hilaris* = *Tibicen hilaris*
T. dorsatus (Say, 1825) = *Cicada dorsata* = *Rihana dorsata* = *Fidicina dorsata* = *Tibicen dorsata* = *Thopha varia* Walker, 1850 = *Cicada varia* = *Fidicina crassa* Walker, 1858
T. duryi Davis, 1917
T. esakii Kato, 1958 = *Tibicen bihamatus* (nec de Motschulsky)
T. esfandiarii Dlabola, 1970 = *Lyristes esfandiarii*
T. figuratus (Walker, 1858) = *Fidicina figurata* = *Cicada figurata* = *Tibicen figurata*
T. flammatus adonis Kato, 1933 = *Tibicen flammatus* f. *adonis* = *Lyristes flammata adonis*
T. flammatus concolor Kato, 1934 = *Tibicen flammatus* ab. *concolor* = *Lyristes flammata concolor*
T. flammatus flammatus (Distant, 1892) = *Cicada flammata* = *Cicada flamma* (sic) = *Tibicen flammata* = *Lyristes flammata* = *Lyristes flammatus* = *Cicada pyropa* Matsumura, 1904
T. flammatus nakamurai Kato, 1940 = *Tibicen flammatus* var. *nakamurai*
T. flammatus viridiflavus Kato, 1939 = *Tibicen flammatus* ab. *viridiflavus*
T. flavomarginatus Hayashi, 1977
T. fuscus Davis, 1934 = *Tibicen fusca*

T. gemellus (Boulard, 1988) = *Lyristes gemellus*
T. heathi Sanborn, 2010
T. hidalgoensis Davis, 1941
T. inauditus Davis, 1917 = *Tibicen inauditis* (sic) = *Lyristes inauditus*
T. intermedius Mori, 1931 = *Tibicen intermedia* = *Lyristes intermedius* = *Lyristes horni* Schmidt, 1932 = *Tibicen horni* = *Tibicen japonica* (nec Kato) = *Lyristes flammata* (nec Distant)
T. isodol (Boulard, 1988) = *Lyristes isodol*
T. jai Ouchi, 1938 = *Lyristes jai* = *Tibicen orientalis* Ouchi, 1938 = *Tibicen wui* Kato, 1934 = *Lyristes wui* = *Lyristes katoi* Liu, 1939
T. japonicus var. ______ Kato, 1933
T. japonicus echigo Kato, 1936 = *Tibicen japonicus* f. *echigo*
T. japonicus hooshianus (Matsumura, 1936) = *Cicada hooshiana* = *Cicada hoshiana* (sic) = *Lyristes japonica hooshiana*
T. japonicus immaculatus Kato, 1933 = *Tibicen japonicus* ab. *immaculatus* = *Tibicen japonicus inmaculatus* (sic)
T. japonicus interruptus Kato, 1943 = *Tibicen japonicus* var. *interruptus*
T. japonicus itoi Kato, 1939 = *Tibicen japonicus* ab. *itoi*
T. japonicus iwaoi Kato, 1939 = *Tibicen japonicus* ab. *iwaoi*
T. japonicus japonicus (Kato, 1925) = *Cicada japonica* = *Tibicen japonica* (sic) = *Tibicen japonensis* (sic) = *Lyristes japonica* = *Lyristes japonicus* = *Cicada flammata* (nec Distant) = *Tibicen dolichoptera* Mori, 1931 = *Tibicen japonicus dolichopterus* = *Tibicen japonica dolichoptera* = *Lyristes japonica dolichoptera* = *Tibicen dorichoptera* (sic) = *Lyristes hooshiana* Matsumura, 1936
T. japonicus kobayashii Kato, 1939 = *Tibicen japonicus* var. *kobayashii*
T. japonicus niger Kato, 1933
T. japonicus nigrofasciatus Kato, 1940 = *Tibicen japonicus* var. *nigrofasciatus* = *Tibicen japonicus ingrofasciatus* (sic)
T. kyushyuensis (Kato, 1926) = *Cicada kyushyuensis* = *Tibicen kyushuensis* (sic) = *Lyristes kyushyuensis* = *Lyristes kyushuensis* (sic) = *Tibicen ishiharai* Kato, 1959 = *Tibicen tsukushiensis* Kato, 1959 = *Tibicen shikokuanus* Kato, 1959
T. latifasciatus (Davis, 1915) = *Cicada pruinosa latifasciata* = *Tibicen latifasciata* (sic) = *Tibicen pruinosus latifasciata* (sic)
T. leechi (Distant, 1890) = *Cicada leechi* = *Tibicen leachi* (sic) = *Lyristes leechi* = *Lyrtistes* (sic) *leechi*
T. linnei (Smith & Grossbeck, 1907) = *Cicada linnei* Smith & Grossbeck 1907 = *Lyristes linnei* = *Tettigonia tibicen* Fabricius 1775 (nec *Cicada tibicen* Linnaeus, 1758) = *Cicada tibicen* (nec Linnaeus) = *Cicada tubicen* (sic) (nec Linnaeus) = *Fidicina tibicen* (nec Linnaeus) = *Rihana tibicen* (nec Linnaeus) = *Tettigonia tibicen* var. ________ Fabricius 1781 = *Tibicen linnei* var. ________
T. longioperculus Davis, 1926 = *Tibicen longiopercula*
T. lyricen engelhardti (Davis, 1910) = *Cicada engelhardti* = *Rihana engelhardti* = *Tibicen engelhardti* = *Tibicen lyricen* var. *engelhardti* = *Tibicen lyricen englehardti* (sic)
T. lyricen lyricen (Degeer, 1773) = *Cicada lyricen* = *Cicada lyricea* (sic) = *Fidicina lyricen* = *Rihana lyricen* = *Tibicen lyricen* = *Lyristes lyricen* = *Tibicen lyricen* var. *lyricen* = *Cicada fulvula* Osborn, 1906 = *Tibicen fulvula* = *Tibicen fulvulus*
T. lyricen virescens Davis, 1935 = *Tibicen lyricen* var. *virescens* = *Tibicen virescens*
T. maculigena (Signoret, 1860) = *Cicada maculigena* = *Cicada stigmosa* Stål, 1866
T. minor Davis, 1934
T. montezuma (Distant, 1881) = *Cicada montezuma*
T. nigriventris (Walker, 1858) = *Cicada nigriventris* = *Rihana vitripennis nigriventris* = *Tibicen nigroventris* (sic) = *Diceroprocta nigriventris*
T. oleacea (Distant, 1891) = *Fidicina oleacea* = *Cicada oleacea*
T. occidentis (Walker, 1850) = *Cicada occidentis* = *Cicada occidentalis* (sic) = *Cicada crassimargo* Spinola, 1852 = *Tettigades crassimargo* = *Tibicina crassimargo* = *Tibicen crassimargo*
T. parallelus Davis, 1923 = *Tibicen parallela*
T. paralleloides Davis, 1934
T. pekinensis (Haupt, 1924) = *Cicada pekinensis* = *Lyristes pekinensis*
T. pieris Kirkaldy, 1909 = *Lyristes pieris*
T. plebejus martorellii Martorell & Pena, 1879, nom. nud.
T. plebejus (Scopoli, 1763) = *Cicada plebeja* = *Tettigonia plebeja* = *Tettigonia plebeia* (sic) = *Cicada plebeia* (sic) = *Cicada plebia* (sic) = *Cicada plebeya* (sic) = *Cicada pelebeja* (sic) = *Cicada (Cicada) plebeja* = *Cicada*

plebja (sic) = *Cicada (= Tibicen) plebia* (sic) = *Rihana plebeja* = *Tibicina plebeja* = *Lyristes plebejus* = *Lyriste* (sic) *plebejus* = *Lyristes plebeja* = *Lyristes plebeius* = *Liristes* (sic) *plebejus* = *Lyristes plebeius* (sic) mph. *plebeius* (sic) = *Tibicen plebeia* (sic) = *Tibicen plebeius* (sic) = *Tibicen pelbeia* (sic) = *Tettigonia orni* Fabricius, 1775 (nec *Cicada orni* Linnaeus, 1758) = *Tibicen orni* (nec Linnaeus) = *Tettigonia fraxini* Fabricius, 1803 = *Cicada fraxini* = *Tettigonia obscura* Fabricius, 1803 = *Cicada plebeja obscura* = *Psaltoda plebeia* Goding & Froggatt, 1904 = *Fidicina africana* Metcalf, 1955 = *Lyristes plebeius* (sic) mph. *castanea* Boulard, 1982 = *Lyristes plebejus* mph. *castanea* = *Lyristes plebejus* var. *castanea*

T. pronotalis pronotalis Davis, 1938 = *Tibicen pronotalis* = *Tibicen marginalis pronotalis* = *Tibicen walkeri* var. *pronotalis* = *Tibicen walkeri pronotalis*

T. pronotalis walkeri Metcalf, 1955 = *Tibicen walkeri* = *Cicada marginata* Say, 1825 (nec *Cicada marginata* Olivier, 1790) = *Tibicen marginata* = *Cicada marginalis* Walker, 1852 (nec *Cicada marginalis* Scopoli, 1763) = *Tibicen marginalis* = *Tibicen marginales* (sic) = *Lyristes marginalis*

T. pruinosus fulvus Beamer, 1924 = *Tibicen pruinosa fulva* = *Tibicen pruinosus fulva* = *Tibicen pruinosus* var. *fulvus*

T. pruinosus pruinosus (Say, 1825) = *Cicada pruinosa* = *Cicada pruinora* (sic) = *Cicada pruinoso* (sic) = *Tibicen pruinosa* = *Lyristes pruinosus* = *Cicada bruneosa* Wild, 1852 = *Cicada brunneosa* (sic)

T. resh (Haldeman, 1852) = *Cicada resh* = *Cicada resch* (sic) = *Cicada robertsonii* Fitch, 1855 = *Cicada robertsoni* (sic) = *Tibicen robertsonii*

T. resonans (Walker, 1850) = *Cicada resonans* = *Lyristes resonans*

T. robinsonianus Davis, 1922 = *Tibicen robinsoniana* = *Lyristes robinsonianus*

T. robustus (Distant, 1881) = *Cicada robusta* = *Rihana robusta*

T. similaris (Smith & Grossbeck, 1907) = *Cicada similaris* = *Rihana similaris* = *Tibicen simularis* (sic)

T. simplex Davis, 1941 = *Tibicen bifida simplex* = *Tibicen bifidus simplex* = *Tibicen bifidus* var. *simplex*

T. slocumi Chen, 1943 = *Lyristes slocumi*

T. sublaqueatus (Uhler, 1903) = *Cicada sublaqueata*

T. sugdeni Davis, 1941

T. superbus (Fitch, 1855) = *Cicada superba* = *Rihana superba* = *Tibicen superba* = *Tibicen superb* (sic)

T. texanus Metcalf, 1963 = *Tibicen tigrina* Davis, 1927 (nec *Tibicen tigrinus* Distant, 1888)

T. tibicen australis (Davis, 1912) = *Rihana sayi australis* = *Rihana sayi* var. *australis* = *Cicada sayi australis* = *Tibicen australis* = *Tibicen chloromera australis* = *Tibicen chloromerus* var. *australis*

T. tibicen tibicen (Linnaeus, 1758) = *Cicada tibicen* = *Tettigonia tibicen* = *Cicada (Tettigonia) tibicen* = *Fidicina tibicen* = *Cicada (Cicada) tibicen* = *Rihana tibicen* = *Diceroprocta tibicen* = *Thopha chloromera* Walker, 1850 = *Tibicen chloromerus* = *Tibicen chloromera* = *Tibicen chloromea* (sic) = *Cicada chloromera* = *Lyristes chloromerus* = *Cicada sayi* Smith & Grossbeck, 1907 = *Rihana sayi* = *Tibicen sayi*

T. tremulus Cole, 2008

T. townsendii (Uhler, 1905) = *Cicada townsendii* = *Cicada townsendi* (sic) = *Rihana townsendi* (sic) = *Tibicen townsendi* (sic)

T. tsaopaonensis Chen, 1943 = *Lyristes tsaopaonensis*

T. variegatus (Fabricius, 1794) = *Tettigonia variegata* = *Tettigonia variata* (sic) = *Cicada variegata* = *Cicada variegata* var. ______ Germar, 1830 = *Tibicen variegata*

T. winnemanna (Davis, 1912) = *Cicada winnemanna* = *Cicada pruinosa winnemanna* = *Tibicen pruinosus winnemanna* = *Tibicen pruinosa winnemanna* = *Lyristes winnemanna*

Cornuplura Davis, 1944

C. curvispinosa (Davis, 1936) = *Tibicen curvispinosa*

C. nigroalbata (Davis, 1936) = *Tibicen nigroalbata*

C. rudis (Walker, 1858) = *Fidicina rudis* = *Cicada rudis* = *Rihana rudis* = *Tibicen rudis*

Hea Distant, 1906 = *Hua* (sic) = *Kinoshitaia* Ouchi, 1938

H. choui Lei, 1992 = *Hua* (sic) *choui*

H. fasciata Distant, 1906 = *Kinoshitaia sinensis* Ouchi, 1938

H. yunnanensis Chou & Yao, 1995

Antankaria Distant, 1904

A. signoreti (Metcalf, 1955) = *Cicada signoreti* = *Cicada punctipes* Signoret, 1860 (nec *Cicada punctipes* Zetterstedt, 1828) = *Tettigia punctipes* = *Cicada madagascariensis* Distant, 1892 = *Antankaria madagascariensis*

Cacama Distant, 1904

C. californica Davis, 1919

C. carbonaria Davis, 1919

C. collinaplaga Sanborn & Heath, 2011 = *Cacama* n. sp. 1 Sanborn and Phillips, 1995

C. crepitans (Van Duzee, 1914) = *Proarna crepitans*

C. dissimilis (Distant, 1881) = *Cicada dissimilis*

C. furcata Davis, 1919

C. longirostris (Distant, 1881) = *Proarna longirostris* = *Ploarna* (sic) *longirostris*

C. maura (Distant, 1881) = *Proarna maura*

C. moorei Sanborn & Heath, 2011 = *Cacama dissimilis* (nec Distant) = *Cacama* n. sp. 2 Sanborn and Phillips, 1995

C. pygmaea Sanborn, 2011

C. valvata (Uhler, 1888) = *Proarna valvata* = *Cacama valvada* (sic) = *Cacama valuate* (sic)

C. variegata Davis, 1919

Orialella Metcalf, 1952 = *Oria* Distant, 1904 (nec *Oria* Huebner, 1821)

O. aerizulae Boulard, 1986

O. boliviana (Distant, 1904) = *Cicada boliviana* = *Oria boliviana*

Cryptotympana Stål, 1861 = *Chryptotympana* (sic) = *Crptotympana* (sic) = *Cryptolympana* (sic) = *Ciyptotympana* (sic) = *Cryptotympanus* (sic) = *Cyrister* (sic) = *Gryptotympana* (sic) = *Cryptotympaua* (sic) = *Criptotympana* (sic) = *Cryplolympana* (sic) = *Crystotympana* (sic) = *Cryptympana* (sic) = *Chryptotympana* (sic) = *Magicicada* (nec Davis)

C. accipiter (Wallker, 1850) = *Fidicina accipiter*

C. acuta (Signoret, 1849) = *Cicada acuta* = *Fidicina acuta* = *Cicada vicinia* Signoret, 1849 = *Fidicina vicinia* = *Fidicina nivifera* Walker, 1850 = *Fidicina vinifera (nivifera)* = *Fidicina bicolor* Walker, 1852 = *Fidicina blicolor* (sic) = *Cryptotympana bicolor*

C. albolineata Hayashi, 1987

C. alorensis Hayashi, 1987

C. aquila (Walker, 1850) = *Fidicina aquila* = *Cryptotympana acquila* (sic) = *Crptotympana* (sic) *aquila* = *Cryptolympana* (sic) *aquila* = *Ciyptotympana* (sic) *aquila*

C. atrata (Fabricius, 1775) = *Tettigonia atrata* = *Cicada atrata* = *Cicada atra* (sic) = *Fidicina atrata* = *Cryptotympana atrat* (sic) = *Cryptotympana atrula* (sic) = *Chryptotympana* (sic) *atrata* = *Tettigonia pustulata* Fabricius, 1787 = *Cryptotympana pustulata* = *Cicada pustulata* = *Cryptotympana pustulala* (sic) = *Cryptotympanus* (sic) *pustulatus* = *Cryptotympana pustaulata* (sic) = *Cryptotympana pusstulafa* (sic) = *Cicada nigra* Olivier, 1790 = *Cryptotympana nigra* = *Fidicina bubo* Walker, 1850 = *Cryptotympana bubo* = *Cryptotympana bubo* = *Cryptotympana sinensis* Distant, 1887 = *Cyrister* (sic) *sinensis* = *Cryptotympana dubia* Haupt, 1917 = *Gryptotympana* (sic) *dubia* = *Cryptotympana coreanus* Kato, 1925 = *Cryptotympana coreana* = *Cryptotympana dubia coreana* = *Cryptotympana santoshonis* Matsumura, 1927 = *Cryptotympana santoshoesnis* (sic) = *Cryptotympana pustulata* (nec Fabricius) = *Cryptotympana wenchewensis* Ouchi, 1938 = *Cryptotypmana pustulata castanea* Liu, 1940 = *Cryptotympana atrata castanea* = *Cryptotympana pustulata fukienensis* Liu, 1940 = *Cryptotympana pustulata fukiensis* (sic) = *Cryptotympana atrata fukienensis*

C. auropilosa Hayashi, 1987

C. brevicorpus Hayashi, 1987 = *Cryptotympana epithesia* (nec Distant)

C. brunnea Hayashi, 1987

C. consanguinea Distant, 1916

C. corvus (Walker, 1850) = *Fidicina corvus* = *Cryptotympana corva* = *Fidicina invarians* Walker, 1858 = *Cryptotympana invarians*

C. demissitia Distant, 1891 = *Cryptotympana demissita* (sic)

C. diomedea (Walker, 1858) = *Fidicina diomedea* = *Cryptotympana dimeda* (sic)

C. distanti Hayashi, 1987 = *Cryptotympana suluensis* (nec Distant)

C. dohertyi Hayashi, 1987

C. edwardsi Kirkaldy, 1902

C. epithesia Distant, 1888

C. exalbida Distant, 1891

C. facialis bimaculata Kato, 1936 = *Cryptotympana facialis* var. *bimaculata*

C. facialis facialis (Walker, 1858) = *Cicada facialis* = *Cicada fascialis* (sic) = *Cryptotympaua* (sic) *facialis* = *Cryptotympana facialis* = *Cryptotympana fascialis* (sic) = *Cryptotympana pustulata* Uhler, 1896 (nec Fabricius) = *Cryptotympana pnstulata* (sic) (nec Fabricius) = *Cryptotympana intermedia* Matsumura, 1907 (nec Signoret) = *Cryptotympana facialis formosana* Kato, 1925 = *Cryptotympana fascialis* (sic) *formosana* = *Cryptotympana japonensis* Kato, 1925 = *Oryptotympana* (sic) *japonensis* = *Criptotympana* (sic) *japonensis* = *Cryptotympana intermedia* (nec Signoret) = *Cryptotympana iutermedia* (sic) (nec Signoret) = *Cryptotympana japonensis riukiuensis* Kato, 1925 = *Cryptotympana riukiuensis* = *Cryptotympana okinawana* Matsumura, 1927 = *Cryptotympana facialis* var. *okinawana* = *Cryptotympana facialis okinawana* = *Cryptotympana facialis* (nec Walker) = *Cryptotympana facialis yonakunina* Ishihara, 1968

C. fumipennis (Walker, 1858) = *Fidicina fumipennis* = *Cryptotympana viridipennis* Distant, 1911 = *Cryptotympana viridipennis infuscata* Moulton & China, 1926 = *Cryptotympana viridipennis* var. *infuscata*

C. gracilis Hayashi, 1987 = *Cryptotympana vesta* (nec Distant)

C. holsti holsti Distant, 1904 = *Cryptotympana mandarina* (nec Distant) = *Cryptotympana vitalasi* Distant, 1917 = *Cryptotympana fusca* Kato, 1925 = *Cryptotympana fusa* (sic) = *Cryptotympana capillata* Kato, 1925 = *Cryptotympana holsti capillata* = *Cryptotympana holsti inornata* Matsumura, 1927 = *Cryptotympana holsti inorrata* (sic) = *Cryptotympana kagiana* Matsumura, 1927 = *Cryptotympana holsti kagiana*

C. holsti takahashii Kato, 1925

C. immaculata (Olivier, 1790) = *Cicada immaculata* = *Fidicina immaculata*

C. insularis Distant, 1887

C. intermedia (Signoret, 1849) = *Cicada intermedia* = *Fidicina immaculata* Walker, 1850 (nec Olivier) = *Cryptotympana immaculata* (nec Olivier) = *Cryptotympana iutermedia* (sic) = *Cryplolympana* (sic) *intermedia* = *Crystotympana* (sic) *inermidia* (sic)

C. izzardi Lallemand & Synave, 1953

C. jacobsoni China, 1926 = *Cryptotympana epithesia* Moulton, 1925 (nec *Cryptotympana epithesia* Distant, 1888)

C. karnyi Moulton, 1923 = *Cryptotympana leopoldi* Lallemand, 1931

C. kotoshoensis Kato, 1925 = *Cryptotympana shirakii* Matsumura, 1927

C. limborgi Distant, 1888 = *Cryptotympana recta* (nec Walker)

C. lombokensis Distant, 1912

C. mandarina Distant, 1891 = *Cryptympana* (sic) *mandarina* = *Fidicina operculata* Walker, 1850 nom. nud. = *Cryptotympana corvus* (nec Walker) = *Cryptotympana mimica* Distant, 1917

C. moultoni Hayashi, 1987 = *Cryptotympana robinsoni* (nec Moulton)

C. niasana Distant, 1909

C. nitidula Hayashi, 1987

C. ochromelas Hayashi, 1987 = *Cryptotympana lombokensis* (nec Distant)

C. pelengensis Hayashi, 1987

C. praeclara Hayashi, 1987

C. recta (Walker, 1850) = *Fidicina recta*

C. robinsoni Moulton, 1923

C. sibuyana Hayashi, 1987

C. socialis Hayashi, 1987

C. suluensis Distant, 1906

C. takasagona Kato, 1925 = *Cryptotympana intermedia* (nec Signoret) = *Cryptotympana intermedia* (nec Signoret) = *Cryptotymapana iutermedia* (nec Signoret) = *Cryptotympana argenteus* Kato, 1925

C. timorica (Walker, 1870) = *Fidicina timorica* = *Cryptotympana heuertzi* Lallemand & Synave, 1953 = *Cryptotympana acuta* (nec Signoret)

C. varicolor Distant, 1904 = *Cryptotympana sumbawensis* Jacobi, 1941

C. ventralis Hayashi, 1987 = *Cryptotympana robinsoni* (nec Moulton)

C. vesta (Distant, 1904) = *Cicada vesta*

C. viridicostalis Hayashi, 1987 = *Cryptotympana acuta* (nec Signoret) = *Cryptotympana? varicolor* (nec Distant)

C. wetarensis Hayashi, 1987

C. yayeyamana Kato, 1925 = *Cryptotympana yaeyamana* (sic) = *Cryptotympana yayeyama* (sic) = *Cryptotympana ishigakiana* Matsumura, 1927

Salvazana Distant, 1913 = *Salvasana* (sic) = *Pulchrocicada* He, 1984

S. mirabilis imperialis Distant, 1918 = *Salvazana imperialis* = *Salvazana mirabilis* var. *imperialis* = *Pulchrocicada guangxiensis* He, 1984

S. mirabilis mirabilis Distant, 1913 = *Salvazana mirabilis* var. *mirabilis* = *Pulchrocicada sinensis* He, 1984

Raiateana Boulard, 1979

R. knowlesi (Distant, 1907) = *(?) Tibicen knowlesi* = *Tibicen knowlesi* = *Lyristes knowlesi* = *Cicada knowlesi*

R. kuruduadua bifasciata Duffels, 1988

R. kuruduadua kuruduadua (Distant, 1881) = *Cicada kuruduadua* = *Tibicen kuruduadua* = *Lyristes kuruduaduus* = *Raiateana kuruduadua taveuniensis* (sic)

R. oulietea Boulard, 1979 = *Raiateana oulietea* morph *poinsignoni* Boulard, 1995 = *Raiateana oulietea* morph *suavis* Boulard, 1995

R. samoensis Duffels, 1988 = *Raiateana kuruduadua samoensis*

Subtribe Heteropsaltriina Distant, 1905 = Heteropsaltriaria

Heteropsaltria Jacobi, 1902

H. aliena Jacobi, 1902

Subtribe Nggelianina Boulard, 1979 = Nggelianaria

Nggeliana Boulard, 1979

N. leveri Boulard, 1979

N. typica Boulard, 1979

Tribe Fidicinini Distant, 1905

Subtribe Fidicinina Distant, 1905

Fidicina Amyot & Audinet-Serville, 1843 = *Cicada (Fidicina)* = *Fidicula* (sic)

F. affinis Haupt, 1918

F. aldegondae Kuhlgatz, 1902

F. christinae Boulard & Martinelli, 1996 = *Fidicina torresi* (nec Boulard & Martinelli)

F. ethelae (Goding, 1925) = *Majeorona ethelae*

F. explanata Uhler, 1903

F. innotabilis (Walker, 1858) = *Cicada innotabilis*

F. mannifera mannifera (Fabricius, 1803) = *Tettigonia mannifera* = *Cicada mannifera* = *Tibicen mannifera* = *Cicada (Fidicina) plebeja mannifera* = *Fidicina vinifera* (sic) = *Fidicina manifera* (sic) = *Cicada cantatrix* Germar, 1821 = *Fidicina cantatrix* = *Fidicina rana* Walker, 1850 = *Fidicina excavata* Walker, 1850 = *Fidicina divisa* Walker, 1858

F. mannifera umbrilinea Walker, 1858 = *Fidicina umbrilinea*

F. muelleri Distant, 1892 = *Fidicina mülleri*

F. obscura Boulard & Martinelli, 1996

F. parvula Jacobi, 1904

F. robini Boulard & Martinelli, 1996

F. rosacordis (Walker, 1850) = *Cicada rosacordis*

F. rubricata Distant, 1892

F. sawyeri Distant, 1912

F. sciras (Walker, 1850) = *Carineta sciras* = *Fidicina mannifera* (nec Fabricius)

F. sp. 1 Pogue, 1997

F. sp. 2 Pogue, 1997

F. sp. 3 Pogue, 1997

F. sp. 4 Pogue, 1997

F. sp. 5 Pogue, 1997

F. sp. 6 Pogue, 1997
F. sp. 7 Pogue, 1997
F. sp. 8 Pogue, 1997
F. sp. 9 Pogue, 1997
F. sp. 10 Pogue, 1997
F. torresi Boulard & Martinelli, 1996 = *Fidicina mannifera* (nec Fabricius) = *Fidicina christinae* (nec Boulard & Martinelli)
F. toulgoeti Boulard & Martinelli, 1996 = *Fidicina toulgoëti*
F. vitellina (Jacobi, 1904) = *Cicada vitellina*

Fidicinoides Boulard & Martinelli, 1996
F. besti Boulard & Martinelli, 1996
F. brunnea Boulard & Martinelli, 1996
F. cachla (Distant, 1899) = *Fidicina cachla*
F. nr. *cachla* (Wolda, 1989) = *Dorisiana* nr. *cachla*
F. carmenae Santos & Martinelli, 2009
F. coffea Sanborn, Moore & Young, 2008 = *Fidicina* "*coffea*" Young, 1977 = *Fidicina* sp. 2, the "coffee cicada" Young, 1984 = *Fidicina coffea* Young, 1984
F. compostela (Davis, 1934) = *Fidicina compostela*
F. descampsi Boulard & Martinelli, 1996
F. determinata (Walker, 1858) = *Fidicina determinata* = *Fidicina pronoe* (nec Walker) = *Fidicinoides picea* (nec Walker)
F. distanti (Goding, 1925) = *Fidicina distanti*
F. dolosa Santos & Martinelli, 2009
F. duckensis Boulard & Martinelli, 1996
F. ferruginosa Sanborn & Heath, 2011, nom. nud.
F. flavibasalis (Distant, 1905) = *Fidicina flavibasalis* = *Cicada* (circa) *rudis* Salazar Escobar, 2005
F. flavipronotum Sanborn, 2007
F. fumea (Distant, 1883) = *Fidicina fumea* = *Dorisiana fumea*
F. guayabana Sanborn, Moore & Young, 2008 = *Fidicina* "*guayabana*" Young, 1977 = *Fidicina* sp. 3, the "guayabana cicada" Young, 1984 = *Fidicina guayabana* Young, 1984
F. jauffreti Boulard & Martinelli, 1996
F. lacteipennis (Distant, 1905) = *Fidicina lacteipennis*
F. opalina (Germar, 1821) = *Cicada opalina* = *Fidicina opalina* = *Fidicina phaeochlora* Walker, 1858 = *Fidicina phoeochlora* (sic) = *Fidicina poeochlora* (sic)
F. passerculus (Walker, 1850) = *Cicada passerculus* = *Cicada lacrines* Walker, 1850 = *Fidicina lacrines*
F. pauliensis Boulard & Martinelli, 1996
F. picea (Walker, 1850) = *Fidicina picea* = *Fidicina pertinax* Stål, 1864
F. poulaini Boulard & Martinelli, 1996
F. pronoe (Walker, 1850) = *Cicada pronoe* = *Fidicina pronoe* = *Fidicina pronae* (sic) = *Fidicina prone* (sic) = *Fidicina vinula* Stål, 1854 = *Cicada compacta* Walker, 1858 = *Fidicina compacta*
F. pseudethelae Boulard & Martinelli, 1996
F. roberti (Distant, 1905) = *Fidicina roberti*
F. rosabasalae Santos & Martinelli, 2009
F. saccifera Boulard & Martinelli, 1996
F. sarutaiensis Santos, Martinelli & Maccagnan, 2010
F. sericans (Stål, 1854) = *Fidicina sericans* = *Fidicina sericana* (sic)
F. spinicosta (Walker, 1850) = *Cicada spinicosta* = *Fidicina spinicosta* = *Fidicina spinocosta* (sic) = *Carinetta* (sic) *spinicosta* = *Carineta* (sic) *spinicosta*
F. steindachneri (Kuhlgatz & Melichar, 1902) = *Fidicina steindachneri*
F. sucinalae Boulard & Martinelli, 1996
F. variegata (Sanborn, 2005) = *Fidicina variegata* = *Fidicina* "*variegata*" Young, 1977 = *Fidicina* n. sp. Young, 1981 = *Fidicina* sp. 1 Young, 1984
F. yavitensis Boulard & Martinelli, 1996

Bergalna Boulard & Martinelli, 1996
B. pullata (Berg, 1879) = *Fidicina pullata* = *Fidicina fullata* (sic) = *Fidicina pulata* (sic)

B. xanthospila (Germar, 1830) = *Cicada xanthospila*

Subtribe Guyalnina Boulard & Martinelli, 1996

Proarna Stål, 1864 = *Proarno* (sic) = *Ploarna* (sic) = *Proarma* (sic) = *Tympanoterpes (Proarua)* (sic) = *Proaria* (sic)
P. alalonga Sanborn & Heath, 2011, nom. nud.
P. bergi (Distant, 1892) = *Tympanoterpes bergi* = *Proarma* (sic) *bergi* = *Proarna bergii* (sic)
P. bufo Distant, 1905
P. cocosensis Davis, 1935
P. dactyliophora Berg, 1879
P. germari Distant, 1905 = *Cicada grisea* Germar, 1821 (nec *Tettigonia grisea* Fabricius, 1775) = *Tympanoterpes grisea* (nec Fabricius)
P. gianucai Sanborn, 2008
P. grisea (Fabricius, 1775) = *Tettigonia grisea* = *Cicada grisea*
P. guttulosa (Walker, 1858) = *Cicada guttulosa*
P. hilaris (Germar, 1834) = *Cicada hilaris* = *Tympanoterpes hilaris* = *Poecilopsaltria hilaris* = *Tympanoterpes (Proarua)* (sic) *hilaris* = *Odopoea hilaris* = *Proalba* (sic) *hilaris* = *Cicada subtincta* Walker, 1850 = *Tympanoterpes subtincta* = *Cicada albiflos* Walker, 1850 = *Tympanoterpes albiflos* = *Cicada tomentosa* Walker, 1858 = *Odopoea tomentosa* = *Tympanoterpes tomentosa*
P. insignis Distant, 1881 = *Proarna alba insignis* = *Proaria* (sic) *insignis*
P. invaria (Walker, 1850) = *Cicada invaria* = *Tympanoterpes invaria* = *Cicada dexithea* Walker, 1850 = *Tympanoterpes dexithea* = *Cicada ovatipennis* Walker, 1858 = *Cicada fulvoviridis* Walker, 1858 = *Proarna* (circa) *insignis* Salazar Escobar, 2005
P. montevidensis Berg, 1882 = *Proarna capistrata* Distant, 1885
P. olivieri Metcalf, 1963 = *Cicada albida* Olivier, 1790 (nec *Cicada albida* Gmelin, 1789) = *Tympanoterpes albida* = *Proarna albida* = *Proarna celbida* (sic)
P. sp. nr. *olivieri* Young, 1981
P. palisoti (Metcalf, 1963) = *Cicada palisoti* = *Cicada bicolor* Palisot de Beauvois, 1813 (nec *Cicada bicolor* Olivier, 1790)
P. parva Sanborn & Heath, 2011, nom. nud.
P. praegracilis Berg, 1881
P. pulverea (Olivier, 1790) = *Cicada pulvera* = *Tympanoterpes pulvera*
P. sallaei Stål, 1864 = *Proarna sallei* (sic) = Species A Sueur, 2002
P. sp. 1 Pogue, 1997
P. sp. 2 Pogue, 1997
P. sp. 3 Pogue, 1997
P. sp. 4 Pogue, 1997
P. sp. 5 Pogue, 1997
P. sp. 6 Pogue, 1997
P. sp. 7 Pogue, 1997
P. sp. 8 Pogue, 1997
P. sp. 9 Pogue, 1997
P. sp. 10 Pogue, 1997
P. squamigera Uhler, 1895
P. strigicollis Jacobi, 1907
P. uruguayensis Berg, 1882

Prasinosoma Torres, 1963
P. fuembuenai Torres, 1963
P. heidemanni (Distant, 1905) = *Proarna heidemanni*
P. inconspicua (Distant, 1906) = *Proarna inconspicua*
P. medialinea Sanborn & Heath, 2011, nom. nud.

Elassoneura Torres, 1964
E. carychrous Torres, 1964

Beameria Davis, 1934
- *B. ansercollis* Sanborn & Heath, 2011 = *Beameria* n. sp. Sanborn and Phillips 1995
- *B. venosa* (Uhler, 1888) = *Prunasis venosa* = *Proarna venosa* = *Beameria vanosa* (sic)
- *B. wheeleri* Davis, 1934

Dorisiana Metcalf, 1952 = *Dorisia* Delétang, 1919 (nec *Dorisia* Moeschler, 1883)
- *D. amoena* (Distant, 1899) = *Fidicina amoena*
- *D. beniensis* Boulard & Martinelli, 2011
- *D. bicolor* (Olivier, 1790) = *Cicada bicolor* = *Fidicina bicolor* = *Fidicina cayennensis* Kirkaldy, 1909
- *D. bogotana* (Distant, 1892) = *Fidicina bogotana* = *Fidicina bgotoma* (sic)
- *D. brisa* (Walker, 1850) = *Cicada brisa* = *Cicada briza* (sic) = *Fidicina brisa* = *Fidicinoides brisa* = *Guyalna brisa* = *Guylana briza* (sic) = *Fidicina amazona* Distant, 1892
- *D. christinae* Boulard & Martinelli, 2011
- *D. crassa* Boulard, 1998
- *D. drewseni* (Stål, 1854) = *Cicada drewseni* = *Fidicina drewseni* = *Fidicina drewensi* (sic) = *Dorisia drewseni* = *Fidicina gastracathophora* Berg, 1879
- *D. glauca* (Goding, 1925) = *Fidicina glauca* = *Fidicinoides glauca*
- *D. noriegai* Sanborn & Heath, 2011, nom. nud.
- *D. panamensis* (Davis, 1939) = *Fidicina panamensis* = *Fidicina panamaensis* (sic)
- *D. semilata* (Walker, 1850) = *Cicada semilata* = *Fidicina semilata* = *Cicada passer* Walker, 1850 = *Cicada brizo* Walker, 1850 = *Fidicina brizo* = *Cicada melisa* Walker, 1850 = *Cicada melina* Walker, 1850 = *Cicada panyases* Walker, 1850 = *Cicada pidytes* Walker, 1850 = *Cicada physcoa* Walker, 1850 = *Cicada braure* Walker, 1850 = *Cicada solennis* Walker, 1850 = *Cicada solemis* (sic)
- *D.* sp. B Wolda, 1989
- *D.* sp. 1 Pogue, 1997
- *D. sutori* Sueur, 2000
- *D. toulgoueti* Boulard & Martinelli, 2011
- *D. viridifemur* (Walker, 1850) = *Cicada viridifemur* = *Fidicina viridifemur* = *Fidicinoides viridifemur*
- *D. viridis* (Olivier, 1790) = *Fidicina viridis* = *Dorisia viridis* = *Dorisiana virides* (sic)

Tympanoterpes Stål, 1861 = *Tympanoterpis* (sic) = *Tympanotherpes* (sic)
- *T. cordubensis* Berg, 1884
- *T. elegans* Berg, 1882
- *T. perpulchra* (Stål, 1854) = *Cicada perpulchra* = *Cicada (Cicada) perpulchra*
- *T. serricosta* (Germar, 1834) = *Cicada serricosta* = *Fidicina pusilla* Berg, 1879 = *Tympanoterpes pusilla*
- *T. xanthogramma* (Germar, 1834) = *Cicada xanthogramma* = *Cicada fuscovenosa* Stål, 1854

Pompanonia Boulard, 1982
- *P. buziensis* Boulard, 1982

Ollanta Distant, 1905
- *O. caicosensis* Davis, 1939
- *O. mexicana* Distant, 1905
- *O. melvini* Ramos, 1983
- *O. modesta* (Distant, 1881) = *Selymbria modesta* = *Diceroprocta belizensis* (nec Distant) = *Conibosa* sp. (nec Distant)

Pacarina Distant, 1905
- *P. championi* (Distant, 1881) = *Proarna championi* = *Conibosa* sp. (nec Distant)
- *P. puella* Davis, 1923 = *Cicada signifera* Walker, 1858 (nec *Cicada signifera* Germar, 1830) = *Proarna signifera* = *Pacarina signifera*
- *P. schumanni* Distant, 1905
- *P. shoemakeri* Sanborn & Heath, 2012 = *Pacarina puella* (nec Davis) = *Pacarina puella* (juniper variety) Sanborn & Phillips, 1995
- *P.* sp. 1 Young, 1984

Ariasa Distant, 1905
- *A. albiplica* (Walker, 1858) = *Fidicina albiplica* = *Tympanoterpes albiplica*

A. alboapicata (Distant, 1905) = *Tympanoterpes alboapicata*
A. arechavaletae (Berg, 1884) = *Tympanoterpes arechavaletae*
A. bilaqueata (Uhler, 1903) = *Cicada bilaqueata* = *Cicada bilagueta* (sic) = *Rihana bilaqueta* (sic) = *Diceroprocta bilaqueata*
A. colombiae (Distant, 1892) = *Tympanoterpes colombiae*
A. diupsilon (Walker, 1850) = *Cicada diupsilon* = *Tympanoterpes diupsilon*
A. marginata (Olivier, 1790) = *Cicada marginata* = *Tympanoterpes marginata* = *Cicada viridis* Stoll, 1788 = *Tettigonia viridis* Fabricius, 1803 (nec *viridis* Stål, 1870, nec *viridis* Distant, 1892, nec *viridis* Moulton, 1923) = *Cicada viridis* Germar, 1830 = *Ariasa brasiliorum* Kirkaldy, 1909
A. nigrorufa (Walker, 1850) = *Fidicina nigrorufa* = *Tympanoterpes nigrorufa*
A. nigrovittata Distant, 1905
A. sp. 1 Pogue, 1997
A. sp. 2 Pogue, 1997
A. sp. 3 Pogue, 1997
A. sp. 4 Pogue, 1997
A. sp. 5 Pogue, 1997
A. sp. 6 Pogue, 1997
A. urens (Walker, 1850) = *Cicada urens* = *Cicada torrida* Walker, 1850 (nec *Cicada torrida* Erichson, 1842) = *Ariasa quaerenda* Kirkaldy, 1909

Guyalna Boulard & Martinelli, 1996
G. atalapae Boulard & Martinelli, 2011
G. bleuzeni Boulard & Martinelli, 2011
G. bonaerensis (Berg, 1879) = *Fidicina bonaerensis* = *Dorisia bonaerensis* = *Dorisiana bonaerensis* = *Dorisiana bonaërensis*
G. bonaerensis bergi Delétang, 1919 nom. nud.
G. bonaerensis dominiquei Delétang, 1919 nom. nud.
G. chlorogena (Walker, 1850) = *Fidicina chlorogena* = *Fidicina basispes* Walker, 1858
G. cuta (Walker, 1850) = *Cicada cuta* = *Fidicina cuta* = *Cicada lucastia* Walker, 1850
G. densusa Boulard & Martinelli, 2011
G. jauffreti Boulard & Martinelli, 2011
G. nigra Boulard, 1999
G. platyrhina Sanborn & Heath, 2011, nom. nud.
G. rufapicalis Boulard, 1998

Hemisciera Amyot & Audinet-Serville, 1843
H. durhami Distant, 1905
H. maculipennis (de Laporte, 1832) = *Cicada maculipennis* = *Fidicina maculipennis* = *Hemisciera pictipennis* (sic) = *Cicada versicolor* Brullé, 1835 = *Hemisciera versicolor* = *Cicada sumptuosa* Blanchars, 1840 = *Hemisciera sumptuosa* = *Fidicina floslofia* Walker, 1858
H. taurus (Walker, 1850) = *Fidicina taurus*

Majeorona Distant, 1905 = *Majerona* (sic) = *Majeroma* (sic)
M. aper (Walker, 1850) = *Fidicina aper* = *Majerona* (sic) *aper*
M. bovilla Distant, 1905
M. durantoni Boulard & Martinelli, 2011
M. ecuatoriana Goding, 1925
M. lutea Distant, 1906 = *Majeroma* (sic) *lutea*
M. orvoueni Boulard & Martinelli, 2011
M. truncata Goding, 1925

Tribe Hyantiini Distant, 1905 = Hyantinii (sic) = Hyantini (sic)

Quesada Distant, 1905 = *Queseda* (sic) = *Quezada* (sic)
Q. gigas (Olivier, 1790) = *Cicada gigas* = *Tympanoterpes gigas* = *Tympanotherpes* (sic) *gigas* = *Queseda* (sic) *gigas* = *Cicada triupsilon* Walker, 1850 = *Cicada triypsilon* (sic) = *Cicada trupsilon* (sic) = *Cicada sonans* Walker, 1850 = *Cicada consonans* Walker, 1850 = *Cicada vibrans* Walker, 1850 = *Tympanoterpes*

grossa (nec Fabricius) = *Cicada grossa* (nec Fabricius) = *Cicada (Cicada) grossa* (nec Fabricius) = *Tympanoterpes sibilarix* Berg, 1879 = *Tympanoterpes sibilantis* (sic) = *Tympanoterpis* (sic) *sibilantes* (sic) = *Tympanotherpes* (sic) *gigas*

Q. sodalis (Walker, 1850) = *Cicada sodalis* = *Tympanoterpes sodalis* = *Fidicina vultur* Walker, 1858

Mura Distant, 1905

M. elegantula Distant, 1905

Hyantia Stål, 1866

H. bahlenhorsti Sanborn, 2011

H. honesta (Walker, 1850) = *Cyclochila honesta*

Tribe Tamasini Moulds, 2005

Tamasa Distant, 1905

T. burgessi (Distant, 1905) = *Abricta burgessi*

T. doddi (Goding & Froggatt, 1904) = *Tibicen doddi* = *Abricta doddi* = *Tamasa tristigma doddi* = *Tamasa tristigma* var. *doddi*

T. rainbowi Ashton, 1912

T. tristigma (Germar, 1834) = *Cicada tristigma* = *Tettigia (Tettigia) tristigma* = *Tettigia tristigma* = *Tibicen kurandae* Goding & Froggatt, 1904 = *Abricta kurundae* = *Tamasa kurandae*

T. sp. nr. *tristigma* Moss, 1988

Parnkalla Distant, 1905

P. muelleri (Distant, 1882) = *Tibicen muelleri* = *Parnkalla mülleri* = *Parnkalla melleri* (sic) = *Parnkalla mueller* (sic) = *Tibicen gregoryi* Distant, 1882 = *Parnkalla gregoryi*

Parnquila Moulds, 2012

P. hillieri (Distant, 1906) (nec *hillieri* Distant, 1907) = *Burbunga hillieri*

P. magna (Distant, 1913) = *Parnkalla magna*

P. unicolor (Ashton, 1921) = *Areopsaltria unicolor*

P. venosa (Distant, 1907) = *Burbunga venosa* = *Burbunga vernosa* (sic)

Tribe Dundubiini Atkinson, 1886 = Dundubini (sic) = Platylomiini Metcalf, 1955

Subtribe Dundubiina Matsumura, 1917 = Dundubiaria = Platylomaria Metcalf, 1955

Dundubia Amyot & Audinet-Serville, 1843 = *Dandubia* (sic) = *Dundbia* (sic) = *Dundula* (sic) = *Dendubia* (sic) = *Dubia* (sic)

D. aerata Distant, 1888

D. andamansidensis (Boulard, 2001) = *Platylomia andamansidensis*

D. annandalei Boulard, 2007 nom. nov. pro *Dundubia intemerata* Boulard, 2003 (nec Walker) = *Dundubia rufivena* (nec Walker) = *Dundubia terpsichore* (nec Walker)

D. ayutthaya Beuk, 1996

D. crepitans Boulard, 2005

D. cochlearata Overmeer & Duffels, 1967

D. dubia Lee, 2009

D. emanatura Distant, 1889

D. ensifera Bloen & Duffels, 1976

D. euterpe Bloem & Duffels, 1976

D. feae (Distant, 1892) = *Cosmopsaltria feae* = *Orientopsaltria feae* = *Dundubia longina* Distant, 1917

D. flava Lee, 2009

D. gravesteini Duffels, 1976 = *Dundubia graveistini* (sic)

D. hainanensis (Distant, 1901) = *Cosmopsaltria hainanensis* = *Platylomia hainanensis* = *Dundubia hainanensi* (sic) = *Dundubia bifasciata* Liu, 1940

D. hastata (Moulton, 1923) = *Cosmopsaltria hastata* = *Orientopsaltria hastata* = *Dundubia spiculata* (nec Nouhalhier)

D. jacoona (Distant, 1888) = *Cosmposaltria jacoona* = *Orientopsaltria jacoona*

D. kebuna Moulton, 1923 = *Dundubia intemerata* (nec Walker)

D. laterocurvata Beuk, 1996 = *Dundubia mannifera* var. *a* (nec Walker) = *Cosmopsaltria hastata* (nec Moulton)

D. myitkyinensis Beuk, 1996

D. nagarasingna Distant, 1881 = *Cosmopsaltria nagarasingna* = *Platylomia nagarasingna* = *Platylomia magarasingna* (sic) = *Platylomia nagalasingna* (sic) = *Dundubia nagarasingha* (sic) = *Dundubia helena* Distant, 1912 = *Cosmopsaltria fratercula* Distant, 1912 = *Orientopsaltria fratercula* = *?Dundubia bifasciata* Ishihara, 1961 = *Platylomia* sp. Gogala, 1995

D. nigripes (Moulton, 1923) = *Dundubia mannifera* var. *a* (nec Walker) = *Dundubia mannifera* (nec Walker) = *Cosmopsaltria nigripes* = *Orientopsaltria nigripes* = *Dundubia rafflesii* (nec Distant)

D. nigripesoides Boulard, 2008

D. oopaga (Distant, 1881) = *Cosmopsaltria oopaga* = *Cosmopsaltria oopaqa* (sic) = *Orientopsaltria oopaga* = *Orientopsaltria oophaga* (sic) = *Cosmopsaltria andersoni* Distant, 1883 = *Terpnosia andersoni* Wu, 1935 (nec Distant, 1892) = *Orientopsaltria andersoni*

D. rafflesii Distant, 1883

D. rhamphodes Bloem & Duffels, 1976

D. rufivena rufivena Walker, 1850 = *Dundubia intemerata* Walker, 1857 = *Dundubia intenerata* (sic) = *Dundubia intermerata* (sic) = *Pomponia intemerata* = *Fidicina confinis* Walker, 1870 = *Dundubia mellea* Distant, 1889

D. rufivena var. *a* Distant, 1889 = *Dundubia mellea* var. *a*

D. simalurensis Overmeer & Duffels, 1967

D. sinbyudaw Beuk, 1996

D. solokensis Overmeer & Duffels, 1967

D. somraji Boulard, 2003

D. sp. 1 Pham & Thinh, 2006

D. spiculata Noualhier, 1896 = *Platylomia spiculata* = *Dundubia speculata* (sic) = *Dundubia siamensis* Haupt, 1918 = *Cosmopsaltria hastata* (nec Moulton) = *Orientopsaltria hastata* (nec Moulton)

D. terpsichore (Walker, 1850) = *Cephaloxys terpsichore* = *Mogannia terpsichore* = *Dundubia intemerata* (nec Walker) = *Dundubia intermerata* (sic) (partim) = *Dundubia mannifera* (nec Linnaeus, 1754 nom. nud.) = *Dundubia manifera* (sic) = *Dundubia mannifera terpsichore* Distant, 1917 = *Dundubia mannifera* var. *a* = *Dundubia vaginata terpischore*

D. vaginata nigrimacula Walker, 1950 = *Dundubia vaginata* var. *nigrimacula* = *Dundubia nigrimacula*

D. vaginata vaginata (Fabricius, 1787) = *Tettigonia vaginata* = *Cicada vaginata* = *Dundubia vaginat* (sic) = *Cicada virescens* Olivier, 1790 = *Dundubia immacula* Walker, 1850 = *Dundubia mannifera immacula* = *Dundubia sobria* Walker, 1850 = *Cicada mannifera* Walker, 1850 nom. nud. = *Dundubia mannifera* = *Dundubia manifera* (sic) = *Fidicina mannifera* (nec Fabricius) = *Mogannia mannifera* = *Dundubia mannifera* var. *a* = *Dundubia mannifera terpsichore* (nec Walker) = *Dundubia mannifera* Banks, 1910 nom. nud. = *Maua* sp. Gogala & Riede 1995

D. vanna Chou, Lei, Li, Lu & Yao, 1997 nom. nud.

Acutivalva Yao, 1985 nom. nud.

A. choui Yao, 1985 nom. nud.

Linguvalva Chou & Yao, 1985 nom. nud.

L. sinensis Chou & Yao, 1985 nom. nud.

Spilomistica Chou & Yao, 1984 nom. nud.

S. sinensis Chou & Yao, 1984 nom. nud. in Hua, 2000

Mata Distant, 1906

M. kama (Distant, 1881) = *Pomponia kama*

M. rama Distant, 1912

M. sp. 1 Pham & Thinh, 2006

Sinosemia Matsumura, 1927 = *Sinosomia* (sic) = *Senosemia* (sic)
- *S. shirakii* Matsumura, 1927 = *Senosemia* (sic) *shirakii* = *Cryptotympana shirakii* = *Senosemia* (sic) sp. 1 Pham & Thinh, 2006

Champaka Distant, 1905
- *C. abdulla* (Distant, 1881) = *Cosmopsaltria abdulla* = *Platylomia abdulla* = *Platylomia distanti* Moulton, 1923 = *Platylomia spinosa distanti*
- *C. aerata* (Distant, 1888) = *Dundubia aerata* = *Platylomia aerata*
- *C. celebensis* Distant, 1913 = *Platylomia celebensis* = *Champaka maculipennis* Haupt, 1917
- *C. constanti* (Lee, 2009) = *Platylomia constanti*
- *C. maxima* (Lee, 2009) = *Platylomia maxima*
- *C. meyeri* (Distant, 1883) = *Cosmopsaltria meyeri* = *Platylomia meyeri* = *Cosmopsaltria majuscula* Distant, 1889 = *Dundubia majuscula* = *Cosmopsaltria maiuscula* (sic) = *Platylomia majuscula*
- *C. nigra* (Distant, 1888) = *Dundubia spinosa* Walker, 1850 (nec Fabricius) = *Cosmopsaltria nigra* = *Platylomia nigra* = *Platylomia albomaculata* Distant, 1905
- *C. spinosa* (Fabricius, 1787) = *Tettigonia spinosa* = *Tettigonia spinnosa* (sic) = *Tettigonia bispinosa* (sic) = *Cicada spinosa* = *Tettigonia 2spinosa* = *Dundubia spinosa* = *Cosmopsaltria spinosa* = *Cosmopsaltria (Cosmopsaltria) spinosa* = *Platylomia spinosa* = *Platylomia umbrata* (nec Distant) = *Platylomia maculata* Liu, 1940
- *C. virescens* (Distant, 1905) = *Platylomia virescens*
- *C. viridimaculata* (Distant, 1889) = *Pomoponia viridimaculata* = *Platylomia viridimaculata* = *Champaka harveyi* Distant, 1912 = *Champaka viridimaculata harveyi*
- *C. wallacei* (Beuk, 1999) = *Platylomia wallacei*

Platylomia Stål, 1870 = *Platlomia* (sic) = *Platytomia* (sic) = *Platylonia* (sic) = *Paltylomia* (sic) = *Cosmopsaltria (Platylomia)*
- *P. amicta* (Distant, 1889) = *Dundubia amicta* = *Cosmopsaltria amicta*
- *P. bivocalis* (Matsumura, 1907) = *Cosmopsaltria bivocalis* = *Comopsaltria* (sic) *bivocalis*
- *P. bocki* (Distant, 1882) = *Platylomia bbccki* (sic) = *Platylomia bccki* (sic) = *Platylomia bock* (sic) = *Dundubia bocki* = *Cosmopsaltria bocki*
- *P. brevis* Distant, 1912
- *P. ficulnea* (Distant, 1892) = *Cosmopsaltria ficulnea*
- *P. flavida* (Guérin-Méneville, 1834) = *Cicada flavida* = *Cosmopsaltria flavida* = *Cosmopsaltria (Platylomia) flavida* = *Dundubia flavida* = *Dundubia flava* (sic)
- *P. hainanensis* (Distant, 1901) = *Cosmopsaltria hainanensis*
- *P. insignis* Distant, 1912
- *P. juno* Distant, 1905 = *Cosmopsaltria juno*
- *P. larus* (Walker, 1858) = *Dundubia larus* = *Cosmopsaltria larus*
- *P. larus* var. _______ (Distant, 1889) = *Cosmopsaltria larus* var. _______
- *P. lemoultii* Lallemand, 1924 = *Playlonia* (sic) *lemoultii*
- *P. malickyi* Beuk, 1998
- *P.* n. sp. Boer & Duffles, 1990
- *P. pendleburyi* Moulton, 1923
- *P. operculata* Distant, 1913 = *Platylomia radha* (nec Distant)
- *P. pendleburyi* Moulton, 1923
- *P. plana* Lei & Li, 1994
- *P. radha* (Distant, 1881) = *Dundubia radha* = *Cosmopsaltria radha* = *Dundubia similis* Distant, 1882 = *Cosmopsaltria similis* = *Platylomia similis*
- *P. stasserae* Boulard, 2005
- *P. strongata* Lei, 1997
- *P.* undescribed species A Sanborn, Phillips & Sites, 2007 = *Platylomia similis* (nec Distant)
- *P. vibrans* (Walker, 1850) = *Dundubia vibrans* = *Cosmopsaltria vibrans*

Sinotympana Lee, 2009
- *S. incomparabilis* Lee, 2009

Ayesha Distant, 1905

A. serva (Walker, 1850) = *Dundubia serva* = *Cosmopsaltria serva* = *Cosmopsaltria (Cosmopsaltria) spathulata* Stål, 1870 = *Cosmopsaltria spathulata* = *Ayesha spathulata* = *Cosmopsaltria operculissima* Distant, 1881 = *Ayesha operculissima* = *Cicada elopurina* Distant, 1888 = *Cosmopsaltria vomerigera* Breddin, 1901 = *Dundubia lelita* Kirkaldy, 1904

Khimbya Distant, 1905 = *Khimbia* (sic)

K. cuneata (Distant, 1897) = *Pomponia sita*
K. diminuta (Walker, 1850) = *Dundubia dimunita* = *Cosmopsaltria dimunita* = *Khimbia* (sic) *diminuta*
K. evanescens (Walker, 1858) = *Dundubia evanescens* = *Pomponia evanescens* = *Khimbia* (sic) *evanescens*
K. immsi Distant, 1912
K. sita (Distant, 1881) = *Cosmopsaltria sita* = *Khimbia* (sic) *sita*

Subtribe Megapomponiina Boulard, 2008

Megapomponia Boulard, 2005

M. atrotunicata Lee & Sanborn, 2009
M. castanea Lee & Sanborn, 2009
M. clamorigravis Boulard, 2005
M. sp. nr. *decem* Boulard, 2008
M. foksnodi Boulard, 2010
M. imperatoria (Westwood, 1842) = *Cicada imperatoria* = *Dundubia imperatoria* = *Pomponia imperatoria* = *Pomponia impteratoria* (sic)
M. intermedia (Distant, 1905) = *Pomponia intermedia* = *Pomponia imperatoria* (nec Westwood)
M. macilenta Lee, 2012
M. merula (Distant, 1905) = *Pomponia merula* = *Pomponia imperatoria* (nec Westwood)
M. pendleburyi (Boulard, 2001) = *Megrapomponia* (sic) *pendleburyi* = *Pomponia pendleburyi* = *Pomponia merula* Gogala and Riede 1995 (nec Distant) = *Pomponia* sp. Gogala and Trilar 2005 = *Pomponia imperatoria* Riede and Kroker 1995 (nec Westwood)
M. sitesi Sanborn & Lee, 2009 = *Megapomponia* undescribed species A Sanborn, Phillips & Sites 2007
M. undescribed species B Sanborn, Phillips & Sites, 2007

Subtribe Macrosemiina Kato, 1925 = Macrosemaria = Macrosemiaria (sic)

Macrosemia Kato, 1925

M. anhweiensis Ouchi, 1938
M. assamensis (Distant, 1905) = *Platylomia assamensis*
M. diana (Distant, 1905) = *Platylomia diana* = *Cosmopsaltria diana*
M. divergens (Distant, 1917) = *Cosmopsaltria divergens* = *Platylomia divergens* = *Orientopsaltria divergens*
M. kareisana (Matsumura, 1907) = *Cosmopsaltria kareisana* = *Platylomia karaisana* (sic) = *Platylomia kareisana* = *Platylomia kareipana* (sic) = *Marosemia karaisana* (sic) = *Platylomia hopponis* Kato, 1925 = *Macrosemia hopponis* = *Macrosemia kareisana hopponis* = *Platylomia karapinensis* Kato, 1925 = *Macrosemia kareisana karapinensis* = *Cosmopsaltria montana* Kato, 1927 = *Orientopsaltria montana*
M. khuanae Boulard, 2001 = *Macrosemia longiterebra* Boulard, 2004
M. kiangsuensis kiangsuensis Kato, 1938 = *Platylomia kingvosana* Liu, 1940 = *Platylomia kinvosana* (sic)
M. kiangsuensis viridescens (Metcalf, 1955) = *Platylomia kingvosana viridescens* = *Platylomia kingvosana virescens* Liu, 1940
M. matsumurai (Kato, 1928) = *Macrosemia matsumura* (sic) = *Macrosemia matsumarai* (sic) = *Platylomia matsumurai*
M. perakana (Moulton, 1923) = *Platylomia saturata perakana*
M. pieli (Kato, 1938) = *Platylomia pieli* = *Platylomia piei* (sic) = *Platylomia pieli elongata* Liu, 1939 = *Platylomia pieli* var. *elongata* = *Platylomia pieli trifuscata* Liu, 1939 = *Platylomia pieli* var. *trifuscata* = *Platylomia chusana* Kato, 1940
M. saturata saturata (Walker, 1858) = *Dundubia saturata* = *Cosmopsaltria saturata* = *Cosmopsaltria saturate* (sic) = *Platylomia saturata* = *Platylomia satura* (sic) = *Paltylomia* (sic) *saturata* = *Dundubia obtecta* Walker, 1850 (nec Fabricius)

M. saturata var. *a* (Distant, 1891) = *Cosmopsaltria saturata* var. *a*
M. saturata var. *b* (Distant, 1891) = *Cosmopsaltria saturata* var. *b*
M. suavicolor Boulard, 2008
M. tonkiniana (Jacobi, 1905) = *Cosmopsaltria tonkiniana* = *Cosmoscarta* (sic) *tonkiniana* = *Platylomia tonkiniana* = *Orientopsaltria tonkiniana*
M. umbrata (Distant, 1888) = *Cosmopsaltria umbrata* = *Platylomia umbrata* = *Macrosemia chantrainei* Boulard, 2003

Subtribe Aolina Distant, 1905 = Aolaria

Haphsa Distant, 1905 = *Aola* Distant, 1905 = *Pomponia (Aola)*
H. bicolora Sanborn, 2009
H. bindusara (Distant, 1881) = *Haphsa bindusura* (sic) = *Pomponia bindusara* = *Aola bindusara*
H. conformis Distant, 1917
H. dianensis Chou, Lei, Li, Lu & Yao, 1997
H. durga (Distant, 1881) = *Cosmopsaltria durga* = *Meimuna durga* = *Dundubia durga*
H. fratercula Distant, 1917
H. jsguillotsi (Boulard, 2005) = *Haphsa jdguillotsi* (sic) = *Meimuna jsguillotsi* = *Meimuna guillotsi* (sic)
H. karenensis Ollenbach, 1928 = *Meimuna nauhkae* Boulard, 2005
H. nana Distant, 1913
H. nicomache (Walker, 1850) = *Dundubia nicomache* = *Cosmopsaltria nicomache* = *Haphsa nicoache* (sic) = *Cicada delineata* Walker, 1858
H. opercularis Distant, 1917
H. scitula (Distant, 1888) = *Pomponia scitula* = *Aola scitula* = *Pomponia (Aola) scitula*
H. stellata Lee, 2009
H. sulaiyai (Boulard, 2005) = *Aola sulaiyai*

Kaphsa Lee, 2012
K. aculeus (Lee, 2009) = *Haphsa aculeus*
K. concordia Lee, 2012

Subtribe Cosmopsaltriina Kato, 1932 = Cosmopsaltriaria

Lethama Distant, 1905
L. locusta (Walker, 1850) = *Cephaloxys locusta* = *Mogannia locusta* = *Dundubia locusta*

Sinapsaltria Kato, 1940
S. annamensis Kato, 1940
S. typica Kato, 1940

Dilobopyga Duffels, 1977
D. alfura alfura (Breddin, 1900) = *Cosmopsaltria alfura* = *Meimuna alfura*
D. alfura alfura var. _______ (Breddin, 1901) = *Cosmopsaltria alfura* var. _______ = *Meimuna alfura* var. _______
D. aprina Lee, 2009
D. breddini (Duffels, 1970) = *Diceropyga breddini* = *Cosmopsaltria opercularis* Breddin, 1901 (nec Walker)
D. chlorogaster (Boisduval, 1835) = *Cicada chlorogaster* = *Dundubia chlorogaster* = *Cosmopsaltria chlorogaster* = *Cosmopsaltria chlorogastri* (sic) = *Diceropyga chlorogaster* = *Dundubia maculosa* Walker, 1858 = *Diceropyga maculosa* = *Dundubia fuliginosa* (nec Walker, 1850)
D. gemina gemina (Distant, 1888) = *Cosmopsaltria gemina* = *Diceropyga gemina* = *Diceropyga minahassae* Distant (partim) = *Dundubia vaginata* Walker, 1868 (nec Fabricius) = *Dundubia vibrans* Walker, 1868 (nec Walker, 1850)
D. gemina toxopei (Schmidt, 1926) = *Diceropyga toxopei* = *Dicercopyga* (sic) *roteri* Schmidt, 1926 = *Diceropyga roteri*
D. janskocki Duffels, 1990

D. margarethae margarethae Duffels, 1977 = *Diceropyga gemina* (nec Distant, 1888) = *Diceropyga chlorogaster* (nec Boisduval)
D. margarethae parvula Duffels, 1977
D. minahassae (Distant, 1888) = *Cosmopsaltria minahasae* (sic) = *Cosmopsaltria minahassae* = *Diceropyga minabassae* (sic) = *Diceropyga minahassae* = *Diceropyga gemina* auct. (nec Distant, 1888)
D. multisignata (Breddin, 1901) = *Cosmopsaltria multisignata* = *Diceropyga* (?) *multisignata*
D. opercularis (Walker, 1858) = *Dundubia insularis* Walker, 1858 = *Cosmopsaltria insularis* = *Diceropyga insularis*
D. ornaticeps (Breddin, 1901) = *Cosmopsaltria ornaticeps* = *Diceropyga ornaticeps*
D. remanei Duffels, 1999
D. similis Duffels, 1977 = *Cosmopsaltria opercularis* Breddin (nec Walker)

Brachylobopyga Duffels, 1982
B. montana Duffels, 1988
B. toradja (Breddin, 1901) = *Cicada toradja* = *Cosmopsaltria toradja* = *Tibicen toradja* = *Tibicen toradjus* = *Brachylobopyga decorata* Duffels, 1982

Cosmopsaltria Stål, 1866 = *Cosmopsaltria (Cosmopsaltria)* = *Cosmopsatria* (sic) = *Cosmoscarta* (sic) = *Dundubia (Cosmopsaltria)* = *Fatima* Distant, 1905 = *Sawda* Distant, 1905
C. aurata Duffels, 1983
C. bloetei Duffels, 1965
C. capitata Distant, 1888 = *Fatima capitata* = *Sawda froggatti* Distant, 1911 = *Cosmopsaltria froggatti*
C. delfae Duffels, 1983
C. doryca (Boisduval, 1835) = *Cicada doryca* = *Cicada dorei* (sic) = *Dundubia doryca* = *Dundubia dorei* (sic) = *Dundubia doryca*
C. emarginata Duffels, 1969
C. gestroei (Distant, 1905) = *Sawda gestroei* = *Sawda gestroi* (sic)
C. gigantea gigantea (Distant, 1897) = *Pomponia gigantea* = *Sawda pratti* Distant, 1905 = *Cosmopsaltria pratti* = *Sawda sharpi* Distant, 1905 = *Cosmopsaltria sharpi* = *Sawda* sp. Brongersma & Venema, 1960
C. gigantea occidentalis Duffels, 1983 = *Cosmopsaltria giganta* (sic) *occidentalis*
C. gracilis Duffels, 1983
C. halmaherae Duffels, 1988
C. huonensis Duffels, 1983
C. inermis Stål, 1870 = *Cosmopsaltria (Cosmopsaltria) inermis*
C. kaiensis Duffels, 1988
C. lata (Walker, 1868) = *Dundubia lata*
C. loriae Distant, 1897 = *Fatima loriae*
C. meeki (Distant, 1906) = *Haphsa meeki*
C. mimica Distant, 1897 = *Sawda mimica*
C. papuensis Duffels, 1983 = *Fatima capitata* Distant, 1912
C. personata Duffels, 1983
C. retrorsa Duffels, 1983
C. satyrus Duffels, 1969
C. signata Duffels, 1983
C. sp. a Duffels, 1986
C. toxopeusi Duffels, 1983
C. vitiensis (Distant, 1906) = *Sawda ? vitiensis* = *Sawda vitiensis* = *(?) Cosmopsaltria vitiensis* = *Cosmopsaltria (?) vitiensis*
C. waine Duffels, 1991

Diceropyga Stål, 1870 = *Cosmopsaltria (Diceropyga)* = *Cosmopsaltria (Diceropygis)* (sic)
D. acutipennis (Walker, 1858) = *Cicada acutipennis*
D. auriculata Duffels, 1977
D. aurita Duffels, 1977
D. bacanensis Duffels, 1988
D. bicornis Duffels, 1977
D. bihamata Duffels, 1977

D. bougainvillensis Duffels, 1977
D. didyma (Boisduval, 1835) = *Cicada didyma* = *Cosmopsaltria didyma* = *Meimuna didyma*
D. gravesteini Duffels, 1977 = *Diceropyga pigafettae* Lallemand (nec Distant)
D. guadalcanalensis Duffels, 1977 = *Diceropyga guadalcanensis* (sic) = *Diceropyga pigafettae* (nec Distant)
D. junctivitta (Walker, 1868) = *Dundubia junctivitta* = *Cosmopsaltria junctivitta* = *Diceropyga junctivitta* = *Diceropyga obtecta* (nec Fabricius) = *Dundubia subapicalis* (nec Walker) = *Cosmopsaltria pigafettae* (nec Distant) = *Diceropyga pigafettae* (nec Distant)
D. major Duffels, 1977
D. malaitensis Duffels, 1977
D. n. sp. Biak Boer, 1995 = *Diceropyga* sp. A Duffels & Turner, 2002
D. n. sp. Nunfer Boer, 1995 = *Diceropyga* sp. B Duffels & Turner, 2002
D. noonadani Duffels, 1977
D. novaebritannicae Duffels, 1977
D. novaeguinae Distant, 1912
D. novaeguinea Duffels, 1977
D. obliterans Duffels, 1977
D. obtecta (Fabricius, 1803) = *Tettigonia obtecta* = *Cicada obtecta* = *Cosmopsaltria obtecta* = *Cosmopsaltria (Diceropyga) obtecta* = *Cosmopsaltria (Diceropygis)* (sic) *obtecta* = *Dundubia obtecta* = *Dundubia bicaudata* Walker, 1858 = *Cosmopsaltria bicaudata* = *Cosmopsaltria (Diceropyga) bicaudata* = *Dundubia subapicalis* (nec Walker) = *Cosmopsaltria pigafettae* (nec Distant)
D. ochrothorax Duffels, 1977
D. pigafettae (Distant, 1888) = *Cosmopsaltria pigafettae*
D. rennellensis Duffels, 1977
D. subapicalis (Walker, 1870) = *Diceropyga apicalis* (sic) = *Cosmopsaltria atra* Distant, 1897 = *Diceropyga atra* = *Diceroprocta obtecta* (nec Fabricius) = *Dundubia subapicalis* (partim)
D. subjuga Duffels, 1977
D. tortifer Duffels, 1977
D. triangulata Duffels, 1977
D. woodlarkensis inexpectata Duffels, 1977
D. woodlarkensis woodlarkensis Duffels, 1977

Rhadinopyga Duffels, 1985
R. acuminata (Duffels, 1977) = *Diceropyga acuminata*
R. duffelsi Sanborn, 2005
R. epiplatys Duffels, 1985
R. impar impar (Walker, 1868) = *Dundubia impar* = *Cosmopsaltria impar* = *Diceropyga impar*
R. impar impar var. *β* (Distant, 1891) = *Cosmopsaltria impar* var. *β* = *Diceropyga impar* var. *β*
R. recedens (Walker, 1870) = *Dundubia recedens* = *Cosmopsaltria recedens* = *Diceropyga recedens* = *Cosmopsaltria lutulenta* Distant, 1888 = *Diceropyga lutulenta*

Inflatopyga Duffels, 1997 = Genus "I" Duffels
I. boulardi Duffels, 1997
I. ewarti Duffels, 1997
I. langeraki Duffels, 1997
I. mouldsi Duffels, 1997
I. verlaani Duffels, 1997
I. webbi Duffels, 1997

Aceropyga Duffels, 1977
A. acuta Duffels, 1988
A. albostriata (Distant, 1888) = *Cosmopsaltria albostriata* = *Diceropyga albostriata*
A. corynetus corynetus Duffels, 1977 = *Aceropyga corynetus*
A. corynetus monacantha Duffels, 1988
A. corynetus ungulata Duffels, 1988 = *Aceropyga corynetus ungulatus* (sic)

A. distans distans (Walker, 1858) = *Dundubia distans* = *Cosmopsaltria distans* = *Diceropyga distans* = *Dundubia subfascia* Walker, 1858
A. distans lineifera (Walker, 1858) = *Dundubia lineifera* = *Cosmopsaltria lineifera* = *Cosmopsaltria (Diceropyga) lineifera*
A. distans taveuniensis Duffels, 1977
A. egmondae Duffels, 1988
A. huireka Duffels, 1988
A. macracantha Duffels, 1988
A. philorites Duffels, 1988 = *Aceropyga philoritis* (sic)
A. poecilochlora (Walker, 1858) = *Dundubia poecilochlora* = *Cosmopsaltria poecilochlora* = *Cosmopsaltria (Diceropyga) poecilochlora* = *Diceropyga poecilochlora* = *Aceropyga poecilochora* (sic) = *Dundubia connata* Walker, 1858 = *Khimbya kusaiensis* Esaki, 1939 = *Diceropyga kusaiensis* = *Khimbya musaiensis* (sic)
A. pterophon Duffels, 1988
A. stuarti pallens Duffels, 1977
A. stuarti stuarti (Distant, 1882) = *Cosmopsaltria stuarti* = *Diceropyga stuarti*

Meimuna Distant, 1905 = *Meinuna* (sic) = *Meinuma* (sic) = *Miemuna* (sic) = *Meimuua* (sic) = *Maimuna* (sic) = *Meimura* (sic)
M. boninensis (Distant, 1905) = *Diceropyga boninensis* = *Cosmopsaltria ogasawarensis* Matsumura, 1906 = *Meimuna ogasawarensis*
M. bonininsulana Kato, 1928 = *Meimuna bonin-insulana*
M. cassandra Distant, 1912 = *Meimuna casandra* (sic)
M. chekianga Kato, 1940
M. chekiangensis Chen, 1940
M. choui Lei, 1994
M. crassa (Distant, 1905) = *Haphsa crassa* = *Haphsa craasa* (sic)
M. duffelsi Boulard, 2005
M. gakokizana Matsumura, 1917
M. gamameda (Distant, 1902) = *Cosmopsaltria gamameda* = *Meimuna ganameda* (sic)
M. goshizana Matsumura, 1917 = *Meimuna gozhizana* (sic)
M. infuscata Lei & Beuk, 1997 = *Meinuna* (sic) *infuscata*
M. iwasakii Matsumura, 1913 = *Meimuna ishigakina* Kato, 1925 = *Meimuna iwasakii ishigakina* = *Meimuna uraina* Kato, 1925 = *Meimuna uraina* var. *a* Kato, 1927 = *Meimuna uraina* var. *b* Kato, 1927 = *Meimuna uraina* var. *c* Kato, 1927 = *Meimuna iriomoteana* Kato, 1936
M. khadiga (Distant, 1904) = *Cosmopsaltria khadiga*
M. kuroiwae Matsumura, 1917 = *Meimuna sakaguchii* Matsumura, 1927 = *Meimuna amamioshimana* Kato, 1928 = *Meimuna amami-oshimana* = *Meimuna sakaguchii* (nec Matsumura) = *Meimuna kikaigashimana* Kato, 1937 = *Meimuna tsuchidai* Kato, 1931 = *Meimuna hentonaensis* Kato, 1960 = *Meimuna oshimensis* (nec Matsumura) = *Meimura* (sic) *oshimensis* (nec Matsumura) = *Meimuna nakaoi* Ishihara, 1968 = *Meimuna kuroiwai* (sic)
M. microdon (Walker, 1850) = *Dundubia microdon* = *Cosmopsaltria microdon* = *Meimuna omeinensis* Chen, 1940
M. mongolica (Distant, 1881) = *Cosmopsaltria mongolica* = *Meimuna suigensis* Matsumura, 1927 = *Meimuna chosensis* Matsumura, 1927 = *Meimuna heijonis* Matsumura, 1927 = *Meimuna santoshonis* Matsumura, 1927 = *Meimuna gallosi* Matsumura, 1927 = *Meimuna galloisi* (sic)
M. multivocalis (Matsumura, 1917) = *Cosmpsaltria multivocalis* = *Orientopsaltria multivocalis*
M. neomongolica Liu, 1940
M. opalifera (Walker, 1850) = *Dundubia opalifera* = *Cosmopsaltria opalifera* = *Dundubia (Cosmopsaltria) opalifera* = *Meinuma* (sic) *opalifera* = *Miemuna* (sic) *opalifera* = *Meimuua* (sic) *opalifera* = *Maimuna* (sic) *opalifera* = *Meimuna oplifera* (sic) = *Meimuna operlifera* (sic) = *Meimuna opaliferia* (sic) = *Meimura* (sic) *opalifera* = *Meimura* (sic) *oparifera* (sic) = *Meimuna opalifer* (sic) = *Meimuna opalifera formosana* Kato, 1925 = *Meimuna opalifera* var. *formosana* = *Meimuna opalifera formosans* (sic) = *Meimuna opalifera nigroventris* Kato, 1927 = *Meimuna opalifera* var. *nigroventris* = *Meimuna opalifera nigoventria* (sic) = *Meimuna opalifera punctata* Kato, 1932 = *Meimuna opalifera* f. *punctata* = *Meimuna opalifera*

ab. *punctata* = *Meimuna opalifera* var. *punctata* = *Meimuna longipennis* Kato, 1937 = *Meimura* (sic) *longipennis*

M. oshimensis (Matsumura, 1906) = *Cosmopsaltria oshimensis* = *Meimura* (sic) *oshimensis* = *Meimuna oshimensis tokunoshimana* Ishihara, 1968

M. pallida Ollenbach, 1929

M. raxa Distant, 1913

M. silhetana (Distant, 1888) = *Cosmopsaltria silhetana* = *Cosmopsalta* (sic) *silhetana*

M. subviridissima Distant, 1913

M. tavoyana (Distant, 1888) = *Dundubia tavoyana* = *Cosmopsaltria tavoyana*

M. tripurasura (Distant, 1881) = *Dundubia tripurasura* = *Cosmopsaltria tripurasura*

M. undescribed species A Sanborn, Phillips & Sites, 2007

M. undescribed species B Sanborn, Phillips & Sites, 2007

M. velitaris (Distant, 1897) = *Cosmopsaltria velitaris* = *Haphsa velitaris*

M. viridissima Distant, 1912

Moana Myers, 1928

M. aluana aluana (Distant, 1905) = *Aceropyga aluana aluana* = *Diceropyga aluana* = *Diceropyga aluana* Lallemand (partim)

M. aluana minima (Duffels, 1977) = *Aceropyga aluana minima* = *Aceropyga aluana minuta* (sic)

M. aluana torquata (Duffels, 1977) = *Aceropyga aluana torquata* = *Diceropyga pigafettae* (nec Distant)

M. expansa Myers, 1928

M. novaeirelandicae (Duffels, 1977) = *Moana irelandica* (sic) = *Aceropyga novaeirelandicae* = *Diceropyga pigafettae* (nec Distant) = *Diceropyga aluana* (nec Distant)

M. obliqua (Duffels, 1977) = *Aceropyga obliqua*

Subtribe Orientopsaltriina Boulard, 2004 = Orientopsaltriaria

Orientopsaltria Kato, 1944

O. agatha (Moulton, 1911) = *Cosmopsaltria agatha* = *Cosmopsaltria* sp. aff. *agatha* Endo & Hayashi, 1979 = *Cosmopsaltria montivaga agatha*

O. alticola (Distant, 1905) = *Cosmopsaltria alticola* = *Cosmopsaltria alticola pontiaka* Kirkaldy, 1913 = *Cosmopsaltria alticola* var. *pontianka* = *Platylomia bangueyensis* Distant, 1912 = *Cosmopsaltria inermis* (nec Stål)

O. angustata Duffels & Zaidi, 2000 = *Cosmopsaltria guttigera* (nec Walker) = *Orientopsaltria guttigera* (nec Walker) = *Orientopsaltria* nr. *guttigera* n. sp. Zaidi & Ruslan 1995

O. bardi Boulard, 2008

O. beaudouini Boulard, 2003

O. brooksi (Moulton, 1923) = *Cosmopsaltria brooksi*

O. cantavis Boulard, 2003

O. ceslaui (Lallemand & Synave, 1953) = *Cosmopsaltria ceslaui*

O. confluens Duffels & Zaidi, 2000

O. duarum (Walker, 1857) = *Dundubia duarum* = *Cosmopsaltria duarum* = *Cosmopsaltria lauta* Distant, 1888 = *Orientopsaltria lauta*

O. endauensis Duffels & Schouten, 2002

O. fangrayae Boulard, 2001

O. fuliginosa (Walker, 1850) = *Dundubia fuliginosa* = *Cosmopsaltria fuliginosa* = *Cosmopsaltria (Cosmopsaltria) fuliginosa* = *Platylomia fuliginosa* = *Cosmopsaltria* sp. aff. *fuliginosa* Endo & Hayashi, 1979 = *Dundubia melpomene* Walker, 1850

O. guttigera (Walker, 1856) = *Dundubia guttigera* = *Cosmopsaltria guttigera*

O. hollowayi Duffels & Zaidi, 2000

O. ida (Moulton, 1911) = *Cosmopsaltria ida*

O. inermis (Stål, 1870) = *Cosmopsaltria inermis*

O. jmalleti Boulard, 2005

O. kinabaluana Duffels & Zaidi, 2000

O. latispina Duffels & Zaidi, 2000

O. lucieni Boulard, 2005

O. maculosa Duffels & Zaidi, 2000 = *Orientopsaltria* n. sp. 1 nr. *guttigera* Zaidi & Ruslan, 1995 = *Orientopsaltria* nr. *guttigera* Zaidi, Ruslan & Mahadir, 1996 = *Orientopsaltria* nr. *guttiger* (sic) = *Orientopsaltria* nr. *guttifer* (sic)
O. montivaga (Distant, 1889) = *Cosmopsaltria montivaga*
O. moultoni (China, 1926) = *Cosmopsaltria moultoni*
O. musicus Boulard, 2005
O. noonadani Duffels & Zaidi, 2000 = *Cosmopsaltria inermis* (nec Stål) = *Cosmopsaltria alticola* (nec Distant)
O. padda (Distant, 1887) = *Cosmopsaltria padda* = *Cosmopsaltria latilinea* (nec Walker) = *Cosmopsaltria duarum vera* Moulton, 1911 = *Cosmopsaltria duarum* var. *vera* = *Cosmopsaltria duarum padda* = *Cosmopsaltria duarum* var. *padda*
O. nr. *padda* Zaidi, Ruslan & Mahadir, 1996
O. palawana Duffels & Zaidi, 2000
O. phaeophila (Walker, 1850) = *Dundubia phaeophila* = *Dundubia phoeopila* (sic) = *Cosmopsaltria phaeophila* = *Cosmopsaltria phoeophila* (sic) = *Cosmopsaltria phaephila* (sic)
O. ruslani Duffels & Zaidi, 1998 = *Cosmopsaltria montivaga* (nec Distant) = *Orientopsaltria montivaga* (nec Distant)
O. saudarapadda Duffels & Zaidi, 1998 = *Cosmopsaltria montivaga* (nec Distant)
O. sp. 1 Zaidi, Ruslan & Azman, 2000
O. sp. 2 Zaidi, Ruslan & Azman, 2000
O. sumatrana (Moulton, 1917) = *Cosmopsaltria sumatrana*
O. vanbreei Duffels & Zaidi, 2000 = *Orientopsaltria sumatrana* (nec Moulton) = *Orientopsaltria* sp. n. 2 (near *sumatrana* (Moulton)) Zaidi & Ruslan, 1995 = *Orientopsaltria* near *sumatrana* Zaidi, Rusland & Mahadir, 1996

Subtribe Crassopsaltriina Boulard, 2008

Crassopsaltria Boulard, 2008
C. giovannae (Boulard, 2005) = *Macrosemia giovannae*

Tribe Tosenini Amyot & Audinet-Serville, 1843 = Tosenaria = Toseniini (sic) = Tacuaria (nec Distant)

Tosena Amyot & Audinet-Serville, 1843 = *Cicada (Tosena)* = *Tossena* (sic) = *Josena* (sic) = *Tosema* (sic)
T. albata Distant, 1878 = *Tosena melanoptera* var. *albata* = *Tosena melanoptera* var. *c* Distant, 1906 = *Tosena albata* var. *c* = *Tosena melanopterix* (sic) var. *albata* = *Tosena melanopterix* (sic) (nec White)
T. depicta Distant, 1888
T. dives (Westwood, 1842) = *Cicada dives* = *Gaeana dives* = *Huechys transversa* Walker, 1858
T. fasciata fasciata (Fabricius, 1787) = *Tettigonia fasciata* = *Cicada fasciata* = *Josena* (sic) *fasciata* = *Tosena faciata* (sic)
T. fasciata var. *a* Distant, 1889
T. fasciata var. *b* Distant, 1889
T. fasciata var. *c* Distant, 1889
T. mearesiana (Westwood, 1842) = *Cicada mearesiana* = *Cicada mareaseana* (sic)
T. melanoptera melanoptera (White, 1846) = *Cicada (Tosena) melanoptera* = *Cicada melanoptera* = *Tosena fasciata* Moulton, 1923 (nec Fabricius) = *Tosena melanopteryx* Kirkaldy, 1909 = *Tosema* (sic) *melanopteryx* = *Tosena fasciata* var. *d* Distant, 1889
T. melanoptera var. *a* Distant, 1889 = *Tosena melanopteryx* var. *a*
T. melanoptera var. *b* Distant, 1889 = *Tosena melanopteryx* var. *b*
T. paviei (Noualhier, 1896) = *Gaeana paviei* = *Gaeana pavici* (sic)
T. sp. 1 Pham & Thinh, 2006
T. sp. 2 Pham & Thinh, 2006

Distantalna Boulard, 2009
D. splendida splendida (Distant, 1878) = *Tosena splendida* = *Tosena splendidula* (sic)
D. splendida var. *a* (Distant, 1889) = *Tosena splendida* var. *a*
D. splendida var. *b* (Distant, 1889) = *Tosena splendida* var. *b*

Trengganua Moulton, 1923 = *Teregganua* (sic)
T. sibylla (Stål, 1863) = *Gaeana sibylla* = *Tosena sibylla* = *Teregganua* (sic) *sibylla*

Ayuthia Distant, 1919
A. spectabile Distant, 1919

Tribe Distantadini Orian, 1963

Distantada Orian, 1963
D. thomasseti Orian, 1963 = *Distantada thomaseti* (sic)

Tribe Sonatini Lee, 2010

Hyalessa China, 1925 = *Sonata* Lee, 2009
H. ella (Lei & Chou, 1997) = *Oncotympana ella* = *Sonata ella*
H. expansa (Walker, 1858) = *Carineta expansa* = *Pomponia expansa* = *Oncotympana expansa* = *Sonata expansa*
H. fuscata (Distant, 1905) = *Oncotympana fuscata* = *Sonata fuscata* = *Pomponia maculaticollis* (nec de Motschulsky) = *Oncotympana coreanus* Kato, 1925 = *Oncotympana coreana* = *Oncotympana nigrodorsalis* Mori, 1931 = *Oncotympana nigridorsalis* (sic) = *Oncotympana coreana nigrodorsalis* = *Oncotympana maculaticollis fuscata* = *Oncotympana maculaticollis pruinosa* Kato, 1927 = *Oncotympana coreana pruinosa* = *Oncotympana coreana* var. *pruinosa*
H. maculaticollis maculaticollis (de Motschulsky, 1866) = *Cicada maculaticollis* = *Pomponia maculaticollis* = *Pomponia maculicollis* (sic) = *Pomponia maculaticollia* (sic) = *Oncotympana maculaticollis* = *Onchotympana* (sic) *maculaticollis* = *Acotympana* (sic) *maculaticollis* = *Oncotympana naculaticollis* (sic) = *Oncotympana maculaticoliis* (sic) = *Oncotympana maculicollis* (sic) = *Oncotympana maculaticollis aka* Kato, 1934 = *Oncotympana maculaticollis* ab. *aka* = *Oncotympana maculaticollis clara* Kato, 1932 = *Oncotympana maculaticollis* ab. *clara* = *Oncotympana macullaticollis* (sic) *clara* = *Oncotympana maculaticollis pygmaea* Kato, 1932 = *Oncotympana maculaticollis* ab. *pygmaea* = *Oncotympana maculaticollis nigroventris* Kato, 1932 = *Oncotympana maculaticollis* f. *nigroventris* = *Oncotympana fuscata* (nec Distant) = *Oncotympana maculaticollis fuscata* = *Sonata maculaticollis* = *Cicada orni* (nec Linnaeus)
H. maculaticollis mikido (Kato, 1925) = *Oncoympana maculaticollis mikado* = *Oncotympana maculaticollis micado* (sic) = *Sonata maculaticollis mikido*
H. mahoni (Distant, 1906) = *Oncotympana mahoni* = *Sonata mahoni*
H. melanoptera (Distant, 1904) = *Pomponia melanoptera* = *Oncotympana melanoptera* = *Sonata melanoptera*
H. obnubila (Distant, 1888) = *Pomponia obnubila* = *Oncotympana obnubila* = *Sonata obnubila*
H. ronshana China, 1925
H. stratoria (Distant, 1905) = *Oncotympana stratoria* = *Sonatoa stratoria*
H. virescens (Distant, 1905) = *Oncotympana virescens* = *Sonata virescens*

Tribe Gaeanini Distant, 1905

Subtribe Gaeanina Distant, 1905 = Gaeanaria

Gaeana Amyot & Audinet-Serville, 1843 = *Geana* (sic) = *Geaena* (sic) = *Gaena* (sic)
G. atkinsoni Distant, 1892
G. cheni Chou & Yao, 1985 = *Gaeana yunnanensis* Chou & Yao, 1985 = *Gaeana maculata sheshuangpana Chou & Yao, 1985 nom. nud.* = *Gaena* (sic) *maculata sheshuangpana* = *Gaeana maculata xishuanbanna* Chou & Yao, 1985 *nom. nud.*
G. chinensis Kato, 1940
G. consors Atkinson, 1884 = *Gaeana maculata consors* = *Gaeana maculata* var. *a* Distant, 1892
G. hainanensis Chou & Yao, 1985
G. maculata barbouri Liu, 1940
G. maculata distanti Liu, 1940

G. maculata maculata (Drury, 1773) = *Cicada maculata* = *Tettigonia maculata* = *Cicada flavomaculata* (sic) = *Geaena* (sic) *maculata* = *Gaeana maculatus* (sic)
G. nigra Lei & Chou, 1997
G. variegata Yen, Robinson & Quicke, 2005 nom. nud.

Sulphogaeana Chou & Yao, 1985
S. dolicha Lei, 1997
S. sulphurea (Westwood, 1839) = *Cicada sulphurea* = *Gaeana sulphurea* = *Sulphogaeana sulphura* (sic) = *Cicada pulchella* Westwood, 1942 = *Cicada pulcilla* (sic)
S. vestita (Distant, 1905) = *Gaeana vestita* = *Callogaeana vestita* = *Gaeana vestita* var. _______ Distant, 1917 = *Sulphogaeana vestita* var. _______ (Distant, 1917)

Ambragaeana Chou & Yao, 1985
A. ambra Chou & Yao, 1985
A. laosensis (Distant, 1917) = *Gaeana laosensis*
A. stellata (Walker, 1858) = *Huechys stellata* = *Gaeana stellata*
A. stellata var. *a* (Distant, 1892) = *Gaeana stellata* var. *a*
A. sticta (Chou & Yao, 1985) = *Gaeana sticta*

Callogaeana Chou & Yao, 1985 = *Callagaeana* (sic) = *Callagaena* (sic)
C. annamensis annamensis (Distant, 1913) = *Callagaena* (sic) *annamensis* = *Gaeana annamensis*
C. annamensis var. *b* Distant, 1892 = *Gaeana annamensis* var. *b* = *Gaeana festiva* var. *b*
C. aurantiaca Chou & Yao, 1985
C. festiva festiva (Fabricius, 1803) = *Tettigonia festiva* = *Cicada festiva* = *Gaeana festiva* = *Gaeana fastiva* (sic) = *Callagaeana* (sic) *festiva* = *Cicada thalassina* Perchon, 1838 (nec *Cicada thalassina* Germar, 1830) = *Gaeana thalassina* = *Cicada percheronii* Guérin-Méneville, 1844 = *Gaeana consobrina* Walker, 1850 nom. nud.
C. festiva var. *a* (Distant, 1892) = *Gaeana festiva* var. *a*
C. guangxiensis Chou & Yao, 1985
C. hageni hageni (Distant, 1889) = *Callagaeana* (sic) *hageni* = *Gaeana hageni*
C. hageni var. *a* (Distant, 1889) = *Gaeana hageni* var. *a*
C. jinghongensis Chou & Yao, 1985
C. sultana Distant, 1913 = *Callagaeana* (sic) *sultana* = *Gaeana sultana*
C. viridula Chou & Yao, 1985
C. vitalisi (Distant, 1913) = *Callagaeana* (sic) *vitalisi* = *Gaeana vitalisi* = *Gaeana vitalis* (sic)

Balinta Distant, 1905
B. auriginea Distant, 1905
B. delinenda (Distant, 1888) = *Gaeana delinenda* = *Huechys octonotata* Walker, 1850 (nec Westwood)
B. flavoterminalia Boulard, 2008
B. kershawi Kirkaldy, 1909 = *Balinta rershawi* (sic)
B. minuta Boulard, 2008
B. octonotata octonotata (Westwood, 1842) = *Cicada 8-notata* = *Huechys octonotata* = *Cicada octonotata* = *Gaeana octonotata* = *Huechys picta* Walker, 1858
B. octonotata var. *a* Distant, 1892
B. octonotata var. *b* Distant, 1892
B. pulchella Distant, 1913
B. sanguiniventris Ollenbach, 1929
B. tenebricosa tenebricosa (Distant, 1888) = *Gaena* (sic) *tenebricosa* = *Gaeana tenebricosa* = *Gaeana tenebriscosa* (sic) = *Gaeana tenbricosa* (sic)
B. tenebricosa var. *a* Distant, 1892

Taona Distant, 1909
T. immaculata Chen, 1940
T. versicolor Distant, 1909

Chloropsalta Haupt, 1920 = *Chloropsaltria* (sic) = *Chloropsatia* (sic)
C. ochreata (Melichar, 1902) = *Cicadatra ochreata* = *Chloropsatia* (sic) *ochreata*
C. smaragdula Haupt, 1920
C. viridiflava (Distant, 1914) = *Psalmocharias viridiflava* = *Cicadetta viridiflava* Horváth nom. nud. = *Cicadetta (Melampsalta) viridiflava* Horváth nom. nud. = *Chloropsaltria* (sic) *viridiflava*

Subtribe Becquartinina Boulard, 2005 = Becquartinaria

Becquartina Kato, 1940 = *Becquardina* (sic) = *Sinopsaltria* Chen, 1943
B. bleuzeni Boulard, 2005
B. bifasciata (Chen, 1943) = *Sinopsaltria bifasciata* Chen, 1943
B. decorata (Kato, 1940) = *Gaeana ? decorata* = *Gaeana decorata* = *Gaeana decolata* (sic)
B. octonotata octonotata (Jacobi, 1902) = *Gaeana electa* = *Gaeana electra* (sic)
B. ruiliensis Chou & Yao, 1985 = *Becquartina uiliensis* (sic)
B. versicolor Boulard, 2005 = *Becquartina versicolor* var. *chantrainei* Boulard, 2005 = *Becquartina versicolor* var. *flammea* Boulard, 2005 = *Becquartina versicolor* var. *fuscalae* Boulard, 2005 = *Becquartina versicolor var. obscura* Boulard, 2005

Tribe Lahugadini Distant, 1905 = Lahudadini (sic)

Lahugada Distant, 1905
L. dohertyi (Distant, 1891) = *Pomponia dohertyi*

Tribe Cicadini Latreille, 1802 = Cicadaria = Psithyristriini Distant, 1905 = Psithyrisrini (sic) = Leptopsaltriini Moulton, 1923 = Leptopsaltrini (sic) = Pomponiini Kato, 1932 = Pomponiaria = Terpnosiina Kato, 1932 = Oncotympanini Ishihara, 1961

Subtribe Cicadina Latreille, 1802 = Cicadaria

Cicada Linnaeus, 1758 = *Cicada (Cicada)* = *Cicda* (sic) = *Gicada* (sic) = *Cicida* (sic) = *Cidada* (sic) = *Tettigia* Kolenati, 1857 = *Cicada (Tettigia)* = *Tettigia (Tettigia)* = *Tettigea* (sic) = *Teltigia* (sic) = *Testigia* (sic) = *Macroprotopus* Costa, 1877
C. albicans Walker, 1858
C. albiceps Stephens, 1829 nom. nud.
C. albida Gmelin, 1789
C. afona Costa, 1834 Of uncertain position
C. albella Billberg, 1820 nom. nud. Of uncertain position
C. alboguttata Goeze, 1778 Of uncertain position
C. albula Zetterstedt, 1828 Of uncertain position
C. ambigua Donovan, 1798 Of uncertain position
C. americana Gmelin, 1789 Of uncertain position
C. amphibia Pontoppidan, 1763 Of uncertain position
C. apicalis Billberg, 1820 nom. nud. Of uncertain position
C. aptera Linnaeus, 1758 Of uncertain position
C. argentea De Villers, 1789 Of uncertain position
C. arvensis Müller, 1776 Of uncertain position
C. asius Walker, 1850 = *Tympanoterpes asius*
C. barbara (Stål, 1866) = *Tettigia barbara*
C. barbara lusitanica Boulard, 1982
C. bella Billberg, 1820 nom. nud. Of uncertain position
C. biguttulus Billberg, 1820 nom. nud. Of uncertain position
C. bilineata De Villers, 1789 Of uncertain position
C. brazilensis Metcalf, 1963 = *Cicada vitrea* Germar, 1830 (nec *Cicada vitrea* Fabricius, 1803) Of uncertain position
C. bupthalmica Gravenhorst, 1807 Of uncertain position
C. cafra Olivier, 1790 Of uncertain position

C. *canaliculata* De Villiers, 1789 Of uncertain position
C. *cancellata* Gmelin, 1789 = *Cicada reticulata* Müller, 1776 (nec *Cicicada reticulata* Linnaeus, 1758) Of uncertain position
C. *carnaria* Stephens, 1829 nom. nud. Of uncertain position
C. *carniolica* Gmelin, 1789 = *Cicada lineata* Scopoli, 1763 (nec *Cicada lineata* Linnaeus, 1758) Of uncertain position
C. *casmatmema* Capanni, 1894
C. *casta* Gmelin, 1789 Of uncertain position
C. *cerisyi* Guérin-Méneville, 1844 = *Cicada* cf. *cerisyi* Schedl, 1993
C. *cinerea* Rossi, 1792
C. *clarisona* Hancock, 1834 Of uncertain position
C. *coeca* de Fourcroy, 1785 Of uncertain position
C. *collaris* Degeer, 1773 Of uncertain position
C. *complex* Walker, 1850
C. *confusa* Metcalf, 1963 = *Cicada rufipes* Germar, 1830 (nec *Cicada rufipes* Fabricius, 1803)
C. *conspurcata* (Fabricius, 1777) = *Tettigonia conspurcata* Of uncertain position
C. *costa* Dohrn, 1859 nom. nud. Of uncertain position
C. *costai* Metcalf, 1963 = *Cicada flava* Costa, 1834 (nec *Cicada flava* Linnaeus, 1758) Of uncertain position
C. *cretensis* Quartau & Simões, 2005
C. *crocata* Gmelin, 1789 Of uncertain position
C. *dinigrata* Gmelin, 1789 Of uncertain position
C. *devilliersi* Metcalf, 1963 = *Cicada maculata* De Villiers, 1789 (nec *Cicada maculata* Drury, 1773) Of uncertain position
C. *dipalliata* Capanni, 1894 = *Cicada dippalliata* (sic)
C. *dwigubskyi* Metcalf, 1963 = *Cicada nebulosa* Dwigubsky, 1802 (nec *Cicada nebulosa* Fabricius, 1781) Of uncertain position
C. *effecta* Walker, 1858 Of uncertain position
C. *egregia* Uhler, 1903
C. *elegans* Billberg, 1820 nom. nud. Of uncertain position
C. *elongella* Zetterstedt, 1840 nom. nud.
C. *erythroptera* Gmelin, 1789 Of uncertain position
C. *fasciaaurata* De Villiers, 1789 Of uncertain position
C. *flava* Linnaeus, 1758 Of uncertain position
C. *flavicollis* Stephens, 1829 nom. nud.
C. *flavifrons* Gmelin, 1789 Of uncertain position
C. *flavofasciata* Goeze, 1778 = *Cicada fasciolata* de Fourcroy, 1785 Of uncertain position
C. *fuschsii* Eversmann, 1837 nom. nud. Of uncertain position
C. *fuscomaculata* Goeze, 1778
C. *fuscomaculosa* Metcalf, 1963 = *Cicada maculosa* Gmelin, 1789 of uncertain position
C. *geographica* Goeze, 1778 Of uncertain position
C. *gibberula* Gmelin, 1789 Of uncertain position
C. *gigantea* Germar, 1830 Of uncertain position
C. *glabra* Müller, 1776 Of uncertain position
C. *gmelini* Metcalf, 1963 = *Cicada flava* Gmelin, 1789 (nec *Cicada flava* Linnaeus, 1758) Of uncertain position
C. *goezei* Metcalf, 1963 = *Cicada maculata* Goeze, 1778 (nec *Cicada maculata* Drury, 1773)
C. *guineensis* Metcalf, 1963 = *Cicada flavescens* Turton, 1802 (nec *Cicada flavescens* Fabricius, 1794) Of uncertain position
C. *guttata* Forster, 1771 Of uncertain position
C. *inexacta* Walker, 1858 Of uncertain position
C. *intacta* Walker, 1852 Of uncertain position
C. *javanensis* Metcalf, 1963 = *Cicada undata* Germar, 1830 (nec *Cicada undata* Degeer, 1773) Of uncertain position
C. *kirkaldyi* Metcalf, 1963 = *Cicada obtusa* Uhler, 1903 (nec *Cicada obtusa* Fabricius, 1787) = *Cicada brasiliensis* Kirkaldy, 1909 (nec *Cicada brasiliensis* Gmelin, 1789)
C. *lactea* Gmelin, 1789 Of uncertain position
C. *leucophaea* Preissler, 1792 Of uncertain position

C. leucothoe Walker, 1852 Of uncertain position
C. lineola von Schrank, 1801
C. lineolata Gmelin, 1789
C. lodosi Boulard, 1979
C. longicornis Poda, 1761 Of uncertain position
C. lugubris Gmelin, 1789 Of uncertain position
C. lutea Gmelin, 1789 Of uncertain position
C. majuscula Germar, 1818 nom. nud. Of uncertain position
C. mannifica Forskal, 1775
C. marginata Sulzer, 1761 Of uncertain position
C. melanaria Germar, 1830 Of uncertain position
C. melanoptera Gmelin, 1789 = *Cicada marginata* Degeer, 1773 (nec *Cicada marginata* Sulzer, 1761) Of uncertain position
C. membranacea de Fourcroy, 1785 Of uncertain position
C. microcephala La Guillou, 1841 Of uncertain position
C. mordoganensis Boulard, 1979 = *Cicada oleae* (partim) = *Cicada sp. nr. orni* Quartau et al., 1998
C. muscilliformis Capanni, 1894
C. nervoptera de Fourcroy, 1785 Of uncertain position
C. nigropunctata Goeze, 1778
C. nodosa (Walker, 1850) = *Cicada nodosa* = *Diceropyga* (?) *nodosa* = *Diceropyga nodosa*
C. notata Scopoli, 1763 Of uncertain position
C. novella Metcalf, 1963 = *Cicada gracilis* Germar, 1830 (nec *Cicada gracilis* Schellenberg, 1802) Of uncertain position
C. oculata de Tigny, 1802 Of uncertain position
C. olivierana Metcalf, 1963 = *Cicada ferruginea* Olivier, 1790 (nec *Cicada ferruginea* Fabricius, 1787) Of uncertain position
C. olivieri Metcalf, 1963 = *Cicada guttata* Olivier, 1790 (nec *Cicada guttata* Foster, 1771) Of uncertain position
C. opercularis Olivier, 1790 = *Fidicina opercularis* = *Cicada tibicen opercularis* = *Cicada variegata opercularis*
C. orni Linnaeus, 1758 = *Tettigonia orni* = *Tibicen orni* = *Cicida* (sic) *orni* = *Cidada* (sic) *orni* = *Cicada ornia* (sic) = *Tetigia (Tettigia) orni* = *Cicada (Tettigia) orni* = *Teltigia* (sic) *orni* = *Testigia* (sic) *orni* = *Cicadetta orni* = *Tettigonia punctata* Fabricius, 1798 = *Tettigia orni punctata* = *Tettigonia orni* var. *b* Billberg, 1820 nom. nud. = *Macroprotopus oleae* Costa, 1877 = *Cicada pallida* var. *a* Petagna, 1787 = *Cicada orni lesbosiensis* Boulard 2000
C. osalbum De Villiers, 1789 Of uncertain position
C. palearctica Metcalf, 1963 = *Cicada hyalina* Gmelin, 1789 (nec *Cicada hyalina* Fabricius, 1775)
C. pallens Gmelin, 1789
C. pallescens Müller, 1776 Of uncertain position
C. pallida Goeze, 1778 = *Cicada pallida* var. *A* de Fourcroy, 1785 = *Cicada pallida* var. *B* de Fourcroy, 1785 = *Cicada pallida* var. *C* de Fourcroy, 1785 Of uncertain origin
C. pennata (Distant, 1881) = *Tettigia pennata*
C. permagna (Haupt, 1917) = *Tettigia permagna*
C. poae Capanni, 1849
C. prasina Eversmann, 1837 nom. nud. Of uncertain position
C. purpurescens Metcalf, 1963 = *Cicada coerulescens* Germar, 1830 (nec *Cicada coerulescens* Fabricius, 1803)
C. pusilla Müller, 1776 Of uncertain position
C. quadrinotata Gmelin, 1789 = *Cicada quadrimaculata* Müller, 1776 (nec *Cicada quadrimaculata* von Schrank, 1776) Of uncertain position
C. quadristrigata Gmelin, 1789 = *Cicada 4-striata* Of uncertain position
C. regalis de Fourcroy, 1785 Of uncertain position
C. regina Graber, 1876
C. rivulosa Gmelin, 1789 Of uncertain position
C. roselii Gravenhorst, 1807 Of uncertain position
C. rubiniae Capanni, 1894 nom. nud.
C. rubroelytrata Goeze, 1778 Of uncertain position
C. rufgescens Stephens, 1829 nom. nud. Of uncertain position

C. sahlbergi Stål, 1854 = *Cicada (Cicada) sahlbergi* Of uncertain position
C. scutata Goeze, 1778 Of uncertain position
C. sexpunctata Gmelin, 1789 Of uncertain position
C. sebai Metcalf, 1963 = *Cicada fusca* Goeze, 1778 (nec *Cicada fusca* Müller, 1776)
C. signata Billberg, 1820 nom. nud. Of uncertain position
C. smaragdina Capanni, 1894
C. squalida Müller, 1776 Of uncertain position
C. strigosa Gmelin, 1789 Of uncertain position
C. superba Billberg, 1820 nom. nud. Of uncertain position
C. thomasseti China, 1924
C. thalassina Germar, 1830 Of uncertain position
C. thlaspi Piller & Mitterpacher, 1783 Of uncertain position
C. torquata de Fourcroy, 1785 Of uncertain position
C. triangulum De Villiers, 1789 Of uncertain position
C. tristis de Fourcroy, 1785 Of uncertain position
C. turtoni Metcalf, 1963 = *Cicada 3-punctata* Turton, 1802 (nec *Cicada tripunctata* Fabricius, 1781)
C. variola Gravenhorst, 1807 Of uncertain position
C. venulosa Turton, 1802 Of uncertain position
C. vertumnus Stephens, 1829 nom. nud.
C. viridiflava McAtee, 1934 nom. nud. Of uncertain position
C. vitrata De Villiers, 1789 Of uncertain position

Cicadalna Boulard, 2006
C. takensis Boulard, 2006 = *Cicadalna takensis* mph. *suffusca* Boulard, 2006

Inthaxara Distant, 1913
I. flexa Lei & Li, 1996
I. olivacea Chen, 1940
I. rex Distant, 1913

Rustia Stål, 1866
R. dentivitta (Walker, 1862) = *Cicada dentivitta* = *Rustia pedunculata* Stål, 1866 = *Tibicen amussitata* Distant, 1892 = *Rustia dentivitta amussitata* = *Rustia dentivitta* var. *amussitata*
R. tigrina (Distant, 1888) = *Tibicen tigrinus*

Subtribe Oncotympanina Ishihara, 1961

Oncotympana Stål, 1870 = *Pomponia (Oncotympana)* = *Octotymana* (sic) = *Onctympana* (sic) = *Onchotympana* (sic) = *Acotympana* (sic) = *Dokuma* Distant, 1905
O. averta Lee, 2010
O. brevis Lee, 2010
O. grandis Lee, 2010
O. nigristigma (Walker, 1850) = *Dundubia nigrostigma* = *Pomponia nigrostigma* = *Dokuma nigristigma*
O. pallidiventris (Stål, 1870) = *Pomponia (Oncotympana) pallidiventris* = *Pomponia pallidiventris* = *Pomponia (Oncotympana) pallidiveniris* (sic)
O. simonae Lee, 2010
O. sp. Lee, 2010 = *Oncotympana viridicincta* Endo & Hayashi, 1979
O. undata Lee, 2010
O. viridicincta (Stål, 1870) = *Onchotympana* (sic) *viridicincta* = *Pomponia (Oncotympana) viridicincta* = *Pomponia viridicincta* = *Dokuma consobrina* Distant, 1906

Neoncotympana Lee, 2010
N. leeseungmoi Lee, 2010

Subtribe Psithyristriina Distant, 1905 = Psithyristriaria = Pomponiaria Kato, 1932 = Pomponariaria (sic) = Terpnosiina Kato, 1932 = Terpnosiaria

Psithyristria Stål, 1870
- *P. albiterminalis* Lee & Hill, 2010
- *P. crassinervis* Stål, 1870
- *P. genesis* Lee & Hill, 2010
- *P. grandis* Lee & Hill, 2010
- *P. incredibilis* Lee & Hill, 2010
- *P. moderabilis* Lee & Hill, 2010
- *P. nodinervis* Stål, 1870
- *P. paracrassis* Lee, 2010
- *P. paraspecularis* Lee & Hill, 2010
- *P. paratenuis* Lee, 2010
- *P. peculiaris* Lee & Hill, 2010
- *P. simplicinervis* Stål, 1870
- *P. specularis* Stål, 1870
- *P. tenuinervis* Stål, 1870

Basa Distant, 1905
- *B. singularis* (Walker, 1858) = *Dundubia singularis* = *Pomponia singularis*

Pomponia Stål, 1866 = *Pomponia (Pomponia)* = *Pomonia* (sic)
- *P. adusta* (Walker, 1850) = *Cicada adusta* = *Cicada buddha* Kirkaldy, 1909 = *Pomponia budda*
- *P. backanensis* Pham & Yang, 2009 = Pham 2004 drawing = *P.* sp. Lee 2008
- *P. bullata* Schmidt, 1924
- *P. bulu* Zaidi & Azman, 2000 = *Pomponia* sp. Zaidi, Ruslan & Azman 1999
- *P. cinctimanus* (Walker, 1850) = *Dundubia cinctimanus* = *Dundubia linearis cinctimanus*
- *P. cyanea* Fraser, 1948
- *P. daklakensis* Sanborn, 2009
- *P. decem* (Walker, 1857) = *Dundubia decem* = *Megapomponia decem* = *Pomponia diffusa* Breddin, 1900
- *P. dolosa* Boulard, 2001 = *Pomponia fusca* (nec Olivier) = *Cicada dolosa*
- *P. fuscoides* Boulard, 2002
- *P. gemella* Boulard, 2005
- *P. kiushiuensis* Kato, 1925 = *Pomponia kiusiuensis* (sic) = *Pomponia huishiuensis* (sic)
- *P. langkawiensis* Zaidi & Azman, 1999
- *P. linearis* (Walker, 1850) = *Dundubia linearis* = *Pomponia fusca* (nec Olivier) = *Pomponia fusca fusca* = *Dundubia linealis* (sic) = *Pomponia linealis* (sic) = *Pomonia* (sic) *fusca*
- *P.* nr. *linearis* n. sp. Zaidi & Ruslan, 1995
- *P. mickwanae* Boulard, 2009 = *Pomponia lactea* (nec Distant)
- *P. namtokola* Boulard, 2005
- *P. nasanensis* Boulard, 2005
- *P. noualhieri* Boulard, 2005
- *P. orientalis* (Distant, 1912) = *Tettigia orientalis* = *Cicada orientalis* = *Pomponia occidentalis* (sic)
- *P. parafusca* Kato, 1944
- *P. piceata* Distant, 1905 = *Pomponia hierogryphica* (sic) Kato, 1938 = *Pomponia hieroglyphica*
- *P. picta* (Walker, 1868) = *Dundubia picta* = *Cicada fusca* Olivier, 1790 (nec *Cicada fusca* Müller, 1776) = *Dundubia fusca* = *Pomponia (Pomponia) fusca* = *Pomponia fusca fusca* = *Pomponia linearis* (nec Walker) = *Pomponia fusca* (nec Olivier)
- *P. ponderosa* Lee, 2009 = *Pomponia linearis* (nec Walker) = *Pomponia fusca* (nec Olivier)
- *P. promiscua* Distant, 1887
- *P. quadrispinae* Boulard, 2002
- *P. rajah* Moulton, 1923 = *Megapomponia rajah*
- *P. ramifera* (Walker, 1850) = *Dundubia ramifera* = *Dundubia linearis ramifera*
- *P. secreta* Hayashi, 1978
- *P. siamensis* China, 1925 nom. nud.
- *P. solitaria* Distant, 1888
- *P.* sp. 1 Kahono, 1993

P. sp. 2 Kahono, 1993
P. subtilita Lee, 2009 = *Pomponia linearis* (nec Walker) = *Pomponia fusca* (nec Olivier)
P. surya Distant, 1904
P. undescribed species A Sanborn, Phillips & Sites, 2007
P. undescribed species B Sanborn, Phillips & Sites, 2007
P. undescribed species C Sanborn, Phillips & Sites, 2007
P. urania (Walker, 1850) = *Dundubia urania*
P. yayeyamana Kato, 1933 = *Pomponia fusca yayeyamana* = *Pomponia tuba* Lee, 2009 = *Pomponia linearis* (nec Walker, 1850) = *Pomponia fusca* (nec Olivier)
P. zakrii Zaidi & Azman, 1998
P. zebra Bliven, 1964

Terpnosia Distant, 1892 = *Terprosia* (sic) = *Ternopsia* (sic) = *Trepnosia* (sic) = *Ierpnosia* (sic)
T. abdullah Distant, 1904 = *Terpnosia adbulla* (sic) = *Terpnosia abdullahi* (sic)
T. andersoni Distant, 1892
T. nr. *andersoni* Zaidi & Ruslan, 1995
T. chapana Distant, 1917
T. clio (Walker, 1850) = *Dundubia clio*
T. collina (Distant, 1888) = *Pomponia collina*
T. confusa Distant, 1905 = *Terpnosia psecas* (nec Walker)
T. elegans (Kirby, 1891) = *Pomponia elegans*
T. ganesa Distant, 1904
T. graecina (Distant, 1889) = *Pomponia graecina* = *Pomponia graecinea* (sic)
T. jenkinsi Distant, 1912
T. jinpingensis Lei & Chou, 1997
T. lactea (Distant, 1887) = *Pomponia lactea* = *Pomponia lacteal* (sic) = *Leptopsaltria lactea*
T. maculipes (Walker, 1850) = *Dundubia maculipes*
T. mawi Distant, 1909 = *Terpnosia nawi* (sic)
T. mega Chou & Lei, 1997
T. mesonotalis Distant, 1917 = *Calcagninus salvazanus* Distant, 1917
T. neocollina Liu, 1940
T. nigella Chou & Lei, 1997
T. nonusaprilis Boulard, 2003
T. oberthuri Distant, 1912
T. obscurana Metcalf, 1955 = *Terpnosia obscura* Liu, 1940 (nec *Terpnosia obscura* Kato, 1938)
T. polei (Henry, 1931) = *Pomponia polei*
T. posidonia Jacobi, 1902 = *Terponia* (sic) *posidonia* = *Cicada stipata* Walker, 1850 (nec *Dundubia stipata* Walker, 1850)
T. psecas (Walker, 1850) = *Dundubia psecas* = *Cicada psecas* = *Yezoterpnosia psecas* = *Ternopsia* (sic) *psecas*
T. pumila (Distant, 1891) = *Pomponia pumila*
T. puriticis Lei & Chou, 1997
T. ransonneti (Distant, 1888) = *Pomponia ransonneti* = *Euterpnosia ransonneti* = *Pomponia ransonetti* (sic) = *Terpnosia ransonetti* (sic) = *Pomponia greeni* Kirby, 1891
T. renschi Jacobi, 1941
T. ridens Pringle, 1955
T. rustica Distant, 1917
T. similis obsoleta (China, 1926) = *Pomponia similis obsoleta*
T. similis similis Schmidt, 1924 = *Pomponia similis* = *Pomponia picta* (nec Walker)
T. simusa (Boulard, 2008) = *Pomponia simusa*
T. n. sp. Zaidi & Ruslan, 1995
T. sp. 1 Zaidi, Ruslan & Mahadir, 1996
T. sp. 1 Pham & Thinh, 2006
T. sp. 2 Zaidi, Ruslan & Mahadir, 1996

T. stipata (Walker, 1950) = *Dundubia stipata* = *Ternopsia* (sic) *stipata* = *Dundubia clonia* Walker, 1850 = *Dundubia chlonia* (sic)
T. translucida (Distant, 1891) = *Pomponia translucida*
T. versicolor Distant, 1912
T. viridissima Boulard, 2005

Semia Matsumura, 1917
S. klapperichi Jacobi, 1944
S. lachna (Lei & Chou, 1997) = *Pomponia lachna*
S. majuscula (Distant, 1917) = *Terpnosia majuscula* = *Terpnosia majuscule* (sic)
S. spinosa Pham, Hayashi & Yang, 2012
S. watanabei (Matsumura, 1907) = *Leptopsaltria watanabei* = *Pomponia watanabei*

Subtribe Minipomponiina Boulard, 2008

Minipomponia Boulard, 2008
M. fuscacuminis (Boulard, 2005) = *Pomponia fuscacuminis*
M. littldollae (Boulard, 2002) = *Pomponia littldollae*

Subtribe Leptopsaltriina Moulton, 1923 = Leptopsaltriaria = Terpnosiaria

Yezoterpnosia Matsumura, 1917 = *Yzoterpnosia* (sic)
Y. fuscoapicalis (Kato, 1938) = *Terpnosia fuscoapicalis*
Y. ichangensis (Liu, 1940) = *Terpnosia ichangensis* = *Terpnosia ichangens* (sic)
Y. nigricosta (de Motschulsky, 1866) = *Cicada nigricosta* = *Terpnosia nigricosta* = *Terpnosia nigrocosta* (sic) = *Yezoterpnosia nigricosta* = *Yezoterpnosia sapporensis* Kato, 1925 = *Terpnosia sapporensis* = *Terpnosia nigricosta sapporensis* = *Terpnosia obscura* (nec Kato)
Y. obscura Kato, 1938 = *Terpnosia obscura* = *Terpnosia obscula* (sic) = *Terpnosia obscurana* (sic)
Y. shaanxiensis (Sanborn, 2006) = *Terpnosia shaanxiensis*
Y. vacua (Olivier, 1790) = *Cicada vacua* = *Terpnosia vacua* = *Terpnosia vacau* (sic) = *Trepnosia* (sic) *vacuna* (sic) = *Terpnosia vacuna* (sic) = *Terpnosia vacva* (sic) = *Terpnosia nacua* (sic) = *Ierpnosia* (sic) *vacua* = *Cicada clara* de Motschulsky, 1866 = *Terpnosia pryeri* Distant, 1892 = *Terpnosia kawamurae* Matsumura, 1913 = *Terpnosia kuramensis* Kato, 1925 = *Terpnosia knramensis* (sic) = *Terpnosia vacua kuramensis* = *Terpnosia vacua nigra* Kato, 1927 = *Terpnosia vacua* var. *nigra* = *Terpnosia vacuna* (sic) *nigra*

Euterpnosia Matsumura, 1917 = *Euterponosia* (sic) = *Futerpnosia* (sic) = *Euternosia* (sic) = *Terpnosia (Euterpnosia)*
E. alipina Chen, 2005
E. ampla Chen, 2006
E. arisana Kato, 1925 = *Euterpnosia arisanus* (sic)
E. chibensis chibensis Matsumura, 1917 = *Leptopsaltria tuberosa* (nec Signoret)
E. chibensis daitoensis Matsumura, 1917 = *Euterpnosia chibensis* var. *daitoensis* = *Euterpnosia chibensis* var. *gotoi* Kato, 1936 = *Euterpnosia chibensis gotoi* = *Leptopsaltria tuberosa* (nec Signoret)
E. chibensis nuaiensis Boulard, 2005
E. chinensis Kato, 1940
E. chishingsana Chen & Shiao, 2008
E. crowfooti (Distant, 1912) = *Terpnosia crowfooti*
E. cucphuongensis Pham, Ta & Yang, 2010
E. gina Kato, 1931
E. hohogura Kato, 1933
E. hoppo Matsumura, 1917 = *Euterpnosia sozanensis* Kato, 1925
E. iwasakii Matsumura, 1913 = *Purana iwasakii* = *Euternosia* (sic) *iwasakii* = *Euterpnosia iwasaku* (sic)
E. koshunensis Kato, 1927
E. kotoshoensis Kato, 1925
E. laii Lee, 2003 = *Euterpnosia viridifrons* var. *a* Kato, 1927 (partim) = *Euterpnosia viridifrons* f. *a* (partim)
E. latacauta Chen & Shiao, 2008

E. madhava (Distant, 1881) = *Pomponia madhava* = *Terpnosia madhava*
E. madawdawensis Chen, 2005
E. okinawana Ishihara, 1968 = *Euterpnosia chibensis okinawana*
E. olivacea Kato, 1927
E. ruida Lei & Chou, 1997 = *Euterpnosia* sp. 1 Pham & Thinh, 2006
E. sinensis hwaisana Chen, 1940 = *Euterpnosia sinensis hwasiana* (sic)
E. sinensis sinesnis Chen, 1940
E. suishana Kato, 1931
E. varicolor Kato, 1926 = *Euterpnosia elongata* Lee, 2003
E. viridifrons Matsumura, 1917 = *Euterpnosia viridifrous* (sic) = *Euternosia* (sic) *viridifrons* = *Euterpnosia viridifrons* var. *a* Kato, 1927 (partim) = *Euterpnosia viridifrons* f. *a* (partim)

Leptosemia Matsumura. 1917 = *Leptosomia* (sic) = *Leptopsalta* (sic) = *Chosenosemia* Doi, 1931
L. huasipana Chen, 1943
L. sakaii (Matsumura, 1913) = *Leptopsaltria sakaii* = *Cicada sakaii* = *Terpnosia fuscolimbata* Schumacher, 1915 = *Terpnosia (Euterpnosia) fuscolimbata* = *Euterpnosia fuscolimbata* = *Leptosemia fuscolimbata* = *Leptosemia conica* Kato, 1926
L. takanonis Matsumura, 1917 = *Cicada takanonis* = *Leptosomia* (sic) *takanonis* = *Leptosemia takanois* (sic) = *Euterpnosia inanulata* Kishida, 1929 = *Leptosemia sakaii* (nec Matsumura) = *Chosenosemia souyoensis* Doi, 1931 = *Leptosemia souyoensis* = *Tana* (sic) *japonensis* (nec Distant)
L. umbrosa Ku. In Wagner, 1968

Onomacritus Distant, 1912
O. sumatranus Distant, 1912

Neocicada Kato, 1932
N. australamexicana Sanborn & Sueur, 2005 = *Neocicada* sp. Sueur 2002 = *Tettigia hieroglyphica* (nec Say)
N. centramericana Sanborn, 2005
N. chisos (Davis, 1916) = *Cicada chisos* = *Tettigia hieroglyphica* (nec Say) = *Neocicada hieroglyphica* (nec Say)
N. hieroglyphica hieroglyphica (Say, 1830) = *Cicada hieroglyphica* = *Tettigia hieroglyphica* = *Tettigea* (sic) *hieroglyphica* = *Tibicen hieroglyphica* = *Cicada hieroblyphica* (sic) = *Cicada hierglyphica* (sic) = *Neocicada hieroglyphica* = *Cicada characterea* Germar, 1830 = *Cicada characteria* (sic) = *Tettigia characteria* (sic)
N. hieroglyphica johannis (Walker, 1850) = *Cicada johannis* = *Tettigia johannis* = *Cicada hieroglyphica johannis* = *Cicada sexguttata* Walker, 1850 = *Tettigia sexguttata*
N. mediamexicana Sanborn, 2005

Puranoides Moulton, 1917
P. abdullahi Azman & Zaidi, 2002
P. goemansi Lee, 2009
P. jaafari Azman & Zaidi, 2002
P. klossi Moulton, 1917
P. cf. *klossi* Schouten, Duffels & Zaidi, 2004

Leptopsaltria Stål, 1866 = *Leptosaltria* (sic)
L. andamanensis Distant, 1888
L. draluobi Boulard, 2003
L. jaesornensis Boulard, 2008 = *Leptopsaltria jaesornensis* Boulard, 2009
L. mickwanae Boulard, 2007
L. phra Distant, 1913
L. samia (Walker, 1850) = *Dundubia samia* = *Pomponia samia* = *Formosemia samia*
L. tuberosa (Signoret, 1847) = *Cicada tuberosa* = *Dundubia tuberosa* = *Leptosaltria* (sic) *tuberosa* = *Leptopsaltria toberosa* (sic) = *Leptopsaltria tuberose* (sic)

Tanna Distant, 1905 = *Tann* (sic) = *Tana* (sic)
T. abdominalis Kato, 1938 = *Neotanna abdominalis* = *Neotanna sinensis* Ouchi, 1938
T. apicalis Chen, 1940 = *Tana* (sic) *apicalis*

T. auripennis Kato, 1930
T. bakeri Moulton, 1923 = *Tana* (sic) *bakeri*
T. bhutanensis Distant, 1912 = *Neotanna bhutanensis*
T. chekiangensis Ouchi, 1938 = *Tanna sinensis* Kato, 1938 = *Tanna japonensis* (nec Distant)
T. harpesi Lallemand & Synave, 1953
T. herzbergi Schmidt, 1932 = *Tanna herrbergi* (sic)
T. infuscata Lee & Hayashi, 2004 = *Tanna infuscate* (sic)
T. insignis Distant, 1906
T. ishigakiana Kato, 1960 = *Tanna japonensis ishigakiana*
T. japonensis japonensis (Distant, 1892) = *Pomponia japonensis* = *Tanna japonensie* (sic) = *Tanna nipponensis* (sic) Matsumura = *Tanna japonnensis* (sic) = *Tanna japonesis* (sic) = *Tanna joponensis* (sic) = *Leptopsaltria japonica* Horváth, 1892 = *Tanna* (?) *sasaii* Matsumura, 1939 = *Tanna japonensis kimotoi* Kato, 1943 = *Tanna japonensis nigrofusca* Kato, 1940 = *Tanna obliqua* Liu, 1940
T. japonensis var. ______ Ishihara, 1939
T. karenkonis Kato, 1939
T. kimtaewooi Lee, 2010
T. minor Hayashi, 1978
T. ornata Kato, 1940
T. ornatipennis Esaki, 1933 = *Tanna (Neotanna) ornatipennis* = *Neotanna ornatipennis*
T. pallida Distant, 1906
T. pseudocalis Lei & Chou, 1997
T. puranoides Boulard, 2008
T. sayurie Kato, 1926
T. sinensis (Ouchi, 1938) = *Neotanna sinensis*
T. sozanensis Kato, 1926
T. sp. 1 Pham & Thinh, 2006
T. taipinensis (Matsumura, 1907) = *Leptopsaltria taipinensis* = *Tanna tapinensis* (sic) = *Tanna taikosana* Kato, 1925 = *Tanna horiensis* Kato, 1926
T. tairikuana Kato, 1940
T. tigroides (Walker, 1858) = *Dundubia tigroides* = *Pomponia tigroides* = *Leptopsaltria tigroides* = *Purana tigroides* = *Neotanna tigroides*
T. undescribed species A Sanborn, Phillips & Sites, 2007
T. ventriroseus Boulard, 2003
T. viridis Kato, 1925 = *Neotanna viridis* = *Neottanna* (sic) *viridis* = *Tanna horishana* Kato, 1925 = *Neotanna horishana* = *Neotanna horishaua* (sic) = *Euterpnosia horishana* = *Tanna viridis niitakaensis* Kato, 1926 = *Neotanna viridis niitakaensis* = *Neotanna tarowanensis* Matsumura, 1927 = *Neotanna rantaizana* Matsumura 1927 = *Neotanna viridis inmaculata* Kato, 1932 = *Tanna viridis immaculata* (sic)
T. yanni Boulard, 2003
T. yunnanensis (Lei & Chou, 1997) = *Neotanna yunnanensis*

Paratanna Lee, 2012
P. parata Lee, 2012

Formocicada Lee & Hayashi, 2004
F. taiwana Lee & Hayashi, 2004

Formosemia Matsumura, 1917 = *Formosia* (sic)
F. apicalis (Matsumura, 1907) = *Leptopsaltria apicalis* = *Purana apicalis* = *Formosemia apicalia* (sic)

Maua Distant, 1905 = *Mau* (sic)
M. affinis Distant, 1905 = *Leptopsaltria quadrituberculata* (nec Signoret) = *Maua quadrituberculata* (nec Signoret)
M. albigutta (Walker, 1857) = *Dundubia albigutta* = *Dundubia albiguttata* (sic) = *Leptopsaltria albigutta* = *Leptopsaltria albiguttata* (sic) = *Purana albigutta* = *Purana albiguttata* (sic) = *Maua albiguttata* (sic) = *Mau* (sic) *albigutta* = *Naua* (sic) *albigutta*
M. albistigma (Walker, 1850) = *Dundubia albistigma* = *Leptopsaltria albistigma*

M. borneensis Duffels, 2009 = *Maua affinis* (nec Distant)
M. fukienensis Liu, 1940 = *Maua quikiensis* (sic) = *Purana fukiensis* (sic)
M. latilinea (Walker, 1868) = *Dundubia latilinea* = *Cosmopsaltria latilinea* = *Cosmopsatria duarum laticincta* (sic) = *Maua dohrni* Schmidt, 1912
M. linggana Moulton, 1923
M. palawanensis Duffels, 2009 = *Maua affinis* (nec Distant)
M. philippinensis Schmidt, 1924
M. platygaster Ashton, 1912
M. quadrituberculata quadrituberculata (Signoret, 1847) = *Cicada quadrituberculata* = *Cicada quadrimaculata* (sic) = *Dundubia quadrituberculata* = *Leptopsaltria quadrituberculata* = *Maua 4-tuberculata* = *Maua quadrimaculata* (sic) = *Maua tuberculata* (sic) = *Maua ackermanni* Schmidt, 1924
M. quadrituberculata tavoyana (Ollenbach, 1929) = *Maua tavoyana* = *Purana tavoyana* = *Maua quadrituberculata khunpaworensis* Boulard, 2007 = *Maua quadrituberculata kpaworensis* (sic) = *Maua 4-drituberculata* (sic) *kpaworensis*

Purana Distant, 1905 = *Paruna* (sic)
P. abdominalis Lee, 2009
P. atroclunes Boulard, 2002
P. australis (Kato, 1944) = *Formosemia australis*
P. barbosae (Distant, 1889) = *Leptopsaltria barbosae* = *Purana carmente barbosae*
P. campanula Pringle, 1955
P. capricornis Kos & Gogala, 2000
P. carmente (Walker, 1850) = *Dundubia carmente* = *Leptopsaltria carmente* = *Chremistica* (sic) *carmente* = *Purana carmente carmente* = *Leptopsaltria nigrescens* Distant, 1889
P. carolettae Esaki, 1936
P. celebensis (Breddin, 1901) = *Leptopsaltria celebensis*
P. chueatae Boulard, 2007
P. clavohyalina Liu, 1939
P. conifacies (Walker, 1858) = *Cicada conifacies* = *Dundubia conifacies*
P. conspicua Distant, 1910
P. davidi Distant, 1905 = *Purana davini* (sic) = *Purana dividi* (sic) = *Purana davidii* (sic) = *Leptopsaltria davidi*
P. dimidia Chou & Lei, 1997
P. doiluangensis Boulard, 2005
P. gemella Boulard, 2007
P. gigas (Kato, 1930) = *Purana clavohyalina* Liu, 1940 = *Formosemia gigas* = *Formosia* (sic) *gigas*
P. guttularis (Walker, 1858) = *Cicada guttularis* = *Leptopsaltria guttularis* = *Purana gutturalis* (sic) = *Purana* sp. 1 Pham & Thinh, 2006
P. nr. *guttularis* Zaidi, Ruslan & Mahadir, 1996
P. hermes Schouten & Duffels, 2002
P. hirundo (Walker, 1850) = *Cicada hirundo*
P. infuscata Schouten & Duffels, 2002
P. jacobsoni Distant, 1912
P. jdmoorei Boulard, 2005 = *Purana jdmoorrei* (sic)
P. johanae Boulard, 2005
P. karimunjawa Duffels & Schouten, 2007
P. khaosokensis Boulard, 2007
P. khuanae Boulard, 2002
P. khuanae Boulard, 2002 = *Purana khunpaworensis* (sic)
P. kpaworensis Boulard, 2007
P. latifascia Duffels & Schouten, 2007
P. metallica Duffels & Schouten, 2007 = *Purana* aff. *tigrina* Gogala, 1995 = *Purana langkawi* (sic)
P. mickhuanae Boulard, 2009
P. montana Kos & Gogala, 2000
P. morrisi (Distant, 1892) = *Leptopsaltria morrisi*
P. mulu Duffels & Schouten, 2007
P. nana Lee, 2009

P. natae Boulard, 2007
P. nebulilinea (Walker, 1868) = *Dundubia nebulilinea* = *Dundubia nebulinea* (sic) = *Leptopsaltria nebulilinea* = *Leptopsaltria nebulinea* (sic) = *Purana tigrina* (nec Walker) = *Purana pryeri* (nec Distant)
P. nr. *nebulilinea* n. sp. Zaidi & Ruslan, 1995
P. niasica Kos & Gogala, 2000
P. notatissima Jacobi, 1944
P. obducta Schouten & Duffels, 2002
P. opaca Lee, 2009
P. parvituberculata Kos & Gogala, 2000
P. phetchabuna Boulard, 2008 = *Purana phechabuna* (sic)
P. pigmentata Distant, 1905
P. pryeri (Distant, 1881) = *Leptopsaltria pryeri* = *Paruna* (sic) *pryeri*
P. ptorti Boulard, 2007
P. sagittata Schouten & Duffels, 2002
P. sp. Lee, 2010
P. taipinensis (Kato, 1944) = *Formosemia taipinensis*
P. tanae Boulard, 2007
P. tigrina (Walker, 1850) = *Dundubia tigrina* = *Leptopsaltria tigrina* = *Purana tigrina mjöbergi* Moulton, 1923
P. cf. *tigrina* Schouten, Duffels & Zaidi, 2004
P. tigrinaformis Boulard, 2007 = *Purana tigrina* (nec Walker)
P. tripunctata Moulton, 1923
P. trui Pham, Schouten & Yang, 2012
P. ubina Moulton, 1923 = *Purana tigrina* (nec Walker)
P. nr. *ubina* Zaidi, Ruslan & Mahadir, 1996
P. undescribed species A Sanborn, Phillips & Sites, 2007
P. usnani Duffels & Schouten, 2007
P. vesperalba Boulard, 2009
P. vindevogheli Boulard, 2008

Kamalata Distant, 1889
K. javanensis Distant, 1905
K. pantherina Distant, 1889 = *Kamalata pantheriana* (sic)

Calcagninus Distant, 1892
C. divaricatus Bliven, 1964
C. nilgirensis (Distant, 1887) = *Leptopsaltria nilgirensis* = *Calcagninus nilgiriensis* (sic)
C. picturatus (Distant, 1888) = *Leptopsaltria picturata* = *Calcagninus pictulatus* (sic)

Qurana Lee, 2009
Q. ggoma Lee, 2009

Nabalua Moulton, 1923 = *Nabula* (sic)
N. borneensis Duffels, 2004 = *Nabula* (sic) *mascula* (nec Distant)
N. maculata Duffels, 2004
N. mascula (Distant, 1889) = *Nabula* (sic) *mascula* = *Leptopsaltria mascula*
N. nr. *mascula* sp. n. Zaidi & Ruslan, 1995
N. neglecta Moulton, 1923 = *Leptopsaltria neglecta*
N. sumatrana Duffels, 2004
N. zaidii Duffels, 2004 = *Nabalua mascula* Zaidi, Ruslan & Mahabir, 1996

Neotanna Kato, 1927 = *Tanna (Neotanna)*
N. conyla Chou & Lei, 1997
N. shensiensis Sanborn, 2006
N. simultaneous Chen, 1940

N. thalia (Walker, 1850) = *Dundubia thalia* = *Pomponia thalia* = *Neotanna thalia* = *Tanna thalia* = *Dubia* (sic) *thalia* = *Pomponia thaila* (sic) = *Cicada sphinx* Walker, 1850 = *Puranoides sphinx* = *Pomponia horsfieldi* Distant, 1893

Taiwanosemia Matsumura, 1917 = *Higurasia* Kato, 1925
T. hoppoensis (Matsumura, 1907) = *Leptopsaltria hoppoensis* = *Tanna hoppoensis* = *Purana hoppoensis* = *Terpnosia bicolor* Kato, 1925 = *Higurasia bicolor*

Subtribe Gudabina Boulard, 2008

Gudaba Distant, 1906
G. apicata Distant, 1906
G. longicauda Lei, 1996
G. maculata Distant, 1912
G. marginatus (Distant, 1897) = *Calcagninus marginatus*

Tribe Cicadatrini Distant, 1905 = Cicadatraria = Moganniini Distant, 1905 = Moganniaria = Moganini (sic) = Mogannini (sic) = Moganiini (sic)

Emathia Stål, 1866
E. aegrota Stål, 1866 = *Tibicen aurengzebe* Distant, 1881

Vagitanus Distant, 1918
V. luangensis Distant, 1918
V. vientianensis Distant, 1918

Triglena Fieber, 1875
T. virescens Fieber, 1876

Taungia Ollenbach, 1929
T. abnormis Ollenbach, 1929
T. albantennae Boulard, 2009 = *Purana* (sic) *albantennae*

Cicadatra Kolenati, 1857 = *Cicada (Cicadatra)* = *Tettigia (Cicadatra)* = *Cicadrata* (sic) = *Cicadatra (Rustavelia)* Horváth, 1912 = *Rustavelia*
C. abdominalis Schumacher, 1923
C. acberi (Distant, 1888) = *Tibicen acberi* = *Sena acberi* = *Psalmocharias acberi*
C. adanai Kartal, 1980 = *Cicadatra adanae* (sic)
C. alhageos (Kolenati, 1857) = *Cicada (Cicadatra) atra alhageos* = *Cicada alhageos* = *Cicadatra atra alhageos* = *Cicadatra alhageos alhageos* = *Cicadatra algheos* (sic) = *Cicadatra olivacea* Melichar, 1913 = *Cicadatra alhageos olivacea* = *Cicadatra stenoptera* Haupt, 1917 = *Cicadatra alhageos stenoptera*
C. anoea (Walker, 1850) = *Cicada anoea* = *Cicada anae* (sic)
C. appendiculata Linnavuori, 1954
C. atra atra (Olivier, 1790) = *Cicada atra* = *Cicadatra atra* = *Tettigia (Cicadatra) atra* = *Tettigia atra* = *Cicadrata* (sic) *atra* = *Cicada arfa* (sic) = *Cicada arta* (sic) *Cicada concinna* (nec Germar) = *Cicadatra atra concinna* = *Cicadatra concinna* = *Cicada transversa* Germar, 1830 = *Cicadatra transversa* = *Cicadatra atra aquila* Fieber, 1876 = *Cicadatra atra pallipes* Fieber, 1876 = *Cicadatra atra tau* Fieber, 1876
C. atra hyalinata (Brullé, 1832) = *Cicadatra hyalinata* = *Tibicen hyalinatus* = *Cicadatra atra* var. *hyalinata*
C. atra putoni Boulard, 1992 = *Cicadatra atra* var. *putoni*
C. atra sufflava Boulard, 1992 = *Cicadatra atra* var. *sufflava*
C. atra vitrea (Brullé, 1832) = *Tibicen vitreus* = *Cicada vitreus* = *Cicadatra atra* var. *vitrea*
C. bistunensis Mozaffarian & Sanborn, 2010
C. erkowitensis Linnavuori, 1973
C. flavicollis Horváth, 1911 = *Psalmocharias flavicollis* = *Cicadatra foveicollis* (sic)
C. genoina Dlabola, 1979

C. gingat China, 1926

C. glycyrrhizae (Kolenati, 1857) = *Cicada (Cicadatra) glycyrrhizae* = *Cicada glycyrrhizae* = *Cicadatra glycyrrhizae* = *Cicadatra atra glycyrrhizae* = *Cicadatra glacyrrhiza* (sic) = *Cicadatra glycyrrhiza* (sic) = *Cicadatra glycirrhizae* (sic) = *Cicadatra glycyrrhizicae* (sic) = *Cicadatra atra glycirrhizae* (sic)

C. gregoryi China, 1925

C. hagenica Dlabola, 1987 = nom. nov. pro *Cicada virens* Hagen, 1856 nec *Cicada virens* Herrich Schäffer, 1835 = *Cicadatra hyalina virens* = *Cicadatra virens*

C. hyalina (Fabricius, 1798) = *Tettigonia hyalina* = *Cicada hyalina* = *Tibicen hyalina* = *Cicada (Cicadatra) hyalina* = *Tettigia (Cicadatra) hyalina* = *Tettigonia hyalina* = *Cicadetta hyalina* = *Cicadatra hyaline* (sic) = *Cicadatra (Rustavelia) burriana* Horváth, 1912 = *Cicadatra burriana* = *Rustavelia burriana* = *Cicada (Cicadatra) geodesma* Kolenati, 1857 = *Cicada geodesma* = *Cicadatra geodesma* = *Cicadatra goedesma* (sic) = *Cicadatra hyalina geodesma* = *Cicada (Cicadrata)* (sic) *hyalina geodesma* = *Cicadatra geodesma geodesma* = *Cicadatra geodesma discrepans* Schumacher, 1923 = *Cicadatra hyalina discrepens* = *Cicadatra geodesma rossica* Schumacher, 1923 = *Cicadatra hyalina rossica* = *Cicadatra taurica* Fieber, 1876 = *Cicadatra hyalina taurica* = *Cicadatra geodesma taurica* = *Cicadatra hyalina virens* Fieber, 1876 = *Cicada virens* = *Cicadatra hyalina vireus* (sic) = *Cicadatra viridis* Haupt, 1917 = *Cicadetta coriaria* (nec Puton, 1875)

C. icari Simões, Sanborn & Quartau, 2012

C. inconspicua Distant, 1912

C. karachiensis Ahmed, Sanborn & Hill, 2010

C. karpathosensis Simões, Sanborn & Quartau, 2012

C. kermanica Dlabola, 1970

C. longipennis Schumacher, 1923

C. loristanica Mozaffarian & Sanborn, 2010

C. mirzayansi Dlabola, 1981 = *Cicadatra mirzyansi* (sic)

C. naja Dlabola, 1979 = *Cicadetta* (sic) *naja*

C. pallisi Schumacher, 1923

C. persica Kirkaldy, 1909 = *Cicadatra lineola* Hagen, 1856 (nec *Cicada lineola* von Schrank, 1801) = *Tettigia (Cicadatra) lineola* = *Cicadatra lineola*

C. platyptera Feiber, 1876 = *Cicadatra atra platyptera* = *Cicadatra decumana* Schumacher, 1923 = *Cicadatra hyalina decumana* = *Cicadatra platypleura livida* Schumacher, 1923 = *Cicadatra livida* = *Cicadatra platyptera livens* (sic) = *Cicadatra platyptera melanaria* Schumacher, 1923

C. raja Distant, 1906

C. ramanensis Linnavouri, 1962

C. reinhardi Kartal, 1980

C. sankara (Distant, 1904) = *Tibicen sankara* = *Cicadatra sankana* (sic)

C. shaluensis China, 1925

C. shapur Dlabola, 1981

C. tenebrosa Fieber, 1876

C. vulcania Dlabola & Heller, 1962

C. walkeri Metcalf, 1963 = *Cicada striata* Walker, 1850 (nec *Cicada striata* Linnaeus, 1758) = *Cicadatra striata*

C. xantes (Walker, 1850) = *Cicada xantes* = *Cicadatra xanthes* (sic) = *Cicada subvenosa* Walker, 1858

C. zahedanica Dlabola, 1970

C. ziaratica Ahmed, Sanborn & Akhter, 2012

Klapperichicen Dlabola, 1957 = *Klapperischicen* (sic) = *Klapparichicen* (sic)

K. acoloratus Dlabola, 1960 = *Klapperichicen accoloratus* (sic) = *Klapperichicen* (sic) *accoloratus* (sic)

K. dubius (Jacobi, 1927) = *Cicadatra dubius*

K. turbatus (Melichar, 1902) = *Klapperichicen* (sic) *turbatus* = *Tibicen turbatus* = *Tibicina turbata*

K. viridissimus (Walker, 1858) = *Cicada viridissima* = *Cicadatra viridissima* = *Chloropsalta viridissima* = *Klapperischen viridissima* (sic) = *Klapperichien* (sic) *viridissima*

Psalmocharias Kirkaldy, 1908 = *Cicadatra (Psalmocharias)* = *Psalmochavias* (sic) = *Sena* Distant, 1905

P. balochii Ahmed & Sanborn, 2010

P. cataphractica (Popov, 1989) = *Cicadatra cataphractica* = *Cicadatra ex.* gr. *querula* Dlabola, 1961

P. chitralensis Ahmed & Sanborn, 2010

P. flava Dlabola, 1970

P. gizarensis Ahmed & Sanborn, 2010
P. japokensis Ahmed & Sanborn, 2010
P. paliur (Kolenati, 1857) = *Cicada (Cicadatra) querula paliuri* = *Cicada paliuri* = *Cicadatra querula paliuri* = *Cicada (Cicadatra) querula paliuri* = *Cicadatra querula palinri* (sic) = *Cicadatra paliuri* = *Psalmocharias querula paliuri* = *Psalmocharias paiiuri* (sic)
P. plagifera (Schumacher, 1922) = *Cicadatra (Psalmocharias) plagifera* = *Cicadatra plagifera*
P. querula (Pallas, 1773) = *Cicada querula* = *Tettigonia querula* = *Cicada concinna querula* Germar, 1830 = *Cicada quaerula* (sic) = *Cicada quercula* (sic) = *Cicada (Cicadatra) querula* = *Tettigia (Cicadatra) querula* = *Cicadatra querula* = *Cicadatra guerula* (sic) = *Cicadatra guerula* (sic) = *Sena querula* = *Sena quaerula* (sic) = *Cicadatra (Psalmocharias) querula* = *Psalmocharias quaerula* (sic) = *Cephaloxys quadrimacula* Walker, 1850 = *Cephaloxys quadrimaculata* (sic) = *Cicadatra quadrimacula* = *Cicadatra quadrimaculata* (sic) = *Mogannia quadrimacula* = *Cicada steveni* Stål, 1854 = *Cicadatra steveni*
P. rugipennis (Walker, 1858) = *Cicada rugipennis* = *Cicadatra rugipennis* = *Sena rugipennis* = *Sena querula* var. ______ Distant, 1906 = *Cicadatra quaerula* (sic) var. ______ = *Psalmochavias* (sic) *rugipennis*

Nipponosemia Kato, 1925 = *Niponosemia* (sic)
N. guangxiensis Chou & Wang, 1993
N. metulata Chou & Lei, 1993
N. terminalis (Matsumura, 1913) = *Abroma terminalis* = *Lemuriana terminalis* = *Cicada terminalis* = *Gicada* (sic) *terminalis* = *Niponosemia* (sic) *terminalis* = *Cicada fuscoplaga* Schumacher, 1915 = *Gicada* (sic) *fuscoplaga* = *Nipponosemia fuscoplaga* = *Cicada terminalis formosana* Kato, 1925 = *Cicada terminalis* var. *formosana* = *Nipponosemia terminalis formosana*
N. virescens Kato, 1926 = *Nipponosemia virecens* (sic)

Shaoshia Wei, Ahmed & Rizvi, 2010
S. zhangi Wei, Ahmed & Rizvi, 2010

Mogannia Amyot & Audinet-Serville, 1843 = *Cephaloxys* Signoret, 1847 = *Mogonnia* (sic) = *Mongania* (sic) = *Moggania* (sic) = *Mognnia* (sic) = *Mogana* (sic)
M. aliena Distant, 1920
M. aurea Fraser, 1942
M. basalis Matsumura, 1913 = *Mogannia flavocapitata* Kato, 1925 = *Mogannia basalis tienmushanensis* Ouchi, 1938 = *Mogannia basalis* var. *tienmushanensis*
M. binotata Distant, 1906 = *Prasia elegans* Kirkaldy, 1913
M. caesar Jacobi, 1902 = *Mogannia cyanea* (nec Walker)
M. conica conica (Germar, 1830) = *Cicada conica* = *Mogannia illustrata* Amyot & Audinet-Serville, 1843 = *Mogannia conica illustrata* = *Cephaloxys hemelytra* Signoret, 1847 = *Mogannia indicans* Walker, 1850 = *Mogannia conica indicans* = *Mogannia incidans* var. *β* Walker, 1850 = *Mogannia ignifera* Walker, 1850 = *Mogannia conica ignifera* = *Mogannia avicula* Walker, 1850 = *Mogannia avicula* (sic) = *Mogannia recta* Walker, 1858 = *Mogannia histrionica* Uhler, 1861
M. conica var. *d* Distant, 1892
M. cyanea Walker, 1858 = *Mogannia nigrocyanea* Matsumura, 1913 = *Mogannia nigocyanea* (sic) = *Mogannia nigrocyanea flavofascia* Kato, 1925 = *Mogannia cyanea flavofascia* = *Mogannia cyanea flavofescia* (sic) = *Mogannia bella* Kato, 1927 = *Mogannia chekingensis* Ouchi, 1938 = *Mogannia chinensis* Ouchi, 1938 (nec *Mogannia chinensis* Stål, 1865) = *Mogannia tienmushana* Chen, 1957 = *Mogannia ouchii* Metcalf, 1963
M. doriae Distant, 1888
M. effecta effecta Distant, 1892 = *Mogannia efecta* (sic) = *Mogannia conica* (nec Germar)
M. effecta var. *a* Distant, 1892
M. effecta var. *b* Distant, 1892
M. flammifera Moulton, 1923
M. nr. *flammifera* Zaidi, Ruslan & Mahadir, 1996
M. formosana Matsumura, 1907 = *Mogannia nasalis* (nec White, 1844) = *Mogannia rubricosta* Matsumura, 1917 = *Mogannia formosana rubricosta* = *Mogannia rubrlcosta* (sic) = *Mogannia pallipes* Matsumura, 1917 = *Mogannia formosana pallipes* = *Mogannia fasciata pallipes* = *Mogannia formosana taikosana* Kato, 1925 = *Mogannia formosana* var. *b* Kato, 1925 = *Mogannia fasciata* Kato, 1925 = *Mogannia formosana fasciata* = *Mogannia fasciata nigroventris* Kato, 1925 = *Mogannia formosana fasciata nigroventris* =

Mogannia fuscomaculata Kato, 1925 = *Mogannia formosana fuscomaculata* = *Mogannia fusculata* (sic) = *Mogannia kashotoensis* Kato, 1925 = *Mogannia kwashotoensis* (sic) = *Mogannia kashotoensis flavoguttata* Kato, 1925 = *Mogannia kwashotoensis* (sic) *flavoguttata* = *Mogannia kasotoensis* (sic) *flavoguttata* = *Mogannia kanoi* Kato, 1925 = *Mogannia formosana elegans* Kato, 1926 = *Mogannia formosana interrupta* Kato, 1926 = *Mogannia formosana rubricosta* Kato, 1930 = *Mogannia formosana uraina* Kato, 1933 = *Mogannia sinensis* Ouchi, 1938

M. funebris funebris Stål, 1865

M. funebris var. *a* Distant, 1892

M. funettae Boulard, 2008

M. grelaka Moulton, 1923

M. guangdongensis Chen, Yang & Wei, 2012

M. hainana Shen, Lei & Yang, 2006

M. hebes (Walker, 1858) = *Cephaloxys hebes* = *Mogonnia* (sic) *hebes* = *Mogannia hebets* (sic) = *Mogannia spurcata* Walker, 1858 = *Mogannia hebes* var. *a* Kato, 1925 = *Mogannia hebes* var. *b* Kato, 1925 = *Mogannia hebes* var. *c* Kato, 1925 = *Mogannia flavescens* Kato, 1925 = *Mogannia hebes flavescens* = *Mogannia delta* Kato, 1925 = *Mogannia hebes* var. *d* Kato, 1925 = *Mogannia ritozana* Matsumura, 1927 = *Mogannia hebes ritozana* = *Mogannia ritozana dorsovitta* Matsumura, 1927 = *Mogannia ritozana dorsovittata* (sic) = *Mogannia ritozana dorsovitta* (sic) = *Mogannia hebes dorsovitta* = *Mogannia hebes dorsivitta* (sic) = *Mogannia katonis* Matsumura, 1927 = *Mogannia hebes katonis* = *Mogannia subfusca* Kato, 1928 = *Mogannia hebes concolor* Kato, 1932 = *Mogannia janea* Hua, 2000

M. horsfieldi Distant, 1905

M. indigotea Distant, 1917

M. malayana Moulton, 1923

M. nr. *malayana* Zaidi, Ruslan & Mahadir, 1996

M. mandarina Distant, 1905 = *Mogannia mandarinea* (sic)

M. minuta Matsumura, 1907 = *Mogannia iwasakii* Matsumura, 1913 = *Mogannia formosana* (nec Matusumura) = *Mogannia iwasakii nigroventris* Kato, 1937

M. montana Moulton, 1923

M. moultoni Distant, 1916

M. nasalis fusca Liu, 1939 = *Mogannia nasalis fuscous* (sic)

M. nasalis nasalis (White, 1844) = *Cicada (Mogannia) nasalis* = *Mogannia chinensis* Stål, 1865 = *Mogannia kikowensis* Ouchi, 1938 = *Mogannia kikowensis* var. *a* Ouchi, 1938 = *Mogannia kikowensis a* = *Mogannia kikowensis* var. *b* Ouchi, 1938 = *Mogannia kikowensis b* = *Mogannia kikowensis* var. *c* Ouchi, 1938 = *Mogannia kikowensis c* = *Mogannia nasalis kikowensis*

M. obliqua Walker, 1858 = *Mogannia obligua* (sic)

M. paae Boulard, 2009 = *Mogannia* sp. Boulard, 2005, 2007

M. rosea Moulton, 1923

M. ruiliensis Chen, Yang & Wei, 2012

M. saucia Noualhier, 1896 = *Mogannia sauciei* (sic) = *Mogannia cyanea yungshienensis* Liu, 1940

M. sesioides Walker, 1870

M. sp. 1 Pham & Thinh, 2006

M. sumatrana Moulton, 1923

M. tienmushana Chen, 1957

M. tumdactylina Chen, Yang & Wei, 2012

M. unicolor (Walker, 1852) = *Cephaloxys unicolor*

M. venustissima venustissima Stål, 1865 = *Mogannia venutissima* (sic) = *Mognnia* (sic) *venutissima* (sic)

M. venustissima var. *a* Stål, 1865

M. venustissima var. *b* Stål, 1865

M. viridis (Signoret, 1847) = *Cephaloxys viridis* = *Cephaloxys rostrata* Walker, 1850 = *Mogannia distinguenda* Distant, 1920

Pachypsaltria Stål, 1863 = *Pachysaltria* (sic)

P. camposi Goding, 1925

P. cinctomaculata (Stål, 1854) = *Cicada cinctomaculata* = *Carineta ciliaris* Walker, 1858

P. cirrha Torres, 1960

P. haematodes Torres, 1960

P. monrosi Torres, 1960
P. phaedima Torres, 1960
P. peristicta Torres, 1960

Species of uncertain position in Cicadidae
Genus A Pogue, 1996
Genus B Pogue, 1996

Subfamily Cicadettinae Buckton, 1889

Tribe Dazini Kato, 1932

Daza Distant, 1905 = *Dazu* (sic)
D. montezuma (Walker, 1850) = *Zammara montezuma* = *Odopoea montezuma* = *Odopaea* (sic) *montezuma*
D. nayaritensis Davis, 1934

Procollina Metcalf, 1963 = *Collina* Distant, 1905 (nec *Collina* Bonarelli, 1893)
P. biolleyi (Distant, 1903) = *Odopoea biolleyi* = *Collina biolleyi*
P. medea (Stål, 1864) = *Odopoea medea* = *Collina medea*
P. obesa (Distant, 1906) = *Collina obesa*
P. queretaroensis Sanborn, 2007

Aragualna Champanhet, Boulard & Gaiani, 2000
A. plenalinea Champanhet, Boulard & Gaiani, 2000

Onoralna Boulard, 1996
O. falcata Boulard, 1996

Tribe Huechysini Distant, 1905

Subtribe Huechysina Distant, 1905 = Huechysaria

Graptotettix Stål, 1866
G. guttatus Stål, 1866
G. rubromaculatus Distant, 1905
G. thoracicus Distant, 1892

Parvittya Distant, 1905 = *Paravittya* (sic) = *Tarvittya* (sic)
P. samara Distant, 1905

Huechys Amyot & Audinet-Serville, 1843 = *Cicada (Huechys)* = *Huechys (Huechys)* = *Huechys (Amplihuechys)* = *Hyechis* (sic) = *Huechis* (sic) = *Heuchys* (sic) = *Hyechys* (sic) = *Huachas* (sic) = *Peuchys* (sic)
H. aenea Distant, 1913
H. beata Distant, 1892 = *Huechys sanguinea* var. *b* Distant, 1892 = *Huechys philaemata* var. *b* = *Huechys sanguinea* (nec Degeer)
H. celebensis celebensis Distant, 1888 = *Huechys (Amplihuechys) celebensis*
H. celebensis var. *a* Breddin, 1901 = *Huechys (Amplihuechys) celebensis* var. *a*
H. celebensis var. *b* Breddin, 1901 = *Huechys (Amplihuechys) celebensis* var. *b*
H. chryselectra Distant, 1888 = *Huechys (Amplihuechys) chryselecrta* = *Huechys chtyselectra* (sic)
H. curvata Haupt, 1924 = *Huechys (Amplihuechys) curvata*
H. dohertyi dohertyi Distant, 1892
H. dohertyi var. *a* Distant, 1892
H. eos Breddin, 1901
H. facialis facialis Walker, 1857 = *Huechys fascialis* (sic) = *Huechys (Amplihuechys) facialis*
H. facialis obscura Haupt, 1924 = *Huechys (Amplihuechys) facialis obscura*

H. fretensis Haupt, 1924 = *Huechys (Huechys) fretensis*
H. funebris Lee, 2011
H. fusca fusca Distant, 1892 = *Huechys (Huechys) fusca*
H. fusca var. *a* Distant, 1892 = *Huechys (Huechys) fusca* var. *a*
H. fusca nigra Lallemand, 1931 = *Huechys (Huechys) fusca nigra*
H. haematica Distant, 1888 = *Huechys fusca haematica* = *Huechys (Huechys) fusca haematica*
H. incarnata aterrima Haupt, 1924 = *Huechys (Huechys) incarnata aterrima*
H. incarnata var. *b* Distant, 1897 = *Huechys (Huechys) incarnata* var. *b*
H. incarnata brunnea Haupt, 1924 = *Huechys (Huechys) incarnata brunnea*
H. incarnata incarnata (Germar, 1834) = *Cicada incarnata* = *Huechis* (sic) *incarnata* = *Huechys sanguinea incarnata* = *Huechys (Huechys) incarnata* = *Cicada sanguinolenta* (nec Fabricius) = *Cicada germari* Guérin-Méneville, 1838 = *Huechys germari*
H. incarnata novaeguineensis Haupt, 1924 = *Huechys (Huechys) incarnata novaeguineensis*
H. incarnata nox Schmidt, 1919 = *Huechys incarnata* var. *a* Distant, 1897 = *Huechys (Huechys) incarnata nox*
H. incarnata testacea (Fabricius, 1787) = *Tettigonia testacea* = *Cicada testacea* = *Cicada sanguinolenta* Germar (partim) = *Huechys (Huechys) incarnata testacea*
H. incarnata venosa Haupt, 1924 = *Huechys (Huechys) incarnata venosa*
H. insularis borneensis Haupt, 1924 = *Huechys (Huechys) insularis borneensis*
H. insularis insularis Haupt, 1924 = *Huechys (Huechys) insularis*
H. insularis lombokensis Haupt, 1924 = *Huechys (Huechys) insularis lombokensis*
H. insularis niasensis Haupt, 1924 = *Huechys (Huechys) insularis niasensis*
H. insularis sumatrana Haupt, 1924 = *Huechys (Huechys) insularis sumatrana*
H. lutulenta Distant, 1892 = *Huechys (Amplihuechys) lutulenta*
H. nigripennis Moulton, 1923
H. parvula Haupt, 1924 = *Huechys (Huechys) parvula*
H. phaenicura (Germar, 1834) = *Cicada phaenicura* = *Cicada phoenicura* (sic) = *Huechys phoenicura* (sic) = *Huechys (Huechys) phaenicura* = *Huechys phaenicura balabakensis* Haupt, 1924 = *Huechys phoenicura* (sic) *balabakensis* = *Huechys (Huechys) phaenicura balabakensis* = *Huechys (Huechys) phoenicura* (sic) f. *balabakensis* = *Huechys phaenicura palawanensis* Haupt, 1924 = *Huechys phoenicura* (sic) *palawanensis* = *Huechys (Huechys) phaenicura palawanensis* = *Huechys (Huechys) phoenicura* (sic) form. *palawanensis*
H. pingenda breddini Haupt, 1924 = *Huechys (Amplihuechys) pingenda breddini*
H. pingenda pingenda Distant, 1888 = *Huechys (Amplihuechys) pingenda* = *Graptotettix pingenda*
H. sanguinea var. _______ Breddin, 1899 = *Huechys (Huechys) sanguinea* var. _______
H. sanguinea var. _______ Walker, 1870 = *Huechys (Huechys) sanguinea* var. _______
H. sanguinea hainanensis Kato, 1931 = *Huechys philaemata hainanensis* = *Huechys sanguinea ocher* Liu, 1940 = *Huechys sanguinea ochra* = *Huechys (Huechys) sanguinea hainanensis*
H. sanguinea philaemata (Fabricius, 1803) = *Tettigonia philaemata* = *Cicada philaemata* = *Cicada incarnata philaematis* = *Cicada philoemata* (sic) = *Huechys philaemata* = *Huechys (Huechys) sanguinea philaemata* = *Huechys sanguinea* var. *philaemata* = *Huechys sanguinea philoemata* (sic) = *Huechys sanguinea* var. *a* Distant, 1892 = *Huechys sanguinea* (nec Degeer)
H. sanguinea sanguinea (Degeer, 1773) = *Cicada sanguinea* = *Heuchys* (sic) *sanguinea* = *Hyechys* (sic) *sanguinea* = *Huachas* (sic) *sanguinea* = *Huechys sanquiena* (sic) = *Huechys sanquinea* (sic) = *Huechys (Huechys) sanguinea* = *Scieroptera sanguinea* = *Tettigonia sanguinolenta* Fabricius, 1775 = *Cicada sanguineolenta* = *Tettigonia sangvinolenta* (sic) = *Tettigonia philaemata* Fabricius, 1803 = *Huechys vesicatoria* Smith, 1871 = *Peuchys* (sic) *vesicatoria* = *Huechys quadrispinosa* Haupt, 1924 = *Huechys sanguinea philaemata* = *Huechys sanguinea philaemata albifascia* Kato, 1927 = *Huechys philaemata albifascia*
H. sanguinea suffusa Distant, 1888 = *Huechys (Huechys) sanguinea suffusa*
H. sanguinea wuchangensis Liu, 1940 = *Huechys (Huechys) sanguinea wuchangensis*
H. sumbana sumbana Jacobi, 1941
H. sumbana var. _______ Jacobi, 1941
H. thoracica Distant, 1879 = *Huechys (Huechys) thoracica*
H. tonkinensis Distant, 1917

H. vidua vidua White, (1846) = *Cicada (Huechys) vidua* = *Cicada vidua* = *Huechys (Huechys) vidua*
H. vidua var. *a* Distant, 1892 = *Huechys (Huechys) vidua* var. *a*

Scieroptera Stål, 1866 = *Schieroptera* (sic) = *Sleroptera* (sic) = *Scieretopra* (sic) = *Scieoptera* (sic) = *Scleroptera* (sic)
S. borneensis Schmidt, 1918
S. crocea crocea (Guérin-Méneville, 1838) = *Cicada crocea*
S. crocea var. *a* Distant, 1892
S. nr. *crocea* Zaidi & Ruslan, 1995
S. delineata Distant, 1917 = *Scieroptera delineate* (sic)
S. distanti Schmidt, 1920 = *Schieroptera* (sic) *distanti* = *Scieroptera distani* (sic)
S. flavipes Schmidt, 1918
S. formosana Schmidt, 1918 = *Scieroptera formosana ater* Kato, 1925 = *Scieroptera formosana atra* = *Scieroptera formosana ater* = *Scieroptera formosana trigutta* Kato, 1926 = *Scieroptera formosana albifascia* Kato, 1926 = *Scieretopra* (sic) *formosana* = *Scieoptera* (sic) *formosana* = *Scieroptera formasana* (sic) = *Scieroptera splendidula* (nec Fabricius) = *Sleroptera* (sic) *splendidula* (nec Fabricius) = *Scieroptera splendidula cuprea* (nec Walker)
S. fulgens Schmidt, 1918
S. fumigata (Stål, 1854) = *Huechys fumigata*
S. fuscolimbata Schmidt, 1918
S. hyalinipennis Schmidt, 1912
S. limpida Schmidt, 1918
S. montana Schmidt, 1918 = *Scieroptera carmente* Schu.
S. niasana Schmidt, 1918
S. orientalis Schmidt, 1918
S. sanaoensis Schmidt, 1924
S. sarasinorum Breddin, 1901
S. n. sp. Zaidi & Ruslan, 1995
S. splendidula cuprea (Walker, 1870) = *Huechys cuprea* = *Scieroptera splendidula* var. *a* Distant, 1982 = *Huechys cuprae* (sic) = *Scieroptera splendidula cuprae* (sic)
S. splendidula splendidula (Fabricius, 1775) = *Tettigonia splendidula* = *Telligonia* (sic) *splendidula* = *Cicada splendidula* = *Huechys splendidula* = *Scleroptera* (sic) *splendidula* = *Scieroptera splendidula cuprea* (partim) = *Scieroptera splendidula vittata* Kato, 1940
S. splendidula var. *c* Distant, 1892
S. splendidula var. *d* Distant, 1892
S. splendidula trabeata (Germar, 1830) = *Huechys trabeata* = *Scieroptera splendidula* var. *b* Distant, 1892
S. sumatrana nigripennis Schmidt, 1918 = *Huechys nigripennis*
S. sumatrana sumatrana Schmidt, 1918 = *Scieroptera splendidula* (nec Fabricius) = *Huechys splendidula* (nec Fabricius)

Tribe Carinetini Distant, 1905 = Carinettini (sic) = Sinosenini Boulard, 1975

Subtribe Carinetiina Distant, 1905 = Carinetaria

Carineta Amyot & Audinet-Serville, 1843 = *Cerinetha* (sic) = *Carinata* (sic) = *Carinetta* (sic) = *Carinita* (sic) = *Carinenta* (sic) = *Clarineta* (sic)
C. aestiva Distant, 1883
C. apicalis Distant, 1883
C. apicoinfuscata Sanborn, 2011
C. aratayensis Boulard, 1986
C. argentea Walker, 1852 = *Carinita* (sic) *argentea* = *Carineta argentata* (sic)
C. basalis Walker, 1850
C. boliviana Distant, 1905
C. boulardi Champanhet, 1999
C. calida Walker, 1858
C. carayoni Boulard, 1986

C. cearana Distant, 1906
C. centralis Distant, 1892
C. cinara Distant, 1883
C. cingenda Distant, 1883
C. congrua Walker, 1858
C. crassicauda Torres, 1948 = *Carinenta* (sic) *crassicauda*
C. criqualicae Boulard, 1986
C. cristalinea Champanhet, 2001
C. crocea Distant, 1883
C. crumena Goding, 1925
C. cyrili Champanhet, 1999
C. detoulgoueti Champanhet, 2001 = *Carineta detoulgouëti*
C. diardi (Guérin-Méneville, 1829) = *Cicada formosa* Germar, 1830 = *Carineta formosa* = *Carinata* (sic) *formosa* = *Cicada polychroa* Perty, 1833 = *Cicada duvaucelii* Guérin-Méneville, 1838 = *Cicada duvancellii* (sic)
C. dolosa Boulard, 1986
C. doxiptera Walker, 1858
C. durantoni Boulard, 1986
C. ecuatoriana Goding, 1925
C. fasciculata (Germar, 1821) = *Cicada fasciculata* = *Carinenta* (sic) *fasciculata* = *Carinetta* (sic) *fasciculata* = *Fidicina fasciculata* = *Cicada bilineosa* Walker, 1858 = *Carineta bilineosa* = *Cicada obtusa* Walker, 1858 = *Carineta obtusa* = *Cicada tenuistriga* Walker, 1858 = *Carineta trenuistriga* = *Carineta diplographa* Berg, 1879
C. fimbriata Distant, 1891
C. garleppi Jacobi, 1907
C. gemella Boulard, 1986
C. genitalostridens Boulard, 1986
C. guianaensis Sanborn, 2011
C. illustris Distant, 1905
C. imperfecta Boulard, 1986
C. indecora (Walker, 1858) = *Cicada indecora*
C. lichiana Boulard, 1986
C. limpida Torres, 1948 = *Clarineta* (sic) *limpida*
C. liturata Torres, 1948
C. lydiae Champanhet, 1999
C. maculosa Torres, 1948 = *Carinenta* (sic) *maculosa* = *Carineta fasciculata* (nec Germar)
C. martiniquensis Davis, 1934
C. matura Distant, 1892
C. modesta Sanborn, 2011
C. naponore Boulard, 1986
C. peruviana Distant, 1905
C. picadae Jacobi, 1907
C. pilifera Walker, 1858
C. pilosa Walker, 1850
C. platensis Berg, 1882
C. porioni Champanhet, 2001
C. postica Walker, 1858 = *Carineta viridicata* (nec Distant)
C. producta Walker, 1858
C. propinqua Torres, 1948 = *Carinenta* (sic) *propinqua*
C. quinimaculata Sanborn, 2011
C. rubricata Distant, 1883
C. rufescens (Fabricius, 1803) = *Tettigonia rufescens* = *Cicada rufescens* = *Carineta oberthuri* Distant, 1881 = *Carineta oberthuerii* (sic)
C. rustica Goding, 1925
C. scripta Torres, 1948 = *Carinenta* (sic) *scripta*
C. socia Uhler, 1875 = *Carineta socio* (sic) = *Carineta* sp. 3 Salazar Escobar, 2005

C. sp. Young, 1981
C. sp. M Wolda & Ramos, 1992
C. sp. 1 Pogue, 1997
C. sp. 2 Pogue, 1997
C. sp. 3 Pogue, 1997
C. sp. 4 Pogue, 1997
C. sp. 5 Pogue, 1997
C. sp. 6 Pogue, 1997
C. sp. 7 Pogue, 1997
C. sp. 1 Salazar Escobar, 2005
C. sp. 2 Salazar Escobar, 2005
C. spoliata (Walker, 1858) = *Cicada spoliata*
C. strigilifera Boulard, 1986
C. submarginata Walker, 1850
C. tetraspila Jacobi, 1907
C. tigrina Boulard, 1986
C. titschacki Jacobi, 1951 = *Carineta titchaki* (sic) = *Carineta titschachi* (sic)
C. tracta Distant, 1892
C. trivittata Walker, 1858 = *Carineta strigimargo* Walker, 1858
C. turbida Jacobi, 1907
C. urostridulens Boulard, 1986
C. ventralis Jacobi, 1907
C. ventrilloni Boulard, 1986
C. verna Distant, 1883
C. viridicata Distant, 1883
C. viridicollis viridicollis (Germar, 1830) = *Cicada viridicollis* = *Carineta viricollis* (sic) = *Carineta viridis* (sic)
C. viridicollis var. *b* Stål, 1862

Toulgoetalna Boulard, 1982
T. tavakiliani Boulard, 1982

Novemcella Goding, 1925
N. ecuatoriana Goding, 1925

Herrera Distant, 1905
H. ancilla (Stål, 1864) = *Carineta ancilla* = *Cicada marginella* Walker, 1858 = *Carineta marginella* = *Herrera marginella*
H. coyamensis Sanborn, 2007
H. humilistrata Sanborn & Heath, 2011, nom. nud.
H. infuscata Sanborn 2009
H. laticapitata Davis, 1938
H. lugubrina compostelensis Davis, 1938 = *Herrera lugubrina* var. *compostelensis* = *Herrera tugubrina* (sic) *compostelensis*
H. lugubrina lugubrina (Stål, 1864) = *Carineta lugubrina*
H. nr. *lugubrina* Wolda & Ramos, 1992
H. sp. F3 Wolda & Ramos, 1992
H. sp. H1 Wolda, 1989
H. sp. Q5 Wolda & Ramos, 1992
H. umbraphila Sanborn & Heath, 2011, nom. nud.

Paranistria Metcalf, 1952 = *Tympanistria* Stål, 1861 (nec *Tympanistria* Reichenbach, 1852)
P. trichiosoma (Walker, 1850) = *Carineta trichiosoma* = *Tympanistria trichiosoma*
P. villosa villosa (Fabricius, 1781) = *Tettigonia villosa* = *Cicada villosa* = *Carineta villosa* = *Tympanistria villosa*
P. villosa var. *a* (Stål, 1866) = *Tympanistria villosa* var. *a*
P. villosa var. *b* (Stål, 1866) = *Tympanistria villosa* var. *b*

P. villosa var. *c* (Stål, 1866) = *Tympanistria villosa* var. *c*

Guaranisaria Distant, 1905
G. bicolor Torres, 1958
G. dissimilis Distant, 1905
G. llanoi Torres, 1964

Subtribe Kareniina Boulard, 2008

Karenia Distant, 1888 = *Sinosenia* Chen, 1943
K. caelatata Distant, 1890 = *Karenia coelata* (sic) = *Karenia coelatata* (sic) = *Sinosena caelatata* = *Sinosena caelatta* (sic)
K. chama Wei & Zhang, 2009
K. hoanglienensis Pham & Yang, 2012 = *Karenia* undescribed species Pham & Yang, 2010
K. ravida Distant, 1888 = *Sinosena ravida*
K. sulcata Lei & Chou, 1997

Tribe Parnisini Distant, 1905 = Parnisiini (sic) = Parnesini (sic)

Quintilia Stål, 1866 = *Tibicen (Quintilia)* = *Quintillia* (sic)
Q. annulivena (Walker, 1850) = *Cicada annulivena* = *Tibicen annulivena*
Q. aurora (Walker, 1850) = *Cicada aurora* = *Tibicen aurora* = *Tibicen (Quintilia) sanguinarius* Stål, 1866 = *Quintilia sanguinaria*
Q. carinata (Thunberg, 1822) = *Quintilia carinatus* (sic) = *Tettigonia carinata* = *Tibicen carinata* = *Tibicen (Quintilia) carinatus* = *Cicada tristis* Germar, 1834 = *Cicada nigroviridis* Walker, 1852 = *Tibicen nigroviridis*
Q. catena (Fabricius, 1775) = *Tettigonia catena* = *Cicada catena* = *Tibicen catenus* = *Cicada catenata* (sic) = *Cicada nebulosa* Olivier, 1790 = *Cicada hyalina* Olivier, 1790 = *Tibicen hyalina* = *Cicada garrula* Olivier, 1790 = *Tibicen garrulus* = *Tettigonia crenata* Thunberg, 1822
Q. diversa (Walker, 1850) = *Cicada diversa*
Q. dorsalis (Thunberg, 1822) = *Tettigonia dorsalis* = *Tibicen dorsalis* = *Cicada maculipennis serricosta* Walker, 1850 nom. nud. = *Cicada serricosta* Germar nom. nud. = *Cicada serricostata* (sic)
Q. maculiventris Distant, 1905
Q. monilifera (Walker, 1850) = *Cicada monilifera* = *Tibicen monilifera* = *Tibicen (Quintilia) monolifera* = *Tibicen (Quintilia) maculinervis*
Q. musca (Olivier, 1790) = *Cicada musca* = *Tettigomyia musca* = *Tibicen musca* = *Cicada tachinaria* Germar, 1830
Q. obliqua (Walker, 1850) = *Cicada obliqua* = *Tibicen obliquus*
Q. pallidiventris (Stål, 1866) = *Tibicen (Quintilia) pallidiventris*
Q. peregrina peregrina (Linnaeus, 1764) = *Cicada peregrina* = *Tibicen peregrina* = *Tibicen (Quintilia) peregrinus* (sic) = *Cicada maculipennis* Germar, 1834
Q. peregrina var. *a* (Stål, 1866) = *Tibicen (Quintilia) peregrinus* var. *a*
Q. cf. *peregrina* Dean, 1992
Q. pomponia Distant, 1912 = *Lycurgus pomponia* = *Lycurgus pomponius*
Q. primitiva (Walker, 1850) = *Cicada primitiva* = *Tibicen primitiva* = *Tibicen (Quintilia) primitiva* = *Tibicen (Quintilia) haematinus* Stål, 1866 = *Quintilia haematinus*
Q. rufiventris (Walker, 1850) = *Cicada rufiventris* = *Cicada holmgreni* Stål, 1856 = *Tibicen holmgreni* = *Tibicen (Quintilia) holmgreni*
Q. semipunctata Karsch, 1890 = *Cicada semipunctata* Walker, 1850 nom. nud. = *Cicada semipunctata* Dohrn, 1859 nom. nud. = *Melampsalta semipunctata*
Q. sp. A Phillips, Sanborn & Villet, 2002
Q. sp. B Phillips, Sanborn & Villet, 2002
Q. umbrosa (Stål, 1866) = *Tibicen (Quintilia) umbrosus* = *Tibicen umbrosus* = *Cicada diaphana* Germar, 1830 = *Cicada maculipennis diaphana* = *Tibicen diaphanus* = *Tibicen (Quintilia) peregrinus* var. *b* Stål, 1866
Q. vitripennis Karsch, 1890 = *Tibicen (Quintilia) vitripennis* = *Lycurgus vitripennis*

Q. cf. *vitripennis* Dean, 1992
Q. vittativentris (Stål, 1866) = *Tibicen (Quintilia) vittativentris*
Q. wallkeri China, 1925
Q. wealei (Distant, 1892) = *Tibicen (Quintilia) wealei*

Rhinopsaltria Melichar, 1908
R. sicardi Melichar, 1908

Jafuna Distant, 1912
J. melichari Distant, 1912

Malgotilia Boulard, 1980
M. sogai Boulard, 1980

Lycurgus China, 1925
L. conspersa (Karsch, 1890) = *Quintilia conspersa*
L. cf. *conspersa* Dean & Milton, 1991
L. frontalis (Karsch, 1890) = *Quintilia frontalis*
L. sinensis Jacobi, 1944
L. subvittus (Walker, 1850) = *Cicada subvitta* = *Tibicen subvitta* = *Tibicen subvittatus* (sic) = *Quintilia subvittatus* (sic) = *Quintilia subvitta* = *Lycurgus subvitta* (sic) = *Cicada strigosa* Walker, 1858

Psilotympana Stål, 1861
P. ruficollis (Thunberg, 1822) = *Tettigonia ruficollis* = *Cicada ruficollis*
P. signifera (Germar, 1830) = *Cicada signifera* = *Cicada elana* Walker, 1850 = *Cicada brevipennis* Schaum nom. nud.
P. varicolor Distant, 1920
P. walkeri Metcalf, 1963 = *Cicada gracilis* Walker, 1850 (nec *Cicada gracilis* Schellenberg, 1802) = *Psilotympana gracilis*

Abagazara Distant, 1905
A. bicolorata (Distant, 1892) = *Callipsaltria bicolorata*
A. omaruruensis Hesse, 1925

Henicotettix Stål, 1858
H. hageni Stål, 1858

Koranna Distant, 1905 = *Korania* (sic)
K. analis Distant, 1905

Masupha Distant, 1892
M. ampliata Distant, 1892
M. delicata Distant, 1892
M. dregei Distant, 1892

Derotettix Berg, 1882
D. mendosensis Berg, 1882 = *Derotettix mendocensis* (sic)
D. wagneri Distant, 1905 = *Derotettix proseni* Torres, 1945

Acyroneura Torres, 1958
A. singularis Torres, 1958

Taipinga Distant, 1905
T. albivenosa (Walker, 1858) = *Cicada albivenosa* = *Psilotympana albivenosa*
T. consobrina Distant, 1906
T. fuscata Distant, 1906

T. fusiformis (Walker, 1850) = *Cicada fusiformis* = *Psilotympana fusiformis*
T. infuscata (Distant, 1892) = *Psilotympana infuscata*
T. luctuosa (Stål, 1855) = *Cicada luctuosa* = *Tibicen luctuosus* = *Tibicen (Quintilia) luctuosus*
T. nana (Walker, 1850) = *Cicada nana* = *Tibicen nanus* = *Tibicen nana*
T. nigricans (Stål, 1855) = *Cicada nigricans* = *Tibicen nigricans* = *Tibicen (Quintilia) nigricans*
T. rhodesi Distant, 1920
T. undulata (Thunberg, 1822) = *Tettigonia undulata* = *Tibicen (Quintilia) undulates* = *Quintilia undulata* = *Tibicen undulates* = *Cicada scripta* Germar, 1834 = *Tibicen scriptus*

Zouga Distant, 1906
Z. apiana Hesse, 1925
Z. beaumonti Boulard, 1994
Z. delalandei Distant, 1914
Z. festiva Distant, 1920
Z. fornicata (Linnaeus, 1758) = *Cicada fornicata* = *Tettigonia fornicata* = *Tympanistria fornicata* = *Paranistria fornicata* = *Cicada fornicata* Olivier, 1790 = *Cicada dimidiata* Olivier, 1790 = *Tettigonia dimidiata* = *Carineta leuconeura* Walker, 1850 = *Tympanistria leuconeura*
Z. heterochroma Boulard, 1994
Z. hottentota Distant, 1914
Z. peringueyi (Distant, 1920) = *Neomuda peringueyi*
Z. typica Distant, 1906

Luangwana Distant, 1914
L. capitata Distant, 1914

Calopsaltria Stål, 1861 = *Callipsaltria* Stål, 1866
C. convexa Haupt, 1918 = *Callipsaltria* (sic) *convexa*
C. hottentota Kirkaldy, 1909 = *Cicada elongata* Stål, 1855 (nec *Cicada elongata* Fabricius, 1781) = *Calopsaltria elongata* = *Callipsaltria elongata*
C. longula (Stål, 1855) = *Cicada longula* = *Callipsaltria longula*
C. nigra (Karsch, 1890) = *Callipsaltria nigra*

Kageralna Boulard, 2012
K. dargei Boulard, 2012

Calyria Stål, 1862 = *Cicada (Calyria)*
C. cuna (Walker, 1850) = *Cicada cuna* = *Cicada (Calyria) blanda* Stål, 1862
C. fenestrata (Fabricius, 1803) = *Tettigonia fenestrata* = *Cicada fenestrata*
C. jacobii Bergroth, 1914
C. mogannioides Jacobi, 1907 = *Calyria mogannioides* (sic)
C. stigma (Walker, 1850) = *Cicada stigma*
C. telifera (Walker, 1858) = *Cicada telifera* = *Calyira* (sic) *telifera* = *Calyria cuna* (nec Walker)

Parnisa Stål, 1862 = *Cicada (Parnisa)* = *Pamisa* (sic) = *Parnissa* (sic)
P. angulata Uhler, 1903 = *Pamisa* (sic) *angulata*
P. demittens (Walker, 1858) = *Cicada demittens*
P. designata (Walker, 1858) = *Cicada designata* = *Cicada casta* Stål, 1854 = *Cicada (Parnisa) casta* = *Parnisa casta*
P. fraudulenta (Stål, 1862) = *Cicada (Parnisa) fraudulenta* = *Cicada fraudulenta*
P. haemorrhagica Jacobi, 1904
P. lineaviridia Sanborn & Heath, 2011, nom. nud.
P. moneta (Germar, 1834) = *Cicada moneta* = *Melampsalta moneta* = *Cicada tricolor* Walker, 1850 = *Parnisa tricolor*
P. proponens proponens (Walker, 1858) = *Cicada proponens* = *Cicada (Parnisa) biplagiata* Stål, 1862
P. proponens var. *β* (Walker, 1858) = *Cicada proponens* var. *β*
P. protracta Uhler, 1903 = *Pamisa* (sic) *protracta*

P. viridis Sanborn & Heath, 2011, nom. nud.

Mapondera Distant, 1905

M. capicola Kirkaldy, 1909 = *Cicada pulchella* Stål, 1855 (nec *Cicada pulchella* Westwood, 1845) = *Psilotympana pulchella* = *Mapondera pulchella*

M. hottentota Kirkaldy, 1909 = *Cicada abdominalis* Stål, 1855 (nec *Cicada abdominalis* Donovan, 1798) = *Tibicen abdominalis* = *Tibicen (Quintilia) abdominalis* = *Mapondera abdominalis*

Bijaurana Distant, 1912

B. sita Distant, 1912

B. typica Distant, 1912

Adeniana Distant, 1905 = *Adenia* Distant, 1905 (nec *Adenia* Robineau-Desvoidy, 1863) = *Hymenogaster* Horváth, 1911

A. decolorata Korsakoff, 1947

A. korsakoffi Villiers, 1943

A. kovacsi (Horváth, 1911) = *Hymenogaster kovacsi* = *Zouga kovasci* (sic)

A. leouffrei Korsakoff, 1947

A. licenti Villiers, 1943

A. longiceps (Puton, 1887) = *Cicadatra longiceps* = *Hymenogaster longiceps* = *Zouga longiceps* = *Hymenogaster planiceps* Horváth, 1917 = *Adeniana planiceps* = *Hymenogaster tabia* Horváth, 1911 = *Zouga tabia* = *Adeniana tabia*

A. mairei (de Bergevin, 1917) = *Hymenogaster mairei* = *Adeniana bourlieri* de Bergevin, 1931

A. maroccana Villiers, 1943

A. minuta Villiers, 1943

A. nigerrima Villiers, 1943

A. nigricans Distant, 1914

A. obokensis Distant, 1914

A. robusta Villiers, 1943

A. safadi Schedl, 2007

A. seitzi Distant, 1914

A. yerburyi Distant, 1905

A. yotvataensis Schedl, 1999

Arcystasia Distant, 1882 = *Arcystacea* (sic)

A. godeffroyi Distant, 1882 = *Arcystasia doddefroyi* (sic)

Prunasis Stål, 1862 = *Cicada (Prunasis)*

P. pulcherrima pulcherrima (Stål, 1854) = *Cicada pulcherrima* = *Cicada (Prunasis) pulcherrima* = *Cicada viridula* Walker, 1850 = *Prunasis viridula*

P. pulcherrima var. *a* (Stål, 1862) = *Cicada (Prunasis) pulcherrima* var. *a* Stål, 1862

P. pulcherrima var. *b* (Stål, 1862) = *Cicada (Prunasis) pulcherrima* var. *b* Stål, 1862

P. pulcherrima var. *c* (Stål, 1862) = *Cicada (Prunasis) pulcherrima* var. *c* Stål, 1862

Crassisternalna Boulard, 1980

C. pauliani Boulard, 1980

Tribe Taphurini Distant, 1905 = Taphusinini (sic) = Eaphurarini (sic)

Subtribe Taphurina Distant, 1905 = Taphuraria

Chalumalna Boulard, 1998

C. martinensis Boulard, 1998

Psallodia Uhler, 1903

P. espinii Uhler, 1903

Chrysolasia Moulds, 2003
C. guatemalena (Distant, 1883) = *Tibicen guatemalenus* = *Tibicen guatemalanus* (sic) = *Abricta guatemalena*

Dorachosa Distant, 1892
D. explicta Distant, 1892

Dulderana Distant, 1905
D. truncatella (Walker, 1850)

Nosola Stål, 1866
N. paradoxa Stål, 1866

Selymbria Stål, 1861 = *Selymbra* (sic) = *Selymbia* (sic)
S. ahyetios Ramos & Wolda, 1985 = *Selymbria achyetios* (sic) = *Selymbria* sp. H2 Wolda, 1977
S. danieleae Sanborn, 2011
S. pandora Distant, 1911
S. pluvialis Ramos & Wolda, 1985 = *Selymbria stigmatica* (nec Germar) = *Selymbria* sp. 1 Wolda, 1977
S. sp. 1 Pogue, 1996
S. stigmatica (Germar, 1834) = *Cicada signifera* Germar, 1830 (nec *Cicada signifera* Germar, 1830) = *Chicada* (sic) *signifera* = *Cicada stigmatica* = *Cicada macrophthalma* Stål, 1854 = *Cicada macropthalma* (sic)
S. subolivacea (Stål, 1862) = *Cicada (Cicada) subolivacea* = *Cicada subolivacea*

Prosotettix Jacobi, 1907
P. sphecoidea Jacobi, 1907

Taphura Stål, 1862 = *Cicada (Taphura)*
T. boulardi Sanborn, 2011
T. charpentierae Boulard, 1971
T. debruni Boulard, 1990
T. egeri Sanborn, 2011
T. hastifera (Walker, 1858) = *Cicada hastifera* = *Cicada frontalis* Walker, 1858 = *Taphura frontalis*
T. maculata Sanborn, 2011
T. minusculus Thouvenot, 2012
T. misella (Stål, 1854) = *Cicada misella* = *Cicada (Taphura) misella* = *Thaphura* (sic) *misella*
T. nitida (Degeer, 1773) = *Cicada nitida* = *Cicada (Taphura) nitida*
T. sauliensis Boulard, 1971
T. sp. C Wolda, 1977

Elachysoma Torres, 1964
E. quadrivittata Torres, 1964 = *Elachysoma quadrimarginatus* (sic)

Imbabura Distant, 1911
I. typica Distant, 1911

Subtribe Tryellina Moulds, 2005

Abricta Stål, 1866 = *Tibicen (Abricta)*
A. brunnea (Fabricius, 1798) = *Tettigonia brunnea* = *Cicada brunnea* = *Tibicen (Abricta) brunneus* = *Tibicen brunneus*
A. ferruginosa (Stål, 1866) = *Tibicen (Abricta) ferruginosa*
A. pusilla (Fabricius, 1803) = *Tettigonia pusilla* = *Cicada pusilla* = *Tibicen pusillus*

Aleeta Moulds, 2003
A. curvicosta (Germar, 1834) = *Cicada curvicosta* = *Tibicen curvicosta* = *Abricta curvicosta* = *Abricta curyvicosii* (sic) = *Cicada tephrogaster* Boisduval, 1835 = *Tibicen (Abricta) tephrogaster* = *Abricta tephrogaster*

Tryella Moulds, 2003
T. adela Moulds, 2003
T. burnsi Moulds, 2003
T. castanea (Distant, 1905) = *Abricta castanea*
T. crassa Moulds, 2003
T. graminea Moulds, 2003
T. infuscata Moulds, 2003
T. kauma Moulds, 2003
T. lachlani Moulds, 2003 = *Abricta* sp. "G" Ewart, 1993 = *Abricta* sp. Zborowski & Storey, 1995 = Heathlands sp. G Ewart
T. noctua (Distant, 1913) = *Abricta noctua* = *Abricta rufonigra* Ashton, 1914
T. occidens Moulds, 2003
T. ochra Moulds, 2003
T. rubra (Goding & Frogatt, 1904) = *Tibicen ruber* = *Abricta ruber* = *Abricta rubra* = *Abricta elseyi* Distant, 1905 = *Abricta kadulua* Burns, nom. nud.
T. stalkeri (Distant, 1907) = *Abricta stalkeri*
T. willsi (Distant, 1882) = *Tibicen willsi* = *Abricta willsi*

Chrysocicada Boulard, 1989
C. franceaustralae Boulard, 1989 = *Chyrsocicada franceaustralasae* Boulard, 1990

Pictila Moulds, 2012
P. occidentalis (Goding & Froggatt, 1904) = *Tibicen occidentalis* = *Abricta occicentalis*

Magicicada Davis, 1925 = *Magicicadas* (sic) = *Magicicadidae* (sic) = *Migicicada* (sic) = *Megacicada* (sic) = *Magicicadae* (sic) = *Cicada* (nec Linnaeus) = *Mogannia* (nec Amyot & Audinet-Serville, 1843)
M. cassinii (Fisher, 1852) = *Cicada cassinii* = *Cicada septendecim cassinii* = *Cicada septendecim cassini* = *Tibicen cassinii* = *Tibicen cassini* (sic) = *Tibicen septendecim cassina* (sic) = *Tibicina septendecim cassini* (sic) = *Tibicina septendecim cassinii* = *Cicada cassinii* = *Cicada septendecim cassinii* = *Tibicena* (sic) *cassinii* = *Magicicada* = *Magicicada septendecim* var. *cassinii septendecim cassinii* = *Magicicada septendecim cassini* (sic) = *Magicicada cassini* (sic) = *Magicicada cassani* (sic) = *Migicicada* (sic) *cassini* (sic) = *Magiciada ccassini* (sic) = *Magicicada casini* (sic) = *Magicicada cassi* (sic)
M. neotredecim Marshall & Cooley, 2000
M. septendecim (Linnaeus, 1758) = *Cicada septendecim* = *Tettigonia septendecim* = *Tettigonia septendedecim* (sic) = *Cicada septemdecim* (sic) = *Tettigonia septemdecim* (sic) = *Tettigonia septendecem* (sic) = *Tettigonia septendecim* = *Tibicen septendecim* = *Cicada 17-decim* = *Tibicen septemdecim* (sic) = *Tibicina septemdecim* = *Tibicim* (sic) *septendecim* = *Cicada septedecim* (sic) = *Tibicena* (sic) *septendecim* = *Ciccada* (sic) *septemdecim* (sic) = *Magicicada septendcim* (sic) = *Megacicada* (sic) *septemdecim* (sic) = *Magicada* (sic) *septendecim* = *Tibicina septendceim* (sic) = *Cicada septednecim* (sic) = *Ciccada* (sic) *septemdecim* (sic) = *Magicicada septendecem* (sic) = *Magicicada septemdecim* (sic) = *Magicicada sependecim* (sic) = *Migicicada* (sic) *septendecim* = *Magicicada septemdecem* (sic) = *Magicicada septendecin* (sic) = *Magicicada septendecium* (sic) = *Magicicada sepiendecim* (sic) = *Magicicada septimdecim* (sic) = *Magicicada septidecim* (sic) = *Magicicada septindecim* (sic) = *Magicicada septemdicim* (sic) = *Magicicada septendecimthe* (sic) = *Magicicada septendedim* (sic) = *Magicicada septemdecima* (sic) = *Tettigonia costalis* Fabricius, 1798 = *Cicada costalis*
M. septendecula Alexander & Moore, 1962 = *Magicicada septendicula* (sic) = *Migicicada* (sic) *septendecula*
M. tredecassini Alexander & Moore, 1962 = *Magicicada trecassini* (sic) = *Magicicada tredecessini* (sic) = *Magicicada tredicassini* (sic)
M. tredecim (Walsh & Riley, 1868) = *Cicada tredecim* = *Cicada septendecim tredecim* = *Cicada tredecem* (sic) = *Tibicen tredecim* = *Cicada trydecim* (sic) = *Tibicen septendecim tredecim* = *Tibicina septendecim tredecim* = *Magicicada septendecim tredecim* = *Magicicada septendecim tridecim* (sic) = *Megacicada* (sic) *tredecim* = *Magicicada tridecim* (sic) = *Magicicada tredecem* (sic) = *Magicicada tredicim* (sic)
M. tredecula Alexander & Moore, 1962

Ueana Distant, 1905 = *Ueana (Tibicen)* = *Uena* (sic) = *Neana* (sic)
U. bergerae Boulard, 1993

U. boudinoti Boulard, 1992
U. bouleti Boulard, 1993 = *Uena* (sic) *bouleti*
U. crepitans Boulard, 1992 = *Uena* (sic) *crepitans*
U. desserti Boulard, 1997 = *Uena* (sic) *desserti*
U. fungigera Boulard, 1991
U. harmonia Kirkaldy, 1905 = *Ueana polymnia* Kirkaldy, 1905
U. hyalinata Boulard, 1994
U. latreillae Boulard, 1993 = *Uena* (sic) *latreillae*
U. letocarti Boulard, 1992
U. lifuana (Montrouzier, 1861) = *Cicada lifuana* = *Tibicen lifuana* = *Neana* (sic) *lifuana*
U. lullae Boulard, 1994
U. maculata Distant, 1906
U. montaguei Distant, 1920
U. procera Boulard, 1994
U. quirosi Boulard, 1994
U. rosacea (Distant, 1892) = *Melampsalta rosacea* = *Uena* (sic) *rosacea* = *Parasemia rosacea*
U. simonae Boulard, 1993
U. speciosa Boulard, 1992
U. spectible Boulard, 1997
U. stasserae Boulard, 1994
U. tintinnabula Boulard, 1991 = *Uena* (sic) *tintinabula*
U. tiyanni Boulard, 1992
U. variegata Boulard, 1993

Nelcyndana Distant, 1906 = *Nelcynda* Stål, 1870 (nec *Nelcynda* Walker, 1862) = *Tibicen (Nelcynda)* = *Necyndana* (sic)
N. borneensis Duffels, 2010
N. madagascariensis (Distant, 1905) = *Nelcynda madagascariensis*
N. mulu Duffels, 2010
N. tener (Stål, 1870) = *Tibicen (Nelcynda) tener* = *Nelcynda tener* = *Tibicen tener* = *Necyndana* (sic) *tener*
N. cf. *tener* Schouten, Duffels & Zaidi, 2004
N. vantoli Duffels, 2010

Trismarcha Karsch, 1891 = *Trimarcha* (sic)
T. angolensis Distant, 1905
T. atrata Distant, 1905 = *Trismarcha atra* (sic)
T. consobrina (Distant, 1920) = *Lemuriana consobrina*
T. decolorata Dlabola, 1960
T. excludens (Walker, 1858) = *Cicada excludens* = *Tibicen excludens* = *Trismarcha sericosta* Karsch, 1891 = *Trismarcha sanguinea*
T. exsul Jacobi, 1904
T. ferruginosa Karsch, 1891 = *Trimarcha* (sic) *ferruginosa* = *Trismarcha fuliginosa* Karsch, 1991
T. flavocostata (Distant, 1905) = *Lemuriana flavocostata*
T. guillaumeti Boulard, 1986
T. guineica Dlabola, 1971
T. heimi Boulard, 1975
T. lobayensis Boulard, 1973
T. nana Dlabola, 1960
T. sikorae (Distant, 1905) = *Lemuriana sikorae*
T. sirius (Distant, 1899) = *Tibicen sirius* = *Lemuriana sirius* = *Limuriana* (sic) *sirius*
T. umbrosa retusa Boulard, 1986 = *Trismarcha umbrosa* var. *retusa*
T. umbrosa umbrosa Karsch, 1891 = *Trismarcha umbrosa*
T. voeltzkowi Jacobi, 1917

Monomotapa Distant, 1879 = *Monomatapa* (sic)
M. insignis Distant, 1897 = *Mogana* (sic) *insignis* = *Monomatapa* (sic) *insignis* = *Mogannia* (sic) *insignis* = *Abricta offirmata* Dlabola, 1962

M. matoposa Boulard, 1980
M. socotrana Distant, 1905

Kanakia Distant, 1892
K. congrua Goding & Froggatt, 1904 = *Kanakia congruens* (sic)
K. flavoannulata (Distant, 1920) = *Abricta flavoannulata* = *Kanakia annulata* (sic)
K. gigas Boulard, 1988
K. parva Boulard, 1991
K. typica Distant, 1892

Abroma Stål, 1866 = *Tibicen (Abroma)* = *Abraoma* (sic)
A. antandroyae Boulard, 2008
A. apicalis Ollenbach, 1929
A. apicifera (Walker, 1850) = *Cicada terminus* Walker, 1850
A. bengalensis Distant, 1906
A. bowringi Distant, 1905 = *Abroma bouringi* (sic)
A. canopea Boulard, 2007
A. cincturae Boulard, 2009
A. egae (Distant, 1892)
A. ferraria (Stål, 1870) = *Tibicen (Abroma) ferrarius* = *Tibicen ferrarius* = *Abroma ferrarius*
A. guerinii (Signoret, 1847) = *Cicada guerinii* = *Tibicen (Abroma) guerinii* = *Abroma guerini* = *Lemuriana sikoriae* (nec Distant)
A. inaudibilis Boulard, 1999
A. maculicollis (Guérin-Méneville, 1838) = *Cicada maculicollis* = *Tibicen (Abroma) maculicollis* = *Abraoma* (sic) *maculicollis*
A. mameti Boulard, 1979
A. minor Jacobi, 1917
A. nubifurca (Walker, 1858) = *Cicada nubifurca* = *Tibicen nubifurca* = *Cicada nubifera* (sic) = *Cicada apicalis* Kirkaldy, 1891 = *Tibicen apicalis* = *Tibicen nubifurca apicalis*
A. orhanti Boulard, 2008
A. philippinensis Distant, 1905
A. probably n. sp. Schouten, Duffels & Zaidi, 2004
A. pumila (Distant, 1892) = *Tibicen pumila*
A. reducta (Jacobi, 1902) = *Tibicen reductus*
A. sp. 1 Zaidi, Ruslan & Mahadir, 1996
A. tahanensis Moulton, 1923
A. temperata (Walker, 1858) = *Cicada temperata* = *Cicada blandula* Walker, 1858
A. vinsoni Boulard, 1979

Allobroma Duffels, 2011
A. kedenburgi Breddin, 1905 = *Lemuriana chandaea* Moulton, 1912 = *Abroma maculicollis* (nec Guérin-Méneville, 1838)
A. portegiesi Duffels, 2011

Sundabroma Duffels, 2011
S. bimaculata Duffels, 2011
S. suffusca Duffels, 2011

Malagasia Distant, 1882 = *Tibicen (Epora)* Stål, 1866 = *Epora*
M. aperta (Signoret, 1860) = *Cicada aperta* = *Tibicen (Epora) apertus*
M. inflata Distant, 1882
M. mariae Boulard, 1980

Malgachialna Boulard, 1980
M. noualhieri Boulard, 1980

Musimoia China, 1929
M. burri China, 1929

Hylora Boulard, 1971
H. bonneti Boulard, 1975
H. differata (Dlabola, 1960) = *Panka differata* = *Hylora centralensis* Boulard, 1971
H. mondziana Boulard, 1973
H. villiers Boulard, 1984

Ligymolpa Karsch, 1890 = *Ligymorpa* (sic)
L. madegassa Karsch, 1890

Lemuriana Distant, 1905 = *Limuriana* (sic)
L. apicalis (Germar, 1830) = *Cicada apicalis* = *Tibicen apicalis* = *Tibicen (Abroma) apicalis* = *Limuriana* (sic) *apicalis* = *Limuriana* (sic) *apicatus* (sic) = *Cicada semicinta* Walker, 1850 = *Tibicen semicinctus*
L. cambodiana Lee, 2010
L. vinhcuuensis Pham & Yang, 2010 = *Lemuriana* undescribed species Pham & Yang, 2010

Unduncus Duffels, 2011
U. connexa (Distant, 1910) = *Lemuriana connexa*

Panka Distant, 1905
P. africana Distant, 1905
P. duartei Boulard, 1975
P. lunguncus Boulard, 1970
P. minimuncus Boulard, 1970
P. parvula Boulard, 1973
P. parvulina Boulard, 1995
P. silvestris Jacobi, 1912
P. simulata Distant, 1905 = *Tibicen nubifurca* Distant, 1892 (nec *Tibicen nubifurca* Walker, 1858)
P. umbrosa Distant, 1920

Nyara Villet, 1999
N. thanatotica Villet, 1999

Oudeboschia Distant, 1920
O. festiva Distant, 1920

Neomuda Distant, 1920
N. abdominalis Distant, 1920
N. trimeni Distant, 1920

Viettealna Boulard, 1980
V. griveaudi Boulard, 1980

Subtribe Anopercalnina Boulard, 2008

Anopercalna Boulard, 2008
A. distanti (Karsch, 1890) = *Malagasia distanti*
A. vadoni (Boulard, 1980) = *Malagasia vadoni*
A. viettei Boulard, 2008
A. virescens (Distant, 1905) = *Malagasia virescens*

Tribe Cicadettini Buckton, 1889 = Melampsaltini Distant, 1905 = Melampsaltaria

Subtribe Cicadettina Buckton, 1889

Auta Distant, 1897
- *A. daymanensis* Boer, 1999
- *A. insignis* Distant, 1897
- *A. tapinensis* Boer, 1999

Nigripsaltria Boer, 1999 = *Nigripsalta* (sic)
- *N. carinata* Boer, 1999 = *Nigripsalta* (sic) *carinata*
- *N. dali* Boer, 1999
- *N. mouldsi* Boer, 1999
- *N. tagulaensis* Boer, 1999

Toxopeusella Schmidt, 1926
- *T. fakfakensis* Boer, 1999
- *T. moluccana* (Kirkaldy, 1909) = *Baeturia moluccana* = *Melampsalta moluccana* = *Cicada stigma* Walker, 1870 (nec *Cicada stigma* Walker, 1850) = *Baeturia stigma* = *Toxopeusella stigma* = *Melampsalta stigma* = *Toxopeusella browni* Boulard, 1981 = *Toxopeusella cheesmanae* Boulard, 1981 = *Toxopeusella smarti* Boulard, 1981
- *T. montana* Boer, 1999

Pagiphora Horváth, 1912 = *Enneaglena* Haupt, 1917
- *P. annulata* (Brullé, 1832) = *Pagiphora anullata* (sic) = *Tibicen annulatus* = *Cicada annulata* = *Cicadetta annulata* = *Cicadetta anulata* (sic) = *Parnisa anulata* (sic) = *Melampsalta annulata* = *Melampsalta (Melampsalta) annulata* = *Pagiphora annullata* (sic)
- *P. aschei* Kartal, 1978
- *P. hauptosa* Boulard, 1981 = *Enneaglena annulata* Haupt, 1917 = *Pagiphora annulata* (nec Horváth)
- *P. maghrebensis* Boulard, 1981
- *P. yanni* Boulard, 1992

Tympanistalna Boulard, 1982
- *T. distincta* (Rambur, 1840) = *Cicada distincta* = *Cicadetta distincta* = *Melampsalta distincta* = *Melampsalta (Melampsalta) distincta*
- *T. gastrica* (Stål, 1854) = *Cicada gastrica* = *Melampsalta gastrica* = *Cicadetta gastrica* = *Melampsalta (Melampsalta) gastrica*

Cicadetta Kolenati, 1857 = *Cicada (Cicadetta)* = *Cacadetta* (sic) = *Cicadette* (sic) = *Sicadetta* (sic) = *Cicadelta* (sic) = *Heptaglena* Horváth, 1911 (nec *Heptaglena* Schmarda, 1849) = *Oligoglena* Horváth, 1912 = *Heptaglena (Oligglena)*
- *C. abscondita* Lee, 2008 = *Takapsalta ichinosawana* (nec Matsumura) = *Takapsalta* sp. Kato, 1956 = *Cicadetta montana* (nec Scopoli) = *Cicadetta* sp. Gogala & Trilar 2004
- *C. afghanistanica* Dlabola, 1964
- *C. albipennis* Fieber, 1876 = *Melampsalta albipennis* = *Melampsalta (Melampsalta) albipennis*
- *C. anapaistica* Hertach, 2011
- *C. brevipennis* Fieber, 1872 = *Cicadetta brevipenis* (sic) = *Cicadetta montana* var. *brevipennis* = *Cicadetta montana brevipennis* = *Melampsalta montana brevipennis* = *Melampsalta (Melampsalta) montana brevipennis* = *Cicadetta montana* var. *brevipennis* = *Cicadetta montana* (nec Scopoli)
- *C. calliope calliope* (Walker, 1850) = *Cicada calliope* = *Melampsalta calliope* = *Cicada parvula* Say, 1825 = *Melampsalta parvula* = *Carineta parvula* = *Cicada pallescens* Germar, 1830 = *Melampsalta pallescens*
- *C. calliope floridensis* (Davis, 1920) = *Melampsalta calliope floridensis* = *Cicadetta calliope* var. *floridensis* = *Melampsalta floridensis*
- *C. camerona* (Davis, 1920) = *Melampsalta camerona*
- *C. cantilatrix* Sueur & Puissant, 2007 = *Cicadetta cerdaniensis* (nec Puissant & Boulard)
- *C. capistrata* (Ashton, 1912) = *Melampsalta capistrata*
- *C. caucasica* Kolenati, 1857 = *Cicada (Tettigetta) tibialis caucasica* = *Tettigetta tibialis caucasica* = *Cicadetta prasina caucasica* = *Cicadetta tibialis caucasica* = *Melampsalta (Pauropsalta) caucasica* = *Pauropsalta tibialis caucasica* = *Melampsalta tibialis caucasica* = *Cicada tibialis imbecilis* Eversman, 1859 = *Cicadetta sareptana* Fieber, 1876 = *Melampsalta sareptana* = *Melampsalta (Melampsalta) sareptana*

C. cerdaniensis Puissant & Boulard, 2000 = *Cicadetta montana* (nec Scopoli)
C. chaharensis (Kato, 1938) = *Melampsalta chaharensis*
C. chinensis Distant, 1905 = *Tibicina chinensis* = *Tibicen chinensis*
C. concinna (Germar, 1821) = *Cicada concinna* = *Cicada coccina* (sic) = *Cicadetta concina* (sic) = *Cicadatra concinna* = *Cicada adusta concinna* = *Melampsalta (Melampsalta) anglica concinna* = *Cicadatra concinna concinna* = "*Cicadatra concinna*" = *Cicadetta concina* (sic) = *Cicada aestivalis* Eversmann, 1837 nom. nud. = *Cicada subapicalis* Walker, 1858 = *Cicadetta subapicalis* = *Melampsalta (Melampsalta) subapicalis* = *Cicada montana* var. *adusta* Hagen, 1856 = *Cicada adusta* = *Cicadetta adusta* = *Melampsalta adusta* = *Cicada (Cicadetta) montana adusta* = *Cicada montana* var. *adusta* = *Melampsalta (Melampsalta) adusta* = *Cicadetta montana adusta* = *Cicada concinna* var. Fieber, 1876
C. diminuta Horváth, 1907 = *Melampsalta diminuta* = *Melampsalta (Melampsalta) diminuta*
C. dirfica Gogala, Trilar & Drosopoulos, 2011
C. fangoana fangoana Boulard, 1976 = *Cicadetta (Cicadetta) fangoana* = *Cicada haematodes* (nec Scopoli, nec Linnaeus) = *Cicadetta montana* var. *dimidiata* (nec Fabricius) = *Cicadetta montana* (nec Scopoli)
C. fangoana rufinervis Boulard, 1980 = *Cicadetta fangoana* var. *rufinervis*
C. fuscoclavalis (Chen, 1943) = *Melampsalta fuscoclavalis* = *Melampsalta fuscoclavalis chungnanshana* Chen, 1943 = *Cicadetta fuscoclavalis chungnanshana*
C. haematophleps Fieber, 1876
C. hageni Fieber, 1872 = *Cicada hageni* = *Melampsalta hageni* = *Melampsalta (Melampsalta) hageni* = *Cicada annulata* Hagen, 1856 (nec *Tibicen annulata* Brullé, 1832) = *Cicadetta annulata* = *Cicada angulata* (sic) = *Tibicen annulatus* (nec Brullé) = *Melampsalta (Melampsalta) annulata*
C. hannekeae Gogala, Drosopoulos & Trilar, 2008
C. hodoharai (Kato, 1940) = *Melampsalta hodoharai*
C. inglisi (Ollenbach, 1929) = *Melampsalta inglisi*
C. inserta Horváth, 1911 = *Melampsalta inserta* = *Melampsalta (Melampsalta) inserta*
C. intermedia (Ollenbach, 1929) = *Melampsalta intermedia*
C. iphigenia Emeljanov, 1996 = *Cicadetta tibicalis* (nec Panzer)
C. juncta (Walker, 1850) = *Cicada juncta* = *Melampsalta juncta*
C. kansa (Davis, 1919) = *Melampsalta kansa* = *Melampsalta parvula* (nec Say)
C. kissavi Gogala, Drosopoulos & Trilar, 2009
C. kollari Fieber, 1876 = *Melampsalta kollari* = *Melampsalta (Melampsalta) kollari* = *Cicadetta collari* (sic) = *Cicada brachyptera* Hagen, 1856 nom. nud. = *Cicada brachyptera* Kolenati, 1857 nom. nud.
C. konoi (Kato, 1937) = *Melampsalta konoi* = *Melampsalta yezoensis* Kato, 1932 (nec *Melampsalta yezoensis* Matsumura, 1898)
C. laevifrons (Stål, 1870) = *Melampsalta laevifrons*
C. limitata (Walker, 1852) = *Cicada limitata* = *Melampsalta limitata* = *Cicada signifera* Waker, 1850 (nec *Cicada signifera* Germar, 1830)
C. macedonica Schedl, 1999 = *Cicadetta montana macedonica*
C. mediterranea Fieber, 1872 = *Melampsalta mediterranea* = *Melampsalta (Melampsalta) mediterranea* = *Euryphara mediterranea*
C. minuta (Ollenbach, 1929) = *Melampsalta minuta*
C. montana montana (Scolpoli, 1772) = *Cicada montana* = *Cicada montata* (sic) = *Tettigonia montana* = *Cicada (Cicadetta) montana* = *Melampsalta montana* = *Melampsalta (Melampsalta) montana* = *Tettigia montana* = *Cicadetta (Melampsalta) montana* = *Cicadetta montana montana* = *Cacadetta* (sic) *montana* = *Cicadette* (sic) *montana* = *Cicadetta montana* form *typica* = *Cicada haematodes* Linnaeus, 1767 (nec Scopoli) = *Tettigonia haematodes* (nec Scopoli) = *Cicada hoematodes* (sic) (nec Scopoli) = *Cicada hematodes* (sic) (nec Scopoli) = *Cicada haematodis* (sic) (nec Scolopli) = *Cicada haematodes* (nec Scopoli, nec Linnaeus) = *Cicada haematoda* (sic) (nec Scopoli) = *Cicada haemalodes* (sic) (nec Scopoli) = *Cicadetta haematodes* (nec Scopoli) = *Cicada haematoides* (sic) (nec Scolopli) = *Melampsalta haematodes* (nec Scolopli) = *Melampsalta (Melampsalta) haematodes* (nec Scopoli) = *Tetigonia* (sic) *secunda* Schaeffer, 1767 = *Tetigonia* (sic) *prima* Schaeffer, 1767 = *Cicada flavofenestrata* Goeze, 1778 = *Cicadetta montana flavofenestrata* = *Cicadetta montana* var. *flavofenestrata* = *Cicadetta montana* form *flavofenestrata* = *Cicada schaefferi* Gmelin, 1789 = *Tettigonia schaefferi* = *Melampsalta (Melampsalta) schaefferi* = *Tettigonia dimidiata* Fabricius, 1803 = *Cicadetta montana* var. *dimidiata* = *Cicadetta montana dimidiata* = *Cicadetta montana* form *dimidiata* = *Cicada anglica* Samouelle, 1824 = *Melampsalta (Melampsalta) anglica* = *Cicada parvula* Walker, 1850 = *Cicada saxonica* Hartwig, 1857 = *Melampsalta (Melampsalta) saxonica* = *Cicadetta megerlei* Fieber, 1872 = *Cicadetta montana megerlei* = *Cicadetta*

montana var. *megerlei* = *Melampsalta megerlei* = *Melampsalta (Melampsalta) megerlei* = *Cicadetta montana longipennis* Fieber, 1876 = *Cicada orni* (nec Linnaeus) = *Cicada tibialis* (nec Panzer) = *Cicadetta tibialis* (nec Panzer) = *Cicadetta montana* var. *longipennis* (partim) = *Cicadetta montana longipenis* (sic) = *Cicadetta pygmea* (nec Olivier) = *Cicadetta pygmaea* (sic) (nec Olivier) = *Tettigonia sanguinea* (nec Fabricius) = *Cicadetta montana* var. *brevipennis* (partim) = *Cicadetta* cf. *montana* Hugel et al. 2008 = *Takapsalta ichinosawana* (nec Matsumura)

C. montana peregrina Gogala, 1998

C. cf. *montana* Gogala & Trilar, 2004

C. cf. *montana* Vincent 2006

C. cf. *montana* Sueur & Puissant 2007

C. cf. *montana* Niedringhaus, Biedermann & Nickel 2010

C. nigropilosa Logvinenko, 1976

C. olympica Gogala, Drosopoulos & Trilar, 2009

C. pellosoma (Uhler, 1861) = *Cicada pellosoma* = *Melampsalta pellosoma* = *Melampalta* (sic) *pellosoma* = *Melampsalta (Melampsalta) pellosoma* = *Melampsalta pellosoma concolor* Kato, 1932 = *Melampsalta pellosoma* var. *concolor* = *Melampsalta pellosoma flava* Kato, 1927 = *Melampsalta pellosoma* var. *flava*

C. petrophila Popov, Aronov & Sergeyeva, 1985

C. petryi Schumacher, 1924 = *Cicadetta montana petryi* = *Cicadetta montana* form *petryi*

C. pieli (Kato, 1940) = *Melampsalta pieli* = *Melampsalta wulsini* Liu, 1940

C. pilosa Horváth, 1901 = *Pauropsalta pilosa*

C. podolica Eichwald, 1830 = *Cicada podolica* = *Melampsalta (Melampsalta) podolica* = *Cicadetta adusta* (nec Hagen) = *Cicadetta concinna* (nec Germar)

C. cf. *podolica* Gogala, Drosopoulos & Trilar, 2008

C. popovi Emeljanov, 1996 = *Cicadetta* ex. gr. *tibialis* Popov, 1985

C. ramosi Sanborn, 2009 = *Cicadetta calliope* (nec Walker)

C. sarasini (Distant, 1914) = *Melampsalta sarasini*

C. shansiensis (Esaki & Ishihara, 1950) = *Melampsalta shansiensis*

C. sibilatrix Horváth, 1901 = *Pauropsalta sibilatrix* = *Melampsalta (Pauropsalta) sibilatrix* = *Melampsalta (Cicadetta) sibilatrix* = *Melampsalta sibilathrix* (sic)

C. sp. n. Drosopoulos, Asche & Hoch 1986

C. surinamensis Kirkaldy 1909 = *Cicada marginella* Olivier, 1790 (nec *Cicada marginella* Fabricius, 1787) = *Melampsalta marginella*

C. texana (Davis, 1936) = *Melampsalta texana*

C. transylvanica hyalinervis Boulard, 1981 = *Cicadetta transylvanica* var. *hyalinervis*

C. transylvanica transylvanica Fieber, 1876 = *Cicadetta transsylvanica* (sic) = *Cicadetta assylvanica* (sic) = *Melampsalta transylvanica* = *Melampsalta (Melampsalta) transsylvanica* (sic) = *Cicadivetta* (sic) *transylvanica*

C. tumidifrons Horváth, 1911 = *Cicadetta (Pauropsalta) tumidifrons* = *Melampsalta (Pauropsalta) tumidifrons* = *Melampsalta tumidifrons*

C. tunisiaca (Karsch, 1890) = *Melampsalta tunisiaca* = *Melampsalta (Melampsalta) tunisiaca*

C. turcica (Schedl, 2001)

C. ventricosa (Stål, 1866) = *Melampsalta ventricosa*

C. walkerella Kirkaldy, 1909 = *Cicada connexa* Walker, 1850: 177 (nec *Cicada connexa* Walker, 1850: 173) = *Melampsalta connexa*

C. yamashitai (Esaki & Ishihara, 1950) = *Melampsalta yamashitai*

Euboeana Gogala, Drosopoulos & Trilar, 2011

E. castaneivaga Gogala, Drosopoulos & Trilar, 2011

Cicadivetta Boulard, 1982 = *Cicadetta (Cicadivetta)*

C. carayoni (Boulard, 1982) = *Tettigetta carayoni*

C. flaveola (Brullé, 1832) = *Tettigetta flaveola* = *Tibicen flaveolus* = *Cicada flaveola* = *Cicadetta flaveola* = *Cicadetta (Cicadivetta) flaveola* = *Melampsalta flaveola* = *Melampsalta (Melampsalta) flaveola*

C. goumenissa Gogala, Drosopoulos & Trilar, 2012

C. parvula (Fieber, 1876) = *Tettigetta parvula* = *Cicadetta parvula* = *Melampsalta parvula* = *Cicadetta fieberi* Oshanin, 1908 = *Cicadetta fieberi* Kirkaldy, 1909 = *Melampsalta (Melampsalta) fieberi* = *Melampsalta fieberi* = *Heptaglena (Oligglena) libanotica* Horváth, 1911 = *Oligglena libanotica*

C. tibialis (Panzer, 1798) = *Tettigonia tibialis* = *Cicada tibialis* = *Cicadetta tibialis* = *Cicada (Tettigetta) tibialis* = *Melampsalta tibialis* = *Pauropsalta tibialis* = *Melampsalta (Pauropsalta) tibialis* = *Melampsalta (Melampsalta) tibialis* = *Melampsalta tibalis* (sic) = *Cicadetta cissylvanica* Haupt, 1835 = *Cicada minor* Eversmann, 1837 = *Cicada tibialis minor* = *Cicadetta tibialis minor* = *Cicadivetta tibialis* var. *minor* = *Heptaglena libanotica* Horváth, 1911 = *Oligoglena libanotica* = *Cicadetta libanotica* = *Cicadetta tibialis acuta* Dlabola, 1961 = *Cicadivetta tibialis acuta*

Dimissalna Boulard, 2007

D. dimissa (Hagen, 1856) = *Cicada dimissa* = *Cicadetta dimissa* = *Melampsalta dimissa* = *Melampsalta (Melampsalta) dimissa* = *Melampsalta dismissa* (sic) = *Tettigetta dimissa* = *Melampsalta hodoharai* (nec Kato)

Curvicicada Chou & Lu, 1997 = *Oedicida* (sic)

C. xinjiangensis Chou & Lu, 1997 = *Curvicicada buerensis* (sic)

Pinheya Dlabola, 1963

P. violacea violacea (Linnaeus, 1758) = *Cicada violacea* = *Tettigonia violacea* = *Tettigonia violecea* (sic) = *Melampsalta violacea* = *Cicada hottentota* Olivier, 1790 = *Cicada hottentotta* (sic) = *Cicadetta hottentota*

P. violacea var. *β* Walker, 1850 = *Cicada hottentota* var. *β* = *Cicadetta violacea* var. *β*

Hilaphura Webb, 1979 = *Hylaphura* (sic)

H. varipes (Waltl, 1837) = *Cicada varipes* = *Cicadetta varipes* = *Melampsalta varipes* = *Melampsalta (Melampsalta) varipes* = *Melapmpsalta varipes* = *Cicada segetum* Rambur, 1840 = *Cicadatra segetum* = *Cicadetta segetum* = *Melampsalta segetum* = *Hilaphura segetum* = *Hylaphura* (sic) *segetum* = *Cicada picta* Germar, 1830 (nec *Cicada picta* Fabricius, 1794) = *Melampsalta picta* (nec Fabricius) = *Cicadetta picta* (nec Fabricius) = *Melampsalta (Melampsalta) picta* (nec Fabricius) = *Cicadetta decorata* Kirkaldy, 1909 = *Melampsalta (Melampsalta) decorata* = *Cicada musiva* var. *caspica* (nec Kolenati)

Euryphara Horváth, 1912

E. cantans (Fabricius, 1794) = *Tettigonia cantans* = *Cicada cantans* = *Melampsalta cantans* = *Melampsalta (Melampsalta) cantans* = *Cicadetta cantans* = *Cicada haematodes* Linnaeus, 1767 (nec *Cicada haematodes* Scopoli, 1763)

E. contentei Boulard, 1982

E. dubia (Rambur, 1840) = *Cicada dubia* = *Melampsalta dubia* = *Melampsalta (Melampsalta) dubia* = *Cicadetta dubia* = *Cicadetta euphorbiae* Fieber, 1876 = *Melampsalta (Melampsalta) euphorbiae*

E. ribauti Boulard, 1982

E. undulata (Waltl, 1837) = *Cicada undulata* = *Cicadetta undulata* = *Melampsalta (Melampsalta) undulata*

E. virens (Herrich-Schäffer, 1835) = *Cicada virens* = *Cicadetta virens* = *Melampsalta (Melampsalta) virens* = *Tettigetta virens*

Saticula Stål, 1866 = *Sacula* (sic)

S. coriaria Stål, 1866 = *Cicadetta coriaria* = *Cicadetta coriacea* (sic) = *Cicadetta coriascea* (sic) = *Melampsalta coriaria* = *Sacula* (sic) *coriaria* = *Cicadetta violacea* Hagen, 1855 (nec *Cicada violacea* Linnaeus, 1758) = *Cicadetta auratiaca* Puton, 1883 = *Cicadetta aurantica* (sic) = *Melampsalta aurantiaca* = *Melampsalta (Melampsalta) aurantiaca* = *Saticula aurantiaca*

S. vayssieresi Boulard, 1988

Xossarella Boulard, 1980

X. giovanninae Boulard, 1980

X. marmottani Boulard, 1980

X. soussensis Boulard, 1980

Takapsalta Matsumura, 1927

T. ichinosawana Matsumura, 1927 = *Cicadetta ichinosawana*

T. komaensis Kato, 1944

Kosemia Matsumura, 1927 = *Leptopsalta* Kato, 1928 = *Leptosalta* (sic) = *Loptopsalta* (sic) = *Karapsalta* Matsumura, 1931

K. admirabilis (Kato, 1927) = *Melampsalta admirabilis* = *Leptopsalta admirabilis* = *Leptopsalta apmirabilis* (sic) = *Leptopsalta abmirabilis* (sic) = *Leptosalta* (sic) *admirabilis* = *Cicadetta admirabilis* = *Leptopsalta admirabilis kishidae* Kato, 1932 = *Leptopsalta admirabilis* var. *kishidai* = *Leptopsalta kishidae* = *Melampsalta radiator* Liu (nec Uhler)

K. bifuscata (Liu, 1940) = *Melampsalta bifuscata* = *Leptopsalta bifuscata*

K. fuscoclavalis (Chen, 1942) = *Melampsalta fuscoclavalis* = *Cicadetta fuscoclavalis* = *Leptopsalta fuscoclavalis* = *Leptopsalta fuscoflavelis* (sic) = *Leptopsalta juscoclavalis* (sic) = *Melampsalta fuscoclavalis chungnanshana* Chen, 1943

K. phaea (Chou & Lei, 1997) = *Leptopsalta phaea*

K. prominentis (Lei & Chou, 1997) = *Leptopsalta prominentis*

K. radiator (Uhler, 1896) = *Melampsalta radiator* = *Melampsaltria* (sic) *radiator* = *Cicadetta radiator* = *Melampsalta (Melampsalta) radiator* = *Tibicen radiator* = *Leptopsalta radiator* = *Loptopsalta* (sic) *radiator*

K. rubicosta (Chou & Lei, 1997) = *Leptopsalta rubicosta*

K. yamashitai (Esaki & Ishihara, 1950) = *Melampsalta yamashitai* = *Leptopsalta yamashitai* = *Leptosalta* (sic) *yamashitai*

K. yezoensis (Matsumura, 1898) = *Melampsaltria* (sic) *yezoensis* = *Melampsalta yezoensis* = *Melampsalta (Melampsalta) yezoensis* = *Melampsalta yezoensis gigas* = *Cicadetta yezoensis* = *Cicadetta jezoensis* (sic) = *Leptopsalta yezoensis* = *Cicadetta sachalinensis* Matsumura, 1917 = *Melampsalta sachalinensis* = *Kosemia sachalinensis* = *Karapsalta sachalinensis* = *Cicada orni* (nec Linnaeus) = *Melampsalta konoi* (nec Kato)

Melampsalta Kolenati, 1857 = *Cicada (Melampsalta)* = *Melampsalta (Melampsalta)* = *Cicadetta (Melampsalta)* = *Cicadetta (Melampsarta)* (sic) = *Melampsaltria* (sic) = *Melampalta* (sic) = *Melampsaltera* (sic) = *Melamosalta* (sic) = *Malampsalta* (sic) = *Melampsata* (sic) = *Parasemia* Matsumura, 1927 (nec *Parasemia* Hübner, 1820) = *Parasemoides* Strand, 1928

M. aethiopica Distant, 1905 = *Cicadetta aethiopica*

M. albeola (Eversmann, 1859) = *Cicada albeola* = *Cicadetta albeola* = *Melampsalta (Melampsalta) albeola* = *Cicadetta (Melampsalta) albeola*

M. artensis (Montrouzier, 1861) = *Cicada artensis*

M. bifuscata Liu, 1940 = *Melampsalta bifasciata* (sic) = *Cicadetta bifasciata* (sic)

M. cadisia (Walker, 1850) = *Cicada cadisia* = *Cicada variegata* Olivier, 1790 = *Melampsalta variegata* = *Cicada pellucida* Germar, 1830 = *Cicada ruficollis* (nec Thunberg) = *Melampsalta ruficollis* (nec Thunberg)

M. caspica (Kolenati, 1857) = *Cicada (Melampsalta) musiva caspica* = *Cicadetta musiva caspica* = *Cicada musiva caspica* = *Melampsalta musiva caspica* = *Melampsalta (Melampsalta) musiva caspica* = *Melampsalta musiva* var. *caspica* = *Cicadetta caspica* = *Melampsalta caspia* (sic) = *Cicadetta musiva* = *Melampsalta (Cicadetta) caspica*

M. charharensis Kato, 1938

M. expansa Haupt, 1918 = *Cicadetta expansa*

M. fraseri China, 1938 = *Cicadetta fraseri*

M. germaini Distant, 1906 = *Cicadetta germaini* = *Germalna* (sic) *germaini*

M. kewelensis Distant, 1907 = *Melampsalta kewellensis* (sic) = *Cicadetta kewelensis* = *Cicadetta kewellensis* (sic)

M. kulingana Kato, 1938

M. leucoptera leucoptera (Germar, 1830) = *Cicada leucoptera* = *Melampsalta leucoptera* = *Melampsaltera* (sic) *leucoptera* = *Cicada fusconervosa* Stål, 1855 = *Cicadetta fusconervosa*

M. leucoptera var. *a* Stål, 1866

M. leucoptera var. *b* Stål, 1866

M. literata (Distant, 1888) = *Cicadetta literata* = *Cicadetta (Melampsalta) literata*

M. lobulata Fieber, 1876 = *Cicadetta lobulata* = *Melampsalta (Melampsalta) lobulata*

M. mogannia (Distant, 1905) = *Quintilia mogannia* = *Lycurgus mogannia* = *Lycurgus monganius* (sic) = *Cicadetta mogannia* = *Quintilia kozanensis* Ouci, 1938

M. mokanshanensis Ouchi, 1938 = *Cicadetta mokanshanensis*

M. neocruentata Liu, 1940 = *Melamosalta* (sic) *neocruentata* = *Cicadetta neocruentata*

M. occidentalis (Schumacher, 1922) = *Cicadetta (Melampsalta) occidentalis* = *Cicadetta occidentalis*

M. sinuatipennis (Oshanin, 1906) = *Cicadetta sinuatipennis* = *Cicadetta sinautipennis* (sic) = *Melampsalta (Melampsalta) sinuatipennis* = *Cicadetta (Melampsalta) sinuatipennis*

M. soulii soulii (Distant, 1905) = *Quintilia soulii* = *Melampsalta soulii* = *Cicadetta soulii* = *Lycurgus soulii*
M. soulii var. *a* (Distant, 1905) = *Quintilia soulii* var. *a* = *Cicadetta soulii* var. *a*
M. soulii var. *b* (Distant, 1905) = *Quintilia soulii* var. *b* = *Cicadetta soulii* var. *b*
M. sp. a Phillips, Sanborn & Villet, 2002
M. sp. b Phillips, Sanborn & Villet, 2002
M. telxiope (Walker, 1850) = *Cicada telxiope* = *Cicadetta telxiope* = *Cicada arche* Walker, 1850 = *Cicadetta arche* = *Cicada duplex* Walker, 1850 = *Melampsalta duplex*
M. uchiyamae (Matsumura, 1927) = *Parasemia uchiyamae* = *Melampsalta rosacea* (nec Distant)
M. yapensis Esaki, 1947
M. yasumatsui Esaki, 1947

Linguacicada Chou & Lu, 1997
L. continuata (Distant, 1888) = *Cicadetta continuata* = *Melampsalta continuata* = *Melampsalta (Melampsalta) continuata* = *Cicadetta (Melampsalta) continuata* = *Curvicicada continuata*

Amphipsalta Fleming, 1969
A. cingulata cingulata (Fabricius, 1775) = *Tettigonia cingulata* = *Cicada cingulata* = *Melampsalta cingulata* = *Cicadetta cingulata* = *Cicada mendosa* Walker, 1858 = *Cicada indivulsa* Walker, 1858 = *Cicada indivisula* (sic) = *Cicadetta indivulsa* = *Cicada flexicosta* Stål, 1859 = *Melampsalta flexicosta* = *Cicada flavicosta* (sic)
A. cingulata var. ______ (Walker, 1852) = *Cicada zealandica* (sic) var. ______
A. strepitans (Kirkaldy, 1891) = *Cicadetta strepitans* = *Melampsalta strepitans* = *Cicada cingulara obscura* Hudson, 1891 = *Melampsalta cingulata obscura* = *Melampsalta obscura* = *Cicadetta obscura* = *Melampsalta cingulata* Kirby (nec Fabricius)
A. zelandica (Boisduval, 1835) = *Cicada zelandica* = *Cicada zealandica* (sic) = *Cicada zeylandica* (sic) = *Melampsalta zelandica* = *Melampsalta zealandica* (sic) = *Cicadetta zealandica* (sic) = *Amphipsalta zealandica* (sic)

Tibeta Lei & Chou, 1997
T. minensis Lei & Chou, 1997
T. nanna Lei & Chou, 1997
T. planarius Lei & Chou, 1997
T. zenobia (Distant, 1912) = *Melampsalta zenobia* = *Cicadetta zenobia* = *Melampsalta kulingana* Kato, 1938

Scolopita Chou & Lei, 1997
S. cupis Chou & Lu, 1997
S. fusca Chou & Lei, 1997
S. lusiplex Chou & Lei, 1997 = *Scolopita* sp. 1 Pham & Thinh, 2006
S. mokanshanensis (Ouchi, 1938) = *Melampsalta mokanshanensis*

Samaecicada Popple & Emery, 2010
S. subolivacea (Ashton, 1912) = *Pauropsalta subolivacea* = *Melampsalta subolivacea*

Gudanga Distant, 1905 = *Paragudanga* Distant, 1913
G. adamsi Moulds, 1996 = *Gudanga* sp. nov. Lithgow, 1988 = *Gudanga* sp. Ewart, 1988 = *Gudanga browni* (nec Distant)
G. sp. nr. *adamsi* Southern Brigalow Blackwing Popple & Strange 2002
G. sp. nr. *adamsi* Watson, Watson, Hu, Brown, Cribb & Myhra, 2010
G. aurea Moulds, 1996
G. boulayi Distant, 1905
G. browni (Distant, 1913) = *Paragudanga browni*
G. kalgoorliensis Moulds, 1996
G. lithgowae Ewart & Popple, 2013 = *Gudanga* sp. Ewart, 1988 = *Gudanga* n. sp. Ewart, 1989 = *Gudanga* sp. A Ewart and Popple, 2001 = *Gudanga adamsi* (nec Moulds)
G. nowlandi Ewart & Popple, 2013 = *Gudanga* sp. B Ewart and Popple, 2001
G. pterolongata Olive, 2007

G. solata Moulds, 1996

Adelia Moulds, 2012

A. borealis (Goding & Froggatt, 1904) = *Tibicen borealis* = *Abricta borealis*

Diemeniana Distant, 1906 = *Diemenia* Distant, 1905 (nec *Diemenia* Spinola, 1850)

D. cincta (Fabricius, 1803) = *Tettigonia cincta* = *Cicada cincta* = *Tibicen cinctus* = *Abricta cincta* = *Diemeniana tillyardi* Hardy, 1918

D. euronotiana (Kirkaldy, 1909) = *Abricta euronotiana* = *Cicada aurata* Walker, 1850 (nec *Cicada aurata* Linnaeus, 1758) = *Tibicen auratus* = *Tibicen(?) auratus* = *Abricta aurata* = *Diemeniana richesi* Distant, 1913

D. frenchi (Distant, 1907) = *Abricta frenchi* = *Cicada coleoptrata* Walker, 1858 (nec *Cicada coleoptrata* Linnaeus, 1758) = *Tibicen coleoptratus* = *Diemeniana coleoptrata* = *Diemeniana tasmani* Kirkaldy, 1909

D. hirsuta (Goding & Froggatt, 1904) = *Tibicen hirsutus* = *Tibicen hirsuta* = *Abricta hirsuta* = *Diemeniana hirsutus* = *Diemeniana turneri* Distant, 1914

D. neboissi Burns, 1958

Uradolichos Moulds, 2012

U. longipennis (Ashton, 1914) = *Urabunana longipennis*

Pauropsalta Goding & Froggatt, 1904 = *Cicadetta (Pauropsalta)* = *Melampsalta (Pauropsalta)*

P. aktites Ewart, 1989 = *Psaltoda* (sic) *aktites*

P. annulata Goding & Froggatt, 1904

P. annulata song type *A* Popple, Walter & Raghu, 2008

P. annulata song type *B* Popple, Walter & Raghu, 2008

P. annulata song type *C* Popple, Walter & Raghu, 2008

P. annulata song type C_N Popple, Walter & Raghu, 2008

P. annulata song type C_S Popple, Walter & Raghu, 2008

P. annulata song type *X* Popple & Walter, 2010

P. sp. nr. *annulata* Southern Red-eyed Squeaker Moss & Popple, 2000

P. sp. nr. *annulata* Inland Sprinkler Squeaker Popple & Strange, 2002

P. sp. nr. *annulata* Maraca Squeaker Popple & Strange, 2002

P. sp. nr. *annulata* Static Squeaker Popple & Strange, 2002

P. sp. nr. *annulata* Emery, Emery, Emery & Popple, 2005

P. aquilus Ewart, 1989 = *Pauropsalta* sp. D Ewart, 1988

P. ayrensis Ewart, 1989

P. borealis Goding & Froggatt, 1904 = *Melampsalta borealis*

P. collina Ewart, 1989 = *Pauropsalta* sp. C Ewart, 1988

P. sp. nr. *collina* Emery, Emery, Emery & Popple, 2005

P. sp. nr. *collina/encaustica* Haywood, 2005

P. corticinus Ewart, 1989 = *Pauropsaltra* sp. A Ewart, 1986 = *Pauropsalta corticina*

P. sp. nr. *corticinus* Emery, Emery, Emery & Popple, 2005

P. dolens (Walker, 1850) = *Cicada dolens*

P. elgneri Ashton, 1912 = *Melampsalta elgneri*

P. encaustica (Germar, 1834) = *Cicada encaustica* = *Melampsalta encaustica* = *Cicada arclus* Walker, 1850 = *Melampsalta arclus* = *Pauropsalta arclus* = *Cicada juvenis* Walker, 1850 = *Cicada dolens* Walker, 1850 = *Pauropsalta dolens*

P. sp. nr. *P. encaustica* Moss, 1989

P. n. sp. #16M nr. *encaustica* Gwynne & Schatral, 1988

P. sp. nr. *encaustica* Sandstone Squeaker M26 Moss & Popple, 2000

P. sp. nr. *encaustica* Montane Squeaker M14 Moss & Popple, 2000

P. exaequata (Distant, 1892) = *Melampsalta exaequata* = *Melampsalta exequata* (sic) = *Pauropsalta exequata* (sic)

P. extensa Goding & Froggatt, 1904

P. extrema (Distant, 1892) = *Melampsalta extrema*

P. near *extrema* Marshall & Hill, 2009

P. fuscata Ewart, 1989 = *Pauorpsalta furcata* (sic) = *Pauropsalta* sp. B Ewart, 1986 = *Melampsalta (Pauropsalta) encaustica* (nec Germar)
P. sp. nr. *fuscata* Montane Grass Squeaker Moss & Popple, 2000
P. sp. nr. *fuscata* Moss & Popple, 2000
P. sp. nr. *fuscata* Haywood, 2005
P. infrasila Moulds, 1987
P. infuscata (Goding & Froggatt, 1904) = *Melampsalta infuscata* = *Cicadetta infuscata*
P. johanae Boulard, 1993
P. judithae Boulard, 1997 = *Pauropsalta juditae* (sic)
P. lineola Ashton, 1914 = *Melampsalta lineola*
P. melanopygia (Germar, 1834) = *Cicada melanopygia* = *Tibicen melanopygius* = *Melampsalta melanopygia* = *Pauropsalta leurensis* Goding & Froggatt, 1904
P. sp. nr. *melanopygia* Moss, 1988
P. mimica Distant, 1907
P. mneme (Walker, 1850) = *Cicada mneme* = *Melampsalta mneme* = *Melampsalta mnemae* (sic) = *Cicadetta (Pauropsalta) mneme* = *Cicada antica* Walker, 1850 = *Pauropsalta leurensis* Goding & Froggatt, 1904
P. nigristriga Goding & Froggatt, 1904 = *Pauropsalta nigristrigata* (sic)
P. opacus Ewart, 1989 = *Pauropsalta opaca*
P. prolongata Goding & Froggatt, 1904
P. rubea (Goding & Froggatt, 1904) = *Melampsalta rubea* = *Cicadetta rubea* = *Melampsalta nebulosa* Goding & Froggatt, 1904 = *Cicadetta nebulosa* = *Melampsalta geisha* Distant, 1915 = *Cicadetta geisha* = *Pauropsalta geisha*
P. rubra Goding & Froggatt, 1904 = *Melampsalta rubra*
P. rubristrigata (Goding & Froggatt, 1904) = *Melampsalta rubristrigata* = *Melampsalta rubrostrigata* (sic) = *Cicadetta rubristrigata*
P. rufifascia (Walker, 1850) = *Cicada rufifascia* = *Melampsalta rufifascia*
P. Sandstone Marshall & Hill, 2009
P. siccanus Ewart, 1989 = *Pauropsalta* sp. F Ewart, 1988
P. sp. nr. *siccanus* Callistris Squeaker Popple & Strange, 2002
P. sp. nov. 1 Lithgow, 1988
P. sp. nov. 2 Lithgow, 1988
P. sp. A Ewart, 2009
P. sp. C Ewart, 1993
P. sp. M Moulds, 2005
P. stigmatica Distant, 1905 = *Melampsalta stigmatica*
P. walkeri Moulds & Owen, 2011
P. near *walkeri* Marshall & Hill, 2009

Graminitigrina Ewart & Marques, 2008
G. bolloni Ewart & Marques, 2008
G. bowensis Ewart & Marques, 2008
G. carnarvonensis Ewart & Marques, 2008
G. karumbae Ewart & Marques, 2008 = *Pauropsalta* sp. D Ewart, 1993 = Heathlands sp. D Ewart, 2005
G. triodiae Ewart & Marques, 2008 = *Urabunana* sp. Dunn, 2002

Palapsalta Moulds, 2012
P. circumdata (Walker, 1852) = *Cicada circumdata* (nom. nov. pro *Cicada marginata* Leach, 1814 nec *Cicada marginata* Olivier, 1790) = *Pauropsalta circumdata* = *Tettigonia marginata* = *Cicada marginata* = *Melampsalta marginata* = *Cicadetta marginata* = *Pauropsalta marginata* = *Cicada themiscura* Walker, 1850 = *Melampsalta themiscura* = *Melampsalta fletcheri* Goding & Froggatt, 1904 = *Cicadetta fletcheri*
P. eyrei (Distant, 1882) = *Pauropsalta eyreyi* = *Pauropsalta eyrey* (sic) = *Melampalta eyrei* = *Cicadetta eyrei*
P. virgulatus (Ewart, 1989) = *Pauropsalta virgulatus*

P. vitellinus (Ewart, 1989) = *Pauropsalta vitellinus* = *Pauropsalta* sp. E Ewart, 1988 = *Pauropsalta* sp. near *melanopygia* Moss, 1988

Nanopsalta Moulds, 2012

N. basalis (Goding & Froggatt, 1904) = *Pauropsalta basalis* = *Melampsalta basalis* = *Pauropsalta endeavourensis* Distant, 1907 = *Melampsalta endeavourensis* = *Cicadetta endeavourensis*

Punia Moulds, 2012

P. minima (Goding & Froggatt, 1904) = *Cicadetta minima* = *Pauropsalta minima* = *Melampsalta minima*

Neopunia Moulds, 2012

N. graminis (Goding & Froggatt, 1904) = *Cicadetta graminis* = *Melampsalta graminis*

Marteena Moulds, 1986

M. rubricincta (Goding & Froggatt, 1904) = *Tibicen rubricinctus* = *Tibicen rubricincta* = *Tibicina rubricinctus* = *Tibicina rubricincta*

Auscala Moulds, 2012

A. spinosa (Goding & Froggatt, 1904) = *Cicadetta spinosa* = *Melampsalta spinosa* = *Kobonga clara* Distant, 1916

Birrima Distant, 1906

B. castanea (Goding & Froggatt, 1904) = *Pauropsalta castanea* = *Melampsalta castanea* = *Cicadetta castanea* = *Melampsalta fulva* Goding & Froggatt, 1904 = *Cicadetta fulva* = *Birrima montrouzieri* Distant, 1906 = *Cicadetta montrouzieri*

B. varians (Germar, 1834) = *Cicada varians* = *Dundubia varians* = *Melampsalta varians*

Yoyetta Moulds, 2012

Y. aaede (Walker, 1850) = *Melampsalta aaede* = *Cicadetta aaede*

Y. abdominalis (Distant, 1892) = *Melampsalta abdominalis* = *Cicadetta abdominalis* = *Pauropsalta castanea* Goding & Froggatt

Y. sp. nr. *abdominalis* sp. 4 Katydid Cicada (Moss & Popple, 2000) = *Cicadetta* sp. nr. *abdominalis* sp. 4 Katydid Cicada

Y. sp. nr. *abdominalis* (Emery, Emery, Emery & Popple, 2005) = *Cicadetta* sp. nr. *abdominalis*

Y. sp. aff. *abdominalis* (Haywood, 2005) = *Cicadetta* sp. aff. *abdominalis*

Y. sp. nr. *abdominalis* (Emery, 2008) = *Cicadetta* sp. nr. *abdominalis*

Y. celis (Moulds, 1988) = *Cicadetta celis*

Y. sp. nr. *celis* (Emery, Emery, Emery & Popple, 2005) = *Cicadetta* sp. nr. *celis*

Y. denisoni (Distant, 1893) = *Melampsalta denisoni* = *Cicadetta denisoni* = *Melampsalta kershawi* Goding & Froggatt, 1904 = *Cicadetta kershawi*

Y. sp. nr. *denisoni* Brown Firetail (Moss & Popple, 2000) = *Cicadetta* sp. nr. *denisoni*

Y. sp. nr. *denisoni* (Emery, Emery, Emery & Popple, 2005) = *Cicadetta* sp. nr. *denisoni*

Y. hunterorum (Moulds, 1988) = *Cicadetta hunterorum*

Y. incepta (Walker, 1850) = *Cicada incepta* = *Melampsalta incepta* = *Melempsalta* (sic) *incepta* = *Cicadetta incepta*

Y. sp. nr. *incepta* (Greenup, 1964) = *Melampsalta* sp. nr. *incepta*

Species complex near *Y. incepta* (Ewart & Popple, 2001) = species complex near *Cicadetta incepta*

Species complex near *Y. incepta* (Adavale) (Ewart & Popple, 2001) = species complex near *Cicadetta incepta* (Advale)

Species complex near *Y. incepta* (Blackall) (Ewart & Popple, 2001) = species complex near *Cicadetta incepta* (Blackall)

Species complex near *Y. incepta* (Charleville) (Ewart & Popple, 2001) = species complex near *Cicadetta incepta* (Charleville)

Y. sp. nr. *incepta* (Ewart, 2009) = *Cicadetta* sp. nr. *incepta*

Y. landsboroughi (Distant, 1882) = *Melampsalta landsboroughi* = *Melampsalta telxiope* (nec Walker) = *Cicadetta landsboroughi*

Y. sp. nr. *landsboroughi* (Ewart, 1986) = *Cicadetta* sp. nr. *landsboroughi*
Y. toowoombae (Distant, 1915) = *Melampsalta toowoombae* = *Cicadetta toowoombae*
Y. tristrigata (Goding & Froggatt, 1904) = *Melampsalta tristrigata* = *Cicadetta tristrigata*
Y. sp. nr. *tristrigata* sp. 5 Fiery Ambertail M43 (Moss & Popple, 2000) = *Cicadetta* sp. nr. *tristrigata* sp. 5 Fiery Ambertail M43
Y. sp. nr. *tristrigata* sp. 16 Robust Ambertail (Moss & Popple, 2000) = *Cicadetta* sp. nr. *tristrigata* sp. 16 Robust Ambertail
Y. sp. nr. *tristrigata* (Ewart & Popple, 2001) = *Cicadetta* sp. nr. *tristrigata*
Y. sp. nr. *tristrigata* (Popple & Ewart, 2002) = *Cicadetta* sp. nr. *tristrigata*
Y. sp. nr. *tristrigata* (Emery, Emery, Emery & Popple, 2005) = *Cicadetta* sp. nr. *tristrigata*
Y. sp. nr. *tristrigata* (Haywood, 2005) = *Cicadetta* sp. nr. *tristrigata*

Taurella Moulds, 2012
T. forresti (Distant, 1882) = *Melampsalta forresti* = *Cicadetta forresti* = *Melampsalta warburtoni* Distant, 1882 = *Cicadetta warburtoni* = *Melampsalta capistrata* Ashton, 1912 = *Cicadetta capistrata*
T. froggatti (Distant, 1907) = *Melampsalta froggatti* = *Cicadetta froggatti* = *Melampsalta sulcata* Distant, 1907 = *Cicadetta sulcata*
T. viridis (Ashton, 1912) = *Melampsalta viridis* = *Cicadetta viridis* = *Cicadetta* sp. nr. *viridis* Ewart, 1998

Urabunana Distant, 1905
U. sp. nov. Lithgow, 1988
U. segmentaria Distant, 1905
U. sericeivitta (Walker, 1862) = *Cicada seiceivitta* = *Melampsalta seicevitta* (sic) = *Pauropsaltria sericeivitta*
U. sp. A Ewart, 1998

Sylphoides Moulds, 2012
S. arenaria (Distant, 1907) = *Melampsalta arenaria* = *Melampsalta orenaria* (sic) = *Cicadetta arenaria* = *Cicadetta areuarai* (sic)

Pyropsalta Moulds, 2012
P. melete (Walker, 1850) = *Cicada melete* = *Melampsalta melete* = *Cicadetta melete* = *Melampsalta rubricincta* Goding & Froggatt, 1904 = *Cicadetta rubricincta*

Physeema Moulds, 2012
P. bellatrix (Ashton, 1914) = *Pauropsalta bellatrix* = *Melampsalta bellatrix*
P. convergens (Walker, 1850) = *Cicada convergens* = *Melampsalta convergens* = *Cicadetta convergens* = *Macrotristria intersecta* (nec Walker) = *Melampsalta landsboroughi convergens* = *Melampsalta cylindrica* Ashton, 1912 = *Cicadetta landsboroughi* var. *ashtoni* Metcalf, 1963 = *Melampsalta landsboroughi* var. *ashtoni*
P. labyrinthica (Walker, 1850) = *Dundubia labyrinthica* = *Melampsalta labyrinthica* = *Cicadetta labyrinthica* = *Cicada interstans* Walker, 1858 = *Melampsalta interstans*
P. sp. aff. *labyrinthica* (Haywood, 2005) = *Pauropsalta* sp. aff *labyrinthica*
P. latorea (Walker, 1850) = *Cicada latorea* = *Melampsalta latorea* = *Cicadetta latorea*
P. n. sp. #75M nr. *latorea* (Gwynne & Schatral, 1988) = *Cicadetta* n. sp. #75M nr. *latoria* (sic)
P. quadricincta (Walker, 1850) = *Cicada quadricincta* = *Melampsalta quadricincta* = *Cicadetta quadricincta*

Gelidea Moulds, 2012
G. torrida torrida (Erichson, 1842) = *Cicada torrida* = *Melampsalta torrida* = *Cicadetta torrida* = *Cicada basiflamma* Walker, 1850 = *Cicada connexa* Walker, 1850 = *Cicada damater* Walker, 1850 = *Melampsalta damater* = *Cicadetta (Tettigeta) torrida* = *Melampsalta spreta* Goding & Froggatt, 1904 = *Cicadetta spreta*
G. torrida var. *β* Walker, 1850 = *Cicada torrida* var. *β* = *Cicadetta torrida* var. *β* = *Cicadetta (Tettigeta) torrida* var. *β*
G. torrida var. *δ* Walker, 1850 = *Cicada torrida* var. *δ* = *Cicadetta torrida* var. *δ* = *Cicadetta (Tettigetta) torrida* var. *δ*

G. torrida var. γ Walker, 1850 = *Cicada torrida* var. γ = *Cicadetta torrida* var. γ = *Cicadetta (Tettigetta) torrida* var. γ

Clinopsalta Moulds, 2012

C. adelaida (Ashton, 1914) = *Melampsalta adelaida* = *Cicadetta adelaida*

C. sp. nr. *adelaida* (Emery, Emery, Emery & Popple, 2005) = *Cicadetta* sp. nr. *adelaida*

C. tigris (Ashton, 1914) = *Melampsalta tigris* = *Cicadetta tigris*

C. sp. nr. *tigris* (Ewart, 1988) = *Cicadetta* sp. nr. *tigris* = Species C Ewart & Popple, 2001

C. sp. nr. *tigris* (Ewart, 2009) = *Cicadetta* sp. nr. *tigris* = *Notopsalta* sp. G Ewart 1998

Galanga Moulds, 2012

G. labeculata (Distant, 1892) = *Melampsalta labeculata* = *Cicadetta labeculata*

G. sp. nr. *labeculata* (Greenup, 1964) = *Melampsalta* sp. nr. *labeculata*

Kobonga Distant, 1906

K. apicans Moulds & Kopestonsky, 2001

K. apicata (Ashton, 1914) = *Cicadetta apicata* = *Melampsalta apicata* = *Kobonga* sp. nr. *Cicadetta apicata* Ewart, 2009

K. sp. nr. *apicata* Callitris Clicker (Popple & Strange, 2002) = *Cicadetta* sp. nr. *apicata* Callitris Clicker

K. froggatti Distant, 1913 = *Kobonga castanea* Ashton, 1914

K. fuscomarginata Distant, 1914 = *Pauropsalta fuscomarginata* = *Pauropsalta fuscomarginatus* (sic) = *Melampsalta fuscomarginata*

K. godingi Distant, 1905 = *Melampsalta godingi* = *Melampsalta umbrimargo* (nec Walker)

K. oxleyi (Distant, 1882) = *Melampsalta oxleyi* = *Cicadetta oxleyi*

K. sp. nr. *Cicadetta oxleyi* (Dunn, 1991)

K. sp. C Ewart, 2009

K. umbrimargo (Walker, 1858) = *Cicada umbrimargo* = *Melampsalta umbrimargo* = *Kobongo umbrimago* (sic)

Notopsalta Dugdale, 1972

N. melanesiana (Myers, 1926) = *Cicadetta melanesiana*

N. sericea sericea (Walker, 1850) = *Cicada sericea* = *Melampsalta sericea* = *Melampsalta cruentata sericea* = *Cicadetta sericea* = *Melamsalta* (sic) *sericea* = *Cicada nervosa* Walker, 1850 = *Melampsalta nervosa* = *Cicadetta nervosa* = *Melampsalta scutellaris* = *Melampsalta indistincta* Myers, 1921 (nec Walker) = *Melampsalta cruentata* (nec Fabricius) = *Melampsalta nervosa* (nec Walker)

N. sericea var. β (Walker, 1850) = *Cicada nervosa* var. β

Rhodopsalta Dugdale, 1972

R. cruentata (Fabricius, 1775) = *Tettigonia cruentata* = *Cicada cruentata* = *Melampsalta cruentata* = *Melampsalta muta cruentata* = *Cicadetta cruentata* = *Cicadetta cincta* Walker, 1850 = *Melampsalta cincta* = *Melampsalta muta cincta* = *Cicadetta cincta* = *Cicada muta minor* Hudson, 1891 = *Cicadetta muta minor* = *Cicadetta minor*

R. leptomera (Myers, 1921) = *Melampsalta leptomera* = *Malampsalta* (sic) *leptomera* = *Cicadetta leptomera*

R. microdora (Hudson, 1936) = *Melampsalta microdora* = *Cicadetta microdora*

Plerapsalta Moulds, 2012

P. incipiens (Walker, 1850) = *Cicada incipiens* = *Melampsalta incipiens* = *Cicadetta incipiens* = *Melampsalta murrayensis* Distant, 1907 = *Melampsalta mureensis* (sic) = *Cicadetta murrayensis* = *Cicadetta murrayiensis* (sic) = *Melampsalta abbreviata* (nec Walker)

P. multifascia (Walker, 1850) = *Cicada multifasciata* = *Melampsalta multifascia* = *Pauropsalta multifascia* = *Cicadetta multifascia* = *Cicada obscurior* Walker, 1850 = *Cicada singula* Walker, 1850 = *Pauropsalta singula* = *Melampsalta singula* = *Cicadetta singula*

Caliginopsalta Ewart, 2005

C. percola Ewart, 2005 = *Notopsalta* sp. D Ewart, 1988

Terepsalta Moulds, 2012

T. infans (Walker, 1850) = *Cicada infans* = *Tibicen infans* = *Quintilia infans* = *Cicada abbreviata* Walker, 1862 = *Melampsalta abbreviata*

Telmapsalta Moulds, 2012

T. hackeri (Distant, 1915) = *Melampsalta hackeri* = *Cicadetta hackeri*

Limnopsalta Moulds, 2012

L. stradbrokensis (Distant, 1915) = *Melampsalta stradbrokensis* = *Cicadetta stradbrokensis*

Heliopsalta Moulds, 2012

H. polita (Popple, 2003) = *Cicadetta polita* = *Cicadetta* sp. G Popple and Strange, 2002

Tettigetta Kolenati, 1857 = *Cicada (Tettigetta)* = *Tettigella* (sic) = *Tettigretta* (sic)

T. dlabolai Mozaffarian & Sanborn, 2010 = *Cicadetta gastrica* (nec Stål)

T. golestani Gogala & Schedl, 2008

T. hoggarensis Boulard, 1980

T. isshikii (Kato, 1926) = *Melampsalta isshikii* = *Cicadetta isshikii* = *Melampsalta isshikii flavicosta* Kato, 1927 = *Cicadetta isshikii flavicosta* = *Leptopsalta radiator* (nec Uhler) = *Leptopsalta radiater* (sic) (nec Uhler) = *Mogannia hebes* (nec Walker) = *Melampsalta pellosoma* (nec Uhler) = *Cicadetta pellosoma* (nec Uhler)

T. maroccana Boulard, 1980

T. megalopercula Mozaffarian & Sanborn, 2012

T. merkli Boulard, 1980

T. musiva (Germar, 1830) = *Cicada musiva* = *Melampsalta musiva* = *Cicadetta musiva* = *Melampsalta (Melampsalta) musiva* = *Melampsalta (Cicadetta) musiva* = *Cicadetta (Melampsalta) musiva* = *Cicada tamarisci* Walker, 1870 = *Cicada tamaricis* (sic) = *Cicada tamarisca* (sic) = *Cicadetta musiva suoedicola* de Bergevin, 1913 = *Cicadetta musiva suaedicola* (sic)

T. omar (Kirkaldy, 1899) = *Melampsalta omar* = *Cicadetta omar*

T. prasina (Pallas, 1773) = *Cicada prasina* = *Cicada brasina* (sic) = *Cicada prosina* (sic) = *Cicada porasina* (sic) = *Tettigonia prasina* = *Tettigetta pallas* (sic) = *Cicada (Tettigetta) prasina* = *Cicadetta prasina* = *Melampsalta prasina* = *Melampsalta (Melampsalta) prasina* = *Cicada lutescens* Olivier, 1790 = *Melampsalta (Melampsalta) lutescens*

T. safavii Mozaffarian & Sanborn, 2012

T. turcica Schedl, 2001

T. yemenensis Schedl, 2004

Tettigettalna Puissant, 2010

T. aneabi (Boulard, 2000) = *Tettigetta aneabi*

T. argentata argentata (Olivier, 1790) = *Cicada argentata* = *Cicadea argentea* (sic) = *Cicadetta argentata* = *Melampsalta argentata* = *Melampsalta (Melampsalta) argentata* = *Tettigia argentata* = *Tettigretta* (sic) *argentata* = *Tettigia* (sic) *argentata* = *Cicadetta (Melampsalta) argentata* = *Cicada sericans* Herrich-Schäffer, 1835 = *Cicadetta sericans* = *Melampsalta (Melampsalta) sericans* = *Melampsalta argentata atra* Gomez-Menor Ortega, 1957 = *Melampsalta argentata* var. *atra* = *Cicadetta argentata atra* = *Tettigetta atra* = *Cicada helianthemi* (nec Rambur)

T. argentata flavescens (Boulard, 1982) = *Tettigetta argentata* mph. *flavescens* = *Tettigetta argentata* var. *flavescens*

T. argentata mauresqua (Boulard, 1982) = *Tettigetta argentata* mph. *mauresqua* = *Tettigetta argentata* var. *mauresqua*

T. argentata nana (Boulard, 2000) = *Tettigetta argentata* var. *nana*

T. armandi Puissant, 2010

T. boulardi Puissant, 2010

T. defauti Puissant, 2010

T. estrellae (Boulard, 1982) = *Tettigetta estrellae* = *Tettigetta septempulsata* Boulard & Quartau, 1991

T. helianthemi galantei Puissant, 2010 = *Tettigettalna argentata* (nec Olivier)

T. helianthemi helianthemi (Rambur, 1840) = *Cicada helianthemi* = *Melampsalta (Melampsalta) helianthemi* = *Cicadetta helianthemi* = *Cicadatra atra* (nec Olivier) = *Cicadatra concinna* (nec Germar) = *Tettigetta argentata* (nec Olivier) = *Tettigetta argentata?* = *Tettigetta manueli* Boulard, 2000

T. josei (Boulard, 1982) = *Tettigetta josei*
T. mariae (Quartau & Boulard, 1995) = *Tettigetta mariae*

Tettigettula Puissant, 2010
T. pygmea (Olivier, 1790) = *Cicada pygmea* = *Cicada pygmaea* (sic) = *Cicada pygmacea* (sic) = *Cicada pygmoea* (sic) = *Melampsalta (Melampsalta) pygmaea* (sic) = *Cicadetta pygmea* = *Tettigetta pygmea* = *Tettigetta pygmaea* (sic) = *Cicadetta brullei* Fieber, 1872 = *Tettigetta brullei* = *Melampsalta brullei* = *Tettigetta brellei* (sic) = *Melampsalta (Melampsalta) brullei* = *Cicadetta transsylvanica* (sic) (nec Fieber) = *Cicadetta tibialis* (nec Panzer) = *Cicadetta montana* (nec Scopoli) = *Melampsalta (Melampsalta) montana* (nec Scopoli)

Tettigettacula Puissant, 2010
T. baenai (Boulard, 2000) = *Tettigetta baenai* = *?Tettigetta brullei* (nec Fieber) = *?Tettigetta pygmea* (nec Olivier)
T. bergevini (Boulard, 1980) = *Tettigetta bergevini*
T. floreae (Boulard, 1987) = *Tettigetta floreae*
T. linaresae (Boulard, 1987) = *Tettigetta linaresae*
T. sedlami (Boulard, 1981) = *Tettigetta sedlami*

Pseudotettigetta Puissant, 2010
P. melanophyrs leunami (Boulard, 2000) = *Tettigetta leunami* = *Cicadetta helianthemi* (nec Rambur) = *Tettigettalna helianthemi* (nec Rambur)
P. melanophrys melanophrys (Horváth, 1907) = *Melampsalta (Melampsalta) melanophrys* = *Cicadetta melanophrys* = *Melampsalta melanophrys*

Gagatopsalta Ewart, 2005
G. auranti Ewart, 2005 = *Notopsalta* sp. H Ewart, 1998
G. obscurus Ewart, 2005 = species B Ewart & Popple, 2001 = *Gagatopsalta obscura*

Simona Moulds, 2012
S. sancta (Distant, 1913) = *Melampsalta sancta* = *Cicadetta sancta* = *Melampsalta subgulosa* Ashton, 1914 = *Melampsalta subglusa* (sic) = *Cicadetta subgulosa*

Chelapsalta Moulds, 2012
C. puer (Walker, 1850) = *Cicada puer* = *Melampsalta puer* = *Melampsata* (sic) *puer* = *Melampsalta pueri* (sic) = *Pauropsalta puer* = *Cicadetta puer* = *Cicadetta (Melampsarta)* (sic) *puer*

Erempsalta Moulds, 2012
E. hermansburgensis (Distant, 1907) = *Melampsalta hermansburgensis* = *Melampsalta hermannsburgensis* (sic) = *Cicadetta hermansburgensis* = *Cicadetta hermannsburgensis* (sic) = Species D Ewart & Popple, 2001

Pipilopsalta Ewart, 2005
P. ceuthoviridis Ewart, 2005 = species E Ewart & Popple, 2001

Dipsopsalta Moulds, 2012
D. signata (Distant, 1914) = *Pauropsalta signata* = *Melampsalta signata*

Paradina Moulds, 2012
P. leichardti (Distant, 1882) = *Melampsalta leichardti* = *Pauropsalta leichardti* = *Urabunana leichardti*

Mugadina Moulds, 2012
M. emma (Goding & Froggatt, 1904) = *Pauropsalta emma* = *Melampsalta emma*
M. festiva (Distant, 1907) = *Urabunana festiva*
M. marshalli (Distant, 1911) = *Urabunana marshalli*
M. sp. nr. *marshalli* Lime Grass-ticker (Popple & Strange, 2002) = *Urabunana* sp. nr. *marshalli* Lime Grass-ticker Popple & Strange
M. sp. nov. Little Green grass-ticker (Popple & Strange, 2002) = *Urabunana* sp. nov. Little Green grass-ticker Popple & Strange

Myopsalta Moulds, 2012

M. atrata (Goding & Froggatt, 1904) = *Melampsalta atrata* = *Cicadetta atrata* = *Notopsalta atrata*
M. sp. nr. *atrata* Montane Grass Buzzer (Moss & Popple, 2000) = *Notopsalta* sp. nr. *atrata*
M. sp. nr. *atrata* Black Acacia Buzzer (Popple & Strange, 2002) = *Notopsalta* sp. nr. *atrata*
M. sp. nr. *atrata* Black Brigalow Buzzer (Popple & Strange, 2002) = *Notopsalta* sp. nr. *atrata*
M. sp. nr. *atrata* (Emery, Emery, Emery & Popple, 2005) = *Notopsalta* sp. nr. *atrata*
M. sp. nr. *atrata* (Haywood, 2005) = *Notopsalta* sp. nr. *atrata*
M. binotata (Goding & Froggatt, 1904) = *Melampsalta binotata* = *Cicadetta binotata*
Species near *M. binotata* (Ewart & Popple, 2001) = Species near *Cicadetta binotata*
M. sp. nr. *binotata* Mitchell Smokey Buzzer (Popple & Strange, 2002) = *Cicadetta* sp. nr. *binotata*
M. sp. nr. *binotata* Moree Smokey Buzzer (Popple & Strange, 2002) = *Cicadetta* sp. nr. *binotata*
M. crucifera (Ashton, 1912) = *Melampsalta crucifera* = *Cicadetta crucifera*
M. sp. nr. *crucifera* (Ewart, 1998) = *Cicadetta* sp. nr. *crucifera*
M. sp. nr. *crucifera* Fishing Reel Buzzer (Popple & Strange, 2002) = *Cicadetta* sp. nr. *Cicadetta crucifera* = *Cicadetta* sp. H Ewart, 2009
M. lactea (Distant, 1905) = *Melampsalta lactea* = *Cicadetta lactea*
M. mackinlayi (Distant, 1882) = *Melampsalta mackinlayi* = *Cicadetta mackinlayi*
M. sp. A (Ewart, 1988) = *Notopsalta* sp. A
M. sp. B (Ewart, 1988) = *Notopsalta* sp. B
M. sp. C (Ewart, 1988) = *Notopsalta* sp. C
M. sp. E (Ewart, 1998) = *Notopsalta* sp. E
M. sp. F (Ewart, 1998) = *Notopsalta* sp. F
M. sp. G (Ewart, 1998) = *Notopsalta* sp. G
M. waterhousei (Distant, 1906) = *Melampsalta waterhousi* = *Cicadetta waterhousei*
M. sp. aff. *waterhousei* (Haywood, 2005) = *Cicadetta* sp. aff. *waterhousei*
M. wollomombii (Coombs, 1995) = *Urabunana wollomombii*

Noongara Moulds, 2012

N. issoides (Distant, 1905) = *Melampsalta issoides* = *Cicadetta issoides*

Froggattoides Distant, 1910 = *Frogattoides* (sic) = *Froggatoides* (sic) = *Larrakeeya* Ashton, 1912

F. pallida (Ashton, 1912) = *Larrakeeya pallida*
F. typicus Distant, 1910 = *Frogattoides* (sic) *typicus*

Clinata Moulds, 2012

C. nodicosta (Goding & Froggatt, 1904) = *Pauropsalta nodicosta* = *Melampsalta nodicosta*
C. sp. aff *nodicosta* (Little Shrub Cicada) (Haywood, 2005) = *Pauropsalta* sp. aff. *nodicosta* (Little Shrub Cicada)
C. sp. aff *nodicosta* (Little Buloke Cicada) (Haywood, 2005) = *Pauropsalta* sp. aff. *nodicosta* (Little Buloke Cicada)

Toxala Moulds, 2012

T. verna (Distant, 1912) = *Urabunana verna* = *Curvicicada verna* = *Oecicicada* (sic) *verna*

Platypsalta Moulds, 2012

P. dubia (Goding & Froggatt, 1904) = *Pauropsalta dubia* = *Melampsalta dubia*
P. sp. nr. *dubia* Broad-wing Scrub-buzzer (Popple & Strange, 2002) = *Pauropsalta* sp. nr. *dubia*
P. mixta (Distant, 1914) = *Pauropsalta mixta* = *Melampsalta mixta* = *Cicadetta mixta*
P. sp. 1 nr. *mixta* Black Scrub-buzzer (Popple & Strange, 2002) = *Cicadetta* sp. 1 nr. *mixta*
P. sp. 2 nr. *mixta* Surat Scrub-buzzer (Popple & Strange, 2002) = *Cicadetta* sp. 2 nr. *mixta*
P. sp. 3 nr. *mixta* Little Grass Buzzer (Popple & Strange, 2002) = *Cicadetta* sp. 3 nr. *mixta* = *Cicadetta mixta* (partim)

Drymopsalta Ewart, 2005

D. crepitum Ewart, 2005 = species F Ewart, 1993 = Healthlands sp. F Ewart, 2005

D. dameli (Distant, 1905) = *Pauropsalta dameli* = *Melampsalta dameli* = *Urabunana daemeli* (sic) = *Drymopsaltra daemeli* (sic) = *Urabunana rufilinea* Ashton, 1914 = *Pauropsalta* sp. F Ewart, 1993

Crotopsalta Ewart, 2005
C. fronsecetes Ewart, 2005 = *Notopsalta* sp. J Ewart, 1998
C. leptotigris Ewart, 2009 = *Crotopsalta* sp. A Ewart, 2009
C. plexis Ewart, 2005 = *Notopsalta* sp. I sp. nr. *Pauropsalta stigmata* Ewart, 1998 = *Notopsalta* sp. I Wilga Ticker
C. poaecetes Ewart, 2005
C. strenulum Ewart, 2005

Ewartia Moulds, 2012
E. brevis (Ashton, 1912) = *Melampsalta brevis* = *Cicadetta brevis*=Heathlands sp. E Ewart, 1993 = *Pauropsalta* sp. E Ewart, 1993
E. cuensis (Distant, 1913) = *Melampsalta cuensis* = *Cicadetta cuensis* = *Melampsalta hermannsburgensis* (sic) (nec *Melampsalta hermansburgensis* Distant, 1907)
E. oldfieldi (Distant, 1883) = *Melampsalta oldfieldi* = *Cicadetta oldfieldi* = *Cicaderna* (sic) *oldfieldi*
E. sp. nr. *oldfieldi* Green Heath-Buzzer (Moss & Popple, 2000) = *Cicadetta* sp. nr. *oldfieldi*
E. sp. nr. *oldfieldi* Western Wattle Cicada (Popple & Strange, 2002) = *Cicadetta* sp. nr. *oldfieldi*

Aestuansella Boulard, 1981
A. aestuans (Fabricius, 1794) = *Tettigonia aestuans* = *Cicada aestuans* = *Melampsalta aestuans* = *Cicadetta aestuans* = *Melampsalta (Pauropsalta) aestuans* = *Pauropsalta aestuans* = *Tettigonia algira* Fabricius, 1803 = *Tibicen algira* = *Cicada algira* = *Cicadetta algira* = *Tettigetta algira* = *Cicada severa* Stål, 1854 = *Melampsalta severa* = *Cicadetta severa*
A. dogueti Boulard, 1981

Ggomapsalta Lee, 2009
G. vernalis (Distant, 1916) = *Pauropsalta vernalis*

Kikihia Dugdale, 1972 = *Melampslata* (nec Kolenati)
K. "acoustica" Marshall, Cooley, Hill & Simon, 2005
K. angusta (Walker, 1850) = *Cicada angusta* = *Melampsalta angusta* = *Melampsalta muta angusta* = *Cicadetta angusta* = *Cicadetta angustata* (sic) = *Cicada muta cinerascens* (sic) = *Melampsalta muta* (nec Fabricius) = *Melampsalta cruentata* (nec Fabricius) = *Melampsalta cruentata angusta* = *Cicadetta muta cinerascens* (sic)
K. "aotea" Marshall & Hill, 2009
K. "aotea east" Marshall, Cooley, Hill & Simon, 2005
K. "aotea west" Marshall, Cooley, Hill & Simon, 2005
K. "astragali" Marshall, Cooley, Hill & Simon, 2005
K. "balaena" Fleming, 1984
K. cauta (Myers, 1921) = *Melampsalta cauta* = *Cicadetta cauta*
K. convicta (Distant, 1892) = *Melampsalta convicta* = *Cicadetta convicta*
K. cutora cumberi Fleming, 1973
K. cutora cutora (Walker, 1850) = *Cicada cutora* = *Melampsalta cuterae* (sic) = *Melampsalta cutora* = *Cicadetta cutora* = *Kikihia cutara* (sic) = *Melampsalta cruentata subalpina* (nec Hudson) = *Melampsalta subalpine* (sic) (nec Hudson) = *Melampsalta muta cutora* = *Cicada muta* var. *cutora* = *Melampsalta muta* var. *c cutora* (nec Walker) = *Melampsalta exulis* = *Melampsalta muta* (nec Fabricius) = *Melampsalta subalpina* (nec Hudson) = *Melampsalta ochrina* (nec Walker) = *Melampsalta cutora* (nec Walker) = *Melampsalta cuterae* Kirby, 1896 (nec Walker)
K. cutora exulis (Hudson, 1950) = *Melampsalta exulis* = *Melampsalta cruentata subalpina* Hudson = *Melampsalta muta* var. *c. cutora* (nec Walker) = *Cicadetta cutora exulis* = *Cicadetta exulis* = *Melampsalta muta cutora* = *Melampsalta muta subalpina* (nec Hudson)
K. dugdalei Fleming, 1984 = *Kikihia* sp. V Fleming, 1975
K. "flemingi" Marshall, Cooley, Hill & Simon, 2005

K. horologium Fleming, 1984 = *Cicada muta flavescens* Hudson, 1891 (nec *C. flavescens* Fabricius, 1794 (Cicadellidae), nec C. Rossi, 1790 nec Turton, 1802 (Cicadidae) *nomen dubium* = *Melampsalta muta* var. a *muta* (nec Fabricius) = *Melampsalta muta* var. *d flavescens* (nec Fabricius, etc.) = *Melampsalta subalpina* (nec Hudson) = *Melampsalta muta flavescens* = *Melampsalta cruentata flavescens* = *Cicadetta muta flavescens* = *Kikihia* sp. 7 Fleming, 1975

K. laneorum Fleming, 1984 = *Kikihia* sp. L Fleming, 1975

K. longula (Hudson, 1950) = *Melampsalta muta longula* = *Cicadetta longula* = *Kikihia muta longula* = *Melampsalta cruentata* (nec Fabricius) = *Cicadetta cruentata muta* (nec Fabricius) = *Cicadetta muta subalpina* (nec Hudson) = *Melampsalta muta muta* (nec Fabricius)

K. "murihikua" Fleming, 1984

K. "muta east" Marshall, Cooley, Hill & Simon, 2005

K. muta muta (Fabricius, 1775) = *Tettigonia muta* = *Cicada muta* = *Melampsalta muta* = *Melampsalta mudta* (sic) = *Melampsalta muta muta* = *Cicadetta muta* = *Cicada bilinea* Walker, 1858 = *Melampsalta bilinea* = *Cicadetta bilinea* = *Cicada muta cinerescens* Hudson, 1891 = *Cicada muta cinerascens* (sic) = *Cicadetta muta cinerescens* = *Cicadetta muta cinerascens* (sic) = *Cicada cruentata* (nec Fabricius) = *Melampsalta creuentata* (nec Fabricius) = *Cicadetta cruentata* var. *muta* = *Melampsalta fuliginosa* Myers, 1921 = *Cicadetta fuliginosa*

K. muta pallida (Hudson, 1950) = *Kikihia muta* var. *pallida* = *Melampsalta muta pallida* = *Melampsalta muta* var. *pallida*

K. "muta west" Simon, 2009

K. "nelsonensis" Fleming, 1984

K. "NWCM" Fleming & Dugdale in Arensburger, Simon and Holsinger, 2004

K. ochrina (Walker, 1858) = *Cicada ochrina* = *Melampsalta ochrina* = *Cicadetta ochrina* = *Cicada orbrina* (sic) = *Cicada aprilina* Hudson, 1891 = *Cicadetta aprilina* = *Kikihia aprilina* = *Melampsalta muta* (nec Fabricius) = *Melampsalta cuterae* (nec Walker) = *Melampsalta cutora* (nec Walker)

K. paxillulae Fleming, 1984

K. "peninsularis" Fleming, 1984

K. rosea (Walker, 1850) = *Cicada rosea* = *Cicada rosa* (sic) = *Melampsalta rosea* = *Cicadetta rosea* = *Melampsalta angusta* (nec Walker) = *Melampsalta muta cruentata* (nec Fabricius) = *Melampsalta cruentata* (nec Fabricius) = *Melampsalta muta muta* (nec Fabricius) = *Melampsalta muta* (nec Fabricius)

K. scutellaris (Walker, 1850) = *Melampsalta scutellaris* = *Cicadetta scutellaris* = *Cicada tristis* Hudson, 1891

K. subalpina (Hudson, 1891) = *Cicada muta* var. *sub-alpina* = *Cicada muta subalpina* = *Cicada muta subalpina* = *Cicadetta subalpina* = *Cicadelta* (sic) *muta subalpina* = *Melampsalta cruentata* var. *sub-alpina* = *Melampsalta muta* var. *subalpina* = *Melampsalta muta subalpina* = *Melampsalta cruentata sub-alpina* = *Melampsalta subalpina* = *Kikihia subalpine* (sic) = *Cicada muta rufescens* Hudson, 1891 = *Melampsalta muta rufescens* = *Cicadetta muta rufescens* = *Cicada muta flavescens* Hudson, 1891 = *Cicadetta muta flavescens* = *Melampsalta muta callista* Hudson, 1950 = *Cicadetta muta callista* = *Melampsalta muta muta* Kirby (nec Fabricius) = *Melampsalta muta* Hutton (nec Fabricius) = *Melampsalta cruentata flavescens* Hudson = *Melampsalta muta* var. b *subalpina*

K. "tasmani" Fleming, 1984

K. "tuta" Marshall, Cooley, Hill & Simon, 2005

K. "westlandica" Simon, 2009

K. "westlandica North" Marshall, Cooley, Hill & Simon, 2005

K. "westlandica South" Marshall, Cooley, Hill & Simon, 2005

Maoricicada Dugdale, 1972 = *Maoripsalta* (sic) = *Maoricicad* (sic) = *Maoripsalta* (sic) = *Melampsalta* (nec Kolenati)

M. alticola Dugdale & Fleming, 1978

M. campbelli (Myers, 1923) = *Melampsalta campbelli* = *Cicadetta campbelli* = *Cicadetta cambellii* (sic) = *Maoripsalta* (sic) *campbelli* = *Pauropsaltra maorica* Myers, 1923 = *Melampsalta maorica* = *Cicadetta maorica*

M. cassiope (Hudson, 1891) = *Cicada cassiope* = *Melampsalta cassiope* = *Cicadetta cassiope* = *Maoripsalta* (sic) *cassiope* = *Melampsalta quadricincta* (nec Walker) = *Melampsalta nervosa* (nec Walker)

M. clamitans Dugdale & Fleming, 1978 = *Maoricicada* sp. C Fleming, 1975

M. hamiltoni (Myers, 1926) = *Cicadetta hamiltoni*
M. iolanthe (Hudson, 1891) = *Cicada iolanthe* = *Melampsalta iolanthe* = *Cicadetta iolanthe* = *Cicadetta planthe* (sic)
M. lindsayi (Myers, 1923) = *Pauropsalta lindsayi* = *Melampsalta linsayi* = *Cicadetta lindsayi* = *Maoricicada mangu lindsayi*
M. mangu celer Dugdale & Fleming, 1978
M. mangu gourlayi Dugdale & Fleming, 1978
M. mangu mangu (White, 1879) = *Melampsalta mangu* = *Cicadetta mangu* = *Melampsalta quadricincta* Myers (nec Walker)
M. mangu multicostata Dugdale & Fleming, 1978
M. myersi (Fleming, 1971) = *Cicadetta myersi* = *Melampsalta iolanthe* Myers (nec Hudson)
M. nigra frigida Dugdale & Fleming, 1978 = *Maoricicada nigra frigide* (sic) = *Maoricicada* sp. F Fleming, 1975
M. nigra nigra (Myers, 1921) = *Melampsalta nigra* = *Cicadetta nigra*
M. oromelaena (Myers, 1926) = *Melampsalta oromelaena* = *Cicadetta oromelaena*
M. otagoensis maceweni Dugdale & Fleming, 1978
M. otagoensis otagoensis Dugdale & Fleming, 1978 = *Maoricicada* sp. O Fleming, 1975
M. phaeoptera Dugdale & Fleming, 1978 = *Maoricicada* sp. P Fleming, 1975
M. tenuis Dugdale & Fleming, 1978 = *Maoriciada* sp. T Fleming, 1975

Mouia Distant, 1920
M. variabilis variabilis Distant, 1920
M. variabilis var. *a* Distant, 1920
M. variabilis var. *b* Distant, 1920
M. variabilis var. *c* Distant, 1920

Buyisa Distant, 1907
B. umtatae Distant, 1907

Fijipsalta Duffels, 1988
F. tympanistria (Kirkaldy, 1907) = *Cicadetta tympanistria* = *(?) Cicadetta tympanistria*

Myersalna Boulard, 1988
M. depicta (Distant, 1920) = *Melampsalta depicta* = *Cicadetta depicta* = *Myersalna bigoti* Boulard, 1988
M. dumbeana Distant, 1920 = *Melampsalta dumbeana* = *Cicadetta dumbeana*
M. flavocaerulea Boulard, 1992

Rouxalna Boulard, 1999
R. rouxi (Distant, 1914) = *Cicadetta rouxi* = *Melampsalta rouxi*

Poviliana Boulard, 1997
P. sarasini (Distant, 1914)
P. vincentiensis Boulard, 1997

Stellenboschia Distant, 1920
S. rotundata (Distant, 1892) = *Melampsalta rotundata* = *Pauropsalta rotundata*

Katoa Ouchi, 1938 = *Katao* (sic) = *Lisu* Liu, 1940
K. chlorotica Chou & Lu, 1997 = *Katoa chlorotiea* (sic) = *Katoa* sp. 1 Pham & Thinh, 2006 = *Katoa* sp. 2 Pham & Thinh, 2006
K. neokanagana (Liu, 1940) = *Lisu neokanagana*
K. paucispina Lei & Chou, 1995
K. paura Chou & Lu, 1997
K. taibaiensis Chou & Lei, 1995
K. tenmokuensis Ouchi, 1938 = *Katao* (sic) *tenmokuensis*

Species of uncertain position in Cicadettini

Cicadetta sp. Soper, 1981
Cicadetta sp. n. Ewart, 1986
Cicadetta sp. nov. 1 Lithgow, 1988
Cicadetta sp. nov. 2 Lithgow, 1988
Cicadetta sp. nov. 3 Lithgow, 1988
Cicadetta sp. nov. 4 Lithgow, 1988
Cicadetta sp. nov. 5 Lithgow, 1988
Cicadetta sp. A, Gidyea Cicada Ewart & Popple, 2001
Cicadetta sp. A, Brown Sandplain Cicada Ewart, 2009
Cicadetta sp. B Ewart, 2009 = Cravens Buzzing Cicada Ewart, 2009
Cicadetta sp. F Ewart, 2009 = Species F (Spinifex rattler) Ewart and Popple, 2001
Cicadetta sp. H Capparis Cicada (Popple & Strange, 2002) = *Cicadetta* sp. H Popple, 2003 = *Cicadetta* sp. K Ewart, 2009
Cicadetta sp. I Pink-wing Cicada Popple & Strange, 2002
Cicadetta sp. I Small Acacia Cicada Ewart, 2009
Cicadetta sp. J Ewart, 2009 = Roaring Senna Cicada Ewart, 2009
Cicadetta sp. K Ewart, 2009 = Capparis Cicada Ewart, 2009
Cicadetta sp. M Ewart, 2009 = *Notopsalta* sp. B Ewart, 1988 = Copper Shrub Buzzer Ewart, 2009
Cicadetta sp. nov. Small Robust Ambertail Popple & Strange, 2002
Cicadetta sp. nov. Small Treetop Ticker Popple & Strange, 2002
Melampsalta alanubilis Goding & Froggatt nom. nud. in Stevens and Carver, 1986
Cicadettini sp. a Phillips, Sanborn & Villet, 2002
Cicadettini sp. b Phillips, Sanborn & Villet, 2002
Cunnamulla Smoky Buzzer Ewart, 2009
Jundah Grass Ticker Ewart, 2009
Melaleuca sp. A Ewart, 2005
N. gen. n. sp. Phillips, Sanborn & Villet, 2002
N. Gen. "Kynuna" Marshall & Hill, 2009
N. Gen. "northwestern" Marshall & Hill, 2009
N. Gen. "Nullarbor wingbanger" Marshall & Hill, 2009
N. Gen. "pale grass cicada" Marshall & Hill, 2009
N. Gen. "near pale grass cicada" Marshall & Hill, 2009
N. Gen. "revving tigris" Marshall & Hill, 2009
N. Gen. "swinging tigris" Marshall & Hill, 2009
N. Gen. "tigris H2" Marshall & Hill, 2009
N. Gen. "troublesome tigris" Marshall & Hill, 2009

Tribe Prasiini Matsumura, 1917 = Prasini (sic) = Prassini (sic)

Subtribe Prasiina Matsumura, 1917

Arfaka Distant, 1905
A. fulva (Walker, 1870) = *Cephaloxys fulva* = *Mogannia fulva*
A. hariola (Stål, 1863) = *Prasia hariola*

Sapatanga Distant, 1905 = *Spatanga* (sic)
S. nutans (Walker, 1850) = *Cephaloxys nutans* = *Selymbria nutans* = *Selymbia* (sic) *nutans*

Jacatra Distant, 1905
J. typica Distant, 1905

Mariekea Jong & Boer, 2004
M. acuta Jong, 2004
M. euharderi Jong, 2004

M. floresiensis Jong, 2004
M. groenendaeli Boer, 2004
M. harderi (Schmidt, 1925) = *Lembeja harderi*
M. major Jong, 2004

Prasia Stål, 1863 = *Drepanopsaltria* Breddin, 1901 = *Drepanopsatra* (sic) *(Lembeja)* = *Parasia* (sic)
P. breddini Jong, 1985
P. faticina Stål, 1863 = *Prasia culta* Distant, 1898 = *Drepanopsaltria culta* = *Drepanopsatra* (sic) *(Lembeja) culta*
P. nigropercula Jong, 1985
P. princeps Distant, 1888 = *Drepanopsaltria princeps* = *Prasia culta* (nec Distant)
P. sarasinorum Jong, 1985
P. senilirata Jong, 1985
P. tuberculata Jong, 1985

Lacetas Karsch, 1890 = *Lacetas (Lacetasiastes)* = *Lacetas (Lacetas)*
L. annulicornis Karsch, 1890 = *Lacetas (Lacetas) annulicornis* = *Lacetas anulicornis* (sic)
L. breviceps Schumacher, 1912 = *Lacetas (Lacetasiastes) breviceps*
L. jacobii Schumacher, 1912 = *Lacetas (Lacetas) jacobii*
L. longicollis longicollis Schumacher, 1912 = *Lacetas (Lacetas) longicollis*
L. longicollis tendaguruensis Schumacher, 1912 = *Lacetas (Lacetas) tendaguruensis*

Lembeja Distant, 1892 = *Lembejam* (sic) = *Perissoneura* Distant, 1883 (nec *Perissoneura* Mac-Lachlan, 1871)
L. brendelli Jong, 1986
L. consanguinea Jong, 1987
L. dekkeri Jong, 1986
L. distanti Jong, 1986
L. elongata Jong, 1986
L. fatiloqua (Stål, 1870) = *Prasia fatiloqua*
L. foliata (Walker, 1858) = *Cephaloxys foliata* = *Prasia foliata* = *Lembeja foliate* (sic)
L. fruhstorferi Distant, 1897 = *Prasia fruhstorferi*
L. hollowayi Jong, 1986
L. incisa Jong, 1986
L. lieftincki Jong, 1987
L. maculosa (Distant, 1883) = *Perissoneura maculosa* = *Prasia maculosa* = *Prasia faticina* (nec Distant)
L. majuscula Jong, 1986
L. minahassae Jong, 1986
L. mirandae Jong, 1986
L. oligorhanta Jong, 1986
L. papuensis Distant, 1897 = *Prasia papuensis* = *Drepanopsaltria russula* Jacobi, 1903 = *Lembeja crassa* Distant, 1909
L. paradoxa (Karsch, 1890) = *Perissoneura paradoxa* = *Prasia paradoxa* = *Perissoneura acutipennis* Karsch, 1890 = *Prasia acutipennis* = *Lembeja acutipennis* = *Lembeja brunneosa* Distant, 1910 = *Lembeja australis* Ashton, 1912
L. parvula Jong, 1987
L. pectinulata Jong, 1986
L. robusta Distant, 1909
L. roehli Schmidt, 1925 = *Prasia fatiloqua* (nec Stål)
L. sangihensis Jong, 1986 = *Prasia foliata* (nec Walker)
L. sanguinolenta Distant, 1909
L. sumbawensis Jong, 1987
L. tincta (Distant, 1909) = *Prasia tincta*
L. vitticollis (Ashton, 1912) = *Prasia vitticollis* = *Prasia viticollis* (sic)
L. wallacei Jong, 1987
L. 1 new species Boer & Duffels, 1996

L. 2 new species Boer & Duffels, 1996
L. 3 new species Boer & Duffels, 1996
L. 4 new species Boer & Duffels, 1996
L. 5 new species Boer & Duffels, 1996

Subtribe Iruanina Boulard, 1993 = Iruanaria

Iruana Distant, 1905
I. brignolii Boulard, 1982
I. meruana Boulard, 1990
I. rougeoti Boulard, 1975
I. sulcata Distant, 1905

Subtribe Bafutalnina Boulard, 1993 = Bafutalnaria

Bafutalna Boulard, 1993
B. mirei Boulard, 1993

Tribe Chlorocystini Distant, 1905 = Chlorocystaria = Chlorocystinae (sic) = Gymnotympanini Boulard, 1979

Dinarobia Mamet, 1957 = *Mauricia* Orian, 1954 (nec *Mauricia* Harris, 1919)
D. claudeae (Orian, 1954) = *Mauricia claudeae*

Owra Ashton, 1912
O. insignis Ashton, 1912

Chlorocysta Westwood, 1851 = *Cystosoma (Chlorocysta)* = *Chlorocysta (Chlorocysta)* = *Mardalana* Distant, 1905 = *Mardarana* (sic) = *Mardalena* (sic)
C. fumea (Ashton, 1914) = *Mardalana fumea*
C. sp. n. Ewart, 1998
C. suffusa (Distant, 1907) = *Mardalana suffusa*
C. vitripennis (Westwood, 1851) = *Cystosoma (Chlorocysta) vitripennis* = *Chlorocysta* (sic) *(Chlorocysta) vitripennis* = *Cystosoma vitripennis* = *Cicada congrua* Walker, 1862 = *Mardalana congrua* = *Mardarana* (sic) *congrua* = *Kanakia congrua* = *Chlorocysta macrula* Stål, 1863

Glaucopsaltria Goding & Froggatt, 1904 = *Glaucocysta* (sic)
G. viridis Goding & Froggatt, 1904 = *Chlorocysta viridis*

Thaumastopsaltria Kirkaldy, 1900 = *Thaumatopsaltria* (sic) = *Thoumastopsaltria* (sic) = *Taumatopsaltria* (sic) = *Thaumastropsaltria* (sic) = *Acrilla* Stål, 1863 (nec *Acrilla* Adams, 1860)
T. adipata (Stål, 1863) = *Acrilla adipata* = *Thoumastopsaltria* (sic) *adipata*
T. globosa (Distant, 1897) = *Acrilla globosa* = *Thaumastopsaltria glauca* Ashton, 1912 = *Thaumatopsaltria* (sic) *glauca*
T. lanceola Boer, 1992
T. pneumatica Boer, 1992
T. sarissa Boer, 1992
T. sicula Boer, 1992
T. smithersi Moulds, 2012
T. spelunca Boer, 1992

Cystosoma Westwood, 1842 = *Cicada (Cystosoma)* = *Cystisoma* (sic) = *Cyrtosoma* (sic) = *Cytosoma* (sic)
C. saundersii (Westwood, 1842) = *Cicada (Cystosoma) saundersii* = *Cystosoma saundersi* (sic) = *Cystosoma laudersii* (sic) = *Cytosoma* (sic) *saundersii*
C. schmeltzi Distant, 1882 = *Cystosoma schmelzi* (sic)

Cystopsaltria Goding & Froggatt, 1904
C. immaculata Goding & Froggatt, 1904

Venustria Goding & Froggatt, 1904 = *Venustra* (sic)
V. superba Goding & Froggatt, 1904

Mirabilopsaltria Boer, 1996
M. globulata Boer, 1996
M. humilis (Blöte, 1960) = *Baeturia humilis*
M. inconspicua Boer, 1996
M. inflata Boer, 1996
M. toxopeusi (Blöte, 1960) = *Baeturia toxopeusi*
M. viridicata (Distant, 1897) = *Baeturia viridicata*

Guineapsaltria Boer, 1993
G. chinai (Blöte, 1960) = *Baeturia chinai*
G. flava (Goding & Froggatt, 1904) = *Tibicen flavus* = *Abricta flava* = *Abricta flavus* = *Baeturia flava* = *Melampsalta flava* = *Cicadetta flava* = *Baeturia modesta* Distant, 1907 = *Baeturia exhausta* (nec Guérin-Méneville) = *Baeturia minuta* Blöte, 1960 = *Baeturia stylata* (nec Blöte)
G. flaveola Boer, 1993
G. pallida (Blöte, 1960) = *Baeturia pallida*
G. pallidula Boer, 1993
G. pennyi Boer, 1993
G. stylata (Blöte, 1960) = *Baeturia stylata*
G. viridula (Blöte, 1960) = *Baeturia viridula*

Scottotympana Boer, 1991
S. biardae Boer, 1991
S. huibregtsae Boer, 1991
S. sahebdivanii Boer, 1991

Gymnotympana Stål, 1861
G. dahli (Kuhlgatz, 1905) = *Tibicen dahli* = *Ueana dahli* = *Ueana (Tibicen) dahli* = *Ueana dahlii* (sic) = *Baeturia valida* Blöte. 1960
G. hirsuta Boer, 1995
G. langeraki Boer, 1995
G. membrana Boer, 1995
G. minoramembrana Boer, 1995
G. montana Boer, 1995
G. nigravirgula Boer, 1995
G. obiensis Boer, 1995
G. olivacea Distant, 1905
G. parvula Boer, 1995
G. phyloglycera Boer, 1995
G. rubricata (Distant, 1897) = *Baeturia rubicata* = *Gymnotympana nenians* Jacobi, 1903 = *Gymnotympana neniens* (sic)
G. rufa (Ashton, 1914) = *Baeturia rufa*
G. stenocephalis Boer, 1995
G. strepitans (Stål, 1861) = *Cicada strepitans*
G. stridens (Stål, 1861) = *Cicada stridens* = *Dundubia significata* Walker, 1870
G. subnotata (Walker, 1868) = *Cicada subnotata* = *Baeturia subnotata* = *Cicada innotabilis* Walker, 1868 (nec *Cicada innotablis* Walker, 1858) = *Fidicina innotabilis* (nec Walker, 1858) = *Gymnotympana innotabilis*
G. varicolor varicolor (Distant, 1907) = *Baeturia varicolor* = *Kanakia congrua* (nec Walker)
G. varicolor var. *a* (Distant, 1907) = *Baeturia varicolor* var. *a*
G. varicolor var. *b* (Distant, 1907) = *Baeturia varicolor* var. *b*
G. verlaani Boer, 1995
G. viridis Boer, 1995

Fractuosella Boulard, 1979
- *F. breoni* Boulard, 1989
- *F. darwini* (Distant, 1905) = *Stagira darwini*
- *F. vinsoni* Boulard, 1979
- *F. virginiae* Boulard, 1979

Papuapsaltria Boer, 1995
- *P. angulata* Boer, 1995
- *P. baasi* Boer, 1995
- *P. bidigitula* Boer, 1995
- *P. bosaviensis* Boer, 2000
- *P. brassi* Boer, 1995
- *P. dioedes* Boer, 1995
- *P. dolabrata* Boer, 1995
- *P. goniodes* Boer, 1995
- *P. lachlani* Boer, 1995
- *P. nana* (Jacobi, 1903) = *Acrilla nana* = *Thaumastopsaltria nana* = *Thaumastropsaltria* (sic) *nana* = *Baeturia famulus* Distant, 1906
- *P. novariae* Boer, 1995
- *P. phyllophora* (Blöte, 1960) = *Baeturia phyllophora*
- *P. plicata* Boer, 1995
- *P. pusilla* Boer, 2000
- *P. spinigera* Boer, 1995
- *P. stoliodes* Boer, 1995
- *P. toxopei* Boer, 1995
- *P. ungula* Boer, 1995
- *P. ustulata* (Blöte, 1960) = *Baeturia ustulata*
- *P. woodlarkensis* Boer, 1995

Baeturia Stål, 1866
- *B. arabuensis* Blöte, 1960
- *B. bemmeleni* Boer, 1994
- *B. bicolorata* Distant, 1892
- *B. bilebanarai* Boer, 1989
- *B. bipunctata* Blöte, 1960 = *Baeturia bipuncta* (sic)
- *B. biroi* Boer, 1994
- *B. bismarkensis* Boer, 1989
- *B. bloetei* Boer, 1989 = *Baeturia bicolorata* Blöte, 1960
- *B. boulardi* Boer, 1989
- *B. brandti* Boer, 1989
- *B. brongersmai* Blöte, 1960
- *B. colossea* Boer, 1994 = *Baeturia gigantea* (sic)
- *B. conviva* (Stål, 1861) = *Cicada conviva* = *Tibicen convivus* = *Cicada parallela* Walker, 1870
- *B. cristovalensis* Boer, 1989
- *B. daviesi* Boer, 1994
- *B. edauberti* Boulard, 1979 = *Baeturia aubertae* Boulard, 1979 = *Baeturia efatensis* Boulard, 1979 = *Baeturia epiensis* Boulard, 1979
- *B. exhausta* (Guérin-Méneville, 1831) = *Cicada exhausta* = *Cicada hastipennis* Walker, 1858 = *Baeturia hastipennis* = *Cicada parallella* Walker, 1868 = *Baeturia parallela* = *Dundubia parabola* Walker, 1858 = *Baeturia parabola*
- *B. fortuini* Boer, 1994
- *B. furcillata* Boer, 1992
- *B. galeata* Boer, 2000
- *B. gibberosa* Boer, 1994
- *B. gressitti* Boer, 1989

B. guttulinervis Blöte, 1960
B. guttulipennis Blöte, 1960
B. hamiltoni Boer, 1994
B. hardyi Boer, 1986
B. hartonoi Boer, 1994
B. inconstans Boer, 1994
B. intermedia Boer, 1982
B. karkarensis Boer, 1992
B. laminifer Blöte, 1960
B. laperousei Boulard, 2005
B. laureli Boer, 1986
B. lorentzi Boer, 1992
B. loriae Distant, 1897 = *Gymnotympana loriae*
B. maai Boer, 1994
B. macgillavryi Boer, 1989 = *Baeturia famulus* (nec Myers) = *Cephaloxys dilectus* nom. nud. = *Baeturia dilectus* nom. nud. = *Cephaloxys cillogata* nom. nud.
B. maddisoni Duffels, 1988 = *Baeturia exhausta* (nec Guérin-Méneville)
B. mamillata Blöte, 1960
B. manusensis Boer, 1989
B. marginata Boer, 1989
B. marmorata Blöte, 1960
B. mendanai Boer, 1989
B. mussauensis Boer, 1989
B. nasuta Blöte, 1960
B. papuensis Boer, 1989
B. parva Blöte, 1960
B. pigrami Boer, 1994
B. polhemi Boer, 2000
B. quadrifida (Walker, 1868) = *Cicada quadrifida* = *Baeturia hirsuta* Blöte, 1960
B. reijnhoudti Boer, 1989 = *Baeturia reynhoudti* (sic) = *Baeturia reinhoudti* (sic)
B. retracta Boer, 1994
B. roonensis Boer, 1994
B. rossi Boer, 1994
B. rotumae Duffels, 1988
B. rufula Blöte, 1960 = *Baeturia rufa* (sic)
B. schulzi Schmidt, 1926
B. sedlacekorum Boer, 1989
B. silveri Boer, 1994
B. sp. 1 Kahono, 1993
B. sp. 2 Kahono, 1993
B. sp. 3 Kahono, 1993
B. splendida Boer, 1994
B. tenuispina Blöte, 1960
B. turgida Boer, 1992
B. uveiensis Boulard, 1996 = *Baeturia uveaiensis* (sic)
B. valida Blöte, 1960
B. vanderhammeni Blöte, 1960
B. versicolor Boer, 1994
B. viridis Blöte, 1960
B. wauensis Boer, 1994
B. wegeneri Boer, 1994

Aedestria Boer, 1990
A. bullata Boer, 1993
A. cheesmanae Boer, 1993

A. cobrops Boer, 1990
A. digitata (Blöte, 1960) = *Baeturia digitata*
A. dilobata Boer, 1993
A. hastulata Boer, 1993
A. kaiensis Boer, 1993
A. latifrons (Blöte, 1960) = *Baeturia latifrons* = *Baeturia beccarii* Distant, 1892 (nec *Baeturia beccarii* Distant, 1888) = *Baeturia famulus* Myers, 1928 = *Baeturia famulus* (nec Distant)
A. moluccensis Boer, 1993
A. obiensis Boer, 1993
A. sepia Boer, 1990
A. waigeuensis Boer, 1993

Decebalus Distant, 1920
D. ugandanus Distant, 1920

Muda Distant, 1897 = *Iwasemia* Matsumura, 1927 = *Nahasemia* Matsumura, 1930
M. kuroiwae (Matsumura, 1913) = *Prasia kuroiwae* = *Parasia* (sic) *kuroiwae* = *Iwasemia kuroiwae* = *Nahasemia kuroiwae* = *Nahasemia knroiwae* (sic) = *Baeturia kuroiwae* = *"Baeturia" kuroiwae*
M. obtusa (Walker, 1858) = *Cephaloxys obtusa* = *Baeturia obtusa* = *Muda concolor* Distant, 1897
M. sp. 3 Zaidi, Azman & Ruslan, 2001
M. sp. 4 Zaidi, Azman & Ruslan, 2001
M. tua Duffels, 2004
M. undescribed species A Sanborn, Phillips & Sites, 2007
M. undescribed species B Sanborn, Phillips & Sites, 2007
M. virguncula (Walker, 1857) = *Cicada virguncula* = *Baeturia beccarii* Distant, 1888 = *Muda beccarii* = *Melampsalta flava* (nec Goding & Froggatt)

Musoda Karsch, 1890
M. flavida Karsch, 1890
M. gigantea Distant, 1914
M. occidentalis Boulard, 1974
M. orientalis Boulard, 1974

Cephalalna Boulard, 2006
C. francimontanum Boulard, 2006

Nablistes Karsch, 1891 = *Nabristes* (sic)
N. heterochroma Boulard, 1986
N. terebrata Karsch, 1891 = *Nabristes* (sic) *terebrata*

Akamba Distant, 1905 = *Aramba* (sic)
A. aethiopica Distant, 1905

Kumanga Distant, 1905
K. sandaracata (Distant, 1888) = *Baeturia sandaracata*

Conibosa Distant, 1905
C. occidentis (Walker, 1858) = *Cephaloxys occidentis* = *Calyria occidentis* = *Conibosa occidentris* (sic) = *Conibosa occidentalis* (sic) = *Conibosa occidentes* (sic) = *Calyria virginea* Stål, 1864
C. sp. nr. *occidentis* Young, 1980
C. sp. D Wolda & Ramos, 1992
C. sp. G Wolda & Ramos, 1992

Durangona Distant, 1911
D. tigrina Distant, 1911

Tribe Tettigomyiini Distant, 1905 = Tettigomyiarini (sic) = Tettigomyini (sic)

Gazuma Distant, 1905
G. barrettae Distant, 1905 = *Gazuma barettae* (sic) = *Gazuma balettae* (sic)
G. delalandei Distant, 1905
G. pretoriae Distant, 1905

Paectira Karsch, 1890 = *Inyamana* Distant, 1905 = *Inymana* (sic) = *Inyanama* (sic)
P. boulardi Heller, 1980
P. bouvieri (Melichar, 1911) = *Inymana* (sic) *bouvieri* = *Inyamana bouvieri*
P. capeneri Boulard, 1992
P. dispar Heller, 1980
P. cf. *dispar* Villet & Noort, 1999
P. dulcis Karsch, 1890 = *Paectira ducalis* (sic)
P. feminavirens Boulard, 1977 = *Paectira femiravirens* (sic)
P. forchhammeri Boulard, 1992
P. fuliginosa (Distant, 1905) = *Xosopsaltria fuliginosa*
P. hemaris (Distant, 1905) = *Inyamana hemaris*
P. hyndi Boulard, 1992
P. jeanuaudi Boulard, 1977
P. lindneri Heller, 1980
P. modesta Heller, 1980
P. ochracea ochracea (Distant, 1905) = *Inyamana ochracea*
P. ochracea viridans Heller, 1980
P. cf. *ochracea* Villet & Noort, 1999
P. oreas (Jacobi, 1905) = *Inyamana oreas*
P. scheveni Heller, 1980
P. ventricosa (Melichar, 1914) = *Inyanama* (sic) *ventricosa* = *Inyamana ventricosa*

Xosopsaltria Kirkaldy, 1904 = *Pydna* Stål, 1861 (nec *Pydna* Walker, 1856)
X. annulata (Germar, 1830) = *Cicada annulata* = *Tettigomyia annulata* = *Pydna annulata* = *Xosopsaltria capicola* Kairkaldy, 1909
X. n. sp. Phillips, Sanborn & Villet, 2002
X. scholandi Distant, 1907
X. scurra (Germar, 1830) = *Cicada scurra* = *Tettigomyia scurra* = *Pydna scurra* = *Cicada lutea* Olivier, 1790 = *Tettigomyia lutea* = *Pydna lutea* = *Xosopsaltria lutea*
X. thunbergi Metcalf, 1955 = *Tettigonia punctata* Thunberg, 1822 (nec *Tettigonia punctata* Fabricius, 1798) = *Pydna punctata* = *Xosopsaltria punctata* = *Cicada tabaniformis* Germar nom. nud. = *Tettygomyia* (sic) *tabaniformis* Walker, 1850 nom. nud. = *Tettigomyia tabaniformis* = *Xosopsaltria tabaniformis*

Tettigomyia Amyot & Audinet-Serville, 1843 = *Cicada (Tettigomyia)* = *Tettigomya* (sic) = *Tettigomia* (sic) = *Tettygomyia* (sic)
T. vespiformis Amyot & Audinet-Serville, 1843

Bavea Distant, 1905 = *Baeva* (sic)
B. concolor (Walker, 1850) = *Cephaloxys concolor*

Stagira Stål, 1861
S. abagnata Villet, 1997
S. acrida (Walker, 1850) = *Cicada acridia*
S. aethlius (Walker, 1850) = *Cicada aethlius*
S. caffrariensis Villet, 1997
S. celsus Villet, 1997
S. chloana Villet, 1997

S. consobrina Distant, 1920
S. consobrinoides Villet, 1997
S. dendrophila Villet, 1997
S. dracomontana Villet, 1997
S. dracomontanoides Villet, 1997
S. elegans Villet, 1997
S. empangeniensis Villet, 1997
S. enigmatica Villet, 1997
S. eshowiensis Villet, 1997
S. furculata Villet, 1997
S. fuscula Villet, 1997
S. microcephala (Walker, 1850) = *Cephaloxys microcephala* = *Cicada cereris* Stål, 1855 = *Stagira cereris* = *Cicada ceresis* (sic)
S. mystica Villet, 1997
S. nasuta Villet, 1997
S. natalensis Villet, 1997
S. ngomiensis Villet, 1997
S. nkandlhaensis Villet, 1997
S. oxyclypea Villet, 1997
S. pondoensis Villet, 1997
S. pseudaethlius Villet, 1997
S. purpurea Villet, 1997
S. segmentaria Karsch, 1890 = *Stagira sanguinea* Distant, 1920 = *Stagira ruficostata* Distant, 1920
S. sexcostata Villet, 1997
S. selindensis Villet, 1997
S. simplex (Germar, 1934) = *Cicada simplex* = *Cicada simplicis* (sic)
S. stygia Villet, 1997
S. sylvia Villet, 1997
S. virescens Kirkaldy, 1909 = *Cicada viridula* Walker, 1858 (nec *Cicada viridula* Walker, 1850) = *Stagira viridula* = *Stagira simplex* (nec Germar)
S. cf. *viridula* Villet, 1992
S. viridoptera Villet, 1997
S. vulgata Villet, 1997
S. xenomorpha Villet, 1997
S. zebrata Villet, 1997
S. zuluensis Villet, 1997

Stagea Villet, 1994
S. platyptera Villet, 1994

Spoerryana Boulard, 1974 = *Spoeryana* (sic)
S. llewelyni Boulard, 1974 = *Spoerryana llewelini* (sic) = *Spoerryana llyewellini* (sic)

New Genus in Tettigoyiini Villet & Noort, 1999

Tribe Hemidictyini Distant, 1905 = Hemidictyaria = Hemidyctyini (sic)

Hemidictya Burmeister, 1905 = *Hemidyctya* (sic) = *Hemididicta* (sic) = *Hemidyctia* (sic)
H. frondosa Burmeister, 1835 = *Hemidyctya* (sic) *frondosa* = *Hemidyctia* (sic) *frondosa*

Hovana Distant, 1905
H. distanti (Brancsik, 1893) = *Hemidictya distanti*

Subfamily Tibicininae Distant, 1905 = Tibicinae (sic) = Tibiciniinae (sic) = Tibiciinae (sic) = Tettigadinae Distant, 1905 = Platypediinae Kato, 1932 = Platypedinae (sic) = Ydiellinae Boulard, 1973

Tribe Tettigadini Distant, 1905 = Tettigadesini = Tettigadesaria

Tettigades Amyot & Audinet-Serville, 1843 = *Tettigodes* (sic) = *Tettigates* (sic)

T. angularis Torres, 1958
T. auropilosa Torres, 1944
T. bosqi Torres, 1942
T. brevicauda Torres, 1958
T. brevivenosa Torres, 1942
T. chilensis Amyot & Audinet-Serville, 1843 = *Tettigades chiliensis* (sic) = *Cicada rubrolineata* Spinola, 1852 = *Tettigades rubrolineata* = *Fidicina crassivena* Walker, 1858 = *Cicada eremophila* Philippi, 1866
T. compacta Walker, 1850
T. crassa Torres, 1958
T. curvicosta Torres, 1958
T. distanti Torres, 1958
T. dumfriesi Distant, 1920 = *Tettigades dumfriesii* (sic)
T. lacertosa Torres, 1944
T. lebruni Distant, 1906 = *Odopoea lebruni* (partim)
T. limbata Torres, 1958
T. lizeri Torres, 1942
T. lizeriana Delétang, 1919
T. major Torres, 1944
T. mexicana Distant, 1881
T. opaca Jacobi, 1907 = *Tettigades porteri* Bréthes, 1920
T. parva Distant, 1892 = *Tettigades parvus* = *Tettigades brevivenosa* Torres, 1942
T. pauxilla Torres, 1958
T. porteri Bréthes, 1920
T. procera Torres, 1958
T. sarcinatrix Torres, 1944 = *Tettigades sarsinatrix* (sic)
T. sordida Torres, 1944
T. ulnaria Distant, 1906 = *Tettigades urnaria* (sic)
T. undata Torres, 1958
T. unipuncta Torres, 1958
T. varivenosa Distant, 1906

Torresia Sanborn & Heath, 2011, nom. nud.

T. lariojaensis Sanborn & Heath, 2011, nom. nud.
T. sanjuanensis Sanborn & Heath, 2011, nom. nud.

Alarcta Torres, 1958

A. albicerata (Torres, 1949) = *Tettigades albicerata*
A. bahiensis (Torres, 1942) = *Tettigades bahiensis*
A. blanchardi (Torres, 1948) = *Tettigades blanchardi*
A. macrogina (Torres, 1949) = *Tettigades macrogina*
A. micromaculata Sanborn & Heath, 2011, nom. nud.
A. minuta (Torres, 1949) = *Tettigades minuta*
A. quadrimacula Torres, 1958
A. terrosa Torres, 1958

Tettigotoma Torres, 1942

T. maculata Torres, 1942

Psephenotettix Torres, 1958

P. grandis Torres, 1958
P. minor Torres, 1958

Babras Jacobi, 1907
B. sonorivox Jacobi, 1907

Calliopsida Torres, 1958
C. cinnabarina (Berg, 1879) = *Tettigades cinnabarina* = *Chonosia cinnabarina*

Coata Distant, 1906
C. facialis facialis Distant, 1906 = *Coata fascialis* (sic)
C. facialis var. ________ Jacobi, 1907
C. fragilis Goding, 1925
C. humeralis (Walker, 1858) = *Cicada humeralis* = *Plautilla humeralis*

Chonosia Distant, 1905
C. atrodorsalis Torres, 1945
C. crassipennis (Walker, 1858) = *Fidicina crassipennis*
C. crassipennis metequei Delétang, 1919 nom. nud.
C. longiopercula Sanborn & Heath, 2011, nom. nud.
C. papa (Berg, 1882) = *Tettigades papa*
C. septentrionala Sanborn & Heath, 2011, nom. nud.
C. trigonocelis Torres, 1945

Acuticephala Torres, 1958
A. alipuncta Torres, 1958

Mendozana Distant, 1906 = *Mendozana (Tettigades)* = *Semaiophora* Haupt, 1918 = *Fadylia* Delétang, 1919
M. antennaria (Jacobi, 1907) = *Tettigades antennaria* = *Mendozana (Tettigades) antennaria* = *Semaiophora dietherti* Schmidt, 1926
M. platypleura Distant, 1906 = *Semaiophora basimaculata* Haupt, 1918 = *Fadylia bruchi* Delétang, 1919

Tribe Lamotialnini Boulard, 1976 = Lamotialnii (sic)

Subtribe Lamotialnina Boulard, 1976 = Lamotialnaria

Lamotialna Boulard, 1976
L. condamini Boulard, 1976
L. couturieri Boulard, 1986

Tribe Platypediini Kato, 1932 = Platypedini (sic)

Platypedia Uhler, 1888
P. affinis Davis, 1939
P. aperta Van Duzee, 1915
P. areolata (Uhler, 1861) = *Cicada areolata* = *Platypedia areoloata* (sic) = *Platypedia aerolata* (sic)
P. australis Davis, 1941
P. balli Davis, 1936
P. barbata Davis, 1920
P. bernardinoensis Davis, 1932 = *Platypedia rufipes bernardinoensis* = *Platypedia rufipes bernardoensis* (sic) = *Platypedia bernardoensis* (sic) = *Platypedia rufipes* var. *angustipennis* Davis, 1932
P. falcata Davis, 1920 = *Platypedid* (sic) *falcata*
P. intermedia Van Duzee, 1915 = *Platypedia areolata intermedia*
P. laticapitata Davis, 1921
P. mariposa Davis, 1935 = *Platypedia plumbea* Van Duzee nom. nud. (partim)
P. middlekauffi Simons, 1953
P. minor Uhler, 1888 = *Platypedia minsi* Johnson & Ledig, 1918 nom. nud. = *Platypedia gressitti* Kato, 1932

P. mohavensis mohavensis Davis, 1920 = *Platypedia mohavensis* = *Platypedia latipennis* Davis, 1921
P. mohavensis rufescens Davis, 1932 = *Platypedia rufescens* = *Platypedia mohavensis* var. *rufescens*
P. putnami keddiensis Davis, 1920 = *Platypedia putnami* var. *keddiensis* = *Platypedia keddiensis*
P. putnami lutea Davis, 1920 = *Platypedia putnami* var. *lutea* = *Platypedia lutea*
P. putnami occidentalis Davis, 1920 = *Platypedia putnami* var. *occidentalis* = *Platypedia occidentalis*
P. putnami putnami (Uhler, 1877) = *Cicada putnami* = *Platypedia putnami* = *Plptypedia* (sic) *putnami*
P. rufipes Davis, 1920
P. scotti Davis, 1935 = *Platypedia sierra* Wymore, 1935
P. similis Davis, 1920 = *Platypedia areolata similis* = *Platypedia plumbea* Van Duzee nom. nud. (partim)
P. sylvesteri Simons, 1953
P. tomentosa Davis, 1942
P. usingeri Simons, 1953
P. vanduzeei Davis, 1920

Neoplatypedia Davis, 1920 = *Neoplaytpedia* (sic)
N. ampliata (Van Duzee, 1915) = *Platypedia ampliata*
N. constricta Davis, 1920

Tribe Ydiellini Boulard, 1973

Subtribe Ydiellina Boulard, 1993 = Ydiellaria

Maroboduus Distant, 1920 = *Ydiella* Boulard, 1973
M. fractus Distant, 1920 = *Ydiella gilloni* Boulard, 1973 = *Maroboduus gilloni*
M. maxicapsulifera (Boulard, 1986) = *Ydiella maxicapsulifera*

Tribe Tibicinini Distant, 1905 = Tibicinaria = Tibicenini (sic) = Tibicini (sic) = Tibicinae (sic)

Subtribe Tibicinina Distant, 1905

Tibicina Kolenati, 1857 = *Cicada (Tibicina)* = *Tibicicina* (sic) = *Tebicina* (sic) = *Tibicea* (sic) = *Tibicena* (sic) = *Tybicina* (sic) = *Tibucina* (sic) = *Tibicen* (nec Latreille) = *Paharia* (nec Distant)
T. casyapae (Distant, 1888) = *Tibicen casyapae* = *Paharia casyapae*
T. cisticola (Hagen, 1855) = *Cicada cisticola* = *Tibicen cisticola* = *Tettigia cisticola*
T. corsica corsica (Rambur, 1840) = *Cicada corsica* = *Cicadetta corsica* = *Melampsalta corsica* = *Melampsalta (Melampsalta) corsica* = *Tibicen corsicus* = *Tibicina corsica* = *Tibicen tomentosus* (nec Olivier) = *Cicada cisticola* (nec Fairmairei)
T. corsica corsica var. *lerestifi* Boulard, 1980 = *Tibicina cisticola* var. *lerestifi* = *Tibicina cisticola lerestifi* = *Tibicina corsica* var. *lerestifi* = *Tibicina corsica larestifi* (sic)
T. corsica fairmairei Boulard, 1984 = *Cicada cisticola* Fairmaire, 1855 = *Tettigia cisticola* = *Tibicen cisticola* = *Tibicina cisticola* = *Tibicina fairmairei* = *Tibicina faimairei* (sic) = *Cicada cisticola* (nec Hagen)
T. fieberi Puissant & Sueur, 2002
T. garricola Boulard, 1983 = *Tibicicina* (sic) *garricola*
T. haematodes flammea Boulard, 1982 = *Tibicina haematodes* mph. *flammea* = *Tibicina haematodes* var. *flammea*
T. haematodes haematodes (Scopoli, 1763) = *Cicada haematodes* = *Cicada haematoides* (sic) = *Cicada (Tettigonia) haematodes* = *Tettigonia haematodes* = *Tibicen haematodes* = *Tibicina haematodes* = *Tebicina* (sic) *haematodes* = *Tibicea* (sic) *haematodes* = *Tibicena* (sic) *haematodes* = *Tybicina* (sic) *haematodes* = *Tibicen haematordes* (sic) = *Tibicina haematedes* (sic) = *Tibicina hematodes* (sic) = *Tibicina haematotes* (sic) = *Tettigonia haematodes* Fabricius, 1794 = *Tettigonia sanguinea* Fabricius, 1803 = *Cicada sanguinea* = *Cicada (Tettigonia) sanguinea* = *Tibicen sanguinea* = *Cicada plebeja* (nec Scopoli) = *Tibicen plebeja* (nec Scopoli)
T. haematodes helveola (Germar, 1821) = *Cicada helvola* = *Cicada helveola* (sic) = *Cicada helveolus* (sic) = *Tibicen helvola* = *Tibicen helvolus*
T. haematodes viridinervis Fieber, 1876 = *Tibicina haematodes* var. *viridinervis*
T. insidiosa Boulard, 1977

T. intermedia Fieber, 1876 = *Tibicen intermedia* = *Tibicen intermedius* (sic) = *Tibicina intermedius* (sic)
T. maldesi Boulard, 1981
T. nigronervosa Fieber, 1876 = *Tibicen nigronervosa* = *Tibicen nigronervosus* = *Tibicina nigronervossa* (sic) = *Tibicina nigrovenosa* (sic) = *Tibicen luctuosa* Costa, 1888 = *Tibicina luctuosa* = *Tibicen luctuosa* (sic) = *Tibicen luctuosus* = *Tibicina tomentosa* (nec Olivier)
T. picta (Fabricius, 1794) = *Tettigonia picta* = *Cicada picta* = *Tibicen picta* = *Tibicen pictus* = *Tibicien* (sic) *pictus* = *Cicadetta picta*
T. pieris (Kirkaldy, 1909) = *Tibicen pieris*
T. quadrisignata (Hagen, 1855) = *Cicada haematodes 4-signata* = *Cicada haematodes quadrisignata* = *Cicada 4-signata* = *Tibicen quadrisignata* = *Tibicen quadrisignatus* = *Tibicina 4-signata* = *Tibicina (Tibicina) quadrisignata* = *Tibicen quadrosignata* (sic) = *Tibicina nigronervosa* (nec Fieber)
T. reticulata Distant, 1888 = *Tibicen reticulatus* = *Paharia reticulata*
T. serhadensis Koçak & Kemal, 2010
T. steveni (Krynicki, 1837) = *Cicada steveni* = *Cicada stevenii* (sic) = *Cicada stevensi* (sic) = *Tibicina steveni* = *Cicada (Tibicina) haematodes stevenii* (sic) = *Tibicina haematodes* var. *steveni* = *Tibicen haematodes steveni* = *Tibicina intermedia* (nec Fieber)
T. tomentosa (Olivier, 1790) = *Cicada tomentosa* = *Cicado* (sic) *tomentosus* (sic) = *Tibicen tomentosa* = *Tibicen tomentosus* = *Tibicina tomentosa* = *Tettigia tomentosa* = *Tibicin* (sic) *tomentosa* = *Tibicina (Tibicina) tomentosa* = *Cicada baetica* (Rambur, 1840) = *Tibicina baetica* = *Tibicina boetica* (sic) = *Tibicen beaticus* = *Tibicen baetica* = *Tibicen picta* (nec Germar) = *Tettigonia picta* (nec Fabricius) = *Tibicen pictus* (nec Fabricius) = *Tibicien* (sic) *pictus* (nec Fabricius) = *Tibicina picta* (nec Fabricius) = *Tibicina cisticola* (nec Hagen)

Subtibicina Lee, 2012
S. tigris Lee, 2012

Paharia Distant, 1905
P. lacteipennis (Walker, 1850) = *Cephaloxys lacteipennis* = *Mogannia lacteipennis* = *Tibicen lacteipennis* = *Tibicina lacteipennis*
P. putoni putoni (Distant, 1892) = *Tibicen putoni* = *Tibicina putoni* = *Tibicina lacteipennis* Puton, 1883 (nec *Cephaloxyx lacteipennis* Walker) *Sena lacteipennis* = *Tibicen lacteipennis* = *Tibicina lacteipennis* = *Psalmocharias lacteipennis*
P. putoni var. ______ Jacobi, 1928
P. semenovi Oshanin, 1906 = *Tibicen semenovi* = *Tibicina semenovi* = *Psalmocharias semenovi*
P. zevara (Kusnezov, 1931) = *Tibicina zevara* = *Tibucina* (sic) *zevara* = *Paharia zevara*

Tibicinoides Distant, 1914 = *Tibicenoides* (sic)
T. cupreosparsa (Uhler, 1889) = *Tibicen cupreosparsa* = *Tibicen cupreosparsus* = *Tibicina cupreosparsa* = *Tibicinoides cupreosparsus* = *Tibicen cupressparsa* (sic)
T. mercedita (Davis, 1915) = *Okanagana mercedita*
T. minuta (Davis, 1915) = *Okanagana minuta*

Okanagana Distant, 1905 = *Okanagama* (sic) = *Okanaga* (sic) = *Okanogana* (sic) = *Ocanagana* (sic) = *Kanagana* (sic)
O. annulata Davis, 1935
O. arboraria Wymore, 1934 = *Okanagana arboraria crocea* Wymore, 1934
O. arctostaphylae Van Duzee, 1915
O. aurantiaca Davis, 1917
O. aurora Davis, 1936
O. balli Davis, 1919
O. bella Davis, 1919 = *Okanagana belli* (sic) = *Okanagana rubrocaudata* Davis, 1925 = *Okanagana bella* var. *rubrocaudata* = *Okanagana bella rubrocaudata* = *Okanagana rubracaudata* (sic)
O. canadensis (Provancher, 1889) = *Cicada canadensis* = *Tibicen canadensis*
O. canescens Van Duzee, 1915 = *Okanagana albibasalis* Wymore, 1934 = *Okanagana alibasalis* (sic)
O. catalina Davis, 1936 = *Okanagana hirsuta* var. *catalina*
O. cruentifera (Uhler, 1892) = *Tibicen cruentifera* = *Tibicina cruentifera* = *Tibicens* (sic) *cruentifera* = *Tibicen cruentiferus* = *Cicada cruentifera*
O. ferrugomaculata Davis, 1936

O. formosa Davis, 1926
O. fratercula Davis, 1915
O. fumipennis Davis1932
O. georgi Heath & Sanborn, 2007
O. gibbera Davis, 1927
O. hesperia (Uhler, 1872) = *Cicada hesperia* = *Tibicen hesperina* (sic) = *Tibicen hesperia* = *Tibicina hesperia* = *Tibicinoides hesperia* = *Tibicinoides hesperius* (sic) = *Okanagana striatipes var. beameri* Davis, 1930 = *Okanagana striatipes beameri*
O. hirsuta Davis, 1915
O. lurida Davis, 1919
O. luteobasalis Davis, 1935 = *Okanagana fratercula* (nec Davis)
O. magnifica Davis, 1919 = *Okanagana magnica* (sic)
O. mariposa mariposa Davis, 1915 = *Okanagana mairposa*
O. mariposa oregonensis Davis, 1939 = *Okanagana mariposa* var. *oregonensis*
O. napa Davis, 1919
O. nigriviridis Davis, 1921 = *Okanagana nigrinervis* (sic)
O. nigrodorsata Davis, 1923
O. noveboracensis (Emmons, 1854) = *Cicada noveboracensis* = *Cicada novaeboracensis* (sic) = *Tibicen noveboracensis* = *Tibicen rimosa noveboracensis* = *Tibicen novaboracensis* (sic) = *Okanagana novaeboracensis* (sic) = *Okanogana* (sic) *noveboracensis*
O. occidentalis (Walker, 1866) = *Cicada occidentalis* = *Tibicen occidentalis*
O. opacipennis Davis, 1926 = *Okanagana arctostaphylae opacipennis*
O. oregona Davis, 1916
O. orithyia Bliven, 1964
O. ornata Van Duzee, 1915
O. pallidula Davis, 1917 = *Okanagana pallidula nigra* Davis, 1938 = *Okanagana pallidula* var. *nigra* = *Okanagana nigra*
O. pernix Bliven, 1964
O. rhadine Bliven, 1964
O. rimosa ohioensis Davis, 1942 = *Okanagana rimosa* var. *ohioensis*
O. rimosa rimosa (Say, 1830) = *Cicada rimosa* = *Tibicen rimosa* = *Okanagana rinosa* (sic) = *Okanaga* (sic) *rimosa* = *Okanagana rimrosa* (sic) = *Ocanagana* (sic) *rimosa* = *Okanagana rumorsa* (sic)
O. rubrovenosa Davis, 1915 = *Okanagana rubrovenosa rubrovenosa* = *Okanagana rubrovenosa rubida* Davis, 1936 = *Okanagana rubrovenosa* var. *rubida* = *Okanagana rubida*
O. salicicola Bliven, 1964
O. schaefferi Davis, 1915 = *Okanagana shaefferi* (sic)
O. sequoiae Bliven, 1964
O. simulata Davis, 1921
O. sperata Van Duzee, 1935
O. striatipes (Haldeman, 1852) = *Cicada striatipes* = *Tibicen striatipes*
O. sugdeni Davis, 1938
O. synodica nigra Davis, 1944 = *Okanagana synodica* var. *nigra*
O. synodica synodica (Say, 1825) = *Cicada synodica* = *Tibicen synodica*
O. tanneri Davis, 1930 = *Okanagana schaefferi tanneri*
O. triangulata Davis, 1915 = *Okanagana triangulata crocea* Wymore, 1934 = *Okanagana triangulata croncina* (sic)
O. tristis rubrobasalis Davis, 1926 = *Okanagana tristis* var. *rubrobasalis* = *Okanagana tristis rubrobasilis* (sic)
O. tristis tristis Van Duzee, 1915 = *Okanagana davisi* Simons, 1953
O. uncinata Van Duzee, 1915
O. utahensis Davis, 1919
O. vanduzeei Distant, 1914 = *Okanagana vanduzei* (sic) = *Okanaga* (sic) *vanduziae* (sic) = *Okagana* (sic) *vanduzeei* = *Okanagana californica* Distant, 1914 = *Okanagana californicus* (sic) = *Okanagana vanduzeei californica* = *Okanagana vanduzeei consobrina* Distant, 1914 = *Okanagana vanduzei* (sic) *consobrina* = *Okanagana consobrina*
O. vandykei Van Duzee, 1915
O. venusta Davis, 1935
O. villosa Davis, 1941

O. viridis Davis, 1918
O. vocalis Bliven, 1964
O. wymorei Davis, 1935 = *Okanagana wymori* (sic)
O. yakimaensis Davis, 1939

Okanagodes Davis, 1919
O. gracilis gracilis Davis, 1919 = *Okanagodes gracilis pallida* Davis, 1932 = *Okanagodes gracilis* var. *pallida*
O. gracilis viridis Davis, 1934 = *Okanagodes gracilis* var. *viridis*
O. terlingua Davis, 1932

Clidophleps Van Duzee, 1915
C. astigma Davis, 1917
C. beameri Davis, 1936
C. blaisdellii (Uhler, 1892) = *Tibicen blaisdellii* = *Tibicen blaisdelli* (sic) = *Okanagana blaisdelli* (sic) = *Clidophleps blaisdelli* (sic) =
C. distanti distanti (Van Duzee, 1914) = *Okanagana dstanti* (sic) = *Okanagana distanti* = *Okanagana distanti pallida* Van Duzee, 1914 = *Clidophleps distanti pallida* = *Clidophleps pallida* = *Okanagana distanti pallidus* (sic) = *Clidophleps distanti* var. *pallidus*
C. distanti truncata (Van Duzee, 1914) = *Okanagana distanti truncatus* (sic) = *Okanagana distanti truncata* (sic) = *Clidophleps distanti* var. *truncatus* (sic) = *Clidophleps truncata* = *Okanagana truncata*
C. rotundifrons (Davis, 1916) = *Okangana rotundifrons*
C. tenuis Davis, 1927
C. vagans Davis, 1925
C. wrighti Davis, 1926

Subpsaltria Chen, 1943 = *Subpsultria* (sic) = *Subpsalta* (sic) = *Subpsaltra* (sic)
S. yangi Chen, 1943 = *S. sienyangensis* Chen, 1943

Ahomana Distant, 1905
A. chilensis Distant, 1905
A. neotropicalis Distant, 1905

BIBLIOGRAPHY

Abaii, M. 2000. *Pests of forest trees & shrubs of Iran*. Ministry of Agriculture, Agricultural Research, Education & Extension Organization, Tehran. 178pp.

Achtziger, R., & U. Nigmann. 2002. Zikaden in Mythologie, Knust und Folklore. *Denisia* 4: 1–16.

Agnello, A.M. 2004. Apple pests and their management. Pages 161–176 *in* J. Capinera, ed. *Encyclopedia of entomology*, volume 1. Springer-Verlag, Heidelberg.

Agnello, A.M. 2008. Apple pests and their management. Pages 244–261 *in* J. Capinera, ed. *Encyclopedia of entomology*, second edition, volume 1. Springer-Verlag, Heidelberg.

Ahern, R.G., S.D. Frank, & M.J Raupp. 2005. Comparison of exclusion and imidacloprid for reduction of oviposition damage to young trees by periodical cicadas (Hemiptera: Cicadidae). *Journal of Economic Entomology* 98: 2133–2136.

Ahmad, A. 1984. A comparative study on flight surface and aerodymanic patterns of insects, birds, and bats. *Indian Journal of Experimental Biology* 22: 270–278.

Ahmed, Z., & S.A. Rizvi. 2010. Review of species of *Cicadatra* (Hemiptera: Cicadidae of Pakistan. *Thirteenth International Auchenorrhyncha Congress and the 7th International Workshop on Leafhoppers and Planthoppers of Economic Significance, 28 June-2 July, Vaison-la-Romaine, France, Abstracts: Talks and Posters*: 25–26.

Ahmed, Z., & A.F. Sanborn. 2010. The cicada fauna of Pakistan including the description of four new species (Hemiptera: Cicadoidea: Cicadidae). *Zootaxa* 2516: 27–48.

Ahmed, Z., A.F. Sanborn, & K.B.R. Hill. 2010. A new species of the cicada genus *Cicadatra* (Hemiptera: Cicadoidea: Cicadidae) from Pakistan. *Zoology in the Middle East* 51: 75–81.

Ahmed, Z., T.F. Siddiqui, M.A. Akhter, & S.A. Rizvi. 2005. Newly recorded species of *Cosmopsaltria saturata* Walker (Homoptera: Cicadidae) from Pakistan with their taxonomic characteristics. *International Journal of Biology and Biotechnology* 2: 843–845.

Ahn, M.Y., B.-S. Hahn, K.S. Ryu, & S.I. Cho. 2002. Effects of insect crude drugs on blood coagulation and fibrinolysis system. *Natural Products Sciences* 8: 66–70.

Ahn, M.Y., K.S. Ryu, Y.W. Lee, & Y.S. Kim. 2000. Cytotoxicity and L-amino acid oxidaase activity of crude insect drugs. *Archives of Pharmical Research* 23: 477–481.

Aidley, D.J. 1989. Structure and function in flight muscle. Pages 31–49 *in* G.J.Goldsworthy & C.H.Wheeler, eds. *Insect flight*. CRC Press, Boca Raton.

Aizu, H., N. Sekita, & M. Yamada. 1984. [Notes on apple fruit damage caused by sucking of the large brown cicada *Graptopsaltria nigrofuscata* Motschulsky.] *Annual Report of the Society of Plant Protection of North Japan* 35: 140–143. [In Japanese, English summary]

Ajesh, K., & K. Sreejith. 2009. Peptide antibiotics: an alternative and effective antimicrobial strategy to circumvent fungal infections. *Peptides* 30: 999–1006.

Akers, D. 2007. …mountains high. On Sunday 21st October IFFA neld a field trip to Bush's Paddock and Mt. Cottrell in association with Pinkerton Landcare and Environmental Group. *Indigenotes* 18: 5–6.

Al Azawi, A.F. 1986. A survey of insect pests of date palms in Qatar. *Date Palm Journal* 4: 247–266.

Albuquerque, F.A., N.M. Martinelli, A.B. Ros, & M. Stülp. 2000. Levantamento de cigarras associadas ao cafeeiro (*Coffea arabica*) na região de Maringá - PR. *XXVI Congresso Brasileiro de Pesquisas Cafeeiras, 2000, Marília. Anais XXVI Congresso Brasileiro de Pesquisas Cafeeiras* 26: 196.

Aldini, R.N. 2008. *Apis amphibia, Cicada, Cimex, Cimices sylvestres, Tipulae*…The insects now known as Hemiptera, in Ulisse Aldrovandi's *De Animalibus Insectis* (1602). *Bulletin of Insectology* 61: 103–105.

Alexander, R.D., D.C. Marshall, & J.R. Cooley. 1997. Evolutionary perspectives on insect mating. Pages 4–31 *in* J.C.Choe & B.J.Crespi, eds.*The evolution of mating systems in insects and arachnids*. Cambridge University Press, Cambridge.

Alexinschi, A. 1955. Contributiumi la cunoasterea faunaei entomologice a nisipurilpr de la Hanul Conachi, cu consideratiuni speciale asupra Ord. Lepidoptera. *Studii si Cercetari Stiintifice, Academia Republicii Populare Romîne, Filiala Iasi* VI: 219–226.

Alfken, J.D. 1903. Beitrag zur Insectenfauna der hawaiischen und neuseelandish Inseln (Ergebnisse einer Reise nach dem Pacific) Schauinsland 1986–97. *Zoologische Jahrbücher Abteilung für Systematik Ökologie und Geographie der Tiere* 19: 561–628.

Almeida, J.E.M. de. 2004. Controle biológico de cigarras-do-cafeeiro. *Anais X Reunião Itinerante de Fitossandidade do Instituto Biológico Café, 10. Mococa, SP, 19 de Outubro 2004*: 101–113.

Almeida, L.M.L., M.D.A. Fialho, M. Schneider, & C.E.G. Pinheiro. 2004. Dinâmica populacional de *Quesada gigas* (Hemiptera, Cicadidae) em Brasília, DF. *XXV Congresso Brasileiro de Zoologia, 2004, Brasília. Livro de Resumos do XXV Congresso Brasileiro de Zoologia* 25: 125–126.

Andersen, D.C. 1987. Below-ground herbivory in natural communities: a review emphasizing fossorial animals. *Quarterly Review of Biology* 62: 261–286.

Andersen, D.C. 1994. Are cicadas (*Diceroprocta apache*) both a "keystone" and a "critical-link" species in lower Colorado River riparian communities? *Southwestern Naturalist* 39: 26–33.

Andersen, D.C., & M.L. Folk. 1993. *Blarina brevicauda* and *Peromyscus leucopus* reduce overwinter survivorship of acorn weevils in an Indiana hardwood forest. *Journal of Mammalogy* 74: 656–664.

Anderson, N.A., M.E. Ostry, & G.W. Anderson. 1979. Insect wounds as infection sites for *Hypoxylon mammatum* on trembling aspen. *Phytopathology* 69: 476–479.

Andrews, J.R.H., & G.W. Gibbs. 1989. The first New Zealand insects collected on Cook's Endeavour voyage. *Pacific Science* 43: 102–114.

Anonymous. 1884. The insects of the year. *Science n.s.* 4: 565–568.

Anonymous. 1911. Proceedings of the Society. *The Glasgow Naturalist* 3: 99–103.

Anonymous. 1932. *Insect Pest Survey Bulletin* 12(10): 424–425.

Anonymous. 1933a. *Insect Pest Survey Bulletin* 13(10): 336–337.

Anonymous. 1933b. *Entomological World* 1: 183–187.

Anonymous. 1936. *Entomological World* 4: 765–767.

Anonymous. 1937. *Entomological World* 5: 259.

Anonymous. 1957. Forest insect survey and life. *Forest Insect Survey Newsletter* 6: 1–32.

Anonymous. 1985. Cane pests. Bureau of Sugar Experiment Stations, Indooroopilly, Queensland, Australia. 85th Annual Report to the Minister for Primary Industries 1985: 17–23.

Anonymous. 1986. The insect with designer muscles. *New Scientist* 109: 23.

Anonymous. 1991. Periodical cicadas may form communal choruses. *American Entomologist* 37: 227.

Anonymous. 1994. *Zhong guo guo shu bing chong zhi (Chinese fruit tree pests)*. China Agriculture Press, Beijing. 1093pp.

Anonymous. 1996a. Naughty? Nice? Or neutral? Periodical Cicadas (*Magicicada* species). *Organic Gardening* 43: 20.

Anonymous. 1996b. Souvenirs "cicado-philatéliques". *Ecole Pratique des Hautes Etudes, Travaux du Laboratoire Biologie et Evolution des Insectes Hemipteroidea* 9: 82.

Anonymous. 1997a. Champions of the insect world. *Explorer* 38: 4–5.

Anonymous. 1997b. Cicadas (Hemiptera). Pages 102–103 *in* D. Smith, G.A.C. Beattie, & R. Broadley, eds. *Citrus pests and their natural enemies: Integrated pest management in Australia*. Horticultural Research and Development Corporation and Queensland Department of Primary Industries, Brisbane.

Anonymous. 1997c. *Planung Vernetzter Biotopsysteme. Bereich Landkreis Südwestpfalz und Kreisfreie Städte Zweibrücken und Pirmasens*. Landesamt für Umweltschutz und Gewerbeaufsicht Rheinland-Pfalz, Oppenheim. 303pp.

Anonymous. 1998. Souvenirs "cicado-philatéliques". *Ecole Pratique des Hautes Etudes, Travaux du Laboratoire Biologie et Evolution des Insectes Hemipteroidea* 10: 135.

Anonymous. 1999a. Periodical Cicada - Brood V. *United States Department of Agriculture, Forest Service, Northeastern Forest Experiment Station*. 2pp.

Anonymous. 1999b. The periodical cicada in West Virginia. *West Virginia Department of Agriculture Leaflet*. 2pp.

Anonymous. 1999c. Corrections and clarifications. *Science* 283: 489.

Anonymous. 2002. Natural exposure. *Australian Geographic* (66): 30.

Anonymous. 2004. The 17-year itch. *Science* 304: 367.

Anonymous. 2007. Warning...periodical cicada will emerge soon Spring of 2008. *The Market Bulletin* 91(July): 3.

Anonymous. 2008. Periodical cicada will emerge soon! *The Market Bulletin* 92(May): 6.

Anufriev, G.A. 2006. [About the fauna of Homoptera Cicadina of Bashkir State Reserve.] *Russian Entomological Journal* 15: 247–251. [In Russian, English summary]

Anufriev, G.A., & A.F. Emelyanov. 1988. [1.Suborder Cicadina (Auchenorrhyncha) - cicadas.] Pages 12–495 *in* P.A. Ler, ed. *Keys to the identification of insects of the Soviet far east. Volume 2: Homoptera and Zheteroptera*. Nauka, Leningrad. [In Russian]

Anufriev, G.A., & N.V. Smirnova. 2009a. [Composition of the fauna and the communities' structure of ther Cicadina (Homoptera) in the lowland trans-Volga woodlands.] *Entomologicheskoe Obozrenie* 88: 529–538. [In Russian, English summary]

Anufriev, G.A., & N.V. Smirnova. 2009b. Composition of the fauna and the communities' structure of ther Cicadina (Homoptera) in the lowland trans-Volga woodlands. *Entomological Review* 89: 784–792.

Aoki, C., F. Santos Lopes, & F.L. de Souza. 2010. Insecta, Hemiptera, Cicadidae, *Quesada gigas* (Olivier, 1790), *Fidicina mannifera* (Fabricius, 1803), *Dorisiana viridis* (Olivier, 1790) and *Dorisiana drewseni* (Stål, 1854): first records for the state of Mato Grosso do Sul, Brazil. *Check List* 6: 162–163.

Arambourg, Y. 1985. The olive's entomological fauna. II. Species of localized economic significance. *Olivae* 5: 14–19.

Archie, J., C. Simon, & D. Wartenberg. 1985. Geographic patterns and population structure in periodical cicadas based on spatial analysis of allozyme frequencies. *Evolution* 39: 1261–1274.

Arensburger, P. 2002. Phylogenetic relationships of New Zealand cicadas. *Dissertation Abstracts International* 63: 5055B.

Arensburger, P., T.R. Buckley, C. Simon, M. Moulds, & K.E. Holsinger. 2004. Biogeography and phylogeny of the New Zealand cicada genera (Hemiptera: Cicadidae) based on nuclear and mitochondrial DNA data. *Journal of Biogeography* 31: 557–569.

Arensburger, P., C. Simon, & K.E. Holsinger. 2004. Evolution and phylogeny of the New Zealand cicada genus *Kikihia* Dugdale (Hemiptera: Cicadidae) with special reference to the origin of the Kermadec and Norfolk Islands' species. *Journal of Biogeography* 31: 1769–1783.

Arias Giralda, A., J.M. Pérez Rodríguez, & F. Gallego Pintor. 1998. La cigarra del olivo en "Tierra de Barros" (Extremadura): Especie dominante, vuelo de adultos, oviposicion y avivamiento. *Revista de Ciências Agraias* 21: 261–269.

Arnett, R.H. 1985. *American insects, a handbook of the insects of America north of Mexico*. Van Nostrand Reinbald, New York. 850pp.

Arnett, R.H. 2000. *American insects, a handbook of the insects of America north of Mexico*. Second edition. CRC Press, Boca Raton, FL. 1003pp.

Arnett, R.H., Jr., & R.L. Jacques, Jr. 1981. *Simon and Schuster's guide to insects*. Simon & Schuster, New York. 511pp.

Artmann, G. 1987. Die fremdsprachige Bergzikade (*Cicadetta montana peregrina*). *Mitteilungen der Entomologischen Gesellschaft Basel* 37: 190–192.

Asahina, S. 1972. [Insects encountered on board a research vessel Hakudo-Maru.] *New Entomologist* 21: 67–71. [In Japanese, English summary]

Ascunce, M.S., J.A. Ernst, A. Clark, & H.N. Nigg. 2008. Mitochondrial nucleotide variability in invasive populations of the root weevil *Diaprepes abbreviatus* (Coleoptera: Curculionidae) of Florida and preliminary assessment of *Diaprepes* from Dominica. *Journal of Economic Entomology* 101: 1443–1454.

Aspöck, U., & H. Aspöck. 2010. Arthropoda – ein Fascinosum. Zur Biodiversität und Systematik der erfolgreichsten Metazoa im Spiegel ihrer medizinischen Bedeutung. *Denisia* 30: 33–80.

Azman, S., & M.I. Zaidi. 2002. Two new species of the genus *Puranoides* Moulton from Malaysia (Homoptera: Cicadidae). *Serangga* 7: 1–14.

Bâbâi, H. 1968. *Cicadatra ocreata* Melichar (Homoptera – Cicadidae). *Entomologie et Phytopathologie Appliquees* 27: 55–62.

Bachmetjew, P. 1907. *Experimentelle Entomologische Studien vom Physikalisch-chemischen Standpunkt Aus*. Zweiter Band. Einfluss der Äusseren Faktoren auf Insekten. Staatsdruckerei, Sophia. 944pp.

Bailey, W.J. 1991a. *Acoustic behavior of insects, an evolutionary perspective*. Chapman and Hall, London. 225pp.

Bailey, W.J. 1991b. Mate finding: selection on sensory cues. Pages 42–74 *in* W.J. Bailey & J. Ridsdill-Smith, eds. *Reproductive behaviour of insects*. Chapman and Hall, London.

Bailey, W.J. 1998. Do large bushcrickets have more sensitive ears? Natural variation in hearing thresholds within populations of the bushcricket *Requena verticalis* (Listroscelidinae: Tettigoniidae). *Physiological Entomology* 23: 105–112.

Bailey, W.J. 2003. Insect duets: underlying mechanisms and their evolution. *Physiological Entomology* 28: 157–174.

Bairyamova, V. 1992. [Studies on cicade fauna (Homoptera, Auchenorrhyncha) of the Rhodopes.] *Acta Zoologica Bulgarica* 44: 44–56. [In Russian, English summary]

Baker, A. 2005. Are there genuine mathematical explanations of physical phenomena? *Mind* 114: 223–238.

Banks, C.S. 1904a. A preliminary report on insects of the Cacao, prepared especially for the benefit of farmers. *Report of the Philippine Commission, 1903, Part 2, War Deptartment, Washington, D.C. (=Department of the Interior, Bureau of Government Laboratories, Biological Laboratory, Entomological Division, Bulletin Number)* 1: 1–52.

Banks, C.S. 1904b. A preliminary report on insects of the Cacao, prepared especially for the benefit of farmers. *Annual Report of the Philippine Commission* 4: 597–620.

Barberoglou, M., V. Zorba, E. Stratakis, E. Spanakis, P. Tzanetakis, S.H. Anastasiadis, & C. Fotakis. 2009. Bio-inspired water repellant surfaces produced by ultrafast laser structuring of silicon. *Applied Surface Sciences* 255: 5425–5429.

Barbosa, P., & M.R. Wagner. 1989. *Introduction to forest and shade tree insects*. Academic Press, Inc., San Diego. 639pp.

Barnes, J.K. 1988. Asa Fitch and the emergence of American entomology with an entomological bibliography and a catalog of taxonomic names and type specimens. *New York State Museum Bulletin* 461: 1–120.

Barratt, B.I.P. 1983. Distribution survey of soil insects at Millers Flat, Central Otago. *New Zealand Journal of Experimental Agriculture* 11: 83–87.

Barratt, B.I.P., & B.H. Patrick. 1987. Insects of snow tussock grassland on the East Otago Plateau. *New Zealand Entomologist* 10: 69–98.

Barrett, B.A. 2001. Periodical cicadas in Missouri. *University of Missouri Agricultural Publication G7259*: 2pp.

Bartholomew, G.A., & M.C. Barnhart. 1984. Tracheal gases, respiratory gas exchange, body temperature and flight in some tropical cicadas. *Journal of Experimental Biology* 111: 131–144.

Bartlett, C.R. 2000. An annotated list of planthoppers (Hemiptera: Fulgoroidea) of Guana Island (British West Indies). *Entomological News* 111: 120–132.

Barton, B.S. 1804. Remarks. *Philadelphia Medical and Physical Journal Collected and Arranged by B.S. Barton* 1: 55–56.

Bashan, M., H. Akbas, & K. Yurdakovoc. 2002. Phospholipid and triacylglycerol fatty acid composition of major life stages of sunn pest, *Eurygaster integriceps* (Heteroptera: Scutelleridae). *Comparative Biochemistry and Physiology. B. Biochemical and Molecular Biology* 132: 375–380.

Bashan, M., & O. Cakmak. 2005. Changes in composition of phospholipid and triacylglycerol fatty acids prepared from prediapausing and diapausing individuals of *Dolycoris baccarum* and *Piezodorus lituratus* (Heteroptera: Pentatomidae). *Annals of the Entomological Society of America* 98: 575–579.

Bator, A. 1952. Über einige xerotherme Lokalitäten Tirols und ihre Fauna. *Wetter und Leben* 4: 198–201.

Bauernfeind, R.J. 2000. 1998 distribution of Brood IV *Magicicada* (Homoptera: Cicadidae) in Kansas. *Journal of the Kansas Entomological Society* 73: 238–241.

Bauman, J., B. Rivera-Orduño, & S.P. Stock. 2004. Interactions between *Fusarium oxysporum* f. sp. *asparagi* (Ascomycota: Pyrenomycetes) and *Heterorhabditis caborca* strain (Heterorhabditidae) in *Galleria mellonella* larvae. *Book of Abstracts, 37th Annual Meeting of the Society for Invertebrate Pathology and 7th International Conference on* Bacillus thuringiensis, *Helsinki, Finland, 1–6 August 2004*: 104.

Bayefsky-Anand, S. 2005. Effect of location and season on the arthropod prey of *Nycteris grandis* (Chiroptera: Nycteridae). *African Zoology* 40: 93–97.

Beaupre, S.J., & K.G. Roberts. 2001. *Agkistrodon contortrix contortrix* (Southern Copperhead)). Chemotaxis, arboreality and diet. *Herpetological Review* 32: 44–45.

Bedick, J.C., H. Tunaz, A.R. Nor Aliza, S.M. Putnam, M.D. Ellis, & D.W. Stanley. 2001. Eicosanoids act in nodulation reactions to bacterial infections in newly emerged adult honeybees, *Apis mellifera*, but not in older forageers. *Comparative Biochemistry and Physiology. Part C. Toxicology and Pharmacology* 130: 107–117.

Beenen, R. 2001. *Leptus mariae* (Acari: Erythraeidae) parasiterend op bladkevers (Coleoptera: Chrysomelidae). *Entomologische Berichten, Amsterdam* 61: 201–202.

Behncke, H. 2000. Periodical cicadas. *Journal of Mathematical Biology* 40: 413–431.

Benko, T.P., & M. Perc. 2006. Deterministic chaos in sounds of Asian cicadas. *Journal of Biological Systems* 14: 555–566.

Bennet-Clark, H.C. 1994. The world's noisiest insects - cicada song as a model of the biophysics of animal sound production. *Verhandlungen der Deutschen Zoologischen Gesellschaft* 87: 165–176.

Bennet-Clark, H.C. 1995. Insect sound production: transduction mechanisms and impedance matching. *Symposia of the Society for Experimental Biology* 49: 199–218.

Bennet-Clark, H.C. 1997. Tymbal mechanics and the control of song frequency in the cicada *Cyclochila australasiae*. *Journal of Experimental Biology* 200: 1681–1694.

Bennet-Clark, H.C. 1998a. Size and scale effects as constraints in insect sound communication. *Philosophical Transactions of the Royal Society of London. B* 353: 407–419.

Bennet-Clark, H.C. 1998b. How cicadas make their noise. *Scientific American* 278(5): 58–61.

Bennet-Clark, H.C. 1999. Resonators in insect sound production: How insects produce loud pure-tone songs. *Journal of Experimental Biology* 202: 3347–3357.

Bennet-Clark, H.C. 2007. The first description of resilin. *Journal of Experimental Biology* 210: 3879–3881.

Bennet-Clark, H.C., & A.G. Daws. 1999. Transduction of mechanical energy into sound energy in the cicada *Cyclochila australasiae*. *Journal of Experimental Biology* 202: 1803–1817.

Bennet-Clark, H.C., & D. Young. 1992. A model of the mechanism of sound production in cicadas. *Journal of Experimental Biology* 173: 123–153.

Bennet-Clark, H.C., & D. Young. 1994. The scaling of song frequency in cicadas. *Journal of Experimental Biology* 191: 291–294.

Bennet-Clark, H.C., & D. Young. 1998. Sound radiation by the bladder cicada *Cystosoma saundersii*. *Journal of Experimental Biology* 201: 701–715.

Bennett, B.G. 1984. Blue, red, and yellow insects. *New Zealand Entomologist* 8: 88–90.

Bennett, P.M., & K.A. Hobson. 2009. Trophic structure of a boreal forest arthropod community revealed by stable isotope ($d^{13}C$, $d^{15}N$) analysis. *Entomological Science* 12: 17–24.

Benson, J. 1987. The cicadas are here. *Agricultural Research, USA* 35: 5.

Bento, J.M.S. 2007. Diagnóstico da situação de pragas de solo em São Paulo. Pages 26–28 *in* C.J. Ávila & R.P.S. Júnior, eds. *10ª Reunião Sul-Brasileira sobre pragas de solo, Pragas-Solo-Sul, Anaise Ata, 25–27 September 2007, Dourados, MS*. Embrapa Agropecuária Oeste, Dourados.

Berenbaum, M. 1995. Ah! Humbug! *American Entomologist* 41: 132–133.

Bergh, C.J., & B.D. Short. 2008. Ecological and life-history notes on syrphid predators of woolly apple aphid in Virginia, with emphasis on *Heringia calcarata*. *BioControl* 53: 773–786.

Bergier, E. 1941. *Peuples entomophages et insectes comestibles: Étude sur les moeurs de l'homme et de l'insecte*. Imprimerie Rullière Fières, Avignon. 231pp.

Bernhardt, G.E., L. Kutschbach-Brohl, B.E. Washburn, R.B. Chipman, & L.C. Francoeur. 2010. Temporal variation in terrestrial invertebrate consumption by laughing gulls in New York. *American Midland Naturalist* 163: 442–454.

Bernhardt, K.G. 1991. Zum Auftreten von *Cicadetta montana* Scopoli, 1772 (Homoptera-Auchenorrhyncha) bei Tecklenburg und Lengerich. *Natur und Heimat* 51: 77–78.

Bernier, C. 2006. Les cigales (Hemiptera, Cicadidae) de l'ouest de la France. Appel à participation dans le cadre de l'enquête nationale sur les cigales. *Lettre de l'Atlas entomologique regional (Nantes)* (19): 1–9.

Bessin, R., & L.H. Townsend. 1992. Apple: Periodical cicada control, 1991. Page 3 *in* A.K.Burditt, Jr., ed. *Insecticide and Acaricide Tests, vol. 17*. Entomological Society of America, Lanham, MD.

Beuk, P.L.T. 1996. The *jacoona* assemblage of the genus *Dundubia* Amyot & Serville (Homoptera: Cicadidae): a taxonomic study of its species and a discussion of its phylogenetic relationships. *Contributions to Zoology* 66: 129–184.

Beuk, P.L.T. 1998. Revision of the *radha* group of the genus *Platylomia* Stål, 1870 (Homoptera, Cicadidae). *Tijdschrift voor Entomologie* 140: 147–176.

Beuk, P.L.T. 1999. Revision of the cicadas of the *Platylomia spinosa* group (Homoptera: Cicadidae). *Oriental Insects* 33: 1–84.

Bey, I. 1913. Note sur *Cicadatra foveicollis* Hov. *Bulletin de la Société Entomologique d'Egypte* 1912–1913: 141–14.

Beyer, W.N., O.H. Pattee, L. Sileo, D.J. Horrman, & B.M. Mulhern. 1985. Metal contamination in wildlife living near two zinc smelters. *Environmental Pollution (Series A, ecological and Biological)* 38: 63–86.

Biedermann, R., R. Achtziger, H. Nickel, & A.J.A. Stewart. 2005. Conservation of grassland leafhoppers: a brief review. *Journal of Insect Conservation* 9: 229–243.

Bigot, L. 1985. Contribution a l'etude des peuplements littoraux et cotiers de la Nouvelle-Caledonie (Grande Terre, ile des Pins) et d'une des iles Loyaute (Ouvea): Premier inventaire entomologique. *Annales de la Société Entomologique de France* 21: 317–329.

Binet, A. 1894. Contributions a l'étude du systéme neuveux sous-intestinal des insects. *Journal de l'Anatome et de la Physiologie* 30: 449–580.

Birkenshaw, C. 2003. Exploitation of the cicada (*Pycna madagascariensis*) for food by black lemurs (*Eulemur macaco macaco*) and brown lemurs (*Eulemur fulvus fulvus*). *Lemur News* 8: 3.

Biswas, B., M. Ghosh, A. Bal, & G.C. Sen. 2004. Insecta: Hemiptera: Homoptera: Fulgoroidea, Cercopoidea, Cicadoidea and Membracoidea. Pages 239–251 *in* J.R.B. Alfred, ed. *State Fauna Series 10: Fauna of Manipur, Part 2, Insects*. Zoological Survey of India, Kolkata.

Blair, M. 1998. Seventeen years waiting. *Kansas Wildlife & Parks* 55: 2–6.

Boer, A.J. de. 1982. The taxonomy and biogeography of the *nasuta* group of the genus *Baeturia* Stål, 1866 (Homoptera, Tibicinidae). *Beaufortia* 32: 57–77.

Boer, A.J. de. 1986. The taxonomy and biogeography of the *conviva* group of the genus *Baeturia* Stål, 1866 (Homoptera, Tibicinidae). *Beaufortia* 36: 167–182.

Boer, A.J. de. 1989. The taxonomy and biogeography of the *bloetei* group of the genus *Baeturia* Stål, 1866 (Homoptera, Tibicinidae). *Beaufortia* 39: 1–43.

Boer, A.J. de. 1990. *Aedeastria*, a new cicada genus from New Guinea, its phylogeny and biogeography (Homoptera, Tibicinidae), preceded by a discussion of the taxonomy of New Guinean Tibicinidae. *Beaufortia* 40: 63–72.

Boer, A.J. de. 1991. *Scottotympana*, a new cicada genus from New Guinea, with the description of three new species, their taxonomy and biogeography (Homoptera, Tibicinidae). *Beaufortia* 42: 1–11.

Boer, A.J. de. 1992a. The taxonomy and biogeography of the genus *Thaumastopsaltria* Kirkaldy, 1900 (Homoptera, Tibicinidae). *Beaufortia* 43: 17–44.

Boer, A.J. de. 1992b. The taxonomy and biogeography of the *viridis* group of the genus *Baeturia* Stål, 1866 (Homoptera, Tibicinidae). *Bijdragen tot de Dierkunde* 61: 163–183.

Boer, A.J. de. 1993a. *Guineapsaltria*, a new genus from the Australian-New Guinean region (Homoptera, Tibicinidae), with notes on its taxonomy and biogeography. *Bijdragen tot de Dierkunde* 63: 15–41.

Boer, A.J. de. 1993b. Ten species added to the genus *Aedeastria* De Boer, 1990, with the description of eight new species and notes on the taxonomy and biogeography (Homoptera, Tibicinidae). *Beaufortia* 43: 140–167.

Boer, A.J. de. 1994a. The taxonomy and biogeography of the *loriae* group of the genus *Baeturia* Stål, 1866 (Homoptera, Tibicinidae). *Tijdschrift voor Entomologie* 137: 1–26.

Boer, A.J. de. 1994b. Four species added to the *Baeturia nasuta* group, with notes on taxonomy and biogeography (Homoptera, Tibicinidae). *Tijdschrift voor Entomologie* 137: 161–172.

Boer, A.J. de. 1994c. The taxonomy and biogeography of the *exhausta* group of the genus *Baeturia* Stål, 1866 (Homoptera, Tibicinidae). *Beaufortia* 44: 127–158.

Boer, A.J. de. 1994d. The taxonomy and biogeography of the *guttulinervis* group of the genus *Baeturia* Stål, 1866 (Homoptera, Tbicinidae). *Bijdragen tot de Dierkunde* 64: 87–100.

Boer, A.J. de. 1995a. The taxonomy and biogeography of the cicada genus *Papuapsaltria* gen. n. (Homoptera, Tibicinidae). *Tijdschrift voor Entomologie* 138: 1–44.

Boer, A.J. de. 1995b. The taxonomy, phylogeny and biogeography of the cicada genus *Gymnotympana* Stål, 1861 (Homoptera, Tibicinidae). *Invertebrate Taxonomy* 9: 1–81.

Boer, A.J. de. 1995c. The phylogeny and taxonomic status of the Chlorocystini (*sensu stricto*) (Homoptera, Tibicinidae). *Contributions to Zoology* 65: 201–231.

Boer, A.J. de. 1995d. Islands and cicadas adrift in the West-Pacific: biogeographic patterns related to plate tectonics. *Tijdschrift voor Entomologie* 138: 169–244.

Boer, A.J. de. 1996. *Mirabilopsaltria*, a new cicada genus from New Guinea, its taxonomy and biogeography (Homoptera Tibicinidae). *Tropical Zoology* 9: 349–379.

Boer, A.J. de. 1997. Phylogeny and biogeography of Australian genera of Chlorocystini (Insecta: Homoptera: Tibicinidae). *Memoirs of the Museum of Victoria* 56: 91–123.

Boer, A.J. de. 1999. Taxonomy and biogeography of the New Guinean Cicadettini (Hemiptera, Tibicinidae). *Mitteilungen aus dem Museum für Naturkunde im Berlin, Deutsche Entomologishe Zeitschrift* 46: 115–147.

Boer, A.J. de. 2000. The cicadas of Mt. Bosavi and the Kikori Basin, southern Papua New Guinea (Homoptera, Tibicinidae). *Bishop Museum Occasional Papers* 61: 1–23.

Boer, A.J. de, & J.P. Duffels. 1996a. Historical biogeography of the cicadas of Wallacea, New Guinea and the West Pacific: a geotectonic explanation. *Palaeogeography, Palaeoclimatology and Palaeoecology* 124: 153–177.

Boer, A.J. de, & J.P. Duffels. 1996b. Biogeography of Indo-Pacific cicadas east of Wallace's Line. Pages 297–330 *in* A. Keast & S.E. Miller, eds. *The origin and evolution of Pacific Island biotas, New Guinea to Eastern Polynesia: patterns and processes*. SPB Academic Publishing, Amsterdam.

Bohart, R.M., & L. Stange. 1976. *Liogorytes joergenseni* (Bréthes), a cicada killer in Argentina (Sphecidae, Nyssoninae). *Pan-Pacific Entomologist* 52: 313.

Bolcatto, P.G., M.C. Medrano, & C. De Santis. 2006. Cuando cantan las chicharras: un contrapunto entre físicos y biólogos. *ECO LOGICA* (14): 6–10.

Bonaccorso, F.J. 1979. Foraging and reproductive ecology in a Panamanian bat community. *Bulletin of the Florida State Museum, Biological Sciences* 24: 359–408.

Bonfils, J., S. Quilici, & B. Reynaud. 1994. Les Hémiptères Auchénorhynques de l'ile de la Réunion. *Bulletin de la Société Entomologique de France* 99: 227–240.

Borror, D.J., & R.E. White. 1970. *A field guide to insects: America north of Mexico*. Houghton Mifflin Company, Boston. 404pp. 16 plates.

Bosik, J.J. 1997. *Common names of insects & related organisms 1997*. Committee on Common Names of Insects, Entomological Society of America, Lanham, MD. 232pp.

Boucias, D.G., J.C. Pendland, & Ö. Kalkar. 2004. *Nomuraea*. Pages 1558–1560 *in* J. Capinera, ed. *Encyclopedia of entomology*, volume 2. Springer-Verlag, Heidelberg.

Boucias, D.G., J.C. Pendland, & Ö. Kalkar. 2008. *Nomuraea*. Pages 2613–2615 *in* J. Capinera, ed. *Encyclopedia of entomology*, second edition, volume 3. Springer-Verlag, Heidelberg.

Boulard, M. 1981a. Homoptères Cicadoidea récoltés en Algerie par M. J.-M. Maldes. *Revue Française d'Entomologie (N.S.)* 3: 37–45.

Boulard, M. 1981b. Materiaux pour une révision de la faune cicadéenne de l'Ouest palearctique (2e note). [Hom.]. *Bulletin de la Société Entomologique de France* 86: 41–53.

Boulard, M. 1981c. Sur trois *Toxopeusella* nouvelles des collections du British Museum (Hom. Cicadoidea). *The Entomologists's Monthly Magazine* 117: 129–138.

Boulard, M. 1981d. Sur une cigale inedite des Comores (Hom.). *Bulletin de la Société Entomologique de France* 86: 103–106.

Boulard, M. 1982a. Taxa nouveaux pour la faune des cigales de France (Hom.). *Bulletin de la Société Entomologique de France* 87: 49–50.

Boulard, M. 1982b. Recherches Zoologiques en Ethiopie: Homoptera Cicadoidea. *Problemi Attutali di Sciencza e di Cultura. Accademia Nazionale dei Lincei, Roma Quadernie* 252: 57–60.

Boulard, M. 1982c. Sur deux cigales nouvelles du Bassin Méditerranéen. *Nouvelle Revue d'Entomologie* 12: 101–105.

Boulard, M. 1982d. Une nouvelle cigale néotropicale, halophile et crépusculaire (Homoptera, Cicadoidea). *Revue Française d'Entomologie (N.S.)* 4: 108–112.

Boulard, M. 1982e. Description d'une cigale guyanaise pourvue de cymbales miniscules (Homoptera, Cicadoidea). *Revue Française d'Entomologie (N.S.)* 4: 179–182.

Boulard, M. 1982f. Les cigales du Portugal, contribution a leur etude (Hom. Cicadidae). *Annales de la Société Entomologique de France* 18: 181–198.

Boulard, M. 1982g. Cigales dans la garrigue. *Dossier pédagogique, Radio et Télévision scolaire, 1981–1982* 9: 37–48.

Boulard, M. 1983a. *Tibicina garricola* n. sp., cigale méconnue du Sud de la France (Hom. Cicadoidea). *L'Entomologiste* 39: 309–312.

Boulard, M. 1983b. La cigale, la guêpe et les fourmis…Note de chasse a Port-Cros. *Travaux Scientifique du Parc National de Port-Cros, France* 8: 119–122.

Boulard, M. 1984a. La cigale des cistes de Fairmaire, son identité scientifique et sa localisation géographique in Gallia primordia patria [Hom. Tibicinidae]. *Entomologica Gallica* 1: 45–48.

Boulard, M. 1984b. *Hylora villiersi*, cigale nouvelle de l'Ouest Africain (Homoptera, Tibicinidae). *Revue Française d'Entomologie (N.S.)* 6: 52–54.

Boulard, M. 1984c. Arguments pour la suppression du nom de genre *Tibicen* et de ses dérivés dans la nomenclature de la superfamille de Cicadoidea. *Bulletin of Zoological Nomenclature* 41: 166–184.

Boulard, M. 1984d. Cigales de France et d'Afrique (Aperçu de leur biologie générale et particulière). *Fueille d'Information, Société dea amis du Muséum National d'Histoire naturelle et du Jardin des Plantes* (Juin 1984): 5–7, Plate.

Boulard, M. 1985a. Signalisations nouvelles pour la faune des cigales lusitaniennes (Homoptera: Cicadoidea). *L'Entomologiste* 41: 33–34.

Boulard, M. 1985b. Cigales in concert. *Sciences et Avenir* 462: 58–65.

Boulard, M. 1985c. Cigales Africaines nouvelles ou mal connues de la famille des Cicadidae (Homoptera, Cicadoidea). *Annales de la Société Entomologique de France* 21: 175–188.

Boulard, M. 1985d. Comment on the proposed conservation of *Tibicina* Amyot 1847 and *Lyristes* Horváth 1926. Z.N. (S.)239. *Bulletin of Zoological Nomenclature* 42: 212–214.

Boulard, M. 1985e. Apparence et mimetisme chez les cigales (Hom. Cicadoidea). *Bulletin de la Société Entomologique de France* 90: 1016–1051.

Boulard, M. 1986a. Nouvelles cigales guyano-amazoniennes du genre *Carineta* (Homoptera, Tibicinidae). *Nouvelle Revue d'Entomologie (N.S.)* 2: 415–429.

Boulard, M. 1986b. Les cigales de Madagascar, point bibliographique et perspectives d'études. *Madagascar Recherches Scientifiques. Bulletin Trimestriel de L'Association pour L'Avancement de la Recherche Scientifique Concernant Madagascar et la Region Malgache* 19: 30–35.

Boulard, M. 1986c. Une singulière évolution morphologique: celle d'un appareil stridulant sur les genitalia des males de *Carineta*. Description de cinq espèces nouvelles (Hom. Tibicinidae). *Annales de la Société Entomologique de France* 22: 191–204.

Boulard, M. 1986d. Cigales de la Forêt de Taï (Côte d'Ivoire) et compléments à la faune cicadéenne afrotropicale [Homoptera, Cicadoidea]. *Revue Française d'Entomologie (N.S.)* 7: 223–240.

Boulard, M. 1986e. *Orialella aerizulae* n. sp., cigale nouvelle de la forêt guyanaise (Hom. Cicadoidea). *L'Entomologiste* 42: 345–347.

Boulard, M. 1987. Cigales nouvelles d'Afrique du Nord (Hom. Cicadoidea). *L'Entomologiste* 43: 215–218.

Boulard, M. 1988a. Biologie et comportement des cigales de France. *Insectes* (69): 7–13.

Boulard, M. 1988b. Sur une deuxième espèce du genre *Saticula* (Homoptera, Cicadoidea, Tibicinidae). *Revue Française d'Entomologie (N.S.)* 10: 41–43.

Boulard, M. 1988c. Homoptères Cicadoidea de Nouvelle-Calédonie. 1. Description d'un genre nouveau et deux espèces nouvelles de Tibicinidae. *in*: S. Tiller, ed. *Zoologia Neocaledonica, I. Memoires du Museum National d'Histoire Naturelle Serie A Zoologie.* 142: 61–66.

Boulard, M. 1988d. La tératologie des cigales. Aperçu bibliographique et signalisation d'anomaliles ou de monstrosités nouvelles. *Bulletin de la Société Entomologique de France* 92: 149–159.

Boulard, M. 1988e. Les *Lyristes* d'Asie Mineure (Hom. Cicadidae). I - Sur deux formes éthospécifiques syntopiques et description de deux espèces nouvelles. *L'Entomologiste* 44: 153–167.

Boulard, M. 1988f. Taxonomie et nomenclature superieures des Cicadoidea. Histoire, problems et solutions. *Ecole Pratique des Hautes Etudes, Travaux du Laboratoire Biologie et Evolution des Insectes Hemipteroidea* 1: 3–89.

Boulard, M. 1988g. Les cigales. *Lou Prouvençau à l'Escolo, Revisto dóu Centre Internaciounau de Recerco e d'Estùdi Prouvençau* (116, 4–1988): 3–5.

Boulard, M. 1989a. Une nouvelle espèce de cigale de la Réunion (Tibicenidae, Gymnotympanini). *Bulletin de la Société Entomologique de France* 93: 129–132.

Boulard, M. 1989b. Non-régénération d'une antenne chez une larve de cigale au dernier stade (Hom. Cicadoidea). *Bulletin de la Société Entomologique de France* 93: 205–208.

Boulard, M. 1989c. Contributions a l'entomologie general et appliquee. 1. Insectes nuisibles ou associes aux plantes perennes cultivees en Afrique tropicale. *Ecole Practique des Hautes Etudes, Travaux du Laboratoire Biologie et Evolution des Insectes Hemipteroidea* 1: 1–54, plates I-!V.

Boulard, M. 1989d. Expedition France-Australe (5 juillet-5 août 1988). Participation entomologique. Compte Rendu de mission en Australie Occidentale. *Ecole Pratique des Hautes Etudes, Travaux du Laboratoire Biologie et Evolution des Insectes Hemipteroidea* 2: 55–82.

Boulard, M. 1990a. A new genus and species of cicada (Insecta, Homoptera, Tibicinindae) from western Australia. *Reserch in Shark Bay. Report of the France-Australe Bicentenary Expedition Committee*: 237–243.

Boulard, M. 1990b. Contributions a l'entomologie general et appliquee. 2. Cicadaires (Homopteres Auchenorhynques). Première partie: Cicadoidea. *Ecole Practique des Hautes Etudes, Travaux du Laboratoire Biologie et Evolution des Insectes Hemipteroidea* 3: 55–245.

Boulard, M. 1991a. L'urine des Homoptères, un matèriau utilisé ou recyclé de façons étonnantes. 2de partie. *Insectes* (81): 7–8.

Boulard, M. 1991b. Quelle espèce-type pour le genre *Melampsalta* Amyot, 1847? (Homoptera, Tibicinidae). *Nouvelle Revue d'Entomologie* 8: 25–28.

Boulard, M. 1991c. Cigales colligées lors des "Voyages de découvertes" conduit par J. Dumont d'Urville. Description de *Poecilipsaltria durvillei*, n. sp. (Hom. Cicadoidea, Cicadidae). *Bulletin de la Société Entomologique de France* 96: 117–124.

Boulard, M. 1991d. A propos de la nuisibilité de certaines cigales Néocalédoniennes. *Bulletin de la Société Entomologique de France* 116: 261–265.

Boulard, M. 1991e. Descriptions de trois espèces nouvelles de cigales néocalédoniennes (Homoptera, Cicadoidea, Tibicinidae). *L'Entomologiste* 47: 259–268.

Boulard, M. 1991f. Le Mimétisme. Caractéristiques et Modalités principales. Éléments de Vocabulaire. *Ecole Pratique des Hautes Etudes, Travaux du Laboratoire Biologie et Evolution des Insectes Hemipteroidea* 4: 1–48.

Boulard, M. 1991g. Sur une nouvelles Cigale néocalédoniennes et son étoonnante cymbalisation (Hom. Tibicinidae). *Ecole Pratique des Hautes Etudes, Travaux du Laboratoire Biologie et Evolution des Insectes Hemipteroidea* 4: 73–82.

Boulard, M. 1992a. Descriptions de six especes nouvelles de Cigales neocaledoniennes: premieres notes ethologiques (Homoptera, Cicadoidea). *Bulletin de la Société Entomologique de France* 97: 119–133.

Boulard, M. 1992b. *Pagiphora yanni*, cigale anatolienne inédite: description et premières informations biologiques (cartes d'identité et d'éthologie sonores) (Homoptera, Cicadoidea, Tibicinidae). *Nouvelle Revue d'Entomologie (N.S.)* 9: 365–374.

Boulard, M. 1992c. Identité et bio-écologie de *Cicadatra atra* (Olivier, 1790), la cigale noire, in gallia primordia patria (Homoptera, Cicadoidea, Cicadidae). *Travaux du Laboratoire de la Biologie et Evolution d'Insectes de l'Ecole Practique des Hautes Etudes* 5: 55–86.

Boulard, M. 1992d. Trois nouvelle cigales *Paectira* d'Afrique orientale et du sud (Homoptera, Cicadoidea, Tibicinidae). *Ecole Pratique des Hautes Etudes, Travaux du Laboratoire Biologie et Evolution des Insectes Hemipteroidea* 5: 105–114.

Boulard, M. 1992e. A propos des tours préimaginales de *Cicadetta montana* (Scopoli). *Ecole Pratique des Hautes Etudes, Travaux du Laboratoire Biologie et Evolution des Insectes Hemipteroidea* 5: 126.

Boulard, M. 1993a. *Pagiphora yanni*, cigale anatolienne inédite. Description et premières informations biologiques (cartes d'identité et d'éthologie sonores) (Homoptera, Cicadoidea, Tibicinidae). *Türkije Entomoloji Dergisi* 17: 1–9.

Boulard, M. 1993b. Description d'une cigale originaire de Nouvelle-Calédonie (*Homoptera, Cicadoidea, Tibicinidae*). *L'Entomologiste* 49: 261–264.

Boulard, M. 1993c. Les métamorphoses animales. Caractéristiques et modalités principales. Éléments de vocabulaire. *Ecole Practique des Hautes Etudes, Travaux du Laboratoire Biologie et Evolution des Insectes Hemipteroidea* 6: 1–72.

Boulard, M. 1993d. *Bafutalna mirei*, n. gen., n. sp. de cigale acymbalique (Homoptera, Cicadoidea, Tibicinidae). *Ecole Pratique des Hautes Etudes, Travaux du Laboratoire Biologie et Evolution des Insectes Hemipteroidea* 6: 87–92.

Boulard, M. 1993e. Quatre nouvelles espèces pour la faune cicadéennes de la Nouvelle-Calédonie (homoptera, Cicadoidea, Tibicinidae). *Ecole Practique des Hautes Etudes, Travaux du Laboratoire Biologie et Evolution des Insectes Hemipteroidea* 6: 109–126.

Boulard, M. 1994a. Sur deux cigales nouvelles de la tribu des Polyneurini (Homoptera, Cicadoidea, Cicadidae). *Revue Française d'Entomologie (N.S.)* 16: 61–65.

Boulard, M. 1994b. *Tacua speciosa*, variété *decolorata* n. var. (*Homoptera, Cicadidae*). *Revue Française d'Entomologie (N.S.)* 16: 66.

Boulard, M. 1994c. Cigales inédites de Nouvelle-Calédonie et du Vanuatu (Homoptera, Cicadoidea, Tibicinidae). *Nouvelle Revue d'Entomologie (N.S.)* 11: 143–149.

Boulard, M. 1994d. Sur deux nouvelles *Zouga* Distant, 1906, originaires du Maroc, et l'étonnante membranisation sous-abdominale des mâles (Homoptera, Cicadoidea, Tibicinidae). *Bulletin de la Société Entomologique Suisse* 67: 457–467.

Boulard, M. 1995a. Postures de cymbalisation, cymbalisations et cartes d'identité acoustique des cigales. 1. Généralités et espèces méditerranéennes (Homoptera Cicadoidea). *Ecole Practique des Hautes Etudes, Travaux du Laboratoire Biologie et Evolution des Insectes Hemipteroidea* 7/8: 1–72.

Boulard, M. 1995b. *Panka parvulina* n. sp., miniscule cigale de la grande forêt Ivoirienne (Homoptera Cicadoidea Tibicinidae). *Ecole Pratique des Hautes Etudes, Travaux du Laboratoire Biologie et Evolution des Insectes Hemipteroidea* 7/8: 121–124.

Boulard, M. 1995c. *Raiateana oulietea*, l'etonnante et énigmatique Cigale de la Polynésie française (Homoptera Cicadoidea Cicadidae). *Ecole Pratique des Hautes Etudes, Travaux du Laboratoire Biologie et Evolution des Insectes Hemipteroidea* 7/8: 161–178.

Boulard, M. 1996a. Sur une cigale wallisienne, crépusculaire et ombrophile (Homoptera, Cicadoidea, Tibicinidae). *Bulletin de la Société Entomologique de France* 101: 151–158.

Boulard, M. 1996b. *Onoralna falcata*, nouveau genre, nouvelle espèce de cigale néotropicale (Cicadomorpha, Cicadoidea, Tibicinidae). *Ecole Pratique des Hautes Etudes, Travaux du Laboratoire Biologie et Evolution des Insectes Hemipteroidea* 9: 3–8.

Boulard, M. 1996c. Sur deux nouvelles cigales malgaches représentant deux genres nouveaux pour Madagascar (Cicadomorpha, Cicadidae, Cicadinae). *Ecole Practique des Hautes Etudes, Travaux du Laboratoire Biologie et Evolution des Insectes Hemipteroidea* 9: 95–104.

Boulard, M. 1996d. Postures de cymbalisation, cymbalisations et cartes d'identité acoustique des cigales. 2. Espèces forestières afro- et néotropicales (Cicadoidea, Cicadidae et Tibicinidae). *Ecole Practique des Hautes Etudes, Travaux du Laboratoire Biologie et Evolution des Insectes Hemipteroidea* 9: 113–158.

Boulard, M. 1997a. Nouvelles cigales remarquables originaires de la Nouvelle-Calédonie (Homoptera, Cicadoidea, Tibicinidae). *in* J. Najt & L. Matile, eds. *Zoologia Neocaledonica*, Vol 4. *Mémoires du Muséum National d'Histoire Naturelle. A* 171: 179–196.

Boulard, M. 1997b. Abécédaire illustré du mimétisme. *Ecole Practique des Hautes Etudes, Travaux du Laboratoire Biologie et Evolution des Insectes Hemipteroidea* 10: 3–77.

Boulard, M. 1997c. Nomenclature et taxonomie supérieures des Cicadoidea ou vraies cigales. Histoire, problèmes et solutions (Rhynchota Homoptera Cicadomorpha). *Ecole Pratique des Hautes Etudes, Travaux du Laboratoire Biologie et Evolution des Insectes Hemipteroidea* 10: 79–129.

Boulard, M. 1998. Aperçu de cicadologie antillo-guyanaise. *123e Congrès national des sociétés historiques et scientifiques, Antilles-Guyane, 1998, Histoire naturelle*: 75–94.

Boulard, M. 1999a. Postures de cymbalisation, cymbalisations et cartes d'identité acoustique des cigales 3.- Espèces tropicales des savanes et milieux ouverts (Cicadoidea, Cicadidae et Tibicinidae). *Ecole Pratique des Hautes Etudes, Travaux du Laboratoire Biologie et Evolution des Insectes Hemipteroidea* 11/12: 77–116.

Boulard, M. 1999b. A propos d'une affichette cigales. *Ecole Pratique des Hautes Etudes, Travaux du Laboratoire Biologie et Evolution des Insectes Hemipteroidea* 11/12: 183–185.

Boulard, M. 2000a. Description de deux cigales malgaches dédiées à Loïc Matile [Auchenorhyncha, Cicadidae, Cryptotympanini]. *Revue Française d'Entomologie (N.S.)* 22: 255–262.

Boulard, M. 2000b. Espèce, milieu et comportement. *Ecole Pratique des Hautes Etudes, Travaux du Laboratoire Biologie et Evolution des Insectes Hemipteroidea* 13: 1–40.

Boulard, M. 2000c. Cymbalisation d'appel et cymbalisation de cour chez quatre cigales thaïlandaises. *Ecole Pratique des Hautes Etudes, Travaux du Laboratoire Biologie et Evolution des Insectes Hemipteroidea* 13: 49–66.

Boulard, M. 2000d. Appareils, productions et communications sonores chez les insectes en générale et chez les cigales enparticulier. *Ecole Pratique des Hautes Etudes, Travaux du Laboratoire Biologie et Evolution des Insectes Hemipteroidea* 13: 75–110.

Boulard, M. 2000e. *Tosena albata* (Distant, 1878) bonne espèce confirmée par l'acoustique (Auchenorhyncha, Cicadidae, Tosenini. *Ecole Pratique des Hautes Etudes, Travaux du Laboratoire Biologie et Evolution des Insectes Hemipteroidea* 13: 119–126.

Boulard, M. 2000f. Description de quatre *Tettigetta* ibériques nouvelles (Auchenorhyncha, Cicadidae, Tibicinae). *Ecole Pratique des Hautes Etudes, Travaux du Laboratoire Biologie et Evolution des Insectes Hemipteroidea* 13: 133–143.

Boulard, M. 2000g. Note rectificative (C.I.A. de *Cicadatra platyptera* et de *C^{atra} glycirrhizae*). *Ecole Pratique des Hautes Etudes, Travaux du Laboratoire Biologie et Evolution des Insectes Hemipteroidea* 13: 144.

Boulard, M. 2000h. *Métamorphoses et transformations animals*. Oblitérations évolutives. Société Nouvelle des Éditions Boubeé, Paris. 311pp.

Boulard, M. 2000i. *Cicadogéographie de la France européenne: premier fichier signalétique et éco-éthologiques et premier atlas des Cigales et Membracoïdés*. Ministère de l'Environment, Direction de la Nature de des Pasages, Paris. 131pp.

Boulard, M. 2001a. Statut acoustique et comportement sonore de quelques cigales thaïlandaises. Description d'une espèce nouvelle (Homoptera, Cicadidae). *Bulletin de la Société entomologique de France* 106: 127–147.

Boulard, M. 2001b. Higher taxonomy and nomenclature of the Cicadoidea or true cicadas: history, problems and solutions (Rhynchota Auchenorhyncha Cicadomorpha). *Ecole Pratique des Hautes Etudes, Travaux du Laboratoire Biologie et Evolution des Insectes Hemipteroidea* 14: 1–48.

Boulard, M. 2001c. Cartes d'identité acoustique et éthologie sonore de cinq espèces de cigales thaïlandaises, don't deux sont nouvelles (Rhynchota, Auchenorhyncha, Cicadidae). *Ecole Pratique des Hautes Etudes, Travaux du Laboratoire Biologie et Evolution des Insectes Hemipteroidea* 14: 49–71.

Boulard, M. 2001d. Éthologie sonore et larvaire de *Pomponia pendleburyi* n. sp. (précédé d'un historique taxonomique concernant les grandes *Pomponia*) (Auchenorhyncha, Cicadidae, Cicadinae, Pomponiini). *Ecole Pratique des Hautes Etudes, Travaux du Laboratoire Biologie et Evolution des Insectes Hemipteroidea* 14: 79–103.

Boulard, M. 2001e. Sur les destineés taxonomiques de <<La Cigale à bords verts>> et de << La Cigale à deux taches>> Stoll, 1788 et sur l'espèce type du genre *Chremistica* Stål, 1870 (Rhynchota, Auchenorhyncha, Cicadidae). *Ecole Pratique des Hautes Etudes, Travaux du Laboratoire Biologie et Evolution des Insectes Hemipteroidea* 14: 113–120.

Boulard, M. 2001f. A propos de deux formes inédites ou mal connues de *Chremistica* malgaches et séchelliennes (Rhynchota, Auchenorhyncha, Cicadidae). *Ecole Pratique des Hautes Etudes, Travaux du Laboratoire Biologie et Evolution des Insectes Hemipteroidea* 14: 129–140.

Boulard, M. 2001g. Note rectificative: *Pomponia dolosa* nom. nov. pro *P. fusca* Boulard, 2000 [non Olivier, 1790] redescription de *P. fusca* (Olvr) (Rhynchota, Auchenorhyncha, Cicadidae). *Ecole Pratique des Hautes Etudes, Travaux du Laboratoire Biologie et Evolution des Insectes Hemipteroidea* 14: 157–163.

Boulard, M. 2002a. Éthologie sonore et cartes d'identité acoustique de dix espèces de cigales thaïlandaises, dont six restées jusqu'ici inédites ou mal connues [Auchenorhyncha, Cicadoidea, Cicadidae]. *Revue francaise d'Entomologie* (N.S.) 24: 35–66.

Boulard, M. 2002b. Diversité des Auchénorhynchques Cicadomorphes Formes, couleurs et comportements (Diversité structurelle ou taxonomique Diversité particulière aux Cicadidés). *Denisia* 4: 171–214.

Boulard, M. 2003a. Éthologie sonore et statut acoustique de quelques cigales thaïlandaises, incluant la description de deux espèces nouvelles (Hemiptera: Auchenorhyncha, Cicadoidea, Cicadidae). *Annales de la Société entomologique de France* 39: 97–119.

Boulard, M. 2003b. Comportement sonore et identité acoustique de six espèces de cigales thaïlandaises, dont une nouvelles espèces (Auchenorhyncha, Cicadidae). *Bulletin de la Société entomologique de France* 108: 185–200.

Boulard, M. 2003c. Contribution à la connaissance des cigales thaïlandaises incluant la description de quarte espèces nouvelles [Rhynchota, Cicadoidea, Cicadidae]. *Revue Française d'Entomologie (N.S.)* 25: 171–201.

Boulard, M. 2003d. *Macrosemia longiterebra*, une nouvelle espèce de cigale asienne exceptionnelle par la longueur de la tarière del femelles (Rhynchota, Cicadoidea, Cicadidae). *L'Entomologist* 59: 187–192.

Boulard, M. 2003e. Statut taxonomique et acoustique de quatre cigales thaïlandaises, dont deux restées inédites jusqu'ici (Rhynchota, Cicadomorpha, Cicadidae). *Nouvelle Revue d'Entomologie (N.S.)* 25: 259–279.

Boulard, M. 2003f. *Tibicen* Latreille, 1825: <<Fatal error>>. *Nouvelle Revue d'Entomologie (N.S.)* 25: 371–372.

Boulard, M. 2004. Description et statut acoustique d'une nouvelle cigale thailandaise (*Chremistica chueatae*). *L'Entomologist* 60: 237–243.

Boulard, M. 2005a. La polychromie des *Mogannia*, petites cigales nonchalantes du sud-est asiatique. *Insectes* (139): 31–33.

Boulard, M. 2005b. Sur le statut acoustique de deux espèces sommitales et ombrophiles d'Asie tropico-continentale, *Macrosemia longiterebra* Blrd et *Pycna concinna* n. sp. (Rhynchota, Homoptera, Cicadidae). *L'Entomologist* 61: 111–120.

Boulard, M. 2005c. A propos de la découverte exceptionnelle d'*Angamiana melanoptera* nouvelle espèce pour la tribu asienne des Polyneurini (Rhynchota, Homoptera, Cicadidae). *Bulletin de la Société entomologique de France* 110: 233–239.

Boulard, M. 2005d. Acoustic signals, diversity and behaviour of cicadas (Cicadidae, Hemiptera). Pages 331–349 *in* S. Drosopoulos & M.F. Claridge, eds. *Insect sounds and communication: physiology, behaviour, ecology and evolution.* CRC Press, Boca Raton.

Boulard, M. 2005e. Taxonomie et statut acoustique de huit cigales thaïlandaises, incluant cinq espèces nouvelles [Rhynchota, Cicadoidea, Cicadidae]. *Revue Française d'Entomologie (N.S.)* 27: 117–143.

Boulard, M. 2005f. *Baeturia laperousei* n.sp., cigale de Vanikoro [Rhynchota, Cicadidae, Tibicininae]. *Revue française d'Entomologie (N.S.)* 27: 189–192.

Boulard, M. 2005g. Description de deux espèces inédites de Cigales asiennes, dont une étonnamment polymorphe (Rhynchota, Cicadoidea, Cicadidae). *Revue Française d'Entomologie (N.S.)* 21: 365–374.

Boulard, M. 2005h. Données statutaires et éthologiques sur des cigales thaïlandaises, incluant la description de huit espèces nouvelles, ou mal connues. *Ecole Pratique des Hautes Etudes, Travaux du Laboratoire Biologie et Evolution des Insectes Hemipteroidea* 15: 5–57.

Boulard, M. 2005i. Sur six nouvelles espèces de cigales d'Asie continento-tropicale (Rhynchota, Cicadoidea, Cicadidae). *Ecole Pratique des Hautes Etudes, Travaux du Laboratoire Biologie et Evolution des Insectes Hemipteroidea* 15: 59–78.

Boulard, M. 2005j. Á propos de la découverte exceptionnelle d'*Angamiana melanoptera* nouvelle espèces pour la tribu asienne des Polyneurini (Rhynchota Homoptera Cicadidae). *Ecole Pratique des Hautes Etudes, Travaux du Laboratoire Biologie et Evolution des Insectes Hemipteroidea* 15: 79–91.

Boulard, M. 2005k. Création du genre *Megapomponia* et description de *Mp. clamorigravis* n. sp. (Rhynchota, Cicadoidea, Cicadidae). *Ecole Pratique des Hautes Etudes, Travaux du Laboratoire Biologie et Evolution des Insectes Hemipteroidea* 15: 93–110.

Boulard, M. 2005l. Note rectificative: *Tosena albata* Distant, 1878 bonne espèce confirmée par l'acoustique (addenda et corrigenda). *Ecole Pratique des Hautes Etudes, Travaux du Laboratoire Biologie et Evolution des Insectes Hemipteroidea* 15: 111–123.

Boulard, M. 2005m. Description d'une espèce méconnue du genre *Angamiana* Distant, 1890 (Rhynchota Cicadoidea Cicadidae). *Ecole Pratique des Hautes Etudes, Travaux du Laboratoire Biologie et Evolution des Insectes Hemipteroidea* 15: 125–127.

Boulard, M. 2005n. Taxonomie et statut acoustique de quatre cigales thaïlandaises, dont deux sont nouvelles (Rhynchota, Cicadoidea, Cicadidae). *Ecole Pratique des Hautes Etudes, Travaux du Laboratoire Biologie et Evolution des Insectes Hemipteroidea* 15: 129–146.

Boulard, M. 2005o. Note technique: Missions cicadologiques en Thaïlande 2000–2004. Chronologie des espèces colligées, publiées ou sous presse (addenda et corrigenda). *Ecole Pratique des Hautes Etudes, Travaux du Laboratoire Biologie et Evolution des Insectes Hemipteroidea* 15: 147–152.

Boulard, M. 2006a. Une cigale sabulicole et buveuse d'eau: un "scoop" cicadologique! *Insectes* (142): 3–4.

Boulard, M. 2006b. *Cephalalna francimontanum* nouveau genre, nouvelle espèce de cigale malgache (Rhynchota, Cicadidae, Tibicininae). *Bulletin de la Société entomologique de France* 111: 129–132.

Boulard, M. 2006c. Compléments biotaxonomiques à la cicadofaune thaïlandaise incluant la description et l'éthologie de trios nouvelles espèces [Rhynchota, Cicadoidea, Cicadidae]. *Revue Française d'Entomologie (N.S.)* 28: 131–143.

Boulard, M. 2006d. Cigales nouvelles, originaires de Thaïlande et appartenant au genre *Purana* Distant, 1905 (Rhynchota, Cicadoidea, Cicadidae). *Nouvelle Revue d'Entomologie (N.S.)* 23: 195–212.

Boulard, M. 2006e. *Cicadalna takensis*, nouvelle espèce d'un genre nouveau découverte en Thaïlande du nord (Rhynchota, Cicadidae). *Bulletin de la Société entomologique de France* 111: 529–534.

Boulard, M. 2006f. Premières donnes sur l'"imaginaison" haute en couleurs de deux cigales asiennes, *Tosena splendida* Distant et *Huechys sanguinea* (De Geer). Mise au point conceptuelle a propos de cet evenement (Insects, Cicadidae). *Lambillionea* CVI: 373–381.

Boulard, M. 2006g. Une exception écologique: une Platypleure cespiticole. Description et premières données sur la biologie de cette espèce d'Asie continento-tropicale restée jusqu'ici inédite (Rhynchota, Auchenorhyncha, Cicadidae). *Lambillionea* CVI: 620–630.

Boulard, M. 2006h. Facultés acoustiques, éthologie sonore des cigales, entomophonateurs par excellence. *Ecole Pratique des Hautes Etudes, Travaux du Laboratoire Biologie et Evolution des Insectes Hemipteroidea* 16: 1–181.

Boulard, M. 2007a. Description et ethologie sonore de sept cigales d'Asie continento-tropicale don't six nouvelles pour la science et une jusqu'ici inedited pour la Thaïlande [Rhynchota, Cicadoidea, Cicadidae]. *Revue Française d'Entomologie (N.S.)* 29: 93–120.

Boulard, M. 2007b. *Dimissalna*, nouveau gene de Cicadette (Rhynchota, Cicadidae). *Bulletin de la Société entomologique de France* 112: 97–98.

Boulard, M. 2007c. Nouvelle image de la biodiversité chez les cigales thaïlandaise (Hem., Cicadidae). *Bulletin de la Société entomologique de France* 112: 238.

Boulard, M. 2007d. Additions biotaxonomique à la faune des cigales thaïlandaises, comprenant la description et la C.I.A. d'une espèce nouvelle et des données inédites concernant trente et une autres espèces [Rhynchota, Auchenorrhyncha, Cicadidae]. *Lambillionea* CVII: 493–510.

Boulard, M. 2007e. *The cicadas of Thailand: general and particular characteristics*. Volume 1. White Lotus Co., Ltd. Bangkok. 103pp., plates 1–48.

Boulard, M. 2007f. Être ou ne pas être les facettes du mimétisme. *Ecole Pratique des Hautes Etudes, Travaux du Laboratoire Biologie et Evolution des Insectes Hemipteroidea* 17: 1–151.

Boulard, M. 2008a. *Baeturia laperousei* Boulard, 2005 statut spécifique corroboré par l'acoustique [Rhynchota, Cicadidae, Cicadettinae, Chlorocystini]. *Revue Française d'Entomologie (N.S.)* 30: 67–70.

Boulard, M. 2008b. *Platylomia operculata* Distant, 1913, a cicada that takes water from hot springs and becomes victim of the people (Rhynchota: Cicadomorpha: Cicadidae). *Acta Entomologica Slovenica* 16: 105–116.

Boulard, M. 2008c. Sur un magasin de cigales asiennes, comprenant quatre espèces nouvelles et quelques autres captures intéressantes (Rhynchota, Cicadoidea, Cicadidae). *Nouvelle Revue d'Entomologique* 24: 351–371.

Boulard, M. 2008d. Statuts et éthologies sonores de quatre cigales malgaches, trois sont nouvelles et deux représentent un nouveau genre et une nouvelle sous-tribu (Rhynchota, Auchenorhyncha Cicadidae). *Lambillionea* CVIII: 161–178.

Boulard, M. 2008e. *Platylomia operculata* Distant, une cigale curiste devenant altruiste malgre elle (Rhynchota, Auchenorhyncha Cicadidae). *Lambillionea* CVIII: 345–357.

Boulard, M. 2008f. Les cigales thaïes. Liste actualisée incluant la description de deux nouveaux genres, de sept espèces nouvelles et les cartes d'identité acoustique (CIA) de *Chremistica siamensis* Bregman et de *Leptopsaltria samia* (Walker). *Ecole Pratique des Hautes Etudes, Travaux du Laboratoire Biologie et Evolution des Insectes Hemipteroidea* 18: 1–112.

Boulard, M. 2009a. *Pomponia mickwanae* n.sp., nouvelle espèce de cigale Thaïe [Rhynchota, Cicadoidea, Cicadidae]. *Revue Française d'Entomologie (N.S.)* 31: 39–43.

Boulard, M. 2009b. Descriptions et éthologies singulières de deux espèces de la tribu des Leptopsaltriini (Rhynchota, Cicadoidea, Cicadidae). *Bulletin de la Société entomologique de France* 114: 47–54.

Boulard, M. 2009c. La cigale persévérante et l'Araugnée mélomane (Rhynchota Cicadidae). *L'Entomologist* 65: 63–64.

Boulard, M. 2009d. Nouvelles cigales colligées en Thaïlande, (Rhynchota, Cicadoidea, Cicadidae). *Lambillionea* CIX: 39–58.

Boulard, M. 2009e. Sur la naissance imaginable de *Formotosena montivaga* (Distant), grande et somptueuse cigale asienne. Clarificaion conceptuelle à propos de cet événement (Rhynchota Cicadomorpha Cicadidae). *L'Entomologiste* 65: 175–180.

Boulard, M. 2009f. Sur une nouvelle espéce de *Mogannia*, (Rhynchota, Cicadoidea, Cicadidae). *Lambillionea* CIX: 250–253.

Boulard, M. 2009g. *Distantalna* genre nouveau pour les Tosenini. *Nouvelle Revue d'Entomologie (n.s.)* 25: 325–328.

Boulard, M. 2010a. Description de *Megapomponia foksnodi* n. sp. (Rhynchota, Cicadidae, Pomponiini). *Bulletin de la Société entomologique de France* 115: 15–16.

Boulard, M. 2010b. Création du genre *Sechellalna* pour *Yanga seychellensis* Distant 1912 Cigale dépurvue d'organe sonore (Rhynchota, Cicadidae). *Lambillionea* CX: 334–337.

Boulard, M., & K. Chueata. 2004. Une cigale asienne périodique? *Gaena festiva* F. *L'Entomologist* 60: 141–143.

Boulard, M., J. Deeming, & L. Matile. 1989. Deux nouvelles espèces de *Pseudogaurax* Malloch associées à l'Araingée *Nephila turneri* Blackwell en Côte d'Ivoire [Diptera, Chloropidae; Araneae, Araneidae]. *Revue Française d'Entomologie (N.S.)* 11: 143–150.

Boulard, M., & N.M. Martinelli. 1996. Révision des Fidicini, nouveau statut de la tribu, espèces connues et nouvelles espèces (Cicadomorpha, Cicadidae, Cicadinae). Premiere partie: Sous-tribu nouvelle des Fidicinina. *Ecole Pratique des Hautes Etudes, Travaux du Laboratoire Biologie et Evolution des Insectes Hemipteroidea* 9: 11–81.

Boulard, M., & B. Mondon. 1996. *Vies et mémoires de cigales*. Equinoxe, Barbentane, France. 159pp.

Boulard, M., & M.S. Moulds. 2001. *Ueana simonae* n. sp. Nouvelle cigale Neo-Caledonienne (Auchenorhyncha, Cicadidae, Tibicininae). *Ecole Pratique des Hautes Etudes, Travaux du Laboratoire Biologie et Evolution des Insectes Hemipteroidea* 14: 73–78.

Boulard, M., & A. Nel. 1990. Sur deux cigales fossiles des terrains tertaires de la France [Homoptera, Cicadoidea]. *Revue Française d'Entomologie (N.S.)* 12: 37–45.

Boulard, M., & J.A. Quartau. 1991. *Tettigetta septempulsata*, nouvelle cigale lusitanienne (Homoptera Cicadoidea Tibicinidae). *Travaux du Laboratoire de la Biologie et Evolution d'Insectes de l'Ecole Practique des Hautes Etudes* 4: 49–56.

Boulard, M., & B. Riou. 1988. *Tibicina gigantea* n. sp., cigale fossile de la Montaigne d'Andance (Homoptera, Tibicinidae). *Nouvelle Revue d'Entomologie* 5: 349–351.

Boulard, M., & B. Riou. 1999. *Miocenoprasia grasseti n. g., n. sp.*, grande cigale fossile du Miocene de la Montagne d'Andance (Cicadomorpha Cicadidae Tibicininae). *Ecole Practique des Hautes Etudes, Travaux du Laboratoire Biologie et Evolution des Insectes Hemipteroidea* 11/12: 135–140.

Boulard, M., & J. Sueur. 1996. Sur deux nouvelles Zammarini originaires du Vénézuéla (Cicadomorpha, Cicadoidea, Cicadidae). *Ecole Pratique des Hautes Etudes, Travaux du Laboratoire Biologie et Evolution des Insectes Hemipteroidea* 9: 105–112.

Bowie, M.H., J.W.M. Marris, R.W. Emberson, I.G. Andrew, J.A. Berry, C.J. Vink, E.G. White, M.A.W. Stufkins, E.H.A. Oliver, J.W. Early, J. Klimaszewski, P.M. Johns, S.D. Wratten, K. Mahlfeld, B. Brown, A.C. Eyles, S.M. Pawson, & R.P. Macfarlane. 2003. A terrestrial invertebrate inventory of Quail Island (Otamahua): towards the restoration of the invertebrate community. *New Zealand Natural Sciences* 28: 81–109.

Brackenbury, J.H. 1992. *Insects in flight*. Blandford Press, London. 192pp.

Braker, H.E., & H.W. Greene. 1994. Population biology: Life histories, abundance, demography, and predator-prey interactions. Pages 244–255 *in* L.A. McDade, K.S. Bawa, H.A. Hespenheide, & G.S. Hartshorn, eds. *La Selva: Ecology and natural history of a Neotropical rain forest*. University of Chicago Press, Chicago.

Brambila, J., & G.S. Hodges. 2004. Bugs (Hemiptera). Pages 354–371 *in* J. Capinera, ed. *Encyclopedia of entomology*, Volume 1. Kluwer Academic Publishers, Dordrecht.

Brambila, J., & G.S. Hodges. 2008. Bugs (Hemiptera). Pages 591–644 *in* J. Capinera, ed. *Encyclopedia of entomology*, second edition, volume 1. Springer-Verlag, Heidelberg.

Branson, B.A. 1994. The loudmouth of the woods. *American Horticulturist* 73: 10–11.

Brandstetter, R. 1999. Notes on *Cicadetta abdominalis* (Distant) at Flowerdale, Victoria. *Victorian Entomologist* 29: 40.

Bregman, R. 1985. Taxonomy, phylogeny and biogeography of the *tridentigera* group of the genus *Chremistica* Stal, 1870 (Homoptera, Cicadidae). *Beaufortia* 35: 37–60.

Britcher, H.W. 1899. Onondaga Academy of Sciences. *Science* 10: 737.

Brittain, R.A., V.J. Meretsky, J.A. Gwinn, J.G. Hammond, & J.K. Riegel. 2009. Northern Saw-whet owl (*Aegolius acadicus*) autumn migration magnitude and demographics in south-central Indiana. *Journal of Raptor Research* 43: 199–209.

Brodskiy, A.K. 1993. Structure, functioning and evolution of the tergum in alate insects. II. Organization features of the terga in different orders of insects. *Entomological Review* 72: 8–28.

Brodsky, A.K. 1992. [Structure, functioning and evolution of the tergum in alate insects. II. Characteristics of the organization of a wing-bearing tergal plate in insects of different orders.] *Entomologicheskoe Obozrenie* 71: 39–59. [In Russian, English summary]

Brodsky, A.K. 1994. *The evolution of insect flight*. Oxford University Press, Oxford. 229pp.

Brown, D. 2001. Black-billed gulls hawking cicadas over shrubland. *Notornis* 48: 111–112.

Brown, D.M., & C.A. Toft. 1991. Molecular systematics and biogeography of the cockatoos (Psittaciformes: Cacatuidae). *Auk* 116: 141–157.

Brown, E.E., & J.D. Brown. 1990. Occurrence of a northern cicada, *Okanagana rimosa* (Homoptera, Cicadidae), in the southern Appalachians. *Brimleyana* 16: 5–7.

Brown, J. 1987. Bladder cicadas (*Cystosoma saundersii* Westwood). *Victorian Entomologist* 17(1): 6.

Brown, K. 1999. Listen, we're different. *New Scientist* 163: 32–35.

Brown, M.W., & J.J. Schmitt. 1994. Population dynamics of woolly apple aphid (Homoptera: Aphididae) in West Virginia apple orchards. *Environmental Entomology* 23: 1182–1188.

Brown, W.P., & M.E. Zuefle. 2009. Does the periodical cicada, *Magicicada septendecim*, prefer to oviposit on native or exotic plant species? *Ecological Entomology* 34: 346–355.

Brua, C., & S. Hugel. 2008. Présence des cigales *Cicadetta montana* (Scopoli, 1772) et *Cicadetta cantilatrix* Sueur & Puissant, 2007 en Alsace (Hemiptera, Cicadidae). *Bulletin de la Société entomologique de Mulhouse* 64: 49–52.

Brunet, B.L. 1986. Some observations on the double drummer cicada (*Thopha saccata* Fabr.). Location: Myall Lakes, New South Wale 16th December to the 24th December 1985. *Royal Zoological Society of New South Wales, Entomology Section Circular* (43): 12–13.

Brunet, B. 1987. Feasting at the Manna gum. Location: Araluen Valley, N.S.W. *Royal Zoological Society New South Wales, Entomology Section Circular* (55): 44–45.

Brunet, B. 1995. Arthropods of the eastern suburbs. *The Society for Insect Studies, Preserve the Habitat, Circular* (48): 18–23.

Buchanan White, F. 1879. List of the Hemiptera of New Zealand. *Entomologist's Monthly Magaszine* 15: 213–220.

Buck, J., & E. Buck. 1968. Mechanisms of rhythmic synchronous flashing of fireflies. *Science* 159: 1319–1327.

Buckley, T.R., P. Arensburger, C. Simon, & G.K. Chambers. 2002. Combined data, Bayesian phylogenetics, and the origin of the New Zealand cicada genera. *Systematic Biology* 51: 4–18.

Buckley, T.R., M. Cordeiro, D.C. Marshall, & C. Simon. 2006. Differentiating between hypotheses of lineage sorting and introgression in new Zealand alpine cicadas (*Maoricicada* Dugdale). *Systematic Biology* 55: 411–425.

Buckley, T.R., & C. Simon. 2007. Evolutionary radiation of the cicada genus *Maoricicada* Dugdale (Hemiptera: Cicadoidea) and the origins of the New Zealand alpine biota. *Biological Journal of the Linnean Society* 91: 419–435.

Buckley, T.R., C. Simon, & G.K. Chambers. 2001a. Exploring among-site rate variation models in a maximum likelihood framework using empirical data: Effects of model assumptions on estimates of topology, branch lengths, and bootstrap support. *Systematic Biology* 50: 67–86.

Buckley, T.R., C. Simon, & G.K. Chambers. 2001b. Phylogeography of the New Zealand cicada *Maoricicada campbelli* based on mitochondrial DNA sequences: ancient clades associated with Cenozoic environmental change. *Evolution* 55: 1395–1407.

Buckley, T.R., C. Simon, P.K. Floor, & B. Misof. 2000. Secondary structure and conserved motifs of the frequently sequenced domains IV and V of the insect mitochondrial large subunit rRNA gene. *Insect Molecular Biology* 9: 565–580.

Buckley, T.R., C. Simon, H. Shimodaira, & G.K. Chambers. 2001. Evaluating hypotheses on the origin and evolution of the New Zealand Alpine cicadas (*Maoricicada*) using multiple-comparison tests of tree topology. *Molecular Biology and Evolution* 18: 223–234.

Bukhvalova, M. 2005. Partitioning of acoustic transmission channels in grasshopper communities. Pages 199–205 *in* S. Drosopoulos & M.F. Claridge, eds. *Insect sounds and communication: physiology, behaviour, ecology and evolution.* CRC Press, Boca Raton.

Bunescu, A. 1959. Contributii la studiul raspindirii geographice a unor animale mediteraneene din R.P.R. Nota I. Artropode. *Probleme de Geografie* 6: 86–107.

Bunker, G.J., M.L. Neckermann, & J.R. Cooley. 2007. New northern records of *Tibicen chloromerus* (Hemiptera: Cicadidae) in Connecticut. *Transactions of the American Entomological Society* 133: 357–361.

Burk, T. 1982. Evolutionary significance of predation on sexually signalling males. *Florida Entomologist* 65: 90–104.

Burk, T. 1988. Acoustic signals, arms races and the costs of honest signalling. *Florida Entomologist* 71: 400–409.

Burton, J.F., & E.D.H. Johnson. 1984. Insect, amphibian or bird? *British Birds* 77: 87–104.

Burwell, C.J. 1991. New distribution records of three Queensland cicadas (Homoptera: Cicadidae). *Australian Entomology Magazine* 18: 124.

Buschbeck, E.K., & M. Hauser. 2009. The visual system of male scale insects. *Naturwissenschaften* 96: 365–374.

Büyükgüzel, K., H. Tunaz, S.M. Putnam, & D.W. Stanley. 2002. Prostaglandin biosynthesis by midgut tissue isolated from the tobacco hornworm, *Manduca sexta*. *Insect Biochemistry and Molecular Biology* 32: 435–443.

Byrne, D.N., S.L. Buchmann, & H.G. Spencer. 1988. Relationship between wing loading, wingbeat frequency and body mass in homopterous insects. *Journal of Experimental Biology* 135: 9–23.

Cade, W. 1985. Insect mating and courtship behaviour. Pages 591–619 *in* G.A. Kerkut & L.I. Gilbert, eds. *Comprehensive insect physiology, biochemistry and pharmacology. Volume 9. Behaviour*. Pergammon Press, Oxford.

Cahoon, G.A., & C.W. Donoho, Jr. 1982. The influence of urea sprays, mulch, and pruning on apple tree decline. *Research Circular Ohio Agricultural Research and Development Center, Wooster, Ohio* 272: 16–19.

Callaham, M.A., Jr. 2000. Ecology of belowground invertebrates in tallgrass prairie. *Dissertation Abstracts International* 61: 3957B.

Callaham, M.A., Jr., J.M. Blair, T.C. Todd, D.J. Kitchen, & M.R. Whiles. 2003. Macroinvertebrates in North American tallgrass prairie soills: effects of fire, mowing and fertilization on density and biomass. *Soil Biology & Biochemistry* 35: 1079–1093.

Callaham, M.A., Jr., M.R. Whiles, & J.M. Blair. 2002. Annual fire, mowing and fertilization effects on two cicada species (Homoptera: Cicadidae) in tallgrass prairie. *American Midland Naturalist* 148: 90–101.

Callaham, M.A., Jr., M.R. Whiles, C.K. Meyer, B.L. Brock, & R.E. Charlton. 2000. Feeding ecology and emergence production of annual cicadas (Homoptera: Cicadidae) in tallgrass prairie. *Oecologia* 123: 535–542.

Campbell, B.C., J.D. Steffen-Campbell, & R.J. Gill. 1994. Evolutionary origin of whiteflies (Hemiptera: Stenorrhyncha: Aleyrodidae) inferred from 18S rDNA sequences. *Insect Molecular Biology* 3: 73–88.

Campbell, B.C., J.D. Steffen-Campbell, J.T. Sorensen, & R.J. Gill. 1995. Paraphyly of Homoptera and Auchenorrhyncha inferred from 18s rDNA nucleotide sequences. *Systematic Entomology* 20: 175–194.

Campos, M.C., J.A.A. Silva, R.L.C. Ferreira, & E.J.R. Dias. 2004. Morfometria de espécies simpátricas de cigarras (Homptera: Cicadidae) em Salvador, BA. *XXV Congresso Brasiliero de Zoologia, Brasília. Livro de Resumos do XXV Congresso Brasileiro de Zoologia* 25: 125.

Campos, P.R.A., & V.M. de Olivieira. 2005. Campos et al. reply. *Physical Review Letters* 95(229802): 1p.

Campos, P.R.A., V.M. de Oliveira, R. Giro, & D.S. Galvão. 2004a. The emergence of prime numbers as the result of evolutionsry strategy. *arXiv:q-bio/0406017v1 [q-bio.PE]*. 11pp.

Campos, P.R.A., V.M. de Oliveira, R. Giro, & D.S. Galvão. 2004b. Emergence of prime numbers as the result of evolutionary strategy. *Physical Review Letters* 95(098107): 4pp.

Cantoreanu, M. 1960. Contributii la cunoasterea elementelor mediteraneene din fauna de Cicadine a R.P.R. *Comunicari Academmia Republicii Populare Romîne* 10: 331–335.

Cantoreanu, M. 1968a. Cicadine (*Homoptera – Auchenorrhyncha*) din reiunea viitorului lac de acumulare de la Portile de Fier (I). *Studii si Cercertari de Biologie, Seria Zoologie* 20: 341–345.

Cantoreanu, M. 1968b. Ord. Homoptera (*Auchenorrhyncha*). *Travaux du Muséum d'Histoire Naturelle "Grigore Antipa"* 9: 127–131.

Cantoreanu, M. 1969. *Homoptera-Auchenorrhyncha*-Arten aus der Dobrogea. *Lucrarile Statiunii de Cercetari Marine "Prof. Ioan Borcea" Agigea* III: 259–265.

Cantoreanu, M. 1972a. Studii sistematice asupra insectelor Homoptera Auchenorrhyncha din bazinup rîului Arges. *Studii si Comunicari Muzeaul Judetean Arges Pitesti* 1972: 109–115.

Cantoreanu, M. 1972b. Studii sistematice asupra cicadinelor (Hom. – Auchen.) in Masivul Ciucas. *Comunicari si Referate Muzeual de Stiinte Naturale, Ploiesti* 1972: 207–217.

Cantoreanu, M. 1975. Homoptera-Auchenorrhyncha. Pages 90–95 *in* M. Ionescu, ed. *Fauna Seria Monografica.* Academia Republicii Socialiste Romania, Portile de Fier.

Capinera, J. 2004. Entomophthorales. Pages 787–789 *in* J. Capinera, ed. *Encyclopedia of entomology*, volume 1. Springer-Verlag, Heidelberg.

Capinera, J. 2008. Entomophthorales. Pages 1344–1345 *in* J. Capinera, ed. *Encyclopedia of entomology*, second edition, volume 2. Springer-Verlag, Heidelberg.

Cardozo, G. de O., D. de A.M.M. Silvestre, & A. Colato. 2007a. Periodical cicadas: a minimal automaton model. *arXiv:q-bio/0611067v5 [q-bio.PE]*. 13pp.

Cardozo, G. de O., D. de A.M.M. Silvestre, & A. Colato. 2007b. Periodical cicadas: a minimal automaton model. *Physica A: Statistical Mechanics and its Applications* 382: 439–444.

Carl, M., P. Huemer, A. Zanettl, & C. Salvadori. 2005. Ecological assessment in alpine forest ecosystems: bioindication with insects (Auchenorrhyncha, Coleoptera (Staphylinidae), Lepidoptera). *Studi Trenti de Scienze Naturali, Acta Biologia* 81 (Supplemento 1): 167–217.

Carnevali, M.D.C., & R. Valvassori. 1981. Z-line in insect muscle: structural and functional diversities. *Bolletino di Zoologia* 48: 1–9.

Carpenter, F.M. 1992. *Treatise on invertebrate paleontology. Part R. Arthropoda 4. Volume 3: Superclass Hexapoda*. The Geological Society of America, Inc. and the University of Kansas, Boulder Colorado, and Lawrence, Kansas. 227pp.

Carpenter, W.B., & W.S. Dallas. 1867. *Zoology: being a systematic account of the general structure, habits, instincts, and uses of the principal families of the animal kingdom, as well as of the chief forms of fossil remains*. Volume II. Bell & Daldy, London. 588pp.

Carus, C.G. 1829. Ueber die stimmwerkzeuge der italienischen zikaden. *Carus' Analekten zur Naturwissenschaft und Heilkunde, gesammelt auf einer Reise durch Italien im Jahre* 1828: 141–168.

Carvalho, R.P.L., N.M. Martinelli, H.N. Lusvarghi, & R.L. Andrade. 1990. Controle de cigarra, *Quesada gigas* em cafeeiro com inseticida granulado Counter. *16° Congresso Brasileiro de Pesquisas Cafeeiras, 1990, Pinhal. Anais do 16° Congresso Brasileiro de Pesquisas Cafeeiras* 16: 39.

Carver, M. 1990. Book Review: *Australian Cicadas. Journal of the Australian Entomological Scoiety* 29: 260.

Carver, M., G.F. Gross, & T.E. Woodward. 1994. Hemiptera (bugs, leafhoppers, cicadas, aphids, scale insects etc.). Pages 316–329 *in* I.D. Naumann, ed. *Systematic and applied entomology: an introduction*. Melbourne University Press, Melbourne.

Carver, M., G.F. Gross, T.E. Woodward, G. Cassis, J.W. Evans, M.J. Fletcher, L. Hill, I. Lansbury, M.B. Malipatil, G.B. Monteith, M.S. Moulds, J.T. Polhemus, J.A. Slater, P. Stys, K.L. Taylor, T.A. Weir, & D.J. Williams. 1991. Hemiptera (Bugs, leafhoppers, cicadas, aphids, scale insects etc.). Pages 429–509 *in* I.D. Naumann, P.B. Carne, J.F. Lawrence, E.S. Nielsen, J.P. Spradbery, R.W. Taylor, M.J. Whitten, & M.J. Littlejohn, eds. *The insects of Australia. A textbook for students and research workers*. Second edition, Volume I. Melbourne University Press, Carleton.

Castellani, O. 1952. Contributo alla fauna emitterologica d'Italia. *Bolletino dell'Associazione Romana di Entomologia* 7: 15–16.

Ceotto, P., G.J. Kergoat, J.-Y. Rasplus, & T. Bourgoin. 2008. Molecular phylogenetics of cixiid planthoppers (Hemiptera: Fulgoromorpha): new insights from combined analyses of mitochondrial and nuclear genes. *Molecular Phylogenetics and Evolution* 48: 667–678.

Chadwick, C. 1984. Notes and exhibits. *Royal Zoological Society New South Wales, Entomology Section Circular* 30: 24.

Chambers, G.K., W.M. Boon, T.R. Buckley, & R.A. Hit. 2001. Using molecular methods to understand Gondwanan affinities of the New Zealand biota: three case studies. *Australian Journal of Botany* 49: 377–387.

Chambers, S. 1975. Catching a new cicada. *New Zealand Farmer* 96: 95.

Chambers, S. 1976. A locust is not a cicada. *New Zealand Farmer* 97: 83.

Chambers, S. 1978. At the close of a cicada summer. *New Zealand Farmer* 99: 69.

Chambers, V.T. 1868. (Stinging cicadas). *American Naturalist* 2: 332–333.

Champanhet, J.-M. 1999. Nouvelles *Carineta* de l'Équateur et de Bolivie (Cicadomorpha, Cicadoidae, Tibicinidae). *Ecole Pratique des Hautes Etudes, Travaux du Laboratoire Biologie et Evolution des Insectes Hemipteroidea* 11/12: 65–76.

Champanhet, J.-M. 2001. Nouvelles *Carineta* originaires de l'Equateur et de Bolivie (Cicadomorpha, Cicadidae, Tibicininae). *Ecole Pratique des Hautes Etudes, Travaux du Laboratoire Biologie et Evolution des Insectes Hemipteroidea* 14: 105–112.

Champanhet, J.-M., M. Boulard, & M. Gaiani. 2000. *Aragualna plenalinea* n. g., n. sp. de cigale néotropicale pour la tribu des Dasini (Auchenorhyncha, Cicadidae, Tibicininae). *Ecole Pratique des Hautes Etudes, Travaux du Laboratoire Biologie et Evolution des Insectes Hemipteroidea* 13: 41–48.

Champlain, A.B., & J.N. Knull. 1923. Notes on Pennsylvania Diptera. *Entomological News* 34: 211–215.

Chapman, G.B. 2005. A light and transmission electron microscope study of some cells and tissues associated with the tymbal muscle of a periodical cicada (Homoptera, Cicadidae). *Invertebrate Biology* 124: 321–331.

Chang, H.M., & P.P.H. But. 1987. Chantui. Pages 775–1320 *in* H.M. Chang, ed. *Pharmacology and applications of Chinese Materia Medica*. Volume 2. World Scientific Publishing, Singapore.

Chari, N., S. Shailaja, & M. Vengal-Reddy. 1995. Studies on flight dynamics of insect fliers: Low frequency flier *Platypleura octogullata* (sic) and high frequency flier *Tessaratoma javanica*. *Indian Journal of Comparative Animal Physiology* 13: 25–28.

Chaudhry, G.-U., M.I. Chaudhry, & S.M. Khan. 1966. *Survey of insect fauna of forests of Pakistan*. Final technical report. Biological Sciences Research Division, Pakistan Forest Institute, Peshwar. 167pp.

Chaudhry, G.-U., M.I. Chaudhry, & N.K. Malik. 1970. *Survey of insect fauna of forests of Pakistan*. Volume II. Biological Sciences Research Division (Forest Entomology Branch), Pakistan Forest Institute, Peshwar. 205pp.

Chavanduka, D.M. 1975. Insects as a source of protein to the African. *Rhodesia Science News* 9: 217–220.

Chawanji, A.S., A.N. Hodgson, & M.H. Villet. 2005a. Sperm morphology in four species of African platypleurine cicadas (Hemiptera: Cicadomorpha: Cicadidae). *Tissue and Cell* 37: 257–267.

Chawanji, A.S., A.N. Hodgson, & M.H. Villet. 2005b. Spermiogenesis in the cicada *Kongota punctigera* (Cicadinae: Platypleurini) and *Monomotapa matoposa* (Tibicinae: Taphurini). *Proceedings of the Microscopy Society of Southern Africa* 35: 66.

Chawanji, A.S., A.N. Hodgson, & M.H. Villet. 2006. Sperm morphology in five species of cicadettine cicadas (Hemiptera: Cicadomorpha: Cicadidae). *Tissue and Cell* 38: 373–388.

Chawanji, A.S., A.N. Hodson, M.H. Villet, A.F. Sanborn, & P.K. Phillips. 2007. Spermiogenesis in three species of cicadas (Hemiptera: Cicadidae). *Acta Zoologica (Stockholm)* 88: 337–348.

Chawanji, A.S., S.C. Pinchuck, A.N. Hodgson, & M.H. Villet. 2005. Morphology of the female reproductive tract in a cicada *Azanicada zuluensis* (Hemiptera: Cicadidae). *Proceedings of the Microscopy Society of Southern Africa* 35: 77.

Chelpakova, Z.M. 1991. [Comparison of cicada arid station of central Kazakhstan and north-east Kyrghyzstan.] *Kyrgyz Respublikasy Ilimder Akademiyasynyn Kabarlarg Khimiya-Tekhnologiya Zhana Biologiya Ilimderi* 1991: 66–68, 97. [In Russian]

Chen C.-H. 2004. [*Cicadas of Taiwan*.] Big Tree Culture Enterprise Co., Ltd., Taipei. 206pp. [In Chinese]

Chen, C.-H., & S.-F. Shiao. 2008. Two new species of the genus *Euterpnosia* Matsumura (Hemiptera: Cicadidae) from Taiwan. *Pan-Pacific Entomologist* 84: 81–91.

Chen, J.-H. 2005. Redescription of *Euterpnosia varicolor* and description of two new species of cicadas from Taiwan (Hemiptera: Cicadidae). *Collection and Research* 18: 37–50.

Chen, J.-H. 2006. *Euterpnosia ampla*, a new species of cicada from Taiwan (Hemiptera: Cicadidae). *BioFormosa* 41: 71–76.

Chen, K.F. 1933. *Entomological World* 1: 358–161.

Chen, K.F. 1942. The Cicadidae (Homoptera) collected by Prof. I. Chou in his expedition to Sikang Province (East Tibet). *Nanking Journal* 11: 143–147.

Chen, P.P., S. Wongsiri, T. Jamyanya, T.E. Rinderer, S. Vongsamanode, M. Matsuke, H.A. Sylvester, & B.P. Oldroyd. 1998. Honey bees and other edible insects used as human food in Thailand. *American Entomologist* 44: 24–28.

Chen, S., S. Bao, & J.-X. Chen. 2005. A model analyzing life cycle of periodical cicadas. *arXiv:q-bio/0510010v1 [q-bio.PE]*. 4pp.

Chen S., & Yang C.T. 1995. The metatarsi of the Fulgoridea (Homoptera: Auchenorrhyncha). *Chinese Journal of Entomology* 15: 257–269.

Chen Z.-Z. 2000. [Outbreak and control of *Cryptotympana atrata*.] *Forest Research, Beijing* 13: 551–553. [In Chinese, English summary]

Chilcote, C.A., & F.W. Stehr. 1984. A new record for *Magicicada septendecim* in Michigan (Homoptera: Cicadidae). *Great Lakes Entomologist* 17: 53–54.

Chinery, M. 1986. *Collins guide to the insects of Britain and western Europe*. Collins, London. 320pp.

Cho, P.S. 1955. The fauna of Dagelet Island (Ulnung-do). *Bulletin of the Sungkyunkwan University* 2: 179–266.

Cho, P.S. 1965. Beiträge zur Kenntnis der Insekten-Fauna Insel Dagelet (Ulung-do). Pages 157–205 in *Commemoration Theses, 60th Anniversary (Natural Science), Korea University = Koryŏ Taehakkyo yuksipchunyŏn kinyŏm nonmunjip.* Koryŏ Taehakkyo, Seoul.

Chou I. 1987. Auchenorrhyncha and Auchenorrhynochologists from China. *Procedings of the 6th Auchenorrhyncha Meeting, Turin, Italy,* 7–11 September 1987: 19–31.

Chou, I., Z. Lei, & J. Wang. 1993. [A study on the classification of the genus *Nipponosemia* Kato (Homoptera: Cicadoidea) from China.] *Entomotaxonomia* 15: 82–86. [In Chinese, English summary]

Chou, I., Z. Lei, L. Li, X. Lu, & W. Yao. 1997. [*The Cicadidae of China (Homoptera: Cicadoidea)*]. Tianze Eldoneio, Hong Kong. 380pp. 16 Plts [In Chinese, English summary]

Chou, I., Z. Lei, & W. Yao. 1992. [Revision of the genus *Paratalainga* He (Homoptera: Cicadidae).] *Entomotaxonomia* 14: 170–178. [In Chinese, English summary]

Chou, I., & W. Yao. 1985. [Studoj pri la tribo Gaeanini el Cineo (Homopteroj: Cikadedoj).] *Entomotaxonomia* 7: 123–140. [In Chinese, Esparanto summary]

Chou, I., & W. Yao. 1986. [Una nova genro kaj ok novaj specioj de Polyneurini el Cinio (Homoptera: Cicadidae).] *Entomotaxonomia* 8: 185–194. [In Chinese, Esperanto summary]

Chrétien, U. 2005. Chilpen: 2 Orchideen- und 2 Zikadenarten. *Pro Natura* 2005(2): 5.

Chrétien, U. 2008. Unbekannte Sänger. *Pro Natura* 2008(1): 1, 3.

Christensen, H., & M.L. Fogel. 2010. Feeding ecology and evidence for amino acid synthesis in the periodical cicada (*Magicicada*). *Journal of Insect Physiology* 57: 211–219.

Chung, A.Y.C. 2010. Edible insects and entomophagy in Borneo. Pages 141–150 *in* P.B. Durst, D.V. Johnson, R.N. Leslie, & K. Shons, eds. *Forest insects as human food: humans bite back.* Food and Agriculture Organization of the United Nations Regional Office for Asia and the Pacific, Bangkok.

Cioara, A., & A.F. Sanborn. 2001. Descriptive analysis of the calling songs of some North American cicadas (Homoptera: Cicadoidea). *Florida Scientist* 64(Supplement 1): 25.

Claridge, M.F. 1983. Acoustic signals and species problems in the Auchenorrhyncha. Pages 111–120 *in* W.J. Knight, N.C. Pant, T.S. Robertson, & M.R. Wilson, eds. *Proceedings of the 1st International Workshop on Biotaxonomy, Classification and Biology of Leafhoppers and Planthoppers (Auchenorrhyncha) of Economic Importance, London, 4–7 October 1982.*

Claridge, M.F. 1985. Acoustic signals in the Homoptera: Behavior, taxonomy, and evolution. *Annual Review of Entomology* 30: 297–317.

Claridge, M.F. 2005. Acoustic signals, species and speciation in Auchenorrhyncha; a historical review. *Abstracts of Talks and Posters, 12th International Auchenorrhyncha Congress, 6th International Workshop on Leafhopper and Planthoppers of Economic Significance, University of California, Berkeley, August 7–12, 2005*: I-4-I-5.

Claridge, M.F., J.C. Morgan, & M.S. Moulds. 1999a. Substrate-transmitted acoustic signals of the primitive cicada, *Tettigarcta crinita* Distant (Hemiptera Cicadoidea, Tettigarctidae). *Journal of Natural History* 33: 1831–1834.

Claridge, M.F., J.C. Morgan, & M.S. Moulds. 1999b. Substrate transmitted acoustic signals in the primitive cicada *Tettigarcta crinita*: an ancestral feature. *Abstracts of Papers and Posters, 10th International Auchenorrhyncha Congress, 6–10 September 1999, Cardiff, Wales.* 1p (no pagination).

Clark, D.R., Jr. 1992. Organochlorines and heavy metals in 17-year cicadas pose no apparent dietary threat to birds. *Environmental Monitoring and Assessment* 20: 47–54.

Clark, W.H. 1983. Cicada nymphs, *Okanagana* sp. (Homoptera: Cicadidae), in nests of the Owyhee harvester ant, *Pogonomyrmex owyhee* (Hymenoptera: Formicidae) in Idaho with a note on mound rebuilding by *Pogonomyrmex*. *Journal of the Idaho Academy of Science* 19: 29–32.

Clarke, A.R., S. Balagawi, B. Clifford, R.A.I. Drew, L. Leblanc, A. Mararuai, D. McGuire, D. Putulan, T. Romig, S. Sar, & D. Tenakanai. 2004. Distribution and biogeography of *Bactrocera* and *Dacus* species (Diptera: Tephritidae) in Papua New Guinea. *Australian Journal of Entomology* 43: 148–156.

Claus, C. 1884. *Elementary text-book of zoology. General part and special part: Protozoa to Insecta*. W. Swan Sonnenschein & Co., London. 615pp.

Clay, K., A.L. Shelton, & C. Winkle. 2009a. Differential susceptibility of tree species to oviposition by periodical cicadas. *Ecological Entomology* 34: 277–286.

Clay, K., A.L. Shelton, & C. Winkle. 2009b. Effects of oviposition by periodical cicadas on tree growth. *Canadian Journal of Forest Research* 39: 1688–1697.

Cloudsley-Thompson, J.L. 1957. La Cigale. *Discovery*, London 18: 341–343.

Cloudsley-Thompson, J.L. 1960. Some aspect of the fauna of the coastal dunes of the Bay of Biscay. *Entomologists Monthly Magazine* 96: 49–53.

Cloudsley-Thompson, J.L. 1970. Terrestrial invertebrates. Pages 15–77 *in* G.C. Whittow, ed. *Comparative physiology of thermoregulation, Vol. I. Invertebrates and nonmammalian vertebrates.* Academic Press, New York.

Cloudsley-Thompson, J.L. 1982. Art of insects. *Nature* 296: 524.

Cloudsley-Thompson, J.L. 1991. *Ecophysiology of desert arthropods and reptiles. Adaptation of Desert Organisms.* Springer Verlag, Berlin. 203pp.

Cloudsley-Thompson, J.L., & J.H.P. Sankey. 1957. Some aspect of the fauna of the district around the Etang-de-Berre, Bouches-du-Rhone, France. *Annals and Magazine of Natural History* (12) 10: 417–424.

Cocroft, R.B., & M. Pogue. 1996. Social behavior and communication in the neotropical cicada *Fidicina mannifera* (Homoptera: Cicadidae). *Journal of the Kansas Entomological Society* 69: 85–97.

Cocroft, R.B., & R.L. Rodríguez. 2005. The behavioral ecology of insect vibrational communication. *BioScience* 55: 323–334.

Coelho, J.R. 1997. Sexual size dimorphism and flight behavior in cicada killers, *Sphecius speciosus*. *Oikos* 79: 371–375.

Coelho, J.R. 2001. Behavioral and physiological thermoregulation in male cicada killers (*Sphecius speciosus*) during territorial behavior. *Journal of Thermal Biology* 26: 109–116.

Coelho, J.R., & C.W. Holliday. 2001. Effects of size and flight performance on intermale mate competition in the cicada killer, *Sphecius speciosus* Drury (Hymenoptera: Sphecidae). *Journal of Insect Behavior* 14: 345–351.

Coelho, J.R., & L.D. Ladage. 1999. Foraging capacity of the great golden differ wasp *Sphex ichneumoneus*. *Ecological Entomology* 24: 480–483.

Coelho, J.R., & K. Wiedman. 1999. Functional morphology of the hind tibial spurs of the cicada killer (*Sphecius speciosus* Drury)(Hymenoptera: Sphecidae). *Journal of Hymenoptera Research* 8: 6–12.

Coffin, J. 2001a. Sur le comportement de ponte et la ponte de *Tibicina quadrisignata* (Hagen, 1855) (Rhynchota, Auchenorhyncha, Cicadidae). *Ecole Pratique des Hautes Etudes, Travaux du Laboratoire Biologie et Evolution des Insectes Hemipteroidea* 14: 121–127.

Coffin, J. 2001b. Une recontre inhabituelle de *Cicada orni* (L.) en altitude. *Ecole Pratique des Hautes Etudes, Travaux du Laboratoire Biologie et Evolution des Insectes Hemipteroidea* 14: 128.

Cohen, B.L., S. Stark, A.B. Gawthrop, M.E. Burke, & C.W. Thayer. 1998. Comparison of articulate brachiopod nuclear and mitochondrial gene trees leads to a clade-based redefinition of protostomes (Protostomozoa) and deuterostomes (Deuterostomozoa). *Proceedings of the Royal Society. Series B. Biological Sciences* 265: 475–482.

Colbourne, R., K. Baird, & J. Jolly. 1990. Relationship between invertebrates eaten by little spotted kiwi, *Apteryx owenii*, and their availability on Kapiti Island, New Zealand. *New Zealand Journal of Zoology* 17: 533–542.

Cole, J.A. 2008. A new cryptic species of cicada resembling *Tibicen dorsatus* revealed by calling song (Hemiptera: Auchenorrhyncha: Cicadoidea). *Annals of the Entomological Society of America* 101: 815–823.

Coleman, J.D. 1977. The foods and feeding of starlings in Canterbury. *Proceedings of the New Zealand Ecological Society* 24: 94–109.

Cominetti, M.C., C. Aoki, & F.L. Souza. 2004. Razão sexual de três espécies de cigarras (Hemiptera: Cicadidae) em Campo Grande, MS. *XXV Congresso Brasileiro de Zoologia, 2004, Brasília. Livro de Resumos do XXV Congresso Brasileiro de Zoologia* 25: 125.

Cook, P.M., M.P. Rowe, & R.W. Van Devender. 1994. Allometric scaling and interspecific differences in the rattling sounds of rattlesnakes. *Herpetologica* 50: 358–368.

Cook, W.M., & R.D. Holt. 2002. Periodical cicada (*Magicicada cassini*) oviposition damage: visually impressive yet dynamically irrelevant. *American Midland Naturalist* 147: 214–224.

Cook, W.M., & R.D. Holt. 2006. Influence of multiple factors on insect colonization of heterogeneous landscapes: a review and case study with periodical cicadas. *Annals of the Entomological Society of America* 99: 809–820.

Cook, W.M., R.D. Holt, & J. Yao. 2001. Spatial variability in oviposition damage by periodical cicadas in a fragmented landscape. *Oecologia* 127: 51–61.

Cooley, J.R. 2001. Long-range acoustic signals, phonotaxis, and risk in the sexual pair-forming behaviors of *Okanagana canadensis* and *O. rimosa* (Hemiptera: Cicadidae). *Annals of the Entomological Society of America* 94: 755–760.

Cooley, J.R. 2004. Asymmetry and mating success in a periodical cicada, *Magicicada septendecim* (Hemiptera: Cicadidae). *Ethology* 110: 745–759.

Cooley, J.R. 2007. Decoding asymmetries in reproductive character displacement. *Proceedings of the Academy of Natural Sciences of Philadelphia* 156: 89–96.

Cooley, J.R., G.S. Hammond, & D.C. Marshall. 1998. Effects of enamel paint on the behavior and survival of the periodical cicada, *Magicicada septendecim* (Homoptera) and the lesser migratory grasshopper, *Melanoplus sanguinipes* (Orthoptera). *Great Lakes Entomologist* 31: 161–168.

Cooley, J.R., G. Kritsky, M.J. Edwards, J.D. Zyla, D.C. Marshall, K.B.R. Hill, R. Krauss, & C. Simon. 2009. The distribution of periodical cicada Brood X in 2004. *American Entomologist* 55: 106–112.

Cooley, J.R., & D.C. Marshall. 2001. Sexual signaling in periodical cicadas, *Magicicada* spp. (Hemiptera: Cicadidae). *Behaviour* 138: 827–855.

Cooley, J.R., & D.C. Marshall. 2004. Thresholds or comparisons: mate choice criteria and sexual selection in a periodical cicada, *Magicicada septendecim* (Hemiptera: Cicadidae). *Behaviour* 141: 647–673.

Cooley, J.R., D.C. Marshall, K.B.R. Hill, & C. Simon. 2006. Reconstructing asymmetrical reproductive character displacement in a periodical cicada contact zone. *Journal of Evolutionary Biology* 19: 855–868.

Cooley, J.R., D.C. Marshall, & C. Simon. 2004. The historical contraction of periodical cicada Brood VII (Hemiptera: Cicadoidea: Cicadidae). *Journal of the New York Entomological Society*112: 198–204.

Cooley, J.R., C. Simon, & D.C. Marshall. 2003. Temporal separation and speciation in periodical cicadas. *BioScience* 53: 151–157.

Cooley, J.R., C. Simon, D.C. Marshall, K. Sloan, & C. Ehrhardt. 2001. Allochronic speciation, secondary contact, and reproductive character displacement in periodical cicadas (Hemiptera: *Magicicada* spp.): genetic, morphological, and behavioural evidence. *Molecular Evolution* 10: 661–671.

Coombs, M. 1993a. New distribution records for three cicadas (Hemiptera: Cicadidae) in south-western New South Wales. A*ustralian Entomologist* 20: 133–134.

Coombs, M. 1993b. Seasonality and reproductive behaviour of *Cicadetta tristrigata* (Goding and Froggatt) (Hemiptera: Cicadidae) at Armidale, N.S.W. A*ustralian Entomologist* 20: 143–144.

Coombs, M. 1995. A new species of *Urabunana* Distant and new locality records for *U. marshalli* Distant (Hemiptera: Cicadidae) from northern New South Wales. *Journal of the Australian Entomological Society* 34: 13–15.

Coombs, M. 1996. Seasonality of cicadas (Hemiptera) on the northern tablelands of New South Wales. *Australian Entomologist* 23: 55–60.

Coombs, M., & E. Toolson. 1991. New distribution record for the double drummer cicada, *Thopha saccata* (Fabricius) (Homoptera: Cicadidae). *Australian Entomology Magazine* 18: 100.

Correia, T., A.I. Moura Santos, & P. Fonseca. 2008. Potassium currents of auditory receptor cells in *Tettigetta josei* (Hemiptera, Cicadoidea). *Abstracts of the XII Invertebrate Sound & Vibration Meeting, Tours, France, 27–30 October 2008*: 74.

Costa Lima, A.M. da. 1922. Relação dos insectos que mais commumente atacam as principaes culturas do Brasil. *A Lavoura* 26: 110–112.

Costa Lima, A.M. da. 1928. Segundo catálogo systematico dos insetos que vivem nas plantas do Brasil e ensaio de bibliographia brasileira. Archivos da Escola Superior de Agricultura e Medicina Veterinaria, Rio de Janiero (1927) 8: 69–301.

Costa-Neto, E.M. 2005. Entomotherapy, or the medicinal use of insects. *Journal of Ethnobiology* 25: 93–114.

Costa-Neto, E.M. 2008. As cigarras (Hemiptera: Cicadidae) na visão dos moradores do povoado de Pedra Branca, Bahia, Brasil. Cicadas (Hemiptera: Cicadidae) as perceived by the inhabitants of the Perda Branca settlement, Bahia, Brazil. *Boletin de la Sociedad Entomológico Aragonesa* 43: 453–457.

Costilla, M.A., & C.E. Pastor. 1986. El efecto del riego y la presencia del hongo *Cordyceps* sp. en el control de ninfas de la chicharra *Proarna bergi* Distant en caña de azucar. *Revista Industrial y Agrícola de Tucumán* 63: 121–130.

Counts, J.W., & S.T. Hasiotis. 2009. Neoichnological experiments with masked chafer beetles (Coleoptera: Scarabaeidae): implications for backfilled continental trace fossils. *Palaios* 24: 74–91.

Courts, S.E. 1998. Dietary strategies of Old World fruit bats (Megachiroptera, Pteropodidae): how do they obtain sufficient protein? *Mammal Review* 28: 185–194.

Cowan, P.E., & A. Moeed. 1987. Invertebrates in the diet of brushtail possums, *Trichosurus vulpecula*, in lowland podocarp/broadleaf forest, Orongoronogo Valley, Wellington, New Zealand. *New Zealand Journal of Zoology* 14: 163–178.

Cox, J. 1997. *New Forest natural area profile*. Jonathan Cox Associates, Lymington. 33pp.

Cox, R.T. 1992. A comment on Pleistocene population bottlenecks in periodical cicadas (Homoptera, Cicadidae, *Magicicada* spp.). *Evolution* 46: 845–846.

Cox, R.T., & C.E. Carlton. 1988. Paleoclimatic influences in the evolution of periodical cicadas (Insecta: Homoptera: Cicadidae: *Magicicada* spp.). *American Midland Naturalist* 120: 183–193.

Cox, R.T., & C.E. Carlton. 1991. Evidence of genetic dominance of the 13-year life cycle in periodical cicadas (Homoptera: Cicadidae: *Magicicada* spp.). *American Midland Naturalist* 125: 63–74.

Cox, R.T., & C.E. Carlton. 1998. A commentary on prime numbers and life cycles of periodical cicadas. *American Naturalist* 152: 162–164.

Cox, R.T., & C.E. Carlton. 2003. A comment on gene introgression versus en masse cycle switching in the evolution of 13-year and 17-year life cycles in periodical cicadas. *Evolution* 57: 428–432.

Craighead, F.C. 1949. Insect enemies of eastern forests. *United States Department of Agriculture Miscellaneous Publications* 657: 1–679.

Cranshaw, W.S., & B.C. Kondratieff. 1991. Pest profiles: Cicadas of Colorado. *Tree Leaves* 15: 10–12.

Cranshaw, W.S., & B.C. Kondratieff. 1995. *Bagging big bugs: How to identify, collect and display the largest and most colorful insects of the Rocky Mountain region*. Fulcrum Publishing, Golden, Colorado. 324pp.

Cranston, P.S. 2009. Biodiversity of Australasian insects. Pages 83–105 *in* R. Foottit & P. Adler, eds., *Insect biodiversity: science and society*. Blackwell Publishing Ltd., Oxford.

Crowhurst, P.S. 1970. Notes on *Neoitamus bulbus* (Walker) (Diptera: Asilidae). *New Zealand Entomologist* 4: 121–124.

Cryan, J.R. 2004. Molecular phylogeny of Cicadomorpha (Insecta: Hemiptera: Cicadoidea, Cercropoidea and Membracoidea): adding evidence to the controversy. *Systematic Entomology* 30: 563–574.

Cryan, J.R., & G.J. Svenson. 2010. Family-level relationships of the spittlebugs and froghoppers (Hemiptera: Cicadomorpha: Cercropoidea). *Systematic Entomology* 35: 393–415.

Cryanoski, D. 2007. Flying insects threaten to deafen Japan. *Nature* 448: 977.

Cullen, M.J. 1974. The distribution of asynchronous muscle in insects with particular reference to the Hemiptera: an electron microscope study. *Journal of Entomology. Series A. General Entomology* 49: 17–41.

Curtis, H.S. 1999. Cicada choirs. *New Scientist* 163: 56.

Dadamirzaev, A., & A. Kholmuminov 1982. [On nutrition of cicindelids on cicadas.] *Uzbekskii Biologicheskii Zhunal* 2: 66–67. [In Russian]

Dajoz, R. 1998. *Les insectes et la forêt: Rôle et diversité des insectes dans milieu forestier*. Editions Technique & Documentation, Paris. 594pp.

Dajoz, R. 2000. *Insects and forests: The role and diversity of insects in the forest environment*. Intercept Ltd., London. 668pp.

Dallas, W.S. 1870. Rhynchota. *Zoological Records* 6: 472–504.

Daniel, H.J. 1996. Erasing his ego with bugs. *Minotaur* #30 4: 14–15.

Daniel, H.J., C. Knight, T.M. Charles, & A.L. Larkins. 1993. Predicting species by call in three species of North Carolina cicadas. *Journal of the Elisha Mitchell Scientific Society* 109: 67–76.

Daniel, M.J. 1972. Bionomics of the ship rat (*Rattus r. rattus*) in a New Zealand indigenous forest. *New Zealand Journal of Science* 15: 313–341.

Daniel, M.J. 1973. Seasonal diet of the ship rat (*Rattus r. rattus*) in lowland forest in New Zealand. *Proceedings of the New Zealand Ecological Society* 20: 21–30.

Daniel, M.J. 1979. The New Zealand short-tailed bat, *Mystacina tuberculata*; a review of present knowledge. *New Zealand Journal of Zoology* 6: 357–370.

Danks, H.V. 1992. Long life cycles in insects. *Canadian Entomologist* 124: 167–187.

Das, S.C., & S.K. Roy. 1980. A cicada from tea in the Dooars. *Two and a Bud* 27: 33.

Davidson, R.H., & W.F. Lyon. 1987. *Insect pests of farm, garden, and orchard.* 8th edition. Wiley, New York. 640pp.

Davis, J.J. 1933. Insects of Indiana for 1932. *Proceedings of the Indiana Academy of Science* 42: 213–225.

Davis, J.J. 1935. Insects of Indiana for 1935. *Proceedings of the Indiana Academy of Science* 45: 257–268.

Davis, J.J. 1954. Insects of Indiana in 1953. *Proceedings of the Indiana Academy of Science* 63: 152–156.

Davis, W.T. 1894. Seventeen-year locust pupae. *Proceedings of the Natural Science Association of Staten Island* 4: 21.

Davis, W.T. 1895. Insects at Watchogue and Beulah land, Staten Island, New York. *Journal of the New York Entomological Society* 3: 140–143.

Davis, W.T. 1906a. The seventeen-year locust on Staten Island in 1906. *Proceedings of the Staten Island Association of Arts and Sciences* 1: 91.

Davis, W.T. 1906b. Observations on *Cicada tibicen* L. and allied forms. *Entomological News* 17: 237–239.

Davis, W.T. 1906c. Several species of cicada (minutes). *Journal of the New York Entomological Society* 14: 240.

Davis, W.T. 1908. The seventeen-year cicada on Staten Island in 1907. *Proceedings of the Staten Island Association of Arts and Sciences* 2: 1–2.

Davis, W.T. 1911a. Miscellaneous notes on collecting in Georgia. *Journal of the New York Entomological Society* 19: 216–219.

Davis, W.T. 1911b. Notes on the seventeen-year cicada in 1911. *Proceedings of the Staten Island Association of Arts and Sciences* 4: 1–3.

Davis, W.T. 1911c. The seventeen-year cicada on Staten Island between the years of 1894 and 1911. *Proceedings of the Staten Island Association of Arts and Sciences* 3: 120–122.

Davis, W.T. 1912. Notes on the seventeen-year cicada in 1911. *Proceedings of the Staten Island Association of Arts and Sciences* 4: 1–3.

Davis, W.T. 1913. The seventeen-year cicada on Staten Island in 1912. *Proceedings of the Staten Island Association of Arts and Sciences* 4: 99.

Davis, W.T. 1919a. *Tibicen inauditus*. *Journal of the New York Entomological Society* 27: 108.

Davis, W.T. 1919b. A belated *Tibicina cassinii*. *Journal of the New York Entomological Society* 27: 341–342.

Davis, W.T. 1919c. Minutes of October 10, 1918. *Bulletin of the Brooklyn Entomological Society* 14: 28.

Davis, W.T. 1920. Mating habits of *Sphecius speciosus*, the cicada killing wasp. *Bulletin of the Brooklyn Entomological Society* 15: 128–129.

Davis, W.T. 1922. Species collected during an automobile trip in Oregon and California. *Bulletin of the Brooklyn Entomological Society* 22: 73.

Davis, W.T. 1927. Natural history records from the Meeting of the Staten Island Nature Club. *Proceedings of the Staten Island Institute of Arts and Sciences* 4: 116–123.

Davis, W.T. 1929. [Review] Studies on the Biology of Kansas Cicadidae. By Raymond H. Beamer. *Journal of the New York Entomological Society* 37: 451–452.

Davis, W.T. 1933. Natural history records from the meeting of the Staten Island Nature Club. *Proceedings of the Staten Island Institute of Arts and Sciences* 7: 25–50.

Davis, W.T. 1936. Natural history records from the meeting of the Staten Island Nature Club. *Proceedings of the Staten Island Institute of Arts and Sciences* 8: 19–57.

Daws, A.G., & R.M. Hennig. 1996. Tuning of the peripheral auditory system of the cicada, *Cyclochila australasiae*. *Zoology - Analysis of Complex Systems* 99: 175–188.

Daws, A.G., R.M. Hennig, & D. Young. 1997. Phonotaxis in the cicadas *Cystosoma saundersii* and *Cyclochila australasiae*. *Bioacoustics* 7: 173–188.

Day, E., D. Pfeiffer, & E. Lewis. 2002. Periodical cicada. *Virginia Cooperative Extension Publication* 444–276W: 3pp.

Dean, W.R.J. 1992. Effects of animal activity on the absorption rate of soils in the southern Karoo, South Africa. *Journal of the Grassland Society of Southern Africa* 9: 178–180.

Dean, W.R.J. 1993. Alpine swifts opportunistically feeding on cicadas. *Ostrich* 64: 42–43.

Dean, W.R.J., & S.J. Milton. 1991. Emergence and oviposition of *Quintilia* cf. *conspersa* Karsch (Homoptera: Cicadidae) in the southern Karoo, South Africa. *Journal of the Entomological Society of Southern Africa* 54: 111–119.

Dean, W.R.J., & S.J. Milton. 1992. Emergence and density of *Quintillia* cf. *vitripennis* Karsch (Homoptera: Cicadidae) in the southern Karoo. *Journal of the Entomological Society of Southern Africa* 55: 71–75.

Dean, W.R.J., & S.J. Milton. 2001. The density and stability of birds in shrubland and drainage line woodland in the southern Karoo, South Africa. *Ostrich* 72: 185–192.

Dean, W.R.J., S.J. Milton, M. du Plessis, & W.R. Siegfried. 1995. Dryland degradation: Symptoms, stages, and hypothetical cures. *General Technical Report Intermountain Research Station, United States Department of Agriculture Forest Service,* No. INT-GTR 315: 178–182.

Dearborn, D.C., L.S. MacDade, S. Robinson, A.D. Dowlng Fink, & M.L. Fink. 2009. Offspring development mode and the evolution of brood parasitism. *Behavioral Ecology* 20: 517–524.

Deay, H.O. 1953. The periodical cicada, *Magicicada septendecim* (L.) in Indiana. *Proceedings of the Indiana Academy of Science* 62: 203–206.

DeFoliart, G.R. 2003. Food, insects as. Pages 431–437 *in* V.H. Resh & R.T. Cardé, eds. *Encyclopedia of insects*. Academic Press, New York.

Deitz, L.L. 1979. Selected references for identifying New Zealand Hemiptera (Homoptera and Heteroptera), with notes on nomenclature. *New Zealand Entomologist* 7: 20–29.

Delaye, L., & A. Moya. 2010. Evolution of reduced prokaryotic genomes and the minimal cell concept: variation on a theme. *BioEssays* 32: 281–287.

Delescaille, L.-M. 2005. La gestion des pelouses sèches en Région wallonne. *Biotechnologie, Agronomie, Société et Environnement* 9: 119–124.

Demichelis, S., A. Manino, C. Sartor, D. Cifuentes, & A. Patetta. 2010. Specific identification of some female Empoascini (Hemiptera: Cicadellidae), using morphological characters of the ovipositor and isozyme and mtCOI sequence analyses. *Canadian Entomologist* 142: 513–531.

Demir, E. 2006. Contributions to the knowledge of Turkish Auchenorrhyncha (Homoptera) with a new record, *Pentastiridius nanus* (Ivanoff, 1885). *Munis Entomology & Zoology* 1: 97–122.

Demir, E. 2007a. Contributions to the knowledge of Turkish Auchenorrhyncha (Homoptera, Fulgoromorpha and Cicadomorpha, excl. Cicadellidae) with a new record, *Setapius klapperichianus* Dlabola, 1988. *Munis Entomology & Zoology* 2: 39–58.

Demir, E. 2007b. Auchenorrhyncha (Homoptera) data from Ankara with two new records to Turkey. *Munis Entomology & Zoology* 2: 481–492.

Demir, E. 2008. The Fulgoromorpha and Cicadomorpha of Turkey. Part I: Mediterranean Region (Hemiptera). *Munis Entomology & Zoology* 3: 447–522.

De Santis, C.L., M.C. Medrano, A.F. Sanborn, & P.G. Bolcatto. 2007. Cicádidos (Insecta: Hemiptera: Cicadidae) del Museo Provincial de Ciencias Naturales "Florentino Ameghino" Santa Fe, Argentina. *Museo Provincial de Ciencias Naturales "Florentino Ameghino" Serie Catálogo* No. 19: 1–19.

De Santis, C.L., R. Urteaga, & P.G. Bolcatto. 2006. Cicada wings as determinant factor for the sound emission: the case of *Quesada gigas*. *arXiv:q-bio/0608011v1 [q-bio.OT]*. 5pp.

Dey, S. 1999. Ultrastructural features of photoreceptor and associated structures in relation to visual adaptation of *Cicadetta montana*: A transmission electron microscopic study and review. *Cytobios* 98: 159–174.

De Zayas, F. 1988. *Entomofauna cubana: Tópicos entomológicos a nivel medio para uso didáctico. Superorden Hemipteroidea. Ordern Homoptera. Orden Heteroptera*. Tomo VII. Editorial Cientifico-Técnica, La Habana, Cuba. 261pp.

Didier, B. 2006. Parlez-vous entomo? *Insectes* (143): 28.

Dietrich, C.H. 1989. Surface sculpturing of the abdominal integument of Membracidae and other Auchenorrhyncha (Homoptera). *Proceedings of the Entomological Society of Washington* 91: 143–152.

Dietrich, C.H. 2002. Evolution of Cicadomorpha (Insecta, Hemiptera). *Denisia* 4: 155–170.

Dietrich, C.H. 2003. Auchenorrhyncha (cicadas, spittlebugs, leafhoppers, treehoppers, and planthoppers). Pages 431–437 *in* V.H. Resh & R.T. Cardé, eds. *Encyclopedia of insects*. Academic Press, New York.

Dietrich, C.H. 2005. Keys to the families of Cicadomorpha and subfamilies and tribes of Cicadellidae (Hemiptera: Auchenorrhyncha). *Florida Entomologist* 88: 502–517.

Dietrich, C.H., D.A. Dmitriev, R.A. Rakitov, D.M. Takiya, & J.N. Zahniser. 2005. Phylogeny of cicadellidae (Cicadomorpha: Membracoidea) based on combined morphological and 28S rDNA sequence data. *Abstracts of Talks and Posters, 12th International Auchenorrhyncha Congress, 6th International Workshop on Leafhopper and Planthoppers of Economic Significance, University of California, Berkeley, August 7–12, 2005*: S-13-S-15.

Dietrich, C.H., R.A. Rakitov, J.L. Holmes, & W.C. Black, IV. 2001. Phylogeny of the major lineages of Membracoidea (Insecta: Hemiptera: Cicadimorpha) based on 28S rDNA sequences. *Molecular Phylogenetics and Evolution* 18: 293–305.

Dingley, J.M. 1974. Vegetable caterpillar and vegetable cicada. *New Zealand Natural Heritage* 2: 535–536.

Distant, W.L. 1881. [Exhibited a very large cicada from Madagascar]. *Proceedings of the Entomological Society of London* 1881: ii-iii.

Distant, W.L. 1882. Contributions to a knowledge of the Rhynchotal fauna of Sumatra. *Entomologist's Monthly Magazine* 19: 156–160.

Distant, W.L. 1906. Recent bibliographical and nomenclatorial notes on the Rhynchota. *Entomologist* 39: 36–37.

Dlabola, J. 1971. Taxonomische und Chorologische Ergänzungen der Zikadenfauna von Anatolien, Iran, Afghanistan und Pakistan (Homptera, Auchenorrhyncha). *Acta Entomologica Bohemoslovaca* 68: 377–396.

Dlabola, J. 1981. Ergebnisse der Tschechoslowakisoh-Iranischen Entomologischen Expedition nach dem Iran.1970–1973). *Acta Entomologica Musie Nationalis Pragae* 40: 127–308.

Dlabola, J. 1987. Neue ostmediterrane und iranische Zikadentaxone (Homoptera, Auchenorrhyncha). *Acta Entomologica Bohemoslovaca* 84: 295–312.

Dmitriev, D.A. 2001a. [Fauna of the (Homoptera, Cicadina) of Voronezh Province.] *Entomologicheskoe Obozrenie* 80: 54–72. [In Russian, English summary]

Dmitriev, D.A. 2001b. Fauna of the Homoptera Cicadina of Voronezh Province. *Entomological Review* 81: 7–25.

Dmitriev, D.A. 2007a. [Zoogeographical analysis and statial distribution of Auchenorrhyncha (Homoptera, Cicadina) in the Central Chernozem region.] *Entomologicheskoe Obozrenie* 86: 807–826. [In Russian, English summary]

Dmitriev, D.A. 2007b. Zoogeographical analysis and statial distribution of Auchenorrhyncha (Homoptera, Cicadina) in the Central Chernozem region. *Entomological Review* 87: 1201–1217.

Dmitriev, D.A. 2010. Homologies of the head of Membracoidea based on nymphal morphology with notes on other groups of Auchenorrhyncha (Hemiptera). *European Journal of Entomology* 107: 597–613.

Dobosz, R. 1993. Nowe stanowiska *Ledra aurita* (L.) i *Cicadetta montana* (Scop.) z Górnego Slaska (*Homoptera: Auchenorrhyncha*). [New records of *Ledra aurita* (L.) and *Cicadetta montana* (Scop.) from Upper Silesia (*Homoptera: Auchenorrhyncha*).] *Acta Entomologica Silensiana* 1: 15.

Dohlen, C.D. von, & N.A. Moran. 1995. Molecular phylogeny of the Homoptera: a paraphyletic taxon. *Journal of Molecular Evolution* 41: 211–223.

Doi, H. 1932. [Miscellaneous notes on insects.] *Journal of the Chosen Natural History Society* (13): 30–49. [In Korean]

Doi, H. 1934. [Miscellaneous notes on insects (4).] *Journal of the Chosen Natural History Society* (17): 64–68. [In Korean]

Dolling, W.R. 1991. *The Hemiptera*. Oxford University Press, Oxford. 274pp.

Doolan, J.M. 1981. Male spacing and the influence of female courtship behaviour in the bladder cicada, *Cystosoma saundersii* Westwood. *Behavioral Ecology and Sociobiology* 9: 269–276.

Doolan, J.M., & R.C. MacNally. 1981. Spatial dynamics and breeding ecology in the cicada *Cystosoma saundersii*: the interaction between distributions of resources and intraspecific behaviour. *Journal of Animal Ecology* 50: 925–940.

Doolan, J.M., & D. Young. 1981. The organization of the auditory organ of the bladder cicada, *Cystosoma saundersii*. *Philosophical Transactions of the Royal Society of London. B* 291: 525–540.

Doolan, J.M., & D. Young. 1989. Relative importance of song parameters during flight phonotaxis and courtship in the bladder cicada *Cystosoma saundersii*. *Journal of Experimental Biology* 141: 113–131.

Donat, P. 1981. Metamorfóza cikády *Tibicina intermedia*. *Ziva* 29: 145.

Douglas, M.J.W. 1970. Foods of harriers in a high country habitat. *Notornis* 17: 92–95.

Dourlot, S. 2009 *Insect museum*. Firefly Books, Buffalo. 256pp.

Drosopoulos, S. 1980. Hemipterological studies in Greece. Part II Heteroptera. A catalogue of the reported species. *Biologia gallo-hellencia* 9: 187–194

Drosopoulos, S., M. Asche, & H. Hoch. 1986. A preliminary list and some notes on the Cicadomorpha (Homoptera-Auchenorrhyncha) collected in Greece. *Proceedings of the 2nd International Congress Concerning the Rhynchota Fauna of the Balkan and Adjacent Regions, Mikrolimni, Greece* 1986: 8–13.

Drosopoulos, S., E. Eliopoulos, & P. Tsakalov. 2005. Is migration responsible for the peculiar geographical distribution and speciation based on acoustic divergence of two cicada species in the Aegean archipelago? Pages 219–226 *in* S. Drosopoulos & M.F. Claridge, eds. *Insect sounds and communication: physiology, behaviour, ecology and evolution*. CRC Press, Boca Raton.

Drugescu, C. 1983. Entomofauna din coronamentul unei paduri de foioase din Valea Cernei. *Lucrarile Celei de a III-a Conferinta de entomologie, Iasi, 20–22 mai 1983*: 189–198.

Dubovsky, G.K. 1982. [On the fauna of cicadas (Auchenorrhyncha) of Repetek Reserve.] *Uzbekskii Biologicheskii Zhurnal* (1): 49–52. [In Russian]

Dudley, R. 2000. *Biomechanics of insect flight: form, function, evolution*. Princeton University Press, Princeton. 476pp.

Duffels, J.P. 1982. *Brachylobopyga decorata* n. gen., n. sp. from Sulawesi, a new taxon of the subtribe Cosmopsaltriaria (Homoptera, Cicadoidea: Cicadidae). *Entomologische Berichten (Amsterdam)* 42: 156–160.

Duffels, J.P. 1983a. Distribution patterns of oriental Cicadoidea (Homoptera) east of Wallace's line and plate tectonics. *in* I.W.B. Thornton, ed. Symposium on Distribution and Evolution of Pacific Insects. *GeoJournal* 7: 491–498.

Duffels, J.P. 1983b. Taxonomy, phylogeny and biogeography of the genus *Cosmopsaltria*, with remarks on the historic biogeography of the subtribe Cosmopsaltriaria (Homoptera: Cicadidae). *Pacific Insects Monographs* 39: 1–127.

Duffels, J.P. 1985. *Rhadinopyga*, n. gen. from the "Vogelkop" of New Guinea and adjacent islands, a new genus of the subtribe Cosmopsaltriaria (Homoptera, Cicadoidea: Cicadidae). *Bijdragen tot de Dierkunde* 55: 275–279.

Duffels, J.P. 1986. Biogeography of Indopacific Cicadoidea, a tentative recognition of areas of endemism. *Cladistics* 2: 318–336.

Duffels, J.P. 1988a. Biogeography of the cicadas of the islands of Bacan, Maluku, Indonesia, with description of *Diceropyga bacanensis* n. sp. (Homoptera, Cicadidae). *Tijdschrift voor Entomologie* 131: 7–12.

Duffels, J.P. 1988b. *Cosmopsaltria halmaherae* n. sp. endemic to Halmahera, Maluku, Indonesia (Homoptera, Cicadidae). *Bijdragen tot de Dierkunde* 58: 20–24.

Duffels, J.P. 1988c. *Cosmopsaltria kaiensis* n. sp., a new cicada from the Kai Islands, Indonesia (Homoptera: Cicadidae). *Entomologische Berichten (Amsterdam)* 48: 187–190.

Duffels, J.P. 1988d. The cicadas of the Fiji, Samoa, and Tonga Islands, their taxonomy and biogeography (Homoptera, Cicadoidea) with a chapter on the geological history of the area by A. Ewart. *Entomonograph* 10: 1–108.

Duffels, J.P. 1989. The Sulawesi genus *Brachylobopyga* (Homoptera: Cicadidae). *Tijdschrift voor Entomologie* 132: 123–127.

Duffels, J.P. 1990a. *Dilobopyga janstocki* n. sp., a new cicada endemic to Sulawesi (Homoptera, Cicadidae). *Bijdragen tot de Dierkunde* 60: 323–326.

Duffels, J.P. 1990b. Biogeography of Sulawesi cicadas (Homoptera: Cicadoidea). Pages 63–72 *in* W.J. Knight & J.D. Holloway, eds. *Insects and the rain forests of South East Asia (Wallacea)*. Royal Entomological Society, London.

Duffels, J.P. 1991a. The eye-catching cicada *Hamza ciliaris* (Linnaeus, 1758) comb. n. in Indonesia and the Pacific: taxonomic status, synonymy, and distribution (Homoptera, Cicadoidea). *Bijdragen tot de Dierkunde* 61: 119–130.

Duffels, J.P. 1991b. Revision of the genus *Champaka* (Homoptera, Cicadidae) from Borneo and Sulawesi. *Tijdschrift voor Entomologie* 134: 177–182.

Duffels, J.P. 1993. The systematic position of *Moana expansa* (Homoptera: Cicadidae), with reference to sound organs and the higher classification of the superfamily Cicadoidea. *Journal of Natural History* 27: 1223–1237.

Duffels, J.P. 1997. *Inflatopyga*, a new cicada genus (Homoptera: Cicadoidea: Cicadidae) endemic to the Solomon islands. *Invertebrate Taxonomy* 11: 549–568.

Duffels, J.P. 1999. *Dilobopyga remanei*: a new species from Sulawesi (Indonesia) (Hemiptera: Auchenorrhyncha: Cicadomorpha: Cicadidae). *Reichenbachia, Staatliches Museum für Tierkunde Dresden* 33: 81–86.

Duffels, J.P. 2004. Revision of the cicada genus *Nabalua* (Hemiptera: Cicadidae) from southeast Asia. *Oriental Insects* 38: 463–490.

Duffels, J.P. 2009. Revision of the cicadas of the genus *Maua* Distant (Hemiptera, Cicadidae) from Sundaland. *Tijdschrift voor Entomologie* 152: 303–332.

Duffels, J.P. 2010. The genus *Nelcyndana* Stål (Hemiptera, Cicadidae, Taphurini) with description of three new species from Borneo. *ZooKeys* 61: 11–31.

Duffels, J.P., & A.J. de Boer. 1990. Areas of endemism and composite areas in East Malesia. Pages 249–272 *in* P. Baas, K. Kalkman, & R. Geesink, eds. *The plant diversity of Malesia. Proceedings of the flora Malesiana symposium commemorating Professor Dr. C.G.G.J. van Steenis, Leiden, August 1989*. Kluwer Academic, Dordrecht.

Duffels, J.P., & A.J. de Boer. 2006. Cicada endemism in Papua. Pages 532–538 *in* A.J. Marshall & B.M. Beehler, eds. *The ecology of Papua, part one*. Periplus Editions, Singapore.

Duffels, J.P., & M. Hayashi. 2006. On the identity of the cicada species *Pomponia picta* (Walker) (=*P. fusca* (Olivier)) and *P. linearis* (Walker) (Hemiptera, Cicadidae). *Tijdschrift voor Entomologie* 149: 189–201.

Duffels, J.P., & M.A. Schouten. 2002. *Orientopsaltria endauensis*, a new cicada species from Endau Pompia National Park, Malaysia (Homoptera, Cicadidae). *Malayan Nature Journal* 56: 49–55.

Duffels, J.P., M.A. Schouten, & M. Lammertink. 2007. A revision of the cicadas of the *Purana tigrina* group (Hemiptera, Cicadidae) in Sundaland. *Tijdschrift voor Entomologie* 150: 367–387.

Duffels, J.P., & H. Turner. 2002. Cladistic analysis and biogeography of the cicadas of the Indo-Pacific subtribe Cosmopsaltriina (Hemiptera: Cicadoidea: Cicadidae). *Systematic Entomology* 27: 235–261.

Duffels, J.P., & H.J.G. van Mastrigt. 1991. Recognition of cicadas (Homoptera, Cicadidae) by the Ekagi people of Irian Jaya (Indonesia), with a description of a new species of *Cosmopsaltria*. *Journal of Natural History* 25: 173–182.

Duffels, J.P., & M.I. Zaidi. 1998. Two new species of the genus of the genus *Orientospaltria* Kato (Homoptera, Cicadidae) from Peninsular Malaysia, a contribution to study of cicada biodiversity. *Serangga* 3: 317–341.

Duffels, J.P., & M.I. Zaidi. 1999. Correction to: Two new species of the genus *Orientopsaltria* Kato (Homoptera, Cicadidae) from Peninsular Malaysia, a contribution to the study of cicada biodiversity *Serangga* 3(2): 317–341. *Serangga* 4: 147–148.

Duffels, J.P., & M.I. Zaidi. 2000. A revision of the cicada genus *Orientopsaltria* Kato (Homoptera, Cicadidae) from Southeast Asia. *Tijdschrift voor Entomologie* 142: 195–297.

Dugdale, J.S. 1988. Charles Alexander Fleming, KBE – 1916–1987 (Life member of the Society 1977). *New Zealand Entomologist* 11: 92–93.

Duke, L., D. Steinkraus, J. English, & K. Smith. 2002. Infectivity of resting spores of *Massospora cicadina* (Entomophthorales: Entomophthoraceae), an entomopathogenic fungus of periodical cicadas (*Magicicada* spp.) (Homptera: Cicadidae). *Journal of Invertebrate Pathology* 80: 1–6.

Dumbleton, L.J. 1966. Genitalia, classification and zoogeography of the New Zealand Hepialidae (Lepidoptera). *New Zealand Journal of Science* 9: 920–981.

Dunn, G.A. 1996. *Insects of the Great Lakes region*. University of Michigan Press, Ann Arbor. 324pp.

Dunn, K.L. 1991. New distribution records for some eastern Australian cicadas. *Victorian Naturalist* 21: 52–53.

Dunn, K.L. 1992. Notes on the silver cicada *Cicadetta celis* Moulds. *Victorian Naturalist* 22: 12–17.

Dunn, K.L. 1998. Butterfly watching in Tasmania – Part II. *Victorian Naturalist* 28: 44–49.

Dunn, K.L. 1999a. Butterfly watching in Tasmania – Part IV. *Victorian Naturalist* 29: 4–9.

Dunn, K.L. 1999b. Butterfly watching in Tasmania – Part V. *Victorian Naturalist* 29: 33–39.

Dunn, K.L. 1999c. Butterfly watching in Tasmania – Part VI. *Victorian Naturalist* 29: 56–64.

Dunn, K.L. 1999d. Notes on *Cicadetta abdominalis* (Distant) at Flowerdale, Victoria. *Victorian Naturalist* 29: 40.

Dunn, K.L. 2002. Observations on mating cicadas in northern Queensland. *Victorian Entomologist* 32: 26–27.

Durin, B. 1981. *Insects etc.: an anthology of arthropods featuring a bounty of beetles*. Hudson Hills Press, New York. 108pp.

D'Urso, V. 1986. On the Auchenorrhyncha (Homoptera) from Aeolian island (Sicily, Italy). *Proceedings of the 2nd International Congress Concerning the Rhynchota Fauna of the Balkan and Adjacent Regions, Mikrolimni, Greece* 1986: 23–27.

D'Urso, V. 2002. The wing-coupling apparatus of Hemiptera Auchenorrhyncha: Structure, function, and systematic value. *Denisia* 4: 401–410.

D'Urso, V., & S. Ippolitto. 1994. Wing-coupling apparatus of Auchenorrhyncha (Insecta, Homoptera). *International Journal of Insect Morphology and Embryology* 23: 211–224.

D'Utra, G. 1908. Cigarras nos cafesaes. *Boletim de agricultura, Secretaria da Agricultura, Commercio e Obras Publicas Estado Sao Paulo*, 9ª Ser. N. 5: 357–365.

Dworakowska, I. 1988. Main veins of the wing of Auchenorrhyncha (Insecta, Rhynchota, Hemelytrata). *Entomologische Abhandlungen Staatliches Museum fur Tierkunde Dresden* 52: 63–108.

Easton, E. 1999. Something old and something rare: the work of one of South China's earliest naturalists. *Porcupine* 20: 28.

Easton, E.R., & P. Wing-Wah. 1999. Observations on twelve families of Homoptera in Macau, Southeastern China, from 1989 to the present. *Proceedings of the Entomological Society of Washington* 101: 99–105.

Eaton, E.R. 1991. Class of '91: Cicada style. *Explorer* 33: 8–10.

Eaton, E.R., & K. Kaufman. 2007. *Kaufman field guide to insects of North America.* Hillstar Editions, L.C., Houghton Mifflin Company, New York. 393pp.

Eberhard, W.G. 2004. Male-female conflict and genitalia: failure to confirm predictions in insects and spiders. *Biological Reviews* 79: 121–186.

Eckert, M., J. Gabriel, H. Birkenbeil, G. Greiner, J. Rapus, & G. Gäde. 1996. A comparative immunocytochemical study using an antiserum against a synthetic analogue of the corpora cardiaca peptide Pea-CAH-I (MI, neurohormone D) of *Periplaneta americana. Cell & Tissue Research* 284: 401–413.

Edwards, M.J., A.E. Faivre, R.C. Crist III, M.I. Sitvarin, & J. Zyla. 2005. Distribution of the 2004 emergence of seventeen-year periodical cicadas (Hemiptera: Cicadidae: *Magicicada* spp., Brood X) in Pennsylvania, U.S.A. *Entomological News* 116: 273–282.

Ellingson, A.R., & D.C. Andersen. 2002. Spatial correlations of *Diceroprocta apache* and its host plants: evidence for a negative impact from *Tamarix* invasion. *Ecological Entomology* 27: 16–24.

Ellingson, A.R., D.C. Andersen, & B.C. Kondratieff. 2002. Obsrvations of the larval stages of *Diceroprocta apache* (Homoptera: Tibicinidae). *Journal of the Kansas Entomological Society* 75: 283–289.

Elliott, L., & W. Hershberger. 2006. *The songs of insects.* Houghton Mifflin Company, Boston. 228pp.

Elven, H. 1993. Litt om sangsikaden *Cicadetta montana* i Norge. *Insekt-Nytt* 18(2): 5–7.

Emeljanov, A.F. 1981a. [Morphology and origin of endopleural structures of metathorax in Auchenorrhyncha (Homoptera).] *Entomologicheskoye Obozreniye* 60: 732–744. [In Russian]

Emeljanov, A.F. 1981b. Morphology and origin of endopleural structures of metathorax in Auchenorrhyncha (Homoptera). *Entomological Review* 60: 13–26.

Emeljanov, A.F. 1984. Cicadidae. Pages 56–57 *in* V.S. Velikan, ed. [*Keys to the identification of harmful and useful insects and mites of fruit and berry plantations in the USSR.*] Kolos, Leningrad. [In Russian]

Emeljanov, A.F. 1996a. [Two new Cicadidae species similar to *Cicadetta tibialis* (Homoptera, Cicadidae).] *Zoologichesky Zhurnal* 75: 1897–1900. [In Russian, English summary]

Emeljanov, A.F. 1996b. Two new cicad species similar to *Cicadetta tibialis* (Homoptera, Cicadidae). *Entomological Review* 76: 1054–1057.

Emeljanov, A.F., & V.I. Kirillova. 1991. [Trends and modes of the karyotype evolution in Cicadina: II. Peculiarities and evolutionary changes of karyotypes in the superfamilies Cercopoidea, Cicadoidea, Fulgoridea and the Cicadina as a whole (Homoptera).] *Entomologicheskoye Obozreniye* 70: 796–817. [In Russian, English summary]

Emeljanov, A.F., & V.I. Kirillova. 1992. Trends and modes of the karyotype evolution in Cicadina: II. Peculiarities and evolutionary changes of karyotypes in the superfamilies Cercopoidea, Cicadoidea, Fulgoridea and the Cicadina as a whole (Homoptera). *Entomological Review* 70: 59–81.

Emery, D. 2008. Phenology and diversity of cicadas in the Sydney region. *Tarsus* (584): 34–35.

Emery, D.L., & S.J. Emery. 2002. First records for the paperback cicada *Cicadetta hackeri* (Distant) and *Cicadetta spinosa* (Goding & Froggatt) (Hemiptera: Cicadidae) from Sydney, New South Wales. *Australian Entomologist* 29: 137–139.

Emery, D.L., S.J. Emery, N.J. Emery, & L.W. Popple. 2005. A phenological study of the cicadas (Hemiptera: Cicadidae) in western Sydney, New South Wales, with notes on plant associations. *Australian Entomologist* 32: 97–110.

Emigh, T.H. 1979. The dynamics of finite haploid populations with overlapping generations. I. Moments, fixation probabilities and stationary distributions. *Genetics* 92: 323–337.

Emmrich, R. 1984. Vorkommen und Verbreitung von *Cicadetta montana* (Scop.) im Gebiet der DDR, unter besonderer Berücksichtigung der Sächsischen Schweiz (Insecta, Homoptera, Auchenorrhyncha, Cicadidae). *Faunistische Abhandlungen, Staatliches Museum für Tierkunde in Dresden* 11: 109–117.

English, L.D., J.J. English, R.N. Dukes, & K.G. Smith. 2006. Timing of 13-year periodical cicadas (Homoptera: Cicadidae) emergence determined 9 months before emergence. *Environmental Entomology* 35: 245–248.

Escalante, G.J.A. 1974. Notes on insects of Alto Urubamba, Cuzco. *Revista Peruana de Entomologia* 17: 120–121.

Esson, M.J. 2007. Sex ratios of emerging cicadas based on nymphal exuviae. *The Weta* 33: 20–22.

Esteban, M., & B. Sanchiz. 1997. Descripción de nuevas especies animales de la Península Ibérica e Islas Baleares 1978–1994): tendencias taxonómicas y listado sistemático. *Graellsia* 53: 111–175.

Ettling, J.A. 1986. Cicada feeding by captive copperheads (*Agkistrodon contortrix*). *Bulletin of the Chicago Herpetological Society* 21: 32.

Evans, H.E. 1966. *The comparative ethology and evolution of the sand wasps*. Harvard University Press, Cambridge. 526pp.

Evans, J.W. 1956. Paleozoic and Mesozoic Hemiptera (Insecta). *Australian Journal of Zoology* 4: 165–258.

Evans, J.W. 1963. The zoogeography of New Zealand leafhoppers and froghoppers (Insecta, Homoptera, Cicadelloidea and Cercopoidea). *Transactions of the Royal Society of New Zealand* 3: 85–91.

Evans, K.L. 2004. The potential for interactions between predation and habitat change to cause population declines of farmland birds. *Ibis* 146: 1–13.

Eve, R. 1991. L'environment acoustique d'un peuplement d'oiseaux en forêt tropical; organisation ou chaos? *Revue d'Ecologie: La Terre et La Vie* 46: 191–229.

Ewart, A. 1986. Cicadas of Kroombit tops. *Queensland Naturalist* 27: 50–57.

Ewart, A. 1988. Cicadas (Homoptera). Pages 180–201 *in* G. Scott, ed. *Lake Broadwater: the natural history of an island lake and its environs*. Darling Downs Institute Press, Toowoomba, Queensland.

Ewart, A. 1989a. Cicada songs - Song production, structures, variation and uniqueness within species. *Entomological Society of Queensland News Bulletin* 17: 75–82.

Ewart, A. 1989b. Revisionary notes on the genus *Pauropsalta* Goding and Froggatt (Homoptera: Cicadidae) with special reference to Queensland. *Memoirs of the Queensland Museum* 27: 289–375.

Ewart, A. 1990. Status of the Germar and Leach types of Australian cicadas (Homoptera) held at the Hope Entomological Collections, Oxford. *Australian Entomological Magazine* 17: 1–5.

Ewart, A. 1993. Cicadas of the Heathlands region, Cape York Peninsula. Pages 135–147 *in Royal Geographical Society of Queensland, Cape York Peninsula Scientific Expedition, Wet season 1992. Report, Vol. 2*. Royal Geographical Society of Queensland, Brisbane.

Ewart, A. 1996. Cicadas. Tracks 80–88 *in* D.C.F.Rentz, Co-ordinator. Nature sounds of Australia, original recordings of Australian wildlife, including birds, frogs, cicadas and crabs. The Sound Heritage Association, Ltd.

Ewart, A. 1998a. Cicadas, and their songs, of the Miles-Chinchilla Region. *Queensland Naturalist* 36: 54–72.

Ewart, A. 1998b. Cicadas of Musselbrook Reserve. *in* Musselbrook Reserve Scientific Study Report. *Geography Monograph Series* 4: 135–138.

Ewart, A. 1999. Accoustic [sic] signalling within smaller cicadas (Homoptera) of Queensland, Australia: song patterns and their possible evolution. *Abstracts of Papers and Posters, 10th International Auchenorrhyncha Congress, 6–10 September 1999, Cardiff, Wales.* 2pp (no pagination).

Ewart, A. 2001a. Emergence patterns and densities of cicadas (Hemiptera: Cicadidae) near Caloundra, South-east Queensland. *Australian Entomologist* 28: 69–84.

Ewart, A. 2001b. Dusk chorusing behaviour in cicadas (Homoptera: Cicadidae) and a mole cricket, Brisbane, Queensland. *Memoirs of the Queensland Museum* 46: 499–510.

Ewart, A. 2005a. Cicadas of the Pennefather River-Weipa areas, October/November 2002, with comparative notes on the cicadas from Heathlands, Cape York peninsula. *Geography Monographs* 10: 169–179.

Ewart, A. 2005b. New genera and species of small ticking and 'chirping' cicadas (Hemiptera: Cicadoidea: Cicadidae) from Queensland, with descriptions of their songs. *Memoirs of the Queensland Museum* 51: 439–500.

Ewart, A. 2009a. *Crotopsalta leptotigris*, a new species of ticking cicada (Hemiptera: Cicadoidea: Cicadidae) from Cravens Peak, southwest Queensland. *Australian Entomologist* 36: 139–151.

Ewart, A. 2009b. Cicadas of the eastern segment of the Cravens Peak Reserve, northeastern Simpson Desert, S.W. Queensland; January/February 2007. Pages 137–170 *in* H. Freemantle, ed. *Scientific Study Report, Geography Monograph Series No. 13*. The Royal Geographical Society of Queensland Inc., Milton, Queensland.

Ewart, A., & D. Marques. 2008. A new genus of grass cicadas (Hemiptera: Cicadoidea: Cicadidae) from Queensland, with descriptions of their songs. *Memoirs of the Queensland Museum* 52: 150–202.

Ewart, A., & L.W. Popple. 2001. Cicadas, and their songs, from south-western Queensland. *Queensland Naturalist* 39: 52–71.

Ewart, A., & L.W. Popple. 2007. Songs and calling behaviour of *Frogattoides typicus* Distant (Hemiptera: Cicadoidea: Cicadidae), a nocturnally singing cicada. *Australian Entomologist* 34: 127–139.

Ewart, T. 1995. Cicadas. Pages 79–88 *in* M. Ryan, ed. *Wildlife of greater Brisbane*. Queensland Museum with the support of the Environment Management Branch of Brisbane City Council, Brisbane.

Ewing, A.W. 1984. Acoustic signals in insect sexual behaviour. Pages 223–240 *in* T. Lewis, ed. *Insect Communication*. Academic Press, London.

Ewing, A.W. 1989. *Arthropod bioacoustics: neurobiology and behaviour*. Comstock Publishing Associates, Cornell University Press, Ithaca. 260pp.

Eyles, H.C. 1960. Insects associated with the major fodder crops in the North Island. *New Zealand Journal of Agricultural Research* 3: 994–1008.

Faber, A. 1928. Die Bestimmung der deutschen Geradflügler (Orthoptera) nach ihren Lautäusserungen. *Zeitschrift für Wissenschaftliche Insektenbiologie* 23: 203–234.

Fain, A., J.-Y. Baugnee, & F. Hidvegi. 1992. Acariens phorétiques ou parasites récoltés sur des hyménoptères et un homoptère dans la région de Treignes, en Belgique. *Bulletin et Annales de la Société Entomologique de Belgique* 128: 335–338.

Faithfull, I. 1990. Some vertebrate predators of the green Monday cicada *Cyclochila australasiae* (Donovan) in Melbourne. *Victorian Entomologist* 20(5): 97–98.

Fan, Y., R. Wu, M.-H. Chen, L. Kuo, & P.O. Lewis. 2010. Choosing among partitioning models in Bayesian phylogenetics. *Molecular Biology and Evolution* 28: 523–532.

Farris, H.E., M.L. Oshinsky, T.G. Forrest, & R.R. Hoy. 2008. Auditory sensitivity of an acoustic parasitoid (*Emblemasoma* sp., Sarcophagidae, Diptera) and the calling behavior of potential hosts. *Brain, Behavior and Evolution* 72: 16–25.

Fattorusso, R., S. Frutos, X. Sun, N.J. Sucher, & M. Pellecchia. 2006. Traditional Chinese medicines with caspase-inhibitory activity. *Phytomedicine* 13: 16–22.

Feltwell, J. 1997. The cicada and the copper underwing. *British Journal of Entomology and Natural History* 10: 102.

Feng D. 1988. [Features of waveforms and spectra of songs in black cicada (*Cryptotympana atrata* Fabricius).] *Acta Biophysica Sinica* 4: 25–29, 77. [In Chinese, English summary]

Feng D. 1990. [Structures of the sound-producing organs of cicada *Cryptotympana atrata* Fabricius: tymbal and songs.] *Acta Biophysica Sinica* 6: 443–448. [In Chinese, English summary]

Feng D.-X., & Jiang J.-C. 1988. [Structures of the sound-producing organs of *Cryptotympana atrata* Fabricius: The median mediastinal membrane.] *Acta Entomologica Sinica* 31: 351–357. [In Chinese, English summary]

Feng D., & Nie S. 1993. [Timbre variations of the calling songs of the cicada *Cryptotympana atrata* Fabricius.] *Acta Biophysica Sinica* 9: 415–420. [In Chinese, English summary]

Feng D., & Shen J. 1990. [Song information and dual octamonic features of its spectra in the cicada *Oncotympana maculaticollis* Motschulsky.] *Acta Biophysica Sinica* 8: 463–469. [In Chinese, English summary]

Feng D., & Shen J. 1992. [Timbre and coding of calling songs of the cicada *Cryptotympana atrata* Fabricius.] *Acta Biophysica Sinica* 8: 8–13. [In Chinese, English summary]

Feng Y., Chen X.-M., Ye S.-D., Wang S.-Y., Chen Y., & Wang Z.-L. 1999. [Records of four species of edible insects in Homoptera and their nutritious elements analysis.] *Forest Research, Beijing* 12: 515–518. [In Chinese, English summary]

Fenton, M.B., D.H.M. Cumming, J.M. Hutton, & C.M. Swanepoel. 1987. Foraging and habitat use by *Nycteris grandis* (Chiroptera: Nycteridae) in Zimbabwe. *Journal of Zoology* 211: 709–716.

Ferenca, R. 1996. [Some rare insect species in the environs of Kaunas.] Pages 147–149 *in* V. Jonaitis, ed. *Lietuvos entomologu darbai. Lietuvos entomologu diagijos 10-meciui*. Lietuvos Entomologu Draugija, Ekologijos Institutos, Vilnius. [In Russian]

Ferreira, R.S., C. Aoki, & F.L. Souza. 2004. Levantamento de plantas hospedeiras de cigarras (Hemiptera: Cicadidae) em Campo Grande, MS. *XXV Congresso Brasileiro de Zoologia, 2004, Brasília. Livro de Resumos do XXV Congresso Brasileiro de Zoologia* 25: 125.

Fialho, M.D.A., L.M.L. Almeida, M. Schneider, & C.E.G. Pinheiro. 2004. Predação de *Quesada gigas* (Hemiptera, Cicadidae) por aves. *XXV Congresso Brasileiro de Zoologia, 2004, Brasília. Livro de Resumos do XXV Congresso Brasileiro de Zoologia* 25: 126.

Fitch, H.S. 1960. Autecology of the copperhead. *University of Kansas Publications Museum of Natural History* 13: 85–288.

Fitzgerald, B.M., & B.J. Karl. 1979. Foods of feral house cat (*Felis catus* L.) in forest of the Orongorongo Valley, Wellington. *New Zealand Journal of Zoology* 6: 110–125.

Fitzgerald, B.M., M.J. Meads, & A.H. Whitaker. 1986. Food of the kingfisher (*Halcyon sancta*) during nesting. *Notornis* 33: 23–32.

Fleming, C.A. 1966. More about cicadas. *Nature (Naturalists' Club Notes – Correspondance School, Department of Education)* 15: 1–7.

Fleming, C.A. 1975a. Cicadas (1). *New Zealand's Nature Heritage* 4: 1568–1572.

Fleming, C.A. 1975b. Cicadas (2). *New Zealand's Nature Heritage* 4: 1591–1595.

Fleming, C.A. 1977. The history of life in New Zealand forests. *New Zealand Journal of Forestry* 22: 249–262.

Fleming, C.A. 1984. The cicada genus *Kikihia* (Hemiptera: Homoptera): I. The New Zealand green foliage cicadas. *National Museum of New Zealand Records* 2: 191–206.

Fletcher, N.H. 1992. *Acoustic systems in Biology*. Oxford University Press, New York. 333pp.

Fletcher, N.H. 2004. A simple frequency-scaling rule for animal communication. *Journal of the Acoustical Society of America* 115: 2334–2338.

Flook, P.K., & C.H.F. Rowell. 1998. Inferences about orthopteroid phylogeny and molecular evolution from small subunit nuclear ribosomal DNA sequences. *Insect Molecular Biology* 7: 163–178.

Flory, S.L., & W.B. Mattingly. 2008. Response of host plants to periodical cicada oviposition damage. *Oecologia* 156: 649–656.

Flowers, D. 1991. Cicadas! Remember the summer of 1990? *Brookfield Zoo Bison* 5: 22–24

Fonseca, F.V., J.R. Silva, R.I. Samuels, R.A. DaMatta, W.R. Terra, & C.P. Silva. 2010. Purification and partial characterization of a midgut membrane-bound α-glucosidase from *Quesada gigas* (Hemiptera: Cicadidae). *Comparative Biochemistry and Physiology. Part B. Biochemistry & Molecular Biology* 155: 20–25.

Fonseca, P.J. 1991. Characteristics of the acoustic signals in nine species of cicadas (Homoptera, Cicadidae). *Bioacoustics* 3: 173–182.

Fonseca, P.J. 1993. Directional hearing in a cicada: biophysical aspects. *Journal of Comparative Physiology. A* 172: 767–774.

Fonseca, P.J. 1996. Sound production in cicadas: Timbal muscle activity during calling song and protest song. *Bioacoustics* 7: 13–31.

Fonseca, P.J., & H.C. Bennet-Clark. 1998. Asymmetry of tymbal action and structure in a cicada: A possible role in the production of complex songs. *Journal of Experimental Biology* 201: 717–730.

Fonseca, P.J., & T. Correia. 2007. Effects of temperature on tuning of the auditory pathway in the cicada *Tettigetta josei* (Hemiptera, Tibicinidae). *Journal of Experimental Biology* 210: 1834–1845.

Fonseca, P.J., & R.M. Hennig. 1996. Phasic action of the tensor muscle modulates the calling song in cicadas. *Journal of Experimental Biology* 199: 1535–1544.

Fonseca, P.J., & R.M. Hennig. 2004. Directional characteristics of the auditory system of cicadas: is the sound producing tymbal an integral part of directional hearing? *Physiological Entomology* 29: 400–408.

Fonseca, P.J., & A. Michelsen. 2008. The mechanisms of frequency discrimination in cicadas (Hemiptera, Cicadoidea): recent insights. *Abstracts of the XII Invertebrate Sound & Vibration Meeting, Tours, France, 27–30 October 2008*: 30.

Fonseca, P.J., D. Münch, & R.M. Hennig. 2000. Auditory perception: How cicadas interpret acoustic signals. *Nature* 405: 297–298.

Fonseca, P.J., & A.V. Popov. 1994. Sound radiation in a cicada: the role of different structures. *Journal of Comparative Physiology. A* 175: 349–361.

Fonseca, P.J., & A.V. Popov. 1997. Directionality of tympanal vibrations in a cicada: a biophysical analysis. *Journal of Comparative Physiology. A* 180: 417–427.

Fonseca, P.J., & M. A. Revez. 2002a. Song discrimination by male cicadas *Cicada barbara lusitanica* (Homoptera, Cicadidae). *Journal of Experimental Biology* 205: 1285–1292.

Fonseca, P.J., & M. A. Revez. 2002b. Temperature dependence of cicada songs (Homoptera, Cicadoidea). *Journal of Comparative Physiology. A. Sensory, Neural and Behavioral Physiology* 187: 971–976.

Fonseca, P.J., E.A. Serrão, F. Pina-Martins, P. Silva, S. Mira, J.A. Quartau, O.S. Paulo, & L. Cancela. 2008. The evolution of cicada songs contrasted with the relationships inferred from mitochondrial DNA (Insecta, Hemiptera). *Bioacoustics* 18: 17–34.

Fontaine, K.M, J.R. Cooley, & C. Simon. 2007. Evidence for paternal leakage in hybrid periodical cicadas (Hemiptera: *Magicicada* spp.). *PLoS ONE* 2(9): e892. 8pp. doi: 10.1371/journal.pone.0000892.

Forbes, S.A. 1888. On the present state of our knowledge concerning insect diseases. *Psyche* 5: 3–12.

Forbush, E.H. 1924. Gulls and terns feeding on the seventeen-year cicada. *Auk* 41: 468–470.

Fornazier, M.J., & N.M. Martinelli. 2000. Ocorrência de cigarras em café arábica na região de montanha do Estado do Espírito Santo. *1° Simpósio de Pesquisa dos Cafés do Brasil, 2000, Poços de Caldas. Anais do 1° Simpósio de Pesquisa dos Cafés do Brasil* 2: 1175–1177.

Fornazier, M. J., & A.C. da Rocha. 2000. Controle da cigarra do cafeeiro em regiões declivosas no estado do Espírito Santo. *1° Simpósio de Pesquisa dos Cafés do Brasil, 2000, Poços de Caldas. Anais do 1° Simpósio de Pesquisa dos Cafés do Brasil* 2: 1167–1170.

Forrest, T.G. 1991. Power output and efficiency of sound production in crickets. *Behavioral Ecology* 2: 327–338.

Forrest, T.G., & R. Raspert. 1994. Models of female choice in acoustic communication. *Behavioral Ecology* 5: 293–303.

Fox, K.J. 1982. Entomology of the Egmont National Park. *New Zealand Entomologist* 7: 286–289.

Freytag, P.H. 2000. Auchenorrhyncha on stamps. *Bio-Philately* 49: 55–56.

Frith, C.B. 1995. Cicadas (Insecta: Cicadidae) as prey of regent bowerbirds *Sericulus chrysocephalus* and other bowerbird species (Ptilonorhynchidae). *Sunbird* 25: 44–48.

Froeschner, R.C. 1952. A synopsis of the Cicadidae of Missouri (Homoptera). *Journal of the New York Entomological Society* 60: 1–13.

Frost, S.W. 1969. Supplement to Florida insects taken in light traps. *Florida Entomologist* 52: 91–101.

Fu G.-Q., Ma P.-C., Wu Q.-X., Wei J., Yan W., & Li L.-J. 2003. Inhibitory effects of ethanolic extracts from 196 kind of Chinese herbs on tyrosinase. *Zhonghua Pifuke Zazhi (Chinese Journal of Dermatology)* 36: 103–106.

Fuhrmann, M. 2010. Zum Vorkommen von Singzikaden (Insecta: Hemiptera, Cicadidae) im NSG Kahle Haardt bei Vöhl (Hessen). *Phillippia* 14: 135–137.

Fujiyama, I. 1982. Some fossil Cicadas from Neogene of Japan. *Bulletin of the National Science Museum, Series C (Geology & Paleontology)* 8: 181–187.

Fujiyama, S. 1996. Annual thermoperiod regulating an eight-year life-cycle of a periodical diplopod, *Parafontaria laminata armigera* Verhoeff (Diplopoda). *Pedobiologia* 40: 541–547.

Fullard, J.H., & B. Heller. 1990. Functional organization of the arctiid moth tymbal (Insecta, Lepidoptera). *Journal of Morphology* 204: 57–65.

Fullard, J.H., J.M. Ratclife, & D.S. Jacobs. 2008. Ignoring the irrelevant: auditory tolerance of audible but innocuous sounds in the bat-detecting ears of moths. *Naturwissenschaften* 95: 241–245.

Furlow, B. 2000. Wake-up call: Wnat to keep track of the passing years? Just follow the trees. *New Scientist* 167: 24.

Gäde, G. 2005. Adipokinetic hormones and the hormonal control of metabolic activity in Hemiptera. *Pestycydy* 4: 49–54.

Gäde, G., L. Auerswald, R. Predel, & H.G. Marco. 2004. Substrate usage and its regulation during flight and swimming in the backswimmer, *Notonecta glauca. Physiological Entomology* 29: 84–93.

Gäde, G., & M. P.-E. Janssens. 1994. Cicadas contain novel members of the AKH/RPCH family peptides with hypertrehalosaemic activity. *Biological Chemistry Hoppe-Seyler* 375: 803–809.

Galacgac, E.S., & C.M. Balisacan. 2009. Traditional weather forecasting for sustainable agroforestry practices in Ilocos Norte Province, Phillippines. *Forest Ecology and Management* 257: 2044–2053.

Gallagher, E., P. O'Hare, R. Stephenson, G. Waite, & N. Vock. 2003. *Macadamia problem solver & bug identifier*. Department of Primary Industries, Brisbane. 215pp.

Ganchev, T., & I. Potamitis. 2007. Automatic acoustic identification of singing insects. *Bioacoustics* 16: 281–328.

Gardenhire, R.Q. 1959. Summary of insect conditions in Iran 1958. *Applied Entomology and Phytopathology* 18: 51–61.

Garrick, R.C. 2010. Montane refuges and topographic complexity generate and maintain invertebrate biodiversity: recurring themes across space and time. *Journal of Insect Conservation* 15: 469–478.

Gebicki, C. 1987. Leafhopper associations (Homoptera, Auchenorrhyncha) in xerothermic communities in the vicinity of Pinczów. *Acta biologica Silesiana* 6: 87–98.

Genaro, J.A., & C. Juarrero de Varona. 1998. Comportamiento de *Sphecius hogardii* durnate la nidificación (Hymenoptera: Sphecidae). *Caribbean Journal of Science* 34: 323–324.

Genov, P.V., & G. Massei. 1997. Larve di cicala (*Cicada orni*) nella dieta del cinghiale (*Sus scrofa*): dati preliminari. *Supplemento Alle Ricerche di Biologia della Selvaggina* 27: 561–564.

Gerhardt, H.C., & F. Huber. 2002. *Acoustic communication in insects and anurans: common problems and diverse solutions.* University of Chicago Press, Chicago. 531pp.

Gess, H. 1981. Cicadas of the Grahamstown area. *Redwing* 1981: 3–10.

Gil, R., A. Latorre, & A. Moya. 2010. Evolution of prokayote-animal symbiosis from a genomics perspective. Pages 207–233 *in* J.H.P. Hackstein, ed. *(Endo)symbiotic Methanogenic Archaea.* Microbiology Monographs 19. Springer-Verlag, Berlin Heidelberg.

Gilbert, C., & C. Klass. 2006. Decrease in geographic range of the finger Lakes Brood (Brood VII) of the periodical cicada (Hemiptera: Cicadidae: *Magicicada* spp.). *Journal of the New York Entomological Society* 114: 78–85.

Gill, B.J., R.G. Powlesland, & M.H. Powlesland. 1983. Laying seasons of three insectivorous song-birds at Kowhai Bush, Kaikoura. *Notornis* 30: 81–85.

Gillett, M., & C. Gillett. 2005. Insects & other arthropods. Pages 169–194 *in* P. Hellyer & S. Aspinall, eds. *The Emirates – a natural history*. Trident Press Ltd., London.

Gillooly, J.F., & A.G. Ophir. 2010. The energetic basis of acoustic communication. *Proceedings of the Royal Society. Series B. Biological Sciences* 277: 1325–1331.

Girgenrath, M., & R.L. Marsh. 1997. *In vivo* performance of trunk muscles in tree frogs during calling. *Journal of Experimental Biology* 200: 3101–3108.

Girgenrath, M., & R.L. Marsh. 1999. Power output of sound-producing muscles in the tree frogs *Hyla versicolor* and *Hyla chrysoscelis*. *Journal of Experimental Biology* 202: 3225–3237.

Glare, T.R., M. O'Callaghan, & P.J. Wigley. 1993. Checklist of naturally occurring entomopathogenic microbes and nematodes. *New Zealand Journal of Zoology* 20: 95–120.

Glick, P.A. 1923. Insects injurious to Arizona crops during 1922. *14th Annual Report of the Arizona Commission of Agriculture and Horticulture* 1921–22: 55–77.

Glinski, R.L., & R.D. Ohmart. 1983. Breeding ecology of the Mississippi Kite in Arizona. *Condor* 85: 200–207.

Glinski, R.L., & R.D. Ohmart. 1984. Factors of reproduction and population densities in the Apache cicada (*Diceroprocta apache*). *Southwestern Naturalist* 29: 73–79.

Goble, T.A., B.W. Price, N.P. Barker, & M.H. Villet. 2008a. Preliminary molecular phylogenetics and inferred historical biogeography of three African cicada genera of the tribe Platypleurini (Hemiptera: Cicadidae). *23rd International Congress of Entomology, Durban, 6–12 July 2008*. Abstract number 673.

Goble, T.A., B.W. Price, N.P. Barker, & M.H. Villet. 2008b. Implications of a preliminary molecular investigation of four African cicada genera in the tribe Parnisini (Hemiptera: Cicadidae). *23rd International Congress of Entomology, Durban, 6–12 July 2008*. Abstract number 674.

Goemans, G. 2010. A historical overview of the classification of the Neotropical tribe Zammarini (Hemiptera, Cicadidae) with a key to the genera. *ZooKeys* 43: 1–13.

Gogala, A., M. Aljancic, M. Gogala, & I. Sivec. 1994. *Insects, the flourishing multitudes*. Prirodoslovni muzej Slovenije, Ljubljana, Slovenia. 71pp. (T.E. Moore, English ed.)

Gogala, M. 1991a. Sedemnajstletni in trinajstletni škržatki. *Proteus* 54: 3–10.

Gogala, M. 1991b. Nove ugotovitbe o sedemnajstletnih in trinajstletnih škržatki. *Proteus* 54: 74–75.

Gogala, M. 1994a. Z mikrofonom, mrežo in fotoaparatom v deželi belega slona. *Proteus* 56: 31–37.

Gogala, M. 1994b. Sosledje zvokov v Belumskem pragozdu. *Proteus* 57: 3–9.

Gogala, M. 1995. Songs of four cicada species from Thailand. *Bioacoustics* 6: 101–116.

Gogala, M. 1998a. Use of acoustic methods to find, locate and recognize singing cicadas in Slovenia, Croatia and Macedonia. *Bioacoustics* 9: 156–157.

Gogala, M. 1998b. Pojoci skrzati Slovenije [The songs of cicadas of Slovenia.] *Proteus* 60: 392–399.

Gogala, M. 2002. Gesänge der Singzikaden aus Südost- und Mittel-Europa. *Denisia* 4: 241–248.

Gogala, M. 2006a. Acoustic website on European singing cicadas. *Razprave IV. Razreda Sazu (Ljubljana)* 47: 155–164.

Gogala, M. 2006b. Neue Erkenntnisse uber die Systematik der *Cicadetta montana* Gruppe (Auchenorrhyncha: Cicadoidea: Tibicinidae). *Beitrage Entomologische Berlin* 56: 369–376.

Gogala, M. 2008. Scopolijeva "Cicada montana" je do danes dobila več kot deset sestrskih vrst škržadov. *Idrijski Razgledi* 53: 53–58.

Gogala, M. 2010. Sporazumevanje živali s pomočjo zvoka, Acoustic communication in animals. Pages 79–83 *in* S. Strgulc-Krajšek & M. Vičar, eds. *Organizmi kot živi sistemi: Mednarodni posvet Biološka znanost in družba. Ljublajana, 21–22 oktober 2010, Zbornik prispevkov*. Zavod RS za šolstvo, Ljubljana.

Gogala, M., & S. Drosopoulos. 2006. Song of *Cicadetta flaveola* Brullé (Auchenorrhyncha: Cicadoidea: Tibicinidae) from Greece. *Russian Entomological Journal* 15: 275–278.

Gogala, M., S. Drosopoulos, & T. Trilar. 2008a. *Cicadetta montana* complex (Hemiptera, Cicadidae) in Greece – a new species and new records based on bioacoustics. *Mitteilungen aus dem Museum für Naturkunde im Berlin, Deutsche Entomologishe Zeitschrift* 55: 91–100.

Gogala, M., S. Drosopoulos, & T. Trilar. 2008b. Present status of mountain cicadas *Cicadetta montana* (sensu lato) in Europe. *Bulletin of Insectology* 61: 123–124.

Gogala, M., S. Drosopoulos, & T. Trilar. 2009. Two mountains, two species: new taxa of the *Cicadetta montana* species complex in Greece (Hemiptera: Cicadidae). *Acta Entomologica Slovenica* 17: 13–28.

Gogala, M., & A. Gogala. 1999. A checklist and provisional atlas of the Cicadoidea fauna of Slovenia (Homoptera: Auchenorrhyncha). *Acta Entomologica Slovenica* 7: 119–128.

Gogala, M., & A.V. Popov. 1997a. Bioacoustics of singing cicadas of the western Palaearctic: *Cicadetta mediterranea* Fieber 1876 (Cicadoidea: Tibicinidae). *Acta Entomologica Slovenica* 5: 11–24.

Gogala, M., & A.V. Popov. 1997b. Verleich der Gesänge der Singzikaden *Cicadivetta tibialis, Cicadetta mediterranea* und *Tettigetta brullei*. *Beiträge zur Zikadenkunde* 1: 34–36.

Gogala, M., & A.V. Popov. 2000. Bioacoustics of singing cicadas of the western Palaearctic: *Tettigetta dimissa* (Hagen) (Cicadoidea: Tibicinidae). *Acta Entomologica Slovenica* 8: 7–20.

Gogala, M., A.V. Popov, & D. Ribaric. 1996. Bioacoustics of singing cicadas of the western Palaearctic: *Cicadetta tibialis* (Panzer) (Cicadoidea: Tibicinidae). *Acta Entomologica Slovenica* 4: 45–62.

Gogala, M., & K. Riede. 1995. Time sharing of song activity by cicadas in Temengor Forest Reserve, Hulu Perak, and in Sabah, Malaysia. *Malayan Nature Journal* 48: 297–305.

Gogala, M., & W. Schedl. 2008. A new *Tettigetta* species from Iran and its song (Hemiptera: Cicadidae, Cicadettinae). *Advances in Arachnology and Developmental Biology Monographs* 12: 395–404.

Gogala, M., & G. Seljak. 2003. Škžadi in škržatki - Auchenorrhyncha. Pages 347–354 *in* B. Sket, M. Gogala, & V. Kuštor, eds. *Živalstvo Slovenije*. 1. natis. Tehniška založba Slovenije, Ljubljana.

Gogala, M., & T. Trilar. 1998a. First record of *Cicadatra persica* Kirkaldy, 1909 from Macedonia, with description of its song. *Acta Entomologica Slovenica* 6: 5–15.

Gogala, M., & T. Trilar. 1998b. Gibt es verschiedene '*Cicadetta montana*' - Arten in Slowenien? Eine bioakustische Untersuchung (Homoptera, Auchenorrhyncha). *Beiträge zur Zikadenkunde* 2: 67–68.

Gogala, M., & T. Trilar. 1999a. The song structure of *Cicadetta montana macedonica* Schedl with remarks on songs of related singing cicadas (Homoptera: Auchenorrhyncha: Cicadomorpha: Tibicinidae). *Reichenbachia, Staatliches Museum für Tierkunde Dresden* 33: 91–97.

Gogala, M., & T. Trilar. 1999b. The importance of acoustic data in the systematics of singing cicadas, with examples from S.E. Europe and S.E. Asia. *Abstracts of Papers and Posters, 10th International Auchenorrhyncha Congress, 6–10 September 1999, Cardiff, Wales.* 1p (no pagination).

Gogala, M., & T. Trilar. 2000. Sound emissions of *Pagiphora annulata* (Homotera: Cicadoidea: Tibicinidae) - A preliminary report. *Acta Entomologica Slovenica* 8: 21–26.

Gogala, M., & T. Trilar. 2001a. Wing clicking in songs of the *Cicadatra* and *Pagiphora* cicadas from SE Europe. *Program and Book of Abstracts, 2nd European Hemiptera Congress, Fiesa, Slovenia, 20–24 June 2001*: 28.

Gogala, M., & T. Trilar. 2001b. Studies of cicada sound production using commercial HS video. *Program & Abstracts, XVIII International Bio-Acoustic Council, Università degli Studi di Pavia, Cogne, 3–6 September 2001*: 14.

Gogala, M., & T. Trilar. 2002. The *Cicadetta montana* complex (Hemiptera, Cicadoidea, Tibicinidae). *Eleventh International Auchenorrhyncha Congress, August 5–9, Potsdam, Germany, Abstracts of Talks and Posters*: 39–40.

Gogala, M., & T. Trilar. 2003. Video analysis of wing clicking in cicadas of the genera *Cicadatra* and *Pagiphora* (Homotera: Auchenorrhyncha; Cicadoidea. *Acta Entomologica Slovenica* 11: 5–15.

Gogala, M., & T. Trilar. 2004a. Bioacoustic investigations and taxonomic consideration on the *Cicadetta montana* complex (Homotera: Cicadoidea: Tibicinidae). *Anais da Academia Brasileira Ciências* 76: 316–324.

Gogala, M., & T. Trilar. 2004b. Present views and news about the Cicadetta montana complex (Auchenorrhyncha: Cicadoidea: Tibicinidae). *Third European Hemiptera Congress, St. Petersburg, June 8–11, 2004, Abstracts*: 35–36.

Gogala, M., & T. Trilar. 2005. Biodiversity of cicadas in Malaysia – a bioacoustic approach. *Serangga* 9: 63–81.

Gogala, M., & T. Trilar. 2006. Kompleks vrst gorskega skrzada (*Cicadetta montana*) v Evropi. Mountain cicada (*Cicadetta montana*) species complex in Europe. *1. Slovenski Entomologki Simpozij, Knjiga Povzetkov/1st Slovenian Entomological Symposium, Book of Abstracts, Ljubljana, 4–5 November 2006*: 16–17.

Gogala, M., & T. Trilar. 2007. Description of the song of *Purana metallica* group from Thailand and *P. latifascia* from Borneo (Hemiptera, Cicadidae). *Tijdschrift voor Entomologie* 150: 389–400.

Gogala, M., & T. Trilar. 2008. Taxomonic value of song parameters – a case of mountain cicadas. *Abstracts of the XII Invertebrate Sound & Vibration Meeting, Tours, France, 27–30 October 2008*: 31.

Gogala, M., & T. Trilar. 2009a. Nova spoznanja o gorskih škržadih *Cicadetta montana* s. lato. New discoveries about mountain cicadas *Cicadetta montana* s. lato. *Drugi Slovenski Entomologški Simpozij, Knjiga Povzetkov/2nd Slovenian Entomological Symposium, Book of Abstracts, Ljubljana, 7–8 February 2009*: 32–33.

Gogala, M., & T. Trilar. 2009b. Distribution of cicadas of the *Cicadetta montana* species complex in Europe. *5th European Hemiptera Congress, 31 August-4 September 2009, Velence, Hungary – Abstracts*: 21.

Gogala, M., T. Trilar, U. Kozina, & H. Duffels. 2004. Frequency modulated song of the cicada *Maua albigutta* (Walker 1856) (Auchenorrhyncha: Cicadoidea) from South East Asia. *Scopolia* No. 54: 1–16.

Gogala, M., T. Trilar, & V.T. Krpač. 2005. Fauna of singing cicadas (Auchenorrhyncha: Cicadoidea) of Macedonia – a bioacoustic survey. *Acta Entomologica Slovenica* 13: 103–126.

Goldberg, J., S.A. Trewick, & A.M. Paterson. 2008. Evolution of New Zealand's terrestrial fauna: a review of molecular evidence. *Philosophical Transactions of the Royal Society of London, Series B. Biological Sciences* 363: 3319–3334.

Goldstein, B. 1929. A cytological study of the fungas *Massospora cicadina*, parasitic on the seventeen-year Cicada, *Magicicada septendecim*. *American Journal of Botany* 16: 394–401.

Goles, E., O. Schulz, & M. Markus. 2000. A biological generator of prime numbers. *Nonlinear Phenomena in Complex Systems* 3: 208–213.

Goles, E., O. Schulz, & M. Markus. 2001. Prime number selection of cycles in a predator-prey model. *Complexity* 6: 33–38.

Gomez-Menor, J. 1951. Homopteros. *Graellsia* 9: 1–13.

Gonçalves, W., & A.M. Faria. 1989. Inseticidas sistêmicos granulados no controle das ninfas móveis das cigarras e seus efeitos na produtividade de cafeeiros. *Bragantia, Campinas* 48: 95–108.

González, M.I., M. Alvarado, J.M. Duran, A. Serrano, & A. De La Rosa. 1998. Estudios sobre *Cicada* sp. (Homoptera: *Cicadidae*) en olivo. *Boletín Sanidad Vegetal y Plagas* 24: 803–816.

González, R.H. 1989. *Insectos y acaros de importancia agrícola y cuarentenaria en Chile*. Editora Ograma, S.A. Universidad de Chile. 310pp.

Goodchild, A.J.P. 1966. Evolution of the alimentary canal in the Hemiptera. *Biological Review* 41: 97–140.

Goode, J. 1980. *Insects of Australia with illustrations from the classic The Insects of Australia and New Zealand by R.J. Tillyard*. Angus & Robertson Publishers, Sydney. 260pp.

Goodwin, S. 1986. Pests of grapes. *Agfacts* No. H7.AE.1: 1–12.

Gordon, D.G. 1998. *The eat-a-bug cookbook*. Ten Speed Press, Berkeley, CA. 101pp.

Goruch, C.S. 1998. Periodical cicadas. *Clemson Cooperative Extension Service, Entomology Insect Information Series, EIIS/TO-10*. 2pp.

Gortner, R.A. 1911a. Studies on Melanin. III. The inhibitory action of certain phenolic substances upon tyrosinase. A suggestion as to the cause of dominant and recessive whites. *Journal of Biological Chemistry* 10: 113–122.

Gortner, R.A. 1911b. A variant in the periodical cicada. *Science* 34: 153.

Gosalbes, M.J., A. Latorre, A. Lamelas, & A. Moya. 2010. Genomics of intracellular symbionts in insects. *International Journal of Medical Microbiology* 300: 271–278.

Gosper, C.R. 1999. Birds foraging for cicadas at Primbee, NSW. *Australian Birds* 31: 142–144.

Gourley, S.A., & Y. Kuang. 2009. Dynamics of a neutral delay equation for an insect population with long larval and short adult phases. *Journal of Differential Equations* 246: 4653–4669.

Grant, J.A., & L.K. Ward. 1992. English Nature Species Recovery Programme – New Forest Cicada (*Cicadetta montana* Scopoli) (Hemiptera: Cicadidae). Institute of Terrestrial Ecology, Natural Environment Research Council, Project T09056K1, Furzebrook Research Station, Wareham. 58pp.

Grant, P. 2005. The priming of periodical cicada life cycles. *Trends in Ecology and Evolution* 20: 169–174.

Grant, P.R. 2006. Opportunistic predation and offspring sex ratios of cicada-killer wasps (*Sphecius speciosus* Drury). *Ecological Entomology* 31: 539–547.

Grant-Taylor, T.L. 1966. Cicadas. Pages 347–349 *in* A.H. McLintock, ed. *Encyclopaedia of New Zealand*, Volume 1. R.E. Owen, Wellington.

Greaves, S.N.J., D.G. Chapple, D.M. Gleeson, C.H. Daugherty, & P.A. Ritchie. 2007. Phylogeography of the spotted skink (*Oligosoma lineoocellatum*) and green skink (*O. chloronoton*) species complex (Lacertilia: Scincidae) in New Zealand reveals pre-Pleistocene divergence. *Molecular Phylogenetics and Evolution* 45: 729–739.

Grechkin, V.P. 1956. [Some important injurious insects in the mountain forests of Tajikistan.] *Zoologicheski Zhurnal* 35: 1476–1492. [In Russian]

Green, O.R. 1977. Field trip to Mt. Egmont. *Weta* 1: 15–16.

Greenfield, M.D. 1994. Cooperation and conflict in the evolution of signal interactions. *Annual Review of Ecology and Systematics* 25: 97–126.

Greenfield, M.D. 2005. Mechanisms and evolution of communal sexual displays in arthropods and anurans. *Advances in the Study of Behavior* 35: 1–62.

Greenfield, M.D., & J. Schul. 2008. Mechanisms and evolution of synchronous chorusing: emergent properties and adaptive functions in *Neoconocephalus* katydids (Orthoptera: Tettigoniidae). *Journal of Comparative Psychology* 122: 289–297.

Grimaldi, D., & M.S. Engel. 2005. *Evolution of the insects*. Cambridge University Press, Cambridge. 755pp.

Guglielmino, A., V. D'Urso, & A. Alma. 2000. Auchenorrhyncha (Insecta, Homoptera) from Sardinia (Italy): a faunistic, ecological and zoogeographical contribution. *Mitteilungen aus dem Museum für Naturkunde im Berlin, Deutsche Entomologishe Zeitschrift* 47: 161–172.

Guindon, S., & O. Gascuel. 2002. Efficient biased estimation of evolutionary distances when substitution rates vary among sites. *Molecular Biology & Evolution* 19: 534–543.

Günthart, H., & R. Mühlenthaler. 2002. Provisorische Checklist der Zikaden der Schweiz (Insecta: Hemiptera, Auchenorrhyncha). *Denisia* 4: 329–338.

Gwynne, D.T. 1987. Sex-biased predation and the risky mate-locating behaviour of male tick-tock cicadas (Homoptera: Cicadidae). *Animal Behaviour* 35: 571–576.

Gwynne, D.T., P. Yeoh, & A. Schatral. 1988. The singing insects of King's Park and Perth Gardens (Australia). *Western Australian Naturalist* 17: 25–71.

Hadley, N.F. 1994. *Water relations of terrestrial arthropods*. Academic Press, San Diego. 356pp.

Hadley, N.F., M.C. Quinlan, & M.L. Kennedy. 1991. Evaporative cooling in the desert cicada: thermal efficiency and water/metabolic costs. *Journal of Experimental Biology* 159: 269–283.

Hadley, N.F., E.C. Toolson, & M.C. Quinlan. 1989. Regional differences in cuticular permeability in the desert cicada, *Diceroprocta apache*: implications for evaporative cooling. *Journal of Experimental Biology* 141: 219–230.

Hagen, H.R. 1929. The fly-insect situation in the Smyrna fig district. *Journal of Economic Entomology* 22: 900–909.

Hahn, D.E. 1962. *A list of the designated type specimens in the Macleay Museum. INSECTA*. The University of Sydney, The Macleay Museum. 184p.

Hahus, S.C., & K.G. Smith. 1990. Food habits of *Blarina*, *Peromyscus*, and *Microtus* in relation to an emergence of periodical cicadas (*Magicicada*). *Journal of Mammalogy* 71: 249–252.

Haitlinger, R. 1990. Four new species of *Leptus* Latrielle, 1796 (Acari, Prostigmata, Erythraeidae) from insects of Australia, New Guinea and Asia. *Wiadomosci Parazytologiczne* 36: 47–53.

Halkka, O. 1959. Chromosome studies on the Hemiptera, Homoptera, Auchenorrhyncha. *Annales Academiae Scientiarum Fennicae. Series A. IV. Biologica* 43: 1–72.

Hamann, H.H.F. 1960. Der Monchgraben vor dem Bau der Autobahn. *Naturkundliches Jahrbuch der Stadt Linz* 6: 113–244.

Hamasaka, Y., C.J. Mohrherr, R. Predel, & C. Wegener. 2005. Chronobiological analysis and mass spectrometric characterization of pigment-dispensing factor in the cockroach *Leucophaea maderae*. *Journal of Insect Science* 5:43. 13pp. Available online: insectscience.org/5.43

Hamilton, K.G.A. 1981. Morphology and evolution of the Rhynchotan head (Insecta: Hemiptera, Homoptera). *Canadian Entomologist* 113: 953–974.

Hamilton, K.G.A. 1985. Comment on the proposed conservation of *Tibicina* Amyot 1847 and *Lyristes* Horváth 1926. Z.N. (S.)239. *Bulletin of Zoological Nomenclature* 42: 211.

Hamilton, K.G.A. 1999. The ground-dwelling leafhoppers Myerslopiidae, new family, and Sagmatiini, new tribe (Homptera: Membracoidea). *Invertebrate Taxonomy* 13: 207–235.

Handke, K. 2004a. Artenliste "Fauna". Pages 148–152 *in* M. Koch & C. Dobes, eds. *Botanisches Geländepraktikum (HP-F): "Wandel mediterraner Vegetation – Istrien/Kroatien"*. Heidelberger Institut für Pflanzenwissenschaften, Biodiversität und Pflanzensystematik, Universität Heidelberg, Heidelberg.

Handke, K. 2004b. Gesamtartenliste "Fauna". Pages 153–159 *in* M. Koch & C. Dobes, eds. *Botanisches Geländepraktikum (HP-F): "Wandel mediterraner Vegetation – Istrien/Kroatien"*. Heidelberger Institut für Pflanzenwissenschaften, Biodiversität und Pflanzensystematik, Universität Heidelberg, Heidelberg.

Hangay, G., & P. German. 2000. *Insects of Australia*. Reed New Holland, Sydney. 128pp.

Hanson, J. 1956. Studies on the cross-striation of the indirect flight myofibrils of the blowfly Calliphora. *The Journal of Biophysical and Biochemical Cytology* 2: 691–710, plates 179–182.

Hara, S., N. Hayashi, S. Hirano, X.-N. Zhong, S. Yasuda, & H. Komae. 1990. Determination of germanium in some plants and animals. *Zeitschrift für Naturforschung Section C Biosciences* 45: 1250–1251.

Harford, B. 1958. Observations of some habits and early development of *Melampsalta cingulata* (Fabr.). Order Homoptera. Family Cicadidae. *New Zealand Entomologist* 2: 4–11.

Harris, A.C. 1993. Unseasonal occurrence of *Kikihia subalpina* (Hudson) (Hemiptera: Homoptera: Cicadidae) adults in Dunedin during June, 1993. *The Weta* 16: 8–9.

Hartbauer, M. 2008. Chorus model of the synchronizing bushcricket species *Mecopoda elongata*. *Ecological Modelling* 213: 105–118.

Hartemink, A.E., & L.S. Kuniata. 1996. Some factors influencing yield trends of sugarcane in Papua New Guinea. *Outlook on Agriculture* 25: 227–234.

Harvey, P.M., & M.B. Thompson. 2006. Energetics of emergence in the cicadas, *Cyclochila australasiae* and *Abricta curvicosta* (Homoptera: Cicadidae). *Journal of Insect Physiology* 52: 905–909.

Harvie, A.E. 1973. Diet of the opossum (*Trichosurus vulpecula* Kerr) on farmland northeast of Waverly, New Zealand. *Proceedings of the New Zealand Ecological Society* 20: 48–52.

Harwood, J.D., & J.J Obrycki. 2004. After 17 years.......Brood X is back. *Antennae* 28: 235–238.

Hasselt, A.W.M. von. 1881. Studien over de klank-organen, den zang en den schreeuw der Cicaden. *Tijdschrift voor Entomologie* 25: 179–212.

Hastings, J. 1986. Provisioning by female Western Cicada killer wasps, *Sphecius grandis* (Hymenoptera: Sphecidae): Influence of body size and emergence time on individual provisioning success. *Journal of the Kansas Entomological Society* 59: 262–268.

Hastings, J. 1989a. Protandry in Western Cicada killer wasps, (*Sphecius grandis*, Hymenoptera: Sphecidae): an empirical study of emergence time and mating opportunity. *Behavioral Ecology and Sociobiology* 25: 255–260.

Hastings, J. 1989b. Thermoregulation in the dog-day cicada, *Tibicen duryi* (Homoptera: Cicadidae). *Transactions of the Kentucky Academy of Sciences* 50: 145–149.

Hastings, J. 1990. Sexual-size dimorphism in western Cicada Killer wasps, *Sphecius grandis* (Hymenoptera: Sphecidae). *Transactions of the Kentucky Academy of Sciences* 51: 1–5.

Hastings, J.M., C.W. Holliday, & J.R. Coelho. 2008. Body size relationship between *Sphecius speciosus* (Hymentoptera: Crabronidae) and their prey: prey size determines wasp size. *Florida Entomologist* 91: 657–663.

Hastings, J.M., C.W. Holliday, A. Long, K. Jones, & G. Rodriguez. 2010. Size-specific provisioning by cicada killers, *Sphecius speciosus*, (Hymenoptera: Crabroniidae) in north Florida. *Florida Entomologist* 93: 412–421.

Hastings, J.M., & E.C. Toolson. 1991. Thermoregulation and activity patterns of two syntopic cicadas, *Tibicen chiricahua* and *T. duryi* (Homoptera: Cicadidae), in central New Mexico. *Oecologia* 85: 513–520.

Hawkeswood, T.J. 2000. Some notes on the occurrence of the Australian beetle *Rhipicera femorata* (Kirby)(Coleoptera: Rhipiceridae). *Mauritiana* 17: 417–419.

Hayakawa, Y. 2007. Parasperm: morphological and functional studies on nonfertile sperm. *Ichthyological Research* 54: 111–130.

Hayashi, M. 1982a. A new species of the Genus *Pycna* from the Himalayas (Homoptera: Cicadidae). *Pacific Insects* 24: 78–83.

Hayashi, M. 1982b. Notes on the distribution of Cicadidae (Insecta, Homoptera) in the Fuji-Hakone-Izu area. *Memoirs of the National Science Museum, Tokyo* 15: 187–194.

Hayashi, M. 1984. [A review of the Japanese Cicadidae.] *Cicada* 5: 25–75. [In Japanese]

Hayashi, M. 1987a. A revision of the Genus *Cryptotympana* (Homoptera, Cicadidae) part I. *Bulletin of the Kitakyushu Museum of Natural History* 6: 119–212.

Hayashi, M. 1987b. A revision of the genus *Cryptotympana* (Homoptera, Cicadidae). Part II. *Bulletin of the Kitakyushu Museum of Natural History* 7: 1–109.

Hayashi, M. 1993. [Cicadidae of SE Asia.]. *Nature & Insects* 28: 13–19. [In Japanese]

Hayashi, M. 1998. [Correct usage to the generic name of '*Baeturia*' *kuroiwae* (Cicadidae, Tibicinidae).] *Cicada* 13: 62. [In Japanese]

Hayashi, M., S. Kamitani, & N. Ohara. 2010. Auchenorhyncan biodiversity in Japan. *Thirteenth International Auchenorrhyncha Congress and the 7th International Workshop on Leafhoppers and Planthoppers of Economic Significance, 28 June-2 July, Vaison-la-Romaine, France, Abstracts: Talks and Posters*: 116–117.

Hayashi, M., & M.S. Moulds. 1999. Observations on the eclosion of the hairy cicada *Tettigarcta crinita* Distant (Homoptera: Cicadoidea: Tettigarctidae). *Australian Entomologist* 25: 97–101.

Hayashi, M., T. Sasaki, & Y. Saisho. 2000. [Additional biological notes on *Euterpnosia chibensis daitoensis*.] *Cicada* 15: 51–54. [In Japanese, English summary]

Hayashi, T., S. Yasui, & K. Sato. 1996. [Predation on nymphs of the evening cicada *Tanna japonensis* by the grey-faced buzzard *Butastur indicus*.] *Japanese Journal of Ornithology* 45: 39–40. [In Japanese, English summary]

Hayes, B. 2004. Bugs that count. *American Scientist* 92: 401–405.

Haynes, K.F., & K.V. Yeargan. 1999. Exploitation of intraspecific communication systems: illicit signalers and receivers. *Annals of the Entomological Society of America* 92: 960–970.

Haywood, B.T. 2005. *Notes on the distribution, lifehistory and conservation status of cicadas (Hemiptera: Homoptera) in the south east of South Australia*. Wildlife Advisory Committee, Department for Environment and Heritage, Adelaide. 60pp.

Haywood, B.T. 2006a. A study of the cicadas (Hemiptera: Homoptera) in the south east of South Australia – Part I. *South Australian Naturalist* 80: 12–23.

Haywood, B.T. 2006b. A study of the cicadas (Hemiptera: Homoptera) in the south east of South Australia – Part II. *South Australian Naturalist* 80: 48–53.

Haywood, B.T. 2007. A study of the cicadas (Hemiptera: Homoptera) in the south east of South Australia – Part III. *South Australian Naturalist* 80: 13–18.

He J. 1984. [Descriptions of two new genera and four new species of Cicadidae from China (Homoptera: Auchenorrhyncha).] *Contributions of the Shanghai Institute of Entomology* 4: 221–228. [In Chinese, English summary]

He, Z., & N. Chen. 1985. [The sound structure of five species of common chirp insects in Beijing.] *Acta Zoologica Sinica* 31: 324–330. [In Chinese, English summary]

Heads, M. 2001. Birds of paradise, biogeography and ecology in New Guinea: a review. *Journal of Biogeography* 28: 893–925.

Heads, M. 2002a. Birds of paradise, vicariance biogeography and terrane tectonics in New Guinea. *Journal of Biogeography* 29: 261–283.

Heads, M. 2002b. Regional patterns of biodiversity in New Guinea animals. *Journal of Biogeography* 29: 285–294.

Heads, M. 2005. Dating nodes on molecular phylogenies: a critique of molecular biogeography. *Cladistics* 21: 62–78.

Heady, S.E., & L.R. Nault. 1991. Acoustic signals of *Graminella nigrifrons* (Homoptera, Cicadellidae). *Great Lakes Entomologist* 24: 9–16.

Hearn, L. 1971. *Shadowings*. Charles E. Tuttle Company, Inc. Rutland, Vermont. 288pp.

Heath, J.E., & M.S. Heath. 1999. Contrasting physiological adaptation to climate by cicadas of Australia and New Zealand. *Abstracts of Papers and Posters, 10th International Auchenorrhyncha Congress, 6–10 September 1999, Cardiff, Wales.* 1p (no pagination).

Heath, J.E., M.S. Heath, A. Sanborn, & F. Noriega. 1990. Thermoregulation in relation to altitude among arid land cicadas. *Abstracts of the Combined Meeting of the 7th International Auchenorrhyncha Congress and the 3rd International Workshop on Leafhoppers and Planthoppers of Economic Importance, Wooster, Ohio, August 13–17.*

Heath, J.E., A. Sanborn, M.S. Heath, & P. Phillips. 1999. Temperature responses and habitat selection by mangrove cicadas in Florida and Queensland. *Abstracts of Papers and Posters, 10th International Auchenorrhyncha Congress, 6–10 September 1999, Cardiff, Wales.* 1p (no pagination).

Heath, M.S. 1983. The semiarid regions of Argentina: a natural laboratory for the study of evolution. Pages 11–15 *in* L.F.C. Mendez & L.W. Bates, eds. *Brazil and Rio de La Plata, challenge and response*. Eastern Illinois University Press, Charleston.

Heath, M.S., & J.E. Heath. 1994. *Okanagana* type localities revisited (Homoptera: Cicadidae). *Tymbal* 20: 16–17.

Heath, M.S., J.E. Heath, & F.G. Noriega. 1990. Geographical distribution of Argentine cicadas. *Abstracts of the Combined Meeting of the 7th International Auchenorrhyncha Congress and the 3rd International Workshop on Leafhoppers and Planthoppers of Economic Importance, Wooster, Ohio, August 13–17.*

Heath, M.S., J.E. Heath, A.F. Sanborn, & F.G. Noriega. 1988. Convergence and parallelism among salt tolerant cicadas of Argentina and the southwestern United States. *Proceedings of the Second International Congress of Comparative Physiology and Biochemistry*: 304.

Heath, M.S., J.E. Heath, A.F. Sanborn, & F.G. Noriega. 1999. Convergence and parallelism among cicadas of Argentina and the Southwestern United States. *Abstracts of Papers and Posters, 10th International Auchenorrhyncha Congress, 6–10 September 1999, Cardiff, Wales.* 1p (no pagination).

Heath, M.S., & A.F. Sanborn. 2007. A new species of cicada of the genus *Okanagana* (Hemiptera: Cicadoidea: Cicadidae) from Arizona. *Annals of the Entomological Society of America* 100: 483–489.

Heckel, P.F., & T.C. Keener. 2007. Sex differences noted in mercury bioaccumulaton in *Magicicada cassini*. *Chemosphere* 69: 79–81.

Heckel, P.F., T.C. Keener. S.-J. Khang, & J.-Y. Lee. 2005. Periodical cicadas as indicators of ecological health. *Proceedings of the Air & Waste Management Association's 98th Annual Conference and Exhibition, Minneapolis, MN, June 21–25, 2005*: 14pp.

Heckel, P.F., J.-Y. Lee, & T.C. Keener. 2007. Comparison of analytical techniques to determine the mercury content of 17-year periodical cicadas. *American Laboratory, News Edition* 39: 16–17.

Heenan, P.B., A.D. Mitchell, P.J. de Lange, J. Keeling, & A.M. Paterson. 2010. Late-Cenozoic origin and diversification of Chatham Islands endemic plant species revealed by analyses of DNA sequence data. *New Zealand Journal of Botany* 48: 83–136.

Heinrich, B. 1979. Keeping a cool head: honeybee thermoregulation. *Science* 205: 1269–1271.

Heinrich, B. 1996. *The thermal warriors: Strategies of insect survival.* Harvard University Press, Cambridge. 221pp.

Heinrich, B. 2003. Thermoregulation. Pages 431–437 *in* V.H. Resh & R.T. Cardé, eds. *Encyclopedia of insects.* Academic Press, New York.

Heinrich, W.O. 1967. Cicada - a coffee pest in Brazil. *World Crops* 19: 43–47.

Helfert, B., & K. Sanger. 1993. Final moulting of *Dundubia vaginata* in the Thale Ban National Park/Thailand [Homoptera: Cicadidae]. *Entomologica Generalis* 18: 37–41.

Heliövaara, K., & R. Väisänen. 1985. Pohjois-Amerikan 13- ja 17-vuotiset kaakaat. *Luonnon Tutkija* 89: 88–93.

Heliövaara, K., R. Väisänen, & C. Simon. 1994. Evolutionary ecology of periodical insects. *Trends in Ecology and Evolution* 9: 475–480.

Heller, F.R. 1987. Eine große Singzikade im Rosensteinpark in Stuttgart. *Mitteilungen Entomologischer Verein Stuttgart* 22: 93–94.

Heller, K.-G. 1986. Warm-up and stridulation in the bushcricket, *Hexacentrus unicolor* Serville (Orthoptera, Conocephalidae, Listroscelidinae). *Journal of Experimental Biology* 126: 97–109.

Helmore, D.W. 1982. Drawings of New Zealand insects. *Entomological Society of New Zealand Bulletin* 8: 1–52.

Helson, G.A.H. 1973. Insect pests: cicada. *New Zealand Journal of Agriculture* 126: 29.

Hempel, A. 1913a. As cigarras do caféeiro. *O Frazendeiro* 6: 139–141.

Hempel, A. 1913b. As cigarras do caféeiro. *O Frazendeiro* 6: 196–197.

Henderson, I.M. 1983. A contribution to the systematics of New Zealand Philopotamidae (Trichoptera). *New Zealand Journal of Zoology* 10: 163–176.

Hendricks, S.B. 1963. Metabolic control of timing. *Science* 141: 21–27.

Henshaw, S. 1903. The Hemiptera described by Philip Reese Uhler. IV. *Psyche* 10: 224–238.

Hennessey, M.K. 1990. Insect type specimens in the Staten Island Institute of Arts and Sciences, New York. *Florida Entomologist* 73: 465–476.

Hennig, R.M., H.-U. Kleindienst, T. Weber, F. Huber, T.E. Moore, & A.V. Popov. 1992. Aspects of sound production and reception in cicadas. Page 180 *in* N. Elsner & D.W. Richter, eds. *Rhythmogenesis in neurons and networks, Proceedings of the 20th Göttingen Neurobiology Congress.* Georg Thieme Verlag, Stuttgart.

Hennig, R.M., T. Weber, F. Huber, H.-U. Kleindienst, T.E. Moore, & A.V. Popov. 1992. Tensor muscle modulates sound amplitude of cicada song. Proceedings of the Third International Congress of Neuroethology, McGill University, Montreal. No. 261.

Hennig, R.M., T. Weber, F. Huber, H.-U. Kleindienst, T.E. Moore, & A.V. Popov. 1993. A new function for an old structure: the "timbal muscle" in cicada females. *Naturwissenschaften* 80: 324–326.

Hennig, R.M., T. Weber, F. Huber, H.-U. Kleindienst, T.E. Moore, & A.V. Popov. 1994. Auditory threshold changes in singing cicadas. *Journal of Experimental Biology* 187: 45–55.

Hennig, R.M., T. Weber, T.E. Moore, F. Huber, H.-U. Kleindienst, & A.V. Popov. 1994. Function of the tensor muscle in the cicada *Tibicen linnei*. *Journal of Experimental Biology* 187: 33–44.

Henríquez M.G. 1998. *Guia ilustrada de los principales ordenes y familias de la calse insects en El Salvador.* Universidad de El Salvador, Facultad de Ciencias Agronomicas, Departamento de Proteccion Vegetal, San Salvador. 231pp.

Hertach, T. 2004. Beitrage zur Klärung der Artkomplexes *Cicadetta montana* – Bergzikade (Hemiptera, Cicadidae): Entdeckung einer Singzikadenart mit ungewissem taxonomischem Status in der Nordschweiz. *Mitteilungen der Entomologischen Gesellschaft Basel* 54: 58–66.

Hertach, T. 2007. Three species instead of only one: distribution and ecology of the *Cicadetta montana* species complex (Hemiptera: Cicadoidea) in Switzerland. *Mitteilungen der Schweizerischen Entomologischen Gesellschaft* 80: 37–61.

Hertach, T. 2008a. A new cicada species for Switzerland: *Tettigetta argentata* (Olivier, 1790)(Hemiptera: Cicadoidea). *Mitteilungen der Schweizerischen Entomologischen Gesellschaft* 81: 209–214.

Hertach, T. 2008b. *Singzikaden (Cicadidae) in den Kantonen Aargau, Baselland, Basel-Stadt und Solothurn: Verbreitung und Schutzempfehlungen*. Otelfingen. 35pp.

Hestir, J.L., & M.D. Cain. 1999. Bird diversity and composition in even-aged loblolly pine stands relative to emergence of 13-year periodical cicadas and vegetation structure. *Proceedings of the Tenth Biennial Southern Silvicultural Research Conference, Shreveport, Louisiana, February 16–18, 1999. General Technical Report Southern Research Station, United States Department of Agriculture Forest Service Number* SRS 30: 243–248.

Hewitt, G.M., R. A. Nichols, & M.G. Ritchie. 1988. 1868 and all that for *Magicicada*. *Nature* 336: 206–207.

Hickey, G., & K. McGhee. 2002. Cicada clocks. *Nature Australia* 27: 12.

Hickson, R.E., C. Simon, & S.W. Perry. 2000. The performance of several multiple-sequence alignment programs in relation to secondary-structure features for an rRNA sequence. *Molecular Biology and Evolution* 17: 530–539.

Hidalgo-Gato González, M. 2001. Notas distribucionales y depredacion de *Uhleroides sagrae* (Homoptera: Cicadidae) por *Nephila clavipes* (Araneae: Tetragnatidae). *Cocuyo* 11: 10.

Hidvegi, F., & J.-Y. Baugnée. 1992. Données nouvelles sur *Cicadetta montana* (Scopoli, 1772) (Homoptera, Cicadoidea, Tibicinidae) abondance, sex-ratio et <<tours>> préimaginales chez une population belge. *Ecole Pratique des Hautes Etudes, Travaux du Laboratoire Biologie et Evolution des Insectes Hemipteroidea* 5: 121–126.

Hill, D.S. 2008. Pests of crops in warmer climates and their control. Springer-Verlag, Heidelberg. i-ix, 1–704.

Hill, D.S., P. Hore, & I.W.B. Thornton. 1982. *Insects of Hong Kong*. Hong Kong University Press, Hong Kong. 503pp.

Hill, K.B.R., & D.C. Marshall. 2008. Cicadas. Pages 523–524 *in* M. Winterbourne, G. Knox, G. Burrows, & I. Marsden, eds. *The Natural History of Canterbury*. Canterbury Univeristy Press, Christchurch.

Hill, K.B.R., & D.C. Marshall. 2009. Confirmation of the cicada *Tibicen pronotalis walkeri* stat. nov. (=*T. walkeri*, Hemiptera: Cicadidae) in Florida: finding singing insects through their songs. *Zootaxa* 2125: 63–66.

Hill, K.B.R., D.C. Marshall, & J.R. Cooley. 2005. Crossing Cook Strait: possible human transportation of two New Zealand cicadas from North Island to South Island (*Kikihia scutellaris* and *K. ochrina*, Hemiptera: Cicadoidea). *New Zealand Entomologist* 28: 71–80.

Hill, K.B.R., C. Simon, & G.K. Chambers. 2005. Phylogeography of a widespread New Zealand subalpine cicada, *Maoricicada campbelli* (Hemiptera, Cicadidae). *Abstracts of Talks and Posters, 12th International Auchenorrhyncha Congress, 6th International Workshop on Leafhopper and Planthoppers of Economic Significance, University of California, Berkeley, August 7–12, 2005*: S-23.

Hill, K.B.R., C. Simon, D.C. Marshall, & G.K. Chambers. 2009. Surviving glacial ages within the Biotic Gap: phylogeography of the New Zealand cicada *Maoricicada campbelli*. *Journal of Biogeography* 36: 675–692.

Hill, K.G., D.B. Lewis, M.E. Hutchins, & R.B. Coles. 1980. Directional hearing in the Japanese quail (*Coturnix coturnix japonica*). *Journal of Experimental Biology* 86: 135–151.

Hill, P.S.M. 2008. *Vibrational communication in animals*. Harvard University Press, Cambridge. 261pp.

Hilton, D. 1997. It's a small world - dog days. *Alberta Naturalist* 27: 25–27.

Hirai, T. 2004. Diet composition of introduced bullfrog, *Rana catesbeiana*, in the Mizorogaike Pond of Kyoto, Japan. *Ecological Research* 19: 375–380.

Hisamitsu, A., Y. Sokabe, T. Terada, Y. Osumi, S. Terada, A. Hirano, M. Sugita, F. Matsuo, R. Katayama, N. Ogino, S. Takami, & Y. Sakuratani. 2010. [Response of animals and plants to the solar eclipse on 22 July 2009 – Observation on the Nara Campus of Kinki University.] *Memoires of the Faculty of Agriculture of Kinki University* 43: 91–104. [In Japanese]

Hitchcock, B. 1983. Cicadas controlled by EDB. *BSES Bulletin* No. 2: 9–10.

Hoback, W.W., R.L. Rana, & D.W. Stanley. 1999. Fatty acid compositions of phospholipids and triacylglycerols of selected tissues, and fatty acid biosynthesis in adult periodical cicadas, *Magicicada septendecim*. *Comparative Biochemistry and Physiology. A* 122: 355–362.

Högmander, J. 2001. *Proposal for the establishment of the Ladoga Skerries National Park*. Karelia Parks Development, Petrozavodsk. 92pp.

Hogmire, H.W., T.A. Baugher, V.L. Crim, & S.I. Walter. 1990. Effects and control of periodical cicada (Homoptera: Cicadidae) oviposition injury on nonbearing apple trees. *Journal of Economic Entomology* 86: 2401–2404.

Hogue, C.L. 1993a. *Insects of the Los Angeles Basin*. Natural History Museum of Los Angeles County, Los Angeles. 446pp.

Hogue, C.L. 1993b. *Latin American Insects and Entomology*. University of California Press, Berkeley. 536pp.

Holden, C. 2007. Debugging Japan's cables. *Science* 317: 1301.

Holliday, C.W., J.M. Hastings, & J.R. Coelho. 2009. Cicada prey of New World cicada killers, *Sphecius* spp. (Dahlbom, 1843) (Hymenoptera: Crabonidae). *Entomological News* 120: 1–17.

Holloway, J.D. 1983. The biogeography of the Macrolepidoptera of south-eastern Polynesia. *GeoJournal* 7: 517–525.

Holloway, J.D. 1997. Sundaland, Sulawesi and eastwards: A zoogeographical perspective. *Malayan Nature Journal* 50: 207–227.

Holloway, J.D. 1998. Geological signal and dispersal noise in two contrasting insect groups in the Indo-Australian tropics: R-mode analysis of pattern in Lepidoptera and cicadas. Pages 291–314 *in* R. Hall & J.D. Holloway, eds. *Biogeography and geological evolution of SE Asia*. Backhuys Publishers, Leiden.

Holloway, J.D., & J.D. Hall. 1998. SE Asian geology and biogeography: an introduction. Pages 1–23 *in* R. Hall & J.D. Holloway, eds. *Biogeography and geological evolution of SE Asia*. Backhuys Publishers, Leiden.

Holman, L., & R.R. Snook. 2006. Spermicide, cryptic female choice and the evolution of sperm form and function. *Journal of Evolutionary Biology* 19: 1660–1670.

Holzinger, W.E. 1995. Bemerkenswerte Zikadenfunde aus Österreich, 2. Teil (Ins.: Homoptera, Auchenorrhyncha). *Linzer Biologische Beitrage* 27: 1123–1127.

Holzinger, W.E. 1996. Kritisches Verzeichnis der Zikaden Österreichs (Ins.: Homoptera, Auchenorrhyncha). *Carinthia II* 186/106: 501–517.

Holzinger, W.E. 1999. Rote Liste der Zikaden Kärnten (Insecta: Auchenorrhyncha). Pages 425–450 *in* T. Rottenburg, C. Wieser, P. Mildner, and W.E. Holzinger, eds. *Rote Listen gefahrdeter Tiere Kärntens, Naturschutz in Kärnten* 15: 1–718.

Holzinger, W.E., W. Fröhlich, H. Günthart, P. Lauterer, H. Nickel, A. Drosz, W. Schedl, & R. Remane. 1997. Vorläufiges Verzeichnis der Zikaden Metteleuropas (Insecta: Auchenorrhyncha). *Beiträge zur Zikadenkunde* 1: 43–62.

Holzinger, W.E., I. Kammerlander, & H. Nickel. 2003. *The Auchenorrhyncha of Central Europe Die Zikaden Mitteleuropas. Volume 1: Fulgoromorpha, Cicadomorpha excl. Cicadellidae*. Brill Academic Publishers, Leiden. 674pp.

Holzinger, W., & B. Komposch. 2008. Von großen und kleinen Minnesängen: die Zikadenfauna der pannonischen Trockenrasen. Pages 139–142 *in* H. Wiesbauer, ed. *Die Steppelebt. Felssteppen und Trockenrasen in Niederösterreich*. Amt der NÖ Landesregierung, St. Pölten.

Hooper, S.L., K.H. Hobbs, & J.B. Thuma. 2008. Invertebrate muscles: thin and thick filament structure; molecular basis of contraction and its regulation, catch and asynchronous muscle. *Progress in Neurobiology* 86: 72–127.

Hopkin, M. 2004. An army in its prime. *Nature News* doi:10.1038/news040510–11.

Hopper, K.R. 1999. Risk-spreading and bet-hedging in insect population biology. *Annual Review of Entomology* 44: 535–560.

Hornig, U. 1994. Ungewöhnliches Verhalten von *Lyristes plebejus* (SCOP.)(Homoptera, Cicadidae). *Entomologische Nachrichten und Berichten* 38: 65–66.

Horváth, G. 1889. Hemipterologiai Közlemények. *Természetrajzi Füzetek* 4: 184–191.

Hostetler, M.E. 1997. *That gunk on your car: a unique guide to insects of North America*. Ten Speed Press, Berkeley. 125pp.

Howard, D.R., & P.S.M. Hill. 2007. The effect of fire on spatial distribution of male mating aggregations in *Gryllotalpa major* Saussure (Orthoptera: Gryllotalpidae) at the Nature Conservancy's Tallgrass Prairie Preserve in Oklahoma: evidence of a fire-dependent species. *Journal of the Kansas Entomological Society* 80: 51–64.

Howard, F.W., D. Moore, R.M. Giblin-Davis, & R.G. Abad. 2001. *Insects on palms*. CABI Publishing, Oxon. 400pp.

Hoy, R.R., & D. Robert. 1996. Tympanal hearing in insects. *Annual Review of Entomology* 41: 433–450.

Hsieh, M.-T., W.-H. Peng, F.-T. Yeh, H.-Y. Tsai, & Y.-S. Chang. 1991. Studies on the anticonvulsive, sedative and hypothermic effects of *Periostracum cicadae* extracts. *Journal of Ethnopharmacology* 35: 83–90.

Hu, D.H., S. Kimura, & K. Maruyama. 1986. Sodium dodecyl sulfate gel electrophoresis studies of connectin-like high molecular weight proteins of various types of vertebrate and invertebrate muscles. *Journal of Biochemistry* 99: 1485–1492.

Hua L.-Z. 2000. *List of Chinese insects*. Volume I. Zhongshan (Sun Yat-sen) Universisty Press, Guangzhou. 448pp. [In Chinese]

Huang, Y.G., J.K. Kang, R.S. Liu, K.W. Oh, C.J. Nam, & H.S. Kim. 1997. Cytotoxic activities of various fractions extracted from some pharmaceutical insect relatives. *Archives of Pharmacological Research (Seoul)* 20: 110–114.

Huang, Y.P., T. Komiya, & K. Maruyama. 1984. Actin isoforms of insect muscles: two-dimensional electrophoresis studies. *Comparative Biochemistry and Physiology. B* 79: 511–514.

Hubbell, S. 1988. The cicada's song. *Wings* 13(3): 7–9.

Huber, F. 1983a. Neural correlates of orthopteran and cicada phonotaxis. Pages 108–135 *in* F. Huber & H. Markl, eds. *Neuroethology and behavioural physiology: roots and growing points*. Springer-Verlag, Berlin.

Huber, F. 1983b. Insekten - Modell-Organismen der Neuroethologie (Aus dem Leben periodisch auftretender Zikaden). *Natur, Religion, Sprache und Universität, 7, Schriftenreiche der Westfälischen Wilhelms-Universitat, Universitatvortäge* 1982/83: 1–37.

Huber, F. 1984. The world of insects: periodical cicadas and their behavior. *Alexander von Humbolt Stiftung Mitteilungen* 43: 24–31.

Huber, F. 1985. Approaches to insect behavior of interest to both neurobiologists and behavioral ecologists. *Florida Entomologist* 68: 52–78.

Huber, F., H.-U. Kleindienst, T.E. Moore, K. Schildberger, & T. Weber. 1990. Acoustic communication in periodical cicadas: neuronal responses to songs of sympatric species. Pages 217–228 *in* F.G. Gribaken, K. Wiese, & A.V. Popov, eds. *Sensory systems and communication in arthropods: including the first comprehensive collection of contributions by Soviet scientists. Advances in Life Sciences*. Birkhäuser, Basel.

Hudson, G.V. 1904. *New Zealand Neuroptera. A popular introduction to the life histories and habits of may-flies, dragon-flies, caddis-flies and allied insects inhabiting New Zealand. Including notes on their relation to angling*. West, Newman & Co., London. 102pp.

Hugel, S., F. Matt, H. Callot, J.-J. Feldtrauer and C. Brua. 2008. Présence de *Cicadetta brevipennis* Fieber, 1876 en Alsace (Hemiptera, Cicadidae). *Bulletin de la Société entomologique de Mulhouse* 64: 5–10.

Hughes, D.R., A.H. Nuttall, R.A. Katz, & G.C. Carter. 2009. Nonlinear acoustics in cicada mating calls enhance sound propagation. *Journal of the Acoustical Society of America* 125: 958–967.

Huis, A. van. 1996. The traditional use of arthropods in sub Saharan Africa. *Proceedings of the 7th Meeting of Experimental and Applied Entomologists in the Netherlands on 15 December 1995. Proceedings of the Section Experimental and Applied Entomology on the Netherlands Entomological Society* 7: 3–20.

Huis, A. van. 2003. Insects as food in sub-Saharan Africa. *Insect Science and its Application* 23: 163–185.

Humer, R.A. 1984. Foundations for an evolutionary classification of the Entomophthorales (Zygomycetes). Pages 166–183 *in* Q. Wheeler & M. Blackwell, eds. *Fungus-Insect relationships: perspectives in ecology and evolution*. Columbia University Press, New York.

Humphreys, G.S. 1989. Earthen structures built by nymphs of the cicada *Cyclochila australasiae* (Donovan)(Homoptera: Cicadidae). *Australian Entomological Magazine* 16: 99–108.

Humphreys, G.S. 1994. Bioturbation, biofabrics and the biomantle: an example from the Sydney Basin. Pages 421–436 *in* A.J. Ringrose-Voase & G.S. Humphreys, eds. *Soil micromorphology: Studies in management and genesis. Proceedings of the IX International Working Meeting on Soil Micromorphology, Townsville, Australia, July 1992. Developments in Soil Science 22*. Elsevier, Amsterdam.

Humphreys, G.S. 2005. Variation in population density of cicadas (Hemiptera: Cicadidae) in the Sydney region: *Psaltoda moerens* (Germar), *Thopha saccata* (Fabricius) and the 'spreta' form of *Cyclochila australasiae* (Donovan). *Australian Entomologist* 32: 67–77.

Humphreys, G.S., & P.B. Mitchell. 1983. A preliminary assessment of the role of bioturbation and rainwash on sandstone hillslopes in the Sydney Basin. Pages 66–80 *in* R.W. Young & G.C. Nanson, eds. *Aspects of Australian sandstone landscapes*. Australian and New Zealand Geomorphology Group Special Publication No. 1.

Hussain, M.E., & M. Amin. 1985. Electroretinogram (ERG) studies on the compound eyes of cicada. *National Academy of Science Letters (India)* 8: 189–191.

Hutchings, M., & B. Lewis. 1983. Insect sound and vibration receptors. Pages 181–205 *in* B. Lewis, ed. *Bioacoustics: a comparative approach*. Academic Press, London.

Hutchins, M., A.V. Evans, R.W. Garrison, & N. Schlager (Eds.). 2004. *Grzimek's Animal Life Encyclopedia*, second edition. Volume 3, Insects. Gale Group, Farmington Hills, MI. 472pp.

Hutchinson, R. 2002. La vie des cigales: un aperçu. *Nouv'Ailes* 12: 6–7.

Hyche, L.L. 1998. Periodical cicadas (The "13-year Locusts") in Alabama. *Alabama Agricultural Experiment Station Bulletin* 635: 1–12.

Hyslop, J.A. 1935. Periodical cicada. *United States Department of Agriculture, Bureau of Entomology and Plant Quarantine E* 364: 1–20.

Hyslop, J.A. 1940. The periodical cicada, Brood XIV. *United States Department of Agriculture, Bureau of Entomology and Plant Quarantine E* 502: 1–15.

Ibanez, M. 2003. *Les cigales*. Cycles de vie, caractéristiques et symbolique. Rapport de stage MST Aménagement et Mise en Valeur Durables des Régions. Observatoire Naturaliste des Écosytèmes Méditerranéens, Prades-Le-Lez. 18pp.

Ibanez, M. 2009. Hémiptères Cicadidae. Pages 208–212 *in* F. Karas, ed. *Etat des lieux des connaissances sur les Invertébrés continentaux des Pays de la Loire, binal final*. Rapport GRETIA pour le Conseil Régional des Pays de la Loire. Groupe d'Etude des Invertébrés Armoricains, Rennes Cedex.

Ibrahim, A.b., & T.M. Leong. 2009. Records of the black and scarlet cicada, *Huechys sanguinea* (De Geer) in Singapore, with notes on its emergence (Homoptera: Cicadidae: Cicadettinae). *Nature in Singapore* 2: 317–322.

Ikeda, K. 1959. Studies on the origin and the pattern of the nimiature electrical oscillation in the insect muscle. *Japanese Journal of Physiology* 9: 484–497.

Inoue, T., H. Goto, S. Makino, K. Okabe, I. Okuchi, K. Hamaguchi, M. Sueyoshi, & E. Shoda-Kagaya. 2007. Cicadas (Hemiptera: Cicadidae) collected with Malaise traps set in natural forests and plantations in the northeastern part of Ibaraki Prefecture. *Bulletin of the Forestry and Forest Products Research Institute* 6: 249–252.

Irie, M., K. Udagawa, M. Nagata, & M. Kubo-Irie. 2004. Three dimensional synergetic movement of cicada sperm linked by the spermatodesm. *Zoological Science* 21: 1282.

Irwin, M.D., & J.R. Coelho. 2000. Distribution of the Iowan brood of periodical cicadas (Homoptera: Cicadidae: *Magicicada* spp.) in Illinois. *Annals of the Entomological Society of America* 93: 82–89.

Isaki, I. 1933. *Entomological World* 1: 380–381. [In Japanese]

Ishii, M. 1990. An observation on the oviposition behavior of the parasitic moth, *Epipomponia nawai* (Dyar) (Lepidoptera, Epipyropidae). *Japanese Journal of Entomology* 58: 441–442.

Ishii, Y. 1953. A new species of Leptus from Kyushu (Acarina: Erythraeidae). *Igaku Kenkyuu (Acta Medica)* 23: 160–163.

Itô, Y. 1976. Status of insect pests of sugarcane in the southwestern islands of Japan. *Japan Agricultural Research Quarterly* 10: 63–69.

Itô, Y. 1978. *Comparative ecology* (edited and translated by J. Kikkawa). Cambridge University Press, Cambridge. 436pp.

Itô, Y. 1982a. [Mass emergence of cicadas: *Magicicada* spp. and *Mogannia minuta*]. *Insectarium (Tokyo)* 19: 108–113. [In Japanese]

Itô, Y. 1982b. [Mass emergence of cicadas: *Magicicada* spp. and *Mogannia minuta*]. *Insectarium (Tokyo)* 19: 136–141. [In Japanese]

Itô, Y. 1998. Role of escape from predators in periodical cicada (Homoptera: Cicadidae) cycles. *Annals of the Entomological Society of America* 91: 493–496.

Itô, Y., & E. Kasuya. 2005. Demography of the Okinawan eusocial wasp *Ropalidia fasciata* (Hymenoptera: Vespidae) I. Survival rate of individuals and colonies, and yearly fluctuations in colony density. *Entomological Science* 8: 41–64.

Itô, Y., & M. Nagamine. 1981. Why a cicada, *Mogannia minuta* Matsumura, became a pest of sugarcane: an hypothesis based on the theory of 'escape'. *Ecological Entomology* 6: 273–283.

Itô, Y., & M. Nagamine. 1998. Slow expansion of the distribution range of the sugarcane cicada, *Mogannia minuta* Matsumura (Homoptera: Cicadidae). *Entomological Science* 1: 533–535.

Iwamoto, H. 2008. 614-Pos. The flight muscle of cicada is synchronous, but is crystalline and stretch-activatable. *Biophysical Journal* 94: 614.

Iwamoto, H., K. Inoue, & N. Yagi. 2006. Evolution of long-range myofibrillar crystallinity in insect flight muscle as examined by X-ray cryomicrodiffraction. *Proceedings of the Royal Society of London. Series B. Biological Sciences* 273: 677–685.

Izzo, A.S., & D.A. Gray. 2004. Cricket song in sympatry: species specificity of song without reproductive character displacement in *Gryllus rubens*. *Annals of the Entomological Society of America* 97: 831–837.

Jäger, U., & D. Frank. 2002. Naturnahe Kalk-Trockenrasen und deren Verbuschungsstadien (Festuco-Brometalia) (*besondere Bestände mit bemerkenswerten Orchideen). *Prioritär zu schützender Lebensraum. Pages 90–101, 294–297 *in* J. Peterson & U. Ruge, eds. *Naturschutz im Land Sachsen-Anhalt, Die Lebensraumtypen nach Anhang I der Fauna-Flora-Habitatrichtlinie im Land Sachsen-Anhalt. 39. Jahrgang 2002 Sonderheft. Landesamt f*ür Umweltschutz, Halle.

Jakubczak, J.L., W.D. Burke, & Eickbushi. 1991. Retrotransposable elements *R1* and *R2* interrupt the rRNA genes of most insects. *Proceedings of the National Academy of Science of the United States of America* 88: 3295–3299.

James, D.A., K.S. Williams, & K.G. Smith. 1986. Survey of 1985 periodical cicada (Homoptera: *Magicicada*) emergence sites in Washington County, Arkansas, with reference to ecological implications. *Proceedings of the Arkansas Academy of Sciences* 40: 37–39.

Jameson, M.L. 1995. Insect musicians. *Museum Notes (University of Nebraska State Museum)* 91: 1–4.

Jami, F., M.N. Kazempour, S.A. Elahinia, & G. Khodakaramian. 2005. First report of *Xanthomonas arbicola* pv. *pruni* on Stone Fruit trees from Iran. *Journal of Phytopathology* 153: 371–372.

Jamieson, B.G.M. 1987. *The ultrastructure and phylogeny of insect spermatozoa*. Cambridge University Press, Cambridge. 320pp.

Jamieson, B.G.M., R. Dallai, & B.A. Afzelius. 1999. *Insects: their spermatozoa and phylogeny*. Wiley, Chichester, USA. 555pp.

Jang, Y., & C. Gerhardt. 2006. Divergence in female calling song discrimination betweern sympatric and allopatric populations of the southerm wood cricket *Gryllus fultoni* (Orthoptera: Gryllidae). *Behavioral Ecology and Sociobiology* 60: 150–158.

Jankovic, L., & R. Papovic. 1981. [New and infrequent species in Yugoslav fauna of cicadas (Homoptera: Auchenorrhyncha).] *Srpska Akademija Nauka I Umetnosti Odeljenje Prirodo-Matematichkikh Nauka Glas* 48: 121–134. [In Russian, English summary]

Janský, V. 2005. Cikády (Auchenorrhyncha). Pages 83–85 *in* O. Majzlan, ed. *Fauna Devínskej Kobyly*. APOP, Bratislava.

Jardak, T., & M. Ksantini. 1996. Key elements of, and economic and environmental need for, a modified approach to olive crop care in Tunisia. *Olivae* No. 61: 24–33.

Jennions, M.D., & M. Petrie. 1997. Variation in mate choice and mating preferences: a review of causes and consequences. *Biological Reviews* 72: 283–327.

Jeon, J.-B., T.-W. Kim, P. Tripotin, & J.-I. Kim. 2002. Notes on a cicada parasitic moth in Korea (Lepidoptera: Epipyropidae). *Korean Journal of Entomology* 32: 239–241.

Jiang J.-C. 1985. [A study of the song characteristics in cicadas at Jinghong in Yunnan Province (China).] *Acta Entomologica Sinica* 28: 257–265. [In Chinese, English summary]

Jiang J.-C. 1987. [Structure of sounding membrane and effects of abdominal moving on calling song in cicada.] *Acta Biophysica Sinica* 3: 106–111. [In Chinese, English summary]

Jiang J.-C. 1989a. [Directivity of songs in black cicada (*C. atrata* Fabricius) and function of third spiracle.] *Acta Biophysica Sinica* 5: 63–69. [In Chinese, English summary]

Jiang J.-C. 1989b. Mechanical characteristics and sounding mechanism of sounding membrane in black cicada (*C. atrata* Fabricius). *Science in China. Series B. Chemistry, Life Sciences, & Earth Sciences* 32: 64–77.

Jiang, J.-C., Chen H., & Hu M.-L. 1990a. Dynamic properties of main sounder in black cicada (*C. atrata* Fabricius). *Science in China. Series B. Chemistry, Life Sciences, & Earth Sciences* 33: 159–169.

Jiang, J.-C., Chen H., & Hu M.-L. 1990b. Dynamic properties of assistant sounder in black cicada (*C. atrata* Fabricius). *Science in China. Series B. Chemistry, Life Sciences, & Earth Sciences* 33: 1181–1191.

Jiang, J.-C., Lin X., & Wang Q. 1987. [Structure of sounding membrane and effects of abdominal moving on calling song in cicada.] *Acta Biophysica Sinica* 3: 106–111. [In Chinese, English summary]

Jiang, J.-C., Liu X., Yang X., Chen H., Xu M., & Tang H. 1994. [Decay of amplitude-modulation property of song and an analysis of sounding mechanism in mottled cicada (*Gaeana maculata* Drury).] *Acta Acustica* 19: 290–299. [In Chinese, English summary]

Jiang, J.-C., Nie S., Liu L., Xu M., Sun W., Liu X., & Wang Q. 1986. Structure and function of the sound apparatus in cicada (*C. atrata* Fabr.). *Chinese Journal of Acoustics* 5: 174–184.

Jiang, J.-C., Wang Q., & Lin X.-Q. 1988. [Characteristics of the structure and timbre of songs of the black cicada *Cryptotympana atrata* Fabricius.] *Acta Entomologica Sinica* 31: 129–139. [In Chinese, English summary]

Jiang, J.-C., Wang Y., Xu M., Zhang H., & Chen H. 1991. Frequency conversion property and dynamic process in songs of mingming cicada (*Oncotympana maculatricollis* Motsch). *Science in China. Series B. Chemistry, Life Sciences, & Earth Sciences* 34: 576–586.

Jiang, J.-C., Wang Y., Xu M., Zhang H., & Chen H. 1992. [Functional property of toning muscle in Mingming cicada (*Oncotympana maculaticollis* Motsch.).] *Shengwu Wuli Xuebao = Acta Biophysica Sinica* 8: 236–244. [In Chinese, English summary]

Jiang J.-C., Xu M., & Chen H. 1990. [Characteristics of song in cricket (*Laxoblemmus doenitzistein*) and effects of call in black cicada (*C. atrata* Fabricius)]. *Acta Biophysica Sinica* 9: 183–190. [In Chinese, English summary]

Jiang, J.-C., Xu M., Chen H., & Zhang H. 1990. A study of sound production acoustics in the black cicada (*C. atrata* Fabricius). *Chinese Journal of Acoustics* 9: 161–169.

Jiang, J.-C., Xu M., & Liu Y. 1991. [The vari-toned complex songs in cicadas and an analysis of the mechanism of their sound production.] *Acta Entomologica Sinica* 34: 159–165. [In Chinese, English summary]

Jiang J., Yang X., & Liu X. 1996. [Two kinds of behavioral songs and analysis of sound-production mechanism in kaempfer cicada.] *Acta Acustica* 21: 313–318. [In Chinese, English summary]

Jiang J., Yang X., Tang H., Xu M., & Chen H. 1995a. Innervation frequency response of sounding muscles in Mingming cicada. *Chinese Science Bulletin* 40: 71–75.

Jiang J., Yang X., Tang H., Xu M., & Chen H. 1995b. [Song property serving the purpose of communication and stimation of flight phonotactic region in cicadas.] *Acta Acustica* 20: 226–231. [In Chinese, English summary]

Jiang, J.-C., Yang X., Wang Y., Xu M., Chen H., & Tang H. 1995. Neural control of sound production in mingming cicada. *Science in China. Series B. Chemistry, Life Sciences, & Earth Sciences* 38: 676–687.

Joermann, G., & H. Schneider. 1987. The songs of four species of cicada in Yugoslavia (Homoptera: Cicadidae). *Zoologischer Anzeiger* 219: 283–296.

Johns, P.M. 1977. The biology of the terrestrial fauna. Pages 311–328 *in* C.J. Burrows, ed. *Cass, history and science in the Cass District, Canterbury, New Zealand.* Department of Botany, University of Canterbury.

Johnsen, T.S., J.D. Hendeveld, J.L. Blank, & K. Yasak. 1996. Epaulet brightness and condition in female red-winged blackbirds. *Auk* 113: 356–362.

Johnson, C.W. 1906. The distribution of the periodical cicada in New England in 1906. *Psyche* 13: 159.

Johnson, J.B., R.H. Hagen, & E.A. Martinko. 2010. Effect of succession and habitat area on wandering spider (Araneae) abundance in an experimental landscape. *Journal of the Kansas Entomological Society* 83: 141–153.

Johnson, L.K., & R.B. Foster. 1986. Associations of large Homoptera (Fulgoridae and Cicadidae) and trees in a tropical forest. *Journal of the Kansas Entomological Society* 59: 415–422.

Johnson, T.S., J.D. Hengeveld, J.L. Blank, K. Yasukawa, & V. Nolan. 1996. Epaulet brightness and condition in female red-winged blackbirds. *The Auk* 113: 356–362.

Johnson, W.T., & H.H. Lyon. 1976. *Insects that feed on trees and shrubs*, First edition. Comstock Publishing, Cornell University Press, Ithaca. 464pp.

Johnson, W.T., & H.H. Lyon. 1988. *Insects that feed on trees and shrubs.* 2nd edition. Cornell University Press, Ithaca. 556pp.

Jolivet, P. 2004. Un mois de mai memorable en América du Nord. *Insectes* (135): 3–7.

Jolivet, P., & K.K. Verma. 2005. *Fascinating insects; some aspects of insect life*. Pensoft Publishers, Sofia. 310pp.

Jong, M.R. de. 1982. The Australian species of the genus *Lembeja* Distant 1892 (Homoptera, Tibicinidae). *Bijdragen tot de Dierkunde* 52: 175–185.

Jong, M.R. de. 1985. Taxonomy and biogeography of Oriental Prasiini. 1. The genus *Prasia* Stål, 1863 (Homoptera, Tibicinidae). *Tijdschrift voor Entomologie* 128: 165–191.

Jong, M.R. de. 1986. Taxonomy and biogeography of Oriental Prasiini. 2. The *foliata* group of the genus *Lembeja* Distant, 1892 (Homoptera, Tibicinidae). *Tijdschrift voor Entomologie* 129: 141–180.

Jong, M.R. de. 1987. Taxonomy and biogeography of Oriental Prasiini. 3. The *fatiloqua* and *parvula* groups of the genus *Lembeja* Distant, 1892 (Homoptera, Tibicinidae). *Tijdschrift voor Entomologie* 130: 177–209.

Jong, M.R. de, & A.J. de Boer. 2004. Taxonomy and biogeography of Oriental Prasiini, 4. The genus *Mariekea* gen. n. (Homoptera, Tibicinidae). *Tijdschrift voor Entomologie* 147: 265–281.

Jong, M.R. de., & J.P. Duffels. 1981. The identity, distribution and synonymy of *Lembeja papuensis* Distant, 1897 (Homoptera, Tibicinidae). *Bulletin Zoologisch Museum, Universiteit van Amsterdam* 8: 53–62.

Jongebloed, W.L., E. Rosenzweig, D. Kalicharan, J.J.L. van der Want, & J.S. Ishay. 1999. Ciliary hair cells and cuticular photoreceptor of the hornet *Vespa orientalis* as components of a gravity detecting system: an SEM/TEM investigation. *Journal of Electron Microscopy* 48: 63–75.

Josephson, R.K. 1973. Contraction kinetics of the fast muscles used in singing by a katydid. *Journal of Experimental Biology* 59: 781–801.

Josephson, R.K. 1981. Temperature and the mechanical performance of insect muscle. Pages 19–44 *in* B. Heinrich, ed. *Insect thermoregulation*. Wiley, New York.

Josephson, R.K. 1984. Contraction dynamics of flight and stridulatory muscles of tettigoniid insects. *Journal of Experimental Biology* 108: 77–96.

Josephson, R.K., & D. Young. 1981. Synchronous and asynchronous muscles in cicadas. *Journal of Experimental Biology* 91: 219–237.

Josephson, R.K., & D. Young. 1985. A synchronous insect muscle with an operating frequency greater than 500 Hz. *Journal of Experimental Biology* 118: 185–208.

Josephson, R.K., & D. Young. 1987. Fiber ultrastructure and contraction kinetics in insect fast muscles. *American Zoologist* 27: 991–1000.

Jowett, I.G., J.W. Hayes, N. Deans, & G.A. Eldon. 1998. Comparison of fish communities and abundance in unmodified streams of Kahurangi National Park with other areas of New Zealand. *New Zealand Journal of Marine and Freshwater Research* 32: 307–322.

Joy, J.B., & B.J. Crespi. 2007. Adaptive radiation of gall-inducing insects within a single host-plant species. *Evolution* 61: 784–795.

Júnior, L.J.Q., J.R.G.S. Almeida, J.T. Lima, X.P. Nunes, J.S. Siqueira, L.E.G. de Oliveira, R.N. Almeida, P.F. de Athayde-Filho, & J.M. Barbosa-Filho. 2008. Plants with anticonvulsant properties – a review. *Revista Brasileira de Farmacognosia* 18 (Supl,): 798–819.

Kadamshoev, M., N. Miralibekov, & A. Abdulnazarov. 2002. [Trophic relations between singing cicadas.] *Zaschita Karantin Rastenii* 5: 45. [In Russian]

Kahono, S. 1993. Fauna Serangga di beberapa daerah di Wamena, Jayawijaya, Irian Jaya. *Prosiding Seminar Hasil Penelitian dan Pengembangan Sumber Daya Hayati, Puslitbang Biologi-LIPI: prosiding, Bogor, 14 Jun 1993*: 199–213.

Kalisz, P.J. 1994. Spatial and temporal patterns of emergence of periodical cicadas (Homoptera: Cicadidae) in a mountainous forest region. *Transactions of the Kentucky Academy of Sciences* 55: 118–123.

Kaltenbach, A., H. Steiner, & L. Aschenbrenner. 1972. Die Tierwelt der Trockenlandschaft. *Naturgeschichte Wiens* 2: 447–494.

Kamijo, N. 1933. On a collection of insects from North Keisho-do, Korea (II). *Journal of the Chosen Natural History Society* (15): 46–63.

Kang, C.H., S. Ferguson-Miller, & E. Margoliash. 1977. Steady state kinetics and binding of eukaryotic cytochromes *c* with yeast cytochrome *c* peroxidase. *The Journal of Biological Chemistry* 252: 919–926.

Kanasugi, T. 2008. [Collecting record of *Tibicen flammatus* (Cicadidae: Hemiptera) at Mt. Akagi and distribution in Gunma Prefecture.] *Bulletin of Gunma Museum of Natural History* 12: 83–84.

Kang, C.H., S. Ferguson-Miller, & E. Margoliash. 1977. Steady state kinetics and binding of eukaryotic cytochromes *c* with yeast cytochrome *c* peroxidase. *The Journal of Biological Chemistry* 252: 919–926.

Kanmiya, K., & S. Nakao. 1988. Acoustic analysis of the calling song of a cicada, *Meimuna kuroiwae* Matsumura, in the Nansei Islands, Japan: Part II. Multivariate analysis of the sound. *Kurume University Journal* 37: 59–82.

Kapla, A., T. Trilar, M. Gogala, & A. Gogala. 2001. Acoustic emissions of the species complex of *Cicadetta montana*. *Program and Book of Abstracts, 2nd European Hemiptera Congress, Fiesa, Slovenia, 20–24 June 2001*: 35–36.

Karalliedde, L.D., & C.T. Kappagoda. 2009. The challenge of traditional Chinese medicines for allopathic practitioners. *American Journal of Physiology –Heart and Circulatory Physiology* 297: H1967-H1969.

Karban, R. 1981a. Effects of local density on fecundity and mating speed for periodical cicadas. *Oecologia* 51: 260–264.

Karban, R. 1981b. Flight and dispersal of periodical cicadas. *Oecologia* 49: 385–390.

Karban, R. 1982a. Experimental removal of 17-year cicada nymphs and growth of host apple trees. *Journal of the New York Entomological Society* 90: 74–81.

Karban, R. 1982b. Increased reproductive success at high densities and predator satiation for periodical cicadas. *Ecology* 63: 321–328.

Karban, R. 1983a. Induced responses of cherry trees to periodical cicada oviposition. *Oecologia* 59: 226–231.

Karban, R. 1983b. Sexual selection, body size and sex-related mortality in the cicada *Magicicada cassini*. *American Midland Naturalist* 109: 324–330.

Karban, R. 1984a. Opposite density effects of nymphal and adult mortality for periodical cicadas. *Ecology* 65: 1656–1661.

Karban, R. 1984b. Variation in development time for cicadas. *International Congress of Entomology Proceedings* 17: 355.

Karban, R. 1985. Addition of periodical cicada nymphs to an oak forest: effects on cicada density, acorn production, and rootlet density. *Journal of the Kansas Entomological Society* 58: 269–276.

Karban, R. 1986. Prolonged development in cicadas. Pages 222–235 *in* F. Taylor & R. Karban, eds. *The evolution of insect life cycles*. Springer-Verlag, New York.

Karban, R. 1997. Evolution of prolonged development: a life table analysis for periodical cicadas. *American Naturalist* 150: 446–461.

Karban, R., C.A. Black, & S.A. Weinbaum. 2000. How 17-year cicadas keep track of time. *Ecology Letters* 3: 253–256.

Karban, R., & G. English-Loeb. 2000. Ants do not affect densities of periodical cicada nymphs (Homoptera: Cicadidae). *Journal of the Kansas Entomological Society* 72: 251–255.

Karlin, A.A., E.C. Stout, L.T. Adams, L.R. Duke, & J.J. English. 1994. Genetic variability in developing periodical cicadas. *Arkansas Academy of Science Proceedings* 48: 89–92.

Karlin, A.A., K.S. Williams, K.G. Smith, & D.W. Sugg. 1991. Biochemical evidence for rapid changes in heterozygosity in a population of periodical cicadas (*Magicicada tredecassini*). *American Midland Naturalist* 125: 213–221.

Kartal, V. 1983. Neue Homoptera aus der Türkei. II. *Marburger Entomologische Publikationen* 1: 235–248.

Kartal, V. 1988. Die Genitaltmorphologie der Arten der Gattung *Chloropsalta* (Homoptera, Auchenorrhyncha, Cicadidae). *Ondokuz Mayis Üniversitesi Fen Dergisi* 1: 9–15.

Kartal, V. 1999. Die Genitalmorphologie der Arten der Gattungen *Hilaphura* und *Cicadetta* aus Spanien (Hemiptera: Auchenorrhyncha: Cicadomorpha: Tibicinidae). *Reichenbachia, Staatliches Museum für Tierkunde Dresden* 33: 99–103.

Kartal, V. 2006. Genital morphology of *Cicadetta albipennis* Fieber, 1876 and *Euryphara cantans* (Fabricius, 1794) (Hemiptera: Auchenorrhyncha: Cicadidae). *Russian Journal of Entomology* 15: 279–280.

Kartal, V., & Ü. Zeybekoglu. 1999a. *Cicadatra persica* Kirkaldy, 1909 (Cicadidae, Homoptera)'nin Essey Organlarinin Morfolojisi ve Yumurtlama Kapasitesi Üzerine Bir Arastirsma. *Turkish Journal of Zoology* 23: 59–62.

Kartal, V., & Ü. Zeybekoglu. 1999b. *Cicadatra hyalina* (Fabricius, 1798)(Homoptera, Cicadidae) Türü Üzerine Taksonomik Bir Arastirsma. *Turkish Journal of Zoology* 23: 63–66.

Kaser, S.A., & J. Hastings. 1981. Thermal physiology of the cicada, *Tibicen duryi*. *American Zoologist* 21: 1016.

Katsuki, Y. 1960. Neural mechanism of hearing in cats and insects. Pages 53–75 *in* Y. Katsuki, ed. *Electrical activity of single cells*. Igakushoin, Hongo, Tokyo.

Kavanagh, M.W. 1987. The efficiency of sound production in two cricket species, *Gryllotalpa australis* and *Teleogryllus commodus* (Orthoptera: Grylloidea). *Journal of Experimental Biology* 130: 107–119.

Kawahara, A.Y. 2007. Thirty-foot telescopic nets, bug-collecting video games, and beetle pests: entomology in modern Japan. *American Entomologist* 53: 160–172.

Kay, M.K. 1980. *Amphisalta zelandica* (Boisduval), *Amphisalta cingulata* (Fabricius), *Amphisalta strepitans* (Kirkaldy) (Hemiptera: Cicadidae): Large cicadas. *Forest and Timber Insects in New Zealand* No. 44: 1–8.

Keener, T.C., S.-J. Khang, J.-Y. Lee, & P.F. Heckel. 2007. Occurrence of mercury in periodical cicadas (*Magicicada: cassini*). *Journal of Entomological Science* 42: 435–436.

Kellner, C.J., K.G. Smith, N.C. Wilkinson, & D.A. James. 1990. Influence of periodical cicadas on foraging behavior of insectivorous birds in an Ozark forest. *Studies in Avian Biology* 13: 375–380.

Kemal, M., & A.Ö. Koçak. 2010. Cicadidae of Turkey and some ecological notes on *Tibicina serhadensis* Koçak & Kemal from Van Province (East Turkey) *(Homoptera)*. *Cesa News* (55): 1–19.

Kevan, D.K.M. 1985. The land of the locusts: being some further verses on grigs and cicadas. Part III. *Lyman Entomological Museum and Research Laboratory Memoir No.* 16. 1–466.

Khamasi, R. 2004. Cicada invasion feeds forest. *Nature News* doi:10.1038/news041122–12.

Khamasi, R. 2005. Cicada invasion feeds forest. Insect broods kick-start forest ecosystems every 17 years. *The Forestry Chronicle* 81: 25.

Kholmuminov, A. 1987. [Changes in Cicadidae fauna in Betpak-Dala (Uzbek SSR, USSR) caused by reclamation.] *Uzbekskii Biologicheskii Zhurnal* (4): 47–49. [In Russian]

Kiho, T., I. Miyamoto, K. Nagai, & S. Ukai. 1988. Minor, protein-containing galactomannans from the insect-body portion of the fungal preparation Chan hua (*Cordyceps cicadae*). *Carbohydrate Research* 181: 207–215.

Kim, H.K. 1958. Insect fauna of Dukchok Islands. *Korean Journal of Applied Zoology* 1(1): 87–102.

Kim, H.-K. 1960. Insect list of Solak Mountain. *Korean Journal of Applied Zoology* 3(1): 21–29.

King, C.M., & J.E. Moody. 1982. The biology of the stoat (*Mustela erminae*) in the National Parks of New Zealand II. Food habits. *New Zealand Journal of Zoology* 9: 57–80.

Kirillova, V.I. 1986a. [Chromosome numbers of world Homoptera Auchenorrhyncha. 1. Fulgoroidea, Cercopodea, and Cicadoidea.] *Entomologicheskoye Obozreniye* 1986: 115–125. [In Russian]

Kirillova, V.I. 1986b. Chromosome numbers of world Homoptera Auchenorrhyncha. 1. Fulgoroidea, Cercopodea, and Cicadoidea. *Entomological Review* 65: 34–47.

Kishida, K. 1929. [Cicadas from Ranan, N.E. Corea.] *Lansania* 1: 109. [In Japanese.]

Klein, U., C. Bock, W.A. Kafka, & T.E. Moore. 1987a. Sensory equipment of the 17-year cicada: structural survey and physiological evidence for olfaction. *Chemical Senses* 12: 210–211.

Klein, U., C. Bock, W.A. Kafka, & T.E. Moore. 1987b. Olfactory sensilla on the antennae of *Magicicada cassini* (Fisher) (Homoptera: Cicadidae): Structure, function and possible behavioral role. *Proceedings of the 6th Auchenorrhyncha Meeting, Turin, Italy, 7–11 September, 1987*: 93–98.

Klein, U., C. Bock, W.A. Kafka, & T.E. Moore. 1988. Antennal sensilla of *Magicicada cassini* (Fisher) (Homoptera: Cicadidae): Fine structure and electrophysiological evidence for olfaction. *International Journal of Insect Morphology and Embryology* 17: 153–167.

Kobayashi, Y. 1951. Notes of fungi. (1) On the newly found genus *Massospora* from Japan. *Shokubutsu Kenkyu Zasshi* 26: 117–119.

Koçak, A.Ö. 1982. List of the genera of Turkish Auchenorrhyncha (Homoptera), with some replacement names for the genera exsisting in the other countries. *Priamus* I: 141–167.

Koçak, A.Ö., & M. Kemal. 2010. List of the species of some pterygot orders recorded in the Province Van (East Turkey) and a description of a new species in the family Cicadidae (Insecta). *Priamus* 12: 130–149.

Kocher, T.D., W.K. Thomas, A. Meyer, S.V. Edwards, S. Pääbo, F.X. Villablance, & A.C. Wilson. 1989. Dynamics of mitochondrial DNA evolution in animals: amplification and sequencing with conserved primers. *Proceedings of the National Academy of Sciences of the United States of America* 86: 6196–6200.

Koenig, W.D., & A.M. Liebhold. 2003. Regional impacts of periodical cicadas on oak radial increment. *Canadian Journal of Forest Research* 33: 1084–1089.

Koenig, W.D., & A.M. Liebhold. 2005. Effects of periodical cicada emergences on abundance and synchrony of avian populations. *Ecology* 86: 1873–1882.

Kohira, M., H. Okada, M. Nakanishi, & M. Yamanaka. 2009. Modeling the effects of human-caused mortality on the brown bear population on the Shiretoko Peninsula, Hokkaido, Japan. *Ursus* 20: 12–21.

Köhler, U., & R. Lakes-Harlan. 2001. Auditory behvaiour of a parasitoid fly (*Emblemasoma auditrix*, Sarcophagidae, Diptera). *Journal of Comparative Physiology. A. Sensory, Neural and Behavioral Physiology* 187: 581–587.

Kölliker, A. 1888. Zur kenntnis der quergestreiften muskelfasern. *Zeitschrift für Wissenschaftliche Zoologie* 47: 689–710.

Kondratieff, B.C., A. Ellington, & D.A. Leatherman. 2002. The cicadas (Homoptera: Cicadidae, Tibicinidae) of Colorado. *Contributions of the C.P. Gillette Museum of Arthropod Diversity. Insects of Western North America 2*. 81pp.

Kondratieff, B.C., J.P. Schmidt, P.A. Opler, & M.C. Garhart. 2005. Survey of selected arthropod taxa of Fort Sill, Comanche County, Oklahoma. III. Arachnida: Ixodidae, Scorpiones, Hexapoda: Ephemeroptera, Hemiptera, Homoptera, Coleoptera, Neuroptera, Trichoptera, Lepidoptera, and Diptera. Pages 70–248 *in* P.A. Opler, ed. *Survey of selected arthropod taxa of Fort Sill, Comanche County, Oklahoma. III.* C.P. Gillette Museum of Arthropod Diversity, Department of Bioagricultural Sciences and Pest Management, Colorado State University, Fort Collins, Colorado.

Kos, M., & M. Gogala. 2000. The cicadas of the *Purana nebulilinea* group (Homoptera, Cicadidae) with a note on their songs. *Tijdschrift voor Entomologie* 143: 1–25.

Kossenko, S.M., & C.H. Fry. 1998. Competition and coexistence of the European bee-eater *Merops apiaster* and the blue-cheeked bee-eater *Merops persicus* in Asia. *Ibis* 140: 1–13.

Kostovski, G., D.J. White, A. Mitchell, A. Mitchell, M.W. Austin, & P.R. Stoddart. 2008. Nanoimprinting on optical fiber end faces for chemical sensing. 19th International Conference on Optical Fibre Sensors, D. Sampson, S. Collins, K. Oh, & R. Yamauchi, eds. *Proceedings of the SPIE 7004*: 70042H-70042H4.

Kostovski, G., D.J. White, A. Mitchell, M.W. Austin, & P.R. Stoddart. 2009. Nanoimprinted optical fibres: biotemplated nanostructures for SERS sensing. *Biosensors and Bioelectronics* 24: 1531–1535.

Kozhevnikova, A.G. 1986. [Cicadidae of cotton fields in the Fergana Valley.] *Uzbekskii Biologicheskii Zhurnal* 1986(5): 45–48. [In Russian, English summary]

Kozhevnikova, A.G. 1996. [Materials on biology of *Cicadatra querula* (Pall.) harming cotton plants in the Republic of Uzbekistan.] *Uzbekskii Biologicheskii Zhurnal* 1996: 97–98. [In Russian]

Krasnoff, S.B., R.F. Reátegui, M.M. Wagenaar, J.B. Gloer, & D.M. Gibson. 2005. Cicadapeptins I and II: new Aib-containing peptides from the entomopathogenic fungus Cordyceps heteropoda. *Journal of Natural Products* 68: 50–55.

Krause, J.M., T.M. Brown, E.S. Bellosi, & J.F. Genise. 2008. Trace fossils of cicadas in the Cenozoic of central Patagonia, Argentina. *Palaeontology* 51: 405–418.

Krell, R.K., E.A. Boyd, J.E. Nay, Y.-L. Park, & T.M. Perring. 2007. Mechanical and insect transmission of *Xylella fastidiosa* to *Vitis vinifera*. *American Journal of Enology and Viticulture* 58: 211–216.

Kritsky, G. 1987. Take two cicadas and call me in the morning. *Bulletin of the Entomological Society of America* 33: 139–141.

Kritsky, G. 1988. The 1987 emergence of the periodical cicada (Homoptera: Cicadidae: *magicicada* spp.: Brood X) in Ohio. *Ohio Journal of Science* 88: 168–170.

Kritsky, G. 1992. The 1991 emergence of the periodical cicadas (Homoptera: Cicadidae: *Magicicada* spp.: Brood XIV) in Ohio. *Ohio Journal of Science* 92: 38–39.

Kritsky, G. 1999. *In Ohio's backyard: periodical cicadas*. Ohio Biological Survey Backyard Series No. 2. Columbus, Ohio. 83pp.

Kritsky, G. 2001a. Periodical revolutions and the early history of the "locust" in American cicada terminology. *American Entomologist* 47: 186–188.

Kritsky, G. 2001b. Cost of horns, fossil formation, quantifying pitfall trap data, cicadas, carnivorous caterpillars, and bugs cooling. *American Entomologist* 47: 118–119.

Kritsky, G. 2004a. *Periodical cicadas: the plague and the puzzle*. Indiana Academy of Sciences, Indianapolis. 147pp.

Kritsky, G. 2004b. John Bartram and the periodical cicadas: a case study. Pages 43–51 *in* J. Van Horne & N. Hoffman, eds. *America's curious botanist: a tercentennial reappraisal of John Bartram, 1699–1777*. American Philosophical Society, Philadelphia.

Kritsky, G. 2004c. They're back! The Mount's periodical visitors are coming. *The Mount Magazine* 2004(Winter): 2–4.

Kritsky, G., A. Hoelmer, M. Noble, & R. Troutman. 2009. Observations on periodical cicadas (Brood XIV) in Indiana and Ohio in 2008 (Hemiptera: Cicadidae: *Magicicada* spp.). *Proceedings of the Indiana Academy of Science* 118: 83–87.

Kritsky, G., & S. Simon. 1996. The unexpected 1995 emergence of periodical cicadas (Homoptera: Cicadidae: *Magicicada* spp.) in Ohio. *Ohio Journal of Science* 96: 27–28.

Kritsky, G., & J. Smith. 2009. Observations on periodical cicadas (Brood XIII) in Indiana in 2007 (Hemiptera: Cicadidae: *Magicicada* spp.). *Proceedings of the Indiana Academy of Science* 118: 81–82.

Kritsky, G., J. Smith, & N.T. Gallagher. 1999. The 1999 emergence of the periodical cicadas in Ohio (Homoptera: Cicadidae: *Magicicada* spp.). *Ohio Biological Survey Notes* 2: 43–47.

Kritsky, G., J. Webb, M. Folsom, & M. Pfiester. 2005. Observations on periodical cicadas (Brood X) in Indiana and Ohio in 2004 (Hemiptera: Cicadidae: *Magicicada* spp.). *Proceedings of the Indiana Academy of Science* 114: 65–69.

Kritsky, G., & F.N. Young. 1991. Observations on periodical cicadas (Brood XIII) in Indiana in 1990. *Proceedings of the Indiana Academy of Sciences* 100: 45–47.

Kritsky, G., & F.N. Young. 1992. Observations on periodical cicadas (Brood XIV) in Indiana in 1991 (Homoptera: Cicadidae). *Proceedings of the Indiana Academy of Sciences* 101: 59–61.

Kroder, S., J. Samietz, D. Schneider, & S. Dorn. 2006. Adjustment of vibratory signals to ambient temperature in a host-searching parasitoid. *Physiological Entomology* 32: 105–112.

Krohne, D.T., T.J. Couillard, & J.C. Riddle. 1991. Population responses of *Peromyscus leucopus* and *Blarina brevicauda* to emergence of periodical cicadas. *American Midland Naturalist* 126: 317–321.

Krombein, K.V., D.R. Smith, & B.D. Burks. 1979. Catalog of Hymenoptera in America North of Mexico, Volume 2, Apocrita (Aculeata), pages 1199–2209. Smithsonian Institution Press, Washington, D.C.

Krpač, V., T. Trilar, & M. Gogala. 2001. Singing cicadas of Macedonia. *Program and Book of Abstracts, 2nd European Hemiptera Congress, Fiesa, Slovenia, 20–24 June 2001*: 29–30.

Kubo-Irie, M., M. Irie, T. Nakazawa, & H. Mohri. 2003. Ultrastructure and function of long and short sperm in Cicadidae (Hemiptera). *Journal of Insect Physiology* 49: 983–991.

Kubota, S., & H. Tanase. 1999. Recent records of the latest droning of *Cryptotympana facialis* (Homoptera, Cicadidae) at Shirahama, Wakayama Prefecture, Japan. *Nanki Seibutu* 41: 64.

Kubota, S., & H. Tanase. 2002. The earliest droning of *Cryptotympana facialis* (Homoptera, Cicadidae) at the seaside in Shirahama Town, Wakayama Prefecture, Japan. *Nanki Seibutu* 44: 114.

Kühnelt, W. 1949. Die Landtierwelt, mit besonderer Berückschtigung des Lunzer Gebietes. *Das Ybbstal, Wein* 1: 89–154.

Kukalová-Peck, J. 2008. Phylogeny of higher taxa in Insecta: finding synapomorphies in the extant fauna and separating them from homoplasies. *Evolutionary Biology* 35: 4–51.

Kuniata, L.S. 1994. *Cordyceps* sp. an important entomopathogenic fungus of cicada nymphs at Ramu, Papua New Guinea. *Papua New Guinea Journal of Agriculture, Forestry and Fisheries* 37: 47–52.

Kuniata, L.S. 2001. Integrated crop production management of sugarcane at the Ramu Sugar Ltd estate. Pages 335–343 *in* R.M. Bourke, M.G. Allen, & J.G. Salisbury, eds. *Food Security for Papua New Guinea. Proceedings of the Papua New Guinea Food and Nutrition 2000 Conference, Papua New Guinea University ot Technology, Lae, 26–30 June 2000*. ACIAR Proceedings No. 99. The Australian Centre for International Agricultural Research, Canberra.

Kuniata, L.S., & H. Nagaraja. 1992. Biology of *Baeturia papuensis* De Boer (Homoptera: Tibicinidae) on sugarcane in Papua New Guinea. *Science in New Guinea* 18: 65–72.

Kuo, Y.C., S.C. Weng, C.J. Chou, T.T. Chang, & W.J. Tsai. 2003. Activation and proliferation signals in primary human T lymphocytes inhibited by ergosterol peroxide isolated from *Cordyceps cicadae*. *British Journal of Pharmacology* 140: 895–906.

Kurahashi, H. 1996. The systematic position of *Nemoraea cicadina* Kato, 1943, a parasite of the cicada *Tanna japonensis* Distant. *Japanese Journal of Systematic Entomology* 2: 109–113.

Kurokawa, H. 1953. A study of the chromosomes in some species of the Cicadidae and Cercopidae (Homoptera). *Japanese Journal of Genetics* 28: 1–5.

Kysela, E. 2002. Zikaden als Schmuck- und Trachtbestandteil in Römischer Kaiserzeit und Völkerwanderungszeit. *Denisia* 4: 21–28.

Lago, P.K., & S. Testa III. 2000. The terrestrial Hemiptera and Auchenorrhynchous Homoptera of Point Clear Island and surrounding marshlands, Hancock County, Mississippi. *Journal of the Mississippi Academy of Sciences* 45: 184–193.

Lakes-Harlan, R., & U. Köhler. 2003. Influence of habitat structure on the phonotactic strategy of a parasitoid fly *Emblemasoma auditrix*. *Ecological Entomology* 28: 758–765.

Lakes-Harlan, R., H. Stölting, & T.E. Moore. 2000. Phonotactic behaviour of a parasitoid fly (*Emblemasoma auditrix*, Diptera, Sarcophagidae) in response to the calling song of its host cicada (*Okanagana rimosa*, Homoptera, Cicadidae). *Zoology* 103: 31–39.

Lakes-Harlan, R., H. Stölting, & A. Stumpner. 1999. Convergent evolution of insect hearing organs from a preadaptive structure. *Proceedings of the Royal Society of London. Series B. Biological Sciences* 266: 1161–1167.

Lamberton, K. 1994. Nature notebook: The cicada. *New Mexico Wildlife* 39: 9.

Lampe, K.-H., D. Rohwedder, & B. Rach. 2006. Insect types in the ZFMK Collection, Bonn: "Homoptera". 34pp. Available online: http://www.zfmk.de/web/2_Downloads/6_Typen/Hym_Typen_ZFMK_Homoptera.pdf

Lane, D. 1990a. An inquiry into suspected hybridization in zones of overlap involving species of the genus *Kikihia* (Homptera: Cicadoidea: Tibicinidae). *Abstracts of the Combined Meeting of the 7th International Auchenorrhyncha Congress and the 3rd International Workshop on Leafhoppers and Planthoppers of Economic Importance, Wooster, Ohio, August 13–17.*

Lane, D. 1990b. The specific mate recognition system in cicadas of the genus *Kikihia* (Homptera: Cicadoidea: Tibicinidae). *Abstracts of the Combined Meeting of the 7th International Auchenorrhyncha Congress and the 3rd International Workshop on Leafhoppers and Planthoppers of Economic Importance, Wooster, Ohio, August 13–17.*

Lane, D. 1993. Can flawed statistics be a substitute for real biology. *New Zealand Journal of Zoology* 20: 51–59.

Lane, D. 1995. The recognition concept of species applied in an analysis of putative hybridization in New Zealand cicadas of the genus *Kikihia* (Insecta: Hemiptera: Tibicinidae). Pages 367–421 *in* D.M. Lambert & H.G. Spencer, eds. *Speciation and the recognition concept: theory and application.* Johns Hopkins University Press, Baltimore.

Lang, A., & H. Walentowski. 2010. *Handbuch der Lebensraumtypen nach Anhang I der Fauna-Flora- Habitat-Richtlinie in Bayern. Bayerisches Landesamt für* Umweltschutz & Bayerisches Landesanstalt *für* Wald und Forstwirtschaft, Augsburg & Freising-Weihenstephan. 220pp.

Langham, N.P.E. 1990. The diet of feral cats (*Felis catus* L.) on Hawke's Bay farmland, New Zealand. *New Zealand Journal of Zoology* 17: 243–255.

Langley, W.M., K. Anderson, D. Stanley, & C. Frey. 2010. Body temperature and responses of periodical cicadas (Homoptera: Cicadidae) to a potential predator. *Journal of the Kansas Entomological Society* 83: 189–191.

Lara, R.I.R., N.W. Perioto, & S. de Freitas. 2010. Diversidade de hemerobiídeos (Neuroptera) e suas associações com presas em cafeeiros. *Pesquisa Agropecuária Brasileira, Brasilia* 45: 115–123.

Lau, M., G. Reels, & J. Fellows. 1996. The biodiversity survey: highlights of the first year. *Porcupine* 15: 33–35.

Lauterer, P. 1985. Comment on the proposed conservation of *Tibicina* Amyot 1847 and *Lyristes* Horváth 1926. Z.N. (S.)239. *Bulletin of Zoological Nomenclature* 42: 214.

Lee, C.-F., M Satô, & M. Sakai. 2005. The Rhipiceridae of Taiwan and Japan (Insecta: Coleoptera: Dascilloidea). *Zoological Studies* 44: 437–444.

Lee, J.M., M. Hatakeyama, & K. Oishi. 2000. A single and rapid method for cloning insect vitellogenin cDNAs. *Insect Biochemistry and Molecular Biology* 30: 189–194.

Lee, J.M., Y. Nishimori, M. Hatakeyama, T.-W. Bae, & K. Oishi. 2000. Vitellogenin of the cicada Graptopsaltria nigrofuscata (Homoptera): Analysis of its primary structure. *Insect Biochemistry and Molecular Biology* 30: 1–7.

Lee, W., M.-K. Jin, W.-C. Yoo, & J.-K. Lee. 2004. Nanostructuring of a polymeric substrate with well-defined nanometer-scale topography and tailored surface wettability. *Langmuir* 20: 7665–7669.

Lee, Y.-F., Y.-H. Lin, & S.-H. Wu. 2010. Spatiotemporal variation in cicada diversity and distribution, and tree use by exuviating nymphs, in East Asian tropical Reef-Karst forests and forestry plantations. *Annals of the Entomological Society of America* 102: 216–226.

Lee, Y.J. 1995. [*The cicadas of Korea.*] Jonah Publications, Seoul. 157pp. 9pls. [In Korean]

Lee, Y.J. 1998. Habitat and habits of *Cicadetta montana* (Homoptera, Cicadidae) in Korea. *Cicada* 13: 59–61.

Lee, Y.J. 1999. A list of Cicadidae (Homoptera) in Korea. *Cicada* 15: 1–16.

Lee, Y.J. 2000. [A note on *Cicadetta admirabilis* (Homoptera, Cicadidae)]. *Lucanus* 1: 4–5. [In Korean, English summary]

Lee, Y.J. 2001a. Cicadas observed in Soya-do Is. of Korea with a note on *Oncotympana fuscata*. *Cicada* 16: 51–52.

Lee, Y.J. 2001b. [Noisy cicadas and how to control them.] *Lucanus (Seoul)* (2): 1–4. [In Korean, English summary]

Lee, Y.J. 2003a. Some morphological variations of *Platypleura kaempferi* (Hemiptera, Cicadidae) in Korea. *Cicada* 17: 47–48.

Lee, Y.J. 2003b. [On the acoustic signals of cicadas (Hemiptera, Cicadidae)]. *Lucanus* (4): 5–8. [In Korean]

Lee, Y.J. 2003c. Some insects observed in Gureopdo Is. of Korea in summer of 2003. *Lucanus* (4): 15–16.

Lee, Y.J. 2005. [*Cicadas of Korea.*] Geobook, Seoul. 191pp. [In Korean, author sometimes cited as Yi, Y.J.]

Lee, Y.J. 2008a. A checklist of Cicadidae (Insecta: Hemiptera) from Vietnam, with some taxonomic remarks. *Zootaxa* 1787: 1–27.

Lee, Y.J. 2008b. Revised synonymic list of Cicadidae (Insecta: Hemiptera) from the Korean peninsula, with the description of a new species and some taxonomic remarks. *Proceedings of the Biological Society of Washington* 121: 445–467.

Lee, Y.J. 2009a. Description of a new cicada in the genus *Dilobopyga* (Hemiptera: Cicadidae). *Florida Entomologist* 92: 87–90.

Lee, Y.J. 2009b. Taxonomic notes on the genus *Haphsa* (Hemiptera: Cicadidae) with descriptions of two new species. *Florida Entomologist* 92: 330–337.

Lee, Y.J. 2009c. Description of two new species of the *Platylomia spinosa* species group (Hemiptera: Cicadidae). *Florida Entomologist* 92: 338–343.

Lee, Y.J. 2009d. A new genus and new species of the subtribe Cicadina (Hemiptera: Cicadidae: Cicadini). *Florida Entomologist* 92: 470–473.

Lee, Y.J. 2009e. A new genus and a new species of Cicadidae (Insecta: Hemiptera) from China. *Proceedings of the Biological Society of Washington* 122: 87–90.

Lee, Y.J. 2009f. Descriptions of two new species of the *Purana abdominalis* species group (Hemiptera: Cicadidae: Cicadini) from the Philippines, with a key to the species groups of *Purana*. *Journal of Natural History* 43: 1487–1504.

Lee, Y.J. 2009g. Three new species of the *Pomponia linearis* species group (Hemiptera: Cicadidae), with a key to the species. *Proceedings of the Biological Society of Washington* 122: 306–316.

Lee, Y.J. 2009h. A checklist of Cicadidae (Insecta: Hemiptera) in Palawan with five new species. *Journal of Natural History* 43: 2617–2639.

Lee, Y.J. 2009i. Cicadas (Hemiptera: Cicadidae) from Panay, Philippines, with a new species and new genus. *Journal of Asia-Pacific Entomology* 12: 293–295.

Lee, Y.J. 2010a. Cicadas (Insecta: Hemiptera: Cicadidae) of Mindanao, Philippines, with the description of a new genus and a new species. *Zootaxa* 2351: 14–28.

Lee, Y.J. 2010b. A checklist of Cicadidae (Insecta: Hemiptera) from Cambodia, with two new species and a key to the genus *Lemuriana*. *Zootaxa* 2487: 19–32.

Lee, Y.J. 2010c. Checklist of cicadas (Hemiptera: Cicadidae) of Luzon, Philippines, with six new species and revised keys to the species of *Oncotympana* Stål and *Psithyristria* Stål. *Zootaxa* 2621: 1–26.

Lee, Y.J. 2010d. New genus and two new species of Oncotympanina stat. nov. (Hemiptera: Cicadidae: Cicadini) and the erection of Sonatini new tribe. *Journal of Asia-Pacific Entomology* 14: 167–171.

Lee, Y.J., H.-J. Choe, M.-L. Lee, & K.S. Woo. 2002. A phylogenetic consideration of the far eastern Palearctic species of the genus *Cicadetta* Kolenati (Homoptera, Cicadidae) based on the mitochondrial COI gene sequences. *Journal of Asia-Pacific Entomology* 5: 3–11.

Lee, Y.J., & M. Hayashi. 2003a. Taxonomic review of Cicadidae (Hemiptera, Auchenorrhyncha) from Taiwan, Part 1. Platypleurini, Tibicenini, Polyneurini, and Dundubiini (Dundubiina). *Insecta Koreana* 20: 149–185.

Lee, Y.J., & M. Hayashi. 2003b. Taxonomic review of Cicadidae (Hemiptera, Auchenorrhyncha) from Taiwan, Part 2. Dundubiini (a part of Cicadina) with two new species. *Insecta Koreana* 20: 359–392.

Lee, Y.J., & M. Hayashi. 2004. Taxonomic review of Cicadidae (Hemiptera, Auchenorrhyncha) from Taiwan, Part 3. Dundubiini (two other genera of Cicadina), Moganiini, and Huechysini with a new genus and two new species. *Journal of Asia-Pacific Entomology* 7: 45–72.

Lee, Y.J., & K.B.R. Hill. 2010. Systematic revision of the genus *Psithyristria* Stål (Hemiptera: Cicadidae) with seven new species and a molecular phylogeny of the genus and higher taxa. *Systematic Entomology* 35: 277–305.

Lee, Y.J., H.-Y. Oh, & S.G. Park. 2004. A new habitat of *Cicadetta pellosoma* and *Cicadetta isshikii* (Hemiptera, Cicadidae) in Korea and their variations in body coloration. *Journal of Asia-Pacific Entomology* 7: 127–131.

Lee, Y.J., & A.F. Sanborn. 2009. Three new species of the genus *Megapomponia* (Hemiptera: Cicadidae) from Indochina, with a key to the species of *Megapomponia*. *Journal of Asia-Pacific Entomology* 13: 31–39.

Leem, J.-Y., D.-S. Park, E.Y. Suh, J.H. Hur, H.-W. Oh, & H.-Y. Park. 2008. Erratum. Isolation and functional analysis of a 24-residue linear α-helical antimicrobial peptide from Korean blackish cicada, *Cryptotympana dubia* (Homoptera). Doo-Sang Park, Jae Yoon Leem, Eun Young Suh, Jang Hyun Hur, Hyun-Woo Oh, & Ho-Yong Park. *Arch Insect Biochem Physiol* 66(4):204–213, 2007. *Archives of Insect Biochemistry and Physiology* 69: 106.

Lehmann-Ziebarth, N., P.P. Heideman, R.A. Shapiro, S.L. Stoddart, C.C.L. Hsiao, G.R. Stephenson, P.A. Milewski, & A.R. Ives. 2005. Evolution of periodicity in periodical cicadas. *Ecology* 86: 3200–3211.

Lei Z. 1994. [One new species and five new records of the genus *Meimuna* (Homoptera: Cicadidae) from China.] *Entomotaxonomia* 16: 184–188. [In Chinese, English summary]

Lei Z. 1996. [A new species and a new record of the genus *Gudaba* from China (Homoptera: Cicadidae).] *Acta Zootaxonomica Sinica* 21: 210–213. [In Chinese, English summary]

Lei Z. 1997. [A new species of the genus *Sulphogaeana* from China (Homoptera: Cicadidae).] *Acta Zootaxonomica Sinica* 22: 72–74. [In Chinese, English summary]

Lei Z., & Chou I. 1995. [Two new species of the genus *Katoa* (Homoptera: Cicadidae) from China.] *Entomotaxonomia* 17: 99–102. [In Chinese, English summary]

Lei Z., Chou I., & Li L. 1994. [The sound characteristics of cicada songs and their significance in classification (Homoptera: Cicadoidea).] *Entomotaxonomia* 16: 51–59. [In Chinese, English summary]

Lei Z., Chou I., Yao W., & Lu X. 1995. [A study of the genus *Hea* Distant (Homoptera: Cicadidae).] *Entomotaxonomia* 17: 201–204. [In Chinese, English summary]

Lei Z., Jiang J., Li L., & Chou I. 1997. [A comparative study of calling songs oin kaempfer cicada at different regions.] *Acta Entomologica Sinica* 40: 349–357. [In Chinese, English summary]

Lei Z., & Li L. 1994. [A new species and two new records of the genus *Platylomia* (Homoptera: Cicadidae) from China.] *Entomotaxonomia* 16: 91–94. [In Chinese, English summary]

Lei Z., & Li L. 1996. [A new species of the genus *Inthaxara* from China (Homoptera: Cicadidae).] *Acta Entomologica Sinica* 39: 410–411. [In Chinese, English summary]

Lei Z., Zhou Y., Chou I., & Li L. 1995. A new species of the genus *Chremistica* from China (Homoptera: Cicadidae). *Entomologia Sinica* 2: 236–238.

Lenhart, P.A., V. Mata-Silva, & J.D. Johnson. 2010. Foods of the Pallid bat, *Antrozous pallidus* (Chiroptera: Vespertilionidae), in the Chihuahuan Desert of western Texas. *Southwestern Naturalist* 55: 110–115.

Lepesme, P. 1947. *Les insectes des palmiers*. Paul Lechevalier, Paris. 904pp.

Lepley, M. M.,Thevenot, C.-P. Guillaume, P. Ponel, & P. Bayle. 2004. Diet of the nominate Southern Grey Shrike *Lanius meridionalis meridionalis* in the north of its range (Mediterranean France). *Bird Study* 51: 156–162.

Lepschi, B.J. 1993. Food of some Australian birds in eastern New South Wales: Additions to Barker and Vestjens. *Emu* 93: 195–199.

Lepschi, B.J. 1997. Food of some Australian birds in eastern New South Wales: Additions to Barker and Vestjens, part 2. *Emu* 97: 84–87.

Le Quesne, W.J. 1965. Hemiptera, Cicadomorpha (excluding Deltocephalinae and Typhlocybinae). *Handbook for the Identification of British Insects* 2(2a): 1–64.

Leskey, T.C., & J.C. Bergh. 2005. Factors promoting infestation of newly planted nonbearing apple orchards by dogwood borer (Lepidoptera: Sesiidae). *Journal of Economic Entomology* 98: 2121–2132.

Levinson, H., & A. Levinson. 2001. Goethes Insekten und Insekten-Nachbildungen in Weiman. *Spixiana Supplement* 27: 11–32.

Li, J., J. Huang, W. Cai, Z. Zhao, W. Peng, & J. Wu. 2009. The vitellogenin of the bumblebee, *Bombus hypocrita*: studies on structural analysis of the cDNA expression of the mRNA. *Journal of Comparative Physiology. B. Biochemical, Systemic and Environmental Physiology* 180: 161–170.

Li X.-F., Chen X.-S., Wang X.-M., & Hou X.-H. 2007. Contents and existing forms of cantharidin in Meloidae (Coleoptera). *Acta Entomologica Sinica* 50: 750–754.

Li, X.-M., Q.-F. Wang, B. Schofield, J. Lin, S.-K. Huang, & Q. Wang. 2009. Modulation of antigen-induced anaphylaxis in mice by a traditional Chines medicine formula, Gua Min Kang. *American Journal of Chinese Medicine* 37: 113–125.

Liebherr, J.K. 2008. *Mecyclothorax kavanaughi* sp. n. (Coleoptera: Carabidae) from the Finisterre Range, Papua New Guinea. *Tijdschrift voor Entomologie* 151: 147–154.

Lin, C.-P., M.S. Cast, T.K. Wood, & M.-Y. Chen. 2007. Phylogenetics and phlyogeography of the oak treehopper *Platycotis vittata* indicate three distinct North American lineages and a neotropical origin. *Molecular Phylogenetics and Evolution* 45: 750–756.

Lin, C.S. 1971. Bionomics of *Stictia carolina* at Lake Texoma, with notes on some neotropical species (Hymenoptera: Sphecidae). *Texas Journal of Science* 23: 275–286.

Linnavuori, R. 1977. Additional notes on the hemipterous fauna of Somalia. *Bolletino della Societa Entomologica Italiana* 109: 61–65.

Lithgow, G. 1988. Insects collected in and around the Chinchilla Shire. Pages 63–72 *in* R. Hando, ed. *Going bush with the Chinchilla Nats. Twenty years of field observations*. Chinchilla Field Naturalists' Club, Chinchilla.

Liu, A.-Y. 1988. A new record of the genus *Metarhyzium*. *Acta Mycologica Sinica* 7: 191–192.

Liu G.Z. 1992. [The history of research in and the utilization of Cicadidae in China.] *Chinese Journal of Entomology, Special Publication No.* 8: 1–43. [In Chinese]

Liu, K.-C. 1990. All about Chinese cicadas. I. Cicada Science in China. *Chinese Journal of Entomology (Zhonghua Kunchong)* 10: 37–47.

Liu, K.-C. 1991. [Development of cicada science in China.] *Chinese Journal of Entomology* 11: 252–259. [In Chinese]

Liu L.-J., & Jiang J.-C. 1992. Ultrastructure and acoustic function of the sounding membrane in black cicada *Cryptotympana atrata* Fabricius. *Acta Entomologica Sinica* 35: 422–427.

Liu L., Jin R., & Xu G. 1999. HPLC determination of uracil xanthine, hypoxanthine and uridine in ten species of animal drugs. *Zhongguo Zhongyao Zazhi* 24: 73–76, 124.

Liu, X., P.K. Chu, & C. Ding. 2010. Surface nano-functionalization of biomaterials. *Materials Science and Engineering* R 70: 275–302.

Lizer y Trelles, C.A. 1955. Ensayos para la destrucción de la "*Fidicina mannifera*" chicharra de la yerba mate. *Ingeniería Agronómica. Publicacion del Centro Argentino de Ingenieros Agronomos* 13: 3–10.

Lizer y Trelles, C.A. 1957. Nuevos ensayos para combatir la chicharra de la yerba mate *Fidicina mannifera* (Fab.). *Anales de la Sociedad Cientifica Argentina* 164: 81–86.

Llewellyn, L.E., P.M. Bell, & E.G. Moczydlowski. 1997. Phylogenetic survey of soluble saxitoxin-binding activity in pursuit of the function and molecular evolution of saxiphilin, a relative of transferrin. *Proceedings of the Royal Society. Series B. Biological Sciences* 264: 891–902.

Lloyd, J.E. 1973. Model for the mating protocol of synchronously flashing fireflies. *Nature* 245: 268–270.

Lloyd, J.E. 1984. Periodical cicadas. *Antenna* 8: 79–91.

Lloyd, J.E. 1986. A successful rearing of 13-year periodical cicadas beyond their present range and beyond that of 17-year cicadas. *American Midland Naturalist* 117: 362–368.

Lloyd, M., & R. Karban. 1983. Chorusing centers of periodical cicadas. *Journal of the Kansas Entomological Society* 56: 299–304.

Lloyd, M., G. Kritsky, & C. Simon. 1983. A simple Mendelian model for 13- and 17- year life cycles of periodical cicadas, with historical evidence of hybridization between them. *Evolution* 37: 1162–1180.

Lloyd, M., & J. White. 1983. Why is one of the periodical cicadas (*Magicicada septendecula*) a comparatively rare species? *Ecological Entomology* 8: 293–303.

Lloyd, M., & J. White. 1987. Xylem feeding by periodical cicada nymphs on pine and grass roots, with novel suggestions for pest control in conifer plantations and orchards. *Ohio Journal of Science* 87: 50–54.

Lloyd, M., J. White, & N. Stanton. 1982. Dispersal of fungus-infected periodical cicadas to new habitat. *Environmental Entomology* 11: 852–858.

Lo, Y.F., & N.Y.L. Fiona. 2002. A new cicada record for Hong Kong – *Meimuna silhetana* (Cicadidae). *Porcupine* 27: 6–7.

Lockley, T.C., & O.P. Young. 1986. *Phidippus audax* (Araneae, Salticidae) predation upon a cicada (*Tibicen* sp.) (Homoptera, Cicadidae). *Journal of Arachnology* 14: 393–394.

Lodos, N., & M. Boulard. 1987. Bazi Cicadidae (Homoptera: Auchenorrhyncha) türlerinin taninmalarinda sesin taksonomik karakter olarak kullanilmasi üzerinde bir araştirma. *Türkiye I. Entomoloji Kongresi, 13–16 Ekim 1987, Izmir, Bildirileri, Entomoloji Derneği Yayınları* (3): 643–648.

Lodos, N., & A. Kalkandelen. 1981. Preliminary list of Auchenorrhyncha with notes on distribution and importance of species in Turkey. V. Families Flatidae, Ricaniidae and Cicadidae. *Türkiye Bitki Koruma Dergisi* 5: 67–82.

Lodos, N., & A. Kalkandelen. 1988. Preliminary list of Auchenorrhyncha with notes on distribution and importance of Turkey. XXVII. (Addenda and corrigenda). *Türkiye Entomoloji Dergisi* 12: 11–22.

Logan, D. 2006. Nymphal development and lifecycle length of *Kikihia ochrina* (Walker) (Homoptera: Cicadidae). *The Weta* 31: 19–22.

Logan, D., & P.A. Alspach. 2007. Negative association between chorus cicada, *Amphipsalta zelandica*, and armillaria root disease in kiwifruit. *New Zealand Plant Protection* 60: 235–240.

Logan, D., & P. Connolly. 2005. Cicadas from kiwifruit orchards in New Zealand and identification of their final instar exuviae (Cicadidae: Homoptera). *New Zealand Entomologist* 28: 33–44.

Logan, D.P., M.G. Hill, P.G. Connolly, B.J. Maher, & S.J. Dobson. 2010. Influences of landscape structure on endemic cicadas in New Zealand kiwifruit orchards. *Agricultural and Forest Entomology* 13: 259–271.

Logan, D., & B.J. Maher. 2009. Options for reducing the number of chorus cicada, *Amphipsalta zelandica* (Boisduval), in kiwifruit orchards. *New Zealand Plant Protection* 62: 268–273.

Londt, J.G.H. 2006. Predation by Afrotropical Asilidae (Diptera): an analysis of 2000 prey records. *African Entomology* 14: 317–328.

Long, C.A. 1993. Evolution in periodical cicadas: a genetical explanation. *Evolutionary Theory* 10: 209–220.

Lorenz, M.W., & G. Gäde. 2009. Hormonal regulation of energy metabolism in insects as a driving force for performance. *Integrative and Comparative Biology* 29: 380–392.

Lovari, S., P. Valier, & M. Ricci Lucchi. 1994. Ranging behaviour and activity of red foxes (*Vulpes vulpes*: mammalia) in relation to environmental variables, in a Mediterranean mixed pinewood. *Journal of Zoology, London* 232: 323–339.

Luken, J.O., & P.J. Kalisz. 1989. Soil disturbance by the emergence of periodical cicadas. *Soil Science Society of America Journal* 53: 310–313.

Luk'yanova, O.N. 1992. [Cicadidae (Homoptera, Auchenorrhyncha) fauna in the "Ramit" reserve.] *Izvestiya Akademii Nauk Respubliki Tadzhikistan Otdelenie Biologicheskikh Nauk* 0(1): 29–32. [In Russian]

Lunz, A.M., R. de Azevedo, M.M. Júnior, O.M. Monteiro, A. Lechinoski, & L.Z. Zaneti. 2010. Método para monitoramento de ninfas de cigarras e controle com inseticidas em reflorestamentos com paricá. *Pesquisa Agropecuária Brasileira, Brasilia* 45: 631–637.

Luthy, P. 1974. *Corynebacterium okanaganae*, an entomopathogenic species of the Corynebacteriaceae. *Canadian Journal of Microbiology* 20: 791–794.

Ma, S.P., R. Qu, & B.Q. Hang. 1989. [Immulogic inhibition and anti-allergic action of *Cryptotympana atrata* Fabricius.] *Chung-Kuo Chung Yao Tsu Chih (China Journal of Chinese Materia Medica)* 14: 490–493. [In Chinese]

Maccagnan, D.H.B., & N.M. Martinelli. 2002. Descrição e caracterização das ninfas de *Quesada gigas* (Olivier, 1790) (Hemiptera: Cicadidae) associadas ao cafeeiro. *XIX Congresso Brasileiro de Entomologia, 2002, Manaus. Anais da Sociedade Entomológica do Brasil* 19: 234.

Maccagnan, D.H.B., & N.M. Martinelli. 2004. Descrição das ninfas de *Quesada gigas* (Olivier) (Hemiptera: Cicadidae) associadas ao cafeeiro. *Neotropical Entomology* 33: 439–446.

Maccagnan, D.H.B., N.M. Martinelli, P.R.R. Prado, & F.M. Sene. 2006. Descrição do som emitido por *Majeorona aper* (Walker, 1850) (Hemiptera: Cicadidae). *XXI Congresso Brasileiro de Entomologia, 2006, Recife - PE. Anais do XXI Congresso Brasileiro de Entomologia* 21: 205.

Maccagnan, D.H.B., N.M. Martinelli, & F.M. Sene. 2007. Padrão de emergência de *Quesada gigas* (Olivier, 1790) (Hemiptera: Cicadidae). Pages 230–233 *in* C.J. Ávila & R.P.S. Júnior, eds. *10ª Reunião Sul-Brasileira sobre pragas de solo, Pragas-Solo-Sul, Anaise Ata, 25–27 September 2007, Dourados, MS.* Embrapa Agropecuária Oeste, Dourados.

Maccagnan, D.H.B., N.M. Martinelli, F.M. Sene, & P.R.R. Prado. 2006. Estudo da emergência de cigarras (Hemiptera: Cicadidae) no Campus da UNESP de Jaboticabal. *XXVI Congresso Brasileiro de Zoologia, 2006, Londrina - PR. Anais do XXVI Congresso Brasileiro de Zoologia.*

Maccagnan, D.H.B., P.R.R. Prado, F.M. Sene, & N.M. Martinelli. 2006a. Etologia sonora de *Fidicina mannifera* e *Dorisiana viridis* (Hemiptera: Cicadidae) em Jaboticabal, SP. *XXVI Congresso Brasileiro de Zoologia, 2006, Londrina - PR. Anais do XXVI Congresso Brasileiro de Zoologia*

Maccagnan, D.H.B., P.R.R. Prado, F.M. Sene, & N.M. Martinelli. 2006b. Etologia sonora de *Quesada gigas* (Olivier, 1790) (Hemiptera: Cicadidae). *XXI Congresso Brasileiro de Entomologia, 2006, Recife - PE. Anais do XXI Congresso Brasileiro de Entomologia* 21: 205.

Macfarlane, R.P. 1979. Notes on insects of the Chatham Islands. *New Zealand Entomologist* 7: 64–70.

Machin, J., J.J.B. Smith, & G.J. Lampert. 1994. Evidence for hydration-dependent closing of pore structures in the cuticle of *Periplaneta americana. Journal of Experimental Biology* 192: 83–94.

MacNally, R.C. 1988a. Simultaneous modelling of distributional patterns in a guild of eastern-Australian cicadas. *Oecologia* 76: 246–253.

MacNally, R.C. 1988b. On the statistical significance of the Hutchinsonian size-ratio parameter. *Ecology* 69: 1974–1982.

MacNally, R.C., & J.M. Doolan. 1982. Comparative reproductive energetics of the sexes in the cicada *Cystosoma saundersii. Oikos* 39: 179–186.

MacNally, R.C., & J.M. Doolan. 1986a. An empirical approach to guild structure: habitat relationships in nine species of eastern-Australian cicadas. *Oikos* 47: 33–46.

MacNally, R.C., & J.M. Doolan. 1986b. Patterns of morphology and behaviour in a cicada guild: a neutral model analysis. *Australian Journal of Ecolgoy* 11: 279–294.

MacNally, R., & D. Young. 1981. Song energetics of the bladder cicada, *Cystosoma saundersii. Journal of Experimental Biology* 90: 185–196.

Macnamara, P. 2005. *Illinois insects and spiders*. The Univeristy of Chicago Press, Chicago. 118pp.

Mader, L. 1922. *Das Insektenleben Österreichs; mit einem Anhang über Gallen und ahnliche Pflanzenverunstaltungen samt deren Erzeuger*. Ein Handbuh und Wegweiser für Naturfreunde jeder Art. Hölder-Pichler-Tempsky A.G., Wien. 216pp.

Maes, J.-M. 1994. *Insectos y ácaros asociados al cultivo de cafeto (Coffee arabica)(Rubiaceae) y sus enemigos naturales*. Memoria de la Reunion Informativa sobre Avances de Investigación. Grupe de Entomologios de Café, Escuela de Ecologia, Universidad Centroaméricana, Managua, Octubre 1994. 29pp.

Maes, J.-M. 1998. *Insectos de Nicaragua.* Volume I. Secretaria, Técnica de Reserva de Biósfera de Nicaragua (BOSAWAS), Cooperación Nicaragua-Alemania: Deutsche Gesellschaft für Zusammenarbeit (GTZ): Ministerio del Ambiente y Recursos Naturales: Unesco. Marena, Managua, Nicaragua. 485pp.

Maes, J.-M., & J. Téllez Robleto. 1988. Catálogo de los insectos y artrópodos terrestres asociados a las principales plantas de importancia económica en Nicaragua. *Revista Nicaragüense Entomología* 5: 1–95.

Maezono, Y., & T. Miyashita. 2000. *Folsomia candida* (Collembola: Isolomidae) living in nest cell of a cicada *Graptopsaltria nigrofuscata* underground. *Edaphologia* 66: 51–57.

Maier, C.T. 1982a. Observations on the seventeen-year periodical cicada, *Magicicada septendecim* (Hemiptera: Homoptera: Cicadidae). *Annals of the Entomological Society of America* 75: 14–23.

Maier, C.T. 1982b. Abundance and distribution of the seventeen-year periodical cicada, *Magicicada septendecim* (Linnaeus) (Hemiptera: Cicadidae - Brood II), in Connecticut. *Proceedings of the Entomological Society of Washington* 84: 430–439.

Maier, C.T. 1985. Brood VI of 17-year periodical cicadas, *Magicicada* spp. (Hemiptera: Cicadidae): New evidence from Connecticut, the hypothetical 4-year deceleration, and the status of the brood. *Journal of the New York Entomological Society* 93: 1019–1026.

Maier, C.T. 1996. Connecticut is awaiting return of the periodical cicada. *Frontiers of Plant Science* 48: 4–6.

Maitland, P., & D. Maitland. 1985. Spineless but not speechless. A selection of invertebrate sound. *Australian Natural History* 21: 331–334.

Májsky, J., & V. Janský. 2006. Dva vzácne druhy veľkých cikád z Tematínskych vrchov. Pages 74–75 *in* K. Rajcová, ed. *Najvzácnejšie prírodné hodnoty Tematínskych vrchov. Zborník výsledkov inventarizačného výskumu územia európskeho významu Tematínske vrchy*. KOZA, Trenčín a Pre Prírodu, Trenčín.

Malaisse, F., & G. Parent. 1997. Chemical composition and energetic value of some edible products provided by hunting or gathering in the open forest (Miombo). *Geo-Eco-Trop* 21: 65–71.

Malherbe, C., M. Burger, & R.D. Stephen. 2004. Rediscovery of the cicada *Pycna sylvia* (Distant, 1899): including behavioural and ecological notes for the species (Hemiptera: Cicadidae). *Annals of the Transvaal Museum* 41: 85–89.

Malhi, C.S., & V.R. Parashad. 1991. Burrow parasitism and killing behaviour in spiny mouse, *Mus platythrix* and its association with cicada (*Platypleura octoguttata*) and toad (*Bufo bufo*). *Zeitschrift für Angewandte Zoologie* 78: 243–245.

Malaisse, F. 1997. *Se nourrir en forêt claire africaine: approche écologique et nutritionelle*. Les Presses Agronomique de Gembloux, Gembloux. 384pp.

Malloch, J.R. 1922. *Panorpa rufescens* feeding on a cicada (Neuroptera). *Bulletin of the Brooklyn Entomological Society* 17: 45.

Mangold, J.R. 1978. Attraction of *Euphasiopteryx ochracea*, *Corethrella* sp. and gryllids to broadcast songs of the southern mole cricket. *Florida Entomologist* 61: 57–61.

Marcello, G.J., S.M. Wilder, & D.B. Meikle. 2008. Population dynamics of a generalist rodent in relation to variability in pulsed food resources in a fragmented landscape. *Journal of Animal Ecology* 77: 41–46.

Marchand, P.J. 2002. Jammming cicadas. *Natural History* 111(5): 36–39.

Marcu, O. 1929. Contributiuni la cunoasterea faunei Olteniei. *Arhivele Olteniei* 45–46: 474–479.

Marden, J.H. 2000. Variability in the size, composition, and function of insect flight muscles. *Annual Review of Physiology* 62: 157–178.

Mariconi, F.A.M. 1958. *Inseticidas e seu emprego no combate às pragas*. Editôra Agronômica "Ceres" Ltda., Sao Paulo. 531pp.

Markus, M., O. Schulz, & E. Goles. 2002. Prey population cycles are stable in an evolutionary modeil if and only if their periods are prime. *ScienceAsia* 28: 199–203.

Marques, O.M., N.M. Martinelli, R.L. Azevedo, M.L. Coutinho, & J.M.L. Serra. 2004. Ocorrência de duas espécies de cigarras (Hemiptera: Cicadidae) no estado da Bahia, Brasil. *Magistra, Cruz das Almas-Bahia* 16: 120–121.

Marshall, A.T. 1983. X-ray microanalysis of the filter chamber of the cicada, *Cyclochila australasiae* Don.: A water-shunting epithelial complex. *Cell and Tissue Research* 231: 215–227.

Marshall, D.C. 2001. Periodical cicada (Homoptera: Cicadidae) life-cycle variations, the historical emergence record, and the geographic stability of brood distributions. *Annals of the Entomological Society of America* 94: 386–399.

Marshall, D.C. 2008. Periodical cicadas, *Magicicada* spp. (Hemiptera: Cicadidae). Pages 2785–2794 *in* J. Capinera, ed. *Encyclopedia of entomology*, second edition, volume 3. Springer-Verlag, Heidelberg.

Marshall, D.C. 2009. Cryptic failure of partitioned Bayesian phylogenetic analyses: lost in the land of long trees. *Systematic Biology* 59: 108–117.

Marshall, D.C., & J.R. Cooley. 2000. Reproductive character displacement and speciation in periodical cicadas, with description of a new species, 13-year *Magicicada neotredecim*. *Evolution* 54: 1313–1325.

Marshall, D.C., J.R. Cooley, R.D. Alexander, & T.E. Moore. 1996. New records of Michigan Cicadidae (Homoptera), with notes on the use of songs to monitor range changes. *Great Lakes Entomologist* 29: 165–169.

Marshall, D.C., J.R. Cooley, K.B.R. Hill, & C. Simon. 2005. Contrasting effects of Pliocene- and Pleistocene-age environmental changes on speciation in New Zealand cicadas. *Abstracts of Talks and Posters, 12th International Auchenorrhyncha Congress, 6th International Workshop on Leafhopper and Planthoppers of Economic Significance, University of California, Berkeley, August 7–12, 2005*: S-21-S-22.

Marshall, D.C., J.R. Cooley, & C. Simon. 2003. Holocene climate shifts, life cycle plasticity, and speciation in periodical cicadas: a reply to Cox and Carlton. *Evolution* 57: 433–437.

Marshall, D.C., & K.B.R. Hill. 2009. Versatile aggressive mimicry of cicadas by an Australian predatory katydid. *PLoS One* 4(1): e4185. 8pp. doi:10.1371/journal.pone.0004185.

Marshall, D.C., K.B.R. Hill, K.M. Fontaine, T.R. Buckley, & C. Simon. 2009. Glacial refugia in a maritime temperatre climate: cicada (*Kikihia subalpina*) mtDNA phylogeography in New Zealand. *Molecular Ecology* 18: 1995–2009.

Marshall, D.C., C. Simon, & T.R. Buckley. 2006. Accurate branch length estimation in partitioned Bayesian analyses requires accommodation of among-partition rate variation and attention to branch length priors. *Systematic Biology* 55: 993–1003.

Marshall, D.C., K. Slon, J.R. Cooley, K.B.R. Hill, & C. Simon. 2008. Steady plio-pleistocene diversification and a 2-million-year sympatry threshold in a New Zealand cicada radiation. *Molecular Phylogenetics and Evolution* 48: 1054–1066.

Marshall, S.A. 2006. *Insects: their natural history and diversity with a photographic guide to insects of eastern North America*. Firefly Books, Buffalo. 718pp.

Marshall, S.A. 2009. *Insects A to Z*. Firefly Books, Ltd., Richmond Hill, Ontario. 32pp.

Martel, R.R., & J.H. Law. 1992. Hemolymph titers, chromophore association and immunological cross-reactivity of an ommochrome-binding protein form the hemolymph of the tobacco hornworm, *Manduca sexta*. *Insect Biochemistry and Molecular Biology* 22: 561–569.

Martin, A.P., & C. Simon. 1988. Anomalous distribution of nuclear and mitochondrial DNA markers in periodical cicadas. *Nature* 336: 237–239.

Martin, A., & C. Simon. 1990a. Differing levels of among-population divergence in the mitochondrial DNA of periodical cicadas related to historical biogeography. *Evolution* 44: 1066–1080.

Martin, A., & C. Simon. 1990b. Temporal variation in insect life cycles: lessons from periodical cicadas. *BioScience* 40: 359–367.

Martin, J.H., & M.D. Webb. 2000. Hemiptera: the true bugs. Pages 54–75 *in* P.C. Barnard, ed. *Identifying British Iinsects and Arachnids: an annotated bibliography of key works*. Cambridge University Press, Cambridge.

Martinelli, N.M. 1990. As cigarras que causam danos à cultura. *Correio Agrícola* 1990(28 Fev.): 11–13.

Martinelli, N.M. 2004. Cigarras do cafeeiro. Pages 517–541 *in* J.R. Salvadori, C.J. Ávila, & M.T. Braga da Silva, eds. *Pragas de Solo do Brasil*. Embrapa Trigo, Passo Fundo.

Martinelli, N.M. 2006. Pragas do solo: cigarras do cafeeiro. 2006. *VI Encontro Sobre "Pragas e Doenças do Cafeeiro."* 4pp.

Martinelli, N.M., M.F.D. Lima, & T.K. Matuo. 2001. Avaliação da eficiência do imidacloprid, aplicado via líquida, para o controle de cigarras do cafeeiro. *XXVII Congresso Brasileiro de Pesquisas Cafeeiras, 2001, Uberaba. Anais do XXVII Congresso Brasileiro de Pesquisas Cafeeiras* 27: 125–126.

Martinelli, N.M., & H.N. Lusvarghi. 1998. Controle de cigarras do cafeeiro com terbufós em duas formulações. *XXIV Congresso Brasileiro de Pesquisas Cafeeiras, 1998, Poços de Caldas. Anais do XXIV Congresso Brasileiro de Pesquisas Cafeeiras* 24: 115.

Martinelli, N.M., & D.H.B. Maccagnan. 2006. Emergência de *Carineta spoliata* (Walker, 1858) (Hemiptera: Cicadoidea: Tibicinidae). *XXI Congresso Brasileiro de Entomologia, 2006, Recife - PE. Anais do XXI Congresso Brasileiro de Entomologia* 21: 246.

Martinelli, N.M., D.H.B. Maccagnan, T. Matuo, S. Silveira Neto, P.R.R. Prado, & F.M. Sene. 2006. Atração de *Quesada gigas* (Olivier, 1790) por som emitido via playback. *XXI Congresso Brasileiro de Entomologia, 2006, Recife - PE. Anais do XXI Congresso Brasileiro de Entomologia* 21: 246.

Martinelli, N.M., D.H.B. Maccagnan, R. Ribiero, M.F.A. Pereira, & R.S. Santos. 2004. Constatação de nova espécie de cigarra (Hemiptera: Cicadidae) associada ao cafeeiro no município de Sarutaiá, SP. *XX Congresso Brasileiro de Entomologia, 2004, Gramado*: 619.

Martinelli, N.M., T. Matuo, & J.H. Fiorelli. 2000. Eficiência do inseticida Premier (Imidacloprid) para o controle de cigarras do cafeeiro. *XXVI Congresso Brasileiro de Pesquisas Cafeeiras, 2000, Marília. Anais XXVI Congresso Brasileiro de Pesquisas Cafeeiras* 26: 102–103.

Martinelli, N.M., T. Matuo, D.R. Martins, & E.B. Malheiros. 2000. Efeito de dosagens de inseticidas granulados aplicados com diferentes métodos para o controle das cigarras do cafeeiro. *Ecossistema* 25: 209–211.

Martinelli, N.M., T.K. Matuo, D.R. Martins, & E.B. Malheiros. 2003. Eficiência de máquina de aplicação para o controle das cigarras do cafeeiro. *Ecossistema* 28: 31–34. 2003.

Martinelli, N.M., T. Matuo, M.R. Yamada, & E.B. Malheiros. 1998. Modo de aplicação e eficiência de inseticidas granulados sistêmicos para o controle de Cigarras (Hemiptera: Cicadidae) do cafeeiro. *Anais da Sociedade Entomológica do Brasil* 27: 133–140.

Martinelli, N.M., R.D. Vieira, & R.A. Zucchi. 1986a. Ocorrência de *Quesada gigas* (Olivier, 1790) (Hom., Cicadidae) em cacaueiros no Estado de São Paulo. *X Congresso Brasileiro de Entomologia, 1986, Rio de Janeiro. Anais do 10º Congresso Brasileiro de Entolomologia* 10: 119.

Martinelli, N.M., R.D. Vieira, & R.A. Zucchi. 1986b. Descrição e ocorrência de *Quesada gigas* (Olivier, 1790) (Hom. Cicadidae) em cacaueiros no Estado de São Paulo. *Ciência Agronômica, Jaboticabal* 1: 5–6.

Martinelli, N.M., & R.A. Zucchi. 1984. Ocorrência de *Fidicina pronoe* (Walker, 1850) (Hom., Cicadidae) em cafeeiros. *IX Congresso Brasileiro de Entomologia, 1984, Londrina. Anais do 9º Congresso Brasileiro de Entomologia* 9: 1.

Martinelli, N.M., & R.A. Zucchi. 1986. Novas constatocoes de especes de cigarras (Hom., Cicadidae) em cafeeiros. *Resumos Congresso Brasileiro de Entomologia* 10: 16.

Martinelli, N.M., & R.A. Zucchi. 1987a. Cigarras associadas ao cafeeiro. I. Gênero *Quesada* Distant, 1905 (Homoptera, Cicadidae, Cicadinae). *Anais da Sociedade Entomológica do Brasil* 16: 51–60.

Martinelli, N.M., & R.A. Zucchi. 1987b. Espécies de cigarras (Hom., Cicadidae - Tibicinidae) associadas ao cafeeiro no Brasil. *XI Congresso Brasileiro de Entomologia, 1987, Campinas. Anais do 11º Congresso Brasileiro de Entomologia* 11: 469.

Martinelli, N.M., & R.A. Zucchi. 1987c. Distribuição geográfica das espécies de cigarras associadas ao cafeeiro no Brasil. *XI Congresso Brasileiro de Entomologia, 1987, Campinas. Anais do 11º Congresso Brasileiro de Entomologia* 11: 470.

Martinelli, N.M., & R.A. Zucchi. 1989a. Cigarras associadas ao cafeeiro. II. Gênero *Fidicina* Amyot & Serville, 1843 (Homoptera, Cicadidae, Cicadinae). *Anais da Sociedade Entomológica do Brasil* 18: 5–12.

Martinelli, N.M., & R.A. Zucchi. 1989b. Cigarras associadas ao cafeeiro. III. Gênero *Dorisiana* Metcalf, 1952 (Homoptera, Cicadidae, Cicadinae). *Anais da Sociedade Entomológica do Brasil* 18(Supl.): 5–12.

Martinelli, N.M., & R.A. Zucchi. 1989c. Cigarras associadas ao cafeeiro. IV. Gênero *Carineta* A. & S., 1843 (Homoptera, Tibicinidae, Tibicininae). *Anais da Sociedade Entomológica do Brasil* 18 (Supl.): 13–22.

Martinelli, N.M., & R.A. Zucchi. 1993a. Caracterização de exúvias de três espécies de cigarras (Hemiptera - Cicadidae). *XIV Congresso Brasileiro de Entomologia, 1993, Piracicaba. Anais do 14° Congresso Brasileiro de Entomologia* 14: 3.

Martinelli, N.M., & R.A. Zucchi. 1993b. Novos registros de hospedeiros para três espécies de cigarras (Hemiptera: Cicadidae). *XIV Congresso Brasileiro de Entomologia, 1993, Piracicaba. Anais do 14° Congresso Brasileiro de Entomologia* 14: 174.

Martinelli, N.M., & R.A. Zucchi. 1997a. Novos registros de *Carineta fasciculata* (Hemiptera - Tibicinidae) em cafezais de três estados brasileiros. *XVI Congresso Brasileiro de Entomologia, 1997, Salvador. Anais do 16° Congresso Brasileiro de Entomologia* 16: 208.

Martinelli, N.M., & R.A. Zucchi. 1997b. Sobre algumas espécies de *Fidicina* depositadas em coleções de instituições brasileiras. *XVI Congresso Brasileiro de Entomologia, 1997, Salvador. Anais do 16° Congresso Brasileiro de Entomologia* 16: 346–347.

Martinelli, N.M., & R.A. Zucchi. 1997c. Cigarras (Hemiptera: Cicadidae: Tibicinidae) associadas ao cafeeiro: Distribuição, hospedeiros e chave para as espécies. *Anais da Sociedade Entomológica do Brasil* 26: 133–143.

Martinelli, N.M., & R.A. Zucchi. 1997d. Primeiros registros de plantas hospedeiras de *Fidicina mannifera, Quesada gigas* e *Dorisiana drewseni* (Hemiptera-Cicadidae). *Revista de Agricultura* 72: 271–281.

Martinelli, N.M., R.A. Zucchi, S. Silveira Neto & J.R.P. Parra. 1989. Ocorrência de *Fidicina pronoe* (Walker, 1850) (Hom.: Cicadidae) em cafeeiros de Araguari, MG. *XII Congresso Brasileiro de Entomologia, 1989, Belo Horizonte. Anais do 12° Congresso Brasileiro de Entomologia* 12: 168.

Matsuda, R. 1960. Morphology of the pleurosternal region of the pterothorax in insects. *Annals of the Entomological Society of America* 53: 712–731.

Mattson, W.J., Jr. 1980. Herbivory in relation to plant nitrogen content. *Annual Review of Ecology and Systematics* 11: 119–161.

Matsushima, A., S. Sato, Y. Chuman, Y. Takeda, S. Yokotani, T. Nose, Y. Tominaga, M. Shimohigashi, & Y. Shimohigashi. 2004. cDNA cloning of the housefly pigment-dispersing factor (PDF) precursor protein and its peptide comparison among the insect circadian neuropeptides. *Journal of Peptide Science* 10: 82–91.

Matsushima, A., S. Yokotani, X. Lui, K. Sumida, T. Honda, S. Sato, A. Kaneki, Y. Takeda, Y. Chuman, M. Ozaki, D. Asai, T. Nose, H. Onoue, Y. Ito, Y. Tominaga, Y. Shimohigashi, & M. Shimohigashi. 2003. Molecular cloning and circadian expression profile of insect neuropeptide PDF in black blowfly, *Phormia regina. Letters in Peptide Science* 10: 419–430.

Matthews, C.E., & D. Hildreth. 1997. Cicada studies: unravelling the mysteries of a unique insect. *The Science Teacher* 64: 34–37.

Matthews, R.W., & J.R. Matthews. 2010. *Insect behavior*, second edition. Springer, Dordrecht Heidelberg London New York. 514pp.

Mattingly, W.B., & S.L. Flory. 2010. Plant architecture affects periodical cicada oviposition behavior on native and non-native hosts. *Oikos* 120: 1083–1091.

Matuo, T.K., C.G. Raetano, T. Matuo, N.M. Martinelli, & G.J. Leite. 2008. Desenvolvimento de equipamento motorizado para aplicação líquida de produtos fitossanitários na cultura do café. *Engenharia Agrícola, Jaboticabal* 28: 543–553.

Maw, H.E.L., R.G. Foottit, K.G.A. Hamilton, & G.G.E. Scudder. 2000. *Checklist of the Hemiptera of Canada and Alaska*. National Research Council Research Press, Ottawa. 220pp.

McCoy, C.E. 1965. A synopsis of the Cicadidae of Arkansas (Homoptera). *Proceedings of the Arkansas Academy of Science* 19: 40–45.

McCutcheon, J.P., B.R. McDonald, & N.A. Moran. 2009a. Origin of an alternative genetic code in the extremely small and GC-rich genome of a bacterial symbiont. *PLoS Genetics* 5(7): e1000565. 11pp. doi: 10.1371/journal.pgen.1000565.

McCutcheon, J.P., B.R. McDonald, & N.A. Moran. 2009b. Convergent evolution of metabolic roles in bacterial co-symbionts of insects. *Proceedings of the National Academy of Science of the United States of America* 106: 15394–15399.

McCutcheon, J.P., & N.A. Moran. 2010. Functional convergence in reduced genomes of bacterial symbionts spanning 200 my of evolution. *Genome Biology and Evolution* 2: 708–718.

McDouall, A. 1998. *Land's End to Minehead Maritime Natural Area. A nature conservation profile*. English Nature, Truro. 81pp.

McGavin, G.C. 1993. *Bugs of the world.* Facts on File, Inc. New York. 193pp.

McGavin, G.C. 2000. *Insects, spiders and other terrestrial arthropods*. Dorling Kindersley, inc., New York. 256pp.

McGhee, K. 2002. Cicada clocks. *Nature Australia* 27: 12.

McKilligan, N.G. 1984. The food and feeding ecology of the cattle egret, *Ardeola ibis*, when nesting in South-East Queensland. *Australian Wildlife Research* 11: 133–144.

McLister, J.D. 2001. Physical factors affecting the cost and efficiency of sound production in the treefrog *Hyla versicolor*. *Journal of Experimental Biology* 204: 69–80.

Meagher, R.L., S.W. Wilson, H.D. Blocker, R.V.W. Eckel, & R.S. Pfannenstiel. 1993. Homoptera associated with sugarcane fields in Texas. *Florida Entomologist* 76: 508–514.

Medler, J.T. 1980. Insects of Nigeria -- Checklist and bibliography. *Memoirs of the American Entomological Institute* 30: 1–919.

Medler, S., & K. Hulme. 2009. Frequency-dependent power output and skeletal muscle design. *Comparative Biochemistry and Physiology, Part A: Molecular & Integrative Physiology* 152: 407–417.

Meelkop, E., L. Temmerman, L. Schoofs, & T. Janssen. 2010. Signalling through pigment dispersing lormone-like peptides in invertebrates. *Progress in Neurobiology* 93: 125–147.

Meer Mohr, J.C. van der. 1965. Insects eaten by the Karo-Batak people (a contribution to entomo-bromatology). *Entomologische Berichten, Amsterdam* 25: 101–107.

Meineke, T., & A. Krügeneer. 2005. Natur und Nutzung im FFH-Gebiet "Dörnberg, Immelburg und Helfenstein" bei Zierenberg im Landkreis Kassel. *Jahrbuch Naturschutz in Hessen* 9: 138–153.

Méléard, S., & V.C. Tran. 2009. Trait substitution sequence process and Canonical Equation for age-structured populations. *Journal of Mathematical Biology* 58: 881–921.

Melville, R.V., & R.W. Sims. 1984. *Tibicina* Amyot, 1847 and *Lyristes* Horvath, 1926 (Insecta, Hemiptera, Homoptera): Proposed conservation by the suppression of *Tibicen* Berthold, 1827. Z.N.(S.) 239. *Bulletin of Zoological Nomenclature* 41: 163–184.

Meng, C., & L.S. Kubatko. 2009. Detecting hybrid speciation in the presence of incomplete lineage sorting using gene tree incongruence: a model. *Theoretical Population Biology* 75: 35–45.

Mengesha, T.E., R.R. Vallance, M. Barraja, & R. Mittal. 2009. Parametric structural modeling of insect wings. *Bioinspiration & Biomimetics* 4: 036004. 15pp. Available online: stacks.iop.org/BB/4/036004

Menninger, H.L., M.A. Palmer, L.S. Craig, & D.C. Richardson. 2008. Periodical cicada detritus impacts stream ecosystem metabolism. *Ecosystems* 11: 1306–1317.

Menon, F. 2005. New record of Tettigarctidae (Insecta, Hemiptera, Cicadidae) from the Lower Cretaceous of Brazil. *Zootaxa* 1087: 53–58.

Merrett, M.F., B.D. Clarkson, & J.L. Bathgate. 2002. Ecology and conservation of *Alseuosmia quercifolia* (Alseuosmiaceae) in the Waikato region, New Zealand. *New Zealand Journal of Botany* 40: 49–63.

Merrick, M.J., & R.J. Smith. 2004. Temperature regulation in burying beetles (*Nicrophorus* spp.) Coleoptera: Silphidae): effects of body size, morphology and environmental temperature. *Journal of Experimental Biology* 207: 723–733.

Messner, B., & J. Adis. 1992. Kutikuläre Waaschsausscheidungen als plastronhaltende Strukturen bei Larven von Schaum- und Singzikaden (Auchenorrhyncha: Cercopidae und Cicadidae). (Cuticular wax secretions as plastron retaining structures in larvae of Spittlebugs and Cicada (Auchenorrhyncha, Cercopidae and Cicadidae)). *Revue Suisse de Zoologie* 99: 713–720.

Metevelis, P. 1992. Martyred childe of God. *Asian Folklore Studies* 51: 183–198.

Mey, E., & H. Steuer. 1985. Vorkommenn von *Cicadetta montana* (Scopoli, 1772) bei Rudeolstadt, Thur. (Insecta, Homoptera, Auchenorrhyncha, Cicadidae). *Faunistische Abhandlungen Staatliches Museum für Tierkunde Dresden* 13: 110.

Meyer-Rochow, V.B. 1973. Edible insects in three different ethnic groups of Papua and New Guinea. *The American Journal of Clinical Nutrition* 26: 673–677.

Michelsen, A., & N. Elsner. 1999. Sound emission and the acoustic far field of a singing acridid grasshopper (*Omocestus viridulus* L.). *Journal of Experimental Biology* 202: 1571–1577.

Michelsen, A., & P.J. Fonseca. 2000. Spherical sound radiation patterns of singing grass cicadas, *Tympanistalna gastrica*. *Journal of Comparative Physiology A. Sensory, Neural, and Behavioral Physiology* 186: 163–168.

Michelsen, A., & O.N. Larsen. 1985. Hearing and sound. Pages495–556 *in* G.A. Kerkut & L.I. Gilbert, eds. *Comprehensive insect physiology, biochemistry and pharmacology. Volume 6. Nervous system: Sensory*. Pergamon Press, Oxford.

Milius, S. 2000. Cicada subtleties, what part of 10,000 cicadas screeching don't you understand? *Science News* 157: 408–410.

Milius, S. 2003. African cicadas warm up before singing. *Science News* 163: 408.

Miller, D. 1952. Insect people of the Maori. *Journal of the Polynesian Society (Wellinton, N.Z.)* 61: 1–61.

Miller, F., & W. Crowley. 1998. Effects of periodical cicada oviposition injury on woody plants. *Journal of Arboriculture* 24: 248–253.

Miller, F.D. 1997. Effects and control of periodical cicada *Magicicada septendecim* and *Magicicada cassini* oviposition injury on urban forest trees. *Journal of Arboriculture* 23: 225–232.

Miller, S.A., & J.B. Harley. 1999. *Zoology*, 4th edition. WCB/McGraw-Hill, Boston. i-xvi, 750pp.

Miller, S.E. 1985. The California Channel Islands -- Past, present, and future: An entomological perspective. Pages 3–27 *in* A.S. Menke & D.R. Miller, eds. *Entomology of the California Channel Islands*. Proceedings of the first symposium. Santa Barbara Museum of Natural History, Santa Barbara.

Milton, S.J., & W.R.J. Dean. 1992. An underground index of rangeland degradation: cicadas in arid southern Africa. *Oecologia* 91: 288–291.

Minoretti, N., & B. Baur. 2006. Among- and within-population variation in sperm quality in the simultaneously hermaphroditic land snail *Arianta arbustorum*. *Behavioral Ecology and Sociobiology* 60: 270–280.

Miranda, X. 2006. Substrate-borne signal repertoire and courtship jamming by adults of *Ennya chrysura* (Hemiptera: membracidae). *Annals of the Entomological Society of America* 99: 374–386.

Mirzayans, H. 1995. *Insects of Iran. The list of Homoptera: Auchenorrhyncha in the insect collection of Plant Pests & Diseases Research Institute*. Plant Pests & Diseases Research Institute, Insects Taxonomy Research Department Publication Number 1. 59pp.

Mirzayans, H., A. Hashemi, H. Borumand, M. Zairi, & GH. Rajabi. 1976. Insect fauna from Province of Fars (Iran) (1). *Journal of Entomological Society of Iran* 3: 109–135.

Mitani, N., S. Mihara, N. Ishii, & H. Koike. 2009. Clues to the cause of the Tsushima leopard cat (*Prionailurus bengalensis euptilura*) decline from isotopic measurements in three species of Carnivora. *Ecological Research* 24: 897–908.

Mizuno, T. 1999. Medicinal effects and utilization of *Cordyceps* (Fr.) link (Ascomycetes) and *Isaria* Fr. (mitosporic fungi) Chinese caterpillar fungi, "Tochukaso"(Review). *International Journal of Medicinal Mushrooms* 1: 251–261.

Mjølhus, E., A. Wikan, & T. Solberg. 2005. On synchronization in semelparous populations. *Journal of Mathematical Biology* 50: 1–21.

Moalla, M., T. Jardak, & M. Ghorbel. 1992. Preliminary observations on a new olive pest in Tunisia: The cicada, *Psalmocharias plagifera* Schum. (Homoptera: Cicadidae). *Olivae* 44: 34–42.

Moeed, A. 1975. Diets of nestling starlings and mynas at Havelock North, Hawke's Bay. *Notornis* 22: 291–294.

Moeed, A. 1980. Diets of adult and nestling starlings (*Sturnus vulgaris*) in Hawke's Bay, New Zealand. *New Zealand Journal of Zoology* 7: 247–256.

Moeed, A., & M.J. Meads. 1983. Invertebrate fauna of four tree species in Orongorongo Valley, New Zealand, as revealed by trunk traps. *New Zealand Journal of Ecology* 6: 39–53.

Moeed, A., & M.J. Meads. 1984. Survey of Hen (Taranga) Island invertebrates as potential food for the little spotted kiwi, *Apteryx oweni*. *TANE* 30: 219–226.

Moeed, A., & M.J. Meads. 1987a. Invertebrate survey of offshore islands in relation to potential food sources for the little spotted kiwi, *Apteryx oweni* (Aves: Apterygidae). *New Zealand Entomologist* 10: 50–64.

Moeed, A., & M.J. Meads. 1987b. Seasonality of arthropods caught in a Malaise trap in mixed lowland forest of the Orongorongo Valley, New Zealand. *New Zealand Journal of Zoology* 14: 197–208.

Moeed, A., & M.J. Meads. 1988. Seasonality and density of emerging insects of a mixed lowland broadleaf-podocarp forest floor, Orongorongo Valley, New Zealand. *New Zealand Journal of Zoology* 14: 447–492.

Mogia, M. 2002. Traditional uses of cicadas by Tabare Sine people in Simbu province of Papua New Guinea. *Denisia* 4: 17–20.

Monteith, S. 1998. Animal oddities: cicada nymphs. *Wildlife Australia* 35: 39.

Moore, T.E. 1981. The evolution of complex acoustical behavior in cicadas. *Entomological News* 92: 167–168.

Moore, T.E. 1990. Comparative acoustic behaviour of cicadas (Homoptera). *7th International Meeting on Insect Sound and Vibration, Piran, Yugoslavia* 1990: 36.

Moore, T.E. 1991. The cicadas of the Fiji, Samoa, and Tonga Islands, their taxonomy and biogeography (Homoptera, Cicadoidea) by J.P. Duffels. (Book review). *Annals of the Entomological Society of America* 84: 213–214.

Moore, T.E. 1993. Acoustic signals and speciation in cicadas (Insecta: Homoptera: Cicadidae). Pages 269–284 *in* D.R. Lees & D. Edwards, eds. *Evolutionary patterns and processes*. Academic Press, London.

Moore, T.E. 1996. Cicadoidea. Pages 221–223 *in* J.E. Llorente Bousquets, A.N. García Aldrete, & E. González Soriano, eds. *Biodiversidad, Taxonomía y Biogeographía de Artrópodos de México: Hacia Una Síntesis de Su Conocimiento*. Cd. Universtiaria, Mexico, Instituto de Biologia, UNAM.

Moore, T.E., F. Huber, H. Kleindienst, K. Schildberger, & T. Weber. 1990. Neuronal responses of periodical cicadas to songs of sympatric species. *Abstracts of the Combined Meeting of the 7th International Auchenorrhyncha Congress and the 3rd International Workshop on Leafhoppers and Planthoppers of Economic Importance, Wooster, Ohio, August 13–17*.

Moore, T.E., F. Huber, T. Weber, U. Klein, & C. Bock. 1993. Interaction between visual and phonotactic orientation during flight in *Magicicada cassini* (Homoptera: Cicadidae). *Great Lakes Entomologist* 26: 199–221.

Moran, N.A., P. Tran, & N.M. Gerardo. 2005. Symbiosis and insect classification: an ancient symbiont of sap-feeding insects from the bacterial phylum *Bacteroidetes*. *Applied and Environmental Microbiology* 71: 8802–8810.

Moreira, C. 1921. Entomologia agricola Brasileria. *Ministero da Agricultura, Industria e Commercio, Instituto Biologico de Defensa Agricola, Boletim* 1: 1–182.

Moreira, C. 1928. Insectas nocivos ao cafeeiro no Brasil. *Revista da Sociedade Rural Brasileira* 8(No. 92): 24–25.

Morikawa, N., K. Matsushita, M. Nishina, & Y. Kono. 2003. High concentrations of trehalose in aphid hemolymph. *Applied Entomology and Zoology* 38: 241–248.

Moriyama, M., & H. Numata. 2006. Induction of egg hatching by high humidity in the cicada *Cryptotympana facialis*. *Journal of Insect Physiology* 52: 1219–1225.

Moriyama, M., & H. Numata. 2008. Diapause and prolonged development in the embryo and their ecological significance in two cicadas, *Cryptotympana facialis* and *Graptopsaltria nigrofuscata*. *Journal of Insect Physiology* 54: 1487–1494.

Moriyama, M., & H. Numata. 2009. Comparison of cold tolerance in eggs of two cicadas, *Cryptotympana facialis* and *Graptopsaltria nigrofuscata*, in relation to climate warming. *Entomological Science* 12: 162–170.

Moriyama, M., & H. Numata. 2010. Dessication tolerance in fully developed embryos of two cicadas, *Cryptotympana facialis* and *Graptopsaltria nigrofuscata*. *Entomological Science* 13: 68–74.

Morris, B. 2004. *Insects and human life*. Berg, Oxford. 317pp.

Morris, C.R., & G.N. Askew. 2010. The mechanical power output of the pectoralis muscle of cockatiel (*Nymphicus hollandicus*): the *in vivo* muscle length trajectory and activity patterns and their implications for power modulation. *Journal of Experimental Biology* 213: 2770–2780.

Moschidis, M.C. 1987. Thin layer chromatographic and infra red spectral evidence for the presence of phospholipids in *Cicada oni*. *Zeitschrift für Naturforschung Section C Biosciences* 42: 1023–1026.

Moss, J.T. St. Leger. 1988. Cicada fauna of Carlisle Island, Queensland, with some notes on habits and songs. *Queensland Naturalist* 29: 29–31.

Moss, J.T. St. Leger. 1989. Notes on the Tasmanian cicada fauna with comments on its uniqueness. *Queensland Naturalist* 29: 102–106.

Moss, J.T. St. Leger. 1990. A survey of the cicadas of Moreton Island, Queensland. *Queensland Naturalist* 30: 68–70.

Moss, J.T. St. Leger. 2000. Notes on the Tasmanian cicada fauna with comments on its uniqueness. *Invertebrata* 16: 6–8.

Moss, J.T. St. Leger, & M.S. Moulds. 2000. A new species of *Psaltoda* Stål, with notes on comparative morphology and song structure. *Australian Entomologist* 27: 47–60.

Moss, J.T. St. Leger, & L.W. Popple. 2000. Cicada, butterfly and moth records from the Gibraltar Range, New South Wales (Hemiptera: Cicadidae; Lepidoptera). *Queensland Naturalist* 38: 53–60.

Motta, P.C. 2003. Cicadas (Hemiptera, Auchenorrhyncha, Cicadidae) from Brasília (Brazil): exuviae of the last instar with key of the species. *Revista Brasiliera de Zoologia* 20: 19–22.

Moulds, M.S. 1983. Summertime is cicada time. *Australian Natural History* 20: 429–435.

Moulds, M.S. 1984. *Psaltoda magnifica* sp. n. and notes on the distribution of *Psaltoda* species (Homoptera: Cicadidae). *General and Applied Entomology* 16: 27–32.

Moulds, M.S. 1985. *Illyria*, a new genus for Australian cicadas currently placed in *Cicada* L. (= *Tettigia* Amyot) (Homoptera: Cicadidae). *General and Applied Entomology* 17: 25–35.

Moulds, M.S. 1986. *Marteena*, a new genus for the cicada *Tibicen rubricinctus* Goding and Froggatt (Homoptera: Tibicinidae). *General and Applied Entomology* 18: 39–42.

Moulds, M.S. 1987a. Australian entomological research: the role of the amateur. (Summary of Zoo Le Souëf Memorial Lecture). *Victorian Entomologist* 17: 22–28.

Moulds, M.S. 1987b. The specific status of *Pauropsalta nigristriga* Goding and Froggatt (Homoptera: Cicadidae) with the description of an allied new species. *Australian Entomological Magazine* 14: 17–22.

Moulds, M.S. 1988. The status of *Cicadetta* and *Melampsalta* (Homoptera: Cicadidae) in Australia with the description of two new species. *General and Applied Entomology* 20: 39–48.

Moulds, M.S. 1990. *Australian cicadas*. New South Wales University Press, Kensington, NSW. 217pp.

Moulds, M.S. 1992. Two new species of *Macrotristria* Stål (Hemiptera: Cicadidae) from Queensland. *Australian Entomology Magazine* 19: 133–138.

Moulds, M.S. 1993. *Henicopsaltria danielsi* sp. n. and a new locality for *H. eydouxii* (Hemiptera: Auchenorrhyncha: Cicadidae). *General and Applied Entomology* 25: 23–26.

Moulds, M.S. 1994. The identity of *Burbunga gilmorei* (Distant) and *B. inornata* Distant (Hemiptera: Cicadidae) with descriptions of two allied new species. *Journal of the Australian Entomological Society* 33: 97–103.

Moulds, M.S. 1996a. Review of the Australian genus *Gudanga* Distant (Hemiptera: Cicadidae) including new species from Western Australia and Queensland. *Australian Journal of Ecology* 35: 19–31.

Moulds, M.S. 1996b. The Australian cicada fauna. Pages 91–94 *in* D.C.F.Rentz, Co-ordinator. Nature sounds of Australia, original recordings of Australian wildlife, including birds, frogs, cicadas and crabs. The Sound Heritage Association, Ltd.

Moulds, M.S. 1999. The higher classification of the Cicadoidea: a cladistic approach. *Abstracts of Papers and Posters, 10th International Auchenorrhyncha Congress, 6–10 September 1999, Cardiff, Wales.* 2pp (no pagination).

Moulds, M.S. 2001. A review of the tribe Thophini Distant (Hemiptera: Cicadoidea: Cicadidae) with the description of a new species of *Thopha* Amyot & Serville. *Insect Systematics & Evolution* 31: 195–203.

Moulds, M.S. 2002. Three new species of *Psaltoda* Stål from eastern Australia (Hemiptera: Cicadoidea: Cicadidae). *Records of the Australian Museum* 54: 325–334.

Moulds, M.S. 2003a. Cicadas. Pages 431–437 *in* V.H. Resh & R.T. Cardé, eds. *Encyclopedia of insects*. Academic Press, New York.

Moulds, M.S. 2003b. An appraisal of the cicadas of the genus *Abricta* Stål and allied genera (Hemiptera: Auchenorrhyncha: Cicadidae). *Records of the Australian Museum* 55: 245–304.

Moulds, M.S. 2005a. Song analyses of cicadas of the genera *Aleeta* Moulds and *Tryella* Moulds (Hemiptera: Cicadidae). *Proceedings of the Linnean Society of New South Wales* 126: 133–142.

Moulds, M.S. 2005b. An appraisal of the higher classification of cicadas (Hemiptera: Cicadoidea) with special reference to the Australian fauna. *Records of the Australian Museum* 57: 375–446.

Moulds, M.S. 2008a. Talcopsaltriini, a new tribe for a genus and species of Australian cicada (Hemiptera: Cicadoidea: Cicadidae). *Records of the Australian Museum* 60: 207–214.

Moulds, M.S. 2008b. *Thopha hutchinsoni*, a new cicada (Cicadoidea: Cicadidae) from Western Australia, with notes on the distribution and colour polymorphism of *Thopha sessiliba* Distant. *Australian Entomologist* 35: 129–140.

Moulds, M.S. 2009a. Insects of the Fly River system. Pages 493–513 *in* B. Bolton, ed. *The Fly River, Papua New Guinea: environmental studies in an impacted tropical river system*. Developments in Earth & Environmental Sciences volume 9. Elsevier, Amsterdam.

Moulds, M.S. 2009b. Cicadas. Pages 163–164 *in* V.H. Resh & R.T. Cardé, eds. *Encyclopedia of insects*, second edition. Academic Press, New York.

Moulds, M.S. 2010. *Platypleura tepperi* Goding & Froggatt, 1904 (Cicadoidea: Cicadidae), a Madagascan cicada erroneously recorded from Australia. *Australian Entomologist* 37: 11–12.

Moulds, M.S., & M. Carver. 1991. Superfamily Cicadoidea. Pages 465–467 *in* CSIRO, ed. *The Insects of Australia: a textbook for students and research workers, Vol. 1*. 2nd edition. Melbourne University Press, Melbourne.

Moulds, M.S., & S.A. Cowan. 2002. Cicadoidea. In: *Zoological Catalogue of Australia*. Australian Biological Resources Study, Canberra. 55pp. Published electronically.

Moulds, M.S., & G. Hangay. 1998. First record of the bladder cicada *Cystosoma saundersii* (Westwood) from Lord Howe Island (Hemiptera: Cicadidae). *Australian Entomologist* 25: 75–76.

Moulds, M.S., & K.A. Kopestonsky. 2001. A review of the genus *Kobonga* Distant with the description of a new species (Hemiptera: Cicadidae). *Proceedings of the Linnean Society of New South Wales* 123: 141–157.

Mozaffarian, F. 2008. Report of *Adeniana longiceps* (Hem.: Auchenorrhyncha: Cicadidae) from Iran. *Journal of Entomological Society of Iran* 27 (Supplement): 9–10.

Mozaffarian, F., & A.F. Sanborn. 2010. The cicadas of Iran with the description of two new species (Hemiptera, Cicadidae). *Mitteilungen aus dem Museum für Naturkunde im Berlin, Deutsche Entomologishe Zeitschrift* 57: 69–84.

Mozaffarian, F., A.F. Sanborn, & P.K. Phillips. 2010. *Cicadatra lorestanica*, a new species of cicada from Iran (Hemiptera: Cicadidae). *Journal of Entomological and Acarological Research* 42(1): 27–37.

Muehleisen, R.T. 2007. Sound recordings from the 2007 emergence of Brood XIII cicada. *Journal of the Acoustical Society of America* 122: 2947.

Muraoka, Y. 2009. *Videoanalyse der Zwergohreule in Unterkärnten: Auswertung von Infrarotaufnahmen aus einem Nistkasten – Brutsaison 2007*. Unveröffentlichter Bericht, erstellt im Auftrag des Amtes der Kärntner Landesregierung, Abteilung 20, Uabt. Naturschutz, Wien. 30pp.

Murphy, M.T. 1986. Temporal components of reproductive variability in eastern kingbirds (*Tyrannus tyrannus*). *Ecology* 67: 1483–1492.

Murty, U.S., K. Jamil, & A. Ahmed. 1993. Aerodynamic parameters and design of flight surface of the mosquito *Culex quinquefasciatus* (Diptera, Culicidae) Say. *Insect Science and its Application* 14: 205–209.

Musgrave, A., & G.P. Whitley. 1931. Nature rambles at Trial Bay. *Australian Museum Magazine* 4: 149–155.

Na, H.-J., H.-Y. Shin, N.-H. Kim, M.-W. Kwon, E.-J. Park, S.-H. Hong, N.-I. Kim, & H.-M. Kim. 2004. Regulatory effects of cytokine production in atopic allergic reaction by gammi-danguieumja. *Inflammation* 28: 291–298.

Nagamine, M., & Y. Itô. 1980. The dynamics of epidemic sugarcane cicada populations. *International Congress of Entomology Proceedings* 16: 132.

Nagamine, M., & Y. Itô. 1982. *Platypleura kuroiwae* Matsumura (Hemiptera: Cicadidae), a cicada which became a pest of sugarcane in the Ryukyus. *Japanese Journal of Applied Entomology and Zoology* 26: 80–83.

Nahirney, P.C., J.C. Forbes, H.D. Morris, S.C. Chock, & K. Wang. 2006. What the buzz was all about: superfast song mucles rattle the tymbals of male periodical cicadas. *The FASEB Journal* 20: 2017–2026.

Nakamura, H., K. Murayama, S. Kubota, & K. Shigemori. 1988. [Feeding behavior, food items and the estimation of total food consumption in ruddy kingfisher.] *Bulletin of the Institute of Nature Education, Shiga Heights* (25): 15–22. [In Japanese, English summary]

Nakao, S. 1952. [On the diurnal rhythm of sound-producing activity in *Meimuna opalifera* Walker (Cicadidae, Homoptera).] *Physiological Ecology* 5: 17–25. [In Japanese, English summary]

Nakao, S. 1987. [Comparative study on the diurnal sound-producing activities of three subspecies of *Cryptotympana facialis* (Walker), (preliminary notes).] *Cicada* 7: 19–23. [In Japanese]

Nakao, S. 1988. [Ecological notes on *Meimuna* (Cicadidae) in the Ryukyu Archipelago.] *Cicada* 7: 43–46. [In Japanese]

Nakao, S. 1990. [*Natural history of cicadas*]. Chûô-Kôronsha, Tokyo. 179pp. [In Japanese]

Nakao, S., & K. Kanmiya. 1988. Acoustic analysis of the calling songs of a cicada, *Meimuna kuroiwae* Matsumura in Nansei Islands, Japan: Part I. Physical properties of the sound. *Kurume University Journal* 37: 25–46.

Namba, T., Y.H. Ma, & K. Inagaki. 1988. Insect-derived crude drugs in the Chinese Song Dynasty. *Journal of Ethnopharmacology* 24: 247–286.

Nanninga, F. 1991. Hemiptera. Plate 3 *in* I.D. Naumann, P.B. Carne, J.F. Lawrence, E.S. Nielsen, J.P. Spradbery, R.W. Taylor, M.J. Whitten, & M.J. Littlejohn, eds. *The insects of Australia. A textbook for students and research workers.* Second edition, Volume I. Melbourne University Press, Carleton.

Narhardiyati, M., & W.J. Bailey. 2005. Biology and natural enemies of the leafhopper *Balclutha incisa* (Matsumura) (Hemiptera: Cicadelllidae: Deltocephalinae) in south-western Australia. *Australlian Journal of Entomology* 44: 104–109.

Naruse, K. 1983. A historical note on the type-specimens of *Platypleura kaempferi* (Fabricius), with description of a new subspecies from Miyoke-Jima Island, Izu Islands, Japan (Homoptera: Cicadidae). *Fragmenta Coleopterologica* 35–37: 151–152.

Nash, R., & J.P. O'Connor. 1990. Insects imported into Ireland 9. Records of Orthoptera, Dictyoptera, Homoptera and Hymenoptera including Roger's ant, *Hypoponera punctatissima* (Roger). *Irish Naturalists' Journal* 23: 255–257.

Nast, J. 1982. Palearctic *Auchenorrhyncha* (*Homoptera*). Part 3. New taxa and replacement names introduced till 1980. *Annales Zoologici* 36: 289–362.

Nast, J. 1987. The *Auchenorrhyncha* (*Homoptera*) of Europe. *Annales Zoologici* 40: 535–661.

Nation, J.L. 2002. *Insect physiology and biochemistry*. CRC Press, Boca Raton, FL. 485pp.

Nation, J.L. 2008. *Insect physiology and biochemistry*, second edition. CRC Press, Boca Raton, FL. 544pp.

Natural England. 2010. *Lost life: England's lost and threatened species.* Natural England publication NE 233. Natural England, Sheffield. 53pp.

Naumann, I.D. 1993. *CSIRO handbook of Australian insect names. Common and scientific names for insects and allied organisms of economic and environmental importance.* 6th edition. CSIRO Publications, Victoria. 200pp.

Navarro, M.E. 1997. Contribucion al estudio del modelo bioacoustico de *Tettigades chilensis* Amyot & Serville (Hemiptera: Auchenorrhyncha: Cicadidae). *Gayana Zoologica* 61: 41–48.

Nel, A. 1996. Un Tettigarctidae fossile du Lias européen (Cicadomorpha, Cicadoidea, Tettigarctidae). *Ecole Practique des Hautes Etudes, Travaux du Laboratoire Biologie et Evolution des Insectes Hemipteroidea* 9: 83–94.

Nel, A., M. Zarbout, G. Barale, & M. Philippe. 1998. *Liassotettigarcta africana* (Auchenorrhyncha: Cicadoidea: Tettigarctidae), the first Mesozoic insect from Tunisia. *European Journal of Entomology* 95: 593–598.

Nelson, L. 2004. Big buzz as cicadas arrive after 17-year gap. *Nature* 429: 233.

New, T.R. 1996. *Name that insect. A guide to the insects of Southeastern Australia.* Oxford University Press, Melbourne. 194pp.

Nichols, P.B., D.R. Nelson, & J.R. Garey. 2006. A family level analysis of tardigrade phylogeny. *Hydrobiologia* 558: 53–60.

Nickel, H. 1994. Wärmeliebende Zikaden (Homoptera, Auchenorrhyncha) im südlichen Niedersachsen. *Braunschweiger naturkundliche Schriften* 4: 533–551.

Nickel, H. 2003. *The leafhoppers and planthoppers of Germany (Hemiptera, Auchenorrhyncha): patterns and strategies in a highly diverse group of phytophagous insects.* Pensoft Series Faunistica No. 28. Pensoft Publishers, Sofia. 429pp.

Nickel, H. 2004. Rote Liste gefährdeter Zikaden (Hemiptera, Auchenorrhyncha) Bayerns. *Schriftenreihe des Bayerischen Landesamtes für Umweltschutz* 166: 59–67.

Nickel, H. 2008. Tracking the elusive: leafhoppers and planthoppers (Insecta: Hemiptera) in tree canopies of European deciduous forests. Pages 175–214 *in* A. Floren & J. Schmidl, eds. *Canopy arthropod research in Europe*. Bioform entomology, Nuremberg.

Nickel, H., W.E. Holzinger, & E. Wachmann. 2002. Mitteleuropäische Lebensräume und ihre Zikadenfauna (Hemiptera: Auchenorrhyncha). *Denisia* 4: 279–328.

Nickel, H., & R. Remane. 2002. Artenliste der Zikaden Deutschlands, mit Angaben zu Nährpflanzen, Nahrungsbreite, Lebenszyklen, Areal und Gefährdung (Hemiptera, Fulgoromorpha et Cicadomorpha). *Beiträge zur Zikadenkunde* 5: 27–64.

Nickel, H., & R. Remane. 2003. Verzeichnis der Zikaden (Auchenorrhyncha) der Bundesländer Deutschlands. *In* B. Klausnitzer, ed. *Entomofauna Germanica*, Band 6. *Entomologische Nachrichten und Berichte, Supplement* 8: 130–154.

Nickel, H., & F.W. Sander. 1996. Kommentiertes Verzeichnis der bisher in Thüringen nachgewiesenen Zikadenarten (Homoptera, Auchenorrhyncha). *Veröffentlichungen des Naturkundemuseum Erfurt* 15: 152–170.

Nickens, T.E. 2000. Insect opera: It's nature's strangest symphony, and if you're lucky, the cicadas will soon be bringing their act to a forest near you. *Audubon* 102: 25–31.

Niedringhaus, R., R. Biedermann, & H. Nickel. 2010. *Verbreitungsatlas der Zikaden des Großherzogtums Luxemburg.* Ferrantia 60. Musée national d'histoire naturelle, Luxembourg. 105pp.

Niehuis, M. 1995. Weitere Nachweise von Röhrenspinne (*Eresus niger*), Gottesanbeterin (*Mantis religiosa*) und Blutaderzikade (*Tibicina haematodes*) in Rheinland-Pfalz. *Fauna und Flora in Rheinland-Pfalz* 8: 33–41.

Niehuis, M., & L. Simon. 1994. Zum Vorkommen von Blutaderzikade –*Tibicina haematodes* (Scop.) – und Bergzikade – *Cicadetta montana* (Scop.) – in Rheinland-Pfalz (Homoptera: Cicadidae). *Fauna und Flora in Rheinland-Pfalz* 7: 253–264.

Niehuis, M., & M. Weitzel. 1995. Die Bergzikade (*Cicadetta montana*) in Rheinland-Pfalz (Insecta: Homoptera: Cicadina). *Fauna und Flora in Rheinland-Pfalz* 8: 439–447.

Nikoh, N., & T. Fukatsu. 2000. Interkingdom host jumping underground: Phylogenetic analysis of entomoparasitic fungi of the genus *Cordyceps*. *Molecular Biology and Evolution* 17: 629–638.

Nkouka, E. 1987. Les insects comestibles dans les sociétés d'Afrique Centrale. *Revue Scientifique et Culturelle du CICIBA, Muntu* 6: 171–178.

Noda, N., S. Kubota, Y. Miyata, & K. Miyahara. 2000. Optically active N-acetyldopamine dimer of the crude drug "Zentai," the cast-off shell of the cicada *Cryptotympana* sp. *Chemical and Pharmaceutical Bulletin* 48: 1749–1752.

Noort, S. van. 1995. An association of *Italochrysa neurodes* (Rambur) (Neuroptera: Chyrsopidae) with *Platypleura capensis* (Linnaeus) (Hemiptera: Cicadidae). *African Entomology* 3: 92–94.

Nosko, L. 1984. Herpetofauna a některé vzácné druhy horné nitry [The herpetofauna and some rare species of evertebrates of region Horná Nitra (Slovakia).] *Fauna Bohemiae Septentrionalis* (9): 55–59.

Novikov, D.V., N.V. Novikova, G.A. Anufriev, & C.H. Dietrich. 2006. Auchenorrhyncha (Hemiptera) of Kyrgyz grasslands. *Russian Journal of Entomology* 15: 303–310.

Novotny, V., & M.R. Wilson. 1997. Why are there no small species among xylem-sucking insects? *Evolutionary Ecology* 11: 419–437.

Nowlin, W.H., M.J. González, M.J. Vanni, M.H.H. Stevens, M.W. Fields, & J.J. Valente. 2007. Allochthonous subsidy of periodical cicadas affects the dynamics and stability of pond communities. *Ecology* 88: 2174–2186.

Nowlin, W.H., M.J. Vanni, & L.H. Yang. 2008. Comparing resource pulses in aquatic and terrestrial ecosystems. *Ecology* 89: 647–659.

Noyes, J.S., & E.W. Valentine. 1989a. Mymaridae (Insecta: Hymenoptera) – introduction, and review of genera. *Fauna of New Zealand* 17: 1–95.

Noyes, J.S., & E.W. Valentine. 1989b. Chalcidoidea (Insecta: Hymenoptera) – introduction, and review of genera in smaller families. *Fauna of New Zealand* 18: 1–96.

Nuorteva, P., M. Lodenius, S. Nagasawa, E. Tulisalo, & S.-L. Nuorteva. 2004. Influence of species, sex, age and food on the accumulation of toxic cadmium and some essential metals in Auchenorrhynchous Homoptera. *Memoranda Societatis pro Fauna et Flora Fennica* 80: 49–63.

Oakley, T.H. 2003. On homology of arthropod compound eyes. *Integrative and Comparative Biology* 43: 522–530.

Oberdörster, U., & P.R. Grant. 2006. Predicting emergence, chorusing, and oviposition of periodical cicadas. *Ecology* 87: 409–418.

Oberdörster, U., & P.R. Grant. 2007a. Predator foolhardiness and morphological evolution in 17-year cicadas (*Magicicada* spp.). *Biological Journal of the Linnean Society* 90: 1–13.

Oberdörster, U., & P.R. Grant. 2007b. Acoustic adaptations of periodical cicadas (Hemiptera: *Magicicada*). *Biological Journal of the Linnean Society* 90: 15–24.

Obrist, M.K., G. Pavan, J. Sueur, K. Riede, D. Liusia, & R. Márquez. 2010. Bioacoustics approaches in biodiversity inventories. *Abc Taxa* 8: 68–99.

O'Connor, J.P., & R. Nash. 1981. Notes on five species of insect (Hemiptera: Coleoptera) imported into Ireland. *Irish Naturalists' Journal* 20: 299–300.

O'Connor, J.P., R. Nash, & R. Anderson. 1983. Insects imported into Ireland 4. Records of Dictoptera, Hemiptera, and Coleoptera (including *Pycnomerus fuliginosus* Erichson). *Irish Naturalists' Journal* 21: 81–83.

O'Geen, A.T., P.A. McDaniel, & A.J. Busacca. 2002. Cicada burrows as indicators of paleosols in the inland Pacific Northwest. *Soil Science Society of America Journal* 66: 1584–1586.

Oh, H., Y. Kim, & N. Kim. 2010. First breeding record of Japanese night heron *Gorsachius goisagi* in Korea. *Ornithological Science* 9: 131–134.

Ohara, K., & M. Hayashi. 2001. Notes on cicadas of the genus *Tibicen* (Homoptera, Cicadidae) in Tokushima Prefecture, Shikoku, Japan. *Bulletin of the Tokushima Prefectural Museum* (11): 1–6.

Ohbayashi, T., H. Sato, & S. Igawa. 1999. An epizootic on *Meimuna boninensis* (Distant)(Homoptera: Cicadidae), caused by *Massospora* sp. (Zygomycetes: Entomophthorales) and *Nomuraea cylindrospora* (Deuterosmycotina: Hyphomycetes) in the Ogasawara (Bonin) Islands, Japan. *Applied Entomology and Zoology* 34: 339–343.

Ohgushi, R. 1954. [Preliminary study on a cicada community at Mt. Nakatsumine.] *Kontyû* 21: 10–18. [In Japanese, English summary]

Ohmart, R.D., B.W. Anderson, & W.C. Hunter. 1988. The ecology of the lower Colorado River from Davis Dam to the Mexico-United States International Boundary: a community profile. *United States Fish and Wildlife Service Biological Report* 85(7.19): 1–296.

Ohno, M. 1982. [Bibliography of the selected important animals occurring in Japan (1c) *Euterpnosia* spp., supplement 3.] *Journal of the Toyo University General Education* (25): 51–56. [In Japanese]

Ohno, M. 1985. [Bibliography of the selected important animals occurring in Japan (1d) *Euterpnosia* spp., supplement 4.] *Journal of the Toyo University General Education* (28): 105–110.

Ohya, E. 2004. Identification of *Tibicen* cicada species by a principle components analysis of their songs. *Anais da Academia Brasileira Ciências* 76: 441–444.

Ojie Ardebilie, M.M., & J. Nozari. 2009. [Study on the effect of digital calling songs of *Cicada orni* and *Okanagana rimosa* (Hemiptera: Cicadidae) for attracting house sparrows *Passer domesticus* (Passeriformes: Passeridae).] *Journal of the Entomological Society of Iran* 29: A13-A21. [In Arabic, English summary]

Okáli, I., & V. Janský. 1998. Cikádovié (Auchenorrhyncha, Cicadidae) Slovenska. *Entomofauna Carpathica* 10: 33–43.

Okane, I., A. Nakagiri, & T. Ito. 1999. Descriptive catalogue of IFO fungus collection XVI. *Research Communications, Institute for Fermentation, Osaka Number* 19: 116–123.

Olive, J.C. 2007. A new species of *Gudanga* Distant (Hemiptera: Cicadidae) from northern Queensland. *Australian Entomologist* 34: 1–6.

Oliver, P. 2009. Eurasian Jay taking peanuts in flight. *British Birds* 102: 146.

O'Neill, S.B., T.R. Buckley, T.R. Jewell, & P.A. Ritchie. 2009. Phylogeographic history of the New Zealand stick insect *Niveaphasma annulata* (Phasmatodea) estimated from mitochondrial and nuclear loci. *Molecular Phylogenetics and Evolution* 53: 523–536.

O'Neill, S.B., D.G. Chapple, C.H. Daugherty, & P.A. Ritchie. 2008. Phylogeography of two New Zealand lizards: McCann's skink (*Oligosoma maccanni*) and the brown skink (*O. zelandicum*). *Molecular Phylogenetics and Evolution* 48: 1168–1177.

Orians, G.H., & H.S. Horn. 1969. Overlap in food and foraging of four species of blackbirds in the Potholes of central Washington. *Ecology* 50: 930–938.

Osaka Museum of Natural History (Ed.). 2008. [*Singing Insects: Stories about their Songs and Acoustic Behavior*.] Osaka shiritsa shizenshi hakubutsukan sosho 4. Tokai University Press, Kanagawa. 331pp. [In Japanese]

Osborn, H. 1896. Tenth annual meeting of the Iowa Academy of Sciences. *Science* 3: 124–126.

O'Shea, R. 1980. A generic revision of the Nematopodini (Heteroptera: Coreidae: Coreinae). *Studies in Neotropical Fauna and Environment* 15: 197–225.

Ossiannilsson, F. 1981. The Auchenorrhyncha (Homoptera) of Fennoscandia and Denmark. Part 2: The families Cicadidae, Cercopidae, Membracidae, and Cicadellidae (excl. Deltocephalinae). *Fauna Entomologica Scandinavica* 7: 223–225.

Ostfeld, R.S., & F. Keesing. 2004. Oh the locusts sang, then they dropped dead. *Science* 306: 1488–1489.

Ostry, M.E., & N.A. Anderson. 1983. Infection of trembling aspen by *Hypoxylon mammatum* through cicada oviposition wounds. *Phytopathology* 73: 1092–1096.

Ostry, M.E., & N.A. Anderson. 1998. *Interactions of insects, woodpeckers, and* Hypoxylon *canker on aspen. Research Paper North Central Forest Experiment Station, United States Department of Agriculture Forest Service,* No. NC331. 17pp.

Ostry, M.E., & N.A. Anderson. 2009. Genetics and ecology of the *Entoleuca mammatum-Populus* pathosystem: implications for aspen improvement and management. *Forest Ecology and Management* 257: 390–400.

Otero, L.S. 1971. *Brazilian insects and their surroundings*. Koyo Shoin Co., Ltd. Kanda, Tokyo. 181pp.

O'Toole, C. 1995. *Alien empire*. BBC Books, London. 224pp.

Ouvrard, D., B.C. Campbell, T. Bourgoin, & K.L. Chan. 2000. 18S rRNA secondary structure and phylogenetic position of Peloridiidae (Insecta, Hemiptera). *Molecular Phylogenetics and Evolution* 16: 403–417.

Owada, M., Y. Arita, U. Jinbo, Y. Kishida, H. Nakajima, M. Ikeda, S. Niitsu, & S. Keino. 2005. [Moths and butterflies of the Akasaka Imperial Gardens, Tokyo, Central Japan.] *Memoires of the National Science Museum, Tokyo* (39): 55–120. [In Japanese]

Özdikmen, H., & A.B. Kury. 2006. Three homonymous generic names in Aranease and Opiliones. *The Journal of Arachnology* 34: 279–280.

Öztürk, N., M.R. Ulusoy, L. Erkiliç, & S. (Ö.) Bayhan. 2004. Malatya ili kayisi bahçelerinde saptanan zararlilar ile avci türler (Pests and predatory species determined in apricot orchards in Malatya province of Turkey). *Bitki Koruma Bülteni* 44: 1–13.

Pachas, P.O. 1966. La chicharra de la yerba mate (*Fidicina mannifera*, Fab., 1803) sur biología e observaciones sobre los métodos de control em Misiones (República Argentina). *Idia* 217: 5–15.

Page, R.D.M., R. Cruickshank, & K.P. Johnson. 2002. Louse (Insecta: Phthiroptera) mitochondrial 12S rRNA secondary structure is highly variable. *Insect Molecular Biology* 11: 361–369.

Paião, F.G., A.M. Meneguim, E.C. Casagrande, & R.P. Leite. 2002. Envolvimento de cigarras (Homoptera: Cicadidae) na transmissão de *Xylella fastidiosa* em cafeeiro. *Fitopatologia brasileira* 27 (Supl.): S67.

Pain, S. 2009. What the katydid next. *New Scientist* 203 (2727): 44–47.

Palma, R.L., P.M. Lovis, & C. Tither. 1989. An annotated list of primary types of the phyla Arthropoda (except Crustacea) and Tardigrada held in the National Museum of New Zealand. *National Museum of New Zealand Miscellaneous Series No.* 20: 1–49.

Palmer, W.A., M.P. Vitelli, & G.R. Donnelly. 2005. The phytophagous insect fauna associated with *Acacia nilotica* ssp. *indica* (Mimosaceae) in Australia. *Australian Entomologist* 32: 173–180.

Panov, A.A., & E.E. Krjutchkova. 1983. The cerebral neurosecretory cells of cicadas (Homoptera: Cicadidae). *Journal für Hirnforschung* 24: 141–148.

Park, D.-S., J.Y. Leem, E.Y. Suh, J.H. Hur, H.-W. Oh, & H.-Y. Park. 2007. Isolation and functional analysis of a 24-residue linear α-helical antimicrobial peptide from Korean blackish cicada, *Cryptotympana dubia* (Homoptera). *Archives of Insect Biochemistry and Physiology* 66: 204–213.

Parker, N.W., & J.R. Wright. 1987. *Bugs*. Mulberry Books, New York. 40pp.

Parry-Jones, K., & M. Augee. 1992. Insects in flying-fox diets. *Bat Research News* 33: 9–11.

Paton, T.R., G.S. Humphreys, & P.B. Mitchell. 1995. *Soils: A new global view*. Yale University Press, New Haven. 234pp.

Patrick, B. 1991. Entomological values of the Rees riverbed near Glenorchy. *The Weta* 14: 3–7.

Patterson, I.J., & P. Cavallini. 1996. The volume of sound as an index to the relative abundance of *Cicada orni* L. (Homoptera: Cicadidae) in different habitats. *Entomologist's Gazette* 47: 206–210.

Patterson, I.J., P. Cavallini, & A. Rolando. 1991. Density, range size and diet of the European Jay *Garrulus glandarius* in the Maremma Natural Park, Tuscany, Italy, in summer and autumn. *Ornis Scandinavica* 22: 79–87.

Patterson, I.J., G. Massei, & P. Genov. 1997. The density of cicadas *Cicada orni* in Mediterranean coastal habitats. *Italian Journal of Zoology* 64: 141–146.

Pechuman, L.L. 1985. The periodical cicada (*Magicicada septendecim*): Brood VII revisited (Homoptera: Cicadidae). *Entomological News* 96: 59–60.

Peck, C.H. 1879. Report of the Botanist. *Thirty-first Report of the New York State Museum of Natural History* 31: 19–60.

Pemberton, R.W., & T. Yamasaki. 1995. Insects: old food in new Japan. *American Entomologist* 41: 227–229.

Peña G., L.E. 1987. *Introducción a los insectos de Chile*. Secundo Edición. Editorial Universitaria, S.A. Santiago de Chile. 256pp.

Peng, H.-Y., C.-C. Chien, & W.-F. Hsiao. 2010. [Biological characteristics of the entomopathogenic fungus, *Nomuraea viridulus*.] *Formosan Entomologist* 30: 145–165. [In Chinese, English summary]

Peng J.W., & Lei Z.R. 1992. Cicadidae. Pages 97–105 *in* Peng J. & Liu Y., eds. *Iconography of forest insects in Hunan China*. Academia Sinica & Hunan Forestry Institute, Hunan, China.

Perepelov, E., A.G. Burgov, & A. Marymanska-Nadachowska. 2002. Constitutive heterochromatin in karyotypes of two Cicadidae species from Japan (Cicadoidea, Hemiptera). *Folia biologica (Kraków)* 50: 217–219.

Perez-Gerlabert, D.E. 2008. Arthropods of Hispaniola (Dominican Republic and Haiti): a checklist and bibliography. *Zootaxa* 1831: 1–530.

Pham, H.T. 2004. [The cicadas genera *Pomponia* Stal, 1866; *Dundubia* Amyot & Serville, 1843 and *Platylomia* Stal, 1870 (Cicadidae: Cicadinae) from some national parks and nature reserves in Vietnam.] *Journal of Biology, Hanoi* 26: 61–65. [In Vietnamese, English summary]

Pham, H.T. 2005a. [Tribe Huechysini (Homoptera: Cicadidae) in Vietnam.] *Proceedings of the 5th Vietnam National Conference on Entomology, Hanoi, 11–12 April 2005*: 216–218. [In Vietnamese, English summary]

Pham, H.T. 2005b. [Key to the species of the genus *Crytotympana* Stal, 1861 (Homoptera, Cicadidae) from Vietnam.] *Proceedings of the 1st National Workshop on Ecology and Biological Resources, Hanoi, 17 May 2005*: 232–235. [In Vietnamese, English summary]

Pham, H.T., & H.T. Thinh. 2005a. [Checklist of family Cicadidae (Homoptera: Auchenorrhyncha) from Vietnam.] *Proceedings of the 1st National Workshop on Ecology and Bio-resources, Hanoi, 17 May 2005*: 236–247. [In Vietnamese, English summary]

Pham, H.T., & H.T. Thinh. 2005b. [Key to identification of the tribes of the cicadas (Cicadidae, Auchenorrhyncha, Homoptera) from Vietnam.] *Problems in Basic Research in Life Sciences. Scientific Reports, 2005 National Conference, Ha Noi Medical University, 3 November 2005*: 287–290. [In Vietnamese, English summary]

Pham, H.T., & H.T. Thinh. 2006. [Study on the composition and distribution of the cicadas (Cicadidae, Homoptera) length across Ho Chi Minh highway from Quangtri, Thuathien-Hue and Quangnam provinces.] *Proceedings of the Scientific Conference on Forward to Agronomy Management Technology for Development of Sustainable Agriculture in Vietnam, Hanoi Agricultural University, 10 October 2006*: 524–528. [In Vietnamese, English summary]

Pham, H.T., H.T. Thinh, & J.-T. Yang. 2010. A new cicada species (Hemiptera: Cicadidae), with a key to the species of the genus *Euterpnosia* Matsumura, 1917 from Vietnam. *Zootaxa* 2512: 63–68.

Pham, H.T., & J.-T. Yang. 2009. A contribution to the Cicadidae fauna of Vietnam (Hemiptera, Auchenorrhyncha), with one new species and twenty new records. *Zootaxa* 2249: 1–19.

Pham, H.T., & J.-T. Yang. 2010a. Biodiversity of Vietnamese cicadas (Hemiptera: Cicadidae). *Thirteenth International Auchenorrhyncha Congress and the 7th International Workshop on Leafhoppers and Planthoppers of Economic Significance, 28 June-2 July, Vaison-la-Romaine, France, Abstracts: Talks and Posters*: 133–134.

Pham, H.T., & J.-T. Yang. 2010b. The genus *Lemuriana* Distant (Hemiptera, Auchenorrhyncha) from Vietnam, with description of a new species. *Oriental Insects* 44: 205–210.

Phillips, P.A. 1992. Minor cicada. Page 251 *in* D.L. Flaherty, L.P. Christensen, W.T. Lanini, J.J. Marois, P.A. Phillips, & L.T. Wilson, eds. *Grape pest management*, 2nd edition. University of California Division of Agriculture and Natural Resources Publication 3343. University of California, Oakland.

Phillips, P.K., & A.F. Sanborn. 2007. Phytogeography influences biogeography of the Cicadidae. *Acta Zoologica Sinica* 53: 454–462.

Phillips, P.K., & A.F. Sanborn. 2010. Evolution of endothermy in the Cicadoidea. *Thirteenth International Auchenorrhyncha Congress and the 7th International Workshop on Leafhoppers and Planthoppers of Economic Significance, 28 June-2 July, Vaison-la-Romaine, France, Abstracts: Talks and Posters*: 74–75.

Phillips, P.K., A.F. Sanborn, & M.H. Villet. 2002. Thermal responses in some African cicadas (Hemiptera: Cicadoidea: Cicadidae). *Eleventh International Auchenorrhyncha Congress, August 5–9, Potsdam, Germany, Abstracts of Talks and Posters*: 31.

Picker, M., C. Griffiths, & A. Weaving. 2002. *Field guide to insects of South Africa.* Struik Publishers, Cape Town. 440pp.

Pickover, C.A. 2009. *The math book. From Pythagorus to the 57th dimension, 250 milestones in the history of mathematics.* Sterling Publishing, New York. 528pp.

Pierantoni, U. 1953. Quelques considérations sur l'histoire des recherches concernant la symbiose physiologique héréditaire. *Bolletino di Zoologia* 20: 35–37.

Pierce, C.M.F., T.J. Gibb, & R.D. Waaltz. 2005. Insects and other arthropods of economic importance in Indiana in 2004. *Proceedings of the Indiana Academy of Science* 114: 105–110.

Pierre, J., & A. Francois. 2006. Une cigale en Bretagne! *Bulletin du GRETIA* (35): 19.

Pigott, C.D. 1991. *Tilia cordata* Miller (*T. europaea* L. pro parte, *T. parvifolia* Ehrh. ex Hoffm., *T. sylvestris* Desf., *T. foemina folio minore* Bauhin). *Journal of Ecology* 79: 1147–1207.

Pigozzi, G. 1991. The diet of the European badger in a Mediterranean coastal area. *Acta Theriologica* 36: 293–306.

Pilarczyk, S., & D. Swierczewski. 2004. Some remarks on the planthoppers and leafhoppers of Poland. *Beiträge zur Entomofaunistik* 5: 146–147.

Pillet, J.M. 1993. Inventaire, écologie et répartition des cigales du Valais (Suisse). (Homoptera, Cicadoidea). *Bulletin de la Murithienne* 111: 95–113.

Pinchen, B.J., & L.K. Ward. 2002. The history, ecology and conservation of the New Forrest cicada. *British Wildlife* 13: 258–266.

Pinchuck, S., & M.H. Villet. 1995. Histology of the tymbal muscles of eastern Cape cicadas (Homoptera: Cicadidae). *Proceedings of the 10th Congress of the Entomological Society of Southern Africa, 3–7 July 1995, Grahamstown*: 112.

Pinto, G.A., J.A. Quartau, J.C. Morgan, & J. Hemingway. 1998. Preliminary data on the sequencing of a fragment of cytochrome *B* gene of mitochondrial DNA in *Cicada orni* L. (Homoptera: Cicadidae) in Portugal. *Boletim da Sociedade Portuguesa de Entomologia* 7: 57–66.

Pinto-Juma, G.A., J.A. Quartau, & M.W. Bruford. 2008. Population structure of *Cicada barbara* Stål (Hemiptera, Cicadoidea) from the Iberian peninsula and Morocco based on mitochondrial DNA analysis. *Bulletin of Entomological Research* 98: 15–25.

Pinto-Juma, G.A., J.A. Quartau, & M.W. Bruford. 2009. Mitochondrial DNA variation and the evolutionsry history of the Mediterranean species of *Cicada* L. (Hemiptera, Cicadoidea). *Zoological Journal of the Linnean Society* 155: 266–288.

Pinto-Juma, G.A., S.G. Seabra, & J.A. Quartau. 2008. Patterns of acoustic variation in *Cicada barbara* Stål (Hemiptera, Cicadoidea) from the Iberian peninsula and Morocco. *Bulletin of Entomological Research* 98: 1–14.

Pinto-Juma, G., P.C. Simões, S.G. Seabra, & J.A. Quartau. 2005. Calling song structure and geographic variation in *Cicada orni* Linnaeus (Hemiptera: Cicadidae). *Zoological Sciences* 44: 81–94.

Piulachs, M.D., K.R. Guidugli, A.R. Barchuk, J. Cruz, Z.L.P. Simões, & X. Bellés. 2002. The vitellogenin of the honey bee, *Apis mellifera*: structural analysis of the cDNA and expression studies. *Insect Biochemistry and Molecular Biology* 33: 459–465.

Pogue, M.G. 1996. Biodiversity of Cicadoidea (Homoptera) of Pakitza, Manu Reserved Zone and Tambopata Reserved Zone, Peru: a faunal comparison. Pages 313–325 *in* D.E. Wilson & A. Sandoval, eds. *Manu, the biodiversity of southeastern Peru.* Smithsonian Institution, Editorial Horizonte, Lima.

Pond, W. 1984. Polynesian classification of small land animals: a description from Niuatoputapu Island in the Kingdom of Tonga. *The Weta* 7: 93–107.

Pons, L. 2006. Cicadas' fungal foe is potential friend of science. *Agricultural Research* 54 (Feb.): 8.

Poole, R.W., R.W. Garrison, W.P. McCafferty, D. Otte, & W.P. Stark. 1997. *Nomina Insecta Nearctica. A checklist of the insects of North America. Volume 4: Non-holometabolous orders.* Entomological Information Services, Rockville, Maryland.

Popov, A.V. 1971. Synaptic transformation in the auditory system of insects. Pages 301–320 *in* G. V. Gersuni, ed. (translated by J. Rose). *Sensory processes at the neuronal and behavioural level.* Academic Press, New York.

Popov, A.V. 1981. Sound production and hearing in the cicada, *Cicadetta sinuatipennis* Osh. (Homoptera, Cicadidae). *Journal of Comparative Physiology. A. Sensory, Neural and Behavioral Physiology* 142: 271–280.

Popov, A.V. 1985. *Akusticheskoe povedenie i slukh nasekomykh (Acoustic behavior and hearing of insects)*. Nauka, Leningrad. 253pp. [In Russian]

Popov, A.V. 1989a. [Species of singing cicadas revealed on the basis of peculiarities of acoustic behaviour. I. *Cicadatra cataphractica* Popov of acoustic behavior 1. (ex gr. *querula*) (Homoptera, Cicadidae).] *Entomologisheskoe Obozrenie* 68: 291–307. [In Russian]

Popov, A.V. 1989b. Species of singing cicadas revealed on the basis of peculiarities of acoustic behaviour. I. *Cicadatra cataphractica* Popov of acoustic behavior 1. (ex gr. *querula*) (Homoptera, Cicadidae). *Entomological Review, Washington* 68: 62–78.

Popov, A.V. 1990a. Co-evolution of sound-production and hearing in insects. Pages 301–304 *in* F.G. Gribakin, K. Wiese, & A.V. Popov, eds. *Sensory systems and communication in arthropods: including the first comprehensive collection of contributions by Soviet scientists. Advances in Life Sciences*. Birkhäuser Verlag, Basel.

Popov, A.V. 1990b. [Species of singing cicadas revealed on the basis of peculiarities of acoustic behaviour. II. *Cicadetta petrophila* Popov (ex gr. *inserta*) (Homoptera, Cicadidae).] *Entomologicheskoe Obozrenie* 69: 579–586. [In Russian]

Popov, A.V. 1991. Species of cicadas identified from features of acoustic behavior. II. *Cicadetta petrophila* Popov (ex gr. *inserta*) (Homoptera, Cicadidae). *Entomological Review* 70: 17–24.

Popov, A.V. 1997a. [Acoustic signals of three morphologically similar species of singing cicadas (Homoptera, Cicadidae).] *Entomologisheskoe Obozrenie* 76: 3–15, 232. [In Russian]

Popov, A.V. 1997b. Acoustic signals of three morphologically similar species of singing cicadas (Homoptera, Cicadidae). *Entomological Review* 77: 1–11.

Popov, A.V. 1998a. [Sibling species of the singing cicadas *Cicadetta prasina* (Pall.) and *C. pellosoma* (Uhler) (Homoptera, Cicadidae).] *Entomologicheskoe Obozrenie* 77: 315–326. [In Russian]

Popov, A.V. 1998b. Sibling species of the singing cicadas *Cicadetta prasina* (Pall.) and *C. pellosoma* (Uhler) (Homoptera, Cicadidae). *Entomological Review* 78: 309–318.

Popov, A.V., I.B. Arnov, & M.V. Sergeeva. 1985a. [Calling songs and hearing in cicadas from Soviet Central Asia.] *Zhurnal Evolutsionnoi Biokhimi I Fiziologii* 21: 451–462. [In Russian, English summary]

Popov, A.V., I.B. Arnov, & M.V. Sergeeva. 1985b. Calling songs and hearing in cicadas from Soviet Central Asia. *Journal of Evolutionary Biochemistry and Physiology* 21: 288–298.

Popov, A.V., I.B. Arnov, & M.V. Sergeeva. 1991. [Frequency characteristics of the tympanal organs and spectrum of the sound communication signals of the singing cicadas (Homoptera, Cicadidae) of Praskoveiskaya Valley. To the problem of matching between soundproduction and soundreception.] *Entomologicheskoe Obozrenie* 70: 281–296. [In Russian]

Popov, A.V., I.B. Arnov, & M.V. Sergeeva. 1992. Frequency characteristics of tympanal organs and spectrum of sound communication signals of singing cicadas (Homoptera, Cicadidae) of the Praskoveiskaya Valley. Contribution to the study of correspondence between sound production and sound reception. *Entomological Review, Washington* 70: 151–166.

Popov, A.V., A. Beganović, & M. Gogala. 1997. Bioacoustics of singing cicadas of the western Palaearctic: *Tettigetta brullei* (Fieber 1876) (Cicadoidea: Tibicinidae). *Acta Entomologica Slovenica* 5: 89–101.

Popov, A.V., & M.V. Sergeeva. 1987. [Sound signalization and hearing in the Baikal cicada, *Cicadetta yezoensis* (Homoptera, Cicadidae).] *Zoologicheski Zhurnal* 66: 681–691. [In Russian, English summary]

Popova, E.A., & A.G. Dubovskaya. 1984. *Species composition of the cicades on food grains in Uzbekistan. Vidovoi sostav tsikadovykh na zermovykh zlakakh v Uzbekistane*. Muhammad Ali Society, Karachi, Pakistan. 8pp.

Popple, L.W. 2003. A new species of *Cicadetta* Amyot (Hemiptera: Cicadidae) from Queensland, with notes on its calling song. *Australian Entomologist* 30: 107–114.

Popple, L.W., & D. Emery. 2010. A new cicada genus and a redescription of *Pauropsalta olivacea* Ashton (Hemiptera: Cicadidae) from eastern Australia. *Australian Entomologist* 37: 147–156.

Popple, L., & A. Ewart. 2002. Cicadas (Hemiptera: Auchenorrhyncha: Cicadidae). Pages 113–118 *in* H. Horton, ed. *A Brisbane Bushland: the history and natural history of Enoggera Reservoir and its environs*. Queensland Naturalist Club, Brisbane.

Popple, L.W., & A.D. Strange. 2002. Cicadas, and their songs, from the Tara and Waroo shires, Southern Central Queensland. *Queensland Naturalist* 40: 15–30.

Popple, L.W., & G.H. Walter. 2010. A spatial analysis of the ecology and morphology of cicadas in the *Pauropsalta annulata* species complex (Hemiptera: Cicadidae). *Biological Journal of the Linnean Society* 101: 553–565.

Popple, L.W., G.H. Walter, & S. Raghu. 2008. The structure of calling songs in the cicada *Pauropsalta annulata* Goding and Froggatt (Hemiptera: Cicadidae): evidence of diverging populations? *Evolutionary Ecology* 22: 203–215.

Potamitis, I., T. Ganchev, & N. Fakotakis. 2007. Automatic acoustic identification of crickets and cicadas. *Proceedings of the Ninth International Symposium for Signal Processing and Its Applications, ISSPA 2007*: 4pp.

Poulton, E.B. 1907. Predaceous insects and their prey. *Transactions of the Entomological Society of London* 54: 323–409.

Powell, J.A., & C.L. Hogue. 1979. *California insects*. University of California Press, Berkeley. 388pp.

Powledge, T.M. 2004. Behold *Magicicada. The Scientist* 18 (July 10): 21.

Prange, H.D. 1996. Evaporative cooling in insects. *Journal of Insect Physiology* 42: 493–499.

Prange, H.D., & B. Pinshow. 1994. Thermoregulation of an unusual grasshopper in a desert environment: the importance of food source and body size. *Journal of Thermal Biology* 19: 75–78.

Pray, C.L., W.H. Nowlin, & M.J. Vanni. 2009. Deposition and decomposition of periodical cicadas (Homoptera: Cicadidae: *Magicicada*) in woodland aquatic ecosystems. *Journal of the North American Benthological Society* 28: 181–195.

Prešern, J., M. Gogala, & T. Trilar. 2004a. Comparison of *Dundubia vaginata* (Auchenorrhyncha: Cicadoidea) songs from Borneo and peninsular Malaysia. *Acta Entomologica Slovenica* 12: 239–248.

Prešern, J., M. Gogala, & T. Trilar. 2004b. Bioacoustics as a systematic tool: Dundubia vaginata from South-East Asia. *Third European Hemiptera Congress, St. Petersburg, June 8–11, 2004, Abstracts*: 62.

Press, M.C., & J.B. Whittaker. 1993. Exploitation of the xylem stream by parasitic organisms. *Philosophical Transactions of the Royal Society of London. B* 341: 101–111.

Preston-Mafham, K., & R. Preston-Mafham. 2000. *The natural world of bugs & insects*. Thunder Bay Press, Advantage Publishers Group, San Diego, CA. 512pp.

Prestwich, K.N. 1994. The energetics of acoustic signalling in anurans and insects. *American Zoologist* 34: 625–643.

Price, B.W., N.P. Barker, & M.H. Villet. 2007. Patterns and processes underlying evolutionary significant units in the *Platypleura stridula* L. species complex (Hemiptera: Cicadidae) in the Cape Floristic Region, South Africa. *Molecular Ecology* 16: 2574–2588.

Price, B.W., N.P. Barker, & M.H. Villet. 2009. A watershed study on gentic diversity: phylogenetic analysis of the *Platypleura plumosa* (Hemiptera: Cicadidae) complex reveals catchment-specific lineages. *Molecular Phylogenetics and Evolution* 54: 617–626.

Price, B.W., T.A. Bellingan, & M.H. Villet. 2008. Phylogeny of the tribe Tettigomyiini (Hemiptera: Cicadidae) in southern Africa. *23rd International Congress of Entomology, Durban, 6–12 July 2008*. Abstract number 1332.

Princen, L.H. 1975. A possible relationship between the eastern mole and the periodical cicada. *Proceedings of the Peoria Academy of Sciences* 8: 26–32.

Pringle, J.W.S. 1957. Insect flight. *Cambridge Monographs in Experimental Biology* 9: 1–132.

Pringle, J.W.S. 1981. The evolution of fibrillar muscle in insects. *Journal of Experimental Biology* 94: 1–14.

Prokop, J., & M. Boulard. 2000. *Tibicina sakalai* n. sp., cigale fossile du Miocène inférieur de Tchécoslovaquie (Auchenorhyncha, Cicadidae Tibicininae). *Ecole Pratique des Hautes Etudes, Travaux du Laboratoire Biologie et Evolution des Insectes Hemipteroidea* 13: 127–131.

Puissant, S. 2001. Eco-ethologie de *Cicadetta montana* (Scopoli, 1772) en France (Auchénorhyncha, Cicadidae, Cicadettini). *Ecole Pratique des Hautes Etudes, Travaux du Laboratoire Biologie et Evolution des Insectes Hemipteroidea* 14: 141–155.

Puissant, S. 2005. Taxonomy, distribution and first eco-ethological data of *Melampsalta varipes* (Waltl, 1837), an unrecognized cicada (Hemiptera, Cicadidae). *Insect Systematics and Evolution* 36: 301–315.

Puissant, S. 2006. *Contribution a la connaissance des cigales de France: geonemie et ecologie des populations (Hemiptera, Cicadidae)*. Association pour la Caractérisation et l'Etude des Entomocénoses, Bedeilhac et Aynat. 193pp.

Puissant, S., & M. Boulard. 2000. *Cicadetta cerdaniensis*, espèce jumelle de *Cicadetta montana* décryptée par l'acoustique (Auchenorhyncha, Cicadidae, Tibicininae). *Ecole Pratique des Hautes Etudes, Travaux du Laboratoire Biologie et Evolution des Insectes Hemipteroidea* 13: 111–117.

Puissant, S., & B. Defaut. 2005. Les synusies de cigales en France (Hemiptera, Cicadidae). *Matériaux Orthoptériques et Entomocénotiques* 10: 115–129.

Puissant, S., & J. Sueur. 2000. Redecouverte en France de *Cicadivetta tibialis* (Panzer, 1798) et nouvelles observations sur la biologie de cette espèce (Cicadidae, Tibicininae, Cicadettini). *Ecole Pratique des Hautes Etudes, Travaux du Laboratoire Biologie et Evolution des Insectes Hemipteroidea* 13: 67–74.

Puissant, S., & J. Sueur. 2001. Contribution à l'étude des cigales de Corse (Hemiptera, Cicadidae). *Bulletin de la Société entomologique de France* 106: 429–436.

Puissant, S., & J. Sueur. 2002. Description of a new cicada species from Turkey [Hemiptera, Cicadoidea]. *Revue française d'Entomologie (N.S.)* 24: 197–200.

Puissant, S., & J. Sueur. 2010. A hotspot for Mediterranean cicadas (Insecta: Hemiptera: Cicadidae): new genera, species and songs from southern Spain. *Systematics and Biodiversity* 8: 555–574.

Punzo, F. 2000. *Desert arthropods: life history variations*. Springer-Verlag, Berlin. 230pp.

Putnam, J.D. 1881. Remarks on the habits of several western Cicadidae. *Proceedings of the Davenport Academy of Natural Science* 3: 67–68.

Quartau, J.A. 1987. On the discrimination between *Cicada orni* Linné, 1758 and *C. barbara lusitanica* Boulard, 1982 by numerical taxonomic methods (Homoptera Cicadidae). *Proceedings of the 6th Auchenorrhyncha Meeting, Turin, Italy, 7–11 September 1987*: 37.

Quartau, J.A. 1988. A numerical taxonomic analysis of interspecific morphological differnces in two closely related species of *Cicada* (Homoptera, Cicadidae) in Portugal. *Great Basin Naturalist Memoirs* 12: 171–181.

Quartau, J.A. 1993. Biosystematic studies of some Auchenorrhyncha in Portugal. *Proceedings of the 8th Auchenorrhyncha Congress, Delphi, Greece, 9–13 August 1993*: 93.

Quartau, J.A. 1995. Cigarras esses insectos quase desconhecidos. *Correio da Natureza* 19: 33–38.

Quartau, J.A. 1998. *Vie et Mémoires de Cigales*. Book Review. *African Entomology* 6: 381.

Quartau, J.A. 2008. Preventative fire procedures in Mediterranean woods are destroying their insect biodiversity: a plea to the EU governments. *Journal of Insect Conservation* 13: 267–270.

Quartau, J.A., G. André, G. Pinto-Juma, S.G. Seabra, & P.C. Simões. 2008. Geographic variation in the calling song of *Cicada orni* L. (Hemiptera: Cicadoidea) in the Mediterranean area. *Boletim do Museu Municipal do Funchal, Suplemento* 14: 127–130.

Quartau, J.A., & M. Boulard. 1995. *Tettigetta mariae* n. sp., nouvelle Cigale lusitanienne (Homoptera Cicadoidea Tibicinidae). *Ecole Pratique des Hautes Etudes, Travaux du Laboratoire Biologie et Evolution des Insectes Hemipteroidea* 7–8: 105–110.

Quartau, J.A., M.M. Coble, A.M. Viegas-Crespo, M.T. Rebelo, M. Ribeiro, G.A. Pinto, P. Simões, G. André, M. Claridge, J. Hemingway, J.C. Morgan, & S. Drosopoulos. 1997. A behavioural, molecular and morphometric approach to population divergence in cicadas (Homoptera, Auchenorrhyncha): Some introductory results. *Proceedings of the 9th International Auchenorrhyncha Congress, Sydney, Australia, 17–21 February 1997*: 47–48.

Quartau, J.A., & P.J. Fonseca. 1988. An annotated check-list of the species of cicadas known to occur in Portugal (Homoptera: Cicadoidea). *Proceedings of the 6th Auchenorrhynchota Meeting, Turin, Italy* 7–11 September 1987: 367–375.

Quartau, J.A., & G.A. Pinto. 1998. Patterns of genetic diversity within *Cicada orni* L. and *C. barbara* (Stål) (Insecta: Homoptera) in Portugal using mtDNA data. *VIII Congresso Ibérico de Entomologia, Évora, Portugal, 7–11 de Setembro 1998*: 106.

Quartau, J.A., G. Pinto, S. Seabra, P.C. Simoes, S. Drosopoulos, M. Bruford, M. Claridge, & J. Morgan. 2002. Studies on species divergence and speciation in cicadas: patterns of morphometric variation (Hemiptera, Cicadoidea). *Eleventh International Auchenorrhyncha Congress, August 5–9, Potsdam, Germany, Abstracts of Talks and Posters*: 21.

Quartau, J.A., & M.T. Rebelo. 1990. The songs of some species of cicadas occuring in Portugal (Homoptera, Auchenorrhyncha, Cicadidae). *Abstracts of the Combined Meeting of the 7th International Auchenorrhyncha Congress and the 3rd International Workshop on Leafhoppers and Planthoppers of Economic Importance, Wooster, Ohio, August 13–17*.

Quartau, J.A., & M.T. Rebelo. 1994. Sinais acústicos em Cicadidae e Cicadellidae (Homoptera, Auchenorrhyncha) que Ocorrem em Portugal. *Actas do I Congresso Nacional de Etologia, Lisboa, 1994*: 137–142.

Quartau, J.A., M.T. Rebelo, & P. Simões. 1999. Cicadídeos (Insectos, Homópteros). Pages 69–74 *in* M. Santos-Reis & A.I. Correia, eds. *Caracterização da Flora e de Fauna do Montado da Herdade da ribeira Abaixo (Grândola-Baixo Alentejo)*. Centro de Biologia Ambiental, Lisboa.

Quartau, J.A., M.T. Rebelo, P.C. Simões, T.M. Fernandes, M.F. Claridge, S. Drosopoulos, & J.C. Morgan. 1999. Acoustic signals of populations of *Cicada orni* L. in Portugal and Greece (Homoptera: Auchenorrhyncha: Cicadomorpha: Cicadidae). *Reichenbachia, Staatliches Museum für Tierkunde Dresden* 33: 71–80.

Quartau, J.A., M. Ribeiro, P. Almeida, & A.M. Crespo. 1996. A pilot electrophoretic study applied to two closely related species of *Cicada* L. (Homoptera, Cicadidae) in Portugal. *VII Congreso Ibérico de Entomología, Santiago de Compostela, España, 19–23 de Septiembre 1996*: 138.

Quartau, J.A., M. Ribiero, P.C. Simões, & M.M. Coelho. 2001. Genetic divergence among populations of two closely related species of *Cicada* Linnaeus (Hemiptera: Cicadoidea) in Portugal. *Insect Systematics & Evolution* 32: 99–106.

Quartau, J.A., M. Ribeiro, P.C. Simões, & A. Crespo. 2000. Taxonomic separation by isozyme electrophoresis of two closely related species of *Cicada* L. (Hemiptera: Cicadoidea) in Portugal. *Journal of Natural History* 34: 1677–1684.

Quartau, J.A., S. Seabra, & A. Sanborn. 2000. Effect of ambient air temperature on some temporal parameters of the calling song of *Cicada orni* Linnaeus, 1758 (Hemiptera: Cicadidae) in Portugal. *Acta Zoologica Cracoviensia* 43: 193–198.

Quartau, J.A., & P.C. Simões. 2003. Bioacoustic and morphological differentiation in two allopatric species of the genus *Tibicina* Amyot (Hemiptera, Cicadoidea) in Portugal. *Mitteilungen aus dem Museum für Naturkunde im Berlin, Deutsche Entomologishe Zeitschrift* 50: 113–119.

Quartau, J.A., & P.C. Simões. 2004. First description of the calling song produced by *Euryphara contentei* Boulard, 1982 (Insecta: Hemiptera, Cicadoidea) in Portugal. *Arquivos do Museu Bocage, Nova Série* 3: 563–572.

Quartau, J.A., & P.C. Simões. 2005a. *Cicada cretensis* sp. n. (Hemiptera, Cicadoidea) from southern Greece. *Biologia, Bratislava* 60: 489–494.

Quartau, J.A., & P.C. Simões. 2005b. Acoustic evolutionary divergence in cicadas: the species of *Cicada* L. in southern Europe. Pages 227–237 *in* S. Drosopoulos & M.F. Claridge, eds. *Insect sounds and communication: physiology, behaviour, ecology and evolution*. CRC Press, Boca Raton.

Quartau, J.A., P.C. Simões, M.T. Rebelo, & G. André. 2001. On two species of the genus *Tibicina* Amyot, 1847 (Hemiptera, Cicadoidea) in Portugal, with one new record. *Arquivos do Museu Bocage, Nova Série* 3: 401–412.

Quartau, J.A., P. Simões, M.T. Rebelo, S. Drosopoulos, M. Claridge, & J.C. Morgan. 1998. Acoustic signals in populations of *Cicada* L. in Greece (Insecta, Cicadidae). *VIII Congresso Ibérico de Entomologia, Évora, Portugal, 7–11 de Setembro 1998*: 66.

Quartau, J.A., P.C. Simões, S.G. Seabra, S. Drosopoulos, M.F. Claridge, & J.C. Morgan. 1999. Acoustic and genetic studies on closely related species of *Cicada* L. (Hemiptera: Cicadoidea) in the Mediterranean area. *Abstracts of Papers and Posters, 10th International Auchenorrhyncha Congress, 6–10 September 1999, Cardiff, Wales.* 2pp (no pagination).

Quek, S.-P., S.J. Davies, T. Itino, & N.E. Pierce. 2004. Codiversification in an ant-plant mutualism: stem texture and the evolution of host use in *Crematogaster* (Formicidae: Myrmicinae) inhabitants of Macaranga (Euphorbiaceae). *Evolution* 58: 554–570.

Quiroga, D., P. Arretz, & J.E. Araya. 1991. Sucking insects damaging jojoba, *Simmondsia chinensis* (Link) Schneider, and their natural enemies, in the North Central and Central regions of Chile. *Crop Protection* 10: 469–472.

Ragge, D.R., & W.J. Reynolds. 1998. *The songs of the grasshoppers and crickets of Western Europe*. Harley Books, London. 591pp.

Raina, A., L. Pannell, J. Kochansky, & H. Jaffe. 1995. Primary structure of a novel neuropeptide isolated from the corpora cardiaca of periodical cicadas having adipokinetic and hypertrehalosemic activities. *Insect Biochemistry and Molecular Biology* 25: 929–932.

Rakitov, R.A. 2002. Structure and function of the malpighian tubules, and related behaviors in juvenile cicadas: evidence of homology with spittlebugs (Hemiptera: Cicadoidea & Cercopoidea). *Zoologischer Anzeiger* 241: 117–130.

Raksakantong, P., N. Meeso, J. Kubola, & S. Siriamornpun. 2010. Fatty acids and proximate composition of eight Thai edible terricolous insects. *Food Research International* 43: 350–355.

Ramallo, N.E.V. de. 1989. El hongo *Cordyceps* sp., parasito natural de ninfas de *Proarna bergi* Distant. *Revista Industrial y Agrícola de Tucumán* 66: 143–147.

Ramos, J.A. 1983. Sinopia de las cigarras de la República Dominicana. *Caribbean Journal of Science* 19: 61–70.

Ramos, J.A. 1988. Zoogeography of the auchenorrhynchous Homoptera of the Greater Antilles (Hemiptera). Pages 61–70 *in* J.K. Liebherr, ed. *Zoogeography of Caribbean insects*. Cornell University Press, Ithaca.

Ramos, J.A., & H. Wolda. 1985. Description and distribution of two new species of *Selymbria* from Panama (Homoptera, Cicadoidea, Tibicinidae). *Caribbean Journal of Science* 21: 177–185.

Ramos Elorduy de Conconi, J., & J.M. Pino Moreno. 1979. Insectos comestibles del Valle del Mezquital y su valor nutritivo. *Anales del Instituto de Biología, Universidad Nacional Autonoma de México, Zoología* 50: 563–574.

Ramos Elorduy de Conconi, J., & J.M. Pino Moreno. 1981. Insectos comestibles del Valle del Mezquital y su valor nutritivo. *Anales del Instituto de Biologia, Universidad Nacional Autonoma de Mexico, Zoologia* 50: 563–574.

Ras, R.H.A., E. Sahramo, J. Malm, J. Raula, & M. Karppinen. 2008. Blocking the lateral film growth at the nanoscale in area-selective atomic layer deposition. *Journal of the American Chemical Society* 130: 11252–11253, S1-S2.

Ratcliff, B.C., & P.C. Hammond. 2002. Insects and the native vegetation of Nebraska. *Transactions of the Nebraska Academy of Sciences* 28: 29–47.

Rau, P. 1922. Ecological and behavior notes on Missouri insects. *Transactions of the Academy of Science of St. Louis* XXIV(7): 1–71, plates V-VIII.

Raven, P.H., & D.K. Yeates. 2007. Australian biodiversity: threats for the present, opportunities for the future. *Australlian Journal of Entomology* 46: 177–187.

Reding, M.E., & S.I. Guttman. 1991. Radionuclide contamination as an influence on the morphology and genetic structure of periodical cicada (*Magicicada cassini*) populations. *American Midland Naturalist* 126: 322–337.

Reger, J.F. 1961. The fine structure of neuromuscular junctions and the sarcoplasmic reticulum of extrinsic eye muscles of *Fundulus heteroclitus*. *The Journal of Biophysical and Biochemical Cytology* 10: 111–121.

Reichholf-Riehm, H. 1984. *Inseckten mit Anhang Spinnentiere*. Mosaik Verlag, München. 288pp.

Reid, B., R.G. Ordish, & M. Harrison. 1982. An analysis of the gizzard contents of 50 North Island brown kiwis, *Apteryx australis mantelli*, and notes on feeding observations. *New Zealand Journal of Ecology* 5: 76–85.

Reinhold, K. 1999. Energetically costly behaviour and the evolution of resting metabolic rate in insects. *Functional Ecology* 13: 217–224.

Reis, P.R., J.C. de Souza, & C. do C.A. Melles. 1984. Pragas do cafeeiro. *Informe Agropecuário, Belo Horizonte* 10: 3–11.

Reis, P.R., J.C. de Souza, & M. Venzon. 2002. Manejo ecológico das principais pragas do cafeeiro. *Informe Agropecuário, Belo Horizonte* 23: 83–99.

Remane, R. 1987. Zum artenbestand der Zikaden (Homoptera: Auchenorrhyncha) auf dem Mainzer Sand. *Mainzer Naturwissenschaftliches Archiv* 25: 273–349.

Remane, R., R. Achtziger, W. Fröhlich, H. Nickel, & W. Witsack. 1998. Rote Liste der Zikaden (Homoptera, Auchenorrhyncha). *In* M. Binot, R. Bless, P. Boye, H. Gruttke, & P. Pretscher, eds. *Rote Liste gefährdeter Tiere Deutschlands. Schriftenreihe für Landschaftspflege und Naturschutz* 55: 243–249.

Remane, R., W. Fröhlich, H. Nickel, W. Witsack, & R. Achtziger. 1997. Rote Liste der Zikaden Deutschlands (Homoptera, Auchenorrhyncha). *Beiträge zur Zikadenkunde* 1: 63–70.

Remane, R., & E. Wachmann. 1993. *Zikaden: kennenlernen - beobacten*. Naturbuch Verlag, Augsburg. 288pp.

Remondino, M., & A. Cappellini. 2005. Periodical insects and prime numbers: an agent based simulation. *Proceedings of the European Simulation and Modelling Conference, ESM 2005, Universidade do Porto, Porto, Portugal, October 24–26, 2005*: 6pp.

Remondino, M., & A. Cappellini. 2006a. Agent based simulation in biology: the case of periodical insect as natural prime number generators. *Vet On-Line – The International Journal of Veterinary Medicine*. 9pp. Available online: www.priory.com/vet/cicada.pdf

Remondino, M., & A. Cappellini. 2006b. An evolutionary selection model based on a biological phenomenon: the periodical Magicicadas. Pages 485–497 *in* S. Nolfi, G. Baldassorre, R. Calabretta, J.C.T. Hallam, D. Marocco, J.-A. Meyer, O. Miglino, & D. Parisi, eds. *From animals to animats 9, 9th International Conference on simulation of Adaptive Behavior, SAB 2006, Rome, Italy, September 25–29, 2006, Proceedings*. Sring-Verlag, Berlin Heildelberg.

Retallack, G.J. 2004. Late Oligocene bunch grassland and early Miocene sod grassland paleosols from central Oregon, USA. *Palaeogeography, Palaeoclimatology, Palaeoecology* 207: 203–237.

Rethwisch, M.D., C.W. McDaniel, & J. Thiessen. 1992. Control of Apache cicada on asparagus, 1991. Page 77 *in* A.K. Burditt, Jr., ed. *Insecticide and Acaricide Tests*, vol. 17. Entomological Society of America, Lanham, MD.

Reynaud, G. 2005. Quelques données supplémentaires pour la faune des cigales (Homoptera, Cicadidae) de France. *Bulletin de la Sociétié Linnéenne de Bordeaux* 33: 141–144.

Ribeiro, M., J.A. Quartau, P. Simões, T.M.A. Fernandes, M.T. Rebelo, A. Crespo, & G. André. 1998. Patterns of morphometric and electrophoretic variation in *Cicada orni* L. and *C. barbara* (Stal) (Insecta: Homoptera) in Portugal. *VIII Congresso Ibérico de Entomologia, Évora, Portugal, 7–11 de Setembro 1998*: 106–107.

Ribeiro, R., M.F.A. Pereira, N.M. Martinelli, & D.H.B. Maccagnan. 2006. Dispersão de *Fidicinoides* sp. (Hemiptera: Cicadidae) em cafeeiro. *Científica (Jaboticabal)* 34: 263–268.

Rice, M.E. 2000. Freid green cicadas and a young African hunter. *American Entomologist* 46: 6–7.

Riede, K. 1995. Räumliche und zeitliche Strukturierung tropischer Rufemeinschaften. *Verhandlungen der deutschen Zoologischen Geselschaft* 88: 45.

Riede, K. 1996. Diversity of sound-producing insects in a Bornean lowland rain forest. Pages 77–84 *in* D.S. Edwards, W.E. Booth, & S.L. Choy, eds. *Tropical rainforest research: current issues*. Kluwer Academic, Dordrecht.

Riede, K. 1997a. Bioacoustic diversity and resource partitioning in tropical calling communities. Pages 275–280 *in* H. Ulrich, ed. *Tropical biodiversity and systematics. Proceedings of the international symposium on biodiversity and systematics in tropical ecosystems, Bonn, 1994*. Zoologisches Forschungsinstitut und Museum Alexander Koenig, Bonn.

Riede, K. 1997b. Bioacoustic monitoring of insect communities in a Bornean rainforest canopy. Pages 442–452 *in* N.E. Stork, J. Adis, & R.K. Didham, eds. *Canopy arthropods*. Chapman & Hall, London.

Riede, K., & A. Kroker. 1995. Bioacoustics and niche differentiation in two cicada species from Bornean lowland forest. *Zoologischer Anzeiger* 234: 43–51.

Riegel, G. 1981. The cicada in Chinese folklore. *Melsheimer Entomological Series* 30: 15–20.

Riley, C.V. 1872. The periodical cicada. *Annual Report on the Noxious, Beneficial and Other Insects of the State of Missouri, Made to the State Board of Agriculture Persuant to an Appropriation for This Purpose From the Legislature of the State. Jefferson City* 4: 30–34.

Riou, B. 1995. *Lyristes renei n. sp.*, cigale fossile du Miocène ardéchois (Homoptera Cicadidae). *Ecole Practique des Hautes Etudes, Travaux du Laboratoire Biologie et Evolution des Insectes Hemipteroidea* 7/8: 73–76.

Ritchie, M.G. 2001. Chronic speciation in periodical cicadas. *Trends in Ecology & Evolution* 16: 59–61.

Rivera-Orduño, B., & S.P. Stock. 2004. Ecological characterization of *Heterorhabditis* sp. (*caborca* strain) (Nematoda: Heterorhabditidae), a natural pathogen of *Diceroprocta ornea* (Homoptera: Cicadidae) from Sonora, Mexico. *Book of abstracts, 37th Annual Meeting of the Society for Invertebrate Pathology and 7th International Conference on* Bacillus thuringiensis, *Helsinki, Finland, 1–6 August 2004*: 98.

Rizza, D. 2010. *Magicicada*, mathematical explanation and mathematical realism. *Erkenntnis* 74: 101–114.

Robert, D., R.N. Miles, & R.R. Hoy. 1999. Tympanal hearing in the Sarcophagid parasitoid fly *Emblemasoma* sp.: the biomechanics of directional hearing. *Journal of Experimental Biology* 202: 1865–1876.

Robertson, M.P., M.H. Villet, & A.R. Palmer. 2004. A fuzzy classification technique for predicting species' distributions: applications using invasive alien plants and indigenous insects. *Diversity and Distributions* 10: 461–474.

Robinson, G.R., P.L. Sibrell, C.J. Boughton, L.H. Yang, & T.C. Hancock. 2005. Biogeochemistry of metals in periodic cicadas. *Eos Transactions American Geophysical Union* 86(18) *Joint Assembly Supplement, Abstract* B31A-06.

Robinson, G.R., Jr., P.L. Sibrell, C.J. Boughton, & L.H. Yang. 2007. Influence of soil chemistry on metal and bioessential element concentrations in nymphal and adult periodical cicadas (*Magicicada* spp.). *Science of the Total Environment* 374: 367–378.

Robinson, W.H. 2005. *Handbook of urban insects and arachnids*. Cambridge University Press, Cambridge. 472pp.

Rodenhouse, N.L., P.J. Bohlen, & G.W. Barrett. 1997. Effects of woodland shape on the spatial distribution and density of 17-year periodical cicadas (Homoptera: Cicadidae). *American Midland Naturalist* 137: 124–135.

Rolando, A. 1998. Factors affecting movements and home ranges in the jay (*Garrulus glandarius*). *Journal of Zoology* 246: 249–257.

Römer, H. 1998. The sensory ecology of acoustic communication in insects. Pages 63–96 *in* R.R. Hoy, A.N. Popper, & R.R. Fay, eds. *Comparative hearing: Insects*. Springer-Verlag, Berlin.

Rosenberg, K.V., R.D. Ohmart, & B.W. Anderson. 1982. Community organization of riparian breeding birds: response to an annual resource peak. *The Auk* 99: 260–274.

Rosenbluth, J., A.G. Szent-Györgyi, & J.T. Thompson. 2010. The ultrastructure and contractile properties of a fast-acting, obliquely striated, myosin-regulated muscle: the funnel retractor of squids. *Journal of Experimental Biology* 213: 2430–2443.

Roth, P. 1949. Les *Sphecius* palearctiques (Hym. Sphecidae). *Annales de la Société Entomologique de France* 118: 79–94.

Roxburgh, L., B. Pinshow, & H.D. Prange. 1996. Temperature regulation by evaporative cooling in a desert grasshopper, *Calliptamus barbarus* (Ramme, 1951). *Journal of Thermal Biology* 21: 331–337.

Ruckstuhl, M. 2010. Förderung der Biodiversität in der Fallätsche. *Grünzeit* 32: 2–3.

Ruffinelli, A. 1970. Contribucion al conocimiento de los Homopteros Auquenorrincos del Uruguay. *Publicação Técnica; Série Zoologia Agrícola* 1: 1–25.

Russell, L.M., & M.B. Stoetzel. 1991. Inquilines in egg nests of periodical cicadas (Homoptera: Cicadidae). *Proceedings of the Entomological Society of Washington* 93: 480–488.

Ryan, M.J. 1985. Energetic efficiency of vocalization by the frog *Physalaemus pustulosus*. *Journal of Experimental Biology* 116: 47–52.

Ryan, M.J. 1988. Energy, calling and selection. *American Zoologist* 28: 885–898.

Rypstra, A.L., & T.G. Gregg. 1986. Facultative carnivory in *Drosophila hydei*. *Ohio Journal of Science* 86: 34.

Saboori, A., & H. Lazarboni. 2008. A new genus and species of Trombidiidae (Acari: Trombidioidea) described from larvae ectoparasitic on *Cicadatra ochreata* Melichar (Hmoptera: Cicadidae) from Iran. *Zootaxa* 1852: 50–58.

Sadof, C.S. 1992. Periodical cicada in Indiana. *Purdue University Cooperative Extension Service* E-47: 1–2.

Sahlberg, C.R. 1842. Cicadae tres novae fennicae. *Acta Societatis Scientarum Fennicae* 1: 85–92.

Sahli, H.F., & S. Ware. 2000. Oviposition sites and emergence habitats of 13-year periodical cicadas (Brood XIX) in eastern Virginia. *Virginia Journal of Science* 51: 187–194.

Saisho, Y. 2001a. [On the positive phototaxis observed in nymphs of *Euterpnosia chibensis chibensis*.] *Cicada* 15: 67–69. [In Japanese, English summary]

Saisho, Y. 2001b. [Observations of *Euterpnosia chibensis daitoensis*.] *Cicada* 15: 69. [In Japanese]

Saisho, Y. 2003. [A report of field researches on *Terpnosia vacua*.] *Cicada* 17: 49–54. [In Japanese, English summary]

Saisho, Y. 2004. [A report of field researches on *Cicadetta radiator*.] *Cicada* 18: 1–4. [In Japanese, English summary]

Saisho, Y. 2005. [A report on *Cryptotympana atrata* observed at Kanazawa, Honshu, in 2002 and 2004.] *Cicada* 18: 43–46. [In Japanese, English summary]

Saisho, Y. 2006a. [Eclosion of *Pomponia linearis*.] *Cicada* 18: 57. [In Japanese]

Saisho, Y. 2006b. [On the relation between the beginning time of male charping and the increasing and decreasing of the number of cicadas.] *Cicada* 18: 58–61. [In Japanese, English summary]

Saisho, Y. 2006c. [Copulation of *Tanna japonensis ishigakiana*.] *Cicada* 18: 62. [In Japanese]

Saisho, Y. 2006d. [Eclosion of *Cryptotympana yaeyamana*.] *Cicada* 18: 62. [In Japanese]

Saisho, Y. 2010a. [Further observations on the eclosion of *Kosemia radiator*.] *Cicada* 19: 57–59. [In Japanese]

Saisho, Y. 2010b. [Eclosion of *Kosemia radiator*.] *Cicada* 19: 59. [In Japanese]

Saisho, Y. 2010c. [Field observations of autumn cicadas on Amami-Oshima Is., Ryukyus.] *Cicada* 19: 60. [In Japanese]

Salazar, J.A., & A. Sanborn. 2009. Nota rectificativa sobre artículo. *Boletím Científico Centro de Museos Museo de Historia Natural* 13(2): 270–271.

Salazar Escobar, J.A. 2005. Algunos cicádidos de Colombia (Homptera: Cicadidae). *Boletín Científico, Centro de Museos, Museo de Historia Natural* 9: 192–204.

Salguero, E. 2003. *Protegiendo el espíritu de Morales*. Fundación para el Ecodesarrollo y la Conservación, Agencia Española de Cooperación International, Sendero Educativo para la Conservación del Agua del Municipio de Morales Izabal. Departamento de Morales, Izabal, Guatemala, C.A. 58pp.

Saljoqi, A.-U.-R., M.S. Rahimi, I.A. Khan, & S. Rehman. 2010. Inseticidal management of cicada, *Tibicen* spp. (Homoptera: Cicadidae) in grape vineyards in northern plains of Afghanistan. *Sarhad Journal of Agriculture* 26: 69–74.

Saljoqi, A.-U.-R., M.S. Rahimi, S. Rehman, & I.A. Khan. 2010. Population density of cicada, *Tibicen* spp. (Homoptera: Cicadidae) in grape vineyards in northern plains of Afghanistan. *Sarhad Journal of Agriculture* 26: 75–78.

Salmah, Y., J.P. Duffels, & M.I. Zaidi. 2005. Taxonomic notes of the Malaysian species of *Chremistica* Stål (Homoptera: Cicadidae). *Journal of Asia-Pacific Entomology* 8: 15–23.

Salmah, Y., & M.I. Zaidi. 2002. The genus *Chremistica* Stal (Homoptera: Cicadidae) from Peninsular Malaysia, with descriptions of two new species, *Chremistica guamusangensis* n. sp. and *Chremistica kecil* n. sp. *Serangga* 7: 225–244.

Salmah, Y., M.I. Zaidi, & J.P. Duffels. 2004. Male genitalia illustrations for genus *Chremistica* Stål (Homoptera: Cicadidae) from Malaysia. *Serangga* 9: 1–8.

Salmon, J.T. 1950. Notes on synonymy among New Zealand insects. 1. *Transactions and Proceedings of the Royal Society of New Zealand* 78: 1–2.

Samietz, J., S. Kroder, D. Schneider, & S. Dorn. 2006. Ambient temperature affects mechanosensory host location in a parasitic wasp. *Journal of Comparative Physiology. A. Neuroethology, Sensory, Neural, and Behavioral Physiology* 192: 151–157.

Samson, R.A., H.C. Evans, & J.-P. Latgé. 1988. *Atlas of entomopathogenic fungi*. Springer-Verlag, Wetenschappelijke uitgeverij Bunge, Utrecht. 187pp.

Sanborn, A.F. 1991. Endothermy in cicadas (Homoptera: Cicadidae). *Dissertation Abstracts International* 51: 5753B.

Sanborn, A.F. 1996. The cicada *Diceroprocta delicata* (Homoptera: Cicadidae) as prey for the dragonfly *Erythemis simplicicollis* (Anisoptera: Libellulidae). *Florida Entomologist* 79: 69–70.

Sanborn, A.F. 1997. Body temperature and the acoustic behavior of the cicada *Tibicen winnemanna* (Homoptera: Cicadidae). *Journal of Insect Behavior* 10: 257–264.

Sanborn, A.F. 1998. Thermal biology of cicadas (Homoptera: Cicadoidea). *Trends in Entomology* 1: 89–104.

Sanborn, A.F. 1999a. Cicada (Homoptera: Cicadidae and Tibicinidae) type material in the collections of the American Museum of Natural History, California Academy of Sciences, Snow Entomological Museum, Staten Island Institute of Arts and Sciences, and the United States National Museum. *Florida Entomologist* 82: 34–60.

Sanborn, A.F. 1999b. Review and type status of North American cicada species. *Abstracts of Papers and Posters, 10th International Auchenorrhyncha Congress, 6–10 September 1999, Cardiff, Wales.* 2pp (no pagination).

Sanborn, A.F. 2000. Comparative thermoregulation of sympatric endothermic and ectothermic cicadas (Homoptera: Cicadidae: *Tibicen winnemanna* and *Tibicen chloromerus*). *Journal of Comparative Physiology. A. Sensory, Neural and Behavioral Physiology* 186(6): 551–556.

Sanborn, A.F. 2001a. Timbal muscle physiology in the endothermic cicada *Tibicen winnemanna* (Homoptera: Cicadidae). *Comparative Biochemistry and Physiology. A. Physiology* 130: 9–19.

Sanborn, A.F. 2001b. A first contribution to a knowledge of the cicada fauna of El Salvador (Homoptera: Cicadoidea). *Florida Entomologist* 84: 449–450.

Sanborn, A.F. 2001c. Distribution of the cicadas (Homoptera: Cicadidae) of the Bahamas. *Florida Entomologist* 84: 651–652.

Sanborn, A.F. 2002a. Periodical cicadas: The magic cicada (Hemiptera, Tibicinidae, *Magicicada* spp.). *Denisia* 4: 225–230.

Sanborn, A.F. 2002b. Cicada thermoregulation (Hemiptera, Cicadoidea). *Denisia* 4: 455–470.

Sanborn, A.F. 2004a. Thermoregulation in the endothermic cicada *Tibicen cultriformis* (Hemiptera: Cicadidae). *The FASEB Journal* 18: A1098.

Sanborn, A.F. 2004b. Thermoregulation and endothermy in the large western cicada *Tibicen cultriformis* (Hemiptera: Cicadidae). *Journal of Thermal Biology* 29: 97–101.

Sanborn, A.F. 2004c. Cicadas (Hemiptera: Cicadoidea). Pages 510–514 *in* J. Capinera, ed. *Encyclopedia of entomology*, Volume 1. Kluwer Academic Publishers, Dordrecht.

Sanborn, A.F. 2004d. Sound production in the Cicadoidea. Pages 2055–2057 *in* J. Capinera, ed. *Encyclopedia of entomology*, Volume 3. Kluwer Academic Publishers, Dordrecht.

Sanborn, A.F. 2004e. Thermoregulation. Pages 2223–2225 *in* J. Capinera, ed. *Encyclopedia of entomology*, Volume 3. Kluwer Academic Publishers, Dordrecht.

Sanborn, A.F. 2004f. Two new cicada *Zammara* species from South America (Hemiptera: Cicadoidea: Cicadidae). *Florida Entomologist* 87: 365–371.

Sanborn, A.F. 2005a. *Fidicina variegata*, a new cicada species from Costa Rica (Hemiptera: Cicadomorpha: Cicadidae). *Annals of the Entomological Society of America* 98: 187–190.

Sanborn, A.F. 2005b. *Rhadinopyga duffelsi*, a new cicada species from New Guinea (Hemiptera: Cicadomorpha: Cicadidae). *Tijdschrift voor Entomologie* 148: 21–26.

Sanborn, A.F. 2005c. Acoustic signals and temperature. Pages 111–125 *in* S. Drosopoulos & M.F. Claridge, eds. *Insect sounds and communication: physiology, behaviour, ecology and evolution*. CRC Press, Boca Raton.

Sanborn, A.F. 2006a. New records for the cicada fauna from four Central American countries (Hemiptera: Cicadoidea: Cicadidae). *Florida Entomologist* 89: 75–79.

Sanborn, A.F. 2006b. New records of cicadas from Mexico (Hemiptera: Cicadoidea: Cicadidae). *Southwestern Naturalist* 51: 255–257.

Sanborn, A.F. 2006c. Two new cicada species from Shaanxi, China (Hemiptera: Cicadomorpha: Cicadidae). *Acta Entomologica Sinica* 49: 829–834.

Sanborn, A.F. 2007a. Additions to the cicada fauna of Venezuela with the description of a new species and checklist of the Venezuelan cicada fauna (Hemiptera: Cicadoidea: Cicadidae). *Zootaxa* 1503: 21–32.

Sanborn, A.F. 2007b. New species, new records and checklist of cicadas from Mexico (Hemiptera: Cicadoidea: Cicadidae). *Zootaxa* 1651: 1–42.

Sanborn, A.F. 2008a. The identity of *Cicada tibicen* Linné [=*Tibicen chloromerus* (Walker, 1850)] (Hemiptera: Cicadoidea: Cicadidae). *Entomological News* 119: 227–231.

Sanborn, A.F. 2008b. Acoustic communication in insects. Pages 33–38 *in* J. Capinera, ed. *Encyclopedia of entomology*, second edition, volume 1. Springer-Verlag, Heidelberg.

Sanborn, A.F. 2008c. Cicadas (Hemiptera: Cicadoidea). Pages 874–877 *in* J. Capinera, ed. *Encyclopedia of entomology*, second edition, volume 1. Springer-Verlag, Heidelberg.

Sanborn, A.F. 2008d. Sound production in the Cicadoidea. Pages 3468–3470 *in* J. Capinera, ed. *Encyclopedia of entomology*, second edition, volume 4. Springer-Verlag, Heidelberg.

Sanborn, A.F. 2008e. Thermoregulation in insects. Pages 3757–3760 *in* J. Capinera, ed. *Encyclopedia of entomology*, second edition, volume 4. Springer-Verlag, Heidelberg.

Sanborn, A.F. 2008f. New records of Brazilian cicadas including the description of a new species (Hemiptera: Cicadoidea, Cicadidae). *Neotropical Entomology* 37: 685–690.

Sanborn, A.F. 2009a. Checklist, new species and key to the cicadas of Cuba (Hemiptera: Cicadoidea: Cicadidae). *Mitteilungen aus dem Museum für Naturkunde in Berlin – Deutsche Entomologische Zeitschrift* 59: 85–92.

Sanborn, A.F. 2009b. Two new species of cicadas from Vietnam (Hemiptera: Cicadoidea: Cicadidae). *Journal of Asia-Pacific Entomology* 12: 307–312.

Sanborn, A.F. 2009c. Redescription and neotype designation for *Okanagana noveboracensis* (Emmons, 1854) (Hemiptera: Cicadidae). *Proceedings of the Entomological Society of Washington* 111: 867–873.

Sanborn, A.F. 2009d. A new species of *Cicadetta* (Hemiptera: Cicadomorpha: Cicadidae) from Hispaniola. *Caribbean Journal of Science* 45: 1–7.

Sanborn, A.F. 2010a. The cicadas of Colombia including new records and the description of a new species (Hemiptera: Cicadidae). *Journal of Natural History* 44: 1577–1607.

Sanborn, A.F. 2010b. Two new species and new records of cicadas from Central America (Hemiptera: Cicadoidea: Cicadidae). *Studies on Neotropical Fauna and Environment* 45(2): 67–76.

Sanborn, A.F., J.H. Breitbarth, J.E. Heath, & M.S. Heath. 2002. Temperature responses and habitat sharing in two sympatric species of *Okanagana* (Homoptera: Cicadoidea). *Western North American Naturalist* 62: 437–450.

Sanborn, A.F., & T.H. Davis. 1992. Wing-loading and minimum flight temperature in two periodical cicadas (Homoptera: Cicadidae: *Magicicada septendecim* and *M. cassini*). *Florida Scientist* 55 (Supplement 1): 18–19.

Sanborn, A.F., & J.E. Heath. 2009. Thermal responses of periodical cicadas: within and between brood parity (Hemiptera: Cicadidae: *Magicicada* spp.). *Open Access Insect Physiology* 1: 13–20.

Sanborn, A.F., J.E. Heath, & M.S. Heath. 1992. Thermoregulation and evapourtive cooling in the cicada *Okanagodes gracilis* (Homoptera: Cicadidae). *Comparative Biochemistry and Physiology. A* 102: 751–757.

Sanborn, A.F., J.E. Heath, & M.S. Heath. 2009. Long-range sound distribution and the calling song of the cicada *Beameria venosa* (Uhler) (Hemiptera: Cicadidae). *Southwestern Naturalist* 54: 24–30.

Sanborn, A.F., J.E. Heath, M.S. Heath, & F.G. Noriega. 1995. Thermoregulation by endogenous heat production in two South American grass dwelling cicadas (Homoptera: Cicadidae: *Proarna*). *Florida Entomologist* 78: 319–328.

Sanborn, A.F., J.E. Heath, M.S. Heath, & P.K. Phillips. 2004. Temperature responses and habitat selection by mangrove cicadas in Florida and Queensland (Hemiptera Cicadoidea). *Tropical Zoology* 17: 65–72.

Sanborn, A.F., J.E. Heath, F.G. Noriega, & M.S. Heath. 1989. Endogenous heat production in two grass dwelling cicadas from South America (Homoptera: Cicadidae). *Physiologist* 32: 154.

Sanborn, A.F., J.E. Heath, P.K. Phillips, M.S. Heath, & F.G. Noriega. 2010. Thermal niche separation and diversity in tropical ecosystems: evidence from cicada thermal responses. *Thirteenth International Auchenorrhyncha Congress and the 7th International Workshop on Leafhoppers and Planthoppers of Economic Significance, 28 June-2 July, Vaison-la-Romaine, France, Abstracts: Talks and Posters*: 71–72.

Sanborn, A.F., M.S. Heath, J.E. Heath, & F.G. Noriega. 1995. Diurnal activity, temperature responses and endothermy in three South American cicadas (Homoptera: Cicadidae: *Dorisiana bonaerensis*, *Quesada gigas* and *Fidicina mannifera*). *Journal of Thermal Biology* 20: 451–460.

Sanborn, A.F., M.S. Heath, J.E. Heath, F.G. Noriega, & P.K. Phillips. 2004. Convergence and parallelism among cicadas of Argentina and the southwestern United States (Hemiptera: Cicadoidea). *Biological Journal of the Linnean Society* 83: 281–288.

Sanborn, A.F., M.S. Heath, J.E. Heath, & P.K. Phillips. 1990. Evaporative cooling in the cicada *Okanagodes gracilis* (Homoptera: Cicadidae). *Physiologist* 33: A106.

Sanborn, A.F., M.S. Heath, J. Sueur, & P.K. Phillips. 2005. The genus *Neocicada* Kato, 1925 (Hemiptera: Cicadomorpha: Cicadidae), with descriptions of three new species. *Systematic Entomology* 30: 191–207.

Sanborn, A.F., & J.F. Maté-Nankervis. 2009. Hexapod diversity on an urban university campus. *Florida Scientist* 72: 66–92.

Sanborn, A.F., & S. Maté Nankervis. 1997. Thermoregulation and the effect of body temperature on call parameters in the cicada *Diceroprocta olympusa* (Homoptera: Cicadidae). *Florida Scientist* 60 (Supplement 1): 9.

Sanborn, A.F., & S. Maté. 2000. Thermoregulation and the effect of body temperature on call temporal parameters in the cicada *Diceroprocta olympusa* (Homoptera: Cicadidae). *Comparative Biochemistry and Physiology. A. Physiology* 125: 141–148.

Sanborn, A.F., T.E. Moore, & A.M. Young. 2008. Two new cicada species from Costa Rica (Hemiptera: Cicadomorpha: Cicadidae) with a key to *Fidicinoides* in Costa Rica. *Zootaxa* 1846: 1–20.

Sanborn, A.F., F.G. Noriega, J.E. Heath, & M.S. Heath. 1988. Homeothermy in the cicada *Dorisiana bonaerensis*. *Proceedings of the Second International Congress of Comparative Physiology and Biochemistry*: 603.

Sanborn, A.F., F.G. Noriega, & P.K. Phillips. 2002a. Is there a thermal benefit to crepitating in cicadas (Homoptera: Cicadoidea)? *Florida Scientist* 65 (Supplement 1): 21–22.

Sanborn, A.F., F.G. Noriega, & P.K. Phillips. 2002b. Thermoregulation in a crepitating cicada (Homoptera: Cicadoidea). *The FASEB Journal* 16(5, Part I): A38.

Sanborn, A.F., F.G. Noriega, & P.K. Phillips. 2002c. Thermoregulation in the cicada *Platypedia putnami* var. *lutea* with a test of a crepitation hypothesis. *Journal of Thermal Biology* 27: 365–369.

Sanborn, A.F., L.M. Perez, C.G. Valdes, & A.K. Seepersaud. 2001. Wing morphology and minimum flight temperature in cicadas (Insecta: Homoptera: Cicadoidea). *The FASEB Journal* 15: A1106.

Sanborn, A.F., & P.K. Phillips. 1992. Observations on the effect of a partial solar eclipse on calling in some desert cicadas (Homoptera: Cicadidae). *Florida Entmologist* 75: 285–287.

Sanborn, A.F., & P.K. Phillips. 1993. The relationship between sound pressure level and body size in some North American cicadas (Homoptera: Cicadidae). *Florida Scientist* 56 (Supplement 1): 29.

Sanborn, A.F., & P.K. Phillips. 1994. Thermal and acoustic biology of Big Bend cicadas. *1993 Research Newsletter, Big Bend National Park*: 45.

Sanborn, A.F., & P.K. Phillips. 1995a. Scaling of sound pressure level and body size in cicadas (Homoptera: Cicadidae; Tibicinidae). *Annals of the Entomological Society of America* 88: 479–484.

Sanborn, A.F., & P.K. Phillips. 1995b. No acoustic benefit to subterranean calling in the cicada *Okanagana pallidula* Davis (Homoptera: Tibicinidae). *Great Basin Naturalist* 55: 374–376.

Sanborn, A.F., & P.K. Phillips. 1996. Thermal responses of the *Diceroprocta cinctifera* species group (Homoptera: Cicadidae). *Southwestern Naturalist* 41: 136–139.

Sanborn, A.F., & P.K. Phillips. 1997. Biogeography of the Big Bend cicadas. *1996 Research-Resource Management Newsletter, Big Bend National Park*: 35–36.

Sanborn, A.F., & P.K. Phillips. 1999. Analysis of acoustic signals produced by the cicada *Platypedia putnami* variety *lutea* (Homoptera: Tibicinidae). *Annals of the Entomological Society of America* 92: 451–455.

Sanborn, A.F., & P.K. Phillips. 2001. Re-evaluation of the *Diceroprocta delicata* (Homoptera: Cicadidae) species complex. *Annals of the Entomological Society of America* 94: 159–165.

Sanborn, A.F., & P.K. Phillips. 2004. Neotype and allotype description of *Tibicen superbus* (Hemiptera: Cicadomorpha: Cicadidae) with description of its biogeography and calling song. *Annals of the Entomological Society of America* 97: 647–652.

Sanborn, A.F., & P.K. Phillips. 2010. Reevaluation of the *Diceroprocta texana* species complex (Hemiptera: Cicadoidea: Cicadidae). *Annals of the Entomological Society of America* 103: 860–865.

Sanborn, A.F., P.K. Phillips, & P. Gillis. 2008. The cicadas of Florida (Hemiptera: Cicadoidea: Cicadidae). *Zootaxa* 1916: 1–43.

Sanborn, A.F., P.K. Phillips, & R.W. Sites. 2007. Biodiversity, biogeography, and bibliography of the cicadas of Thailand (Hemiptera: Cicadoidea: Cicadidae). *Zootaxa* 1413: 1–46.

Sanborn, A.F., P.K. Phillips, & M.H. Villet. 2003a. Thermal responses in some Eastern Cape African cicadas (Hemiptera: Cicadidae). *Journal of Thermal Biology* 28: 347–351.

Sanborn, A.F., P.K. Phillips, & M.H. Villet. 2003b. Analysis of the calling songs of *Platypleura hirtipennis* (Germar, 1834) and *P. plumosa* (Germar, 1834)(Hemiptera: Cicadidae). *African Entomology* 11: 291–296.

Sanborn A.F., M.H. Villet, & P.K. Phillips. 2002. Endothermy in some African Platypleurine cicadas (Hemiptera: Cicadoidea: Cicadidae: *Platypleura* spp.). *Eleventh International Auchenorrhyncha Congress, August 5–9, Potsdam, Germany, Abstracts of Talks and Posters*: 32.

Sanborn A.F., M.H. Villet, & P.K. Phillips. 2003. Hot-blooded singers: endothermy facilitates crepuscular signaling in African platypleurine cicadas (Hemiptera: Cicadidae: *Platypleura* spp.) *Naturwissenschaften* 90: 305–308.

Sanborn A.F., M.H. Villet, & P.K. Phillips. 2004a. Endothermy in African platypleurine cicadas: the influence of body size and habitat (Hemiptera: Cicadidae: *Platypleura* spp.) *Physiological and Biochemical Zoology* 77: 816–823.

Sanborn A.F., M.H. Villet, & P.K. Phillips. 2004b. Body size and habitat influence endothermy in African platypleurine cicadas (Hemiptera: Cicadidae). *The FASEB Journal* 18: A1097.

Sanborn, A.F., & M.D. Webb. 2001. Recognition of the types of *Okanagana occidentalis* (Walker, 1866) (Hemiptera: Cicadoidea: Tibicinidae): Lectotype designation, type locality and species identity. *Florida Entomologist* 84: 451–453.

Sander, F.W. 1985. Zur Kenntnis der Entomofauna von Bulgarien – Zikaden (Homoptera-Auchenorrhyncha). *Wissenschaftliche Zeitschrift Friedrich-Schiller-Universität Jena, Naturwissenschaftliche Reihe* 34: 585–607.

Sander, F.W. 1993. Rote Liste der Zikaden (Homoptera: Auchenorrhyncha) Thüringens. *Naturschutzreport* 5: 70–74.

Sanders, M.G. 2004. Notes on a mass aggregation of *Illyria burkei* (Goding & Froggatt)(Hemiptera: Cicadidae) in central Queensland. *Australian Entomologist* 31: 79–80.

Santos, B.B. dos. 1988. Artropodos asociados as plantas cultivadas no estado do Parana. II. Cafeeiro e mandioca. *Revista do Setor de Ciências Agrárias* 10: 27–28.

Santos, R.S., & N.M. Martinelli. 2004a. Espécies de *Fidicinoides* (Boulard & Martinelli, 1996) (Hemiptera: Cicadidae) do Brasil. *XX Congresso Brasileiro de Entomologia, 2004, Gramado*: 630.

Santos, R.S., & N.M. Martinelli. 2004b. Primeiro Registro de *Fidicinoides pauliensis* (Boulard & Martinelli, 1996) (Hemiptera: Cicadidae) em Cafeeiro. *XX Congresso Brasileiro de Entomologia, 2004, Gramado*: 630.

Santos, R.S., & N.M. Martinelli. 2007. Ocorrência de *Fidicinoides pauliensis* Boulard & Martinelli, 1996 (Hemiptera: Cicadidae) em cafeeiro em Tapiratiba, SP. *Revista de Agricultura (Piracicaba)* 82: 311–314.

Santos, R.S., & N.M. Martinelli. 2009a. Primeiro registro de *Fidicinoides picea* (Walker, 1850) e *Fidicinoides poulaini* Boulard & Martinelli, 1996 no Brasil. *Ciência Rural, Santa Maria* 39: 559–562.

Santos, R.S., & N.M. Martinelli. 2009b. Descrição de novas espécies de *Fidicinoides* Boulard & Martinelli (Hemiptera: Cicadidae) do Brasil. *Neotropical Entomology* 38: 638–642.

Santos, R.S., N.M. Martinelli, D.H.B. Maccagnan, A.F. Sanborn, & R. Ribeiro. 2010. Description of a new cicada species associated with the coffee plant and an identification key to the species of *Fidicinoides* (Hemiptera: Cicadidae) from Brazil. *Zootaxa* 2602: 48–56.

Santos, R.S., N.A. Pereira, & N.M. Martinelli. 2004. Primeira ocorrência de *Fidicinoides piceae* e *Fidicinoides poulaini* (Hemiptera: Cicadidae) no Brasil. *XX Congresso Brasileiro de Entomologia, 2004, Gramado*: 627.

Sato, K., K. Yamada, & T. Tani. 1982. [Occurrence and control of the large brown cicada (*Graptopsaltria nigrofuscata* Motschulsky) in Japanese pear orchards.] *Proceedings of the Association for Plant Protection of Kyushu* 28: 197–199. [In Japanese]

Sato, S., Y. Chuman, A. Matsushima, Y. Shimohigashi, Y. Tominaga, & M. Shimohigashi. 2001. The PERIOD clock protein of the last summer cicada *Meimuna opalifera*: cDNA cloning and structural analysis. Abstracts from the 12th Annual Meeting of the Japanese Society for Comparative Physiology and Biochemistry, Fukuoka, japan, July 14–16, 2001. *Comparative Biochemistry and Physiology. Part A. Molecular and Integrative Physiology* 130: 874–875.

Sato, S., Y. Chuman, A. Matsushima, Y. Tominaga, Y. Shimohigashi, & M. Shimohigashi. 2002. A circadian neuropeptide, pigment-dispersing factor-PDF, in the last-summer cicada *Meimuna opalifera*: cDNA cloning and immunocytochemistry. *Zoological Science* 19: 821–828.

Saulich, A.Kh. 2010. Long life cycles in insects. *Entomological Review* 90: 1127–1152.

Savinov, A.B. 1992. [Morphology of thoracic skeleton of the nymph and adult *Cicadetta montana* Scop. (Homoptera, Cicadidae).] *Entomologicheskoe Obozrenie* 71: 267–283, 500. [In Russian, English summary]

Savinov, A.V. 1993. Morphology of the nymphal and adult thoracic skeleton of *Cicadetta montana* Scop. (Homoptera, Cicadidae). *Entomological Review* 72: 1–15.

Sazima, I. 2009. Insect cornucopia: various bird types prey on the season's first giant cicadas in an urban park in southeastern Brazil. *Biota Neotropica* 9: 259–262.

Scarpellini, J.R. 2004. Controle conjunto de cigarras, broca, bicho mineiro e ferrugem do cafeeiro. *Anais X Reunião Itinerante de Fitossandidade do Instituto Biológico Café, 10. Mococa, SP, 19 de Outubro 2004*: 114–128.

Scarpellini, J.R., A.P. Takematsu, T. Jocys, J.C.C. Santos, & F.L.S. Dias Neto. 1994. Utilizaçao de insecticidas granulodos sistêmicos no controle de ninfas de cigarras do cafeeiro (e seus efeitos na produtividade). *Ecossistema* 19: 54–60.

Schabel, H.G. 2006. *Forest entomology in East Africa: forest insects of Tanzania*. Springer, Dordrect, The Netherlands. 328pp.

Schedl, W. 1986a. Zur Verbreitung, Biologie und Oekologie der Zingzikaden von Istrien und dem angrenzenden Kuestenland (Homoptera: Cicadidae und Tibicinidae). *Zoologische Jahrbücher Abteilung für Systematik Ökologie und Geographie der Tiere* 113: 1–27.

Schedl, W. 1986b. Zur morphologie, Ökologie und Verbreitung der singzikade *Cicadetta podolica* (Eichw.)(Homoptera: Auchenorrhyncha, Tibicinidae). *Annalen des Naturhistorischen Museums in Wien* 88/89B: 579–585.

Schedl, W. 1989. Beobachtungen uber Singzikaden auf den Brioni-Inselan (Istrien) (Homoptera: Cicadidae und Tibicinidae). *Zeitschrift der Arbeitegemeinschaft Österreichischer Entomologen* 40: 108–112.

Schedl, W. 1993. Ein Beitrag zur Singzikaden-Fauna Ägyptens (Homoptera: Cicadidae et tibicinidae). *Linzer Biologische Beiträge* 25: 795–803.

Schedl, W. 1999a. Eine neue Unterart der Bergsingzikade im Balkan, *Cicadetta montana macedonica* ssp. n. (Hemiptera: Auchenorrhyncha: Cicadomorpha: Tibicinidae). *Reichenbachia, Staatliches Museum für Tierkunde Dresden* 33: 87–90.

Schedl, W. 1999b. Contribution to the singing cicadas of Israel and adjacent countries (Homoptera, Auchenorrhyncha: Cicadidae et Tibicinidae). *Linzer Biologische Beitrage* 31: 823–837.

Schedl, W. 2000. Taxonomie, Biologie und Verbreitung der Singzikaden Mitteleuropas (Insecta: Homoptera: Cicadidae et Tibicinidae). *Berichte des Naturwissenschaftlich-medizinischen Vereins im Innsbrück* 60: 79–94.

Schedl, W. 2001a. Eine neue Singzikaden aus der Türkei, *Tettigetta turcica* nov. spec.(Cicadoidea: Tibicinidae). *Linzer Biologische Beitrage* 33: 1287–1290.

Schedl, W. 2001b. Taxonomy, biology and distribution of singing cicadas of central Europe (Cicadidae et Tibicinidae). *Program and Book of Abstracts, 2nd European Hemiptera Congress, Fiesa, Slovenia, 20–24 June 2001*: 27.

Schedl, W. 2002. Die Verbreitung der fünf Singzikaden-Arten in Österreich (Hemiptera: Cicadoidea). *Denisia* 4: 231–240.

Schedl, W. 2003. Zur Morphologie, Taxonomie und Verbreitung westpalaearktischer *Klapperichicen*-Arten (Hemiptera, Cicadoidea: Tibicinidae). *Linzer Biologische Beitrage* 35: 423–432.

Schedl, W. 2004a. Ein Beitrag zur Kenntnis der Singzikadenfauna des Jemen (Homoptera: Auchenorrhyncha: Cicadidae et Tibicinidae). *Linzer Biologische Beitrage* 36: 11–18.

Schedl, W. 2004b. Die Singzikaden des Burgenlandes (Österreich)(Insecta: Homoptera, Cicadoidea). *Linzer Biologische Beitrage* 36: 913–917.

Schedl, W. 2007. Order Auchenorrhyncha, families Cicadidae and Tibicinidae. Pages 153–158 *in* A. van Harten, ed. *Arthropod Fauna of the UAE*, Volume 1. Dar Al Ummah Printing, Publishing, Distribution & Advertising, Abu Dhabi.

Scheele, S. 1997. *Insect pests and diseases on harakeke*. Manaaki Whenua Press, Landcare Research, Lincoln, New Zealand. 27pp.

Schiemenz, H. 1987. Beiträge zur Insektenfauna der DDR: Homoptera-Auchenorrhyncha (Cicadina)(Insecta). Teil I: Überfamilie Cicadoidea excl. Typhlocybinae et Deltocephalinae. *Faunistische Abhandlungen Staatliches Museum für Tierkunde Dresden* 15: 41–108.

Schiemenz, H. 1988. Beiträge zur Insektenfauna der DDR: Homoptera-Auchenorrhyncha (Cicadina)(Insecta). Teil II: Überfamilie Cicadoidea excl. Typhlocybinae et Deltocephalinae. *Faunistische Abhandlungen Staatliches Museum für Tierkunde Dresden* 16: 37–93.

Schildberger, K., H.-U. Kleindienst, T.E. Moore, & F. Huber. 1986. Auditory thresholds and acoustic signal processing in the CNS of periodical cicadas. Page 126 *in* N. Elsner & W. Rathmayer, eds. *Sensomotorik: Identifizierte Neurone*. Beiträge zur 14. Göttinger Neurobiologentagung, Georg Thieme Verlag, Stuttgart.

Schmid, C.K., & D.P.A. Sands. 1991. The healing powers of cicada "flowers". *Verhandlungen der Naturforschenden Gesellschaft in Basel* 101: 67–74.

Schmidt, K.A., & C.J. Whelan. 1999. Nest predation on woodland songbirds: when is nest predation density dependent? *Oikos* 87: 65–74.

Schmidt, O. 2005. Revision of *Scotocyma* Turner (Lepidoptera: Geometridae: Larentiinae). *Australian Journal of Entomology* 44: 257–278.

Schnappauf, W., & E. Müller. 2005. *Rote Liste der gefährdeten Tiere und Gefäßpflanzen Bayerns. Kurfassung*. Bayerisches Staatsministerium für Umwelt, Gesundheit und Verbraucherschutz, München. 186pp.

Schneemann, K.-U. 2008. Wundertüte Natur. *Pro Natura* 2008(1): 2.

Schneider, D. 1996. Insects of Generation X. *Scientific American* 275(2): 28.

Schneiderkötter, K., & R. Lakes-Harlan. 2002. Attack and defense – parasitization behaviour of the parasitoid fly *Emblemasoma auditrix* and the host cicada *Okanagana rimosa*. *Zoology* 105 (Supplement V): 8.

Schneiderkötter, K., & R. Lakes-Harlan. 2004. Infection behavior of a parasitoid fly, *Emblemasoma auditrix*, and its host cicada *Okanagana rimosa*. *Journal of Insect Science* 4:36. 7pp. Available online: insectscience.org/4.36

Schönefeld, P., & U. Göllner-Scheiding. 1984. Die Zikaden-Typen Jacobis der Sunde-Expedition Reusch 1927 im Zoologischen Museum Berlin (Auchenorrhyncha, Homoptera). *Mitterlungen aus dem Zoologischen Museum in Berlin* 60: 69–76.

Schönitzer, K., & U. Oesterling. 1998. Die bayerischen Zikaden der Zoologischen Staatssammlung München, ein Beitrag zur Faunistik der Homoptera. Teil 1: Cixidae, Delphacidae, Issidae, Tettigometridae, Cicadidae, Cercopidae, Membracidae. *Nachrichtenblatt der Bayerischen Entomologen* 47: 30–36.

Schouten, M.A., & J.P. Duffels. 2002. A revision of the cicadas of the *Purana carmente* group (Homoptera, Cicadidae) from the Oriental region. *Tijdschrift voor Entomologie* 145: 29–46.

Schouten, M.A., J.P. Duffels, & M.I. Zaidi. 2004. A checklist of the cicada fauna (Homoptera, Cicadoidea) of Endau Rompin Natinal Park, Malaysia, with description of a new species. *Malayan Nature Journal* 56: 369–386.

Schouteden, H. 1907. Notice sur quelques Hémiptères de L'ile Maurice. *Annales de la Société Entomologique de Belgique* 51: 285–288.

Schouteden, H. 1925. Hemitera. -6. Cicadides. *Revue de Zoologie et de Botanique Africaines* 17: 83–84.

Schuder, D.L. 1985. Periodical cicada in Indiana. *Purdue University Cooperative Extension Service* E-47: 1–2.

Schuder, D.L. 1987. Periodical cicada in Indiana. *Purdue University Cooperative Extension Service* E-47: 1–2.

Seabra, S.G., G. André, & J.A. Quartau. 2008. Variation in the acoustic properties of the calling songs of *Cicada barbara* and *C. orni* (Hemiptera: Cicadidae) at the individual and population levels. *Zoological Studies* 47: 1–10.

Seabra, S.G., G. Pinto-Juma, & J.A. Quartau. 2006. Calling songs of sympatric and allopatric populations of *Cicada barbara* and *C. orni* (Hemiptera: Cicadidae) on the Iberian peninsula. *European Journal of Entomology* 103: 843–852.

Seabra, S., P.C. Simões, S. Drosopoulos, & J.A. Quartau. 2000. Genetic variability and differentiation in two allopatric species of the genus *Cicada* L. (Hemiptera, Cicadoidea) in Greece. *Mitteilungen aus dem Museum für Naturkunde im Berlin, Deutsche Entomologishe Zeitschrift* 47: 143–145.

Seabra, S., J.A. Quartau, & M.W. Bruford. 2009. Spatio-temporal genetic variation in sympatric and allopatric Mediterranean *Cicada* species (Hemiptera, Cicadidae). *Biological Journal of the Linnean Society* 96: 249–265.

Seabra, S.G., H.R. Wilcock, J.A. Quartau, & M.W. Bruford. 2002. Microsatellite loci isolated from the Mediterranean species *Cicada barbara* (Stål) and *Cicada orni* L. (Hemiptera, Cicadoidea). *Molecular Ecology Notes* 2: 173–175.

Secomb, D. 1992. Adult barred cuckoo-shrike feeds cicadas to juvenile. *Australian Birds* 26: 29–30.

Seibt, U., & W. Wickler. 2000. 'Sympathetic song': the silent and the overt vocal repertoire, exemplified with a dueting pair of the African Slate-coloured Boubou, *Laniarius funebris*. *Ethology* 106: 795–809.

Seki, T., S. Fujishita, M. Ito, N. Matsuoka, & K. Tsukida. 1987. Retinoid composition in the compound eyes of insects. *Experimental Biology (Berlin)* 47: 95–103.

Seki, T., K. Isono, M. Ito, & Y. Katsuta. 1994. Flies in the group Cyclorrhapha use (3S)-3-hydroxyretinal as a unique visual pigment chromophore. *European Journal of Biochemistry* 226: 691–696.

Seki, Y. 1933. [On the cicada huts of *Meimuna opalifera* Walker.] *Entomological World* 1: 102–112. [In Japanese]

Senter, P. 2008. Voices of the past: a review of Paleozoic and Mesozoic animal sounds. *Historical Biology* 20: 255–287.

Shafroth, P.B., J.R. Cleverly, T.L. Dudley, J.P. Taylor, C. van Riper, III, E.P. Weeks, & J.N. Stuart. 2005. Control of *Tamarix* in ther western United States: implications for water salvage, wildlife use, and riparian restoration. *Environmental Management* 35: 231–246.

Shcherbakov, D.E. 1981a. [Diagnostics of the families of Auchenorrhyncha (Homoptera) on wings. I. Fore wing.] *Revue d'Entomologie de l'URSS* 60: 828–843. [In Russian]

Shcherbakov, D.E. 1981b. Diagnostics of the families of Auchenorrhyncha (Homoptera) on wings. I. Forewings. *Entomological Review, Washington* 60: 64–81.

Shcherbakov, D.E. 1982a. [Diagnostics of the families of Auchenorrhyncha (Homoptera) on wings. II. Hind wings.] *Revue d'Entomologie de l'URSS* 61: 523–536. [In Russian]

Shcherbakov, D.E. 1982b. Diagnostics of the families of Auchenorrhyncha (Homoptera) on wings. II. Hindwings. *Entomological Review, Washington* 61: 70–78.

Shcherbakov, D.E. 2002. The 270 million year history of Auchenorrhyncha (Homoptera). *Denisia* 4: 29–36.

Shcherbakov, D.E. 2009. Review of the fossil and extant genera of the cicada family Tettigarctidae (Hemiptera: Cicadoidea). *Russian Entomological Journal* 17: 343–348.

Shekarian, B., & A. Rezwani. 2001. An investigation on the bioecology of *Psalmocharias alhageos* (Kol.)(Hom.: Cicadidae) in Lorestan Province of Iran. *Applied Entomology and Phytopathology* 69: 25–26, 109–118.

Shelley, T.E., & T.S. Whittier. 1997. Lek behavior of insects. Pages 273–293 *in* J.C. Choe & B.J. Crespi, eds. *The evolution of mating systems in insects and arachnids*. Cambridge University Press, Cambridge.

Shen, J.-X. 1988. An electrophysiological study on hearing of the cicada *Cryptotympana atrata* Fabricius (Homoptera: Cicadidae). *Chinese Journal of Acoustics* 7: 164–170.

Shen, J.-X. 1993. [Coding of cicada auditory nerve responses to sound signal time index.] *Chinese Science Bulletin* 38: 363–366. [In Chinese]

Shen, J.-X. 1994. Coding of temporal parameters of sound signals by auditory interneurons in the cicada *Cryptotympana atrata*. *Chinese Science Bulletin* 39: 772–777.

Shen, J.-X., P. Gao, & H. Tang. 1992. [An electrophysiological study on coding of acoustic signals by auditory interneurons in cicada *Cryptotympana atrata*.] *Chinese Journal of Acoustics* 17: 382–386. [In Chinese, English summary]

Shen Y., Lei Z., & Yang J. 2006. [A new species of the genus *Mogannia* (Hemiptera: Cicadidae) from China.] *Entomotaxonomia* 28: 175–178. [In Chinese, English summary]

Shimazu, M. 1989. *Metarrhizium cylindrosporea* Chen et Guo (Deuteromycotina: Hyphomycetes), the causative agent of an epizootic on *Graptopsaltria nigrofuscata* Motchulski (Homoptera: Cicadidae). *Applied Entomology and Zoology* 24: 430–434.

Shin, T.Y., J.H. Park, & H.M. Kim. 1999. Effect of *Cryptotympana atrata* extract on compound 48/80-induced anaphylactic reactions. *Journal of Ethnopharmacology* 66: 319–325.

Shiyake, S. 2007. *World Cicadas 200*. Osaka Museum of Natural History, Osaka. 126pp. [In Japanese]

Showalter, W.J. 1935. Insect rivals of the rainbow. Pages 29–90 *in* G.G. Grosvenor, ed. *Our insect friends and foes and spiders*. The National Geographic Society, Washington, D.C.

Silva, M.A.T. da, A. Moino, Jr., I.M. de Oliveira, C.S. Ferreira, & E.A.N. Rosa. 2007. Avaliação do nematóide entomopatogênico *Heterorhabditis* sp. (JPM4) em condições de semi-campo, visando ao controle da cigarra-do-cafeeiro. Pages 127–130 *in* C.J. Ávila & R.P.S. Júnior, eds. *10ª Reunião Sul-Brasileira sobre pragas de solo, Pragas-Solo-Sul, Anaise Ata, 25–27 September 2007, Dourados, MS*. Embrapa Agropecuária Oeste, Dourados.

Simmons, L.W. 1994. Reproductive energetics of the role reversing bushcricket, *Kawanaphila nartee* (Orthoptera, Tettigoniidae, Zaprochilinae). *Journal of Evolutionary Biology* 7: 189–200.

Simões, P.C., M. Boulard, & J.A. Quartau. 2006. On the taxonomic status of *Cicada orni* Linnaeus (Hemiptera: Cicadidae) from Lesbos island in Greece. *Zootaxa* 1105: 17–25.

Simões, P.C., M. Boulard, M.T. Rebelo, S. Drosopoulos, M.F. Claridge, J.C. Morgan, & J.A. Quartau. 2000. Differences in the male calling songs of two sibling species of *Cicada* (Hemiptera: Cicadoidea) in Greece. *European Journal of Entomology* 97: 437–440.

Simões, P.C., & J.A. Quartau. 2006. Selective responsiveness in males of *Cicada orni* to conspecific and allospecific calling songs (Hemiptera: Cicadidae). *Entomologia Generalis* 29: 47–60.

Simões, P.C., & J.A. Quartau. 2007. On the dispersal of males of *Cicada orni* in Portugal (Hemiptera: Cicadidae). *Entomologia Generalis* 30: 245–252.

Simões, P.C., & J.A. Quartau. 2008. Distribution patterns and calling song variation in species of the genus *Cicada* Linnaeus, 1758 (Hemiptera, Cicadidae) in the Aegean Sea area. *Italian Journal of Zoology* 75: 135–146.

Simões, P.C., & J.A. Quartau. 2009. Patterns of morphometric variation among species of the genus *Cicada* (Hemiptera: Cicadidae) in the Mediterranean area. *European Journal of Entomology* 106: 393–403.

Simões, P.C., & J.A. Quartau. 2010. Wing shape variation in genus *Cicada* L. (Hemiptera: Cicadidae) in the Mediterranean area: a landmark based geometric morphometric analysis. *Thirteenth International Auchenorrhyncha*

Congress and the 7th International Workshop on Leafhoppers and Planthoppers of Economic Significance, 28 June-2 July, Vaison-la-Romaine, France, Abstracts: Talks and Posters: 138–140.

Simon, C. 1983a. Morphological differentiation in wing venation among broods of 13- and 17-year periodical cicadas. *Evolution* 37: 104–115.

Simon, C. 1983b. A new coding procedure for morphometrie data with an example from periodical cicada wing veins. Pages 378–382 *in* J. Felsenstein, ed. *Numerical taxonomy*. NATO Advanced Studies Institute Series, Volume G1, Springer Verlag, Berlin.

Simon, C. 1988. Evolution of 13-year and 17-year periodical cicadas (Homoptera: Cicadidae: *Magicicada*). *Bulletin of the Entomological Society of America* 34: 163–176.

Simon, C. 1990. Discriminant analysis of year classes of periodical cicada based on wing morphometric data enhanced by molecular information. Pages 309–323 *in* J.T. Sorensen & R.G. Footit, eds. *Ordinations in the study of morphology, evolution, and systematics: applications and quantitative genetic rationales*. Elsevier, Amsterdam.

Simon, C. 1991. Molecular systematics at the species boundary: Exploiting conserved and variable regions of the mitochondrial genome of animals via direct sequencing from amplified DNA. Pages 33–71 *in* G.M. Hewitt, A.W.B. Johnston, & J.P.W. Young, eds. *Molecular techniques in taxonomy. NATO Advanced Studies Institute Series, Vol. H57*. Springer Verlag, Berlin.

Simon, C. 2009. Using New Zealand examples to teach Darwin's 'Origin of Species': lessons from molecular phylogenetic studies of cicadas. *New Zealand Science Review* 66: 102–112.

Simon, C., T. Buckley, P. Arensburger, & G. Chambers. 1999. Systematics of New Zealand cicadas and their Australian and New Caledonian relatives. *Abstracts of Papers and Posters, 10th International Auchenorrhyncha Congress, 6–10 September 1999, Cardiff, Wales.* 2pp (no pagination).

Simon, C., A. Franke, & A. Martin. 1991. The polymerase chain reaction: DNA extraction and amplification. Pages 329–355 *in* G.M. Hewitt, A.W.B. Johnston, & J.P.W. Young, eds. *Molecular techniques in taxonomy. NATO Advanced Studies Institute Series, Vol. H57*. Springer Verlag, Berlin.

Simon, C., F. Frati, A. Beckenbach, B. Crespi, H. Liu, & P. Flocok. 1994. Evolution, weighting, and phylogenetic utility of mitochondrial gene sequences and a compilation of conserved polymerase chain reaction primers. *Annals of the Entomological Society of America* 87: 651–701.

Simon, C., R. Karban, & M. Lloyd. 1981. Patchiness, density, and aggregative behavior in sympatric allochronic populations of 17-year cicadas. *Ecology* 62: 1525–2535.

Simon, C., & M. Lloyd. 1982. Disjunct synchronic populations of 17-year periodical cicadas: Relicts or evidence of polyphyly? *Journal of the Entomological Society of New York* 90: 275–301.

Simon, C., C. McIntosh, & J. Deniega. 1993. Standard restriction fragment length analysis of the mitochondrial genome is not sensitive enough for phylogenetic analysis or identification of 17-year periodical cicada broods (Hemiptera: Cicadidae): the potential for a new technique. *Annals of the Entomological Society of America* 86: 228–238.

Simon, C., L. Nigro, J. Sullivan, K. Holsinger, A. Martin, A. Grapputo, A. Franke, & C. McIntosh. 1996. Large differences in substitutional pattern and evolutionary rate of 12S ribosomal RNA genes. *Molecular Biology & Evolution* 13: 923–932.

Simon, C., S. Pääbo, T.D. Kocher, & A.C. Wilson. 1990. Evolution of mitochondrial ribosomal RNA in insects as shown by the polymerase chain reaction. Pages 235–244 *in* M.T. Clegg & S.J. O'Brien, eds. *Molecular evolution. UCLA Symposium on Molecular & Cellular Biology, New Series, Vol. 122*. Liss, New York.

Simon, C., J. Tang, S. Dalwadi, G. Staley, J. Deniega, & T.R. Unnasch. 2000. Genetic evidence for assortative mating between 13-year cicadas and sympatric "17-year cicadas with 13-year life cycles" provides support for allopatric speciation. *Evolution* 54: 1326–1336.

Sims, T.W., A.N. Palazotto, & A. Norris. 2010. A structural dynamic analysis of a *Manduca sexta* forewing. *International Journal of Micro Air Vehicles* 2: 119–140.

Sismondo, E. 1990. Synchronous, alternating, and phase-locked stridulation by a tropical katydid. *Science* 249: 55–58.

Skaife, S.H. 1979. *African Insect Life*. Second Edition. C. Struick Publishers, Johannesburg. 279pp.

Smiley, J., & D. Giuliani. 1991. Common insects and other arthropods. Pages 245–271 *in* C.A. Hall, Jr., ed. *Natural history of the White-Inyo Range, eastern California*. University of California Press, Berkeley.

Smit, B. 1964. *Insects in South Africa: How to control them*. Oxford University Press, Cape Town. 399pp.

Smith, D.M., J.F. Kelly, & D.M. Finch. 2006. Cicada emergence in southwestern riparian forest: influence of wildfire and vegetation composition. *Ecological Applications* 16: 1608–1618.

Smith, D.M., J.F. Kelly, & D.M. Finch. 2007. Avian nest box selection and nest success in burned and unburned southwestern riparian forest. *Journal of Wildlife Management* 71: 411–421.

Smith, D.S. 1960. Innervation of the fibrillar flight muscle of an insect: *Tenebrio molitor* (Coleoptera). *The Journal of Biophysical and Biochemical Cytology* 8: 447–466.

Smith, D.S. 1961. The organization of the flight muscle in a dragonfly, *Aeshna* sp. (Odonata). *The Journal of Biophysical and Biochemical Cytology* 11: 119–145.

Smith, D.S. 1965. The organization of flight muscle in an aphid, *Megoura viciae* (Homoptera). *The Journal of Cell Biology* 27: 379–393.

Smith, D.S. 1966. The organization and function of the sarcoplasmic reticulum and T-system of muscle cells. *Progress in Biophysical and Molecular Biology* 16: 107–142.

Smith, D.S. 1984. The structure of insect muscle. Pages 111–150 *in* R.C. King & H. Akai, eds. *Insect ultrastructure*, vol. 2. Plenum Press, New York.

Smith, D.S., B.L. Gupta, & U. Smith. 1966. The organization and myofilament array of insect visceral muscles. *Journal of Cell Science* 1: 49–57.

Smith, J.J., & S.T. Hasiotis. 2008. Traces and burrowing behaviors of the cicada nymph *Cicadetta calliope*: neoichnology and paleoecological signifcance of extant soil-dwelling insects. *Palaios* 23: 503–513.

Smith, K.G. 1985. Periodical cicadas. *Arkansas Naturalist* 3: 1–9.

Smith, K.G. 2009. Cornucopia: birds and periodical cicadas. *Bird Observer* 37: 161–169.

Smith, K.G., N.C. Wilkinson, K.S. Williams, & V.B. Steward. 1987. Predation by spiders on periodical cicadas (Homoptera, *Magicicada*). *Journal of Arachnology* 15: 277–279.

Smith, K.G., K.S. Williams, & F. M. Stephen. 1991. Emergence of 13-year periodical cicadas (Homoptera: Cicadidae). *Bulletin of the Ecological Society of America* 72(Supplement): 253.

Smith, R.C., E.G. Kelly, G.A. Dean, H.R. Bryson, R.L Parker, D.E. Gates, L.L. Peters, S.C. White, & G.A. Salsbury. 2000. *Insect in Kansas*. Kansas Department of Agriculture, Topeka. 521pp.

Smith, R.D., & K.A. Smith. 2003. *Country study for biodiversity of the Republic of Macedonia (first national report)*. Ministry of Environment and Physical Planning, Skopje. 213pp.

Smits, A., J. Cooley, & E. Westerman. 2010. Twig to root: egg-nest density and underground nymph distribution in a periodical cicada (Hemiptera: *Magicicada septendecim* (L.)). *Entomologica Americana* 116: 73–77.

Smolinske, S.C. 2005. Herbal product contamination and toxicity. *Journal of Pharmacy Practice* 18: 188–208.

Snipp, R., & N. Bemment. 2004. *The British & Irish Association of Zoos & Aquariums Joint Management of Species Programmes Annual Report*. The British & Irish Association of Zoos & Aquariums, London. 107pp.

Snodgrass, R.E. 1919. The seventeen-year locust. *Annual Report of the Smithsonian Institution* 74: 381–409.

Snodgrass, R.E. 1921. The mouth parts of the cicada. *Proceedings of the Entomological Society of Washington* 23: 1–15.

Snodgrass, R.E. 1923. Insect musicians, their music, and their instruments. *Annual Report of the Smithsonian Institution* 78: 405–452.

Snodgrass, R.E. 1927. The head and mouth parts of the cicada. *Proceedings of the Entomological Society of Washington* 29: 1–16.

Snodgrass, R.E. 1930. Insects: Their ways and means of living. *Smithsonian Scientific Series* 5: 1–362.

Snodgrass, R.E. 1935. *Principles of insect morphology*. McGraw-Hill Book Company, Inc. New York. 667pp.

Snodgrass, R.E. 1938. The loral plates and the hypopharynx of Hemiptera. *Proceedings of the Entomological Society of Washington* 40: 228–236.

Snow, F.H. 1904. Lists of Coleoptera, Lepidoptera, Diptera and Hemiptera collected in Arizona by the entomological expeditions of the University of Kansas in 1902 and 1903. *Kansas University Science Bulletin* 2: 323–350.

Soares, V.P., L.Z. Zaneti, N.T. Santos, & H.G. Leite. 2008. Análise especial da distribuição de cigarras (*Quesada gigas* Oliver) em povoamentos de paricá (*Schizolobium amazonicum* Huber ex Ducke) na região de Dom Eliseu, PA. *Revista Árvore, Viçosa-MG* 32: 251–258.

Socha, R., & D. Kodrík. 1999. Differences in adipokinetic response of *Pyrrhocoris apterus* (Heteroptera) in relation to wing dimorphism and diapause. *Physiological Entomology* 24: 278–284.

Söderman, G. 2007. *Taxonomy, distribution, biology and conservation status of Finnish Auchenorrhyncha (Hemiptera: Fulgoromorpha et Cicadomorpha)*. The Finnish Environment 7. Finnish Environment Institute, Helsinki. 101pp.

Solier, M. 1837. Observations sur quelques particularites de la stridulation des insectes, et en particulier sur le chant de la cigale. *Annales de la Société Entomologique de France* 6: 199–217.

Song, F., K.L. Lee, A.K. Soh, F. Zhu, & Y.L. Bai. 2004. Experimental studies of the material properties of the forewing of cicada (Homóptera, Cicàdidae). *Journal of Experimental Biology* 207: 3035–3042.

Soós, A. 1956. Revision und Ergänzungen zum Homopteren-Teil des Werkes "Fauna Regni Hungaricae". III. 7. Cicadidae. 8. Membracidae. 9. Ulopidae. *Rovartani Közlemények Folia Entomologica Hungarica (Series Nova)* 9: 411–421.

Soper, R.S. 1974. The genus *Massospora* Peck entomopathogenic for cicadas. Ph.D. dissertation, Cornell University. *Dissertation Abstracts International* 35: 2801B.

Soper, R.S. 1981. New cicada pathogens: *Massospora cicadettae* from Australia and *Massospora pahariae* from Afghanistan. *Mycotaxonomy* 13: 50–58.

Sorensen, J.T., B.C. Campbell, R.J. Gill, & J.D. Steffen-Campbell. 1995. Non-monophyly of Auchenorrhyncha ("Homoptera"), based upon 18s rDNA phylogeny: Eco-evolutionary and cladistic implications within pre-Heteropterodea Hemiptera (S.L.) and a proposal for new monophyletic suborders. *Pan-Pacific Entomologist* 71: 31–60.

Sörensen, W. 1884. Traek af nogle sydamerikanske insecters biologi. *Entomologisk Tidskrift* 5: 1–25.

Sorenson, C.E., & L. Mills. 1989. The Apache cicada in Nevada. *Fact Sheet, College of Agriculture, University of Nevada Reno, Nevada Cooperative Extension* 89–78): 1–4.

Southcott, R.V. 1972. Revision of the larvae of the tribe Callidosomatini (Acarina: Erythraeidae) with observations on post-larval instars. *Australian Journal of Zoology Supplement Series* 13: 1–84.

Southcott, R.V. 1988. Two new larval mites (Acarina: Erythraeidae) ectoparasitic on north Queensland [Australia] cicadas. *Records of the South Australian Museum (Adelaide)* 22: 103–116.

Souza, J.C. de, R.R. Reis, & C. do C.A. Melles. 1983. Cigarras do cafeeiro: historico, reconheaimento, biologia, prejuízos e controle. *Belo Horizonte, Empresa de Pesquisa de Minas Gerais Boletim Technico* 5: 1–28.

Souza, J.C. de, R.R. Reis, & C. do C.A. Melles. 1984. Prejuízos causados pela cigarra do cafeeiro *Quesada gigas* (Olivier, 1854) (Homoptera-Cicadidae) em Minas Gerias. *Congresso Brasileiro de Pesquisa Cafeeiras, 11., Londrina. Resumos. Rio de Janeiro, Instituto Brasileiro do Café*: 152–153.

Spatafora, J.W., G.-H. Sung, J.-M. Sung, N.L. Hywell-Jones, & J.F. White. 2007. Phylogenetic evidence for an animal pathogen origin of ergot and the grass endophytes. *Molecular Ecology* 16: 1701–1711.

Speare, A.T. 1919. The fungus parasite of the periodical cicada. *Science* 50: 116–117.

Speare, A.T. 1921. *Massospora cicadina* Peck, a fungous parasite of the periodical cicada. *Mycologia* 13: 72–82.

Speer, J.H., K. Clay, G. Bishop, & M. Creech. 2010. The effect of periodical cicadas on growth of five tree species in Midwestern deciduous forests. *American Midland Naturalist* 164: 173–186.

Spence, R.H. 1851. [On the apparition of *Cicada septendecim*.] *Transactions of the Entomological Society of London* (2) 1: 103–104.

Spichiger, S.-E. 2004a. Coming soon. *Forest Pests Management News (Pennsylvania Department of Conservation and Natural Resources)* 22(1): 3–4.

Spichiger, S.-E. 2004b. Brood X!? *Forest Pests Management News (Pennsylvania Department of Conservation and Natural Resources)* 22(2): 1–2.

Srivastava, R.K., & R.S. Bhandari. 2004. Cicadas in flame – diminishing background music of the forest. *The Indian Forester* 130: 1478–1479.

Stadler, H. 1922. *Tettigonia haematodes* in Unterfranken. *Archiv für Naturgeschichte* 88: 160.

Stanley, D.W. 2000. *Eicosoanoids ininvertebrate signal transduction systems.* Princeton University Press, Princeton, NJ. i-xi, 1–277.

Stanley, D.W., & R.W. Howard. 1998. The biology of prostaglandins and related eicosanoids in invertebrates: Cellular, organismal and ecological actions. *American Zoologist* 38: 369–381.

Stanley-Samuelson, D.W., R.W. Howard, & E.C. Toolson. 1990. Phospholipid fatty acid composition and arachidonic acid uptake and metabolism by the cicada *Tibicen dealbatus* (Homoptera: Cicadidae). *Comparative Biochemistry and Physiology. B* 97: 285–290.

Steinbauer, M.J. 1997. Notes on *Psaltoda moerens* (Germar) (Hemiptera: Cicadidae) in Tasmania. *Australian Entomologist* 24: 169–171.

Steinhaus, E.A. 1941. A study of the bacteria associated with thirty species of insects. *Journal of Bacteriology* 42: 757–790.

Stephen, F.M., G.W. Wallis, & K.G. Smith. 1990. Bird predation on periodical cicadas in Ozark forests: Ecological release for other canopy arthropods? *Studies in Avian Biology* 13: 369–374.

Stevens, M.M., & M. Carver. 1985. Type-specimens of Hemiptera (Insecta) transferred from Macleay Museum, University of Sydney, to the Australian National Insect Collection, Canberra. *Proceedings of the Linnean Society of New South Wales* 108: 263–266.

Steward, V.B., K.G. Smith, & F.M. Stephen. 1988. Red-winged blackbird predation on periodical cicadas (Cicadidae: *Magicicada* spp.): bird behavior and cicada responses. *Oecologia* 76: 348–352.

Stewart, R.D., V.C. Mallik, & L. Redmond. 1999. *Importation of Longan fruit with stems (Dimocarpus longan) from China into the United States: a qualitative, pathway-initiated pest risk assessment.* United States Department of Agriculture, Animal and Plant Health Inspection Service, Plant Protection and Quasrantine, Scientific Services, Riverdale, Maryland. 31pp.

Stock, S.P., B. Rivera-Orduño, & Y. Flores-Lara. 2009. *Heterorhabditis sonorensis* n. sp. (Nematoda: Heterorhabditidae), a natural pathogen of the seasonal cicada *Diceroprocta ornea* (Walker) (Homoptera: Cicadidae) in the Sonoran desert. *Journal of Invertebrate Pathology* 100: 175–184.

Stoddart, P.R., P.J. Cadusch, T.M. Boyce, R.M. Erasmus, & J.D. Comins. 2006. Optical properties of chitin: surface-enhanced Raman scattering substrates based on antireflection structures on cicada wings. *Nanotechnology* 17: 680–686.

Stoetzel, M.B., & L.M. Russell. 1991. Observations on brood X of periodical cicadas: 1987–1990 (Homoptera: Cicadidae). *Proceedings of the Entomological Society of Washington* 93: 471–479.

Stokes, D.R., & R.K. Josephson. 1999. Power and control muscles of cicada song. *American Zoologist* 39: 72A-73A.

Stokes, D.R., & R.K. Josephson. 2004. Power and control muscles of cicada song: structural and contractile heterogeneity. *Journal of Comparative Physiology. A. Neuroethology, Sensory, Neural and Behavioral Physiology* 190: 279–290.

Stölting, H., T.E. Moore, & R. Lakes-Harlan. 2002. Substrate vibrations during acoustic signalling in the cicada *Okanagana rimosa*. *Journal of Insect Science* 2:2. 7pp. Available online: insectscience.org/2.2

Stölting, H., T.E. Moore, & R. Lakes-Harlan. 2004. Acoustic communication in *Okanagana rimosa* (Say)(Homoptera: Cicadidae). *Zoology* 107: 243–257.

Storm, J.J., & J.O. Whitaker, Jr. 2007. Food habits of mammals during an emergence of 17-year cicadas (Hemiptera: Cicadidae: *Magicicada* spp.). *Proceedings of the Indiana Academy of Sciences* 116: 196–199.

Storm, J.J., & J.O. Whitaker, Jr. 2008. Prey selection of big brown bats (*Eptesicus fuscus*) during an emergence of 17-year cicada (*Magicicada* spp.). *American Midland Naturalist* 160: 350–357.

Strandine, E.J. 1940. A quantitative study of the periodical cicada with respect to soil of three forests. *American Midland Naturalist* 24: 177–183.

Strange, L.A. 2000. The cicada killers of Florida (Hymenoptera: Sphecidae). *Florida Department of Agriculture & Consumer Services, Division of Plant Industry Entomological Circular* (402): 1–2.

Strange, L.A. 2003. Cicada killer, giant ground hornet, *Sphecius hogardii* (Latreille) and *Sphecius speciosus* (Drury) (Insecta: Hymenoptera: Sphecidae). Document EENY-295 (573) Featured Creatures from the Entomology and Nematology Department, Florida Cooperatuve Extension Service, Institute of Food and Agricultureal Sciences, Univeristy of Florida. 3pp. Available online: edis.ifas.ufl.edu

Strausfeld, N.J., I. Sinakevitch, S.M. Brown, & S.M. Farris. 2009. Ground plan of the insect mushroom body: functional and evolutionary implications. *Journal of Comparative Neurology* 513: 2665–291.

Strauß, J., & R. Lakes-Harlan. 2006. Embryonic development of pleuropodia of the cicada, *Magicicada cassini. Journal of Insect Science* 6:27. 6pp. Available online: insectscience.org/6.27

Strauß, J., & R. Lakes-Harlan. 2009. Postembryonic development of the auditory system of the cicada *Okanagana rimosa* (Homoptera: Auchenorrhyncha: Cicadidae). *Zoology* 112: 305–315.

Strecker, H. 1879. The cicada in Texas. *Science News* 1: 256.

Strehl, C.E. 1981. Breeding success of the red-winged blackbird as it is affected by a superabundance of food. *Dissertation Abstracts International B. Sciences and Engineering* 42: 2216.

Strehl, C.E., & J. White. 1986. Effects of superabundant food on breeding success and behavior of the red-winged blackbird. *Oecologia (Berlin)* 70: 178–186.

Strobl, G. 1900. Steirische Hemipteren. *Mittheilungen des Naturwissenschaftlichen Vereines für Steiermark* 36: 170–224.

Strümpel, H. 1983. Homoptera (Pflanzensauger). *Handbuch der Zoologie (Berlin)* 4(28): 1–222.

Stucky, B.J. 2009. Discovery of the cicada *Diceroprocta azteca* in Kansas, with notes on its ecology (Hemiptera: Cicadidae). *Transactions of the Kansas Academy of Science* 112: 123–126.

Stumpner, A., & D. von Helversen. 2001. Evolution and function of auditory systems in insects. *Naturwissenschaften* 88: 159–170.

Sudo, S., K. Tsuyuki, & K. Kanno. 2002. Wing characteristics and flapping behavior of flying insects. *Journal of the Japanese Society for Experimental Mechanics* 2: 257–263.

Sudo, S., K. Tsuyuki, & K. Kanno. 2005. Wing characteristics and flapping behavior of flying insects. *Experimental Mechanics* 45: 550–555.

Sueur, J. 1999. Les cigales dans la culture vietnamienne: de la biologie aux usages culinaires, médicinaux et symboliques. *Ecole Pratique des Hautes Etudes, Travaux du Laboratoire Biologie et Evolution des Insectes Hemipteroidea* 11/12: 55–63.

Sueur, J. 2000. Une nouvelle espèce de cigale du Mexique (Los Tuxtlas, Veracruz), et étude de son émission sonore (Homoptera, Auchenorhyncha, Cicadoidea). *Bulletin de la Société entomologique de France* 105: 217–222.

Sueur, J. 2001. Audiospectrographical analysis of cicada sound production: a catalogue (Hemiptera, Cicadidae). *Mitteilungen aus dem Museum für Naturkunde im Berlin, Deutsche Entomologishe Zeitschrift* 48: 33–51.

Sueur, J. 2002a. Cicada acoustic communication: potential sound partitioning in a multispecies community from Mexico (Hemiptera: Cicadomorpha: Cicadidae). *Biological Journal of the Linnean Society* 75: 379–394.

Sueur, J. 2002b. Les manifestations sonores des cigales: historique des études pionnières (Hemiptera, Cicadomorpha, Cicadidae). *Bulletin de la Société entomologique de France* 107: 109–134.

Sueur, J. 2002c. Choeurs de cigales. *Sciences et Avenir Hors-Série* 131: 50–55.

Sueur, J. 2003. Indirect and direct acoustic aggression in cicadas: first observations in the Palaearctic genus *Tibicina* Amyot (Hemiptera: Cicadomorpha: Cicadidae). *Journal of Natural History* 37: 2931–2948.

Sueur, J. 2005. Insect species and their songs. Pages 207–217 *in* S. Drosopoulos & M.F. Claridge, eds. *Insect sounds and communication: physiology, behaviour, ecology and evolution*. CRC Press, Boca Raton.

Sueur, J., & T. Aubin. 2002a. The decoding process in the acoustic communication of the Palaearctic red cicada: how males reply to synthetic calling songs (Cicadomorpha, Cicadidae, *Tibicina haematodes*). *Eleventh International Auchenorrhyncha Congress, August 5–9, Potsdam, Germany, Abstracts of Talks and Posters*: 37.

Sueur, J., & T. Aubin. 2002b. Acoustic communication in the Palaearctic red cicada, *Tibicina haematodes*: chorus organisation, calling-song structure, and signal recognition. *Canadian Journal of Zoology* 80: 126–136.

Sueur, J., & T. Aubin. 2003a. Is microhabitat segregation between two cicada species (*Tibicina haematodes* and *Cicada orni*) due to calling song propagation constraints? *Naturwissenschaften* 90: 322–326.

Sueur, J., & T. Aubin. 2003b. Specificity of cicada calling songs in the genus *Tibicina* (Hemiptera: Cicadidae). *Systematic Entomology* 28: 481–492.

Sueur, J., & T. Aubin. 2004. Acoustic signals in cicada courtship behaviour (order Hemiptera, genus *Tibicina*). *Journal of Zoology, London* 262: 217–224.

Sueur, J., S. Janique, C. Simonis, J.F.C. Windmill, & M. Baylac. 2010. Cicada ear geometry: species and sex effects. *Biological Journal of the Linnean Society* 101: 922–934.

Sueur, J., S. Pavoine, O. Hamerlynck, & S. Duvail. 2008. Rapid acoustic survey for biodiversity appraisal. *PLoS One* 3(12): e4065. 9pp. doi:10.1371/journal.pone.0004065.

Sueur, J., & S. Puissant. 2000. Émission sonores d'une population française de *Cicadivetta tibialis* (Panzer, 1798) (Hemiptera: Cicadoidea: Cicadidae). *Annales de la Société Entomologique de France (N.S.)* 36: 261–268.

Sueur, J., & S. Puissant. 2001. Une vie de cigale. *Le Courier de la Nature* (191): 18–23.

Sueur, J., & S. Puissant. 2002. Spatial and ecological isolation in cicadas: first data from *Tibicina* (Hemiptera: Cicadoidea) in France. *European Journal of Entomology* 99: 477–484.

Sueur, J., & S. Puissant. 2003. Analysis of sound behaviour leads to new synonymy in Mediterranean cicadas (Hemiptera, Cicadidae, *Tibicina*). *Mitteilungen aus dem Museum für Naturkunde im Berlin, Deutsche Entomologishe Zeitschrift* 50: 121–127.

Sueur, J., & S. Puissant. 2007a. Biodiversity eavesdropping: bioacoustics confirms the presence of *Cicadetta montana* (Insecta: Hemiptera: Cicadidae). *Annales de la Société Entomologique de France (N.S.)* 43: 126–128.

Sueur, J., & S. Puissant. 2007b. Similar look but different song: a new *Cicadetta* species in the *montana* complex (Insecta, Hemiptera, Cicadidae). *Zootaxa* 1442: 55–68.

Sueur, J., S. Puissant, & J.-M. Pillet. 2003. An eastern Mediterranean cicada in the west: first record of *Tibicina steveni* (Krynicki, 1837) in Switzerland and France [Hemiptera, Cicadidae, Tibicininae]. *Revue française d'Entomologie (N.S.)* 25: 105–111.

Sueur, J., S. Puissant, P.C. Simões, S. Seabra, M. Boulard, & J.A. Quartau. 2004. Cicadas from Portugal: revised list of species with eco-ethological data (Hemiptera: Cicadidae). *Insect Systematics & Evolution* 35: 177–187.

Sueur, J., & A.F. Sanborn. 2003. Ambient temperature and sound power of cicada calling songs (Hemiptera: Cicadidae: *Tibicina*). *Physiological Entomology* 28: 340–343.

Sueur, J., D. Vanderpool, C. Simon, D. Ouvrard, & T. Bourgoin. 2007. Molecular phylogeny of the genus *Tibicina* (Hemiptera, Cicadidae): rapid radiation and acoustic behaviour. *Biological Journal of the Linnean Society* 91: 611–626.

Sueur, J., J.F.C. Windmill, & D. Robert. 2006. Tuning the drum: the mechanical basis for frequency discrimination in a Mediterranean cicada. *Journal of Experimental Biology* 209: 4115–4128.

Sueur, J., J.F.C. Windmill, & D. Robert. 2008a. Sexual dimorphism in auditory mechanics: tympanal vibrations of *Cicada orni*. *Journal of Experimental Biology* 211: 2379–2387.

Sueur, J., J.F.C. Windmill, & D. Robert. 2008b. Cicada tympanal mechanics: diversity across species and sex. *Abstracts of the XII Invertebrate Sound & Vibration Meeting, Tours, France, 27–30 October 2008*: 59.

Sueur, J., J.F.C. Windmill, & D. Robert. 2010. Sound emission and reception tuning in three cicada species sharing the same habitat. *Journal of the Acoustical Society of America* 127: 1681–1688.

Sugden, A.M. 2000. Clocks and alternative lifestyles. *Science* 289: 1435.

Sugden, J.W. 1940. Characteristics of certain western cicadas. *Journal of the New York Entomological Society* 48: 117–125.

Sun, T., L. Feng, X.-F. Gao, & L. Jiang. 2005. Bioinspired surfaces with special wettability. *Accounts of Chemical Research* 38: 644–652.

Sun, M., G.S. Watson, Y. Zheng, J.A. Watson, & A. Liang. 2009. Wetting properties on nanostructured surfaces of cicada wings. *Journal of Experimental Biology* 212: 3148–3155.

Surface, H.A. 1907. The study of insects. *The Zoological Bulletin of the Division of Zoology, Pennsylvania State Department of Agriculture* 5: 71–74.

Sutherland, W.J., & M.C.M. De Jong. 1991. The evolutionary stable strategy for secondary sexual characters. *Behavioral Ecology* 2: 16–20.

Sweet, P.R. 1993. The insect prey of the laughing gull. *Kingbird* 43: 27.

Swinton, A.H. 1877. On stridulation in the Cicadidae. *Entomologist's Monthly Magazine* 14: 78–81.

Swinton, A.H. 1879. Audition of the Cicadidae. *Entomologist's Monthly Magazine* 16: 79–82.

Swinton, A.H. 1880. *Insect variety: Its propagation and distribution*. Cassell, Petter, Galpin & Co., London. 326pp.

Swinton, A.H. 1908. The vocal and instrumental music of insects. *The Zoologist (4th Series)* 12: 376–389.

Syme, D.A., & R.K. Josephson. 2002. How to build fast muscles: synchronous and asynchronous designs. *Integrative and Comparative Biology* 42: 762–770.

Synave, H. 1966. Contribution à l'étude de la faune de la Basse Casamance, *Homoptera*: *Cicadidae*, *Cercopidae*, *Fulgoridae*. *Bulletin de l'Institute Fondamental d'Afrique Noire, Séries A, Sciences Naturelles* 28,A,3: 1030–1036.

Szilády, Z. 1870. Magyarországi rovargyüjtésem jegyzéke. *Rovartani Lapok* 7: 113–121.

Taber, S.W., & S.B. Fleenor. 2003. *Insects of the Texas Lost Pines*. Texas A&M University Press, College Station. 283pp.

Taber, S.W., & S.B. Fleenor. 2005. *Invertebrates of central Texas wetlands*. Texas Tech University Press, Lubbock. 322pp.

Takakura, K.I., & K. Yamazaki. 2007. Cover dependence of predator avoidance alters the effect of habitat fragmentation on two cicadas (Hemiptera: Cicadidae). *Annals of the Entomological Society of America* 100: 729–735.

Tan J.J., Zhang Y.W., Wang S.Y., Deng Z.J., & Zhu C.X. 1995. [Investigation on the natural resources and utilization of the Chinese medicinal beetles – Meloidae.] *Acta Entomologica Sinica* 38: 324–331. [In Chinese]

Tanaka, Y., J. Yoshimura, C. Simon, J.R. Cooley, & K.-i. Tainaka. 2009. Allee effect in the selection for prime-numbered cycles in periodical cicadas. *Proceedings of the National Academy of Science of the United States of America* 106: 8975–8979.

Tautz, J. 1990. Mechanorezeptoren bei Invertebraten - Filter verhaltensrelevanter Signale. *Verhandlungen der Deutschen Zoologischen Gesellschaft* 83: 97–108.

Terradas, I. 1999. Les cigales et le rythme des jours. La symolique cicadéenne des îles Andaman (Inde) (avec une comparaison centrée sur celle des Pays méditerranéens). *Ecole Practique des Hautes Etudes, Travaux du Laboratoire Biologie et Evolution des Insectes Hemipteroidea* 11/12: 19–54.

Theron, J.G. 1985. Order Hemiptera (bugs,leafhoppers, cicadas, aphids, scale insects, etc.). Suborder Homoptera. Pages 152–164 *in* C.H. Scholtz & E. Holms, eds. *Insects of Southern Africa*. Butterworths, Durhban.

Thoennes, G. 1941. Effects of injuries caused by the cicada, *Magicicada septendecim*, on the later growth of trees. *Plant Physiology* 16: 827–830.

Thomas, B.W. 1987. Some observations on predation and scavenging by the introduced wasps *Vespula germanica* and *V. vulgaris*. *The Weta* 10: 59–61.

Thomas, D. 1832. Remarks on the American locust (*Cicada septendecim*). *American Journal of Science and Arts* 21: 188–189.

Thouvenot, M. 2007a. Une *Zammara* nouvelle pour la Guyane (Homoptera Cicadidae). *L'Entomologiste* 63: 13–14.

Thouvenot, M. 2007b. Cigales de Guyane: quelques exuvies et leurs pattes-pelleteuses (Homoptera Cicadoidea). *L'Entomologiste* 63: 281–284.

Tian R.-g., Yuan F., & Zhang Y.-a. 2006. Male reproductive system and spermatogenesis in Homoptera (Insecta: Hemiptera). *Entomotaxonomia* 28: 241–253.

Tinkham, E.R. 1941. Biological and faunistic notes on the Cicadidae of the Big Bend Region of Trans-Pecos, Texas. *Journal of the New York Entomological Society* 49: 165–183.

Tinkham, E.R. 1942. Notes on the Cicadidae of Alberta. *Canadian Entomologist* 74: 155–156.

Tishechkin, D.Y. 1988. [The Cicadidae (Homoptera, Cicadinea) of Moscow region.] Pages 3–19 *in* A.P. Rasnitsyn, ed. *Nasekomye Moskovskoi Oblasti I Problemy Kadastra I Okhrany*. Nauka, Moskva. [In Russian]

Tishechkin, D.Y. 2003. Vibrational communication in Cercopoidea and Fulgoroidea (Homoptera: Cicadina) with notes on classification of higher taxa. *Russian Entomological Journal* 12: 129–181.

Toda, M., H. Takahashi, N. Nakagawa, & N. Surigara. 2010. Ecology and control of the green anole (*Anolis carolinensis*), an invasive alien species on the Ogasawara Islands. Pages 145–152 *in* K. Kawakami & I. Okochi, eds. *Restoring oceanic island ecosystems: impact and management of invasive alien species in the Bonin Islands*. Springer, Tokyo Berlin Heidelberg New York.

Toledo, A.V., A.M. Marino de Remes Lenicov, & C.C. López Lastra. 2007. Pathogenicity of fungal isolates (Ascomycota: Hypocreales) against *Peregrinus maidis*, *Delphacodes kuscheli* (Hemiptera: Delphacidae), and *Dalbulus maidis* (Hemiptera: Cicadelllidae), vectors of corn diseases. *Mycopathologia* 163: 225–232.

Tolley, K.A., J.S. Makokha, D.T. Houniet, B.L. Swart, & C.A. Matthee. 2009. The potential for predicted climate shifts to impact genetic landscapes of lizards in the South African Cape Floristic Region. *Molecular Phylogenetics and Evolution* 51: 120–130.

Tomley, A. 1995. Dieback of prickly *Acacia*, with suggestions for further investigations. Page 13 *in* N. March, ed. *Exotic weeds and their control in northwest Queensland*. Queensland Department of Lands, Brisbane.

Toms, S.V. 1991. [Water evaporation for thermal regulation in *Damalacantha vacca sinica* (Orthoptera, Tettigonioidea).] *Zoologichesky Zhurnal* 70: 30–39. [In Russian, English summary]

Toolson, E.C. 1984. Interindividual variation in epicuticular hydrocarbon composition and water loss rates of the cicada, *Tibicen dealbatus* (Homoptera: Cicadidae). *Physiological Zoology* 57: 550–556.

Toolson, E.C. 1985. Evaporative cooling in the desert cicada, *Diceroprocta apache*. *American Zoologist* 25: 144A.

Toolson, E.C. 1987. Water profligacy as an adaptation to hot deserts: water loss rates and evaporative cooling in the Sonoran desert cicada, *Diceroprocta apache* (Homoptera: Cicadidae). *Physiological Zoology* 60: 379–385.

Toolson, E.C. 1993. In the Sonoran Desert, cicadas court, mate, and waste water. *Natural History* 102: 37–39.

Toolson, E.C. 1998. Comparative thermal physiological ecology of syntopic populations of *Cacama valvata* and *Tibicen bifidus* (Homoptera: Cicadidae): Modelling fitness consequences of temperature variation. *American Zoologist* 38: 568–582.

Toolson, E.C., P.D. Ashby, R.W. Howard, & D.W. Stanley-Samuelson. 1994. Eicosanoids mediate control of thermoregulatory sweating in the cicada, *Tibicen dealbatus* (Insecta: Homoptera). *Journal of Comparative Physiology. B. Biochemical, Systemic and Environmental Physiology* 164: 278–285.

Toolson, E.C., & N.F. Hadley. 1987. Energy-dependent facilitation of transcuticular water flux contributes to evaporative cooling in the Sonoran Desert cicada, *Diceroprocta apache* (Homoptera: Cicadidae). *Journal of Experimental Biology* 131: 439–444.

Toolson, E.C., & E.K. Toolson. 1991. Evaporative cooling and endothermy in the 13-year periodical cicada, *Magicicada tredecim* (Homoptera, Cicadidae). *Journal of Comparative Physiology. B. Biochemical, Systemic and Environmental Physiology* 161: 109–115.

Towns, D.R. 2002. Korupaki Island as a case study for restoration of insular ecosystems in New Zealand. *Journal of Biogeography* 29: 593–607.

Trilar, T. 2006. Frequency modulated song of the cicada *Kalabita operculata* (Auchenorrhyncha: Cicadoidea) from Borneo. *Russian Entomological Journal* 15: 341–346.

Trilar, T., & M. Gogala. 1998. Bioakustische Untersuchungen der Singzikaden in Macedonien (Homoptera, Auchenorrhyncha). *Beiträge zur Zikadenkunde* 2: 69–70.

Trilar, T., & M. Gogala. 2002a. Description of the song of *Purana sagittata* Schouten & Duffels (Homoptera, Cicadidae) from peninsular Malaysia. *Tijdschrift voor Entomologie* 145: 47–55.

Trilar, T., & M. Gogala. 2002b. Songs of genus *Purana* from SE Asia (Homoptera, Cicadoidea, Cidadidae). *Eleventh International Auchenorrhyncha Congress, August 5–9, Potsdam, Germany, Abstracts of Talks and Posters*: 41.

Trilar, T., & M. Gogala. 2004. Songs of the cicadas of the genus *Pomponia* from SE Asia (Cicadoidea, Cicadidae, Cicadinae). *Third European Hemiptera Congress, St. Petersburg, June 8–11, 2004, Abstracts*: 81.

Trilar, T., & M. Gogala. 2007. The song structure of *Cicadetta podolica* (Eichwald 1830) (Hemiptera: Cicadidae). *Acta Entomologica Slovenica* 15: 5–20.

Trilar, T., & M. Gogala. 2008. New data on singing cicadas (Hemiptera: Cicadidae) of Romania. *Entomologica Romanica* 13: 29–33.

Trilar, T., & M. Gogala. 2009. Skrzadi iz rodu *Pagiphora* (Hemiptera, Cicadidae) pojejo pri presenetljivo nizkih frekvencah. The cicadas from the genus *Pagiphora* (Hemiptera, Cicadidae) are singing in surprisingly low frequencies. *Drugi Slovenski Entomologki Simpozij, Knjiga Povzetkov/2nd Slovenian Entomological Symposium, Book of Abstracts, Ljubljana, 7–8 November 2009*: 74–75.

Trilar, T., & M. Gogala. 2010. *Tettigetta carayoni* Boulard (Hemiptera: Cicadidae) from Crete, faunistic data and first description of its song. *Acta Entomologica Slovenica* 18: 5–18.

Trilar, T., M. Gogala, & V. Popa. 2006. Contributions to the knowledge of the singing cicadas (Auchenorrhyncha: Cicadoidea) of Romania. *Acta Entomologica Slovenica* 14: 175–182.

Trilar, T., M. Gogala, & J. Szwedo. 2006. Pyrenean Mountain cicada *Cicadetta cerdaniensis* Puissant et Boulard (Hemiptera: Cicadomorpha: Cicadidae) found in Poland. *Polskie Pismo Entomologiczne* 75: 313–320.

Trilar, T., & T. Hertach. 2008. Three species of mountain cicadas *Cicadetta montana* (sensu lato) found in northern Italy. *Bulletin of Insectology* 61: 185–186.

Trilar, T., & W.E. Holzinger. 2004. Bioakustische Nachweise von drei Arten des *Cicadetta montana*-Komplexes aus Österreich (Insecta: Hemiptera: Cicadoidea). *Linzer Biologische Beitrage* 36: 1383–1386.

Tse, W.-P., C.-T. Che, K. Liu, & Z.-X. Lin. 2006. Evaluation of the anti-proliferative properties of selected psoriasis-treating Chinese medicines on cultured HaCaT cells. *Journal of Ethnopharmacology* 108: 133–141.

Tsuyuki, K., S. Sudo, & J. Tani. 2006. Morphology of insect wings and airflow produced by flapping insects. *Journal of Intelligent Material Systems and Structures* 17: 743–751.

Tucker, E.S. 1907. Contributions toward a catalogue of the insects of Kansas. Additions to the Kansas list of Hemiptera-Homoptera. *Transactions of the Kansas Academy of Sciences* 20: 190–201.

Tufail, M., & M. Takeda. 2008. Molecular characteristics of insect vitellogenins. *Journal of Insect Physiology* 54: 1447–1458.

Tunaz, H., J.C. Bedick, J.S. Miller, W.W. Hoback, R.L. Rana, & D.W. Stanley. 1999. Eicosanoids mediate nodulation reactions to bacterial infections in adults of two 17-year periodical cicadas, *Magicicada septendecim* and *M. cassini*. *Journal of Insect Physiology* 45: 923–931.

Tunaz, H., R.A. Jurenka, & D.W. Stanley. 2001. Prostaglandin biosynthesis by fat body from true armyworns, *Pseudaletia unipuncta*. *Insect Biochemistry and Molecular Biology* 31: 435–444.

Turner, H., P. Hovenkamp, & P.C. van Welzen. 2001. Biogeography of southeast Asia and the west Pacific. *Journal of Biogeography* 28: 217–230.

Tyrrell, D., & M.A. Weston. 1983. Isolation and culture of *Massospora levispora*: A pathogen of the adult stage of the cicada *Okanagana rimosa*, biological control. *Canadian Forest Service Research Notes* 3: 1.

Tzean, S.S., L.S. Hsieh, J.L. Chen, & W.J. Wu. 1992. *Nomuraea viridulus*, a new entomogenous fungus from Taiwan. *Mycologia* 84: 781–786.

Tzean, S.S., L.S. Hsieh, J.L. Chen, & W.J. Wu. 1993. *Nomuraea cylindrospora*, comb. nov. *Mycologia* 85: 514–519.

Uchiyama, S., & S.-i. Udagawa. 2002. *Cordyceps owariensis* f. *viridescens* and its new *Nomuraea* anamorph. *Mycoscience* 43: 135–141.

Urban, J.M., & J.R. Cryan. 2007. Evolution of the plathoppers (Insects: Hemiptera: Fulgoroidea). *Molecular Phylogenetics and Evolution* 42: 556–572.

Usuda, N., K. Johkura, T. Hachiya, & A. Nakazawa. 1999. Immunoelectron microscopy of peroxisomes employing the antibody for the SKL sequence PTS1 C-terminus common to peroxisomal enzymes. *Journal of Histochemistry & Cytochemistry* 47: 1119–1126.

Valdes, C.G., & A.F. Sanborn. 2001. Wing-loading and minimum flight temperature in cicadas (Homoptera: Cicadoidea). *Florida Scientist* 64(Supplement 1): 24–25.

Valdes, C.G., A.F. Sanborn, L.M. Perez, & M.R.M. Gebaide. 2002. The relationship between wing morphology and the minimum flight temperature in cicadas (Homoptera: Cicadoidea). *Florida Scientist* 65 (Supplement 1): 21.

Valier, P., M.R. Lucchi, & S. Lovari. 1992. Alimentazione, spozio vitale e attivita della volpe (*Vulpes vulpes* L., 1758) in una puneta costiera. *Supplemento Alle Ricerche di Biologia della Selvaggina* 19: 639–640.

Vandegrift, K.J., & P.J. Hudson. 2009. Response to enrichment, type and timing: small mammals vary in their response to a springtime cicada but not a carbohydrate pulse. *Journal of Animal Ecology* 78: 202–209.

van der Zwet, T., E.W. Brown, & P. Estabrook. 1997. Effect of periodical cicada injury and degree of fire blight severity on Asian pear cultivars. *Fruit Varieties Journal* 51: 35–39.

Van Duzee, E.P. 1893. Corrections and notes to the preeding catalogue in Reprint of Fitch's Catalogue with references and descriptions of the insects collected and arranged for the State Cabinet of Natural History. *Report of the New York State Museum of Natural History* 46: 410.

Van Etten, J.L., L.C. Lane, & D.D. Dunigan. 2010. DNA viruses: the really big ones (Giruses). *Annual Review of Microbiology* 64: 83–99.

Van Pelt, A. 1995. *An annotated inventory of the insects of Big Bend National Park, Texas*. Big Bend Natural History Association, Big Bend National Park, Texas. 151pp.

Van Rensselaer, J. 1828. On the habits of the locust, *Cicada septendecim*. *American Journal of Science and Arts* 13: 224–229.

Vayssière, P. 1955. Les animaux parasites du caféier. Pages 1–381 *in* R. Coste, ed. *Les caféiers et les cafés dans le monde*, Tome 1. Editións Larose, Paris.

Veenstra, J.A., & H.H. Hagedorn. 1995. Isolation of two AKH-related peptides from cicadas. *Archives of Insect Biochemistry and Physiology* 29: 391–396.

Veiga, L.M., & S.F. Ferrari. 2006. Predation of arthropods by Southern Bearded Sakis (*Chiropotes satanas*) in eastern Brazilian Amazonia. *American Journal of Primatology* 68: 209–215.

Verdú, J.R., A. Diaz, & E. Galante. 2004. Thermoregulatory strategies in two closely related sympatric *Scarabaeus* species (Coleoptera: Scarabaeinae). *Physiological Entomology* 29: 32–38.

Vernier, R. 1996. Un chanteur isolé de *Lyristes plebejus* (Scop.)(Hemiptera, Cicadidae) à l'Allondon. *Bulletin Romand d'Entomologie* 14: 147–152.

Vezinet, P. 2003. Le chevêche et les cigales. *Insectes* (131): 13–14.

Vignal, C., & D. Kelley. 2007. Significance of temporal and spectral acoustic cues for sexual recognition in *Xenopus laevis*. *Proceedings of the Royal Society. Series B. Biological Sciences* 274: 479–488.

Vikberg, V. 2008. *Eupelmus (Episolindelia) fuscipennis* Förster kasvatettu vuorilaulukaskaan *Cicadetta montana* (Scopoli) munastosta Etelä-Suomessa (Hymenoptera: Chalcidoidea: Eupelmidae) [*Eupelmus (Episolindelia) fuscipennis* Förster reared from egg batch of *Cicadetta montana* (Scopoli) in southern Finland (Hymenoptera: Chalcidoidea: Eupelmidae)]. *Sahlbergia* 14: 60–67.

Villani, M.G., L.L. Allee, A. Díaz, & P.S. Robbins. 1999. Adaptive strategies of edaphic arthropods. *Annual Review of Entomology* 44: 233–256.

Villet, M. 1987a. Sound pressure levels of some African cicadas (Homoptera: Cicadoidea). *Journal of the Entomological Society of Southern Africa* 50: 269–274.

Villet, M. 1987b. Three new platypleurine cicadas (Homoptera: Cicadidae) from Natal, South Africa. *Journal of the Entomological Society of Southern Africa* 50: 209–215.

Villet, M. 1988. Calling songs of some South African cicadas (Homoptera: Cicadidae). *South African Journal of Zoology* 23: 71–77.

Villet, M. 1989a. Systematic status of *Platypleura stridula* L. and *Platypleura capensis* L. (Homoptera, Cicadidae). *South African Journal of Zoology* 24: 329–332.

Villet, M. 1989b. New taxa of South African platypleurine cicadas (Homoptera: Cicadidae). *Journal of the Entomological Society of Southern Africa* 52: 51–70.

Villet, M. 1989c. Notes on the cicadas (Homoptera: Cicadidae) in the *Insecta Transvaalensia*. *Journal of the Entomological Society of Southern Africa* 52: 325–329.

Villet, M. 1992. Responses of free-living cicadas (Homoptera: Cicadidae) to broadcasts of cicada songs. *Journal of the Entomological Society of Southern Africa* 55: 93–97.

Villet, M. 1993. The cicada genus *Bavea* Distant 1905 (Homoptera Tibicinidae): redescription, distribution and phylogenetic affinities. *Tropical Zoology* 6: 435–440.

Villet, M. 1994a. The cicada genus *Stagea* n. gen. (Homoptera Tibicinidae): systematics. *Tropical Zoology* 7: 293–297.

Villet, M. 1994b. The cicada genus *Tugelana* Distant 1912 (Homoptera Cicadoidea): systematics and distribution. *Tropical Zoology* 7: 87–92.

Villet, M. 1995. Intraspecific variability in SMRS signals: some causes and implications in acoustic signalling systems. Pages 422–439 *in* D.M. Lambert & H.G. Spencer, eds. *Speciation and the recognition concept; theory and application*. Johns Hopkins University Press, Baltimore.

Villet, M. 1997a. Redescription of three species of the genus *Platypleura* Amyot & Serville 1843 (Hemiptera Cicadidae). *Tropical Zoology* 10: 321–332.

Villet, M. 1997b. The cicada genus *Stagira* Stål 1861 (Homoptera Tibicinidae): systematic revision. *Tropical Zoology* 10: 347–392.

Villet, M. 1999a. On two species of *Paectira* Karsch (Hemiptera: Cicadidae) from Kenya. *African Entomology* 7: 151–153.

Villet, M. 1999b. A new cicada of the genus *Severiana* Boulard (Hemiptera: Homoptera: Cicadidae) from northern Namibia. *Cimbebasia* 15: 149–153.

Villet, M. 1999c. The cicada genus *Nyara* n. gen. (Homoptera Cicadidae): systematics, behaviour and conservation status. *Tropical Zoology* 12: 157–163.

Villet, M. 1999d. Re-evaluation of Ashton's types of African cicadas (Homoptera Cicadidae). *Tropical Zoology* 12: 209–218.

Villet, M. 1999e. Cicada biodiversity in Africa. *Abstracts of Papers and Posters, 10th International Auchenorrhyncha Congress, 6–10 September 1999, Cardiff, Wales.* 2pp (no pagination).

Villet, M.H., N.P. Barker, & N. Lunt. 2005. Mechanisms generating biological diversity in the genus *Platypleura* Amyot & Serville, 1843 (Hemiptera: Cicadidae) in southern Africa: implications of a preliminary molecular phylogeny. *South African Journal of Science* 100: 589–594.

Villet, M.H., & I.R. Capitao. 1995. Cicadas (Homoptera: Cicadidae) as indicators of habitat and habitat condition of Valley Bushveld in the Fish River Valley. *Proceedings of the 10th Congress of the Entomological Society of Southern Africa, 3–7 July 1995, Grahamstown*: 140.

Villet, M.H., & I.R. Capitao. 1996. Cicadas (Homoptera: Cicadidae) as indicators of habitat and veld condition in valley bushveld in the Great Fish River Valley. *African Entomology* 4: 280–284.

Villet, M.H., & P.E. Reavell. 1989. Annotated provisional checklist and key to the platypleurine cicadas (Homoptera: Cicadidae) of coastal Natal. *South African Journal of Zoology* 24: 333–336.

Villet, M.H., A.F. Sanborn, & P.K. Phillips. 2002. Calling behaviour in the cicada *Pycna semiclara* (Germar, 1834) (Hemiptera: Cicadoidea: Cicadidae). *Eleventh International Auchenorrhyncha Congress, August 5–9, Potsdam, Germany, Abstracts of Talks and Posters*: 38.

Villet, M.H., A.F. Sanborn, & P.K. Phillips. 2003a. Singers with hot bodies: African cicadas reveal new tricks. Online journal *Science in Africa* (24): www.scienceinafrica.co.za/2003/february/cicadas.htm.

Villet, M.H., A.F. Sanborn, & P.K. Phillips. 2003b. Endothermy, umbraphily and aggression: signaling behaviour in the African cicada *Pycna semiclara* (Hemiptera: Cicadidae). *Proceedings of the 14th Entomological Congress, Entomological Society of Southern Africa, Pretoria, 6–9 July 2003*: 78–79.

Villet, M.H., A.F. Sanborn, & P.K. Phillips. 2003c. Endothermy and chorusing behaviour in the African platypleurine cicada *Pycna semiclara* (Hemiptera: Cicadidae). *Canadian Journal of Zoology* 81: 1437–1444.

Villet, M.H., & S. van Noort. 1999. Cicadas (Hemiptera, Homoptera: Cicadoidea) of Mkomazi. Pages 223–234 *in* M.J. Coe, N.C. McWilliam, G.N. Stone, & M.J. Packer, eds. *Mkomazi: the ecology, biodiversity and conservation of a Tanzania savanna*. Royal Geographical Society (with the Institute of British Geographers), London.

Villet, M.H., & M. Zhao. 1999. The Cicadidae of China (Homoptera: Cicadoidea). *African Entomology* 7: 321–323.

Villiers, A. 1977. *Atlas des Hémiptères*. Société Nouvelle des Édition Boubée & Cie, Paris. 301pp.

Vincent, T. 2006. La petite cigales des montagnes, *Cicadetta* cf. *montana* (Scopoli, 1772) (Insecta, Hemiptera, Cicadidae, Tibicininae), en Haute-Normandie (France). Données anciennes et récentes 1850–2004) et répartition géographique. *Bulletin de la Société Linnéenne de Normandie* 119: 63–73.

Visser, B., & J. Ellers. 2008. Lack of lipogenesis in parasitoids: a review of physiological mechanisms and evolutionary implications. *Journal of Insect Physiology* 54: 1315–1322.

Vogel, R. 1938. Weiteres uber Verbreitung und Lebensweise der Blutroten Singzikade (*Tibicen haematodes*). *Jahreshefte des Vereins für vateriandische Naturkunde in Württemberg* 93: 116–122.

Vokoun, J.C. 2000. Shortnose gar (*Lepisosteus platostomus*) foraging on periodical cicadas (*Magicicada* spp.): Territorial defense of profitable pool positions. *American Midland Naturalist* 143: 261–265.

Vries, T. de, & R. Lakes-Harlan. 2005. Phonotaxis of female parasitoid *Emblemasoma auditrix* (Diptera, Sarcophagidae) in relation to number of larvae and age. *Zoology* 108: 239–246.

Vries, T. de, & R. Lakes-Harlan. 2007. Prenatal cannibalism in an insect. *Naturwissenschaften* 94: 477–482.

Vsevolodova-Perel, T.S., & M.L. Sizemskaya. 2007. Spatial structure of soil macrofauna in the northern Caspian clay semidesert. *Biology Bulletin* 34: 629–634.

Wagner, E. 1952. Contributo alla conoscenza della fauna emitterilogica Italiana. A. Bemerkenswerte *TINGIDAE* und *REDUVIIDAE* aus Italien. *Bollettino della Associazione Romana di Entomologia* 7: 13–16.

Wakakuwa, M., K. Ozaki, & K. Arikawa. 2004. Immunohistochemical localization of *Papilio* RBP in the eye of butterflies. *Journal of Experimental Biology* 207: 1479–1486.

Waldbauer, G. 2000. *Millions of monarchs, bunches of beetles: how bugs find strength in numbers*. Harvard University Press, Cambridge. 264pp.

Walker, D.H., & A.R. Pittaway. 1987. *Insects of Eastern Arabia*. MacMillan, London. 175pp.

Walker, J.J. 1920. The president's address. The fringes of butterfly life. *Transactions of the Entomological Society of London* 1919: xci-cxiii.

Walker, T.J. 1969. Acoustic synchrony: two mechanisms in the snowy tree cricket. *Science* 166: 891–894.

Walker, T.J. 1983. Diel patterns of calling in nocturnal Orthoptera. Pages 45–72 *in* D.T. Gwynne & G.K. Morris, eds. *Orthopteran mating systems: sexual selection in a diverse group of insects*. Westview Press, Boulder, Colorado.

Walker, T.J. 1986. Monitoring flights of field crickets (*Gryllus* spp.) and a tachinid fly (*Euphasiopteryx ochracea*) in North Florida. *Florida Entomologist* 69: 678–685.

Walker, T.J., & T.E. Moore. 2004. Cicadas (of Florida), *Neocicada hieroglyphica* (Say), *Tibicen*, *Diceroprocta* and *Melampsalta* spp. (Insecta: Hemiptera: Cicadidae). Document EENY-327 Featured Creatures from the Entomology and Nematology Department, Florida Cooperatuve Extension Service, Institute of Food and Agricultureal Sciences, Univeristy of Florida. 5pp. Available online: edis.ifas.ufl.edu

Wallis, S. 1841. *Cicada haematodes*. *Entomologist* 1: 203.

Walsh, B.D. 1866. Answers to correspondents. *Practical Entomologist* 2: 33.

Walsh, B.D. 1867. Answers to correspondents. *Practical Entomologist* 2: 56.

Walsh, B.D. 1868. The bug hunter in Egypt. A journal of an entomological tour into south Illinois by the senior editor. Voyage down the Mississippi - The so called mormon flies - locusts. *American Entomologist* 1: 6–11.

Walsh, B.D. 1870. On the group Eurytomides of the Hymenopterous family Chalchididae: with remarks on the theory of species, and a description of Antigaster, a nerw and very anomolous genus of Chalcididae. *American Entomologist* 2: 329–335.

Walsh, B.D., & C.V. Riley. 1868a. The sting of the 17-year cicada. *American Entomologist* 1: 36–37.

Walsh, B.D., & C.V. Riley. 1868b. The periodical cicada. *American Entomologist* 1: 63–72.

Walsh, B.D., & C.V. Riley. 1869a. The periodical cicada. *American Entomologist* 1: 202.

Walsh, B.D., & C.V. Riley. 1869b. Belated individuals of the periodical cicada. *American Entomologist* 1: 217.

Walsh, B.D., & C.V. Riley. 1869c. Insects named. *American Entomologist* 1: 251.

Walter, O., & N. Bemment. 2003. *The Federation of Zoological Gardens of Great Britain and Ireland Joint Management of Species Programmes Annual Report*. The The Federation of Zoological Gardens of Great Britain and Ireland, London. 100pp.

Walter, S., R. Emmrich, & H. Nickel. 2003. *Rote Liste der Zikaden Sachsens. Stand 2003. - Materialien zu Naturschutz und Landschaftspflege 2003*. Freistaat Sachsen, Landesamt für Umwelt und Geologie. 28pp.

Wang B., Zhang H., Fang Y., & Zhang Y. 2008. A revision of Palaeontinidae (Insecta: Hemiptera: Cicadomorpha) from the Jurassic of China with descriptions of new taxa and new combinations. *Geological Journal* 43: 1–18.

Wang, C., & R.J. St. Leger. 2005. Developmental and transcriptional responses to host and nonhost cuticles by the specific locust pathogen *Metarhizium anisopliae* var. *acridum*. *Eukaryotic Cell* 4: 937–947.

Wang, C., & R.J. St. Leger. 2006. A collogenous protective coat enables *Metarhizium anisolpiae* to evade insect immune responses. *Proceedings of the National Academy of Science of the United States of America* 103: 6647–6652.

Wang H., Liu H., Wang K., Chang C., Shih P., & Chu C. 1986. [A study on the Chinese medicine prepared from *Cryptotympana atrata*.] *Bulletin of Chinese Materia Medica* 11: 217–220. [In Chinese]

Wang, H., & T.B. Ng. 2002. Isolation of cicadin, a novel and potent antifungal peptide from dried juvenile cicadas. *Peptides* 23: 7–11.

Wang L.Y. 1988. [The bionomics of a parasitic moth, *Epipomponiis oncotympana* Yang (Lepidoptera: Epipyropidae).] *Sinozoologia* 6: 229–234. [In Chinese, English summary]

Wang, Y., J. Ding, G.S. Wheeler, M.F. Purcell, & G. Zhang. 2009. *Heterapoderopsis bicallosicollis* (Coleoptera: Attelabidae): a potential biological control agent for *Triadica sebifera*. *Environmental Entomology* 38: 1135–1144.

Wang Y., Ren D., Liang J.-H., Liu Y.-S., & Wang Z.-H. 2006. [The fossil Homoptera of China: a review of present knowledge.] *Acta Zootaxonomia Sinica* 31: 294–303. [In Chinese, English summary]

Wang Z.Q. & Zhang X.J. 1987. [Distinguish between some common species of cicadas.] *Kun Chong Zhi Shi = Kunchong Zhishi = Entomological Knowledge* 24: 294–296. [In Chinese]

Wang Z.Q. & Zhang X.J. 1992. [Homoptera: Cicadoidea.] Pages 188–191 *in* Chen S., ed. *Insects of the Hengduan Mountains region*. Volume 1. The comprehensive scientific expedition to the Qinghai-Xizang Plateau, Chinese Academy of Sciences. Science Press, Beijing. [In Chinese, English summary]

Wang Z.Q. & Zhang X.J. 1993. [Homoptera: Cicadidae.] Pages 119–122 *in* Huang F., ed. *Insects of Wuling Mountains area, southwestern China*. Science Press, Beijing. [In Chinese, English summary]

Wang Z.-N. 1988. [Imperfect state of three *Cordyceps* fungi from Taiwan and their pathogenicity to some insects.] *Report of the Taiwan Sugar Research Institute* (121): 17–27. [In Chinese]

Wang Z.-N., & C.E. Pastor. 1988. [Identification of unknown entomogenous fungus on the nymph of *Proarna bergi* from Argentina.] *Transactions of the Mycological Society of the Republic of China* 3: 101–108. [In Chinese, English summary]

Ward, L.F. 1885. Premature appearance of the periodical cicada. *Science* 5: 476.

Warder, R.H. 1869. [*Cicada septendecim* ovipositing in evergreens.] *American Entomologist* 1: 117.

Watanabe, S., & H. Kobayashi. 2006. Sound recording of vocal activity of animals inhabiting subtropical forest on Iriomote Island in the southern Ryukyus, Japan. *Razprave IV. Razreda Sazu (Ljubljana)* 47: 213–228.

Waterhouse, D.F. 1997. The major invertebrate pests and weeds of agriculture and plantation forestry in the southern and western Pacific. *The Australian Centre for International Agricultural Research Monograph* (44): i-vi, 1–99.

Watson, G.S., S. Myhra, B.W. Cribb, & J.A. Watson. 2006. Imaging and investigation of function(s) and efficiency of ordered cuticular nanostructures. Pages 175–198 *in* E.P. Ivanova, ed. *Nanoscale structure and properties of microbial cell surfaces*. Nova Science Publishers, New York.

Watson, G.S., S. Myhra, B.W. Cribb, & J.A. Watson. 2008. Putative functions and functional efficiency of ordered cuticular nanoarrays on insect wings. *Biophysical Journal* 94: 3352–3360.

Watson, G.S., & J.A. Watson. 2004. Natural nano-structures on insects – possible functions of ordered arrays characterized by atomic force microscopy. *Applied Surface Science* 235: 139–144.

Watson, G.S., J.A. Watson, S. Hu, C.L. Brown, B.W. Cribb, & S. Myhra. 2010. Micro and nanostructures found on insect wings – designs for minimising adhesion and friction. *International Journal of Nanomanufacturing* 5: 112–128.

Watts, R.J., & K. Williams. 1992. Host use patterns of adult chaparral cicadas (Homoptera: Cicadidae) in San Diego County, Califormia. *Bulletin of the Ecological Society of America* 73 (Supplement): 380–381.

Way, J.G. 2008. Eastern coyotes, *Canis latrans*, observed feeding on periodical cicadas, *Magicicada septendecim*. *Canadian Field-Naturalist* 122: 221–272.

Weaver, J.E. 1995. Life history, habits, and control of the cicada killer wasp in West Virginia. *West Virginia Agricultural and Forestry Experiment Station Circular* 161: 1–14.

Weaver, R. 1832. Five specimens of *Cicada haematodes*, captured in the New Forest. *Magazine of Natural History and Journal of Zoology, Botany, Minerology, Geology and Meterology* 5: 668–669.

Webb, G.F. 2001. The prime number periodical cicada problem. *Discreet and Continuous Dynamical Systems – Series B* 1: 387–399.

Weber, H. 1931. Homoptera. Pflanzensauger Cicadina - Zikaden. *Biologie der Tiere Deutschlands* 31: 71–208.

Weber, H. 1968. Biologie der Hemipteren. Eine Naturgeschichte der Schnabelkerfe. A. Asher & Company, Amsterdam. 543p.

Weber, T., T.E. Moore, F. Huber, & U. Klein. 1988. Sound production in periodical cicadas (Homoptera: Cicadidae: *Magicicada septendecim, M. cassini*). *Proceedings of the 6th Auchenorrhynchota Meeting, Turin, Italy* 7–11 September 1987: 329–336.

Webster, F.M. 1892. Modern geographical distribution of insects in Indiana. *Proceedings of the Indiana Academy of Sciences* 1892: 81–89.

Webster, F.M. 1898. Distribution of broods XXII, V and VIII, of *Cicada septendecim*, in Indiana. *Proceedings of the Indiana Academy of Sciences* 8: 225–227.

Webster, R.L. 1912. Insects of the year 1912 in Iowa. *Journal of Economic Entomology* 5: 469–472.

Wei C., Z. Ahmed, & S.A. Rizvi. 2010. *Shaoshia*, an unusual new cicada genus from Pakistan with the description of a new species (Hemiptera: Cicadidae). *Zootaxa* 2421: 28–34.

Wei C., Tang G., He H., & Yalin Z. 2009. Review of the cicada genus *Karenia* (Hemiptera: Cicadidae), with a description of one new species trapped by clapping hands and its entomogenous fungus. *Systematics and Biodiversity* 7: 337–343.

Wei C., & Zhang Y. 2010. Geographical distribution of several taxa of Cicadomorpha: where to draw the demarcation line between the Palaearctic and Oriental Regions in China? *Thirteenth International Auchenorrhyncha Congress and the 7th International Workshop on Leafhoppers and Planthoppers of Economic Significance, 28 June-2 July, Vaison-la-Romaine, France, Abstracts: Talks and Posters*: 37–39.

Weis-Fogh, T. 1964. Diffusion in insect wing muscle, the most active tissue known. *Journal of Experimental Biology* 41: 229–256.

Welbourn, W.C. 1983. Potential use of trombidioid and erythraeoid mites as biological control agents of insect pests. Pages 103–140 *in* M.A. Hoy, G.L. Cunningham, & L. Knutson, eds. *Biological control of pests by mites. Agriculture Experiment Station, Division of Agricultural Natural Resources, Special Publication 3304*. University of California, Berkeley.

West-Eberhard, M.J. 2005. Developmental plasticity and the origin of species differences. *Proceedings of the National Academy of Science of the United States of America* 102 (Supplement 1): 6543–6549.

Wetmore, A. 1916. Birds of Porto Rico. *United States Department of Agriculture Bulletin* 326: 1–140.

Whalley, P.E.S. 1983. A survey of recent and fossil cicadas (Insecta, Hemiptera-Homoptera) in Britain. *Bulletin of the British Museum of Natural History (Geology)* 37: 139–147.

Wheeler, A.G., Jr., & G.L. Miller. 2006. Simon Snyder Rathvon: popularizer of agricultural entomology in mid-19th Century America. *American Entomologist* 52: 36–47.

Wheeler, G.L., K.S. Williams, & K.G. Smith. 1992. Role of periodical cicadas (Homoptera; Cicadidae: *Magicicada*) in forest nutrient cycles. *Forest Ecology and Management* 51: 339–346.

Wheeler, W.C., R.T. Schuh, & R. Bang. 1993. Cladistic relationships among higher groups of Heteroptera: congruence between morphological and molecular data sets. *Entomologica Scandinavica* 24: 121–137.

Wheeler, W.M. 1889. Homologues in embryo Hemiptera of the appendages of the first abdominal segment of other insect embryos. *American Naturalist* 23: 644–645.

Whiles, M.R., M.A. Callaham, Jr., C.K. Meyer, B.L. Brock, & R.E. Charlton. 2001. Emergence of periodical cicadas from a Kansas riparian forest: Densities, biomass and nitrogen flux. *American Midland Naturalist* 145: 176–187.

White, D.J., T.M. Boyce, & P.R. Stoddart. 2004. A surface-enhanced Raman scattering substrate of biological origin. *Proceedings of the XIX International Conference on Raman Spectroscopy, Gold Coast, Queensland, Australia, 8–13 August* 2: 292–293.

White, E.G. 1987. Ecological time frames and the conservation of grassland insects. *New Zealand Entomologist* 10: 146–152.

White, E.G., & J.R. Sedcole. 1993a. A study of the abundance and patchiness of cicada nymphs (Homoptera: Tibicinidae) in a New Zealand subalpine shrub-grassland. *New Zealand Journal of Zoology* 20: 38–51.

White, E.G., & J.R. Sedcole. 1993b. Spatio-temporal cicada ecology: a rebuttal of alleged flaws. *New Zealand Journal of Zoology* 20: 59–62.

White, J. 1981. Flagging: host defences versus oviposition strategies in periodical cicadas (*Magicicada* spp., Cicadidae, Homoptera). *Canadian Entomologist* 113: 727–738.

White, J.A., P. Ganter, R. McFarland, N. Stanton, & M. Lloyd. 1983. Spontaneous, field tested and tethered flight in healthy and infected *Magicicada septendecim* L. *Oecologia* 57: 281–286.

White, J., & M. Lloyd. 1981. On the stainability and mortality of periodical cicada eggs. *American Midland Naturalist* 106: 219–228.

White, J., & M. Lloyd. 1983. A pathogenic fungus, *Massospora cicadina* Peck (Entomophthorales), in emerging nymphs of periodical cicadas (Homoptera: Cicadidae). *Environmental Entomology* 12: 1245–1252.

White, J., & M. Lloyd. 1985. Effect of habitat size on nymphs in periodical cicadas (Homoptera: Cicadidae: *Magicicada* spp.). *Journal of the Kansas Entomological Society* 58: 605–610.

White, J., M. Lloyd, & R. Karban. 1982. Why don't periodical cicadas normally live in coniferous forests? *Environmental Entomology* 11: 475–482.

Whitford, W.G., & E. Jackson. 2007. Seed harvester ants (*Pogonomyrmex rugosus*) as "pulse"' predators. *Journal of Arid Environments* 70: 549–552.

Whitten, M.I. 1965. Chromosome numbers in some Australian leaf-hoppers (Homoptera, Auchenorrhyncha). *Proceedings of the Linnean Society of New South Wales* 90: 78–86.

Wiesenborn, W.D. 2005. Biomass of Arthropod trophic levels on *Tamarix ramosissima* (Tamaricaceae) branches. *Environmental Entomology* 34: 656–663.

Wild, P. 1852. (Sur les moeurs de la *Cicada septendecim*). *Annales de la Société Entomologique de France* (2) 10: xviii-xix.

Wiley, R.H., & D.G. Richards. 1982. Adaptations for acoustic communication in birds: sound transmission and signal detection. Pages 131–181 *in* D.E. Kroodsma & E.H. Miller, eds. *Acoustic communication in birds. Production, perception, and design features of sound*, Volume 1. Academic Press, New York.

Williams, G. 2002. A taxonomic and biogeographic review of the invertebrates of the Central Eastern Rainforest Reserves of Australia (CERA) World Heritage Area, and adjacent regions. *Tehnical Reports of the Australian Museum Number* 16: 1–208.

Williams, J.R., & D.J. Williams. 1988. Homoptera of the Mascarene Islands - an annotated catalogue. *Entomology Memoir, Department of Agriculture and Water Supply, Republic of South Africa* 72: 1–98.

Williams, K.S. 1987. Responses of persimmon trees to periodical cicada oviposition damage. Page 424 *in* V. Labeyrie, G. Fabres, & D. Lachise, eds. *Insects-Plants: Proceedings of the 6th International Symposium on Insect-Plant Relationships, Pau, 1986*. Junk, Dordrecht.

Williams, K.S., & C. Simon. 1995. The ecology, behavior, and evolution of periodical cicadas. *Annual Review of Entomology* 40: 269–295.

Williams, K.S., & K.G. Smith. 1986. Spatial and temporal distribution of periodical cicadas (*Magicicada* spp.) and their avian predators. *Proceedings of the IV International Congress of Ecology, Syracuse, New York, 10–16 August 1986*: 353.

Williams, K.S., & K.G. Smith. 1991. Dynamics of periodical cicada chorus centers (Homoptera: Cicadidae: *Magicicada*). *Journal of Insect Behavior* 4: 275–291.

Williams, K.S., & K.G. Smith. 1992. Host plant choices of periodical cicadas, *Magicicada tredecassini*. *Bulletin of the Ecological Society of America, Supplement* 73: 389.

Williams, K.S., & K.G. Smith. 1993. Host plant choices of periodical cicadas, *Magicicada tredecassini*. *Bulletin of the Ecological Society of America* 74 (Supplement): 489.

Williams, K.S., K.G. Smith, & F.M. Stephen. 1993. Emergence of 13-year periodical cicadas (Cicadidae: *Magicicada*): Phenology, mortality, and predator satiation. *Ecology* 74: 1143–1152.

Williams, L.A.D. 1991. Host preferences and seasonal fluctuations of the coffee bug *Odopoea dilata* (Homoptera: Cicadoidea) populations on a coffee farm in the Nassau Valley, Jamaica. *Jamaican Journal of Science and Technology* 2: 25–29.

Wilson, J.W. 1930. *Tibicen davisi* Smith and Grossbeck (Cicadidae), a new pest of economic importance. *Florida Entomologist* 14: 61–65.

Wilson, M.R. 1987. A faunistic review of Auchenorrhyncha on sugarcane. *Proceedings of the 6th Auchenorrhyncha Meeting, Turin, Italy, 7–11 September 1987*: 485–492.

Wilson, M.R. 2001. The need for priority revision of certain Auchenorrhyncha genera: economic and conservation issues. *Program and Book of Abstracts, 2nd European Hemiptera Congress, Fiesa, Slovenia, 20–24 June 2001*: 14.

Wilson, M.R. 2009. A checklist of Fiji Auchenorrhyncha (Hemiptera). *Fiji Arthropods XII. Bishop Museum Occasional Papers* 102: 33–48.

Wilson, M.R., & D.J. Hilburn. 1991. Annotated list of the auchenorrhynchous Homoptera (Insecta) of Bermuda. *Annals of the Entomological Society of America* 84: 412–419.

Windmill, J., D. Robert, & J. Sueur. 2009. Picometre scale mechanics in the cicada ear. *Comparative Biochemistry and Physiology. Part A. Molecular & Integrative Physiology* 153: S130.

Windmill, J.F.C., J. Sueur, & D. Robert. 2009. The next step in cicada audition: measureing pico-mechanics in the cicada's ear. *Journal of Experimental Biology* 212: 4079–4083.

Witsack, W. 1999. Bestandssituation der Zikaden. Pages 422–431 *in* D. Frank & V. Neumann, eds. *Bestandssituation der Pflanzen und Tiere Sachsen-Anhalts*. Ulmer, Stuttgart.

Witsack, W. 2008. Zikaden (Auchenorrhyncha). In: Arten- und Biotopschutzprogramm Sachsen-Anhalt, Biologische Vielfalt und FFH-Management im Landschaftsraum Saale-Unstrut-Triasland, Teil 1. Herausgegeben durch das Lansesamt für Umweltschutz Sachsen-Anhalt. *Berichte des Landesamtes für Umweltschutz Sachsen-Anhalt, Sonderheft 1/2008* 1: 262–267.

Witsack, W., & H. Nickel. 2004. Rote Liste der Zikaden (Hemiptera, Auchenorrhyncha) des Landes Sachsen-Anhalt. *Berichte des Landesamtes für Umweltschutz Sachsen-Anhalt* 39: 228–236.

Wolcott, G.N. 1927. Chapitre XXV. Les Homoptères. Pages 187–193 *in* Entomologie d'Haiti. La Direction du Service Technique du Départment de l'Agriculture et de l'Enseignement Professionel, Port-Au-Prince. 440p.

Wolcott, G.N. 1948. The insects of Puerto Rico. *Puerto Rico University Journal of Agriculture* 32: 1–224.

Wolda, H. 1977. Fluctuations in abundance of some Homoptera in a neotropical forest. *Geo-Eco-Trop* 3: 229–257.

Wolda, H. 1978. Seasonal fluctuations in rainfall, food, and abundance of tropical insects. *Journal of Animal Ecology* 47: 369–381.

Wolda, H. 1983a. "Longterm" stability of tropical insect populations. *Researches on Population Ecology, Supplement* 3: 112–126.

Wolda, H. 1983b. Spatial and temporal variation in abundance in tropical animals. Pages 93–105 *in* S.L. Sutton, T.C. Whitmore, & H.C. Chadwick, eds. *Tropical rain forest: ecology and management*. Blackwell Scientific, Oxford.

Wolda, H. 1984. Diversity and seasonality of Panamanian cicadas. *Mitterlungen der Schweizerischen Entomologischen Gesellschaft* 57: 451.

Wolda, H. 1988. Insect seasonality: Why? *Annual Review of Ecology and Systematics* 19: 1–18.

Wolda, H. 1989. Seasonal cues in tropical organisms. Rainfall? Not necessarily! *Oecologia* 80: 437–442.

Wolda, H. 1993. Diel and seasonal patterns of mating calls in some neotropical cicadas. Acoustic interference? *Proceedings of the Koninklijke Nederlandse Akademie van Wetenschappen. Biological, Chemical, Geological, Physical and Medical Sciences* 96: 369–381.

Wolda, H., & J.A. Ramos. 1992. Cicadas in Panama, their distribution, seasonality and diversity. Pages 271–279 *in* D. Quintero & A. Aiello, eds. *The insects of Panama and Mesoamerica: selected studies*. Oxford University Press, New York.

Won, P.-O., H.-C. Woo, K.-M. Ham, & M.-Z. Chun. 1968. [Chick food analysis of some Korean birds (III).] *Yamashina Chorui Kenkyujo Kenyu Hokoku (Journal of the Yamaashina Institute for Ornithology)* 5: 363–369. [In Japanese, English summary]

Woodall, P.F. 1994. Regent bowerbird feeding on a greengrocer cicada. *Sunbird* 24: 44.

Woodall, P.F. 1996. Cicada predated by ant. *Queensland Naturalist* 34: 32.

Wooton, A. 1984. *Insects of the world*. Facts on File Publications, New York. 224pp.

Wooton, A. 2002. *Insects of the world*. Facts on File Publications, New York. 224pp.

Wooton, R. 1968. The evolution of *Cicadoidea* (*Homoptera*). *Proceedings of the XIII International Congress of Entomology (Moscow)* 1: 318–319.

Woyke, T., D. Tighe, K. Mavromatis, A. Clum, A. Copeland, W. Schackwitz, A. Lapidus, D. Wu, J.P. McCutcheon, B.R. McDonald, N.A. Moran, J. Bristow, & J.-F. Cheng. 2010. One bacterial cell, one complete genome. *PLoS ONE* 5(4): e10314. 8pp. doi: 10.1371/journal.pone.0010314.

Wright, A.G. 1951. Common Illinois insects and why they are interesting. *Illinois State Museum Story of Illinois* No. 8: 1–32.

Wu, C.F. 1935. Catalogus insectorum sinensium (Catalogue of Chinese Insects). Volume II: 1–634.

Wu, G.C., J.C Wright, D.L. Whitaker, & A.N. Ahn. 2010. Kinematic evidence for superfast locolotory muscle in two species of teneriffiid mites. *Journal of Experimental Biology* 213: 2551–2556.

Xia, L., L.-P. Bai, L. Yi, B.-B. Liu, C. Chu, Z.-T. Liang, P. Li, Z.-H. Jiang, & Z.-Z. Zhao. 2007. Authentication of the 31 species of toxic and potent Chinese Materia Medica (T/PCMM) by microscopic technique, part 1: three kinds of toxic and potent animal CMM. *Microscopy Research and Technique* 70: 960–968.

Xu M.-Z., W.S. Lee, J.-M. Han, H.-W. Oh, D.-S. Park, G.-R. Tian, T.S. Jeong, & H.-Y. Park. 2006. Antioxidant and anti-inflammatory activities of *N*-acetyldopamine dimmers from Periodtracum Cicadae. *Bioorganic & Medicinal Chemistry* 14: 7826–7834.

Xu T.-s., Wang H.-j., Xu Q., Hua Z.-y., & Lin C.-c. 2001. Biological characteristics of *Platylomia pieli*. *Forest Research* 14: 396–402.

Xu Y., Ye M., & Chen X. 1995. [Homoptera: Cicadidae and Membracidae.] Pages 84–85 *in* Wu H., ed. *Insects of Baishanzu Mountains, eastern China*. China Forestry Publishing House, Beijing. [In Chinese, English summary]

Yaakop, S., J.P. Duffels, & H. Visser. 2005. The cicada genus *Chremistica* Stål (Hemiptera: Cicadidae) in Sundaland. *Tijschrift voor Entomologie* 148: 247–306.

Yaginuma, K. 2002. *Paecilomyces cicadae* Samson isolated from soil and cicada, and its virulence to the Peach Fruit Moth, *Cuposina sasakii* Matsumura. *Japanese Journal of Applied Entomology and Zoology* 46: 225–231.

Yaginuma, T. 1987. [Encounter with 17-year cicadas.] *Insectarium* 24: 270–272. [In Japanese]

Yamada, K., & T. Tutumi. 1989. [Control of the large brown cicada (*Graptopsaltria nigrofuscata* Motschulsky) by pyrethroids in Japanese pear (Pyrus pyrifolia) orchards.] *Proceedings of the Association for Plant Protection of Kyushu* 35: 154–156. [In Japanese]

Yamanaka, T., & A.M. Liebhold. 2009. Mate-location failure, the Allee effect, and the establishment of invading populations. *Population Ecology* 51: 337–340.

Yamazaki, K. 2007. Cicadas "dig wells" that are used by ants, wasps and beetles. *European Journal of Entomology* 104: 347–349.

Yamazaki, K. 2009. Plant-mediated herbivore-herbivore interactions – some recent progress – . *Journal of Urban living and Health Association (Seikatsu Eisei)* 53: 79–89.

Yang, L.H. 2004. Periodical cicadas as resource pulses in North American forests. *Science* 306: 1565–1567.

Yang, L.H. 2005. Interactions between a detrital resource pulse and a detritivore community. *Oecologia* 147: 522–532.

Yang, L.H. 2006. Periodical cicadas use light for oviposition site selection. *Proceedings of the Royal Society. Series B. Biological Sciences* 273: 2993–3000.

Yang, L.H. 2008. Pulses of dead periodical cicadas increase herbivory of American bellflowers. *Ecology* 89: 1497–1502.

Yang, L.H., & R. Karban. 2009. Long-term habitat selection and chronic root herbivory: explaining the relationship between periodical cicada density and tree growth. *American Naturalist* 173: 105–112.

Yang X., Chen H., Xu M., & Jiang J. 1998. Auditory responses and intrinsic link with sound-production in singing male cicada. *Science in China. Series C. Life Sciences* 41: 576–583.

Yang X., & Jiang J. 1994. [Bilattice structure of the myofibril of sounding muscle and analysis of the original process of sound-production in Mingming cicada (*Oncotympana maculaticollis* Motsch.).] *Acta Biophysica Sinica* 10: 258–262. [In Chinese, English summary]

Yang X., & Jiang J.-C. 1995. [Observation on sounding muscle structure in Mingming cicada.] *Acta Entomologica Sinica* 38: 173–178. [In Chinese, English summary]

Yang X., Jiang J.-C., Chen H., & Xu M. 1994. [A study on action properties of sounding muscle structure in Mingming cicada (*Oncotympana maculaticollis* Motsch).] *Acta Acustica* 19: 216–226. [In Chinese, English summary]

Yasukawa, K. 2010. Yellow-billed Cuckoo hatched and fed by a Red-winged Blackbird. *Wilson Journal of Ornithology* 122: 402–405.

Yaylayan, V.A., J.R.J. Paré, G. Matni, & J.M.R. Bèlanger. 2001. MAP: microwave-assisted extraction of fatty acids and Py/GC/MS analysis of selected insects. *Natural Product Letters* 15: 187–195.

Ye, X.J., & T.B. Ng. 2009. A novel lectin with highly potent antiproliferative and HIV-1 reverse transcriptase inhibitory activities from cicada (*Cicada flammata*). *Applied Microbiology and Biotechnology* 86: 1409–1418.

Yen, S.-H., G.S. Robinson, & D.L.J. Quicke. 2005a. The phylogenetic relationships of Chalcosiinae (Lepidoptera, Zygaenoidea, Zygoenida). *Zoological Journal of the Linnean Society* 143: 161–341.

Yen, S.-H., G.S. Robinson, & D.L.J. Quicke. 2005b. Phylogeny, systematics and evolution of mimetic wing patterns of *Eterusia* moths (Lepidoptera, Zygoenidae, Chalcosiinae). *Systematic Entomology* 30: 358–397.

Yi, Y.J. 2005. [Cicadas of Korea.] [In Korean] See Lee, Y.J. 2005.

Yoshida, A., M. Motoyama, A. Kosaku, & K. Miyamoto. 1996. Nanoprotuberance array in the transparent wing of a hawkmoth, *Cephonodes hylas*. *Zoological Science* 13: 525–526.

Yoshida, T. 2009. Report on the 20th Annual Meeting of the Japanese Association for Development and Comparative Immunology (JADCI), which was held from August 25–27, 2008, at University of Tokyo Medical and Dental School, Tokyo, Japan (Local organizer: Takeshi Yoshida). *Developmental and Comparative Immunology* 33: 948–957.

Yoshimura, J. 1997. The evolutionary origins of periodical cicadas during ice ages. *American Naturalist* 149: 112–124.

Yoshimura, J., T. Hayashi, Y. Tanaka, K.-i. Tainaka, & C. Simon. 2009. Selection for prime-number intervals in a numerical model of periodical cicada evolution. *Evolution* 63: 288–294.

Yoshimura, M., & I. Okochi. 2005. A decrease in endemic odonates in the Ogasawara Islands, Japan. *Bulletin of the Forestry and Forest Products Research Institute* 4: 45–51.

Yoshioka, M., N. Narai, A. Shinkai, T. Tokuda, K. Saito, M. Shioya, N. Tokoro, & Y. Kono. 1994. Analyses of metal ions in spider venoms in relation to insecticidal activity of Clavamine. *Biological & Pharmaceutical Bulletin* 17: 472–475.

Yoshizawa, K., & T. Saigusa. 2001. Phylogenetic analysis of paraneopteran orders (Insecta: Neoptera) based on forewing base structure, with comments on monophyly of Auchenorrhyncha (Hemiptera). *Systematic Entomology* 26: 1–13.

Young, A.M. 1979. A Blue Jay extricates an emerging cicada from its nymphal skin in Wisconsin. *The Passenger Pigeon* 41: 149.

Young, A.M. 1981a. Notes on seasonality and habitat associations of tropical cicadas (Homoptera: Cicadidae) in premontane and montane tropical moist forest in Costa Rica. *Journal of the New York Entomological Society* 89: 123–142.

Young, A.M. 1981b. Notes on the population ecology of cicadas (Homoptera: Cicadidae) in the Cuesta Angel forest ravine of northeastern Costa Rica. *Psyche* 88: 175–195.

Young, A.M. 1981c. Seasonal adult emergences of cicadas (Homoptera: Cicadidae) in northwestern Costa Rica. *Milwaukee Public Museum Contributions in Biology and Geology* 40: 1–29.

Young, A.M. 1981d. Temporal selection for communicatory optimization: The dawn-chorus as an adaptation in tropical cicadas. *American Naturalist* 117: 826–829.

Young, A.M. 1982. *Population biology of tropical insects*. Plenum Press, New York. 511pp.

Young, A.M. 1983a. *Fidicina mannifera* (Chicharra, Sundown cicada). Pages 724–726 *in* D.H. Janzen, ed. *Costa Rican natural history*. University of Chicago Press, Chicago.

Young, A.M. 1983b. Ecological studies of tropical cicadas. *Field Museum of Natural History Bulletin* 54: 20–25.

Young, A.M. 1984a. On the evolution of cicada x host-tree associations in Central America. *Acta Biotheoretica* 33: 163–198.

Young, A.M. 1984b. *Tibicen* cicada (Homoptera: Cicadidae) associated with a potted ornamental palm tree. *The Pan-Pacific Entomologist* 60: 358–359.

Young, A.M. 1991. *Sarapiqui chronicles: a naturalist in Costa Rica*. Smithsonian Institution Press, Washington. 361pp.

Young, D. 1990. Do cicadas radiate sound through their ear-drums? *Journal of Experimental Biology* 151: 41–56.

Young, D. 1996. Sound-production in cicadas. Pages 95–104 *in* D.C.F.Rentz, Co-ordinator. Nature sounds of Australia, original recordings of Australian wildlife, including birds, frogs, cicadas and crabs. The Sound Heritage Association, Ltd.

Young, D. 2003. Tuning in to cicadas. *Nature Australia* 27: 28–35.

Young, D., & H.C. Bennet-Clark. 1995. The role of the tymbal in cicada sound production. *Journal of Experimental Biology* 198: 1001–1019.

Young, D., & R.K. Josephson. 1983a. Mechanisms of sound-production and muscle contraction kinetics in cicadas. *Journal of Comparative Physiology. A* 152: 183–195.

Young, D., & R.K. Josephson. 1983b. Pure-tone songs in cicadas with special reference to the genus *Magicicada*. *Journal of Comparative Physiology. A* 152: 197–207.

Young, D., & R.K. Josephson. 1984. 100Hz is not the upper limit of synchronous muscle contraction. *Nature* 309: 286–287.

Young, F.N., & G. Kritsky. 1987. Observations on periodical cicadas (Brood X) in Indiana in 1987 (Homoptera: Cicadidae). *Proceedings of the Indiana Academy of Sciences* 97: 323–329.

Young, F.N., & G. Kritsky. 1990. Observations on periodical cicadas (Brood XXIII) in Indiana in 1989 (Homoptera-Cicadidae). *Proceedings of the Indiana Academy of Sciences* 99: 25–28.

Yuan, L. 1987. [A preliminary observation of the harm of oviposition by *Cryptotympana atrata* Fabricius in cotton field.] *Kun Chong Zhi Shi = Kunchong Zhishi = Entomological Knowledge* 24: 150–151. [In Chinese]

Yukawa, J., & S. Yamane. 1985. Odonata and Hemiptera collected from the Krakataus and surrounding islands, Indonesia. *Kontyû* 53: 690–698.

Zaidi, M.I. 1993. Riang-riang (Homoptera: Cicadidae) kawasan bandaran Bintulu, Sarawak: tinjauan awal. *Penyelidikan & Pembangunan Sains dan Teknologi, Universiti Kebangsaan Malaysia*: 957–960.

Zaidi, M.I. 1996. Preliminary survey or cicadas within the campus area of the Science Teachers Training College, Bintulu, Sarawak. *Serangga* 1: 97–107.

Zaidi, M.I. 1997. Cicadas within the town centre area of Bintulu, Sarawak: An additional survey. *Serangga* 2: 109–118.

Zaidi, M.I., & S. Azman. 1998. A new species of *Pomponia* Stal (Homoptera: Cicadidae) from peninsular Malaysia. *Serangga* 3: 161–167.

Zaidi, M.I., & S. Azman. 1999. *Pomponia langkawiensis*, a new species of cicada from Malaysia (Homoptera: Cicadidae). *Serangga* 4: 291–297.

Zaidi, M.I., & S. Azman. 2000. *Pomponia bulu*, a new species of cicada from Sabah, Malaysia (Homoptera: Cicadidae). *Serangga* 5: 189–196.

Zaidi, M.I., & S. Azman. 2003. Cicada fauna (Homoptera: Cicadidae) of Sabah: with special reference to six selected protected areas. *Serangga* 8: 95–106.

Zaidi, M.I., S. Azman, & W. Nordin. 2005. Cicada (Homoptera: Cicadidae) fauna of Trus Madi Range, Sabah: a survey and overview. *Journal of Tropical Biology and Conservation* 1: 27–29.

Zaidi, M.I., S. Azuman, & M.Y. Ruslan. 2001a. Cicada (Homoptera: Cicadoidea) fauna of Langkowi Island, Kedah. *Serrangga* 6: 109–122.

Zaidi, M.I., S. Azuman, & M.Y. Ruslan. 2001b. Cicadas from the Samusam Wildlife Sanctuary, Sarawak (Homoptera: Cicadidae). *Serrangga* 6: 123–129.

Zaidi, M.I., & A.A. Hamid. 1996. Preliminary survey of cicadas in Lanjak Entimau Wildlife Sanctuary, Sarawak. *Serangga* 1: 49–58.

Zaidi, M. I., M. S. Mohammedsaid, & O. Yaakob, 1990. Riang-riang kawasan Hutan Simpan Bangi. Pages 259–268 *in* A. Latiff, ed. *Ekologi dan biologi Hutan Simpan Bangi*. Universiti Kebangsaan Malaysia, Bangi.

Zaidi, M.I., B.M. Noramly, & M.Y. Ruslan. 2000a. Cicadas (Homoptera: Cicadidae) from Tibow, Sabah. *Serangga* 5: 319–333.

Zaidi, M.I., B.M. Noramly, & M.Y. Ruslan. 2000b. Cicada (Homoptera: Cicadidae) fauna of Tawau Hills Park: An additional survey and overview. *Serangga* 5: 335–341.

Zaidi, M.I., & W. Nordin. 2001. Cicadas of Crocker Range Park, Sabah (Homoptera: Cicadoidea): an additional survey. *Serrangga* 6: 181–204.

Zaidi, M.I., W. Nordin, M. Maryata, A. Wahab, M.F. Norashikin, K. Catherine, & A. Fatimah. 2002. Cicada (Homoptera: Cicadoidea) fauna of Crocker Range Park, Sabah. Online journal *ASEAN Review of Biodiversity and Environmental Conservation (ARBEC)*, July-September: 13pp. http://www.arbec.com.my/pdf/art19julysep02.pdf.

Zaidi, M. I., & M. Y. Ruslan, 1994. Riang-riang Bukit Fraser: Satu tinjauan awal. Pages 424–430 *in* R. Ahmad, A. Abdullah, M.K. Ayuba, & S.P. Al-Junid, eds. *Kesatuan dalam kepelbagaian Penyelidikan Biologi*. Kumpulan Kertaskerja 27, Universiti Kebangsaan Malaysia, Bangi.

Zaidi, M. I., & M. Y. Ruslan, 1995a. Cicadas of Bario, Sarawak. Pages 169–177 *in* G. Ismail & L.B. Din, eds. *Bario, the Kelabit Highlands of Sarawak: A scientific journey through Borneo*. Pelanduk Publications, Selangor Darul Ehsan, Malaysia.

Zaidi, M. I., & M. Y. Ruslan, 1995b. Cicadas of Sayap-Kinabalu Park, Sabah. Pages 217–223 *in* G. Ismail & L.B. Din, eds. *Sayap-Kinabalu Park, Sabah: A scientific journey through Borneo*. Pelanduk Publications, Selangor Darul Ehsan, Malaysia.

Zaidi, M. I., & M. Y. Ruslan, 1995c. Cicadas (Homoptera: Cicadoidea) in the Kuala Lompat sector of Krau Wildlife Reserve, Pahang. *The Journal of Wildlife and Parks* 14: 62–70.

Zaidi, M. I., & M. Y. Ruslan, 1995d. Cicadas of Tawau Hills Park, Sabah: A preliminary survey. Pages 197–204 *in* G. Ismail, S. Omar, & L.B. Din, eds. *Tawau Hills Park, Sabah: A scientific journey through Borneo*. Pelanduk Publications, Selangor Darul Ehsan, Malaysia.

Zaidi, M. I., & M. Y. Ruslan, 1997. Notes on cicadas (Homoptera: Cicadoidea) in the Zoological Reference Collection, National University of Singapore. *Serangga* 2: 217–233.

Zaidi, M. I., & M. Y. Ruslan, 1998a. Notes on cicadas (Homoptera: Cicadoidea) in the reference collection of Sarawak Museum. *Serangga* 3: 343–371.

Zaidi, M. I., & M. Y. Ruslan, 1998b. Cicadas of Bario, Sarawak. Online journal *ASEAN Review of Biodiversity and Environmental Conservation (ARBEC),* August 1998: 6pp. www.arbec.com.my/pdf/art2sepoct99.pdf.

Zaidi, M. I., & M. Y. Ruslan, 1999. Cicadas of Bario, Sarawak. Online journal *ASEAN Review of Biodiversity and Environmental Conservation (ARBEC),* September-October 1999: 7pp. www.arbec.com.my/pdf/art2sepoct99.pdf.

Zaidi, M.I., M.Y. Ruslan, & S. Azman. 1999. Cicadas (Homoptera: Cicadoidea) in the insect reference collection of Sabah Parks, Malaysia. *Serangga* 4: 299–320.

Zaidi, M.I., M.Y. Ruslan, & S. Azman. 2000. Cicadas (Homoptera: Cicadoidea) in the insect reference collection of Sabah forestry Research Institute, Sandakan. *Serangga* 5: 197–220.

Zaidi, M.I., M.Y. Ruslan, & A.M. Mahadir. 1996. List of cicadas (Homoptera: Cicadoidea) from peninsular Malaysia in the repository of the Centre for Insect Systematics, Universiti Kebangsaan Malaysia. *Serangga* 1: 59–62.

Zamanian, H., M. Mehdipour, & N. Ghaemi. 2008. The study and analysis of the mating behavior and sound production of male cicada *Psalmocharias alhageos* (Kol.) (Homoptera: Cicadidae) to make disruption in mating. *Pakistan Journal of Biological Sciences* 11: 2062–2072.

Zambon, S., K. Nakayama, & O. Nakano. 1980. Verificação do efeito de alguns inseticidas aplicados no solo sobre as ninfas da cigarra *Quesada* spp. (Homoptera-Cicadidae) em cultura de café. *Congresso Brasileiro de Pesquisa Cafeeiras, 11., Campos do Jordão. Resumos. Rio de Janeiro, Instituto Brasileiro do Café*: 255–256.

Zanuncio, J.C., F.F. Pereira, T.V. Zanuncio, N.M. Martinelli, T.B. Moreira Pinon, & E.M. Guimarães. 2004. Occurrence of *Quesada gigas* on *Schizolobium amazonicum* trees in Maranhão and Pará States, Brazil. *Pesquisa Agropecuária Brasileira, Brasília* 39: 943–945.

Zborowski, P., & R. Storey. 1995. *A field guide to insects in Australia*. Reed Books, Chatswood. 207pp.

Zborowski, P., & R. Storey. 2003. *A field guide to insects in Australia*. Reed Books, Sydney. 208pp.

Zeuner, F.E. 1941. The Eocene insects of the Ardtun Beds, Isle of Mull, Scotland. *Annals and Magazine of Natural History* (11)7: 82–100.

Zeuner, F.E. 1944. Notes on Eocene Homoptera from the Isle of Mull, Scotland. *Annals and Magazine of Natural History* (11)11: 110–117.

Zhang, G. J. Zhang, G. Xie, Z. Liu, & H. Shao. 2006. Cicada wings; a stamp from nature for nanoimprint lithography. *Small* 2: 1440–1443.

Zhang J., Sun B., & Zhang X. 1994. [*Miocene insects and spiders from Shanwang, Shandong*.] Science Press, Beijing. 298pp., plates I-XLIV. [In Chinese]

Zhang J.F., & Zhang X.Y. 1990. [Fossil insects of cicada (Homoptera) and true bugs (Heteroptera) from Shanwong, Shandong.] *Acta Palaeontologica Sinica* 29: 337–348, plates I-III. [In Chinese, English summary]

Zhang, S.-W., & L.-J. Xuan. 2007. Five aromatics bearing a 4-*O*-methylglucose unit from *Cordyceps cicadae*. *Helvetica Chimica Acta* 90: 404–410.

Zhang, Y., J. Zhang, Z. Jiang, G. Wang, Z. Luo, Y. Fan, Z. Wu, & Y. Pei. 2010. Requirement of a mitogen-activated protein kinase for appressorium formation and penetration of incest cuticle by the entomopathogenic fungus *Beauveria bassiana*. *Applied and Environmental Microbiology* 76: 2262–2270.

Zhao Z. 2004. *An illustrated Chinese Materia Medica in Hong Kong*. School of Chinese Medicine, Hong Kong Baptist University, Hong Kong. 538pp.

Zheng, D., Z. Zhang, & Q. Wang. 2010. Total and methyl mercury contents and distribution characteristics in cicada, *Cryptotympana atrata* (Fabricius). *Bulletin of Environmental Contamination and Toxicology* 84: 749–753.

Zinn, Y.L., R. Lal, J.M. Bigham, & D.V.S. Resck. 2007. Edaphic controls on soil organic carbon retention in the Brazilian Cerrado: soil structure. *Soil Science Society of America Journal* 71: 1215–1224.

Zoratto, F., D. Santucci, & E. Alleva. 2009. Theories commonly adopted to explain anti-predatory benefits of ther group life: the case of starling (*Sturnus vulgaris*). *Rendiconti Lincei* 20: 163–176.

Zorba, V., E. Stratakis, M. Barberoglou, E. Spanakis, P. Tzanetakis, S.H. Anastasiadis, & C. Fotakis. 2008a. Biomimetic artificial surfaces quantitatively reproduce the water repellency of a lotus leaf. *Advanced Materials* 20: 4049–4054.

Zorba, V., E. Stratakis, M. Barberoglou, E. Spanakis, P. Tzanetakis, S.H. Anastasiadis, & C. Fotakis. 2008b. Tailoring the wetting response of silicon surfaces via fs laser structuring. *Applied Physics. A. Materials Science & Processing* 93: 819–825.

Zyla, J.D. 2004a. First report of the 13-year periodical cicada, *Magicicada tredecim* (Walsh and Riley) (Hemiptera: Cicadidae) in Maryland. *Proceedings of the Entomological Society of Washington* 106: 485–487.

Zyla, J.D. 2004b. Reports of four year accelerated occurrences of the 2004 emergence of periodical cicadas, *Magicicada* spp. (Hemiptera: Cicadidae) Brood X in Maryland, Virginia, and the District of Columbia. *Proceedings of the Entomological Society of Washington* 106: 488–490.

INDEX OF CICADA GENUS-GROUP NAMES

Valid names are *italicized*. Junior synonyms, homonyms and misspellings are in normal font.

D

E

F

G

N

O

P

Q

R

S

T

U

V

X

Y

Z

INDEX OF CICADA SPECIES-GROUP, SUBSPECIES AND INFRASUBSPECIFIC NAMES

Currently valid taxa are listed in *italics* with the authority and year of publication in parentheses if the taxon was originally described in a different genus. Junior synonyms, homonyms and misspellings are in normal font.

A

B

C

D

E

G

H

I

J

K

L

N

O

P

Q

R

T

U

V

W

X

Y

Z

Printed and bound by CPI Group (UK) Ltd, Croydon, CR0 4YY
08/05/2025
01865027-0001